Reprinted in 1989 and 1990 by
Routledge
a division of Routledge, Chapman and Hall
11 New Fetter Lane, London EC4P 4EE

Published in the USA by
Routledge
in association with Routledge, Chapman and Hall, Inc.
29 West 35th Street, New York NY 10001

Printed in Great Britain by Unwin Brothers Ltd.

ISBN 0 415 04309 3 (vol 1)
ISBN 0 415 02768 3 (vol 2)
ISBN 0 415 03654 2 (the set)

KETTRIDGE'S
TECHNICAL DICTIONARY

DICTIONNAIRE TECHNIQUE
PAR
KETTRIDGE

French-English and English-French

DICTIONARY

of technical terms and phrases used in

CIVIL, MECHANICAL, ELECTRICAL, AND MINING ENGINEERING, AND ALLIED SCIENCES AND INDUSTRIES

including

GEOLOGY, PHYSICAL GEOGRAPHY, PETROLOGY, MINERALOGY, CRYSTALLOGRAPHY, METALLURGY, CHEMISTRY, PHYSICS, GEOMETRY, ABBREVIATIONS AND SYMBOLS, WEIGHTS AND MEASURES, COMPOUND CONVERSION FACTORS, ETC.

and a method of

TELEGRAPHIC CODING

by which any entry in the dictionary can be expressed by
a 10-letter cipher word with indicator and check

BY

J. O. KETTRIDGE

Late Officier d'Académie, F.S.A.A., etc.

WITH A SUPPLEMENT BY

YVES R. ARDEN

The CIVIL ENGINEERING terms comprise those used in the Surveying and Construction of Railways, Waterways, Roads, Bridges, etc.

The MINING terms comprise those used in Getting Metals, Coal, Oil, and other Minerals, and Quarrying, Exploration, Surveying, Boring, Excavation, Sinking, Drilling, Ventilation, Lighting, Haulage, Hoisting, Assaying, Ore-Dressing, Management, etc.

The MECHANICAL terms comprise those used in Mechanical Science, Mechanisms, Specifications of Machinery, Plant and Tools, Strength of Materials, Smelting, Refining, Founding and Working of Metals, Building, Carpentry and Joinery, Lighting and Heating, Steam, Hydraulic, Oil and Gas-Engineering, Rail-Locomotion, Machine-Tool Work, Scientific Research, etc.

containing

The Translations of One Hundred Thousand Words, Terms, and Phrases
Illustrated by numerous instructive examples and explanations
The whole arranged on original plans
in progressive alphabetical order
in the readiest form for rapid reference

VOLUME I

ROUTLEDGE
LONDON AND NEW YORK

DICTIONNAIRE

français-anglais, anglais-français,

de termes et locutions techniques de

GÉNIE CIVIL, MÉCANIQUE, ÉLECTRICITÉ, MINES, SCIENCES ET INDUSTRIES CONNEXES

comprenant

GÉOLOGIE, GÉOGRAPHIE PHYSIQUE, PÉTROLOGIE, MINÉRALOGIE, CRISTALLOGRAPHIE, MÉTALLURGIE, CHIMIE, PHYSIQUE, GÉOMÉTRIE, ABRÉVIATIONS ET SYMBOLES, POIDS ET MESURES, FACTEURS DE CONVERSION COMPOSÉS, ETC.

et une méthode de

CODIFICATION TÉLÉGRAPHIQUE

par laquelle tout article du dictionnaire peut être exprimé par un mot conventionnel de 10 lettres avec indicateur et contrôle

PAR

J. O. KETTRIDGE

Feu Officier d'Académie, F.S.A.A., etc.

AUGMENTÉ PAR

YVES R. ARDEN

Les termes de GÉNIE CIVIL comprennent ceux qui sont usités dans les Études et la Construction des Chemins de fer, des Voies navigables, des Routes, des Ponts, etc.

Les termes de MINES comprennent ceux qui sont usités dans l'Exploitation des Métaux, de la Houille, du Pétrole, et d'autres Minéraux, ainsi que des Carrières, Exploration, Levés de Plans, Sondage, Excavation, Fonçage, Perforation, Aérage, Éclairage, Roulage, Extraction, Essais, Préparation mécanique des minerais, Administration, etc.

Les termes de MÉCANIQUE comprennent ceux qui sont usités en Science mécanique, Mécanismes, Devis descriptifs de Machines, de Matériels, et d'Outillage, Résistance des matériaux, Fonte, Affinage, Fonderie et Travail des Métaux, Construction civile, Charpenterie et Menuiserie, Éclairage et Chauffage, Machines à vapeur, à eau, à pétrole, et à gaz, Locomotion sur rails, Travaux de Machines-outils, Recherches scientifiques, etc.

contenant

Les Traductions de Cent Mille Mots, Termes, et Locutions
Illustrés par de nombreux exemples et explications instructifs
Le tout disposé sur des plans originaux
par ordre alphabétique progressif
dans la meilleure forme pour consultation rapide

TOME I

ROUTLEDGE
LONDON AND NEW YORK

PREFACE

This dictionary is not the product of the united labour of a number of collaborators, but the work of one man. It is not a compilation from existing French-English dictionaries, but a new and original production from radical and authentic sources. It is not a random glossary of lifeless words, but a cohesive record of living speech.

The author has striven to produce a work which may be regarded as authoritative. He has also striven to create the ideal dictionary—comprehensive, relevant, accurate, precise, clear, explanative, illustrative, interesting, original, modern, well-arranged, and well-displayed.

Illustrative examples and phrases

A wealth of examples and phrases is given. Examples not only fix the use and illustrate the meaning of the vocabulary word, but attest the correctness of the translations ; they show also how the word is used in connection with other words, thus teaching how to read and write the language. Moreover it will be found that many of the examples impart some useful or interesting piece of knowledge or information.

Definitional matter

It has been the custom with most foreign-language dictionaries, where a word has several or many meanings, to mass the translations together without any indication of the distinctions existing between them, thus presupposing a knowledge of both languages on the part of the user, or leaving him to conjecture which word is the proper one. All that is required of the user of this dictionary is a knowledge of either of the two languages, English or French. Whenever a vocabulary word has several distinct meanings, or whenever reasonable doubt may exist as to its meaning or application, each meaning forms the subject of a separate vocabulary entry, the word being defined in its own language, or some other clear indication given to establish its identity, before proceeding to its translation.

New contributions to technical lexicography

Numbers of technical terms in common use, not to be found in any existing French or English unilingual dictionaries, or even in text-books, appear in the dictionary, forming notable additions to lexically recorded technical language.

Inversion

The English-French section of this dictionary is the inverted counterpart of the French-English section, the genius of both languages being respected with reverent regard.

Arrangement

A new method of vocabulary arrangement has been devised, and is explained in the following pages. It is claimed that great speed and absolute certainty of reference are attained by this method. It is a logical solution of the hitherto unsolved problem of arranging continuous words, hyphened compound words, unhyphened pairs or groups of words, apostrophic contractions, and phrases of any length or kind, in one alphabetical and consistent order throughout.

ARRANGEMENT OF WORK

The vocabulary is arranged in natural progressive order: oil-well will be found under **oil**; main rod under **main**; no thoroughfare under **no**; to get up steam under **get**.

In the English language, ideas are expressed by proceeding from the particular to the general, whereas in French the reverse is the case; thus, **centrifugal pump** becomes in French **pompe centrifuge.** It results, therefore, that although in the English section of the dictionary the different kinds of pumps, for instance, are distributed in the vocabulary, owing to the variations in the initial word, in the French section they all come together, forming therein a subject index.

****

Continuous words are arranged in strictly alphabetical succession.

Hyphened compound words, unhyphened pairs or groups of words, apostrophic contractions, and phrases, are not arranged as though the whole assemblage were one word regardless of breaks of continuity, but are grouped progressively under the common factor; thus,

contre-clavette	iron	travail à l'eau
contre-dépouille	iron and steel constructional work	travail à l'entreprise
contre-échelle	iron bar	travail à la main
contre-latte	iron countersunk wood-screw	travail à la poudre
contre-limon	iron cross twin	travail au feu
ꞓontre-manivelle	iron-founder	travail au tour
contre-plateau	iron or steel bridge	travail aux explosifs
contre-pointe	iron ore	travail d'art
contre-rail	iron wire	travail d'exploitation
contre-vapeur	iron-wire rope	travail de nuit
contrée	ironclad	travail de tour
contrefort	ironman	travail des eaux
contremaître	ironmaster	travail des métaux
contrepoids	ironstone	travail du fond
contrevent	ironwork	travail du jour

faire de l'eau	in a vacuum
faire de l'instantané	in all weathers
faire défaut	in an unminerlike manner
faire déflagrer du salpêtre	in an unworkmanlike manner
faire l'ascension d'une montagne	in-and-out bolt
faire l'impossible	in-and-out calipers
faire la fourche	in place
faire la tare d'un ressort	in-place deposit
faire un joint	in places
faire un levé	in wet weather
faire une découverte	in working order
faire une prise d'eau à une rivière	inaccessibility

Where, by reason of their continuity, compound words are thrown forward in the vocabulary, reference thereto is made in the places which the words would have occupied if discontinuous; thus,

contre-clavette	iron
contre-dépouille	iron and steel constructional work
contre-échelle	iron bar
contrefort See below	ironclad V. ci-après
contre-latte	iron countersunk wood-screw
contre-limon	iron cross twin
contremaître See below	iron-founder
contre-manivelle	ironman, ironmaster V. ci-après
contre-plateau	iron or steel bridge
contrepoids See below	iron ore
contre-pointe	ironstone V. ci-après
contre-rail	iron wire
contre-vapeur	iron-wire rope
contrevent See below	ironwork V. ci-après
contrée	ironclad
contrefort	ironman
contremaître	ironmaster
contrepoids	ironstone
contrevent	ironwork

When various parts of speech are spelt the same way, they are placed in the following order : adjective, adverb, noun, preposition, verb transitive, verb intransitive, verb reflexive ; thus,

<table>
<tr><td>skew (adj.)</td><td>ferme (adj.)</td></tr>
<tr><td>skew (adv.)</td><td>ferme (n.)</td></tr>
<tr><td>skew (n.)</td><td>fermer (v.t.)</td></tr>
<tr><td>skew (v.t.)</td><td>fermer (v.i.)</td></tr>
<tr><td>skew (v.i.)</td><td>fermer (se) (v.r.)</td></tr>
</table>

****

When a continuous word is divided at the end of a line, a hyphen is placed after the part of the word on the first line and another hyphen in front of the remaining part on the next line, in order to distinguish it from a hyphened compound word ; thus,   iron-
-work.

****

The gender and number given at the end of a French phrase refer to the first noun ; thus, in the phrase gaz qui s'échappent dans l'atmosphère (m.pl.), the letters m.pl. indicate that the noun gaz is masculine in gender and plural in number.

****

Synonyms following one another, whether in the translations or in the vocabulary, are placed in order of generality of use, commencing with the commonest.

****

Author's Note

Peculiar difficulty is sometimes met with in the translation of certain technical terms, and it is fallacious to suppose that there is necessarily a corresponding term in a foreign language. The reason is generally that the peculiar process, method, or system to which the terms relate has no application in the other country. For instance, many of the coal-seams in France are highly inclined, distorted, and broken up, and many are very thick. In England the seams are mostly flat, regular, and not very thick. There are consequently a number of French technical terms relating to the former classes of mines which are not required in English practice, and conversely the same may be said of many English terms. Again, the cable-system of drilling is used in countries where the strata are horizontal and easy to work, but in other countries these conditions may be rare, or may not exist at all, and in consequence many of the terms used in this system may be wanting. Similarly, in a country where mountains and glaciers abound, the language is richer in technical terms relating to these subjects than in a country where they are uncommon or non-existent, and where consequently the interest taken in them is more or less casual.

PLAN DE L'OUVRAGE

Le vocabulaire est arrangé par ordre progressif naturel : **oil-well** se trouvera sous **oil**; **main rod** sous **main**; **no thoroughfare** sous **no**; **to get up steam** sous **get**.

Dans la langue anglaise, les idées s'expriment en procédant du particulier au général, tandis qu'en français le contraire a lieu ; ainsi, **centrifugal pump** devient en français **pompe centrifuge**. Il en résulte, que quoique dans la section anglaise du dictionnaire les diverses sortes de pompes, par exemple, soient réparties dans le vocabulaire à cause des variations du mot initial, dans la section française elles se trouvent toutes ensemble, formant ainsi un répertoire raisonné.

✳✳✳✳✳✳✳✳✳✳✳✳✳✳✳✳✳✳✳✳✳✳✳✳✳✳✳✳✳✳

Les mots continus sont arrangés strictement par ordre alphabétique.

Les mots composés séparés par des traits d'union, les paires ou groupes de mots non-séparés par des traits d'union, les contractions apostrophiques, et les phrases, ne sont pas disposés comme si tout l'assemblage était un seul mot sans égard aux solutions de continuité, mais sont groupés progressivement sous le facteur commun ; ainsi,

contre-clavette	iron	travail à l'eau
contre-dépouille	iron and steel constructional work	travail à l'entreprise
contre-échelle	iron bar	travail à la main
contre-latte	iron countersunk wood-screw	travail à la poudre
contre-limon	iron cross twin	travail au feu
contre-manivelle	iron-founder	travail au tour
contre-plateau	iron or steel bridge	travail aux explosifs
contre-pointe	iron ore	travail d'art
contre-rail	iron wire	travail d'exploitation
contre-vapeur	iron-wire rope	travail de nuit
contrée	ironclad	travail de tour
contrefort	ironman	travail des eaux
contremaître	ironmaster	travail des métaux
contrepoids	ironstone	travail du fond
contrevent	ironwork	travail du jour

faire de l'eau	In a vacuum
faire de l'instantané	in all weathers
faire défaut	in an unminerlike manner
faire déflagrer du salpêtre	in an unworkmanlike manner
faire l'ascension d'une montagne	in-and-out bolt
faire l'impossible	in-and-out calipers
faire la fourche	in place
faire la tare d'un ressort	in-place deposit
faire un joint	in places
faire un levé	in wet weather
faire une découverte	in working order
faire une prise d'eau à une rivière	inaccessibility

Où, en conséquence de leur continuité, les mots composés sont reportés dans le vocabulaire, référence y est faite aux endroits que les mots auraient occupés s'ils avaient été discontinus ; ainsi,

contre-clavette	iron
contre-dépouille	iron and steel constructional work
contre-échelle	iron bar
contrefort See below	ironclad V. ci-après
contre-latte	iron countersunk wood-screw
contre-limon	iron cross twin
contremaître See below	iron-founder
contre-manivelle	ironman, ironmaster V. ci-après
contre-plateau	iron or steel bridge
contrepoids See below	iron ore
contre-pointe	ironstone V. ci-après
contre-rail	iron wire
contre-vapeur	iron-wire rope
contrevent See below	ironwork V. ci-après
contrée	ironclad
contrefort	ironman
contremaître	ironmaster
contrepoids	ironstone
contrevent	ironwork

Lorsque les diverses parties du discours ont la même orthographe, elles sont placées dans l'ordre suivant :—adjectif, adverbe, nom, préposition, verbe transitif, verbe intransitif, verbe réfléchi ; ainsi,

skew (*adj.*)	ferme (*adj.*)
skew (*adv.*)	ferme (*n.*)
skew (*n.*)	fermer (*v.t.*)
skew (*v.t.*)	fermer (*v.i.*)
skew (*v.i.*)	fermer (se) (*v.r.*)

****

Quand un mot continu est divisé à la fin d'une ligne, un trait d'union est placé après la partie du mot sur la première ligne et un autre trait d'union devant le reste du mot sur la ligne suivante, afin de le distinguer d'un mot composé séparé par un trait d'union ; ainsi, iron--work.

****

Le genre et le nombre donnés à la fin d'une phrase en français se rapportent au premier nom ; ainsi, dans la phrase gaz qui s'échappent dans l'atmosphère (*m.pl.*), les lettres *m.pl.* indiquent que le nom gaz est du genre masculin et au pluriel.

****

Les synonymes se suivant, soit dans les traductions ou dans le vocabulaire, sont placés dans l'ordre de la généralité d'emploi, en commençant par les plus ordinaires.

****

Note de l'auteur

Des difficultés particulières se rencontrent quelquefois dans la traduction de certains termes techniques et c'est une erreur de supposer qu'il existe nécessairement un terme correspondant dans une langue étrangère. En général la raison en est que le procédé, la méthode, ou le système particulier auquel les termes se rapportent n'a aucune application dans l'autre pays. Par exemple, beaucoup des couches de houille en France sont fortement inclinées (*dressants*), tourmentées, et morcelées, et beaucoup sont d'une grande puissance. En Angleterre les couches sont pour la plupart horizontales (*plateures*), régulières, et non pas très épaisses. Par conséquent, il se trouve nombre d'expressions techniques françaises se rapportant aux classes de mines mentionnées en premier lieu qui n'ont pas d'utilité dans la pratique anglaise, et réciproquement l'on peut dire de même de beaucoup d'expressions anglaises. De plus, le système de sondage à la corde est employé dans les pays où les couches sont horizontales et faciles à exploiter, mais ces conditions peuvent être rares dans d'autres pays, ou ne pas exister du tout, et par conséquent beaucoup d'expressions dont on se sert dans ce système peuvent manquer. De même, dans un pays où les montagnes et les glaciers abondent, la langue est plus riche en fait d'expressions techniques se rapportant à ces sujets que dans un pays où ils ne sont pas très fréquents ou n'existent pas, d'où il s'en suit que l'intérêt qui y est attaché est plus ou moins accidentel.

TELEGRAPHIC CODE

Any entry in this dictionary can be coded in one cipher word of ten letters by the simple method explained below. These words can either be used alone to form a complete message, or along with words from any other code, the code words emanating from the dictionary being identifiable by the prefixal indicator. Inasmuch as the dictionary is for the most part technical, it will be found to be of telegraphic service chiefly as a code supplemental to a general code.

Since each entry in the dictionary consists of both English and French, it will be seen that the dictionary can be used as—

(a) An English-French code, by taking the entry as it stands, or

(b) An English code, by disregarding the French translation, or

(c) A French code, by disregarding the English translation.

In other words, it can be used by two persons, one speaking English and the other French ; or by two English-speaking persons ; or by two French-speaking persons.

The procedure is as follows :—

(1) Begin every code word which is to represent a dictionary entry with the 3-letter indicator given in **Key 1** on page xiv.

(2) Code the **number of the page** in the dictionary containing the entry to be telegraphed into 3 letters from **Keys 2a and 2b.**

(3) Code the **number of the entry** to be telegraphed, given in brackets against the entry in the dictionary, into 2 letters from **Key 3.**

> NOTE.—If an entry is contained partly on one page and partly on the next, the number of the page containing the last part of the entry, i.e., that on which the entry number appears, must be used, not the number of the page containing the commencement of the entry.

(4) Add together individually the figures of the page number and entry number. The total forms the **check figure.** Code this figure into 2 letters from **Key 4.**

The code word is then complete.

Following are specimen codings of dictionary entries :—

EXAMPLE 1

It is desired to code the words :

swing-jib radial drill, machine à percer radiale à potence

These words are contained on page 1050, entry number 61.

Begin the code word with the indicator, which, according to Key 1, is .. .. **KET**

The thousand and hundred figures of the page No. are 10, which, according to Key 2a, are represented by .. .. **T**

The ten and digit figures of the page No. are 50, which, according to Key 2b, are represented by .. .. **IN**

The entry number is 61, which, according to Key 3, is represented by .. .. .. **OB**

The check figure, viz., the total of these figures added together individually, is 13 (viz., $1+0+5+0+6+1$), which, according to Key 4, is represented by .. .. **AR**

Code word :—**KETTINOBAR**

EXAMPLE 2

Entry to be coded :

arbre porte-foret en acier tournant dans des coussinets en bronze à rattrapage de jeu, steel drilling-spindle running in gun-metal bearings with adjustment for taking up wear

Commencing on page 30, and ending on page 31. 31 is used as page number. The entry number is 1.

Indicator	**KET**
Thousand and /or hundred figure(s) of page No. — none	**B**
Ten and digit figures of page No. — 31	**EP**
Entry No. — 1	**AB**
Check figure — 5 (viz., 3+1+1)	**AG**

Code word :—**KETBEPABAG**

EXAMPLE 3

This is an example of the use of another code in conjunction with the dictionary, that is to say, the ordinary language in the example, such as " ship by quickest route," " 8 inches," etc., is supposed to be taken from a general code. The technical language, such as " all-gear single-pulley lathe," " height of centres," etc., is taken from the dictionary.

The codings from the dictionary begin with the indicator **KET**.

The code words for the parts of the message marked with an asterisk would be taken from whatever general code is used.

Code Word	*Translation*
*	ship by quickest route
KETHITOXOL	all-gear single-pulley lathe, for use with high-speed cutting-steels, tour pour l'emploi des aciers à coupe rapide, avec commande par monopoulie et boîte de vitesses
KETLUJEWON	height of centres, hauteur des pointes
*	8 inches
KETLOBALOC	gap-bed, banc rompu
KETNELIHOK	length, longueur
*	10 feet
KETNICANAT	loose head-stock which can be set over for taper turning, contre-pointe pouvant s'excentrer pour tourner conique
KETLIROCOC	four-tool turret with graduated base capable of being swivelled to any angle, porte-outil revolver à quatre faces à diviseur et à orientation quelconque
KETLIRIKOH	four-jaw independent chuck, with reversible jaws, plateau à quatre griffes indé--pendantes et réversibles
KETJIYAFOD	counter-shaft for either floor or ceiling, renvoi de mouvement se fixant indifférem--ment au sol ou au plafond
KETNIFEFOB	lubricating-pump and pipe-connections, pompe de lubrification et sa tuyauterie
KETTIXERAY	taper-turning attachment, appareil à tourner conique

CODE TÉLÉGRAPHIQUE

Tout article figurant dans ce dictionnaire peut être désigné en un seul mot conventionnel de dix lettres par la simple méthode expliquée ci-après. Ces mots peuvent s'employer seuls pour former une dépêche complète, ou bien conjointement avec des mots de tout autre code, les mots codiques provenant du dictionnaire se reconnaissant à l'indicateur préfixe. Le dictionnaire étant pour la majeure partie d'ordre technique, son emploi pour le service télégraphique est surtout indiqué comme code supplémentaire d'un code général.

Comme chaque article du dictionnaire est exprimé en français et en anglais, on s'apercevra que le dictionnaire peut servir à la fois—

a) de code français-anglais, en prenant l'article tel qu'il est énoncé, ou

b) de code français, en passant sur la traduction anglaise, ou bien encore

c) de code anglais, en passant sur la traduction française.

En d'autres termes, il peut être employé par deux personnes, l'une parlant français et l'autre anglais ; ou par deux personnes ne parlant que français ; ou enfin par deux correspondants ne parlant qu'anglais.

Marche à suivre :—

(1) On commencera chaque mot devant représenter un article du dictionnaire par l'**indicateur** à 3 lettres désigné à la **Clef 1**, qui se trouve à la page xiv.

(2) On désignera en trois lettres, suivant les **Clefs 2a et 2b**, le **numéro de la page** du dictionnaire où figure l'article à télégraphier.

(3) On désignera en deux lettres, suivant la **Clef 3**, le **numéro de l'article** à télégraphier. Ce numéro est indiqué entre parenthèses, en regard de l'article.

> **Note.—** Si l'article se trouve en partie sur une page et en partie sur la suivante, on devra employer le numéro de la page où se trouvent la dernière partie de l'article et le numéro d'ordre de celui-ci, et non pas le numéro de la page où se trouve le commencement de l'article.

(4) Additionner les chiffres pris individuellement du numéro de la page et de celui de l'article. Le total forme le **chiffre de contrôle**. Désigner ce chiffre en deux lettres suivant la **Clef 4**.

Le mot codique est alors complet.

Voici quelques exemples de formation de mots codiques, selon les articles du dictionnaire :—

EXEMPLE 1

On désire rendre par un mot l'expression :

> **swing-jib radial drill,** machine à percer radiale à potence

Ces mots se trouvent page 1050, article 61.

Commencer le mot codique par l'indicateur, qui, suivant la Clef 1, est ..   ..   ..   **KET**

Les chiffres des mille et centaines du numéro de la page sont 10, qui, suivant la Clef 2a, sont exprimés par ..   ..   **T**

Les chiffres des dizaines et unités du numéro de la page sont 50, qui, suivant la Clef 2b, sont exprimés par   ..   ..   **IN**

Le numéro de l'article est 61, qui, suivant la Clef 3, est exprimé par   ..   ..   ..   **OB**

Le chiffre de contrôle, c.-à-d. le total de ces chiffres additionnés individuellement, est 13 (c.-à-d. $1+0+5+0+6+1$), qui, suivant la Clef 4, est exprimé par   ..   ..   **AR**

Mot codique :—**KETTINOBAR**

EXEMPLE 2

Article à rendre en un mot codique :

arbre porte-foret en acier tournant dans des coussinets en bronze à rattrapage de jeu, steel
drilling-spindle running in gun-metal bearings with adjustment for taking up wear

Commençant page 30, et finissant page 31. Ce dernier numéro servira de numéro de page.
Le numéro de l'article est 1.

Indicateur	**KET**
Mille et/ou centaines du n° de la page — nul	**B**
Dizaines et unités du n° de la page — 31	**EP**
N° de l'article — 1	**AB**
Chiffre de contrôle — 5 (c.-à-d. 3 + 1 + 1)	**AG**

Mot codique :—**KETBEPABAG**

EXEMPLE 3

Voici un exemple de l'usage d'un autre code en liaison avec le dictionnaire, c'est-à-dire un
texte ordinaire comme "expédiez par la voie la plus rapide," "200 millimètres," etc., censé être
extrait d'un code général. Les expressions techniques, telles que "tour pour l'emploi des aciers à
coupe rapide," "hauteur des pointes," etc., sont extraites du dictionnaire.

Les formations codiques des articles du dictionnaire commencent par l'indicateur **KET**.

Les mots codiques des parties de la dépêche marquées d'un astérisque seraient prises de
n'importe quel code général en usage.

Mot codique	*Traduction*
*	expédiez par la voie la plus rapide
KETGUXEXON	tour pour l'emploi des aciers à coupe rapide, avec commande par monopoulie et boîte de vitesses, all-gear single-pulley lathe, for use with high-speed cutting-steels
KETDOKEJOF	hauteur des pointes, height of centres
*	200 millimètres
KETBINAFAL	banc rompu, gap-bed
KETFADEZAL	longueur, length
*	3 mètres
KETCEFIYAZ	contre-pointe pouvant s'excentrer pour tourner conique, tail-stock which can be set over for taper turning
KETFUPITOD	porte-outil revolver à quatre faces à diviseur et à orientation quelconque, four-tool turret with base graduated in degrees so that tools may be set to any angle
KETFUBILOF	plateau à quatre griffes indépendantes et réversibles, independent four-jaw chuck with reversible jaws
KETGERACAN	renvoi de mouvement se fixant indifféremment au sol ou au plafond, counter-shaft for either floor or ceiling
KETFULOBOH	pompe de lubrification et sa tuyauterie, pump and connections
KETBEKEYOB	appareil à charioter conique, taper-turning attachment

Key 1
Clef 1

Indicator — 1st, 2nd & 3rd letters of code word :—

Indicateur — 1ʳᵉ, 2ᵉ & 3ᵉ lettres du mot codique :—

KET

Key 2a
Clef 2a

Page number — hundreds and thousands — 4th letter of code word :—

Numéro de la page — centaines et mille — 4ᵉ lettre du mot codique : —

0 --	B (To be used when there are no hundreds or thousands in the page number ; i.e., when coding page numbers 1 to 99). (À employer lorsqu'il n'y a pas de centaines ni de mille dans le numéro de la page ; c.-à-d. dans la codification des numéros de pages **1 à 99**).
1 --	C
2 --	D
3 --	F
4 --	G
5 --	H
6 --	J
7 --	L
8 --	N
9 --	P
10 --	T
11 --	V
12 --	X

Key 2b
Clef 2b

Page number — digits and tens — 5th & 6th letters of code word :—

Numéro de la page — unités et dizaines — 5ᵉ & 6ᵉ lettres du mot codique :—

- 00	AB	- 20	EB	- 40	IB	- 60	OB	- 80	UB
- 01	AC	- 21	EC	- 41	IC	- 61	OC	- 81	UC
- 02	AD	- 22	ED	- 42	ID	- 62	OD	- 82	UD
- 03	AF	- 23	EF	- 43	IF	- 63	OF	- 83	UF
- 04	AG	- 24	EG	- 44	IG	- 64	OG	- 84	UG
- 05	AH	- 25	EH	- 45	IH	- 65	OH	- 85	UH
- 06	AJ	- 26	EJ	- 46	IJ	- 66	OJ	- 86	UJ
- 07	AK	- 27	EK	- 47	IK	- 67	OK	- 87	UK
- 08	AL	- 28	EL	- 48	IL	- 68	OL	- 88	UL
- 09	AM	- 29	EM	- 49	IM	- 69	OM	- 89	UM
- 10	AN	- 30	EN	- 50	IN	- 70	ON	- 90	UN
- 11	AP	- 31	EP	- 51	IP	- 71	OP	- 91	UP
- 12	AR	- 32	ER	- 52	IR	- 72	OR	- 92	UR
- 13	AS	- 33	ES	- 53	IS	- 73	OS	- 93	US
- 14	AT	- 34	ET	- 54	IT	- 74	OT	- 94	UT
- 15	AV	- 35	EV	- 55	IV	- 75	OV	- 95	UV
- 16	AW	- 36	EW	- 56	IW	- 76	OW	- 96	UW
- 17	AX	- 37	EX	- 57	IX	- 77	OX	- 97	UX
- 18	AY	- 38	EY	- 58	IY	- 78	OY	- 98	UY
- 19	AZ	- 39	EZ	- 59	IZ	- 79	OZ	- 99	UZ

Key 3
Clef 3

Entry number — 7th & 8th letters of code word :—
Numéro de l'article — 7e & 8e lettres du mot codique :—

1	AB	21	EB	41	IB	61	OB	81	UB	101	BA
2	AC	22	EC	42	IC	62	OC	82	UC	102	BE
3	AD	23	ED	43	ID	63	OD	83	UD	103	BI
4	AF	24	EF	44	IF	64	OF	84	UF	104	BO
5	AG	25	EG	45	IG	65	OG	85	UG	105	BU
6	AH	26	EH	46	IH	66	OH	86	UH	106	CA
7	AJ	27	EJ	47	IJ	67	OJ	87	UJ	107	CE
8	AK	28	EK	48	IK	68	OK	88	UK	108	CI
9	AL	29	EL	49	IL	69	OL	89	UL	109	CO
10	AM	30	EM	50	IM	70	OM	90	UM	110	CU
11	AN	31	EN	51	IN	71	ON	91	UN	111	DA
12	AP	32	EP	52	IP	72	OP	92	UP	112	DE
13	AR	33	ER	53	IR	73	OR	93	UR	113	DI
14	AS	34	ES	54	IS	74	OS	94	US	114	DO
15	AT	35	ET	55	IT	75	OT	95	UT	115	DU
16	AV	36	EV	56	IV	76	OV	96	UV	116	FA
17	AW	37	EW	57	IW	77	OW	97	UW	117	FE
18	AX	38	EX	58	IX	78	OX	98	UX	118	FI
19	AY	39	EY	59	IY	79	OY	99	UY	119	FO
20	AZ	40	EZ	60	IZ	80	OZ	100	UZ	120	FU

Key 4
Clef 4

Check figure -- 9th & 10th letters of code word :—
Chiffre de contrôle — 9e & 10e lettres du mot codique :—

1	AB	16	AV	31	ON
2	AC	17	AW	32	OP
3	AD	18	AX	33	OR
4	AF	19	AY	34	OS
5	AG	20	AZ	35	OT
6	AH	21	OB	36	OV
7	AJ	22	OC	37	OW
8	AK	23	OD	38	OX
9	AL	24	OF	39	OY
10	AM	25	OG	40	OZ
11	AN	26	OH	41	UB
12	AP	27	OJ	42	UC
13	AR	28	OK	43	UD
14	AS	29	OL	44	UF
15	AT	30	OM	45	UG

ABBREVIATIONS
USED IN THIS DICTIONARY

adj.	*adjective.*
adv.	*adverb.*
Arch.	Architecture.
Archeol.	Archeology.
Astron.	Astronomy.
Build.	Building.
Carp.	Carpentry.
Cf.	Compare (*Confer*).
Chem.	Chemistry.
Civ. Engin.	Civil Engineering.
Crystall.	Crystallography.
Elec.	Electricity.
Engin.	Engineering.
f.	*feminine.*
f.pl.	*feminine plural.*
Geog.	Geography.
Geol.	Geology.
Geom.	Geometry.
Hydraul.	Hydraulics.
Hydraul. Engin.	Hydraulic Engineering.
Hydrog.	Hydrography.
Hydrol.	Hydrology.
Hydros.	Hydrostatics.
i.e.	that is (*id est*).
invar.	*invariable.*
m.	*masculine.*
m.pl.	*masculine plural.*
Mach.	Machinery ; Machine.
Math.	Mathematics.
Mech.	Mechanics.
Metall.	Metallurgy.
Meteor.	Meteorology.
Metrol.	Metrology.
n.	*noun.*
n.f.	*noun feminine.*
n.f.pl.	*noun feminine plural.*
n.m.	*noun masculine.*
n.m.pl.	*noun masculine plural.*
Oceanog.	Oceanography.
opp. to	opposed to.
Opt.	Optics.
p.p.	*past participle.*
pers.	person.
Persp.	Perspective.
Petrol.	Petrology (*science of rocks*).
Photog.	Photography.
Photom.	Photometry.
Phys.	Physics.
Phys. Geog.	Physical Geography.
pl.	*plural.*
prep.	*preposition.*
q.v.	which see (*quod vide*).
Rly.	Railways.
sing.	*singular.*
Steam-Engin.	Steam-Engineering.
Surv.	Surveying.
Teleg.	Telegraphy.
Teleph.	Telephony.
Thermochem.	Thermochemistry.
Topog.	Topography.
U.S.A.	United States of America.
v.	*verb.*
v.i.	*verb intransitive* (or *neuter*).
v.r.	*verb reflexive.*
v.t.	*verb transitive* (or *active*).
viz.	namely (*videlicet*).

The sign ' before an initial h denotes that the h is aspirate.

ABRÉVIATIONS
EMPLOYÉES DANS CE DICTIONNAIRE

adj.	*adjectif.*
adv.	*adverbe.*
Arch.	Architecture.
Archéol.	Archéologie.
Arpent.	Arpentage.
Astron.	Astronomie.
c.-à-d.	c'est-à-dire.
Cf.	Conférer.
Ch. de f.	Chemins de fer.
Charp.	Charpenterie.
Chim.	Chimie.
Constr.	Construction.
Cristall.	Cristallographie.
Élec.	Électricité.
f.	*féminin.*
f.pl.	*féminin pluriel.*
Géod.	Géodésie.
Géogr.	Géographie.
Géogr. phys.	Géographie physique.
Géol.	Géologie.
Géom.	Géométrie.
Hydraul.	Hydraulique.
Hydrogr.	Hydrographie.
Hydrol.	Hydrologie.
Hydros.	Hydrostatique.
invar.	*invariable.*
m.	*masculin.*
m.pl.	*masculin pluriel.*
Mach.	Machines ; Machine.
Maçonn.	Maçonnerie.
Math.	Mathématiques.
Méc.	Mécanique.
Menuis.	Menuiserie.
Météor.	Météorologie.
n.	*nom.*
n.f.	*nom féminin.*
n.f.pl.	*nom féminin pluriel.*
n.m.	*nom masculin.*
n.m.pl.	*nom masculin pluriel.*
Océanogr.	Océanographie.
opp. à	par opposition à.
Opt.	Optique.
p.p.	*participe passé.*
pers.	personne.
Persp.	Perspective.
Pétrol.	Pétrologie (*science des roches*).
Photogr.	Photographie.
Photom.	Photométrie.
Phys.	Physique.
pl.	*pluriel.*
prép.	*préposition.*
Serrur.	Serrurerie.
sing.	*singulier.*
Syn. de	Synonyme de (*Same as*).
Télégr.	Télégraphie.
Téléph.	Téléphonie.
Thermochim.	Thermochimie.
Topogr.	Topographie.
Trav. publ.	Travaux publics.
V. ou *V.*	Voir (*See*).
v.	*verbe.*
V. ci-après	Voir ci-après (*See below*).
V. ci-avant	Voir ci-avant (*See above*).
v.i.	*verbe intransitif* (ou *neutre*).
v.r.	*verbe réfléchi* (ou *pronominal*).
v.t.	*verbe transitif* (ou *actif*).

Le signe ' placé devant un h initial indique que l'h est aspiré.

FRENCH - ENGLISH

A

à bras, à main, à droite, à gauche, à raison de, etc.
See **bras** (à), **main** (à), **droite** (à), **gauche** (à),
raison de (à), etc.

à-coup (*n.m.*), jerk ; shock : (1)
éviter les à-coups aux **démarrages,** to avoid
jerks on starting. (2)
batterie d'accumulateurs employée comme
régulateur, qui absorbe tous les à-coups (*f.*),
battery of accumulators used as a regulator,
which absorbs all the shocks. (3)

aa (lave scoriacée) (contrasté avec **pahoéhoé**)
(Géol.) (*n.m.*), a-a ; aa. (4) (contrasted with
pahoehoe).

abaca (*n.m.*), abaca ; Manila hemp. (5)

abaissement (*n.m.*), lowering ; letting down ;
fall ; falling : (6)
abaissement de l'élinde, lowering, letting
down, the dredging-ladder. (7)
abaissement de la température, du baromètre,
du mercure dans le thermomètre, fall of the
temperature, of the barometer, of the
mercury in the thermometer. (8 ou 9 ou 10)
abaissement de potentiel (Élec.), fall of poten-
-tial ; drop of potential ; pressure-drop ;
pressure-loss. (11)
abaissement des teneurs (se dit du minerai, etc.),
falling off in grade. (12)

abaisser (rendre plus bas) (*v.t.*), to lower : (13)
abaisser la température de détonation, to
lower the temperature of detonation. (14)

abaisser (faire aller en bas) (*v.t.*), to lower ; to
let down ; to let fall ; to drop : (15)
abaisser la coulisse (distribution par coulisse),
to lower the link (link-gear). (16)
abaisser un store, to lower, to let down, a
blind. (17)
abaisser une perpendiculaire (opp. à *élever une
perpendiculaire*) (Géom.), to drop a perpen-
-dicular. (18) (opp. to *to erect a perpen-
-dicular*).

abaisser (diminuer, réduire) (*v.t.*), to lessen ;
to diminish ; to reduce : (19)
abaisser le coût de la production, to reduce the
cost of production. (20)

abaisser (s') (*v.r.*), to be lowered ; to go down ;
to decline ; to fall : (21)
plus la fonte est carburée, plus son point de
fusion s'abaisse, the more cast iron is car-
-burized, the more its melting-point goes
down. (22)

abandon *ou* **abandonnement** (*n.m.*), abandon-
-ment : (23)
abandon de débris par les glaces flottantes,
abandonment of debris by floe-ice. (24)

abandon de massifs (méthode d'exploitation de
mines) (*m.*), leaving pillars as permanent
supports of the roof ; chambers-and-per-
-manent-pillars method ; incomplete
removal of the mineral. (25)

abandonner (*v.t.*). See examples : (26)
abandonner la garde d'un moteur, to leave an
engine unattended. (27)
abandonner les galets (Placers & Mines), to
run the boulders to waste. (28)
abandonner les massifs pour soutenir le toit
(Mines), to leave pillars to support the roof.
(29)
abandonner son poste, to leave, to quit, one's
post. (30)
abandonner un procédé, to abandon, to discard,
a process. (31)
abandonner une concession, to abandon, to
throw up, a claim. (32)
chantiers abandonnés (Mines) (*m.pl.*), aban-
-doned workings. (33)
carrière abandonnée (*f.*), disused quarry. (34)
une ville minière abandonnée, a deserted
mining town. (35)

abaque (auge pour laver le minerai d'or) (*n.m.*),
abacus major. (36)

abaque (graphique) (Math.) (*n.m.*), diagram :
(37)
abaque pour le calcul des ressorts à boudin,
diagram for the calculation of spiral springs.
(38)

abat-jour [abat-jour *pl.*] (pour lampes) (*n.m.*),
shade ; lamp-shade. (39)

abat-jour (pour châssis à glace dépolie) (Photogr.)
(*n.m.*), hood (for focussing-screen). (40)

abatage (Mines) (*n.m.*). See examples : (41)
abatage à l'eau, hydraulic mining ; hydrau-
-licking ; piping ; spatter-work. (42)
abatage à la main, hand-stoping. (43)
abatage à la pioche et à la pelle, pick-and-
shovel work ; hand-mining. (44)
abatage au feu, fire-setting. (45)
abatage au moyen de coins, wedging down.
(46)
abatage au moyen de pinces, barring down.
(47)
abatage au pic, breaking ground with the pick.
(48)
abatage combiné (montant et descendant),
combined stoping. (49)
abatage de front, breast-stoping ; breast-work.
(50)
abatage descendant, underhand stoping ;
bottom-stoping. (51)

1

abatage du charbon, breaking down coal; getting down coal; coal-mining. (1)

abatage du minerai, breaking down the ore; stoping the ore; cutting out the ore; cutting-out stoping. (2)

abatage en carrières, quarrying. (3)

abatage en filon, lode-mining. (4)

abatage en gradins, stoping (stoping proper,— i.e., excavating the ore stepwise). (5)

abatage en gradins droits, underhand stoping; bottom-stoping. (6)

abatage en gradins renversés, overhand stop-ing; overhead stoping; overstoping; back-stoping. (7)

abatage en remontage, raise-stoping; working towards the rise. (8)

abatage en taille chassante, drift-stoping. (9)

abatage en traçage, development stoping. (10)

abatage hydraulique, hydraulic mining; hydraulicking; piping; spatter-work. (11)

abatage latéral, side stoping. (12)

abatage mécanique, machine-stoping; stop-ing by machinery; machine-mining. (13)

abatage mécanique du charbon, machine coal-mining. (14)

abatage montant, overhand stoping; over-stoping; overhead stoping; back-stoping. (15)

abatage par gradins latéraux, side stoping. (16)

abatage par les explosifs, breaking ground by explosives; mining by blasting; shooting off the solid. (17)

abatage (proprement dit) (opp. à traçage) (Exploi-tation des couches de houille) (n.m.), work-ing of the broken; working in the broken; broken working; second working; pillar-working; drawing back; cutting away the pillars; removing pillars; robbing. (18)

abatage suivant immédiatement le traçage, working of the broken immediately following the whole; following up the whole with the broken; removing the pillars immediately after the first working. (19)

abatage d'arbres (m.), felling, hewing down, cutting down, trees. (20)

abattant (n.m.), flap. (21)

abattre (Mines) (v.t.). See examples: (22)

abattre le charbon par havage, to mine coal by undercutting. (23)

abattre le minerai, to break down the ore; to stope the ore; to stope; to cut out the ore. (24)

abattre par la méthode hydraulique, to hydraulic: (25)

abattre l'alluvion par la méthode hydraulique, to hydraulic the alluvial. (26)

abattre la poussière, to lay the dust. (27)

abattre un arbre, to fell, to hew down, to cut down, a tree: (28)

abattre un arbre à coups de hache, to fell a tree with an axe. (29)

abattre un wagon afin de le visiter et le réparer, to tip up a wagon in order to examine and repair it: to turn a wagon on its side for examination and repair. (30)

abbcite (Explosif) (n.f.), abbcite. (31)

abée (d'un moulin) (n.f.), flume, leat, (of a mill). (32)

aberration (Opt.) (n.f.), aberration. (33)

aberration chromatique ou aberration de réfrangi-bilité (f.), chromatic aberration; colour-aberration: (34)

l'aberration de réfrangibilité se corrige au moyen de lentilles achromatiques, chro-matic aberration is corrected by means of achromatic lenses. (35)

aberration de courbure du champ (f.), aberration of curvature of field. (36)

aberration de sphéricité (f.), spherical aberration. (37)

aberration latérale (f.), lateral aberration. (38)

aberration longitudinale (f.), longitudinal aberration. (39)

abich (n.m.), a small mortar, used in mineralogical laboratories. (40)

abîme (Géol.) (n.m.), abyss; gulf; chasm. (41)

ablation (Géol.) (n.f.), ablation. (42)

ablation de terre ferme (f.), land ablation (erosion by the sea). (43)

ablation glaciaire (f.), glacial ablation. (44)

abondamment (adv.), abundantly; plentifully; amply; copiously: (45)

être abondamment pourvu (-e) d'eau, to be plentifully supplied with water. (46)

abondance (n.f.), abundance; plentifulness; plenty: (47)

abondance de bons ouvriers, plenty of good workmen. (48)

abondant, -e (adj.), abundant; plentiful; ample; copious: (49)

une provision abondante de bois, an abundant supply of wood; a plentiful supply of timber. (50)

pluie abondante (f.), copious rain. (51)

abonder (v.i.), to abound; to be well supplied (with); to be plentiful; to be abundant: (52)

la staurotide abonde dans certains schistes, cristallins ou argileux, staurolite abounds in certain crystalline or argillaceous schists. (53)

pays qui abonde en forêts (m.), country which abounds with forests. (54)

abord (approche; accès) (n.m.), approach; access: (55)

les abords d'une île, the approaches to an island. (56)

abordable (adj.), approachable; accessible. (57)

abornement (n.m.), pegging; pegging off; pegging out; staking; staking off; staking out. (58)

aborner (v.t.), to peg; to peg off; to peg out; to stake; to stake off; to stake out: (59)

aborner une concession, to peg out, to stake off, a claim. (60)

aboucher des tuyaux, to join up pipes. (61)

about (Charp., etc.) (n.m.), butt; butt-end. (62)

about (en), abutting: (63)

assemblage en about (m.), abutting joint. (64)

aboutement (n.m.), abutment; abutting; butting; butt-jointing. (65)

abouter (v.t.), to abut; to butt; to butt-joint: (66)

abouter deux pièces de bois, to abut, to butt-joint, two pieces of wood. (67)

abouter (s') (v.r.), to abut; to butt. (68)

aboutir à une poche richement minéralisée, to open out into a rich pocket of ore. (69)

abrasif, -ive (adj.), abrasive: (70)

quartz abrasif (m.), abrasive quartz. (71)

abrasif (n.m.), abrasive; abradant. (72)

abrasion (*n.f.*), abrasion. (1)

abrasion marine (*f.*), marine abrasion. (2)

abreuver un tonneau, to season a cask (to fill it with water to swell the wood). (3)

abreuver une pompe, to prime a pump. (4)

abri (*n.m.*), shelter ; refuge ; refuge-hole ; manhole ; place of safety : (5)

se retirer à l'abri, to retire to a place of safety. (6)

abri (d'une locomotive) (*n.m.*), cab (of a loco-motive). (7)

abriter (*v.t.*), to shelter ; to protect : (8)

vallée abritée contre les vents du nord par un rempart de montagnes (*f.*), valley sheltered from the north winds by a barrier of moun-tains. (9)

abriter les machines contre les intempéries, to protect the machinery from the weather. (10)

abriter (s') (*v.r.*), to take shelter : (11)

s'abriter sous un arbre pendant un orage, to take shelter under a tree during a storm. (12)

abrupt, -e (*adj.*), abrupt ; precipitous ; sheer. (13)

abruptement (*adv.*), abruptly ; precipitously ; sheer. (14)

abscisse (Math.) (*n.f.*), abscissa : (15)

abscisse d'un point, abscissa of a point. (16)

absence (manque) (*n.f.*), absence : lack : (17)

absence de bonnes routes, absence, lack, of good roads. (18)

absolu, -e (Phys., Chim., Méc.) (*adj.*), absolute : (19)

température absolue (*f.*), absolute tempera-ture. (20)

alcool absolu (*m.*), absolute alcohol. (21)

vitesse absolue (*f.*), absolute velocity. (22)

absorbable (*adj.*), absorbable. (23)

absorbant, -e (*adj.*), absorbent ; absorbing. (24)

absorbant (*n.m.*), absorbent. (25)

absorbant (base pour dynamite) (*n.m.*), dope ; absorbent. (26)

absorbant actif (*m.*), active dope. (27)

absorbant inerte (*m.*), inert dope. (28)

absorber (*v.t.*), to absorb : (29)

une éponge absorbe l'eau, a sponge absorbs water. (30)

absorber un à-coup, to absorb a shock. (31)

absorptif, -ive (*adj.*), absorptive. (32)

absorptiomètre (*n.m.*), absorptiometer. (33)

absorption (*n.f.*), absorption : (34)

absorption d'oxygène par la houille, des gaz par les substances poreuses, absorption of oxygen by coal, of gases by porous substances. (35 *ou* 36)

absorptivité (*n.f.*), absorptivity ; absorptiveness. (37)

abstrich (Métall.) (*n.m.*), abstrich. (38)

abyssal, -e, -aux (Géol.) (*adj.*), abyssal ; plutonic : (39)

roches abyssales (*f.pl.*), abyssal (*or* plutonic) rocks. (40)

abyssal, -e, -aux *ou* **abyssique** (Océanogr.) (*adj.*), abyssal ; abyssic : (41)

dans la région abyssale, la faune varie fort peu, in the abyssal region, the fauna varies very little. (42)

des profondeurs abyssales (*f.pl.*), abyssal depths. (43)

abysse (Océanogr.) (*n.m.*), abyss. (44)

abzug (Métall.) (*n.m.*), abzug. (45)

acacia (*n.m.*), acacia. (46)

acajou (*n.m.*), mahogany. (47)

acanthikon (Minéral.) (*n.m.*), akanticone. (48)

acanthite (Minéral.) (*n.f.*), acanthite. (49)

accablant, -e (*adj.*), overwhelming ; overpower-ing ; oppressive : (50)

chaleur accablante (*f.*), overpowering heat. (51)

accélérateur (opp. à **retardateur** ou **modérateur**) (Photogr.) (*n.m.*), accelerator. (52)

accélération (*n.f.*), acceleration ; speeding up : (53)

accélération de la pesanteur (Phys.), accelera-tion of gravity ; constant of gravitation ; gravitation constant ; intensity of gravity ; gravity. (54)

accélération de vitesse, acceleration of velocity; acceleration of speed ; increase of speed. (55)

accélération de vitesse d'un corps tombant, acceleration of velocity of a falling body. (56)

accélération (Méc.) (*n.f.*), acceleration. (57)

accélération angulaire (Méc.) (*f.*), angular acceleration. (58)

accélération négative (Méc.) (*f.*), negative acceleration ; minus acceleration ; retarded acceleration ; retardation. (59)

accélération positive (Méc.) (*f.*), positive accelera-tion ; acceleration. (60)

accélération uniforme (Méc.) (*f.*), uniform accel-eration. (61)

accélérer (*v.t.*), See examples : (62)

accélérer la vitesse d'une poulie, to accelerate, to increase, the speed of a pulley. (63)

accélérer les progrès, to accelerate progress. (64)

accélérer les travaux, to speed up, to push on with, the work. (65)

accepter (*v.t.*), to accept : (66)

accepter une offre pour l'outillage, livraison des marchandises, to accept an offer for the plant, delivery of the goods. (67 *ou* 68)

accès (*n.m.*), access ; approach ; entrance ; entry : (69)

accès facile aux chantiers (Mines), ready access to the stopes. (70)

accessibilité (*n.f.*), accessibility ; approach-ability. (71)

accessible (*adj.*), accessible ; approachable. (72)

accessoire (*adj.*), accessory. (73)

accessoires (*n.m.pl.*), accessories ; fittings ; appurtenances. (74)

accessoires de chaudières *ou* **accessoires pour générateurs** (*m.pl.*), boiler-fittings ; boiler accessories ; boiler appurtenances. (75)

accessoires de la voie (Ch. de f.) (*m.pl.*), track accessories ; accessories to track. (76)

accessoires de machines hydrauliques (*m.pl.*), hydraulic accessories ; hydraulic fittings. (77)

accessoires pour perforatrices (*m.pl.*), rock-drill accessories. (78)

accident (événement fortuit, ordinairement fâcheux et inattendu) (*n.m.*), accident ; mishap : (79)

accident au moteur d'extraction, accident, mishap, to the winding-engine. (80)

accident mortel *ou* accident suivi de mort, fatal accident ; fatality ; casualty. (81)

accident non-suivi de mort, non-fatal accident. (82)

accidents de mines, mining accidents. (83)

accidents de personnes, personal accidents. (1)

accidents de translation (Mines), accidents whilst ascending or descending by machinery. (2)

accidents du travail, accidents to workmen. (3)

accident (Géogr. phys.) (*n.m.*), accident; irregularity; unevenness: (4)

les accidents du sol, the accidents, the uneven--ness, of the ground. (5)

il faut parcourir le glacier pour voir ses beaux accidents, one must travel over the glacier to see its delightful irregularities. (6)

accident (rejet; cran) (Géol. & Mines) (*n.m.*), throw; jump; leap; check. (7)

accidenté, -e (*adj.*), undulating; hilly; broken; uneven: (8)

pays accidenté (*m.*), undulating (*or* hilly) (*or* broken) country. (9)

accidentel, -elle (*adj.*), accidental: (10)

trouvaille accidentelle (*f.*), accidental dis--covery. (11)

accidentellement (*adv.*), accidentally. (12)

acclimaté, -e (*adj.*), acclimatized. (13)

acclivité (*n.f.*), acclivity. (14)

accord (Phys.) (*n.m.*), tuning; syntony. (15)

accorder (Phys.) (*v.t.*), to tune; to syntonize: (16)

accorder un récepteur, to tune, to syntonize, a receiver. (17)

accorder (concéder) (*v.t.*), to grant; to allow: (18)

accorder à quelqu'un la permission de visiter la mine, to grant someone permission to visit, to allow someone to see over, the mine. (19)

accorder la théorie avec la pratique, to reconcile theory with practice. (20)

accotement (Ch. de f.) (*n.m.*), side-space; side--long ground. (21)

accouplement (action) (Méc.) (*n.m.*), coupling: (22)

l'accouplement de deux arbres, coupling two shafts. (23)

accouplement (dispositif) (Méc.) (*n.m.*), coupling. (24) See also **manchon.**

accouplement (Élec.) (*n.m.*), connection; con--necting; coupling; grouping: (25)

accouplement des éléments de pile, connecting, connection of, coupling, grouping, battery-cells. (26)

accouplement à débrayage (Méc.) (*m.*), clutch-coupling. (27)

accouplement à fiche (Élec.) (*m.*), plug-connec--tion. (28)

accouplement à manchon (Méc.) (*m*), sleeve-coupling; box-coupling; muff-coupling. (29)

accouplement à plateaux (Méc.) (*m.*), flange-coupling; flanged coupling; plate-coupling; face-plate coupling. (30)

accouplement d'arbres (Méc.) (*m.*), shaft-coupling. (31)

accouplement d'Oldham (Méc.) (*m.*), Oldham coupling. (32)

accouplement direct (à), direct-coupled: (33)

pompe à accouplement direct (*f.*), direct-coupled pump. (34)

accouplement élastique (*m.*), flexible coupling; compensating-coupling. (35)

accouplement en batterie *ou* **accouplement en parallèle** *ou* **accouplement en quantité** *ou* **accouplement en surface** *ou* **accouplement en dérivation** (Élec.) (*m.*), multiple connec--tion; multiple circuit; connecting in multiple; connecting in parallel; connect--ing in quantity; connecting in bridge; coupling in multiple; multiple grouping; abreast connection. (36)

accouplement en série *ou* **accouplement en tension** *ou* **accouplement en cascade** (*m.*), series connection; connecting in series; coupling in series; cascade; cascade con--nection; concatenated connection; con--catenation; tandem connection. (37)

accouplement mixte *ou* **accouplement en séries parallèles** *ou* **accouplement en quantité et en tension** (*m.*), multiple-series connection; parallel-series connection; connecting in multiple series. (38)

accoupler (Méc., etc.) (*v.t.*), to couple; to couple up: (39)

accoupler les roues d'une locomotive, to couple the wheels of a locomotive. (40)

accoupler (Élec.) (*v.t.*), to connect; to couple; to group: (41)

accoupler en quantité, to connect in multiple (*or* in parallel); to bridge. (42) For the numerous synonymous renderings see **accouplement en quantité.**

accoupler (s') (*v.r.*), to be coupled; to be con--nected; to be grouped: (43)

les dynamos peuvent s'accoupler comme des éléments de piles (*f.pl.*), dynamos can be coupled like battery-cells. (44)

accrochage (action) (Mines) (*n.m.*), onsetting; hooking on; hitching. (45)

accrochage (recette) (Mines) (*n.m.*), plat; platt; station; shaft-station; landing; landing-station; landing-stage; onsetting-station; pit-landing; lodge. (46)

accrochage (dans un haut fourneau ou cubilot) (*n.m.*), scaffold; scaffolding; hanging; hang: (47)

les accrochages résultent souvent d'un charge--ment irrégulier, scaffolds often result from irregular charging. (48)

accrochage à l'infini (Photogr.) (*m.*), infinity-catch. (49)

accrochage du fond (Mines) (*m.*), bottom landing; bottom plat; bottom onsetting-station; bottom station. (50)

accrochage du jour (Mines) (*m.*), bank; top landing. (51)

accrochage intermédiaire (Mines) (*m.*), inter--mediate station; intermediate plat. (52)

accroche-coulisse [accroche-coulisses *pl.*] (Son--dage) (*n.m.*), jar-latch; bootjack; boot-latch. (53)

accroche-tube [accroche-tubes *ou* accroche-tube *pl.*] (Sondage) (*n.m.*), casing-grab; pipe-grab. (54)

accrocher (*v.t.*), to hook; to hook on; to hitch; to attach: (55)

accrocher un wagon au câble, to hook, to hitch, a truck to the rope. (56)

accrocher (s') (*v.r.*), to hook; to hook on. (57)

accrocher (s') (hauts fourneaux, etc.) (*v.r.*), to hang. (58) Cf. **accrochage.**

accrocheur (Sondage) (*n.m.*), grab; grab-iron; dog; hoisting-dog. (59)

accrocheur (Mines) (pers.) (*n.m.*), hanger-on ; onsetter ; hooker ; hooker-on ; hitcher ; shackler ; platman. (1)

accrocheur du fond (Mines) (*m.*), bottomer ; bottomman. (2)

accroissement (*n.m.*), increase : (3)
accroissement de pression pouvant amener l'explosion d'une chaudière, increase of pressure which can bring about a boiler explosion. (4)
accroissement de vitesse, increase of speed ; acceleration. (5)
accroissement des réserves, increase in the reserves. (6)

accroître (*v.t.*), to increase : (7)
accroître le rendement d'une mine, to increase the output of a mine. (8)

accroître (*v.i.*), to increase. (9)

accroître (s') (*v.r.*), to increase. (10)

accueillir favorablement une proposition, to entertain a proposal. (11)

accul (*n.m.*), blind alley ; cul-de-sac ; dead-end. (12)

accumulateur (Élec.) (*n.m.*), accumulator ; storage-battery ; secondary battery. (13)

accumulateur (Hydraul., etc.) (*n.m.*), accumu-lator. (14)

accumulateur à caisse de contrepoids (Hydraul.) (*m.*), weight-case accumulator. (15)

accumulateur à contrepoids par anneaux (Hydraul.) (*m.*), weight-ring accumulator. (16)

accumulateur au plomb (Élec.) (*m.*), lead accumu-lator. (17)

accumulateur de vapeur (*m.*), steam-accumu-lator. (18)

accumulateur Edison (*m.*), Edison accumulator ; Edison's storage-battery. (19)

accumulateur hydraulique (*m.*), hydraulic accumulator. (20)

accumulateur, modèle à pression de sens con-traire (Hydraul.) (*m.*), inverted pattern accumulator. (21)

accumulateur, modèle indépendant (Hydraul.) (*m.*), self-contained accumulator. (22)

accumulation (*n.f.*), accumulation ; gathering : (23)
accumulation de grisou dans les vieux chantiers, de gaz dans les remblais, accumulation (*or* gathering) of fire-damp in old workings, of gas in the goaf. (24 *ou* 25)

accumuler (*v.t.*), to accumulate : (26)
accumuler de la force, to accumulate power. (27)

accumuler (s') (*v.r.*), to accumulate. (28)

accuser (*v.t.*), to show ; to indicate ; to mark ; to betray ; to reveal : (29)
résultats qui accusent une amélioration considérable (*m.pl.*), results which show a considerable improvement. (30)
poids trop faibles pour être accusés par la balance la plus sensible (*m.pl.*), weights too light to be indicated by the most delicate balance. (31)
la sursaturation de certains lacs salés est tellement accusée que le corps de ceux qui s'y baignent se recouvre aussitôt de sel, the supersaturation of certain salt lakes is so marked that the bodies of those that bathe in them immediately become covered with salt. (32)
les cendres accusent la présence du fer (*f.pl.*), the ashes betray the presence of iron. (33)

le thermomètre accuse une élévation de tem-pérature, the thermometer reveals a rise of temperature. (34)

acérage (*n.m.*), steeling. (35)

acérain, -e (*adj.*), steely : (36)
fer acérain (*m.*), steely iron. (37)

acerdèse (Minéral) (*n.f.*), acerdese ; manganite ; gray manganese ore. (38)

acéré, -e (*adj.*), steeled : (39)
lame acérée (*f.*), steeled blade. (40)

acérer (*v.t.*), to steel : (41)
acérer un burin, to steel a chisel. (42)

acétate (*n.m.*), acetate. (43)

acétylène (*n.m.*), acetylene. (44)

acétylénique (*adj.*), acetylenic. (45)

achat (*n.m.*), purchase ; buying : (46)
achat de minerai, purchase of ore. (47)
achat illicite d'or [Afrique du Sud], illicit gold-buying ; I.G.B. (48)
achat illicite de diamants [Afrique du Sud], illicit diamond-buying ; I.D.B. (49)

acheter (*v.t.*), to buy ; to purchase : (50)
acheter des approvisionnements pour l'hiver, to buy supplies for the winter. (51)

acheteur, -euse (pers.) (*n.*), buyer ; purchaser. (52)

achèvement (*n.m.*), finishing ; completion : (53)
achèvement d'un chemin de fer, completion of a railway. (54)
achèvement des objets coulés, finishing cast-ings. (55)

achever (*v.t.*), to finish ; to complete. (56)

achrématite (Minéral) (*n.f.*), achrematite. (57)

achroïte (Minéral) (*n.f.*), achroite. (58)

achromatique (*adj.*), achromatic : (59)
lentille achromatique (*f.*) *ou* objectif achro-matique (*m.*), achromatic lens. (60)

achromatisation (*n.f.*), achromatization. (61)

achromatiser (*v.t.*), to achromatize. (62)

achromatisme (*n.m.*), achromatism : (63)
achromatisme d'une lentille, achromatism of a lens. (64)

aciculaire *ou* **aciculé, -e** *ou* **aciculiforme** (*adj.*), acicular ; aciculate ; aciculated ; aciculine ; aciculiform ; aciform ; needle-shaped : (65)
cristal aciculaire (*m.*), acicular crystal. (66)

aciculite (Minéral) (*n.f.*), aciculite. (67)

acide (Chim.) (*adj.*), acid ; acidic : (68)
sel acide (*m.*), acid salt. (69)

acide (Géol.) (*adj.*), acid ; acidic : (70)
roches acides (*f.pl.*), acid (*or* acidic) rocks. (71)

acide (*n.m.*), acid. (72)

acide arsénieux (Minéral) (*m.*), arsenolite ; arsenite. (73)

acide azotique (*m.*), nitric acid. (74)

acide borique (*m.*), boric acid. (75)

acide carbonique (Chim.) (*m.*), carbonic acid ; carbonic-acid gas ; carbon dioxide. (76)

acide carbonique (grisou) (*m.*), choke-damp ; black-damp ; carbonic acid ; mephitic air. (77)

acide chlorhydrique (*m.*), hydrochloric acid ; muriatic acid. (78)

acide chromique (*m.*), chromic acid. (79)

acide fluorhydrique (*m.*), hydrofluoric acid. (80)

acide humique (*m.*), humic acid. (81)

acide minéral (*m.*), mineral acid. (82)

acide molybdique (*m.*), molybdic acid. (83)

acide muriatique (*m.*), muriatic acid ; hydro-chloric acid. (84)

acide nitrique (*m.*), nitric acid. (85)

acide phosphorique (*m.*), phosphoric acid. (1)

acide silicique (*m.*), silicic acid. (2)

acide sulfhydrique (*m.*), sulfhydric acid ; hydrogen sulphide ; sulphureted hydrogen. (3)

acide sulfureux (SO²)(*m.*), sulphur dioxide. (4)

acide sulfurique (*m.*), sulphuric acid ; oil of vitriol ; vitriol ; vitriolic acid. (5)

acide titanique (*m.*), titanic acid. (6)

acide tungstique (*m.*), tungstic acid. (7)

acide vitriolique (*m.*). Same as acide sulfurique.

acidifère (*adj.*), acidiferous. (8)

acidifiable (*adj.*), acidifiable. (9)

acidifiant, -e (*adj.*), acidifiant ; acidific. (10)

acidifiant (*n.m.*), acidifiant ; acidifier. (11)

acidification (*n.f.*), acidification. (12)

acidifier (*v.t.*), to acidify. (13)

acidifier (s') (*v.r.*), to acidify. (14)

acidimètre (*n.m.*), acidimeter ; acidometer. (15)

acidimétrie (*n.f.*), acidimetry. (16)

acidimétrique (*adj.*), acidimetric ; acidimetrical. (17)

acidité (*n.f.*), acidity : (18)
    acidité des tailings, acidity of the tailings. (19)

acidulant, -e (*adj.*), acidulating. (20)

acidule (*adj.*), acidulous ; acidulent. (21)

acidulé, -e (*adj.*), acidulated : (22)
    eau acidulée (*f.*), acidulated water. (23)

aciduler (*v.t.*), to acidulate : (24)
    aciduler une solution, to acidulate a solution. (25)

acier (*n.m.*), steel. (26)

acier à coupe rapide (*m.*), high-speed steel ; high-speed cutting-steel ; high-speed tool-steel. (27)
    aciers à coupe rapide (*m.pl.*), high-speed steels. (28)

acier à haute résistance (*m.*), high-tensile steel. (29)

acier à l'aluminium (*m.*), aluminium steel. (30)

acier à outils (*m.*), tool-steel. (31)

acier à ressort (*m.*), spring-steel. (32)

acier à rivets (*m.*), rivet-steel. (33)

acier au bore (*m.*), boron steel. (34)

acier au carbone (*m.*), carbon steel. (35)

acier au carbone à haute teneur (*m.*), high-carbon steel. (36)

acier au chrome (*m.*), chrome steel ; chromium steel. (37)

acier au chrome-vanadium (*m.*), chrome-vanadium steel. (38)

acier au creuset (*m.*). Same as acier fondu au creuset.

acier au manganèse (*m.*), manganese steel. (39)

acier au molybdène (*m.*), molybdenum steel. (40)

acier au nickel (*m.*), nickel steel. (41)

acier au nickel à haute teneur (*m.*), high-nickel steel. (42)

acier au nickel-chrome (*m.*), nickel-chrome steel ; nickel-chromium steel. (43)

acier au platine (*m.*), platinum steel. (44)

acier au silicium (*m.*), silicon steel. (45)

acier au tungstène (*m.*), tungsten steel ; wolf--ram steel. (46)

acier au vanadium (*m.*), vanadium steel. (47)

acier auto-trempant (*m.*), self-hardening steel ; air-hardening steel. (48)

acier Bessemer (*m.*), Bessemer steel. (49)

acier cémenté (*m.*), cement steel ; cemented steel ; cementation steel ; blister steel. (50)

acier chromé (*m.*), chromium steel ; chrome steel. (51)

acier comprimé (*m.*), compressed steel. (52)

acier corroyé (*m.*), shear-steel. (53)

acier coulé (*m.*), cast steel ; run steel ; steel of molten origin. (54)

acier coulé en coquille (*m.*), chilled steel. (55)

acier d'affinage (*m.*), refining-steel. (56)

acier de carburation (*m.*). Same as acier cémenté.

acier de construction (*m.*), structural steel. (57)

acier de forge (*m.*), steel of smithing quality ; forge-steel ; forging-steel. (58)

acier de haute tension (*m.*), high-tensile steel. (59)

acier de moulage (*m.*), casting-steel. (60)

acier deux fois corroyé (*m.*), double-shear steel. (61)

acier doux (*m.*), mild steel ; low steel ; soft steel. (62)

acier dur (*m.*), hard steel. (63)

acier électrique (*m.*), electric steel. (64)

acier en lingot (*m.*), ingot steel. (65)

acier extra-doux (*m.*), ingot iron. (66)

acier feuillard (*m.*), strip ; strip-steel. (67)

acier fondu au creuset *ou simplement* acier fondu (*m.*), crucible cast steel ; crucible steel ; pot-steel ; cast steel. (68)

acier fondu sur sole *ou* acier Martin (*m.*), open-hearth steel ; Siemens-Martin steel. (69)

acier inoxydable (*m.*), stainless steel ; rustless steel. (70)

acier marchand (*m.*), merchant steel. (71)

acier moulé (*m.*), moulded steel. (72)

acier poule (*m.*), blister steel. (73)

acier pour aéroplanes (*m.*), aeroplane-steel. (74)

acier pour bordages *ou* acier destiné au façonnage de brides (*m.*), flange-steel. (75)

acier pour fleurets (de perforatrices) (*m.*), drill-steel (for rock-drills). (76)

acier pour foyers (*m.*), fire-box steel. (77)

acier pour la construction (*m.*), structural steel. (78)

acier pour mèches (*m.*) *ou* aciers pour mèches (*m.pl.*) *ou* acier pour forets (*m.*), drill-steel ; drill-steels (metal-working drills). (79)

acier pour outils rapides (*m.*), high-speed tool-steel. (80)

acier pour tubes (*m.*), tube-steel. (81)

acier raffiné à deux marques (*m.*), double-shear steel. (82)

acier raffiné à une marque (*m.*), single-shear steel. (83)

acier rapide (*m.*), high-speed steel. (84)

acier sauvage (*m.*), wild steel ; fiery steel. (85)

acier soudant *ou* acier soudé (*m.*), weld steel ; wrought steel. (86)

acier spécial (*m.*), special steel. (87)

acier sur sole (*m.*), open-hearth steel ; Siemens-Martin steel. (88)

acier sur sole acide (*m.*), acid open-hearth steel. (89)

acier sur sole basique (*m.*), basic open-hearth steel. (90)

acier ternaire (*m.*), ternary steel. (91)

acier trempé (*m.*), hardened steel. (92)

acier une fois corroyé (*m.*), single-shear steel. (93)

aciérage (*n.m.*). Same as aciération.

aciération (conversion du fer en acier par cémen--tation) (*n.f.*), acieration ; case-hardening ; steeling : (94)

une dose de carbone déterminée est nécessaire à l'aciération, the addition of a certain proportion of carbon is requisite for aciera--tion. (1)

aciération (soudure de l'acier sur du fer) (n.f.), steeling. (2)

aciération (recouvrement d'acier par galvano--plastie) (n.f.), acierage ; steeling. (3)

aciérer (convertir du fer en acier par cémentation) (v.t.), to acierate ; to case-harden ; to steel : (4)

le carbone acièra le fer, carbon acierates iron. (5)

aciérer le fer par cémentation avec du charbon, to acierate, to case-harden, to steel, iron by cementation with charcoal. (6)

aciérer (souder de l'acier sur du fer) (v.t.), to steel : (7)

aciérer une hache, un burin, to steel an axe, a chisel. (8 ou 9)

aciérer (recouvrir d'acier par galvanoplastie) (v.t.), to steel : (10)

aciérer une plaque de cuivre, to steel a copper plate. (11)

aciéreux, -euse (adj.), steely : (12)

fer aciéreux (m.), steely iron. (13)

aciérie (n.f.), steel-works. (14)

aciforme (adj.).   Same as aciculiforme.

acmite (Minéral) (n.f.), acmite ; ægirite ; ægirine. (15)

acoustique (adj.), acoustic ; acoustical : (16)

téléphone acoustique (m.), acoustic telephone. (17)

acoustique (science) (n.f.), acoustics. (18)

acoustique (tube acoustique) (n.f.), speaking-tube. (19)

acquéreur, -eure ou **-euse** (pers.) (n.), purchaser; buyer. (20)

acquérir (v.t.), to acquire ; to secure ; to pur--chase ; to buy : (21)

acquérir le droit de couper du bois, to acquire the right to cut timber. (22)

acquérir une propriété à un prix raisonnable, to acquire, to secure, a property at a reason--able price. (23)

acquisition (n.f.), acquisition; purchase; buying: (24)

acquisition du droit d'exploiter les gîtes minéraux, acquisition of mineral rights. (25)

acquitter (s') de son devoir, to do one's duty. (26)

acre (mesure agraire anglaise) (n.f.), acre = 0·40468 hectare. (27)

acrobatique (Méc.) (adj.), lifting : (28)

machines acrobatiques (f.pl.), lifting machin--ery. (29)

acte de concession (Mines) (m.), licence ; claim-licence. (30)

acte de concession de mines (m.), mining-licence. (31)

actinique (Phys.) (adj.), actinic ; actinical : (32)

rayons actiniques (m.pl.), actinic rays. (33)

actinisme (n.m.), actinism. (34)

actinium (Chim.) (n.m.), actinium. (35)

actinolithe ou **actinolite** (Minéral) (n.f.), actino--lite ; actinote. (36)

actinolithique (adj.), actinolitic. (37)

actinomètre (n.m.), actinometer. (38)

actinométrie (n.f.), actinometry. (39)

actinométrique (adj.), actinometric ; actino--metrical. (40)

actinoschiste (n.m.), actinolite-schist. (41)

actinote (Minéral) (n.f.), actinolite ; actinote. (42)

actinoteux, -euse ou **actinotique** (adj.), actino--litic. (43)

action (n.f.), action ; effect ; work ; agency : (44)

action chimique, chemical action. (45)

action d'un levier sur une masse, effect of a lever on a mass. (46)

action des agents atmosphériques (Géol.), weathering ; action, work, of atmospheric agents : (47)

roches qui sont exposées à l'action des agents atmosphériques (f.pl.), rocks which are exposed to weathering. (48)

action des eaux courantes, des glaces, de la neige, du vent (Géol.), action, work, agency, of running water, of ice, of snow, of wind. (49 ou 50 ou 51 ou 52)

action oxydante de l'air sur le charbon, oxidizing effect of the air on coal. (53)

actionné, -e (p.p.), driven ; actuated : (54)

actionné (-e) par balancier, beam-driven : (55)

machine actionnée par balancier (f.), beam-driven engine. (56)

actionné (-e) par chaîne, driven by chain ; chain-driven. (57)

actionné (-e) par courroie, belt-driven ; driven by belt. (58)

actionné (-e) par électricité, electrically driven ; driven by electricity ; operated by electricity. (59)

actionné (-e) par force hydraulique, driven by water-power ; hydraulically driven. (60)

actionné (-e) par l'air comprimé, driven by compressed air. (61)

actionné (-e) par la vapeur, steam-driven ; driven by steam. (62)

actionné (-e) par le vent, wind-driven ; driven by wind-power. (63)

actionné (-e) par moteur, engine-driven ; motor-driven ; driven by power ; power-driven. (64)

actionner (v.t.), to drive ; to actuate ; to propel; to run ; to work ; to operate : (65)

actionner les machines par la vapeur, to drive the machinery by steam. (66)

actionner une grue par l'électricité, to operate a crane by electricity ; to work a crane electrically. (67)

actionner une perforatrice par air comprimé, to run, to actuate, a drill by compressed air. (68)

activer (v.t.). See examples : (69)

activer la chauffe, to fire up. (70)

activer la combustion par une machine soufflante, to accelerate, to quicken, to urge, combustion by a blower. (71)

activer la construction d'un chemin de fer, to press forward, to urge forward, to push forward, the construction of a railway. (72)

activer le feu, to stimulate, to urge, to rouse, to arouse, to stir up, the fire. (73)

activer le tirage, to quicken the draught : (74)

l'échappement de vapeur des locomotives sert à activer le tirage (m.), the exhaust from locomotives serves to quicken the draught. (75)

activer les travaux dans la limite du possible, to push on with the work as quickly as possible ; to speed up the work as much as possible. (76)

activité (*n.f.*), activity : (1)
  activité volcanique, volcanic activity. (2)

actuel, -elle (réel) (*adj.*), actual : (3)
  réserves actuelles de minerai (*f.pl.*), actual reserves of ore. (4)

actuellement (*adv.*), actually. (5)

acuité (*n.f.*) *ou* **acutesse** (*n.f.*), sharpness ; acute-ness ; keenness : (6)
  acuité d'une pointe, sharpness, acuteness, keenness, of a point. (7)
  acuité du tranchant d'un outil, sharpness, keenness, of the cutting edge of a tool. (8)

acutangle *ou* **acutangulaire** *ou* **acutangulé, -e** (Géom.) (*adj.*), acute-angled ; acutangular : (9)
  triangle acutangle (*m.*), acute-angled triangle. (10)

adamantin, -e (Minéralogie) (*adj.*), adamantine : (11)
  éclat adamantin (*m.*), adamantine lustre. (12)

adamine (Minéral) (*n.f.*), adamite ; adamine. (13)

adamsite (Minéral) (*n.f.*), adamsite. (14)

adaptable (*adj.*), adaptable. (15)

adaptation (*n.f.*), adaptation ; adaptability ; fitting. (16)

adaptation (Géogr. phys.) (*n.f.*), adjustment. (17)

adapter (*v.t.*), to adapt ; to fit ; to suit : (18)
  adapter un ajutage à l'extrémité d'un tuyau, to fit a nozzle on the end of a pipe. (19)
  procédé adapté (*m.*), suitable process. (20)

adapter (s') (*v.r.*), to fit ; to adapt itself : (21)
  ces télé-objectifs s'adaptent sur tous objectifs (*m.pl.*), these telephoto lenses fit on any lens. (22)

adapter (s') (Géogr. phys.) (*v.r.*), to adjust itself : (23)
  cours d'eau et vallées qui s'adaptent aux conditions nouvelles, streams and valleys which adjust themselves to the new condi-tions. (24)

adapteur (Photogr.) (*n.m.*), adapter. (25)

adapteur de plaques (*m.*) *ou* **adapte-plaques** [adapte-plaques *pl.*] (*n.m.*), plate-adapter. (26)

adapteur pour film-pack (*m.*), film-pack adapter. (27)

addition (Métall.) (*n.f.*), addition ; finishing-metal. (28)

addition de fondant (Métall.) (*f.*), fluxing. (29)

additionner de fondant (Métall.) (*v.t.*), to flux. (30)

adduction d'eau par la gravité (*f.*), conduction of water by gravity. (31)

adelpholite *ou* **adelpholithe** (Minéral) (*n.f.*), adelpholite. (32)

adent (Charp.) (*n.m.*), indent. (33)

adenter (Charp.) (*v.t.*), to indent. (34)

adhérence (*n.f.*), adherence ; adhesion ; adhe-siveness ; sticking : (35)
  adhérence des diamants à la graisse, adherence of diamonds to grease. (36)
  adhérence entre le béton et l'armature, adhe-sion between the concrete and the reinforce-ment. (37)

adhérence (Ch. de f.) (*n.f.*), adhesion : (38)
  adhérence des roues sur la surface des rails, adhesion of the wheels to the surface of the rails. (39)

  chemin de fer à adhérence (*m.*), adhesion-railway. (40)

adhérent, -e (*adj.*), adherent. (41)

adhérer (*v.i.*), to adhere ; to stick : (42)
  deux surfaces qui adhèrent l'une à l'autre, two surfaces which adhere to each other. (43)
  adhérer à la langue (se dit de certains miné-raux, particulièrement les argiles), to adhere, to stick, to the tongue. (44)

adhésif, -ive (*adj.*), adhesive. (45)

adhésion (*n.f.*), adhesion ; adherence ; adhesive-ness ; sticking. (46)

adhésion (Phys.) (*n.f.*), adhesion : (47)
  on nomme adhésion l'attraction moléculaire qui se manifeste entre les surfaces des corps en contact, adhesion is the molecular attraction exerted between the surfaces of bodies in contact. (48)

adhésivement (*adv.*), adhesively. (49)

adiabatique (Phys.) (*adj.*), adiabatic : (50)
  ligne adiabatique (*f.*), adiabatic line. (51)

adiabatique (impénétrable à la chaleur) (*adj.*), non-conducting ; non-conductive. (52)

adiabatique (*n.f.*), adiabatic ; adiabatic line ; adiabat. (53)

adiabatiquement (*adv.*), adiabatically. (54)

adipocérite *ou* **adipocire** (Minéral) (*n.f.*), adipo-cerite ; mineral adipocere. (55)

adjacence (*n.f.*), adjacency ; adjacence ; con-tiguity : (56)
  adjacence de deux angles, adjacence of two angles. (57)

adjacent, -e (*adj.*), adjacent ; adjoining ; con-tiguous : (58)
  rues adjacentes (*f.pl.*), adjoining streets. (59)

adjacent, -e (Géom.) (*adj.*), adjacent : (60)
  angles adjacents supplémentaires (*m.pl.*), supplementary adjacent angles. (61)

adjoint, -e (*adj.*), assistant : (62)
  inspecteur adjoint (*m.*), assistant inspector. (63)

admetteur (d'une machine compound) (*n.m.*), high-pressure cylinder (of a compound engine). (64)

admettre (*v.t.*), to admit ; to admit of : (65)
  admettre la vapeur au cylindre, to admit steam to the cylinder. (66)
  admettre une hypothèse, to admit an hy-pothesis. (67)
  réparation qui n'admet point de retard (*f.*), repair which admits of no delay. (68)

administrateur, -trice (pers.) (*n.*), director. (69)

administrateur délégué *ou* **administrateur directeur** *ou* **administrateur gérant** (*m.*), managing director. (70)

administration (action) (*n.f.*), management ; administration. (71)

administration des mines (*f.*), mining-bureau. (72)

administrer (*v.t.*), to manage ; to direct ; to administer. (73)

admission (action) (*n.f.*), admission ; intake : (74)
  admission de vapeur (au cylindre), admission of steam, steam intake, (to the cylinder). (75)

admission (lumière d'admission de cylindre à vapeur) (*n.f.*), steam-port ; induction-port. (76)

admission anticipée (tiroir) (*f.*), preadmission ; early admission (slide-valve). (77)

adolescence (Géogr. phys.) (*n.f.*), adolescence. (1)

adossé, -e (*adj.*), back-to-back : (2)
tailles adossées (Mines) (*f.pl.*), back-to-back rooms. (3)

adosser un appentis contre un mur, to back a shed against a wall. (4)

adoucir (Métall.) (*v.t.*), to soften : (5)
adoucir la fonte avec de l'oxyde de fer, to soften cast iron with iron oxide. (6)

adoucir une courbe (Ch. de f.), to ease a curve. (7)

adoucir une surface avec de la toile d'émeri, to rub down, to smooth, a surface with emery-cloth. (8)

adoucissage *ou* **adoucissement** (Métall.) (*n.m.*), softening. (9)

adoucissement de l'entrée et de la sortie des courbes (Ch. de f.) (*m.*), easement of the entrance to and departure from curves. (10)

adresser aux autorités une demande pour obtenir une atténuation des règlements, to make an application to the authorities for relaxation of the regulations. (11)

adulaire (Minéral) (*n.f.*), adularia. (12)

adultération (*n.f.*), adulteration ; faking. (13)

adultérer (*v.t.*), to adulterate ; to fake : (14)
adultérer les résultats des analyses, to fake the results of analyses. (15)

ægirine *ou* **ægyrine** (Minéral) (*n.f.*), acmite ; ægirite ; ægyrine ; ægyrite. (16)

ænigmatite (Minéral) (*n.f.*), ænigmatite. (17)

aérable (*adj.*), ventilable. (18)

aérage (*n.m.*), ventilation ; ventilating ; airing ; draught : (19)
aérage des mines, mine-ventilation ; ventilat--ing mines. (20)

aérage artificiel (*m.*), artificial ventilation. (21)

aérage ascensionnel (*m.*), ascensional ventilation. (22)

aérage diagonal (*m.*), diagonal ventilation. (23)

aérage direct (*m.*), direct ventilation. (24)

aérage distinct (*m.*), separate ventilation. (25)

aérage intensif (*m.*), strong ventilation. (26)

aérage mécanique (*m.*), mechanical ventilation. (27)

aérage naturel (*m.*), natural ventilation. (28)

aérage négatif (*m.*), exhaust-draught ; induced draught ; vacuum method of ventilation. (29)

aérage positif (*m.*), forced draught ; pressure-draught ; plenum method of ventilation. (30)

aérage renversé *ou* **aérage rétrograde** (*m.*), reversed ventilation. (31)

aérage sans machines (*m.*), ventilation without machinery ; natural ventilation. (32)

aérant, -e *ou* **aérateur, -trice** (*adj.*), ventilating ; ventilation (*used as adj.*) : (33)
machines aérantes (*f.pl.*), ventilating machinery. (34)
ouverture aérante (*f.*), ventilation-hole. (35)

aération (ventilation) (*n.f.*). Same as **aérage**.

aération (exposition à l'air) (*n.f.*), aeration : (36)
aération de la pulpe, aeration of the pulp. (37)

aéré, -e (ventilé) (*adj.*), ventilated ; aired. (38)

aéré, -e (exposé à l'air) (*adj.*), aerated. (39)

aérer (ventiler) (*v.t.*), to air ; to ventilate : (40)

aérer un appartement, to air a room. (41)
aérer un atelier, to ventilate a workshop. (42)

aérer (exposer à l'air) (*v.t.*), to aerate : (43)
aérer de l'eau, to aerate water. (44)

aérien, -enne (*adj.*), aerial ; overhead : (45)
chemin de fer aérien (*m.*), aerial railway. (46)
réseau électrique aérien (*m.*), overhead electric system. (47)

aériennement (*adv.*), aerially. (48)

aérifère (*adj.*), air (*used as adj.*) : (49)
tube aérifère (*m.*), air-tube. (50)

ærifère (*adj.*), copper-bearing ; copper (*used as adj.*) : (51)
mine ærifère (*f.*), copper-mine. (52)

aérification (Chim. & Phys.) (*n.f.*), aerification ; aerifaction. (53)

aérifier *ou* **aériser** (*v.t.*), to aerify. (54)

aériforme (*adj.*), aeriform ; gaseous. (55)

aérocondenseur (*n.m.*), aerocondenser. (56)

aérolithe (*n.m.*), aerolite ; aerolith ; air-stone ; meteorite. (57)

aérolithique (*adj.*), aerolitic ; meteoritic. (58)

aéromètre (*n.m.*), aerometer. (59)

aérométrie (*n.f.*), aerometry. (60)

aérométrique (*adj.*), aerometric. (61)

aéromoteur (*n.m.*), wind-engine ; windmill. (62)

aéromoteur pour puits (*m.*), windmill for pump--ing ; wind-engine. (63)

aérophore (*n.m.*), aerophore. (64)

aérosite (Minéral) (*n.f.*), aerosite ; pyrargyrite. (65)

æschynite (Minéral) (*n.f.*), eschynite ; æschynite. (66)

aétite (*n.f.*), aetites ; eaglestone. (67)

afeldspathique (*adj.*), feldspar-free : (68)
roches afeldspathiques (*f.pl.*), feldspar-free rocks. (69)

affaiblir (*v.t.*), to weaken : (70)
affaiblir une planche en l'entaillant, to weaken a board by cutting into it. (71)
affaiblir une solution, to weaken a solution. (72)

affaiblir (Photogr.) (*v.t.*), to reduce : (73)
affaiblir un cliché dur, to reduce a hard negative. (74)

affaiblissement (*n.m.*), weakening. (75)

affaiblissement (opp. à **renforcement**) (Photogr.) (*n.m.*), reduction ; reduction in density. (76)

affaiblisseur (opp. à **renforçateur**) (Photogr.) (*n.m.*), reducer. (77)

affaire (entreprise) (*s'emploie aussi au pluriel*) (*n.f.*), business ; piece of business ; concern ; enterprise ; proposition ; affair : (78)
l'arrêt des affaires, the stoppage of business. (79)
affaire de spéculation *ou* affaire spéculative, speculative enterprise. (80)
affaire rémunératrice, paying proposition. (81)
affaire roulante, going concern. (82)
une affaire de filons (Mines), a lode proposition. (83)
une affaire de placers, a placer proposition. (84)

affaissement (*n.m.*), subsidence ; subsiding ; sinking : (85)
affaissement d'un édifice, d'une montagne, subsidence of a building, of a mountain. (86 *ou* 87)

affaissements du sol produits par l'exploita--tion du sel, subsidence of the ground caused by salt-mining. (1)

affaissement (Résistance des matériaux) (*n.m.*), deflection (Strength of materials). (2)

affaissement correspondant à la charge (Résistance des ressorts) (*m.*), deflection, set, for a given load (Strength of springs). (3)

affaisser (*v.t.*), to sink ; to cause to subside : (4)
les grandes pluies affaissent les terres (*f.pl.*), heavy rains cause the earth to subside. (5)

affaisser (s') (*v.r.*), to subside ; to sink : (6)
toit qui s'affaisse peu à peu (*m.*), roof which sinks little by little. (7)

affecter (exercer de l'influence sur) (*v.t.*), to affect : (8)
circonstances qui affectent fâcheusement les opérations (*f.pl.*), circumstances which affect operations adversely. (9)
atmosphère qui affecte la respiration (*f.*), at--mosphere which affects the breathing. (10)
être affecté (-e) par la crise ouvrière, to be affected by the labour crisis ; to suffer from labour troubles. (11)
district non-affecté par des crises ouvrières (*m.*), district free from labour troubles. (12)

affecter (prendre une forme) (*v.t.*), to affect : (13)
le plomb affecte dans sa cristallisation la figure cubique, lead, when crystallizing, affects the cubical form. (14)
minéral qui affecte des nuances variées (*m.*), mineral which affects various shades. (15)

affecter (destiner à un usage) (*v.t.*), to intend ; to design ; to destine ; to allot : (16)
la clef à douille est spécialement affectée aux écrous à six pans, placés dans une telle position que l'on ne peut pas les serrer ou desserrer avec une clef ordinaire, the box-spanner is specially intended for hexagonal nuts, placed in such a position that they cannot be screwed up or unscrewed with an ordinary spanner. (17)
outil spécialement affecté à un certain usage (*m.*), tool specially designed for a certain purpose. (18)
dans une installation électrique de quelque importance, il est bon d'affecter un local spécial aux dynamos, in an electric installa--tion of any importance, it is a good thing to allot a special place to the dynamos. (19)

afficher une copie des règlements à l'entrée de la mine, to post a copy of the regulations at the mine entrance. (20)

affilage *ou* **affilement** (*n.m.*), setting : (21)
l'affilage a pour but d'ôter le morfil, setting has for its object the removal of the wire-edge. (22)

affiler (donner le fil à) (*v.t.*), to set : (23)
affiler un ciseau, to set a chisel. (24)

affiler (passer dans la filière) (*v.t.*), to draw ; to wiredraw : (25)
affiler l'or, to draw gold. (26)

affiloir (*n.m.*), sharpener. (27)

affiloire (*n.f.*), whetstone ; hone ; honestone ; oilstone. (28)

affinage (*n.m.*), refining : (29)
affinage au creuset, refining in the crucible ; crucible refining. (30)
affinage au moyen de l'électricité *ou* affinage électrolytique *ou* affinage par électrolyse, electrorefining ; electrolytic refining. (31)

affinage de la fonte au four à réverbère, refining cast iron in the reverberatory furnace. (32)
affinage sur sole, open-hearth refining. (33)

affinement (*n.m.*), refinement ; refining : (34)
l'affinement des métaux précieux, the refine--ment of the precious metals ; refining precious metals. (35)

affiner (*v.t.*), to refine ; to fine : (36)
affiner de l'or, to refine, to fine, gold. (37)
affiner du cuivre brut en le fondant dans un four, to refine raw copper by smelting it in a furnace. (38)
affiner au moyen de l'électricité *ou* affiner par électrolyse (*v.t.*), to electrorefine : (39)
affiner en grand le cuivre par électrolyse, to electrorefine copper on a large scale. (40)

affiner (s') (*v.r.*), to be refined ; to refine : (41)
l'or et l'argent ne s'affinent plus à la coupelle, gold and silver are no longer refined in the cupel. (42)

affinerie (*n.f.*), refinery ; refining-works. (43)

affineur (pers.) (*n.m.*), refiner. (44)

affinité (*n.f.*), affinity : (45)
affinité chimique, chemical affinity. (46)

affleurement (Géol.) (*n.m.*), outcrop ; outcrop--ping ; crop ; cropping ; cropping out ; basset ; blow ; outburst. (47)

affleurement altéré par l'action des agents atmosphériques (*m.*), weathered outcrop ; blossom. (48)

affleurement de quartz (*m.*), quartz outcrop (*or* blow). (49)

affleurement masqué (*m.*), concealed outcrop. (50)

affleurement rocheux (*m.*), rock (*or* rocky) outcrop. (51)

affleurer (*v.t.*), to flush ; to flush up ; to make flush ; to level : (52)
affleurer un joint, to flush up a joint. (53)

affleurer (Géol.) (*v.i.*), to outcrop ; to crop out ; to crop : (54)
couches qui affleurent à la surface du sol (*f.pl.*), strata which outcrop on the surface of the ground. (55)
une veine de charbon affleure sur le flanc de la colline, a coal-vein crops out on the hill--side. (56)
le gypse n'affleure nulle part, car il est soluble, gypsum never outcrops, because it is soluble. (57)

affleurer (s') (*v.r.*), to be flush : (58)
deux surfaces qui s'affleurent, two surfaces which are flush. (59)

affluence (en parlant des eaux) (*n.f.*), concourse ; affluence : (60)
l'affluence des eaux provenant de la fonte des neiges fait déborder les rivières, the con--course of waters resulting from the melting of the snows causes the rivers to overflow. (61)

affluent, -e (*adj.*), tributary ; affluent : (62)
cours d'eau affluent (*m.*), tributary stream ; affluent stream. (63)

affluent (*n.m.*), tributary ; affluent : (64)
la Marne est un affluent de la Seine, the Marne is a tributary of the Seine. (65)

affluer (*v.i.*), to flow ; to run ; to fall : (66)
fleuve qui afflue dans la mer (*m.*), river which flows (*or* runs) (*or* falls) into the sea. (67)

affouage (droit de prendre du bois) (*n.m.*), right to cut timber. (68)

affouage (entretien en bois d'une usine) (*n.m.*), supply of wood necessary for a works :

especially, the wood required for making
the charcoal used in blast-furnaces and
refining-works. (1)

affouillable (*adj.*), liable to wash away ; liable
to be undermined : (2)

le sol des pentes accentuées est souvent affouil-
-lable, the soil of pronounced slopes is often
liable to wash away. (3)

affouillement (*n.m.*), undermining ; washing ;
wash : (4)

construire un radier de béton afin d'éviter
tout affouillement par les eaux, to construct
a concrete floor in order to prevent any
undermining by water. (5)

affouillement des rives d'un fleuve par les eaux
courantes, undermining, washing, of the
banks of a river by running water. (6)

affouiller (*v.t.*), to undermine ; to wash : (7)

eaux qui affouillent les berges d'une rivière
(*f.pl.*), waters which undermine (*or* water
which washes) the banks of a river. (8)

affranchir les bouts (des barres de fer, d'acier),
to crop the ends. (9)

affranchissement des bouts (*m.*), cropping the
ends. (10)

affût (*n.m.*), stand ; standard ; rest : (11)

affût pour théodolite, pour appareil photo-
-graphique, stand for theodolite, for camera.
(12 *ou* 13)

affût (pour perforatrices) (Mines) (*n.m.*), bar
(rock-drill mounting). (14)

affût-colonne [**affûts-colonnes** *pl.*] (perforatrices)
(*n.m.*), column ; post. (15)

affût pour creusement de tunnels (perforatrices)
(*m.*), tunnel-bar. (16)

affût pour fonçage de puits (perforatrices) (*m.*),
shaft-sinking bar. (17)

affût-trépied [**affûts-trépieds** *pl.*] (perforatrices)
(*n.m.*), tripod. (18)

affûtage (action d'affûter) (*n.m.*), sharpening ;
grinding : (19)

affûtage d'outils, tool-grinding ; tool-sharpen-
-ing : (20)

l'affûtage des outils se fait à la meule en grès
ou à la meule en émeri, tool-grinding is
done on the grindstone or on the emery-
wheel. (21)

affûtage de lames de scies, sharpening saw-
blades. (22)

affûtage de limes par projection de sable,
sharpening files by sand-blasting. (23)

affûtage de mèches *ou* affûtage de forets, drill-
grinding ; drill-sharpening. (24)

affûtage (collection d'outils nécessaires à un
menuisier) (*n.m.*), bench-planes (set of planes,
etc., especially the pair of planes consisting
of a jack-plane and a trying-plane). (25)

affûter (*v.t.*), to sharpen ; to grind : (26)

alésoirs affûtés mécaniquement après la trempe
(*m.pl.*), reamers ground mechanically after
hardening. (27)

affûteur (pers.) (*n.m.*), sharpener ; tool-sharpener ;
grinder. (28)

affûteuse (*n.f.*), sharpening-machine ; sharpener ;
grinding-machine ; grinder (of the tool,
cutter or drill-sharpening type). (29) See
machine à affûter for varieties.

agalite (Minéral) (*n.f.*), agalite. (30)

agalmatolite *ou* **agalmatolithe** (Minéral) (*n.m.*),
agalmatolite ; figure-stone ; bildstein. (31)

agaphite (*n.f.*), agaphite ; oriental turquoise.
(32)

agarice (*n.f.*) *ou* **agaric minéral** (*m.*) *ou* **agarie
fossile** (*m.*), agaric mineral. (33)

agate (Minéral) (*n.f.*), agate. (34)

agate arborisée *ou* **agate herborisée** (*f.*), dendritic
agate ; arborized agate ; tree-agate. (35)

agate enhydre (*f.*), enhydrous agate. (36)

agate jaspée (*f.*), jasper-agate. (37)

agate mousseuse (*f.*), moss-agate. (38)

agate noire (*f.*), obsidian ; volcanic glass. (39)

agate œillée (*f.*), eye agate. (40)

agate rubanée *ou* **agate zonaire** (*f.*), banded
agate ; zoned agate. (41)

agate tachetée (*f.*), clouded agate ; mottled
agate. (42)

agaté, -e (*adj.*), agaty : (43)

jaspe agaté (*m.*), agaty jasper. (44)

agatifère (*adj.*), agatiferous : (45)

terrain agatifère (*m.*), agatiferous ground. (46)

agatifier *ou* **agatiser** (*v.t.*), to agatize. (47)

agatifier (s') *ou* **agatiser** (s') (*v.r.*), to agatize ;
to become transformed into agate. (48)

agatin, -e (*adj.*), agatine. (49)

agatoïde (*adj.*), agatoid. (50)

âge (*n.m.*), age : (51)

l'âge d'un arbre, des roches, the age of a tree,
of rocks. (52 *ou* 53)

âge de la pierre (Archéol.) (*m.*), stone age ; age
of stone ; flint age. (54)

âge néolithique *ou* **âge de la pierre polie** (Archéol.)
(*m.*), neolithic age ; polished-stone age. (55)
[Note.—Neolithic is frequently written with
an initial capital letter ; as, the Neolithic
age.]

âge paléolithique *ou* **âge de la pierre taillée**
(Archéol.) (*m.*), paleolithic age ; chipped-
stone age. (56)
[Note.—Paleolithic is frequently written with
an initial capital letter ; as, the Paleolithic
age.]

âge primaire *ou* **âge paléozoïque** (Géol.) (*m.*),
Primary age ; Paleozoic age. (57)

âge quaternaire (Géol.) (*m.*), Quaternary age.
(58)

âge secondaire *ou* **âge mésozoïque** (Géol.) (*m.*),
Secondary age ; Mesozoic age. (59)

âge tertiaire (Géol.) (*m.*), Tertiary age. (60)

agence (Commerce, etc.) (*n.f.*), agency. (61)

agence de brevets *ou* **agence de brevets d'inven-
-tion** (*f.*), patent agency. (62)

agencement (*n.m.*), arrangement ; putting into
working order ; fitting : (63)

la structure ou texture d'une roche est déter-
-minée par l'agencement des minéraux qui
prennent part à sa composition, the struc-
-ture or texture of a rock is determined by
the arrangement of the minerals of which
it is composed. (64)

agencement (*n.m.*) *ou* **agencements** (*n.m.pl.*)
(objets mobiliers que l'on ne pourrait déplacer
sans les rendre à peu près inutilisables ou
sans commettre des détériorations impor-
-tantes [casiers, rayons de magasin, installa-
-tions de gaz, etc.]), fittings ; fittings and
fixtures ; fixtures and fittings : (65)

mobilier et agencement [mobilier (*n.m.*)],
furniture and fittings ; furniture, fittings, and
fixtures. (66)

agencer (*v.t.*), to arrange ; to put into working
order ; to fit : (67)

les chaudières actuelles sont toutes constituées
de corps et de tubes cylindriques, diverse-
-ment agencés (*f.pl.*), modern boilers are all

formed of barrels and cylindrical tubes, differently arranged (or fitted). (1)

agencer les divers organes d'un appareil, to put the different parts of an apparatus into proper working order. (2)

agencer (s') (v.r.), to be arranged ; to be put into working order; to be fitted; to fit: (3)

toutes les pièces d'une machine doivent s'agencer avec la même justesse (f.pl.), all the parts of a machine should be arranged (or fitted) (or should fit) with the same precision. (4)

agent (celui qui fait les affaires d'autrui) (n.m.), agent : (5)

agent de brevets ou agent de brevets d'inven-tion, patent agent. (6)

agent exclusif pour la France et ses Colonies, sole agent for France and Colonies. (7)

agent (tout ce qui agit, opère ; force) (n.m.), agent ; agency : (8)

agent atmosphérique, atmospheric agent ; atmospheric agency. (9) Cf. action des agents atmosphériques.

agents d'intempérisme (Géol.), weathering agencies ; weathering agents : (10)

roches modifiées sous l'influence des agents d'intempérisme (f.pl.), rocks modified under the influence of weathering agencies (or agents). (11)

agent de dénudation ou agent dénudant (Géol.), denuding agent ; denuding agency : (12)

l'eau, le gel, et le vent, sont de puissants agents de dénudation, water, frost, and wind, are powerful denuding agents. (13)

agent fertilisant, fertilizing agent ; fertilizing agency ; fertilizer. (14)

agent réducteur (Chim.), reducing agent ; reducer. (15)

agent voyer [agents voyers pl.] (surveillance des travaux publics) (pers.) (m.), surveyor. (16)

agglomérant (n.m.), agglomerant. (17)

agglomérat (Géol.) (n.m.), agglomerate. (18)

agglomératif, -ive (adj.), agglomerative ; agglom-eratic : (19)

agents agglomératifs (m.pl.), agglomeratic agents. (20)

agglomération (n.f.), agglomeration ; agglom-erate : (21)

une agglomération de sable, de glace, et de dents d'éléphants, an agglomeration of sand, ice, and elephants' teeth. (22)

aggloméré, -e (adj.), agglomerated : (23)

gravier aggloméré (m.), agglomerated gravel. (24)

aggloméré (de houille) (n.m.), briquette ; briquet. (25)

agglomérer (v.t.), to agglomerate. (26)

agglomérer (Métall.) (v.t.), to sinter. (27)

agglomérer (s') (v.r.), to agglomerate : (28)

la poudre d'asphalte s'agglomère sous l'action de pilons chauffés, asphalt powder agglom-erates under the action of heated rammers. (29)

agglomérer (s') (Métall.) (v.r.), to sinter. (30)

agglutinant, -e (adj.), agglutinant. (31)

agglutinant (n.m.), agglutinant. (32)

agglutinant (d'une meule en émeri) (n.m.), bond (of an emery-wheel). (33)

agglutination (n.f.), agglutination. (34)

agglutiner (v.t.), to agglutinate. (35)

agglutiner (s') (v.r.), to agglutinate : (36)

'houilles grasses qui s'agglutinent (f.pl.), bituminous coals which agglutinate. (37)

agir (v.i.), to act : (38)

les acides agissent sur les métaux (m.pl.), acids act on (or upon) metals. (39)

agir selon le cas, to act according to circum-stances. (40)

agir une sonnerie, to set a bell ringing. (41)

agissant, -e (adj.), active : (42)

volcan agissant (m.), active volcano. (43)

agitateur, -trice (adj.), agitating : (44)

mouvement agitateur (m.), agitating motion. (45)

agitateur (Mach.) (n.m.), agitator. (46)

agitateur (Chim.) (n.m.), stirring-rod ; stirrer. (47)

agitateur en verre (Chim.) (m.), glass stirring-rod ; glass rod. (48)

agitation (n.f.), agitation ; shaking ; stirring. (49)

agitation (fermentation) (n.f.), unrest ; trouble ; disturbance : (50)

agitation dans la classe ouvrière, unrest among the labouring classes ; labour trouble ; labour disturbance. (51)

agiter (v.t.), to agitate ; to shake ; to stir. (52)

aglomération (n.f.), aglomérer (v.t.), etc. Same as **agglomération, agglomérer,** etc.

agnotozoïque (Géol.) (adj.), Agnotozoic. (53)

agnotozoïque (n.m.), Agnotozoic : (54)

l'agnotozoïque, the Agnotozoic. (55)

agoge (Mines) (n.m.), drain ; gutter ; drainage-channel in a mine to conduct water to the sump. (56)

agrafe (crochet) (n.f.), hook ; clasp ; fastener ; snap. (57)

agrafe (crampon) (n.f.), cramp ; clamp. (58)

agrafe (crampon pour pierres) (Arch.) (n.f.), cramp ; cramp-iron. (59)

agrafe (Serrur.) (n.f.), casement-fastener. (60)

agrafe (pour oreilles de châssis, etc.) (Fonderie) (n.f.), dog (for flask-lugs, etc.). (61)

agrafe (du chariot d'une scie à débiter les bois en grume) (n.f.), timber-clip (of the log-carriage of a log-sawing machine). (62)

agrafe à griffes pour courroies (f.), steel belt-lacing. (63)

agrafe à scellement (Arch.) (f.), cramp with stone-hook. (64)

agrafe à T (Arch.) (f.), T cramp. (65)

agrafe à talon (Arch.) (f.), cramp with turned down ends. (66)

agrafe de collets (Sondage) (f.), collar-grab. (67)

agrafe de manœuvre (Sondage) (f.), clamp. (68)

agrafe et porte [agrafe (n.f.) ; porte (n.f.)], hook and eye. (69)

agrafe pour courroies (f.), belt-fastener. (70)

agrafer (v.t.), to hook ; to clasp ; to fasten : (71)

agrafer une porte, to hook a door. (72)

agrafer (s') (v.r.), to hook ; to clasp ; to fasten. (73)

agrandir (v.t.), to enlarge ; to make larger : (74)

agrandir une usine, to enlarge a works. (75)

agrandir une photographie, to enlarge a photo-graph. (76)

agrandir (s') (v.r.), to enlarge ; to grow bigger ; to increase in size : (77)

trou qui s'agrandit par l'usure (m.),   hole which enlarges by wear. (1)

agrandissement (n.m.), enlargement ; enlarging ; increase. (2)

agrandissement (action) (Photogr.) (n.m.), enlargement ; enlarging : (3)
l'agrandissement d'une photographie,   the enlargement of a photograph ; enlarging a photograph. (4)

agrandissement (photographie agrandie) (n.m.), enlargement : (5)
appareils destinés à faire des agrandissements (m.pl.), apparatus for making enlargements. (6)

agrandisseur (Photogr.) (n.m.), enlarger. (7)

agréer (v.t.), to accept ; to approve : (8)
exploseurs d'un type agréé par le ministre des mines (m.pl.), blasting-machines of a type approved by the minister of mines. (9)
lampes de sûreté agréées (f.pl.), approved safety-lamps. (10)

agréeur (pers.) (n.m.), wiredrawer. (11)

agrégat (Géol. & Minéralogie) (n.m.), aggregate : (12)
agrégat cristallin, crystalline aggregate. (13)

agrégatif, -ive (adj.), aggregative. (14)

agrégation (n.f.), aggregation : (15)
les corps se composent de l'agrégation de particules de matières infiniment petites (m.pl.), bodies are composed of the aggrega-tion of infinitely small particles of matter. (16)

agrégé, -e (Géol.) (adj.), aggregate. (17)

agréger (v.t.), to aggregate : (18)
la nature agrège les grains de sable en cristaux, nature aggregates sand grains into crystals. (19)

agréger (s') (v.r.), to aggregate. (20)

agrès (n.m.pl.), tackle ; hoisting-tackle : (21)
apporter les agrès nécessaires au déplacement d'une machine (palans, cordes, leviers, rouleaux, cales, maillets, etc.), to bring the necessary tackle for shifting a machine (pulley-blocks, ropes, levers, rollers, wedges, mallets, etc.). (22)

agrès de chèvre ou simplement **agrès** (n.m.pl.), gin-tackle. (23)

agréyeur (pers.) (n.m.), wiredrawer. (24)

agricolite ou **agricolithe** (Minéral) (n.f.), agric-olite. (25)

agripper (v.t.), to lay hold of ; to clutch ; to grab ; to grip : (26)
agripper un outil tombé dans un trou de sonde, to clutch, to grab, a tool dropped in a bore-hole. (27)

aide (n.f.), aid ; help ; assistance : (28)
venir en aide à quelqu'un, to come to some-one's assistance. (29)

aide (pers.) (n.m.), assistant ; helper ; mate. (30)

aide-galibot [aides-galibots pl.] (Mines) (n.m.), boy ; lad ; pit-boy ; foal. (31)

aide-maçon [aides-maçons pl.] (n.m.), brick-layers' labourer. (32)

aide-mémoire [aide-mémoire pl.] (n.m.), hand-book ; manual. (33)

aide-mémoire de poche de l'ingénieur-mécanicien (m.), mechanical engineers' pocket-book. (34)

aigre (cassant) (adj.), short : (35)
fer aigre (m.), short iron. (36)
argile aigre (f.), short clay. (37)

aigu, -ë (adj.), acute ; sharp ; pointed. (38)

aigu, -ë (Géom.) (adj.), acute : (39)
angle aigu (m.), acute angle. (40)

aigue-marine [aigues-marines pl.] (Minéral) (n.f.), aquamarine. (41)

aiguillage (action) (Ch. de f.) (n.m.), switching ; switching off ; turning off (into a siding) : (42)
l'aiguillage d'un train, switching a train. (43)

aiguillage (changement de voie) (Ch. de f.) (n.m.), switch ;   points ;   turnout ;   turn-off ; switches. (44)

aiguillage à deux voies symétriques (m.), two-way symmetrical switch ; two-way symmetrical turnout ; right and left-hand turn-off. (45)

aiguillage à droite (m.),   right-hand switch ; right-hand turnout ; right-hand turn-off. (46)

aiguillage à gauche (m.),   left-hand switch ; left-hand turnout ; left-hand turn-off. (47)

aiguillage à trois voies avec croisements (pose symétrique) (m.),   three-way switches and crossings with symmetrical points ; 3-way right and left-hand turn-off with crossings. (48)

aiguille (petite verge de métal) (n.f.), needle. (49)

aiguille (d'un cadran) (n.f.), needle, hand, pointer, index, (of a dial). (50)

aiguille (d'une balance) (n.f.),   pointer, needle, index, tongue, (of a balance or scale). (51)

aiguille (d'une boussole) (n.f.),   needle (of a compass). (52)

aiguille (d'une horloge) (n.f.), hand, pointer, (of a clock or timepiece) : (53)
petite aiguille, hour-hand. (54)
grande aiguille, minute-hand. (55)
aiguille trotteuse, second-hand. (56)

aiguille (Travail aux explosifs) (n.f.),   needle ; blasting-needle ; shooting-needle ; priming-needle ; pricker ; picker ; nail ; aiguille. (57)

aiguille (lame d'aiguille) (Ch. de f.) (n.f.),   point ; point-rail ; switch-rail ; slide-rail ; tongue ; switch-tongue ; latch. (58)

aiguille (Constr. hydraul.) (n.f.), needle : (59)
aiguilles de barrage, dam-needles. (60)

aiguille (n.f.) ou **aiguille de réglage** (d'un injec-teur), spindle (of an injector). (61)

aiguille (Géogr. phys.) (n.f.), needle ; aiguille : (62)
une aiguille de rocher, a needle of rock. (63)
les aiguilles se rencontrent surtout dans le massif du Mont-Blanc,   aiguilles occur principally in the Mont Blanc group of mountains. (64)

aiguille (Géol.) (n.f.), needle : (65)
l'aiguille que la mer a isolée des falaises est un curieux exemple d'érosion,   the needle that the sea has cut off from the cliffs is a curious example of erosion. (66)

aiguille (Minéralogie) (n.f.), needle : (67)
aiguilles cristallines, crystalline needles. (68)

aiguille (aiguillage) (Ch. de f.) (n.f.), switch. (69)

aiguilles (Ch. de f.) (n.f.pl.),   points ; switch ; switch-points ; switches. (70)

aiguilles à distance (Ch. de f.) (f.pl.), distance-points. (71)

aiguille à emballer ou **aiguille d'emballage** (f.), packing-needle. (72)

aiguille à retirer les pièces battues (outil de mouleur) (f.),   draw-spike ;   draw-hook ; draw-stick. (73)

aiguille à tracer (f.), scriber. (1)

aiguille à trous d'air (Fonderie) (f.), vent-wire ; pricker ; piercer. (2)

aiguille aimantée (f.), magnetic needle ; magnetized needle. (3)

aiguille astatique (Phys.) (f.), astatic needle. (4)

aiguille automatique (Ch. de f.) (f.), automatic switch ; self-acting switch ; jumping-switch. (5)

aiguille - coin [aiguilles-coins pl.] (Mines) (n.f.). Same as aiguille infernale.

aiguille d'essai (f.), touch-needle ; test needle ; proof needle. (6)

aiguille d'inclinaison (Phys.) (f.), dipping-needle ; inclinatory needle. (7)

aiguille de mire (d'un viseur) (Photogr.) (f.), sighting-pin, sight-bar, sighter, (of a view-finder). (8)

aiguille de déraillement (Ch. de f.) (f.), catch-point ; throw-off point ; derailing-switch ; runaway-switch. (9)

aiguille désaimantée (f.), demagnetized needle. (10)

aiguilles et croisement (Ch. de f.) [aiguille (n.f.) ; croisement (n.m.)], points and crossing ; switch and crossing. (11)

aiguille fixe (Ch. de f.) (f.), fixed point. (12)

aiguille infernale (Mines) (f.), plug and feather ; plug-and-feather wedge ; plug and feathers ; stob and feathers ; feathers and stob. (13)

aiguille mobile (Ch. de f.) (f.), movable point. (14)

aiguilles prises en pointe ou aiguilles abordées en pointe (Ch. de f.) (f.pl.), facing-switch ; facing point-switch. (15)

aiguilles prises par le talon ou aiguilles abordées en talon (Ch. de f.) (f.pl.), trailing-switch ; trailing point-switch. (16)

aiguille rabotée (Ch. de f.) (f.), planed point. (17)

aiguillé, -e (adj.), needle-shaped ; acicular ; aciculate. (18)

aiguiller (Ch. de f.) (v.t.), to switch ; to switch off ; to turn off : (19)

aiguiller un train, to switch, to switch off, to turn off, a train. (20)

aiguiller (s') (v.r.), to switch ; to switch off ; to turn off : (21)

courbe qui s'aiguille en deux parties (f.), curve which switches into two parts. (22)

aiguillette (Géogr. phys.) (n.f.), needle ; small point : (23)

une aiguillette de rocher, a needle of rock ; a small point of rock. (24)

aiguilleur (Ch. de f.) (n.m.), pointsman ; switcher ; switchman ; latchman. (25)

aiguisage ou aiguisement (n.m.), sharpening ; grinding. (26)

aiguisé, -e (adj.), sharp ; ground. (27)

aiguiser (v.t.), to sharpen ; to grind : (28)

aiguiser une hache, un couteau, to sharpen, to grind, an axe, a knife. (29 ou 30).

aiguiser (s') (v.r.), to be sharpened ; to be ground : (31)

la plupart des outils s'aiguisent à la meule, most tools are ground on a wheel. (32)

aiguiserie (n.f.), grinding-works. (33)

aiguiseur (pers.) (n.m.), sharpener ; grinder ; tool-sharpener. (34)

aiguisoir (instrument) (n.m.), sharpener ; knife-sharpener. (35)

aiguïté (n.f.), acuteness ; sharpness : (36)

aiguïté d'un angle, acuteness, sharpness, of an angle. (37)

aikinite (Minéral) (n.f.), aikinite. (38)

aile (d'un bâtiment) (n.f.), wing (of a building). (39)

aile (d'une poutre, d'un fer à T, d'un fer à L, ou analogue) (n.f.), flange, wing, (of a girder, of a T iron, of an L iron, or the like) : (40 ou 41 ou 42 ou 43)

poutre à larges ailes (f.), broad-flange girder. (44)

aile (d'un moulin à vent) (n.f.), wing (of a wind--mill). (45)

aile (d'un ventilateur) (n.f.), blade, vane, wing, (of a fan [or of a ventilator] [or of a ventilating-fan]). (46)

aile (d'une hélice) (n.f.), blade (of a propeller). (47)

aile (d'une vis ailée) (n.f.), wing (of a thumb-screw). (48)

aileron (roue hydraulique) (n.m.), paddle ; paddle-board ; float ; floatboard. (49)

ailette (n.f.). Diminutive of "aile," i.e., small wing, blade, etc. : (50)

les ailettes d'un ventilateur de table, the blades of a table-fan. (51)

aimant (Phys.) (n.m.), magnet. (52)

aimant armé (m.), armed magnet. (53)

aimant artificiel (Phys.) (m.), artificial magnet. (54)

aimant en fer à cheval (m.), horseshoe-magnet ; U magnet. (55)

aimant en forme de barreau (m.), bar magnet. (56)

aimant naturel (Phys.) (m.), natural magnet. (57)

aimant naturel (Minéral) (m.), natural magnet ; loadstone ; lodestone. (58)

aimant permanent (m.), permanent magnet. (59)

aimantation (n.f.), magnetization ; magnetizing : (60)

l'aimantation du fer, the magnetization of iron ; magnetizing iron. (61)

aimantation par double touche (f.), magnetization by double touch. (62)

aimantation par influence (f.), magnetization by influence. (63)

aimantation par l'électricité (f.), magnetization by electricity. (64)

aimantation par simple touche (f.), magnetization by single touch. (65)

aimantation par touche séparée (f.), magnetiza--tion by separate touch. (66)

aimanté, -e (adj.), magnetized ; magnetic : (67) aiguille aimantée (f.), magnetic needle. (68)

aimanter (v.t.), to magnetize : (69)

aimanter une barre d'acier à saturation, to magnetize a bar of steel to saturation. (70)

aimanter (s') (v.r.), to magnetize : (71)

le fer doux s'aimante mieux que l'acier, soft iron magnetizes better than steel. (72)

ainalite (Minéral) (n.f.), ainalite. (73)

air (n.m.), air. (74)

air chaud (m.), hot air. (75)

air comprimé (m.), compressed air. (76)

air de famille (m.), family likeness : (77)

l'air de famille des venues éruptives d'une même province pétrographique, the family likeness of the eruptive occurences of a selfsame petrographical province. (78)

air frais (m.), fresh air. (79)

air grisouteux (*m.*), foul air ; air containing fire-damp. (1)

air impropre à la respiration *ou* **air irrespirable** (*m.*), air unfit for respiration ; irrespirable air. (2)

air libre (*m.*), free air ; open air : (3)

opération qui se fait à l'air libre (*f.*), operation which takes place in the open air. (4)

air liquide (*m.*), liquid air. (5)

air lourd (*m.*), oppressive air. (6)

air pur (*m.*), pure air ; fresh air. (7)

air respirable (*m.*), respirable air. (8)

air vicié (*m.*), foul air ; bad air ; vitiated air. (9)

airage (galerie d'aérage) (Mines) (*n.m.*), airway ; windway ; air-heading ; air-head. (10)

aire (Géom.) (*n.f.*), area : (11)

l'aire d'un carré, d'un cercle, the area of a square, of a circle. (12 *ou* 13)

unité d'aire (*f.*), unit of area. (14)

aire (superficie) (*n.f.*), area : (15)

l'aire d'un bâtiment, the area of a building. (16)

aire (domaine) (*n.f.*), area : (17)

aire d'alluvions, alluvial area. (18)

aire (plancher) (*n.f.*), floor : (19)

l'aire d'un pont, d'un réservoir, the floor of a bridge, of a reservoir. (20 *ou* 21)

aire de lavage, washing-floor. (22)

aire de séchage, drying-floor. (23)

aire (d'une enclume) (*n.f.*), face, crown, (of an anvil). (24)

aire (d'un marteau) (*n.f.*), face (of a hammer). (25)

aire continentale (Géol.) (*f.*), continental area. (26)

aire d'alimentation *ou* **aire de drainage** (bassin hydrographique) (Géogr. phys.) (*f.*), intake ; watershed ; drainage-basin ; drainage-area ; catchment-basin. (27)

aire de coulée (haut fourneau) (*f.*), pig-bed (blast-furnace). (28)

airure (Mines) (*n.f.*), pinch-out ; pinch. (29)

aisance (*n.f.*), ease : (30)

porter avec aisance un pesant fardeau, to carry a heavy burden with ease. (31)

aisé, -e (*adj.*), easy. (32)

aisément (*adv.*), easily. (33)

ajointer (*v.t.*), to join up ; to join end to end : (34)

ajointer des tuyaux, to join up pipes. (35)

ajoupa (*n.m.*), native hut. (36)

ajouré, -e (*adj.*), perforated ; pierced ; latticed. (37)

ajoutage *ou* **ajoutoir** (*n.m.*). Same as **ajutage.**

ajust (*n.m.*), granny-knot ; granny's-knot ; granny's-bend. (38)

ajustage *ou* **ajustement** (*n.m.*), adjustment ; adjusting ; fitting ; setting up : (39)

l'ajustement d'une balance, the adjustment of a balance ; adjusting a scale. (40)

ajuster (*v.t.*), to adjust ; to fit ; to set up : (41)

ajuster les pièces d'une machine, to adjust the parts of a machine. (42)

ajuster un couvercle à une boîte, to fit a lid to a box. (43)

ajuster une machine à vapeur, to set up a steam-engine. (44)

ajuster (s') (*v.r.*), to fit : (45)

clef qui s'ajuste à une serrure (*f.*), key which fits a lock. (46)

ajusteur (pers.) (*n.m.*), fitter ; artificer ; bench-hand. (47)

ajusteur-monteur [ajusteurs-monteurs *pl.*] (*n.m.*), erector ; fitter. (48)

ajut (*n.m.*). Same as **ajust.**

ajutage (*n.m.*), nozzle ; nozle ; nose ; nose-piece ; snout ; jet ; ajutage ; adjutage. (49)

ajutage (Chim.) (*n.m.*), connector. (50)

ajutage à jet de sable (*m.*), sand-blast nozzle. (51)

ajutage à vapeur (d'un injecteur) (*m.*), steam-nozzle, steam-cone, (of an injector). (52)

ajutage convergent (Hydraul.) (*m.*), convergent nozzle ; convergent mouthpiece. (53)

ajutage convergent (d'un injecteur) (*m.*), com-bining-nozzle, combining-tube, combining-cone, (of an injector). (54)

ajutage cylindrique (Hydraul.) (*m.*), cylindrical mouthpiece ; short pipe ; orifice in thick wall ; opening in thick wall. (55) Cf. écoulement.

ajutage divergent (Hydraul.) (*m.*), divergent nozzle ; divergent mouthpiece. (56)

ajutage divergent (d'un injecteur) (*m.*), diverging tube, diverging nozzle, diverging cone, delivery-tube, delivery-nozzle, delivery-cone, (of an injector). (57)

ajutage rentrant de Borda (Hydraul.) (*m.*), reentrant short pipe ; Borda mouthpiece. (58)

ajutoir (*n.m.*). Same as **ajutage.**

akanthicone (Minéral) (*n.m.*), akanticone. (59)

akérite (Pétrol.) (*n.f.*), akerite. (60)

alabandine *ou* **alabandite** (Minéral) (*n.f.*), alaban-dite ; alabandine. (61)

alabastrin, -e (*adj.*), alabastrine. (62)

alabastrite (Minéral) (*n.f.*), gypseous alabaster ; modern alabaster ; alabaster. (63)

alaisage (*n.m.*), **alaiser** (*v.t.*), **alaisoir** (*n.m.*). Same as **alésage, aléser, alésoir.**

alalite (Minéral) (*n.f.*), alalite. (64)

alambic (*n.m.*), still. (65)

alarme (signal) (*n.f.*), alarm : (66)

donner l'alarme, to give the alarm. (67)

alaskaïte (Minéral) (*n.f.*), alaskaite. (68)

alaskite (Pétrol.) (*n.f.*), alaskite. (69)

albâtre (Minéral) (*n.m.*), alabaster. (70)

albâtre calcaire *ou* **albâtre oriental (d'Egypte)** (*m.*), calcareous alabaster ; Oriental alabaster ; Egyptian alabaster ; alabaster-stone ; alabaster ; alabastrites. (71)

albâtre gypseux *ou* **albâtre blanc vulgaire** (*m.*), gypseous alabaster ; modern alabaster ; alabaster. (72)

albâtre onychite (*m.*), Algerian onyx. (73)

albertite (Minéral) (*n.f.*), albertite ; Albert coal. (74)

albionite (Explosif) (*n.f.*), albionite. (75)

albite (Minéral) (*n.f.*), albite ; white schorl. (76)

albraque (Mines) (*n.m.*), lodge (a lower level in a mine in which water collects). (77)

album à coller (Photogr.) (*m.*), paste-on album. (78)

album à passe-partout (Photogr.) (*m.*), slip-in album. (79)

album-classeur pour pellicules [albums-classeurs *pl.*] (*m.*), film-negative storage-album ; negative-album. (80)

album-tarif [albums-tarifs *pl.*] (*n.m.*), illustrated price-list. (81)

alcalescence (*n.f.*), alkalescence ; alkalescency. (82)

alcalescent, -e (*adj.*), alkalescent. (83)

alcali (*n.m.*), alkali. (84)

alcalicité (*n.f.*), alkalinity. (85)

alcalifère (*adj.*), alkaliferous. (86)

alcalifiable (*adj.*), alkalifiable ; alkalizable. (87)

alcalifiant, -e (*adj.*), alkalifying ; alkalizing : (88)

principe alcalifiant (m.), alkalifying principle. (1)

alcaligène (adj.), alkaligenous. (2)

alcalimètre (n.m.), alkalimeter; alcalimeter. (3)

alcalimétrie (n.f.), alkalimetry; alcalimetry. (4)

alcalimétrique (adj.), alkalimetric; alkalimetrical. (5)

alcalimétriquement (adv.), alkalimetrically. (6)

alcalin, -e (adj.), alkaline: (7)
sel alcalin (m.), alkaline salt. (8)

alcalinisation (n.f.), alkalization. (9)

alcaliniser (v.t.), to alkalinize; to alkalize; to alkalify. (10)

alcalinité (n.f.), alkalinity. (11)

alcalino-terreux, -euse (adj.), of or belonging to the alkaline earths: (12)
métaux alcalino-terreux (m.pl.), metals of the alkaline earths. (13)

alcalisation (n.f.), alcalization. (14)

alcaliser (v.t.), to alkalize; to alkalify; to alkalinize. (15)

alcaliser (s') (v.r.), to alkalify. (16)

alcaloïde (adj.), alkaloid; alkaloidal. (17)

alcaloïde (n.m.), alkaloid. (18)

alcool (n.m.), alcohol; spirit; spirits. (19)

alcool à brûler ou **alcool dénaturé** (m.), methylated spirit; denatured alcohol. (20)

alcool absolu ou **alcool enhydre** (m.), absolute alcohol. (21)

alcool vinique (m.), wine-alcohol; spirits of wine. (22)

alcoolique (adj.), alcoholic: (23)
fermentation alcoolique (f.), alcoholic fermentation. (24)

aléa (n.m.), chance; risk: (25)
les aléas d'une entreprise, the risks of an undertaking; the chances of a venture. (26)

alène (n.f.), awl. (27)

alène de cordonnier (f.), shoemakers' awl. (28)

aléné, -e (adj.), awl-shaped. (29)

alésage (agrandissement d'un trou ou l'action de le rendre conique au moyen de l'outil à aléser) (Méc.) (n.m.), reaming; reaming out; riming; broaching: (30)
l'alésage des trous dans le métal, reaming out holes in metal. (31)

alésage (opération qui consiste à rendre parfaitement régulière la surface intérieure d'un cylindre au moyen de la machine à aléser) (Méc.) (n.m.), boring; boring out: (32)
alésage de l'intérieur d'un corps de pompe, boring the inside of a pump-barrel. (33)
l'alésage des centres de roues, boring out wheel-centres. (34)

alésage (action) (Sondage) (n.m.), reaming; broaching: (35)
l'alésage d'un trou de sonde, reaming a bore-hole. (36)

alésage (diamètre intérieur d'un cylindre) (n.m.), bore: (37)
l'alésage en centimètres d'un cylindre, the bore of a cylinder in centimetres. (38)

aléser (agrandir un trou ou le rendre conique) (Méc.) (v.t.), to ream; to ream out; to rime; to broach: (39)
aléser un trou, to ream, to ream out, to rime, to broach, a hole. (40)

aléser (rendre parfaitement régulière la surface intérieure d'un cylindre) (Méc.) (v.t.), to bore; to bore out: (41)
aléser un cylindre, to bore, to bore out, a cylinder. (42)

aléser (Sondage) (v.t.), to ream; to broach. (43)

aléseuse (n.f.), boring-machine. (44)

alésoir (outil à aléser) (Méc.) (n.m.), reamer; rimer; broach; opening-bit. (45) See also équarrissoir.

alésoir (machine à aléser) (n.m.), boring-machine. (46)

alésoir (Sondage) (n.m.), reamer; broaching-bit. (47)

alésoir à bout fileté pour l'amorçage (m.), self-feeding reamer. (48)

alésoir à fines rainures (m.), fine-fluted rimer. (49)

alésoir à fourche (Sondage) (m.), forked reamer. (50)

alésoir à lames rapportées ou **alésoir à lames mobiles** ou **alésoir extensible** (m.), adjustable reamer; adjustable-blade reamer; expansion-reamer; expanding reamer. (51)

alésoir à main à — cannelures torses ou **alésoir à main à — rainures torses** ou **alésoir à main à — tranchants en hélice** (m.), —-flute twist hand-reamer; hand-reamer with — twist flutes; hand-reamer with — spiral cutting edges. (52)

alésoir à — pans (m.), —-sided broach; broach with — sides. (53)

alésoir à queue (m.), chucking-reamer. (54)

alésoir conique (m.), taper reamer. (55)

alésoir-fraise creux [alésoirs-fraises creux pl.] (m.), shell-reamer. (56)

alésoir horizontal (machine) (m.), horizontal boring-machine. (57)

alésoir long à vilebrequin, à — cannelures droites (m.), long-set bit-stock reamer, with — straight grooves. (58)

alésoir pour cônes Morse (m.), Morse-taper reamer. (59)

alésoir pour trous de goupilles coniques (m.), taper-pin reamer. (60)

alésoir vertical (machine) (m.), vertical boring-machine. (61)

alexandrite (Minéral) (n.f.), alexandrite. (62)

algodonite (Minéral) (n.f.), algodonite. (63)

algonkien, -enne (Géol.) (adj.), Algonkian; Algonquian. (64)

algonkien (n.m.), Algonkian; Algonquian: (65)
l'algonkien, the Algonkian. (66)

alichen ou **alichon** (de roue de moulin à eau) (n.m.), paddle, paddle-board, float, floatboard, (of wheel of water-mill). (67)

alidade (n.f.), alidade. (68)

alidade à lunette (f.), telescopic alidade; alidade with telescope. (69)

alidade à lunette à stadia; arc de cercle à vernier double donnant la minute (f.), alidade with telescope with stadia-diaphragm; arc reading by two verniers to single minutes. (70)

alidade à pinnules (f.), sighted alidade; alidade with sights. (71)

alidade à pinnules à charnières, règle cuivre à biseau, boîte noyer (f.), alidade with folding sights, bevelled brass scale, in walnut case. (72)

alidade nivelatrice (f.), levelling-alidade. (73)

alignement (n.m.), alignment; alinement: (74)
l'alignement d'un mur, d'une rangée d'arbres, the alignment of a wall, of a row of trees. (75 ou 76)

alignement de la direction (d'un filon) (m.), line of strike (of a lode). (77)

alignement droit ou simplement **alignement** (n.m.) (Ch. de f.), straight (a straight stretch of line): (78)

une ligne de chemin de fer, en plan, présente des alignements droits et des courbes, a line of railway, in plan, contains straights and curves. (1)

aligner (*v.t.*), to align ; to aline : (2)

aligner des jalons, to align stakes. (3)

alignoir *ou* **alignonet** (*n.m.*), slate-knife. (4)

alimentaire (*adj.*), feed, feeding (*used as adjs*) : (5)

pompe alimentaire (*f.*), feed-pump. (6)

tuyau alimentaire (*m.*), feed-pipe ; feeding-pipe. (7)

alimentateur (*n.m.*), feeder. (8)

alimentateur de minerai (*m.*), ore-feeder. (9)

alimentation (*n.f.*), feeding ; feed ; supply : (10)

alimentation automatique, automatic feed ; automatic feeding. (11)

alimentation d'air comprimé, supply of com--pressed air. (12)

alimentation d'un réseau à 3 fils (Élec.), feeding a 3-wire system. (13)

alimentation des chaudières, feeding boilers ; boiler-feeding. (14)

alimentation en eau d'une machine, watering an engine ; supplying an engine with water. (15)

alimentation en eau potable, supply of drinkable water. (16)

alimenter (*v.t.*), to feed : (17)

alimenter d'eau une chaudière, to feed a boiler with water. (18)

alimenter un fourneau avec du minerai, to feed a furnace with ore. (19)

alimenter (s') (*v.r.*), to be fed. (20)

alios (Pétrol.) (*n.m.*), a kind of rough sandstone of a dark-brown colour, found in S.W. France. (21)

aliotique (*adj.*), of *or* pertaining to alios. (22) See this word in vocabulary.

allactite (Minéral) (*n.f.*), allactite. (23)

allagite (Minéral) (*n.f.*), allagite. (24)

allaise (dans le lit d'une rivière) (*n.f.*), sand-bank. (25)

allanite (Minéral) (*n.f.*), allanite ; cerine ; orthite. (26)

allée de desserte (Mines de houille) (*f.*), gate-road ; gateway ; gate ; stall-road. (27)

allée de roulage (Mines) (*f.*), drawing-road. (28)

allège (tender de locomotive) (*n.f.*), tender. (29)

alléger (rendre plus léger) (*v.t.*), to lighten : (30)

dispositif qui allège le poids d'un appareil (*m.*), device which lightens the weight of an apparatus. (31)

alléger (rendre plus mince) (*v.t.*), to reduce ; to thin ; to thin down : (32)

alléger une pièce de bois, de métal, to reduce a piece of wood, of metal. (33 ou 34)

alléger (s') (*v.r.*), to be lightened ; to be reduced ; to thin down. (35)

allemontite (Minéral) (*n.f.*), allemontite. (36)

aller (s'adapter) (*v.i.*), to fit : (37)

clef qui va à une serrure (*f.*), key which fits a lock. (38)

aller à cheval à la ville voisine, to ride to the next town. (39)

aller aux informations sur quelqu'un, to make enquiries about someone. (40)

aller dans les détails, to go into details. (41)

aller en amont (d'une rivière), to go up-river ; to go up-stream. (42)

aller en aval, to go down-river, to go down-stream. (43)

aller en diminuant, to decrease ; to diminish ; to decline : (44)

débit d'un puits qui va en diminuant (*m.*), flow of a well which declines. (45)

aller en diminuant (s'effiler), to taper : (46)

outil qui va en diminuant vers la pointe (*m.*), tool which tapers towards the point. (47)

aller en voiture d'une ville à une autre, to drive from one town to another. (48)

aller haut le pied (se dit d'un train en marche qui est vide, ou d'une machine locomotive qui circule sans convoi), to run light. (49)

alliage (action) (*n.m.*), alloying ; alloyage. (50)

alliage (métal que l'on combine avec autres, ou résultat de la combinaison) (*n.m.*), alloy ; mixed metal. (51)

alliage antifriction (*m.*), antifriction alloy. (52)

alliage fusible (*m.*), fusible metal ; fusible alloy. (53)

alliage or-argent-cuivre (*m.*), gold-silver-copper alloy. (54)

allier (*v.t.*), to alloy : (55)

allier l'or à (*ou* avec) l'argent, to alloy gold with silver. (56)

allier (s') (*v.r.*), to alloy : (57)

le zinc s'allie facilement avec les métaux, sauf le plomb et le bismuth, zinc alloys readily with metals other than lead and bismuth. (58)

allocation (*n.f.*), allocation ; allowance : (59)

allocation pour nourriture, food allowance ; chop allowance. (60)

allochroïte (Minéral) (*n.f.*), allochroite. (61)

allomorphe (*adj.*), allomorphic. (62)

allomorphie (*n.f.*), allomorphism. (63)

allomorphite (Minéral) (*n.f.*), allomorphite. (64)

allonge (barre, tige, etc.) (*n.f.*), lengthening-piece ; lengthening-rod ; extension-piece. (65)

allonge (tube) (*n.f.*), lengthening-tube : (66)

une allonge en tôle, a sheet-iron lengthening-tube. (67)

allonge (Chim.) (*n.f.*), adapter : (68)

adapter une allonge au col d'une cornue, to fit an adapter to the neck of a retort. (69)

allongement (*n.m.*), lengthening ; elongation : (70)

allongement de la flamme (lampes de sûreté), elongation, spiring, of the flame. (71)

allongement (Méc.) (*n.m.*), elongation ; stretch : (72)

allongement d'un fil de cuivre par traction, elongation, stretch, of a copper wire under tension. (73)

allonger (*v.t.*), to lengthen ; to elongate : (74)

allonger une galerie de mine, to lengthen a mine-level. (75)

allonger (s') (*v.r.*), to lengthen ; to grow longer ; to elongate : (76)

les rails de chemin de fer s'allongent pendant les chaleurs (*m.pl.*), railway-rails elongate during the hot weather. (77)

allopalladium (Minéral) (*n.m.*), allopalladium. (78)

allophane (Minéral) (*n.f.*), allophane. (79)

allotriomorphe (Pétrol.) (*adj.*), allotriomorphic ; xenomorphic. (80)

allotropie (*n.f.*) *ou* **allotropisme** (*n.m.*) (Chim.), allotropy ; allotropism. (81)

allotropique (*adj.*), allotropic ; allotropical. (82)

allotropiquement (*adv.*), allotropically. (83)

alluaudite (Minéral) (*n.f.*), alluaudite. (84)

alluchon (dent mortaisée dans le pourtour d'un rouet) (*n.m.*), cog. (1)

allumage (communication de la lumière) (*n.m.*), lighting : (2)

l'allumage des lampes, lighting lamps. (3)

allumage (moteurs à combustion interne) (*n.m.*), ignition ; sparking. (4)

allumage (Travail aux explosifs) (*n.m.*), lighting ; firing ; ignition : (5)

l'allumage de la charge, lighting the charge ; the ignition of the charge. (6)

allumage électrique, electric firing. (7)

allumage simultané, firing a number of shots simultaneously. (8)

allumer (*v.t.*), See examples : (9)

allumer du feu, to strike fire. (10)

allumer le feu d'une locomotive, to light the fire of a locomotive. (11)

allumer les lampes, to light the lamps. (12)

allumer les mèches (Travail aux explosifs), to light the fuses. (13)

allumer un fourneau de mine, to fire a shot ; to ignite a powder-mine. (14)

allumer une allumette, to light a match ; to strike a light. (15)

allumer une pompe, to prime a pump ; to fetch a pump ; to start a pump in operation. (16)

allumer (s') (*v.r.*), to light ; to ignite ; to catch fire ; to take fire : (17)

le bois sec s'allume aisément, dry wood easily catches fire. (18)

allumeur (*n.m.*), lighter ; igniter. (19)

allumeur (Travail aux explosifs) (pers.) (*n.m.*), shotfirer ; shotlighter ; shotman ; blaster. (20)

allumeur de lampes (pers.) (*m.*), lamp-lighter. (21)

allumeur de sûreté (*m.*), safety-lighter. (22)

allumeur électrique (*m.*), electric igniter. (23)

allumeur-extincteur à poussoir, forme bouton [allumeurs-extincteurs *pl.*] (Élec.) (*m.*), button-switch ; push-button switch. (24)

allumeur-extincteur à poussoir, forme poire (*m.*), push-pattern pear switch. (25)

allumeur intérieur (lampes) (*m.*), internal igniter. (26)

allumoir (pour becs de gaz, etc.) (*n.m.*), lighter ; torch (a mechanical device for lighting gas-jets, etc.) (27)

allumoir électrique (*m.*), electric lighter ; electric torch. (28)

allure (façon de marcher) (*n.f.*), gait ; pace ; *hence, in speaking of a machine,* speed : (29)

machine qui marche à une allure très rapide (*f.*), engine which runs at a very high speed. (30)

ventilateur à allure lente (*m.*), slow-speed fan ; low-speed fan. (31)

allure (tempérament) (*n.f.*), character : (32)

allure du filon en profondeur, character of the lode in depth. (33)

allure en chapelet d'un filon *ou* allure amyg-daline d'un filon, amygdaloidal character of a lode [literally—like a string of beads]. (34)

allure irrégulière des couches, irregular char-acter of the strata. (35)

allure (d'un fourneau) (*n.f.*), working : (36)

régler l'allure d'un fourneau, to regulate the working of a furnace. (37)

réduire la charge de minerai jusqu'à ce que le haut fourneau soit remis en allure normale, to reduce the charge of ore until the blast-furnace is brought back to normal working. (38)

allure chaude (haut fourneau) (*f.*), hot working (39)

allure froide (haut fourneau) (*f.*), cold working. (40)

allure rapide (haut fourneau) (*f.*), driving ; rapid working. (41)

alluvial, -e, -aux *ou* alluvien, -enne (*adj.*), alluvial : (42)

argile alluviale (*f.*), alluvial clay. (43)

or alluvien (*m.*), alluvial gold. (44)

alluvion (*n.f.*) *ou* alluvions (*n.f.pl.*), alluvial ; alluvials ; alluvion ; alluvium ; alluviums ; alluvia. (45)

alluvion à laver (Exploitation des placers) (*f.*), wash-dirt ; wash-gravel ; wash-stuff ; washing-stuff. (46)

alluvions anciennes (*f.pl.*), Old alluvium. (47)

alluvions anciennes aurifères (*f.pl.*), old gold-bearing alluvia. (48)

alluvion aurifère (*f.*), gold-bearing alluvium ; placer-dirt ; dirt. (49)

alluvions d'eaux douces (*f.pl.*), fresh-water alluviums. (50)

alluvions gelées (*f.pl.*), frozen alluvium ; frozen alluvion. (51)

alluvion glaciaire (*f.*), glacial alluvium ; allu--vium of the ice period. (52)

alluvion inférieure (*f.*), deep alluvium. (53)

alluvions marines (*f.pl.*), marine drift. (54)

alluvion pliocène (*f.*), Pliocene alluvium. (55)

alluvions post-glaciaires (*f.pl.*), postglacial alluvia. (56)

alluvions préglaciaires (*f.pl.*), preglacial alluvium. (57)

alluvions récentes (*f.pl.*), recent alluvium ; alluvium of a late period. (58)

alluvion stannifère (*f.*), tin-bearing alluvial. (59)

alluvion verticale (cheminée diamantifère) (Géol.) (*f.*), pipe : diamond pipe. (60)

alluvionnaire (*adj.*), alluvial : (61)

or alluvionnaire (*m.*), alluvial gold. (62)

alluvionnement (*n.m.*), alluviation : (63)

dépôt formé par l'alluvionnement progressif de sables et d'argile (*m.*), deposit formed by the progressive alluviation of sand and clay. (64)

alluvionnien, -enne (*adj.*), alluvial : (65)

or alluvionnien (*m.*), alluvial gold. (66)

almandine (*n.f.*) *ou* almandin (*n.m.*) (Minéral.), almandite ; almandine ; almondine. (67)

alnoïte (Minéral.) (*n.f.*), alnoite ; alnoeite. (68)

aloès (*n.m.*), aloe : (69)

câble d'aloès (*m.*), aloe rope. (70)

alpe (*n.f.*), alp. (71)

Alpes (*n.f.pl.*), Alps : (72)

les Alpes Bernoises, the Bernese Alps. (73)

alpestre *ou* alpin, -e (*adj.*), alpine : (74)

la beauté des régions alpestres, the beauty of the alpine regions. (75)

la prairie alpine, the alpine grass-land. (76)

alphabets et chiffres à frapper [alphabet (*n.m.*) ; chiffre (*n.m.*)], letter and figure stamps for engineers. (77)

alphabets et chiffres à jour, stencil alphabets and figures. (78)

alpin, -e *ou* alpique (qui se rapporte aux Alpes) (*adj.*), Alpine : (79)

les divisions orographiques du système alpin (*f.pl.*), the orographic divisions of the Alpine system. (80)

la chaîne alpique, the Alpine chain. (81)

alquifoux (*n.m.*), alquifou ; arquifoux ; potters' lead ; potters' ore. (82)

alstonite (Minéral) (*n.f.*), alstonite. (1)

altaïte (Minéral) (*n.f.*), altaite. (2)

altazimut (*n.m.*), altazimuth. (3)

altérable (Minéralogie & Géol.) (*adj.*), alterable : (4)

métaux altérables (*m.pl.*), alterable metals. (5)

altération (*n.f.*), alteration ; change : (6)

altération chimique, chemical change. (7)

altération physique, physical change. (8)

altération (Minéralogie) (*n.f.*), alteration : (9)

la kaolinisation du feldspath est un exemple d'altération, the kaolinization of feldspar is an instance of alteration. (10)

on connaît des pseudomorphoses par altération d'orthose en kaolin ou muscovite, avec séparation de quartz, there are instances of pseudomorphoses by alteration of orthoclase into kaolin or muscovite, with separation of quartz. (11)

altération par l'action des agents atmosphériques *ou simplement* **altération** (*n.f.*) (Géol.), weathering ; alteration by the influence of atmospheric agents. (12)

altéré, -e (Minéralogie) (*adj.*), altered : (13)

nombre de minéraux altérés ont été considérés longtemps comme espèces minérales (*m.*), many altered minerals were for a long time regarded as mineral species. (14)

altéré (-e) par l'action des agents atmosphériques *ou simplement* **altéré, -e** (*adj.*) (Géol.), weathered ; altered by the influence of atmospheric agents ; altered : (15)

affleurement altéré par l'action des agents atmosphériques (*m.*), weathered outcrop. (16)

croûte altérée (*f.*), weathered crust. (17)

altérer (Minéralogie) (*v.t.*), to alter. (18)

altérer (Géol.) (*v.t.*), to weather ; to alter. (19)

altérer (s') (Minéralogie) (*v.r.*), to alter. (20)

altérer (s') (Géol.) (*v.r.*), to weather ; to alter : (21)

roche qui s'altère sphéroïdalement (*f.*), rock which weathers spheroïdally. (22)

altérer (s') (se détériorer) (*v.r.*), to deteriorate ; to become impaired : (23)

les lentilles sont susceptibles de s'altérer plus ou moins à la longue (*f.pl.*), lenses are apt to deteriorate more or less in time. (24)

alternance (*n.f.*), alternation : (25)

alternance du jour et de la nuit, alternation of day and night. (26)

alternances de gel et de dégel, alternations of frost and thaw. (27)

alternance des formations marines et d'eau douce (Géol.), alternation of marine and fresh-water formations. (28)

alternant, -e (*adj.*), alternant ; alternating. (29)

alternateur (Élec.) (*n.m.*), alternator ; alternat-ing-current dynamo ; alternating dynamo ; alternating-current generator. (30)

alternateur à courant diphasé (*m.*), two-phase alternator. (31)

alternateur à courant monophasé (*m.*), single-phase alternator ; one-phase generator. (32)

alternateur à courant polyphasé (*m.*), polyphase alternator ; multiphase generator. (33)

alternateur à courant triphasé (*m.*), three-phase alternator. (34)

alternateur à induit fixe et à inducteurs fixes (*m.*), inductor type of alternator ; inductor-alternator. (35)

alternateur à induit fixe et à inducteurs mobiles (*m.*), revolving-field type of alternator. (36)

alternateur à induit mobile et à inducteurs fixes (*m.*), revolving-armature type of alternator. (37)

alternatif, -ive (qui a lieu tour à tour) (*adj.*), alternate ; alternating ; alternative : (38)

l'action alternative de la pluie et du soleil (*f.*), the alternate action of rain and sun. (39)

méthode alternative de transport (*f.*), alterna-tive method of transport. (40)

alternatif, -ive (Méc.) (*adj.*), reciprocating ; alternating : (41)

mouvement alternatif (*m.*), reciprocating motion. (42)

scie alternative (*f.*), reciprocating saw, alter-nating saw. (43)

alternatif, -ive (Élec.) (*adj.*), alternating : (44)

courant alternatif (*m.*), alternating current. (45)

alternation (*n.f.*), alternation : (46)

alternation d'un mouvement (Méc.), alterna-tion of a motion. (47)

alternativité (Élec.) (*n.f.*), alternativity ; half-period. (48)

alterne (Géom.) (*adj.*), alternate : (49)

angles alternes (*m.pl.*), alternate angles. (50)

alterné, -e (*adj.*), alternate ; alternating : (51)

couches alternées de talc et de schiste (*f.pl.*), alternating layers of talc and schist. (52)

alterner (*v.i.*), to alternate : (53)

le jour alterne avec la nuit, day alternates with night. (54)

alternomoteur (*n.m.*), alternating-current motor. (55)

altimètre (*n.m.*), altimeter. (56)

altimétrie (*n.f.*), altimetry. (57)

altimétrique (*adj.*), altimetric ; altimetrical. (58)

altimétriquement (*adv.*), altimetrically. (59)

altitude (*n.f.*), altitude ; elevation ; height : (60)

altitude au-dessus du niveau de la mer, alti-tude, elevation, height, above sea-level. (61)

altitudinal, -e, -aux (*adj.*), altitudinal. (62)

alumiane (Minéral) (*n.f.*), alumian. (63)

aluminate (Chim.) (*n.m.*), aluminate : (64)

aluminate de sodium, de potassium, aluminate of sodium, of potassium. (65 *ou* 66)

alumine (Chim.) (*n.f.*), alumina ; aluminium oxide. (67)

alumine cristallisée (*f.*), crystallized alumina ; corundum. (68)

aluminer (*v.t.*), to aluminate ; to aluminize ; to alum. (69)

aluminerie (fabrique d'alun) (*n.f.*), alum-works. (70)

alumineux, -euse (*adj.*), aluminous : (71)

terre alumineuse (*f.*), aluminous earth. (72)

aluminiate (Chim.) (*n.m.*), aluminate. (73)

aluminico- (*préfixe*), alumino ; aluminio : (74)

aluminico-silicate (*n.m.*), aluminosilicate. (75)

aluminière (mine d'alun) (*n.f.*), alum-mine. (76)

aluminière (fabrique d'alun) (*n.f.*), alum-works. (77)

aluminifère (*adj.*), aluminiferous ; aluniferous : (78)

schiste aluminifère (*m.*), aluminiferous schist ; aluniferous shale. (79)

aluminique (Chim.) (*adj.*), aluminic. (80)

aluminite (Minéral) (*n.f.*), aluminite ; websterite. (81)

aluminium (*n.m.*), aluminium ; aluminum. (82)

aluminium du commerce (*m.*), commercial aluminium. (83)

alun (*n.m.*), alum. (84)

alun d'ammoniaque *ou* alun ammoniacal (*m.*), ammonia-alum. (1)

alun de chrome (*m.*), chrome-alum ; chromium alum. (2)

alun de fer (*m.*), iron alum. (3)

alun de plume (*m.*), feather-alum ; alum-feather ; plume-alum. (4)

alun de potasse *ou* alun potassique (*m.*), potash-alum ; potassium alum. (5)

alun de roche *ou* alun de Rome (*m.*), rock alum ; roche alum ; Roman alum. (6)

alun de soude *ou* alun de sodium (*m.*), soda-alum ; sodium alum. (7)

alun ordinaire (*m.*), common alum. (8)

alunage (*n.m.*) *ou* alunation (*n.f.*), alumination ; aluminization. (9)

aluner (*v.t.*), to alum ; to aluminate ; to alu--minize. (10)

alunerie (*n.f.*), alum-works. (11)

aluneux, -euse (*adj.*), aluminous : (12)
terrain aluneux (*m.*), aluminous ground. (13)

alunière (mine) (*n.f.*), alum-mine. (14)

alunière (fabrique) (*n.f.*), alum-works. (15)

alunifère (*adj.*), aluniferous ; aluminiferous : (16)
schiste alunifère (*m.*), aluniferous schist ; aluminiferous shale. (17)

alunite (Minéral) (*n.f.*), alunite ; alum-rock. (18)

alunogène (Minéral) (*n.m.*), alunogen. (19)

alurgite (Minéral) (*n.f.*), alurgite. (20)

alvéolaire (Géol.) (*adj.*), honeycomb (*used as adj.*): (21)
structure alvéolaire (usure éolienne, etc.) (*f.*), honeycomb structure. (22)

alvite (Minéral) (*n.f.*), alvite. (23)

am-mètre (Élec.) (*n.m.*). Same as ampèremètre.

amaigrir (rendre plus mince) (*v.t.*), to reduce ; to thin ; to thin down : (24)
amaigrir une planche, to reduce, to thin down, a board. (25)

amaigrir (s') (*v.r.*), to be reduced ; to thin down. (26)

amaigrissement (*n.m.*), reduction ; reducing ; thinning down. (27)

amalgamable (*adj.*), amalgamable. (28)

amalgamateur (appareil) (*n.m.*), amalgamator. (29)

amalgamateur (pers.) (*n.m.*), amalgamator. (30)

amalgamateur à cuve (*m.*), pan-amalgamator. (31)

amalgamation (*n.f.*), amalgamation ; amalga--mating. (32)

amalgamation au tonneau (*f.*), barrel-amalgama--tion. (33)

amalgamation aux cuves (*f.*), pan-amalgamation. (34)

amalgame (union du mercure avec un autre métal) (*n.m.*), amalgam. (35)

amalgame (Minéral) (*n.m.*), amalgam ; argental mercury. (36)

amalgame d'argent (*m.*), silver amalgam. (37)

amalgame d'or et de mercure (*m.*), gold and quick--silver amalgam. (38)

amalgame de sodium (*m.*), sodium amalgam. (39)

amalgamer (*v.t.*), to amalgamate : (40)
amalgamer l'or, to amalgamate gold. (41)

amalgamer (s') (*v.r.*), to amalgamate : (42)
l'or s'amalgame bien avec le mercure (*m.*), gold amalgamates readily with mercury. (43)

amalgameur (pers.) (*n.m.*), amalgamator. (44)

amarrage (*n.m.*), mooring : (45)
amarrage d'un câble, mooring a cable. (46)

amarrer (*v.t.*), to moor ; to make fast ; to belay : (47)
amarrer une drague, to moor a dredge. (48)
câble métallique dont les extrémités sont solidement amarrées sur le sol (*m.*), wire-rope whose ends are firmly moored on the ground. (49)
amarrer le bout d'un cordage, to make fast, to belay, the end of a rope. (50)

amas (*n.m.*), heap ; pile : (51)
amas de pierres, heap, pile, of stones. (52)
amas de sable, heap of sand. (53)

amas (Géol. & Mines) (*n.m.*), mass : (54)
amas couché (Géol.), recumbent mass. (55)
amas de minerai, mass of ore ; ore-mass ; bunny. (56)
amas debout (Géol.), erect mass. (57)
amas quartzeux, mass of quartz. (58)

amas (en) (Géol. & Mines), massive : (59)
gisements en amas (*m.pl.*), massive deposits. (60)

amasite (Explosif) (*n.f.*), amasite. (61)

amasser (mettre en tas) (*v.t.*), to heap up ; to heap ; to pile up : (62)
amasser de la terre, to heap up, to pile up, earth. (63)

amasser (assembler) (*v.t.*), to collect ; to gather. (64)

amasser (s') (se mettre en tas) (*v.r.*), to heap up ; to pile up ; to bank up. (65)

amasser (s') (s'assembler) (*v.r.*), to collect ; to gather : (66)
les eaux pluviales s'amassent dans les citernes (*f.pl.*), rain-water collects in cisterns. (67)

amateur (*adj. m.*), amateur : (68)
photographe amateur (*m.*), amateur photog--rapher. (69)

amateur (pers.) (*n.m.*), amateur : (70)
l'amateur et le professionnel, the amateur and the professional. (71)
tour d'amateur (*m.*), amateur's lathe. (72)

amausite (Pétrol.) (*n.f.*), amausite ; petrosilex. (73)

amazonite (Minéral) (*n.f.*), amazonite ; Amazon stone. (74)

amblygone (*adj.*), amblygon ; amblygonal ; obtuse-angled. (75)

amblygonite (Minéral) (*n.f.*), amblygonite. (76)

ambre (*n.m.*) *ou* ambre jaune (*m.*), amber ; yellow amber ; succinite ; succin. (77)

ambrin, -e (*adj.*), amberous ; ambery. (78)

ambrite *ou* ambérite (Minéral) (*n.f.*), ambrite ; amberite. (79)

ambulance (*n.f.*), ambulance. (80)

ambulant, -e (*adj.*), locomotive : (81)
moteur ambulant (*m.*), locomotive engine. (82)

âme (d'un câble métallique, d'un câble électrique, d'une corde) (*n.f.*), core (of a wire rope, of an electric cable, of a rope). (83 *ou* 84 *ou* 85)

âme (d'une poutre, d'un fer à T, d'un rail de chemin de fer) (*n.f.*), web (of a girder, of a T iron, of a railway-rail). (86 *ou* 87 *ou* 88)

amélioration (*n.f.*), improvement ; betterment ; amelioration : (89)
amélioration dans tout l'ensemble, all-round improvement. (90)
amélioration du rendement, betterment of the yield. (91)
amélioration marquée *ou* amélioration sensible, marked improvement. (92)

amélioration (plus-value : opp. à dépréciation *ou* moins-value) (*n.f.*), appreciation. (93)

améliorer (*v.t.*), to improve ; to better ; to ameliorate ; to appreciate : (1)
améliorer l'aérage d'une mine, to improve the ventilation of a mine. (2)

améliorer (s') (*v.r.*), to improve ; to better ; to ameliorate ; to appreciate : (3)
s'améliorer au fur et à mesure de l'avancement, to improve as advance is made. (4)

amenage (avancement) (Méc.) (*n.m.*), feed : (5)
machine à dégauchir avec amenage auto-matique (*f.*), surface-planer with automatic feed. (6)

amenage par crémaillère (*m.*), rack-feed. (7)

amenage par cylindres (*m.*), roller-feed. (8)

aménagement (disposition ; distribution avec ordre) (*n.m.*), arrangement ; arranging ; laying out : (9)
aménagement d'un atelier, arrangement of a workshop ; arranging a shop. (10)
aménagement des sièges d'extraction (Mines), laying out the workings. (11)

aménagement de pente (*m.*), grading. (12)

aménagement des chutes d'eau (*m.*), harnessing waterfalls. (13)

aménagement des eaux de la mine (*m.*), dealing with the water in the mine. (14)

aménagement du courant d'air (Mines) (*m.*), coursing ; coursing the air ; conducting and guiding the air-current. (15)

aménager (*v.t.*). See examples : (16) Cf. **aménagement**.
aménager du bois, to cut up wood. (17)
aménager la pente d'une route, to grade a road. (18)
aménager une mine, un puits, to lay out a mine, a shaft. (19 *ou* 20)

amenée des eaux captées d'une source jusque dans un réservoir (*f.*), conduction of, con-ducting, bringing, water caught from a spring to a reservoir. (21)

amener (*v.t.*). See examples : (22)
amener des remblais du jour (Mines), to bring in stowing from the surface. (23)
amener l'eau sur une propriété, to conduct, to bring, water to a property. (24)
amener la bulle d'air entre ses repères (Nivelle-ment), to bring the air-bubble to the centre of its run. (Levelling) (25)
amener un cheval par la bride, to lead a horse by the bridle. (26)

amenuisement (*n.m.*), reduction ; reducing ; thinning ; thinning down. (27)

amenuiser (*v.t.*), to reduce ; to thin ; to thin down : (28)
amenuiser une planche, to reduce, to thin down, a board. (29)

amenuiser (s') (*v.r.*), to be reduced ; to thin down. (30)

amésite (Minéral) (*n.f.*), amesite. (31)

améthyste (*n.f.*), amethyst. (32)

améthyste orientale (*f.*), Oriental amethyst. (33)

améthystin, -e (*adj.*), amethystine. (34)

ameublir la terre avec une pioche, to loosen the earth with a pick. (35)

amiante (Minéral) (*n.m.*), amianthus ; amiantus ; amiant ; amianth ; asbestos. (36)

amiantinite (Minéral) (*n.f.*), amianthinite. (37)

amiantoïde (*adj.*), amiantoid ; amianthoidal ; amiantoidal. (38)

amiantoïde (Minéral) (*n.f.*), amiantoid ; amian-thoid. (39)

amincir (*v.t.*), to thin ; to thin down ; to reduce : (40)
amincir une plaque en la passant au laminoir, to thin down a plate by passing it through the rolls ; to reduce a plate by passing it through the rolling-mill. (41)

amincir (s') (*v.r.*), to become thin ; to become thinner ; to thin down ; to thin out ; to pinch : (42)
le filon s'amincit quelque peu, the lode is thinning down somewhat. (43)

amincissement (*n.m.*), thinning ; thinning down ; thinning out ; reduction ; reducing ; pinch-ing. (44)

ammiolite (Minéral) (*n.f.*), ammiolite. (45)

ammonal (Explosif) (*n.m.*), ammonal. (46)

ammonia-dynamite (*n.f.*), ammonia dynamite. (47)

ammoniacal, -e, -aux (*adj.*), ammoniacal ; ammoniac : (48)
sel ammoniacal (*m.*), ammoniacal salt. (49)

ammoniaque (Chim.) (*n.f.*), ammonia. (50)

ammonique (*adj.*), ammonic ; ammonical. (51)

ammonite (Explosif) (*n.f.*), ammonite. (52)

ammonium (Chim.) (*n.m.*), ammonium. (53)

amoindrir (*v.t.*), to diminish ; to lessen ; to decrease. (54)

amoindrir (s') (*v.r.*), to diminish ; to lessen ; to decrease : (55)
filon qui s'amoindrit en largeur et en richesse (*m.*), lode which lessens in width and value. (56)

amoindrissement (*n.m.*), diminution ; lessening ; decrease. (57)

amollir (*v.t.*), to soften : (58)
le feu amollit les métaux, fire softens metals. (59)

amollir (s') (*v.r.*), to soften. (60)

amollissement (*n.m.*), softening : (61)
amollissement de l'asphalte, softening of asphalt. (62)

amonceler (*v.t.*), to heap up ; to heap ; to pile up ; to bank ; to bank up ; to drift : (63)
amonceler de la terre, to heap up, to pile up, earth. (64)
les vents amoncellent le sable (*m.pl.*), the winds drift the sand. (65)

amonceler (s') (*v.r.*), to heap up ; to pile up ; to bank up ; to drift : (66)
les nuages s'amoncellent au ciel (*m.pl.*), the clouds are banking up in the sky. (67)

amoncellement (*n.m.*), heaping up ; piling up ; banking up ; drifting : (68)
amoncellement des neiges, drifting of the snow. (69)

amont (d'une rivière) (*n.m.*), upper part (of a river). (70)

amont (d') *ou elliptiquement* **amont**, up-steam ; upstream ; up-river ; upriver ; upper ; head (*used as adj.*) : (71)
le niveau d'amont au-dessus d'un barrage, the up-stream level above a dam. (72)
l'extrémité amont d'une île, the upper extremity of an island. (73)
le bief d'amont d'un moulin, the headrace of a mill. (74)

amont (en) *ou* **amont** (à l'), up-stream ; upstream ; up-river ; upriver ; up : (75)
aller en amont, to go up-stream ; to go upriver. (76)
la vitesse à l'amont, the velocity upstream. (77)

amont de (en) *ou* **amont de** (à l'), above; up: (1)
usine qui reçoit l'eau par un canal d'amenée
situé en amont des chutes (*f.*), works which
receives the water by a feeder situated above
the falls. (2)

les eaux accumulées en amont d'un barrage
(*f.pl.*), the water accumulated above a
dam. (3)

des signaux placés en amont et en aval des
stations (*m.pl.*), signals placed above and
below stations; signals placed up and down
the line. (4)

coup de bélier qui se produit à l'amont d'un
robinet (*m.*), water-hammer which occurs
above a tap. (5)

amont pendage (Mines) (opp. à **aval pendage**) (*m.*),
back; backs. (6) (opp. to **bottom**):

l'amount pendage d'une galerie, d'un chantier
d'abatage, the back of a level, of a stope.
(7 *ou* 8)

la galerie possède environ — mètres d'amont
pendage, the level has about — metres of
backs. (9)

amorçage (Travail aux explosifs) (*n.m.*), priming:
(10)

amorçage d'un coup de mine, priming a shot.
(11)

amorçage (Soudure) (*n.m.*), scarfing (Welding).
(12)

amorçage (Magnétisme) (*n.m.*), energization: (13)
les dynamos sont auto-excitatrices, c'est-à-
dire que le courant nécessaire à l'amorçage
est emprunté à la machine elle-même (*f.pl.*),
dynamos are self-exciting, that is to say
the current required for energization is
taken from the machine itself. (14)

amorçage d'un arc (Élec.) (*m.*), striking an arc.
(15)

amorçage d'un injecteur (*m.*), starting of an
injector. (16)

amorçage de l'écoulement *ou* **amorçage du débit**
(d'un puits) (*m.*), inducing flow. (17)

amorçage du trou de sonde (*m.*), starting the
bore-hole. (18)

amorce (détonateur) (Travail aux explosifs) (*n.f.*),
cap; blasting-cap; detonator; exploder. (19)

amorce (poudre pour enflammer une charge)
(*n.f.*), priming. (20)

amorce (certaine quantité d'eau qu'on verse
dans une pompe) (*n.f.*), priming. (21)

amorce (d'une galerie, d'un travers-banc)
(Mines) (*n.f.*), mouth (of a drift, of a cross-
cut). (22 *ou* 23)

amorce (Soudure) (*n.f.*), scarf (Welding). (24)

amorce (pierre d'attente) (*n.f.*), toothing-stone;
tooth. (25)

amorce à allumage retardé (*f.*), delay-action
detonator. (26)

amorce à friction (*f.*), friction-fuse. (27)

amorce à la chaux vive (*f.*), lime cartridge (28)

amorce de coulée (dans un moule) (Fonderie) (*f.*),
gate; ingate; git; geat (an opening in a
mould through which the melted metal
enters in casting). (29)

amorce de quantité *ou* **amorce à fil** *ou* **amorce à
incandescence** (*f.*), quantity-fuse; battery-
fuse; glow-fuse. (30)

amorce de tension *ou* **amorce à étincelle** (*f.*),
tension-fuse; machine-fuse; spark-fuse. (31)

amorce électrique (*f.*), electric fuse; electric
blasting-cap; electric detonator. (32)

amorce fulminante (*f.*), fulminating cap. (33)

amorcer (*v.t.*). See examples: (34)

amorcer les deux bouts à souder, to scarf the
two ends to be welded. (35)

amorcer un arc (Élec.), to strike an arc. (36)

amorcer un injecteur, to start an injector. (37)

amorcer un siphon, to start the flow of water
in a siphon. (38)

amorcer un trou dans une pièce de métal, to
start a hole in a piece of metal. (39)

amorcer une cartouche (Travail aux explosifs),
to prime a cartridge. (40)

amorcer une dynamo (Magnétisme), to energize
a dynamo. (41)

amorcer une galerie, un montage (Mines),
to start a level, a raise. (42 *ou* 43)

amorcer une machine à vapeur, to bar a steam-
engine. (44)

amorcer une pompe, to prime a pump;
to fetch a pump; to start a pump in
operation. (45)

amorcer (s') (se dit d'un injecteur) (*v.r.*), to start;
to catch on: (46)

injecteur qui s'amorce instantanément,
injector which starts instantaneously (*or*
which catches on instantly). (47)

amorcer (s') (Magnétisme) (*v.r.*), to energize: (48)
dynamo qui s'amorce à circuit ouvert (*f.*),
dynamo which energizes on open circuit. (49)

amorceur (robinet d'amorçage) (*n.m.*), priming-
cock. (50)

amorçoir (Charp.) (*n.m.*), boring-bit. (51)

amorphe (*adj.*), amorphous; amorphic; amor-
-phose; structureless: (52)

phosphore amorphe (*m.*), amorphous phos-
-phorus. (53)

minéraux amorphes (*m.pl.*), amorphous
minerals (uncrystallized). (54)

roches amorphes (*f.pl.*), amorphous rocks;
structureless rocks. (55)

amorphie (*n.f.*) *ou* **amorphisme** (*n.m.*), amor-
-phism; amorphia. (56)

amortir (Élec.) (*v.t.*), to damp. (57)

amortir un choc, to deaden, to absorb, a shock.
(58)

amortissement (Élec.) (*n.m.*), damping. (59)

amortisseur (*n.m.*) *ou* **amortisseur de choc** (Méc.),
shock-absorber. (60)

amortisseur (Élec.) (*n.m.*), damping-grid. (61)

amortisseur (Aérage de mines) (*n.m.*), diffuser;
moderator. (62)

amortisseur à friction (*m.*), friction draft-gear.
(63)

amovible (*adj.*), removable; detachable: (64)
chaudière avec foyer amovible (*f.*), boiler with
removable furnace. (65)

ampère (Élec.) (*n.m.*), ampere. (66)

ampère-heure [**ampères-heure** *pl.*] (*n.m.*),
ampere-hour. (67)

ampère-tour [**ampères-tours** *pl.*] (*n.m.*), ampere-
turn. (68)

ampèremètre (*n.m.*), ammeter; ampere-meter;
amperometer; ampere-gauge. (69)

ampèremètre à aimant mobile (*m.*), movable-
iron ammeter. (70)

ampèremètre à courant mobile (*m.*), movable-
coil ammeter; suspended-coil ammeter. (71)

ampèremètre enregistreur (*m.*), recording
ammeter. (72)

ampèremètre thermique (*m.*), hot-wire ammeter;
thermic amperemeter. (73)

ampères-mètre [**ampères-mètres** *pl.*] (*n.m.*),
Same as **ampèremètre**.

amphibole (Minéral) (*n.f.*), amphibole. (1)
amphibolifère (*adj.*), amphiboliferous. (2)
amphibolique (*adj.*), amphibolic. (3)
amphibolite (Pétrol.) (*n.f.*), amphibolite. (4)
amphiboloïde (*adj.*), amphiboloid. (5)
amphiboloschiste (*n.m.*), hornblende schist ;
  hornblende slate. (6)
amphigène (Minéral) (*n.m.*), amphigene ; leucite ;
  white garnet. (7)
amphithalite (Minéral) (*n.f.*), amphithalite. (8)
amphithéâtral, -e, -aux (*adj.*), amphitheatrical :
  (9)
  dépression amphithéâtrale (*f.*), amphitheatrical
  depression. (10)
amphithéâtre (*n.m.*), amphitheatre : (11)
  un amphithéâtre de collines, an amphitheatre
  of hills. (12)
amphodélite (Minéral) (*n.f.*), amphodelite. (13)
ample (*adj.*), ample ; full : (14)
  ample provision de combustible, ample, full,
  supply of fuel. (15)
amplifiant, -e *ou* amplificatif, -ive (Opt.) (*adj.*),
  amplifying ; amplificative ; amplificatory ;
  magnifying : (16)
  la puissance amplificative d'un microscope,
  the amplifying power of a microscope. (17)
  le pouvoir amplifiant d'une loupe, the magnify-
  -ing power of a lens. (18)
amplificateur (Photogr.) (*n.m.*), enlarger. (19)
amplification (Opt.) (*n.f.*), amplification ;
  magnification. (20)
amplifier (Opt.) (*v.t.*), to amplify ; to magnify :
  (21)
  le microscope amplifie les petits objets, the
  microscope magnifies small objects. (22)
amplitude (Phys.) (*n.f.*), amplitude : (23)
  amplitude des oscillations d'un pendule,
  amplitude of the oscillations of a pendulum.
  (24)
  amplitude des vibrations, amplitude of
  vibrations. (25)
amplitude de battage (Sondage) (*f.*), length of
  stroke of the chisel. (26)
ampoule (d'un thermomètre, d'une lampe élec-
  -trique à incandescence) (*n.f.*), bulb (of a
  thermometer, of an incandescent electric
  lamp). (27 *ou* 28)
ampoule (Verrerie de laboratoire) (*n.f.*), flask.
  (29). See also ballon and flacon.
ampoule (boursouflure qui se trouve sur l'acier
  de cémentation) (*n.f.*), blister. (30)
ampoule de Crookes (*f.*), Crookes tube. (31)
amvis (Explosif) (*n.m.*), amvis. (32)
amygdale (Géol.) (*n.f.*), amygdule. (33)
amygdaloïde *ou* amygdalaire *ou* amygdalin, -e
  (Géol.) (*adj.*), amgydaloid ; amygdaloidal :
  (34)
  structure amygdaloïde (*f.*), amygdaloid
  structure. (35)
an (*n.m.*), year : (36)
  une fois l'an, once a year. (37)
anachromatique (*adj.*), anachromatic ; soft-
  focus ; diffused-focus : (38)
  objectif anachromatique (*m.*), soft-focus lens ;
  diffused-focus lens ; anachromatic lens. (39)
anaclinal, -e, -aux (Géol.) (*adj.*), anaclinal. (40)
analcime (Minéral) (*n.f.*), analcite ; analcime.
  (41)
anallatique (Opt.) (*adj.*), anallatic : (42)
  lunette anallatique (*f.*), anallatic telescope.
  (43)
anallatisme (*n.m.*), anallatism. (44)

analogue (Cristall.) (*adj.*), analogous : (45)
  pôle analogue (*m.*), analogous pole. (46)
analysabilité (*n.f.*), analyzability ; analysability.
  (47)
analysable (*adj.*), analyzable ; analysable. (48)
analyse (Chim.) (*n.f.*), analysis : (49)
  analyse des gaz, de l'eau, analysis of gases,
  of water. (50 *ou* 51)
  analyse faite sur les retours d'air, analysis of
  air-returns ; testing the quality of the air-
  returns. (52)
analyse (Chimie minérale ou métallurgique) (*n.f.*),
  analysis ; analyzing ; assay ; assaying : (53)
  See also essai.
  analyse des métaux par électrolyse, analysis
  of, assay of, assaying, metals by electrol-
  -ysis. (54)
analyse au chalumeau (*f.*), blowpipe-analysis ;
  blowpipe-assay ; blowpipe-assaying ; blow-
  -piping. (55)
analyse calorimétrique (*f.*), calorimetrical analysis.
  (56)
analyse chimique (*f.*), chemical analysis. (57)
analyse contradictoire (*f.*), check-analysis ;
  check-assay. (58)
analyse électrolytique (*f.*), electroanalysis ;
  electrolytic analysis. (59)
analyse élémentaire (*f.*), elementary analysis. (60)
analyse eudiométrique (*f.*), eudiometric analysis.
  (61)
analyse gravimétrique (*f.*), gravimetric analysis.
  (62)
analyse immédiate (*f.*), proximate analysis. (63)
analyse inorganique *ou* analyse minérale (*f.*),
  inorganic analysis ; mineral analysis. (64)
analyse microchimique (*f.*), microchemical
  analysis. (65)
analyse organique (*f.*), organic analysis. (66)
analyse par la voie humide *ou* analyse par voie
  humide (*f.*), wet analysis ; wet assay. (67)
analyse par la voie sèche *ou* analyse par voie
  sèche (*f.*), dry analysis ; dry assay. (68)
analyse polariscopique (*f.*), polariscopic analysis.
  (69)
analyse pyrognostique (*f.*), pyrognostic analysis.
  (70)
analyse qualitative *ou* analyse qualificative (*f.*),
  qualitative analysis. (71)
analyse quantitative (*f.*), quantitative analysis.
  (72)
analyse spectrale (*f.*), spectrum analysis. (73)
analyse ultime (*f.*), ultimate analysis. (74)
analyse volumétrique (*f.*), volumetric analysis.
  (75)
analyser (Chim.) (*v.t.*), to analyze ; to analyse :
  (76)
  analyser un composé chimique, to analyze a
  chemical compound. (77)
  analyser l'air de la mine pour déterminer la
  teneur en grisou, to analyse the air of the
  mine in order to ascertain the content of
  fire-damp ; to test the air for fire-damp.
  (78)
analyser (Chimie minérale ou métallurgique) (*v.t.*),
  to analyze ; to analyse ; to assay : (79)
  analyser un minerai, to analyse, to assay,
  an ore. (80)
analyser (s') (*v.r.*), to be analyzed : (81)
  substance qui ne peut s'analyser (*f.*), substance
  which cannot be analyzed. (82)
analyseur (d'un polariscope) (Phys.) (*n.m.*),
  analyzer (of a polariscope). (83)

analyste (Chim.) (pers.) (*n.m.*), analyst. (1)

analyste (Chimie minérale ou métallurgique) (*n.m.*), analyst ; assayer. (2)

analytique (*adj.*), analytic ; analytical : (3) chimie analytique (*f.*), analytical chemistry. (4)

analytiquement (*adv.*), analytically. (5)

anamésite (Pétrol.) (*n.f.*), anamesite. (6)

anamorphisme (Géol.) (*n.m.*), anamorphism. (7)

anastigmate *ou* anastigmat *ou* anastigmatique (*adj.*), anastigmat ; anastigmatic ; stig--matic : (8)
objectif anastigmat (*m.*), anastigmat lens ; stigmatic lens. (9)

anastigmat (*n.m.*), anastigmat ; anastigmat lens ; anastigmatic lens. (10)

anatase (Minéral) (*n.f.*), anatase ; octahedrite. (11)

anazotique (*adj.*), non-nitrogenous. (12)

anches (d'une chèvre, d'une bigue) (*n.f.pl.*), shears ; sheers ; shear-legs ; sheer-legs. (13 *ou* 14)

ancien, -enne (*adj.*), old ; past ; former ; ancient: (15)
anciens cours d'eau (*m.pl.*), old streams. (16)
ancien propriétaire (*m.*), former owner. (17)
anciens travaux (Mines) (*m.pl.*), old workings ; ancient workings. (18)
ancienne plage (Géogr. phys.) (*f.*), ancient beach. (19)

ancrage (Constr.) (*n.m.*), anchoring ; fixing. (20)

ancrage de la cheminée (Mines) (*m.*), choking, clogging, of chute. (21)

ancre (Marine) (*n.f.*), anchor. (22)

ancre (Constr.) (*n.f.*), anchor : (23)
ancre en forme d'un S, d'un X, anchor in the form of an S, of an X ; S plate, X plate. (24 *ou* 25).

ancre (Télégr., etc.) (*n.f.*), stay-anchor : (26)
ancre qui reçoit le tirage d'un hauban, stay-anchor which receives the pull of a guy. (27)

ancre (d'une lampe électrique à incandescence) (*n.f.*), anchor (of an electric incandescent lamp). (28)

ancrer (Constr.) (*v.t.*), to anchor ; to fix : (29)
ancrer une solive de plancher à un mur, to anchor, to fix, a floor-joist to a wall. (30)

andalousite (Minéral) (*n.f.*), andalusite. (31)

andésine (Minéral) (*n.f.*), andesine ; andesite. (32)

andésite (Pétrol.) (*n.f.*), andesite ; andesyte. (33)

andésitique (*adj.*), andesitic. (34)

andorite (Minéral) (*n.f.*), andorite (*f.*). (35)

andradite (Minéral) (*n.f.*), andradite ; melanite. (36)

andrewsite (Minéral) (*n.f.*), andrewsite. (37)

âne (étau d'établi) (*n.m.*), bench-vice. (38)

anélectrique (*adj.*), anelectric. (39)

anémie des mineurs (*f.*), miners' anemia ; miners' anæmia ; miners' disease. (40)

anémographe (*n.m.*), anemograph. (41)

anémographe enregistreur (*m.*), recording anemograph ; self-registering anemograph. (42)

anémographie (*n.f.*), anemography. (43)

anémographique (*adj.*), anemographic. (44)

anémomètre (*n.m.*), anemometer ; wind-gauge. (45)

anémomètre à moulinet (*m.*), vane-anemometer. (46)

anémomètre de pression (*m.*), pressure-plate anemometer. (47)

anémomètre de rotation *ou* anémomètre rotatif (*m.*), rotary anemometer ; velocity-ane--mometer. (48)

anémomètre différentiel (*m.*), differential anemometer. (49)

anémomètre enregistreur (*m.*), recording ane--mometer. (50)

anémomètre pendulaire (*m.*), pendulum ane--mometer. (51)

anémométrie (*n.f.*), anemometry. (52)

anémométrique (*adj.*), anemometric ; anemo--metrical. (53)

anémométriquement (*adv.*), anemometrically. (54)

anémométrographe (*n.m.*), anemometrograph. (55)

anémométrographique (*adj.*), anemometro--graphic. (56)

anémoscope (*n.m.*), anemoscope. (57)

anéroïde (*adj.*), aneroid : (58)
baromètre anéroïde (*m.*), aneroid barometer. (59). See baromètre anéroïde for varieties.

anéroïde (*n.m.*), aneroid. (60)

anfractueux, -euse (*adj.*), rugged ; rough ; craggy ; uneven : (61)
rochers anfractueux (*m.pl.*), rugged (*or* craggy) rocks. (62)

anfractuosité (*s'emploie souvent au pluriel*) (*n.f.*), ruggedness ; roughness ; cragginess ; cragged--ness ; unevenness : (63)
anfractuosités des rochers, des côtes de la mer, ruggedness, cragginess, of the rocks, of the seacoast. (64 *ou* 65).
anfractuosités d'un chemin, roughness of a way ; unevenness of a road. (66)

angle (*n.m.*), angle. (67)

angle aigu (*m.*), acute angle. (68)

angle curviligne (*m.*), curvilinear angle. (69)

angle d'avance (tiroir) (l'excès de l'angle de calage sur 90°) (*m.*), angle of advance ; angular advance (slide-valve). (70)

angle d'avance à l'admission (tiroir) (*m.*), angle of lead ; angular lead. (71)

angle d'avance à l'échappement (tiroir) (*m.*), angle of prerelease ; angular prerelease. (72)

angle d'incidence (*m.*), angle of incidence. (73)

angle d'incidence (machine-outil) (*m.*), angle of clearance ; angle of relief ; clearance-angle ; relief-angle ; relief ; bottom rake. (74)

angle d'incidence limite (Opt.) (*m.*), critical angle ; angle of total reflection. (75)

angle d'inclinaison (*m.*), angle of inclination : (76)
angle d'inclinaison de la voie d'un chemin de fer de montagne, angle of inclination of the track of a mountain-railway. (77)

angle de calage (de l'excentrique du tiroir) (Méc.) (*m.*), angle of keying (of the valve-eccentric). (78)

angle de calage (Élec.) (*m.*), angle of lead. (79)

angle de calage des balais (Elec.) (*m.*), angle of lead of brushes. (80)

angle de champ (Photogr.) (*m.*), angle of view. (81)

angle de contact *ou* angle de contingence (*m.*), angle of contact. (82)

angle de coupe (machine-outil) (*m.*), cutting angle. (83)

angle de décalage (Élec.) (*m.*), angle of lag. (84)

angle de dégagement (machine-outil) (*m.*), angle of rake ; top rake. (85)

angle de frottement (*m.*), angle of friction; friction-angle. (1)

angle de pendage (d'un filon) (*m.*), angle of dip (of a lode). (2)

angle de polarisation (*m.*), angle of polarization; polarizing angle. (3)

angle de pression (Engrenages) (*m.*), pressure-angle; angle of obliquity (Gearing). (4)

angle de recouvrement (tiroir) (*m.*), angle of lap (slide-valve). (5)

angle de réflexion (*m.*), angle of reflection. (6)

angle de réfraction (*m.*), angle of refraction. (7)

angle de repos (*m.*), angle of repose; natural slope. (8)

angle de sortie (turbines) (*m.*), exit-angle. (9)

angle de taillant (machine-outil) (*m.*), tool-angle. (10)

angle dièdre (*m.*), dihedral angle. (11)

angle droit (*m.*), right angle. (12)

angle droit (à), at right angles: (13)
à angle droit avec une ligne tirée entre — et —, at right angles to a line drawn between — and —. (14)

angle du croisement (Ch. de f.) (*m.*), crossing-angle; angle of crossing. (15)

angle externe (*m.*), exterior angle; external angle; outward angle. (16)

angle inscrit (*m.*), inscribed angle. (17)

angle interne (*m.*), interior angle; internal angle. (18)

angle limite (Opt.) (*m.*), critical angle; angle of total reflection. (19)

angle obtus (*m.*), obtuse angle. (20)

angle optique (*m.*), optic angle. (21)

angle plan *ou* **angle rectiligne** (*m.*), plane angle. (22)

angle polyèdre (*m.*), polyhedral angle. (23)

angle réfringent d'un prisme (*m.*), refracting angle of a prism. (24)

angle rentrant (*m.*), reentrant angle; reentering angle. (25)

angle saillant (*m.*), salient angle. (26)

angle solide (*m.*), solid angle. (27)

angle sphérique (*m.*), spherical angle. (28)

angle trièdre (*m.*), trihedral angle. (29)

angle visuel (Opt.) (*m.*), visual angle. (30)

angles adjacents (*m.pl.*), adjacent angles. (31)

angles adjacents supplémentaires (*m.pl.*), supple-mentary adjacent angles. (32)

angles alternes (*m.pl.*), alternate angles. (33)

angles alternes externes (*m.pl.*), exterior alter-nate angles. (34)

angles alternes internes (*m.pl.*), interior alter-nate angles. (35)

angles complémentaires (*m.pl.*), complementary angles. (36)

angles correspondants (*m.pl.*), corresponding angles. (37)

angles supplémentaires (*m.pl.*), supplementary angles; supplemental angles. (38)

angles trièdres symétriques (*m.pl.*), symmetrical trihedral angles. (39)

anglésite (Minéral) (*n.f.*), anglesite; lead-vitriol. (40)

angloir (*n.m.*), bevel; bevel-square. (41)

anglomètre (*n.m.*), angle-meter; angulometer. (42)

angulaire (*adj.*), angular: (43)
vitesse angulaire (Méc.) (*f.*), angular velocity. (44)

angulairement (*adv.*), angularly. (45)

angularité (*n.f.*), angularity; angularness. (46)

anhydre (*adj.*), anhydrous; anhydric: (47)
sel anhydre (*m.*), anhydrous salt. (48)

anhydride (Chim.) (*n.m.*), anhydride. (49)

anhydride arsénieux (*m.*), arsenious anhydride; arsenious oxide; white arsenic; arsenic. (50)

anhydride sulfocarbonique (*m.*), carbon di-sulphide; carbon bisulphide. (51)

anhydride sulfureux (*m.*), sulphur dioxide; sulphurous anhydride; sulphurous oxide. (52)

anhydride sulfurique (*m.*), sulphur trioxide; sulphuric anhydride. (53)

anhydrite (Minéral) (*n.f.*), anhydrite; karstenite. (54)

animal, -e, -aux (*adj.*), animal: (55)
traction animale (*f.*), animal traction. (56)

animal (*n.m.*), animal. (57)

animal de trait (*m.*), draught animal. (58)

animikite (Minéral) (*n.f.*), animikite. (59)

anion (Élec.) (*n.m.*), anion. (60)

anisométrique (Cristall.) (*adj.*), anisometric. (61)

anisotrope (Phys.) (*adj.*), anisotropic; aniso-tropal; anisotrope; anisotropical; aniso-tropous: (62)
cristal anisotrope (*m.*), anistropic crystal. (63)

anisotropie (*n.f.*), anisotropy; anisotropism. (64)

ankérite (Minéral) (*n.f.*), ankerite. (65)

ankylostomiase (Pathologie) (*n.f.*), ankylosto-miasis; miners' anemia. (66)

annabergite (Minéral) (*n.f.*), annabergite; nickel-ochre. (67)

anneau (*n.m.*), ring. (68)

anneau (d'une chaîne) (*n.m.*), link (of a chain). (69)

anneau (d'une clef) (*n.m.*), bow (of a key). (70)

anneau de cylindre (de broyeur à cylindres) (*m.*), roll-shell (of crushing-rolls). (71)

anneau de garde (d'un électromètre) (*m.*), guard-ring (of an electrometer). (72)

anneau de graissage (*m.*), oil-ring. (73)

anneau de manœuvre (Sondage) (*m.*), ring and wedges; spider and slips. (74)

anneau de Pacinotti *ou* **anneau de Gramme** (Élec.) (*m.*), Pacinotti ring; Gramme ring. (75)

anneau oculaire (Opt.) (*m.*), eye-ring. (76)

anneaux colorés (Phys.) (*m.pl.*), coloured rings. (77)

anneaux de Newton (Phys.) (*m.pl.*), Newton's rings. (78)

anneaux de Nobili (Phys.) (*m.pl.*), Nobili's rings; electric rings; electrical rings. (79)

année (*n.f.*), year; twelvemonth. (80)

annérödite (Minéral) (*n.f.*), annerodite. (81)

annite (Minéral) (*n.f.*), annite. (82)

annonciateur (Téléph.) (*n.m.*), indicator. (83)

annuel, -elle (*adj.*), annual; yearly: (84)
précipitation annuelle (*f.*), annual rainfall. (85)

annuellement (*adv.*), annually; yearly. (86)

annulaire (*adj.*), annular; ring-shaped: (87)
soupape annulaire (*f.*), annular valve. (88)

annulation (*n.f.*), cancellation; cancelling. (89)

annuler (*v.t.*), to cancel; to annul: (90)
annuler une commande, to cancel an order. (91)
ce catalogue annule les précédents, this cata-logue cancels previous ones. (92)

anode (Phys.) (*n.f.*), anode; anelectrode. (93)

anomite (Minéral) (*n.f.*), anomite. (94)

anormal, -e, -aux (*adj.*), abnormal : (1) effort anormal (Méc.) (*m.*), abnormal stress. (2)

anorthite (Minéral) (*n.f.*), anorthite. (3)

anorthose (Minéral) (*n.f.*), anorthose ; anortho- -clase. (4)

anorthosite (Pétrol.) (*n.f.*), anorthosite. (5)

anse (d'un panier, d'un seau, ou analogue) (*n.f.*), handle (of a basket, of a pail, or the like). (6 *ou* 7 *ou* 8)

anse (d'une poche de fonderie) (*n.f.*), bail, bull, bull-handle, bow, bow-handle, (of a foundry- ladle). (9)

anse (d'une benne) (*n.f.*), lifting-bow (of a hoisting-bucket). (10)

anse (d'un cadenas) (*n.f.*), bow, shackle, (of a padlock). (11)

anse (Géogr. phys.) (*n.f.*), cove. (12)

anse de panier (*f.*), basket handle : (13) voûte en anse de panier (Arch.) (*f.*), basket- handle arch. (14)

anspect (*n.m.*), handspike ; lever. (15)

anspect ferré (*m.*), iron-shod lever. (16)

antarctique (*adj.*), antarctic : (17) pôle antarctique (*m.*), antarctic pole. (18)

antécédent, -e (Géol.) (*adj.*), antecedent : (19) cours d'eau antécédent (*m.*), antecedent stream. (20)

anthophyllite *ou* antholite (Minéral) (*n.f.*), anthophyllite ; antholite. (21)

anthosidérite (Minéral) (*n.f.*), anthosiderite. (22)

anthracifère (*adj.*), anthraciferous : (23) terrain anthracifère (*m.*), anthraciferous ground. (24)

anthracite (*n.m.*), anthracite ; anthracite coal ; hard coal ; stone-coal ; glance-coal ; blind coal. (25)

anthraciteux, -euse (*adj.*), anthracitous ; semian- -thracitic. (26)

anthracitique (*adj.*), anthracitic. (27)

anthracosis (Pathologie) (*n.f.*), anthracosis ; blacklung : (28) l'anthracosis est une maladie qui attaque les houilleurs, anthracosis is a disease that attacks colliers. (29)

anti-halo (Photogr.) (*adj.*), anti-halo ; non- halation ; non-halative ; antihalation : (30) plaque anti-halo (*f.*), anti-halo plate ; non- halation plate ; non-halative plate. (31)

anticathode (Phys.) (*n.f.*), anticathode. (32)

anticlinal, -e, -aux (opp. à synclinal, -e, -aux) (Géol.) (*adj.*), anticlinal (33) (opp. to syn- -clinal) : pli anticlinal (*m.*), anticlinal fold. (34)

anticlinal (Géol.) (*n.m.*), anticline ; anticlinal. (35)

anticlinal en forme de selle (Mines) (*m.*), saddle ; saddle-reef. (36)

antidote (*n.m.*), antidote ; counterpoison : (37) en cas d'empoisonnement par l'hydrogène sulfuré, l'antidote indiqué est le chlore (faire respirer au patient du chlorure de chaux imbibé de vinaigre), in case of hydrogen sulphide poisoning, the antidote is chlorine (make the patient breathe chloride of lime soaked in vinegar). (38) antidote pour cyanure *ou* antidote au cyanure *ou* antidote contre l'empoisonnement par le cyanure, antidote for cyanide ; antidote to cyanide ; antidote against cyanide poisoning. (39)

antifriction (*adj. invar.*), antifriction : (40) alliage antifriction (*m.*), antifriction alloy. (41)

antifriction (*n.f.*), antifriction metal ; white metal ; babbitt metal ; babbitt : (42) palier garni d'antifriction (*m.*), bearing lined with antifriction metal ; babbitted bearings. (43)

antigorite (Minéral) (*n.f.*), antigorite. (44)

antilogue (Cristall.) (*adj.*), antilogous : (45) pôle antilogue (*m.*), antilogous pole. (46)

antimagnétique (*adj.*), antimagnetic. (47)

antimoine (Chim.) (*n.m.*), antimony. (48)

antomoine cru (*m.*), crude antimony. (49)

antimoine métallique (*m.*), metallic antimony. (50)

antimoine oxydé (Minéral) (*m.*), antimony- bloom ; valentinite. (51)

antimoine sulfuré (Minéral) (*m.*), antimony- glance ; gray antimony ; stibnite ; anti- -monite. (52)

antimonial, -e, -aux (*adj.*), antimonial : (53) argent antimonial (*m.*), antimonial silver. (54)

antimoniate (Chim.) (*n.m.*), antimonate ; anti- -moniate. (55)

antimonié, -e (*adj.*), antimoniated : (56) plomb antimonié (*m.*), antimoniated lead. (57)

antimonieux, -euse (Chim.) (*adj.*), antimonious ; stibious. (58)

antimonifère (*adj.*), antimoniferous. (59)

antimonique (Chim.) (*adj.*), antimonic ; stibic. (60)

antimonite (Chim.) (*n.m.*), antimonite. (61)

antimoniure (Chim.) (*n.m.*), antimonide. (62)

antimonocre (Minéral) (*n.m.*), antimony ochre ; cervantite. (63)

antiparallèle (Géom.) (*adj.*), antiparallel. (64)

antiparallèle (*n.m.*), antiparallel. (65)

antitartrique (*n.m.*), anti-incrustator. (66)

antozonite (Minéral) (*n.f.*), antozonite. (67)

apatélite (Minéral) (*n.f.*), apatelite. (68)

apatite (Minéral) (*n.f.*), apatite. (69)

apériodique (*adj.*), aperiodic ; dead-beat : (70) galvanomètre apériodique (*m.*), aperiodic galvanometer ; dead-beat galvanometer. (71)

apex (*n.m.*), apex. (72)

aphanèse *ou* aphanésite (Minéral) (*n.f.*), aphane- -site ; clinoclasite. (73)

aphanite (Pétrol.) (*n.f.*), aphanite. (74)

aphanitique (*adj.*), aphanitic. (75)

aphosite (Explosif) (*n.f.*), aphosite. (76)

aphrite (Minéral) (*n.f.*), aphrite ; foam-spar. (77)

aphrizite (Minéral) (*n.f.*), aphrizite. (78)

aphrodite (Minéral) (*n.f.*), aphrodite. (79)

aphrosidérite (Minéral) (*n.f.*), aphrosiderite. (80)

aphtalose *ou* aphthalose *ou* aphthitalite (Minéral) (*n.f.*), aphthitalite ; aphthalose. (81)

aphthonite (Minéral) (*n.f.*), aphthonite. (82)

apjohnite (Minéral) (*n.f.*), apjohnite. (83)

aplanat (Opt.) (*n.m.*), aplanat. (84)

aplaner (*v.t.*), to plane ; to planish ; to smooth : (85) aplaner une douve, to plane, to planish, a stave. (86)

aplanétique (Opt.) (*adj.*), aplanatic ; aplanetic : (87) objectif aplanétique (*m.*), aplanatic lens. (88)

aplanétisme (*n.m.*), aplanatism : (89) aplanétisme des miroirs, des lentilles, aplan- -atism of mirrors, of lenses. (90 *ou* 91)

aplaneur (pers.) (*n.m.*), planisher. (1)
aplanir (*v.t.*), to level ; to smooth : (2)
  aplanir un chemin, to level a path. (3)
aplanir (s') (*v.r.*), to become level ; to become smooth. (4)
aplanissement (*n.m.*), levelling ; smoothing : (5)
  aplanissement des terres, levelling the ground. (6)
aplatir (*v.t.*), to flatten ; to flat : (7)
  aplatir une pièce de métal, to flatten, to flat, a piece of metal. (8)
  aplatir une courbe, to flatten a curve. (9)
aplatir (s') (*v.r.*), to flatten ; to become flat. (10)
aplatissement (action) (*n.m.*), flattening; flatting : (11)
  l'aplatissement du sphéroïde terrestre, the flattening of the terrestrial spheroid. (12)
  aplatissement d'une barre de fer en la faisant passer entre des cylindres, flatting an iron bar by passing it between rolls. (13)
aplatissement (état) (*n.m.*), flatness. (14)
aplatisserie (atelier) (*n.f.*), flatting-works ; flatting-mill ; flattening-mill. (15)
aplatisseur (pers.) (*n.m.*), flatter. (16)
aplatissoir (marteau) (*n.m.*), flatter. (17)
aplatissoire (laminoir) (*n.f.*), flatting-mill ; flattening-mill. (18)
aplite (Pétrol.) (*n.f.*), aplite ; haplite. (19)
aplomb (*n.m.*), perpendicularity ; plumb : (20)
  l'aplomb d'une tour, the perpendicularity of a tower. (21)
  'hors d'aplomb, out of plumb. (22)
aplomb (fil à plomb) (*n.m.*), plumb-bob ; plumb-line ; plummet ; plummet-line. (23)
  See **fil à plomb** for varieties.
aplomb (d') *ou* **à plomb**, plumb ; perpendicularly : (24)
  la maison se tient à plomb, the house stands plumb. (25)
  jalon planté d'aplomb (*m.*), stake set perpendicularly. (26)
aplome (Minéral) (*n.m.*), aplome ; haplome. (27)
apochromatique (Opt.) (*adj.*), apochromatic : (28)
  objectif apochromatique (*m.*), apochromatic lens. (29)
apophyllite (Minéral) (*n.f.*), apophyllite. (30)
apophyse (Géol.) (*n.f.*), apophysis ; offshoot : (31)
  apophyses de roches éruptives, apophyses of eruptive rocks. (32)
apothème (Géom.) (*n.m.*), apothem. (33)
appareil (*n.m.*) *ou* **appareils** (*n.m.pl.*), apparatus ; appliance ; appliances ; plant ; machinery ; gear ; tackle. (34)
appareil (Photogr.) (*n.m.*). See **appareil photographique**.
appareil (disposition des briques, des pierres, dans un mur) (Maçonn.) (*n.m.*), bond : (35)
  appareil en boutisses, heading bond. (36)
appareil à air chaud (pour hauts fourneaux) (*m.*), hot-blast stove for blast-furnaces. (37)
appareil à braser les scies à ruban (*m.*), band-saw brazing apparatus. (38)
appareil à charioter conique (tour) (*m.*), taper-turning attachment (lathe). (39)
appareil à fraiser à grande vitesse (*m.*), high-speed milling-attachment. (40)
appareil à fraiser circulairement (*m.*), circular milling-attachment. (41)
appareil à pulvérisation d'eau (*m.*), spray ; pulverizer. (42)

appareil à redresser et mettre au rond les meules en grès (*m.*), grindstone truing device ; grindstone-trimmer; grindstone-dresser. (43)
appareil à trousser (Fonderie) (*m.*), rig for loam-work ; spindle and sweep. (44)
appareil capteur de poussières (*m.*), dust-catcher. (45)
appareil chauffe-colle (*m.*), glue-heating apparatus ; glue-heater. (46)
appareil d'alimentation (*m.*), feed-gear. (47)
appareils d'analyse *ou* **appareils d'essais** (*m.pl.*), assaying-appliances ; assay-plant. (48)
appareil d'arrêt (*m.*), stop-gear ; stopping-gear. (49)
appareil d'arrêt et de mise en marche automatique (*m.*), automatic stop-and-starting gear ; automatic stopping-and-starting gear. (50)
appareil d'éclairement (Microscopie) (*m.*), illuminating apparatus. (51)
appareil d'extraction (Mines) (*m.*), hoisting-gear ; hoisting-appliances ; winding-gear. (52)
appareils d'extraction (Mines) (*m.pl.*), hoisting-plant ; winding-plant. (53)
appareil de battage (Sondage) (*m.*), percussion-rig. (54)
appareil de changement de marche (*m.*), reversing-gear. (55)
appareil de chargement distributeur (haut fourneau) (*m.*), stock-distributor. (56)
appareils de chloruration (*m.pl.*), chlorination-plant. (57)
appareil de choc (*m.*), drop-test machine ; drop-testing machine. (58)
appareils de concentration (*m.pl.*), concentration-plant ; concentrating-plant ; concentrates-plant. (59)
appareils de cyanuration (*m.pl.*), cyanide-plant ; cyaniding-plant. (60)
appareil de démarrage (*m.*), starting-gear. (61)
appareil de distribution de vapeur (*m.*), valve-gear ; valve-motion. (62)
appareils de distribution du gaz (*m.pl.*), gas-fittings. (63)
appareils de laboratoires (*m.pl.*), chemical and scientific apparatus. (64)
appareils de lavage (*m.pl.*), washing-plant. (65)
appareil de levage (*m.*), lifting-apparatus ; hoist ; tackle. (66)
appareils de levage (*m.pl.*), lifting-machinery ; lifting-gear ; lifting-tackle ; hoisting-tackle ; tackle. (67)
appareil de manœuvre d'aiguilles (Ch. de f.) (*m.*), switch-stand. (68)
appareils de manutention de charbon (*m.pl.*), coal-handling plant. (69)
appareil de mise en marche (*m.*), starting-gear ; (70)
appareil de pointage (contrôle de présence des employés) (*m.*), time-recorder ; time-clock. (71)
appareil de projection (*m.*), magic lantern ; projection-lantern. (72)
appareil de propulsion (*m.*), propelling-gear. (73)
appareil de réception (des cages, — puits de mine) (*m.*), landing-apparatus. (74)
appareil de sauvetage (incendie) (*m.*), rescue-apparatus ; life-saving appliance ; fire-escape. (75)
appareil de sauvetage (appareil respiratoire) (*m.*), rescue-apparatus ; life-saving apparatus. (76)
appareil de sondage (*m.*), drilling-rig. (77)

appareil de sûreté (m.), safety-apparatus; safety-appliance. (1)

appareil de suspension à arc (m.), bow hanger. (2)

appareil de suspension à barre (m.), bar hanger. (3)

appareil de suspension à étrier (m.), stirrup hanger. (4)

appareil de translation (Télégr.) (m.), translator; repeater. (5)

appareil détecteur de grisou (m.), gas-detector; gas-indicator; fire-damp detector; warner. (6)

appareil diviseur (pour fraiser entre pointes) (m.), dividing-heads; index-centres (milling-machine). (7)

appareil diviseur pour écrous à 6 et 8 pans (m.), dividing-apparatus for hexagonal and octagonal nuts. (8)

appareil enregistreur (m.), recording instrument; self-registering apparatus; self-recording apparatus. (9)

appareil frigorifique (m.), freezing-machine; refrigerating-machine. (10)

appareils frigorifiques (m.pl.), refrigeration-plant; refrigerating-plant; freezing-machinery. (11)

appareil mécanique (m.), mechanical appliance; mechanical contrivance; mechanical device. (12)

appareil photographique ou simplement appareil (n.m.), photographic camera; camera. (13) See also chambre noire.
  appareil à main, appareil d'atelier, appareil pliant, etc. Same as chambre à main, chambre d'atelier, chambre pliante, etc. See under chambre noire.

appareil pour fraiser verticalement (m.), vertical milling-attachment. (14)

appareil pour signaux (m.), signalling-apparatus. (15)

appareil protecteur de scie circulaire (m.), circular-saw guard. (16)

appareil protecteur pour scies (m.), saw-guard. (17)

appareils ralentisseurs de vitesse (m.pl.), speed-checking appliances. (18)

appareil respiratoire (m.), breathing-apparatus. (19)

appareil respiratoire indépendant (m.), self-contained breathing-apparatus. (20)

appareil sécheur de vapeur (m.), steam-drier. (21)

appareil télégraphique (m.), telegraphic apparatus; telegraphic instrument. (22)

appareil tendeur (m.), tension-gear. (23)

appareillage (garnitures) (n.m.), fittings: (24)
  appareillage pour lumière électrique, electric-light fittings. (25)
  appareillage de moteurs, engine-fittings. (26)

appareillage (Maçonn.) (n.m.), bonding: (27)
  les ponts biais sont à éviter à cause de la difficulté d'appareillage (m.pl.), skew bridges should be avoided on account of the difficulty of bonding. (28)

appareiller (Maçonn.) (v.t.), to bond. (29)

appareilleur (pers.) (n.m.), fitter. (30)

appareilleur à gaz (m.), gas-fitter; gas-man. (31)

apparence (n.f.), appearance; look: (32)
  apparence spongieuse de l'or, spongy appearance of the gold. (33)

apparent, -e (adj.), apparent; evident; obvious. (34)

apparent, -e (Opt., etc.) (adj.), apparent: (35)
  grandeur apparente d'un objet (f.), apparent magnitude of an object. (36)

apparenté, -e (se dit des minéraux, etc.) (adj.), related. (37)

apparition (venue; arrivée) (n.f.), appearance; advent: (38)
  les tramways firent leur apparition en Angle-terre en 1860 (m.pl.), tramways made their appearance in England in 1860. (39)
  une apparition de roche à la surface, an appearance of rock on the surface. (40)
  l'apparition de l'image (Photogr.), the appearance of the image. (41)
  l'apparition d'un chemin de fer, the advent of a railway. (42)

appauvrissement (n.m.), impoverishment. (43)

appel (Téléph. & Télégr.) (n.m.), call: (44)
  appel téléphonique, telephone-call. (45)

appel d'air (m.), draught; indraught; indraught of air; intake of air: (46)
  jet de vapeur qui provoque un appel d'air (m.), jet of steam which creates a draught. (47)

appeler (v.t.), to call: (48)
  appeler au secours ou appeler à l'aide, to call for help. (49)
  appeler l'attention sur un fait, to call attention to a fact. (50)
  appeler un poste téléphonique, to call a telephone-exchange. (51)
  appeler la bulle d'air entre ses repères (Nivelle-ment), to bring the air-bubble to the centre of its run. (Levelling) (52)

appentis (n.m.), lean-to; penthouse; pentice; shed. (53)

applique (n.f.), bracket: (54)
  applique en col de cygne à rotule, laiton poli, montée sur rosace en porcelaine, saillie — c/m., goose-necked swiveling bracket, polished brass, mounted on porcelain rose, projecting — centimetres. (55)

applique (d'), bracket; wall (used as adjs): (56)
  treuil d'applique (m.), bracket-crab; wall-crab. (57)

appliqué, -e (opp. à rationnel ou pur) (adj.), applied; practical: (58)
  chimie appliquée (f.), applied chemistry; practical chemistry. (59)

appointage (n.m.), pointing: (60)
  appointage d'un crampon, pointing a spike. (61)

appointements (n.m.pl.), salary: (62)
  appointements du directeur, salary of the manager; manager's salary. (63)

appointer (v.t.), to point: (64)
  appointer un bâton, to point a stick. (65)

appointir (v.t.), to point: (66)
  appointir une aiguille, to point a needle. (67)

appointissage (n.m.), pointing. (68)

appointement (n.m.), wharf. (69)

apport (Métall.) (n.m.), addition; finishing-metal. (70)

apport (n.m.) ou apports (n.m.pl.) (Géol.), drift: (71)
  apport des glaciers, glacial drift. (72)
  apports fluvio-glaciaires, fluvioglacial drift. (73)
  apport des fleuves ou apports des rivières, river-drift. (74)
  apports torrentiels, torrential drift. (75)

apport des vents,   wind-drift. (1)

apport de sable,   drift of sand. (2)

apports de ruissellement (Géol.),   wash. (3)

apporter (Géol.) (*v.t.*),   to drift :   (4)

les sédiments apportés par les grands fleuves (*m.pl.*),   the sediments drifted by the great rivers. (5)

appréciable (*adj.*),   appreciable :   (6)

grandeur appréciable (*f.*),   appreciable size. (7)

appréciation (évaluation) (*n.f.*),   valuation ; valuing ; estimating the value *or* amount ; estimate :   (8)

faire l'appréciation de marchandises,   to make a valuation of goods. (9)

appréciation du gîte,   estimating the amount of mineral available. (10)

appréciation réservée,   conservative estimate. (11)

apprécier (*v.t.*),   to value ; to estimate the value of ; to estimate :   (12)

apprécier des marchandises,   to value goods. (13)

apprenti (pers.) (*n.m.*),   apprentice. (14)

apprentissage (*n.m.*),   apprenticeship. (15)

apprêter (*v.t.*),   to prepare ; to make ready. (16)

apprêter (s') (*v.r.*),   to prepare ; to get oneself ready. (17)

approche (*n.f.*),   approach :   (18)

l'approche de l'hiver,   the approach of winter. (19)

approcher (*v.i. & v.t.*),   to approach :   (20)

mine qui approche de la fin de son exploitation (*f.*),   mine which is approaching exhaustion. (21)

l'heure approche,   the hour approaches. (22)

approfondir (*v.t.*),   to deepen :   (23)

approfondir un puits de mine,   to deepen a mine-shaft. (24)

approfondir (s') (*v.r.*),   to deepen ; to become deeper ; to grow deeper :   (25)

le lit des cours d'eau s'approfondit ou s'exhausse,   the bed of streams becomes deeper or shal- -lower ; the bed of watercourses deepens or shallows. (26)

approfondissement (*n.m.*),   deepening :   (27)

approfondissement de puits déjà creusés,   deepening pits already sunk. (28)

approprié, -e (*adj.*),   suitable ; appropriate ; proper :   (29)

une situation appropriée à une usine,   a suitable location for a works. (30)

souvent d'excellents outils sont défectueux parce qu'ils ne sont pas appropriés au travail qu'on exige d'eux,   excellent tools are often wanting because they are not suitable for the work required of them. (31)

approvisionnement (action) (*n.m.*),   supplying ; provisioning. (32)

approvisionnement (provisions ; choses ras- -semblées) (*n.m.*),   supply ; store ; provision : (33)

approvisionnement d'eau assuré,   assured water- supply. (34)

approvisionnements de matières premières,   supplies of raw materials. (35)

approvisionnements disponibles en magasin,   stores in hand ; stores on hand. (36)

approvisionner (*v.t.*),   to supply ; to provision : (37)

approvisionner de bois une mine,   to supply a mine with timber. (38)

approvisionner de charbon un navire,   to coal a ship. (39)

approvisionner en combustible un foyer,   to stoke a furnace. (40)

approvisionner (s') (*v.r.*).   See examples :   (41)

s'approvisionner chez un marchand de la localité,   to obtain one's supplies from a local store ; to get one's supplies from a local dealer. (42)

s'approvisionner de vivres,   to supply oneself (*or* to provide oneself) with provisions ; to lay in a stock (*or* to take in a supply) (*or* to get in a supply) of provisions. (43)

approximatif, -ive (*adj.*),   approximate ; rough : (44)

valeur approximative (*f.*),   approximate value. (45)

calcul approximatif (*m.*),   rough calculation. (46)

approximativement (*adv.*),   approximately ; roughly. (47)

appui (*n.m.*),   support ; rest :   (48)

l'appui d'une voûte, d'un pont,   the support of an arch, of a bridge. (49 *ou* 50)

appui (Charp.) (*n.m.*),   sill (a sill placed between uprights or posts at some distance from the ground, as in a window or in the doorway of a frame or partition). (51)

à hauteur d'appui,   elbow-high. (52)

appui (d'un levier) (Méc.) (*n.m.*),   fulcrum (of a lever). (53)

appui (d'un rail) (*n.m.*),   foot (of a rail) :   (54)

l'appui du rail Vignole s'appelle patin,   the foot of the Vignoles rail is called the flange. (55)

appui de fenêtre (*m.*),   window-sill ; elbow-board. (56)

appui de porte (*m.*),   door-sill. (57)

appui télégraphique (*m.*),   telegraph-post ; tele- -graph-pole. (58)

appuyer (*v.t.*),   to support ; to hold up ; to lean ; to rest :   (59)

appuyer une galerie par des piliers,   to support, to hold up, a gallery by pillars. (60)

appuyer une échelle contre un mur,   to lean, to rest, a ladder against a wall. (61)

appuyer (reposer) (*v.i.*),   to rest ; to lean ; to bear :   (62)

poutre qui appuie sur un mur (*f.*),   beam which rests on a wall. (63)

appuyer sur un bouton,   to press a button. (64)

âpre au toucher,   rough to the touch :   (65)

certains minéraux sont âpres au toucher (*m.pl.*),   certain minerals are rough to the touch. (66)

aptitude (*n.f.*),   aptitude :   (67)

aptitude à la détonation,   detonating aptitude. (68)

aquatique (plein d'eau ; marécageux) (*adj.*),   watery ; marshy :   (69)

terrain aquatique (*m.*),   watery ground ; marshy ground. (70)

aqueduc (*n.m.*),   aqueduct. (71)

aqueux, -euse (*adj.*),   aqueous ; watery :   (72)

fusion aqueuse (*f.*),   aqueous fusion ; watery fusion. (73)

aquifère (*adj.*),   water-bearing ; watery :   (74)

couche géologique aquifère (*f.*),   water-bearing geological bed. (75)

terrain aquifère (*m.*),   water-bearing strata ; watery ground. (76)

aquo-igné, -e (Géol.) (*adj.*),   aqueo-igneous ; hydrothermal :   (77)

fusion aquo-ignée (*f.*), aqueo-igneous fusion ; hydrothermal fusion. (1)

aquosité (*n.f.*), wateriness. (2)

aragonite (Minéral) (*n.f.*), aragonite. (3)

araignée (crochet de puits) (*n.f.*), well-creeper. (4)

arasement (*n.m.*), striking ; striking off ; cutting off. (5)

araser (*v.t.*), to strike ; to strike off ; to cut off : (6)

araser à la règle l'excès de sable de la face supérieure d'un châssis de fonderie, to strike the surplus sand from the top face of a moulding-box with the straight-edge. (7)

araser les têtes des pieux de façon qu'elles se trouvent toutes à un même niveau, to cut off, to strike off, the heads of piles so that they are all at one level. (8)

arbalète (*n.f.*), file-carrier. (9)

arbalétrier (Constr.) (*n.m.*), principal rafter ; chief rafter ; principal. (10)

arborisation (*n.f.*), arborization : (11)
cristaux disposés en arborisation coralloïde (*m.pl.*), crystals occurring in coralloid arborization. (12)

arborisé, -e (*adj.*), arborized ; dendritic : (13)
agate arborisée (*f.*), arborized agate ; dendritic agate ; tree-agate. (14)

arbre (*n.m.*), tree : (15)
une avenue d'arbres, an avenue of trees. (16)
arbres à racines traçantes plantés sur les berges d'une rivière pour préserver les terres contre l'action de l'eau, trees with running roots planted on the banks of a river to protect the earth against the action of the water. (17)

arbre (Méc.) (*n.m.*), shaft ; axle ; spindle ; arbor ; mandrel. (18) See also mandrin.

arbre (Transmission) (*n.m.*), shaft. (19)

arbre (d'un tour) (*n.m.*), spindle, mandrel, (of a lathe) : (20)
arbre en acier percé dans toute sa longueur tournant dans des coussinets en bronze à longue portée *ou* arbre en acier percé de part en part et tournant dans des coussinets en bronze à longue portée, steel spindle bored through its entire length and running in long gun-metal bearings ; hollow steel mandrel revolving in long gun-metal bearings ; steel spindle with thoroughfare hole turning in long gun-metal bearings. (21)

arbre (d'une molette) (Mines) (*n.m.*), gudgeon, shaft, (of a winding-pulley). (22)

arbre (d'une grue) (*n.m.*), post, jib-post, (of a crane). (23)

arbre à came(s) (*m.*), cam-shaft ; camshaft ; tumbling-shaft ; wiper-shaft. (24 *ou* 25)

arbre à deux manivelles (*m.*), two-throw crank-shaft. (26)

arbre à excentrique(s) (*m.*), eccentric-shaft. (27 *ou* 28)

arbre à manivelle (*m.*), crank-shaft ; cranked shaft. (29)

arbre à manivelle rapportée (*m.*), built crank-shaft. (30)

arbre à mouvement planétaire (*m.*), planet-spindle ; planet-action spindle. (31)

arbre à plateau-manivelle (*m.*), disc crank-shaft. (32)

arbre à trois manivelles (*m.*), three-throw crank-shaft. (33)

arbre à trousser (Fonderie) (*m.*), spindle (of a rig for loam-work). (34)

arbre à vilebrequin (*m.*), crank-shaft ; cranked shaft. (35)

arbre à volant (*m.*), fly-wheel shaft. (36)

arbre chablis (*m.*), fallen tree. (37)

arbre commandé *ou* arbre conduit (*m.*), driven shaft. (38)

arbre coudé (*m.*), crank-shaft ; cranked shaft. (39)

arbre creux (*m.*), hollow shaft. (40)

arbre d'attaque (*m.*), driving-shaft. (41)

arbre de changement de marche (distribution) (*m.*). Same as arbre de relevage.

arbre de commande *ou* arbre de couche (*m.*), driving-shaft : (42)
excentrique calé sur l'arbre de couche de la machine (*m.*), eccentric keyed on the driving-shaft of the engine. (43)

arbre de Diane (Chim.) (*m.*), arbor Dianæ ; silver-tree. (44)

arbre de haute futaie *ou* arbre de haute venue (*m.*), timber-tree ; lofty tree. (45)

arbre de Jupiter (Chim.) (*m.*), arbor Jovis ; tin-tree. (46)

arbre de la came (*m.*). Same as arbre à came.

arbre de la poupée fixe (tour) (*m.*), live-spindle (lathe). (47)

arbre de la poupée mobile (tour) (*m.*), dead-spindle ; tail-spindle. (48)

arbre de ligne (*m.*), line-shaft ; line. (49)

arbre de Mars (Chim.) (*m.*), arbor Martis ; iron-tree. (50)

arbre de meule, monté de paliers et plateaux (*m.*), grindstone-spindle, with plummer-block bearings and side-plates. (51)

arbre de relevage (distribution) (*m.*), reversing-shaft ; reverse-shaft ; tumbling-shaft ; lifting-shaft (link-motion). (52)

arbre de relevage (d'un bocard) (*m.*), jack-shaft (of a stamp-mill). (53)

arbre de renvoi (*m.*), counter-shaft ; counter-shaft (the shaft in a counter-driving motion). (54)

arbre de Saturne (Chim.) (*m.*), arbor Saturni : lead-tree. (55)

arbre de transmission (*m.*), shaft ; transmission-shaft. (56) Cf. arbres de transmission.

arbre de transmission horizontal (*m.*), horizontal shaft ; lying shaft. (57)

arbre de transmission vertical (*m.*), vertical shaft ; upright shaft. (58)

arbre déraciné (*m.*), uprooted tree. (59)

arbre des cames (*m.*). Same as arbre à cames.

arbre du frein (*m.*), brake-shaft. (60)

arbre du tambour (*m.*), drum-shaft. (61)

arbre du tiroir (*m.*), valve-stem ; valve-spindle ; stem of the slide-valve ; slide-rod. (62)

arbre en état (*m.*), standing tree. (63)

arbre équivalent (Méc.) (*m.*), equivalent shaft. (64)

arbre flexible (*m.*), flexible shaft. (65)

arbre mandrin (tour) (*m.*), live-spindle (lathe). (66)

arbre-manivelle [arbres-manivelles *pl.*] (*n.m.*), crank-shaft ; cranked shaft. (67)

arbre menant (*m.*), driving-shaft. (68)

arbre mené (*m.*), driven shaft. (69)

arbre moteur (*m.*), driving-shaft. (70)

arbre porte-came(s) (*m.*). Same as arbre à came(s).

arbre porte-foret (*m.*), drilling-spindle. (71)

arbre porte-foret en acier tournant dans des coussinets en bronze à rattrapage de jeu (*m.*), steel drilling-spindle running in gun-metal bearings with adjustment for taking up wear. (1)

arbre porte-foret équilibré avec butée sur billes (*m*), counterbalanced drilling-spindle with the thrust taken on a ball thrust-bearing. (2)

arbre porte-fraise (*m.*), milling-machine arbor; milling-machine cutter-arbor; cutter-arbor. (3)

arbre porte-mèche (*m.*). Same as **arbre porte-foret.**

arbre porte-meule (*m.*), wheel-spindle; grinding-spindle; emery-wheel spindle. (4)

arbre porte-outil (alésoir) (*m.*), cutter-spindle; boring-spindle; cutter-mandrel (boring-machine). (5)

arbre porte-scie (*m.*), saw-spindle; saw-arbor; saw-mandrel. (6)

arbre principal (*m.*), main shaft; first-motion shaft. (7)

arbre secondaire (*m.*), counter-shaft; counter-shaft; intermediate shaft; second-motion shaft; jack-shaft. (8)

arbre sur pied (*m.*), standing tree. (9)

arbres de transmission *ou* **arbres pour trans-mission** *ou simplement* **arbres** (*n.m.pl.*), shafting; transmission-shafting: (10)
poulies et arbres de transmission, pulleys and shafting. (11)
arbres de transmission en acier comprimé, compressed-steel shafting. (12)
une ligne d'arbres, a line of shafting. (13)
Cf. **arbre de transmission.**

arbreux, -euse (*adj.*), wooded; woody: (14)
pays arbreux (*m.*), wooded country. (15)

arc (*n.m.*), arc; bow. (16)

arc (Géom.) (*n.m.*), arc: (17)
un arc de cercle, an arc of a circle; a circular arc. (18)
l'arc d'une ellipse, d'une courbe quelconque, the arc of an ellipse, of any curve. (19 *ou* 20)

arc (Élec.) (*n.m.*), arc. (21)

arc (Arch.) (*n.m.*), arch, i.e., the curve or bow described by an arch, as distinguished from the structure itself, which is termed in French **voûte**, or, in the case of a bridge, **arche.** (22)
See these words in vocabulary.

arc (de tourneur; de perceur) (*n.m.*), bow (turners'; drillers'). (23)

arc à charbon (Élec.) (*m.*), carbon arc. (24)

arc à courant alternatif (Élec.) (*m.*), alternating-current arc. (25)

arc à courant continu (Élec.) (*m.*), direct-current arc; continuous-current arc. (26)

arc à flamme (Élec.) (*m.*), flame-arc; flaming arc. (27)

arc à l'air *ou* **arc à feu libre** (Élec.) (*m.*), open arc. (28)

arc à trois centres (Arch.) (*m.*), three-centred arch. (29)

arc à vapeur de mercure (Élec.) (*m.*), mercury-vapour arc. (30)

arc à vapeurs lumineuses (Élec.) (*m.*), luminous-vapour arc. (31)

arc aplati *ou* **arc à quartre centres** (Arch.) (*m.*), four-centred arch. (32)

arc au mercure (Élec.) (*m.*), mercury-arc. (33)

arc bombé (Arch.) (*m.*), segmental arch; segment arch. (34)

arc-boutant [arcs-boutants *pl.*] (Arch.) (*n.m.*), flying buttress; arch-buttress; buttress. (35)

arc-boutant (contre-fiche d'une ferme) (Constr.) (*n.m.*), strut; spur; brace. (36)

arc-boutant (étai incliné) (Étaiement des murs) (Constr.) (*n.m.*), raking shore; raker. (37)

arc-boutement (Engrenages) (*n.m.*), interference (Gearing). (38)

arc-boutement en voûte (*m.*), arching. (39)

arc-bouter (Arch.) (*v.t.*), to buttress: (40)
arc-bouter un mur, to buttress a wall. (41)

arc-bouter (s') **en voûte**, to arch: (42)
minerai abattu qui s'arc-boute en voûte dans une cheminée et refuse de descendre (*m.*), broken ore which arches in a chute and refuses to descend. (43)

arc carbo-minéral (Élec.) (*m.*), mineralized-carbon arc. (44)

arc d'engrènement *ou* **arc de prise** (Engrenages) (*m.*), pitch-arc: (45)
les engrenages à denture droite présentent l'inconvénient qu'il est impossible d'obtenir avec eux un arc d'engrènement assez long pour que deux dents au moins soient com-plètement en prise, spur-gears have the disadvantage that it is impossible to obtain with them a pitch-arc long enough to permit of two teeth at least being completely in mesh. (46)

arc de cercle (Géom.) (*m.*). See under **arc** (Géom.).

arc de développante *ou* **arc de développante de cercle** (Géom.) (*m.*), involute arc. (47)

arc de plein cintre (*m.*). See **arc plein cintre.**

arc de suspension (*m.*), bow hanger. (48)

arc électrique (*m.*), electric arc. (49)

arc elliptique *ou* **arc ellipsoïdal** (Arch.) (*m.*), elliptical arch. (50)

arc en anse de panier (Arch.) (*m.*), basket-handle arch. (51)

arc-en-ciel [arcs-en-ciel *pl.*] *ou simplement* **arc** (*n.m.*), rainbow; bow: (52)
premier arc, primary rainbow; primary bow. (53)
second arc, secondary rainbow; secondary bow. (54)
arc surnuméraire, supernumerary rainbow; spurious rainbow. (55)

arc-en-ciel blanc (*m.*), white rainbow. (56)

arc en décharge *ou* **arc de décharge** (Arch.) (*m.*), arch of discharge; discharging-arch; safety-arch. (57)

arc en fer à cheval *ou* **arc outrepassé** (Arch.) (*m.*), horseshoe arch. (58)

arc en segment de cercle (Arch.) (*m.*), segmental arch; segment arch. (59)

arc-en-terre [arcs-en-terre *pl.*] (*n.m.*), dew-bow. (60)

arc enfermé (Élec.) (*m.*), enclosed arc. (61)

arc exhaussé (Arch.) (*m.*), stilted arch. (62)

arc extradossé (Arch.) (*m.*), extradosed arch. (63)

arc lumineux (Élec.) (*m.*), luminous arc. (64)

arc métallique (Élec.) (*m.*), metallic arc. (65)

arc plein cintre *ou* **arc de plein cintre** *ou* **arc en plein cintre** *ou* **arc romain** (Arch.) (*m.*), round arch; semicircular arch; Roman arch. (66)

arc rampant (Arch.) (*m.*), rampant arch; rising arch. (67)

arc renversé (Arch.) (*m.*), inverted arch; reversed arch; inflected arch. (68)

arc surbaissé (opp. à **arc surhaussé**) (Arch.) (*m.*), diminished arch; imperfect arch;

segmental arch ; segment arch ; scheme arch ; skene arch ; skeen arch. (1)

arc surhaussé *ou* **arc surmonté** *ou* **arc surélevé** (Arch.) (*m.*), stilted arch. (2)

arc triangulaire (Arch.) (*m.*), triangular arch. (3)

arc voltaïque (*m.*), voltaic arc. (4)

arcade (*n.f.*), arcade ; arch. (5)

arcade aveugle *ou* **arcade feinte** (*f.*), blind arch. (6)

arcanite (Minéral) (*n.f.*), arcanite. (7)

arcanseur (*n.m.*), pinch-bar ; pinching-bar ; pinch. (8)

arceau (*n.m.*), arch ; archway. (9)

arche (*n.f.*), arch (of a bridge) : (10)
les arches se classent en arches de plein cintre, en anse de panier, elliptiques, etc., arches are classified under round, basket-handle, elliptical arches, etc. (11)
pont à arches surbaissées (*m.*), bridge with diminished arches. (12)
See **arc** for other varieties.

arche biaise (*f.*), skew arch ; skewed arch ; oblique arch. (13)

arche de décharge pour les hautes eaux (*f.*), flood arch. (14)

arche naturelle (Géogr. phys.) (*f.*), natural arch. (15)

archéen, -enne (Géol.) (*adj.*), Archæan ; Archean ; Archaian ; Archæic ; Archeic : (16)
roches archéennes (*f.pl.*), Archæan rocks. (17)

archéen (*n.m.*), Archæan ; Archean ; Archaian ; Archæic ; Archeic : (18)
l'archéen, the Archæan. (19)

archet (*n.m.*), bow ; drill-bow. (20)

Archimède (d'), Archimedean ; Archimedes' : (21)
vis d'Archimède (*f.*), Archimedean screw ; Archimedes' screw. (22)
principe d'Archimède (Hydrostatique) (*m.*), Archimedean principle ; Archimedes' principle. (23)

archipel (*n.m.*), archipelago. (24)

architecte (pers.) (*n.m.*), architect. (25)

architectural, -e, -aux (*adj.*), architectural : (26)
l'art architectural (*m.*), the architectural art. (27)

architecture (*n.f.*), architecture : (28)
l'architecture moderne en France, modern architecture in France. (29)
l'architecture de la croûte terrestre, the architecture of the earth's crust. (30)

architecture hydraulique (*f.*), hydraulic engineer-ing. (31)

arctic carbonite (Explosif) (*f.*), arctic carbonite. (32)

arctique (*adj.*), arctic : (33)
la longue nuit de l'hiver arctique, the long night of the arctic winter. (34)

ardennite (Minéral) (*n.f.*), ardennite. (35)

ardent, -e (*adj.*), burning ; hot ; fiery ; blazing ; live : (36)
charbon ardent (*m.*), burning coal ; live coal. (37)
soleil ardent (*m.*), burning sun ; blazing sun. (38)

ardeur (*n.f.*), heat ; burning heat ; great heat ; excessive heat : (39)
l'ardeur du soleil, du feu, the heat of the sun, of the fire. (40 *ou* 41)

ardoise (Pétrol.) (*n.f.*), slate. (42)

ardoise (feuille d'ardoise) (*n.f.*), slate : (43)

on emploie les ardoises pour la couverture des bâtiments, slates are used for roofing buildings. (44)
ardoises d'échantillon, sized slates. (45)

ardoisé, -e (*adj.*), slate-coloured. (46)

ardoiser (*v.t.*), to slate : (47)
ardoiser un toit, to slate a roof. (48)

ardoiserie (*n.f.*), slate-industry. (49)

ardoisier, -ère *ou* **ardoiseux, -euse** (*adj.*), slaty ; slate (*used as adj.*) : (50)
roche ardoisière (*f.*), slaty rock. (51)
formation ardoisière (*f.*), slate formation. (52)

ardoisier (pers.) (*n.m.*), slate-quarryman ; slate-quarrier. (53)

ardoisière (*n.f.*), slate-quarry. (54)

ardu, -e (raide) (*adj.*), steep ; arduous ; up-hill : (55)
sentier ardu (*m.*), steep path ; arduous path ; up-hill path. (56)

ardu, -e (pénible) (*adj.*), hard ; arduous ; up-hill : (57)
travail ardu (*m.*), hard work ; arduous work ; up-hill work. (58)

are (*n.m.*), are = 119·60 square yards. (59)

aréage (*n.m.*), measuring land in *ares*. (60)

arénacé, -e (*adj.*), arenaceous ; sandy. (61)

arénacé, -e (Géol.) (*adj.*), arenaceous : (62)
roche arénacée (*f.*), arenaceous rock. (63)

arendalite (Minéral) (*n.f.*), arendalite. (64)

arène (Pétrol.) (*n.f.*), sand. (65)

arène (Mines) (*n.f.*), drain ; gutter ; drainage-channel. (66)

aréner (*v.i.*), to subside ; to sink in the sand. (67)

arénical, -e, -aux (*adj.*), arenose ; sandy ; mixed with sand. (68)

arénière (*n.f.*), sand-pit. (69)

arénifère (*adj.*), sand-bearing. (70)

aréniforme *ou* **arénulacé, -e** *ou* **arénuleux, -euse** (*adj.*), arenaceous ; arenose ; arenulous ; sandy. (71)

aréomètre (*n.m.*), areometer ; aræometer. (72)

aréométrie (*n.f.*), areometry ; aræometry. (73)

aréométrique (*adj.*), areometric ; aræometric. (74)

arête (saillie) (*n.f.*), ridge ; edge. (75)

arête (angle saillant que forment deux faces droites ou courbes d'une pierre, d'une pièce de bois, etc.) (*n.f.*), arris ; edge : (76)
l'arête d'une poutre, the arris (*or* the edge) of a beam. (77)

arête (d'un tiroir, d'une lumière de cylindre) (*n.f.*), edge (of a slide-valve, of a cylinder-port) : (78 *ou* 79)
arête extérieure, steam-edge ; outer edge. (80)
arête intérieure, exhaust-edge ; inner edge. (81)

arête (angle que forme une voûte avec une autre voûte) (*n.f.*), groin : (82)
voûte d'arête (*f.*), groined vault. (83)

arête (Géol. & Géogr. phys.) (*n.f.*), edge ; ridge ; arête : (84)
l'arête d'un rocher, the edge (*or* the ridge) of a rock. (85)
les arêtes de la pyramide du Weisshorn, the arêtes of the pyramid of the Weisshorn. (86)
une arête de neige, an arête of snow. (87)

arête (Géom.) (*n.f.*), edge : (88)
l'arête d'un prisme, d'un polyèdre, the edge of a prism, of a polyhedron. (89 *ou* 90)

arête (Cristall.) (*n.f.*), edge. (91)

arête (Constr.) (*n.f.*), hip. (92)

arête anticlinale (Géol.) (*f.*), anticlinal ridge. (93)

arête de partage (Géogr. phys.) (*f.*), dividing ridge. (94)

arête de rebroussement (Géom.) (*f.*), line of striction. (1)
arête latérale (d'un cristal) (*f.*), lateral edge (of a crystal). (2)
arête réfringente (d'un prisme) (*f.*), refracting edge (of a prism). (3)
arête tranchante (d'un outil) (*f.*), cutting edge (of a tool). (4)
arête tronquée (Cristall.) (*f.*), truncated edge. (5)
arête vive (*f.*), sharp edge ; sharp arris : (6)
l'arête vive du couteau d'une balance, the sharp edge of the knife-edge of a balance. (7)
poutre à arêtes vives (*f.*), beam with sharp arrises. (8)
arête vive (à) *ou* **arêtes vives** (à) *ou* **vive arête** (à) *ou* **vives arêtes** (à), sharp ; sharp-edged : (9)
sable à arêtes vives (*m.*), sharp sand. (10)
stries d'acier à vives arêtes (*f.pl.*), sharp ridges of steel. (11)
arêteux, -euse (*adj.*), ridgy ; ridged. (12)
arêtier (angle) (Constr.) (*n.m.*), hip. (13)
arêtier (arbalétrier) (*n.m.*), hip-rafter ; hip ; angle-rafter. (14)
arêtière (*n.f.*), hip-tile. (15)
arfvedsonite (Minéral) (*n.f.*), arfvedsonite. (16)
argent (*n.m.*), silver. (17)
argent allemand (alliage) (*m.*), German silver. (18)
argent amalgamé (Minéral) (*m.*), amalgam ; argental mercury. (19)
argent antimonial (Minéral) (*m.*), antimonial silver ; dyscrasite. (20)
argent arsenical (Minéral) (*m.*), arsenical silver blende ; light-red silver ore ; light-ruby silver ore ; proustite. (21)
argent au titre (*m.*), standard silver ; silver of standard fineness. (22)
argent cémentatoire (*m.*), cement-silver. (23)
argent corné (Minéral) (*m.*), horn-silver ; horn-ore ; cerargyrite ; kerargyrite. (24)
argent de coupelle (*m.*), cupelled silver. (25)
argent en barre *ou* **argent en barres** (*m.*), bar silver. (26)
argent fin (*m.*), fine silver. (27)
argent fulminant (*m.*), fulminating silver. (28)
argent natif (*m.*), native silver. (29)
argent noir (Minéral) (*m.*), black silver. (30)
argent rouge (Minéral) (*m.*), red silver ; ruby silver. (31)
argent rouge antimonial (Minéral) (*m.*), dark-red silver ore ; dark-ruby silver ; pyrargyrite ; argyrythrose. (32)
argent rouge arsenical *ou* **argent rouge clair** (Minéral) (*m.*), arsenical silver blende ; light-red silver ore ; light-ruby silver ; proustite. (33)
argent sulfuré (Minéral) (*m.*), argentite. (34)
argent telluré (Minéral) (*m.*), telluric silver ; hessite. (35)
argent tenant or, silver containing gold ; gold-silver. (36)
argent vierge (*m.*), virgin silver. (37)
argentage (*n.m.*). Same as **argenture**.
argental, -e, -aux (*adj.*), argental : (38)
mercure argental (*m.*), argental mercury. (39)
argentate (Chim.) (*n.m.*), argentate. (40)
argentaurum (*n.m.*), argentaurum. (41)
argenté, -e (qui a l'éclat blanc de l'argent) (*adj.*), silvery ; silver : (42)
gris argenté (*m.*), silvery gray ; silver grey. (43)

argenté, -e (recouvert d'une solution d'argent) (*adj.*), silvered ; silver-plated : (44)
plaques argentées (*f.pl.*), silvered plates. (45)
argenter (*v.t.*), to silver ; to silver-plate ; to plate : (46)
argenter sur métaux, to silver on metals. (47)
argenter une glace, to silver a mirror. (48)
argenter par électrolyse *ou* **argenter par le procédé électro-chimique,** to electrosilver ; to silver by the electrochemical process. (49)
argenteur, -euse (*adj.*), silvering : (50)
sel argenteur (*m.*), silvering salt. (51)
argenteur (pers.) (*n.m.*), silver-plater ; silverer. (52)
argentifère (*adj.*), argentiferous ; silver-bearing ; silver : (53)
plomb argentifère (*m.*), argentiferous lead ; silver lead. (54)
minéraux argentifères (*m.pl.*), silver-bearing minerals. (55)
argentine (Minéral) (*n.f.*), argentine. (56)
argentique (Chim.) (*adj.*), argentic. (57)
argentite (Minéral) (*n.f.*), argentite ; argyrite ; argyrose ; silver-glance. (58)
argentopyrite (Minéral) (*n.f.*), argentopyrite. (59)
argenture (*n.f.*), silvering ; silver-plating ; plating. (60)
argenture au trempé *ou* **argenture par immersion** (*f.*), silvering by dipping. (61)
argenture galvanique (*f.*), electrosilvering. (62)
argilacé, -e (*adj.*), argillaceous ; argillous ; clayey. (63)
argile (*n.f.*), clay. (64)
argile à blocaux (Géol.) (*f.*), boulder-clay ; bowlder-clay ; till ; glacial till ; drift ; glacial drift. (65)
argile à briques (*f.*), brick-clay. (66)
argile à modeler (*f.*), modeling-clay ; modelling-clay. (67)
argile à porcelaine (*f.*), porcelain-clay. (68)
argile à poterie (*f.*), potters' clay ; potters' earth ; pot-clay ; pot-earth ; argil. (69)
argile à silex (*f.*), flint clay. (70)
argile alluviale (*f.*), alluvial clay. (71)
argile bleue, grise, jaune, rouge, etc. (*f.*), blue, gray, yellow, red clay, etc. (72 *ou* 73 *ou* 74 *ou* 75)
argile de Londres (Géol.) (*f.*), London clay. (76)
argile entraînée par les eaux (*f.*), washed-down clay. (77)
argile feuilletée (*f.*), shale. (78)
argile figuline (*f.*), figuline ; potters' clay. (79)
argile fluvio-glaciaire (*f.*), glacio-aqueous clay. (80)
argile fusible (*f.*), fusible clay. (81)
argile grasse (*f.*), fat clay ; fatty clay. (82)
argile maigre (*f.*), lean clay. (83)
argile ocreuse (*f.*), ochreous clay ; ocherous clay. (84)
argile plastique (*f.*), plastic clay ; fatty clay. (85)
argile réfractaire (*f.*), fire-clay ; refractory clay. (86)
argile schisteuse (*f.*), shale. (87)
argile smectique (*f.*), smectite ; fullers' earth ; fullers' chalk. (88)
argileux, -euse (*adj.*), argillaceous ; argillous ; clayey : (89)
minerai de fer argileux (*m.*), argillaceous iron ore. (90)
argilifère (*adj.*), argilliferous. (91)
argilite *ou* **argillite** (Pétrol.) (*n.f.*), argillite ; argillyte. (92)

argillitique (*adj.*), argillitic. (1)

argilo-arénacé, -e (*adj.*), argilloarenaceous. (2)

argilo-calcaire (*adj.*), argillocalcareous. (3)

argilo-ferrugineux, -euse (*adj.*), argilloferrugi--nous. (4)

argilo-magnésien, -enne (*adj.*), argillomagnesian. (5)

argilo-sablonneux, -euse (*adj.*), argilloarenaceous. (6)

argilloïde (*adj.*), argilloid. (7)

argon (Chim.) (*n.m.*), argon. (8)

argue (*n.f.*), drawing-bench; draw-bench; drag-bench; wiredrawing-bench; wiredrawer (for drawing gold or silver wire). (9)

arguer (*v.t.*), to draw; to wiredraw : (10)
arguer l'or, to draw, to wiredraw, gold. (11)

argyrite (Minéral) (*n.f.*). Same as argyrose.

argyrodite (Minéral) (*n.f.*), argyrodite. (12)

argyropyrite (Minéral) (*n.f.*), argyropyrite. (13)

argyrose (Minéral) (*n.f.*), argyrite; argyrose; argentite; silver-glance. (14)

argyrythrose (Minéral) (*n.f.*), argyrythrose; pyrargyrite; dark-ruby silver; dark-red silver ore. (15)

aride (*adj.*), arid. (16)

aridité (*n.f.*), aridity; aridness. (17)

aristiforme (*adj.*), ridgy. (18)

arkansite (Minéral) (*n.f.*), arkansite. (19)

arkite (Explosif) (*n.f.*), arkite. (20)

arkose (Pétrol.) (*n.f.*), arkose. (21)

arksutite (Minéral) (*n.f.*), arksutite. (22)

arlequin (Minéral) (*n.m.*), harlequin opal. (23)

armature (assemblage de liens en bois ou en métal soutenant les parties d'un ouvrage de charpente) (*n.f.*), trussing; bracing : (24)
l'armature d'une poutre, the trussing of a girder. (25)
ferme en bois à armature en fer (*f.*), timber truss with iron bracing. (26)

armature (lien en fer ou étrier destiné à renforcer une charpente) (*n.f.*), fastening; strap; strap-bolt; stirrup; stirrup-strap; stirrup-bolt; stirrup-piece. (27)

armature (bande métallique servant à consolider une autre pièce) (*n.f.*), strap : (28)
raidir une plaque par des armatures, to stiffen a plate with straps. (29)

armature (pour béton) (*n.f.*), reinforcement, reenforcement, (for concrete) : (30)
l'adhérence entre le béton et l'armature (*f.*), the adhesion between the concrete and the reinforcement. (31)

armature (d'un four) (*n.f.*), plating (of a furnace). (32)

armature (d'un câble) (*n.f.*), armour, armor, sheathing, (of a cable). (33)

armature (contact) (Magnétisme) (*n.f.*), armature; keeper; lifter; arming : (34)
des armatures de fer doux, armatures of soft iron. (35)

armature (induit) (Élec.) (*n.f.*), armature : (36)
l'armature d'une dynamo, the armature of a dynamo. (37)

armature (d'une bouteille de Leyde) (*f.*), arma--ture (of a Leyden jar). (38)

armature de châssis *ou* armature de moule (Fonderie) (*f.*), flask-grid; mould-grating; spider; tie-rod. (39)

armature de noyau (Fonderie) (*f.*), core-iron; core-grid; grid; grate; grating; spider; tie-rod; rod-reinforcing. (40)

armature de pompe (*f.*), pump-gear. (41)

armature du dessous *ou* armature du fond (d'un moule en terre) (Fonderie) (*f.*), bed-plate, bottom plate, foundation-plate, loam-plate, (of a loam mould). (42)

armature du dessus (d'un moule en terre) (*f.*), cope-ring. (43)

armature en anneau (Élec.) (*f.*), ring armature. (44)

armature en disque (Élec.) (*f.*), disc armature. (45)

armature en tambour (Élec.) (*f.*), drum armature. (46)

armature hydraulique (*f.*), hydraulic fitting. (47)

armaturer (*v.t.*), to reinforce; to reenforce : (48)
armaturer l'autel d'un fourneau par une plaque métallique, to reinforce the bridge of a furnace by a metal plate. (49)

armé, -e (se dit des poutres) (*adj.*), trussed; braced : (50)
poutre armée (*f.*), trussed beam; braced beam. (51)

armé, -e (se dit du béton et du ciment) (*adj.*), reinforced; reenforced; armoured; armored; ferro (*prefix*) : (52)
béton armé (*m.*), reinforced concrete; armoured concrete; ferroconcrete. (53)

armé, -e (se dit des câbles, etc.) (*adj.*), armoured; armored; sheathed : (54)
câble armé (*m.*), armoured cable. (55)

armé (se dit d'un aimant) (*adj.*), armed : (56)
aimant armé (*m.*), armed magnet. (57)

armé, -e (aciéré) (*adj.*), steeled : (58)
lame armée (*f.*), steeled blade. (59)

armement (équipement) (*n.m.*), equipment : (60)
fonçage et armement du puits principal, sink--ing and equipment of the main shaft. (61)

armer (*v.t.*). See examples : (62)
armer le béton, to reinforce, to reenforce, to armour, to armor, concrete. (63)
armer un aimant, to arm a magnet. (64)
armer un câble, to armour, to sheath, a cable. (65)
armer un obturateur (Photogr.), to set a shutter. (66) Cf. obturateur toujours armé.
armer un puits pour l'extraction (Mines), to equip a shaft for winding. (67)
voie de chemin de fer armée de rails dissy--métriques (*f.*), railway-track equipped with bull-headed rails. (68)
armer une hache, un burin, to steel an axe, a chisel. (69 *ou* 70)
armer une poutre, to truss a girder; to brace a beam. (71)

armure (contact) (Magnétisme) (*n.f.*), armature; keeper; lifter; arming : (72)
des armures de fer doux, armatures of soft iron. (73)

armure (d'un câble) (*n.f.*), armour, armor, sheathing, (of a cable). (74)

arpentage (*n.m.*), measuring; land-measuring; land-surveying; surveying; survey. (75)

arpenter (*v.t.*), to measure; to survey : (76)
arpenter un champ, to measure, to survey, a field. (77)

arpenter (s') (*v.r.*), to be measured; to be surveyed). (78)

arpenteur (pers.) (*n.m.*), land-measurer; sur--veyor. (79)

arquer (*v.t.*), to arch; to curve; to bend : (80)
arquer une pièce de bois, to bend a piece of wood. (81)

arquer (*v.i.*), to arch ; to curve ; to bend : (1)
poutre qui arque (*f.*), beam which bends. (2)

arquérite (Minéral) (*n.f.*), arquerite. (3)

arquifoux (*n.m.*), arquifoux ; alquifou. (4)

arrachage des clous (*m.*), pulling out, drawing, extraction of, nails. (5)

arrachage du feu (Mines) (*m.*), digging the fire out. (6)

arrachage du tubage d'isolement (*m.*), pulling, drawing, casing. (7)

arrache-clou [arrache-clous *ou* arrache-clou *pl.*] (*n.m.*), nail-extractor ; nail-puller. (8)

arrache-pieux [arrache-pieux *pl.*] (*n.m.*), pile-drawer. (9)

arrache-sonde [arrache-sondes *ou* arrache-sonde *pl.*] (*n.m.*), drill-rod grab. (10)

arrache-tuyau [arrache-tuyaux *ou* arrache-tuyau *pl.*] (Sondage) (*n.m.*), casing-spear ; casing-dog. (11)

arrache-tuyau à déclic (*m.*), trip casing-spear. (12)

arrachement (action d'arracher) (*n.m.*). Same as **arrachage**.

arrachement (Maçonn.) (*n.m.*), toothing. (13)

arracher (*v.t.*). See examples : (14)
arracher des arbres, to tear up trees. (15)
arracher le feu (Mines), to dig the fire out. (16)
arracher le filet d'un écrou, d'un boulon, to strip, to tear off the thread from, a nut, a bolt. (17 *ou* 18)
arracher un clou, to pull out, to draw, to extract, a nail. (19)

arrangement (*n.m.*), arrangement : (20)
l'arrangement moléculaire de la matière calcaire dans les restes organiques, the molecular arrangement of the calcareous matter in organic remains. (21)

arranger (*v.t.*), to arrange. (22)

arrastre (Préparation mécanique des minerais) (*n.m.*), arrastra ; arrastre ; arastra. (23)

arrastre mexicain (*m.*), Mexican arrastra. (24)

arrêt (cessation) (*n.m.*), stoppage ; stopping ; stop ; suspension ; standstill ; hitch : (25)
arrêt d'urgence, emergency stop : (26)
arrêt d'urgence en fin de cordée (Mines), emergency stop at end of wind. (27)
arrêt de l'aérage, stoppage of, stopping, the ventilation. (28)
arrêt des affaires, stoppage of business. (29)
arrêt des travaux pendant les vacances, stoppage of work during holidays. (30)
arrêt momentané, temporary stoppage ; momentary standstill. (31)
arrêt inopiné par suite d'avaries, break-down. (32)
arrêt pour procéder à des réparations *ou* arrêt pour réparations, stoppage for repairs. (33)

arrêt (Signaux de ch. de f.), stop ; danger : (34)
signal à l'arrêt, signal at stop ; signal at danger. (35)

arrêt (Méc.) (*n.m.*), stop ; block ; catch. (36)

arrêt (règlement administratif) (*n.m.*), edict ; decree : (37)
un arrêt du gouvernement, a government edict ; a Government decree. (38)

arrêt-barrage [arrêts-barrages *pl.*] (*n.m.*), dam. (39)

arrêt-barrage (Aérage de mines) (*n.m.*), stopping ; barrier. (40)

arrêt-barrage d'eau (pour arrêter la propagation des explosions de poussières de houille)

(Mines) (*m.*), damp-sheet ; water-curtain : water-stopping ; water-barrier. (41)

arrêt-barrage de poussières incombustibles (Mines) (*m.*), stone-dust stopping ; rock-dust barrier. (42)

arrêt-barrage des tailings (*m.*), tailings-dam. (43)

arrêt de chanfrein (Menuis.) (*m.*), chamfer-stop ; stop for chamfer. (44)

arrêt de ligne (Télégr., etc.) (*m.*), line terminal. (45)

arrêt sur l'infini (Photogr.) (*m.*), infinity-catch. (46)

arrête-convoi [arrête-convois *pl.*] (plan incliné) (Mines) (*n.m.*), set stopping device. (47)

arrêté (*n.m.*), by-law ; regulation : (48)
un arrêté de police, a police regulation. (49)

arrêté, -e (être), to be stopped ; to be at a stand-still : (50)
être arrêté par suite du manque de combustible, to be stopped for want of fuel. (51)
les travaux sont arrêtés maintenant (*m.pl.*), work is now at a standstill. (52)

arrêter (*v.t.*), to stop ; to cease ; to suspend ; to arrest : (53)
arrêter l'aérage, to stop the ventilation ; to stop ventilating. (54)
arrêter l'exploitation, to stop mining ; to cease operations. (55)
arrêter la circulation dans la galerie de roulage, to suspend the traffic on the haulage-road. (56)
arrêter les travaux, to stop work ; to cease work ; to cease working ; to suspend operations. (57)
arrêter le fléau d'une balance, to arrest the beam of a balance. (58)

arrêter les services de quelqu'un, to retain someone's services. (59)

arrêter une planche avec des clous, to fix a board with nails. (60)

arrêter (s') (*v.r.*), to stop ; to come to a stand-still : (61)
s'arrêter court, to stop short. (62)

arrêter (s') par suite d'avaries, to break down. (63)

arrhénite (Minéral) (*n.f.*), arrhenite. (64)

arrière (partie postérieure) (*n.m.*), back ; back part ; rear : (65)
l'arrière d'une charrette, d'une chambre noire photographique, the back of a cart, of a photographic camera. (66 *ou* 67)

arrière (d') *ou elliptiquement* **arrière**, back ; hind : (68)
essieu d'arrière *ou* essieu arrière (*m.*), back axle ; hind axle. [des essieux d'arrière *ou* des essieux arrière *pl.*]. (69)

arrière (côté du piston opposé à celui où se trouve la manivelle) (machine fixe horizontale) (opp. à **avant**) (*on écrit par abréviation* **AR** *ou* **A R**), head-end : (70)
admission arrière (*f.*), head-end admission. (71)
échappement AR (*m.*), head-end release. (72)
point mort A R (*m.*), head-end dead-centre. (73)

arrière *ou* **AR** *ou* **A R** (côté du piston où se trouve la manivelle) (locomotive), crank-end ; back : (74)
admission arrière (*f.*), crank-end admission. (75)
point mort arrière (*m.*), back dead-centre. (76)

arrière-bec [arrière-becs *pl.*] (*n.m.*) (opp. à avant-bec), down-stream cutwater (of a bridge-pier). (77)

arrière-bief [arrière-biefs *pl.*] (canal) (Navigation) (*n.m.*), aft-bay ; tail-bay. (1)
arrière-cadre réversible [arrière-cadres réversibles *pl.*] (d'un appareil photographique) (*m.*), reversing-frame. (2)
arrière-cylindre (*n.m.*) *ou* arrière du cylindre (*m.*) *ou* AR-cylindre (*n.m.*) (machine fixe horizon-tale), head-end of cylinder. (3)
arrière-cylindre (*n.m.*) *ou* arrière du cylindre (*m.*) *ou* AR-cylindre (*n.m.*) (locomotive), crank-end of cylinder. (4)
arrière-plan [arrière-plans *pl.*] (*n.m.*), back-ground. (5)
arrière-train [arrière-trains *pl.*] (d'un véhicule à quatre roues) (*n.m.*), hind-carriage (of a four-wheeled vehicle). (6)
arrivée (*n.f.*), arrival : (7)
l'arrivée de la cage à la surface, the arrival of the cage at the surface. (8)
arrivée d'air (orifice) (*f.*), air-inlet. (9)
arrivée d'eau (*f.*), water-inlet. (10)
arrivée de la vapeur (*f.*), steam-inlet ; inlet for steam. (11)
arrivée du vent (d'une machine soufflante) (*f.*), air-inlet (of a blowing-engine). (12)
arriver (*v.i.*), to arrive ; to come. (13)
arriver à, to arrive at ; to come to ; to reach ; to attain : (14)
arriver à sa destination, to arrive at, to reach, one's destination. (15)
arriver à un arrangement, to come to an arrangement (*or* to terms). (16)
arriver à une conclusion, to arrive at, to come to, to reach, a conclusion. (17)
arriver à la fin de son exploitation (se dit d'une mine), to approach exhaustion. (18)
arriver à ses fins, to achieve one's object ; to succeed. (19)
arrondir (*v.t.*), to round ; to round off : (20)
arrondir les coins d'une table, to round the corners of a table. (21)
galets arrondis (*m.pl.*), rounded pebbles. (22)
arrondissement (action) (*n.m.*), rounding. (23)
arrondissement (rondeur) (*n.m.*), roundness. (24)
arrosage *ou* arrosement (*n.m.*), watering ; sprinkling : (25)
arrosage des voies de roulage (Mines), watering the haulage-roads ; sprinkling the haulage-ways. (26)
arroser (*v.t.*), to water ; to sprinkle : (27)
de nombreuses rivières arrosent le Chili (*f.pl.*), numerous rivers water Chile. (28)
région bien arrosée (*f.*), well-watered region. (29)
arroser (s') (*v.r.*), to be watered ; to be sprinkled : (30)
les galeries doivent s'arroser méthodiquement (Mines), the roads should be watered systematically. (31)
arroseur (*n.m.*), sprinkler ; watering-apparatus. (32)
arrosion (*n.f.*), corrosion ; erosion : (33)
l'arrosion de la rouille, the corrosion of rust. (34)
arrosoir (*n.m.*), watering-can ; watering-pot. (35)
arrugie (*n.f.*), drain ; gutter ; drainage-channel in a mine to conduct water to the sump. (36)
arséniate (Chim.) (*n.m.*), arsenate ; arseniate : (37)
arséniate de plomb, arsenate of lead ; lead arsenate. (38)
arsenic (corps simple) (As) (*n.m.*), arsenic. (39)

arsenic (*n.m.*) *ou* arsenic blanc (As^{2}O^3 *ou* As^{4}O^6), arsenic ; white arsenic ; arsenious anhydride ; arsenious oxide ; arsenic trioxide. (40)
arsenic amorphe (*m.*), amorphous arsenic. (41)
arsenic natif (*m.*), native arsenic. (42)
arsenic sublimé (*m.*), sublimed arsenic. (43)
arsenic sulfuré rouge (Minéral) (*m.*), red arsenic ; ruby-arsenic ; ruby-sulphur ; ruby of arsenic ; ruby of sulphur ; realgar. (44)
arsenical, -e, -aux (*adj.*), arsenical : (45)
intoxication arsenicale (*f.*), arsenical poison-ing. (46)
arséniciase (*n.f.*) *ou* arsenicisme (*n.m.*) (Toxi-cologie), arseniasis ; arsenicism ; arsenical-ism. (47)
arsénié, -e (*adj.*), arseniated ; arseniureted : (48)
hydrogène arsénié (*m.*), arseniureted hydrogen. (49)
arsénieux (*adj.m.*), arsenious : (50)
anhydride arsénieux (*m.*), arsenious anhydride. (51)
arsénifère (*adj.*), arseniferous. (52)
arséniopléite (Minéral) (*n.f.*), arseniopleite. (53)
arséniosidérite (Minéral) (*n.f.*), arseniosiderite. (54)
arséniosulfure (*n.m.*), sulpharsenide. (55)
arsénique (Chim.) (*adj.*), arsenic. (56)
arsénite (Chim.) (*n.m.*), arsenite. (57)
arsénite (Minéral) (*n.f.*), arsenite ; arsenolite. (58)
arséniure (Chim.) (*n.m.*), arsenide ; arseniuret : (59)
arséniure de nickel, arsenide of nickel ; nickel arsenide. (60)
arséniuré (*adj.m.*), arseniureted. (61)
arsennickel (Minéral) (*n.m.*), arsenical nickel ; copper-nickel ; kupfernickel ; niccolite ; nickelite ; nickeline. (62)
arsénolite (Minéral) (*n.f.*), arsenolite ; arsenite. (63)
arsénopyrite (Minéral) (*n.f.*), arsenopyrite ; arsenical pyrites ; mispickel ; white mundic. (64)
art (*n.m.*), art : (65)
l'art de bâtir, the art of building. (66)
l'art de l'ébéniste, the cabinetmaker's art. (67)
les arts mécaniques, the mechanical arts. (68)
art de l'ingénieur (*m.*), engineering. (69)
art de la mécanique (*m.*), mechanical engineering. (70)
art de la topographie (*m.*), topographic engineer-ing. (71)
art des mines (*m.*), mining engineering. (72)
artère (Élec.) (*n.f.*), feeder ; feed-wire ; feeder-cable ; feeding-conductor : (73)
l'artère et le réseau, the feeder and the network. (74)
artésien, -enne (*adj.*), Artesian : (75)
puits artésien (*m.*), Artesian well. (76)
eaux artésiennes (*f.pl.*), Artesian waters. (77)
articulation (*n.f.*), articulation ; joint. (78)
articulation (Géol.) (*n.f.*), joint. (79)
articulation à genouillère (*f.*), knuckle-joint. (80)
articulation à la Cardan (*f.*), Cardan joint ; cardan joint. (81)
articuler (*v.t.*), to articulate ; to joint ; to link : (82)
locomotive compound articulée système Mallet (*f.*), Mallet articulated compound loco-motive. (83)

règle articulée (*f.*), jointed rule. (1)

articuler (s') (*v.r.*), to articulate ; to joint ; to link. (2)

artificiel, -elle (*adj.*), artificial : (3)
pierre artificielle (*f.*), artificial stone. (4)

artificiellement (*adv.*), artificially. (5)

asbeste (*n.m.*), asbestos. (6)

asbestiforme (*adj.*), asbestiform. (7)

asbestin, -e (*adj.*), asbestine. (8)

asbestoïde (*adj.*), asbestoid ; asbestoidal. (9)

asbolane *ou* **asbolite** (Minéral) (*n.f.*), asbolite ; asbolane ; asbolan ; earthly cobalt ; slaggy cobalt. (10)

ascendant, -e (*adj.*), ascending ; upward : (11)
mouvement ascendant (*m.*), upward motion. (12)

ascenseur (*n.m.*), lift ; passenger-lift ; elevator. (13)

ascenseur à air comprimé (*m.*), pneumatic lift ; pneumatic elevator. (14)

ascenseur à gaz (*m.*), gas-elevator. (15)

ascenseur à piston (*m.*), piston-elevator. (16)

ascenseur à piston plongeur (*m.*), plunger-elevator. (17)

ascenseur électrique (*m.*), electric lift ; electric passenger-lift ; electric elevator. (18)

ascenseur électrique ; modèle à tambour, com--mande par le bas, contrôle par interrupteur dans la cabine (*m.*), electric passenger-lift ; drum type, underdriven, controlled by switch in car. (19)

ascenseur électrique ; modèle à tambour, mécanisme en surplomb, contrôle à câble (*m.*), electric passenger-lift, drum type ; gear overhead, rope control. (20)

ascenseur hydraulique (*m.*), hydraulic lift ; hydraulic elevator. (21)

ascenseur hydraulique à haute pression (*m.*), high-pressure hydraulic passenger-lift. (22)

ascenseur hydropneumatique (*m.*), hydropneu--matic lift. (23)

ascension (action de monter) (*n.f.*), ascent ; rising : (24)
l'ascension d'une montagne, the ascent of a mountain. (25)
l'ascension de l'eau dans le corps de pompe, the rising of the water in the pump-barrel. (26)

ascension (Géol.) (*n.f.*), ascension : (27)
l'ascension de matières à l'état de fusion ignée, the ascension of matter in a state of igneous fusion. (28)

ascension capillaire (Phys.) (*f.*), capillary ascent. (29)

ascensionnel, -elle (*adj.*), ascensional : (30)
aérage ascensionnel (*m.*), ascensional ventila--tion. (31)

aschiste (opp. à diaschiste) (Pétrol.) (*adj.*), aschistic. (32) (opp. to diaschistic) :
roches aschistes (*f.pl.*), aschistic rocks. (33)

asmanite (Minéral) (*n.f.*), asmanite. (34)

aspect (*n.m.*), aspect ; appearance ; look ; outlook ; showing : (35)
aspect commercial d'un sujet, commercial aspect of a subject. (36)
aspect du front de taille (Mines), appearance of the working-face. (37)

aspérité (*n.f.*), roughness ; asperity ; ruggedness ; unevenness : (38)
aspérité du sol, roughness, unevenness, of the ground. (39)

aspérolite (Minéral) (*n.f.*), asperolite. (40)

asphaltage (*n.m.*), asphalting. (41)

asphalte (*n.m.*), asphalt ; asphaltum. (42)

asphaltène (*n.m.*), asphaltene. (43)

asphalter (*v.t.*), to asphalt: (44)
asphalter un trottoir, to asphalt a footpath. (45)

asphaltique (*adj.*), asphaltic : (46)
calcaire asphaltique (*m.*), asphaltic limestone. (47)

asphyxiant, -e (*adj.*), asphyxiating ; suffocating : (48)
gaz asphyxiant (*m.*), asphyxiating gas ; suffocating gas. (49)

asphyxie (*n.f.*), asphyxia ; asphyxiation ; suffocation ; gassing : (50)
asphyxie par les gaz naturels, suffocation by natural gases. (51)
asphyxie produite par les gaz de la poudre, gassing from powder fumes : (52)
la mort fut causée par asphyxie produite par les gaz de la poudre, death was due to gassing from powder fumes. (53)

asphyxier (*v.t.*), to asphyxiate ; to suffocate ; to gas : (54)
nombre d'ouvriers furent asphyxiés par les gaz pendant qu'ils combattaient l'incendie, a number of workmen were gassed while fighting the fire. (55)

aspidolite (Minéral) (*n.f.*), aspidolite. (56)

aspirail (*n.m.*), air-hole ; air-vent ; vent ; flue ; air-flue : (57)
ménager des aspiraux, to make air-holes. (58)

aspirant, -e (Méc.) (*adj.*), suction (*used as adj.*) ; sucking ; aspiring : (59)
pompe aspirante (*f.*), suction-pump ; sucking pump ; aspiring pump. (60)

aspirant (crépine de pompe) (*n.m.*), strainer ; wind-box ; snore-piece ; tail-piece ; rose. (61)

aspirateur (*n.m.*), exhauster ; aspirator ; sucker. (62)

aspirateur d'air (*m.*), air-exhauster. (63)

aspirateur de poussières (*m.*), dust-exhaust fan ; dust-exhauster. (64)

aspirateur hydraulique (Mines) (*m.*), water-blast. (65)

aspiration (*n.f.*), exhaustion ; exhausting ; suction : (66)
aspiration de l'air (opp. à *refoulement*) (Aérage de mines), exhausting the air. (67) (in contradistinction to *forcing* or *forcing down*).
aspiration de l'eau par une pompe, suction of water by a pump. (68)

aspirer (*v.t.*), to suck ; to draw ; to exhaust ; to aspirate : (69)
piston qui aspire de l'air (*m.*), piston which draws (*or* which exhausts) (*or* which aspirates) air. (70)

assainir (*v.t.*), to make healthy ; to cleanse ; to purify : (71)
le desséchement des marais assainit un pays, the draining of marshes makes a country healthy. (72)
assainir le ballast en pratiquant des drains de distance en distance, to cleanse the ballast by making drains at intervals. (73)
assainir un chantier contaminé (Mines), to cleanse polluted workings ; to purify gassy workings ; to clear away the gases from the workings ; to disperse the gases in the workings. (74)

assainir (s') (*v.r.*), to become healthy ; to be cleansed ; to be purified. (75)

assainissement (*n.m.*), making healthy ; cleans-
-ing ; purifying. (1)

asseau (*n.m.*), slaters' hammer. (2)

asséchement (*n.m.*), draining ; drainage ; un-
-watering ; dewatering ; drying : (3)
asséchement d'un marais, drying a swamp ;
draining a marsh. (4)
asséchement des travaux (Mines), unwatering,
dewatering, draining, the workings. (5)

assécher (*v.t.*), to drain ; to unwater ; to
dewater ; to dry up ; to dry : (6)
assécher une mine, to unwater, to dewater,
to drain, to pump out, to fork, a mine.
(7)
assécher une mine par un écoulement spontané,
to unwater a mine by natural drainage ;
to unwater a mine naturally through an
adit. (8)

assécher (s') (*v.r.*), to run dry ; to dry up : (9)
torrent qui s'assèche pendant l'été (*m.*),
torrent which dries up in summer. (10)

assemblage (action) (*n.m.*), assembling ; assem-
-blage ; joining ; jointing : (11)
assemblage des pièces d'une machine, assem-
-bling the parts of a machine. (12)

assemblage (réunion d'objets assemblés) (*n.m.*),
assemblage ; collection. (13)

assemblage (manière de joindre ensemble des
pièces de bois, de métal) (*n.m.*), joint ;
jointing. (14). See also **joint**.

assemblage à boulons (*m.*), bolted joint. (15)

assemblage à brides *ou* **assemblage à brides
boulonnées** (*m.*), flange-joint. (16)

assemblage à charnière (*m.*), hinge-joint ; folding
joint. (17)

assemblage à clef (*m.*), key-joint. (18)

assemblage à clin (Rivetage) (*m.*), lap-joint. (19)

assemblage à collier (*m.*), collar-joint. (20)

assemblage à cornières (*m.*), angle-iron joint.
(21)

assemblage à couvre-joints (*m.*), butt-strap
joint ; welted joint ; fish-joint ; fished
joint. (22)

assemblage à double tenon (*m.*), double-tenon
joint. (23)

assemblage à douille (*m.*), socket-joint. (24)

assemblage à emboîtement (tuyaux) (*m.*), spigot-
joint ; spigot-and-faucet joint ; faucet-joint.
(25)

assemblage à embrèvement (*m.*), joggle-joint.
(26)

assemblage à embrèvements anglais *ou* **assem-
-blage à embrèvements séparés par un plat
joint** (*m.*), bridle-joint ; notch-and-bridle
joint. (27)

assemblage à empattements boulonnés (*m.*),
bolted-tee joint. (28)

assemblage à encastrement (*m.*), housed joint.
(29)

assemblage à enfourchement (*m.*), slot-mortise
joint ; slip-mortise joint. (30)

assemblage à entaille (*m.*), notch-joint ; notched
joint ; notching ; single-notch joint. (31)

assemblage à entailles *ou* **assemblage à double
entaille** (*m.*), double-notch joint. (32)

assemblage à enture (*m.*), scarf-joint. (33)

assemblage à enture à vis (*m.*), screw-and-
socket joint. (34)

assemblage à étrier (*m.*), stirrup-joint. (35)

assemblage à fausse languette (*m.*), slip-tongue
joint ; loose-tongue joint ; filleted joint ;
ploughed and feathered joint ; ploughed and

tongued joint ; grooved and feathered joint ;
feather-tongued joint ; feather-joint. (36)

assemblage à feuillure (*m.*), rabbeted joint ;
rebated joint ; fillistered joint. (37)

assemblage à franc-bord (Rivetage) (*m.*), butt-
joint ; butting-joint. (38)

assemblage à franc-bord à double couvre-joint
(*m.*), butt-joint with two welts. (39)

assemblage à franc-bord avec couvre-joint simple
(*m.*), butt-joint with one welt. (40)

assemblage à grain-d'orge (*m.*), bevelled tongue-
and-groove joint ; angular grooved and
tongued joint. (41)

assemblage à languette rapportée (*m.*). Same
as **assemblage à fausse languette**.

assemblage à lanterne (*m.*), turnbuckle-joint.
(42)

assemblage à manchon (*m.*), sleeve-joint ; fer-
-rule-joint. (43)

assemblage à manchon taraudé (*m.*), sleeve-nut
joint ; screw ferrule-joint. (44)

assemblage à mi-bois (*m.*), halved joint ; halv-
-ing ; half-lap ; half-lap joint ; lap-joint ;
lapped joint ; overlap-joint ; overlapping
joint ; step-joint. (45)

assemblage à mi-fer (*m.*). Same as preceding
applied to wrought-iron work. (46)

assemblage à mi-fonte (*m.*). Same as preceding
applied to cast-iron work. (47)

assemblage à onglet (*m.*), mitre-joint ; miter-
joint ; bevel-joint ; chamfered joint. (48)

assemblage à onglet en sifflet (*m.*), splayed mitre-
joint. (49)

assemblage à plat *ou* **assemblage à plat joint**
(*m.*), square joint ; straight joint ; butt-
joint ; butting-joint. (50)

assemblage à queue d'aronde *ou* **assemblage à
queue d'hironde** *ou* *simplement* **assemblage
à queue** (*m.*), dovetail joint ; fantail joint ;
swallow-tail joint. (51)

assemblage à queue d'aronde à mi-bois (*m.*),
dovetail halved joint. (52)

assemblage à queue d'aronde à recouvrement (*m.*),
dovetail lap-joint. (53)

assemblage à rainure et languette (*m.*), tongue-
and-groove joint ; grooved and tongued
joint ; ploughed and tongued joint ; feather-
joint. (54)

assemblage à recouvrement (*m.*), lap-joint ;
lapped joint ; overlap-joint ; overlapping
joint ; step-joint. (55)

assemblage à sifflet (*m.*). Same as **assemblage
en sifflet**.

assemblage à tenon (*m.*), tenon-joint. (56)

assemblage à tenon avec chaperon et renfort
(*m.*), tusk-tenon joint (a tenon-joint with
a bevelled haunch above and one tusk
beneath—the usual form). (57)

assemblage à tenon avec cheville (*m.*), pinned
tenon-joint. (58)

assemblage à tenon avec renfort carré (*m.*),
haunched-tenon joint. (59)

assemblage à tenon et à mortaise (*m.*), mortise-
and-tenon joint ; mortise-joint ; tenon-
joint ; tenoned and housed joint. (60)

assemblage à tenon et mortaise doubles (*m.*),
double-tenon joint. (61)

assemblage à tenon, mortaise, et cale (*m.*),
wedged mortise-and-tenon joint. (62)

assemblage à tenon passant avec clef (*m.*),
through-tenon joint with key ; pegged
tenon-joint. (63)

assemblage à tenon renforcé *ou* **assemblage à tenon avec renfort** (*m.*), tusk-tenon joint; tenon-and-tusk joint. (1)
assemblage à trait de Jupiter (*m.*), splayed indent scarf. (2)
assemblage à trait de Jupiter horizontal (*m.*), tabled scarf. (3)
assemblage à vis (*m.*), screw-joint. (4)
assemblage bout à bout (Charp.) (*m.*), heading-joint. (5)
assemblage d'angle (*m.*), angle-joint. (6)
assemblage d'onglet (*m.*).   Same as **assemblage à onglet.** (7)
assemblage de charpente (*m.*), carpenters' joint. (7)
assemblage de charpente en fer (*m.*), smiths' joint (girder work and the like). (8)
assemblage de chaudronnerie (*m.*), copper-smiths' joint. (9)
assemblage de la fonte (*m.*), cast-iron joint. (10)
assemblage de menuiserie (*m.*), joiners' joint. (11)
assemblage de plomberie (*m.*), plumbers' joint. (12)
assemblage de rallonges (Sondage) (*m.*), drill-joint; drill-rod joint; drill-pole joint. (13)
assemblage de tuyauterie (*m.*), pipe-joint. (14)
assemblage du fer (*m.*), wrought-iron joint. (15)
assemblage en about (*m.*), abutting joint. (16)
assemblage en onglet (*m.*).   Same as **assemblage à onglet.**
assemblage en queue d'aronde (*m.*).   Same as **assemblage à queue d'aronde.**
assemblage en sifflet (*m.*), splayed joint. (17)
assemblage en sifflet désabouté (*m.*), splayed butt scarf. (18)
assemblage oblique (*m.*), oblique joint. (19)
assemblage oblique à tenon et mortaise avec embrèvement (*m.*), mortise-and-tenon heel-joint. (20)
assemblage par clavette en coin (*m.*), cottered joint. (21)
assemblage rivé *ou* **assemblage par rivets** (*m.*), riveted joint; rivet-joint. (22)
assemblage suspendu (*m.*), suspended joint. (23)
assembler (*v.t.*), to assemble; to join; to joint: (24)
assembler les pièces d'une machine, to assemble the parts of a machine. (25)
assembler les panneaux d'une porte, to joint the panels of a door. (26)
assembler à queue d'aronde (*v.t.*), to dovetail. (27)
asseoir (*v.t.*), to seat; to bed: (28)
asseoir les fondements d'un édifice, to seat, to bed, the foundations of a building. (29)
assette (*n.f.*), slaters' hammer. (30)
assez (suffisamment) (*adv.*), enough: (31)
maison qui n'est pas assez grande (*f.*), house which is not large enough. (32)
assez bon, bonne, fair; fairly good: (33)
assez bonne moyenne (*f.*), fair average. (34)
assez bonnes espérances (*f.pl.*), fair prospects. (35)
assiette (situation stable, solide, d'un corps) (*n.f.*), set; firmness; steadiness; stability: (36)
l'assiette d'une poutre, d'une pierre, the set of a beam, of a stone. (37 *ou* 38)
le ballast est indispensable à l'assiette de la voie, ballast is absolutely necessary to ensure the set of the permanent way. (39)
assiette (base; fondement) (*n.f.*), foundation; base; bed; seat; seating: (40)

machine qui repose sur une assiette solide (*f.*), engine which rests on a solid foundation (*or* on a firm seating). (41)
traverse qui offre une assiette parfaitement plane au patin du coussinet (*f.*), sleeper which presents a perfectly even seat to the foot of the chair. (42)
assiette (position topographique) (*n.f.*), situation; location; site; position: (43)
ville qui a une assiette favorable (*f.*), town which has a favourable location; town which is favorably situated. (44)
assiette pour le pied (*f.*), footing; foothold. (45)
assise (masse formant lit) (*n.f.*), bed; foundation: (46)
plaque d'assise (*f.*), bed-plate; foundation-plate. (47)
assise (Géol.) (*n.f.*), bed; assise. (48)
assise (Maçon.) (*n.f.*), course: (49)
une assise de briques, a course of bricks. (50)
assise de boutisses (Maçonn.) (*f.*), heading-course; heading. (51)
assise de carreaux (Maçonn.) (*f.*), stretching-course. (52)
assise de niveau (Maçonn.) (*f.*), level course. (53)
assise de Pilton (Géol.) (*f.*), Pilton bed; Pilton formation. (54)
assise de retombée (d'une voûte) (*f.*), springing-course (of an arch). (55)
assise houillère (Géol.) (*f.*), coal-bed. (56)
assistance (*n.f.*), assistance; aid; help: (57)
assistance expérimentée, skilled assistance. (58)
association (*n.f.*), association: (59)
association du platine natif à des roches péridotiques, association of native platinum with peridotic rocks. (60)
minerai qui contient le nickel en association avec la pyrite de fer magnétique (*m.*), ore which contains nickel in association with magnetic iron pyrites. (61)
association (réunion de personnes) (*n.f.*), association; partnership. (62)
association des éléments de pile (*f.*), connecting, connection of, coupling, grouping, battery-cells. (63). See **accouplement,** which is synonymous with **association** in this sense, for various modes of connection.
associé, -e (pers.) (*n.*), associate; partner. (64)
associer (*v.t.*), to associate: (65)
espèce minérale qui se rencontre associée au grenat (*f.*), mineral species which is found associated with garnet. (66)
associer (Élec.) (*v.t.*), to connect; to couple; to group. (67)
assorti, -e (*adj.*), assorted: (68)
dimensions assorties (*f.pl.*), assorted sizes. (69)
assortiment (action) (*n.m.*), assorting; assort-ment. (70)
assortiment (collection) (*n.m.*), assortment; set: (71)
un assortiment d'outils de charpentier, an assortment of carpenter's tools; a set of carpenters' tools. (72)
assortir (*v.t.*), to assort. (73)
assortir des minerais, to blend ores. (74)
assujétir *ou* **assujettir** (*v.t.*), to fix; to fasten; to set: (75)
assujétir une poutre, to fix a beam. (76)

assurance (excédent de matière que l'on emploie dans une fonte sur le poids présumé des pièces à couler) (Fonderie) (*n.f.*), spare. (1)

assurer (rendre sûr) (*v.t.*), to assure ; to ensure ; to secure : (2)

assurer la sécurité, to ensure safety. (3)

assurer une bonne fondation, to secure a good foundation. (4)

assurer une main-d'œuvre suffisante, to secure sufficient labour. (5)

assurer (rendre stable) (*v.t.*), to strengthen : (6)

assurer une muraille, to strengthen a wall. (7)

assurer le bout d'un cordage, to make fast, to belay, the end of a rope. (8)

assurer (s') (*v.r.*) *ou* assurer (s') de (s'en garantir la possession), to secure ; to get : (9)

s'assurer le contrôle d'une propriété, to secure control of a property. (10)

s'assurer une main-d'œuvre suffisante, to secure sufficient labour. (11)

assurer (s') du bien-fondé d'un bruit, to ascertain whether there is any truth in a report. (12)

astaki (résidu de pétrole) (*n.m.*), astatki. (13)

astatique (Phys.) (*adj.*), astatic : (14)

aiguille astatique (*f.*), astatic needle. (15)

astérie (Minéral) (*n.f.*), asteriated opal. (16)

astérie (à) *ou* **astéries (à)** (Cristall.), asteriated ; star (*used as adj.*) : (17)

quartz à astéries (*m.*), asteriated quartz ; star quartz. (18)

astérisme (Cristall.) (*n.m.*), asterism. (19)

astigmate (Opt.) (*adj.*), astigmatic. (20)

astigmatisme (*n.m.*), astigmatism : (21)

astigmatisme d'une lentille, astigmatism of a lens. (22)

astragale (moulure) (*n.m.*), astragal. (23)

astrakanite (Minéral) (*n.f.*), astrakanite. (24)

astrophyllite (Minéral) (*n.f.*), astrophyllite. (25)

asymétrie (*n.f.*), asymmetry. (26)

asymétrique (*adj.*), asymmetric : asymmetrical ; dissymmetric ; dissymmetrical ; unsymmetrical. (27)

asynchrone (*adj.*), asynchronous : (28)

moteur asynchrone (*m.*), asynchronous motor. (29)

asynchronisme (*n.m.*), asynchronism. (30)

atacamite *ou* **atakamite** (Minéral) (*n.f.*), atacamite (31)

atélestite (Minéral) (*n.f.*), atelestite. (32)

atelier (*n.m.*), workshop ; shop ; mill ; works ; house ; studio. (33)

atelier (chantier) (Mines) (*n.m.*), working-place ; working : (34)

ateliers de traçage, development working-places. (35)

atelier d'ajustage (*m.*), fitters' shop ; fitting-shop. (36)

atelier d'ébarbage (Fonderie) (*m.*), fettling-shop. (37)

atelier d'enrichissement (Préparation mécanique des minerais) (*m.*), dressing-works. (38)

atelier d'essai (*m.*), testing-shop. (39)

atelier de chaudronnerie (*m.*), boiler-works. (40)

atelier de construction de locomotives (*m.*), loco-motive-works. (41)

atelier de dessablage (Fonderie) (*m.*), dressing-shop ; cleaning-shop. (42)

atelier de forge *ou* **atelier de forgeron** (*m.*), blacksmith's shop ; smiths' shop ; smithy ; smithery ; forge. (43)

atelier de lavage (*m.*), wash-house ; washing-house ; washery. (44)

atelier de maréchal ferrant (*m.*), blacksmiths' shop ; smiths' shop ; smithy ; smithery ; farriery. (45)

atelier de modelage (Fonderie) (*m.*), pattern-shop. (46)

atelier de montage (*m.*), erecting-shop. (47)

atelier de montage de locomotives (*m.*), loco-motive erecting-shop. (48)

atelier de moulage (*m.*), moulding-shop ; moulders' shop. (49)

atelier de parachèvement (*m.*), finishing-shop. (50)

atelier de photographe (*m.*), photographer's studio ; photographic studio. (51)

atelier de préparation mécanique *ou simplement* **atelier** (*n.m.*) (minerais), mill ; dressing-works : (52)

atelier de préparation mécanique des minerais, ore-dressing works. (53)

atelier de réparation (*m.*), repairing-shop ; repair-shop ; repairs-shop. (54)

atelier de réparation de locomotives (*m.*), loco-motive repair-shop. (55)

atelier de réparation de wagons (*m.*), car-shop ; car repair-shop. (56)

atelier de travail des boues (minerais) (*m.*), slimes-concentration mill. (57)

atelier de travail des fines (minerais) (*m.*), fine-concentration mill. (58)

atelier de travail des grenailles et sables (minerais) (*m.*), coarse-concentration mill ; jig-mill. (59)

atelier de travail des gros (minerais) (*m.*), sorting and picking-house. (60)

atelier de vorscheidage (minerais) (*m.*), ragging-floor. (61)

atelier des ajusteurs (*m.*), fitters' shop ; fitting-shop. (62)

atelier des bocards (*m.*), battery-house. (63)

atelier des machines (*m.*), machine-shop. (64)

atélite (Minéral) (*n.f.*), atelite. (65)

athermal, -e, -aux (se dit des eaux minérales froides) (*adj.*), cold. (66) (opp. to thermal).

athermane (Phys.) (*adj.*), athermanous ; ather-mous. (67)

athermanéité (*n.f.*), athermancy. (68)

athermique (*adj.*), athermic. (69)

atlasite (Minéral) (*n.f.*), atlasite. (70)

atmosphère (*n.f.*), atmosphere : (71)

l'atmosphère entourant le globe terrestre, the atmosphere surrounding the terrestrial globe. (72)

atmosphère chargée de poussière, dust-laden atmosphere. (73)

atmosphère de fumée, smoky atmosphere. (74)

atmosphère irrespirable, irrespirable atmos-phere ; atmosphere unfit for respiration. (75)

atmosphère (unité de pression) (Méc.) (*n.f.*), atmosphere ; (76) atmo : (77)

machine qui travaille à — atmosphères (*f.*), machine which works at — atmospheres. (78)

Note.—The word **atmo** in English signifies the French unit=the pressure of 760 millimetres of mercury at 0°C. in the latitude of Paris, at sea-level, equivalent to about 14·7 lbs to the sq. in. **atmosphere** is the English and U.S. standard unit=mercury-column 30 inches high, at the sea-level at London, at a temperature of 0°C.

atmosphère (Phys.) (*n.f.*), medium (the ether through which light and heat pass). (1)

atmosphérique (*adj.*), atmospheric; atmos- -pherical; air (*used as adj.*): (2)
pression atmosphérique (*f.*), atmospheric pressure; air-pressure. (3)

atoll (Géogr. phys.) (*n.m.*), atoll; lagoon-island; lagoon-reef. (4)

atome (Chim. & Phys.) (*n.m.*), atom: (5)
une molécule d'eau est formée de deux atomes d'hydrogène combinés avec un atome d'oxygène, a molecule of water is made up of two atoms of hydrogen combined with one atom of oxygen. (6)

atomicité (*n.f.*), atomicity; atomic value: (7)
atomicité des corps simples et des radicaux, atomicity of elements and of radicals. (8)

atomique (*adj.*), atomic; atomical: (9)
poids atomique d'un corps simple, atomic weight of an element. (10)

atomiquement (*adv.*), atomically. (11)

atopite (Minéral) (*n.f.*), atopite. (12)

âtre (*n.m.*), fire-place; hearth: (13)
âtre d'une forge, d'un fourneau, hearth of a forge, of a furnace. (14 *ou* 15)

attache (lien) (*n.f.*), tie; fastening; binder; bond; clip; band. (16)

attache de câble d'extraction (Mines) (*f.*), winding-rope cappel. (17)

attache-rail [**attache-rails** *pl.*] (*n.m.*), rail-clip. (18)

attachement (*n.m.*), attachment: (19)
attachement des bennes, attachment of tubs. (20)

attacher (*v.t.*), to attach; to tie; to fasten: (21)
attacher de l'importance à une découverte, to attach importance to a discovery. (22)

attacolite (Minéral) (*n.f.*), attacolite. (23)

attaquable (*adj.*), attackable. (24)

attaque (*n.f.*), attack: (25)
attaque des bancs de gravier, attack of the gravel-banks. (26)

attaque (Méc.) (*n.f.*), driving: (27)
méthode d'attaque des machines-outils (*f.*), method of driving machine-tools. (28)

attaque (d') (Méc.), driving: (29)
poulie d'attaque (*f.*), driving-pulley. (30)

attaque (Téléph. & Télégr.) (*n.f.*), call. (31)
Cf. **attaquer**.

attaque de front (Mines) (*f.*), breast-stoping; breast-work. (32)

attaque directe (Mines) (*f.*), outwards attack. (33)

attaque par le mur (Mines) (*f.*), resuing; strip- -ping. (34)

attaque rétrograde (Mines) (*f.*), homewards attack. (35)

attaquer (*v.t.*), to attack: (36)
attaquer un filon par tranches, to attack a lode in slices. (37)

attaquer (Méc.) (*v.t.*), to drive: (38)
crémaillère attaquée par un pignon (*f.*), rack driven by a pinion. (39)

attaquer (Téléph. & Télégr.) (*v.t.*), to call: (40)
attaquer un poste téléphonique, to call a telephone-exchange. (41)
le poste de pompiers, ainsi attaqué, prévient par une sonnerie que l'appel a été entendu, the fire-station, called in this manner, signals by a bell that the call has been heard. (42)

attaquer (s') (*v.r.*), to be attacked: (43)

les perles s'attaquent très facilement par les acides (*f.pl.*), pearls are very easily attacked by acids. (44)

atteindre (*v.t.*), to reach; to attain; to get to: (45)
atteindre le stade de maturité, le but de son voyage, to reach the stage of maturity, the end of one's journey. (46 *ou* 47)
atteindre une profondeur de —, to attain, to reach, to get to, a depth of —. (48)
atteindre une nappe pétrolifère, to reach, to strike, oil. (49)
atteindre le bed-rock (Mines & Placers), to bottom; to strike bed-rock. (50)

attelage (action) (*n.m.*), attachment; hooking on; coupling: (51)
attelage de la cage au câble, attachment of the cage to the rope. (52)
attelage des tenders aux locomotives, coupling tenders to locomotives. (53)

attelage (dispositif) (Ch. de f.) (*n.m.*), coupler; coupling; connector. (54)

attelage (bêtes de somme) (*n.m.*), team: (55)
un attelage de bœufs, a team of oxen. (56)

attelage à chaîne (*m.*), chain-coupling. (57)

attelage de wagons (*m.*), car-coupler; car-coupling. (58)

atteler (*v.t.*), to attach; to couple; to couple up: (59)
atteler une remorque à un tracteur, to attach a trailer to a tractor; to couple a trail-car to a tractor. (60)

atteler des chevaux à une voiture, to yoke horses to a carriage. (61)

attenant, -e (*adj.*), adjoining; contiguous; adjacent; abutting: (62)
une prairie attenante à la rivière, a field adjoining the river; a field contiguous to the river. (63)

attenant (*prép.*), adjoining; contiguous to; adjacent to; abutting on: (64)
une prairie attenant la rivière, a field adjoining the river; a field contiguous to the river. (65)

attendre (*v.t.*), to wait for; to wait; to await. (66)

attendre (s') (*v.r.*), to expect; to rely; to depend: (67)
s'attendre à rencontrer le filon, to expect to strike the lode. (68)
s'attendre à une promesse, to rely, to depend, on a promise. (69)

attente (*n.f.*), expectation; expectations: (70)
attente concernant une propriété, expectations regarding a property. (71)

attente (pierre d'attente) (Maçonn.) (*n.f.*), tooth; toothing-stone. (72)

attention (Signaux de ch. de f.), caution. (73)

attention au train (avis), beware of the trains. (74)

attention aux hommes (signal d'avertissement) (Remontage) (Mines), men. (75)

attention, s.v.p., à la peinture (avis), wet paint. (76)

atténuation (opp. à renforcement) (Photogr.) (*n.f.*), reduction; reduction in density. (77)

atténuer (*v.t.*), to lessen; to diminish; to attenuate; to reduce: (78)
atténuer le frottement, to lessen friction. (79)
atténuer un cliché dur (Photogr.), to reduce a hard negative. (80)

atterrir (*v.i. & v.t.*), to land: (81)

détritus qui atterrissent sur les bords d'un glacier (*m.pl.*), detritus which lands on the margins of a glacier. (1)

atterrir (Mines) (*v.t.*), to land ; to bank : (2)
atterrir la benne, to land the kibble. (3)
atterrir la cage, to bank the cage. (4)

atterrissage (Mines) (*n.m.*), landing ; banking : (5)
atterrissage de la benne, landing of the kibble. (6)
atterrissage de la cage, banking the cage. (7)

atterrissement (Géol.) (*n.m.*), alluvium. (8)

attirable (*adj.*), attractable. (9)

attirail (*n.m.*) [*le pluriel* attirails *est peu usité*], appliances ; apparatus ; implements : (10)
l'attirail d'un photographe, the apparatus (*or* the appliances) of a photographer. (11)

attirail d'outils (Sondage) (*m.*), string of tools. (12)

attirant, -e (*adj.*), attractive : (13)
la force attirante de l'aimant, the attractive force of the magnet. (14)

attirer (*v.t.*), to attract ; to draw : (15)
l'aimant attire le fer et quelques autres métaux (*m.*), the magnet attracts iron and some other metals. (16)
l'ambre frotté attire les corps légers non-métalliques, amber, if rubbed, will attract light non-metallic bodies. (17)
attirer l'attention, to attract attention. (18)
attirer l'attention sur un fait, to draw attention to a fact. (19)

attirer (s') (*v.r.*), to attract each other ; to attract one another : (20)
les pôles de nom contraire s'attirent (*m.pl.*), unlike poles attract each other. (21)
corps qui s'attirent réciproquement (*m.pl.*), bodies which mutually attract one another. (22)

attisage *ou* **attisement** (*n.m.*), poking ; stirring. (23)

attiser (*v.t.*), to poke ; to stir : (24)
attiser le feu, to poke, to stir, the fire. (25)

attisoir *ou* **attisonnoir** (*n.m.*), poker ; prick-bar ; pricker. (26)

attoll (Géogr. phys.) (*n.m.*), atoll ; lagoon-island ; lagoon-reef. (27)

attouchement (*n.m.*), touching : (28)
mort occasionnée par l'attouchement d'un fil électrique (*f.*), death caused by touching an electric wire. (29)

attractif, -ive (*adj.*), attractive : (30)
l'aimant a une force attractive, the magnet has an attractive force. (31)

attraction (*n.f.*), attraction. (32)

attraction capillaire (*f.*), capillary attraction. (33)

attraction électrique (*f.*), electrical attraction. (34)

attraction magnétique (*f.*), magnetic attraction. (35)

attraction moléculaire (*f.*), molecular attraction. (36)

attrempage (*n.m.*), tempering. (37)

attremper (*v.t.*), to temper : (38)
attremper l'acier, to temper steel. (39)
attremper un four, un creuset *ou* pot (les chauffer par degrés), to temper a furnace, a crucible *or* pot. (40 *ou* 41)

attribuer (*v.t.*), to attribute ; to assign ; to ascribe : (42)

la valeur attribuée au minerai est de — fcs, the value assigned to the ore is fcs —. (43)

attrition (Phys.) (*n.f.*), attrition : (44)
attrition d'une masse mouvante, attrition of a moving mass. (45)

au début de l'industrie minière, in the early days of mining. (46)

au cas d'incendie, in case of fire. (47)

au commencement de l'été, early in the summer. (48)

au-dessous de, below : (49)
au-dessous de la moyenne, below the average. (50)
au-dessous du niveau de l'eau, below the water-level. (51)

au-dessus de, above : (52)
au-dessus de la moyenne, above the average. (53)
au-dessus du niveau de la mer, above sea-level. (54)

au fond (Mines), belowground ; underground : (55)
travailler au fond, to work underground. (56)

au hasard, at random ; haphazard. (57)

au jour (Mines), at the surface ; aboveground : (58)
travail au jour (*m.*), surface-work. (59)

au jour le jour, from day to day : (60)
un carnet de sondage devrait être tenu au jour le jour par le sondeur, a bore-holing journal should be kept from day to day by the driller. (61)

au premier abord, at first sight. (62)

au premier plan, in the foreground. (63)

au stade primitif, in the early stage. (64)

au titre, standard ; of standard fineness : (65)
or au titre (*m.*), standard gold ; gold of standard fineness. (66)

aube (d'une roue hydraulique) (*n.f.*), blade, paddle, (of a water-wheel) : (67)
aubes en rayons, radial paddles. (68)
aubes en tangentes, tangent paddles. (69)

aube (d'une turbine, d'un ventilateur) (*n.f.*), blade, vane, (of a turbine, of a fan). (70 *ou* 71)

aube courbe (turbine, etc.) (*f.*), curved vane ; curved blade. (72)

aube directrice *ou* **aube fixe** (turbine) (*f.*), guide-vane ; guide-blade ; stationary vane ; stationary blade. (73)

aube mobile *ou* **aube réceptrice** (turbine) (*f.*), runner-vane ; runner-blade ; moving vane ; moving blade. (74)

aubier (*n.m.*), alburnum ; sap-wood. (75)

auge (récipient ; boîte) (*n.f.*), trough. (76)

auge (rigole) (*n.f.*), flume ; channel. (77)

auge (godet d'une roue hydraulique) (*n.f.*), bucket. (78)

auge (pli synclinal ; fond de bateau) (opp. à voûte *ou* selle) (Géol.) (*n.f.*), trough. (79) (opp. to arch *or* saddle).

auge de forgeron (*f.*), smiths' water-trough ; slake-trough. (80)

auge en fonte pour meule en grès (*f.*), cast-iron trough for grindstone. (81)

augélite (Minéral) (*n.f.*), augelite. (82)

auger (plier en forme de gouttière) (*v.t.*), to trough. (83)

auget (*n.m.*), bucket : (84)
auget d'une roue hydraulique, bucket of a water-wheel. (85)

auget d'élévateur (*m.*), elevator-bucket ; elevator-cup. (86)

augette (Mines) (n.f.), augette ; dish ; small pan. (1)

augite (Minéral) (n.f.), augite. (2)

augitique (Minéralogie) (adj.), augitic. (3)

augitite (Pétrol.) (n.f.), augitite. (4)

augitophyre (Pétrol.) (n.m.), augitophyre. (5)

augitophyrique (Pétrol.) (adj.), augitophyric. (6)

augmentation (n.f.), increase ; augmentation ; enlargement ; rise : (7)
augmentation brusque de section (Hydraul.), sudden enlargement of cross-section. (8)
augmentation de valeur, increase of (or in) value. (9)
augmentation soutenue de la production, steady rise in production. (10)

augmenter (v.t.), to increase ; to augment ; to enlarge ; to raise : (11)
augmenter la production d'une mine, to increase the output of a mine. (12)
augmenter le prix, to raise the price. (13)

augmenter (v.i.), to increase ; to augment ; to enlarge ; to rise : (14)
augmenter de prix, to rise in price. (15)
angmenter de valeur, to rise, to increase, to improve, in value. (16)

augurer (v.i.), to auger ; to predict : (17)
augurer bien, mal, d'une chose, to augur well, ill, of a thing. (18 ou 19)
bien augurer de la richesse d'une mine, to predict a successful future for a mine. (20)

aune ou aulne (arbre) (n.m.), alder. (21)

auralite (Minéral) (n.f.), auralite. (22)

auréole (n.f.) ou auréole métamorphique (Géol.), aureole. (23)

auréole (produite dans une lampe en présence du grisou) (n.f.), cap ; gas-cap ; cap-flame ; blue cap ; show ; testing-flame ; non-luminous flame. (24)

aureux (adj. m.), aurous : (25)
oxyde aureux (m.), aurous oxide. (26)

aurichalcite (Minéral) (n.m.), aurichalcite. (27)

aurifère (adj.), auriferous ; gold-bearing ; gold (used as adj.) : (28)
sable aurifère (m.), auriferous sand ; gold-bearing sand. (29)
champ aurifère (m.), gold-field. (30)

aurique (adj.), auric : (31)
oxyde auri~ue (m.), auric oxide. (32)

auro-argentifère (adj.), auro-argentiferous ; auri-argentiferous : (33)
minerai auro-argentifère (m.), auro-argentif-erous ore. (34)

aurochlorure (Chim.) (n.m.), aurochloride ; aurochlorate ; chloraurate. (35)

aurocyanure (Chim.) (n.m.), aurocyanide. (36)

auroferrifère (adj.), auroferriferous. (37)

auroplombifère (adj.), auroplumbiferous. (38)

aurosulfite (n.m.), sulphaurite. (39)

aurotellurite (Minéral) (n.f.), graphic gold ; graphic tellurium ; graphic ore ; sylvanite ; sylvan. (40)

aussière (n.f.), hawser. (41)

austénite (Métall.) (n.f.), austenite. (42)

austral, -e, -als ou -aux (adj.), austral ; south ; southern. (43)

autel (n.m.), fire-bridge ; flame-bridge ; furnace-bridge ; fire-stop ; bridge-wall ; bridge : (44)
l'autel d'un four à réverbère est la partie qui sépare la sole de la chauffe, the fire-bridge of a reverberatory furnace is the part which separates the hearth from the fire-chamber. (45)

l'autel d'un foyer de chaudière empêche que le combustible ne soit entraîné dans les carneaux, the fire-bridge of a boiler-furnace prevents the fuel being carried into the flues. (46)

auto-excitation (Élec.) (n.f.), self-excitation. (47)

auto-excitatrice (adj. f.), self-exciting : (48)
les dynamos sont auto-excitatrices, c'est-à-dire que le courant nécessaire à l'amorçage est emprunté à la machine elle-même (f.pl.), dynamos are self-exciting, that is to say, the current required for energization is taken from the machine itself. (49)

auto-excitatrice (n.f.), self-exciter. (50)

auto-induction (Élec.) (n.f.), self-induction. (51)

autoclave (adj.), autoclave : (52)
joint autoclave (m.), autoclave joint. (53)

autoclave (n.m.), autoclave. (54)

autoconduction (Phys.) (n.f.), autoconduction. (55)

autodynamique (adj.), autodynamic. (56)

autogène (adj.), autogenous : (57)
soudure autogène (f.), autogenous welding ; autogenous soldering. (58)

automatique (adj.), automatic ; self-acting ; self- : (59)
frein automatique (m.), automatic brake ; self-acting brake. (60)
palier à graissage automatique (m.), self-oiling plummer-block. (61)

automatiquement (adv.), automatically. (62)

automne (n.m.), autumn. (63)

automobile (adj.), self-propelling ; automobile ; self-moving. (64)

automobile ou simplement auto (n.m.), motor-car ; motor ; car ; automobile. (65)

automorphe (Pétrol.) (adj.), automorphic ; auto-morphous ; idiomorphic ; idiomorphous. (66)

automoteur, -trice (adj.), automotive ; self-propelling ; self-acting : (67)
plan automoteur (m.), self-acting incline ; gravity-incline. (68)

autorégulateur, -trice (adj.), self-regulating. (69)

autorégulateur (n.m.), automatic regulator ; self-acting regulator. (70)

autorégulation (n.f.), automatic regulation. (71)

autorisation (n.f.), authorization ; permission ; permit. (72)

autoriser (v.t.), to authorize ; to permit. (73)

autorité (n.f.), authority. (74)
autorités locales (f.pl.), local authorities. (75)

autotransformateur (Élec.) (n.m.), autotrans-former. (76)

autunite (Minéral) (n.f.), autunite ; uranite ; lime uranite. (77)

auvent (n.m.), hood ; penthouse ; pentice ; porch-roof. (78)

aux risques et périls des destinataires, at owner's risk : (79)
marchandises qui voyagent aux risques et périls des destinataires (f.pl.), goods which are sent at owner's risk. (80)

auxiliaire (adj.), auxiliary : (81)
moteur auxiliaire (m.), auxiliary engine. (82)

aval (d'une rivière) (n.m.), lower part (of a river). (83)

aval (d') ou elliptiquement aval, down-stream ; downstream ; down-river ; lower ; tail (used as adj.) ; aft : (84)
le niveau d'aval au-dessous d'un barrage, the down-stream level below a dam. (85)

l'extrémité aval d'une île, the lower extremity of an island. (1)

la porte d'aval d'une écluse, the tail-gate, the aft-gate, of a lock. (2)

aval (en) *ou* **aval (à l')**, down-stream; down-stream; down-river: (3)

aller en aval, to go down-stream (*or* down-river). (4)

la vitesse à l'aval, the velocity downstream. (5)

aval de (en) *ou* **aval de (à l')**, below; down: (6)

des signaux placés en amont et en aval des stations (*m.pl.*), signals placed above and below stations (*or* placed up and down the line). (7)

aval pendage (Mines) (opp. à **amont pendage**) (*m.*), bottom. (8) (opp. to **back** *or* **backs**):

l'aval pendage d'une galerie, d'un chantier d'abatage, the bottom of a level, of a stope. (9 *ou* 10)

avalaison (*n.f.*), freshet; spate. (11)

avalanche (*n.f.*) *ou* **avalanche de neige**, avalanche; snowslide. (12)

avalanche de pierres (*f.*), stoneslide. (13)

avalanche de rochers (*f.*), rockslide. (14)

avalanche de terre (*f.*), landslide. (15)

avalasse (*n.f.*), freshet; spate. (16)

avaler (Mines) (*v.t.*), to sink: (17)

avaler un puits, to sink a shaft. (18)

avaler (*v.i.*), to go down-river; to go down-stream. (19)

avaleresse (Mines) (*n.f.*), shaft in water-bearing ground (in course of excavation): (20)

fonçage des avaleresses (*m.*), sinking shafts in water-bearing ground. (21)

avalite (Minéral) (*n.f.*), avalite. (22)

avals (*n.m.pl.*). Same as **aval pendage**.

avance (Travail au tour) (*n.f.*), feed; traverse feed (Lathe-work): (23)

l'avance de l'outil est le chemin parcouru par l'outil pendant une révolution de la pièce, the feed of the tool is the path travelled over by the tool during one revolution of the work. (24)

avance (du tiroir) (*n.f.*), lead (of the slide-valve): (25)

donner au tiroir de l'avance et du recouvrement, to give to the slide-valve lead and lap. (26)

avance à l'admission (quantité d'ouverture) (tiroir) (*f.*), outside lead; linear lead. (27)

avance à l'admission (phase de la distribution) (opp. à **retard à l'admission**) (*f.*), pread-mission; early admission. (28)

avance à l'allumage (moteur à combustion interne) (*f.*), lead of the ignition. (29)

avance à l'échappement *ou* **avance à l'émission** (quantité d'ouverture) (tiroir) (*f.*), inside lead; exhaust-lead. (30)

avance à l'échappement *ou* **avance à l'émission** (phase de la distribution) (opp. à **retard à l'échappement**) (*f.*), prerelease; early release. (31)

avance angulaire (l'excès de l'angle de calage sur 90°) (tiroir) (*f.*), angular advance; angle of advance. (32)

avance angulaire à l'admission (*f.*), angular lead; angle of lead. (33)

avance angulaire à l'échappement (*f.*), angular prerelease; angle of prerelease. (34)

avance linéaire (du tiroir) (*f.*), linear advance. (35)

avance linéaire à l'admission (*f.*), linear lead; outside lead. (36)

avance linéaire à l'échappement (*f.*), inside lead; exhaust-lead. (37)

avancé, -e (*adj.*), advanced; late: (38)

maturité avancée (Géogr. phys.) (*f.*), late maturity. (39)

avancée (d'un glacier, de la mer) (*n.f.*), advance (of a glacier, of the sea). (40 *ou* 41)

avancement (*n.m.*) *ou* **avancée** (*n.f.*) (partie extrême d'une galerie de mine que l'on creuse), heading; head; advance heading. (42)

avancement (action) (Exploitation de mines) (*n.m.*), heading: (43)

les avancements en galeries éboulées sont très lents, headings in caved-in levels are very slow. (44)

avancement journalier, daily heading; daily progress. (45)

avancement (amenage; entraînement) (Méc.) (*n.m.*), feed; feeding: (46)

machine à raboter à avancement automatique (*f.*), planing-machine with automatic feed. (47)

avancement (serrage; profondeur de passe) (machines-outils) (*n.m.*), depth of cut. (48)

avancement (d'un édifice en construction) (*n.m.*), forwardness (of a building under construction). (49)

avancement (d'une jetée) (*n.m.*), jutting out (of a jetty). (50)

avancement d'une drague (*m.*), stepping a dredge ahead. (51)

avancer (pousser; faire faire des progrès) (*v.t.*), to urge forward; to push on; to speed up: (52)

avancer les travaux, to urge forward, to push on with, to speed up, the work. (53)

avancer (faire saillie) (*v.i.*), to jut out; to project: (54)

les toits doivent avancer un peu pour protéger les murs (*m.pl.*), roofs should project a little to protect the walls. (55)

avancer (Exploitation de mines) (*v.i.*), to head. (56)

avancer (s') (*v.r.*), to jut out; to project: (57)

des deux côtés du port, un vaste roc s'avance, from either side of the harbour, a huge rock juts out. (58)

avancer une drague, to step a dredge ahead. (59)

avant (*n.m.*), front; front part; forepart: (60)

l'avant d'une voiture, d'une chambre noire photographique, the front of a carriage, of a photographic camera. (61 *ou* 62)

avant (d') *ou elliptiquement* **avant**, front; fore: (63)

essieu d'avant *ou* essieu avant (*m.*), front axle; fore axle [des essieux d'avant *ou* des essieux avant *pl.*]. (64)

avant (en), ahead: (65)

train en avant, train ahead. (66)

avant (côté du piston où se trouve la manivelle) (machine fixe horizontale) (opp. à **arrière**) (*on écrit souvent par abréviation* **AV** *ou* **A V**), crank-end: (67)

admission avant (*f.*), crank-end admission. (68)

échappement AV (*m.*), crank-end release. (69)

point mort A V (*m.*), crank-end dead-centre. (70)

avant *ou* **AV** *ou* **A V** (côté du piston opposé à celui où se trouve la manivelle) (locomotive) head-end; forward: (71)

admission avant (*f.*), head-end admission. (1)

point mort avant (*m.*), forward dead-centre. (2)

avant-bec [avant-becs *pl.*] (d'une pile de pont) (*n.m.*) (opp. à **arrière-bec**), up-stream cut-water (of a bridge-pier). (3)

avant-clou [avant-clous *pl.*] (*n.m.*), gimlet. (4)

avant-corps [avant-corps *pl.*] (*n.m.*), projection. (5)

avant-creuset [avant-creusets *pl.*] (de haut fourneau, etc.) (*n.m.*), forehearth (of blast-furnace, etc.). (6)

avant-creuset (de cubilot à avant-creuset) (*n.m.*), receiver (of cupola with receiver). (7)

avant-cylindre (*n.m.*) *ou* avant du cylindre (*m.*) *ou* **AV-cylindre** (*n.m.*) (machine fixe horizon-tale), crank-end of cylinder. (8)

avant-cylindre (*n.m.*) *ou* avant du cylindre (*m.*) *ou* **AV-cylindre** (*n.m.*) (locomotive), head-end of cylinder. (9)

avant-pieu [avant-pieux *pl.*] (*n.m.*), dolly; punch; set. (10)

avant-plan [avant-plans *pl.*] (*n.m.*), foreground. (11)

avant porte-objectif (Photogr.) (*m.*), lens-front. (12)

avant porte-objectif en U pouvant se déplacer (*ou* **pouvant se décentrer**) **en hauteur et en largeur** *ou* **porte-objectif en U à décentre-ment vertical et latéral** (*ou* **à double décentre-ment**) (*ou* **à décentrement dans les deux sens**) (*m.*), stirrup rising-and-cross lens-front; stirrup lens-front with rising-and-cross movements. (13)

avant-projet [avant-projets *pl.*] (*n.m.*), draft; draft scheme; proposal plan: (14)

avant-projet d'un chemin de fer, draft scheme for a railway. (15)

avant-puits [avant-puits *pl.*] (Procédé Kind-Chaudron, etc.) (*n.m.*), guide-pit; guide-bore pit. (16)

avant-train [avant-trains *pl.*] (d'un véhicule à quatre roues) (*n.m.*), fore-carriage. (17)

avantage (*n.m.*), advantage: (18)

les avantages et inconvénients des scies circulaires, the advantages and disadvan-tages of circular saws. (19)

avarie (*n.f.*) *ou* **avaries** (*n.f.pl.*), damage; deterio-ration: (20)

rupture d'une roue qui occasionne de grandes avaries (*f.*), breakage of a wheel which causes a lot of damage. (21)

avaries de bandages et de roues, deterioration of tires and wheels. (22)

avarié, -e (*adj.*), damaged: (23)

marchandises avariées (*f.pl.*), damaged goods. (24)

avarier (*v.t.*), to damage; to deteriorate. (25)

avarier (s') (*v.r.*), to damage; to deteriorate. (26)

aven (*n.m.*), limestone cave; limestone cavern (a cave or cavern under a limestone sink). (27)

avènement d'un chemin de fer (*m.*), advent of a railway. (28)

avenir (*n.m.*), future; prospects; outlook: (29)

avenir d'une mine, future, prospects, of a mine: (30)

mine qui a devant elle un bel avenir (*f.*), mine which has a fine future before it. (31)

avenir plein de promesses, promising future (*or* outlook). (32)

avenir (d'), future: (33)

les perspectives d'avenir d'une contrée (*f.pl.*), the future prospects of a country. (34)

aventure minière (*f.*), mining venture. (35)

aventurier (pers.) (*n.m.*), adventurer. (36)

aventurine (Minéral) (*n.f.*), aventurine (the quartz variety). (37)

aventurine orientale (*f.*), aventurine; aventurine feldspar; sunstone. (38)

aventuriné, -e (*adj.*), aventurine (*used as adj.*): (39)

quartz aventuriné (*m.*), aventurine quartz. (40)

averse (*n.f.*), shower; shower of rain: (41)

averse soudaine, sudden shower. (42)

avertir (*v.t.*), to warn; to give warning. (43)

avertissement (*n.m.*), notice; warning: (44)

avertissement avant le tirage de coups de mines, warning before firing. (45)

avertisseur (*adj. m.*), alarm (*used as adj.*); warning: (46)

signal avertisseur (*m.*), alarm-signal; warning signal. (47)

avertisseur (appareil quelconque qui avertit) (*n.m.*), alarm; indicator. (48)

avertisseur (sifflet d'alarme, manomètre, ou instrument analogue) (*n.m.*), alarm; alarm-gauge. (49)

avertisseur (marteau avertisseur) (Mines) (*n.m.*), rapper. (50)

avertisseur automatique (*m.*), automatic alarm; automatic warning-apparatus. (51)

avertisseur d'incendie (*m.*), fire-alarm. (52)

avertisseur de niveau d'eau (*m.*), water-level alarm. (53)

avertisseur électrique du feu (*m.*), electric fire-alarm. (54)

aveugle (*adj.*), blind: (55)

chevaux aveugles (chevaux de mines) (*m.pl.*), blind horses. (56)

arcade aveugle (*f.*), blind arch. (57)

aveugler une fuite (*ou* **une voie d'eau**), to stop a leak. (58)

avide d'eau, absorbent (*or* greedy) of water; absorptive; thirsty; imbibitory: (59)

l'argile desséchée se fendille: elle est alors très avide d'eau et adhère fortement à la langue (*f.*), dried clay cracks: it is then highly absorbent (*or* very greedy) of water (*or* very thirsty) and adheres strongly to the tongue. (60)

avis (avertissement) (*n.m.*), advice; notice: (61)

avis par écrit, notice in writing. (62)

avis (*n.m.*) *ou* **avis** (*n.m.pl.*), advice. (63)

avisse (*n.f.*), screw-piece. (64)

aviver une surface que l'on veut souder, to tin a surface that one wishes to solder. (65)

avoir (*v.t.*), to have: (66)

avoir une bonne saison, to have a good season. (67)

avoir (être d'une dimension de) (*v.t.*), to be (*v.i.*); to measure (*v.i.*): (68)

la tour Eiffel a trois cents mètres de haut, the Eiffel tower is three hundred metres high. (69)

avoir besoin de, to have need of; to be in need of; to need: (70)

avoir grand besoin de réparations, to be greatly in need of repair. (71)

avoir du mou (se dit des courroies, des câbles, des chaînes), to slack; to be slack; to be loose. (72)

avoir l'espoir de retirer vivants les ouvriers qui manquent, to entertain hope of the missing men being rescued alive. (73)

avoir lieu, to take place ; to occur : (1)
une amélioration sensible a eu lieu, a marked
improvement has taken place. (2)

avoir recours à, to have recourse to : (3)
avoir recours à la ventilation artificielle, to
have recourse to artificial ventilation. (4)

avoir un tirant d'eau de — pieds (se dit d'un
bateau), to have a draught of, to draw, —
feet of water. (5)

avoir (n') pas besoin de commentaires, to call for
no comment. (6)

avoisinant, -e (*adj.*), neighbouring ; adjoining ;
near-by. (7)

avortement d'une entreprise (*m.*), failure of an
enterprise ; miscarriage of an undertaking.
(8)

avoyage des dents d'une scie (*m.*), setting, jumping,
the teeth of a saw. (9)

avoyer une lame de scie, to set, to jump, a saw-
blade. (10)

awaruite (Minéral) (*n.f.*), awaruite. (11)

axe (Géom., etc.) (*n.m.*), axis. (12)

axe (arbre) (Méc.) (*n.m.*), spindle ; shaft ; axle ;
pin ; axis. (13)

axe anticlinal (Géol.) (*m.*), anticlinal axis. (14)

axe binaire (Cristall.) (*m.*), axis of binary sym-
-metry. (15)

axe d'élasticité (Cristall.) (*m.*), axis of elasticity.
(16)

axe d'hémitropie (Cristall.) (*m.*), twinning-axis.
(17)

axe d'oscillation d'un pendule (*m.*), axis of
oscillation of a pendulum. (18)

axe d'un cercle (Géom.) (*m.*), axis of a circle. (19)

axe d'un cône, d'une pyramide régulière (*m.*),
axis of a cone, of a regular pyramid. (20
ou 21)

axe d'un couple (Méc.) (*m.*), axis of a couple. (22)

axe d'un cristal (*m.*), axis of a crystal. (23)

axe d'un cylindre (Géom.) (*m.*), axis of a cylinder.
(24)

axe d'un cylindre (Méc.) (*m.*), spindle of a roller.
(25)

axe d'un édifice (Arch.) (*m.*), axis of an edifice.
(26)

axe d'un fourneau (*m.*), axis of a furnace. (27)

axe d'une lentille (*m.*), axis of a lens. (28)

axe d'une lunette, d'un microscope (*m.*), axis of a
telescope, of a microscope. (29 *ou* 30).

axe d'une roue, d'une poulie (*m.*), axis of a wheel,
of a pulley. (31 *ou* 32).

axe de cristallisation (*m.*), axis of crystallization.
(33)

axe de la terre (*m.*), axis of the earth. (34)

axe de noyau (Fonderie) (*m.*), core-arbor. (35)

axe de polarisation (Opt.) (*m.*), axis of polariza-
-tion. (36)

axe de réfraction (Opt.) (*m.*), axis of refraction.
(37)

axe de révolution (*m.*), axis of revolution. (38)

axe de rotation (*m.*), axis of rotation. (39)

axe de soulèvement (Géol.) (*m.*), axis of elevation.
(40)

axe de suspension du fléau d'une balance (*m.*),
axis of a balance. (41)

axe de symétrie (*m.*), axis of symmetry ;
symmetry-axis. (42)

axe de zone (Cristall.) (*m.*), zone-axis. (43)

axe des abscisses (Géom.) (*m.*), axis of abscissas.
(44)

axe des ordonnées (Géom.) (*m.*), axis of ordinates.
(45)

axe du coulisseau (Mach.) (*m.*), link-block pin.
(46)

axe du filon (*m.*), axis of the lode. (47)

axe en axe (d'). See under **D**.

axe instantané de rotation (Méc.) (*m.*), in-
-stantaneous axis of rotation. (48)

axe isoclinal (Géol.) (*m.*), isoclinal axis. (49)

axe neutre (*m.*), neutral axis. (50)

axe optique (*m.*), optic axis : (51)
axe optique d'une lunette, optical axis of a
telescope. (52)

axe porte-meule (*m.*), grinding-spindle. (53)

axe principal (Opt., etc.) (*m.*), principal axis :
(54)
axe principal d'un miroir sphérique, d'une
lentille, principal axis of a spherical mirror,
of a lens. (55 *ou* 56)
axe principal d'un cristal, principal axis of a
crystal. (57)

axe principal (arbre) (Méc.) (*m.*), main shaft ;
first-motion shaft. (58)

axe secondaire (Opt., etc.) (*m.*), secondary axis :
(59)
axe secondaire d'un miroir sphérique, d'une
lentille, secondary axis of a spherical mirror,
of a lens. (60 *ou* 61)

axe secondaire (arbre) (Méc.) (*m.*), counter-
shaft ; countershaft ; intermediate shaft ;
second-motion shaft ; jack-shaft. (62)

axe synclinal (Géol.) (*m.*), synclinal axis. (63)

axes des coordonnées (Géom.) (*m.pl.*), axes of
coordinates ; coordinate axes. (64)

axial, -e, -aux (*adj.*), axial : (65)
ligne axiale (*f.*), axial line. (66)

axiforme (*adj.*), axiform. (67)

axinite (Minéral) (*n.f.*), axinite. (68)

axuel, -elle (*adj.*), axial : (69)
ligne axuelle (*f.*), axial line. (70)

azimut (*n.m.*), azimuth : (71)
l'azimut d'une étoile, du soleil, the azimuth
of a star, of the sun. (72 *ou* 73)

azimut du plan de polarisation (Opt.) (*m.*),
azimuth of the plane of polarization. (74)

azimut magnétique (*m.*), magnetic azimuth.
(75)

azimutal, -e, -aux (*adj.*), azimuthal : (76)
angles azimutaux (*m.pl.*), azimuthal angles.
(77)

azoïque (Géol.) (*adj.*), azoic. (78)

azorite (Minéral) (*n.f.*), azorite. (79)

azotate (*n.m.*), nitrate. (80)

azotate d'ammoniaque (*m.*), ammonium nitrate.
(81)

azotate d'argent (*m.*), nitrate of silver ; silver
nitrate. (82)

azotate de potassium (*m.*), potassium nitrate ;
nitrate of potash ; nitre ; saltpetre. (83)

azotate de sodium *ou* **azotate de soude** (*m.*),
sodium nitrate ; nitrate of soda. (84)

azote (*n.m.*), nitrogen. (85)

azoté, -e (*adj.*), nitrogenous ; nitrogeneous ;
nitrogenic. (86)

azoter (*v.t.*), to nitrogenize : (87)
azoter de l'hydrogène, to nitrogenize hydrogen.
(88)

azoteux, -euse (Chim.) (*adj.*), nitrous : (89)
acide azoteux (*m.*), nitrous acid. (90)

azotique (Chim.) (*adj.*), nitric : (91)
acide azotique (*m.*), nitric acid. (92)

azurite (Minéral) (*n.f.*), azurite ; chessylite ;
Chessy copper ; blue malachite ; blue copper
carbonate. (93)

B

babingtonite (Minéral) (*n.f.*), babingtonite. (1)

bac (bateau servant à passer un cours d'eau) (*n.m.*), ferry-boat ; ferry ; trail-bridge. (2)

bac (cuve) (*n.m.*), vat ; trough ; tub. (3)

bac (chariot de mine) (*n.m.*), truck ; tub. (4)

bac à eau pour forgerons (*m.*), smiths' water-trough ; slake-trough. (5)

bac à piston (Préparation mécanique des minerais) (*m.*), fixed-sieve jig. (6)

bac avec agitation mécanique *ou* bac muni d'agitateurs (Préparation mécanique des minerais) (*m.*), agitation-vat. (7)

bac collecteur (Préparation mécanique des minerais) (*m.*), collecting-vat. (8)

bac d'éléments *ou simplement* bac (*n.m.*) (récipient destiné à contenir les éléments d'un accumulateur électrique) (*m.*), cell-jar ; jar ; container ; containing vessel ; vessel. (9)

bac de cyanuration (*m.*), cyaniding-vat. (10)

bac de filtration (*m.*), filtration-vat ; filter-vat. (11)

bac de précipitation (*m.*), precipitation-vat. (12)

bac ménagé sous le banc (tour) (*m.*), trough, tray, drop-pan, under the bed (lathe). (13)

bac transbordeur (*m.*), ferry-bridge. (14)

bâche (réservoir) (*n.f.*), tank. (15)

bâche (grosse toile) (*n.f.*), awning ; tarpaulin. (16)

bâche (grue hydraulique) (Ch. de f.) (*n.f.*), water-pillar ; water-crane. (17)

bâche à eau (*f.*), water-tank. (18)

bâche d'alimentation (*f.*), feed-tank. (19)

bâche d'aspiration (Hydraul.) (*f.*), suction-tank. (20)

bâche fermée (pour le moteur d'une turbine) (*f.*), enclosed casing ; closed turbine-chamber. (21)

bâcher un puits (Mines), to crib a shaft. (22)

bacnure (Mines) (*n.f.*), stone-drift ; rock-drift ; gallery in dead ground. (23)

badine (de forgeron) (*n.f.*), side-mouth tongs ; crook-bit tongs : elbow-tongs. (24)

bague (*n.f.*), ring. (25)

bague (virole pour tubes de chaudière) (*n.f.*), ferrule ; thimble. (26)

bague (anneau métallique monté sur un arbre de transmission) (*n.f.*), collar. (27)

bague (buselure ; coquille) (Méc.) (*n.f.*), bush ; bushing. (28)

bague (calibre à bague) (Méc.) (*n.f.*), external cylindrical gauge ; ring-gauge ; ring : (29) (often occurs in the expression **bagues et bouchons,** which is usually rendered in the reverse order in English, i.e., **internal and external cylindrical gauges ; plug and ring-gauges ; plugs and rings**). (30)

bague (d'un presse-étoupe) (*n.f.*), bush, bushing, (of a stuffing-box). (31)

bague à billes (*f.*), ball-ring. (32)

bague à cames (*f.*), cam-wheel. (33)

bague à tolérance (calibre) (*f.*), limit external gauge. (34)

bague-butoir [bagues-butoirs *pl.*] (*n.f.*), knock-off link. (35)

bague d'amarrage (*f.*), ring-bolt. (36)

bague d'arrêt *ou* bague de butée (pour arbres de transmission, etc.) (*f.*), set-collar ; stop-collar. (37)

bague d'arrêt en deux pièces (*f.*), split set-collar. (38)

bague d'excentrique (*f.*), eccentric-strap ; eccentric-collar ; eccentric-hoop. (39)

bague de bielle d'accouplement (locomotive) (*f.*), rod-bush ; bush of side-rod. (40)

bague de cylindre (de broyeur à cylindre) (*f.*), roll-shell (of crushing-rolls). (41)

bague de graissage (*f.*), oil-ring. (42)

bague de piston (*f.*), piston-ring. (43)

bague fendue (*f.*), split ring. (44)

baguer (*v.t*), to ring ; to collar ; to ferrule ; to bush ; (45)

baguer les tubes d'une chaudière, to ferrule the tubes of a boiler. (46)

douille baguée en bronze (*f.*), socket bushed with gun-metal. (47)

baguette (verge ; petit bâton menu) (*n.f.*), rod ; stick : (48)

baguette en verre, glass rod. (49)

baguette de charbon (Élec.), stick of carbon. (50)

baguette (moulure servant à cacher un joint) (Menuis.) (*n.f.*), reed. (51)

baguette (moulure ronde poussée au bord d'une pièce de menuiserie) (*n.f.*), bead. (52)

baguette divinatoire (*f.*), divining-rod ; divining-staff ; dowsing-rod ; douser : (53)

dans les campagnes, on croit encore à la découverte des sources au moyen de la baguette de coudrier, in country places, they still believe in the discovery of springs by means of the hazel rod. (54)

baïart (*n.m.*), hand-barrow. (55)

baie (Géogr.) (*n.f.*), bay. (56)

baie (dans un mur) (*n.f.*), opening (in a wall). (57)

baie de fenêtre (*f.*), window-opening. (58)

baie de porte (*f.*), door-opening ; doorway. (59)

baïkalite (Minéral) (*n.f.*), baikalite. (60)

bail [baux *pl.*] (*n.m.*), lease. (61)

bail minier (*m.*), mining-lease. (62)

bailer (jélonka ; tube à boulet) (Exploitation des puits à pétrole) (*n.m.*), bailer. (63)

bâillement (*n.m.*), fissure ; crack ; rent. (64)

bailleur, -eresse (pers.) (*n.*), lessor. (65)

bain (*n.m.*), bath. (66)

bain (matière en fusion) (Métall.) (*n.m.*), bath : (67)

désoxyder le bain, to deoxidize the bath. (68)

bain (état de fusion parfaite d'un métal) (*n.m.*), fusion : (69)

métal au bain (*m.*), metal in fusion. (70)

bain d'huile (*m.*), oil bath : (71)

engrenages tournant dans un bain d'huile (*m.pl.*), gears running in an oil bath. (72)

bain de développement *ou* bain révélateur (Photogr.) (*m.*), developing-bath. (73)

bain de fixage *ou* bain fixateur (Photogr.) (*m.*), fixing-bath. (74)

bain de trempe (*m.*), tempering-bath ; hardening-bath. (75)

bain de virage-fixage *ou* bain de virage-fixage combiné (Photogr.) (*m.*), toning-and-fixing bath ; combined toning-and-fixing bath ; combined bath. (76)

bain-marie [bains-marie *pl.*] (Chim.) (*n.m.*), water-bath. (77)

bain métallique (*m.*), metal bath. (1)
baissage (opp. à renforcement) (Photogr.) (*n.m.*), reduction ; reduction in density. (2)
baisse (*n.f.*), fall ; drop ; reduction ; reducing ; subsidence : (3)
baisse d'une rivière (décrue), fall, subsidence, of a river. (4)
baisse de potentiel (Élec.), drop of potential ; fall of potential ; pressure-drop ; pressure-loss. (5)
baisse de prix, fall, drop, reduction, in price. (6)
baisse de valeur, fall, falling off, in value : (7)
une baisse sensible de valeur, a marked falling off in value. (8)
baisse du mercure dans le thermomètre, fall of the mercury in the thermometer. (9)
baisser (*v.t.*), to lower ; to reduce : (10)
baisser le prix de certaines marchandises, to lower, to reduce, the price of certain goods. (11)
baisser un cliché dur (Photogr.), to reduce a hard negative. (12)
baisser (*v.i.*), to fall ; to drop ; to subside ; to recede ; to go down : (13)
la température de l'atmosphère baisse à mesure qu'on atteint des régions plus élevées, the temperature of the atmosphere falls as higher regions are reached. (14)
le fleuve a baissé de — mètres, the river has fallen — metres. (15)
baisser de valeur, to fall in value. (16)
bajoyer (d'une écluse, du coursier d'une roue hydraulique) (*n.m.*), side-wall (of a canal-lock, of the breasting of a water-wheel). (17 *ou* 18)
bajoyer (d'une rivière) (*n.m.*), river-wall. (19)
baladeurs (engrenages baladeurs) (*n.m.pl.*), sliding gears ; balladeur gears. (20)
baladeuse (lampe baladeuse) (Élec.) (*n.f.*), hang-ing lamp. (21). See **lampe baladeuse** for extended example.
balai (*n.m.*), broom : (22)
balai de bouleau, birch broom. (23)
balai (Élec.) (*n.m.*), brush : (24)
le balai d'une dynamo, the brush of a dynamo. (25)
balance (instrument pour peser) (*n.f.*), balance ; scales ; scale ; weighing-scale ; pair of scales. (26)
balance (Élec.) (*n.f.*), balance : (27)
la balance Thomson, the Kelvin balance. (28)
balance (équilibre) (*n.f.*), balance ; equilibrium. (29)
balance à colonne (*f.*), pillar-balance ; pillar-scales. (30)
balance à court fléau (*f.*), short-beam balance. (31)
balance à fléau (*f.*), beam-balance ; beam-scales ; beam and scales. (32)
balance à levier coudé (*f.*), bent-lever balance. (33)
balance à long fléau (*f.*), long-beam balance. (34)
balance à plateaux (*f.*), pan-scales ; scale with pans. (35)
balance à poids (soupape de sûreté) (*f.*), steam-balance (safety-valve). (36)
balance à ressort (*f.*), spring-balance valve. (37)
balance actinique (*f.*), actinic balance. (38)
balance aréométrique de Mohr (*f.*), Mohr's specific-gravity balance. (39)

balance d'analyse (*f.*), analytical balance. (40)
balance d'eau (*f.*), water-balance ; hydraulic balance. (41)
balance d'essai *ou* **balance d'essayeur** (*f.*), assay-scales ; assay-balance. (42)
balance d'induction voltaïque (*f.*), induction-balance. (43)
balance d'une force de — grammes, sensible au milligramme (*f.*), balance to weigh up to — grammes, sensible to 1 milligramme ; balance to carry — grammes, turning to 1 milligramme. (44)
balance danoise (*f.*), Danish balance. (45)
balance de Jolly (*f.*), Jolly balance. (46)
balance de laboratoire (*f.*), chemical balance. (47)
balance de locomotive (ressort contrôlant la soupape de sûreté d'une chaudière de loco-motive) (*f.*), locomotive-balance. (48)
balance de précision, sous cage verre (*f.*), balance of precision, precision balance, in glass case. (49)
balance de Roberval (*f.*), Roberval's balance. (50)
balance de torsion (*f.*), torsion balance. (51)
balance de Wheatstone (Élec.) (*f.*), Wheatstone's bridge ; Wheatstone's balance ; electric balance ; electrical bridge. (52)
balance fausse (opp. à balance juste) (*f.*), unjust balance ; unjust scale. (53)
balance hydraulique (*f.*), hydraulic balance ; water-balance. (54)
balance hydrostatique (*f.*), hydrostatic balance ; specific-gravity balance. (55)
balance juste (*f.*), just balance ; just scale. (56)
balance pour produits chimiques (*f.*), chemical scales. (57)
balance pour raffinerie (*f.*), refinery scales. (58)
balance romaine (*f.*), steelyard ; scale-beam ; lever scales ; Roman balance. (59)
balance sèche (Mines) (*f.*), gravity-hoist. (60)
balance sèche à double effet (*f.*), double gravity-hoist ; gravity-hoist with cage at each end. (61)
balance sèche à simple effet (*f.*), single gravity-hoist ; gravity-hoist with cage and counter-poise. (62)
balance sensible (*f.*), sensitive balance ; sensible balance ; delicate scale. (63)
balance-trébuchet [**balances-trébuchets** *pl.*] (*n.f.*), physical balance ; balance (light scales on stand for delicate work, such as chemical analysis, assaying, physical experiments, etc.) : (64)
balance-trébuchet de précision, sur socle acajou, à tiroir, precision balance on mahog-any stand, with drawer. (65)
balancement (*n.m.*), balancing ; rocking ; swaying ; swinging ; swing : (66)
le balancement d'une locomotive est une cause de déraillement, the swaying of a locomotive is a cause of derailment. (67)
balancement des marches d'escalier (*m.*), dancing the steps of stairs. (68)
balancer (*v.t.*), to balance ; to swing : to rock : to oscillate ; to sway : (69)
balancer une soupape de sûreté, to balance a safety-valve. (70)
le vent balance la cime des arbres, the wind sways the tops of the trees. (71)
balancer (*v.i.*), to balance ; to oscillate ; to swing ; to rock ; to sway ; to sway about : (72)
locomotive qui balance (*f.*), engine which sways about. (73)

balancer (se) (*v.r.*), to balance ; to oscillate ; to swing ; to rock ; to sway ; to sway about: (1)

lampe qui se balance au plafond (*f.*), lamp which swings on the ceiling. (2)

balancer une marche d'escalier, to dance a stair-step. (3)

balancier (Mach.) (*n.m.*), beam ; bob ; walking-beam ; working-beam ; logging-head. (4)

balancier (Fonderie) (*n.m.*), beam ; sling-beam ; lifting-beam. (5)

balancier (d'une porte d'écluse) (*n.m.*), balance-bar (of a lock-gate). (6)

balancier (*n.m.*) *ou* **balancier de répartition** *ou* **balancier de suspension** (wagon ou locomotive de ch. de f.), equalizing-bar ; equalizing-beam ; equalizing-lever ; equalizer ; com-pensating-beam ; rock-lever ; rocking-lever. (7)

balancier (découpoir à la main) (*n.m.*), fly-press. (8)

balancier (pers.) (*n.m.*), scale-maker. (9)

balancier à boules (*m.*), fly-bar with two balls. (10)

balancier à contrepoids (pour maîtresse-tige de pompe) (Mines) (*m.*), balance-bob (of pump spear-rod). (11)

balancier d'équerre (pompe de mine) (*m.*), angle-bob ; V bob ; quadrant. (12)

balancier de battage (Sondage) (*m.*), walking-beam ; working-beam. (13)

balancier de renvoi (distribution de vapeur) (*m.*), rock-shaft ; rocking-shaft ; rocker-shaft ; rocker. (14)

balayage (*n.m.*), sweeping ; sweeping up : (15)

balayage des poussières, sweeping up the dust. (16)

balayer (*v.t.*), to sweep ; to sweep up : (17)

balayer les poussières, to sweep up the dust. (18)

courant d'air balayant les fronts de taille (Mines) (*m.*), air-current sweeping the working-faces. (19)

balayeur, -euse (pers.) (*n.*), sweeper. (20)

balayeuse mécanique (*f.*), mechanical sweeper. (21)

balayures (*n.f.pl.*), sweepings. (22)

balcon (Fonderie) (*n.m.*), fin ; burr. (23)

baleine (pont de service) (*n.f.*), service-bridge. (24)

balène *ou* **balètre** (Fonderie) (*n.f.*), fin ; burr. (25)

balèvre (Fonderie) (*n.f.*), scale ; burr ; scab. (26)

balisage *ou* **balisement** (action) (*n.m.*), beaconing. (27)

balisage (ensemble de balises) (*n.m.*), beaconage. (28)

balise (*n.f.*), beacon. (29)

baliser (*v.t.*), to beacon : (30)

baliser le tracé d'un chemin de fer, to beacon the alignment of a railway. (31)

baliveau (Constr.) (*n.m.*), scaffold-pole ; stand-ard. (32)

ballage (Métall.) (*n.m.*), balling. (33)

ballast (*n.m.*), ballast ; ballasting ; bottom. (34)

ballast en pierre cassée (*m.*) broken-stone ballast. (35)

ballast en sable (*m.*), sand ballast. (36)

ballastage (*n.m.*), ballasting : (37)

ballastage de la voie (voie ferrée), ballasting the road. (38)

ballaster (*v.t.*), to ballast : (39)

ballaster une voie (Ch. de f.), to ballast a road (*or* a permanent way). (40)

ballastière (*n.f.*), borrow-pit ; borrow. (41)

balle (petit corps arrondi) (*n.f.*), ball. (42)

balle (de marchandises) (*n.f.*), bale (of goods). (43)

balle (*n.f.*) *ou* **balle de puddlage** (Métall.), ball ; puddle-ball ; loup ; loupe. (44)

baller (Métall.) (*v.i.*), to ball ; to ball up. (45)

ballon (Mines) (*n.m.*), fire-damp pocket. (46)

ballon (Verrerie de laboratoire) (*n.m.*), balloon ; flask ; globular flask ; bulb. (47)

ballon à long col (*m.*), bolthead ; matrass ; receiver. (48)

ballon de Dumas, pour densité de vapeurs (*m.*), Dumas' vapour-density bulb. (49)

ballon fond plat (*m.*), flask with flat bottom. (50)

ballon ordinaire (*m.*), flask with round bottom. (51)

ballon pour distillation fractionnée (*m.*), frac-tional-distillation flask. (52)

ballon tubulé (*m.*), tubulated flask. (53)

ballot (de marchandises) (*n.m.*), bale, ballot, (of goods). (54)

ballottement (*n.m.*), rattling. (55)

ballottement (Fonderie) (*n.m.*), rapping. (56)

ballotter (Fonderie) (*v.t.*), to rap. (57)

ballotter (remuer ; être secoué) (*v.i.*), to rattle : (58)

fenêtre qui ballotte (*f.*), window which rattles. (59)

balourd (*n.m.*), want of balance ; unbalance : (60)

le moindre balourd des meules en émeri provoque des trépidations de l'arbre, the least want of balance of emery-wheels causes vibrations of the shaft. (61). Cf. **présenter un balourd.**

baltimorite (Minéral) (*n.f.*), baltimorite. (62)

balustre (*n.m.*), baluster. (63)

bambou [bambous *pl.*] (*n.m.*), bamboo. (64)

banc (Géol. & Géogr. phys.) (*n.m.*), bank ; bed ; reef. (65)

banc (établi) (*n.m.*), bench. (66)

banc (d'une machine-outil) (*n.m.*), bed (of a machine-tool) : (67)

le banc d'un tour, the bed of a lathe. (68)

banc à couper (*m.*), cutting-bench. (69)

banc à étirer *ou* **banc à tirer** (*m.*). Same as banc d'étirage.

banc à rectifier (*m.*), straightening-machine ; straightener. (70)

banc calcaire (*m.*), limestone bed. (71)

banc d'âne (*m.*), shaving-horse ; draw-horse ; drawing-bench. (72)

banc d'épreuve (*m.*), testing-bench. (73)

banc d'étirage (*m.*), draw-bench ; drawing-bench ; drag-bench. (74)

banc de cassage (Mines) (*m.*), breaker-station. (75)

banc de ciel (Mines ou Carrières) (*m.*), roof. (76)

banc de graviers (*m.*), bank of gravel ; gravel-bank. (77)

banc de menuisier (*m.*), joiners' bench. (78)

banc de moulage (*m.*), moulding-bench. (79)

banc de redressage (*m.*), straightening-machine ; straightener. (80)

banc de sable (*m.*), sand-bank. (81)

banc de tréfilerie (*m.*), wiredrawing-bench ; wiredrawer. (82)

banc demi-rompu (tour) (*m.*), half-gap bed (lathe). (83)

banc droit en fonte, raboté et dressé, monté sur pieds (tour) (*m.*), straight cast-iron bed, planed and finished, mounted on standards. (1)

banc monté sur pieds faisant armoire (tour) (*m.*), bed standing on cabinet feet; bed mounted on standards formed as cabinets (for storing tools, etc.); bed mounted on cupboard base. (2)

banc robuste reposant sur le sol dans toute sa longueur (tour) (*m.*), strong bed resting on the floor the whole length of base. (3)

banc rompu (tour) (*m.*), gap-bed. (4)

bandage (*n.m.*) *ou* bandage de roue, tire; tyre; wheel-tire. (5)

bandage (Fonderie) (*n.m.*), iron band used to strengthen a mould against the casting pressure. (6)

bandage d'acier (*m.*), steel tire. (7)

bandage de cylindre (de broyeur à cylindres) (*m.*), roll-shell (of crushing-rolls). (8)

bandage dépourvu de mentonnet (*m.*), flangeless tire; blank tire; bald tire; blind tire; plain tire. (9)

bandage fixé par un cercle engagé dans une rainure (*m.*), tire secured by a retaining-ring in a groove. (10)

bandage pneumatique (*m.*), pneumatic tyre. (11)

bande (*n.f.*), band; strip; belt. (12)

bande (Spectroscopie) (*n.f.*), band: (13)
bandes brillantes *ou* bandes lumineuses, bright bands. (14)
bandes d'absorption, absorption bands. (15)
bandes obscures *ou* bandes sombres, dark bands. (16)

bande (d'un ressort) (*n.f.*), tension (of a spring). (17)

bande (de la table des lumières du cylindre) (*n.f.*), bridge (of the port-face of the cylinder). (18)

bande (d'un bateau) (*n.f.*), list (of a boat). (19)

bande boueuse *ou* bande de boue (Glaciologie) (*f.*), dirt-band; dirt-line. (20)

bande de contrée (*f.*), stretch, belt, tract, of country: (21)
bande de contrée métallifère, metalliferous belt of country. (22)

bande de gravier cimenté (*f.*), band of cemented gravel. (23)

bande de minerai (*f.*), ore-streak; streak of ore. (24)

bande de recouvrement (couvre-joint) (*f.*), covering-plate; cover; welt; wrapper; butt-strap; strap; fish-plate. (25)

bande de terrain *ou* bande de terre (*f.*), strip, stretch, of ground: (26)
bande de terrain maigre *ou* bande de terrain pauvre, stretch of lean ground; strip of poor ground. (27)

bande du frein (*f.*), brake-band. (28)

bande sans fin (*f.*), endless belt. (29)

bande souple de transport (*f.*), conveyor-belt; travelling apron; apron. (30)

bander (tendre) (*v.t.*), to stretch; to make tense; to tense; to strain to stiffness: (31)
bander un câble, un arc, un ressort, to stretch a rope, a bow, a spring. (32 *ou* 33 *ou* 34)

bander une roue, to tire, to tyre, to shoe, a wheel. (35)

banket (*n.m.*) *ou* banket aurifère (conglomé--rat aurifère), banket; auriferous banket. (36)

banne (*n.f.*). Same as benne.

banquette (Ponts et Chaussées) (*n.f.*), banquette; bench; berm; berme. (37)

banquette de halage (le long d'un canal) (*f.*), tow-path; towing-path (along a canal). (38)

banquise (*n.f.*), pack; ice-pack. (39)

baquet de forgeron (*m.*), smiths' water-trough; slake-trough. (40)

baquettes (*n.f.pl.*), draw-tongs. (41)

baquettes à main (*f.pl.*), hand draw-tongs. (42)

baquettes mécaniques (*f.pl.*), bench draw-tongs. (43)

bar (civière) (*n.m.*), hand-barrow. (44)

bar (Météor.) (*n.m.*), bar. (45)

baraque (*n.f.*) *ou* baraque d'outils, shed; tool-shed. (46)

baraque (*n.f.*) *ou* baraquement (*n.m.*) (pour hommes), bunk-house; living-quarters; barrack. (47)

baraquer (*v.t.*), to barrack. (48)

baraquer (*v.i.*), to barrack. (49)

baraquer (se) (*v.r.*), to barrack. (50)

barbe (*n.f.*) *ou* barbe de pêne, notch (in a lock-bolt, to receive the bit of the key). (51)

barbes (d'une pièce de métal) (*n.f.pl.*), burr (on a piece of metal). (52)

barboche (*n.f.*), frame-saw file. (53)

barboteur (*n.m.*), washer with paddles; patouillet. (54)

barbure (*n.f.*) *ou* barbures (*n.f.pl.*), burr; scale. (55)

bard (*n.m.*), hand-barrow. (56)

bardeau (planchette en forme de tuile) (Constr.) (*n.m.*), shingle. (57)

barette (d'un four) (*n.f.*), damper (of a furnace). (58)

barge (*n.f.*), barge; sailing-barge. (59)

baricaut (*n.m.*), keg. (60)

baril (*n.m.*), barrel; cask; keg. (61)

baril (de vilebrequin) (*n.m.*), socket (of carpenters' brace). (62)

baril d'amalgamation (*m.*), amalgamation-barrel; amalgamating-barrel. (63)

baril de chloruration (*m.*), chlorination-barrel. (64)

barillet (petit baril) (*n.m.*), keg. (65)

barillet (d'un indicateur de pression, ou instru--ment analogue) (*n.m.*), drum (of a steam-engine indicator, or the like). (66 *ou* 67)

barillet (d'une pompe) (*n.m.*), barrel, cylinder, (of a pump). (68)

barillet (d'un objectif) (*n.m.*), tube, barrel, body-tube, (of a lens). (69)

bariolé, -e (*adj.*), variegated; mottled; party-coloured; parti-colored: (70)
marne bariolée (*f.*), variegated marl. (71)

baritel (manège) (*n.m.*), horse-power; horse-gear; animal driving-gear; animal-gear. (72)

barkévicite (Minéral) (*n.f.*), barkevikite. (73)

barle (Géol. & Mines) (*n.f.*), fault. (74)

barnhardtite (Minéral) (*n.f.*), barnhardtite. (75)

baromètre (*n.m.*), barometer; glass; weather-glass. (76)

baromètre à cadran (*m.*), wheel-barometer. (77)

baromètre à cuvette (*m.*), cup-barometer; cistern-barometer. (78)

baromètre à eau (*m.*), water-barometer. (79)

baromètre à mercure (*m.*), mercurial barometer. (80)

baromètre à réserve auxiliaire (*m.*), auxiliary barometer. (81)

baromètre à siphon (*m.*), siphon-barometer. (1)

baromètre altimétrique (*m.*), mountain-barom-eter ; orometer ; hypsobarometer. (2)

baromètre anéroïde (*m.*), aneroid barometer. (3)

baromètre anéroïde à cadran à jour, avec thermomètre courbé (*m.*), open-face aneroid barometer, with curved thermometer. (4)

baromètre anéroïde à cadran en métal argenté (*m.*), aneroid barometer with silvered metal dial. (5)

baromètre anéroïde avec index tournant servant de repère, divisions orométriques pour — mètres ; loupe à fort grossissement disposée pour permettre la lecture tout antour du cercle ; dans un solide écrin en cuir à courroie (*m.*), aneroid barometer with revolving altitude scale reading to — metres ; strong reading-lens arranged to traverse the whole circle ; in solid leather sling case. (6)

baromètre anéroïde de nivellement, compensé en température (*m.*), surveying aneroid barometer, compensated for temperature. (7)

baromètre anéroïde enregistreur (*m.*), recording aneroid barometer ; self-recording aneroid barometer ; aneroidograph. (8)

baromètre anéroïde forme montre, avec boussole derrière (*m.*), watch-form aneroid barometer, with compass on reverse ; watch-aneroid with compass on back ; watch-barometer with compass on back. (9)

baromètre de Bourdon (*m.*), Bourdon's barom-eter. (10)

baromètre de Fortin (*m.*), Fortin's barometer. (11)

baromètre de Gay-Lussac (*m.*), Gay-Lussac's barometer. (12)

baromètre de mine (*m.*), mine-barometer ; pit-barometer. (13)

baromètre de montagne (*m.*), mountain-barom-eter ; orometer ; hypsobarometer. (14)

baromètre de poche (*m.*), pocket barometer. (15)

baromètre enregistreur (*m.*), recording barom-eter ; self-recording barometer ; self-registering barometer. (16)

baromètre enregistreur de poche, de nivellement, révolutionnant en 24 heures, la plume pointant toutes les 2 minutes, avec série de feuilles de diagrammes et accessoires, en boîte (*m.*), pocket-size recording barometer for surveying, with chart running for 24 hours, and recording every 2 minutes, with set of charts and accessories, in case. (17)

baromètre étalon (*m.*), standard barometer. (18)

baromètre holostérique (*m.*), holosteric barom-eter. (19)

baromètre métallique (*m.*), aneroid barometer. (20)

baromètre normal (*m.*), normal barometer. (21)

baromètre orométrique (*m.*), mountain-barom-eter ; orometer ; hypsobarometer. (22)

baromètre-témoin [baromètres-témoins *pl.*] (*n.m.*), auxiliary barometer. (23)

barométrie (*n.f.*), barometry. (24)

barométrique (*adj.*), barometric ; barometrical : (25)

pression barométrique (*f.*), barometric pressure. (26)

barométriquement (*adv.*), barometrically. (27)

barométrographe (*n.m.*), barograph ; baro-metrograph. (28)

barométrographie (*n.f.*), barometrography. (29)

baroque (*adj.*), baroque : (30)

les perles baroqu.s sont celles qui ne sont ni rondes ni en forme de poire (*f.pl.*), baroque pearls are those which are neither round nor pear-shaped. (31)

baroscope (*n.m.*), baroscope. (32)

barothermographe *ou* barothermomètre (*n.m.*), barothermograph ; barothermometer. (33)

barrage (action) (*n.m.*), damming ; barrage ; stopping ; closing : (34)

barrage des vallées hautes, damming high valleys. (35)

barrage d'une vallée fluviale par une coulée de laves, barrage of a river-valley by a flow of lava. (36)

barrage d'une rue, closing a street. (37)

barrage (construction) (Hydraul.) (*n.m.*), dam ; barrage ; weir. (38)

barrage (Aérage de mines) (*n.m.*), stopping. (39)

barrage (Géol.) (*n.m.*), barrier of rock crossing the course of a stream. [Natural barrier or dam presented by rocky beds which outcrop across certain streams and which, on account of their hardness, offer greater resistance to erosion than the other parts of the stream-bed. These barriers produce falls and rapids]: (40)

une cascade est provoquée par un barrage, a cascade is caused by a barrier of rock crossing the course of a stream. (41)

barrage à aiguilles *ou* barrage à fermettes (*m.*), needle-dam. (42)

barrage à encoffrement en charpente (barrage formé par des cadres de bois rondins super-posés horizontalement, avec remplissage de rocailles et terre) (*m.*), cribwork-dam. (43)

barrage de retenue (*m.*), retaining dam. (44)

barrage de retenue d'eau (*m.*), dam for retaining a water-supply. (45)

barrage de retenue des débris *ou* barrage pour retenir les débris (*m.*), dam for holding back waste. (46)

barrage de retenue des tailings *ou* barrage pour retenir les tailings (*m.*), tailings-dam. (47)

barrage en maçonnerie (*m.*), masonry dam. (48)

barrage mobile (*m.*), movable dam. (49)

barrage provisoire (*m.*), temporary dam ; pro-visional dam. (50)

barragiste (pers.) (*n.m.*), weir-keeper. (51)

barranco (*n.m.*) *ou* barranca (*n.f.*) (Géogr. phys.), barranco ; barranca. (52)

barrandite (Minéral) (*n.f.*), barrandite. (53)

barre (tige) (*n.f.*), bar ; rod ; rail : (54)

une barre de fer, a bar of iron ; an iron bar. (55)

barres en cuivre rouge pour entretoises et rivets, copper bars for stay-bolts and rivets. (56)

barre (Géol. & Mines) (*n.f.*), bar : (57)

une barre de basalte, de grès, de terre dure, a bar of basalt, of sandstone, of hard ground. (58 *ou* 59 *ou* 60)

barre (Hydrogr.) (*n.f.*), bar : (61)

la barre à l'embouchure d'un fleuve, à l'entrée d'un port, the bar at the mouth of a river, at the entrance to a harbour. (62 *ou* 63)

barre à mine (*f.*), miner's bar ; bar ; jumper ; jumper-drill ; jumping-drill. (64)

barre à rivets (*f.*), rivet-bar ; rivet-rod. (65)

barre d'alésage *ou* barre à aléser (*f.*), boring-bar ; cutter-bar. (66)

barre d'appui (*f.*), window-bar (a fixed bar to prevent a person falling out or entering a window). (1)

barre d'attelage (Ch. de f.) (*f.*), draw-bar ; draught-iron ; draft-iron ; draw-link ; drag-bar ; drag-link. (2)

barre d'eau (*f.*), bore ; tidal bore. (3)

barre d'ébranlage (Fonderie) (*f.*), rapping-bar ; loosening-bar. (4)

barre d'excentrique (*f.*), eccentric-rod. (5)

barre d'excentrique de marche arrière (*f.*), back--ward eccentric-rod. (6)

barre d'excentrique de marche avant (*f.*), forward-eccentric-rod. (7)

barre de cabestan (*f.*), capstan-bar. (8)

barre de carrière (*f.*), quarry-bar ; jumper ; baby drill. (9)

barre de chariotage *ou* barre de commande de chariotage (tour) (*f.*), feed-shaft (lathe). (10)

barre de débrayage (renvoi de mouvement), (*f.*), fork-bar ; strap-bar (countershaft). (11)

barre de fermeture (*f.*), window-bar (a movable bar to secure a shutter). (12)

barre de mineur (*f.*), miners' bar ; bar ; jumper ; jumper-drill ; jumping-drill. (13)

barre de porte (*f.*), door-bar ; door-brand. (14)

barre de puddlage (*f.*), paddle. (15)

barre de rallonge (*f.*), lengthening-piece ; lengthening-rod ; extension-piece. (16)

barre de relevage (distribution de vapeur) (*f.*), reversing-lever rod ; reversing-rod ; reach-rod. (17)

barre de renvoi (*f.*), flat rod ; string-rod ; transmission-rod. (18)

barre de rivière (*f.*), river-bar. (19)

barre de sonde (*f.*), boring-bar. (20)

barre de surcharge (Sondage à la corde) (*f.*), sinker-bar. (21)

barre de traction (*f.*). Same as barre d'attelage.

barre étirée (*f.*), drawn bar. (22)

barre omnibus [barres omnibus *pl.*] (Élec.) (*f.*), bus-bar ; omnibus-bar. (23)

barre porte-couteaux (d'une haveuse à barre coupante) (*f.*), cutter-arm (of a bar coal-cutting machine). (24)

barreau (*n.m.*), bar ; rail : (25)

soumettre un barreau d'acier à des charges graduellement croissantes, to subject a bar of steel to gradually increasing stresses. (26)

barreau (d'une échelle) (*n.m.*), rung, round, step, (of a ladder). (27)

barreau à chevrons (grille de foyer) (*m.*), herring-bone bar (furnace-grate). (28)

barreau à pique (barreau de grille, — clôture) (*m.*), spear-headed railing. (29)

barreau aimanté *ou simplement* barreau (*n.m.*), bar needle ; bar magnet ; magnetic bar : (30)

barreau à chape d'agate (boussole), bar needle with agate centre (compass). (31)

barreau creux (grille à eau) (*m.*), hollow bar (water-grate). (32)

barreau d'essai (éprouvette) (*m.*), test-bar ; coupon. (33)

barreau de battement (grille de clôture) (*m.*), shutting-post. (34)

barreau de côtière (*m.*), hinge-post ; hinging-post ; swinging-post. (35)

barreau de fenêtre (*m.*), window-bar. (36)

barreau de grille (foyer) (*m.*), grate-bar ; fire-bar ; furnace-bar ; grid-bar. (37)

barreau de grille (clôture) (*m.*), railing ; rail. (38)

barreau de jette-feu (*m.*), drop-grate bar. (39)

barreau dormant (grille de foyer) (*m.*), bearer. (40)

barrer (*v.t.*). See examples : (41)

barrer le chemin, to bar the way. (42)

barrer un cours d'eau, to dam a stream. (43)

barrer une galerie (Mines), to fence off a road. (44)

barrer une porte, to bar a door (*or* a gate). (45)

barrer une route, to close, to obstruct, a road. (46)

barrette (petite barre) (*n.f.*), small bar ; bar. (47)

barrette (de châssis de fonderie) (*n.f.*), bar, cross-bar, stay, (of a foundry-flask). (48)

barrette (du tiroir) (*n.f.*), face-flange (of the slide-valve). (49)

barricade (*n.f.*), barricade. (50)

barricader (*v.t.*), to barricade. (51)

barrière (*n.f.*), barrier ; fence ; gate. (52)

barrière (d'un passage à niveau) (Ch. de f.) (*n.f.*), gate (of a level crossing). (53)

barrière à bascule *ou* barrière oscillante (*f.*), lift-gate ; lifting-gate. (54)

barrière à guillotine (Mines) (*f.*), guillotine gate. (55)

barrière à pivot (*f.*), swing-gate. (56)

barrière automatique (Mines) (*f.*), self-closing gate. (57)

barrière de cage (Mines) (*f.*), cage-fence ; cage-gate. (58)

barrière de sûreté (*f.*), safety-fence. (59)

barrière géographique entre deux pays (*f.*), geographical barrier between two countries. (60)

barrière roulante (*f.*), sliding gate. (61)

barriquaut (*n.m.*), keg. (62)

barrique (*n.f.*), cask ; barrel. (63)

barye (unité C.G.S.) (Phys.) (*n.f.*), barie. (64)

barylite (Minéral) (*n.f.*), barylite. (65)

barysphère (Géol.) (*n.f.*), barysphere. (66)

baryte (Chim.) (*n.f.*), baryta ; heavy earth. (67)

barytine *ou* barytite (Minéral) (*n.f.*), barite ; heavy spar ; terra ponderosa. (68)

barytine crêtée (*f.*), crested barite. (69)

barytique (*adj.*), barytic. (70)

barytocalcite (Minéral) (*n.f.*), barytocalcite. (71)

barytocélestine (Minéral) (*n.f.*), barytocelestite. (72)

baryum (Chim.) (*n.m.*), barium. (73)

bas, basse (*adj.*), low ; base ; shallow : (74)

basse fréquence (Élec.) (*f.*), low frequency. (75)

basse mer (*f.*), low tide. (76)

bas métal (*m.*), base metal. (77)

basse pression (*f.*), low pressure ; low tension. (78)

basse pression (à), low-pressure ; low-tension : (79)

cylindre à basse pression, low-pressure cylinder. (80)

basse teneur (d'un minerai) (*f.*), low grade, baseness, (of an ore). (81)

basse tension (*f.*), low tension ; low pressure. (82)

basses terres (*f.pl.*), lowlands. (83)

bas voltage (*m.*), low voltage ; low tension ; low pressure : (84)

lampe à bas voltage (*f.*), low-voltage lamp. (85)

écrou bas (*m.*), shallow nut. (86)

bas (*n.m.*), bottom ; foot ; lower part : (87)

bas d'une colline, bottom, foot, of a hill. (88)

bas d'une rivière, lower part of a river. (1)
le bas Rhin, le bas Danube, the lower Rhine, the lower Danube. (2 *ou* 3)
bas de l'eau (*m.*), low water. (4)
bas-fond [**bas-fonds** *pl.*] (terrain bas) (*n.m.*), flat : (5)
un bas-fond alcalifère, an alkali flat. (6)
bas-fond (dans une rivière ou dans la mer) (*n.m.*), flat ; shoal. (7 *ou* 8).
bas foyer (*m.*), low hearth ; forge-hearth ; hearth (of a Catalan forge, or the like). (9)
bas-joyer (*n.m.*). Same as **bajoyer**.
basalte (*n.m.*), basalt. (10)
basalte à leucite (*m.*), leucite-basalt. (11)
basalte à néphéline (*m.*), nepheline-basalt. (12)
basaltiforme (*adj.*), basaltiform. (13)
basaltine (Minéral) (*n.f.*), basaltine. (14)
basaltique (*adj.*), basaltic ; basaltine : (15)
lave basaltique (*f.*), basaltic lava. (16)
basaltoïde (*adj.*), basaltoid. (17)
basanite (Pétrol.) (*n.f.*), basanite. (18)
basculage (*n.m.*), seesawing ; swinging ; rocking ; rotation ; tipping ; tilting ; dumping : (19)
basculage d'une benne, dumping a bucket. (20)
basculaire (*adj.*), seesaw ; swinging ; balancing : (21)
mouvement basculaire (*m.*), seesaw motion ; swinging movement ; balancing motion. (22)
basculant, -e (*adj.*). See **bascule** (à).
bascule (levier) (*n.f.*), bascule ; balanced lever. (23)
bascule (machine à peser) (*n.f.*), weighing-machine ; scale ; scales. (24)
bascule (planches placées presque en équilibre sur les boulins) (Échafaudage) (*n.f.*), trap : (25)
avoir bien soin d'éviter les bascules, to be very careful to avoid traps. (26)
bascule (à) *ou* **basculant, -e** (*adj.*), tipping ; dumping ; tilting : (27)
wagon à bascule (*m.*), tipping wagon ; dump-ing wagon. (28)
four basculant (*m.*), tilting furnace. (29)
bascule à l'arrière (appareil photographique) (*f.*), swing-back. (30)
bascule à l'avant (appareil photographique) (*f.*), swing-front. (31)
bascule à romaine (*f.*), platform scales. (32)
bascule automatique à cadran (*f.*), automatic weighing-machine. (33)
bascule de battage (Sondage) (*f.*), walking-beam ; working-beam. (34)
bascule de sonnette (*f.*), bell-crank ; bell-crank lever. (35)
bascule en l'air (*f.*), suspended weighing-machine ; suspension-scales. (36)
basculement (*n.m.*). Same as **basculage**.
basculer (*v.i.*), to seesaw ; to swing ; to rotate ; to rock ; to tip ; to tip up ; to tilt ; to dump : (37)
poutre qui bascule (*f.*), beam which seesaws. (38)
appareil qui bascule sur un pivot (*m.*), appara-tus which swings (*or* rotates) on a pivot. (39)
wagon qui bascule (*m.*), wagon which tips ; car which dumps. (40)
basculeur (*n.m.*), dump ; tip ; tipper ; tippler ; tilter ; tumbler. (41)
basculeur en bout (*m.*), end-dump. (42)
base (partie inférieure d'un corps) (*n.f.*), base ; bottom ; foot : (43)
base d'une colline, base, bottom, foot, of a hill. (44)

base (Arch.) (*n.f.*), base : (45)
la base d'un mur, d'une colonne, the base of a wall, of a column. (46 *ou* 47)
base (*n.f.*) *ou* **base d'opérations** (Topogr.) base ; base-line ; datum-line : (48)
la base d'un levé trigonométrique, the base-line for a trigonometrical survey. (49)
de la précision de la mesure de la base dépend l'exactitude de tout le levé, the exactness of the whole survey depends on the precision of the measurement of the datum-line. (50)
base (Chim.) (*n.f.*), base. (51) See examples in vocabulary.
base (Cristall.) (*n.f.*), base : (52)
la base d'un cristal, the base of a crystal. (53)
base (Géom.) (*n.f.*), base : (54)
la base d'un triangle, d'un tronc de cône, the base of a triangle, of a frustum of cone. (55 *ou* 56)
base (principe ; fondement) (*n.f.*), basis ; ground-work : (57)
base d'un système, basis, groundwork, of a system. (58)
base (de chambre noire) (*n.f.*), base-board, bed, (of camera). (59)
base alcaline (Chim.) (*f.*), alkaline base. (60)
base d'opérations (*f.*), base of operations. (61)
Cf. **base** *ou* **base d'opérations** (Topogr.).
base de contrôle (Triangulation) (*f.*), base of verification. (62)
base salifiable (Chim.) (*f.*), salifiable base. (63)
baser (*v.t.*), to base : (64)
baser ses calculs sur des faits, to base one's calculations on facts. (65)
basicité (*n.f.*), basicity : (66)
basicité des acides, basicity of acids. (67)
basification (Chim.) (*n.f.*), basification. (68)
basile (*n.m.*), angle of the bit of a plane. (69)
basique (Chim.) (*adj.*), basic : (70)
sel basique (*m.*), basic salt. (71)
basique (Géol.) (*adj.*), basic : (72)
roches basiques (*f.pl.*), basic rocks. (73)
basique (Métall.) (*adj.*), basic : (74)
garnissage basique (*m.*), basic lining. (75)
basique (Cristall.) (*adj.*), basal : (76)
clivage basique (*m.*), basal cleavage. (77)
basse (*f. de bas adj.*). See **bas, basse.**
basse-riche [**basses-riches** *pl.*] (*n.f.*), a black, shelly stone, found at Mont-dore (France), used for making vases, artistic pedestals, etc. (78)
bassicot (pour voie à câble aérien) (*n.m.*), bucket (for aerial ropeway). (79)
bassin (récipient) (*n.m.*), basin ; bowl. (80)
bassin (d'une balance) (*n.m.*), pan, scale, bowl, (of a balance). (81)
bassin (Géol.) (*n.m.*), basin ; rock-basin. (82)
bassin (Géogr. phys.) (*n.m.*), basin. (83)
bassin (réservoir) (*n.m.*), basin ; reservoir ; pond. (84)
bassin (dock) (*n.m.*), dock ; basin. (85)
bassin (d'un four à réverbère) (*n.m.*), sloping hearth (of a reverberatory furnace). (86)
bassin alluvial (*m.*), alluvial basin. (87)
bassin carbonifère (*m.*), coal-basin. (88)
bassin d'éboulements (*m.*), rock-fall basin. (89)
bassin de chasse (Hydraul.) (*m.*), flush-pond. (90)
bassin de coulée *ou* *simplement* **bassin** (*n.m.*) (Aciérie), casting-pit. (91)
bassin de coulée (réservoir établi en amont de l'orifice de coulée d'un moule) (Fonderie) (*m.*), pouring-basin ; runner-basin. (92)

bassin de dépôt (*m.*), settling-pond. (1)
bassin de dépôt des schlamms (*m.*), slimes-pit. (2)
bassin de filtration (*m.*), filter-bed. (3)
bassin de réception pour les scories (*m.*), slag-receiver. (4)
bassin éolien (Géol.) (*m.*), eolian basin ; æolian basin. (5)
bassin épurateur (Fonderie) (*m.*), skim-basin. (6)
bassin fluvial (*m.*), river-basin. (7)
bassin géologique (*m.*), geological basin ; rock-basin : (8)
les bassins de la Seine et du Pô sont à la fois des bassins hydrographiques et géologiques, the basins of the Seine and the Po are both drainage-basins and geological basins. (9)
bassin glaciaire (*m.*), glacial basin. (10)
bassin houiller (*m.*), coal-basin. (11)
bassin hydrographique (Géogr. phys.) (*m.*), drain-age-basin ; catchment-basin ; hydrographi-cal basin ; drainage-area ; watershed ; intake. (12)
bassin lacustre (*m.*), lake-basin. (13)
bassin océanique (*m.*), ocean-basin. (14)
bassin pétrolifère (*m.*), oil-basin. (15)
bassin salifère (*m.*), salt bottom. (16)
bassin tectonique (Géol.) (*m.*), tectonic basin. (17)
bassin volcanique (*m.*), volcanic basin. (18)
bastite (Minéral) (*n.f.*), bastite ; schiller-spar. (19)
bastnaésite (Minéral) (*n.f.*), bastnasite. (20)
bât (*n.m.*), pack-saddle. (21)
bâtard, -e (*adj.*), bastard : (22)
lime bâtarde (*f.*), bastard file. (23)
batardeau (*n.m.*), coffer-dam ; bartardeau. (24)
bateau (*n.m.*), boat. (25)
bateau à rames (*m.*), rowboat. (26)
bateau à vapeur (*m.*), steamboat. (27)
bateau à voiles (*m.*), sailing-ship ; sailing-vessel. (28)
bateau caboteur (*m.*), coasting-vessel ; coaster. (29)
bateau charbonnier (*m.*), coal-ship ; collier. (30)
bateau charbonnier à voiles (*m.*), sailing-collier. (31)
bateau-citerne [bateaux-citernes *pl.*] (*n.m.*), tank-ship ; tank-vessel ; tanker. (32)
bateau-citerne à vapeur (*m.*), tank-steamer. (33)
bateau-citerne à voiles (*m.*), tank sailing-vessel. (34)
bateau de commerce (*m.*), merchantman ; trad-ing-vessel. (35)
bateau de passage (*m.*), ferry-boat. (36)
bateau-drague [bateaux-dragues *pl.*] (*n.m.*) *ou* bateau dragueur [bateaux dragueurs *pl.*] (*m.*), dredge-boat. (37)
bateau fluvial (*m.*), river-boat. (38)
bateau pétrolier (*m.*), oil-ship. (39)
bateau-rabot [bateaux-rabots *pl.*] (*n.m.*), dredge-boat. (40)
bateau remorqueur (*m.*), tug ; tug-boat. (41)
batée (*n.f.*) *ou* batée de lavage, batea ; dish ; washing-dish ; pan. (42)
batée de terre aurifère (*f.*), dish of dirt. (43)
batée sibérienne (*f.*), Siberian batea. (44)
batelage (*n.m.*), lighterage. (45)
batelée (*n.f.*), boat-load. (46)
batholithe (Géol.) (*n.m.*), batholith ; batholite ; bathylith ; bathylite. (47)
batholithique (Géol.) (*adj.*), batholithic. (48)
bathygraphique (*adj.*), bathygraphic. (49)

bathymètre *ou* bathomètre (*n.m.*), bathymeter ; bathometer. (50)
bathymétrie *ou* bathométrie (Océanogr.) (*n.f.*), bathymetry ; bathometry. (51)
bathymétrique *ou* bathométrique (*adj.*), bathy-metric ; bathymetrical ; bathometric ; batho-metrical : (52)
carte bathymétrique des océans (*f.*), bathy-metrical chart of the oceans. (53)
bâti (charpente en fer ou en bois) (*n.m.*), frame ; framing ; framework. (54)
bâti (d'une machine à vapeur, ou analogue) (*n.m.*), frame (of a steam-engine, or the like. (55 *ou* 56) Cf. bâti *ou* bâti d'assise.
bâti (*n.m.*) *ou* bâti d'assise (d'une machine à vapeur, ou analogue), bed-plate (of a steam-engine, or the like). (57 *ou* 58)
bâti à nervures *ou* bâti nervuré (*m.*), ribbed frame. (59)
bâti-baïonnette [bâtis-baïonnettes *pl.*] (d'une machine à vapeur) (*n.m.*), girder-type frame (of a steam-engine). (60)
bâti creux monté sur haute colonne avec pied à fourche (machine à percer) (*m.*), hollow frame mounted on tall pillar with U-shaped base-plate (drilling-machine). (61)
bâti de bocard (*m.*), battery-frame ; battery-framework (ore-stamp). (62)
bâti de croisée *ou* bâti dormant (*m.*), window-frame. (63)
bâti de forge (*m.*), smiths' hearth. (64)
bâti de forge, bâti à deux feux (*m.*), smiths' hearth, double hearth. (65)
bâti de forge, bâti à un seul feu, avec ou sans hotte (*m.*), smiths' hearth, single hearth, with or without hood. (66 *ou* 67)
bâti de porte *ou* bâti dormant (châssis scellé dans le mur) (*m.*), door-casing ; door-case ; door-pocket ; door-frame ; casing ; frame. (68)
bâti de porte (châssis qui reçoit les panneaux) (*m.*), door-frame. (69)
bâti en forme de caisson (*m.*), box-form frame ; box-section frame. (70)
bâti en pyramide (*m.*), A frame. (71)
bâti formant armoire et réservoir d'huile (tour, etc.) (*m.*), frame forming a cupboard and an oil-reservoir (lathe, etc.) (72)
bâtiment (*n.m.*), building ; house. (73)
bâtiment (navire) (*n.m.*), ship ; vessel. (74)
bâtiment de générateur *ou* bâtiment des chaudières (*m.*), boiler-house. (75)
bâtiment de la machine *ou* bâtiment des machines *ou* bâtiment des moteurs (*m.*), engine-house ; power-house. (76)
bâtiment des pompes (*m.*), pump-house. (77)
bâtiments de l'exploitation (Mines) (*m.pl.*), mine-buildings. (78)
bâtir (*v.t.*), to build ; to construct : (79)
bâtir une maison, un mur, un barrage, to build a house, a wall, a dam. (80 *ou* 81 *ou* 82)
batitures (*n.f.pl.*) *ou* batitures de fer, scale ; iron-scale ; hammer-scale ; forge-scale ; hammer-slag ; cinder ; anvil-dross ; nill. (83)
bâton (*n.m.*), stick : (84)
bâton de soudure, stick of solder. (85)
bâton de soufre, stick, roll, cane, of sulphur. (86)
bâton-pilote [bâtons-pilotes *pl.*] (Ch. de f.) (*n.m.*), staff ; train-staff. (87)
battage (martelage) (*n.m.*), beating ; hammering : (88)
battage d'or, gold-beating. (89)

battage à froid, cold hammering. (1)

battage (Sondage) (*n.m.*), percussion : (2)
battage au trépan, percussion with the bit. (3)

battage à la masse (*m.*), sledging. (4)

battage au large (entaillage des épontes) (Exploitation de mines) (*m.*), brushing the roof and the floor ; broaching ; lameskirting. (5)

battage au large (Travail aux explosifs) (*m.*), blasting from the centre to the out-circle. (6)

battage contre-vapeur (*m.*), reversing steam. (7)

battage de pieux *ou* **battage de pilots** (*m.*), driving piles ; pile-driving. (8)

battage des aiguilles infernales (Mines) (*m.*), plug and feathering. (9)

battage du sable autour d'un modèle (Fonderie) (*m.*), ramming the sand round a pattern. (10)

battant (d'une porte) (*n.m.*), leaf (of a door). (11)

battant (d'un loquet de porte) (*n.m.*), lift (of a gate-latch). (12)

battant neuf (*m.*), **battant neuve** (*f.*), brand new. (13)

batte (maillet) (*n.f.*), mallet. (14)

batte (de mouleur) (Fonderie) (*n.f.*), rammer. (15)

batte (de plombier) (*n.f.*), dresser ; lead dresser ; beater ; bat. (16)

batte (des chemins de fer ou de terrassier) (*n.f.*), beater ; beater-pick ; beating-pick ; packer ; tamping-pick. (17)

batte à bourre (*f.*), tamping-bar ; tamping-iron ; tamping-rod ; tamper. (18)

batte plate (Fonderie) (*f.*), flat rammer. (19)

batte-plate [**battes-plates** *pl.*] (*n.f.*). Same as **batte** (de plombier).

battée (*n.f.*). Same as **batée**.

battement (*n.m.*) *ou* **battements** (*n.m.pl.*), banging; slamming ; jarring ; jar. (20)

battement de piston (*m.*), stroke of piston ; piston-stroke. (21)

battements (d'une lampe synchronisante) (Élec.) (*n.m.pl.*), pulsations (of a synchronizing lamp). (22)

batterand (pour casser les pierres) (*n.m.*), stone-breakers' hammer. (23)

batterand (marteau de carrier) (*n.m.*), quarry-man's hammer (for driving wedges in the stone). (24)

batterie (série ; jeu) (*n.f.*), battery ; set. (25)

batterie (Élec.) (*n.f.*), battery. (26). See also **pile** (*n.f.*).

batterie (mise en). See under **mise**.

batterie d'accumulateurs (Élec.) (*f.*), battery of accumulators ; storage-battery ; secondary battery. (27)

batterie d'étais (Constr.) (*f.*), set of shores ; system of shoring ; frame. (28)

batterie de bocards *ou* **batterie de pilons** (pour minerai) (*f.*), stamp-battery. (29)

batterie de chaudières (*f.*), battery of boilers. (30)

batterie de cylindres (de laminoir) (*f.*), set, battery, of rolls (of rolling-mill). (31)

batterie de laminoirs (*f.*), train of rolls ; roll-train ; rolling-mill train ; set of rolls ; battery of rolls. (32)

batterie de piles (Élec.) (*f.*), battery. (33)

batterie de — pilons (pour minerai) (*f.*), —-head battery ; —-stamp mill : (34)
battery de 5 pilons, 5-head battery ; 5-stamp mill. (35)

batterie électrique (*f.*), Leyden-jar battery ; Leyden battery ; battery of Leyden jars. (36)

batterie en cascade (Élec.) (*f.*), battery connected in series (*or* in cascade). (37)

batterie en surface (Élec.) (*f.*), battery connected in multiple (*or* in parallel). (38)

batterie-tampon [**batteries-tampons** *pl.*] (Élec.) (*n.f.*), buffer-battery. (39)

batteur d'or (pers.) (*m.*), gold-beater. (40)

battitures (*n.f.pl.*). Same as **batitures**.

battrant (*n.m.*). Same as **batterand**.

battre (*v.t.*). See examples : (41)
battre à froid, to cold-hammer. (42)
battre à la masse, to sledge. (43)
battre au large (entailler les épontes) (Exploitation de mines), to brush the roof and the floor ; to broach. (44)
battre au large (Travail aux explosifs), to blast from the centre to the out-circle. (45)
battre contre-vapeur, to reverse steam : (46)
mécanicien qui bat contre-vapeur (*m.*), engine-driver who reverses steam. (47)
battre de l'or, to beat gold. (48)
battre des aiguilles infernales (Mines), to plug and feather. (49)
battre des pieux, to drive piles. (50)
battre en retraite (Exploitation de mines), to work on the retreat ; to work home ; to work homewards ; to retreat. (51)
battre la ligne (cordeau blanchi à la craie), to snap the line. (52)
battre le fer sur l'enclume, to hammer iron on the anvil. (53)
battre le pays, to scour, to explore, the country. (54)
battre le rappel (se dit des mineurs ensevelis), to knock to attract attention ; to attract attention by knocking. (55)
battre le sable dans un châssis de fonderie, to ram the sand in a foundry-flask. (56)

baudet (de scieur de long) (*n.m.*), pit-saw horse ; pit-sawyers' trestle. (57)

baudruche (*n.f.*), gold-beaters' skin. (58)

baudrucheur (pers.) (*n.m.*), gold-beater. (59)

baulite (Pétrol.) (*n.f.*), baulite. (60)

baume du Canada (*m.*), Canada balsam. (61)

bauxite (Minéral) (*n.f.*), bauxite ; beauxite. (62)

bavure (Fonderie) (*n.f.*), fin ; burr : (63)
une bavure est une trace saillante laissée sur une pièce moulée à l'endroit des joints du moule, a fin is a ridge left on a casting at the junction of the parts of a mould. (64)

bavure (morfil) (*n.f.*), wire-edge ; burr. (65)

bayart (*n.m.*), hand-barrow. (66)

béant, -e (*adj.*), gaping ; yawning ; open : (67)
fissure béante (*f.*), gaping (*or* yawning) (*or* open) fissure. (68)

beau (**bel** *devant une voyelle*) **belle** (*adj.*), beautiful ; fine : (69)
pierre susceptible d'un beau poli (*f.*), stone capable of receiving a fine polish. (70)
mine qui a devant elle un bel avenir (*f.*), mine which has a fine future before it. (71)
la belle fluorine violette de Cornouailles, the beautiful violet fluorite of Cornwall. (72)

beau (*n.m.*) *ou* **beau temps** (Barométrie), fair : (73)
le baron être est au beau, the barometer is at fair. (74)

beau fixe (Barométrie) (*m.*), set fair. (75)

beau temps (*m.*), fine weather ; fair weather ; beautiful weather ; good weather. (76)

beaumontite (Minéral) (*n.f.*), beaumontite. (1)

beauxite (Minéral) (*n.f.*), bauxite ; beauxite. (2)

bec (de tuyau, etc.) (*n.m.*), nozzle ; nozle ; nose ; nose-piece ; snout. (3)

bec (d'un outil) (*n.m.*), nose (of a tool) : (4) le bec d'un burin, d'un outil de tour, d'une pince, the nose of a chisel, of a lathe-tool, of a pliers. (5 *ou* 6 *ou* 7)

bec (d'un compas à coulisse, d'une clef) (*n.m.*), jaw (of a sliding caliper-gauge, of a spanner). (8 *ou* 9)

bec (d'une burette à huile) (*n.m.*), spout (of an oil-can). (10)

bec (d'un vase, d'une éprouvette) (*n.m.*), spout, lip, (of a beaker, of a test-glass). (11 *ou* 12)

bec (d'une pile de pont) (*n.m.*), cutwater (of a bridge-pier). (13)

bec (Géogr.) (*n.m.*), point ; beak ; promontory. (14)

bec (brûleur à gaz) (*n.m.*), burner ; jet : (15) bec — bougies, — - candle-power burner. (16)

bec à gaz (*m.*), gas-burner ; gas-jet. (17)

bec à incandescence par le gaz, avec veilleuse (*m.*), incandescent-gas burner, with by-pass. (18)

bec à queue de poisson (gaz) (*m.*), fish-tail burner. (19)

bec Bunsen (*m.*), Bunsen burner ; Bunsen's lamp. (20)

bec Bunsen, avec virole à air, robinet à gaz et veilleuse (*m.*), Bunsen burner, with air-regulator and by-pass stop-cock. (21)

bec courbe (d'un robinet à bec courbe (*m.*), bib-nozzle (of a bib-cock). (22)

bec-d'âne [becs-d'âne *pl.*] (*n.m.*). Same as **bédane**.

bec d'étain (Cristall.) (*m.*), beak of tin. (23)

bec-de-cane [becs-de-cane *pl.*] (pince plate) (*n.m.*), flat-nose pliers ; flat-nosed pliers. (24)

bec-de-cane (Serrur.) (*n.m.*), lever handle (of a shop-door latch). (25)

bec-de-cane (organe de l'ancienne distribution, servant à obtenir le changement de marche) (*n.m.*), gab ; gab-hook ; eccentric-hook ; hook-gear : (26) la coulisse supplanta le bec-de-cane, the link supplanted the gab-hook. (27)

bec-de-corbeau [becs-de-corbeau *pl.*] (pinces) (*n.m.*), long-nose cutting-pliers. (28)

bec de coulée (d'un convertisseur Bessemer) (*m.*), nose (of a Bessemer converter). (29)

bec de gaz (*m.*), gas-burner ; gas-jet. (30)

bec de prise d'eau (alimentation en marche de soute du tender) (*m.*), water-scoop. (31)

bec de terre (*m.*), point, beak, of land ; promon-tory. (32)

bec de tuyère (*m.*), nozzle-tip ; tuyère-nozzle. (33)

bec droit (d'un robinet à bec droit) (*m.*), straight nose (of a straight-nose cock). (34)

bec droit (opp. à bec renversé) (gaz) (*m.*), upright burner ; erect burner. (35)

bec papillon (gaz) (*m.*), butterfly-burner. (36)

bec renversé (gaz) (*m.*), inverted burner. (37)

bécasse (haut fourneau) (*n.f.*), stock-indicator ; charge-gauge ; test-rod ; gauge. (38)

bêche (*n.f.*), spade. (39)

becherglas (verre à filtrations chaudes) (Verrerie de laboratoire) (*n.m.*), beaker. (40)

bêchevet (Charp.) (*n.m.*), splayed joint. (41)

bed-rock (roche de fond) (Géol.) (*n.m.*), bed-rock. (42)

bed-rock véritable (*m.*), true bed-rock. (43)

bédane (de charpentier, de menuisier) (*n.m.*), mortise-chisel ; mortising-chisel ; framing-chisel ; heading-chisel. (44)

bédane (pour machine à mortaiser le bois) (*n.m.*), mortise-chisel, mortising-chisel, (for wood-mortising machine). (45)

bédane (*n.m.*) *ou* **bédane à froid** (de mécanicien ; de serrurier), cold-chisel (cross-cut) ; engineers' chipping-chisel (cross-cut) ; cross-cut chisel. (46)

bédane à joues (machine à mortaiser le bois) (*m.*), self-coring chisel. (47)

bédane double (*m.*), double mortise-chisel. (48)

bédaner (*v.t.*), to mortise ; to mortice. (49)

bée (d'un moulin à eau) (*n.f.*), flume, leat, (of a water-mill). (50)

bée de fenêtre (*f.*), window-opening. (51)

bée de porte (*f.*), door-opening ; doorway. (52)

beffroi (d'une drague) (*n.m.*), gantry (of a dredge). (53)

béguettes (*n.f.pl.*), short-nose pliers ; short-nosed pliers. (54)

bel-outil [beaux-outils *pl.*] (*n.m.*), jewellers' anvil. (55)

bélier (*n.m.*), ram. (56)

bélier aspirateur (*m.*), suction-ram. (57)

bélier hydraulique (*m.*), hydraulic ram ; hydraulic-impulse ram ; water-ram ; ram-pump. (58)

belle-fleur (chevalement de puits d'extraction) [Belgique] (*n.f.*), head-frame ; pit-head frame ; pit-frame ; pulley-frame ; poppet-head ; puppet-head ; head-stock ; gallows frame ; shaft-tackle. (59)

bellite (Explosif) (*n.f.*), bellite. (60)

bélonite (Minéral) (*n.f.*), belonite. (61)

bénarde (Serrur.) (*n.f.*), pinned-key lock. (62)

bénéfice (*n.m.*), profit. (63)

bénéfice brut (*m.*), gross profit. (64)

bénéfice net (*m.*), net profit. (65)

bénéficier (*v.t.*), to work at a profit : (66) un minerai difficile à bénéficier, an ore difficult to work at a profit. (67)

bénéficier (*v.i.*), to make a profit ; to profit : (68) bénéficier sur une marchandise, to make a profit on a line of goods. (69)

bénéficier (se) (*v.r.*), to be worked at a profit : (70) mine qui se bénéficie aisément (*f.*), mine which is easily worked at a profit. (71)

benne (Mines) (*n.f.*), bucket ; kibble ; tub. (72)

benne à bascule *ou* **benne basculante** (*f.*), tipping-bucket ; dumping-bucket. (73)

benne à eau (*f.*), water-kibble ; water-bucket. (74)

benne à fond mobile (*f.*), bucket with drop-bottom. (75)

benne à patins (*f.*), sledge-bucket ; sledge-kibble ; sledge-tub ; sledge-corve. (76)

benne automatique (*f.*), grab-bucket ; grab ; clam-shell bucket. (77)

benne d'épuisement (*f.*), water-kibble ; water-bucket. (78)

benne d'extraction (*f.*), hoisting-bucket ; mining-bucket. (79)

benne flottante (opp. à cage guidée) (*f.*), loose kibble. (80)

benne preneuse *ou* **benne piocheuse** (*f.*), grab-bucket ; grab ; clam-shell bucket. (81)

benne roulante (*f.*), wheeled kibble ; wheeled bucket. (82)

benne suspendue (*f.*), hanging bucket ; suspended bucket. (1)

benzène (*n.m.*), benzene. (2)

benzine (*n.f.*), benzine. (3)

benzol (*n.m.*), benzol. (4)

benzoline (*n.f.*), benzoline. (5)

béquille (*n.f.*), crutch handle ; T handle. (6)

béquille (d'un robinet) (*n.f.*), crutch key (of a cock). (7)

berceau (Arch.) (*n.m.*), cradle vault. (8)

berceau (Mines) (*n.m.*), cradle ; rocker ; cradle-rocker. (9)

berceau (*n.m.*) *ou* **berceau cylindrique** (de la boîte à feu) (locomotive), roof-sheet, wrapper-plate, (of fire-box). (10)

berceau lavoir (*m.*), washing-cradle. (11)

bercer (*v.t.*), to rock. (12)

bercer (se) (*v.r.*), to rock. (13)

berge (*n.f.*), bank : (14)
berge d'un fleuve, d'un canal, d'un chemin de fer, d'une fouille, bank of a river, of a canal, of a railway, of a cut. (15 *ou* 16 *ou* 17 *ou* 18)

berge (d'une montagne) (*n.f.*), side, flank, (of a mountain). (19)

bergmannite (Minéral) (*n.f.*), bergmannite. (20)

béril (Minéral) (*n.m.*). Same as **béryl**.

berline *ou* **berlaine** (Mines) (*n.f.*), tram ; tub ; corf ; corve ; rolley ; colliery-wagon ; mine-car. (21)

berme (Trav. publ.) (*n.f.*), berm ; berme ; bench ; banquette. (22)

bérouche (Mines) (*n.f.*), kibble ; bucket. (23)

berthiérite (Minéral) (*n.f.*), berthierite. (24)

bertrandite (Minéral) (*n.f.*), bertrandite. (25)

béryl (Minéral) (*n.m.*), beryl. (26)

béryl doré (Minéral) (*m.*), golden beryl. (27)

béryllium (Chim.) (*n.m.*), beryllium ; glucinum ; glucinium. (28)

béryllonite (Minéral) (*n.f.*), beryllonite. (29)

berzélianite (Minéral) (*n.f.*), berzelianite. (30)

berzéliite (Minéral) (*n.f.*), berzeliite. (31)

besaiguë (*n.f.*). Same as **bisaiguë**.

besogne (*n.f.*), task. (32)

besoin (*n.m.*), need ; requirement : (33)
besoin pressant de facilités de transport, pressing need of transportation facilities. (34)
besoins actuels, present needs ; actual requirements. (35)

bête de somme (*f.*), beast of burden. (36)

bête de trait (*f.*), draught-animal. (37)

bétoire (*n.f.*), limestone cave ; limestone cavern. (38)

béton (*n.m.*), concrete. (39)

béton armé (*m.*), reinforced concrete ; armoured concrete ; forroconcrete. (40)

béton Coignet (*m.*), beton ; béton ; beton Coignet. (41)

béton de ciment (*m.*), cement concrete. (42)

béton de ciment armé (*m.*), reinforced cement concrete. (43)

béton fretté (*m.*), hooped concrete ; stirruped concrete. (44)

béton gras (*m.*), rich concrete. (45)

béton immergé (*m.*), submerged concrete. (46)

béton maigre (*m.*), poor concrete. (47)

bétonnage (*n.m.*), concreting ; concrete-work. (48)

bétonner (*v.t.*), to concrete. (49)

bétonnière (*n.f.*), concrete-mixer. (50)

beudantite (Minéral) (*n.f.*), beudantite. (51)

beurre de montagne *ou* **beurre de pierre** *ou* **beurre de roche** (Minéral) (*m.*), stone-butter ; rock-butter. (52)

beurre des tourbières (*m.*), bog-butter ; butyrellite. (53)

beurtia (Mines) (*n.m.*), blind shaft ; blind pit ; internal shaft ; staple ; staple-shaft ; staple-pit. (54)

beyrichite (Minéral) (*n.f.*), beyrichite. (55)

biais, -e (*adj.*), skew ; skewed ; oblique : (56)
voûte biaise (*f.*), skew (*or* skewed) (*or* oblique) arch. (57)

biais (en), on the skew ; obliquely : (58)
éviter de poser les traverses en biais, to avoid laying the sleeper on the skew. (59)

biais (*n.m.*), skew ; obliquity : (60)
le biais d'un pont, d'une voûte, the skew of a bridge, of an arch. (61 *ou* 62)

biaisement (*n.m.*), skewing. (63)

biaiser (*v.i.*), to skew. (64)

biatomique (Chim.) (*adj.*), diatomic. (65)

biaxe (*adj.*), biaxial ; biaxal : (66)
cristal biaxe (*m.*), biaxial crystal. (67)

bibasique (Chim.) (*adj.*), bibasic ; dibasic. (68)

bicâble (*n.m.*), double-rope cable-tramway. (69)

bicarbonate (Chim.) (*n.m.*), bicarbonate. (70)

biconcave (*adj.*), biconcave ; double-concave : (71)
lentille biconcave (*f.*), biconcave (*or* double-concave) lens. (72)

biconvexe (*adj.*), biconvex ; double-convex : (73)
lentille biconvexe (*f.*), biconvex (*or* double-convex) lens. (74)

bicylindrique (*adj.*), two-cylinder ; double-cylinder ; double ; duplex-cylinder ; duplex : (75)
machine bicylindrique (*f.*), two-cylinder engine ; double-cylinder engine ; double engine ; duplex-cylinder engine ; duplex engine. (76)

bidet (*n.m.*), plain ungeared lathe. (77)

bidon (*n.m.*), tin ; can. (78)

bidon à essence (*m.*), petrol-tin. (79)

bidon à huile (*m.*), oil-can (tin for holding oil). (80)

bidon à pétrole (*m.*), oil-can (tin for holding petroleum). (81)

bidon pour carbure (*m.*), carbide-tin. (82)

biebérite (Minéral) (*n.f.*), bieberite ; cobalt-vitriol. (83)

bief (espace compris entre deux écluses, sur un canal navigable) (*n.m.*), pond ; reach ; level. (84)

bief (d'un moulin à eau) (*n.m.*), race, raceway, (of a water-mill). (85)

bief d'amont (partie d'un canal qui commence en dessus du sas) (*m.*), head-bay ; head-crown. (86)

bief d'amont (d'un moulin) (*m.*), headrace ; forebay. (87)

bief d'aval (partie d'un canal en dessous du sas) (*m.*), tail-bay ; aft-bay. (88)

bief d'aval *ou* **bief de fuite** (d'un moulin) (*m.*), tail-race ; mill-tail. (89)

bief de moulin (*m.*), mill-race ; mill-run ; mill-course. (90)

bielle (Mach.) (*n.f.*), connecting-rod ; rod ; pitman ; connection. (91)

bielle (d'une ferme de comble en fer) (*n.f.*), strut, brace, (of an iron roof-truss). (92)

bielle articulée (*f.*), knuckle-jointed connecting-rod. (93)

bielle avec tête à cage fermée *ou* **bielle à cage fermée** (*f.*), box-end connecting-rod. (94)

bielle avec tête à cage ouverte et à chapeau *ou* **bielle à chapeau** (*f.*), marine-pattern con--necting-rod. (95)

bielle avec tête à chape *ou* bielle à chape (*f.*), strap-end connecting-rod ; strap-head con--necting-rod. (1)

bielle avec tête à fourche *ou* bielle à fourche (*f.*), fork-end connecting-rod. (2)

bielle d'accouplement (des roues d'une locomotive) (*f.*), side-rod ; parallel rod ; coupling-rod. (3)

bielle d'accouplement à bagues (*f.*), side-rod with solid eye ends and bushes. (4)

bielle d'accouplement à clavettes de réglage (*f.*), side-rod with cottered ends, cotters secured in position by set-screws. (5)

bielle d'excentrique (*f.*), eccentric-rod. (6)

bielle d'excentrique de marche en arrière (loco--motive) (*f.*), backward eccentric-rod ; back-up eccentric-rod. (7)

bielle d'excentrique de marche en avant (loco--motive) (*f.*), forward eccentric-rod ; go-ahead eccentric-rod. (8)

bielle de machine à balancier (*f.*), pitman of beam-engine. (9)

bielle de relevage (distribution) (*f.*), shifting-rod (valve-gear). (10)

bielle de suspension de l'entretoise des sabots (*f.*), brake-hanger; brake-beam hanger. (11)

bielle de suspension de la coulisse (*f.*), link-hanger ; link-support. (12)

bielle du tiroir (*f.*), valve-rod ; rod of the slide-valve ; slide-rod. (13)

bielle du tiroir (distribution Walschaerts) (*f.*), radius-rod ; radius-bar (Walschaerts valve-gear). (14)

bielle infinie *ou* bielle de longueur infinie (*f.*), connecting-rod of infinite length : (15)

l'influence de l'obliquité de la bielle du tiroir est à peu près nulle, car sa longueur est très grande par rapport à l'excentricité. On suppose donc les bielles infinies, the influence of the angularity of the valve-rod is practically nil, because its length is very great in relation to the throw. Connecting-rods are therefore supposed to be of infinite length. (16)

bielle motrice (*f.*), main rod ; main connecting-rod ; driving-rod. (17)

bielle motrice à cages fermées sur les deux têtes (*f.*), main rod with box ends. (18)

bielle motrice avec petite tête à fourche et à chapes rapportées, et avec grosse tête à cage ouverte et à chapeau (*f.*), main rod with fork little end and straps secured by bolts, and with marine-type big end. (19)

bielle motrice avec petite tête simple et à bague, et avec grosse tête à chape rapportée (*f.*), main rod with solid-eye-and-bush little end, and bolted-strap big end. (20)

bien-être (*n.m.*), well-being : (21)

se préoccuper du bien-être matériel et moral du personnel et des ouvriers, to pay attention to the material and moral well-being of the staff and the workmen. (22)

biez (*n.m.*). Same as bief.

bifilaire (*adj.*), bifilar ; two-wire : (23)

suspension bifilaire (Phys.) (*f.*), bifilar suspen--sion. (24)

réseau bifilaire (Élec.) (*m.*), two-wire system. (25)

bifurcation (*n.f.*), bifurcation ; forking ; fork ; junction. (26)

bifurcation (Ch. de f.) (*n.f.*), junction : (27)

un train doit ralentir sa marche aux bifurca--tions importantes, a train should slow down at important junctions. (28)

bifurcation en gare (*f.*), junction within station limits ; junction inside station limits. (29)

bifurcation en pleine voie (*f.*), junction outside station limits. (30)

bifurquer (*v.i.*), to bifurcate ; to fork : (31)

rivet bifurqué (*m.*), bifurcated rivet. (32)

bifurquer (se) (*v.r.*), to bifurcate ; to fork : (33)

parfois, un dyke se bifurque en deux ou plusieurs autres plus petits, occasionally, a dyke forks into two or more smaller ones. (34)

bigarré, -e (*adj.*), variegated ; mottled ; party-coloured ; parti-colored : (35)

grès bigarré (*m.*), variegated sandstone. (36)

bigorne (*n.f.*) *ou* bigorne à queue, beak-iron ; beck-iron ; bickern ; beakhorn-stake ; stake. (37)

bigorne (d'une enclume) (*n.f.*), beak, horn, beak-iron, (of an anvil). (38)

bigorne à gouttière (*f.*), groove-punch. (39)

bigorne à pince (*f.*), beak-iron with two taper ends. (40)

bigue (l'appareil) (*n.f.*), shears ; sheers ; shear-legs ; sheer-legs. (41)

bigues (les deux montants, ou hanches, de l'appareil) (*n.f.pl.*), shears ; sheers ; shear-legs ; sheer-legs (42).

bijou (*n.m.*), jewel. (43)

bijouterie (*n.f.*), jewelry ; jewellery. (44)

bijoutier (pers.) (*n.m.*), jeweler ; jeweller. (45)

bilan calorifique (Métall.) (*m.*), heat balance-sheet ; thermal balance-sheet ; heat-sheet : (46)

dresser le bilan calorifique d'un haut fourneau, to draw up the heat balance-sheet of a blast-furnace. (47)

bilan chimique (*m.*), chemical balance-sheet. (48)

bildstein (Minéral.) (*n.m.*), bildstein ; agalmato--lite ; figure-stone. (49)

billage (de métaux) (*n.m.*), ball-testing. (50)

bille (bloc de bois non-travaillé) (*n.f.*), saw-log ; log. (51)

bille (Ch. de f.) (*n.f.*). In Belgium.—a railway-sleeper. (52)   In France.—a piece of wood 5·40 metres long intended to be cut in half to make 2 sleepers. (53)

bille (pour coussinets) (*n.f.*), ball (for bearings) : (54)

billes en acier spécial fondu au creuset, trempées à cœur, parfaitement polies et calibrées, special crucible steel balls, hardened through--out, accurately ground and polished. (55)

biller (*v.t.*), to ball-test : (56)

biller un métal, to ball-test a metal. (57)

billette (Métall.) (*n.f.*), billet. (58)

billot d'enclume (*m.*), anvil-block. (59)

bimétallique (*adj.*), bimetallic : (60)

fil bimétallique (*m.*), bimetallic wire. (61)

binaire (*adj.*), binary : (62)

composé binaire (Chim.) (*m.*), binary com--pound. (63)

le laiton est un alliage binaire de cuivre et de zinc, brass is a binary alloy of copper and zinc. (64)

binard *ou* binart (*n.m.*), lorry ; trolley. (65)

biotite (Minéral.) (*n.f.*), biotite. (66)

bioxyde (Chim.) (*n.m.*), dioxide. (67)

bipolaire (Phys.) (*adj.*), two-pole ; double-pole ; bipolar : (68)

interrupteur bipolaire (*m.*), two-pole switch ; double-pole switch. (69)

dynamo à courant continu bipolaire (*f.*), bipolar direct-current dynamo. (70)

bipolarité (*n.f.*), bipolarity. (1)
bipoulie (*n.f.*), two-step cone; two-lift cone-pulley. (2)
biprisme (Opt.) (*n.m.*), biprism. (3)
bipyramidal, -e, -aux *ou* **bipyramidé -e,** (*adj.*), bipyramidal: (4)
    cristal de quartz bipyramidé (*m.*), bipyramidal quartz crystal. (5)
bipyramide (Cristall.) (*n.f.*), bipyramid. (6)
biquartz (*n.m.*), biquartz. (7)
biréfringence (Opt.) (*n.f.*), birefringence. (8)
biréfringent, -e (*adj.*), birefringent: (9)
    cristal biréfringent (*m.*), birefringent crystal. (10)
bisaiguë (*n.f.*), mortise-axe (double-ended). (11)
biscuit (Minéralogie) (*n.m.*), streak plate: (12)
    la molybdénite donne un trait bleu-verdâtre sur le biscuit, molybdenite gives a greenish blue streak on the streak plate. (13)
bise (*n.f.*), north wind; cold wind. (14)
bise glaciale (*f.*), icy wind. (15)
biseau (*n.m.*), bevel. (16)
biseau (Cristall.) (*n.m.*), bevelment. (17)
biseau (outil de tourneur en bois) (*n.m.*), side chisel; skew chisel. (18)
biseau (en), bevelled: (19)
    bord en biseau (*m.*), bevelled edge. (20)
    cristal de quartz en biseau (*m.*), bevelled quartz crystal. (21)
biseautage (*n.m.*), bevelling. (22)
biseauter (*v.t.*), to bevel: (23)
    biseauter une glace, to bevel a mirror. (24)
bismite (Minéral) (*n.f.*), bismite. (25)
bismuth (Chim.) (*n.m.*), bismuth. (26)
bismuthide (*adj.*), bismuthal. (27)
bismuthifère (*adj.*), bismuthiferous. (28)
bismuthine *ou* **bismuthinite** (Minéral) (*n.f.*), bismuthine; bismuthinite. (29)
bismuthique (Chim.) (*adj.*), bismuthic. (30)
bismuthite *ou* **bismutite** (Minéral) (*n.f.*), bismutite; bismuthite. (31)
bismuthocre (Minéral) (*n.m.*), bismuth ochre. (32)
bissecter (*v.t.*), to bisect: (33)
    bissecter un angle, to bisect an angle. (34)
bissecteur, -trice (*adj.*), bisecting. (35)
bissection (*n.f.*), bisection. (36)
bissectrice (*n.f.*), bisector; bisectrix. (37)
bissel (de locomotive) (Ch. de f.) (*n.m.*), pony truck; pony; two-wheeled truck. (38)
bissel arrière (*m.*), trailing-truck. (39)
bisulfate (Chim.) (*n.m.*), bisulphate; disulphate. (40)
bisulfite (Chim.) (*n.m.*), bisulphite. (41)
bisulfure (Chim.) (*n.m.*), bisulphide; disulphide. (42)
bit (Sondage) (*n.m.*), bit. (43)
bit creux (*m.*), hollow bit. (44)
bit plein (*m.*), solid bit. (45)
bittern (eau mère) (*n.m.*), bittern; mother-liquor. (46)
bittersalz (Minéral) (*n.m.*), bitter salt; epsomite. (47)
bitterspath (Minéral) (*n.m.*), bitter-spar; dolomite. (48)
bitumage (*n.m.*), bitumination; bituminization; asphalting; tarring. (49)
bitume (*n.m.*), bitumen; pitch; tar. (50)
bitume de Judée (*m.*), Judean bitumen; bitumen of Judea; Jew's pitch; Jews' pitch. (51)
bitume glutineux (Minéral) (*m.*), maltha; pissasphalt. (52)
bitume minéral (*m.*), mineral pitch. (53)

bitumé, -e (*adj.*), **bitumer** (*v.t.*), **bitumeux, -euse** (*adj.*). Same as **bituminé, -e, bituminer, bitumineux, -euse.**
bituminé, -e (*adj.*), bituminated; bituminized; tarred: (54)
    papier bituminé (*m.*), bituminized paper. (55)
    carton bituminé (*m.*), tarred felt. (56)
bituminer (*v.t.*), to bituminate; to bituminize; to asphalt; to tar: (57)
    bituminer un trottoir, to asphalt a path. (58)
bitumineux, -euse (*adj.*), bituminous; asphaltic; tarry: (59)
    'houille bitumineuse (*f.*), bituminous coal. (60)
bituminifère (*adj.*), bituminiferous. (61)
bituminisation (transformation en bitume) (*n.f.*), bituminization. (62)
bituminiser (*v.t.*), to bituminize. (63)
bituminiser (se) (*v.r.*), to bituminize. (64)
bituminite (Explosif) (*n.f.*), bituminite. (65)
bivalence (Chim.) (*n.f.*), bivalence. (66)
bivalent, -e (*adj.*), bivalent. (67)
bizaiguë (*n.f.*). Same as **bisaiguë.**
black-band (fer carbonaté lithoïde) (Minéral) (*n.m.*), blackband; black-band; black-band ore. (68)
blaireau (de mouleur) (*n.m.*), swab; swab brush. (69)
blaireau plat pour épousseter les plaques (Photogr.) (*m.*), flat plate-dusting brush; flat plate-duster. (70)
blanc, -che (*adj.*), white. (71)
blanc (couleur) (*n.m.*), white. (72)
blanc (pers.) (*n.m.*), white; white man. (73)
blanc (pièce non-travaillée) (*n.m.*), blank: (74)
    vis en blanc (*f.*), screw-blank. (75)
blanc (à) (marche des machines, des machines-outils) (opp. à **en charge** *ou* **chargé, -e**), light; on no load; empty; idle; non-cutting: (76)
    machine marchant à blanc (*f.*), machine running light; machine running on no load; empty machine. (77)
    course à blanc (*f.*), idle stroke; non-cutting stroke. (78)
    Cf. charge (en) *ou* chargé, -e in vocabulary.
blanc d'argent *ou* **blanc de céruse** (*m.*), white lead; ceruse. (79)
blanc de chaux (*m.*), whitewash; limewash. (80)
blanc de plomb (*m.*), white lead; ceruse. (81)
blanc éblouissant (coloration du fer) (*m.*), dazzling white. (82)
blanc-soudant (*adj. invar.*), welding; sparkling: (83)
    chaude blanc-soudant (*f.*), welding-heat; sparkling heat. (84)
blanc-soudant (*n.m.*), welding-heat; sparkling heat. (85)
blanchâtre (*adj.*), whitish; whity. (86)
blanche-bleue [blanches-bleues *pl.*] (*n.f.*), a kind of bluish white slate. (87)
blancheur (*n.f.*), whiteness. (88)
blanchiment (*n.m.*) *ou* **blanchiment à la chaux,** whitewashing; limewashing. (89)
blanchir (*v.t.*), to whiten: (90)
    le soufre blanchit la fonte, sulphur whitens pig iron. (91)
blanchir (*v.i.*), to whiten; to become white: (92)
    l'or blanchit dès qu'il est touché par le mercure (*m.*), gold whitens as soon as it comes in contact with mercury. (93)
blanchir (*v.t.*) *ou* **blanchir à la chaux,** to white-wash; to limewash: (94)

blanchir un mur, un plafond, to whitewash, to limewash, a wall, a ceiling. (1 ou 2)

blanchir le bois, to clean up wood. (3)

blanchir une pièce de fonderie à la meule en émeri, to clean up a casting on the emery-wheel. (4)

blanchir (se) (*v.r.*), to whiten. (5)

blauspath (Minéral) (*n.m.*), blue-spar. (6)

bleispath (Minéral) (*n.m.*), lead-spar; cerusite; cerussite; ceruse. (7)

bleivitriol (Minéral) (*n.m.*), lead-vitriol; angle-site. (8)

blende (Minéral) (*n.f.*), blende; zinc blende; false galena; pseudogalena; mock lead; mock ore; blackjack; rosinjack; jack; lead marcasite; sphalerite. (9)

blendeux, -euse (*adj.*), blendous; blendy: (10) plomb blendeux (*m.*), blendous lead. (11)

blesser (*v.t.*), to injure: (12) effondrements qui peuvent se produire inopiné-ment et blesser ou tuer des ouvriers (*m.pl.*), falls which may occur unexpectedly and injure or kill workmen. (13)

blessures reçues par les ouvriers (*f.pl.*), injuries to workmen. (14)

bleu, -e (*adj.*), blue. (15)

bleu (couleur) (*n.m.*), blue. (16)

bleu (épreuve en traits blancs sur fond bleu) (*n.m.*), blue-print. (17)

bleuâtre (*adj.*), bluish. (18)

bleui, -e (*adj.*), blued: (19) acier bleui (*m.*), blued steel. (20)

blindage (Trav. publ.) (*n.m.*), sheeting. (21)

blindage (des galeries de mines) (*n.m.*), steel timbering. (22)

blinder (Trav. publ.) (*v.t.*), to sheet: (23) blinder une tranchée, to sheet a trench. (24)

blinder un puits de mine (avec un blindage métallique), to line a mine-shaft with metal. (25)

blinder une galerie de mine (avec un blindage en acier), to steel-timber a mine-level. (26)

bloc (masse) (*n.m.*), block; mass; lump; clump; piece: (27)
bloc de béton, concrete block. (28)
bloc de marbre, block of marble. (29)
bloc de rocher, piece, clump, of rock. (30)
bloc de sel, lump, clump, of salt. (31)

bloc (à), chock-a-block; block and block; home; right home; hard up; hard on: (32)
'hisser à bloc, to hoist chock-a-block; to hoist block and block. (33)
serrer une vis à bloc, to drive a screw right home; to tighten a screw hard up. (34)
serrer les freins à bloc, to put the brakes hard on. (35)

bloc charrié ou bloc de **charriage** (*m.*) (*s'emploie surtout au pluriel*) (Géol.), floating reef. (36)

bloc continental (Géogr. phys.) (*m.*), continental block. (37)

bloc d'enclume (*m.*), anvil-block. (38)

bloc de battage (tubage de puits à pétrole) (*m.*), drive-block; driving-block. (39)

bloc de fondation du mortier (bocard) (*m.*), mortar-block (ore-stamp). (40)

bloc de tête de tubage (tubage de puits à pétrole) (*m.*), drive-head. (41)

bloc erratique (Géol.) (*m.*), erratic block; erratic boulder; erratic; boulder; bowlder. (42)

bloc-film [blocs-films *pl.*] (Photogr.) (*n.m.*), film-pack. (43)

bloc perché (bloc de rocher ou gros galet déposé par un glacier dans une position isolée) (*m.*), perched block; perched rock. (44)

bloc perché (pyramide coiffée) (Géol.) (*m.*), erosion column; chimney-rock. (45)

blocage (débris de pierres, etc.) (*n.m.*), rubble; rubble-stone; filling. (46)

blocage (Maçonn.) (*n.m.*), rubblework. (47)

blocage (assujétissement) (*n.m.*), clamping; locking. (48)

blocage (organes de fixation) (*n.m.*), clamping mechanism; locking mechanism: (49)
un blocage énergique permet de fixer la tourelle en place, the turret is fixed in place by a strong locking mechanism. (50)

blocageux, -euse (Constr.) (*adj.*), rubbly. (51)

blocaille (*n.f.*), rubble; rubble-stone; ballast. (52)

blocailleux, -euse (*adj.*), rubbly. (53)

blochet (pièce de bois reliant une jambe de force à la sablière d'un comble) (*n.m.*), tie; tie-brace (between a strut and the plate of a roof). (54)

blochet (pièce de bois d'un assez fort équarrissage, remplacent l'entrait dans certains combles) (*n.m.*), tie-beam. (55)

block-system (Ch. de f.), (*n.m.*), block system; block-signal system. (56)

block-system absolu (*m.*), absolute block-signal system. (57)

block-system permissif (*m.*), permissive block-signal system. (58)

blœdite ou **bloedite** (Minéral) (*n.f.*), bloedite; blödite. (59)

bloom (Métall.) (*n.m.*), bloom. (60)

bloomerie (Métall.) (*n.f.*), bloomery. (61)

blooming (Métall.) (*n.m.*), blooming. (62)

bloquer (assujétir) (*v.t.*), to clamp; to lock: (63)
bloquer la tourelle d'un tour, to clamp, to lock, the turret of a lathe. (64)

bloquer la section (Ch. de f.), to close the section. (65)

bloquer les freins, to put the brakes hard on; to skid the wheels. (66)

bloquer (se) (*v.r.*), to be clamped; to lock. (67)

blue-billy (oxyde de fer) (*n.m.*), blue-billy. (68)

blue-lias (Géol.) (*n.m.*), blue lias. (69)

blutage (*n.m.*), bolting; sifting. (70)

bluter (*v.t.*), to bolt; to sift: (71) bluter du sable, to bolt sand. (72)

bobbinite (Explosif) (*n.f.*), bobbinite. (73)

bobéchon (*n.m.*), workman's candlestick. (74)

bobierrite (Minéral) (*n.f.*), bobierrite. (75)

bobinage (*n.m.*), coiling; winding; spooling; reeling. (76)

bobine (*n.f.*), reel; spool: (77)
bobine de pellicules (Photogr.), spool of films (78)

bobine (Élec.) (*n.f.*), coil; bobbin. (79)

bobine d'extraction ou simplement **bobine** (*n.f.*) (pour câble plat) (Mines), hoisting-reel, winding-reel, reel, (for flat rope). (80)

bobine d'induction (Élec.) (*f.*), induction-coil. (81)

bobine de dérivation (Élec.) (*f.*), shunt coil. (82)

bobine de réactance ou **bobine de self** (Élec.) (*f.*), choking-coil; choke-coil; reactance-coil; reactive coil; reactor; impedance-coil; economy-coil. (83)

bobine de résistance (Élec.) (*f.*), resistance-coil. (84)

bobine de Ruhmkorff (Élec.) (*f.*), Ruhmkorff coil. (1)

bobine de tirerie (Tréfilerie) (*f.*), drawing-block ; wiredrawing block. (2)

bobine folle (*f.*), loose reel. (3)

bobine inductrice (Élec.) (*f.*), induction-coil. (4)

bobine primaire (Élec.) (*f.*), primary coil. (5)

bobine secondaire (Élec.) (*f.*), secondary coil. (6)

bobine suceuse (Élec.) (*f.*), sucking-coil. (7)

bobiner (*v.t.*), to coil ; to wind ; to spool ; to reel : (8)

bobiner des fils, to coil wires. (9)

bobiner des fils autour du noyau d'un induit, to wind wires round the core of an armature. (10)

bocage (*n.m.*) *ou* **bocages** (*n.m.pl.*) (Fonderie), scrap. (11)

bocage de bâtiment (*m.*), heavy scrap. (12)

bocage de fonte (*m.*), cast scrap ; foundry-scrap. (13)

bocage de poterie (*m.*), hardware scrap ; pot-scrap ; muck-scrap. (14)

bocage mécanique (*m.*), machine-scrap. (15)

bocal (Verrerie de laboratoire) (*n.m.*), bottle, very wide mouth ; bottle, extra wide mouth. (16)

bocard (*n.m.*), stamp ; stamp-mill ; stamping-mill ; mill. (17) See also **moulin.**

bocard à chute libre (*m.*), free-falling stamp-mill ; gravity-stamp mill ; gravitation-stamp mill. (18)

bocard à eau (*m.*), wet mill ; wet stamping-mill. (19)

bocard à grande vitesse (*m.*), high-speed stamp-mill. (20)

bocard à mineral (*m.*), ore-stamp. (21)

bocard à — pilons (*m.*), — -stamp mill ; — -head battery : (22)

bocard à 5 pilons, 5-stamp mill ; five-head battery. (23)

bocard à sec (*m.*), dry mill ; dry stamping-mill. (24)

bocard à simple pilon (*m.*), single-stamp mill. (25)

bocard australien (*m.*), Australian stamp. (26)

bocard californien (*m.*), Californian stamp. (27)

bocard chinois (*m.*), Chinese stamp. (28)

bocard pneumatique (*m.*), pneumatic stamp. (29)

bocardage (*n.m.*), milling ; stamping ; stamp-milling : (30)

bocardage du minerai, milling, stamping, ore. (31)

bocardage à l'eau, wet stamping. (32)

bocardage à sec, dry stamping. (33)

bocarder (*v.t.*), to mill ; to stamp : (34)

bocarder du minerai, to mill, to stamp, ore. (35)

bocardeur (pers.) (*n.m.*), millman ; batteryman. (36)

bocfil (*n.m.*), fret-saw ; fret-saw frame. (37)

bog-manganèse (Minéral) (*n.m.*), bog-manganese ; bog-ore ; bog-mine ; bog-mine ore. (38)

boghead (*n.m.*), bog-head ; boghead-coal. (39)

bogie (*n.m.*) *ou* **bogie de locomotive**, bogie ; bogey ; truck ; engine-truck ; locomotive-truck ; four-wheeled truck. (40)

bogie à menottes (*m.*), swing-motion truck. (41)

bogie pivotant (*m.*), swivelling bogie ; swivelling truck. (42)

bois (*n.m.*), wood ; timber ; lumber : (43)

un morceau de bois, a piece of wood. (44)

la conservation des bois, the preservation of timber. (45)

un bois mal assujetti peut céder, a badly fixed timber may give way. (46)

chantier de bois (*m.*), timber-yard ; lumber-yard. (47)

bois (lieu planté d'arbres) (*n.m.*), wood. (48)

bois (fût d'un rabot) (*n.m.*), stock (of a plane). (49)

bois à façon (*m.*), timber supplied to a contractor (the value being deducted from his contract price). (50)

bois à limer (*m.*), vice-cap ; wood vice-clamp. (51)

bois à œuvre (*m.*). Same as **bois d'œuvre.**

bois agatisé (*m.*), agatized wood. (52)

bois blancs (*m.pl.*), the class of woods consisting of the following : poplar, alder, birch, lime, acacia, sycamore, willow, etc. (53)

bois cani (*m.*), druxy timber ; druxey timber. (54)

bois carié (*m.*), rotten timber. (55)

bois carré (*m.*), square timber. (56)

bois chablis (*m.pl.*), windfalls ; wind-blown timber. (57)

bois combustible (*m.*), wood fuel ; fire-wood. (58)

bois compact (*m.*), close-grained wood. (59)

bois compensé (*m.*), made-up wood. (60)

bois contreplaqué *ou* **bois de contreplacage** *ou* **bois croisé et contreplaqué** *ou* **bois composite** (*m.*), ply-wood. (61)

bois d'ébène (*m.*), ebony. (62)

bois d'échantillon (*m.*), dimension-timber. (63)

bois d'étais (*m.*), propwood. (64)

bois d'œuvre (*m.*), timber suitable for conversion into market forms (as distinguished from *fire-wood*). (65)

bois de charpente (*m.*), timber ; lumber ; wood suitable for use in buildings, carpentry, etc. (66)

bois de chauffage (*m.*), fire-wood ; wood fuel. (67)

bois de chêne (*m.*), oak timber. (68)

bois de construction (*m.*), builders' timber ; constructional timber ; wood suitable for use in buildings, etc. (69)

bois de contreplacage (*m.*). See **bois contre-plaqué.**

bois de gaïac (*m.*), lignum-vitæ. (70)

bois de grume (*m.pl.*), logs ; wood in the log. (71)

bois de l'année (*m.*), this year's timber. (72)

bois de mine (*m.*), mine-timber ; mining-timber ; timber for mining purposes ; pit-wood. (73)

bois de montagne (Minéral) (*m.*), mountain-wood. (74)

bois de refend (*m.*), lathwood. (75)

bois de sciage (*m.*), saw-timber. (76)

bois de soutènement (*m.*), propwood. (77)

bois de stère (*m.*), fire-wood sold by the stère or cubic metre (analogous to the English term **cord-wood**). (78)

bois débité (*m.*), broken-down timber ; sawn timber. (79)

bois déjeté *ou* **bois déversé** (*m.*), warped timber. (80)

bois domaniaux *ou* **bois de l'État** (*m.pl.*), State timber ; Government timber ; Crown timber (England). (81)

bois dur (*m.*), hardwood ; hard-wood. (82)

bois éclatable (*m.*), splintery timber. (83)

bois en état *ou* **bois en étant** (*m.*), standing timber. (84)

bois en grume (*m.pl.*), logs ; wood in the log. (85)

bois équarris (*m.pl.*), squared timber ; baulks ; balks. (1)

bois flache *ou* **bois flacheux** (*m.*), waney timber. (2)

bois flotté (*m.*), floated timber. (3)

bois forestier (*m.*), forest timber. (4)

bois fossile (*m.*), fossil wood. (5)

bois gauchi (*m.*), warped timber. (6)

bois gisant (*m.*), windfall ; fallen tree. (7)

bois incombustible (*m.*), fire-proof wood ; unin--flammable wood ; incombustible wood. (8)

bois léger (*m.*), light wood. (9)

bois lourd (*m.*), heavy wood. (10)

bois mort (*m.*), dead wood. (11)

bois mouliné (*m.*), worm-eaten timber. (12)

bois noueux (*m.*), knot-wood ; knotty timber. (13)

bois opalisé (*m.*), opalized wood. (14)

bois parfait (*m.*), heart-wood ; duramen. (15)

bois pelard (*m.*), barked timber (bark removed). (16)

bois pétrifié (*m.*), petrified wood ; wood-stone. (17)

bois plaqué triplé (*m.*), 3-ply wood ; three-ply wood. (18)

bois plein (*m.*), close-grained wood. (19)

bois pouilleux (*m.*), foxed timber. (20)

bois pourri (*m.*), rotten timber. (21)

bois résineux (*m.pl.*), resinous woods (fir, pine, larch, cedar, cypress, and yew). (22)

bois silicifié (*m.*), silicified wood. (23)

bois taillis (*m.*), underwood ; copsewood. (24)

bois tendre (*m.*), soft wood ; softwood. (25)

bois vermoulu (*m.*), worm-eaten timber. (26)

bois vert (*m.*), green wood. (27)

boisage (*n.m.*), timbering. (28)

boisage (de puits de mine) (*n.m.*), timbering, cribbing, (of mine-shaft). (29)

boisage armé (galerie de mine) (*m.*), herring-bone timbering ; double timbering ; reinforced timbering. (30)

boisage complet (Mines) (*m.*), full timbering. (31)

boisage de faîte (galerie de mine) (*m.*), roof timbering. (32)

boisage de tailles (Mines) (*m.*), room-timbering. (33)

boisage de voies maîtresses (Mines) (*m.*), entry-timbering. (34)

boisage du puits par la méthode ascendante (*m.*), shaft-timbering ascending method. (35)

boisage du puits par la méthode descendante (*m.*), shaft-timbering descending method. (36)

boisage en acier (blindage) (Mines) (*m.*), steel timbering. (37)

boisage jointif *ou* **boisage serré** (Mines) (*m.*), close timbering ; close cribbing. (38)

boisage parallélépipédique (Mines) (*m.*), square sets ; square timbering ; Nevada timbering. (39)

boisage sans sole (galerie de mine) (*m.*), timber--ing without sills. (40)

boisage suivant de très près l'abatage (Mines) (*m.*), timbering carried close to excavation. (41)

boiser (garnir de bois) (*v.t.*), to timber. (42)

boiser (un puits de mine) (*v.t.*), to timber, to crib, (a mine-shaft). (43)

boiser (garnir d'arbres) (*v.t.*), to wood ; to timber : (44)

pays bien boisé (*m.*), well-wooded country. (45)

boiser (se) (*v.r.*), to be timbered : (46)

les galeries de mine se boisent au moyen de cadres, mine-levels are timbered with frames (*or* sets). (47)

boiserie (*n.f.*), woodwork. (48)

boiseur (Mines) (pers.) (*n.m.*), timberman. (49)

boisseau (d'un robinet) (*n.m.*), dome (of a cock *or* tap). (50)

boisseau (mesure anglaise de capacité) (*n.m.*), bushel = 3·637 decalitres. (51)

boitard (Mach.) (*n.m.*), vertical floor-bearing (bearing for vertical shaft passing through ceiling and floor). (52)

boîte (*n.f.*), box. (53)

boîte (d'une serrure) (*n.f.*), case (of a lock). (54)

boîte à air (*f.*), air-box. (55)

boîte à amalgame fermant à clef (*f.*), amalgam-safe. (56)

boîte à bourrage (Mach.) (*f.*), stuffing-box. (57)

boîte à concentrés (Préparation mécanique des minerais) (*f.*), concentrates-box. (58)

boîte à débourbage (sluices) (*f.*), mud-box. (59)

boîte à clapet (*f.*), clack-box ; valve-chest ; valve-box ; valve-chamber ; valve-case ; valve-casing. (60)

boîte à engrenages (*f.*), gear-box. (61)

boîte à étoupe (Mach.) (*f.*), stuffing-box. (62)

boîte à feu (locomotive) (*f.*), fire-box ; outer fire-box. (63)

boîte à feu étroite (plongeant entre les longerons) (*f.*), narrow fire-box (extending down between the frames). (64)

boîte à feu large (débordant au-dessus des roues) (*f.*), wide fire-box (extending out over the wheels). (65)

boîte à forets (porte-foret à archet) (*f.*), bow-drill ; piercer. (66)

boîte à forets (douille pour forets) (*f.*), drill-socket. (67)

boîte à fumée (locomotive) (*f.*), smoke-box ; smoke-arch ; arch ; front end. (68)

boîte à fumée allongée (*f.*), extended smoke-box. (69)

boîte à galets (*f.*), roller-bearing box ; friction-roller box. (70)

boîte à graisse *ou* **boîte à huile** (*f.*). Same as **boîte d'essieu.**

boîte à mousse (Mines) (*f.*), moss-box. (71)

boîte à noyau *ou* **boîte à noyaux** (Fonderie) (*f.*), core-box ; core-stock. (72)

boîte à noyaux en deux parties (*f.*), split core-box. (73)

boîte à onglet (*f.*), mitre-box ; mitre-board. (74)

boîte à outils (*f.*), tool-chest ; tool-box. (75)

boîte à piles (Élec.) (*f.*), battery-box. (76)

boîte à plongeur (*f.*), expansion-box ; slip-box. (77)

boîte à réparations (*f.*), repair-outfit. (78)

boîte à sable (locomotive) (*f.*), sand-box ; sander. (79)

boîte à slimes (Préparation mécanique des minerais) (*f.*), slimes-box. (80)

boîte à soupape (*f.*), valve-chest ; valve-box ; valve-case ; valve-casing ; valve-chamber. (81)

boîte à vapeur *ou* **boîte à tiroir** (*f.*), steam-chest ; valve-chest ; steam-box ; valve-box ; steam-case ; valve-case ; valve-casing ; steam-chamber ; valve-chamber ; slide-box. (82)

boîte à vent (d'un cubilot, d'un haut fourneau) (*f.*), wind-box, air-box, (of a cupola, of a blast-furnace). (83 *ou* 84) boîte à zinc (*f.*), zinc-box ; zinc precipitation box. (1)

boîte alimentaire *ou* boîte d'alimentation (*f.*), feed-box. (2)

boîte-classeur [boîtes-classeurs *pl.*] (*n.f.*) *ou* boîte pour conserver les clichés *ou* boîte à plaques, plate-box ; box for negatives. (3)

boîte d'artère (Élec.) (*f.*), feeder-box. (4)

boîte d'échappement (*f.*), silencer ; muffler. (5)

boîte d'engrenage (*f.*), gear-box. (6)

boîte d'essieu [boîtes des essieux *pl.*] (Ch. de f.) (*f.*), axle-box ; journal-box ; grease-box ; oil-box ; pedestal-box ; housing-box ; box. (7)

boîte d'essieu moteur (locomotive) (*f.*), driving-box ; driving axle-box. (8)

boîte d'onglet (*f.*), mitre-box ; mitre-board. (9)

boîte de changement de vitesse (*f.*), change-speed gear-box ; speed-box ; gear-box. (10)

boîte de changement de vitesse à monopoulie (*f.*), change-speed gear-box with constant-speed pulley. (11)

boîte de dilatation (*f.*), expansion-box ; slip-box. (12)

boîte de distribution de vapeur (*f.*).   Same as boîte à vapeur.

boîte de filières, garnie de coussinets (*f.*), case of stocks and dies. (13)

boîte de graissage (*f.*). Same as boîte d'essieu.

boîte de joint (*f.*), joint-box. (14)

boîte de jonction (*f.*), junction-box. (15)

boîte de manœuvre des aiguilles (Ch. de f.) (*f.*), lever-box ; lever-box for switches. (16)

boîte de précipitation (*f.*), precipitation-box. (17)

boîte de pression (*f.*), pressure-box. (18)

boîte de queue (sluices) (*f.*), tail-box ; end-box. (19)

boîte de recouvrement de tringles (Ch. de f.) (*f.*), switch-rod covering. (20)

boîte de regard (*f.*), inspection-fitting. (21)

boîte de résistance (Élec.) (*f.*), resistance-box. (22)

boîte de roue (*f.*), axle-box (of a wagon or carriage). (23)

boîte de sluice (*f.*), sluice-box. (24)

boîte de sonde (*f.*), pole-box. (25)

boîte de sûreté et vis pour laminoir [boîte (*n.f.*) ; vis (*n.f.*)], box and pin for rolling-mill. (26)

boîte de tête (sluices) (*f.*), head-box. (27)

boîte de vitesses (*f.*). Same as boîte de change-ment de vitesse.

boîte des coussinets (Mach.) (*f.*), journal-box ; pedestal-box. (28)

boîte en bois, avec séparations (*f.*), wooden box, with divisions. (29)

boîte en carton (*f.*), cardboard box. (30)

boîte et vis pour étau [boîte (*n.f.*) ; vis (*nf.*)], box and pin for vice. (31)

boîte et vis pour tiges de sonde, box and pin for drill-rods ; pole pin and box. (32)

boîte fermant à clef (*f.*), lock-box. (33)

boîte inférieure de décharge (sluices) (*f.*), drop-box. (34)

boîte métallique pour tenir les explosifs (*f.*), canister for keeping explosives. (35)

boîte patent (*f.*), patent axle-box. (36)

boîte pointue (Préparation mécanique des minerais) (*f.*), pointed box ; funnel-box ; V vat ; spitzkasten. (37)

boîte réfrigérante (de four métallurgique) (*f.*), water-box, water-bosh, (of metallurgical furnace). (38)

boîte réfrigérante (de tuyère) (*f.*), cooler, jumbo, (of tuyère). (39)

bol (argile) (*n.m.*), bole. (40)

bolaire (*adj.*), bolar ; clayey. (41)

boléite (Minéral) (*n.f.*), boleite. (42)

bolomètre (*n.m.*), bolometer. (43)

bombe (*n.f.*) *ou* bombe volcanique (Géol.), bomb ; volcanic bomb. (44)

bombe à torsion (*f.*), lava-ball. (45)

bombe calorimétrique (Phys.) (*f.*), calorimetric bomb ; bomb. (46)

bombe-chandelle [bombes-chandelles *pl.*] (Phys.) (*n.f.*), Prince Rupert's drop ; Rupert's drop ; Rupert's ball ; Rupert's tear ; detonating bulb. (47)

bombe craquelée *ou* bombe en croûte de pain (Géol.) (*f.*), bread-crust bomb. (48)

bombe d'olivine (Géol.) (*f.*), olivine bomb. (49)

bombement (action) (*n.m.*), bulging ; swelling. (50)

bombement (convexité ; renflement) (*n.m.*), bulge ; swell ; camber ; crown ; belly. (51)

bombement (d'un mur) (*n.m.*), bulging, bulge, belly, (of a wall). (52)

bombement (d'une route) (*n.m.*), camber, crown, (of a road). (53)

bombement (d'une poulie bombée) (*n.m.*), crown, crowning, (of a crown-faced pulley). (54)

bomber (*v.t.*), to bulge ; to swell ; to swell out ; to camber : (55)
les chaussées sont bombées pour faciliter l'écoulement des eaux, roadways are cambered to facilitate the drainage of the water. (56)

bomber (*v.i.*), to bulge ; to swell ; to camber : (57)
mur qui bombe (*m.*), wall which bulges. (58)
un tonneau bombe au centre, a cask swells in the middle. (59)

bomber (se) (*v.r.*), to bulge ; to swell ; to camber. (60)

bombonne à gaz (*f.*), gas-bottle. (61)

bon, bonne (*adj.*), good : (62)
bon ajustage (*m.*), good fit. (63)
bon conducteur (*m.*), good conductor : (64)
le cuivre est bon conducteur de l'électricité, copper is a good conductor of electricity. (65)
bon mur (Mines) (*m.*), good bottom. (66)
bon toit (Mines) (*m.*), good roof. (67)
bonnes espérances (*f.pl.*), good prospects. (68)

bonanza (*n.f.*), bonanza ; rich strike of ore. (69)

bonbonne à gaz (*f.*), gas-bottle. (70)

bonde (*n.f.*), plug : (71)
la bonde d'un réservoir, the plug of a reservoir. (72)

bonde (bouchon de tonneau) (*n.f.*), bung. (73)

bonde (trou dans un tonneau, pour y verser le liquide) (*n.f.*), bung-hole. (74)

bonde (d'un étang) (*n.f.*), shut off, shut, (of a pond). (75)

bondieu (*n.m.*), pit-sawyers' wedge. (76)

bondon (de tonneau) (*n.m.*), bung. (77)

bone-bed (Géol.) (*n.m.*), bone-bed. (78)

bonnet carré (Sondage) (*m.*), four-wing star bit. (79)

bonnette à portrait (Photogr.) (*f.*), portrait attachment. (80)

boort (égrisé) (*n.m.*), bort (for polishing diamonds). (81)

boracite (Minéral) (*n.f.*), boracite. (82)

borate (Chim.) (*n.m.*), borate : (83)
borate de sodium *ou* borate de soude, borate of sodium ; sodium borate. (84)

borax (*n.m.*), borax. (1)

bord (partie qui termine le contour) (*n.m.*), border ; edge ; margin ; rim. (2)

bord (d'une table) (*n.m.*), edge (of a table). (3)

bord (d'un glacier) (*n.m.*), margin (of a glacier). (4)

bord (d'un tiroir, d'une lumière de cylindre) (*n.m.*), edge (of a slide-valve, of a cylinder-port) : (5 *ou* 6).

    bord extérieur, steam-edge ; outer edge. (7)

    bord intérieur, exhaust-edge ; inner edge. (8)

bord (d'un fleuve) (*n.m.*), bank (of a river). (9)

bord de l'eau (*m.*), water's edge ; riverside. (10)

bord de la mer (*s'emploie surtout au pluriel*) (*m.*), seashore ; seacoast ; shore of the sea : (11)

    les bords de la mer Baltique, the shores of the Baltic ; the Baltic coast. (12)

bord en biseau (*m.*), bevelled edge. (13)

bordage (façonnage de brides) (Méc.) (*n.m.*), flanging. (14)

bordage (Hydraul.) (*n.m.*), cleading : (15)

    bordage d'une porte d'écluse, d'un batardeau, cleading of a lock-gate, of a coffer-dam. (16 *ou* 17)

border (*v.t.*), to border ; to edge ; to fringe : (18)

    le continent asiatique est bordé au nord par une immense plate-forme sous-marine qui s'étend sur une largeur atteignant par places 7°, the Asiatic continent is bordered on the north by an immense submarine shelf which stretches out over a breadth reaching 7° in places. (19)

border (faire une bride à, des brides à) (*v.t.*), to flange : (20)

    border une tôle de chaudière, to flange a boiler-plate. (21)

bordure (*n.f.*), border ; edging ; fringe ; rim : (22)

    bordure d'un gisement (Géol.), fringe of a seam. (23)

bordure (de trottoir) (*n.f.*), kerb ; kerb-stone ; curb ; curbstone. (24)

bordures du placer (*f.pl.*), rim-gravels. (25)

bore (Chim.) (*n.m.*), boron. (26)

boréal, -e, -aux (*adj.*), boreal ; north ; northern. (27)

borinage (*n.m.*), a coal-mining district in Belgium. (28)

borique (*adj.*), boric : (29)

    acide borique (*m.*), boric acid. (30)

bornage (*n.m.*), bounding ; limiting ; setting landmarks ; setting boundaries ; pegging ; pegging off ; pegging out ; staking ; staking off ; staking out : (31)

    bornage illégal de concessions, illegal pegging of claims. (32)

borne (*n.f.*), boundary ; limit ; landmark ; boundary-mark ; boundary-stone ; boundary-post. (33)

borne (Élec.) (*n.f.*), terminal : (34)

    borne d'un élément de pile, terminal of a battery-cell. (35)

borne à vis (Élec.) (*f.*), screw-terminal ; binding-post ; binding-screw. (36)

borne kilométrique (*f.*), kilometric stone (the metric equivalent of a mile-stone). (37)

borne légale (Mines) (*f.*), legal post. (38)

borne-signal [bornes-signaux *pl.*] (Topogr.) (*n.f.*), monument : (39)

    placer des bornes-signaux sur la méridienne, to place monuments on the meridian. (40)

borner (*v.t.*), to bound ; to limit ; to peg ; to peg off ; to peg out ; to stake ; to stake off ; to stake out : (41)

    concession bornée au nord, au sud, à l'est, à l'ouest, par une rivière (*f.*), claim bounded on the north, on the south, on the east, on the west, by a river. (42 *ou* 43 *ou* 44 *ou* 45)

    borner une concession, to peg out, to stake off, a claim. (46)

borner (se) au strict nécessaire, to limit oneself to the strict necessary. (47)

bornine [$Bi^6Te^6S^3$] (Minéral) (*n.f.*), tetradymite ; telluric bismuth. (48)

bornite *ou* bornine [Cu^3FeS^3] (Minéral) (*n.f.*), bornite ; erubescite ; horse-flesh ore ; peacock ore ; peacock copper ; variegated copper ore ; variegated pyrites ; purple copper. (49)

borosilicate (*n.m.*), borosilicate ; silicoborate. (50)

bort (égrisé) (*n.m.*), bort (for polishing diamonds). (51)

borure (Chim.) (*n.m.*), boride. (52)

bossage (*n.m.*), boss : (53)

    bossage d'un corps de manivelle, boss of a crank-web. (54)

bossage de l'excentrique (*m.*), eccentric-disc ; eccentric-sheave. (55)

bosse (*n.f.*), boss. (56)

bosseyage (battage des aiguilles infernales) (Mines) (*n.m.*), plug and feathering. (57)

bosseyer (*v.t.*), to plug and feather. (58)

bosseyeuse (*n.f.*), plug-and-feathering machine ; bosseyeuse. (59)

botryogène (Minéral) (*n.m.*), botryogen. (60)

botryoïde (*adj.*), botryoidal. (61)

botte de feuillards (*f.*), bundle of strip (*or* of hoop-iron). (62)

botte de fil de fer (*f.*), coil of iron wire. (63)

bouc (poulie à pignon) (*n.m.*), sprocket-wheel ; rag-wheel. (64)

bouchage (action) (*n.m.*), stopping ; stopping up ; stoppage ; obstruction ; plugging ; corking ; sealing : (65)

    le bouchage des bouteilles, stopping (*or* cork-ing) bottles. (66)

    le bouchage d'un puits de mine (en cas d' incendie), sealing a mine-shaft (in case of fire). (67)

bouchage (ce qui sert à boucher) (*n.m.*), stopping : (68)

    un bouchage hermétique, an air-tight stopping. (69)

bouchage pour arrêt momentané (haut fourneau) (*m.*), damping down (blast-furnace). (70)

bouche (orifice) (*n.f.*), mouth : (71)

    la bouche d'un fleuve, d'une caverne, d'un four, the mouth of a river, of a cave, of a furnace. (72 *ou* 73 *ou* 74)

bouche d'eau *ou simplement* bouche (*n.f.*), (hy--drante), water-hydrant ; water-plug ; hy--drant ; plug. (75)

bouche d'eau (Ch. de f.) (*f.*), water-pillar ; water-crane. (76)

bouche d'entrée de l'air (*f.*), air-inlet nozzle (77)

bouche d'incendie (*f.*), fire-hydrant ; fire-plug. (78)

bouche de sortie de l'air (*f.*), air-discharge nozzle. (79)

bouche du puits (*f.*), mouth of the shaft ; pit-mouth ; pit-head. (80)

bouche-trou [bouche-trous *pl.*] (*n.m.* & *adj.*), stop-gap. (1)
bouché (-e) à l'émeri *ou* bouché (-e) émeri, stoppered : (2)
flacon bouché à l'émeri (*m.*), stoppered bottle. (3)
fiole bouchée émeri (*f.*), stoppered flask. (4)
boucher (*v.t.*), to stop up ; to stop ; to obstruct ; to plug ; to plug up ; to cork ; to seal ; to seal up : (5)
boucher hermétiquement un vase, to seal up a retort hermetically. (6)
boucher le trou de coulée (Fonderie), to plug up, to stop up, to bott up, the tapping-hole ; to boat up the furnace. (7)
boucher les puits (puits de mine—en cas d'incendie), to seal the shafts. (8)
boucher un passage, to stop up, to obstruct, a passage. (9)
boucher un trou (*ou* une ouverture), to stop up a hole (*or* a gap) ; to close an opening. (10)
boucher un trou dans un puits, to plug a hole in a well. (11)
boucher une bouteille, to stop, to cork, a bottle. (12)
boucher une voie d'eau, to stop a leak. (13)
boucher (se) (*v.r.*), to be stopped up ; to choke : (14)
les tuyaux de sablière se bouchent facilement (*m.pl.*), sand-pipes choke easily. (15)
bouchon (*n.m.*), stopper ; cork ; plug ; cap. (16)
bouchon (tampon d'obturation de trou de coulée) (Fonderie) (*n.m.*), bott ; plug ; boat. (17)
bouchon (calibre à bouchon) (Méc.) (*n.m.*), internal cylindrical gauge ; plug-gauge ; plug : (18)
Often occurs in the expression **bagues et bouchons,** which is usually rendered in the reverse order in English, i.e., **internal and external cylindrical gauges ; plug and ring-gauges ; plugs and rings.** (19)
bouchon à l'émeri (verre) (*m.*), ground stopper. (20)
bouchon à tolérance (calibre) (*m.*), limit internal gauge. (21)
bouchon d'objectif (Photogr.) (*m.*), lens-cap. (22)
bouchon de bouteille (*m.*), cork ; bottle-stopper. (23)
bouchon de lavage (*m.*), washout-plug. (24)
bouchon de remplissage (*m.*), filling-plug. (25)
bouchon de tonneau *ou simplement* **bouchon**(*n.m.*), bung. (26)
bouchon de vidange (*m.*), emptying-plug. (27)
bouchon en bois (*m.*), wood plug. (28)
bouchon en liège (*m.*), cork. (29)
bouchon en verre (*m.*), glass stopper. (30)
bouchon fusible (chaudière) (*m.*), fusible plug ; safety-plug. (31)
boucle (anneau) (*n.f.*), buckle ; ring ; shackle. (32)
boucle (d'une courroie) (*n.f.*), buckle (of a strap). (33)
boucle (Élec.) (*n.f.*), ring ; loop : (34)
montage en boucle (*m.*), ring-connection ; loop-connection. (35)
boucle (d'une rivière) (*n.f.*), sweep (of a river). (36)
boucle (d'un cordage) (*n.f.*), loop, eye, bight, (of a rope). (37)
boucle à nœud coulant (*f.*), running loop. (38)
boucle d'amarrage (*f.*), ring-bolt. (39)
boucle de tiroir (*f.*), drawer-handle. (40)
boucle de vapeur (*f.*), steam-loop. (41)

boucler (bomber) (*v.i.*), to bulge : (42)
mur qui boucle (*m.*), wall which bulges. (43)
bouclier (Percement de tunnels) (*n.m.*), shield. (44)
bouclier (Mines) (*n.m.*), breast-boards. (45)
bouclier Greathead (*m.*), Greathead shield. (46)
boudin (*n.m.*) *ou* **boudin de roue,** flange ; wheel-flange. (47)
boudin (moulure) (Menuis.) (*n.m.*), ovolo ; Grecian ovolo. (48)
boudin (bouvet) (Menuis.) (*n.m.*), ovolo-plane. (49)
boudin à carré (Menuis.) (*m.*), quirk ovolo. (50)
boue (*n.f.*) (*s'emploie souvent au pluriel*), mud ; slime ; sludge. (51)
boue de meule (*f.*), wheel-swarf ; swarf ; grit from grindstone. (52)
boue glaciaire (Géol.) (*f.*), glacier-mud ; glacier-silt. (53)
boue salifère (*f.*), salt mud. (54)
boues (slimes ; schlamms) (Traitement des minerais) (*n.f.pl.*), slimes ; slime ; sludge. (55)
boues minérales (limon que déposent certaines eaux minérales) (*f.pl.*), mineral mud. (56)
boueux, -euse (*adj.*), muddy : (57)
eau boueuse (*f.*), muddy water. (58)
bougeoir (*n.m.*), candlestick ; candle-holder. (59)
bougie (*n.f.*), candle. (60)
bougie (Photom.) (*n.f.*), candle : (61)
la bougie anglaise (de spermaceti) brûle 120 grains, ou 7^{gr}, 8, à l'heure. Elle équi--vaut aux 0, 9 environ de la bougie française stéarique dite de l'Étoile, the English candle (of spermaceti) burns 120 grains, or 7·8 grammes, per hour. It is equivalent to about ·9 of the French stearine candle called star candle. (62)
bougie anglaise (Photom.) (*f.*), English candle ; British candle ; standard candle. (63)
bougie d'allumage (*f.*), sparking-plug ; spark-plug ; sparker. (64)
bougie de blanc de baleine *ou* **bougie de spermaceti** (*f.*), spermaceti candle. (65)
bougie de cire (*f.*), wax candle. (66)
bougie de paraffine (*f.*), paraffin candle. (67)
bougie décimale (*f.*), bougie décimale ; decimal candle ; pyr : (68)
la bougie décimale (l'unité française d'intensité lumineuse) vaut un vingtième de l'étalon Violle, the bougie décimale, or decimal candle (the French light-unit), has the value of one-twentieth of the Violle standard. (69)
See **Violle** in vocabulary.
bougie électrique *ou* **bougie Jablochkoff** (*f.*), electric candle ; Jablochkoff's candle. (70)
bougie-mètre [bougies-mètre *pl.*] (Photom.) (*n.f.*), metre-candle ; candle-metre. (71)
— **bougies** (intensité en bougies), — -candle-power : (72)
bec — bougies (*m.*), — -candle-power burner. (73)
bouillant, -e (*adj.*), boiling : (74)
de l'eau bouillante (*f.*), boiling water. (75)
bouilleurs (d'une chaudière à bouilleurs) (*n.m.pl.*), the horizontal cylindrical subsidiary shells of a French boiler, below the main shell, and connected with it by water-legs. (76)
bouillie d'argile (Fonderie) (*f.*), claywash. (77)
bouillir (*v.i.*), to boil : (78)
l'eau bout à 100° C. (*ou* 212° F.), water boils at 100° C. (*or* 212° F.). (79)
bouilloire à suif (*f.*), tallow-kettle. (80)

bouillon (bulle gazeuse) (*n.m.*), bubble. (1)

bouillon (dans une masse de métal fondu)(*n.m.*), blow-hole ; blow ; air-hole. (2)

bouillonnant, -e (*adj.*), bubbling ; boiling. (3)

bouillonnement (*n.m.*), bubbling ; bubbling over ; boiling. (4)

bouillonner (*v.i.*), to bubble ; to bubble over ; to boil ; to boil over. (5)

boulangérite (Minéral) (*n.f.*), boulangerite. (6)

boulbène (*n.f.*), an argilloarenaceous earth. (7)

boule (*n.f.*), ball : (8)

boule d'un régulateur (Mach.), ball of a governor. (9)

boule (*n.f.*) *ou* boule de puddlage (Métall.), ball ; puddle-ball ; loup ; loupe. (10)

boule de condensation (*f.*), condenser-bell ; condenser-chamber. (11)

bouleau (arbre) (*n.m.*), birch. (12)

boulets (pour moulins à boulets) (*n.m.pl.*), balls (for ball-mills) : (13)

boulets d'acier forgé, forged steel balls. (14)

bouleversement (Géol.) (*n.m.*), disruption ; convulsion. (15)

boulin (échafaudage) (*n.m.*), putlog ; putlock. (16)

boulon (*n.m.*), bolt ; pin : (17)

boulons et écrous, bolts and nuts. (18)

boulon à clavette (*m.*), cotter-bolt ; key-bolt ; forelock-bolt. (19)

boulon à croc *ou* boulon à crochet (*m.*), hook-bolt. (20)

boulon à écrou (*m.*), screw-bolt. (21)

boulon à émerillon (*m.*), swivel-bolt. (22)

boulon à ergot (*m.*), snug-head bolt ; feather-necked bolt ; lip-bolt. (23)

boulon à œil (*m.*), eye-bolt ; eye-headed bolt. (24)

boulon à oreilles (*m.*), wing-bolt ; thumb-bolt. (25)

boulon à T *ou* boulon à tête de marteau (*m.*), T bolt ; T-headed bolt. (26)

boulon à tête carrée (*m.*), square-head bolt ; square-headed bolt. (27)

boulon à tête en goutte-de-suif (*m.*), button-head bolt ; mushroom-head bolt. (28)

boulon à tête et écrou 6 pans, acier doux decolleté, fileté à moitié, *ou* fileté entière--ment, longueur en m/m tête non-com--prise —, hexagon-head bolt and hexagon nut, mild steel, machine-made, screwed half way up, *or* screwed to head, length in millimetres under head to point —. (29 *ou* 30)

boulon à tête fraisée (*m.*), countersunk bolt ; countersunk-head bolt ; countersunk-headed bolt. (31)

boulon à tête hémisphérique *ou* boulon à tête ronde (*m.*), round-head bolt ; round-headed bolt. (32)

boulon à tête noyée *ou* boulon à tête perdue (*m.*), flush bolt. (33)

boulon à tête ronde fendue (*m.*), slotted round-head bolt. (34)

boulon d'ancrage (*m.*), anchor-bolt ; anchoring-bolt. (35)

boulon d'assemblage (*m.*), assembling-bolt. (36)

boulon d'attelage (Ch. de f.) (*m.*), coupling-pin ; draw-bolt ; draw-pin ; drag-bolt. (37)

boulon d'éclisse (Ch. de f.) (*m.*), fish-bolt. (38)

boulon de chaîne (*m.*), chain-bolt. (39)

boulon de courroie (*m.*), belt-bolt. (40)

boulon de fixation (*m.*), holding-down bolt. (41)

boulon de fondation (*m.*), foundation-bolt. (42)

boulon de scellement à crans (*m.*), rag-bolt ; jag-bolt ; jagged bolt ; barb-bolt ; barbed bolt ; sprig-bolt ; bay-bolt ; bat-bolt. (43)

boulon de scellement ordinaire (*m.*), fang-bolt. (44)

boulon de suspension (locomotive ou wagon) (*m.*), spring-hanger pin ; hanger-pin. (45)

boulon libre (*m.*), through bolt ; thorough bolt ; in-and-out bolt. (46)

boulon mécanique (*m.*), machine-bolt. (47)

boulon pour assemblage de la pointe et de la branche de pointe (Ch. de f.) (*m.*), splice-bolt. (48)

boulon pour courroies (*m.*), belt-bolt. (49)

boulon prisonnier (*m.*), stud-bolt ; stud ; stand--ing-bolt. (50)

boulon taraudé (*m.*), threaded bolt ; tap-bolt. (51)

boulonnage (*n.m.*), bolting. (52)

boulonner (*v.t.*), to bolt : (53)

boulonner une machine sur son bâti, to bolt an engine on its bed-plate. (54)

boulonner une porte, to bolt a door. (55)

boulonnerie (*n.f.*), bolt and nut works. (56)

boulonnière (pour former les têtes de boulons) (*n.f.*), bolt-header. (57)

boulonnière (tarière) (*n.f.*), bolt-auger. (58)

bouniou (Mines) (*n.m.*), sump ; sink-hole ; standage. (59)

bouquet d'arbres (*m.*), clump of trees. (60)

bouquet de lampes électriques (*m.*), cluster of electric lamps. (61)

bourbe (*n.f.*), mud ; mire. (62)

bourbeux, -euse (*adj.*), muddy ; miry. (63)

bourdonneau (*n.m.*), hook (of a double-strap gate-hinge). (64)

bourdonnement des fils (Téléph.) (*m.*), buzzing of wires. (65)

bourdonner (*v.i.*), to buzz. (66)

bourdonnière (opp. à bourdonneau) (*n.f.*), strap (of a double-strap gate-hinge). (67)

bourdonnière (d'un chardonnet) (opp. à pivot) (*n.f.*), pintle (of a hanging-post). (68) (opposed to pivot,—at heel of post).

bournonite (Minéral) (*n.f.*), bournonite ; cog-wheel ore. (69)

bourrage (action) (*n.m.*), ramming ; tamping ; stemming : (70)

bourrage des fourneaux de mine, stemming holes. (71)

bourrage (matière) (*n.m.*), tamping ; ram. (72)

bourrage (d'une boîte à bourrage *ou* d'un presse-étoupe) (*n.m.*), packing, stuffing, (of a stuffing-box). (73)

bourrage (garniture du piston hydraulique d'une pompe) (*n.m.*), packing. (74)

bourrage de la cheminée (Mines) (*m.*), choking, clogging, of chute. (75)

bourrage de piston (*m.*), piston-packing. (76)

bourrage des traverses (Ch. de f.) (*m.*), packing the sleepers. (77)

bourre d'argile (*f.*), ram of clay ; clay-tamping. (78)

bourre de coton (*f.*), cotton-waste ; waste. (79)

bourre de laine (*f.*), wool-waste. (80)

bourre de sûreté (Travail aux explosifs) (*f.*), safety-ram ; safety-tamping. (81)

bourre en bois (*f.*), wooden tamping. (82)

bourre en caoutchouc (*f.*), india-rubber tamping. (83)

bourrelet (d'une roue de chemin de fer, d'un champignon de rail) (*n.m.*), flange (of a railway-wheel, of a rail-head). (84 *ou* 85)

bourrer (tasser, après avoir placé une bourre) (*v.t.*), to stem ; to tamp ; to ram ; to ram home : (1)

bourrer un trou de mine, to stem a shot-hole. (2)

bourrer une charge de mine, to ram home a charge. (3)

bourrer le piston hydraulique d'une pompe, to pack the water-piston of a pump. (4)

bourrer les traverses (Ch. de f.), to pack the sleepers. (5)

bourrer un presse-étoupe, to pack, to stuff, a stuffing-box. (6)

bourreur (pers.) (*n.m.*), tamper. (7)

bourriquet (*n.m.*), windlass. (8)

bourroir (de mineur) (*n.m.*), tamper ; tamping-bar ; tamping-rod ; tamping-iron ; stemmer ; bar. (9)

bourroir (de poseur de voie) (*n.m.*), beater ; beater-pick ; beating-pick ; packer ; tamp--ing-pick. (10)

boursault *ou* **bourseau** (de plombier) (*n.m.*), dresser ; lead dresser ; beater ; bat. (11)

boursouflement (de la sole d'une galerie de mine) (*n.m.*), heave, heaving, creep, creeping, (of the floor of a mine-level). (12)

boursoufler (se) (Mines) (*v.r.*), to heave ; to creep. (13)

bouse de vache (employée dans la fonderie) (*f.*), cow-dung. (14)

boussole (*n.f.*), compass ; magnetic compass ; dial. (15)

boussole à cadran (*f.*), compass-dial. (16)

boussole à pinnules, à charnière, en bronze, suspension, barreau à chape d'agate, en boîte acajou (*f.*), sight-compass, folding sights, in gun-metal, bar needle, agate centre, with lifter, in mahogany case. (17)

boussole d'arpenteur (*f.*), surveyor's compass ; surveyors' dial ; surveying-compass. (18)

boussole d'arpenteur montée sur genou articulé (*f.*), surveyors' compass, ball and socket mounting (Jacob's-staff). (19)

boussole d'inclinaison (*f.*), dipping compass ; inclination compass ; dip-circle ; inclinom--eter. (20)

boussole de déclinaison (*f.*), declination compass. (21)

boussole de géologue (*f.*), geological compass. (22)

boussole de mine (*f.*), mine-dial ; miners' dial ; miners' compass. (23)

boussole de poche (*f.*), pocket magnetic compass ; pocket-compass. (24)

boussole de poche, dite savonnette *ou* **boussole savonnette** (*f.*), pocket magnetic compass, in hunter case. (25)

boussole de variation (*f.*), variation-compass. (26)

boussole des sinus (*f.*), sine galvanometer. (27)

boussole des tangentes (*f.*), tangent galvanom--eter. (28)

boussole forme montre (*f.*), watch-form compass. (29)

boussole marine (*f.*), mariners' compass ; compass. (30)

boussole nivelante (*f.*), levelling-compass. (31)

boussole suspendue (*f.*), hanging compass. (32)

boussole topographique, à prisme (*f.*), prismatic compass. (33)

bout (*n.m.*), end : (34)

le bout d'un bâton, d'une rue, d'un tunnel, the end of a stick, of a street, of a tunnel. (35 *ou* 36 *ou* 37)

bout à bout, end to end ; endwise : (38)

poser des rails bout à bout, to lay rails end to end. (39)

bout affranchi (d'une barre de fer ou d'acier) (*m.*), crop-end. (40)

bout de course (d'un outil, d'un piston) (*m.*), end of stroke, end of travel, (of a tool, of a piston). (41 *ou* 42)

bout rejeté (de bois) (*m.*), baff-end (of timber). (43)

boutefeu [**boutefeux** *pl.*] (pers.) (*n.m.*), shotfirer ; shotlighter ; blaster ; shotman. (44)

bouteille (*n.f.*), bottle. (45)

bouteille à gaz (*f.*), gas-bottle. (46)

bouteille de Lane (*f.*), Lane's unit jar ; Lane's electrometer. (47)

bouteille de Leyde (*f.*), Leyden jar ; Leyden bottle ; Leyden phial ; Leyden vial ; electric jar. (48)

bouteille de mercure (*f.*), bottle of mercury ; flask of mercury : (49)

le mercure est généralement évalué en bouteilles (*ou* flasques) de kg. 34,65, mercury is generally reckoned in bottles (or flasks) of kg. 34·65. (50)

bouter le minerai dans la voie de fond, to drop, to dump, to shoot, the ore into the bottom level. (51)

bouterolle (*n.f.*), rivet-snap ; rivet-set ; rivetting-set ; snap ; snap-tool ; snap-head. (52)

bouterolle à main (*f.*), hand rivet-snap. (53)

bouterolle à œil (*f.*), eyed rivet-snap. (54)

bouteroller (*v.t.*), to snap (to form by a snap-tool) : (55)

bouteroller un rivet, to snap a rivet. (56)

boutique (ensemble des outils d'un artisan) (*n.f.*), set of tools ; tools ; implements : (57)

une boutique de menuisier, a set of joiners' tools. (58)

boutisse (brique ou pierre placée selon sa longueur dans un mur, de manière à laisser voir ses deux bouts,—opp. à **carreau**) (Maçonn.) (*n.f.*), header. (59)

appareil en boutisses (*m.*), heading-bond. (60)

bouton (*n.m.*), button ; knob. (61)

bouton (de sonnerie, d'interrupteur de courant, ou analogue) (*n.m.*), button ; push-button ; push. (62 *ou* 63 *ou* 64)

bouton à friction (d'un palmer) (*m.*), ratchet-stop (of a micrometer). (65)

bouton à olive (*m.*), oval knob. (66)

bouton d'appel (sonnerie électrique) (*m.*), call-button. (67)

bouton d'essai *ou* **bouton de fin** *ou* *simplement* **bouton** (*n.m.*), button ; assay-button. (68)

bouton de manivelle (Méc.) (*m.*), crank-pin ; wrist-pin ; wrist. (69)

bouton de manivelle de l'essieu moteur (locomotive) (*m.*), main crank-pin ; main-pin. (70)

bouton de mise au point (appareil photographi--que) (*m.*), focussing-knob. (71)

bouton de porte (*m.*), door-knob. (72)

bouton de sonnerie (*m.*), bell-push. (73)

bouton moleté (*m.*), milled head ; milled knob. (74)

bouton or-argent (Essais) (*m.*), gold-silver button. (75)

bouton pour courroies (*m.*), button fastener (for belts). (76)

bouveau (Mines) (*n.m.*), stone-drift ; rock-drift ; gallery in dead ground. (77)

bouvet à approfondir *ou* **bouvet de deux pièces** (*m.*), plough ; plow ; plough-plane. (78)

bouvet à joindre (*m.*), match-plane. (1)

bouvet à languette *ou* bouvet mâle (*m.*), tonguing-plane. (2)

bouvet à noix (*m.*), cock-bead plane. (3)

bouvet à rainure *ou* bouvet femelle (*m.*), grooving-plane. (4)

bouvet en deux morceaux (la paire) (*m.*), match-planes ; tonguing-and-grooving planes. (5)

bouvet en un morceau *ou* bouvet à fourchement (*m.*), double-ended match-plane. (6)

bouveter (*v.t.*), to match ; to plough : (7) bouveter des planches, to match boards. (8)

bow-string [bow-string *pl.*] (*n.m.*), bow-string bridge ; tension bridge. (9)

bowénite (Minéral) (*n.f.*), bowenite. (10)

bowete (*n.f.*). Same as bouveau.

bowlingite (Minéral) (*n.f.*), bowlingite. (11)

boy (jeune serviteur indigène dans les colonies) (*n.m.*), boy. (12)

boyau (*n.m.*) *ou* boyau de chat, gut ; catgut. (13)

boyau (tuyau flexible) (*n.m.*), hose. (14)

boyau d'accouplement de l'intercommunication pneumatique (matériel roulant de ch. de f.) (*m.*), air-signal hose ; signal-pipe hose ; signal-hose. (15)

boyau d'accouplement du frein à air comprimé (*m.*), air-brake hose ; brake-hose ; train-pipe hose. (16)

boyau de mine (*m.*), break-through ; break-through (17)

boyau en caoutchouc (*m.*), rubber hose ; india-rubber hose. (18)

boyau en cuir (*m.*), leather hose. (19)

boyau en toile (*m.*), canvas hose. (20)

brachistocrone *ou* brachystocrone (Géom.) (*adj.*), brachistocronic. (21)

brachistocrone *ou* brachystocrone (*n.f.*), brachis-tocrone. (22)

brachyanticlinal (opp. à brachysynclinal *ou* cuvette *ou* cuvette synclinale) (Géol.) (*n.m.*), quaquaversal fold ; dome. (23)

brachydiagonal, -e, -aux (Cristall.) (*adj.*), brachy-diagonal. (24)

brachydiagonale (*n.f.*), brachydiagonal. (25)

brachydôme (Cristall.) (*n.m.*), brachydome. (26)

brachypinacoïde (Cristall.) (*n.m.*), brachy-pinacoid. (27)

brachyprisme (Cristall.) (*n.m.*), brachyprism. (28)

brachypyramide (Cristall.) (*n.f.*), brachy-pyramid. (29)

brachysynclinal (Géol.) (*n.m.*), centroclinal fold ; basin. (30)

bracon (d'une porte d'écluse) (*n.m.*), diagonal brace (of a lock-gate). (31)

[bracons (*n.m.pl.*), diagonal bracing.] (32)

bradysisme (*n.m.*), bradyseismism. (33)

bradysismique (*adj.*), bradyseismal ; bradyseis-mic ; bradyseismical. (34)

brai (*n.m.*), pitch. (35)

braise (combustible) (*n.f.*), breeze. (36)

brame (Métall.) (*n.f.*), slab. (37)

bran de scie (*m.*), sawdust. (38)

brancard (ambulances) (*n.m.*), stretcher. (39)

brancard (Ch. de f.) (*n.m.*), side-sill ; sill : (40) le châssis d'un wagon de chemin de fer se compose de deux brancards ou longerons et de deux traverses de tête, the underframe of a railway-wagon is composed of two side-sills and two end-sills. (41)

brancard du bogie (*m.*), truck side-frame. (42)

branche (d'un arbre) (*n.f.*), branch, bough, (of a tree). (43)

branche (d'un compas) (*n.f.*), leg, branch, (of a compass). (44)

branche (d'un pied à trois branches, d'un trépied) (*n.f.*), leg (of a tripod). (45)

branche (d'une fourche) (*n.f.*), prong, tooth, tine, (of a fork). (46)

branche (d'une clef) (Serrur.) (*n.f.*), stem, shank, (of a key). (47)

branche (d'un cadenas) (*n.f.*), bow, shackle, (of a padlock). (48)

branche (d'une penture à T, d'un couplet) (Serrur.) (*n.f.*), strap, leaf, (of a butt-and-strap hinge, of a strap-hinge). (49 *ou* 50)

branche (d'une tenaille) (*n.f.*), handle (of a pincers, nippers, or tongs) : (51) tenaille à mors coupants, branches noires, mâchoires polies (*f.*), cutting-nippers, black handles, bright jaws. (52)

branche (d'une tenaille de forgeron) (*s'emploie généralement au pluriel*) (*n.f.*), rein (of a blacksmiths' tongs). (53)

branche à coulisse *ou* branche coulissante (trépied) (*f.*), sliding leg ; telescopic leg. (54)

branche articulée (trépied) (*f.*), fold-over leg. (55)

branche de pointe (Ch. de f.) (*f.*), splice-rail ; short point rail ; short rail ; side-point : (56) la pointe et la branche de pointe, the point-rail and the splice-rail ; the main point and the side-point. (57)

branche droite (trépied) (*f.*), rigid leg. (58)

branchement (*n.m.*), branching. (59)

branchement (Tuyauterie) (*n.m.*), branch ; branch-pipe. (60)

branchement (opp. à conducteur principal) (Élec.) (*n.m.*), branch ; tap. (61) (opp. to main *or* lead).

branchement (Ch. de f.) (*n.m.*), turnout ; turn-off. (62)

branchement d'abonné (Élec.) (*m.*), service-lead. (63)

branchement de tailles (Mines) (*m.*), room turning ; turning off rooms ; setting off stalls ; branching off breasts. (64)

branchement double (Tuyauterie) (*m.*), double V branch. (65)

brancher (*v.t.*), to branch : (66) brancher des tuyaux, to branch pipes. (67)

brancher des tailles sur une voie maîtresse (Mines), to turn off, to set off, to branch off, rooms from an entry. (68)

brandisite (Minéral) (*n.f.*), brandisite. (69)

branlant, -e (se dit des sièges de soupapes, etc.) (*adj.*), fluttering ; dancing ; loose : (70) siège branlant (*m.*), fluttering (*or* dancing) (*or* loose) seat. (71)

branles (d'un étau) (*n.m.pl.*), jaws (of a vice). (72)

branloire (d'un soufflet de forge) (*n.f.*), breakstaff, rocker, (of a smiths' bellows). (73)

bras (*n.m.*), arm. (74)

bras (d'une machine à percer radiale) (*n.m.*), arm (of a radial drilling-machine). (75)

bras (du fléau d'une balance) (*n.m.pl.*), arms (of the beam of a balance). (76)

bras (d'un moulin à vent) (*n.m.*), arm, vane, (of a windmill). (77)

bras (d'une roue) (*n.m.*), arm, spoke, (of a wheel). (78)

bras (d'une poulie) (*n.m.*), arm (of a pulley). (79)

bras (d'une bobine pour câble plat d'extraction) (Mines) (*n.m.*), arm, horn, (of a reel for flat winding-rope). (80)

bras (à), by hand ; by hand-power ; hand (*used as adj.*) ; manual : (1)
machine marchant à bras (*f.*), machine driven by hand-power. (2)
pompe à bras (*f.*), hand-pump. (3)
traction à bras d'hommes (*f.*), manual haulage. (4)

bras de la mer (*m.*), arm of the sea. (5)

bras de levier (*m.*), lever-arm ; arm of a force ; leverage of a force ; mechanical advantage : (6)
l'effet d'une force appliquée à un levier croît proportionnellement à la longueur du bras de levier sur lequel elle agit (*m.*), the effect of a force applied to a lever increases pro- -portionally to the length of the lever-arm on which it acts. (7)

bras de manivelle (*m.*), crank-web ; crank-arm ; crank-cheek. (8)

bras mort (délaissé) (Géogr. phys.) (*m.*), cut-off ; ox-bow lake. (9)

bras porte-profil (appareil à trousser) (Fonderie) (*m.*), spindle-arm (rig for loam-work). (10)

bras radial (Méc.) (*m.*), radial arm. (11)

brasage *ou* **brasement** (*n.m.*), brazing ; hard soldering. (12)

braser (*v.t.*), to braze ; to hard solder : (13)
braser une scie à ruban, to braze, to hard solder, a band-saw. (14)

braser (se) (*v.r.*), to braze : (15)
la fonte se brase très mal avec la fonte, mais très bien avec le fer ou l'acier doux, cast iron brazes very badly with cast iron, but very well with wrought iron or soft steel. (16)

brasero [braseros *pl.*] *ou* **brasier** (*n.m.*), fire- basket ; fire-devil. (17)

brasquage (Métall.) (*n.m.*), brasquing ; brasking. (18)

brasque (*n.f.*), brasque ; brask ; steep. (19)

brasquer (*v.t.*), to brasque ; to brask : (20)
brasquer un creuset, une coupelle, un fourneau, to brasque a crucible, a cupel, a furnace. (21 *ou* 22 *ou* 23)

brassage (*n.m.*), stirring ; stirring up : (24)
brassage du courant d'air, stirring up the air-current. (25)

brassage (Métall.) (*n.m.*), rabbling ; stirring. (26)

brasser (*v.t.*), to stir ; to stir up : (27)
brasser le courant d'air, to stir up the air- current. (28)

brasser (Métall.) (*v.t.*), to rabble ; to stir : (29)
brasser le fer puddlé avec le crochet, to rabble the puddled iron with the staff ; to stir the puddled iron with the rabble. (30)

brasure (action) (*n.f.*), brazing ; hard soldering : (31)
brasure des métaux, brazing metals. (32)

brasure (composition) (*n.f.*), brazing-solder ; brass solder ; hard solder. (33)

brasure dure (*f.*), hard brazing-solder ; hard brass solder. (34)

brasure tendre *ou* **brasure douce** (*f.*), soft brazing- solder ; soft brass solder. (35)

braunite (Minéral) (*n.f.*), braunite. (36)

braunspath (Minéral) (*n.m.*), brown spar. (37)

braye (*n.f.*) *ou* **brayer** (*n.m.*), sling ; rope-sling. (38)

brayer (*v.t.*), to sling : (39)
brayer une pierre, to sling a stone. (40)

brèche (*n.f.*), breach ; gap ; opening. (41)

brèche (Géol.) (*n.f.*), breccia ; brockram. (42)

brèche calcaire (*f.*), calcareous breccia. (43)

brèche de faille (*f.*), fault-breccia ; fault-rock ; rock-rubble. (44)

brèche de friction *ou* **brèche de dislocation** (*f.*), friction-breccia ; crush-breccia. (45)

brèche éruptive (*f.*), flow breccia ; volcanic breccia. (46)

brèche osseuse (*f.*), bone breccia ; osseous breccia. (47)

brèche volcanique (*f.*), volcanic breccia. (48)

bréchiforme *ou* **bréchoïde** (*adj.*), breccial ; brecciated : (49)
calcaire bréchiforme (*m.*), brecciated lime- -stone. (50)

bréguets (élingue à pattes) (*n.m.pl.*), can-hooks ; sling-dogs ; sling ; dog-hook sling. (51)

breithauptite (Minéral) (*n.f.*), breithauptite ; antimonial nickel. (52)

breunérite (Minéral) (*n.f.*), breunerite. (53)

brevet d'invention *ou* **simplement** **brevet** (*n.m.*), patent ; letters patent. (54)

brevet de perfectionnement (*m.*), patent of improvement. (55)

breveté, -e (*adj.*), patent : (56)
dispositif breveté de sûreté (*m.*), patent safety- device. (57)

breveté (pers.) (*n.m.*), patentee. (58)

breveté S.G.D.G., patented. (*Note.*—the letters S.G.D.G. are the abbreviated form of " sans garantie du gouvernement " which means " no government guarantee.") (59)

breveter (*v.t.*), to patent ; to grant a patent to : (60)
breveter une invention, to patent an inven- -tion. (61)

brewstérite (Minéral) (*n.f.*), brewsterite. (62)

bricole (*n.f.*), breast-strap ; byard. (63)

bridage (*n.m.*), clamping ; fastening together. (64)

bride (saillie circulaire à l'extrémité d'un tuyau, ou analogue) (*n.f.*), flange : (65)
bride d'un tuyau, d'un tube, d'un cylindre, flange of a pipe, of a tube, of a cylinder. (66 *ou* 67 *ou* 68)

bride (à) *ou* **brides (à)**, flanged ; flange (*used as adj.*) : (69)
tuyau à bride (*m.*), flanged pipe ; flange- pipe. (70)

bride (collier) (*n.f.*), collar. (71)

bride (*n.f.*) *ou* **bride de serrage**, clamp ; cramp. (72)

bride à capote (*f.*), G cramp ; G clamp ; C cramp ; C clamp ; cramp ; clamp. (73)

bride de sabotage (Ch. de f.) (*f.*), chairing-gauge. (74)

bride du ressort (pour ressort à lames super- -posées) (*f.*), spring-band, spring-buckle, (for leaf-spring). (75)

brider (serrer avec une bride) (*v.t.*), to clamp ; to clamp together ; to fasten together : (76)
brider des tuyaux avec des boulons, to clamp, to fasten, pipes together with bolts. (77)

brider l'air (Aérage de mines), to regulate the quantity of air passing in each split to the desired amount. (78)

brider une pierre (Carrières), to sling a stone (to place it in a sling for hoisting). (79)

brider (se) (*v.r.*), to be clamped : (80)
les tubes de fonte se brident avec des boulons (*m.pl.*), cast-iron tubes are clamped with bolts. (81)

brigade de nivellement (*f.*), levelling-party. (82)

brigadier-poseur [brigadiers-poseurs *pl.*] (Ch. de f.) (pers.) (*n.m.*), permanent-way inspector ; roadmaster ; trackmaster. (1)

brillant, -e (*adj.*), brilliant ; bright ; shiny ; glossy : (2)

viseur brillant (Photogr.) (*m.*), brilliant view-finder. (3)

de brillantes couleurs (*f.pl.*), bright colours. (4)

papier brillant (Photogr.) (*m.*), glossy paper. (5)

brillant (lustre ; éclat) (*n.m.*), brilliance ; brilliancy ; brightness ; glossiness : (6)

le brillant d'une pierre précieuse, the brilliance of a precious stone (*or* of a gem). (7)

le brillant de l'acier, the brightness of steel. (8)

brillant (diamant) (*n.m.*), brilliant. (9)

brin (*n.m.*), strand : (10)

cordage composé de — brins (*m.*), rope made up of — strands. (11)

palan à — brins (*m.*), — -strand pulley. (12)

brin libre (d'une corde, d'une chaîne de palan) (*m.*), free end (of a rope, of a pulley-block chain). (13 *ou* 14)

brin mou *ou* **brin lâche** *ou* **brin conduit** *ou* **brin mené** (d'une courroie, d'un câble de trans--mission) (*m.*), slack side, slack length, slack portion, driven side, idle side, (of a belt, of a transmission-rope). (15 *ou* 16)

brin tendu *ou* **brin conducteur** *ou* **brin moteur** *ou* **brin menant** (d'une courroie, d'un câble de transmission) (*m.*), tight side ; tight length ; taut portion ; driving side. (17 *ou* 18)

briolette (Joaillerie) (*n.f.*), briolette. (19)

brique (*n.f.*), brick. (20)

brique à couteau (*f.*), feather-edged brick ; gauge-brick ; arch-brick. (21)

brique creuse (*f.*), hollow brick ; perforated brick. (22)

brique crue (*f.*), dried brick ; sun-dried brick. (23)

brique cuite (*f.*), burned brick. (24)

brique de bois (*f.*), wood brick. (25)

brique de dolomie (*f.*), dolomite brick. (26)

brique de laitier (*f.*), slag brick. (27)

brique de magnésite (*f.*), magnesite brick. (28)

brique émaillée *ou* **brique vernissée** (*f.*), glazed brick ; enamelled brick. (29)

brique pleine (*f.*), solid brick. (30)

brique réfractaire (*f.*), fire-brick. (31)

briquet (Menuis.) (*n.m.*), flap-hinge. (32)

briquetage (*n.m.*), brickwork. (33)

briqueter (*v.t.*), to brick. (34)

briqueterie (*n.f.*), brick-field. (35)

briqueteur (pers.) (*n.m.*), bricklayer. (36)

briquetier (pers.) (*n.m.*), brick-maker. (37)

briqueton (*n.m.*), bat. (38)

briquette (*n.f.*), briquette ; briquet. (39)

briquette de houille (*f.*), coal briquette. (40)

bris (*n.m.*), breakage ; breaking : (41)

le bris de certaines pièces d'une machine résultant du grippement du piston, the breakage of certain parts of an engine resulting from seizing of the piston. (42)

brisance (d'un explosif) (*n.f.*), disruptive effect, rending effect, (of an explosive). (43)

brisant, -e (en parlant des explosifs) (*adj.*), rend--ing ; disruptive : (44)

explosifs brisants (*m.pl.*), rending (*or* disruptive) explosives. (45)

brise-glace *ou* **brise-glaces** [brise-glace *ou* brise-glaces *pl.*] (*n.m.*), ice-breaker. (46)

brise-lames [brise-lames *pl.*] (*n.m.*), breakwater. (47)

briser (*v.t.*), to break ; to smash : (48)

en cas de sinistre, briser la glace (avertisseur d'incendie), in case of fire, break the glass (fire-alarm). (49)

briser en éclats (*v.t.*), to splinter. (50)

briser (se) (*v.r.*), to break ; to break up : (51)

le verre se brise aisément, glass breaks easily. (52)

brisis (angle, endroit de contact) (Constr.) (*n.m.*), break : (53)

le brisis d'un comble brisé, the break of a curb-roof. (54)

brisou (*n.m.*), fire-damp ; gas ; pit-gas , fire. (55)

brisure (*n.f.*), break ; rent : (56)

les brisures des rochers, the rents in the rocks. (57)

brisure (d'une charnière) (joint ; endroit de contact) (*n.f.*), break (of a hinge). (58)

britonite (Explosif) (*n.f.*), britonite. (59)

brocatelle (marbre) (*n.f.*), brocatelle. (60)

brochage (Méc.) (*n.m.*), drifting : (61)

brochage des trous de rivets, drifting rivet-holes. (62)

brochantite (Minéral) (*n.f.*), brochantite. (63)

broche (long clou) (*n.f.*), spike ; spike-nail. (64)

broche (petit levier) (Méc.) (*n.f.*), tommy ; tommy-bar ; tommy-lever ; tommy-rod : (65)

vis à tête ronde percée de trous pour l'emploi d'une broche (*f.*), round-headed screw pierced with holes for use with a tommy. (66)

broche (arbre porte-mèche d'une machine à percer, arbre porte-taraud d'une machine à tarauder, ou analogue) (*n.f.*), spindle : (67 *ou* 68 *ou* 69)

machine à tarauder à — broches (*f.*), — spindle tapping-machine. (70)

broche (d'une serrure) (*n.f.*), drill-pin, pin, pintle, broach, stem, (of a lock). (71)

broche (d'une charnière) (*n.f.*), pin, pintle, (of a hinge). (72)

broche (d'une plaque de fond d'un moule en terre) (Fonderie) (*n.f.*), prod, pricker, dabber, (of a loam-plate of a loam mould). (73)

broche (mandrin) (*n.f.*), drift ; driftpin ; driver. (74)

broche conique (*f.*), taper drift ; taper driftpin. (75)

broche conique lisse (*f.*), smooth taper drift ; smooth taper driftpin. (76)

broche d'assemblage (*f.*), rivet-drift. (77)

broche d'évent (Fonderie) (*f.*), riser-pin ; riser-stick. (78)

broche de coulée (Fonderie) (*f.*), runner-pin ; runner-stick ; gate-pin ; gate-stick ; sprue. (79)

broche de déchassage (*f.*), drift ; centre-key (for loosening and removing taper-shank drills from sockets). (80)

broche porte-fraise (*f.*), milling-machine arbor ; cutter-arbor. (81)

broche porte-meule (*f.*), grinding-spindle. (82)

brocher (Méc.) (*v.t.*), to drift : (83)

brocher un trou de rivet, to drift a rivet-hole. (84)

broiement *ou* **broîment** (*n.m.*). Same as **broyage.**

bromargyrite (*n.f.*) *ou* **bromargyre** (*n.m.*) (Minéral), bromyrite ; bromargyrite ; bromite. (85)

bromate (Chim.) (*n.m.*), bromate. (1)
brome (Chim.) (*n.m.*), bromine. (2)
bromique (Chim.) (*adj.*), bromic. (3)
bromite (Minéral) (*n.f.*). Same as **bromargyrite.**
bromlite (Minéral) (*n.f.*), bromlite ; alstonite. (4)
bromocyanuration (*n.f.*), bromocyanidation. (5)
bromocyanure (*n.m.*), bromocyanide. (6)
bromocyanurer (*v.t.*), to bromocyanide. (7)
bromure (Chim.) (*n.m.*), bromide : (8)
　bromure d'argent, bromide of silver ; silver bromide. (9)
bromyrite (Minéral) (*n.f.*). Same as **bromargyrite.**
broncal (*n.m.*), pump-drill (upright drill-stock with bob and cord). (10)
bronzage (*n.m.*), bronzing ; (11)
　bronzage de la fonte, de l'acier, du zinc, bronz-ing cast iron, steel, zinc. (12 ou 13 ou 14)
bronze (*n.m.*), bronze. (15)
bronze (*n.m.*) *ou* bronze industriel, gun-metal : (16)
　coussinets en bronze (*m.pl.*), gun-metal bear-ings. (17)
bronze à canon (*m.*), gun-metal ; cannon-metal. (18)
bronze au manganèse *ou* bronze manganésé (*m.*), manganese bronze ; manganese copper. (19)
bronze au zinc (*m.*), zinc-bronze. (20)
bronze d'aluminium (*m.*), aluminium bronze ; aluminium gold. (21)
bronze d'art *ou* bronze statuaire (*m.*), statuary bronze. (22)
bronze de cloches (*m.*), bell-metal. (23)
bronze phosphoreux *ou* bronze phosphuré (*m.*), phosphor-bronze ; phosphorus-bronze. (24)
bronze pour robinetterie (*m.*), cock-metal ; cock-brass. (25)
bronze silicié (*m.*), silicon-bronze. (26)
bronzer (*v.t.*), to bronze : (27)
　bronzer une pièce venue de fonte, to bronze a casting. (28)
bronzite (Minéral) (*n.f.*), bronzite. (29)
brookite (Minéral) (*n.f.*), brookite. (30)
brossage (*n.m.*), brushing. (31)
brossailles (*n.f.pl.*). Same as **broussailles.**
brosse (*n.f.*), brush. (32)
brosse à caractères à jour *ou* brosse pour vignettes (*f.*), stencil-brush. (33)
brosse à goudronner (*f.*), tar-brush. (34)
brosse à tubes *ou* brosse à écouvillon (*f.*), tube-brush ; flue-brush. (35)
brosse à verres de lampes (*f.*), lamp-glass brush. (36)
brosse circulaire pour polisseurs, à — rangs (*f.*), circular polishing-brush, — -row. (37)
brosse de lampisterie *ou* brosse pour lampes (*f.*), lamp-brush. (38)
brosse dite moignon (pour polisseurs) (*f.*), mop-end brush. (39)
brosse en fil de fer (*f.*), wire brush ; iron-wire brush. (40)
brosse pour fontes brutes (plate à manche, avec sangle, à manche courbe, ou en bout) (*f.*), foundry-brush ; casting-brush (flat with handle, with strap, with curved handle, or round). (41 ou 42 ou 43 ou 44)
brosse pour limes (*f.*), file-brush. (45)
brouettage (*n.m.*), barrowing. (46)
brouette (*n.f.*), barrow ; hand-barrow ; wheel-barrow. (47)
brouettée (*n.f.*), barrowful ; barrowload ; barrow. (48)

brouetter (*v.t.*), to barrow : (49)
　brouetter de la terre, to barrow earth. (50)
brouetteur (pers.) (*n.m.*), barrowman. (51)
brouillage (Géol.) (*n.m.*), jumbling. (52)
brouillard (*n.m.*), fog. (53)
brouillé, -e (Géol.) (*adj.*), jumbled : (54)
　roches brouillées (*f.pl.*), jumbled rocks. (55)
broussailles (*n.f.pl.*), brush ; brushwood ; under-wood ; scrub. (56)
broussailleux, -euse (*adj.*), brushy ; bushy. (57)
brousse (*n.f.*), bush (thick undergrowth). (58)
broutage *ou* broutement (*n.m.*), chattering. (59)
brouter (*v.i.*), to chatter : (60)
　outil de tour qui broutte (*m.*), lathe-tool which chatters. (61)
broyable (*adj.*), crushable. (62)
broyage (*n.m.*), crushing ; grinding ; milling : (63)
　broyage à l'eau, wet crushing (*or* grinding). (64)
　broyage à mort, comminution ; pulverization. (65)
　broyage à sec, dry crushing (*or* grinding). (66)
　broyage au bocard, stamp-milling. (67)
　broyage d'échantillons, sample-grinding. (68)
　broyage d'essai, trial crushing. (69)
　broyage de galets, crushing boulders. (70)
　broyage de minerai à façon, customs crushing, public crushing, crushing for the public, of ore. (71)
　broyage de mortier, grinding mortar. (72)
　broyage fin, fine crushing (*or* grinding). (73)
　broyage grossier, coarse crushing (*or* grinding). (74)
　broyage moyen, medium crushing (*or* grinding). (75)
　broyage par cylindres, crushing by rolls. (76)
　broyage par la voie humide, wet crushing (*or* grinding). (77)
　broyage par la voie sèche, dry crushing (*or* grinding). (78)
broyer (*v.t.*), to crush ; to grind ; to mill : (79)
　broyer à sec, to dry-grind. (80)
　broyer du quartz, to crush quartz. (81)
　broyer l'amalgame, to grind the amalgam. (82)
broyeur, -euse (*adj.*), crushing ; grinding ; milling : (83)
　une machine broyeuse, a crushing (*or* a grinding) machine. (84)
　cylindres broyeurs (*m.pl.*), crushing rolls. (85)
broyeur (*n.m.*), crusher ; crushing-machine ; grinder ; grinding-machine ; breaker ; rock-breaker ; rock-crusher ; mill. (86). See also **moulin.**
broyeur à boulets (*m.*), ball-crusher ; ball-mill. (87)
broyeur à ciment (*m.*), cement-mill. (88)
broyeur à cylindres (*m.*), crushing rolls ; rolls. (89)
broyeur à force centrifuge (*m.*), centrifugal mill. (90)
broyeur à galets (*m.*), pebble-mill. (91)
broyeur à mâchoires (*m.*), jaw-crusher ; jaw-breaker. (92)
broyeur à meules horizontales (*m.*), roller-mill. (93)
broyeur à meules verticales (*m.*), edge-mill. (94)
broyeur à mortier (*m.*), mortar-mill. (95)
broyeur à noix (*m.*), gyratory crusher ; gyratory rock-breaker. (96)
broyeur à pendule (*m.*), swinging crusher. (97)
broyeur à quartz (*m.*), quartz-mill ; quartz-crusher ; quartz-battery. (98)

broyeur à sable (*m.*), sand-mill. (1)
broyeur à sec (*m.*), dry crusher. (2)
broyeur d'échantillons (*m.*), sample-grinder. (3)
broyeur des fins (*m.*), fine crusher. (4)
broyeur des gros (*m.*), coarse crusher. (5)
broyeur des mixtes (*m.*), middles crusher. (6)
broyeur giratoire (*m.*), gyratory crusher; gyratory rock-breaker. (7)
broyeur malaxeur *ou* broyeur mélangeur (*m.*), mixing-mill. (8)
broyeur pour minerais (*m.*), ore-crusher. (9)
broyeur pour quartz (*m.*), quartz-mill; quartz-crusher; quartz-battery. (10)
broyeur tamiseur (*m.*), combined grinder and sieve. (11)
brucelles (*n.f.pl.*), tweezers. (12)
brucite (Minéral) (*n.f.*), brucite. (13)
bruit (*n.m.*), noise: (14)
le bruit d'une détonation, the noise of a detonation. (15)
brûlage du sol (destruction par le feu des herbes sèches, des broussailles) (*m.*), burning the ground. (16)
brûlant, -e (*adj.*), burning: (17)
des maisons brûlantes (*f.pl.*), burning houses. (18)
brûlement (*n.m.*), burning: (19)
brûlement des extrémités de pieux, burning the ends of posts. (20)
brûler (*v.t.*), to burn: (21)
brûler du bois, to burn wood fuel. (22)
brûler la brousse (*ou* le sol), to burn the bush (*or* the ground). (23)
brûler (*v.i.*), to burn: (24)
le soufre brûle avec une flamme bleue, sulphur burns with a blue flame. (25)
brûler la station (Ch. de f.), to run past, to pass beyond, the station. (26)
brûler les mèches (Travail aux explosifs), to light the fuses. (27)
brûleur (*n.m.*), burner. (28)
brûleur à anneau (gaz) (*m.*), ring burner. (29)
brûleur à champignon (Bunsen) (*m.*), mushroom burner. (30)
brûleur à couronne (Bunsen) (*m.*), rose burner; rosette burner. (31)
brûleur à gaz (*m.*), gas-burner; gas-jet. (32)
brûleur à gaz Argand (*m.*), Argand gas-burner. (33)
brûleur à incandescence (*m.*), incandescent burner. (34)
brûleur à pétrole (d'une locomotive chauffée au pétrole) (*m.*), oil-burner (of an oil-fired locomotive). (35)
brûleur Bunsen (*m.*), Bunsen burner; Bunsen's lamp. (36)
brûleur Bunsen avec virole à air, robinet à gaz et veilleuse (*m.*), Bunsen burner, with air-regulator and by-pass stop-cock. (37)
brûlure (*n.f.*), burn. (38)
brun, -e (*adj.*), brown. (39)
brun (*n.m.*), brown. (40)
brunâtre (*adj.*), brownish. (41)
bruni (*n.m.*), burnish. (42)
brunir (rendre brun) (*v.t.*), to brown: (43)
brunir l'acier, to brown steel. (44)
brunir (polir) (*v.t.*), to burnish: (45)
brunir de l'or, to burnish gold. (46)
brunissage (*n.m.*), burnishing. (47)
brunisseur (pers.) (*n.m.*), burnisher. (48)
brunissoir (*n.m.*), burnisher. (49)
brunissoir en agate (*m.*), agate burnisher. (50)

brunissure (poli) (*n.f.*), burnish. (51)
brushite (Minéral) (*n.f.*), brushite. (52)
brusque (*adj.*), sudden: (53)
variations brusques de température (*f.pl.*), sudden changes of temperature. (54)
brusquement (*adv.*), suddenly. (55)
brut, -e (en parlant de poids) (*adj.*), gross: (56)
poids brut (*m.*), gross weight. (57)
brut, -e (en parlant de matériaux grossiers) (*adj.*), raw; crude; rough: (58)
matière brute (*f.*), raw material. (59)
pétrole brut (*m.*), crude petroleum. (60)
diamant brut (*m.*), rough diamond. (61)
fonte brute de moulage et toutes pièces de fonderie brutes ou travaillées, raw pig-iron and all kinds of rough or machined castings. (62)
pièces de forge brutes, dégrossies ou finies, de tous genres, poids et dimensions (*f.pl.*), forgings, rough, roughed or finished, of all kinds, weights and sizes. (63)
brut (-e) d'estampage, rough stamped; as stamped: (64)
soupapes brutes d'estampage (*f.pl.*), rough stamped valves; valves as stamped. (65)
brut (-e) de fonte *ou* brut (-e) de fonderie *ou* brut (-e) de coulée, rough cast; as cast: (66)
engrenages bruts de fonte (*m.pl.*), rough cast gears; gears as cast. (67)
roues d'engrenages à chevrons brutes de fonderie en fonte (*f.pl.*), rough cast herring-bone gear-wheels, of cast iron. (68)
brut (-e) de forge, rough forged; as forged: (69)
outils bruts de forge (*m.pl.*), rough forged tools; tools as forged. (70)
brut (-e) de laminage, rough rolled; as rolled: (71)
l'essai de pliage se fait sur une éprouvette à froid et sur une éprouvette trempée sur chaque tôle ou barre brute de laminage (Cahier des charges unifié), one cold-bend and one temper-bend test from each plate or bar as rolled. (Standard specification). (72)
brutal, -e, -aux (grossier) (*adj.*), rough; brutal: (73)
manipulations brutales (*f.pl.*), rough handling. (74)
bûcheron (pers.) (*n.m.*), wood-cutter: woodman; lumberman. (75)
bucholzite (Minéral) (*n.f.*), bucholzite. (76)
buddle (Préparation mécanique des minerais) (*n.m.*), buddle. (77)
buffle (*n.m.*), buff. (78)
buis (arbre) (*n.m.*), box. (79)
buis (bois) (*n.m.*), box; boxwood. (80)
bulle (*n.f.*), bubble. (81)
bulle d'air (*f.*), air bubble. (82)
bulle de gaz *ou* bulle gazeuse (*f.*), gas bubble. (83)
bulletin (rapport qui publie quelque chose d'officiel) (*n.m.*), bulletin; statement: (84)
bulletin officiel du conseil des mines, official statement (*or* bulletin) of the chamber of mines. (85)
bulletin (reconnaissance; certificat) (*n.m.*), certificate. (86)
bulletin d'analyse (*m.*), analytical certificate. (87)
bulletin de l'essayeur de commerce (*m.*), certificate of assay; assayer's certificate. (88)
bunsénine (Minéral) (*n.f.*), krennerite. (89)
bunsénite (Minéral) (*n.f.*), bunsenite. (90)

buratite (Minéral) (*n.f.*), buratite. (1)

bure (Mines) (*n.f.*), blind shaft; blind pit; internal shaft; staple; staple-shaft; staple-pit. (2)

bureau (personnel des dignitaires d'une assemblée) (*n.m.*), board: (3)
　bureau des **examinateurs**, board of examiners. (4)

bureau (endroit où s'expédient les affaires) (*n.m.*), office; bureau. (5)

bureau d'essais (*m.*), assay-office. (6)

bureau d'expédition (*m.*), forwarding-office. (7)

bureau de dessin (*m.*), drawing-office. (8)

Bureau des mines (*m.*), Bureau of Mines. (9)

Bureau météorologique (*m.*), Meteorological Office; Weather Bureau. (10)

bureau télégraphique (*m.*), telegraph-office. (11)

bureau téléphonique (*m.*), telephone call-office. (12)

burette (Verrerie de laboratoire) (*n.f.*), burette: (13)
　burette avec pince, burette with clip. (14)
　burette avec poire en caoutchouc, burette with india-rubber blowing-ball. (15)
　burette avec robinet, burette with stop-cock. (16)
　burette avec tube d'affluence, burette with side-tube for filling. (17)
　burette de Mohr, Mohr's burette. (18)

burette (*n.f.*) *ou* **burette à huile** *ou* **burette à graisse** *ou* **burette à graisser** *ou* **burette de graissage**, oil-can; oiler; oil-feeder: (19)
　burette à huile, à bec long, à bec court, oil-can, with long spout, with short spout. (20 *ou* 21)
　burette à huile, à bec droit, à bec cintré, oil-can, with straight spout, with bent spout. (22 *ou* 23)
　burette à huile, à bec fixe, à bec interchange-able, oil-can, with fixed spout, with inter-changeable spout. (24 *ou* 25)
　burette à huile, en fer-blanc, en tôle étamée, en cuivre poli, en cuivre rouge poli, oil-can, in tin, in tinned steel, in polished brass, in polished copper. (26 *ou* 27 *ou* 28 *ou* 29)
　burette à **huile**, contenance . . ., oil-can, capacity . . . . (30)
　burette à pompe, oil-can with force-pump; oil-squirt. (31)
　burette à suif, tallow-pot. (32)
　burette "chemin de fer," railway oil-can. (33)

burin (*n.m.*) *ou* **burin à froid** (de mécanicien; de serrurier), chisel; cold-chisel; flat chisel; engineers' chipping chisel (flat). (34)

burin (de menuisier) (*n.m.*), mortise-chisel. (35)

burin (de mineur) (*n.m.*), chisel; drill; borer. (36)

burin bédane (de mécanicien) (*m.*), chipping-chisel (cross-cut); cold-chisel (cross-cut); cross-cut chisel; cape-chisel. (37)

burin en pointe de diamant (de mécanicien) (*m.*), chipping-chisel (diamond); cold-chisel (diamond-point); diamond-nose chisel; diamond-point chisel. (38)

burin grain-d'orge (de mécanicien) (*m.*), round-nosed cold-chisel; chipping-chisel (round-nosed). (39)

burinage (*n.m.*), chiseling; chiselling; chipping: (40)
　burinage d'une rainure de cale, chiseling a keyway. (41)
　burinage des pièces venues de fonte, chipping castings. (42)

buriner (*v.t.*), to chisel; to chip: (43)
　glacier qui a buriné les parois d'une vallée (*m.*), glacier which has chiseled the sides of a valley. (44)

burineur (pers.) (*n.m.*), chiseler; chipper. (45)

busc (d'une écluse de canal) (*n.m.*), mitre-sill (of a canal-lock). (46)

busc d'amont (*m.*), head mitre-sill. (47)

busc d'aval (*m.*), tail mitre-sill. (48)

buse (conduit qui amène l'eau sur une roue hydraulique) (*n.f.*), flume; leat. (49)

buse (tubage de puits artésien) (*n.f.*), Artesian-well casing. (50)

buse (Aérage de mines) (*n.f.*), air-pipe. (51)

buse (ajutage) (*n.f.*), nose; nose-piece; nozzle; snout. (52)

buse (*n.f.*) *ou* **busillon** (*n.m.*) (du porte-vent d'une tuyère), nozzle, blast-nozzle, nose, (of the blowpipe of a tuyère). (53)

buse à jet de sable (*f.*), sand-blast nozzle. (54)

buse en béton (*f.*), concrete pipe. (55)

buselure (d'un palier, ou analogue) (*n.f.*), bush, bushing, (of a bearing, or the like): (56 *ou* 57)
　buselures en bronze, gun-metal bushes (*or* bushings). (58)

bustamite (Minéral) (*n.f.*), bustamite. (59)

butée (Méc.) (*n.f.*), thrust-bearing; thrust-block; thrust; stop. (60)

butée (culée de pont) (*n.f.*), abutment. (61)

butée (étai) (Constr.) (*n.f.*), shore. (62)

butée (déclanchement) (*n.f.*), knock-off; kick-up: (63)
　came de butée (*f.*), knock-off cam. (64)

butée (effort) (*n.f.*), thrust. (65)

butée à billes (*f.*), thrust ball bearing; thrust ball bearings. (66)

butée à billes à double effet (*f.*), double-thrust ball bearings. (67)

butée à l'arrière (tour) (*f.*), end thrust-bearing; tail-pin thrust (lathe). (68)

butée des terres (Ponts & Chaussées) (*f.*), thrust of the ground. (69)

butée permettant de régler la profondeur des trous à percer (*f.*), depth-stop; adjustable stop for depth of holes. (70)

butée réglable (*f.*), adjustable stop. (71)

buter (appuyer contre) (*v.t.*), to butt. (72)

buter (étayer) (*v.t.*), to prop; to shore; to shore up: (73)
　buter un mur, to shore up a wall. (74)

buter (s'appuyer contre) (*v.i.*), to butt: (75)
　poutre qui bute contre un mur (*f.*), beam which butts against a wall. (76)

butoir (Ch. de f.) (*n.m.*), bumping-post; buffer: (77)
　le butoir d'une voie de garage, the bumping-post of a siding. (78)
　le butoir d'une voie principale, the buffer of a main-line track. (79)

butoir (d'un cabestan) (*n.m.*), pawl-head (of a capstan). (80)

butoir (Géol.) (*n.m.*), horst. (81). See **horst** for example of use.

butte (éminence de terrain) (*n.f.*), mound; knoll; hillock. (82)

butte (bois droit et isolé; étai; chandelle) (Mines) (*n.f.*), prop; tree; post; shore. (83)

butte métallique (Mines) (*f.*), cast-iron prop. (84)

butte-témoin [**buttes-témoins** *pl.*] (opp. à **fenêtre**) (Géol.) (*n.f.*) outlier; butte (85) (**opp.** to **inlier**):

les monticules isolés qui échappent à la destruction par ablation sont connus sous le .nom de buttes-témoins. Ils sont souvent le meilleur indice de l'ancienne extension d'une couche aujourd'hui presque entièrement disparue, the isolated mounds which escape destruction by ablation are known as outliers. They are often the best indication of the ancient extension of a bed which to-day has almost entirely disappeared. (1)

buttée (*n.f.*). Same as **butée**.
butter (*v.t.*). Same as **buter**.
buttoir (*n.m.*). Same as **butoir**.
butyrite *ou* **butyrellite** (*n.f.*), butyrellite ; bog-butter. (2)
buvable (*adj.*), drinkable ; potable : (3)
eau buvable (*f.*), drinkable water. (4)
by-pass (tube de dégagement) (*n.f.*), by-pass. (5)
byssolite (Minéral) (*n.f.*), byssolite. (6)
bytownite (Minéral) (*n.f.*), bytownite. (7)

C

cabane (*n.f.*), cabin ; hut ; house. (8)
cabane de bois (*f.*), log cabin ; log hut ; log house. (9)
cabestan (*n.m.*), capstan. (10)
cabestan à bras (*m.*), hand-capstan. (11)
cabestan à cheval (*m.*), horse-whim ; whim ; whin ; gin. (12)
cabestan à vapeur (*m.*), steam-capstan. (13)
cabestan double (*m.*), double capstan. (14)
cabestan électrique (*m.*), electric capstan. (15)
cabestan simple (*m.*), single capstan. (16)
cabine (d'une grue) (*n.f.*), cabin, cab, house, (of a crane). (17)
cabine (d'un ascenseur) (*n.f.*), car (of a passenger-lift) ; passenger-car (of a lift). (18)
cabine à signaux (*f.*), signal-box ; signal-cabin ; signal-tower. (19)
cabine téléphonique (*f.*), telephone-box ; tele-phone-booth. (20)
cabinet noir (Photogr.) (*m.*), dark room. (21)
câblage (*n.m.*), cabling. (22)
câble (cordage) (*n.m.*), rope ; cable. (23)
câble (Élec.) (*n.m.*), cable. (24)
câble à section décroissante (Extraction) (Mines) (*m.*), tapering rope ; taper rope. (25)
câble à signaux (*m.*), signalling-wire. (26)
câble aérien (*m.*), aerial ropeway ; ropeway ; wire-ropeway ; wireway ; wire-tramway ; cableway ; rope-railway. (27)
câble armé (*m.*), armoured cable. (28)
câble armé sous plomb (*m.*), lead-covered cable ; lead-sheathed cable ; leaded cable. (29)
câble au-dessous des wagons (*m.*), rope below the wagons. (30)
câble au-dessus des wagons (*m.*), rope above the wagons. (31)
câble-chaîne [**câbles-chaînes** *pl.*] (*n.m.*), cable-chain ; chain cable. (32)
câble-chaîne à étais *ou* **câble-chaîne étançonnée** *ou* **câble-chaîne à mailles étançonnées** (*m.*), stud-link cable-chain ; studded-link chain cable. (33)
câble-chaîne à maillons courts *ou* **câble-chaîne à mailles serrées** (*m.*), short-link cable-chain. (34)
câble concentrique (Élec.) (*m.*), concentric cable. (35)
câble conducteur (traction électrique,—système par fil aérien) (*m.*), trolley-wire. (36)
câble conique (Extraction) (Mines) (*m.*), tapering rope ; taper rope. (37)
câble d'aloès (*m.*), aloe rope. (38)
câble d'équilibre (Extraction) (Mines) (*m.*), tail-rope ; counterbalancing-rope ; balance-rope ; load-rope. (39)

câble d'extraction (Mines) (*m.*), winding-rope ; hoisting-rope. (40)
câble de chanvre (*m.*), hemp rope. (41)
câble de contrepoids (*m.*). Same as **câble d'équilibre**.
câble de distribution électrique (*m.*), electric dis-tribution-main ; electric supply-main. (42)
câble de forage (*m.*), drill-rope ; drilling-rope ; drilling-cable. (43)
câble de haubanage (*m.*), guy-rope ; stay-rope. (44)
câble de levage (d'un transporteur aérien à voie unique) (*m.*), hoisting-rope, fall-rope, (of a cableway). (45)
câble de queue (Roulage) (Mines) (opp. à **câble de tête** *ou* **câble-tête**) (*m.*), tail-rope. (46)
câble de remorque (*m.*), tow-rope ; tow-line ; towing-rope. (47)
câble de renvoi (*m.*), return-rope. (48)
câble de sondage (*m.*), drill-rope ; drilling-rope ; drilling-cable. (49)
câble de sûreté (*m.*), safety-rope. (50)
câble de terre (Élec.) (*m.*), earth-cable. (51)
câble de tête (Roulage) (Mines) (*m.*), main rope. (52)
câble de traction *ou* **câble de tirage** (*m.*), hauling-rope ; traction-rope ; pulling-rope. (53)
câble de traction souterraine (*m.*), underground hauling-rope. (54)
câble diminué (Extraction) (Mines) (*m.*), tapering rope ; taper rope. (55)
câble électrique (*m.*), electric cable. (56)
câble en acier *ou* **câble en fils d'acier** (*m.*), steel-wire rope. (57)
câble en fer *ou* **câble en fils de fer** (*m.*), iron-wire rope. (58)
câble en fibres végétales (*m.*), vegetable-fibre rope. (59)
câble épissé (*m.*), spliced rope. (60)
câble flottant (Roulage) (Mines) (*m.*), endless rope above wagons. (61)
câble-guide [**câbles-guides** *pl.*] (*n.m.*), guide-rope. (62)
câble hélicoïdal (Abatage en carrières) (*m.*), wire-saw. (63)
câble métallique (*m.*), wire rope ; stranded wire. (64)
câble métallique sans fin (*m.*), endless wire rope. (65)
câble mou (*m.*), slack rope. (66)
câble muni d'une armature de fer *ou* **câble revêtu d'une cuirasse de fils de fer** (*m.*), armoured cable ; wire-sheathed cable. (67)
câble-noria [**câbles-norias** *pl.*] (*n.m.*), single-rope cable-tramway. (68)

câble nu (Élec.) (*m.*), bare cable. (1)

câble plat (*m.*), flat rope. (2)

câble porteur (*m.*), carrying-rope ; carrier-rope ; track-rope ; supporting-rope. (3)

câble-queue [câbles-queues *pl.*] (Roulage) (Mines) (opp. à câble-tête) (*n.m.*), tail-rope. (4)

câble rond (*m.*), round rope. (5)

câble sans armature extérieure (*m.*), unarmoured cable. (6)

câble sans fin (*m.*), endless rope ; endless cable. (7)

câble sous caoutchouc (*m.*), rubber-insulated cable ; rubber cable. (8)

câble sous papier et plomb (*m.*), paper-insulated and lead-covered cable. (9)

câble sous plomb (*m.*), lead-covered cable ; lead-sheathed cable ; leaded cable. (10)

câble souterrain à armature d'asphalte (*m.*), underground cable with coating of asphalt. (11)

câble télédynamique *ou* câble télodynamique (*m.*), teledynamic cable ; telodynamic rope. (12)

câble-tête [câbles-têtes *pl.*] (Roulage) (Mines) (*n.m.*), main rope. (13)

câble tors (*m.*), twisted rope. (14)

câble tracteur (*m.*), hauling-rope ; traction-rope ; pulling-rope. (15)

câble tracteur et porteur à la fois (*m.*), combined hauling and carrying rope. (16)

câble traînant (Roulage) (Mines) (*m.*), endless rope below wagons. (17)

câbler (*v.t.*), to cable. (18)

cabochon (Joaillerie) (*n.m.*), cabochon : (19) pierre taillée en cabochon (*f.*), stone cut en cabochon. (20)

cabotage (*n.m.*), coasting-trade ; coastwise trade. (21)

caboteur (*n.m.*), coasting-vessel ; coaster. (22)

cabrérite (Minéral) (*n.f.*), cabrerite. (23)

cabrouet (*n.m.*), truck (two-wheeled hand-truck). (24)

cache *ou* cachette (*n.f.*), hiding-place ; cache. (25)

cache-cadre pour le tirage des épreuves (*m.*), [cache-cadre *pl.*] *ou simplement* cache (*n.m.*) (Photogr.), printing-mask ; mask. (26)

cache-entrée [cache-entrée *pl.*] (*n.m.*), drop ; key-drop ; keyhole-guard. (27)

cacher (*v.t.*), to hide ; to conceal. (28)

cacholong (Minéral) (*n.m.*), cacholong ; pearl-opal. (29)

cacoxène (Minéral) (*n.m.*), cacoxenite. (30)

cadenas (*n.m.*), padlock. (31)

cadenas à combinaisons (*m.*), combination pad--lock. (32)

cadenasser (*v.t.*), to padlock : (33) cadenasser une porte, une chaîne, to padlock a door, a chain. (34 *ou* 35)

cadmie (*n.f.*) *ou* cadmie des fourneaux, cadmia. (36)

cadmifère (*adj.*), cadmiferous. (37)

cadmique (*adj.*), cadmic. (38)

cadmium (Chim.) (*n.m.*), cadmium. (39)

cadmium sulfuré (Minéral) (*m.*), cadmium-blende ; cadmium ochre ; greenockite. (40)

cadran (*n.m.*) dial ; dial-plate : (41) le cadran d'un baromètre anéroïde, the dial of an aneroid barometer. (42)

cadran à aiguille (*m.*), needle-dial. (43)

cadran à boussole (*m.*), compass-dial. (44)

cadran à métal argenté (*m.*), silvered-metal dial. (45)

cadran gradué *ou* cadran divisé (d'instrument de précision) (*m.*), graduated dial ; divided dial. (46)

cadran gradué (de machine-outil) (*m.*), graduated index disc. (47)

cadran solaire *ou simplement* cadran (*n.m.*), sun-dial. (48)

cadranure (*n.f.*) *ou* cadrannure (*n.f.*) *ou* cadran (*n.m.*) (dans un tronc d'arbre), star shake. (49)

cadre (*n.m.*), frame ; framing ; framework. (50)

cadre (Boisage de mines) (*n.m.*), frame ; set ; frame-set ; framing ; durn. (51)

cadre à étiquettes (*m.*), card-rack. (52)

cadre à oreilles (Fonçage de puits de mine) (*m.*), sinking-frame ; curb ; crib. (53)

cadre à un seul montant (Boisage de galeries de mine) (*m.*), post-and-bar ; post-and-cap ; two-stick set ; half set. (54)

cadre à verre dépoli (*ou* à glace dépolie) pour la mise au point (Photogr.) (*m.*), focussing-screen ; focussing-frame ; ground-glass focussing-back. (55)

cadre arrière réversible (appareil photographique) (*m.*), reversing-frame. (56)

cadre avec tendard (galerie de mine) (*m.*), set with stretcher-piece. (57)

cadre-cache [cadres-caches *pl.*] (Photogr.) (*n.m.*), mask ; printing-mask. (58)

cadre coffrant (galerie de mine) (*m.*), false set. (59)

cadre colleté au ferme (puits de mine) (*m.*), wedging-curb ; wedging-crib. (60)

cadre complet (galerie) (*m.*), full set ; four-stick set : (61) un cadre complet est formé de quatre pièces, à savoir : un chapeau, deux montants, et une semelle, a full set consists of four sticks, viz. : a head-piece, two uprights, and a sill : a four-stick set is composed of four pieces, i.e. : a cap, two legs, and a sole-piece. (62)

cadre de boisage (*m.*), set of timber. (63)

cadre de puits (*m.*), shaft-frame ; shaft-set. (64)

cadre de superficie (puits de mine) (*m.*), head-frame ; sinking head-frame ; shaft-collar. (65)

cadre du foyer (locomotive) (*m.*), foundation-ring ; mud-ring. (66)

cadre intermédiaire (pour châssis négatifs) (Photogr.) (*m.*), carrier (for dark slides). (67)

cadre ordinaire (un chapeau et deux montants) (galerie) (*m.*), ordinary set ; three-stick set ; double timber (a head-piece and two uprights). (68)

cadre porteur (puits) (*m.*), sinking-frame ; curb ; crib ; supporting-curb ; supporting-frame. (69)

cadre porteur (galerie) (*m.*), bearer-set ; carrying-set. (70)

cadres jointifs (galerie) (*m.pl.*), close sets ; skin-to-skin timbering. (71)

cænozoïque (Géol.) (*adj.* & *n.m.*), Cenozoic ; Cænozoic ; Cainozoic. (72)

cæsium (Chim.) (*n.m.*), cæsium ; cesium. (73)

cafre (*adj.*), Kafir ; Kaffir (*used as adjs*) : (74) ouvrier cafre (*m.*), Kafir labourer ; Kaffir boy. (75)

Cafre (pers.) (*n.m.*), Kafir ; Kaffir. (76)

cage (Mines, etc.) (*n.f.*), cage. (77)

cage (d'un laminoir) (*n.f.*), housing, holster, standard, bearer, (of a rolling-mill). (1)

cage (d'une balance de précision) (*n.f.*), case : (2)

balance de précision, sous cage verre (*f.*), balance of precision, in glass case. (3)

cage à billes (*f.*), ball-cage. (4)

cage à deux étages (Mines) (*f.*), two-decker cage. (5)

cage à double voie (Mines) (*f.*), double-track cage. (6)

cage à eau (Mines) (*f.*), water-hoist tank. (7)

cage à plaque (Filetage) (*f.*), screw-plate with holding-down plate for loose dies. (8)

cage d'écureuil (Élec.) (*f.*), squirrel-cage : (9) rotor à cage d'écureuil (*m.*), squirrel-cage rotor. (10)

cage d'escalier (*f.*), staircase ; stairway. (11)

cage d'extraction (Mines) (*f.*), hoisting-cage. (12)

cage en fil de fer (*f.*), wire guard. (13)

cage guidée (Mines) (*f.*), guided cage ; cage working between guides. (14)

cahier des charges (*m.*), specification : (15)

cahiers des charges unifiés des grandes compagnies de chemins de fer français, des chemins de fer de l'État belge, standard specifications of the great French railway companies, of the Belgian State railways. (16 *ou* 17)

cahot (*n.m.*), jolt. (18)

cahotage *ou* cahotement (*n.m.*), jolting. (19)

cahotant, -e *ou* cahoteux, -euse (*adj.*), jolting ; jolty. (20)

cahoter (*v.t. & v.i.*), to jolt. (21)

caillou (*n.m.*), pebble ; pebble-stone ; stone ; boulder ; bowlder. (22)

caillou à facettes (Géol.), glyptolith ; gibber ; dreikanter. (23)

caillou du Rhin, Rhine quartz. (24)

caillou façonné par l'usure éolienne (Géol.), wind-worn stone. (25)

cailloux impressionnés (Géol.), impressed pebbles. (26)

cailloux roulés (Géol.), drift boulders. (27)

cailloux roulés par les eaux (Géol.), water-rolled pebbles. (28)

cailloux striés (Géol.), striated boulders ; striated pebbles. (29)

cailloutage (pavage en cailloux) (*n.m.*), pebble paving. (30)

caillouteux, -euse (*adj.*), pebbly. (31)

cailloutière (*n.f.*), gravel-pit. (32)

cailloutis (Trav. publ.) (*n.m.*), roadstones. (33)

cailloutis fluvio-glaciaire (Géol.) (*m.*), fluvio-glacial drift-stones. (34)

caïnite (Minéral) (*n.f.*), kainite. (35)

caisse (*n.f.*), case ; box. (36)

caisse (d'une voiture, d'un wagon) (*n.f.*), body (of a carriage, of a wagon). (37 *ou* 38)

caisse à claire-voie (*f.*), crate ; skeleton case. (39)

caisse à débourber (Mines) (*f.*), trunk. (40)

caisse à eau (de tender de locomotive) (*f.*), tank, water-tank, (of locomotive tender). (41)

caisse à sable (*f.*), sand-box. (42)

caisse à vent (de cubilot, de haut fourneau, etc.) (*f.*), wind-box, air-box, (of cupola, of blast-furnace, etc.). (43 *ou* 44)

caisse allemande *ou* caisse à tombeau (Préparation mécanique des minerais) (*f.*), German chest. (45)

caisse d'emballage (*f.*), packing-case. (46)

caisse de cémentation (de four à cémenter) (*f.*), converting-pot (of cementation-furnace). (47)

caisse de criblage (Mines) (*f.*), screening-box. (48)

caisse de marchandises (*f.*), case of goods. (49)

caisse de précipitation par le zinc *ou* caisse à zinc (*f.*), zinc precipitation box ; zinc-box. (50)

caisse de provisions (*f.*), box, case, of provisions ; chop-box. (51)

caisse du tender (locomotive) (*f.*), tender-tank. (52)

caisse guidée (Mines) (*f.*), skip ; gunboat ; skep. (53)

caisse pleine (opp. à caisse à claire-voie) (*f.*), close case. (54)

caisse pointue (Préparation mécanique des minerais) (*f.*), pointed box ; funnel-box ; V vat ; spitzkasten. (55)

caisson (Trav. publ.) (*n.m.*), caisson. (56)

caisson à minerai (*m.*), ore-bin ; ore-bunker. (57)

caisson allemand *ou* caisson à tombeau (Préparation mécanique des minerais) (*m.*), German chest. (58)

caisson d'aérage (*m.*), air-box. (59)

calage (*n.m.*). See examples : (60) See also caler (*v.t.*) for further examples.

calage d'un instrument de nivellement, clamping a levelling-instrument. (61)

calage d'une poulie sur l'arbre, keying a pulley on the shaft. (62)

calage d'une soupape, jamming of a valve. (63)

calage des balais (Élec.), lead of brushes. (64)

calage des roues sur les essieux, fixing wheels on axles. (65)

calage des wagonnets dans la cage (Mines), scotching, chocking, spragging, blocking, cogging, the trucks in the cage. (66)

calaminaire (*adj.*), calamine-bearing : (67) gîte calaminaire (*m.*), calamine-bearing deposit. (68)

calamine ($H_2Zn_2SiO_5$) (Minéral) (*n.f.*), calamine ; electric calamine ; siliceous calamine ; hemiomorphite. (69)

calamine ($ZnCO_3$) (Minéral) (*n.f.*), calamine ; smithsonite ; zinc-spar. (70)

calamite (Minéral) (*n.f.*), calamite. (71)

calavérite (Minéral) (*n.f.*), calaverite. (72)

calcaire (*adj.*), calcareous ; limestone-bearing ; limestone (used as adj.) : (73) roche calcaire (*f.*), calcareous rock ; limestone rock. (74)

calcaire (*n.m.*), limestone ; lime-rock. (75)

calcaire à ciment (*m.*), cement-stone. (76)

calcaire à crinoïdes (*m.*), crinoidal limestone. (77)

calcaire à entroques (*m.*), entrochal limestone. (78)

calcaire à nummulites (*m.*), nummulitic limestone ; nummulite-limestone. (79)

calcaire carbonifère (Géol.) (*m.*), Carboniferous limestone ; mountain-limestone. (80)

calcaire charbonneux (*m.*), carbonaceous limestone. (81)

calcaire coquillier (*m.*), shell-limestone ; shelly limestone. (82)

calcaire corallien (*m.*), coral limestone ; coralline limestone. (83)

calcaire de montagne (Géol.) (*m.*) mountain-limestone ; Carboniferous limestone. (84)

calcaire dolomitique (*m.*), dolomitic limestone. (85)

calcaire gras (*m.*), rich limestone. (1)

calcaire gréseux (*m.*), cornstone. (2)

calcaire grossier (Géol.) (*m.*), calcaire grossier. (3)

calcaire lacustre (*m.*), lacustrine limestone. (4)

calcaire lithographique (*m.*), lithographic stone; lithographic slate. (5)

calcaire magnésien (*m.*), magnesian limestone; dolomite. (6)

calcaire oolithique (*m.*), oolitic limestone; rag--stone. (7)

calcaire pisolithique (*m.*), pisolitic limestone. (8)

calcaréo-argileux, -euse (*adj.*), calcareo-argilla--ceous. (9)

calcaréo-bitumineux, -euse (*adj.*), calcareobitu--minous. (10)

calcaréo-ferrugineux, -euse (*adj.*), calcareofer--ruginous. (11)

calcaréo-magnésien, -enne (*adj.*), calcareomag--nesian. (12)

calcaréo-sableux, -euse (*adj.*), arenaceocalcareous. (13)

calcaréo-siliceux, -euse (*adj.*), calcareosilicious. (14)

calcareux, -euse (*adj.*), calcareous. (15)

calcarifère (*adj.*), calcariferous. (16)

calcarone [calcaroni *pl.*] [Sicile] (*n.m.*), calcarone; sulphur-kiln. (17)

calccélestine (Minéral) (*n.f.*), calciocelestite. (18)

calcédoine (Minéral) (*n.f.*), chalcedony; calced--ony. (19)

calcédoine enhydre (*f.*), enhydrous chalcedony. (20)

calcédoine saphirine (*f.*), sapphirine chalcedony. (21)

calcédonieux, -euse (*adj.*), chalcedonous. (22)

calcédonique (*adj.*), chalcedonic. (23)

calcédonix (Minéral) (*n.m.*), chalcedonyx. (24)

calcifère (*adj.*), calciferous. (25)

calcification (*n.f.*), calcification. (26)

calcifié, -e (*adj.*), calcified. (27)

calcin (incrustation dans les chaudières) (*n.m.*), scale; boiler-scale; fur; furring; incrusta--tion. (28)

calcinable (*adj.*), calcinable. (29)

calcination (*n.f.*), calcination; calcining; roasting : (30)

   calcination du minerai, calcination of ore; calcining the ore. (31)

calciné, -e (*adj.*), calcined; roasted. (32)

calciner (*v.t.*), to calcine; to roast : (33)

   calciner des minerais, to calcine ores. (34)

calciner (se) (*v.r.*), to calcine. (35)

calciocélestine (Minéral) (*n.f.*), calciocelestite. (36)

calcioferrite (Minéral) (*n.f.*), calcioferrite. (37)

calciothorite (Minéral) (*n.f.*), calciothorite. (38)

calcique (*adj.*), calcic; limy. (39)

calcite (Minéral) (*n.f.*), calcite. (40)

calcium (Chim.) (*n.m.*), calcium. (41)

calcoferrite (Minéral) (*n.f.*), calcioferrite. (42)

calcovolborthite (Minéral) (*n.f.*), calciovol--borthite. (43)

calcschiste (*n.m.*), micaceous calcareous schist. (44)

calcul (*n.m.*), calculation; reckoning; com--putation : (45)

   un calcul précis, a precise calculation; an exact reckoning. (46)

calculer (*v.t.*), to calculate; to reckon; to compute. (47)

caldeira (Géol.) (*n.f.*), caldera; chaldera. (48)

cale (*n.f.*) *ou* cale d'appui (fragment de bois, etc., que l'on place sous un objet pour lui donner l'assiette, le niveau, ou une certaine inclinaison), shim; packing-piece; wedge. (49)

cale (fragment de bois, etc., que l'on place sous un objet, une roue, pour l'empêcher de rouler) (*n.f.*), scotch; block; chock; sprag; wedge. (50)

cale (pièce de bois scellée dans un mur et contre laquelle bute la tête d'une contre-fiche) (Étaiement) (*n.f.*), needle. (51)

cale (coin) (Méc.) (*n.f.*), key; wedge. (52)

cale (pour coussinets de rails) (Ch. de f.) (*n.f.*), key; rail-key; plug. (53)

cale (placée dans la sous-cave) (Boisage de mines) (*n.f.*), sprag; holing-prop; punch-prop; gib. (54)

cale (d'un navire) (*n.f.*), hold (of a ship). (55)

cale à joints (Pose de voie) (Ch. de f.) (*f.*), expan--sion-gauge; expansion-shim; grasshopper shim; spider shim. (56)

cale d'arbre (*f.*), shaft-key. (57)

cale en V pour traçage (*f.*), draughtman's V block; engineers' vee block : (58)

   cale en V à une entaille, à — entailles, V block, with one vee, with — vees. (59 *ou* 60)

calebasse (*n.f.*), calabash. (61)

calédonite (Minéral) (*n.f.*), caledonite. (62)

caléfaction (*n.f.*), calefaction. (63)

caler (*v.t.*). See examples : (64)

   caler la tête d'une hache, to wedge an axe-head. (65)

   caler les balais (Élec.), to advance the brushes. (66)

   caler les roues d'un wagonnet de mine, to scotch, to chock, to sprag, to block, to cog, the wheels of a mine-car. (67)

   caler un instrument dans sa boîte, to pack an instrument in its case. (68)

   caler un instrument de nivellement, to clamp a levelling-instrument. (69)

   caler une pierre, un rail de chemin de fer, to shim a stone, a railway-rail. (70 *ou* 71)

   caler une poulie, les excentriques sur l'arbre, to key a pulley, the eccentrics on the shaft. (72 *ou* 73)

   caler une roue sur l'essieu, to fix a wheel on the axle. (74)

   caler une soupape, to jam a valve. (75)

calfait (*n.m.*), caulking-chisel; calking-iron; yarning-chisel. (76)

calfat (pers.) (*n.m.*), caulker; calker. (77)

calfatage (*n.m.*), caulking; calking. (78)

calfater (*v.t.*), to caulk; to calk. (79)

calibrage (*n.m.*), gauging; calipering : (80)

   calibrage des tubes, gauging (*or* calipering) tubes. (81)

calibrage du tube d'un thermomètre (*m.*), calibra--tion of the tube of a thermometer. (82)

calibre (diamètre) (*n.m.*), calibre; diameter. (83)

calibre (pièce préparée pour servir d'étalon) (*n.m.*), gauge; gage. (84) See also jauge.

calibre (pièce préparée pour servir de mesure) (*n.m.*), templet; template; pattern; patron. (85)

calibre (singe) (*n.m.*), former; jig; template. (86)

calibre (compas d'épaisseur) (*n.m.*), caliper; calipers; caliper-gauge. (87)

calibre à bague (*m.*), ring-gauge. (88)

calibre à bouchon (*m.*), plug-gauge. (89)

calibre à coulisse (*m.*), slide-calipers ; sliding calipers ; sliding caliper-gauge ; caliper-gauge ; caliper-square ; beam-caliper ; beam caliper-gauge. (1)

calibre à découper, en glace forte (Photogr.) (*m.*), plate-glass cutting-shape ; glass cutting-shape. (2)

calibre à lames (*m.*), feeler-gauge ; feeler ; thickness-gauge. (3)

calibre à tige coulissante et vis micrométrique (*m.*), internal micrometer. (4)

calibre à vis micrométrique (*m.*), micrometer-caliper ; micrometer-calipers ; micrometer-gauge ; micrometer. (5)

calibre conique (*m.*), taper gauge. (6)

calibre d'affûtage (pour la vérification de l'angle de coupe des forets) (*m.*), drill-gauge. (7)

calibre d'épaisseur (jauge pour fines tôles) (*m.*), sheet-gauge ; plate-gauge. (8)

calibre de détail (*m.*), detail-gauge. (9)

calibre de forage (*m.*), drilling-jig ; drilling-template. (10)

calibre de forme (*m.*), former ; jig. (11)

calibre de fraisage (*m.*), milling-jig. (12)

calibre de profondeur (*m.*), depth-gauge. (13)

calibre de tolérance *ou* calibre à tolérances *ou* calibre à limites (*m.*), limit-gauge. (14)

calibre étalon *ou* calibre de référence (*m.*), stand-ard gauge ; reference gauge. (15)

calibre mère *ou* calibre d'ensemble (*m.*), master-gauge. (16)

calibre pour écrous (*m.*), nut-gauge. (17)

calibre pour la vérification des vis (*m.*), screw-gauge ; screw-thread gauge ; thread-gauge. (18)

calibre prismatique (*m.*), measuring-block ; tool-setting gauge. (19)

calibre reproducteur (*m.*), former ; jig. (20)

calibre simple, femelle (*m.*), caliper-gauge, external ; fixed caliper-gauge, female. (21)

calibre simple, mâle (*m.*), caliper-gauge, internal ; fixed caliper-gauge, male. (22)

calibrer (*v.t.*), to gauge ; to gage ; to caliper : (23) calibrer un tube, to gauge, to caliper, a tube. (24)

calibrer le tube d'un thermomètre, to calibrate the tube of a thermometer. (25)

calibrer (se) (*v.r.*), to be gauged ; to be calipered. (26)

calibreur (*n.m.*), tube-gauge ; tube-caliper. (27)

caliche (nitrate de soude) (*n.m.*), caliche. (28)

caliner l'acier (Métall.), to kill the steel. (29)

calomel (Minéral) (*n.m.*), calomel ; horn-quick-silver. (30)

calorie (Phys.) (*n.f.*), calory ; calorie = 3·968 British thermal units. (31)

calorifiant, -e (*adj.*), heating ; calorific : (32) l'action calorifiante du soleil (*f.*), the heating action of the sun. (33)

calorifique (*adj.*), calorific ; thermic ; thermal ; heating : (34) pouvoir calorifique (*m.*), calorific (*or* heating) power. (35)

calorifuge (*adj.*), non-conducting ; non-conduc-tive ; insulating (i.e., non-conductive of heat) : (36) enveloppe calorifuge (*f.*), non-conducting lagging ; insulating jacketing. (37)

calorifuge (*n.m.*), non-conductor : (38) garnir une chaudière, un cylindre, d'un calori-fuge, to lag a boiler, a cylinder, with a non-conductor. (39 *ou* 40)

calorimètre (*n.m.*), calorimeter. (41)

calorimètre à eau (*m.*), water-calorimeter. (42)

calorimètre à glace (*m.*), ice-calorimeter. (43)

calorimétrie (*n.f.*), calorimetry. (44)

calorimétrique (*adj.*), calorimetric ; calori-metrical : (45) analyse calorimétrique (*f.*), calorimetric analysis. (46)

calorimétriquement (*adv.*), calorimetrically. (47)

calorique (*n.m.*), heat. (48)

calorique latent (*m.*), latent heat. (49)

calorique spécifique (*m.*), specific heat. (50)

calotte (Géom.) (*n.f.*), cap. (51)

calotte (d'un convertisseur Bessemer) (*n.f.*), nose (of a Bessemer converter). (52)

calotte glaciaire *ou* calotte de glace (*f.*), ice-cap ; ice-sheet ; glacial sheet : (53) la calotte glaciaire du Mont-Blanc, the ice-cap of Mont Blanc. (54) les immenses calottes de glace qui couvraient à l'époque quaternaire toute la surface de l'Europe septentrionale et les deux tiers de l'Amérique du Nord avaient leur origine, comme les glaciers continentaux du Groen-land et de l'Antarctique à l'époque actuelle, dans des régions élevées, the immense ice-sheets which covered in the Quaternary epoch the whole surface of northern Europe and two-thirds of North America had their origin, like the continental glaciers of Greenland and the Antarctic in the present epoch, in lofty regions. (55)

calotte glaciaire continentale (Glaciologie) (*f.*), continental ice-sheet ; continental glacier. (56)

calotte glaciaire locale (Glaciologie) (*f.*), local ice-cap. (57)

calotte sphérique (Géom.) (*f.*), spherical cap ; spherical segment. (58)

calquage (action) (*n.m.*), tracing. (59)

calque (*n.m.*), tracing : (60) calque d'un plan de mine, tracing of a mine-plan. (61)

calquer (*v.t.*), to trace ; to make a tracing of : (62) calquer un dessin, to trace a drawing. (63)

calqueur, -euse (pers.) (*n.*), tracer. (64)

camarade (compagnon de travail) (*n.m.*), mate. (65)

camarade d'atelier (*m.*), shopmate. (66)

camarades de travail (*m.pl.*), fellow workmen ; fellow-workmen ; mates. (67)

cambrage (*n.m.*), cambering. (68)

cambre (*n.f.*), camber. (69)

cambrer (*v.t.*), to camber. (70)

cambrer (se) (*v.r.*), to camber : (71) poutre qui commence à se cambrer (*f.*), beam which begins to camber. (72)

cambrien, -enne (Géol.) (*adj.*), Cambrian : (73) les roches du système cambrien (*f.pl.*), the rocks of Cambrian system. (74)

cambrien (*n.m.*), Cambrian : (75) le cambrien, the Cambrian. (76)

cambrite (Explosif) (*n.f.*), cambrite. (77)

cambrure (*n.f.*), camber : (78) la cambrure d'une pièce de bois, the camber of a piece of wood. (79)

came (Méc.) (*n.f.*), cam ; wiper ; lifter. (80)

came à cylindre (*f.*), cylinder-cam ; drum-cam ; barrel-cam. (81)

came à cylindre à rainure (*f.*), grooved cylinder-cam. (82)

came à développante (*f.*), involute cam. (1)

came à montagne russe (*f.*), waved wheel ; wave-wheel. (2)

came à tambour (*f.*), drum-cam ; cylinder-cam ; barrel-cam. (3)

came à tambour à languettes (*f.*), strake drum-cam. (4)

came de butée (*f.*), knock-off cam. (5)

came désaxée (*f.*), eccentric cam. (6)

came en cœur (*f.*), heart cam ; heart-shaped cam ; heart. (7)

camera lucida (*f.*), camera lucida. (8)

camion (*n.m.*), lorry. (9)

camion à vapeur (*m.*), steam-lorry. (10)

camion automobile (*m.*), motor-lorry ; autotruck. (11)

camion-citerne [camions-citernes *pl.*] (*n.m.*), tank-wagon. (12)

camionnage (*n.m.*), cartage ; carting. (13)

camp (*n.m.*), camp. (14)

camp de concentration (*m.*), compound ; con-centration-camp. (15)

camp minier (*m.*), mining camp. (16)

campagne (durée de temps exigée pour l'accom-plissement d'une opération industrielle) (*n.f.*), campaign ; run. (17)

campagne de bocard (minerai) (*f.*), mill-run. (18)

campagne de haut fourneau (*f.*), blast-furnace campaign ; run of blast-furnace. (19)

campan (marbre) (*n.m.*), Campan. (20)

campement (action) (*n.m.*), camping. (21)

camper (*v.i.*), to camp : (22)

camper dans une forêt, to camp in a forest. (23)

camptonite (Pétrol.) (*n.f.*), camptonite. (24)

campylite (Minéral) (*n.f.*), campylite. (25)

canal (Navigation) (*n.m.*), canal. (26)

canal (Trav. hydraul.) (*n.m.*), race ; ditch ; channel ; sluice ; raceway ; sluiceway ; canal. (27)

canal (conduit) (*n.m.*), pipe ; passage : (28)

canal pour le gaz, gas-pipe. (29)

canal pour la vapeur, steam-pipe. (30)

canal d'émission (de la vapeur), exhaust-passage ; exhaust-steam passage ; steam-passage. (31)

canal à bief de partage *ou* canal à point de partage (Navigation) (*m.*), summit-canal. (32)

canal à écluses et à lac intérieur (*m.*), lock and inland-lake canal (such as the Suez canal). (33)

canal à niveau (*m.*), level canal ; ditch-canal. (34)

canal collecteur (d'une pompe centrifuge) (*m.*), volute chamber, whirlpool-chamber, (of a centrifugal pump). (35)

canal d'alimentation (canal placé entre les rails pour l'alimentation en eau des locomotives pendant leur marche) (*m.*), track-tank ; feed-trough. (36)

canal d'amenée (*m.*), feeder ; lead ; leat ; flume ; penstock ; pentstock ; pentrough. (37)

canal d'assèchement *ou* canal d'écoulement *ou* canal d'épuisement (*m.*), drainage-channel. (38)

canal d'écoulement naturel (*m.*), natural drain-age-channel. (39)

canal d'évacuation *ou* canal d'échappement *ou* canal de décharge (*m.*), tail-race. (40)

canal d'irrigation (*m.*), irrigation-ditch ; catch-feeder. (41)

canal de coulée (Fonderie) (*m.*). Same as chenal de coulée.

canal de dérivation (*m.*). Same as canal d'amenée.

canal de desséchement (à la surface, pour dessécher un terrain bas ou humide) (*m.*), drainage-channel ; drainage-ditch. (42)

canal de fuite (*m.*), tail-race. (43)

canal de graissage (*m.*), oil-channel ; oil-groove. (44)

canal de navigation (*m.*), navigation. (45)

canal de prise (*m.*). Same as canal d'amenée.

canal découvert (opp. à conduite forcée) (Hydraul.) (*m.*), open channel (46) (contradistin-guished from full pipe) :

écoulement dans les canaux découverts (*m.*), flow in open channels. (47)

canal distributeur (*m.*), delivery-race ; delivery-channel ; distributing-canal. (48)

canal éclusé (*m.*), locked canal. (49)

canal en bois *ou* canal en planches (supporté par chevalets) (Méthode hydraulique) (*m.*), flume. (50)

canal latéral (Navigation) (*m.*), lateral canal. (51)

canal maritime (*m.*), ship-canal. (52)

canalisation (action de canaliser, ou établisse-ment de canaux) (*n.f.*), canalization ; canalizing : (53)

la canalisation d'une rivière, the canalization of a river ; canalizing a river. (54)

la canalisation du gaz, de la vapeur, the canalization of gas, of steam. (55 ou 56)

canalisation (système de canaux, ou réseau de conduites) (*n.f.*), canalization ; system of pipes ; piping ; pipage ; pipe-line ; line ; ditch-line : (57)

la canalisation qui amène la vapeur à la machine, the piping which brings the steam to the engine. (58)

canalisation (Élec.) (*n.f.*), transmission system ; canalization : (59)

influence de la nature du courant sur le rende-ment de la canalisation (*f.*), influence of the kind of the current on the efficiency of the transmission system. (60)

canalisation à pétrole (*f.*), petroleum pipe-line ; oil pipe-line ; oil-line. (61)

canalisation aérienne (Élec.) (*f.*), overhead transmission system. (62)

canalisation d'air (*f.*), air-line ; air pipe-line. (63)

canalisation d'air comprimé (*f.*), compressed-air line. (64)

canalisation d'amenée (*f.*), flume ; fluming. (65)

canalisation d'eau (*f.*), water-system. (66)

canalisation de gaz (*f.*), gas pipe-line. (67)

canalisation de tuyaux distributeurs (*f.*), canaliza-tion of distributing-pipes ; system of distributing-pipes. (68)

canalisation de vapeur (*f.*), steam pipe-line. (69)

canalisation électrique (*f.*), electric transmission system ; electric canalization : (70)

une canalisation électrique de transport de force et d'éclairage, an electric transmission system for the conveyance of power and light. (71)

canalisation pour le pétrole (*f.*). Same as canalisation à pétrole.

canalisation principale (*f.*), main ; main pipe-line. (72)

canalisation souterraine (Élec.) (*f.*), underground transmission system ; underground canal-ization. (73)

canaliser (couper de canaux) (*v.t.*), to canalize : (74)

canaliser une contrée, to canalize a country. (1)

canaliser (rendre navigable) (*v.t.*), to canalize : (2) canaliser une rivière, to canalize a river. (3)

canaliser (envoyer par canalisation) (*v.t.*), to pipe ; to pipe-line ; to send by pipe-line : (4) canaliser le gaz, l'eau d'une source, to pipe gas, water from a spring. (5 *ou* 6)

canaliser le pétrole, to pipe oil ; to pipe-line oil ; to send oil by pipe-line. (7)

canar *ou* canard (Aérage de mines) (*n.m.*), air-pipe. (8)

canari (*n.m.*), canary : (9) canaris employés pour indiquer la présence de gaz toxiques, canaries used to detect the presence of foul gases. (10)

cancrinite (Minéral) (*n.f.*), cancrinite. (11)

candel-coal *ou* candle-coal (*n.m.*). Same as cannel-coal.

candite (Minéral) (*n.f.*), candite ; ceylonite ; ceylanite. (12)

canette (Travail à la poudre) (*n.f.*), squib ; german ; match. (13)

canevas (*n.m.*), canvas. (14)

cani (*adj.m.*), druxy ; druxey : (15) du bois cani, druxy timber. (16)

cani (*n.m.*), druxy timber. (17)

canillon (d'un robinet) (*n.m.*), plug (of a cock *or* tap). (18)

caniveau (*n.m.*), gutter. (19)

caniveau souterrain (ligne de tramway électrique) (*m.*), conduit ; underground conduit. (20)

cannel-coal *ou simplement* cannel (*n.m.*), cannel-coal ; cannel ; candle-coal ; jet-coal ; kennel-coal. (21)

cannelé, -e (*adj.*), channeled ; fluted ; grooved ; corrugated : (22) cylindres cannelés (*m.pl.*), grooved rolls ; corrugated rolls ; fluted rollers. (23)

canneler (*v.t.*), to channel ; to flute ; to groove : to corrugate. (24)

cannelle (d'un robinet) (la partie fixe ; le cylindre extérieur) (*n.f.*), body (of a cock *or* tap). (25)

cannelle (robinet pour fûts) (*n.f.*), butt-cock. (26)

cannelure (*n.f.*), channeling ; channelling ; fluting ; groove ; flute ; channel ; corruga-tion. (27)

cannelure (Géol.) (*n.f.*), fault-fissure. (28)

cannelure (rainure pratiquée dans un cylindre de laminoir) (*n.f.*), groove : (29) cylindre à cannelures (*m.*), grooved roll. (30)

cannelure (ouverture formée entre une paire de cylindres à cannelures et par laquelle passe la barre) (*n.f.*), pass. (31)

cannelure carrée (laminoir) (*f.*), box pass. (32)

cannelure fermée *ou* cannelure emboîtée (laminoir) (*f.*), closed pass. (33)

cannelure finisseuse (laminoir) (*f.*), finishing-pass. (34)

cannelure glaciaire (Géol.) (*f.*), glacial grooving. (35)

cannelure ogive (laminoir) (*f.*), Gothic pass. (36)

cannelure ouverte *ou* cannelure roulante (laminoir) (*f.*), open pass. (37)

cannette (*n.f.*), butt-cock. (38)

canon (d'une clef) (la partie forée) (*n.m.*), pipe, tube, barrel, (of a key). (39)

canon (d'une serrure) (partie qui reçoit la tige de la clef) (*n.m.*), pipe, tube, barrel, (of a lock). (40)

canon (de la poupée fixe, de la poupée mobile, d'un tour) (*n.m.*), barrel (of the fixed head-stock, of the loose head-stock, of a lathe). (41 *ou* 42)

canon (d'une seringue) (*n.m.*), barrel (of a syringe). (43)

canon (d'une vis d'Archimède) (*n.m.*), barrel (of an Archimedean screw). (44)

canon de soufre (*m.*), stick, roll, cane, of sulphur. (45)

cañon (*n.m.*), cañon ; canyon : (46) le grand cañon du Colorado, the Grand Cañon of the Colorado. (47)

cantonite (Minéral) (*n.f.*), cantonite. (48)

canule (*n.f.*), butt-cock. (49)

caolin (*n.m.*), kaolin ; china-clay ; porcelain-clay. (50)

caoutchouc (*n.m.*), india-rubber ; rubber ; caoutchouc. (51)

caoutchouc de Para *ou* caoutchouc para (*m.*), Para rubber. (52)

caoutchouc durci (*m.*), hard rubber ; vulcanite ; ebonite. (53)

caoutchouc minéral *ou* caoutchouc fossile (*m.*), mineral caoutchouc ; elastic bitumen ; elastic mineral pitch ; elaterite. (54)

caoutchouc pneumatique (de bicyclette) (*m.*), pneumatic tyre. (55)

caoutchouc vulcanisé (*m.*), vulcanized rubber. (56)

caoutchouté, -e (*adj.*), rubberized : (57) toile caoutchoutée (*f.*), rubberized canvas. (58)

caoutchouter (*v.t.*), to rubberize. (59)

caoutchouteux, -euse (*adj.*), rubbery. (60)

cap (Géogr.) (*n.m.*), cape. (61)

capable (pouvant contenir, embrasser) (*adj.*), capable : (62) wagon capable de transporter jusqu'à — tonnes (*m.*), truck capable of carrying up to — tons. (63)

capable (habile ; expert) (*adj.*), capable ; com-petent ; skilful. (64)

capablement (*adv.*), capably. (65)

capacité (contenance ; aptitude) (*n.f.*), capacity : (66) capacité calorifique (Phys.), calorific capacity ; specific heat. (67)

capacité d'emmagasinage d'eau, water-storage capacity. (68)

capacité d'un accumulateur, capacity of an accumulator. (69)

capacité de filtration des tailings, leaching capacity of the tailings. (70)

capacité de la soute à combustible (de locomo-tive), fuel-capacity. (71)

capacité de la soute à eau (de locomotive), tank-capacity. (72)

capacité de travail, working capacity. (73)

capacité du cylindre, capacity of cylinder. (74)

capacité électrique, electric capacity : (75) l'unité de capacité électrique est le farad (*f.*), the unit of electric capacity is the farad. (76)

capacité inductrice (Élec.), inductive capacity. (77)

capacité limitée (*ou* restreinte) d'une raffinerie, limited capacity of a refining-works. (78)

capacité (talent) (*n.f.*), capacity ; ability ; efficiency : (79) manquer de capacité pour les affaires, to lack capacity for business ; to lack business ability. (80)

un homme d'une grande capacité, a man of great ability. (1)

capacité dans les travaux miniers, mining ability. (2)

la capacité professionnelle des ouvriers, the efficiency of the workmen. (3)

capelage des câbles (*m.*), capping ropes. (4)

capeler un câble, to cap a rope. (5)

capillaire (Phys.) (*adj.*), capillary : (6)

attraction capillaire (*f.*), capillary attraction. (7)

capillarité (*n.f.*), capillarity. (8)

capnite (Minéral) (*n.f.*), monheimite. (9)

caponnière (*n.f.*), shelter ; refuge ; refuge-hole ; manhole ; place of safety. (10)

capot en tôle pour moteur (*m.*), bonnet of sheet iron for engine ; sheet-iron hood for motor. (11)

capote en tôle pour cheminée (*f.*), cowl, hood, bonnet, of sheet iron for chimney. (12)

caprice (Mines) (*n.m.*), shoot ; dropper. (13)

capricieux, -euse (*adj.*), capricious ; fitful ; freak : (14)

source capricieuse (*f.*), capricious (*or* fitful) (*or* freak) spring. (15)

capsule (vase de laboratoire) (*n.f.*), capsule. (16)

capsule (d'une bouteille) (*n.f.*), capsule (of a bottle). (17)

capsule (Travail aux explosifs) (*n.f.*), cap ; blasting-cap ; detonator ; exploder ; capsule. (18)

capsule renforcée (Travail aux explosifs) (*f.*), strong detonator. (19)

captage (*n.m.*). See examples : (20)

captage d'eau, catchment of water ; catching water ; collection of water. (21)

captage d'une nappe pétrolifère, catchment of an oil-sheet ; catching an oil-sheet. (22)

captage de l'amalgame, catching the amal-gam. (23)

captage de l'or, gold-saving ; recovery of gold. (24)

captage du métal précieux, precious-metal saving ; recovery of precious metal. (25)

captation (*n.f.*). Same as **captage**.

capter (*v.t.*). See examples : (26)

capter l'or (dans le traitement), to save, to recover, gold. (27)

capter le métal sur les couvertures, to catch the metal on the blankets. (28)

capter les eaux, to catch, to collect, water. (29)

capture (Géogr. phys.) (*n.f.*), capture ; piracy : (30)

capture de cours d'eau, capture of streams ; stream piracy. (31)

capturer (Géogr. phys.) (*v.t.*), to capture. (32)

capuchon (*n.m.*), hood : (33)

capuchons pour lampes, pour cheminées de loco-motives, pour meules d'émeri, pour châssis à glace dépolie, hoods for lamps, for loco-motive smoke-stacks, for emery-wheels, for focussing-screens. (34 *ou* 35 *ou* 36 *ou* 37)

caracole (Sondage) (*n.f.*), crow's-foot ; pole-hook ; fishing-hook. (38)

caractère (*n.m.*), character ; nature ; property : (39)

caractère de la roche encaissante, character of the wall-rocks ; nature of the country-rock. (40)

caractères physiques et chimiques de l'or, physical and chemical properties of gold. (41)

caractères à jour, lettres et chiffres (*m.pl.*), stencil letters and figures. (42)

caractériser (*v.t.*), to characterize : (43)

métal caractérisé par un haut poids atomique (*m.*), metal characterized by a high atomic weight. (44)

caractéristique (*adj.*), characteristic : (45)

fossiles caractéristiques d'une période géologi-que (*m.pl.*), characteristic fossils of a geological period. (46)

caractéristique (*n.f.*), characteristic ; feature : (47)

caractéristiques des graviers aurifères, char-acteristics of the gold-bearing gravel. (48)

principales caractéristiques géologiques, main geological features. (49)

carat (unité de poids pour diamants) (*n.m.*), carat : (50)

diamant de — carats (*m.*), diamond of — carats. (51)

carat (collectivement, diamants très petits) (*n.m.*), carat-goods : (52)

le carat est fort demandé, carat-goods are much in demand. (53)

carat (*n.m.*) *ou* **carat de fin,** carat ; carat fine : (54)

or à — carats *ou* or à — carats de fin (*m.*), — -carat gold ; gold — carats fine. (55)

carat international (pierres précieuses) (*m.*), international carat = 200 milligrammes. (56)

carat métrique (pierres précieuses) (*m.*), metric carat = 200 milligrammes. (57)

carbo-minéral, -e, -aux (Élec.) (*adj.*), mineralized-carbon : (58)

arcs carbo-minéraux (*m.pl.*), mineralized-carbon arcs. (59)

charbon carbo-minéral (*m.*), mineralized carbon ; impregnated carbon. (60)

carbonado (*n.m.*), carbonado ; carbon ; bort ; black diamond. (61)

carbonatation (Chim.) (*n.f.*), carbonation ; carbonatation. (62)

carbonate (*n.m.*), carbonate : (63)

carbonate de chaux, carbonate of lime. (64)

carbonate de fer, carbonate of iron ; iron carbonate. (65)

carbonate de soude, carbonate of soda ; sodium carbonate. (66)

carbonate hydraté de cuivre, hydrous car-bonate of copper. (67)

carbonater (Chim.) (*v.t.*), to carbonate. (68)

carbone (Chim.) (*n.m.*), carbon. (69)

carbone de cémentation *ou* **carbone de carbure** (*m.*), cement carbon ; carbide carbon ; combined carbon. (70)

carbone de recuit (*m.*), temper carbon ; anneal-ing carbon. (71)

carbone de trempe (*m.*), hardening carbon. (72)

carbone fixe (*m.*), fixed carbon. (73)

carbone volatil (*m.*), volatile carbon. (74)

carboné, -e (*adj.*). Same as **carburé, -e.**

carboneux, -euse (*adj.*), carbonous. (75)

carbonifère (Minéralogie) (*adj.*), carboniferous ; coal-bearing : (76)

une formation carbonifère, a carboniferous (*or* a coal-bearing) formation. (77)

carbonifère *ou* **carboniférien, -enne** (Géol.) (*adj.*), Carboniferous : (78)

période carbonifère *ou* période carboniférienne (*f.*), Carboniferous period. (79)

carbonifère *ou* **carboniférien** (Géol.) (*n.m.*), Carboniferous. (80)

carbonification (*n.f.*), carbonization. (1)
carbonique (*adj.*), carbonic : (2)
  acide carbonique (*m.*) *ou* gaz carbonique (*m.*), carbonic acid ; carbonic-acid gas. (3)
carbonisation (*n.f.*), carbonization ; charring ; charing : (4)
  carbonisation du bois, carbonization of wood ; charring wood. (5)
carbonisé, -e (*adj.*), carbonized ; charred ; chared. (6)
carboniser (*v.t.*), to carbonize ; to char : (7)
  carboniser du bois, to carbonize, to char, wood. (8)
  carboniser la houille dans un four à coke, to carbonize, to char, coal in a coke-oven. (9)
carboniser (se) (*v.r.*), to carbonize ; to char : (10)
  bois qui se carbonise (*m.*), wood which car--bonizes (*or* chars). (11)
carbonite (Explosif) (*n.f.*), carbonite. (12)
carborundum (*n.m.*), carborundum. (13)
carburateur, -trice (*adj.*), carburating ; carburet--ing ; carburizing. (14)
carburateur (de moteur à essence) (*n.m.*), car--bureter (of petrol-motor). (15)
carburateur (pour gaz d'éclairage) (*n.m.*), car--bureter ; gas-machine. (16)
carburation (*n.f.*), carburization ; carbureting ; carburetting ; carburation : (17)
  carburation du fer par cémentation avec du charbon, carburization of wrought iron by cementation with charcoal. (18)
carbure (Chim.) (*n.m.*), carbide. (19)
carbure (*n.m.*) *ou* **carbure de calcium**, carbide ; carbide of calcium ; calcium carbide. (20)
  carbure tout venant, ungraded carbide. (21)
carbure (*n.m.*) *ou* **carbure d'hydrogène** (hydro--carbure) (Chim.), hydrocarbon : (22)
  carbures aromatiques, aromatic hydrocarbons. (23)
  carbures homologues, homologous hydro--carbons. (24)
  carbures saturés, saturated hydrocarbons. (25)
  carbures volatils, volatile hydrocarbons. (26)
carbure de fer (*m.*), carbide of iron. (27)
carburé, -e (*adj.*), carburized ; carbureted ; carburated. (28)
carburer (*v.t.*), to carburize ; to carburet ; to carburate. (29)
carcas (Métall.) (*n.m.*), horse ; old horse ; bear ; sow ; salamander ; shadrach ; freeze. (30)
carcasse (Charp., etc.) (*n.f.*), carcass ; carcase ; skeleton ; frame : (31)
  on arme le béton en le coulant sur une carcasse métallique, concrete as reinforced by pouring it upon a metal skeleton. (32)
carcasse (d'un moteur électrique) (*n.f.*), carcass, carcase, frame, (of an electric motor). (33)
carcel (Photom.) (*n.m.*), carcel. (34)
carde à limes *ou* **carde pour limes** (*f.*), file-card ; file-cleaning card. (35)
cargaison (*n.f.*), cargo. (36)
cargaison d'aller (*f.*), outward cargo. (37)
cargaison de retour (*f.*), homeward cargo. (38)
cargneule (Géol.) (*n.f.*), rauchwacke. (39)
carie du bois (*f.*), timber-rot ; rotting of timber ; decay of timber. (40)
carie humide (*f.*), wet rot. (41)
carie sèche (*f.*), dry rot. (42)
carié, -e (*adj.*), rotten ; decayed. (43)

carier (*v.t.*), to rot ; to decay. (44)
carier (se) (*v.r.*), to rot ; to decay. (45)
carnallite (Minéral) (*n.f.*), carnallite. (46)
carneau (*n.m.*), flue. (47)
carneau à air (*m.*), air-flue. (48)
carneau à fumées *ou* **carneau de fumée** (*m.*), smoke-flue. (49)
carneau à gaz (*m.*), gas-flue. (50)
carneau de chaudière (*m.*), boiler-flue. (51)
carneau de four (*m.*), furnace-flue. (52)
carnet (registre) (*n.m.*), note-book ; book. (53)
carnet d'aérage (Mines) (*m.*), airway ; air-sollar. (54)
carnet d'opérations (Topogr.) (*m.*), field-book. (55)
carnet de levé (*m.*), survey-book. (56)
carnet de levé à la boussole *ou* **carnet de mine** (*m.*), dialing-book. (57)
carnet de levé au théodolite (*m.*), transit-book. (58)
carnet de nivellement (*m.*), levelling-book ; level-book. (59)
carnet de sondage (*m.*), bore-holing journal ; drillers' log. (60)
carniole (Géol.) (*n.f.*), rauchwacke. (61)
carnotite (Minéral) (*n.f.*), carnotite. (62)
carotte *ou* **carotte-témoin** [carottes-témoins *pl.*] (Sondage) (*n.f.*), core ; bore-core ; drill-core. (63)
carotte (cale pour enrayer les roues d'un wagon sur rails) (*n.f.*), sprag. (64)
carotte-échantillon [carottes-échantillons *pl.*] (*n.f.*), core-sample : (65)
  carottes-échantillons de roches, core-samples of rocks. (66)
carpholite (Minéral) (*n.f.*), carpholite. (67)
carrare (*n.m.*), Carrara marble. (68)
carré, -e (*adj.*), square : (69)
  mètre carré (*m.*), square metre. (70)
  boulon à tête carrée (*m.*), square-headed bolt. (71)
carré (Géom.) (*n.m.*), square. (72)
carré (fer, acier, etc.) (*s'emploie le plus souvent au pluriel*) (*n.m.*), square. (73)
carré (surface plane entre deux moulures, ou à l'emplacement d'une baguette) (Menuis.) (*n.m.*), quirk. (74)
carré (palier d'escalier) (*n.m.*), landing. (75)
carreau (plaque carrée ou rectangulaire) (*n.m.*), square : (76)
  carreau de verre, square, pane, of glass. (77)
carreau (espèce de pavé plat) (*n.m.*), tile : (78)
  carreau de carrelage, flooring-tile. (79)
carreau (brique ou pierre disposée de telle sorte que sa face la plus longue soit en parement,—opp. à boutisse) (Maçonn.) (*n.m.*), stretcher. (80)
carreau (lime carrée) (*n.m.*), square file. (81)
carreau (*n.m.*) (d'une carrière) *ou* **carreau de la carrière**, head (of a quarry) ; quarry-head. (82)
carreau (d'une mine) (*n.m.*) *ou* **carreau de la mine**, grass ; bank ; bank-head ; pit-bank : (83)
  minerai sur le carreau de la mine (*m.*), ore on the bank ; ore at grass. (84)
carrefour (*n.m.*), crossroads : (85)
  le carrefour d'une forêt, the crossroads of a forest. (86)
carrelage (en carreaux) (*n.m.*), tile flooring ; tile floor. (87)
carrelage en briques (*m.*), brick paving ; brick pavement. (88)

carreler (*v.t.*), to floor ; to pave. (1)

carrelet (Dessin) (*n.m.*), square ruler. (2)

carrelet (*n.m.*) *ou* carrelette (*n.f.*) (lime), square file (small). (3)

carrément (*adv.*), squarely. (4)

carrer (*v.t.*), to square : (5)
carrer un bloc de marbre, to square a block of marble. (6)

carrer (se) (*v.r.*), to be squared ; to square : (7)
pierre qui ne peut pas se carrer (*f.*), stone which cannot be squared. (8)
le bout et le côté doivent se carrer l'un avec l'autre, the end and the side should square with each other. (9)
le cercle ne peut se carrer, the circle cannot be squared. (10)

carrier (pers.) (*n.m.*), quarryman ; quarrier. (11)

carrière (*n.f.*), quarry ; pit ; stone-quarry ; stone-pit. (12)

carrière à ciel ouvert (*f.*), open quarry. (13)

carrière abandonnée (*f.*), abandoned quarry ; disused quarry. (14)

carrière d'argile (*f.*), clay-pit. (15)

carrière de craie (*f.*), chalk-pit. (16)

carrière de grès (*f.*), sandstone-quarry. (17)

carrière de marbre (*f.*), marble-quarry. (18)

carrière de pierre à chaux *ou* carrière de pierre calcaire (*f.*), limestone-quarry ; lime-quarry ; lime-pit. (19)

carrière de pierre à plâtre (*f.*), gypsum-quarry. (20)

carrière de pierre de taille (*f.*), cut-stone quarry. (21)

carrière de remblais (*f.*), quarry worked at the surface for supplying gobbing-material. (22)

carte (représentation du globe ou d'une de ses parties) (*n.f.*), map ; chart. (23)

carte à grande échelle (*f.*), large-scale map. (24)

carte bathymétrique des océans (*f.*), bathy-metrical chart of the oceans. (25)

carte de circulation (Ch. de f.) (*f.*), pass (over a railroad). (26)

carte de l'état-major (*f.*), ordnance map. (27)

carte en courbes de niveau (*f.*), contour-map. (28)

carte géographique (*f.*), geographical map ; map. (29)

carte géologique (*f.*), geological map. (30)

carte hydrographique (*f.*), hydrographic map ; hydrograph. (31)

carte plane (*f.*), plane-chart. (32)

carte pluviométrique (*f.*), rain-chart ; rain-map. (33)

carte routière (*f.*), route-map ; road-map. (34)

carte topographique (*f.*), topographical map. (35)

carter (*n.m.*) *ou* carter des engrenages, gear-case ; gear-casing ; gear-cover ; gear-guard ; gear box ; wheel-guard. (36)

cartogramme (*n.m.*), cartogram. (37)

cartographe (pers.) (*n.m.*), cartographer ; mapper. (38)

cartographie (*n.f.*), cartography ; mapping. (39)

cartographique (*adj.*), cartographic ; carto-graphical. (40)

carton (*n.m.*), cardboard : (41)
boîte en carton (*f.*), cardboard box. (42)

carton bitumé (*m.*), tarred felt ; roofing-felt. (43)

carton d'amiante (*m.*), asbestos millboard ; amianthus. (44)

carton de montagne *ou* carton fossile *ou* carton minéral (Minéral) (*m.*), mountain-paper ; fossil paper. (45)

carton pour coller les épreuves (Photogr.) (*m.*), paste-on mount. (46) Cf. passe-partout.

cartouche (*n.f.*), cartridge. (47)

cartouche à chemise d'eau (*f.*), water-cartridge. (48)

cartouche-amorce [cartouches-amorces *pl.*] (*n.f.*), primer. (49)

cartouche de dynamite (*f.*), dynamite cartridge. (50)

cartouche de sûreté (*f.*), safety-cartridge. (51)

cartouche éclairante (Photogr.) (*f.*), flash-candle. (52)

cascade (*n.f.*), cascade ; waterfall ; fall ; falls. (53)

cascade de séracs (Glaciologie) (*f.*), ice-cascade ; ice-fall. (54)

cascader (*v.i.*), to cascade : (55)
eau qui cascade de rocher en rocher (*f.*), water which cascades from rock to rock. (56)

cascalho (*n.m.*), cascalho ; mellan. (57)

case (*n.f.*) *ou* casier (*n.m.*), bin ; bunker. (58)

case à minerai (*f.*), ore-bin ; ore-bunker. (59)

casque (*n.m.*), helmet. (60)

casque de pompier (*m.*), fireman's helmet. (61)

casque de sauvetage *ou* casque respiratoire (*m.*), smoke-helmet. (62)

cassable (*adj.*), breakable. (63)

cassage (*n.m.*), breaking : (64)
cassage des gros blocs de minerai au chantier, breaking the big lumps of ore in the stope. (65)
cassage des gueuses de fonte à la masse, breaking iron pigs with the sledge. (66)

cassant, -e (*adj.*), brittle : (67)
le verre non-recuit est très cassant, unannealed glass is very brittle. (68)

cassant (*n.m.*), breakableness : (69)
le cassant du verre, the breakableness of glass. (70)

casse (action de casser) (*n.f.*), breaking ; break-age : (71)
pour éviter la casse du manche, on creuse un trou au moyen d'un foret, to avoid breaking the handle, a hole should be bored with a drill. (72)

casse (objets cassés) (*n.f.*), breakages : (73)
payer la casse, to pay for breakages. (74)

casse (coupelle) (*n.f.*), gold-refining cupel ; mellan. (75)

casse (cône d'éboulis) (Géol.) (*n.f.*), talus. (76)

casse-coke [casse-ccke *pl.*] (*n.m.*), coke-breaker ; coke-breaking machine. (77)

casse-cou (*adj.*), breakneck. (78)

casse-cou [casse-cou *pl.*] (*n.m.*), breakneck. (79)

casse-fer [casse-fer *pl.*] (*n.m.*), hardy ; hardie ; anvil-cutter. (80)

casse-fonte [casse-fonte *pl.*] (*n.m.*), castings-breaker. (81)

casse-gueuses [casse-gueuses *pl.*] (*n.m.*), pig-iron breaker. (82)

casse-pierre [casse-pierre *pl.*] (*n.m.*). Same as casse-pierres.

casse-pierres [casse-pierres *pl.*] (machine) (*n.m.*), stone-breaker ; stone-crusher ; stone-mill ; rock-breaker. (83)

casse-pierres (masse) (*n.m.*), stone-hammer. (84)

casse-pierres (trépan de sonde) (*n.m.*), four-wing star bit. (85)

casser (*v.t.*), to break : (86)
casser un verre, to break a glass. (87)

casser (v.i.), to break. (1)

casser (se) (v.r.), to break. (2)

cassette d'instruments de mathématiques (f.), magazine-case of mathematical instruments. (3)

casseur de pierres (pers.) (m.), stone-breaker. (4)

cassitérite (Minéral) (n.f.), cassiterite ; tinstone ; tin ore. (5)

cassitérotantalite (Minéral) (n.f.), cassitero-tantalite. (6)

cassure (action) (n.f.), breaking ; fracture ; rending : (7)
la cassure d'un verre, the breaking of a glass. (8)

cassure (solution de continuité ; fente) (n.f.), break ; fracture ; rent. (9)

cassure (Géol.) (n.f.), fracture ; rent : (10)
l'étude des lois qui président à la répartition géographique des grandes cassures de la lithosphère (f.), the study of the laws which govern the geographical distribution of the great fractures of the lithosphere. (11)
une cassure de terrain, a rent in the ground. (12)

cassure (Minéralogie & Métall.) (n.f.), fracture. (13)

cassure à grains (Minéralogie & Métall.) (f.), granular fracture. (14)

cassure à nerf (Métall.) (f.), fibrous fracture. (15)

cassure argentée (Métall.) (f.), silvery fracture. (16)

cassure brillante (Minéralogie) (f.), brilliant fracture. (17)

cassure céroïde (Minéralogie) (f.), splintery fracture. (18)

cassure cireuse (Minéralogie) (f.), waxy fracture (fracture having the dull whitish or yellowish colour of wax). (19)

cassure compacte (Minéralogie) (f.), compact fracture. (20)

cassure conchoïdale ou cassure conchoïde (Minéralogie) (f.), conchoidal fracture. (21)

cassure conique ou cassure conoïde (Minéralogie) (f.), conical fracture. (22)

cassure écailleuse ou cassure esquilleuse (Minéralogie) (f.), splintery fracture. (23)

cassure fibreuse (Minéralogie & Métall.) (f.), fibrous fracture. (24)

cassure fraîche (f.), fresh fracture. (25)

cassure grenue (Minéralogie & Métall.) (f.), granular fracture. (26)

cassure hachée ou cassure crochue (Minéralogie) (f.), hackly fracture. (27)

cassure inégale (Minéralogie) (f.), uneven fracture. (28)

cassure lamellaire ou cassure lamelleuse (Minéralogie) (f.), lamellated fracture. (29)

cassure laminaire (Minéralogie) (f.), laminar fracture. (30)

cassure mate (Minéralogie) (f.), lustreless fracture. (31)

cassure nerveuse (Métall.) (f.), fibrous fracture. (32)

cassure nette (Minéralogie) (f.), clean fracture. (33)

cassure plate (Minéralogie) (f.), plane fracture. (34)

cassure raboteuse (Minéralogie) (f.), uneven fracture. (35)

cassure résineuse (Minéralogie) (f.), resinous fracture. (36)

cassure saccharoïde (Minéralogie) (f.), saccha-roidal fracture. (37)

cassure schistoïde (Minéralogie) (f.), schistoid fracture. (38)

cassure simple (Géol.) (f.), fissure ; diaclase (a simple fracture of strata without disloca-tion, distinguished from fault or paraclase). (39)

cassure terne (Minéralogie) (f.), dull fracture. (40)

cassure unie (Minéralogie) (f.), even fracture. (41)

cassure vitreuse (Minéralogie) (f.), vitreous fracture. (42)

cassure vive (Minéralogie) (f.), bright fracture. (43)

castine (n.f.), limestone flux. (44)

castor (n.m.) ou castorite (n.f.) (Minéral), castor-ite ; castor. (45)

catacaustique (Opt.) (adj.), catacaustic. (46)

catacaustique (n.f.), catacaustic ; caustic by reflection. (47)

cataclastique (Géol.) (adj.), cataclastic : (48)
roches cataclastiques (f.pl.), cataclastic rocks. (49)

cataclinal, -e, -aux (Géol.) (adj.), cataclinal. (50)

cataclysme (Géol.) (n.m.), cataclysm. (51)

cataclysmique (adj.), cataclysmic ; cataclysmal ; cataclysmatic. (52)

catalogue (n.m.), catalogue : (53)
catalogues illustrés et devis sur demande, illustrated catalogues and estimates on application. (54)

catamaran (n.m.), catamaran : (55)
drague en catamaran (f.), catamaran dredge. (56)

catamorphisme (Géol.) (n.m.), catamorphism. (57)

catapléite (Minéral) (n.f.), catapleite. (58)

cataracte (chute d'eau) (n.f.), cataract. (59)

cataracte (régulateur hydraulique) (Mach.) (n.f.), cataract. (60)

catastrophe (n.f.), catastrophe. (61)

catégorie (n.f.), category ; class : (62)
catégorie de travail, class of work. (63)

cathétomètre (n.m.), cathetometer. (64)

cathode (Phys.) (n.f.), cathode ; kathode. (65)

cathodique (adj.), cathodic ; cathodical ; kathodic. (66)

cation (Phys.) (n.m.), cation. (67)

catlinite (argile) (n.f.), catlinite. (68)

cause (n.f.), cause : (69)
la cause d'un accident, the cause of an accident. (70)

causer (v.t.), to cause ; to occasion : (71)
causer beaucoup d'ennuis, to cause a lot of annoyance ; to give a lot of trouble ; to occasion much trouble. (72)
causer l'arrêt des travaux, to cause a stoppage of work. (73)

causticité (n.f.), causticity. (74)

caustique (adj.), caustic : (75)
soude caustique (f.), caustic soda. (76)

caustique (corrosif) (n.m.), caustic. (77)

caustique (Opt.) (n.f.), caustic. (78)

caustique par réflexion (Opt.) (f.), caustic by reflection ; catacaustic. (79)

caustique par réfraction (Opt.) (f.), caustic by refraction ; diacaustic. (80)

cavalier (d'un tour à fileter) (n.m.), quadrant, quadrant-plate, radial arm, radial plate, swing-frame, tangent plate, (of a screw-cutting lathe). (81)

cavalier (d'une balance de précision) (*n.m.*), rider (of a balance of precision). (1)

cavalier (clou à deux pointes) (*n.m.*), staple ; wire staple. (2)

cavalier de déblais (*m.*), spoil-bank ; spoil-heap. (3)

cave (*n.f.*), cellar ; vault. (4)

caver (*v.t.*), to hollow ; to hollow out ; to under--mine ; to wear away ; to wear hollow : (5) la mer cave les rochers, the sea hollows the rocks. (6) l'eau cave lentement la pierre (*f.*), water slowly wears away stone. (7)

caver (se) (*v.r.*), to hollow ; to hollow out ; to become hollow ; to wear away ; to wear hollow. (8)

caverne (*n.f.*), cave ; cavern. (9)

caverne à ossements (*f.*), bone-cave. (10)

caverneux, -euse (*adj.*), cavernous ; hollow : (11) rochers caverneux (*m.pl.*), cavernous rocks. (12)

cavernosité (*n.f.*), hollowness : (13) la cavernosité d'un rocher, the hollowness of a rock. (14)

cavité (*n.f.*), cavity ; hollow : (15) cavité rocheuse, rock cavity. (16) les cavités intérieures de la terre, the internal cavities of the earth. (17)

céder (*v.i.*), to cede ; to yield ; to give in ; to give way : (18) un bois mal assujetti peut céder, a badly fixed timber may give way. (19)

ceinture de sûreté (*f.*), safety-belt. (20)

céladonite (Minéral) (*n.f.*), celadonite. (21)

célestine (*n.f.*) *ou* célestin (*n.m.*) (Minéral), celestite ; celestine. (22)

cellulaire (*adj.*), cellular : (23) la texture cellulaire de certaines roches, the cellular texture of certain rocks. (24)

cellule (*n.f.*), cell : (25) les cellules d'une éponge, the cells of a sponge. (26)

celluloïd *ou* celluloïde (*n.m.*), celluloid. (27)

celtite (Explosif) (*n.f.*), celtite. (28)

cément (Métall.) (*n.m.*), cement. (29)

cémentation (*n.f.*), cementation ; case-hardening. (30)

cémentatoire (*adj.*), cementatory : (31) poudre cémentatoire (*f.*), cementatory powder. (32)

cémenter (*v.t.*), to cement ; to case-harden : (33) cémenter le fer pour le transformer en acier malléable, to cement iron in order to trans--form it into malleable steel. (34)

cémenter (se) (*v.r.*), to cement ; to become cemented : (35) le fer se cémente facilement, iron cements easily. (36)

cémenteux, -euse (*adj.*), cementing : (37) matière cémenteuse (*f.*), cementing material. (38)

cémentite (Métall.) (*n.f.*), cementite. (39)

cendre (*n.f.*) *ou* cendres (*n.f.pl.*), ash ; ashes. (40)

cendre bleue native (Minéral) (*f.*), blue ashes ; ash-blue. (41)

cendre d'os (*f.*), bone-ash. (42)

cendre de bois (*f.*), wood-ash. (43)

cendre de houille (*f.*), coal-ash. (44)

cendres volcaniques (Géol.) (*f.pl.*), volcanic ashes ; volcanic ash. (45)

cendrier (de locomotive) (*n.m.*), ash-pan. (46)

cendrier (de chaudière à vapeur fixe) (*n.m.*), ash-pit. (47)

cendrier (vase, fosse pour recevoir les cendres) (*n.m.*), ash-bin ; ash-pit. (48)

cendrure (défaut dans le métal) (*n.f.*), cinder-pit. (49)

centiare (*n.m.*), centiare ; centare = 10·7639 sq. feet *or* 1·1960 sq. yards. (50)

centibar (Météor.) (*n.m.*), centibar. (51)

centigrade (*adj.*), centigrade : (52) échelle centigrade (*f.*), centigrade scale. (53)

centigramme (*n.m.*), centigramme = 0·154 grain. (54)

centilitre (*n.m.*), centilitre = 0·070 gill. (55)

centimètre (*n.m.*), centimetre = 0·3937 inch. (56)

centimètre carré (*m.*), square centimetre = 0·15500 square inch. (57)

centimètre cube (*m.*), cubic centimetre = 0·0610 cubic inch. (58)

centrage (Méc.) (*n.m.*), centering ; centring : (59) le centrage d'un cylindre creux, centering a hollow cylinder. (60)

central, -e, -aux (*adj.*), central : (61) chauffage central (*m.*), central heating. (62)

centrale électrique *ou simplement* centrale (*n.f.*), central electric station ; central power-station ; central generating station. (63)

centralement (*adv.*), centrally. (64)

centre (*n.m.*), centre ; center ; middle : (65) le centre d'un cercle, the centre of a circle. (66)

centre (Sismologie) (*n.m.*), centrum ; centre ; focus. (67)

centre d'oscillation (*m.*), centre of oscillation. (68)

centre de dépression (Météor.) (*m.*), centre of depression. (69)

centre de figure (Math.) (*m.*), centre of figure. (70)

centre de gravité (Méc.) (*m.*), centre of gravity ; centre of mass ; centre of inertia : (71) le centre de gravité du fléau d'une balance, the centre of gravity of a balance-beam. (72)

centre de percussion (*m.*), centre of percussion. (73)

centre de poussée (Hydros.) (*m.*), centre of buoyancy ; centre of displacement ; centre of immersion. (74)

centre de pression (Hydros.) (*m.*), centre of pressure. (75)

centre de roue (*m.*), wheel-centre. (76)

centre de symétrie (*m.*), centre of symmetry. (77)

centre des forces parallèles (*m.*), centre of parallel forces. (78)

centre houiller (*m.*), coal-mining centre. (79)

centre instantané de rotation (Méc.) (*m.*), instan--taneous centre of rotation. (80)

centre minier (*m.*), mining centre. (81)

centre optique des lentilles (*m.*), optical centre of lenses. (82)

centrer (*v.t.*), to centre ; to center : (83) centrer une tige pour montage entre pointes de tour, to centre a rod for mounting between lathe-centres. (84) une pièce à tourner doit toujours être centrée convenablement, a part to be turned should always be properly centered (*or* centred). (85) centrer une lentille, to centre a lens. (86)

centrifuge (*adj.*), centrifugal : (87) force centrifuge (*f.*), centrifugal force. (88)

centripète (*adj.*), centripetal : (89) force centripète (*f.*), centripetal force. (90)

centripétence (*n.f.*), centripetence ; centrip-
-etency. (1)
cérargyrite (*n.f.*) *ou* cérargyre (*n.m.*) (Minéral),
cerargyrite ; , kerargyrite ; horn-silver ;
horn-ore. (2)
cerceau (*n.m.*), hoop : (3)
cerceau d'un tonneau, hoop of a cask. (4)
cerceau de Delezenne (Élec.) (*m.*), Delezenne's
circle. (5)
cerclage (*n.m.*), hooping : (6)
cerclage de tonneaux, hooping casks. (7)
cercle (Géom.) (*n.m.*), circle. (8)
cercle (cerceau) (*n.m.*), hoop ; band ; ring ;
circle. (9)
cercle à calcul (*m.*), circular slide-rule ; pocket-
calculator ; watch-calculator. (10)
cercle d'aberration *ou* cercle d'aberration chroma-
-tique (Opt.) (*m.*), circle of aberration. (11)
cercle d'alignement (*m.*), transit-instrument ;
transit. (12)
cercle d'échanfrinement *ou* cercle d'échanfreine-
-ment (Engrenages) (*m.*). Same as cercle
de tête (Gearing).
cercle d'évidement (Engrenages) (*m.*). Same as
cercle de pied.
cercle de base (tracé d'un engrenage à dévelop-
-pante) (*m.*), base-circle. (13)
cercle de contact (Engrenages) (*m.*). Same as
cercle primitif.
cercle de couronne (Engrenages) (*m.*). Same as
cercle de tête.
cercle de déclinaison (Astron.) (*m.*), declination-
circle ; circle of declination. (14)
cercle de division (Engrenages) (*m.*). Same as
cercle primitif.
cercle de fixation de l'enveloppe (locomotive)
(*m.*), jacket-band. (15)
cercle de galets (plaque tournante, etc.) (*m.*),
ring of rollers. (16)
cercle de pied *ou* cercle de racine (Engrenages)
(*m.*), root-circle ; dedendum-circle (Gearing).
(17)
cercle de renfort de l'avant de la boîte à fumée
(locomotive) (*m.*), smoke-box front ring.
(18)
cercle de roue (*m.*), wheel-tire ; wheel-tyre. (19)
cercle de roulement (*m.*), circular track. (20)
cercle de tête (Engrenages) (*m.*), addendum-
circle ; point-circle ; outside circle ; blank-
circle. (21)
cercle de tonneau (*m.*), cask-hoop. (22)
cercle divisé *ou* cercle gradué (*m.*), divided circle ;
graduated circle. (23)
cercle extérieur (Engrenages) (*m.*). Same as
cercle de tête.
cercle générateur (*m.*), generating circle. (24)
cercle horaire (*m.*), horary circle ; hour-circle.
(25)
cercle horizontal (d'un théodolite, ou instrument
analogue) (*m.*), horizontal circle (of a theodo-
-lite, or the like). (26 *ou* 27)
cercle limite (Engrenages) (*m.*), clearance-circle.
(28)
cercle méridien (instrument) (*m.*), meridian-
circle ; transit-circle ; transit. (29)
cercle moleté d'un oculaire (*m.*), milled ring of
an eyepiece. (30)
cercle oculaire (Opt.) (*m.*), eye-ring. (31)
cercle polaire (*s'emploie généralement au pluriel*)
(*m.*), polar circle. (32)
cercle polaire antarctique (*m.*), antarctic circle.
(33)

cercle polaire arctique (*m.*), arctic circle. (34)
cercle primitif (Engrenages) (*m.*), pitch-circle ;
pitch-line ; primitive circle : dividing-
circle ; rolling circle. (35)
cercle vertical (d'un théodolite, ou instrument
analogue) (*m.*), vertical circle (of a theodolite,
or the like). (36 *ou* 37)
cerclé (-e) de fer, iron-bound ; hooped with iron.
(38)
cercler (*v.t.*), to hoop ; to bind ; to encircle ;
to circle : (39)
cercler un tonneau, to hoop a cask. (40)
cérérite (Minéral) (*n.f.*), cererite ; cerite. (41)
cérésine (*n.f.*), ceresin ; ceresine ; fossil wax ;
mineral wax. (42)
céreux, -euse (Chim.) (*adj.*), cerous : (43)
oxyde céreux (*m.*), cerous oxide. (44)
cérides (*n.m.pl.*), cerium metals ; metals of the
cerium family. (45)
cérifère (Minéralogie) (*adj.*), cerium-bearing ;
cerium-producing. (46)
cérine (Minéral) (*n.f.*), cerine ; allanite ; orthite.
(47)
cérique (Chim.) (*adj.*), ceric : (48)
oxyde cérique (*m.*), ceric oxide. (49)
terres cériques (*f.pl.*), ceric earths. (50)
cérite (Minéral) (*n.f.*), cerite ; cererite. (51)
cérium (Chim.) (*n.m.*), cerium. (52)
cérolite (Minéral) (*n.f.*), cerolite. (53)
certificat (*n.m.*), certificate. (54)
certificat d'amélioration (*m.*), certificate of
improvement. (55)
certificat d'épreuve (*m.*), test-certificate. (56)
certificat de franc-mineur (*m.*), free-miner's
certificate. (57)
céruse (blanc d'argent) (*n.f.*), white lead ;
ceruse. (58)
cérusite *ou* cérussite *ou* céruse (Minéral) (*n.f.*),
cerusite ; cerussite ; ceruse ; lead-spar. (59)
cervantine (Minéral) (*n.f.*), cervantite ; antimony
ochre. (60)
césium (Chim.) (*n.m.*), cesium ; cæsium. (61)
cessation (*n.f.*), cessation ; stoppage ; suspension :
(62)
cessation de travail, d'affaires, cessation of
work, of business. (63 *ou* 64)
cesser (*v.i.*), to cease ; to discontinue ; to stop :
(65)
trains qui cessent de circuler (*m.pl.*), trains
which cease running. (66)
cesser de se servir d'une machine, to discontinue
the use of a machine. (67)
la pluie a cessé, the rain has stopped. (68)
cesser le travail (se mettre en grève), to stop
work ; to down tools ; to strike. (69)
ceylonite *ou* ceylanite (Minéral) (*n.f.*), ceylonite ;
ceylanite ; candite. (70)
chabasie *ou* chabacite (Minéral) (*n.f.*), chabazite ;
chabasite ; chabasie. (71)
chablis *ou* chable (*n.m.*), windfall ; wind-blown
tree. (72)
chabotte (d'un marteau-pilon) (*n.f.*), anvil-
block (of a power-hammer). (73)
chadouf (*n.m.*), shadoof. (74)
chaînage (Arpent.) (*n.m.*), chaining ; chainage.
(75)
chaîne (*n.f.*), chain. (76)
chaîne (de collines, de montagnes) (*n.f.*), chain
(of hills, of mountains). (77 *ou* 78)
chaîne (Chim.) (*n.f.*), chain. (79)
chaîne (mesure anglaise de longueur) (*n.f.*),
chain = 20·1168 metres. (80)

chaîne à blocs (*f.*), block-chain. (1)

chaîne à fuseau (*f.*), stud-chain; pintle-chain; pin-chain. (2)

chaine a godets *ou* **chaîne à augets** (*f.*), bucket-chain. (3)

chaîne à la Vaucanson (*f.*), ladder chain. (4)

chaîne à maillons (*f.*), link-chain. (5)

chaîne à maillons courts (*f.*), short-link chain. (6)

chaîne à maillons longs (*f.*), long-link chain. (7)

chaîne à maillons ordinaires (*f.*), ordinary-link chain. (8)

chaîne à maillons ouverts (*f.*), open-link chain. (9)

chaîne à maillons pleins (*f.*), block-chain. (10)

chaîne à rouleaux (*f.*), roller-chain. (11)

chaîne à rouleaux et à manchons (*f.*), bush-roller chain. (12)

chaîne à ruban d'acier (Arpent.) (*f.*), steel band-chain ; band-chain. (13)

chaîne blanche (opp. à chaîne noire) (*f.*), bright chain. (14)

chaîne-câble [chaînes-câbles *pl.*] (*n.f.*), chain cable ; cable-chain. (15)

chaîne-câble à étais *ou* chaîne-câble étançonnée *ou* chaîne-câble à mailles étançonnées (*f.*), stud-link chain cable ; studded-link cable-chain. (16)

chaîne-câble à maillons courts *ou* chaîne-câble à mailles serrées (*f.*), short-link chain cable. (17)

chaîne calibrée (*f.*), pitched chain ; pitch-chain. (18)

chaîne d'allongement (Sondage) (*f.*), drill-chain. (19)

chaîne d'arpentage *ou* chaîne d'arpenteur (*f.*), surveying-chain ; surveyors' chain ; land-chain ; measuring-chain ; land-measuring chain. (20)

[Note.—The length of the chaîne d'arpentage in France = 1 decametre. (21) In England and the United States of America the surveyors' chain *or* Gunter's chain = 4 rods, or 66 feet, each link being 7·92 inches. (22) Ramsden's chain has 100 links of 1 foot each, as has also the engineers' chain. (23)]

chaîne d'attelage (*f.*), coupling-chain ; drag-chain. (24)

chaîne de commande du frein à main (*f.*), brake-chain. (25)

chaîne de contrepoids (*f.*), tail-chain ; balance-chain. (26)

chaîne de Galle (*f.*), sprocket-chain. (27)

chaîne de levage *ou* chaîne de charge (d'un palan, etc.) (*f.*), lifting-chain ; load-chain. (28)

chaîne de manœuvre (d'un palan, d'un monte-charge, ou analogue) (opp. à chaîne de levage *ou* chaîne de charge) (*f.*), hand-chain (of a pulley-block, hoist, or the like). (29 *ou* 30 *ou* 31)

chaîne de sûreté (*f.*), safety-chain ; bridle-chain. (32)

chaîne de suspension (*f.*), chain sling ; sling-chain ; sling. (33)

chaîne de suspension à grappin (*f.*), dog-hook sling-chain. (34)

chaîne de Vaucanson (*f.*), ladder chain. (35)

chaîne dragueuse (*f.*), dredge-chain. (36)

chaîne droite (opp. à chaîne torse) (*f.*), straight-link chain. (37)

chaîne en acier sans soudure (*f.*), weldless-steel chain. (38)

chaîne en gerbe (*f.*), indented chain. (39)

chaîne en S (*f.*), twisted chain; curbed chain. (40)

chaîne éprouvée (*f.*), tested chain. (41)

chaîne fermée (Chim.) (*f.*), closed chain ; ring. (42)

chaîne flottante (Roulage) (Mines) (*f.*), endless chain above wagons ; car-haul chain over tubs. (43)

chaîne-Galle (*n.f.*) *ou* chaîne Galle [chaînes-Galle *ou* chaînes Galle *pl.*], sprocket-chain. (44)

chaîne galvanisée (*f.*), galvanized chain. (45)

chaîne lâche (*f.*), long-link chain. (46)

chaîne " ne varietur " (*f.*), silent chain ; noise-less chain. (47)

chaîne noire (*f.*), black chain. (48)

chaîne ouverte (Méc.) (*f.*), open-link chain. (49)

chaîne ouverte (Chim.) (*f.*), open chain. (50)

chaîne pendante (*f.*), hanging chain. (51)

chaîne plate (*f.*), flat-link chain. (52)

chaîne polie (opp. à chaîne noire) (*f.*), bright chain. (53)

chaîne porte-couteaux (d'une haveuse à chaîne) (*f.*), cutter-chain (of a chain coal-cutting machine). (54)

chaîne principale (montagnes) (*f.*), main chain. (55)

chaîne sans fin (*f.*), endless chain. (56)

chaîne secondaire (montagnes) (*f.*), secondary chain. (57)

chaîne serrée (*f.*), short-link chain. (58)

chaîne silencieuse (*f.*), silent chain ; noiseless chain. (59)

chaîne soudée (*f.*), welded chain. (60)

chaîne torse (*f.*), twisted chain ; curbed chain. (61)

chaîne traînante (Roulage) (Mines) (*f.*), endless chain below wagons ; car-haul chain under tubs ; creeper-chain. (62)

chaîne traînante à maillons d'accrochage (*f.*), creeper-chain with hooks. (63)

chaîne traînante avec chariot d'accrochage (*f.*), creeper-chain with grip. (64)

chaînée (mesure prise à la chaîne d'arpenteur) (*n.f.*), chain. (65)

chaîner (Arpent.) (*v.t.*), to chain. (66)

chaînetier (pers.) (*n.m.*), chain-maker ; chain-smith. (67)

chaînette (petite chaîne) (*n.f.*), small chain ; chain. (68)

chaînette (Math.) (*n.f.*), catenary ; catenary curve ; funicular curve. (69)

chaîneur (Arpent.) (pers.) (*n.m.*), chainman. (70)

chaînier *ou* chaîniste (pers.) (*n.m.*), chain-maker ; chainsmith. (71)

chaînon (Méc.) (*n.m.*), link : (72)

les chaînons d'une chaîne, d'un câble-chaîne, the links of a chain, of a cable-chain. (73 *ou* 74)

chaînon (montagnes) (*n.m.*), secondary chain. (75)

chair fossile *ou* chair minérale (Minéral) (*f.*), mountain-flesh. (76)

chaise (support) (Méc.) (*n.f.*), seat : (77)

la chaise d'un coussinet, the seat of a bearing. (78)

chaise (de laminoir) (*n.f.*), chock (of rolling-mill). (79)

chaise (*n.f.*) *ou* chaise pour transmission, hanger ; shafting-hanger ; shaft-hanger. (80)

chaise à deux jambages (f.) ou chaise-lyre (n.f.) [chaises-lyres pl.], double-sided hanger; sling-hanger. (1)

chaise à deux jambages, palier graisseur à rotule et à bague (f.), double-sided self-oiling swivel-hanger with ring; self-oiling sling-hanger, with swivel-bearings and ring. (2)

chaise à un seul jambage ou chaise à une seule jambe ou chaise pendante en col de cygne ou chaise en J (f.), open-sided hanger; open-side hanger; J hanger. (3)

chaise à un seul jambage et à palier ordinaire ou chaise à une seule jambe et à palier ordinaire (f.), ordinary J hanger. (4)

chaise à un seul jambage, palier graisseur à rotule (f.), open-sided self-oiling swivel-hanger; self-oiling open-side hanger, with swivel-bearings. (5)

chaise-console [chaises-consoles pl.] ou chaise-applique [chaises-appliques pl.] (n.f.), wall-bracket; bracket-hanger. (6)

chaise de sol ou chaise sur le sol (f.), A standard; standard; floor-hanger. (7)

chaise en bout (f.), end wall-bracket; end-bracket; angle-bracket. (8)

chaise-niche [chaises-niche pl.] (n.f.), wall-box. (9)

chaise-palier [chaises-paliers pl.] (n.f.), hanger with bearings. (10)

chaise-palier pendante à rotule, portée — fois le diamètre, coussinets bronze (f.), hanger with swivel-bearings, length of brasses equal to — diameters, gun-metal bearings. (11)

chaise pendante (f.), ceiling-hanger; pendant bracket; hanger. (12)

chaise pendante à deux jambes ou chaise en U (f.), sling-hanger. (13)

chaland ou chalan (n.m.), barge; lighter; scow. (14)

chaland-citerne [chalands-citernes pl.] (n.m.), tank-barge. (15)

chalcanthite (Minéral) (n.f.), chalcanthite; cyanose; cyanosite. (16)

chalcédoine (Minéral) (n.f.), chalcedony; calcedony. (17)

chalcédonieux, -euse (adj.), chalcedonous. (18)

chalcédonique (adj.), chalcedonic. (19)

chalcédonix (Minéral) (n.m.), chalcedonyx. (20)

chalcocite (Minéral) (n.f.). Same as chalcosine.

chalcolite (Minéral) (n.f.), chalcolite; torbernite; torberite; uranite; copper uranite. (21)

chalcomorphite (Minéral) (n.f.), chalcomorphite. (22)

chalcophanite (Minéral) (n.f.), chalcophanite. (23)

chalcophyllite (Minéral) (n.f.), chalcophyllite; copper-mica. (24)

chalcopyrite (Minéral) (n.f.), chalcopyrite; copper pyrites; yellow copper; yellow copper ore; yellow pyrites; yellow ore. (25)

chalcopyrrhotine (Minéral) (n.f.), chalcopyr-rhotite. (26)

chalcosidérite (Minéral) (n.f.), chalcosiderite. (27)

chalcosine (Minéral) (n.f.), chalcocite; chal-cosine; copper-glance; glance-copper; gray ore. (28)

chalcostibite (Minéral) (n.f.), chalcostibite; wolfsbergite. (29)

chalcotrichite (Minéral) (n.f.), chalcotrichite; plush-copper. (30)

chaleur (Phys.) (n.f.), heat: (31)

la chaleur du feu, du soleil, the heat of the fire, of the sun. (32 ou 33)

chaleur (n.f.) ou chaleurs (n.f.pl.) (temps chaud), heat; hot weather: (34)

pendant les chaleurs, during the hot weather. (35)

les grandes chaleurs, the great heat; the very hot weather. (36)

chaleur atomique (f.), atomic heat. (37)

chaleur blanche (f.), white heat. (38)

chaleur cachée (f.). Same as chaleur latente.

chaleur centrale du globe (f.), central heat of the earth. (39)

chaleur de combinaison (Thermochim.) (f.), heat of formation. (40)

chaleur de combustion (Thermochim.) (f.), heat of combustion. (41)

chaleur de vaporisation (f.). Same as chaleur latente de vaporisation.

chaleur insoutenable ou chaleur insupportable (f.), unbearable (or insupportable) heat. (42)

chaleur latente (f.), latent heat. (43)

chaleur latente de fusion (f.), latent heat of fusion. (44)

chaleur latente de vaporisation (f.), latent heat of vaporization: heat of vaporization; vaporization latent temperature. (45)

chaleur moléculaire (f.), molecular heat. (46)

chaleur perdue (f.) ou chaleurs perdues (f.pl.), waste heat. (47)

chaleur radiante ou chaleur rayonnante (f.), radiant heat. (48)

chaleur rouge (f.), red heat. (49)

chaleur sensible (f.), sensible heat. (50)

chaleur spécifique (f.), specific heat; calorific capacity. (51)

chaleur spécifique de vaporisation (f.), specific heat of vaporization; vaporization specific temperature. (52)

chalumeau (n.m.), blowpipe. (53)

chalumeau à acétylène (m.), acetylene-blow-pipe. (54)

chalumeau à bouche (m.), mouth-blowpipe. (55)

chalumeau à gaz ou chalumeau à gaz d'éclairage (m.), gas-blowpipe. (56)

chalumeau à pointe de platine de Plattner (m.), Plattner's platinum-tip blowpipe. (57)

chalumeau électrique (m.), electric blowpipe. (58)

chalumeau oxhydrique (m.), oxyhydrogen blow-pipe; compound blowpipe. (59)

chalumeau oxy-acétylénique (m.), oxyacetylene blowpipe. (60)

chalybite (Minéral) (n.f.), chalybite. (61)

chambranle (n.m.), jamb-lining. (62)

chambre (n.f.), chamber; room. (63)

chambre (d'un injecteur) (n.f.), chamber (of an injector). (64)

chambre (Photogr.) (n.f.). See chambre noire.

chambre à air (de pompe, etc.) (f.), air-chamber (of pump, etc.). (65)

chambre à air (de cubilot ou haut fourneau) (f.), wind-box (of cupola or blast-furnace). (66 ou 67)

chambre à air (d'un bandage pneumatique) (f.), inner tube (of a pneumatic tyre). (68)

chambre à eau (d'une chaudière) (f.), water-space (of a boiler). (69)

chambre à feu (d'une chaudière à vapeur) (f.), fire-chamber (of a steam-boiler). (70)

chambre barométrique (f.), Torricellian vacuum. (71)

chambre claire (*f.*), camera lucida. (1)

chambre d'accrochage *ou* chambre d'envoyage (Mines) (*f.*), landing-station ; onsetting-station ; plat ; platt ; lodge ; station ; shaft-station ; pit-landing. (2)

chambre d'air (dans un trou de mine) (Travail aux explosifs) (*f.*), air-space (in a shot-hole) (Blasting). (3)

chambre d'aspiration (*f.*), suction-box. (4)

chambre d'eau (d'une chaudière) (*f.*), water-space (of a boiler). (5)

chambre d'eau (pour le moteur d'une turbine) (*f.*), flume ; turbine-chamber ; turbine-pit ; forebay. (6)

chambre d'eau ouverte (turbine)(*f.*), open flume ; open turbine-chamber. (7)

chambre d'écluse (canal) (*f.*), lock-chamber. (8)

chambre d'emprunt (balastière) (*f.*), borrow-pit ; borrow. (9)

chambre d'épuration (dans un moule) (Fonderie) (*f.*), skim-gate ; skimming-chamber ; whirling runner. (10)

chambre d'explosion (de moteur à combustion interne) (*f.*), combustion-chamber (of internal-combustion engine). (11)

chambre de chauffe (*f.*), boiler-room. (12)

chambre de combustion (de foyer de chaudière à vapeur) (*f.*), combustion-chamber (of steam-boiler furnace). (13)

chambre de compression (*f.*), compression-chamber. (14)

chambre de dessablage au jet de sable (*f.*), sand-blast cleaning-room. (15)

chambre de distribution (tiroir) (*f.*), steam-chest ; valve-chest ; steam-box ; valve-box ; steam-case ; valve-case ; valve-casing ; steam-chamber ; valve-chamber ; slide-box (slide-valve). (16)

chambre de grillage du minerai (*f.*), ore-roasting chamber ; stall. (17)

chambre de la machine (*f.*), engine-room. (18)

chambre de mine (fourneau de mine) (*f.*), blast-hole ; shot-hole ; mine-chamber ; mining-hole ; powder-mine. (19)

chambre de plomb (*f.*), leaden chamber. (20)

chambre de séchage (*f.*), drying-room ; drying-chamber. (21)

chambre de travail (d'un caisson) (*f.*), working-chamber (of a caisson). (22)

chambre de vapeur (d'une chaudière) (*f.*), steam-space (of a boiler). (23)

chambre des pompes (*f.*), pump-room. (24)

chambre des portes (écluse) (*f.*), lock-chamber. (25)

chambre du moteur (*f.*), engine-room. (26)

chambre noire (cabinet noir ; laboratoire obscur) (Photogr.) (*f.*), dark room. (27)

chambre noire *ou simplement et ordinairement* chambre (*n.f.*) (appareil photographique), camera : (28)

chambre photographique *ou* chambre noire photographique, photographic camera. (29)

chambre à magasin, magazine-camera. (30)

chambre à main, hand-camera. (31)

chambre à main à foyer fixe, fixed-focus hand-camera. (32)

chambre à mise au point (opp. à *chambre à foyer fixe*), focussing-camera. (33)

chambre à obturateur de plaque, focal-plane camera. (34)

chambre à pellicules *ou* chambre pour pelli-cules, film-camera. (35)

chambre à (*ou* pour) pellicules en bobines (*ou* en rouleaux) se chargeant en plein jour, daylight-loading roll-film camera. (36)

chambre à (*ou* pour) plaques, plate-camera. (37)

chambre à soufflet, bellows camera. (38)

chambre d'agrandissement, enlarging-camera. (39)

chambre d'atelier, studio camera. (40)

chambre de forme rectangulaire, box-form camera. (41)

chambre de voyage *ou* chambre de touriste *ou* chambre touriste [chambres touriste *pl.*] *ou* chambre à pied, field-camera ; stand-camera ; view-camera. (42)

chambre demi-plaque de voyage, half-plate field-camera. (43)

chambre détective, detective camera. (44)

chambre pliante *ou* chambre folding, folding camera. (45)

chambre reflex, reflex camera. (46)

chambre se mettant dans la poche du gilet, vest-pocket camera. (47)

chambre stéréoscopique, stereoscopic camera. (48)

chambre obscure (*f.*).   Same as **chambre noire**.

chambré, -e (*adj.*), chambered : (49)

noyau chambré (Fonderie) (*m.*), chambered core. (50)

mine chambrée (Travail aux explosifs) (*f.*), chambered hole. (51)

chambres isolées (panneau) (Exploitation des mines de charbon) (*f.pl.*), panel. (52)

chambrière (reculoire de chariot de mine) (*n.f.*), backstay, safety-dog, dog, drag, (of mine-car). (53)

chameau (*n.m.*), camel : (54)

transport à dos de chameau (*m.*), camel transport. (55)

courroies en poil de chameau (*f.pl.*), camel-hair belting. (56)

chamois (peau de chamois) (*n.m.*), wash-leather ; shammy ; chamois. (57)

chamoisite *ou* chamosite (Minéral) (*n.f.*), chamo-site ; chamoisite. (58)

champ (étendue de terre) (*n.m.*), field. (59)

champ (Opt.) (*n.m.*), field : (60)

le champ d'une lunette, d'un microscope, the field of a telescope, of a microscope. (61 *ou* 62)

champ (Phys.) (*n.m.*), field. (63)

champ (côté d'une pièce équarrie le plus étroit dans le sens de la longueur) (*n.m.*), edge : (64)

barre plate à champs ronds (*f.*), flat bar with round edges. (65)

planche placée sur champ (*f.*), board placed on edge. (66)

champ (de), edgeways ; edgewise : (67)

placer les poutres de champ, to place the beams edgeways. (68)

champ aurifère *ou* champ d'or (*m.*), gold-field. (69)

champ d'exploitation (Mines) (*m.*), winning : (70)

les limites du champ d'exploitation (*f.pl.*), the boundaries of the winning. (71)

champ d'investigation (*m.*), field of investigation. (72)

champ de cuivre (*m.*), copper-field. (73)

champ de dragage (*m.*), dredging-field. (74)

champ de force (Phys.) (*m.*), field of force. (75)

champ de fractures (Géol.) (*m.*), group of faults. (76)

champ de gaz (*m.*), gas-field. (1)

champ de glace (*m.*), ice-field : (2)
d'immenses champs de glace couvrent, en Norvège, les plateaux qui s'élèvent au-dessus de la limite des neiges éternelles, in Norway, immense ice-fields cover the plateaux which rise above the snow-line. (3)

champ de lave (*m.*), lava-field. (4)

champ de neige (*m.*), snow-field. (5)

champ de pétrole (*m.*), oil-field ; petroleum-field. (6)

champ de vision *ou* champ de vue (Opt.) (*m.*). Same as champ visuel.

champ diamantifère (*m.*), diamond-field. (7)

champ électrique (*m.*), electric field. (8)

champ magnétique (*m.*), magnetic field ; field of magnetic force. (9)

champ pétrolifère (*m.*), oil-field ; petroleum-field. (10)

champ plan (Photogr.) (*m.*), flat field. (11)

champ salicole *ou* champ salifère (*m.*), salt-field. (12)

champ tournant (Élec.) (*m.*), rotary field. (13)

champ visuel (Opt.) (*m.*), field of view ; field of vision ; visual field ; field of regard. (14)

champignon (de cheminée) (*n.m.*), cowl (of chimney). (15)

champignon d'essieu (*m.*), axle-collar. (16)

champignon de rail (*m.*), rail-head ; head of railway-rail. (17)

championite (Explosif) (*n.f.*), championite. (18)

chance (*n.f.*), chance ; prospect ; luck : (19)
chances de trouver du charbon, chances (*or* prospects) of finding coal. (20)
bonne chance, good luck. (21)
mauvaise chance, bad luck. (22)

chandelier (*n.m.*), candlestick ; candle-holder. (23)

chandelier (pour supporter le corps cylindrique d'une chaudière) (*n.m.*), pedestal. (24)

chandelle (bougie) (*n.f.*), candle. (25)

chandelle (étai vertical) (Constr.) (*n.f.*), vertical shore ; dead-shore. (26)

chandelle (bois droit et isolé ; butte ; étai) (Mines) (*n.f.*), prop; tree; post; shore. (27)

chandelle de glace (*f.*), icicle. (28)

chandelle de suif (*f.*), tallow candle. (29)

chandeller (*v.t.*), to prop ; to prop up ; to shore ; to shore up : (30)
chandeller le toit d'une galerie de mine, to prop, to shore, the roof of a mine-level. (31)

chanfrein (*n.m.*), chamfer. (32)

chanfrein arrêté (Menuis.) (*m.*), stopped chamfer. (33)

chanfreiner *ou* chanfreindre *ou* chanfraindre (*v.t.*), to chamfer : (34)
chanfreiner un écrou, une tôle de chaudière, to chamfer a nut, a boiler-plate. (35 *ou* 36)

chanfreineuse (*n.f.*), plate-edge planing-machine. (37)

changeant, -e (*adj.*), changing ; changeable ; variable. (38)

changement (*n.m.*), change ; alteration ; varia-tion : (39)
un changement de vitesse, a change of speed. (40)

changement (*n.m.*) *ou* changement de voie (Ch. de f.), switch ; points ; turnout ; turn-off : (41)
introduction d'un changement dans une voie déjà posée (*f.*), introducing a switch into an existing track. (42)

changement à aiguilles (*m.*), split-switch ; point-switch. (43)

changement à deux voies symétriques *ou* changement simple (pose symétrique) (*m.*), two-way symmetrical switch ; two-way symmetrical turnout ; right and left-hand turn-off. (44)

changement à droite (*m.*), right-hand switch ; right-hand turnout ; right-hand turn-off. (45)

changement à gauche (*m.*), left-hand switch ; left-hand turnout ; left-hand turn-off. (46)

changement à rails mobiles (*m.*), stub-switch ; blunt switch. (47)

changement à trois voies avec croisements (pose symétrique) *ou* changement triple avec croisements (pose symétrique) *ou* changement double avec croisements (pose symétrique) (*m.*), three-way switches and crossings with symmetrical points ; 3-way right and left-hand turn-off with crossings. (48)

changement avec croisement, switch and cros-sing ; points and crossing. (49)

changement de marche (action) (Méc.) (*m.*), reversing ; reversing the motion. (50)

changement de marche (appareil) (*m.*), revers-ing-gear ; reversing-motion. (51)

changement de marche à vis (*m.*), screw reversing-gear. (52)

changement de marche du mouvement de pression (*m.*), feed-reversing gear. (53)

changement de marche pour fileter à droite ou à gauche (tour) (*m.*), reversing-motion for cutting right or left-hand threads ; screw-cutting reverse (lathe). (54)

changement de vitesse (Engrenages) (*m.*), change-speed gear ; change-gear. (55)

changement de voie (Ch. de f.) (*m.*). See changement *ou* changement de voie.

changer (*v.t.*), to change ; to alter : (56)
changer ses projets, to change one's plans. (57)

changer (*v.i.*), to change ; to become changed ; to alter : (58)
changer d'aspect, to change in appearance. (59)
changer sous l'action de la chaleur, to change under the influence of heat ; to become changed by the action of heat. (60)

changer de mains (*ou* de propriétaire), to change hands (*or* ownership) (*or* owners). (61)

changer de place (*v.t.*), to shift : (62)
changer une machine de place, to shift a machine. (63)

changer le sens de la marche, to reverse. (64)

changer (se) (*v.r.*), to change ; to alter. (65)

chanlate *ou* chanlatte (*n.f.*), eaves-board ; eaves-catch ; eaves-lath. (66)

chant du grisou (Mines) (*m.*), singing, sing, of the fire-damp. (67)

chantier (*n.m.*) *ou* chantier de travail (Mines), working ; workings ; working-place. (68)

chantier (*n.m.*) *ou* chantier d'abatage (Mines), stope. (69)

chantier (d'une fonderie) (*n.m.*), floor (of a foundry). (70)

chantier à ciel ouvert (Mines) (*m.*), open-cast ; open-cut ; open-cast workings ; openwork. (71)

chantier à ciel ouvert (Placers) (*m.*), open-cast ; open-cut ; paddock. (72)

chantiers à mi-pente (*m.pl.*), mid-workings. (73)

chantiers abandonnés (*m.pl.*), abandoned workings. (1)

chantiers anciens (*m.pl.*), old workings ; former workings ; ancient workings. (2)

chantier barré (*m.*), fenced-off working-place. (3)

chantier chassant (*m.*), drift-stope. (4)

chantier d'avancement (*m.*), heading-stope. (5)

chantier de bois (*m.*), timber-yard ; lumber-yard. (6)

chantier de décharge des débris (*m.*), dump ; dumping-ground ; spoil-heap ; spoil-bank. (7)

chantier de dépôt *ou* chantier de versage (*m.*), dumping-ground ; dump ; tip. (8)

chantier de houille (*m.*), coal-yard. (9)

chantier de lavage (Placers) (*m.*), washing ; washings : (10)

chantier de lavage des sables aurifères, gold-washings. (11)

chantier de moulage (Fonderie) (*m.*), moulding-floor. (12)

chantier de moulage en sable (Fonderie) (*m.*), sand-floor. (13)

chantier de moulage en terre (Fonderie) (*m.*), loam-work floor. (14)

chantier de pierre (*m.*), stone-yard. (15)

chantiers du fond (*m.pl.*), bottom workings. (16)

chantier en butte (Dragage) (*m.*), bank-working (a working-place on the bank of a river). (17)

chantiers en cloche (*m.pl.*), bell-shaped works. (18)

chantier en gradins (Mines) (*m.*), stope. (19)

chantier en mort-terrain (*m.*), dead-workings. (20)

chantier en remonte (*m.*), raise-stope. (21)

chantier en vallée (Dragage) (*m.*), bed-working (a working-place in the bed of a river). (22)

chantiers épuisés (*m.pl.*), exhausted workings ; stoped out workings ; worked out stopes. (23)

chantier étriqué *ou* chantier étroit (*m.*), narrow working-place ; scanty working-place. (24)

chantiers inférieurs (*m.pl.*), lower workings. (25)

chantiers les plus éloignés (*m.pl.*), farthest-in workings. (26)

chantier profond (*m.*), deep workings. (27)

chantiers sous-marins (*m.pl.*), submarine work-ings ; undersea workings. (28)

chantiers souterrains (*m.pl.*), underground workings. (29)

chantiers supérieurs (*m.pl.*), upper workings. (30)

chantiers très éloignés (*m.pl.*), far distant work-ings. (31)

chantignole *ou* chantignolle (Constr.) (*n.f.*), purlin-cleat. (32)

chantournage (*n.m.*), jig-sawing. (33)

chantourner (*v.t.*), to jig-saw. (34)

chanvre (*n.m.*), hemp. (35)

chanvre de Manille (*m.*), Manila hemp ; Manila ; manila ; manilla. (36)

chaos (Géogr. phys.) (*n.m.*), chaos : (37)

un chaos de rochers, de blocs, a chaos of rocks, of blocks (*or* of boulders). (38 *ou* 39)

chaotique (*adj.*), chaotic. (40)

chape (couvercle) (*n.f.*), cover ; cap ; lid ; head. (41)

chape (enduit) (*n.f.*), coating : (42)

une chape en asphalte recouvrant l'extrados de la voûte d'un pont, a coating of asphalt covering the back of the arch of a bridge. (43)

chape (*n.f.*) *ou* chape de châssis (Fonderie), cheek ; middle part ; middle. (44)

chape (d'un moule en terre) (Fonderie) (*n.f.*), cope (of a loam mould). (45)

chape (d'une balance) (*n.f.*), bearings (of a balance) : (46)

couteaux du fléau d'une balance reposant dans une chape d'agate (*m.pl.*), knife-edges of a balance-beam resting in agate bearings. (47)

chape (d'une aiguille de boussole) (*n.f.*), cap, centre, (of a compass-needle). (48)

chape (d'une articulation à genouillère) (*n.f.*), jaw, fork, (of a knuckle-joint). (49)

chape d'agate (boussole) (*f.*), agate cap ; agate centre : (50)

barreau à chape d'agate (*m.*), bar needle with agate cap. (51)

chape de bielle *ou simplement* chape (*n.f.*), rod-end strap (connecting-rod *or* side-rod) ; strap. (52)

chape de poulie (*f.*), pulley-shell. (53)

chape de suspension des fermes (chaudière de locomotive) (*f.*), sling-stay. (54)

chape en pierre dure (rubis *ou* saphir) (boussole) (*f.*), jewelled cap (ruby *or* sapphire). (55 *ou* 56)

chapeau (*n.m.*), cap ; capping. (57)

chapeau (pièce de bois posée horizontalement sur la partie la plus élevée d'une charpente) (Constr.) (*n.m.*), cap. (58)

chapeau (d'un palier) (*n.m.*), cap, pedestal-cover, (of a bearing-block *or* of a plummer-block). (59)

chapeau (d'un pieu) (*n.m.*), cap, hood, (of a pile). (60)

chapeau (d'une soupape) (*n.m.*), cap (of a valve). (61)

chapeau (d'une vanne) (*n.m.*), cap-sill (of a sluice-gate). (62)

chapeau (d'un presse-étoupe) (*n.m.*), gland, follower, (of a stuffing-box). (63)

chapeau (d'une poutre composée) (*n.m.*), flange-plate, flange, boom-plate, boom, (of a built-up girder). (64)

chapeau (Boisage des galeries de mine) (*n.m.*), crown-piece ; crown ; head-piece ; cap-piece ; cap ; collar. (65)

chapeau (lampe) (*n.m.*), cap ; hood ; bonnet. (66)

chapeau de châssis *ou simplement* chapeau (*n.m.*) (Fonderie), cope ; top part of flask. (67)

chapeau de fer (Géol.) (*m.*), iron-hat ; hat ; ironstone blow ; iron-gossan ; gossan. (68)

chapeau de filon (Géol. & Mines) (*m.*), cap ; capping ; cap-rock ; rock-capping. (69)

chapelet d'îles (*m.*), string of islands. (70)

chapelet d'outils (Sondage) (*m.*), string of tools. (71)

chapelet hydraulique (*m.*), chain-pump ; chapelet ; paternoster pump. (72)

chapelle (Méc.) (*n.f.*), a box or chamber contain-ing some mechanism. (73)

chapelle (boîte à clapet) (*n.f.*), clack-box ; valve-box ; valve-chest ; valve-chamber ; valve-case ; valve-casing. (74)

chapelle de refoulement (*f.*), delivery-box. (75)

chapelle du tiroir (*f.*), steam-chest ; valve-chest ; steam-box ; valve-box ; valve-case ; valve-case ; valve-casing ; steam-chamber ; valve-chamber ; slide-box. (76)

chaperon (Arch.) (*n.m.*), coping. (77)

chaperonner (Arch.) (v.t.), to cope : (1)
chaperonner un mur, to cope a wall. (2)

chapiteau (d'un alambic) (n.m.), head, cap, (of
a still). (3)

chapiteau de cheminée (de locomotive) (m.),
cap of smoke-stack ; smoke-stack top. (4)

chaque (adj.), each : (5)
pilons de — kilos chaque, stamps of — kilos
each. (6)

char à bennes (Mines) (m.), lorry. (7)

charbon (de bois) (n.m.), charcoal ; wood-
charcoal ; wood-coal. (8)

charbon (de terre) (n.m.), coal. (9)   See also
houille.

charbon (Élec.) (n.m.), carbon : (10)
charbons pour lampes à arc (m.pl.), carbons
for arc-lamps. (11)
balais en charbon (m.pl.), carbon brushes. (12)

charbon à coke (m.), coking coal. (13)

charbon à gaz (m.), gas-coal. (14)

charbon à mèche (Élec.) (m.), cored carbon. (15)

charbon à souder (m.), soldering-charcoal ;
soldering-block ; charcoal block. (16)

charbon bitumineux (m.). Same as charbon
gras.

charbon carbo-minéral [charbons carbo-miné-
-raux pl.] (Élec.) (m.), mineralized carbon ;
impregnated carbon. (17)

charbon cuivré (Élec.) (m.), coppered carbon.
(18)

charbon de bois (m.), charcoal ; wood-charcoal ;
wood-coal. (19)

charbon de cornue (m.), gas-carbon. (20)

charbon de forge (m.), smithing coal ; black-
-smith coal. (21)

charbon de laverie (m.), washery-coal. (22)

charbon de ménage (m.), house-coal ; domestic
coal. (23)

charbon de Paris (m.), briquettes composed of
tar and various carbonized fuels (sawdust,
bark, spent tan, peat, coke and coal-dust).
(24)

charbon de terre (m.), coal. (25)

charbon demi-gras (m.), semibituminous coal.
(26)

charbon du pays de Galles (m.), Welsh coal. (27)

charbon extrait (Production houillère) (m.),
mineral gotten ; mineral extracted ; mineral
raised. (28)

charbon feuilleté (m.), laminated coal ; foliated
coal. (29)

charbon fossile (m.), fossil charcoal ; mineral
charcoal ; fibrous coal ; mother-of-coal.
(30)

charbon gras (m.), bituminous coal ; soft coal ;
fat coal ; smoking coal ; flaming coal. (31)

charbon gros ou charbon en roche (m.), large
coal ; round coal ; lump-coal. (32)

charbon imprégné (Élec.) (m.), impregnated
carbon ; mineralized carbon. (33)

charbon lavé (m.), washed coal. (34)

charbon maigre (m.), lean coal. (35)

charbon menu (m.), small coal ; small ; coal-
slack ; slack ; slack coal. (36)

charbon minéral (m.), coal. (37)

charbon minéralisé (Élec.) (m.), mineralized
carbon ; impregnated carbon. (38)

charbon moulé (m.), briquette ; briquet. (39)

charbon négatif (Élec.) (m.), negative carbon.
(40)

charbon non-lavé (m.), unwashed coal. (41)

charbon ordinaire (Élec.) (m.), solid carbon. (42)

charbon polaire (Élec.) (m.), electrode-carbon.
(43)

charbon positif (Élec.) (m.), positive carbon. (44)

charbon pour arc à flamme (Élec.) (m.), flame-
carbon. (45)

charbon pour fours métallurgiques (m.), furnace-
coal. (46)

charbon pour production de vapeur (m.), steam-
coal ; coal for steam-raising. (47)

charbon pulvérisé (m.), powdered coal ; pul-
-verized coal. (48)

charbon sans fumée (m.), smokeless coal. (49)

charbon tout venant (m.), run-of-mine coal ;
through-and-through coal ; unsorted coal.
(50)

charbon transporté par voie d'eau (m.), water-
borne coal. (51)

charbon végétal (m.), vegetable charcoal. (52)

charbonnage (exploitation de la houille) (n.m.),
coal-mining. (53)

charbonnage (houillère) (n.m.), colliery ; coal-
mine. (54)

charbonnaille (charbon menu) (n.f.), small coal ;
small ; fines ; coal-slack. (55)

charbonné, -e (adj.), charred ; chared ; carbon-
-ized. (56)

charbonner (v.t.), to char ; to carbonize ; to
coal : (57)
charbonner du bois, to char, to carbonize, to
coal, wood. (58)

charbonner (v.i.) ou charbonner (se) (v.r.), to
char ; to carbonize : (59)
bois qui charbonne (ou se charbonne) (m.),
wood which chars (or carbonizes). (60)

charbonneux, -euse (adj.), carbonaceous ;
coaly : (61)
calcaire charbonneux (m.), carbonaceous
limestone. (62)
roches charbonneuses (f.pl.), carbonaceous
rocks. (63)
matière charbonneuse (f.), coaly matter. (64)

charbonnier, -ère (qui a rapport à l'industrie, au
commerce du charbon) (adj.), coal (used as
adj.) : (65)
industrie charbonnière (f.), coal trade. (66)

charbonnier (bateau charbonnier) (n.m.), collier ;
coal-ship. (67)

chardonnet (armature ou poutre latérale d'une
porte à bourdonnière) (n.m.), hanging-
post. (68)

chardonnet (poteau tourillon d'une porte d'écluse)
(n.m.), heel-post. (69)

chardonnet (partie courbe de la maçonnerie d'un
sas d'écluse contre laquelle s'appuie le
poteau tourillon d'une porte d'écluse) (n.m.),
hollow quoin. (70)

charge (ce que peut porter une voiture, etc.) (n.f.),
load ; capacity : (71)
charge de wagon, wagon-load ; car-load ;
truck-load. (72)
charge normale (d'un wagon de ch. de f., ou
analogue), normal load ; normal capacity.
(73 ou 74)
charge limite, maximum load ; maximum
capacity. (75)
charge remorquée par une locomotive, load
drawn by a locomotive. (76)

charge (force totale) (Résistance des matériaux)
(n.f.), load (Strength of materials) : (77)
charge d'épreuve ou charge d'essai, test load.
(78)
charge de rupture, breaking-load. (79)

charge de sécurité, safe load. (1)

charge pratique *ou* charge de travail, working-load : (2)

  charge pratique (*ou* charge de travail) d'une chaîne, working-load of a chain. (3)

charge statique, static load : (4)

  la charge statique d'une chaîne, the static load of a chain. (5)

charge (intensité de la charge par unité de surface) (Résistance des matériaux) (*n.f.*), stress ; intensity of stress ; unit stress : (6)

charge à la rupture par flexion, breaking-stress under bending. (7)

charge de rupture en kg : mm², breaking-stress in kilos per sq. millimetre. (8)

charge de cisaillement, shearing-stress ; intensity of shear-stress. (9)

charge de sécurité, safe stress. (10)

charge de sécurité par flexion, safe stress under bending. (11)

charge déterminant l'écrasement par centi--mètre carré de section, crushing-stress per square centimetre of section. (12)

charge pratique, working stress. (13)

charge (d'un fourneau métallurgique) (*n.f.*), charge, stock, (of a metallurgical furnace) : (14)

charge d'allumage, bed-charge ; bed-fuel. (15)

charge (Travail aux explosifs) (*n.f.*), charge (Blasting) : (16)

charge d'amorçage, priming-charge. (17)

charge de mine, blasting-charge. (18)

charge (Élec.) (*n.f.*), charge : (19)

  la charge d'une bouteille de Leyde, the charge of a Leyden jar. (20)

charge résiduelle, residual charge ; electrical residue. (21)

charge (d'une soupape) (*n.f.*), load (of a valve). (22)

charge (*n.f.*) *ou* **charge d'eau** (Hydraul.), pressure ; water-pressure ; head ; head of water ; pressure-head ; static head : (23)

  l'épaisseur des parois doit être inversement proportionnelle à leur résistance et directe--ment proportionnelle à la charge (*f.*), the thickness of the walls should be inversely proportional to their strength and directly proportional to the pressure. (24)

écoulement à travers un orifice de — pouce(s) carré(s) sous une charge d'eau de — pouce(s) au-dessus du bord supérieur de l'orifice (*m.*), flow through a hole — inch(es) square, under a head — inch(es) above the top edge of the hole. (25)

charge de la vitesse, velocity-head. (26)

charge (Mines) (*n.f.*), weight ; weighting ; pressure : (27)

  la charge sur les piliers, the weight on the pillars. (28)

se faire une idée de la charge maximum à supporter par les piliers, to form an idea of the maximum weighting to be supported by the pillars. (29)

charge (en) (sous tension) (Élec.) (opp. à **hors courant**), live : (30)

conducteur en charge (*m.*), live conductor. (31)

charge (en) *ou* **chargé, -e** (marche des moteurs, des machines-outils, etc.) (opp. à **à vide** *ou* **à blanc**), under load ; loaded : (32)

moteur qui démarre en charge (*m.*), engine which starts under load. (33)

machine marchant chargée (*f.*), machine running under load ; loaded machine. (34) Cf. **vide (à)** and **blanc (à)** in vocabulary.

charge due à la neige (*f.*), snow-load. (35)

charge due au vent (*f.*), wind-load. (36)

charge limite d'élasticité (Méc.) (*f.*), elastic limit ; elastic strength ; limit of elasticity : (37)

300 kilogrammes par centimètre carré est la charge limite d'élasticité du cuivre, 300 kilogrammes per square centimetre is the elastic limit of copper. (38)

charge utile (*f.*), live-weight : (39)

le poids mort des wagonnets de mines représente en moyenne 40 0/0 de la charge utile, the dead-weight of mine-cars represents on an average 40 % of the live-weight. (40)

chargement (*n.m.*), charging ; loading ; filling : (41)

chargement d'un cubilot, charging a cupola. (42)

chargement en plein jour (Photogr.), day--light loading. (43)

chargement en wagonnets du minerai abattu, loading the broken ore into trucks ; filling the trucks with broken ore. (44)

chargement (ensemble des objets formant une charge) (*n.m.*), load : (45)

le chargement d'une voiture, d'un bâtiment, d'une bête de somme, the load of a carriage, of a building, of a beast of burden. (46 *ou* 47 *ou* 48)

chargement (Travail aux explosifs) (*n.m.*), charging ; loading : (49)

chargement des coups de mine, charging shots ; loading holes. (50)

charger (*v.t.*), to charge ; to load ; to fill : (51)

charger un accumulateur, to charge an accumu--lator. (52)

charger un fourneau avec du minerai, to charge a furnace with ore. (53)

charger un fourneau de mine, to charge a shot-hole ; to load a hole. (54)

charger un wagonnet, to load, to fill, a truck. (55)

charger du minerai dans des véhicules, to load ore into trucks. (56)

colonne chargée (Résistance des matériaux) (*f.*), loaded column. (57)

charger (donner des instructions à) (*v.t.*), to instruct. (58)

charger (se dit du toit d'une galerie de mine) (*v.i.*), to squeeze ; to undergo squeeze : (59)

le toit charge, the roof is undergoing squeeze. (60)

charger (se) (*v.r.*). See examples : (61)

se charger (recevoir la charge), to be loaded ; to be charged ; to be filled ; to fill : (62)

fusil qui se charge par la culasse (*m.*), gun which is loaded at the breech. (63)

appareil photographique qui peut se charger en plein jour (*m.*), photographic camera which can be loaded in daylight. (64)

se charger de faire un rapport sur une propriété, to undertake to report on a property. (65)

se charger du personnel, to take charge of the staff. (66)

charger avec des poids, to weight ; to weight down : (67)

charger un moule avec des poids suffisants pour résister au soulèvement du châssis,

to weight a mould sufficiently to resist the lifting of the box. (1)

chargette (Travail aux explosifs) (*n.f.*), charging-spoon. (2)

chargeur (*n.m.*), charger ; loader ; filler. (3)

chargeur à la taille (Exploitation des mines de houille) (*pers.*) (*m.*), coal-filler ; loader. (4)

chargeur de bennes (Mines) (*pers.*) (*m.*), kibble-filler. (5)

chargeur de haut fourneau (*pers.*) (*m.*), blast-furnace filler ; top-filler. (6)

chargeur mécanique (foyer) (*m.*), mechanical stoker ; stoker ; fire-feeder. (7)

chargeur mécanique à alimentation par en dessous (*m.*), underfeed stoker. (8)

chargeur mécanique à alimentation par le dessus (*m.*), overfeed stoker. (9)

chargeuse (dispositif pour la manutention de matériaux, etc.) (*n.f.*), loader. (10)

chargeuse (*n.f.*) *ou* **chargeuse mécanique** (de four), charger ; charging-machine. (11)

chargeuse-défourneuse de brames [chargeuses-défourneuses *pl.*] (*f.*), slab-charging machine. (12)

chariot (voiture) (*n.m.*), car ; truck ; trolley ; wagon ; lorry. (13)

chariot (Mach.) (*n.m.*), carriage ; carrier. (14)

chariot (d'un tour) (*n.m.*), carriage (of a lathe). (15)

chariot (de chambre noire) (*n.m.*), base-board, bed, (of camera). (16)

chariot à agrafes (scierie) (*m.*), log-carriage with timber-clips. (17)

chariot à charnière (de chambre photographique) (*m.*), folding base-board (of camera). (18)

chariot automobile (câble aérien) (*m.*), self-propelling carrier. (19)

chariot butoir (Mines) (*m.*), barney. (20)

chariot contrepoids *ou* **chariot de contrepoids** (plans inclinés) (Mines) (*m.*), balance-truck ; counterbalance-carriage ; counterbalancing-carriage. (21)

chariot d'étuve à noyaux (*m.*), core-oven truck ; core-oven car ; core-carriage. (22)

chariot de mine (*m.*), mine-car ; mine-truck ; wagon ; trolley ; trawley. (23)

chariot de roulement (de pont roulant ou de câble aérien) (*m.*), runner ; truck ; trolley ; travelling runner ; travelling trolley ; travelling carriage ; monkey carriage ; jenny ; jinny ; block-carriage. (24 ou 25)

chariot de service (Ch. de f.) (*m.*), traverse-table ; traverse-carriage ; traverser ; slide-rail ; railway-slide. (26)

chariot de suspension (*m.*), (câble aérien) (*m.*), carrier ; carrying-block. (27)

chariot porte-broche équilibré par contrepoids et se déplaçant automatiquement et à la main dans le sens vertical avec butée per--mettant le réglage de la course (machine à fraiser verticale) (*m.*), spindle-carriage balanced by counterweight and moving up and down automatically or by hand, adjustable to the required travel by stops (vertical milling-machine). (28)

chariot porte-meule (machine à meuler) (*m.*), wheel-carriage (grinding-machine). (29)

chariot porte-outil (*m.*), tool-carriage. (30)

chariot porte-outil (d'un étau-limeur) (*m.*), ram (of a shaper). (31)

chariot porte-palan (1) **roulant sur l'aile inféri--eure du fer, ou** (2) **roulant sur l'aile supérieure**

du fer (*m.*), lifting-block fitted with travelling trolley (1) arranged to run on the bottom flange of the joist, or (2) arranged to run on the top flange of the joist. (32 *ou* 33)

chariot porte-tourelle (tour) (*m.*), turret-slide (lathe). (34)

chariot porteur (Mines) (*m.*), carriage for tub on inclines ; slope-carriage. (35)

chariot porteur à patins pour dressants (*m.*), sledge-carriage for highly inclined seams. (36)

chariot roulant (Ch. de f.) (*m.*). Same as **chariot de service**.

chariot roulant (de pont roulant ou de câble aérien) (*m.*). Same as **chariot de roulement**.

chariot tendeur *ou* **chariot tenseur** (*m.*), tension-carriage ; tightener. (37)

chariot transbordeur (Ch. de f.) (*m.*). Same as **chariot de service**.

chariotage (Travail au tour) (*n.m.*), traversing ; traverse (Lathe-work). (38)

chariotage longitudinal (Travail au tour) (*m.*), sliding ; longitudinal traverse. (39)

chariotage transversal (Travail au tour) (*m.*), cross-traverse ; surfacing. (40)

charioter (Travail au tour) (*v.i.*), to traverse. (41)

charme (arbre) (*n.m.*), hornbeam. (42)

charmouthien (Géol.) (*n.m.*), Charmouthian stage. (43)

charnière (*n.f.*), hinge ; butt-hinge ; butt. (44)

charnière(s) (à), hinged ; hinge (*used as adj.*) : (45 ou 46)

couvercle à charnière (*m.*), hinged lid. (47)

assemblage à charnière (*m.*), hinge-joint. (48)

charnière (d'un pli) (Géol.) (*n.f.*), bend (of a fold). (49)

charnière anticlinale (Géol.) (*f.*), anticlinal crest ; crest of an anticline. (50)

charnière et gond *ou* **charnière et pivot** [charnière (*n.f.*) ; gond (*n.m.*) ; pivot (*n.m.*)], hinge and pin. (51)

charnière synclinale (Géol.) (*f.*), synclinal trough ; trough of a syncline. (52)

charnière universelle (Mach. et Méc.) (*f.*), univer--sal joint ; Cardan joint ; cardan joint. (53)

charnon (d'une charnière) (*n.m.*), knuckle (of a hinge). (54)

charpente (assemblage de pièces de bois ou de métal) (*n.f.*), frame-work ; framing ; frame : (55)

la charpente d'une grue, the framework of a crane. (56)

la charpente d'un cintre de pont, the framing of a bridge-centre. (57)

une charpente de chêne, de fer, a frame of oak, of iron. (58 ou 59)

charpente à croisillons (*f.*), lattice bracing. (60)

charpente de batterie (bocard) (*f.*), battery-frame ; battery-framework (ore-stamp). (61)

charpente de comble (*f.*), roof-frame : (62)

une charpente de comble se compose de fermes, de pannes, de chevrons, de lattis, etc., a roof-frame is composed of trusses, purlins, rafters, lathing, etc. (63)

charpente de pont (*f.*), bridge-frame. (64)

charpente du chevalement *ou* **charpente exté--rieure de puits** (Mines) (*f.*), shaft-head frame. (65)

charpente en bois (art) (*f.*), carpentry. (66)

charpente en fer (*f.*), ironwork ; iron construc--tional work : (67)

traité de charpente en fer (*m.*), treatise on ironwork. (1)

charpente métallique (*f.*), iron and steel construc-tional work. (2)

charpenterie (en bois) (*n.f.*), carpentry. (3)

charpenterie en fer (*f.*), ironworking. (4)

charpentier (pers.) (*n.m.*), carpenter. (5)

charpentier à tout faire (*m.*), general carpenter. (6)

charpentier en fer (*m.*), ironworker (a worker in heavy iron, such as trusses, girders, or the like). (7)

charretée (*n.f.*), cartload ; cartful. (8)

charretier (pers.) (*n.m.*), carman. (9)

charrette (*n.f.*), cart. (10)

charrette à bras (*f.*), hand-cart. (11)

charriage (*n.m.*), carting ; cartage ; conveyance ; carriage. (12)

charriage (transport) (Géol.) (*n.m.*), drifting ; travelling : (13)
dépôts accumulés par charriage (*m.pl.*), deposits accumulated by drifting. (14)

charriage (chevauchement) (Géol.) (*n.m.*), thrust ; overthrust : (15)
charriage de nappes, de plis couchés, thrust, overthrust, of rock-sheets, of recumbent folds. (16 *ou* 17)

charrier (entraîner) (*v.t.*), to carry along ; to wash down ; to drift : (18)
matières terreuses charriées par les eaux (*f.pl.*), earthy matter carried along by the water ; earthy material washed down by the water. (19)
plusieurs fleuves charrient de l'or dans leur sable (*m.pl.*), some rivers drift gold in their sands. (20)

charroi (*n.m.*), carting ; cartage ; transport by wagon ; road-transport. (21)

charron (pers.) (*n.m.*), wheelwright. (22)

charronnage (*n.m.*), wheelwrights' work. (23)

charronnerie (*n.f.*), wheelwrights' trade. (24)

charrue (*n.f.*), plough ; plow. (25)

chas (d'une aiguille) (*n.m.*), eye (of a needle). (26)

chassage (action) (Mines) (*n.m.*), drifting ; driving. (27)

chassage (*n.m.*) *ou* **chassante** (*n.f.*) (galerie de mine), drift ; drifting-level ; driftway ; drive (parallel in direction to the strike) ; level running along the strike. (28)

chasse (jeu) (Méc.) (*n.f.*), play ; clearance. (29)

chasse (courbure des dents d'une scie) (*n.f.*), set. (30)

chasse à coins (Pose de voie) (Ch. de f.) (*f.*), keying-hammer (Plate-laying). (31)

chasse à parer (outil de forgeron) (*f.*), flattener ; flatter (blacksmiths' tool). (32)

chasse-bestiaux [chasse-bestiaux *pl.*] (*n.m.*), cow-catcher ; pilot. (33)

chasse-boulon [chasse-boulon *ou* chasse-boulons *pl.*] (*n.m.*), driftbolt ; drive-bolt ; teeming-punch. (34)

chasse carrée (pour forgerons) (*f.*), set-hammer. (35)

chasse-clavette [chasse-clavette *ou* chasse-clavettes *pl.*] *ou* chasse-clef [chasse-clef *ou* chasse-clefs *pl.*] (*n.m.*), key-drift ; drift ; driver. (36)

chasse-clou [chasse-clou *ou* chasse-clous *pl.*] (*n.m.*), Same as chasse-pointe.

chasse d'air (*f.*), air-blast ; rush of air : (37)
cage de mine qui est lancée dans le chevale-ment par une violente chasse d'air (*f.*),

mine-cage which is hurled up into the head-frame by a violent air-blast. (38)

chasse d'eau *ou simplement* **chasse** (*n.f.*), (Hydraul.), flush ; flushing ; scour ; scour-ing. (39)

chasse-goupille [chasse-goupille *ou* chasse-goupilles *pl.*] (*n.m.*), pin-punch ; pin-drift. (40)

chasse-neige [chasse-neige *pl.*] (*n.m.*), snow-plough. (41)

chasse-pierres *ou* **chasse-pierre** [chasse-pierres *ou* chasse-pierre *pl.*] (de locomotive, etc.) (*n.m.*), guard-iron ; rail-guard : (42)
chasse-pierres formant supports en cas de rupture de roues (grues-chevalets, etc.), guard-irons forming supports in case of breakage of wheels. (43)

chasse-pointe [chasse-pointe *ou* chasse-pointes *pl.*] (*n.m.*), nail-set ; brad-punch ; nail-punch ; set. (44)

chasse-rivet [chasse-rivet *ou* chasse-rivets *pl.*] (*n.m.*), rivet-snap ; rivet-set ; rivetting-set ; snap ; snap-tool ; snap-head. (45)

chasse-tampon [chasse-tampons *pl.*] (*n.m.*), plugging-bar. (46)

chasser (pousser en avant) (*v.t.*), to drive : (47)
le vent chasse la pluie, the wind drives the rain. (48)
être chassé (-e) des chantiers par une venue d'eau, to be driven out of the workings by an inrush of water ; to be drowned (*or* to be flooded) out of the workings. (49)
chasser l'eau (d'une substance), to drive off the water (from a substance). (50)
chasser le mauvais air, to drive out the bad air. (51)
chasser les produits volatils, to expel the volatile matter. (52)
chasser un clou à coups de marteau, to drive in a nail with a hammer. (53)
chasser une clavette, to drive out a key. (54)

chasser (Serrur.) (*v.t.*), to shoot : (55)
la clef chasse le pêne de la serrure, the key shoots the bolt of the lock. (56)

chasser (Mines) (*v.i.*), to drive ; to drift : (57)
chasser sur le filon à droite et à gauche, to drive on the lode right and left. (58)
chasser le long du mur, to drift along the foot-wall. (59)

châssis (cadre) (*n.m.*), frame. (60)

châssis (Ch. de f.) (*n.m.*). See **châssis de loco-motive, châssis de wagon**, etc.

châssis (Mines) (*n.m.*). See **châssis de mine**, etc.

châssis (Fonderie) (*n.m.*). See **châssis de fonderie**, etc.

châssis (Photogr.) (*n.m.*). See **châssis négatif**, etc.

châssis (de fenêtre) (Constr.) (*n.m.*). See **châssis de fenêtre**, etc.

châssis (d'un automobile) (*n.m.*), chassis (of a motor-car). (61)

châssis à brisures (Photogr.) (*m.*), roller-blind slide. (62)

châssis à charnière (Photogr.) (*m.*), book-form dark slide ; book-form plate-holder. (63)

châssis à colonnes (Fonderie) (*m.*), column-box. (64)

châssis à deux parties (dessous, dessus) (Fonderie) (*m.*), two-part flask, two-parted box (drag, cope). (65)

châssis à flches (fenêtre) (*m.*), casement ; French sash. (66)

châssis à glace dépolie pour la mise au point avec capuchon *ou* châssis à verre dépoli pour la mise au point avec abat-jour (Photogr.) (*m*.), focussing-screen with hood ; ground-glass focussing-screen with hood ; focussing-frame with hood ; hooded focussing-back. (1)

châssis à guillotine (fenêtre) (*m*.), sliding sash. (2)

châssis à magasin (Photogr.) (*m*.), changing-box ; plate-changing box ; plate-magazine. (3)

châssis à molettes (Mines) (*m*.), pulley-frame ; head-frame ; pit-head frame ; pit-frame ; poppet-head ; puppet-head ; head-stock ; gallows frame ; shaft-tackle. (4)

châssis à pivot (fenêtre) (*m*.), pivot-hung sash. (5)

châssis à plaques (Photogr.) (*m*.), plate-holder ; dark slide. (6)

châssis à rideau (Photogr.) (*m*.), roller-blind slide. (7)

châssis à rouleaux (Photogr.) (*m*.), spool-holder ; roll-holder. (8)

châssis à trois parties (dessous, chape, dessus) (Fonderie) (*m*.), three-part flask, three-parted box, (drag, cheek, cope). (9)

châssis à tuyaux (Fonderie) (*m*.), pipe-box. (10)

châssis à vitroses (Photogr.) (*m*.), cut-film holder. (11)

châssis à volet(s) *ou* châssis à tirette(s) (Photogr.) (*m*.), plate-holder. (12 *ou* 13)

châssis-adaptateur [châssis-adapteurs *pl*.] (Photogr.) (*n.m*.), adapter. (14)

châssis articulé pour moulage en mottes (Fonderie) (*m*.), snap-flask ; snap. (15)

châssis de charpente (*m*.), frame ; outer frame ; skeleton (the timbering—posts, sill, and transom—surrounding an inner frame, or any open framework supporting the parts of a structure). (16)

châssis de fenêtre *ou* châssis dormant *ou* simple-*ment* châssis (*n.m*.), window-frame ; frame. (17)

châssis de fenêtre *ou* châssis mobile *ou* simple-*ment* châssis (*n.m*.), window-sash ; sash ; sash-frame. (18)

châssis de fonderie *ou* châssis de moulage *ou* simplement châssis (*n.m*.), foundry-flask ; flask ; moulding-flask ; casting-box ; molding-box ; box. (19)

châssis de locomotive *ou* simplement châssis (*n.m*.), engine-frame ; frame ; main frame : (20)

un châssis de locomotive se compose de deux longerons et plusieurs entretoises, an engine-frame is composed of two main side-frames and several cross-ties. (21)

châssis de mine *ou* simplement châssis (*n.m*.) (Boisage des galeries de mines), frame ; set ; frame-set ; framing ; durn. (22)

châssis de série (Fonderie) (*m.pl*.), flasks for repetitive work. (23)

châssis de suspension (câble aérien) (*m*.), carry--ing-block ; carrier. (24)

châssis de wagon *ou* simplement châssis (*n.m*.) (Ch. de f.), underframe ; undercarriage : (25)

le châssis d'un wagon de chemin de fer se compose de deux brancards ou longerons et de deux traverses de tête, the under--frame of a railway-wagon is composed of two side-sills and two end-sills. (26)

châssis double (Photogr.) (*m*.), double slide. (27)

châssis du bissel (locomotive) (*m*.), pony-truck frame. (28)

châssis du bogie (*m*.), bogie-frame ; truck-frame ; engine-truck frame. (29)

châssis faux (Boisage des galeries de mine) (*m*.), false set. (30)

châssis-magasin [châssis-magasins *pl*.] (Photogr.) (*n.m*.), changing-box ; plate-changing box ; plate-magazine. (31)

châssis négatif *ou* châssis pour plaques *ou* simplement châssis (*n.m*.) (Photogr.), dark slide ; plate-holder ; slide ; holder. (32)

châssis négatif s'ouvrant par le milieu, dit châssis anglais *ou* châssis-livre [châssis-livres *pl*.] (*n.m*.), book-form dark slide ; book-form plate-holder. (33)

châssis passe-vues va-et-vient *ou* châssis porte-vues va-et-vient (*m*.), lantern-slide carrier. (34)

châssis porte-plaque(s) (Photogr.) (*m*.), plate-holder ; holder ; dark slide ; slide. (35 *ou* 36)

châssis porte-scie (*m*.), saw-frame ; saw-gate. (37)

châssis pour film-pack *ou* châssis pour bloc-film (*m*.), film-pack adapter. (38)

châssis-presse [châssis-presses *pl*.] (*n.m*.) *ou* châssis positif (Photogr.), printing-frame ; pressure-frame : (39)

des châssis-presses demi-plaque, half-plate printing-frames. (40)

châssis simple (Photogr.) (*m*.), single slide. (41)

châssis vitré (fenêtre) (*m*.), glazed frame ; glazed sash. (42)

chat (fonte qui s'échappe du haut fourneau par suite de la rupture du tampon du trou de coulée) (*n.m*.), breakout (of molten metal from a blast-furnace). (43)

châtaignier (arbre) (*n.m*.), chestnut. (44)

château d'eau (*m*.), water-tower. (45)

chatoiement *ou* chatoîment (*n.m*.), chatoyancy : (46)

chatoiement d'un minéral, chatoyancy of a mineral. (47)

chatoyant, -e (*adj*.), chatoyant : (48)

lustre chatoyant (*m*.), chatoyant lustre. (49)

chatoyante (Minéral) (*n.f*.), chatoyant. (50)

chaud, -e (*adj*.), hot ; warm : (51)

eau chaude (*f*.), hot water ; warm water. (52)

chaud (à), hot (in the sense of *when hot* or *while heated*) : (53)

substance qui est plus soluble à chaud qu'à froid (*f*.), substance which is more soluble hot than cold. (54)

chaud (*n.m*.), heat ; warmth : (55)

craindre le chaud et le froid, to be unable to bear heat and cold. (56)

chaude (Métall.) (*n.f*.), heat : (57)

tôle laminée en une seule chaude (*f*.), plate rolled at one heat. (58)

forgeage qui exige plusieurs chaudes (*m*.), forging which requires several heats. (59)

chaude blanc-soudant (*f*.), welding-heat ; sparkling heat. (60)

chaude blanche (*f*.), white heat. (61)

chaude grasse (*f*.), white-flame heat. (62)

chaude rouge (*f*.), red heat. (63)

chaude rouge sang (*f*.), blood-red heat. (64)

chaude suante (*f*.), sweating heat. (65)

chaudière (*n.f*.), boiler. (66)

chaudière à bouilleurs (*f*.), French boiler ; elephant boiler. (67)

chaudière à bouts bombés (*f.*), dish-ended boiler. (1)

chaudière à carneau(x) intérieur(s) (*f.*), flue-boiler. (2 *ou* 3)

chaudière à chauffage extérieur *ou* chaudière à foyer extérieur (*f.*), externally fired boiler ; outside-fired boiler. (4)

chaudière à chauffage intérieur *ou* chaudière à foyer intérieur (*f.*), internally fired boiler ; inside-fired boiler. (5)

chaudière à circulation (*f.*), circulating-boiler. (6)

chaudière à deux tubes-foyers (*f.*), two-flue boiler. (7)

chaudière à eau chaude (*f.*), hot-water boiler. (8)

chaudière à faible volume d'eau *ou* chaudière à petits éléments (*f.*), small water-capacity boiler ; sectional boiler (such as the water-tube, multitubular boiler, etc.). (9)

chaudière à flamme directe (*f.*), direct-draught boiler. (10)

chaudière à flamme en retour *ou* chaudière à retour de flamme (*f.*), return-tube boiler; return tubular boiler; return-flue boiler. (11)

chaudière à foyer(s) intérieur(s) cylindrique(s) *ou* chaudière à tube(s)-foyer(s) (*f.*), cylin-drical-flue boiler ; flue-boiler. (12 *ou* 13)

chaudière à grand volume d'eau *ou* chaudière à gros éléments (*f.*), large water-capacity boiler (such as the cylindrical, Cornish, Lancashire, Galloway, French, semitubular boiler, etc.). (14)

chaudière à tubes d'eau *ou* chaudière aquatu-bulaire (*f.*), water-tube boiler. (15)

chaudière à tubes de fumée *ou* chaudière à tubes de flammes (*f.*), fire-tube boiler. (16)

chaudière à un tube-foyer (*f.*), single-flue boiler. (17)

chaudière à vapeur (*f.*), steam-boiler. (18)

chaudière à vapeur verticale (*f.*), vertical steam-boiler. (19)

chaudière à vaporisation instantanée (*f.*), flash-boiler ; flasher. (20)

chaudière cylindrique (*f.*), cylindrical boiler ; cylinder boiler. (21)

chaudière cylindrique simple (*f.*), plain cylin-drical boiler. (22)

chaudière de bitumier (*f.*), asphalt or tar-boiler. (23)

chaudière de Cornouailles *ou* chaudière à un foyer intérieur cylindrique *ou* chaudière à un tube-foyer (*f.*), Cornish boiler. (24)

chaudière de Cornouailles à tubes Galloway (*f.*), Cornish boiler with Galloway tubes. (25)

chaudière de Lancashire *ou* chaudière à deux foyers intérieurs cylindriques *ou* chaudière à deux tubes-foyers (*f.*), Lancashire boiler. (26)

chaudière de locomotive (*f.*), locomotive-boiler ; fire-box boiler. (27)

chaudière demi-fixe *ou* chaudière mi-fixe (*f.*), semiportable boiler. (28)

chaudière en tombeau (*f.*), wagon boiler ; caravan boiler. (29)

chaudière fixe *ou* chaudière placée à demeure (*f.*), stationary boiler. (30)

chaudière Galloway (*f.*), Galloway boiler. (31)

chaudière horizontale (*f.*), horizontal boiler. (32)

chaudière ignitubulaire (*f.*), fire-tube boiler. (33)

chaudière locomobile (chaudière à vapeur) (*f.*), portable boiler ; non-stationary boiler. (34)

chaudière marine *ou* chaudière de marine (*f.*), marine boiler. (35)

chaudière multitubulaire *ou* chaudière inexplo-sible (*f.*), multitubular boiler ; multiple boiler ; safety-boiler. (36)

chaudière roulante (pour chauffer l'asphalte, le goudron, etc.) (*f.*), portable boiler. (37)

chaudière sectionnelle (*f.*), sectional boiler. (38)

chaudière semi-tubulaire (*f.*), semitubular boiler. (39)

chaudière terrestre (opp. à chaudière marine) (*f.*), land-boiler. (40)

chaudière tubulaire (*f.*), tubular boiler ; multiflue boiler. (41)

chaudière tubulaire à retour de flamme (*f.*), return tubular boiler; return-tube boiler; return-flue boiler. (42)

chaudière tubuleuse *ou* chaudière tubulée (*f.*), tubulous boiler. (43)

chaudière verticale (*f.*), vertical boiler ; upright boiler. (44)

chaudron (*n.m.*), cauldron ; caldron. (45)

chaudronnerie (art) (*n.f.*), boiler-making ; boiler-work. (46)

chaudronnerie (atelier) (*n.f.*), boiler-works. (47)

chaudronnier (pers.) (*n.m.*), boiler-maker ; boiler-smith ; coppersmith. (48)

chauffage (action de rendre chaud) (*n.m.*), heating ; warming. (49)

chauffage (élévation de température qui se produit dans les organes d'une machine, par suite d'un défaut de graissage) (*n.m.*), heating : (50) le chauffage d'une boîte à graisse, the heating of an axle-box. (51)

chauffage (action d'approvisionner en combustible, etc.) (*n.m.*), stoking ; firing : (52) bien conduire le chauffage d'un four, to pay proper attention to the firing of a furnace. (53)

chauffage à air chaud (*m.*), hot-air heating. (54)

chauffage à eau chaude (*m.*), hot-water heating. (55)

chauffage à la main (*m.*), hand-stoking ; firing by hand. (56)

chauffage à vapeur (*m.*), steam-heating. (57)

chauffage au gaz (*m.*), gas-heating. (58)

chauffage au pétrole (chauffage d'appartements, etc.) (*m.*), oil-heating ; heating by petroleum. (59). Cf. chauffé (-e) au pétrole.

chauffage central (*m.*), central heating. (60)

chauffage mécanique (*m.*), mechanical stoking ; mechanical firing. (61)

chauffage mixte (charbon et pétrole) (*m.*), mixed heating. (62)

chauffe (*n.f.*), stoking ; firing : (63) donner une chauffe, to attend to the stoking ; to look after the firing. (64)

chauffe (lieu où se brûle le combustible dans un fourneau) (*n.f.*), fire-chamber : (65) l'autel d'un four à réverbère est la partie qui sépare la sole de la chauffe (*m.*), the fire-bridge of a reverberatory furnace is the part which separates the hearth from the fire-chamber. (66)

chauffe-colle [chauffe-colle *pl.*] (*n.m.*), glue-heater ; glue-heating apparatus. (67)

chauffé (-e) à blanc, white-hot ; at a white heat. (68)

chauffé (-e) au charbon, coal-fired : (69) locomotive chauffée au charbon (*f.*), coal-fired locomotive. (70)

chauffé (-e) au coke, coke-fired : (71)

étuve à noyaux, chauffée au coke (*f.*), coke-fired core-oven. (1)

chauffé (-e) au gaz, gas-fired : (2)

four chauffé au gaz (*m.*), gas-fired furnace. (3)

chauffé (-e) au pétrole, oil-fired : (4)

locomotive chauffée au pétrole (*f.*), oil-fired locomotive. (5)

chauffé (-e) au rouge, red-hot ; at a red heat. (6)

chauffer (rendre chaud) (*v.t.*), to heat ; to warm : (7)

chauffer à blanc, to heat to whiteness ; to make white hot : (8)

chauffer du métal à blanc, to heat metal to whiteness ; to make metal white hot. (9)

chauffer au rouge, to heat to redness ; to make red hot. (10)

chauffer de l'eau, to heat water. (11)

chauffer un appartement, to warm a room. (12)

chauffer (*v.i.*), to heat ; to be heated ; to get hot ; to run hot : (13)

lorsqu'un coussinet chauffe, when a bearing heats. (14)

les chaudières chauffent au charbon de terre (*f.pl.*), boilers are heated with coal. (15)

le four chauffe, the furnace is getting hot. (16)

moteur qui chauffe (*m.*), engine which is running hot. (17)

chauffer (allumer ; entretenir le feu de) (*v.t.*), to stoke ; to fire ; to light the fire of : (18)

chauffer un four, to stoke a furnace. (19)

chauffer un four au coke, une locomotive au pétrole, to fire a furnace with coke, a loco-motive with oil. (20 *ou* 21)

chauffer une chaudière, to stoke, to fire, to light the fire of, to tend, to serve, a boiler. (22)

chauffer le moulin au bois, to run the mill on wood fuel. (23)

chauffer (se) (*v.r.*), to heat ; to warm ; to get hot : (24)

l'eau se chauffe lentement (*f.*), the water heats slowly. (25)

chaufferie (salle des chaudières) (*n.f.*), boiler-room. (26)

chaufferie (Métall.) (*n.f.*), chafery. (27)

chauffeur (pers.) (*n.m.*), stoker ; fireman. (28)

chauffeur conducteur d'un générateur de vapeur (*m.*), fireman in charge of a steam-boiler. (29)

chauffeur de locomotive (*m.*), locomotive fireman. (30)

chauffeur de rivets (pers.) (*m.*), rivet-heater. (31)

chauffeur-mécanicien [chauffeurs-mécaniciens *pl.*] (*n.m.*), engineman. (32)

chaufour (*n.m.*), lime-kiln ; lime-pit. (33)

chaufournerie (*n.f.*), lime-burning ; lime-burning industry. (34)

chaufournier (pers.) (*n.m.*), lime-burner. (35)

chaulage (*n.m.*), liming. (36)

chaussée (*n.f.*), road ; roadway. (37)

chaussée (d'une route) (opp. au trottoir) (*n.f.*), roadway (of a road) (38) (as distinguished from the path).

chaussée empierrée (*f.*), metalled road. (39)

chaux (*n.f.*), lime. (40)

chaux anhydre (*f.*). Same as chaux vive.

chaux carbonatée (*f.*), carbonate of lime. (41)

chaux éteinte *ou simplement* chaux (*n.f.*), slaked lime ; slacked lime. (42)

chaux éteinte à l'air *ou* chaux délitée *ou* chaux amortie (*f.*), air-slaked lime. (43)

chaux fluatée (*f.*), fluor spar. (44)

chaux grasse (*f.*), fat lime ; fatty lime. (45)

chaux hydratée *ou simplement* chaux (*n.f.*), hydrated lime. (46)

chaux hydraulique (*f.*), hydraulic lime. (47)

chaux incuite (*f.*), unburnt lime. (48)

chaux-limite (*n.f.*), Portland cement. (49)

chaux maigre (*f.*), poor lime ; quiet lime. (50)

chaux sulfatée (*f.*), sulphate of lime. (51)

chaux vive (*f.*), quicklime ; unslaked lime ; unslacked lime. (52)

chavirement (*n.m.*), capsizing ; turning over ; tipping ; dumping ; shooting. (53)

chavirer (*v.t.*), to tip ; to dump ; to shoot. (54)

chavirer (*v.i.*), to capsize ; to turn over : (55)

wagon qui chavire (*m.*), wagon which turns over. (56)

cheddite (Explosif) (*n.f.*), cheddite. (57)

chef d'atelier (*m.*), shop-foreman. (58)

chef d'équipe *ou* chef d'escouade (*m.*), foreman ; gang-foreman ; ganger ; gaffer ; shift-boss ; shifter ; chargeman. (59)

chef de base (d'une ardoise) (Constr.) (*m.*), tail (of a slate). (60)

chef de chauffe (*m.*), chief stoker ; head stoker. (61)

chef de district (*m.*), district manager. (62)

chef de drague *ou* chef dragueur (*m.*), dredge-master. (63)

chef de fonderie (*m.*), foundry-foreman. (64)

chef de gare (Ch. de f.) (*m.*), station-master. (65)

chef de gare adjoint (*m.*), assistant station-master. (66)

chef de l'usine (*m.*), works' foreman. (67)

chef de manœuvre (Ch. de f.) (*m.*), shunting-foreman. (68)

chef de pose (Ch. de f.) (*m.*), ganger ; track-fore-man. (69)

chef de pose (Télégr., etc.) (*m.*), chief linesman ; chief lineman. (70)

chef de poste (*m.*). Same as chef d'équipe.

chef de section (*m.*), section-manager. (71)

chef de tête (d'une ardoise) (Constr.) (*m.*), head (of a slate). (72)

chef du gisement (Géol. & Mines) (*m.*), cap ; capping ; cap-rock ; rock capping. (73)

chef du mouvement (Ch. de f.) (*m.*), superintend-ent of the line. (74)

chef du service des ateliers (*m.*), works' manager. (75)

chef-mineur [chefs-mineurs *pl.*] (*n.m.*), mine-captain ; mine-master. (76)

chef-pompier [chefs-pompiers *pl.*] (exhaure) (*n.m.*), chief pumpman. (77)

chef sondeur (*m.*), drilling-foreman ; drill-fore-man ; boring-master ; master borer. (78)

chemin (*n.m.*), way ; road ; path ; track. (79)

chemin de bois (Mines, etc.) (*m.*), wooden-rail track. (80)

chemin de clavette (Méc.) (*m.*), keyway ; key-seat ; key-seating ; key-groove. (81)

chemin de fer (*m.*), railway ; railroad. (82)

chemin de fer à adhérence (opp. à chemin de fer à crémaillère) (*m.*), adhesion-railway. (83)

chemin de fer à crémaillère (*m.*), rack-railway ; cogway. (84)

chemin de fer à voie de 1 mètre (*m.*), metre-gauge railway. (85)

chemin de fer à voie étroite (*m.*), narrow-guage railway ; light railway ; light rail-road. (86)

chemin de fer à voie large (*m.*),   broad-gauge railway ; wide-gauge railway. (1)

chemin de fer à voie normale (*m.*),   standard-gauge railway. (2)

chemin de fer aérien (*m.*), aerial railway. (3)

chemin de fer atmosphérique (*m.*),   atmospheric railway ; pneumatic railroad. (4)

chemin de fer d'intérêt local (*m.*), local railway. (5)

chemin de fer de l'État (*m.*), State railway. (6)

chemin de fer de mines (*m.*),   mines-railway. (7)

chemin de fer de montagne (*m.*),   mountain railway ; mountain railroad. (8)

chemin de fer de pénétration (*m.*), railway into the interior. (9)

chemin de fer Decauville (*m.*),   Decauville railway. (10)

chemin de fer économique (*m.*),   economic railway ; light railway. (11)

chemin de fer électrique (*m.*), electric railway. (12)

chemin de fer électrique souterrain (*m.*), under-ground electric railway. (13)

chemin de fer funiculaire (*m.*),   cable-railway ; cable-railroad ; cable-road ; rope-railway ; funicular railway. (14)

chemin de fer industriel (*m.*), industrial railway. (15)

chemin de fer portatif (*m.*),   portable railway ; portable railroad. (16)

chemin de fer secondaire (*m.*),   secondary railway. (17)

chemin de fer souterrain (*m.*),   underground railway. (18)

chemin de fer suspendu (monorail transporteur) (*m.*), runway ; overhead runway ; overhead track. (19)

chemin de fer télédynamique (*m.*),   teledynamic railway. (20)

chemin de halage (*m.*),   towing-path ; tow-path. (21)

chemin de roulement (*m.*),   runway ; race ; roller-track ; roller-path. (22)

chemin de roulement pour billes (*m.*), ball-race. (23)

chemin suivi (*ou* parcouru) par le bouton de manivelle (*m.*), crank-path. (24)

cheminée (pour fumée) (*n.f.*),   chimney ; stack ; chimney-stack ;   smoke-stack ;   shaft ; funnel. (25)

cheminée (Géol.) (*n.f.*),   chimney ; vent ; neck ; pipe. (26)

cheminée (d'un injecteur) (*n.f.*), chamber (of an injector). (27)

cheminée (*n.f.*) *ou* cheminée à minerai (Mines), chute ; chute-raise ; shoot ; shute ; pass ; chimney ;   mill ;   mill-hole ;   ore-chute ; ore-pass : (28)
une cheminée dans un massif de minerai, dans le remblai,   a chute in an ore-body, in the waste. (29 *ou* 30)

cheminée (*n.f.*) *ou* cheminée à charbon (Mines), chute ; shoot ; shute ; coal-chute. (31)

cheminée à gaz (*f.*), gas-stove. (32)

cheminée d'aérage (Mines) (*f.*),   air-stack ; ventilation-stack. (33)

cheminée d'ascension (Géol.) (*f.*), neck. (34)

cheminée d'usine (*f.*),   factory chimney ; chimney-stack. (35)

cheminée de fées (Géol.) (*f.*),   chimney-rock ; erosion column. (36)

cheminée de forge (*f.*),   forge chimney ; forge smoke-stack. (37)

cheminée de haut fourneau (*f.*),   blast-furnace shaft. (38)

cheminée de lampe (*f.*),   lamp-chimney ; lamp-glass. (39)

cheminée de locomotive (*f.*),   engine smoke-stack ;   locomotive-chimney ;   funnel of locomotive. (40)

cheminée de remblai (*f.*),   waste-chute. (41)

cheminée diamantifère (Géol.) (*f.*),   diamond pipe ; pipe. (42)

cheminée évasée (d'un ventilateur de mine) (*f.*), évasé chimney (of a mine-fan). (43)

cheminée volcanique (*f.*),   volcanic chimney ; volcanic vent ;   volcanic neck ;   volcanic pipe. (44)

cheminement (*n.m.*), creeping ; creep : (45)
le cheminement des rails qui sont entraînés par les trains, principalement dans les pentes, et auquel il convient de s'opposer dans la mesure du possible, the creeping of rails which are drawn along by the trains, principally on down grades, and which it is advisable to check as far as possible. (46)

cheminement (Géol.) (*n.m.*),   creep ; creeping. (47)

cheminement (Topogr.) (*n.m.*),   meandering : (48)
cheminement au théodolite,   meandering with the theodolite. (49)

cheminer (*v.i.*),   to creep ; to move along ; to advance : (50)
rails qui cheminent (*m.pl.*),   rails which creep. (51)
les sables marins cheminent le long des côtes sous l'action des courants de marées (*m.pl.*), sea-sands creep along the coasts under the action of tidal currents. (52)
moraine médiane qui chemine avec le glacier (*f.*),   medial moraine which moves along with the glacier. (53)

chemise (*n.f.*),   jacket ; jacketing ; lining ; cleading ; clothing ; sheathing. (54)

chemise (d'un four métallurgique) (*n.f.*),   lining (of a metallurgical furnace). (55)

chemise (d'un haut fourneau) (*n.f.*),   lining, shirt, (of a blast-furnace). (56)

chemise (d'une pompe) (*n.f.*), lining (of a pump). (57)

chemise d'air (*f.*), air-casing. (58)

chemise d'eau *ou* chemise à circulation d'eau (*f.*), water-jacket. (59)

chemise de vapeur (*f.*),   steam-jacket. (60)

chemise en briques réfractaires (*f.*),   lining of fire-bricks. (61)

chemise extérieure (haut fourneau) (*f.*),   mantle ; outer stack ; building. (62)

chemise intérieure (haut fourneau) (*f.*),   lining ; shirt. (63)

chenal (*n.m.*),   channel. (64)

chenal à huile (*m.*).   See chenaux à huile.

chenal d'ancien cours d'eau *ou* chenal d'ancienne rivière (*m.*),   old stream-channel ; ancient river-channel. (65)

chenal de coulée (de cubilot ou d'autre fourneau) (*m.*),   tapping-shoot (of cupola or other furnace). (66 *ou* 67)

chenal de coulée (dans l'aire de coulée d'un haut fourneau) (*m.*),   runner (in the pig-bed of a blast-furnace). (68)

chenal de coulée (dans un moule) (Fonderie) (*m.*), runner, gate, main runner, main gate, (in a mould). (1)

chenal de coulée à talon (moule) (*m.*), side-runner ; side-gate. (2)

chenal de coulée en chute directe (moule) (*m.*), direct-pouring gate ; drop-runner ; plump-gate ; pop-gate. (3)

chenal de coulée en source (moule) (*m.*), foun--tain-runner ; horn-gate. (4)

chenal de cours d'eau *ou* chenal de rivière (*m.*), stream-channel ; streamway ; channelway ; river-channel. (5)

chenal filonien (*m.*), lode-channel. (6)

chenal souterrain (ancien chenal de rivière rempli d'alluvion aurifère) (*m.*), deep lead. (7)

chenaux à huile (pattes d'araignée) (*m.pl.*), oil-channels. (8)

chêne (*n.m.*), oak. (9)

chéneau (Constr.) (*n.m.*), gutter. (10)

chéneau à l'anglaise (*m.*), parapet gutter. (11)

chéneau encaissé (*m.*), parallel gutter ; trough-gutter ; box-gutter. (12)

cher, -ère (*adj.*), dear ; expensive. (13)

chercher (*v.t.*), to search ; to search for ; to look for ; to seek ; to seek for : (14)

chercher du travail (*ou* un emploi) dans les mines, to seek employment in the mines. (15)

chercher constamment son niveau, to try to find its own level : (16)

l'eau cherche constamment son niveau (*f.*), water tries to find its own level. (17)

chercher la valeur de l'inconnue (Math.), to find the value of the unknown quantity. (18)

chercheur (Exploitation des gisements alluvion--naires) (pers.) (*n.m.*), digger ; alluvial digger. (19)

chercheur d'aventures minières (*m.*), adventurer ; mining-adventurer. (20)

chercheur d'or (*m.*), gold-digger. (21)

chercheur de grisou (Mines) (*m.*), fireman. (22)

chert (Pétrol.) (*n.m.*), chert ; phthanite ; hornstone ; rock-flint. (23)

chessylite (Minéral) (*n.f.*), chessylite ; Chessy copper ; azurite ; blue malachite ; blue copper carbonate. (24)

cheval (animal) (*n.m.*), horse. (25)

cheval (Méc.) (*n.m.*), horse-power : (26)

une machine de — chevaux, a — horse-power engine. (27)

[Note however, the **French horse-power** or **metric horse-power** or **cheval** = 4,500 kilogrammetres a minute : equivalent to 32,549 foot-pounds a minute, or about ·9863 of the English or ordinary horse-power, which equals 33,000 foot-pounds of work per minute]. (28 *ou* 29)

[cheval, cheval-vapeur and force de cheval are often used in English to signify French horse-power, and conversely le horse-power, or H.P., or HP, are used in French.]

cheval aciéré (*m.*), roughshod horse. (30)

cheval alimentaire (Mach.) (*m.*), donkey feed-pump. (31)

cheval-an [chevaux-ans *pl.*] (Méc.) (*n.m.*), horse-power year. (32)

cheval de bât (*m.*), pack-horse. (33)

cheval de force (Méc.) (*m.*), horse-power : (34)

un automobile de — chevaux de force, a — horse-power car. (35)

See notes under **cheval** (Méc.).

cheval de mine (*m.*), mine-horse ; pit-pony. (36)

cheval de selle (*m.*), saddle-horse. (37)

cheval de terre (cavité remplie d'une matière terreuse) (Mines) (*m.*), horse of ground. (38)

cheval de trait (*m.*), draught-horse. (39)

cheval dynamique (Méc.) (*m.*), horse-power. (40) See notes under **cheval** (Méc.).

cheval effectif (Méc.) (*m.*), brake horse-power ; effective horse-power ; actual horse-power ; net horse-power. (41)

cheval électrique (*m.*), electric horse-power. (42)

cheval ferré à glace (*m.*), roughshod horse. (43)

cheval-heure [chevaux-heure *pl.*] (Méc.) (*n.m.*), horse-power hour. (44)

cheval-heure effectif [chevaux-heure effectifs *pl.*] (*m.*), brake horse-power hour ; effective horse-power hour ; actual horse-power hour. (45)

cheval-heure électrique (*m.*), electric horse-power hour. (46)

cheval-heure indiqué [chevaux-heure indiqués *pl.*] (*m.*), indicated horse-power hour. (47)

cheval indiqué (Méc.) (*m.*), indicated horse-power ; gross horse-power. (48)

cheval nominal *ou* cheval-vapeur nominal [chevaux nominaux *ou* chevaux-vapeur nominaux *pl.*] (*m.*), nominal horse-power. (49)

cheval-vapeur [chevaux-vapeur *pl.*] (Méc.) (*n.m.*), horse-power. (50). See notes under **cheval** (Méc.).

chevalement (au-dessus du puits d'extraction) (Mines) (*n.m.*), head-frame ; pit-head frame ; pit-frame ; pulley-frame ; poppet-head ; puppet-head ; head-stock ; gallows frame ; shaft-tackle. (51)

chevalement (Étaiement) (Constr.) (*n.m.*), trestle-shore. (52)

chevalement à palan (Mines) (*m.*), whip and derry ; whipsey-derry ; whipsiderry. (53)

chevalement-abri [chevalements-abris *pl.*] (Mines) (*n.m.*), shaft-house ; head-house. (54)

chevalement de sondage *ou* simplement chevale--ment (*n.m.*), derrick. (55)

chevalement de sondage de puits à pétrole (*m.*), oil-well derrick ; carpenters' rig. (56)

chevalet (*n.m.*), horse ; trestle. (57)

chevalet d'extraction (Mines) (*m.*). Same as chevalement.

chevalet de scieur (*m.*), sawhorse ; sawbuck ; jack ; horse ; buck. (58)

chevalet de scieur de long (*m.*), pit-saw horse ; pit-sawyers' trestle. (59)

chevauchant, -e (*adj.*), overlapping ; straddling : (60)

des ardoises chevauchantes, overlapping slates. (61)

chevauché, -e (Géol.) (*adj.*), overthrust ; over-cast. (62)

chevauchement (*n.m.*), overlapping ; overlap ; straddling ; spanning : (63)

le chevauchement des tuiles, overlapping of tiles. (64)

chevauchement (Géol.) (*n.m.*), thrust ; over--thrust : (65)

chevauchement de nappes, de plis couchés, thrust, overthrust, of rock-sheets, of recum--bent folds. (66 *ou* 67)

chevaucher (*v.t.*), to overlap ; to straddle ; to span : (68)

grue à portique chevauchant — voies de chemin de fer (*f.*), gantry-crane spanning — sets of rails. (1)

chevaucher (*v.i.*), to overlap ; to straddle : (2) tuiles qui ne chevauchent pas régulièrement (*f.pl.*), tiles which do not overlap in a regular manner. (3)

chevaucher (Géol.) (*v.t. & v.i.*), to overthrust. (4)

chevel (*n.m.*), portable vice. (5)

chever une pièce de métal, to groove out a piece of metal. (6)

chevet (d'un filon) (*n.m.*), foot-wall, ledger-wall, ledger, (of a lode). (7)

chevêtre (*n.m.*), trimmer ; trimmer-beam. (8)

chevêtre sous la marche palière (escalier) (*m.*), landing-trimmer. (9)

chevêtrier (*n.m.*), trimming-joist. (10)

cheveux de Pélé (filaments de lave) (*m.pl.*), Pele's hair. (11)

cheveux de Vénus (Minéral) (*m.pl.*), Venus's hairstone ; Veneris crinis ; flèches d'amour ; sagenite ; sagenitic quartz ; rutilated quartz. (12)

cheville (*n.f.*), pin ; peg ; bolt. (13)

cheville (d'une charnière) (*n.f.*), pin, pintle, (of a hinge). (14)

cheville (pour assujétir un tenon dans une mortaise) (*n.f.*), draw-bore pin. (15)

cheville (tire-fond) (*n.f.*), spike. (16)

cheville d'attelage (Ch. de f.) (*f.*), coupling-pin ; draw-pin ; draw-bolt ; drag-bolt. (17)

cheville de bois (*f.*), peg ; wood pin ; treenail. (18)

cheville fusible (chaudière) (*f.*), fusible plug. (19)

cheville ouvrière (de bogie, de wagon) (*f.*), king-bolt ; king-pin ; centre-pin ; main-pin ; bogie-pin ; pintle. (20 *ou* 21)

cheviller (*v.t.*), to pin ; to peg ; to bolt : (22) cheviller un assemblage à tenon et mortaise, to pin a mortise-and-tenon joint. (23)

chèvre (appareil propre à élever des fardeaux) (*n.f.*), gin. (24)

chèvre (support pour scier, etc.) (*n.f.*), **horse** ; sawhorse ; sawbuck ; jack. (25)

chèvre à hauban (*f.*), shear-legs and guy. (26)

chèvre à trois pieds (*f.*), shear-legs ; sheer-legs ; gin. (27)

chèvre d'échafaudage (*f.*), trestle. (28)

chèvre de carrossier (*f.*), carriage-jack (notched lever and horse type). (29)

chèvre verticale *ou simplement* **chèvre** (*n.f.*), derrick-crane ; derrick. (30)

chevron (*n.m.*), rafter. (31)

chevron d'arêtier *ou* **chevron arêtier** (*m.*), angle-rafter ; hip-rafter ; hip. (32)

chevron de long pan (*m.*), common rafter (of a hip-roof). (33)

chevron intermédiaire (*m.*), common rafter. (34)

chevron principal (*m.*), principal rafter ; chief rafter ; principal. (35)

chevronnage (*n.m.*), raftering. (36)

chevronner (*v.t.*), to rafter : (37) chevronner un comble, to rafter a roof. (38)

chiastolite (Minéral) (*n.f.*), chiastolite ; macle. (39)

chicane (*n.f.*), baffle ; baffle-plate ; baffler : (40) planches en chicane (*f.pl.*), baffle-boards. (41) cuve à chicanes (*f.*), baffle-tank. (42)

chien (cliquet) (*n.m.*), **dog** ; pawl ; ratchet ; catch ; latch ; click ; trigger ; detent. (43)

chien (Tréfilerie) (*n.m.*), wire-dog ; bench draw-tongs ; draw-bench tongs ; lion's claw ; devil's claw. (44)

chien de mine (*m.*), dog (a mine-wagon running upon two longitudinals). (45)

chiffre (*n.m.*), figure : (46) en chiffres ronds, in round figures. (47) chiffres et lettres à chaud [chiffre (*n.m.*) ; lettre (*n.f.*)], figures and letters for brand-ing ; figures and letters for wood. (48) chiffres et lettres à froid, figures and letters for stamping ; letter and figure stamps for engineers. (49)

chiffrer (se) (*v.r.*), to work out ; to figure out : (50) économie de combustible qui se chiffre par — tonnes chaque jour (*f.*), saving of fuel which works out at — tons each day. (51)

childrénite (Minéral) (*n.f.*), childrenite. (52)

chiléite (Minéral) (*n.f.*), chileite. (53)

chimico-analytique (*adj.*), chemico-analytical. (54)

chimico-électrique (*adj.*), chemico-electrical. (55)

chimico-métallurgique (*adj.*), chemicometallur-gical. (56)

chimico-minéralogique (*adj.*), chemicomineral-ogical. (57)

chimico-physique (*adj.*), chemicophysical. (58)

chimie (*n.f.*), chemistry. (59)

chimie analytique (*f.*), analytical chemistry. (60)

chimie appliquée (*f.*), applied chemistry ; practical chemistry. (61)

chimie de l'or (*f.*), chemistry of gold. (62)

chimie industrielle (*f.*), industrial chemistry ; technical chemistry. (63)

chimie inorganique (*f.*), inorganic chemistry. (64)

chimie métallurgique (*f.*), metallurgical chemistry. (65)

chimie minérale (*f.*), mineralogical chemistry. (66)

chimie organique (*f.*), organic chemistry. (67)

chimie physique (*f.*), physical chemistry. (68)

chimie pure (*f.*), pure chemistry ; theoretical chemistry. (69)

chimique (*adj.*), chemical : (70) la composition chimique de l'atmosphère, the chemical composition of the atmosphere. (71)

chimiquement (*adv.*), chemically : (72) corps chimiquement combinés (*m.pl.*), chemi-cally combined substances. (73) minéral presque chimiquement pur (*m.*), mineral almost chemically pure. (74)

chimiste (pers.) (*n.m.*), chemist. (75)

chimiste-analyste [chimistes-analystes *pl.*] (*n.m.*), analytical chemist. (76)

chimiste-conseil [chimistes-conseils *pl.*] (*n.m.*), consulting-chemist. (77)

chimiste-essayeur [chimistes-essayeurs *pl.*] (*n.m.*), analytical chemist ; assayer chemist. (78)

chimiste-métallurgiste [chimistes-métallurgistes *pl.*] (*n.m.*), metallurgical chemist. (79)

china-clay (*n.m.*), china-clay ; kaolin ; china-stone. (80)

Chinois (pers.) (*n.m.*), Chinaman ; Chinese ; Chinese boy. (81)

chiolite (Minéral) (*n.f.*), chiolite. (82)

chiviatite (Minéral) (*n.f.*), chiviatite. (83)

chloanthite (Minéral) (*n.f.*), chloanthite. (84)

chlorastrolite (Minéral) (*n.f.*), chlorastrolite. (85)

chlorate (Chim.) (*n.m.*), chlorate : (86)

chlorate de potasse *ou* chlorate de potassium, chlorate of potash; potassium chlorate. (1)

chlore (Chim.) (*n.m.*), chlorine. (2)

chloreux, -euse (Chim.) (*adj.*), chlorous. (3)

chlorhydrique (*adj.*), hydrochloric. (4)

chlorique (Chim.) (*adj.*), chloric. (5)

chlorite (Chim.) (*n.m.*), chlorite. (6)

chlorite (Minéral) (*n.f.*), chlorite: (7)
une chlorite, les chlorites, le genre chlorite, a chlorite, the chlorites, the chlorite group. (8 *ou* 9 *ou* 10)

chlorité, -e *ou* **chloriteux, -euse** *ou* **chloritique** (*adj.*), chloritic; chloritous. (11)

chloritoïde (*n.f.*) *ou* **chloritspath** (*n.m.*) (Minéral), chloritoid. (12)

chloritoschiste (*n.m.*), chlorite schist; chlorite slate. (13)

chlorocalcite (Minéral) (*n.f.*), chlorocalcite. (14)

chloromélanite (Minéral) (*n.f.*), chloromelanite. (15)

chloropale (Minéral) (*n.f.*), chloropal. (16)

chlorophane (Minéral) (*n.f.*), chlorophane. (17)

chloroschiste (*n.m.*), chlorite schist; chlorite slate. (18)

chlorospinelle (Minéral) (*n.m.*), chlorospinel. (19)

chloruration (*n.f.*), chlorination; chlorinating: (20)
chloruration de l'or, chlorination of gold; chlorinating the gold. (21)

chlorure (Chim.) (*n.m.*), chloride: (22)
chlorure cuivrique, cupric chloride. (23)
chlorure d'ammonium, chloride of ammonium; sal ammoniac. (24)
chlorure d'argent, chloride of silver; silver chloride. (25)
chlorure d'or, chloride of gold; gold chloride. (26)
chlorure de calcium, chloride of calcium; calcium chloride. (27)
chlorure de chaux, chloride of lime. (28)
chlorure de potassium, potassium chloride. (29)
chlorure de sodium, sodium chloride. (30)

chlorurer (*v.t.*), to chlorinate; to chlorinize: (31)
chlorurer un métal, to chlorinate a metal. (32)

choc (*n.m.*), shock; impact; concussion; percussion: (33)
le choc d'une collision de trains, the shock of a collision of trains. (34)

choc (d'une roue contre le rail) (Ch. de f.) (*n.m.*), hammer-blow (of a wheel against the rail). (35)

choc électrique (*m.*), electric shock. (36)

choc en retour (Phys.) (*m.*), return-shock; return-stroke. (37)

choisir (*v.t.*), to choose; to select: (38)
choisir un nouvel emplacement pour une usine, to select a new site for a works. (39)

choix (*n.m.*), choice; choosing; selection: (40)
choix de l'emplacement du moulin, choosing the site of the mill; choice of the site of the mill. (41)

chômage (*n.m.*), idleness; standing; shutting down; closing; stoppage: (42)
chômage d'une usine, shutting down of a works; closing of a factory. (43)
chômage de la mine, standing of the mine. (44)
chômage du dimanche, Sunday closing. (45)

chômage forcé, enforced idleness. (46)
chômage momentané, temporary stoppage. (47)
chômage prolongé, prolonged stoppage; long-continued idleness. (48)

chômage (en) (se dit d'une mine), standing; inactive: (49)
mine en chômage (*f.*), standing mine; inactive mine. (50)

chômage (en) (se dit des machines, etc.), idle; standing. (51)

chômer (*v.i.*), to stand; to stand idle; to lie idle; to be idle; to shut down; to close down; to stop work: (52)
mine qui chôme (*f.*), mine which is standing. (53)
ouvriers qui chôment (*m.pl.*), workmen who are idle. (54)
lorsque les eaux sont basses, les moulins chôment, when the water is low, the mills shut down. (55)

chômer d'ouvrage, to be out of work. (56)

chômeur (pers.) (*n.m.*), idle workman: (57)
la ville est pleine de chômeurs, the town is full of idle workmen. (58)

chondroarsénite (Minéral) (*n.f.*), chondrarsenite. (59)

chondrodite (Minéral) (*n.f.*), chondrodite. (60)

chott (lac salé plus ou moins desséché) (Géol.) (*n.m.*), salt-pan; salt bottom. (61)

christianite (zéolite) (Minéral) (*n.f.*), christianite; phillipsite; lime harmotome. (62)

christianite (anorthite) (Minéral) (*n.f.*), christianite. (63)

chromate (Chim.) (*n.m.*), chromate: (64)
chromate de fer, chromate of iron; iron chromate. (65)

chromatique (*adj.*), chromatic; chromatical; colour (*used as adj.*): (66)
aberration chromatique (*f.*), chromatic aberration; colour-aberration. (67)

chrome (Chim.) (*n.m.*), chromium; chrome. (68)

chromé, -e (*adj.*), chrome; chromium (*used as adjs*): (69)
acier chromé (*m.*), chrome steel; chromium steel. (70)

chromeux, -euse (Chim.) (*adj.*), chromous. (71)

chromifère (*adj.*), chromiferous. (72)

chromique (Chim.) (*adj.*), chromic: (73)
acide chromique (*m.*), chromic acid. (74)

chromite (Minéral) (*n.f.*), chromite. (75)

chromocre (Minéral) (*n.m.*), chrome ochre. (76)

chronomètre (*n.m.*), chronometer. (77)

chronométrie (*n.f.*), chronometry. (78)

chronométrique (*adj.*), chronometric; chrono-metrical. (79)

chronométriquement (*adv.*), chronometrically. (80)

chrysobéryl (Minéral) (*n.m.*), chrysoberyl. (81)

chrysocolle (Minéral) (*n.f.*), chrysocolla. (82)

chrysolithe *ou* **chrysolite** (Minéral) (*n.f.*), chrysolite. (83)

chrysolithique *ou* **chrysolitique** (*adj.*), chrysolitic. (84)

chrysoprase (Minéral) (*n.f.*), chrysoprase. (85)

chrysotile (Minéral) (*n.m.*), chrysotile. (86)

chuck (mandrin de tour) (Méc.) (*n.m.*), chuck. (87)

chute (*n.f.*), fall; falling; drop. (88)

chute (*n.f.*) *ou* **chute d'eau** (masse d'eau qui tombe d'une certaine hauteur), fall; water-fall; water-jump: (89)

les chutes du Rhin, the falls of the Rhine. (1)

chute (*n.f.*) *ou* **chute d'eau** (différence de niveau) (Hydraul.), fall ; head ; head of water. (2)

chute brute (Hydraul.) (*f.*), total fall ; total head. (3)

chute dans un puits (*f.*), falling into (*or* down) a shaft. (4)

chute de barre (bout affranchi)(*f.*), crop-end. (5)

chute de comble (*f.*), roof-pitch ; pitch of roof. (6)

chute de la cage (Mines) (*f.*), fall of cage ; run of the cage. (7)

chute de neige (*f.*), fall of snow ; snowfall. (8)

chute de pierres dans le puits (*f.*), fall of stones in shaft. (9)

chute de potentiel *ou* **chute de tension** (Élec.) (*f.*), fall of potential ; drop of potential ; pressure-drop ; pressure-loss. (10)

chute de sur un échafaudage (*f.*), falling off a scaffold. (11)

chute des corps (*f.*), fall of bodies ; falling of bodies. (12)

chute des corps dans le vide (*f.*), fall of bodies in vacuo. (13)

chute des corps dans les fluides (*f.*), fall of bodies in fluids. (14)

chute effective *ou* **chute disponible** (Hydraul.) (*f.*), working fall ; working head ; effective head ; available fall. (15)

chute inductive (Élec.) (*f.*), inductive drop. (16)

chute ohmique (Élec.) (*f.*), ohmic drop. (17)

ciel (*n.m.*), sky : (18)

un ciel étoilé, a starry sky. (19)

ciel (d'un foyer) (*n.m.*), roof, crown, (of a furnace or fire-box). (20)

ciel (d'une galerie de mine) (*n.m.*), roof (of a mine-level). (21)

ciel ouvert (à) (Carrières, Mines, etc.), open-cast ; open-cut ; open ; daylight (*used as adj.*) : (22)

des carrières exploitées à ciel ouvert (*f.pl.*), quarries worked open-cast. (23)

exploitation à ciel ouvert (*f.*), open-cut mining. (24)

carrière à ciel ouvert (*f.*), open quarry. (25)

'houillère à ciel ouvert (*f.*), daylight-colliery. (26)

cime (*n.f.*), top ; summit ; peak : (27)

cime d'une montagne, top, summit, peak, of a mountain : (28)

l'acuité d'une cime de montagne, the sharp-ness of a mountain peak. (29)

ciment (*n.m.*), cement. (30)

ciment (Géol.) (*n.m.*), cement. (31)

ciment à prise lente (*m.*), slow-setting cement. (32)

ciment à prise rapide (*m.*), quick-setting cement. (33)

ciment armé (*m.*), reinforced cement. (34)

ciment de laitier (*m.*), slag cement. (35)

ciment de Portland (*m.*), Portland cement. (36)

ciment hydraulique (*m.*), hydraulic cement ; water-cement. (37)

ciment pour courroies (*m.*), belting-cement. (38)

ciment pouzzolanique (*m.*), pozzuolanic cement. (39)

ciment romain (*m.*), Roman cement. (40)

cimentage (*n.m.*), cementing ; cementation : (41)

cimentage des forages pour empêcher l'enva-hissement par les eaux, cementing bore-holes to prevent flooding. (42)

cimentage des puits à pétrole, cementing oil-wells. (43)

cimentaire (*adj.*), cementing ; cementatory : (44)

matière cimentaire (*f.*), cementing material. (45)

cimentation (*n.f.*), cementation ; cementing : (46)

cimentation des terrains dans la traversée des niveaux aquifères, cementation of the ground in negotiating water-bearing strata. (47)

cimenté, -e (Géol., etc.) (*adj.*), cemented : (48)

gravier cimenté (*m.*), cemented gravel. (49)

cimenter (*v.t.*), to cement : (50)

cimenter des pierres, to cement stones. (51)

cimentier (pers.) (*n.m.*), cement-maker ; cement-manufacturer. (52)

ciminite (Pétrol.) (*n.f.*), ciminite. (53)

cimolite (Minéral) (*n.f.*), cimolite. (54)

cinabarin, -e (*adj.*), cinnabarine ; cinnabaric. (55)

cinabre (Minéral) (*n.m.*), cinnabar ; cinabar. (56)

cinabrifère (*adj.*), cinnabarine ; cinnabaric. (57)

cinématique (*adj.*), kinematic ; kinematical ; cinematic ; cinematical. (58)

cinématique (*n.f.*), kinematics ; cinematics : (59)

la cinématique des mécanismes, the kinematics of mechanisms. (60)

cinématiquement (*adv.*), kinematically ; cine-matically. (61)

cinémomètre (*n.m.*), kinemometer. (62)

cinérite (Géol.) (*n.f.*), cinereous tuff. (63)

cinétique (*adj.*), kinetic ; kinetical : (64)

énergie cinétique (*f.*), kinetic energy. (65)

cinétique (*n.f.*), kinetics : (66)

la cinétique des gaz, the kinetics of gases. (67)

cinétiquement (*adv.*), kinetically. (68)

cinglage (Métall.) (*n.m.*), shingling ; knobbling. (69)

cingler (Métall.) (*v.t.*), to shingle ; to knobble : (70)

cingler le fer, to shingle, to knobble, iron. (71)

cingler la ligne (avec un cordeau blanchi à la craie) (Charp.), to snap the line. (72)

cingleur (Métall.) (*n.m.*), shingler. (73)

cingleur rotateur (*m.*), rotary squeezer. (74)

cintrage (*n.m.*), arching ; bending ; curving : (75)

cintrage du toit, arching the roof. (76)

cintrage des tôles, des tuyaux, des rails, bending plates, pipes, rails. (77 *ou* 78 *ou* 79)

cintre (courbure) (*n.m.*), arch ; curve : (80)

le cintre d'un tunnel, the arch of a tunnel. (81)

le cintre d'une voûte, the curve of an arch ; the arch of a vault. (82)

cintre (arcade de bois sur laquelle on bâtit les voûtes en pierre) (*n.m.*), centre ; centering : (83)

cintre pour une voûte en plein cintre, centre for a round arch ; centering for a semi-circular arch. (84)

cintrer (*v.t.*), to arch ; to bend ; to curve : (85)

cintrer une pièce de bois, to bend a piece of wood. (86)

cintreuse (*n.f.*), bender ; bending-machine. (87)

See **machine à cintrer** for varieties.

cipolin (marbre) (*n.m.*), cipolin. (88)

circonférence (*n.f.*), circumference ; girth : (89)

la circonférence d'un cercle, the circumference of a circle. (90)

circonférence primitive (Engrenages) (*f.*), pitch-circumference. (91)

circonférentiel, -elle (*adj.*), circumferential : (1) pas circonférentiel (*m.*), circumferential pitch. (2)

circonscription (*n.f.*) *ou* **circonscription territoriale,** area ; district : (3) circonscription administrative, administrative area. (4) la circonscription territoriale des opérations de la société, the area of the company's operations. (5)

circonscrire (Géom.) (*v.t.*), to circumscribe : (6) circonscrire un polygone à un cercle, to cir- -cumscribe a polygon about a circle. (7) courbe circonscrite à un polygone (*f.*), curve circumscribed about a polygon. (8)

circonscrire un dérangement (Télégr.), to locate a fault. (9)

circuit (*n.m.*), circuit. (10)

circuit d'aérage (Mines) (*m.*), ventilation-circuit. (11)

circuit de retour (*m.*), return-circuit. (12)

circuit dérivé (Élec.) (*m.*), shunt circuit ; derived circuit. (13)

circuit des faîtes (Paratonnerres) (*m.*), ridge-circuit. (14)

circuit électrique (*m.*), electric circuit. (15)

circuit extérieur (Élec.) (*m.*), external circuit. (16)

circuit fermé (Élec.) (*m.*), closed circuit ; made circuit. (17)

circuit inducteur (Élec.) (*m.*), inductive circuit. (18)

circuit induit (Élec.) (*m.*), induced circuit. (19)

circuit intérieur (Élec.) (*m.*), internal circuit. (20)

circuit magnétique (*m.*), magnetic circuit. (21)

circuit magnétique fermé (*m.*), closed magnetic circuit. (22)

circuit magnétique ouvert (*m.*), open magnetic circuit. (23)

circuit ouvert (Élec.) (*m.*), open circuit ; broken circuit. (24)

circuit total (Élec.) (*m.*), complete circuit ; full circuit. (25)

circulaire (*adj.*), circular ; round : (26) scie circulaire (*f.*), circular saw. (27)

circulairement (*adv.*), circularly. (28)

circulation (*n.f.*), circulation ; travelling ; traffic : (29) la circulation de l'air dans les mines, de l'eau dans les tuyaux, the circulation of air in mines, of water in pipes. (30 *ou* 31) circulation ascensionnelle du courant d'air, ascensional travelling of the air-current. (32) entraver la circulation, to impede the traffic. (33)

circulation (des trains) (Ch. de f.) (*n.f.*), running, working, traffic, (of trains) : (34) des tableaux dits graphiques réglementent la circulation des trains (*m.pl.*), representa- -tions called diagrams regulate the working of trains. (35)

circulation à contre-voie (Ch. de f.) (*f.*), running on wrong line. (36)

circulation à double voie (Ch. de f.) (*f.*), double-line working. (37)

circulation à voie fermée (Ch. de f.) (*f.*), absolute block-signal system. (38)

circulation à voie ouverte (Ch. de f.) (*f.*), per- -missive block-signal system. (39)

circulation à voie unique (Ch. de f.) (*f.*), single-line working. (40)

circuler (*v.i.*), to circulate ; to run ; to travel : (41) air qui circule autour des fronts de taille (Mines) (*m.*), air which circulates around the working-faces. (42) les trains ont cessé de circuler par suite de l'abondance de la neige, trains have ceased to run owing to the heavy fall of snow. (43)

cire (*n.f.*), wax. (44)

cire d'abeilles (*f.*), beeswax. (45)

cire fossile *ou* **cire minérale** (*f.*), fossil wax ; mineral wax ; ceresin ; ceresine. (46)

cireux, -euse (*adj.*), waxy : (47) éclat cireux (*m.*), waxy lustre. (48)

cirque (Géol.) (*n.m.*), cirque ; circ ; corrie ; comb : (49) région alpestre célèbre par ses escarpements gigantesques, ses cimes dentelées, ses cirques sauvages, que les névés et les glaciers rayent de taches blanches, alpine region celebrated for its titanic cliffs, its jagged peaks, its wild cirques streaked with white patches by the névés and the glaciers. (50)

cirrholite *ou* **cirrolite** (Minéral) (*n.f.*), cirrolite ; cirrholite. (51)

cisaille (*n.f.*), shear. (52). See cisailles.

cisaillement (*n.m.*), shearing ; clipping : (53) le cisaillement des tôles, shearing plates. (54)

cisaillement (Résistance des matériaux) (*n.m.*), shearing ; shear ; slide (Strength of materials) : (55) rivet qui est soumis à la fois à des efforts de traction et de cisaillement (*m.*), rivet which is at the same time subjected to tensile and shearing stresses. (56) rivet qui travaille au double cisaillement, rivet which is in double shear. (57)

cisaillement double, double shear. (58)

cisaillement simple, single shear. (59)

cisailler (*v.t.*), to shear ; to clip : (60) cisailler une tôle, des fers cornières, to shear a plate, angle-iron. (61 *ou* 62)

cisailler à froid, to cold-shear. (63)

cisailler (Résistance des matériaux) (*v.t.*), to shear. (64)

cisailler (se) (Résistance des matériaux) (*v.r.*), to shear. (65)

cisailles (*n.f.pl.*), shears, shear ; shearing-machine. (66)

cisailles à barres (*f.pl.*), bar-shears. (67)

cisailles à billettes (*f.pl.*), billet-shears. (68)

cisailles à blooms (*f.pl.*), bloom-shears. (69)

cisailles à brames (*f.pl.*), slab-shears. (70)

cisailles à câbles (Sondage à la corde) (*f.pl.*), rope-knife. (71)

cisailles à chaud de lingots (*f.pl.*), hot-ingot shear. (72)

cisailles à goupilles (*f.pl.*), pin-shears. (73)

cisailles à guillotine (*f.pl.*), guillotine shears ; guillotine shearing-machine. (74)

cisailles à levier à contrepoids (*f.pl.*), lever shearing-machine with counterweight. (75)

cisailles à levier pour cisailler des tôles par le milieu (*f.pl.*), lever shearing-machine to cut plates through the middle. (76)

cisailles à lingots (*f.pl.*), ingot-shear. (77)

cisailles à main (*f.pl.*), hand-shears. (78)

cisailles à métaux (*f.pl.*), metal-shears. (79)

cisailles à tôles (*f.pl.*), plate-shears. (80)

cisailles circulaires pour métaux (*f.pl.*), straight-cutting circular shearing-machine. (81)

cisailles d'établi (*f.pl.*), bench-shears. (82)

cisailles de ferblantier (*f.pl.*), tinmen's shears; tinmen's snips. (1)

cisailles marchant à bras (*f.pl.*), hand-lever shearing-machine. (2)

cisailleur (pers.) (*n.m.*), shearman. (3)

ciseau (*n.m.*), chisel. (4)

ciseau (Sondage) (*n.m.*), chisel; bit. (5). See **trépan**, the more usual word, for varieties.

ciseau à chanfrein *ou* **ciseau à biseau** (*m.*), bevelled-edge chisel. (6)

ciseau à déballer (*m.*), case-opener. (7)

ciseau à froid (*m.*), cold-chisel. (8)

ciseau à gouge (*m.*), gouge; firmer-gouge. (9)

ciseau à larder *ou* **ciseau bédane** (*m.*), mortise-chisel. (10)

ciseau de calfat (*m.*), caulking-chisel; calking-iron. (11)

ciseau de tour, nez carré, 1 biseau (*m.*), turning-chisel, square nose, single bevel. (12)

ciseau long (*m.*), paring-chisel. (13)

ciseau ordinaire *ou* **ciseau à planche** *ou* **ciseau à panne** (Charp. ou Menuis.) (*m.*), firmer-chisel. (14)

ciseau renforcé (*m.*), strong firmer-chisel. (15)

ciseaux (instrument à deux branches tranchantes) (*n.m.pl.*), scissors. (16)

ciseaux (mécanisme extenseur en forme de plusieurs X) (*n.m.pl.*), lazy-tongs. (17)

ciselant, -e (*adj.*), chiseling; chiselling: (18) l'action ciselante des glaciers (*f.*), the chiseling action of glaciers. (19)

ciseler (*v.t.*), to chisel: (20) ciseler la pierre, to chisel stone. (21)

ciseleur (pers.) (*n.m.*), chiseler. (22)

ciselure (*n.f.*), chiseling; chiselling. (23)

citerne (*n.f.*), cistern; tank; water-tank. (24)

citrine (Minéral) (*n.f.*), citrine; citrine quartz; false topaz. (25)

civière (*n.f.*), hand-barrow. (26)

claie (*n.f.*), screen. (27)

claie à sable (*f.*), sand-screen; harp. (28)

claim (concession) (Mines) (*n.m.*), claim; con-cession. (29) See also **concession**.

claim alluvionnaire (*m.*), alluvial claim. (30)

claim de découverte (*m.*), discovery-claim. (31)

claim de quartz aurifère (*m.*), reef-claim. (32)

claim minier (*m.*), mining claim. (33)

claim pour exploitation hydraulique (*m.*), hydraulic-mining claim. (34)

clain (Mines) (*n.m.*). Same as **claim**.

clair, -e (se dit de l'eau, du feu, du temps, etc.) (*adj.*), clear. (35)

clair, -e (se dit des couleurs) (*adj.*), light; pale: (36) couleur claire (*f.*), light colour; pale colour. (37)

clair, -e (coloration du fer) (*adj.*), bright: (38) température du rouge clair (*f.*), bright-red heat. (39)

clair (se dit du fil: par opposition à **noir**) (*adj.*), bright (40) (said of wire: contradis-tinguished from **black**).

claire-voie [**claires-voies** *pl.*] (clôture) (*n.f.*), paling. (41)

claire-voie (à), open; open-worked; lattice (*used as adj.*); skeleton: (42) chevalement à claire-voie (*m.*), lattice head-gear. (43) caisse à claire-voie (*f.*), crate; skeleton case. (44)

clairement (*adv.*), clearly. (45)

clairer du minerai, to wash ore. (46)

clameaux (*n.m.pl.*) *ou* **clampe** (*n.f.*), cramp; cramp-iron; dog-iron; joint-cramp; joiners' dogs; timber dogs; clamp. (47)

clapet (*n.m.*), clack-valve; clack; flap-valve; clapper; clapper-valve; valve. (48). See also **valve** and **soupape**.

clapet à charnière (*m.*), hinged valve; leaf-valve. (49)

clapet d'admission (*m.*), inlet-valve. (50)

clapet d'aspiration (*m.*), suction-valve. (51)

clapet d'échappement (*m.*), exhaust-valve. (52)

clapet de démarrage (locomotive compound) (*m.*), intercepting-valve. (53)

clapet de pied (*m.*), foot-valve. (54)

clapet de refoulement (*m.*), delivery-valve. (55)

clapet de retenue (*m.*), check-valve; retaining valve; back-pressure valve; non-return valve. (56)

clapet de sûreté (soupape de trop-plein) (*m.*), relief-valve; escape-valve. (57)

clapet de trop-plein (*m.*), overflow-valve. (58)

clapet du cendrier (locomotive) (*m.*), ash-pan damper. (59)

clapet sphérique (*m.*), ball valve. (60)

clapier (talus d'éboulis) (Géol.) (*n.m.*), talus. (61)

claquement (Mach.) (*n.m.*), rattling; chattering. (62)

claquer (*v.i.*), to rattle; to chatter. (63)

clarificateur, -trice (*adj.*), clarifying: (64) filtre clarificateur (*m.*), clarifying filter. (65)

clarificateur (*n.m.*), clarifier. (66)

clarification (*n.f.*), clarification. (67)

clarifier (*v.t.*), to clarify: (68) clarifier un liquide, to clarify a liquid. (69)

clarite (Minéral) (*n.f.*), clarite. (70)

clarté (*n.f.*), clearness: (71) la clarté de l'eau, the clearness of water. (72)

classe (*n.f.*), class; sort; grade: (73) la classe recommandée de matériel de sondage, the class of drilling-plant recommended. (74) la classe laborieuse, the labouring class. (75) les classes ouvrières, the working classes. (76)

classement (*n.m.*), classing; classifying; sort-ing; grading. (77)

classement (Ch. de f.) (*n.m.*), marshalling; drilling: (78) classement des wagons de marchandises, marshalling goods-wagons. (79)

classement en grosseur *ou* **classement par grosseur** (*m.*), sizing. (80)

classer (*v.t.*), to class; to classify; to sort; to grade: (81) classer par densité, to classify, to sort, by density. (82) classer, suivant la grandeur des grains, les minerais broyés, to grade the crushed ore according to grain-size. (83)

classer des wagons (Ch. de f.), to marshal trucks; to drill cars. (84)

classer par dimension *ou* **classer par grosseur** *ou* **classer suivant grandeur** *ou* *simplement* **classer** (*v.t.*), to size: (85) classer le minerai suivant grosseur avec des tôles perforées, to size the ore with punched plates. (86)

classeur (*adj.*), classifying; sorting; sizing: (87) crible classeur (*m.*), sizing-screen. (88)

classeur (pour minerai, etc.) (*n.m.*), classifier; sizer: (89)

le trommel est un appareil robuste et commode, mais c'est un mauvais classeur, the trommel is a strong and convenient apparatus, but it is a bad sizer. (1)

classeur à schlamms (*m.*), slimes-classifier. (2)

classeur conique (*m.*), cone-classifier. (3)

classeur hydraulique (*m.*), hydraulic classifier. (4)

classeur-trieur [classeurs-trieurs *pl.*] (*n.m.*). Same as **classeur**.

classificateur (*n.m.*). Same as **classeur**.

classification (*n.f.*), classification; classing; sorting. (5)

classifier (*v.t.*), to classify; to class; so sort: (6) classifier par densité, to classify, to sort, by density. (7)

classique (*adj.*), standard: (8) formule classique (*f.*), standard formula. (9)

clastique (Géol.) (*adj.*), clastic: (10) roches clastiques (*f.pl.*), clastic rocks. (11)

claudétite (Minéral) (*n.f.*), claudetite. (12)

clause pour cas de grève (*f.*), strike-clause. (13)

clausthalite (Minéral) (*n.f.*), clausthalite. (14)

clavage (d'une voûte) (*n.m.*), keying, keying up, keying in, (an arch). (15)

claveau (Arch.) (*n.m.*), voussoir; arch-stone; quoin. (16)

clavetage *ou* **clavettage** (Méc.) (*n.m.*), keying; cottering; forelocking: (17) clavetage qui a pris du jeu, keying which has worked loose. (18)

claveter (Méc.) (*v.t.*), to key; to cotter; to forelock: (19) claveter une poulie sur l'arbre, to key a pulley on the shaft. (20)

clavette (*n.f.*), key; cotter; pin; forelock. (21)

clavette à mentonnet *ou* **clavette à tête** (*f.*), gib-head key. (22)

clavette d'arbre *ou* **clavette de calage** (*f.*), shaft-key; steady-pin. (23)

clavette d'essieu (*f.*), axle-pin; linchpin. (24)

clavette et contre-clavette (*f.pl.*), gib and cotter; gib and key. (25)

clavette évidée *ou* **clavette creuse** *ou* **clavette à friction** (*f.*), saddle-key; hollow key. (26)

clavette fendue (*f.*), split pin; split cotter-pin; split key; spring-cotter; spring-forelock; spring-key. (27)

clavette linguiforme (clavette d'arbre) (*f.*), feather; feather-key; spline. (28)

clavette noyée (*f.*), sunk key. (29)

clavette plate (*f.*), flat key. (30)

clavette posée à plat *ou* **clavette à méplat** (*f.*), key on flat. (31)

clavette tangentielle (*f.*), tangent-key. (32)

clavetter (*v.t.*). Same as **claveter**.

claya (intercalé au milieu des schistes houillers) (*n.m.*), clay-band; ramble: (33) les clayas, se décollant très facilement du toit, obligent les mineurs à consolider les boisages, afin d'éviter la chute des faux toits qu'ils forment, clay-bands, coming away very easily from the roof, compel the miners to strengthen the timbering, in order to prevent the fall of the false roofs which they form. (34)

clayonnage (Constr. hydraul.) (*n.m.*), mat; mattress. (35)

clayonner (*v.t.*), to mat: (36) clayonner le talus d'un canal, d'une rivière, to mat the bank of a canal, of a river. (37 *ou* 38)

clé (*n.f.*). Same as **clef**.

clef (Serrur.) (*n.f.*), key: (39) la clef d'une porte, the key of a door. (40)

clef (outil destiné à serrer ou desserrer les écrous) (*n.f.*), spanner; wrench; key. (41)

clef (Charp.) (*n.f.*), key: (42) assemblage à tenon passant avec clef, through-tenon joint with key. (43)

clef (pièce qui entre dans le boisseau d'un robinet) (*n.f.*), plug; spigot: (44) robinet à clef (*m.*), plug-cock. (45)

clef à béquille (*f.*), box-spanner with T handle; box-key with crutch handle. (46)

clef à béquille pour tire-fonds (1) **à trou carré, ou** (2) **à trou à six pans** (*f.*), box-spanner for coach-screws (1) with square opening, or (2) with hexagonal opening. (47 *ou* 48)

clef à chaîne pour tubes (*f.*), chain pipe-wrench. (49)

clef à clavette (*f.*), wedge-spanner. (50)

clef à cliquet, noire, ou polie (*f.*), ratchet-spanner; ratchet-wrench, black, or bright. (51 *ou* 52)

clef à crémaillère (*f.*), rack-spanner. (53)

clef à deux mâchoires *ou* **clef à double mâchoire** (*f.*), two-jaw spanner; double-jaw spanner. (54)

clef à douille (*f.*), box-spanner; box-wrench; box-key; socket-wrench; socket-spanner. (55)

clef à écrous (*f.*), nut-wrench. (56)

clef à fer creux (*f.*), pipe-wrench; tube-wrench; cylinder-wrench. (57)

clef à fer creux, articulée (*f.*), parrot-nose pipe-wrench. (58)

clef à fourche (*f.*), fork-wrench; gap-spanner. (59)

clef à gobelet pour tire-fonds (*f.*), box-spanner for coach-screws. (60)

clef à griffes en bout (*f.*), pin-spanner; pin-wrench. (61)

clef à griffes sur le côté (*f.*), face-spanner. (62)

clef à manche bois (*f.*), wood-handled spanner. (63)

clef à marteau (*f.*), screw-hammer; wrench-hammer. (64)

clef à molette *ou* **clef à mâchoires mobiles** (*f.*), monkey-wrench; adjustable spanner; shifting-spanner; screw-wrench; turn-screw. (65)

clef à molette façon Clyburn *ou* **clef à molette à une seule mâchoire mobile** (*f.*), Clyburn spanner; Clyburn wrench. (66)

clef à rochet, avec douille à 6 pans (*f.*), ratchet-wrench; with hexagon opening; ratchet-spanner, for hexagon nuts. (67)

clef à 6 pans (*f.*), spanner for hexagon nuts. (68)

clef à tubes (*f.*), pipe-wrench; pipe-twister; tube-wrench; cylinder-wrench. (69)

clef à vis (*f.*), screw-key; screw-wrench. (70)

clef anglaise (*f.*), coach-wrench. (71)

clef anglaise, double à marteau *ou* **clef à écrous, double à marteau, dite anglaise** (*f.*), double-bar coach-wrench. (72)

clef anglaise, simple à marteau (*f.*), single-bar coach-wrench. (73)

clef bénarde (Serrur.) (*f.*), pin-key; pinned key. (74)

clef coudée (*f.*), bent spanner. (75)

clef de bicyclette (*f.*), bicycle spanner; bicycle wrench. (76)

clef de cadenas (*f.*), padlock-key. (77)

clef de calibre (*f.*), ring spanner. (1)
clef de rallonge (*f.*), extension piece ; lengthen-
-ing bar (for box-spanners). (2)
clef de relevée (Sondage) (*f.*), hoisting-plug.
(3)
clef de retenue (Sondage) (*f.*), catch-wrench ;
rod-support ; supporting-fork ; resting-
fork ; lye-key ; tiger. (4)
clef de robinet (*f.*), cock-key ; faucet-key ;
spanner for cocks. (5)
clef de scie (*f.*), winding-bar (for stretching the
cord of a frame or bow saw). (6)
clef de voiture (*f.*), coach-wrench. (7)
clef de voûte (*f.*), keystone, key, (of an arch).
(8)
clef double *ou* clef double à fourches (*f.*), double-
ended spanner ; double-headed wrench. (9)
clef droite (*f.*), straight spanner. (10)
clef en acier estampé (*f.*), drop-forged steel
spanner. (11)
clef en deux pièces pour tubes (*f.*), grip pipe-
wrench. (12)
clef en S *ou* clef cintrée en S (*f.*), S-shaped
spanner ; curved spanner ; S wrench. (13)
clef fermée (*f.*), ring spanner. (14)
clef forée (Serrur.) (*f.*), piped key ; pipe-key.
(15)
clef jaspée (*f.*), spanner, mottled finish. (16)
clef pour automobiles (*f.*), motor-spanner. (17)
clef pour écrous ronds, à encoches (*f.*), pin-
spanner ; pin-wrench. (18)
clef pour essieux (*f.*), axle-spanner ; axle-
wrench. (19)
clef pour tubes *ou* clef pour tuyaux (*f.*), pipe-
wrench ; pipe-twister ; tube-wrench ;
cylinder-wrench. (20)
clef serre-tubes à chaîne (*f.*), chain pipe-wrench.
(21)
clef simple *ou* clef simple à fourche (*f.*), single-
ended spanner. (22)
clef simple à fourche, à nervures (*f.*), single-
ended ribbed spanner ; light single-ended
spanner. (23)
clefs de service (*f.pl.*), spanners ; all necessary
spanners ; screw-keys necessary for working
the machine. (24)
clenche *ou* clenchette (de loquet de porte) (*n.f.*),
bar (of catch for gate-latch). (25)
cléveite (Minéral) (*n.f.*), cleveite. (26)
clichage (de plan incliné) (Mines) (*n.m.*), tub-
controller. (27)
clichages (de puits de mine) (*n.m.pl.*), keps ;
keeps ; props ; rests ; cage-shuts ; cage-
sheets ; landing-chairs ; landing-dogs ;
dogs ; catches ; wings ; fangs. (28). See
taquets (synonymous word) for varieties.
cliché (Photogr.) (*n.m.*), negative. (29)
cliché (photogravure dans un catalogue) (*n.m.*),
illustration ; cut ; plate. (30)
cliché doux (*m.*), soft negative. (31)
cliché dur (*m.*), hard negative. (32)
cliché faible (*m.*), weak negative ; thin negative.
(33)
cliché heurté (*m.*), contrasty negative. (34).
See example under heurté, -e.
cliché opaque *ou* cliché dense *ou* cliché intense
(*m.*), dense negative. (35)
cliché pur (*m.*), clear negative. (36)
cliché sans contrastes (*m.*), flat negative ; nega-
-tive without contrasts. (37)
cliché sous-exposé *ou* cliché manquant de pose
(*m.*), underexposed negative. (38)

cliché surexposé *ou* cliché trop posé (*m.*), over-
-exposed negative. (39)
cliché trop développé (*m.*), overdeveloped
negative. (40)
cliché trop peu développé (*m.*), underdeveloped
negative. (41)
cliché vigoureux (*m.*), plucky negative ; vigour-
-ous negative ; strong negative. (42)
cliché voilé (*m.*), fogged negative. (43)
clicheur (Mines) (pers.) (*n.m.*), hanger-on ;
onsetter ; hooker ; hooker-on ; platman.
(44)
cliffite (Explosif) (*n.f.*), cliffite. (45)
climat (*n.m.*), climate. (46)
climat insalubre *ou* climat malsain (*m.*), un-
-healthy climate. (47)
climat rigoureux (*m.*), severe climate ; rigorous
climate. (48)
climat salubre (*m.*), healthy climate. (49)
climat sous-tropical *ou* climat subtropical (*m.*),
subtropical climate. (50)
climat tempéré (*m.*), temperate climate. (51)
climat tropical (*m.*), tropical climate. (52)
climat variable (*m.*), changeable climate. (53)
climatérique *ou* climatique (*adj.*), climatic ;
climatical : (54)
le régime climatérique d'une vallée, the
climatic conditions prevailing in a valley.
(55)
le mécanisme des variations climatiques, the
mechanism of climatic changes. (56)
climatologie (*n.f.*), climatology. (57)
climatologique (*adj.*), climatological ; climatologic.
(58)
climatologiquement (*adv.*), climatologically. (59)
clinche (de loquet de porte) (*n.f.*), bar (of catch
for gate-latch). (60)
clinochlore (Minéral) (*n.m.*), clinochlore. (61)
clinoclasite *ou* clinoclase (Minéral) (*n.f.*),
clinoclasite ; clinoclase. (62)
clinodiagonale (Cristall.) (*n.f.*), clinodiagonal.
(63)
clinodôme (Cristall.) (*n.m.*), clinodome. (64)
clinohumite (Minéral) (*n.f.*), clinohumite. (65)
clinomètre (*n.m.*), clinometer ; inclinometer ;
clinoscope. (66)
clinométrie (*n.f.*), clinometry. (67)
clinométrique (*adj.*), clinometric , clinometrical.
(68)
clinopinacoïde (Cristall.) (*n.m.*), clinopinacoid.
(69)
clinoprisme (Cristall.) (*n.m.*), clinoprism. (70)
clinorhombique (*adj.*), clinorhombic. (71)
clinoscope (*n.m.*), clinoscope ; clinometer ;
inclinometer. (72)
clinquant (*n.m.*), foil. (73)
clinquant d'étain (*m.*), tin-foil. (74)
clintonite (Minéral) (*n.f.*), clintonite ; seybertite.
(75)
clintonites (*n.f.pl.*) *ou* genre clintonite (*m.*)
(Minéralogie), clintonites ; clintonite group ;
brittle micas. (76)
cliquet (*n.m.*), click ; catch ; trigger ; pawl ;
dog ; keeper ; latch ; ratchet. (77)
cliquet (*n.m.*) *ou* cliquet simple (clef à cliquet),
ratchet-spanner ; ratchet-wrench. (78)
cliquet à canon *ou* *simplement* cliquet (*n.m.*)
(porte-foret), ratchet-brace ; engineers'
ratchet-brace ; ratchet drill-brace ; ratchet-
drill ; ratchet. (79)
cliquet à canon, à pans (*m.*), engineers' ratchet-
brace, with long hexagonal nut. (80)

cliquet à clef (*m.*), ratchet-brace with tommy-bar. (1)

cliquet à manchon court (*m.*), short-head ratchet-brace. (2)

cliquet d'arrêt *ou simplement* cliquet (*n.m.*) (d'un treuil, ou analogue), pawl (of a winch, or the like). (3 *ou* 4)

cliquet du frein à main (*m.*), brake-pawl; brake-dog. (5)

cliquet, façon anglaise (*m.*), English pattern ratchet-brace. (6)

clitographe (*n.m.*), inclinometer. (7)

clivable (*adj.*), cleavable : (8)
roche clivable (*f.*), cleavable rock. (9)

clivage (*n.m.*), cleavage; cleaving : (10)
clivage basique, basal cleavage. (11)
clivage des minéraux, cleavage of minerals; mineral cleavage. (12)
clivage des roches, slaty cleavage : (13)
on appelle clivage des roches, la structure qui les rend susceptibles de se séparer en plaquettes et en lames dans une direction indépendante de la stratification, what is known as slaty cleavage is the structure which renders a rock capable of being split into plates or laminæ in a direction independent of the planes of bedding. (14)
clivage du diamant, diamond cleaving. (15)

cliver (*v.t.*), to cleave : (16)
cliver un cristal, to cleave a crystal. (17)

cliver (se) (*v.r.*), to cleave : (18)
pierres qui ne se clivent pas (*f.pl.*), stones which will not cleave. (19)
cristaux qui se clivent aisément (*m.pl.*), crystals which cleave readily. (20)

cloche (*n.f.*), bell. (21)

cloche (d'isolateur) (Élec.) (*n.f.*), petticoat (of insulator). (22)

cloche (support de rails de ch. de f.) (*n.f.*), pot-sleeper. (23)

cloche (*n.f.*) *ou* cloche de verre (Verrerie de laboratoire), bell jar; bell glass; receiver. (24)

cloche (*n.f.*) *ou* cloche de curage (Sondage), shell-pump; sand-pump; sludger; sludge-pump; wimble. (25)

cloche (d'un cabestan) (*n.f.*), barrel, drum, (of a capstan). (26)

cloche (Mines) (*n.f.*), cauldron; pot-hole. (27)

cloche (d'une carrière) (*n.f.*), opening (of a quarry). (28)

cloche (fondis) (*n.f.*), subsidence (of surface) (owing to underground caving). (29)

cloche à air *ou* cloche d'air (d'une pompe) (*f.*), air-chamber (of a pump). (30)

cloche à boulet (Sondage) (*f.*), sand-pump, sludger, sludge-pump, with ball valve. (31)

cloche à clapets (Sondage) (*f.*), sand-pump, sludger, sludge-pump, with clack-valves. (32)

cloche à échantillon (Sondage) (*f.*), core-extractor; core-lifter; core-catcher. (33)

cloche à écrou (Sondage) (*f.*), screw-bell; screw-grab; bell-box; bell-screw; bell-mouth socket. (34)

cloche à plongeur (Trav. publ.) (*f.*), diving-bell. (35)

cloche à soupape (Sondage) (*f.*), sludger with valve. (36)

cloche d'accrocheur (*f.*). Same as cloche à écrou.

cloche de réservoir (*f.*), tank-cover. (37)

cloche du gueulard (haut fourneau) (*f.*), bell; cone. (38)

cloche électrique (*f.*), electric bell (large dome-shaped gong for giving one or more strokes, as used on railway-tracks). (39)

cloche taraudée (*f.*). Same as cloche à écrou.

cloison (*n.f.*), partition; division. (40)

cloison (Mines) (*n.f.*), brattice; partition. (41)

cloison d'aérage (*f.*), air-brattice. (42)

cloison d'eau (d'une chaudière) (*f.*), film of water (between two plates in a boiler). (43)

cloison de planches *ou* cloison en bois (*f.*), wood partition; boarding. (44)

cloison en brique (*f.*), brick partition. (45)

cloison en décharge *ou* cloison en porte-à-faux (Constr.) (*f.*), self-supporting partition. (46)

cloison en madriers (*f.*), plank partition. (47)

cloison en toile grossière (Mines) (*f.*), canvas brattice. (48)

cloison étanche (*f.*), bulkhead; water-tight bulkhead. (49)

cloison médiane (d'un puits de mine) (*f.*), mid-wall (of a mine-shaft). (50)

cloisonnage (*n.m.*), partitioning. (51)

cloisonnage (Mines) (*n.m.*), bratticing; partition-ing. (52)

cloisonner (*v.t.*), to partition; to partition off; to divide up : (53)
cloisonner une maison, to partition a house. (54)

cloisonner (Mines) (*v.t.*), to brattice; to partition : (55)
cloisonner un puits, une galerie, to brattice a shaft, a level. (56 *ou* 57)

clore (fermer) (*v.t.*), to close; to shut : (58)
clore une porte, une fenêtre, to shut a door, a window. (59 *ou* 60)

clos, -e (*adj.*), closed; shut; close : (61)
trouver porte close, to find the door shut. (62)
essai en vase clos (*m.*), close test. (63)

clôture (*n.f.*). enclosure; fence; fencing. (64)

clôture de bornage (*f.*), boundary-fence. (65)

clôture de l'orifice du puits (*f.*), fencing the pit-top. (66)

clôture en fil de fer (*f.*), wire-fence. (67)

clôture murée (*f.*), walled enclosure. (68)

clôturer (*v.t.*), to enclose; to fence; to fence in. (69)

clou (*n.m.*), nail. (70)

clou à bateaux (*m.*), boat-nail. (71)

clou à cheval *ou* clou à chevaux (*m.*), horseshoe-nail. (72)

clou à crochet (*m.*), tenterhook. (73)

clou à deux pointes (*m.*), staple; wire-staple. (74)

clou à latter (*m.*), lath-nail. (75)

clou à parquet (*m.*), flooring-nail. (76)

clou à patte (*m.*), holdfast; wall-holdfast. (77)

clou à river (*m.*), rivet; boiler-rivet. (78)

clou à tête de diamant (*m.*), diamond nail. (79)

clou à tête plate (*m.*), flat nail. (80)

clou barbelé (*m.*), barbed-wire nail. (81)

clou de mouleur (*m.*), hanger; gagger; lifter. (82)

clou de Paris (*m.*), French nail; wire nail. (83)

clou découpé (*m.*), cut nail. (84)

clou en zinc (*m.*), zinc nail. (85)

clou fondu (*m.*), cast nail. (86)

clou pour noyaux (Fonderie) (*m.*), core-nail. (87)

clouage *ou* clouement (*n.m.*), nailing. (88)

clouer (le bois, etc.) (*v.t.*), to nail; to nail up : (89)

clouer une caisse, to nail up a case. (1)

clouer une planche, to nail a board. (2)

clouer (le métal) (*v.t.*), to rivet : (3)

clouer une tôle, to rivet a plate. (4)

clouer (se) (*v.r.*), to be nailed ; to be riveted : (5)

les tôles des chaudières se clouent à froid, boiler-plates are riveted cold. (6)

cloup (Géol.) (*n.m.*), sink ; sink-hole ; limestone sink ; swallow ; swallow-hole ; dolina. (7)

clouterie (fabrication) (*n.f.*), nail-making. (8)

clouterie (usine) (*n.f.*), nail-works. (9)

cloutier (pers.) (*n.m.*), nail-maker. (10)

clouure (*n.f.*), nailing ; riveting. (11)

cluse (Géol.) (*n.f.*), transverse valley ; cross-valley. (12)

clydite (Explosif) (*n.f.*), clydite. (13)

coagulabilité (*n.f.*), coagulability. (14)

coagulable (*adj.*), coagulable. (15)

coagulant, -e (*adj.*), coagulant. (16)

coagulant (*n.m.*), coagulant ; coagulator. (17)

coagulateur, -trice (*adj.*), coagulative ; coagulatory. (18)

coagulation (*n.f.*), coagulation. (19)

coaguler (*v.t.*), to coagulate. (20)

coaguler (se) (*v.r.*), to coagulate. (21)

coagulum [coagulums *pl.*] (*n.m.*), coagulum. (22)

coaltar (*n.m.*), coal-tar ; gas-tar ; tar. (23)

coaltarement (*n.m.*) *ou* **coaltarisation** (*n.f.*), tarring. (24)

coaltarer *ou* **coaltariser** (*v.t.*), to tar. (25)

coassociation (*n.f.*), copartnership. (26)

coassocié, -e (pers.) (*n.*), copartner ; joint partner. (27)

cobalt (Chim.) (*n.m.*), cobalt. (28)

cobalt arséniaté (Minéral) (*m.*), cobalt-bloom ; red cobalt ; erythrite. (29)

cobalt arsenical (Minéral) (*m.*), gray cobalt ; smaltite ; speiss cobalt ; speiskobalt. (30)

cobalt gris (Minéral) (*m.*). Same as **cobaltine**.

cobalt oxydé noir (Minéral) (*m.*), earthy cobalt ; slaggy cobalt ; asbolite ; asbolane ; asbolan. (31)

cobalteux, -euse (Chim.) (*adj.*), cobaltous. (32)

cobaltifère (*adj.*), cobaltiferous. (33)

cobaltine (Minéral) (*n.f.*), cobaltite ; cobaltine ; cobalt-glance ; glance-cobalt. (34)

cobaltique (Chim.) (*adj.*), cobaltic. (35)

cobaltocre (Minéral) (*n.m.*), cobalt-ochre ; cobalt-crust. (36)

coccolite *ou* **coccolithe** (Minéral) (*n.f.*), coccolite. (37)

cochon (mélange de métal et de scories qui obstrue les fourneaux) (*n.m.*), horse ; old horse ; sow ; bear ; salamander ; shadrach ; freeze. (38)

code (*n.m.*), code. (39)

code des signaux (*m.*), code of signals. (40)

code employé (*m.*), code used. (41)

code minier (*m.*), mining code. (42)

code privé (*m.*), private code. (43)

code télégraphique (*m.*), telegraphic code. (44)

codirecteur (pers.) (*n.m.*), joint manager. (45)

coefficient, -e (*adj.*), coefficient. (46)

coefficient (*n.m.*), coefficient. (47)

coefficient d'aimantation (*m.*), coefficient of induced magnetization ; magnetic suscepti-bility. (48)

coefficient d'écoulement (Hydrog.) (*m.*), coefficient of run-off : (49)

le coefficient d'écoulement est le rapport de l'écoulement avec les précipitations, the coefficient of run-off is the ratio between the run-off and the rainfall. (50)

coefficient d'écrasement (*m.*), modulus of com-pression. (51)

coefficient d'effet utile (Méc.) (*m.*), coefficient of efficiency ; coefficient of useful effect ; commercial coefficiency ; efficiency : (52)

le coefficient d'effet utile d'une machine est le rapport du travail utile produit au travail moteur absorbé, the coefficient of efficiency of a machine is the ratio of the useful work performed to the mechanical energy expended ; the efficiency of an engine is the ratio of the useful (or available) energy to the mechanical energy required to drive it. (53)

coefficient d'élasticité (*m.*), coefficient of elastic-ity ; modulus of elasticity ; stretch-modulus ; Young's modulus. (54)

coefficient d'hystérésis (*m.*), coefficient of hysteresis. (55)

coefficient d'induction mutuelle (*m.*), coefficient of mutual induction ; mutual inductance. (56)

coefficient de compressibilité des liquides (*m.*), coefficient of compressibility of liquids. (57)

coefficient de contraction (Hydraul.) (*m.*), coefficient of contraction ; contraction coefficent. (58)

coefficient de dépense (Hydraul.) (*m.*), coefficient of discharge ; coefficient of efflux. (59)

coefficient de dilatation (*m.*), coefficient of expansion ; expansion coefficient : (60)

les coefficients de dilatation des solides, des liquides, des gaz, the coefficients of expansion of solids, of liquids, of gases. (61 *ou* 62 *ou* 63)

coefficient de frottement (*m.*), coefficient of friction ; friction-coefficient ; friction-factor. (64)

coefficient de frottement intérieur (*m.*), coefficient of viscosity. (65)

coefficient de glissement *ou* **coefficient de frotte-ment de glissement : fer sur fer, laiton sur fonte, etc.** (*m.*), coefficient of sliding friction : wrought iron on wrought iron, brass on cast iron, etc. (66 *ou* 67)

coefficient de rendement (Méc.) (*m.*). Same as **coefficient d'effet utile.**

coefficient de résistance (*m.*), coefficient of resistance. (68)

coefficient de roulement *ou* **coefficient de frotte-ment de roulement** (*m.*), coefficient of friction of rolling motion : (69)

coefficient de roulement : gaïac sur gaïac, orme sur gaïac, fer sur fer, acier sur acier, coefficient of friction of rolling motion : lignum vitæ on lignum vitæ, elm on lignum vitæ, iron on iron, steel on steel. (70 *ou* 71 *ou* 72 *ou* 73)

coefficient de rupture (Résistance des matériaux) (*m.*), modulus of rupture. (74)

coefficient de self-induction (Élec.) (*m.*), coeffi-cient of self-induction ; self-inductance. (75)

coefficient de viscosité (*m.*), coefficient of viscosity. (76)

coercibilité (*n.f.*), coercibility : (77)

la coercibilité des fluides, the coercibility of fluids. (78)

coercible (*adj.*), coercible : (79)

les liquides et les gaz sont coercibles, liquids and gases are coercible. (80)

coercitif, -ive (*adj.*), coercive ; coercitive : (81)

force coercitive (Magnétisme) (*f.*), coercive (or coercitive) force. (82)

cœsium (Chim.) (*n.m.*), cesium ; cæsium. (83)

cœur de chêne (*m.*), heart of oak. (84)

cœur de croisement (Ch. de f.) (*m.*), point of crossing; tongue of crossing; frog-point. (1)

cœur de l'été (*m.*), height of summer; mid--summer. (2)

cœur de l'hiver (*m.*), depth of winter; mid--winter. (3)

cœur du bois (*m.*), heart-wood. (4)

coffrage (*n.m.*), coffering: (5)
le coffrage d'un puits de mine, coffering a mine-shaft. (6)

coffrage (moule pour la fabrication d'ouvrages en béton) (*n.m.*), framing; form. (7)

coffre (*n.m.*), chest; box; bin. (8)

coffre (d'une serrure) (*n.m.*), case (of a lock). (9)

coffre à minerai (*m.*), ore-bin; ore-bunker. (10)

coffre d'écluse (*m.*), lock-chamber; coffer. (11)

coffre-fort [coffres-forts *pl.*] *ou simplement* **coffre** (*n.m.*), safe; strong box. (12)

coffrer (Mines, etc.) (*v.t.*), to coffer: (13)
coffrer un puits, to coffer a shaft. (14)

coffret de pharmacie (*m.*), medicine-chest. (15)

cognée de bûcheron (*f.*), felling-axe. (16)

cogner (Mach.) (*v.i.*), to knock: (17)
moteur qui cogne (*m.*), engine which knocks. (18)

cogner à une porte, to knock at a door. (19)

cogner un clou, to drive in, to hammer in, a nail. (20)

cohérence (*n.f.*), coherence. (21)

cohérent, -e (*adj.*), coherent. (22)

cohéreur (Phys.) (*n.m.*), coherer. (23)

cohésif, -ive (*adj.*), cohesive. (24)

cohésion (*n.f.*), cohesion; cohesiveness: (25)
la cohésion est forte dans les solides, faible dans les liquides, et probablement absente dans les gaz, cohesion is strong in solids, weak in liquids, and probably absent in gases. (26)
la cohésion du gisement (Géol. ou Mines), the cohesion of the seam. (27)

cohésivement (*adv.*), cohesively. (28)

coiffe (*n.f.*), cap: (29)
une coiffe de fer, an iron cap. (30)

coiffe (d'une chèvre) (*n.f.*), cap (of a sheer-legs). (31)

coiffer (*v.t.*), to cap: (32)
une pyramide terreuse coiffée d'une pierre plate, an earthy pyramid capped with a flat stone. (33)
coiffer un puits à pétrole, to cap an oil-well. (34)

coin (angle) (*n.m.*), corner: (35)
les coins d'une chambre, the corners of a room. (36)
un coin étranglé, a tight corner. (37)

coin (pièce prismatique) (*n.m.*), wedge; key; coin; quoin. (38)

coin (pour coussinets de rail) (*n.m.*), key; rail-key; plug. (39)

coin (Mines & Carrières) (*n.m.*), wedge; gad. (40)

coin (Géom.) (*n.m.*), wedge. (41)

coin à fendre le bois (*m.*), timber-splitting wedge. (42)

coin à pierre (*m.*), stone-wedge; plug. (43)

coin de calage (*m.*), keying-wedge. (44)

coin de calfat (*m.*), reaming-iron. (45)

coin de desserrage (*m.*), loosening-wedge. (46)

coin de rattrapage de jeu (machines-outils) (*m.*), slip for taking up wear. (47)

coin de serrage (*m.*), tightening-wedge. (48)

coin demi-rond (*m.*), feather; plug (one of the two external wedges of a plug-and-feather wedge). (49)

coin hydraulique (Mines) (*m.*), hydraulic wedge. (50)

coin multiple (Mines) (*m.*), multiple wedge; compound wedge. (51)

coin prisonnier (*m.*), set-key; fitting-key. (52)

coinçage (*n.m.*), wedging; keying; jamming; binding: (53)
coinçage des rails, keying, wedging, rails. (54)
coinçage du trépan dans le forage, wedging, jamming, of the bit in the bore-hole. (55)

coincement (*n.m.*), wedging; jam; jamming; binding: (56)
coincement du trépan (Sondage), wedging of the bit. (57)
coincement dans le puits (Mines), jam in shaft. (58)
coincement du piston d'indicateur, binding of indicator-piston. (59)

coincer (*v.t.*), to wedge; to jam; to jam in; to key: (60)
coincer des rails, to key, to wedge, rails. (61)

coincer (se) (*v.r.*), to wedge; to jam; to bind: (62)
burin qui se coince dans la rainure qu'il creuse (*m.*), chisel which binds in the groove it cuts. (63)

coistresse (Mines) (*n.f.*), subdrift; counter-level; counter-gangway; counter. (64)

coke (*n.m.*), coke. (65)

coke de gaz *ou* **coke d'usine à gaz** (*m.*), gas-coke; retort-coke. (66)

coke distillé en vase clos (*m.*), retort-coke. (67)

coke métallurgique (*m.*), oven-coke; foundry-coke; furnace-coke. (68)

coke naturel (*m.*), native coke. (69)

cokéfaction *ou* **cokéfication** (*n.f.*), coking: (70)
cokéfaction des combustibles, coking of fuels. (71)

cokéfiant, -e (*adj.*), coking: (72)
combustible cokéfiant (*m.*), coking fuel. (73)

cokéfier (*v.t.*), to coke. (74)

cokéfier (se) (*v.r.*), to coke: (75)
'houille qui se cokéfie bien (*f.*), coal which cokes well. (76)

col (d'une bouteille) (*n.m.*), neck (of a bottle). (77)

col (d'une cornue) (*n.m.*), neck, beak, (of a retort). (78)

col (Géogr. phys.) (*n.m.*), pass; col; saddle: (79)
le col de la Bernina, the Bernina pass. (80)

col d'oie *ou* **col de cygne** (*m.*), gooseneck; swan--neck: (81)
crochet en col de cygne (*m.*), goosenecked (*or* goose-necked) hook. (82)
robinet à col de cygne (*m.*), swanneck cock. (83)

colcotar (*n.m.*), colcothar. (84)

colémanite (Minéral) (*n.f.*), colemanite. (85)

colimaçon (escalier tournant) (*n.m.*), spiral stairs; winding stairs; corkscrew stairs. (86)

colique de plomb *ou* **colique saturnine** (*f.*), lead-colic. (87)

colis (*n.m.*), package; parcel. (88)

collage (*n.m.*), gluing; sticking. (89)

collant, -e (en parlant du charbon, de la houille) (*adj.*), caking: (90)
'houille collante (*f.*), caking coal. (91)
'houille non-collante, non-caking coal. (92)

colle (*n.f.*), glue. (93)

colle de poisson (*f.*), fish glue. (1)
collecteur, -trice (*adj.*), collecting : (2)
bac collecteur (*m.*) *ou* cuve collectrice (*f.*), collecting vat. (3)
collecteur, -trice (principal, — se dit des drains, des égouts, etc.) (*adj.*), main : (4)
drain collecteur (*m.*), main drain. (5)
collecteur (*n.m.*), collector. (6)
collecteur (d'une machine dynamo-électrique) (*n.m.*), collector (of a dynamo-electric machine). (7)
collecteur (conduite principale) (*n.m.*), main. (8)
collecteur d'eau (Mines) (*m.*), sump ; sink-hole ; standage. (9)
collecteur de poussières (*m.*), dust-collector. (10)
collecteur de surchauffeur (*m.*), superheater-header ; superheater-manifold. (11)
collectif, -ive (*adj.*), collective ; joint : (12)
rapport collectif (*m.*), joint report. (13)
collection (*n.f.*), collection : (14)
collection de minéraux, collection of minerals. (15)
collectionnement (*n.m.*), collecting. (16)
collectionner (*v.t.*), to collect : (17)
collectionner des minéraux, des fossiles, to collect minerals, fossils. (18 *ou* 19)
collectionneur (pers.) (*n.m.*), collector. (20)
coller (*v.t.*), to glue ; to stick. (21)
coller (*v.i.*), to stick ; to adhere. (22)
coiler (en parlant de la houille) (*v.i.*), to cake : (23)
'houille qui colle, qui ne colle pas (*f.*), coal which cakes, which does not cake. (24 *ou* 25)
coller (se) (*v.r.*), to stick ; to adhere. (26)
collerette (d'un tube) (*n.f.*), flange (of a tube). (27)
collet (Méc.) (*n.m.*), collar ; collet. (28)
collet (d'un ciseau) (*n.m.*), neck (of a chisel). (29)
collet (d'un cylindre de laminoir) (*n.m.*), neck, journal, (of a rolling-mill roll). (30)
collet (d'une lunette) (*n.m.*), collar (of a telescope). (31)
collet (partie formant saillie ; rebord ; bourrelet) (*n.m.*), flange : (32)
rabattre un collet sur une tôle, to turn up a flange on a sheet. (33)
collet rabattu *ou* collet tombé, turned up flange. (34)
collet de la fusée (*m.*), axle-collar. (35)
collet du palier de butée (*m.*), thrust-collar. (36)
colletage (Mines) (*n.m.*), wedging ; wedge-driving. (37)
colleter (Mines) (*v.t.*), to wedge : (38)
colleter le boisage, to wedge the timbering. (39)
colliénation (*n.f.*), collineation ; collimation. (40)
collier (*n.m.*), collar. (41)
collier d'arbre (*m.*), shaft-collar. (42)
collier d'arrêt (*m.*), stop-collar ; set-collar. (43)
collier d'embrayage (*m.*), clutch-collar. (44)
collier d'excentrique (*m.*), eccentric-strap ; eccentric-collar ; eccentric-hoop. (45)
collier d'excentrique de marche en arrière (loco-motive) (*m.*), backward eccentric-strap. (46)
collier d'excentrique de marche en avant (loco-motive) (*m.*), forward eccentric-strap. (47)
collier de butée (*m.*), thrust-collar. (48)
collier de retenue (pour la descente des tubages de puits) (*m.*), casing-clamps, pipe-clamps, pipe-clambs, (for lowering well-casing). (49)
collier de tuyau (*m.*), pipe-collar. (50)
collimater (*v.t.*), to collimate ; to collineate. (51)
collimateur (*n.m.*), collimator. (52)
collimation (*n.f.*), collimation ; collineation. (53)
colline (*n.f.*), hill : (54)

colline escarpée, steep hill. (55)
collision (*n.f.*), collision : (56)
collision de deux trains, collision of two trains. (57)
colloïdal, -e, -aux (*adj.*), colloidal : (58)
poudre colloïdale (*f.*), colloidal powder. (59)
colloïde (*adj.*), colloid : (60)
l'opale est colloïde, c'est-à-dire réfractaire à la cristallisation (*f.*), opal is colloid, that is to say, refractory to crystallization. (61)
colloïde (*n.m.*), colloid. (62)
collyrite (Minéral) (*n.f.*), collyrite. (63)
colombate (Chim.) (*n.m.*), columbate ; niobate. (64)
colombite (Minéral) (*n.f.*), columbite. (65)
colombium (Chim.) (*n.m.*), columbium ; niobium. (66)
colon (pers.) (*n.m.*), colonist. (67)
colonial, -e, -aux (*adj.*), colonial. (68)
colonie (*n.f.*), colony : (69)
une colonie de mineurs, a colony of miners. (70)
les colonies d'outre-mer, the oversea colonies. (71)
colonne (*n.f.*), column ; pillar. (72)
colonne (d'un laminoir) (*n.f.*), housing, holster, standard, bearer, (of a rolling-mill). (73)
colonne barométrique (*f.*), barometric column. (74)
colonne cannelée (*f.*), fluted pillar ; fluted column. (75)
colonne coiffée (Géol.) (*f.*), erosion column ; chimney-rock. (76)
colonne couchée (de minerai) (*f.*), course of ore. (77)
colonne creuse en fonte (*f.*), hollow cast-iron pillar. (78)
colonne d'eau (*f.*), column of water ; water-column. (79)
colonne d'échappement (locomotive) (*f.*), nozzle-stand. (80)
colonne d'exhaure (puits de mine) (*f.*), rising-main. (81)
colonne de fumée (*f.*), column of smoke : (82)
volcan qui projette une colonne de fumée dans l'espace (*m.*), volcano which projects a column of smoke into space. (83)
colonne de mercure (*f.*), column of mercury. (84)
colonne de richesse (Géol. & Mines) (*f.*), shoot of ore ; ore-shoot ; ore-chute ; ore-shute ; chimney of ore ; ore-chimney ; pipe of ore ; pipe-vein. (85)
colonne de tubage *ou* **colonne de tubes** (*f.*), casing-line ; tube-column ; string of casing. (86)
colonne lisse (Arch.) (*f.*), plain column. (87)
colonne montante (Constr.) (*f.*), riser ; riser-pipe ; rising-main. (88)
colophonite (Minéral) (*n.m.*), colophonite. (89)
coloradoïte (Minéral) (*n.f.*), coloradoite. (90)
colorant, -e (*adj.*), colouring ; coloring : (91)
matière colorante (*f.*), colouring matter. (92)
colorant (*n.m.*), colouring matter ; colorant. (93)
coloration (*n.f.*), colouration ; coloration ; colouring ; coloring : (94)
coloration de la flamme (Essais au chalumeau), colouration of the flame (Blowpipe-assaying). (95)
coloration de recuit (Métall.) (*f.*), heat-tint. (96)
colorer (*v.t.*), to colour ; to color : (97)
pierre colorée par l'oxyde de fer (*f.*), stone coloured with iron oxide. (98)
colorimètre (*n.m.*), colorimeter ; chromometer. (99)

coltinage (Ch. de f., etc.) (*n.m.*), carrying : (1)
le coltinage des traverses, carrying sleepers. (2)
coltinage du charbon (*m.*), coal-backing. (3)
coltiner (Ch. de f.) (*v.t.*), to carry : (4)
coltiner des traverses ou des rails, et les déposer
à leur place, to carry sleepers or rails
and lay them in their place. (5 *ou* 6)
coltineur (Ch. de f.) (pers.) (*n.m.*), carrier. (7)
columbite (Minéral) (*n.f.*), columbite. (8)
columbium (Chim.) (*n.m.*), columbium ; niobium.
(9)
coma (Opt.) (*n.m.*), coma. (10)
combattre (*v.t.*), to fight ; to fight against ; to
combat : (11)
combattre la température excessive des roches
profondes, to fight against the extreme heat
of the deep rocks. (12)
combattre un incendie, to fight, to combat, a
fire. (13)
combe (Géol.) (*n.f.*), anticlinal valley. (14)
combinable (*adj.*), combinable : (15)
objectif combinable (Photogr.) (*m.*), combin-
-able lens. (16)
combinaison (assemblage, arrangement dans un
certain ordre) (*n.f.*), combination : (17)
une combinaison de couleurs, de sons, d'engre-
-nages, a combination of colours, of sounds, of
gears. (18 *ou* 19 *ou* 20)
combinaison (Chim.) (*n.f.*), combination : (21)
séparer l'or de ses combinaisons, to separate
gold from its combinations. (22)
combinaison (mesures prises pour assurer le
succès d'une entreprise) (*n.f.*), contrivance ;
scheme ; device. (23)
combiné, -e (Chim., etc.) (*adj.*), combined : (24)
l'eau combinée entrant dans la composition
des minéraux (*f.*), the combined water
entering into the composition of minerals.
(25)
l'eau qui se trouve à l'état combiné dans
certains minéraux (*f.*), the water which
occurs in a combined state in certain
minerals. (26)
l'élément le plus répandu dans les roches est
la silice, à l'état libre ou combiné (*m.*),
the most widely distributed element in
rocks is silica, in the free or combined state.
(27)
or combiné (*m.*), combined gold. (28)
combiné (Chim.) (*n.m.*), combination ; compound :
(29)
l'eau est un combiné d'oxgyène et d'hydrogène
(*f.*), water is a compound of oxygen and
hydrogen. (30)
combiner (*v.t.*), to combine : (31)
combiner de l'hydrogène et de l'oxygène au
moyen de l'étincelle électrique, to combine
hydrogen and oxygen by means of the electric
spark. (32)
combiner deux éléments chimiques, to combine
two chemical elements. (33)
combiner (calculer ; disposer) (*v.t.*), to devise ;
to contrive : (34)
combiner un nouveau projet, to devise a new
scheme ; to contrive a new plan. (35)
combiner (se) (*v.r.*), to combine : (36)
métal qui se combine au soufre (*m.*), metal
which combines with sulphur. (37)
l'arsenic se combine avec le soufre, avec le
fluor, le chlore, le brome, et l'iode (*m.*),
arsenic combines with sulphur, with fluorine,
chlorine, bromine, and iodine. (38)

comble (d'un bâtiment) (*n.m.*), roof (of a building).
(39)
comble à deux égouts *ou* comble à deux pentes
ou comble à deux rampes *ou* comble à deux
versants *ou* comble à deux longs pans (*m.*),
span-roof ; ridge-roof ; ridged roof ; double-
pitch roof ; pitch-roof. (40)
comble à deux longs pans avec croupes (*m.*),
hipped ridge-roof ; hip-and-ridge roof ;
hip-roof with ridge. (41)
comble à deux longs pans sans fermes (*m.*), couple-
roof. (42)
comble à la Mansard *ou* comble à la française *ou*
comble brisé (*m.*), Mansard roof ; French
roof ; curb-roof. (43)
comble à une seule pente *ou* comble à un versant
(*m.*), single-pitch roof. (44)
comble avec avant-toit (*m.*), umbrella-roof. (45)
comble en appentis (*m.*), lean-to roof ; pent
roof ; penthouse-roof. (46)
comble en croupe (*m.*), hip-roof ; hipped roof. (47)
comble en dôme (*m.*), dome roof. (48)
comble en pavillon *ou* comble pyramidal *ou*
comble à autant de versants que de côtés (*m.*),
pavilion roof ; pyramidal roof. (49)
comble retroussé (*m.*), collar-beam roof ; collar-
tie roof. (50)
comble sans ferme (*m.*), untrussed roof. (51)
comble-shed [combles-sheds *pl.*] (*n.m.*) *ou* comble
en dent de scie, square-to roof ; shed-
roof ; saw-tooth roof. (52)
comble sur fermes (*m.*), trussed roof. (53)
comble sur pignon *ou* comble sur pignons (*m.*),
gable-roof. (54)
comblement (*n.m.*), filling up ; filling : (55)
comblement d'un fossé, d'une dépression,
d'un lac de montagne par des matières
détritiques, filling up a ditch, a depression,
of a mountain lake by detrital matter.
(56 *ou* 57 *ou* 58).
combler (*v.t.*), to fill up ; to fill. (59)
combler (se) (*v.r.*), to fill up ; to fill ; to be filled :
(60)
comme le lac Majeur, le lac de Côme se
comble rapidement par les apports des
soixante torrents alpestres qui y débouchent,
like Lake Maggiore, the Lake of Como is
filling up rapidly with the drift of the sixty
alpine torrents which fall into it. (61)
combles s'intersectant (*m.pl.*), hip-and-valley
roof. (62)
comburant, -e (*adj.*), comburent. (63)
combustibilité (*n.f.*), combustibility ; combusti-
-bleness : (64)
la combustibilité d'un gaz, the combustibility
of a gas. (65)
combustible (*adj.*), combustible ; fuel (*used as
adj.*) : (66)
matière combustible (*f.*), combustible material.
(67)
gaz combustible (*m.*), fuel-gas. (68)
combustible (*n.m.*), fuel ; firing ; combustible.
(69)
combustible de gueulard (*m.*), furnace-gas. (70)
combustible de ménage (*m.*), household fuel.
(71)
combustible fossile (*m.*), fossil fuel. (72)
combustible liquide (*m.*), liquid fuel. (73)
combustible minéral (*m.*), mineral fuel. (74)
combustible pulvérisé (*m.*), powdered fuel ;
pulverized fuel. (75)
combustion (*n.f.*), combustion. (76)

combustion interne (*f.*), internal combustion : (1)
  moteur à combustion interne (*m.*), internal-combustion engine. (2)
combustion lente (*f.*), slow combustion. (3)
combustion spontanée (*f.*), spontaneous com--bustion. (4)
combustion vive (*f.*), quick combustion. (5)
commande (Commerce) (*n.f.*), order : (6)
  recevoir une forte commande de marchandises, to receive a big order for goods. (7)
commande (action) (Méc. & Mach.) (*n.f.*), driving. (8)
commande (organes transmettant le mouvement) (Mach.) (*n.f.*), drive ; driving-gear. (9)
commande (de) (opp. à **commandé, -e**) (Méc. & Mach.), driving : (10)
  arbre de commande (*m.*), driving-shaft. (11)
commande (d'une soupape) (*n.f.*), mechanical operation (of a valve). (12)
  soupape commandée (*f.*), mechanically operated valve. (13)
commande de vapeur directe (à), direct steam-driven. (14)
commande directe (à), direct-driven ; directly driven : (15)
  ventilateur à commande directe (*m.*), direct-driven fan. (16)
commande électrique (*f.*), electric drive ; motor-drive. (17)
commande par balancier (*f.*), beam-drive. (18)
commande par boîte de vitesses à monopoulie (*f.*). Same as **commande par monopoulie et boîte de vitesses.**
commande par câble (*f.*), rope-drive. (19)
commande par chaîne (*f.*), chain-drive. (20)
commande par cône *ou* **commande par cône-poulie** (*f.*), cone-pulley drive ; step-cone drive. (21)
commande par cônes de friction (*f.*), friction-cone drive ; cone-drive ; cone-gear. (22)
commande par courroie (*f.*), belt-drive. (23)
commande par engrenages (*f.*), gear-drive. (24)
commande par le bas (à) (Méc.), underdriven : (25)
  ascenseur à commande par le bas (*m.*), under--driven passenger-lift. (26)
commande par monopoulie et boîte de vitesses (*f.*), all-gear single-pulley drive ; single-pulley drive through a change-speed gear-box ; constant-speed pulley drive through a change-speed gear-box. (27)
commande par poulies (*f.*), belt-drive. (28)
commande par tambour (*f.*), drum-drive. (29)
commande, -e (opp. à **de commande**) (Méc. & Mach.) (*adj.*), driven : (30)
  arbre commandé (*m.*), driven shaft. (31)
commandé, -e (Mach.) (*p.p.*), driven : (32)
  commandé (-e) par balancier, beam-driven. (33)
  commandé (-e) par courroie, belt-driven ; driven by belt. (34)
  commandé (-e) par l'air comprimé, driven by compressed air. (35)
  commandé (-e) par moteur, engine-driven ; motor-driven ; driven by power ; power-driven. (36)
commander (Mach.) (*v.t.*), to drive : (37)
  l'excentrique commande le tiroir (*m.*), the eccentric drives the slide-valve. (38)
commander des marchandises, to order goods. (39)
comme l'indique la gravure *ou* **comme le montre la gravure,** as shown in the illustration : (40)

comme l'indique la gravure, celle machine possède deux supports, etc., as shown in the illustration, this machine has two rests, etc. (41)
commençant, -e (*adj.*), commencing ; early : (42)
  maturité commençante (Géogr. phys.) (*f.*), early maturity. (43)
commencement (*n.m.*), commencement ; begin--ning ; start. (44)
commencement de l'admission (vapeur) (*m.*), admission-point ; admission. (45)
commencement de l'émission (vapeur) (*m.*), release-point ; release. (46)
commencement de la compression (vapeur) (*m.*), compression-point ; compression. (47)
commencement de la détente (vapeur) (*m.*), expansion-point ; expansion. (48)
commencer (*v.t.* & *v.i.*), to commence ; to begin ; to start. (49)
commerçant, -e (pers.) (*n.*), trader ; merchant. (50)
commerce (*n.m.*), trade ; commerce. (51)
commerce caboteur (*m.*), coasting-trade ; coastwise trade. (52)
commerce d'exportation (*m.*), export-trade. (53)
commerce d'importation (*m.*), import-trade. (54)
commerce (du) *ou* **commerce (de),** commercial (made or put up for market) : (55)
  fer du commerce (*m.*), commercial iron. (56)
  sel du commerce (*m.*), commercial salt. (57)
  tôles de commerce recuites (*f.pl.*), annealed commercial sheets. (58)
commercial, -e, -aux (*adj.*), commercial : (59)
  transactions commerciales (*f.pl.*), commercial transactions. (60)
commercialement (*adv.*), commercially. (61)
comminution (*n.f.*), comminution. (62)
commis de bureau *ou simplement* **commis** (*n.m.*) (pers.), clerk. (63)
commissaire de transport (pers.) (*m.*), forward--ing-agent. (64)
commission d'examen (*f.*), board of examiners. (65)
commode (*adj.*), convenient. (66)
commun, -e (*adj.*), common ; ordinary. (67)
communément (*adv.*), commonly ; ordinarily. (68)
communicateur, -trice (*adj.*), communicating ; connecting : (69)
  le fil communicateur, the connecting wire. (70)
communicateur de mouvement (Méc.) (*m.*), transmitter of motive power ; organ of transmission. (71)
communication (*n.f.*), communication ; con--nection : (72)
  communication entre deux puits, communica--tion, connection, between two shafts. (73)
  communication entre un conducteur et la terre (Élec.), connection between a con--ductor and the earth. (74)
  communication reçue du fond (Mines), com--munication received from below. (75)
communiquer (*v.t.*), to communicate : (76)
  communiquer le mouvement au moyen d'une manivelle, to communicate motion by means of a crank. (77)
commutateur (appareil servant à changer la direction ou l'intensité d'un courant électrique) (*n.m.*), commutator : (78)
  commutateur des pôles d'une pile, commutator of the poles of a battery. (79)

commutateur (de dynamo, de moteur) (n.m.), commutator. (1 ou 2)

commutateur (appareil servant à établir ou à supprimer la communication entre deux ou plusieurs points d'un circuit électrique) (n.m.), switch. (3) See also interrupteur.

commutateur à bain d'huile (m.), oil-switch. (4)

commutateur à cheville (m.), pin-switch. (5)

commutateur à double rupture (m.), double-break switch. (6)

commutateur à glissement (m.), sliding switch. (7)

commutateur à manivelle ou commutateur à manette (m.), handle switch; lever switch. (8)

commutateur à ressort (m.), lock-switch; snap-switch. (9)

commutateur automatique (m.), automatic switch; self-acting switch. (10)

commutateur bipolaire (m.), double-pole switch. (11)

commutateur conjoncteur (m.), circuit-closer. (12)

commutateur-coupleur [commutateurs-coupleurs pl.] (n.m.), grouping-switch. (13)

commutateur de sûreté (m.), safety-switch. (14)

commutateur de voltmètre (m.), voltmeter-switch. (15)

commutateur disjoncteur (m.), circuit-breaker; disjunctor; break-circuit. (16)

commutateur immergé dans un bain d'huile (m.), switch resting in an oil bath. (17)

commutateur inverseur (m.), reversing-switch. (18)

commutateur monopolaire (m.), single-pole switch. (19)

commutateur permutateur (m.), change-over switch. (20)

commutateur unipolaire (m.), single-pole switch. (21)

commutatrice (de dynamo, de moteur) (n.f.), commutator. (22 ou 23)

compacité (n.f.), compactness: (24)
compacité des mortiers hydrauliques, compactness of hydraulic mortars. (25)

compact, -e (adj.), compact; close-grained. (26)

compagnie (n.f.), company. (27)

compagnie de chemins de fer (f.), railway-company. (28)

compagnie du gaz (f.), gas-company. (29)

compagnons de travail (m.pl.), fellow workmen; fellow-workmen. (30)

comparaison (n.f.), comparison: (31)
comparaison des échelles thermométriques centigrade et Fahrenheit, comparison of the centigrade and Fahrenheit thermometric scales. (32)

comparateur (Phys.) (n.m.), comparator. (33)

comparatif, -ive (adj.), comparative. (34)

compartiment (n.m.), compartment. (35)

compartiment d'extraction (d'un puits de mine) (m.), hoisting-compartment, winding-compartment, (of a mine-shaft). (36)

compartiment de l'exhaure (puits de mine) (m.), pump-compartment. (37)

compartiment des échelles (puits de mine) (m.), ladderway; climbingway; ladder-road. (38)

compartimentage d'un puits (m.), dividing a shaft into compartments; partitioning a shaft. (39)

compartimentage de la mine (m.), division of the mine into separate districts. (40)

compas (instrument servant à tracer des cir--conférences, etc.) (n.m.), compasses; compass; pair of compasses. (41)

compas (instrument servant à prendre des mesures extérieures ou intérieures) (n.m.), calipers; callipers; calibre-compass; caliper-compasses. (42)

compas (boussole marine) (n.m.), compass; mariners' compass. (43)

compas à balustre (m.). Same as compas-balustre.

compas à cheveu (m.), hair compass. (44)

compas à coulisse (m.), sliding caliper-gauge; slide-calipers; sliding calipers; caliper-gauge; caliper-square; beam-caliper; beam caliper-gauge. (45)

compas à crémaillère (m.), rack-compass. (46)

compas à lame tranchante (m.), washer-cutter. (47)

compas à pièces de rechange à aiguille, tire-ligne, porte-crayon, et rallonge (m.), pillar-compasses with needle points, pen and pencil points, and lengthening-bar. (48)

compas à pointes sèches ou compas à diviser (m.), dividers; divider; pair of dividers. (49)

compas à pompe (m.). Same as compas-balustre à pompe.

compas à porte-crayon (m.), compasses with pencil point; pencil-compasses. (50)

compas à ressort (m.), spring-dividers. (51)

compas à tire-ligne (m.), compasses with pen point. (52)

compas à trois branches (m.), triangular com--pass. (53)

compas à verge (m.), beam compasses; trammel; tram. (54)

compas à verge, à vis de rappel (m.), beam compasses, with adjusting-screw. (55)

compas-balustre [compas-balustre pl.] (n.m.), bow compass; bow compasses; bow; bows. (56)

compas-balustre à pincettes (m.), bow spring compasses; spring bow compasses. (57)

compas-balustre à pincettes, à bascule (m.), bow spring compasses, with knee joint. (58)

compas-balustre à pincettes, à pointes sèches (m.), bow spring dividers; spring bow divider. (59)

compas-balustre à pincettes, à porte-crayon (m.), bow spring compasses, with pencil point. (60)

compas-balustre à pincettes, à tire-ligne (m.), bow spring compasses, with pen point. (61)

compas-balustre à pointes sèches (m.), bow dividers. (62)

compas-balustre à pompe (m.), pump spring bow; rotational compass. (63)

compas-balustre à pompe, changeant à tire-ligne, à porte-crayon, et à pointes sèches (m.), pump spring bow, set of 3, ink, pencil, and divider. (64)

compas d'ellipse (m.), elliptic compass; oval compass. (65)

compas d'épaisseur (m.), outside calipers. (66)

compas d'épaisseur à ressort, avec écrou rapide (m.), outside spring-calipers, with spring-nut. (67)

compas d'épaisseur, avec quart de cercle (m.), outside calipers, with wing. (68)

compas d'extérieur (m.), outside calipers. (69)

compas d'intérieur (m.), inside calipers. (70)

compas de calibre (m.), calipers; callipers; calibre-compass; caliper-compasses. (71)

compas de charpentier (m.), carpenters' com--passes. (72)

compas de poche (pour tracer) (*m.*), pocket-compasses ; compasses with turn-in points. (1)

compas de proportion (*m.*), proportional com--passes. (2)

compas de réduction (*m.*), reduction-com--passes. (3)

compas dit 1/2 8 de chiffre (*m.*), egg calipers. (4)

compas dit 8 de chiffre (*m.*), figure-of-eight calipers ; hour-glass calipers. (5)

compas dit maître à danser *ou* **compas maître de danse** (*m.*), outside-and-inside calipers ; in-and-out calipers ; double callipers. (6)

compas droit *ou* **compas droit à pointes** (*m.*), dividers ; divider ; pair of dividers. (7)

compas droit à pointes à ressort (*m.*), spring dividers. (8)

compas elliptique (*m.*), elliptic compass ; oval compass. (9)

compas quart de cercle *ou* **compas droit, 1/4 de cercle** (*m.*), quadrant compasses ; wing-compasses. (10)

compas sphérique (*m.*), bow calipers. (11)

compas universel (*m.*), universal compass. (12)

compasser les feux (Travail aux explosifs), to prepare a volley. (13)

compensateur (Phys., etc.) (*n.m.*), compensator. (14)

compensateur (Signaux de ch. de f.) (*n.m.*), compensator ; wire-compensator. (15)

compensateur (pendule compensateur) (*n.m.*), compensation pendulum ; compensated pendulum ; compensating pendulum. (16)

compensateur magnétique (*m.*), magnetic compensator. (17)

compensation (*n.f.*), compensation : (18)

compensation de l'usure (Méc.), compensa--tion for wear ; taking up the wear. (19)

compensation en température (en parlant de baromètres anéroïdes, etc.), compensation for temperature. (20)

compensation pour affaissement du sol, compensation for subsidence. (21)

compensatrice (Réseau à 3 fils) (Élec.) (*n.f.*), (22)

compenser (*v.t.*), to compensate : (23)

compenser l'usure (Méc.), to compensate for wear ; to take up the wear. (24)

compenser un baromètre anéroïde pour varia--tions de température, to compensate an aneroid barometer for variations of tempera--ture. (25)

compétence (*n.f.*), competence ; competency. (26)

compétent, -e (*adj.*), competent. (27)

complémentaire (*adj.*), complementary : (28)

angles complémentaires (*m.pl.*), complementary angles. (29)

couleurs complémentaires (*f.pl.*), comple--mentary colours. (30)

roches complémentaires (*f.pl.*), complementary rocks. (31)

complet, -ète (*adj.*), complete ; finished. (32)

complètement (*adv.*), completely ; wholly ; entirely ; totally. (33)

compléter (*v.t.*), to complete ; to finish : (34)

compléter les réparations d'un puits, to complete the repairs to a shaft. (35)

complexe (*adj.*), complex : (36)

minerai complexe (*m.*), complex ore. (37)

complexe (Géom. & Géol.) (*n.m.*), complex. (38)

complexité (*n.f.*), complexity : (39)

complexité de la structure (Géol.), complexity of the structure. (40)

complication (*n.f.*), complication. (41)

compliquer (*v.t.*), to complicate : (42)

système qui a le grave inconvénient de compli--quer les pièces du mécanisme (*m.*), system which has the serious disadvantage of complicating the parts of the mechanism. (43)

comporter (*v.t.*), to admit of ; to contain ; to call for ; to require : (44)

dans certains cas la machine comporte trois poulies, in certain cases the machine admits of three pulleys. (45)

machine qui comporte tous les derniers perfectionnements (*f.*), machine which contains all the latest improvements. (46)

morceaux de minerai qui comportent une attention spéciale ou des traitements particuliers (*m.pl.*), pieces of ore which call for (*or* require) special attention or particular treatment. (47)

composant, -e (*adj.*), component ; constituent : (48)

forces composantes (*f.pl.*), component forces. (49)

composant (*n.m.*), component ; constituent : (50)

les composants d'un alliage, the components of an alloy. (51)

les composants de l'eau, the constituents of water. (52)

composante (Méc.) (*n.f.*), component. (53)

composé, -e (*adj.*), compound ; composite : (54)

poutre composée (*f.*), compound beam. (55)

trépan composé (*m.*), compound bit ; com--posite borer. (56)

composé (*n.m.*), compound : (57)

l'eau est un composé d'hydrogène et d'oxygène (*f.*), water is a compound of hydrogen and oxygen. (58)

composé binaire (*m.*), binary compound. (59)

composé chimique (*m.*), chemical compound. (60)

composé nitré *ou* **composé nitrogéné** (*m.*), nitro-compound. (61)

composé ternaire (*m.*), ternary compound. (62)

composer une poutre, to build up a girder : (63)

on compose parfois les poutres au moyen d'une tôle verticale qui forme l'âme, et de cornières qui forment les semelles, sans plates-bandes horizontales, girders are sometimes built up by means of a vertical plate which forms the web, and of angles which form the treads, without horizontal flange-plates. (64)

composition (*n.f.*), composition : (65)

la composition chimique de l'atmosphère, the chemical composition of the atmosphere. (66)

composition des forces (Méc.) (*f.*), composition of forces. (67)

composition inattaquable par le pétrole (*f.*), petroleum-resisting composition. (68)

compound (machines à vapeur) (*adj. invar.*), compound : (69)

machine compound (*f.*), compound engine. (70)

compound (Élec.) (*adj. invar.*), compound ; compound-wound : (71)

enroulement compound (*m.*), compound winding. (72)

dynamo compound (*f.*), compound-wound dynamo. (73)

compoundage (machines à vapeur & Élec.) (*n.m.*), compounding. (1)

compounder (machines à vapeur & Élec.) (*v.t.*), to compound. (2)

comprendre (renfermer en soi) (*v.t.*), to comprise; to consist of; to include; to cover: (3)
la propriété comprend plusieurs concessions, the property comprises several concessions. (4)
le matériel doit comprendre les pièces suivantes, plant to consist of the following. (5)
les prix ne comprennent pas la pose (*m.pl.*), prices do not include fixing. (6)
le minerai compris dans le bulletin de l'essayeur, the ore covered by the assayer's certificate. (7)

compresseur (*n.m.*), compressor. (8)

compresseur à colonne d'eau (*m.*), water-column compressor. (9)

compresseur à faible vitesse (*m.*), slow-speed compressor. (10)

compresseur à grande vitesse (*m.*), high-speed compressor. (11)

compresseur à injection *ou* **compresseur à injec--tion d'eau** (*m.*), injection-compressor; spray-compressor. (12)

compresseur à piston liquide (*m.*), water-piston compressor. (13)

compresseur compound *ou* **compresseur étagé** (*m.*), compound compressor; stage-compressor (air-cylinders compounded or staged). (14)

compresseur d'air (*m.*), air-compressor. (15)

compresseur humide (*m.*), wet compressor. (16)

compresseur hydraulique (*m.*), hydraulic com--pressor. (17)

compresseur monocylindrique *ou* **compresseur simple** (*m.*), single-stage compressor (single-stage air-cylinder). (18)

compresseur monocylindrique double (*m.*), duplex compressor (two single-stage air-cylinders, driven by two steam-cylinders set tandem). (19)

compresseur monocylindrique simple *ou* **com--presseur à groupe unique de cylindres en tandem** (*m.*), straight-line compressor (one single-stage air-cylinder, driven by a steam-cylinder set tandem). (20)

compresseur sec (*m.*), dry compressor. (21)

compressibilité (*n.f.*), compressibility; com--pressibleness: (22)
la compressibilité des gaz, the compressibility of gases. (23)

compressible (*adj.*), compressible. (24)

compressif, -ive (*adj.*), compressive. (25)

compression (*n.f.*), compression: (26)
la compression de la vapeur dans le cylindre, the compression of steam in the cylinder. (27)
la compression d'un ressort par un poids, the compression of a spring by a weight. (28)

compression (Résistance des matériaux) (*n.f.*), compression; crushing (Strength of materials): (29)
essai à la compression (*m.*), compression-test; crushing-test. (30)
pièce prismatique qui travaille à la compression (*f.*), prismatic bar which is in compression. (31)
poutre soumise à des efforts de compression (*f.*), beam subjected to compressive stresses. (32)

compression adiabatique (*f.*), adiabatic com--pression. (33)

compression compound *ou* **compression étagée** (d'air) (*f.*), compound compression; stage compression. (34)

compression en deux étages (d'air) (*f.*), two-stage compression. (35)

compression isotherme (*f.*), isothermal com--pression. (36)

compression sans étage (d'air) (*f.*), stageless compression. (37)

compression simple (d'air) (*f.*), single-stage compression. (38)

comprimable (*adj.*), compressible. (39)

comprimant, -e (*adj.*), compressing: (40)
force comprimante (*f.*), compressing force. (41)

comprimé, -e (*adj.*), compressed: (42)
air comprimé (*m.*), compressed air. (43)

comprimer (*v.t.*), to compress: (44)
comprimer des gaz, to compress gases. (45)
comprimer en briquettes des poussiers de houille, to compress coal-dust into briquettes. (46)

comprimer (se) (*v.r.*), to be compressed; to be able to be compressed: (47)
certains gaz se compriment indéfiniment sans passer à l'état liquide, certain gases can be compressed indefinitely without passing into the liquid state. (48)

compromettre la sécurité des ouvriers, to endanger, to imperil, to jeopardize, the safety of the workmen. (49)

compte (*n.m.*), account; computation; reckoning. (50)

compte d'établissement de prix de revient (*m.*), cost-account. (51)

compte de vente (*m.*), account sales. (52)

compte-gouttes [compte-gouttes *pl.*] (graissage) (*n.m.*), drip-feed lubricator. (53)

compte-gouttes [compte-gouttes *pl.*] (flacon) (*n.m.*), drop-bottle; dropping-bottle. (54)

compte-gouttes capillaire, à tube de caoutchouc (*m.*), dropping-tube, dropper, dripping-tube, with india-rubber top. (55)

compte-pas [compte-pas *pl.*] (*n.m.*), paceometer; passimeter; passometer; pedometer. (56)

compte rendu des études faites et des travaux accomplis (*m.*), report of researches made and of work done. (57)

compte-tours [compte-tours *pl.*] (*n.m.*). Same as **compteur de tours.**

compter (calculer) (*v.t.*), to count; to compute; to reckon. (58)

compter (regarder comme certain) (*v.i.*), to expect: (59)
compter reprendre les travaux, to expect to resume work. (60)
compter sur quelqu'un, to rely, to count, to depend, on someone. (61)

compteur (*n.m.*), counter; meter. (62)

compteur à eau (*m.*), water-meter. (63)

compteur à gaz (*m.*), gas-meter. (64)

compteur à paiement préalable (*m.*), slot-meter. (65)

compteur d'air (*m.*), air-meter. (66)

compteur d'électricité (*m.*), electricity-meter; electric meter. (67)

compteur de tours (*m.*), speed-counter; revolu--tion-counter: (68)
compteur de tours, enregistrant jusqu'à — tours, remise à 0 rapide *ou* compteur de tours, — guichets, ramenage à 0 rapide, speed-counter, counting to — revolutions, sets rapidly to zero; speed-counter, — figures, quickly resets to zero. (69)

Note.—A speed-counter counting to 100,000 revs is otherwise expressed as a speed-counter 5 figures (viz., 99,999), one counting to 10,000 revs as a 4 figures, and so on.
compteur de tours de poche (*m.*), pocket speed-counter. (1)
compteur de vapeur (*m.*), steam-meter. (2)
compteur de vitesse linéaire (*m.*), surface-speed indicator. (3)
compteur Venturi (Hydraul.) (*m.*), Venturi meter. (4)
concassage (*n.m.*), breaking; crushing: (5)
concassage du minerai au concasseur, breaking ore in the rock-breaker; crushing ore in the crusher. (6)
concassage à la main, breaking by hand; hand-breaking. (7)
concasser (*v.t.*), to break; to crush: (8)
concasser à la main, to break by hand. (9)
concasseur (*n.m.*), breaker; crusher; rock-breaker; rock-crusher; stone-breaker; stone-crusher; stone-mill. (10)
concasseur à charbon (*m.*), coal-breaker; coal-cracker; cracker. (11)
concasseur à mâchoires (*m.*), jaw-breaker; jaw-crusher. (12)
concasseur à minerais (*m.*), ore-crusher. (13)
concasseur giratoire (*m.*), gyratory crusher; gyratory rock-breaker. (14)
concave (*adj.*), concave: (15)
surface concave (*f.*), concave surface. (16)
miroir concave (*m.*), concave mirror. (17)
concavité (*n.f.*), concavity: (18)
la concavité d'une lentille, the concavity of a lens. (19)
concavo-concave *ou* **concave-concave** (*adj.*), con-cavo-concave; biconcave; double-concave. (20)
concavo-convexe *ou* **concave-convexe** (*adj.*), concavo-convex. (21)
concéder (*v.t.*), to concede; to grant. (22)
concentrateur, -trice (*adj.*), concentrating: (23)
appareil concentrateur (*m.*), concentrating apparatus. (24)
concentrateur (*n.m.*), concentrator. (25)
concentrateur à boues *ou* **concentrateur pour slimes** (*m.*), slimes-concentrator. (26)
concentrateur magnétique (*m.*), magnetic concentrator. (27)
concentration (*n.f.*), concentration; concen-trating: (28)
la concentration de la chaleur, des rayons solaires, the concentration of heat, of the sun's rays. (29 *ou* 30)
concentration à l'eau *ou* **concentration par voie humide** (du minerai) (*f.*), water-concentra-tion; wet concentration. (31)
concentration à sec *ou* **concentration par voie sèche** (*f.*), dry concentration. (32)
concentration ignée (*f.*), igneous concentration. (33)
concentration par flottement (*f.*), concentration by flotation. (34)
concentré, -e (*adj.*), concentrated. (35)
concentré (*n.m.*), concentrate: (36)
concentré d'étain, tin concentrate. (37)
concentrés de plomb argentifère, silver-lead concentrates. (38)
concentrer (*v.t.*), to concentrate: (39)
concentrer des minerais par lavage, to con-centrate ores by washing. (40)
concentrer (se) (*v.r.*), to concentrate: (41)

minerai qui se concentre facilement (*m.*), ore which concentrates readily. (42)
concentrique (*adj.*), concentric; self-centring: (43)
cercles concentriques (*m.pl.*), concentric circles. (44)
mandrin à serrage concentrique (*m.*), concentric chuck; self-centring chuck. (45)
concentriquement (*adv.*), concentrically. (46)
concessible (*adj.*), concessible. (47)
concession (*n.f.*), concession; claim; grant; licence. (48). See also **claim**.
concession à perpétuité (*f.*), perpetual con-cession. (49)
concession alluvionnaire (*f.*). Same as **con-cession de terrains d'alluvion**.
concession d'eau (*f.*), water-privilege; water-concession. (50)
concession de chemin de fer (*f.*), railway-con-cession; concession for a railway. (51)
concession de découverte (*f.*), discovery-claim; reward-claim. (52)
concession de lit de cours d'eau (*f.*), bed-claim. (53)
concession de mines (*f.*), mining-concession; mining-claim; mineral claim; mining-licence. (54)
concession de placer aurifère (*f.*), placer gold-claim; gold-claim. (55)
concession de quartz aurifère (*f.*), quartz-claim; reef-claim. (56)
concession de terrain (*f.*), concession of land. (57)
concession de terrains d'alluvions (*f.*), alluvial claim; diggers' claim. (58)
concession filonienne (*f.*), lode-claim. (59)
concession forestière (*f.*), timber-concession. (60)
concession minière (*f.*), mining-concession; mining-claim; mineral claim; mining-licence. (61)
concession riveraine (Mines) (*f.*), bank-claim. (62)
concessionnaire (*adj.*), concessionary: (63)
société concessionnaire (*f.*), concessionary company. (64)
concessionnaire (pers.) (*n.m. ou f.*), claimholder; concessionaire; concessioner; concessionary; grantee; licence-holder. (65)
conchoïdal, -e, -aux *ou* **conchoïde** (*adj.*), con-choidal; conchoid: (66)
cassure conchoïdale (*f.*), conchoidal fracture. (67)
conchylien, -enne (*adj.*), conchitic. (68)
concluant, -e (*adj.*), conclusive: (69)
des expériences concluantes (*f.pl.*), conclusive experiments. (70)
conclure (*v.t.*), to conclude. (71)
conclusion (*n.f.*), conclusion. (72)
concordance (Géol.) (*n.f.*), conformability; conformity. (73)
concordant, -e (Géol.) (*adj.*), conformable: (74)
stratification concordante (*f.*), conformable stratification. (75)
concourant, -e (*adj.*), concurrent: (76)
forces concourantes (*f.pl.*), concurrent forces. (77)
concréfier (*v.t.*), to concrete. (78)
concréfier (se) (*v.r.*), to concrete. (79)
concret, -ète (*adj.*), concrete: (80)
exemple concret (*m.*), concrete example. (81)
concret (béton) (*n.m.*), concrete. (82)
concrétion (*n.f.*), concretion: (83)
concrétion nodulaire, nodular concretion. (84)

concrétionné, -e (Géol.) (*adj.*), concretionary ; concretional : (1)
roches concrétionnées (*f.pl.*), concretionary rocks. (2)
condamnation pour une infraction aux règlements sur les mines (*f.*), conviction for an offence against the mining regulations. (3)
condamner à une amende (*v.t.*), to fine. (4)
condensabilité (*n.f.*), condensability : (5)
la condensabilité des gaz, the condensability of gases. (6)
condensable (*adj.*), condensable. (7)
condensant, -e *ou* condensateur, -trice (*adj.*), condensing : (8)
électroscope condensateur (*m.*), condensing electroscope. (9)
condensateur (Phys.) (*n.m.*), condenser. (10)
condensateur (Élec.) (*n.m.*), condenser. (11)
condensateur (Opt.) (*n.m.*), condenser. (12)
condensation (*n.f.*), condensation ; condensing : (13)
la condensation de la vapeur, de l'électricité, de la lumière, the condensation of steam, of electricity, of light. (14 *ou* 15 *ou* 16)
condensation par mélange (de la vapeur) (*f.*), direct-contact condensation. (17)
condensation par surface (de la vapeur) (*f.*), surface-condensation. (18)
condenser (*v.t.*), to condense : (19)
condenser la vapeur d'échappement, to con-dense exhaust-steam. (20)
condenser (se) (*v.r.*), to condense. (21)
condenseur (des machines à vapeur) (*n.m.*), condenser. (22)
condenseur (pour purifier le gaz d'éclairage) (*n.m.*), condenser. (23)
condenseur à siphon (*m.*), siphon-condenser. (24)
condenseur *avec* pompe à air concentrique (*m.*), condenser with tail air-pump. (25)
condenseur avec pompe à air séparée (*m.*), con-denser with independent air-pump. (26)
condenseur barométrique *ou* condenseur avec pompe à air seul (*m.*), barometric condenser. (27)
condenseur par éjection *ou* condenseur à trompe (*m.*), ejector-condenser. (28)
condenseur par injection *ou* condenseur à injec-tion *ou* condenseur à jet (*m.*), injection-condenser ; jet-condenser. (29)
condenseur par mélange *ou* condenseur à mélange (*m.*), direct-contact condenser. (30)
condenseur par surface *ou* condenseur à surface *ou* condenseur à sec (*m.*), surface-condenser. (31)
condenseur par surface au moyen de l'air (*m.*), air-cooled surface-condenser. (32)
condenseur par surface au moyen de l'eau (*m.*), water-cooled surface-condenser. (33)
condenseur rectangulaire par surface (*m.*), rectangular surface-condenser. (34)
condition (situation) (*n.f.*), condition ; state : (35)
conditions atmosphériques, weather conditions. (36)
condition heureuse des affaires *ou* conditions à souhait, welcome state of affairs. (37)
conditions de travail, working conditions. (38)
condition (clause restrictive) (*n.f.*), condition : (39)
respecter les clauses et conditions d'un con-trat, to respect the terms and conditions of an agreement. (40)

condition (à) *ou* condition (sous) (Commerce), on approval : (41)
envoyer des marchandises à condition, to send goods on approval. (42)
conductance (Élec.) (*n.f.*), conductance. (43)
conducteur, -trice (Phys.) (*adj.*), conducting ; conductive ; conductor (*used as adj.*) : (44)
fil conducteur (Élec.) (*m.*), conducting wire ; conductor-wire. (45)
conducteur, -trice (opp. à conduit, -e) (Mach.) (*adj.*), driving (46) (opp. to driven) :
poulie conductrice (*f.*), driving pulley. (47)
conducteur (Élec.) (*n.m.*), conductor : (48)
le cuivre est bon conducteur de l'électricité, copper is a good conductor of electricity. (49)
conducteur (filon guide) (Mines) (*n.m.*), leader : (50)
s'apercevoir, un beau jour, qu'un petit filon n'est que le conducteur d'une masse filoni-enne beaucoup plus importante, to discover, one fine day, that a little lode is only the leader of a much more important lode mass. (51)
conducteur auxiliaire (Élec.) (*m.*), branch ; tap. (52)
conducteur d'énergie (Élec.) (*m.*), power-conductor. (53)
conducteur d'équilibre (Élec.) (*m.*). Same as conducteur neutre.
conducteur de chevaux (Mines) (pers.) (*m.*), pony-driver. (54)
conducteur de lingots (Laminage) (*m.*), travelling table. (55)
conducteur de réseau (Élec.) (*m.*), distributing-main ; distribution-main ; supply-main. (56)
conducteur des travaux (Constr.) (pers.) (*m.*), foreman of job ; works' foreman ; gaffer. (57)
conducteur électrique principal (*m.*), electric main. (58)
conducteur en charge (Élec.) (*m.*), live conductor. (59)
conducteur neutre (Système de distribution à fils multiples) (Élec.) (*m.*), neutral conductor ; neutral wire. (60)
conducteur nu (Élec.) (*m.*), bare conductor. (61)
conducteur principal (opp. à conducteur secon-daire, conducteur auxiliaire *ou* branchement) (Élec.) (*m.*), main ; lead ; main conductor. (62)
conducteur principal (opp. à conducteur neutre) (Système de distribution à trois fils) (Élec.) (*m.*), outer conductor ; outer. (63)
conducteur ramifié (Élec.) (*m.*), branched lead. (64)
conducteur secondaire (Élec.) (*m.*), branch ; tap. (65)
conducteur tubulaire (Sondage) (*m.*), guide-tube ; conductor-pipe ; drill-pipe. (66)
conductibilité (*n.f.*), conductivity ; conducti-bility : (67)
conductibilité du sol, conductibility of the ground. (68)
conductibilité électrique du cuivre, electric conductivity of copper. (69)
conductibilité moléculaire (*f.*), molecular con-ductivity. (70)
conductibilité spécifique (Élec.) (*f.*), specific conductivity ; specific conductance. (71)
conductibilité thermique *ou* conductibilité calori-fique (*f.*), thermal conductivity. (72)
conductibilité thermique extérieure (*f.*), external conductivity ; emissivity. (73)

conductible (*adj.*), conductible. (1)
conduction (Phys.) (*n.f.*), conduction. (2)
conductivité (Phys.) (*n.f.*), conductivity. (3)
conduire (*v.t.*). See examples : (4)
  conduire des marchandises d'une ville à une autre, to convey goods from one town to another. (5)
  conduire des travaux, to conduct, to direct, to supervise, operations. (6)
  conduire l'eau au moyen de conduites, to conduct the water by means of pipes. (7)
  conduire la chauffe, to stoke ; to attend to the firing. (8)
  conduire les feux d'une chaudière, to look after the fires of a boiler. (9)
  bien conduire le chauffage d'un four, to pay proper attention to the firing of a furnace. (10)
  conduire quelqu'un sur une propriété, to conduct someone to a property. (11)
  conduire un bateau, un automobile, to steer a boat, a motor-car. (12 *ou* 13)
  conduire un canal jusqu'à la mer, to carry a canal as far as the sea. (14)
  conduire un cheval, to drive, to lead, a horse. (15)
  conduire une expérience, to conduct an experiment. (16)
  conduire une locomotive, to drive a locomo-tive. (17)
conduit, -e (opp. à **conducteur, -trice**) (Mach.) (*adj.*), driven (18) (opp. to **driving**) :
  poulie conduite (*f.*), driven pulley. (19)
conduit (*n.m.*), pipe ; conduit. (20). See also **conduite** and **tuyau.**
conduit à vent *ou* **conduit aérifère** *ou* **conduit d'air** (*m.*), air-conduit ; air-pipe. (21)
conduit d'échappement (*m.*), exhaust-pipe ; blast-pipe. (22)
conduit de fumée (*m.*), smoke-pipe. (23)
conduit de gaz (*m.*), gas-pipe. (24)
conduit en fonte (*m.*), cast-iron pipe. (25)
conduit souterrain (*m.*), underground conduit. (26)
conduite (action) (*n.f.*), conduct ; conducting ; leading ; driving ; steering ; etc. : (27)
  la conduite des affaires, the conduct of affairs. (28)
  See the verb **conduire** for applications of other meanings.
conduite (tuyau) (*n.f.*), pipe ; conduit. (29). See also **conduit** and **tuyau.**
conduite à emboîtement *ou* **conduite à joints à emboîtements** (*f.*), socket-pipe. (30)
conduite circulaire de vent (haut fourneau) (*f.*), circular blast-main ; horseshoe main ; bustle-pipe (blast-furnace). (31)
conduite d'air (*f.*), air-pipe ; air-hose. (32)
conduite d'amenée (*f.*), supply-pipe ; head-pipe. (33).
conduite d'amenée d'eau (*f.*), water supply-pipe. (34)
conduite d'amenée d'eau fraîche (*f.*), fresh-water supply-pipe. (35)
conduite d'arrivée (*f.*), inlet-pipe. (36)
conduite d'eau (*f.*), water-pipe. (37)
conduite de distribution d'eau (*f.*), water-main. (38)
conduite de gaz (*f.*), gas-pipe. (39)
conduite de l'intercommunication pneumatique (locomotive) (*f.*), signal-pipe. (40)
conduite de refoulement (d'une pompe) (*f.*), delivery-pipe. (41)

conduite de vapeur (*f.*), steam-pipe. (42)
conduite de vent (haut fourneau) (*f.*), blast-main (blast-furnace). (43)
conduite de vent (cubilot) (*f.*), blast-pipe ; blast-main (cupola). (44)
conduite de vent chaud (haut fourneau) (*f.*), hot-blast main. (45)
conduite électrique (*f.*), electric conduit. (46)
conduite en tôle d'acier rivée (*f.*), riveted steel pipe. (47)
conduite forcée (opp. à **conduite libre** *ou* **canal découvert**) (Hydraul.) (*f.*), full pipe. (48) (contradistinguished from **open channel**).
conduite générale du frein (*f.*), train-pipe ; brake-pipe. (49)
conduite libre (Hydraul.) (*f.*), open channel. (50)
conduite maîtresse *ou simplement* **conduite** (*n.f.*), main. (51)
conduite maîtresse d'air (*f.*), air-main. (52)
conduite maîtresse de gaz (*f.*), gas-main. (53)
conduite maîtresse de vapeur (*f.*), main steam-pipe. (54)
conduite mixte (*f.*), compound pipe (varying diameters). (55)
conduite principale (*f.*). Same as **conduite maîtresse.**
conduite principale d'air, de gaz, etc. Same as **conduite maîtresse d'air, de gaz, etc.**
conduite souple (*f.*), hose ; flexible pipe. (56)
condurrite (Minéral) (*n.f.*), condurrite. (57)
cône (Géom., etc.) (*n.m.*), cone. (58)
cône (conicité) (*n.m.*), taper ; (59)
  cône Morse, Morse taper. (60)
  goupilles coniques en acier, cône — p. 100 (*f.pl.*), steel taper pins, taper — %. (61)
cône (de haut fourneau) (*n.m.*), cone, bell, (of blast-furnace) : (62)
  le cup et cône, the cup-and-cone ; the bell-and-hopper. (63)
cône (*n.m.*) *ou* **cône-poulie** (*n.m.*) [**cônes-poulies** *pl.*] *ou* **cône de transmission** *ou* **cône de vitesse**, cone ; cone-pulley ; stepped cone ; step-cone ; stepped pulley ; speed-cone ; speed-pulley ; speed-rigger ; speeder. (64)
cône à corde (*m.*), cone-sheave ; grooved cone. (65)
cône à courroie (*m.*), belt-cone ; speed-cone for flat strap. (66)
cône à — gradins *ou* **cône à — étages** (*m.*), — -step cone ; — -speed cone-pulley : (67)
  un cône à 3 gradins, a 3-step cone ; a 3-speed cone-pulley. (68)
cône adventif (volcanique) (Géol.) (*m.*), parasitic cone ; subordinate cone ; lateral cone. (69)
cône convergent (d'un injecteur) (*m.*), combining-cone, combining-tube, combining-nozzle, (of an injector). (70)
cône d'avancement (Travail aux explosifs dans le fonçage d'un puits) (*m.*), sink ; wedge-shaped centre-cut (depression in shaft made by centre-blast). (71)
cône d'éboulement *ou* **cône d'éboulis** (Géol.) (*m.*), talus. (72)
cône d'éruption (Géol.) (*m.*), cone of eruption. (73)
cône de débris (volcanique) (Géol.) (*m.*), debris-cone. (74)
cône de déjection (Géogr. phys.) (*m.*), fan delta ; cone delta ; delta ; alluvial fan ; alluvial cone. (75)
cône de division (Engrenages) (*m.*), pitch-cone (Gearing). (76)

cône de forgeron (*m.*), smiths' cone ; smiths' mandrel. (1)

cône de friction (Méc.) (*m.*), friction-cone. (2)

cône de lave (*m.*), lava cone. (3)

cône de lumière (Opt.) (*m.*), cone of rays. (4)

cône de renvoi (*m.*), cone for overhead motion ; overhead cone ; overhead cone-pulley ; counter-shaft cone. (5)

cône de révolution (Géom.) (*m.*), cone of revolu-tion. (6)

cône de scories (volcaniques) (Géol.) (*m.*), cinder-cone ; scoria-cone. (7)

cône de vitesse (*m.*). See cône *ou* cône de vitesse.

cône divergent (d'un injecteur) (*m.*), diverging cone, diverging tube, diverging nozzle, delivery-cone, delivery-tube, delivery-nozzle, (of an injector). (8)

cône droit (Géom.) (*m.*), right cone. (9)

cône droit circulaire (*n.*) (*m.*), right circular cone. (10)

cône femelle (*m.*), female cone. (11)

cône mâle (*m.*), male cone. (12)

cône lavique (*m.*), lava cone. (13)

cône mixte (volcanique) (*m.*), composite cone. (14)

cône Morse (*m.*), Morse taper. (15)

cône oblique (Géom.) (*m.*), oblique cone. (16)

cône parasitaire (Géol.) (*m.*), parasitic cone ; subordinate cone ; lateral cone. (17)

cône-poulie [cônes-poulies *pl.*] (*n.m.*). See cône *ou* cône-poulie.

cône primitif (Engrenages) (*m.*), pitch-cone (Gearing). (18)

cône taraudé (Sondage) (*m.*), bell-screw ; bell-box ; bell-mouth socket ; screw-bell ; screw-grab. (19)

cône tronqué (*m.*), truncated cone. (20)

cône volcanique (*m.*), volcanic cone. (21)

cônes alternes *ou* cônes complémentaires (Méc.) (*m.pl.*), belt-speeder. (22)

confection (*n.f.*), making ; making up : (23)
confection d'une route, making a road. (24)
confection des joints, making joints. (25)
confection des noyaux (Fonderie), core-making. (26)
matériaux employés dans la confection du béton armé (*m.pl.*), materials used in making reinforced concrete. (27)

confectionner (*v.t.*), to make ; to make up : (28)
l'ouvrier serrurier devrait savoir confectionner ses outils (*m.*), the metal-worker should know how to make his tools. (29)
confectionner une cartouche, to make up a cartridge. (30)

confiance (*n.f.*), confidence ; faith : (31)
avoir une confiance sans bornes dans une propriété, to have unbounded faith in a property. (32)

confier à quelqu'un la direction d'une mine, to entrust the management of a mine to someone. (33)

configuration (*n.f.*), configuration ; lay ; lie : (34)
configuration du pays, configuration of the country ; lay of the country ; lie of the land. (35)
configuration du terrain, configuration of the ground. (36)

confiner (*v.t.*), to confine ; to limit : (37)
confiner les travaux dans les étages supérieurs, to confine the work to the upper levels ; to limit work to the upper levels. (38)

confirmer (*v.t.*), to confirm : (39)

confirmer une nouvelle, to confirm a report. (40)

confiscation (*n.f.*), confiscation. (41)

confisquer (*v.t.*), to confiscate. (42)

conflagration (*n.f.*), conflagration. (43)

confluent (*n.m.*), confluence ; junction ; meeting : (44)
le confluent de deux rivières, the confluence (*or* the junction) of two rivers. (45)

confluer (*v.i.*), to meet ; to join ; to conflow : (46)
le Rhône et la Saône confluent à Lyon, the Rhone and the Saône meet at Lyons. (47)

conformation (*n.f.*), conformation. (48)

conforme (*adj.*), conformable. (49)

conformer (se) (*v.r.*), to conform : (50)
se conformer aux plans de schistosité (Géol.), to conform to the planes of schistosity. (51)
se conformer aux règlements, to conform to, to comply with, the regulations. (52)

conformité (Géol.) (*n.f.*), conformity ; conform-ability. (53)

confus, -e (*adj.*), confused : (54)
amas confus de débris (*m.*), confused mass of debris. (55)

conge (Mines) (*n.m.*), ore-basket. (56)

conge à patins (*m.*), sledge-corve. (57)

congé (renvoi d'une personne à gages) (*n.m.*), dismissal : (58)
motif de congé (*m.*), reason for dismissal. (59)

congé (autorisation de s'absenter) (*n.m.*), leave of absence. (60)

congé (*n.m.*) *ou* congé de raccordement, fillet ; neck-moulding ; neck-mold ; hollow moulding ; hollow curvature ; congé ; apophyge. (61)

congé (rabot de menuisier) (*n.m.*), neck-moulding plane. (62)

congé (quart de rond creux qui raccorde le boudin et la surface de roulement d'une roue de ch. de f.) (*n.m.*), throat. (63)

congédier (*v.t.*), to dismiss ; to discharge : (64)
congédier des ouvriers, to discharge workmen. (65)

congélabilité (*n.f.*), congealableness. (66)

congelable (*adj.*), congealable ; freezable. (67)

congélation (*n.f.*), congelation ; congealment ; gelation ; freezing : (68)
congélation du sol (procédé de fonçage), freezing the ground. (69)

congeler (*v.t.*), to congeal ; to freeze. (70)

congeler (se) (*v.r.*), to congeal ; to freeze. (71)

conglomérat (Géol.) (*n.m.*), conglomerate ; cemented gravel. (72)

conglomérat aurifère (*m.*), auriferous conglom-erate ; auriferous banket ; banket. (73)

conglomérat de base (*m.*), basal conglomerate. (74)

conglomérat de faille (*m.*), fault-conglomerate. (75)

conglomérat de friction (*m.*), crush-conglom-erate. (76)

conglomérat de quartz (*m.*), quartz conglomer-ate. (77)

conglomération (*n.f.*), conglomeration. (78)

conglomératique (*adj.*), conglomeratic. (79)

congloméré, -e (*adj.*), conglomerate ; conglom-erated. (80)

conicité (*n.f.*), conicity ; conicality ; taper : (81)
la conicité d'une roue d'angle, the conicity of a bevel-wheel. (82)
la conicité de la surface de roulement d'un bandage de roue de locomotive, the taper of the tread of a locomotive wheel-tire. (83)

une conicité de 1 sur —, a taper of 1 in —. (1)

conique (*adj.*), conical ; conic ; cone-shaped ; taper ; tapering : (2)

surface conique (*f.*), conical surface. (3)

goupille conique (*f.*), taper pin. (4)

conjoncteur (Élec.) (*n.m.*), circuit-closer. (5)

conjoncteur-disjoncteur [**conjoncteurs-disjonc-teurs** *pl.*] (Élec.) (*n.m.*), make-and-break. (6)

conjoncture (*n.f.*), juncture ; conjuncture. (7)

conjugué, -e (Méc., etc.) (*adj.*), twin : (8)

machines conjuguées (*f.pl.*), twin engines. (9)

conjugué, -e (Géom.) (*adj.*), conjugate ; conju-gated : (10)

lignes conjuguées sur une surface (*f.pl.*), con-jugate lines on a surface. (11)

conjugué, -e (Opt.) (*adj.*), conjugate ; con-jugated : (12)

foyers conjugués (*m.pl.*), conjugate foci. (13)

miroirs conjugués (*m.pl.*), conjugate mirrors. (14)

conjugué, -e (Chim.) (*adj.*), conjugate ; con-jugated. (15)

connecter (*v.t.*), to connect : (16)

accouplement employé pour connecter deux pièces (*m.*), coupling used to connect two parts. (17)

connexion (*n.f.*), connection : (18)

connexions entre conducteurs, connections between conductors. (19)

conoïde (*adj.*), conoid. (20)

consacrer (*v.t.*), to devote ; to sanction : (21)

ateliers qui sont exclusivement consacrés à la fabrication des appareils de haute précision (*m.pl.*), workshops which are exclusively devoted to the manufacture of apparatus of great precision. (22)

synonymies que l'usage a consacrées (*f.pl.*), synonymies that use has sanctioned. (23)

consanguinité des laves (Géol.) (*f.*), consanguinity of lavas. (24)

conscience (plaque à conscience pour drilles) (*n.f.*), breast-plate ; palette ; conscience. (25)

conseil (*n.m.*) *ou* **conseils** (*n.m.pl.*), advice ; counsel. (26)

conseil d'administration (*m.*), board ; board of directors. (27)

conseil général des mines (*m.*), chamber of mines. (28)

conseillable (*adj.*), advisable ; recommendable. (29)

conseiller (pers.) (*n.m.*), adviser : (30)

conseiller technique, technical adviser. (31)

conseiller (*v.t.*), to advise : (32)

conseiller l'emploi de l'énergie électrique, to advise the employment of electric power. (33)

conséquent, -e (*adj.*), consequent : (34)

pôles conséquents *ou* points conséquents (Magnétisme) (*m.pl.*), consequent poles ; consequent points. (35)

cours d'eau conséquent (Géogr. phys.) (*m.*), consequent stream. (36)

conservation (*n.f.*), conservation ; keeping ; preservation ; care : (37)

conservation d'eau, conservation of water. (38)

conservation de l'énergie *ou* conservation des forces vives (Méc.), conservation of energy. (39)

conservation des bois, preservation of timber. (40)

conservation des explosifs, care of explosives. (41)

conserver (*v.t.*), to conserve ; to keep ; to pre-serve : (42)

conserver des galeries dans les remblais (Mines), to keep the gob open ; to reserve passages in the stowing. (43)

conserver l'eau en excédent, to conserve the spare water. (44)

conserver la direction, to keep the direction. (45)

conserver (se) (*v.r.*), to keep : (46)

la solution ne se conserve pas, the solution will not keep. (47)

considérable (*adj.*), considerable ; extensive ; large. (48)

consignataire (pers.) (*n.m.*), consignee. (49)

consignateur, -trice (pers.) (*n.*), consignor. (50)

consignation (marchandises) (*n.f.*), consignment. (51)

consigner les cotes sur des carnets de nivelle-ment, to record the readings in levelling-books. (52)

consistance (*n.f.*), consistency ; consistence ; firmness : (53)

la consistance de l'argile, de la graisse, the consistency of clay, of grease. (54 *ou* 55)

consistant, -e (*adj.*), consistent ; firm ; set : (56)

terrain consistant (*m.*), firm ground. (57)

graisse consistante (*f.*), set grease. (58)

consister (*v.i.*), to consist : (59)

la majeure partie de l'atmosphère consiste en azote, the major part of the atmosphere consists of nitrogen. (60)

console (*n.f.*), bracket. (61)

console-équerre [**consoles-équerres** *pl.*] (*n.f.*), knee bracket ; knee. (62)

console murale (*f.*), wall-bracket. (63)

consolidation (*n.f.*), consolidation ; strengthen-ing : (64)

consolidation du puits, strengthening the shaft. (65)

consolider (*v.t.*), to consolidate ; to strengthen : (66)

consolider le boisage, to strengthen the timber-ing. (67)

consommation (*n.f.*), consumption : (68)

consommation journalière de charbon, daily consumption of coal. (69)

consommation de vapeur par cheval-heure indiqué, consumption of steam per indicated horse-power hour. (70)

consommer (*v.t.*), to consume. (71)

constance en richesse du minerai en profondeur (*f.*), persistence of high values of ore in depth. (72)

constant, -e (*adj.*), constant. (73)

constante (Phys.) (*n.f.*), constant. (74)

constante calorifique (*f.*), heat-constant. (75)

constante capillaire (Phys.) (*f.*), capillary con-stant ; constant of capillarity. (76)

constante diélectrique *ou* **constante de diélectri-cité** (*f.*), dielectric constant ; dielectric capacity ; specific inductive capacity ; permittivity. (77)

constante électrique (*f.*), electrical constant. (78)

constante lumineuse (*f.*), light-constant. (79)

constater (*v.t.*), to ascertain ; to prove ; to establish : (80)

constater l'étendue des dégâts, to ascertain the extent of the damage. (81)

constater l'existence d'un filon, to prove, to establish, the existence of a lode. (82)

constituant, -e (*adj.*), constituent ; component : (1)
partie constituante (*f.*), constituent part ; component part. (2)
constitution (*n.f.*), constitution : (3)
la constitution chimique de l'acier, the chemical constitution of steel. (4)
constructeur (pers.) (*n.m.*), builder ; maker ; constructor. (5)
constructeur d'instruments de mathématiques (*m.*), mathematical-instrument maker. (6)
constructeur de locomotives (*m.*), locomotive-builder. (7)
constructeur de machines (*m.*), engine-maker ; engine-builder. (8)
constructeur de maisons (*m.*), builder ; house-builder. (9)
construction (action) (*n.f.*), building ; construc-tion ; constructing ; erection ; erecting ; making : (10)
matériaux employés dans la construction du matériel roulant des chemins de fer (*m.pl.*), materials used in the construction of railway rolling-stock. (11)
construction (édifice construit) (*n.f.*), building ; construction ; erection ; structure : (12)
construction en bois, wooden building. (13)
construction hydraulique (art) (*f.*), hydraulic engineering. (14)
construire (*v.t.*), to construct ; to build ; to erect ; to put up ; to make : (15)
construire une maison, to build a house. (16)
construire une route, to construct, to make, a road. (17)
construire une usine métallurgique, to erect a smelting-works ; to put up a smelter. (18)
construire une voie ferrée, to build a railroad ; to construct a railway. (19)
consumable (*adj.*), consumable. (20)
consumer (*v.t.*), to consume. (21)
consumptible (*adj.*), consumptible ; consumable. (22)
contact (*n.m.*), contact. (23)
contact (Géol. & Mines) (*n.m.*), contact : (24)
métamorphisme de contact (*m.*), contact metamorphism. (25)
minéral de contact (*m.*), contact-mineral. (26)
contact (Élec.) (*n.m.*), contact ; connection. (27)
contact (touche) (Élec.) (*n.m.*), contact : (28)
contact en platine, platinum contact. (29)
contact (armature) (Magnétisme) (*n.m.*), keeper ; armature ; arming ; lifter : (30)
des contacts de fer doux, keepers of soft iron. (31)
contact à frottement (*m.*), frictional contact. (32)
contact à la terre (Élec.) (*m.*), earth-connection. (33)
contact de roulement (*m.*), rolling contact. (34)
contact électrique (*m.*), electric contact ; electric connection. (35)
contact fixe (Ch. de f.) (*m.*), automatic stop (on railway-line to operate a signal in cab of locomotive). (36)
contact glissant (*m.*), sliding contact. (37)
contamination (*n.f.*), contamination ; pollution. (38)
contaminer (*v.t.*), to contaminate ; to pollute. (39)
contemporain, -e (*adj.*), contemporaneous. (40)
contenance (*n.f.*), capacity ; content : (41)
la contenance d'un tonneau, the capacity of a cask. (42)

la contenance d'un champ, the content of a field. (43)
contenir (*v.t.*), to contain ; to hold ; to carry : (44)
tonneau qui contient — litres (*m.*), cask which contains (*or* holds) — litres. (45)
minerai qui contient — grammes d'argent par tonne (*m.*), ore which carries — grammes of silver per ton. (46)
contenu (*n.m.*), contents ; content : (47)
le contenu d'une boîte, the contents of a box. (48)
contestation (*n.f.*), dispute : (49)
les contestations qui pourraient s'élever entre l'acheteur et le fournisseur sur l'exécution d'un marché, the disputes which may arise between the buyer and the supplier on the carrying out of a contract. (50)
contigu, -ë (*adj.*), contiguous ; adjoining ; conterminous. (51)
contiguïté (*n.f.*), contiguity. (52)
continent (*n.m.*), continent ; mainland. (53)
continental, -e, -aux (*adj.*), continental : (54)
glacier continental (*m.*), continental glacier. (55)
continu, -e (*adj.*), continuous. (56)
continuation (*n.f.*), continuation ; continuance. (57)
continuel, -elle (*adj.*), continual : (58)
faire de continuels efforts, to make continual efforts. (59)
continuellement (*adv.*), continually. (60)
continuer (*v.t. & v.i.*), to continue. (61)
continuer (se) (*v.r.*), to continue. (62)
continuité (*n.f.*), continuity : (63)
la continuité d'une chaîne de montagnes, the continuity of a chain of mountains. (64)
contorsion (*n.f.*), contortion : (65)
contorsions des couches, contortions of the strata. (66)
contorsionné, -e (*adj.*), contorted. (67)
contour (*n.m.*), contour ; outline : (68)
le contour d'une colonne, d'un dôme, the contour of a pillar, of a dome. (69 *ou* 70)
contourné, -e (*adj.*), contorted. (71)
contracter (*v.t.*), to contract ; to shrink ; to narrow : (72)
le froid contracte les métaux, cold contracts metals. (73)
contracter (se) (*v.r.*), to contract ; to shrink ; to narrow : (74)
lorsque le rivetage est fait à chaud, les rivets se contractent en se refroidissant, when riveting is done hot, the rivets contract in cooling. (75)
lorsque la fonte se solidifie, elle se contracte au fur et à mesure que la température baisse, when cast iron solidifies, it shrinks according as the temperature falls. (76)
contraction (*n.f.*), contraction ; shrinking ; shrinkage ; narrowing : (77)
contraction de la croûte terrestre par refroi-dissement du foyer central, contraction of the earth's crust by cooling of the central fire. (78)
contraction de la masse, shrinkage of the mass. (79)
contraction de la veine *ou* contraction de la veine liquide (écoulement de l'eau par orifices) (Hydraul.), contraction of the jet ; contraction of the water-vein (flow of water through orifices). (80)

contraction latérale (écoulement de l'eau par déversoir) (Hydraul.) (*f.*), end-contraction (flow of water over weir). (1)

contradictoire (*adj.*), check ; control (*used as adjs*) : (2)

essai contradictoire (*m.*), check-assay ; control-assay. (3)

contrarier (*v.t.*), to interfere with : (4)

contrarier les plans de quelqu'un, to interfere with someone's plans. (5)

cendres qui contrarient le tirage d'un four (*f.pl.*), ashes which interfere with the draught of a furnace. (6)

contraste (*n.m.*), contrast : (7)

un excès de pose donne un cliché gris sans contrastes, overexposure gives a grey negative without contrasts. (8)

contrasté, -e (Photogr.) (*adj.*), contrasty : (9)

avec le papier contrasté on obtient, des clichés faibles, des épreuves contrastées, with contrasty paper, contrasty prints are obtained from weak negatives. (10) Cf. heurté, -e.

contrat (*n.m.*), contract. (11)

contrat d'entreprise (Exploitation des mines de houille) (*m.*), bargain-contract. (12)

contravention (*n.f.*), contravention ; violation. (13)

contre-balancer (*v.t.*), to counterbalance : (14)

contre-balancer la déperdition d'énergie, to counterbalance the loss of energy. (15)

contre-balancer (se) (*v.r.*), to balance ; to balance each other. (16)

contre-balancier [contre-balanciers *pl.*] (de maîtresse-tige de pompe) (Mines) (*n.m.*), balance-bob (of pump spear-rod). (17)

contre-câble d'équilibre (Extraction) (Mines) (*m.*), counterbalancing-rope. (18)

contre-clavette [contre-clavettes *pl.*] (Mach.) (*n.f.*), gib : (19)

clavette et contre-clavette, gib and cotter ; gib and key. (20)

contre-clef [contre-clefs *pl.*] (*n.f.*), each of the voussoirs placed on either side of the keystone of an arch. (21)

contre-cœur [contre-cœur *ou* contre-cœurs *pl.*] (Ch. de f.) (*n.m.*), wing-rail (of a crossing or frog). (22)

contre-cône [contre-cônes *pl.*] (*n.m.*), counter-shaft cone ; cone for overhead motion ; overhead cone ; overhead cone-pulley. (23)

contre-coup [contre-coups *pl.*] (*n.m.*). See examples : (24)

contre-coup d'une explosion, recoil, backlash, of an explosion ; backlashing. (25)

contre-coup d'une locomotive après une collision, recoil of a locomotive after a collision. (26)

contre-coup des chocs des roues contre les rails (Ch. de f.), reaction of the hammer-blows of the wheels against the rails. (27)

contre-courant [contre-courants *pl.*] (*n.m.*), counter-current. (28)

contre-coussinet [contre-coussinets *pl.*] (Mach.) (*n.m.*), top brass. (29)

contre-dépouille (Fonderie, etc.) (*n.f.*), undercut. (30)

contre-dépouille (être en) (Fonderie, etc.), to be undercut : (31)

partie d'un modèle qui est en contre-dépouille (*f.*), part of a pattern which is undercut. (32)

contre-échelle [contre-échelles *pl.*] (Dessin) (*n.f.*), diagonal scale. (33)

contre-écrou [contre-écrous *pl.*] (*n.m.*), lock-nut ; check-nut ; jam-nut ; pinch-nut ; pinching-nut ; grip-nut ; set-nut ; safety-nut ; keeper. (34)

contre-fer [contre-fers *pl.*] (d'un rabot) (*n.m.*), break-iron, back iron, (of a plane). (35)

contre-fiche [contre-fiches *pl.*] (d'une ferme) (Constr.) (*n.f.*), strut, spur, brace, (of a truss). (36)

contre-fiche (étai incliné) (Étaiement des murs) (*n.f.*), raking shore ; raker. (37)

contre-fil (à) (en parlant du bois, etc.), against the grain : (38)

travailler du bois à contre-fil, to work wood against the grain. (39)

contre-fil de l'eau (Hydraul.) (*m.*), counter stream-line. (40)

contrefort (*n.m.*). See below.

contre-foulement [contre-foulements *pl.*] (Hydraul.) (*n.m.*), backset. (41)

contre-lattage (Constr.) (*n.m.*), counter-lathing. (42)

contre-latte [contre-lattes *pl.*] (*n.f.*), counter-lath. (43)

contre-latter (*v.t.*), to counter-lath : (44)

contre-latter des chevrons, to counter-lath rafters. (45)

contre-limon [contre-limons *pl.*] (d'escalier) (*n.m.*), wall-string. (46)

contremaître (*n.m.*). See below.

contre-manivelle [contre-manivelles *pl.*] (*n.f.*), fly-crank. (47)

contre-marche [contre-marches *pl.*] (hauteur de chaque marche d'escalier) (*n.f.*), rise. (48)

contre-marche (planche qui forme le devant vertical d'une marche d'escalier) (*n.f.*), riser. (49)

contre-marche palière (*f.*), landing-riser. (50)

contre-plaque pour chaises (transmission) [contre-plaques *pl.*] (*f.*), wall-plate for brackets (shafting). (51)

contre-plateau pour le montage des mandrins (de tour) [contre-plateaux *pl.*] (*m.*), back-plate for mounting chucks ; chuck-back (of lathe). (52)

contrepoids (*n.m.*). See below.

contre-poinçon [contre-poinçons *pl.*] (*n.m.*), counter-punch ; bolster. (53)

contre-poinçonner (*v.t.*), to counter-punch. (54)

contre-pointe [contre-pointes *pl.*] (de tour, de fraiseuse horizontale) (*n.f.*), loose head-stock ; loose head ; tail-stock ; tailstock ; tail-block ; loose poppet-head ; sliding puppet ; sliding headstock ; deadhead ; back-head ; back-puppet ; foot-stock. (55 *ou* 56)

contre-pointe à fourreau (*f.*), cylinder tail-stock ; cylinder poppet-head. (57)

contre-pointe avec corps déporté *ou* **contre-pointe dégagée sur le devant** *ou* **contre-pointe évidée à l'avant** (*f.*), tail-stock cut away at the front. (58)

contre-pointe pouvant s'excentrer pour tourner conique *ou* **contre-pointe se déplaçant sur sa semelle pour tourner conique** (*f.*), tail-stock which can be set over for taper turning ; loose head-stock adjustable for taper turning ; loose head to slide for taper turning. (59)

contrepoison (*n.m.*). See below.

contre-porte de foyer [contre-portes de foyers *pl.*] (*f.*), fire-door shield ; fire-door liner. (1)

contre-poupée [contre-poupées *pl.*] (*n.f.*). Same as contre-pointe.

contre-pression [contre-pressions *pl.*] (*n.f.*), counterpressure. (2)

contre-pression (pression existant sur la face du piston opposée à celle sur laquelle s'exerce l'effort de la vapeur venant du tiroir) (*n.f.*), back pressure. (3)

contre-proposition [contre-propositions *pl.*] (*n.f.*), counter-proposition ; counter-proposal. (4)

contre-rail [contre-rails *pl.*] (voie ferrée) (*n.m.*), guard-rail ; check-rail ; wing-rail ; wing ; edge-rail ; guide-rail ; rail-guard ; side-rail ; safety-rail. (5)

contre-rivure [contre-rivures *pl.*] (*n.f.*), riveting-burr ; burr ; rivet-washer. (6)

contretemps (*n.m.*). See below.

contre-tige de piston (*f.*), tail-rod of piston ; extended piston-rod. (7)

contre-vapeur (*n.f.*), reversed steam : (8)
marche à contre-vapeur (*f.*), running with reversed steam. (9)

contrevenir (*v.i.*). See below.

contrevent (*n.m.*), contreventement (*n.m.*), contreventer (*v.t.*). See below.

contrée (*n.f.*), country. (10) See also pays.

contrée accidentée (*f.*), undulating country ; uneven country ; rolling country. (11)

contrée boisée *ou* contrée arbreuse (*f.*), wooded country ; woody country. (12)

contrée d'avenir (*f.*), promising country ; coming country. (13)

contrée densément boisée (*f.*), densely wooded country. (14)

contrée marécageuse (*f.*), marshy country ; swampy country. (15)

contrée minière (*f.*), mining country. (16)

contrée montagneuse (*f.*), mountainous country. (17)

contrée ondulée (*f.*), undulating country ; rolling country. (18)

contrée pétrolifère (*f.*), oil region ; oil-bearing territory. (19)

contrée peu sûre (*f.*), unsettled country. (20)

contrée plate (*f.*), flat country ; plain country. (21)

contrée stérile (*f.*), barren country ; sterile country. (22)

contrée sûre (*f.*), settled country. (23)

contrefort (Arch.) (*n.m.*), buttress ; close buttress : (24)
le contrefort d'une muraille, d'un pont, the buttress of a wall, of a bridge. (25 *ou* 26)

contrefort (Géogr. phys.) (*n.m.*), spur ; counter-fort : (27)
les contreforts méridionaux des Alpes Centrales, the southern spurs of the Central Alps. (28)

contremaître (pers.) (*n.m.*), foreman. (29)

contremaître de fonderie (*m.*), foundry-foreman. (30)

contremaître de mine (*m.*), mine-foreman. (31)

contremaître du fond (Mines) (*m.*), underground foreman ; foreman of the underground workings ; pit-bottom foreman. (32)

contremaître du jour (Mines) (*m.*), surface-foreman. (33)

contremaître mécanicien (*m.*), foreman mechanic. (34)

contremarche (*n.f.*). Same as contre-marche.

contrepoids (*n.m.*), counterweight ; counter-balance ; counterpoise ; balance-weight. (35)

contrepoids (remplissage métallique entre les rayons d'une roue motrice de locomotive) (*n.m.*), counterbalance ; counterbalance-weight ; balance-weight ; locomotive-balance. (36)

contrepoids de la cage (Mines) (*m.*), balance-weight of the cage. (37)

contrepoids de queue (*m.*), back balance ; tail-weight. (38)

contrepoids du levier de manœuvre des aiguilles (Ch. de f.) (*m.*), counterweight of the switch-lever. (39)

contrepoids hydraulique (*m.*), hydraulic counter-poise. (40)

contrepoids tendeur (*m.*), balanced tension-block. (41)

contrepoison (*n.m.*), antidote ; counterpoison : (42)
en cas d'empoisonnement par l'hydrogène sulfuré, le contrepoison indiqué est le chlore (faire respirer au patient du chlorure de chaux imbibé de vinaigre), in case of hydrogen sulphide poisoning, the antidote is chlorine (make the patient breathe chloride of lime soaked in vinegar). (43)

contretemps (*n.m.*), contretemps ; unfortunate occurrence ; awkward incident. (44)

contrevenir aux ordres qu'on a reçus, to act contrary to instructions. (45)

contrevenir aux règlements, to contravene (*or* to infringe) (*or* to violate) the regulations (*or* the rules). (46)

contrevent (Constr.) (*n.m.*), wind-brace. (47)

contreventement (*n.m.*), wind-bracing ; bracing against wind-pressure. (48)

contreventer (*v.t.*), to brace against wind-pressure to wind-brace : (49)
contreventer un comble, une ferme, to brace a roof, a truss, against wind-pressure. (50 *ou* 51)

contribuer (*v.t. & v.i.*), to contribute : (52)
contribuer à une dépense, to contribute to an expense. (53)
contribuer pour un tiers, pour un quart, to contribute a third, a quarter ; to contribute one-third, one-quarter. (54 *ou* 55)

contribution (*n.f.*), contribution. (56)

contrôle (*n.m.*), control ; controlling ; checking ; supervision ; inspection : (57)

contrôle à câble, rope control. (58)

contrôle à la recette (Mines), checking the ascent and descent of workmen at the pit-head. (59)

contrôle de présence, timekeeping. (60)

contrôle du débit *ou* contrôle du jaillissement (Puits à pétrole), controlling the outflow. (61)

contrôle numérique des lampes (Mines), checking the lamps (to ascertain whether all the men are out of the mine) ; handing in of lamps before leaving the pit-head. (62)

contrôler (*v.t.*), to control ; to supervise ; to inspect ; to check ; to tally. (63)

contrôleur (pers.) (*n.m.*), controller ; examiner ; inspector ; checker ; tallyman ; time-keeper. (64)

contrôleur (instrument) (*n.m.*), controller ; telltale. (65)

contrôleur (Élec.) (*n.m.*), controller. (66)

contrôleur d'aiguilles (Ch. de f.) (*m.*), switch-controller. (67)

contrôleur d'incendie (*m.*), automatic fire-alarm. (1)

contrôleur de niveau (*m.*), telltale ; telltale pipe. (2)

contrôleur de ronde *ou* **contrôleur de ronde de nuit** (*m.*), watchman's clock ; telltale ; telltale clock ; time-recorder ; time-clock. (3)

contrôleur de vitesse (*m.*), speed-controller ; speed-reducer ; speeder. (4)

convection (*n.f.*), convection. (5)

convenable (*adj.*), suitable ; convenient : (6)
moyens convenables (*m.pl.*), suitable means. (7)

convenablement (*adv.*), suitably ; conveniently ; properly. (8)

convenir aux besoins de quelqu'un, to suit, to serve, someone's purpose. (9)

convenir d'un prix, to agree on a price ; to come to terms. (10)

conventionnel, -elle (*adj.*), conventional : (11)
signe conventionnel (*m.*), conventional sign. (12)

convergence (*n.f.*), convergence ; convergency : (13)
convergence de deux lignes, convergence of two lines. (14)
convergence des rayons réfléchis par un miroir concave, convergence of the rays reflected by a concave mirror. (15)

convergent, -e (*adj.*), converging ; convergent : (16)
rayons convergents (*m.pl.*), converging rays. (17)

convergent (d'un injecteur) (*n.m.*), combining-tube, combining-nozzle, combining-cone, (of an injector). (18)

converger (*v.i.*), to converge : (19)
la plupart des chemins de fer français convergent sur Paris, most of the French railways converge on Paris. (20)
rayons qui convergent vers le foyer (*m.pl.*), rays which converge towards the focus. (21)

conversion (*n.f.*), conversion ; transformation : (22)
conversion du fer en acier, conversion of iron into steel. (23)

convertibilité (*n.f.*), convertibility. (24)

convertir (*v.t.*), to convert ; to transform : (25)
convertir de la glace en eau, to convert ice into water. (26)

convertir (Élec.) (*v.t.*), to convert : (27)
convertir les courants alternatifs en courant continu, to convert alternating currents into direct current. (28)

convertissable (*adj.*), convertible : (29)
le silicate est convertissable en verre, silicate is convertible into glass. (30)

convertisseur (Métall.) (*n.m.*), converter. (31)

convertisseur (Élec.) (*n.m.*), converter. (32)

convertisseur Bessemer (*m.*), Bessemer converter. (33)

convertisseur pour le cuivre en mattes (*m.*), converter for copper matte. (34)

convertisseur tournant (Élec.) (*m.*), rotary converter. (35)

convexe (*adj.*), convex : (36)
lentille convexe (*f.*), convex lens. (37)

convexion (*n.f.*), convection. (38)

convexité (*n.f.*), convexity : (39)
convexité d'une lentille, convexity of a lens. (40)

convexo-concave (*adj.*), convexo-concave. (41)

convexo-convexe (*adj.*), convexo-convex. (42)

convoi (Ch. de f., etc.) (*n.m.*), train. (43)

convoi à marchandises (*m.*), goods-train ; freight-train. (44)

convoi de ballast (*m.*), ballast-train. (45)

convoi de bêtes de somme (*m.*), pack-train. (46)

convoi de wagons (Mines) (*m.*), train, trip, set, of cars. (47)

convoyeur (*n.m.*), conveyer ; conveyor ; trans-porter. (48)   See also **transporteur.**

convoyeur à godets (*m.*), bucket-conveyor. (49)

convoyeur à palettes (*m.*), push-plate conveyor ; flight-conveyor ; scraper-conveyor ; trough-conveyor. (50)

convoyeur à plateaux (*m.*), tray-conveyor. (51)

convoyeur à vis sans fin (*m.*), screw-conveyor ; spiral conveyor. (52)

convoyeur de grumes (*m.*), log-conveyor. (53)

coolie (pers.) (*n.m.*), coolie. (54)

coordonnée (Math.) (*n.f.*), coordinate : (55)
coordonnées d'un point, coordinates of a point. (56)
coordonnées rectilignes, rectilinear coordinates. (57)

copal (*n.m.*) *ou* **copale** (*n.f.*), copal. (58)

copal fossile (*m.*) *ou* **copaline** (*n.f.*) (Minéral), fossil copal ; copalin ; copaline ; copalite ; Highgate resin. (59)

copeau (*n.m.*), shaving ; chip ; cutting. (60)

copeaux de bois (*m.pl.*), wood shavings. (61)

copeaux de métal (*m.pl.*), metal chips ; metal cuttings ; swarf : (62)
la vis mère est située à l'intérieur du banc et à l'abri des copeaux, the lead-screw is placed inside the bed and protected from chips. (63)

copeaux de zinc (*m.pl.*), zinc shavings ; zinc cuttings. (64)

copiapite (Minéral) (*n.f.*), copiapite. (65)

copperasine (Minéral) (*n.m.*), copperasine. (66)

coprolithe *ou* **coprolite** (*n.m.*), coprolite ; copro-lith. (67)

copropriétaire (*n.m. ou f.*), coproprietor ; joint owner ; part-owner. (68)

coque de la drague (*f.*), hull of the dredge ; hull of dredger. (69)

coquillage (trempe en coquille) (Fonderie) (*n.m.*), chilling ; chill-hardening : (70)
le coquillage de la fonte, chilling cast iron. (71)

coquille (*n.f.*), shell. (72)

coquille (d'un palier, ou analogue) (*n.f.*), bush, bearing (of a bearing, or the like) : (73 *ou* 74)
coquilles en bronze, gun-metal bushes ; gun-metal bushings. (75)

coquille (moule pour fabriquer les moulages en fonte trempée) (*n.f.*), chill ; chill-mould. (76)

coquiller (Fonderie) (*v.t.*), to chill ; to chill-harden. (77)

coquilles (d'un moule) (Fonderie) (*n.f.pl.*), parts (of a mold). (78)

coquilleux, -euse *ou* **coquillier, -ère** (*adj.*), con-chiferous ; shelly ; shell (*used as adj.*) : (79)
calcaire coquillier (*m.*), shelly limestone ; shell-limestone. (80)

coquins (Géol.) (*n.m.pl.*), phosphatic nodules. (81)

coracite (Minéral) (*n.f.*), coracite. (82)

corail [coraux *pl.*] (*n.m.*), coral. (83)

corail noir (*m.*), black coral. (84)

corail rouge (*m.*), red coral. (85)

corallien, -enne (*adj.*), coralline ; coral : (86)

calcaire corallien (*m.*), coralline limestone ; coral limestone. (1)

corallifère (*adj.*), coralliferous. (2)

coralliforme (*adj.*), coralliform. (3)

coralligène (*adj.*), coralligenous. (4)

coralligère (*adj.*), coralligerous ; coralliferous. (5)

coralrag (Géol.) (*n.m.*), coral-rag. (6)

corbeau (Arch.) (*n.m.*), corbel. (7)

corbeau (pièce de bois fixée à la maîtresse-tige d'une pompe de mine pour l'arrêter en cas de rupture) (*n.m.*), catch-piece ; bang-piece. (8)

corbeille (*n.f.*), basket. (9)

corbeille de coulée (pour moulage en mottes) (Fonderie) (*f.*), slip-jacket (for snap-mould-ing). (10)

corbeille de retenue de rivets (*f.*), rivet-catcher. (11)

cordage (*n.m.*), rope ; cordage. (12)

cordage blanc (*m.*), white rope. (13)

cordage goudronné *ou* **cordage noir** (*m.*), tarred rope. (14)

corde (*n.f.*), cord ; line ; rope ; cable ; band. (15)

corde (Géom.) (*n.f.*), chord : (16)

la corde d'un arc, the chord of an arc. (17)

corde à boutons (transporteur aérien à voie unique) (*f.*), button-rope (cableway). (18)

corde à boyau *ou* **corde en boyau de chat** (*f.*), gut band. (19). The plural **cordes en boyaux** may be translated as **gut bands** or **gut banding**.

corde-courroie [**cordes-courroies** *pl.*] (*n.f.*), rope belt. (20)

corde d'amiante (*f.*), asbestos plaited packing. (21)

corde de curage (Sondage des puits à pétrole) (*f.*), sand-line. (22)

corde de fil métallique (*f.*), telegraph-wire. (23)

corde de levage (d'un transporteur aérien à voie unique) (*f.*), hoisting-rope, fall-rope, (of a cableway). (24)

corde de manœuvre (d'un palan, d'un monte-charge, ou analogue) (*f.*), hand-rope (of a pulley-block, hoist, or the like). (25 *ou* 26 *ou* 27)

corde de piano (*f.*), piano-wire. (28)

corde de tension (*f.*), stretching-cord ; tighten-ing-cord. (29)

corde de tirage (*f.*), pull-rope. (30)

corde en cuir (pour machines-outils) (*f.*), round leather band, round leather belt, (for machine-tools). (31)

Note.—The plural **cordes en cuir** can be translated, round leather banding or round leather belting.

corde en foin (Fonderie) (*f.*), hay rope ; hay band. (32)

corde en paille (*f.*), straw rope ; straw band. (33)

corde sans fin (*f.*), endless cord ; endless line. (34)

corde-signal [**cordes-signaux** *pl.*] (*n.f.*), signal-cord ; signal-rope ; bell-cord. (35)

cordeau (*n.m.*), line ; string. (36)

cordeau (*n.m.*) *ou* **cordeau blanchi à la craie**, chalk-line. (37)

cordeau Bickford (*m.*), Bickford fuse ; common fuse ; safety-fuse. (38)

cordeau détonant (*m.*), detonating-fuse. (39)

cordée (Mines) (*n.f.*), wind ; hoist ; lift ; journey ; trip : (40)

arrêt en fin de cordée (*m.*), stopping at the end of the wind. (41)

cordiérite (Minéral) (*n.f.*), cordierite. (42)

cordon (fausse cannelure) (Laminage) (*n.m.*), false pass. (43)

cordon (Géol. & Mines) (*n.m.*), string ; stringer : (44)

un cordon de nodules, a string of nodules. (45)

cordon de quartz, quartz stringer. (46)

cordon littoral (Géogr. phys.) (*m.*), offshore bar : (47)

le cordon littoral exerce le rôle protecteur d'un brise-lames, the offshore bar performs the protective function of a breakwater. (48)

corindon (Minéral) (*n.m.*), corundum ; diamond-spar. (49)

corindon adamantin (*m.*), adamantine spar. (50)

cornaline (Minéral) (*n.f.*), cornelian ; carnelian. (51)

cornbrash (Géol.) (*n.m.*), Cornbrash : (52)

le cornbrash, the Cornbrash. (53)

corne (*n.f.*), horn : (54)

spatule en corne (*f.*), horn spatula. (55)

corne polaire (dynamo) (*f.*), pole-horn ; pole-tip. (56)

corné, -e (Pétrol.) (*adj.*), horny : (57)

une pierre cornée, a horny stone. (58)

orthose corné (*m.*), horny orthoclase. (59)

cornet d'essai (*m.*), cornet ; cornette. (60)

corniche (Arch.) (*n.f.*), cornice. (61)

corniche (de neige) (*n.f.*), cornice (of snow). (62)

corniche (de rocher) (*n.f.*), ledge (of rock). (63)

cornier, -ère (*adj.*), corner (*used as adj.*) : (64)

poteau cornier (*m.*), corner-post. (65)

tuile cornière (*f.*), corner-tile. (66)

cornière (*n.f.*) *ou* **cornières** (*n.f.pl.*), angle-iron ; L iron ; angles. (67)

cornières à angles vifs (*f.pl.*), square root and edge angles. (68)

cornières égales *ou* **cornières à branches égales** *ou* **cornières à ailes égales** (*f.pl.*), even-sided angles ; equal-sided angles. (69)

cornières inégales *ou* **cornières à branches inégales** *ou* **cornières à ailes inégales** (*f.pl.*), uneven-sided angles ; unequal-sided angles. (70)

cornish-stone (*n.f.*), Cornish stone. (71)

cornue (*n.f.*), retort ; still. (72)

cornue à gaz (*f.*), gas-retort. (73)

cornue à goudron (*f.*), tar-still. (74)

cornue à schiste (*f.*), shale-retort. (75)

cornue de distillation d'amalgame (*f.*), gold-retort ; pot retort. (76)

cornue en verre (*f.*), glass retort. (77)

corocoro (Métall.) (*n.m.*), corocoro. (78)

coron (Houillères) (*n.m.*), mining village. (79)

corps (solidité ; consistance) (*n.m.*), body : (80)

métal sans corps (*m.*), metal without body. (81)

corps (substance) (*n.m.*), body ; substance. (82)

corps (d'un marteau) (tête ou masse métallique) (opp. au **manche**) (*n.m.*), head (of a hammer) (83) (in contradistinction to the **handle**).

corps (d'un pointeau) (*n.m.*), barrel (of a centre-punch). (84)

corps composé (Chim.) (*m.*), compound : (85)

l'eau est un corps composé, water is a com-pound. (86)

corps cylindrique (d'une chaudière) (*m.*), barrel, shell, (of a boiler). (87)

corps d'ambulanciers (*m.*), ambulance corps. (88)

corps d'arrière (de chambre noire) (*m.*), back (of camera). (1)

corps d'avant (de chambre noire) (*m.*), front. (2)

corps d'essieu (*m.*), axle-centre. (3)

corps d'inspecteurs (*m.*), body of inspectors ; inspectorate. (4)

corps de bielle (*m.*), rod-body (connecting-rod). (5)

corps de châssis *ou simplement* corps (*n.m.*) (Fonderie), drag ; nowel ; bottom part of flask. (6)

corps de cylindre (*m.*), cylinder-barrel. (7)

corps de manivelle (*m.*), crank-web ; crank-arm ; crank-cheek. (8)

corps de minerai (*m.*), body of ore ; ore-body. (9)

corps de palier (*m.*), pillow of plummer-block. (10)

corps de piston (*m.*), piston-head ; piston. (11)

corps de pompe (*m.*), pump-barrel ; pump-cylinder. (12)

corps de pompiers (*m.*), fire-brigade ; fire-fighting corps. (13)

corps de roue (*m.*), wheel-centre. (14)

corps de roue à centre plein (*m.*), disc wheel-centre. (15)

corps de roue à rayons (*m.*), spoke wheel-centre. (16)

corps de roue à toile (*m.*), plate wheel-centre. (17)

corps de sauvetage (*m.*), rescue-corps. (18)

corps de sonde (*m.*), rods (the assemblage of rods in the bore-hole, including the top rod and the jar). (19)

corps diamagnétique (*m.*), diamagnetic substance ; diamagnetic ; diamagnet : (20)
le bismuth est le meilleur exemple d'un corps diamagnétique, bismuth is the best example of a diamagnetic. (21)

corps diaphane (Opt.) (*m.*), diaphanous body ; transparent body. (22)

corps éclairé (opp. à corps lumineux) (Opt.) (*m.*), illuminated body. (23)

corps en mouvement (*m.*), body in motion ; moving body. (24)

corps en repos (*m.*), body at rest : (25)
la statique est la science des corps en repos, statics is the science of bodies at rest. (26)

corps étranger (*m.*), foreign substance. (27)

corps flottants (*m.pl.*), floating bodies. (28)

corps graves (Phys.) (*m.pl.*), heavy bodies. (29)

corps lumineux (Opt.) (*m.*), luminous body. (30)

corps magnétique (*m.*), magnetic substance ; magnetic. (31)

corps noir (Phys.) (*m.*), black body ; ideal black body. (32)

corps paramagnétique (*m.*), paramagnetic substance ; paramagnetic ; paramagnet. (33)

corps simple (Chim.) (*m.*), element : (34)
l'hydrogène est un corps simple (*m.*), hydrogen is an element. (35)

corps simple métallique (*m.*), metallic element. (36)

corps translucide (Opt.) (*m.*), translucent body. (37)

corps transparent (Opt.) (*m.*), transparent body ; diaphanous body. (38)

corrasion (Géol.) (*n.f.*), corrasion ; erosion. (39)

correct, -e (*adj.*), correct. (40)

correctement (*adv.*), correctly. (41)

correction (*n.f.*), correction : (42)

correction de l'astigmatisme, des aberrations chromatiques (Opt.), correction for astigma-tism, for chromatic aberrations. (43 *ou* 44)

correction de la capillarité (Barométrie), cor-rection for capillarity. (45)

correction de la température (Barométrie), correction for temperature. (46)

correction de niveau (Barométrie), correction for level. (47)

corrélation (*n.f.*), correlation : (48)
la corrélation qui existe entre les effondrements et les phénomènes volcaniques, entre les mouvements orogéniques et les tremble-ments de terre, the correlation which exists between subsidences and volcanic phenomena, between orogenic movements and earthquakes. (49 *ou* 50)

correspondance (appareil de correspondance électrique) (Signaux de ch. de f.) (*n.f.*), annunciator. (51)

correspondance à guichets (*f.*), drop-annuncia-tor. (52)

corridor (*n.m.*), corridor. (53)

corriger (*v.t.*), to correct : (54)
corriger la différence, moitié par les vis calantes, moitié par les vis de réglage de la bulle (mode de réglage d'un niveau), to correct half the error with the foot-screws, and the other half with the bubble-nuts. (55)

corroborer (*v.t.*), to corroborate : (56)
corroborer les assertions de quelqu'un, to corroborate somebody's statements. (57)

corrodant, -e (*adj.*), corroding ; corrosive. (58)

corrodant (*n.m.*), corrodent ; corrosive. (59)

corroder (*v.t.*), to corrode : (60)
les acides corrodent les métaux (*m.pl.*), acids corrode metals. (61)

corroder (Géol.) (*v.t.*), to corrade ; to erode : (62)
les rivières corrodent leur lits (*f.pl.*), rivers corrade their beds. (63)

corroder (se) (*v.r.*), to corrode. (64)

corrosif, -ive (*adj.*), corrosive ; corroding. (65)

corrosif (*n.m.*), corrosive ; corrodent. (66)

corrosion (*n.f.*), corrosion : (67)
la corrosion des tubes de chaudière, de tubage, the corrosion of boiler-tubes, of casing. (68 *ou* 69)

corrosion (Géol.) (*n.f.*), corrasion ; erosion : (70)
les berges des fleuves sont soumises à des corrosions dues au glissement continuel des eaux contre leur parois, river-banks are subjected to erosion due to the continual sliding of the water against their faces. (71)

corrosivement (*adv.*), corrosively. (72)

corrosiveté (*n.f.*), corrosiveness ; corrodibility. (73)

corroyage (Métall.) (*n.m.*), welding : (74)
corroyage de l'acier, du fer puddlé, welding steel, puddled iron. (75 *ou* 76)

corroyage (Menuis.) (*n.m.*), trimming ; dressing. (77)

corroyer (Métall.) (*v.t.*), to weld. (78)

corroyer (Menuis.) (*v.t.*), to trim ; to dress : (79)
corroyer du bois, to trim wood. (80)

corroyer (se) (Métall.) (*v.r.*), to weld. (81)

corsite (Pétrol.) (*n.f.*), corsite. (82)

corundellite (Minéral) (*n.f.*), corundellite. (83)

corundophyllite *ou* corundophilite (Minéral) (*n.f.*), corundophilite. (84)

corynite (Minéral) (*n.f.*), corynite. (85)

cosalite (Minéral) (*n.f.*), cosalite. (86)

cosinus (Math.) (*n.m.*), cosine. (87)

cosismal, -e, -aux *ou* cosiste (*adj.*), coseismal ; coseismic ; homoseismal ; homoseismic. (1)

cosiste (*n.f.*), coseismal ; homoseismal. (2)

cosse (d'une ardoisière) (*n.f.*), top, overlay, (of a slate-quarry). (3)

cosse (d'un câble) (*n.f.*), thimble, eyelet, (of a rope). (4)

cosse de l'estrope (câble) (*f.*), eye of the loop. (5)

cosse du câble (*f.*), rope-thimble ; eye of the rope. (6)

cosse ovale (*f.*), egg-shaped thimble. (7)

cossyrite (Minéral) (*n.f.*), cossyrite. (8)

costière (Mines) (*n.f.*), drift ; drifting-level ; driftway ; drive (parallel in direction to the strike) ; level running along the strike. (9)

costresse (Mines) (*n.f.*), subdrift ; counter-level ; counter-gangway ; counter. (10)

cote (Géod.) (*n.f.*), reading : (11)
carnet qui contient les cotes lues sur la mire (*m.*), note-book which contains the rod-readings. (12)
consigner les cotes sur des carnets de nivelle--ment, to record the readings in levelling-books. (13)
surface qui est — mètres au-dessous de la cote zéro (*f.*), surface which is — metres below the zero reading. (14)
cote de profondeur, depth-reading. (15)

coté, -e (*adj.*), showing dimensions or values ; dimensioned : (16)
croquis coté (*m.*), dimensioned sketch. (17)

côte (penchant) (*n.f.*), slope : (18)
monter une côte, to climb a slope. (19)

côte (Marine) (*s'emploie souvent au pluriel*) (*n.f.*), coast ; seacoast ; shore ; seaboard ; coast-line : (20)
les côtes d'Angleterre, the English coast. (21)
les côtes de l'Atlantique, the Atlantic sea--board. (22)

côte (escarpement de faille) (Géol.) (*n.f.*), fault-escarpment ; fault-scarp. (23)

côte à côte, side by side : (24)
cylindres placés côte à côte (*m.pl.*), side-by-side cylinders. (25) (opp. à cylindres en tandem, opposed to tandem cylinders).

côté (*n.m.*), side : (26)
le côté d'une boîte, d'une route, the side of a box, of a road. (27 *ou* 28)

côté (Géom.) (*n.m.*), side : (29)
le côté d'un carré, d'un triangle, the side of a square, of a triangle. (30 *ou* 31)

côté chair (d'une courroie) (*m.*), flesh side (of a belt). (32)

côté cuir (d'une courroie) (*m.*), grain side (of a belt). (33)

coteau (versant d'une colline) (*n.m.*), hillside ; slope. (34)

coteau (colline) (*n.m.*), hill. (35)

coter (*v.t.*), to show dimensions or values ; to dimension. (36) See coté, -e for example.

coticule (Minéral) (*n.f.*), novaculite ; novaculyte ; Turkey stone ; Turkey slate. (37)

côtier, -ère (*adj.*), coastal : (38)
fleuve côtier (*m.*), coastal river. (39)

coton (*n.m.*), cotton. (40)

coton-collodion (Explosif) (*n.m.*), pyroxyline ; pyroxyle. (41)

coton minéral (*m.*), mineral wool ; mineral cotton ; rock-wool ; slag-wool ; cinder-wool. (42)

coton nitré (*m.*) *ou* coton-poudre (*n.m.*), nitro-cotton ; guncotton. (43)

cotunnite (Minéral) (*n.f.*), cotunnite. (44)

couchant (*n.m.*), west. (45)

couche (substance appliquée sur une autre) (*n.f.*), bed ; layer ; sheet : (46)
une couche de béton, a bed of concrete. (47)
une couche de poussière, a layer of dust. (48)
une couche de glace, a sheet of ice. (49)

couche (de caoutchouc, de caoutchouc para, de caoutchouc vulcanisé) (Isolement des fils électriques) (*n.f.*), lap, lapping (of rubber, of Para rubber, of vulcanized rubber). (50 *ou* 51 *ou* 52)

couche (enduit) (*n.f.*), coat ; coating ; covering : (53)
une couche de peinture, a coat of paint. (54)
une couche infinitésimale d'or, an infinitesimal coating of gold. (55)

couche (pièce de bois qui se place sous le pied d'un étai pour former empattement) (Constr.) (*n.f.*), footing-block ; sole-piece. (56)

couche (Géol.) (*n.f.*), bed ; seam ; layer ; stratum ; deposit. (57) See also gîte, gisement and lit :
couches à gaz, gas-strata. (58)
couches aquifères, water-bearing strata. (59)
couche d'argile, bed of clay ; clay-seam. (60)
couche de faible pendage, slightly dipping seam. (61)
couche de fort pendage, highly inclined seam. (62)
couche de houille, coal-bed ; coal-seam. (63)
couche de moyenne puissance, seam of medium thickness. (64)
couche de pendage modéré, seam of moderate dip. (65)
couche de terre, de gravier, bed, layer, of earth, of gravel. (66 *ou* 67)
couche de terres de couverture, layer of overburden. (68)
couche en plateure, flat seam ; slightly dipping seam. (69)
Note.—So-called flat seams usually have a more or less decided dip, often rising to 3 %, in one direction or another.
couche grisouteuse, fiery seam ; gassy seam. (70)
couche inexploitable, unworkable seam. (71)
couche inférieure, substratum ; understra--tum ; lower bed. (72)
couche lacustre, lacustrine deposit ; lacustrine bed. (73)
couche mince, thin seam. (74)
couche pétrolifère, oil-bearing strata ; oil--strata. (75)
couche puissante, thick seam. (76)
couche salifère, bed of salt ; salt-bed. (77)
couche sous-jacente *ou* couche subjacente, subjacent bed ; underlying seam ; under-layer. (78)
couche supérieure, superstratum ; upper bed. (79)
couche surjacente *ou* couche susjacente, over--lying bed. (80)
couches traversées (Sondage ; Fonçage), strata passed through. (81)

couche d'isolement *ou* couche isolante imperméable (Constr.) (*f.*), damp-proof course. (82)

couche d'impression (Peinture) (*f.*), priming-coat. (83)

couché, -e (opp. à droit, -e *ou* debout) (Géol.) (*adj.*), recumbent. (84) (opp. to erect) :
pli couché (*m.*), recumbent fold. (85)

couchis (pièce de bois appuyée contre un mur, qui reçoit la tête d'une contre-fiche, ou sur laquelle vient buter un étrésillon) (Constr.) (*n.m.*), wall-piece. (1)

couchis (pièce de bois qui se place sous le pied d'une contre-fiche, d'un étai, pour former empattement) (Constr.) (*n.m.*), footing-block ; sole-piece. (2)

coude (*n.m.*), elbow ; bend ; crank. (3)

coude (d'une rivière, d'une route) (*n.m.*), bend, elbow, (of a river, of a road). (4 *ou* 5)

coude (bout de tuyau coudé)(*n.m.*), elbow ; bend : (6)

un coude en fonte, a cast-iron elbow (*or* bend). (7)

coude circulaire (*m.*), disc crank ; wheel crank. (8)

coude d'équerre (*m.*), right-angled bend. (9)

coude d'essieu (*m.*), axle-crank. (10)

coude de raccord (*m.*), union-elbow. (11)

coude du porte-vent (haut fourneau) (*m.*), goose-neck ; leg-pipe ; tuyère-stock (blast-furnace). (12)

couder (*v.t.*), to bend ; to crank : (13)

couder une barre de fer, to bend a bar of iron. (14)

couder un essieu, to crank an axle. (15)

couder (se) (*v.r.*), to bend ; to elbow. (16)

couffe à charbon (*f.*), coal-basket. (17)

coufflée (Géol. & Mines) (*n.f.*), throw ; jump ; leap ; check. (18)

coulage (action de couler, en parlant d'une matière en fusion ou inconsistante) (*n.m.*), running : (19)

le coulage d'une bougie, the running of a candle. (20)

boisage qui s'oppose au coulage d'un sol ébouleux (*m.*), timbering which resists the running of loose ground. (21)

coulage (perte d'un liquide, etc.) (*n.m.*), leakage : (22)

un coulage d'eau, a leakage of water. (23)

coulage (échappement de métal fondu d'un moule) (Fonderie) (*n.m.*), run-out ; break-out. (24)

coulage (action de jeter un métal en fusion dans un moule) (*n.m.*), pouring ; teeming ; casting : (25)

coulage à noyau, core-casting. (26)

coulage plein, solid casting. (27)

coulant, -e (*adj.*), flowing ; running : (28)

un ruisseau coulant, a running brook. (29)

coulant (anneau mobile ; curseur) (*n.m.*), runner ; slider. (30)

coulant (tube à tirage d'un instrument d'optique) (*n.m.*), draw-slide ; draw-tube ; sliding-tube. (31)

coulants (*n.m.pl.*) *ou* **coulantage** (*n.m.*) (guides de puits de mine), guides ; slides; cage-guides; pit-guides. (32)

coulé (ouvrage jeté en moule) (*n.m.*), casting. (33)

coulée (flot de matière) (*n.f.*), flow ; stream ; run : (34)

coulée de lave *ou* coulée de laves *ou* coulée lavique, flow of lava ; stream of lava ; lava-flow ; lava-stream ; coulée. (35)

coulée boueuse *ou* coulée de boue (Géol.), mud-flow ; mud-flood. (36)

une coulée de terres, a run of ground. (37)

coulée (action de jeter dans un moule) (*n.f.*), pouring ; teeming ; casting ; running : (38)

coulée des moules, pouring, casting, running, moulds. (39)

coulée d'un lingot, casting, teeming, an ingot. (40)

coulée des gueuses en halle, casting pigs under cover. (41)

coulée en première fusion, direct casting. (42)

coulée à la descente *ou* coulée en chute directe, top pouring ; top casting. (43)

coulée à talon, side pouring ; side casting. (44)

coulée en source *ou* coulée en source directe, bottom pouring ; bottom casting ; bottom-fed casting. (45)

coulée (trou de coulée dans un moule) (Fonderie) (*n.f.*), runner ; gate ; git ; geat ; sprue ; cast-gate ; pouring-gate ; running-gate ; pouring-hole ; down-runner ; down-gate ; upright runner ; sprue-hole ; funnel ; jet : (46)

pratiquer une coulée dans un moule, to make a runner in a mould. (47)

coulée (partie métallique qui reste attenante après la pièce coulée) (*n.f.*), gate ; git ; geat ; sprue ; header ; runner ; jet : (48)

la coupe des coulées et bavures, cutting off the gates and fins. (49)

coulée (broche de coulée) (Fonderie) (*n.f.*), runner-pin ; runner-stick ; gate-pin ; gate-stick ; sprue : (50)

retirer les coulées, to withdraw the runner-pins. (51)

coulée (action de faire une prise à un fourneau) (*n.f.*), tapping ; running off : (52)

coulée du métal en fusion d'un haut fourneau, tapping, running off, the molten metal from a blast-furnace. (53)

la coulée de la fonte, tapping the metal (cast iron). (54)

coulée du laitier, flushing, running off, the slag. (55)

coulée (quantité de métal qui échappe à une seule opération) (*n.f.*), tap ; cast : (56)

une coulée de — kilogrammes de métal, a tap (*or* a cast) of — kilogrammes of metal. (57)

coulée (trou de coulée d'un fourneau) (*n.f.*), tap-hole ; tapping-hole ; draw-hole ; mouth. (58)

coulée (d'un creuset) (*n.f.*), lip (of a crucible). (59)

coulée d'éboulis mouvants (Géol.) (*f.*), rock-glacier ; talus-glacier. (60)

coulée de minerai (colonne de richesse) (*f.*), shoot of ore ; ore-shoot ; ore-chute ; ore-shute ; chimney of ore ; ore-chimney ; pipe of ore ; pipe-vein. (61)

coulée sur métal (soudure par la fonte liquide) (*f.*), burning ; casting on. (62)

couler (faire écouler) (*v.t.*), to run off : (63)

couler l'eau d'un réservoir, to run off the water from a tank. (64)

couler (jeter en moule) (*v.t.*), to pour ; to teem ; to cast ; to run : (65)

couler du métal en fusion dans un moule, to pour molten metal (*or* to run melted metal) into a mold. (66)

couler un lingot, to cast, to teem, an ingot. (67)

on coulait pleins autrefois les cylindres de laminoirs, rolling-mill rolls used to be cast solid. (68)

une roue coulée d'une seule pièce, a wheel cast in one piece. (69)

couler une couche de bitume autour d'un tuyau, to run a coating of pitch round a pipe. (1)

couler à découvert, to cast in open sand. (2)

couler à vert, to cast in green sand. (3)

couler (faire une prise à un fourneau) (*v.t.*), to tap ; to run off : (4)

couler le métal d'un four, to tap, to run off, the metal from a furnace. (5)

couler le laitier d'un haut fourneau, to flush, to run off, the slag from a blast-furnace. (6)

couler (immerger) (*v.t.*), to sink : (7)

couler un bateau, to sink a boat. (8)

couler (*v.i.*), to flow ; to run : (9)

fleuve qui coule dans la mer (*m.*), river which flows (*or* runs) into the sea. (10)

rivière qui coule au travers d'une propriété (*f.*), river which runs through a property. (11)

puits qui coule lentement (*m.*), well which flows slowly. (12)

couler (glisser) (se dit du terrain) (Mines) (*v.i.*), to run. (13)

couler (fondre et s'épancher) (*v.i.*), to run : (14)

bougie qui coule (*f.*), candle which runs. (15)

couler (s'échapper du moule par quelque fente) (*v.i.*), to run ; to break out : (16)

cette pièce a coulé, this casting has run (*or* broken out). (17)

couler (laisser échapper) (*v.i.*), to leak : (18)

cuve qui coule (*f.*), vat which leaks. (19)

couler (s'avancer sans effort) (*v.i.*), to work : (20)

outil qui coule bien (*m.*), tool which works well. (21)

piston qui ne coule pas (*m.*), piston which does not work. (22)

couler (glisser) (*v.i.*), to slide ; to slip : (23)

se laisser couler le long d'une corde, to slide down a rope. (24)

couler (se) (être jeté en moule) (*v.r.*), to be poured ; to be cast : (25)

les tuyaux de fonte se coulent de trois manières : horizontalement, en plan incliné, et debout (*m.pl.*), iron pipes are cast in three ways : on the flat, on the bank, and on end. (26)

couler bas *ou* couler à fond (*v.i.*), to sink : (27)

bateau qui coule bas (*m.*), boat which sinks. (28)

couleur (Fonderie) (pers.) (*n.m.*), pourer ; teemer ; caster. (29)

couleur (*n.f.*), colour ; color. (30)

couleur (de), coloured ; colored ; of colour : (31)

pierre de couleur (*f.*), coloured stone. (32)

personne de couleur (*f.*), coloured person ; person of colour. (33)

couleur (trace d'or dans la batée) (*n.f.*), colour ; color. (34)

couleur (peinture) (*n.f.*), paint ; colour ; color : (35)

couleur minérale, mineral paint. (36)

couleur claire *ou* couleur pâle (*f.*), light colour ; pale colour. (37)

couleur de la poussière (Minéralogie) (*f.*), streak. (38)

couleur de recuit *ou* couleur du revient (Métall.) (*f.*), tempering-colour. (39)

couleur éclatante *ou* couleur vive (*f.*), bright colour. (40)

couleur faible (or) (*f.*), weak colour (gold). (41)

couleur foncée (*f.*), dark colour. (42)

couleur forte (or) (*f.*), strong colour (gold). (43)

couleurs primitives (*f.pl.*), primary colours ; primitive colours ; fundamental colours. (44)

couleurs primitives du spectre (violet, indigo, bleu, vert, jaune, orangé, rouge) (*f.pl.*), primitive colours of the spectrum (violet, indigo, blue, green, yellow, orange, red). (45)

couleurs prismatiques (*f.pl.*), prismatic colours. (46)

couleurs spectrales (*f.pl.*), spectral colours. (47)

coulissant, -e (*adj.*) *ou* coulisse (à), sliding ; slide (*used as adj.*) : (48)

banc coulissant (*m.*), sliding bench. (49)

pied à coulisse (*m.*), sliding caliper-gauge ; slide-calipers. (50)

coulisse (pièce mobile qui glisse dans une rainure, etc.) (*n.f.*), slide ; slider. (51)

coulisse (rainure dans laquelle on fait glisser une pièce mobile) (*n.f.*), slideway : (52)

les coulisses d'un tiroir, the slideways of a slide-valve. (53)

coulisse (*la coulisse proprement dite*—organe d'une distribution par coulisse) (*n.f.*), link : (54)

la coulisse de Stephenson, the Stephenson link. (55)

la coulisse peut être courbe ou droite, mobile ou fixe, the link can be curved or straight, shifting or stationary. (56)

coulisse (*par ellipse pour* distribution par coulisse—appareil de changement de marche et de variation de la détente) (*n.f.*), link-motion ; link-gear ; valve-gear ; valve motion ; link valve-motion. (57) See also distribution.

coulisse (de sonde) (*n.f.*), jar ; drill-jar. (58)

coulisse à barres croisées *ou* coulisse à barres fermées (*f.*), link-motion with crossed rods. (59)

coulisse à barres ouvertes *ou* coulisse à barres droites (*f.*), link-motion with open rods. (60)

coulisse à crémaillère pour la mise au point (Photogr.) (*f.*), focussing-rack. (61)

coulisse à flasques (distribution) (*f.*), plate-link. (62)

coulisse à river (*f.*), riveting-slide. (63)

coulisse d'Allan (*f.*), Allan's link-motion. (64)

coulisse de changement de marche (*f.*), reversing-link ; reverse-link. (65)

coulisse de Gooch *ou* coulisse renversée *ou* coulisse retournée (*f.*), Gooch's link-motion. (66)

coulisse de pêchage (Sondage) (*f.*), fishing-jar. (67)

coulisse de Stephenson (*f.*), Stephenson's link-motion. (68)

coulisse découpée (distribution) (*f.*), slotted link. (69)

coulisse fixe (*f.*), stationary link. (70)

coulisse Joy (*f.*), Joy's valve-gear. (71)

coulisse mobile (*f.*), shifting link. (72)

coulisse Walschaerts (*f.*), Walschaerts valve-gear. (73)

coulisseau (petite coulisse) (*n.m.*), slide ; slider. (74)

coulisseau (d'une distribution à coulisse) (*n.m.*), link-block ; slide-block ; die, (of a link-motion). (75)

coulisseau (de crosse de piston) (*n.m.*), slipper-block ; block ; body, (of piston cross-head). (76)

coulisseau porte-outil (machine à mortaiser, etc.) (*m.*), tool-slide ; tool-holding slide ; tool-ram (slotting-machine, etc.). (77)

coulisser (*v.t.*), to provide with slides. (78)

coulisser (*v.i.*), to slide : (79)

étau qui coulisse entre deux barres parallèles (*m.*), vice which slides between two parallel bars. (1)

curseur qui coulisse le long d'une tige (*m.*), runner which slides along a rod. (2)

couloir (*n.m.*), corridor ; passage ; passageway : (3)

couloirs d'un édifice, corridors, passages, of a building. (4)

wagon à couloir (Ch. de f.) (*m.*), corridor-carriage. (5)

couloir (Géol.) (*n.m.*), couloir. (6)

couloir (Mines) (*n.m.*), chute ; shoot ; shute. (7)

couloir à charbon (*m.*), coal-chute ; coal-shoot. (8)

couloir à minerai (*m.*), ore-chute ; ore-shoot. (9)

couloir de chargement (*m.*), loading-chute. (10)

couloir en tôle d'acier (*m.*), steel chute. (11)

couloir oscillant (*m.*), swinging chute. (12)

coulomb (Élec.) (*n.m.*), coulomb. (13)

coulombmètre (Élec.) (*n.m.*), coulomb-meter ; coulometer. (14)

coulure (Fonderie) (*n.f.*), run-out ; breakout. (15)

coup (choc donné par un corps en mouvement) (*n.m.*), blow. (16)

coup (chacun des mouvements d'un corps qui doivent se répéter) (*n.m.*), stroke : (17)

un coup de piston dans une pompe, a stroke of the piston in a pump. (18)

coup (*n.m.*) *ou* **coup de lunette** (Levés & Nivelle--ment), sight ; observation : (19)

cote de chaque coup de lunette (*f.*), reading of each sight (*or* observation). (20)

coup arrière [coups arrière *pl.*] (Nivellement) (*m.*), backsight ; back observation ; + sight ; plus sight. (21)

coup avant [coups avant *pl.*] (Nivellement) (*m.*), foresight ; fore observation ; — sight ; minus sight. (22)

coup d'air (*m.*), rush of air : (23)

coups d'air résultant de l'allumage des gaz lors de la coulée, rushes of air resulting from the ignition of the gases when casting. (24)

coup d'eau (*m.*), rush of water ; inrush of water ; water-inflow ; outburst of water : (25)

la résistance d'un barrage aux coups d'eau qu'il aura à supporter par suite des inonda--tions subites, the resistance of a dam to the rushes of water that it will have to sustain consequent on sudden floods. (26)

coup d'échappement (*m.*), blast of the exhaust. (27)

coup d'empiétage (Travail aux explosifs) (*m.*), centre-blast. (28)

coup de bélier (*m.*), water-hammering ; water-hammer : (29)

coups de bélier dans les conduites forcées, water-hammering in full pipes. (30)

coup de feu (fusion de minerai, etc.) (*m.*), hot blast : (31)

donner un coup de feu sur le minerai grillé, to pass a hot blast over the roasted ore. (32)

coup de feu (Mines) (*m.*), explosion (of fire-damp or coal-dust). (33 *ou* 34)

coup de feu à une chaudière (*m.*), local over--heating of a boiler. (35)

coup de fortune (*m.*), stroke of fortune. (36)

coup de foudre (*m.*), thunderbolt ; stroke of lightning. (37)

coup de fouet (du câble d'extraction) (Mines) (*m.*), whip, whipping, flapping, flopping, surging, (of the winding-rope). (38)

coup de fouet (au commencement de la décharge d'un accumulateur) (Élec.) (*m.*), rapid drop in voltage. (39)

coup de frein (*m.*), sudden application of the brake. (40)

coup de froid (Météor.) (*m.*), cold-snap. (41)

coup de grisou (*m.*), fire-damp explosion ; gas-explosion. (42)

coup de lunette (*m.*). See **coup** *ou* **coup de lunette**.

coup de mine *ou simplement* **coup** (*n.m.*), shot ; blast. (43)

coup de mine ayant débourré *ou* **coup débourré** *ou* **coup de mine ayant fait canon** *ou* **coup de mine qui fait canon** (*m.*), blown-out shot. (44)

coup de molette (Mines) (*m.*). Same as **coup de fouet**.

coup de piston (*m.*), stroke of piston ; piston-stroke. (45)

coup-de-poing [coups-de-poing *pl.*] (graisseur) (*n.m.*), hand-pump lubricator. (46)

coup de pointeau (*m.*), punch-mark ; centre-punch mark ; centre-pop : (47)

repérer les centres par des coups de pointeau, to mark the centres by punch-marks. (48)

coup de poussières (Mines) (*m.*), coal-dust explosion ; dust-explosion. (49)

coup de relevage (Travail aux explosifs) (*m.*), bottom-shot. (50)

coup de sifflet (*m.*), blast of a whistle. (51)

coup de soleil (Pathologie) (*m.*), sunstroke ; insolation. (52)

coup de sonde (*m.*), bore-hole. (53)

coup de timbre (*m.*), stroke of bell. (54)

coup de vent (*m.*), gust of wind ; squall ; gale. (55)

coup raté (Travail aux explosifs) (*m.*), miss-fire shot. (56)

coup sec (*m.*), dead blow. (57)

coupage (*n.m.*), cutting ; cutting off : (58)

coupage des jets de coulée (Fonderie), cutting off gates. (59)

coupage à la patte du câble (Mines), cutting off the end of the rope. (60)

coupant, -e (*adj.*), cutting ; edge (*used as adj.*) : (61)

outils coupants (*m.pl.*), cutting-tools ; edge-tools. (62)

coupant (*n.m.*), cutting edge ; edge : (63)

le coupant d'un couteau, the edge of a knife. (64)

coupe (action de couper) (*n.f.*), cutting : (65)

la coupe des métaux par le jet d'oxygène, cutting metals by the oxygen jet. (66)

coupe (entaille) (*n.f.*), cut : (67)

coupe en travers, cross-cut. (68)

profondeur de coupe (*f.*) (machines-outils), depth of cut. (69)

coupe (havage) (Exploitation du charbon) (*n.f.*), cut ; cutting. (70)

coupe (représentation graphique) (*n.f.*), section : (71)

coupe d'un puits de mine, section of a mine-shaft. (72)

la machine figurée en coupe, the machine shown in section. (73)

coupe (équipe d'ouvriers) (*n.f.*), shift ; gang. (74)

coupe à charbon (opp. à **coupe à terre**) (Houillères) (*f.*), working-shift. (75)

coupe à mi-bois (Charp.) (*f.*), halving ; halved joint ; lap-joint ; half-lap ; overlap-joint ; step-joint. (76)

coupe à terre (Houillères) (*f.*), repairing-shift. (1)

coupe-boulons [coupe-boulons *pl.*] (*n.m.*), bolt-cropper; bolt-clipper; bolt-cutter. (2)

coupe-cercle [coupe-cercle *ou* coupe-cercles *pl.*] (*n.m.*), washer-cutter. (3)

coupe-circuit [coupe-circuit *pl.*] (Élec.) (*n.m.*), cut-out. (4)

coupe-circuit à fusible (Élec.) (*m.*), fuse. (5)

coupe-coulées [coupe-coulées *pl.*] (Fonderie) (*n.m.*), gate-cutter; git-cutter; sprue-cutter. (6)

coupe en long (*f.*), lengthwise section; lon-gitudinal section. (7)

coupe en travers (*f.*), cross-section; transverse section. (8)

coupe-épreuves [coupe-épreuves *pl.*] (Photogr.) (*n.m.*), print-trimmer. (9)

coupe-épreuves à molette (*m.*), wheel print-trim-mer. (10)

coupe géologique (*f.*), geological section. (11)

coupe horizontale (*f.*), horizontal section. (12)

coupe-net [coupe-net *pl.*] (*n.m.*), wire-netting cutter. (13)

coupe oblique (*f.*), oblique section. (14)

coupe-rondelle [coupe-rondelle *ou* coupe-ron-delles *pl.*] (*n.m.*), washer-cutter. (15)

coupe schématique (*f.*), diagrammatic section; schematic section. (16)

coupe transversale (*f.*), cross-section; trans-verse section: (17)

pont représenté en coupe transversale (*m.*), bridge shown in cross-section. (18)

coupe-tubage [coupe-tubage *pl.*] (Sondage) (*n.m.*), casing-cutter. (19)

coupe-tube *ou* coupe-tubes [coupe-tube *ou* coupe-tubes *pl.*] (*n.m.*), tube-cutter. (20)

coupe-tubes à 3 molettes (*m.*), three-wheel tube-cutter. (21)

coupe-tubes avec pince (*m.*), tube-cutter and tube-wrench combined. (22)

coupe-tuyaux [coupe-tuyaux *pl.*] (*n.m.*), pipe-cutter. (23)

coupe-vent [coupe-vent *pl.*] (*n.m.*), wind-cutter. (24)

coupe verticale (*f.*), vertical section: (25)

coupe verticale des terrains traversés, vertical section of the strata passed through. (26)

coupellation (*n.f.*), cupellation: (27)

coupellation de l'argent, du plomb d'œuvre, cupellation of silver, of work-lead. (28 *ou* 29)

coupelle (*n.f.*), cupel. (30)

coupeller (*v.t.*), to cupel: (31)

coupeller de l'or, to cupel gold. (32)

coupement (*n.m.*), cutting. (33)

coupement (*n.m.*) *ou* coupement de voie (traversée oblique) (Ch. de f.), diamond crossing. (34)

couper (*v.t.*), to cut; to cut off: (35)

couper du bois, to cut timber; to timber; to wood. (36)

couper en travers *ou* couper de travers, to cross-cut: (37)

couper du bois de travers, to cross-cut wood. (38)

couper l'eau, to cut off, to turn off, the water. (39)

couper la communication téléphonique, to cut off telephonic communication. (40)

couper la vapeur, to shut off, to cut off, steam. (41)

couper le courant (Élec.), to switch off the current. (42)

couper riche le filon (Mines), to strike it rich. (43)

couper un arbre, to cut down, to fell, a tree. (44)

couper un circuit électrique, to disconnect, to break, to open, an electric circuit. (45)

couper un tube à la longueur voulue, to cut a tube to the required length. (46)

couper une pierre, to cut a stone. (47)

couper (se) (*v.r.*), to cut; to be cut. (48)

couper (se) (s'entrecroiser) (*v.r.*), to cut each other; to cross each other; to intersect each other: (49)

routes qui se coupent à angle droit (*f.pl.*), roads which cut each other at right angles. (50)

couperose (*n.f.*), copperas. (51)

couperose blanche (*f.*), white copperas. (52)

couperose bleue (*f.*), blue copperas; blue vitriol; bluestone; sulphate of copper; copper sulphate; copper vitriol. (53)

couperose verte (*f.*), green vitriol. (54)

coupeuse (pour havage circulaire) (Mines) (*n.f.*), heading-machine; header. (55)

coupeuse de rails (*f.*), rail-cutting machine. (56)

coupeuse rotative (Mines) (*f.*), rotary heading-machine. (57)

couplage (*n.m.*), coupling; connecting: (58)

le couplage des roues d'une locomotive, coupling the wheels of a locomotive. (59)

couplage (Élec.) (*n.m.*), connection; connecting; coupling; grouping: (60)

le couplage des éléments de pile, the connection of battery-cells; connecting (*or* coupling) (*or* grouping) battery-cells. (61)

couplage en batterie *ou* couplage en parallèle *ou* couplage en quantité *ou* couplage en surface *ou* couplage en dérivation (Élec.) (*m.*), multiple connection; multiple circuit; connecting in multiple; connecting in parallel; connecting in quantity; connect-ing in bridge; coupling in multiple; multiple grouping; abreast connection. (62)

couplage en série *ou* couplage en tension *ou* couplage en cascade (*m.*), series connection; connecting in series; cascade; cascade connection; concatenated connection; concatenation; tandem connection. (63)

couplage mixte *ou* couplage en séries parallèles *ou* couplage en quantité et en tension (*m.*), multiple-series connection; parallel-series connection; connecting in multiple series. (64)

couple (*n.m.*) *ou* couple moteur (Méc.), torque; couple: (65)

couple de démarrage *ou* couple au démarrage, starting-torque. (66)

l'axe d'un couple (*m.*), the axis of a couple. (67)

couple (*n.m.*) *ou* couple voltaïque, cell; element; couple; voltaic couple; galvanic element. (68)

couple Leclanché (*m.*), Leclanché cell. (69)

couple réversible (*m.*), reversible cell; reversible element. (70)

coupler (*v.t.*), to couple; to couple up; to connect: (71)

coupler les fourneaux de mine, to couple up shots. (72)

coupler (Élec.) (*v.t.*), to connect; to couple; to group: (73) Cf. couplage.

coupler en quantité, to connect in multiple; to connect in parallel; to bridge. (74)

couplet (*n.m.*), strap-hinge. (1)

coupleur (Élec.) (*n.m.*), make-and-break. (2)

coupon (*n.m.*), coupon ; section ; short length ; portion ; test-bar ; test-piece : (3)
coupon de rail, coupon of rail ; short length (*or* portion) (*or* section) of rail. (4)

coupure (*n.f.*), cut ; cutting : (5)
faire une coupure à travers un marais, to make a cutting across a marsh. (6)

coupure (du banc de tour) (*n.f.*), gap (of lathe-bed). (7)

coupure de méandre (Géogr. phys.) (*f.*), cut-off. (8)

coupure transversale (Géol.) (*f.*), transverse valley ; cross-valley. (9)

courant, -e (qui est en cours) (*adj.*), current ; present : (10)
mois courant (*m.*), current (*or* present) month. (11)

courant, -e (qui court, — se dit des eaux vives) (*adj.*), running : (12)
un ruisseau d'eau courante, a brook of running water. (13)

courant (*n.m.*), current. (14)

courant (*n.m.*) *ou* courant de palan, fall ; tackle-fall ; rope-fall ; purchase-fall. (15)

courant à haute fréquence (Élec.) (*m.*), high-frequency current. (16)

courant à moyenne tension (Élec.) (*m.*), medium-pressure current. (17)

courant alternatif (Élec.) (*m.*), alternating current. (18)

courant bipolaire (Élec.) (*m.*), bipolar current ; two-pole current. (19)

courant constant (Élec.) (*m.*), constant current. (20)

courant continu (Élec.) (*m.*), direct current ; continuous current ; unidirectional current. (21)

courant d'air (*m.*), current of air ; air-current ; blast ; air-blast ; draught ; draught of air. (22)

courant d'air ascendant (Mines) (*m.*), upcast. (23)

courant d'air chaud (*m.*), hot blast ; hot air-blast. (24)

courant d'air descendant (*m.*), down-draught. (25)

courant d'air descendant (Mines) (*m.*), downcast. (26)

courant d'air entrant (Mines) (*m.*), ingoing air-current. (27)

courant d'air forcé (*m.*), forced draught : (28)
soufflage d'un courant d'air forcé à travers un tamis (*m.*), blowing a forced draught through a sieve. (29)

courant d'air principal (Mines) (*m.*), main air-current. (30)

courant d'air sortant (Mines) (*m.*), outgoing air-current. (31)

courant d'excitation (Élec.) (*m.*), exciting current ; leakage current. (32)

courant d'induction (Élec.) (*m.*), induction-current. (33)

courant de conduction (Élec.) (*m.*), conduction-current. (34)

courant de convection (Élec.) (*m.*), convection-current. (35)

courant de flamme (d'une chaudière) (*m.*), path of flame (in a boiler). (36)

courant de fleuve (*m.*), river-current. (37)

courant de Foucault *ou* courant de remous (Élec.) (*m.*), eddy current ; Foucault current ; local current. (38)

courant de ligne (Élec.) (*m.*), line-current. (39)

courant de marée (*m.*), tidal current. (40)

courant de phase (Élec.) (*m.*), phase-current. (41)

courant de régime (Élec.) (*m.*), normal current. (42)

courant de retour (Élec.) (*m.*), return-current. (43)

courant dérivé (Élec.) (*m.*), shunt current. (44)

courant dérivé (sluices) (*m.*), underflow ; under-current. (45)

courant déwatté (Élec.) (*m.*), wattless current ; reactive current ; inactive current ; quad-rature current ; magnetizing current ; ninety-degrees current ; idle current. (46)

courant diphasé (Élec.) (*m.*), two-phase current ; diphase current ; biphase current. (47)

courant électrique (*m.*), electric current. (48)

courant harmonique (Élec.) (*m.*), harmonic current. (49)

courant inducteur (Élec.) (*m.*), inductive current. (50)

courant induit (Élec.) (*m.*), induced current. (51)

courant monophasé (Élec.) (*m.*), single-phase current ; one-phase current ; monophase current ; uniphase current. (52)

courant ondulatoire (Élec.) (*m.*), undulatory current. (53)

courant oscillatoire (Élec.) (*m.*), oscillating current. (54)

courant parasite (Élec.) (*m.*). Same as courant de Foucault.

courant partiel (Aérage de mines) (*m.*), split (Mine-ventilation) : (55)
fractionner le courant principal en — courants partiels, to divide the main current into — splits. (56)

courant périodique (Élec.) (*m.*), periodic current. (57)

courant polyphasé (Élec.) (*m.*), polyphase current ; multiphase current. (58)

courant primaire (Élec.) (*m.*), primary current. (59)

courant pulsatoire (Élec.) (*m.*), pulsating current ; pulsatory current. (60)

courant redressé (Élec.) (*m.*), rectified current ; commuted current. (61)

courant secondaire (Élec.) (*m.*), secondary current. (62)

courant sinusoïdal (Élec.) (*m.*), sinusoidal current. (63)

courant tourbillonnaire (Élec.) (*m.*). Same as courant de Foucault.

courant triphasé (Élec.) (*m.*), three-phase current ; triphase current. (64)

courant uniphasé (Élec.) (*m.*). Same as courant monophasé.

courant vagabond (Élec.) (*m.*), stray current ; vagabond current. (65)

courant watté (Élec.) (*m.*), active current ; effort current ; working current. (66)

courbable (*adj.*), bendable. (67)

courbage (*n.m.*), curving ; bending. (68)

courbe (*adj.*), curve ; curved : (69)
ligne courbe (*f.*), curve line. (70)

courbe (*n.f.*), curve ; bend. (71)

courbe à double courbure (*f.*), tortuous curve ; twisted curve. (72)

courbe à grand rayon *ou* courbe aplatie (courbure) (*f.*), long-radius curve ; flat curve. (73)

courbe à grand rayon (tuyau) (*f.*), long-radius bend. (74)

courbe à petit rayon *ou* courbe de faible rayon (*f.*), short-radius curve ; sharp curve. (1)

courbe adiabatique (Phys.) (*f.*), adiabatic curve ; adiabat. (2)

courbe caractéristique (*f.*), characteristic curve. (3)

courbe d'échauffement (Métall.) (*f.*), heating-curve. (4)

courbe d'équilibre (Métall.) (*f.*), equilibrium-curve. (5)

courbe d'équilibre (Géogr. phys.) (*f.*). Same as courbe de lit.

courbe de consommation (Élec.) (*f.*), load-curve. (6)

courbe de contact (Engrenages) (*f.*), curve of contact (Gearing). (7)

courbe de détente (diagramme) (*f.*), expansion-curve. (8)

courbe de fusibilité (Métall.) (*f.*), freezing-curve ; fusibility-curve. (9)

courbe de la cloche (*f.*), probability-curve ; curve of frequency of error ; frequency-curve. (10)

courbe de lit (Géogr. phys.) (*f.*), grade : (11) rivière qui atteint, qui possède, sa courbe de lit (*f.*), river which reaches, which is at, grade. (12 *ou* 13)

courbe de magnétisme (*f.*), magnetization-curve ; magnetization-graph. (14)

courbe de niveau (Topogr.) (*f.*), contour-line ; isohyp : (15) la configuration du terrain est indiquée par des courbes de niveau espacées à des inter--valles convenables, dits équidistances, the configuration of the ground is shown by contour-lines spaced at convenient intervals, called contour-intervals. (16)

courbe de poursuite (*f.*), curve of pursuit. (17)

courbe de raccordement (Ch. de f.) (*f.*), transi--tion curve ; transition spiral ; easement-curve ; tapering curve. (18)

courbe de refroidissement (Métall.) (*f.*), cooling-curve. (19)

courbe des points de fusion (Métall.) (*f.*), melting-point curve. (20)

courbe en œuf (épure de distribution) (*f.*), valve-ellipse. (21)

courbe funiculaire (*f.*), funicular curve ; cate--nary curve ; catenary. (22)

courbe gauche (*f.*), tortuous curve ; twisted curve. (23)

courbe irrégulière (*f.*). See courbes irrégulières.

courbe magnétique (*f.*). See courbes magnéti--ques.

courbe piézométrique (Hydraul.) (*f.*), pressure-curve. (24)

courbe plane (*f.*), plane curve. (25)

courbe raide *ou* courbe vive (*f.*), sharp curve. (26)

courbé, -e (*adj.*), bent ; curved. (27)

courbement (*n.m.*), bending ; curving : (28) le courbement des bois, bending wood. (29)

courber (*v.t.*), to bend ; to curve : (30) courber un bâton, to bend a stick. (31)

courber (*v.i.*), to bend ; to curve : (32) poutre qui courbe sous une charge (*f.*), beam which bends under a load. (33)

courber (se) (*v.r.*), to bend ; to curve. (34)

courbes irrégulières dites pistolets (Dessin) (*f.pl.*), French curves. (35)

courbes magnétiques (*f.pl.*), magnetic curves ; magnetic fantom ; magnetic spectrum. (36)

courbure (*n.f.*), curvature ; bending : (37) courbure de la jante d'une roue, curvature of the rim of a wheel. (38)

courbure des bois, bending wood. (39)

courbure du champ (Opt.), curvature of field. (40)

coureur de jour (*m.*), outcropping seam of coal. (41)

courir (*v.i.*), to run : (42) le bail a encore — ans à courir, the lease has still — years to run. (43) chaîne de montagnes qui court du nord au sud (*f.*), chain of mountains which runs from north to south. (44)

courir un risque, to run, to incur, a risk. (45)

couronne (*n.f.*), crown. (46)

couronne (d'un fourneau) (*n.f.*), crown, dome, (of a furnace). (47)

couronne (d'une poulie, d'une roue d'engrenage) (*n.f.*), rim (of a pulley, of a gear-wheel). (48 *ou* 49)

couronne (Sondage) (*n.f.*), boring-head ; boring-crown. (50)

couronne (cercle de fer dont on garnit la tête d'un pieu) (*n.f.*), ferrule. (51)

couronne (de galets, de billes) (*n.f.*), ring (of rollers, of balls). (52 *ou* 53)

couronne (de fil) (*n.f.*), coil (of wire). (54)

couronne (*n.f.*) *ou* couronne circulaire (Géom.), annulus. (55)

couronne (d'une turbine) (*n.f.*), ring (of a turbine). (56)

couronne à barres *ou simplement* couronne (*n.f.*) (d'un cabestan), drumhead (of a capstan). (57)

couronne à pointes de diamant (*f.*), diamond boring-crown ; boring-crown set with diamonds. (58)

couronne de renfort (Moulage en terre) (*f.*), building-ring (Loam-moulding). (59)

couronne des linguets (cabestan) (*f.*), pawl-ring ; pawl-rim (capstan). (60)

couronne directrice *ou* couronne fixe (turbine) (*f.*), guide-ring. (61)

couronne en toile remplie de graine de lin (Sondage) (*f.*), seed-bag. (62)

couronne mobile (d'une turbine) (*f.*), runner, wheel, (of a turbine). (63)

couronne mobile (d'une pompe centrifuge) (*f.*), impeller, runner, (of a centrifugal pump). (64)

couronne sans diamants (Sondage) (*f.*), unset crown. (65)

couronnement (Arch.) (*n.m.*), crowning ; coping. (66)

couronnement (des bajoyers d'une écluse) (*n.m.*), coping (of the side walls of a lock). (67)

couronnement pour brûleur Bunsen (*m.*), rose for Bunsen burner. (68)

couronner (*v.t.*), to crown ; to cap : (69) basaltes qui couronnent les collines (*m.pl.*), basalts which cap the hills. (70) couronner le faîte d'un comble, to cap the ridge of a roof ; to ridge a roof. (71)

courroie (*n.f.*), belt ; band ; strap. (72)

courroie à boucle (*f.*), buckle-strap ; strap with buckle. (73)

courroie balata (*f.*), balata belt. (74)

courroie croisée (*f.*), crossed belt ; cross-belt ; halved belt. (75)

courroie de commande (*f.*), driving-belt ; driv--ing-band. (76)

courroie de transmission (*f.*), belt; driving-belt; driving-band. (1) Cf. **courroies de transmission.**

courroie demi-tordue (*f.*). Same as **courroie croisée.**

courroie en caoutchouc (*f.*), india-rubber belt; rubber belt. (2)

courroie en coton (*f.*), cotton belt. (3)

courroie en cuir (*f.*), leather belt; leather machine-band. (4)

courroie en toile caoutchoutée (*f.*), rubberized canvas belt. (5)

courroie ouverte (*f.*), open belt. (6)

courroie sans fin (*f.*), endless belt; continuous belt. (7)

courroie sans fin à godets (*f.*), endless bucket-belt. (8)

courroie tordue d'un demi-tour (*f.*). Same as **courroie croisée.**

courroie tordue d'un quart *ou* courroie tordue au quart *ou* courroie semi-croisée (*f.*), quarter-twist belt; quarter-turn belt; quartering-belt; quartered belt. (9)

courroie transporteuse (*f.*), conveyor-belt; travelling apron; apron. (10)

courroies (*n.f.pl.*) *ou* **courroies de transmission,** belting; belts; driving-belts: (11)

courroies et poulies de transmission (*f.pl.*), belting and pulleys. (12)

courroies à trois plis *ou* courroies en 3 épaisseurs (*f.pl.*), 3-ply belting; three-ply belting. (13)

courroies doubles en cuir (*f.pl.*), double leather belting. (14)

courroies en cuir articulé (*f.pl.*), leather link belting. (15)

courroies en poil de chameau (*f.pl.*), camel-hair belting. (16)

courroies rondes en cuir (*f.pl.*), round leather banding; round leather belting. (17)

cours (d'une rivière — son mouvement) (*n.m.*), flow (of a river). (18)

cours (d'une rivière — sa longueur) (*n.m.*), course (of a river). (19)

cours (des sluices) (*n.m.*), run (of sluices) (length of sluices). (20)

cours (du filon) (*n.m.*), run (of the lode). (21)

cours (Commerce) (*n.m.*), price: (22)

cours du marché, market-price. (23)

cours élevé de l'étain, high price of tin. (24)

cours d'eau (*m.*), stream; watercourse. (25)

cours d'eau à droit de passage (*m.*), right-of-way stream. (26)

cours d'eau de montagne (*m.*), mountain stream. (27)

cours d'eau permanent (*m.*), perennial stream. (28)

cours d'eau souterrain (*m.*), underground stream. (29)

cours d'instruction de premiers secours, pour travaux de sauvetage (*m.*), course of training for, classes for instruction in, first-aid work, rescue work. (30 *ou* 31)

cours inférieur (d'une rivière) (*m.*), lower part (of a river). (32)

cours supérieur (d'une rivière) (*m.*), upper part (of a river). (33)

course (d'une rivière, d'un fleuve) (*n.f.*), course (of a river). (34)

course (mouvement rectiligne d'un organe mécanique) (*n.f.*), stroke; travel; length of stroke. (35)

course à vide *ou* course à blanc (machines-outils *ou* analogues) (*f.*) (opp. à **course utile**), idle stroke; non-cutting stroke. (36)

course aller *ou* course directe *ou* course avant (du piston, etc.) (*f.*), outstroke; forward stroke. (37)

course arrière (*f.*). Same as **course retour.**

course de l'outil (*f.*), travel of the tool; length of stroke. (38)

course de l'outil réglable à volonté, même en marche (*f.*), length of stroke adjustable at will, even while tool is in motion. (39)

course descendante (*f.*), downstroke. (40)

course du chariot (machine-outil) (*f.*), travel of carriage. (41)

course du pilon (*f.*), lift of the stamp. (42)

course du piston (*f.*), piston-stroke (length of stroke); stroke of the piston; play of the piston. (43)

course du piston (piston plongeur) (*f.*), stroke of ram. (44)

course du tiroir (*f.*), valve-travel; travel of valve; stroke of slide-valve. (45)

course montante (*f.*), upstroke. (46)

course retour *ou* course rétrograde (du piston, etc.) (*f.*), instroke; back-stroke; return-stroke. (47)

course transversale du chariot (tour) (*f.*), trans-verse movement of carriage (lathe). (48)

course utile (machines-outils) (*f.*), cutting stroke. (49)

course verticale de l'arbre porte-foret (*f.*), vertical movement of drilling-spindle; vertical feed of spindle; stroke of spindle; fall of spindle; traverse of spindle. (50)

coursier (Hydraul.) (*n.m.*), breasting; soleing. (51)

court, -e (*adj.*), short: (52)

une courte distance, a short distance. (53)

une échelle courte, a short ladder. (54)

court (être à) *ou* court, -e (être), to be short; to run short; to run out: (55)

être à court d'approvisionnements, de com-bustible, to be short, to run short, to run out, of supplies, of fuel. (56 *ou* 57)

être à court de main-d'œuvre, to be short of labour; to be shorthanded. (58)

court-circuit [courts-circuits *pl.*] (Élec.) (*n.m.*), short circuit; short. (59)

court-circuité, -e (*adj.*), short-circuited; shorted. (60)

court-circuiter (*v.t.*), to short-circuit; to short. (61)

court foyer (Phys.) (*m.*), short focus. (62)

courte dérivation (*f.*) *ou* court-shunt (*n.m.*) (Élec.), short shunt; (63)

excitation compound à courte dérivation (*f.*), short-shunt compound excitation. (64)

courte distance de la rivière (à) *ou* courte proxi-mité de la rivière (à), within easy reach of the river. (65)

coussinet (de palier, ou analogue) (Mach.) (*n.m.*), bearing, bearing-bush, brass, pillow, carriage, (of plummer-block, or the like). (66 *ou* 67)

coussinet (Filetage) (*n.m.*), die; screwing-die (Screw-cutting). (68)

coussinet (pour rails de ch. de f.) (*n.m.*), chair; railway-chair; rail-chair; carriage. (69)

coussinet (premier claveau reposant à plat sur le pied-droit d'une voûte) (*n.m.*), springer; springing-stone; springing. (70)

coussinet à billes (*m.*), ball bearing. (71)

coussinets à galets *ou* coussinets à roules (*m.pl.*), roller-bearings. (72)

coussinet à la machine (m.), machine-die. (1)

coussinets à rapprochement concentrique (m.pl.), self-centring dies. (2)

coussinet aléseur (m.), reaming die. (3)

coussinet d'une seule pièce (m.), one-part die. (4)

coussinet de changement (Ch. de f.) (m.), switch-chair. (5)

coussinet de croisement (Ch. de f.) (m.), crossing-chair. (6)

coussinet de fusée ou **coussinet de boîte d'essieu** (m.), journal-bearing; axle-box bearing; journal-brass. (7)

coussinet de glissement (Ch. de f.) (m.), slide-chair. (8)

coussinet de joint (Ch. de f.) (m.), joint-chair. (9)

coussinets de l'essieu du bogie (m.pl.), brasses of engine-truck axle. (10)

coussinet de la boîte d'essieu moteur (m.), driving-box brass; driving-box bearing; crown-brass. (11)

coussinets de la grosse tête de bielle (m.pl.), big end brasses. (12)

coussinet de talon (Ch. de f.) (m.), heel-chair. (13)

coussinet de voiture (Ch. de f.) (m.), carriage-bearing; car-bearing. (14)

coussinets en acier trempé (de palier) (m.pl.), hardened steel bearings. (15)

coussinets en bronze (m.pl.), gun-metal bearings. (16)

coussinets en bronze dur (m.pl.), hard gun-metal bearings. (17)

coussinets en bronze phosphoreux à longue portée garnis de graisseurs automatiques, à bagues et réservoirs d'huile (m.pl.), long self-oiling phosphor-bronze bearings, with rings and oil-reservoirs. (18)

coussinets en bronze phosphoreux cylindro-coniques à rattrapage de jeu (m.pl.), cylindro--conical phosphor-bronze bearings for taking up wear; phosphor-bronze bearings with taper ends for adjusting wear. (19)

coussinet en deux pièces (m.), two-part die; die in two halves. (20)

coussinets en fonte garnis d'antifriction (m.pl.), babbitted cast-iron bearings. (21)

coussinet fendu (m.), split die. (22)

coussinet inférieur ou **simplement coussinet** (n.m.) (de palier) (opp. à **coussinet supérieur** ou **contre-coussinet**), bottom brass. (23)

coussinet – lunette [coussinets-lunettes pl.] (pour filière à coussinets) (n.m.), solid die (for engineers' screw-stock). (24)

coussinet réglable (Filetage) (m.), adjustable die. (25)

coussinets réglables (de palier) (m.pl.), adjustable bearings. (26)

coussinet supérieur (de palier) (m.), top brass. (27)

coût (n.m.), cost; costs. (28)

coût d'entretien (m.), maintenance costs. (29)

coût d'exploitation (m.), working-cost. (30)

coût de la vie (m.), cost of living. (31)

coût de manutention du gravier (m.), cost of handling the gravel. (32)

coût, fret, et assurance, cost, freight, and insurance. (33)

couteau (n.m.), knife; cutter. (34)

couteau (d'une balance) (n.m.), knife-edge (of a balance or scale). (35)

couteau (d'une mèche à trois pointes) (n.m.), router (of a centre-bit). (36)

couteau à démastiquer (m.), hacking-knife. (37)

couteau à l'attache des plateaux (balance) (m.), extreme knife-edge; knife-edge on which pans are suspended. (38)

couteau à mastiquer (m.), putty-knife; stopping-knife. (39)

couteau à palette (m.), palette-knife. (40)

couteau central ou **couteau du fléau** (balance) (m.), centre knife-edge; middle knife-edge. (41)

couteau de déclenche (m.), trip-catch. (42)

couteau fendeur (scierie) (m.), splitter; splitter-wheel; spreader; spreading-wheel. (43)

coûter (v.i.), to cost. (44)

coûteux, -euse (adj.), expensive; costly. (45)

coutume (n.f.), custom; practice: (46)

une coutume du pays, a custom of the country. (47)

couture (ligne de jonction des pièces) (n.f.), seam: (48)

la couture d'un tuyau, the seam of a pipe. (49)

une couture entre des tôles de chaudière, a seam between boiler-plates. (50)

couver (v.i.), to breed; to smoulder; to smolder: (51)

laisser couver le feu d'un haut fourneau pendant 24 à 36 heures, to allow the fire of a blast-furnace to breed for 24 to 36 hours. (52)

feu qui couve (Mines) (m.), breeding fire. (53)

couvercle (n.m.), cover; lid; cap. (54)

couvercle (d'une boîte) (n.m.), lid (of a box). (55)

couvercle à charnière (m.), hinged lid. (56)

couvercle à glissière (m.), pull-off lid. (57)

couvercle de cylindre (machine fixe horizontale) (m.), front cylinder-head; front cylinder-cover; front-head of cylinder. (58)

couvercle de la boîte à huile (m.), axle-box cover; axle-box lid; journal-box cover; journal-box lid. (59)

couvercle de la boîte à vapeur (m.), cap, cover, of the steam-chest. (60)

couvercle du dôme (dôme de vapeur) (m.), dome-cap; dome-head. (61)

couvercle du piston (m.), follower; junk-ring. (62)

couvert, -e (Météor.) (adj.), overcast; cloudy. (63)

couverture (n.f.), cover; covering. (64)

couverture (toiture) (n.f.), roof; roofing. (65) See also **toit** and **toiture.**

couverture (Préparation mécanique des minerais) (n.f.), blanket: (66)

recueillir sur les couvertures les sulfures contenus dans la pulpe, to recover on the blankets the sulphides contained in the pulp. (67)

couverture de tuiles (f.), roof of tiles; tile roofing. (68)

couverture en verre (f.), glass roof. (69)

couvre-boîtes à graisse (m.), wheel-guard. (70)

couvre-engrenages [couvre-engrenages pl.] (n.m.), gear-case; gear-cover; gear-guard; gear-box; wheel-guard. (71)

couvre-joint [couvre-joints pl.] (n.m.), butt strap; butt strip; strap; welt; cover; covering-plate; junction-plate; fish; fish-plate; wrapper. (72)

couvre-meule [couvre-meules pl.] (n.m.), wheel-guard; cover for emery-wheel; hood for emery-wheel. (73)

couvre-objet [couvre-objets *pl.*] (Microscopie) (*n.m.*), cover-glass ; cover-slip. (1)

couvre-roue [couvre-roues *pl.*] (*n.m.*), wheel-guard. (2)

couvre-soleil [couvre-soliel *pl.*] (d'une lunette) (*n.m.*), sunshade, ray-shade, (of a telescope). (3)

couvreur (pers.) (*n.m.*), slater ; roofer. (4)

couvrir (*v.t.*), to cover : (5)
  neige couvrant la terre (*f.*), snow covering the ground. (6)

couvrir (munir d'un toit) (*v.t.*), to roof : (7)
  couvrir une maison, to roof a house. (8)

couvrir d'ardoises (*v.t.*), to slate : (9)
  couvrir d'ardoises un toit, to slate a roof. (10)

couvrir la voie (Signaux de ch. de f.), to block the line. (11)

couvrir le train (Signaux de ch. de f.), to protect the train. (12)

couvrir les dépenses, to cover, to pay, expenses. (13)

covelline (Minéral) (*n.f.*), covelline ; covellite. (14)

cow-catcher (*n.m.*), cow-catcher ; pilot. (15)

coyau (d'un comble) (Charp.) (*n.m.*), furring ; furring-piece, (of a roof). (16)

cracher (projeter des parcelles) (*v.i.*), to spit : (17)
  roche qui crache (*f.*), rock which spits. (18)

cradle (berceau) (Mines) (*n.m.*), cradle ; washing-cradle ; rocker ; cradle-rocker. (19)

crag (Géol.) (*n.m.*), crag. (20)

crag blanc (*m.*), white crag. (21)

crag corallin (*m.*), coralline crag. (22)

craie (*n.f.*), chalk. (23)

craie à silex *ou* **craie marneuse** (*f.*), chalk with flints ; chalk rock. (24)

craie blanche (*f.*), white chalk. (25)

craie chloritée *ou* **craie glauconieuse** (*f.*), chloritic marl ; chalk marl ; greensand marl. (26)

craie de Briançon (*f.*), French chalk. (27)

craie phosphatée (*f.*), phosphatic chalk. (28)

craie sans silex (*f.*), chalk without flints. (29)

crain (Géol. & Mines) (*n.m.*), throw ; jump ; leap ; check. (30)

craindre (*v.t.*), to fear : (31)
  on n'a pas à craindre les effets de la gelée, the effects of frost need not be feared. (32)

crainte (*n.f.*), fear : (33)
  craintes au sujet de la sécurité des ouvriers, fears as to the safety of the workmen. (34)

crampe (*n.f.*), staple. (35)

crampon (*n.m.*), cramp ; cramp-iron ; dog-iron ; clamp ; holdfast. (36)

crampon (à deux pointes) (*n.m.*), staple. (37)

crampon (fixage de rails) (*n.m.*), spike ; dog-spike ; dog-nail. (38)

crampon (pour chevaux) (*n.m.*), frost-nail ; caulk ; calk. (39)

crampon à scellement (*m.*), cramp-iron with stone-hook. (40)

crampon de fermeture (*m.*), window-fastener ; sash-fastener ; window-catch ; window-lock. (41)

cramponner (*v.t.*), to cramp ; to clamp ; to fasten : (42)
  cramponner des pierres, les pièces d'une charpente, to cramp stones, the members of a frame. (43 *ou* 44)

cran (entaille) (Méc.) (*n.m.*), notch ; catch : (45)
  les crans d'un secteur denté, the notches of a quadrant. (46)

cran (dans un métal) (*n.m.*), flaw ; fault. (47)

cran (Géol. & Mines) (*n.m.*), throw ; jump ; leap ; check. (48)

cran de détente (secteur denté) (*m.*), expansion-notch (quadrant). (49)

cran de point mort (secteur denté) (*m.*), dead-centre notch (quadrant). (50)

cran de sûreté *ou* **cran de repos** (Méc.) (*m.*), safety-catch. (51)

crapaud (fixage de rails) (*n.m.*), clip ; sleeper-clip. (52)

crapaud (dans un diamant) (*n.m.*), flaw ; feather. (53)

crapaudine (pièce métallique percée de trous) (*n.f.*), strainer ; grating. (54)

crapaudine (bouchon au fond d'un réservoir, ou analogue) (*n.f.*), plug. (55)

crapaudine (palier de pied) (*n.f.*), footstep ; footstep-bearing ; step-bearing ; step-box ; step ; shaft-step. (56)

crapaudine (partie évidée dans laquelle entre le gond d'une porte) (*n.f.*), crapaudine. (57)

crapaudine (partie évidée dans laquelle s'appuie, pour pivoter, l'axe d'un appareil quelconque) (*n.f.*), centre-casting ; centre-plate ; main casting. (58)

crapaudine du bogie (locomotive) (*f.*), truck centre-casting ; engine-truck centre-casting ; centre-plate of bogie. (59)

craque (cavité pleine de cristaux, dans une roche) (*n.f.*), vug ; vugh ; geode ; druse. (60)

craquelures dues au soleil (Géol.) (*f.pl.*), sun-cracks ; shrinkage-cracks. (61)

craquement (*n.m.*), cracking ; crackling ; creaking. (62)

craquer (*v.i.*), to crack ; to crackle ; to creak. (63)

crasse (*n.f.*) *ou* **crasses** (*n.f.pl.*), scum ; dross ; slag ; cinder. (64)

crasse de fonte (scories provenant du cubilot) (Fonderie) (*f.*), iron-slag ; cinder ; scoria ; clinker. (65)

crasse de fonte (scories dans les poches de coulée ou dans les moules) (Fonderie) (*f.*), sullage. (66)

crasses des chaudières (*f.pl.*), fur ; furring ; scale ; boiler-scale ; incrustation. (67)

crassier (*n.m.*), slag-dump ; slag-tip ; slag-heap ; cinder-bank. (68)

cratère (*n.m.*), crater. (69)

cratère adventif (*m.*), parasitic crater ; lateral crater. (70)

cratère central (opp. à **cratère adventif**) (*m.*), central crater. (71)

cratère d'effondrement (*m.*), pit-crater. (72)

cratère d'explosion (*m.*), explosion-crater. (73)

cratère de soulèvement (*m.*), elevation-crater ; crater of elevation. (74)

cratère-lac [cratères-lacs *pl.*] (*n.m.*), crater-lake. (75)

cratériforme (*adj.*), crateriform. (76)

craterlet (Géol.) (*n.m.*), craterlet ; crater-basin ; craterkin. (77)

crayer (*v.t.*), to chalk. (78)

crayère (*n.f.*), chalk-pit. (79)

crayeux, -euse (*adj.*), chalky ; cretaceous : (80)
  terrain crayeux (*m.*), chalky (*or* cretaceous) ground. (81)

crayon (*n.m.*), pencil. (82)

crayon (de charbon) (Élec.) (*n.m.*), pencil (of carbon). (83)

crémaillère (Méc.) (*n.f.*), rack : (84)
  crémaillère et pignon, rack and pinion. **(85)**

crémaillère (d'un chemin de fer à crémaillère) (*n.f.*), rack ; rack-rail ; cog-rail. (1)

crémaillère (limon à crémaillère) (d'escalier) (*n.f.*), cut string ; open string. (2)

crémaillère à vis sans fin (*f.*), worm-rack. (3)

crénelage (*n.m.*), castellation ; toothing. (4)

crénelé, -e (*adj.*), castellated ; toothed : (5)
écrou crénelé (*m.*), castellated nut. (6)
secteur crénelé, toothed sector ; quadrant. (7)

créneler une roue, to tooth, to ratch, a wheel. (8)

créosotage (*n.m.*), creosoting : (9)
créosotage du bois, creosoting timber. (10)

créosote (*n.f.*), creosote. (11)

créosoter (*v.t.*), to creosote : (12)
créosoter des traverses de chemin de fer, to creosote railway-sleepers. (13)

crépine (*n.f.*) *ou* **crépine d'aspiration** (pompe), strainer ; wind-bore ; snore-piece ; tail-piece ; rose. (14)

crépiter (*v.i.*), to crepitate ; to crackle : (15)
le sel crépite dans le feu, salt crepitates in fire. (16)

crétacé, -e (de la nature de la craie) (*adj.*), creta--ceous ; chalky : (17)
terrain crétacé (*m.*), cretaceous ground ; chalky ground. (18)

crétacé, -e *ou* **crétacique** (Géol.) (*adj.*), Cretaceous ; Cretacic (*used as adj.*) : (19)
la période crétacée, the Cretaceous period ; the chalk period. (20)

crétacé (Géol.) (*n.m.*), Cretaceous ; chalk period : (21)
le crétacé, the Cretaceous ; the chalk period. (22)

crête (*n.f.*), crest ; ridge ; comb. (23)

crête (d'une colline) (*n.f.*), crest, comb, ridge, crown, (of a hill). (24)

crête (d'une montagne) (*n.f.*), crest, ridge, (of a mountain). (25)

crête (d'un toit) (*n.f.*), ridge, crest, comb, (of a roof). (26)

crête (d'un barrage, d'un déversoir) (*n.f.*), crest (of a dam, of a weir). (27 *ou* 28)

crête anticlinale (Géol.) (*f.*), anticlinal ridge. (29)

creusé, -e (se dit des lames de scies circulaires, etc.) (*adj.*), hollow-ground. (30)

creusement *ou* **creusage** (*n.m.*), digging ; sinking ; excavation : (31)
creusement de fossés (*ou* de tranchées), digging ditches ; digging trenches ; trenching. (32)
creusement par dragage rotatif (Mines), sinking by rotary dredging. (33)
creusement des vallées (Géol.) (*m.*), hollowing out of valleys ; excavation of valleys : (34)
le ruissellement des eaux participe au creuse--ment des vallées en portant aux thalwegs les matériaux meubles des pentes, the run--ning off of water over the surface shares in the hollowing out of valleys by carrying down the loose material from the slopes to the thalwegs. (35)

creuser (faire une cavité) (*v.t.*), to dig ; to sink ; to excavate : (36)
creuser un fossé (*ou* une tranchée), to dig a ditch ; to dig a trench ; to trench. (37)
creuser un puits de mine, une descenderie, to sink a mine-shaft, a winze. (38 *ou* 39)
creuser une fouille dans le sol d'une fonderie, to dig a pit in the floor of a foundry. (40)

creuser (rendre creux) (*v.t.*), to hollow ; to hollow out ; to scoop out ; to excavate ; to wear hollow : (41)

creuser une pierre, to hollow out a stone. (42)
bandage de roue dont la surface de roulement est creusée (*m.*), wheel-tire whose tread is worn hollow. (43)

creuser (rendre plus profond) (*v.t.*), to deepen : (44)
creuser un puits qui ne donne pas assez d'eau, to deepen a well which does not give enough water. (45)

creuser (se) (*v.r.*), to wear hollow ; to hollow. (46)

creuset (*n.m.*), crucible ; pot ; melting-pot ; melter. (47)

creuset (d'un haut fourneau) (*n.m.*), crucible, hearth, well, (of a blast-furnace). (48)
Note.—creuset is the part below the tuyère-zone ; distinguished from ouvrage, the part between the tuyère-zone and the boshes. There is no such distinction in England, the whole being known as the **crucible, hearth** or **well**.

creuset en argile (*m.*), clay crucible. (49)

creuset en grès (*m.*), stoneware crucible. (50)

creuset en platine (*m.*), platinum crucible. (51)

creuset en plombagine *ou* **creuset en mine de plomb** *ou* **creuset en graphite** (*m.*), plumbago crucible ; black-lead crucible ; graphite crucible ; blue-pot. (52)

creuset en terre réfractaire (*m.*), fire-clay crucible. (53)

creuset réducteur (*m.*), reduction-crucible. (54)

creux, -euse (qui a une cavité intérieure) (*adj.*), hollow : (55)
piston creux (*m.*), hollow piston. (56)

creux, -euse (profond) (*adj.*), deep : (57)
puits creux (*m.*), deep well. (58)

creux, -euse (qui est en contre-bas) (*adj.*), sunken : (59)
chemin creux (*m.*), sunken road. (60)

creux (*n.m.*), hollow ; cavity ; space ; hole ; depression ; hollowness : (61)
dans le creux d'un rocher, in the hollow of a rock. (62)
tomber dans un creux, to fall into a hole. (63)

creux (d'une vis) (opp. à plein) (*n.m.*), groove (of a screw). (64) (opp. to **tooth**).

creux (opp. à plein *ou* dent) (Engrenages) (*n.m.*), space. (65) (opp. to **tooth**) (Gearing) :
creux entre les dents, space between teeth. (66)

creux (cannelure de laminoir) (*n.m.*), pass. (67)

creux (moule) (*n.m.*), mould ; mold. (68)

creux (doline) (Géol.) (*n.m.*), sink ; sink-hole ; limestone sink ; swallow ; swallow-hole ; dolina. (69)

creux de la coquille (tiroir) (*m.*), exhaust-cavity (slide-valve). (70)

crevaison (*n.f.*), bursting : (71)
crevaison des tubes de chaudière, bursting of boiler-tubes. (72)

crevasse (fente) (*n.f.*), crevice : (73)
les crevasses dans les vieilles murailles, the crevices in old walls. (74)

crevasse (*n.f.*) *ou* **crevasse glaciaire**, crevasse. (75)

crevasse séismique (*f.*), seismic fissure ; gap caused by an earthquake. (76)

crevasser (*v.t.*), to crevice ; to crevasse. (77)

crever (*v.t.*), to burst : (78)
crever un sac à force de le remplir, to burst a sack by overfilling it. (79)

crever (*v.i.*), to burst. (80)

crever un pilier (dépilage) (Exploitation de la houille), to split a pillar. (81)

cri de l'étain (*m.*), cry of tin ; tin-cry. (82)

criblage (*n.m.*), screening; griddling; riddling; sifting; jigging: (1)
le criblage du charbon, screening coal. (2)
crible (*n.m.*), screen; griddle; riddle; sieve. (3)
crible (*n.m.*) *ou* crible hydraulique (pour minerai ou pour charbon), jig; jigger; jigging-machine; hotching-machine. (4 *ou* 5)
crible à grille filtrante (*m.*), filter-screen. (6)
crible à grille fixe *ou* crible à piston *ou* crible hydraulique à piston (*m.*), fixed-sieve jig; plunger-jig. (7)
crible à grille mobile (*m.*), movable-sieve jig. (8)
crible à mouvement différentiel (*m.*), differential-motion jig. (9)
crible à percussion (*m.*), percussion-sieve; impact-screen. (10)
crible à secousses (*m.*), shaking-screen; bump-ing screen; push-screen. (11)
crible classeur (*m.*), sizing-screen. (12)
crible du Hartz (*m.*), Hartz jig. (13)
crible en toile métallique (*m.*), wire-cloth screen; wire-wove screen. (14)
crible en tôle perforée (*m.*), punched-plate screen. (15)
crible oscillant (*m.*), oscillating screen; swing-ing screen. (16)
crible pour le sable (*m.*), sand-screen. (17)
crible rotatif (*m.*), rotary screen. (18)
cribler (passer à travers un crible; isoler au moyen du crible) (*v.t.*), to screen; to griddle; to riddle; to sift; to jig. (19)
cribler (percer en beaucoup d'endroits) (*v.t.*), to riddle; to honeycomb: (20)
région criblée de trous de sonde (*f.*), district honeycombed with bore-holes. (21)
cribler (se) (*v.r.*), to sift. (22)
cribleur (pers.) (*n.m.*), screenman; screener; jigger. (23)
cribleuse (de fonderie) (*n.f.*), sand-sifter; sand-riddler. (24)
cric (*n.m.*), jack; lifting-jack; hoisting-jack. (25)
Note.—cric is properly a jack of the rack type, as distinguished from **vérin**, a jack of the screw type, but the distinction is not always observed.
cric à crémaillère (*m.*), rack-and-pinion jack. (26)
cric à crémaillère avec fût en bois à simple engrenage (*m.*), single-purchase rack-and-pinion jack with wood body. (27)
cric à double engrenage (*m.*), double-purchase jack. (28)
cric à levier (*m.*), lever-jack; lifting-jack. (29)
cric complètement métallique (*m.*), all-metal jack. (30)
cric de charpentier (*m.*), timber-jack. (31)
cric pour locomotives (*m.*), locomotive-jack. (32)
cric relève-rails (*m.*), rail-jack; railway-jack; railway track-lifter. (33)
cric-tenseur [crics-tenseurs *pl.*] (*n.m.*), wire-stretcher, wire-strainer; strainer; wire-straining ratchet; raidisseur. (34)
crichtonite (Minéral) (*n.f.*), crichtonite. (35)
crin (filet de quartz, etc.) (Mines) (*n.m.*), string; stringer (36)
crique (fente) (*n.f.*), crack. (37)
crique (Géogr. phys.) (*n.f.*), creek. (38)
crique bonanza (*f.*), bonanza-creek. (39)
crise ouvrière (*f.*), labour crisis. (40)
crispite (Minéral) (*n.f.*), rutile. (41)

cristal (Chim. & Minéralogie) (*n.m.*), crystal. (42)
cristal (verre blanc, très pur, et très limpide) (*n.m.*), glass; crystal; crystal glass. (43)
cristal à deux axes [cristaux à deux axes *pl.*] *ou* cristal biaxe (*m.*), biaxial crystal; biaxal crystal. (44)
cristal à un axe [cristaux à un axe *pl.*] (*m.*), uniaxial crystal; uni.,axal crystal; mono--axal crystal. (45)
cristal aciculaire (*m.*), acicular crystal. (46)
cristal armé (*m.*), wire-glass; ferroglass: (47)
protecteur de niveau d'eau en cristal armé (*m.*), water-level protector, with wire-glass tube. (48)
cristal attractif (*m.*), positive crystal. (49)
cristal de montagne (*m.*). Same as cristal de roche.
cristal de quartz (*m.*), quartz crystal. (50)
cristal de quartz bipyramidé (*m.*), bipyramidal quartz crystal. (51)
cristal de quartz comprimé (*m.*), quartz crystal, rhombohedron and prism. (52)
cristal de quartz en biseau (*m.*), bevelled quartz crystal. (53)
cristal de quartz maclé (*m.*), twin quartz crystal; macled quartz crystal. (54)
cristal de quartz oblique (*m.*), oblique quartz crystal. (55)
cristal de quartz prismé (*m.*), prismatic quartz crystal; quartz crystal, bipyramid and prism. (56)
cristal de roche *ou simplement* cristal (*n.m.*), rock-crystal; mountain-crystal; berg-crystal; crystal. (57)
cristal droit (*m.*), right-handed crystal. (58)
cristal en forme de trémie (*m.*), hopper crystal. (59)
cristal fusiforme (*m.*), fusiform crystal. (60)
cristal gauche (*m.*), left-handed crystal. (61)
cristal liquide (*m.*), liquid crystal. (62)
cristal maclé *ou* cristal hémitrope (*m.*), twin crystal; twinned crystal; compound crystal; macled crystal; hemitrope crystal. (63). See macle (*n.f.*) for varieties of twins.
cristal négatif *ou* cristal en creux (*m.*), negative crystal (in a mineral mass, a cavity having the form of a crystal). (64)
cristal négatif *ou* cristal répulsif (*m.*), negative crystal (a crystal exhibiting negative double refraction). (65)
cristal positif *ou* cristal attractif (*m.*), positive crystal. (66)
cristal uniaxe *ou* cristal monoaxifère (*m.*), uni--axial crystal; uniaxal crystal; monoaxal crystal. (67)
cristallière (*n.f.*), mine of rock-crystal. (68)
cristallifère (*adj.*), crystalliferous: (69)
géode cristallifère (*f.*), crystalliferous geode. (70)
cristallin, -e (*adj.*), crystalline: (71)
roches cristallines (*f.pl.*), crystalline rocks. (72)
cristallinité (*n.f.*), crystallinity. (73)
cristallisabilité (*n.f.*), crystallizability. (74)
cristallisable (*adj.*), crystallizable: (75)
sel cristallisable (*m.*), crystallizable salt. (76)
cristallisant, -e (*adj.*), crystallizing: (77)
propriétés cristallisantes (*f.pl.*), crystallizing properties. (78)
cristallisation (*n.f.*), crystallization; crystallizing: (79)
cristallisation du fer dans un champ magnétique, crystallization of iron in a magnetic field. (80)

cristallisation du sel, crystallization of salt. (1)

cristallisation par dissolution et évaporation, crystallization by solution and vaporization. (2)

cristallisation par dissolution et refroidisse--ment, crystallization by solution and cooling. (3)

cristallisation par fusion, crystallization by fusion. (4)

cristallisation par les courants électriques, crystallization by electric currents. (5)

cristallisation par sublimation, crystallization by sublimation. (6)

cristalliser (*v.t.*), to crystallize : (7)
cristalliser du sel, to crystallize salt. (8)

cristalliser (*v.i.*), to crystallize : (9)
tous les grenats cristallisent dans le système cubique, all garnets crystallize in the cubic system. (10)

cristalliser (se) (*v.r.*), to crystallize. (11)

cristallisoir *ou* **cristalliseur** (appareil) (*n.m.*), crystallizer. (12)

cristallite (Pétrol.) (*n.f.*), crystallite. (13)

cristallitique (*adj.*), crystallitic. (14)

cristallogénie (*n.f.*), crystallogeny. (15)

cristallogénique (*adj.*), crystallogenic ; crystallo--genical. (16)

cristallographe (pers.) (*n.m.*), crystallographer. (17)

cristallographie (*n.f.*), crystallography : (18)
la cristallographie du fer, the crystallography of iron. (19)

cristallographie chimique (*f.*), chemical crystal--lography. (20)

cristallographie géométrique (*f.*), geometrical crystallography. (21)

cristallographie physique (*f.*), physical crystal--lography. (22)

cristallographique (*adj.*), crystallographic ; crystallographical : (23)
notation cristallographique (*f.*), crystallo--graphic notation. (24)

cristallographiquement (*adv.*), crystallographi--cally : (25)
minéral qui est cristallographiquement et optiquement identique avec un autre (*m.*), mineral which is crystallographically and optically identical with another. (26)

cristalloïde (*adj.*), crystalloid. (27)

cristallométrie (*n.f.*), crystallometry. (28)

cristallométrique (*adj.*), crystallometric. (29)

cristallophyllien, -enne (*adj.*), schistose crystalline; foliated crystalline ; phyllocrystalline : (30)
roches cristallophylliennes (*f.pl.*), schistose crystalline rocks ; foliated crystalline rocks ; phyllocrystalline rocks. (31)

cristobalite (Minéral) (*n.f.*), cristobalite. (32)

critique (Phys.) (*adj.*), critical : (33)
point critique (*m.*), critical point. (34)

croc (*n.m.*), hook ; eye-hook. (35)

croc à ciseaux (*m.*), sister-hooks ; match-hooks ; clip-hooks. (36)

croc à cosse (*m.*), eye-hook and thimble. (37)

croc à émerillon (*m.*), swivel-hook. (38)

croc à puits (*m.*), well-creeper. (39)

crochet (*n.m.*), hook. (40)

crochet (outil de tour) (*n.m.*), cranked tool ; hanging tool ; hook tool. (41)

crochet (ringard de puddlage) (*n.m.*), rabble ; rabbler ; rabbling tool ; staff. (42)

crochet à ancre (*m.*). Same as **crochet à tête de bélier**.

crochet à feu (*m.*), rake. (43)

crochet à queue de cochon (*m.*), pigtail hook. (44)

crochet à ramasser (outil de mouleur) (*m.*), lifter. (45)

crochet à ressort (*m.*), spring-hook ; safety-hook ; clip-hook ; clevis ; clevice. (46)

crochet à talon pour lisser (outil de mouleur) (*m.*), cleaner. (47)

crochet à tête de bélier (*m.*), ram's-horn crane-hook ; ram's-horn ; double crane-hook ; double hook ; change-hook. (48)

crochet à vis (*m.*), screw-hook. (49)

crochet d'assemblage (*m.*), joint-cramp ; cramp ; cramp-iron ; dog-iron. (50)

crochet d'attelage (*m.*), coupling-hook. (51)

crochet de balancier (Fonderie) (*m.*), beam-hook. (52)

crochet de boîte (*m.*), box-hook. (53)

crochet de fonderie (*m.*), hanger ; gagger ; lifter. (54)

crochet de gouttière (1) **à queue en pointe** (2) **pour être fixé à vis** (3) **pour être fixé sur chevron** (*m.*), gutter-bracket (1) to drive (2) to screw on (3) for attaching to rafter. (55 ou 56 ou 57)

crochet de grue (*m.*), crane-hook. (58)

crochet de levage (*m.*), lifting-hook. (59)

crochet de mouleur (pointe de mouleur) (*m.*), moulders' nail, hook head ; moulders' point, hook head. (60)

crochet de palan (*m.*), tackle-hook ; pulley-block hook. (61)

crochet de puits dit araignée (*m.*), well-creeper. (62)

crochet de sûreté (pour bennes, cuffats, etc.) (*m.*), safety-hook ; spring-hook ; clip-hook; clevis ; clevice. (63)

crochet de sûreté (évite-molettes) (Mines) (*m.*), detaching-hook ; safety detaching-hook ; safety-hook. (64)

crochet de suspension (*m.*), hanger ; suspension-hook ; hook. (65)

crochet de suspension des boyaux (*m.*), hose-hanger. (66)

crochet de suspension du récepteur (téléphone) (*m.*), receiver-hook. (67)

crochet de traction (*m.*), draw-hook ; drag-hook. (68)

crochet double (*m.*). Same as **crochet à tête de bélier**.

crochet en queue de cochon (*m.*), pigtail hook. (69)

crochet en S (*m.*), S hook ; S-shaped hook. (70)

crochet méplat (*m.*), flat hook. (71)

crochet pour câble (*m.*), rope-hook. (72)

crochet pour établis de menuisiers (*m.*), bench-stop. (73)

crochet pour la pose des courroies (*m.*), belt-mounter ; belt-shifter ; belt-shipper. (74)

crochet pour tuyaux (*m.*), pipe-hook. (75)

crochet tournant (*m.*), swivel-hook. (76)

crocheter une serrure, to pick a lock. (77)

crochets pour cordes en boyaux (*m.pl.*), hooks and eyes for gut bands. (78)

crochon (Géol. & Mines) (*n.m.*), bend (of the seam). (79)

crochu, -e (cassure) (Minéralogie) (*adj.*), hackly. (80)

crocidolite (Minéral) (*n.f.*), crocidolite. (81)

crocodile (concasseur à minerai) (*n.m.*), alligator. (82)

crocodile (contact fixe) (Ch. de f.) (*n.m.*), auto-
-matic stop (on railway-line to operate a
signal in cab of locomotive). (1)

crocoïse *ou* **crocoïte** (Minéral) (*n.f.*), crocoite ;
crocoisite ; lead chromate ; red lead ore.
(2)

crocus (*n.m.*), crocus. (3)

croisée (point où deux choses se croisent) (*n.f.*),
crossing : (4)
le point de croisée de deux fils, the crossing-
point of two wires. (5)

croisée (fenêtre) (*n.f.*), window. (6)

croisement (intersection) (*n.m.*), crossing : (7)
les tableaux de distribution devraient être
étudiés de manière à éviter autant que
possible le croisement de fils (*m.pl.*), dis-
-tributing-boards should be designed in
such a way as to avoid as far as possible
the crossing of wires. (8)

croisement (*n.m.*) *ou* **croisement de voie** *ou*
croisement de voies (endroit où deux
voies se coupent), crossing ; track-crossing.
(9)

croisement (*n.m.*) *ou* **croisement de voie** *ou*
croisement de voies (l'appareil à pointe de
cœur que l'on établit au point où deux voies
se croisent) (Ch. de f.), crossing ; frog. (10)

croisement à niveau (Ch. de f.) (*m.*), grade
crossing ; crossing at grade ; level crossing.
(11)

croisement à pattes de lièvre mobiles (*m.*),
movable-wing frog. (12)

croisement à pointes mobiles (*m.*), movable-
point crossing ; movable-point frog. (13)

croisement à ressort (*m.*), spring-rail frog. (14)

croisement de changement *ou* **croisement aigu**
(*m.*), acute crossing ; V crossing ; common
crossing. (15)

croisement de rues (*m.*), street crossing. (16)

croisement de traversée (*m.*), obtuse crossing ;
elbow-crossing ; K crossing ; crossing-
frog ; cross-frog. (17)

croisement de voies d'aérage (Mines) (*m.*), air-
crossing ; air-bridge. (18)

croisement rigide (Ch. de f.) (*m.*), rigid frog. (19)

croiser (couper en travers) (*v.t.*), to cross ; to
intersect : (20)
sentier qui croise la route (*m.*), path which
crosses the road. (21)
croiser des fils, to cross wires. (22)

croiser (se) (*v.r.*), to cross ; to cross each other ;
to intersect : (23)
routes qui se croisent (*f.pl.*), roads which
cross each other. (24)

croisette (Minéral) (*n.f.*), cross-stone. (25)

croiseur (filon croiseur) (*n.m.*), cross-lode ;
cross-vein ; cross-course ; crossing ; counter-
lode ; counter ; contra-lode ; caunter-lode ;
caunter ; bar. (26)

croisillon (objet en croix) (*n.m.*), cross (any
object in the form of a cross). (27)

croisillon (d'une poutrelle à croisillons, d'une
poutre en treillis) (*n.m.*), brace. (28)

croisillon à poignées *ou* *simplement* **croisillon**
(*n.m.*), star-wheel ; spoke-wheel. (29)

croix (*n.f.*), cross : (30)
un objet en forme de croix, an object in the
form of a cross. (31)

croquis (*n.m.*), sketch. (32)

croquis à l'encre *ou* **croquis à la plume** (*m.*), pen-
and-ink sketch. (33)

croquis au crayon (*m.*), pencil sketch. (34)

croquis coté (*m.*), dimensioned sketch. (35)

croquis en couleurs (*m.*), coloured sketch. (36)

croquiseur (pers.) (*n.m.*), sketcher. (37)

crosse *ou* **crossette** (de piston) (*n.f.*), cross-head,
crosshead, slipper-block, (of piston). (38)

crottin de cheval (*m.*), horse-dung ; horse-
manure. (39)

croupe (Géogr. phys.) (*n.f.*), ridge : (40)
le trait le plus curieux du relief sous-marin de
l'océan Atlantique est la longue croupe qui
s'étend du Nord au Sud depuis l'Islande
jusqu'au Cap, séparant deux cuvettes
allongées où les profondeurs atteignent
6 000 mètres, the most curious feature of the
submarine relief of the Atlantic ocean is the
long ridge which stretches from north to
south from Iceland to the Cape, separating
two long basins where the depths reach
6,000 metres. (41)
croupe de partage, dividing ridge. (42)

croupe (Arch.) (*n.f.*), hip (a truncated roof or
gable). (43)

croûte (*n.f.*), crust. (44)

croûte (de la fonte) (*n.f.*), skin (of cast iron). (45)

croûte altérée (Géol.) (*f.*), weathered crust. (46)

croûte terrestre (*f.*), earth's crust ; crust of the
earth. (47)

croûtes de coke (traces laissées par les coups de
poussières) (Mines) (*f.pl.*), coked coal-dust ;
caked coal-dust (traces left by dust-
explosions). (48)

crown-glass *ou* *simplement* **crown** (*n.m.*), crown
glass. (49)

cru, -e (*adj.*), raw ; crude ; hard : (50)
huile de lin crue (*f.*), raw linseed oil. (51)
métal cru (*m.*), crude metal ; raw metal. (52)
antimoine cru (*m.*), crude antimony. (53)
eau crue (*f.*), hard water. (54)

cruche (*n.f.*), jar. (55)

crue (d'un cours d'eau) (*n.f.*), rise ; rising ;
flood ; spate : (56)
la crue de la Seine, the rising of the Seine. (57)
les crues du Nil, the floods of the Nile. (58)
mesurer le débit d'un cours d'eau en temps de
crue, to measure the flow of a stream in
time of flood. (59)

crue (d'un glacier) (*n.f.*), advance (of a glacier).
(60)

cryolite *ou* **cryolithe** (Minéral) (*n.f.*), cryolite ;
Greenland spar. (61)

crypton (Chim.) (*n.m.*), crypton ; krypton. (62)

cubage (*n.m.*), cubic measurement ; measure-
-ment ; cubature ; cubage ; yardage : (63)
cubage des placers, cubic measurement,
measurement, yardage, of placers. (64)

cubane *ou* **cubanite** (Minéral) (*n.f.*), cubanite. (65)

cube (*adj.*), cubic : (66)
mètre cube (*m.*), cubic metre. (67)

cube (*n.m.*), cube. (68)

cube maritime (*m.*), shipping dimensions. (69)

cuber (*v.t.*), to cube ; to measure (solids) ;
to gauge : (70)
cuber des bois, des pierres, des gîtes minéraux,
to measure wood, stones, mineral deposits.
(71 *ou* 72 *ou* 73)
cuber la production d'un puits à pétrole, to
gauge the production of an oil-well. (74)

cubilot (Fonderie) (*n.m.*), cupola ; cupola-
furnace. (75)

cubilot à avant-creuset (*m.*), cupola with receiver.
(76)

cubilotier (pers.) (*n.m.*), cupolaman. (77)

cubique (*adj.*), cubic ; cubical ; cubiform ; cube-shaped. (1)

cubizite *ou* **cuboïte** (Minéral) (*n.f.*), analcite. (2)

cuboïde (*adj.*), cuboid ; cuboidal. (3)

cuffat *ou* **cufat** (Mines) (*n.m.*), kibble ; bucket ; hoisting-bucket ; tub. (4)

cuffat à eau *ou* **cuffat d'épuisement** (*m.*), water-kibble ; water-bucket ; bailing-tank ; baling-tank. (5)

cuffat à eau à vidange automatique (*m.*), self-discharging water-bucket. (6)

cuiller (pour métaux en fusion, etc.) (*n.f.*), ladle. (7)

cuiller (outil de mouleur) (*n.f.*), spoon ; spoon-tool. (8)

cuiller (Sondage) (*n.f.*), surface-auger ; clay-auger ; earth-auger ; earth-borer ; scoop ; shell ; shell-auger ; wimble. (9)

cuiller à boulet, à gobelet et ciseau (Sondage) (*f.*), combination bit and sludger with ball valve. (10)

cuiller à boulet, à gobelet et mouche de tarière (*f.*), combination auger and sludger with ball valve. (11)

cuiller à boulet et gobelet (*f.*), sludger with ball valve. (12)

cuiller à essais (*f.*), assay-spoon ; éprouvette ; provett. (13)

cuiller à fondants (*f.*), flux-spoon ; éprouvette ; provett. (14)

cuiller à fondre (*f.*), casting-ladle (of the spoon type). (15)

cuiller à mouche ordinaire (Sondage) (*f.*), closed scoop. (16)

cuiller à vapeur (Trav. publ.) (*f.*), steam-shovel ; steam-digger ; steam-excavator ; steam-navvy ; navvy. (17)

cuiller de coulée (*f.*), casting-ladle (of the spoon type). (18)

cuiller de plombier (*f.*), lead-ladle. (19)

cuiller écumoire (*f.*), skimming-ladle ; skimmer. (20)

cuiller fermée (Sondage) (*f.*), closed scoop. (21)

cuiller ouverte (Sondage) (*f.*), open scoop. (22)

cuiller ouverte à mouche de tarière (*f.*), scoop with auger ; wimble with auger. (23)

cuillère (*n.f.*). Same as **cuiller.**

cuir (peau tannée) (*n.m.*), leather : (24)
courroie en cuir (*f.*), leather belt. (25)

cuir (pièce, article en cuir) (*n.m.*), leather : (26)
les cuirs d'une pompe, d'une soupape, the leathers of a pump, of a valve. (27 *ou* 28)

cuir de montagne (Minéral) (*m.*), mountain-leather. (29)

cuir embouti (*m.*), U leather ; cup-leather. (30)

cuir tanné à l'écorce de chêne (*m.*), oak-bark tanned leather. (31)

cuir vert *ou* **cuir cru** *ou* **cuir d'abatis** (*m.*), raw hide : (32)
pignon en cuir vert (*m.*), raw-hide pinion. (33)

cuirasse (d'un tour, ou autre machine-outil) (le traînard) (*n.f.*), saddle (of a lathe, or other machine-tool) : (34 *ou* 35)
cuirasse à chariot horizontal mobile fonction-nant automatiquement et à la main dans les deux sens (machine à aléser et à percer horizontale), saddle with sliding table with automatic and hand longitudinal and cross-feed motions (horizontal boring and drilling-machine). (36)

cuirasse (d'un câble) (*n.f.*), sheathing, armour, (of a cable). (37)

cuirasse (lampe) (*n.f.*), shield ; bonnet. (38)

cuirasse du traînard (d'un tour) (le tablier) (*f.*), apron (of a lathe). (39)

cuirassé, -e (armé d'une cuirasse en fer) (*adj.*), ironclad : (40)
moteur cuirassé (Élec.) (*m.*), ironclad motor. (41)
électro-aimant cuirassé (*m.*), ironclad electro-magnet. (42)

cuire (calciner) (*v.t.*), to bake ; to burn ; to kiln ; to fire : (43)
cuire du plâtre, to burn plaster. (44)

cuissard (d'une chaudière) (*n.m.*), water-leg, leg, (of a boiler). (45)

cuisson *ou* **cuite** (*n.f.*), baking ; burning ; kilning ; firing : (46)
la cuisson des briques, baking (*or* burning) (*or* kilning) bricks. (47)
la cuisson du calcaire donne des chaux, the burning of limestone produces limes. (48)
le kaolin ne subit aucun retrait à la cuisson, china-clay does not shrink during the process of firing. (49)

cuivrage (*n.m.*), coppering : (50)
cuivrage du fer, coppering iron. (51)

cuivre (*n.m.*), copper. (52)

cuivre (cuivre jaune) (*n.m.*), brass. (53)

cuivre (d'un fer à souder) (*n.m.*), copper-bit, bit, (of a soldering-iron). (54)

cuivre blanc (alliage) (*m.*), white copper. (55)

cuivre brut (*m.*), raw copper. (56)

cuivre carbonaté bleu (Minéral) (*m.*), blue copper carbonate ; blue malachite ; azurite ; chessylite ; Chessy copper. (57)

cuivre cémentatoire (*m.*), cement-copper. (58)

cuivre de rosette (*m.*), rose copper ; rosette copper. (59)

cuivre électrolytique (*m.*), electrolytic copper. (60)

cuivre en feuilles *ou* **cuivre en planches** (*m.*), sheet copper. (61)

cuivre en paillettes (*m.*), float copper. (62)

cuivre gris (Minéral) (*m.*), gray copper. (63)

cuivre gris antimonial (Minéral) (*m.*), panabase. (64)

cuivre gris arsenical (Minéral) (*m.*), tennantite. (65)

cuivre jaune (*m.*), brass. (66)

cuivre natif (*m.*), native copper. (67)

cuivre noir (cuivre non-purifié) (*m.*), black copper. (68)

cuivre panaché (Minéral) (*m.*), peacock copper ; peacock ore ; horse-flesh ore ; purple copper ; variegated copper ore ; variegated pyrites ; bornite ; erubescite. (69)

cuivre pur (*m.*), pure copper. (70)

cuivre pyriteux (Minéral) (*m.*), copper pyrites ; chalcopyrite ; yellow copper ore ; yellow copper ; yellow pyrites ; yellow ore. (71)

cuivre rosette (*m.*), rose copper ; rosette copper. (72)

cuivre rouge (*m.*), copper ; pure copper. (73)

cuivre-type (*n.m.*) *ou* **cuivre standard**, standard copper. (74)

cuivre vitreux rouge (Minéral) (*m.*), red copper ; red copper ore ; ruby-copper ; ruby-copper ore ; cuprite. (75)

cuivrer (*v.t.*), to copper. (76)

cuivrerie (usine) (*n.f.*), copper-works. (77)

cuivrerie (ouvrages en cuivre jaune) (*n.f.*), brass-work : (78)
la cuivrerie dans l'intérieur de l'abri d'une locomotive, the brasswork in the inside of the cab of a locomotive. (79)

cuivreux, -euse (*adj.*), coppery ; cupreous : (1)
  pyrite de fer cuivreuse (*f.*), cupreous iron
  pyrites. (2)
cuivreux (Chim.) (*adj. m.*), cuprous : (3)
  oxyde cuivreux (*m.*), cuprous oxide. (4)
cuivrique (Chim.) (*adj.*), cupric : (5)
  chlorure cuivrique (*m.*), cupric chloride. (6)
cul (d'une bouteille) (*n.m.*), bottom (of a bottle).
  (7)
cul (d'une poulie) (*n.m.*), arse (of a pulley-block).
  (8)
cul de poche (Fonderie) (*m.*), scull ; skull. (9)
cul-de-sac [**culs-de-sac** *pl.*] (*n.m.*),   cul-de-sac ;
  dead-end ; blind alley. (10)
culasse (d'un aimant) (*n.f.*), yoke (of a magnet).
  (11)
culbutage (*n.m.*), tipping ; tilting ; dumping ;
  shooting ; discharge : (12)
  wagon à culbutage automatique (*m.*), self-
  tipping wagon ; self-dumping truck. (13)
culbuter (*v.t.*), to tip ; to tilt ; to dump ; to
  shoot : (14)
  culbuter un wagon, to dump a wagon ; to tip
  a truck. (15)
  culbuter un wagon entièrement, to tip a wagon
  right over ; to turn a truck right over. (16)
  culbuter le minerai dans une cheminée, to
  dump the ore into a chute ; to shoot the ore
  down a pass. (17)
culbuter (*v.i.*), to tip ; to tip up ; to tilt ; to
  dump ; to topple over : (18)
  wagon qui culbute dans le sens latéral (*m.*),
  wagon which tips sideways. (19)
culbuter (se) (*v.r.*), to tip ; to tip up ; to tilt ; to
  dump ; to topple over. (20)
culbuteur (*n.m.*), dump ; tipper ; tippler ; tip ;
  tipping-apparatus ; tilter. (21)
culée (Arch.) (*n.f.*), abutment : (22)
  la culée d'un pont, the abutment of a bridge.
  (23)
culminant, -e (*adj.*), culminating : (24)
  le mont Blanc est le point culminant des
  Alpes, Mont Blanc is the culminating
  point of the Alps. (25)
culmination (Astron.) (*n.f.*),   culmination ;
  transit : (26)
  culmination de la Polaire, culmination, transit,
  of Polaris. (27)
culmination inférieure (Astron.) (*f.*),   lower
  culmination ; lower transit. (28)
culmination supérieure (*f.*), upper culmination ;
  upper transit. (29)
culot (de lampe à huile) (*n.m.*), bowl (of oil-lamp) ;
  oil-vessel ; vessel ; fount. (30)
culot (de lampe électrique à incandescence) (*n.m.*),
  cap (of incandescent electric lamp). (31)
culot (d'une cartouche) (*n.m.*),   base (of a
  cartridge). (32)
culot (d'une crapaudine) (*n.m.*), centre bearing-
  plate (of a centre-plate). (33)
culot (plateau que l'on interpose entre le creuset
  et le feu) (*n.m.*), baffle-plate ; baffle-brick.
  (34)
culot (massif d'injection) (Géol.) (*n.m.*), boss :
  (35)
  un culot granitique, a granitic boss. (36)
culot (d'une cheminée volcanique) (Géol.) (*n.m.*),
  plug (of a volcanic neck). (37)
culot (d'un trou de mine) (Mines) (*n.m.*), socket
  (of a shot-hole). (38)
culot à baïonnette (lampe à incandescence) (*m.*),
  bayonet-cap. (39)

culot à vis (lampe à incandescence) (*m.*), screw-
  cap. (40)
culot de plomb (Coupellation) (*m.*), lead button.
  (41)
culotte (tuyau bifurqué) (*n.f.*),   Y branch ;
  breeches-pipe. (42)
culotte (cuissard de chaudière) (*n.f.*), water-leg ;
  leg. (43)
culotte de bifurcation du tuyau de prise de vapeur
  (locomotive) (*f.*),   T head ; niggerhead ;
  steam-head ; branch-pipe ; steam-pipe tee
  piece. (44)
culottes de la cheminée   (chaudière)   (*f.pl.*),
  breeching. (45)
cumengite *ou* **cumengéite** (Minéral) (*n.f.*), cumen-
  -gite ; cumengeite. (46)
cummingtonite (Minéral) (*n.f.*), cummingtonite.
  (47)
cumulo-volcan [**cumulo-volcans** *pl.*] (*n.m.*),
  cumulo-volcano. (48)
cunette (*n.f.*), gutter. (49)
cup (haut fourneau) (*n.m.*),   cup ; hopper
  (blast-furnace). (50)
cup et cône (haut fourneau),   cup-and-cone ;
  bell-and-hopper : (51)
  le cup et cône, the cup-and-cone : the bell-and-
  hopper. (52)
cuprifère (*adj.*), cupriferous ; copper-bearing :
  (53)
  l'horizon cuprifère (*m.*),   the cupriferous
  horizon. (54)
  grès cuprifère (*m.*), copper-bearing sandstone.
  (55)
cuprique (*adj.*), cupric : (56)
  acide cuprique (*m.*), cupric acid. (57)
cuprite (Minéral) (*n.f.*), cuprite ; red copper ;
  red copper ore ; ruby-copper ; ruby-copper
  ore. (58)
cuproapatite (Minéral) (*n.f.*), cuproapatite. (59)
cuprodescloizite (Minéral) (*n.f.*), cuprodescloizite.
  (60)
cupromagnésite (Minéral) (*n.f.*), cupromagnesite.
  (61)
cupromanganèse (Alliage) (*n.m.*), cupromanga-
  -nese. (62)
cuproplombite (Minéral) (*n.f.*),   cuproplumbite.
  (63)
cuproschéelite (Minéral) (*n.f.*),   cuproscheelite.
  (64)
cuprotungstite (Minéral) (*n.f.*), cuprotungstite.
  (65)
curage (*n.m.*),   cleansing ; cleaning ; cleaning
  out ; flushing ; scraping : (66)
  curage d'une rivière, d'un fossé, cleaning out,
  cleansing, a river, a ditch. (67 *ou* 68)
  curage d'un trou de sonde, cleaning, flushing,
  a bore-hole. (69)
curcuma (*n.m.*), turmeric ; curcuma : (70)
  papier de curcuma (*m.*), turmeric-paper. (71)
cure-feu [**cure-feu** *pl.*] (*n.m.*), prick-bar ; pricker ;
  poker. (72)
curer (*v.t.*), to cleanse ; to clean ; to clean out ;
  to flush ; to scrape. (73)
curette (*n.f.*), scraper ; cleaner ; clearer. (74)
curette (Travail aux explosifs) (*n.f.*),   cleaner ;
  drag. (75)
curseur (*n.m.*), runner ; slider ; traveller. (76)
curseur (d'un instrument de mathématiques)
  (*n.m.*), cursor (of a mathematical instrument).
  (77)
curseur (d'une cuve d'amalgamation) (*n.m.*),
  spider (of an amalgamating-pan). (78)

curtisite (Explosif) (*n.f.*), curtisite. (1)
curviligne (*adj.*), curvilinear ; curvilineal : (2)
mouvement curviligne (*m.*), curvilinear motion. (3)
curvimètre (*n.m.*), curvometer ; curvimeter. (4)
cut-off [**cut-offs** *pl.*] (méandre recoupé) (Géogr. phys.) (*n.m.*), cut-off : (5)
les cut-offs du Mississipi, the cut-offs of the Mississippi. (6)
cuve (*n.f.*), vat ; tank ; tub ; pan. (7)
cuve (d'un haut fourneau) (*n.f.*), tunnel, fire-room, shaft, (of a blast-furnace). (8)
cuve (en fonte) (d'une plaque tournante) (*n.f.*), basin (of cast iron) (of a turntable). (9)
cuve à chicanes (*f.*), baffle-tank. (10)
cuve à mercure (d'un baromètre) (*f.*), mercury-cup. (11)
cuve collectrice (*f.*), collecting-vat. (12)
cuve-cylindre [**cuves-cylindres** *pl.*] (*n.f.*), drum (for mixing, washing, etc.). (13)
cuve d'amalgamation *ou* **cuve de broyage** (*f.*), amalgamating-pan ; amalgamation-pan ; pan ; pan-mill ; grinding-pan ; grinder. (14)
cuve de cyanuration (*f.*), cyaniding-vat. (15)
cuve de décantation (*f.*), dewaterer. (16)
cuve de filtration *ou* **cuve filtrante** (*f.*), filtration-vat ; filter-vat ; leaching-vat ; lixiviation-tank. (17)
cuve de lavage (*f.*), washing-vat. (18)
cuve de lavage, dite settler (*f.*), settler. (19)
cuve de précipitation (*f.*), precipitation-tank. (20)
cuve de repos (*f.*), settling-tank. (21)
cuve pour développement (Photogr.) (*f.*), develop-ing-tank ; development-tank. (22)
cuve pour développer en plein jour les pellicules en rouleaux (*f.*), daylight roll-film develop-ing-tank ; tank for developing roll films in daylight. (23)
cuve pour développer les film-packs (*f.*), film-pack developing-tank. (24)
cuve pour le traitement des résidus aurifères et l'épuration d'amalgame (*f.*), clean-up pan. (25)
cuveau à patins (Mines) (*m.*), sledge-tub ; sledge-corve. (26)
cuvelage (d'un puits de mine) (*n.m.*), tubbing (of a mine-shaft) (27)
cuvelage (d'un puits à pétrole) (*n.m.*), casing (of an oil-well). (28)
cuvelage en bois (puits de mine) (*m.*), wooden tubbing. (29)
cuvelage en fonte (*m.*), cast-iron tubbing. (30)
cuvelage en pierre (*m.*), stone tubbing. (31)
cuveler (*v.t.*), to tub ; to case : (32)
cuveler un puits de mine, to tub a mine-shaft. (33)
cuveler un puits à pétrole, to case an oil-well. (34)
cuvellement (*n.m.*). Same as **cuvelage.**
cuvette (*n.f.*), basin ; dish ; tray : (35)
une cuvette en verre (Photogr., etc.), a glass dish. (36)
cuvette (d'un baromètre) (*n.f.*), cistern, mercury-cup, cup, (of a barometer). (37)
cuvette (d'une boussole, d'un compas) (*n.f.*), bowl (of a compass). (38)
cuvette (Géogr. phys.) (*n.f.*), basin : (39)
les grandes cuvettes océaniques, the great oceanic basins. (40)
cuvette lacustre, lake-basin. (41)

cuvette (*n.f.*) *ou* **cuvette synclinale** (opp. à dôme *ou* **brachyanticlinal** (Géol.), basin ; centro-clinal fold. (42) (opp. to dome *or* **quaqua-versal fold**).
cuvette (entonnoir placé à la tête du tuyau de descente pour recevoir les eaux des gouttières) (Constr.) (*n.f.*), hopper head ; head. (43)
cuvette (à), dished : (44)
meule à cuvette (meule en émeri) (*f.*), dished wheel. (45)
cuvette à billes (roulement à billes) (*f.*), ball-cage (ball bearings). (46)
cuvette d'égouttage (*f.*), drip-cup ; drain-cup. (47)
cuvette pour le développement des bobines de pellicules (*f.*), roll-film developing-dish. (48)
cyanhydrique (*adj.*), hydrocyanic. (49)
cyanicide (*n.m.*), cyanicide ; cyanide killer. (50)
cyanique (Chim.) (*adj.*), cyanic. (51)
cyanite (Minéral) (*n.f.*), cyanite ; kyanite ; disthene. (52)
cyanogène (Chim.) (*n.m.*), cyanogen. (53)
cyanose (Minéral) (*n.f.*), cyanosite ; cyanose ; chalcanthite. (54)
cyanotrichite (Minéral) (*n.f.*), cyanotrichite ; lettsomite. (55)
cyanuration (*n.f.*), cyanidation ; cyaniding. (56)
cyanure (Chim.) (*n.m.*), cyanide. (57)
cyanure d'argent (*m.*), cyanide of silver ; silver cyanide. (58)
cyanure d'or (*m.*), cyanide of gold ; gold cyanide. (59)
cyanure de potassium (*m.*), cyanide of potassium ; potassium cyanide. (60)
cyanurer (*v.t.*), to cyanide : (61)
cyanurer des concentrés, des sables, des slimes, des tailings, to cyanide concentrates, sands, slimes, tailings. (62 *ou* 63 *ou* 64 *ou* 65)
cyanurer (se) (*v.r.*), to cyanide. (66)
cycle (*n.m.*), cycle. (67)
cycle d'érosion (Géogr. phys.) (*m.*), cycle of erosion. (68)
cycle d'hystérésis (Phys.) (*m.*), hysteresis cycle ; hysteretic cycle ; hysteresis loop. (69)
cycle de Carnot (Thermodynamique) (*m.*), Carnot's cycle. (70)
cycle de Rankine (Thermodynamique) (*m.*), Rankine cycle ; Rankine's cycle. (71)
cycle théorique (machines à vapeur) (*m.*), ideal cycle : (72)
pour étudier méthodiquement l'action de la vapeur, on imagine une évolution simple, débarrassée de toutes les complications accessoires de la réalité, suivant ce qu'on appelle le cycle théorique de la machine à vapeur, in order to study the action of steam methodically, a simple evolution is imagined, relieved of all the accessory complications of reality, according to what is called the ideal cycle of the steam-engine. (73)
cycloïdal, -e, -aux (Géom.) (*adj.*), cycloidal : (74)
engrenage cycloïdal (*m.*), cycloidal gear. (75)
cycloïde (Géom.) (*n.f.*), cycloid. (76)
cyclonal, -e, -aux *ou* **cyclonique** (*adj.*), cyclonal ; cyclonic : (77)
tourbillon cyclonal (*m.*), cyclonal whirlwind. (78)
cyclone (*n.m.*), cyclone. (79)
cylindrage (action de passer sous ou entre des rouleaux) (*n.m.*), rolling : (80)
cylindrage des chaussées empierrées, **rolling** metalled roads. (81)

cylindrage des tôles, rolling plates. (1)
cylindrage (Travail de tour) (*n.m.*), straight turning; cylindrical turning; rounding: (2) cylindrage des rouleaux en bois, rounding wooden rollers (in the lathe); turning wooden rollers. (3)
cylindrage à chaud (*m.*), hot-rolling. (4)
cylindrage à froid (*m.*), cold-rolling. (5)
cylindre (Géom.) (*n.m.*), cylinder. (6)
cylindre (tube cylindrique) (*n.m.*), cylinder. (7)
cylindre (rouleau) (*n.m.*), roll; roller. (8)
cylindre (d'une pompe) (*n.m.*), barrel, cylinder, (of a pump). (9)
cylindre (d'un indicateur de pression) (*n.m.*), drum (of a steam-engine indicator). (10)
cylindre à air (*m.*), air-cylinder. (11)
cylindre à basse pression (*m.*), low-pressure cylinder. (12)
cylindre à cannelures (*m.*), grooved roll; cor-rugated roll; fluted roller. (13)
cylindre à cintrer (*m.*), bending-roll. (14)
cylindre à eau (*m.*), water-cylinder. (15)
cylindre à fer marchand (*m.*), merchant roll; finishing-roll. (16)
cylindre à frein (d'un frein à air comprimé) (*m.*), brake-cylinder (of a compressed-air brake). (17)
cylindre à gradins (*m.*), stepped roll. (18)
cylindre à grande vitesse *ou* cylindre à marche rapide (*m.*), high-speed roll. (19)
cylindre à haute pression (*m.*), high-pressure cylinder. (20)
cylindre à moyenne pression (*m.*), intermediate-pressure cylinder; intermediate cylinder. (21)
cylindre à tôle (*m.*), plate-roll. (22)
cylindre à vapeur (*m.*), steam-cylinder. (23)
cylindre à vent (d'une machine soufflante) (*m.*), air-cylinder, blast-box, (of a blowing-engine). (24)
cylindre broyeur (*m.*), crushing-roll; crusher-roll; roll. (25)
cylindre cannelé (*m.*), grooved roll; corrugated roll; fluted roller. (26)
cylindre compresseur (pour fer puddlé) (*m.*), rotary squeezer; cam squeezer; squeezer. (27)
cylindre compresseur *ou* cylindre de compression (pour chaussées) (*m.*), roller; road-roller. (28)
cylindre compresseur à vapeur (pour chaussées) (*m.*), steam-roller. (29)
cylindre creux (Géom.) (*m.*), hollow cylinder. (30)
cylindre d'amenage *ou* cylindre d'entraînement (*m.*), feed-roller; feed-roll. (31)
cylindre de détente (*m.*), low-pressure cylinder. (32)
cylindre de friction (*m.*), friction-wheel. (33)
cylindre de laminoir (*m.*), rolling-mill roll. (34)
cylindre de machine (*m.*), engine-cylinder. (35)
cylindre de pompe (*m.*), pump-barrel; pump-cylinder. (36)
cylindre de révolution (Géom.) (*m.*), cylinder of revolution. (37)
cylindre dégrossisseur *ou* cylindre de dégrossisse-ment (*m.*), roughing-roll; roughing-down roll; rough-down roll; breaking-down roll. (38)
cylindre dégrossisseur (laminage du fer) (*m.*), roughing-roll; roughing-down roll; rough-down roll; breaking-down roll; muck-roll; puddle-roll. (39)
cylindre droit (Géom.) (*m.*), right cylinder. (40)
cylindre du blooming (laminoir) (*m.*), blooming-roll (rolling-mill). (41)

cylindre du dessous (laminoir) (*m.*), bottom roll; lower roll (rolling-mill). (42)
cylindre du dessus (laminoir) (*m.*), top roll; upper roll (rolling-mill). (43)
cylindre ébaucheur (*m.*). Same as cylindre dégrossisseur.
cylindre en fonte trempée (*m.*), chilled-iron roll; chilled cast-iron roll. (44)
cylindre en gradins (*m.*), stepped roll. (45)
cylindre en porte-à-faux (*m.*), overhanging cylinder. (46)
cylindre enrouleur (*m.*), winding-drum. (47)
cylindre entraîneur (*m.*), feed-roll; feed-roller. (48)
cylindre espatard (*m.*), planishing-roll. (49)
cylindre femelle (laminoir) (*m.*), bottom roll; lower roll (rolling-mill). (50)
cylindre finisseur (*m.*), finishing-roll. (51)
cylindre finisseur (laminage du fer) (*m.*), finish-ing-roll; merchant roll. (52)
cylindre forgeur (*m.*), forge-roll. (53)
cylindre intermédiaire (machine à vapeur) (*m.*), intermediate cylinder; intermediate-pres-sure cylinder. (54)
cylindre lamineur (*m.*), rolling-mill roll. (55)
cylindre lisse (*m.*), plain roll; smooth roller. (56)
cylindre malaxeur (*m.*), mixing-cylinder. (57)
cylindre mâle (laminoir) (*m.*), top roll; upper roll (rolling-mill). (58)
cylindre oblique (Géom.) (*m.*), oblique cylinder. (59)
cylindre oscillant (machine à vapeur) (*m.*), oscillating cylinder. (60)
cylindre oscillant (broyage) (*m.*), swinging roll. (61)
cylindre polisseur (*m.*), planishing-roll; polishing-cylinder; polishing-drum. (62)
cylindre rouleur (*m.*), roller (a cylinder used to carry something). (63)
cylindre tronqué (Géom.) (*m.*), truncated cylinder. (64)
cylindrée (*n.f.*), cylinder (the contents of a cylinder): (65)
une cylindrée de vapeur, d'air, a cylinder of steam, of air. (66 *ou* 67)
cylindrer (passer sous ou entre des rouleaux) (*v.t.*), to roll: (68)
cylindrer une barre de fer, to roll a bar of iron. (69)
cylindrer une route, to roll a road. (70)
cylindrer (donner la forme d'un cylindre à) (*v.t.*), to round; to make cylindrical; to turn: (71)
cylindrer une pièce de bois, to round a piece of wood. (72)
cylindrer à chaud, to hot-roll. (73)
cylindrer à froid, to cold-roll. (74)
cylindres en tandem (*m.pl.*), tandem cylinders. (75)
cylindres placés côte à côte (*m.pl.*), side-by-side cylinders. (76)
cylindreur (pers.) (*n.m.*), roller. (77)
cylindricité (*n.f.*), cylindricity. (78)
cylindrique (*adj.*), cylindrical; cylindric; cylindriform: (79)
surface cylindrique (*f.*), cylindrical surface. (80)
cylindriquement (*adv.*), cylindrically. (81)
cylindro-conique (*adj.*), cylindroconic; cylindro-conical: (82)
tambour cylindro-conique (*m.*), cylindroconical drum. (83)
cylindro-spiraloïde (*adj.*), cylindrospiral: (84)
tambour cylindro-spiraloïde (*m.*), cylindro-spiral drum. (85)
cylindroïde (*adj.*), cylindroid; cylindroidal. (86)
cymogène (pétrole) (*n.m.*), cymogen. (87)
cymophane (Minéral) (*n.f.*), cymophane. (88)
cyprine (Minéral) (*n.f.*), cyprine. (89)

D

d'axe en axe, from centre to centre : (1)
d'axe en axe des fusées (essieu), from centre to centre of journals (axle). (2)
d'axe en axe des bogies (wagon de ch. de f.), truck-centres. (3)
dacite (Pétrol.) (*n.f.*), dacite. (4)
dahménite (Explosif) (*n.f.*), dahmenite. (5)
dallage (action) (*n.m.*), flagging ; paving. (6)
dallage (pavé) (*n.m.*), flagging ; flagstone pave--ment ; pavement. (7)
dallage de marbre (*m.*), marble pavement. (8)
dallage strié (Géol.) (*m.*), striated pavement. (9)
dalle (pavé) (*n.f.*), flag ; flagstone. (10)
dalle (Géol.) (*n.f.*), flag ; flagstone : (11)
dalles à lingules, lingula flags. (12)
dalle de verre (*f.*), patent light. (13)
daller (*v.t.*), to flag ; to pave with flagstones. (14)
dalleur (pers.) (*n.m.*), flagger ; pavior. (15)
dalot (Ponts & Chaussées) (*n.m.*), box culvert ; flat-top culvert. (16)
damage (action) (*n.m.*), ramming ; tamping. (17)
damage (ouvrage) (*n.m.*), tamp-work. (18)
dame (de paveur) (*n.f.*), beetle ; rammer ; earth-rammer. (19)
dame (petit mur incliné qui forme la partie antérieure du creuset d'un haut fourneau) (*n.f.*), dam. (20)
dame (plaque de fonte sur laquelle s'écoulent les laitiers, — haut fourneau) (*n.f.*), dam-plate. (21)
dame (Géol.) (*n.f.*), erosion column ; chimney-rock. (22)
damer (pilonner) (*v.t.*), to ram ; to tamp : (23)
damer un pieu, to ram, to tamp, a pile. (24)
damer une allée, to ram, to tamp, a path. (25)
damer (taluter) (*v.t.*), to slope ; to batter : (26)
damer une terrasse, to slope a terrace. (27)
damourite (Minéral) (*n.f.*), damourite. (28)
danaïde (Hydraul.) (*n.f.*), danaide ; danaïde. (29)
danaïte (Minéral) (*n.f.*), danaite. (30)
danalite (Minéral) (*n.f.*), danalite. (31)
danburite (Minéral) (*n.f.*), danburite. (32)
danger (*n.m.*), danger : (33)
danger de contact (Élec.), danger resulting from contact. (34)
danger des poussières de charbon, danger arising from coal-dust. (35)
dangereusement (*adv.*), dangerously. (36)
dangereux, -euse (*adj.*), dangerous ; unsafe ; insecure ; risky : (37)
quartier dangereux barré (Mines), dangerous place fenced off. (38)
dannemorite (Minéral) (*n.f.*), dannemorite. (39)
dans l'ensemble, on the whole. (40)
dans la pratique, in practice. (41)
dans le cas où une nappe de pétrole sera atteinte, in the event of oil being reached. (42)
dans le vide, in a vacuum ; in vacuo : (43)
les corps tombent parallèlement dans le vide (*m.pl.*), bodies fall parallelly in a vacuum. (44)
chute des corps dans le vide (*f.*), fall of bodies in vacuo. (45)
dans les conditions actuelles, under existing conditions. (46)
dans les premiers jours de l'été, early in the summer. (47)
dans les temps humides, in wet weather. (48)

dans œuvre, inside ; in the clear : (49)
mesure dans œuvre (*f.*), inside measurement. (50)
un puits de mine de — mètres de diamètre dans œuvre, a mine-shaft — metres in diameter in the clear. (51)
dans peu de jours, in (*or* within) a few days. (52)
dans tous les temps, in all weathers : (53)
point nettement visible dans tous les temps (*m.*), point clearly visible in all weathers. (54)
dans un avenir prochain, in the near future. (55)
dans un rayon de — kilomètre(s), within a radius of — kilometre(s). (56)
danseuses (Ch. de f.) (*n.f.pl.*), yielding sleepers : (57)
les danseuses sont des traverses qui ballottent au passage des trains, yielding sleepers are sleepers which jump up and down on the passing of trains. (58)
dartre (Fonderie) (*n.f.*), scab. (59)
dartrer (Fonderie) (*v.i.*), to scab. (60)
dash-pot (Méc.) (*n.m.*), dash-pot. (61)
data (*n.m.pl.*), data. (62)
datolite *ou* **datholite** (Minéral) (*n.f.*), datolite ; datholite. (63)
dauphin (extrémité inférieure recourbée d'un tuyau de descente des eaux pluviales, qui sert à rejeter les eaux dans un caniveau) (Constr.) (*n.m.*), shoe ; boot. (64)
davyne (Minéral) (*n.f.*), davyne. (65)
de centre à centre, from centre to centre. (66)
de fond en comble, from top to bottom. (67)
de jour en jour, from day to day. (68)
de la surface jusqu'au bed-rock, from grass-roots to bed-rock. (69)
de prime abord, at first sight. (70)
de temps en temps *ou* **de temps à autre**, from time to time. (71)
de tout repos, safe ; reliable : (72)
estimation de tout repos (*f.*), safe estimate ; reliable estimate. (73)
dé (bocards) (*n.m.*), die (ore-stamps) : (74)
dé en acier spécial, special-steel die. (75)
dé (coussinet) (*n.m.*), bearing ; bearing-bush ; brass ; pillow ; carriage : (76)
dé en bronze, gun-metal bearing. (77)
débâcle (des glaces) (*n.f.*), breaking up, debacle, (of the ice) : (78)
les glaçons d'une débâcle (*m.pl.*), the floes of a debacle. (79)
déballage (*n.m.*), unpacking. (80)
déballer (*v.t.*), to unpack : (81)
les marchandises doivent être déballées à l'arrivée et mises en magasin dans un local sec (*f.pl.*), the goods should be unpacked on arrival and stored in a dry place. (82)
débarcadère (*n.m.*), landing-place ; wharf. (83)
débarder une cargaison, to discharge, to unload, a cargo. (84)
débarquement (*n.m.*), disembarking ; debarking ; landing ; discharging ; unloading. (85)
débarquer (*v.t.*), to disembark ; to debark ; to land ; to discharge ; to unload : (86)
débarquer des marchandises, to land, to unload, goods. (87)
débarquer une cargaison, to discharge, to unload, a cargo. (88)

débarquer (*v.i.*), to land ; to disembark ; to debark. (1)

débarrasser (*v.t.*), to free ; to clear ; to rid : (2)
débarrasser un minerai de la gangue adhérente par le bocardage et le lavage, to free an ore from the adhering gangue by stamping and washing. (3)
puits de mine qui est débarrassé d'eau (*m.*), mine-shaft which is clear of (*or* free from) water. (4)

débarrasser (se) **de**, to get rid of ; to scrap : (5)
se débarrasser des résidus, to get rid of the tailings. (6)

débaucher les ouvriers, to induce the men to strike. (7)

débillardement (Charp. & Menuis.) (*n.m.*), wreath- -ing. (8)

débillarder (*v.t.*), to wreath : (9)
limon débillardé (escalier) (*m.*), wreathed string (stairs). (10)

débit (Hydraul.) (*n.m.*), flow ; discharge : (11)
mesurer le débit d'un cours d'eau, to measure the flow of a stream ; to gauge the discharge of a stream. (12)
débit d'étiage, flow at low water. (13)

débit (d'un déversoir) (*n.m.*), discharge (of a weir). (14)

débit (d'une source, d'un puits à pétrole) (*n.m.*), flow, yield (of a spring, of an oil-well). (15 *ou* 16)
débit qui baisse, failing yield. (17)

débit (d'une pompe) (*n.m.*), flow, capacity, (of a pump). (18)

débit (d'un injecteur, d'un graisseur) (*n.m.*), feed (of an injector, of a lubricator) (the fluid ejected). (19 *ou* 20)

débit (du vent par minute) (fourneau) (*n.m.*), feed (of air per minute). (21)

débit (de gaz, d'électricité) (*n.m.*), output (of gas, of electricity). (22 *ou* 23)

débit de rouage (Méc.) (*m.*), rim-speed. (24)

débit des bois (*m.*), breaking down timber ; breaking up timber ; lumbering. (25)

débit des bois en grume *ou* **débit des grumes** (*m.*), breaking down logs ; breaking up logs. (26)

débitage (du bois) (*n.m.*), breaking down ; break- -ing up ; sawing. (27)

débitage (de la pierre) (*n.m.*), sawing. (28)

débiter (*v.t.*). See examples : (29)
débiter aux dimensions voulues une longue barre de fer, to cut up a long bar of iron into the requisite sizes. (30)
débiter de l'ardoise en feuillets, to split slate into thin sheets. (31)
débiter de la vapeur, to discharge steam : (32)
la quantité de vapeur débitée par une soupape de sûreté, the quantity of steam discharged by a safety-valve. (33)
débiter du bois, to break down timber ; to break up timber ; to lumber ; to saw wood. (34)
débiter le marbre en plaques, to saw marble into slabs. (35)
débiter un tronc d'arbre en planches, to break down, to saw up, a tree-trunk into boards. (36)
débiter une moyenne de — litres par minute, (se dit d'une source, d'un cours d'eau), to yield an average of —litres per minute. (37)

débiter (se) (*v.r.*), to be broken down ; to saw ; to be sawn ; to be split : (38)

la pierre se débite bien, the stone saws easily. (39)
le marbre se débite à la scie sans dents, marble is sawn with the stone-saw. (40)
l'ardoise se débite en feuillets, slate is split into thin sheets. (41)

débitoir à taillant double (*m.*), spalling-hammer ; spaul-hammer. (42)

débituminisation (*n.f.*), debituminization : (43)
débituminisation du coke, debituminisation of coke. (44)

débituminiser (*v.t.*), to debituminize : (45)
débituminiser le charbon de terre pour le convertir en coke, to debituminize coal to convert it into coke. (46)

déblai (action) (*n.m.*), cutting ; excavation. (47)

déblai (ouvrage) (*n.m.*), cutting ; cut ; excava- -tion. (48)

déblai (de voie ferrée) (*n.m.*), cutting ; cut ; railway-cutting. (49)

déblais (*n.m.pl.*), waste ; waste stuff ; rubbish. (50)

déblais (Génie civil) (*n.m.pl.*), spoil. (51)

déblais de mine (*m.pl.*), attle ; deads ; waste ; waste stuff. (52)

déblayement *ou* **déblaiement** (*n.m.*), clearing ; clearing out ; clearing away : (53)
déblayement d'un éboulement (de terre), clearing a fall. (54)

déblayer (*v.t.*), to clear ; to clear out ; to clear away ; to cut : (55)
déblayer et reboiser un puits, to clear out and retimber a shaft. (56)
déblayer le front de taille, to clear the working-face. (57)
déblayer un banc de sable, to clear away a sand-bank. (58)
déblayer une recette (Mines), to cut a plat. (59)

débloquer la section (Signaux de ch. de f.), to open the section. (60)

déboisage (Mines) (*n.m.*), untimbering ; drawing timber ; timber-drawing ; removal of timber ; removal of the timbering ; prop-drawing. (61)

déboiser (Mines) (*v.t.*), to untimber ; to remove the timbering ; to draw the props : (62)
déboiser un puits, to untimber a shaft. (63)

déboiser une contrée, to denude a region of timber. (64)

déboiseur (Mines) (pers.) (*n.m.*), timber-drawer. (65)

débordement (*n.m.*), overflowing ; overflow : (66)
débordement d'un fleuve, overflow of a river. (67)

déborder (*v.i.*), to overflow ; to run over : (68)
verre qui déborde (*m.*), glass which runs over. (69)

déborder (se dit d'une rivière) (*v.i.*), to overflow ; to overflow its banks ; to overrun its banks. (70)

déborder (se) (*v.r.*), to overflow ; to run over. (71)

débouché (*n.m.*), outlet : (72)
le débouché d'une mine, the outlet of a mine. (73)

débouché (d'un ravin, d'un défilé) (*n.m.*), debou- -chure, débouchure, opening out, (of a ravine). (74)

débouché (d'un pont) (*n.m.*), waterway (of a bridge). (75)

débouché à l'industrie (*m.*), outlet for trade ; market ; débouché. (76)

débouché d'air (*m.*), air-outlet. (1)

déboucher (ôter le bouchon de) (*v.t.*), to unstop ; to uncork ; to draw the plug from : (2) déboucher un tuyau, to unstop a pipe. (3) déboucher une bouteille, to uncork a bottle. (4) déboucher le trou de coulée (four métallurgi- -que), to draw the plug from the tap-hole. (5)

déboucher (*v.i.*) See examples : (6) déboucher d'un bois, to emerge, to debouch, from a wood. (7) déboucher dans la mer (se dit d'un fleuve), to flow, to fall, to run, to debouch, into the sea. (8)

débouchure (disque de métal enlevé par un poinçon) (*n.f.*), punching. (9)

déboulonnage *ou* **déboulonnement** (*n.m.*), un- -bolting. (10)

déboulonner (*v.t.*), to unbolt. (11)

débourbage (curage) (*n.m.*), cleaning out; cleansing. (12)

débourbage (du minerai, du charbon) (*n.m.*), washing (ore, coal). (13 *ou* 14)

débourbage à bras (*m.*), washing by hand. (15)

débourbage mécanique (*m.*), machine-washing. (16)

débourber (un fossé, un étang) (*v.t.*), to clean out, to cleanse, (a ditch, a pond). (17 *ou* 18)

débourber (du minerai, du charbon) (*v.t.*), to wash (ore, coal). (19 *ou* 20)

débourbeur, -euse (Mines) (*adj.*), washing : (21) trommel débourbeur (*m.*), washing-trommel. (22)

débourbeur (Mines) (pers.) (*n.m.*), washerman. (23)

débourrage (Travail aux explosifs) (*n.m.*), un- -ramming ; unstemming : (24) débourrage des mines ratées, unramming, unstemming, miss-holes. (25)

débourrer (*v.t.*), to unram ; to unstem. (26)

débourrer (faire canon) (*v.i.*), to blow out : (27) coup de mine qui débourre (*m.*), shot which blows out. (28)

débours (*n.m.pl.*), disbursements ; outlay ; expenses ; out-of-pocket expenses. (29)

débourser une somme considérable d'argent, to lay out a large sum of money ; to spend a considerable amount of money. (30)

debout (*adv.*), on end ; erect (*adj.*): (31) mettre un tonneau debout, to put a cask on end. (32)

déboutonnage (de rivets) (*n.m.*), flying. (33)

déboutonner (se) (Rivetage) (*v.r.*), to fly : (34) il est toujours à craindre que les rivets fraisés ne se déboutonnent, it is always to be feared that countersunk rivets may fly. (35)

débrayable (*adj.*), releasable ; which can be thrown out of action : (36) mouvement débrayable à volonté par levier placé à la portée de l'ouvrier (*m.*), motion which can be thrown out of action at will by a lever placed within reach of the work- -man. (37)

débrayage (action) (*n.m.*), throwing out of gear ; throwing out of action ; throwing out of operation ; throwing out of feed ; putting out of gear ; releasing ; release ; tripping ; disconnecting ; unclutching : (38) débrayage automatique du mouvement de pression, automatic tripping of the feed ; automatic release of the feed-motion. (39)

débrayage automatique du renvoi de mouve- -ment à chaque extrémité de course du plateau, automatic throwing out of action of the counter-motion at each end of the travel of the table. (40)

débrayage d'une courroie, throwing a belt off ; forking a belt off ; laying a belt off. (41)

débrayage instantané, instantaneous release. (42)

débrayage par excentrique, release by cam. (43)

débrayage sur le devant, disengaging-gear in front ; front release ; stop-motion in front. (44)

débrayage (dispositif servant à débrayer une courroie) (*n.m.*). Same as **débrayeur.**

débrayer (*v.t.*), to throw out of gear ; to throw out of action ; to throw out of operation ; to throw out of feed ; to put out of gear ; to release ; to trip ; to disconnect ; to unclutch : (45)

débrayer l'arbre porte-foret, to release the drilling-spindle. (46)

débrayer le mouvement de filetage, to throw the screw-cutting motion out of action. (47)

débrayer une courroie, to throw a belt off ; to fork a belt off ; to lay a belt off ; to unship a belt. (48)

débrayeur (*n.m.*), belt-shifter ; belt-shipper ; belt-fork ; striker ; striking-gear ; strike- gear ; belt-striking gear ; belt-shifting apparatus. (49)

débrider deux pièces de bois, to unclamp two pieces of wood. (50)

débrider une pierre (Carrières), to unsling a stone. (51)

débrideur (pers.) (*n.m.*), lander ; unloader. (52)

débris (*n.m.pl.*), debris ; débris ; waste ; rubbish. (53)

débris (de métal) (*n.m.pl.*), scrap. (54)

débris de fer (*m.pl.*), wrought-iron scrap ; scrap-iron. (55)

débris de fonte (*m.pl.*), cast scrap ; foundry-scrap ; scrap. (56)

débris de fonte de bâtiment (*m.pl.*), heavy scrap. (57)

débris de fonte de poterie (*m.pl.*), hardware scrap ; pot-scrap ; muck-scrap. (58)

débris de fonte mécanique (*m.pl.*), machine-scrap. (59)

débris fossiles (*m.pl.*), fossil debris. (60)

débris métalliques (*m.pl.*), scrap-metal. (61)

débrouiller du fil, to disentangle wire. (62)

débroussaillement (*n.m.*), clearing of brushwood. (63)

débroussailler le terrain, to clear away the bush ; to clear the ground of brush and undergrowth. (64)

début (*n.m.*), start ; commencement. (65)

décadenasser (*v.t.*), to unpadlock : (66) décadenasser une porte, to unpadlock a door. (67)

décagement des bennes (Mines) (*m.*), taking, drawing, tubs out of the cage ; clearing the cage. (68)

décager les bennes, to take, to draw, tubs out of the cage ; to clear the cage. (69)

décagramme (*n.m.*), decagramme = 5·644 drams. (70)

décalage (Méc.) (*n.m.*), unkeying ; driving out keys ; withdrawing wedges. (71)

décalage des balais (Élec.) (*m.*), lag of brushes. (72)

décalcification (*n.f.*), decalcification. (1)

décaler (*v.t.*), to unkey ; to drive out the keys ; to withdraw the wedges : (2)
décaler l'excentrique, to unkey the eccentric. (3)

décaler les balais (Élec.), to displace the brushes. (4)

décalitre (*n.m.*), decalitre = 2·200 gallons. (5)

décamètre (*n.m.*), decametre = 10·936 yards. (6)

décamètre (*n.m.*) *ou* décamètre à ruban, measur--ing-tape (10 metres long) ; tape-line. (7)

décamètre avec ruban en acier (*m.*), measure with steel tape. (8)

décantation (*n.f.*), decantation. (9)

décanter (*v.t.*), to decant : (10)
décanter un liquide, to decant a liquid. (11)

décanteur (appareil) (*n.m.*), decanter. (12)

décapage (mécanique,—avec la pierre ponce, etc.) (*n.m.*), scouring. (13)

décapage (*n.m.*) *ou* décapage à l'acide (chimique), pickling : (14)
décapage des pièces de fonte, pickling castings. (15)

décapelage (enlèvement des terrains de couverture) (Mines) (*n.m.*), removing the overburden ; stripping ; baring. (16)

décapeler le gîte (Mines), to remove the over--burden ; to strip. (17)

décaper (nettoyer avec la pierre ponce, etc.) (*v.t.*), to scour : (18)
décaper les plaques d'amalgamation, to scour the amalgamation-plates. (19)

décaper *ou* décaper à l'acide (*v.t.*), to pickle : (20)
décaper une pièce de fonte dans une solution nitrique, to pickle a casting in a nitric solution. (21)

décapitation (de rivières) (Géogr. phys.) (*n.f.*), beheading (of rivers). (22)

décapiter (Géogr. phys.) (*v.t.*), to behead : (23)
cours d'eau décapité (*m.*), beheaded river. (24)

decapod (locomotive) (*n.f.*), decapod. (25)

décapolaire (Élec.) (*adj.*), ten-pole. (26)

décarbonater *ou* décarboniser (*v.t.*), to decarbon--ate. (27)

décarburant, -e *ou* décarburateur, -trice (*adj.*), decarburating ; decarbonizing. (28)

décarburation (*n.f.*), decarburization ; decarbon--ization : (29)
décarburation partielle de la fonte, partial decarburization of cast iron. (30)

décarburer (*v.t.*), to decarburize ; to decarbonize : (31)
décarburer de la fonte, to decarbonize, to decarburize, cast iron. (32)

décarburer (se) (*v.r.*), to decarbonize. (33)

Decauville (*n.m.*), Decauville railway : (34)
un Decauville, a Decauville railway. (35)

déceler (*v.t.*), to reveal : (36)
appareil qui décèle la présence du grisou (*m.*), apparatus which reveals the presence of fire--damp. (37)

décentrage *ou* décentrement (*n.m.*) *ou* décentration (*n.f.*), decentration ; throwing off centre. (38) Cf. avant porte-objectif en U.

décentrer (*v.t.*), to decentre ; to throw off centre : (39)
décentrer une lentille, to decentre a lens. (40)
décentrer la contre-pointe d'un tour, to throw the tail-stock of a lathe off centre. (41)

décharge (*n.f.*), discharge ; outflow : (42)
une décharge d'air, de gaz, a discharge of air, of gas. (43 *ou* 44)
une décharge d'eau, an outflow of water. (45)

décharge (d'un accumulateur) (*n.f.*), discharge (of an accumulator). (46)

décharge (d'un pan de bois) (Charp.) (*n.f.*), brace (of a timber frame). (47)

décharge automatique (à), self-discharging ; self-tipping ; self-dumping : (48)
wagon à décharge automatique (*m.*), self-tipping wagon. (49)

décharge conductive (Élec.) (*f.*), conductive discharge. (50)

décharge convective (Élec.) (*f.*), convective discharge. (51)

décharge disruptive (Élec.) (*f.*), disruptive discharge. (52)

décharge électrique (*f.*), electric discharge. (53)

décharge en aigrettes (Élec.) (*f.*), brush dis--charge. (54)

décharge oscillante (Élec.) (*f.*), oscillatory discharge. (55)

déchargement (*n.m.*), discharging ; unloading ; emptying : (56)
déchargement de la cage (Mines), unloading, clearing, the cage. (57)

décharger (*v.t.*), to discharge ; to unload ; to empty : (58)
décharger un chaland, to discharge a lighter. (59)
décharger des marchandises, to unload goods. (60)

décharger (se) (*v.r.*), to empty ; to empty itself : (61)
rivière qui se décharge dans un lac (*f.*), river which empties itself into a lake. (62)

décharger (se) (se dit d'un wagonnet) (*v.r.*), to empty itself ; to dump its load. (63)

déchargeur (parafoudre) (*n.m.*), lightning-arrester ; lightning-discharger ; lightning-protector. (64). See also parafoudre.

déchargeur à écoulement (*m.*), water-column lightning-arrester. (65)

déchargeur à écoulement en gouttes (*m.*), water-dropping lightning-arrester. (66)

déchargeur à jet d'eau (*m.*), water-jet lightning-arrester. (67)

déchausser le front de taille à l'aide de petits coups de mine, to dislodge the face by small blasts. (68)

déchausser un caillou de son alvéole, to dislodge a stone from its setting. (69)

déchausser un poteau télégraphique, to lay bare the foot of a telegraph-pole. (70)

déchéance de droits de concession (*f.*), forfeiture of claims. (71)

déchénite (Minéral) (*n.f.*), dechenite. (72)

déchet (*s'emploie très souvent au pluriel*) (*n.m.*), waste ; refuse ; scrap. (73)

déchet de coton (*m.*), cotton-waste. (74)

déchet de métal (*m.*), scrap of metal. (75)

déchets (Mines) (*n.m.pl.*), waste ; waste stuff ; refuse ; attle ; deads. (76)

déchets de criblage (*m.pl.*), screenings. (77)

déchets de fonderie (*m.pl.*), foundry-scrap ; cast scrap ; scrap. (78)

déchets métalliques (*m.pl.*), scrap-metal. (79)

déchirer (*v.t.*), to tear ; to rend : (80)
tarière qui déchire le bois (*f.*), auger which tears the wood. (81)
fissures qui déchirent les flancs des volcans (*f.pl.*), fissures which rend the flanks of vocanoes. (82)

déchirure (Géol.) (*n.f.*), rent. (83)

déchu (-e) de (être), to forfeit : (84)

être déchu de ses droits à une concession, to forfeit a claim. (1)

décibar (Météor.) (*n.m.*), decibar. (2)

décigramme (*n.m.*), decigramme = 1·543 grains. (3)

décilitre (*n.m.*), decilitre = 0·176 pint. (4)

décimètre (*n.m.*), decimetre = 3·937 inches. (5)

décimètre carré (*m.*), square decimetre = 15·500 square inches. (6)

décimètre cube (*m.*), cubic decimetre = 61·024 cubic inches. (7)

décintrage *ou* **décintrement** (d'une voûte, d'une arche) (*n.m.*), striking the centring (of an arch). (8)

décintrer une voûte, to strike the centre of an arch. (9)

décintroir à talus *ou* **décintroir de talus** (*m.*), pick-mattock ; mattock. (10)

déclanche (*n.f.*), **déclanchement** (*n.m.*), **déclancher** (*v.t.*), **déclancheur** (*n.m.*). Same as **déclenche, déclenchement, etc.**

déclarer (se) (en parlant d'un sinistre) (*v.r.*), to break out : (11)

un incendie s'est déclaré, a fire has broken out. (12)

déclenchement (*n.m.*) *ou* **déclenche** (*n.f.*) (action), tripping ; release ; disengagement : (13)

déclenchement automatique du mouvement de pression (machines à percer, etc.), auto--matic tripping of the feed ; automatic release of the feed-motion ; automatic disengagement of the feed-motion. (14)

déclenchement par butée réglable, tripping the feed by adjustable stop. (15)

déclenchement pneumatique (de l'obturateur) *ou* déclenche à la poire (Photogr.), pneu--matic release (of the shutter) ; bulb release. (16)

déclenchement au doigt (Photogr.), finger release ; trigger release. (17)

déclenchement (*n.m.*) *ou* **déclenche** (*n.f.*), (mé--canisme), trip ; tripper ; trip-gear ; release. (18)

déclencher (Méc.) (*v.t.*), to trip ; to trip the feed ; to throw out ; to disengage ; to release : (19)

déclencher un obturateur (Photogr.), to release a shutter. (20)

déclencher (se) (Méc.) (*v.r.*), to trip ; to disengage ; to come apart. (21)

déclencheur (*n.m.*) *ou* **déclencheur d'obturateur** (Photogr.), release ; shutter-release. (22)

déclencheur automatique *ou* **déclencheur à dis--tance** (Photogr.) (*m.*), self timer ; auto--matic timer. (23)

déclencheur métallique *ou* **déclencheur Antioüs** (Photogr.) (*m.*), wire release ; cable release ; Antinous release. (24)

déclic (*n.m.*), trigger ; click ; catch ; trip ; tripper ; trip-dog ; tripping-device ; release. (25)

déclic (d'une sonnette) (*n.m.*), tripping-device, releasing-hook, (of a pile-driver). (26)

déclic d'attelage (évite-molettes) (Mines) (*m.*), trip ; detaching-hook : (27)

évite-molettes à déclic d'attelage (*m.*), trip overwind-gear. (28)

déclin (d'un volcan) (*n.m.*), waning (of a volcano). (29)

déclinaison (*n.f.*), declination ; angle of declina--tion. (30)

déclinaison magnétique *ou* **déclinaison de l'aiguille aimantée** *ou* *simplement* **déclinaison** (*n.f.*),

magnetic declination ; declination of the needle ; declination ; variation ; variation of the compass. (31)

déclinateur (instrument pour déterminer la déclinaison) (*n.m.*), declinator. (32)

déclinatoire *ou* **déclinateur** (boussole de forme rectangulaire) (*n.m.*), trough-compass ; long compass. (33)

décliner (*v.i.*), to decline. (34)

décliquer (*v.t.*), to trip ; to release : (35)

décliquer le mouton d'une sonnette, to trip, to release, the monkey of a pile-driver. (36)

déclive (*adj.*), sloping : (37)

la partie déclive d'une toiture, the sloping part of a roof. (38)

décliver (*v.i.*), to slope. (39)

déclivité (*n.f.*), declivity ; slope : (40)

déclivité d'une montagne, declivity of a mountain. (41)

déclivité d'un terrain, slope of a piece of ground. (42)

décochage (Fonderie) (*n.m.*), stripping : (43)

après le décochage, les pièces de fonte sont râpées, les noyaux vidés, tout le sable adhérent brûlé est enlevé, etc., after strip--ping, the castings are rasped, the cores knocked out, all the adhering burnt sand is removed, etc. (44)

décocher (*v.t.*), to strip. (45)

décoincement (*n.m.*), unkeying ; driving out keys ; withdrawing wedges. (46)

décoincer (*v.t.*), to unkey ; to drive out the keys ; to withdraw wedges. (47)

décoincer les tiges (Sondage), to jar the rods slack ; to disengage the rods. (48)

décoincer (se) (*v.r.*), to become unkeyed ; to work loose. (49)

décollage *ou* **décollement** (cessation d'adhérence *ou* de contact) (*n.m.*), parting ; coming off ; coming away : (50)

le décollement d'un massif sous-cavé, d'un pan de rocher qui s'éboule, the parting of a holed block, of a slab of rock which falls. (51 *ou* 52)

décollage *ou* **décollement** (plan de séparation) (*n.m.*), parting : (53)

le cassage des gros blocs de rocher se fait au moyen de coins s'ils ont des lits ou des décollages bien marqués, the breaking of big lumps of rock is done with wedges if they have well-marked beds or partings. (54)

décoller un bloc de rocher, to dislodge a piece of rock. (55)

décoller (se) (*v.r.*), to part ; to come off ; to come away. (56)

décolletage (*n.m.*), cutting off. (57)

décolleté, -e *ou* **décolleté (-e) dans la barre** (se dit des vis, des boulons, des écrous, des rondelles, etc.) (*adj.*), machine-made ; cut from bar : (58)

boulons décolletés *ou* boulons décolletés dans la barre (*m.pl.*), machine-made bolts ; bolts cut from bar. (59)

décolleter (*v.t.*), to cut off. (60)

décolorant, -e (*adj.*), discolouring. (61)

décoloration (*n.f.*), discolouration. (62)

décoloré, -e (*adj.*), discoloured. (63)

décolorer (*v.t.*), to discolour. (64)

décolorer (se) (*v.r.*), to discolour ; to become discoloured : (65)

le permanganate de potasse se décolore sous l'influence d'une solution acide de protoxyde de fer, permanganate of potash discolours under the influence of an acid solution of protoxide of iron. (1)

déco:nbler un puits, to clear out a shaft. (2)

décomposable (*adj.*), decomposable. (3)

décomposé, -e (*adj.*), decomposed. (4)

décomposer (*v.t.*), to decompose : (5)

décomposer un composé chimique, une substance par la chaleur, la lumière blanche du soleil par le prisme, to decompose a chemical compound, a substance by heat, the white light of the sun by the prism. (6 *ou* 7 *ou* 8)

décomposer (se) (*v.r.*), to decompose ; to become decomposed. (9)

décomposition (résolution d'un corps en ses principes) (*n.f.*), decomposition : (10)

décomposition de l'eau, decomposition of water. (11)

décomposition des forces (Méc.), decomposition of forces. (12)

décomposition électrolytique, electrolytic decomposition. (13)

décomposition (altération) (*n.f.*), decomposition ; decay ; alteration : (14)

décomposition des matières végétales, decom--position of vegetal substances. (15)

décomposition des bois de mine, decay, decomposition, of mine-timber. (16)

décomposition du prix de revient en ses principaux éléments (*f.*), analysis of the cost-price into its chief components. (17)

déconcerter les plans de quelqu'un, to upset someone's plans. (18)

déconstruction d'une machine (*f.*), taking a machine to pieces. (19)

déconstruire une machine, to take a machine to pieces. (20)

découler (*v.i.*), to flow out ; to run out : (21)

métal fondu qui découle du fourneau (*m.*), molten metal which runs out of the furnace. (22)

découpage (*n.m.*), cutting out ; cutting off ; cutting up ; cutting : (23)

découpage de la pierre par câble hélicoïdal (Abatage en carrières), cutting stone with the wire-saw (Quarrying). (24)

découpage des métaux au moyen du jet d'oxygène, cutting metals with the oxygen jet. (25)

découpage (taillage en suivant les contours d'un dessin) (*n.m.*), fret-cutting ; fret-sawing. (26)

découpage (à l'emporte-pièce) (*n.m.*), punching, punching out, (with the hollow punch). (27)

découpage en massifs d'abatage (Mines) (*m.*), blocking out. (28)

découper (*v.t.*), to cut out ; to cut off ; to cut up ; to cut : (29)

découper une éprouvette dans le sens du laminage, to cut out a test-piece in the direction of rolling. (30)

découper des barreaux d'essais à la machine-outil, to cut off test-bars by machine-tool. (31)

découper (tailler en suivant les contours d'un dessin) (*v.t.*), to fret-saw. (32)

découper (à l'emporte-pièce) (*v.t.*), to punch ; to punch out : (33)

découper un trou dans une plaque, to punch a hole in a plate. (34)

découper en massifs d'abatage (Mines), to block out ; to cut up into pillars : (35)

découper un étage en massifs d'abatage, to block out a level. (36)

découper un gîte en massifs, to cut up a bed into pillars. (37)

découper (se) (*v.i.*), to cut : (38)

le grès se découpe facilement, sandstone cuts easily. (39)

découpeur de carottes *ou* **découpeur de témoins** (Sondage) (*m.*), core-cutter ; core-bit. (40)

découpeuse (à métaux, etc.) (*n.f.*), punching-machine (hollow-punch type). (41)

découplement (*n.m.*), uncoupling : (42)

découplement de wagons, uncoupling wagons. (43)

découpler (*v.t.*), to uncouple. (44)

découpoir (emporte-pièce) (*n.m.*), punch ; hollow punch ; socket-punch. (45)

découpoir à col de cygne (*m.*), swanneck fly-press ; swanneck screw-press. (46)

découpoir à la main (balancier) (*m.*), fly-press. (47)

découpure (*n.f.*), fretwork. (48)

décourageant, -e (*adj.*), discouraging ; unpromis--ing. (49)

découronnement (*n.m.*), uncapping. (50)

découronner (*v.t.*), to uncap : (51)

volcans, découronnés par le temps, qui sont aujourd'hui entaillés de profondes vallées, jadis occupées par des glaciers (*m.pl.*), volcanoes, uncapped by the weather, which are to-day grooved with deep valleys, formerly occupied by glaciers. (52)

découvert, -e (*adj.*), uncovered ; unprotected ; exposed ; open : (53)

pays découvert (*m.*), open country. (54)

découvert (tiroir) (*n.m.*), negative lap ; clearance (slide-valve). (55)

découvert extérieur (tiroir) (*m.*), negative outside lap ; outside clearance. (56)

découvert intérieur (tiroir) (*m.*), negative inside lap ; inside clearance. (57)

découverte (*n.f.*), discovery ; find : (58)

la découverte d'une mine d'or, the discovery of a gold-mine. (59)

découverte fortuite, chance discovery ; adventitious find. (60)

découverte (Métall.) (*n.f.*), uncovering. (61)

découverture (Mines) (*n.f.*), stripping ; unsoiling ; untopping ; baring ; removing the over--burden : (62)

découverture de la surface pour obtenir le dégelage naturel du sol, stripping the surface for natural thawing of the ground. (63)

découvrir (trouver ce qui était inconnu) (*v.t.*), to discover ; to find : (64)

découvrir une mine, to discover a mine. (65)

découvrir (ôter ce qui couvrait ; mettre au jour) (*v.t.*), to uncover ; to expose ; to disclose : (66)

découvrir le filon sur une longueur de — mètres, to uncover the vein for a distance of — metres ; to expose the lode over a length of — metres. (67)

découvrir (Métall.) (*v.i.*), to uncover : (68)

bouton, bain, qui découvre (*m.*), button, bath, which uncovers. (69 *ou* 70)

découvrir (se) (Métall.) (*v.r.*), to uncover. (71)

décramponner (*v.t.*), to unspike. (72)

décrassage (d'un foyer) (*n.m.*), slicing, pricking, (a grate). (73)

décrassage (Métall.) (*n.m.*), stirring the molten metal and skimming off the dross. (74)

décrassage mécanique (gazogène) (*m.*), mechani-
-cal poking (gas-producer). (1)

décrasse-meule avec 2 garnitures de molettes
[**décrasse-meules** *pl.*] (*m.*), emery-wheel
dresser, with 2 sets of cutters. (2)

décrasser la grille d'un foyer, to slice a furnace-
grate ; to prick the fire-bars. (3)

décrasseur (gazogène) (*n.m.*), poker ; stirrer. (4)

décrépitation (*n.f.*), decrepitation. (5)

décrépiter (*v.t.*), to decrepitate : (6)
décrépiter du sel, to decrepitate salt. (7)

décrépiter (*v.i.*), to decrepitate : (8)
l'anthracite décrépite lorsqu'on le chauffe (*m.*),
anthracite decrepitates when heated. (9)
la disomose décrépite au feu et donne une
forte odeur d'ail, nickel-glance decrepitates
in the fire and gives off a strong smell of
garlic. (10)

décret du gouvernement (*m.*), government
decree ; Government order. (11)

décrire (*v.t.*), to describe : (12)
décrire une courbe, to describe a curve. (13)

décrochage *ou* **décrochement** (*n.m.*), unhooking ;
hooking off ; detachment : (14)
décrochement fortuit, accidental detachment.
(15)

décrochement horizontal *ou* **décrochement
transversal** *ou* *simplement* **décrochement**
(*n.m.*) (Géol.), transcurrent fault ; trans-
-verse thrust. (16)

décrocher (*v.t.*), to unhook ; to take off (the hook);
to detach : (17)
décrocher le récepteur (Téléph.), to take off
the receiver. (18)

décrocher (se) (*v.r.*), to become unhooked. (19)

décroissance (*n.f.*) *ou* **décroissement** (*n.m.*),
decrease ; decline ; dwindling. (20)

décroître (*v.i.*), to decrease ; to diminish ;
to decline ; to dwindle : (21)
teneur décroissante du minerai (*f.*), dwindling
grade of the ore. (22)

décroûtage (Fonderie) (*n.m.*), skinning : (23)
décroûtage au jet de sable des pièces coulées,
skinning castings by sand-blasting. (24)

décroûter (*v.t.*), to skin. (25)

décrue (d'un fleuve) (*n.f.*), fall, falling, sub-
-sidence, (of a river) (26) (as opposed to
the **rise** *or* **rising**).

décrue (d'un glacier) (*n.f.*), retreat (of a glacier).
(27)

dedans (*n.m.*), inside ; interior : (28)
le dedans de la voie (Ch. de f.), the inside of
the road. (29)
les dedans d'un édifice, the interior of a
building. (30)

dédier à, to name after : (31)
la struvite est dédiée au ministre russe Struve,
struvite is named after the Russian minister
Struve. (32)

dédommagement (*n.m.*), compensation : (33)
recevoir une somme en dédommagement
d'un accident, to receive a sum as com-
-pensation for an accident. (34)

dédommager (*v.t.*), to compensate : (35)
dédommager quelqu'un de ses pertes, to
compensate someone for his losses. (36)

déduction (*n.f.*), deduction. (37)
déduction faite des frais divers, after deduct-
-ing sundry charges. (38)

déduire (*v.t.*), to deduct : (39)
déduire ses frais, to deduct one's expenses. (40)

défaire un joint, to unmake a joint. (41)

défausser (*v.t.*), to true ; to straighten : (42)
défausser une tringle, to true, to straighten, a
rod. (43)

défausser (se) (*v.r.*), to straighten. (44)

défaut (absence) (*n.m.*), want ; lack : (45)
défaut de jugement, want of judgment. (46)
défaut d'homogénéité dans l'acier, lack of
homogeneity in the steel. (47)

défaut (défectuosité) (*n.m.*), defect ; fault ;
flaw ; blemish : (48)
défaut dû à mauvaise construction, defect
due to faulty workmanship. (49)
diamant qui a un léger défaut (*m.*), diamond
which has a slight flaw. (50)

défaut (Élec.) (*n.m.*), fault ; defect : (51)
recherche des défauts (*f.*), locating faults. (52)
défaut d'isolement, fault in insulation ; defect
in insulation. (53)
défaut de contact, contact-fault. (54)

défavorable (*adj.*), unfavourable : (55)
dans les conditions défavorables, under
unfavourable conditions. (56)

défectueusement (*adv.*), defectively ; imperfectly ;
faultily. (57)

défectueux, -euse (*adj.*), defective ; imperfect ;
faulty. (58)

défectuosité (*n.f.*), defect ; imperfection ; fault ;
flaw : (59)
défectuosité passée inaperçue, unperceived
flaw. (60)

défendre (prohiber ; interdire) (*v.t.*), to prohibit ;
to forbid. (61)

défendre (protéger ; mettre à l'abri) (*v.t.*), to
protect : (62)
défendre la surface d'un talus par un revête-
-ment en gazon ou en maçonnerie, to protect
the surface of a slope by a facing of turf or
masonry. (63 *ou* 64)
défendre quelque chose contre le froid (*ou* du
froid), to protect something against the cold
(*or* from the cold). (65)

défense (prohibition ; interdiction) (*n.f.*), pro-
-hibition ; restriction : (66)
défense de sous-traiter sans autorisation, pro-
-hibition to sublet without permission. (67)
défense absolue de toucher aux fils (avis) (Élec.),
keep off the wires ; keep clear of the wires.
(68)
défense d'entrer (avis), no admittance. (69)
défense d'entrer sans autorisation (avis), no
admittance except on business. (70)
défense de fumer (avis), no smoking ; no
smoking allowed ; smoking forbidden ;
smoking prohibited. (71)
défense de passer (avis), no thoroughfare ;
trespassers will be prosecuted. (72)
défense (*ou* défense absolue) de traverser les
voies, passage souterrain et passerelle (avis)
(Ch. de f.), passengers are forbidden (*or*
are strictly forbidden) to cross the line
except by the subway or foot-bridge. (73)

défense (mise à l'abri) (*n.f.*), protection ;
protecting : (74)
défense et conservation des rives d'une rivière,
protection and preservation of the banks of
a river. (75)

déferrer un chemin de fer, to take up the rails of
a railway. (76)

défilé (Géogr. phys.) (*n.m.*), defile. (77)

définir (*v.t.*), to define ; to mark : (78)
massif de quartz bien défini (*m.*), well-defined
body of quartz. (79)

particularités bien définies (*f.pl.*), well-marked peculiarities. (1)

définitif, -ive (*adj.*), final; definitive; ultimate: (2)

étude définitive d'un chemin de fer (*f.*), final survey of a railway. (3)

déflagrant, -e (*adj.*), deflagrating. (4)

déflagrateur (*n.m.*), deflagrator. (5)

déflagration (*n.f.*), deflagration: (6)

déflagration du salpêtre, deflagration of nitre. (7)

déflagrer (*v.i.*), to deflagrate: (8)

les azotates déflagrent sur les charbons ardents (*m.pl.*), nitrates deflagrate on live coal. (9)

déflation (Géol.) (*n.f.*), deflation. (10)

défléchir (*v.t.*), to deflect: (11)

défléchir une aiguille, to deflect a needle. (12)

défléchir (*v.i.*), to deflect: (13)

rayons qui défléchissent (*m.pl.*), rays which deflect. (14)

déflecteur (*n.m.*), deflector: (15)

le déflecteur du foyer d'une locomotive, the deflector of the fire-box of a locomotive. (16)

déflexion (*n.f.*), deflection: (17)

déflexion de la lumière, deflection of light. (18)

défoncement (*n.m.*), recessing; recess. (19)

défoncer (former un creux, un retrait) (*v.t.*), to recess: (20)

défoncer un panneau, to recess a panel. (21)

défonceuse (*n.f.*), recessing-machine. (22)

déformation (*n.f.*), deformation: (23)

déformation de la voie d'un chemin de fer, deformation of a railway-track. (24)

déformation (de pièces coulées) (Fonderie) (*n.f.*), buckling, warping, (of castings). (25)

déformation (Méc.) (Résistance des matériaux) (*n.f.*), strain; deformation; set. (26)

déformation due à la compression (*f.*), compressive strain. (27)

déformation due à la torsion (*f.*), torsional strain; twist. (28)

déformation due à la traction (*f.*), tensile strain. (29)

déformation due au cisaillement (*f.*), shearing-strain; shear-strain. (30)

déformation momentanée *ou* **déformation élasti-que** (*f.*), temporary set; elastic deformation; elastic lag; lag. (31)

déformation permanente *ou* **déformation plastique** (*f.*), permanent set; permanent deforma-tion; plastic deformation: (32)

déformation permanente d'un métal provo-quée par des efforts mécaniques, permanent set of a metal caused by mechanical stresses. (33)

déformation tangentielle (*f.*), tangential strain. (34)

déformer (*v.t.*), to deform; to disfigure; to injure: (35)

machine qui cisaille les fers à L et à T sans déformer les profils (*f.*), machine which shears L and T iron without injuring the sections. (36)

déformer (Méc.) (*v.t.*), to strain; to deform. (37)

déformer (se) (Méc.) (*v.r.*), to strain; to be deformed: (38)

tous les corps se déforment plus ou moins quand ils sont soumis à l'action des forces extérieures (*m.pl.*), all bodies strain more or less when subjected to the action of external forces. (39)

défournement (*n.m.*), drawing the charge; drawing, taking, out of the furnace, oven, or kiln: (40)

défournement des briques, drawing, taking, bricks out of the kiln. (41)

défournement du feu (Mines) (*m.*), digging the fire out. (42)

défourner (*v.t.*), to draw the charge; to draw, to take, out of the furnace, oven, or kiln. (43)

défourner le feu (Mines), to dig the fire out. (44)

défrichement *ou* **défrichage** (*n.m.*), clearing; surface-clearing; grubbing: (45)

défrichement d'un bois, du terrain, clearing a wood, the ground. (46 *ou* 47)

défricher (*v.t.*), to clear; to grub. (48)

dégagement (action de retirer ce qui est engagé dans quelque chose) (*n.m.*), disengagement: (49)

dégagement de la tête porte-outil, disengage-ment of the tool-head. (50)

dégagement (émanation) (*n.m.*), giving off; disengagement; emission; evolution; discharge: (51)

dégagements de gaz provenants du charbon, disengagement of gas from the coal. (52)

dégagement instantané de grisou, outburst of fire-damp. (53)

dégagement (dépouille) (Méc.) (*n.m.*), relief; clearance; backing off. (54)

dégager (retirer ce qui était engager) (*v.t.*), to disengage: (55)

dégager l'hydrogène de l'eau (Chim.), to disengage hydrogen from water. (56)

dégager (produire une émanation) (*v.t.*), to give off; to emit; to evolve; to liberate; to disengage: (57)

l'hydrogène sulfuré est dégagé librement des fumerolles volcaniques et de quelques sources minérales (*m.*), sulphureted hydro-gen is given off freely from volcanic fu-maroles and from some mineral springs. (58)

toute fermentation dégage de l'acide carboni-que (*f.*), every fermentation disengages carbonic acid gas. (59)

dégager (débarrasser de ce qui encombrait) (*v.t.*), to free; to clear; to open up: (60)

dégager un passage, to clear a passage. (61)

dégager le minerai sur trois faces, to open up the ore on three sides. (62)

dégager (dépouiller) (Méc.) (*v.t.*), to relieve; to back off; to clear: (63)

taraud dégagé mécaniquement (*m.*), machine-relieved tap. (64)

dégager (déporter; évider) (*v.t.*), to cut away: (65)

contre-pointe (de tour) dégagée sur le devant (*f.*), tailstock (of lathe) cut away in front. (66)

dégager (se) (en parlant des vapeurs) (*v.r.*), to be given off; to be liberated; to be evolved; to be disengaged; to escape. (67)

dégager (se) (Méc.) (*v.r.*), to disengage; to clear itself: (68)

outil qui se dégage dans son mouvement de retour (*m.*), tool which clears itself in the return motion. (69)

dégarnir de ballast les bouts des traverses, to clear away the ballast from the ends of the sleepers. (70)

dégât (*n.m.*) (*s'emploie très souvent au pluriel*), damage: (71)

dégâts occasionnés par la chute de pierres, damage done by falling stones. (72)

dégauchir (*v.t.*), to true; to true up; to straighten; to take out of wind; to take out of winding; to surface: (1)

dégauchir une tringle, to true, to straighten, a rod. (2)

dégauchir une planche, to take a board out of wind; to surface a board; to true up a board to a plane surface. (3)

dégauchir (se) (*v.r.*), to straighten. (4)

dégauchissage *ou* **dégauchissement** (*n.m.*), truing; truing up; straightening; taking out of wind; surface-planing; surfacing. (5)

dégauchisseuse (*n.f.*), surface-planing machine; surface-planer; surfacer. (6)

dégel (*n.m.*), thaw. (7)

dégelage *ou* **dégèlement** (*n.m.*), thawing; dege-lation: (8)

dégelage des alluvions aurifères, thawing the gold-bearing alluvia. (9)

dégeler (*v.t.*), to thaw: (10)

dégeler les terrains, to thaw the ground. (11)

dégeler (*v.i.*), to thaw. (12)

dégorgement (*n.m.*), outfall; discharge: (13)

dégorgement d'air, discharge of air. (14)

dégorgement (tuyau) (*n.m.*), discharge-pipe. (15)

dégorgeoir (endroit où les eaux se dégorgent) (*n.m.*), outlet; spout. (16)

dégorgeoir (pour forgerons) (*n.m.*), fuller; creaser; creasing-tool. (17)

dégorgeoir (burin grain-d'orge) (*n.m.*), round-nosed cold-chisel; chipping-chisel (round-nosed). (18)

dégorgeoir à mortaises (*m.*), mortise boring-bit. (19)

dégorgeoir de dessous (*m.*), bottom fuller. (20)

dégorgeoir de dessus (*m.*), top fuller. (21)

dégorger (Travail de forge) (*v.t.*), to fuller. (22)

dégoutter (*v.t.*), to drip; to drop. (23)

dégoutter (*v.i.*), to drip; to drop: (24)

l'eau dégoutte des stalactites, the water drips from stalactites. (25)

dégradateur (Photogr.) (*n.m.*), vignetter. (26)

dégradation (*n.f.*), dilapidation: (27)

l'humidité amène la dégradation des murs (*f.*), damp brings about the dilapidation of walls. (28)

dégradation (Géol.) (*n.f.*), degradation: (29)

dégradation continue du sommet d'une montagne par l'action répétée du gel et du dégel, continual degradation of the summit of a mountain by the repeated action of frost and thaw. (30)

dégradation des falaises de craie attribuable à l'érosion opérée par les eaux marines sur les côtes, degradation of chalk cliffs due to erosion caused by the action of sea-water on the coasts. (31)

dégradation de l'énergie (Phys.) (*f.*), degrada-tion of energy; dissipation of energy. (32)

dégradation de la température (*f.*), fall of the temperature. (33)

dégradation des eaux (*f.*), wash, washing, scour, of water: (34)

préserver la maçonnerie du travail et des dégradations incessantes des eaux, to pro-tect the masonry from the continual action and wash (*or* scour) of the water. (35)

dégradation des rives d'un fleuve par les eaux courantes, washing (*or* scour) of the banks of a river by running water. (36)

dégrader (*v.t.*), to damage; to deface; to dilapi-date; to wash; to scour. (37)

dégrader (se) (*v.r.*), to delapidate. (38)

dégraissage (*n.m.*), scouring; cleaning. (39)

dégraisser (*v.t.*), to scour; to clean: (40)

dégraisser les plaques d'amalgamation, to scour the amalgamation-plates. (41)

dégraisser une lime, to clean a file. (42)

dégraisseur de vapeur (*m.*), oil-separator; steam-separator. (43)

degré (*n.m.*), degree: (44)

degrés du thermomètre, degrees of the ther-mometer. (45)

degrés au-dessous de zéro, degrees below zero. (46)

degrés au-dessus de zéro, degrees above zero. (47)

degrés centigrades, degrees centigrade. (48)

degrés de froid, degrees of frost. (49)

degrés Fahrenheit, degrees Fahrenheit. (50)

degrés Réaumur, degrees Réaumur. (51)

degré de longitude, de latitude, degree of longitude, of latitude. (52 *ou* 53)

degrés d'un cercle, degrees of a circle: (54)

le cercle est divisé en trois cent soixante degrés, the circle is divided into three hundred and sixty degrees. (55)

degré (marche d'escalier) (*n.m.*), step. (56)

degré de dureté (d'une meule en émeri) (*m.*), grade, degree of hardness, (of an emery-wheel): (57)

degré de dureté dur, hard grade. (58)

degré de dureté dur moyen, medium hard grade. (59)

degré de dureté moyen, medium grade. (60)

degré de dureté tendre, soft grade. (61)

degré de dureté tendre moyen, medium soft grade. (62)

dégringoler (*v.i.*), to tumble down; to fall down; to topple over: (63)

dégringoler de l'escalier, to fall down the stairs. (64)

dégrossi, -e (*adj.*), rough-hewn: (65)

marbre dégrossi (*m.*), rough-hewn marble. (66)

dégrossir (*v.t.*), to rough down; to rough-hew: (67)

dégrossir du bois, to rough down timber. (68)

dégrossir un bloc de marbre, to rough-hew a block of marble. (69)

dégrossissage *ou* **dégrossissement** (*n.m.*), rough-ing; roughing down. (70)

dégrossisseur (*n.m.*), roughing-roll; roughing-down roll; rough-down roll; breaking-down roll. (71)

dégrossisseur (laminage du fer) (*n.m.*), roughing-roll; roughing-down roll; rough-down roll; breaking-down roll; muck-roll; puddle-roll. (72)

dehors (*n.m.*), outside; exterior: (73)

le dehors d'une maison, the outside (*or* exterior) of a house. (74)

déhouillement des piliers *ou simplement* **déhouille-ment** (*n.m.*), robbing; getting the coal out of the pillars; cutting away the pillars; second working; broken working; working in the broken; pillar-working; pillar-drawing; drawing back; removing pillars; pulling pillars; extracting pillars. (75)

déhouiller les piliers, to rob the pillars; to get the coal out of the pillars; to cut away the pillars; to take out the pillars; to draw the pillars; to pull the pillars; to extract the pillars; to work in the broken. (76)

déhouilleuse (*n.f.*), coal-cutting machine; coal-cutter; cutter; undercutting-machine; undercutter; holing-machine; mining-machine; ironman. (1) See **haveuse** for varieties.

déhourdage (d'une cheminée) (Mines) (*n.m.*), start-ing, unkeying, loosening, (a clogged chute). (2)

déhourder une cheminée, to start, to unkey, to loosen, a clogged chute. (3)

déjections volcaniques (*f.pl.*), volcanic ejecta. (4)

déjeter (en parlant du bois) (*v.t.*), to warp; to spring: (5)

l'humidité déjette les bois (*f.*), damp warps wood. (6)

déjeter (en parlant du métal) (*v.t.*), to buckle; to bend. (7)

déjeter (se) (en parlant du bois) (*v.r.*), to warp; to spring. (8)

déjeter (se) (en parlant du métal) (*v.r.*), to buckle; to bend. (9)

déjettement (du bois) (*n.m.*), warping, springing, (of wood). (10)

déjettement (du métal) (*n.m.*), buckling; buckle: (11)

déjettement d'une tôle de chaudière, buckling of a boiler-plate. (12)

délabré, -e (*adj.*), dilapidated; tumble-down; ramshackle. (13)

délabrement (*n.m.*), dilapidation. (14)

délai (Commerce) (*n.m.*), time (required or given for delivery, etc.): (15)

le délai accordé pour l'exécution d'une com-mande, the time granted for the execution of an order. (16)

délaissé (Géogr. phys.) (*n.m.*), cut-off; ox-bow lake. (17)

délégué mineur [**délégués mineurs** *pl.*] (pers.) (*m.*), deputy. (18)

delessite (Minéral) (*n.f.*), delessite. (19)

délétère (*adj.*), deleterious. (20)

délié, -e (*adj.*), slender; thin: (21)

un fil délié, a slender wire. (22)

délimitation (*n.f.*), delimitation: (23)

délimitation de frontières, delimitation of boundaries. (24)

délimiter (*v.t.*), to delimit. (25)

déliquescence (*n.f.*), deliquescence. (26)

déliquescent, -e (*adj.*), deliquescent: (27)

sel déliquescent (*m.*), deliquescent salt. (28)

délit (fissure dans la masse) (Carrières) (*n.m.*), rift. (29)

délit (en) (Constr.), bed out: (30)

poser une pierre en délit, to set a stone bed out. (31)

déliter (placer une pierre de taille dans un sens qui n'est pas celui de son lit de carrière) (*v.t.*), to surbed. (32)

déliter (se) (*v.r.*), to crumble; to crumble up; to disintegrate; to scale off: (33)

se déliter en une poudre fine, to crumble into a fine powder. (34)

se déliter par exposition à l'air, to crumble by exposure to the air; to disintegrate on exposure to the air. (35)

délivrer (*v.t.*), to deliver: (36)

délivrer de la marchandise, to deliver goods. (37)

delphinite (Minéral) (*n.f.*), delphinite. (38)

delta (Géogr. phys.) (*n.m.*), delta: (39)

les deltas du Nil, du Pô, du Rhône, du Gange, the deltas of the Nile, of the Po, of the Rhone, of the Ganges. (40 *ou* 41 *ou* 42 *ou* 43)

delta (système triphasé) (Élec.) (*n.m.*), delta; mesh: (44)

montage en delta (*m.*) *ou* groupement en delta (*m.*), delta-connection; mesh-group-ing. (45)

delta (alliage métallique) (*n.m.*), delta-metal. (46)

delta lacustre (*m.*), lacustrine delta. (47)

delta marin (*m.*), marine delta. (48)

déluge (*n.m.*), deluge; flood. (49)

déluge de boue (Géol.) (*m.*), mud-flood; mud-flow. (50)

délutage (*n.m.*), unluting. (51)

déluter (*v.t.*), to unlute. (52)

démanchage *ou* **démanchement** (*n.m.*), un-handling: (53)

démanchage accidentel rendu impossible, accidental unhandling rendered impossible. (54)

démancher (*v.t.*), to unhandle. (55)

démancher (se) (*v.r.*), to become unhandled: (56)

lime qui se démanche (*f.*), file which becomes unhandled. (57)

demande (*n.f.*), application; demand; request; claim: (58)

demande de permis de recherches, application for prospecting-licence. (59)

demande en concession de chemin de **fer**, application for railway-concession. (60)

demande en concession de mines, application for mining-licence. (61)

prix sur demande (*m.pl.*), prices on application. (62)

demandé, -e (*p.p.*), in demand; sought after: (63)

la main-d'œuvre européenne est très demandée, European labour is very much in demand. (64)

demander (*v.t.*), to request; to ask for; to require; to apply for; to call for; to claim; to demand: (65)

demander l'autorisation (*ou* la permission) de visiter la mine, to request to look over the mine; to ask for (*or* to apply for) permis-sion to visit the mine. (66)

demander de légères réparations, to require, to need, some slight repairs. (67)

galerie de mine qui demande à être boisée (*f.*), mine-level which requires timbering. (68)

demander une concession de mines, to apply for, to make application for, a mining-licence. (69)

ne pas demander de commentaires, to call for no comment. (70)

demander un rabais, to claim a rebate. (71)

demander dans les journaux des soumissions, to advertise for tenders. (72)

demandez le catalogue, ask for catalogue. (73)

demandeur en concession (pers.) (*m.*), applicant for licence; applicant for concession. (74)

démantèlement (*n.m.*), dismantling. (75)

démanteler (*v.t.*), to dismantle: (76)

démanteler une drague, to dismantle a dredge. (77)

démantoïde (Minéral) (*n.f.*), demantoid; Uralian emerald. (78)

démarcation (*n.f.*), demarcation: (79)

démarcation de concessions, demarcation of claims. (80)

démarcation superficielle, surface-demarcation. (81)

démarrage (*n.m.*), starting; unmooring. (82)

démarrage au rhéostat (*m.*), rheostat-starting. (1)

démarrer (*v.t.*), to unmoor : (2)
démarrer une drague, to unmoor a dredge. (3)

démarrer (*v.i.*), to start (to commence to move) : (4)
train qui démarre (*m.*), train which starts. (5)
les moteurs asynchrones peuvent démarrer sur le courant alternatif directement : ils ne requièrent pas d'excitatrice (*m.pl.*), asynchronous motors can start directly on the alternating current : they do not require an exciter. (6)

démarreur (Élcc.) (*n.m.*), starter ; motor-starter ; starting-rheostat ; starting-resistance ; motor-starting resistance ; starting-box. (7)

démasquer (*v.t.*), to uncover : (8)
l'obturateur d'une chambre noire est destiné à masquer ou à démasquer l'objectif pour la pose (*m.*), the shutter of a camera is intended to cover or uncover the lens for exposure. (9)

demeure (à), fixed ; permanent ; stationary ; set : (10)
établir un châssis à demeure, to erect a permanent frame. (11)
chaudière placée à demeure (*f.*), stationary boiler. (12)
goupille à demeure (*f.*), set-pin. (13)

demi, -e (*adj.*), half. (14)

demi (*n.m.*), half. (15)

demi-boisage (*n.m.*), half-timbering. (16)

demi-boîte (*n.f.*) *ou* **demi-boîte à noyau** (Fonderie), half-box ; half core-box. (17)

demi-calcaire (*adj.*), semicalcareous. (18)

demi-cercle (Géom.) (*n.m.*), semicircle. (19)

demi-cercle (Géod.) (*n.m.*), demi-circle ; graphometer. (20)

demi-circulaire (*adj.*), semicircular. (21)

demi-coupe (Dessin) (*n.f.*), half-section. (22)

demi-course (tiroir, etc.) (*n.f.*), half-travel (slide-valve, etc.). (23)

demi-coussinet (*n.m.*), brass (bearing-bush). (24)

demi-coussinet inférieur (*m.*), bottom brass. (25)

demi-coussinet supérieur (*m.*), top brass. (26)

demi-cylindrique (*adj.*), semicylindrical. (27)

demi-ferme (Constr.) (*n.f.*), half-truss. (28)

demi-fixe (*adj.*), semiportable : (29)
chaudière demi-fixe (*f.*), semiportable boiler. (30)

demi-grandeur naturelle (*f.*), half natural size. (31)

demi-gras, -grasse (se dit du charbon, de la houille) (*adj.*), semibituminous. (32)

demi-kilomètre (*n.m.*), half a kilometre. (33)

demi-long, -longue (*adj.*), of medium length. (34)

demi-métallique (*adj.*), semimetallic. (35)

demi-nœud (*n.m.*), overhand knot ; simple knot ; single knot. (36)

demi-noyau (*n.m.*), half-core. (37)

demi-opacité (*n.f.*), semiopacity. (38)

demi-opale (*n.f.*), semiopal. (39)

demi-opaque (*adj.*), semiopaque. (40)

demi-pente (galerie de mine) (*n.f.*), slant ; run. (41)

demi-pincette (*n.f.*), half-elliptic spring ; semi-elliptic spring. (42)

demi-plaque (Photogr.) (*n f.*), half-plate : (43)
chambre noire demi-plaque de voyage (*f.*), half-plate field-camera. (44)

châssis-presses demi-plaque (*m.pl.*), half-plate printing-frames. (45)

demi-pont pour le rompu (tour) (*m.*), half bridge-piece for the gap (lathe). (46)

demi-pose (Photogr.) (*n.f.*), bulb exposure. (47)
See pose, demi-pose et l'instantané *or* obturateur toujours armé.

demi-rompu (tour) (*n.m.*), half-gap. (48)

demi-rond, -e *ou* **1/2 rond, -e** (*adj.*), half-round : (49)
lime demi-ronde *ou* lime 1/2 ronde (*f.*), half-round file. (50)

demi-rond *ou* **demi-rondin** (*n.m.*), half-round. (51)

demi-solide (*adj.*), semisolid. (52)

demi-stratifié, -e (*adj.*), semistratified. (53)

demi-tordu (*n.m.*), half-twist. (54)

demi-tour (d'une serrure) (*n.m.*), latch-bolt, spring-bolt, (of a lock). (55)

demi-translucide (*adj.*), semitranslucent. (56)

demi-transparence (*n.f.*), semitransparency. (57)

demi-transparent, -e (*adj.*), semitransparent. (58)

demi-varlope (*n.f.*), jack-plane. (59)

demie (*n.f.*), half. (60)

demoiselle (de paveur) (*n.f.*), beetle ; rammer ; earth-rammer. (61)

demoiselle (Géol.) (*n.f.*), erosion column ; chimney-rock. (62)

démontable (*adj.*), capable of being taken to pieces ; made in sections ; sectional ; removable ; parting : (63)
machine facilement démontable (*f.*), machine which can easily be taken to pieces. (64)
machine démontable pour le transport à dos de mulets (*f.*), machine made in sections for mule-back transportation. (65)
concasseur démontable (*m.*), sectional rock-breaker. (66)
table démontable (d'une machine) (*f.*), removable table. (67)
poulie démontable (*f.*), parting pulley. (68)

démontage (*n.m.*), taking to pieces ; taking down ; unmaking : (69)
démontage de la sonde, unmaking the joints of the rods. (70)

démonter (*v.t.*), to take to pieces ; to take down ; to dismount ; to unmake : (71)
démonter une machine, to take a machine to pieces ; to dismount a machine. (72)
démonter un joint, to unmake a joint. (73)

démontrable (*adj.*), demonstrable ; provable. (74)

démontrer (*v.t.*), to demonstrate ; to prove ; to show ; to point to. (75)

démoulage des modèles (Fonderie) (*m.*), withdrawal, withdrawing, drawing, of patterns. (76)

démoulage des pièces coulées (*m.*), stripping castings : (77)
le démoulage de la pièce coulée s'opère le lendemain du jour de la coulée, the stripping of the casting is done the day after the day of casting. (78)

démouler un modèle (Fonderie), to withdraw, to draw, a pattern. (79)

démouler une pièce coulée, un lingot, to strip a casting, an ingot. (80 *ou* 81)

démultiplication (Engrenages) (*n.f.*), reduction : (82)
démultiplication obtenue au moyen d'engrenages droits, reduction obtained by means of spur-gears. (83)

dendriforme (*adj.*), dendriform. (84)

dendrite (Minéralogie) (*n.f.*), dendrite : (1)
cristaux qui se groupent en dendrites (*m.pl.*),
crystals which are grouped in dendrites. (2)
dendritique (*adj.*), dendritic. (3)
dendroïde (*adj.*), dendroid. (4)
dendrolithe (*n.m.*), dendrolite. (5)
déniveler (*v.t.*), to throw out of level. (6)
dénivellation (*n.f.*) *ou* **dénivellement** (*n.m.*),
oscillations of level ; dislevelment ; deflection
from the level ; drop : (7)
dénivellation des masses montagneuses, oscil-
-lations of level of mountain masses. (8)
dénivellation de la voie (Ch. de f.), oscillations
of level of the road. (9)
une dénivellation brusque du terrain, a sudden
drop in the ground. (10)
dénivellation piézométrique (Hydraul.) (*f.*),
pressure-drop ; pressure-loss ; loss of head ;
loss of pressure. (11)
dénomination (*n.f.*), denomination ; name ;
names : (12)
dénomination des différentes parties d'un
engrenage cylindrique, names of the various
parts of a spur-gear. (13)
dénoyage (d'une mine) (*n.m.*), unwatering,
bailing out, (a mine). (14)
dénoyer (*v.t.*), to unwater ; to bail out. (15)
dense (*adj.*), dense : (16)
métal dense (*m.*), dense metal. (17)
densement (*adv.*), densely. (18)
densimètre (*n.m.*), densimeter. (19)
densimétrie (*n.f.*), densimetry. (20)
densité (*n.f.*), density. (21)
densité (Phys.) (*n.f.*), density ; specific gravity :
(22)
densité des solides, des liquides, des gaz,
density, specific gravity, of solids, of liquids,
of gases. (23 *ou* 24 *ou* 25)
densité de flux *ou* **densité magnétique** (*f.*), flux
density ; magnetic flux density ; magnetic
density ; magnetic induction. (26)
densité de vapeur (Chim.) (*f.*), vapour-density.
(27)
densité du courant (Élec.) (*f.*), current-density ;
density of the current. (28)
densité gravimétrique (*f.*), gravimetric density.
(29)
densité superficielle (Élec.) (*f.*), surface-density.
(30)
dent (*n.f.*), tooth ; cog : (31)
dents chevronnées (Engrenages), herring-bone
teeth (Gearing). (32)
dents couchées (scie), ripping teeth (saw). (33)
dents crochues *ou* dents à crochets (scie),
hook teeth ; hooked teeth. (34)
dents d'engrenage, gear-teeth ; wheel-teeth ;
cogs. (35)
dents d'une fourche, prongs, teeth, tines,
of a fork. (36)
dents d'une roue d'engrenage, d'une crémaillère,
teeth, cogs, of a gear-wheel, of a rack. (37
ou 38)
dents d'une scie, d'une lime, teeth of a saw,
of a file. (39 *ou* 40)
dents dégagées. Same as **denture à dépouille.**
dents droites (scie), peg-teeth ; fleam-teeth.
(41)
dents pour couper en travers (scie), cross-cut
teeth. (42)
dents taillées à la machine, machine-cut teeth.
(43)
denté, -e (*adj.*), toothed ; cogged : (44)

roue dentée (*f.*), toothed wheel. (45)
dentelé, -e (*adj.*), indented ; notched ; jagged :
(46)
une cime de montagne dentelée, a jagged
mountain peak. (47)
denteler (*v.t.*), to indent ; to notch ; to jag. (48)
dentelure (*n.f.*), indentation. (49)
denter (*v.t.*), to tooth ; to ratch ; to cog : (50)
denter une roue, to tooth, to ratch, to cog,
a wheel. (51)
denture (*n.f.*), teeth : (52)
série de roues de filetage à denture divisée et
taillée à la machine (*f.*), set of change-
wheels with machine-divided and machine-
cut teeth. (53)
denture à dépouille (fraises, etc.) (*f.*), relieved
teeth ; backed off teeth ; cleared teeth
(milling-cutters, etc.). (54)
dénudation (*n.f.*), denudation. (55)
dénudation (Géol.) (*n.f.*), denudation ; dilapida-
-tion : (56)
dénudation des blocs de grès, denudation of
blocks of sandstone. (57)
dénuder (*v.t.*), to denude : (58)
dénuder une forêt, to denude a forest. (59)
dénuder un arbre, to bark a tree. (60)
départ (Chimie métallurgique) (*n.m.*), parting :
(61)
appliquer le procédé du départ par l'acide
azotique à la séparation de l'or et de l'argent
dans leurs alliages, to apply the process of
parting by nitric acid to the separation of
gold and silver in their alloys. (62)
départ avec inquartation d'argent, de cadmium,
parting with inquartation of silver, of
cadmium. (63 *ou* 64)
dépasser (*v.t.*), to exceed : (65)
dépasser la limite d'élasticité, to exceed the
limit of elasticity. (66)
dépasser un signal d'arrêt, to run past a stop-
signal ; to pass beyond a danger-signal. (67)
dépècement du massif (Mines) (*m.*), cutting up
the seam. (68)
dépendances (*n.f.pl.*), outbuildings. (69)
dépense (*n.f.*) *ou* **dépenses** (*n.f.pl.*), expense ;
expenses ; expenditure ; outgoings ; cost :
(70)
dépenses d'épuisement, pumping costs. (71)
dépense (de combustible, de vapeur) (*n.f.*),
consumption (of fuel, of steam). (72 *ou* 73)
dépense (quantité de liquide passant par un orifice
dans un temps donné) (*n.f.*), efflux : (74)
le coefficient de dépense, the coefficient of
efflux. (75)
dépenser (*v.t.*), to spend ; to expend : (76)
dépenser une somme d'environ — francs, to
spend about (*or* something like) fcs —. (77)
déperdition (*n.f.*), loss ; waste ; wastage ;
leakage : (78)
déperdition de calorique (*ou* de chaleur) par
rayonnement, loss of heat by radiation. (79)
déperdition de liquide (pompes), loss of liquid ;
waste of liquid ; slip. (80)
déperdition d'essence de pétrole par évaporation,
wastage of petroleum spirit by evaporation.
(81)
déperdition d'électricité, leakage of electricity.
(82)
déphasage (Élec.) (*n.m.*), dephasing. (83)
déphaser (Élec.) (*v.t.*), to dephase. (84)
déphlegmateur (*n.m.*), dephlegmator. (85)
déphosphoration (*n.f.*), dephosphorization : (86)

déphosphoration de l'acier, dephosphorization of steel. (1)

déphosphorer (*v.t.*), to dephosphorize. (2)

dépilage (*n.m.*) *ou* **dépilage des piliers** (Mines), robbing the pillars ; removing the pillars ; extracting the pillars ; cutting away the pillars ; pulling pillars ; drawing back ; pillar-drawing ; pillar-working ; second working ; broken working ; working in the broken. (3)

dépilage d'une jambe de charbon (*m.*), drawing an entry. (4)

dépilage en long avec planche protectrice (*m.*), rib-and-pillar method. (5)

dépilage en retour *ou* **dépilage en retraite** (*m.*), working on the retreat ; working home ; working homewards ; homewards working ; mining-retreating. (6)

dépilage peu à peu des piliers (*m.*), cutting away the pillars gradually ; thinning of the pillars. (7)

dépilement (*n.m.*). Same as **dépilage**.

dépiler les piliers (Mines), to rob, to remove, to cut away, to draw, to pull, to extract, to take out, the pillars ; to work in the broken. (8)

déplacement (changement de place) (*n.m.*), dis-placement ; movement : (9)

le déplacement du piston, du tiroir, the displacement of the piston, of the slide-valve. (10 *ou* 11)

déplacements des lignes de rivage (Géogr. phys.), displacements of shore-lines. (12)

déplacement longitudinal du porte-foret à la main sur crémaillère, longitudinal move-ment of the drill-holder by hand on a rack. (13)

déplacement (transport d'une place dans une autre) (*n.m.*), removing ; removal; shifting : (14)

déplacement de bornes, removing landmarks. (15)

déplacement de failles (Géol.), shifting of faults. (16)

déplacement des machines, removal of, removing, shifting, the machinery. (17)

déplacement latéral (d'une faille) (Géol. & Mines) (*m.*), heave. (18)

déplacer (*v.t.*), to displace ; to remove ; to move ; to shift : (19)

déplacer le chariot à la main (Méc.), to move the carriage by hand. (20)

déplacer la coulisse (distribution par coulisse), to shift the link (link-gear). (21)

déplacer (se) (*v.r.*), to shift ; to get out of place. (22)

dépolarisant (*n.m.*), depolarizer. (23)

dépolarisation (Opt. & Élec.) (*n.f.*), depolariza-tion : (24)

dépolarisation de la lumière, depolarization of light. (25)

dépolariser (Opt. & Élec.) (*v.t.*), to depolarize. (26)

déporter (évider ; dégager) (*v.t.*), to cut away : (27)

contre-pointe (de tour) avec corps déporté (*f.*), tailstock (of lathe) cut away in front. (28)

déposer (*v.t.*), to deposit : (29)

eau qui dépose de la vase (*f.*), water which deposits mud. (30)

déposer (*v.i.*), to deposit ; to settle. (31)

déposer par électrolyse *ou* **déposer électrolitique-ment** (*v.t.*), to electrodeposit : (32)

déposer de l'or électrolytiquement, to electro-deposit gold. (33)

déposer une demande de concession de mines, to make an application for, to apply for, a mining-licence. (34)

déposer (se) (*v.r.*), to deposit ; to settle : (35)

enlever le carbonate de chaux qui se dépose dans les chaudières, to remove the carbonate of lime which deposits (*or* settles) in boilers. (36)

dépôt (action) (*n.m.*), depositing ; deposition ; settling ; settlement : (37)

dépôt du cuivre, depositing copper. (38)

dépôt électrolytique de l'argent, electrolytic deposition of silver. (39)

dépôt des boues, settling of slimes. (40)

dépôt (*n.m.*) *ou* **dépôts** (*n.m.pl.*) (matières solides qu'abandonne un liquide au repos), deposit ; sediment ; settlings. (41)

dépôt (*n.m.*) *ou* **dépôts** (*n.m.pl.*) (sels d'incrusta-tions dans les chaudières à vapeur), scale ; boiler-scale ; fur ; furring ; incrustation. (42)

dépôt (*n.m.*) *ou* **dépôts** (*n.m.pl.*) (Géol.), deposit ; bed. (43)

dépôt (magasin) (*n.m.*), depot ; store ; store-house ; storeroom ; warehouse ; shed ; yard. (44)

dépôt alluvien (*m.*), alluvial deposit. (45)

dépôt d'eau douce (Géol.) (*m.*), fresh-water deposit. (46)

dépôt d'estuaire (Géol.) (*m.*), estuarine deposit ; estuarine bed. (47)

dépôt de bois (*m.*), timber-yard. (48)

dépôt de dynamite (*m.*), dynamite-store. (49)

dépôt de machines (Ch. de f.) (*m.*), engine-shed ; locomotive-shed. (50)

dépôt de marchandises (*m.*), goods-yard ; goods-depot. (51)

dépôt de mer (*m.*), marine bed ; marine deposit. (52)

dépôt de wagons (*m.*), wagon-shed ; car-shed. (53)

dépôt des cendres (*m.*), ash-dump. (54)

dépôt des déblais (action) (*m.*), dumping the waste : (55)

l'emplacement choisi pour le dépôt des déblais (*m.*), the site chosen for dumping the waste. (56)

dépôt des déblais (chantier) (*m.*), dump ; dump-ing-ground ; tip : (57)

l'emplacement choisi pour le dépôt des déblais, the site chosen for the dump. (58)

dépôt détritique (Géol.) (*m.*), detrital deposit. (59)

dépôt erratique (argile à blocaux) (Géol.) (*m.*), drift ; glacial drift ; till ; glacial till ; boulder-clay ; bowlder-clay. (60)

dépôt fluviatile *ou* **dépôt d'origine fluviale** (*m.*), fluviatile deposit. (61)

dépôt fluvio-marin (*m.*), fluviomarine deposit. (62)

dépôt galvanoplastique (*m.*), electrodeposit ; electroplating. (63)

dépôt houiller (*m.*), coal deposit. (64)

dépôt lacustre (*m.*), lacustrine deposit. (65)

dépôt littoral (*m.*), littoral deposit. (66)

dépôt marin (*m.*), marine bed ; marine deposit. (67)

dépôt meuble (Géol.) (*m.*), loose deposit. (68)
Cf. dépôts meubles sur les pentes.

dépôt souterrain de dynamite (*m.*), underground dynamite-store. (69)

dépôts meubles sur les pentes (Géol.) (*m.pl.*), slide. (1)

dépouille (dégagement) (Méc.) (*n.f.*), relief ; clearance ; backing off. (2)

dépouille (augmentation donnée à certaines parties d'un modèle en vue de faciliter sa sortie du moule) (Fonderie) (*n.f.*), draught ; draft ; draw-taper ; taper ; delivery ; strip. (3)

dépouille (découverte) (Métall.) (*n.f.*), uncover-ing. (4)

dépouillement (dénudation) (*n.m.*), stripping : (5)

perdre la meilleure partie de la recette par un dépouillement insuffisant du fond (Dragage en rivière), to lose the best part of the yield by insufficient stripping of the bottom. (6)

dépouiller (dénuder) (*v.t.*), to strip : (7)

dépouiller un arbre de son écorce, un fil recou-vert de sa matière isolante, to strip a tree of its bark, a covered wire of its insulating material. (8 *ou* 9)

dépouiller le gîte (Mines), to strip the over-burden. (10)

dépouiller (dégager) (Méc.) (*v.t.*), to relieve ; to back off ; to clear : (11)

dépouiller un foret, un taraud, un alésoir, une fraise, to relieve, to back off, to clear, a drill, a tap, a reamer, a milling-cutter. (12 *ou* 13 *ou* 14 *ou* 15)

dépouiller (Fonderie) (*v.t.*), to taper : (16)

dépouiller un modèle, to taper a pattern. (17)

dépouiller (se) (Métall.) (*v.r.*), to uncover. (18)

dépourvu, -e (*p.p*), devoid ; destitute ; barren ; free ; —less (*suffix*) : (19)

dépourvu (-e) d'arbres, treeless : (20)

région dépourvue d'arbres (*f.*), treeless region. (21)

dépourvu (-e) d'eau, waterless. (22)

dépourvu (-e) d'or (se dit d'un pays, d'une région), devoid of gold ; destitute of gold ; goldless ; gold-free. (23)

dépourvu (-e) de feldspath, feldspar-free : (24)

roches dépourvues de feldspath (*f.pl.*), feld-spar-free rocks. (25)

dépourvu (-e) de mentonnet, flangeless : (26)

roue dépourvue de mentonnet (*f.*), flangeless wheel. (27)

dépourvu (-e) de pluie (se dit d'un pays, d'une région), rainless. (28)

dépoussiérage (*n.m.*), removing the dust ; sweeping up the dust. (29)

dépoussiérer (*v.t.*), to sweep up the dust. (30)

dépréciation (*n.f.*), depreciation ; falling off in value : (31)

dépréciation du matériel, depreciation of plant. (32)

déprécier (*v.t.*), to depreciate. (33)

déprécier (se) (*v.r.*), to depreciate ; to fall in value. (34)

dépressif, -ive (*adj.*), depressive. (35)

dépression (*n.f.*), depression : (36)

dépression barométrique, barometric depres-sion. (37)

dépression capillaire du baromètre, capillary depression of the barometer. (38)

dépression du sol, depression in the ground. (39)

dépression océanique, oceanic depression. (40)

dépression (*n.f.*) *ou* **dépression motrice** (Ventila-tion), depression : (41)

la méthode la meilleure pour créer la dépression nécessaire à la circulation des gaz dans les générateurs est le ventilateur, the best method of creating the necessary depression for the circulation of the gases in boilers is the fan. (42)

déprimer (*v.t.*), to depress ; to cause to sink : (43)

la capillarité déprime la hauteur de la colonne barométrique, capillarity depresses the height of the barometric column. (44)

fleuve qui déprime le sol (*m.*), river which causes the ground to sink. (45)

dépuration (*n.f.*), purification. (46)

dépurer (*v.t.*), to purify : (47)

dépurer un métal, to purify a metal. (48)

dépurer (se) (*v.r.*), to purify ; to become clean : (49)

l'eau se dépure par la distillation, water purifies by distillation. (50)

déraillement (*n.m.*), derailment ; going off the line ; running off the rails ; leaving the metals ; jumping the track. (51)

dérailler (*v.i.*), to derail ; to go off the line ; to jump the metals ; to leave the track ; to run off the rails. (52)

Cf. **faire dérailler** and **train déraillé**.

dérangement (*n.m.*), derangement ; disarrange-ment ; disturbance ; unsettled state : (53)

dérangement d'une machine, derangement of a machine. (54)

dérangement des affaires, disturbance, unsettled state, of business. (55)

dérangement du temps, unsettled state, inclemency, of the weather. (56)

dérangement (Géol.) (*n.m.*), disturbance ; fault-ing ; faultage. (57)

dérangement (Élec.) (*n.m.*), fault : (58)

recherche des dérangements (*f.*), locating faults. (59)

déranger (*v.t.*), to derange ; to put out of order ; to disturb ; to disarrange : (60)

déranger une serrure, le rouage d'une machine, to derange a lock, the wheelwork of a machine. (61 *ou* 62)

déranger les plans de quelqu'un, to disturb someone's calculations ; to interfere with someone's plans. (63)

déranger (se) (*v.r.*), to become deranged ; to get out of order. (64)

dérapage (*n.m.*), skidding. (65)

déraper (*v.i.*), to skid. (66)

déréglé, -e (*p.p.*), out of order. (67)

dérivateur (pour lampe à arc) (Élec.) (*n.m.*), shunter ; cut-out. (68)

dérivation (détournement) (*n.f.*), diversion ; diverting : (69)

dérivation d'un cours d'eau, diversion of a stream ; diverting a stream. (70)

dérivation (canal de dérivation) (Hydraul.) (*n.f.*), feeder ; lead ; leat ; flume ; penstock ; pentstock ; pentrough. (71)

dérivation (Ch. de f.) (*n.f.*), loop ; loop-line. (72)

dérivation (Géol.) (*n.f.*), offshoot ; apophysis. (73)

dérivation (Aérage de mines) (*n.f.*), split (the workings ventilated by a divided air-cur-rent) : (74)

la dépression décroît en raison inverse du cube du nombre des dérivations, the depression decreases inversely as the cube of the number of splits. (75)

dérivation (Élec.) (*n.f.*), shunt : (1)
une dérivation prise sur le circuit général, alimente quelques lampes à incandescence, qui servent à l'éclairage de la voiture, a shunt taken on the general circuit, feeds several incandescent lamps, which serve to light the car. (2)

dérivation (en) (Élec.), in shunt ; shunt (*used as adj.*) : (3)
une lampe en dérivation sur un réseau d'éclairage, a lamp in shunt on a lighting system. (4)
enroulement en dérivation (*m.*), shunt winding. (5)

dérive (Ch. de f.) (*n.f.*), running back (of wagons, —down an incline) ; running wild ; breaking away ; breakaway. (6)

dérivé (Élec.) (*adj.m.*), shunt (*used as adj.*) ; derived : (7)
circuit dérivé (*m.*), shunt circuit ; derived circuit. (8)
courant dérivé (*m.*), shunt current. (9)

dérivé (Chim.) (*adj. m.*), derived : (10)
un produit dérivé, a derived product. (11)

dérivé (*n.m.*), derivative ; derived product : (12)
un sel est un dérivé de l'acide correspondant, a salt is a derivative of the corresponding acid. (13)
dérivé du pétrole, product derived from petroleum. (14)

dériver (détourner de son cours) (*v.t.*), to divert : (15)
dériver un cours d'eau, to divert a stream. (16)

dériver (Élec.) (*v.t.*), to shunt : (17)
dériver un courant électrique, to shunt an electric current. (18)

dériver (défaire la rivure de) (*v.t.*), to unhead : (19)
dériver un clou, to unhead a rivet. (20)

dériver (défaire ce qui est rivé) (*v.t.*), to unrivet : (21)
dériver une tôle de chaudière, to unrivet a boiler-plate. (22)

dériver (se) (perdre sa rivure) (*v.r.*), to become unrivetted. (23)

dernbachite (Minéral) (*n.f.*), dernbachite. (24)

dernier, -ère (*adj.*), last ; latest : (25)
dernier degré de perfection (*m.*), highest degree of perfection. (26)
dernier prix (*m.*), lowest price ; bed-rock price ; rock-bottom price. (27)
dernière main (*f.*), finishing touches : (28)
mettre la dernière main à un ouvrage, to put the finishing touches to a work. (29)
dernière recette du fond (Mines) (*f.*), bottom station. (30)
derniers mois de l'année (*m.pl.*), latter part of the year. (31)
derniers perfectionnements (*m.pl.*), latest improvements. (32)

dérocher le métal (le séparer de sa gangue), to separate metal from its gangue. (33)

dérouillement (*n.m.*), rubbing off rust. (34)

dérouiller (*v.t.*), to rub the rust off : (35)
dérouiller un couteau, to rub the rust off a knife. (36)

déroulement (*n.m.*), unwinding ; uncoiling ; unrolling ; unfolding. (37)

dérouler (*v.t.*), to unwind ; to uncoil ; to unroll ; to unfold : (38)
dérouler la corde d'un treuil, to unwind the rope from a winch. (39)

dérouler ses plans à quelqu'un, to unfold one's plans to someone. (40)

dérouler (se) (*v.r.*), to unwind ; to uncoil : (41)
fil qui se déroule de la bobine (*m.*), wire which unwinds from the spool. (42)

derrick (Exploitation des puits à pétrole, etc.) (*n.m.*), derrick. (43)

derrick (chèvre verticale) (Constr.) (*n.m.*), derrick ; derrick-crane. (44)

derrick à trois pieds (*m.*), tripod derrick. (45)

derrick de puits à pétrole (*m.*), oil-well derrick ; carpenters' rig. (46)

désaccoupler (*v.t.*), to uncouple. (47)

désagrégation (*n.f.*), disintegration ; crumbling ; crumbling away : (48)
la désagrégation de maçonneries de briques, the crumbling of brickwork. (49)

désagrégation (Géol.) (*n.f.*), disintegration ; weathering : (50)
le sable est une matière minérale, pulvérulente, composée de grains généralement fins, provenant de la désagrégation des roches siliceuses ou calcaires, sand is a mineral matter, powdery, composed of grains generally fine, resulting from the disintegration of siliceous or limestone rocks. (51)

désagrégeable (*adj.*), disintegrable. (52)

désagrégeant, -e (*adj.*), disintegrating ; weathering. (53)

désagréger (*v.t.*), to disintegrate ; to weather. (54)

désagréger (se) (*v.r.*), to disintegrate ; to crumble ; to weather. (55)

désaimantation (*n.f.*), demagnetization. (56)

désaimanter (*v.t.*), to demagnetize : (57)
désaimanter un barreau d'acier, to demagnetize a steel bar. (58)

désaimanter (se) (*v.r.*), to become demagnetized. (59)

désamorçage (Magnétisme) (*n.m.*), failure to generate ; loss of excitation ; deenergization. (60)

désamorçage d'un injecteur (opp. à amorçage) (*m.*), stopping, failure, of an injector. (61) (opp. to starting).

désamorcer (Magnétisme) (*v.t.*), to deenergize. (62)

désamorcer un injecteur, to stop an injector. (63)

désamorcer une cartouche, to unprime a cartridge. (64)

désamorcer une pompe, to drain, to dewater, a pump : (65)
on désamorce les pompes quand il gèle, pour éviter que l'eau qu'elles contiennent ne fasse éclater les parois en se congelant, pumps are drained in frosty weather, to prevent the water they contain bursting the walls in congealing. (66)

désamorcer (se) (Magnétisme) (*v.r.*), to fail to generate ; to deenergize : (67)
dynamo qui se désamorce lorsque la résistance du circuit total dépasse une certaine limite (*f.*), dynamo which fails to generate when the resistance of the full circuit exceeds a certain limit. (68)

désamorcer (se) (injecteurs) (*v.r.*), to stop ; to fail : (69)
les injecteurs en général, ont tous le défaut de se désamorcer dès qu'ils sont exposés à des chocs ou secousses, comme celà arrive sur les locomotives, etc. (*m.pl.*), injectors in

general, all have the defect of stopping (or failing) as soon as they are subjected to shocks or jars, as happens on locomotives, etc. (1)

désancrage (d'une cheminée) (Mines) (n.m.), starting, unkeying, loosening, (a clogged chute). (2)

désancrer une cheminée, to start, to unkey, to loosen, a clogged chute. (3)

désancreur (Mines) (n.m.), cannon (for loosening a clogged chute). (4)

désappointer (casser la pointe de) (v.t.), to break the point of : (5)
désappointer une aiguille, to break the point of a needle. (6)

désappointer (émousser la pointe de) (v.t.), to blunt ; to dull : (7)
désappointer un outil, to blunt, to dull, a tool. (8)

désargentage (n.m.) ou **désargentation** (n.f.), desilvering ; desilverization : (9)
désargentation des minerais de plomb, du cuivre argentifère, desilverization of lead ores, of argentiferous copper. (10 ou 11).
désargentation du plomb d'œuvre par pattin-sonage, desilverization of work-lead by pattinsonation. (12)
désargentation par imbibition, desilverization by soaking. (13)

désargenter (v.t.), to desilverize ; to desilver : (14)
désargenter un amalgame d'or et d'argent, to desilverize a gold-silver amalgam. (15)
désargenter le minerai d'argent, to reduce silver-ore. (16)

désargenteur (n.m.), desilverizer. (17)

désassemblage (n.m.), disassembling ; taking to pieces ; disjointing ; disconnecting ; un-coupling. (18)

désassembler (v.t.), to disassemble ; to take to pieces ; to disjoint ; to disconnect ; to uncouple : (19)
désassembler les planches d'une cloison, to disjoint the boards of a partition. (20)
désassembler les rallonges (Sondage), to disconnect, to uncouple, the rods. (21)

désastre (n.m.), disaster ; catastrophe. (22)

désastreusement (adv.), disastrously. (23)

désastreux, -euse (adj.), disastrous. (24)

désattelage (n.m.), uncoupling ; detachment. (25)

désatteler (v.t.), to uncouple ; to detach : (26)
désatteler des wagons, to uncouple trucks. (27)

désavantage (n.m.), disadvantage. (28)

désavantagé (-e) par suite du manque de moyens de transport (être), to be handicapped for want of transport facilities. (29)

désaxage (n.m.), setting over ; throwing off centre. (30)

désaxé, -e (adj.), set over ; thrown off centre ; eccentric : (31)
came désaxée (f.), eccentric cam. (32)

désaxer (v.t.), to set over ; to throw off centre : (33)
désaxer la contre-pointe d'un tour, to set over the tailstock of a lathe ; to throw the loose head of a lathe off centre. (34)

descendant, -e (adj.), descending ; downgoing ; falling ; downward ; down : (35)
benne descendante (f.), descending bucket, downgoing bucket. (36)

marée descendante (f.), falling tide. (37)
mouvement descendant (m.), downward motion. (38)
train descendant (m.), down train. (39)

descenderie (opp. à montage, remontage, ou remontée) (Mines) (n.f.), winze ; way-shaft ; winds. (40)

descenderie de remblai ou **descenderie de remblais** (f.), filling-raise ; waste-raise. (41)

descendre (parcourir de haut en bas) (v.t.), to go down ; to descend : (42)
descendre une montagne, une rivière, to go down, to descend, a mountain, a river. (43 ou 44)

descendre (mettre ou porter plus bas) (v.t.), to lower ; to drop ; to take down : (45)
descendre les ouvriers, les chevaux (Mines), to lower workmen, horses. (46 ou 47)
descendre le minerai à un étage inférieur, to drop the ore to a lower level. (48)

descendre (aller de haut en bas) (v.i.), to come down ; to go down ; to descend : (49)
descendre d'une montagne, to come down from a mountain. (50)

descendre (s'étendre jusqu'en bas) (v.i.), to go down : (51)
le puits descend à — mètres environ, the shaft goes down about — metres. (52)

descendre (baisser) (v.i.), to fall ; to go down ; to go out : (53)
le thermomètre descend, the thermometer is falling. (54)
la mer descend, the sea is going down. (55)
la marée descend, the tide is going out. (56)

descendre (être disposé en pente) (v.i.), to go downhill : (57)
chemin qui descend (m.), road which goes downhill. (58)

descendre à terre (débarquer) (v.i.), to land. (59)

descente (action d'aller d'un point élevé à un autre plus bas) (n.f.), descent ; descending ; fall : (60)
descente dans la benne, dans une boucle du câble (Mines), descent in the kibble, in a loop at the end of the rope. (61 ou 62)
descente de la charge (fourneau), descent of the charge. (63)
descente automatique du porte-outil, auto-matic fall of the tool-holder. (64)

descente (action de mettre ou porter une chose en bas) (n.f.), lowering : (65)
descente des ouvriers, des chevaux, des bois, des remblais, de pompes dans le puits, lowering workmen, horses, timber, gobbing-material, pumps into the shaft. (66 ou 67 ou 68 ou 69 ou 70)

descente (pente) (n.f.), slope ; incline ; descent : (71)
une descente douce, a gentle slope ; an easy incline. (72)
une descente escarpée, a precipitous descent. (73)

descente (descenderie) (Mines) (n.f.), winze ; way-shaft ; winds. (74)

descente (n.f.) ou **descente d'eau** (tuyau de descente des eaux pluviales), downpipe ; rain-water downpipe ; downcomer ; leader ; stack-pipe. (75)

descloizite (Minéral) (n.f.), descloizite. (76)

descriptif, -ive (adj.), descriptive : (77)
géométrie descriptive (f.), descriptive geom-etry. (78)

description (*n.f.*), description. (1)

désembrayage (*n.m.*), désembrayer (*v.t.*). Same as débrayage, débrayer.

désengageur (Ch. de f.) (*n.m.*), disengaging-lever. (2)

désengrenage (*n.m.*), throwing out of gear; putting out of gear. (3)

désengrené, -e (*adj.*), out of gear. (4)

désengrener (*v.t.*), to throw out of gear; to put out of gear; to disengage: (5)
désengrener les roues d'une machine, to throw the wheels of a machine out of gear. (6)

désengrener (se) (*v.r.*), to be thrown out of gear. (7)

désensabler le tubage (Exploitation des puits à pétrole), to free the casing from sand. (8)

désensableur (Sondage) (*n.m.*), sand-pump; sludger, sludge-pump. (9)

désensibilisateur (Photogr.) (*n.m.*), desensitizer. (10)

désensibilisation (*n.f.*), desenitization. (11)

désensibiliser (*v.t.*), to desensitize: (12)
désensibiliser une plaque photographique, to desensitize a photographic plate. (13)

déséquilibrer (*v.t.*), to disequilibrate; to throw out of balance. (14)

déséquiper une drague, to dismantle a dredge. (15)

désert, -e (*adj.*), deserted; abandoned: (16)
une contrée déserte, a deserted country. (17)

désert (*n.m.*), desert. (18)

désertique (*adj.*), desert: (19)
région désertique (*f.*), desert region. (20)

déshuilheur de vapeur (*m.*), oil-separator; steam-separator. (21)

déshydratation (*n.f.*), dehydration; anhydra-tion: (22)
déshydratation des sels, dehydration, anhydra-tion, of salts. (23)

déshydrater (*v.t.*), to dehydrate; to anhydrate: (24)
déshydrater de la chaux, to dehydrate, to anhydrate, lime. (25)

désincrustant (*n.m.*), disincrustant. (26)

désincrustation (*n.f.*), scaling; furring; removing incrustation; preventing incrustation. (27)

désincruster (*v.t.*), to scale; to remove scale from; to fur: (28)
désincruster les parois d'une chaudière, to scale the walls of a boiler; to fur the inside of a boiler. (29)

désintégrateur (*n.m.*), disintegrator. (30)

désintégration (*n.f.*), disintegration. (31)

désintégrer (*v.t.*), to disintegrate. (32)

désintégrer (se) (*v.r.*), to disintegrate. (33)

désintéressé, -e (*adj.*), disinterested; unbiassed. (34)

desmine (Minéral) (*n.f.*), desmine; stilbite. (35)

désobstruer (*v.t.*), to clear; to clear from obstruction. (36)

désolé, -e (*adj.*), desolate: (37)
une région désolée, a desolate region. (38)

désordre (*n.m.*), disorder. (39)

désordres ouvriers (*m.pl.*), labour disturbances; labour troubles. (40)

désorganisateur, -trice (*adj.*), disorganizing. (41)

désorganisation (*n.f.*), disorganization. (42)

désorganiser (*v.t.*), to disorganize. (43)

désoufrage (*n.m.*), desulphurization; de-sulphuration. (44)

désoufrer (*v.t.*), to desulphurize; to desulphurate. (45)

désoxydant, -e (*adj.*), deoxidizing. (46)

désoxydant (*n.m.*), deoxidizer. (47)

désoxydation (*n.f.*), deoxidization: (48)
désoxydation d'un métal, deoxidation of a metal. (49)

désoxyder (*v.t.*), to deoxidize: (50)
désoxyder le bain (Métall.), to deoxidize, to reduce, the bath. (51)

dessablage (des pièces coulées) (Fonderie) (*n.m.*), cleaning, dressing, (castings) (removing the sand): (52)
dessablage des petites pièces au grattoir, à la brosse, au tonneau dessableur, au jet de sable, cleaning small castings with the scraper, with the brush, in the tumbling-barrel, by sand-blast. (53 *ou* 54 *ou* 55 *ou* 56)

dessablage au tonneau (Fonderie) (*m.*), tumbling; rattling; cleaning in the tumbling-barrel: (57)
le dessablage mécanique au tonneau convient particulièrement aux petites pièces, mechanical tumbling (*or* rattling) is specially suitable for small castings. (58)

dessabler (*v.t.*), to clean; to dress. (59)

dessabler au tonneau, to tumble; to rattle; to clean in the tumbler. (60)

dessableur (Fonderie) (pers.) (*n.m.*), cleaner; dresser. (61)

dessalure (*n.f.*), freshening: (62)
endroits où des fleuves ont anené une dessalure des eaux marines (*m.pl.*), places where rivers have brought about a freshening of sea-water. (63)

desséchant, -e (*adj.*), drying: (64)
vents desséchants (*m.pl.*), drying winds. (65)

desséchement (*n.m.*), drying up; draining: (66)
desséchement du sol, drying up of the ground. (67)
desséchement d'un marais, draining a marsh. (68)

dessécher (*v.t.*), to dry up; to drain: (69)
dessécher un étang, to drain a pond. (70)

dessécher les bois, to season timber. (71)

dessécher (se) (*v.r.*), to dry up. (72)

dessécheur (*n.m.*), dryer; drier. (73)

dessécheur de vapeur (*m.*), steam-drier. (74)

desserrage (action) (*n.m.*), loosening; slacking. (75)

desserrage (état) (*n.m.*), looseness; slackness: (76)
desserrage d'un écrou, looseness, slackness, of a nut. (77)

desserrage (Freinage) (Ch. de f.), brakes off. (78)

desserrage à fond (Ch. de f.), brakes full off. (79)

desserré, -e (*adj.*), loose; slack: (80)
écrou desserré (*m.*), loose nut; slack nut. (81)

desserrer (*v.t.*), to loosen; to slack: (82)
desserrer un écrou, to loosen, to slack, a nut. (83)

dessertir (*v.t.*), to unset. (84)

dessertir (se) (*v.r.*), to become unset: (85)
sondage qui traverse des terrains hétérogènes, dans lesquels les diamants risquent de se dessertir (*m.*), bore which passes through heterogeneous strata, in which the diamonds are liable to become unset. (86)

desservir (faire le service de) (*v.t.*), to serve: (87)
ligne de chemin de fer qui dessert le Centre et le Midi (*f.*), railway-line which serves the Midlands and the South. (88)
champs d'exploitation desservis par plusieurs puits (*m.pl.*), winnings served by several shafts. (89)

dessiccateur (*n.m.*), dryer ; desiccator. (1)

dessiccatif, -ive (*adj.*), drying. (2)

dessiccation (*n.f.*), desiccation ; drying. (3)

dessiccation des bois (*f.*), seasoning timber. (4)

dessin (art) (*n.m.*), drawing : (5)
  apprendre le dessin, to learn drawing. (6)

dessin (représentation) (*n.m.*), drawing ; design ; sketch. (7)

dessin à la plume (*m.*), pen-drawing. (8)

dessin à main levée (*m.*), free-hand drawing. (9)

dessin au crayon (*m.*), pencil-drawing. (10)

dessin au trait (*m.*), sketch ; sketch drawn in outline ; outline. (11)

dessin d'architecture (*m.*), architectural drawing. (12)

dessin de coupe (*m.*), section-drawing. (13)

dessin de machine *ou* dessin industriel (*m.*), machine-drawing. (14)

dessin en demi-grandeur naturelle *ou* dessin 1/2 nature (*m.*), half-size drawing. (15)

dessin en grandeur naturelle *ou* dessin en vraie grandeur *ou* dessin nature (*m.*), full-size drawing. (16)

dessin géométrique (*m.*), geometrical drawing ; mechanical drawing. (17)

dessin graphique (*m.*), graphic drawing ; diagrammatic sketch. (18)

dessin lavé (*m.*), wash drawing. (19)

dessin linéaire (*m.*), linear drawing ; lineal design. (20)

dessin ombré (*m.*), shaded drawing. (21)

dessin 1/4 nature (*m.*), quarter-size drawing. (22)

dessin schématique (*m.*), diagrammatic sketch ; diagram ; scheme : (23)
  dessin schématique du fonctionnement d'une bielle, diagrammatic sketch of the working of a connecting-rod. (24)

dessinateur (pers.) (*n.m.*), draughtsman ; designer. (25)

dessiner (*v.t.*), to draw ; to design ; to sketch. (26)

dessiner à grands traits, to make a rough sketch of ; to sketch (rapidly or roughly). (27)

dessiner au trait, to draw in outline ; to sketch (in outline). (28)

dessoudé, -e (soudure à l'étain, etc.) (*adj.*), unsoldered. (29)

dessoudé, -e (soudure à chaud) (*adj.*), unwelded. (30)

dessoudure (*n.f.*), unsoldering. (31)

dessous d'étampe (outil de forgeron) (*m.*), bottom swage ; bottom rounding-tool. (32)

dessous de boîte (boîte d'essieu) (*m.*), axle-box cellar ; axle-box sponge-box. (33)

dessous de châssis *ou simplement* dessous (*n.m.*) (Fonderie), drag ; nowel ; bottom part of flask. (34)

dessus d'étampe (outil de forgeron) (*m.*), top swage ; top rounding-tool. (35)

dessus de châssis *ou simplement* dessus (*n.m.*) (Fonderie), cope ; top part of flask. (36)

destinataire (pers.) (*n.m. ou f.*), consignee. (37)

destiner (*v.t.*), to destine ; to intend ; to design : (38)
  les machines électriques sont des appareils destinés à transformer l'énergie mécanique en énergie électrique (*f.pl.*), electric machines are apparatuses intended to convert mechanical energy into electrical energy. (39)

destructeur, -trice *ou* destructif, -ive (*adj.*), destructive : (40)

le feu est un élément destructeur, fire is a destructive element. (41)

destruction (*n.f.*), destruction. (42)

destruction des bancs de graviers (*f.*), breaking down the gravel banks. (43)

désulfurant, -e (*adj.*), desulphurizing. (44)

désulfuration (*n.f.*), desulphurization ; desulphuration. (45)

désulfurer (*v.t.*), to desulphurize ; to desulphurate : (46)
  désulfurer de l'acier, to desulphurize steel. (47)

détachable (*adj.*), detachable. (48)

détacher (*v.t.*), to detach ; to dislodge : (49)
  détacher un bloc de rocher, to dislodge a piece of rock. (50)

détacher (se) (*v.r.*), to become detached ; to come away : (51)
  un gros bloc se détacha et tomba dans le puits, a heavy stone came away and fell down the shaft. (52)

détails (*n.m.pl.*), details ; particulars. (53)

détartrage (*n.m.*), scaling ; furring ; removing scale. (54)

détartrer (*v.t.*), to scale ; to remove scale from ; to fur : (55)
  détartrer une chaudière, to scale, to fur, a boiler. (56)

détartreur (*n.m.*), detartarizer. (57)

détecteur de grisou (*m.*), gas-detector ; fire-damp detector ; gas-indicator ; warner. (58)

détective (*n.m.*), detective camera. (59)

dételage (*n.m.*), uncoupling ; hooking off ; detachment : (60)
  dételage de wagons, uncoupling wagons. (61)

dételer (*v.t.*), to uncouple ; to hook off ; to detach. (62)

détendeur (*n.m.*) *ou* détendeur de pression, relief-valve ; escape-valve : (63)
  détendeur de vapeur, steam relief-valve ; escape-valve for steam. (64)

détendeur (cylindre à basse pression) (*n.m.*), low-pressure cylinder. (65)

détendre (*v.t.*), to slack ; to slacken ; to loosen : (66)
  détendre une corde, to slack a cord ; to slacken a rope. (67)

détendre (se) (se dit des vapeurs) (*v.r.*), to expand : (68)
  en sortant du générateur, la vapeur se détend avant de venir en contact avec les aubes de la roue-turbine, on leaving the boiler, the steam expands before impinging upon the vanes of the turbine-wheel. (69)

détente (Méc.) (*n.f.*), detent ; keeper ; pawl ; click ; trigger ; dog. (70)

détente (expansion) (*n.f.*), expansion : (71)
  la détente d'un gaz, de la vapeur, the expansion of a gas, of steam. (72 *ou* 73)

détente adiabatique (Phys.), adiabatic expansion. (74)

détente (organe pour couper la vapeur afin de permettre la détente) (*n.f.*), expansion-gear ; valve-gear ; valve-motion ; link-motion. (75)

détente automatique (*f.*), automatic expansion-gear. (76)

détente Corliss (*f.*), Corliss valve-gear. (77)

détente d'Allan (*f.*), Allan's link-motion. (78)

détente de Gooch (*f.*), Gooch's link-motion. (79)

détente de Stephenson (*f.*), Stephenson's link-motion. (80)

détente Meyer (*f.*), Meyer valve-gear; Meyer expansion-gear. (1)

détérioration (*n.f.*), deterioration; damage. (2)

détériorer (*v.t.*), to deteriorate; to damage. (3)

détériorer (se) (*v.r.*), to deteriorate; to damage. (4). (*Also used without the reflexive pronoun, as*—laisser détériorer les marchandises, to allow the goods to deteriorate.)

détermination (*n.f.*), determination; fixing: (5)

détermination de la teneur en métal d'un minerai, determination of the metal contents of an ore. (6)

détermination expérimentale du centre de gravité des corps solides, experimental determination of the centre of gravity of solid bodies. (7)

déterminer (*v.t.*), to determine; to fix; to ascertain: (8)

déterminer l'inclinaison des strates, la direction du filon, to determine, to ascertain, the dip of the strata, the strike of the lode. (9 *ou* 10)

déterminer l'emplacement d'un puits, to determine the site of a shaft; to locate a shaft; to fix upon a site for a shaft. (11)

déterminer l'emplacement des outils perdus dans le trou de sonde, to locate tools lost in the bore-hole. (12)

déterrer (*v.t.*), to dig up; to unearth. (13)

détirefonner un rail, to unspike a rail. (14)

détonant, -e (*adj.*), detonating; detonable; explosive. (15)

détonateur (*n.m.*), detonator; exploder; cap; blasting-cap. (16)

détonation (*n.f.*), detonation; explosion; report: (17)

détonation incomplète, incomplete detonation. (18)

détonation violente, violent explosion; loud report. (19)

détoner (*v.i.*), to detonate. (20)

détordre (*v.t.*), to untwist: (21)

détordre une corde, to untwist a cord. (22)

détors, -e (*adj.*), untwisted. (23)

détorsion (*n.f.*), untwisting: (24)

détorsion du câble, untwisting of the rope. (25)

détour (*n.m.*), winding; turn; detour: (26)

les détours d'un souterrain, the windings of an underground passage. (27)

détournement (*n.m.*), diversion; diverting: (28)

détournement d'un cours d'eau, diverting a stream. (29)

détourner (dévier) (*v.t.*), to divert: (30)

détourner une rivière de son cours, un cours d'eau au moyen d'un fossé, to divert a river from its course, a stream by means of a ditch. (31 *ou* 32)

détourner l'eau dans un fossé, to turn water through a ditch. (33)

détourner (défaire ce qui était tourné) (*v.t.*), to wind off: (34)

détourner une corde roulée sur un tambour, to wind off a cord rolled round a drum. (35)

détrempe (Métall.) (*n.f.*), softening (destroying the temper by reheating). (36)

détremper (*v.t.*), to soften: (37)

détremper l'acier, to soften steel. (38)

détret (*n.m.*), hand-vice; pin-vice. (39)

détritique (Géol.) (*adj.*), detritic; detritic: (40)

dépôt détritique (*m.*), detrital deposit. (41)

détritus (*n.m.*) *ou* **détritus** (*n.m.pl.*), detritus. (42)

détritus charriés (Géol.) (*m.pl.*), drift detritus. (43)

détroit (Géogr. phys.) (*n.m.*), strait; straits; sound; narrows: (44)

le détroit de Gibraltar, the Straits of Gibraltar. (45)

détruire (*v.t.*), to destroy. (46)

deutéroprisme (Cristall.) (*n.m.*), deuteroprism; prism of the second order; second order prism. (47)

deutéropyramide (Cristall.) (*n.f.*), deuteropyra-mid; pyramid of the second order; second order pyramid. (48)

deutogène (Géol.) (*adj.*), clastic; fragmentary; fragmental; detrital: (49)

roches deutogènes (*f.pl.*), clastic rocks; fragmentary rocks; fragmental rocks; detrital rocks. (50)

deuxième *ou* **2ᵉ** (*adj.*), second; 2nd: (51)

prisme de deuxième espèce (*m.*), second order prism. (52)

devant de la marche (escalier) (*m.*), riser (stairs). (53)

devanture (d'un édifice) (*n.f.*), front, facade, (of an edifice). (54)

devanture (d'une chaudière, d'un foyer, etc.) (*n.f.*), front (of a boiler, of a furnace, etc.) (55 *ou* 56)

devanture de foyer, furnace-front. (57)

développable (*adj.*), developable: (58)

surface développable (Géom.) (*f.*), develop-able surface. (59)

développable (Géom.) (*n.f.*), developable; developable surface. (60)

développante (Géom.) (*n.f.*), involute; involute curve: (61)

développante de cercle, involute of a circle. (62)

développante de cercle (à) *ou simplement* **à déve-loppante,** involute: (63)

engrenage à développante de cercle *ou* engre-nage à développante (*m.*), involute gear; involute gearing. (64)

développateur (Photogr.) (*n.m.*), developer. (65)

Note.—*développateur* is the developer proper; the mixture used for developing is called *révélateur.*

développement (*n.m.*), development; develop-ing; opening up; growth: (66)

développement de l'industrie minière du charbon, development, growth, of the coal-mining industry. (67)

développement (Géom.) (*n.m.*), development; stretch out: (68)

développement du cylindre droit, du cône, de la courbure de la came, development of the right cylinder, of the cone, of cam-curve. (69 *ou* 70 *ou* 71)

développement d'un limon débillardé (escalier), stretch out, development, of a wreathed string (stairs). (72)

développement (Photogr.) (*n.m.*), development; developing: (73)

développement d'une photographie, develop-ment of a photograph; developing a photograph. (74)

développer (*v.t.*), to develop; to open up: (75)

développer les ressources d'un pays neuf, to develop the resources of, to open up, a new country. (76)

développer (Géom.) (*v.t.*), to develop: (77)

développer une développable sur un plan, to develop a developable on a plane. (1)

développer (Photogr.) (*v.t.*), to develop : (2) développer un cliché, to develop a negative. (3)

développer (se) (*v.r.*), to develop. (4)

déverrouillage d'aiguilles (Ch. de f.) (*m.*), un-locking switches. (5)

déverrouiller une aiguille (Ch. de f.), to unlock a switch. (6)

dévers, -e (*adj.*), out of plumb ; out of true. (7)

dévers (défaut d'aplomb) (*n.m.*), cant ; inclina-tion. (8)

dévers (Ch. de f.) (*n.m.*), superelevation ; elevation of the outer rail ; cant : (9) le dévers augmente à mesure que le rayon de la courbe diminue, the elevation of the outer rail increases as the radius of the curve diminishes. (10)

déversé, -e (se dit des bois de mines) (*adj.*), under-set : (11) montant déversé (*m.*), underset prop. (12)

déversement (de wagons) (*n.m.*), discharge ; shooting ; tipping. (13)

déversement (Hydraul.) (*n.m.*), discharge : (14) déversement du trop-plein d'un étang, dis-charge of the overflow of a pond. (15)

déversement (inclinaison) (*n.m.*), inclination ; bias : (16) le déversement d'un mur, the inclination of a wall. (17)

déversement (des montants) (Mines) (*n.m.*), underset, underlie, (of props). (18)

déversement (gauchissement) (*n.m.*), warping : (19) le déversement d'une planche, the warping of a board. (20)

déversement (Géol.) (*n.m.*), inversion ; overturn. (21)

déverser (faire couler) (*v.t.*), to discharge ; to shoot ; to tip : (22) cours d'eau qui déverse ses eaux dans un lac (*m.*), stream which discharges its waters into a lake. (23) wagon qui déverse sa charge (*m.*), truck which discharges its load. (24) déverser directement le charbon des wagonets dans des bateaux, to shoot the coal directly from the trucks into boats. (25)

déverser (incliner) (*v.t.*), to incline : (26) déverser un poteau de la perpendiculaire, to incline a post from the perpendicular. (27)

déverser (gauchir) (*v.t.*), to warp : (28) déverser une pièce de bois, to warp a piece of wood. (29)

déverser (perdre son aplomb) (*v.i.*), to incline : (30) mur qui déverse un peu à gauche, à droite (*m.*), wall which inclines a little to the left, to the right. (31 *ou* 32)

déverser (gauchir) (*v.i.*), to warp ; to wind. (33)

déverser (se) (perdre son aplomb) (*v.r.*), to incline. (34)

déverser (se) (se gauchir) (*v.r.*), to warp ; to wind. (35)

déversoir (d'un canal) (*n.m.*), weir ; waste-weir. (36)

déversoir (*n.m.*) *ou* **déversoir de jaugeage** (Hydraul.), weir. (37)

déversoir (établi sur un moule) (Fonderie) (*n.m.*), flow-off gate ; cut-off. (38)

déversoir à crête épaisse *ou* **déversoir sur seuil** (Hydraul.) (*m.*), broad-crested weir ; flat-crested weir. (39)

déversoir à nappe libre (*m.*), weir with free nappe. (40)

déversoir à nappe noyée *ou* **déversoir noyé** (*m.*), drowned weir ; submerged weir. (41)

déversoir en mince paroi *ou* **déversoir sur mince paroi** *ou* **déversoir à arête vive** (*m.*), sharp-crested weir ; thin-edged weir. (42)

déviation (*n.f.*), deviation ; deflection : (43) déviation d'un trou de sonde de la verticale, deviation of a bore-hole from the vertical ; deflection of a bore-hole from verticality ; drift of a bore-hole. (44)

dévidage (*n.m.*), reeling ; winding ; spooling. (45)

dévider (*v.t.*), to reel ; to wind ; to spool. (46)

dévider (se) (*v.r.*), to wind. (47)

dévidoir (*n.m.*), reel ; spool ; winder. (48)

dévidoir à tuyaux (*m.*), hose-reel. (49)

dévier (*v.t.*), to deflect ; to deviate : (50) dévier une aiguille sur un cadran, to deflect a needle on a dial. (51)

dévier (*v.i.*), to deviate ; to deflect ; to run out of true : (52) dévier de la ligne verticale, to run out of the vertical. (53)

dévier (se) (*v.r.*), to deviate ; to deflect. (54)

devis (*n.m.*) *ou* **devis estimatif**, estimate : (55) devis d'une installation complète pour éclairage électrique, estimate of a complete installa-tion for electric lighting. (56) devis estimatif pour matériel neuf, estimate for new plant. (57) devis sur demande, estimates on application. (58)

devis descriptif (*m.*), specification : (59) devis descriptif des travaux pour une maison d'habitation, specification of works for a dwelling-house. (60)

dévissage (*n.m.*), unscrewing : (61) dévissage des lampes de sûreté, unscrewing, unlocking, safety-lamps. (62)

dévisser (*v.t.*), to unscrew. (63)

dévisser (se) (*v.r.*), to unscrew ; to come un-screwed. (64)

devoir (*n.m.*), duty ; obligation : (65) s'acquitter de son devoir, to do one's duty. (66) devoirs des concessionnaires vis-à-vis des inventeurs, des explorateurs, des proprié-taires du sol, et envers l'État, obligations of claimholders to discoverers, explorers, landowners, and to the State. (67)

dévoltage (Élec.) (*n.m.*), lowering the voltage ; dropping the voltage. (68)

dévolter (Élec.) (*v.t.*), to lower, to drop, the volt-age. (69)

dévolteur (Élec.) (*n.m.*), negative booster ; sucking booster. (70)

dévonien, -enne (Géol.) (*adj.*), Devonian : (71) la période dévonienne, the Devonian period. (72)

dévonien (Géol.) (*n.m.*), Devonian : (73) le dévonien, the Devonian. (74)

dévorer la fumée (se dit d'un foyer fumivore), to consume smoke (said of a smoke-consum-ing furnace). (75)

deweylite (Minéral) (*n.f.*), deweylite. (76)

dextrorsum (*adj. invar.*), dextrorsal ; dextrorse ; right-handed : (77)

hélice dextrorsum (*f.*), dextrorsal helix ; right-handed spiral. (1)

dextrorsum (*adv.*), dextrorsally ; dextrorsely ; right-handedly : (2)
les pas de vis ordinaires et les fils des bobines d'induction sont enroulés dextrorsum (*m.pl.*), ordinary screw-threads and wires of induction-coils are wound dextrorsally. (3)

dézincifier (*v.t.*), to dezincify. (4)

diabase (Pétrol.) (*n.f.*), diabase. (5)

diabase à olivine (*f.*), olivine diabase. (6)

diable (charrette à quatre roues fort basses) (*n.m.*), trolley ; trolly. (7)

diable *ou* **diable-brouette** [diables-brouettes *pl.*] (*n.m.*), hand-truck ; truck (two-wheeled hand-truck). (8)

diacaustique (Opt.) (*adj.*), diacaustic. (9)

diacaustique (*n.f.*), diacaustic ; caustic by refraction. (10)

diaclase (fracture du sol non-accompagnée de rejet) (opp. à **paraclase**) (Géol.) (*n.f.*), diaclase ; fissure ; joint. (11)

diaclasite (Minéral) (*n.f.*), diaclasite. (12)

diaclinal, -e, -aux (Géol.) (*adj.*), diaclinal. (13)

diadochite (Minéral) (*n.f.*), diadochite. (14)

diagénèse (Minéralogie & Chim.) (*n.f.*), diagenesis. (15)

diagénétique (*adj.*), diagenetic. (16)

diagonal, -e, -aux (*adj.*), diagonal. (17)

diagonal (voie diagonale) (Ch. de f.) (*n.m.*), cross-over ; crossover-road ; diagonal road. (18)

diagonale (ligne) (*n.f.*), diagonal. (19)

diagonale (galerie de mine) (*n.f.*), slant ; run. (20)

diagonalement (*adv.*), diagonally. (21)

diagramme (*n.m.*), diagram. (22)

diagramme (d'un baromètre enregistreur) (*n.m.*), chart of a recording barometer). (23)

diagramme d'indicateur (*m.*), indicator-diagram ; indicator-card. (24)

diagramme de distribution (tiroir) (*m.*), valve-diagram ; distribution-diagram. (25)

diagramme de la distribution Corliss (*m.*), diagram of Corliss valve-gear. (26)

diagramme de la production (*m.*), diagram of output. (27)

diagramme de montage (Élec.) (*m.*), diagram of connections. (28)

diagramme de puissance (*m.*), power-diagram. (29)

diagramme défectueux (*m.*), distorted card : (30)
diagramme défectueux dû au coincement du piston de l'indicateur, distorted card due to binding of indicator-piston. (31)

diagramme des efforts *ou* **diagramme des forces** (*m.*), stress-diagram ; stress-sheet. (32)

diagramme entropique (*m.*), entropy diagram : (33)
diagramme entropique de la vapeur laminée, entropy diagram of wiredrawn steam. (34)

diagramme indiquant l'effet de la perte de travail résultant d'une détente incomplète (*m.*), diagram showing the effect of lost motion resulting from incomplete expansion. (35)

diagramme théorique (*m.*), theoretical diagram ; ideal indicator-card. (36)

diagramme totalisé *ou* **diagramme rankinisé** (machine polycylindrique) (*m.*), combined diagram (multicylinder engine). (37)

diallage (Minéral) (*n.f.*), diallage. (38)

diallage verte (Minéral) (*f.*), green diallage ; smaragdite. (39)

diallagique (*adj.*), diallagic. (40)

diallagite (Pétrol.) (*n.f.*), diallagite. (41)

dialogite *ou* **diallogite** (Minéral) (*n.f.*), dialogite ; diallogite ; rhodochrosite ; rhodochroisite. (42)

diamagnétique (*adj.*), diamagnetic : (43)
le bismuth est fortement diamagnétique, c'est-à-dire qu'il est repoussé par l'aimant, à l'inverse du fer, bismuth is strongly diamagnetic, that is to say, it is repelled by the magnet, the opposite to iron. (44)

diamagnétiquement (*adv.*), diamagnetically. (45)

diamagnétisme (*n.m.*), diamagnetism. (46)

diamagnétomètre (*n.m.*), diamagnetometer. (47)

diamant (*n.m.*), diamond. (48)

diamant à pointes naïves (*m.*), point diamond. (49)

diamant alluvien (*m.*), alluvial diamond. (50)

diamant brillant *ou* **diamant taillé en brillant** (*m.*), brilliant ; brilliant-cut diamond. (51)

diamant brut (*m.*), rough diamond. (52)

diamant de vitrier (*m.*), glaziers' diamond ; glass-cutters' diamond ; diamond pencil ; diamond point ; diamond tool. (53)

diamant du Brésil (*m.*), Brazilian diamond. (54)

diamant du Cap (*m.*), Cape diamond. (55)

diamant en rose *ou* **diamant taillé en rose** (*m.*), rose diamond ; rose-cut diamond. (56)

diamant en table *ou* **diamant taillé en table** (*m.*), table diamond ; table-cut diamond. (57)

diamant noir (*m.*), black diamond ; carbon ; carbonado ; bort. (58)

diamant non-taillé (*m.*), uncut diamond. (59)

diamant poli (*m.*), polished diamond. (60)

diamant savoyard (*m.*), brown diamond. (61)

diamant taillé (*m.*), cut diamond. (62)

diamantaire (*adj.*), diamantoid : (63)
pierres diamantaires (*f.pl.*), diamantoid stones. (64)

diamantaire (pers.) (*n.m.*), diamond-cutter ; lapidary. (65)

diamantifère (*adj.*), diamondiferous ; diamantif-erous ; diamond-bearing ; diamond-pro-ducing : (66)
argile diamantifère (*f.*), diamondiferous clay ; diamond-bearing clay. (67)
les terrains diamantifères du Transvaal (*m.pl.*), the diamond-producing grounds of the Transvaal. (68)

diamantin, -e (*adj.*), diamantine. (69)

diamétral, -e, -aux (*adj.*), diametric ; diametrical; diametral : (70)
pas diamétral (*m.*), diametrical pitch ; dia-metral pitch. (71)

diamétralement (*adv.*), diametrically ; dia-metrally : (72)
pôles diamétralement opposés (*m.pl.*), dia-metrically opposite poles. (73)

diamètre (*n.m.*), diameter. (74)

diamètre de serrage (d'un mandrin) (*m.*), hold-ing capacity (of a chuck). (75)

diamètre définitif du puits (*m.*), finished diameter of the shaft. (76)

diamètre extérieur (*m.*), external diameter ; outside diameter. (77)

diamètre initial du sondage (*m.*), initial diameter of the bore-hole. (78)

diamètre intérieur (*m.*), internal diameter ; inside diameter. (79)

diamètre maximum admis *ou* **diamètre maxi-mum admissible** (tour) (*m.*), swing (lathe) : (80)

diamètre maximum admis au-dessus des chariots, swing over saddle ; swing of the rest. (1)

diamètre maximum admis au-dessus du banc, swing over bed ; swing of the bed. (2)

diamètre maximum admis dans la coupure *ou* diamètre maximum admissible dans le rompu, swing in gap. (3)

diamètre primitif (Engrenages) (*m.*), pitch-diameter (Gearing). (4)

diaphane (*adj.*), diaphanous : (5)

l'albâtre gypseux est moins diaphane que l'albâtre calcaire (*m.*), gypseous alabaster is less diaphanous than calcareous alabaster. (6)

corps diaphane (Opt.) (*m.*), diaphanous body. (7)

diaphorite (Minéral) (*n.f*). diaphorite. (8)

diaphragmation (Opt.) (*n.f.*), diaphragming ; stopping ; stopping down. (9)

diaphragme (*n.m.*), diaphragm. (10)

diaphragme (Photogr.) (*n.m.*), diaphragm ; stop. (11)

diaphragme (d'une lunette) (*n.m.*), diaphragm, cross-wire ring, (of a telescope). (12)

diaphragme iris (Photogr.) (*m.*), iris-diaphragm. (13)

diaphragme tournant (Opt.) (*m.*), revolving diaphragm ; rotating diaphragm. (14)

diaphragmer (*v.t.*), to diaphragm ; to stop ; to stop down : (15)

diaphragmer un objectif, to stop down, to stop, to diaphragm, a lens. (16)

diapositive (*n.f.*) *ou* **diapositif** (*n.m.*) (Photogr.), slide ; diapositive. (17)

diapositive pour la projection (*f.*), lantern-slide. (18)

diapositive pour la stéréoscopie (*f.*), stereoscopic slide. (19)

diaschiste (opp. à **aschiste**) (Pétrol.) (*adj.*), diaschistic. (20) (opp. to **aschistic**) :

roches diaschistes (*f.pl.*), diaschistic rocks. (21)

diaspore (Minéral) (*n.m.*), diaspore. (22)

diastrophique (Géol.) (*adj.*), diastrophic. (23)

diastrophisme (Géol.) (*n.m.*), diastrophism. (24)

diathermane *ou* **diathermique** (*adj.*), diathermanous ; diathermal ; diathermic ; diather-mous. (25)

diathermanéité *ou* **diathermansie** (*n.f.*), diather-mancy ; diathermance ; diathermaneity. (26)

diatomique (Chim.) (*adj.*), diatomic. (27)

diatrème (cheminée d'ascension) (Géol.) (*n.m.*), neck. (28)

dibasique (Chim.) (*adj.*), dibasic ; bibasic. (29)

dichroïque (*adj.*), dichroic. (30)

dichroïsme (*n.m.*), dichroism. (31)

dichroïte (Minéral) (*n.f.*), dichroite. (32)

dichroscope (Cristall.) (*n.m.*), dichroscope. (33)

dichroscopique (*adj.*), dichroscopic. (34)

dickinsonite (Minéral) (*n.f.*), dickinsonite. (35)

diclinique (Cristall.) (*adj.*), diclinic. (36)

didyme *ou* **didymium** (Chim.) (*n.m.*), didymium. (37)

dièdre (Géom.) (*adj.*), dihedral ; diedral : (38)

angle dièdre (*m.*), dihedral angle. (39)

dièdre (*n.m.*), dihedron. (40)

dief (Houillères) (*n.m.*), clay-dyke. (41)

diélectricité (*n.f.*), dielectricity. (42)

diélectrique (*adj.*), dielectric : (43)

hystérésis diélectrique (*f.*), dielectric hysteresis. (44)

diélectrique (*n.m.*), dielectric : (45)

tout diélectrique est nécessairement un corps isolant, mais tout corps isolant peut ne pas être un bon diélectrique, every dielectric is necessarily an insulating substance, but every insulating substance need not be a good dielectric. (46)

diélectriquement (*adv.*), dielectrically. (47)

diève (Houillères) (*n.f.*), clay-dyke. (48)

différence (*n.f.*), difference. (49)

différence de phase (Élec.) (*f.*), phase-difference. (50)

différence de potentiel (Élec.) (*f.*), difference of potential ; potential difference : (51)

différence de potentiel aux bornes, difference of potential at the terminals. (52)

différenciation (*n.f.*), differentiation : (53)

différenciation des roches, differentiation of rocks. (54)

différencier (*v.t.*), to differentiate. (55)

différends entre patrons et ouvriers (*m.pl.*), differences between employers and work-men. (56)

différent, -e (*adj.*), different ; dissimilar ; various. (57)

différentiel, -elle (*adj.*), differential : (58)

engrenage différentiel (*m.*), differential gear. (59)

différentiel (Méc.) (*n.m.*), differential ; differential gear ; differential gearing ; balance-gear ; compensating-gear ; equalizing-gear. (60)

difficile (*adj.*), difficult. (61)

difficilement (*adv.*), with difficulty ; difficultly. (62)

difficulté (*n.f.*), difficulty : (63)

difficulté à se procurer des provisions, diffi-culty in obtaining provisions. (64)

difficultés d'ordre technique, technical diffi-culties ; difficulties of a technical nature. (65)

difficultés inséparables du travail des pionniers, difficulties incident to pioneer work. (66)

diffracter (*v.t.*), to diffract : (67)

diffracter les rayons lumineux, to diffract light-rays. (68)

diffractif, -ive (*adj.*), diffractive. (69)

diffraction (*n.f.*), diffraction. (70)

diffus, -e (*adj.*), diffused : (71)

lumière diffuse (*f.*), diffused light. (72)

diffuser (*v.t.*), to diffuse : (73)

diffuser la lumière, to diffuse light. (74)

diffuseur (*n.m.*), diffuser. (75)

diffusibilité (*n.f.*), diffusibility. (76)

diffusible (*adj.*), diffusible. (77)

diffusif, -ive (*adj.*), diffusive. (78)

diffusion (*n.f.*), diffusion. (79)

digne d'étude, worthy of investigation. (80)

digression (Astron.) (*n.f.*), elongation ; digression. (81)

digue (*n.f.*), dike ; dam. (82)

digue en crochet *ou* **digue recourbée en crochet** (Géogr. phys.) (*f.*), hook ; recurved spit. (83)

digue en épi (Géogr. phys.) (*f.*), spit. (84)

dilatabilité (*n.f.*), expansibility ; dilatability. (85)

dilatable (*adj.*), expansible ; dilatable. (86)

dilatation (*n.f.*), expansion ; dilatation : (87)

dilatation des gaz, des solides, expansion of gases, of solids. (88 *ou* 89)

dilatation du mercure, de l'eau, expansion of mercury, of water. (90 *ou* 91) dilatation d'un corps sous l'influence de la chaleur, des rails de chemins de fer pendant les chaleurs, expansion of a substance under the influence of heat, of railway-rails during the hot weather. (1 *ou* 2)

dilatation électrique (*f.*), electric expansion. (3)

dilatation négative (*f.*), negative dilatation. (4)

dilatation positive (*f.*), positive dilatation. (5)

dilater (*v.t.*), to expand ; to dilate : (6)
la chaleur dilate les gaz, heat expands gases. (7)
dilater un ballon, to dilate a balloon. (8)

dilater (se) (*v.r.*), to expand ; to dilate : (9)
les liquides se dilatent beaucoup plus que les solides (*m.pl.*), liquids expand much more than solids. (10)

dilatomètre (*n.m.*), dilatometer. (11)

dilatométrique (*adj.*), dilatometric. (12)

dilué, -e (*adj.*), diluted ; dilute. (13)

diluer (*v.t.*), to dilute : (14)
diluer d'eau un acide, to dilute an acid with water. (15)

diluvien, -enne *ou* **diluvial, -e, -aux** (Géol.) (*adj.*), diluvial : (16)
dépôt diluvien (*m.*), diluvial deposit. (17)

diluvium (Géol.) (*n.m.*), diluvium. (18)

dimension (*n.f.*), dimension ; size ; measure--ment : (19)
tout corps a trois dimensions : longueur, largeur, et profondeur (*m.*), every body has three dimensions : length, breadth, and thickness. (20)
dimensions assorties, assorted sizes. (21)
dimensions types, standard sizes. (22)
dimensions superficielles, superficial measure--ments ; surface measurements. (23)

diminuer (*v.t.*), to dimish ; to decrease ; to lessen ; to reduce : (24)
diminuer les ressources d'un pays, to diminish the resources of a country. (25)
diminuer la pression, to lighten, to reduce, the pressure. (26)
diminuer la longueur d'une planche, to reduce the length of a board. (27)

diminuer (*v.i.*), to diminish ; to decrease ; to lessen ; to fall ; to fall off : (28)
diminuer de grandeur, to diminish, to decrease, in size. (29)
diminuer de valeur, to dimish, to fall, in value ; to depreciate. (30)

diminution (*n.f.*), diminution ; decrease ; decline ; falling off ; falling away ; reduction : (31)
diminution de la production, decrease of out--put ; decline of production. (32)

dimorphe (Cristall.) (*adj.*), dimorphous ; di--morphic : (33)
minéral dimorphe (*m.*), dimorphous mineral. (34)

dimorphisme (*n.m.*) *ou* **dimorphie** (*n.f.*) (Cristall.), dimorphism ; dimorphy. (35)

dinantien, -enne (Géol.) (*adj.*), Dinantian. (36)

dinantien (Géol.) (*n.m.*), Dinantian ; Culm : (37)
le dinantien, the Dinantian ; the Culm. (38)

diopside (Minéral) (*n.m.*), diopside. (39)

dioptase (Minéral) (*n.f.*), dioptase ; emerald copper. (40)

diorite (Pétrol.) (*n.f.*), diorite. (41)

diorite augitique *ou* **diorite à pyroxène** (*f.*), augite-diorite. (42)

diorite micacée (*f.*), mica-diorite. (43)

diorite orbiculaire (*f.*), orbicular diorite. (44)

diorite quartzifère (*f.*), quartz-diorite. (45)

dioritique (*adj.*), dioritic. (46)

diphasé, -e (Élec.) (*adj.*), two-phase ; diphase ; biphase : (47)
courant diphasé (*m.*), two-phase current. (48)

diplôme (*n.m.*), certificate : (49)
diplôme de capacité dans les travaux de sauvetage à l'intérieur des mines, certificate of competency in mine-rescue work. (50)

diplômer (*v.t.*), to certificate : (51)
diplômer un candidat, to certificate a candi--date. (52)

dipyre (Minéral) (*n.m.*), dipyre. (53)

dire (*v.t.*), to say ; to tell ; to state : (54)
une roche telle que la craie est dite maigre au toucher, a rock like chalk is said to be meagre to the touch. (55)
Cf. **dit, -e.**

dire d'experts (à), at a valuation : (56)
reprendre le matériel à dire d'experts, to take over the plant at a valuation. (57)

direct, -e (*adj.*), direct ; straight. (58)

directement (*adv.*), directly ; straight ; due : (59)
le filon se dirige directement au nord, au sud, à l'est, à l'ouest, the lode runs due north, south, east, west. (60 *ou* 61 *ou* 62 *ou* 63)

directeur (pers.) (*n.m.*), manager. (64)

directeur de mine (*m.*), mine-manager. (65)

directeur diplômé (*m.*), certificated manager. (66)

directeur en résidence (*m.*), resident manager. (67)

directeur général (*m.*), general manager. (68)

direction (conduite des affaires) (*n.f.*), manage--ment ; conduct ; direction ; charge. (69)

direction (action de diriger un bateau, un auto--mobile, ou analogue) (*n.f.*), steering. (70)

direction (organes de direction d'un automobile, ou analogue) (*n.f.*), stearing-gear. (71)

direction (ligne de mouvement) (*n.f.*), direction ; drift ; course : (72)
direction d'un courant, d'une galerie de mine, direction, drift, course, of a current, of a mine-level. (73 *ou* 74)
direction d'une force (Méc.), direction of a force. (75)

direction (d'un filon) (Géol. & Mines) (*n.f.*), strike, trend, bearing, level-bearing, course, level-course, run, (of a lode). (76)

direction (en) (Géol. & Mines), parallel in direction to the strike ; along the strike ; along the line of strike ; following the line of strike ; on the line of the strike : (77)
mère-galerie en direction (*f.*), main road parallel in direction to the strike. (78)
en direction sur le filon, on the course (*or* on the run) of the lode. (79)

direction de solution (Cristall.) (*f.*), solution-plane. (80)

directrice (Géom.) (*n.f.*), director. (81)

directrice (d'une turbine) (*n.f.*), guide (of a turbine). (82)

diriger (*v.t.*). See examples : (83)
diriger une entreprise, to direct an enterprise ; to manage an undertaking. (84)
diriger une lunette sur un point éloigné, to direct a telescope on a distant point. (85)
diriger un bateau, un automobile, to steer a boat, a motor-car. (86 *ou* 87)

diriger (se) (en parlant d'un filon, etc.) (*v.r.*), to run ; to strike ; to follow a course ; to take a direction : (88)

le filon se dirige vers le nord, the lode runs north ; the lode strikes north ; the lode follows a northerly course ; the lode takes a northerly direction. (1)

discernement de la graisse employée comme trieur (*m.*), discrimination of grease as a sorter. (2)

discontinu, -e (*adj.*), discontinuous. (3)

discontinuer (*v.t.* & *v.i.*), to discontinue : (4) discontinuer de travailler pendant l'hiver, to discontinue to work during the winter. (5)

discordance (Géol.) (*n.f.*), unconformability ; unconformity ; discordance ; transgression : (6) discordance de stratification, unconform-ability, discordance, transgression, of strati-fication. (7)

discordant, -e (Géol.) (*adj.*), unconformable ; discordant ; transgressive. (8)

disette (*n.f.*), scarcity ; shortage ; want ; dearth : (9) disette de charbon, scarcity, shortage, want, dearth, of coal. (10)

disilicate (Chim.) (*n.m.*), disilicate. (11)

disilicique (*adj.*), disilicic. (12)

disjoncteur (Élec.) (*n.m.*), circuit-breaker ; break-circuit ; disjunctor. (13)

disjoncteur à maximum (*m.*), maximum circuit-breaker. (14)

disjoncteur à minimum (*m.*), minimum circuit-breaker. (15)

disjoncteur retardé (*m.*), delay-action circuit-breaker. (16)

dislocation (*n.f.*), dislocation ; dislodgment : (17) dislocation des pierres d'un mur, dislodgment of the stones of a wall. (18)

dislocation (Géol.) (*n.f.*), dislocation : (19) dislocations de la croûte terrestre par refroi-dissement du foyer central, dislocations of the earth's crust by cooling of the central fire. (20)

disloquer (*v.t.*), to dislocate ; to dislodge. (21)

disomose (Minéral) (*n.f.*), nickel-glance ; gers-dorffite. (22)

disparaître (*v.i.*), to disappear. (23)

disparition (*n.f.*), disappearance. (24)

dispendieux, -euse (*adj.*), expensive : (25) procédé qui est très dispendieux (*m.*), process which is very expensive. (26)

dispense (*n.f.*), exemption. (27)

disperser (*v.t.*), to disperse ; to scatter : (28) disperser des débris, to scatter debris. (29)

disperser (Opt.) (*v.t.*), to disperse : (30) disperser la lumière en ses couleurs spectrales composantes, to disperse light into its component spectral colours. (31)

disperser (se) (*v.r.*), to disperse ; to scatter. (32)

disperser (se) (Opt.) (*v.r.*), to disperse : (33) rayons qui se dispersent après avoir traversé un prisme (*m.pl.*), rays which disperse after having passed through a prism. (34)

dispersif, -ive (*adj.*), dispersive. (35)

dispersion (*n.f.*), dispersion : (36) dispersion de la lumière blanche en rayons de diverses couleurs au moyen d'un prisme, dispersion of white light into rays of different colours by means of a prism. (37) dispersion des axes optiques (Cristall.), dis-persion of the optic axes. (38)

disponible (*adj.*), available ; on hand ; in hand : (39)

chute disponible (Hydraul.) (*f.*), available head. (40)

approvisionnements disponibles en magasin (*m.pl.*), stores on (*or* in) hand. (41)

disposé (-e) en couches (être), to occur in beds : (42) du minerai disposé en couches, ore occurring in beds. (43)

disposé (-e) en gradins *ou* **disposé (-e) en cascades,** stepped : (44) collines disposées en gradins (*f.pl.*), stepped hills. (45)

dispositif (*n.m.*), device ; arrangement ; con-trivance. (46)

dispositif breveté de sûreté (*m.*), patent safety-arrangement ; patent safety-device. (47)

dispositif de changement de vitesse (*m.*), speed-changing device ; speed-changing mechanism. (48)

dispositif de fortune *ou* **dispositif de circonstance** (*m.*), makeshift. (49)

dispositif de sécurité empêchant l'embrayage simultané de deux mouvements automatiques (*m.*), safety-device preventing two auto-matic movements being put into operation at the same time. (50)

dispositif mécanique (*m.*), mechanical contrivance ; mechanical device. (51)

dispositifs de mines (*m.pl.*), preparatory work ; preliminary work ; dead-work. (52)

disposition (*n.f.*), arrangement ; disposition : (53) la disposition des chaudières, the arrangement of the boilers. (54) disposition d'ensemble, general arrangement. (55)

disque (*n.m.*), disc ; disk (56)

disque (Ch. de f.) (*n.m.*), disc signal ; signal. (57)

disque à distance (Ch. de f.) (*m.*), distant signal. (58)

disque à pression (pour roulements à billes) (*m.*), bearing-plate (for ball bearings). (59)

disque d'arrêt (Ch. de f.) (*m.*), stop signal ; danger signal. (60)

disque d'arrêt absolu (Ch. de f.) (*m.*), absolute stop signal. (61)

disque d'excentrique (*m.*), eccentric-disc ; eccentric-sheave. (62)

disque de carton divisé en secteurs colorés (Opt.) (*m.*), colour-wheel. (63)

disque de Newton (Opt.) (*m.*), Newton's disc. (64)

disque de noyau (Élec.) (*m.*), core-disc ; core-plate ; core-lamination. (65)

disque en buffle (*m.*), buff-wheel. (66)

disque en drap (*m.*), calico mop ; rag-wheel. (67)

disque indicateur d'aiguillage (Ch. de f.) (*m.*), switch-signal ; switch-target ; ground-disc. (68)

disque oculaire (Opt.) (*m.*), eye-ring. (69)

disque-scie [disques-scies *pl.*] (*n.m.*), disc saw ; disk saw. (70)

disque-signal [disques-signaux *pl.*] (Ch. de f.) (*n.m.*), disc signal. (71)

disruptif, -ive (Élec.) (*adj.*), disruptive : (72) décharge disruptive (*f.*), disruptive discharge. (73)

disséminer (*v.t.*), to disseminate ; to scatter : (74) minéral disséminé à travers le gravier (*m.*), mineral disseminated through the gravel. (75)

dissimuler (*v.t.*), to conceal ; to hide ; to cover up : (76)

dissimuler une fente dans un ouvrage de menuiserie, to conceal a crack in a piece of joinery work. (1)

dissipation (*n.f.*), dissipation : (2)
dissipation de la chaleur par rayonnement, par convection, dissipation of heat by radiation, by convection. (3 *ou* 4)

dissiper (*v.t.*), to dissipate. (5)

dissiper (se) (*v.r.*), to dissipate. (6)

dissociable (Chim.) (*adj.*), dissociable. (7)

dissociation (Chim.) (*n.f.*), dissociation : (8)
la dissociation de chlorure ammoniacal, the dissociation of ammonium chloride. (9)

dissocier (Chim.) (*v.t.*), to dissociate. (10)

dissolubilité (*n.f.*), dissolubility. (11)

dissoluble (*adj.*), dissoluble. (12)

dissolution (action) (*n.f.*), dissolution ; dissolv--ing ; solution : (13)
dissolution par les eaux souterraines de roches calcaires ou autres, dissolution by under--ground water of limestone or other rocks. (14)

dissolution (liquide contenant un corps dissous) (*n.f.*), solution. (15)

dissolution cyanurée (*f.*), cyanide solution. (16)

dissolution saturée (*f.*), saturated solution. (17)

dissolution sursaturée (*f.*), supersaturated solution. (18)

dissolvant, -e (*adj.*), dissolvent ; solvent. (19)

dissolvant (*n.m.*), dissolvent ; solvent ; men--struum. (20)

dissoudre (*v.t.*), to dissolve : (21)
dissoudre du sel dans l'eau, to dissolve salt in water. (22)

dissoudre (se) (*v.r.*), to dissolve (23)
dans les acides, la turquoise se dissout sans effervescence et sans résidu, en offrant les colorations du cuivre, in acids, turquoise dissolves without effervescing and without residue, exhibiting the colourations of copper. (24)

dissous, -oute (*adj.*), dissolved : (25)
sel dissous (*m.*), dissolved salt. (26)

dissymétrie (*n.f.*), dissymmetry ; asymmetry : (27)
dissymétrie du relief terrestre, dissymmetry of the relief of the earth's surface. (28)

dissymétrique (*adj.*), dissymmetric ; dissym--metrical ; unsymmetrical ; asymmetrical. (29)

distance (*n.f.*), distance. (30)

distance des centres (*f.*), distance between centres ; centre distance : (31)
distance des centres d'une roue d'engrenage et d'un pignon, distance between centres of a gear-wheel and of a pinion. (32)

distance entre pointes (tour) (*f.*), distance between centres (lathe). (33)

distance focale (Opt.) (*f.*), focal distance ; focal length. (34)

disthène (Minéral) (*n.m.*), disthene ; cyanite ; kyanite. (35)

disthénique (*adj.*), disthenic. (36)

distillable (*adj.*), distillable. (37)

distillation (action) (*n.f.*), distillation ; retorting : (38)
distillation de la houille en vases clos, distilla--tion of coal in closed retorts ; retorting coal. (39)
distillation dans le vide, vacuum distillation. (40)

distillation fractionnée, fractional distillation. (41)
distillation sèche, dry distillation ; destructive distillation. (42)

distillation (produit) (*n.f.*), distillate ; distilla--tion. (43)

distillatoire (*adj.*), distillatory. (44)

distiller (*v.t.*), to distil ; to retort : (45)
distiller l'amalgame, to retort the amalgam. (46)
distiller les schistes, to retort, to distil, shales. (47)

distiller (*v.i.*), to distil. (48)

distiller (se) (*v.r.*), to distil ; to be distilled : (49)
substances qui se distillent en alambics (*f.pl.*), substances which are distilled in stills. (50)

distordre (*v.t.*), to distort. (51)

distorsion (*n.f.*), distortion : (52)
distorsion dans les objectifs photographiques, distortion in photographic lenses. (53)

distorsion en barillet (*f.*), barrel-shaped distor--tion. (54)

distorsion en croissant *ou* **distorsion en coussin** (*f.*), pincushion distortion. (55)

distribuer (*v.t.*), to distribute : (56)
distribuer l'eau au moyen de conduites, to distribute the water by means of pipes. (57)

distributeur (appareil servant à distribuer la matière soumise à l'action des machines) (*n.m.*), distributer ; distributor. (58)

distributeur (*n.m.*) *ou* **distributeur de vapeur**, valve ; steam-valve (a valve for opening or closing a steam-port). (59)

distributeur (d'une turbine) (*n.m.*), guide-ring (of a turbine). (60)

distributeur automatique (*m.*), slot-machine. (61)

distributeur Corliss (*m.*), Corliss valve. (62)

distributeur oscillant *ou* **distributeur tournant** *ou* **distributeur glissant à robinets** (*m.*), rocking valve ; rocking slide-valve ; swinging valve. (63)

distributeur soulevant *ou* **distributeur à soupape** (*m.*), poppet-valve ; drop-valve. (64)

distribution (*n.f.*), distribution : (65)
le problème des causes de la distribution géographique des volcans peut être aujourd' hui considéré comme à peu près résolu, the problem of the geographical distribution of volcanoes can be considered to-day as nearly solved. (66)
distribution d'air dans les chantiers (Mines), distribution of air into the workings. (67)
distribution d'énergie électrique, distribution of electrical power. (68)

distribution (manière dont la vapeur se répartit) (*n.f.*), distribution : (69)
la distribution de vapeur par tiroirs, the dis--tribution of steam by slide-valves. (70)

distribution (appareil de distribution de vapeur) (*n.f.*), valve-gear ; valve-motion. (71)

distribution à coulisse *ou* **distribution par coulisse** *ou* **simplement** distribution (*n.f.*), link-motion ; link-gear ; link valve-motion. (72)

distribution à déclic (*f.*), trip valve-gear. (73)

distribution à soupape (*f.*), poppet valve-gear ; drop valve-gear. (74)

distribution à tiroir (*f.*), slide valve-gear. (75)

distribution Corliss (*f.*), Corliss valve-gear. (76)

distribution d'Allan (*f.*), Allan's link-motion. (77)

distribution d'eau *ou* **distribution des eaux** (*f.*), water service; water-supply; distribution of water: (1)
la distribution d'eau d'une ville, the water service of a town. (2)
distribution de Gooch (*f.*), Gooch's link-motion; Gooch valve-gear. (3)
distribution de locomotive par coulisse (*f.*), locomotive link-motion. (4)
distribution de Stephenson (*f.*), Stephenson's link-motion. (5)
distribution en dérivation *ou* **distribution en quantité** (Élec.) (*f.*), parallel distribution. (6)
distribution en série *ou* **distribution en tension** (Élec.) (*f.*), series distribution. (7)
distribution Joy (*f.*), Joy's valve-gear. (8)
distribution Marshall (*f.*), Marshall's valve-gear. (9)
distribution Meyer (*f.*), Meyer valve-gear. (10)
distribution mixte (Élec.) (*f.*), parallel-series distribution. (11)
distribution par coulisse (*f.*), link-motion. (12)
distribution par coulisse d'Allan, de Gooch, de Stephenson (*f.*). Same as **distribution d'Allan, de Gooch, de Stephenson.**
distribution par soupapes (*f.*), drop valve-gear; poppet valve-gear. (13)
distribution radiale (*f.*), radial valve-gear. (14)
distribution Walschaerts (*f.*), Walschaerts valve-gear. (15)
district (*n.m.*), district; field; region. (16)
district aurifère (*m.*), gold-bearing district; gold-field. (17)
district dépourvu de charbon *ou* **district non-carbonifère** (*m.*), coalless district. (18)
district éloigné (*m.*), outlying district. (19)
district houiller (*n.m.*), coal-district; coal-field. (20)
district minier (*m.*), mining-district; mining-region; mine-field; minery. (21)
district pétrolifère (*m.*), oil-district; oil-field; petroleum-field. (22)
dit, -e (appelé communément) (*adj. & p.p.*), called: (23)
la mèche ordinaire, dite mèche de sûreté, the common fuse, called safety-fuse. (24)
il est quelquefois nécessaire d'épingler certaines parties du moule avec des pointes spéciales, dites de mouleur, it is sometimes necessary to pierce certain parts of the mould with special nails, called moulders' nails. (25)
The word is also commonly used in French technical expressions to indicate that the words following are curious or arbitrary, but sanctioned by usage; thus, **compas dit " maître à danser "** (outside-and-inside calipers); **boussole de poche dite savon-nette** (pocket magnetic compass, in hunter case). Such expressions should preferably be used entire, but the word or words follow-ing *dit* or *dite* are sufficient. Thus we may say **compas dit maître à danser** or simply **maître à danser**; **boussole de poche dite savonnette** or simply **savonnette.** The quotation-marks **" maître à danser "** are optional. In printing, the words *dit* and *dite* are often italicized.
In English the word *called* is not, or is very rarely, used in such connections, the usual practice being to precede such oddities by the word *or*; thus, **thickness-gauge or " feeler "** (jauge d'épaisseur). Here also quotation-marks are optional. Cf. **dire** (*v.t.*).

ditroïte (Pétrol.) (*n.f.*), ditroite; ditroyte. (26)
diurne (*adj.*), diurnal: (27)
le mouvement diurne du soleil, the diurnal motion of the sun. (28)
divalence (Chim.) (*n.f.*), divalence. (29)
divalent, -e (*adj.*), divalent. (30)
divergence (*n.f.*), divergence. (31)
divergent, -e (*adj.*), divergent; diverging: (32)
ménisque divergent (*m.*), diverging meniscus. (33)
divergent (d'un injecteur) (*n.m.*), diverging tube, diverging nozzle, diverging cone, delivery-tube, delivery-nozzle, delivery-cone, (of an injector). (34)
divers, -e (*adj.*), different; diverse; various; sundry; miscellaneous: (35)
couleurs diverses (*f.pl.*), different colours. (36)
divers (*n.m.pl.*), sundries. (37)
divisé (-e) à la machine, machine-divided: (38)
engrenages divisés à la machine (*m.pl.*), machine-divided gears. (39)
diviser (*v.t.*), to divide: (40)
diviser un cercle en 360 degrés, une droite en un nombre quelconque de parties égales, to divide a circle into 360 degrees, a straight line into any number of equal parts. (41 *ou* 42)
diviser en deux parties égales, to bisect: (43)
diviser un angle en deux parties égales, to bisect an angle. (44)
diviser (se) (*v.r.*), to divide: (45)
minéral qui a une tendance à se diviser en lames (*m.*), mineral which has a tendency to divide into laminæ. (46)
division (*n.f.*), division; dividing: (47)
division d'une propriété, division of a property; dividing up a property. (48)
division d'un massif en tranches (Mines de charbon), dividing a seam into slices. (49)
division du courant d'air principal en plusieurs courants partiels (Aérage de mines), division of the main air-current into several splits. (50)
djongle (*n.f.*), jungle. (51)
docile (*adj.*), docile: (52)
minerai docile (*m.*), docile ore. (53)
docimasie *ou* **docimastique** (*n.f.*), docimasy. (54)
docimastique (*adj.*), docimastic. (55)
dock (bassin) (*n.m.*), dock. (56)
dock (entrepôt) (*n.m.*), dock; dock-warehouse. (57)
dodécaèdre *ou* **dodécaédrique** (Géom.) (*adj.*), dodecahedral; dodecahedric; duodecahedral. (58)
dodécaèdre (*n.m.*), dodecahedron; duodecahedron. (59)
dodécagonal, -e, -aux (Géom.) (*adj.*), dodecagonal. (60)
dodécagone (*n.m.*), dodecagon. (61)
doguin (de tour) (*n.m.*), dog; lathe-dog; carrier; lathe-carrier; turning-carrier; driver. (62)
doigt (*n.m.*), finger; trigger: (63)
déclenchement au doigt (obturateur photo-graphique) (*m.*), finger release; trigger release. (64)
doigt (*n.m.*) *ou* **doigt de retenue** *ou* **doigt de relevage** (pour suspendre les pilons d'un bocard), finger, finger-bar (to hang up the stamps of a stamp-mill). (65)
doigt d'encliquetage (*m.*), pawl; ratchet; click; dog. (66)

doigt d'entraînement (du plateau-toc d'un tour) (*m.*), catch-pin (of the catch-plate of a lathe). (1)

doigtier (*n.m.*), finger-stall. (2)

dolérite (Pétrol.) (*n.f.*), dolerite. (3)

doléritique (*adj.*), doleritic. (4)

dolérophane *ou* **dolérophanite** (Minéral) (*n.f.*), dolerophanite. (5)

doline (Géol.) (*n.f.*), dolina ; sink-hole ; sink ; limestone sink ; swallow-hole ; swallow. (6)

dolomie *ou* **dolomite** (Minéral) (*n.f.*), dolomite ; magnesian limestone ; bitter-spar. (7)

dolomie vacuolaire (Géol.) (*f.*), rauchwacke. (8)

dolomitique (*adj.*), dolomitic. (9)

dolomitisation (*n.f.*), dolomitization : (10)
dolomitisation des calcaires magnésiens, dolomitization of magnesian limestones. (11)

domaine (*n.m.*), property ; area : (12)
domaine minier, mining area. (13)

dôme (Arch.) (*n.m.*), dome. (14)

dôme (opp. à **cuvette** ou **cuvette synclinale** ou **brachysynclinal**) (Géol.) (*n.m.*), dome ; quaquaversal fold. (15) (opp. to **basin** or **centroclinal fold**).

dôme (d'un fourneau) (*n.m.*), dome (of a furnace). (16)

dôme de prise de vapeur *ou* **dôme de vapeur** (*m.*), steam-dome ; dome. (17)

dôme de soulèvement (Géol.) (*m.*), upheaval dome. (18)

domeykite (Minéral) (*n.f.*), domeykite. (19)

dominant, -e (*adj.*), dominating ; principal ; chief : (20)
minéral qui se présente sous deux formes dominantes, l'octaèdre et le dodécaèdre (*m.*), mineral which occurs in two dominant forms, the octahedral and the dodecahedral. (21)

dominer (*v.t.*), to dominate ; to overlook : (22)
le mont Cervin domine la vallée de Zermatt, the Matterhorn dominates the Valley of Zermatt. (23)

dominite (Explosif) (*n.f.*), dominite. (24)

domite (Pétrol.) (*n.f.*), domite. (25)

dommage (*s'emploie très souvent au pluriel*) (*n.m.*), damage : (26)
orage de grêle qui cause de grands dommages (*m.*), hail-storm which does a lot of damage. (27)

dommage (Belgique) (*n.m.*), In Belgium, the ground adjoining the buildings attached to a coal-mine. (28)

donarite (Explosif) (*n.f.*), donarite. (29)

donné, -e (*adj.*), given : (30)
vitesse à un instant donné (*f.*), velocity at a given moment. (31)
point donné (*m.*), given point. (32)
quantités données (*f.pl.*), given quantities. (33)

donnée (point incontestable ou admis comme tel) (*n.f.*), datum : (34)
on manque de données techniques pour déterminer la question, there are not sufficient technical data to determine the question. (35)

donnée (*généralement au pl.*) (Math.) (*n.f.*), datum : (36)
les données d'un problème de géométrie, the data of a geometrical problem. (37)

donner (*v.t.*), to give. (38)

donner (Mines) (*v.i.*), to sag ; to squeeze : (39)
lorsque le toit donne, ou que la sole gonfle, when the roof sags, or the floor creeps ;

when the roof squeezes, or the pavement heaves. (40)

donner à bail, to lease ; to let on lease. (41)

donner avis par écrit de terminer un contrat, to give notice in writing to terminate an agreement. (42)

donner bon espoir, to give good promise ; to promise well ; to look very promising. (43)

donner coup *ou* **donner charge** (Mines) (*v.i.*), to weight : (44)
laisser les terrains donner coup sur le boisage, avant d'établir une voûte de maçonnerie, to allow the ground to weight on the timbering before building a masonry arch. (45)

donner de bons rendements, to yield good returns. (46)

donner de l'air à une chambre, to air a room. (47)

donner de la voie à une scie, to set, to jump, a saw. (48)

donner des traces d'or (se dit d'un échantillon), to show, to return, traces of gold. (49)

donner droit à une concession de découverte, to entitle to a reward-claim. (50)

donner du fruit à un mur, to batter a wall. (51)

donner du mou au câble, to slack, to slacken, the rope. (52)

donner du travail à — ouvriers, to give employment to — men ; to find work for — men. (53)

donner l'entreprise pour les tous travaux du jour, to let a contract for all surface-work. (54)

donner l'impulsion à une machine, to propel a machine. (55)

donner les premiers secours aux ouvriers blessés, to give first aid to injured men. (56)

donner naissance à des explosions, to give rise to explosions. (57)

donner un bénéfice (*ou* **un profit**), to pay ; to show a profit. (58)

donner un coup de feu sur le mineral grillé, to pass a hot blast over the roasted ore. (59)

donner un rendement de — 0/0 d'or, to give a return of — % of gold. (60)

donner une chasse à, to flush ; to scour : (61)
donner une chasse à un égout, to flush a sewer ; to scour a drain. (62)

donner une chauffe, to look after the fires ; to attend to the firing ; to stoke. (63)

donner une commande pour des marchandises, to give an order for goods. (64)

donner une forte odeur (se dit de certaines substances lorsqu'on les chauffe), to give off a strong smell. (65)

donner une grande impulsion à l'industrie minière en général, to give a great impetus to mining generally. (66)

donner une moyenne de, to average : (67)
le minerai donne une moyenne de — par tonne, the ore averages — per ton. (68)

dopplérite (Minéral) (*n.f.*), dopplerite. (69)

dorer (*v.t.*), to gild. (70)

dorer par électrolyse *ou* **dorer par le procédé électrochimique,** to electrogild ; to gild by the electrochemical process. (71)

doreur (pers.) (*n.m.*), gilder. (72)

dormant (*n.m.*) *ou* **dormant de porte,** casing ; frame ; door-casing ; door-case ; door-pocket ; door-frame. (73)

dormant (*n.m.*) *ou* **dormant de fenêtre,** casing ; frame ; window-frame. (74)

dormant (d'un palan) (*n.m.*), standing end (of a tackle). (1)

dortoir pour les ouvriers (*m.*), sleeping-quarters for the men. (2)

dorure (*n.f.*), gilding. (3)

dorure électrochimique *ou* **dorure électrolytique** *ou* **dorure galvanoplastique** (*f.*), electro-gilding. (4)

dorure par immersion *ou* **dorure au trempé** (*f.*), gilding by dipping. (5)

dos (*n.m.*), back : (6)

le dos d'un couteau, the back of a knife. (7)

dos d'âne [**dos d'âne** *pl.*] (Géol., etc.) (*m.*), hog-back ; hog's-back ; horseback. (8)

dos d'âne (en), hogbacked : (9)

pont en dos d'âne (*m.*), hogbacked bridge. (10)

dosage (*n.m.*), action or process of combining, making up, or measuring a compound or solution, or of adding or introducing an ingredient ;   admixture ;   making up ; measuring ; dosing : (11)

le dosage des alcalis dans les chaux et ciments, the admixture of alkalis in limes and cements. (12)

dosage d'une solution, making up a solution. (13)

dosage de la perte au feu (Chim.), measuring the ignition loss. (14)

tube pour dosage (Chim.) (*m.*), measuring-tube. (15)

dose (*n.f.*), the addition of that proportion, or the proportion itself, of an element or com-pound which, in combination with other elements or compounds, will produce a state of chemical equilibrium in the mass ; addition of a certain proportion ; proportion ; percentage ; certain proportion *or* percentage ; amount ; measure ; dose : (16)

une dose de carbone déterminée est nécessaire à l'aciération, the addition of a certain proportion of carbon is requisite for aciera-tion. (17)

la dose de cyanure dans la solution, the proportion of cyanide in the solution. (18)

acide injecté à la dose de — grammes au litre (*m.*), acid injected in the proportion of — grammes to the litre. (19)

introduire une dose de zinc dans du bronze, to introduce a certain proportion (*or* a measure) of zinc into bronze. (20)

une dose de soufre neutralise l'effet d'une dose cinq à dix fois plus forte de silicium, a dose of sulphur neutralizes the effect of a dose of silicon five or ten times stronger. (21)

doser (*v.t.*), to add *or* introduce a certain pro-portion ; to make up ; to proportion ; to measure ; to dose : (22)

quand on a mal dosé le zinc, le bronze est cassant, when zinc has been introduced in wrong proportion, the bronze is brittle. (23)

doser la solution pour la cyanuration, to make up the cyaniding solution. (24)

doser l'acide carbonique contenu dans l'air, to measure the carbonic acid contained in the air. (25)

doser l'acier avec du manganèse pour le rendre doux, to dose steel with manganese to render it soft. (26)

dosse (de bois) (*n.f.*), slab (of timber). (27)

doter (pourvoir) (*v.t.*), to equip ; to provide : (28)

doter une mine d'un ou de plusieurs groupes électrogènes, to equip a mine with one or more generating sets. (29)

douane (*n.f.*), customs. (30)

doublage (action de rendre double) (*n.m.*), doubling. (31)

doublage (action de garnir d'une doublure) (*n.m.*), lining : (32)

doublage des emballages d'exportation, lining export cases. (33)

double (*adj.*), double ; twofold : (34)

double bascule à l'avant (appareil photo-graphique) (*f.*), double swing-front. (35)

double cisaillement (Résistance des matériaux) (*m.*), double shear. (36)

double duo (laminoir) (*m.*), four-high rod-mill. (37)

double effet (à),   double-acting ;   double-action : (38)

pompe à double effet (*f.*), double-acting pump ; double-action pump. (39)

double équerre (*f.*), double square. (40)

double fond (*m.*), false bottom. (41)

double harnais d'engrenages enfermé dans des carters *ou* double harnais d'engrenages recouverts de gaines protectrices contre les accidents (*m.*),   double encased gearing ; double gear guarded by covers. (42)

double jet (*m.*) *ou* double travée (*f.*) (Épuise-ment), double lift. (43)

double pesée (Phys.) (*f.*),   double weighing. (44)

double quartier tournant (escalier)(*m.*),   half space. (45) Cf. escalier.

double réfraction (Opt.) (*f.*), double refraction : (46)

double réfraction du spath d'Islande, double refraction of Iceland spar. (47)

double T (en parlant des fers profilés), H ; I : (48)

poutre en double T (*f.*), H beam ; I girder. (49)

double tirage (appareil photographique) (*m.*), double extension. (50)

double touche (Magnétisme) (*f.*), double touch. (51)

palan double (*m.*),   double tackle ; twofold purchase. (52)

doublement (*adv.*), doubly. (53)

doublement (*n.m.*), doubling. (54)

doubler (rendre double) (*v.t.*), to double : (55)

doubler le rendement (*ou* la production),   to double the yield (*or* the output). (56)

doubler (garnir d'une doublure) (*v.t.*), to line : (57)

doubler de fer-blanc une boîte, to line a box with tin. (58)

doubler (devenir double) (*v.i.*), to double : (59)

arbre qui a doublé en hauteur (*m.*), tree which has doubled in height. (60)

terrain qui a doublé de valeur (*m.*), ground which has doubled in value. (61)

doublet de Wollaston (Opt.) (*m.*), Wollaston's doublet. (62)

doucement (*adv.*), softly ; mildly ; smoothly ; gently ; easily. (63)

douceur de la température (*f.*), mildness of the weather. (64)

douceur de roulement (Mach.) (*f.*), smoothness of running. (65)

doucine (moulure) (Menuis.) (*n.f.*), ogee ; O.G. (66)

doucine (rabot) (*n.f.*), ogee-plane. (1)

doucine à carré (*f.*), quirk ogee. (2)

doucine renversée (*f.*), reversed ogee. (3)

doucir (*v.t.*) *ou* **doucir à la pierre à l'huile**, to set ; to set on the oilstone ; to oilstone : (4)

les outils de tour doivent être soigneusement doucis à la pierre à l'huile (*m.pl.*), lathe-tools should be carefully set on the oilstone (*or* should be carefully oilstoned). (5)

doucissage (*n.m.*), setting. (6)

douelle (d'une voûte) (*n.f.*), intrados, soffit, (of an arch). (7)

douille (*n.f.*), socket. (8)

douille (d'une pelle) (*n.f.*), socket (of a shovel). (9)

douille (pour lampes électriques à incandescence) (*n.f.*), lamp-holder ; holder ; socket. (10)

douille à baïonnette (*f.*), bayonet lamp-holder ; bayonet-holder ; bayonet-socket. (11)

douille à clef (*f.*), key lamp-holder ; key-holder ; key-socket. (12)

douille à vis *ou* **douille filetée** (Méc.) (*f.*), screw-socket. (13)

douille à vis (lampes) (*f.*), screw lamp-holder ; screw-holder ; screw-socket. (14)

douille de secours (outil de repêchage) (Sondage) (*f.*), horn socket. (15)

douille destinée à recevoir les queues des forets aux cônes Morse (*f.*), socket for Morse taper shank drills. (16)

douille du câble (*f.*), rope-socket. (17)

douille pour câble métallique (*f.*), wire-rope socket. (18)

douille pour coulisse (Sondage) (*f.*), jar-socket. (19)

douille pour forets à queue carrée (*f.*), socket for square-shank drills. (20)

douille sans clef d'allumage (*f.*), keyless lamp-holder ; keyless holder ; keyless socket. (21)

douve (*n.f.*), stave : (22)

les douves d'un tonneau, the staves of a cask. (23)

doux, douce (*adj.*), soft ; mild ; smooth ; gentle ; easy : (24)

fer doux (*m.*), soft iron. (25)

acier doux (*m.*), mild steel. (26)

l'amiante est doux au toucher (*m.*), asbestos is soft to the touch. (27)

lime à taille douce (*f.*), smooth-cut file. (28)

marche douce (Mach.) (*f.*), smooth running. (29)

pente douce (*f.*), gentle slope ; easy gradient. (30)

down (Géogr. phys.) (*n.m.*), down. (31)

drachme (poids anglais) (*n.f.*), dram = 1·772 grammes. (32)

dragage (*n.m.*), dredging. (33)

dragage aurifère (*m.*), gold-dredging. (34)

dragage de rivières (*m.*), river-dredging. (35)

dragage en butte (*m.*), banks-dredging ; dredg-ing the banks of streams. (36)

dragage en vallée (*m.*), dredging the beds of streams. (37)

dragage stannifère (*m.*), tin-dredging. (38)

dragonite (Explosif) (*n.f.*), dragonite. (39)

draguage (*n.m.*). Same as **dragage**.

drague (*n.f.*), dredge ; dredger ; dredging-machine. (40)

drague à aspiration (*f.*), suction-dredge. (41)

drague à couloir (*f.*), dredge with shoot. (42)

drague à cuiller (*f.*), spoon-dredger. (43)

drague à élinde *ou* **drague à échelle** (*f.*), ladder-dredge. (44)

drague à étain (*f.*), tin-dredger ; tin-dredge. (45)

drague à filet (*f.*), net-dredger ; net-dredge. (46)

drague à godets (*f.*), bucket-dredge. (47)

drague à mâchoires (*f.*), grab-dredge ; grab-dredger ; clam-shell bucket dredging-machine (48)

drague à or (*f.*), gold-dredge ; gold-dredger ; gold-boat. (49)

drague à sable (*f.*), sand-dredge. (50)

drague à sac (*f.*), bag and spoon. (51)

drague à succion (*f.*), suction-dredge. (52)

drague à vapeur (*f.*), steam-dredge ; steam-dredger. (53)

drague aspirante (*f.*), suction-dredge. (54)

drague avec élévateur en queue (*f.*), dredge with tail elevator. (55)

drague californienne (*f.*), Californian dredge. (56)

drague de prospection (*f.*), prospecting-dredge. (57)

drague électrique (*f.*), electric dredge ; electric dredger. (58)

drague en catamaran (*f.*), catamaran dredge. (59)

drague fluviale (*f.*), river-dredge. (60)

drague hydraulique (*f.*), hydraulic dredge. (61)

drague-pompe [dragues-pompes *pl.*] (*n.f.*), pump-dredger ; pump-dredge. (62)

drague pour placers (*f.*), placer-dredge. (63)

drague suceuse (*f.*), suction-dredge. (64)

draguer (*v.t.*), to dredge : (65)

draguer une rivière, une passe, to dredge a river, a channel. (66 *ou* 67)

dragueur (pers.) (*n.m.*), dredger. (68)

drain (*n.m.*), drain. (69)

drain collecteur (*m.*), main drain. (70)

drainable (*adj.*), drainable. (71)

drainage (*n.m.*), drainage ; draining : (72)

drainage des terrains humides, drainage of wet ground ; draining wet ground. (73)

drainage du grisou, draining fire-damp. (74)

drainer (*v.t.*), to drain : (75)

drainer le grisou, une mine, to drain fire-damp, a mine. (76 *ou* 77)

drainer (se) (*v.r.*), to drain ; to drain away. (78)

dravite (Minéral) (*n.f.*), dravite. (79)

dressage (*n.m.*). See examples. (80) (See also the verb **dresser**) :

dressage d'un échafaud, erecting a scaffold. (81)

dressage d'une pierre, dressing, squaring, a stone. (82)

dressage d'une tige, d'une barre, straighten-ing a rod, a bar. (83 *ou* 84)

dressage de la voie (Ch. de f.), lining the road. (85)

dressant (*n.m.*) *ou* **dressants de houille** (*m.pl.*) (Mines de charbon), highly inclined seam ; seam of high dip ; edge-seam ; edge-coals ; rearer-coals. (86)

dressant (Mines métallifères) (*n.m.*), steep vein ; steeply inclined lode ; steeply dipping lode ; rearer. (87)

dresser (*v.t.*). See examples : (88)

dresser à la meule d'émeri une surface plane, to true, to rectify, a flat surface with an emery-wheel. (89)

dresser la carte d'un district, to prepare, to lay down, the map of a district ; to map a district. (90)

dresser la voie (Ch. de f.), to line the road. (1)

dresser le bilan calorifique d'un haut fourneau, to draw up the heat balance-sheet of a blast-furnace. (2)

dresser le bois (le marquer d'un trait au moyen du cordeau), to line out wood. (3)

dresser le bois (le corroyer), to dress wood ; to trim timber. (4)

dresser le champ d'une planche, to shoot the edge of a board. (5)

dresser procès-verbal d'un accident suivi de mort, to report on a fatal accident. (6)

dresser sur face, to face : (7)

bride dressée sur face (f.), faced flange. (8)

dresser un devis, un rapport, to draw up, to make out, an estimate, a report. (9 ou 10)

dresser un mât, to erect a mast. (11)

dresser un niveau, to level the ground. (12)

dresser une barre de fer, to straighten an iron bar. (13)

dresser une meule d'émeri, to true an emery-wheel. (14)

dresser une pierre, to dress, to square, a stone. (15)

dresser une surface plane (Travail au tour), to face a plane surface (Lathe-work). (16)

drift (argile à blocaux) (Géol.) (n.m.), drift ; glacial drift ; till ; glacial till ; boulder-clay ; bowlder-clay. (17)

drille (n.f.), Archimedean drill ; lightning-brace. (18)

drille à engrenages à plaque conscience (f.), breast-drill ; breast-brace ; breast drill-brace. (19)

drille à main (f.), hand-drill. (20)

driller (v.t.), to drill (with a hand-drill). (21)

droit, -e (qui n'est pas courbe) (adj.), straight : (22)

ligne droite (f.), straight line. (23)

droit, -e (perpendiculaire) (adj.), upright; erect; plumb : (24)

un mur droit, a plumb wall. (25)

droit, -e (opposé à gauche) (adj.), right ; right-hand ; right-handed : (26)

la rive droite d'une rivière, the right bank of a river. (27)

cristal droit (m.), right-handed crystal. (28)

droit, -e (Géom.) (adj.), right : (29)

angle droit (m.), right angle. (30)

cône droit (m.), right cone. (31)

droit (n.m.) ou **droits** (n.m.pl.) (privilège), right ; rights ; privilege. (32)

droit (n.m.) ou **droits** (n.m.pl.) (imposition), tax ; taxes ; taxation ; duty ; duties ; dues. (33)

droit ad valorem (Douane) (m.), ad valorem duty. (34)

droit d'enregistrement (m.), registration-duty. (35)

droits d'entrée (m.pl.), import-duty. (36)

droits d'exploiter le pétrole (m.pl.), oil-rights. (37)

droit d'exploiter les pierres précieuses (m.), right to mine for precious stones. (38)

droits d'exportation (m.pl.), export-duty. (39)

droits d'importation (m.pl.), import-duty. (40)

droits d'un propriétaire (m.pl.), rights of an owner. (41)

droits de captation d'eau (m.pl.), water-rights. (42)

droits de douane (m.pl.), customs duty ; customs dues. (43)

droit de faire des sondages (m.), boring-rights. (44)

droit de faire du bois (m.), right to cut timber ; right to timber ; right to wood ; timber-rights. (45)

droit de faire du bois et de l'eau pour travaux miniers (m.), right to wood and water for mining purposes. (46)

droit de faire une prise d'eau à un lac ou **droit de prendre l'eau d'un lac** (m.), right to take water from a lake. (47)

droits de fouiller sous le terrain d'autrui (m.pl.), undermining-rights. (48)

droit de passage (m.), right of way ; way-leave. (49)

droit de propriété (m.), proprietary rights ; proprietorship ; ownership ; right of posses-sion : (50)

avoir le droit de propriété moyennant rede-vance, to have the right of possession on royalty. (51)

droits de sortie (m.pl.), export-duty. (52)

droits de superficie (m.pl.), surface-rights. (53)

droits de vue (m.pl.), right of light ; ancient lights. (54)

droit de wharf (m.), wharfage ; wharf-dues ; quayage. (55)

droit exclusif (m.), sole right. (56)

droit fixe (droit d'enregistrement) (m.), fixed duty. (57)

droits miniers (m.pl.), mining-rights ; mineral rights ; ore-leave. (58)

droit proportionnel (droit d'enregistrement) (m.), ad valorem duty. (59)

droit spécifique (Douane) (m.), specific duty. (60)

droite (côté droit ; partie droite) (n.f.), right ; right hand ; right side. (61)

droite (ligne droite) (n.f.), straight ; straight line : (62)

droites parallèles, parallel straight lines. (63)

droite (à) (locution adjective), right-hand ; right-handed : (64)

vis à droite (f.), right-hand screw ; right-handed screw. (65)

droite (à) (locution adverbiale), right ; to the right ; right-handedly : (66)

à droite et à gauche du puits, right and left of the shaft. (67)

droite d'échanfrinement (Engrenages) (f.), addendum-line (Gearing). (68)

droite d'évidement (Engrenages) (f.), root-line ; dedendum-line. (69)

droite primitive (Engrenages) (f.), pitch-line : (70)

droite primitive d'une crémaillère, pitch-line of a rack. (71)

drop (appareil pour le chargement des navires) (n.m.), drop. (72)

drumlin (Géol.) (n.m.), drumlin. (73)

druse (Géol.) (n.f.), druse ; geode ; vug ; vugh. (74)

drusiforme (adj.), drusiform. (75)

drusique (adj.), drusy. (76)

dû, due (adj.), due : (77)

rupture due à la torsion (f.), breakage due to torsion. (78)

ductile (adj.), ductile : (79)

métaux ductiles (m.pl.), ductile metals. (80)

ductilité (*n.f.*), ductility. (1)
dudgeon (extendeur système Dudgeon) (*n.m.*), tube-expander ; Dudgeon's pattern tube-expander. (2)
dudgeonnage (*n.m.*), expanding tubes. (3)
dudgeonner un tube, to expand a tube. (4)
dufrénite (Minéral) (*n.f.*), dufrenite. (5)
dufrénoysite (Minéral) (*n.f.*), dufrenoysite. (6)
dumortiérite (Minéral) (*n.f.*), dumortierite. (7)
dune de sable (*f.*), sand-dune, sand-hill ; sand-drift. (8)
dunite (Pétrol.) (*n.f.*), dunite. (9)
duo (laminoir) (*n.m.*), two-high mill ; two-high rolls ; two-high train. (10)
duplex (*adj. invar.*), duplex : (11)
télégraphe duplex (*m.*), duplex telegraph. (12)
dur, -e (*adj.*), hard : (13)
bois dur (*m.*), hard wood. (14)
roche dure (*f.*), hard rock. (15)
durabilité (*n.f.*), durability. (16)
durable (*adj.*), durable ; lasting. (17)
duramen (*n.m.*), duramen ; heart-wood. (18)
durangite (Minéral) (*n.f.*), durangite. (19)
durci, -e (*adj.*), hardened ; indurated ; tempered : (20)
argile durcie (*f.*), hardened clay ; indurated clay. (21)
verre durci (*m.*), hardened glass ; tempered glass. (22)
durcir (*v.t.*), to harden ; to indurate ; to temper : (23)
la gelée durcit le sol, frost hardens the ground. (24)
la chaleur durcit l'argile, heat indurates clay. (25)
durcir (*v.i.*), to harden ; to become hard ; to indurate : (26)
l'acier durcit sous l'action de la trempe (*m.*), steel hardens under the influence of temper-ing. (27)
durcir (se) (*v.r.*), to harden ; to become hard ; to indurate ; to be tempered : (28)
les bois se durcissent en séchant (*m.pl.*), woods harden in drying. (29)
durcissement (*n.m.*), hardening ; induration ; tempering : (30)
durcissement des ciments, hardening of cements. (31)
durée (*n.f.*), duration ; length ; life : (32)
durée d'une épreuve, duration of a test. (33)
durée d'un bail, length of a lease. (34)
durée d'une lampe à incandescence, d'un câble, du matériel, life of a glow-lamp, of a rope, of the plant. (35 ou 36 ou 37)
durée d'une mine, life, lifetime, of a mine. (38)
durée de travail dans les mines, working-hours in mines. (39)
durée de pose (Photogr.) (*f.*), exposure. (40)
durer (*v.i.*), to last. (41)
dureté (*n.f.*), hardness : (42)
la dureté du diamant, de l'acier, d'une meule d'émeri, the hardness of diamond, of steel, of an emery-wheel. (43 ou 44 ou 45)
la dureté du diamant = 10 ou diamant : D. = 10 (Minéralogie), the hardness of diamond = 10 ; diamond : H. = 10. (46)
dureté du verre (Métall.) (*f.*), glass-hardness. (47)
dyas (Géol.) (*n.m.*), Dyas. (48)
dyasique ou **dyssique** (Géol.) (*adj.*), Dyassic. (49)
dyke (Géol. & Mines) (*n.m.*), dyke ; dike. (50)
dyke de grès (*m.*), sandstone dyke. (51)

dyke de roche ignée (*m.*), dyke of igneous rock. (52)
dynamètre (*n.m.*), **dynamétrie** (*n.f.*), **dynamétrique** (*adj.*). Same as **dynamomètre**, etc.
dynamique (*adj.*), dynamic ; dynamical : (53)
électricité dynamique (*f.*), dynamic electricity. (54)
dynamique (*n.f.*), dynamics : (55)
dynamique des forces, dynamics of forces. (56)
dynamiquement (*adv.*), dynamically. (57)
dynamite (*n.f.*), dynamite. (58)
dynamite à base active (*f.*), active-dope dynamite. (59)
dynamite à base inerte (*f.*), inert-dope dynamite. (60)
dynamite ammoniacale (*f.*), ammonia dynamite. (61)
dynamite-gomme [dynamites-gommes *pl.*] (*n.f.*), gum-dynamite. (62)
dynamiter (*v.t.*), to dynamite. (63)
dynamitière (*n.f.*), dynamite-store. (64)
dynamitière souterraine (*f.*), underground dynamite-store. (65)
dynamitière superficielle (*f.*), aboveground dynamite-store. (66)
dynamo (*n.f.*), dynamo ; generator. (67)
dynamo à anneau (*f.*), ring dynamo. (68)
dynamo à courant alternatif (*f.*), alternating-current dynamo ; alternating dynamo ; alternating-current generator ; alternator. (69)
dynamo à courant continu (*f.*), direct-current dynamo ; continuous-current generator. (70)
dynamo à courant continu bipolaire (*f.*), bipolar direct-current dynamo. (71)
dynamo à courant continu multipolaire (*f.*), multipolar direct-current dynamo. (72)
dynamo à disque (*f.*), disc dynamo. (73)
dynamo à excitation indépendante (*f.*), separate-ly excited dynamo. (74)
dynamo à pôles (*f.*), pole dynamo. (75)
dynamo à tambour (*f.*), drum dynamo. (76)
dynamo auto-excitatrice (*f.*), self-exciting dynamo. (77)
dynamo compound [dynamos compound *pl.*] (*f.*), compound-wound dynamo ; compound dynamo. (78)
dynamo de compensation (*f.*), balancer. (79)
dynamo-électrique (*adj.*), dynamo-electric ; dynamo-electrical : (80)
machine dynamo-électrique (*f.*), dynamo-electric machine. (81)
dynamo multipolaire-série [dynamos multi-polaires-série *pl.*] (*f.*), series-multipolar dynamo. (82)
dynamo-série [dynamos-série *pl.*] (*n.f.*) ou **dynamo en série,** series-dynamo ; series-wound dynamo. (83)
dynamo-shunt [dynamos-shunt *pl.*] (*n.f.*) ou **dynamo-dérivation** [dynamos-dérivation *pl.*] (*n.f.*) ou **dynamo en dérivation,** shunt-wound dynamo ; shunt dynamo. (84)
dynamo unipolaire (*f.*), unipolar dynamo ; homopolar dynamo. (85)
dynamographe (*n.m.*), dynamograph. (86)
dynamométamorphisme (Géol.) (*n.m.*), dynamo-metamorphism ; dynamic metamorphism. (87)
dynamomètre (*n.m.*), dynamometer. (88)
dynamomètre d'absorption (*m.*), absorption dynamometer. (89)

dynamomètre de transmission (*m.*), transmission dynamometer. (1)
dynamométrie (*n.f.*), dynamometry. (2)
dynamométrique (*adj.*), dynamometric ; dynamo--metrical : (3)
  frein dynamométrique (*m.*), dynamometrical brake. (4)
dyne (C.G.S.) (*n.f.*), dyne. (5)
dysanalyte (Minéral) (*n.f.*), dysanalyte. (6)

dysclasite (Minéral) (*n.f.*), dyclasite ; okenite. (7)
dyscrase *ou* **dyscrasite** (Minéral) (*n.f.*), dyscrasite ; antimonial silver. (8)
dysluite (Minéral) (*n.f.*), dysluite. (9)
dysodile *ou* **dysodyle** (Minéral) (*n.m.*), dysodile ; papyraceous coal. (10)
dysprosium (Chim.) (*n.m.*), dysprosium. (11)

E

eau (*s'emploie souvent au pluriel*) (*n.f.*), water. (12)
eau (des pierres précieuses) (*n.f.*), water : (13)
  diamant d'une belle eau (*m.*), diamond of a fine water. (14)
eau à souder (*f.*), soldering-liquid. (15)
eaux basses (étiage) (*f.pl.*), low water. (16)
eau boueuse (*f.*), muddy water. (17)
eau bouillante (*f.*), boiling water. (18)
eau buvable (*f.*), drinkable water. (19)
eau chaude (*f.*), hot water ; warm water. (20)
eau claire (*f.*), clear water. (21)
eau courante (*f.*) *ou* **eaux courantes** (*f.pl.*), running water. (22)
eau crue (*f.*), hard water. (23)
eau d'alimentation (*f.*), feed-water. (24)
eau d'alimentation des chaudières (*f.*), boiler-feed water. (25)
eau d'amont (Hydraul.) (*f.*), headwater. (26)
eau d'aval (Hydraul.) (*f.*), tail-water. (27)
eau d'hydratation (*f.*), water of hydration. (28)
eau d'imbibition (*f.*), soakage-water. (29)
eau d'infiltration (*f.*), percolating water ; seepage-water. (30)
eau de carrière (*f.*), quarry-water. (31)
eau de chaux (*f.*), lime water. (32)
eau de constitution (*f.*), water of constitution ; inseparable water of crystallization. (33)
eau de cristallisation (*f.*), water of crystalliza--tion. (34)
eau de fontaine (*f.*), spring-water. (35)
eaux de fonte *ou* **eaux de fusion** (des neiges, de la glace) (*f.pl.*), water produced by melting, water from melting, (of the snows, of ice). (36)
eau de lavage (*f.*), wash-water. (37)
eau de mer (*f.*), sea-water. (38)
eau de pluie (*f.*), rain-water. (39)
eau de puits (*f.*), well-water ; pump-water. (40)
eau de roche (*f.*), rock-water. (41)
eaux de ruissellement (Géogr. phys.) (*f.pl.*), water which has run off over the surface. (42) Cf. **ruissellement.**
eau de saturation (*f.*), water of saturation. (43)
eau de savon (*f.*), suds ; soap-suds : (44)
  pompe refoulante à eau de savon (*f.*), sud-pump. (45)
eau de source (*f.*), spring-water. (46)
eaux de surface (*f.pl.*), surface-water ; day-water (47)
eau de ville (*f.*), town water. (48)
eaux des nappes (Géol.) (*f.pl.*), sheet-water. (49)
eau distillée (*f.*), distilled water. (50)
eau dormante (*f.*), still water ; standing water ; stagnant water. (51)

eau douce (*f.*), fresh water ; soft water. (52)
eau douce (d'), fresh-water : (53)
  mollusques d'eaux douces (*m.pl.*), fresh-water molluscs. (54)
eaux du jour (*f.pl.*), day-water ; surface-water. (55)
eau entraînée (chaudières) (*f.*), entrained water. (56)
eau ferrugineuse (*f.*), ferruginous water ; chalybeate water ; chalybeate. (57)
eau fluviale (*f.*), river-water. (58)
eaux folles (*f.pl.*), day-water ; surface-water. (59)
eau-forte (*n.f.*), aqua fortis. (60)
eau froide (*f.*), cold water. (61)
eau limpide (*f.*), clear water. (62)
eaux marines (*f.pl.*), sea-water. (63)
eaux mères (*f.pl.*), mother-water ; mother-liquor ; mother-liquid ; mother-lye ; mothers. (64)
eaux météoriques (*f.pl.*), meteoric waters. (65)
eau minérale *ou* **eau minérale naturelle** (*f.*), mineral water. (66)
eau minérale sulfatée (*f.*), sulphatic mineral water. (67)
eau morte (*f.*), still water ; stagnant water ; standing water. (68)
eau non-buvable *ou* **eau non-potable** (*f.*), undrinkable water. (69)
eau peu profonde (*f.*), shallow water. (70)
eaux pluviales (*f.pl.*), rain-water. (71)
eau potable (*f.*), drinkable water. (72)
eau profonde (opp. à eau peu profonde) (*f.*), deep water. (73)
eaux profondes (opp. à eaux de surface) (*f.pl.*), underground water. (74)
eau régale (*f.*), aqua regia. (75)
eau salée *ou* **eau saline** (*f.*), brine ; salt-water ; saline water. (76)
eau saumâtre (*f.*), brackish water. (77)
eaux sauvages (*f.pl.*), surface-water ; day-water. (78)
eau schlammeuse (Préparation mécanique des minerais) (*f.*), slime-water. (79)
eau sous pression (*f.*), water under pressure. (80)
eaux souterraines (*f.pl.*), underground water ; subterranean waters. (81)
eau stagnante (*f.*), stagnant water ; standing water ; still water. (82)
eaux superficielles (*f.pl.*), surface-water ; day-water. (83)
eaux terrestres (*f.pl.*), earth-waters. (84)
eau thermale (*f.*), thermal water. (85)
eau tiède (*f.*), lukewarm water ; tepid water. (86)
eau tranquille (*f.*), still water. (87)

eau **trouble** (*f.*), cloudy water. (1)
eau **trouble**(Préparation mécanique des minerais) (*f.*), slime-water. (2)
eau **vive** (*f.*) *ou* **eaux vives** (*f.pl.*), running water. (3)
ébarbage (*n.m.*), taking off the burr ; trimming ; fettling : (4)
ébarbage des pièces coulées (Fonderie), fettling, trimming, castings. (5)
ébarber (*v.t.*), to take the burr off ; to trim ; to trim off ; to trim the edges of ; to trim off rough edges ; to fettle : (6)
ébarber un joint (Plomberie), to wipe a joint. (7)
ébarber un tube intérieurement, to trim off the burr inside a tube. (8)
ébarber une pièce coulée (Fonderie), to fettle, to trim, a casting. (9)
ébarber une pièce de métal, to take the burr off a piece of metal. (10)
ébarbeur (Fonderie) (pers.) (*n.m.*), fettler; chipper. (11)
ébarbeuse (*n.f.*), grinding-machine (of the fettling or trimming type) ; fettling-machine ; trimming-machine. (12)
ébarbure (*n.f.*), burr ; fin. (13)
ébardoir (*n.m.*), shave-hook. (14)
ébauchage (action de donner une première façon à un objet) (*n.m.*), roughing ; roughing out ; roughing-out work ; roughing down. (15)
ébauche (dessin sans détails) (*n.f.*), sketch ; rough sketch. (16)
ébauché de puddlage (*m.*), puddle-bar ; muck-iron ; muck-bar. (17)
ébaucher (*v.t.*), to rough ; to rough out ; to rough down ; to sketch ; to sketch out : (18)
ébaucher le bois, to rough down timber. (19)
ébaucher un plan, to rough out, to sketch, to sketch out, a plan. (20)
ébaucheur (Laminage du fer) (*n.m.*), roughing-roll ; roughing-down roll ; rough-down roll ; breaking-down roll ; muck-roll ; puddle-roll. (21)
ébauchoir à mortaises (*m.*), mortise boring-bit. (22)
ébauchoir de charpentier (*m.*), carpenters' chisel (all iron, steeled edge). (23)
ébène (bois ou arbre) (*n.f.*), ebony. (24)
ébénier (arbre) (*n.m.*), ebony. (25)
ébéniste (pers.) (*n.m.*), cabinetmaker. (26)
ébénisterie (*n.f.*), cabinetmaking ; cabinetwork. (27)
ébonite (*n.f.*), ebonite ; vulcanite ; hard rubber. (28)
ébotter (*v.t.*), to cut the head off : (29)
ébotter un clou, to cut the head off a nail. (30)
éboulement (*n.m.*), falling in ; fall ; caving in ; caving ; run ; rush together : (31)
éboulement de rocher, fall of rock ; rock-fall. (32)
éboulement de terrain (*ou* de terre), fall of ground ; fall of earth ; run of ground. (33)
éboulement des parois d'une carrière, caving in of the sides of a quarry. (34)
éboulement de terres *ou simplement* **éboulement** (*n.m.*) (Géol.), landslip ; landslide ; landfall ; earth-fall ; éboulement. (35)
ébouler (s') (*v.r.*), to fall ; to fall in ; to cave in. (36)
ébouleux, -euse (*adj.*), loose ; running : (37)
terrain ébouleux (*m.*), loose ground ; running ground. (38)

éboulis (*n.m.*) *ou* **éboulis de roches** (Géol.), fall of rock ; rock-rubble. (39)
éboulis (matières éboulées) (Mines) (*n.m.pl.*), falls. (40)
éboulis (aire d'éboulement) (Mines) (*n.m.pl.*), caved area. (41)
ébranlage (Fonderie) (*n.m.*), rapping. (42)
ébranler (Fonderie) (*v.t.*), to rap : (43)
ébranler un modèle pour faciliter sa sortie du moule, to rap a pattern to facilitate its delivery from the mould. (44)
ébullition (*n.f.*), boiling ; ebullition. (45)
écaillage (*n.m.*), scaling ; flaking ; peeling off in flakes : (46)
écaillage du marbre, scaling, flaking, of marble. (47)
écaille (*n.f.*), scale ; flake : (48)
écailles de laminage, roll-scale. (49)
murs qui s'effritent et s'en vont par écailles (*m.pl.*), walls which crumble and disappear in flakes. (50)
écaille (*n.f.*) *ou* **écailles de fer** (*f.pl.*) (batitures), scale ; iron-scale ; hammer-scale ; forge-scale ; hammer-slag ; cinder ; anvil-dross ; nill. (51)
écaille (éclat) (*n.f.*), chip : (52)
écaille de marbre, chip of marble. (53)
écaillement (*n.m.*). Same as **écaillage**.
écailler (s') (*v.r.*), to scale ; to flake ; to peel off in flakes. (54)
écailleux, -euse (*adj.*), scaly ; flaky. (55)
écart (variation) (*n.m.*), variation ; difference : (56)
écarts de teneur, variations, differences, in grade. (57)
écarté, -e (*adj.*), remote ; out-of-the-way : (58)
endroit écarté (*m.*), out-of-the-way place. (59)
écartement de voie *ou simplement* **écartement** (*n.m.*) (Ch. de f.), gauge ; railway-gauge ; track-width ; distance between rails. (60)
écartement des barreaux d'une grille (*m.*), distance between the bars of a grate. (61)
écartement des cylindres réglable par vis (*m.*), distance apart of rollers adjustable by screw. (62)
écartement des essieux (*m.*), wheel-base. (63)
écartement des essieux couplés (locomotive) (*m.*), driving-wheel base. (64)
écartement des essieux extrêmes (*m.*), total wheel-base. (65)
écartement des pivots des bogies (wagon de ch. de f.) (*m.*), truck-centres. (66)
écartement des rivets d'axe en axe (*m.*), pitch of rivet-centres. (67)
écartement du foret à la colonne (*m.*), distance of drill from column. (68)
écartement normal (Ch. de f.) (*m.*), standard gauge ; normal gauge = 1·435 metres or 4 feet 8½ inches. (69) (Varies slightly in France and other countries, some tracks measuring 1·44 m. = 4'8$\frac{11}{16}$", others 1·45 m. = 4'9$\frac{3}{32}$", etc.).
écarter (*v.t.*), to turn aside ; to swing aside out of the way ; to swing clear ; to push aside : (70)
écarter le plateau d'une machine à percer lors du perçage de pièces placées sur la plaque d'assise, to turn aside the table of a drilling-machine when drilling articles placed on the bed-plate ; to swing the table clear of any work placed on the bed of the machine. (71)

écarter un obstacle, to push an obstacle aside. (1)

écarter (s') (dévier) (v.r.), to deviate ; to turn aside ; to run out : (2)

s'écarter de la verticale, to deviate from, to run out of, the vertical. (3)

échafaud ou échafaudage (n.m.), scaffold ; scaffolding ; stage ; staging. (4)

échafaud de maçon ou échafaud ordinaire ou échafaud simple, bricklayers' scaffold ; common scaffold. (5)

échafaudage en bascule ou échafaud à bascule ou échafaud en encorbellement, flying scaffold. (6)

échafaudage suspendu, suspended scaffold. (7)

échafaudage volant ou échafaud volant (Construction civile), travelling cradle ; cradle ; boat ; hanging scaffold ; hanging stage. (8)

échafaudage volant (Mines), walling scaffold ; walling stage ; walling platform ; cradle ; hanging scaffold ; hanging stage. (9)

échafaudage (action) (n.m.), scaffolding ; erect- -ing a scaffold ; erecting scaffolds. (10)

échafauder (v.t.), to scaffold. (11)

échafauder (v.i.), to erect a scaffold; to erect scaffolds. (12)

échancrer (v.t.), to hollow out ; to notch ; to indent ; to groove : (13)

trou échancré (d'une filière) (m.), notched hole (of a screw-plate). (14)

échancrure (n.f.), hollowing out ; indentation ; notch : (15)

échancrures d'une côte, indentations of a coast. (16)

échancrure (Hydraul.) (n.f.), notch. (17)

échange (n.m.), exchange ; interchange : (18)

échange de trafic, exchange of traffic. (19)

échanger (v.t.), to exchange ; to interchange : (20)

échanger des signaux, to interchange, to exchange, signals. (21)

échantignole ou échantignolle (n.f.), purlin-cleat. (22)

échantillon (n.m.), sample : (23)

un échantillon de minerai, d'air, a sample of ore, of air. (24 ou 25)

échantillon (type de certains matériaux de construction) (n.m.), dimension ; size ; section : (26)

bois d'échantillon (m.), dimension timber. (27)

ardoises d'échantillon (f.pl.), sized slates. (28)

les ardoises modèles anglais, de quelque échantillon qu'elles soient, s'emploient par superposition (f.pl.), English pattern slates, of whatever size they may be, are laid one upon another. (29)

une barre de fer d'échantillon uniforme, an iron bar of uniform section. (30)

fers de faible échantillon (m.pl.), light-section iron. (31)

les fers marchands comprenant les plats, carrés, et ronds, sont divisés suivant leurs dimensions en classes de valeur différente d'autant plus élevée que l'échantillon de la section est plus petit (m.pl.), merchant bars including flats, squares, and rounds, are divided according to their dimensions into classes of different value, the smaller the size of the section, the higher the value. (32)

échantillon (partie des ardoises non recouverte par les ardoises superposées) (n.m.), bare. (33)

échantillon auquel on ne peut se fier (m.), un- -reliable sample. (34)

échantillon choisi ou échantillon de choix (m.), picked sample. (35)

échantillon d'actionnaires ou échantillon salé (m.), salted sample. (36)

échantillon prélevé consciencieusement (m.), fairly drawn sample. (37)

échantillon pris au hasard (m.), random sample ; grab-sample. (38)

échantillon témoin (m.), check-sample. (39)

échantillon type ou échantillon sérieux (m.), typical sample ; representative sample : (40)

échantillon type du minerai sur la halde, typical sample of the ore on the dump. (41)

échantillonnage (n.m.), sampling. (42)

échantillonnage d'une mine d'or, sampling a gold-mine. (43)

échantillonnage au hasard, random sampling ; grab sampling. (44)

échantillonner (v.t.), to sample : (45)

échantillonner du minerai, to sample ore. (46)

échantillonneur (pers.) (n.m.), sampler. (47)

échantillonneuse (n.f.), sampling-machine ; sampler. (48)

échappée (hauteur de passage) (n.f.), headroom. (49)

échappée de beau temps (f.), spell of fine weather. (50)

échappement (fuite) (n.m.), escape ; leakage : (51)

échappement de gaz, escape, leakage, of gas. (52)

échappement d'eau, leakage of water. (53)

échappement (Méc.) (action) (n.m.), exhaust ; release : (54)

l'échappement de vapeur des locomotives sert à activer le tirage, the exhaust of steam from locomotives serves to quicken the draught. (55)

tirage qui est activé par l'échappement des gaz brûlés (moteur à combustion interne) (m.), draught which is quickened by the exhaust of the burnt gases. (56)

échappement (conduite) (n.m.), exhaust-pipe ; blast-pipe. (57)

échappement anticipé (tiroir) (m.), prerelease ; early release (slide-valve). (58)

échapper (s') (v.r.), to escape ; to leak : (59)

vapeur qui s'échappe dans l'atmosphère (f.), steam which escapes into the atmosphere. (60)

échapper (s') par la tangente, to fly off at a tangent. (61)

écharde (n.f.), splinter. (62)

écharpe (d'un pan de bois) (n.f.), brace (of a timber frame). (63)

échasse d'échafaud (f.), scaffold-pole ; standard. (64)

échaudage (n.m.), limewashing. (65)

échauffant, -e (adj.), heating. (66)

échauffement (n.m.), heating : (67)

échauffement des coussinets, d'une lame de scie, des conducteurs électriques, heating of bearings, of a saw-blade, of electric con- -ductors. (68 ou 69 ou 70)

échauffement du ciment pendant la prise, heating of cement during setting. (71)

échauffement des remblais (Mines), heating in the goaf. (72)

échauffer (v.t.), to warm ; to heat. (73)

échauffer (s') (v.r.), to get warm ; to get hot ; to heat : (74)

corps qui s'échauffent lentement (*m.pl.*), substances which heat slowly. (1)

échec (*n.m.*), check ; failure : (2)
échecs dus à une prospection insuffisante, failures due to insufficient prospecting. (3)

échelle (engin) (*n.f.*), ladder. (4)

échelle (dimensions relatives) (*n.f.*), scale : (5)
plan rapporté à l'échelle de 1 : — (*m.*), plan drawn to the scale of 1 in —. (6)

échelle (série de divisions sur un instrument) (*n.f.*), scale. (7)

échelle (moyen de comparaison) (*n.f.*), scale : (8)
travailler sur une échelle modérée, to work on a moderate scale. (9)

échelle à coulisse (*f.*), extension-ladder. (10)

échelle centigrade (*f.*), centigrade scale. (11)

échelle d'étiage (Hydraul.) (*f.*), water-gauge (for determining the level of the water in a river, or other body of water). (12)

échelle d'ivoire de Plattner pour mesurer les boutons (*f.*), Plattner's ivory button-scale. (13)

échelle de Baumé (*f.*), Baumé's scale. (14)

échelle de Celsius (*f.*), Celsius scale. (15)

échelle de corde (*f.*), rope ladder. (16)

échelle de dureté de Mohs (Minéralogie) (*f.*), Mohs scale of hardness. (17)

échelle de fusibilité (Minéralogie) (*f.*), scale of fusibility. (18)

échelle de Gunter (*f.*), Gunter's scale ; plane scale. (19)

échelle de mise au point (Photogr.) (*f.*), focus-sing-scale ; focussing-index. (20)

échelle de perroquet (Mines et Carrières) (*f.*), centipede ladder. (21)

échelle de profondeur de foyer (Photogr.) (*f.*), depth-of-focus scale. (22)

échelle de proportion (Dessin) (*f.*), proportional scale ; proportionate scale. (23)

échelle de puits (Mines) (*f.*), shaft-ladder ; footway. (24)

échelle de sauvetage (*f.*), fire-escape ; fire-ladder ; firemen's ladder ; aerial ladder. (25)

échelle des salaires (*f.*), scale of wages ; wage-scale. (26)

échelle double (*f.*), steps ; pair of steps ; step-ladder. (27)

échelle en bois (*f.*), wood ladder. (28)

échelle en chaîne (*f.*), chain ladder. (29)

échelle Fahrenheit (*f.*), Fahrenheit scale. (30)

échelle logarithmique (*f.*), logarithmic scale. (31)

échelle mécanique *ou* **échelle mobile** (Mines) (*f.*), man-engine ; man-machine ; mining-engine ; travelling ladderway ; movable ladder. (32)

échelle mobile (*f.*), sliding scale : (33)
échelle mobile des salaires, sliding scale of wages. (34)

échelle normale de températures (*f.*), normal scale of temperatures. (35)

échelle proportionnelle (Dessin) (*f.*), propor-tional scale ; proportionate scale. (36)

échelle thermo-électrique (*f.*), thermoelectric series. (37)

échelle thermométrique (*f.*), thermometric scale. (38)

échelle transversale (Dessin) (*f.*), diagonal scale. (39)

échelon (d'une échelle) (*n.m.*), rung, round, step, (of a ladder). (40)

échelonné, -e (*adj.*) *ou* **échelons (en)**, stepped : (41)

engrenage échelonné (*m.*), stepped gear. (42)

fente en échelons (*f.*), stepped fissure. (43)

éclair (*n.m.*), flash ; flash of light : (44)
l'éclair d'un diamant, the flash of a diamond. (45)

lumière-éclair (Photogr.) (*f.*), flash-light. (46)

éclair (Météor.) (*n.m.*), lightning ; flash of light-ning. (47)

éclair (Métall.) (*n.m.*), brightening ; blick ; fulguration. (48)

éclair arborescent *ou* **éclair ramifié** (*m.*), forked lightning. (49)

éclair de chaleur (*m.*), heat-lightning ; summer lightning. (50)

éclair diffus *ou* **éclair en nappes** (*m.*), sheet lightning. (51)

éclair en boule (*m.*), ball lightning ; globe lightning ; globular lightning. (52)

éclair en chapelet (*m.*), beaded lightning ; chapleted lightning ; pearl lightning. (53)

éclair en zigzag (*m.*), zigzag lightning. (54)

éclair fulminant *ou* **éclair en sillons** (*m.*), fillet lightning ; ribbon lightning. (55)

éclair sinueux (*m.*), chain lightning. (56)

éclairage (action) (*n.m.*), lighting ; illumination ; illuminating : (57)
l'éclairage au moyen de bougies est peu employé aujourd'hui, lighting by means of candles is little used nowadays. (58)
éclairage des chantiers (Mines), lighting the workings. (59)

éclairage (effet) (*n.m.*), light ; illumination : (60)
les lampes à arc donnent un bon éclairage (*f.pl.*), arc-lamps give a good light. (61)

éclairage à feu nu (Mines) (*m.*), open-light lighting ; naked-light lighting. (62)

éclairage à incandescence par le gaz (*m.*), incandescent-gas lighting. (63)

éclairage à l'acétylène (*m.*), acetylene-lighting. (64)

éclairage artificiel (*m.*), artificial lighting ; artificial light. (65)

éclairage au gaz (*m.*), gas-lighting. (66)

éclairage au gaz par incandescence (*m.*), incan-descent-gas lighting. (67)

éclairage électrique (*m.*), electric lighting. (68)

éclairage magnésique (*m.*), magnesium-light. (69)

éclairage par le gaz de pétrole (*m.*), oil-gas lighting. (70)

éclairant, -e (*adj.*), lighting ; illuminating : (71)
pouvoir éclairant (*m.*), lighting-power ; illuminating power. (72)

éclaircir la grille d'un foyer, to prick the fire-bars of a furnace ; to slice the furnace-grate ; to clean the fire. (73)

éclairement (Photom.) (*n.m.*), illumination. (74)

éclairer (*v.t.*), to light ; to light up ; to illumi-nate ; to illumine : (75)
éclairer une chambre avec des bougies, to light a room with candles. (76)

éclairer la marche, to reconnoitre. (77)

éclat (fragment) (*n.m.*), splinter ; chip ; chip-ping ; fragment : (78)
éclat de bois, splinter of wood. (79)
éclat de pierre, splinter of stone ; chip of stone ; stone chipping : (80)
on emploie fréquemment des éclats de pierre dans la fabrication du béton, stone chip-pings are frequently used in the manu-facture of concrete. (81)
éclat de roche, splinter of rock. (82)

éclats de fonte, chippings of cast iron. (1)
éclat (dans le bois) (*n.m.*), shake (in timber);
split (in wood): (2)
l'éclat est fréquent dans les bois employés
verts, shakes frequently occur in wood
used green. (3)
éclat (brillant) (*n.m.*), brilliancy; brightness;
lustre: (4)
l'éclat d'une pierre précieuse, the brilliancy
of a precious stone; the lustre of a gem.
(5)
éclat (Photom.) (*n.m.*), brilliancy; brightness.
(6)
éclat (Minéralogie) (*n.m.*), lustre. (7)
éclat adamantin (*m.*), adamantine lustre. (8)
éclat argenté *ou* **éclat** argentin (*m.*), silvery
lustre. (9)
éclat cireux (*m.*), waxy lustre. (10)
éclat de perle (*m.*), pearly lustre. (11)
éclat demi-métallique (*m.*), semimetallic lustre.
(12)
éclat gras (*m.*), greasy lustre. (13)
éclat intrinsèque (Photom.) (*m.*), intrinsic
brilliancy; intrinsic brightness. (14)
éclat mat (Minéralogie) (*m.*), dull lustre. (15)
éclat métallique (*m.*), metallic lustre. (16)
éclat métalloïde (*m.*), metalloidal lustre. (17)
éclat nacré *ou* **éclat** perlé (*m.*), pearly lustre. (18)
éclat résineux (*m.*), resinous lustre. (19)
éclat sous-métallique (*m.*), submetallic lustre.
(20)
éclat soyeux (*m.*), silky lustre. (21)
éclat total (Photom.) (*m.*), total brilliancy. (22)
éclat vitreux (Minéralogie) (*m.*), vitreous lustre.
(23)
éclatable (*adj.*), splintery: (24)
bois éclatable (*m.*), splintery timber. (25)
éclatant, -e (*adj.*), bright; brilliant: (26)
lumière éclatante (*f.*), bright light. (27)
poli éclatant (*m.*), brilliant polish. (28)
éclatement (*n.m.*), bursting: (29)
l'éclatement d'une conduite d'eau, des tubes
de chaudière, d'une meule d'émeri, the
bursting of a water-pipe, of boiler-tubes,
of an emery-wheel. (30 *ou* 31 *ou* 32)
éclater (faire explosion) (*v.i.*), to explode. (33)
éclater (se dit d'une chaudière) (*v.i.*), to burst;
to explode. (34)
éclater (se dit d'un tuyau) (*v.i.*), to burst. (35)
éclater (se dit d'un incendie) (*v.i.*), to break out.
(36)
éclater (se dit du bois) (*v.i.*), to split: (37)
le bois vert est sujet à éclater, green wood is
liable to split. (38)
éclimètre (*n.m.*), eclimeter. (39)
éclipse (Astron.) (*n.f.*), eclipse. (40)
éclipse annulaire (*f.*), annular eclipse. (41)
éclipse de lune *ou* **éclipse de la lune** *ou* **éclipse
lunaire** (*f.*), eclipse of the moon; lunar
eclipse. (42)
éclipse de soleil *ou* **éclipse du soleil** *ou* **éclipse
solaire** (*f.*), eclipse of the sun; solar eclipse.
(43)
éclipse partielle (*f.*), partial eclipse. (44)
éclipse totale (*f.*), total eclipse. (45)
éclipser (*v.t.*), to eclipse: (46)
la lune éclipse le soleil, the moon eclipses the
sun. (47)
éclissage (Ch. de f.) (*n.m.*), fishing; fish-plating:
(48)
éclissage du rail dissymétrique, fishing the
bull-headed rail. (49)

éclisse (pour rails de chemins de fer) (*n.f.*),
fish-plate; fish; splice-piece; splice-bar.
(50)
éclisser (*v.t.*), to fish; to fish-plate: (51)
éclisser la voie (*ou* les rails), to fish, to fish-
plate, the rails. (52)
éclogite (Pétrol.) (*n.f.*), eclogite. (53)
éclusage d'un bateau (*m.*), lockage of a boat;
locking a boat. (54)
écluse (*n.f.*) *ou* **écluse de canal**, lock; canal-
lock. (55)
écluse (porte d'écluse) (*n.f.*), lock-gate; canal
lock-gate. (56)
écluse (d'un caisson) (*n.f.*), lock, air-lock, (of a
caisson). (57)
écluse (Fonderie) (*n.f.*), shut; gate-shutter. (58)
écluse à sas *ou* **écluse double** (*f.*), lift-lock. (59)
éclusée (quantité d'eau nécessaire pour écluser
un bateau) (*n.f.*), feed. (60)
écluser le métal en fusion (Fonderie), to shut
off the molten metal. (61)
écluser un bateau, to lock a boat; to pass a boat
through a lock. (62)
écoin (Boisage des galeries de mine) (*n.m.*),
lid; bonnet; headboard; blocking-board;
clog. (63)
écoine (*n.f.*), **écoiner** (*v.t.*), **écoinette** (*n.f.*).
Same as **écouane, écouaner, écouanette**.
école (*n.f.*), school. (64)
école des mines (*f.*), school of mines. (65)
école industrielle (*f.*), technical school. (66)
école professionnelle de mineurs (*f.*), miners'
training-school. (67)
économie (*n.f.*), economy; saving: (68)
économie dans le coût de la manutention, sav-
-ing in handling-costs. (69)
économie de combustible, economy in fuel
consumption. (70)
économie de main-d'œuvre et de temps,
saving of time and labour. (71)
économique (*adj.*), economic; economical. (72)
économiquement (*adv.*), economically. (73)
économiser (*v.t.*), to economize; to save; to
husband: (74)
économiser de la force motrice, to economize
power. (75)
économiser la main-d'œuvre (*ou* le travail), to
save labour: (76)
mode de travail qui économise la main-
d'œuvre (*m.*), method of working which
saves labour. (77)
machines économisant le travail (*f.pl.*), labour-
saving machinery. (78)
économiser son argent, to economize, to save,
one's money; to husband one's resources.
(79)
économiseur (pour le chauffage des eaux d'alimen-
-tation de chaudières) (*n.m.*), economizer.
(80)
économiseur d'eau (machine hydraulique) (*m.*),
water-saving apparatus; water-saving gear.
(81)
écoperche (*n.f.*), scaffold-pole; standard. (82)
écorce (d'un arbre) (*n.f.*), bark (of a tree). (83)
écorce terrestre (*f.*), earth's crust; crust of the
earth. (84)
écorcer un arbre, to bark a tree. (85)
écouane *ou* **écouenne** (*n.f.*), cabinet-file. (86)
écouaner (*v.t.*), to file (with a cabinet-file). (87)
écouanette (*n.f.*), small cabinet-file. (88)
écoulement (*n.m.*), flow; flowing; flowage;
outflow; drainage; draining: (89)

écoulement d'électricité, flow of electricity. (1)

écoulement du courant d'air, flow of the air-current. (2)

écoulement de la glace d'un glacier, flowage of the ice of a glacier. (3)

écoulement naturel, natural drainage. (4)

les chaussées sont bombées pour faciliter l'écoulement des eaux (*f.pl.*), roads are cambered to facilitate the drainage of the water. (5)

écoulement (Hydraul.) (*n.m.*), flow : (6)

écoulement à gueule-bée *ou* écoulement à plein tuyau *ou* écoulement par orifice avec ajutage cylindrique, flow through orifice in thick wall ; flow through short pipe. (7)

écoulement à travers un orifice de — pouce(s) carré(s), sous une charge d'eau de — pouce(s) au-dessus du bord supérieur de l'orifice, flow through a hole — inch(es) square, under a head — inch(es) above the top of the hole. (8)

écoulement dans les canaux découverts, flow in open channels. (9)

écoulement de l'eau dans les tuyaux, flow of water in pipes. (10)

écoulement de l'eau par orifices, flow of water through orifices. (11)

écoulement en mince paroi *ou* écoulement par orifice percé en mince paroi, flow through sharp-edged orifice ; flow through orifice in thin plate. (12)

écoulement par barrages et déversoirs, flow over dams and weirs. (13)

écoulement (Hydrog.) (*n.m.*), run-off. (14)

écoulement de la matière (Méc.) (*m.*), flowage. (15)

écouler(s') (*v.r.*), to flow ; to flow out ; to flow away ; to run off ; to run out ; to drain away : (16)

l'eau s'écoule avec une vitesse proportionnelle à la pente, water flows with a velocity proportional to the slope. (17)

vapeur qui s'écoule du petit cylindre dans le grand cylindre, steam which flows from the high-pressure cylinder into the low-pressure cylinder. (18)

écoute-s'il-pleut [écoute-s'il-pleut *pl.*] (*n.m.*), water-mill which is more or less dependent on a fall of rain to run it. (19)

écouteur (d'un récepteur téléphonique) (*n.m.*), earpiece (of a telephone-receiver). (20)

écoutille (d'un bateau) (*n.f.*), hatchway, hatch, (of a boat). (21)

écouvillon (*n.m.*), tube-brush ; flue-brush. (22)

écran (*n.m.*), screen. (23)

écran (Élec.) (*n.m.*), screen. (24)

écran (*n.m.*) *ou* écran compensateur (Photogr.), screen ; filter. (25)

écran (d'une locomotive) (*n.m.*), cab-side (of a locomotive) (i.e., the side of the cab forming the screen). (26)

écran de ciel (Photogr.) (*m.*), sky-shade. (27)

écran ignifuge *ou simplement* écran (*n.m.*), fire-screen ; fire-proof screen ; screen. (28)

écran jaune (Photogr.) (*m.*), yellow screen. (29)

écran orthochromatique (Photogr.) (*m.*), ortho--chromatic screen ; colour-screen ; light-filter ; ray-filter. (30)

écran pour la projection (*m.*), lantern-screen. (31)

écran se montant à friction sur le barillet (Photogr.) (*m.*), push-on screen ; slip-on filter. (32)

écran se vissant à l'intérieur du barillet (Photogr.) (*m.*), screw-in screen. (33)

écrasable (*adj.*), crushable : (34)

des pierres écrasables (*f.pl.*), crushable stones. (35)

écrasage (*n.m.*), crushing. (36)

écrasant, -e (*adj.*), crushing : (37)

la puissance écrasante d'une machine, the crushing power of a machine. (38)

écrasé (-e) par une locomotive (être), to be run over by a locomotive. (39)

écrasée (*n.f.*), subsidence ; collapse ; fall ; falling in ; caving ; caving in ; crush : (40)

l'écrasée d'un toit de mine, the collapse of a mine-roof. (41)

écrasement (*n.m.*), crushing. (42)

écrasement (d'une chaudière) (*n.m.*), implosion (of a boiler). (43)

écrasement (Résistance des matériaux) (*n.m.*), crushing ; compression (Strength of mate--rials) : (44)

essai d'écrasement (*m.*), crushing-test ; compression-test. (45)

écraser (*v.t.*), to crush : (46)

écraser du minerai, du quartz, to crush ore, quartz. (47 *ou* 48)

pression du toit qui tend à écraser le charbon et à produire beaucoup de menu (*f.*), roof-pressure which tends to crush the coal and to produce a lot of slack. (49)

écrémage (*n.m.*), skimming ; skimming off ; drossing. (50)

écrémer (*v.t.*), to skim ; to skim off ; to dross : (51)

écrémer les scories dans une poche de fonderie, to skim the sullage in a foundry-ladle. (52)

écrémer les placers, to waste the placers ; to gopher ; to coyote. (53)

écrémer une mine, to rob, to pick the eyes out of, a mine. (54)

écrémoir (Fonderie) (*n.m.*), skimmer ; skimming-ladle. (55)

écrevisse (*n.f.*), self-adjusting stone dogs ; block-clip ; lever-grip tongs. (56)

écrin (*n.m.*), case : (57)

solide écrin en cuir à courroie, solid leather sling case. (58)

écriteau (*n.m.*), notice ; notice-board ; board ; warning : (59)

mettre un écriteau à une porte, to put a notice on a door. (60)

écriteau signalant un quartier grisouteux (Mines de houille) (*m.*), fire-board ; fireboard. (61)

écrou (*n.m.*), nut : (62)

écrous et boulons, bolts and nuts. (63)

écrou à chapeau (*m.*), box-nut. (64)

écrou à créneaux *ou* écrou crénelé (*m.*), castle nut ; castellated nut. (65)

écrou à entailles *ou* écrou à rainures *ou* écrou à dents (*m.*), slotted nut ; notched nut. (66)

écrou à huit pans *ou* écrou à 8 pans (*m.*), octagonal nut. (67)

écrou à molette (*m.*), milled nut ; knurled nut ; nurled nut. (68)

écrou à oreilles (*m.*), wing-nut ; thumb-nut ; finger-nut ; fly-nut ; butterfly-nut ; hand-nut. (69)

écrou à six pans *ou* écrou à 6 pans (*m.*), hex-agonal nut. (1)

écou à trous (*m.*), holed nut. (2)

écrou bas (*m.*), shallow nut. (3)

écrou carré (*m.*), square nut. (4)

écrou cylindrique (*m.*), cylindrical nut ; round nut. (5)

écrou d'essieu (*m.*), axle-nut. (6)

écrou de raccord (*m.*), coupling-nut. (7)

écrou de réglage (*m.*), regulating-nut. (8)

écrou de serrage (écrou ordinaire alors qu'il est employé conjointement avec un contre-écrou) (*m.*), maiden nut (ordinary nut when used in conjunction with a lock-nut). (9)

écrou de sûreté (*m.*), safety-nut (10)

écrou décolleté (*m.*), machine-made nut ; nut cut from bar. (11)

écrou desserré (*m.*), loose nut ; slack nut. (12)

écrou fendu (*m.*), elastic nut. (13)

écrou fileté (*m.*), threaded nut. (14)

écrou godronné *ou* écrou moleté (*m.*), milled nut ; knurled nut ; nurled nut. (15)

écrou haut (*m.*), deep nut. (16)

écrou indesserrable (*m.*), lock-nut. (17)

écrou non-taraudé (*m.*), untapped nut ; blank nut. (18)

écrou noyé (*m.*), flush nut. (19)

écrou ordinaire (*m.*), standard nut. (20)

écrou papillon (*m.*). Same as écrou à oreilles.

écrou taraudé (*m.*), tapped nut. (21)

écrou taraudeur (*m.*), die-nut. (22)

écrouir (battre un métal à froid) (*v.t.*), to hammer-harden ; to cold-hammer : (23)

écrouir un métal le rend cassant, to cold-hammer a metal makes it brittle. (24)

écrouir (passer par la filière) (*v.t.*), to cold-draw ; to draw cold : (25)

écrouir le fil de fer, to cold-draw iron wire. (26)

écrouir (passer par le laminoir) (*v.t.*), to cold-roll : (27)

écrouir de la tôle, to cold-roll sheet iron. (28)

écrouissage (*n.m.*) *ou* écrouissage à froid (*m.*) *ou* écrouissement (*n.m.*), hammer-hardening ; cold-hammering ; cold-drawing ; cold-rolling. (29)

écroulement (*n.m.*), falling in ; falling down ; tumbling down ; giving way ; collapse : (30)

l'écroulement d'une muraille, the collapse of a wall. (31)

écrouler (s') (*v.r.*), to fall in ; to fall down ; to give way ; to collapse : (32)

pont qui s'écroule (*m.*), bridge which gives way. (33)

Note.—*In certain cases the reflexive pronoun is suppressed; as*, laisser écrouler une maison, to allow a house to fall in.

écuelle (*n.f.*), dish ; bowl ; augette ; pan : (34)

écuelle de lavage, washing-dish. (35)

écuelle de bois, wooden bowl. (36)

écuelle (d'une vis) (opp. à plein) (*n.f.*), groove (of a screw). (37) (opp. to tooth).

écumage (*n.m.*), skimming ; scumming : (38)

écumage des métaux en fusion, skimming molten metals. (39)

écume (*n.f.*), froth ; scum ; dross ; skim ; skimmings. (40)

écume de manganèse (Minéral) (*f.*), bog-man-ganese ; bog-ore ; bog-mine ; bog-mine ore. (41)

écume de mer (Minéral) (*f.*), meerschaum ; sepiolite. (42)

écumer (*v.t.*), to skim ; to skim off ; to scum ; to dross : (43)

écumer l'étain fondu, to skim melted tin. (44)

écumer la scorie qui surnage, to skim off the slag which floats on the top. (45)

écumer (*v.i.*), to scum ; to froth. (46)

écumeux, -euse (*adj.*), frothy ; drossy. (47)

écumoire (*n.f.*), skimmer ; skimming-ladle. (48)

écumoire (crépine) (*n.f.*), strainer ; rose. (49)

écurer (*v.t.*), to clean ; to clean out ; to cleanse : (50)

écurer un puits, to clean out a well. (51)

écureuil (roue de volée) (*n.m.*), man-power ; manual gear. (52)

écurie (*n.f.*), stable. (53)

écusson (Serrur.) (*n.m.*), scutcheon ; escutcheon , keyhole-escutcheon ; key-plate. (54)

édénite (Minéral) (*n.f.*), edenite. (55)

édifier (*v.t.*), to build ; to erect. (56)

édingtonite (Minéral) (*n.f.*), edingtonite. (57)

effacé, -e (Signaux de ch. de f.) (*adj.*), blind ; open : (58)

le signal effacé, c'est-à-dire disposé parallèle-ment à la voie, indique que la voie est libre, the blind target or open switch, i.e., set parallelly to the road, indicates that the road is clear. (59)

effectif, -ive (*adj.*), effective. (60)

effectif, -ive (Méc.) (*adj.*), effective ; actual ; net ; brake (*used as adj.*) : (61)

puissance effective en chevaux (*f.*), effective horse-power ; actual horse-power ; net horse-power ; brake horse-power. (62)

effectif de la main-d'œuvre (*m.*), force of men employed. (63)

effectivement (*adv.*), effectively. (64)

effectuer (*v.t.*), to effect : (65)

effectuer des améliorations importantes, to effect considerable improvements. (66)

effectuer une expérience, to conduct an experi-ment. (67)

effectuer la tombée du front de taille (Mines), to bring down the face. (68)

effectuer la tombée du front de taille au moyen de coins, to wedge down the face. (69)

effectuer la tombée du front de taille au moyen de pinces, to bar down (*or* to gad down) the face. (70)

effervescence (*n.f.*), effervescence ; effervescency : (71)

effervescence d'un carbonate avec un acide, effervescence of a carbonate with an acid. (72)

faire effervescence, to effervesce (*v.i.*) : (73)

les alcalis font effervescence avec les acides (*m.pl.*), alkalis effervesce with acids. (74)

effervescent, -e (*adj.*), effervescent. (75)

effet (*n.m.*), effect ; action : (76)

effets mécaniques des explosions, mechanical effects of explosions. (77)

machine à simple effet (*f.*), single-acting engine ; single-action engine. (78)

pompe à double effet (*f.*), double-acting pump ; double-action pump. (79)

effet (puissance transmise par une force, par une machine) (Méc.) (*n.m.*), effect : (80)

l'effet d'une locomotive, the effect of a loco-motive. (81)

effet (Élec.) (*n.m.*), effect. (82)

effet brisant (d'un explosif) (*m.*), rending effect, disruptive effect, (of an explosive). (83)

effet Ferranti (Élec.) (*m.*), Ferranti effect. (84)

effet Joule (Élec.) (*m.*), joulean effect; Joule effect. (1)

effet Thomson (Élec.) (*m.*), Thomson effect. (2)

effet utile (Méc.) (*m.*), efficiency; coefficient of efficiency; coefficient of useful effect: (3) l'effet utile d'une locomotive, the efficiency of a locomotive. (4) l'effet utile d'une machine est le rapport du travail utile produit au travail moteur absorbé, the coefficient of efficiency of a machine is the ratio of the useful work produced to the mechanical energy expended. (5)

effet-volant [**effets-volants** *pl.*] (Méc.) (*n.m.*), fly-wheel effect. (6)

efficace (*adj.*), efficacious; effectual; efficient. (7)

efficacement (*adv.*), effectually; efficiently. (8)

efficacité (*n.f.*), efficiency: (9) efficacité de l'aérage, efficiency of the ventilation. (10)

effilé, -e (*adj.*), slender; thin; tapering; drawn out. (11)

effilement progressif (*m.*), tapering: (12) l'effilement progressif des masses continentales au sud de l'équateur, the tapering of the continental masses south of the equator. (13)

effiler (*v.t.*), to thin; to taper; to draw out: (14) tube en verre effilé à son extrémité (*m.*), glass tube drawn out at end. (15)

effiler (s') (*v.r.*), to become thin; to taper; to be drawn out. (16)

effleurir (*v.i.*), to effloresce. (17)

effleurir (s') (*v.r.*), to effloresce: (18) certains sels s'effleurissent à l'air (*m.pl.*), certain salts effloresce in the air. (19)

efflorescence (*n.f.*), efflorescence; efflorescency. (20)

efflorescent, -e (*adj.*), efflorescent. (21)

effluve (*n.m.*), effluvium: (22) effluves grisouteux, fire-damp effluvia. (23)

effondrement (*n.m.*), falling in; fall; caving in; collapse; subsidence: (24) effondrement du puits (Mines), falling in, fall, caving in, collapse, of the shaft. (25) effondrement du toit (Mines), falling in, fall, caving in, collapse, subsidence, of the roof. (26) effondrement provoqué par le travail des eaux souterraines, subsidence due to the action of underground streams. (27)

effondrer (*v.t.*), to weigh down; to cause to sink: (28) effondrer un plancher en le surchargeant, to weigh down a floor by overloading it. (29)

effondrer (s') (*v.r.*), to fall in; to cave in; to give way; to collapse; to subside: (30) toit qui s'effondre (*m.*), roof which gives way. (31) Note.—*The reflexive pronoun is suppressed in certain cases*; as, faire effondrer le sol, to cause the ground to subside.

effort (déploiement volontaire de force physique) (*n.m.*), effort: (32) faire effort pour soulever un fardeau, to make an effort to lift a burden. (33)

effort (force mécanique) (*n.m.*), force: (34) les efforts du vent, the force of the wind. (35) l'effort irrésistible, qui arrive, quoi qu'on puisse faire, à détruire, parfois même dans un temps très court, l'édifice souterrain, the irresistible force, which succeeds, in spite of all one can do, in destroying, sometimes even in a very short time, the underground fabric. (36)

effort (force qui tend à allonger, à raccourcir, à infléchir, à tordre, ou à couper en le cisaillant, un corps quelconque) (Résistance des matériaux) (*n.m.*), stress (Strength of materials). (37)

effort (force de résistance développée par la pièce) (travail) (Résistance des matériaux) (*n.m.*), stress. (38)

effort anormal (*m.*), abnormal stress. (39)

effort de cisaillement (*m.*), shearing-stress; shear-stress: (40) rivet qui est soumis à la fois à des efforts de traction et de cisaillement (*m.*), rivet which is at the same time subjected to tensile and shearing-stresses. (41)

effort de compression (*m.*), compressive stress; crushing-stress; positive stress. (42)

effort de dilatation (*m.*), expansion-stress. (43)

effort de flexion (*m.*), bending-stress; cross-breaking stress; deflective stress. (44)

effort de torsion (*m.*), torsional stress. (45)

effort de traction (Résistance des matériaux) (*m.*), tensile stress. (46)

effort de traction (traction par moteur) (*m.*), tractive effort; pull: (47) effort de traction d'une locomotive, tractive effort, pull, of a locomotive. (48)

effort sur le crochet de traction (locomotive) (*m.*), draw-bar pull. (49)

effort tangentiel (*m.*), tangential stress. (50)

effort tranchant (*m.*). Same as **effort de cisaillement**.

effort transversal (*m.*). Same as **effort de flexion**.

effritement (*n.m.*), crumbling; disintegration. (51)

effriter (*v.t.*), to crumble; to render friable; to disintegrate: (52) le gel effrite les roches les plus dures, frost disintegrates the hardest rocks. (53)

effriter (s') (*v.r.*), to crumble; to crumble to pieces; to crumble into dust; to disintegrate. (54)

égal, -e, -aux (le même en nature, en quantité, en qualité) (*adj.*), equal: (55) un diamètre est égal à deux rayons, a diameter is equal to two radii. (56)

égal, -e, -aux (qui ne varie pas) (*adj.*), even; equal; equable: (57) une température égale, an even (*or* an equal) (*or* an equable) temperature. (58)

égal, -e, -aux (uni; de niveau) (*adj.*), even; level: (59) chemin égal (*m.*), level path. (60) route bien égale (*f.*), very even road. (61)

également (*adv.*), equally. (62)

égaler (être égal à) (*v.t.*), to equal: (63) 5 multiplié par 4 égale 20, 5 multiplied by 4 equals 20. (64)

égalisation (*n.f.*), equalization. (65)

égaliser (rendre égal) (*v.t.*), to equalize: (66) égaliser les chances, to equalize the chances. (67)

égaliser (unir; aplanir) (*v.t.*), to level: (68) égaliser un terrain, to level a plot of ground. (69)

égaliser (s') (devenir uni) (*v.r.*), to find its own level; to become level; to get level: (70)

chemin qui commence à s'égaliser (m.), path which is beginning to get level. (1)

égaliseur de potentiel (Élec.) (m.), equalizer. (2)

égalité (n.f.), equality ; evenness : (3)
égalité de deux lignes, de deux nombres, equality of two lines, of two numbers. (4 ou 5)
égalité d'une surface, equality, evenness, of a surface. (6)
égalité d'un terrain, evenness of a piece of ground. (7)

égarer (s') (v.r.), to lose one's way ; to go astray ; to lose oneself : (8)
s'égarer dans un bois, to lose one's way in a wood. (9)

égohine ou **égoïne** (n.f.), small hand-saw. (10)

égout (conduit pour l'écoulement des eaux sales, des immondices) (n.m.), sewer ; drain. (11)

égout (rangées d'ardoises ou de tuiles faisant saillie hors d'un toit) (n.m.), eaves. (12)

égout (pente d'un toit) (n.m.), slope (of a roof). (13)

égout collecteur (m.), main sewer. (14)

égouttage ou **égouttement** (n.m.), draining ; dripping ; dropping. (15)

égoutter (v.t.), to drain : (16)
égoutter des terres, to drain land. (17)

égoutter (s') (v.r.), to drip ; to drop; to drain. (18)

égouttoir (Photogr.) (n.m.), draining-rack. (19)

égrisé (n.m.) ou **égrisée** (n.f.), diamond-dust ; diamond-powder ; bort. (20)

égrugeures de minerai (f.pl.), crushed ore. (21)

eisenglimmer (Minéral) (n.m.), iron-glance. (22)

eisengymnite (Minéral) (n.f.), iron-gymnite. (23)

eisenspath (Minéral) (n.m.), siderite. (24)

eisspath (Minéral) (n.m.), ice-spar. (25)

éjecteur (n.m.), ejector. (26)

éjecteur à vapeur (m.), steam-ejector. (27)

éjecto-condenseur [éjecto-condenseurs pl.] ou **éjecteur-condenseur** [éjecteurs-condenseurs pl.] (n.m.), ejector-condenser. (28)

ekaaluminium (Chim.) (n.m.), ekaluminum ; ekaluminium. (29)

ekabore (Chim.) (n.m.), ekaboron. (30)

ekasilicium (Chim.) (n.m.), ekasilicon. (31)

ekebergite (Minéral) (n.f.), ekebergite. (32)

élaboration (n.f.), elaboration : (33)
l'élaboration des matières brutes, the elaboration of raw materials. (34)

élaborer (v.t.), to elaborate. (35)

élæolite (Minéral) (n.f.), elæolite ; eleolite. (36)

élæolitique (adj.), elæolithic ; eleolitic. (37)

élæomètre ou **élaïomètre** (n.m.), oleometer ; oilometer ; oil-gauge. (38)

élagage (n.m.), lopping : (39)
élagage des arbres plantés en bordure des voies publiques, lopping trees planted alongside highways. (40)

élaguer (v.t.), to lop. (41)

élargir (v.t.), to enlarge ; to widen : (42)
élargir un puits (Mines), to enlarge, to cut down, a shaft. (43)
élargir une galerie d'aérage, to enlarge an air-way. (44)
elargir une route, to widen a road. (45)

élargir (Sondage) (v.t.), to underream : (46)
élargir un trou tubé au-dessous de la colonne de tubes, to underream a cased hole beneath the casing-column. (47)

élargir (s') (v.r.), to enlarge ; to widen ; to widen out. (48)

élargissement (n.m.), enlarging ; enlargement ; widening ; widening out : (49)
élargissement brusque de section (Hydraul.), sudden enlargement of cross-section. (50)
élargissement d'un puits (Mines), enlarging, cutting down, a shaft. (51)
élargissement des galeries (Mines), enlarging roads ; widening levels ; lameskirting. (52)

élargissement (Sondage) (n.m.), underreaming. (53)

élargisseur (appareil servant à élargir les trous tubés au-dessous de la colonne de tubes) (Sondage) (n.m.), underreamer. (54)

élargisseur (appareil servant à élargir le tubage) (Sondage) (n.m.), expander ; casing-expander ; drift. (55)

élargisseur à couronne de diamants (m.), under-reamer set with diamonds. (56)

élargisseur à couteaux (m.), underreamer with cutters. (57)

élasmose (nagyagite) (Minéral) (n.f.), nagyagite ; black tellurium ; foliated tellurium ; leaf tellurium. (58)

élasmose (altaïte) (Minéral) (n.f.), altaite. (59)

élasmosine (Minéral) (n.f.). Same as élasmose.

élasticité (n.f.), elasticity ; spring ; springiness : (60)
élasticité des solides, des liquides, des gaz, elasticity of solids, of liquids, of gases. (61 ou 62 ou 63)

élasticité de flexion (Phys.) (f.), elasticity of flexure. (64)

élasticité de torsion (f.), torsional elasticity ; rotational elasticity. (65)

élasticité de traction (f.), elasticity of elonga-tion. (66)

élastique (adj.), elastic ; springy ; spring (used as adj.) : (67)
limite élastique (Phys.) (f.), elastic limit. (68)
tension élastique pour lame de scie (f.), spring tension for saw-blade. (69)

élatérite (Minéral) (n.f.), elaterite ; mineral caoutchouc ; elastic bitumen ; elastic mineral pitch. (70)

élatéromètre (n.m.), elatrometer. (71)

Eldorado (n.m.), Eldorado ; El Dorado. (72)

électricien (pers.) (n.m.), electrician. (73)

électricité (n.f.), electricity. (74)

électricité atmosphérique (f.), atmospheric electricity. (75)

électricité dynamique (f.), dynamic electricity. (76)

électricité négative (f.), negative electricity. (77)

électricité positive (f.), positive electricity. (78)

électricité statique (f.), static electricity ; fric-tional electricity. (79)

électricité voltaïque (f.), voltaic electricity ; voltaelectricity ; galvanic electricity. (80)

électrification (n.f.), electrification : (81)
électrification d'un chemin de fer, electrifica-tion of a railway. (82)

électrifier (v.t.), to electrify. (83)

électrique (adj.), electric ; electrical : (84)
lumière électrique (f.), electric light. (85)
unité électrique (f.), electrical unit. (86)

électriquement (adv.), electrically : (87)
lignes de chemins de fer exploitées électrique-ment (f.pl.), railway-lines worked electri-cally. (88)

électrisable (adj.), electrifiable ; electrizable. (89)

électrisant, -e (adj.), electrifying. (90)

électrisation (n.f.), electrification ; electrization. (1)

électriser (v.t.), to electrify ; to electrize : (2) électriser un corps, to electrify a substance ; to electrize a body. (3)

électriser (s') (v.r.), to electrify : (4) l'ambre s'électrise très facilement par le frotte- -ment (m.), amber electrifies readily by friction. (5)

électro-aimant [électro-aimants pl.] (quelquefois abrégé électro) (n.m.), electromagnet ; magnet. (6)

électro-aimant à plongeur (m.), plunger-electro- -magnet ; plunger-magnet. (7)

électro-aimant boiteux (m.), club-footed magnet. (8)

électro-aimant cuirassé (m.), ironclad electro- -magnet. (9)

électro-aimant porteur ou électro-aimant de levage (m.), lifting-magnet. (10)

électrochimie (n.f.), electrochemistry. (11)

électrochimique (adj.), electrochemical. (12)

électrode (n.f.), electrode. (13)

électrode négative (f.), negative electrode ; cathode. (14)

électrode positive (f.), positive electrode ; anode ; anelectrode. (15)

électrodynamique (adj.), electrodynamic ; electrodynamical. (16)

électrodynamique (n.f.), electrodynamics. (17)

électrodynamisme (n.m.), electrodynamism. (18)

électrodynamomètre (n.m.), electrodynamometer. (19)

électrodynamométrique (adj.), electrodynamo- -metric ; electrodynamometrical. (20)

électrogène (adj.), generating ; electric gener- -ating : (21) groupe électrogène (m.), generating set ; electric generating set. (22)

électrogénérateur (n.m.), electric generator ; dynamo. (23)

électrolysable (adj.), electrolyzable. (24)

électrolysation (n.f.), electrolyzation. (25)

électrolyse (n.f.), electrolysis. (26)

électrolyser (v.t.), to electrolyze : (27) électrolyser un corps, to electrolyze a sub- -stance. (28)

électrolyte (n.m.), electrolyte. (29)

électrolytique (adj.), electrolytic ; electrolytical : (30) cuivre électrolytique (m.), electrolytic copper. (31) décomposition électrolytique (f.), electrolytical decomposition. (32)

électrolytiquement (adv.), electrolytically. (33)

électromagnétique (adj.), electromagnetic ; electromagnetical : (34) induction électromagnétique (f.), electro- -magnetic induction. (35)

électromagnétiquement (adv.), electromagneti- -cally. (36)

électromagnétisme (n.m.), electromagnetism. (37)

électrométallurgie (n.f.), electrometallurgy : (38) électrométallurgie du fer, electrometallurgy of iron. (39)

électrométallurgique (adj.), electrometallurgical. (40)

électrométallurgiste (pers.) (n.m.), electrometal- -lurgist. (41)

électromètre (n.m.), electrometer. (42)

électromètre à quadrants (m.), quadrant electrom- -eter. (43)

électromètre absolu (m.), absolute electrometer. (44)

électromètre capillaire (m.), capillary electrom- -eter. (45)

électrométrie (n.f.), electrometry. (46)

électrométrique (adj.), electrometric ; electro- -metrical. (47)

électromoteur, -trice (adj.), electromotive : (48) force électromotrice (f.), electromotive force. (49)

électromoteur (Phys.) (n.m.), electromotor. (50)

électromoteur (moteur électrique) (n.m.), electro- -motor ; electric motor ; electromagnetic machine. (51)

électromotographe (n.m.), electromotograph. (52)

électron (n.m.), electron. (53)

électronégatif, -ive (adj.), electronegative. (54)

électronite (Explosif) (n.f.), electronite. (55)

électropompe (n.f.), electropump. (56)

électropositif, -ive (adj.), electropositive. (57)

électroscope (n.m.), electroscope. (58)

électroscope à feuilles d'or ou électroscope à pailles d'or (m.), gold-leaf electroscope ; gold-leaf electrometer. (59)

électroscope condensateur (m.), condensing electroscope. (60)

électroscopique (adj.), electroscopic. (61)

électrosémaphore (n.m.), electrosemaphore. (62)

électrostatique (adj.), electrostatic ; electro- -statical : (63) induction électrostatique (f.), electrostatic induction. (64)

électrostatiquement (adv.), electrostatically. (65)

électrostriction (n.f.), electrostriction. (66)

électrotechnique (adj.), electrotechnic ; electro- -technical. (67)

électrotechnique (n.f.), electrotechnics ; electro- -technology. (68)

électrothermique (adj.), electrothermic ; electro- -thermal ; electrothermotic. (69)

électrotrieuse (pour minerais) (n.f.), magnetic separator ; electromagnetic separator. (70)

électrum (Minéral) (n.m.), electrum. (71)

élément (un des quatre éléments,—l'air, le feu, la terre, et l'eau) (n.m.), element : (72) l'élément liquide, the watery element. (73) le feu est un élément destructeur, fire is a destructive element. (74)

élément (principe constitutif d'un objet matériel quelconque) (n.m.), element : (75) substance contenant des éléments étrangers (f.), substance containing foreign elements. (76)

élément (principe actif, ou moyen d'action) (n.m.), element : (77) éléments de perturbation, disturbing elements. (78)

élément (corps simple) (Chim.) (n.m.), element : (79) l'hydrogène est un élément, hydrogen is an element. (80)

élément (chacun des morceaux de métal, ou autres substances, formant un couple voltaïque) (n.m.), element : (81) les deux éléments d'un couple voltaïque, the two elements of a voltaic couple. (82)

élément (n.m.) ou élément voltaïque (couple d'une pile voltaïque), cell ; element ; voltaic element ; galvanic element ; couple. (83)

élément (Géom.) (n.m.), element : (84) les éléments d'une ligne, d'une surface, d'un solide, the elements of a line, of a surface, of a solid. (85 ou 86 ou 87) élément (d'une chaudière à petits éléments) (n.m.), element, unit, section, (of a sectional boiler). (1)

élément chimique (m.), chemical element. (2)

élément d'accumulateur (Élec.) (m.), accumu-lator-cell ; storage-cell ; secondary cell. (3)

élément de pile (m.), battery-cell. (4)

élément Leclanché (m.), Lechanché cell. (5)

élément métallique (m.), metallic element. (6)

élément métalloïdique (m.), metalloidal element ; non-metallic element. (7)

élément surchauffeur (m.), superheater-unit ; superheater-element. (8)

élémentaire (adj.), elementary : (9)
analyse élémentaire (f.), elementary analysis. (10)
la composition élémentaire du pétrole brut, the elementary composition of crude petroleum. (11)

éléments (premiers principes) (n.m.pl.), elements : (12)
les éléments de géométrie, de physique, the elements of geometry, of physics. (13 ou 14)

éléolite (Minéral) (n.f.), eleolite ; elæolite. (15)

éléolitique (adj.), eleolitic ; elæolitic. (16)

éléomètre (n.m.), oleometer ; oilometer ; oil-gauge. (17)

élévateur (n.m.), elevator ; lift ; lifter. (18)

élévateur à air comprimé (m.), compressed-air elevator ; compressed-air lift. (19)

élévateur à chaîne (m.), chain-elevator. (20)

élévateur à godets (m.), bucket-elevator. (21)

élévateur à sec (pour matières sèches) (m.), dry elevator. (22)

élévateur à vis sans fin (m.), screw-elevator. (23)

élévateur d'eau (m.), water-elevator ; water-lifter. (24)

élévateur de marchandises (m.), goods-lift ; freight-elevator. (25)

élévateur de tailings (m.), tailings-elevator ; stacker. (26)

élévateur de tubage (m.), casing-elevator. (27)

élévateur électrique (m.), electric elevator ; electric lift. (28)

élévateur en queue (drague) (m.), tail-elevator. (29)

élévateur flottant (m.), floating elevator. (30)

élévateur hydraulique (m.), hydraulic elevator ; hydraulic lift. (31)

élévateur pneumatique (m.), pneumatic elevator ; pneumatic lift. (32)

élévation (exhaussement) (n.f.), raising : (33)
l'élévation d'un mur, raising a wall. (34)

élévation (hausse) (n.f.), rise : (35)
élévation de température (f.), rise of tempera-ture. (36)
élévation du prix des denrées, rise in the price of provisions. (37)
l'élévation incessante des salaires, the continual rise in wages. (38)

élévation (soulèvement d'un corps) (n.f.), eleva-tion ; elevating ; raising ; lifting : (39)
élévation du minerai, elevation of the ore ; raising the ore. (40)

élévation (éminence) (n.f.), eminence ; elevation ; height ; rise : (41)
gravir une petite élévation, to climb a small eminence. (42)

élévation (distance en hauteur) (n.f.), elevation ; height ; altitude : (43)
atteindre une grande élévation, to reach a great height. (44)

élévation (représentation d'une façade de bâtiment, d'une construction quelconque) (n.f.), elevation ; raised plan : (45)
l'élévation d'un bâtiment, the elevation of a building. (46)

élévatoire (adj.), elevating ; lifting : (47)
machines élévatoires (f.pl.), elevating machinery ; lifting machinery. (48)

élève (pers.) (n.m.), learner ; pupil. (49)

élèvement (n.m.), raising ; elevation ; elevating. (50)

élever (rendre plus haut) (v.t.), to raise : (51)
élever une maison d'un étage, to raise a house by a storey. (52)

élever (hausser) (v.t.), to raise : (53)
élever un prix, to raise a price. (54)

élever (remonter) (v.t.), to elevate ; to raise ; to lift : (55)
élever les tailings, to elevate, to raise, the tailings. (56)
élever des fardeaux à l'aide d'une grue, to lift weights with a crane. (57)

élever (ériger ; construire) (v.t.), to erect : (58)
élever un bâtiment, to erect a building. (59)

élever (opp. à abaisser) (Géom.) (v.t.), to erect (60) (opp. to to drop) :
élever une perpendiculaire à l'extrémité d'une droite donnée, to erect a perpendicular at the end of a given straight line. (61)

élever (s') (v.r.), to rise : (62)
pression qui s'élève au-dessus de celle de l'atmosphère (f.), pressure which rises above that of the atmosphere. (63)
le terrain s'élève au nord, au sud, à l'est, à l'ouest, the ground rises to the north, to the south, to the east, to the west. (64 ou 65 ou 66 ou 67)

éliminateur d'hyposulfite (Photogr.) (m.), hypo-eliminator ; hypo-killer. (68)

élimination (n.f.), elimination ; expulsion : (69)
élimination du soufre, du phosphore, contenu dans la fonte, elimination of the sulphur, of the phosphorus, contained in pig iron. (70 ou 71)
élimination du soufre des produits sidérurgiques, expulsion of sulphur from siderurgical products. (72)

éliminer (v.t.), to eliminate ; to expel : (73)
éliminer la cause d'une erreur, to eliminate the cause of an error. (74)
éliminer au cinglage la scorie renfermée dans la balle de fer sortant du four à puddler, to eliminate the slag inclusions in the ball of iron coming from the puddling-furnace during the process of shingling. (75)

élinde (n.f.) ou élinde de drague, ladder ; dredging-ladder ; digging-ladder ; bucket-ladder. (76)

élinde de l'élévateur (f.), stacker-ladder. (77)

élingue (n.f.), sling ; rope-sling. (78)

élingue à pattes (f.), dog-hook sling ; sling-dogs ; can-hook ; can-hooks ; sling. (79)

élingue de barrique (f.), barrel-sling. (80)

élingue sans fin (f.), endless sling. (81)

élinguer (v.t.), to sling : (82)
élinguer une barrique, to sling a barrel. (83)
barrique élinguée (f.), slung barrel. (84)

ellipse (n.f.), ellipse. (85)

ellipse du tiroir (f.), valve-ellipse. (86)

ellipsoïdal, -e, -aux ou ellipsoïde (adj.), ellipsoi-dal ; ellipsoid. (87)

ellipsoïde (n.m.), ellipsoid. (88)

ellipsoïde allongé (*m.*), prolate ellipsoid. (1)
ellipsoïde aplati (*m.*), oblate ellipsoid. (2)
ellipsoïde de révolution (*m.*), ellipsoid of revolu-
-tion. (3)
ellipsoïde des déformations (*m.*), strain-ellipsoid ;
ellipsoid of expansion. (4)
elliptique (*adj.*), elliptic ; elliptical. (5)
elliptiquement (*adv.*), elliptically. (6)
éloigné, -e (*adj.*), distant ; outlying ; far off ;
remote ; away : (7)
objet éloigné (*m.*), distant object. (8)
district éloigné (*m.*), outlying district. (9)
les chantiers sont éloignés du puits de—mètres,
the workings are — metres away from the
shaft. (10)
éloignement (*n.m.*), remoteness ; distance : (11)
l'éloignement de deux villes, the distance (i.e.,
the considerable distance) between two
towns. (12)
élongation (Astron.) (*n.f.*), elongation ; digression.
(13)
éluvial, -e, -aux (Géol.) (*adj.*), eluvial. (14)
éluvium (Géol.) (*n.m.*), eluvium. (15)
elvan (Pétrol.) (*n.m.*), elvan ; elvanite. (16)
elvanitique (*adj.*), elvanitic. (17)
émail (*n.m.*), enamel. (18)
émaillage (*n.m.*), enameling ; enamelling : (19)
émaillage de la tôle, enameling sheet iron. (20)
émailler (*v.t.*), to enamel : (21)
émailler de la tôle, to enamel iron. (22)
émanation (*n.f.*), emanation ; vapour ; gas :
(23)
émanations de radium, emanations of radium.
(24)
émanations thermales, thermal vapours ;
thermal gases. (25)
émanations volcaniques, volcanic vapours.
(26)
emballage (*n.m.*), packing : (27)
emballage en caisse pleine, en caisse à claire-
voie, en vrac, packing in close cases, in
crates, in bulk. (28 *ou* 29 *ou* 30)
emballage maritime *ou* emballage pour transport
outre-mer, packing for shipment. (31)
emballement (Mach.) (*n.m.*), racing. (32)
emballer (*v.t.*), to pack : (33)
emballer des marchandises, to pack goods. (34)
emballer (s') (Mach.) (*v.r.*), to race : (35)
accroissement brusque de vitesse produit
lorsque le moteur s'emballe (*m.*), sudden
increase of speed created when the engine
races. (36)
embarcadère (*n.m.*), wharf ; landing-place. (37)
embarquement (*n.m.*), shipping ; shipment ;
embarcation. (38)
embarquer (*v.t.*), to ship ; to put on board ;
to embark. (39)
embarrasser (obstruer) (*v.t.*), to obstruct ; to
stop up ; to block up : (40)
embarrasser une rue, to obstruct, to block up,
a road. (41)
embarrasser (gêner) (*v.t.*), to embarrass ; to
hinder. (42)
embase (d'un arbre de transmission) (*n.f.*), collar
(annular enlargement of a shaft). (43)
embase (d'un couteau, d'un ciseau) (*n.f.*),
shoulder, bolster, (of a knife, of a chisel).
(44 *ou* 45)
embase de la cheminée (locomotive) (*f.*), stack-
base. (46)
**embase en bois pour recevoir une rosace, un
interrupteur électrique** (*f.*), wood base,

wood block, to receive a rose, an electric
switch. (47 *ou* 48)
embatage (*n.m.*), **embatre** (*v.t.*). Same as
embattage, embattre.
embattage (*n.m.*), tiring ; tyring ; shoeing (a
wheel) : (49)
embattage des roues de locomotives, tiring
locomotive-wheels. (50)
embattre (*v.t.*), to tire ; to tyre ; to shoe : (51)
embattre une roue de voiture, to tire, to shoe,
a carriage-wheel. (52)
embauchage des ouvriers (*m.*), engaging workmen ;
taking on hands. (53)
embaucher des ouvriers, to engage workmen ;
to take on hands. (54)
emboîtement (*n.m.*), boxing ; housing ; socketing ;
jointing ; fitting in ; fitting into ; nesting :
(55)
emboîtement des matrices et des tenons d'une
charpente, boxing the housings and tenons
of a frame ; housing the mortices and tenons
of a frame. (56)
emboîtement d'une pièce dans une autre,
fitting of one piece into another. (57)
emboîter (*v.t.*), to box ; to house ; to socket ;
to joint ; to fit in ; to fit into ; to nest :
(58)
emboîter un tenon dans une mortaise, to box
(*or* to house) a tenon in a mortice. (59)
emboîter des tuyaux, to socket pipes ; to joint
tubes ; to fit pipes into each other. (60)
la lunette se compose de tubes emboîtés, etc.,
the telescope is composed of nested tubes,
etc. (61)
emboîter des caisses les unes dans les autres,
to nest boxes. (62)
emboîter (s') (*v.r.*), to fit into ; to fit in : (63)
pièce qui s'emboîte directement dans l'arbre
d'une machine (*f.*), part that fits directly
into the spindle of a machine. (64)
roue hydraulique qui s'emboîte entre deux
pans de mur verticaux (*f.*), water-wheel
which fits in between two vertical faces of
wall. (65)
embolite (Minéral) (*n.f.*), embolite. (66)
embouage (Mines) (*n.m.*), flushing ; slushing.
(Injection of a mixture of water and
clay, mud, silt, culm, or other refuse, into a
mine to quench a fire). (67)
embouchure (d'un fleuve) (*n.f.*), mouth (of a
river). (68)
embouchure (d'un concasseur) (*n.f.*), mouth
(of a rock-breaker). (69)
embouchure (d'un chalumeau) (*n.f.*), mouth-
-piece (of a blowpipe). (70)
embouchure (d'un transmetteur téléphonique)
(*n.f.*), mouthpiece (of a telephone-trans-
-mitter). (71)
embouchure convergente (Hydraul.) (*f.*), con-
-vergent mouthpiece ; convergent nozzle.
(72)
embouchure divergente (Hydraul.) (*f.*), divergent
mouthpiece ; divergent nozzle. (73)
embouer (Mines) (*v.t.*), to flush ; to slush. (74).
See embouage.
embout (d'un tube acoustique) (*n.m.*), mouth-
-piece (of a speaking-tube). (75)
emboutir (courber par choc ou par pression) (*v.t.*),
to shape ; to stamp ; to press ; to swage ;
to swedge ; to beat (into shape with a
hammer, or the like) : (76)
emboutir une chaudière, to shape a boiler. (77)

emboutir un godet, to stamp a bucket. (1)

emboutir (revêtir d'une garniture métallique) (*v.t.*), to cap (a rod, or the like, with a metal cap or ferrule) ; to ferrule ; to tip. (2)

emboutissage (*n.m.*), shaping ; stamping ; swaging ; swedging ; beating (into shape) : (3) emboutissage des objets en métal mince, stamping thin metal ware. (4)

emboutisseur (pers.) (*n.m.*), stamper ; shaper. (5)

emboutisseuse (*n.f.*) *ou* **emboutissoir** (*n.m.*) (machine pour emboutir les métaux par chocs ou pressions), shaping-machine ; shaper ; stamping-machine ; stamping-press ; stamper ; stamp (machine for swaging sheet metal between dies to the requisite form). (6)

emboutissoir (Sondage) (*n.m.*), casing-swedge ; drift ; swage. (7). See **redresseur** (synonymous word) for varieties.

embranchement (division en plusieurs branches) (*n.m.*), branching ; branching off : (8) l'embranchement d'un filon, the branching of a lode. (9)

embranchement (réunion de chemins qui se croisent) (*n.m.*), junction. (10)

embranchement (subdivision d'une voie ferrée en voies secondaires) (*n.m.*), junction : (11) embranchement en gare, junction within (or inside) station limits. (12) embranchement en pleine voie, junction out--side station limits. (13)

embranchement (ramification de tuyaux) (*n.m.*), branch ; branch-pipe. (14)

embranchement (chaîne secondaire de montagnes ou de collines) (*n.m.*), secondary chain (of mountains or hills). (15 *ou* 16)

embrancher (*v.t.*), to branch : (17) embrancher des tuyaux, to branch pipes. (18)

embrancher (s') (*v.r.*), to branch ; to branch off. (19)

embraser (*v.t.*), to set on fire ; to fire : (20) embraser une maison, to set a house on fire. (21)

embrasser (*v.t.*), to embrace : (22) courroie qui embrasse un certain arc (*f.*), belt which embraces a certain arc. (23) propriété qui embrasse plusieurs concessions (*f.*), property which embraces several claims. (24)

embrasure (*n.f.*), recess. (25)

embrayage (action) (*n.m.*), engaging ; connecting ; throwing into gear ; throwing into action ; throwing into feed ; throwing into operation ; putting into gear ; coming into gear : (26) embrayage de deux arbres, connecting two shafts (by means of a clutch, or the like). (27) embrayage simultané de deux mouvements, throwing two motions into action at the same time. (28) embrayage d'une courroie, throwing a belt on ; forking a belt on ; laying a belt on. (29)

embrayage (manchon d'embrayage) (*n.m.*), clutch. (30)

embrayage à coins (*m.*), V-groove clutch. (31)

embrayage à cône (*m.*), cone-clutch. (32)

embrayage à disque (*m.*). Same as **embrayage à plateau**.

embrayage à enroulement (*m.*). Same as **embrayage à spirale**.

embrayage à friction (*m.*), friction-clutch. (33)

embrayage à griffes (*m.*), claw-clutch. (34)

embrayage à mâchoires (*m.*), jaw-clutch. (35)

embrayage à plateau (*m.*), plate-clutch ; disc-clutch. (36)

embrayage à plateaux multiples (*m.*), multiple-plate clutch. (37)

embrayage à ruban (*m.*), band-clutch ; strap-clutch. (38)

embrayage à spirale (*m.*), scroll-clutch ; spiral clutch ; coil-clutch. (39)

embrayage électrique (*m.*), electric clutch. (40)

embrayage hydraulique (*m.*), hydraulic clutch. (41)

embrayage magnétique (*m.*), magnetic clutch. (42)

embrayage métallique (*m.*), metal-to-metal clutch. (43)

embrayer (*v.t.*), to engage ; to connect ; to throw into gear ; to throw into action ; to throw into operation ; to throw into feed ; to put into gear ; to trip in : (44) embrayer le mouvement de filetage, to throw the screw-cutting motion into action. (45) embrayer une courroie, to throw a belt on ; to fork a belt on ; to lay a belt on. (46)

embrayer (s') (*v.r.*), to engage ; to come into gear. (47)

embrayeur (pour courroies) (*n.m.*), belt-shifter ; belt-shipper ; belt-fork ; striker ; striking-gear ; strike-gear ; belt-striking gear ; belt-shifting apparatus. (48)

embrayeur (d'un manchon d'embrayage) (*n.m.*), fork (of a clutch). (49)

embrèvement (Charp.) (*n.m.*), joggle : (50) assemblage à embrèvement (*m.*), joggle-joint. (51)

embrever (*v.t.*), to joggle : (52) en général, pour obvier aux inconvénients de l'assemblage oblique à tenon et mortaise, on embrève tous les assemblages obliques, in general, to obviate the disadvantages of the oblique mortise-and-tenon joint, all oblique joints are joggled. (53)

embut (*n.m.*), limestone cave ; limestone cavern. (54)

émeraude (*n.f.*), emerald. (55)

émeraude brute *ou* **émeraude moraillon** (*f.*), rough emerald. (56)

émeraude du Brésil (*f.*), Brazilian emerald. (57)

émeraude orientale (*f.*), Oriental emerald. (58)

émergence (*n.f.*), emergence ; emersion. (59)

émergent, -e (*adj.*), emergent : (60) rayons émergents (Phys.), emergent rays. (61) rochers émergents (à marée basse) (*m.pl.*), emergent rocks. (62)

émerger (*v.i.*), to emerge. (63)

émeri (*n.m.*), emery. (64)

émerillon (*n.m.*), swivel. (65)

émersion (*n.f.*), emersion ; emergence : (66) émersion d'un rocher à la marée basse, emer--gence of a rock at low tide. (67) émersion des rayons lumineux qui traversent une lentille, emergence of the luminous rays which traverse a lens. (68)

émerylite (Minéral) (*n.f.*), emerylite. (69)

émetteur, -trice (*adj.*), transmitting : (70) station émettrice (*f.*), transmitting station. (71)

émettre (*v.t.*), to emit ; to give off : (72) émettre des rayons, to emit rays. (73) émettre des vapeurs, to give off fumes ; to emit vapours : (74)

les vapeurs émises par une substance au contact du fer rougi (*f.pl.*), the fumes given off by a substance on contact with red-hot iron. (1)

émeulage (*n.m.*), grinding : (2)
émeulage des surfaces planes, grinding plane surfaces. (3)

émeuler (*v.t.*), to grind. (4)

émiettement (*n.m.*), crumbling : (5)
émiettement d'une roche, crumbling of a rock. (6)

émietter (*v.t.*), to crumble. (7)

émietter (s') (*v.r.*), to crumble. (8)

éminemment (*adv.*), eminently ; in a high degree ; highly ; admirably : (9)
matière éminemment inflammable (*f.*), highly inflammable material. (10)
roche éminemment siliceuse (*f.*), highly siliceous rock. (11)
terrain éminemment propre au dragage (*m.*), ground admirably suited to dredging ; ground eminently suited for dredging. (12)

émissaire (Géogr. phys.) (*n.m.*), emissary : (13)
les émissaires d'un lac, the emissaries of a lake. (14)

émissif, -ive (*adj.*), emissive : (15)
pouvoir émissif (*m.*), emissive power. (16)

émission (*n.f.*), emission : (17)
émission de calorique, emission of heat. (18)

émission (de la vapeur du cylindre) (*n.f.*), release (of steam from the cylinder). (19)

emmagasinage *ou* **emmagasinement** (*n.m.*), stor- -ing ; storage ; warehousing : (20)
emmagasinage des explosifs, storing explosives ; storage of explosives. (21)
emmagasinage de marchandises, warehousing goods. (22)

emmagasiner (*v.t.*), to store ; to store up ; to put into store ; to warehouse : (23)
le magma fluide emmagasiné au-dessous de l'écorce terrestre, the fluid magma stored beneath the earth's crust. (24)
emmagasiner la quantité d'eau nécessaire aux besoins journaliers, to store up sufficient water for daily needs. (25)
emmagasiner du travail, to store up energy. (26)

emmanché, -e (muni d'un manche) (*adj.*), handled : (27)
pelles emmanchées (*f.pl.*), handled shovels. (28)

emmanchement (action d'adapter un manche) (*n.m.*), handling (providing with a handle) : (29)
emmanchement d'un outil, handling a tool. (30)

emmanchement à chaud (*m.*), shrinking ; shrinking on : (31)
l'emmanchement à chaud d'un piston sur la tige, shrinking a piston on the rod. (32)

emmanchement de tiges de sonde (*m.*), drill-rod coupling. (33)

emmancher (munir d'un manche) (*v.t.*), to handle : (34)
emmancher un balai, to handle a broom. (35)

emmancher (enchâsser) (*v.t.*), to fix ; to mount ; to couple ; to slip : (36)
le piston doit être solidement emmanché sur la tige, the piston should be firmly fixed on the rod. (37)
emmancher une poulie sur l'arbre, to mount a pulley on the shaft. (38)

emmancher des tiges de sonde, to couple drill-rods. (39)
les poteaux tubulaires se composent le plus souvent de tubes de fer de diamètres décroissants, emmanchés l'un dans l'autre (*m.pl.*), tubular poles are generally built up of wrought-iron tubes of decreasing diameter, slipped one into the other. (40)

emmancher à chaud (*v.t.*), to shrink on : (41)
emmancher à chaud un bandage, to shrink on a tire. (42)
emmancher un bandage à chaud sur une roue, to shrink a tire on a wheel. (43)

emmancher par effort (*v.t.*), to force on : (44)
emmancher une roue de locomotive sur la portée de calage de l'essieu par l'effort d'une presse hydraulique, to force a locomotive-wheel on the axle-seat with a hydraulic press. (45)

emmarchement (largeur de l'escalier) (*n.m.*), length of step. (46)

emmortaiser (*v.t.*), to mortise (to join by a tenon and mortise). (47)

émorfiler (*v.t.*), to take off the wire-edge : (48)
émorfiler un ciseau, to take the wire-edge off a chisel. (49)

émoudre (*v.t.*), to grind ; to whet ; to sharpen : (50)
émoudre un couteau, to grind a knife. (51)

émoulage (*n.m.*), grinding ; sharpening. (52)

emouleur (pers.) (*n.m.*), grinder. (53)

émoulu, -e (*adj.*), ground ; sharpened. (54)

émoussé, -e (*adj.*), blunt ; dull. (55)

émoussement (*n.m.*), blunting ; dulling. (56)

émousser (*v.t.*), to blunt ; to dull ; to take off the edge of : (57)
émousser un outil, to blunt, to dull, a tool. (58)

émousser (s') (*v.r.*), to blunt ; to dull ; to lose its edge : (59)
outil qui s'émousse vite (*m.*), tool which blunts quickly (*or* dulls rapidly). (60)

empannon *ou* **empanon** (Constr.) (*n.m.*), jack- rafter. (61)

empâté, -e (Géol.) (*pp.* & *adj.*), pasted ; pasty : (62)
morceaux qui sont empâtés dans la masse de certaines roches (*m.pl.*), fragments which are pasted in the mass of certain rocks. (63)
texture empâtée (*f.*), pasty texture. (64)

empattement (Constr.) (*n.m.*), footing. (65)

empattement (écartement des essieux) (*n.m.*), wheel-base. (66)

empattement des essieux couplés (locomotive) (*m.*), driving-wheel base. (67)

empattement total (locomotive) (*m.*), total wheel- base. (68)

empêchement (*n.m.*), hindrance ; obstacle. (69)

empêcher (*v.t.*), to prevent ; to avert ; to hinder ; to impede ; to stop : (70)
empêcher un bateau d'être entraîné par le courant, to prevent a boat being carried away by the current. (71)

empenoir (*n.m.*), lock-mortise chisel. (72)

empierrement (Ponts et Chaussées) (*n.m.*), metalling ; metaling. (73)

empierrer (*v.t.*), to metal : (74)
empierrer une route, to metal a road. (75)

empiétement (*n.m.*), encroachment : (76)
empiétement de la mer sur la terre ferme, encroachment of the sea on the land. (77)

empiétement involontaire, unintentional encroachment. (1)

empléter (*v.i.*), to encroach : (2)

empiéter sur une concession avoisinante, to encroach on a neighbouring concession. (3)

empilage *ou* empilement (*n.m.*), piling ; piling up ; stacking : (4)

empilement du bois, piling lumber ; stacking wood. (5)

empilement de blocs dénudés, de plis couchés (Géol.), piling up of denuded blocks, of recumbent folds. (6 *ou* 7)

empiler (*v.t.*), to pile ; to pile up ; to stack : (8)

la pile voltaïque est constituée par une série de disques empilés les uns sur les autres, the voltaic pile is formed by a series of discs piled the ones upon the others. (9)

empiler les tailings, to stack the tailings. (10)

empiler (s') (*v.r.*), to pile up. (11)

empirique (*adj.*), empiric ; empirical : (12)

formule empirique (*f.*), empirical formula. (13)

empiriquement (*adv.*), empirically : (14)

graduer une échelle empiriquement, to graduate a scale empirically. (15)

emplacement (*n.m.*), site ; location ; position ; place ; spot : (16)

emplacement d'un trou de sonde, d'un puits à pétrole, location, position, of a bore-hole, of an oil-well. (17 *ou* 18)

emplacement d'usine *ou* emplacement de moulin, site for works ; mill-site. (19)

emplacement de forage *ou* emplacement de sondage, boring-site ; drilling-site. (20)

emplacement des travaux, location of workings. (21)

emplacement isolé, out-of-the-way place ; isolated spot. (22)

emplectite (Minéral) (*n.f.*), emplectite. (23)

emplir (*v.t.*), to fill ; to fill up. (24)

emplir (s') (*v.r.*), to fill ; to fill up. (25)

emplissage (*n.m.*), filling ; filling up : (26)

emplissage d'un haut fourneau, filling a blast-furnace. (27)

emploi (action) (*n.m.*), use ; employment : (28)

emploi des explosifs dans les mines grisouteuses, use, employment, of explosives in gassy mines. (29)

emplois industriels du diamant noir, industrial uses of the black diamond. (30)

emploi (charge ; fonction ; occupation) (*n.m.*), employment ; employ ; work ; situation ; place ; post ; berth : (31)

chercher de l'emploi, to seek employment ; to look for work ; to look out for a situation (or a place) (or a berth). (32)

employé (pers.) (*n.m.*), employee ; employé. (33)

employé de bureau *ou simplement* employé (*n.m.*), clerk. (34)

employé de bureau indigène (*m.*), native clerk. (35)

employé du gaz (*m.*), gas-worker. (36)

employer (*v.t.*), to employ ; to use : (37)

les filles et les femmes, chez nous, ne peuvent pas être employées aux travaux souterrains. Les enfants ne peuvent être employés à un travail quelconque, soit au fond, soit au jour, qu'à partir de — ans révolus (Législation minière), with us, girls and women may not be employed in underground work. Children may only be employed in work of any kind,

either below or above ground, from above the age of — years. (Mining legislation). (38)

employer la force, to use force. (39)

employer l'énergie de la vapeur, to employ steam-power. (40)

employer la vapeur lorsque l'eau vient à manquer, to use steam when water fails. (41)

employer le pétrole comme combustible, to use oil as fuel. (42)

employer plusieurs jours à l'examen d'une propriété, to spend several days examining a property. (43)

empointage (*n.m.*), pointing. (44)

empointer (*v.t.*), to point : (45)

empointer une aiguille, une épingle, to point a needle, a pin. (46 *ou* 47)

empoise (de laminoir) (*n.f.*), chock (of rolling-mill). (48)

empoisonné, -e (*adj.*), poisoned. (49)

empoisonnement (*n.m.*), poisoning : (50)

empoisonnement par l'oxyde de carbone, carbon monoxide poisoning ; carbonic oxide poisoning ; C.O. poisoning. (51)

empoisonnement par le cyanure de potassium, poisoning by potassium cyanide. (52)

empoisonner (*v.t.*), to poison. (53)

emporte-pièce [emporte-pièce *pl.*] (Méc.) (*n.m.*), punch ; hollow punch ; socket-punch. (54)

emporte-pièce (de témoins) (Sondage) (*n.m.*), core-extractor ; core-lifter ; core-catcher ; core-breaker. (55)

emporter (enlever de vive force ; entraîner) (*v.t.*), to sweep away ; to carry away ; to wash away : (56)

inondations qui emportent un barrage (*f.pl.*), floods which sweep away a dam. (57)

emposieu (Géol.) (*n.m.*), sink ; sink-hole ; lime-stone sink ; swallow ; swallow-hole ; dolina. (58)

empoter (Boisage de mines) (*v.t.*), to hitch. (59)

empreinte (*n.f.*), impression ; print ; mark : (60)

empreinte d'un modèle dans le sable, d'une portée dans un moule (Fonderie), impression, print, of a pattern in the sand, of a core-print in a mould. (61 *ou* 62)

empreinte du fond du trou de sonde, impression of the bottom of the bore-hole. (63)

empreintes de pas (fossiles), footprints. (64)

empreintes de plantes et d'animaux (fossiles), impressions of plants and animals. (65)

empreintes de pluie (Géol.), rain-prints ; rain-pits. (66)

empreintes digitales *ou* empreintes des doigts, finger-marks. (67)

emprunter (s'aider d'un secours étranger) (*v.t.*), to make use of ; to use : (68)

canalisation aérienne qui emprunte un ouvrage d'art pour franchir une voie ferrée (*f.*), overhead transmission system which makes use of a permanent structure to cross a railway. (69)

voies publiques empruntées pour une distribu-tion d'énergie électrique (*f.pl.*), highways used for an electric-power distribution. (70)

émulseur (*n.m.*), emulsifier. (71)

émulsion (*n.f.*), emulsion. (72)

en amont, en aval, en hauteur, en profondeur, etc. See amont (en), aval (en), hauteur (en), profondeur (en), etc.

en attendant de plus amples recherches, pending further research (or investigation). (1)

en bon mineur, in a minerlike manner : (2)
travailler en bon mineur, to work in a miner-like manner. (3)

en bon ouvrier, in a workmanlike manner. (4)

en cas d'incendie ou en cas de sinistre, in case of fire. (5)

en chiffres ronds, in round figures. (6)

en construction, under construction. (7)

en deux pièces ou en deux parties (se dit des poulies, des manchons d'assemblage, etc.), split ; two-part : (8)
poulie en deux pièces (f.), split pulley. (9)
manchon cylindrique en deux pièces (m.), split muff-coupling. (10)
coussinet en deux pièces (m.), two-part die. (11)

en direction (Géol. & Mines), parallel in direction to the strike ; along the strike ; along the line of strike ; following the line of strike ; on the line of strike : (12)
mère-galerie en direction (f.), main road parallel in direction to the strike. (13)

en direction sur le filon, on the course of the lode ; on the run of the lode ; on the line of strike of the lode. (14)

en droite ligne, in a straight line. (15)

en équilibre, in equlibrio. (16)

en état de fonctionner (se dit des machines), in working order. (17)

en exploitation normale, in working order : (18)
temps nécessaire pour mettre la mine en exploitation normale (m.), time required to put the mine in working order. (19)

en feu, on fire. (20)

en forme d'entonnoir, funnel-shaped : (21)
cratère en forme d'entonnoir (m.), funnel-shaped crater. (22)

en forme de bassin, basin-shaped. (23)

en forme de cloche, bell-shaped. (24)

en forme de cœur, heart-shaped. (25)

en forme de coin, wedge-shaped. (26)

en gare (Ch. de f.), within station limits ; inside station limits : (27)
bifurcation en gare (f.), junction within (or inside) station limits. (28)

en grand, on a large scale : (29)
exploitation en grand (f.), working on a large scale. (30)

en inclinaison (Géol. & Mines), along the dip ; following the dip ; on the line of the dip. (31)

en ligne droite, in a straight line. (32)

en mauvais état, in bad condition ; in a bad state ; out of repair. (33)

en mauvais mineur, in an unminerlike manner : (34)
travailler en mauvais mineur, to work in an unminerlike manner. (35)

en mauvais ouvrier, in an unworkmanlike manner. (36)

en moyenne, on an average ; on the average ; upon an average : (37)
la mèche de sûreté brûle à raison d'un mètre par minute, en moyenne, safety-fuse burns at the rate of one metre a minute, on an average. (38)

en palier, on the level ; level : (39)
locomotives roulant en palier (f.pl.), engines running on the level. (40)

galerie de roulage en palier (f.), level haulage-way. (41)

en peu de jours, in a few days. (42)

en place (Géol.), in place ; in situ : (43)
gisement en place (m.), deposit in situ ; in-place deposit. (44)

en plein filon, right in the lode : (45)
travailler en plein filon, to work right in the lode. (46)

en pleine activité ou en pleine exploitation ou en pleine marche, in full working order ; in full swing. (47)

en pleine voie (Ch. de f.), outside station limits : (48)
bifurcation en pleine voie (f.), junction outside station limits. (49)

en pression (en parlant d'une machine à vapeur), under steam : (50)
la visite des soupapes, quand la machine est en pression, présente quelque danger, the inspection of valves, when the engine is under steam, is somewhat dangerous. (51)

en profondeur (Mines), in depth ; at depth : (52)
tracer de nouveaux niveaux en profondeur, to develop new levels in depth. (53)

en raison inverse de, in inverse ratio to ; inversely as. (54). See raison inverse for examples of use.

en règle générale, as a general rule. (55)

en réparation, under repair ; undergoing repairs. (56)

en tout temps, at any time. (57)

en une pièce ou en une seule pièce, whole ; solid ; one-part : (58)
poulie en une pièce (f.), whole pulley ; solid pulley. (59)
coussinet en une seule pièce (m.), one-part die. (60)

en vrac ou en vrague, in bulk : (61)
expédier en vrac, to ship in bulk. (62)

énargite (Minéral) (n.f.), enargite. (63)

encagement (Mines) (n.m.), caging ; loading the cage ; decking ; pushing (wagons) on the cage. (64)

encager les wagons, to cage the trucks ; to deck the wagons ; to push the cars on the cage. (65)

encageur (pers.) (n.m.), cager ; loader. (66)

encaissant, -e (Géol.) (adj.), enclosing ; country (used as adj.) ; wall (used as adj.) : (67)
la roche encaissante d'un minerai, the enclosing rock of an ore. (68)
roche encaissante, country-rock ; enclosing rock ; wall-rock. (69)

encastré, -e (adj.), fixed ; encastré : (70)
poutre encastrée (f.), fixed beam ; encastré beam. (71)

encastrement (Charp., etc.) (n.m.), housing : (72)
le champignon inférieur d'un rail de chemin de fer n'a plus d'autre raison d'être que d'assurer un bon encastrement dans le coussinet, the foot of a railway-rail has no other object than to ensure a good housing in the chair. (73)

encastrer (Charp., etc.) (v.t.), to house : (74)
encastrer une poutre dans un poteau, to house a beam in a post. (75)

encastrer (Arch.) (v.t.), to tail ; to tail in ; to tail into ; to tail on : (76)
encastrer une poutre dans un mur, to tail a beam into a wall. (77)
encastrer un bois, to tail in a timber. (78)

encastrer (s') (Arch.) (*v.r.*), to tail; to tail in; to tail on: (1)
solive qui s'encastre (*f.*), joist which tails in. (2)

enceindre (*v.t.*), to enclose; to surround; to fence: (3)
enceindre une terre d'un mur, to enclose a piece of ground with a wall. (4)

enceinte (*n.f.*), enclosure; wall; fencing: (5)
enceinte de palissades, enclosure, fencing, of palisades. (6)
enceinte de montagnes, wall of mountains. (7)

enchaîneur (Mines) (pers.) (*n.m.*), hanger-on; onsetter; hooker-on; hitcher; shackler. (8)

enchâsser (*v.t.*), to set; to mount: (9)
enchâsser un diamant dans une bague, to set, to mount, a diamond in a ring. (10)
enchâsser un diamant dans la couronne (Sondage), to set a diamond in the crown. (11)

enchère (*n.f.*) *ou* **enchères** (*n.f.pl.*), auction: (12)
vendre une maison à l'enchère (*ou* aux enchères), to sell a house by (*or* at) auction. (13)

enchérir (*v.t.*), to raise the price of. (14)
enchérir (*v.i.*), to rise in price. (15)
enchérissement des matières premières (*m.*), rise in the price of raw material. (16)

enchevêtrement (*n.m.*), entanglement; con--fusion. (17)

enchevêtrer (*v.t.*), to entangle; to confuse: (18)
enchevêtrer du fil, to entangle wire. (19)
masse vitreuse formée d'une multitude de cristaux enchevêtrés (*f.*), vitreous mass formed of a multitude of confused crystals. (20)

enchevêtrer des solives (Constr.), to trim joists. (21)

enclanchement (*n.m.*), interlocking: (22)
enclanchement des aiguilles (Ch. de f.), interlocking of the points. (23)
enclanchement des portes d'aérage (Mines), interlocking of the air-doors. (24)

enclancher (*v.t.*), to interlock. (25)
enclave (d'une écluse) (*n.f.*), gate-chamber, offset, (of a canal-lock). (26)
enclave de roche éruptive (Pétrol.) (*f.*), enclave; xenolith. (27)

enclaver deux poutres l'une dans l'autre, to fit two beams one into the other. (28)

enclenchement (*n.m.*), **enclencher** (*v.t.*). Same as **enclanchement, enclancher.**

encliquetage (*n.m.*), ratchet; ratchet-motion. (29)

encliquetage à arc-boutement *ou* **encliquetage à frottement** (*m.*), strut-action pawl-motion. (30)

encliquetage à levier (*m.*), lever ratchet-motion. (31)

encliquetage à rochet (*m.*), ratchet-and-pawl motion; pawl-and-ratchet motion. (32)

encliqueter (*v.t.*), to ratch. (33)
enclore (*v.t.*), to enclose; to fence in. (34)
enclos (*n.m.*), enclosure. (35)
enclos muré (*m.*), walled enclosure. (36)
enclume (*n.f.*), anvil. (37)
enclume à surface aciérée (*f.*), steel-faced anvil. (38)
enclume de maréchal *ou* **enclume pour maréchaux** (*f.*), farriers' anvil. (39)

enclumeau (*n.m.*) *ou* **enclumette** (*n.f.*) *ou* **enclumot** (*n.m.*), stake anvil; stake; square stake; baby anvil. (40)

encoche (*n.f.*), notch; nick; slot: (41)
les encoches d'un secteur denté, the notches of a quadrant. (42)
les encoches d'une douille à baïonnette, the slots of a bayonet lamp-holder. (43)

encoche (à) *ou* **encoché, -e** (*adj.*), notched; slotted. (44)

encochement (*n.m.*), notching; nicking; slotting. (45)

encocher (*v.t.*), to notch; to nick; to slot: (46)
encocher un morceau de bois, to notch a piece of wood. (47)

encoffrement en charpente (cadre de bois rondins, superposés horizontalement, avec remplissage de pierres, rocailles, terre, ou matières analogues, employé dans les constructions hydrauliques et autres travaux publics) (*m.*), crib; cribwork. (48)

encollage (Soudure) (*n.m.*), shutting up; shutting together (Welding): (49)
encollage des amorces d'une soudure, shutting up the scarfs of a weld. (50)

encoller (Soudure) (*v.t.*), to shut up; to shut together. (51)

encombrement (*n.m.*), obstruction; block; blocking up; blockage: (52)
encombrement des puits par la tuyauterie de vapeur, blocking up of the shafts by the steam-piping. (53)
encombrement du trafic des voies ferrées, congestion of railway traffic; congested state of the traffic on the railway; block on a railway. (54)

encombrement (*n.m.*) *ou* **encombrement sur le plancher,** space occupied; space taken up; floor-space; floor-space occupied: (55)
appareil photographique d'un encombrement minime (*m.*), camera which takes up (*or* which occupies) the smallest possible space (*or* the minimum amount of space). (56)
un des grands avantages des moteurs électriques réside dans leur faible encombrement, one of the great advantages of electric motors consists in the small floor-space occupied. (57)

encombrer (*v.t.*), to obstruct; to stop up; to block up; to block; to fill up: (58)
encombrer une rue, to obstruct, to block up, a street. (59)

encorbellement (*n.m.*), cantilever. (60)

encourageant, -e (*adj.*), encouraging; promis--ing: (61)
trouvaille encourageante (*f.*), promising find. (62)

encourir (*v.t.*), to incur; to run: (63)
encourir un risque, to incur, to run, a risk. (64)
encourir une dépense, une responsabilité, to incur an expense, a responsibility. (65 *ou* 66)

encrassement (*n.m.*), fouling; getting dirty; clogging; gumming; gumming up: (67)
encrassement de la toile métallique (lampes de mine), clogging of the gauze; gauze getting dirty. (68)

encrasser (*v.t.*), to foul; to dirty; to clog; to gum; to gum up: (69)
combustible qui encrasse les grilles (*m.*), fuel which clogs the grates. (70)

encrasser (s') (*v.r.*), to clog ; to get dirty ; to gum ; to gum up : (1)
meule d'émeri qui s'encrasse (*f.*), emery-wheel which gums up. (2)

encre de Chine (*f.*), Indian ink ; India ink ; Chinese ink ; China ink. (3)

endent (Charp.) (*n.m.*), indent. (4)

endentement (*n.m.*), toothing. (5)

endenter (Charp.) (*v.t.*), to indent. (6)

endenter (garnir de dents) (*v.t.*), to tooth ; to ratch ; to cog : (7)
endenter une roue d'engrenage, to tooth, to ratch, a gear-wheel. (8)

endigage *ou* **endiguement** (*n.m.*), dyking ; embanking ; banking ; damming. (9)

endiguer (*v.t.*), to dyke ; to embank ; to bank ; to dam. (10)

endogène (Géol.) (*adj.*), endogenous : (11)
roches endogènes (*f.pl.*), endogenous rocks. (12)

endommager (*v.t.*), to damage ; to injure. (13)

endomorphe (Minéralogie) (*adj.*), endomorphic ; endomorphous ; perimorphic ; perimorphous. (14)

endomorphe (*n.m.*), endomorph ; perimorph. (15)

endomorphisme (*n.m.*), endomorphism ; peri-morphism. (16)

endosmose (Phys.) (*n.f.*), endosmose ; endosmosis. (17)

endosmotique (*adj.*), endosmotic ; endosmic. (18)

endothermique (*adj.*), endothermic. (19)

endroit (*n.m.*), place. (20)

enduire (*v.t.*), to coat ; to cover ; to do over. (21)

enduit (*n.m.*), coat ; coating : (22)
enduit de goudron, coat of tar. (23)

endurance (*n.f.*), endurance : (24)
endurance des aciers aux efforts alternatifs répétés, endurance of steels against repeated alternating stresses. (25)

enduri, -e (*adj.*), hardened ; indurated : (26)
argile endurcie (*f.*), hardened clay ; indurated clay. (27)

endurcir (*v.t.*), to harden ; to indurate : (28)
tremper du fer pour l'endurcir, to temper iron to harden it. (29)
la chaleur endurcit l'argile, heat indurates clay. (30)

endurcir (s') (*v.r.*), to harden ; to indurate. (31)

endurcissement (*n.m.*), hardening ; induration. (32)

énergie (*n.f.*), energy ; power : (33)
l'énergie d'un gaz, d'un réactif, the energy of a gas, of a reagent. (34 *ou* 35)

énergie actuelle *ou* **énergie cinétique** (Phys.) (*f.*), actual energy ; active energy ; kinetic energy ; motive energy. (36)

énergie calorifique (*f.*), heat-energy ; thermal energy ; calorific energy. (37)

énergie chimique (*f.*), chemical energy. (38)

énergie de la vapeur (*f.*), steam-power. (39)

énergie de rayonnement (*f.*), radiant energy. (40)

énergie électrique (*f.*), electric energy ; electric power. (41)

énergie emmagasinée (*f.*), stored up energy. (42)

énergie hydro-électrique (*f.*), hydroelectric energy. (43)

énergie mécanique (*f.*), mechanical energy. (44)

énergie potentielle (*f.*), potential energy. (45)

énergie potentielle de self-induction (*f.*), potential energy of self-induction. (46)

énergique (*adj.*), energetic ; vigourous ; strong : (47)
serrage énergique (*m.*), strong grip. (48)

énergiquement (*adv.*), energetically ; vigourously ; strongly. (49)

enfance (Géogr. phys.) (*n.f.*), infancy. (50)

enfer (place de travail d'un foyer) (*n.m.*), stoke-hole. (51)

enfermer (*v.t.*), to shut up ; to enclose. (52)

enferrer (Carrières) (*v.t.*), to plug ; to plug off. (53)

enferrure (Carrières) (*n.f.*), plugging. (54)

enfilage (Boisage de mines) (*n.m.*), forepoling ; spilling ; spiling ; poling ; lagging. (55)

enfiler des palplanches entre les flandres (Boisage de mines), to spill, to insert spills, between the crowntrees. (56)

enflammer (*v.t.*), to set on fire ; to set fire to ; to fire ; to ignite ; to light : (57)
émission d'étincelles pouvant enflammer le grisou (*f.*), emission of sparks which may ignite the fire-damp. (58)

enflammer (s') (*v.r.*), to ignite ; to become ignited ; to be ignited ; to take fire ; to catch fire : (59)
'houille qui s'enflamme facilement (*f.*), coal which ignites easily. (60)

enflement (*n.m.*), swelling ; inflation. (61)

enfler (*v.t.*), to swell ; to inflate : (62)
la fonte des neiges enfle les fleuves, the melting of the snows swells the rivers. (63)
enfler un ballon avec du gaz, to inflate a balloon with gas. (64)

enfler (*v.i.*), to swell ; to swell out ; to inflate. (65)

enfler (s') (*v.r.*), to swell ; to swell out ; to inflate. (66)

enfoncement (*n.m.*), driving ; driving in : (67)
enfoncement de pilotis, driving piles. (68)
enfoncement par choc, par pression (tubage des trous de sonde), driving by percussion, by pressure (casing bore-holes). (69 *ou* 70)
enfoncement d'un clou, driving in a nail. (71)

enfoncement (défoncement) (*n.m.*), recess ; recessing. (72)

enfoncer (*v.t.*), to drive ; to drive in ; to drive into : (73)
enfoncer des pilotis, to drive piles. (74)
enfoncer un coin à coups de masse, to drive in a wedge with a sledge-hammer. (75)
enfoncer un clou dans un mur, un piquet en terre, to drive a nail into a wall, a peg into the ground. (76 *ou* 77)
enfoncer une porte, to break open a door. (78)

enfoncer (former un creux, un retrait) (*v.t.*), to recess. (79)

enfourchement (Menuis.) (*n.m.*), slot-mortise ; slit-mortise. (80)

enfournement (*n.m.*) *ou* **enfournage** (*n.m.*) *ou* **enfournée** (*n.f.*), charging : (81)
porte qui est utilisée pour l'enfournement du fer (*f.*), door which is used for charging the iron. (82)

enfourner (*v.t.*), to charge : (83)
enfourner du minerai, des briques, to charge ore, bricks. (84 *ou* 85)

enfourneur (pers.) (*n.m.*), charger ; filler. (86)

enfourneuse (*n.f.*), charger ; charging-machine. (87)

enfourneuse-défourneuse de lingots et blooms [enfourneuses-défourneuses *pl.*] (*f.*), ingot-and-bloom charger. (88)

enfreindre les règlements, to infringe the regu-lations. (89)

enfumé, -e (*adj.*), smoked ; smoky : (1)
verre enfumé (*m.*), smoked glass. (2)
quartz enfumé (*m.*), smoky quartz. (3)
engagement (*n.m.*), engagement. (4)
engager (*v.t.*), to engage : (5)
être engagé (-e) dans les affaires minières, to
be engaged in mining. (6)
engager un personnel d'employés, to engage a
staff. (7)
engendrement (*n.m.*), generation ; generating. (8)
engendrer (*v.t.*), to generate : (9)
une chaudière engendre de la vapeur, a boiler
generates steam. (10)
un demi-cercle tournant autour d'un diamètre
engendre une sphère, a semicircle revolving
round a diameter generates a sphere. (11)
engendrer (s') (*v.r.*), to be generated : (12)
chaudière dans laquelle s'engendre la vapeur
(*f.*), boiler in which steam is generated. (13)
engin (*n.m.*), appliance ; contrivance ; gear ;
tackle : (14)
fournir les aides et les engins nécessaires à la
manœuvre des pièces, to supply the necessary
assistance and appliances (*or* tackle) for
handling the machinery. (15)
engin de sauvetage (incendie) (*m.*), fire-escape.
(16)
engin extérieur (Mines) (*m.*), head-work. (17)
engins de levage (*m.pl.*), hoisting-appliances ;
hoisting-gear ; hoisting-tackle ; tackle ;
lifting-gear. (18)
engins de mine (*m.pl.*), mining-appliances. (19)
engins de sauvetage (*m.pl.*), rescue-appliances.
(20)
engins de sûreté (*m.pl.*), safety-appliances. (21)
engorgement (*n.m.*), choking ; obstruction ;
stoppage : (22)
engorgement d'un tuyau, choking of a pipe. (23)
engorgement de la cheminée (Mines), choking,
clogging, of chute. (24)
engorgement (d'un haut fourneau, d'un cubilot)
(*n.m.*), choking, bunging up, gobbing up,
(of a blast-furnace, of a cupola). (25 *ou* 26)
engorgement (d'un crible, d'un tamis) (*n.m.*),
blinding (of a screen, of a sieve). (27 *ou* 28)
engorger (*v.t.*), to choke ; to choke up ; to stop
up. (29)
engoujonnage (action) (*n.m.*), registering : (30)
engoujonnage qui est fait au moyen de broches,
registering which is done by means of pins.
(31)
engoujonnage (appareil) (*n.m.*), register. (32)
engoujonnage à billes (*m.*), ball register. (33)
engoujonner (*v.t.*), to register : (34)
engoujonner les deux parties d'un châssis de
fonderie, to register the two parts of a
moulding-box. (35)
engoujonner (s') (*v.r.*), to register ; to be in
register : (36)
partie de dessus du châssis qui s'engoujonne
avec le dessous, top part of the box which
registers with the bottom ; cope which is
in register with the drag. (37)
engrenage (action) (*n.m.*), gearing ; engaging ;
engagement ; throwing into gear ; putting
into gear ; coming into gear ; meshing ;
mesh ; pitching. (38)
engrenage (roues dentées) (*n.m.*), gear ; gearing ;
gear-wheels. (39)
engrenage(s) (à), geared : (40 *ou* 41)
poche à engrenage (Fonderie) (*f.*), geared
ladle. (42)

étau-limeur à engrenages (*m.*), geared shaping-
machine. (43)
engrenage à chaîne (*m.*), chain-gear. (44)
engrenage à chevrons (*m.*), herring-bone gear ;
double-helical gear ; V gear. (45)
engrenage à crémaillère (*m.*), rack-and-pinion
gear. (46)
engrenage à denture croisée (*m.*), stepped gear ;
step-tooth gear. (47)
engrenage à denture droite (*m.*), straight-tooth
gearing. (48)
engrenage à développante de cercle *ou simplement*
engrenage à développante (*m.*), involute
gear ; involute gearing. (49)
engrenage à flancs (*m.*), flank-gear. (50)
engrenage à flancs rectilignes (*m.*), straight-
flank gear. (51)
engrenage à joues (*m.*), shrouded gear ; shroud-
-ing gear ; flanged gear. (52)
engrenage à lanterne *ou* engrenage à fuseaux
(*m.*), lantern-gear ; lantern-gearing ; cog
and round. (53)
engrenage à 45 degrés *ou* engrenage à onglet
(*m.*), mitre-gear. (54)
engrenage à retour (*m.*), reversible gear ;
recriprocal gearing. (55)
engrenage à vis globique *ou* engrenage à vis
sans fin globique *ou* engrenage globoïde *ou*
engrenage à vis à filets convergents (*m.*),
hour-glass screw-gear ; Hindley's screw-
gear ; Hindley's worm-gear ; curved-worm
gear ; globoid gear. (56)
engrenage à vis sans fin (*m.*), worm-gear ;
worm-gearing. (57)
engrenage conique *ou* engrenage d'angle (*m.*),
bevel-gear ; bevel-gearing ; conical gear.
(58)
engrenage cycloïdal (*m.*), cycloidal gear. (59)
engrenage cylindrique (*m.*), circular gear ;
circular gearing. (60)
engrenage de commande *ou* engrenage d'attaque
(*m.*), driving-gear. (61)
engrenage de fatigue (*m.*), heavy-duty gear ;
gear subject to heavy duty. (62)
engrenage de rechange (*m.*), change-gear. (63)
engrenage de rechange pour le filetage des vis
(*m.*), change-gear for screw-cutting. (64)
engrenage de renvoi (*m.*), counter-gear. (65)
engrenage de transmission (*m.*), transmission-
gear ; intermediate gear. (66)
engrenage démultiplicateur (*m.*) *ou* engrenage
de réduction de vitesse (*m.*) *ou* engrenages
réducteurs (*m.pl.*), reducing-gear ; reduction-
gear ; reducing-wheels. (67)
engrenage différentiel (*m.*), differential gear ;
differential gearing ; differential ; equaliz-
-ing-gear ; compensating-gear ; balance-gear.
(68)
engrenage droit (*m.*), spur-gear ; spur-gearing.
(69)
engrenage droit (à), spur-geared : (70)
palan à engrenage droit (*m.*), spur-geared
pulley-block. (71)
engrenage échelonné (*m.*), stepped gear ; step-
tooth gear. (72)
engrenage elliptique (*m.*), elliptical gear. (73)
engrenage épaulé *ou* engrenage gardé (*m.*).
Same as engrenage à joues.
engrenage épicycloïdal (*m.*), epicycloidal gear.
(74)
engrenage extérieur (*m.*), external gear ; outside
gear. (75)

engrenage hélicoïdal (*m.*), spiral gear; spiral gearing; helical gear; helical gearing; screw-gear; screw-gearing. (1)

engrenage hélicoïdal double (*m.*), double-helical gear; herring-bone gear; V gear. (2)

engrenage hyperboloïde (*m.*), hyperbolical gear; skew bevel gear; skew-gear; skew-gearing. (3)

engrenage hypocycloïdal (*m.*), hypocycloidal gear. (4)

engrenage intérieur (*m.*), internal gear; inside gear; annular gear. (5)

engrenage menant (*m.*), driving-gear. (6)

engrenage mené (*m.*), driven gear; following-gear. (7)

engrenage multiplicateur (*m.*), multiplying-gear; multiplying-gearing; multiplying-wheels. (8)

engrenage partiellement denté (*m.*), mutilated gear. (9)

engrenage planétaire (*m.*), planet-gear; planet-gearing; sun-and-planet motion. (10)

engrenage réciproque (*m.*), reciprocal gear; reversible gearing. (11)

engrenages (*n.m.pl.*), gears; gearing; gear; gear-wheels. (12)

engrenages à denture divisée et taillée à la machine et recouverts de gaines protectrices contre les accidents (*m.pl.*), gear with machine-divided and machine-cut teeth, encased with guards to prevent accidents. (13)

engrenages baladeurs (*m.pl.*), sliding gears; balladeur gears. (14)

engrenages enfermés dans des carters *ou* **engrenages recouverts de gaines protectrices** (*m.pl.*), enclosed gear; guarded gears. (15)

engrenages taillés à la machine *ou simplement* **engrenages taillés** (*m.pl.*), machine-cut gearing; cut gears. (16)

engrenages taillés dans la masse (*m.pl.*), gears cut from the solid; gear-wheels cut from solid blanks. (17)

engrenant, -e (*adj.*), gearing; engaging; mesh--ing. (18)

les roues engrenantes (*f.pl.*), the gearing wheels. (19)

engrènement (*n.m.*), gearing; engaging; engagement; meshing; mesh; pitching; coming into gear: (20)

engrènement de deux roues, gearing, engaging, meshing, pitching, of two wheels. (21)

engrener (*v.t.*), to gear; to throw into gear; to put into gear; to engage; to mesh: (22)

engrener un pignon dans une roue, to gear a pinion in a wheel. (23)

roue qui engrène une vis sans fin (*f.*), wheel which engages a worm. (24)

engrener (*v.i.*) *ou* **engrener (s')** (*v.r.*), to gear; to come into gear; to engage; to mesh; to pitch: (25)

un pignon est une roue dentée d'un petit diamètre, s'engrenant sur une roue plus grande, a pinion is a toothed wheel of small diameter, gearing with a larger wheel. (26)

roues qui s'engrènent l'une dans l'autre (*f.pl.*), wheels which gear one in the other. (27)

pignon qui engrène sur une roue (*m.*), pinion which engages on a wheel. (28)

pignon qui engrène avec une crémaillère (*m.*), pinion which meshes (*or* pitches) with a rack. (29)

enhydre (*adj.*), enhydrous: (30)

agate enhydre (*f.*), enhydrous agate. (31)

enhydre *ou* **enhydros** (Minéral) (*n.m.*), enhydrite; enhydros. (32)

enlevage (opp. à la patte du câble) (Mines) (*n.m.*), lifting-point (of the rope on the pit-head pulleys.) (33) (in contradistinction to the **end of the rope** [attached to the cages]):

les coupages à la patte du câble ont l'avantage de changer l'enlevage (*m.pl.*), cutting off the end of the rope has the advantage of changing the lifting-point. (34)

enlèvement (*n.m.*), removing; removal; taking away; clearing away: (35)

enlèvement d'un gîte métallifère, removal of a mineral deposit. (36)

enlèvement des cosses (Ardoisières), untopping; unsoiling; quarry-stripping. (37)

enlèvement des débris (Mines), clearing away, taking away, the debris. (38)

enlèvement des échelles avant le tirage des coups de mines, removal of ladders before firing. (39)

enlèvement des piliers (Mines), removing the pillars; extracting the pillars; taking out the pillars; cutting away the pillars; drawing back the pillars; pulling pillars; pillar-drawing; pillar-working; second working; broken working; working in the broken; robbing. (40)

enlèvement des piliers à la fin de l'exploitation, removing the pillars in the final stage of the process of mining. (41)

enlèvement des poussières, sweeping up the dust; removing the dust. (42)

enlèvement des salbandes (Mines), removing the selvage; bulking. (43)

enlèvement des terrains de couverture (*ou* des terrains stériles superposés), removing the overburden; removal of the overburden; stripping; baring; unsoiling; untopping. (44)

enlèvement du boisage, drawing timber; removal of the timbering; removal of timber. (45)

enlèvement du massif par enlevures horizon--tales successives en ordre montant, working away the mass by horizontal slices in ascending order. (46)

enlèvement du minerai des chantiers, removal, withdrawal, of the ore from the stopes. (47)

enlever (lever en haut) (*v.t.*), to lift; to raise; to hoist; to wind: (48)

enlever des fardeaux à l'aide d'une grue, to raise weights with the help of a crane; to lift weights with a crane. (49)

enlever en vitesse (Mines), to wind, to hoist, to draw, at full speed. (50)

enlever (emporter; arracher) (*v.t.*), to carry away: (51)

les eaux enlèvent quelquefois les ponts (*f.pl.*), the waters sometimes carry away the bridges. (52)

enlever (faire disparaître) (*v.t.*), to remove; to take away; to clear away: (53)

enlever le minerai abattu des chantiers, to clear the broken ore from the stopes. (54)

enlever le " mou " avant le " dur " (Mines), to remove the selvage; to hulk. (55)

enlever les cosses (Ardoisières), to untop; **to** unsoil. (56)

enlever les étais, to withdraw the props ; to draw the props ; to unprop. (1)

enlever les piliers (Mines), to remove the pillars ; to extract the pillars ; to take out the pillars ; to cut away the pillars ; to draw the pillars ; to pull the pillars ; to work in the broken. (2)

enlever les poussières, to sweep up the dust. (3)

enlever les rails d'un chemin de fer, to take up the rails of a railway. (4)

enlever les salbandes (Mines), to remove the selvage ; to hulk. (5)

enlever les terrains de recouvrement (ou les terres de couverture), to remove the over--burden; to strip the overburden ; to strip; to unsoil ; to untop. (6)

enlever un couvercle, to take off a lid ; to remove a cover. (7)

enlevure (Exploitation des couches de houille) (n.f.), lift ; slice ; jud ; judd. (8)

enquête (n.f.), enquiry ; inquiry : (9)

enquête scientifique, scientific enquiry. (10)

enraiement ou **enrayage** ou **enrayement** (n.m.), dragging ; skidding ; locking ; spragging · braking ; applying the brakes. (11)

enrayer (entraver le mouvement des roues) (v.t.), to drag ; to skid ; to lock ; to sprag ; to brake ; to apply the brake. (12)

enrayer (garnir de ses rais) (v.t.), to spoke : (13)

enrayer une roue, to spoke a wheel. (14)

enregistrement (n.m.), registration ; registering ; recording. (15)

enregistrer (v.t.), to register ; to record : (16)

enregistrer le titre de propriété, to register the title to a property. (17)

enregistreur (adj. m.), registering ; recording ; self-registering ; self-recording : (18)

baromètre enregistreur (m.), recording ba--rometer ; self-registering barometer. (19)

enregistreur (n.m.), self-registering apparatus ; register ; recorder. (20)

enregistreur de pression de la vapeur (m.), registering steam-gauge. (21)

enregistreur de vitesse (m.), speed-recorder; tachograph. (22)

enregistreur des pressions atmosphériques (m.), barograph ; self-registering barometer. (23)

enregistreur des variations de température (m.), registering thermometer. (24)

enregistreur des variations du vent (m.), self-registering anemograph. (25)

enregistreur des variations électriques (m.), electrometer. (26)

enregistreur du niveau des liquides (m.), rain-gauge ; pluviometer ; pluviameter ; udom--eter ; ombrometer. (27)

enrichir (v.t.), to enrich : (28)

enrichir le minerai à la teneur de — par tonne, to enrich, to dress up, the ore to a content of — per ton. (29)

enrichir (s') (v.r.), to become richer. (30)

enrichir (s') (se dit d'un filon) (v.r.), to improve in grade or width. (31 ou 32)

enrichissement (n.m.), enrichment : (33)

enrichissement par flottage, par secousses (du minerai), enrichment by flotation, by joggling (of ore). (34 ou 35)

enrobant, -e (adj.), encasing ; enclosing : (36)

gangue enrobante (f.), enclosing matrix. (37)

enrober (v.t.), to encase ; to enclose : (38)

minéral enrobé dans une gangue dure (m.), mineral encased in a hard matrix. (39)

enrochement (n.m.), enrockment ; riprap : (40)

enrochement des piles d'un pont, enrockment of the piles of a bridge. (41)

enrocher (v.t.), to enrock ; to riprap : (42)

enrocher un barrage, to riprap a dam. (43)

enroulage ou **enroulement** (n.m.), winding ; coiling ; wrapping ; rolling up : (44)

Note.—**coiling** is rendered more specifically in French by the expression **enroulement en couronne**.

enroulement du fil, coiling wire. (45)

enroulement des barres autour d'un cylindre de laminoir, wrapping of bars round a rolling-mill roll. (46)

enroulement (Élec.) (n.m.), winding : (47)

enroulement à anneau, ring-winding. (48)

enroulement compound, compound winding. (49)

enroulement différentiel, differential winding. (50)

enroulement en dérivation, shunt-winding. (51)

enroulement en série, series-winding. (52)

enroulement en tambour, drum-winding. (53)

enroulement multiple, multiple winding. (54)

enrouler (v.t.), to wind ; to coil ; to wrap ; to roll up ; to roll : (55)

Note.—**to coil** is rendered more specifically in French by the expression **enrouler en couronne** or **rouler en couronne**, thus, commercial lead piping is coiled, les tuyaux en plomb du commerce sont roulés en conronne (m.pl.).

enrouler un câble sur un tambour, to wind a rope round a drum ; to coil a rope on a drum. (56)

enrouler des fils autour du noyau d'un induit, to wind wires round the core of an armature. (57)

enrouler du papier autour d'un cylindre, to roll paper round a drum. (58)

enrouler (s') (v.r.), to wind ; to coil ; to wrap ; to roll : (59)

Note.—**to coil** is rendered more specifically in French by the expression **s'enrouler en couronne**.

câble qui s'enroule autour du tambour (m.), cable which winds round the drum ; rope which coils round the barrel. (60)

enrouleur, -euse (adj.), winding ; coiling : (61)

cylindre enrouleur (m.), winding-drum. (62)

ensablement ou **ensablage** (action) (n.m.), sanding up ; choking with sand : (63)

ensablement qui a ruiné un port, sanding up which has ruined a harbour. (64)

ensablement du tubage, sanding up of the casing ; obstruction of the tubing by sand. (65)

ensablement (amas de sable) (n.m.), sand-bank. (66)

ensabler (v.t.), to sand up ; to sand ; to fill with sand ; to choke with sand ; to cover with sand : (67)

les barres ensablent l'embouchure des fleuves (f.pl.), bars sand up the mouths of rivers. (68)

les inondations ensablent souvent les terres (f.pl.), floods often cover the ground with drift-sand. (69)

ensabler (s') (v.r.), to sand up : (70)

puits qui s'ensable (m.), well which sands up. (71)

ensachement (n.m.), sacking ; putting into sacks ; bagging ; putting into bags. (72)

ensacher (*v.t.*), to bag ; to sack ; to put into bags : (1)

ensacher le minerai, to bag the ore ; to put the ore into bags. (2)

ensacheur (pers.) (*n.m.*), sacker ; bagger. (3)

ensemble (d'), general : (4)

disposition d'ensemble (*f.*), general arrange-ment. (5)

vue d'ensemble (*f.*), general view. (6)

ensemble d'outils (*m.*), set, outfit, of tools. (7)

ensemble de concessions de mines contiguës (*m.*), block of mining-claims. (8)

enseveli (-e) dans la mine (être), to be entombed in the mine. (9)

enseveli (-e) sous un éboulement de terrain (être), to be buried by a fall of ground. (10)

enstatite (Minéral) (*n.f.*), enstatite. (11)

entablement (d'un marteau-pilon, d'une presse à forger) (*n m.*), entablature (of a power-hammer, of a forging-press). (12 *ou* 13)

entaillage (*n.m.*), notching ; nicking ; grooving ; slotting ; cutting. (14)

entaillage à l'herminette *ou simplement* entaillage (*n.m.*), adzing : (15)

l'entaillage des traverses de chemins de fer, adzing railway-sleepers. (16)

entaille (*n.f.*), notch ; nick ; groove ; slot ; cut : (17)

pratiquer une entaille dans une pièce de bois, to make a notch in a piece of wood. (18)

entaille (d'une lime) (*n.f.*), cut (of a file). (19)

entaille (du banc de tour) (*n.f.*), gap (of lathe-bed). (20)

entaille à affûter les scies (*f.*), saw-clamp. (21)

entaille à mi-bois (*f.*), halving ; halved-joint ; half-lap ; lap-joint ; overlap-joint ; step-joint. (22)

entaillé, -e (*adj.*), notched ; nicked ; grooved ; slotted ; cut. (23)

entailler (*v.t.*), to notch ; to nick ; to groove ; to slot ; to cut : (24)

entailler une barre au burin, à la tranche, to nick a bar with the chisel, with the set. (25 *ou* 26)

entailler à l'herminette *ou simplement* entailler (*v.t.*), to adze. (27)

entailler les épontes (Mines), to brush the roof and the floor ; to broach. (28)

entamer (couper ; faire une incision) (*v.t.*), to cut ; to cut into ; to make an impression upon : (29)

l'acier au tungstène entame l'acier trempé (*m.*), tungsten steel cuts hardened steel. (30)

entartrage (*n.m.*), incrustation ; scaling. (31)

entartrer (*v.t.*), to incrust ; to scale ; to fur : (32)

l'eau de chaux entartre les chaudières (*f.*), lime water furs boilers. (33)

entartrer (s') (*v.r.*), to become incrusted ; to become scaled ; to scale ; to fur. (34)

entassement (*n.m.*), heaping up ; banking up ; piling up ; stacking. (35)

entasser (*v.t.*), to heap up ; to bank up ; to pile up ; to stack : (36)

entasser des pierres, to heap up stones. (37)

entasser les tailings, to stack the tailings. (38)

entasser (s') (*v.r.*), to heap up ; to pile up ; to bank up. (39)

enter (Charp.) (*v.t.*), to scarf : (40)

enter deux pièces de bois, to scarf two pieces of wood. (41)

enterrer (engloutir sous les décombres) (*v.t.*), to bury : (42)

éboulement qui enterre les mineurs, fall of earth which buries the miners. (43)

entibois (*n.m.*), vice-cap ; wood vice-clamp. (44)

entonnoir (ustensile ayant la forme d'un cône évasé) (*n.m.*), funnel. (45)

entonnoir (cône très évasé) (*n.m.*), hollow : (46)

village situé au fond d'un entonnoir (*m.*), village situated at the bottom of a hollow. (47)

entonnoir (cirque) (Géol.) (*n.m.*), corrie ; comb ; bowl ; cirque ; circ. (48)

entonnoir à filtre (*m.*), filter-funnel. (49)

entonnoir à robinet et bouché à l'émeri (*m.*), funnel with stopper and stopcock. (50)

entonnoir à séparation (*m.*), separating-funnel ; separation-funnel ; separatory funnel ; separator. (51)

entonnoir de dissolution (Géol.) (*m.*), dissolution-basin. (52)

entonnoir de retassement (vide central dans un lingot d'acier) (*m.*), pipe. (53)

entonnoir en fer-blanc, avec tamis métallique cuivre (*m.*), tin funnel, with brass gauze strainer. (54)

entonnoir en verre (*m.*), glass funnel. (55)

entonnoir pour filtrer à chaud *ou* entonnoir pour filtrer au bain-marie (*m.*), hot-filtration funnel. (56)

entourer (*v.t.*), to surround ; to enclose. (57)

entr'axe (*n.m.*), distance between centres ; from centre to centre : (58)

entr'axe des fusées, distance between centres of journals. (59)

entrailles de la terre (*f.pl.*), entrails of the earth. (60)

entraînement (*n.m.*), carrying along ; washing down ; entrainment. (61) See the verb entraîner for other renderings and applica-tions.

entraînement (amenage ; avancement) (Méc.) (*n.m.*), feeding ; feed : (62)

machine à raboter avec entraînement auto-matique par cylindres (*f.*), planing-machine with automatic roller-feed. (63)

entraînement positif (tour), positive feed (lathe). (64)

entraînement (d') (Méc.), feed (*used as adj.*) ; feeding ; driving : (65)

cylindre d'entraînement (*m.*), feed-roll. (66)

pignon d'entraînement (*m.*), driving-pinion. (67)

entraîner (traîner avec soi) (*v.t.*), to carry along with, up, down, into, out, *or* away (*according to sense, as illustrated in the following examples*) ; to draw along, in, etc. ; to wash away, *or* down ; to entrain ; to drift : (68)

matières terreuses entraînées par les eaux (*f.pl.*), earthy material (*or* matter) carried along (*or* washed down) (*or* drifted) by the water. (69)

les eaux de pluie entraînent avec elles les parties les plus meubles du terrain (*f.pl.*), rain-water carries along with it the looser particles of ground. (70)

du combustible entraîné dans les carneaux, fuel carried into (*or* drawn into) the flues. (71)

les sables marins cheminent le long des côtes sous l'action des courants de marées et sont entraînés au large, quelquefois à des distances assez considérables du rivage (*m.pl.*),

sea-sands creep along the coasts under the action of tidal currents and are carried out to sea, sometimes to fairly considerable distances from the shore. (1)

empêcher un bateau d'être entraîné par le courant, to prevent a boat being carried away by the current. (2)

le métal se trouve entraîné par des rouleaux, the metal is drawn in by rollers. (3)

l'atmosphère est entraîné avec la terre dans son mouvement de rotation diurne (*f.*), the atmosphere is drawn round with the earth in its diurnal rotatory motion. (4)

locomotive qui entraîne un lourd convoi (*f.*), locomotive which draws a heavy load. (5)

en réalité, la vapeur entraîne toujours avec elle une certaine quantité de liquide que les bulles gazeuses projettent en crevant à la surface. Cet entraînement d'eau s'appelle primage, in point of fact, steam always entrains with it a certain quantity of liquid which the gassy bubbles throw up when bursting on the surface. This entrain--ment of water is called priming. (6)

entraîner (avoir pour résultat) (*v.t.*), to involve ; to entail ; to give rise to ; to cause ; to bring about : (7)

installation qui entraîne de grands frais (*f.*), installation which involves a big expenditure. (8)

reculer devant la dépense que devait entraîner un changement de système, to shrink from incurring the expense which a change of system would entail. (9)

l'incrustation des chaudières peut entraîner des explosions (*f.*), the incrustation of boilers can give rise to explosions. (10)

entrait (d'une ferme de comble en bois) (*n.m.*), tie-beam, stringer, stretcher, binding-beam, (of a wooden roof-truss). (11)

entrait (d'une ferme de comble en fer) (*n.m.*), tie-rod, tie-bar, (of an iron roof-truss). (12)

entrait retroussé (*m.*), collar-beam ; collar-tie ; collar ; span-piece. (13)

entraver (*v.t.*), to fetter ; to impede ; to hinder ; to interfere with : (14)

entraver la circulation, to impede the traffic. (15)

malheureusement, le poids des accumulateurs au plomb a été jusqu'ici un inconvénient sérieux qui a entravé la généralisation de leur emploi, unfortunately, the weight of lead accumulators has been up to now a serious drawback which has hindered their coming into general use. (16)

circonstances qui entravent la bonne marche d'un service (*f.pl.*), circumstances which interfere with the smooth running of a service. (17)

entre-deux [entre-deux *pl.*] (*n.m.*), parting : (18)

galeries conjuguées séparées par des entre-deux et réunies par des recoupes d'aérage (*f.pl.*), parallel entries separated by partings and connected by cross-headings. (19)

entre-deux de gore (Mines), clay-parting. (20)

entre-deux (bande de la table des lumières du cylindre) (*n.m.*), bridge (of the port-face of the cylinder). (21)

entre-pointes (tour) (*n.m.*), distance between centres (lathe). (22)

entre-rail [entre-rails *pl.*] (Ch. de f.) (*n.m.*), space between rails. (23) In England the

entre-rail of the standard-gauge road is called the **four-foot** way or simply the **four-foot** (although the actual space between the rails is more than 4 ft). Cf. **entre-voie.**

entre-voie [entre-voies *pl.*] (Ch. de f.) (*n.f.*), space between running tracks. (24) Named in English according to the actual distance between the two pairs of rails, thus in England the **entre-voie** of the standard-gauge road is called the **six-foot way** or **6-ft way** or simply the **six-foot** : in America it is called the **7-ft way.** In France, the **entre-voie** of the standard-gauge track is 2 metres wide.

entrecouper (*v.t.*), to intersect ; to cut ; to cross : (25)

entrecouper une prairie par de nombreuses rigoles, to intersect a field by numerous ditches. (26)

entrecouper (s') (*v.r.*), to cut each other ; to intersect. (27)

entrecroiser (s') (*v.r.*), to cross each other. (28)

entredeux (*n.m.*). Same as **entre-deux.**

entrée (*n.f.*), entrance ; entry ; mouth ; inlet ; gate. (29)

entrée (d'une caverne) (*n.f.*), entrance, mouth, (of a cave). (30)

entrée (d'un tunnel) (*n.f.*), entrance (to or of a tunnel) ; portal (of a tunnel). (31)

entrée (d'un sac) (*n.f.*), mouth (of a sack). (32)

entrée (droits d'entrée) (Douane) (*n.f.*), import-duty (Customs). (33)

entrée d'air (orifice) (*f.*), air-inlet. (34)

entrée d'air (appel d'air) (*f.*), indraught of air ; indraught ; draught. (35)

entrée de serrure (ouverture) (*f.*), keyhole. (36)

entrée de serrure (écusson) (*f.*), key-plate ; escutcheon ; keyhole-escutcheon ; scutcheon. (37)

entrée des cannelures (l'excédent de largeur d'une cannelure sur celle qui la précède immédiatement) (Laminage des métaux) (*f.*), draught. (38)

entrée latérale (cage de mine) (*f.*), side-gate. (39)

entrée pour voitures (*f.*), carriage-entrance. (40)

entrefer (circuit magnétique ou électrique) (*n.m.*), air-gap ; interferric space. (41)

entremêler (*v.t.*), to intermix ; to mix up. (42)

entreposage (*n.m.*), warehousing. (43)

entreposer (*v.t.*), to warehouse. (44)

entreposeur (pers.) (*n.m.*), warehouse-keeper. (45)

entrepôt (*n.m.*), warehouse. (46)

entreprendre (prendre la résolution de faire une chose et la commencer) (*v.t.*), to undertake ; to take in hand : (47)

entreprendre un travail, un voyage, to undertake a task, a journey. (48 *ou* 49)

entreprendre *ou* **entreprendre à forfait** (s'engager à faire ou à fournir) (*v.t.*), to undertake ; to contract for : (50)

entreprendre des travaux à forfait, to undertake work on contract ; to contract for work. (51)

entreprendre une fourniture de vivres, to contract for a supply of provisions. (52)

entrepreneur (pers.) (*n.m.*), contractor ; under--taker ; entrepreneur. (53)

entrepreneur de bâtiments *ou* **entrepreneur de constructions** (*m.*), building-contractor ; builder. (54)

entrepreneur de roulage (*m.*), cartage-con--tractor ; haulage-contractor ; carrier. (55)

entrepreneur de sondages (*m.*), boring-con--tractor ; drilling-contractor. (56)

entrepreneur de travaux de mine (*m.*), mining-contractor. (1)

entreprise (*n.f.*), enterprise ; concern ; business ; undertaking : (2)

entreprise particulière, private concern (*or* enterprise) (*or* undertaking). (3)

entreprise rémunératrice, paying concern ; remunerative undertaking ; profitable business. (4)

entreprise (ce qu'on s'est chargé de faire à forfait) (*n.f.*), contract : (5)

l'entreprise d'un pont, the contract for a bridge. (6)

entreprise (à l') *ou* **entreprise** (**par**), on contract ; by contract : (7)

travailler à l'entreprise, to work on contract. (8)

faire exécuter des travaux par entreprise, to get work done on contract. (9)

entrer (*v.i.*), to enter ; to go in ; to go into ; to come in : (10)

entrer dans les détails, to enter, to go, into details. (11)

entrer en ébullition, to begin to boil. (12)

entrer en fusion, to begin to melt. (13)

entretenir (*v.t.*), to maintain ; to keep in repair ; to keep up ; to keep ; to supply : (14)

entretenir une voie ferrée, to maintain a railroad. (15)

entretenir un bâtiment, to keep a building in repair. (16)

entretenir le feu d'un fourneau, to keep up the fire of a furnace. (17)

entretenir une toiture à étanche d'eau, to keep a roof water-tight. (18)

entretenir en bois une mine, to supply a mine with timber. (19)

entretien (soin) (*n.m.*), maintenance ; upkeep ; keeping in repair : (20)

entretien d'un chemin, maintenance of a road. (21)

entretien de la voie (Ch. de f.), maintenance of permanent way ; track maintenance. (22)

entretien d'un bâtiment, upkeep of a building ; keeping a building in repair. (23)

entretien (ensemble des choses nécessaires) (*n.m.*), supply ; upkeep : (24)

entretien en bois d'une usine, supply of wood required by a works. (25)

entretien d'eau d'une mine, amount or volume of mine-water collected in 24 hours. (26)

entretoisage (*n.m.*), bracing ; strutting ; staying. (27)

entretoise (*n.f.*), brace ; strut ; spur ; cross-piece. (28)

entretoise (entre les solives de plancher) (*n.f.*), strut, strutting-piece, bridging-piece, (between floor-joists). (29) *If in the plural* —entretoises—struts ; strutting ; stiffening ; bridging. (30)

entretoise (tige d'écartement pour chaudières) (*n.f.*), stay-bolt ; stay-rod ; stay ; brace : (31)

entretoises de foyers de locomotives, stay-bolts of fire-boxes of locomotives. (32)

entretoise (cale pour maintenir l'écartement de l'ornière) (Ch. de f.) (*n.f.*), flangeway-filling ; filler ; filler-block ; distance-block ; block-ing-piece. (33)

entretoise (entre les barreaux d'une grille) (*n.f.*), distance-piece (between the bars of a grate). (34)

entretoise (d'une porte d'écluse) (*n.f.*), cross-piece (of a lock-gate). (35)

entretoise de derrick (*f.*), derrick-girt. (36)

entretoise de plaque de garde (Ch. de f.) (*f.*), horn-plate stay ; pedestal-binder ; pedestal-cap. (37)

entretoise des sabots (*f.*), brake-bar ; brake-beam. (38)

entretoisement (*n.m.*), bracing ; strutting ; staying. (39)

entretoiser (*v.t.*), to brace ; to strut : (40)

entretoiser une charpente, to brace a frame ; to strut a framing. (41)

entretoiser (*v.t.*) *ou* **entretoiser par des boulons** (écarter au moyen de tiges métalliques) (*v.t.*), to stay ; to stay-bolt : (42)

entretoiser la boîte à fumée d'une locomotive, to stay, to stay-bolt, the smoke-box of a locomotive. (43)

entretoiser par des tubes-tirants, to stay-tube. (44)

entrevoie [entrevoies *pl.*] (*n.f.*). Same as **entrevoie**.

entropie (Phys.) (*n.f.*), entropy. (45)

enture (Charp.) (*n.f.*), scarf ; scarf-joint ; scarf-jointing ; splice-joint. (46)

enture à goujon *ou* **enture à simple tenon** (*f.*), plug-tenon joint ; spur-tenon joint. (47)

enture à mi-bois avec abouts carrés *ou* **enture en paume** (*f.*), lapped scarf. (48)

enture à mi-bois avec tenons d'about et clef (*f.*), lipped table-scarf with key. (49)

enture à trait de Jupiter (*f.*), splayed indent scarf. (50)

enture en sifflet (*f.*), splayed scarf. (51)

envahi (-e) par l'eau (*ou* par les eaux), flooded ; water-logged : (52)

mine envahie par l'eau (*f.*), flooded mine. (53)

vieux chantiers envahis par les eaux (*m.pl.*), water-logged old workings. (54)

envahir (*v.t.*), to invade ; to break into ; to rush in ; to encroach upon : (55)

région qui a été envahie par une transgression marine (*f.*), region which has been invaded by an encroachment of the sea. (56)

eaux qui envahissent un puits à pétrole (*f.pl.*), water which breaks into an oil-well. (57)

envahissement (*n.m.*), invasion ; breaking in ; inrush ; encroachment : (58)

en Hollande, les digues protègent le pays contre l'envahissement de la mer, in Holland, the dikes protect the country from the invasion of the sea. (59)

envahissement de la mer sur les terres, encroach-ment of the sea on the land. (60)

envasé, -e (*adj.*), silty. (61)

envasement (envahissement par la vase) (*n.m.*), silting up ; silting ; filling up with mud ; choking with mud : (62)

envasement d'un canal, silting up of a canal. (63)

envasement (dépôt de vase) (*n.m.*), silt. (64)

envaser (*v.t.*), to silt up ; to choke with mud. (65)

envaser (s') (*v.r.*), to silt up ; to become choked up with mud. (66)

enveloppe (*n.f.*), covering ; cover ; case ; casing ; jacket ; jacketing ; lagging ; clead-ing ; clothing ; sheath ; sheathing ; dead-ing. (67)

enveloppe (d'un câble électrique) (*n.f.*), covering (of an electric cable). (68)

enveloppe (d'un bandage pneumatique) (*n.f.*), cover (of a pneumatic tyre). (69)

enveloppe (d'une turbine, d'une pompe centrifuge, d'un ventilateur) (*n.f.*), casing (of a turbine, of a centrifugal pump, of a fan). (1 *ou* 2 *ou* 3)

enveloppe calorifuge (*f.*), non-conducting lagging ; non-conducting jacketing ; insulating cover--ing. (4)

enveloppe d'eau (*f.*), water-jacket. (5)

enveloppe de chaudière (*f.*), boiler-jacket ; boiler-lagging ; boiler-deading ; boiler-clothing ; boiler-cleading. (6)

enveloppe de cylindre (*f.*), cylinder-jacket ; cylinder-lagging. (7)

enveloppe de tôle isolant la chaudière (*f.*), boiler-jacket ; jacket ; clothing-plate. (8)

enveloppe de vapeur (*f.*), steam-jacket. (9)

enveloppe du dôme (dôme de vapeur) (*f.*), dome-casing ; dome-cover. (10)

enveloppe du fond de cylindre (locomotive) (*f.*), cylinder-head casing ; cylinder-head cover. (11)

enveloppe du foyer (locomotive) (*f.*), fire-box casing. (12)

enveloppe en bois du corps du cylindre (*f.*), cylinder-lagging. (13)

enveloppe en spirale (d'une turbine, d'une pompe centrifuge, d'un ventilateur) (*f.*), spiral casing. (14 *ou* 15 *ou* 16)

enveloppe métallique (*f.*), metallic sheath. (17)

envelopper (*v.t.*), to cover ; to case ; to jacket ; to lag : (18)

tube de lunette enveloppé de maroquin (*m.*), telescope-tube covered with morocco. (19)

envelopper une chaudière, un cylindre, d'un calorifuge, to jacket, to lag, a boiler, a cylinder, with a non-conductor. (20 *ou* 21)

environnant, -e (*adj.*), surrounding : (22)

les lieux environnants, the surrounding places. (23)

environner (*v.t.*), to surround. (24)

environs (*n.m.pl.*), environs ; surrounding country ; neighbourhood. (25)

envoi (action) (*n.m.*), sending ; despatch : dispatch ; forwarding : (26)

envoi du minerai à la fonderie, sending ore to the smelting-works ; despatch of ore to the smeltery. (27)

envoi (chose envoyée) (*n.m.*), consignment ; shipment : (28)

envoi à titre d'essai, trial shipment. (29)

envoi aux molettes (Mines) (*m.*), overwinding ; overrunning ; pulleying ; drawing up against pulley. (30)

envoiler (s') (*v.r.*), to warp. (31)

envoyage (Mines) (*n.m.*), onsetting-station ; plat ; platt ; landing ; landing-station ; pit-landing : station ; shaft-station ; lodge. (32)

envoyage du fond (*m.*), bottom onsetting-station ; bottom landing ; bottom station ; bottom platt. (33)

envoyer (*v.t.*), to send ; to despatch ; to dispatch ; to forward : (34)

envoyer des marchandises, to send, to despatch, to forward, goods. (35)

envoyer le pétrole par canalisation, to send oil by pipe-line ; to pipe-line the oil ; to pipe oil. (36)

envoyer aux molettes (Mines) (*v.t.*), to overwind ; to overrun ; to pulley ; to draw up against the pulley. (37)

envoyeur (expéditeur) (pers.) (*n.m.*), sender ; consignor ; forwarder. (38)

envoyeur (accrocheur) (Mines) (pers.) (*n.m.*), hanger-on ; onsetter ; hooker ; hooker-on. (39)

éocène (Géol.) (*adj.*), Eocene ; Eocenic : (40)

la série éocène, the Eocene series. (41)

éocène (*n.m.*), Eocene : (42)

l'éocène, the Eocene. (43)

éogène (Géol.) (*n.m.*), Paleogene. (44)

éolien, -enne (Géol.) (*adj.*), eolian ; æolian; wind (*used as adj.*) : (45)

roches éoliennes (*f.pl.*), eolian rocks ; æolian rocks. (46)

érosion éolienne (*f.*), wind erosion. (47)

éolien, -enne (qui est mû par le vent) (*adj.*), wind (*used as adj.*) : (48)

moteur éolien (*m.*), wind-engine. (49)

éolienne (*n.f.*), windmill for pumping ; wind-engine. (50)

éolithique (Archéol. & Géol.) (*adj.*), eolithic. (51)

éozoïque (Géol.) (*adj.*), Eozoic. (52)

éozoïque (*n.m.*), Eozoic. (53)

épais, -aisse (*adj.*), thick ; dense : (54)

un mur épais, a thick wall. (55)

un brouillard épais, a thick fog ; a dense fog. (56)

épais, -aisse (évaluation) (*adj.*), thick : (57)

une planche épaisse de — centimètres, a board — centimetres thick. (58)

épais (*n.m.*), thickness : (59)

mur qui a — pied(s) d'épais (*m.*), wall which is — foot (feet) in thickness (*or* which is — foot (feet) thick). (60)

épaissement (*adv.*), thickly. (61)

épaisseur (*n.f.*), thickness : (62)

épaisseur de la croûte terrestre, thickness of the earth's crust. (63)

épaisseur d'un liquide, thickness of a liquid. (64)

épaisseur de la dent au primitif (Engrenages), thickness of tooth at pitch-circle (Gearing). (65)

épaisseur du stérile (Mines), thickness of the overburden. (66)

épaisseur (pli) (*n.f.*), ply : (67)

courroies en trois (*ou* en 3) épaisseurs (*f.pl.*), three-ply belting ; 3-ply belting. (68)

épaisseur (fausse pièce) (Fonderie) (*n.f.*), thick--ness ; thickness piece : (69)

une épaisseur de terre, a thickness of loam. (70)

épaissir (*v.t.*), to thicken. (71)

épaissir (*v.i.*), to thicken ; to get thick. (72)

épaissir (s') (*v.r.*), to thicken ; to get thick. (73)

épaississement (*n.m.*), thickening. (74)

épanchement (Géol.) (*n.m.*), extrusion ; effusion ; outpouring : (75)

épanchements volcaniques par lesquels la masse interne reparaît au jour, volcanic extrusions by which the internal mass re--appears on the surface. (76)

roches d'épanchement (*f.pl.*), extrusive rocks ; effusive rocks. (77)

épancher (*v.t.*), to pour out : (78)

épancher de l'eau, to pour out water. (79)

épancher (s') (Géol.) (*v.r.*), to extrude ; to effuse ; to be poured out : (80)

roches ignées qui se sont épanchées à la surface du sol (*f.pl.*), igneous rocks which have been poured out at the surface of the ground. (81)

épanouissement polaire (dynamo) (*m.*), pole-shoe. (82)

éparpiller (*v.t.*), to scatter ; to disperse ; to disseminate : (83)

des grains d'or éparpillés au travers du gravier, grains of gold disseminated through the gravel. (1)

épaulement (*n.m.*), shoulder : (2)
l'épaulement d'un tenon, d'une tige de sonde, the shoulder of a tenon, of a boring-rod. (3 *ou* 4)

épauler (*v.t.*), to shoulder : (5)
épauler une poutre, to shoulder a beam. (6)

épée de vanne (Hydraul.) (*f.*), gate-stem. (7)

épéirogénie (Géol.) (*n.f.*), epeirogeny ; epirogeny. (8)

épéirogénique *ou* **épéirogénétique** (*adj.*), epeiro-genic ; epeirogenetic ; epirogenic ; epiro-genetic. (9)

éperon (Arch.) (*n.m.*), buttress ; close buttress : (10)
l'éperon d'un pont, d'une muraille, the buttress of a bridge, of a wall. (11 *ou* 12)

éperon (Géogr. phys.) (*n.m.*), spur : (13)
un éperon des Alpes, des Pyrénées, a spur of the Alps, of the Pyrenees. (14 *ou* 15)

épicentral, -e, -aux (Sismologie) (*adj.*), epicentral. (16)

épicentre (Sismologie) (*n.m.*), epicentre ; epi-centrum ; epifocal point. (17)

épicycloïdal, -e, -aux (Géom.) (*adj.*), epicycloidal : (18)
engrenage épicycloïdal (*m.*), epicycloidal gear. (19)

épicycloïde (Géom.) (*n.f.*), epicycloid. (20)

épidote (Minéral) (*n.m.*), epidote. (21)

épigène (Cristall.) (*adj.*), epigene. (22)

épigénétique (Géol.) (*adj.*), epigenetic ; super-imposed : (23)
vallée épigénétique (*f.*), epigenetic valley ; superimposed valley. (24)

épigénie (*n.f.*), epigenesis. (25)

épinglage (Fonderie) (*n.m.*), piercing : (26)
épinglage des moules, des noyaux, piercing moulds, cores. (27 *ou* 28)

épingle (*n.f.*), pin. (29)

épingler (Fonderie) (*v.t.*), to pierce. (30)

épinglette (Travail aux explosifs) (*n.f.*), pricker ; picker ; needle ; shooting-needle ; blasting-needle ; priming-needle ; nail ; aiguille. (31)

épirogénie (Géol.) (*n.f.*), epirogeny ; epeirogeny. (32)

épirogénique *ou* **épirogénétique** (*adj.*), epirogenic ; epirogenetic ; epeirogenic ; epeirogenetic. (33)

épisser (*v.t.*), to splice : (34)
épisser un câble (*ou* un cordage), to splice a rope. (35)

épissure (*n.f.*), splice. (36)

épissure à œillet (*f.*), eye-splice ; ring-splice. (37)

épistilbite (Minéral) (*n.f.*), epistilbite. (38)

éponge (*n.f.*), sponge. (39)

éponge d'or (*f.*), gold-sponge ; sponge gold. (40)

éponge de platine (*f.*), platinum sponge. (41)

éponte (d'un filon, d'un dyke) (Géol. & Mines) (*n.f.*), wall (of a lode, of a dyke) : (42 *ou* 43)
dans les filons, le toit et le mur portent la qualification collective d'épontes, in lodes, the hanging wall and the foot-wall bear the collective name of walls. (44)

époque (*n.f.*), epoch. (45)

époque glaciaire (Géol.) (*f.*), Glacial epoch ; Glacial period ; ice period ; ice age ; boulder period ; Bowlder period. (46)

épouser la forme de *ou simplement* **épouser** (*v.t.*), to correspond in shape to ; to conform in shape to ; to affect the form of ; to assume the shape of : (47)
coussinet qui épouse la forme du rail (*m.*), chair which corresponds in shape to the rail. (48)
les fils électriques sont très souples et peuvent épouser sans la moindre difficulté et sans la moindre gêne pour la circulation, les parois des galeries (de mine) les plus tortueuses (*m.pl.*), electric wires are very flexible and can conform in shape to the walls of the most tortuous roadways without the least difficulty and without the least hindrance to the traffic. (49)
l'antimoine ainsi préparé est pur et épouse la forme cristalline (*m.*), antimony thus prepared is pure and affects the crystalline form. (50)

époussetage (*n.m.*), dusting. (51)

épousseter (*v.t.*), to dust : (52)
épousseter une plaque photographique, to dust a photographic plate. (53)

épreuve (*n.f.*), test ; testing ; proof ; trial. (54)
See also **essai**.

épreuve (Photogr.) (*n f.*). See **épreuve positive**, etc.

épreuve à l'eau (*f.*), water-test ; hydraulic test. (55)

épreuve à la vapeur (*f.*), steam-test ; steam-trial. (56)

épreuve de (à l') proof : (57)
à l'épreuve du feu, fire-proof ; fireproof. (58)
à l'épreuve des intempéries, weather-proof. (59)
à l'épreuve des maladresses, fool-proof. (60)

épreuve de chaudière (*f.*), boiler-test. (61)

épreuve de fonctionnement (*f.*), working-trial ; working-test. (62)

épreuve de réception (*f.*), acceptance-test. (63)

épreuve de soudabilité (*f.*), welding-test. (64)

épreuve de texture (métaux) (*f.*), homogeneity-test. (65)

épreuve de trempe (*f.*), temper-test. (66)

épreuve du câble (*f.*), testing the rope. (67)

épreuve en traits blancs (*f.*), white-line print. (68)

épreuve en traits blancs sur fond bleu (*f.*), blue-print. (69)

épreuve en traits noirs (*f.*), black-line print. (70)

épreuve hydraulique (*f.*), hydraulic test ; water-test : (71)
épreuve hydraulique des chaudières, hydraulic test of boilers. (72)

épreuve négative (Photogr.) (*f.*), negative. (73)

épreuve par agrandissement (opp. à **épreuve par contact**) (Photogr.) (*f.*), enlargement-print. (74)

épreuve par choc (*f.*), falling-weight test ; impact-test ; drop-test. (75)

épreuve par contact (Photogr.) (*f.*), contact-print. (76)

épreuve par rabattement (métaux) (*f.*), ram's-horn test ; plating-out test. (77)

épreuve positive *ou simplement* **épreuve** (*n.f.*) (Photogr.), positive ; print : (78)
tirage des épreuves, printing positives. (79)
virer une épreuve, to tone a print. (80)
épreuve trop virée, overtoned print. (81)

épreuve statique à la déformation (*f.*), static strain test. (82)

éprouver (essayer) (*v.t.*), to test ; to prove ; to try : (1)
éprouver le câble, to test the rope. (2)
éprouver un canon de fusil, to prove a gun-barrel. (3)
éprouver (être exposé à ; ressentir ; subir) (*v.t.*), to experience ; to meet with ; to sustain ; to entertain : (4)
éprouver un désastre sérieux, to meet with, to sustain, a serious disaster. (5)
éprouver des difficultés à se procurer une main-d'œuvre suffisante, to experience difficulty in obtaining sufficient labour. (6)
éprouver des craintes au sujet de la sécurité des ouvriers, to entertain fears as to the safety of the workmen. (7)
éprouvette (vase) (Verrerie de laboratoire) (*n.f.*), test-glass. (8)
éprouvette (tube) (Verrerie de laboratoire) (*n.f.*), test-tube. (9)
éprouvette (cuiller) (Métall.) (*n.f.*), éprouvette ; provett ; assay-spoon. (10)
éprouvette (barreau, morceau d'essai) (Méc.) (*n.f.*), test-bar ; test-piece ; coupon. (11)
éprouvette à pied, et à bec (*f.*), test-glass on foot, and with spout ; test-tube with foot and lip. (12)
éprouvette de pliage (Méc.) (*f.*), bend-test piece. (13)
éprouvette de traction (Méc.) (*f.*), tensile-test piece. (14)
éprouvette graduée (*f.*), graduated test-tube. (15)
éprouvette-type [**éprouvettes-types** *pl.*] (Méc.) (*n.f.*), standard test-piece. (16)
epsomite (Minéral.) (*n.f.*), epsomite. (17)
épuisant, -e (*adj.*), exhausting : (18)
travail épuisant (*m.*), exhausting work. (19)
climat épuisant (*m.*), exhausting climate. (20)
épuise (*n.f.*), water-elevator. (21)
épuise-volante [**épuises-volantes** *pl.*] (*n.f.*), wind-mill for pumping ; wind-engine. (22)
épuisé, -e (Chim.) (*adj.*), spent : (23)
acide épuisé (*m.*), spent acid. (24)
épuisement (affaiblissement considérable ou suppression complète de rendement) (*n.m.*), exhaustion : (25)
l'épuisement d'une mine, d'une carrière, the exhaustion of a mine, of a quarry. (26 *ou* 27)
épuisement (exhaure) (*n.m.*), draining ; drainage ; pumping ; pumping out ; unwatering ; dewatering ; bailing ; bailing out : (28)
épuisement d'une galerie de mine, draining, unwatering, dewatering, pumping out, a mine-level. (29)
épuisement des sables coulants, bailing quick-sands. (30)
épuisement en répétitions (Mines), stage pumping ; multistage pumping ; pumping in successive lifts. (31)
épuisement en un seul jet (Mines), single-stage pumping ; pumping in one lift. (32)
épuisement par cuffats, unwatering with water-buckets ; bailing with kibbles. (33)
épuiser (*v.t.*). See examples : (34)
épuiser les chantiers d'abatage (Mines), to exhaust, to work out, the stopes ; to spend ground. (35)
épuiser les réserves de minerai, to exhaust the reserves of ore. (36)
épuiser une citerne, un tonneau, to empty a cistern, a cask. (37 *ou* 38)

épuiser une mine (la mettre à sec), to drain, to unwater, to dewater, to pump out, to fork, a mine. (39)
épuiser (s') (*v.r.*), to become exhausted ; to come to an end. (40)
épurateur (Mach.) (*n.m.*), purifier. (41)
épurateur de gaz (*m.*), gas-purifier. (42)
épuration (*n.f.*), purification ; purifying ; refining : (43)
épuration des eaux, du gaz d'éclairage, purifi-cation of water, of lighting-gas. (44 *ou* 45)
épuration des pétroles, refining petroleum ; oil-refining. (46)
épure (*n.f.*), working plan ; working drawing ; diagram : (47)
tracer une épure, to lay out a working plan ; to plot a diagram. (48)
épure de distribution *ou* **épure de régulation** (du tiroir) (*f.*), valve-diagram, distribution-diagram, (slide-valve). (49)
épure de la distribution Corliss (*f.*), diagram of Corliss valve-gear. (50)
épure Zeuner du tiroir Meyer (*f.*), Zeuner diagram of Meyer valve. (51)
épurer (*v.t.*), to purify ; to refine : (52)
épurer des métaux, des huiles, des gaz, to purify metals, oils, gases. (53 *ou* 54 *ou* 55)
épurer (s') (*v.r.*), to purify ; to refine. (56)
équarrir (rendre carré) (*v.t.*), to square : (57)
équarrir un bloc de marbre, des grumes, to square a block of marble, logs. (58 *ou* 59)
équarrir (tailler à angle droit) (*v.t.*), to square : (60)
équarrir une poutre, to square a beam. (61)
équarrir (agrandir un trou en faisant usage de l'équarrissoir) (*v.t.*), to broach ; to ream ; to ream out : (62)
équarrir un trou, to broach a hole. (63)
équarrissage (action de rendre carré ou de tailler à angle droit) (*n.m.*), squaring : (64)
équarrissage du bois, squaring timber. (65)
équarrissage (état de ce qui est équarri) (*n.m.*), squareness. (66)
équarrissage (dimensions du bois équarri) (*n.m.*), scantling : (67)
montant qui a un équarrissage de — sur — (*m.*), upright which has a scantling of — by —. (68)
équarrissage (agrandissement d'un trou au moyen de l'équarrissoir) (*n.m.*), broaching ; ream-ing ; reaming out. (69)
équarrissement (*n.m.*). Same as **équarrissage.**
équarrissoir (Méc.) (*n.m.*), broach ; reamer ; rimer ; opening-bit. (70) See also **alésoir.**
équarrissoir (Sondage) (*n.m.*), reamer. (71)
équarrissoir à — pans (*m.*), — -sided broach ; broach with — sides. (72)
équarrissoir à 8 pans (*m.*), octagonal broach ; octagonal reamer. (73)
équarrissoir conique (*m.*), taper broach. (74)
équarrissoir cylindrique (*m.*), parallel broach. (75)
équateur (*n.m.*), equator. (76)
équateur magnétique (*m.*), magnetic equator. (77)
équatorial, -e, -aux (*adj.*), equatorial. (78)
équatorialement (*adv.*), equatorially. (79)
équerre (instrument) (*n.f.*), square. (80)
équerre (*n.f.*) *ou* **équerres** (*n.f.pl.*) (fer en T), T iron ; tee ; tees. (81)
équerre (*n.f.*) *ou* **équerres** (*n.f.pl.*) (fer en L), angle-iron ; L iron ; angles. (82)

équerre à coulisse (*f.*), caliper-square ; sliding caliper-gauge ; sliding-calipers ; slide-calipers ; caliper-gauge ; beam-caliper ; beam caliper-gauge. (1)

équerre à 8 pans (*f.*), octagonal-nut angle-gauge. (2)

équerre à lame d'acier (*f.*), try-square ; trying-square. (3)

équerre à onglet (*f.*), mitre-square. (4)

équerre à 45°, en poirier (*f.*), set-square 45°, pear wood. (5)

équerre à réflexion (*f.*), optical square. (6)

équerre à 6 pans (*f.*), hexagonal-nut angle-gauge. (7)

équerre à 60°, en ébène et à filets cuivre (*f.*), set-square 60°, ebony, brass edges. (8)

équerre-applique [**équerres-appliques** *pl.*] (pour transmissions) (*n.f.*), end wall-bracket ; end bracket ; angle-bracket (for shafting). (9)

équerre assemblée à jour (*f.*), set-square, open centre ; framed set-square. (10)

équerre d'angle (fer) (*f.*), angle-plate ; corner-band. (11)

équerre d'arpenteur *ou simplement* **équerre** (*n.f.*), cross-staff ; cross-staff head ; cross ; surveyors' cross : (12)

équerre octogone, octagonal cross-staff. (13)

équerre cylindrique, cylindrical cross-staff head ; circular cross. (14)

équerre en T *ou* **équerre en té** *ou* **équerre à T** (*f.*), T square. (15)

équidistance (Math.) (*n.f.*), equidistance. (16)

équidistance (Topogr.) (*n.f.*), contour-interval : (17)

la configuration du terrain est indiquée par des courbes de niveau espacées à des intervalles convenables, dits équidistances, the con-figuration of the ground is shown by contour-lines spaced at convenient intervals, called contour-intervals. (18)

carte où l'équidistance est de — mètres (*f.*), map on which the contour-interval is — metres. (19)

équidistant, -e (*adj.*), equidistant. (20)

équilatéral, -e, -aux (*adj.*), equilateral : (21)

triangle équilatéral (*m.*), equilateral triangle. (22)

équilibrage (*n.m.*) *ou* **équilibration** (*n.f.*), equili-bration ; counterbalancing : (23)

équilibrage de la sonde, counterbalancing the rods. (24)

équilibre (*n.m.*), equilibrium ; equipoise ; poise ; balance. (25)

équilibre (en), in equilibrio. (26)

équilibre des forces (*m.*), equilibrium of forces. (27)

équilibre indifférent (*m.*), neutral equilibrium ; mobile equilibrium ; indifferent equilibrium. (28)

équilibre instable (*m.*), unstable equilibrium : (29)

un corps en surfusion est en état d'équilibre instable, a body in surfusion is in a state of unstable equilibrium. (30)

équilibre stable (*m.*), stable equilibrium. (31)

équilibrer (*v.t.*), to equilibrate ; to balance ; to equipoise ; to poise ; to counterpoise ; to counterbalance : (32)

équilibrer une balance, to equilibrate, to counterpoise, a balance. (33)

équilibrer une soupape de sûreté, to balance a safety-valve. (34)

équilibrer (s') (*v.r.*), to balance ; to equilibrate ; to poise ; to counterbalance : (35)

des poids qui s'équilibrent, weights which balance. (36)

équinoxe (*n.m.*), equinox. (37)

équinoxial, -e, -aux (*adj.*), equinoctial ; equi-noxial : (38)

ligne équinoxiale (*f.*), equinoctial line. (39)

équipage (*n.m.*), outfit ; rig ; set. (40)

équipage d'engrenages (*m.*), gear-train ; train of gear-wheels ; multiple of gearing ; cluster of wheels ; nest of wheels. (41)

équipage d'outils (*m.*), outfit, set, of tools. (42)

équipage de laminoir (*m.*), roll-train ; rolling-mill train ; rolling-mill rolls. (43)

équipage de sonde (*m.*), drilling-outfit. (44)

équipage dégrossisseur (laminoir) (*m.*), muck-train ; puddle-train ; puddle-bar train ; puddle-rolls ; roughing-rolls. (45)

équipage finisseur (laminoir) (*m.*), finishing-mill ; finishing-rolls. (46)

équipe (d'ouvriers) (*n.f.*), shift ; gang ; crew ; squad ; spell. (47) See also **poste.**

équipe d'entretien (Ch. de f.) (*f.*), maintenance gang. (48)

équipe de 8 heures (*f.*), eight-hour shift. (49)

équipe de jour (*f.*), day-shift. (50)

équipe de mineurs (*f.*), shift, gang, of miners ; mining crew. (51)

équipe de nuit (*f.*), night-shift. (52)

équipe de pompiers (Mines) (*f.*), fire-fighting corps ; fire-company. (53)

équipe de poseurs (Ch. de f.) (*f.*), plate-laying gang ; track-laying gang. (54)

équipe de relève (*f.*), relieving gang. (55)

équipe de sauvetage *ou* **équipe de sauveteurs** (*f.*), rescue-party ; rescue-corps. (56)

équipe de secours (*f.*), breakdown gang. (57)

équipe du matin (*f.*), morning shift. (58)

équipe du soir (*f.*), night-shift. (59)

équipe réduite (*f.*), short shift. (60)

équipe volante (*f.*), flying squad. (61)

équipe voltigeante (Ch. de f.) (*f.*), floating gang. (62)

équipement (*n.m.*), equipment ; equipping ; fitting out ; outfitting ; outfit : (63)

équipement d'une expédition, equipment of, fitting out, an expedition. (64)

équiper (*v.t.*), to equip ; to fit out ; to rig out ; to rig up ; to rig. (65)

équipotentiel, -elle (Phys.) (*adj.*), equipotential : (66)

ligne équipotentielle (*f.*), equipotential line. (67)

équitombant, -e (*adj.*), equal-falling : (68)

grains équitombants (*m.pl.*), equal-falling grains. (69)

équivalence (*n.f.*), equivalence. (70)

équivalent, -e (*adj.*), equivalent : (71)

orifice équivalent d'une mine (Aérage) (*m.*), equivalent orifice of a mine. (72)

équivalent (*n.m.*), equivalent. (73)

équivalent mécanique de la chaleur (*m.*), mechanical equivalent of heat ; Joule's equivalent. (74)

érable (*n.m.*), maple. (75)

erbium (Chim.) (*n.m.*), erbium. (76)

erbue (opp. à castine) (Métall.) (*n.f.*), clay flux. (77) (distinguished from **limestone flux**.)

erdcobalt (Minéral) (*n.m.*), earthy cobalt ; slaggy cobalt ; asbolite ; asbolane ; asbolan. (78)

ère (*n.f.*), era : (1)
  ère paléozoïque (Géol.), Paleozoic era. (2)
érection (*n.f.*), erection. (3)
erg (C.G.S.) (*n.m.*), erg. (4)
ergmètre (*n.m.*), ergmeter. (5)
ergot (*n.m.*), snug ; lug. (6)
ergot (du pêne d'une serrure à gorge) (*n.m.*),
  stump, fence, stub, (of the bolt of a tumbler-
  lock). (7)
ergot (du culot d'une lampe à baïonnette) (Élec.)
  (*n.m.*), pin (of the cap of a bayonet-lamp).
  (8)
ériger (*v.t.*), to erect ; to put up : (9)
  ériger une fonderie, to erect a smelter ; to put
  up a smeltery. (10)
érinite (Minéral) (*n.f.*), erinite. (11)
erminette (*n.f.*), adze. (12) See **herminette**
  for varieties.
éroder (*v.t.*), to erode : (13)
  la rouille érode le fer, rust erodes iron. (14)
éroder (Géol.) (*v.t.*), to erode ; to corrade : (15)
  l'eau érode le fond du lit des rivières (*f.*), water
  erodes the bottom of river-beds ; water
  corrades the bottoms of the beds of rivers.
  (16)
éroder (s') (*v.r.*), to erode. (17)
érosif, -ive (*adj.*), erosive : (18)
  l'action érosive des acides, the erosive action
  of acids. (19)
érosion (*n.f.*), erosion. (20)
érosion (Géol.) (*n.f.*), erosion ; corrasion : (21)
  le grand cañon du Colorado est l'exemple le
  plus grandiose d'érosion profonde, the
  Grand Cañon of the Colorado is the most
  imposing example of deep erosion. (22)
  érosion éolienne, wind erosion. (23)
  érosion fluviale, river erosion. (24)
  érosion glaciaire, glacial erosion. (25)
  érosion opérée par les eaux marines sur les
  côtes, erosion caused by the action of sea-
  water on the coasts. (26)
  érosion pluviale, rain erosion. (27)
  érosion régressive *ou* érosion remontante,
  head-erosion ; headwater erosion ; re-
  -trogressive erosion. (28)
  érosion subaérienne, subaerial erosion. (29)
erratique (Géol.) (*adj.*), erratic : (30)
  bloc erratique (*m.*), erratic block. (31)
erreur (*n.f.*), error ; mistake : (32)
  erreur d'appréciation, error in valuation. (33)
  erreur de signaux, error in signalling. (34)
érubescite (Minéral) (*n.f.*), erubescite ; bornite ;
  horse-flesh ore ; peacock ore ; peacock
  copper ; purple copper ; variegated copper
  ore ; variegated pyrites. (35)
éruptif, -ive (*adj.*), eruptive ; volcanic : (36)
  roche éruptive (*f.*), eruptive rock. (37)
  roche d'origine éruptive, rock of volcanic
  origin. (38)
éruption (*n.f.*), eruption ; outburst : (39)
  éruption boueuse, mud eruption ; outburst of
  mud. (40)
  éruption de lave, eruption of lava. (41)
  éruption volcanique, volcanic eruption. (42)
érythrine (cobalt arséniaté) (Minéral) (*n.f.*),
  erythrite ; cobalt-bloom ; red cobalt. (43)
érythrite (orthose) (Minéral) (*n.f.*), erythrite. (44)
escaille (Mines) (*n.m.*), clay (running through
  coal-deposits) : (45)
  un lit d'escaille, a bed of clay. (46)
escalier (*n.m.*), stairs ; stair ; staircase ; stair-
  -way ; steps. (47)

escalier à double quartier tournant (*m.*), half-
  turn stairs. (48)
escalier à limons superposés *ou* **escalier sans jour
  médian** (*m.*), dogleg stairs. (49)
escalier à marches mobiles (*m.*), moving stair-
  -case ; moving stairway ; travelling stair-
  -case ; travelling stairs. (50)
escalier à noyau creux *ou* **escalier à noyau évidé**
  ou **escalier à jour** (*m.*), open-newel stair ;
  hollow-newel stair ; open-well stair. (51)
escalier à noyau plein (*m.*), solid-newel stair. (52)
escalier à quartier tournant (*m.*), quarter-turn
  stair. (53)
escalier à rampe droite *ou* **escalier droit** (*m.*),
  straight flight of stairs ; flyers ; fliers. (54)
escalier en 8 (*m.*), figure-of-eight stairs. (55)
escalier rompu en paliers (*m.*), stairs interrupted
  by landings. (56)
escalier suspendu *ou* **escalier en encorbellement**
  (*m.*), hanging stairs ; hanging steps. (57)
escalier tournant *ou* **escalier en vis** *ou* **escalier à
  vis** *ou* **escalier hélicoïdal** *ou* **escalier en hélice**
  ou **escalier en spirale** *ou* **escalier en colimaçon**
  ou **escalier en limaçon** *ou* **escalier en escargot**
  (*m.*), spiral stairs ; winding stairs ; cork-
  -screw stairs. (58)
escarbille (*n.f.*), cinder ; coal cinder : (59)
  locomotive qui sème son chemin d'escarbilles
  (*f.*), locomotive which strews its path with
  cinders. (60)
escarboucle (Minéral) (*n.f.*), precious garnet. (61)
escargot (escalier tournant) (*n.m.*), spiral stairs ;
  winding stairs ; corkscrew stairs. (62)
escarpé, -e (*adj.*), precipitous ; steep : (63)
  descente escarpée (*f.*), precipitous descent.
  (64)
  colline escarpée (*f.*), steep hill. (65)
escarpement (état de ce qui est escarpé) (*n.m.*),
  precipitousness ; steepness. (66)
escarpement (pente abrupte) (Géogr. phys.) (*n.m.*),
  cliff ; escarpment ; scarp ; bluff : (67)
  vallée dominée par de beaux escarpements
  calcaires (*f.*), valley dominated by fine
  limestone cliffs. (68)
escarpement de faille (Géol.) (*m.*), fault-escarp-
  -ment ; fault-scarp. (69)
eschynite (Minéral) (*n.f.*), eschynite ; æschynite.
  (70)
escouade (d'ouvriers) (*n.f.*), gang ; crew ; squad.
  (71)
espace (étendue superficielle et limitée) (*n.m.*),
  space : (72)
  travailler dans un espace resserré, to work in
  a confined space. (73)
espace (étendue indéfinie ; milieu sans bornes)
  (*n.m.*), space : (74)
  volcan qui projette une colonne de fumée
  dans l'espace (*m.*), volcano which projects a
  column of smoke into space. (75)
espace (intervalle libre) (*n.m.*), space : (76)
  l'espace entre deux murs, the space between
  two walls. (77)
espace (entre le piston et le fond de cylindre)
  (*n.m.*), space (between piston and cylinder-
  head). (78)
espace creux (*m.*), cavity ; hollow space. (79)
espace intermédiaire (machine à détente) (*m.*),
  receiver-space (expansion-engine). (80)
espace nuisible *ou* **espace mort** *ou* **espace neutre**
  ou **espace libre** (de cylindre à vapeur) (*m.*),
  clearance ; clearance-space ; piston-clear-
  -ance. (81)

espacement (*n.m.*), spacing : (1)
espacement des traverses (Ch. de f.), spacing of sleepers. (2)
espacer (*v.t.*), to space. (3)
espatard (*n.m.*), planishing-roll. (4)
espèce (*n.f.*), kind ; sort ; species : (5)
espèce de matière, kind of matter. (6)
espèce minérale, mineral species : (7)
l'acmite est une espèce minérale apparte-nant au genre pyroxène (*f.*), acmite is a mineral species belonging to the pyroxene group. (8)
espérance (*n.f.*), hope ; expectation : (9)
espérances relatives à une propriété, expecta-tions regarding a property. (10)
espérer (*v.t. & v.i.*), to hope. (11)
espoir (*n.m.*), hope : (12)
espoir de sauver la vie des ouvriers, hope of rescuing the men alive. (13)
esprit (*n.m.*), spirit ; spirits. (14)
esprit de bois (*m.*), wood-spirit ; wood-naphtha. (15)
esprit de sel décomposé (*m.*), killed spirit. (16)
esprit-de-vin (*n.m.*), spirits of wine ; wine-alcohol. (17)
esquisse (*n.f.*), sketch ; rough sketch. (18)
essai (épreuve) (*n.m.*), test ; testing ; trial : (19). See also épreuve.
l'essai d'une machine, testing a machine ; the trial of an engine. (20)
essai (analyse) (Chim. minérale ou métallurgique) (*n.m.*), assay ; assaying ; analysis ; testing : (21). See also analyse.
essai des métaux, assay of metals ; assaying metals ; analysis of metals. (22)
essai d'argent par voie humide, assaying silver by the wet method. (23)
essai par coupellation, assaying by cupellation. (24)
essai à basse tension (*m.*), low-tension test. (25)
essai à froid (essais des métaux, ou épreuve hydraulique des chaudières) (*m.*), cold-test. (26 *ou* 27)
essai à haute tension (*m.*), high-tension test. (28)
essai à l'eau de mer sur les ciments (*m.*), sea-water test on cements. (29)
essai à la compression (*m.*), compression-test ; crushing-test. (30)
essai à la flexion (*m.*), bend-test ; bending-test ; deflection-test. (31)
essai à la perle (*m.*), bead-test. (32)
essai à la torsion (*m.*), torsional test ; twisting-test. (33)
essai à la traction (*m.*), tensile test. (34)
essai au borax (*m.*), borax-bead test. (35)
essai au carbonate de soude *ou* essai au soude (*m.*), sodium-carbonate bead-test. (36)
essai au chalumeau (épreuve) (*m.*), blowpipe test. (37)
essai au chalumeau (*m.*) *ou* essais au chalumeau (*m.pl.*) (analyse), blowpipe assay ; blowpipe assaying ; blowpipe analysis ; blowpiping. (38)
essai au choc (*m.*), falling-weight test ; impact-test ; drop-test. (39)
essai au creuset (*m.*), crucible-test. (40)
essai au dynamomètre (*m.*), dynamometer-test. (41)
essai au four (*m.*), furnace-test. (42)
essai au fourneau à chalumeau (*m.*), blowpipe-furnace test. (43)

essai au moulin (minerais) (*m.*), mill-test ; mill-trial ; mill-run (ores). (44)
essai au sel de phosphore (*m.*), salt of phos-phorus bead-test. (45)
essai au spectroscope (*m.*), spectroscope-test. (46)
essai au touchau (*m.*), touch-needle test. (47)
essai au tube fermé (essais au chalumeau) (*m.*), closed-tube test. (48)
essai au tube ouvert (*m.*), open-tube test. (49)
essai aux acides *ou* essai avec les acides (*m.*), acid-test. (50)
essai avec récipient à pétrole fermé (*m.*), closed-cup test. (51)
essai avec récipient à pétrole ouvert (*m.*), open-cup test. (52)
essai comparatif *ou* essai comparé (*m.*), com-parative test. (53)
essai contradictoire (*m.*), check-analysis ; check-assay ; control-assay. (54)
essai d'aplatissement (*m.*), flattening-test. (55)
essai d'arbitre (*m.*), umpire's assay. (56)
essai d'écrasement (*m.*), crushing-test ; com-pression-test. (57)
essai d'emboutissage (*m.*), stamping-test. (58)
essai de billage (métaux) (*m.*), ball-test. (59)
essai de chargement (poutres, etc.) (*m.*), loading-test. (60)
essai de cintrage (*m.*), bend-test ; bending-test. (61)
essai de coloration (essais au chalumeau) (*m.*), flame-reaction ; flame-test. (62)
essai de cyanuration (*m.*), cyanide test. (63)
essai de décantation (tailings) (*m.*), decantation-test. (64)
essai de dureté (*m.*), hardness test. (65)
essai de filtration (tailings) (*m.*), percolation-test. (66)
essai de fonctionnement (*m.*), working-test ; working-trial ; running-test. (67)
essai de forgeage (*m.*), forging-test. (68)
essai de fusion (*m.*), smelting-trial. (69)
essai de laboratoire (*m.*), laboratory-test. (70)
essai de pliage (*m.*), bend-test ; bending-test ; doubling-over test. (71)
essai de pliage à froid (*m.*), cold bend-test. (72)
essai de rabattement des collerettes (tubes) (*m.*), flanging-test. (73)
essai de réception (*m.*), acceptance-test. (74)
essai de résistance au choc (*m.*), falling-weight test ; impact-test ; drop-test. (75)
essai de scorification (*m.*), scorification-assay. (76)
essai de soudabilité (*m.*), welding-test. (77)
essai de surtension (*m.*), overpressure test ; excess-pressure test. (78)
essai du froid (huiles) (*m.*), cold-test. (79)
essai jusqu'à la rupture (*m.*), destruction test. (80)
essai par la coloration de la flamme (Essais au chalumeau) (*m.*), flame-reaction ; flame-test. (81)
essai par voie humide (*m.*), wet assaying. (82)
essai par voie sèche (*m.*), dry assaying. (83)
essai pyrognostique (*m.*), fire assay. (84)
essai radiographique (*m.*), radiographic test. (85)
essai sous pression (*m.*), steam-test ; steam-trial. (86)
essai spectroscopique (*m.*), spectroscopic test. (87)
essai sur le charbon (Minéralogie) (*m.*), charcoal-test. (88)
essayage (*n.m.*), testing. (89)

essayer (éprouver) (*v.t.*), to test ; to try : (1)
essayer au pliage, à la compression (Méc.),
to test for bending, for compression. (2
ou 3)
essayer le câble, to test the rope. (4)
essayer un procédé, to try, to test, a process.
(5)
essayer (Chim. minérale ou métallurgique) (*v.t.*),
to assay ; to analyze ; to test : (6)
essayer de l'or, to assay gold. (7)
essayer de l'or avec la pierre de touche, to
test gold with the touchstone. (8)
essayer de (*v.i.*), to try ; to endeavour ; to
attempt ; to make an attempt : (9)
essayer de tracer le filon, to try, to endeavour,
to attempt, to make an attempt, to trace
the lode. (10)
essayerie (*n.f.*), assay-office. (11)
essayeur (pers.) (*n.m.*), assayer ; analyst ; tester.
(12)
essayeur de commerce (*m.*), public assayer ;
assayer. (13)
esse (quelque chose en forme d'esse) (*n.f.*), ess ;
S : (14)
en cet endroit, la rivière décrit une esse, here,
the river describes an S. (15)
esse (crochet en forme d'S) (*n.f.*), S hook. (16)
esse (clavette d'essieu) (*n.f.*), linchpin ; axle-
pin. (17)
esse (jauge) (*n.f.*), iron-wire gauge. (18)
esselier (*n.m.*), angle-brace ; angle-tie ; raking
strut ; strut. (19)
essence (Chim.) (*n.f.*), spirit. (20)
essence (*n.f.*) *ou* **essence de pétrole** *ou* **essence**
minérale *ou* **essence pour automobiles** *ou*
essence pour autos, petrol ; petroleum spirit ;
motor-spirit ; gasoline. (21)
essence de bois (*f.*), kind of wood ; species of
wood : (22)
le chêne est la meilleure essence de bois pour
la confection des traverses de chemin de
fer, oak is the best kind of wood for making
railway-sleepers. (23) Cf. **essences d'arbres.**
essence de térébenthine (*f.*), spirit of turpentine.
(24)
essences d'arbres (*f.pl.*), kind or species of trees
predominating in a district : (25)
forêt en essences de chêne (*f.*), oak forest. (26)
essences résineuses, resinous trees. (27)
essentiel, -elle (*adj.*), essential : (28)
données essentielles (*f.pl.*), essential data. (29)
minéraux essentiels (*m.pl.*), essential minerals.
(30)
essentiellement (*adv.*), essentially. (31)
esser le fil de fer, to gauge iron wire. (32)
esseret (*n.m.*), carpenters' auger. (33)
essieu (*n.m.*), axle ; axletree : (34)
essieu d'un wagon, axle, axletree, of a truck
or wagon. (35)
essieux droits en acier fondu sur sole, bruts de
forge, ébauchés, ou finis, pour locomotives
et tenders, straight axles of open-hearth
steel, rough-forged, rough-machined, or
finished-machined, for locomotives and
tenders. (36 *ou* 37 *ou* 38)
essieu à patente (*m.*), patent axletree. (39)
essieu arrière [essieux arrière *pl.*] *ou* **essieu**
d'arrière [essieux d'arrière *pl.*] (*m.*), back
axle ; hind axle. (40)
essieu avant [essieux avant *pl.*] *ou* **essieu d'avant**
[essieux d'avant *pl.*] (*m.*), front axle ; fore
axle. (41)

essieu bissel (locomotive) (*m.*), pony-truck axle.
(42)
essieu coudé (*m.*), crank-axle. (43)
essieu du bogie (*m.*), bogey-axle ; engine-truck
axle. (44)
essieu fixe (*m.*), dead axle. (45)
essieu moteur (*m.*), driving-axle. (46)
essieu patent *ou* **essieu patenté** (*m.*), patent
axletree. (47)
essieu porteur (de locomotive) (opp. à **essieu**
moteur) (*m.*), carrying-axle. (48)
essieu porteur (de bogie, de bissel) (locomotive)
(*m.*), truck-axle. (49 *ou* 50)
essieu porteur d'arrière *ou* **essieu porteur à**
l'arrière (*m.*), trailing-axle ; trailing-truck
axle. (51)
essieu porteur d'avant *ou* **essieu porteur à l'avant**
(*m.*), leading-axle ; front-truck axle. (52)
essieu surbaissé (*m.*), dropped axle. (53)
essieu tournant (*m.*), live axle. (54)
essonite (Minéral) (*n.f.*), essonite ; hessonite.
(55)
essor (*n.m.*), progress : (56)
l'industrie minière a pris un grand essor (*f.*),
the mining industry has made great progress.
(57)
est (*adj.*), east ; eastern ; easterly. (58)
est (*n.m.*), east. (59)
est-nord-est (*n.m.*), east-northeast. (60)
est-sud-est (*n.m.*), east-southeast. (61)
estacade (plate-forme de chargement) (*n.f.*),
elevated runway ; tip ; tipple. (62)
estacade (Construction hydraulique) (*n.f.*),
stockade. (63)
estampage (*n.m.*), **estamper** (*v.t.*), **estampeur**
(*n.m.*), **estampeuse** (*n.f.*). Same as **étampage,**
étamper, etc.
estau (dans une galerie de mine) (*n.m.*), arch
(portion of lode left standing to support
hanging wall). (64)
estau (dans un puits de mine) (*n.m.*), pentice
(roof of rock or slice of ground left pro-
-visionally between the bottom of a shaft
and its extension beneath). (65)
estibois (*n.m.*), vice-cap ; wood vice-clamp. (66)
estimateur (pers.) (*n.m.*), valuer ; valuator. (67)
estimatif, -ive (*adj.*), estimated : (68)
réserves estimatives de minerai (*f.pl.*), esti-
-mated ore reserves. (69)
estimation (*n.f.*), valuing ; valuation ; estimat-
-ing ; estimate : (70)
estimation de diamants, valuing diamonds.
(71)
estimation de la teneur en or des alluvions,
estimating the gold contents of alluvials. (72)
estimation de tout repos, reliable estimate ;
safe estimate. (73)
estimation prudente, conservative estimate.
(74)
estimer (*v.t.*), to estimate ; to value : (75)
estimer la valeur du minerai en réserve, to
estimate the value of the ore in reserve.
(76)
estimer le minerai à — par tonne *ou* estimer le
minerai — par tonne, to value the ore at
— per ton. (77)
estimer une propriété au-dessous de sa valeur,
to underestimate, to underrate, the value of
a property. (78)
estival, -e, -aux (*adj.*), summer : (79)
température estivale (*f.*), summer temperature.
(80)

estomac (d'une enclume) (*n.m.*), web, body, (of an anvil). (1)

estomac (d'une bobine pour câble plat d'extrac- -tion) (Mines) (*n.m.*), hub (of a reel for flat winding-rope). (2)

estomac (plaque conscience) (*n.m.*), breast-plate ; conscience ; palette. (3)

estomac d'excentrique (*m.*),   eccentric-disc ; eccentric-sheave. (4)

estouffée *ou* **estoupée** (Mines) (*n.f.*), dam (in a mine-level). (5)

estrope (de poulie, de moufle) (*n.f.*),   strap, strop, (of pulley-block, of tackle-block) : (6 *ou* 7)

moufle à estrope double (*f.*),   double-strapped block. (8)

estrope en fer de poulie (*f.*), iron binding, for tackle-block. (9)

estuaire (*n.m.*), estuary. (10)

établi (*n.m.*), bench. (11)

établi de charpentier (*m.*),   carpenters' bench. (12)

établi de mouleur (*m.*), moulders' bench. (13)

établi roulant pour étaux (*m.*),   portable vice- stand ; portable vice-bench. (14)

établir (*v.t.*).   See examples :   (15)

établir de meilleures conditions hygiéniques, to establish better sanitary conditions. (16)

établir l'existence d'un filon,   to prove, to establish, the existence of a lode. (17)

établir la cause du désastre,   to ascertain the cause of the disaster. (18)

établir les grandes lignes d'un projet de développement futur,   to outline a plan for future development. (19)

établir un barrage dans un cours d'eau,   to construct a dam in a stream. (20)

établir une canalisation dans,   to lay a pipe- line in ; to pipe :   (21)

établir une canalisation dans une maison, to pipe a house. (22)

établir une machine,   to set up, to erect, a machine. (23)

établissement de prix de revient (*m.*), costing. (24)

étage (d'une maison) (*n.m.*), story, storey, floor, (of a house). (25)

étage (Géol.) (*n.m.*), stage :   (26)

le néocomien est un étage du système crétacé, the Neocomian is a stage of the Cretaceous system. (27)

étage (Mines) (*n.m.*), stage , lift ; level : (28)

partager un gîte en étages et sous-étages, to divide up a bed into stages and sub- -stages. (29)

un puits ou un flanc de montagne est divisé en étages : chaque étage ou niveau est caractérisé par sa voie de fond,   a shaft or mountainside is divided into stages : each stage, lift, or level, has its own bottom level (i.e., bottom or main roadway). (30)

tracer de nouveaux étages en profondeur, to develop new levels in depth.   (31)

le —ᵉ étage, the —th level. (32)

l'étage de — mètres *ou* l'étage —,   the — - metre level. (33)

étage (d'un cône, d'un cône-poulie) (*n.m.*), step, lift, (of a cone, of a cone-pulley) : (34)

cône à 3 étages (*m.*), 3-step cone ; 3-lift cone- pulley ; 3-speed cone. (35)

étage de pression (turbines) (*m.*), pressure-stage. (36).   (*In the plural*—**étages de pression**— pressure-stages *or* pressure-staging (37),

thus, au lieu d'**étages de vitesse**, on peut faire usage d'**étages de pression**, in place of velocity-staging, pressure-staging can be resorted to.)

étage de vitesse (*m.*), velocity-stage : (38)

turbine à étages de vitesse (*f.*), velocity-stage turbine. (39)

étage du fond (Mines) (*m.*), bottom level. (40)

étage houiller (Géol.) (*m.*), coal-measures. (41)

étagé, -e (*adj.*),   stage (*used as adj.*) ; staged : (42)

compression étagée (d'air) (*f.*),   stage-com- -pression. (43)

compresseur étagé (*m.*),   stage-compressor. (44)

étai (*n.m.*), shore ; prop ; post ; stay ; support. (45)

étai (pour supporter un mur, etc., menaçant ruine) (*n.m.*), shore. (46)

étai (bois droit et isolé ;   chandelle ;   butte) (Mines) (*n.m.*),   prop ;   tree ;   post ;   shore. (47)

étai de mine (*m.*), pit-prop ; pit-post. (48)

étai incliné (*m.*), raking shore ; raker. (49)

étai vertical (*m.*),   vertical shore ; dead-shore. (50)

étaiement (*n.m.*), shoring ; propping ; staying : (51)

étaiement d'un mur par contrefiches,   shoring a wall by rakers. (52)

étaiement (combinaison de pièces de bois de charpente appelées étais) (Constr.) (*n.m.*), shoring. (53)

étain (*n.m.*), tin. (54)

étain alluvionnaire *ou* **étain d'alluvion** (*m.*), alluvial tin. (55)

étain de bois (Minéral) (*m.*), wood-tin. (56)

étain de roche (*m.*), lode-tin ; mine-tin. (57)

étain en larmes (*m.*),   grain-tin ;   drop-tin ; feathered tin. (58)

étain en saumon (*m.*), block tin ; bar tin. (59)

étain métallique (*m.*), metallic tin. (60)

étain noir (*m.*), black tin. (61)

étain oxydé (Minéral) (*m.*), cassiterite ; tinstone. (62)

étain pyriteux (Minéral) (*m.*),   tin pyrites ; stannite ; bell-metal ore. (63)

étalage (*n.m.*), bosh :   (64)

les étalages d'un haut fourneau, d'un cubilot, the boshes of a blast-furnace, of a cupola. (65 *ou* 66)

étaler (*v.t.*), to lay out ; to spread ; to spread out :   (67)

étaler un jet d'eau, to spread a jet of water. (68)

étalon (*n.m. employé adjectivement*), standard : (69)

mesures étalons (*f.pl.*),   standard measures. (70)   (Note.—*Sometimes* **étalon** *is hyphened to the noun qualified.*)

étalon (*n.m.*), standard. (71)

étalon d'éclairement (Photom.) (*m.*), standard of illumination. (72)

étalon de lumière (*m.*), light-standard ; standard of light. (73)

étalon industriel (*m.*), commercial standard. (74)

étalon Violle (Photom.) (*m.*), Violle standard ; Violle's standard ; violle : (75)

l'étalon Violle correspond à la lumière émise par un centimètre de platine à la température de solidification. Il équivaut à environ 18½ bougies anglaises,   the Violle standard corresponds to the light afforded by a square centimetre of platinum at solidification-point. It is equivalent to about 18½ English candles. (1)

étalonnage (des lampes) (Photom.) (*n.m.*), rating (of lamps). (2)

étalonnage (d'un galvanomètre) (*n.m.*), standard--ization (of a galvanometer). (3)

étalonner (Photom.) (*v.t.*), to rate : (4)
les lampes anglaises sont étalonnées en candles, les lampes françaises en bougies décimales, les lampes allemandes en hefners (*f.pl.*), English lamps are rated in candles, French lamps in decimal candles, German lamps in hefners. (5)

étalonner un galvanomètre, to standardize a galvanometer. (6)

étamage (*n.m.*), tinning : (7)
étamage de la tôle, tinning sheet iron. (8)

étamer (*v.t.*), to tin : (9)
étamer le fer, to tin iron. (10)

étampage (*n.m.*), stamping ; pressing ; swaging ; swedging ; drop-forging. (11)

étampe (*n.f.*) *ou* **étampe pour fers ronds** (outil de forgeron), swage ; rounding-tool ; dolly. (12)

étampe de dessous *ou* **étampe de dessous pour fers ronds** (*f.*), bottom swage ; bottom rounding-tool. (13)

étampe de dessus *ou* **étampe de dessus pour fers ronds** (*f.*), top swage ; top rounding-tool. (14)

étamper (*v.t.*), to stamp ; to press ; to swage ; to swedge ; to drop-forge. (15)

étamper à chaud, to hot-stamp. (16)

étamper à froid, to cold-stamp. (17)

étampeur (pers.) (*n.m.*), stamper. (18)

étampeuse (*n.f.*), stamping-machine ; stamping-press ; stamper ; stamp. (19)

étampeuse à main (*f.*), hand-stamp. (20)

étamure (couche d'étain) (*n.f.*), tinning ; coating of tin. (21)

étamure (métal pour étamer) (*n.f.*), tinning-metal. (22)

étanche (*adj.*), tight ; water-tight ; fluid-tight ; steam-tight ; gas-tight ; impervious ; stanch : (23)
tonneau étanche (*m.*), tight cask ; water-tight cask. (24)
bien peu de tunnels sont étanches, very few tunnels are water-tight. (25)
chaudière étanche (*f.*), steam-tight boiler. (26)
terrain étanche (*m.*), impervious ground. (27)

étanche d'eau (à), water-tight : (28)
entretenir une toiture à étanche d'eau, to keep a roof water-tight. (29)

étanchéité (*n.f.*), tightness ; water-tightness ; imperviousness ; stanchness : (30)
étanchéité d'un joint, tightness, water-tight--ness, imperviousness, stanchness, of a joint. (31)

étancher un tonneau, to make a cask water-tight. (32)

étancher une voie d'eau, to stop a leak. (33)

étançon (*n.m.*), shore ; stanchion ; prop ; stay ; support. (34)

étançonnement (*n.m.*), shoring ; propping. (35)

étançonner (*v.t.*), to shore ; to shore up : to prop : (36)
étançonner un mur, to shore up a wall. (37)

étang (*n.m.*), pond. (38)

étape (*n.f.*), stage : (39)
procéder par étapes successives, to proceed by successive stages. (40)
la dernière étape d'un voyage, the last stage of a journey. (41)

état (condition) (*n.m.*), state ; condition : (42)
état d'une mine, state, condition, of a mine. (43)
état du temps, state of the weather. (44)

état (exposé ; relevé) (*n.m.*), statement ; account ; return. (45)

État (nation) (*n.m.*), state, government (often written with initial capital letter : prend souvent une majuscule) : (46)
chemin de fer de l'État (*m.*), State railway. (47)

état brut (à l'), in the crude state ; in the rough : (48)
à l'état brut, tel que sortant de la carrière, in the rough, as quarried. (49)

état critique (Phys.) (*m.*), critical state. (50)

état estimatif (*m.*), estimate. (51)

état naissant (Chim.) (*m.*), nascent state ; nascent condition. (52)

état natif (à l'), in a native state ; in the native state ; native : (53)
or à l'état natif (*m.*), gold in the native state ; native gold. (54)

état sphéroïdal (Phys.) (*m.*), spheroidal state. (55)

étau (*n.m.*), vice ; vise. (56)

étau à agrafes d'établi (*m.*), bench-vice, table-vice, with clamp. (57)

étau à barres parallèles (*m.*), vice sliding between parallel bars. (58)

étau à base fixe pour établis, à serrage ordinaire ; corps fonte, mordaches en acier (*m.*), fixed bench-vice, with ordinary screw-grip ; body of cast iron fitted with detachable steel jaws. (59)

étau à base tournante pour établis, à serrage instantané (*m.*), instantaneous-grip swivel bench-vice ; quick-grip revolving bench-vice ; sudden-grip rotary bench-vice. (60)

étau à chaîne (pour tubes) (*m.*), chain-vice. (61)

étau à chaud (*m.*), hot-vice. (62)

étau à crémaillère (*m.*), rack vice ; machine-vice with slots on top of slide. (63)

étau à enclumette (*m.*), anvil-vice. (64)

étau à griffes (*m.*), vice with clamp. (65)

étau à mâchoires rapportées (*m.*), vice with detachable jaws ; vice with inserted jaws. (66)

étau à main *ou* **étau à main, à vis** (*m.*), hand-vice. (67)

étau à main, à queue (*m.*), pin-vice ; tail-vice. (68)

étau à main, avec queue percée de part en part (*m.*), pin-vice, with hole through handle ; pin-vice, with thoroughfare hole. (69)

étau à main, filet carré, poli partout, manche bois verni (*m.*), hand-vice with square thread, bright all over, polished wood handle. (70)

étau à main parallèle (*m.*), parallel hand-vice. (71)

étau à pied (*m.*), leg-vice ; staple-vice. (72)

étau à pied roulant (*m.*), portable vice-stand, with leg-vice. (73)

étau à serrage concentrique (*m.*), self-centring vice. (74)

étau à table (*m.*), standing-vice. (75)

étau à tubes (*m.*), tube-vice ; pipe-vice. (1)

étau à tubes, à charnière (*m.*), hinged tube-vice ; hinged pipe-vice. (2)

étau à tubes, ouvert (*m.*), tube-vice, open type. (3)

étau à vis cachée (*m.*), vice with protected screw. (4)

étau d'affûtage et d'avoyage pour scies (*m.*), saw-filers' vice ; saw-vice ; saw-clamp. (5)

étau de menuisier (*m.*), joiners' vice. (6)

étau de tuyauteur (*m.*), pipe-fitters' vice ; pipe-vice. (7)

étau-limeur [étaux-limeurs *pl.*] (*n.m.*), shaping-machine ; shaper ; shaping-planer. (8)

étau-limeur à commande par bielle (*m.*), crank shaping-machine. (9)

étau-limeur à commande par crémaillères avec triple harnais d'engrenages, retour rapide, course réglable pendant la marche, tête porte-outil à secteur denté, descente verticale automatique du porte-outil, table d'équerre à rainures se déplaçant automatiquement et à la main dans le sens transversal et à la main dans le sens vertical (*m.*), triple-geared rack-actuated shaping-machine, quick-return motion, stroke adjustable whilst running, quadrant head, self-acting down feed to tool-box, slotted box table with self-acting and hand cross-feed and hand vertical feed. (10)

étau-limeur à course variable (*m.*), adjustable-stroke shaper. (11)

étau-limeur à engrenages (*m.*), geared shaping-machine. (12)

étau-limeur à outil mobile (*m.*), traversing-head shaping-machine ; traverse shaper. (13)

étau-limeur à table mobile (*m.*), pillar-shaper. (14)

étau-limeur double (*m.*), double-headed shaping-machine. (15)

étau parallèle (*m.*), parallel vice. (16)

étau parallèle à base tournante avec mors en acier trempé (*m.*), parallel swivel-vice with hardened steel jaws. (17)

étau parallèle roulant (*m.*), portable vice-bench, with parallel vice. (18)

étau-plateau [étaux-plateaux *pl.*] (pour machines à percer) (*n.m.*), vice-plate ; combined parallel vice and circular table. (19)

étau roulant (*m.*), vice-bench ; vice-stand. (20)

étau roulant, à table (*m.*), vice-bench, with table. (21)

étau tendeur *ou* étau tenseur *ou* étau raidisseur (pour la télégraphie) (*m.*), draw-vice ; wire-straining vice. (22)

étau tendeur à cliquet (*m.*), draw-vice with ratchet. (23)

étau-tiroir [étaux-tiroirs *pl.*] (*n.m.*), sliding parallel vice ; machine-vice (working in parallel slides). (24)

étau-tiroir à tige (*m.*), machine-vice on post. (25)

étau-tiroir s'inclinant sous tout angle (*m.*), machine-vice tilting to any angle. (26)

étau tournant à mâchoires parallèles pour machines à fraiser (*m.*), swivel-vice with parallel jaws for milling-machines. (27)

étau tournant parallèle américain pour établis, à serrage instantané (*m.*), instantaneous-grip rotary American parallel bench-vice. (28)

étayage *ou* étayement (*n.m.*), shoring ; propping ; staying. (29)

étayer (*v.t.*), to shore ; to shore up ; to prop ; to stay ; to support : (30)

étayer un mur qui a perdu son aplomb et menace ruine, to shore up a wall which has departed from the perpendicular and threatens to fall. (31)

pilier qui étaie un plafond (*m.*), pillar which supports a roof. (32)

été (*n.m.*), summer. (33)

éteindre (étouffer) (*v.t.*), to extinguish ; to put out : (34)

éteindre un incendie, to extinguish, to put out, a fire. (35)

éteindre le gaz, to put out the gas. (36)

éteindre (s') (cesser de brûler) (*v.r.*), to go out. (37)

éteindre de la chaux, to slake, to slack, to kill, lime. (38)

éteindre le fer dans de l'eau (trempe), to quench iron in water. (39)

éteint, -e (*adj.*), extinct : (40)

volcan éteint (*m.*), extinct volcano. (41)

éteint, -e (se dit de la chaux, etc.) (*adj.*), slaked ; slacked. (42)

éteint (-e) à l'air, air-slaked : (43)

chaux éteinte à l'air (*f.*), air-slaked lime. (44)

étendre (*v.t.*), to extend ; to spread ; to stretch ; to stretch out ; to widen : (45)

étendre les limites de sa propriété, to extend the boundaries of one's property. (46)

étendre une feuille d'or en la battant, to spread a leaf of gold by beating it. (47)

étendre (affaiblir, en ajoutant de l'eau) (*v.t.*), to dilute : (48)

étendre d'eau un acide, to dilute an acid with water ; to water an acid. (49)

étendre (s') (*v.r.*), to extend ; to spread ; to stretch ; to stretch out ; to widen : (50)

inondations qui s'étendent sur — kilomètre(s) environ (*f.pl.*), floods which extend for about — kilometre(s). (51)

incendie qui s'étend (*m.*), fire which spreads. (52)

plaine qui s'étend à perte de vue (*f.*), plain which stretches out of sight. (53)

étendu, -e (considérable) (*adj.*), extended ; extensive ; wide ; broad. (54)

étendu, -e (se dit des liquides) (*adj.*), diluted ; dilute ; watered. (55)

étendue (dimension en superficie) (*n.f.*), expanse ; stretch ; spread ; area ; tract : (56)

étendue de pays, expanse, stretch, spread, tract, of country. (57)

étendue (portée) (*n.f.*), extent ; range ; reach : (58)

étendue d'un filon, extent, range, of a lode. (59)

étendue de l'explosion, extent of the explosion. (60)

étendue de la vue, reach, range, of vision. (61)

étendue superficielle, superficial extent ; area. (62)

éther (*n.m.*), ether. (63)

éther de pétrole (*m.*), petroleum ether. (64)

éthylène (*n.m.*), ethylene. (65)

étiage (d'une rivière) (*n.m.*), low water ; low-water mark : (66)

débit d'étiage (*m.*), flow at low water. (67)

la vitesse du Rhône, qui est de 0 m. 50 à 1 m. à l'étiage, atteint de 4 à 5 m. aux hautes eaux, the speed of the Rhone, which is from

0·5 m. to 1 m. at low water, reaches from 4 to 5 m. at high water. (1)

le Rhône atteint son étiage en hiver, la Seine en été, the Rhone reaches its low-water mark in winter, the Seine in summer. (2)

étincelant, -e (*adj.*), sparkling ; glittering ; glistening : (3)

pierreries étincelantes (*f.pl.*), sparkling gems. (4)

étinceler (*v.i.*), to sparkle ; to glitter ; to glisten. (5)

étincelle (*n.f.*), spark. (6)

étincelle électrique (*f.*), electric spark. (7)

étincellement (*n.m.*), sparkling ; glittering ; glistening. (8)

étirable (*adj.*), drawable : (9)

fer étirable (*m.*), drawable iron. (10)

étirage (Tréfilerie) (*n.m.*), drawing ; wire-draw--ing. (11)

étirage à chaud (*m.*), hot-drawing. (12)

étirage à froid (*m.*), cold-drawing. (13)

étirement (*n.m.*), drawing ; drawing out : (14)

l'étirement du noyau d'un pli (Géol.), the drawing out of the core of a fold. (15)

étirer (*v.t.*), to draw ; to draw out : (16)

pli avec flanc inverse étiré (Géol.) (*m.*), fold with trough limb drawn out. (17)

étirer (Tréfilerie) (*v.t.*), to draw ; to draw out ; to wiredraw : (18)

étirer des métaux par la filière, to draw metals through the draw-plate. (19)

étirer des tubes, to draw tubes. (20)

étirer à chaud, to hot-draw. (21)

étirer à froid, to cold-draw : (22)

tubes en acier sans soudure, étirés à froid (*m.pl.*), cold-drawn weldless-steel tubes. (23)

étireur (pers.) (*n.m.*), drawer ; wiredrawer. (24)

étireuse (*n.f.*) *ou* **étireur** (*n.m.*), draw-bench ; drawing-bench ; drag-bench ; wiredrawing-bench ; wiredrawer. (25)

étoile (*n.f.*), star. (26)

étoile (Élec.) (*n.f.*), star ; Y : (27)

montage en étoile (*m.*), star-connection ; Y-connection. (28)

montage en étoile-triangle (*m.*), star-delta connection. (29)

étoile polaire *ou* **Etoile du Nord** (*f.*), pole-star ; polar star ; North Star ; Polaris. (30)

étonner la roche *ou* **étonner les roches par le feu**, to disintegrate the rock by fire. (31)

étouffant, -e (*adj.*), suffocating ; stifling ; oppressive : (32)

atmosphère étouffante (*f.*), stifling atmosphere. (33)

étouffement (*n.m.*), smothering : (34)

étouffement de l'incendie, smothering the fire. (35)

étouffer (*v.t.*), to smother : (36)

étouffer un commencement d'incendie, to smother the beginnings of a fire. (37)

étouffoir de surchauffeur (*m.*), superheater-damper. (38)

étoupe (*n.f.*), tow ; junk : (39)

garniture d'étoupe (*f.*), tow packing ; junk packing. (40)

étoupe de chanvre (*f.*), hemp tow. (41)

étoupille (Travail aux explosifs) (*n.f.*), fuse. (42). See also **mèche** and **amorce**.

étoupille à friction (*f.*), friction-fuse. (43)

étoupille à percussion (*f.*), percussion-fuse. (44)

étoupille de sûreté (*f.*), safety-fuse ; common fuse ; Bickford fuse. (45)

étoupiller (*v.t.*), to fuse : (46)

étoupiller un trou de mine, to fuse a shot-hole. (47)

étranger, -ère (*adj.*), foreign : (48)

substance étrangère (*f.*) *ou* corps étranger (*m.*), foreign substance. (49)

étranglement (Géogr. phys.) (*n.m.*), narrow : (50)

lacs séparés les uns des autres par des étranglements et des chutes (*m.pl.*), lakes separated from each other by narrows and falls. (51)

étranglement (*n.m.*) *ou* **étranglement de terre** (Mines), pinch ; balk ; nip. (52)

étranglement (Mach.) (*n.m.*), throttling. (53)

étrangler (Mach.) (*v.t.*), to throttle : (54)

étrangler la vapeur, to throttle steam. (55)

étrangleur (*n.m.*), throttle ; throttle-valve. (56)

être à court (de) *ou* **être court (-e) (de)**, to be short (of) ; to run short (of) ; to run out (of) : (57)

être à court de combustible, to be short, to run short, to run out, of fuel. (58)

être à l'arrêt (Procédé Bessemer), to drop : (59)

lorsque le bain est à l'arrêt, when the flame drops. (60)

être à même de faire face à une demande, to be in a position, to be able, to be ready, to meet a demand. (61)

être dans le filon, to be in ore : (62)

le tunnel est dans le filon sur toute sa longueur, the tunnel is in ore all the way. (63)

être de mesure, to be to size : (64)

pièces de bois qui ne sont pas de mesure (*f.pl.*), pieces of wood which are not to size. (65)

être de vente facile, to command a ready sale ; to find a ready market ; to sell well. (66)

être en contravention avec les règlements, to contravene the regulations. (67)

être en face d'une difficulté, to be faced, to be confronted, with a difficulty. (68)

être en même de commencer le broyage, to be just about to start crushing. (69)

être entre deux airs, to be in a draught. (70)

être refusé (-e) par le crible, to pass over the screen : (71)

matières refusées par le crible (*f.pl.*), stuff passing over the screen. (72)

être soumis (-e) à des efforts de traction, de compression, de cisaillement, etc. (Résistance des matériaux), to be subjected to tensile stress, to compressive stress, to shearing-stress, etc. (Strength of materials) : (73 *ou* 74 *ou* 75)

rivet qui est à la fois soumis à des efforts de traction et de cisaillement (*m.*), rivet which is at the same time subjected to tensile and shearing-stresses. (76)

être tributaire de l'étranger, to depend, to be dependent, on (*or* upon) foreign supplies. (77)

étrécir (*v.t.*), to narrow ; to shrink ; to contract. (78)

étrécir (s') (*v.r.*), to narrow ; to shrink ; to contract. (79)

étrécissement (*n.m.*), narrowing ; shrinking ; shrinkage ; contraction : (80)

étrécissement du lit d'une rivière, narrowing, contraction, of the bed of a river. (81)

étreinte (Mines) (*n.f.*), pinch ; balk ; nip. (82)

étrésillon (*n.m.*), strut ; brace. (83)

étrésillon (pièce de bois posée entre les parois d'une tranchée pour empêcher l'éboulement des terres) (*n.m.*), strut ; brace. (84)

étrésillon (pièce de bois qu'on pose entre deux murs menaçant ruine) (*n.m.*), flying shore ; horizontal shore. (1)

étrésillon (pièce de bois posée entre les solives de plancher) (*n.m.*) strut ; strutting-piece ; bridging-piece. (2) (*If in the plural*— **étrésillons**—struts ; strutting, stiffening *or* bridging.) (3)

étrésillonnement (*n.m.*), strutting ; bracing. (4)

étrésillonner (*v.t.*), to strut ; to brace : (5)
étrésillonner les berges d'une fouille, to strut, to brace, the banks of a cut. (6)

étrier (*n.m.*), stirrup ; U bolt ; strap ; strap-bolt ; stirrup-strap ; stirrup-bolt ; stirrup-piece ; bow shackle. (7)

étrier (d'une bielle) (*n.m.*), strap (of a con-necting-rod). (8)

étrier (d'une balance de précision) (*n.m.*), link (of a balance of precision). (9)

étrier (de tour, d'étau-limeur, de raboteuse, ou analogue) (*n.m.*), tool-post, tool-stock, (of lathe, of shaper, of planer, or the like). (10 *ou* 11 *ou* 12 *ou* 13)

étrier à touret (*m.*), swivel-stirrup. (14)

étrier d'échafaudage (*m.*), cradle-iron ; cradle-stirrup. (15)

étriqué, -e (*adj.*), scanty ; narrow : (16)
chantier étriqué (*m.*), scanty working-place. (17)

étroit, -e (*adj.*), narrow : (18)
vallée étroite (*f.*), narrow valley. **(19)**

étroitement (*adv.*), narrowly. (20)

étroitesse (*n.f.*), narrowness : (21)
étroitesse d'un passage, narrowness of a passage. (22)

étude (application d'esprit pour apprendre ou approfondir) (*n.f.*), study : (23)
étude des conditions actuelles, study of existing conditions. (24)

étude (travaux qui précèdent l'exécution d'un projet) (*n.f.*), survey ; surveying : (25)
étude d'un chemin de fer, d'un canal, survey of a railway, of a canal. (26 *ou* 27)
étude de l'emplacement d'une usine, survey of site for works. (28)
étude définitive, final survey. (29)
étude géologique des terrains traversés (Sondage & Mines), geological survey of strata passed through. (30)
étude hydrographique, hydrographic survey. (31)
étude préliminaire, preliminary survey. (32)
étude sur le terrain, field-survey. (33)

étudier (*v.t.*), to study : (34)
étudier un problème, to study a problem. (35)

étudier (faire des travaux pour l'exécution d'un projet) (*v.t.*), to survey : (36)
piqueter la direction générale du tracé à étudier, to stake out the general direction of the alignment to be surveyed. (37)

étui (*n.m.*), case. (38)

étui de cartouche (*m.*), cartridge-case. (39)

étui de mathématique (*m.*), case of mathematical instruments. (40)

étui en cuir à courroie *ou* **étui en cuir à courroie, pour porter en bandoulière** (*m.*), leather case with sling strap ; leather sling case. (41)

étuvage *ou* **étuvement** (*n.m.*), stoving ; baking. (42)

étuve (*n.f.*), stove ; oven ; drying-house. (43)

étuve à chauffer l'air (*f.*), hot-air stove. (44)

étuve à noyaux (Fonderie) (*f.*), core-oven ; core-stove. (45)

étuve de séchage (*f.*), drying-stove ; drying-oven. (46)

étuver (*v.t.*), to stove ; to bake : (47)
étuver des noyaux, des moules (Fonderie), to stove, to bake, cores, molds. (48 *ou* 49)

eucaïrite (Minéral) (*n.f.*), eucairite ; eukairite. (50)

eucalyptus (*n.m.*), eucalyptus. (51)

euchroïte (Minéral) (*n.f.*), euchroite. (52)

euclase (Minéral) (*n.f.*), euclase. (53)

eucolite (Minéral) (*n.f.*), eucolite. (54)

eudialyte (Minéral) (*n.f.*), eudialyte. (55)

eudiomètre (Chim.) (*n.m.*), eudiometer. (56)

eudiométrie (*n.f.*), eudiometry. (57)

eudiométrique (*adj.*), eudiometric ; eudiometrical : (58)
analyse eudiométrique (*f.*), eudiometrical analysis. (59)

eudiométriquement (*adv.*), eudiometrically. (60)

eudnophite (Minéral) (*n.f.*), eudnophite. (61)

eukairite (Minéral) (*n.f.*), eukairite ; eucairite. (62)

eulysite (Pétrol.) (*n.f.*), eulysite. (63)

eulytine (Minéral) (*n.f.*), eulytite ; eulytine. **(64)**

euosmite (Minéral) (*n.f.*), euosmite. (65)

euphotide (Pétrol.) (*n.f.*), euphotide. (66)

eurite (Pétrol.) (*n.f.*), eurite. (67)

euritique (*adj.*), euritic : (68)
roches euritiques (*f.pl.*), euritic rocks. (69)

eustatique (Géol.) (*adj.*), eustatic. (70)

eusynchite (Minéral) (*n.f.*), eusynchite. (71)

eutaxite (Pétrol.) (*n.f.*), eutaxite. (72)

euxénite (Minéral) (*n.f.*), euxenite. (73)

évacuation (*n.f.*), evacuation ; clearing out ; removal ; withdrawal ; discharge : (74)
évacuation du minerai hors des chantiers, removal, withdrawal, of the ore from the stopes. (75)

évacuer (faire sortir ; faire écouler) (*v.t.*), to evacuate ; to clear out ; to remove ; to withdraw ; to carry off ; to drain off : (76)
évacuer les tailings, to evacuate the tailings. (77)
une travée étant foncée, avec ou sans boisage provisoire, les mineurs évacuent le puits, qui est livré aux maçons, a section being sunk, with or without temporary timbering, the miners evacuate the shaft, which is given over to the masons. (78)
évacuer l'eau d'un fossé au moyen d'un canal, to carry off, to drain off, the water from a ditch by means of a channel. (79)

évaluation (*n.f.*), valuation ; valuing ; estimate : (80)
évaluation de diamants, valuation of diamonds ; valuing diamonds. (81)
évaluation prudente du minerai en réserve, conservative estimate of the ore in reserve. (82)

évaluer (*v.t.*), to value ; to make a valuation of ; to estimate the value of ; to estimate : (83)
évaluer le minerai actuellement en réserve, to value, to estimate the value of, to make a valuation of, the ore actually in reserve. (84)
évaluer la production des puits à pétrole, to estimate the production of the oil-wells. (85)

évaporation (*n.f.*), evaporation ; evaporating : (86)

évaporation de l'eau, evaporation of water. (1)

évaporer (*v.t.*), to evaporate : (2)

évaporer de l'eau salée pour en précipiter le sel, to evaporate brine to precipitate the salt. (3)

évaporer à siccité (*ou* à sec) (Chim.), to evaporate to dryness. (4)

évaporer (**s'**) (*v.r.*), to evaporate : (5)

l'eau et toute sorte de liquide s'évaporent naturellement, soit par la seule action de l'air, soit par la chaleur du soleil (*f.*), water and every kind of liquid evaporate naturally, either under the action of the air alone, or under the action of the heat of the sun. (6)

évasé, -e (*adj.*), bell-mouthed ; bell-mouth (*used as adj.*) ; wide-mouthed ; flaring ; évasé ; rounded-approach (*used as adj.*) : (7)

trou évasé (*m.*), bell-mouthed hole ; bell-mouth hole ; wide-mouthed hole. (8)

la partie évasée d'un entonnoir, the flaring part of a funnel. (9)

cheminée évasée (d'un ventilateur de mine) (*f.*), évasé chimney (of a mine-fan). (10)

orifice évasé (Hydraul.) (*m.*), rounded-approach orifice. (11)

évasement (*n.m.*), bellmouth ; widening at the mouth ; flaring ; flaring enlargement ; flare : (12)

une fraisure est un évasement pratiqué à l'orifice d'un trou destiné à recevoir une vis ou autre objet de forme analogue, a counter-sink is a flare made at the mouth of a hole to receive a screw or other object of like form. (13)

évaser (*v.t.*), to bellmouth ; to widen at the mouth ; to flare : (14)

évaser un trou, to bellmouth a hole ; to widen a hole at the mouth. (15)

évaser l'orifice du trou de coulée d'un moule, to flare the mouth of the pouring-gate of a mould. (16)

évaser (**s'**) (*v.r.*), to flare ; to widen at the mouth. (17)

événement (*n.m.*), event ; occurrence ; incident : (18)

un événement dangereux, a dangerous occurrence. (19)

évent (air libre ; grand air) (*n.m.*), open air : (20)

opération qui se fait à l'évent (*f.*), operation which takes place in the open air. (21)

évent (canal pour renouveler l'air) (*n.m.*), vent ; vent-hole ; air-vent ; air-hole. (22)

évent (orifice créé à la partie supérieure des moules) (Fonderie) (*n.m.*), riser ; rising-head : (23)

les évents servent à la sortie des gaz chassés par la fonte, risers serve as outlets for the gases driven out by the molten metal. (24)

évent (partie métallique qui reste attenante après la pièce coulée) (*n.m.*), riser ; rising-head : (25)

l'évent que l'on coupe après coulée, the riser that is cut off after casting. (26)

évent (broche d'évent) (Fonderie) (*n.m.*), riser-pin ; riser-stick. (27)

éventail (Géol.) (*n.m.*), fan. (28)

éventail (**en**), fan-shaped : (29)

structure en éventail (Géol.) (*f.*), fan-shaped structure. (30)

éventail composé (Géol.) (*m.*), anticlinorium ; anticlinore. (31)

éventail composé inverse (Géol.) (*m.*), syncli-norium ; synclinore. (32)

éventer une carrière, to open up a quarry. (33)

évidage *ou* **évidement** (*n.m.*), hollowing out ; chambering ; scooping out ; cutting away ; channeling out ; grooving ; fluting. (34). See the verb for examples.

évidence (*n.f.*), evidence : (35)

évidence de minéralisation à la surface, evidence of mineralization on the surface. (36)

évident, -e (*adj.*), evident. (37)

évider (*v.t.*), to hollow out ; to chamber ; to scoop out ; to cut away ; to channel out ; to groove ; to flute : (38)

évider une pierre pour en faire une auge, to hollow out a stone to make a trough. (39)

excentrique évidé (*m.*), chambered eccentric. (40)

contre-pointe (de tour) évidée à l'avant (*f.*), tailstock (of lathe) cut away in front. (41)

évider le corps d'une bielle, to channel out the body of a connecting-rod. (42)

évider des alésoirs, des tarauds, des mèches hélicoïdales, to groove, to flute, reamers, taps, twist-drills. (43 *ou* 44 *ou* 45)

évite-molettes (Mines) (*n.m.pl.* & *sing.*), over-winding-gear ; overwind-gear ; overwinder ; catcher ; automatic apparatus to prevent overwinding. (46)

évite-molettes à coincement des guides (*m.pl.*), grip overwind-gear ; overwinding-gear with grips clutching the guides. (47)

évite-molettes à déclic d'attelage (*m.pl.*), trip overwind-gear. (48)

évite-molettes modérateurs de vitesse (*m.pl.*), overspeed and overwinding-gear ; over-winder and overspeeder. (49)

évitement (Ch. de f.) (*n.m.*), passing-place ; passing-track ; turnout ; pass-by ; passing ; shunting-loop. (50)

éviter (*v.t.*), to avoid ; to prevent : (51)

éviter l'arrêt des travaux, to avoid stoppage of work. (52)

éviter un obstacle, to avoid an obstacle. (53)

éviter un accident, to prevent an accident. (54)

évolution (*n.f.*), evolution : (55)

évolution des roches, evolution of rocks. (56)

ex-radio (Chim.) (*n.m.*), exradio. (57)

exact, -e (*adj.*), exact ; correct ; accurate : (58)

proportion exacte (*f.*), exact proportion ; correct proportion. (59)

carte exacte (*f.*), accurate map. (60)

exactement (*adv.*), exactly ; correctly ; accurately. (61)

exactitude (*n.f.*), exactness ; exactitude ; correct-ness ; accuracy. (62)

exagération (*n.f.*), exaggeration. (63)

exagération de pose (Photogr.) (*f.*), overexposure. (64)

exagération des courbes (Ch. de f.) (*f.*), sharpness of curves. (65)

exagérer (*v.t.*), to exaggerate : (66)

exagérer une difficulté, to exaggerate a difficulty. (67)

examen (*n.m.*), examination ; investigation : (68)

examen microscopique, microscopic examina-tion. (69)

examinateur (pers.) (*n.m.*), examiner. (70)

examiner (*v.t.*), to examine ; to investigate ; to look into ; to inquire into : (71)

examiner les retours d'air d'un quartier (Mines), to examine the air-returns from a district. (1)

examiner une question, to look into, to inquire into, to investigate, a matter. (2)

examiner à nouveau, to reexamine. (3)

excavateur (*n.m.*), excavator, digging-machine; digger. (4)

excavateur à air comprimé (*m.*), pneumatic excavator. (5)

excavateur à godets (*m.*), bucket-excavator. (6)

excavateur à section entière (Mines) (*m.*), full-cut heading-machine. (7)

excavateur à vapeur (*m.*), steam-excavator; steam-digger; steam-shovel; steam-navvy; navvy. (8)

excavation (action) (*n.f.*), excavation; excavating; digging: (9)

excavation d'un puits, excavating a shaft. (10)

excavation à la main, excavating by hand; digging by hand. (11)

excavation (fouille) (*n.f.*), excavation; pit; cut; cutting: (12)

les cavernes sont des excavations naturelles (*f.pl.*), caverns are natural excavations. (13)

une excavation à la main, a hand-pit. (14)

excaver (*v.t.*), to excavate; to cut: (15)

excaver un tunnel, to excavate a tunnel. (16)

excaver une recette (Mines), to cut a station (*or* a plat). (17)

excédent (*n.m.*), surplus; excess: (18)

excédent d'eau, surplus water. (19)

excéder (*v.t.*), to exceed: (20)

dépense qui excède la recette (*f.*), expenditure which exceeds the receipts. (21)

excellite (Explosif) (*n.f.*), excellite. (22)

excentration (Tournage) (*n.f.*), setting over; throwing off centre. (23)

excentrer (*v.t.*), to set over; to throw off centre: (24)

excentrer la contre-pointe d'un tour, to set over the tailstock of a lathe; to throw the loose head of a lathe off centre. (25)

excentrer (s') (*v.r.*), to be set over; to be thrown off centre: (26)

contre-pointe pouvant s'excentrer pour tourner conique (*f.*), tailstock which can be set over for taper turning. (27)

excentricité (*n.f.*), eccentricity; throw: (28)

excentricité de l'excentrique, eccentricity, throw, of the eccentric. (29)

excentrique (*adj.*), eccentric: (30)

mandrin excentrique (*m.*), eccentric chuck. (31)

excentrique (*n.m.*), eccentric; cam. (32) See also **came**.

excentrique (mandrin excentrique) (*n.m.*), eccentric chuck. (33)

excentrique à cadre (*m.*), cam and yoke. (34)

excentrique à collier (*m.*), eccentric and strap. (35)

excentrique à développement (*m.*), involute cam. (36)

excentrique à galets (*m.*), cam and followers. (37)

excentrique de marche en arrière (locomotive) (*m.*), back eccentric; backward eccentric; back-up eccentric; reverse-eccentric. (38)

excentrique de marche en avant (locomotive) (*m.*), fore eccentric; forward eccentric; go-ahead eccentric. (39)

excentrique du tiroir *ou* **excentrique de commande du tiroir** (*m.*), valve-eccentric. (40)

excentrique en cœur (*m.*), heart cam; heart-shaped cam; heart. (41)

excentrique en triangle *ou* **excentrique triangulaire** (*m.*), triangular cam. (42)

excentrique évidé (*m.*), chambered eccentric. (43)

excentrique fictif (*m.*), equivalent eccentric. (44)

excentriquement (*adv.*), eccentrically. (45)

exception (*n.f.*), exception: (46)

à de rares exceptions près *ou* à bien peu d'exceptions près *ou* sauf de rares exceptions, with very few exceptions. (47)

exceptionnel, -elle (*adj.*), exceptional; unusual: (48)

nature exceptionnelle du minerai (*f.*), unusual nature of the ore. (49)

exceptionnellement (*adv.*), exceptionally; unusually. (50)

excès (*n.m.*), excess. (51)

excès de pose (Photogr.) (*m.*), overexposure. (52)

excès de pression (*m.*), excess pressure; over-pressure. (53)

excessif, -ive (*adj.*), excessive. (54)

excessivement (*adv.*), excessively. (55)

excitateur (Élec.) (*n.m.*), exciter (a static exciter). (56) Cf. **excitatrice**.

excitation (Élec.) (*n.f.*), excitation: (57)

excitation des dynamos, excitation of dynamos. (58)

excitation compound (*f.*), compound excitation. (59)

excitation en dérivation *ou* **excitation en shunt** (*f.*), shunt excitation. (60)

excitation en série (*f.*), series excitation. (61)

excitation séparée *ou* **excitation indépendante** (*f.*), separate excitation; independent excitation. (62)

excitatrice (Élec.) (*n.f.*), exciter (an exciting dynamo). (63). Cf. **excitateur**.

exciter (Élec.) (*v.t.*), to excite: (64)

exciter une dynamo, to excite a dynamo. (65)

exciter (s') (Élec.) (*v.r.*), to excite: (66)

un alternateur ne peut pas s'exciter lui-même, puisqu'il produit du courant alternatif et que l'excitation exige du courant continu, an alternator cannot excite itself, seeing that it produces alternating current and that excitation requires continuous current. (67)

excroissance (*n.f.*), excrescence. (68)

excursion (*n.f.*), trip; excursion; tour: (69)

excursion de prospection, prospecting tour. (70)

exécutable (*adj.*), workable; feasible. (71)

exécuter (*v.t.*), to execute; to carry out; to perform; to do: (72)

exécuter des travaux obligatoires, to execute obligatory work. (73)

exécuter un projet, to carry out a project. (74)

exécuter beaucoup de travaux sur une propriété, to do a lot of work on a property. (75)

exemplaire (*n.m.*), example; specimen: (76)

un bel exemplaire d'un genre de minéraux, a fine specimen of a group of minerals. (77)

exempt, -e (*adj.*), exempt; free: (78)

exempt (-e) de tous impôts, exempt from all taxes; free from (*or* of) all taxes. (79)

endroit exempt de poussières (*m.*), place free from dust. (80)

les bateaux en bois ne sont jamais complètement exempts des voies d'eau (*m.pl.*), wooden boats are never completely free from leaks. (81)

exempter (*v.t.*), to exempt; to free: (1)
exempter une compagnie de certaines res-
-trictions, to exempt a company from
certain restrictions. (2)
exemption (*n.f.*), exemption: (3)
exemption d'exploiter une concession, exemp-
-tion from working a claim. (4)
exercé, -e (expérimenté) (*adj.*), trained; skilled:
(5)
hommes exercés pour les travaux de sauvetage
(*m.pl.*), trained men for rescue-work. (6)
opérateur exercé (*m.*), skilled operator. (7)
exercer (*v.t.*), to exercise; to act; to exert:
(8)
exercer un droit, to exercise a right. (9)
exercer les fonctions de boutefeu sans être
titulaire d'un diplôme, to act as shotfirer
without having a certificate. (10)
exercer une pression (sur), to exert a pressure
(on); to bring pressure to bear (upon). (11)
exercices de sauvetage (*m.pl.*), fire-drill. (12)
exfoliation (*n.f.*), exfoliation: (13)
exfoliation d'ardoises, exfoliation of slates.
(14)
exfolier (*v.t.*), to exfoliate: (15)
exfolier une roche, to exfoliate a rock. (16)
exfolier (s') (*v.r.*), to exfoliate: (17)
minéral qui, placé sur un charbon ardent,
devient blanc et s'exfolie (*m.*), mineral
which, placed on a hot coal, becomes white
and exfoliates. (18)
exhalaison (*n.f.*), exhalation: (19)
exhalaisons des marécages, exhalations from
swamps. (20)
exhaure (*n.f.*), pumping; pumping out; un-
-watering; dewatering: (21)
exhaure des avaleresses, unwatering shafts in
water-bearing ground. (22)
exhaussement (*n.m.*), raising: (23)
exhaussement d'un mur, raising a wall. (24)
exhausser (*v.t.*), to raise: (25)
exhausser une maison d'un étage, to raise a
house by a story. (26)
exhausteur (*n.m.*), exhauster. (27)
exhaustion (*n.f.*), exhaustion: (28)
exhaustion d'un gaz contenu dans un récipient,
exhaustion of a gas contained in a receiver.
(29)
exigence (nécessité) (*n.f.*), requirement; need;
demand. (30)
exiger (*v.t.*), to require; to need; to exact;
to demand: (31)
forgeage qui exige plusieurs chaudes (*f.*),
forging which requires several heats. (32)
existence (*n.f.*), existence; presence: (33)
existence d'un gîte établie de façon indubitable,
presence of a seam proved beyond a doubt.
(34)
existence en magasin (*f.*), stock in hand; stock
on hand. (35)
exitèle (Minéral) (*n.f.*), valentinite; antimony-
bloom. (36)
exogène (Géol.) (*adj.*), exogenous: (37)
roches exogènes (*f.pl.*), exogenous rocks. (38)
exomorphe (Géol.), (*adj.*), exomorphic. (39)
exomorphisme (*n.m.*), exomorphism. (40)
exosmose (Phys.) (*n.f.*), exosmose; exosmosis.
(41)
exosmotique (*adj.*), exosmotic; exosmic. (42)
exothermique *ou* **exotherme** (*adj.*), exothermic;
exothermous. (43)
exotique (*adj.*), exotic. (44)

expansibilité (*n.f.*), expansibility. (45)
expansible (*adj.*), expansible. (46)
expansif, -ive (*adj.*), expansive. (47)
expansion (*n.f.*), expansion: (48)
expansion des gaz, de la vapeur d'eau, ex-
-pansion of gases, of steam. (49 *ou* 50)
expédier (envoyer à destination) (*v.t.*), to despatch;
to dispatch; to send; to forward; to ship:
(51)
expédier des marchandises, to send, to despatch,
to forward, goods. (52)
expédier du minerai en Amérique, to ship ore
to America. (53)
expédier en vrac, to ship in bulk. (54)
expédier (faire promptement) (*v.t.*), to expedite;
to despatch; to dispatch: (55)
expédier des affaires, to expedite matters;
to despatch business. (56)
expéditeur (pers.) (*n.m.*), sender; consignor;
forwarder; shipper. (57)
expédition (action d'envoyer à destination) (*n.f.*),
despatch; dispatch; despatching; sending;
forwarding. (58)
expédition (chose expédiée) (*n.f.*), consignment;
shipment. (59)
expédition (excursion) (*n.f.*), expedition; tour;
trip: (60)
expédition explorative (*ou* exploratrice), ex-
-plorative (*or* exploratory) expedition (*or*
tour). (61)
expérience (habileté) (*n.f.*), experience; skill.
(62)
expérience (essai) (*n.f.*), experiment; test: (63)
expériences de laboratoire, laboratory ex-
-periments. (64)
expériences sur l'inflammabilité des poussières
charbonneuses, experiments on the in-
-flammability of coal-dust. (65)
expérimental, -e, -aux (*adj.*), experimental: (66)
détermination expérimentale du centre de
gravité des corps solides (*f.*), experimental
determination of the centre of gravity of
solid bodies. (67)
physique expérimentale (*f.*), experimental
physics. (68)
expérimentalement (*adv.*), experimentally. (69)
expérimentation (essai d'application) (*n.f.*),
experiment. (70)
expérimenté, -e (*adj.*), skilled; experienced:
(71)
ouvrier expérimenté (*m.*), skilled workman;
experienced man. (72)
expérimenter (*v.t.*), to experiment upon; to try;
to test: (73)
expérimenter un gaz, to experiment upon a
gas. (74)
expérimenter un nouveau procédé, to test,
to try, a new process. (75)
expérimenter (*v.i.*), to experiment; to make
experiments: (76)
expérimenter sur une plus grande échelle,
to experiment on a larger scale. (77)
expert, -e (*adj.*), expert. (78)
expert (pers.) (*n.m.*), expert: (79)
expert diplômé, qualified expert. (80)
expert en pétroles, petroleum expert; oil
expert. (81)
expert minier, mining-expert. (82)
expertise (visite et opération des experts) (*n.f.*),
survey; examination by an expert. (83)
expertise (rapport d'expert) (*n.f.*), expert's report.
(84)

expiration (*n.f.*), expiration: (1)
l'expiration d'un bail, the expiration of a
lease. (2)

expirer (*v.i.*), to expire. (3)

explication (*n.f.*), explanation. (4)

expliquer (*v.t.*), to explain: (5)
expliquer le mode d'opérer, to explain the
mode of operation. (6)

exploitabilité (*n.f.*), workability; workableness;
gettability; gettableness; workable nature;
payable nature: (7)
exploitabilité d'un gisement de houille, work-
-ability, gettability, workable nature, payable
nature, of a coal-deposit. (8)

exploitable (Mines, Placers, etc.) (*adj.*),
workable; mineable; gettable; exploit-
-able; payable; pay (*used as adj.*): (9)
'houille exploitable (*f.*), workable coal; mine-
-able coal; gettable coal. (10)
gravier exploitable (Placers) (*m.*), payable
gravel; pay-gravel. (11)

exploitant (d'une mine) (pers.) (*n.m.*), owner or
agent, operator, (of a mine): (12)
le nom du directeur est porté par l'exploitant
à la connaissance de l'inspecteur des mines,
notice of the name of the manager shall be
sent by the owner or agent to the inspector
of mines. (13)

exploitation (action d'exploiter des biens) (*n.f.*),
working; work; operation; exploitation:
(14)
l'exploitation d'un brevet, working a patent.
(15)
l'exploitation d'une forêt, the exploitation of
a forest. (16)
exploitation en grand, working, work, operation,
exploitation, on a large scale. (17)

exploitation (action d'exploiter des mines) (*n.f.*),
working; work; mining; getting; opera-
-tion; exploitation: (18)
exploitation d'une couche de houille déjà
reconnue par sondages, working a bed of
coal already proved by boring; mining a
coal-seam already proved by bore-holing.
(19)
exploitation pratique d'une mine de charbon,
practical working of a coal-mine. (20)
exploitation rémunératrice de minerai à basse
teneur, profitable working of low-grade ore.
(21)

exploitation (chantier, lieu où l'on exploite)
(Mines) (*n.f.*), workings; mine. (22)

exploitation (chantier placérien) (*n.f.*), workings;
diggings. (23)

exploitation à ciel ouvert *ou* **exploitation au jour**
(action) (*f.*), open-cast mining; open-cut
mining; openwork mining; surface-mining;
surface-working; paddocking. (24)

exploitation à ciel ouvert *ou* **exploitation au jour**
(chantier) (*f.*), open-cast; open-cut; open-
cast workings; openwork; surface-mine;
surface-working; paddock. (25)

exploitation à ciel ouvert des gîtes minéraux (*f.*),
surface-working of ore deposits. (26)

exploitation à ciel ouvert et à bras d'homme (*f.*),
surface-mining by hand. (27)

exploitation à ciel ouvert par pelles à vapeur (*f.*),
open-cut mining by steam-shovels. (28)

exploitation aurifère (*f.*), gold-workings; gold-
diggings; gold-mine. (29)

exploitation d'alluvions (*f.*), alluvial diggings;
alluvial workings. (30)

exploitation d'alluvions diamantifères (*f.*), al-
-luvial-diamond diggings. (31)

exploitation de l'étain alluvionnaire (*f.*), alluvial-
tin mining. (32)

exploitation de l'or (*f.*), gold-mining. (33)

exploitation de la houille (*f.*), coal-mining;
getting coal. (34)

exploitation de pétrole (*f.*), petroleum-workings.
(35)

exploitation de placers (*f.*), placer-workings;
placer-diggings; placer-mine. (36)

exploitation des alluvions (*f.*), alluvial digging;
alluvial mining; working alluvials. (37)

exploitation des alluvions au moyen de dragues
(*f.*), dredge-mining. (38)

exploitation des alluvions aurifères (*f.*), gold-
digging. (39)

exploitation des alluvions gelées (*f.*), working
the frozen alluvia. (40)

**exploitation des alluvions immergées au fond de
cours d'eau** (*f.*), river-mining. (41)

exploitation des bas métaux (*f.*), base-metal
mining. (42)

exploitation des bois (*f.*), logging. (43)

exploitation des carrières (*f.*), quarrying. (44)

exploitation des carrières d'ardoise (*f.*), slate-
quarrying. (45)

exploitation des carrières d'argile (*f.*), clay-
mining. (46)

exploitation des carrières par galeries souterraines
(*f.*), working quarries by tunnels. (47)

exploitation des chenaux souterrains (anciens
chenaux de rivières remplis d'alluvion
aurifère) (*f.*), deep-lead mining. (48)

exploitation des couches (*f.*), bed-mining. (49)

exploitation des filons (*f.*), lode-mining; vein-
mining. (50)

exploitation des gisements (*f.*), bed-mining. (51)

exploitation des gisements pétrolifères (*f.*), oil-
mining; petroleum-mining. (52)

exploitation des minerais (*f.*), ore-mining. (53)

exploitation des minerais à basse teneur (*f.*),
low-grade ore-mining. (54)

exploitation des mines (*f.*), mining. (55)

exploitation des mines à ciel ouvert (*f.*), open-
cast mining; open-cut mining. (56)

exploitation des mines d'étain (*f.*), tin-mining.
(57)

exploitation des mines d'or (*f.*), gold-mining.
(58)

exploitation des mines de diamants (*f.*), diamond-
mining. (59)

exploitation des pierres précieuses (*f.*), gem-
mining. (60)

exploitation des placers (*f.*), placer-mining;
placer-working; placer-work. (61)

exploitation des placers stannifères (*f.*), placer
tin-mining; alluvial tin-mining. (62)

exploitation des puits à pétrole (*f.*), oil-mining;
petroleum-mining. (63)

exploitation des sables de mer métallifères (*f.*),
beach-combing. (64)

exploitation du charbon (*f.*), coal-mining;
getting coal. (65)

exploitation du pétrole (*f.*), getting petroleum;
oil-mining; petroleum-mining. (66)

exploitation du sel (*f.*), salt-mining. (67)

exploitation en avant et en retour (*f.*), advancing
and retreating mining. (68)

exploitation en battant en retraite (*f.*). Same as
exploitation en retour.

exploitation en gradins (Mines) (*f.*), stoping. (69)

exploitation en gradins droits (*f.*), underhand stoping ; bottom-stoping. (1)

exploitation en gradins renversés (*f.*), overhand stoping ; overhead stoping ; overstoping ; back-stoping. (2)

exploitation en profondeur (*f.*), deep mining ; deep working. (3)

exploitation en retour *ou* exploitation en retraite (*f.*), working on the retreat ; working home ; working homewards ; homewards working ; mining retreating. (4)

exploitation filonienne (*f.*), lode-mining ; vein-mining. (5)

exploitation hydraulique (*f.*), hydraulic mining ; hydraulicking ; piping ; spatter-work : (6) exploitation hydraulique de l'or, hydraulic gold-mining ; hydraulicking for gold. (7)

exploitation minière (*f.*), mining. (8)

exploitation minière des métaux (*f.*), metal-mining. (9)

exploitation minière des minéraux (*f.*), mineral-mining. (10)

exploitation minière du quartz (*f.*), quartz-mining ; quartz-reefing. (11)

exploitation par cantonnement (Ch. de f.) (*f.*), block system ; block-signal system. (12)

exploitation par chambres descendantes sous voûtes (*f.*), underground glory-hole method ; underground milling. (13)

exploitation par chambres isolées (Mines) (*f.*), panel-work ; panel-working ; panelling. (14)

exploitation par éboulement (*f.*), caving system of mining ; subsidence-of-the-roof method. (15)

exploitation par la méthode directe (Mines) (*f.*), working out ; working outwards. (16)

exploitation par la méthode rétrograde (*f.*), working on the retreat ; mining retreating ; working home ; working homewards ; homewards working. (17)

exploitation par ouvrages en travers (*f.*), cross-cutting. (18)

exploitation par quartiers indépendants (Mines) (*f.*), panel work ; panel working ; panelling. (19)

exploitation placérienne (action) (*f.*), placer-working ; placer-mining ; placer-work. (20)

exploitation placérienne (lieu) (*f.*), placer-work--ings ; placer-diggings ; placer-mine. (21)

exploitation profonde (*f.*), deep mining. (22)

exploitation souterraine (action) (*f.*), underground work ; closed work ; working underground. (23)

exploitation souterraine (chantier) (*f.*), under--ground workings. (24)

exploitation souterraine des gisements aurifères *ou* exploitation souterraine des graviers cimentés (*f.*), drift-mining. (25)

exploiter (mettre en œuvre ; faire valoir) (*v.t.*), to work ; to run ; to operate ; to exploit : (26) exploiter le moulin à profit, à perte, to work, to run, to operate, the mill at a profit, at a loss. (27 *ou* 28)

exploiter un brevet, to work a patent. (29)

exploiter une fonderie à plein rendement, to operate a smelter at full capacity. (30)

exploiter une forêt, to exploit a forest. (31)

exploiter une propriété en location, to work a property under lease. (32)

exploiter (Mines) (*v.t.*), to work ; to mine ; to get ; to operate ; to exploit : (33) exploiter à ciel ouvert, to work open-cast ; to paddock. (34)

exploiter en gradins, to stope. (35)

exploiter la houille, to mine coal ; to mine for coal ; to get coal. (36)

exploiter le pétrole, to get petroleum ; to mine for oil ; to exploit for petroleum. (37)

exploiter le terrain moyennant une redevance de — 0/0, to work the ground on a royalty of — %. (38)

exploiter par la méthode directe, to work out. (39)

exploiter une couche de houille déjà reconnue par sondages, to mine a bed of coal already proved by boring ; to work a coal-seam already proved by bore-holing. (40)

exploiter une mine, une carrière, to work, to exploit, to operate, a mine, a quarry. (41 *ou* 42)

explorable (*adj.*), explorable. (43)

explorateur, -trice *ou* exploratif, -ive (*adj.*), explorative ; exploratory ; exploring. (44)

explorateur (pers.) (*n.m.*), explorer. (45)

exploration (*n.f.*), exploration ; exploring : (46) exploration d'une contrée vierge, exploring a virgin country. (47) exploration d'une mine après une explosion, exploring a mine after an explosion. (48)

explorativement (*adv.*), exploratively. (49)

explorer (*v.t.*), to explore : (50) explorer un souterrain, to explore a cave. (51)

exploser (*v.i.*), to explode : (52) la dynamite explose facilement, dynamite explodes easily. (53)

exploseur (*n.m.*), blasting-machine ; firing-machine ; exploder. (54)

exploseur à manivelle (*m.*), crank blasting-machine. (55)

exploseur à poignée (*m.*), push-down blasting-machine. (56)

exploseur dynamo-électrique (*m.*), dynamo-electrical blasting-machine. (57)

exploseur électrique à basse tension (*m.*), low-tension electric blasting-machine. (58)

exploseur électrique à haute tension (*m.*), high-tension electric blasting-machine. (59)

exploseur électrostatique (*m.*), frictional blasting-machine. (60)

exploseur magnéto-électrique (*m.*), magneto blasting-machine. (61)

explosible (*adj.*), explosible ; detonable. (62)

explosif, -ive (*adj.*), explosive ; detonating : (63) onde explosive (*f.*), explosive wave. (64)

explosif (*n.m.*), explosive. (65)

explosifs à base de coton nitré *ou* explosifs au coton nitré (*m.pl.*), nitro-cotton explosives. (66)

explosifs à base de nitrate d'ammoniaque *ou* explosifs à l'azotate d'ammoniaque (*m.pl.*), ammonium-nitrate explosives. (67)

explosifs à base de nitrobenzine *ou* explosifs à la nitrobenzine (*m.pl.*), nitrobenzene explosives. (68)

explosifs à base de nitroglycérine *ou* explosifs à la nitroglycérine (*m.pl.*), nitroglycerine explosives. (69)

explosifs à base de nitronaphtaline *ou* explosifs à la nitronaphtaline (*m.pl.*), nitronaphtha--lene explosives. (70)

explosifs autorisés (*m.pl.*), permitted explosives ; permissible explosives. (71)

explosifs brisants (*m.pl.*), rending explosives ; disruptive explosives. (72)

explosifs chloratés (*m.pl.*), chlorate mixtures. (1)

explosifs-couche (*n.m.pl.*), explosives for coal-mines ; coal-mining explosives. (2)

explosifs d'amorce (*m.pl.*), priming-explosives. (3)

explosifs de grande puissance (*m.pl.*), high explosives. (4)

explosifs de mines (*m.pl.*), mining-explosives. (5)

explosifs de sûreté pour mines grisouteuses (*m.pl.*), safety-explosives for fiery mines ; flameless explosives for gassy mines. (6)

explosifs déflagrants (*m.pl.*), deflagrating explosives. (7)

explosifs détonants (*m.pl.*), shattering explosives. (8)

explosifs gelés (*m.pl.*), frozen explosives. (9)

explosifs interdits (*m.pl.*), prohibited explosives. (10)

explosifs lents (*m.pl.*), slow explosives. (11)

explosifs nitrés (*m.pl.*), nitrate mixtures. (12)

explosifs-roche (*n.m.pl.*), explosives for rock-work ; rock-work explosives. (13)

explosion (*n.f.*), explosion; detonation; report: (14)
explosion violente, violent explosion ; violent detonation ; loud report. (15)

explosion de chaudière (*f.*), boiler-explosion. (16)

explosion de grisou (*f.*), fire-damp explosion ; explosion of fire-damp. (17)

explosion de poussières de charbon (*f.*), coal-dust explosion. (18)

explosion intempestive (*ou* prématurée) d'un coup de mine (*f.*), premature explosion of a shot ; premature blast. (19)

explosion retardée (*f.*), delayed explosion. (20)

explosionner (*v.i.*), to explode. (21)

exportable (*adj.*), exportable. (22)

exportateur (pers.) (*n.m.*), exporter. (23)

exportation (*n.f.*), exportation ; export: (24)
exportations d'Angleterre en Amérique, exports from England to America. (25)

exporter (*v.t.*), to export: (26)
exporter des marchandises, to export goods. (27)

exposé (*n.m.*), statement ; account ; returns. (28)

exposer (mettre en vue) (*v.t.*), to expose : (29)
exposer de grandes réserves de minerai, to expose large ore-reserves. (30)

exposer (soumettre à l'influence de) (*v.t.*), to expose : (31)
exposer à la lumière une plaque sensible, to expose a sensitized plate to the light. (32)

exposer (mettre en péril) (*v.t.*), to expose ; to endanger ; to jeopardize ; to imperil : (33)
exposer sa vie, to expose, to endanger, to jeopardize, to imperil, one's life. (34)

exposer (orienter du côté de) (*v.t.*), to face : (35)
maison exposée au midi (*f.*), house facing south. (36)

exposition (*n.f.*), exposure : (37)
exposition à l'air, aux intempéries, exposure to the air, to the weather. (38 *ou* 39)
exposition d'une plaque photographique, ex-posure of a photographic plate. (40)

exprimer (faire sortir par la pression) (*v.t.*), to squeeze out ; to press out ; to express : (41)

exprimer le mercure de l'amalgame, to squeeze, to press, the mercury out of the amalgam ; to express the quicksilver from the amalgam. (42)

exprimer (manifester) (*v.t.*), to express : (43)
exprimer une opinion, to express an opinion. (44)

expulser (*v.t.*), to expel ; to drive out : (45)
expulser les matières volatiles, to expel the volatile matter. (46)

exsudation (*n.f.*), exudation. (47)

exsuder (*v.t.*), to exude : (48)
arbre qui exsude de la résine (*m.*), tree which exudes resin. (49)

exsuder (*v.i.*), to exude. (50)

extendeur (mandrin à arrondir les tubes de chaudières) (*n.m.*), tube-expander ; boiler-tube expander. (51)

extendeur fonctionnant à broche (*m.*), tube-expander, with round-head mandrel, operated by tommy. (52)

extendeur système Dudgeon, fonctionnant par tourne-à-gauche (*m.*), Dudgeon's pattern tube-expander, with square-head mandrel, operated by tap-wrench. (53)

extensibilité (*n.f.*), extensibility ; tensility ; tensibility : (54)
extensibilité du câble, extensibility of the rope. (55)
extensibilité du caoutchouc, tensibility of india-rubber. (56)

extensible *ou* extensile (*adj.*), extensible ; tensile : (57)
les métaux sont extensibles à divers degrés, metals are tensile to different degrees. (58)

extension (*n.f.*), extension : (59)
extension d'un bail, extension of a lease. (60)
extension d'une plaque de métal au marteau, au laminoir, expansion, spreading, of a metal plate with the hammer, in the rolling-mill. (61 *ou* 62)

extérieur, -e (*adj.*), exterior ; external ; outer ; outside. (63)

extérieur (*n.m.*), exterior ; outside : (64)
l'extérieur d'une maison, the exterior (*or* the outside) of a house. (65)

extérieurement (*adv.*), externally ; on the ex-terior ; outside ; exteriorly. (66)

externe (*adj.*), external. (67)

extincteur (*n.m.*) *ou* extincteur d'incendie, extinguisher ; fire-extinguisher. (68)

extinction (*n.f.*), extinction ; extinguishing ; putting out ; quenching : (69)
extinction d'un incendie, extinction of, extinguishing, putting out, a fire. (70)
extinction du fer chaud dans l'eau (Trempe), quenching hot iron in water (Tempering). (71)

extinction de la chaux (*f.*), slaking, slacking, lime. (72)

extinction du mercure (*f.*), flouring, sickening, of mercury. (73)

extra-courant [extra-courants *pl.*] (Élec.) (*n.m.*), extra current ; current of self-induction. (74)

extra-latéral, -e, -aux (*adj.*), extralateral. (75)

extracteur (*n.m.*), extractor. (76)

extracteur de carottes (Sondage) (*m.*), core-extractor. (77)

extraction (action de retirer) (*n.f.*), extraction ; extracting ; taking out ; winning : (78)

extraction de l'or d'une mine, de l'or de ses minerais, extraction of gold from a mine, of gold from its ores ; winning gold from a mine, gold from its ores. (1 ou 2)

extraction d'un clou, extraction of, extracting, pulling out, drawing, a nail. (3)

extraction du charbon, winning, getting, coal. (4)

extraction (remontée) (Mines) (*n.f.*), hoisting ; winding ; raising ; drawing : (5)

extraction du minerai, hoisting, winding, ore. (6)

extraction du pétrole, raising oil ; drawing petroleum. (7)

extraction à vapeur (*f.*), steam-hoisting. (8)

extraction de pierre d'une carrière (*f.*), quarrying stone. (9)

extraction du marbre (*f.*), marble-quarrying ; quarrying marble. (10)

extraction du mercure d'un minerai (*f.*), ex--traction of mercury from an ore ; mercurification of an ore. (11)

extraction électrique (*f.*), electric hoisting. (12)

extraction hydraulique (*f.*), hydraulic hoisting. (13)

extraction par l'air comprimé *ou* **extraction pneumatique** (*f.*), hoisting by compressed air ; air-hoisting ; pneumatic hoisting. (14)

extraction par moteurs animés (*f.*), hand or animal-power hoisting. (15)

extrados (d'une voûte) (*n.m.*), extrados, back, (of an arch.) (16)

extradossé, -e (*adj.*), extradosed : (17)

voûte extradossée (*f.*), extradosed arch. (18)

extraire (retirer) (*v.t.*), to extract ; to take out ; to win : (19)

extraire l'or d'une mine, to extract, to win, gold from a mine. (20)

extraire le charbon, to win, to get, coal. (21)

extraire le mercure d'un minerai, to extract mercury from an ore ; to mercurify an ore. (22)

extraire le métal des minerais, l'or du quartz, to extract, to win, metal from ores, gold from quartz. (23 ou 24)

extraire du minerai par empiétement, to take out ore under trespass. (25)

extraire un clou, to extract a nail ; to pull out a nail ; to pull a nail out ; to draw a nail. (26)

extraire (remonter) (Mines) (*v.t.*), to hoist ; to wind ; to raise ; to draw : (27)

extraire le charbon hors de la mine, to raise, to wind, to hoist, to draw, coal from the mine. (28)

extraire le pétrole, to raise oil ; to draw petroleum. (29)

extraire de la pierre d'une carrière, to quarry stone. (30)

extraire l'ardoise, to quarry slate. (31)

extraire (s') (*v.r.*), to be extracted : (32)

le sélénium s'extrait de la zorgite, selenium is extracted from zorgite. (33)

extrait, -e (Production des mines) (*p.p.*), gotten ; extracted ; raised : (34)

charbon extrait (*m.*) *ou* 'houille extraite (*f.*), mineral gotten ; mineral extracted ; mineral raised. (35)

minerai extrait (Mines métallifères) (*m.*), mineral gotten ; mineral extracted ; ore raised. (36)

extrait (abrégé d'un ouvrage plus étendu) (*n.m.*), extract : (37)

extrait certifié de l'enregistrement d'une demande en concession de mines, certified extract of the registration of an application for mining-licence. (38)

extrait d'un rapport, extract from a report. (39)

extrait (Chim.) (*n.m.*), extract. (40)

extraordinaire (*adj.*), extraordinary : (41)

rayon extraordinaire (Opt.) (*m.*), extraordinary ray. (42)

extraordinairement (*adv.*), extraordinarily. (43)

extrême (*adj.*), extreme : (44)

chaleur extrême (*f.*), extreme heat. (45)

extrêmement (*adv.*), extremely. (46)

extrémité (*n.f.*), extremity ; end ; tip : (47)

extrémité d'une ligne, extremity, end, of a line. (48)

extrémité de la came (bocard), tip of the cam (ore-stamp). (49)

extrusion (Géol.) (*n.f.*), extrusion : (50)

une extrusion de lave, an extrusion of lava. (51)

F

fabricant (pers.) (*n.m.*), maker ; manufacturer. (52)

fabrication (*n.f.*), make ; making ; manufacture ; work : (53)

marchandises de fabrication anglaise (*f.pl.*), goods of English make. (54)

fabrication directe du fer, making wrought iron direct from the ore. (55)

fabrication de l'acide sulfurique, manufacture of sulphuric acid. (56)

fabrication en série, gang work ; repetition work ; repetitive work. (57)

fabrique (*n.f.*), manufactory ; factory ; works ; mill. (58)

fabriquer (*v.t.*), to make ; to manufacture. (59)

façade (*n.f.*), front ; façade : (60)

façade d'un édifice, front, façade, of a building. (61)

façade de chaudière, boiler-front. (62)

face (*n.f.*), face ; front. (63)

face avant de la boîte à fumée (locomotive) (*f.*), smoke-arch front ; smoke-box front. (64)

face d'un marteau (*f.*), face of a hammer. (65)

face d'une dent d'engrenage (*f.*), face of a gear-tooth. (66)

face de glissement (Cristall.) (*f.*), gliding-face ; slip-face. (67)

face de groupement (hémitropie) (Cristall.) (*f.*), composition face. (68)

face du tiroir (*f.*), valve-face. (69)

faces d'un coin (*f.pl.*), faces of a wedge (*or* of a key). (70)

faces d'un cristal (*f.pl.*), faces of a crystal. (71)

faces d'un cube (*f.pl.*), faces of a cube. (72)

faces de clivage (*f.pl.*), faces of cleavage. (73)

facette (*n.f.*), facet : (74)

facettes d'un diamant, facets of a diamond. (1)

facettes (à), faceted. (2)

facetter (*v.t.*), to facet : (3)

facetter une pierre précieuse, to facet a precious stone. (4)

facies [facies *pl.*] (*n.m.*), facies : (5)

facies d'un groupe de strates, facies of a group of strata. (6)

facile (*adj.*), easy ; ready. (7)

facilement (*adv.*), easily ; readily. (8)

facilité (*n.f.*), facility ; ease ; easiness : (9)

facilités d'exploitation économique, facilities for economic working. (10)

facilités de transport par chemin de fer, railway-transport facilities. (11)

facilité de manœuvre d'une machine, ease of operation of a machine. (12)

faciliter (*v.t.*), to facilitate : (13)

ébranler un modèle pour faciliter sa sortie du moule, to rap a pattern to facilitate its delivery from the mould. (14)

façon (manière dont une chose est faite) (*n.f.*), workmanship ; make ; pattern : (15)

la façon d'un meuble, the workmanship of a piece of furniture. (16)

cliquet façon anglaise (*m.*), English pattern ratchet-brace. (17)

façon (main-d'œuvre) (*n.f.*), making ; labour : (18)

la façon coûte plus que la matière, the making costs more than the materials. (19)

payer tant pour la façon, to pay so much for the labour. (20)

façon de se comporter (*f.*), behaviour : (21)

matériaux si distincts dans leur façon de se comporter (*m.pl.*), materials so distinct in their behaviour. (22)

façon dont se comporte la fonte sous l'outil, behaviour of the iron under the tool. (23)

façonnage *ou* **façonnement** (*n.m.*), shaping ; forming. (24)

façonnage de brides (*m.*), flanging. (25)

façonner (*v.t.*), to shape ; to form : (26)

façonner un bloc de marbre, to shape a block of marble. (27)

facteur (Math.) (*n.m.*), factor. (28)

facteur d'impédance (Élec.) (*m.*), impedance-factor. (29)

facteur d'utilisation (Élec.) (*m.*), load-factor. (30)

facteur de puissance (Élec.) (*m.*), power-factor ; lag-factor. (31)

facteur de sécurité (*m.*), factor of safety ; coefficient of safety. (32)

facteur humain (*m.*), human element. (33)

factice (*adj.*), imitation ; artificial ; factitious : (34)

pierre factice (*f.*), imitation stone. (35)

facture (*n.f.*), invoice ; bill. (36)

facturer (*v.t.*), to invoice ; to bill : (37)

facturer les caisses d'emballage à leur entière valeur, to invoice the packing-cases at their full value. (38)

facturer (se) (*v.r.*), to be invoiced. (39)

se facturer une longueur de 25 m/m en plus de leur longueur réelle (se dit des prix des limes), to advance 1 inch : (40)

les limes à deux soies se facturent une longueur de 25 m/m en plus de leur longueur réelle (*f.pl.*), two-tanged files advance 1 in. (41)

facultatif, -ive (*adj.*), optional. (42)

faculté (propriété d'un corps) (*n.f.*), property ; power ; capability : (43)

l'aimant a la faculté d'attirer le fer (*m.*), the magnet has the property of attracting iron. (44)

faculté de recevoir un poli (se dit des pierres, des métaux, etc.), capability of receiving a polish. (45)

faculté hygrométrique de certains sels, hygrometric power of certain salts. (46)

faculté (Commerce & Droit) (*n.f.*), option ; right. (47)

fagot (*n.m.*), fagot ; faggot. (48)

fahlbande (Géol.) (*n.f.*), fahlband. (49)

fahlerz (Minéral) (*n.m.*), fahlerz ; fahl-ore. (50)

fahlunite (Minéral) (*n.f.*), fahlunite. (51)

fahrkunst [fahrkunst *pl.*] (Mines) (*n.m.*), man-engine ; man-machine ; mining-engine ; travelling ladderway ; movable ladder. (52)

faible (*adj.*), weak ; feeble ; low ; slow ; slight ; light ; small ; thin : (53)

chaîne trop faible pour supporter un poids (*f.*), chain too weak to support a weight. (54)

acide faible (*m.*), weak acid. (55)

une résistance faible, a feeble (*or* a weak) resistance. (56)

cliché faible (Photogr.) (*m.*), weak negative ; thin negative. (57)

faible épaisseur d'un gîte (*f.*), thinness of a seam. (58)

faible passe (Travail aux machines-outils) (*f.*), light cut. (59)

faible pourcentage d'humidité (*m.*), low percentage of moisture. (60)

faible quantité (*f.*), small quantity. (61)

faible teneur (*f.*), low grade ; low tenor. (62)

faible vitesse (*f.*), low speed ; slow speed. (63)

faibles variations de température (*f.pl.*), slight variations of temperature. (64)

faiblement (*adv.*), weakly ; feebly ; slightly : (65)

corps faiblement magnétiques (*m.pl.*), feebly magnetic substances. (66)

faiblisseur (opp. à renforçateur) (Photogr.) (*n.m.*), reducer. (67)

faille (Géol.) (*n.f.*), fault. (68)

failles à rejet compensateur (*f.pl.*), trough-and-ridge faults. (69)

faille à rejets multiples (*f.*), multithrow fault. (70)

faille complexe (*f.*), complex fault. (71)

faille conforme (*f.*), conformable fault. (72)

failles conjuguées (*f.pl.*), conjugate faults. (73)

faille contraire (*f.*), unconformable fault. (74)

faille d'effondrement (*f.*), slip-fault. (75)

faille de plongement *ou* **faille de plongée** (*f.*), dip-fault ; cross-fault. (76)

faille en direction *ou* **faille parallèle à la direction** *ou* **faille longitudinale** (*f.*), strike-fault ; longitudinal fault. (77)

faille en gradins *ou* **faille en escalier** (*f.*), step-fault ; distributive fault. (78)

faille fermée (*f.*), closed fault. (79)

faille horizontale (*f.*), horizontal fault. (80)

faille inverse (*f.*), reverse fault ; reversed fault ; overlap-fault. (81)

faille normale (*f.*), normal fault ; gravity-fault ; ordinary fault. (82)

faille oblique *ou* **faille diagonale** (*f.*), oblique fault ; cross-fault. (83)

faille ouverte *ou* **faille béante** *ou* **faille disjonctive** (*f.*), open fault. (84)

faille **périphérique** (*f.*), peripheral fault. (1)
faille **radiale** (*f.*), radial fault. (2)
faille **ramifiée** (*f.*), branched fault ; branching fault. (3)
faille **simple** (*f.*), simple fault. (4)
faille **transversale** (*f.*), transcurrent fault ; transverse thrust. (5)
faille **verticale** (*f.*), vertical fault. (6)
faillé, -e (Géol.) (*adj.*), faulted : (7)
région faillée (*f.*), faulted region. (8)
faillir (faire défaut) (*v.i.*), to fail ; to give out ; to run out : (9)
le jour commençait a faillir, daylight was beginning to fail. (10)
faire (*v.t.*), to make ; to do. (11)
faire aciérer un cheval, to rough, to roughen, to frost-nail, to caulk the shoes of, a horse. (12)
faire attention, to pay attention ; to take care ; to be careful ; to look out. (13)
faire augurer (*v.i.*), to augur : (14)
tout celà fait bien augurer de l'avenir du jeune territoire, all this augurs well for the future of the young territory. (15)
faire balancer une marche d'escalier, to dance a stair-step. (16)
faire basculer (*v.t.*), to seesaw ; to swing ; to rotate ; to tip ; to dump ; to tilt : (17)
faire basculer un convertisseur sur ses tourillons, to swing, to rotate, a converter on its trunnions. (18)
faire basculer un wagon, to tip a wagon ; to dump a car. (19)
faire bonne figure, to make a good showing. (20)
faire bouillir (*v.t.*), to boil : (21)
faire bouillir de l'eau, to boil water. (22)
faire breveter une invention, to patent an invention. (23)
faire canon (Travail aux explosifs)(*v.i.*), to blow out : (24)
la bourre est destinée à empêcher le coup de faire canon, etc., the tamping is intended to prevent the shot blowing out, etc. (25)
coup de mine ayant fait canon (*m.*), blown out shot. (26)
faire circuler de l'air frais dans les chantiers, to circulate fresh air through the working-places. (27)
faire corps (*v.i.*), to corporate : (28)
roues qui font corps avec leurs essieux (*f.pl.*), wheels which corporate with their axles. (29)
faire côte (*v.i.*), to run ashore : (30)
navire qui fait côte (*m.*), ship which runs ashore. (31)
faire crever (*v.t.*), to burst : (32)
la gelée a fait crever les tuyaux, the frost has burst the pipes. (33)
faire de grands travaux sur une propriété, to do a lot of work on a property. (34)
faire de l'eau (*v.i.*), to take in water ; to water : (35)
locomotive qui fait de l'eau (*f.*), engine which takes in water ; locomotive which waters. (36)
faire de l'instantané (Photogr.), to snap-shot ; to snap : (37)
faire de l'instantané par temps sombre, to snap-shot in dull weather. (38)
faire défaut (*v.i.*), to fail ; to run out ; to give out ; to be wanting ; to be lacking : (39)

l'approvisionnement d'eau a fait défaut (*m.*), the water-supply has failed ; the supply of water has run out (*or* has given out). (40)
étage géologique qui fait défaut en Europe (*m.*), geological stage which is wanting in Europe. (41)
sujet sur lequel les renseignements font défaut (*m.*), subject on which information is lack-ing. (42)
faire déflagrer du salpêtre, to deflagrate nitre. (43)
faire démarrer un moteur, to start an engine. (44)
faire déposer (*v.t.*), to deposit : (45)
faire déposer un métal au moyen d'une pile, to deposit a metal by means of a battery. (46)
faire déposer électriquement *ou* **faire déposer électriquement une couche sur** (*v.t.*), to electrodeposit ; to electroplate ; to plate : (47)
faire déposer électriquement sur un objet une couche de métal préalablement dissous dans un liquide, to electroplate an object with a metal ; to electrodeposit on an object a coating of metal previously dissolved in a liquid. (48)
faire dérailler (*v.t.*), to derail : (49)
l'exagération des courbes fait dérailler les trains (*f.*), sharp curves derail trains. (50)
faire dériver les eaux d'un fleuve, to divert the waters of a river. (51)
faire des affaires, to transact, to do, business. (52)
faire des économies, to economize ; to save ; to husband one's resources. (53)
faire des expériences, to make experiments ; to experiment. (54)
faire des progrès, to make progress (*or* head-way) ; to progress. (55)
faire des sondages, to bore ; to bore-hole : (56)
faire des sondages pour trouver de l'eau, to bore for water. (57)
faire des travaux pour trouver du charbon, to mine for coal. (58)
faire détoner (*v.t.*), to detonate : (59)
faire détoner de la dynamite, to detonate dynamite. (60)
faire du bois (*v.i.*), to cut timber ; to timber ; to wood : (61)
droit de faire du bois pour travaux miniers (*m.*), right to wood for mining purposes. (62)
faire du charbon (*v.i.*), to take in coal ; to coal : (63)
navire qui fait du charbon (*m.*), ship which takes in coal ; ship which coals. (64)
faire du feu. Same as **faire feu.**
faire eau (*v.i.*), to make water ; to leak : (65)
bateau qui fait eau (*m.*), boat which makes water (*or* which leaks). (66)
faire échec à, to put a check on : (67)
les conditions troublées du pays ont fait échec à la production (*f.pl.*), the disturbed con-ditions of the country have put a check on the production. (68)
faire éclater (*v.t.*). Same as **faire crever.**
faire écouler (*v.t.*), to run off ; to drain off : (69)
faire écouler l'eau d'un réservoir, to run off, to drain off, the water from a tank. (70)
faire effervescence (*v.i.*), to effervesce : (71)
les alcalis font effervescence avec les acides (*m.pl.*), alkalies effervesce with acids. (72)

faire égoutter (*v.t.*), to drain ; to drain off. (1)

faire évaporer (*v.t.*), to evaporate ; to dry off : (2)

faire évaporer le mercure, to evaporate, to dry off, the mercury. (3)

faire exécuter des travaux par entreprise, to get work done on contract. (4)

faire exploser (*v.t.*), to explode : (5)

faire exploser la poudre, to explode powder. (6)

faire explosion (*v.i.*), to explode : (7)

poudre qui fait explosion (*f.*), powder which explodes. (8)

faire face à (pourvoir à), to meet ; to keep pace with : (9)

faire face à une demande, to meet a demand. (10)

faire face au débit d'un puits à pétrole, to keep pace with the flow of an oil-well. (11)

faire face à (être placé vis-à-vis de), to face: (12)

maison qui fait face à la mer (*f.*), house which faces the sea. (13)

faire fausse route, to lose one's way ; to go astray. (14)

faire feu (*v.t.*), to strike fire : (15)

matières non susceptibles de faire feu avec le fer ou le quartz (*f.pl.*), materials not liable to strike fire with iron or quartz. (16)

faire flamber (*v.t.*), to buckle : (17)

faire flamber la tige (Sondage), to buckle the rods. (18)

faire fonctionner (*v.t.*), to run ; to drive ; to work : (19)

faire fonctionner un moteur au gaz de ville, to run an engine on town gas. (20)

faire fonctionner une dynamo par une turbine, to drive a dynamo by a turbine. (21)

faire glisser (*v.t.*), to slide ; to slip : (22)

faire glisser la réglette d'une règle à calcul, to slide the slide of a slide-rule. (23)

faire joint étanche, to make a tight joint. (24)

faire jouer la mine (Travail aux explosifs), to fire the blast. (25)

faire jouer un fusible *ou* **faire fondre un fusible** (Élec.), to blow, to blow out, to melt, a fuse : (26)

terre trop faible pour faire jouer un fusible (*f.*), earth too weak to blow out a fuse. (27)

faire l'appréciation de marchandises, to make a valuation of goods. (28)

faire l'ascension d'une montagne, to make the ascent of, to ascend, a mountain. (29)

faire l'évaluation des pertes occasionnées par un incendie, to make a valuation of the loss occasioned by a fire. (30)

faire l'impossible, to do one's very utmost. (31)

faire la fourche (*v.i.*), to fork : (32)

chemin qui fait la fourche (*m.*), road which forks. (33)

faire la tare de (Dynamométrie), to calibrate ; to scale ; to measure : (34)

faire la tare d'un ressort, to calibrate, to scale, a spring. (35)

faire la visite des chantiers d'une mine, to make an inspection of the mine-workings. (36)

faire le balancement d'une marche d'escalier, to dance a stair-stop. (37)

faire le chemin (enferrer les roches) (Carrières), to plug ; to plug off. (38)

faire le levé d'une mine, to survey a mine. (39)

faire le plein de, to fill ; to fill up : (40)

faire le plein d'un réservoir, to fill up a tank. (41)

faire le pourcentage des frais généraux, to work out the percentage of standing expenses. (42)

faire le relevé de, to plot : (43)

faire le relevé du terrain, to plot the ground. (44)

faire le service, to run ; to ply : (45)

des navires à vapeur font le service entre Marseille et Gênes (*m.pl.*), steamers run (*or* ply) between Marseilles and Genoa. (46)

faire le ventre. Same as **faire ventre.**

faire le vide, to create, to produce, a vacuum : (47)

faire le vide dans un récipient, to create, to produce, a vacuum in a receiver. (48)

faire long feu (Travail aux explosifs) (*v.i.*), to hang fire. (49)

faire machine en arrière (conduite des loco--motives), to reverse ; to back. (50)

faire osciller (*v.t.*), to oscillate : (51)

faire osciller un balancier, to oscillate a balance-beam. (52)

faire précipiter (*v.t.*), to precipitate ; to deposit ; to bed : (53)

faire précipiter de l'or en solution, to pre--cipitate, to bed, gold in solution. (54)

faire prise (se dit du ciment, etc.) (*v.i.*), to set. (55)

faire provision pour imprévus, to make pro--vision, to provide, for contingencies. (56)

faire rendre son maximum à l'outillage, to get the most out of the plant. (57)

faire ressortir (*v.t.*), to show ; to show up : (58)

compte qui fait ressortir une perte (*m.*), account which shows a loss. (59)

faire revenir (*v.t.*), to draw the temper of ; to let down the temper of ; to temper : (60)

faire revenir un burin, to draw the temper of, to let down the temper of, to temper, a chisel. (61)

pièce qui doit être trempée et revenue jaune paille (*f.*), part which should be hardened and tempered straw yellow. (62)

Cf. **revenu** *ou* **revenu après trempe.**

faire révolutionner complètement (en parlant d'un cercle d'alignement, d'un théodolite à lunette centrale) (*v.t.*), to transit : (63)

faire révolutionner complètement une lunette, to transit a telescope. (64)

faire rompre l'équilibre, to upset the equilibrium. (65)

faire saillie (*v.i.*), to jut out ; to stand out ; to project. (66)

faire sauter (*v.t.*), to blast ; to blow up : (67)

faire sauter la roche par la dynamite, to blast, to blow up, the rock with dynamite. (68)

faire table rase, to make a clean sweep. (69)

faire — tours à la minute, to run at — revo--lutions a minute. (70)

faire tous ses efforts pour augmenter la pro--duction, to make every endeavour to increase the output. (71)

faire travailler (*v.t.*), to work ; to run : (72)

faire travailler un moteur à plein rendement, to work, to run, an engine to its full capacity. (73)

faire travailler (Résistance des matériaux) (*v.t.*), to stress (Strength of materials) : (74)

dans la pratique, il est d'usage de faire travailler les câbles métalliques au sixième environ de leur résistance à la rupture, et même quelque--fois au dixième, suivant le degré de sécurité dont on veut s'entourer, in practice it is usual to stress wire ropes to about one-sixth of their breaking-strength, and sometimes even to one-tenth, according to the degree of safety adopted. (1)

faire un devis estimatif pour matériel neuf, to estimate for new plant. (2)

faire un joint, to make a joint. (3)

faire un levé *ou* **faire un levé de plans,** to make a survey ; to survey : (4)

faire un levé des terrains, to make a survey of, to survey, the land. (5)

faire un nouveau levé d'une mine, des terrains, to resurvey a mine, the ground. (6 *ou* 7)

faire un prix, to quote, to make, a price. (8)

faire un rapport, to make a report ; to report : (9)

faire un rapport sur une mine, to make a report, to report, on a mine. (10)

faire une découverte, to make a discovery. (11)

faire une prise d'eau à une rivière, to tap a river. (12)

faire une prise sur un câble électrique, to tap an electric cable. (13)

faire une révolution complète (en parlant d'un cercle d'alignement, d'un théodolite à lunette centrale) (*v.i.*), to transit : (14)

lunette qui fait une révolution complète autour de son axe horizontal (*f.*), telescope which transits about its horizontal axis. (15)

faire une tentative pour tracer le filon, to make an attempt to trace the lode. (16)

faire varier la course d'un tiroir, to vary the travel of a slide-valve. (17)

faire venir de fonte (*v.t.*), to cast ; to cast on : (18)

faire venir de fonte une console avec un palier, to cast a bracket in one piece with a plummer-block. (19)

faire ventre (*v.i.*), to belly ; to belly out ; to bulge : (20)

mur qui fait ventre (*m.*), wall which bulges. (21)

faisable (*adj.*), feasible : (22)

faisable mais pas à conseiller, feasible but hardly advisable. (23)

faisceau aimanté (*m.*), bunch of magnets. (24)

faisceau de chaleur (*m.*), beam of heat. (25)

faisceau de failles (Géol.) (*m.*), group of faults. (26)

faisceau de lampes électriques (*m.*), cluster of electric lamps. (27)

faisceau de rayons *ou* **faisceau de lumière** *ou* **faisceau lumineux** (Opt.) (*m.*), pencil of rays ; pencil of light ; luminous pencil ; beam ; beam of light. (28)

faisceau de ressorts (*m.*), nest of springs ; cluster of springs ; cluster spring. (29)

faisceau de triage (Ch. de f.) (*m.*), gridiron track ; ladder-track ; ladder. (30)

faisceau de verges, de cannes, de piquets (*m.*), bundle, sheaf, of rods, of sticks, of pickets. (31 *ou* 32 *ou* 33)

faisceau tubulaire (*m.*), nest of boiler-tubes. (34)

fait (-e) à la main, made by hand ; hand-made. (35)

fait (-e) à la mécanique, made by machine ; made by machinery ; machine-made. (36)

faîtage (*n.m.*) *ou* **faîte** (*n.m.*) (arête au haut d'un comble entre les deux versants opposés), ridge, crest, comb, (of a roof). (37)

faîtage (*n.m.*) *ou* **faîte** (*n.m.*) *ou* **faîtière** (*n.f.*) *ou quelquefois même* **faît** (*n.m.*) *ou* **faîtier** (*n.m.*) (panne faîtière), ridge-pole ; ridge-piece ; ridge-plate ; ridge-tree ; ridge-beam. (38)

faîtage (plomb ou suite de tuiles au haut d'un toit) (*n.m.*), ridge-capping ; ridge-tiles. (39)

faîte (d'un comble) (*n.m.*). See **faîtage** *ou* **faîte.**

faîte (d'un arbre) (*n.m.*), top (of a tree). (40)

faîte (d'un filon) (*n.m.*), apex (of a lode) ; ore-apex. (41)

faîte (d'une galerie de mine) (*n.m.*), roof (of a mine-level). (42)

faîtière (tuile faîtière) (*n.f.*), ridge-tile. (43)

faîtière (panne faîtière) (*n.f.*). See **faîtage.**

falaise (*n.f.*), cliff : (44)

falaise de craie, chalk cliff. (45)

falaise marine, sea-cliff. (46)

falaise (en), cliffed : (47)

rivage en falaise (*m.*), cliffed shore. (48)

falsification (*n.f.*), falsification ; fake. (49)

falsifier (*v.t.*), to falsify ; to fake. (50)

falun (Géol.) (*n.m.*), falun. (51)

famatinite (Minéral) (*n.f.*), famatinite. (52)

famille (*n.f.*), family : (53)

familles de minéraux, families of minerals. (54)

la famille de la silice, des feldspaths, the family of the silicas, of the feldspars. (55 *ou* 56)

fanal (*n.m.*), light (as in front of a locomotive). (57)

fanal de tête (*m.*), headlight. (58)

fantôme magnétique (*m.*), magnetic phantom ; magnetic fantom ; magnetic spectrum ; magnetic curves. (59)

farad (Élec.) (*n.m.*), farad. (60)

fardeau (*n.m.*), load ; burden. (61)

fardier (*n.m.*), lorry ; trolley. (62)

farine de bocard *ou* **farine minérale** (*f.*), powdered ore. (63)

farine de sondage *ou simplement* **farine** (*n.f.*) *ou* **farines** (*n.f.pl.*), bore-meal ; drilling ; drillings. (64)

farine fossile (*f.*), fossil farina ; fossil meal ; rock meal ; bergmehl. (65)

fascine (*n.f.*), fascine : (66)

revêtement en fascines (*m.*), revetment formed of fascines. (67)

fassaïte (Minéral) (*n.f.*), fassaïte ; fassite. (68)

fathom (mesure linéaire anglaise) (*n.m.*), fathom = 1·8288 metres. (69)

fatigue (*n.f.*), fatigue ; tiredness ; stress ; strain ; straining ; overstrain ; excessive duty ; heavy duty : (70)

fatigue de la voie (Ch. de f.), fatigue of the track. (71)

fatigue des métaux, fatigue of metals. (72)

fatigue du câble, fatigue of the rope ; stress, strain, on the rope. (73)

placer une poutre à travers une autre sans que l'une impose la moindre fatigue à l'autre, to place a beam across another without the one laying the slightest stress on the other. (74)

wagons très longs qui peuvent franchir sans fatigue des courbes de faible rayon (*m.pl.*), very long cars which can take sharp curves without strain. (75)

engrenages ayant de la fatigue (*m.pl.*), gear subject to heavy duty. (76)

fatiguer (*v.t.*), to fatigue ; to stress ; to strain ; to overstrain ; to prove heavy on : (77)

circulation qui fatigue la voie (*f.*), traffic which fatigues the track (*or* which proves heavy on the rails). (1)

la fibre la plus fatiguée (Résistance des matériaux), the most stressed fibre. (2)

les poulies en bois sont légères et ne fatiguent pas les transmissions (*f.pl.*), wood pulleys are light and do not stress the shafting. (3)

faujasite (Minéral) (*n.f.*), faujasite. (4)

faune (*n.f.*), fauna. (5)

faune fossile (*f.*), fossil fauna. (6)

fausérite (Minéral) (*n.f.*), fauserite. (7)

fausse (*adj. f.*). See **faux, fausse**.

faussé, -e (*adj.*), bent; deflected; buckled; out of true : (8)

redresser un essieu **faussé**, to straighten a bent axle. (9)

fausser (plier ; tordre) (*v.t.*), to bend ; to deflect ; to buckle ; to twist out of shape ; to kink out of line : (10)

fausser un essieu, une clef, to bend an axle, a key. (11 *ou* 12)

fausser (déranger ; pervertir) (*v.t.*), to derange ; to upset ; to foul : (13)

fausser une serrure, le rouage d'une machine, to derange, to upset, a lock, the wheelwork of a machine. (14 *ou* 15)

cage de mine qui **fausse** les guides (*f.*), mine-cage which fouls the guides. (16)

fausser un filet (filet de vis), to cross a thread (screw-thread). (17)

fausser (se) (se plier) (*v.r.*), to bend ; to deflect ; to buckle ; to kink out of line : (18)

rails qui se **faussent** (*m.pl.*), rails which kink out of line. (19)

fausser (se) (se déranger ; se pervertir) (*v.r.*), to become deranged ; to get out of order ; to foul : (20)

instrument qui se **fausse** par suite d'un choc violent (*m.*), instrument which becomes deranged owing to a violent shock. (21)

absence de pièces pouvant se **fausser** (*f.*), absence of parts which can get out of order. (22)

fausset (*n.m.*), spigot ; vent-peg ; vent-plug. (23)

faux, fausse (*adj.*), false : (24)

faux anticlinal (Géol.) (*m.*), false anticlinal. (25)

faux bed-rock (*m.*), false bed-rock. (26)

faux bois (*m.*), sap-wood ; alburnum. (27)

faux busc (d'écluse) (*m.*), false mitre-sill (canal-lock). (28)

fausse cannelure (Laminage) (*f.*), false pass. (29)

fausse cartouche (*f.*), dummy cartridge. (30)

faux comble (Constr.) (*m.*), false roof. (31)

fausse crémaillère (d'escalier) (*f.*), cut wall-string ; open wall-string. (32)

faux entrait (Constr.) (*m.*), collar-beam; collar-tie ; collar ; span-piece. (33)

fausse équerre (sauterelle) (*f.*), bevel ; bevel-square. (34)

fausse fenêtre (*f.*), false window ; blank window ; blind window. (35)

faux fond (*m.*), false bottom. (36)

faux-fuyant [faux-fuyants *pl.*] (*n.m.*), byway ; bypath ; by-road. (37)

fausse galène (*f.*), false galena ; pseudogalena ; mock lead ; mock ore ; blende ; zinc blende ; blackjack ; rosinjack ; jack ; lead marcasite ; sphalerite. (38)

fausse grille (*f.*), false grate. (39)

fausse languette (Charp.) (*f.*), slip-tongue ; loose tongue ; fillet ; feather tongue ; feather. (40)

faux limon (d'escalier) (Constr.) (*m.*), wall-string (staircase). (41)

fausse maille (*f.*), connecting-link ; mending-link ; repair-link. (42)

faux mur (Mines) (*m.*), false foot-wall. (43)

faux nez (pour recevoir un outil) (*m.*), false nose. (44)

fausse pièce (Fonderie) (*f.*), thickness ; thick-ness piece. (45)

faux pieu (*m.*), false pile. (46)

faux plateau pour le montage des mandrins (tour) (*m.*), back-plate for mounting chucks ; chuck-back (lathe). (47)

faux poids (*m.*) (opp. à *poids juste*), unjust weight. (48) (opp. to *just weight.*)

fausse stratification (Géol.) (*f.*), false bedding ; cross-bedding ; current-bedding ; cross-lamination ; oblique lamination ; diagonal stratification. (49)

faux synclinal (Géol.) (*m.*), false synclinal. (50)

faux toit (Mines) (*m.*), false hanging wall ; false roof. (51)

fausse topaze (*f.*), false topaz ; citrine quartz ; citrine. (52)

fausse turquoise (*f.*), bone turquoise ; odon-tolite turquoise. (53)

balance **fausse** (*f.*) (opp. à *balance juste*), unjust balance ; unjust scale. (54) (opp. to *just balance or just scale.*)

porte **fausse** (*f.*), false door ; blank door ; blind door. (55)

favorable (*adj.*), favourable ; favorable. (56)

favorable (Mines) (*adj.*), kind ; kindly : (57)

strates favorables (*f.pl.*), kind strata ; kindly strata. (58)

favorablement (*adj.*), favourably ; favorably. (59)

fayalite (Minéral) (*n.f.*), fayalite. (60)

fédération syndicale ouvrière (*f.*), trade-union ; trades-union. (61)

federerz (Minéral) (*n.m.*), feather-ore ; plumosite; heteromorphite. (62)

feeder (Élec.) (*n.m.*), feeder ; feed-wire ; feeder-cable ; feeding-conductor. (63)

feldspath (*n.m.*), feldspar ; felspar. (64)

feldspath alcalin (*m.*), alkali feldspar. (65)

feldspath argiliforme (*m.*), kaolin ; china-clay ; china-stone ; porcelain-clay. (66)

feldspath aventuriné (*m.*), aventurine feldspar ; aventurine ; sunstone. (67)

feldspath barytique (*m.*), baryta feldspar. (68)

feldspath calcaire (*m.*), lime feldspar. (69)

feldspath nacré (*m.*), argentine ; moonstone. (70)

feldspath potassique (*m.*), potash feldspar. (71)

feldspath sodico-calcique *ou* **feldspath** calcoso-dique (*m.*), soda-lime feldspar. (72)

feldspath sodique (*m.*), soda feldspar. (73)

feldspath vert (*m.*), Amazon stone ; amazonite. (74)

feldspath vitreux (*m.*), glassy feldspar ; sanidine. (75)

feldspathide (*n.m.*), feldspar proper, *as distin-guished from* feldspathoid. (76)

feldspathique (*adj.*) *ou* **feldspathiforme** (*adj.*) *ou* **feldspaths** (à), feldspathic ; felspathic ; felspathose : (77)

roches à feldspaths (*f.pl.*), feldspathic rocks. (78)

feldspathoïde (*n.m.*), feldspathoid. (79)

fêler (*v.t.*), to crack : (1)
 fêler un verre, to crack a glass. (2)
fêler (se) (*v.r.*), to crack. (3)
felsite (Pétrol.) (*n.f.*), felsite. (4)
felsitique (*adj.*), felsitic. (5)
felsophyre (Pétrol.) (*n.m.*), felsophyre ; feld-
 -sparphyre. (6)
fêlure (*n.f.*), crack. (7)
femelle (Méc.) (*adj.*), female ; internal ; interior ;
 inside : (8)
 vis femelle (*f.*), female screw ; internal screw;
 interior screw ; inside screw. (9)
fendage (*n.m.*), splitting ; cleaving : (10)
 fendage du diamant, diamond cleaving ;
 diamond splitting. (11)
fendage (d'une tête de vis) (*n.m.*), slotting, nick-
 -ing, slitting, (a screw-head). (12)
fendante (*n.f.*), slot-file ; slotting-file ; cotter-
 file ; cottering-file. (13)
fendeur (d'ardoise, de bois, etc.) (pers.) (*n.m.*),
 splitter. (14 *ou* 15)
fendeur (de diamants) (pers.) (*n.m.*), cleaver. (16)
fendillement (*n.m.*), fissuration ; cracking. (17)
fendiller (*v.t.*), to fissure ; to crack. (18)
fendiller (se) (*v.r.*), to fissure ; to crack : (19)
 l'argile desséchée se fendille (*f.*), dried clay
 cracks. (20)
fendre (*v.t.*), to split ; to cleave ; to crack : (21)
 fendre du bois avec des coins, to split wood
 with wedges. (22)
 fendre un diamant suivant ses faces de clivage,
 to cleave, to split, a diamond according to
 its cleavage-faces. (23)
 la gelée fend les pierres, frost cracks stones. (24)
fendre une tête de vis, to slot, to nick, to slit, a
 screw-head. (25)
fendre (se) (*v.r.*), to split ; to crack. (26)
fendu, -e (*adj.*), split ; cleft : (27)
 goupille fendue (*f.*), split pin. (28)
 bague fendue (*f.*), split ring. (29)
fendue (galerie débouchant au jour) (Mines) (*n.f.*),
 slope ; day-drift ; day-hole ; day-level ;
 surface-drive. (30)
fendue (cheminée) (Mines) (*n.f.*), chute-raise ;
 chimney. (31)
fenêtre (Arch.) (*n.f.*), window. (32)
fenêtre (d'un vernier) (*n.f.*), window (of a vernier).
 (33)
fenêtre (opp. à [1] **butte-témoin,** **témoin** *ou*
 butoir *ou* [2] **lambeau de recouvrement,**
 massif exotique, paquet exotique *ou* **klippe**)
 (Géol.) (*n.f.*), inlier. (34) (opp. to **outlier**.)
fenêtre à bascule *ou* **fenêtre basculante** (*f.*),
 centre-hung window ; pivot-hung window
 (pivoted horizontally). (35)
fenêtre à guillotine *ou* **fenêtre à coulisses** (*f.*),
 sash-window. (36)
fenêtre à pivot *ou* **fenêtre pivotante** (*f.*), centre-
 hung window ; pivot-hung window (pivoted
 vertically). (37)
fenêtre feinte (*f.*), blank window ; blind window ;
 false window. (38)
fenêtre glissante *ou* **fenêtre roulante** (*f.*), York-
 -shire light. (39)
fenêtre ordinaire (*f.*), French window ; case-
 -ment ; casement window ; French case-
 -ment. (40)
fenêtre ouvrant en dehors (*f.*), window opening
 outwards. (41)
féniestre (Houillères) (*n.m.*), air-door. (42)
fente (*n.f.*), crack ; rent ; cleft ; split ; slit ;
 fissure ; crevice. (43)

fente (dans une tête de vis) (*n.f.*), slot, nick,
 slit, (in a screw-head). (44)
fente (d'une équerre d'arpenteur, d'un collimateur)
 (*n.f.*), slit (of a cross-staff head, of a colli-
 -mator). (45 *ou* 46)
fente de remplissage (Géol.) (*f.*), fissure-vein. (47)
fente en échelons (Géol.) (*f.*), stepped fissure.
 (48)
fer (métal) (*n.m.*), iron. (49)
 Note.—As a trade term for rolled sections **fer**
 or **fers** is often to be translated as **bar** or
 bars, plate, etc., and sometimes as **sections,**
 as will be observed from the following
 entries.
 Compare also **fer forgé** *ou simplement* **fer.**
fer (instrument en fer, etc.) (*n.m.*), iron ; bit. (50)
fer à bandages de roues (*m.*) *ou* **fers à bandages
 de roues** (*m.pl.*), tire-bar ; tyre-bars ; tyre-
 iron. (51)
fer à barreaux de grille (*m.*) *ou* **fers à barreaux
 de grille** (*m.pl.*), (1) *pour grille de foyer.*—
 fire-bar ; fire-bars ; fire-bar iron ; grate-
 bar ; grate-bars ; furnace-bar ; furnace-
 bars ; grid-bar ; grid-bars. (52)
 (2) *pour grille de claie.*—screen-bar ; screen-
 bars. (53)
 Note.—**fers à barreaux de grille,** as a general
 trade term, may be translated as **fire and
 screen-bars,** many of the same sections being
 used for either purpose. (54)
fer à boudin (*m.*) *ou* **fers à boudin** (*m.pl.*), bulb
 plate ; bulb plates ; bulb iron. (55)
fer à boudin à patin (*m.*) *ou* **fers à boudin à
 patin** (*m.pl.*), bulb tee ; bulb tees ; bulb T
 iron. (56)
fers à bouveter (*m.pl.*), tonguing and grooving
 irons. (57)
fer à cheval (*m.*), horseshoe. (58)
fer à côte (*m.*) *ou* **fers à côte** (*m.pl.*), ribbed
 plate ; ribbed plates. (59)
fer à double T *ou* **fer à ⊏⊐** *ou* **fer à I** (*m.*). Same
 as **fer double T.**
fer à glace (pour chevaux) (*m.*), roughened
 shoe ; frost-nailed shoe ; caulked shoe ;
 sharpened shoe. (60)
fer à grains (opp. à **fer nerveux**) (*m.*), granular
 iron. (61)
fer à nervure (*m.*) *ou* **fers à nervure** (*m.pl.*),
 ribbed plate ; ribbed plates. (62)
fer à olive (*m.*) *ou* **fers à olive** (*m.pl.*), oval iron ;
 oval bar ; ovals. (63)
fer à raboter (*m.*), plane-iron ; plane-bit. (64)
fer à souder (*m.*), soldering-iron ; soldering-
 bit ; soldering-copper ; copper-bit. (65)
fer à souder à chauffage automatique (*m.*),
 self-heating soldering-bit. (66)
fer à souder au gaz (*m.*), gas soldering-bit ;
 gas-heated soldering-iron. (67)
fer à souder électrique (*m.*), electric soldering-
 bit. (68)
fer à souder fonctionnant à l'essence minérale
 (*m.*), benzoline soldering-bit. (69)
fer à T (*m.*) *ou* **fers à T** (*m.pl.*), T iron ; T bar ;
 T bars ; tee ; tees. (70)
fer à vitrage (*m.*) *ou* **fers à vitrage** (*m.pl.*),
 sash-iron ; sash-bar ; sash-bars ; sash-bar
 iron. (71)
fer à Z (*m.*) *ou* **fers à Z** (*m.pl.*), Z iron ; Z bar ;
 zed bar ; zed ; zeds ; zees. (72)
fer acérain (*m.*) *ou* **fer aciéreux** (*m.*), steely
 iron. (73)
fer affiné (*m.*), malleable iron. (74)

fer aigre (*m.*), short iron. (1)
fer alpha *ou* fer α (*m.*), alpha iron ; α iron. (2)
fer arsenical (Minéral) (*m.*), mispickel ; arseno-
-pyrite ; arsenical pyrites ; white mundic. (3)
fer au bois *ou* fer au charbon de bois (*m.*), charcoal iron. (4)
fer au coke (*m.*), coke iron. (5)
fer ballé (*m.*), balled iron. (6)
fer battu (*m.*), hammered iron. (7)
fer bêta *ou* fer β (*m.*), beta iron ; β iron. (8)
fer biseauté (*m.*), bevel-edged flat (rolled section). (9)
fer-blanc [fers-blancs *pl.*] (*n.m.*), tin-plate ; tin ; tinned iron ; tinned sheet iron ; white iron. (10)
fer brut (*m.*), muck-iron ; muck-bar ; puddle-bar. (11)
fer carbonaté *ou* fer carbonaté lithoïde (Minéral) (*m.*), blackband ; black-band ; black-band ore. (12)
fer carré (*m.*) *ou* fers carrés (*m.pl.*), square bar ; square bars ; squares. (13)
fer cassant à chaud (*m.*), hot-short iron ; red-short iron ; hot-short. (14)
fer cassant à froid (*m.*), cold-short iron. (15)
fer cavalier (*m.*) *ou* fers cavaliers (*m.pl.*), shoe-
-ing bars ; horseshoe iron ; horseshoe sections. (16)
fer chromé (Minéral) (*m.*), chrome iron ; chrome iron ore ; chromite. (17)
fer cornière (*m.*) *ou* fers cornières (*m.pl.*), angle iron ; L iron ; angles. (18)
See cornières for varieties.
fer corroyé (*m.*), fagot-iron ; faggot-iron ; fagotted iron. (19)
fer coudé (fer à souder) (*m.*), hatchet bit (solder-
-ing-iron). (20)
fer d'angle (*m.*) *ou* fers d'angle (*m.pl.*). Same as fer cornière.
fer de bouvet double (*m.*), tonguing-iron. (21)
fer de bouvet simple (*m.*), grooving-iron. (22)
fer de construction (*m.*), structural iron. (23)
fer de couleur (*m.*), hot-short iron ; red-short iron ; hot-short. (24)
fer de forge (*m.*), wrought iron of smithing quality ; forging-iron. (25)
fer de guillaume (*m.*), rabbet-iron ; rebate-iron. (26)
fers de petit échantillon *ou* fers de faible échantillon (*m.pl.*), light-section iron. (27)
fer de rabot (*m.*), plane-iron ; plane-bit. (28)
fer de rabot double, plat ou oblique, acier fondu (*m.*), double plane-iron, flat or skew, cast steel. (29 *ou* 30)
fer de rabot simple (*m.*), single plane-iron. (31)
fer de roche (*m.*), hard iron. (32)
fer de Suède (*m.*), Swedish iron. (33)
fer demi-rond (*m.*) *ou* fers demi-ronds (*m.pl.*), half-round iron ; half-round bar ; half-rounds. (34)
fer des marais (Minéral) (*m.*), bog-iron ; bog-iron ore ; bog-ore ; bog-mine ; bog-mine ore ; swamp-ore ; morass-ore ; meadow-ore. (35)
fers divers (*m.pl.*), miscellaneous sections ; sundry sections. (36)
fer double T (*m.*), I iron ; I bar ; H iron ; H bar. (37)
fer doux (*m.*), soft iron. (38)
fer droit (fer à souder) (*m.*), straight bit (solder-
-ing-iron). (39)

fer du commerce (*m.*), commercial iron. (40)
fer ductile (*m.*), malleable iron. (41)
fer dur (*m.*), hard iron. (42)
fer ébauché (*m.*), muck-iron ; muck-bar ; puddle-bar. (43)
fer en bande (*m.*), flat bar ; flat bars. (44)
fer en barre *ou* fer en barres (*m.*), bar iron. (45)
fer en croix (*m.*), cross-section iron ; cruciform iron. (46)
fer en équerre *ou* fer en L (*m.*). Same as fer cornière.
fer en lame (*m.*), sheet iron. (47)
fer en rondin (*m.*), rod-iron ; rods. (48)
fer en T (*m.*). Same as fer à T.
fer en U (*m.*), U iron ; U bar ; channel-iron. (49)
fer en ⌐____⌐ (*m.*) *ou* fers en ⌐____⌐ (*m.pl.*), channel-iron ; channel-bar ; channels. (50)
fer en verge (*m.*), rod-iron ; rods. (51)
fer feuillard (*m.*). Same as feuillard de fer.
fer fondu (*m.*), ingot iron. (52)
fer forgé *ou* *simplement* fer (*n.m.*) (opp. à fonte), wrought iron : (53)
la fonte truitée est très propre à être convertie en fer forgé, mottled iron is very suitable for conversion into wrought iron. (54)
le puddlage est une opération qui a pour but d'affiner la fonte en la décarburant et la transformant en fer, the object of puddling is to refine cast iron by decarbonizing it and converting it into wrought iron. (55)
tuyau de fer (*m.*), wrought-iron pipe. (56)
fer galvanisé (*m.*), galvanized iron. (57)
fer gamma *ou* fer γ (*m.*), gamma iron ; γ iron. (58)
fer homogène (*m.*), ingot iron. (59)
fer laminé (*m.*) *ou* fers laminés (*m.pl.*), rolled iron ; rolled sections. (60)
fer lithoïde (*m.*). Same as fer carbonaté lithoïde.
fer marchand (*m.*), merchant iron ; merchant bar ; bar iron ; finished iron. (61)
fer martelé (*m.*), hammered iron. (62)
fer méplat (*m.*), flat bar ; flat bars. (63)
fer métallique (*m.*), metallic iron. (64)
fer météorique (*m.*), meteoric iron. (65)
fer métis (*m.*), medium iron (between hard and soft). (66)
fer mouluré (*m.*) *ou* fers moulurés (*m.pl.*), moulding iron ; mouldings. (67)
fer natif (*m.*), native iron. (68)
fer nerveux (opp. à fer à grains) (*m.*), fibrous iron. (69)
fer noir (*m.*), black iron. (70)
fer octogone (*m.*) *ou* fers octogones (*m.pl.*), octagon bar ; octagon iron ; octagons. (71)
fer oligiste (Minéral) (*m.*), oligist iron ; oligist. (72)
fer ovale (*m.*) *ou* fers ovales (*m.pl.*), oval iron ; oval bar ; ovals. (73)
fer ouvré (*m.*), hammered ironwork. (74)
fer oxydé hydraté (Minéral) (*m.*), brown hematite ; brown iron ore ; limonite. (75)
fer oxydulé (Minéral) (*m.*), black oxide of iron ; magnetic oxide of iron ; magnetic iron ore ; magnetic iron ; ferrosoferric oxide ; mag-
-netite. (76)
fer pailleux (*m.*), flawy iron. (77)
fer pisiforme (Minéral) (*m.*), pea iron ore. (78)
fer plat (*m.*) *ou* fers plats (*m.pl.*), flat bar ; flat bars ; flats. (79)

fers profilés (*m.pl.*), sectional iron ; sections ; shapes. (1)

fer puddlé (*m.*), puddled iron. (2)

fer rond (*m.*) *ou* **fers ronds** (*m.pl.*), round bar ; round bars ; rounds. (3)

fer rouge (*m.*), red-hot iron. (4)

fer rouverain *ou* **fer rouverin** (*m.*), hot-short iron ; red-short iron ; hot-short. (5)

fer soudant *ou* **fer soudé** (*m.*), weld-iron. (6)

fer spathique (Minéral) (*m.*), spathic iron ; sparry iron. (7)

fers spéciaux (*m.pl.*), special sections : (8) machine à percer pour cornières et fers spéciaux (*f.*), drilling-machine for angles and special sections. (9)

fer spéculaire (Minéral) (*m.*), specular iron ; specular iron ore ; looking-glass ore. (10)

fer sulfaté rouge (Minéral) (*m.*), botryogen. (11)

fer tellurique (opp. à **fer météorique**) (*m.*), telluric iron. (12)

fer tendre (*m.*), cold-short iron. (13)

fer titané (Minéral) (*m.*), titanic iron ore ; titaniferous iron ore ; titaniferous oxide of iron ; ilmenite. (14)

ferblanterie (métier) (*n.f.*), tin-plate working. (15)

ferblanterie (commerce) (*n.f.*), tinsmithing ; tinsmiths' trade. (16)

ferblanterie (boutique) (*n.f.*), tin-shop. (17)

ferblantier (pers.) (*n.m.*), tin-plate worker ; tinsmith ; tinman ; tinner ; whitesmith : dealer in tinware. (18)

fergusonite (Minéral) (*n.f.*), fergusonite. (19)

ferme (*adj.*), firm ; solid. (20)

ferme (terrain ferme) (*n.m.*), solid : (21) percer une voie dans le ferme (Mines), to drive a road in the solid. (22)

ferme (exploitation agricole) (*n.f.*), farm : (23) ferme ouverte au public pour la prospection, farm open to prospecting. (24)

ferme (Constr.) (*n.f.*), truss ; girder. (25)

ferme à faux entrait (*f.*), collar-beam truss ; collar-truss. (26)

ferme à tirant (*f.*), couple-close ; close-couple truss. (27)

ferme-circuit [**ferme-circuit** *pl.*] (Élec.) (*n.m.*), circuit-closer. (28)

ferme de comble (Constr.) (*f.*), roof-truss. (29)

ferme de croupe (*f.*), hip-truss. (30)

ferme de pont (*f.*), bridge-truss. (31)

ferme du ciel du foyer (locomotive) (*f.*), crown-bar. (32)

fermement (*adv.*), firmly ; solidly. (33)

fermentation (*n.f.*), fermentation ; heating : (34) fermentation des schistes, heating of the shales. (35)

fermentation (des bois) (*n.f.*), decomposition, decay, (of timber). (36)

fermentation ouvrière (*f.*), labour disturbances ; labour troubles ; unrest among the labouring classes. (37)

fermenter (*v.i.*), to ferment ; to heat. (38)

fermer (*v.t.*). See examples : (39) fermer à clef *ou* fermer à la clef (*v.t.*), to lock : (40) fermer une porte à clef, to lock a door. (41) fermer à clef *ou* fermer à la clef (*v.i.*), to lock : (42) armoire qui ne ferme pas à clef (*f.*), cupboard which will not lock. (43) fermer l'eau, to turn off, to shut off, the water. (44)

fermer la vapeur *ou* fermer l'introduction, to shut off steam. (45)

fermer le courant (Elec.), to switch off the current. (46)

fermer le moulin, to shut, to shut up, to shut down, to close, to close up, to close down, the mill. (47)

fermer un circuit électrique, to close, to make, an electric circuit. (48)

fermer un moule (Fonderie), to close a mould. (49)

fermer un robinet, to turn off a tap. (50)

fermer un trou, to stop up, to block up, a hole. (51)

fermer un tube à un de ses bouts (*ou* par une de ses extrémités), to close a tube at one end. (52)

fermer une lampe de sûreté, to lock, to fasten, a safety-lamp. (53)

fermer une porte, to shut, to close, a door. (54)

fermer (*v.i.*), to close ; to close down ; to shut ; to fasten : (55) porte qui ferme bien (*f.*), door which shuts properly. (56) fermer pour l'hiver, to close down for the winter. (57)

fermer (se) (*v.r.*), to close ; to shut : (58) soupape qui s'ouvre et se ferme (*f.*), valve which opens and shuts. (59)

fermeté (*n.f.*), firmness. (60)

fermeture (action) (*n.f.*), closing ; shutting ; locking ; fastening ; fencing ; covering : (61)

fermeture à la clef, locking. (62)

fermeture de l'orifice du puits (Mines), closing the shaft-top ; covering the pit-top. (63)

fermeture de la navigation, closing of navigation. (64)

fermeture de la cage, des recettes (Mines), fencing of the cage, of the loading-stations. (65 *ou* 66)

fermeture des lampes de sûreté, locking, fastening, safety-lamps. (67)

fermeture (des travaux) (*n.f.*), stoppage ; shutdown. (68)

fermeture (d'un circuit électrique) (*n.f.*), closing ; making ; make. (69)

fermeture (dispositif) (*n.f.*), fastening ; fastener ; lock : (70) une fermeture solide, a solid fastening. (71)

fermeture (porte) (Mines, etc.), (*n.f.*), door ; closing-device. (72)

fermeture à clef (lampes de sûreté) (*f.*), key-lock. (73)

fermeture à rivet de plomb (lampes) (*f.*), lead-rivet fastening. (74)

fermeture de l'admission (de vapeur dans le cylindre) (*f.*), cut-off. (75)

fermeture électromagnétique (lampes) (*f.*), magnetic lock. (76)

fermoir (agrafe) (*n.m.*), clasp ; fastener ; snap. (77)

fermoir (ciseau à deux biseaux) (*n.m.*), double-bevelled chisel (cutting edge doubly bevelled). (78)

fermoir de tour (*m.*), double-bevelled turning-chisel. (79)

fermoir de tour nez rond *ou* **fermoir de tour néron** (*m.*), skew double-bevelled turning-chisel ; side chisel. (80)

féroélite (Minéral) (*n.f.*), mesole. (81)

ferraille (*n.f.*), old iron ; scrap-iron. (1)
ferré, -e (*adj.*), iron-shod ; bound with iron. (2)
ferrer (*v.t.*), to iron ; to bind with iron ; to metal ;
to shoe ; to tip : (3)
ferrer un wagon, to iron a wagon. (4
ferrer une route, to metal a road. (5)
ferrer un cheval, to shoe a horse. (6)
ferrer à glace un cheval, to rough, to roughen,
to frost-nail, to caulk the shoes of, a horse. (7)
ferrerie (commerce) (*n.f.*), iron-trade. (8)
ferrerie (gros ouvrages de fer) (*n.f.*), ironwork.
(9)
ferret (noyau dur dans les pierres) (Minéralogie)
(*n.m.*), hard core (in stones or rocks). (10)
ferret (*n.m.*) *ou* ferret d'Espagne (Minéral), red
hematite. (11)
ferreux, -euse (Métall.) (*adj.*), ferrous : (12)
métaux ferreux (*m.pl.*) (opp. à *métaux autres
que le fer*), ferrous metals (13) (opp. to
non-ferrous metals).
ferreux, -euse (Chim.) (*adj.*), ferrous : (14)
oxyde ferreux (*m.*), ferrous oxide. (15)
ferreux, -euse (Minéralogie) (*adj.*), ferriferous ;
iron-bearing ; iron : (16)
minerais ferreux (*m.pl.*), iron ores. (17)
ferricyanure (Chim.) (*n.m.*), ferricyanide. (18)
ferrifère (*adj.*), ferriferous ; iron-bearing : (19)
sol ferrifère (*m.*), iron-bearing ground. (20)
ferrique (Chim.) (*adj.*), ferric : (21)
oxyde ferrique (*m.*), ferric oxide. (22)
ferrite (Pétrol.) (*n.m.*), ferrite. (23)
ferrite (Métall.) (*n.f.*), ferrite. (24)
ferro-alliage (Métall.) (*n.m.*), ferro-alloy. (25)
ferro-aluminium (Métall.) (*n.m.*), ferro-aluminum ;
ferro-aluminium. (26)
ferrocalcite (Minéral) (*n.f.*), ferrocalcite. (27)
ferrochrome (Métall.) (*n.m.*), ferrochrome ;
ferrochromium. (28)
ferrocobaltite (Minéral) (*n.f.*), ferrocobaltite. (29)
ferrocyanate (Chim.) (*n.m.*), ferrocyanate. (30)
ferrocyanique (Chim.) (*adj.*), ferrocyanic. (31)
ferrocyanure (Chim.) (*n.m.*), ferrocyanide. (32)
ferromagnésien, -enne (Minéralogie) (*adj.*),
ferromagnesian. (33)
ferromagnétique (*adj.*), ferromagnetic ; sidero-
-magnetic. (34)
ferromanganèse (Métall.) (*n.m.*), ferromanganese.
(35)
ferromolybdène (Métall.) (*n.m.*), ferromolyb-
-denum. (36)
ferronickel (Métall.) (*n.m.*), ferronickel. (37)
ferrosilicium (Chim.) (*n.m.*), ferrosilicon. (38)
ferrotantalite (Minéral) (*n.f.*), tantalite. (39)
ferrotellurite (Minéral) (*n.f.*), ferrotellurite. (40)
ferrotitane (Métall.) (*n.m.*), ferrotitanium. (41)
ferrotungstène (Métall.) (*n.m.*), ferrotungsten.
(42)
ferrovanadium (Métall.) (*n.m.*), ferrovanadium.
(43)
ferrugineux, -euse (*adj.*), ferruginous : (44)
quartz ferrugineux (*m.*), ferruginous quartz.
(45)
eau ferrugineuse (*f.*), ferruginous water ;
chalybeate water. (46)
ferruginosité (*n.f.*), ferrugination. (47)
ferrure (*n.f.*) *ou* ferrures (*n.f.pl.*), ironwork ;
fittings ; mountings : (48)
les ferrures d'un wagon, the ironwork of a
wagon. (49)
appareil photographique à ferrures en cuivre
(*m.*), camera with brass fittings (*or
mountings*). (50)

ferrure (d'un cheval) (action) (*n.f.*), shoeing (a
horse). (51)
ferrure d'isolateur (Élec.) (*f.*), insulator-pin. (52)
fête publique (*f.*), public holiday ; general
holiday. (53)
fétu (Travail à la poudre) (*n.m.*), squib ; straw ;
straw squib ; rush ; reed ; spire ; german.
(54)
feu (développement simultané de chaleur et de
lumière) (*n.m.*), fire ; flame ; light : (55)
le feu est un élément destructeur, fire is a
destructive element. (56)
feu (incendie) (*n.m.*), fire. (57)
feu (fanal) (*n.m.*), light ; lamp. (58)
feux (d'un diamant) (*n.m.pl.*), fire (of a diamond).
(59)
feu brisou (Mines) (*m.*), fire-damp ; gas ;
pit-gas ; fire. (60)
feu central (Géol.) (*m.*), central fire. (61)
feu d'arrière (Ch. de f.) (*m.*), tail-lamp ; tail-
light. (62)
feu d'avant (Ch. de f.) (*m.*), headlight. (63)
feu d'oxydation (Essais au chalumeau) (*m.*),
oxidizing flame. (64)
feu dans les remblais (Mines) (*m.*), gob-fire. (65)
feu de couleur (*m.*), coloured light ; coloured
lamp : (66)
signal de chemin de fer muni de feux de couleur
(*m.*), railway-signal provided with coloured
lights. (67)
feu de mine (*m.*), mine-fire. (68)
feu de réduction (Essais au chalumeau) (*m.*),
reducing flame. (69)
feu de réverbère (*m.*), reverberatory flame.
(70)
feu du chalumeau (*m.*), blowpipe-flame. (71)
feux du prisme (*m.pl.*), prismatic lights. (72)
feu grieux (Mines) (*m.*), fire-damp ; gas ; pit-
gas ; fire. (73)
feu modéré *ou* feu retenu (*m.*), moderate fire.
(74)
feux nus (*m.pl.*), naked lights ; open lights. (75)
feu poussé (*m.*), blown fire. (76)
feu qui couve (*m.*), breeding-fire. (77)
feu terrou (Mines) (*m.*), fire-damp ; gas ;
pit-gas ; fire. (78)
feuillard (*n.m.*) *ou* feuillard de fer (bande large,
plate, et mince) (*s'emploie aussi au pluriel*),
strip ; strip-iron. (79)
feuillard (*n.m.*) *ou* feuillard d'acier, strip ; strip-
steel. (80)
feuillard (*n.m.*) *ou* feuillard de fer (bande de fer
mince et étroite que l'on emploie pour
consolider les angles d'ouvrages en bois),
hoop-iron ; strap-iron ; hoop ; hoops ;
strap ; straps. (81)
feuille (*n.f.*), sheet : (82)
une mince feuille de caoutchouc, a thin sheet
of india-rubber. (83)
des feuilles de tôle ondulée, corrugated-iron
sheets ; corrugated-iron sheeting. (84)
feuille d'étain (*f.*), tin-foil. (85)
feuille de paie *ou* feuille des salaires (*f.*), pay-
roll ; pay-sheet ; wages-sheet. (86)
feuille de placage (*f.*), veneer. (87)
feuille de présence (*f.*), time-sheet. (88)
feuille de quinzaine (*f.*), fortnightly pay-roll.
(89)
feuille de route (Transport) (*f.*), way-bill. (90)
feuille-de-sauge [feuilles-de-sauge *pl.*] (*n.f.*),
cross-file ; crossing-file ; double half-round
file. (91)

feuille de scie (*f.*), saw-blade. (1)

feuille de verre (*f.*), sheet of glass. (2)

feuille du charbon (*f.*), joint-planes, joints₂ cleat, of the coal. (3)

feuiller (Charp.) (*v.t.*), to rabbet ; to rebate : (4) feuiller une planche, to rabbet, to rebate, a board. (5)

feuilleret (*n.m.*), fillister-plane ; fillister. (6)

feuillet (*n.m.*), thin sheet ; thin plate ; lamina. (7)

feuillet magnétique (*m.*), magnetic shell. (8)

feuilleté, -e (*adj.*), foliated ; laminated : (9) structure feuilletée (*f.*), foliated structure. (10) charbon feuilleté (*m.*), laminated coal. (11)

feuillure (Charp., etc.) (*n.f.*), rabbet ; rebate. (12)

feuillure (d'un châssis de fenêtre) (*n.f.*), fillister (of a window-sash). (13)

feutre (*n.m.*), felt ; felting. (14)

fibre (*n.f.*), fibre ; fiber. (15)

fibre élémentaire (Résistance des matériaux) (*f.*), elementary fibre. (16)

fibre moyenne (Résistance des matériaux) (*f.*), mean fibre. (17)

fibre vulcanisée (*f.*), vulcanized fibre. (18)

fibreux, -euse (*adj.*), fibrous : (19) cassure fibreuse (*f.*), fibrous fracture. (20)

fibroferrite (Minéral) (*n.f.*), fibroferrite. (21)

fibrolaminaire (*adj*), fibrolamellar. (22)

fibrolite (Minéral) (*n.f.*), fibrolite. (23)

ficelle (*n.f.*), string ; twine. (24)

ficelle d'amiante (*f.*), asbestos string ; asbestos twine ; asbestos thread. (25)

ficelle de lin (*f.*), flax twine. (26)

fiche (cheville) (*n.f.*), peg ; pin ; stake. (27)

fiche (Élec.) (*n.f.*), plug ; plug-key ; key ; peg. (28)

fiche (Arpent.) (*n.f.*), arrow ; chain-pin. (29)

fiche (Microscopie) (*n.f.*), slide ; slider. (30)

fiche (charnière) (*n.f.*), hinge (a cabinet hinge, usually of brass). (31)

fiche à barrages (*f.*), line-pin for roping in street excavations. (32)

fiche à bouton (*f.*), hinge with knobbed pin. (33)

fiche à nœud (*f.*), loose-pin hinge. (34)

fiche de commutateur (Élec.) (*f.*), switch-plug ; switch-key ; switch-peg. (35)

fiche plombée (Arpent.) (*f.*), drop arrow. (36)

ficher (*v.t.*), to drive in ; to drive into : (37) ficher un clou dans un mur, des pieux en terre, to drive a nail into a wall, stakes into the earth. (38 *ou* 39)

fichtélite (Minéral) (*n.f.*), fichtelite ; fichtellite. (40)

fièvre (*n.f.*), fever. (41)

fièvre de l'or (*f.*), gold-fever ; yellow fever ; rush to the gold-fields. (42)

fièvre des jungles (*f.*), jungle-fever. (43)

fièvre des marais *ou* fièvre miasmatique *ou* fièvre paludéenne (*f.*), marsh-fever ; ma- -larial fever ; paludal fever. (44)

fièvre du pétrole (*f.*), oil-fever ; rush to the oil- fields. (45)

fièvre jaune (Pathologie) (*f.*), yellow fever. (46)

figer (*v.t.*), to congeal ; to coagulate. (47)

figer (se) (*v.r.*), to congeal ; to coagulate. (48)

figulin, -e (*adj.*), figuline. (49)

figure (Géom. et Dessin) (*n.f.*), figure. (50)

figure d'interférence (Opt.) (*f.*), interference figure. (51)

figure de corrosion (Cristall.) (*f.*), etching-figure ; corrosion-figure. (52)

figure de percussion *ou* figure de décollement (Cristall.) (*f.*), percussion-figure. (53)

figure de pression (Cristall.) (*f.*), pressure-figure. (54)

figures de Widmanstætten (Cristall.) (*f.pl.*), Widmanstätten figures ; Widmanstättian figures. (55)

figure plane (Géom.) (*f.*), plane figure. (56)

figurer (*v.t.*), to figure ; to represent ; to show : (57) machine figurée en coupe (*f.*), machine shown in section. (58)

fil (métal étiré) (*n.m.*), wire. (59)

fil (du diaphragme d'une lunette) (*n.m.*), cross- wire, cross-hair, (of the diaphragm or cross- wire ring of a telescope). (60) See next entry for plural rendering : fils (*n.m.pl.*) *ou* fils du réticule *ou* fils d'araignée, cross-wires ; cross-hairs ; spider-lines ; web ; reticle ; reticule : (61) les fils, the cross-wires. (62) fil horizontal *ou* fil axial, horizontal cross- wire ; horizontal cross-hair. (63) fil vertical *ou* fil collimateur, vertical cross- wire ; vertical cross-hair. (64)

fil (d'un outil) (*n.m.*), edge, cutting edge, (of a tool) : (65) le fil d'un couteau, the edge, the cutting edge, of a knife. (66) un fil tranchant, a keen edge. (67)

fil (des lampes à incandescence) (Élec.) (*n.m.*), filament (of incandescent lamps). (68)

fil (du bois) (*n.m.*), grain (of wood) : (69) prendre du bois contre le fil, to work wood against the grain. (70)

fil à plomb (*m.*), plumb-line ; plummet-line ; plumb-bob ; plummet ; plumb. (71)

fil à plomb à lampe (pour travaux souterrains) (*m.*), lamp-plummet, plummet-lamp, (for underground work). (72)

fil à plomb, en cuivre, à pointe d'acier (*m.*), plumb-bob, in brass, steel pointed ; brass plummet with steel point. (73)

fil à plomb en fonte de fer (*m.*), cast-iron plum- -met. (74)

fil clair (opp. à fil noir) (*m.*), bright wire. (75)

fil conducteur (Élec.) (*m.*), conducting-wire ; conductor-wire. (76)

fil cuivré (*m.*), coppered wire. (77)

fil d'acier (*m.*), steel wire. (78)

fil d'acier en clair (*m.*), bright steel wire. (79)

fil d'amenée *ou* fil d'aller (opp. à fil de retour) (Élec.) (*m.*), feed-wire. (80)

fil d'araignée (Opt.) (*m.*). See under fil (du diaphragme d'une lunette).

fil d'équilibre (Élec.) (*m.*), balance-wire. (81)

fil de branchement d'abonné (Élec.) (*m.*), service- lead. (82)

fil de compensation (Élec.) (*m.*). Same as fil neutre.

fil de cuivre (*m.*), copper wire. (83)

fil de cuivre recouvert de soie (*m.*), silk-covered copper wire. (84)

fil de dérivation (Élec.) (*m.*), shunt wire. (85)

fil de fer (*m.*), iron wire. (86)

fil de fer barbelé (*m.*), barbed wire ; barb-wire. (87)

fil de fer galvanisé (*m.*), galvanized iron wire. (88)

fil de fer pour clôture (*m.*), fence-wire. (89)

fil de l'eau (Hydrodynamique) (*m.*), stream- line. (90)

fil de laiton (*m.*), brass wire. (91)

fil de ligne (Élec.) (*m.*), line-wire. (1)
fil de ligne souterraine (*m.*), underground line-wire. (2)
fil de magnésium (*m.*), magnesium wire. (3)
fil de pierre (Carrières) (*m.*), grain. (4)
fil de pile (Élec.) (*m.*), battery-wire. (5)
fil de platine (*m.*), platinum wire. (6)
fil de retour (Élec.) (*m.*), return-wire. (7)
fil de stadia (*m.*), stadia-wire ; stadia-hair. (8)
fil de terre (Élec.) (*m.*), earth-wire ; ground-wire. (9)
fil de trolley (Élec.) (*m.*), trolley-wire. (10)
fil en charge (Élec.) (*m.*), live wire. (11)
fil étamé (*m.*), tinned wire. (12)
fil fin (opp. à fil gros) (*m.*), fine wire. (13)
fil fusible (*m.*), fusible wire. (14)
fil galvanisé (*m.*), galvanized wire. (15)
fil gros (*m.*), thick wire. (16)
fil hors courant (Élec.) (*m.*), dead wire. (17)
fil isolé (Élec.) (*m.*), insulated wire. (18)
fil neutre *ou* fil intermédiaire (Système de dis--tribution à fils multiples) (Élec.) (*m.*), neutral wire ; neutral conductor. (19)
fil noir (*m.*), black wire. (20)
fil nu (Élec.) (*m.*), bare wire. (21)
fil omnibus (Élec.) (*m.*), bus-wire ; omnibus-wire. (22)
fil-pilote [fils-pilotes *pl.*] (Élec.) (*n.m.*), pilot-wire. (23)
fil recouvert (Élec.) (*m.*), covered wire. (24)
fil recuit (*m.*), annealed wire. (25)
fil souple (Élec.) (*m.*), flexible wire ; flex. (26)
fil stadimétrique (*m.*), stadia-wire ; stadia-hair. (27)
fil télégraphique (*m.*), telegraph-wire. (28)
fil téléphonique (*m.*), telephone-wire. (29)
fil-témoin [fils-témoins *pl.*] (Élec.) (*n.m.*), pilot-wire. (30)
filament (*n.m.*), filament : (31)
les filaments de l'asbeste, the filaments of asbestos. (32)
filament étiré (pour lampes électriques à incandes--cence) (*m.*), drawn filament. (33)
filament graphité (*m.*), graphitized filament ; metallized filament. (34)
filament métallique (*m.*), metallic filament. (35)
filament passé par pression à la filière (*m.*), squirted filament. (36)
filamenteux, -euse (*adj.*), filamentous ; filamen--tose. (37)
filasse de montagne (Minéral) (*f.*), asbestos. (38)
file de rails (*f.*), line, string, of rails : (39)
la voie d'un chemin de fer est formée le plus souvent de deux files de rails parallèles, the road of a railway is generally formed of two parallel lines of rails. (40)
filer (tirer à la filière) (*v.t.*), to draw : (41)
filer l'or, l'argent, to draw gold, silver. (42 *ou* 43)
filet (*n.m.*), net : (44)
filet de protection (Canalisation aérienne), guard-net (Overhead transmission system). (45)
filet (Géol.) (*n.m.*), veinlet ; venule ; stringer ; string ; thread : (46)
filet aurifère, auriferous veinlet. (47)
filet de minerai, de quartz, stringer, string, thread, of ore, of quartz. (48 *ou* 49)
filet (d'une vis, d'un écrou) (*n.m.*), thread, worm, fillet, (of a screw, of a nut) : (50 *ou* 51)
filet à droite, right-hand thread. (52)
filet à gauche, left-hand thread. (53)

filet carré *ou* filet rectangulaire, square thread : (54)
vis à filet carré *ou* vis à filet rectangulaire, square-thread screw ; square-threaded screw. (55)
filet de vis, screw-thread. (56)
filet extérieur, outside thread. (57)
filet intérieur, inside thread. (58)
filet rond, round thread. (59)
filet trapézoïdal, trapezoidal thread ; buttress thread. (60)
filet triangulaire, V thread. (61)
filet (émission peu abondante, mais formant un écoulement continu) (*n.m.*), stream : (62)
un maigre filet d'eau, a thin stream of water. (63)
l'huile arrive par mince filet dans la cornue (*f.*), the oil enters the retort in a thin stream. (64)
filet (*n.m.*) *ou* filet fluide (Phys.), stream ; fila--ment : (65)
le mouvement d'un filet d'air, the motion (*or* the movement) of a stream (*or* of a filament) of air. (66)
filet (*n.m.*) *ou* filet liquide (Phys.), stream ; fila--ment : (67)
le frottement des filets liquides contre la paroi intérieure des tuyaux, the friction of the streams (*or* the filaments) against the interior surface of pipes. (68)
les filets d'eau qui touchent la crête d'un déversoir, the streams (*or* filaments) of water which touch the crest of a weir. (69)
filet de tourbillon (Phys.) (*m.*), vortex-filament. (70)
filetage (de vis ou d'écrous) (*n.m.*), threading ; screw-cutting ; screwing : (71)
filetage de tuyaux, pipe-threading ; pipe-screwing. (72)
filetage sur le tour, screw-cutting on the lathe. (73)
filetage en plusieurs passes, screwing in several cuts. (74)
filetage (pas de vis) (*n.m.*), thread ; pitch : (75)
filetage normal anglais (Whitworth), Whitworth (British standard) thread ; Whitworth's standard thread. (76)
For other varieties of thread, see pas, the more common word.
filetage (étirage du fil métallique) (*n.m.*), drawing ; wiredrawing. (77)
fileter (faire un filet de vis ou d'écrou) (*v.t.*), to thread ; to worm ; to screw ; to cut a thread on ; to cut a screw on : (78)
fileter un boulon, to thread, to worm, to screw, to cut a thread on, a bolt. (79)
fileter un tuyau, to thread, to screw, a pipe. (80)
fileter (passer du fil à la filière) (*v.t.*), to draw ; to draw through the draw-plate ; to wire--draw : (81)
fileter des métaux, to draw metals ; to draw metals through the draw-plate ; to wire--draw metals. (82)
filière (*n.f.*) *ou* filière à étirer (pièce d'acier destinée à étirer les fils métalliques), draw-plate ; die ; die-plate : (83)
passer de l'or à la filière, to draw gold through the draw-plate. (84)
filière à trous ronds, à trous carrés, draw-plate with round holes, with square holes. (85 *ou* 86)

filière (pièce d'acier pour fileter en vis) (*n.f.*), die ; die-plate ; screwing-die. (1)

filière (*n.f.*) *ou* **filière à truelle** *ou* **filière à palette** *ou* **filière simple** *ou* **filière plate** (simple plaque d'acier trempée, dans laquelle ont été forés, et taraudés, des trous au diamètre voulu), screw-plate ; tap-plate (the screw-plate proper, i.e., a hardened steel plate with a series of threaded holes, graded in size). (2) For varieties, see **filière à truelle**, below.

filière (*n.f.*) *ou* **filière à cage** *ou* **filière à coussinets** (fût composé généralement d'une cage, terminée par deux bras de levier), stock ; screw-stock ; screwing-stock ; screw-plate ; die-stock ; engineers' screw-stock. (3)

filière (jauge pour fils métalliques) (*n.f.*), wire-gauge. (4)

filière (panne) (Constr.) (*n.f.*), purlin ; purline ; side-timber ; side-waver. (5)

filière (dans le toit) (Mines) (*n.f.*), slip (in the roof). (6)

filière (Échafaudage) (Constr.) (*n.f.*), ledger. (7)

filière à anneau et à coussinets (*f.*), screw-plate, with loose dies, ring end, and screw-head. (8)

filière à bois (*f.*), box screw. (9)

filière à cliquet (*f.*), ratchet screwing-stock. (10)

filière à déclenchement automatique (d'une machine à tarauder) (*f.*), self-opening diehead, self-opening screwing-head, (of a screwing-machine). (11)

filière à deux vis de côté (*f.*), screw-stock with two screws, one on each side of cage. (12)

filière à gaz (*f.*), gas-stock. (13)

filière à guide (*f.*), guide-stock. (14)

filière à la machine (*f.*), machine die-plate. (15)

filière à lunettes *ou* **filière à coussinets-lunettes** (*f.*), solid-die stock. (16)

filière à lunettes et à guides pour fers creux, taraudant d'une seule passe, avec coussinets-lunettes, guides, et tarauds à gaz coniques et cylindriques (*f.*), solid-die guide-stock for screwing iron gas-tube at once going over, with dies, guides, and taper and plug gas-taps. (17)

filière à lunettes rondes (*f.*), circular-die stock ; frame for circular dies. (18)

filière à peignes (*f.*), chaser-die screwing-stock. (19)

filière à plaque (*f.*), screw-plate with holding-down plate for loose dies. (20)

filière à truelle (*f.*), screw-plate ; tap-plate. (21) See **filière** *ou* **filière à truelle**.

filière à truelle, manche coudé (*f.*), screw-plate, bent handle. (22)

filière à truelle, manche droit, trous échancrés (*f.*), screw-plate, straight handle, notched holes. (23)

filière à une vis (*f.*), screw-stock with one handle screwed in cage. (24)

filière double (distingué de **filière simple**) (*f.*), two-part screw-plate. (Any screw-plate or die-stock of which the screw-cutting plate or die is in two halves). (25)

filière droite (distingué de **filière oblique**) (*f.*), straight-pattern screw-stock (double-hand-ed engineers' screw-stock with sides of cage parallel with handles). (26)

filière garnie de coussinets *ou* *simplement* **filière garnie** (*f.*), stock and dies. (27)

filière oblique *ou* **filière inclinée** (distingué de **filière droite**) (*f.*), skew-pattern screw-stock (double-handed engineers' screw-stock with sides of cage oblique to handles). (28)

filière oblique, cage noircie et manches polis (*f.*), skew-pattern screw-stock, black cage and bright handles. (29)

filière pour bijouterie, pas très fin (*f.*), watch screw-plate, very fine pitch. (30)

filière pour horlogerie (*f.*), clock screw-plate. (31)

filière pour tubes en fer, pas du gaz (*f.*), stock for iron gas-tubes ; gas-thread pipe-stock. (32)

filière simple (distingué de **filière double**) (*f.*), one-part screw-plate. (Any screw-plate or die-stock of which the screw-cutting plate or die is in one piece, plain or undivided. The steel plate with a series of threaded holes, graded in size, is in this category and is what is generally meant by **filière simple**.) (33)

film (Photogr.) (*n.m.*), film. (34)

film-pack [**film-packs** *pl.*] (Photogr.) (*n.m.*), film-pack. (35)

film rigide (*m.*), stiff film ; cut film ; flat film ; flat cut film. (36)

filon (Géol. & Mines) (*n.m.*), lode ; vein ; lead. (37)

filon aveugle (filon sans affleurement) (*m.*), blind lode ; blind lead. (38)

filon-couche [**filons-couches** *pl.*] (*n.m.*), bed-vein ; sill. (39)

filon croisé (*m.*), intersected lode. (40)

filon croiseur (*m.*), cross-lode ; cross-vein ; cross-course ; crossing ; counter-lode ; counter ; contra-lode ; caunter-lode ; caunter ; bar. (41)

filon cuprifère (*m.*), copper-bearing lode. (42)

filon d'étain (*m.*), tin-lode. (43)

filon d'exsudation (*m.*), exudation-vein. (44)

filon d'incrustation (*m.*), infiltration-vein. (45)

filon d'injection (*m.*), dyke ; dike. (46)

filon de contact (*m.*), contact lode. (47)

filon de fracture (*m.*), fissure-vein. (48)

filon de quartz aurifère (*m.*), auriferous quartz vein. (49)

filon de ségrégation (*m.*), segregation-vein. (50)

filon de sublimation (*m.*), sublimation-vein. (51)

filon en gradins (*m.*), stepped lode ; lob. (52)

filon épuisé (*m.*), worked out lode ; dead lode. (53)

filon guide (*m.*), leader. (54)

filon intercepté (*m.*), intercepted lode. (55)

filon mère *ou* **filon principal** (*m.*), mother lode ; master-lode ; main lode ; champion lode. (56)

filon nodulaire (*m.*), ball vein. (57)

filon pourri (*m.*), rotten lode ; rotten vein. (58)

filon qui fait le ventre *ou* **filon renflé** (*m.*), swelled lode. (59)

filon ramifié (*m.*), branched lode. (60)

filon stannifère (*m.*), tin-lode ; tin-bearing lode. (61)

filon stérile (*m.*), barren lode ; hungry lode. (62)

filonien, -enne (*adj.*), lode ; vein (*used as adjs*) : (63)

exploitation filonienne (*f.*), lode-mining ; vein-mining. (64)

or filonien (*m.*), lode-gold ; vein-gold. (65)

filonnet (*n.m.*), vein ; small lode. (66)

filtrage (*n.m.*), filtering ; filtration ; straining. (67)

filtrage artificiel des eaux destinées à l'alimenta--tion publique ou privée, artificial filtration of water intended for public or private consumption. (1)

filtrage des eaux destinées à l'alimentation des chaudières à vapeur, filtering water for use in steam-boilers; filtration of boiler-feed water. (2)

filtrage naturel des eaux à travers les bancs sableux formant le sous-sol du terrain, natural filtration of water through sandy banks forming the subsoil. (3)

filtrant, -e (*adj.*), filtering: (4)
matière filtrante (*f.*), filtering-material. (5)

filtration (*n.f.*), filtration; filtering; straining; percolation; percolating: (6)
filtration à travers le sable, filtering through sand. (7)
eaux de filtration (Géol.) (*f.pl.*), percolating water. (8)

filtre (*n.m.*), filter. (9)
filtre à charbon de bois (*m.*), charcoal filter. (10)
filtre à eau (*m.*), water-filter. (11)
filtre à pression *ou* filtre sous pression (*m.*), pres--sure-filter. (12)
filtre à sable (*m.*), sand-filter. (13)
filtre d'amalgame (*m.*), amalgam-filter. (14)
filtre en papier (*m.*), paper filter. (15)
filtre-presse [filtres-presses *pl.*] (*n.m.*), filter-press. (16)

filtrer (*v.t.*), to filter; to strain; to percolate; to leach: (17)
filtrer de l'eau, une solution, to filter water, a solution. (18 *ou* 19)
filtrer un liquide, to filter, to strain, a liquid. (20)
filtrer (*v.i.*), to filter; to strain; to percolate; to leach: (21)
eaux qui filtrent à travers le sol (*f.pl.*), water which filters (*or* percolates) (*or* leaches) through the ground (*or* soil). (22)

fin, fine (délié et menu) (*adj.*), fine: (23)
sable fin (*m.*), fine sand. (24)
fil fin (*m.*), fine wire. (25)
pluie fine (*f.*), fine rain. (26)

fin, fine (précieux par la qualité) (*adj.*), fine: (27)
or fin (*m.*), fine gold. (28)
fins diamants (*m.pl.*), fine diamonds. (29)
perles fines (*f.pl.*), fine pearls. (30)

fin (*n.f.*), end; close; termination: (31)
fin de l'année, end, close, of the year. (32)
fin (de), fine: (33)
pièce d'or qui contient neuf dixièmes de fin (*f.*), gold coin which is nine-tenths fine (*or* 900 fine). (34)
or à — carats de fin (*m.*), gold — carats fine. (35)

fin de l'admission et commencement de la détente (vapeur) [fin (*n.f.*), commencement (*n.m.*)], cut-off. (36)

finage (de la fonte) (*n.m.*), refining (of cast iron). (37)

final, -e, -als (*adj.*), final: (38)
inspection finale (*f.*), final inspection. (39)
fine-métal (*n.m.*), plate metal. (40)
finement (*adv.*), finely. (41)
finerie (pour la fonte) (*n.f.*), refinery; finery; finery-furnace. (42)
fines (houille ou minerai) (*n.f.pl.*), fines (coal or ore). (43 *ou* 44)
fines tôles (fer ou acier) (*f.pl.*), sheets; thin plates (iron or steel). (45 *ou* 46)

finesse (*n.f.*), fineness; thinness; slenderness; sharpness; keenness: (47)
finesse d'une poudre, fineness of a powder. (48)
finesse du tranchant d'un couteau, sharpness, keenness, of the edge of a knife. (49)

fini, -e (Math.) (*adj.*), finite: (50)
l'expérience montre que la lumière ne se propage pas instantanément, mais avec une vitesse finie (*f.*), experiment shows that light does not propagate instantaneously, but with a finite speed. (51)
fini (*n.m.*), finish: (52)
qualité et fini du travail, quality and finish of the work. (53)
finir (*v.t.*), to finish; to end; to terminate; to put an end to. (54)
finir (*v.i.*), to finish; to end; to terminate; to come to an end. (55)
finissage (*n.m.*), finishing: (56)
finissage des objets coulés, finishing castings. (57)
finisseur, -euse (*adj.*), finishing: (58)
cylindre finisseur (*m.*), finishing-roll. (59)
cannelure finisseuse (*f.*), finishing-pass. (60)
finisseur (pers.) (*n.m.*), finisher. (61)
fins (charbon ou minerai) (*n.m.pl.*), fines (coal or ore). (62 *ou* 63)
fiole (Verrerie de laboratoire) (*n.f.*), flask; vial; phial. (64)
fiole (*n.f.*) *ou* fiole d'arpentage, bubble-tube. (65)
fiole conique en verre épais, tubulée, pour filtrer à la trompe (*f.*), filtering-flask, filter-flask, conical, stout glass, with side tube for con--nection to filter-pump. (66)
fiole en verre d'Iéna, fond rond, non-bouchée (*f.*), flask of Jena glass, round-bottomed, unstoppered. (67)
fiole en verre de Bohême, fond plat, bouchée à l'émeri (*f.*), flask of Bohemian glass, flat-bottomed, stoppered. (68)
fiole jaugée (*f.*), flask with mark on neck. (69)
fiole rodée (Arpent.) (*f.*), ground bubble-tube. (70)
fiord (*n.m.*), fiord; fjord; firth. (71)
fiorite (Minéral) (*n.f.*), fiorite. (72)
fissile (*adj.*), fissile: (73)
ardoise fissile (*f.*), fissile slate. (74)
fissilité (*n.f.*), fissility: (75)
fissilité de certains schistes, fissility of certain schists. (76)
fissuration (*n.f.*), fissuration: cracking. (77)
fissure (*n.f.*), fissure; crack; cleft; crevice; rent: (78)
fissure dans un mur, crack in a wall. (79)
fissure minéralisée (Géol.), mineralized fissure; fissure-vein. (80)
fissures de retrait dues au soleil (Géol.), sun-cracks. (81)
fissuré, -e (*adj.*), fissured; fissury; cracked; rent; split. (82)
fissurer (*v.t.*), to fissure; to crack; to crevice. (83)
fissurer (se) (*v.r.*), to fissure; to crack. (84)
fixage (*n.m.*), fixing; fastening: (85)
fixage des rails sur les coussinets, fixing rails on the chairs. (86)
fixage des rails sur les traverses de bois au moyen de crampons, fastening rails down on timber sleepers by dog-spikes. (87)
fixage (action) (Photogr.) (*n.m.*), fixing: (88)
fixage acide, acid fixing. (89)
fixage *ou* **fixateur** (Photogr.) (*n.m.*), fixing-solution; fixing: (90)

fixage acide, acid fixing-solution; acid fixing. (1)

fixation (*n.f.*), fixing; fixation: (2)

fixation à vis, screw-fixing. (3)

fixation de l'emplacement d'un sondage, fixing upon, location of, the site of a bore-hole. (4)

fixation d'une huile (Chim.), fixation of an oil. (5)

fixation d'une photographie, fixing a photo--graph. (6)

fixe (*adj.*), fixed; fast; stationary: (7)

point fixe (*m.*), fixed point. (8)

poulie fixe (*f.*), fast pulley. (9)

chaudière fixe (*f.*), stationary boiler. (10)

fixer (*v.t.*), to fix; to fasten; to secure; to set: (11)

fixer l'emplacement d'un puits à pétrole, to locate the site of, to locate, to fix upon a site for, an oil-well. (12)

fixer un volet qui bat, to fasten, to secure, a banging shutter. (13)

fixer un rivet à demeure, to set a rivet. (14)

fixer (Chim. & Photogr.) (*v.t.*), to fix: (15)

fixer le mercure, to fix the mercury. (16)

fixer une plaque, une épreuve, dans un bain d'hyposulfite, to fix a plate, a print, in a hyposulphite bath. (17 *ou* 18)

fixité (*n.f.*), fixity: (19)

fixité des roues aux essieux, fixity of wheels on the axles. (20)

fjord (*n.m.*), fjord; fiord; firth. (21)

flache *ou* **flacheux, -euse** (*adj.*), waney: (22)

bois flache (*ou* flacheux) (*m.*), waney timber. (23)

flache (*n.f.*), wane. (24)

flacon (*n.m.*), flask; bottle. (25)

flacon à densité (*m.*), specific-gravity bottle; specific-gravity flask; pycnometer. (26)

flacon à deux, à trois tubulures (*m.*), bottle with two, with three tubulures. (27 *ou* 28)

flacon à toucher (Essais) (*m.*), bottle for reagents. (29)

flacon-bocal [flacons-bocaux *pl.*] (*n.m.*), bottle, very wide mouth; bottle, extra wide mouth. (30)

flacon bouché à l'émeri (*m.*), stoppered bottle. (31)

flacon compte-gouttes (*m.*), drop-bottle; drop--ping-bottle. (32)

flacon de Mariotte (*m.*), Mariotte's bottle; Mariotte's flask. (33)

flacon, étiquette vitrifiée (*m.*), bottle, enamelled label, burnt in. (34)

flacon, étroite ouverture, col moulé *ou* **flacon, étroite ouverture, goulot moulé** (*m.*), bottle, narrow mouth, moulded neck. (35)

flacon, large ouverture (*m.*), bottle, wide mouth. (36)

flacon laveur (*m.*), washing-flask; washing-bottle. (37)

flacon non-bouché (*m.*), unstoppered bottle. (38)

flacon tubulé (*m.*), tubulated bottle. (39)

flambage (conservation des bois) (*n.m.*), charring. (40)

flambage (Fonderie) (*n.m.*), skin-drying; torch--ing: (41)

flambage des moules à vert, skin-drying green-sand moulds. (42)

flambage (Résistance des matériaux) (Méc.) (*n.m.*), buckling; buckle (Strength of materials). (43)

flambant, -e (*adj.*), flaming; blazing: (44)

feu flambant (*m.*), blazing fire. (45)

flambeau (*n.m.*), torch. (46)

flambement (*n.m.*). Same as **flambage**.

flamber (essai au borax) (Minéralogie) (*n.m.*), flaming. (47)

flamber (Fonderie) (*v.t.*), to skin-dry; to torch: (48)

flamber à la poêle un moule, to skin-dry a mould with the kettle. (49)

flamber (Méc.) (*v.i.*), to buckle: (50)

trop gros, les pieux opposent à l'enfoncement une résistance énorme; trop minces, ils sont sujets à flamber: on leur donne en moyenne un diamètre égal au trentième de leur hauteur, too big, piles offer an enor--mous resistance to driving; too slender, they are liable to buckle: they are given on an average a diameter equal to a thirtieth of their height. (51)

l'expérience a montré que, lorsque les rivets sont mis en place à chaud, leur longueur ne doit pas dépasser 4 diamètres: au delà de cette limite, ils flambent sous l'action du choc du marteau, experience has shown that when rivets are put in place hot, their length should not exceed 4 diameters: beyond this limit, they buckle under the action of the impact of the hammer. (52)

flamme (*n.f.*), flame: (53)

la flamme d'une bougie, the flame of a candle. (54)

flamme Bunsen (*f.*), Bunsen flame. (55)

flamme du chalumeau (*f.*), blowpipe-flame. (56)

flamme nue (*f.*), naked light; open light. (57)

flamme oxydante (*f.*), oxidizing flame. (58)

flammes perdues (de fourneau) (*f.pl.*), waste gases. (59)

flamme réductrice (*f.*), reducing flame. (60)

flammèche (*n.f.*), spark. (61)

flanc (*n.m.*), side; flank. (62)

flanc (d'une colline, d'une montagne, d'une vallée) (*n.m.*), side, flank, (of a hill, of a mountain, of a valley). (63 *ou* 64 *ou* 65)

flanc (d'une dent d'engrenage) (*n.m.*), flank (of a gear-tooth). (66)

flanc (d'un pli) (Géol.) (*n.m.*), limb, leg, side, flank, (of a fold). (67)

flanc de coteau *ou* **flanc de colline** (*m.*), hillside. (68)

flanc de montagne (*m.*), mountainside. (69)

flanc inverse *ou* **flanc renversé** (d'un pli) (Géol.) (opp. à flanc normal) (*m.*), trough limb (of a fold). (70)

flanc normal (d'un pli) (Géol.) (*m.*), arch limb. (71)

flandre (Boisage des galeries de mine) (*n.f.*), crowntree. (72)

flaque (*n.f.*) *ou* **flaque d'eau**, pool; pond: (73)

flaque d'eau stagnante, stagnant pool; stand--ing pond. (74)

flasque de mercure (*f.*), flask, bottle, of mercury: (75)

le mercure est généralement évalué en flasques (*ou* bouteilles) de kg. 34,65, mercury is generally reckoned in flasks (*or* bottles) of kg. 34·65. (76)

flasques (d'un tour) (*n.m.pl.*), shears, sheers, cheeks, sides, bed-bars, (of a lathe). (77)

flasques (des maillons d'une chaîne à rouleaux) (*n.m.pl.*), side-plates, side-bars, guide-plates, (of the links of a roller-chain). (78)

flasques (d'un treuil à manivelle) (*n.m.pl.*), sides (of a crab-winch). (79)

flasques (d'un soufflet) (*n.m.pl.*), boards (of a bellows). (80)

flasques de coulisse (distribution à coulisse) (*m.pl.*), link-plates. (1)

fléau de balance (*m.*), balance-beam; scale-beam. (2)

flèche (signe servant à désigner une direction) (*n.f.*), arrow: (3)

    direction qui s'indique par une flèche (*f.*), direction which is indicated by an arrow. (4)

    arbre qui tourne dans le sens de la flèche (*m.*), shaft which turns in the direction of the arrow. (5)

flèche (d'une grue) (*n.f.*), jib, boom, (of a crane). (6)

flèche (d'une porte d'écluse) (*n.f.*), balance-bar (of a lock-gate). (7)

flèche (d'une voûte) (*n.f.*), rise, height above impost-level, (of an arch). (8)

flèche (d'un manège) (*n.f.*), sweep (of an animal driving-gear or horse-power). (9)

flèche (du pilon) (bocard) (*n.f.*), stem (of the stamp) (ore-stamp). (10)

flèche (Méc.) (*n.f.*), deflection; set; sag; dip: (11)

    flèche correspondant à la charge (Résistance des ressorts), deflection, set, for a given load (Strength of springs). (12)

    flèche d'une ligne aérienne, sag, dip, of an overhead line. (13)

flèche de pierre (Minéralogie) (*f.*), arrow-stone. (14)

flèche littorale (Géogr. phys.) (*f.*), spit. (15)

flèches d'amour (Minéral.) (*f.pl.*), flèches d'amour; Venus's hairstone; Veneris crinis; rutilated quartz; sagenite; sagenitic quartz. (16)

fléchir (*v.t.*), to flex; to bend; to bow; to sag. (17)

fléchir (*v.i.*), to flex; to bend; to bow; to sag; to yield; to give way: (18)

    les planches à trousser doivent être suffisamment épaisses pour ne pas fléchir à l'usage (*f.pl.*), strickle-boards should be sufficiently thick so as not to bend in use. (19)

    toit de mine qui flèche sous la charge (*m.*), mine-roof which sags under pressure. (20)

fléchissement (*n.m.*), deflection; bending; yielding; sagging. (21)

flectomètre (*n.m.*), deflectometer. (22)

fleur d'émeri (*f.*), flour-emery; flour of emery. (23)

fleur de (à), level with; even with; flush with; on a level with: (24)

    à fleur du sol, flush (*or* even) (*or* level) with the ground. (25)

    à fleur d'eau, on a level with the water. (26)

fleur de soufre (*f.*), flowers of sulphur; flour of sulphur. (27)

fleuret (long ciseau dont le mineur se sert pour percer des trous dans le roc) (*n.m.*), chisel; drill; borer. (28)

fleuret (d'une haveuse à pic) (*n.m.*), bit, cutting-bit, pick, puncher-pick, (of a pick coal-cutting machine). (29)

fleuret (d'une perforatrice, d'un marteau perforateur) (*n.m.*), bit, drill-bit, (of a rock-drill, of a hammer-drill). (30 *ou* 31)

fleuret à archet (Méc.) (*m.*), drill-bow; bow. (32)

fleuve (*n.m.*), river. (33) (a river discharging into the sea: distinguished from **rivière**, a stream discharging into a lake, or into another stream). See also **rivière**.

fleuve à marée *ou* **fleuve soumis à la marée** (*m.*), tidal river. (34)

fleuve côtier (*m.*), coastal river (a river with a short course and feeble flow, discharging into the sea) (35)

fleuve de glace (*m.*), ice-river; glacier. (36)

fleuve de montagne (*m.*), mountain river. (37)

fleuve de plaine (*m.*), river of the plain. (38)

fleuve navigable (*m.*), navigable river. (39)

flexibilité (*n.f.*), flexibility; flexibleness; suppleness; pliability; pliancy: (40)

    flexibilité d'un ressort, flexibility of a spring. (41)

flexible (*adj.*), flexible; supple; pliant; pliable: (42)

    arbre flexible (*m.*), flexible shaft. (43)

flexible (Méc.) (*n.m.*), flexible shaft. (44)

flexion (*n.f.*), deflection; bending; yielding; sagging: (45)

    flexion d'un ressort, deflection of a spring. (46)

flexion (de pièces coulées) (Fonderie) (*n.f.*), buckling, warping, (of castings). (47)

flexure (Géol.) (*n.f.*), flexure; fold. (48) See also **pli**.

flexure anticlinale (*f.*), anticlinal flexure; anti-clinal fold. (49)

flexure du type Park (*f.*), Park type of flexure. (50)

flexure du type Uinta *ou* **flexure du type dinarique** (*f.*), Uinta type of flexure. (51)

flexure monoclinale (*f.*), monoclinal flexure; uniclinal fold. (52)

flexure synclinale (*f.*), synclinal flexure; synclinal fold. (53)

flint (pierre à fusil) (*n.m.*), flint; flintstone. (54)

flint-glass *ou* **flint** (*n.m.*), flint glass; crystal glass; heavy glass. (55)

flore (*n.f.*), flora. (56)

flore fossile (*f.*), fossil flora. (57)

flos-ferri (Minéral.) (*n.m.*), flos ferri. (58)

floss (Métall.) (*n.m.*), floss. (59)

flot (onde; vague) (*n.m.*), wave: (60)

    les flots de la mer, the waves of the sea. (61)

flot (de la marée) (*n.m.*), rise, flow, flux, flood, (of the tide): (62)

    flot et jusant de la mer, rise and fall, flow and ebb, flux and reflux, of the sea. (63)

flottable (*adj.*), floatable. (64)

flottage (Préparation mécanique des minerais) (*n.m.*), flotation; floatation (Ore-dressing): (65)

    procédé par flottage (*m.*), flotation-process. (66)

    enrichissement par flottage (*m.*), enrichment by floatation. (67)

flottage des bois (*m.*), floating logs; running logs (down a stream). (68)

flottant, -e (*adj.*), floating: (69)

    corps flottants (*m.pl.*), floating bodies. (70)

flotter (*v.t.*), to float: (71)

    flotter des bois, to float, to run, logs. (72)

    flotter les schlamms (minerai), to float the slimes (ore). (73)

flotter (*v.i.*), to float. (74)

flotteur (*n.m.*), float; floater; ball. (75)

flotteur à clapet (*m.*), ball valve. (76)

flotteur d'alarme (chaudière) (*m.*), boiler-float. (77)

flotteur en liège (*m.*), cork float. (78)

flou, -e (Photogr.) (*adj.*), fuzzy: (79)

    image floue (*f.*), fuzzy image. (80)

flou (*adv.*), fuzzily. (81)

flou (*n.m.*), fuzziness: (82)

l'objectif anachromatique donne flou par défaut de correction chromatique (*m.*), the soft-focus lens gives fuzziness by lack of chromatic correction. (1)

fluctuation (*n.f.*), fluctuation : (2)
les fluctuations de la mer, the fluctuations of the sea. (3)

fluellite (Minéral) (*n.f.*), fluellite. (4)

fluide (*adj.*), fluid : (5)
les corps fluides se divisent en corps liquides et corps gazeux (*m.pl.*), fluid substances are divided into liquid substances and gaseous substances. (6)

fluide (*n.m.*), fluid. (7)

fluide électrique (*m.*), electric fluid. (8)

fluide moteur (*m.*), motive fluid. (9)

fluidifier (*v.t.*), to fluidify. (10)

fluidité (*n.f.*), fluidity ; fluidness : (11)
fluidité de l'eau, fluidity of water. (12)

fluocérite *ou* **fluocérine** (Minéral) (*n.f.*), fluo--cerite ; fluocerine. (13)

fluor (Chim.) (*n.m.*), fluorine. (14)

fluoré, -e (Chim.) (*adj.*), fluorous. (15)

fluorescence (*n.f.*), fluorescence. (16)

fluorescent, -e (*adj.*), fluorescent. (17)

fluorhydrique (*adj.*), hydrofluoric : (18)
acid fluorhydrique (*m.*), hydrofluoric acid. (19)

fluorine *ou* **fluorite** (Minéral) (*n.f.*), fluor ; fluor spar ; fluorite. (20)

fluoritique (Minéralogie) (*adj.*), fluoritic. (21)

fluorure (Chim.) (*n.m.*), fluoride : (22)
fluorure d'argent, fluoride of silver ; silver fluoride. (23)

flusspath (Minéral) (*n.m.*), fluor ; fluor spar ; fluorite. (24)

fluvial, -e, -aux (*adj.*), fluvial ; river (*used as adj.*) : (25)
apports fluviaux (*m.pl.*), fluvial drift ; river-drift. (26)

fluviatile (*adj.*), fluviatile ; fluviatic : (27)
sable fluviatile (*m.*), fluviatile sand. (28)

fluvio-glaciaire (*adj.*), fluvioglacial ; glacio-aqueous ; aqueoglacial : (29)
apports fluvio-glaciaires (*m.pl.*), fluvioglacial drift. (30)
argile fluvio-glaciaire (*f.*), glacio-aqueous clay : aqueoglacial clay. (31)

fluvio-marin, -e (*adj.*), fluviomarine : (32)
dépôts fluvio-marins (*m.pl.*), fluviomarine deposits. (33)

flux (Métall., Chim., Phys.) (*n.m.*), flux. (34)

flux (de la marée) (*n.m.*), rise, flux, flow, flood, (of the tide) : (35)
flux et reflux de la mer, rise and fall, flux and reflux, flow and ebb, of the sea. (36)

flux acide (Métall.) (*m.*), acid flux. (37)

flux basique (Métall.) (*m.*), basic flux. (38)

flux d'induction (Élec.) (*m.*), induction-flux. (39)

flux de force *ou* **flux de force magnétique** *ou* **flux magnétique** (*m.*), magnetic flux. (40)

flux lumineux *ou* **flux de lumière** (Photom.) (*m.*), luminous flux ; flux of light ; light-flux. (41)

fluxmètre (Élec.) (*n.m.*), fluxmeter. (42)

flysch *ou* **flys** (Géol.) (*n.m.*), Flysch : (43)
le flysch, the Flysch. (44)

focal, -e, -aux (*adj.*), focal : (45)
distance focale (*f.*), focal distance ; focal length. (46)
plan focal (*m.*), focal plane. (47)

focalisation (*n.f.*), focalization ; focusing ; focussing. (48)

focet (*n.m.*), spigot ; vent-peg ; vent-plug. (49)

focimètre *ou* **focomètre** (*n.m.*),   focimeter ; focometer. (50)

foi (*n.f.*), faith ; confidence. (51)

foisonnement (*n.m.*), swell ; swelling ; expan--sion ; expansion in bulk ; increase in bulk : (52)
foisonnement de la terre extraite d'une fouille, swell of earth taken out of a trench. (53)
le foisonnement moyen dans les roches ordi--nairement employées pour le remblayage de mines (grès, schistes, etc.) varie en général de 50 à 60 0/0, the average swell of the rocks ordinarily used for filling mines (sandstones, schists, etc.) varies in general from 50 to 60 %. (54)
foisonnement de la chaux vive à la suite de son extinction, expansion of quicklime as a result of slaking. (55)

foisonner (*v.i.*), to swell ; to increase in bulk : to expand ; to expand in bulk : (56)
terrain susceptible de foisonner par le contact de l'air humide (*m.*), ground liable to swell on contact with damp air. (57)

folding (chambre pliante) (Photogr.) (*n.f.*), folding camera. (58)

foliacé, -e (*adj.*), foliaceous : (59)
structure foliacée (*f.*), foliaceous structure. (60)

folié, -e (*adj.*), foliated. (61)

folle (*adj. f.*). See fou, folle.

fonçage (Mines) (*n.m.*), sinking. (62)

fonçage (Carrières d'ardoise) (*n.m.*), excavating a lift. (63)

fonçage à niveau bas *ou* **fonçage à niveau vide** *ou* **fonçage des avaleresses en niveau bas** (*m.*), sinking through watery strata with continuous pumping. (64)

fonçage à niveau plein (*m.*), sinking through watery strata without any pumping. (65)

fonçage au poussage (*m.*), sinking by piling. (66)

fonçage au trépan (*m.*), sinking by boring. (67)

fonçage de puits (mines métallifères) (*m.*), shaft-sinking. (68)

fonçage de puits (eau, pétrole, etc.) (*m.*), well-sinking. (69)

fonçage de puits à grande profondeur (*m.*), deep-shaft sinking. (70)

fonçage de puits au moyen de la congélation *ou* **fonçage par congélation des puits de mine** (*m.*), sinking shafts by artificial freezing process. (71)

fonçage de puits en terrains inconsistants et aquifères (*m.*), shaft-sinking in loose and water-bearing ground. (72)

fonçage de puits par injection de ciment (*m.*), sinking shafts by cement-injection process. (73)

fonçage en terrains ébouleux (*m.*), sinking through running ground. (74)

fonçage en terrains solides (*m.*), sinking through firm ground. (75)

fonçage sous stot (*m.*), sinking beneath pentice ; sinking beneath a slice of ground left provisionally at the bottom of the winding-shaft. (76)

foncé, -e (*adj.*), dark ; dull : (77)
vert foncé (*m.*), dark green. (78)
orange foncé (coloration du fer) (*m.*), dull orange. (79)

foncée (gradin dans une carrière d'ardoise) (*n.f.*), lift. (80)

foncement (*n.m.*). Same as **fonçage**.

foncer (*v.t.*), to sink : (1)
foncer à niveau plein, to sink through watery strata without any pumping. (2)
foncer à niveau vide, to sink through watery strata with continuous pumping. (3)
foncer un pieu, to sink a post. (4)
foncer un puits (Mines métallifères), to sink a shaft. (5)
foncer un puits (eau, pétrole, etc.), to sink a well. (6)
foncet (d'une serrure) (*n.m.*), cap (of a lock). (7)
fonceur (pers.) (*n.m.*), sinker. (8)
fonceur chef d'équipe (*m.*), foreman sinker ; chargeman sinker. (9)
fonceur de puits (Mines) (*m.*), shaft-sinker ; pit-sinker. (10)
fonceur de puits (eau, pétrole, etc.) (*m.*), well-sinker. (11)
fonction (*n.f.*), function. (12)
fonction de forces (Math.) (*f.*), force-function. (13)
fonctionnement (d'une machine) (*n.m.*), running ; working. (14)
fonctionner (*v.i.*), to run ; to work : (15)
machine qui fonctionne sans à-coups (*f.*), machine which runs without jerks. (16)
fond (l'endroit le plus bas d'une chose creuse) (*n.m.*), bottom : (17)
le fond d'un puits de mine, d'un trou de sonde, d'une vallée, the bottom of a mine-shaft, of a bore-hole, of a valley. (18 *ou* 19 *ou* 20)
wagon à fond mobile (*m.*), drop-bottom truck. (21)
fond (partie solide sur laquelle on trouve une grande masse d'eau) (*n.m.*), bottom ; floor : (22)
rivière à fond sableux, à fond pierreux (*f.*), river with sandy bottom, with stony bottom. (23 *ou* 24)
fond sous-marin *ou* fond de la mer, ocean bottom ; ocean floor ; sea-floor ; bottom of the sea : (25)
l'océanographie étudie la topographie des fonds sous-marins, etc. (*f.*), oceanography treats of the topography of the ocean bottoms, etc. (26)
fond (lieu le plus reculé, le plus éloigné) (*n.m.*), bottom ; depth : (27)
les conditions physiques dans les grands fonds de l'océan (*f.pl.*), the physical conditions in the great depths of the ocean. (28)
fond (d'une serrure) (*n.m.*), plate (of a lock). (29)
fond de bateau (pli synclinal ; auge) (opp. à **voûte** *ou* **selle**) (Géol.) (*m.*), trough. (30) (opp. to arch *or* saddle).
fond de cylindre (*m.*), cylinder-head ; cylinder-cover : (31)
les boulons fixant les fonds au cylindre, the bolts fixing the heads to the cylinder. (32)
fond de cylindre (plus particulièrement, le plateau opposé au couvercle) (machine fixe horizon-tale) (*m.*), back cylinder-head ; back cylinder-cover ; back-head of cylinder. (33)
fond de cylindre arrière *ou* **fond arrière du cylindre** (locomotive) (*m.*), back cylinder-head ; back cylinder-cover ; back-head of cylinder. (34)
fond de cylindre avant du cylindre (locomotive) (*m.*), front cylinder-head ; front cylinder-cover ; front-head of cylinder. (35)

fond de poche (Fonderie) (*m.*), scull ; skull. (36)
fondamental, -e, -aux (*adj.*), fundamental : (37)
roches fondamentales (*f.pl.*), fundamental rocks. (38)
fondant, -e (*adj.*), melting ; dissolving : (39)
glace fondante (*f.*), melting ice. (40)
fondant (*n.m.*), flux. (41)
fondant acide (Métall.) (*m.*), acid flux. (42)
fondant basique (Métall.) (*m.*), basic flux. (43)
fondation (*n.f.*), foundation ; base ; bed : (44)
fondations d'un bâtiment, foundations of a building. (45)
fondations sous l'eau, foundations under water. (46)
fondations sur pieux *ou* fondations sur pilotis, foundations on piles. (47)
fondation du mortier (bocard), mortar-bed (ore-stamp). (48)
fondement (*n.m.*), foundation ; base : (49)
fondements d'une maison, foundations of a house. (50)
fondements d'une montagne, base of a mountain. (51)
fonder (*v.t.*), to found ; to base ; to rest : (52)
fonder une maison sur le roc, to found a house upon a rock. (53)
fonderie (art du fondeur) (*n.f.*), founding : (54)
les progrès de l'électrométallurgie en fonderie d'acier (*m.pl.*), the progress of electro-metallurgy in steel-founding. (55)
fonderie (établissement où l'on fabrique des objets en métal fondu) (*n.f.*), foundry : (56)
l'outillage d'une fonderie d'acier (*m.*), the plant of a steel-foundry. (57)
fonderie de cuivre, brass-foundry. (58)
fonderie de fonte *ou* fonderie de fer, iron-foundry ; ironworks. (59)
fonderie (usine métallurgique) (*n.f.*), smelting-works ; smeltery ; smelter ; works : (60)
fonderie d'étain, tin-smelting works ; tin-smeltery ; tin-works. (61)
fonderie de cuivre, copper-smelting works ; copper-works. (62)
fonderie de fer, ironworks. (63)
fonderie de plomb, lead-smelting works. (64)
fondeur (celui qui dirige une fabrique d'objets en métal fondu, ou ouvrier qui jette en fonte les divers produits de l'industrie) (*n.m.*), founder ; caster. (65)
fondeur (celui qui dirige une usine où l'on fond et raffine les métaux) (*n.m.*), smelter. (66)
fondeur (ouvrier de hauts fourneaux, qui donne issue à la fonte pour la couler en gueuses) (*n.m.*), keeper. (67)
fondeur (d'un four à sole, d'un four électrique) (*n.m.*), melter (of an open-hearth furnace, of an electric furnace). (68 *ou* 69)
fondeur (marchand de fonte) (*n.m.*), iron-merchant. (70)
fondeur affecté aux laitiers (*m.*), slag-man ; cinder-snapper. (71)
fondeur de cuivre (*m.*), brass-founder. (72)
fondeur de fonte *ou* **fondeur en fer** (*m.*), iron-founder. (73)
fondis (*n.m.*), subsidence (of surface) (owing to underground caving). (74)
fondis à jour (*m.*), day-hole (due to subsidence). (75)
fondre (amener à l'état liquide ; dissoudre) (*v.t.*), to melt ; to dissolve : (76)
fondre de la neige, to melt snow. (77)

fondre du sel dans l'eau, to dissolve salt in water. (1)

fondre (liquéfier : en parlant des métaux, des minerais) (v.t.), to melt ; to melt down ; to smelt ; to fuse : (2)

fondre des métaux, to melt, to fuse, metals. (3)

fondre du minerai, de l'étain dans un four à manche, to smelt ore, tin in a shaft-furnace. (4 ou 5)

fondre (confectionner en métal fondu) (v.t.), to cast ; to found : (6)

fondre un banc de tour, to cast, to found, a lathe-bed. (7)

une roue fondue en une seule pièce, a wheel cast in one piece. (8)

fondre (v.i.), to melt ; to melt down ; to dissolve ; to smelt ; to fuse : (9)

la glace fond à 0°, ice melts at 0° C. (10)

le sel fond dans l'eau, salt dissolves in water. (11)

minerai qui fond facilement (m.), ore which smelts easily. (12)

le fer ne fond qu'à une très haute température, iron fuses only at a very high temperature. (13)

minéral qui fond au chalumeau (m.), mineral which fuses in the blowpipe. (14)

fondre (Élec.) (v.i.), to melt ; to blow ; to blow out ; to go : (15)

un plomb a fondu, a fuse has melted (or has blown) (or has gone). (16)

fondre (se) (v.r.), to melt ; to dissolve ; to smelt ; to fuse. (17)

fondrière (crevasse dans le sol) (n.f.), pit ; hollow : (18)

des fondrières causées par les pluies, hollows caused by the rains. (19)

fondrière (terrain marécageux) (n.f.), bog ; morass ; quagmire. (20)

fondrière (minière) (n.f.), surface-working ; surface-mine ; daylight-mine. (21)

fonds (propriété ; domaine) (n.m.), property ; estate. (22)

fondu, -e (liquéfié) (adj.), melted ; molten ; dissolved ; smelted ; fused : (23)

plomb fondu (m.), melted lead ; molten lead. (24)

fondu, -e (coulé) (adj.), cast ; founded : (25)

laiton fondu (m.), cast brass. (26)

fontaine (n.f.), fountain ; spring ; source ; well. (27) See also **source** and **puits**.

fontaine ardente (Géol.) (f.), fire-well. (28)

fontaine d'arrosage (pour mouiller une meule d'émeri, etc.) (f.), water-can. (29)

fontaine de boue (f.), mud spring. (30)

fontaine de gaz ou **fontaine gazeuse** (Géol.) (f.), gas-spring ; gas-well ; natural gas-well ; gasser. (31)

fontaine de tour (f.), lathe-can. (32)

fontaine jaillissante (f.), spouter ; gusher ; spout-well. (33)

fontaine périodique (f.), periodical fountain. (34)

fontainier (fonceur de puits) (pers.) (n.m.), well-sinker. (35)

fonte (transformation d'un corps qui se liquéfie) (n.f.), melting ; dissolving : (36)

les pluies activent la fonte des neiges d'hiver, the rains accelerate the melting of the winter snows. (37)

fonte (action de liquéfier des métaux) (n.f.), melting ; smelting ; fusing. (38)

fonte (l'art, le travail du fondeur) (n.f.), casting ; founding : (39)

fonte d'acier, casting steel ; steel-founding. (40)

fonte en coquille, chill-casting. (41)

fonte (produit d'une fusion) (n.f.), melt : (42)

procédé qui permet de faire une fonte de — kilos de métal (m.), process which allows of a melt of — kilos of metal. (43)

fonte (produit immédiat du traitement des minerais de fer par le charbon) (n.f.), cast iron ; iron ; pig iron ; pig ; metal. (44)

fonte à facettes (f.). Same as **fonte spéculaire**.

fonte à grain moyen (f.), medium-grained pig iron. (45)

fonte à grain serré (f.), close-grained pig iron. (46)

fonte à grains (f.), granular iron. (47)

fonte à gros grain (f.), coarse-grained pig iron. (48)

fonte à l'air chaud (f.), hot-blast pig ; hot-blast iron ; warm-air pig. (49)

fonte à l'air froid (f.), cold-blast pig ; cold-blast iron. (50)

fonte améliorante (f.), improver pig. (51)

fonte au bois ou **fonte au charbon de bois** (f.), charcoal iron ; charcoal pig. (52)

fonte au bois suédoise (f.), Swedish charcoal iron. (53)

fonte au coke (f.), coke iron ; coke pig. (54)

fonte Bessemer (f.), Bessemer iron ; Bessemer pig. (55)

fonte blanche (f.), white iron. (56)

fonte brûlée (f.), burnt iron ; perished metal. (57)

fonte brute de moulage (f.), raw pig iron. (58)

fonte chromée (f.), ferrochrome ; ferrochromium. (59)

fonte d'acier (f.), cast steel ; steel of molten origin. (60)

fonte d'addition (f.), finishing metal ; addition. (61)

fonte d'affinage (f.), conversion pig ; converter-pig. (62)

fonte de Berlin (f.), Berlin iron. (63)

fonte de fer (f.), cast iron ; iron. (64)

fonte de marque (f.), branded pig ; marked pig. (65)

fonte de moulage (f.), foundry-iron. (66)

fonte de première, de seconde, de troisième, de quatrième fusion (f.), cast iron of first, of second, of third, of fourth melting. (67 ou 68 ou 69 ou 70)

fonte de Suède au charbon de bois (f.), Swedish charcoal iron. (71)

fonte douce (f.), soft iron ; soft pig iron ; soft cast iron. (72)

fonte dure (f.), hard iron ; hard pig iron ; hard cast iron. (73)

fonte écossaise ou **fonte d'Écosse** (f.), Scotch pig iron. (74)

fonte en fusion (f.), molten pig ; molten iron ; molten metal. (75)

fonte en gueuses ou **fonte en saumons** (f.), pig iron : (76)

les fontes en gueuses du commerce, commercial pig irons. (77)

fonte épurée (f.), washed metal. (78)

fonte finée (f.), plate metal (refined iron). (79)

fonte grise (f.), gray pig ; gray iron. (80)

fonte hématite (f.), hematite pig. (81)

fonte malléable (*f.*), malleable cast iron ; malleableized cast iron ; malleable pig iron ; malleable iron. (1)

fonte manganésée (*f.*), manganese pig. (2)

fonte marchande (*f.*), merchant pig. (3)

fonte mazée (*f.*), plate metal. (4)

fonte miroitante (*f.*). Same as fonte spéculaire.

fonte phosphoreuse (*f.*), phosphoric pig. (5)

fonte refondue (*f.*), remelted cast iron. (6)

fonte spéciale (*f.*), special iron. (7)

fonte spéculaire *ou* fonte spiegel (*f.*), specular cast iron ; specular pig iron ; spiegeleisen ; spiegel iron ; spiegel. (8)

fonte Thomas (*f.*), Thomas-Gilchrist pig ; basic-Bessemer pig. (9)

fonte trempée (trempée en coquille) (*f.*), chilled cast iron ; chilled iron. (10)

fonte truitée (*f.*), mottled iron ; mottled pig. (11)

fontenier (fonceur de puits) (pers.) (*n.m.*), well-sinker. (12)

fontis (*n.m.*), subsidence (of surface) (owing to underground caving). (13)

forage (sondage) (*n.m.*), boring ; bore-holing ; drilling : (14)

forage de puits (eau, pétrole, etc.), boring wells ; well-drilling. (15)

forage de puits à grand diamètre (Mines), boring shafts of large diameter. (16)

forage des roches, drilling rocks ; rock-boring. (17)

forage pour recherches de pétrole, boring for oil ; drilling for petroleum. (18)

forage (trou de sonde) (*n.m.*), bore-hole ; drill-hole ; bore ; hole. (19)

forage (action de percer un trou dans une pièce métallique) (*n.m.*), drilling. (20)

forage à chute libre (*m.*), free-fall boring ; free-fall drilling. (21)

forage à grande profondeur (*m.*), deep boring ; deep drilling. (22)

forage à l'entreprise (*m.*), contract boring. (23)

forage à la barre (*m.*), jump-drilling. (24)

forage à la grenaille d'acier (*m.*), shot-boring ; shot-drilling. (25)

forage à la main (*m.*), hand-drilling. (26)

forage au diamant (*m.*), diamond-drilling. (27)

forage d'essai (*m.*), trial-boring. (28)

forage mécanique (*m.*), machine-drilling. (29)

forage par rôdage (*m.*), rotary drilling ; boring by rotation. (30)

forage par rôdage avec courant d'eau (*m.*), hydraulic rotary drilling. (31)

force (*n.f.*), force ; power ; strength : (32)

force d'une explosion, d'un explosif, des gaz qui s'échappent, de l'eau, force of an explo-sion, of an explosive, of escaping gases, of water. (33 *ou* 34 *ou* 35 *ou* 36)

force d'un courant d'eau, strength of a current of water. (37)

force d'un mur, strength of a wall. (38)

force d'une chute d'eau (pression), force of a waterfall. (39)

force d'une chute d'eau (capacité de travail), power of a waterfall. (40)

force du vent (pression), force of the wind. (41)

force du vent (force motrice), wind-power. (42)

forces de la nature, forces of nature. (43)

force des choses, force of circumstances. (44)

forces perturbatrices, disturbing forces. (45)

forces souterraines, subterranean forces. (46)

force (d'un liquide) (*n.f.*), strength (of a liquid) : (47)

la force d'un acide, the strength of an acid. (48)

force (jambe de force) (*n.f.*), force-piece. (49)

force (terme de mineur) (*n.f.*), deoxidation of the air. (50)

force absorbée en chevaux (*f.*), horse-power required. (51)

force accélératrice (*f.*), accelerating force. (52)

force centrale (Phys. & Méc.) (*f.*), central force. (53)

force centrifuge (*f.*), centrifugal force. (54)

force centripète (*f.*), centripetal force. (55)

force coercitive (Magnétisme) (*f.*), coercive force ; coercitive force. (56)

forces concourantes (*f.pl.*), concurrent forces. (57)

force constante *ou* force continue (opp. à force impulsive *ou* force d'impulsion) (Méc.) (*f.*), constant force. (58)

force contre-électromotrice de self-induction (*f.*), counter electromotive force ; back electro-motive force ; spurious resistance. (59)

force d'impulsion (*f.*), impulsive force. (60)

force d'inertie (*f.*), inertia ; vis inertiæ. (61)

force de bras (*f.*), hand-power : (62)

à force de bras, by hand ; by hand-power. (63)

force de cheval (*f.*), horse-power. (64) See note under cheval.

force de frottement (*f.*), force of friction. (65)

force de l'homme (*f.*), man-power. (66)

force de levage (*f.*), lifting power ; lifting capacity. (67)

force de levier (*f.*), leverage. (68)

force de percussion (*f.*), percussive force. (69)

force élastique de la vapeur (Phys.) (*f.*), vapour-pressure ; vapour-tension ; tension of vapor. (70)

force électrique (*f.*), electric force. (71)

force électromotrice (*f.*), electromotive force. (72)

force élévatoire (*f.*), lifting power ; lifting capacity. (73)

force en chevaux (*f.*), horse-power. (74) See note under cheval.

force en chevaux exigée (*f.*), horse-power required. (75)

force hydraulique (*f.*), water-power ; hydraulic power. (76)

force impulsive (*f.*), impulsive force. (77)

force locomotrice (*f.*), locomotive power ; power of locomotion. (78)

force magnétique (*f.*), magnetic force. (79)

force magnétisante (*f.*), magnetizing force ; intensity of field ; field intensity ; strength of field ; field strength ; field density ; magnetic intensity ; magnetomotive gradient. (80)

force magnétomotrice (*f.*), magnetomotive force. (81)

force morte (*f.*), dead force ; vis mortua. (82)

force motrice (*f.*), motive power ; power ; motive force : (83)

force motrice d'une chute d'eau, motive power of a waterfall. (84)

force motrice de la vapeur, steam-power. (85)

force motrice transmise par un moteur au jour, power transmitted from an engine at the surface. (86)

forces parallèles (*f.pl.*), parallel forces. (1)
force percutante (*f.*), percussive force. (2)
force polaire (Phys.) (*f.*), polar force. (3)
force portante (d'un aimant) (*f.*), portative force, lifting capacity, (of a magnet). (4)
force propulsive (*f.*), propelling force; pro-pulsive force. (5)
force thermo-électrique (*f.*), thermoelectric force. (6)
force tirante *ou* force tractive (*f.*), tractive force. (7)
force vive (*f.*), living force; vis viva: (8)
transformer la force vive des eaux courantes en travail utile au moyen d'une roue hydrau-lique, to transform the living force of running water into useful work by means of a water-wheel. (9)
forcement (*n.m.*), forcing; forcing open : (10)
forcement d'une serrure, forcing a lock. (11)
forcer (*v.t.*). See examples : (12)
forcer la production, to force the production ; to crowd the output. (13)
forcer le tirage, to force the draught. (14)
forcer les travaux, to crowd the work. (15)
forcer quelqu'un à faire une chose, to force, to compel, someone to do a thing. (16)
forcer un cheval, to overwork a horse. (17)
forcer un obstacle, to overcome an obstacle. (18)
forcer une clef, to force a key. (19)
forcer une porte, to force, to force open, a door. (20)
forer (Sondage) (*v.t.*), to bore ; to drill : (21)
forer un puits, un trou de sonde, to drill, to bore, a well, a bore-hole. (22 *ou* 23)
forer à la barre, to jump ; to bore with the jumping-drill. (24)
forer (Méc.) (*v.t.*), to drill : (25)
forer un trou dans une pièce de métal, to drill a hole in a piece of metal. (26)
forerie (*n.f.*), drilling-machine ; drill ; driller. (27) See also machine à percer.
forerie à colonne, pour percer au vilebrequin (*f.*), drilling-pillar. (28)
forerie à levier à colonne (*f.*), lever-feed pillar drilling-machine. (29)
forerie murale à double engrenage (*f.*), double-geared wall drilling-machine. (30)
forerie portative à engrenage, pour établi (*f.*), bench-drill. (31)
forerie portative à étau (*f.*), bench-drill, with vice. (32)
forerie pour percer au vilebrequin (*f.*), drilling-cramp. (33)
forerie sur colonne pour établis (*f.*), bench pillar drilling-machine. (34)
forestier, -ère (*adj.*), forest (*used as adj.*) : (35)
bois forestier (*m.*), forest-timber. (36)
foret (*n.m.*), drill ; bit. (37) See also mèche.
foret à arçon (*m.*), bow-drill ; piercer. (38)
foret à canon (*m.*), half-round drill ; hog-nose drill. (39)
foret à centre (*m.*), centre-bit. (40)
foret à centre, à queue carrée (*m.*), centre brace-bit ; centre-bit, for bit-stock. (41)
foret à centrer (*m.*), centre-drill ; combined drill and countersink. (42)
foret à cuiller (*m.*), spoon-bit. (43)
foret à hélice (*m.*), twist-drill. (44)
foret à hélice dit américain (*m.*), American twist-drill. (45)

foret à langue d'aspic (*m.*), flat drill ; arrow-headed drill. (46)
foret à queue carrée (*m.*), square-shank drill. (47)
foret à rainures droites (*m.*), straight-fluted drill ; straightway drill ; fluted drill. (48)
foret à repos (*m.*), stop-drill. (49)
foret à spire (*m.*), twist-drill. (50)
foret à teton *ou* foret à téton (*m.*), centre-bit. (51)
foret à teton cylindrique (*m.*), pin-drill ; plug centre-bit ; counterbore. (52)
foret demi-rond (*m.*), half-round drill ; hog-nose drill. (53)
foret hélicoïdal (*m.*), twist-drill. (54)
foret hélicoïdal à queue carrée (*m.*), taper square shank twist-drill. (55)
foret hélicoïdal à queue carrée pour vilebrequins (*m.*), bit-stock twist-drill. (56)
foret hélicoïdal à queue conique (*m.*), taper-shank twist-drill. (57)
foret hélicoïdal à queue conique au cône Morse, avec trait de centre pour affûtage régulier (*m.*), Morse taper shank twist-drill, with centre grinding-line. (58)
foret hélicoïdal à queue cylindrique (*m.*), straight-shank twist-drill ; parallel-shank twist-drill. (59)
foret hélicoïdal à tubes d'huile (*m.*), oil-tube twist-drill. (60)
foret hélicoïdal, cylindrique dans toute sa longueur (*m.*), twist-drill cylindrical throughout its entire length. (61)
foret hélicoïdal en acier rapide, (1) tournant à droite ou (2) tournant à gauche (*m.*), high-speed twist-drill, (1) right-hand or (2) left-hand. (62 *ou* 63)
foret hélicoïde (*m.*). Same as foret hélicoïdal.
foret pour tour (*m.*), chuck-drill. (64)
foret pour vilebrequin (*m.*), bit-stock drill ; brace-bit. (65)
forêt (*n.f.*), forest. (66)
forêt de haute futaie (*f.*), forest of timber-trees. (67)
forêt en essences de chêne (*f.*), oak-forest. (68)
forêt primitive (*f.*), primeval forest. (69)
forêt vierge (*f.*), virgin forest. (70)
foreur (pers.) (*n.m.*), driller. (71)
foreur de puits (*m.*), well-driller. (72)
foreur indigène (*m.*), drill boy. (73)
foreuse (pour roches) (*n.f.*), boring-machine ; drilling-machine ; drill ; driller ; rock-drill ; rock-borer. (74) See also sondeuse, sonde, and perforatrice.
foreuse (pour métal) (*n.f.*), drilling-machine ; drill ; driller. (75) See also machine à percer.
foreuse à carottes *ou* foreuse à témoins (*f.*), core-drill. (76)
foreuse à grenaille d'acier (*f.*), shot-drill. (77)
foreuse à pointes de diamant (*f.*), diamond-drill. (78)
foreuse mécanique (pour métal) (*f.*), drilling-machine. (79)
forfait (*n.m.*), contract : (80)
entreprendre des travaux à forfait, to under-take work on contract. (81)
installations complètes à forfait (*f.pl.*), complete installations on contract. (82)
forge (atelier où l'on travaille les métaux au feu et au fourneau) (*n.f.*), forge. (83)

forge (établissement où l'on transforme la fonte en fer malléable) (n.f.), forge. (1)

forge (atelier de forgeron) (n.f.), forge ; smithy ; smith's shop ; blacksmiths' shop ; smithery. (2)

forge (fourneau pour forger) (n.f.), forge. (3)

forge à braser (f.), brazing-forge. (4)

forge à double vent (f.), double-blast forge. (5)

forge à fer ouvré ou forge à main ou forge de maréchalerie (f.), blacksmiths' forge ; shoeing-forge. (6)

forge à la catalane ou forge catalane (f.), Catalan forge ; Catalan furnace ; Biscayan forge ; forge. (7)

forge à simple vent (f.), single-blast forge. (8)

forge à tuyère centrale (f.), bottom-blast forge. (9)

forge à tuyère en bout (f.), back-blast forge. (10)

forge à ventilateur ou forge avec ventilateur (f.), fan-forge. (11)

forge-atelier [forges-ateliers pl.] (n.f.), combined forge, vice, and drilling-stand. (12)

forge pliante (f.), folding forge. (13)

forge portative ou forge de campagne ou forge volante (f.), portable forge. (14)

forge portative à soufflet (f.), portable forge with bellows. (15)

forgé, -e (adj.), forged : (16)
manivelle forgée (f.), forged crank. (17)

forgeable (adj.), forgeable : (18)
la fonte n'est pas forgeable, cast iron is not forgeable. (19)

forgeage ou forgement (n.m.), forging ; smithing ; smithery. (20)

forger (v.t.), to forge : (21)
forger une barre de fer, to forge a bar of iron. (22)

forger à chaud, to forge hot ; to work hot. (23)

forger à froid, to forge cold ; to work cold. (24)

forger (se) (v.r.), to forge ; to be forged : (25)
tandis que le fer se forge, la fonte ne peut généralement pas se forger, whereas wrought iron forges, cast iron cannot generally be forged. (26)

forgerie (n.f.), smithing ; smithery ; forging. (27)

forgeron (pers.) (n.m.), smith ; blacksmith ; ironsmith. (28)

forgeur (pers.) (n.m.), forger ; forgeman. (29)

formation (n.f.), formation : (30)
la formation des dunes de sable est due à l'action des vents, the formation of sand dunes is due to the action of the winds. (31)
la formation de la houille, de l'écorce terrestre, the formation of coal, of the earth's crust. (32 ou 33)

formation (des trains de marchandises) (Ch. de f.) (n.f.), making up (goods-trains). (34)

formation calcaire (Géol.) (f.), limestone formation ; calcareous formation. (35)

formation carbonifère (f.), caboniferous formation ; coal-bearing formation ; coal-formation ; coal-measures. (36)

formation cimentée ou formation conglomérée (f.), cement formation ; conglomerate formation ; banket formation. (37)

formation corallienne (f.), corallian formation. (38)

formation d'eau douce (f.), fresh-water formation. (39)

formation éolienne (f.), eolian formation ; æolian formation. (40)

formation filonienne (f.), lode formation. (41)

formation géologique (f.), geological formation. (42)

formation granitique (f.), granitic formation. (43)

formation gréseuse (f.), sandstone formation. (44)

formation houillère (f.), coal-formation ; coal-measures. (45)

formation lacustre (f.), lacustrine formation. (46)

formation marine (f.), marine formation. (47)

formation métallifère (f.), ore-bearing formation ; metalliferous formation ; ore-formation. (48)

formation pétrolifère (f.), petroliferous formation ; oil-bearing formation. (49)

formation pneumatolytique (f.), pneumatolytic formation. (50)

formation saliférienne (f.), salt formation. (51)

formation schisteuse (f.), schistose formation ; shale formation. (52)

formation volcanique (f.), volcanic formation. (53)

forme (n.f.), form ; shape : (54)
forme cristalline, crystal form. (55)
forme des palettes du ventilateur, form of fan-blades. (56)
formes de puits, forms of shaft. (57)
forme des massifs de minerai, shape of ore-bodies. (58)

formène (Chim.) (n.m.), formene. (59)

former (v.t.), to form ; to shape. (60)

former (se) une opinion favorable, to form a favourable opinion. (61)

formule (n.f.), formula : (62)
formules des corps composés, formulas, formulæ, of compounds. (63)

formule chimique (f.), chemical formula. (64)

formule classique (f.), standard formula. (65)

formule de Bazin (Hydraul.) (f.), Bazin's formula. (66)

formule empirique ou formule brute (f.), empirical formula ; composition formula. (67)

formule rationnelle ou formule de constitution (f.), rational formula ; constitutional formula ; graphic formula ; structural formula. (68)

forstérite (Minéral.) (n.f.), forsterite. (69)

fort, -e (adj.), strong ; powerful ; heavy ; large ; big ; high : (70)
acide fort (m.), strong acid. (71)
forte baisse dans le prix de l'étain (f.), big drop in the price of tin. (72)
forte passe (Travail aux machines-outils) (f.), heavy cut. (73)
forte pente (f.), high (or steep) gradient. (74)
forte proportion d'humidité (f.), large, high, heavy, percentage of moisture. (75)

fortage (n.m.), surface-royalty ; royalty payable to surface-owner. (76)

fortement (adv.), strongly ; highly : (77)
eau fortement acide (f.), strongly acid water. (78)

fortex (Explosif) (n.m.), fortex. (79)

fortifier (v.t.), to strengthen ; to fortify : (80)
fortifier un mur, to strengthen a wall. (81)

fortuit, -e (adj.), accidental ; casual ; fortuitous ; chance : (82)
découverte fortuite (f.), chance discovery. (83)

fortuitement (adv.), accidentally ; casually ; fortuitously ; by chance ; by accident. (84)

fosse (n.f.), hole ; pit ; depression. (85)

fosse (creusée dans le sol d'une fonderie) (n.f.), pit ; hole. (1)

fosse (puits à charbon) (n.f.), pit ; coal-pit. (2)

fosse (n.f.) ou fosse profonde (Océanogr.), deep ; deep trough ; antiplateau : (3)

la fosse du Tuscarora, the Tuscarora deep. (4)

l'opposition d'une chaîne montagneuse élevée et d'une fosse profonde est un fait très fréquent sur les bords de l'océan (f.), the opposition of a high mountain range and of a deep trough is of very frequent occurrence on the margins of the ocean. (5)

fosse à ciel ouvert (f.), open pit. (6)

fosse à piquer le feu ou fosse à visiter (f.), engine-pit ; cleaning-pit ; inspection-pit. (7)

fosse de coulée (Fonderie) (f.), casting-pit. (8)

fosse de moulage (Fonderie) (f.), moulding-hole. (9)

fosse de remplissage (Mines) (f.), shaft-pocket ; pocket. (10)

fosse de renmoulage (Fonderie) (f.), setting-up pit. (11)

fosse de repos (f.), settling-pit. (12)

fossé (n.m.), ditch ; trench : (13)

le fossé d'une tranchée de chemin de fer, the ditch of a railway-cutting ; the trench of a railway-cut. (14)

fossé (Géol.) (n.m.), graben ; rift-valley. (15)
See opposing term horst for example of use.

fossé central (Travail aux explosifs)(Mines) (m.), centre-cut. (16)

fosset (n.m.), spigot ; vent-peg ; vent-plug. (17)

fossile (adj.), fossil : (18)

bois fossile (m.), fossil wood. (19)

fossile (n.m.), fossil. (20)

fossiles caractéristiques (m.pl.), index fossils ; guide fossils ; type fossils. (21)

fossilifère (adj.), fossiliferous : (22)

strates fossilifères (f.pl.), fossiliferous strata. (23)

fossilisateur, -trice (adj.), fossilizing : (24)

sels fossilisateurs (m.pl.), fossilizing salts. (25)

fossilisation (n.f.), fossilization ; fossilification. (26)

fossilisé, -e (adj.), fossilized. (27)

fossiliser (v.t.), to fossilize ; to fossilify. (28)

fossiliser (se) (v.r.), to fossilize ; to fossilify : (29)

la houille est constituée par des végétaux qui se sont fossilisés, coal is formed by vegetable substances which have fossilized. (30)

fou ou fol, folle (Méc.) (adj.), loose : (31)

manchon qui est fou sur l'arbre (m.), sleeve which is loose on the shaft. (32)

poulie folle (f.), loose pulley. (33)

foudre (n.f.), lightning ; thunderbolt ; bolt : (34)

protéger les lignes électriques contre la foudre, to protect electric lines against lightning. (35)

foudroyage (Méthode d'exploitation de mines) (n.m.), caving ; subsidence of the roof : (36)

méthode de foudroyage (f.), caving system of mining ; subsidence-of-the-roof method. (37)

fouettement (du câble d'extraction) (Mines) (n.m.), whipping, flapping, flopping, surging, (of the winding-rope). (38)

fouettement (des tiges dans le sondage) (n.m.), lashing (of the rods in the bore-hole). (39)

fouille (action) (n.f.), excavation ; digging ; trenching : (40)

fouille de puits, well-digging. (41)

fouille (excavation faite) (n.f.), excavation ; cut ; trench ; pit. (42)

fouille (Dragage, Exploitation des placers, etc.) (n.f.), cut : (43)

l'alluvion est traitée sur place dans un lavoir mobile et les résidus stériles sont rendus aussitôt à la fouille (f.), the alluvial is treated on the spot in a movable washer and the tailings are immediately thrown back into the cut. (44)

fouille (fosse creusée dans le sol d'une fonderie) (n.f.), pit ; hole. (45) See also fosse.

fouille (pour une plaque tournante) (n.f.), pit (for a turntable). (46)

fouille à ciel ouvert (f.), open pit. (47)

fouille à la surface (f.), surface-cut. (48)

fouille de renmoulage (Fonderie) (f.), setting-up pit. (49)

fouille des ouvriers (pour allumettes) (Mines) (f.), search of workmen (for matches). (50)

fouiller (v.t. & v.i.) ou fouiller la terre, to mine ; to dig or burrow in or below the surface of the earth ; to excavate. (51)

suivant la loi française tout propriétaire d'un fonds peut ouvrir une carrière sur son terrain, mais il ne peut fouiller sous le terrain d'autrui, according to French law, the owner of a property may open a quarry on his own land, but he must not mine under the land of others. (52)

droits de fouiller sous le terrain d'autrui (m.pl.), undermining rights. (53)

foulant, -e (adj.), force (used as adj.) ; forcing : (54)

pompe foulante (f.), force-pump ; forcing-pump. (55)

foulement (Fonderie) (n.m.), ramming. (56)

fouler (Fonderie) (v.t.), to ram : (57)

fouler le sable dans un châssis de moulage, to ram the sand in a moulding-box. (58)

fouloir (de mouleur) (Fonderie) (n.m.), rammer. (59)

fouloir à air comprimé ou fouloir pneumatique (m.), pneumatic rammer. (60)

fouloir à bras (m.), floor-rammer. (61)

fouloir à main (m.), hand-rammer. (62)

fouloir à manche en bois (m.), rammer with wooden shaft. (63)

fouloir d'établi (m.), bench-rammer. (64)

fouloir électrique (m.), electric rammer. (65)

four (n.m.), furnace ; kiln ; oven. (66) See also fourneau.

four à affiner (m.), refining-furnace. (67)

four à air soufflé ou four à air forcé (m.), forced-draught furnace. (68)

four à arc (m.), arc-furnace. (69)

four à briques (m.), brick-kiln. (70)

four à calcination (m.), calcining-furnace ; calcining-kiln. (71)

four à cémenter (m.), cementation-furnace ; cementing-furnace. (72)

four à chaux (m.), lime-kiln ; lime-pit. (73)

four à chemise d'eau (m.), water-jacket furnace. (74)

four à coke (m.), coke-oven ; pit-kiln. (75)

four à coupellation ou four à coupelles (m.), cupellation-furnace ; cupelling-furnace ; cupel-furnace. (76)

four à creuset(s) (m.), crucible-furnace ; pot-furnace. (77 ou 78) four à cristaux (cavité dans la roche tapissée de cristaux) (Géol.) (*m.*), vug ; vugh ; geode ; druse. (1)

four à cuve (*m.*), shaft-furnace. (2)

four à électrodes (*m.*), electrode-furnace. (3)

four à fritter (*m.*), calcining-furnace. (4)

four à gaz (*m.*), gas-furnace ; gas-fired furnace. (5)

four à griller (*m.*). Same as four de grillage.

four à induction (Élec.) (*m.*), induction-furnace. (6)

four à loupes (*m.*), balling-furnace. (7)

four à manche (*m.*), shaft-furnace. (8)

four à manche pour la fusion de l'étain (*m.*), shaft-furnace for tin-smelting. (9)

four à minerai de cuivre (*m.*), copper-furnace ; copper-smelting furnace. (10)

four à moufle (*m.*), muffle-furnace ; blind roaster. (11)

four à ouvreaux et courant d'air naturel aspiré (*m.*), natural-draught furnace. (12)

four à plâtre (*m.*), gypsum-furnace. (13)

four à puddler (*m.*), puddling-furnace ; puddler. (14)

four à puddler rotatif (*m.*), rotary puddling-furnace. (15)

four à pyrites (*m.*), desulphurizing-furnace ; pyrites-furnace. (16)

four à réchauffer (*m.*), reheating-furnace. (17)

four à recuire (*m.*), annealing-furnace. (18)

four à récupérateur *ou* four à régénérateur *ou* four à régénération (*m.*), regenerating-furnace ; regenerative furnace ; regeneratory furnace. (19)

four à résistance (Élec.) (*m.*), resistance-furnace. (20)

four à réverbère (*m.*), reverberatory furnace ; reverberatory. (21)

four à réverbère à bassin (*m.*), sloping-hearth reverberatory furnace. (22)

four à réverbère à sole plate (*m.*), level-hearth reverberatory furnace. (23)

four à ruche (fabrication du coke) (*m.*), beehive oven. (24)

four à sole (*m.*), open-hearth furnace ; Siemens-Martin furnace. (25)

four à soufre (*m.*), sulphur-burner ; sulphur-kiln ; sulphur-furnace. (26)

four à tremper (*m.*), hardening-furnace ; tempering-furnace. (27)

four à tuyères et courant d'air forcé (*m.*), forced-draught furnace. (28)

four à vent (*m.*), air-furnace ; wind-furnace. (29)

four basculant (*m.*), tilting-furnace. (30)

four belge (réduction du zinc) (*m.*), Belgian furnace. (31)

four chauffé au gaz (*m.*), gas-fired furnace ; gas-furnace. (32)

four chauffé au pétrole (*m.*), oil-fired furnace ; oil-fuel furnace ; petroleum-furnace. (33)

four d'affinage (*m.*), refining-furnace. (34)

four d'affinage du cuivre (*m.*), copper-refining furnace. (35)

four de calcination (*m.*), calcining-furnace ; calcining-kiln. (36)

four de cémentation (*m.*), cementation-furnace. (37)

four de fusion (*m.*), smelting-furnace ; smelter. (38)

four de galère (*m.*), gallery-furnace. (39)

four de grillage (*m.*), roasting-furnace ; roasting-kiln ; roasting-oven ; roaster ; calcining-furnace ; calciner ; burning-house. (40)

four de laboratoire (*m.*), laboratory-furnace. (41)

four de liquation (*m.*), liquation-furnace ; sweating-furnace. (42)

four de puddlage (*m.*), puddling-furnace ; puddler. (43)

four de puddlage mécanique (*m.*), mechanical puddler ; puddling-machine. (44)

four de réchauffage (*m.*), reheating-furnace. (45)

four de recuite (*m.*), annealing-furnace. (46)

four de réduction (*m.*), reduction-furnace ; reducing-furnace. (47)

four de séchage (*m.*), drying-furnace ; drying-kiln ; dryer. (48)

four de séchage des moules (Fonderie) (*m.*), mould-dryer. (49)

four électrique (*m.*), electric furnace ; electric oven. (50)

four électrique pour la fabrication de l'acier (*m.*), electric steel furnace. (51)

four gallois (*m.*), Welsh furnace. (52)

four liégeois (réduction du zinc) (*m.*), Belgian furnace. (53)

four Martin *ou* four Martin-Siemens (*m.*), open-hearth furnace ; Siemens-Martin furnace. (54)

four Martin fixe (*m.*), fixed open-hearth furnace. (55)

four métallurgique (*m.*), metallurgical furnace. (56)

four oscillant (*m.*), rolling furnace. (57)

four pit (Métall.) (*m.*), soaking-pit. (58)

four potager (Métall.) (*m.*), pot-furnace ; crucible-furnace. (59)

four pour l'extraction du bismuth (*m.*), bismuth-furnace. (60)

four pour la fusion du plomb (*m.*), lead-smelting furnace. (61)

four pour le traitement des minerais d'antimoine (*m.*), antimony-furnace. (62)

four rotatif *ou* four rotatoire *ou* four tournant (*m.*), rotary furnace ; revolving furnace. (63)

four rotatif à soufflerie centrale, approprié pour le coke, pour ateliers de boulonnerie, ferronnerie, estampage, etc. (*m.*), revolving coke furnace with central air-blast, for bolt-makers, smiths, makers of stamped goods, etc. (64)

four sécheur (*m.*). Same as four de séchage.

four silésien (*m.*), Silesian furnace ; Silesian zinc furnace. (65)

four soufflé au vent chaud (*m.*), hot-blast furnace. (66)

fourche (*n.f.*), fork. (67)

fourche à coke (*f.*), coke-fork. (68)

fourche de manœuvre des courroies *ou* fourche de débrayage (*f.*), belt-fork ; strap-fork. (69)

fourcher (*v.i.*), to fork : (70)

chemin qui fourche (*m.*), road which forks. (71)

fourchette (d'un manchon d'embrayage) (*n.f.*), fork (of a clutch). (72)

fourchette (trébuchet qui soulève le fléau d'une balance) (*n.f.*), beam-support. (73)

fourchette (trébuchet qui soulève les plateaux d'une balance) (*n.f.*), pan-support. (74)

fourchette d'accrochage *ou simplement* fourchette (*n.f.*) *ou* fourche (*n.f.*) (Traction par câble sans fin) (Mines), gripping-fork ; fork. (75)

fourgon (attisoir) (*n.m.*), poker ; prick-bar ; pricker. (1)

fourgon (Ch. de f.) (*n.m.*), brake-van ; brake ; guard's van ; guard's brake-van ; van. (2)

fourgon à bagages (*m.*), luggage-van. (3)

fourgon à bestiaux (*m.*), cattle-box. (4)

fourgon de queue (*m.*), rear brake-van. (5)

fourgon de tête (*m.*), front brake-van. (6)

fourgon pour train à marchandises (*m.*), goods-brake. (7)

fourgonner le feu, to poke, to stir, to prick, the fire. (8)

fourneau (*n.m.*), furnace ; stove. (9) See also four.

fourneau à chalumeau (*m.*), blowpipe-furnace. (10)

fourneau à charbon (*m.*), charcoal-pit. (11)

fourneau à chauffer les rivets (*m.*), rivet-heating furnace. (12)

fourneau à essais (*m.*), assay-furnace ; test-furnace. (13)

fourneau à gaz (cheminée à gaz) (*m.*), gas-stove. (14)

fourneau à pétrole (1) avec mèche ou (2) sans mèche (*m.*), oil-stove, petroleum-oil stove, (1) with wick or (2) without wick or wickless. (15 ou 16)

fourneau de mine (*m.*), blast-hole ; shot-hole ; mine-chamber ; mining-hole ; powder-mine. (17)

fournir (*v.t.*), to furnish ; to provide ; to supply ; to contribute : (18)

fournir les fonds nécessaires, to furnish the necessary funds ; to provide the money required. (19)

fournir à une dépense, to contribute to an expense. (20)

fournir (se) chez un marchand de la localité, to get one's supplies from a local dealer. (21)

fournisseur (pers.) (*n.m.*), supplier ; merchant. (22)

fourniture (*n.f.*), supply ; supplying ; provision ; store : (23)

fourniture de vivres, supply of provisions ; supply of food. (24)

fournitures (*n.f.pl.*), supplies ; requisites ; stores : (25)

fournitures de bureau et de dessin, office and drawing requisites. (26)

fournitures de mine, mining supplies ; mining requisites. (27)

fournitures pour usines, engineers' stores. (28)

fourrage (nourriture pour les animaux) (*n.m.*), fodder ; provender. (29)

fourré (*n.m.*), thicket ; jungle. (30)

fourreau (*n.m.*), cylinder ; sleeve : (31)

contre-pointe à fourreau (*f.*), cylinder tail-stock. (32)

broche coulissant dans un fourreau (*f.*), spindle sliding in a sleeve. (33)

fourreau (d'une pompe) (*n.m.*), priming-pipe (of a pump). (34)

fourrure (certain nombre de tours de câble accumulés sur l'estomac, sans les dérouler à chaque cordée) (*n.f.*), spare laps of rope on the drum. (35)

fourrure (couvre-joint) (*n.f.*), welt ; cover ; covering-plate ; fish ; fish-plate ; wrapper. (36)

fourrure (cale pour maintenir l'écartement de l'ornière) (Ch. de f.) (*n.f.*), flangeway-filling ; filler ; filler-block ; distance-block ; blocking-piece. (37)

fowlérite (Minéral) (*n.f.*), fowlerite. (38)

foyaïte (Pétrol.) (*n.f.*), foyaite. (39)

foyer (lieu où on fait le feu) (*n.m.*), hearth : (40)

la cendre du foyer, the cinders on the hearth. (41)

foyer (*n.m.*) ou foyer de chaudière, furnace ; boiler-furnace : (42)

le foyer d'une chaudière à vapeur, the furnace of a steam-boiler. (43)

foyer (d'une locomotive) (*n.m.*), fire-box, inner fire-box, (of a locomotive). (44)

foyer (d'un four à réverbère) (*n.m.*), fire-chamber (of a reverberatory furnace). (45)

foyer (d'un incendie) (*n.m.*), seat (of a fire). (46)

foyer (Phys.) (*n.m.*), focus : (47)

rayons qui convergent vers le foyer (*m.pl.*), rays which converge towards the focus. (48)

le foyer d'une lentille, d'un prisme, the focus of a lens, of a prism. (49 ou 50)

foyer catalan (*m.*), Catalan forge ; Catalan furnace ; Biscayan forge ; forge. (51)

foyer central (Géol.) (*m.*), central fire : (52)

la contraction de la croûte terrestre par refroidissement du foyer central, the con-traction of the earth's crust by cooling of the central fire. (53)

foyer d'aérage (Mines) (*m.*), ventilating-furnace ; ventilation-furnace ; underground furnace ; air-furnace ; dumb furnace ; furnace. (54)

foyer en porte-à-faux (locomotive) (*m.*), over-hung fire-box ; overhanging fire-box. (55)

foyer fixe (Photogr.) (*m.*), fixed focus. (56)

foyer fumivore (*m.*), smoke-consuming furnace. (57)

foyer intérieur cylindrique (*m.*), cylindrical flue ; flue (as of Cornish or Lancashire boiler). (58)

foyer mécanique (*m.*), mechanical stoker ; stoker ; fire-feeder. (59)

foyer mécanique à alimentation par en dessous (*m.*), underfeed stoker. (60)

foyer mécanique à alimentation par le dessus (*m.*), overfeed stoker. (61)

foyer réel (Opt.) (*m.*), real focus. (62)

foyer virtuel (Opt.) (*m.*), virtual focus. (63)

foyers conjugués (Opt.) (*m.pl.*), conjugate foci. (64)

fraction (*n.f.*), fraction. (65)

fractionnement (des huiles) (*n.m.*), fractionation (of oils). (66)

fractionner (*v.t.*), to fractionate. (67)

fracture (*n.f.*), fracture ; breaking : (68)

fracture d'une serrure, breaking of a lock. (69)

fracture (Géol.) (*n.f.*), fracture : (70)

fracture de l'écorce terrestre, fracture of the earth's crust. (71)

fracturer (*v.t.*), to fracture ; to break. (72)

fracturer (se) (*v.r.*), to fracture ; to break. (73)

fracturite (Explosif) (*n.f.*), fracturite. (74)

fragile (*adj.*), fragile ; frail ; weak ; brittle. (75)

fragilité (*n.f.*), fragility ; brittleness : (76)

fragilité du verre, brittleness of glass. (77)

fragment (*n.m.*), fragment ; piece ; bit ; scrap ; chip : (78)

fragments angulaires de quartz, angular frag-ments of quartz. (79)

fragmentaire (*adj.*), fragmentary ; fragmental. (1)

fragmentation (*n.f.*), breakage ; breakage into fragments ; breaking up : (2)
fragmentation du charbon dans les couloirs, breakage of the coal in the chutes. (3)

fragmenter (*v.t.*), to break ; to break up : (4)
opération qui présente l'inconvénient de fragmenter le charbon et d'augmenter la proportion de menu (*f.*), operation which has the disadvantage of breaking up the coal and increasing the percentage of slack. (5)

fragmenter (**se**) (*v.r.*), to break ; to break up : (6)
toit de mine qui se fragmente en tombant (*m.*), mine-roof which breaks in falling. (7)

frais, fraîche (*adj.*), fresh : (8)
cassure fraîche (*f.*) fresh fracture. (9)
eau fraîche (*f.*), fresh water. (10)
nouvelles fraîches (*f.pl.*), fresh news. (11)

frais (*n.m.pl.*), expenses ; cost ; costs ; charges : (12)
frais d'emmagasinage *ou* frais de magasinage, storage-charges. (13)
frais d'entretien des routes, cost of upkeep of roads. (14)
frais d'exploitation, working-expenses ; work--ing-costs. (15)
frais de débarquement, landing-charges. (16)
frais de transformation, smelting-costs ; returning-charges. (17)
frais de transport, transport-charges. (18)
frais de voyage *ou* frais de déplacement, travelling-expenses. (19)
frais divers, sundry charges. (20)
frais élevés d'épuisement, high pumping-costs ; heavy pumping-costs. (21)
frais mensuels, monthly expenses. (22)

fraisage (action d'évaser l'orifice d'un trou déjà percé) (*n.m.*), countersinking. (23)

fraisage (action de travailler, d'entailler le métal) (*n.m.*), milling : (24)
fraisage des dents d'engrenage, milling gear-teeth. (25)

fraisage d'angle (*m.*), angle-milling ; angular milling. (26)

fraisage d'après calibre (*m.*), jig-milling. (27)

fraisage de côté (*m.*), side-milling. (28)

fraisage de face (*m.*), face-milling. (29)

fraisage de forme (*m.*), form-milling. (30)

fraisage des pièces en série (*m.*), gang-milling ; multiple milling. (31)

fraisage des rainures (*m.*), slot-milling. (32)

fraisage en bout (*m.*), end-milling. (33)

fraisage en plan (*m.*), plane-milling ; surface-milling. (34)

fraise (*n.f.*) *ou* fraise pour fraiser les trous (outil en forme de cône renversé, servant à évaser l'orifice d'un trou déjà percé), countersink : countersink-bit : (35)
fraise à bois, wood-countersink. (36)
fraise pour vilebrequins, countersink for ratchet-braces. (37)
fraise à couteau, snail countersink ; snail-horn countersink-bit. (38)
fraise taillée, rose countersink ; rose-head countersink-bit. (39)

fraise (petite roue dentée qui sert à couper les métaux, à refendre les roues d'engrenages, etc.) (*n.f.*), milling-cutter ; mill ; cutter ; milling-tool. (40)

fraise à champ demi-rond (*f.*), half-round cutter. (41)

fraise à champ rond (*f.*), round-edge cutter ; round-nose cutter. (42)

fraise à défoncer (*f.*), side-and-face cutter. (43)

fraise à dents dégagées *ou* fraise à denture à dépouille (*f.*), relieved cutter ; backed off cutter. (44)

fraise à denture droite (*f.*), milling-cutter with straight teeth ; straight mill. (45)

fraise à denture hélicoïdale (*f.*), milling-cutter with spiral teeth ; spiral mill. (46)

fraise à denture interrompue *ou* fraise à denture recoupée (*f.*), nicked-tooth mill. (47)

fraise à fine denture, en acier rapide (*f.*), high-speed cutter, with fine teeth. (48)

fraise à lames rapportées *ou* fraise à dents rapportées *ou* fraise à lames amovibles (*f.*), inserted-blade milling-cutter ; milling-cutter with inserted teeth. (49)

fraise à moyeu (*f.*), cutter with hub. (50)

fraise à quart de rond (*f.*), quarter-round cutter. (51)

fraise à queue *ou* fraise à manche (opp. à fraise à trou) (*f.*), cutter with shank. (52)

fraise à queue conique (*f.*), taper-shank mill. (53)

fraise à queue cylindrique (*f.*), straight-shank mill ; parallel-shank mill. (54)

fraise à tailler les engrenages (*f.*), gear-cutter ; gear-tooth cutter. (55)

fraise à tailler les engrenages à développante (*f.*), involute gear-cutter. (56)

fraise à tailler les fraises (*f.*), cutter for mills. (57)

fraise à tarauds (*f.*), cutter for taps ; tap-cutter. (58)

fraise à trois tailles (*f.*), side-and-face cutter. (59)

fraise à trou (*f.*), cutter with hole. (60)

fraise à trou taraudé (*f.*), cutter with screwed hole. (61)

fraise avec denture à droite (*f.*), right-hand cutter. (62)

fraise avec denture à gauche (*f.*), left-hand cutter. (63)

fraise axiale (*f.*), face-cutter ; facing-mill. (64)

fraise composée (*f.*), interlocking milling-cutter. (65)

fraise concave (*f.*), concave cutter. (66)

fraise convexe (*f.*), convex cutter. (67)

fraise creuse (*f.*), shell mill ; hollow cutter ; running down cutter. (68)

fraise cylindrique (*f.*), parallel milling-cutter. (69)

fraise d'angle (*f.*), angle-cutter ; angular cutter ; angle-mill ; bevel cutter. (70)

fraise d'angle, avec l'une des faces en bout (*f.*), single-angle cutter ; single angular cutter. (71)

fraise d'angle, avec les deux faces inclinées (*f.*), double-angle cutter ; double angular cutter. (72)

fraise de bout (*f.*), end-mill. (73)

fraise de côté (*f.*), side-mill. (74)

fraise de face (*f.*), face-cutter ; face-mill ; facing-mill. (75)

fraise de forme (*f.*), formed cutter ; formed milling-cutter. (76)

fraise destinée à fraiser les forets hélicoïdaux (*f.*), cutter for making twist-drills ; twist-drill cutter. (77)

fraise ébarbeuse, mâle, ou femelle (*f.*),   tube burr remover ; coning-tool ; pipe coning-tool ; burring-reamer ; tube-reamer; male, or female. (1 *ou* 2)

fraise en bout (*f.*), end-mill. (3)

fraise pour congés, droite, gauche, ou droite et gauche (*f.*), corner-rounding cutter, right-hand, left-hand, or double. (4 *ou* 5 *ou* 6)

fraise pour dégrossir les rainures (*f.*), roughing slot-mill. (7)

fraise pour finir les rainures (*f.*), finishing slot-mill. (8)

fraise pour peignes (*f.*), hob. (9)

fraise pour rainures *ou* **fraise pour faire les rainures et les mortaises** (*f.*), slotting-cutter ; slot-mill. (10)

fraise pour rainures à T (*f.*), T-slot cutter. (11)

fraise pour tailler les alésoirs (*f.*), cutter for grooving reamers ; reamer-cutter. (12)

fraise pour tailler les queues-d'aronde (*f.*), dove-tail-cutter. (13)

fraise radiale (*f.*), end-mill. (14)

fraisement (*n.m.*). Same as **fraisage**.

fraiser (évaser l'orifice d'un trou pour recevoir la tête d'une vis) (*v.t.*), to countersink : (15)

fraiser un trou pour une vis, to countersink a hole for a screw. (16)

fraiser (travailler, entailler le métal) (*v.t.*), to mill : (17)

fraiser une rainure de cale, les rainures d'une mèche hélicoïdale, to mill a keyway, the grooves of a twist-drill. (18 *ou* 19)

fraiser entre pointes, to mill between centres. (20)

fraiseur (pers.) (*n.m.*), miller. (21)

fraiseuse (*n.f.*), miller ; milling-machine. (22)

fraiseuse genre raboteuse [fraiseuses genre raboteuses *pl.*] (*f.*), slabbing miller ; slabber ; planomiller. (23)

fraiseuse horizontale (*f.*), horizontal milling-machine. (24)

fraiseuse universelle (*f.*), universal milling-machine. (25)

fraiseuse verticale (*f.*), vertical milling-machine. (26)

See **machine à fraiser** for other varieties.

fraisil (*n.m.*), breeze ; coal-cinders. (27)

fraisure (*n.f.*), countersink : (28)

la fraisure d'un trou, d'un rivet, the counter-sink of a hole, of a rivet. (29 *ou* 30)

franc, franche (*adj.*), free : (31)

marchandise franche de tout droit (*f.*), duty-free goods. (32)

franc de port. Same as **franco de port.**

franc-mineur [francs-mineurs *pl.*] (pers.) (*n.m.*), free miner. (33)

franchir (*v.t.*). See examples : (34)

franchir des rapides (Navigation fluviale), to shoot rapids. (35)

franchir le seuil d'une porte, to cross the threshold of a door. (36)

franchir les aiguilles (en parlant des wagons d'un train de ch. de f.), to pass over, to clear, the points. (37)

franchir les bornes du devoir, to overstep the bounds of duty. (38)

franchir toutes sortes de difficultés, to sur-mount all sorts of difficulties ; to overcome all kinds of difficulties. (39)

franchir un fossé, to jump over, to leap over, to clear, a ditch. (40)

franchir un signal d'arrêt, to run past a stop-signal ; to pass beyond a danger-signal. (41)

franchir une barrière, to pass, to go beyond, to break through, a barrier. (42)

franchir une courbe (se dit de matériel roulant), to pass through, to traverse, to take, to negotiate, a curve : (43)

wagons très longs qui peuvent franchir sans fatigue des courbes de faible rayon (*m.pl.*), very long cars which can take sharp curves without strain. (44)

franchir une vallée (se dit d'un pont), to bridge, to bridge over, to span, a valley. (45)

franchissable (*adj.*), passable : (46)

une rivière franchissable, a passable river. (47)

franchissement (*n.m.*), the act expressed by the verb **franchir**, in any of its senses ; as, le franchissement d'un fossé, jumping over a ditch. (48)

franco (*adv.*) *ou* **franco de port**, free ; carriage free ; carriage paid ; post free : (49)

marchandises rendues franco (*f.pl.*), goods delivered free. (50)

expédier un paquet franco de port, to send a package carriage paid. (51)

une instruction jointe à chaque instrument est envoyée gratis et franco sur demande, instructions for each instrument are sent gratis and post free on application. (52)

franco à bord *ou* **franco bord**, free on board ; F.O.B. ; f.o.b. (53)

franco gare *ou* **franco sur wagon**, free on rail ; F.O.R. ; f.o.r. : (54)

les prix s'entendent franco gare Londres (*m.pl.*), the prices are quoted F.O.R. London. (55)

franco pied d'œuvre, free on site ; delivered site. (56)

frange (Opt.) (*n.f.*), fringe. (57)

frange d'interférence (*f.*), interference fringe. (58)

frangibilité (*n.f.*), frangibility ; frangibleness ; breakableness ; brittleness. (59)

frangible (*adj.*), frangible ; breakable ; brittle. (60)

franklinite (Minéral) (*n.f.*), franklinite. (61)

frappant, -e (*adj.*), striking : (62)

l'analogie frappante qui existe entre les météorites et les éléments terrestres d'origine profonde (*f.*), the striking analogy which exists between meteorites and terrestrial elements of deep origin. (63)

une particularité frappante, a striking peculiarity. (64)

frappe (*n.f.*) *ou* **frappe supérieure** (de marteau-pilon) (opp. à **tas** *ou* **tas inférieur**), tup-pallet, top pallet, pallet for tup, die for tup, top pallet, top die, (of power-hammer). (65)

frappe-devant [frappe-devant *pl.*] (*n.m.*), sledge-hammer ; sledge ; about-sledge. (66) (blacksmiths' : for use with both hands and swung at arm's length over the head : distinguished from **marteau à main**, uphand sledge).

frapper (donner un ou plusieurs coups à) (*v.t.*), to strike ; to tap ; to knock : (67)

frapper avec un marteau, to strike, to tap, with a hammer. (68)

frapper un coup sur, to tap : (69)

frapper un coup sur un vase contenant un corps en surfusion, to tap a vessel contain-ing a body in surfusion. (70)

frapper à une porte, to knock at a door. (1)

frapper (donner une empreinte à) (*v.t.*), to stamp : (2)

frapper la marque du fournisseur sur une pièce de forge, to stamp the brand of the manufacturer on a forging. (3)

frapper le chiffre de son poids sur un lingot, to stamp the weight of the metal on an ingot. (4)

frappeur (marteau de frappeur) (*n.m.*), striker. (5)

frappeur (aide de forgeron) (pers.) (*n.m.*), striker. (6)

frappeuse pour boulons, rivets, crampons, etc. (*f.*), bolt-forging machine, for making bolts, rivets, spikes, etc. (7)

frase (*n.f.*). Same as fraise.

frasil (*n.m.*). Same as fraisil.

frayer (se) (*v.r.*). See examples : (8)

se frayer sa voie sans guide ni assistance, to grope one's way without guide or assistance. (9)

se frayer un passage au travers des vapeurs toxiques, to fight one's way through poisonous fumes. (10)

torrent qui se fraye un chemin à travers les débris morainiques anciennement déposés par un glacier (*m.*), torrent which forces a passage through morainic debris formerly deposited by a glacier. (11)

freibergite (Minéral) (*n.f.*), freibergite. (12)

freieslébénite (Minéral) (*n.f.*), freieslebenite. (13)

frein (*n.m.*), brake. (14)

frein à air comprimé (*m.*), compressed-air brake ; air-brake ; atmospheric brake. (15)

frein à bande *ou* **frein à ruban** *ou* **frein à sangle** *ou* **frein à enroulement** (*m.*), band-brake ; ribbon-brake ; strap-brake. (16)

frein à corde (*m.*), rope-brake. (17)

frein à huile (*m.*), oil dash-pot. (18)

frein à levier (*m.*), lever-brake. (19)

frein à main (*m.*), hand-brake. (20)

frein à palettes (*m.*), fan-brake. (21)

frein à pied (*m.*), foot-brake. (22)

frein à sabot (*m.*), brake with shoe. (23)

frein à vapeur (*m.*), steam-brake. (24)

frein à vide (*m.*), vacuum brake. (25)

frein à vide automatique (*m.*), automatic vacuum brake. (26)

frein à vide de Smith (*m.*), Smith's vacuum brake. (27)

frein à vide duplex automatique (*m.*), duplex automatic vacuum brake. (28)

frein à vis (*m.*), screw-brake. (29)

frein automatique (*m.*), automatic brake ; self-acting brake. (30)

frein automatique continu (*m.*), continuous automatic brake. (31)

frein continu (*m.*), continuous brake. (32)

frein continu à air comprimé système Wenger (*m.*), Wenger continuous atmospheric brake. (33)

frein continu automatique à air comprimé Westinghouse (*m.*), Westinghouse continuous automatic atmospheric brake. (34)

frein d'écrou (*m.*), nut-lock. (35)

frein d'urgence (*m.*), emergency brake. (36)

frein de Prony (*m.*), Prony brake. (37)

frein de retenue (pour la descente des tubages de puits) (*m.*), casing-clamps, pipe-clamps, pipe-clambs, (for lowering well-casing). (38)

frein du bogie (locomotive) (*m.*), truck-brake. (39)

frein dynamométrique (*m.*), dynamometrical brake ; absorption dynamometer. (40)

frein électrique (*m.*), electric brake. (41)

frein hydraulique (*m.*), hydraulic brake. (42)

frein modérable (*m.*), overspeed-brake. (43)

freinage (action) (*n.m.*), applying the brake ; putting on the brake ; braking. (44)

freinage (système de freins) (*n.m.*), brake-gear : (45)

un freinage puissant, a powerful brake-gear. (46)

freinage brusque *ou* **freinage brusqué** (*m.*), applying the brake suddenly ; sudden application of the brake ; jamming on the brake. (47)

freinage modéré en fin de cordée (Mines) (*m.*), overspeed braking at end of hoist. (48)

freiner (*v.t.*), to brake ; to apply the brake to : (49)

freiner un wagon, to brake a wagon. (50)

freiner (*v.i.*), to brake ; to apply the brake ; to put on the brake : (51)

freiner à la descente, to put on the brake going downhill. (52)

freineur (pers.) (*n.m.*), brakeman ; brakesman ; braker. (53)

freineur de plan incliné (*m.*), incline-braker. (54)

freinter (Mines) (*v.t.*). Same as freiner.

freinteur (Mines) (*n.m.*). Same as freineur.

frêne (arbre) (*n.m.*), ash. (55)

fréquence (Élec.) (*n.f.*), frequency : (56)

courant à haute fréquence (*m.*), high-frequency current. (57)

vague à basse fréquence (*f.*), low-frequency wave. (58)

fréquence du grisou dans les chantiers (*f.*), prevalence of fire-damp in the workings. (59)

fréquent, -e (*adj.*), frequent. (60)

fret (*n.m.*), freight. (61)

fret d'aller (*m.*), outward freight. (62)

fret de retour (*m.*), homeward freight. (63)

frettage (*n.m.*), hooping ; ferruling ; banding ; binding ; binding with hoop-iron. (64)

frette (*n.f.*), hoop ; collar ; band ; ring ; ferrule. (65)

frette autour d'un manchon (*f.*), collar round a coupling ; ring round a sleeve. (66)

frette de moyeu (*f.*), nave-band ; nave-hoop. (67)

frette de pieu (*f.*), pile-hoop ; pile-ferrule. (68)

fretter (*v.t.*), to hoop ; to collar ; to band ; to bind ; to bind with hoop-iron ; to ring ; to ferrule : (69)

fretter une roue hydraulique, to hoop a water-wheel. (70)

pilot fretté d'un cercle de fer (*m.*), pile bound with an iron hoop. (71)

béton fretté (*m.*), hooped concrete ; stirruped concrete. (72)

fretter les deux bouts d'une planche d'échafaud, to bind the two ends of a scaffold-board with hoop-iron. (73)

fretter un pilotis, to ferrule a pile. (74)

friabilité (*n.f.*), friability : (75)

friabilité d'un minéral, friability of a mineral. (76)

friable (*adj.*), friable : (77)

terre friable (*f.*), friable earth. (78)

friction (*n.f.*), friction. (79). See also frotte-ment.

friedélite (Minéral) (*n.f.*), friedelite. (80)

frieséite (Minéral) (*n.f.*), frieseite. (81)

frigorifique (*adj.*), refrigerating ; freezing : (82)

appareil frigorifique (*m.*), refrigerating appa-
-ratus ; freezing apparatus. (1)
frittage (*n.m.*), roasting ; calcination. (2)
fritter (*v.t.*), to roast ; to calcine. (3)
froid, -e (*adj.*), cold : (4)
eau froide (*f.*), cold water. (5)
froid (à), cold (in the sense of *when cold*) ; when
cold : (6)
substance qui est plus soluble à chaud qu'à
froid (*f.*), substance which is more soluble
hot than cold. (7)
le verre, très mauvais conducteur à froid,
devient conducteur au rouge, glass, a very
bad conductor when cold, becomes a conductor
when red-hot. (8)
froid (absence de chaleur) (*n.m.*), cold. (9)
Cf. degrés de froid.
froid (temps froid) (*n.m.*), cold ; cold weather.
(10)
froid humide (Météor.) (*m.*), damp cold. (11)
froid sec (Météor.) (*m.*), dry cold. (12)
froideur (*n.f.*), coldness. (13)
front (Mines, etc.) (*n.m.*), face. (14) Cf.
front de taille.
front (d'une colline) (*n.m.*), brow (of a hill). (15)
front d'avancement (Mines) (*m.*), heading-face.
(16)
front de dragage (*m.*), dredging-face ; digging-
face. (17)
front de gravier (Dragage) (*m.*), gravel-face.
(18)
front de taille *ou* **front d'attaque** *ou* **front
d'abatage** *ou simplement* **front** (*n.m.*) (Mines),
working-face ; face under attack ; stope-
face ; face ; breast ; forebreast ; forefield ;
bank. (19)
front de taille du charbon (*m.*), coal-face ; coal-
wall. (20)
front de taille en surplomb *ou* **front de taille en
porte-à-faux** (*m.*), overhanging face. (21)
front de taille léché par le courant d'air (*m.*),
working-face swept by the air-current. (22)
frontal, -e, -aux (*adj.*), frontal. (23)
frontal (marteau frontal) (*n.m.*), frontal hammer ;
frontal helve. (24)
frontière (*n.f.*), frontier : boundary ; border.
(25)
frottage (*n.m.*), rubbing. (26)
frottant, -e (*adj.*), rubbing ; frictionable : (27)
surface frottante (*f.*), rubbing-surface. (28)
frottement (*n.m.*), rubbing ; friction. (29)
frottement (Méc. & Phys.) (*n.m.*), friction : (30)
frottement d'un pivot, d'un collier, des essieux,
friction of a pivot, of a collar, of axles. (31
ou 32 *ou* 33)
frottement de l'air dans les galeries de mine,
de l'eau dans les tuyaux, friction of air in
mine-galleries, of water in pipes. (34 *ou*
35)
frottement à vide (*m.*), friction running light. (36)
frottement au départ (*m.*), starting friction ;
friction of rest ; friction at starting from
rest ; static friction. (37)
frottement de glissement *ou* **frottement de la
première espèce** *ou simplement* **frottement**
(*n.m.*), sliding friction. (38)
frottement de roulement *ou* **frottement de la
seconde espèce** (*m.*), rolling friction. (39)
frottement en charge (*m.*), friction under load.
(40)
frottement extérieur (*m.*), external friction. (41)
frottement intérieur (*m.*), internal friction. (42)

frottement pendant le mouvement *ou* **frottement
pendant la marche** *ou* **frottement en mouve-
-ment** (*m.*), friction of motion. (43)
frotter (*v.t.*), to rub : (44)
frotter d'huile une surface, to rub a surface
with oil. (45)
frotteur (ch. de f. électrique) (*n.m.*), shoe ;
collecting-shoe. (46)
frotteur (dynamo) (*n.m.*), current-collector ;
brush. (47)
frotteur souterrain (tramway électrique) (*m.*),
plow ; plough ; underground trolley ; shoe.
(48)
fructueux, -euse (*adj.*), fruitful ; profitable ;
successful. (49)
Frue vanner [**Frue vanners** *pl.*] (Préparation
mécanique des minerais) (*m.*), Frue vanner.
(50)
fruit (Maçonn.) (*n.m.*), batter : (51)
fruit d'un mur, batter of a wall. (52)
fuchsite (Minéral) (*n.f.*), fuchsite. (53)
fuir (laisser échapper) (*v.i.*), to leak : (54)
tonneau qui fuit (*m.*), cask which leaks. (55)
fuir (s'échapper) (*v.i.*), to leak out ; to escape :
(56)
eau qui fuit (*f.*), water which leaks out (*or*
escapes). (57)
gaz qui fuit (*m.*), gas which escapes (*or* leaks
out). (58)
fuite (échappement) (*n.f.*), escape ; leakage ;
leak : (59)
fuite d'eau, leakage, leak, escape, of water.
(60)
fuite de gaz, de vapeur, escape, leakage, leak,
of gas, of steam. (61 *ou* 62)
fuite (fissure) (*n.f.*), leak : (63)
fuite dans un tuyau, dans un tonneau, dans
un fossé, leak in a pipe, in a cask, in a ditch.
(64 *ou* 65 *ou* 66)
fuite électrique (*f.*), electric leakage. (67)
fuites de fonte (*f.pl.*), spillings. (68)
fulguration (Métall.) (*n.f.*), fulguration ; brighten-
-ing ; blick. (69)
fulgurite (*n.m.*), fulgurite ; lightning-tube. (70)
fulmicoton (*n.m.*), guncotton ; nitro-cotton. (71)
fulminant, -e (*adj.*), fulminating : (72)
amorce fulminante (*f.*), fulminating cap. (73)
fulminate (*n.m.*), fulminate. (74)
fulminate de mercure (*m.*), fulminate of mercury ;
mercury fulminate ; fulminating mercury.
(75)
fulmination (*n.f.*), fulmination. (76)
fulminer (*v.i.*), to fulminate : (77)
l'or fulmine avant d'être chauffé jusqu'au
rouge (*m.*), the gold fulminates before
becoming red hot. (78)
fulminique (Chim.) (*adj.*), fulminic : (79)
acide fulminique (*m.*), fulminic acid. (80)
fumant, -e (*adj.*), smoking : (81)
fumée (*n.f.*), smoke ; fume : (82)
fumée du charbon de terre, de la poudre,
smoke from coal, from gunpowder. (83
ou 84)
fumées produites par les explosifs, fumes of
explosives. (85)
fumer (*v.i.*), to smoke : (86)
charbon qui fume beaucoup (*m.*), coal which
smokes a lot. (87)
fumerolle *ou* **fumerole** (Géol.) (*n.f.*), fumarole.
(88)
fumivore (*adj.*), smoke-consuming ; fumivorous :
(89)

foyer fumivore (m.), smoke-consuming furnace. (1)

fumivore (n.m.), smoke-consumer; smoke-burner; smoke-preventer. (2)

fumivorité (n.f.), consumption of smoke (in smoke-consuming apparatus): (3)
fumivorité des locomotives, consumption of smoke in locomotives. (4)

funiculaire (adj.), funicular: (5)
polygone funiculaire (m.), funicular polygon. (6)

funiculaire (chemin de fer funiculaire) (n.m.), cable-railway; cable-railroad; cable-road; funicular railway. (7)

furlong (mesure linéaire anglaise) (n.m.), furlong =201·168 metres. (8)

fusain (n.m.), wood-coal (a variety of lignite having the appearance of charcoal). (9)

fuseau (d'une chaîne à fuseau) (n.m.), pintle, stud, (of a pintle-chain, of a stud-chain). (10)

fuseaux (d'une lanterne) (n.m.pl.), trundles (of a lantern-wheel); rundles (of a lantern); rounds (of a wallower). (11)

fusée (Mach.) (n.f.), journal; neck: (12)
fusée d'un essieu, journal, neck, of an axle. (13)

fusée (Travail à la poudre) (n.f.), fuse. (14)
See also **mèche, étoupille**, and **amorce**.

fusée à combustion lente (f.), slow-burning fuse; slow-match. (15)

fusée de sûreté (f.), safety-fuse; common fuse; Bickford fuse. (16)

fusée percutante (f.), percussion-fuse. (17)

fuser (se fondre par l'action de la chaleur) (v.i.), to melt away; to burn away: (18)
bougie qui fuse trop vite (f.), candle which burns away too quickly. (19)

fuser (se réduire en poudre) (v.i.), to slake: (20)
quand ils sont très calcaires, les laitiers fusent, ou tombent en poussière après refroidissement, when high in lime, slags slake, or crumble to dust after cooling. (21)

fuser (déflagrer) (v.i.), to deflagrate: (22)
tous les azotates fusent lorsqu'on les jette sur des charbons ardents (m.pl.), all nitrates deflagrate when thrown on live coal. (23)

fusibilité (n.f.), fusibility; fusibleness; fluxibility; fluxibleness; fluxility: (24)
fusibilité d'un métal, des laitiers, fusibility of a metal, of the slags. (25 ou 26)

fusible (adj.), fusible; fusile; fluxible; fluxile; meltable: (27)
alliage fusible (m.), fusible alloy. (28)

fusible (Élec.) (n.m.), fuse; fuse-wire; safety-fuse: (29)
terre trop faible pour faire jouer un fusible (f.), earth too weak to blow out a fuse. (30)

fusible (bouchon fusible) (chaudière) (n.m.), fusible plug; safety-plug. (31)

fusiforme (adj.), fusiform; spindle-shaped: (32)
cristal fusiforme (m.), fusiform crystal. (33)

fusion (n.f.), melting; smelting; fusing; fusion: (34) Cf. **fondre**.
fusion des neiges d'hiver, melting of the winter snows. (35)
fusion et affinage de la fonte et des riblons chargés à froid, melting and refining of cast iron and swarf charged cold. (36)
fusion de l'étain, tin-smelting. (37)
métal en fusion (m.), molten metal; metal in fusion. (38)

fusion (de concessions) (n.f.), amalgamation (of claims). (39)

fusion aqueuse (f.), aqueous fusion; watery fusion: (40)
la chaleur fait subir à l'alun la fusion aqueuse, heat causes alum to undergo aqueous fusion. (41)

fusion aquo-ignée (Géol.) (f.), aqueo-igneous fusion; hydrothermal fusion. (42)

fusion directe (Métall.) (f.), direct fusion; direct smelting. (43)

fusion électrique (f.), electric smelting; electro-smelting. (44)

fusion électrothermique (f.), electrothermic smelting. (45)

fusion ignée (f.), igneous fusion. (46)

fusion plombeuse (Métall.) (f.), lead fusion. (47)

fusionner plusieurs concessions, to amalgamate several concessions. (48)

fût (tonneau) (n.m.), barrel; cask; drum: (49)
fût à huile ou pétrole, tôle galvanisée, galvanized-iron oil-drum. (50)

fût (d'une colonne) (Arch.) (n.m.), shaft (of a column or pillar). (51)

fût (d'une grue) (n.m.), post, jib-post, (of a crane). (52)

fût (d'un rivet) (n.m.), shank (of a rivet). (53)

fût (d'un rabot) (n.m.), stock (of a plane). (54)

fût pour forerie (m.), crank-brace. (55)

fût-pylône (n.m.), trellis-post. (56)

futaie (arbre) (n.f.), timber-tree. (57)

futaille (n.f.), cask; barrel. (58)

futur, -e (adj.), future: (59)
le rendement futur, the future yield. (60)

futur (n.m.), future. (61)

G

gabare (n.f.), lighter; barge. (62)

gabariage (n.m.), gauging; gaging. (63)

gabarier (pers.) (n.m.), lighterman; bargeman; bargee. (64)

gabarier (v.t.), to gauge; to gage: (65)
gabarier une plaque métallique, to gauge a metal plate. (66)

gabarit ou **gabari** (n.m.), gauge; gage; templet; template; pattern; former; patron. (67)

gabarit (Moulage en terre) (Fonderie) (n.m.), template; templet; strickle; strike; sweep; former. (68)

gabarit à bandages (m.), wheel-gauge (for gauging treads of car-wheels). (69)

gabarit d'écartement ou **gabarit d'écartement de voie** ou **gabarit de voie** ou **simplement gabarit** (n.m.) (Ch. de f.), rail-gauge; permanent-way gauge; track-gauge; gauge. (70)

gabarit de chargement ou **simplement gabarit** (n.m.) (Ch. de f.), loading-gauge; gauge. (71)

gabarit de sabotage (Ch. de f.) (m.), chairing-gauge. (72)

gabarit passe-partout (gabarit de chargement) (Ch. de f.) (m.), master-gauge. (73)

gabarit type (*m.*), standard gauge. (1)

gabarit universel (*m.*), universal gauge. (2)

gabbro (Pétrol.) (*n.m.*), gabbro. (3)

gâchage (Constr.) (*n.m.*), tempering. (4)

gâche (pour recevoir un pêne de serrure) (*n.f.*), lock-staple ; box-staple ; rim-staple ; staple ; keeper ; box ; nosing ; strike-plate ; striking-plate. (5)

gâcher (Constr.) (*v.t.*), to temper : (6)
gâcher du mortier, du plâtre, du ciment, to temper, mortar, plaster, cement. (7 *ou* 8 *ou* 9)

gâcher un travail, to spoil a piece of work ; to scamp a job. (10)

gâchette (*n.f.*), trigger. (11)

gâchis (Constr.) (*n.m.*), mortar (made of plaster, lime, sand and cement). (12)

gadolinite (Minéral) (*n.f.*), gadolinite. (13)

gadolinium (Chim.) (*n.m.*), gadolinium ; gado--linum. (14)

gagner (*v.t.*), to gain ; to earn ; to get : (15)
gagner du terrain, to gain ground ; to spread ; to make headway : (16)
méthode qui perd plutôt du terrain qu'elle n'en gagne (*f.*), method which is rather losing than gaining ground. (17)
gagner le dessus, to get the upper hand. (18)
gagner sa vie, to earn one's living ; to earn one's livelihood. (19)
gagner un bon salaire, to earn good wages ; to get good pay. (20)

gahnite (Minéral) (*n.f.*), gahnite ; zinc-spinel. (21)

gaïac (bois) (*n.m.*), lignum-vitæ. (22)

gailleterie *ou* gaillette (*n.f.*), cobbles ; cob coal. (23)

gain (*n.m.*), gain ; earnings ; profit. (24)

gaine (*n.f.*), case ; cover ; guard ; casing ; sheath : (25)
gaine en cuir à courroie *ou* gaine en cuir à courroie, pour porter en bandoulière, leather case with sling strap ; leather sling case. (26)
gaine métallique, metallic sheath. (27)
gaine protectrice contre les accidents, cover to prevent accidents ; gear-case ; gear-casing ; gear-cover ; gear-guard ; gear-box ; wheel-guard. (28)

gaine (Géol.) (*n.f.*), gangue ; matrix. (29)

gainer (*v.t.*), to case ; to cover ; to sheath : (30)
appareil photographique gainé de cuir (*m.*), camera covered with leather. (31)

gainerie (*n.f.*) *ou* gainage (*n.m.*) (d'un appareil photographique), covering (of a camera). (32)

gaize (Pétrol.) (*n.f.*), gaize. (33)

gaize (Sondage) (*n.f.*), extremely hard strata met with in earth-boring. (34)

galandage (*n.m.*), partition. (35)

galène (Minéral) (*n.f.*), galena ; galenite ; lead-glance. (36)

galénobismuthite (Minéral) (*n.f.*), galenobis--mutite. (37)

galénocératite (Minéral) (*n.f.*), phosgenite. (38)

galère (four) (*n.f.*), gallery-furnace. (39)

galère (demi-varlope) (Charp.) (*n.f.*), jack-plane. (40)

galerie (*n.f.*), gallery : (41)
galerie filtrante (Trav. publ.), filter-gallery. (42)

galerie (Mines) (*n.f.*), level ; drive ; drivage ; drift ; road ; roadway ; way ; passageway ; gallery. (43) See also **voie.**

galerie à boisage complet (*f.*), full-timbered level. (44)

galerie à boisage jointif *ou* galerie à boisage serré (*f.*), close-timbered level. (45)

galerie à demi-boisage (*f.*), half-timbered level. (46)

galerie à flanc de coteau (*f.*), adit. (47)

galerie au charbon (*f.*), coal-road. (48)

galerie au rocher (*f.*), rock-drift ; stone-drift ; metal-drift ; gallery in dead ground. (49)

galerie barrée (*f.*), fenced-off gallery ; fenced-off road. (50)

galerie chassante (*f.*). Same as **galerie en direction.**

galerie costresse (*f.*), subdrift ; counter-level ; counter-gangway ; counter. (51)

galerie d'aérage (*f.*), airway ; windway ; air-level. (52)

galerie d'allongement (*f.*). Same as **galerie en direction.**

galerie d'appel d'air (*f.*), intake ; intake-airway ; intake-drift ; intake-heading. (53)

galerie d'asséchement (*f.*). Same as **galerie de drainage.**

galerie d'avancement (*f.*), heading ; head. (54)

galerie d'écoulement (*f.*). Same as **galerie de drainage.**

galerie d'évacuation d'air (*f.*), return-airway. (55)

galerie d'évacuation d'eau (*f.*). Same as **galerie de drainage.**

galerie dans les remblais (*f.*), gob-road. (56)

galerie de chassage (*f.*). Same as **galerie en direction.**

galerie de découpage du gîte (*f.*), blocking-out level. (57)

galerie de desserte *ou* galerie desservant la taille (*f.*), gate-road ; gateway ; gate ; stall-road. (58)

galerie de direction (*f.*). Same as **galerie en direction.**

galerie de drainage (*f.*), drainage-level ; water-level ; watercourse-level ; water-adit ; off--take. (59)

galerie de fond (*f.*), bottom level. (60)

galerie de forme triangulaire (*f.*), triangular passageway. (61)

galerie de navigation (*f.*), boat-level. (62)

galerie de niveau (*f.*), level. (63)

galerie de niveau de mine *ou simplement* galerie de mine (*f.*), mine-level. (64)

galerie de niveau débouchant au jour (*f.*), day-drift ; day-level ; day-hole. (65)

galerie de recherches *ou* galerie de prospection (*f.*), prospecting-level ; exploration-level ; pioneer-level ; pioneer-heading ; drift ; monkey-drift. (66)

galerie de refoulement d'air (*f.*). Same as **galerie d'appel d'air.**

galerie de retour d'air (*f.*), return-airway. (67)

galerie de roulage (*f.*), haulageway ; haulage-road ; haulage-level ; tram-level ; wagon--way ; wagon-road ; rolleyway ; hutch-road. (68)

galerie de roulage à double voie (*f.*), double-track haulage-road. (69)

galerie de roulage à la cale (*f.*), sprag-road. (70)

galerie de roulage à voie unique (*f.*), single-track haulage-road. (71)

galerie de roulage en palier (*f.*), level haulage--way. (72)

galerie de traçage (*f.*), development-headway ; winning-headway ; winning-level ; fore--winning-heading. (73)

galerie de traînage (*f.*), drawing-road. (74)

galerie en couche (*f.*), level in the seam. (1)

galerie en cul-de-sac (*f.*), blind level ; dumb drift. (2)

galerie en direction (*f.*), drift ; drifting-level ; driftway ; drive (parallel in direction to the strike) ; level running along the strike. (3)

galerie en rocher (*f.*). Same as **galerie au rocher**.

galerie inclinée (*f.*), inclined level ; inclined road ; slant. (4)

galerie interdite (*f.*), fenced-off gallery ; fenced-off road. (5)

galerie latérale (*f.*), side entry. (6)

galerie maîtresse *ou* **galerie principale** (*f.*), main road ; mainway ; main level ; gangway ; entry. (7)

galerie principale d'aérage (*f.*), main airway. (8)

galerie principale de roulage (*f.*), main haulage--way. (9)

galerie simple (*f.*), single entry. (10)

galerie transversale (*f.*), cross-heading. (11)

galeries jumelles *ou* **galeries conjuguées** (*f.pl.*), parallel entries. (12)

galeries jumelles doubles *ou* **galeries conjuguées doubles** (*f.pl.*), double entries. (13)

galeries jumelles triples *ou* **galeries conjuguées triples** (*f.pl.*), triple entries. (14)

galet (petit caillou) (*n.m.*), pebble ; pebble-stone. (15) *In the plural*, **galets**, pebbles ; pebble-stones ; pebble-stone ; shingle ; *thus*, **grève de galets** (*f.*), shingle beach ; pebble beach.

galet (gros galet) (*n.m.*), boulder ; bowlder. (16)

galet (*n.m.*) *ou* **galet mécanique**, roller ; wheel ; runner. (17)

galet (*n.m.*) *ou* **galet de roulement** (pour chariot de pont roulant, ou analogue), travelling wheel ; running-wheel ; runner ; rail-wheel (of runner of overhead travelling crane, or the like). (18 *ou* 19)

galet à deux joues (pour chariot de pont roulant) (*m.*), double-flanged travelling wheel. (20)

galet à gorge (Méc.) (*m.*), grooved roller. (21)

galet à une joue (pour chariot de pont roulant) (*m.*), single-flanged travelling wheel. (22)

galet de diable (*m.*), truck-wheel (hand-truck). (23)

galet de friction *ou simplement* **galet** (*n.m.*), friction-roller ; roller. (24)

galet de la came (*m.*), cam-follower ; cam-roller. (25)

galet de renvoi *ou* **galet-guide** (*n.m.*) [**galets-guides** *pl.*] *ou simplement* **galet** (*n.m.*), idler ; belt-idler ; idler-pulley ; idle pulley ; idler-wheel ; guide-pulley ; guide ; leading-on pulley ; runner. (26)

galet de silex (*m.*), flint pebble ; flint ball. (27)

galet pivotant (*m.*), caster ; castor. (28)

galet plein (pour chariot de pont roulant) (opp. à **galet à joues**) (*m.*), plain travelling wheel ; flat runner. (29)

galet presseur (*m.*), pressure-roller. (30)

galet tendeur (*m.*), tension-roller ; tension-pulley ; jockey-roller ; jockey-pulley ; jockey-wheel ; jockey ; belt-tightener ; tightening-pulley ; binding-pulley ; idle pulley ; idler ; belt-idler ; idler-pulley. (31)

galets abandonnés (Mines) (*m.pl.*), boulders run to waste. (32)

galets charriés (Géol.) (*m.pl.*), drift-boulders. (33)

galets de houille (Géol.) (*m.pl.*), coal-balls. (34)

galets impressionnés (Géol.) (*m.pl.*), impressed pebbles. (35)

galibot (Mines) (pers.) (*n.m.*), boy ; lad ; pit-boy ; foal. (36)

gallinace (*n.f.*), obsidian ; volcanic glass. (37)

gallium (Chim.) (*n.m.*), gallium. (38)

gallon (mesure anglaise de capacité) (*n.m.*), gallon = 4·5459631 litres. (39)

gallon (mesure de capacité pour les liquides en usage dans les États-Unis) (*n.m.*), gallon = 3·785310 litres. (40)

galoche (poulie coupée) (*n.f.*), snatch-block ; return-block. (41)

galoche (support sur lequel repose le sommier d'une grille de foyer) (*n.f.*), bearer-bracket ; bearer-cradle. (42)

galop (mouvement d'une locomotive) (*n.m.*) hunting. (43)

galvanique (*adj.*), galvanic : (44)

pile galvanique (*f.*), galvanic pile. (45)

galvaniquement (*adv.*), galvanically. (46)

galvanisation (*n.f.*), galvanization ; galvanizing ; zincing ; zinking ; zincification : (47)

galvanisation de la tôle, galvanization of sheet iron ; zincing sheet iron. (48)

galvanisé, -e (*adj.*), galvanized ; zinked : (49)

fil de fer galvanisé (*m.*), galvanized-iron wire. (50)

galvaniser (*v.t.*), to galvanize ; to zinc ; to zincify : (51)

galvaniser du fer, to galvanize, to zinc, to zincify, iron. (52)

galvanomètre (*n.m.*), galvanometer. (53)

galvanomètre à aimant mobile (*m.*), movable-iron galvanometer. (54)

galvanomètre à courant mobile (*m.*), suspended-coil galvanometer ; movable-coil galva--nometer. (55)

galvanomètre à mercure (*m.*), mercury-galva--nometer. (56)

galvanomètre absolu (*m.*), absolute galva--nometer. (57)

galvanomètre apériodique (*m.*), aperiodic gal--vanometer ; dead-beat galvanometer. (58)

galvanomètre astatique (*m.*), astatic galva--nometer. (59)

galvanomètre balistique (*m.*), ballistic galva--nometer. (60)

galvanomètre des tangentes (*m.*), tangent galvanometer. (61)

galvanomètre différentiel (*m.*), differential galvanometer. (62)

galvanomètre Thomson (lord Kelvin) (*m.*), Lord Kelvin's mirror galvanometer. (63)

galvanoplastie (reproduction d'objets par élec--trolyse) (*n.f.*), galvanoplasty ; galvano--plastics. (64)

galvanoplastie (dépôt sur un objet d'une mince couche de métal par électrolyse) (*n.f.*), electroplating ; plating ; electrodeposition. (65)

galvanoplastique (qui a rapport à la reproduction d'objets par électrolyse) (*adj.*), galvano--plastic. (66)

galvanoplastique (qui a rapport au dépôt d'une couche de métal par électrolyse) (*adj.*), electroplating ; plating : (67)

procédé galvanoplastique (*m.*), electroplating process. (68)

bain galvanoplastique (*m.*), plating bath. (69)

gamelle (*n.f.*), pan ; dish ; batea. (70)

gamme de vitesses (*f.*), range, set, of speeds. (71)

gangue (*n.f.*), gangue ; matrix ; gangue-matter. (72)

gangue encaissante *ou* **gangue enrobante** (*f.*), enclosing matrix. (73)

gangue fusible (Métall.) (*f.*), self-fluxing gangue. (1)

gangue rocheuse (*f.*), rock matrix. (2)

gangue terreuse (*f.*), earthy gangue. (3)

ganister (*n.m.*), ganister; gannister : (4)
　　revêtement de ganister d'un cubilot (*m.*), ganister lining of a cupola-furnace. (5)

gants en caoutchouc (*m. pl.*), india-rubber gloves. (6)

garage (Ch. de f.) (*n.m.*), shunting; turning off (into a siding); switching; switching off. (7)

garage (pour automobiles) (*n.m.*), garage (for motor-cars). (8)

garant (d'un palan) (*n.m.*), hauling end, running end, (of a tackle). (9)

garanti, -e (*adj.*), guaranteed; warranted : (10)
　　machine garantie pour — ans (*f.*), machine warranted for — years. (11)

garantie (*n.f.*), guarantee; warranty. (12)

garantir (*v.t.*), to guarantee; to warrant. (13)

garde (Serrur.) (*n.f.*), ward : (14)
　　les gardes d'une serrure, the wards of a lock. (15)

garde (pour empêcher les barres de s'enrouler autour des cylindres de laminoir) (*n.f.*), guard. (16)

garde (pers.) (*n.m.*), watchman; keeper; tender; caretaker. (17)

garde-barrière [gardes-barrière *ou* gardes-barrières *pl.*] (*n.m.*), gatekeeper; gateman. (18)

garde de nuit (*m.*), night-watchman. (19)

garde-feu [gardes-feu *ou* gardes-feux *pl.*] (*n.m.*), fire-tender. (20)

garde-fou [garde-fous *pl.*] *ou* **garde-corps** [garde-corps *pl.*] (*n.m.*), hand-rail; guard-rail; railing. (21)

garde-fourneau [gardes-fourneau *ou* gardes-fourneaux *pl.*] (*n.m.*), helper (man who assists the keeper of a blast-furnace). (22)

garde-fraisil [garde-fraisil *pl.*] *ou* **garde-frasil** [garde-frasil *pl.*] (forge) (*n.m.*), coal-box. (23)

garde-frein [gardes-frein *ou* gardes-freins *pl.*] (*n.m.*), brakeman; brakesman; braker. (24)

garde-ligne [gardes-ligne *ou* gardes-lignes *pl.*] (Ch. de f.) (*n.m.*), lineman; trackman; track-walker. (25)

garde-magasin [gardes-magasin *ou* gardes-magasins *pl.*] (*n.m.*), storekeeper. (26)

garde-mines [gardes-mines *pl.*] (*n.m.*), assistant mining-engineer. (27)

garder quelqu'un à son service, to retain someone in one's service. (28)

gardien (pers.) (*n.m.*), watchman; keeper; caretaker. (29)

gardien de nuit (*m.*), night-watchman. (30)

gardien de porte (*m.*), gatekeeper; gateman. (31)

gare (*n.f.*), station; railway-station; principal station (as opposed to an intermediate station or way station). (32)

gare (*interjection*), look out (*imperative*). (33)

gare d'arrivée (câble aérien) (*f.*), discharging-station; off-loading station; knock-out station. (34)

gare d'embranchement *ou* **gare de bifurcation** (*f.*), junction. (35)

gare d'évitement (*f.*), sidings. (36)

gare de départ (câble aérien) (*f.*), loading-station. (37)

gare de marchandises *ou* **gare de transbordement** (*f.*), goods-station. (38)

gare de passage (*f.*), intermediate station; way station. (39)

gare de triage (*f.*), yard; switching-yard; railroad-yard; marshalling-yard; drill-yard; yard for making up trains. (40)

gare la mine, fire (the warning cry used when a blast is to be fired). (41)

gare terminus *ou* **gare de tête de ligne** (*f.*), terminus; terminal. (42)

garer (Ch. de f.) (*v.t.*), to shunt; to turn off (into a siding); to run off (into a siding); to switch; to switch off; to side-track : (43)
　　garer un train, to shunt, to switch off, to side-track, a train. (44)

garer (se) (Ch. de f.) (*v.r.*), to shunt; to turn off (into a siding); to switch; to switch off : (45)
　　tamponner un train qui n'a pas eu le temps de se garer, to collide with a train which has not had time to shunt. (46)
　　voie secondaire où un convoi peut se garer pour laisser la voie principale libre pour un autre (*f.*), side line where a train can switch off in order to leave the main line clear for another. (47)

gareur (pers.) (*n.m.*), shunter. (48)

gargouille (pour la gouttière d'un toit) (*n.f.*), waterspout (for the gutter of a roof). (49)

gargouille (d'un fourneau) (*n.f.*), waste-gas main (of a furnace). (50)

gargouille (d'un puits de mine) (*n.f.*), water-ring, garland, garland-curb, curb-ring, (of a mine-shaft). (51)

garniérite (Minéral) (*n.f.*), garnierite; noumeite. (52)

garnir (fournir des choses nécessaires) (*v.t.*), to provide; to furnish; to supply; to fit up; to stock : (53)
　　garnir d'aubes une roue-turbine, to provide a turbine-wheel with blades; to furnish a turbine-wheel with vanes; to blade a turbine-wheel. (54)
　　garnir de marchandises un magasin, to stock a warehouse with goods. (55)

garnir (revêtir) (*v.t.*), to line : (56)
　　garnir d'antifriction un palier, to line a bearing-block with antifriction metal; to babbitt a bearing. (57)
　　garnir une poche de fonderie, to line a foundry-ladle; to daub a casting-ladle. (58)

garnir (envelopper) (*v.t.*), to lag; to jacket : (59)
　　garnir une chaudière, un cylindre, d'un calorifuge, to lag, to jacket, a boiler, a cylinder, with a non-conductor. (60 *ou* 61)

garnir (bourrer) (*v.t.*), to pack; to stuff : (62)
　　garnir un presse-étoupe, to pack a stuffing-box. (63)

garnir de palplanches, to sheet-pile. (64)

garnir de planches, to board; to line with boards; to cover with boards. (65)

garnir de tubes une chaudière, to tube a boiler. (66)

garnir la chaudière, to stoke the boiler. (67)

garnir les parois d'un puits, to lag the walls of a shaft. (68)

garnir une lampe, to trim a lamp. (69)

garnissage (revêtement) (action ou matière) (*n.m.*), lining. (70)

garnissage (revêtement de poches de fonderie, etc.) (action ou matière) (*n.m.*), lining; daubing. (71)

garnissage (bourrage) (action) (*n.m.*), packing ; stuffing. (1)

garnissage (bourrage) (matière) (*n.m.*), packing ; stuffing ; gasket. (2)

garnissage (Boisage de mines) (*n.m.*), lagging. (3)

garnissage acide (convertisseur) (*m.*), acid lining (converter). (4)

garnissage basique (convertisseur) (*m.*), basic lining. (5)

garnissage de lampes (*m.*), trimming lamps. (6)

garnissage de tubes d'une chaudière (*m.*), tubing a boiler. (7)

garniture (ce qui sert à garnir un objet, les diverses parties d'une construction, etc.) (*n.f.*) (*s'emploie aussi au pluriel*), fittings ; fit-ments ; furniture ; mountings ; requisites : (8)

appareil photographique à garniture en cuivre (*m.*), camera with brass fittings. (9)

serrure à garniture en cuivre (*f.*), lock with brass furniture. (10)

garniture (enveloppe qui forme joint hermétique autour de divers organes de machines) (*n.f.*) (*s'emploie souvent au pluriel*), packing ; packer ; stuffing ; gasket : (11)

la garniture (*ou* les garnitures) d'un presse-étoupe, the packing of a stuffing-box. (12)

garniture (enveloppe pour éviter la déperdition de chaleur par rayonnement) (*n.f.*) (*s'emploie souvent au pluriel*), lagging ; lining ; jacket-ing ; jacket ; deading ; cleading ; clothing ; sheathing. (13)

garniture (assortiment complet) (*n.f.*), set : (14)

décrasse-meule, avec 2 garnitures de molettes (*m.*), emery-wheel dresser, with two sets of cutters. (15)

garniture (d'une pompe) (*n.f.*), mountings (of a pump). (16)

garniture (garde d'une serrure) (*n.f.*) (*s'emploie généralement au pluriel*), ward (of a lock). (17)

garniture à labyrinthe (*f.*), labyrinth-packing. (18)

garniture d'étoupe (*f.*), tow packing ; junk packing. (19)

garniture de chanvre (*f.*), hemp packing. (20)

garniture de chaudière (*f.*), boiler-lagging ; boiler-deading ; boiler-cleading ; boiler-clothing ; boiler-jacket ; boiler-jacketing. (21)

garniture de piston (*f.*), piston-packing ; piston-ring ; piston packing-ring. (22)

garniture de vapeur (*f.*), steam-packing. (23)

garniture en caoutchouc (*f.*), india-rubber pack-ing ; india-rubber packer ; rubber gasket. (24)

garniture en cuir (*f.*), leather packer ; leather gasket. (25)

garniture étanche (*f.*), fluid-tight packing ; water-packer ; packer. (26)

garniture étanche de tubage (*f.*), casing-packer. (27)

garniture métallique (*f.*), metallic packing. (28)

garniture réfractaire (*f.*), refractory lining. (29)

garrot de scie (*m.*), winding-bar (for stretching the cord of a frame or bow saw). (30)

gaspillage (*n.m.*), wasting ; waste : (31)

gaspillage des placers, wasting the placers ; gophering ; coyoting. (32)

gaspiller (*v.t.*), to waste : (33)

gaspiller de la force, to waste power. (34)

gaspiller du temps et de l'argent, to waste time and money. (35)

gaspiller les placers, to waste the placers ; to gopher ; to coyote. (36)

gâteau (*n.m.*), cake : (37)

gâteaux de soufre, cakes of sulphur. (38)

gauche (*adj.*), left ; left-hand ; left-handed. (39)

gauche (*n.f.*), left ; left hand ; left side. (40)

gauche (à) (*locution adjective*), left-hand ; left-handed : (41)

vis à gauche (*f.*), left-hand screw ; left-handed screw. (42)

gauche (à) (*locution adverbiale*), left ; to the left ; left-handedly : (43)

à droite et à gauche, right and left. (44)

gauchir (en parlant de tiges, de tubes, etc.) (*v.i.*), to buckle ; to bend. (45)

gauchir (en parlant du bois) (*v.i.*), to wind ; to warp : (46)

planche qui gauchit (*f.*), board which winds. (47)

gauchir (se) (tiges, tubes, etc.) (*v.r.*), to buckle ; to bend. (48)

gauchir (se) (bois) (*v.r.*), to wind ; to warp. (49)

gauchissement (*n.m.*), buckling ; buckle ; bend-ing ; running out of true ; winding ; warp-ing : (50)

gauchissement de pièces coulées (Fonderie), warping, buckling, of castings. (51)

gault (Géol.) (*n.m.*), gault ; galt. (52)

gauss (Élec.) (*n.m.*), gauss. (53)

gave (*n.m.*), torrent. (54)

gay-lussite (Minéral) (*n.f.*), gaylussite. (55)

gaz [gaz *pl.*] (*n.m.*), gas. (56)

gaz à l'air (*m.*), air-gas. (57)

gaz à l'eau (*m.*), water-gas. (58)

gaz acétylène (*m.*), acetylene gas. (59)

gaz azote (*m.*), nitrogen gas. (60)

gaz carbonique (*m.*), carbonic-acid gas ; carbonic acid ; carbon dioxide. (61)

gaz combustible (*m.*), fuel-gas ; power-gas. (62)

gaz d'air (*m.*), air-gas. (63)

gaz d'eau (*m.*), water-gas. (64)

gaz d'échappement (*m.*), exhaust-gas. (65)

gaz d'éclairage (*m.*), lighting-gas ; illuminating-gas. (66)

gaz d'huile (*m.*), oil-gas. (67)

gaz de four à coke (*m.*), coke-oven gas. (68)

gaz de gazogène (*m.*), producer-gas ; generator-gas. (69)

gaz de haut fourneau (*m.*), blast-furnace gas. (70)

gaz de houille (*m.*), coal-gas. (71)

gaz de pétrole (*m.*), petroleum-gas ; oil-gas. (72)

gaz de ville (*m.*), town gas : (73)

faire fonctionner un moteur au gaz de ville, to run an engine on town gas. (74)

gaz dégagés par la poudre (*m.pl.*), powder-fumes. (75)

gaz délétères dégagés par le charbon (*m.pl.*), deleterious gases given off by coal. (76)

gaz des marais (*m.*), marsh-gas. (77)

gaz Dowson (*m.*), Dowson gas. (78)

gaz hydrogène (*m.*), hydrogen gas. (79)

gaz inerte (*m.*), inert gas. (80)

gaz inflammable (*m.*), inflammable gas ; fiery gas. (81)

gaz intoxicant (*m.*), poisonous gas. (82)

gaz méphitique (*m.*), mephitic gas. (83)

gaz mixte (*m.*). Same as gaz pauvre.

gaz naturel (*m.*), natural gas ; rock-gas. (84)

gaz nitrogène (*m.*), nitrogen gas. (1)
gaz occlus (*m.pl.*), occluded gases. (2)
gaz oléfiant *ou* gaz oléifiant (*m.*), olefiant gas. (3)
gaz oxhydrique (*m.*), oxyhydrogen gas. (4)
gaz oxygène (*m.*), oxygen gas. (5)
gaz parfait (*m.*), perfect gas ; ideal gas. (6)
gaz pauvre *ou* gaz pauvre de gazogène (*m.*), producer-gas ; generator-gas. (7)
gaz pauvre de haut fourneau (*m.*), blast-furnace gas. (8)
gaz pauvre par aspiration (*m.*), suction-gas. (9)
gaz pauvre par pression (*m.*), pressure-gas. (10)
gaz perdus (*m.pl.*), waste gases. (11)
gaz portatif (*m.*), portable gas. (12)
gaz provenant des explosifs (*m.pl.*), fumes of explosives. (13)
gaz puant (*m.*), sulphureted hydrogen ; hydrogen sulphide ; sulfhydric acid. (14)
gaz riche (*m.*), oil-gas. (15)
gaz toxique (*m.*), poisonous gas ; foul gas. (16)
gazéifiable (*adj.*), gasifiable. (17)
gazéificateur (*n.m.*), flare-lamp ; flare. (18)
gazéification (*n.f.*), gasification : (19)
   gazéification de l'eau, de la houille, gasification of water, of coal. (20 ou 21)
gazéifier (*v.t.*), to gasify : (22)
   gazéifier l'eau, to gasify water. (23)
gazéifier (se) (*v.r.*), to gasify. (24)
gazéiforme (*adj.*), gasiform. (25)
gazéité (*n.f.*), gaseous nature. (26)
gazeux, -euse (*adj.*), gaseous ; gassy : (27)
   état gazeux (*m.*), gaseous state. (28)
   eau gazeuse (*f.*), gassy water. (29)
gazier (pers.) (*n.m.*), gas-fitter ; gas-man. (30)
gazifère (*adj.*), gas-making : (31)
   appareil gazifère (*m.*), gas-making apparatus (32)
gazogène (*n.m.*), producer ; gas-producer ; generator ; gasogene ; gazogene. (33)
gazogène à aspiration *ou* gazogène aspiré (*m.*), suction gas-producer ; suction-producer. (34)
gazogène à fond à grille *ou* gazogène à grille (*m.*), fire-bar bottom producer ; grate gas-producer. (35)
gazogène à fond hydraulique (*m.*), water-bottom producer. (36)
gazogène à fond plein (*m.*), solid-bottom pro--ducer. (37)
gazogène à tirage naturel (*m.*), natural-draught producer. (38)
gazogène à vent soufflé *ou* gazogène soufflé (*m.*), pressure gas-producer ; pressure-producer ; forced-draught producer. (39)
gazoline (*n.f.*) *ou* gazolène (*n.m.*) *ou* gazoléine (*n.f.*), gasoline ; gasolene ; petrol ; petroleum spirit ; motor-spirit. (40)
gazomètre (*n.m.*), gasometer ; gas-holder ; gas-tank. (41)
gazométrie (*n.f.*), gasometry. (42)
gazométrique (*adj.*), gasometric ; gasometrical. (43)
gazoscope (*n.m.*), gasoscope. (44)
géant, -e (*adj.*), giant : (45)
   les sommets géants des Andes (*m.pl.*), the giant peaks of the Andes. (46)
géant (lance hydraulique) (Exploitation des placers) (*n.m.*), giant ; monitor ; monitor-nozzle. (47)
géanticlinal, -e, -aux (Géol.) (*adj.*), geanticlinal. (48)

géanticlinal (*n.m.*), geanticline ; geanticlinal. (49)
gédrite (Minéral) (*n.f.*), gedrite. (50)
gehlénite (Minéral) (*n.f.*), gehlenite. (51)
geiser (*n.m.*). Same as geyser.
gel (*n.m.*), freezing ; frost : (52)
   le phénomène du gel profond du terrain est connu depuis longtemps en Sibérie, the phenomenon of the deep freezing of the ground has long been known in Siberia. (53)
   le gel et le dégel, freezing and thawing ; frost and thaw. (54)
gelable (*adj.*), freezable. (55)
gélatine (*n.f.*), gelatine. (56)
gélatine détonante (*f.*), blasting-gelatine ; explosive gelatine. (57)
gélatine-dynamite [gélatine-dynamites *pl.*] (*n.f.*), gelatine dynamite ; nitrogelatine. (58)
gélatineux, -euse (*adj.*), gelatinous. (59)
gelé, -e (*adj.*), frozen : (60)
   explosifs gelés (*m.pl.*), frozen explosives. (61)
gelée (*n.f.*), frost. (62)
gelée à glace *ou* gelée noire (*f.*), black frost. (63)
gelée blanche (*f.*), white frost ; hoar frost. (64)
geler (*v.t.*), to freeze. (65)
geler (*v.i.*), to freeze : (66)
   l'esprit-de-vin ne gèle jamais, spirits of wine never freeze. (67)
geler (se) (*v.r.*), to freeze. (68)
gélignite (Explosif) (*n.f.*), gelignite. (69)
gélivure (dans un tronc d'arbre) (*n.f.*), frost-shake. (70)
géloxite (Explosif) (*n.f.*), geloxite. (71)
gemme (*adj.*), gem (*used as adj.*) ; gemmeous : (72)
   pierre gemme (*f.*), gem-stone. (73)
   cristaux gemmes (*m.pl.*), gem-crystals. (74)
gemme (*n.f.*), gem ; gem-stone. (75)
gemmifère (*adj.*), gemmiferous ; gem-bearing. (76)
gêner (*v.t.*), to hinder ; to hamper ; to embarrass ; to interfere with : (77)
   être gêné (-e) par le manque d'eau, to be hampered by want of water. (78)
gêner (se) mutuellement, to get in each other's way : (79)
   l'atelier doit être suffisamment vaste pour permettre aux ouvriers de travailler à l'aise, sans se gêner mutuellement (*m.*), the work--shop should be sufficiently roomy to permit of the men working at ease, without getting in each other's way. (80)
général, -e, -aux (*adj.*), general. (81)
généralement (*adv.*), generally. (82)
généraliser (se) (*v.r.*), to become general : (83)
   le pavé de bois se généralise, wood-paving is becoming general. (84)
générateur, -trice (*adj.*), generating : (85)
   machine génératrice d'électricité (*f.*), electric generating machine. (86)
générateur (*n.m.*), generator. (87)
générateur (*n.m.*) *ou* génératrice (*n.f.*) (machine dynamo-électrique), generator ; dynamo ; generatrix. (88)
générateur (*n.m.*) *ou* génératrice (*n.f.*) (opp. à récepteur *ou* réceptrice) (Élec.), generator ; source. (89) (opp. to load *or* sink) :
   toute installation ayant pour but de transporter l'énergie sous forme d'électricité comprend une ou plusieurs génératrices du courant, des conducteurs pour l'amener aux lieux

d'emploi, parfois des transformateurs pour en modifier la nature, enfin des réceptrices qui l'utilisent en la convertissant (*f.*), every installation having for its object to convey power in the form of electricity comprises one or more current-generators, conductors to bring it to the places of employ- -ment, sometimes transformers to modify the nature of it, finally loads which utilize it by converting it. (1)

générateur à courant alternatif (*m.*), alternating-current generator ; alternating-current dynamo ; alternating dynamo ; alternator. (2)

générateur à courant continu (*m.*), direct-current generator ; continuous-current dynamo. (3)

générateur à courant triphasé (*m.*), three-phase generator ; triphase generator. (4)

générateur d'acétylène (*m.*), acetylene-generator. (5)

générateur d'oxygène (*m.*), oxygen-generator. (6)

générateur de courant (Élec.) (*m.*), current-generator. (7)

générateur de gaz (*m.*), gas-generator. (8)

générateur de vapeur *ou simplement* **générateur** (*n.m.*), steam-boiler ; boiler. (9)

Note.—For the many varieties of steam-boilers, see the synonymous word **chaudière**.

générateur pyromagnétique d'électricité (*m.*), pyromagnetic generator. (10)

génératrice (Élec.) (*n.f.*). Same as **générateur**.

génératrice (Géom.) (*n.f.*), generatrix ; generator. (11)

génératrice à courant alternatif, continu, etc. Same as **générateur à courant alternatif, continu, etc.**

génératrice de courant (*f.*), current-generator. (12)

genèse (*n.f.*), genesis : (13)

genèse des roches, des dépôts stannifères, genesis of rocks, of tin-bearing deposits. (14 *ou* 15)

génétique (*adj.*), genetic ; genetical. (16)

génétiquement (*adv.*), genetically. (17)

génie civil (*m.*), civil engineering. (18)

genou (Méc.) (*n.m.*), knee ; elbow-joint. (19)

genou de Cardan (*m.*), Cardan joint ; cardan joint ; universal joint. (20)

genouillère (*n.f.*), knuckle ; knuckle-joint. (21)

genre (sorte) (*n.m.*), kind ; sort ; type : (22)

les leviers se divisent en trois genres (*m.pl.*), levers are divided into three kinds. (23)

tours genre américain (*m.pl.*), lathes of the American type. (24)

genre (catégorie composée d'espèces) (Minéralogie) (*n.m.*), group : (25)

minéral qui est une espèce du genre grenat (*m.*), mineral which is a species of the garnet group. (26)

genre chlorite, chlorite group. (27)

genre clintonite, clintonite group ; brittle micas. (28)

genre humite, humite group. (29)

genre pyroxène, pyroxene group. (30)

genre wernérite, scapolite group. (31)

genthite (Minéral) (*n.f.*), genthite ; nickel-gymnite. (32)

gentilhomme (haut fourneau) (*n.m.*), dam-plate (blast-furnace). (33)

géochimie (*n.f.*), geochemistry. (34)

géochimique (*adj.*), geochemical. (35)

géocratique (opp. à **hydrocratique**) (Géol.) (*adj.*), geocratic. (36) (opposed to **hydrocratic**).

géode (rognon tapissé de cristaux) (Géol.) (*n.f.*), geode. (37)

géode (druse ; poche à cristaux) (Géol.) (*n.f.*), geode ; druse ; vug ; vugh. (38)

géodésie (*n.f.*), geodesy ; geodetics. (39)

géodésien *ou* **géodésiste** (pers.) (*n.m.*), geodesist ; geodesian. (40)

géodésique (*adj.*), geodesic ; geodesical ; geodetic ; geodetical : (41)

ligne géodésique (*f.*), geodesic line ; geodetic line. (42)

géodésiquement (*adv.*), geodetically. (43)

géodique (Géol.) (*adj.*), geodic ; geodal. (44)

géodynamique (*adj.*), geodynamic ; geodynamical. (45)

géodynamique (*n.f.*), geodynamics. (46)

géogénie (*n.f.*), geogeny. (47)

géogénique (*adj.*), geogenic ; geogenical. (48)

géognosie (*n.f.*), geognosy. (49)

géognostique (*adj.*), geognostic ; geognostical. (50)

géographe (pers.) (*n.m.*), geographer. (51)

géographie (*n.f.*), geography. (52)

géographie commerciale (*f.*), commercial geography. (53)

géographie mathématique (*f.*), mathematical geography. (54)

géographie physique (*f.*), physical geography. (55)

géographier (*v.t.*), to map ; to lay down a map of : (56)

géographier le bassin d'un fleuve, to map the basin of a river ; to lay down a map of a river-basin. (57)

géographique (*adj.*), geographic ; geographical : (58)

carte géographique (*f.*), geographical map. (59)

géographiquement (*adv.*), geographically. (60)

géologie (*n.f.*), geology. (61)

géologie appliquée (*f.*), applied geology. (62)

géologie appliquée aux mines (*f.*), geology applied to mining. (63)

géologie comparée (*f.*), comparative geology. (64)

géologie expérimentale (*f.*), experimental geology. (65)

géologie stratigraphique (*f.*), stratigraphic geology. (66)

géologie sur le terrain (*f.*), field-geology. (67)

géologique (*adj.*), geologic ; geological : (68)

étude géologique des terrains traversés (*f.*), geological survey of the strata passed through. (69)

géologiquement (*adv.*), geologically. (70)

géologue (pers.) (*n.m.*), geologist. (71)

géologue-conseil [géologues-conseils *pl.*] (*n.m.*), consulting-geologist. (72)

géologue sur le terrain (*m.*), field-geologist. (73)

géomètre (pers.) (*n.m.*), geometrician ; geometer. (74)

géomètre du cadastre (*m.*), Ordnance surveyor. (75)

géométrie (*n.f.*), geometry : (76)

la géométrie, cette suprême expression de l'ordre naturel—*Lapparent*, geometry, that supreme expression of natural order. (77)

une géométrie ne saurait être plus vraie qu'une autre : elle est seulement plus ou moins commode—*H. Poincaré,* one geometry cannot be more true than another : it is only more or less convenient. (1)

géométrie à trois dimensions (*f.*), solid geometry ; geometry of three dimensions. (2)

géométrie analytique (*f.*), analytical geometry. (3)

géométrie cinématique (*f.*), kinematic geometry. (4)

géométrie dans l'espace (*f.*), geometry of figures in space. (5)

géométrie descriptive (*f.*), descriptive geometry. (6)

géométrie élémentaire (*f.*), elementary geometry. (7)

géométrie euclidienne (*f.*), Euclidean geometry. (8)

géométrie non-euclidienne (*f.*), non-Euclidean geometry. (9)

géométrie plane (*f.*), plane geometry. (10)

géométrie projective (*f.*), projective geometry ; graphic geometry. (11)

géométrie sphérique (*f.*), spherical geometry. (12)

géométrie supérieure (*f.*), higher geometry. (13)

géométrie synthétique (*f.*), synthetic geometry. (14)

géométrique (*adj.*), geometric ; geometrical : (15)

plan géométrique (*m.*), geometric plane. (16)

dessin géométrique (*m.*), geometrical drawing. (17)

géométriquement (*adv.*), geometrically. (18)

géomorphie (*n.f.*), geomorphy. (19)

géomorphique (*adj.*), geomorphic. (20)

géomorphogénie (*n.f.*), geomorphogeny. (21)

géomorphogénique (*adj.*), geomorphogenic. (22)

géonomie (*n.f.*), geonomy. (23)

géonomique (*adj.*), geonomic. (24)

géophysique (*n.f.*), geophysics. (25)

géostatique (*adj.*), geostatic. (26)

géostatique (*n.f.*), geostatics. (27)

géosynclinal, -e, -aux (Géol.) (*adj.*), geosynclinal. (28)

géosynclinal (*n.m.*), geosyncline ; geosynclinal. (29)

géothermal, -e, -aux *ou* **géothermique** (*adj.*), geothermal ; geothermic : (30)

degré géothermique (*m.*), geothermic degree. (31)

géothermie (*n.f.*), geothermy. (32)

géothermomètre (*n.m.*), geothermometer. (33)

géothermométrique (*adj.*), geothermometric. (34)

gérance (*n.f.*), management. (35)

gérant (pers.) (*n.m.*), manager. (36)

gerbe d'étincelles (*f.*), shower of sparks : (37)

convertisseur qui donne une gerbe d'étincelles (*m.*), converter which gives off a shower of sparks. (38)

gercer (*v.t.*), to crack : (39)

le soleil gerce la terre, the sun cracks the ground. (40)

gerçure (*n.f.*), crack ; surface crack. (41)

gerçure (dans le bois) (*n.f.*), shake (in timber). (42)

gérer (*v.t.*), to manage. (43)

germanium (Chim.) (*n.m.*), germanium. (44)

gersdorffite (Minéral) (*n.f.*), gersdorffite ; nickel-glance. (45)

gestion (*n.f.*), management. (46)

geyser (*n.m.*), geyser. (47)

geysérien, -enne (*adj.*), geyseric ; geyserine. (48)

geysérite (Minéral) (*n.f.*), geyserite. (49)

gibbsite (Minéral) (*n.f.*), gibbsite. (50)

gibet de puisatier (*m.*), shear-legs ; sheer-legs. (51)

gieseckite (Minéral) (*n.f.*), gieseckite. (52)

giffard (*n.m.*), Giffard's injector. (53)

gigantolite (Minéral) (*n.f.*), gigantolite. (54)

gilbert (unité C.G.S.) (Élec.) (*n.m.*), gilbert. (55)

gilbertite (Minéral) (*n.f.*), gilbertite. (56)

gill (mesure anglaise de capacité) (*n.m.*), gill = 1·42 decilitres. (57)

gill (mesure de capacité pour les liquides en usage dans les États-Unis) (*n.m.*), gill = 1·18291 decilitres. (58)

globertite (Minéral) (*n.f.*), giobertite ; magnesite. (59)

girasol (*n.m.*) *ou* **girasolle** (*n.f.*) (Minéral), girasol ; girasole. (60)

giration (*n.f.*), gyration. (61)

giratoire (*adj.*), gyratory ; gyrating : (62)

mouvement giratoire (*m.*), gyratory motion. (63)

giron (d'une marche d'escalier) (*n.m.*), tread (of a stair-step). (64)

girouette (*n.f.*), vane ; weather-vane ; wind-vane ; weathercock. (65)

gisement (Géol.) (*n.m.*), bed ; seam ; deposit. (66) See also **gîte, couche,** and **lit.**

gisement d'argile (*m.*), deposit of clay ; clay-seam ; bed of clay. (67)

gisement d'eau douce (*m.*), fresh-water deposit. (68)

gisement de gaz (*m.*), gas-strata. (69)

gisement de gypse (*m.*), gypsum-deposit. (70)

gisement de soude (*m.*), soda-deposit. (71)

gisement ébouleux (*m.*), seam that runs [seams that run *pl.*]. (72)

gisement en amas, en couche, etc. See gise--ments en amas, en couche, etc.

gisement en place (*m.*), deposit in place ; in-place deposit. (73)

gisement houiller (*m.*), coal-bed ; coal-deposit ; coal-seam. (74)

gisement marin (*m.*), marine deposit. (75)

gisement minéral *ou* **gisement minier** (*m.*), mineral deposit. (76)

gisement pétrolifère (*m.*), oil-bearing strata. (77)

gisements alluvionnaires (Placers) (*m.pl.*), dig--gings ; alluvial diggings ; placer-deposits. (78)

gisements alluvionnaires d'or (*m.pl.*), gold-diggings. (79)

gisements en amas (*m.pl.*), massive deposits. (80)

gisements en couche (*m.pl.*), bedded deposits. (81)

gisements en filons (*m.pl.*), lode deposits. (82)

gismondine (Minéral) (*n.f.*), gismondite. (83)

gîte (Géol.) (*n.m.*), bed ; seam ; deposit. (84) See also **gisement, couche,** and **lit.**

gîte alluvionnaire (*m.*), alluvial deposit. (85)

gîte aurifère (*m.*), auriferous deposit ; gold-bearing deposit. (86)

gîte d'eau (*m.*), water-logged bed ; water-logged strata. (87)

gîte d'émanation (*m.*), sublimation vein. (88)

gîte d'épaisseur moyenne (*m.*), seam of medium thickness. (89)

gîte d'étain (*m.*), tin-deposit. (1)

gîte d'exsudation (*m.*), exudation-vein. (2)

gîte d'imprégnation (*m.*), impregnation. (3)

gîte de contact (*m.*), contact-bed. (4)

gîte de cuivre (*m.*), copper-deposit. (5)

gîte de fer (*m.*), iron-deposit. (6)

gîte de graviers cimentés (Placers) (*m.*), drift-claim. (7)

gîte de pétrole (*m.*), oil-bearing strata. (8)

gîte de plomb (*m.*), lead-deposit. (9)

gîte de sécrétion (*m.*), secretion-vein. (10)

gîte de substitution (*m.*), irregular ore-formation. (11)

gîte de surface (*m.*), surface-deposit. (12)

gîte de zinc (*m.*), zinc-deposit. (13)

gîte en amont pendage (*m.*), hanging wall seam. (14)

gîte en aval pendage (*m.*), foot-wall seam. (15)

gîte exploitable (*m.*), workable deposit. (16)

gîte filonien (*m.*), lode ; vein. (17)

gîte gazeux (*m.*), gas-strata. (18)

gîte houiller (*m.*), coal-bed ; coal-deposit ; coal-seam. (19)

gîte mince (*m.*), thin seam ; scanty bed. (20)

gîte minéral (*m.*), mineral deposit. (21)

gîte pétrolifère (*m.*), oil-bearing strata. (22)

gîte puissant (*m.*), thick deposit ; thick seam. (23)

gîte stannifère (*m.*), tin-deposit : (24)
les gîtes stannifères de Cornouailles, the tin-deposits of Cornwall. (25)

glaçage (*n.m.*), glazing ; glossing : (26)
glaçage des meules d'émeri, glazing of emery-wheels. (27)
glaçage des épreuves photographiques, glazing, glossing, photographic prints. (28)

glace (eau congelée) (*n.f.*), ice. (29) (sometimes used in plural in French where only singular can be used in English. See glace flottante for example).

glace (température de congélation de l'eau) (*n.f.*), freezing ; freezing-point : (30)
le thermomètre est à la glace, the thermometer is at freezing. (31)

glace (miroir) (*n.f.*), mirror. (32)

glace (tache dans une pierre précieuse) (*n.f.*), flaw. (33)

glace de banquise (*f.*), pack-ice. (34)

glace de glacier (*f.*), glacial ice. (35)

glace de réflexion (*f.*), reflecting mirror. (36)

glace de vitrage (*f.*), plate glass. (37)

glace du tiroir *ou* glace de distribution *ou* glace du cylindre (*f.*), valve-seat ; port-face ; cylinder port-face. (38)

glace flottante (*f.*) *ou* glaces flottantes (*f.pl.*), floe-ice. (39)

glace fondante (*f.*), melting ice. (40)

glace fossile (*f.*), fossil ice. (41)

glacé (*n.m.*), glaze ; gloss. (42)

glacer (congeler) (*v.t.*), to freeze. (43)

glacer (vernir d'une couche transparente ; couvrir d'une croûte) (*v.t.*), to glaze ; to gloss : (44)
glacer une épreuve (Photogr.), to glaze, to gloss, a print. (45)

glacer (se) (se congeler) (*v.r.*), to freeze. (46)

glacer (se) (se couvrir d'une croûte) (*v.r.*), to glaze : (47)
les meules en émeri à grain fin sont plus sujettes à se glacer que celles à grain gros (*f.pl.*), fine-grained emery-wheels are more liable to glaze than coarse-grained wheels. (48)

glaceur (plaque pour glacer les épreuves) (Photogr.) (*n.m.*), glazer ; glazing-pad ; print-glazer. (49)

glaciaire (Géol.) (*adj.*), glacial ; glacier (*used as adj.*) : (50)
érosion glaciaire (*f.*), glacial erosion. (51)
boue glaciaire (*f.*), glacier-mud. (52)

glacial, -e, -als (*adj.*), glacial ; icy ; frigid : (53)
vent glacial (*m.*), icy wind. (54)

glaciation (*n.f.*), glaciation. (55)

glacier (*n.m.*), glacier ; ice-river. (56)

glacier alpin (*m.*), Alpine glacier. (57)

glacier composé *ou* glacier polysynthétique (*m.*), compound glacier. (58)

glacier continental (*m.*), continental glacier ; continental ice-sheet. (59)

glacier d'écoulement (*m.*), ice-tongue. (60)

glacier de cirque (*m.*), cliff-glacier ; corrie-glacier. (61)

glacier de piedmont *ou* glacier du pied des monts *ou* glacier de type alaskien (*m.*), piedmont glacier ; malaspina glacier. (62)

glacier de plateau (*m.*), plateau-glacier. (63)

glacier de vallée (*m.*), valley-glacier. (64)

glacier local (*m.*), local glacier. (65)

glacier remanié (*m.*), recemented glacier ; regenerated glacier ; reconstructed glacier. (66)

glacier suspendu (*m.*), hanging glacier. (67)

glaciologie (*n.f.*), glaciology. (68)

glaciologiste (pers.) (*n.m.*), glaciologist. (69)

glaçon (*n.m.*), floe : (70)
les glaçons d'une débâcle, the floes of a debacle. (71)

glaisage (*n.m.*), claying ; pugging ; puddling : (72)
glaisage d'un trou de mine, claying a shot ; bulling a hole. (73)

glaise (*n.f.*), clay ; clay soil. (74)

glaise (Hydraul.) (*n.f.*), pug ; puddle ; pugging ; puddling ; puddled clay : (75)
bassin enduit de glaise (*m.*), reservoir lined with puddle. (76)

glaiser (*v.t.*), to clay ; to pug ; to puddle : (77)
glaiser les berges d'un canal, to pug, to puddle, the banks of a canal. (78)
glaiser un trou de mine, to clay a shot ; to bull a hole ; to pug a hole. (79)

glaiseux, -euse (*adj.*), clayey ; clay (*used as adj.*) : (80)
sol glaiseux (*m.*), clay soil. (81)

glaisière (*n.f.*), clay-pit. (82)

glaner des renseignements, to glean informa-tion. (83)

glasérite (Minéral) (*n.f.*), glaserite. (84)

glaubérite (Minéral) (*n.f.*), glauberite. (85)

glaucodot *ou* glaucodote (Minéral) (*n.m.*), glauco-dot. (86)

glaucolite (Minéral) (*n.f.*), glaucolite. (87)

glauconie *ou* glauconia (Minéral) (*n.f.*), glauconite. (88)

glauconieux, -euse (*adj.*), glauconitic. (89)

glauconifère (*adj.*), glauconiferous. (90)

glaucophane (Minéral) (*n.f.*), glaucophane. (91)

glèbe (Mines) (*n.f.*), glebe. (92)

glène (*n.f.*), coil (of rope). (93)

gléner un cordage, to coil a rope. (94)

glissant, -e (*adj.*), slippery : (95)
rails glissants (*m.pl.*), slippery rails. (96)

glissement (action) (*n.m.*), slipping ; slip ; sliding : (97)
glissement d'une courroie, slipping, slip, creep, of a belt. (98)

glissement du terrain (*ou* des terrains), slipping, slump, of the ground. (1)

glissement du toit (Mines), slip of the roof. (2)

glissement (éboulement) (*n.m.*), slip ; slide ; slump ; fall : (3)

glissement de terrains (*ou* de terres), landslide ; landslip ; landfall ; earth-fall ; slippings. (4)

glissement (opp. à roulement) (Méc.) (*n.m.*), sliding. (5) (opp. to rolling) :

frottement de glissement (*m.*), sliding friction. (6)

glissement (cisaillement) (Résistance des matériaux) (*n.m.*), slide ; shear. (7)

glissement (Élec.) (*n.m.*), slip. (8)

glissement (Cristall.) (*n.m.*), gliding ; slip. (9)

glisser (*v.i.*), to slip ; to slide ; to slump : (10)

courroie qui glisse (*f.*), belt which slips. (11)

le piston doit glisser librement dans le cylindre, the piston should slide freely in the cylinder. (12)

pierres qui glissent sur une pente (*f.pl.*), stones which slide down a slope. (13)

glissette (Géom.) (*n.f.*), glissette. (14)

glissière (de machine-outil, etc.) (*n.f.*), slide (of machine-tool, etc.). (15)

glissière de crosse *ou* glissière de la tête de piston (*f.*), cross-head guide ; guide-bar ; guide ; slide-bar ; slide-rod. (16)

glissière de détente (de tiroir de détente) (*f.*), expansion-plate, expansion-slide, cut-off plate, (of expansion-valve, of cut-off valve). (17)

glissière de plaque de garde *ou* glissière de la boîte à huile (*f.*), axle-guide ; axle-box guide ; pedestal-horn ; journal-box guide. (18)

globe (*n.m.*), globe. (19)

globe opalin (Verrerie) (*m.*), opal globe. (20)

globulaire (*adj.*), globular ; guttate. (21)

globule (*n.m.*), globule. (22)

globuleux, -euse (*adj.*), globulous. (23)

glossecollite (Minéral) (*n.f.*), glossecolite. (24)

glu marine (*f.*), marine glue. (25)

glucine (oxyde de glucinium) (*n.f.*), glucina ; glucine. (26)

glucinium *ou* glucium (Chim.) (*n.m.*), glucinum ; glucinium ; beryllium. (27)

glutineux, -euse (*adj.*), glutinous. (28)

glyptogénèse (Géol.) (*n.f.*), glyptogenesis ; land-sculpture. (29)

gmélinite (Minéral) (*n.f.*), gmelinite. (30)

gneiss [gneiss *pl.*] (Pétrol.) (*n.m.*), gneiss ; bastard granite. (31)

gneiss à amphibole (*m.*), hornblende-gneiss ; hornblendic gneiss. (32)

gneiss à pyroxène (*m.*), augite-gneiss. (33)

gneiss chloriteux (*m.*), chloritic gneiss ; chlorite-gneiss. (34)

gneiss fibreux (*m.*), fibrous gneiss. (35)

gneiss fondamental *ou* gneiss gris (*m.*), funda-mental gneiss ; basement complex ; basal complex ; fundamental complex. (36)

gneiss granitoïde (*m.*), granitoid gneiss. (37)

gneiss graphiteux *ou* gneiss graphitique (*m.*), graphitic gneiss ; graphite-gneiss. (38)

gneiss œillé *ou* gneiss œilleté *ou* gneiss glandu-leux (*m.*), eye-gneiss ; augen-gneiss. (39)

gneisseux, -euse (*adj.*), gneissose ; gneissoid. (40)

gneissique (*adj.*), gneissic ; gneissitic ; gneissy. (41)

gobelet gradué (Photogr.) (*m.*), graduated measure, tumbler form. (42)

godet (auget de roue hydraulique, d'élévateur, etc.) (*n.m.*), bucket ; cup. (43 *ou* 44)

godet à huile (*m.*), oil-cup ; oiler. (45)

godet d'élévateur (*m.*), elevator-bucket ; eleva-tor-cup. (46)

godet de drague (*m.*), dredge-bucket. (47)

godet en tôle d'acier emboutie (*m.*), pressed-steel bucket. (48)

godet en verre à tige (*m.*), needle-lubricator. (49)

godet graisseur (*m.*), oil-cup ; grease-cup ; grease-lubricator ; oiler. (50)

godronnage (*n.m.*), knurling ; nurling ; milling. (51)

godronner (*v.t.*), to knurl ; to nurl ; to mill : (52)

godronner la tête d'une vis pour l'empêcher de glisser dans les doigts, to knurl the head of a screw to prevent it slipping in the fingers. (53)

godronnoir (de tourneur) (*n.m.*), knurl ; knurling-tool ; nurl ; nurling-tool ; milling-tool. (54)

gœthite (Minéral) (*n.f.*), goethite. (55)

golfe (Géogr. phys.) (*n.m.*), gulf. (56)

gomme élastique (*f.*), india-rubber ; rubber ; caoutchouc. (57)

gomme laque (*f.*), lac. (58)

gond (*n.m.*) *ou* gond de porte, hook ; gate-hook. (59)

gond et penture [gond (*n.m.*) ; penture (*n.f.*)], hook and hinge ; hook and ride. (60)

gond et penture, avec gond à écrou, hook and hinge, with hook to bolt through post, screwed and nutted ; hook and hinge, with bolt-hook. (61)

gond et penture, avec gond à patte, hook and hinge, with hook on flap. (62)

gond et penture, avec gond à pointe, hook and hinge, with hook to drive. (63)

gond et penture, avec gond à repos, hook and hinge, with hook on plate ; hook and hinge, with plate-hook. (64)

gond et penture, avec gond à scellement, hook and hinge, with stone-hook. (65)

gond et penture, avec gond à vis, hook and hinge, with hook to screw into post ; hook and hinge, with coach-screw hook. (66)

gonflement (*n.m.*), swelling ; inflation. (67)

gonflement (de la sole d'une galerie de mine) (*n.m.*), creep, creeping, heave, heaving, (of the floor of a mine-level). (68)

gonfler (*v.t.*), to swell ; to inflate : (69)

la pluie gonfle les torrents, the rain swells the torrents. (70)

gonfler un ballon avec du gaz, to inflate a balloon with gas. (71)

gonfler (*v.i.*), to swell ; to inflate. (72)

gonfler (Mines) (*v.i.*), to creep ; to heave : (73)

lorsque le toit donne, ou que la sole gonfle, when the roof sags, or the floor creeps ; when the roof squeezes, or the pavement heaves. (74)

gonfler (se) (*v.r.*), to swell ; to inflate. (75)

goniomètre (*n.m.*), goniometer. (76)

goniomètre à réflexion (*m.*), reflecting goniom-eter. (77)

gore (*n.f.*) *ou* gord (*n.m.*) (Mines de houille), clay (running through coal-deposits) ; bat : (78)

un entre-deux de gore, a clay parting. (79)

gorge (ravine) (*n.f.*), gorge. (80)

gorge (rainure) (*n.f.*), groove : (1)
la gorge d'une poulie à gorge, the groove of a grooved pulley. (2)

gorge (*n.f.*) *ou* gorge mobile (d'une serrure à gorge, d'une serrure à gorges mobiles) (*s'emploie généralement au pluriel*), tumbler (of a tumbler-lock). (3)

goslarite (Minéral) (*n.f.*), goslarite. (4)

goudron (*n.m.*), tar. (5)

goudron de bois (*m.*), wood-tar. (6)

goudron de houille *ou* goudron de gaz (*m.*), coal-tar ; gas-tar. (7)

goudron minéral (*m.*), mineral tar. (8)

goudronnage (*n.m.*), tarring : (9)
goudronnage du bois, tarring wood. (10)

goudronner (*v.t.*), to tar : (11)
cordage goudronné (*m.*), tarred rope. (12)

goudronneuse (*n.f.*), tar-sprinkler. (13)

goudronneux, -euse (*adj.*), tarry : (14)
matière goudronneuse (*f.*), tarry matter. (15)

gouffre (*n.m.*), gulf ; abyss ; chasm. (16)

gouge (*n.f.*), gouge. (17)

gouge (de mouleur) (*n.f.*), bead ; bead-tool. (18)

gouge de forgeron (*f.*), blacksmiths' gouge. (19)

gouge de tour *ou* gouge de tourneur (*f.*), turning-gouge. (20)

gouge pleine (burin grain-d'orge) (*f.*), round-nosed cold-chisel ; chipping-chisel, round-nosed. (21)

gouger (*v.t.*), to gouge. (22)

goujon (*n.m.*) *ou* goujon prisonnier (cheville de fer qui sert à assembler deux pierres), joggle ; gudgeon. (23)

goujon (*n.m.*) *ou* goujon prisonnier (cheville de bois qui sert à assembler deux planches, ou autres pièces en bois), dowel ; dowel-pin ; set-pin ; coak ; joggle. (24)

goujon (*n.m.*) *ou* goujon prisonnier (cheville de fer, taraudée aux deux extrémités, servant à lier certaines pièces de machines), stud-bolt ; stud ; standing-bolt. (25)

goujon (tourillon qui sert à assujettir la partie inférieure d'un poteau, etc.) (*n.m.*), plug-tenon ; spur-tenon ; coak ; joggle. (26)

goujon (d'un arbre) (*n.m.*), gudgeon (of a shaft). (27)

goujon (d'une charnière) (*n.m.*), pin, pintle, (of a hinge). (28)

goujon (de châssis de moulage) (*n.m.*), pin, steady-pin, (of moulding-box) ; flask-pin. (29)

goujon (de boîte à noyaux) (*n.m.*), dowel, dowel-pin, (of core-box). (30)

goujonner (*v.t.*), to joggle ; to dowel ; to coak ; to stud : (31)
goujonner des pierres, to joggle stones. (32)
goujonner des planches, to dowel boards. (33)
goujonner des poutres, to coak, to joggle, beams. (34)
goujonner le couvercle d'un cylindre, to stud the back-head of a cylinder. (35)

goule (*n.f.*), limestone cave ; limestone cavern. (36)

goulet (*n.m.*), gully. (37)

goulot (d'une bouteille) (*n.m.*), neck (of a bottle) (narrow neck). (38)

goulotte (*n.f.*), spout. (39)

goupille (*n.f.*), pin ; cotter-pin. (40)

goupille au cône Morse (*f.*), Morse taper pin. (41)

goupille conique (*f.*), taper pin. (42)

goupille de calage *ou* goupille à demeure (*f.*), set-pin. (43)

goupille fendue (*f.*), split pin ; split cotter-pin. (44)

goupille pleine (*f.*), solid pin ; solid cotter-pin. (45)

goupiller (*v.t.*), to pin. (46)

gousset (*n.m.*), gusset. (47)

goût fin (Mines) (*m.*), fire-stink. (48)

goutte (*n.f.*), drop ; drip : (49)
goutte de pluie, drop of rain ; rain-drop. (50)
acide versé goutte à goutte dans une liqueur alcaline (*m.*), acid poured drop by drop into an alkaline solution. (51)
gouttes de scories (Métall.), slag inclusions. (52)
gouttes froides (Fonderie), cold shot. (53)

goutte-de-suif [gouttes-de-suif *pl.*] (*n.f.*) *ou* goutte de suif, button ; dome-shaped head : (54)
punaise à tête en goutte-de-suif (*f.*), drawing-pin with dome-shaped head. (55)

goutte-de-suif (Élec.) (*n.f.*), button ; contact. (56)

goutte-de-suif (à tête en) (se dit des rivets, des boulons, des vis), button-head ; button-headed ; mushroom-head ; raised-head : (57)
rivet à tête en goutte-de-suif (*m.*), button-head rivet. (58)
boulon à tête en goutte-de-suif (*m.*), button-headed bolt ; mushroom-head bolt. (59)
vis à tête en goutte-de-suif (*f.*), raised-head screw. (60)

goutter (*v.i.*), to drop ; to drip. (61)

gouttière (*n.f.*), gutter. (62)

gouttière en dessus (Constr.) (*f.*), parapet gutter. (63)

gouttière pendante (Constr.) (*f.*), eaves-gutter ; eaves-trough. (64)

gouvernement (*n.m.*), government. (65)

gouverneur (pers.) (*n.m.*), governor : (66)
gouverneur d'une colonie, governor of a colony. (67)

gouverneur (marqueur) (Mines) (*n.m.*), tallyman ; checker. (68)

goyot (Mines) (*n.m.*), upcast compartment (of a two-compartment air-shaft). (69)

graben (opp. à horst) (Géol.) (*n.m.*), graben ; rift-valley. (70)

gradation (*n.f.*), gradation. (71)

gradin (*n.m.*), step : (72)
collines disposées en gradins (*f.pl.*), stepped hills. (73)

gradin (d'un cône, d'un cône-poulie) (*n.m.*), step, lift, (of a cone, of a cone-pulley) : (74)
cône à 3 gradins, 3-step cone ; 3-lift cone-pulley ; 3-speed cone. (75)

gradin (Mines & Carrières) (*n.m.*), stope ; lift ; bench. (76)

gradin chassant (Mines) (*m.*), drift-stope. (77)

gradins couchés (Mines) (*m.pl.*), bench-work. (78)

gradins droits (Mines) (*m.pl.*), underhand stopes ; bottom-stopes. (79)

gradins renversés (Mines) (*m.pl.*), overhand stopes ; back-stopes ; step-faced overhand stopes. (80)

graduateur (*n.m.*), graduator. (81)

graduation (*n.f.*), graduation ; graduating : (82)
graduation d'une échelle, graduation of, graduating, a scale. (83)

graduel, -elle (*adj.*), gradual : (1)
  pente graduelle (*f.*), gradual slope. (2)
graduellement (*adv.*), gradually. (3)
graduer (*v.t.*), to graduate : (4)
  graduer une échelle empiriquement, to graduate
    a scale empirically. (5)
grain (petite parcelle) (*n.m.*),   grain ;   speck ;
    particle : (6)
  grain de sable, grain of sand. (7)
  grain de poussière,   speck, particle, of dust.
    (8)
grain (degré de rugosité ou texture) (*n.m.*),
    grain : (9)
  grain d'un marbre, grain of a marble. (10)
  grain fin, fine grain. (11)
  grain fin (à) *ou* grains fins (à),   fine-grained :
    (12)
    variété à grain fin de granit (*f.*),   fine-grained
      variety of granite. (13)
    meule à grain fin (émeri, ou analogue) (*f.*),
      fine-grained wheel. (14)
  grain grossier *ou* gros grain *ou* grain gros,
    coarse grain ; rough grain. (15)
  grain grossier (à) *ou* gros grain (à) *ou* gros grains
    (à) *ou* grain gros (à),   coarse-grained ; rough-
    grained : (16)
    fonte à gros grain (*f.*),   coarse-grained pig
      iron. (17)
  grain serré, close grain : (18)
    l'acier a le grain plus serré que le fer,   steel
      has a closer grain than iron. (19)
  grain serré (à), close-grained. (20)
  grains (à) *ou* grains (en), granular ; granulated :
    (21)
    cassure à grains (*f.*), granular fracture. (22)
    or natif en grains (*m.*),   granular native gold ;
      granulated native gold. (23)
grain (poids anglais) (*n.m.*),   grain = 0·0648
    gramme. (24)
grain (d'un presse-étoupe) (*n.m.*),   bush (of a
    stuffing-box) : (25)
  le grain du chapeau, the follower-bush. (26)
grain d'acier *ou* **grain en acier** (d'une crapaudine)
    (*m.*),   centre bearing-plate (of a centre-
    plate). (27)
grain-d'orge [grains-d'orge *pl.*] (burin) (*n.m.*),
    round-nosed   cold-chisel ;   chipping-chisel,
    round-nosed. (28)
grain-d'orge (outil de tour à métaux) (*n.m.*),
    round-nose tool. (29)
grain-d'orge de tour (tour à bois) (*m.*),   V-shaped
    turning-tool. (30)
grain de fin (bouton de fin) (*m.*),   button ;
    assay-button. (31)
graissage (*n.m.*),   greasing ;   lubrication ;
    lubricating : (32)
  graissage des roues d'une voiture,   greasing
    the wheels of a carriage. (33)
graissage à l'huile *ou simplement* **graissage** (*n.m.*),
    oiling ; lubrication ; lubricating : (34)
  graissage d'un palier, lubrication of, lubricat-
    -ing, oiling, a bearing. (35)
  graissage automatique, automatic lubrication ;
    automatic oiling. (36)
  graissage automatique (à),   self-oiling ; self-
    lubricating : (37)
    palier à graissage automatique (*m.*),   self-
      oiling plummer-block. (38)
  graissage sous pression, force-feed lubrication.
    (39)
graisse (*n.f.*), grease ; lubricating-oil. (40)
graisse consistante (*f.*), set grease. (41)

graisse pour engrenages (*f.*),   gear-grease. (42)
graisser (*v.t.*), to grease ; to lubricate : (43)
  graisser un essieu, to grease an axle. (44)
graisser à l'huile *ou simplement* **graisser** (*v.t.*),
    to oil ; to lubricate : (45)
  graisser à l'huile une machine, to oil a machine.
    (46)
graisseur, -euse (*adj.*), self-oiling ; self-lubricating ;
    grease (*used as adj.*) : (47)
  palier graisseur (*m.*),   self-oiling plummer-
    block. (48)
  robinet graisseur (*m.*), grease-cock. (49)
graisseur (Mach.) (*n.m.*),   lubricator ;   oil-cup ;
    grease-cup ;   grease-lubricator ;   greaser ;
    oiler. (50)
graisseur (pers.) (*n.m.*), greaser ; oiler. (51)
graisseur à casque *ou* **graisseur à chapeau mobile**
    (*m.*), helmet-cap lubricator. (52)
graisseur à compression (*m.*),   spring grease-
    lubricator. (53)
graisseur à condensation   (*m.*),   condensing
    lubricator. (54)
graisseur à couvercle élastique (*m.*),   spring-lid
    oil-cup. (55)
graisseur à épinglette *ou* **graisseur à tige** (*m.*),
    needle-lubricator. (56)
graisseur à graisse consistante (*m.*),   set-grease
    cup. (57)
graisseur à pointeau (*m.*), needle-valve lubricator.
    (58)
graisseur à ressort (*m.*), spring grease-lubricator.
    (59)
graisseur à vase verre (*m.*),   glass lubricator.
    (60)
graisseur cache-poussière avec chapeau à charnière
    (*m.*),   dust-proof oil-cup with hinged cap.
    (61)
graisseur compte-gouttes *ou* **graisseur débit**
    d'huile (*m.*),   drip-feed lubricator. (62)
graisseur compte-gouttes à débit visible (*m.*),
    sight-feed lubricator,   spot-by-spot flow ;
    telltale lubricator. (63)
graisseur coup-de-poing (*m.*),   hand-pump
    lubricator. (64)
graisseur de la pompe à air (*m.*),   air-pump
    lubricator. (65)
graisseur de roues (pers.) (*m.*),   wheel-greaser ;
    wheel-oiler. (66)
graisseur de wagons (pers.) (*m.*),   car-greaser.
    (67)
graisseur du cylindre (*m.*),   cylinder-lubricator.
    (68)
graisseur tournant (*m.*),   swivel-cap lubricator.
    (69)
graisseux, -euse (*adj.*), greasy ; oily. (70)
grammatite (Minéral) (*n.f.*),   grammatite ;
    tremolite. (71)
gramme (*n.m.*),   gramme = 15·432 grains or
    0·03215 oz. troy. (72)
gramme-masse [grammes-masse *pl.*] (*n.m.*),
    gramme in mass. (73)
granatique (*adj.*), garnetiferous. (74)
grand, -e (*adj.*), great ; large ; big : (75)
  les grandes forêts du Brésil (*f.pl.*),   the great
    forests of Brazil. (76)
  grand air (*m.*), open air : (77)
    travail au grand air (*m.*), open-air work. (78)
  grande calorie (Phys.) (*f.*),   great calory ;
    greater calory ; kilogramme calory ; kilo-
    -calory ; kilocalorie. (79)
  grandes chaleurs (temps très chaud) (*f.pl.*),
    great heat ; very hot weather. (80)

grand cylindre (d'une machine à vapeur compound) (*m.*), low-pressure cylinder (of a compound steam-engine). (1)

grands explosifs (*m.pl.*), high explosives. (2)

grande ligne *ou* grande artère (Ch. de f.) (*f.*), main line ; trunk line : (3)

les grandes lignes partant de Paris, the main lines out of Paris. (4)

les grandes artères du réseau français, the trunk lines of the French system. (5)

grande multiplication (Méc.) (*f.*), high gear. (6)

grande pluie (*f.*), heavy rain ; downpour ; rainpour ; copious rainfall. (7)

grande pluie (Barométrie), much rain. (8)

grand rayon (d'une courbe de ch. de f.) (*m.*), outer radius (of a railway-curve). (9)

grand ressort (*m.*), main spring. (10)

grande route (*f.*), main road ; highway ; high--road ; wagon-road. (11)

grande terre (*f.*), continent ; mainland. (12)

grande vitesse (Méc., etc.) (*f.*), high speed. (13)

grande vitesse (à), high-speed : (14)

machine à percer à grande vitesse (*f.*), high-speed drilling-machine. (15)

grandeur (dimension relative) (*n.f.*), size ; magnitude. (16)

grandeur apparente d'un objet (Opt.) (*f.*), apparent magnitude of an object. (17)

grandeur naturelle (*f.*), natural size ; full size : (18)

dessin en grandeur naturelle (*m.*), full-size drawing. (19)

grandeur réelle d'un objet (Opt.) (*f.*), real magnitude of an object. (20)

grandir (faire paraître plus grand) (*v.t.*), to magnify ; to enlarge : (21)

le microscope grandit les petits objets, the microscope magnifies small objects. (22)

grandissement (*n.m.*), magnification ; enlarge--ment : (23)

le grandissement d'une photographie, the enlargement of a photograph. (24)

granit *ou* **granite** (*n.m.*), granite. (25)

granit gris (*m.*), gray granite. (26)

granit rose (*m.*), red granite. (27)

granitelle (Pétrol.) (*n.m.*), granitell ; granitelle. (28)

granitique *ou* **graniteux, -euse** (*adj.*), granitic ; granitical : (29)

formation granitique (*f.*), granitic formation. (30)

granitite (Pétrol.) (*n.f.*), granitite. (31)

granitoïde (*adj.*), granitoid : (32)

structure granitoïde (*f.*), granitoid structure. (33)

granophyre (Pétrol.) (*n.m.*), granophyre. (34)

granulaire (*adj.*), granular : (35)

structure granulaire (*f.*), granular structure. (36)

granulairement (*adv.*), granularly. (37)

granulation (*n.f.*), granulation. (38)

granule (*n.m.*), granule. (39)

granulé, -e (*adj.*), granulated. (40)

granuler (*v.t.*), to granulate : (41)

granuler de l'étain, to granulate tin. (42)

granuler (se) (*v.r.*), to granulate. (43)

granuleux, -euse (*adj.*), granulous ; granulated. (44)

granuliforme (*adj.*), granuliform. (45)

granulite (Pétrol.) (*n.f.*), granulite. (46)

granulitique (*adj.*), granulitic : (47)

roche granulitique (*f.*), granulitic rock. (48)

granulophyre (Pétrol.) (*n.m.*), microgranulitic porphyry. (49)

graphique (Minéralogie) (*adj.*), graphic ; graphical : (50)

pegmatite graphique (*f.*), graphic pegmatite. (51)

graphique (Dessin) (*adj.*), graphic ; graphical ; diagrammatic : (52)

dessin graphique (*m.*), graphic drawing ; diagrammatic sketch. (53)

graphique (art du dessin appliqué aux sciences) (*n.m.*), graphics. (54)

graphique (diagramme) (*n.m.*), diagram ; chart : (55)

graphique adiabatique, adiabatic diagram. (56)

graphique d'un baromètre enregistreur, chart of a recording barometer. (57)

graphique de la production, diagram of out--put. (58)

graphique des trains (Ch. de f.), train-diagram. (59)

graphiquement (*adv.*), graphically ; diagram--matically ; diagramically. (60)

graphite (Minéral) (*n.m.*), graphite ; plumbago ; black lead. (61)

graphiteux, -euse *ou* **graphitique** (*adj.*), graphitic ; plumbaginous ; plumbago-bearing : (62)

gneiss graphiteux *ou* gneiss graphitique (*m.*), graphitic gneiss. (63)

graphitisation (*n.f.*), graphitization. (64)

graphiter (*v.t.*), to graphitize : (65)

filament graphité (*m.*), graphitized filament. (66)

graphomètre (*n.m.*), graphometer ; demi-circle. (67)

gras, grasse (*adj.*), fat ; fatty ; greasy ; unctuous : (68)

argile grasse (*f.*), fat clay ; fatty clay. (69)

éclat gras (*m.*), greasy lustre. (70)

rails gras (cause de mauvaise adhérence) (*m.pl.*), greasy rails. (71)

grattage (*n.m.*), scratching ; scraping. (72)

gratte-boessage *ou* **gratte-bossage** (*n.m.*), scratch-brushing. (73)

gratte-boesse [gratte-boesses *pl.*] *ou* **gratte-bosse** [gratte-bosses *pl.*] (*n.f.*), scratch-brush. (74)

gratte-boesser *ou* **gratte-bosser** (*v.t.*), to scratch-brush. (75)

gratter (*v.t.*), to scratch ; to scrape : (76)

gratter la terre (*ou* le sol), to scratch the ground. (77)

grattoir (instrument ; outil) (*n.m.*), scraper. (78)

grattoir (charrue pour gratter le sol) (Trav. publ., etc.) (*n.m.*), scraper. (79)

grattoir triangulaire (*m.*), shave-hook. (80)

grattures (*n.f.pl.*), scrapings : (81)

grattures de cuivre, copper scrapings. (82)

gratuit, -e (*adj.*), free ; gratuitous. (83)

grauwacke (Pétrol.) (*n.f.*), graywacke ; grey--wacke ; grauwacke. (84)

grave (*adj.*), serious ; grave : (85)

inconvénient grave (*m.*), serious disadvantage. (86)

graveleux, -euse (*adj.*), gravelly ; gritty. (87)

gravier (*n.m.*), gravel ; grit. (88)

gravier aurifère (*m.*) (*s'emploie souvent au pluriel*) (Placers), auriferous gravel ; gold-bearing gravel ; dirt ; placer dirt. (89)

gravier bleu (*m.*), blue gravel. (1)

gravier coquillier (*m.*), shell gravel. (2)

gravier de rivière *ou* gravier fluviatile (*m.*), river-gravel ; fluviatile gravel. (3)

gravier exploitable *ou* gravier payant *ou* gravier rémunérateur (Placers) (*m.*), pay-gravel ; payable gravel ; pay-dirt. (4)

gravier jaune (*m.*), yellow gravel. (5)

gravière (*n.f.*), gravel-pit. (6)

gravimètre (*n.m.*), gravimeter. (7)

gravimétrie (*n.f.*), gravimetry. (8)

gravimétrique (*adj.*), gravimetric ; gravimetrical : (9)

analyse gravimétrique (*f.*), gravimetric analysis. (10)

gravimétriquement (*adv.*), gravimetrically. (11)

gravir (*v.t.*), to climb ; to climb up ; to clamber ; to clamber up ; to scale : (12)

gravir une montagne, to climb a mountain. (13)

gravir (*v.i.*), to climb ; to climb up ; to clamber ; to clamber up : (14)

gravir sur les rochers, to climb up on the rocks ; to clamber over the rocks. (15)

gravitation (*n.f.*), gravitation : (16)

les lois de la gravitation (*f.pl.*), the laws of gravitation. (17)

gravité (*n.f.*), gravity : (18)

centre de gravité (*m.*), centre of gravity. (19)

graviter (*v.i.*), to gravitate. (20)

gravure (dans un catalogue) (*n.f.*), illustration ; engraving ; cut. (21)

greenockite (Minéral) (*n.f.*), greenockite ; cadmium-blende ; cadmium ochre. (22)

greenovite (Minéral) (*n.f.*), greenovite. (23)

grelin (*n.m.*), hawser. (24)

grenaille (Métall.) (*n.f.*), shot. (25)

grenaille d'acier trempé (*f.*), chilled shot. (26)

grenaille de plomb (*f.*), lead shot. (27)

grenaillement (*n.m.*), shotting. (28)

grenailler (*v.t.*), to shot. (29)

grenailler (se) (*v.r.*), to shot. (30)

grenailles (Préparation mécanique des minerais) (*n.f.pl.*), coarse sand middlings. (31)

grenat (Minéral) (*n.m.*), garnet. (32)

grenat alumineux (*m.*), aluminium garnet. (33)

grenat alumino-calcareux (*m.*), calcium-alu-minium garnet. (34)

grenat alumino-ferreux (*m.*), iron-aluminium garnet. (35)

grenat alumino-magnésien (*m.*), magnesium-aluminium garnet. (36)

grenat cabochon (*m.*), carbuncle. (37)

grenat chromifère (*m.*), chromium garnet. (38)

grenat chromo-calcareux (*m.*), calcium-chromium garnet. (39)

grenat commun (*m.*), common garnet. (40)

grenat de Bohême (*m.*), Bohemian garnet ; pyrope. (41)

grenat ferreux (*m.*), iron garnet. (42)

grenat ferro-calcareux (*m.*), calcium-iron garnet. (43)

grenat noble *ou* grenat oriental (*m.*), noble garnet ; precious garnet ; Oriental garnet ; oriental garnet. (44)

grenat syrien (*m.*), grenat syrien ; rubino di rocca ; rock-ruby. (45)

grenatique (*adj.*), garnetiferous. (46)

grenatite (Pétrol.) (*n.f.*), garnet-rock. (47)

grener (*v.t.*), to granulate ; to grain. (48)

grener (se) (en parlant du sel) (*v.r.*), to grain. (49)

grenouillère (de pompe) (*n.f.*), strainer ; snore-piece ; tail-piece ; wind-box ; wind-bore ; rose. (50)

grenu, -e (*adj.*), granular ; grained : (51)

cassure grenue (*f.*), granular fracture. (52)

grenu (*n.m.*), granularity : (53)

le grenu d'une roche, d'un marbre, the granularity of a rock, of a marble. (54 *ou* 55)

grès (Pétrol.) (*n.m.*), sandstone ; grit ; sandrock. (56)

grès (poterie) (*n.m.*), stoneware. (57)

grès à pavés (*m.*), paving-stone. (58)

grès bigarré (*m.*), variegated sandstone. (59)

grès bitumineux (*m.*), bituminous sandstone. (60)

grès calcaire (*m.*), calcareous grit. (61)

grès flexible (*m.*), flexible sandstone. (62)

grès houiller (*m.*), carboniferous sandstone. (63)

grès lustré (*m.*), glossy sandstone. (64)

grès meulier *ou* grès molaire (*m.*), millstone grit. (65)

grès paf (*m.*), sandstone suitable for flagging. (66)

grès pif (*m.*), sandstone suitable for flagging, but very hard. (67)

grès pouf (*m.*), sandstone unsuitable for flagging, on account of its softness. (68)

grès psammite (*m.*), psammitic sandstone. (69)

grès rouge (Géol.) (*m.*), red sandstone. (70)

grès rouge récent (Géol.) (*m.*), New Red Sandstone. (71)

grès tacheté (*m.*), mottled sandstone. (72)

grès vert (*m.*), greensand. (73)

grès vert inférieur (Géol.) (*m.*), Lower Green-sand. (74)

grès vert supérieur (*m.*), Upper Greensand. (75)

gréseux, -euse (*adj.*), sandstone (*used as adj.*) ; gritty ; arenilitic : (76)

formation gréseuse (*f.*), sandstone formation. (77)

grésière *ou* gréserie (*n.f.*), sandstone-quarry. (78)

grésillon (menu charbon) (*n.m.*), small coal. (79)

grève (plage) (*n.f.*), beach ; shore : (80)

grève de galets, shingle beach ; pebble beach. (81)

grève de sable, sandy beach. (82)

grève (ligue de personnes qui se coalisent pour faire cesser le travail) (*n.f.*), strike ; walk-out : (83)

se mettre en grève, to go on strike ; to strike. (84)

gréviste (pers.) (*n.m.*), striker. (85)

griefs des ouvriers (*m.pl.*), grievances of the men. (86)

griffe (*n.f.*) *ou* griffe de serrage, grip ; clip ; clutch. (87)

griffe (d'un plateau de tour) (*n.f.*), jaw (of a lathe-chuck). (88)

griffe (mandrin de tour à trois pointes) (*n.f.*), prong-chuck ; fork-chuck ; spur-chuck ; three-pronged chuck. (89)

griffe (outil de forgeron) (*n.f.*), hook-spanner ; hook-wrench. (90)

griffe (de l'agrafe du chariot d'une scie à débiter les bois en grume) (*n.f.*), dog, clamping-dog, gripping-dog, (of the timber-clip of the log-carriage of a log-sawing machine). (91)

griffe (appareillage électrique) (*n.f.*), gallery ; glass-holder. (92)

griffe de blocage (de tour, etc.) (*f.*), clamping-handle (of lathe, etc.). (1)

griffe de guidage (cage de mine) (*f.*), guide-shoe. (2)

griffer la grume à scier sur le chariot, to secure the log to be sawn on the log-carriage. (3)

gril (faisceau de triage) (Ch. de f.) (*n.m.*), grid-iron track; ladder-track; ladder. (4)

gril (Hydraul.) (*n.m.*), grating: (5)
 gril établi en amont d'une vanne pour arrêter les bois et détritus charriés par les eaux, grating placed above a sluice-gate to stop the wood and rubbish carried down by the water. (6)

grillage (calcination) (Métall.) (*n.m.*), roasting; calcining; burning: (7)
 grillage de minerais pyriteux, roasting pyritous ores. (8)

grillage (assemblage de barreaux) (*n.m.*), grating. (9)

grillage (treillis) (*n.m.*), netting; wire netting. (10)

grillage à mort (*m.*), dead roasting; sweet roasting. (11)

grillage en tas (Métall.) (*m.*), heap-roasting; heap-roast; open-heap roasting. (12)

grille (assemblage de barreaux fermant une ouverture) (*n.f.*), grating; grid. (13)

grille (d'un accumulateur) (Élec.) (*n.f.*), grid (of an accumulator). (14)

grille (clôture) (*n.f.*), railing. (15)

grille (claie) (*n.f.*), screen. (16)

grille (*n.f.*) *ou* grille à barreaux (pour séparer les minerais bruts du menu de mines), grizzly; ore-grizzley; bar-screen. (17)

grille (châssis métallique qui soutient le charbon dans un fourneau) (*n.f.*), grate; grid: (18)
 grille d'un foyer de chaudière, d'un four à réverbère, d'un gazogène, grate, grid, of a boiler-furnace, of a reverberatory furnace, of a gas-producer. (19 *ou* 20 *ou* 21)

grille à barreaux basculants (grille de foyer) (*f.*), rocking grate. (22)

grille à barreaux mobiles (*f.*). Same as grille à secousse.

grille à chaîne (*f.*), chain-grate. (23)

grille à eau (*f.*), water-grate. (24)

grille à flammèches (locomotive) (*f.*), smoke-box netting; netting; spark-netting; spark-catcher; spark-arrester; spark-consumer. (25)

grille à gradins *ou* grille à étages (*f.*), step-grate; stepped grate. (26)

grille à secousse *ou* grille à secousses (oscillations longitudinales) (grille de foyer) (*f.*), shaking grate. (27)

grille à secousse *ou* grille à secousses (oscillations horizontales) (grille de foyer) (*f.*), rocking grate. (28)

grille de fenêtre (*f.*), window-grating. (29)

grille de foyer (*f.*), furnace-grate; fire-grate. (30)

grille de protection en fil de fer (*f.*), wire guard. (31)

grille extensible (barrière) (*f.*), collapsible gate. (32)

grille inclinée (foyer) (*f.*), inclined grate. (33)

grille oscillante (foyer) (*f.*). Same as grille à secousse.

grille rotative (*f.*), rotary grate. (34)

grille sans fin (*f.*), travelling-grate stoker. (35)

grille tournante (foyer) (*f.*), revolving grate. (36)

griller (calciner) (*v.t.*), to roast; to calcine: (37)
 griller du minerai, to roast ore. (38)

griller (munir d'une grille) (*v.t.*), to grate: (39)
 griller une fenêtre, to grate a window. (40)

grimper (*v.i.*), to climb; to climb up; to clamber; to clamber up. (41)

griotte (marbre) (*n.f.*), griotte. (42)

grip (Roulage) (*n.m.*), grip; clip; haulage-clip; rope-clip; jigger. (43)

grip à mâchoires (*m.*), jaw-grip: jaw-clip. (44)

grippage *ou* grippement (Méc.) (*n.m.*), seizing: (45)
 grippage d'un piston, seizing of a piston. (46)

gripper (Méc.) (*v.i.*), to seize: (47)
 coussinets qui grippent (*m.pl.*), bearings which seize. (48)

gris, -e (*adj.*), gray; grey. (49)

gris (*n.m.*), gray; grey. (50)

gris d'acier, tirant sur le blanc d'argent (*m.*), steel-grey, inclining to silver-white. (51)

gris de fer (*m.*), iron gray. (52)

grisou (*n.m.*), fire-damp; gas: pit-gas; fire. (53)

grisoumètre (*n.m.*), fire-damp delicate indicator; gas-verifier; warner. (54)

grisoumétrie (*n.f.*), study and science of fire-damp. (55)

grisoumétrique (*adj.*), of or relating to fire-damp measurement: (56)
 mesures grisoumétriques (*f.pl.*), fire-damp measurement; fire-damp testing. (57)
 laboratoire grisoumétrique (*m.*), fire-damp laboratory. (58)

grisounite (Explosif) (*n.f.*), grisounite. (59)

grisounite-couche (*n.f.*), grisounite for coal-mines. (60)

grisounite-roche (*n.f.*), grisounite for rock-work. (61)

grisouteux, -euse (*adj.*), fiery; gassy; gaseous: (62)
 mine grisouteuse (*f.*), fiery (*or* gassy) (*or* gaseous) mine. (63)

grondements souterrains (*m.pl.*), subterranean rumblings. (64)

groroïlite (Minéral) (*n.f.*), groroilite. (65)

gros, grosse (*adj.*), big; large: (66)
 un gros arbre, a big tree. (67)
 une grosse somme d'argent, a large sum of money. (68)
 gros fil (opp. à *fil fin*) (*m.*), thick wire. (69) (opp. to *fine wire*).

gros galet (*m.*), boulder; bowlder. (70)

gros grain (*m.*), coarse grain; rough grain. (71)

gros grain (à) *ou* gros grains (à), coarse-grained; rough-grained: (72)
 fonte à gros grain (*f.*), coarse-grained pig-iron. (73)

gros morceau (*m.*), lump. (74)

gros morceaux de minerai (*m.pl.*), lumps of ore; lump ore; large; coarse ore. (75)

gros murs (d'un bâtiment) (*m.pl.*), main walls (of a building). (76)

gros profilés (acier ou fer) (*m.pl.*), heavy sections. (77 *ou* 78)

gros sables (Préparation mécanique des minerais) (*m.pl.*), coarse sands; coarse sand. (79)

gros train de cylindres à fer marchand (*m.*), merchant train; merchant-bar train; merchant mill; bar-mill. (80)

grosse lime (*f.*), rough file. (81)

grosse serrurerie (*f.*), heavy ironwork. (82)

263

grosse tête de bielle (*f.*), connecting-rod big end ; big end, large end, stub-end, butt, of connecting-rod. (1)

grosse venue d'eau (Mines) (*f.*), heavy make of water. (2)

grosses pièces (Fonderie) (*f.pl.*), heavy castings. (3)

grosses tôles (*f.pl.*), heavy plates. (4)

gros (minerai) (*n.m. sing. & pl.*), large ; lumps of ore ; lump ore ; coarse ore. (5)

gros (charbon) (*n.m. sing. & pl.*), large ; round coal ; round ; lump coal ; lumps (of coal). (6)

grosse (12 douzaines) (*n.f.*), gross. (7)

grosseur (dimensions) (*n.f.*), size : (8)

grosseur naturelle, natural size. (9)

grossier, -ère (*adj.*), coarse ; rough : (10)

variété à grain grossier de granit (*f.*), coarse-grained (*or* rough-grained) variety of granite. (11)

grossièrement (*adv.*), coarsely ; roughly : (12)

esquisser grossièrement un plan, to sketch a plan roughly. (13)

estimer grossièrement, to estimate roughly. (14)

grossièreté (*n.f.*), coarseness ; roughness. (15)

grossir (*v.t.*), to make bigger ; to swell : (16)

la fonte des neiges grossit les fleuves, the melting of the snows swells the rivers. (17)

grossir (faire paraître plus grand) (*v.t.*), to magnify; to amplify ; to enlarge : (18)

les microscopes grossissent les petits objets, microscopes magnify small objects. (19)

grossir (*v.i.*), to swell ; to enlarge ; to rise : (20)

la rivière grossit, the river is rising. (21)

grossissant, -e (*adj.*), magnifying ; amplifying; enlarging : (22)

verre grossissant (*m.*), magnifying-glass. (23)

grossissement (*n.m.*), magnification ; amplification ; enlargement ; enlarging. (24)

grosso modo, roughly : (25)

séparer grosso modo les gros cailloux, to separate the boulders roughly. (26)

grossulaire *ou* **grossularite** (Minéral) (*n.f.*), grossular ; grossularite ; gooseberry-stone. (27)

grotte (*n.f.*), grotto : (28)

grotte stalactifère, stalactited grotto. (29)

groupe (*n.m.*), group : (30)

groupe d'éléments de pile, group of battery-cells. (31)

groupe de mines, group of mines. (32)

groupe de mineurs, group, party, of miners. (33)

groupe (Géol.) (*n.m.*), group. (34)

groupe (ensemble de machines électriques) (*n.m.*), set : (35)

groupe convertisseur, converter set. (36)

groupe de secours, emergency set ; standby set ; breakdown set. (37)

groupe électrogène *ou* groupe générateur, generating set ; electric generating set. (38)

groupe moteur-générateur, motor-generator set. (39)

groupe transformateur, transformer set. (40)

groupe transformateur de démarrage, starting transformer set. (41)

groupement (*n.m.*), grouping : (42)

groupement de trommels, grouping of trommels. (43)

groupement des cristaux maclés, grouping of twin crystals. (44)

groupement (Élec.) (*n.m.*), grouping ; coupling ; connecting ; connection : (45)

groupement en boucle, ring-connection ; loop-connection. (46)

groupement en dérivation *ou* groupement en parallèle, grouping in multiple ; multiple connection ; connecting in parallel ; cou-pling in multiple ; abreast connection. (47)

groupement en étoile, star-connection ; star-grouping ; Y-connection. (48)

groupement en étoile-triangle, star-delta connection ; star-delta grouping. (49)

groupement en série *ou* groupement en chaîne *ou* groupement en cascade, grouping in series ; series connection ; concatenated connection ; cascade connection ; tandem connection. (50)

groupement en triangle *ou* groupement en delta, delta-connection ; delta-grouping ; mesh-connection. (51)

groupement mixte, grouping in multiple series ; multiple-series connection ; parallel-series connection. (52)

grouper (*v.t.*), to group : (53)

les cristaux de la desmine sont ordinairement groupés à plusieurs en forme de gerbes (*m.pl.*), the crystals of desmine are usually grouped in sheaf-like aggregates. (54)

grouper (Élec.) (*v.t.*), to group ; to couple ; to connect : (55)

grouper des éléments de pile, to group, to couple, to connect, battery-cells. (56)

grue (*n.f.*), crane. (57)

grue à colonne (*f.*), pillar-crane ; post-crane ; column-crane. (58)

grue à demi-portique (*f.*), semigantry crane ; semiportal crane. (59)

grue à demi-portique électrique, chevauchant deux voies de chemin de fer (*f.*), semiportal electric crane, spanning two sets of rails. (60)

grue à flèche (*f.*), jib-crane. (61)

grue à grappin (*f.*), grab-crane. (62)

grue à pivot (*f.*). Same as **grue pivotante**.

grue à portique (*f.*), gantry-crane ; gauntry-crane. (63)

grue à portique électrique (*f.*), electric gantry-crane. (64)

grue à potence *ou* **grue à potance murale** (*f.*), wall-crane. (65)

grue à rotation complète (*f.*), all-round swing crane. (66)

grue à vapeur (*f.*), steam-crane. (67)

grue à volée (*f.*), jib-crane. (68)

grue à volée variable *ou* **grue à volée basculante** (*f.*), derricking crane ; luffing crane. (69)

grue centrale *ou* **grue de milieu** (*f.*), independent crane. (70)

grue-chevalet [grues-chevalets *pl.*] (*n.f.*) *ou* **grue-chevalet sur roues**, goliath crane. (71)

grue-chevalet à main, sur roues (*f.*), hand goliath crane. (72)

grue-chevalet à rallonges (*f.*), goliath crane with overhanging ends. (73)

grue-chevalet électrique, sur roues, à un seul jambage (*f.*), electric goliath crane, with one leg. (74)

grue d'alimentation (Ch. de f.) (*f.*), water-pillar : water-crane. (75)

grue d'applique (*f.*), wall-crane. (76)

grue d'atelier (*f.*), shop-crane. (77)

grue d'entrepôt (*f.*), warehouse-crane. (78)

grue d'une puissance de levage de — tonnes (f.), crane to lift — tons ; — -ton crane. (1)

grue de fonderie (f.), foundry-crane. (2)

grue de manœuvre pour les trépans (f.), derrick-crane (a crane used on an oil-well derrick). (3)

grue de quai (f.), wharf-crane. (4)

grue derrick [grues derrick pl.] (f.), derrick-crane ; derrick. (5)

grue derrick volante (f.), portable derrick-crane. (6)

grue électrique (f.), electric crane. (7)

grue flottante (f.), floating crane. (8)

grue hydraulique (appareil de levage) (f.), hydraulic crane ; water-crane. (9)

grue hydraulique (bâche ; bouche d'eau) (Ch. de f.) (f.), water-pillar ; water-crane. (10)

grue locomobile à vapeur (f.), portable steam-crane. (11)

grue locomotive (f.), locomotive crane. (12)

grue locomotive à vapeur (f.), locomotive steam-crane. (13)

grue murale (f.), wall-crane. (14)

grue pivotante (f.), swing-crane. (15)

grue pivotante (pouvant virer un cercle complet) (f.), rotary crane. (16)

grue pivotante fixe à main, puissance — kilos, portée — mètres, hauteur de levage — mètres (f.), fixed hand-crane, to lift — kilos ; radius — metres, lift — metres. (17)

grue pivotante fixe électrique (f.), electric rotary fixed crane. (18)

grue pivotante murale (f.), rotary wall-crane. (19)

grue pivotante roulante à main (f.), hand-power rotary travelling crane. (20)

grue pivotante roulante électrique (f.), electric rotary travelling crane. (21)

grue pour forges (f.), forge-crane. (22)

grue roulante ou grue roulante aérienne (f.), travelling crane ; traveling crane ; traveller ; overhead travelling crane ; overhead crane ; overhead traveller. (23). See also pont roulant.

grue roulante à portique (f.), travelling gantry-crane. (24)

grue roulante manœuvrée à la main (f.), port-able hand-crane. (25)

grue-terrassier [grues-terrassiers pl.] (n.f.), steam-shovel ; steam-digger ; steam-excavator ; steam-navvy ; navvy. (26)

grue titan (f.), titan crane. (27)

grue vélocipède sur un rail de translation (f.), walking-crane ; single-rail crane. (28)

grume (n.f.), log : (29)
équarrir des grumes, to square logs. (30)

grünauite (Minéral) (n.f.), grunauite. (31)

grünérite (Minéral) (n.f.), grünerite. (32)

gué (n.m.), ford. (33)

guéable (adj.), fordable : (34)
rivière guéable (f.), fordable river. (35)

guérite (d'une grue) (n.f.), cabin, cab, house, (of a crane). (36)

guérite (d'un wagon à guérite) (Ch. de f.) (n.f.), lookout, cupola, (of a lookout-car, of a cupola-car). (37)

guérite à signaux (f.), signal-box; signal-cabin ; signal-tower. (38)

guérite en encorbellement pour poste de signaux (f.), offset signal-box. (39)

gueulard (d'un haut fourneau) (n.m.), mouth, throat, (of a blast-furnace). (40)

gueulard (d'un cubilot) (n.m.), charging-hole (of a cupola furnace). (41)

gueule (d'un four, d'un convertisseur) (n.f.), mouth (of a furnace, of a converter). (42 ou 43)

gueule (d'un concasseur) (n.f.), mouth (of a rock-breaker). (44)

gueule-bée [gueules-bées pl.] ou gueule bée [gueules bées pl.] (Hydraul.) (n.f.), orifice in thick wall ; opening in thick wall ; short pipe ; cylindrical mouthpiece. (45) Cf. écoulement.

gueule de brochet (Sondage à la corde) (f.), rope-grab. (46)

gueule-de-loup [gueules-de-loup pl.] (de cheminée) (n.f.), turncap (of chimney). (47)

gueule-de-loup (locomotive) (n.f.), exhaust-muffler ; muffler. (48)

gueule-de-loup (Menuis.) (n.f.), groove (to receive a cock-bead). (49)

gueusat ou gueuset (Métall.) (n.m.), small pig. (50)

gueuse (masse de fonte) (n.f.), pig. (51)

gueuse (moule pratiqué dans le sable pour recevoir la fonte) (n.f.), pig-mould. (52)

gueuse de fonte (f.), iron pig ; pig of iron ; pig iron. (53)

gueuse des mères (f.), sow. (54)

guhr (Géol.) (n.m.), guhr. (55)

guichet (n.m.) ou guichet mobile (d'une porte à guichet) (Aérage de mines), shutter, sliding shutter, (of a regulating-door). (56)

guidage (Mines) (n.m.). Same as guidonnage.

guidages (d'un tour) (n.m.pl.), ways, track, (of a lathe). (57)

guide (n.m.), guide. (58)

guide (coulant de puits de mine) (n.m.) (s'emploie généralement au pluriel), guide ; slide ; cage-guide ; pit-guide. (59)

guide (scierie) (n.m.), fence. (60)

guide (lanterne-guide) (Sondage) (n.m.), winged substitute. (61)

guide-âne [guide-ânes pl.] (n.m.), manual ; hand-book ; pocket-book : (62)
un guide-âne pour les mineurs, a manual for miners ; a miners' pocket-book. (63)

guide-chaîne [guide-chaîne pl.] (d'un palan, ou analogue) (n.m.), chain-guide (of a pulley-block, or the like). (64 ou 65)

guide-coussinets [guide-coussinets pl.] (de filière) (n.m.), die-guide. (66)

guide d'épaisseur (m.), thickness guide. (67)

guide d'onglet (scierie) (m.), mitre-fence. (68)

guide de la tête de piston (m.), cross-head guide ; guide-bar ; guide ; slide-bar ; slide-rod. (69)

guide de pilon (bocards) (m.), stamp-guide (ore-stamps). (70)

guide-lame [guide-lame pl.] (scierie) (n.m.), saw-guide. (71)

guide pour filière à lunettes (m.), guide for solid-die stock. (72)

guide rectiligne (scierie) (m.), parallel fence ; straight fence. (73)

guider (v.t.), to guide : (74)
guider le courant d'air, to guide the air-current. (75)

guidon (Géol. & Mines) (n.m.), marker ; float ; float-mineral ; float-ore ; shoad ; shoad-stone ; shode ; shode-stone. (76)

guidonnage (action) (n.m.), guiding : (77)
guidonnage du câble, guiding the rope. (78)

guidonnage (système) (*n.m.*), guides ; guiding system : (1)
   guidonnage de la maîtresse tige d'une pompe d'épuisement, guides of the main rod of a mine-pump. (2)

guidonnage (de puits de mine) (*n.m.*), guides ; slides ; cage-guides ; pit-guides : (3)
   le guidonnage (*ou* guidage) est destiné à empêcher le tournoiement et les rencontres des cages, guides are intended to prevent spinning and collisions of cages. (4)

guillaume (*n.m.*), rabbet-plane ; rebate-plane ; rabbet. (5)

guillaume de côté (*m.*), side rabbet-plane. (6)

guillaume de fil *ou* guillaume de bout (*m.*), square rabbet-plane ; square-mouthed re-bate-plane. (7)

guillaume en navette (*m.*), compass-rabbet. (8)

guillaume oblique (*m.*), skew rabbet-plane ; skew-mouthed rabbet. (9)

guillocher (*v.t.*), to chequer ; to checker. (10)

guillochure (*n.f.*), chequering ; checking ; checkering : (11)

les guillochures sur le corps d'un pointeau, the chequerings on the barrel of a centre-punch. (12)

guipage (de coton) (isolement des fils électriques) (*n.m.*), lap, lapping, (of cotton). (13)

Gulf-Stream (*n.m.*), Gulf Stream. (14)

gummite (Minéral) (*n.f.*), gummite. (15)

gutta-percha *ou* gutta-perca (*n.f.*), gutta-percha. (16)

gymnite (Minéral) (*n.f.*), gymnite. (17)

gypse (*n.m.*), gypsum ; plaster-rock ; plaster-stone. (18)

gypseux, -euse (*adj.*), gypseous : (19)
   albâtre gypseux (*m.*), gypseous alabaster. (20)

gypsifère (*adj.*), gypsiferous ; gypsum-bearing : (21)
   calcaire gypsifère (*m.*), gypsiferous limestone. (22)

gyromètre (*n.m.*), gyrometer. (23)

gyroscope (*n.m.*), gyroscope. (24)

gyroscopique (*adj.*), gyroscopic : (25)
   toupie gyroscopique (*f.*), gyroscopic top. (26)

H

'haarkies (Minéral) (*n.f.*), hair-pyrites ; capil-lary pyrites ; millerite. (27)

habile (*adj.*), able ; clever ; skilful ; skilled. (28)

habileté (*n.f.*), ability ; cleverness ; skill ; skilfulness. (29)

habitation (*n.f.*), dwelling. (30)

habitation ouvrière (*f.*), workman's dwelling. (31)

habitude (*n.f.*), habit ; custom ; use ; way ; practice. (32)

habitude (Cristall., etc.) (*n.f.*), habit. (33)

habituel, -elle (*adj.*), usual ; habitual ; customary : (34)
   les accessoires habituels fournis comprennent . . . (*m.pl.*), the usual accessories supplied include . . . (35)

habituellement (*adv.*), usually ; habitually. (36)

'hachard (*n.m.*), shears ; hand-shears ; tin-men's shears ; tinmen's snips. (37)

'hache (*n.f.*), axe. (38)

'hache à fendre (*f.*), splitting-axe. (39)

'hache à main (*f.*), hatchet. (40)

'hache américaine (*f.*), American axe. (41)

'hache d'ouvrage (*f.*), slate-axe ; sax. (42)

'hache de bûcheron (*f.*), felling-axe. (43)

'hachette (*n.f.*), hatchet. (44)

'hachure (Topogr.) (*n.f.*), hachure. (45)

'hafnium (Chim.) (*n.m.*), hafnium. (46)

'haie (*n.f.*), hedge. (47)

'haie vive (*f.*), quick-hedge ; quickset ; quickset hedge. (48)

'halage (*n.m.*), towing : (49)
   halage d'un bateau sur un canal par des chevaux, towing a boat on a canal by horses. (50)

'halde (Mines) (*n.f.*), dump ; dump-heap. (51)

'halde de déchets *ou* 'halde de déblais *ou* simple-*ment* 'halde (*n.f.*), dump ; dump-heap ; pit-heap ; waste-heap ; spoil-heap ; spoil-bank ; barrow ; burrow. (52)

'halde de minerai (*f.*), ore-dump ; paddock. (53)

'halde de scories (*f.*), slag-dump ; slag-tip ; slag-heap. (54)

'haler (*v.t.*), to tow ; to haul ; to pull : (55)
   haler un chaland le long d'un canal, to tow a barge along a canal. (56)
   haler un câble, to pull, to haul, a rope. (57)
   haler une haveuse le long du front de taille, to pull a coal-cutting machine along the face. (58)

halite (Minéral) (*n.f.*), halite ; rock salt ; fossil salt ; sal gemmæ. (59)

'halle de coulée (haut fourneau) (*f.*), cast-house (blast-furnace). (60)

'halle de fonderie (*f.*), foundry-house. (61)

'halle de montage (*f.*), erecting-shop. (62)

halloysite (Minéral) (*n.f.*), halloysite. (63)

'halo (Photogr.) (*n.m.*), halation. (64)

halogène (Chim.) (*adj.*), halogenous. (65)

halogène (Chim.) (*n.m.*), halogen. (66)

'halotrichite (Minéral) (*n.f.*), halotrichite. (67)

'hanches (d'une chèvre, d'une bigue) (*n.f.pl.*), shears ; sheers ; shear-legs ; sheer-legs. (68)

'hangar (*n.m.*), shed ; outhouse. (69)

haplome (Minéral) (*n.m.*), haplome ; aplome. (70)

'happe (clameaux ; clampe) (*n.f.*), cramp-iron ; cramp ; dog-iron ; joiners' dogs ; joint-cramp. (71)

'happe (tenaille de fondeur) (*n.f.*), crucible-tongs. (72)

'happe (presse à vis pour menuisiers, etc.) (*n.f.*), G cramp ; G clamp ; C clamp. (73)

'happe à gros ventre, vis à oreilles à filets carrés, pouvant serrer — m/m (*ou* serrage de — m/m) (*f.*), deep-pattern square-thread G cramp with thumb-screw, to take in — millimetres. (74)

'happe à nervure (*f.*), ribbed G cramp. (75)

'happer à la langue (se dit de certains minéraux, particulièrement les argiles), to adhere, to stick, to the tongue : (76)
   l'argile sèche happe à la langue (*f.*), dry clay sticks to the tongue. (77)

'hardénite (Métall.) (*n.f.*), hardenite. (1)
'harkise (Minéral) (*n.f.*), hair-pyrites ; capillary pyrites ; millerite. (2)
harmonique (*adj.*), harmonic : (3)
courant harmonique (Élec.) (*m.*), harmonic current. (4)
harmoniques (Élec.) (*n.f.pl.*), harmonics. (5)
harmotome (*n.m.*) *ou* harmotome barytique (Minéral), harmotome ; barium harmotome. (6)
harmotome calcaire (*m.*), lime harmotome ; christianite ; phillipsite. (7)
'harnais (pour chevaux) (*n.m.*), harness (for horses). (8)
'harnais d'engrenages (Méc.) (*m.*), gearing ; gear : (9)
poupée fixe à double harnais d'engrenages (*f.*), double-geared fast head-stock. (10)
'harnais d'engrenages placé à l'intérieur du cône *ou* 'harnais d'engrenages logé dans le cône (*m.*), gearing placed inside the cone. (11)
'harpe (grille pour séparer les minerais bruts du menu de mines) (*n.f.*), grizzly ; grizzley ; ore-grizzly. (12)
'harpe (pierre d'attente) (Maçonn.) (*n.f.*), tooth-ing-stone ; tooth. (13)
'harpon (Sondage) (*n.m.*), spear ; harpoon. (14)
'harpon (scie) (*n.m.*), two-handed saw. (15)
'harpon pour câbles (Sondage à la corde) (*m.*), rope-spear. (16)
'harpon pour câbles métalliques (*m.*), wire-rope spear. (17)
'hartine (Minéral) (*n.f.*), hartin. (18)
'hartite (Minéral) (*n.f.*), hartite. (19)
'hasard (*n.m.*), hazard ; chance ; accident. (20)
'hasarder (*v.t.*), to hazard ; to venture ; to risk. (21)
'hasardeux, -euse (*adj.*), hazardous ; risky ; venturous. (22)
'hatchettine (Minéral) (*n.f.*), hatchettite ; hatchettine ; mineral tallow ; mountain tallow. (23)
'hauban (gros cordage qui maintient une chèvre, une grue, etc., dressée) (*n.m.*), guy ; stay ; tie-back. (24)
'haubanage (*n.m.*), guying ; staying. (25)
'haubaner (*v.t.*), to guy ; to stay : (26)
haubaner un poteau télégraphique, to guy a telegraph-pole. (27)
'hauérite (Minéral) (*n.f.*), hauerite. (28)
'hausmannite (Minéral) (*n.f.*), hausmannite. (29)
'hausse (*n.f.*), rise : (30)
hausse dans le cours de l'étain, rise in the price of tin. (31)
'hausse (d'un barrage, d'une vanne) (*n.f.*), flashboard, flushboard, stop-plank, (of a dam, of a sluice-gate). (32 *ou* 33)
'hausser (*v.t.*), to raise ; to lift : (34)
hausser une maison d'un étage, to raise a house by a story. (35)
hausser le prix des marchandises, to raise the price of the goods. (36)
'hausser (*v.i.*), to rise : (37)
fleuve qui hausse rapidement (*m.*), river which rises rapidly. (38)
l'étain hausse (*m.*), tin is rising (in price). (39)
'haussière (*n.f.*), hawser. (40)
'haut, -e (*adj.*), high : (41)
haut voltage (*m.*), high voltage ; high tension ; high pressure. (42)

haute fréquence (Élec.) (*f.*), high frequency : (43)
courant à haute fréquence (*m.*), high-frequency current. (44)
haute pression (*f.*), high pressure ; heavy pressure ; high tension. (45)
haute pression (à), high-pressure ; high-tension : (46)
cylindre à haute pression (*m.*), high-pressure cylinder. (47)
haute teneur (*f.*), high grade ; high tenor : (48)
haute teneur d'un minerai, high grade of an ore. (49)
haute teneur de soufre, high content of sulphur. (50)
haute teneur (à), high-grade : (51)
minerai à haute teneur (*m.*), high-grade ore. (52)
haute teneur d'or (à), highly auriferous : (53)
minerai à haute teneur d'or (*m.*), highly auriferous ore. (54)
haute tension (*f.*), high tension ; high intensity ; high pressure. (55)
haute tension (à), high-tension ; high-pressure : (56)
ligne de transport de force à haute tension (*f.*), high-tension power-transmission line. (57)
hautes eaux (*f.pl.*), high water. (58)
hautes terres (*f.pl.*), highlands. (59)
'haut (sommet) (*n.m.*), top ; summit : (60)
haut d'une montagne, top, summit, of a mountain. (61)
haut de course du piston, top of stroke of piston. (62)
'haut (d'une rivière) (*n.m.*), upper part (of a river). (63)
le haut Rhin, the upper Rhine. (64)
'haut de *ou* de haut, high ; in height : (65)
un mur haut de — mètres *ou* un mur de — mètres de haut, a wall — metres high ; a wall — metres in height. (66)
'haut de l'eau (*m.*), high water. (67)
'haut-fond [hauts-fonds *pl.*] (*n.m.*), shoal ; flat. (68)
'haut fourneau [hauts fourneaux *pl.*] (*m.*), blast-furnace. (69)
'haut fourneau au charbon de bois (*m.*), charcoal blast-furnace. (70)
'haut fourneau au coke (*m.*), coke blast-furnace. (71)
'haut fourneau électrique (*m.*), electric shaft-furnace. (72)
'haut-le-pied *ou* 'haut le pied (*invar.*) (Ch. de f.), light (without wagons or carriages attached, —said of an engine ; or, empty,—said of a train) : (73)
machine haut-le-pied (*f.*), light engine. (74)
aller haut le pied, to run light. (75)
marche haut-le-pied (*f.*), light running. (76)
'hautement (*adv.*), highly : (77)
gravier hautement aurifère (*m.*), highly aurif-erous gravel. (78)
roche hautement minéralisée (*f.*), highly mineralized rock ; heavily mineralized rock. (79)
'hauteur (élévation verticale) (*n.f.*), height ; elevation ; altitude : (80)
la hauteur d'un mât, the height of a mast. (81)
'hauteur (de) *ou* 'hauteur (en), high ; in height : (82)

le mât **a** — mètres de hauteur, the mast is — metres high ; the mast is — metres in height. (1)

'hauteur (Géom.) (*n.f.*), altitude : (2)
   la hauteur d'un triangle, d'un cône, the altitude of a triangle, of a cone. (3 *ou* 4)

'hauteur (éminence) (*n.f.*), height ; elevation ; hill. (5)

'hauteur (*n.f.*) *ou* 'hauteur d'élévation *ou* 'hauteur manométrique (pompes), head ; lift ; head pumped against ; static head ; static lift ; manometric head. (6)

'hauteur (*n.f.*) *ou* 'hauteur de chute *ou* 'hauteur d'eau (Hydraul.), head ; head of water ; fall. (7)

'hauteur au-dessous du primitif (Engrenages)(*f.*), depth below pitch-line ; dedendum. (Gear--ing). (8)

'hauteur au-dessus du niveau de la mer (*f.*), height, altitude, elevation, above sea-level. (9)

'hauteur au-dessus du primitif (Engrenages)(*f.*), height above pitch-line ; addendum. (10)

'hauteur barométrique (*f.*), barometric height. (11)

'hauteur d'appui (à) (Constr.), elbow-high. (12)

'hauteur d'aspiration (opp. à hauteur de re--foulement) (pompes) (*f.*), suction-head ; suction-lift ; static suction-lift. (13)

'hauteur d'eau (Hydraul.) (*f.*). See 'hauteur *ou* 'hauteur d'eau.

'hauteur d'eau à marée basse (*f.*), depth of water at low tide. (14)

'hauteur d'échappée (*f.*), headroom. (15)

'hauteur d'élévation (*f.*). See 'hauteur *ou* 'hauteur d'élévation.

'hauteur de chute (Hydraul.) (*f.*). See 'hauteur *ou* 'hauteur de chute.

'hauteur de chute des pilons (Bocardage des minerais) (*f.*), drop of stamps ; drop given to stamps. (16)

'hauteur de chute du trépan (Sondage) (*f.*), length of stroke of the chisel. (17)

'hauteur de démoulage (d'une machine à démouler) (Fonderie) (*f.*), drop, draft, (of a pattern-withdrawing machine). (18)

'hauteur de l'instrument en station (Topogr.) (*f.*), height of instrument. (19)

'hauteur de l'œil *ou* 'hauteur des yeux (*f.*), eye-level. (20)

'hauteur de la dent (distance entre le cercle de pied et le cercle de tête) (Engrenages) (*f.*), working-depth of tooth. (21)

'hauteur de levage (d'une grue) (*f.*), lift (of a crane). (22)

'hauteur de marche (escalier) (*f.*), rise (stairs). (23)

'hauteur de passage (*f.*), headroom. (24)

'hauteur de refoulement (pompes) (*f.*), discharge-head ; discharge-lift ; static discharge-head ; delivery-head ; delivery-lift. (25)

'hauteur de sécurité (espace surabondant dans le chevalement) (Mines) (*f.*), overwinding allowance ; clearance. (26)

'hauteur des pointes *ou* HDP (tour) (*f.*), height of centres (lathe). (27)

'hauteur du pas (d'une vis)(*f.*), lead (of a screw). (28)

'hauteur manométrique (*f.*). See 'hauteur *ou* 'hauteur manométrique.

'hauteur maximum à scier *ou* 'hauteur maximum du trait de scie (*f.*), depth of cut. (29)

'hauteur moyenne de l'année (Barométrie) (*f.*), mean yearly height. (30)

hauteur moyenne des continents (Géogr. phys.) (*f.*), average elevation of the continents. (31)

'hauteur moyenne du jour (Barométrie) (*f.*), mean daily height. (32)

'hauteur piézométrique (Hydraul.) (*f.*), pressure-head ; static-head ; head. (33)

'hauteur sous clef (d'une voûte) (*f.*), height above impost-level, rise, (of an arch). (34)

'hauteur stérile (Mines) (*f.*), thickness of the overburden. (35)

'hauteur totale (Hydraul.) (*f.*), total head ; total fall. (36)

'hautcur totale *ou* 'hauteur d'élévation totale (pompes) (*f.*), total head ; total lift ; total head pumped against. (37)

'hauteur totale de dragage (*f.*), full dredging-depth. (38)

'hauteur totale de la dent (Engrenages) (*f.*), whole depth of tooth. (39)

'hauteur variable (à), rising and falling : rising ; which can be raised or lowered : (40)
   table à hauteur variable (*f.*), table which can be raised or lowered ; rising table. (41)

'hauteur verticale (hauteur d'étage) (opp. à relevée) (Mines) (*f.*), vertical depth ; lift (vertical distance between two levels). (42) (distinguished from **inclined depth**).

haüyne (Minéral) (*n.f.*), hauynite ; hauyne. (43)

'havage (action) (Mines) (*n.m.*), cutting ; under--cutting ; holing ; underholing ; kirving ; kerving ; jadding ; slotting : (44)
   havage mécanique du charbon, coal-cutting by machinery. (45)
   havage du front de taille, cutting, under--cutting, holing, underholing, slotting, the face. (46)

'havage (entaille) (*n.m.*), cut ; undercut ; holing ; kerf ; jad. (47)

'havé, -e (*adj.*), undercut ; holed ; underholed ; slotted. (48)

'haver (*v.t.*), to cut ; to undercut ; to hole ; to underhole ; to kirve ; to kerve ; to jad ; to slot : (49)
   haver un massif de houille, to cut, to undercut, to hole, to underhole, to slot, a mass of coal. (50)

'haveur (pers.) (*n.m.*), cutter ; coal-cutter ; undercutter ; holer. (51)

'haveuse (*n.f.*), coal-cutting machine ; coal-cutter ; cutter ; undercutting-machine ; undercutter ; holing-machine ; mining-machine ; ironman. (52)

'haveuse à barre coupante (*f.*), bar coal-cutting machine. (53)

'haveuse à chaîne (*f.*), chain coal-cutting machine. (54)

'haveuse à disque (*f.*), disc coal-cutting machine. (55)

'haveuse à pic *ou* 'haveuse à percussion (*f.*), pick coal-cutting machine ; puncher coal-cutting machine ; percussive coal-cutter. (56)

'haveuse à pic sur affût-colonne (*f.*), post-puncher. (57)

'haveuse aux grands fronts de taille (*f.*), long-wall coal-cutting machine. (58)

'haveuse ripante (*f.*), short-wall coal-cutting machine. (59)

'haveuse-rouilleuse [haveuses-rouilleuses *pl.*] (*n.f.*), undercutting-and-nicking machine ; holing-and-shearing machine. (60)

'havrits (Exploitation des couches de houille) (*n.m.pl.*), fakes; rashings. (1)

'hayésine (Minéral) (*n.f.*), hayesine. (2)

'haylite (Explosif) (*n.f.*), haylite. (3)

hebdomadaire (*adj.*), weekly. (4)

hebdomadairement (*adv.*), weekly. (5)

héberger un inspecteur durant sa visite d'une propriété, to entertain an inspector during his visit to a property. (6)

hébronite (Minéral) (*n.f.*), hebronite. (7)

hectare (*n.m.*), hectare = 2·4711 acres. (8)

hectogramme (*n.m.*), hectogramme = 3·527 ozs avoirdupois. (9)

hectolitre (*n.m.*), hectolitre = 2·75 bushels. (10)

hectomètre (*n.m.*), hectometre = 109·36 yards. (11)

hectowatt (Élec.) (*n.m.*), hectowatt. (12)

hectowatt-heure [hectowatts-heure *pl.*] (*n.m.*), hectowatt-hour. (13)

hédenbergite (Minéral) (*n.f.*), hedenbergite. (14)

hédyphane (Minéral) (*n.m.*), hedyphane. (15)

hefner (Photom.) (*n.m.*), hefner; hefner-unit. (16)

hélice (Géom.) (*n.f.*), helix; spiral: (17)

hélice dextrorsum *ou* hélice à droite, dextrorsal helix; right-handed spiral. (18)

hélice sinistrorsum *ou* hélice senestrorsum *ou* hélice à gauche, sinistrorsal helix; left-handed spiral. (19)

hélice (en) *ou* hélice (à), helical; spiral; twist (*used as adj.*): (20)

escalier en hélice (*m.*), spiral staircase. (21)

alésoir à main en hélice (*m.*), twist hand-reamer. (22)

hélice (d'un bateau) (*n.f.*), screw, propeller, screw-propeller, (of a boat): (23)

hélices jumelles, twin screws; twin propellers. (24)

hélicoïdal, -e, -aux *ou* hélicoïde (*adj.*), helicoid; helicoidal; helical; spiral; twist (*used as adj.*); screw (*used as adj.*): (25)

engrenage hélicoïdal (*m.*), spiral gear; helical gear; screw-gearing. (26)

escalier hélicoïdal (*m.*), spiral stairs. (27)

mèche hélicoïdale (*f.*), twist-drill. (28)

hélicoïde (*n.m.*), helicoid; spiral. (29)

héliotrope (Minéral) (*n.m.*), heliotrope; blood-stone. (30)

hélium (Chim.) (*n.m.*), helium. (31)

helvine *ou* helvite (Minéral) (*n.f.*), helvite; helvine. (32)

hématite (Minéral) (*n.f.*), hematite; specularite. (33)

hématite brune (*f.*), brown hematite; brown iron ore; limonite. (34)

hématite rouge (*f.*), red hematite; red iron ore; red iron. (35)

hématolite (Minéral) (*n.f.*), hematolite. (36)

hématurie (Pathologie) (*n.f.*), hematuria; black-water fever. (37)

hémièdre *ou* hémiédrique (Cristall.) (*adj.*), hemihedral; hemihedric; hemisymmetric; hemisymmetrical. (38)

hémièdre (*n.m.*), hemihedron. (39)

hémiédrie (*n.f.*), hemihedrism; hemihedry; hemisymmetry. (40)

hémimorphe (Cristall.) (*adj.*), hemimorphic; hemimorphous. (41)

hémimorphisme (Cristall.) (*n.m.*), hemimorphism. (42)

hémiomorphite (Minéral) (*n.f.*), hemiomorphite; calamine; electric calamine; siliceous calamine. (43)

hémisphère (*n.m.*), hemisphere. (44)

hémisphère austral *ou* hémisphère méridional *ou* hémisphère sud (*m.*), southern hemisphere. (45)

hémisphère boréal *ou* hémisphère septentrional *ou* hémisphère nord (*m.*), northern hemi-sphere. (46)

hémisphérique (*adj.*), hemispheric; hemi-spherical; round: (47)

boulon à tête hémisphérique (*m.*), round-headed bolt; round-head bolt. (48)

hémisphéroïdal, -e, -aux *ou* hémisphéroïde (*adj.*), hemispheroidal. (49)

hémisphéroïde (*n.m.*), hemispheroid. (50)

hémitropie (Cristall.) (*adj.*), twin; twinned; hemitrope; hemitropic; macled: (51)

cristal hémitrope (*m.*), twin crystal; hemitrope crystal. (52)

hémitropie (Cristall.) (*n.f.*), twinning; hemitropy; hemitropism; twin cristallization. (53)

henry [henrys *pl.*] (Élec.) (*n.m.*), henry: (54)

l'henry est l'unité pratique de self-induction, the henry is the practical unit of self-induction. (55)

hépatite (Minéral) (*n.f.*), hepatite. (56)

herborisé, -e (*adj.*), arborized; dendritic: (57)

agate herborisée (*f.*), arborized agate; dendritic agate; tree-agate. (58)

herbue (distingué de castine) (Métall.) (*n.f.*), clay flux. (59) (distinguished from lime-stone flux).

'herchage (*n.m.*). Same as 'herschage.

'herche (*n.f.*), tram; mine-wagon. (60)

'hercher (*v.t.*), 'hercheur (*n.m.*). Same as 'herscher, 'herscheur.

hercynite (Minéral) (*n.f.*), hercynite. (61)

'hérisson (*n.m.*), sprocket-wheel; rag-wheel. (62)

herméticité (*n.f.*), tightness; water-tightness; imperviousness: (63)

herméticité d'un joint, tightness, water-tightness, imperviousness, of a joint. (64)

hermétique (*adj.*), hermetic; hermetical; air-tight. (65)

hermétiquement (*adv.*), hermetically: (66)

porte qui doit fermer hermétiquement (*f.*), door which should close hermetically. (67)

herminette (*n.f.*), adze. (68)

herminette à gouge (*f.*), rounding-adze. (69)

herminette à tête (*f.*), poll-adze. (70)

'herschage (Mines) (*n.m.*), haulage; hauling; tramming; drawing; wheeling; putting: (71)

herschage souterrain, underground haulage; underground tramming. (72)

'herscher (*v.t.*), to haul; to tram; to draw; to wheel; to put. (73)

'herscheur (pers.) (*n.m.*), trammer; drawer; wheeler; putter; haulageman. (74)

hertzien, -enne (*adj.*), Hertzian: (75)

ondes hertziennes (*f.pl.*), Hertzian waves. (76)

hessite (Minéral) (*n.f.*), hessite; telluric silver. (77)

hessonite (Minéral) (*n.f.*), hessonite; essonite. (78)

hétérogène (*adj.*), heterogeneous: (79)

roches hétérogènes (*f.pl.*), heterogeneous rocks. (80)

hétérogénéité (*n.f.*), heterogeneity. (81)

hétérogénite (Minéral) (*n.f.*), heterogenite. (82)

hétérologue (Chim.) (*adj.*), heterologous: (83)

série hétérologue (*f.*), heterologous series. (84)

hétéromorphe (*adj.*), heteromorphic. (1)
hétéromorphisme (*n.m.*), heteromorphism. (2)
hétéromorphite (Minéral) (*n.f.*), heteromorphite ; plumosite ; feather-ore. (3)
hétérosite (Minéral) (*n.f.*), heterosite. (4)
'hêtre (arbre) (*n.m.*), beech. (5)
'heulandite (Minéral) (*n.f.*), heulandite. (6)
heure (*n.f.*), hour ; time. (7)
heure d'éclairage (*f.*), lighting-hour. (8)
heure d'été (changement d'heure) (*f.*), summer time. (9)
heure d'hiver (*f.*), winter time. (10)
heure du lieu (*f.*), local time. (11)
heure-lampe [heures-lampe *pl.*] (Élec.) (*n.f.*), lamp-hour. (12)
heures de travail (*f.pl.*), working-hours ; hours of labour. (13)
heures supplémentaires *ou* heures supplémentaires de travail (*f.pl.*), overtime. (14)
'heurté, -e (Photogr.) (*adj.*), contrasty : (15) un cliché heurté provient d'un manque de pose et d'un excès de développement, a contrasty negative arises from an underexposure and an overdevelopment. (16) Cf. contrasté, -e.
'heurtequin (d'un essieu) (*n.m.*), collar (of an axle). (17)
'heurtoir (d'une voie de garage) (Ch. de f.) (*n.m.*), bumping-post (of a siding). (18)
'heurtoir (d'une voie principale) (*n.m.*), buffer (of a main-line track). (19)
hexaèdre *ou* hexaédrique (*adj.*), hexahedral. (20)
hexaèdre (*n.m.*), hexahedron. (21)
hexagonal, -e, -aux *ou* hexagone (*adj.*), hexagonal ; hexagon (*used as adj.*) : (22)
prisme hexagonal (*m.*), hexagonal prism. (23)
tourelle hexagonale (tour) (*f.*), hexagon turret (lathe). (24)
hexagone (*n.m.*), hexagon. (25)
hexane (Chim.) (*n.m.*), hexane. (26)
hexapolaire (Élec.) (*adj.*), six-pole. (27)
hexatomique (Chim.) (*adj.*), hexatomic. (28)
hiddénite (Minéral) (*n.f.*), hiddenite ; lithia emerald. (29)
'hie (de paveur) (*n.f.*), beetle ; rammer ; earth-rammer. (30)
'hie (sonnette) (*n.f.*), pile-driver ; post-driver ; spile-driver ; gin. (31)
'hierschage, 'hierscher, 'hierscheur. Same as 'herschage, etc.
'hissage (*n.m.*), hoisting ; winding ; raising ; pulling up. (32)
'hisser (*v.t.*), to hoist ; to wind ; to raise ; to pull up : (33)
hisser à bloc, to hoist chock-a-block ; to hoist block and block. (34)
hisser un bloc de pierre du fond d'une carrière, to raise a block of stone from the bottom of a quarry. (35)
hisser un tonneau, to hoist a cask. (36)
hisser un wagon sur un plan incliné, to pull a wagon up an incline. (37)
hiver (*n.m.*), winter. (38)
hiver bénin *ou* hiver doux (*m.*), mild winter. (39)
hiver rigoureux (*m.*), severe winter ; hard winter. (40)
hiver tardif (*m.*), late winter. (41)
hivernage (saison) (*n.m.*), winter season. (42)
hivernal, -e, -aux (*adj.*), winter : (43)
neige hivernale (*f.*), winter snow. (44)
hiverner sur la propriété, to winter on the property. (45)
hohlspath (Minéral) (*n.m.*), hollow-spar. (46)

holmium (Chim.) (*n.m.*), holmium. (47)
holoaxe (Cristall.) (*adj.*), holoaxial. (48)
holocristallin, -e (*adj.*), holocrystalline : (49)
roches holocristallines (*f.pl.*), holocrystalline rocks. (50)
holoèdre *ou* holoédrique (Cristall.) (*adj.*), holohedral ; holohedric ; holosymmetric ; holosymmetrical. (51)
holoèdre (*n.m.*), holohedron. (52)
holoédrie (*n.f.*), holohedrism ; holosymmetry. (53)
holostérique (*adj.*), holosteric : (54)
baromètre holostérique (*m.*), holosteric barometer. (55)
homilite (Minéral) (*n.f.*), homilite. (56)
homme (*n.m.*), man. (57)
homme au basculeur (*m.*), tipman ; tipper. (58)
homme au lavage (*m.*), washerman. (59)
homme au treuil (*m.*), winchman. (60)
homme de confiance *ou* homme d'un commerce sûr *ou* homme sur lequel on peut compter (*m.*), reliable man. (61)
homme de couleur (*m.*), man of colour. (62)
homme de pétrole (*m.*), oilman (a man interested in the petroleum industry) : (63)
champ qui est l'Eldorado des hommes de pétrole (*m.*), field which is the Eldorado of oilmen. (64)
homme de travail *ou* homme de peine (*m.*), working man ; labouring man ; labourer ; common labourer. (65)
homme manivelle (*f.*), force equivalent to 7 kilogrammetres. (66)
homme rangé (*m.*), steady man. (67)
homogène (*adj.*), homogeneous : (68)
objectif à immersion homogène (*m.*), homogeneous immersion lens. (69)
homogénéité (*n.f.*), homogeneity : (70)
l'homogénéité du béton et sa résistance permettent de diminuer les dimensions attribuées ordinairement aux constructions, the homogeneity of concrete and its strength allow of the dimensions ordinarily allotted to buildings being reduced. (71)
homogènement (*adv.*), homogeneously. (72)
homologue (*adj.*), homologous : (73)
série homologue (*f.*), homologous series. (74)
homomorphe (*adj.*), homomorphous ; homomorphic. (75)
homomorphisme (*n.m.*), homomorphism. (76)
homosiste (*adj.*), homoseismal ; homoseismic ; coseismal ; coseismic. (77)
homosiste (*n.f.*), homoseismal ; coseismal. (78)
honigstein (Minéral) (*n.m.*), honeystone ; mellite. (79)
honoraires (*n.m.pl.*), fee ; fees : (80)
honoraires de l'ingénieur-conseil, consulting-engineer's fees. (81)
hopéite (Minéral) (*n.f.*), hopeite ; hopite. (82)
horaire (*adj.*), hourly ; horary : (83)
débit horaire (d'une pompe) (*m.*), hourly flow (of a pump). (84)
horizon (*n.m.*), horizon. (85)
horizon apparent *ou* horizon visible (*m.*), apparent horizon ; local horizon ; sensible horizon ; visible horizon. (86)
horizon carbonifère (Géol.) (*m.*), coal-horizon. (87)
horizon cuprifère (*m.*), cupriferous horizon ; copper-bearing zone. (88)
horizon géologique (*m.*), geological horizon. (89)
horizon pétrolifère (*m.*), oil-horizon. (90)

horizontal, -e, -aux (*adj.*), horizontal : (1)
ligne horizontale (*f.*), horizontal line. (2)
horizontalement (*adv.*), horizontally. (3)
horizontalité (*n.f.*), horizontality : (4)
horizontalité du fléau d'une balance, horizon-
-tality of the beam of a balance. (5)
horloge (*n.f.*), clock. (6)
'**hornblende** (Minéral) (*n.f.*), hornblende. (7)
'**hornblendite** (Pétrol.) (*n.f.*), hornblendite. (8)
'**hornstein** (*n.m.*), horn-flint. (9)
'**hors courant** (Élec.) (*adj.*), dead : (10)
fil hors courant (*m.*), dead wire. (11)
'**hors d'aplomb**, out of plumb ; off plumb. (12)
'**hors d'équerre**, out of square. (13)
'**hors d'œuvre** (en saillie), projecting. (14)
'**hors d'œuvre** (mesure), outside : (15)
mesure hors d'œuvre (*f.*), outside measure-
-ment. (16)
'**hors d'œuvre** (se dit des pierres précieuses),
unmounted ; unset. (17)
'**hors d'usage** *ou* '**hors de service**, out of use ;
out of action. (18)
'**horst** (Géol.) (*n.m.*), horst : (19)
les Vosges et la Forêt-Noire sont deux grands
horsts dissymétriques, qui se font face de
part et d'autre de la vallée du Rhin. Le
Sinaï est un horst en forme de coin,
situé entre les deux fossés du golfe
de Suez et du golfe d'Akaba, the Vosges
and the Black Forest are two great
dissymmetrical horsts, which face each
other on both sides of the Valley of the Rhine.
Sinai is a wedge-shaped horst, situated
between the two grabens of the Gulf of Suez
and of the Gulf of Akabah. (20)
hôte (Minéralogie) (*n.m.*), host : (21)
parasites et leur hôte (*m.pl.*), parasites and
their host. (22)
hôtel de la Monnaie *ou* **hôtel des Monnaies** (*m.*),
mint. (23)
'**hotte** (d'une forge, d'un laboratoire) (*n.f.*),
hood (of a forge, of a laboratory). (24 *ou* 25)
'**hotte** (cuvette placée à la tête du tuyau de
descente) (Constr.) (*n.f.*), hopper-head ;
head. (26)
'**houage** (*n.m.*), cubic measurement of a coal-
deposit. (27)
'**houille** (*n.f.*), coal. (28) See also **charbon.**
'**houille à courte flamme** (*f.*), short-flaming coal.
(29)
'**houille à gaz** (*f.*), gas-coal. (30)
'**houille à longue flamme** (*f.*), long-flaming coal.
(31)
'**houille anthraciteuse** (*f.*), semianthracite. (32)
'**houille blanche** (*f.*), white coal ; white coals ;
water-power. (33)
'**houille bleue** (force de la marée) (*f.*), tide-
power. (34)
'**houille collante** (*f.*), caking coal. (35)
'**houille de chaudière** (*f.*), steam-coal. (36)
'**houille demi-grasse** (*f.*), semibituminous coal.
(37)
'**houille extraite** (Production houillère) (*f.*),
mineral gotten ; mineral extracted ; mineral
raised. (38)
'**houille flambante** (*f.*), splint-coal. (39)
'**houille grasse** (*f.*), bituminous coal ; soft coal ;
fat coal ; smoking coal ; flaming coal. (40)
'**houille grasse à courte flamme** (*f.*), steam-coal.
(41)
'**houille grasse à longue flamme** (*f.*), cherry coal.
(42)

'**houille grasse maréchale** *ou* '**houille maréchale**
(*f.*), smithing coal ; blacksmith coal. (43)
'**houille grise** (force du vent) (*f.*), wind-power.
(44)
'**houille maigre** (*f.*), lean coal. (45)
'**houille maigre à courte flamme** *or* '**houille
maigre anthraciteuse** (*f.*), semianthracite.
(46)
'**houille miroitante** (*f.*), peacock coal. (47)
'**houille non-collante** (*f.*), non-caking coal. (48)
'**houille papyracée** (*f.*), papyraceous coal ;
dysodile. (49)
'**houille schisteuse** (*f.*), shaly coal. (50)
'**houille sèche** (*f.*), dry coal. (51)
'**houille sèche à longue flamme** (*f.*), splint-coal.
(52)
'**houille verte** (*f.*), stream-power ; power derived
from rivers or streams. (53)
'**houiller, -ère** (*adj.*), of or pertaining to coal or
coal-mining ; carboniferous : (54)
terrain houiller (*m.*) *ou* formation houillère
(*f.*), coal-measures ; coal-formation. (55)
district houiller (*m.*), coal-mining district.
(56)
production houillère en milliers de tonnes
(*f.*), production of coal in thousands of tons.
(57)
grès houiller (*m.*), carboniferous sandstone.
(58)
'**houiller** (Géol.) (*n.m.*), Carboniferous : (59)
le houiller, the Carboniferous. (60)
'**houiller inférieur** (Géol.) (*m.*), Lower Carbonif-
-erous. (61)
'**houiller supérieur** (Géol.) (*m.*), Upper Carbonif-
-erous. (62)
'**houillère** (*n.f.*), colliery ; coal-mine. (63)
'**houillère à ciel ouvert** (*f.*), daylight-colliery.
(64)
'**houilleur** (pers.) (*n.m.*), collier ; coal-miner ;
pitman. (65)
'**houilleuse** (*n.f.*). Same as '**haveuse.**
'**houilleux, -euse** (*adj.*), coaly ; carboniferous :
(66)
roche houilleuse (*f.*), carboniferous rock. (67)
'**houillification** (*n.f.*), carbonization ; conversion
into coal : (68)
houillification de matières végétales, carboni-
-zation of vegetable matter. (69)
'**houillifier** (*v.t.*), to carbonize ; to convert into
coal : (70)
tronc d'arbre houillifié (*m.*), carbonized tree-
trunk. (71)
'**hourdage** *ou* '**hourdis** (couche de gros plâtre sur
un lattis) (*n.m.*), pugging ; deadening ;
deafening. (72)
'**hourder** (*v.t.*), to pug ; to deaden ; to deafen :
(73)
hourder un plancher, une cloison, to pug, to
deaden, to deafen, a floor, a partition. (74
ou 75)
hübnérite (Minéral) (*n.f.*), hübnerite. (76)
'**huche** (pour le moteur d'une turbine) (*n.f.*),
enclosed casing ; closed turbine-chamber.
(77)
'**huche** (cuve à laver le minerai) (*n.f.*), hutch.
(78)
huilage (*n.m.*), oiling : (79)
huilage de machines, oiling machinery. (80)
huilage (trempe à l'huile) (*n.m.*), oil-tempering.
(81)
huile (*n.f.*), oil. (82)
huile à cylindre (*f.*), cylinder-oil. (83)

huile à graisser (*f.*), lubricating-oil. (1)

huile à mécanisme *ou* huile à mouvement (*f.*), machine-oil. (2)

huile animale (*f.*), animal oil. (3)

huile bitumineuse (*f.*), asphaltic oil. (4)

huile brute (*f.*), crude oil. (5)

huile d'éclairage (*f.*), illuminating-oil ; lamp-oil. (6)

huile de baleine (*f.*), whale-oil. (7)

huile de bras *ou* huile de coude *ou* huile de poignets (*expression populaire*) (*f.*), elbow-grease. (8)

huile de colza épurée (*f.*), refined colza-oil. (9)

huile de faible densité (*f.*), light-density oil. (10)

huile de graissage (*f.*), lubricating-oil. (11)

huile de houille (*f.*), coal-naphtha ; coal-tar naphtha. (12)

huile de lin *ou* huile de graine de lin (*f.*), linseed-oil. (13)

huile de lin bouillie *ou* huile de lin cuite (*f.*), boiled linseed-oil. (14)

huile de lin crue (*f.*), raw linseed-oil. (15)

huile de navette (*f.*), rape-oil. (16)

huile de phoque (*f.*), seal-oil. (17)

huile de ricin (*f.*), castor-oil. (18)

huile de roche *ou* huile de pierres (*f.*), rock-oil ; stone-oil ; earth-oil ; fossil oil ; coal-oil ; mineral naphtha ; petroleum. (19)

huile de schiste (*f.*), shale-oil. (20)

huile de vitriol (*f.*), oil of vitriol ; sulphuric acid ; vitriolic acid. (21)

huile fixe *ou* huile grasse (*f.*), fixed oil ; fatty oil. (22)

huile lampante (*f.*), lamp-oil. (23)

huile légère (*f.*), light oil. (24)

huile lourde (*f.*), heavy oil. (25)

huile minérale (*f.*), mineral oil. (26)

huile paraffinée (*f.*), paraffin-oil ; paraffin. (27)

huile pour cylindres (*f.*), cylinder-oil. (28)

huile pour organes de machines (*f.*), engine-oil. (29)

huile végétale (*f.*), vegetable oil. (30)

huile volatile (*f.*), volatile oil. (31)

huilement (*n.m.*), oiling. (32)

huiler (*v.t.*), to oil : (33)
huiler les rouages d'une machine, to oil the wheelwork of a machine. (34)

huileux, -euse (*adj.*), oily : (35)
substance huileuse (*f.*), oily substance. (36)

huisserie (*n.f.*), door-framing (a door-frame and the timbers around it, as in a partition or a frame-building). (37)

'huit de chiffre *ou simplement* 'huit (*n.m.*) (compas d'épaisseur), figure-of-eight calipers ; hour-glass calipers. (38)

huître perlière (*f.*), pearl oyster. (39)

humboldtilite (Minéral) (*n.f.*), humboldtilite ; melilite. (40)

humboldtine (Minéral) (*n.f.*), humboldtine. (41)

humide (*adj.*), damp ; humid ; moist ; wet : (42)
temps humide (*m.*), wet weather ; damp weather. (43)

humidification (*n.f.*), damping ; humidification ; humidifying ; moistening ; wetting. (44)

humidifier (*v.t.*), to damp ; to humidify ; to moisten ; to wet. (45)

humidité (*n.f.*), dampness ; humidity ; damp ; moisture ; moistness ; wet ; wetness : (46)
humidité de l'atmosphère, dampness, humidity, moistness, of the atmosphere. (47)

humidité des murs, dampness of walls. (48)

humique (*adj.*), humic. (49)

'humite (Minéral) (*n.f.*), humite. (50)

'hundredweight (*en abrégé s'écrit* cwt) [Angleterre] (*n.m.*), hundredweight=112 lbs *or* 50·80 kilogrammes. (51)

'hundredweight [Amérique] (*n.m.*), hundredweight =100 lbs *or* 45·36 kilogrammes. (52)

'hureaulite (Minéral) (*n.f.*), hureaulite. (53)

'hutte (*n.f.*), hut. (54)

hyacinthe (Minéral) (*n.f.*), hyacinth ; hyacinth-stone. (55)

hyalin, -e (*adj.*), hyaline : (56)
quartz hyalin (*m.*), hyaline quartz. (57)

hyalite (Minéral) (*n.f.*), hyalite ; water-opal. (58)

hyalomélane (Pétrol.) (*n.f.*), hyalomelan. (59)

hyalomicte (Pétrol.) (*n.f.*), hyalomicte ; greisen. (60)

hyalophane (Minéral) (*n.f.*), hyalophane. (61)

hyalotourmalite (Pétrol.) (*n.f.*), schorl-rock ; shorl-rock. (62)

hydracide (Chim.) (*n.m.*), hydracid. (63)

hydrante (*n.m.*), hydrant ; plug. (64)

hydrargilite *ou* hydrargillite (Minéral) (*n.f.*), hydrargillite. (65)

hydrargyrisme (Pathologie) (*n.m.*), hydrargyri--asis ; hydrargyrism. (66)

hydratation (*n.f.*), hydration ; hydratation. (67)

hydrate (*n.m.*), hydrate : (68)
hydrate de calcium, calcium hydrate. (69)
hydrate de chlore, chlorine hydrate. (70)

hydraté, -e (*adj.*), hydrous ; hydrated : (71)
carbonate hydraté de cuivre (*m.*), hydrous carbonate of copper. (72)
silicate d'aluminium hydraté (*m.*), hydrous aluminum silicate. (73)

hydrater (*v.t.*), to hydrate. (74)

hydrater (s') (*v.r.*), to hydrate. (75)

hydraulicien (pers.) (*n.m.*), hydraulician ; hydraulic engineer. (76)

hydraulicité (*n.f.*), hydraulicity : (77)
hydraulicité des ciments, hydraulicity of cements. (78)

hydrauliquage (abatage à l'eau) (Mines) (*n.m.*), hydraulicking ; hydraulic mining. (79)

hydraulique (*adj.*), hydraulic ; water (*used as adj.*) : (80)
force hydraulique (*f.*), hydraulic power ; water-power. (81)

hydraulique (*n.f.*), hydraulics ; hydraulic science ; hydraulic engineering. (82)

hydrauliquement (*adv.*), hydraulically : (83)
pression obtenue hydrauliquement (*f.*), pressure obtained hydraulically. (84)

hydrauliquer (*v.t.*), to hydraulic ; to pipe ; to nozzle : (85)
hydrauliquer le front de taille, to hydraulic the working-face. (86)

hydrauliste (pers.) (*n.m.*), hydraulist ; hydraulic engineer. (87)

hydrique (Chim.) (*adj.*), hydric. (88)

hydro-électricité (*n.f.*), hydroelectricity. (89)

hydro-électrique (*adj.*), hydroelectric : (90)
énergie hydro-électrique (*f.*), hydroelectric energy. (91)

hydrocarboné, -e (*adj.*), hydrocarbonaceous ; hydrocarbonic ; hydrocarbonous. (92)

hydrocarbure (*n.m.*), hydrocarbon. (93)

hydrocarbures aromatiques (*m.pl.*), aromatic hydrocarbons. (94)

hydrocarbures gazeux (*m.pl.*), gaseous hydro-carbons. (1)
hydrocarbures insaturés (*m.pl.*), unsaturated hydrocarbons. (2)
hydrocarbures liquides (*m.pl.*), liquid hydro-carbons. (3)
hydrocarbures saturés (*m.pl.*), saturated hydro-carbons. (4)
hydrocarbures solides (*m.pl.*), solid hydrocarbons. (5)
hydrocarbures volatils (*m.pl.*), volatile hydro-carbons. (6)
hydroclasseur (*n.m.*), hydraulic classifier. (7)
hydrocratique (opp. à géocratique) (Géol.) (*adj.*), hydrocratic. (8) (opp. to geocratic).
hydrocyanite (Minéral) (*n.f.*), hydrocyanite. (9)
hydrodolomite (Minéral) (*n.f.*), hydrodolomite. (10)
hydrodynamique (*adj.*), hydrodynamic. (11)
hydrodynamique (*n.f.*), hydrodynamics. (12)
hydrofère (*adj.*), water-bearing. (13)
hydrofluorique (*adj.*), hydrofluoric. (14)
hydrofuge (*adj.*), water-resisting. (15)
hydrogénation (*n.f.*), hydrogenation. (16)
hydrogène (*n.m.*), hydrogen. (17)
hydrogène libre (*m.*), free hydrogen. (18)
hydrogène sulfuré (*m.*), hydrogen sulphide ; sulphureted hydrogen ; sulfhydric acid. (19)
hydrogéner (*v.t.*), to hydrogenize. (20)
hydrogénifère (*adj.*), hydrogeniferous. (21)
hydrogéologie (*n.f.*), hydrogeology. (22)
hydrogéologique (*adj.*), hydrogeological. (23)
hydrographe (pers.) (*n.m.*), hydrographer. (24)
hydrographie (*n.f.*), hydrography. (25)
hydrographique (*adj.*), hydrographic ; hydro-graphical : (26)
étude hydrographique (*f.*), hydrographic survey. (27)
hydrologie (*n.f.*), hydrology : (28)
hydrologie des bassins de montagne, hydrology of mountain basins. (29)
hydrologique (*adj.*), hydrologic ; hydrological. (30)
hydrologiquement (*adv.*), hydrologically. (31)
hydrologue (pers.) (*n.m.*), hydrologist. (32)
hydromagnésite (Minéral) (*n.f.*), hydromagnesite. (33)
hydrométallurgie (*n.f.*), hydrometallurgy. (34)
hydrométallurgique (*adj.*), hydrometallurgic ; hydrometallurgical. (35)
hydrométallurgiquement (*adv.*), hydrometal-lurgically. (36)
hydromètre (*n.m.*), hydrometer. (37)
hydrométrie (*n.f.*), hydrometry. (38)
hydrométrique (*adj.*), hydrometric ; hydro-metrical. (39)
hydrominéral, -e, -aux (*adj.*), hydromineral. (40)
hydrophane (*adj.*), hydrophanous : (41)
opale hydrophane (*f.*), hydrophanous opal. (42)
hydrophane (Minéral) (*n.f.*), hydrophane. (43)
hydropneumatique (*adj.*), hydropneumatic : (44)
ascenseur hydropneumatique (*m.*), hydro-pneumatic lift. (45)
hydroscope (pers.) (*n.m.*), water-diviner ; hydros-copist ; dowser ; douser. (46)
hydroscopie (*n.f.*), water-divining ; dowsing ; dousing. (47)
hydrosilicate (*n.m.*), hydrosilicate. (48)
hydrosiliceux, -euse (*adj.*), hydrosiliceous. (49)
hydrosphère (Géogr. phys.) (*n.f.*), hydrosphere. (50)

hydrostatique (*adj.*), hydrostatic ; hydro-statical : (51)
balance hydrostatique (*f.*), hydrostatic balance. (52)
hydrostatique (*n.f.*), hydrostatics. (53)
hydrosulfurique (*adj.*), hydrosulphuric. (54)
hydrotamis (*n.m.*), jig. (55)
hydrotechnique (*adj.*), hydrotechnic ; hydro-technical. (56)
hydrotechnique (*n.f.*), hydrotechny. (57)
hydrothermal, -e, -aux (*adj.*), hydrothermal : (58)
émission hydrothermale (*f.*), hydrothermal emission. (59)
hydrozincite (Minéral) (*n.f.*), hydrozincite ; zinc-bloom. (60)
hydrure (Chim.) (*n.m.*), hydride. (61)
hygiène (*n.f.*), hygiene ; sanitation. (62)
hygiénique (*adj.*), hygienic ; sanitary. (63)
hygromètre (*n.m.*), hygrometer. (64)
hygrométricité (*n.f.*), hygrometricity. (65)
hygrométrie (*n.f.*), hygrometry. (66)
hygrométrique (*adj.*), hygrometric ; hygrometrical: (67)
la faculté hygrométrique de certains sels, the hygrometrical power of certain salts. (68)
hygrométriquement (*adv.*), hygrometrically. (69)
hygroscope (*n.m.*), hygroscope. (70)
hygroscopicité (*n.f.*), hygroscopicity : (71)
hygroscopicité des pierres, hygroscopicity of stones. (72)
hygroscopie (*n.f.*), hygroscopy. (73)
hygroscopique (*adj.*), hygroscopic ; hygroscopical. (74)
hypabyssal, -e, -aux (Géol.) (*adj.*), hypabyssal : (75)
roches hypabyssales (*f.pl.*), hypabyssal rocks. (76)
hyperbole (Géom.) (*n.f.*), hyperbola. (77)
hyperbolique (*adj.*), hyperbolic ; hyperbolical : (78)
courbe hyperbolique (*f.*), hyperbolic curve. (79)
hyperboloïde (*adj.*), hyperboloidal. (80)
hyperboloïde (*n.m.*), hyperboloid. (81)
hyperboloïde de révolution (*m.*), hyperboloid of revolution. (82)
hypercompoundage (Élec.) (*n.m.*), overcom-pounding. (83)
hypercompounder (Élec.) (*v.t.*), to overcom-pound. (84)
hyperfocal, -e, -aux (Photogr.) (*adj.*), hyperfocal : (85)
distance hyperfocale (*f.*), hyperfocal distance. (86)
hypérite (Pétrol.) (*n.f.*), hyperite. (87)
hypersthène (Minéral) (*n.m.*), hypersthene ; Labrador hornblende. (88)
hypersthénique (*adj.*), hypersthenic. (89)
hypersthénite (Pétrol.) (*n.f.*), hypersthenite. (90)
hypoazoteux (*adj.m.*), hyponitrous. (91)
hypoazotite (*n.m.*), hyponitrite. (92)
hypocristallin, -e (*adj.*), hypocrystalline : (93)
roches hypocristallines (*f.pl.*), hypocrystalline rocks. (94)
hypocycloïdal, -e, -aux (Géom.) (*adj.*), hypo-cycloidal : (95)
engrenage hypocycloïdal (*m.*), hypocycloidal gear. (96)
hypocycloïde (Géom.) (*n.f.*), hypocycloid. (97)
hypogé, -e (Géol.) (*adj.*), hypogeal. (98)

hypogène (Géol.) (*adj.*), hypogene. (1)
hyposulfite (*n.m.*), hyposulphite : (2)
   hyposulfite de soude *ou simplement* hypo-
   -sulfite, hyposulphite of soda ; hypo ;
   sodium thiosulphate. (3)
hypothèse (*n.f.*), hypothesis ; theory : (4)
   hypothèse du feu central (Géol.), hypothesis
   of the central fire. (5)
   hypothèses sur la formation du diamant dans
   la nature, hypotheses on the formation of
   the diamond in nature. (6)
   hypothèse vulcanienne *ou* hypothèse pluto-
   -nienne (Géol.), Vulcanian theory ; Plutonic
   theory. (7)
hypsographie (*n.f.*), hypsography. (8)

hypsographique (*adj.*), hypsographic ; hypso-
   -graphical. (9)
hypsomètre (*n.m.*), hypsometer ; hypsother-
   -mometer ; boiling-point thermometer. (10)
hypsométrie (*n.f.*), hypsometry. (11)
hypsométrique (*adj.*), hypsometric ; hypso-
   -metrical. (12)
hystérésis (Phys.) (*n.f.*), hysteresis. (13)
hystérésis diélectrique (*f.*), dielectric hysteresis ;
   electrostatic hysteresis. (14)
hystérésis magnétique (*f.*), magnetic hysteresis.
   (15)
hystérésis visqueuse (*f.*), viscous hysteresis ;
   time-lag of magnetism. (16)
hystéromorphe (*adj.*), hysteromorphous. (17)

I

iceberg (*n.m.*), iceberg. (18)
icosaèdre (*adj.*), icosahedral ; icosihedral. (19)
icosaèdre (*n.m.*), icosahedron ; icosihedron. (20)
idéal, -e, -aux (*adj.*), ideal : (21)
   conditions idéales (*f.pl.*), ideal conditions.
   (22)
idéalement (*adv.*), ideally. (23)
idio-électrique (*adj.*), idioelectric. (24)
idiomorphe (Pétrol.) (*adj.*), idiomorphic ; idio-
   -morphous ; automorphic ; automorphous.
   (25)
idocrase (Minéral) (*n.f.*), idocrase ; vesuvianite ;
   vesuvian. (26)
idrialite (Minéral) (*n.f.*), idrialite. (27)
igné, -e (*adj.*), igneous : (28)
   roches ignées (*f.pl.*), igneous rocks. (29)
ignescent, -e (*adj.*), ignescent : (30)
   pierres ignescentes (*f.pl.*), ignescent stones.
   (31)
ignifuge (*adj.*), fire-proof ; fireproof : (32)
   écran ignifuge (*m.*), fire-proof screen. (33)
ignifuge (*n.m.*), fireproofing ; fire-proof material.
   (34)
ignifuger (*v.t.*), to fireproof : (35)
   ignifuger des objets combustibles, to fireproof
   combustible articles. (36)
ignition (*n.f.*), ignition. (37)
ignorance des règlements (*f.*), ignorance of the
   regulations. (38)
ijolite (Pétrol.) (*n.f.*), ijolite. (39)
île (*n.f.*), island. (40)
île continentale (*f.*), continental island. (41)
île corallienne (*f.*), coral island. (42)
île d'origine volcanique (*f.*), island of volcanic
   origin. (43)
île de boue (Géol.) (*f.*), mud-lump : (44)
   les îles de boue du Mississipi, the mud-lumps
   of the Mississippi. (45)
île pélagique (*f.*), oceanic island. (46)
illicite (*adj.*), illicit. (47)
illimité, -e (*adj.*), unlimited. (48)
illusion d'optique (*f.*), optical illusion. (49)
ilménite (Minéral) (*n.f.*), ilmenite ; titanic iron
   ore ; titaniferous iron ore ; titaniferous
   oxide of iron. (50)
ilménium (Chim.) (*n.m.*), ilmenium. (51)
îlot (*n.m.*), small island ; islet. (52)
ilsemannite (Minéral) (*n.f.*), ilsemannite. (53)
ilvaïte (Minéral) (*n.f.*), ilvaite. (54)
image (*n.f.*), image ; picture. (55)
image latente (Photogr.) (*f.*), latent image. (56)

image nette (Opt.) (*f.*), clear image ; sharp
   image. (57)
image réelle (Opt.) (*f.*), real image. (58)
image virtuelle (Opt.) (*f.*), virtual image. (59)
imaginer un appareil ingénieux, to devise an
   ingenious apparatus ; to invent an ingenious
   device. (60)
imbibé (-e) d'eau, soaked with water ; water-
   logged : (61)
   terrains imbibés d'eau (*m.pl.*), water-logged
   strata. (62)
imbiber (mouiller intérieurement) (*v.t.*), to soak :
   (63)
   une pluie abondante imbibe la terre, a heavy
   rain soaks the ground. (64)
   déchet de coton imbibé d'huile (*m.*), cotton-
   waste soaked in oil. (65)
imbiber (attirer par imbibition) (*v.t.*), to imbibe :
   (66)
   l'eau que la terre imbibe, the water that the
   earth imbibes. (67)
imbibition (*n.f.*), imbibition ; soaking : (68)
   imbibition d'eau par un corps poreux,
   imbibition of water by a porous substance.
   (69)
imbibition (désargentation) (Métall.) (*n.f.*),
   soaking. (70)
imbibition des bois (*f.*), impregnation of timber.
   (71)
imbrifuge (*adj.*), rain-proof ; rain-tight ; water-
   -proof. (72)
imbriqué, -e (*adj.*), imbricate ; imbricated :
   (73)
   structure imbriquée (Géol.) (*f.*), imbricate
   (*or* imbricated) structure. (74)
imbrisable (*adj.*), unbreakable. (75)
imbrûlable (*adj.*), unburnable. (76)
imbuvable (*adj.*), undrinkable : (77)
   eau imbuvable (*f.*), undrinkable water. (78)
imitation (*n.f.*), imitation. (79)
imitation (pierres fausses) (*n.f.*), imitation ;
   imitation stone ; paste. (80)
imitation phoque (*m.*), imitation sealskin. (81)
immaniable (*adj.*), unmanageable ; unweildy ;
   unhandy. (82)
immerger (*v.t.*), to immerge ; to immerse ; to
   plunge : to dip : (83)
   immerger le fer dans l'eau (trempe), to plunge
   iron into water. (84)
immerger (recouvrir avec du métal par trempage)
   (*v.t.*), to dip. (85)

immersion (*n.f.*), immersion ; dipping. (1)
immersion (recouvrement du métal par trempage) (*n.f.*), dipping : (2)
dorure par immersion (*f.*), gilding by dipping. (3)
immobile (*adj.*), immovable ; immobile. (4)
immunité contre la fièvre paludéenne (*f.*), immunity from malaria. (5)
impact (*n.m.*), impact : shock. (6)
impalpabilité (*n.f.*), impalpability : (7)
impalpabilité d'une poudre, impalpability of a powder. (8)
impalpable (*adj.*), impalpable : (9)
poudre impalpable (*f.*), impalpable powder. (10)
impaludisme (*n.m.*), impaludism ; paludism ; marsh-fever. (11)
imparfait, -e (*adj.*), imperfect : (12)
la lignite est une houille imparfaite, lignite is an imperfect coal. (13)
imparfaitement (*adv.*), imperfectly. (14)
impartial, -e, -aux (*adj.*), impartial : (15)
rapport impartial (*m.*), impartial report. (16)
impartialement (*adv.*), impartially. (17)
impartialité (*n.f.*), impartiality. (18)
impassable (*adj.*), impassable. (19)
impasse (*n.f.*), dead-end ; blind alley ; cul-de-sac. (20)
impédance (Élec.) (*n.f.*), impedance ; virtual resistance : (21)
impédance d'un circuit, impedance of a circuit. (22)
impénétrable (*adj.*), impenetrable : (23)
forêt impénétrable (*f.*), impenetrable forest. (24)
imperceptible (*adj.*), imperceptible. (25)
imperceptiblement (*adv.*), imperceptibly. (26)
imperfection (*n.f.*), imperfection. (27)
imperméabilisation (*n.f.*), waterproofing. (28)
imperméabiliser (*v.t.*), to render impermeable ; to waterproof. (29)
imperméabilité (*n.f.*), impermeability. (30)
imperméable (*adj.*), impermeable ; impervious ; waterproof ; rain-proof ; rain-tight : (31)
roche imperméable (*f.*), impermeable rock. (32)
bottes imperméables (*f.pl.*), waterproof boots. (33)
imperméable à l'air, air-tight. (34)
imperméable à l'eau, water-tight ; waterproof. (35)
imperméable à la chaleur, impervious to heat. (36)
impluviosité (*n.f.*), rainlessness : (37)
impluviosité des déserts, rainlessness of deserts. (38)
impolarisable (Élec.) (*adj.*), impolarizable : (39)
pile impolarisable (*f.*), impolarizable battery. (40)
imporosité (*n.f.*), imporosity. (41)
importable (*adj.*), importable. (42)
importance (*n.f.*), importance. (43)
importance du gisement (Mines) (*f.*), size of deposit. (44)
important, -e (*adj.*), important ; large : (45)
découverte importante de minerai (*f.*), important discovery of ore. (46)
venue d'eau importante (*f.*), large inflow of water. (47)
importateur (pers.) (*n.m.*), importer. (48)
importation (*n.f.*), importation ; import : (49)
importations d'Amérique en Angleterre, imports into England from America. (50)

importer (*v.t.*), to import : (51)
importer des marchandises, to import goods. (52)
importer la main-d'œuvre d'un autre district, to import, to bring, labour from another district. (53)
impossible à se procurer à aucun prix, unobtainable at any price. (54)
imposte (Arch.) (*n.f.*), impost. (55)
impôt (*n.m.*), tax : (56)
impôt sur le minerai, ore-tax. (57)
impôts et contributions [impôt (*n.m.*), contribution (*n.f.*)], rates and taxes. (58)
impotable (*adj.*), undrinkable ; impotable : (59)
eau impotable (*f.*), undrinkable water. (60)
impraticabilité (*n.f.*), impracticability. (61)
impraticable (*adj.*), impracticable. (62)
imprégnation (*n.f.*), impregnation : (63)
imprégnation des bois, impregnation of timber. (64)
imprégné, -e (*adj.*), impregnated. (65)
imprégné (-e) d'oxyde de fer, iron-stained : (66)
quartz imprégné d'oxyde de fer (*m.*), iron-stained quartz. (67)
imprégner (*v.t.*), to impregnate. (68)
imprégner (s') (*v.r.*), to become impregnated. (69)
impréparé, -e (*adj.*), unprepared ; unready. (70)
impression (*n.f.*), impression : (71)
impression d'un objet sur la cire, impression of an object in wax. (72)
impression (Peinture) (*n.f.*), priming. (73)
impression (action) (Photogr.) (*n.f.*), printing. (74)
impression (*n.f.*) *ou* imprimé (*n.m.*) (épreuve) (Photogr.), print ; positive. (75)
imprévu, -e (*adj.*), unforeseen ; unexpected : (76)
circonstances imprévues (*f.pl.*), unforeseen circumstances. (77)
imprévu (*n.m.*), unforeseen ; unexpected ; contingency : (78)
faire provision pour imprévus, to provide for contingencies. (79)
imprimer (Peinture) (*v.t.*), to prime. (80)
imprimer le mouvement à, to impart motion to ; to propel : (81)
imprimer un mouvement de rotation à un arbre, to impart a rotary motion to a shaft. (82)
imprimure (Peinture) (*n.f.*), priming. (83)
improductif, -ive (*adj.*), unproductive ; non-productive. (84)
impropre (*adj.*), unsuitable ; unfit : (85)
pierres impropres à la taille (*f.pl.*), stones unsuitable for cutting. (86)
impropre au but, unsuitable for the purpose. (87)
air impropre à la respiration (*m.*), air unfit for respiration. (88)
improviser (*v.t.*), to improvise : (89)
improviser un bateau, to improvise a boat. (90)
improviste (à l'), unexpectedly ; unawares ; suddenly. (91)
imprudence (*n.f.*), carelessness ; imprudence. (92)
imprudent, -e (*adj.*), careless ; imprudent. (93)
impulsif, -ive (*adj.*), impulsive ; propulsive. (94)
impulsion (*n.f.*), impulse ; impetus ; impulsion ; propulsion : (95)
impulsion d'une force (Méc.), impulse of a force. (96)

impur, -e (*adj.*), impure : (1)
eau impure (*f.*), impure water. (2)
impureté (*n.f.*), impurity ; impureness : (3)
impureté d'un métal, impurity of a metal.
(4)
in situ (Géol.), in situ ; in place. (5)
inabordable (*adj.*), unapproachable ; inaccessible.
(6)
Inaccessibilité (*n.f.*), inaccessibility. (7)
inaccessible (*adj.*), inaccessible : (8)
point inaccessible (*m.*), inaccessible point. (9)
longtemps réputé inaccessible, le mont Cervin
fut gravi pour la première fois en 1865 par
une caravane de sept personnes, dont quatre
périrent à la descente, for a long time deemed
inaccessible, the Matterhorn was climbed
for the first time in 1865 by a party of seven
persons, of whom four perished in the descent.
(10)
inacclimaté, -e (*adj.*), unacclimatized. (11)
inaccoutumé, -e (*adj.*), unaccustomed ; unused ;
unusual : (12)
être inaccoutumé (-e) au travail, to be un-
-accustomed to work. (13)
montrer un zèle inaccoutumé, to show unusual
zeal. (14)
inachevé, -e (*adj.*), unfinished ; incomplete. (15)
inachèvement (*n.m.*), incompletion : (16)
inachèvement d'un travail, incompletion of a
work. (17)
inactif, -ive (*adj.*), inactive ; idle ; standing : (18)
machine inactive (*f.*), idle machine ; standing
engine. (19)
inactinique (*adj.*), inactinic. (20)
inactinisme (*n.m.*), inactinism. (21)
inaction (*n.f.*), inaction ; idleness : (22)
une usine dans l'inaction, a works lying idle. (23)
inactivité (*n.f.*), inactivity. (24)
inallié, -e (*adj.*), unalloyed : (25)
métaux inalliés (*m.pl.*), unalloyed metals. (26)
inaltérable (*adj.*), inalterable ; unalterable : (27)
substance inaltérable à l'air (*f.*), substance
unalterable in the air. (28)
inaltéré, -e (Géol.) (*adj.*), unaltered ; un-
-weathered : (29)
zone inaltérée (*f.*), unweathered zone. (30)
inanalysable (*adj.*), unanalyzable. (31)
inanalysé, -e (*adj.*), unanalyzed. (32)
inattaquable par les acides, unattacked by acids ;
impervious to the action of acids. (33)
inattendu, -e (*adj.*), unexpected ; startling. (34)
inattention (*n.f.*), inattention ; negligence ;
carelessness. (35)
incalciné, -e (*adj.*), uncalcined. (36)
incandescence (*n.f.*), incandescence ; glow. (37)
incandescent, -e (*adj.*), incandescent ; glowing.
(38)
incapable (*adj.*), incapable. (39)
incapacité (*n.f.*), incapacity ; inefficiency : (40)
incapacité professionnelle de la main-d'œuvre,
inefficiency of labour. (41)
incassable (*adj.*), unbreakable. (42)
incendie (*n.m.*), fire ; conflagration. (43)
incendie de mine (*m.*), mine-fire. (44)
incendie maîtrisé (*m.*), subdued fire ; mastered
fire ; fire under control. (45)
incendie qui couve (*m.*), breeding-fire. (46)
incendie souterrain (*m.*), underground fire. (47)
incidence (Phys.) (*n.f.*), incidence : (48)
incidence d'un rayon de lumière sur une surface
réfléchissante, incidence of a ray of light
on a reflecting surface. (49)

incidence brewstérienne (Opt.), angle of
polarization ; polarizing angle. (50)
incident, -e (*adj.*), incident : (51)
rayons incidents (*m.pl.*), incident rays. (52)
inclinable (*adj.*), inclinable ; tilting ; canting :
(53)
statif inclinable (d'un microscope) (*m.*), in-
-clinable stand (of a microscope). (54)
table inclinable (Méc.) (*f.*), tilting table. (55)
inclinaison (*n.f.*), inclination ; tilting ; gradient ;
slope ; pitch ; slant ; cant ; dip ; dipping.
(56)
inclinaison (attitude, autre qu'horizontale, que
présente une couche géologique) (*n.f.*), dip.
(57)
inclinaison (par extension—angle que fait la
ligne de plus grande pente d'une couche
géologique avec l'horizon) (*n.f.*), dip ; angle
of dip. (58)
inclinaison (d'une couche géologique considérée
d'en bas) (Mines) (*n.f.*), rise. (59)
inclinaison (en) (Mines), along the dip ; follow-
-ing the dip ; on the line of the dip. (60)
inclinaison de comble (Arch.) (*f.*), roof-pitch ;
pitch of roof. (61)
inclinaison magnétique *ou* **inclinaison de l'aiguille
aimantée** *ou simplement* **inclinaison** (*n.f.*),
magnetic dip ; dip of the needle ; inclination
of the needle ; dip ; inclination ; angle of
dip ; angle of inclination. (62)
incliner (*v.t.*), to incline ; to slope ; to slant ;
to cant ; to tilt. (63)
incliner (*v.i.*). Same translations as **incliner** (s').
incliner (s') (*v.r.*), to incline ; to slope ; to slope
away ; to pitch ; to slant ; to dip : (64)
le terrain s'incline vers l'ouest, the ground
slopes away to the west. (65)
incliner (s') (Géol. & Mines) (*v.r.*), to dip : (66)
filon qui s'incline vers le nord (*m.*), lode which
dips northward. (67)
incliner (s') (Méc.) (*v.r.*), to cant ; to tilt : (68)
table s'inclinant sous tout angle (*f.*), table
canting to any angle. (69)
inclinomètre (*n.m.*), inclinometer. (70)
inclusion (Minéralogie) (*n.f.*), inclusion ; en-
-closure : (71)
le corindon renferme de nombreuses inclusions
gazeuses et liquides, corundum contains
numerous gaseous and liquid inclusions. (72)
inclusion fluide (*f.*), fluid inclusion. (73)
inclusion gazeuse (*f.*), gaseous inclusion ; gas
enclosure. (74)
inclusion liquide (*f.*), liquid inclusion. (75)
inclusion solide (*f.*), solid inclusion. (76)
inclusion vitreuse (*f.*), vitreous inclusion ; glass-
inclusion. (77)
incohésif, -ive (*adj.*), incohesive. (78)
incohésion (*n.f.*), incohesion. (79)
incolore (*adj.*), colourless : (80)
l'eau pure est incolore (*f.*), pure water is
colourless. (81)
incoloré, -e (*adj.*), uncoloured. (82)
incombustibilité (*n.f.*), incombustibility : (83)
incombustibilité d'un gaz, incombustibility
of a gas. (84)
incombustible (*adj.*), fire-proof ; fireproof ;
incombustible ; uninflammable : (85)
bois incombustible (*m.*), fire-proof wood ;
incombustible wood ; uninflammable wood.
(86)
plancher incombustible (*m.*), fire-proof floor.
(87)

incompétence (*n.f.*), incompetence; incom-
-petency. (1)
incompétent, -e (*adj.*), incompetent. (2)
incomplet, -ète (*adj.*), incomplete; unfinished: (3)
détonation incomplète (*f.*), incomplete deto-
-nation. (4)
incompressibilité (*n.f.*), incompressibility. (5)
incompressible (*adj.*), incompressible. (6)
incomprimé, -e (*adj.*), uncompressed. (7)
inconducteur, -trice (*adj.*), non-conducting (said
of heat or electricity). (8)
inconducteur (*n.m.*), non-conductor. (9)
incongelable (*adj.*), uncongealable; unfreezable:
(10)
liquide incongelable (*m.*), unfreezable liquid.
(11)
inconnue (Math.) (*n.f.*), unknown quantity;
unknown: (12)
chercher la valeur de l'inconnue, to find the
value of the unknown quantity. (13)
inconsistance (*n.f.*), inconsistency; inconsis-
-tence; looseness: (14)
inconsistance de la vase, inconsistency of the
mud; inconsistence of the silt. (15)
inconsistance du terrain, looseness of the
ground. (16)
inconsistant, -e (*adj.*), inconsistent; lacking
consistency; loose; running: (17)
vase inconsistante (*f.*), mud lacking con-
-sistency. (18)
terrain inconsistant (*m.*), loose ground;
running ground. (19)
incontaminé, -e (*adj.*), uncontaminated. (20)
incontrôlable (*adj.*), uncontrollable. (21)
inconvénient (*n.m.*), inconvenience; dis-
-advantage; drawback; trouble: (22)
l'inconvénient des hautes tensions, c'est la
nécessité d'un isolement parfait des lignes,
the disadvantage of high tensions is the
necessity of a perfect insulation of the lines.
(23)
les avantages et inconvénients des scies
circulaires, the advantages and disadvan-
-tages of circular saws. (24)
le blanc de céruse a le grave inconvénient de
brunir par le contact des vapeurs sulfureuses,
white lead has the serious drawback of
turning dark by contact with sulphurous
fumes. (25)
incorrect, -e (*adj.*), incorrect. (26)
incorrectement (*adv.*), incorrectly. (27)
incristallisable (*adj.*), uncrystallizable: (28)
sel incristallisable (*m.*), uncrystallizable salt.
(29)
incrustant, -e (*adj.*), incrustating; incrusting.
(30)
incrustation (*n.f.*), incrustation: (31)
l'incrustation de sel sur les roches, the in-
-crustation of salt on rocks. (32)
incrustation (des chaudières, etc.) (*n.f.*), incrus-
-tation; scale; boiler-scale; fur; furring.
(33)
incruster (*v.t.*), to incrust; to scale; to fur:
(34)
l'eau de chaux incruste les chaudières (*f.*),
lime water incrusts boilers. (35)
incruster (s') (*v.r.*), to become incrusted; to
become scaled; to scale; to fur: (36)
tubes qui s'incrustent de calcaire (*m.pl.*),
tubes which fur with lime. (37)
incurie (*n.f.*), want of care; carelessness;
negligence. (38)

indécomposable (*adj.*), undecomposable; in-
-decomposable. (39)
indécomposé, -e (*adj.*), undecomposed. (40)
indécouvert, -e (*adj.*), undiscovered. (41)
indécouvrable (*adj.*), undiscoverable; untrace-
-able. (42)
indéfini, -e (*adj.*), indefinite. (43)
indemniser (*v.t.*), to indemnify; to recoup;
to compensate: (44)
indemniser un ouvrier pour blessures, to
compensate a workman for injuries received.
(45)
indemnité (*n.f.*), indemnity; compensation;
penalty: (46)
indemnité payée à un ouvrier par son patron,
compensation of a workman by his employer.
(47)
indemnité pour affaissement du sol, compensa-
-tion for subsidence. (48)
indemnité pour retard de livraison, penalty for
delay in delivery. (49)
indentations (*ou* **indentures**) **d'une côte** (*f.pl.*),
indentations of a coast. (50)
indépendance (*n.f.*), independence. (51)
indépendant, -e (*adj.*), independent; self-
contained: (52)
plateau à quatre griffes indépendantes (*m.*),
independent four-jaw chuck. (53)
appareil respiratoire indépendant (*m.*), self-
contained breathing-apparatus. (54)
indéréglable (*adj.*), which cannot get out of order:
(55)
appareil indéréglable (*m.*), apparatus which
cannot get out of order. (56)
indicateur (*n.m.*), indicator; gauge; gage. (57)
indicateur à sifflet d'alarme (*m.*), low-water
indicator, with alarm; low-water alarm;
boiler-alarm. (58)
indicateur d'aiguillage (Ch. de f.) (*m.*), switch-
indicator. (59)
indicateur de grisou (*m.*), gas-indicator; gas-
detector; fire-damp delicate indicator;
warner. (60)
indicateur de grisou de Chesneau (*m.*), Chesneau
fire-damp delicate indicator. (61)
indicateur de niveau d'eau (chaudières, réservoirs,
etc.) (*m.*), water-gauge; water-indicator.
(62)
indicateur de niveau d'eau (Mines) (*m.*), water-
level indicator. (63)
indicateur de pentes et de rampes (Ch. de f.) (*m.*),
gradient-post; grade-post; indicator. (64)
indicateur de position de la cage (*m.*), indicator
showing the engineman the position of the
cage in the shaft. (65)
indicateur de pression (*m.*), steam-engine
indicator. (66)
indicateur de pression du vent (*m.*), blast-gauge;
wind-gauge. (67)
indicateur de profondeur (*m.*), depth-indicator.
(68)
indicateur de tension (Élec.) (*m.*), voltage-
indicator. (69)
indicateur de terre *ou* **indicateur de terres** *ou*
indicateur de défauts d'isolement (Élec.)
(*m.*), leakage-indicator; ground-indicator;
ground-detector; earth-indicator; earth-
detector. (70)
indicateur de tirage (*m.*), draught-gauge. (71)
indicateur de vitesse (*m.*), speed-indicator;
revolution-indicator; speed-gauge; tachom-
-eter. (72)

indicateur des pressions (*m.*),   pressure-gauge.
    (1)
indicateur du vide (*m.*), vacuum-gauge. (2)
indicateur dynamométrique (*m.*), steam-engine
    indicator. (3)
indication (*n.f.*), indication ; sign ; show : (4)
    indications à la surface, surface indications ;
    surface shows ; signs on the surface. (5)
    indications de l'existence d'or, indications of
    the existence of gold. (6)
indication (renseignement) (*n.f.*), information ;
    particular : (7)
    donner une fausse indication, to give false
    information. (⁀)
indice (signe apparent) (*n.m.*), indication ; sign ;
    show ; index : (9)
    indices de la présence du pétrole *ou* indices de
    pétrole, indications of oil ; signs of oil ;
    shows of oil. (10)
indice (Math.) (*n.m.*), index. (11)
indice absolu (Opt.) (*m.*),   absolute refractive
    index. (12)
indice de réfraction (Opt.) (*m.*), index of refrac-
    -tion ; refractive index. (13)
indice relatif (Opt.) (*m.*),   relative refractive
    index. (14)
indicolite *ou* **indigolite** (Minéral) (*n.f.*), indicolite ;
    indigolite. (15)
indigène (*adj.*), native ; indigenous : (16)
    main-d'œuvre indigène (*f.*), native labour.
    (17)
indigène (pers.) (*n.m.*),   native ; native boy ;
    boy. (18)
indigo (couleur) (*n.m.*), indigo. (19)
indiqué, -e (Méc.) (*adj.*), indicated ; gross : (20)
    puissance indiquée en chevaux (*f.*), indicated
    horse-power ; gross horse-power. (21)
indiquer (*v.t.*), to indicate ; to show ; to mark :
    (22)
    affleurement qui indique la présence du charbon
    (*m.*), outcrop which indicates the presence
    of coal. (23)
    lieu indiqué sur les cartes (*m.*), place marked
    (*or* shown) on the maps. (24)
    comme l'indique la gravure, cette machine
    possède deux supports, etc., as shown in
    the illustration, this machine has two rests,
    etc. (25)
    indiquer le chemin, to show the way. (26)
indiquer (s') (*v.r.*), to be indicated ; to be shown :
    (27)
    direction qui s'indique par une flèche (*f.*),
    direction which is indicated by an arrow.
    (28)
indirect, -e (*adj.*), indirect : (29)
    pompe à action indirecte (*f.*), indirect-action
    pump. (30)
indiscernable à l'œil nu, invisible to the naked
    eye. (31)
indispensable (*adj.*), indispensable : (32)
    l'oxygène est indispensable à la vie (*m.*),
    oxygen is indispensable to life. (33)
indissoluble (*adj.*), indissoluble ; insoluble. (34)
indium (Chim.) (*n.m.*), indium. (35)
inductance (Élec.) (*n.f.*), inductance : (36)
    inductance et résistance dans les circuits à
    courant continu, inductance and resistance
    in direct-current circuits. (37)
inductance mutuelle (*f.*), mutual inductance ;
    coefficient of mutual induction. (38)
inducteur, -trice *ou* **inductif, -ive** (Élec.) (*adj.*),
    inductive : (39)

courant inducteur (*m.*),   inductive current.
    (40)
chute inductive (*f.*), inductive drop. (41)
inducteur (Élec.) (*n.m.*), inductor. (42)
inductile (*adj.*), inductile. (43)
inductilité (*n.f.*), inductility. (44)
induction (*n.f.*), induction. (45)
induction électromagnétique (*f.*),   electromag-
    -netic induction. (46)
induction électrostatique (*f.*), electrostatic in-
    -duction. (47)
induction magnétique (*f.*), magnetic induction ;
    magnetic flux density ; flux density ;
    magnetic density. (48)
induction magnéto-électrique (*f.*),   magneto-
    electric induction. (49)
induction mutuelle (*f.*), mutual induction. (50)
induction propre (*f.*), self-induction. (51)
inductivité (*n.f.*), inductivity. (52)
inductomètre (*n.m.*), inductometer. (53)
induire (Phys.) (*v.t.*), to induce. (54)
induire en erreur (*v.t.*), to mislead. (55)
induit, -e (Élec.) (*adj.*), induced : (56)
    courant induit (*m.*), induced current. (57)
induit (Élec.) (*n.m.*), armature. (58)
induit à cage d'écureuil (*m.*),   squirrel-cage
    armature. (59)
induit denté (*m.*),   toothed armature ; slotted
    armature ; punched armature. (60)
induit en anneau *ou* **induit à anneau** (*m.*), ring
    armature. (61)
induit en disque *ou* **induit à disque** (*m.*), disc
    armature. (62)
induit en tambour *ou* **induit à tambour** (*m.*),
    drum armature. (63)
induit Gramme (*m.*), Gramme armature. (64)
induit lisse (*m.*), smooth-core armature. (65)
industrie (*n.f.*), industry ; trade. (66)
industrie aurifère *ou* **industrie de l'or** (*f.*), gold-
    mining industry. (67)
industrie des huiles de schiste (*f.*),   shale-oil
    industry. (68)
industrie du bâtiment (*f.*), building-trade. (69)
industrie minière (*f.*), mining industry. (70)
industrie minière diamantifère (*f.*),   diamond-
    mining industry. (71)
industrie minière du charbon (*f.*), coal-mining
    industry. (72)
industrie minière minérale (*f.*), mineral-mining
    industry. (73)
industrie pétrolifère *ou* **industrie pétroléenne** *ou*
    industrie pétrolière (*f.*),   oil-industry ;
    petroleum-industry. (74)
industrie salicole (*f.*), salt-industry. (75)
industriel, -elle (*adj.*), industrial ; commercial ;
    technical : (76)
    centre industriel (*m.*), industrial centre. (77)
    valeur industrielle d'une roche (*f.*), commercial
    value of a rock. (78)
    école industrielle (*f.*), technical school. (79)
industriel (pers.) (*n.m.*), manufacturer. (80)
industriellement (*adv.*),   industrially ; com-
    -mercially. (81)
ineffectif, -ive (*adj.*), ineffective. (82)
inefficace (*adj.*), ineffectual ; inefficient ; in-
    -effective. (83)
inefficacement (*adv.*), ineffectually ; inefficiently.
    (84)
inefficacité (*n.f.*), inefficiency. (85)
inégal, -e, -aux (qui n'est point égal à autre
    chose) (*adj.*), unequal. (86)
inégal, -e, -aux (raboteux) (*adj.*), uneven. (87)

inégalité (irrégularité d'une surface) (n.f.), in-equality ; unevenness : (1)
les inégalités d'une voie, the inequalities of a track ; the unevenness of a road. (2)
inélasticité (n.f.), inelasticity. (3)
inélastique (adj.), inelastic. (4)
inépuisable (adj.), inexhaustible ; unfailing ; never-failing : (5)
approvisionnement inépuisable d'eau (m.), inexhaustible water-supply ; never-failing supply of water. (6)
inépuisé, -e (adj.), unexhausted. (7)
inépuré, -e (adj.), unrefined. (8)
inerte (adj.), inert : (9)
gaz inerte (m.), inert gas. (10)
inertie (n.f.), inertia : (11)
vaincre l'inertie au démarrage, to overcome the inertia on starting. (12)
inespéré, -e (adj.), unhoped for ; unexpected. (13)
inétirable (Tréfilerie) (adj.), undrawable. (14)
inexact, -e (adj.), inexact ; incorrect ; in-accurate. (15)
inexactement (adv.), inexactly ; incorrectly ; inaccurately. (16)
inexactitude (n.f.), inexactness ; incorrectness ; inaccuracy. (17)
inexercé, -e (adj.), unskilled. (18)
inexpérience (n.f.), inexperience. (19)
inexpérimenté, -e (adj.), inexperienced ; un-skilled ; raw : (20)
ouvriers inexpérimentés (m.pl.), unskilled workmen ; raw hands. (21)
inexploitable (adj.), unworkable : (22)
couche inexploitable (f.), unworkable seam. (23)
inexploité, -e (adj.), unworked. (24)
inexploré, -e (adj.), unexplored. (25)
inexplosible (adj.), non-explosive ; inexplosive. (26)
inférence (n.f.), inference ; conclusion ; deduc-tion. (27)
inférieur, -e (situé au-dessous) (adj.), lower ; bottom ; under. (28)
inférieur, -e (qui est au-dessous d'un autre par la valeur, ou de mauvaise qualité) (adj.), inferior. (29 ou 30)
infériorité (n.f.), inferiority : (31)
infériorité de niveau, inferiority of level. (32)
infiltration (n.f.), infiltration ; percolation ; seepage : (33)
infiltration des eaux sauvages à travers les terrains perméables ou meubles, infiltration, percolation, seepage, of surface-water through permeable or loose ground. (34)
infiltrer (v.t.), to infiltrate ; to percolate : (35)
l'eau infiltre le sable (f.), the water infiltrates the sand. (36)
infiltrer (s') (v.r.), to infiltrate ; to percolate ; to seep : (37)
l'eau s'infiltre dans le sable (f.), water per-colates (or infiltrates) into the sand. (38)
infini, -e (adj.), infinite : (39)
l'espace est infini (m.), space is infinite. (40)
infini (n.m.), infinity : (41)
visée à l'infini (f.), sighting at infinity. (42)
mettre au point un appareil photographique sur (ou à) l'infini, to focus a camera on infinity. (43)
infinitésimal, -e, -aux ou infinitésime (adj.), infinitesimal. (44)
infinitésimalement (adv.), infinitesimally. (45)

inflammabilité (n.f.), inflammability ; igniti-bility ; ignitability ; flashing ; flash : (46)
inflammabilité des poussières charbonneuses, inflammability, ignitibility, of coal-dust. (47)
inflammable (adj.), inflammable ; ignitible ; ignitable : (48)
poussière inflammable (f.), ignitible dust. (49)
inflammation (n.f.), ignition : (50)
inflammation des poussières de charbon, coal-dust ignition. (51)
inflammation du grisou, ignition of fire-damp. (52)
inflexibilité (n.f.), inflexibility : (53)
inflexibilité d'un essieu, inflexibility of an axle. (54)
inflexible (adj.), inflexible ; unbending. (55)
inflexion (n.f.), inflection ; bending ; bend ; curvature : (56)
inflexion d'une verge de fer, inflection, bending, of a rod of iron. (57)
influence (n.f.), influence : (58)
dilatation d'un corps sous l'influence de la chaleur (f.), expansion of a substance under the influence of heat. (59)
influence (Phys.) (n.f.), influence : (60)
aimantation par influence (f.), magnetization by influence. (61)
information (n.f.), enquiry ; inquiry : (62)
prendre des informations sur quelqu'un, to make enquiries about someone. (63)
informe (qui n'a pas de forme arrêtée) (adj.), shapeless ; formless : (64)
masse informe (f.), shapeless mass. (65)
infra-rouge (adj.), infra-red ; ultra-red : (66)
rayons infra-rouges (m.pl.), infra-red rays ; ultra-red rays. (67)
infracrétacé, -e (Géol.) (adj.), Lower Cretaceous : (68)
la flore infracrétacée, the Lower Cretaceous flora. (69)
infracrétacé (Géol.) (n.m.), Lower Cretaceous. (70)
infraction aux règlements (f.), infraction of the rules ; breach of the regulations. (71)
infrajurassique (Géol.) (n.m.), Black Jura. (72)
infranchissable (adj.), impassable ; insurmount-able ; unsurmountable : (73)
rivière infranchissable (f.), impassable river. (74)
infrastructure (n.f.), substructure ; underframe : (75)
infrastructure d'un pont roulant, substructure of an overhead travelling crane ; underframe of an overhead traveller. (76)
infrastructure de la voie (Ch. de f.), substructure of the permanent way. (77)
infriable (adj.), unfriable. (78)
infructueux, -euse (adj.), unprofitable ; unpro-ductive ; unsuccessful ; fruitless ; vain : (79)
travail infructueux (m.), unprofitable work. (80)
efforts infructueux (m.pl.), unsuccessful (or fruitless (or vain) efforts. (81)
infusibilité (n.f.), infusibility : (82)
infusibilité du platine, infusibility of platinum. (83)
infusible (adj.), infusible : (84)
le talc est infusible au chalumeau, talc is infusible in the blowpipe. (85)
ingénieur (pers.) (n.m.), engineer. (86)
ingénieur-chimiste [ingénieurs-chimistes pl.] (n.m.), chemical engineer. (87)

ingénieur civil (*m.*), civil engineer. (1)
ingénieur civil des mines (*m.*), mining-engineer. (2)
ingénieur-conseil [ingénieurs-conseils *pl.*] (*n.m.*), consulting-engineer. (3)
ingénieur-conseil de mines (*m.*), consulting mining-engineer. (4)
ingénieur-constructeur [ingénieurs-constructeurs *pl.*] (*n.m.*), engineer (maker). (5)
ingénieur de fosse (Mines) (*m.*), pit-engineer. (6)
ingénieur de hauts fourneaux (*m.*), blast-furnace engineer. (7)
ingénieur de la voie (Ch. de f.) (*m.*), permanent-way engineer. (8)
ingénieur des mines (*m.*), mining-engineer. (9)
ingénieur des mines intérimaire (*m.*), acting mining-engineer. (10)
ingénieur des ponts et chaussées (*m.*), engineer of roads, railways, bridges, canals, etc. (11)
ingénieur des services du jour (Mines) (*m.*), surface-engineer. (12)
ingénieur des voies ferrées (*m.*), railway-engi-neer. (13)
ingénieur-dragueur [ingénieurs-dragueurs *pl.*] (*n.m.*), dredge-mining engineer. (14)
ingénieur-électricien [ingénieurs-électriciens *pl.*] (*n.m.*), electrical engineer. (15)
ingénieur en chef (*m.*), engineer-in-chief; chief engineer. (16)
ingénieur en résidence (*m.*), resident engineer. (17)
ingénieur-géographe [ingénieurs-géographes *pl.*] (*n.m.*), geographical engineer. (18)
ingénieur-hydraulicien [ingénieurs-hydrauliciens *pl.*] (*n.m.*), hydraulic engineer; hydraulician; hydraulist. (19)
ingénieur-hydrographe [ingénieurs-hydrographes *pl.*] (*n.m.*), hydrographical engineer. (20)
ingénieur-mécanicien [ingénieurs-mécaniciens *pl.*] (*n.m.*), mechanical engineer. (21)
ingénieur principal (*m.*), chief engineer; engi-neer-in-chief. (22)
ingénieur-topographe [ingénieurs-topographes *pl.*] (*n.m.*), topographic engineer. (23)
ingénieux, -euse (*adj.*), ingenious: (24)
un dispositif ingénieux, an ingenious device. (25)
ingravissable (*adj.*), unclimbable. (26)
ingrédient (*n.m.*), ingredient. (27)
inguéable (*adj.*), unfordable: (28)
rivière inguéable (*f.*), unfordable river. (29)
inhabitable (*adj.*), uninhabitable. (30)
pays inhabitable pour les blancs (*m.*), country uninhabitable for white men. (31)
inhabité, -e (*adj.*), uninhabited: (32)
île inhabitée (*f.*), uninhabited island. (33)
inhalation (*n.f.*), inhalation: (34)
inhalation d'oxygène dans le cas d'empoisonne-ment par l'oxyde de carbone, inhalation of oxygen in case of carbon monoxide poison-ing. (35)
inhalation des poussières, inhalation of dust. (36)
ininflammabilité (*n.f.*), uninflammability. (37)
ininflammable (*adj.*), uninflammable; non-inflammable; fire-proof; fireproof. (38)
ininterrompu, -e (*adj.*), uninterrupted; unbroken; steady: (39)
une succession ininterrompue de malheurs, an uninterrupted succession of misfortunes. (40)

initial, -e, -aux (*adj.*), initial: (41)
vitesse initiale (Méc.) (*f.*), initial velocity. (42)
initialement (*adv.*), initially. (43)
injecter (*v.t.*), to inject: (44)
injecter la vapeur d'eau, to inject steam. (45)
injecteur (*n.m.*), injector. (46)
injecteur à mise en marche automatique (*m.*), auto-matic injector; self-acting injector; re-starting injector. (47)
injecteur à vapeur d'échappement (*m.*), exhaust-steam injector. (48)
injecteur à vapeur vive (*m.*), live-steam injector. (49)
injecteur aspirant (*m.*), lifting-injector; inspira-tor. (50)
injecteur Giffard (*m.*), Giffard's injector. (51)
injecteur non-aspirant (*m.*), non-lifting injector. (52)
injection (*n.f.*), injection: (53)
injection de ciment à l'air comprimé, injection of cement by compressed air. (54)
injection de masses d'air pur, injection of large quantities of fresh air. (55)
injection (des bois) (*n.f.*), impregnation (of timber). (56)
injection (Géol.) (*n.f.*), injection; intrusion: (57)
une injection de porphyre, a porphyry intru-sion. (58)
injection lit par lit (Géol.) (*f.*), lit-par-lit injection; lit-par-lit intrusion. (59)
inlandsis [inlandsis *pl.*] (Glaciologie) (*n.m.*), continental ice-sheet; continental glacier. (60)
innavigable (*adj.*), unnavigable: (61)
rivière innavigable (*f.*), unnavigable river. (62)
inoccupé, -e (*adj.*), unoccupied; idle: (63)
la plupart des machines sont inoccupées faute de main-d'œuvre, most of the machines are idle for want of labour. (64)
inodore (*adj.*), odourless. (65)
inoffensif, -ive (*adj.*), inoffensive; innoxious. (66)
inofficiel, -elle (*adj.*), unofficial. (67)
inofficiellement (*adv.*), unofficially. (68)
inondation (*n.f.*), flood; inundation. (69)
inonder (*v.t.*), to flood; to inundate: (70)
rivière qui inonde périodiquement ses bords (*f.*), river which floods its banks periodically. (71)
inopiné, -e (*adj.*), unexpected; unforeseen. (72)
inopinément (*adv.*), unexpectedly. (73)
inorganique (*adj.*), inorganic; inorganical: (74)
chimie inorganique (*f.*), inorganic chemistry. (75)
inorganiquement (*adv.*), inorganically. (76)
inoxydable (*adj.*), inoxidizable; unoxidizable; non-corrodible; incorrodible; rustless; stainless: (77)
l'aluminium est inoxydable à l'air sec ou humide (*m.*), aluminium is inoxidizable in dry or damp air. (78)
acier inoxydable (*m.*), stainless steel; rustless steel. (79)
inoxydé, -e (*adj.*), unoxidized: (80)
zone inoxydée (Géol.) (*f.*), unoxidized zone. (81)
inquartation (*n.f.*) *ou* **inquart** (*n.m.*) (Métall.), inquartation; inquarting; quartation: (82)
départ avec inquartation d'argent, de cadmium (*m.*), parting with inquartation of silver, of cadmium. (83 *ou* 84)
inquartation (*n.f.*) *ou* **inquart** (*n.m.*) (Échantillon-nage du minerai), quartering. (85)
inquarter (*v.t.*), to inquartate. (86)

Insalubre (*adj.*), unhealthy; insalubrious. (1)
Insalubrité (*n.f.*), unhealthiness; insalubrity. (2)
Insaturable (*adj.*), unsaturable. (3)
Insaturé, -e (*adj.*), unsaturated: (4)
hydrocarbures insaturés (*m.pl.*), unsaturated hydrocarbons. (5)
Inscription en courbe (*f.*), negotiating curves; taking curves. (6)
Inscrire (s') en courbe, to negotiate curves; to take curves: (7)
bogie de locomotive qui s'inscrit bien en courbe et se prête aux inégalités de la voie (*m.*), engine-truck which negotiates curves well and accommodates itself to the inequalities of the road. (8)
Insécurité (*n.f.*), insecurity. (9)
Insolation (exposition aux rayons du soleil) (*n.f.*), insolation. (10)
Insolation (Pathologie) (*n.f.*), insolation; sun-stroke. (11)
Insoler (*v.t.*), to insolate. (12)
Insolubiliser (*v.t.*), to insolubilize: (13)
calciner un minéral pour insolubiliser la silice, to calcine a mineral to insolubilize the silica. (14)
Insolubilité (*n.f.*), insolubility. (15)
Insoluble (*adj.*), insoluble: (16)
substance insoluble dans l'eau (*f.*), substance insoluble in water. (17)
Insonore (*adj.*), insonorous; unsonorous. (18)
Insoutenable (*adj.*), unbearable: (19)
chaleur insoutenable (*f.*), unbearable heat. (20)
Inspecter (*v.t.*), to inspect; to examine; to make an inspection *or* examination of: (21)
inspecter chaque jour les galeries de roulage, to make a daily examination of the roads; to inspect the haulageways daily. (22)
Inspecteur (pers.) (*n.m.*), inspector; examiner. (23)
Inspecteur des mines (*m.*), inspector of mines; mining inspector; surveyor of mines. (24)
Inspecteur des wagons (Ch. de f.) (*m.*), carriage-examiner; car-inspector. (25)
Inspecteur du travail (*m.*), factory-inspector. (26)
Inspection (*n.f.*), inspection; examination: (27)
inspection fréquente et minutieuse du câble, frequent minute examination of the rope. (28)
Inspectorat (*n.m.*), inspectorate; inspectorship. (29)
Instabilité (*n.f.*), instability; unstableness: (30)
instabilité d'un corps en équilibre, insta-bility of a body in equilibrium. (31)
Instable (*adj.*), unstable: (32)
équilibre instable (*m.*), unstable equilibrium. (33)
Installation (action) (*n.f.*), installation; equip-ment: (34)
installation d'appareils mécaniques, installa-tion of mechanical appliances. (35)
installation d'un siège de charbonnage, equip-ment of a coal-works. (36)
Installation (objets nécessaires à un travail) (*n.f.*), installation; plant; equipment: (37)
installation d'essai *ou* installation à titre d'essai, trial-installation. (38)
installation de forage *ou* installation de sondage, drilling-plant; boring-plant; bore-holing plant. (39)
installation de lumière électrique, electric-light installation. (40)

installation de mines, mining-plant; mining-equipment. (41)
installation de surface (Mines), surface-plant; surface-equipment. (42)
installation isolée (opp. à *usine centrale*) (Élec.), isolated plant. (43) (distinguished from *central power-station*).
Installer (*v.t.*), to install; to equip: (44)
installer l'éclairage électrique dans une usine, to install electric lighting in a works. (45)
laboratoire bien installé, mal installé (*m.*), well-equipped, badly equipped, laboratory. (46 *ou* 47)
Instant (*n.m.*), instant; moment: (48)
vitesse à un instant donné (*f.*), velocity at a given moment. (49)
Instantané, -e (*adj.*), instantaneous: (50)
centre instantané de rotation (Méc.) (*m.*), instantaneous centre of rotation. (51)
Instantané (vue instantanée) (Photogr.) (*n.m.*), snap; snap-shot. (52)
Instantané (pose instantanée) (*n.m.*), instanta-neous exposure: (53)
obturateur faisant la pose ou l'instantané (*m.*), shutter for time or instantaneous exposures. (54)
Instantanéité (*n.f.*), instantaneity; instanta-neousness. (55)
Instantanément (*adv.*), instantaneously: (56)
la lumière ne se propage pas instantanément, mais avec une vitesse finie, light does not propagate instantaneously, but with a finite speed. (57)
Instruction (enseignement) (*n.f.*), instruction; training: (58)
instruction concernant les travaux de sauvetage, instruction, training, in rescue work. (59)
Instruction (manuel d'instructions) (*n.f.*), instruc-tion-book. (60)
Instructions (ordres) (*n.f.pl.*), instructions; orders: (61)
instructions pour s'en servir, instructions for use. (62)
Instruire (enseigner) (*v.t.*), to instruct; to train: (63)
instruire les candidats dans l'emploi des appareils de sauvetage, to train candidates in the use of life-saving apparatus. (64)
Instrument (*n.m.*), instrument; implement: (65)
instrument de recherches, explorer. (66)
instrument universel (théodolite), universal instrument. (67)
instruments d'optique, optical instruments. (68)
instruments de dessin, drawing-instruments. (69)
instruments de mathématiques, mathematical instruments. (70)
instruments de mesure, measuring-instru-ments. (71)
instruments de précision, scientific instru-ments; instruments of precision. (72)
instruments tranchants, cutting implements. (73)
Instrumental, -e, -aux (*adj.*), instrumental: (74)
erreurs instrumentales (*f.pl.*), instrumental errors. (75)
Insubordination des indigènes (*f.*), insubordina-tion of natives. (76)
Insuccès (*n.m.*), insuccess; failure: (77)
insuccès dans le développement des plaques photographiques, failures in the develop-ment of photographic plates. (78)
Insuffisamment (*adv.*), insufficiently. (79)

insuffisance (*n.f.*), insufficiency; inadequacy; shortage: (1)

insuffisance de poids, shortage in weight. (2)

insuffisance de pose (Photogr.) (*f.*), underexposure. (3)

insuffisant, -e (*adj.*), insufficient; inadequate; short. (4)

insuffler de l'air dans un fourneau, to blow air into a furnace. (5)

insupportable (*adj.*), insupportable. (6)

insurmontable (*adj.*), insurmountable; unsur--mountable; insuperable. (7)

insurrection des indigènes (*f.*), insurrection of the natives; native rising. (8)

intact, -e (*adj.*), intact. (9)

intarissable (*adj.*), unfailing; perennial; in--exhaustible: (10)

source intarissable (*f.*), unfailing spring; per--ennial spring. (11)

une mine intarissable, an inexhaustible mine. (12)

intempérie (*n.f.*) (*s'emploie souvent au pluriel*), weather; inclemency of the weather: (13)

exposition aux intempéries (*f.*), exposure to the weather. (14)

intempérisme (Géol.) (*n.m.*), weathering: (15)

roches qui sont exposées à l'intempérisme (*f.pl.*), rocks which are exposed to weather--ing. (16)

roches modifiées sous l'influence des agents d'intempérisme (*f.pl.*), rocks modified under the influence of weathering agencies. (17)

intempestif, -ive (*adj.*), premature: (18)

explosion imtempestive (*f.*), premature explo--sion. (19)

intempestivement (*adv.*), prematurely. (20)

intense (*adj.*), intense. (21)

intensif, -ive (*adj.*), intensive; vigorous. (22)

intensité (*n.f.*), intensity. (23)

intensité calorifique (*f.*), calorific intensity; heat-intensity. (24)

intensité d'aimantation (*f.*), intensity of magnet--ization. (25)

intensité de champ *ou* **intensité de champ magné--tique** (*f.*), intensity of field; field intensity; strength of field; field strength; field density; magnetic intensity; magneto--motive gradient; magnetizing force. (26)

intensité de courant (Élec.) (*f.*), intensity of current; current-intensity; strength of current; current-strength. (27)

intensité de la frappe (Sondage) (*f.*), percussive intensity. (28)

intensité de la pesanteur (Phys.) (*f.*), intensity of gravity; acceleration of gravity; constant of gravitation; gravitation constant; gravity. (29)

intensité de pôle (Magnétisme) (*f.*), strength of pole; pole strength. (30)

intensité du froid (*f.*), intensity of the cold. (31)

intensité en bougies (*f.*), candle-power intensity; candle-power. (32)

intensité lumineuse (Photom.) (*f.*), luminous intensity; intensity of light: (33)

l'intensité lumineuse d'un arc, the luminous intensity of an arc. (34)

intensité moyenne hémisphérique (Photom.) (*f.*), mean hemispherical intensity; mean hemi--spherical candle-power. (35)

intensité moyenne horizontale (*f.*), mean hori--zontal intensity; mean horizontal candle--power. (36)

intensité moyenne sphérique (*f.*), mean spherical intensity; mean spherical candle-power. (37)

intercalation (*n.f.*), intercalation; insertion: (38)

intercalation d'un rhéostat sur un induit, insertion of a rheostat on an armature. (39)

intercaler (*v.t.*), to intercalate; to insert: (40)

un banc gréseux ou calcaire intercalé dans des schistes, a sandstone or limestone bed intercalated in shales. (41)

intercaler (s') (*v.r.*), to become intercalated. (42)

intercepter (*v.t.*), to intercept: (43)

intercepter des rayons lumineux, to intercept light-rays. (44)

interception (*n.f.*), interception. (45)

interchangeable (*adj.*), interchangeable: (46)

machine à pièces interchangeables (*f.*), machine with interchangeable parts. (47)

intercristallisation (*n.f.*), intercrystallization. (48)

interdiction (*n.f.*), prohibition. (49)

interdire (*v.t.*), to prohibit; to forbid. (50)

interdit, -e (*adj.*), forbidden; prohibited: (51)

explosifs interdits (*m.pl.*), prohibited explosives. (52)

interférence (Phys.) (*n.f.*), interference. (53)

interférentiel, -elle (*adj.*), interferential: (54)

réfractomètre interférentiel (*m.*), interferential refractometer. (55)

interférer (*v.i.*), to interfere: (56)

rayons qui interfèrent (*m.pl.*), rays which interfere. (57)

interglaciaire (Géol.) (*adj.*), interglacial. (58)

intérieur, -e (*adj.*), interior; internal; inner; inside; inland. (59)

intérieur (*n.m.*), interior; inside: (60)

intérieur de la terre, d'une contrée, interior of the earth, of a country. (61 *ou* 62)

intérieur d'une boîte, inside of a box. (63)

intérimaire (qui a lieu par intérim) (*adj.*), interim: (64)

rapport intérimaire (*m.*), interim report. (65)

intérimaire (qui remplit un intérim) (*adj.*), acting: (66)

ingénieur des mines intérimaire (*m.*), acting mining-engineer. (67)

intermédiaire (*adj.*), intermediate. (68)

intermédiaire (pour châssis négatifs) (Photogr.) (*n.m.*), carrier (for dark slides). (69)

intermittence (*n.f.*), intermittence: (70)

intermittence de certaines sources, inter--mittence of certain springs. (71)

intermittent, -e (*adj.*), intermittent; periodical: (72)

source intermittente (*f.*), intermittent spring; periodical fountain. (73)

interne (*adj.*), internal; interior; inside; inner. (74)

interpolaire (*adj.*), interpolar. (75)

interrompre (*v.t.*), to interrupt; to stop; to suspend: (76)

interrompre l'aérage, to interrupt the ventila--tion; to stop ventilating. (77)

interrompre la circulation dans la galerie de roulage, to suspend the traffic on the haulage--road. (78)

interrompre un circuit électrique, to interrupt an electric circuit. (79)

interrupteur (appareil qui a pour fonction de fermer ou d'ouvrir un courant électrique) (*n.m.*), switch. (80) See also **commuta--teur.**

Interrupteur (appareil pour produire les ferme-
-tures et les ouvertures de circuits électriques
avec une très grande rapidité) (n.m.),
interrupter ; quick make-and-break ; circuit-
breaker. (1)

Interrupteur à air (Élec.) (m.), air-interrupter ;
air-break switch. (2)

Interrupteur à double rupture (m.), double-break
switch. (3)

Interrupteur à huile (m.), oil-interrupter ; oil-
break switch. (4)

Interrupteur à jet de mercure (m.), mercury-jet
interrupter ; jet interrupter. (5)

Interrupteur à maximum (m.), maximum circuit-
breaker. (6)

Interrupteur à mercure (m.), mercury-interrupter ;
mercury-break switch. (7)

Interrupteur à minimum (m.), minimum circuit-
breaker. (8)

Interrupteur à trembleur (m.), trembler-switch.
(9)

Interrupteur bipolaire (m.), double-pole switch ;
two-pole switch. (10)

Interrupteur conjoncteur (m.), circuit-closer. (11)

Interrupteur d'urgence ou interrupteur de secours
(m.), emergency switch. (12)

Interrupteur de courant (m.). Same as inter-
-rupteur.

Interrupteur de Foucault (m.), Foucault inter-
-rupter. (13)

Interrupteur de Wehnelt (m.), Wehnelt interrup-
-ter. (14)

Interrupteur disjoncteur (m.), circuit-breaker ;
disjunctor ; break-circuit. (15)

Interrupteur isolé à la fibre (m.), fibre-insulated
switch. (16)

Interrupteur modèle " tumbler " (m.), tumbler-
switch. (17)

Interrupteur monopolaire ou interrupteur unipo-
-laire (m.), single-pole switch. (18)

Interruption (n.f.), interruption ; stoppage ;
stopping ; suspension : (19)
interruption d'un circuit électrique, interrup-
-tion of an electric circuit. (20)
interruption de l'aérage, stoppage of, stopping,
the ventilation. (21)

Intersecter (v.t.), to intersect. (22)

Intersecter (s') (v.r.), to intersect. (23)

Intersection (n.f.), intersection : (24)
intersection de deux plans, intersection of
two planes. (25)

Interstice (n.m.), interstice : (26)
interstices des rochers, interstices of the
rocks. (27)

Interstitiel, -elle (adj.), interstitial. (28)

Interstratification (n.f.), interstratification ;
interbedding. (29)

Interstratifié, -e (Géol.) (adj.), interstratified ;
interbedded : (30)
roches interstratifiées (f.pl.), interbedded
rocks. (31)

Intervalle (n.m.), interval ; space ; gap : (32)
intervalle entre deux murs, space between two
walls. (33)

Intervalle (largeur du vide) (Engrenages) (n.m.),
width of space (Gearing). (34)

Intervalle (machine dynamo-électrique) (n.m.),
clearance ; clearance-space. (35)

Intervalle d'air (Élec.) (m.), air-gap ; interferric
space. (36)

Intime (adj.), intimate : (37)
mélange intime (m.), intimate mixture. (38)

Intoxicant, -e (adj.), poisonous : (39)
gaz intoxicant (m.), poisonous gas. (40)

Intoxication (n.f.), poisoning. (41)

Intoxication arsenicale (f.), arsenical poisoning.
(42)

Intoxication chronique par le plomb ou intoxica-
-tion saturnine (f.), lead-poisoning ; saturn-
-ism ; saturnismus. (43)

Intoxiquer (v.t.), to poison. (44)

Intrados (d'une voûte) (n.m.), intrados, soffit,
(of an arch). (45)

Intraglaciaire (Géol.) (adj.), intraglacial ; engla-
-cial : (46)
cavité intraglaciaire (f.), intraglacial (or
englacial) cavity. (47)

Intraitable (adj.), untreatable. (48)

Intransparence (n.f.), opacity ; opaqueness. (49)

Intransparent, -e (adj.), untransparent ; opaque.
(50)

Intratellurique (Géol.) (adj.), intratelluric ;
intratellural. (51)

Intrinsèque (adj.), intrinsic : (52)
valeur intrinsèque (f.), intrinsic value. (53)

Intrinsèquement (adv.), intrinsically. (54)

Introduction (n.f.), introduction. (55)

Introduire (faire adopter) (v.t.), to introduce :
(56)
introduire l'emploi des ventilateurs dans une
mine, to introduce the use of fans in a mine.
(57)

Introduire (faire entrer) (v.t.), to insert ; to intro-
-duce ; to put in ; to put into : (58)
introduire une clef dans une serrure, to insert,
to introduce, to put, a key into a lock. (59)

Intrusif, -ive (adj.), intrusive. (60)

Intrusion (Géol.) (n.f.), intrusion : (61)
l'intrusion de magma fondu entre les strates,
the intrusion of molten magma between
the strata. (62)

Intrusion en forme de nappe (f.), sheet-like
intrusion. (63)

Intrusion ignée (f.), igneous intrusion. (64)

Intrusion lit par lit (Géol.) (f.), lit-par-lit intru-
-sion ; lit-par-lit injection. (65)

Intrusion rocheuse (f.), rock-intrusion. (66)

Intrusion volcanique (f.), volcanic intrusion. (67)

Inutile (adj.), useless ; unuseful ; unnecessary ;
needless : (68)
un meuble inutile, a useless piece of furniture.
(69)
inutile de dire que . . ., needless to say
that . . . (70)

Inutilement (adv.), uselessly ; unusefully ;
needlessly. (71)

Inutilisable (adj.), unutilizable ; inutilizable ;
unserviceable. (72)

Inutilisé, -e (adj.), unutilized ; inutilized. (73)

Inutilité (n.f.), inutility. (74)

Invar (alliage très peu dilatable) (n.m.), invar. (75)

Invasion d'eau (f.), inrush, invasion, of water.
(76)

Inventaire (n.m.), inventory ; list ; schedule. (77)

Inventer (v.t.), to invent ; to devise : (78)
inventer un procédé, to invent a process. (79)
inventer un expédient, to devise an expedient.
(80)

Inventeur (pers.) (n.m.), inventor : (81)
inventeur d'un procédé, inventor of a process.
(82)

Inventeur (d'une mine, ou exploitation analogue)
(pers.) (n.m.), discoverer (of a mine, or the
like) : (83 ou 84)

l'inventeur d'un nouveau champ aurifère, d'un placer, the discoverer of a new gold-field, of a placer. (1 *ou* 2)

inventif, -ive (*adj.*), inventive : (3)
l'esprit inventif des Américains (*m.*), the inventive spirit of Americans. (4)

invention (*n.f.*), invention : (5)
invention du téléphone, invention of the tele--phone. (6)

inventorier (*v.t.*), to inventory ; to make a list of : (7)
inventorier des marchandises, to inventory, to make a list of, goods. (8)

inverse (*adj.*), inverse ; reciprocal ; reverse ; reversed : (9)
l'intensité de la lumière est en raison inverse du carré de la distance du foyer, the intensity of light is in inverse ratio to the square of the distance of the focus. (10)
la polarité inverse du bismuth, the inverse polarity of bismuth. (11)
faille inverse (Géol.) (*f.*), reverse fault ; reversed fault. (12)

inverse (*n.m.*), inverse ; reciprocal : (13)
en électricité, la conductivité est l'inverse de la résistivité, comme la conductance est l'inverse de la résistance, in electricity, conductivity is the reciprocal of resistivity, as conductance is the reciprocal of resistance. (14)

inversement (*adv.*), inversely ; reciprocally : (15)
quantités inversement proportionnelles (*f.pl.*), inversely proportional quantities. (16)

inverser (*v.t.*), to invert ; to reverse : (17)
inverser le sens d'un courant électrique, to reverse the direction of an electric current. (18)

inverseur (Élec.) (*n.m.*), reverser. (19)

inverseur (d'un tour) (*n.m.*), reversing-gear, reversing-motion, (of a lathe). (20)

inversion (*n.f.*), inversion ; reversing : (21)
inversion du sens de la marche (Méc.), reversing. (22)
inversion du sens de rotation, reversing the direction of rotation. (23)

invertir (*v.t.*), to invert ; to reverse : (24)
invertir le sens d'un courant électrique, to reverse the direction of an electric current. (25)

investigation (*n.f.*), investigation. (26)

investison (Mines) (*n.f.*), barrier-pillar ; barrier ; boundary-pillar. (27)

invisible (*adj.*), invisible : (28)
invisible à l'œil nu, invisible to the naked eye. (29)

invitré, -e (*adj.*), unvitrified : (30)
sable invitré (*m.*), unvitrified sand. (31)

involontaire (*adj.*), unintentional ; involuntary : (32)
empiétement involontaire (*m.*), unintentional encroachment. (33)

iodargyrite (*n.f.*) *ou* **iodargyre** (*n.m.*) (Minéral), iodyrite ; iodargyrite. (34)

iodate (Chim.) (*n.m.*), iodate. (35)

iode (Chim.) (*n.m.*), iodine. (36)

iodifère (*adj.*), iodiferous. (37)

iodique (Chim.) (*adj.*), iodic. (38)

iodobromite (Minéral) (*n.f.*), iodobromite. (39)

iodure (Chim.) (*n.m.*), iodide : (40)
iodure d'argent, iodide of silver ; silver iodide. (41)
iodure de méthylène, methylene iodide. (42)

iolite (Minéral) (*n.f.*), iolite. (43)

ion (Élec.) (*n.m.*), ion. (44)

ionisation (*n.f.*), ionization. (45)

ioniser (*v.t.*), to ionize. (46)

ionium (Chim.) (*n.m.*), ionium. (47)

iridescence (*n.f.*), iridescence. (48)

iridescent, -e (*adj.*), iridescent. (49)

iridifère (*adj.*), iridiferous ; iridium-bearing. (50)

iridium (Chim.) (*n.m.*), iridium. (51)

iridosmine (*n.f.*) *ou* **iridosmium** (*n.m.*) (Minéral), iridosmine ; iridosmium ; osmiridium. (52)

iris (*n.m.*), iris ; play of colours ; rainbow colours ; pavonine. (53)

irisation (*n.f.*), irisation ; iridescence. (54)

irisé, -e (*adj.*), irisated ; iridescent. (55)

irradiateur, -trice (Opt.) (*adj.*), irradiative. (56)

irradiation (*n.f.*), irradiation. (57)

irradier (*v.i.*), to irradiate. (58)

irrécouvrable (*adj.*), irrecoverable ; unrecover--able. (59)

irréductibilité (*n.f.*), irreducibility. (60)

irréductible (*adj.*), irreducible ; unreducible. (61)

irrégularité (*n.f.*), irregularity ; unevenness : (62)
irrégularité d'une surface, irregularity, uneven--ness, of a surface. (63)

irrégulier, -ère (*adj.*), irregular : (64)
allure irrégulière des couches (*f.*), irregular character of the strata. (65)

irrégulièrement (*adv.*), irregularly. (66)

irréparable (*adj.*), irreparable. (67)

irréparablement (*adv.*), irreparably. (68)

irrespirable (*adj.*), irrespirable ; unfit for respira--tion : (69)
air irrespirable (*m.*), irrespirable air ; air unfit for respiration. (70)

irrotationnel, -elle (Phys.) (*adj.*), irrotational : (71)
mouvement irrotationnel (*m.*), irrotational motion. (72)

irruption (*n.f.*), irruption ; inburst ; inrush : (73)
irruption d'eau, irruption, inrush, of water. (74)

isobare *ou* **isobarique** *ou* **isobarométrique** (*adj.*), isobaric ; isobarometric : (75)
lignes isobares (*f.pl.*), isobaric lines. (76)

isobare (*n.f.*), isobar. (77)

isobase (Géol.) (*n.f.*), isobase. (78)

isobathe (*adj.*), isobath ; isobathic. (79)

isochromatique (*adj.*), isochromatic. (80)

isochrone *ou* **isochronique** (*adj.*), isochronous ; isochronal ; isochronic ; isochrone : (81)
régulateur isochrone (*m.*), isochronous governor. (82)

isoclinal, -e, -aux (Géol.) (*adj.*), isoclinal ; isoclinic : (83)
pli isoclinal (*m.*), isoclinal fold. (84)
structure isoclinale (*f.*), isoclinal structure. (85)

isoclinal (Géol.) (*n.m.*), isoclinal ; isocline ; isoclinic. (86)

isocline (Phys.) (*adj.*), isoclinal ; isoclinic : (87)
lignes isoclines (*f.pl.*), isoclinal lines ; isoclinic lines. (88)

isohypse (*adj.*), isohypsometric. (89)

isohypse (*n.f.*), isohyp ; contour-line. (90)

isolant, -e *ou* **isolateur, -trice** (Phys.) (*adj.*), insulating : (91)
ruban isolant (*ou* isolateur) (*m.*), insulating tape. (92)
matière isolante (*f.*), insulating material. (93)

isolant (*n.m.*), insulator ; insulating material : (94)
le verre est un isolant, glass is an insulator. (95)

isolateur (appareil) (Phys.) (*n.m.*), insulator. (96)

Isolateur à cloche (*m.*) *ou* isolateur-cloche [isolateurs-cloches *pl.*] (*n.m.*), petticoat insulator ; mushroom insulator. (1)

Isolateur à cloche double (*m.*), double-petticoat insulator. (2)

Isolateur à cloche simple (*m.*), single-petticoat insulator. (3)

Isolateur à haute tension (*m.*), high-tension insulator ; high-voltage insulator. (4)

Isolateur à l'huile (*m.*), oil-insulator. (5)

Isolateur d'angle (*m.*), shackle-insulator. (6)

Isolateur d'arrêt (*m.*), terminal insulator. (7)

Isolateur d'entrée en bâtiment (*m.*), wall-entrance insulator. (8)

Isolateur d'étincelles (*m.*), spark-box ; sparker-box ; spark-insulating box. (9)

Isolateur en forme d'ombrelle (*m.*), umbrella-insulator. (10)

Isolateur en porcelaine (*m.*), porcelain insulator. (11)

Isolateur pour alignements droits (*m.*), straight-line insulator. (12)

Isolation (*n.f.*). Same as isolement.

Isolement (séparation des objets environnants) (*n.m.*), isolation ; isolating ; shutting off : (13)

isolement au moyen de zones humides (Mines), isolation by watered spaces. (14)

isolement des eaux d'un puits, shutting off the water from a shaft. (15)

Isolement (Phys.) (*n.m.*), insulation : (16)

isolement des câbles électriques, insulation of electric cables. (17)

Isoler (séparer) (*v.t.*), to isolate ; to shut off : (18)

isoler les chantiers dans le cas d'incendie, to isolate the workings in case of fire. (19)

isoler les nappes aquifères, to shut off the water-bearing strata. (20)

Isoler (Phys.) (*v.t.*), to insulate : (21)

isoler un conducteur, to insulate a conductor. (22)

Isoler (Chim.) (*v.t.*), to isolate : (23)

isoler un corps simple, to isolate an element. (24)

Isomère *ou* isomérique (*adj.*), isomeric ; isomeri-cal. (25)

Isomérie (*n.f.*) *ou* isomérisme (*n.m.*), isomerism ; isomery. (26)

Isométrique (*adj.*), isometric ; isometrical : (27)

perspective isométrique (*f.*), isometric pro-jection ; isometric perspective. (28)

Isomorphe (*adj.*), isomorphous ; isomorphic : (29)

cristaux isomorphes (*m.pl.*), isomorphous crystals. (30)

minéral qui est isomorphe avec un autre (*m.*), mineral which is isomorphous with another. (31)

Isomorphisme (*n.m.*), isomorphism. (32)

Isosiste *ou* isoséiste *ou* isosismique *ou* isoséismique (*adj.*), isoseismal ; isoseismic. (33)

Isosiste *ou* isoséiste (*n.f.*), isoseismal ; isosist. (34)

Isotherme *ou* isothermique (*adj.*), isothermal ; isothermic ; isothermical ; isothermous : (35)

zones isothermes (*f.pl.*), isothermal zones. (36)

Isotherme (*n.f.*), isotherm. (37)

Isothermiquement (*adv.*), isothermically. (38)

Isotrope (Phys.) (*adj.*), isotropic ; isotrope ; isotropous : (39)

cristal isotrope (*m.*), isotropic crystal. (40)

Isotropie (*n.f.*), isotropy ; isotropism. (41)

Issue (*n.f.*), outlet ; exit : (42)

issue d'un tunnel, outlet of a tunnel. (43)

Isthme (*n.m.*), isthmus. (44)

Itabirite (Pétrol.) (*n.f.*), itabirite ; specular schist. (45)

Itacolumite (Pétrol.) (*n.f.*), itacolumite. (46)

Itinéraire (*n.m.*), itinerary ; route : (47)

tracer un itinéraire, to mark out a route. (48)

J

Jacinthe (Minéral) (*n.f.*), jacinth. (49)

Jack (Téléph. & Télégr.) (*n.m.*), jack. (50)

Jade (Minéral) (*n.m.*), jade ; jade-stone ; ne-phritic stone. (51)

Jade de Saussure (*m.*), saussurite. (52)

Jade océanique (*m.*), jade oceanic. (53)

Jadéite (Minéral) (*n.f.*), jadeite. (54)

Jaillir (*v.i.*), to gush ; to gush out ; to spout ; to spurt out : (55)

pétrole qui jaillit d'un puits (*m.*), oil which gushes from a well. (56)

Jaillir (Élec.) (*v.i.*), to jump ; to leap : (57)

étincelle qui jaillit entre deux électrodes (*f.*), spark which jumps between two electrodes. (58)

Jaillissant, -e (*adj.*), gushing ; spouting. (59)

Jaillissement (*n.m.*), gushing ; gushing out ; gush ; spurting out ; spouting : (60)

jaillissement de pétrole, gushing, gush, of oil. (61)

Jais (Minéral) (*n.m.*), jet. (62)

Jalon (petit) (*n.m.*), peg ; stake ; picket. (63)

Jalon (grand) (Géod.) (*n.m.*), range-pole ; ranging-pole ; ranging-rod ; picket : (64)

jalon en bois, douille acier, peint rouge et blanc, ranging-rod in wood, steel shoe, painted red and white. (65)

Jalon-mire [jalons-mires *pl.*] (*n.m.*), target levelling-rod. (66)

Jalonnement (*n.m.*), pegging ; pegging out ; pegging off ; staking ; staking off ; staking out. (67)

Jalonner (*v.t.*), to peg ; to peg out ; to peg off ; to stake ; to stake off ; to stake out ; to dot : (68)

jalonner une concession, to peg out (*or* to stake off) a claim. (69)

ligne de fracture de l'écorce terrestre jalonnée par des volcans (*f.*), line of fracture of the earth's crust, dotted with volcanoes. (70)

Jalpaïte (Minéral) (*n.f.*), jalpaite. (71)

Jambage (d'une baie de porte, d'une baie de fenêtre) (*n.m.*), jamb, jamb-post, (of a doorway, of a window-opening). (72 *ou* 73)

Jambage (d'une cheminée) (*n.m.*), jamb (of a fire-place). (1)

Jambage en pierre (*m.*), jamb-stone. (2)

Jambages (d'une grue-chevalet) (*n.m.pl.*), legs (of a goliath crane). (3)

Jambages (d'un marteau-pilon) (*n.m.pl.*), stand-ards (of a power-hammer). (4)

Jambe (de charbon) (Exploitation des couches de houille) (*n.f.*), entry-pillar; entry-stump; stump. (5)

Jambe de force (d'une ferme de comble) (*f.*), strut (of a roof-truss) (a diagonal compression member between a straining-beam and a tie-beam). (6)

Jambe de force (Boisage de mines) (*f.*), strut; force-piece. (7)

Jambette (Constr.) (*n.f.*), purlin-post. (8)

Jambette (de chaudière) (*n.f.*), water-leg (of boiler). (9)

Jamesonite (Minéral) (*n.f.*), jamesonite. (10)

Jante (d'une roue) (*n.f.*), felloe, felly, rim, (of a wheel). (11)

Jante (d'une poulie, d'une roue d'engrenage) (*n.f.*), rim (of a pulley, of a gear-wheel). (12 *ou* 13)

Jantille (d'une roue hydraulique) (*n.f.*), paddle (of a water-wheel). (14)

Jargon (Minéral) (*n.m.*), jargoon. (15)

Jarre électrique (*f.*), electric jar; Leyden jar. (16)

Jaspe (Minéral) (*n.m.*), jasper. (17)

Jaspe noir (*m.*), lydite; Lydian stone; touch-stone. (18)

Jaspe opale (*m.*), opal jasper; jasper opal. (19)

Jaspe rubané (*m.*), ribbon jasper. (20)

Jaspe sanguin (*m.*), bloodstone; heliotrope. (21)

Jaspé, -e (*adj.*), jaspery; jasperated. (22)

Jaspé, -e (en parlant des outils, etc.) (*adj.*), mottled finish : (23)

clef jaspée (*f.*), spanner, mottled finish. (24)

Jaspique (*adj.*), jasperous. (25)

Jaspoïde (*adj.*), jaspoid. (26)

Jauge (*n.f.*), gauge; gage. (27) See also calibre.

Jauge à billes (*f.*), ball-gauge. (28)

Jauge alphabétique pour fils d'acier (*f.*), steel-wire letter-gauge. (29)

Jauge circulaire pour fils métalliques (*f.*), circular wire-gauge. (30)

Jauge d'épaisseur (calibre à lames) (*f.*), thickness-gauge; feeler-gauge; feeler. (31)

Jauge de Paris (*f.*), the French wire-gauge : (32)
en France, les fils de fer se mesurent suivant la jauge de Paris, divisée en 30 dimensions dont la plus forte est de 10 millimètres, in France, iron wire is measured by the "jauge de Paris," divided into 30 sizes, the largest of which is 10 millimetres. (33)

Jauge de pas (*f.*), screw-pitch gauge. (34)

Jauge étalon pour fils de fer (*f.*), standard iron-wire gauge. (35)

Jauge pour fils métalliques *ou* jauge de tréfilerie (*f.*), wire-gauge. (36)

Jauge pour fines tôles *ou simplement* jauge pour tôles (*f.*), sheet-gauge; sheet-iron gauge. (37)

Jauge pour fortes tôles *ou simplement* jauge pour tôles (*f.*), plate-gauge. (38)

Jauge pour mèches (*f.*), drill-gauge. (39)

Jauge pour mèches hélicoïdales et fils d'acier (*f.*), twist-drill and steel-wire gauge. (40)

Jauge Stubs (*f.*), Stubs' gauge. (41)

Jaugeage (*n.m.*), gauging; gaging : (42)

jaugeage d'un cours d'eau, gauging the volume of water in a stream. (43)

Jaugeage en déversoir (Hydraul.) (*m.*), notch-gauging. (44)

Jauger (*v.t.*), to gauge; to gage : (45)
jauger le débit d'une pompe, to gauge the flow of a pump. (46)
jauger le fil de fer, to gauge iron wire. (47)
jauger un tonneau, to gauge a cask. (48)

Jaunâtre (*adj.*), yellowish. (49)

Jaune (*adj.*), yellow. (50)

Jaune (couleur) (*n.m.*), yellow. (51)

Jaune paille (*m.*), straw-yellow. (52)

Jayet (Minéral) (*n.m.*), jet. (53)

Jélonka (Extraction du pétrole) (*n.f.*), bailer; Russian bailer; baler. (54)

Jenkinsite (Minéral) (*n.f.*), jenkinsite. (55)

Jérémélévite (Minéral) (*n.f.*), jeremejevite. (56)

Jet (action de jeter, de lancer) (*n.m.*), throwing; casting; cast : (57)
le jet d'une pierre, throwing (*or* casting) a stone. (58)

Jet (action de couler en moule) (Fonderie) (*n.m.*), pouring; running; teeming; casting : (59)
jet de la fonte, pouring the metal. (60)

Jet (émission d'un fluide) (*n.m.*), jet; stream : (61)
jet d'eau sortant d'un ajutage, jet, stream, of water issuing from a nozzle. (62)

Jet (ajutage) (*n.m.*), jet; nozzle; spout : (63)
lance avec jet à éventail (*f.*), branch-pipe with spreader jet. (64)
jet de pompe, pump-spout. (65)

Jet (travée) (Épuisement) (Mines) (*n.m.*), lift; stage : (66)
épuisement en un seul jet (*m.*), pumping in one lift; single-stage pumping. (67)

Jet d'eau (*m.*), jet of water; water-jet. (68)

Jet d'eau (de fenêtre) (*m.*), weathering; wash. (69)

Jet de coulée (trou de coulée dans un moule) (Fonderie) (*m.*), runner; gate; git; geat; sprue; cast-gate; pouring-gate; running-gate; pouring-hole; down-runner; down-gate; upright runner; sprue-hole; funnel; jet. (70)

Jet de coulée *ou simplement* jet (*n.m.*) (partie métallique qui reste attenante après la pièce coulée), gate; git; geat; sprue; header; runner; jet : (71)
la coupe des jets et bavures, cutting off the gates and fins. (72)
jets et débris provenant de la coulée (Fonderie) [jets (*n.m.pl.*); débris (*n.m.pl.*)], home-scrap; returns. (73)

Jet de gaz (*m.*), gas-jet. (74)

Jet de pelle (*m.*), shovel's cast. (75)

Jet de pierre (*m.*), stone's throw; stone's cast : (76)
deux maisons situées à un jet de pierre l'une de l'autre, two houses situated a stone's throw from each other. (77)

Jet de sable (*m.*), sand-blast; sand-jet. (78)

Jet de vapeur (*m.*), steam-jet. (79)

Jet sur banquette (Fouilles) (*m.*), cast after cast. (80)

Jetée (*n.f.*), jetty; pier. (81)

Jeter (*v.t.*), to throw; to cast : (82)
jeter de l'eau par la fenêtre, to throw water out of the window. (83)
jeter de la terre dans une brouette, to cast, to throw, earth into a wheelbarrow. (84)
jeter de côté, to cast aside. (85)

jeter un pont sur une rivière, to throw a bridge over a river. (1)

jeter (couler) (Fonderie) (*v.t.*), to pour ; to run ; to teem ; to cast : (2)

jeter du métal en fusion dans un moule, to pour, to run, molten metal into a mould. (3)

jeter bas les feux *ou* **jeter le feu**, to draw, to dump, the fire. (4)

jeter les fondations d'une maison, to lay the foundations of a house. (5)

jeton (Houillères) (*n.m.*), tally. (6)

jette-feu [**jette-feu** *pl.*] (*n.m.*), drop-grate ; dumping-grate ; dump-grate ; tip-grate. (7)

jeu (fonctionnement) (Méc.) (*n.m.*), working ; action ; movement of parts : (8)

machine qui est en jeu (*f.*), machine which is working (*or* which is in action). (9)

regard servant à constater le jeu d'un méca--nisme (*m.*), peep-hole for watching the working (*or* the action) of a mechanism ; door for the purpose of observing the movement of the parts of a mechanism. (10)

jeu (facilité de se mouvoir) (Méc.) (*n.m.*), play ; slack ; clearance ; lash : (11)

le léger jeu entre la vis et l'écrou, the slight play between the screw and the nut. (12)

donner du jeu à un essieu, to give play to an axle. (13)

porte qui a trop de jeu (*f.*), door which has too much clearance. (14)

jeu au fond des dents (Engrenages), top and bottom clearance (Gearing). (15)

jeu de la voie (Ch. de f.), gauge-clearance ; clearance-space between the wheel-flanges and the rails. (16)

jeu en bout, end play. (17)

jeu entre les dents (Engrenages), side clearance. (18)

jeu inutile *ou* jeu de la denture *ou simplement et ordinairement* jeu, backlash. (19)

jeu latéral, side play ; side lash ; side clearance. (20)

jeu (course) (Méc.) (*n.m.*), play ; stroke ; length of stroke : (21)

le jeu du piston, du tiroir, the play of the piston, of the slide-valve. (22 *ou* 23)

jeu (d'un fusible) (Élec.) (*n.m.*), blowing, blowing out, (of a fuse). (24)

jeu (assortiment) (*n.m.*), set ; assortment ; battery ; train : (25)

jeu de pièces de rechange, de roues à fileter, d'accessoires pour générateurs, set of spare parts, of change-wheels, of boiler-fittings. (26 *ou* 27 *ou* 28)

jeu de laminoirs, train of rolls ; roll-train ; rolling-mill train ; set of rolls ; battery of rolls. (29)

jeu (de pompes de mine) (*n.m.*), set, lift, (of mine-pumps) : (30)

jeu de fonçage, sinking-lift. (31)

jeu de fond, bottom lift. (32)

jeu foulant (opp. à *jeu soulevant*), plunger-set ; plunger-lift. (33)

jeu soulevant, bucket-set ; bucket-lift. (34)

jeu posé *ou* jeu fixe (opp. à *jeu volant*), fixed lift ; fixed set. (35)

jeu volant, suspended lift ; suspended set. (36)

jeu de la lumière (*m.*) *ou* **jeux de la lumière** (*m.pl.*), play of light. (37)

jeune (Géogr. phys.) (*adj.*), young ; youthful : (38)

vallée jeune (*f.*), young valley ; youthful valley. (39)

jeune garçon (*m.*), boy ; lad. (40)

jeunesse (Géogr. phys.) (*n.f.*), youth. (41)

jig (Préparation mécanique des minerais) (*n.m.*), jig ; jigger ; jigging-machine ; movable-sieve jig. (42)

jig à air *ou* **jig pneumatique** (*m.*), air-jig ; pneumatic jig. (43)

joindre (*v.t.*), to join ; to connect ; to couple : (44)

joindre deux morceaux de bois en les collant, to join two pieces of wood by sticking them together. (45)

joindre les deux bouts *ou* joindre les deux bouts de l'année, to make both ends meet. (46)

joindre (*v.i.*), to join. (47)

joindre (se) (*v.r.*), to join. (48)

joint (*n.m.*), joint ; connection ; coupling. (49) See also **assemblage.**

joint (d'une charnière) (*n.m.*), joint (of a hinge). (50)

joint (d'un moule) (Fonderie) (*n.m.*), joint, parting, (of a mould). (51)

joint (Géol.) (*n.m.*), joint ; fissure ; diaclase. (52)

joint à baïonnette (*m.*), bayonet-joint. (53)

joint à boudin *ou* **joint à bride** (*m.*), flange-joint. (54)

joint à calotte sphérique (*m.*). Same as **joint à rotule sphérique.**

joint à chute libre (Sondage) (*m.*), free-fall. (55)

joint à éclisses (*m.*), fish-joint. (56)

joint à emboîtement (tuyaux) (*m.*), spigot-joint ; spigot-and-faucet joint ; faucet-joint. (57)

joint à genou (*m.*). Same as **joint à rotule sphérique.**

joint à insertion (*m.*), inserted joint. (58)

joint à longue manche (*m.*), long sleeve coupling. (59)

joint à plat (*m.*), square joint ; straight joint ; butt-joint ; butting-joint. (60)

joint à recouvrement (*m.*). Same as **joint de recouvrement.**

joint à rotule (*m.*), swivel-joint. (61)

joint à rotule sphérique *ou simplement* **joint à rotule** (*m.*), ball-and-socket joint ; cup-and-ball joint ; cup-joint ; globe-joint ; socket-joint. (62)

joint à vis (*m.*), screw-joint. (63)

joint anglais (assemblage à embrèvements séparés par un plat joint) (Charp.) (*m.*), bridle-joint ; notch-and-bridle joint. (64)

joint appuyé (rails de ch. de f.) (*m.*), supported joint. (65)

joint au mastic de fonte (*m.*), rust-joint. (66)

joint au minium (*m.*), putty-joint. (67)

joint autoclave (*m.*), autoclave joint. (68)

joint brisé (*m.*), break-joint. (69)

joint calfaté (*m.*), caulked joint. (70)

joint chevauché (*m.*), overlap-joint. (71)

joint d'eau (*m.*), water-joint. (72)

joint d'Oldham (*m.*), Oldham coupling. (73)

joint de bout (Charp.) (*m.*), heading-joint. (74)

joint de Cardan *ou* **joint brisé** (*m.*), Cardan joint ; cardan joint ; universal joint. (75)

joint de dilatation (*m.*), expansion-joint ; slip-joint. (76)

joint de paume (*m.*), splayed joint. (77)

joint de paume à pointes coupées *ou* **joint de paume désabouté** (*m.*), splayed butt scarf. (78)

joint de plomb (*m.*),   lead-joint ;   solder-joint ; soldering-joint.   (1)

joint de rail (*m.*),   rail-joint.   (2)

joint de recouvrement (tôles) (*m.*),   lap-joint ; overlap-joint ;   overlapping   joint ;   step-joint.   (3)

joint éclissé (rails) (*m.*),   fished joint.   (4)

joint en porte à faux (rails de ch. de f.) (*m.*), suspended   joint ;   unsupported   rail-joint. (5)

joint en sifflet (*m.*),   splayed joint.   (6)

joint étanche (*m.*),   tight   joint ;   close   joint ; stanch joint.   (7)

joint hollandais (*m.*).   Same as **joint de Cardan.**

joint hydraulique (*m.*),   hydraulic joint.   (8)

joint invisible (*m.*),   invisible joint.   (9)

joint lisse (*m.*),   flush joint.   (10)

joint par torsade (fils) (*m.*),   twist-joint.   (11)

joint parallèle à l'inclinaison (Géol.) (*m.*),   dip-joint.   (12)

joint parallèle à la direction (Géol.) (*m.*),   strike-joint.   (13)

joint principal (Géol.) (*m.*),   master-joint.   (14)

joint refeuillé (*m.*),   bevelled tongue-and-groove joint ; angular grooved and tongued joint. (15)

joint rivé (*m.*),   riveted joint.   (16)

joint télescopique (*m.*),   telescope-joint.   (17)

joint tournevis (*m.*),   Oldham coupling.   (18)

joint universel (*m.*),   universal joint.   (19)

jointif, -ive (*adj.*),   close (with edges touching) : (20)

lattage jointif (*m.*),   close lathing.   (21)

jointout (*n.m.*),   trying-plane ;   try-plane.   (22)

jointure (*n.f.*),   joining ;   joint.   (23)

jointure sphérique (*f.*),   ball-and-socket joint ; cup-and-ball   joint ;   cup-joint ;   globe-joint ;   socket-joint.   (24)

jonc (*n.m.*),   rush.   (25)

jonction (*n.f.*),   junction ;   joining :   (26)

jonction de deux filons,   junction of two lodes. (27)

jonction et ligature de câbles,   joining and bind-ing cables.   (28)

jonction (voie de jonction) (Ch. de f.) (*n.f.*), crossover ;   crossover-road.   (29)

jongle (*n.f.*),   jungle.   (30)

joue (*n.f.*),   cheek ;   flange :   (31)

joues d'un coussinet (coussinet de palier), cheeks of a bearing.   (32)

joues d'un coussinet (Ch. de f.),   cheeks of a railway-chair.   (33)

joues d'une mortaise,   cheeks of a mortise.   (34)

joues d'une poulie,   cheeks of a pulley-block. (35)

joues d'un engrenage à joues,   flanges, flanging, shrouding, of a flanged gear, of a shrouded gear.   (36)

joues du tambour d'un treuil,   flanges of the barrel of a winch.   (37)

joue d'un galet à joues pour chariot de pont roulant,   flange of a flanged travelling wheel of a runner of an overhead travelling crane. (38)

galet à deux joues (*m.*),   double-flanged travelling wheel.   (39)

jouée (Arch.) (*n.f.*),   reveal.   (40)

jouées (d'une mortaise) (*n.f.pl.*),   cheeks (of a mortise).   (41)

jouer (se dit des fusibles) (Élec.) (*v.i.*),   to blow ; to blow out ;   to go :   (42)

en cas de court-circuit ces fusibles joueront presque immédiatement,   in case of short

circuit these fuses will blow out almost immediately.   (43)

jouer un rôle important,   to play an important part ;   to be an important factor :   (44)

en affaires, l'argent joue un grand rôle,   money plays a very important part in business. (45)

le rôle de l'eau, en géologie, est considérable, in geology, water is an important factor.   (46)

joug (Méc.) (*n.m.*),   yoke.   (47)

jouissance (*n.f.*),   privilege ;   right to use ;   use. (48)

jouissance d'eau (*f.*),   water-privilege ;   water-right ;   use of water.   (49)

jouissance de passage (*f.*),   right of way ;   way-leave.   (50)

joule (unité d'énergie électrique) (*n.m.*),   joule ; volt-coulomb ;   coulomb-volt.   (51)

joule par seconde (*m.*),   joule per second ;   volt-ampere ;   ampere-volt ;   watt.   (52)

jour (*n.m.*),   day.   (53)

jour (opp. à fond) (Mines) (*n.m.*),   day ;   surface ; grass :   (54)

eaux du jour (*f.pl.*),   day-water ;   surface-water.   (55)

amener des remblais du jour,   to bring in stowing from the surface.   (56)

remonter le minerai au jour,   to bring ore to the surface (*or* to grass).   (57)

jour (*n.m.*) *ou* **jour de l'escalier** (Constr.),   well ; well-hole.   (58)

jour de fête *ou* **jour férié** (*m.*),   holiday ;   public holiday.   (59)

jour de la semaine (*m.*),   week-day.   (60)

jour de paye (*m.*),   pay-day.   (61)

jour ouvrable *ou* **jour ouvrier** (*m.*),   work-day ; working-day.   (62)

journal de sondage (*m.*),   bore-holing journal. (63)

journal du bord (Dragage) (*m.*),   log-book.   (64)

journalier, -ère (*adj.*),   daily :   (65)

consommation journalière de charbon (*f.*), daily consumption of coal.   (66)

journalier (pers.) (*n.m.*),   day-labourer.   (67)

journée (*n.f.*),   day.   (68)

journée de huit heures (*f.*),   eight-hour day ; 8-hour day.   (69)

journée de travail (*f.*),   working-day.   (70)

journée de travail de huit heures (*f.*),   working-day of eight hours ; 8-hour working-day. (71)

joyau (*n.m.*),   jewel.   (72)

juger sur l'apparence,   to judge by appearances. (73)

jumeau, -elle *ou* **jumelé, -e** (*adj.*),   twin :   (74)

machines jumelles (*f.pl.*),   twin engines.   (75)

jumelle (*n.f.*),   binocular ;   binocular glass ; glass ;   glasses.   (76)

jumelle à prismes *ou* **jumelle prismatique** (*f.*), prism-binocular.   (77)

jumelle de campagne (*f.*),   field-glass ;   field-glasses.   (78)

jumelles (d'une machine à raboter) (*n.f.pl.*), uprights, standards, housings, (of a planing-machine).   (79)

jumelles (d'une sonnette) (*n.f.pl.*),   guide-poles (of a pile-driver).   (80)

jungle (*n.f.*),   jungle.   (81)

jura (Géol.) (*n.m.*),   Jura ;   Jurassic :   (82)

le jura,   the Jura ;   the Jurassic.   (83)

jura blanc (*m.*),   White Jura.   (84)

jura brun (*m.*),   Brown Jura.   (85)

jura noir (*m.*),   Black Jura.   (86)

jurassique (Géol.) (*adj.*),   Jurassic :   (87)

système jurassique (*m.*),   Jurassic system. (88)

Jurassique (*n.m.*), Jurassic ; Jura : (1)
le jurassique, the Jurassic ; the Jura. (2)
Jusant (de la marée) (*n.m.*), fall, ebb, reflux, (of the tide). (3)
Juste (*adj.*), just ; correct ; accurate ; precise ; exact ; proper : (4)
balance juste (*f.*) (opp. à *balance fausse*), just balance ; just scale. (5) (opp. to *unjust balance* or *unjust scale*).
poids juste (*m.*)(opp. à *faux poids*), just weight. (6) (opp. to *unjust weight*).
juste proportion (*f.*), exact proportion ; precise proportion. (7)

justement (*adv.*), justly ; just ; correctly ; accu-rately ; precisely ; exactly ; properly. (8)
justesse (*n.f.*), justness ; correctness ; accuracy ; precision ; exactness ; exactitude : (9)
justesse d'une balance, justness, correctness, accuracy, (of a balance or scale). (10)
justifier (*v.t.*), to justify ; to warrant : (11)
résultats qui justifient des dépenses supplé-mentaires (*m.pl.*), results which justify (*or* warrant) further expenditure. (12)
Jute (*n.m.*), jute. (13)
juxtaposer (*v.t.*), to juxtapose. (14)
juxtaposition (*n.f.*), juxtaposition. (15)

K

kaïnite (Minéral) (*n.f.*), kainite. (16)
kainozoïque (Géol.) (*adj. & n.m.*), Cenozoic ; Cænozoic ; Cainozoic. (17)
kammererite (Minéral) (*n.f.*), kammererite. (18)
kaneelstein (Minéral) (*n.m.*), cinnamon-stone. (19)
kaolin (*n.m.*), kaolin ; china-clay ; china-stone ; porcelain-clay ; kaolinite. (20)
kaolinique (*adj.*), kaolinic. (21)
kaolinisation (*n.f.*), kaolinization : (22)
kaolinisation du feldspath, kaolinization of feldspar. (23)
kaoliniser (*v.t.*), to kaolinize. (24)
kaolinite (Minéral) (*n.f.*). Same as kaolin.
karat (*n.m.*), carat. (25) See carat (*n.m.*).
karst (Géol.) (*n.m.*), Karst : (26)
le nom de karst (*ou* Carso), qui désigne la région calcaire entre la Carniole et l'Istrie est devenu classique pour désigner les pays calcaires ; on parle de régions de karst, de modelé karstique, d'hydrographie karstique. Toute région calcaire n'est pourtant pas au sens propre un karst, the name of Karst which designates the limestone region between Carniola and Istria has become classic to denote calcareous countries ; one speaks of regions of Karst, of Karstic relief, of Karstic hydrography. Every limestone region is not however properly a Karst. (27)
karsténite (Minéral) (*n.f.*), karstenite ; anhydrite. (28)
karstique (Géol.) (*adj.*), Karstic. (29)
kauchteux, -euse (*adj.*), abounding in coal ; rich in coal : (30)
mine kauchteuse (*f.*), rich coal-mine. (31)
keilhauite (Minéral) (*n.f.*), keilhauite ; yttro-titanite. (32)
kérargyre (Minéral) (*n.m.*), kerargyrite ; cerargy-rite ; horn-silver ; horn-ore. (33)
kermésite (*n.f.*) *ou* kermès minéral (*m.*) (Minéral), kermesite ; kermes mineral ; antimony-blende ; red antimony. (34)
kernrostung (Métall.) (*n.m.*), heap-roasting ; heap-roast ; open-heap roasting. (35)
kérosène (*n.m.*), kerosene. (36)
kersantite (*n.f.*) *ou* kersanton (*n.m.*) (Pétrol.), kersantite. (37)
kieselguhr (*n.m.*), kieselguhr. (38)
kiesérite (Minéral) (*n.f.*), kieserite. (39)
killas (schiste) [Cornouailles] (*n.m.*), killas. (40)

kilocalorie (Phys.) (*n.f.*), kilocalory ; kilo-calorie ; kilogramme calory ; great calory ; greater calory. (41)
kilogramme *ou* kilo [kilos *pl.*] (*n.m.*), kilo-gramme ; kilo = 2·2046223 lbs *or* 15432·3564 grains. (42)
kilogrammètre (*n.m.*), kilogrammetre = 7·233 foot-pounds. (43)
kilomètre (*n.m.*), kilometre = 0·62137 mile. (44)
kilométrique (*adj.*), kilometric ; kilometrical. (45)
kilovolt (Élec.) (*n.m.*), kilovolt. (46)
kilovolt-ampère [kilovolts-ampères *pl.*] (*n.m.*), kilovolt-ampere. (47)
kilowatt (Élec.) (*n.m.*), kilowatt. (48)
kilowatt-heure [kilowatts-heure *pl.*] (*n.m.*), kilowatt-hour. (49)
kilowatt-minute [kilowatts-minute *pl.*] (*n.m.*), kilowatt-minute. (50)
kilowatt-seconde [kilowatts-seconde *pl.*] (*n.m.*), kilowatt-second. (51)
kimberlite (Pétrol.) (*n.f.*), kimberlite ; blue-stuff. (52)
kimméridgien, -enne *ou* kiméridgien,-enne (Géol.) (*adj.*), Kimmeridgian ; Kimeridgian. (53)
kimméridgien (*n.m.*), Kimmeridgian. (54)
kiosque de transformation (Élec.) (*m.*), trans-former-box ; transformer-pillar. (55)
klaprothine (Minéral) (*n.f.*), lazulite. (56)
klaubage (du minerai) (*n.m.*), picking ; hand-picking. (57)
klauber (*v.t.*), to pick ; to hand-pick. (58)
klaubés (*n.m.pl.*), picked ore ; hand-picked ore. (59)
klaubeur, -euse (pers.) (*n.*), picker ; hand-picker. (60)
klippe [klippes *ou* klippen *pl.*] (Géol.) (*n.f.*), outlier. (61)
knébélite (Minéral) (*n.f.*), knebelite. (62)
könleinite *ou* könlite (Minéral) (*n.f.*), koenlite ; koenleinite. (63)
kopje (*n.m.*), kopje. (64)
kotschubéite (Minéral) (*n.f.*), kotschubeite. (65)
köttigite (Minéral) (*n.f.*), koettigite. (66)
krémersite (Minéral) (*n.f.*), kremersite. (67)
krennérite (Minéral) (*n.f.*), krennerite. (68)
krypton (Chim.) (*n.m.*), krypton ; crypton. (69)
kupferblende (Minéral) (*n.f.*), copper-blende. (70)
kupfernickel (Minéral) (*n.m.*), kupfernickel ; copper-nickel ; arsenical nickel ; niccolite ; nickelite ; nickeline. (71)
kynite (Explosif) (*n.f.*), kynite. (72)

L

labeur (*n.m.*), labour ; work. (1)
laboratoire (Chim.) (*n.m.*), laboratory. (2)
laboratoire (d'un four à réverbère) (puddlage) (*n.m.*), iron-chamber ; charge-chamber ; reverberatory chamber ; laboratory. (3)
laboratoire d'essais (*m.*), sampling-works ; test--ing-station. (4)
laboratoire de l'usine (*m.*), works' laboratory. (5)
laboratoire industriel (*m.*), technical laboratory. (6)
laboratoire obscur (Photogr.) (*m.*), dark room. (7)
labradorite (*n.f.*) *ou* **labrador** (*n.m.*) (Minéral), labradorite ; Labrador feldspar ; Labrador stone. (8)
labyrinthe (Préparation mécanique du minerai) (*n.m.*), labyrinth. (9)
lac (*n.m.*), lake. (10)
lac de barrage (*m.*), barrier-lake. (11)
lac de barrage glaciaire *ou* **lac de glacier** (*m.*), glacier-lake : (12)
  le lac de Märjelen, barré par le glacier d'Aletsch, dans les Alpes Bernoises, est l'exemple le plus connu d'un lac de barrage glaciaire. L'existence d'un pareil lac est à la merci des variations du glacier barreur, the Lake of Märjelen, dammed by the Aletsch Glacier, in the Bernese Alps, is the best known example of a glacier-lake. The existence of such a lake is at the mercy of the variations of the damming glacier. (13)
lac de barrage morainique *ou* **lac de vallée glaciaire** (*m.*), morainic lake ; glacial lake (formed by morainal damming). (14)
lac de barrage volcanique (*m.*), volcanic lake. (15)
lac de bitume de la Trinité (*m.*), pitch lake of Trinidad. (16)
lac de cirque (*m.*), tarn ; corrie-lake. (17)
lac intérieur (*m.*), inland lake. (18)
lac salé (*m.*), salt lake. (19)
laccolithe (*n.m.*) *ou* **laccolite** (*n.f.*) (Géol.), laccolith ; laccolite. (20)
laccolithique (Géol.) (*adj.*), laccolithic ; lacco--litic. (21)
lacer une courroie, to lace a belt. (22)
laceret (tarière) (*n.m.*), auger. (23)
laceret (piton à vis) (*n.m.*), screw-eye. (24)
laceret à cuiller (*m.*), spoon auger. (25)
laceret à vrille (*m.*), gimlet auger. (26)
lacet (d'une charnière) (*n.m.*), pin, pintle, (of a hinge). (27)
lacet (mouvement d'une locomotive, d'un wagon) (*n.m.*), nosing. (28)
lacet (tarière) (*n.m.*), Same as **laceret**.
lacets (d'une route) (*n.m.pl.*), windings (of a road). (29)
lâche (*adj.*), loose ; slack : (30)
  nœud lâche (*m.*), loose knot. (31)
  le brin lâche d'une courroie, the slack side of a belt. (32)
lâcher les eaux d'un réservoir, to let the water out of a tank. (33)
lâcher prise, to let go. (34)
lâcher un robinet, to turn on a tap ; to open a cock. (35)
lâcher une corde, to loosen, to slacken, to slack, to let out, to relax, a cord. (36)

lacis (*n.m.*), network : (37)
  lacis de fibres, network of fibers. (38)
lacunaire (*adj.*), lacunar. (39)
lacune (*n.f.*), lacuna : (40)
  minéral plein de lacunes (*m.*), mineral full of lacunæ. (41)
lacustre *ou* **lacustral, -e, -aux** (*adj.*), lacustrine ; lacustral ; lake (*used as adj.*) : (42)
  dépôts lacustres (*m.pl.*), lacustrine deposits. (43)
  bassin lacustre (*m.*), lake-basin. (44)
lagunaire (*adj.*), lagoonar. (45)
lagune (*n.f.*), lagoon. (46)
laine (*n.f.*), wool. (47)
laine de bois (*f.*), wood-wool. (48)
laine de scorie *ou* **laine de la salamandre** (*f.*), slag-wool ; cinder-wool ; mineral wool ; mineral cotton ; rock-wool. (49)
lais (Géol.) (*n.m.*), warp ; silt. (50)
laisse (Marine & Navigation fluviale) (*n.f.*), water-mark. (51)
laisse de basse mer (*f.*), low-water mark. (52)
laisse de haute mer (*f.*), high-water mark. (53)
laisser cuire le bain (Métall.), to allow the bath to stand. (54)
laisser déposer (*v.t.*), to deposit : (55)
  dissolution qui laisse déposer à l'évaporation des cristaux (*f.*), solution which deposits crystals on evaporation. (56)
laisser du mou au câble, to slack, to slacken, the rope. (57)
laisser échapper le moment favorable, to let the favourable moment go by ; to let an opportunity slip. (58)
laisser fort à désirer, to leave much to be desired. (59)
laisser le gaz s'échapper, to permit, to allow, the gas to escape. (60)
laisser reposer les chantiers (Mines), to give the workings a rest. (61)
laisser revenir (Trempe), to draw the temper of ; to let down the temper of ; to temper : (62)
  laisser revenir un burin, to draw the temper of, to let down the temper of, to temper, a chisel. (63)
laisser se déposer au fond du vase les matières solides qui se trouvent en suspension dans le liquide, to allow the solid matter in suspension in the liquid to settle on the bottom of the vessel. (64)
laisser tomber (*v.t.*), to drop ; to allow to fall ; to let fall : (65)
  laisser tomber une pierre, to drop a stone. (66)
  laisser tomber le toit (Mines), to allow the roof to fall. (67)
laisser (se) couler le long d'une corde, to slide down a rope. (68)
lait de chaux (*m.*), whitewash ; limewash. (69)
lait de lune *ou* **lait de montagne** (Minéral) (*m.*), mountain-milk. (70)
lait de roche (Minéral) (*m.*), rock-milk. (71)
laitier (*n.m.*) *ou* **laitiers** (*n.m.pl.*) (Métall.), slag ; cinder ; scoria ; scoriæ. (72)
  NOTE.—Particularly, **laitiers** are silicates con--taining earthy bases, as distinguished from **scories,** which are silicates retaining metallic oxides.
laitier acide (*m.*), acid slag. (73)

laitier alumineux *ou* laitier à alumine (*m.*), aluminous slag. (1)

laitier basique (*m.*), basic slag. (2)

laitier calcaire *ou* laitier à chaux (*m.*), calcareous slag. (3)

laitier des volcans (*m.*), glassy lava. (4)

laitier extra-calcaire (*m.*), extra-limy slag. (5)

laitier neutre (*m.*), neutral slag. (6)

laitier réfractaire (*m.*), refractory slag. (7)

laitier siliceux *ou* laitier à silice (*m.*), siliceous slag. (8)

laitier ultra-acide (*m.*), highly acid slag ; strongly acid slag. (9)

laitier ultra-basique (*m.*), highly basic slag ; strongly basic slag. (10)

laitiers de hauts fourneaux (*m.pl.*), blast-furnace slag. (11)

laiton (*n.m.*), brass. (12)

laiton d'aluminium (*m.*), aluminium brass. (13)

laiton fondu (*m.*), cast brass. (14)

laitonnage (*n.m.*), brassing. (15)

laitonner (*v.t.*), to brass. (16)

lambeau de recouvrement (Géol.) (*m.*), outlier. (17)

lambourde (pièce de bois pour soutenir les bouts des solives) (Constr.) (*n.f.*), wall-plate. (18)

lambourde (Boisage des puits de mines) (*n.f.*), lagging (behind wedging-curb). (19)

lame (morceau de métal ou d'autre matière, plat et très mince) (*n.f.*), sheet ; thin sheet ; plate : (20)

une lame de plomb, a sheet of lead. (21)

lame (mince feuillet qui s'exfolie) (Minéralogie, etc.) (*n.f.*), lamina ; flake ; plate : (22)

l'apophyllite a une tendance à se diviser en lames (*f*), apophyllite has a tendency to divide into laminæ. (23)

lame (fer d'un instrument propre à couper) (*n.f.*), blade ; cutter : (24)

lame d'un couteau, d'un canif, blade of a knife, of a penknife. (25 *ou* 26)

lame (nappe d'eau souterraine) (Mines) (*n.f.*), underground sheet of water. (27)

lame à aléser (*f.*), boring-cutter. (28)

lame à cisailler *ou* lame de cisailles (*f.*), shear-blade. (29)

lame d'accumulateur (*f.*), accumulator-plate. (30)

lame d'aiguille [lames d'aiguille *pl.*] (Ch. de f.) (*f.*), switch-rail ; point-rail ; point ; tongue-rail ; tongue ; switch-tongue ; slide-rail ; latch. (31)

lame d'eau (d'une chaudière) (*f.*), film of water (between two plates in a boiler). (32)

lame d'eau au-dessus du seuil d'un déversoir (Hydraul.) (*f.*), sheet of water above the crest of a weir. (33)

lame de collecteur (Élec.) (*f.*), commutator-bar. (34)

lame de pile (*f.*), battery-plate. (35)

lame de raboteuse (*f.*), plane-iron. (36)

lame de sabre (d'une distribution Corliss, ou analogue) (*f.*), lifting-rod (of a Corliss valve-gear, or the like). (37)

lame de scie (*f.*), saw-blade. (38)

lame de scie à métaux (*f.*), hack-saw blade. (39)

lame de scie circulaire (*f.*), circular saw. (40)

lame de scie circulaire creusée (*f.*), hollow-ground circular saw. (41)

lame de tournevis pour vilebrequins (*f.*), turn-screw bit ; screwdriver-bit. (42)

lame mince (Phys., Microscopie, etc.) (*f.*), section ; thin section ; thin plate. (43)

lame négative (élément de pile) (*f.*), negative plate (battery-cell). (44)

lame porte-objet [lames porte-objets *pl.*] (Micro-scopie) (*f.*), slide ; slider. (45)

lame positive (élément de pile) (*f.*), positive plate (battery cell). (46)

lamellaire (*adj.*), lamellar. (47)

lamellation (*n.f.*), lamellation. (48)

lamelle (*n.f.*), lamina ; folium : (49) lamelles de talc, folia of talc. (50)

lamelle couvre-objet [lamelles couvre-objets *pl.*] (Microscopie) (*f.*), cover-glass ; cover-slip. (51)

lamellé, -e (*adj.*), lamellated ; lamellate : (52) l'ardoise est lamellée (*f.*), slate is lamellated. (53)

lamelleux, -euse (*adj.*), lamellose ; lamellous ; lamellate ; lamellated. (54)

laminage (*n.m.*), lamination ; laminating. (55)

laminage (de métaux) (*n.m.*), rolling : (56) laminage en — passages, rolling in — passes. (57)

laminage à chaud, hot-rolling. (58)

laminage à froid, cold-rolling. (59)

laminage (de la vapeur) (*n.m.*), wiredrawing (of steam). (60)

laminé, -e (*adj.*), laminated ; flaky. (61)

laminer (Métall.) (*v.t.*), to roll : (62) laminer du métal en feuilles minces, to roll metal into thin sheets. (63)

laminer le fer, to roll iron. (64)

laminer à chaud, to hot-roll ; to roll hot. (65)

laminer à froid, to cold-roll ; to roll cold. (66)

laminer (vapeur) (*v.t.*), to wiredraw : (67) laminer la vapeur pendant l'admission, par suite de la faible ouverture des lumières, to wiredraw the steam during admission, owing to the small opening of the ports. (68)

laminer (se) (Métall.) (*v.r.*), to roll : (69) le métal se lamine facilement, the metal rolls easily. (70)

laminer (se) (vapeur) (*v.r.*), to wiredraw : (71) vapeur qui se lamine pendant l'admission (*f.*), steam which wiredraws during admission. (72)

laminerie (atelier) (*n.f.*), rolling-mill. (73)

lamineur (pers.) (*n.m.*), roller. (74)

lamineux, -euse (*adj.*), laminiferous. (75)

laminoir (*n.m.*), rolling-mill ; mill ; rolls ; train. (76) See also train.

laminoir à bandages (*m.*), tire-mill. (77)

laminoir à billettes (*m.*), billetting-rolls. (78)

laminoir à blindages (*m.*), armour-plate mill. (79)

laminoir à blooms (*m.*), blooming-mill ; cogging-mill. (80)

laminoir à fines tôles (*m.*), sheet-mill. (81)

laminoir à grosses tôles (*m.*), heavy-plate mill. (82)

laminoir à guides (*m.*), guide-mill. (83)

laminoir à mouvement alternatif (*m.*), recipro-cating mill. (84)

laminoir à plomb (*m.*), lead-mill. (85)

laminoir à profils *ou* laminoir à profilés (*m.*), section-mill. (86)

laminoir à rails (*m.*), rail-mill ; rail-train. (87)

laminoir à tôle (grosses tôles ou tôles moyennes) (*m.*), plate-mill. (88)

laminoir à tôles moyennes (*m.*), medium-plate mill. (89)

laminoir de puddlage *ou* laminoir ébaucheur (*m.*), puddle-train ; puddle-bar train ; puddle-rolls ; muck-train ; roughing-rolls. (1)

laminoir double duo (*m.*), four-high rod-mill. (2)

laminoir duo [laminoirs duos *pl.*] (*m.*), two-high mill ; two-high rolls ; two-high train. (3)

laminoir finisseur (*m.*), finishing-mill ; finishing-rolls. (4)

laminoir marchand (*m.*), merchant mill ; bar-mill ; merchant train ; merchant-bar train. (5)

laminoir moyen (*m.*), medium-sized bar-mill. (6)

laminoir réversible (*m.*), reversing mill. (7)

laminoir trio [laminoirs trios *pl.*] (*m.*), three-high mill ; three-high rolls ; three-high train. (8)

laminoir universel (*m.*), universal mill ; slabbing-mill. (9)

lampadite (Minéral) (*n.f.*), lampadite. (10)

lampe (*n.f.*), lamp. (11)

lampe à accumulateurs (*f.*), storage-battery lamp ; accumulator-lamp. (12)

lampe à acétylène (*f.*), acetylene-lamp. (13)

lampe à alcool (*f.*), alcohol-lamp ; spirit-lamp. (14)

lampe à arc *ou* lampe à arc électrique (*f.*), arc-lamp. (15)

lampe à arc à charbon (*f.*), carbon-arc lamp. (16)

lampe à arc à flamme (*f.*), flame-arc lamp. (17)

lampe à arc à l'air (*f.*), open-arc lamp. (18)

lampe à arc dans le vide (*f.*), vacuum-arc lamp. (19)

lampe à arc enfermé (*f.*), enclosed-arc lamp. (20)

lampe à bas voltage, forme poire en verre clair, filament métallique, intensité en bougies— (*f.*), low-voltage lamp, pear-shaped, clear glass, metallic filament, — candle-power. (21)

lampe à braser (*f.*), brazing-lamp ; blow-lamp ; blow-torch. (22)

lampe à double toile (Mines) (*f.*), double-gauze lamp. (23)

lampe à esprit-de-vin (*f.*), spirit-lamp ; alcohol-lamp. (24)

lampe à feu nu (*f.*), naked-light lamp ; open-light lamp. (25)

lampe à feu rouge (*f.*), lamp showing a red light. (26)

lampe à fil à plomb (pour travaux souterrains, pour alignements dans les mines) (*f.*), plummet-lamp, lamp-plummet, (for under-ground work, for mine surveying). (27)

lampe à filament de carbone (*f.*), carbon-filament lamp. (28)

lampe à filament de tantale (*f.*), tantalum-filament lamp ; tantalum-lamp. (29)

lampe à filament de tungstène (*f.*), tungsten-filament lamp ; wolfram-lamp. (30)

lampe à filament métallique (*f.*), metallic-filament lamp ; wire lamp. (31)

lampe à flamme nue (*f.*), naked-light lamp ; open-light lamp. (32)

lampe à gaz, à incandescence (*f.*), incandescent-gas lamp. (33)

lampe à haut voltage, forme sphérique en verre dépoli, — bougies (*f.*), high-voltage lamp, round, frosted glass, — candle-power. (34)

lampe à huile (*f.*), oil-lamp. (35)

lampe à huile à feu nu *ou* lampe à huile à flamme nue (Mines) (*f.*), naked-light oil-lamp ; open-light oil-lamp. (36)

lampe à incandescence (*f.*), incandescent lamp ; glow-lamp. (37)

lampe à incandescence dans l'air (*f.*), open incandescent lamp. (38)

lampe à incandescence dans le vide (*f.*), vacuum incandescent lamp. (39)

lampe à luminescence (*f.*), luminescence-lamp. (40)

lampe à magnésium (*f.*), magnesium-lamp. (41)

lampe à main (*f.*), hand-lamp. (42)

lampe à osmium (*f.*), osmium-lamp. (43)

lampe à pétrole (*f.*), oil-lamp ; petroleum-lamp. (44)

lampe à souder (*f.*), soldering-lamp ; blow-lamp ; blow-torch. (45)

lampe à suif (Essais au chalumeau) (*f.*), grease-lamp. (46)

lampe à toile métallique (Mines) (*f.*), wire-gauze lamp. (47)

lampe à tube à vide (*f.*), vacuum-tube lamp. (48)

lampe à tube de quartz (*f.*), quartz-lamp. (49)

lampe à vapeur de mercure (*f.*), mercury-vapour lamp ; mercury-lamp. (50)

lampe au carbone (*f.*), carbon-lamp. (51)

lampe au néon (*f.*), neon-lamp. (52)

lampe au tantale (*f.*), tantalum-lamp. (53)

lampe au tungstène (*f.*), tungsten-lamp ; wolfram-lamp. (54)

lampe avec chapeau (*f.*), bonneted lamp. (55)

lampe baladeuse avec douille baïonnette à interrupteur, grille de protection en fil de fer étamé, pour garage et ateliers (*f.*), hanging lamp with bayonet-socket and switch, tinned wire guard, for garage and shop use. (56)

lampe baladeuse avec manche formant pied (*f.*), hanging and standing lamp. (57)

lampe Chesneau (*f.*), Chesneau fire-damp delicate indicator ; Chesneau lamp for detecting fire-damp. (58)

lampe d'Argand (*f.*), Argand lamp. (59)

lampe Davy (Mines) (*f.*), Davy lamp ; miners' friend. (60)

lampe de couleur (*f.*), coloured lamp ; coloured light. (61)

lampe de mineur (*f.*), miners' lamp. (62)

lampe de peintre en bâtiment (*f.*), paint-burning lamp ; blow-lamp ; blow-torch. (63)

lampe de poche (Élec.) (*f.*), pocket-lamp ; flash-lamp ; electric torch : (64)

lampe de poche, garniture nickelée, corps en métal oxydé, lentille ronde, avec pile donnant — heures d'éclairage intermittent, ampoule à filament métallique étiré, flash-lamp, nickeled fittings, body of oxidized metal, round lens, with battery giving — hours of intermittent light, bulb with drawn metal filament. (65)

lampe de sûreté (*f.*), safety-lamp : (66)

lampe de sûreté à rallumeur, safety-lamp with relighter. (67)

lampe de sûreté agréée, approved safety-lamp. (68)

lampe de sûreté munie de rallumeurs à amorces fulminantes, safety-lamp provided with fulminating-cap relighters. (69)

lampe-éclair [lampes-éclairs pl.] (Photogr.) (n.f.), flash-lamp. (1)

lampe électrique à incandescence (f.), incan--descent electric lamp. (2)

lampe électrique portative (f.), electric hand-lamp. (3)

lampe essayée (f.), tested lamp. (4)

lampe étalon (f.), standard lamp. (5)

lampe fil à plomb (f.). Same as lampe à fil à plomb.

lampe grisoumétrique (Mines) (f.), warning-lamp. (6)

lampe Harcourt au pentane (Photom.) (f.), Harcourt pentane-lamp. (7)

lampe-heure [lampes-heure pl.] (Élec.) (n.f.), lamp-hour. (8)

lampe non-protégée (Mines) (f.), unshielded lamp. (9)

lampe-phare [lampes-phares pl.] (n.f.), flare-lamp ; flare. (10)

lampe portative (f.), hand-lamp ; portable lamp. (11)

lampe protégée (Mines) (f.), shielded lamp. (12)

lampe sans chapeau (f.), unbonneted lamp. (13)

lampe-témoin [lampes-témoins pl.] (Élec.) (n.f.), pilot-lamp ; pilot-light. (14)

lampe type (f.), standard lamp. (15)

lampiste (pers.) (n.m.), lampman. (16)

lampisterie (n.f.), lamp-room ; lamp-station ; lamp-cabin. (17)

lamprophyre (Pétrol.) (n.m.), lamprophyre ; mica-trap. (18)

lanarkite (Minéral) (n.f.), lanarkite. (19)

lance (n.f.) ou lance d'eau (tube métallique adapté à l'extrémité d'un tuyau pour diriger le jet), branch ; branch-pipe ; nozzle. (20)

lance (ringard) (n.f.), prick-bar ; pricker ; poker. (21)

lance à incendie avec jet à éventail (f.), fire branch-pipe with spreader jet. (22)

lance d'arrosage (f.), branch-pipe (for spraying or watering). (23)

lance d'incendie (f.). Same as lance à incendie.

lance de sonde (f.), probe ; pricker. (24)

lancement (d'une affaire) (n.m.), launching, floating, (an enterprise). (25)

lancement (d'une pierre) (n.m.), throwing, casting, (a stone). (26)

lancer de l'air, to inject air : (27)

dans les anciens hauts fourneaux, la tuyère lançait de l'air froid, in the old blast-furnaces, the tuyère injected cold air. (28)

lancer de la terre dans une brouette, to throw, to cast, earth into a wheelbarrow. (29)

lancer en l'air, to hurl up, to throw up, into the air : (30)

cage de mine qui est lancée dans le chevalement par une violente chasse d'air (f.), mine-cage which is hurled up into the head-frame by a violent air-blast. (31)

lancer un pont sur une rivière, to throw a bridge over a river. (32)

lande (n.f.), moor ; moorland. (33)

landeux, -euse (adj.), moorland ; moory : (34)

région landeuse (f.), moorland region. (35)

langelotte (Préparation mécanique des minerais) (n.m.), log-washer ; trough-washer ; puddling-trough ; disintegrating-trough. (36)

langite (Minéral) (n.f.), langite. (37)

langue (n.f.), tongue : (38)

langue de feu, tongue of flame. (39)

langue de terre, tongue of land. (40)

langue glaciaire, ice-tongue. (41)

langue (d'une balance) (n.f.). Same as languette (d'une balance).

langue-d'aspic [langues-d'aspic pl.] (n.f.), arrow-head : (42)

foret à langue-d'aspic (m.), arrow-headed drill ; flat drill. (43)

languette (d'une balance) (n.f.), pointer, index, tongue, needle, (of a balance). (44)

languette (clavette linguiforme) (Méc.) (n.f.), feather ; feather-key ; spline. (45)

languette (n.f.) ou languette venue de bois (opp. à languette rapportée) (Charp.), tongue ; feather : (46)

assemblage à rainure et languette (m.), tongue-and-groove joint ; feather-joint. (47)

languette (n.f.) ou languette rapportée ou fausse languette (Charp.), slip-tongue ; loose tongue ; fillet ; feather tongue ; feather : (48)

assemblage à languette rapportée (m.), slip-tongue joint ; loose-tongue joint ; filleted joint ; feather-tongued joint ; feather-joint. (49)

languette de bois (f.), strip of wood ; cleat. (50)

lanières pour courroies (f.pl.), belt-laces. (51)

lanterne (lampe) (n.f.), lantern ; lamp. (52)

lanterne (pour la confection de noyaux) (Fonderie) (n.f.), core-barrel ; lantern. (53)

lanterne (d'une pompe) (n.f.), strainer ; wind-box ; snore-piece ; tail-piece ; rose. (54)

lanterne (Engrenages) (n.f.), lantern ; lantern-wheel ; lantern-pinion ; trundle ; wallower. (55)

lanterne (n.f.) ou lanterne de serrage (unissant deux tendeurs) (Méc.), turnbuckle ; right-and-left coupling ; right-and-left screw-link. (56)

lanterne à trois feux (rouge, vert, et blanc) (f.), tricolored lamp (red, green, and white). (57)

lanterne-applique [lanternes-appliques pl.] (n.f.), bracket-lamp. (58)

lanterne d'agrandissement (Photogr.) (f.), enlarg--ing-lantern. (59)

lanterne de décharge (locomotive) (f.), exhaust-muffler ; muffler. (60)

lanterne de laboratoire (Photogr.) (f.), dark-room lamp. (61)

lanterne de projection ou lanterne à projections ou lanterne magique (f.), magic lantern ; projection-lantern. (62)

lanterne-guide [lanternes-guides pl.] ou simple--ment lanterne (Sondage) (n.f.), winged substitute. (63)

lanterne pliante (f.), folding lantern. (64)

lanterne signal de position d'aiguilles (Ch. de f.) (f.), target-lantern ; target-lamp ; switch-lantern ; switch-box with revolving signal-lamp ; ground-disc, with lamp. (65)

lanterne sourde (f.), dark lantern ; bull's-eye lantern. (66)

lanterne-tempête [lanternes-tempête pl.] (n.f.), hurricane-lamp ; tornado-lamp ; tornado-lantern ; tubular lantern. (67)

lanthane (Chim.) (n.m.), lanthanum ; lanthanium. (68)

lanthanite (Minéral) (n.f.), lanthanite. (69)

lapidaire (adj.), lapidary. (70)

lapidaire (machine à polir les pierres précieuses, etc.) (n.m.), lapidary lathe ; lapidary mill. (71)

lapidaire (machine à meule d'émeri) (*n.m.*), horizontal emery grinding-machine. (1)

lapidaire (meule lapidaire en émeri) (*n.m.*), face-wheel. (2)

lapidaire (pers.) (*n.m.*), lapidary. (3)

lapidescence (*n.f.*), lapidescence ; lapidescency. (4)

lapidescent, -e (*adj.*), lapidescent. (5)

lapideux, -euse (*adj.*), lapideous ; lapidose. (6)

lapidification (*n.f.*), lapidification. (7)

lapidifier (*v.t.*), to lapidify. (8)

lapidifier (se) (*v.r.*), to lapidify. (9)

lapidifique (*adj.*), lapidific ; lapidifical. (10)

lapilli [lapillus *sing.*] (Géol.) (*n.m.pl.*), lapilli : (11)

les tufs volcaniques sont formés de cendres, de lapilli, et boues volcaniques (*m.pl.*), volcanic tuffs are formed of ashes, of lapilli, and volcanic muds. (12)

lapilliforme (*adj.*), lapilliform. (13)

lapis (*n.m.*) *ou* lapis-lazuli (*n.m.*) *ou* lapis oriental (*m.*), lapis lazuli ; lazuli ; Armenian stone. (14)

laque (*n.f.*), lac. (15)

laque en écailles *ou* laque en plaques *ou* laque en feuilles *ou* laque plate (*f.*), shellac. (16)

lardite (Minéral) (*n.f.*), lardite. (17)

lardoire (sabot de pieu) (*n.f.*), pile-shoe. (18)

large (*adj.*), broad ; wide ; large : (19)

une large rivière, a broad river ; a wide river. (20)

larges tôles (*f.pl.*), large plates. (21)

large de *ou* de large, wide ; in width ; broad ; in breadth : (22)

une galerie large de — mètres *ou* une galerie de — mètres de large, a level — metres wide. (23)

largeur (*n.f.*), width ; breadth : (24)

tout corps a trois dimensions : longueur, largeur, et profondeur, every body has three dimensions : length, breadth, and thickness. (25)

largeur (de) *ou* largeur (en), in width ; wide ; in breadth ; broad : (26)

la galerie a — mètres de largeur, the gallery is — metres in width ; the gallery is — metres wide. (27)

largeur d'abatage (*f.*), stoping-width. (28)

largeur de l'escalier (*f.*), length of step. (29)

largeur de la dent (Engrenages) (*f.*), width of tooth (Gearing). (30)

largeur de la voie (Ch. de f.) (*f.*), gauge ; rail-way-gauge ; track-width ; distance between rails. (31)

largeur de lumière (cylindre) (Mach.) (*f.*), port width. (32)

largeur de marche *ou* largeur de giron (distance horizontale entre deux contre-marches d'escalier) (*f.*), going. (33)

largeur du vide (Engrenages) (*f.*), width of space (Gearing). (34)

larme batavique (Phys.) (*f.*), Prince Rupert's drop ; Rupert's drop ; Rupert's ball ; Rupert's tear ; detonating bulb. (35)

larmes volcaniques (*f.pl.*), volcanic drops (masses of vitreous matter met with on volcanoes). (36)

larmier (*n.m.*), eaves. (37)

lasseret (*n.m.*), Same as laceret.

lasulith (Minéral) (*n.m.*), lazulite. (38)

lasurfeldspath (Minéral) (*n.m.*), blue spar. (39)

latent, -e (*adj.*), latent : (40)

chaleur latente (*f.*), latent heat. (41)

latéral, -e, -aux (*adj.*), lateral ; side : (42)

mouvement latéral (*m.*), lateral motion ; side motion. (43)

latéralement (*adv.*), laterally (44)

latérite (argile) (*n.f.*), laterite. (45)

latéritique (*adj.*), lateritic. (46)

latitude (*n.f.*), latitude. (47)

latitude australe *ou* latitude sud (*f.*), latitude south ; south latitude. (48)

latitude boréale *ou* latitude nord (*f.*), latitude north ; north latitude. (49)

latitude géographique (*f.*), geographical latitude. (50)

latitudinal, -e, -aux (*adj.*), latitudinal. (51)

lattage (action ou ouvrage) (*n.m.*), lathing ; lagging : (52)

lattage d'un plafond, lathing a ceiling. (53)

lattage espacé, spaced lathing. (54)

lattage jointif, close lathing. (55)

latte (*n.f.*), lath. (56)

latte à ardoises *ou* latte volige *ou* latte volisse (*f.*), slate-lath ; batten. (57)

latte à rainures (moulure pour pose de fils) (Élec.) (*f.*), casing ; moulding. (58)

latter (*v.t.*), to lath : (59)

latter une cloison, to lath a partition. (60)

lattis (ouvrage) (*n.m.*), lathing ; lagging. (61)

lattis espacé (*m.*), spaced lathing. (62)

lattis jointif (*m.*), close lathing. (63)

laumonite *ou* laumontite (Minéral) (*n.f.*), lau-monite ; laumontite. (64)

laurionite (Minéral) (*n.f.*), laurionite. (65)

lavable (*adj.*), washable. (66)

lavage (*n.m.*), washing ; washing out : (67)

lavage à la batée (des graviers aurifères, etc.), panning ; pan-washing. (68)

lavage à la main, hand-washing ; washing by hand. (69)

lavage à sec par le vent *ou* lavage dans l'air *ou* lavage à l'air (des alluvions, des minerais), dry washing. (70)

lavage au bac à piston, jigging (fixed-sieve jig). (71)

lavage au jig, jigging (movable-sieve jig). (72)

lavage aux sluices, sluicing. (73)

lavage d'épreuves photographiques, washing photographic prints. (74)

lavage d'or, gold-washing. (75)

lavage de chaudières, washing out boilers. (76)

lavage du charbon, coal-washing. (77)

lavage du minerai, washing ore. (78)

lave (volcanique) (Géol.) (*n.f.*), lava. (79)

lave (coulée de boue) (Géol.) (*n.f.*), mud-flow ; mud-flood. (80)

lave acide (*f.*), acid lava. (81)

lave basaltique (*f.*), basaltic lava. (82)

lave basique (*f.*), basic lava. (83)

lave boueuse (*f.*), mud-lava. (84)

lave cordée (*f.*), ropy lava. (85)

lave neutre (*f.*), intermediate lava. (86)

lave obsidienne (*f.*), obsidian lava. (87)

lave trachytique (*f.*), trachytic lava. (88)

lavée (minerai concentré après lavage) (*n.f.*), washing : (89)

une lavée de minerai, a washing of ore. (90)

laver (*v.t.*), to wash ; to wash out : (91)

laver à la batée (*v.t.*), to pan ; to pan out ; to pan off. (92)

laver à la main, to wash by hand. (93)

laver au bac à piston, to jig (fixed-sieve jig). (94)

laver au jig, to jig (movable-sieve jig). (1)

laver au sluice, to sluice : (2)

laver au sluice du gravier aurifère, to sluice gold-bearing gravel. (3)

laver du minerai d'étain, to wash tin ore. (4)

laver une chaudière, to wash out a boiler. (5)

laver (se) (*v.r.*), to wash. (6)

laverie (usine où s'opère le lavage du charbon, des minerais) (*n.f.*), washery ; washing-house ; wash-house. (7)

laverie à charbon (*f.*), coal-washery. (8)

laveur (pers.) (*n.m.*), washer ; washerman : (9)

laveur d'or, gold-washer. (10)

laveur (Mach.) (*n.m.*), washer ; washing-machine. (11)

laveur à charbon (*m.*), coal-washer ; coal-washing machine. (12)

laveur à feldspath (*m.*), felspar washer. (13)

laveur à la batée (Mines) (pers.) (*m.*), panner. (14)

laveur à slimes (Mach.) (*m.*), slimes-washer. (15)

laveur à tambour (*m.*), drum-washer. (16)

laveur circulaire (*m.*), circular washer. (17)

laveur de gaz (*m.*), gas-washer. (18)

laveur mécanique (*m.*), washing-machine. (19)

laveur rotatif (*m.*), rotary washer ; rotary washing-machine. (20)

lavique (*adj.*), lavic ; lavatic. (21)

lavis (Dessin) (*n.m.*), wash : (22)

plan au lavis (*m.*), wash-plan. (23)

lavoir (Mach.) (*n.m.*). Same as **laveur.**

laxmannite (Minéral) (*n.f.*), laxmannite. (24)

lazulite (Minéral) (*n.f.*), lazulite. (25)

lazurite (Minéral) (*n.f.*). lazurite. (26)

leadhillite (Minéral) (*n.f.*), leadhillite. (27)

lécheur (graissage) (Mach.) (*n.m.*), licker. (28)

lecture (*n.f.*), reading : (29)

la lecture des divisions du vernier se fait à l'aide d'une loupe, the reading of the divisions of the vernier is done with a lens. (30)

légal, -e, -aux (*adj.*), legal : (31)

borne légale (*f.*), legal boundary. (32)

légende (d'une carte, d'un plan) (*n.f.*), legend (of a map, of a plan). (33 ou 34)

léger, -ère (*adj.*), light ; slight ; gentle : (35)

un fardeau léger, a light burden. (36)

une légère amélioration, a slight improvement. (37)

une légère inclinaison, a slight (or a gentle) dip. (38)

légèrement (*adv.*), lightly ; slightly ; gently. (39)

légèreté (*n.f.*), lightness ; slightness : (40)

la légèreté d'un gaz, the lightness of a gas. (41)

législation (*n.f.*), legislation ; laws : (42)

législation minière, mining legislation ; mining laws. (43)

lehm (Géol.) (*n.m.*), lehm. (44)

lent, -e (*adj.*), slow : (45)

travail lent (*m.*), slow work. (46)

combustion lente (*f.*), slow combustion. (47)

lentement (*adv.*), slowly. (48)

lenteur (*n.f.*), slowness. (49)

lenticulaire ou **lenticulé, -e** (*adj.*), lenticular : (50)

veine lenticulaire (*f.*), lenticular vein. (51)

lentiforme (*adj.*), lentiform. (52)

lentille (Opt.) (*n.f.*), lens. (53) See also **objectif.**

lentille (Géol.) (*n.f.*), lens ; lentille ; lentil : (54)

une lentille de minerai, a lens of ore. (55)

lentille à portrait (*f.*), portrait attachment. (56)

lentille achromatique (*f.*), achromatic lens. (57)

lentille actinique (*f.*), actinic lens. (58)

lentille anallatique (*f.*), anallatic lens. (59)

lentille biconcave ou **lentille concavo-concave** (*f.*), biconcave lens ; double-concave lens ; concavo-concave lens. (60)

lentille biconvexe ou **lentille convexo-convexe** (*f.*), biconvex lens ; double-convex lens ; convexo-convex lens. (61)

lentille concave (*f.*), concave lens. (62)

lentille concave-convexe ou **lentille concavo-convexe** ou **lentille concave-convexe à bord épais** (*f.*), concavo-convex lens ; diverging concavo-convex lens ; diverging meniscus. (63)

lentille convexe (*f.*), convex lens. (64)

lentille convexe-concave ou **lentille convexo-concave** ou **lentille concave-convexe à bord mince** (*f.*), convexo-concave lens ; converging concavo-convex lens ; converging meniscus. (65)

lentille de pendule (*f.*), pendulum-bob ; pendulum-ball. (66)

lentille dioptrique (d'un viseur direct) (Photogr.) (*f.*), magnifying-lens, eyepiece, (of a direct view-finder). (67)

lentille du levier de manœuvre des aiguilles (Ch. de f.) (*f.*), switch-lever counterweight. (68)

lentille en crown-glass (*f.*), crown-glass lens. (69)

lentille en flint-glass (*f.*), flint-glass lens. (70)

lentille objective (*f.*), object-lens. (71)

lentille périscopique (*f.*), periscopic lens. (72)

lentille plan-concave (*f.*), plano-concave lens. (73)

lentille plan-convexe (*f.*), plano-convex lens ; convexo-plane lens. (74)

lentille supplémentaire (*f.*), supplementary lens. (75)

lenzinite (Minéral) (*n.f.*), lenzinite. (76)

lépidocrocite (Minéral) (*n.f.*), lepidocrocite. (77)

lépidolite (Minéral) (*n.m.*), lepidolite. (78)

lépidomélane (Minéral) (*n.m.*), lepidomelane. (79)

leptochlorite (Minéral) (*n.f.*), leptochlorite. (80)

leptinite (Pétrol.) (*n.f.*), leptinite ; leptynite ; leptynyte. (81)

leptynolite (Pétrol.) (*n.f.*), leptinolite. (82)

lessivage (*n.m.*), leaching ; lixiviation : (83)

lessivage au cyanure de potassium des minerais d'or, leaching gold ores by cyanide of potassium. (84)

lessiver (*v.t.*), to leach ; to lixiviate. (85)

lessiver (se) (*v.r.*), to leach. (86)

lettre d'avis (*f.*), advice-note. (87)

lettres et chiffres en cuivre pour mouleurs [**lettre** (*n.f.*), **chiffre** (*n.m.*)], letters and figures for moulders, brass. (88)

lettsomite (Minéral) (*n.f.*), lettsomite ; cyano-trichite. (89)

leuchtenbergite (Minéral) (*n.f.*), leuchtenbergite. (90)

leucite (Minéral) (*n.f.*), leucite ; amphigene ; white garnet. (91)

leucitique (Pétrol.) (*adj.*), leucitic. (92)

leucitite (Pétrol.) (*n.f.*), leucitite. (93)

leucitophyre (Pétrol.) (*n.m.*), leucitophyre. (94)

leucocrate (Pétrol.) (*adj.*), leucocratic : (95)

roches leucocrates (*f.pl.*), leucocratic rocks. (96)

leucophane (Minéral) (*n.m.*), leucophanite. (97)

leucopyrite (Minéral) (*n.f.*), leucopyrite. (98)

leucoxène (Minéral) (*n.m.*), leucoxene. (99)

levage (action de lever) (*n.m.*), lifting ; hoisting ; raising. (100)

levant (*n.m.*), east. (1)

levé (*n.m.*) *ou* levé de plan *ou* levé de plans *ou* levé des plans, survey ; surveying. (2)

levé à la boussole *ou* levé de plan à la boussole (Mines) (*m.*), dialing ; latching ; compass-survey. (3)

levé à la boussole suspendue (Mines) (*m.*), surveying with the hanging compass. (4)

levé à la planchette (*m.*), plane-table survey. (5)

levé de détail (*m.*), detailed survey. (6)

levé de mines *ou* levé des plans de mines (*m.*), mine-survey ; mine-surveying. (7)

levé de reconnaissance (*m.*), reconnaissance survey. (8)

levé de terrains (*m.*), land-surveying ; land-survey. (9)

levé des plans au stadia (*m.*), stadia-surveying. (10)

levé des plans de surface *ou* levé au jour (*m.*), surface-survey ; surface-surveying. (11)

levé par cheminement (*m.*), meander-survey. (12)

levé photographique (*m.*), photographic survey-ing ; photographic survey. (13)

levé souterrain (*m.*), underground survey ; surveying underground. (14)

levé topographique (*m.*), topographic survey. (15)

levé trigonométrique (*m.*), trigonometrical survey. (16)

lève-roues [lève-roues *pl.*] (*n.m.*), carriage-jack ; wheel-jack ; lifting-jack. (17)

lève-roues à crémaillère (*m.*), carriage-jack of the rack-and-lever type. (18)

lève-roues à vis (*m.*), carriage-jack of the screw-type. (19)

levée (Ponts & Chaussées) (*n.f.*), causeway. (20)

levée de galets (Géogr. phys.) (*f.*), spit. (21)

levée de la production (récolte de l'or) (Mines) (*f.*), clean-up. (22)

levée du clapet (*f.*), lift of the clack-valve. (23)

levée du levier de battage (Sondage) (*f.*), stroke of the walking-beam. (24)

levée du pilon (*f.*), lift of the stamp. (25)

levée du piston (*f.*), play of the piston ; piston-stroke. (26)

lever (hausser) (*v.t.*), to lift ; to hoist ; to raise : (27)

    lever une trappe, to lift a trap-door. (28)

lever (tracer sur le papier) (*v.t.*), to plot ; to lay down ; to lay off ; to draw : (29)

    lever les profils en long et les profils en travers nécessaires pour rédiger le projet définitif d'un chemin de fer, to plot the longitudinal and cross-sections required for the drawing up of the final plan of a railway. (30)

    lever un terrain, to plot a piece of ground. (31)

    lever un plan, to plot a map ; to lay down a plan. (32)

lever (se) (*v.r.*), to lift ; to rise : (33)

    les soupapes de sûreté sont disposées pour se lever dès que la pression aura atteint une certaine limite (*f.pl.*), safety-valves are arranged to lift as soon as the pressure reaches a certain limit. (34)

lever des plans (*m.*). Same as levé de plans.

leverrierite (Minéral) (*n.f.*), leverrierite. (35)

levier (Méc.) (*n.m.*), lever. (36)

levier (barre de fer ou de bois propre à soulever les fardeaux) (*n.m.*), lever ; handspike ; crowbar ; prise. (37)

levier à bascule (*m.*), balanced lever. (38)

levier à chaîne (*m.*), chain-lever. (39)

levier à cliquet (*m.*), ratchet-lever. (40)

levier à contrepoids (*m.*), swing-bob lever. (41)

levier à pied-de-biche et à pointe (*m.*), pick-and-claw crowbar. (42)

levier à poignée (*m.*), handle lever. (43)

levier arithmétique (*m.*), arithmetical lever. (44)

levier composé (*m.*), compound lever. (45)

levier coudé (*m.*), bent lever ; bell-crank ; bell-crank lever. (46)

levier d'aiguille (Ch. de f.) (*m.*), switch-lever ; pointer. (47)

levier d'avance (distribution Walschaerts) (*m.*), lap-and-lead lever, combination lever, (Walschaerts valve-gear). (48)

levier d'entraînement (*m.*). Same as levier porte-train.

levier de battage (Sondage) (*m.*), walking-beam ; working-beam. (49)

levier de changement de marche (*m.*), reversing-lever ; reverse-lever. (50)

levier de débrayage (*m.*), disengaging-lever. (51)

levier de distribution (tiroir) (*m.*), valve-lever (slide-valve). (52)

levier de freinage *ou* levier de commande du frein (*m.*), brake-lever. (53)

levier de l'arbre de relevage (*m.*), reversing-shaft lever ; reverse-shaft lever. (54)

levier de manœuvre (*m.*), operating-lever. (55)

levier de manœuvre de la sablière (locomotive) (*m.*), sand-box lever ; sand-lever. (56)

levier de manœuvre des aiguilles (Ch. de f.) (*m.*), switch-lever ; pointer. (57)

levier de mise en marche (*m.*), starting-lever. (58)

levier de renvoi (*m.*), reversing-lever ; reverse-lever. (59)

levier de signal (Ch. de f.) (*m.*), signal-lever. (60)

levier de treuil de curage (*m.*), sand-reel handle. (61)

levier droit (*m.*), straight lever. (62)

levier du frein (*m.*), brake-lever. (63)

levier du frein à vapeur (*m.*), steam-brake lever. (64)

levier du premier genre *ou* levier inter-appui [leviers inter-appuis *pl.*] (*m.*), lever of the first kind : (65)

    un levier du premier genre est celui dans lequel le point d'appui est entre la résistance et la puissance, a lever of the first kind is that in which the fulcrum is between the power and the resistance (*or* weight). (66)

levier du régulateur (*m.*), throttle-lever. (67)

levier du second genre *ou* levier du deuxième genre *ou* levier inter-résistant [leviers inter-résistants *pl.*] (*m.*), lever of the second kind (celui dans lequel la résistance est entre le point d'appui et la puissance, that in which the resistance is between the power and the fulcrum). (68)

levier du troisième genre *ou* levier inter-puissant [leviers inter-puissants *pl.*] (*m.*), lever of the third kind (celui dans lequel la puissance est entre le point d'appui et la résistance, that in which the power is between the resist-ance and the fulcrum). (69)

levier en bois ferré (*m.*), iron-shod lever. (70)

levier équilibré par contrepoids (*m.*), counter-balanced lever. (71)

levier hydraulique (Mines) (*m.*), hydraulic wedge. (72)

levier inverseur (Méc.) (*m.*), reversing-lever. (73)

levier inverseur (Élec.) (*m.*), reversing-key. (1)

levier placé bien à la portée de l'ouvrier (*m.*), conveniently placed lever. (2)

levier porte-train [leviers porte-trains *pl.*] (d'un train épicycloïdal) (Engrenages) (*m.*), arm (of an epicyclic train) (Gearing). (3)

lévigation (*n.f.*), levigation. (4)

léviger (*v.t.*), to levigate. (5)

lèvre (*n.f.*), lip : (6)
lèvre d'un foret (*ou* d'une mèche), lip of a drill. (7)
lèvre d'un godet, lip of a bucket. (8)

lèvre (d'une faille) (Géol.) (*n.f.*), side (of a fault): (9)
lèvre affaissée, thrown side ; downthrow side ; downcast side. (10)
lèvre soulevée, heaved side ; upthrow side ; upcast side. (11)

lévyne (Minéral) (*n.f.*), levynite ; levine ; levyne. (12)

lézarde (*n.f.*), crevice ; crack ; chink. (13)

lézarder (*v.t.*), to crack ; to chink ; to crevice : (14)
mur que le feu a lézardé (*m.*), wall which the fire has cracked. (15)

lézarder (se) (*v.r.*), to crack ; to chink. (16)

lherzolite (Pétrol.) (*n.f.*), lherzolite. (17)

liais (*n.m.*), a hard, compact limestone—a kind of Portland stone. (18)

liaison (Constr.) (*n.f.*), bonding. (19)

liaison de la chaudière au châssis (locomotive) (*f.*), attachment of the boiler to the frame. (20)

liaisonner (Constr.) (*v.t.*), to bond : (21)
liaisonner un joint de maçonnerie, to bond a masonry joint. (22)

liant, -e (*adj.*), supple ; pliant ; pliable ; flexible. (23)

liant (*n.m.*), suppleness ; pliancy ; pliableness ; flexibility : (24)
le liant de l'acier, the flexibility of steel. (25)

lias (Géol.) (*n.m.*), Lias : (26)
le lias, the Lias. (27)

lias bleu (roche) (*m.*), blue lias. (28)

liasique *ou* liassique (Géol.) (*adj.*), Liassic. (29)

libelle (Minéralogie) (*n.f.*), the air bubble in a liquid inclusion. (30)

libérer (*v.t.*), to liberate ; to release ; to free. (31)

libéthénite (Minéral) (*n.f.*), libethenite. (32)

libre (*adj.*), free ; open : (33)
air libre (*m.*), free air ; open air. (34)

libre (Chim.) (*adj.*), free ; uncombined : (35)
hydrogène libre (*m.*), free hydrogen. (36)
or libre (*m.*), free gold. (37)
l'élément le plus répandu dans les roches est la silice, à l'état libre ou combiné (*m.*), the most widely distributed element in rocks is silica, either in the free or combined state. (38)

librement (*adv.*), freely : (39)
gaz qui est dégagé librement (*m.*), gas which is given off freely. (40)
une aiguille aimantée librement suspendue, a magnetic needle freely suspended. (41)

lice (main courante) (*n.f.*), hand-rail ; hand-railing. (42)

licence (*n.f.*), licence ; license ; permit : (43)
licence pour importer des marchandises prohibées, licence, permit, to import pro-hibited goods. (44)

licencié, -e (pers.) (*n.*), licence-holder. (45)

liebénérite (Minéral) (*n.f.*), liebenerite. (46)

liège (*n.m.*), cork. (47)

liège de montagne *ou* liège fossile (Minéral) (*m.*), mountain-cork ; rock-cork ; fossil cork. (48)

lien (attache) (*n.m.*), tie ; binder ; bond ; strap ; fastening. (49)

lien (ce qui unit, ce qui met en rapport intime) (*n.m.*), binder ; bond : (50)
dans le sable de fonderie, l'argile constitue un lien, in foundry-sand, clay constitutes a binder ; in foundry-sand clay acts as a bond. (51)

lien de faîte (Constr.) (*m.*), ridge-pole ; ridge-piece ; ridge-plate ; ridge-tree ; ridge-beam. (52)

lien en fer à U *ou simplement* lien (*n.m.*) (Char-pente), strap ; strap-bolt ; stirrup ; stirrup strap ; U bolt ; fastening. (53)

lier (*v.t.*), to bind ; to tie up ; to fasten : (54)
lier des pierres avec du mortier, to bind stones with mortar. (55)
lier un fagot, to tie up a fagot. (56)

lierne (Constr.) (*n.f.*), rail (a bar, as one extend-ing from post to post, used as a member of a fence, or other structure). (57)

lierne (entre les solives de plancher) (*n.f.*), strut ; strutting-piece ; bridging-piece. (58) *In the plural,* liernes, struts ; strutting ; stiffening ; bridging.

lieu (*n.m.*), place. (59)

lieu apparent d'un objet (Opt.) (*m.*), apparent place of an object. (60)

lieu de refuge (*m.*), place of safety ; shelter ; refuge. (61)

lieu écarté (*m.*), out-of-the-way place. (62)

lieu vrai d'un objet (Opt.) (*m.*), true place of an object. (63)

liévrite (Minéral) (*n.f.*), lievrite ; ilvaite. (64)

lift (*n.m.*), lift ; passenger-lift ; elevator. (65)

ligature (de câbles) (*n.f.*), binding, lashing, (of cables). (66)

ligaturer (*v.t.*), to bind ; to lash : (67)
ligaturer deux fils, to bind, to lash, two wires. (68)

ligne (*n.f.*), line. (69)

ligne à double voie (Ch. de f.) (*f.*), double-track line. (70)

ligne à faible trafic (Ch. de f.) (*f.*), light-traffic line. (71)

ligne à grand trafic (Ch. de f.) (*f.*), heavy-traffic line. (72)

ligne à voie unique (Ch. de f.) (*f.*), single-track line. (73)

ligne adiabatique (*f.*), adiabatic line ; adiabatic ; adiabat. (74)

ligne aérienne de transport d'énergie électrique (*f.*), overhead electric power-transmission line. (75)

ligne atmosphérique (diagramme d'indicateur) (*f.*), atmospheric line. (76)

ligne-barre [lignes-barres *pl.*] (Élec.) (*n.f.*), bus-bar ; omnibus-bar. (77)

ligne brisée (*f.*), broken line. (78)

ligne courbe (*f.*), curve line. (79)

ligne d'arbres (Méc.) (*f.*), line of shafting. (80)

ligne d'artère (Ch. de f.) (*f.*), main line. (81)

ligne d'emmarchement (escalier) (*f.*), walking-line (stairs). (82)

ligne d'engrènement (Engrenages) (*f.*), pitch-line (Gearing). (83)

ligne d'horizon (Perspective) (*f.*), horizontal line ; horizon. (84)

ligne d'induction magnétique (*f.*), line of magnetic induction. (85)

ligne d'opérations (Topogr.) (*f.*), datum-line ; datum ; base-line ; base. (1)

ligne de charpentier (*f.*), carpenters' line. (2)

ligne de chemin de fer (*f.*), railway-line ; line of railway. (3)

ligne de collimation (*f.*), line of collimation ; axis of collimation. (4)

ligne de conduite (*f.*), policy ; line of conduct. (5)

ligne de contact (*f.*), line of contact. (6)

ligne de côte [lignes de côtes *pl.*] (Géogr. phys.) (*f.*), coast-line. (7)

ligne de courbure (*f.*), line of curvature. (8)

ligne de couronne (Engrenages) (*f.*), addendum-line (Gearing). (9)

ligne de couronnement (d'un comble) (*f.*), ridge-line (of a roof.) (10)

ligne de démarcation (*f.*), line of demarcation. (11)

ligne de direction (Géol. & Mines) (*f.*), line of strike ; strike. (12)

ligne de division (Engrenages) (*f.*). Same as ligne primitive.

ligne de faiblesse (*f.*), line of weakness. (13)

ligne de faille (Géol.) (*f.*), fault-line. (14)

ligne de faîte (d'un comble) (Constr.) (*f.*), ridge-line (of a roof). (15)

ligne de faîte (Géogr. phys.) (*f.*). Same as ligne de partage.

ligne de foi (Opt. & Géod.) (*f.*) (*f.*), fiducial line. (16)

ligne de force (*f.*), line of force ; force-line ; line of induction : (17)
lignes de force d'un champ magnétique, lines of force of a magnetic field. (18)

ligne de foulée *ou* ligne de giron (escalier) (*f.*), walking-line (stairs). (19)

ligne de fracture de l'écorce terrestre (*f.*), line of fracture of the earth's crust. (20)

ligne de fuite (Élec.) (*f.*), leakage path. (21)

ligne de glissement (Cristall.) (*f.*), slip-band ; slip-line. (22)

ligne de moindre résistance (*f.*), line of least resistance : (23)
on trouve les volcans sur les lignes de moindre résistance de l'écorce terrestre, volcanoes occur on the lines of least resistance of the earth's crust. (24)

ligne de niveau (*f.*), line of level. (25)

ligne de nulle pression (diagramme d'indicateur) (*f.*), no-pressure line ; absolute line. (26)

ligne de partage *ou* ligne de partage des eaux (Géogr. phys.) (*f.*), divide ; divide-line ; water-parting ; shed-line ; watershed : (27)
série de hauteurs qui sert de ligne de partage des eaux entre deux bassins fluviaux (*f.*), series of heights which serves as a divide between two river-basins. (28)

ligne de plus grande pente (*f.*), line of swiftest descent. (29)

ligne de poussée *ou* ligne de pression (Engrenages) (*f.*), line of action ; common normal to the curves of the teeth in contact (Gearing). (30)
See normale commune des profils for example.

ligne de racine (Engrenages) (*f.*), root-line ; dedendum-line (Gearing). (31)

ligne de soulèvement (Géol.) (*f.*), line of up-heaval. (32)

ligne de striction (Géom.) (*f.*), line of striction. (33)

ligne de tangence (*f.*), line of tangency : (34)
ligne de tangence de deux cylindres, line of tangency of two cylinders. (35)

ligne de telphérage (*f.*), telpher-line ; telpher-way (36)

ligne de terre (Élec.) (*f.*), earth-line ; ground-line. (37)

ligne de terre (Géométrie descriptive & Per-spective) (*f.*), ground-line ; base-line. (38)

ligne de tourbillon (Phys.) (*f.*), vortex-line. (39)

ligne de tramway (*f.*), tramway-line ; tram-line. (40)

ligne de transmission (Méc.) (*f.*), line of shafting ; transmission-line. (41)

ligne de transmission à longue distance (Élec.) (*f.*), long-distance transmission-line. (42)

ligne de transmission d'énergie électrique (*f.*), electric-power transmission-line. (43)

ligne de transport de force à haute tension (*f.*), high-tension power-transmission line. (44)

ligne de vapeur (Phys.) (*f.*), steam-line. (45)

ligne de visée (d'une lunette) (*f.*), line of sight (of a telescope). (46)

ligne des centres (*f.*), line of centres. (47)

ligne des naissances (d'une voûte) (*f.*), springing-line (of an arch). (48)

ligne des niveaux piézométriques (Hydraul.) (*f.*), hydraulic gradient. (49)

ligne droite (*f.*), straight line. (50)

ligne du nez de marche (escalier) (*f.*), nosing-line (stairs). (51)

ligne du rivage (Géogr. phys.) (*f.*), shore-line ; beach-line. (52)

ligne électrique aérienne (*f.*), overhead electric line. (53)

ligne électrique souterraine (*f.*), underground electric line. (54)

ligne équinoxiale (*f.*), equinoctial line. (55)

ligne ferrée (*f.*), railway-line ; line of railway. (56)

ligne frontière (*f.*), boundary-line. (57)

ligne fuyante (Perspective) (*f.*), vanishing-line. (58)

ligne géodésique (*f.*), geodetic line. (59)

ligne interrompue (Dessin) (*f.*), broken-section line. (60)

ligne latérale (*f.*), side line. (61)

ligne limite (Engrenages) (*f.*), clearance-line (Gearing). (62)

ligne médiane (Géom.) (*f.*), median line. (63)

ligne mixte (Géom.) (*f.*), mixed line. (64)

ligne nette de démarcation (*f.*), clear line of demarcation. (65)

ligne neutre (Phys.) (*f.*), neutral line. (66)

ligne neutre d'un barreau aimanté (*f.*), neutral line of a bar magnet ; equator of a magnet. (67)

ligne oblique (*f.*), oblique line. (68)

ligne perpendiculaire (*f.*), perpendicular line. (69)

ligne pleine (opp. à ligne pointillée *ou* ligne ponctuée *ou* ligne interrompue) (*f.*), full line ; solid line. (70)

ligne pointillée *ou* ligne ponctuée (*f.*), dotted line. (71)

ligne primitive (Engrenages) (*f.*), pitch-line (Gearing): (72)
ligne primitive d'une crémaillère, pitch-line of a rack. (73)

ligne secondaire (Ch. de f.) (*f.*), branch-line. (74)

ligne télégraphique (*f.*), telegraph-line ; tele-graphic line. (75)

ligne téléphonique (*f.*), telephone-line ; tele-phonic line. (76)

ligne tremblée (*f.*), wavy line. (77)

ligne visuelle (Opt.) (*f.*), line of sight ; line of vision. (1)

ligner (*v.t.*), to line : (2)
ligner une pierre avec une ficelle frottée de rouge, to line a stone with a cord rubbed with red. (3)

lignes artificielles (Géom.) (*f.pl.*), artificial lines. (4)

ligneux, -euse (*adj.*), ligneous ; woody. (5)

lignite (*n.m.*), lignite ; brown coal. (6)

lignitifère (*adj.*), lignitiferous ; lignite-bearing. (7)

ligroïne (distillé de pétrole) (*n.f.*), ligroin ; ligroine. (8)

ligurite (Minéral) (*n.f.*), ligurite. (9)

limace (*n.f.*) *ou* limaçon (*n.m.*), Archimedean screw ; Archimedes' screw. (10)

limaçon (escalier tournant) (*n.m.*), spiral stairs ; winding stairs ; corkscrew stairs. (11)

limage (*n.m.*), filing : (12)
limage des dents d'une scie, filing the teeth of a saw. (13)

limaille (*n.f.*), filings. (14)

limaille de fer (*f.*), iron filings. (15)

limaille de zinc (*f.*), zinc-dust. (16)

limbe (d'un instrument de mathématiques) (*n.m.*), limb (of a mathematical instrument). (17)

limbe gradué (*m.*), graduated limb. (18)

limbe horizontal (d'un théodolite) (*m.*), horizontal limb (of a theodolite). (19)

limbe vertical (d'un théodolite) (*m.*), vertical limb. (20)

limburgite (Pétrol.) (*n.f.*), limburgite. (21)

lime (*n.f.*), file. (22)

lime à aiguille (*f.*), needle file. (23)

lime à biseau *ou* lime à barrettes (*f.*), cant file ; barrette file. (24)

lime à bouter (*f.*), warding-file ; key-file. (25)

lime à côté lisse *ou* lime à côtés lisses (*f.*), safe-edge file : (26)
lime avec un côté lisse, avec deux côtés lisses, file with one safe edge, with two safe edges. (27 *ou* 28)

lime à côtés ronds *ou* lime à champs ronds *ou* lime à bords arrondis (*f.*), round-edge file. (29)

lime à couteau (*f.*), knife-file ; knife-edge file. (30)

lime à deux soies *ou* lime à deux queues (*f.*), two-tanged file. (31)

lime à double taille *ou* lime à taille croisée (*f.*), double-cut file ; cross-cut file. (32)

lime à égalir *ou* lime à égaliser (*f.*), equalizing-file ; equalizing-file. (33)

lime à fendre *ou* lime à losange *ou* lime à dossière (*f.*), slitting-file ; featheredge file ; screw-head file. (34)

lime à garnir (*f.*), warding-file ; key-file. (35)

lime à grosse taille *ou* lime à taille rude (*f.*), rough-cut file. (36)

lime à lanterne (barboche) (*f.*), frame-saw file. (37)

lime à scies (*f.*), saw-file. (38)

lime à simple taille *ou* lime à taille simple (*f.*), single-cut file ; float-cut file ; float. (39)

lime à taille demi-douce *ou* lime à taille mi-douce (*f.*), second-cut file ; 2nd-cut file. (40)

lime à taille moyenne (*f.*), middle-cut file. (41)

lime agricole (*f.*), reaper file. (42)

lime bâtarde (*f.*), bastard file. (43)

lime carrée *ou* lime 4/4 (*f.*), square file. (44)

lime conique (*f.*), taper file. (45)

lime-couteau [limes-couteaux *pl.*] (*n.f.*), knife-file ; knife-edge file. (46)

lime d'entrée (*f.*), entering-file. (47)

lime demi-ronde *ou* lime 1/2 ronde (*f.*), half-round file. (48)

lime douce (*f.*), smooth file. (49)

lime extra-douce (*f.*), extra-smooth file. (50)

lime feuille-de-sauge *ou* lime double demi-ronde (*f.*), cross-file ; crossing-file ; double half-round file. (51)

lime-fraiseuse à main [limes-fraiseuses à main *pl.*] (*f.*), milling-file ; circular-cut file. (52)

lime grosse (*f.*), rough file. (53)

lime olive *ou* lime queue-de-rat ovale (*f.*), oval file. (54)

lime parallèle (*f.*), parallel file. (55)

lime plate-à-main [limes plates-à-main *pl.*] (*f.*), hand-file ; arm-file. (56)

lime plate pointue [limes plates pointues *pl.*] (*f.*), flat file. (57)

lime pointue (*f.*), taper file. (58)

lime pour l'affûtage des scies (*f.*), saw-file. (59)

lime queue-de-rat (*f.*), rat-tail file ; rat-tailed file. (60)

lime ronde (*f.*), round file. (61)

lime ronde à scies (*f.*), gulleting saw-file. (62)

lime superdouce (*f.*), superfine file. (63)

lime tiers-point [limes tiers-points *pl.*] *ou* lime 3/4 (*f.*), three-square file ; tri-square file ; triangular file. (64)

lime 3/4 pointue, une taille demi-douce, à angles arrondis (*f.*), taper saw-file, 2nd cut single, blunt. (65)

lime très douce (*f.*), dead smooth file. (65)

lime triangulaire (*f.*). Same as lime tiers-point.

limer (*v.t.*), to file : (67)
limer un chanfrein, to file a chamfer. (68)

limets (de la houille) (*m.pl.*), cleat, cleats, joint-planes, joints, (of the coal). (69)

limets parallèles à l'inclinaison (*m.pl.*), end cleats ; butt cleats. (70)

limets parallèles à la direction (*m.pl.*), face cleats. (71)

limeuse (*n.f.*), shaping-machine ; shaper ; shaping-planer. (72). See étau-limeur for varieties.

limitation (*n.f.*), limitation : (73)
limitation de l'explosion, limitation of the explosion. (74)

limite (*n.f.*), limit. (75)

limite (borne ; frontière) (*n.f.*), boundary ; limit ; border : (76)
limites d'une concession, boundaries, limits, of a concession (*or* of a claim). (77)
limites du champ d'exploitation (Mines), boundaries of the winning. (78)

limite (tolérance) (Méc.) (*n.f.*), limit ; margin ; tolerance : (79)
calibre à limites (*m.*), limit-gauge. (80)

limite d'élasticité *ou* limite élastique (*f.*), limit of elasticity ; elastic limit ; elastic strength : (81)
atteindre la limite d'élasticité, to reach the limit of elasticity. (82)
dépasser la limite élastique, to exceed the elastic limit. (83)

limite d'exploitabilité (Mines) (*f.*), pay-line ; grade-limit at which working is not payable. (84)

limite d'inflammabilité du grisou (*f.*), fire-damp ignition-limit. (85)

limite de charge (d'un wagon de ch. de f., ou analogue) (*f.*), maximum load ; maximum capacity. (1 *ou* 2)

limite de fatigue (Méc.) (*f.*), stress-limit. (3)

limite de la zone forestière (*f.*), timber-line. (4)

limite de proportionnalité (Méc.) (*f.*), limit of proportionality. (5)

limite de travail (Méc.) (*f.*), stress-limit. (6)

limite des neiges éternelles *ou* **limite des neiges persistantes** (*f.*), snow-line ; snow-limit. (7)

limiter (restreindre) (*v.t.*), to limit ; to confine ; to restrict : (8)
 limiter les travaux au traçage, to limit, to confine, to restrict, operations to develop-ment. (9)

limiter (former la limite) (*v.t.*), to bound : (10)
 concession limitée au nord, au sud, à l'est, à l'ouest, par une rivière (*f.*), claim bounded on the north, on the south, on the east, on the west, by a river. (11 *ou* 12 *ou* 13 *ou* 14)

limitrophe (*adj.*), adjoining ; adjacent ; con-tiguous ; bordering. (15)

limnologie (*n.f.*), limnology. (16)

limnologique (*adj.*), limnologic ; limnological. (17)

limon (boue) (*n.m.*), mud ; slime ; ooze ; silt. (18)

limon (d'escalier) (Constr.) (*n.m.*), string ; string-piece ; string-board ; stringer ; outside string ; face-string. (19)
 [Note.—The **wall-string** is called **contre-limon** *or* **faux limon**].

limon à crémaillère *ou* **limon à l'anglaise** (*m.*), cut string ; open string. (20)

limon de ruissellement (Géol.) (*m.*), wash. (21)

limon débillardé (escalier) (*m.*), wreathed string (stairs). (22)

limon droit *ou* **limon à la française** (*m.*), close string ; closed string ; housed string. (23)

limoneux, -euse (*adj.*), muddy ; slimy ; oozy ; silty. (24)

limonite (Minéral) (*n.f.*), limonite ; brown hematite ; brown iron ore. (25)

limoniteux, -euse (*adj.*), limonitic. (26)

limpide (*adj.*), clear ; limpid ; crystal-clear ; transparent : (27)
 une variété limpide de feldspath, a transparent variety of felspar (28)

limpidité (*n.f.*), clearness ; limpidity ; trans-parency : (29)
 limpidité de l'atmosphère, clearness, limpidity, of the atmosphere. (30)
 limpidité de l'eau, clearness of the water. (31)
 limpidité du diamant, transparency of the diamond. (32)

lin fossile (*m.*), fossil flax. (33)

linarite (Minéral) (*n.f.*), linarite. (34)

linéaire *ou* **linéal, -e, -aux** (*adj.*), linear ; lineal : (35)
 mesure linéaire (*f.*), linear measure. (36)
 dessin linéaire (*m.*), lineal design. (37)

linéalement (*adv.*), lineally ; linearly. (38)

lingot (*n.m.*), ingot : (39)
 lingot d'or, d'argent, d'acier, de cuivre, d'étain, ingot of gold, of silver, of steel, of copper, of tin. (40 *ou* 41 *ou* 42 *ou* 43 *ou* 44)

lingotière (*n.f.*), ingot-mould. (45)

linguet (*n.m.*), pawl. (46)

linnéite (Minéral) (*n.f.*), linnæite ; cobalt pyrites. (47)

linoléum (*n.m.*), linoleum. (48)

linteau (*n.m.*), lintel ; transom. (49)

liparite (Pétrol.) (*n.f.*), liparite ; rhyolite. (50)

liquater (Métall.) (*v.t.*), to liquate ; to eliquate : (51)
 liquater un minerai, to liquate, to eliquate, an ore. (52)

liquation (*n.f.*), liquation ; eliquation. (53)

liquéfaction (*n.f.*), liquefaction : (54)
 liquéfaction des gaz, liquefaction of gases. (55)

liquéfiable (*adj.*), liquefiable. (56)

liquéfiant, -e (*adj.*), liquefactive. (57)

liquéfier (*v.t.*), to liquefy. (58)

liquéfier (se) (*v.r.*), to liquefy. (59)

liquescence (*n.f.*), liquescence ; liquescency. (60)

liquescent, -e (*adj.*), liquescent. (61)

liqueur (*n.f.*), liquor ; solution. (62)

liqueur alcaline (*f.*), alkaline solution. (63)

liqueur cyanurée (*f.*), cyanide liquor. (64)

liqueur normale *ou* **liqueur titrée** (*f.*), normal solution ; standard solution : (65)
 l'analyse volumétrique repose sur l'emploi de liqueurs titrées (*f.*), volumetric analysis is based on the use of standard solutions. (66)

liquide (*adj.*), liquid : (67)
 air liquide (*m.*), liquid air. (68)

liquide (*n.m.*), liquid. (69)

liquide incolore (*m.*), colourless liquid. (70)

liquide réfrigérant (*m.*), refrigerating-liquid ; freezing-liquid. (71)

liquidité (*n.f.*), liquidity : (72)
 liquidité de l'eau, liquidity of water. (73)

lire (*v.t.*), to read : (74)
 lire un baromètre, la température sur une échelle thermométrique, to read a barom-eter, the temperature on a thermometric scale. (75 *ou* 76)

liroconite (Minéral) (*n.f.*), liroconite. (77)

lise (*n.f.*), quicksand. (78)

lisière (bord) (*n.f.*), border ; edge ; margin ; fringe ; outskirts : (79)
 lisière d'un bois, outskirts, border, of a wood. (80)
 lisière d'une plaine, edge of a plain. (81)
 lisière maritime des régions désertiques sub-tropicales, maritime fringe of the sub-tropical desert regions. (82)

lisière (Géol. & Mines) (*n.f.*), selvage ; clay-selvage ; clay-course ; gouge ; dig ; pug ; flucan ; selfedge ; salband ; saalband. (83)

lissage (*n.m.*), smoothing. (84)

lisse (*adj.*), smooth ; plain ; flush : (85)
 surface lisse (*f.*), smooth surface ; plain surface. (86)
 palier lisse (*m.*), plain bearings. (87)
 tuyau à joint lisse (*m.*), flush-joint pipe. (88)

lisse (*n.m.*), smoothness : (89)
 lisse du verre, smoothness of glass. (90)

lisse (main courante) (*n.f.*), hand-rail ; hand-railing. (91)

lisser (*v.t.*), to smooth. (92)

lisser (Fonderie) (*v.t.*), to slick ; to sleek ; to smooth : (93)
 lisser un moule, to slick a mould. (94)

lissoir (outil de mouleur) (Fonderie) (*n.m.*), slick ; slicker ; sleeker ; slaker ; smoother. (95)

lissoir à brides (*m.*), flange-smoother. (96)

lissoir à champignon (*m.*), 'bacca-box smoother ; button sleeker. (97)

lissoir à congé (*m.*), round-edge smoother ; fillet-slick. (98)

lissoir à tuyau *ou* lissoir à tuyaux (*m.*), pipe-slick ; pipe-smoother. (1)

lissoir à tuyaux à chapeau (*m.*), safe-end pipe-smoother. (2)

lissoir d'équerre *ou* lissoir équerre (*m.*), corner-slick ; corner-smoother. (3)

lissoir équerre, à congé (*m.*), round-edge corner-smoother. (4)

lissoir équerre, vif (*m.*), square corner-smoother. (5)

liste (*n.f.*), list. (6)

liste nominative (*f.*), list of names. (7)

lit (meuble sur lequel on se couche) (*n.m.*), bed : (8)
lit de camp, camp-bed. (9)

lit (couche de matière) (*n.m.*), bed ; layer : (10)
un lit de béton, a bed of concrete. (11)
barres de fer disposées par lits (*f.pl.*), bars of iron arranged in layers. (12)

lit (Géol.) (*n.m.*), bed : (13)
un lit d'argile, de gravier, a bed of clay, of gravel. (14 *ou* 15). See also couche, gîte, and gisement.

lit (fond sur lequel repose une grande masse d'eau) (*n.m.*), bed : (16)
le lit d'une rivière, de la mer, the bed of a river, of the sea. (17 *ou* 18)

lit (chacune des deux faces par lesquelles se touchent les pierres de taille superposées dans une construction) (*n.m.*), bed. (19)

lit d'ancien cours d'eau (*m.*), ancient river-bed. (20)

lit de carrière (*m.*), natural bed ; quarry-face : (21)
poser une pierre dans le sens de son lit de carrière, to set a stone in the sense of its natural bed. (22)

lit de coke (*m.*), coke-bed. (23)

lit de fusion (Métall.) (*m.*), burden : (24)
on appelle lit de fusion la masse de minerais et de fondants chargée dans le haut fourneau en proportions déterminées, the burden is the mass of ores and fluxes charged in the blast-furnace in determined proportions. (25)

lit de glacier (*m.*), glacier-bed. (26)

lit de houille (*m.*), bed of coal ; coal-bed. (27)

lit de mâchefer (*m.*), cinder-bed. (28)

lit isolant (Constr.) (*m.*), damp-proof course. (29)

lit majeur (d'une rivière) (*m.*), flood-bed. (30)

lit mineur (d'une rivière) (*m.*), normal bed. (31)

lité, -e (*adj.*), bedded. (32)

litharge (*n.f.*), litharge. (33)

lithionite (Minéral) (*n.f.*), lithionite. (34)

lithium (Chim.) (*n.m.*), lithium. (35)

lithoclase (Géol.) (*n.f.*), fracture. (36)

lithofracteur (Explosif) (*n.m.*), lithofracteur. (37)

lithogénèse (*n.f.*), lithogenesis. (38)

lithogénésie (*n.f.*), lithogenesy. (39)

lithogénétique (*adj.*), lithogenetic. (40)

lithoïde (*adj.*), lithoid ; lithoidal. (41)

lithoïdite (Pétrol.) (*n.f.*), lithoidite. (42)

lithologie (*n.f.*), lithology. (43)

lithologique (*adj.*), lithologic ; lithological. (44)

lithologiquement (*adv.*), lithologically. (45)

lithologue (pers.) (*n.m.*), lithologist. (46)

lithomarge (Minéral) (*n.f.*), lithomarge. (47)

lithosphère (*n.f.*), lithosphere. (48)

litier (Métall.) (*n.m.*). Same as laitier.

litre (*n.m.*), litre =1·75980 pints. (49)

littoral, -e, -aux (*adj.*), littoral : (50)
dépôts littoraux (*m.pl.*), littoral deposits. (51)

littoral (*n.m.*), littoral ; coast ; shore ; seacoast ; seaboard : (52)
littoral atlantique, Atlantic seaboard. (53)

livraison (*n.f.*), delivery : (54)
livraison des marchandises, delivery of goods. (55)

livre (*n.m.*), book. (56)

livre (avoirdupois) (unité anglaise de poids) (*n.f.*), pound =0·45359243 kilogramme. (57)

livre de commandes (*m.*), order-book. (58)

livre de pointage (*m.*), time-book. (59)

livrer (*v.t.*), to deliver : (60)
livrer des marchandises à réception de com-mande, to deliver goods on receipt of order. (61)

livreuse (*n.f.*), delivery-van ; parcel-van ; van. (62)

lixiviation (*n.f.*), lixiviation ; leaching. (63)

lixivier (*v.t.*), to lixiviate ; to leach. (64)

loam (Géol.) (*n.m.*), loess ; löss. (65)

lobe (*n.m.*), lobe : (66)
lobe glaciaire, glacial lobe. (67)

local, -e, -aux (*adj.*), local : (68)
usages locaux (*m.pl.*), local customs. (69)

local (*n.m.*), place ; premises ; room. (70)

localement (*adv.*), locally. (71)

localisation (*n.f.*), localization ; localizing ; location ; locating ; limiting : (72)
la théorie de la localisation des volcans dans le voisinage des côtes, the theory of the location of volcanoes in the vicinity of coasts. (73)

localiser (*v.t.*), to localize ; to locate ; to limit : (74)
localiser un incendie, to localize, to limit, a fire. (75)
localiser un défaut (Télégr., etc.), to locate a fault. (76)

localité (*n.f.*), locality. (77)

locataire (pers.) (*n.m. ou f.*), tenant ; renter ; hirer : (78)
locataire trimestriel, quarterly tenant. (79)

locataire à bail (*m. ou f.*), leaseholder ; lessee. (80)

locateur, -trice (pers.) (*n.*), lessor. (81)

location (louage) (*n.f.*), hiring ; hire ; renting : (82)
location de chalands, barge-hire. (83)
location de wagons, truck-hire. (84)

location (prix du loyer) (*n.f.*), rent ; rental ; hire. (85)

location à bail (*f.*), leasing. (86)

lochet (*n.m.*), mining-shovel. (87)

lock-out (coalition de patrons) (*n.m.*), lock-out. (88)

locking-bar (Ch. de f.) (*n.f.*), locking-bar ; detector-bar. (89)

locomobile (*adj.*), portable : (90)
chaudière locomobile (*f.*), portable boiler. (91)

locomobile (*n.f.*). Same as locomotive routière.

locomoteur, -trice *ou* locomotif, -ive (*adj.*), loco-motive : (92)
force locomotrice (*f.*), locomotive power. (93)
machine locomotive (*f.*), locomotive engine. (94)

locomotion (*n.f.*), locomotion. (95)

locomotion à vapeur (*f.*), steam-locomotion. (96)

locomotion sur rails (*f.*), rail-locomotion. (97)

locomotive (*n.f.*), locomotive ; engine ; railway-engine. (98)

locomotive à air comprimé (*f.*), compressed-air locomotive. (99)

locomotive à benzine (*f.*). Same as locomotive à essence.

locomotive à bogie (f.), bogy-locomotive. (1)
locomotive à crémaillère (f.) rack locomotive ; rack-rail locomotive. (2)
locomotive à cylindres extérieurs (f.), outside-cylinder locomotive. (3)
locomotive à cylindres intérieurs (f.), inside-cylinder locomotive. (4)
locomotive à deux cylindres (f.), two-cylinder locomotive. (5)
locomotive à deux essieux couplés (f.), four-coupled locomotive. (6)
locomotive à essence ou locomotive à hydro-carbures volatils (f.), petrol-engine ; benzine-locomotive ; gasoline-locomotive. (7)
locomotive à marchandises (f.), goods-engine ; freight-locomotive. (8)
locomotive à pétrole (f.), oil-locomotive. (9)
locomotive à quatre cylindres et à grande vitesse (f.), high-speed four-cylinder locomotive. (10)
locomotive à — roues (f.), — -wheeled locomo-tive : (11)
locomotive à 8 roues (f.), 8-wheeled locomotive. (12)
locomotive à trois essieux couplés (f.), six-coupled locomotive. (13)
locomotive à un seul essieu moteur (f.), single-driver locomotive ; bicycle locomotive. (14)
locomotive à vapeur (f.), steam-locomotive. (15)
locomotive à vapeur sans foyer (f.). See loco-motive sans foyer.
locomotive à voie étroite (f.), narrow-gauge locomotive. (16)
locomotive à voie normale (f.), standard-gauge locomotive. (17)
locomotive à voyageurs (f.), passenger-engine. (18)
locomotive articulée (f.), articulated locomotive. (19)
locomotive atlantic (f.), Atlantic locomotive. (20)
locomotive chauffée au charbon (f.), coal-fired locomotive. (21)
locomotive chauffée au pétrole (f.), oil-fired locomotive. (22)
locomotive compound (f.), compound locomotive. (23)
locomotive compound à cylindres en tandem (f.), tandem compound locomotive. (24)
locomotive compound articulée système Mallet (f.), Mallet articulated compound locomotive. (25)
locomotive consolidation (f.), consolidation locomotive. (26)
locomotive de manœuvre ou locomotive de gare (f.), shunting engine ; yard engine ; switch locomotive. (27)
locomotive de mines ou locomotive minière (f.), mine-locomotive ; mining-locomotive. (28)
locomotive de secours (f.), relief engine ; engine standing by. (29)
locomotive decapod (f.), decapod locomotive. (30)
locomotive électrique (f.), electric locomotive ; electromotive. (31)
locomotive électrique à accumulateurs (f.), electric storage-battery locomotive. (32)
locomotive électrique à tambour (Mines) (f.), electric cable-reel locomotive. (33)
locomotive électrique à trolley (f.), electric trolley locomotive. (34)

locomotive express compound à 4 cylindres et tandem (f.), four-cylinder tandem com--pound express locomotive. (35)
locomotive mikado (f.), Mikado locomotive. (36)
locomotive mogul (f.), Mogul locomotive. (37)
locomotive pacific (f.), Pacific locomotive. (38)
locomotive-pilote [locomotives-pilotes pl.] (n.f.), pilot-engine. (39)
locomotive prairie (f.), Prairie locomotive. (40)
locomotive routière (f.), traction-engine ; road-engine ; road-locomotive ; locomobile. (41)
locomotive sans foyer ou locomotive à vapeur sans foyer ou locomotive sans feu ou locomo--tive à provision de vapeur ou locomotive à vapeur emmagasinée (f.), fireless loco--motive ; steam-storage locomotive ; thermal storage engine ; hot-water engine. (42)
locomotive-tender [locomotives-tenders pl.] (n.f.), tank-locomotive ; tank-engine. (43)
locomotive-tender à six roues accouplées com--pound à deux cylindres (f.), six-coupled compound two-cylinder tank-locomotive. (44)
locomotive-tender de — tonnes à vide et — tonnes en charge (f.), tank-engine, — tons empty, — tons charged. (45)
locomotive type baltic (f.), Baltic type locomotive. (46)
lœllingite (Minéral) (n.f.), loellingite ; löllingite. (47)
lœss (Géol.) (n.m.), loess ; löss. (48)
logarithme (n.m.), logarithm. (49)
logarithmique (adj.), logarithmic ; logarithmical : (50)
échelle logarithmique (f.), logarithmic scale. (51)
logement (d'une clavette d'arbre, ou autre pièce mécanique analogue) (n.m.), housing (of a shaft-key, or the like). (52 ou 53)
logements des barres (d'un cabestan) (m.pl.), poppet-holes, bar-holes, sockets, (of a capstan). (54)
loi (n.f.), law ; act : (55)
loi concernant les responsabilités des accidents dont les ouvriers sont victimes dans leur travail (f.), workmen's compensation act. (56)
loi d'Ohm (Élec.) (f.), Ohm's law. (57)
loi de Hooke (Méc.) (f.), Hooke's law. (58)
loi de l'albite (Cristall.) (f.), albite law. (59)
loi de l'offre et de la demande (f.), law of supply and demand. (60)
loi de répartition des erreurs (f.), law of errors ; law of frequency of errors. (61)
loi des moyennes (f.), law of averages. (62)
loi des proportions définies (Chim.) (f.), law of definite proportions. (63)
loi des proportions multiples (Chim.) (f.), law of multiple proportions. (64)
loi du péricline (Cristall.) (f.), pericline law. (65)
loi périodique (des éléments chimiques) (f.), periodic law (of the chemical elements) ; Mendeleeff's law. (66)
loi réglementant l'exploitation des mines (f.), mines regulation act. (67)
loi réglementant l'exploitation des mines de charbon (f.), coal-mines regulation act. (68)
loi réglementant l'exploitation des mines métalli--fères (f.), metalliferous mines regulation act. (69)
loi sur les carrières (f.), quarries act. (70)

loi sur les mines de charbon (f.), coal-mines act. (1)

lois de la physique (f.pl.), laws of physics ; physical laws. (2)

lois du frottement (f.pl.), laws of friction. (3)

lois mécaniques (f.pl.), mechanical laws. (4)

löllingite (Minéral) (n.f.), loellingite ; löllingite. (5)

long, longue (adj.), long : (6)
une longue rue, a long street. (7)

long de ou de long (locution adjective), long ; in length : (8)
une galerie long de—mètres, a gallery — metres long. (9)
un tunnel de — mètres de long, a tunnel — metres in length. (10)

long (de) ou long (en) (locution adverbiale), length--wise ; lengthways : (11)
scier de long, to saw lengthwise. (12)
coupe en long (f.), lengthwise section. (13)

long-courrier [long-courriers pl.] (n.m.), ocean-going ship. (14)

long-feu [longs-feux pl.] (Travail aux explosifs) (n.m.), hang-fire ; hung shot. (15)

long-shunt (n.m.) ou longue dérivation (f.), (Élec.), long shunt : (16)
excitation compound à longue dérivation (f.), long-shunt compound excitation. (17)

long-tom [long-toms pl.] (Préparation mécanique des minerais) (n.m.), long-tom. (18)

longer (v.t.), to run along ; to run alongside ; to skirt : (19)
chemin qui longe une rivière (m.), road which runs alongside a river ; path which skirts a river. (20)

longeron (n.m.), main beam ; longitudinal beam ; longitudinal girder ; longitudinal ; stringer. (21)

longeron (d'un châssis de locomotive) (n.m.), main side-frame ; main frame ; engine-frame ; frame : (22)
un châssis de locomotive se compose de deux longerons et plusieurs entretoises, an engine-frame is composed of two main side-frames and several cross-ties. (23)

longeron (d'un wagon de chemin de fer) (n.m.), side-sill ; sill : (24)
le châssis d'un wagon de chemin de fer se compose de deux longerons ou brancards et et de deux traverses de tête, the underframe of a railway-wagon is composed of two side-sills and two end-sills. (25)

longeron d'arrière (locomotive) (m.), back frame ; rear frame ; engine frame, back section. (26)

longeron d'avant (m.), front frame ; engine-frame, front section. (27)

longeron du bogie (m.), truck side-frame. (28)

longeron principal (locomotive) (m.), main frame. (29)

longévité (n.f.), life ; lifetime : (30)
longévité d'une mine, life, lifetime, of a mine. (31)

longitude (n.f.), longitude. (32)

longitude occidentale ou longitude ouest (f.), longitude west ; west longitude. (33)

longitude orientale ou longitude est (f.), longitude east ; east longitude. (34)

longitudinal, -e, -aux (adj.), longitudinal. (35)

longitudinalement (adv.), longitudinally. (36)

longrinage (boisage armé) (galerie de mine) (n.m.), herring-bone timbering ; double timbering ; reinforced timbering. (37)

longrine (n.f.), stringer ; longitudinal beam ; longitudinal girder ; longitudinal : (38)
sur les ponts métalliques, sur les fosses à piquer le feu des stations et des dépôts, la voie est ordinairement posée sur des pièces de bois appelées longrines placées dans le sens de la longueur de la ligne, on iron bridges, on engine-pits at stations and sheds, the road is ordinarily laid on pieces of wood called stringers or longitudinals placed in the direction of the length of the line. (39)

longue (adj. f.). See under long.

longueur (n.f.), length. (40)

longueur (de) ou longueur (en), in length ; long : (41)
un tunnel de — mètres de longueur, a tunnel — metres in length ; a tunnel — metres long. (42)

longueur (en) (dans le sens de la longueur), lengthways ; lengthwise : (43)
refendre une planche en longueur, to rip a board lengthways. (44)

longueur d'emmarchement (escalier) (f.), length of step (stairs). (45)

longueur d'onde (Phys.) (f.), wave-length. (46)

longueur de course du piston (f.), length of piston-stroke. (47)

longueur de portée une fois et demie le diamètre (palier) (f.), length of bearing equal to $1\frac{1}{2}$ diameters (plummer-block). (48)

longueur du banc (tour, etc.) (f.), length of bed (lathe, etc.) (49)

longueur entre pointes (tour) (f.), distance between centres. (50)

longueur focale (f.), focal length ; focal distance. (51)

longueur, largeur, et profondeur, length, breadth, and thickness : (52)
tout corps a trois dimensions : longueur, largeur, et profondeur, every body has three dimensions : length, breadth, and thickness. (53)

longueur sous tête ou longueur tête non-comprise (vis et boulons) (f.), length under head to point : (54)
les vis à tête ronde et les vis à tête cylindrique se mesurent longueur sous tête (f.pl.), round-headed screws and cheese-headed screws are measured length under head to point. (55)

longueur tête comprise (vis et boulons) (f.), length over all : (56)
les vis à tête fraisée et les vis à tête goutte de suif se mesurent tête comprise (f.pl.), countersunk screws and raised-head screws are measured length over all. (57)

lopin (Mines) (n.m.), pillar ; post ; stoop. (58)

lopin de terre (m.), plot, patch, piece, of ground. (59)

loquet (n.m.), latch. (60)

loquet à bouton ou loquet à bouton simple (m.), lift-latch. (61)

loquet à poucier (m.), thumb-latch. (62)

loquet de porte (m.), gate-latch. (63)

loqueteau (n.m.), catch (to fasten a window, or the like). (64)

lory [lories pl.] (petit véhicule qui sert à visiter la voie) (Ch. de f.) (n.m.), push-car. (65)

losange (Géom.) (n.m.), lozenge ; rhomb ; rhombus ; diamond. (66)

losangique (adj.), lozenged ; lozenge-shaped ; diamond-shaped. (67)

lot (n.m.), lot ; parcel : (68)

lot de minerai, lot, parcel, of ore. (1)
lot de terrain, lot, parcel, plot, piece, block, of land. (2)
louage (n.m.), letting ; hiring ; renting. (3)
loubette (n.f.), lewis ; lewisson. (4)
louchet de canal (m.), grafting-tool. (5)
louchet de tourbier (m.), peat-spade. (6)
louer (v.t.), to let ; to hire ; to rent : (7)
louer un ouvrier à la journée, to hire a man by the day. (8)
loueur, -euse (pers.) (n.), hirer. (9)
loup (agglomération de matière mal fondue qui se forme dans le minerai en fusion) (n.m.), horse ; old horse ; bear ; sow ; salamander ; shadrach ; freeze. (10)
loupe (Opt.) (n.f.), lens ; magnifying-glass ; glass ; loupe : (11)
le pouvoir amplifiant d'une loupe, the magnify-ing power of a lens. (12)
loupe (n.f.) ou loupe de puddlage (Métall.), ball ; puddle-ball ; loup ; loupe. (13)
loupe à lire (f.), reading-glass ; reading-lens. (14)
loupe à lire à fort grossissement (f.), strong reading-glass. (15)
loupe de mise au point (Photogr.) (f.), focussing-magnifier ; focussing-glass. (16)
lourd, -e (adj.), heavy ; weighty : (17)
une lourde pierre, a heavy stone. (18)
lourdement (adv.), heavily ; weightily. (19)
louve (n.f.) ou louve à pierres, lewis ; lewisson. (20)
louve à genouillère ou louve à pince (f.), self-adjusting stone-dogs ; block-clip ; lever-grip tongs. (21)
louve en trois pièces (f.), double lewis. (22)
louver (v.t.), to lewis : (23)
louver une pierre, to lewis a stone. (24)
lover un cordage, to coil a rope. (25)
löwéite (Minéral.) (n.f.), loeweite ; löweite. (26)
loyer (n.m.), rent ; rental. (27)
lubrifiant, -e (adj.), lubricating. (28)
lubrifiant (n.m.), lubricant. (29)
lubrification (n.f.), lubrication ; greasing. (30)
lubrifier (v.t.), to lubricate ; to grease : (31)
lubrifier les rouages d'une machine, to lubricate the wheels of a machine. (32)
luisant, -e (adj.), bright ; glittering. (33)
lumachelle (marbre) (n.f.), lumachelle. (34)
lumen [lumens pl.] (Photom.) (n.m.), lumen. (35)
lumenmètre (n.m.), lumenmeter ; globe pho-tometer. (36)
lumière (n.f.), light. (37)
lumière (d'un cylindre à vapeur) (n.f.), port, port-hole, (of a steam-cylinder). (38)
lumière (d'une pinnule) (Opt.) (n.f.), hole (of a sight). (39)
lumière (d'un coussinet) (n.f.), oil-hole (of a bearing). (40)
lumière (d'une pompe) (n.f.), spout-hole (of a pump). (41)
lumière (d'un rabot) (n.f.), throat, mouth, (of a plane). (42)
lumière artificielle (f.), artificial light. (43)
lumière d'admission ou lumière d'entrée (cylindre à vapeur) (f.), steam-port ; induction-port. (44)
lumière d'échappement ou lumière de décharge (cylindre à vapeur) (f.), exhaust-port ; eduction-port. (45)
lumière de l'arc voltaïque (f.), arc-light. (46)
lumière diffuse (f.), diffused light ; scattered light. (47)

lumière du gaz (f.), gaslight. (48)
lumière du jour (f.), daylight. (49)
lumière du soleil (f.), sunlight. (50)
lumière-éclair (Photogr.) (n.f.), flash-light. (51)
lumière électrique (f.), electric light. (52)
lumière réfléchie (f.), reflected light. (53)
lumière réfractée (f.), refracted light. (54)
lumière solaire réfléchie (f.), reflected solar light. (55)
lumière sûre (Photogr.) (f.), safe light. (56)
lumière transmise (f.), transmitted light. (57)
lumineux, -euse (adj.), luminous ; light (used as adj.) : (58)
l'intensité lumineuse d'un arc (f.), the luminous intensity of an arc. (59)
rayons lumineux (m.pl.), luminous rays ; light-rays. (60)
luminosité (n.f.), luminosity : (61)
luminosité des lampes, luminosity of lamps. (62)
lunaire (adj.), lunar : (63)
éclipse lunaire (f.), lunar eclipse. (64)
lune (n.f.), moon. (65)
lunette (n.f.) ou lunette d'approche (Opt.), telescope ; refracting telescope : (66)
le télescope diffère de la lunette par la substi-tution d'un miroir concave à la lentille objective, the reflecting telescope differs from the refracting telescope by the substi-tution of a concave mirror for the object-lens. (67)
lunette (d'une locomotive) (n.f.), cab-window (of a locomotive). (68)
lunette (raccord de tuyaux) (n.f.), pipe-con-nection. (69)
lunette (d'un tour) (n.f.), steady-rest, steady, centre-rest, back-rest, stay, collar-plate, cat-head, (of a lathe). (70)
lunette (n.f.) ou lunette à fileter ou lunette à tarauder, die ; solid die (for engineers' screw-stocks). (71)
lunette à réticule (f.), telescope with web diaphragm. (72)
lunette à révolution complète (f.), transit-telescope. (73)
lunette à stadia (f.), telescope with stadia-diaphragm ; telescope with stadia-reading diaphragm. (74)
lunette à suivre (tour) (f.), following steady-rest ; follower-rest ; follow-rest ; travelling stay. (75)
lunette à trois touches réglables (tour) (f.), three-jaw steady-rest ; three-paw steady. (76)
lunette anallatique (f.), anallatic telescope. (77)
lunette de regard (d'un fourneau, d'un cubilot) (f.), eyepiece (of a furnace, of a cupola). (78 ou 79)
lunette de repère (f.), auxiliary telescope ; finder. (80)
lunette fixe (tour) (f.), fixed steady-rest ; stationary stay. (81)
lunette ronde d'une seule pièce (Filetage) (f.), round die in one piece. (82)
lunette viseur (f.), sighting-telescope. (83)
lunettes (de mécanicien, de casseur de pierres, etc.) (n.f.pl.), goggles. (84)
lunnite (Minéral.) (n.f.), pseudomalachite ; phosphorochalcite ; phosphochalcite. (85)
lussatite (Minéral.) (n.f.), lussatite. (86)
lustre (n.m.), lustre ; gloss ; glossiness ; glaze. (87)
lustré, -e (adj.), lustrous ; glossy. (88)

lustrer (*v.t.*), to gloss ; to glaze. (1)
lut (*n.m.*), lute. (2)
lutation (*n.f.*), lutation ; luting. (3)
lutécine *ou* lutécite (Minéral) (*n.f.*), lutecin ; lutecite. (4)
lutécium (Chim.) (*n.m.*), lutecium ; lutetium. (5)
luter (*v.t.*), to lute : (6)
  luter une cornue avec de l'argile, to lute a retort with clay. (7)
lutétien, -enne (Géol.) (*adj.*), Lutetian. (8)
lutétien (*n.m.*), Lutetian. (9)
lutte (*n.f.*), fight ; battle : (10)
  lutte contre les feux souterrains, fight against underground fires. (11)

lutter (*v.i.*), to fight; to contend; to battle: (12)
  lutter contre une difficulté, to contend, to battle, against a difficulty. (13)
lux [lux *pl.*] (Photom.) (*n.f.*), lux = 0·929 foot-candles or candle-feet. (14)
lydite (Minéral) (*n.f.*), lydite ; Lydian stone ; touchstone. (15)
lyre (compensateur de dilatation) (Tuyauterie) (*n.f.*), expansion-bend. (16)
lyre (d'un tour à fileter) (*n.f.*), quadrant, quad--rant-plate, radial arm, radial plate, swing-frame, tangent plate, (of a screw-cutting lathe). (17)

M

macadam (*n.m.*), macadam. (18)
macadamisage (*n.m.*) *ou* macadamisation (*n.f.*), macadamizing ; macadamization. (19)
macadamiser (*v.t.*), to macadamize : (20)
  macadamiser une route, to macadamize a road. (21)
macaron en régule (*m.*), hole filled with babbitt. (22)
maccalube (Géol.) (*n.f.*), maccaluba ; mud-volcano (volcanic) ; mud-cone. (23)
mâchefer (*n.m.*), clinker ; furnace-clinker. (24)
machine (*n.f.*), machine. (25)
machine (moteur) (*n.f.*), engine ; motor. (26) See also moteur.
machine (locomotive) (*n.f.*), engine ; railway-engine ; locomotive. (27) See also loco--motive.
machine à affûter (*f.*), sharpening-machine ; sharpener ; grinding-machine ; grinder (of the tool, cutter, or drill-sharpening type). (28)
machine à affûter les forets *ou* machine à affûter les mèches (*f.*), drill-grinder ; drill-sharpener. (29)
machine à affûter les forets hélicoïdaux *ou* machine à affûter les mèches hélicoïdales (*f.*), twist-drill grinder. (30)
machine à affûter les fraises (*f.*), milling-cutter sharpening-machine ; cutter-grinder. (31)
machine à affûter les lames de couteaux (*f.*), knife-grinding machine. (32)
machine à affûter les lames de raboteuses (*f.*), plane-iron grinder. (33)
machine à affûter les lames de scies (*f.*), saw-sharpening machine ; saw-grinder. (34)
machine à affûter les outils (*f.*), tool-sharpener ; tool-grinder. (35)
machine à affûter les outils, travaillant à l'eau (*f.*), wet tool-grinder ; water tool-grinder. (36)
machine à affûter les outils, travaillant à sec (*f.*), dry tool-grinder. (37)
machine à air (*f.*), air-engine. (38)
machine à air chaud (*f.*), hot-air engine ; caloric engine. (39)
machine à air comprimé (*f.*), compressed-air engine. (40)
machine à alcool (*f.*), alcohol-engine ; spirit-engine. (41)
machine à aléser (*f.*), boring-machine. (42)
machine à aléser et à percer horizontale, à hauteur variable (*f.*), horizontal boring and drilling-machine, with sliding head and upright. (43)
machine à aléser les boîtes à noyaux (*f.*), core-box boring-machine. (44)
machine à aléser verticale (*f.*), vertical boring-machine. (45)
machine à assembler et à démotter (Fonderie) (*f.*), snap-moulding machine. (46)
machine à avoyer *ou* machine à avoyer les scies (*f.*), setting-machine ; saw-setting machine. (47)
machine à balancier (*f.*), beam-engine. (48)
machine à basse pression (*f.*), low-pressure engine. (49)
machine à border (*f.*), flanging-machine ; flanger. (50)
machine à bouveter (*f.*), matching-machine. (51)
machine à bras (*f.*), hand-power machine. (52)
machine à braser, à affûter, et à avoyer les lames des scies à ruban (*f.*), band-saw brazing, sharpening, and setting machine. (53)
machine à canneler portative (*f.*), keyway-cutting machine. (54)
machine à centrer (*f.*), centring-machine. (55)
machine à chanfreiner (*f.*), plate-edge planing-machine. (56)
machine à charger (pour fours métallurgiques) (*f.*), charging-machine ; charger. (57)
machine à charger les lingots et blooms (*f.*), ingot-and-bloom charger. (58)
machine à cingler (*f.*), shingler. (59)
machine à cintrer (*f.*), bending-machine ; bender. (60)
machine à cintrer les cercles de roues (*f.*), tire-bending machine. (61)
machine à cintrer les fers cornières (*f.*), angle-iron bending-machine ; angle-iron bender. (62)
machine à cintrer les rails, à rouleaux horizontaux (*f.*), rail-bending machine, with horizontal rollers ; rail-bender, with horizontal rolls. (63)
machine à cintrer les tôles (*f.*), plate-bending rollers ; plate-bending rolls. (64)
machine à cintrer les tuyaux (*f.*), pipe-bender. (65)
machine à cintrer pour cercles de roues, frettes, et fers à T, L, et H (*f.*), machine for bending T, L, and H iron, tyres, and circle plates. (66)

machine à cisailler (f.), shearing-machine; shears; shear. (1) See cisailles for many varieties.

machine à cisailler à molettes (f.), revolving-cutter machine (for cutting off tubes, etc.) (2)

machine à cisailler les barres (f.), bar-shearing machine; bar-shears. (3)

machine à commande électrique (f.), electri-cally driven machine; motor-driven machine. (4)

machine à compression (machine à mouler) (Fonderie)(f.), squeezer; press. (5)

machine à condensation (f.), condensing engine. (6)

machine à couper d'onglet (f.), mitre-cutting machine; mitre-machine. (7)

machine à couper les coulées (Fonderie) (f.), gate-cutter; git-cutter; sprue-cutter. (8)

machine à couper les rails (f.), rail-cutting machine. (9)

machine à courroie émerisée (f.), emery-belt polishing-machine. (10)

machine à cylindre oscillant (f.), oscillating engine. (11)

machine à cylindre renversé (f.), inverted-cylinder engine. (12)

machine à cylindre unique (f.), single-cylinder engine; one-cylinder engine. (13)

machine à défoncer (f.), recessing-machine. (14)

machine à dégauchir (f.), surface-planing machine; surface-planer; surfacer. (15)

machine à démouler (Fonderie) (f.), pattern-drawing machine; draft-machine. (16)

machine à démouler à levier (f.), lever draft-machine. (17)

machine à démouler à plaque rotative (f.), turn-over table draft-machine. (18)

machine à démouler à renversement (f.), roll-over draft-machine; rock-over drop-machine. (19)

machine à dessabler à table rotative (Fonderie) (f.), rotary-table sand-blast machine. (20)

machine à détente (f.), expansion-engine. (21)

machine à disque émerisé (f.), disc-grinder. (22)

machine à diviser (f.), dividing-engine; divid-ing-machine; graduating-engine. (23)

machine à diviser les cercles (f.), dividing-engine for circles. (24)

machine à diviser les lignes droites (f.), dividing-engine for straight lines. (25)

machine à donner de la voie aux scies (f.), saw-setting machine. (26)

machine à double effet (f.), double-acting engine. (27)

machine à double expansion (f.), double-expan-sion engine. (28)

machine à draguer (f.), dredging-machine. (29)

machine à eau chaude (f.), hot-water engine; thermal storage engine; steam-storage locomotive; fireless locomotive. (30)

machine à ébarber (f.), grinding-machine (of the fettling or trimming type); fettling-machine; trimming-machine. (31)

machine à encocher (f.), notching-machine. (32)

machine à essayer (f.), testing-machine. (33)

machine à essayer au choc (f.), drop-test machine. (34)

machine à étamper, forger, cisailler, et poinçonner (f.), combined stamping, forging, shearing, and punching-machine. (35)

machine à expansion (f.), expansion-engine. (36)

machine à faire des entretoises (f.), stay-bolt turning and threading-lathe. (37)

machine à faire les joints (f.), jointing-machine; jointer. (38)

machine à faire les moulures (bois) (f.), mould-ing-machine. (39)

machine à faire les rainures de graissage (f.), oil-channeling machine. (40)

machine à faire les rainures et languettes (f.), tonguing-and-grooving machine. (41)

machine à forger les écrous carrés et à six pans (f.), nut-making machine, for making square and hexagon nuts. (42)

machine à fraiser (f.), milling-machine; miller. (43). See also fraiseuse.

machine à fraiser horizontale avec appareil à fraiser verticalement (f.), horizontal milling-machine with vertical milling-attachment. (44)

machine à fraiser horizontale, avec mouvement longitudinal automatique de la table (f.), horizontal milling-machine, with automatic longitudinal movement of the table. (45)

machine à fraiser les rainures de cales (f.), key-way-cutting machine, key-seating machine, key-seat cutter, key-seater, key-grooving machine, (of the milling type). (46)

machine à fraiser universelle (f.), universal milling-machine. (47)

machine à fraiser verticale avec mouvement de reproduction pour fraiser suivant gabarit (f.), vertical milling-machine with profiling attachment. (48)

machine à frapper les boulons, rivets, crampons, etc. (f.), bolt-forging machine, bolt-heading machine, for making bolts, rivets, spikes, etc. (49)

machine à gaz (f.), gas-engine; gas-motor. (50) See moteur à gaz for varieties.

machine à haute pression (f.), high-pressure engine. (51)

machine à lapider (f.), horizontal emery grind-ing-machine. (52)

machine à lime à affûter et donner la voie aux scies à ruban (f.), band-saw filing and setting-machine. (53)

machine à maîtresse-tige (Mines) (f.), bull-engine. (54)

machine à meule(s) d'émeri (f.), emery grind-ing-machine; emery-grinder; emery-machine. (55 ou 56)

machine à meule d'émeri travaillant sur face (f.), emery face-wheel grinding-machine. (57)

machine à meule d'émeri travaillant sur péri-phérie (f.), emery edge-wheel grinding-machine. (58)

machine à meuler (f.), grinding-machine; grinder (as a general term, or specifically, a grinding-machine of the roughing type). (59 ou 60) Cf. machine à ébarber, machine à rectifier, and machine à affûter.

machine à meuler double (f.), double-wheel grinding-machine; two-wheel grinder. (61)

machine à meuler les surfaces planes extérieures (f.), surface-grinder; surface-grinding machine. (62)

machine à meuler simple (f.), single-wheel grinding-machine; one-wheel grinder. (63)

machine à molettes (Mines) (f.), head-gear; pit-head gear. (64)

machine à mortaiser (bois) (f.), mortising-machine. (65)

machine à mortaiser (métal) (*f.*), slotting-machine; slotter. (1)

machine à mortaiser à engrenages, avec retour rapide et avec tous les mouvements auto--matiques (*f.*), geared slotting-machine, completely automatic and with quick return motion. (2)

machine à mortaiser et percer (bois) (*f.*), mortis--ing and boring-machine. (3)

machine à mortaiser les rainures de cales (*f.*), keyway-cutting machine, key-seating machine, key-seat cutter, key-seater, key-grooving machine, (of the slotting type). (4)

machine à mouler (Fonderie) (*f.*), moulding-machine. (5)

machine à mouler à plaque rotative (*f.*), turn-over table moulding-machine. (6)

machine à mouler d'établi (*f.*), bench moulding-machine. (7)

machine à moulurer (bois) (*f.*), moulding-machine. (8)

machine à moulurer dite toupie (*f.*), spindle moulding-machine. (9)

machine à multiple expansion (*f.*), multiple-expansion engine. (10)

machine à noyauter (Fonderie) (*f.*), core-machine; core-making machine; coring-machine. (11)

machine à percer (*f.*), drilling-machine; drill; driller. (12)

machine à percer à commande électrique directe (*f.*), direct motor-driven drilling-machine. (13)

machine à percer à — forets (*f.*), — -spindle drilling-machine. (14)

machine à percer à grande vitesse à mono--poulie (*f.*), single-pulley high-speed drilling-machine. (15)

machine à percer à grande vitesse avec com--mande par boîte de vitesses à monopoulie (*f.*), all-gear single-pulley high-speed drilling-machine; all-geared single-pulley high-speed drill; high-speed drilling-machine driven by a constant-speed pulley through a change-speed gear-box. (16)

machine à percer à pédale (*f.*), treadle drilling-machine. (17)

machine à percer, aléser, et fraiser horizontale (*f.*), horizontal drilling, boring, and milling-machine. (18)

machine à percer, avec pression du porte-foret à la main par levier (*f.*), lever-feed drilling-machine. (19)

machine à percer d'établi (*f.*), bench drilling-machine. (20)

machine à percer et à tarauder sur place les trous d'entretoises des foyers de locomotives (*f.*), portable drilling and tapping-machine for stay-bolt holes in locomotive fire-boxes. (21)

machine à percer horizontale, pour cornières et fers spéciaux (*f.*), horizontal drilling-machine, for angles and special sections. (22)

machine à percer marchant à bras ou au moteur (*f.*), drilling-machine for hand or power; hand or power-drill. (23)

machine à percer montée sur bâti *ou* machine à percer sur bâti *ou* machine à percer à bâti (*f.*), drilling-machine on upright frame. (24)

machine à percer montée sur colonne *ou* machine à percer sur colonne *ou* machine à

percer à colonne (*f.*), pillar drilling-machine; upright drilling-machine of the pillar type; drill-press. (25)

machine à percer montée sur colonne, avec collier tournant à plateau et étau (*f.*), pillar drilling-machine, with table and vice revolving on pillar. (26)

machine à percer montée sur colonne, avec plateau-étau tournant coulissant par volant, pignon et crémaillère; monte et baisse par crémaillère et pignon manœuvrés par vis sans fin et roue à vis sans fin (*f.*), pillar drilling-machine, with revolving vice-plate on arm sliding by hand-wheel, rack and pinion; raising and lowering by rack and pinion worked by worm and worm-wheel. (27)

machine à percer multiple *ou* machine à percer à broches multiples (*f.*), multiple drilling-machine; multiple-spindle drill; multi--spindle drill; gang-drill. (28)

machine à percer portative (*f.*), portable drilling-machine. (29)

machine à percer pouvant se fixer à un mur ou à un poteau (*f.*), drilling-machine for wall or post. (30)

machine à percer radiale (*f.*), radial drilling-machine. (31)

machine à percer radiale à potence (*f.*), swing-jib radial drill. (32)

machine à percer radiale de précision, per--mettant l'emploi des mèches en acier rapide et acier ordinaire (*f.*), precision radial drill, for use with high-speed steel and ordinary steel drills. (33)

machine à percer radiale murale (*f.*), radial wall drilling-machine. (34)

machine à percer radiale universelle (*f.*), uni--versal radial drill. (35)

machine à percer sensitive à plateaux de friction (*f.*), sensitive friction-drill. (36)

machine à percer sensitive de précision (*f.*), sensitive precision drill. (37)

machine à percer sensitive multiple (*f.*), sensi--tive gang-drill. (38)

machine à pétrole (*f.*), petroleum-engine; oil-engine. (39)

machine à picots (Mines) (*f.*), wedge-driving machine. (40)

machine à planer (*f.*), planishing-machine; planisher. (41)

machine à plier (*f.*), folding-machine. (42)

machine à plier et couder les tôles (*f.*), plate-folding and bending machine. (43)

machine à plusieurs cylindres (*f.*), multicylinder engine. (44)

machine à poinçonner et à cisailler (*f.*), punching and shearing-machine. (45) See poin--çonneuse-cisaille for many varieties.

machine à polir (*f.*), polishing-machine. (46)

machine à poncer (*f.*), sandpapering-machine. (47)

machine à poncer à courroie (*f.*), belt sand--papering-machine. (48)

machine à poncer à deux plateaux (*f.*), double-disc sandpapering-machine. (49)

machine à poncer à plateau (*f.*), disc sand--papering-machine. (50)

machine à quadruple expansion (*f.*), quadruple-expansion engine. (51)

machine à quintuple expansion (*f.*), quintuple-expansion engine. (52)

machine à raboter (f.), planing-machine ; planer. (1)

machine à raboter à commande par pignon et crémaillère (f.), rack-and-pinion planer. (2)

machine à raboter à fosse (f.), pit-planer. (3)

machine à raboter à retour rapide, à un ou deux porte-outils, et avec mouvements auto--matiques dans tous les sens (f.), planing-machine, with one or two heads ; self-acting in vertical, horizontal, and angular cuts ; quick return motion. (4 ou 5)

machine à raboter latérale, à course variable et à retour rapide (f.), side-planing machine, with variable feed motion and quick return. (6)

machine à raboter marchant à bras (f.), hand planing-machine ; hand-power planer. (7)

machine à raboter ouverte sur le côté (f.), open-side planing-machine. (8)

machine à raboter tirant les bois d'épaisseur (f.), planing and thicknessing machine. (9)

machine à raboter travaillant dans les deux sens de marche (f.), double-cutting planing-machine. (10)

machine à rainer (bois, etc.) (f.), grooving-machine. (11)

machine à rainer (métal) (f.), slot-drilling machine. (12)

machine à rainurer les poulies (f.), pulley key-seating machine. (13)

machine à rectifier (f.), grinding-machine ; grinder (of the truing type). (14)

machine à rectifier à main (f.), hand-grinder. (15)

machine à rectifier dite universelle (f.), universal grinding-machine. (16)

machine à rectifier, en l'air, les surfaces cylindri--ques (f.), vertical cylinder-grinding machine. (17)

machine à rectifier les bagues (f.), ring-grinder. (18)

machine à rectifier les cames (f.), cam-grinder. (19)

machine à rectifier les coulisses de changement de marche (f.), link-grinder. (20)

machine à rectifier les surfaces cylindriques extérieures (f.), plain grinder. (21)

machine à rectifier les surfaces extérieures (f.), external grinder. (22)

machine à rectifier les surfaces intérieures (f.), internal grinder. (23)

machine à rectifier les surfaces planes extérieures (f.), surface-grinder ; surface-grinding machine. (24)

machine à rectifier les surfaces profilées (f.), profile-grinder. (25)

machine à redresser (f.), straightening-machine ; straightener. (26)

machine à refouler et à souder à table courbe (ou à table droite) (f.), shrinking and weld--ing-machine with dished table (or with straight table). (27 ou 28)

machine à refouler et à souder les cercles de roues (f.), tire-welding, upsetting, and contracting-machine. (29)

machine à refouler la denture des scies (f.), saw-setting machine. (30)

machine à refouler les cercles de roues (f.), tire-upsetter ; upsetter. (31)

machine à refouler, souder, couder, et contre--couder (f.), combined bending, shrinking, and welding-machine. (32)

machine à remouler (Fonderie) (f.), setting-up machine. (33)

machine à river ou machine à riveter (f.), rivet--ing-machine ; riveter ; rivetter. (34) See also riveuse.

machine à river hydraulique fixe (f.), stationary hydraulic rivetter. (35)

machine à river hydraulique portative pour cercles de fondation et couronnes des portes de foyer de locomotives ; type à action directe (f.), portable hydraulic rivetter for locomotive fire-door and foundation rings ; direct-acting class. (36)

machine à roder (f.), lapping-machine. (37)

machine à rouler à trois cylindres pour fer--blantiers (f.), tinmen's three-roll bending-machine. (38)

machine à rouler et à cintrer les tôles (f.), plate-flattening and bending machine. (39)

machine à saigner (f.), cutting-off machine. (40)

machine à scier (f.), sawing-machine ; sawmill ; saw-bench. (41) See scierie for varieties.

machine à scier alternative au moteur (f.), power hack-saw. (42)

machine à scier et à percer les rails (f.), rail-sawing and drilling machine. (43)

machine à scier les métaux (f.), metal-sawing machine. (44)

machine à scier les métaux à froid (f.), cold sawing-machine. (45)

machine à sculpter (f.), carving-machine. (46)

machine à simple effet (f.), single-acting engine. (47)

machine à simple expansion (f.), single-expan--sion engine. (48)

machine à souder électrique (f.), electric welding-machine. (49)

machine à tailler les engrenages (f.), gear-cutting machine ; gear-cutter. (50)

machine à tailler les engrenages droits (f.), spur-gear cutting-machine. (51)

machine à tailler les fraises (f.), cutter-making machine. (52)

machine à tailler les fraises de forme suivant gabarit (f.), cutter-making machine for making formed milling-cutters according to copy ; pantographic machine. (53)

machine à tailler les pignons (f.), pinion-cutting machine. (54)

machine à tarauder (f.), tapping-machine ; screwing-machine ; screw-machine ; screw--ing and tapping-machine. (55)

machine à tarauder à — broches pour le taraudage de tiges et d'écrous (f.), — -spindle bolt-screwing and nut-tapping machine. (56)

machine à tarauder et à tronçonner les tubes et à les ébarber intérieurement (f.), tube-screwing and cutting-off machine, with interior burr-trimming arrangement. (57)

machine à tarauder les tiges et les tubes (f.), screwing-machine for tubes or bolts. (58)

machine à tarauder les tubes (f.), tube-screwing machine ; pipe-threader. (59)

machine à tenons (f.), tenoning-machine. (60)

machine à trancher et dresser les bois en bout (f.), trimmer ; wood-trimmer ; wood-trimming machine ; paring-machine. (61)

machine à trancher et dresser les bois en bout, fonctionnant à bras ou machine à trancher à levier (f.), draw-stroke trimmer. (62)

machine à trancher les bois en feuilles de placage (f.), veneer-cutter. (63)

machin? à triple expansion (f.), triple-expansion engine. (1)

machine à trois cylindres (f.), three-cylinder engine; triple-cylinder engine; triplex engine. (2)

machine à tronçonner (f.), cutting-off machine. (3)

machine à tunnels (f.), tunnelling-machine. (4)

machine à un seul cylindre (f.), one-cylinder engine; single-cylinder engine. (5)

machine à vapeur (f.), steam-engine; steam-motor. (6)

machine à vapeurs combinées (f.), binary heat engine; binary engine. (7)

machine à volant (f.), fly-wheel engine. (8)

machine aspirante (f.), suction-pump; sucking pump; aspiring pump; aspirating pump. (9)

machine atmosphérique (f.), atmospheric engine. (10)

machine au jet de sable (f.), sand-blast; sand-jet. (11)

machine automatique à tailler les engrenages à denture droite (f.), automatic spur-gear cutting machine. (12)

machine auxiliaire (f.), auxiliary engine; assistant engine. (13)

machine bicylindrique (f.), two-cylinder engine; double-cylinder engine; double engine; duplex-cylinder engine; duplex engine. (14)

machine combinée à dégauchir, raboter, moulurer, et mortaiser (bois) (f.), combined surfacing, planing, moulding, and slot-mortising machine. (15)

machine compound (f.), compound engine. (16)

machine compound à condensation (f.), compound condensing engine. (17)

machine Corliss (f.), Corliss engine. (18)

machine d'aérage (f.), ventilating-machine; air-machine. (19)

machine d'alimentation (f.), donkey-pump; boiler-feed pump. (20)

machine d'Atwood (Phys.) (f.), Atwood's machine. (21)

machine d'épuisement ou machine d'exhaure (Mines) (f.), pumping-engine; draining-engine. (22)

machine d'épuisement compound à double effet (f.), double-acting compound pumping-engine. (23)

machine d'épuisement souterraine à triple expansion (f.), triple-expansion under-ground pumping-engine. (24)

machine d'essai (f.), testing-machine. (25)

machine d'extraction (Mines) (f.), winding-engine; winder; hoisting-engine; hoist; drawing-engine; draught-engine; draft-engine. (26)

machine d'extraction à vapeur (f.), steam-winding-engine; steam-winder; steam hoisting-engine; steam-hoist. (27)

machine d'extraction compound bicylindrique (f.), two-cylinder, duplex, double-cylinder, compound winding-engine. (28)

machine d'extraction compound jumelle-tandem à quatre cylindres [machines d'extraction compound jumelles-tandems à quatre cylindres pl.] (f.), four-cylinder twin-tandem compound winding-engine. (29)

machine d'extraction électrique (f.), electric winding-engine; electric winder; electric hoisting-engine; electric hoist. (30)

machine de banlieue (Ch. de f.) (f.), suburban engine. (31)

machine de carrière (f.), quarrying-machine. (32)

machine de Cornouailles (f.), Cornish engine; Cornish pumping-engine. (33)

machine de gare ou machine de manœuvre (Ch. de f.) (f.), shunting engine; yard locomotive; switch locomotive. (34)

machine de Holtz (Élec.) (f.), Holtz machine. (35)

machine de marine (f.), marine engine. (36)

machine de renfort (Ch. de f.) (f.), bank-engine; banking engine; banker; pusher. (37)

machine de secours (Ch. de f.) (f.), helper; helper engine. (38)

machine demi-fixe (f.), semiportable engine. (39)

machine démouleuse (Fonderie) (f.), stripping-machine. (40)

machine dynamo-électrique (f.), dynamo-electric machine. (41)

machine électrique (f.), electric machine; electrical machine. (42)

machine électromotrice (f.), electromotor. (43)

machine exécutant du travail marchand (f.). Same as machine marchant chargée.

machine fixe (f.), stationary engine. (44)

machine frigorifique (f.), freezing-machine. (45)

machine génératrice d'électricité (f.), electric generating machine. (46)

machine haut-le-pied (Ch. de f.) (f.), light engine (i.e., locomotive running without wagons or carriages attached) (47)

machine horizontale (f.), horizontal engine. (48)

machine hydraulique (f.), hydraulic engine; water-motor; water-engine. (49)

machine locomobile (f.), portable engine. (50)

machine locomotive (f.), locomotive engine. (51)

machine magnéto-électrique (f.), magneto-electric machine. (52)

machine marchant à vide ou machine marchant à blanc (opp à machine marchant chargée ou machine exécutant du travail marchand) (f.), machine running light; machine running on no load; empty machine. (53)

machine marchant chargée (f.), machine running under load; loaded machine. (54)

machine mi-fixe (f.), semiportable engine. (55)

machine monocylindrique (f.), single-cylinder engine; one-cylinder engine. (56)

machine-outil [machines-outils pl.] (n.f.), machine-tool: (57)

machines-outils à travailler les métaux, à travailler le bois, machine-tools for metal-working, for wood-working. (58 ou 59)

machines-outils de précision, precision machine-tools. (60)

machine-pilon [machines-pilons pl.] (vapeur) (n.f.), overhead engine. (61)

machine-pilote [machines-pilotes pl.] (Ch. de f.) (n.f.), pilot-engine. (62)

machine pneumatique (Phys.) (f.), air-pump; pneumatic pump; aspiring pump. (63)

machine polycylindrique (f.), multicylinder engine. (64)

machine portative à rainer l'intérieur des moyeux (f.), portable keyway cutter, for cutting keyways in the bores of hubs. (65)

machine pour la montée des rampes (Ch. de f.) (f.), bank-engine; banking engine; banker; pusher. (66)

machine rotative (f.), rotary engine. (1)

machine routière (f.), traction-engine ; road-engine. (2)

machine sans condensation (f.), non-condensing engine. (3)

machine sans détente (f.), non-expansion engine. (4)

machine soufflante (f.), blowing-engine ; blower ; piston-blower ; blast-engine. (5)

machine-tender [machines-tenders pl.] (Ch. de f.), (n.f.), tank-engine ; tank-locomotive. (6)

machine tri-compound (f.), triple-expansion engine. (7)

machine universelle à rectifier les surfaces cylindriques et coniques (f.), universal grinding-machine for rectifying cylindrical and conical surfaces. (8)

machine verticale (f.), vertical engine ; upright engine. (9)

machine Woolf (f.), Woolf engine. (10)

machinerie (ensemble de machines) (n.f.), machinery : (11)
la machinerie d'un moulin, the machinery of a mill. (12)

machinerie (endroit où sont les machines) (n.f.), engine-room ; engine-house. (13)

machines conjuguées (f.pl.), twin engines. (14)

machines couplées (f.pl.), coupled engines. (15)

machines de mines (f.pl.), mining-machinery. (16)

machines économisant la main-d'œuvre (f.pl.), labour-saving machinery. (17)

machines jumelles (f.pl.), twin engines. (18)

machines pour travaux miniers (f.pl.), machinery for mining purposes. (19)

machiniste (pers.) (n.m.), machinist. (20)

machiniste (mécanicien) (n.m.), engineman. (21)

machiniste de grue (m.), craneman. (22)

mâchoire (n.f.), jaw. (23)

mâchoire à tendre (Télégr., etc.) (f.), draw-vice ; wire-straining vice. (24)

mâchoire à tordre (Télégr., etc.) (f.), twisting-vice. (25)

mâchoires (d'un étau) (n.f.pl.), jaws, cheeks, chaps, (of a vice). (26)

mâchoires (d'un étau de menuisier) (n.f.pl.), chops (of a joiners' vice). (27)

mâchoires (d'une tenaille) (n.f.pl.), jaws (of a pincers). (28)

mâchoires (d'une clef) (n.f.pl.), jaws (of a spanner). (29)

mâchoires (d'une poulie à gorge) (n.f.pl.), flanges (of a grooved pulley). (30)

mâchoires (d'une lampe à arc) (Élec.) (n.f.pl.), clutch (of an arc-lamp). (31)

mâchoires à ressort (f.pl.), saw-clamp. (32)

mâchoires de serrage (f.pl.), gripping-jaws. (33)

macle (Minéral.) (n.f.), macle ; chiastolite. (34)

macle (Cristall.) (n.f.), twin ; macle ; hemitrope. (35)

macle à charnière multiple (f.), lattice twin. (36)

macle de Baveno (f.), Baveno twin. (37)

macle de Carlsbad (f.), Carlsbad twin. (38)

macle de deux cubes (f.), two-cube twin. (39)

macle de l'albite (f.), albite twin. (40)

macle de la croix de fer (f.), iron cross twin. (41)

macle de la sperkise ou macle de la speerkies (f.), spear-shaped twin. (42)

macle de Manebach ou macle de Four-la-Brouque (f.), Manebach twin. (43)

macle des spinelles (f.), spinel twin. (44)

macle du péricline (f.), pericline twin. (45)

macle en chevron (f.), herring-bone twin. (46)

macle en cœur (f.), heart-shaped twin. (47)

macle en crête de coq (f.), coxcomb twin. (48)

macle en croix (f.), cruciform twin ; cross-shaped twin. (49)

macle en croix de Saint-André (f.), cruciform twin, crossing at 60°. (50)

macle en croix grecque (f.), cruciform twin, crossing at 90°. (51)

macle en étoile (f.), stellate twin. (52)

macle en fer de lance (f.), arrow-head twin ; swallow-tail twin. (53)

macle en genou (f.), knee-shaped twin ; genic-ulating twin ; elbow-shaped twin. (54)

macle en gouttière (f.), grooved twin. (55)

macle en papillon (f.), butterfly twin. (56)

macle en visière (f.), visor twin. (57)

macle par accolement ou macle par juxtaposition (f.), contact twin ; juxtaposition twin. (58)

macle par pénétration ou macle par entrecroise-ment (f.), penetration twin. (59)

maclé, -e (Cristall.) (adj.), twinned ; twin ; macled ; hemitrope ; hemitropic. (60)

macler (Cristall.) (v.t.), to twin. (61)

macler (se) (Cristall.) (v.r.), to twin. (62)

maclifère (adj.), macled : (63)
schiste maclifère (m.), macled shale. (64)

maçon (pers.) (n.m.), mason ; stone-mason ; stonemason ; bricklayer. (65)

maçonnage (n.m.), masonry ; masonry-work ; stonework ; brickwork. (66)

maçonnage des noyaux (Moulage en terre) (m.), bricking up cores (Loam-moulding). (67)

maçonner (v.t.), to mason ; to mason up ; to brick ; to brick up : (68)
maçonner les parois d'un réservoir, to mason up the walls of a reservoir. (69)
maçonner un noyau (Fonderie), to brick up a core. (70)

maçonnerie (n.f.), masonry ; masonry-work ; stonework. (71)

maçonnerie (n.f.) ou maçonnerie de brique, brickwork. (72)

maçonnerie de béton (f.), concrete-masonry. (73)

maçonnerie en liaison (f.), bonded masonry. (74)

macque (Métall.) (n.f.), alligator squeezer ; crocodile squeezer ; saurian squeezer ; squeezer. (75)

macquer (Métall.) (v.t.), to squeeze : (76)
macquer le fer puddlé, to squeeze puddled iron. (77)

macrodiagonal, -e, -aux (Cristall.) (adj.), macro-diagonal. (78)

macrodiagonale (n.f.), macrodiagonal. (79)

macrodôme (Cristall.) (n.m.), macrodome. (80)

macropinacoïde (Cristall.) (n.m.), macropin-acoid. (81)

macroprisme (Cristall.) (n.m.), macroprism. (82)

macropyramide (Cristall.) (n.f.), macropyramid. (83)

macrosisme (n.m.), macroseism. (84)

macrosismique (adj.), macroseismic ; macro-seismical. (85)

madrier (n.m.), plank ; deal. (86)

magasin (n.m.), store ; storehouse ; storeroom ; warehouse ; stock : (87)
pièces de rechange toujours disponibles en magasin (f.pl.), spare parts always in stock. (88)

magasin (châssis-magasin) (Photogr.) (*n.m.*), changing-box ; plate-changing box ; plate-magazine ; magazine. (1)

magasin à poudre (*m.*), powder-magazine ; powder-house. (2)

magasinage (*n.m.*), storage ; storing ; ware-housing. (3)

magasiner (*v.t.*), to store ; to warehouse. (4)

magasinier (pers.) (*n.m.*), storekeeper ; ware-houseman. (5) More specifically storekeeper is rendered in French by the expression magasinier d'approvisionnements. (6)

magistral (Métall.) (*n.m.*), magistral. (7)

magma (Géol.) (*n.m.*), magma. (8)

magma de consolidation (Géol.) (*m.*), paste. (9)

magma éruptif (*m.*), eruptive magma. (10)

magma fondu (*m.*), molten magma. (11)

magmatique (Géol.) (*adj.*), magmatic. (12)

magnéferrite *ou* magnésioferrite (Minéral) (*n.f.*), magnesioferrite ; magnoferrite ; magnesia ferrite. (13)

magnésie (Chim.) (*n.f.*), magnesia. (14)

magnésien, -enne (*adj.*), magnesian : (15)
calcaire magnésien (*m.*), magnesian limestone. (16)

magnésique (*adj.*), magnesic. (17)

magnésite (silicate) (Minéral) (*n.f.*), meerschaum ; sepiolite. (18)

magnésite (giobertite) (Minéral) (*n.f.*), magnesite ; giobertite. (19)

magnésium (Chim.) (*n.m.*), magnesium. (20)

magnétipolaire (*adj.*), magnetipolar. (21)

magnétique (*adj.*), magnetic : (22)
force magnétique (*f.*), magnetic force. (23)

magnétiquement (*adv.*), magnetically. (24)

magnétisable (*adj.*), magnetizable : (25)
minerai magnétisable (*m.*), magnetizable ore. (26)

magnétisme (tout ce qui regarde les propriétés de l'aimant) (*n.m.*), magnetism. (27)

magnétisme (partie de la physique dans laquelle on étudie les propriétés des aimants) (*n.m.*), magnetism ; magnetics. (28)

magnétisme résiduel *ou* magnétisme rémanent (*m.*), residual magnetism ; remanent magnetism ; remanence. (29)

magnétite (Minéral) (*n.f.*), magnetite ; magnetic iron ore ; magnetic iron ; black oxide of iron ; magnetic oxide of iron ; ferrosoferric oxide. (30)

magnéto (*n.f.*), magneto. (31)

magnéto d'allumage (*f.*), ignition-magneto. (32)

magnéto-électricité (*n.f.*), magneto-electricity. (33)

magnéto-électrique (*adj.*), magneto-electric ; magneto-electrical : (34)
machine magnéto-électrique (*f.*), magneto-electric machine. (35)

magnétographe (*n.m.*), magnetograph. (36)

magnétoïde (*adj.*), magnetoid. (37)

magnétomètre (*n.m.*), magnetometer. (38)

magnétomoteur, -trice (*adj.*), magnetomotive : (39)
force magnétomotrice (*f.*), magnetomotive force. (40)

magnifique (*adj.*), magnificent ; splendid : (41)
temps magnifique (*m.*), splendid weather. (42)

magnoferrite (Minéral) (*n.f.*). Same as magné-ferrite.

maigre (*adj.*), lean ; poor ; meagre : (43)
argile maigre (*f.*), lean clay. (44)

béton maigre (*m.*), poor concrete. (45)

une roche telle que la craie est dite maigre au toucher, a rock like chalk is said to be meagre to the touch. (46)

maillage (Mines) (*n.m.*), airway. (47)

maille (espace vide dont les fils forment le contour) (*n.f.*), mesh : (48)
pierres qui passent à travers les larges mailles d'une claie (*f.pl.*), stones which pass through the coarse meshes of a screen. (49)

maille (d'un câble-chaîne) (*n.f.*), link (of a cable-chain). (50)

maille à étai *ou* maille étançonnée (*f.*), stud-link. (51)

maillechort (*n.m.*), German silver ; nickel silver ; maillechort. (52)

maillet (*n.m.*), mallet. (53)

maillet de calfat (*m.*), caulking-mallet. (54)

maillet de charpentier (*m.*), carpenters' mallet. (55)

maillet de mouleur (*m.*), casters' mallet. (56)

maillet en bois (*m.*), wooden mallet. (57)

mailloche (*n.f.*), mallet ; maul. (58)

maillon (*n.m.*), link : (59)
les maillons d'une chaîne, d'un câble-chaîne, the links of a chain, of a cable-chain. (60 *ou* 61)

maillon à crémaillère (*m.*), cranked link. (62)

maillon d'attache *ou* maillon de jonction (*m.*), shackle ; clevis ; clevice. (63)

main (*n.f.*), hand. (64)

main (à) (*used as adj.*) : (65)
scie à main (*f.*), hand-saw. (66)

main (à la), by hand ; hand (*used as adj.*) ; manual : (67)
travail à la main (*m.*), hand-work ; manual labour. (68)

main courante [mains courantes *pl.*] *ou* main coulante [mains coulantes *pl.*] (garde-fou ; lisse) (*f.*), hand-rail ; hand-railing : (69)
main courante d'un escalier, hand-rail, hand-railing, of a stairs. (70)

main courante (de la chaudière d'une locomotive) (*f.*), hand-rail, side-rail, (of the boiler of a locomotive). (71)

main courante *ou* main de guidage (de cage de mine) (*f.*), guide-shoe. (72)

main courante de la boîte à fumée (locomotive) (*f.*), arch hand-rail. (73)

main d'arrêt (valet d'arrêt d'une plaque tour-nante) (*f.*), latch (of a turntable). (74)

main-d'œuvre [mains-d'œuvre *pl.*] (*n.f.*), labour ; hand-labour ; manual labour : (75)
le prix de main-d'œuvre et celui des matières premières, the price of labour and that of raw materials. (76)

main-d'œuvre (façon) (*n.f.*), workmanship : (77)
le personnel dont nous disposons nous permet d'assurer une main-d'œuvre irréprochable, the staff at our command enables us to guarantee irreproachable workmanship. (78)

main-d'œuvre à la surface (Mines) (*f.*), surface-labour. (79)

main-d'œuvre blanche (*f.*), white labour. (80)

main-d'œuvre chinoise (*f.*), Chinese labour. (81)

main-d'œuvre indigène (*f.*), native labour. (82)

main-d'œuvre noire (*f.*), black labour. (83)

maintenage (Mines) (*n.m.*), overhand stope ; overhead stope ; back-stope. (84)

maintenir (*v.t.*), to maintain ; to keep : (85)
maintenir la direction, to keep the direction. (86)

maintenir le boisage rapproché du front de taille, to carry the timbering close to the face. (1)

maintenir le niveau bas (Fonçage), to keep the water down by pumping. (2)

maintenir un rendement minimum de — tonnes par jour, to maintain an output of at least — tons per day. (3)

maintien (*n.m.*), maintenance ; maintaining ; keeping : (4)

maintien de la direction, keeping the direction. (5)

maintien de la production, maintaining the output. (6)

maintien du boisage près du front de taille, carrying the timbering close to the face. (7)

maison (*n.f.*), house. (8)

maison d'habitation (*f.*), dwelling-house. (9)

maison éclusière (*f.*), lock-house. (10)

maison en pans de bois (*f.*), frame house. (11)

maître, -esse (*adj.*), main ; principal ; master : (12)

(Note.—**maître, -esse**, when it precedes the qualified noun, is often connected to it by a hyphen, but the use of the hyphen is not compulsory) :

maître couloir (Mines) (*m.*), main chute. (13)

maître drain (*m.*), main drain. (14)

maîtresse-lame [maîtresses-lames *pl.*] *ou* maîtresse-feuille [maîtresses-feuilles *pl.*] (d'un ressort à lames superposées) (*n f.*), top plate (of a leaf spring). (15)

maîtresse pièce (d'une charpente) (*f.*), principal member (of a frame). (16)

maîtresse-poutre [maîtresses-poutres *pl.*] (*n.f.*), main girder ; main beam. (17)

maîtresse-tige [maîtresses-tiges *pl.*] (pompe de mine) (*n.f.*), main rod ; spear rod. (18)

rivière maîtresse (*f.*), master-river. (19)

maître (pers.) (*n.m.*), master ; employer. (20)

maître à danser *ou* **maître de danse** (compas) (*m.*), outside-and-inside calipers ; double callipers ; in-and-out calipers. (21)

maître de forges (*m.*), ironmaster. (22)

maître-dragueur [maîtres-dragueurs *pl.*] (*n.m.*), dredgemaster. (23)

maître-foreur (*n.m.*). Same as **maître-sondeur**.

maître-laveur (*n.m.*), washery-foreman. (24)

maître-mineur (*n.m.*), overman. (25)

maître-sondeur (*n.m.*), boring-master ; master borer ; drilling-foreman ; drill-foreman ; foreman driller. (26)

maîtresse (*adj. f.*). See **maître, -esse**.

maîtriser (*v.t.*), to master ; to overmaster ; to overpower ; to cope with ; to control ; to bring under control : (27)

maîtriser le débit d'un puits à pétrole, to control, to overmaster, to overpower the flow of an oil-well. (28)

maîtriser le gaz dégagé dans les chantiers d'abatage, to cope with the gas given off in the workings. (29)

maîtriser un puits, to bring a well under control. (30)

maîtriser une venue d'eau, to master the incoming water ; to cope with an inflow of water. (31)

majeure partie (*f.*), major part ; greater part : (32)

la majeure partie de l'atmosphère consiste en azote, the major part of the atmosphere consists of nitrogen. (33)

mal'aria (*n.f.*), malaria ; malarial fever. (34)

mal défini, -e, ill-defined. (35)

mal gérer (*v.t.*), to mismanage. (36)

mal réussir (*v.i.*), to fail ; to turn out badly. (37)

malachite (Minéral) (*n.f.*), malachite ; green mineral. (38)

malacolite (Minéral) (*n.f.*), malacolite. (39)

malacon *ou* **malakon** (Minéral) (*n.m.*), malacon ; malacone. (40)

maladie (*n.f.*), disease ; illness ; sickness ; dis--order : (41)

maladies des mineurs, des bois de construction, des machines électriques, diseases of miners, of constructional timber, of electrical machinery. (42 *ou* 43 *ou* 44)

maladie des mineurs (Pathologie) (*f.*), miners' disease ; miners' anemia ; miners' anæmia. (45)

malaria (*n.f.*), malaria ; malarial fever. (46)

malaxage (*n.m.*) *ou* **malaxation** (*n.f.*), mixing ; malaxage. (47)

malaxer (*v.t.*), to mix ; to malaxate : (48)

malaxer du mortier, to mix mortar. (49)

malaxeur (Mach.) (*n.m.*), mixer ; malaxator. (50)

malaxeur à mortier (*m.*), mortar-mixer ; malax--ator. (51)

malaxeur à terres (Fonderie) (*m.*), loam-mixer. (52)

maldonite (Minéral) (*n.f.*), maldonite. (53)

mâle (Méc.) (*adj.*), male ; external ; exterior ; outside : (54)

vis mâle (*f.*), male screw ; external screw ; exterior screw ; outside screw. (55)

malfaçon (*n.f.*) *ou* **malfaçons** (*n.f.pl.*), bad workmanship ; bad work ; faulty workman--ship : (56)

construction gâtée par les malfaçons (*f.*), build--ing spoiled by bad workmanship. (57)

malfil (*n.m.*), wire-edge. (58)

mallardite (Minéral) (*n.f.*), mallardite. (59)

malléabiliser (*v.t.*), to malleableize. (60)

malléabilité (*n.f.*), malleability ; malleableness. (61)

malléable (*adj.*), malleable : (62)

l'or est le plus malléable de tous les métaux (*m.*), gold is the most malleable of all metals. (63)

malsain, -e (*adj.*), unhealthy ; insanitary ; unsanitary. (64)

malthacite (Minéral) (*n.f.*), malthacite. (65)

malthe (bitume glutineux) (*n.m.*), maltha. (66)

mamelon (Plomberie, etc.) (*n.m.*), nipple. (67)

mamelon (Géogr. phys.) (*n.m.*), mamelon. (68)

mamelonné, -e (Géogr. phys.) (*adj.*), mame--lonated. (69)

mamillaire (*adj.*), mammillated ; mammillary ; mammillar. (70)

manche (d'un outil) (*n.m.*), handle (of a tool). (71)

manche (d'un marteau, d'une hache) (*n.m.*), handle, helve, shaft, (of a hammer, of an axe). (72 *ou* 73)

manche (d'un couteau) (*n.m.*), handle, haft, (of a knife). (74)

manche (conduit en toile, en cuir, en métal) (*n.f.*), hose. (75)

manche à béquille (de pelle, etc.) (*m.*), crutch-handle (of shovel, etc.) (76)

manche à frapper devant (*m.*), sledge-handle. (77)

manche à œil (d'une pelle) (*m.*), eye-handle, D handle, (of a shovel). (78)

manche à pans (*m.*), polygon-sided handle. (79)

manche de ciseau (*m.*), chisel-handle. (80)

manche de lime (m.), file-handle. (1)

manche de manœuvre (Sondage) (m.), tiller; turning-tool; rod-turning tool; pole-turner; brace-head; brace-key. (2)

manche de manœuvre à vis (m.), screw-tiller. (3)

manche de pelle (m.), shovel-handle. (4)

manche de pic (m.), pick-handle; pick-hilt. (5)

manche en toile à voile (filtrage de l'amalgame) (f), canvas filter; bag-filter. (6)

manchon (Méc.) (n.m.), coupling; box; sleeve; muff; bush; bushing. (7)

manchon (gaine cylindrique de verre, de glaise, de terre cuite, etc.) (n.m.), mantle. (8)

manchon à boulons noyés ou manchon avec boulons à têtes noyées (m.), shrouded coup-ling; recessed coupling. (9)

manchon à embrayage (m.), clutch. (10). See embrayage for varieties.

manchon à frettes (m.), collared coupling. (11)

manchon à friction (m.), friction-coupling. (12)

manchon à incandescence (m.), incandescent-gas mantle; gas-mantle. (13)

manchon à plateaux (m.), flange-coupling; flanged coupling; plate-coupling; face-plate coupling. (14)

manchon à plateaux à boulons noyés (m.), shrouded flange-coupling; recessed flange-coupling; pulley-coupling. (15)

manchon à tocs (m.), pin-coupling. (16)

manchon à vis (m.), screw-coupling; screw-box. (17)

manchon-bride [manchons-brides pl.] (n.m.), cappel; capel; wire-rope clamp. (18)

manchon chargeur (Photogr.) (m.), changing-bag. (19)

manchon conique (pour forets) (m.), taper socket. (20)

manchon cylindrique (m.), muff-coupling. (21)

manchon cylindrique en deux pièces (m.), split muff-coupling. (22)

manchon cylindrique en fonte avec boulons à têtes noyées (m.), cast-iron shrouded muff-coupling. (23)

manchon d'accouplement ou manchon d'assem--blage (m.), coupling; coupling-box. (24)

manchon d'accouplement d'arbres ou manchon d'assemblage d'arbres (m.), shaft-coupling. (25)

manchon d'accouplement élastique (m.), flexible coupling; compensating-coupling. (26)

manchon d'embrayage (m.), clutch. (27). See embrayage for varieties.

manchon d'embrayage à dents (m.), claw-clutch. (28)

manchon de bielle (m.), rod-bearing; rod-brass. (29)

manchon de couplage (m.), coupling-box; coupling. (30)

manchon de cristal (m.), glass mantle. (31)

manchon de dilatation (m.), expansion-coupling. (32)

manchon de poulie (m.), pulley-bush; pulley-bushing. (33)

manchon de raccord (m.), connecting-box. (34)

manchon de refroidissement (haut fourneau) (m.), cooler; jumbo (blast-furnace). (35)

manchon de verre (m.), glass mantle. (36)

manchon destiné à recevoir les queues des forets aux cônes Morse (m.), sleeve for Morse taper shank drills. (37)

manchon élastique (m.), flexible coupling; compensating-coupling. (38)

manchon fou (m.), loose sleeve. (39)

manchon pour les mèches à queues coniques (m.), socket for taper-shank drills. (40)

manchon pour tuyaux (m.), pipe-coupling; pipe-connection. (41)

manchon pour tuyaux à incendie (m.), fire-hose coupling. (42)

manchon taraudé ou manchon fileté (m.), sleeve-nut; screw-ferrule. (43)

manchon vissé (m.), screwed coupling. (44)

manchonnage (action) (n.m.), coupling; bushing: (45)

manchonnage de deux arbres, coupling two shafts. (46)

manchonnage d'une poulie, bushing a pulley. (47)

manchonnage (dispositif) (n.m.). Same as manchon.

manchonner (v.t.), to couple; to bush. (48)

mandrin (pièce sur laquelle le tourneur assujettit son ouvrage) (n.m.), chuck. (49). See also plateau.

mandrin (arbre) (Méc.) (n.m.), mandrel; spindle; arbor. (50) See also arbre.

mandrin (broche) (Rivetage, etc.) (n.m.), drift; driftpin; driver. (51)

mandrin à arrondir les tubes de chaudière (m.), boiler-tube expander. (52)

mandrin à combinaisons (m.), combination chuck. (53)

mandrin à coussinets (m.), die-chuck. (54)

mandrin à deux mâchoires ou mandrin à deux mors (m.), two-jaw chuck. (55)

mandrin à gobelet (m.), cup-chuck. (56)

mandrin à portée conique à rattrapage de jeu concentrique (m.), mandrel with conical neck for taking up wear. (57)

mandrin à queue de cochon (m.), screw-chuck; taper-screw chuck. (58)

mandrin à serrage concentrique (m.), concentric chuck; self-centring chuck. (59)

mandrin à spirale (m.), scroll-chuck. (60)

mandrin à toc (m.), catch-plate; driver-plate; driver-chuck; dog-chuck; point-chuck; take-about chuck. (61)

mandrin à trois pointes ou mandrin à tulipe (m.), prong-chuck; fork-chuck; spur-chuck; three-pronged chuck. (62)

mandrin à vis (m.), bell-chuck; shell-chuck. (63)

mandrin à 8 vis, eight-screw bell-chuck. (64)

mandrin américain à 3 mâchoires à serrage concentrique (m.), 3-jaw American self-centring chuck. (65)

mandrin carré (m.), square drift; square drift-pin. (66)

mandrin conique (m.), taper drift; taper drift-pin. (67)

mandrin conique lisse (m.), smooth taper drift. (68)

mandrin d'évent (Fonderie) (m.), riser-pin; riser-stick. (69)

mandrin de coulée (Fonderie) (m.), runner-pin; runner-stick; gate-pin; gate-stick; sprue. (70)

mandrin de forgeron (m.), smiths' mandrel. (71)

mandrin en acier percé de part en part et tour--nant dans des coussinets en bronze à longue portée ou mandrin percé dans toute sa longueur tournant dans des coussinets en bronze à longue portée (m.), steel spindle bored through its entire length and running in long gun-metal bearings; hollow steel mandrel revolving in long gun-metal bearings; steel spindle with thoroughfare hole, turning in long gun-metal bearings. (1)

mandrin lisse (*m.*), smooth drift; smooth drift-pin. (2)

mandrin porte-fraise (*m.*), milling-machine arbor; cutter-arbor. (3)

mandrin porte-mèche *ou* **mandrin porte-foret** (*m.*), drill-chuck. (4)

mandrin porte-outils (alésoir) (*m.*), cutter-spindle; boring-spindle; cutter-mandrel. (5)

mandrin porte-taraud (*m.*), tapping-chuck. (6)

mandrin pour tour à bois (*m.*), wood-chuck. (7)

mandrin principal (tour) (*m.*), main spindle; live-spindle. (8)

mandrin taillé (opp. à **mandrin lisse**) (*m.*), cutting-drift. (9)

mandrin universel (*m.*), universal chuck. (10)

mandrinage (Travail au tour) (*n.m.*), chucking. (11)

mandrinage (brochage) (*n.m.*), drifting. (12)

mandrinage des tubes de chaudières (*m.*), expanding boiler-tubes. (13)

mandriner (fixer sur ou dans un mandrin) (Travail au tour) (*v.t.*), to chuck: (14)
mandriner une pièce que l'on veut travailler au tour, to chuck a part that one wishes to turn in the lathe. (15)

mandriner (brocher) (*v.t.*), to drift: (16)
mandriner un trou de rivet, to drift a rivet-hole. (17)

mandriner un tube (au dudgeon), to expand a tube (with the tube-expander). (18)

manège (*n.m.*), horse-power; horse-gear; animal driving-gear; animal-gear. (19)

maneton (bouton de manivelle) (Mach.) (*n.m.*), crank-pin; wrist-pin; wrist. (20)

maneton (partie d'une manivelle qu'on fait tourner avec la main) (*n.m.*), handle (of a hand-crank). (21)

manette (*n.f.*), handle; lever. (22)

manette d'interrupteur *ou simplement* **manette** (*n.f.*) (Élec.), switch-handle; switch-lever; key. (23)

manette de mise en marche (*f.*), starting-handle; starting-lever. (24)

manette du régulateur (*f.*), throttle-lever handle. (25)

manganate (Chim.) (*n.m.*), manganate. (26)

manganèse (Chim.) (*n.m.*), manganese. (27)

manganèse oxydé hydraté barytifère (Minéral) (*m.*), manganese hydrate; psilomelane. (28)

manganésé, -e (*adj.*), manganiferous; manganese (*used as adj.*): (29)
bronze manganésé (*m.*), manganese bronze. (30)

manganésien, -enne (*adj.*), manganesian. (31)

manganésifère (*adj.*), manganiferous. (32)

manganeux (Chim.) (*adj. m.*), manganous. (33)

manganique (Chim.) (*adj.*), manganic. (34)

manganite (Minéral) (*n.f.*), manganite; acerdese; gray manganese ore. (35)

manganocalcite (Minéral) (*n.f.*), manganocalcite. (36)

maniement (*n.m.*), handling. (37)

manier (*v.t.*), to handle: (38)
manier un outil, to handle a tool. (39)

manière (*n.f.*), manner; way. (40)

manière de se comporter (*f.*), behaviour: (41)

matériaux si distincts dans leur manière de se comporter (*m.pl.*), materials so distinct in their behaviour. (42)

manière dont se comportent les lampes dans les mines, behaviour of the lamps in under-ground workings. (43)

manille d'assemblage (*f.*), shackle; connecting-link; clevis; clevice. (44)

manille d'attelage (Ch. de f.) (*f.*), coupling-link; attachment-link; shackle. (45)

maniment (*n.m.*). Same as **maniement**.

manipulateur (Télégr.) (*n.m.*), transmitter: sender; manipulator. (46)

manipulation (*n.f.*), handling; manipulation: (47)
manipulation des explosifs, handling of explo-sives. (48)
manipulations brutales, rough handling. (49)

manipuler (*v.t.*), to handle; to manipulate. (50)

manivelle (à main) (*n.f.*), crank; handle; winch; cranked handle: (51)
tourner une manivelle, to turn a crank (*or* a handle). (52)

manivelle (Mach.) (*n.f.*), crank: (53)
manivelle de l'essieu d'une locomotive, crank of the axle of a locomotive. (54)

manivelle (Sondage) (*n.f.*), tiller; turning-tool; rod-turning tool; pole-turner; brace-head; brace-key. (55)

manivelle à contrepoids (*f.*), balance-crank. (56)

manivelle à main *ou* **manivelle à bras** (*f.*), hand-crank. (57)

manivelle à plateau (*f.*), disc crank; wheel crank. (58)

manivelle de machine (*f.*), engine-crank. (59)

manivelle de mise en marche (*f.*), starting-crank. (60)

manivelle double (*f.*), two-throw crank; duplex crank. (61)

manivelle en porte-à-faux (*f.*), overhanging crank; overhung crank. (62)

manivelle motrice (*f.*), driving-crank. (63)

manivelle multiple (*f.*), multithrow crank. (64)

manivelle simple (*f.*), single-throw crank. (65)

manivelle triple (*f.*), three-throw crank. (66)

manne à charbon (*f.*), coal-basket. (67)

manœuvre (action de diriger) (*n.f.*), working; manipulating; manipulation; operating; operation: (68)
manœuvre d'un ascenseur, d'une drague, working a lift, a dredge. (69 *ou* 70)
manœuvre des aiguilles à distance (Ch. de f.), manipulating distance-points. (71)

manœuvre (Ch. de f.) (*n.f.*), shunting: (72)
la manœuvre des trains en gare, the shunting of trains in stations. (73)

manœuvre (pers.) (*n.m.*), labourer. (74)

manœuvre (cordage) (Marine) (*n.f.*), rigging; rope. (75)

manœuvre à la surface (Mines) (*m.*), surface-labourer; outside labourer. (76)

manœuvre au treuil (*m.*), winchman. (77)

manœuvre courante (*f.*), running rigging. (78)

manœuvre dormante (*f.*), standing rigging. (79)

manœuvre préposé à la conduite de la grue (*m.*), craneman. (80)

manœuvre réduite (*f.*), reduced number of hands. (81)

manœuvrer (*v.t.*), to work; to manipulate; to operate: (82)
manœuvrer un levier, to work, to manipulate, a lever. (83)

manœuvrer une drague, to work, to operate, a dredge. (1)

manœuvrer une pompe, to work a pump. (2)

manœuvrer (Ch. de f.) (v.t.), to shunt : (3)

manœuvrer des wagons, to shunt trucks. (4)

manomètre (n.m.), pressure-gauge ; manometer ; gauge ; gage. (5)

manomètre à air (m.), air-gauge. (6)

manomètre à mercure (m.), mercurial pressure-gauge. (7)

manomètre à mercure à air comprimé (m.), mercurial pressure-gauge with closed tube. (8)

manomètre à mercure à air libre (m.), mercurial pressure-gauge with open tube. (9)

manomètre à piston (m.), piston pressure-gauge. (10)

manomètre à plaque (m.), diaphragm pressure-gauge. (11)

manomètre à ressort (m.), spring pressure-gauge. (12)

manomètre à tube (m.), tube pressure-gauge. (13)

manomètre de Bourdon (m.), Bourdon gauge. (14)

manomètre de pression de vapeur ou simplement manomètre (n.m.), steam pressure-gauge ; steam-gauge. (15)

manomètre enregistreur (m.), recording pressure-gauge ; registering pressure-gauge. (16)

manomètre étalon (m.), standard pressure-gauge. (17)

manomètre indicateur de pression (m.), pressure-gauge. (18)

manométrique (adj.), manometric ; manometrical : (19)

le rendement manométrique d'un ventilateur de mine, the manometric efficiency of a mine-fan. (20)

manquant, -e (adj.), missing ; short. (21)

manque (n.m.), want ; lack ; shortage ; short--ness ; scarcity ; dearth ; insufficiency ; absence : (22)

manque d'eau, de pluie, de combustible, de fonds, want, lack, shortage, scarcity, dearth, of water, of rain, of fuel, of funds. (23 ou 24 ou 25 ou 26)

manque de pose (Photogr.), underexposure. (27)

manqué, -e (adj.), spoiled ; defective : (28)

cliché manqué (Photogr.) (m.), spoiled negative. (29)

manquer (faillir ; tomber en défaut ; faire défaut) (v.i.), to fail ; to be wanting : (30)

l'eau manque souvent pendant la saison sèche (f.), the water often fails in the dry season. (31)

manquer d'observer les règlements, to fail to comply with the regulations. (32)

étage géologique qui manque en Europe (m.), geological stage which is wanting in Europe. (33)

manquer (être de moins) (v.i.), to be missing : (34)

un incendie fait rage dans la fosse—un homme manque, a fire is raging in the pit—one man is missing. (35)

manquer (rater) (v.i.), to misfire ; to miss fire ; to fail : (36)

fusil qui manque (m.), gun which misfires (or which misses fire). (37)

coup de mine qui manque de faire explosion (m.), shot which fails to explode. (38)

manquer de (être dépourvu ; ne pas avoir), to lack ; to run short of ; to run out of ; to be out of : (39)

manquer d'expérience, de capacité pour les affaires, to lack experience, business ability. (40 ou 41)

manquer de charbon, to run short (or to run out) (or to be out) of coal. (42)

manteau (n.m.), mantle : (43)

un manteau de détritus, de neige, de végéta--tion forestière, a mantle of detritus, of snow, of forestine vegetation. (44 ou 45 ou 46)

mantonnet (n.m.). Same as mentonnet.

manuel, -elle (adj.), manual : (47)

travail manuel (m.), manual labour. (48)

manuel (n.m.), manual ; hand-book ; pocket-book ; book : (49)

manuel d'instructions, instruction-book. (50)

manuellement (adv.), manually. (51)

manufacture (établissement) (n.f.), manufactory ; factory ; works ; mill. (52)

manufacturer (v.t.), to manufacture. (53)

manufacturier, -ère (pers.) (n.), manufacturer. (54)

manutention (n.f.), handling : (55)

manutention des wagons (Ch. de f.), car-handling. (56)

manutention du minerai dans les chantiers, handling of mineral in stopes. (57)

manutention mécanique des minerais, mechan--ical handling of ore. (58)

appareils de manutention du charbon (m.pl.), coal-handling plant. (59)

manutentionner (v.t.), to handle : (60)

matériel pour manutentionner — tonnes par jour (m.), plant to handle — tons per day. (61)

mappemonde (n.f.), map of the world. (62)

maque (n.f.), maquer (v.t.). Same as macque, macquer.

marais (n.m.), marsh ; swamp ; bog. (63)

marais salant (m.), salina ; salt marsh. (64)

marais tourbeux (m.), peat-bog. (65)

maraudage (Mines & Placers) (n.m.), fossicking. (66)

marauder (v.i.), to fossick. (67)

maraudeur (pers.) (n.m.), fossicker ; tramp miner. (68)

marbre (pierre calcaire) (n.m.), marble. (69)

marbre (n.m.) ou marbre à dresser ou marbre d'ajusteur, surface-plate ; engineers' sur--face-plate ; plane table ; plane : (70)

marbre en fonte, à nervures, et à poignées, raboté et gratté, cast-iron surface-plate, ribbed, and with handles, planed and scraped. (71)

marbre à paysages (de Cathane) (m.), landscape-marble. (72)

marbre de Carrare (m.), Carrara marble. (73)

marbre de Languedoc (m.), Languedoc marble. (74)

marbre de Paros (m.), Parian marble. (75)

marbre onyx (m.), onyx marble. (76)

marbre rose (m.), red marble. (77)

marbre serpentin (m.), serpentine marble. (78)

marbre statuaire (m.), statuary marble. (79)

marbre veiné (m.), veined marble. (80)

marbre vert antique (m.), verd antique. (81)

marbrier, -ère (adj.), marble (used as adj.) : (82)

industrie marbrière (f.), marble industry. (83)

marbrière (n.f.), marble-quarry. (84)

marcasite *ou* **marcassite** (Minéral) (*n.f.*), marcasite; marcassite; white pyrites; white iron pyrites. (1)
marchand, -e (*adj.*), merchantable; merchant; saleable; sale (*used as adj.*); marketable: (2)
fer marchand (*m.*), merchant iron; merchant bar. (3)
fonte marchande (*f.*), merchant pig. (4)
valeur marchande (*f.*), marketable value; sale-value. (5)
marchandise (*n.f.*) (*s'emploie souvent au pluriel*), goods; merchandise. (6)
marchandise franche de tout droit (*f.*) *ou* **mar-chandises reçues en franchise** (*f.pl.*), duty-free goods. (7)
marchant à bras (Méc.), driven by hand-power; hand-power (*used as adj.*): (8)
machine à raboter marchant à bras (*f.*), hand-power planing-machine. (9)
marchant au moteur,   power-driven;   power (*used as adj.*); driven by power; engine-driven; motor-driven: (10)
tour marchant au moteur (*m.*),   power-lathe. (11)
marchant au pied *ou* **marchant à pédale**, driven by foot-power; foot (*used as adj.*); treadle (*used as adj.*); pedal (*used as adj.*): (12)
tour marchant au pied *ou* tour marchant à pédale (*m.*),   foot-lathe;   treadle-lathe; pedal-lathe. (13)
marchant par courroies, belt-driven. (14)
marche (mouvement) (*n.f.*),   running; run; going; movement; march; course; speed: (15)
la marche des trains, the running of trains. (16)
la marche d'un glacier, des dunes de sable, des alluvions dans les rivières à fond mobile, the movement, the march, of a glacier, of sand-dunes, of alluvials in rivers with a shifting bottom. (17 *ou* 18 *ou* 19)
marche rétrograde, backward movement. (20)
marche à suivre, procedure; course to be followed. (21)
marche d'une expérience, course of an experi-ment. (22)
marche des opérations, procedure; course of the operations. (23)
See also specific phrases in vocabulary below.
marche (fonctionnement) (*n.f.*),   working; running: (24)
obstacles qui entravent la bonne marche d'un service (*m.pl.*), obstacles which interfere with the smooth working of a service. (25)
marche idéale du haut fourneau, ideal working of the blast-furnace. (26)
marche (d'un escalier, d'une échelle) (*n.f.*), step (of a stairs, of a ladder). (27 *ou* 28)
marche (d'une vis d'Archimède) (*n.f.*),   flange (of an Archimedean screw). (29)
marche à contre-vapeur (*f.*),   running with reversed steam. (30)
marche à la volée (transmission par courroie) (*f.*), belt-drive: (31)
passer de la marche par engrenages à la marche à la volée, to change over from gear-drive to belt-drive. (32)
marche à régulateur fermé (locomotive) (*f.*), running with the throttle closed. (33)
marche à vide *ou* **marche à blanc** (Mach.) (*f.*), light running; running light; running on no load. (34)
Cf. marcher à vide *ou* marcher à blanc.

marche arrière (conduite de locomotives) (*f.*), running backward; backing. (35)
marche au point mort de la distribution (loco-motive) (*f.*), running on the dead-point of the valve-gear. (36)
marche avant (conduite des locomotives) (*f.*), running forward; going ahead. (37)
marche balancée *ou* **marche balançante** (escalier) (*f.*), dancing step; balanced step (stairs). (38)
marche carrée *ou* **marche droite** (escalier) (*f.*), flyer; flier. (39)
marche courbe *ou* **marche cintrée** (escalier) (*f.*), commode step. (40)
marche d'essai (*f.*),   trial run; experimental run; test run: (41)
marche d'essai du matériel neuf, experimental run, trial run, test run, of new plant. (42)
marche dansante (escalier) (*f.*), winder (a step in a circular part of a stair with one end wider than the other). (43)
marche de départ (escalier) (*f.*), first step. (44)
marche des rayons (Opt.) (*f.*), path of rays. (45)
marche en charge (Mach.) (*f.*), running under load. (46)
marche en surcharge (*f.*), running on overload; overload running. (47)
marche haut-le-pied (Ch. de f.) (*f.*), light run-ning; running light (i.e., without wagons or carriages attached,—said of an engine; or, empty,—said of a train). (48)
marche-palier [**marches-paliers** *pl.*] (*n f.*) *ou* **marche palière** (*f.*)(escalier), landing-step. (49)
marche par engrenages (*f.*), gear-drive. (50)
marche rapide (*f.*), high speed; speed: (51)
tour à marche rapide (*m.*), high-speed lathe; speed-lathe. (52)
marche rayonnante *ou* **marche tournante** *ou* **marche gironnée** (escalier) (*f.*),   winder (a step radiating from a centre, as one of the steps winding round a newel in a regular circular stairs). (53)
marche régressive de l'érosion torrentielle (*f.*), working back of torrential errosion. (54)
Cf. **érosion régressive.**
marche silencieuse (d'un moteur, d'une machine) (*f.*), silent running; quiet running; noise-less running. (55)
marché (endroit public) (*n.m.*), market: (56)
marché des métaux, metal-market. (57)
marché (convention)(*n.m.*), bargain; contract:(58)
marché à la tâche, piece-work bargain. (59)
marché de sondage, boring-contract; drilling-contract. (60)
marché pour une fourniture de charbon, contract for a supply of coal. (61)
marchepalier (*n.f.*). Same as **marche-palier.**
marchepied [**marchepieds** *pl.*] (échelle double) (*n.m.*), step-ladder; steps; pair of steps. (62)
marchepied (de locomotive) (*n.m.*), foot-plate; foot-board. (63)
marchepied (d'un wagon de ch. de f.) (*n.m.*), foot-board. (64)
marchepied du frein (*m.*), brake-step. (65)
marcher (fonctionner; se mouvoir) (*v.i.*),   **to** work; to be driven; to run: (66)
machine qui marche à vapeur (*f.*), machine which works by steam. (67)
scie construite pour marcher au moteur (*f.*), saw constructed to be driven by power. (68)
machine disposée pour marcher électriquement (*f.*), machine arranged to be driven electri-cally. (69)

machine qui marche sans trépidation (*f.*), machine which runs without tremor. (1)

marcher à — kilomètres à l'heure (se dit des trains, etc.), to run, to travel, at — kilo-metres an hour. (2)

marcher à — tours à la minute (se dit des arbres de machines, etc.), to run at — revolutions per minute. (3)

marcher à une allure très rapide, to run at a very high speed. (4)

marcher à vide *ou* marcher à blanc (se dit des moteurs, des machines-outils, etc.), to run light ; to run on no load : (5)

si le moteur marche à vide, il n'absorbe que la puissance nécessaire pour vaincre les frottements, la résistance de l'air, etc., en un mot, les résistances passives, if the motor runs light, it only absorbs the power neces-sary to overcome friction, the resistance of the air, etc., in a word, the resistances to motion. (6)

marcher au ralenti (Mach.), to run slow. (7)

marcher en charge (opp. à *marcher à vide*), to run under load. (8)

marcher en surcharge, to run on overload. (9)

marcher haut le pied (Ch. de f.), to run light (i.e., without wagons or carriages attached, —said of an engine ; or, empty,—said of a train). (10)

mare (*n.f.*), pool ; pool of water ; pond. (11)

mare de pétrole (*f.*), pool of petroleum. (12)

marécage (*n.m.*), marsh ; swamp ; bog. (13)

marécageux, -euse (*adj.*), marshy ; swampy ; boggy. (14)

maréchal *ou* **maréchal ferrant** [maréchaux fer-rants *pl.*] (pers.) (*n.m.*), farrier ; shoeing-smith ; blacksmith ; ironsmith ; smith. (15)

maréchale (houille maréchale) (*n.f.*), smithing coal ; blacksmith coal. (16)

maréchalerie (art ou profession du maréchal ferrant) (*n.f.*), farriery. (17)

maréchalerie (atelier de maréchal ferrant) (*n.f.*), farriery ; smithy ; smithery ; smiths' shop ; blacksmiths' shop. (18)

marée (*n.f.*), tide. (19)

marée (à) *ou* **marée (de)**, tidal : (20)

mers à marées (*f.pl.*), tidal seas. (21)

courant de marée (*m.*), tidal current. (22)

marée basse (*f.*), low tide ; low water. (23)

marée de mortes eaux (*f.*), neap tide. (24)

marée de vives eaux (*f.*), spring-tide. (25)

marée descendante (*f.*), ebb-tide ; falling tide. (26)

marée haute (*f.*), high tide ; high water. (27)

marée montante (*f.*), flood-tide ; rising tide. (28)

marfil (*n.m.*), wire-edge. (29)

margarite (Minéral) (*n.f.*), margarite ; pearl-mica. (30)

margaritifère (*adj.*), margaritiferous ; pearl-bearing ; pearly. (31)

margarodite (Minéral) (*n.f.*), margarodite. (32)

marge (*n.f.*), margin : (33)

marge d'un fleuve, margin of a river. (34)

marginal, -e, -aux (*adj.*), marginal : (35)

rayons marginaux (Opt.) (*m.pl.*), marginal rays. (36)

mariage de deux cordages (*m.*), marriage of two ropes. (37)

marialite (Minéral) (*n.f.*), marialite. (38)

marier des cordages, to marry ropes. (39)

marin, -e (*adj.*), marine ; sea (*used as adj.*) : (40)

formation marine (*f.*), marine formation. (41)

sable marin (*m.*), sea-sand. (42)

maritime (*adj.*), maritime : (43)

lisière maritime des régions désertiques sub-tropicales (*f.*), maritime fringe of the subtropical desert regions. (44)

droit maritime (*m.*), maritime law. (45)

marmite (*n f.*) *ou* **marmite torrentielle** (Géol.), pot-hole ; kettle. (46)

marmite de géants (Géol.) (*f.*), giant-kettle. (47)

marmolite (Minéral) (*n.f.*), marmolite. (48)

marmoréen, -enne (*adj.*), marmoraceous : (49)

roche marmoréenne (*f.*), marmoraceous rock. (50)

marmorisation (*n.f.*), marmarosis ; marmorosis. (51)

marmoriser (*v.t.*), to marmarize ; to marmorize. (52)

marne (*n.f.*), marl. (53)

marne à ciment (*f.*), cement marl. (54)

marne bariolée (*f.*), variegated marl. (55)

marneux, -euse (*adj.*), marly ; marlaceous. (56)

marnière (*n.f.*), marl-pit. (57)

maroquin (cuir) (*n.m.*), morocco : (58)

tube de lunette enveloppé de maroquin (*m.*), telescope-tube covered with morocco. (59)

marque (*n.f.*), mark. (60)

marque à brûler *ou* **marque à chaud** (*f.*), branding-iron. (61)

marque à la craie (*f.*), chalk-mark. (62)

marque de fabrication *ou* **marque de fabrique** *ou* **simplement marque** (*n.f.*), trade-mark ; brand : (63)

marque de fabrique déposée *ou* marque déposée, registered trade-mark. (64)

marque (de), branded ; marked ; by well-known maker(s) ; by leading maker(s) ; high-class : (65)

fonte de marque (*f.*), branded pig ; marked pig. (66)

appareils de marque (*m.pl.*), apparatus by well-known maker (*or* by leading makers) ; high-class apparatus. (67)

marque des explosifs (*f.*), mark of explosives. (68)

marque tracée sur le câble (Extraction) (Mines) (*f.*), mark on the rope. (69)

marquer (*v.t.*), to mark : (70)

les dimensions marquées d'un astérisque sont en magasin (*f.pl.*), the sizes marked with an asterisk are in stock. (71)

accidents géologiques marqués par des venues d'eaux thermales (*m.pl.*), geological accidents marked by occurrences of thermal water. (72)

marquer (Grisoumétrie) (Mines) (*v.i.*), to show : (73)

le grisou marque, dans les lampes de sûreté, dès qu'il y en a 3 à 4 %, the fire-damp shows, in safety-lamps, so soon as there is 3 to 4 % present. (74)

marquer à la craie (*v.t.*), to mark with chalk ; to chalk : (75)

marquer à la craie sur les rails la position exacte des traverses, to chalk on the rails the exact position of the sleepers. (76)

marqueur (Mines) (pers.) (*n.m.*), tallyman ; checker. (77)

marteau (*n.m.*), hammer. (78)

marteau à ardoise (*m.*), slaters' hammer. (79)

marteau à bascule (*m.*), tilt-hammer ; tilting-hammer ; trip-hammer ; shingling-hammer. (80)

marteau à buriner (*m.*), chipping-hammer. (81)

marteau à charbon (*m.*), coal-hammer. (1)

marteau à déballer *dit* à fruits secs (*m.*), packing-hammer ; grocers' hammer. (2)

marteau à dent (*m.*), claw-hammer. (3)

marteau à devant (*m.*), sledge-hammer ; sledge ; about-sledge. (4) (blacksmiths', for use with both hands and swung at arm's length over the head : distinguished from marteau à main).

marteau à devant, à deux têtes (*m.*), double-faced sledge-hammer. (5)

marteau à devant, panne en long (*m.*), straight-peen sledge-hammer. (6)

marteau à devant, panne en travers (*m.*), cross-peen sledge-hammer. (7)

marteau à ébarber (Fonderie) (*m.*), fettling-hammer. (8)

marteau à forger (*m.*), forge-hammer ; forging-hammer. (9)

marteau à frapper devant (*m.*). Same as marteau à devant.

marteau à main (*m.*), hand-hammer. (10)

marteau à main (de forgeron) (*m.*), black-smiths' hammer ; sledge-hammer ; sledge ; uphand sledge. (11) (for use with one hand only : distinguished from marteau à devant).

marteau à panne (*m.*), peen-hammer ; pean-hammer. (12)

marteau à panne bombée (*m.*), ball-peen ham-mer ; ball-pean hammer ; ball pane hammer. (13)

marteau à panne en long (*m.*), straight-peen hammer ; straight-pean hammer ; straight-pane hammer. (14)

marteau à panne en travers (*m.*), cross-peen hammer ; cross-pean hammer ; cross-pane hammer. (15)

marteau à panne fendue (*m.*), claw-hammer. (16)

marteau à pioche (*m.*), pick-hammer. (17)

marteau à piquer *ou* marteau à piquer les chaudières (*m.*), scaling-hammer ; boiler-scaling hammer. (18)

marteau à plaquer (*m.*), veneering-hammer. (19)

marteau à river (*m.*), riveting-hammer. (20)

marteau à soulèvement *ou* marteau allemand (*m.*), tilt-hammer ; tilting-hammer ; trip-hammer ; shingling-hammer. (21)

marteau à têtes rapportées (*m.*), iron-bound mallet ; metal-bound mallet. (22)

marteau à vapeur (*m.*), steam-hammer. (23)

marteau chasse-coins (Pose de voie) (Ch. de f.) (*m.*), keying-hammer (Plate-laying). (24)

marteau chasse-crampons (Pose de voie) (*m.*), spiking-hammer. (25)

marteau d'ajusteur (*m.*), fitters' hammer. (26)

marteau d'eau (Phys.) (*m.*), water-hammer. (27)

marteau de carrier (*m.*), quarryman's hammer ; striking-hammer. (28)

marteau de couvreur (*m.*), slaters' hammer. (29)

marteau de géologue (*m.*), geologists' hammer. (30)

marteau de maçon (*m.*), masons' hammer. (31)

marteau de mineur (*m.*), miners' hammer. (32)

marteau de paveur (*m.*), paviors' hammer. (33)

marteau de poche (*m.*), pocket-hammer. (34)

marteau de scheidage (pour minerais) (*m.*), cobbing-hammer ; bucking-hammer ; buck-ing iron ; bucker ; spalling-hammer ; sorting-hammer. (35)

marteau de tailleurs de pierres (*m.*), stonemasons' hammer. (36)

marteau de vorscheidage (minerais) (*m.*), rag-ging-hammer. (37)

marteau en métal tendre (*m.*), soft-metal hammer. (38)

marteau frontal (*m.*), frontal hammer ; frontal helve. (39)

marteau-matoir [marteaux-matoirs *pl.*] (*n.m.*), caulking-hammer. (40)

marteau perforateur (Mines) (*m.*), hammer-drill ; percussion-drill ; percussive drill. (41)

marteau perforateur pneumatique (*m.*), pneu-matic hammer-drill. (42)

marteau-pilon [marteaux-pilons *pl.*] *ou simple-ment* marteau (*n.m.*), power-hammer ; forge-hammer ; forging-hammer ; hammer. (43)

marteau-pilon à double effet asservi (*m.*), double-acting power-hammer, taking steam above and below the piston. (44)

marteau-pilon à friction (*m.*), drop-hammer ; friction-hammer. (45)

marteau-pilon à ressort(s) (*m.*), spring power-hammer. (46 *ou* 47)

marteau-pilon à simple effet (*m.*), single-acting power-hammer. (48)

marteau-pilon à vapeur (*m.*), steam-hammer. (49)

marteau-pilon atmosphérique (*m.*), pneumatic power-hammer ; atmospheric hammer. (50)

marteau pour ajusteurs (*m.*), fitters' hammer. (51)

marteau pour chaudronniers (*m.*), boilermakers' hammer. (52)

marteau pour piquer les chaudières (*m.*), boiler-scaling hammer ; scaling-hammer. (53)

marteau rivoir (*m.*), riveting-hammer. (54)

martelage *ou* martèlement (*n.m.*), hammering ; malleation : (55)

martelage à froid, cold-hammering. (56)

martèlement de la voie (par une locomotive mal équilibrée), hammering of the road (by an ill-balanced locomotive). (57)

martelé, -e (*adj.*), hammered : (58)

fer martelé (*m.*), hammered iron. (59)

marteler (*v.t.*), to hammer : (60)

marteler le fer sur l'enclume, to hammer iron on the anvil. (61)

marteler à froid, to cold-hammer. (62)

marteleur (pers.) (*n.m.*), hammerman. (63)

martelures (*n.f.pl.*), hammer-scale ; hammer-slag ; scale ; iron-scale ; forge-scale ; cinder ; anvil-dross ; nill. (64)

martensite (Métall.) (*n.f.*), martensite. (65)

martinet (*n.m.*), tilt-hammer ; tilting-hammer ; trip-hammer ; shingling-hammer. (66)

martineur (pers.) (*n.m.*), hammerman. (67)

martite (Minéral.) (*n.f.*), martite. (68)

mascaret (*n.m.*), bore ; tidal bore ; mascaret : (69)

le mascaret de la Seine, the bore of the Seine. (70)

masonite (Minéral.) (*n.f.*), masonite. (71)

masquer (*v.t.*), to cover ; to conceal : (72)

l'obturateur d'une chambre noire est destiné à masquer et à démasquer l'objectif pour la pose (*m.*), the shutter of a camera is intended to cover and uncover the lens for exposure. (73)

lumière de cylindre qui est masquée par le tiroir (*f.*), cylinder-port which is covered by the slide-valve. (74)

réparations ayant pour but de masquer les
défauts (*f.pl.*), repairs intended to conceal
flaws. (1)

masse (*n.f.*), mass ; lump ; heap : (2)
masse de plomb, mass, lump, of lead. (3)
masse de métal incandescent, glowing mass
of metal. (4)
masse de pierres, heap of stones. (5)

masse (Géol.) (*n.f.*), mass ; bed : (6)
masse rocheuse, rock-mass. (7)
masse de gypse, bed of gypsum. (8)

masse (Géol. & Minéralogie) (*n.f.*), mass. (9).
Cf. **masses (en).**

masse (bloc ou morceau de métal) (*n.f.*), solid ;
solid blank : (10)
engrenages taillés dans la masse (*m.pl.*), gears
cut from the solid ; gear-wheels cut from
solid blanks. (11)

masse (d'un pont) (*n.f.*), pier (of a bridge). (12)

masse (Phys.) (*n.f.*), mass : (13)
la masse est une des trois qualités fondamentales
(longueur, masse, et temps) sur lesquelles
sont basées toutes les mesures physiques,
mass is one of the three fundamental qualities
(length, mass, and time) on which all physical
measurements are based. (14)
masse magnétique, magnetic mass. (15)

masse (à la) (Élec.), earthed ; connected to
frame : (16)
magnéto qui a un pôle à la masse (*f.*), magneto
which has an earthed pole. (17)

masse (gros marteau) (*n.f.*), sledge-hammer ;
sledge ; hammer. (18)

masse de mineur (*f.*), miners' hammer. (19)

masse surface d'un terrain (*f.*), total area of a
piece of ground. (20)

masselotte (Fonderie) (*n.f.*), shrinkage-head ;
shrink-head , sinking-head ; feed-head ;
feeding-head ; feeder ; deadhead ; sullage-
piece ; sullage-head ; head ; head metal. (21)

masses (en) (Géol. & Minéralogie), in masses ;
massive : (22)
la cérusite se présente en cristaux, en masses
compactes, en stalactites, etc., cerusite
occurs in crystals, in compact masses, in
stalactites, etc. (23)
minéral qui se présente en masses (*m.*), mineral
which occurs massive. (24)

massette (*n.f.*), sledge-hammer ; sledge ;
hammer. (25)

massette de casseur de pierres (*f.*), stonebreakers'
hammer. (26)

massette de mineur (*f.*), miners' hammer. (27)

massette de scheidage (*f.*), cobbing-hammer ;
bucking-hammer ; bucking-iron ; bucker ;
spalling-hammer ; sorting-hammer. (28)

massicot (Minéral) (*n.m.*), massicot ; lead ochre ;
plumbic ochre. (29)

massif, -ive (*adj.*), massive : (30)
un pilier massif, a massive pillar. (31)
pont de structure massive (*m.*), bridge of
massive structure. (32)

massif (*n.m.*), mass ; block ; body ; large body :
(33)
'haver un massif de houille, to undercut a
mass of coal. (34)
massif d'ancrage, anchorage-block. (35)
massif de fondation, foundation-block. (36)
massif de machine, engine-foundation. (37)
massif de maçonnerie, block of masonry. (38)
massif de minerai, body of ore ; ore-body ;
large body of ore. (39)

massif de recouvrement (Mines), large body
of overburden. (40)

massif (d'arbres) (*n.m.*), clump (of trees). (41)

massif (*n.m.*) *ou* **massif montagneux** (Géol.),
massif ; group of mountains : (42)
le massif du Mont-Blanc, the massif of Mont
Blanc ; the Mont Blanc group of mountains.
(43)
le massif de la Bernina, the Bernina massif. (44)

massif (pilier) (Mines) (*n.m.*), pillar ; post ;
stoop. (45)

massif d'injection (culot) (Géol.) (*m.*), boss. (46)

massif de protection (Mines) (*m.*), barrier-pillar ;
barrier ; boundary-pillar ; pillar. (47)

massif de protection de galerie (*m.*), chain-pillar.
(48)

massif de protection de puits (*m.*), shaft-pillar ;
bottom pillar. (49)

massif exotique (Géol.) (*m.*), outlier. (50)

massif réservé *ou* **massif en réserve** (Mines) (*m.*),
reserve-pillars. (51)

massivement (*adv.*), massively. (52)

mastic (*n.m.*), putty. (53)

mastic à la chaux (*m.*), lime putty ; putty.
(54)

mastic de fer (*m.*), iron-putty. (55)

mastic de fonte (*m.*), cast-iron cement ; rust
cement ; rust ; beaumontague ; beau-
-montague ; fake. (56)

mastic de minium (*m.*), red-lead putty. (57)

mastic de vitrier (*m.*), putty ; glaziers' putty.
(58)

masticage (*n.m.*), puttying. (59)

mastiquer (*v.t.*), to putty ; to stop up with putty :
(60)
mastiquer des carreaux de vitre, to putty
window-panes. (61)
mastiquer un trou, to stop up a hole with
putty. (62)

mat, -e (*adj.*), mat ; matt ; lustreless ; dull : (63)
métal mat (*m.*), mat metal. (64)
papier mat lisse (Photogr.) (*m.*), matt smooth
paper. (65)
cassure mate (*f.*), lustreless fracture. (66)

mat (*n.m.*), mat : (67)
le mat et le bruni, the mat and the burnish.
(68)

mât (*n.m.*), mast ; pole ; post. (69)

mât de signaux (*m.*), signal-post ; signal-mast.
(70)

mât-grue [mâts-grues *pl.*] (*n.m.*), mast-crane.
(71)

matage (*n.m.*), caulking ; calking ; jagging : (72)
le matage des tôles de chaudières aveugle les
fuites, the caulking of boiler-plates stops
leaks. (73)
le matage du plomb dans les joints, calking
lead in joints. (74)
matage autour des têtes de rivets, jagging
around rivet-heads ; caulking round the
heads of rivets. (75)

matelas (de vapeur dans le cylindre) (*n.m.*),
cushion (of steam in the cylinder). (76)

matelasser (*v.t.*), to cushion. (77)

mater (rendre mat) (*v.t.*), to mat. (78)

mater (serrer, battre avec le matoir) (*v.t.*), to
caulk ; to calk ; to jag : (79)
mater une tôle de chaudière en frappant le
chanfrein, to caulk a boiler-plate by striking
the chamfer. (80)
mater la tête d'un rivet, to jag the head of a
rivet ; to caulk a rivet-head. (81)

matériaux (*n.m.pl.*) (*le singulier* matériau *s'em-ploie aussi*), materials ; material ; stuff : (1) matériaux employés dans la confection du béton armé, materials used in making reinforced concrete. (2)

matériaux de construction, materials of con-struction ; building-materials. (3)

le bois est un matériau sujet à pourrir, wood is a material liable to rot. (4)

la section d'une marche pleine (*marche prise dans un seul morceau de bois*) est la même que celle d'une marche en pierre, seul le matériau est différent, the section of a solid step (*step made out of one piece of wood*) is the same as that of a stone step, only the material is different. (5)

matériaux d'empierrement (Ponts & Chaussées) (*m.pl.*), metal. (6)

matériaux d'empierrement pour routes (*m.pl.*), road-metal. (7)

matériaux d'éruption (Géol.) (*m.pl.*), volcanic ejecta. (8)

matériel [matériels *pl.*] (*n.m.*), plant ; appliances ; apparatus ; material ; equipment ; outfit ; stock. (9)

matériel d'éclairage électrique (*m.*), electric-lighting plant. (10)

matériel d'énergie électrique (*m.*), electric-power plant. (11)

matériel d'épuisement *ou* matériel d'exhaure (Mines) (*m.*), pumping-plant. (12)

matériel d'extraction (Mines) (*m.*), hoisting-plant ; winding-plant. (13)

matériel de chemins de fer (*m.*), railway-material ; railway-plant. (14)

matériel de chemins de fer à voie étroite (*m.*), light-railway material. (15)

matériel de forage (*m.*), boring-plant ; bore-holing plant ; drilling-plant. (16)

matériel de mines (*m.*), mining-plant ; mining-appliances ; mining-outfit. (17)

matériel de roulage (*m.*), haulage-plant. (18)

matériel de sauvetage (*m.*), rescue-appliances. (19)

matériel du jour (Mines) (*m.*), surface-plant ; surface-equipment. (20)

matériel électrique (*m.*), electric plant. (21)

matériel fixe (*m.*), fixed plant ; stationary plant. (22)

matériel flottant (*m.*), floating plant. (23)

matériel générateur (*m.*), generating plant. (24)

matériel lourd (matériel roulant) (Ch. de f.) (*m.*), heavy stock. (25)

matériel mobile (*m.*), loose plant ; portable plant. (26)

matériel photographique (*m.*), photographic apparatus. (27)

matériel pour mines (*m.*), mining-plant ; mining-appliances ; mining-outfit. (28)

matériel pour mines d'or (*m.*), gold-mining plant. (29)

matériel pour puits à pétrole (*m.*), oil-well appliances. (30)

matériel remorqueur (opp. à matériel roulant) (Ch. de f.) (*m.*), hauling-stock. (31)

matériel roulant (Ch. de f.) (*m.*), rolling stock ; rolling plant. (32)

mathématicien (pers.) (*n.m.*), mathematician. (33)

mathématique (*adj.*), mathematical : (34)

sciences mathématiques (*f.pl.*), mathematical sciences. (35)

mathématiquement (*adv.*), mathematically. (36)

mathématiques (*n.f.pl.*), mathematics. (37)

mathématiques élémentaires (*f.pl.*), elementary mathematics. (38)

mathématiques mixtes *ou* mathématiques appli-quées (*f.pl.*), applied mathematics ; mixed mathematics. (39)

mathématiques pures (*f.pl.*), pure mathematics ; abstract mathematics. (40)

mathématiques spéciales (*f.pl.*), higher mathe-matics. (41)

matière (*n.f.*) (*s'emploie souvent au pluriel*), matter ; material ; stuff. (42)

matière brute (*f.*), raw material. (43)

matière colorante (*f.*), colouring-matter. (44)

matière étrangère (*f.*), foreign matter : (45)

scorifier les matières étrangères contenues dans un métal, to scorify the foreign matter contained in a metal. (46)

matière filonienne (*f.*), vein-matter ; vein-material ; veinstuff ; lode-matter ; lode-stuff. (47)

matière inorganique (*f.*), inorganic matter. (48)

matière organique (*f.*), organic matter. (49)

matière première (*f.*), raw material. (50)

matière rocheuse (*f.*), rocky matter ; rock-material. (51)

matière terreuse (*f.*), earthy matter ; earthy material. (52)

matière volatile (*f.*), volatile matter. (53)

matières d'or et d'argent *ou simplement* ma-tières (*n.f.pl.*), bullion ; gold and silver bullion. (54)

matières d'or, gold bullion. (55)

matières d'argent, silver bullion. (56)

matières détritiques (Géol.) (*f.pl.*), detrital matter ; detritus. (57)

matières draguées (*f.pl.*), dredged material. (58)

matières en suspension dans un liquide (*f.pl.*), matter in suspension in a liquid. (59)

matières refusées par le crible (*f.pl.*), oversize ; oversize stuff ; stuff passing over the screen ; riddlings ; skimpings. (60)

matières tamisées (*f.pl.*), siftings : siftage. (61)

matières traversant le crible (*f.pl.*), undersize ; undersize stuff ; stuff passing through the screen ; fell. (62)

matir (*v.t.*). Same as mater.

matlockite (Minéral) (*n.f.*), matlockite. (63)

matoir (*n.m.*), caulking-chisel ; calking-iron ; fullering-tool. (64)

matras (Verrerie de laboratoire) (*n.m.*), matrass ; bolthead ; receiver. (65)

matriçage (Méc.) (*n.m.*), dying. (66)

matrice (Minéralogie) (*n.f.*), matrix ; gangue ; gangue-matter. (67)

matrice (Charp.) (*n.f.*), housing ; mortise ; mortice. (68)

matrice (*n.f.*) *ou* matrice à découper, die ; die-plate ; bolster. (69)

matrice (taraud mère) (*n.f.*), master tap ; hob ; hob-tap. (70)

matricer (Méc.) (*v.t.*), to die. (71)

matte (Métall.) (*n.f.*), matte ; matt. (72)

matte de cuivre *ou* matte cuivreuse (*f.*), copper matte. (73)

matte de nickel (*f.*), nickel matte. (74)

matte de plomb *ou* matte plombeuse (*f.*), lead matte. (75)

matter (battre avec le matoir) (*v.t.*). Same as mater.

maturité (Géogr. phys.) (*n.f.*), maturity. (76)

mauvais, -e (*adj.*), bad : (1)
mauvais ajustage (*m.*), bad fit. (2)
mauvais conducteur (*m.*), bad conductor :
(3)
le caoutchouc est mauvais conducteur de
l'électricité, india-rubber is a bad con-
-ductor of electricity. (4)
mauvais mur (Mines) (*m.*), bad bottom. (5)
mauvais temps (*m.*), bad weather. (6)
mauvais toit (Mines) (*m.*), bad roof. (7)
mauvaise construction (*f.*), bad construction ;
faulty workmanship : (8)
défaut dû à mauvaise construction (*m.*),
defect due to faulty workmanship. (9)
mauvaise gestion (*f.*), bad management ;
mismanagement. (10)
mauvaises terres (Géol.) (*f.pl.*), bad lands. (11)
maximum [maxima *ou* maximums *pl.*] (*adj.*)
(*invariable pour les deux genres*), maximum :
(12)
poids maximum (*m.*), maximum weight. (13)
ouverture maximum de la lumière (*f.*) (cylindre
à vapeur), maximum port-opening. (14)
altitudes maxima *ou* altitudes maximums
(*f.pl.*), maximum altitudes. (15)
charge maximum (*f.*) *ou* maximum de charge
(*m.*), maximum load ; peak-load ; peak
of the load. (16)
maximum [maxima *ou* maximums *pl.*] (*n.m.*),
maximum. (17)
maximum et minimum (calibres à tolérances),
go and not go ; go on and not go on ; go in
and not go in : (18)
le plus souvent, les calibres à tolérances portent
les indications *maximum* et *minimum*.
Pour que les pièces usinées se trouvent dans
la tolérance, elles doivent passer dans
l'ouverture maximum et ne le doivent pas
dans celle minimum, limit-gauges are
usually marked *go* and *not go* (or *go on*
and *not go on*) (or *go in* and *not go in*). For
machined parts to be within limits, they
must pass in the go opening but not in the not
go. (19)
maxite (Minéral.) (*n.f.*), maxite. (20)
maxwell (unité C.G.S.) (Élec.) (*n.m.*), maxwell.
(21)
mazéage de la fonte (*m.*), refining cast iron. (22)
mazer la fonte, to refine cast iron. (23)
mazerie (*n.f.*), refinery ; finery ; finery-furnace.
(24)
mazout (*n.m.*), masut. (25)
méandre (Géogr. phys.) (*n.m.*), meander ;
winding : (26)
les méandres d'un fleuve, the meanders of a
river. (27)
méandre divagant (*m.*), flood-plain meander.
(28)
méandre encaissé (*m.*), incised meander ;
intrenched meander. (29)
méandre recoupé (*m.*), cut-off. (30)
mécanicien (personne qui invente ou construit
des machines) (*n.m.*), mechanic : (31)
machine qui est l'ouvrage d'un très habile
mécanicien (*f.*), machine which is the work
of a very clever mechanic. (32)
mécanicien *ou* **mécanicien-chauffeur** [mécani-
-ciens-chauffeurs *pl.*] (personne qui dirige une
machine) (*n.m.*), engineman. (33)
mécanicien *ou* **mécanicien-conducteur** [mécani-
-ciens-conducteurs *pl.*] (de tramway
électrique) (*n.m.*), driver ; motorman. (34)

mécanicien (*n.m.*) *ou* **mécanicien conducteur de
locomotive**, driver ; engine-driver ; engineer ;
engine-runner ; running engineer ; runner ;
locomotive-engineer. (35) Note.—**driver**
and **engine-driver** are the English terms ;
the others are American.
mécanicien constructeur (*m.*), fitter ; artificer.
(36)
mécanicien d'extraction (Mines) (*m.*), winding
engineman ; hoisting engineman. (37)
mécanique (*adj.*), mechanical ; machine (*used
as adj.*); power (*used as adj.*) : (38)
arts mécaniques (*m.pl.*), mechanical arts. (39)
rivetage mécanique (*m.*), machine-riveting.
(40)
scie mécanique (*f.*), power-saw ; sawing-
machine. (41)
mécanique (science) (*n.f.*), mechanics. (42)
mécanique (combinaison d'organes) (*n.f.*),
mechanism : (43)
la mécanique d'une soupape, the mechanism
of a valve. (44)
mécanique appliquée (*f.*), applied mechanics ;
practical mechanics. (45)
mécanique de précision (*f.*), scientific instru-
-ments. (46)
mécanique rationnelle (*f.*), rational mechanics ;
abstract mechanics ; pure mechanics ;
theoretical mechanics. (47)
mécaniquement (*adv.*), mechanically. (48)
mécanisme (*n.m.*), mechanism ; gear : (49)
mécanisme d'une soupape, mechanism of a
valve. (50)
mécanisme d'avancement et de retour rapide,
mechanism for feeding and quick return.
(51)
mécanisme de manœuvre, working-mecha-
-nism ; rig ; rigging : (52)
mécanisme de manœuvre de la grille, grate-
shaking rig ; grate-shaking rigging. (53)
mécanisme de vannage (Hydraul., etc.), gate-
gear. (54)
mèche (Méc.) (*n.f.*), bit ; drill. (55) See also
foret.
mèche (Travail aux explosifs) (*n.f.*), fuse. (56)
mèche (d'une lampe) (*n.f.*), wick (of a lamp).
(57)
mèche (d'une corde) (*n.f.*), core (of a rope).
(58)
mèche (d'une machine à mortaiser) (*n.f.*), core-
driver (of a mortising-machine). (59)
mèche (d'un cabestan) (*n.f.*), spindle (of a
capstan). (60)
mèche à canon (*f.*), half-round drill ; hog-nose
drill. (61)
mèche à centre (*f.*), centre-bit. (62)
mèche à centre, à queue carrée (*f.*), centre
brace-bit ; centre-bit, for bit-stock. (63)
mèche à centrer (*f.*), centre-drill ; combined
drill and countersink. (64)
mèche à combustion lente (*f.*), slow-burning
fuse ; slow-match. (65)
mèche à cuiller (*f.*), spoon bit. (66)
mèche à langue d'aspic, à queue cylindrique (*f.*),
flat drill, with parallel shank ; arrow-headed
drill, with straight shank. (67)
mèche à talon (*f.*), wobble drill. (68)
mèche à trois pointes *ou* **mèche anglaise** (*f.*),
centre-bit. (69)
mèche avec queue carrée (*f.*), square-shank
drill. (70)
mèche creuse (*f.*), hollow drill. (71)

mèche de sûreté (*f.*), safety-fuse ; common fuse ; Bickford fuse. (1)

mèche demi-ronde (*f.*), half-round drill ; hog-nose drill. (2)

mèche évidée (*f.*), fluted drill ; straight-fluted drill ; straightway drill. (3)

mèche extensible (*f.*), extension-bit. (4)

mèche hélicoïdale (*f.*), twist-drill. (5) See foret hélicoïdal for varieties.

mèche ordinaire (*f.*). Same as mèche de sûreté.

mèche plate (*f.*). Same as mèche à langue d'aspic.

mèche pour lampes (*f.*), lamp-wick. (6)

mèche pour terrains humides (*f.*), fuse for wet ground. (7)

mèche pour terrains secs (*f.*), fuse for dry ground. (8)

mèche pour vilebrequin (*f.*), bit-stock drill ; brace-bit. (9)

mèche soufrée (Mines) (*f.*), sulphur squib. (10)

mèche styrienne (*f.*), German pod-bit. (11)

mèche torse (*f.*), auger bit. (12)

mécompte (*n.m.*), miscalculation ; mistake. (13)

médian, -e (*adj.*), median ; medial ; middle : (14)
  ligne médiane (Géom.) (*f.*), median line. (15)
  moraine médiane (Géol.) (*f.*), medial moraine. (16)

médiocre (*adj.*), middling ; poor in quality ; moderate ; mediocre. (17)

médiojurassique (Géol.) (*n.m.*), Brown Jura. (18)

mégadyne (C.G.S.) (*n.f.*), megadyne. (19)

mégerg (Élec.) (*n.m.*), megaerg ; megerg. (20)

mégohm (Élec.) (*n.m.*), megohm. (21)

méionite (Minéral) (*n.f.*), meionite ; mionite. (22)

mélaconise (Minéral) (*n.m.*), melaconite ; melac--onisa ; melaconise. (23)

mélange (*n.m.*), mixing ; mixture : (24)
  mélange des fontes (Fonderie), mixing cast iron ; blending. (25)
  mélange explosif *ou* mélange détonant (Travail aux explosifs), explosive mixture ; detonating mixture. (26)
  mélange intime, intimate mixture. (27)
  mélange tonnant *ou* mélange détonant (moteur à combustion interne), explosive mixture. (28)

mélanger (*v.t.*), to mix : (29)
  mélanger deux solutions par parties égales, to mix two solutions in equal parts. (30)

mélanger (se) (*v.r.*), to mix. (31)

mélangeur (*n.m.*), mixer. (32)

mélangeur (d'un injecteur) (*n.m.*), chamber (of an injector). (33)

mélangeur actif (*m.*), active mixer. (34)

mélangeur inactif (*m.*), inactive mixer. (35)

mélanite (Minéral) (*n.f.*), melanite ; andradite. (36)

mélanocrate (Pétrol.) (*adj.*), melanocratic : (37)
  roches mélanocrates (*f.pl.*), melanocratic rocks. (38)

mélantérie *ou* mélantérite (Minéral) (*n.f.*), melanterite ; melanteria. (39)

mélaphyre (Pétrol.) (*n.m.*), melaphyre. (40)

mêler (*v.t.*), to mix. (41)

mêler (se) (*v.r.*), to mix. (42)

mélèze (arbre) (*n.m.*), larch. (43)

mélilite (Minéral) (*n.f.*), melilite ; humboldtilite. (44)

mélinite (Minéral) (*n.f.*), melinite. (45)

mélinophane (*n.m.*) *ou* méliphanite (*n.f.*) (Minéral), melinophanite. (46)

mélinose (Minéral) (*n.f.*), wulfenite ; yellow lead ore. (47)

mellite (Minéral) (*n.f.*), mellite ; honeystone. (48)

menacer (*v.t.*), to threaten : (49)
  voûte qui menace ruine (*f.*), arch which threatens to collapse. (50)

ménagement de pente (*m.*), grading. (51)

ménager la pente de, to grade : (52)
  ménager la pente d'une route, to grade a road. (53)
  sluice à pente ménagée (*m.*), graded sluice. (54)

ménager son argent, to make good use of one's money ; to husband one's resources. (55)

menant, -e (opp. à mené, -e) (Méc.) (*adj.*), driving : (56)
  roue menante (*f.*), driving-wheel ; driver. (57)

mendipite (Minéral) (*n.f.*), mendipite. (58)

mendozite (Minéral) (*n.f.*), mendozite. (59)

mené, -e (Méc.) (*adj.*), driven : (60)
  roue menée (*f.*), driven wheel ; follower. (61)

mener (*v.t.*), to lead ; to drive : (62)
  mener un cheval par la bride, to lead a horse by the bridle. (63)

mener (Géom.) (*v.t.*), to draw : (64)
  mener une ligne droite entre deux points, to draw a straight line between two points. (65)

meneur (Mines) (pers.) (*n.m.*), trammer ; drawer ; wheeler ; putter ; haulageman. (66)

meneur (porte-lame) (Méc.) (*n.m.*), cutter-bar. (67)

meneur à la machine à aléser (*m.*), boring-bar ; cutter-bar for boring-machine. (68)

meneur de tour (*m.*), lathe cutter-bar. (69)

ménilite (Minéral) (*n.f.*), menilite. (70)

ménisque (Phys. & Opt.) (*n.m.*), meniscus. (71)

ménisque concave (capillarité) (Phys.) (*m.*), concave meniscus. (72)

ménisque convergent (lentille) (Opt.) (*m.*), con--verging meniscus ; converging concavo-convex lens. (73)

ménisque convexe (capillarité) (Phys.) (*m.*), convex meniscus. (74)

ménisque divergent (lentille) (Opt.) (*m.*), diverg--ing meniscus ; diverging concavo-convex lens. (75)

menotte de suspension (bogie) (*f.*), swing-hanger ; swing-link ; swing-link hanger. (76)

menstrue (Chim.) (*n.m.*), menstruum ; solvent ; dissolvent. (77)

mensuel, -elle (*adj.*), monthly. (78)

mensuellement (*adv.*), monthly ; per month. (79)

mentonnet (pièce saillante pour déterminer un arrêt) (*n.m.*), catch ; catch-pin ; latch-pin. (80)

mentonnet (pièce qui reçoit l'impulsion d'une came) (*n.m.*), tappet. (81)

mentonnet (boudin d'une roue) (*n.m.*), flange. (82)

mentonnet (d'un loquet) (*n.m.*), keeper (of a gate-latch) ; latch-catch ; nosing. (83)

mentonnet (du pêne d'une serrure à gorge) (*n.m.*), stump, fence, stub, (of the bolt of a tumbler-lock). (84)

menu, -e (*adj.*), small ; fine : (85)
  menus morceaux (*m.pl.*), small bits. (86)
  menu gravier (*m.*), fine gravel. (87)

menu (*n.m.*) *ou* **menus de houille** (*m.pl.*), small ; smalls ; small coal ; slack ; coal-slack. (1)

menu (*n.m.*) *ou* **menus de minerai** (*m.pl.*), small ; smalls ; small ore. (2)

menuiserie (*n.f.*), joinery. (3)

menuisier (pers.) (*n.m.*), joiner. (4)

méphitique (*adj.*), mephitic ; noxious. (5)

méphitisme (*n.m.*), mephitis ; mephitism. (6)

méplat, -e (*adj.*), flat (having broad surface and little thickness ; thicker one way than the other) : (7)

rail méplat (*m.*), flat rail. (8)

mer (*n.f.*), sea. (9)

mer bordière (*f.*), epicontinental sea. (10)

mer fermée *ou* **mer intérieure** (*f.*), inland sea. (11)

mer libre (*f.*), open sea. (12)

mercure (*n.m.*), mercury ; quicksilver. (13)

mercure argental (Minéral) (*m.*), argental mercury ; amalgam. (14)

mercureux (Chim.) (*adj. m.*), mercurous. (15)

mercuriel, -elle (*adj.*), mercurial : (16)

vapeurs mercurielles (*f.pl.*), mercurial vapours. (17)

mercurifère (*adj.*), quicksilver-bearing ; mercuriferous. (18)

mercurification (*n.f.*), mercurification. (19)

mercurique (Chim.) (*adj.*), mercuric. (20)

mère (*adj.*), mother ; master ; main : (21)

la roche mère du diamant, the mother rock of diamond. (22)

filon mère (*m.*), mother lode ; master-lode ; main lode ; champion lode. (23)

mère-galerie [mères-galeries *pl.*] (Mines) (*n.f.*), main road ; mainway ; main level ; gang-way ; entry ; mother-gate : (24)

mère-galerie en direction, main road parallel in direction to the strike ; mainway following the strike. (25)

mère *ou* **mère-gueuse** [mères-gueuses *pl.*] (rigole principale par laquelle s'écoule la fonte en fusion venant du haut fourneau) (*n.f.*), sow. (26)

mère d'émeraude (*f.*), mother-of-emerald. (27)

méridien, -enne (*adj.*), meridian. (28)

méridien (*n.m.*), meridian. (29)

méridien magnétique (*m.*), magnetic meridian. (30)

méridional, -e, -aux (*adj.*), meridional ; south ; southern. (31)

mériédrie (Cristall.) (*n.f.*), merohedrism. (32)

mériédrique (Cristall.) (*adj.*), merohedral ; merohedric. (33)

merlin (hache pour fendre le bois) (*n.m.*), cleaver. (34)

méroxène (Minéral) (*n.m.*), meroxene. (35)

mesa (Géogr. phys.) (*n.f.*), mesa. (36)

mésestimation (*n.f.*), underestimation ; under-valuation. (37)

mésestimer (*v.t.*), to underestimate ; to under-rate ; to undervalue. (38)

mésocrate (Pétrol.) (*adj.*), mesocratic : (39)

roches mésocrates (*f.pl.*), mesocratic rocks. (40)

mésole (Minéral) (*n.m.*), mesole. (41)

mésolite (Minéral) (*n.f.*), mesolite. (42)

mésotype (Minéral) (*n.f.*), mesotype. (43)

mésozoïque (Géol.) (*adj.*), Mesozoic ; Secondary. (44)

mésozoïque (*n.m.*), Mesozoic ; Secondary : (45)

le mésozoïque, the Mesozoic ; the Secondary. (46)

mesurage (*n.m.*), measurement ; measuring : (47)

mesurage des courants d'air, measurement of, measuring, air-currents. (48)

mesure (évaluation) (*n.f.*), measure ; measure-ment. (49)

mesure (unité) (*n.f.*), measure : (50)

le mètre est une mesure de longueur, the metre is a measure of length. (51)

mesure (dimension) (*n.f.*), measurement ; size ; dimension : (52)

prendre la mesure d'un champ, to take the measurement of a field. (53)

pièces de bois qui ne sont pas de mesure (*f.pl.*), pieces of wood which are not to size. (54)

mesure (moyen) (*n.f.*), measure : (55)

mesures préventives contre la poussière, pre-cautionary measures, precautions, against dust. (56)

mesures de sécurité, safety-measures. (57)

mesure à coulisse (*f.*), sliding caliper-gauge ; slide-calipers ; sliding calipers ; caliper-gauge ; caliper-square ; beam-caliper ; beam caliper-gauge. (58)

mesure à coulisse pour mesurer les profondeurs (*f.*), depth-gauge. (59)

mesure à ruban (*f.*), tape measure ; measuring-tape. (60)

mesure agraire (*f.*), land measure. (61)

mesure dans œuvre (*f.*), inside measurement. (62)

mesure de capacité (*f.*), measure of capacity. (63)

mesure de capacité pour les liquides (*f.*), liquid measure. (64)

mesure de capacité pour les matières sèches (*f.*), dry measure. (65)

mesure de longueur *ou* **mesure linéaire** (*f.*), linear measure ; long measure ; measure of length. (66)

mesure de poids *ou* **mesure pondérale** (*f.*), weight : (67)

un tableau de mesures pondérales, a table of weights. (68)

mesure de surface *ou* **mesure de superficie** (*f.*), square measure ; measure of area. (69)

mesure de volume (*f.*), cubic measure ; solid measure ; measure of solidity. (70)

mesure graduée (Verrerie de laboratoire) (*f.*), graduated measure. (71)

mesure hors d'œuvre (*f.*), outside measurement. (72)

mesure métrique (*f.*), metric measure. (73)

mesurer (*v.t.*), to measure : (74)

mesurer le débit d'un cours d'eau, l'intensité d'un courant électrique, to measure the flow of a stream, the strength of an electric current. (75 *ou* 76)

mesurer des terres, to measure land. (77)

mesurer au mètre, to measure by the metre. (78)

mesurer (avoir en dimension) (*v.i.*), to measure (*v.i.*) : (79)

table qui mesure — m.×— m. (*f.*), table which measures — m. × — m. (80)

mesurer (se) (*v.r.*), to be measured : (81)

les vis à tête fraisée et les vis à tête goutte de suif se mesurent longueur tête comprise. Les vis à tête ronde et les vis à tête cylindrique se mesurent longueur sous tête, countersunk screws and raised-head screws are measured length over all. Round-headed screws and cheese-headed screws are measured length under head to point. (1)

mesureur des courants (Élec.) (*m.*), current-meter. (2)

mesureur des pressions (*m.*), pressure-register. (3)

métachlorite (Minéral) (*n.f.*), metachlorite. (4)

métacinnabarite (Minéral) (*n.f.*), metacinnabarite. (5)

métal (*n.m.*), metal. (6)

métal (fonte) (terme de fonderie) (*n.m.*), metal ; cast iron. (7)

métal à coussinets (*m.*), bearing-metal ; lining-metal ; bush-metal ; bushing-metal ; box-metal. (8)

métaux alcalino-terreux (*m.pl.*), metals of the alkaline earths. (9)

métaux alcalins (*m.pl.*), alkaline metals. (10)

métal anglais (*m.*), Britannia metal ; britannia metal ; britannia. (11)

métal antifriction (*m.*), antifriction metal ; white metal ; babbitt metal ; babbitt. (12)

métal antifriction Magnolia (*m.*), Magnolia antifriction metal. (13)

métal au bain (*m.*), molten metal. (14)

métaux autres que le fer (opp. à **métaux ferreux**) (*m.pl.*), non-ferrous metals. (15)

métal Babbitt (*m.*), Babbitt metal ; Babbitt's metal. (16)

métal blanc (alliage de divers métaux qui ressemblent à de l'argent métallique) (*m.*), white metal. (17)

métal blanc (métal antifriction) (*m.*), white metal ; antifriction metal ; babbitt metal ; babbitt. (18)

métal Britannia (*m.*), Britannia metal ; bri-tannia metal ; britannia. (19)

métal brut *ou* **métal cru** (*m.*), raw metal ; crude metal. (20)

métaux cassants (*m.pl.*), brittle metals. (21)

métal d'apport (pour soudure autogène) (*m.*), filling-metal (for autogenous welding). (22)

métal de cloche (*m.*), bell-metal. (23)

métaux de la famille du cérium (*m.pl.*), cerium metals ; metals of the cerium family. (24)

métaux de la mine de platine (*m.pl.*), platinum metals ; metals of the platinum group. (25)

métal delta (*m.*), delta-metal. (26)

métal déployé (*m.*), expanded metal. (27)

métal en fusion (*m.*), molten metal ; metal in fusion. (28)

métal en gueuse *ou* **métal en saumon** (*m.*), pig-metal. (29)

métal en lingot (*m.*), ingot-metal. (30)

métaux ferreux (*m.pl.*), ferrous metals. (31)

métaux légers (*m.pl.*), light metals. (32)

métaux lourds (*m.pl.*), heavy metals. (33)

métal mère [**métaux mères** *pl.*] (*m.*) *ou* **métal-mère** [**métaux-mères** *pl.*] (*n.m.*), mother metal : (34)

la fonte est le métal mère des métaux dérivés du fer, pig iron is the mother metal of the metals derived from iron. (35)

métaux misés (*m.pl.*), fagoted metals. (36)

métal mou (*m.*), soft metal : (37)

le plomb et l'étain sont des métaux mous, lead and tin are soft metals. (38)

métal Müntz (*m.*), Muntz metal ; Muntz's metal. (39)

métal natif *ou* **métal nu** (*m.*), native metal. (40)

métal noble (*m.*), noble metal ; precious metal. (41)

métal oxydé (*m.*), oxidized metal. (42)

métal pauvre (*m.*), base metal. (43)

métal pour coussinets (*m.*). Same as **métal à coussinets**.

métal précieux (*m.*), precious metal ; noble metal. (44)

métal tendre (*m.*), soft metal : (45)

marteau en métal tendre (*m.*), soft-metal hammer. (46)

métal vierge (*m.*), virgin metal. (47)

métal vil (*m.*), base metal. (48)

métalepsie (Chim.) (*n.f.*), metalepsy ; metalepsis ; metathesis ; substitution. (49)

métalléité (*n.f.*), metallicity ; metalleity. (50)

métallifère (*adj.*), metalliferous ; metal-bearing ; ore-bearing : (51)

mine métallifère (*f.*), metalliferous mine. (52)

roche métallifère (*f.*), metalliferous rock ; metal-bearing rock ; ore-bearing rock. (53)

métallin, -e (*adj.*), metalline. (54)

métallique (*adj.*), metallic ; metal (*used as adj.*) : (55)

corps simple métallique (*m.*), metallic element. (56)

bain métallique (*m.*), metal bath. (57)

Note.—métallique is often used in the general sense of iron (either wrought or cast), or steel, or iron and steel, as opposed to wood, stone, or some other material ; or wire, as opposed to vegetable fibre, etc., as, un pont métallique, an iron bridge. buttes métalliques (Mines) (*f.pl.*), cast-iron props. pylône métallique (*m.*), steel tower or trellis-post. traité de charpente métallique (*m.*), treatise on iron and steel construc-tional work. câble métallique (*m.*), wire rope.

métallisage (*n.m.*) *ou* **métallisation** (*n.f.*), metal-lization. (58)

métalliser (*v.t.*), to metallize ; to metallify. (59)

métallographe (pers.) (*n.m.*), metallographist ; metallographer. (60)

métallographie (*n.f.*), metallography. (61)

métallographique (*adj.*), metallographic ; metallo-graphical. (62)

métalloïde (qui se ressemble à un métal) (*adj.*), metalloid ; metalloidal : (63)

éclat métalloïde (*m.*), metalloidal lustre. (64)

métalloïde (*n.m.*), metalloid ; non-metal. (65)

metalloïdique (qui se rapporte aux métalloïdes) (*adj.*), metalloid ; metalloidal ; non-metallic : (66)

élément métalloïdique (*m.*), metalloidal element ; non-metallic element. (67)

métallurgie (*n.f.*), metallurgy : (68)

métallurgie du fer, metallurgy of iron. (69)

métallurgique (*adj.*), metallurgic ; metallurgical : (70)

four métallurgique (*m.*), metallurgical furnace. (71)

métallurgiquement (*adv.*), metallurgically. (72)

métallurgiste (pers.) (*n.m.*), metallurgist. (73)

métamorphique (*adj.*), metamorphic ; meta-morphous : (74)

roches métamorphiques (*f.pl.*), metamorphic rocks. (75)

métamorphisme (*n.m.*), metamorphism. (76)

métamorphisme de contact (Géol.) (*m.*), con-tact metamorphism ; local metamorphism. (77)

métamorphisme général (Géol.) (*m.*), general metamorphism; regional metamorphism. (1)

métamorphoser (*v.t.*), to metamorphose. (2)

metasilicate (Chim.) (*n.m.*), bisilicate; meta--silicate. (3)

métasilicique (*adj.*), bisilicic; metasilicic. (4)

métasomatique (Géol.) (*adj.*), metasomatic. (5)

métasomatose (Géol.) (*n.f.*), metasomatosis. (6)

métaxite (Minéral) (*n.f.*), metaxite. (7)

météore (*n.m.*), meteor. (8)

météorique (*adj.*), meteoric: (9)
fer météorique (*m.*), meteoric iron. (10)
eaux météoriques (*f.pl.*), meteoric waters. (11)

météoriquement (*adv.*), meteorically. (12)

météorite (*n.m.*), meteorite; aerolite; aerolith; air-stone. (13)

météoritique (*adj.*), meteoritic; aerolitic. (14)

météorologie (*n.f.*), meteorology. (15)

météorologique (*adj.*), meteorologic; meteoro--logical; weather (*used as adj.*): (16)
Bureau météorologique (*m.*), Meteorological Office; Weather Bureau. (17)

méthane (*n.m.*), methane; methane gas. (18)

méthanomètre (*n.m.*), methanometer. (19)

méthode (*n.f.*), method; way; system; process; plan; rule. (20)

méthode d'analyse chimique par la voie humide (*f.*), wet method, humid way, moist process, of chemical analysis. (21)

méthode d'analyse chimique par la voie sèche (*f.*), dry method of chemical analysis. (22)

méthode d'exploitation (*f.*), method of working; working method; system of working: (23)
méthode d'exploitation des gîtes minéraux, method of working mineral deposits. (24)

méthode d'exploitation des mines (*f.*), method of mining; mining method. (25)

méthode d'exploitation du charbon (*f.*), method of working the coal; coal-mining method. (26)

méthode d'oxydation par le minerai (opp. à la **méthode par dilution**) (acier) (*f.*), ore-process. (27) (contrasted with **scrap-process**).

méthode de forage à sec (*f.*), dry method of drilling. (28)

méthode de forage avec courant d'eau (*f.*), wet method of drilling. (29)

méthode de foudroyage (Mines) (*f.*), caving method; subsidence-of-the-roof method. (30)

méthode de l'abandon de massifs (Mines) (*f.*), chambers-and-permanent-pillars method; pillars left as permanent supports of the roof. (31)

méthode de travail (*f.*), working method; plan of working; system of working. (32)

méthode des massifs courts *ou* **méthode des piliers et galeries** (Exploitation de la houille) (*f.*), bord-and-pillar system; board-and-pillar system; pillar-and-stall system; post-and-stall system; stoop-and-room system. (33)

méthode des massifs longs (Exploitation de la houille) (*f.*), room-and-pillar system; pillar-and-room system. (34)

méthode directe (Exploitation des mines de houille) (opp. à **méthode rétrograde**) (*f.*), direct method; outward method. (35)

méthode du recoupement (Levés) (*f.*), method of intersections (Surveying). (36)

méthode empirique (*f.*), empirical method; rule of thumb. (37)

méthode en long (Mines) (*f.*), flat-back stoping-method. (38)

méthode galloise de fusion du cuivre (*f.*), Welsh method of copper-smelting. (39)

méthode grossière et expéditive (*f.*), rough and ready method. (40)

méthode par dilution (acier) (*f.*), scrap-process. (41)

méthode rétrograde (Exploitation des mines de houille) (*f.*), retreating-method; mining-retreating method; homewards method. (42)

méthodique (*adj.*), methodic; methodical. (43)

méthodiquement (*adv.*), methodically. (44)

métier (*n.m.*), trade; craft; business; pro--fession; work: (45)
métier de maçon, mason's craft. (46)
métier de serrurier, locksmith's trade. (47)

métrage (action) (*n.m.*), measurement (in metres); measuring (in metres); surveying; quantity surveying: (48)
le métrage d'un travail de maçonnerie, the measurement of a piece of masonry work; surveying a piece of masonry work. (49)

métrage (longueur) (*n.m.*), measurement: (50)
pièce qui a un métrage de — mètres (*f.*), piece which has a measurement of — metres; part which measures — metres. (51)

mètre (*n.m.*), metre; meter = 39·370113 inches *or* 3·280843 feet *or* 1·0936143 yards. (52)

mètre (objet servant à mesurer et ayant la longueur d'un mètre) (*n.m.*), any measuring instrument 1 metre in length. (53)

mètre à retrait (Fonderie) (*m.*), shrinkage-rule; shrink-rule; contraction-rule. (54)

mètre à ruban (*m.*), measuring-tape (a metre long); tape measure. (55)

mètre-canne [mètres-cannes *pl.*] (*n.m.*), metre-stick (a "yardstick" a metre long). (56)

mètre carré (*m.*), square metre = 10·7639 square feet *or* 1·1960 square yards. (57)

mètre courant (*m.*), running metre. (58)

mètre cube (*m.*), cubic metre = 35·3148 cubic feet *or* 1·307954 cubic yards. (59)

mètre cube d'air (*m.*), cubic metre of air. (60)

mètre en buis (*m.*), metre boxwood rule. (61)

mètre étalon (*m.*), standard metre. (62)

mètre linéaire (*m.*), lineal metre; running metre. (63)

métrer (*v.t.*), to measure (by the metre); to survey: (64)
métrer le travail, to measure the work done; to survey the work. (65)

métreur (pers.) (*n.m.*), surveyor; quantity surveyor: (66)
honoraires des architectes, experts, et métreurs (*m.pl.*), fees of architects, valuers, and surveyors. (67)

métrique (*adj.*), metric; metrical: (68)
système métrique (*m.*), metric system. (69)

mettre (*v.t.*), to put. (70) See also **mise**.

mettre à bord (*v.t.*), to put on board; to embark; to ship. (71)

mettre à contribution les réserves, to draw on the reserves. (72)

mettre à découvert (*v.t.*), to uncover; to expose; to disclose: (73)
mettre à découvert des vieux chantiers (Mines), to uncover old workings. (74)

mettre la veine à découvert sur une longueur de — mètres, to uncover the vein for a distance of — metres. (1)

mettre à dimensions (*v.t.*), to size : (2)
mettre à dimensions un trou, to size a hole. (3)

mettre à exécution (*v.t.*), to carry into execution ; to carry into effect ; to carry out : (4)
mettre un projet à exécution, to carry a plan into execution ; to carry a plan into effect ; to carry out a project. (5)

mettre à feu un haut fourneau, to blow in a blast-furnace. (6)

mettre à l'abri (*v.t.*), to put under cover ; to shelter ; to house. (7)

mettre à la terre (Élec.) (*v.t.*), to earth ; to ground : (8)
mettre à la terre le fil neutre d'un réseau à trois fils, to earth, to ground, the neutral of a three-wire system. (9)

mettre à nu *ou* **mettre à** (*ou* **au**) **jour,** to denude ; to uncover ; to strip ; to expose : (10)
mettre à nu l'avancement (Mines), to expose the heading-face. (11)

mettre à terre (*v.t.*), to put ashore ; to land. (12)

mettre au point (*v.t.*), to focus ; to focalize ; to adjust ; to bring into position for use ; to position : (13)
mettre au point un microscope, un appareil photographique sur (*ou* à) l'infini, to focus a microscope, a camera on infinity. (14 *ou* 15)
mettre au point un microscope au moyen d'une vis micrométrique, to adjust a micro-scope by means of a micrometer-screw. (16)
mettre au point un foret, to bring a drill into position for use ; to position a drill. (17)

mettre au raide (*v.t.*), to take up the slack : (18)
mettre un câble au raide, to take up the slack of a rope. (19)

mettre au rebut (*v.t.*), to scrap ; to throw back into the waste ; to back. (20)

mettre au travail (*v.t.*), to set to work : (21)
mettre un ouvrier au travail, to set a man to work. (22)

mettre aux enchères (*v.t.*), to put up to auction. (23)

mettre d'épaisseur (*v.t.*), to thickness : (24)
mettre d'épaisseur une planche, to thickness a board. (25)

mettre dans le circuit (Élec.), to make, to close, a circuit. (26)

mettre en branle une sonnerie, to set a bell ringing : (27)
avertisseur d'incendie qui met en branle une sonnerie (*m.*), fire-alarm which sets a bell ringing. (28)

mettre en chantier (Fonderie) (*v.t.*), to bed in ; to plant. (29)

mettre en circuit (Élec.) (*v.t.*), to connect. (30)

mettre en court-circuit (Élec.) (*v.t.*), to short-circuit ; to short : (31)
mettre en court-circuit les pôles d'une dynamo, to short-circuit, to short, the poles of a dynamo. (32)

mettre en danger la sécurité des ouvriers, to endanger, to imperil, to jeopardize, to jeopard, the safety of the workmen. (33)

mettre en évidence un cube de minerai, to expose an ore-body. (34)

mettre en feu un haut fourneau, to blow in a blast-furnace. (35)

mettre en liberté (*v.t.*), to liberate ; to release. (36)

mettre en marche *ou* **mettre en mouvement** *ou* **mettre en train** *ou* **mettre en route** (*v.t.*), to start ; to start up ; to set in motion : (37)
mettre une machine en route, to start (*or to* start up) an engine ; to set an engine in motion. (38)

mettre en œuvre (s'en servir de) (*v.t.*), to make use of : (39)
diamants si petits qu'a peine les peut-on mettre en œuvre (*m.pl.*), diamonds so small that one can hardly make use of them. (40)

mettre en œuvre une machine, to set a machine to work. (41)

mettre en opération (*v.t.*), to put into operation. (42)

mettre en péril (*v.t.*). Same as **mettre en danger.**

mettre en place (*v.t.*), to place in position : (43)
mettre en place le moteur d'extraction, to place the hoisting-engine in position. (44)

mettre en poudre (*v.t.*), to reduce to powder. (45)

mettre en pratique (*v.t.*), to put into practice. (46)

mettre en pression. Same as **mettre sous pression.**

mettre en production (*v.t.*), to bring into pro-duction : (47)
mettre un puits à pétrole en production, to bring an oil-well into production. (48)

mettre en route (*v.t.*). See **mettre en marche.**

mettre en sacs (*v.t.*), to put into sacks ; to sack ; to bag : (49)
mettre en sacs le minerai, to bag the ore ; to put the ore into bags. (50)

mettre en station (*v.t.*), to set up : (51)
mettre une lunette en station, to set up a telescope. (52)

mettre en tas (*v.t.*), to stack ; to heap up ; to pile up. (53)

mettre en train. See **mettre en marche.**

mettre en valeur (*v.t.*), to turn to account. (54)

mettre en vente (*v.t.*), to put up for sale ; to offer for sale ; to market. (55)

mettre fin à, to put an end to : (56)
mettre fin à toute équivoque, to put an end to all equivocation. (57)

mettre hors circuit (Élec.) (*v.t.*), to disconnect ; to break a circuit ; to open a circuit. (58)

mettre hors feu un haut fourneau, to blow out a blast-furnace. (59)

mettre la dernière main à un ouvrage, to put, to give, the finishing touches to a piece of work. (60)

mettre la mine en exploitation normale, to put the mine in working order. (61)

mettre le feu à (*v.t.*), to set fire to ; to fire ; to light : (62)
dans les mines grisouteuses, il faut des précautions spéciales pour éviter de mettre le feu au grisou, in gassy mines special precautions are necessary to avoid setting fire to the fire-damp. (63)
mettre le feu à un fourneau de mine, to fire a shot. (64)
mettre le feu aux mèches, to light the fuses. (65)

mettre obstacle à (*v.t.*), to interfere with ; to impede : (66)

mettre obstacle au fonctionnement normal d'une mine, to interfere with the regular working of a mine. (1)

mettre obstacle aux opérations, to impede operations. (2)

mettre sous pression (chaudière), to get up steam. (3)

mettre (se) à l'abri, to take shelter. (4)

mettre (se) en court-circuit (Élec.), to short-circuit ; to short. (5)

mettre (se) en grève, to go on strike ; to strike ; to down tools. (6)

mettre (se) en marche ou mettre (se) en train ou mettre (se) en route (v.r.), to start : (7) locomotive qui se met en marche (f.), loco-motive which starts. (8)

meuble (adj.), loose ; running : (9) terrain meuble (m.), loose ground ; running ground. (10)

meuble (n.m.), furniture ; piece of furniture : (11) meubles de bureau ou meuble de bureau, office-furniture. (12) un meuble inutile, a useless piece of furniture. (13)

meulage (n.m.), grinding : (14) meulage des grosses pièces de forge, grinding heavy forgings. (15)

meule (n.f.), wheel. (16)

meule (Géol.) (n.f.), a stone that is whirled round in a pot-hole or kettle. (17)

meule à affûter ou meule à aiguiser ou simplement meule (n.f.), grinding-wheel ; sharpening-wheel. (18)

meule à cuvette (f.), dished wheel. (19)

meule à dégrossir (f.), roughing-wheel. (20)

meule à ébarber (f.), trimming-wheel. (21)

meule à grain fin (f.), fine-grained wheel. (22)

meule à gros grains (f.), coarse-grained wheel. (23)

meule à noyau rentrant (f.), cup-wheel ; tub-wheel. (24)

meule à polir (f.), polishing-wheel. (25)

meule à rectifier (f.), truing-wheel. (26)

meule à user (f.), abrading-wheel ; reducing-wheel. (27)

meule affûteuse (f.), grinding-wheel ; sharpening-wheel. (28)

meule aléseuse (f.), internal-grinding wheel. (29)

meule-boisseau [meules-boisseaux pl.] (n.f.), cup-wheel ; tub-wheel. (30)

meule d'émeri ou meule en émeri (f.), emery-wheel : (31) meules en émeri, travaillant à sec ou à l'eau, emery-wheels, for dry or wet grinding. (32 ou 33)

meule de remouleur (f.), grinding-wheel ; knife-grinder. (34)

meule dure (émeri) (f.), hard wheel. (35) Cf. degré de dureté.

meule en carborundum (f.), carborundum-wheel. (36)

meule en corindon (f.), corundum-wheel. (37)

meule en grès (f.), grindstone. (38)

meule lapidaire ou meule travaillant sur face (f.), face-wheel. (39)

meule mouillée (f.), wet wheel. (40)

meule tendre (émeri) (f.), soft wheel. (41)

meule verticale (f.), edge-runner ; edge-wheel ; edgestone. (42)

meuler (v.t.), to grind : (43) meuler un outil, to grind a tool. (44)

meulier, -ère (adj.), millstone (used as adj.) : (45) grès meulier (m.), millstone grit. (46)

meulière (n.f.), millstone grit. (47)

meurtiat en pierres sèches (Mines) (m.), pack ; pack-wall ; building. (48)

mho (Élec.) (n.m.), mho. (49)

mi-chemin (à), midway ; half-way : (50) bois situé à mi-chemin de deux villages (m.), wood situated midway between two villages. (51)

mi-course (tiroir, etc.) (n.f.), half-travel (slide-valve, etc.). (52)

mi-dur, -e (adj.), half-hard. (53)

mi-hauteur (à) ou mi-côte (à), half-way up ; half-way up the hill : (54) une station génératrice d'énergie située à mi-hauteur, sur le flanc de la montagne, a power-generating station situated half-way up, on the side of the mountain. (55)

miargyrite (Minéral) (n.f.), miargyrite. (56)

miarolithique (Pétrol.) (adj.), miarolitic. (57)

miascite (Pétrol.) (n.f.), miaskite ; miascite ; miascyte. (58)

miasmatique (adj.), miasmatic ; miasmatical. (59)

miasme (n.m.), miasma. (60)

mica (Minéral) (n.m.), mica ; glimmer. (61)

mica blanc (m.), muscovite. (62)

mica ferreux (m.), iron mica. (63)

mica ferromagnésien (m.), magnesium-iron mica ; iron-magnesia mica. (64)

mica lithinifère (m.), lithium mica ; lithia mica. (65)

mica magnésien (m.), magnesium mica ; magnesia mica. (66)

mica noir (m.), biotite. (67)

mica potassique (m.), potassium mica ; potash mica. (68)

mica sodique (m.), sodium mica ; soda mica. (69)

mica triangulaire (m.), penninite ; pennine. (70)

micacé, -e (adj.), micaceous ; micacious : (71) roche micacée (f.), micaceous rock. (72)

micaschiste (Géol.) (n.m.), mica-schist ; mica-slate. (73)

micaschisteux, -euse (adj.), mica-schistous ; mica-schistose. (74)

micro-électrique (adj.), microelectric. (75)

microchimie (n.f.), microchemistry. (76)

microchimique (adj.), microchemical. (77)

microchimiquement (adv.), microchemically. (78)

microcline (Minéral) (n.m.), microcline. (79)

microcristallin, -e (adj.), microcrystalline. (80)

microfarad (Élec.) (n.m.), microfarad. (81)

microgranit (Pétrol.) (n.m.), microgranite. (82)

microgranulitique (adj.), microgranulitic. (83)

microgrenu, -e (adj.), microgranular : (84) structure microgrenue (f.), microgranular structure. (85)

microhm (Élec.) (n.m.), microhm. (86)

microlithe (Pétrol.) (n.m.), microlith ; microlite. (87)

microlithique (adj.), microlithic ; microlitic : (88) structure microlithique (f.), microlithic struc-ture. (89)

micromètre (n.m.), micrometer. (90) See also palmer.

micromètre avec vis d'arrêt (m.), micrometer with clamp-screw. (91)

micromètre-objectif [micromètres-objectifs pl.] (n.m.), eyepiece micrometer. (92)

micrométrie (*n.f.*), micrometry. (1)

micrométrique (*adj.*), micrometric; micro-
-metrical; micrometer (*used as adj.*) : (2)
vis micrométrique (*f.*), micrometer-screw;
micrometric screw. (3)

micrométriquement (*adv.*), micrometrically. (4)

micron *ou* micromillimètre (*n.m.*), micron;
micromillimetre; micromil=·001 millimetre
or ·000039 inch. (5)

micropegmatite (Pétrol.) (*n.f.*), micropegmatite;
micropegmatyte. (6)

microperthite (Minéral) (*n.f.*), microperthite. (7)

microphotographie (*n.f.*), microphotography. (8)

microscope (*n.m.*), microscope. (9)

microscope à bascule *ou* microscope à genouillère
(*m.*), microscope on jointed stand. (10)

microscope polarisant (*m.*), polarization-micro-
-scope. (11)

microscopie (*n.f.*), microscopy. (12)

microscopique (*adj.*), microscopic; microscopical : (13)

examen microscopique (*m.*), microscopic
examination. (14)

poussières microscopiques (*f.pl.*), micro-
-scopical dust. (15)

microscopiquement (*adv.*), microscopically. (16)

microséisme (*n.m.*), microseism. (17)

microsismique (*adj.*), microseismic; microseis-
-mical. (18)

microsphérolithique (Pétrol.) (*adj.*), micro-
-spherulitic. (19)

microstructure (*n.f.*), microstructure. (20)

microvolt (Élec.) (*n.m.*), microvolt. (21)

midi (sud) (*n.m.*), south. (22)

migration des lignes de faîte *ou* migration des
lignes de partage (Géogr. phys.) (*f.*), shifting
of divides; migration of divides. (23)

migration des mineurs vers d'autres districts (*f.*),
migration of miners to other fields. (24)

milarite (Minéral) (*n.f.*), milarite. (25)

milieu (*n.m.*), middle; centre. (26)
milieux miniers, mining circles. (27)
milieu du courant, midstream. (28)

milieu (Phys.) (*n.m.*), medium : (29)
produire un ébranlement en un point d'un
milieu élastique, to produce a disturbance
at a point in an elastic medium. (30)
milieu ambiant *ou* milieu environnant, sur-
-rounding medium. (31)
milieux réfractifs, refractive media (*or*
mediums). (32)

mille (mesure linéaire anglaise) (*n.m.*), mile
=1·6093 kilometres. (33)

mille carré (*m.*), square mile =259·00 hectares. (34)

millérite (Minéral) (*n.f.*), millerite; hair-pyrites;
capillary pyrites. (35)

milliampère (Élec.) (*n.m.*), milliampere. (36)

millibar (Météor.) (*n.m.*), millibar. (37)

milligramme (*n.m.*), milligramme=0·015 grain. (38)

millimètre (*n.m.*), millimetre=0·03937 inch. (39)

millivolt (Élec.) (*n.m.*), millivolt. (40)

millstonegrit (*n.m.*), millstone grit. (41)

mimétèse *ou* mimétite *ou* mimétène (Minéral)
(*n.f.*), mimetite; mimetene; mimetese;
mimetesite. (42)

mince (*adj.*), thin; slender : (43)
une mince feuille de métal, a thin sheet of
metal. (44)

mincement (*adv.*), thinly; slenderly. (45)

minceur (*n.f.*), thinness; slenderness. (46)

mine (*n.f.*), mine. (47)

mine (minerai) (*n.f.*), ore; stone. (48)

mine (coup de mine) (*n.f.*), shot; blast : (49)
l'allumage de la mine se fait soit à l'aide de
mèches de sûreté, soit à l'aide d'amorces
électriques, lighting the shot is done either
by safety-fuses, or by electric fuses; firing
the blast is effected either by safety-fuses,
or by electric fuses. (50)

mine (trou de mine) (*n.f.*), shot-hole; blast-hole;
hole; mine-chamber; mining-hole; powder-
mine : (51)
débourrage de mines ratées (*m.*), unstemming
miss-holes. (52)

mine à ciel ouvert (*f.*), surface-mine; daylight-
mine. (53)

mine à grande profondeur (*f.*), deep-level mine. (54)

mine à grisou (*f.*), fiery mine; gassy mine;
gaseous mine. (55)

mine bonanza (*f.*), bonanza-mine. (56)

mine chambrée (Travail aux explosifs) (*f.*),
chambered hole. (57)

mine d'émeraudes (*f.*), emerald-mine. (58)

mine d'empiétage (disposition des trous de mine)
(*f.*), cut-hole; centre-cut hole; keyhole;
breaking-in hole; centre-shot; breaking
shot; bursting shot; busting shot; buster
shot. (59)

mine d'or (*f.*), gold-mine : (60)
les mines d'or du Transvaal, the Transvaal
gold-mines. (61)

mine de charbon (*f.*), coal-mine; colliery. (62)

mine de couronne (disposition des trous de
mine) (*f.*), top hole; back hole. (63)

mine de dégraissage (disposition des trous de
mine) (*f.*), enlarging-hole; shoulder hole;
easer. (64)

mine de diamants (*f.*), diamond-mine. (65)

mine de fer (exploitation) (*f.*), iron-mine. (66)

mine de fer (minerai) (*f.*), iron ore; ironstone. (67)

mine de fer spéculaire (*f.*), specular iron ore;
specular iron; looking-glass ore. (68)

mine de gravier (exploitation aurifère) (*f.*),
gravel-mine. (69)

mine de houille (*f.*), coal-mine; colliery. (70)

mine de maizières (disposition des trous de mine)
(*f.*), side hole; skimmer. (71)

mine de mercure (*f.*), quicksilver-mine. (72)

mine de pierres précieuses (*f.*), gem-mine. (73)

mine de plomb (exploitation) (*f.*), lead-mine. (74)

mine de plomb (plombagine) (*f.*), black lead;
plumbago; graphite. (75)

mine de relevage (disposition des trous de mine)
(*f.*), bottom-hole; lifter. (76)

mine de sel (*f.*), salt-mine. (77)

mine de sel gemme (*f.*), rock-salt mine. (78)

mine dépourvue d'eau (*f.*), dry mine. (79)

mine douce (*f.*), soft iron ore. (80)

mine en chômage (*f.*), standing mine; inactive
mine. (81)

mine en pleine activité (*f.*), mine in full working
order. (82)

mine en rapport (*f.*), producing mine. (83)

mine envahie par les eaux (*f.*), flooded mine. (84)

mine épuisée (*f.*), worked out mine. (85)

mine exploitée par puits (*f.*), shaft-mine. (86)

mine grisouteuse (*f.*), fiery mine ; gassy mine ' gaseous mine. (1)
mine humide (*f.*), wet mine. (2)
mine hydraulique (*f.*), hydraulic mine. (3)
mine improductive (*f.*), unproductive mine ; non-producing mine. (4)
mine métallifère (*f.*), metalliferous mine. (5)
mine métallique (*f.*), metal-mine. (6)
mine non-grisouteuse (*f.*), non-fiery mine ; non-gaseous mine. (7)
mine noyée (*f.*), flooded mine. (8)
mine poussiéreuse (*f.*), dusty mine. (9)
mine productive (*f.*), producing mine. (10)
mine ratée (Travail aux explosifs) (*f.*), miss-hole : (11)
débourrage des mines ratées (*m.*), unstemming miss-holes. (12)
mine reconnue (*f.*), proved mine. (13)
mine sous-marine (*f.*), undersea mine ; sub--marine mine. (14)
miner *ou* miner par le bas (*v.t.*), to undermine ; to mine ; to sap : (15)
fleuve qui mine ses bords (*m.*), river which undermines its banks. (16)
l'écroulement des falaises, que l'eau mine par le bas (*m.*), the collapse of cliffs, which the water undermines. (17)
minerai (*n.m.*), ore ; mineral ; stone ; mine-stone ; mine-stuff ; minestuff ; metal. (18)
minerai à basse teneur *ou* minerai à bas titre (*m.*), low-grade ore ; base ore. (19)
minerai à gangue fusible (*m.*), self-fluxing ore ; self-going ore ; fluxing-ore. (20)
minerai à haute teneur (*m.*), high-grade ore. (21)
minerai à la surface (*m.*), ore at grass ; stone at grass. (22)
minerai à teneur constante (*m.*), ore constant in grade. (23)
minerai à teneur variable (*m.*), ore varying in grade. (24)
minerai abattu (*m.*), ore broken down ; broken ore. (25)
minerai argileux (*m.*), argillaceous ore ; clayey ore ; clayed ore. (26)
minerai bocardé *ou* minerai broyé (*m.*), milled ore ; crushed ore. (27)
minerai brut (*m.*), raw ore ; crude ore. (28)
minerai carbonaté (*m.*), carbonate ore. (29)
minerai complexe (*m.*), complex ore. (30)
minerai concassé (*m.*), broken ore. (31)
minerai contenant du métal à l'état libre (*m.*), free-milling ore. (32)
minerai d'argent à basse teneur (*m.*), low-grade silver ore. (33)
minerai de basse teneur *ou* minerai de faible teneur (*m.*), low-grade ore ; base ore. (34)
minerai de fer (*m.*), iron ore ; ironstone. (35)
minerai de fer argileux (*m.*), argillaceous iron ore ; clay iron ore ; clay ironstone. (36)
minerai de fer des marais (*m.*), bog-iron ore ; bog-iron ; bog-ore ; bog-mine ; bog-mine ore ; swamp-ore ; morass-ore ; meadow-ore. (37)
minerai de fer micacé (*m.*), micaceous iron ore. (38)
minerai de filon (*m.*), lode-ore. (39)
minerai de haute teneur (*m.*), high-grade ore. (40)
minerai de mercure (*m.*), mercury ore ; quick--silver ore. (41)

minerai de plomb argentifère (*m.*), argentiferous lead ore ; silver-lead ore. (42)
minerai de rebut (*m.*), waste ore. (43)
minerai de scheidage (*m.*), cobbed ore ; bucked ore ; spalls ; hand-sorted ore. (44)
minerai découpé (*m.*), blocked out ore ; ore blocked out. (45)
minerai des tourbières (*m.*). Same as minerai de fer des marais.
minerai docile (*m.*), docile ore. (46)
minerai dont le métal est totalement amalgamable (*m.*). free-milling ore. (47)
minerai en place (*m.*), mineral in place ; reef in situ ; unbroken ore. (48)
minerai en réserve (*m.*), ore in reserve. (49)
minerai en transit (*m.*), ore in transit. (50)
minerai en vue (*m.*), ore in sight. (51)
minerai extrait (*m.*), mineral gotten ; mineral extracted ; mineral raised ; ore raised. (52)
minerai ferreux (*m.*), iron ore ; ironstone. (53)
minerai friable (*m.*), friable ore. (54)
minerai non-abattu (*m.*), unbroken ore ; reef in situ. (55)
minerai non-payant *ou* minerai non-rémunérateur (*m.*), unpayable ore ; non-payable stone. (56)
minerai non-triable (*m.*), unsortable ore. (57)
minerai oxydé (*m.*), oxide ore ; oxidized ore. (58)
minerai pauvre (*m.*), lean ore ; poor stone ; low-grade ore ; base ore. (59)
minerai payant (*m.*), payable ore ; pay-ore ; payable stone. (60)
minerai pisiforme (*m.*), pea ore ; pisiform ore. (61)
minerai possible (*m.*), possible ore. (62)
minerai probable (*m.*), probable ore. (63)
minerai rebelle *ou* minerai réfractaire (*m.*), rebellious ore ; stubborn ore ; refractory ore. (64)
minerai rémunérateur (*m.*), payable ore ; pay-ore ; payable stone. (65)
minerai rocheux (*m.*), rocky ore. (66)
minerai simple (*m.*), simple ore. (67)
minerai sulfuré (*m.*), sulphide ore. (68)
minerai sur la halde (*m.*), ore on the dump. (69)
minerai telluré (*m.*), telluride ore. (70)
minerai tenace (*m.*), tough ore. (71)
minerai terreux *ou* minerai sablonneux (*m.*), earthy ore ; muddy ore ; sandy ore. (72)
minerai tout venant (*m.*), run of mine ; unsorted ore. (73)
minerai trié à la main (*m.*), hand-sorted ore. (74)
minéral, -e, -aux (*adj.*), mineral : (75)
huile minérale (*f.*), mineral oil. (76)
minéral (*n.m.*), mineral : (77)
minéral de contact, contact-mineral. (78)
minéraux accessoires, accessory minerals. (79)
minéraux apparentés, related minerals. (80)
minéraux essentiels, essential minerals. (81)
minéraux originels, original minerals. (82)
minéraux secondaires, secondary minerals. (83)
minéraux simples, simple minerals. (84)
minéralisable (*adj.*), mineralizable. (85)
minéralisateur, -trice (*adj.*), mineralizing , ore-carrying : (86)
fluide minéralisateur (*m.*), mineralizing fluid. (87)
solutions minéralisatrices (*f.pl.*), mineralizing solutions ; ore-carrying solutions. (88)

les propriétés minéralisatrices du soufre (*f.pl.*), the mineralizing properties of sulphur. (1)

minéralisateur (*n.m.*), mineralizer. (2)

minéralisation (*n.f.*), mineralization. (3)

minéralisé, -e (*adj.*), mineralized ; mineral-bearing ; mineral : (4)

zone minéralisée (*f.*), mineralized zone ; mineral-bearing zone ; mineral belt. (5)

minéraliser (*v.t.*), to mineralize. (6)

minéraliser (se) (*v.r.*), to mineralize ; **to become** mineralized. (7)

minéralogie (*n.f.*), mineralogy. (8)

minéralogie chimique (*f.*), chemical mineralogy. (9)

minéralogie descriptive (*f.*), descriptive min-eralogy. (10)

minéralogie déterminative (*f.*), determinative mineralogy. (11)

minéralogie économique (*f.*), economic min-eralogy. (12)

minéralogie physique (*f.*), physical mineralogy. (13)

minéralogique (*adj.*), mineralogic ; mineralogical : (14)

collection minéralogique (*f.*), mineralogical collection. (15)

minéralogiquement (*adv.*), mineralogically. (16)

minéralogiste *ou* **minéralogue** (pers.) (*n.m.*), mineralogist. (17)

minerie (*n.f.*), rock-salt mine. (18)

minette (Pétrol.) (*n.f.*), minette. (19)

minette (minerai) (*n.f.*), a low grade iron ore. (20)

mineur (pers.) (*n.m.*), miner. (21)

mineur à l'entreprise (*m.*), contract-miner ; bargainman. (22)

mineur au charbon (*m.*), coal-miner ; collier. (23)

mineur au rocher (*m.*), stoneman ; metal-man ; drifter ; mullocker. (24)

mineur d'or (*m.*), gold-miner. (25)

mineur travaillant à la pioche et à la pelle (*m.*), hand-miner. (26)

minier, -ère (*adj.*), mining (*used as adj.*) : (27)

droits miniers (*m.pl.*), mining-rights. (28)

région minière (*f.*), mining-district. (29)

minière (mine peu profonde) (*n.f.*), surface-mine ; surface-working ; diggings. (30)

minière (gangue) (*n.f.*), gangue ; matrix. (31)

minière à ciel ouvert (*f.*), daylight-mine ; open-diggings. (32)

minière à flanc de coteau (*f.*), hillside-workings ; hill-diggings. (33)

minière de minerais de fer d'alluvion (*f.*), iron-stone-quarry ; iron-placer. (34)

minime (*adj.*), minute ; small. (35)

minimum [**minima** *ou* **minimums** *pl.*] (*adj.*) (*invariable pour les deux genres*), minimum : (36)

poids minimum *ou* minimum de poids (*m.*), minimum weight. (37)

pression minimum (*f.*), minimum pressure. (38)

altitudes minima *ou* altitudes minimums (*f.pl.*), minimum altitudes. (39)

minimum [**minima** *ou* **minimums** *pl.*] (*n.m.*), minimum. (40)

minimum (calibres à tolérances), not go ; not go on ; not go in. (41) See **maximum et minimum** for example.

ministre des mines (*m.*), minister of mines. (42)

minite (Explosif) (*n.f.*), minite. (43)

minium (Minéral) (*n.m.*), minium. (44)

minium (article de commerce) (*n.m.*), red lead : (45)

mastic de minium (*m.*), red-lead putty. (46)

minuscule (*adj.*), minute. (47)

minute (unité de temps ou de mesure angulaire) (*n.f.*), minute. (48)

minutieusement (*adv.*), minutely. (49)

minutieux, -euse (*adj.*), minute. (50)

miocène (Géol.) (*adj.*), Miocene ; Miocenic ; Meiocene ; Meiocenic. (51)

miocène (*n.m.*), Miocene ; Meiocene : (52)

le miocène, the Miocene. (53)

mionite (Minéral) (*n.f.*), mionite ; meionite. (54)

mirabilite (Minéral) (*n.f.*), mirabilite ; Glauber's salt ; sal mirabile. (55)

mire (*n.f.*), levelling-staff ; levelling-rod ; levelling-pole ; station-staff ; station-rod ; station-pole ; staff ; rod ; pole. (56)

mire à voyant (*f.*), target levelling-rod. (57)

mire parlante (*f.*), self-reading rod ; speaking rod. (58)

mire parlante en trois parties s'emboîtant l'une dans l'autre *ou* **mire à emboîtement** (*f.*), telescopic levelling-staff. (59)

mire stadia *ou* **mire pour lecture à la stadia** *ou* **mire forestière** (*f.*), stadia-rod. (60)

miroir (*n.m.*), mirror. (61)

miroir ardent (Phys.) (*m.*), burning-mirror. (62)

miroir concave (*m.*), concave mirror. (63)

miroir hyperbolique (*m.*), hyperbolic mirror. (64)

miroir parabolique (*m.*), parabolic mirror. (65)

miroir plan (*m.*), plane mirror. (66)

miroir réfléchissant *ou* **miroir réflecteur** (*m.*), reflecting mirror. (67)

miroir sphérique (*m.*), spherical mirror. (68)

miroirs de faille *ou* **miroirs de filon** (Géol.) (*m.pl.*), slickensides. (69)

miroirs de Fresnel (*m.pl.*), Fresnel's mirrors. (70)

mis (-e) à la terre (Élec.), earthed ; grounded : (71)

conducteur mis à la terre (*m.*), earthed con-ductor ; grounded conductor. (72)

mis (-e) en court-circuit (Élec.), short-circuited ; shorted. (73)

miscibilité (*n.f.*), mixability ; mixibility ; miscibility : (74)

miscibilité de deux métaux, mixability of two metals. (75)

miscible (*adj.*), mixable ; mixible ; miscible. (76)

mise (*n.f.*), putting. (77) See also **mettre**.

mise (d'un paquet) (Métall.) (*n.f.*), bar (of a pile) (one of the bars composing a pile, fagot, or packet) : (78)

mises en long et en travers, bars placed lengthwise and crosswise ; bars in one layer placed across those of the layer beneath. (79)

mises en long avec plein sur joint, bars crossing joint ; bars breaking joint. (80)

mise à dimensions (*f.*), sizing : (81)

mise à dimensions d'un trou, sizing a hole. (82)

mise à feu (Travail aux explosifs) (*f.*), lighting ; firing. (83)

mise à feu d'un haut fourneau (*f.*), blowing in a blast-furnace. (84)

mise à l'abri (*f.*), putting under cover ; shelter-ing ; housing. (85)

mise à la terre (Élec.) (*f.*), earthing ; grounding : (86)

mise à la terre des conducteurs, earthing, grounding, conductors. (1)

mise à nu (*f.*), denudation; stripping; exposure; exposing: (2)
mise à nu du minerai dans une galerie, exposure of ore in a level. (3)

mise à terre de marchandises (*f.*), landing of goods. (4)

mise à terris (Mines) (*f.*), dumping: (5)
mise à terris des stériles, dumping the wastes. (6)

mise au net d'un travail (*f.*), cleaning up a piece of work. (7)

mise au point (Opt., etc.) (*f.*), focusing; focussing; focalization; adjustment; bring-ing into position for use; positioning: (8)
mise au point d'un microscope, d'une lunette, d'un appareil photographique, focussing a microscope, a telescope, a camera. (9 *ou* 10 *ou* 11)
mise au point sur un objet plus ou moins éloigné, focussing on a more or less distant object. (12)
mise au point d'un foret, bringing a drill into position for use; positioning a drill. (13)
mise au point grossière par crémaillère, coarse adjustment by rack and pinion. (14)
mise au point définitive par vis micrométrique *ou* mise au point de précision au moyen d'une vis micrométrique, fine adjustment by micrometer-screw. (15)

mise aux molettes (Mines) (*f.*), overwinding; overrunning; pulleying; drawing up against pulley. (16)

mise du feu (Travail aux explosifs) (*f.*), lighting; firing. (17)

mise en batterie instantanée *ou* **mise en service immédiate** (se dit d'un appareil photo-graphique, ou analogue) (*f.*), ready for instant use; ready for use in a moment. (18)

mise en chantier (Fonderie) (*f.*), bedding in; floor bedding; planting: (19)
mise en chantier du modèle sur une couche en sable ou en plâtre, bedding in the pattern on a bed of sand or plaster. (20)

mise en circuit (Élec.) (*f.*), connecting. (21)

mise en court-circuit (Élec.) (*f.*), short-circuit-ing; shorting. (22)

mise en feu d'un haut fourneau (*f.*), blowing in a blast-furnace. (23)

mise en liberté du gaz par des trous de sonde percés en avance des travaux (*f.*), releasing gas by bore-holes kept in advance. (24)

mise en marche (*f.*), starting. (25)

mise en pression (vapeur) (*f.*), getting up steam: (26)
mise en pression d'une chaudière, getting up steam in a boiler. (27)

mise en sacs (*f.*), putting into sacks; sacking; bagging: (28)
mise en sacs du minerai, bagging the ore; putting the ore into sacks; sacking ore. (29)

mise en station (*f.*), setting up; set-up: (30)
mise en station d'une lunette, setting up a telescope. (31)

mise en train (action) (*f.*), starting. (32)

mise en train (appareil) (*f.*), starting-gear. (33)

mise en valeur (*f.*), turning to account. (34)

mise en vente (*f.*), putting up for sale; offering for sale; marketing. (35)

mise en vitesse (*f.*), getting up speed. (36)

mise en voie des dents d'une scie (*f.*), setting the teeth of a saw. (37)

mise hors circuit (Élec.) (*f.*), disconnecting; disconnection; breaking a circuit; opening a circuit. (38)

mise hors feu d'un haut fourneau (*f.*), blowing out a blast-furnace. (39)

mise sous pression (vapeur) (*f.*), getting up steam: (40)
mise sous pression d'une chaudière, getting up steam in a boiler. (41)

misé, -e (Métall.) (*adj.*), fagoted: (42)
métaux misés (*m.pl.*), fagoted metals. (43)

mispickel *ou* **misspickel** (Minéral) (*n.m.*), mis-pickel; mispikel; arsenopyrite; arsenical pyrites; white mundic. (44)

missourite (Pétrol.) (*n.f.*), missourite. (45)

mitraille (*n.f.*), scrap-iron; scrap. (46)

mitre (d'une cheminée) (*n.f.*), cowl (of a chimney). (47)

mixte (*adj.*), mixed; composite: (48)
chauffage mixte (charbon et pétrole) (*m.*), mixed heating (coal and oil). (49)
poutre mixte (bois et fer) (*f.*), composite beam (wood and iron). (50)

mixte (Élec.) (*adj.*), multiple series; parallel series: (51)
groupement mixte (*m.*), grouping in multiple series; multiple-series connection; parallel-series connection. (52)

mixtes (Préparation mécanique des minerais) (*n.m.pl.*), middlings; middles: (53)
la teneur en métal des mixtes, the metal contents of the middlings. (54)

mixtiligne (Géom.) (*adj.*), mixtilinear; mixtilinial: (55)
triangle mixtiligne (*m.*), mixtilinear triangle. (56)

mizzonite (Minéral) (*n.f.*), mizzonite. (57)

mobile (*adj.*), movable; removable; portable; mobile. (58)

mobile (corps en mouvement) (*n.m.*), moving body; body in motion: (59)
la force d'impulsion d'un mobile, the impulsive force of a moving body. (60)

mobile (force motrice) (*n.m.*), motive power; motive force; moving power; mover; prime mover: (61)
la vapeur est un puissant mobile, steam is a powerful motive force. (62)

mobilier de bureau (*m.*), office-furniture. (63)

mobilier et agencement, furniture and fittings; furniture, fittings, and fixtures. (64)

mobilité (*n.f.*), mobility; movableness. (65)

mode (*n.m.*), mode; method; system; manner; plan. (66) See also **méthode**.

mode d'accouplement *ou* **mode d'attachement** *ou* **mode d'attelage** (*m.*), mode of connection; attachment system; coupling method. (67)

mode d'emploi (*m.*), directions for use. (68)

mode d'exploitation (*m.*), working method. (69)

mode d'opérer (*m.*), mode of operation; modus operandi. (70)

mode de rencontre d'un minéral, du pétrole (*m.*), mode, manner, of occurrence of a mineral, of petroleum. (71 *ou* 72)

mode de soutènement (*m.*), system of support. (73)

mode de travail (*m.*), working method; plan of working. (74)

modelage (*n.m.*), modeling; modelling; pattern-making. (75)

modelage (Fonderie) (*n.m.*), pattern-making. (1)
modèle (parfait en son genre) (*adj.*), model : (2)
un atelier modèle, a model workshop. (3)
modèle (*n.m.*), model ; pattern ; type : (4)
modèle d'une machine, model of a machine. (5)
modèle perfectionné, improved pattern (*or* type). (6)
modèle (Fonderie) (*n.m.*), pattern : (7)
modèle pour la fonte, pattern for casting. (8)
modelé (Géogr. phys.) (*n.m.*), form ; relief : (9)
dans les régions tempérées, le modelé de la surface terrestre est l'œuvre combinée des agents atmosphériques et des eaux courantes, in temperate regions, the form (*or* the relief) of the earth's surface is the combined work of atmospheric agents and running water. (10)
modeler (*v.t.*), to model ; to shape ; to form. (11)
modeleur (Fonderie) (pers.) (*n.m.*), pattern-maker. (12)
modérateur, -trice (*adj.*), moderating ; govern-ing. (13)
modérateur (*n.m.*), moderator ; governor. (14)
modérateur (régulateur de vitesse) (Mach.) (*n.m.*), governor. (15) See **régulateur**, the synonymous and more usual word, for varieties.
modérateur (prise de vapeur de locomotive) (*n.m.*), throttle ; throttle-valve. (16)
modérateur (opp. à **accélérateur**) (Photogr.) (*n.m.*), restrainer ; retarder. (17)
modérateur de vitesse (Freinage) (*m.*), over--speed-gear ; overspeeder. (18)
modéré, -e (*adj.*), moderate : (19)
feu modéré (*m.*), moderate fire. (20)
modérément (*adv.*), moderately. (21)
modérer (*v.t.*), to moderate ; to check : (22)
modérer la vitesse d'une machine, to check the speed of a machine. (23)
moderne (*adj.*), modern ; up-to-date : (24)
méthodes modernes d'exploitation de mines (*f.pl.*), modern (*or* up-to-date) mining methods. (25)
moderniser (*v.t.*), to modernize. (26)
modificateur instantané (Méc.) (*m.*), trip. (27)
modification (*n.f.*), modification ; alteration. (28)
modifier (*v.t.*), to modify ; to alter. (29)
modifier (se) (*v.r.*), to alter ; to change. (30)
module (Math., Méc., & Phys.) (*n.m.*), modulus ; coefficient. (31)
module (Hydraul.) (*n.m.*), module. (32)
module (Engrenages) (*n.m.*), module (Gearing). (33)
module d'élasticité (Phys.) (*m.*), modulus of elasticity ; coefficient of elasticity ; stretch-modulus ; Young's modulus. (34)
module de résistance (*m.*), modulus of resistance. (35)
module de rupture (*m.*), modulus of rupture. (36)
moelle (Botanie) (*n.f.*), pith. (37)
moelle de roche (Minéral) (*f.*), rock-marrow. (38)
moelle de rocher (Minéral) (*f.*), amianthus. (39)
moellon (*n.m.*), rubble ; rubble-stone. (40)
moellon de roche (pour travaux hydrauliques) (*m.*), rock-rubble. (41)
moellonage (*n.m.*), rubble-work ; rubble. (42)
mofette (*n.f.*), damp ; mofette ; mephitic air. (43)

mofette inflammable (*f.*), fire-damp ; gas ; pit-gas ; fire. (44)
moignon (brosse de polisseur) (*n.m.*), mop-end brush. (45)
moindre résistance (*f.*), least resistance : (46)
on trouve les volcans sur les lignes de moindre résistance de l'écorce terrestre, volcanoes occur on the lines of least resistance of the earth's crust. (47)
moins-value [**moins-values** *pl.*] (*n.f.*), deprecia--tion ; decrease of (*or* in) value : (48)
moins-value du matériel, depreciation of plant. (49)
mois (*n.m.*), month. (50)
mois (prix convenu pour un mois de travail) (*n.m.*), month's pay ; month's salary ; month's wages ; monthly allowance : (51)
toucher son mois, to draw a month's pay. (52)
mois d'été (*m.pl.*), summer months. (53)
mois d'hiver (*m.pl.*), winter months. (54)
moisage (Construction hydraulique) (*n.m.*), waling : (55)
moisage des pieux, waling piles. (56)
moisage (Constr.) (*n.m.*), bracing. (57)
moise (Construction hydraulique) (*n.f.*), wale-piece ; wale ; waling. (58)
moise (Boisage des puits de mines) (*n.f.*), bunton ; dividing ; divider. (59)
moise (Constr.) (*n.f.*), brace : (60)
les moises d'une batterie d'étais, the braces of a system of shoring. (61)
moise (échafaudage) (Constr.) (*n.f.*), ledger. (62)
moiser (Constr. hydraul.) (*v.t.*), to wale. (63)
moiser (Constr.) (*v.t.*), to brace : (64)
moiser ensemble des étais pour éviter le flam--bage, to brace shores together to prevent buckling. (65)
moitié (*n.f.*), half. (66)
moitié chemin (à), half-way ; midway : (67)
rester à moitié chemin, to stop half-way. (68)
mol, molle (*adj.*). See **mou**.
molarite (*n.f.*), millstone grit. (69)
molasse (Géol.) (*n.f.*), molasse. (70)
môle (*n.m.*), mole ; breakwater. (71)
môle (Géol.) (*n.m.*), horst. (72)
moléculaire (*adj.*), molecular : (73)
le poids moléculaire de l'eau, the molecular weight of water. (74)
molécule (*n.f.*), molecule. (75)
moletage (*n.m.*), **moleter** (*v.t.*). Same as **molettage, moletter**.
molettage (*n.m.*), milling ; knurling ; nurling. (76)
molette (écrou taillé en scie sur le champ) (*n.f.*), milled nut ; knurled nut ; nurled nut. (77)
molette (de tourneur) (*n.f.*), milling-tool ; knurl--ing-tool ; nurling-tool ; knurl ; nurl. (78)
molette (*n.f.*) *ou* **molette coupante**, cutter ; cutter-wheel ; cutting-wheel ; revolving cutter : (79)
molettes pour décrasse-meules, cutters for emery-wheel dressers. (80)
molettes pour coupe-tuyaux, cutter-wheels for tube-cutter. (81)
molette (grande poulie fixée au-dessus d'un puits de mine, et sur laquelle passe le câble d'extraction) (*n.f.*), winding-pulley ; hoist--ing-pulley ; pit-head pulley ; pulley. (82)
molette (poulie supérieure d'un derrick) (*n.f.*), crown-pulley. (83)
molette à gorge (*f.*), sheave ; grooved wheel. (84)

molette à vis (*f.*), screw-pulley. (1)

moletter (*v.t.*), to mill ; to knurl ; to nurl : (2)
moletter la tête d'une vis pour l'empêcher de glisser dans les doigts, to mill, to knurl, to nurl, the head of a screw to prevent it slipping in the fingers. (3)

mollasse (Géol.) (*n.f.*), molasse. (4)

molybdate (Chim.) (*n.m.*), molybdate. (5)

molybdène (Chim.) (*n.m.*), molybdenum ; molybdena. (6)

molybdénite (Minéral) (*n.f.*), molybdenite. (7)

molybdeux (Chim.) (*adj. m.*), molybdenous ; molybdous. (8)

molybdine *ou* **molybdite** *ou* **molybdénocre** (Minéral) (*n.f.*), molybdite ; molybdine ; molybdic ochre. (9)

molybdique (Chim.) (*adj.*), molybdic ; molybdenic : (10)
acide molybdique (*m.*), molybdic acid. (11)

molysite (Minéral) (*n.f.*), molysite. (12)

moment (*n.m.*), moment ; instant : (13)
opération qui ne dure qu'un moment (*f.*), operation which only lasts a moment. (14)
à un moment donné, at a given moment. (15)

moment (produit de l'intensité d'une force par la distance de la droite suivant laquelle elle est appliquée à un point donné) (Méc.) (*n.m.*), moment. (16)

moment (quantité de mouvement d'un corps dans le premier instant qui suit la rupture de l'équilibre) (Méc.) (*n.m.*), momentum. (17)

moment d'inertie (*m.*), moment of inertia ; moment of rotation ; rotational inertia : (18)
le moment d'inertie de la section considérée par rapport à l'axe neutre, the moment of inertia of the section about the neutral axis. (19)

moment d'inertie polaire (*m.*), polar moment of inertia (20)

moment d'un couple (*m.*), moment of a couple ; twisting moment ; turning moment. (21)

moment d'une force (*m.*), moment of a force ; statical moment. (22)

moment d'une force par rapport à un axe (*m.*), moment of a force with respect to a line ; moment of a force with regard to an axis. (23)

moment d'une force par rapport à un point (*m.*), moment of a force with respect to a point. (24)

moment de stabilité (*m.*), moment of stability. (25)

moment des quantités de mouvement (*m.*), moment of momentum. (26)

moment fléchissant *ou* **moment de flexion** (*m.*), bending moment ; moment of flexure. (27)

moment magnétique (*m.*), magnetic moment. (28)

monadnock (Géogr. phys.) (*n.m.*), monadnock. (29)

monazite (Minéral) (*n.f.*), monazite. (30)

monceau (*n.m.*), heap ; pile. (31)

monde (*n.m.*), world. (32)

mondial, -e, -aux (*adj.*), world (*used as adj.*) ; world's : (33)
à production mondiale d'or, the world's production of gold. (34)

monheimite (Minéral) (*n.f.*), monheimite. (35)

monitor (lance hydraulique) (*n.m.*), monitor ; monitor-nozzle ; giant. (36)

Monnaie (hôtel des Monnaies) (*n.f.*), mint. (37)

monoatomique (Chim.) (*adj.*), monatomic. (38)

monoaxifère (Cristall.) (*adj.*), monoaxal ; uniaxial ; uniaxal : (39)
cristal monoaxifère (*m.*), monoaxal crystal. (40)

monobase (Cristall.) (*adj.*), monobasal. (41)

monobasique (Chim.) (*adj.*), monobasic. (42)

monocâble (*n.m.*), single-rope cable-tramway. (43)

monoclinal, -e, -aux (Géol.) (*adj.*), monoclinal ; monoclinous ; uniclinal : (44)
pli monoclinal (*m.*), monoclinal fold. (45)

monoclinal (Géol.) (*n.m.*), monocline ; mono-clinal. (46)

monoclinique (Cristall.) (*adj.*), monoclinic. (47)

monocylindrique (*adj.*), single-cylinder ; one-cylinder : (48)
moteur monocylindrique (*m.*), single-cylinder engine ; one-cylinder engine. (49)

monogénique (Géol.) (*adj.*), monogenetic : (50)
roche monogénique (*f.*), monogenetic rock. (51)

monophasé, -e (Élec.) (*adj.*), single-phase ; one-phase ; monophase ; uniphase : (52)
courant monophasé (*m.*), single-phase current. (53)

monophote (Élec.) (*adj.*), monophote ; mono-photal : (54)
régulateur monophote (*m.*), monophote regulator. (55)

monopole (*n.m.*), monopoly. (56)

monopoulie (*n.f.*), single pulley ; constant-speed pulley. (57)

monorail (*n.m.*), monorail. (58)

monorail transporteur *ou* **monorail aérien** (*m.*), runway ; overhead runway ; overhead track. (59)

monoréfringence (Opt.) (*n.f.*), monorefringence. (60)

monoréfringent, -e (*adj.*), monorefringent : (61)
cristal monoréfringent (*m.*), monorefringent crystal. (62)

monovalence (Chim.) (*n.f.*), monovalence ; monovalency ; univalence ; univalency. (63)

monovalent, -e (*adj.*), monovalent ; univalent. (64)

montage (action de porter de bas en haut) (*n.m.*), raising : (65)
montage des pierres au moyen d'une grue, raising stones by means of a crane. (66)

montage (action de disposer toutes les parties d'un ensemble pour qu'il soit en état de faire le travail auquel il est destiné) (*n.m.*), mounting ; setting ; setting up ; rigging up ; fitting ; fitting up ; erecting : (67)
montage d'un mandrin de tour sur un faux plateau, mounting a lathe-chuck on a back-plate. (68)
montage d'une pierre précieuse, mounting, setting, a precious stone. (69)
montage des photographies, mounting photo-graphs. (70)
montage des machines, erection of machinery ; setting up machines. (71)

montage (garniture ; armature) (*n.m.*), fitting : (72)
un montage mural, a wall fitting. (73)

montage (ouvrage montant) (opp. à descenderie) (Mines) (*n.m.*), rise ; raise ; riser ; rising ; upraise. (74)

montage (Élec.) (*n.m.*), connection ; connecting ; grouping ; coupling : (75)

montage des dynamos en série, une pile en quantité, connecting, connection of, group-ing, coupling, dynamos in series, a battery in multiple. (1 ou 2)

montage à force (Méc.) (*m.*), force fit. (3)

montage à frottement doux (*m.*),   push fit ; snug fit. (4)

montage à frottement dur (*m.*),   driving-fit ; drive fit. (5)

montage à glissement (*m.*), running fit. (6)

montage de Scott (Élec.) (*m.*), Scott connection. (7)

montage en boucle (Élec.) (*m.*),   ring-connec-tion ; loop-connection. (8)

montage en étoile (Élec.) (*m.*), star-connection ; star-grouping ; Y-connection. (9)

montage en étoile-triangle (Élec.) (*m.*),   star-delta connection ; star-delta grouping. (10)

montage en l'air (tour) (*m.*), face-plate mounting ; chucking independent of support from back centre. (11)

montage en triangle ou montage en delta (Élec.) (*m.*),   delta-connection ;   delta-grouping ; mesh-connection. (12)

montage entre pointes (tour) (*m.*),   mounting between centres ; chucking between centres. (13)

montage et démontage de la sonde (*m.*), making and unmaking the joints of the rods. (14)

montagne (*n.f.*), mountain. (15)

montagneux, -euse (*adj.*), mountainous : (16) pays montagneux (*m.*), mountainous country. (17)

montant, -e (*adj.*), rising ; ascending ; upgoing ; upward ; up : (18)

marée montante (*f.*), rising tide. (19)

benne montante (*f.*),   ascending bucket ; upgoing bucket. (20)

mouvement montant (*m.*),   upward motion. (21)

train montant (*m.*), up train. (22)

montant (pièce posée verticalement) (*n.m.*), upright ; post : (23)

deux montants et une traverse, two uprights and a crosspiece ; two posts and a cross-beam. (24)

montant (d'un bâti de porte ou d'un châssis de fenêtre) (*n.m.*), stile, style, (of a door-frame [frame that receives the panels] or of a window-sash). (25 ou 26)

montant (d'un dormant de porte ou de fenêtre) (*n.m.*),   jamb ; jamb-post ; post, (of a door or window-frame [frame fixed in the wall]). (27 ou 28)

montant (de cadre) (Boisage de galeries de mine) (*n.m.*),   post, prop, upright, arm, leg, leg-piece, (of frame or set) (Timbering mine-levels). (29)

montant (d'une machine à raboter) (*n.m.*), upright, standard, housing, (of a planing-machine). (30)

montant (d'une échelle) (*n.m.*),   side, side-bar, upright, (of a ladder). (31)

montant (total d'un compte) (*n.m.*), amount. (32)

montant d'angle (*m.*),   corner-post ; corner-pillar. (33)

montant de battement (*m.*), shutting-post. (34)

montant de busc (d'une porte d'écluse) (*m.*), mitre-post, meeting-post, (of a lock-gate). (35)

montant de chevalement (Mines) (*m.*),   poppet-leg. (36)

montant de côtière (*m.*),   hinge-post ; hinging-post ; swinging-post. (37)

montant de derrick (*m.*),   derrick-post. (38)

montant de porte (*m.*),   door-post ; gate-post. (39)

montant déversé (Mines) (*m.*),   underset prop. (40)

monte (action de porter de bas en haut) (*n.f.*), raising : (41)

monte et baisse par manivelle et vis sans fin agissant sur pignons et crémaillère,   raising and lowering by handle, endless screw and rack and pinion. (42)

monte-charge [monte-charge *pl.*] (*n.m.*),   goods-lift ; lift ; lifter ; elevator ; freight-elevator ; hoist. (43)

monte-charge (pour fourneau) (*n.m.*),   furnace-hoist. (44)

monte-charge à bras (*m.*),   hand-lift ; hand-power goods-lift. (45)

monte-charge à câble de traction (*m.*),   rope-lift. (46)

monte-charge à cages équilibrées (*m.*),   lift with balanced cages. (47)

monte-charge à frein de sûreté à friction, la charge restant suspendue dès que l'on cesse d'actionner le volant (*m.*),   self-sustaining friction-brake hoist. (48)

monte-charge à vis sans fin à transmission par courroie ; modèle à commande par le bas (*m.*),   belt-driven worm-gear lift ; under-driven type. (49)

monte-charge au moteur (*m.*),   power-lift. (50)

monte-charge électrique (*m.*),   electric lift ; electric elevator. (51)

monte-charge hydraulique à haute pression (*m.*), high-pressure hydraulic goods-lift. (52)

monte-coulée ou monte-coulées [monte-coulée ou monte-coulées *pl.*] (Fonderie) (*n.m.*), gate-spool. (53)

monte-courroie [monte-courroie *pl.*] (*n.m.*), belt-mounter ; belt-shipper ; belt-shifter. (54)

monte-sac [monte-sac *pl.*] (*n.m.*), sack-hoist. (55)

monte-wagon [monte-wagon *pl.*] (*n.m.*),   wagon-lift ; wagon-elevator ; car-elevator. (56)

montebrasite (Minéral) (*n.f.*),   montebrasite. (57)

montée (action) (*n.f.*),   ascent ; rising ; rise : (58) montée du pétrole, ascent, rising, rise, of the oil. (59)

montée (lieu, chemin qui va en montant) (*n.f.*), ascent ; rise ; acclivity : (60)

montée douce,   gentle rise ;   easy ascent ; gradual acclivity. (61)

montée (de la sole d'une galerie de mine) (*n.f.*), creep, creeping, heave, heaving, (of the floor of a mine-level). (62)

montée (ouvrage montant) (Mines) (*n.f.*).   Same as montage.

montée (d'une voûte) (*n.f.*),   rise, height above impost-level, (of an arch.) (63)

monter (*v.t.*), to ascend ; to go up ; to mount ; to climb : (64)

monter une colline,   to ascend, to go up, to mount, to climb, a hill. (65)

monter un fleuve, to go up a river. (66)

monter (*v.i.*),   to rise ; to go up ; to ascend ; to mount : (67)

route qui monte rapidement (*f.*),   road which rises rapidly. (68)

la rivière monte,   the river is rising. (69)

marchandise qui monte de prix (*f.*),   goods which rise in price. (70)

le thermomètre monte, the glass is going up. (1)

monter (transporter en un lieu plus élevé) (v.t.), to raise ; to hoist ; to elevate : (2)

monter les ouvriers mineurs au jour, to raise, to bring, the miners to the surface. (3)

monter (ajuster ; assembler) (v.t.), to mount ; to set up ; to rig up ; to fit up ; to erect : (4)

monter entre pointes une pièce cylindrique à tourner, to mount a cylindrical part to be turned between centres. (5)

monter une machine sur roues, un microscope sur son statif, to mount a machine on wheels, a microscope on its stand. (6 ou 7)

monter une machine à vapeur, to erect, to set up, a steam-engine. (8)

monter (enchâsser dans une garniture) (v.t.), to set ; to mount ; to fit : (9)

monter un diamant, to set, to mount, a diamond. (10)

monter un objectif sur un appareil photographique, to mount, to set, to fit, a lens on a camera. (11)

monter une photographie, to mount a photograph. (12)

monter (bander les ressorts) (v.t.), to wind ; to wind up : (13)

monter une montre, un ressort, to wind, to wind up, a watch, a spring. (14 ou 15)

monter (Élec.) (v.t.), to connect ; to couple ; to group. (16). See the noun montage for examples of use.

monter (percer en montant) (Mines) (v.i.), to rise ; to raise ; to put up a rise ; to put up a raise. (17)

monter (se) (être enchâssé ; s'adapter) (v.r.), to be mounted, set, fitted ; to fit : (18)

ces objectifs se montent sur tous appareils photographiques (m.pl.), these lenses fit on any camera. (19)

monteur (de machines) (pers.) (n.m.), fitter ; erector. (20)

monteur (de pierres précieuses) (n.m.), setter ; mounter. (21)

monteur de diamants (m.), diamond-setter. (22)

monteur-électricien [monteurs-électriciens pl.] (n.m.), electrical fitter. (23)

monticellite (Minéral) (n.f.), monticellite. (24)

monticule (n.m.), hillock ; mound ; knoll. (25)

montmorillonite (Minéral) (n.f.), montmorillonite. (26).

montre (n.f.), watch. (27)

montrer (v.t.), to show ; to exhibit : (28)

montrer une amélioration dans tout l'ensemble, to show an all-round improvement. (29)

montrer une tendance à augmenter de prix, to show a tendency to advance in price ; to exhibit a tendency to increase in price. (30)

montueux, -euse (adj.), hilly ; mountainous : (31)

pays montueux (m.), hilly country. (32)

monture (travail) (n.f.), mounting : setting ; fitting : (33)

monture d'une pierre précieuse, mounting, setting, a precious stone. (34)

monture (garniture) (n.f.), mounting ; mountings ; mount ; setting ; fitting : (35)

monture d'une pierre précieuse, mounting, setting, of a precious stone. (36)

monture d'un objectif photographique, mount, mounting, setting, of a photographic lens. (37)

monture d'un télescope, mounting of a telescope. (38)

monture de scie circulaire, circular saw spindle and pedestals ; mountings of a circular saw. (39)

monture hélicoïdale (objectif) (f.), focussing-mount ; focussing-setting. (40)

monture normale (objectif) (f.), ordinary mount (or setting). (41)

monture rentrante (objectif) (f.), sunk mount (or setting) ; countersunk mount (or setting). (42)

monzonite (Pétrol.) (n.f.), monzonite. (43)

moraillon (Serrur.) (n.m.), hasp. (44)

moraine (n.f.), moraine ; moraine-stuff ; till ; glacial till ; drift ; glacial drift. (45)

moraine de fond (f.), ground-moraine ; subglacial till ; subglacial drift. (46)

moraine déposée (f.), deposited drift. (47)

moraine frontale (f.), terminal moraine ; end-moraine. (48)

moraine inférieure (f.), basal moraine. (49)

moraine interne (f.), englacial till ; englacial drift. (50)

moraine latérale (f.), lateral moraine. (51)

moraine marginale (f.), marginal moraine. (52)

moraine médiane (f.), medial moraine : (53)

certains glaciers présentent à leur surface cinq ou six moraines médianes indépendantes (m.pl.), certain glaciers have on their surface five or six independent medial moraines. (54)

moraine mouvante (f.), moving drift. (55)

moraine profonde (f.), ground-moraine of retreat. (56)

moraine rempart [moraines remparts pl.] (f.), boulder-wall ; bowlder-wall. (57)

moraine riveraine (f.), lateral moraine of retreat. (58)

moraine superficielle (f.), superficial moraine ; surface-moraine ; superglacial till ; superglacial drift. (59)

morainique (adj.), morainic ; morainal ; morainial : (60)

dépôts morainiques (m.pl.), morainic deposits. (61)

morasse (n.f.), morass. (62)

morceau (n.m.), piece ; bit ; fragment : (63)

morceau de tuyau, piece of pipe ; bit of piping. (64)

morceau de terrain, piece of land ; patch of ground. (65)

morceau d'essai (éprouvette) (m.), test-piece ; coupon. (66)

morceaux de ferraille (m.pl.), scrap-iron. (67)

morceler (v.t.), to split up ; to divide up ; to break up ; to cut up. (68)

morceler (se) (v.r.), to split up ; to divide up ; to break up ; to be cut up. (69)

morcellement (n.m.), splitting up ; dividing up ; breaking up ; cutting up : (70)

morcellement d'une concession, dividing up a claim. (71)

morcellement des aires continentales (Géol.), breaking up of continental areas. (72)

mord (n.m.). Same as mors.

mordache (pour un étau) (n.f.), vice-cap ; vice-clamp ; gripping-pad ; vice-jaw : (73)

mordaches en plomb, lead vice-jaws ; lead vice-clamps. (74)

mordage (n.m.), biting : (75)

mordage des cylindres d'un laminoir (opp. à patinage), biting of the rolls of a rolling-mill. (1) (opp. to slipping).

mordâne (Charp.) (n.m.), bevelled haunch. (2)

mordant (n.m.), bite ; mordant : (3)
le mordant d'une lime, du sable, the bite of a file, of sand. (4 ou 5)
l'alun est employé en teinture comme mordant (m.), alum is used in dyeing as a mordant. (6)

mordre (entamer ; user) (v.t.), to bite : (7)
la lime mord le métal, the file bites the metal. (8)

mordre (s'enfoncer dans) (v.t.), to bite : (9)
fer à cheval garni de crampons qui mordent sur la glace et empêchent l'animal de glisser (m.), horseshoe provided with frost-nails which bite on ice and prevent the animal slipping. (10)
vis qui n'a pas mordu le bois (f.), screw which has not bitten the wood. (11)

mordre (serrer en laissant une empreinte) (v.t.), to bite ; to bite into : (12)
étau qui a fortement mordu le fer (m.), vice which has bitten into the iron badly. (13)

mordre (attaquer) (v.t.), to bite ; to act on ; to act upon : (14)
les acides mordent les métaux (m.pl.), acids bite metals. (15)

mordre (v.i.), to bite : (16)
une lime mord, a file bites. (17)
la vis mord, the screw bites. (18)

mordre (engrener) (Méc.) (v.i.), to interlock ; to engage ; to gear : (19)
pignon qui ne mord pas assez (m.), pinion which does not interlock properly ; pinion which does not engage closely enough. (20)

morénosite (Minéral) (n.f.), morenosite. (21)

morfil (n.m.), wire-edge. (22)

morfiler (v.t.), to remove the wire-edge from : (23)
morfiler le tranchant d'un couteau, to remove the wire-edge from the edge of a knife. (24)

morfil (n.m.), wire-edge. (25)

morillons (n.m.pl.), rough emeralds. (26)

morion (Minéral) (n.m.), morion. (27)

morphologie (n.f.), morphology. (28)

morphologique (adj.), morphologic ; morpho--logical. (29)

morphologiquement (adv.), morphologically. (30)

mors (n.m.), jaw : (31)
mors d'un étau, jaws, cheeks, chaps, of a vice. (32)
mors d'une tenaille, jaws of a pincers. (33)

mort (n.f.), death : (34)
mort produite par suffocation, death by suffocation. (35)

mort-fil [morts-fils pl.] (n.m.), wire-edge. (36)

mort mur. See morts murs.

mort-terrain (n.m.) ou morts-terrains (n.m.pl.) (Mines & Géol.), dead-ground : (37)
mort-terrain aquifère, water-bearing dead-ground. (38)
morts-terrains de recouvrement (Géol. & Mines), overburden ; overplacement ; burden ; soil-cap ; strippings ; muck. (39)

mortaisage (bois) (n.m.), mortising ; morticing. (40)

mortaisage (métal) (n.m.), slotting. (41)

mortaise (bois) (n.f.), mortise ; mortice. (42)

mortaise (métal) (n.f.), slot. (43)

mortaise (d'un palan) (n.f.), mortise (of a tackle-block). (44)

mortaise (logement d'une barre de cabestan) (n.f.), socket ; bar-hole. (45)

mortaise (d'un rabot) (n.f.), mouth, throat, (of a plane). (46)

mortaise aveugle (f.), stub-mortise. (47)

mortaise et tenon [mortaise (n.f.) ; tenon (n.m.)], mortise and tenon. (48)

mortaise passante (f.), through-mortice. (49)

mortaiser (bois) (v.t.), to mortise ; to mortice : (50)
mortaiser un montant, to mortise a post. (51)

mortaiser (métal) (v.t.), to slot : (52)
mortaiser une rainure de cale, to slot a key-way. (53)

mortaiseur, -euse (bois) (adj.), mortising ; mor--ticing : (54)
un outil mortaiseur, a mortising tool. (55)

mortaiseur, -euse (métal) (adj.), slotting : (56)
un outil mortaiseur, a slotting tool. (57)

mortaiseuse (bois) (n.f.), mortising-machine. (58)

mortaiseuse (métal) (n.f.), slotter ; slotting-machine. (59) See also machine à mortaiser.

mortalité (n.f.), death-rate ; mortality : (60)
mortalité par 1 000 ouvriers du fond, death-rate, mortality, per 1,000 persons employed underground. (61)

mortel, -elle (adj.), fatal ; mortal : (62)
accident mortel (m.), fatal accident. (63)

mortellement (adv.), fatally ; mortally : (64)
être blessé (-e) mortellement, to be fatally injured. (65)

mortes eaux (f.pl.), neap tide (66)

mortier (mélange de chaux, de sable, et d'eau) (n.m.), mortar. (67)

mortier (vase où l'on écrase les matières) (n.m.), mortar : (68)
mortier et pilon, pestle and mortar. (69)

mortier à double décharge (1) pour broyage à sec, ou (2) pour broyage à l'eau (m.), double-discharge mortar (1) for dry crushing, or (2) for wet crushing. (70 ou 71)

mortier avec embase en fonte (m.), mortar with anvil-block. (72)

mortier avec large pied pour fondation en béton (m.), wide-base mortar for setting on concrete. (73)

mortier de bocard (m.), stamp-mortar ; mortar-box. (74)

mortier de chaux (m.), lime mortar. (75)

mortier hydraulique (m.), hydraulic mortar. (76)

mortier sans amalgamation intérieure ou mortier avec paroi postérieure droite (m.), concen--tration-mortar ; straight-back mortar. (77)

morts murs d'un four de fusion (m.pl.), walls of a smelting-furnace. (78)

morts-terrains (n.m.pl.). See mort-terrain.

morvénite (Minéral) (n.f.), morvenite. (79)

mosandrite (Minéral) (n.f.), mosandrite. (80)

mot télégraphique ou mot de code ou mot codique (m.), code word. (81)

moteur, -trice (adj.), motive ; driving : (82)
force motrice (f.), motive force. (83)
roue motrice (f.), driving-wheel. (84)

moteur (appareil qui engendre l'énergie) (n.m.), engine ; motor. (85) See also machine.

moteur (force qui engendre ou qui transmet le mouvement) (Méc.) (n.m.), motor ; mover ; prime mover. (86)

moteur à air (m.), air-engine. (87)

moteur à air chaud (*m.*), hot-air engine ; caloric engine. (1)

moteur à air comprimé (*m.*), compressed-air engine. (2)

moteur à air raréfié (*m.*), atmospheric engine. (3)

moteur à alcool (*m.*), alcohol-engine ; spirit-engine. (4)

moteur à basse pression (*m.*), low-pressure engine. (5)

moteur à champ tournant (Élec.) (*m.*), rotary-field motor. (6)

moteur à combustion interne (*m.*), internal-com--bustion engine. (7)

moteur à condensation (*m.*), condensing engine. (8)

moteur à courant alternatif (Élec.) (*m.*), alternat--ing-current motor. (9)

moteur à courant continu (*m.*), direct-current motor ; continuous-current motor (10)

moteur à courant triphasé (*m.*), three-phase motor. (11)

moteur à cylindre unique (*m.*), single-cylinder engine ; one-cylinder engine. (12)

moteur à détente (*m.*), expansion-engine. (13)

moteur à double effet (*m.*), double-acting engine ; double-action engine. (14)

moteur à eau (*m.*), water-motor ; water-engine. (15)

moteur à essence *ou* **moteur à essence de pétrole** (*m.*), petrol-engine ; petrol-motor ; gasoline-engine ; gasolene-motor. (16)

moteur à explosions (*m.*), explosion engine ; explosion motor. (17)

moteur à flammes perdues (*m.*), waste-gas engine. (18)

moteur à gaz (*m.*), gas-engine ; gas-motor. (19)

moteur à gaz de hauts fourneaux (*m.*), blast-furnace gas engine. (20)

moteur à gaz pauvre (*m.*), producer-gas engine. (21)

moteur à gaz pauvre par aspiration (*m.*), suction gas-engine. (22)

moteur à gaz pauvre par pression (*m.*), pressure gas-engine. (23)

moteur à gaz perdus (*m.*), waste-gas engine. (24)

moteur à gazoline (*m.*), gasoline-engine ; gaso--lene-motor ; petrol-engine ; petrol-motor. (25)

moteur à grande vitesse (*m.*), high-speed engine. (26)

moteur à haute pression (*m.*), high-pressure engine. (27)

moteur à huile lourde (*m.*), heavy-oil engine. (28)

moteur à la surface (Mines) (*m.*), surface-engine ; bank-engine. (29)

moteur à mouvement alternatif (*m.*), reciprocat--ing engine. (30)

moteur à moyenne vitesse (*m.*), medium-speed engine. (31)

moteur à petite vitesse (*m.*), low-speed engine. (32)

moteur à pétrole (*m.*), oil-engine ; petroleum-engine. (33)

moteur à pétrole lampant (*m.*), paraffin-engine ; lamp-oil motor. (34)

moteur à refroidissement d'air (*m.*), air-cooled motor. (35)

moteur à refroidissement d'eau (*m.*), water-cooled motor. (36)

moteur à répulsion (Élec.) (*m.*), repulsion-motor. (37)

moteur à simple effet (*m.*), single-acting engine ; single-action engine. (38)

moteur à un seul cylindre (*m.*), one-cylinder engine ; single-cylinder engine. (39)

moteur à vapeur (*m.*), steam-engine ; steam-motor. (40)

moteur à vent (*m.*), wind-engine ; windmill. (41)

moteur à vitesse constante (*m.*), constant-speed motor. (42)

moteur ambulant (*m.*), locomotive engine. (43)

moteur animé (*m.*). See **moteurs animés.**

moteur asservi (*m.*), servo-motor ; relay. (44)

moteur asynchrone (Élec.) (*m.*), asynchronous motor. (45)

moteur atmosphérique (*m.*), atmospheric engine. (46)

moteur au jour (Mines) (*m.*), surface-engine ; bank-engine : (47)

moteur au jour commandant des pompes souterraines, surface-engine driving under--ground pumps. (48)

moteur auxiliaire (*m.*), auxiliary engine ; assist--ant engine. (49)

moteur bicylindrique (*m.*), two-cylinder engine ; double-cylinder engine ; double engine ; duplex-cylinder engine ; duplex engine. (50)

moteur compound (Élec.) (*m.*), compound motor ; compound-wound motor. (51)

moteur cuirassé (Élec.) (*m.*), totally enclosed motor ; ironclad motor. (52)

moteur d'épuisement (*m.*), pumping-engine ; draining-engine. (53)

moteur d'extraction (Mines) (*m.*), winding-engine ; winder ; hoisting-engine ; hoist ; drawing-engine ; draught-engine ; draft-engine. (54) See also **machine d'extraction.**

moteur d'extraction à vapeur (*m.*), steam winding-engine ; steam-winder ; steam hoisting-engine ; steam-hoist. (55)

moteur d'extraction électrique (*m.*), electric winding-engine ; electric winder ; electric hoisting-engine ; electric hoist. (56)

moteur d'induction (Élec.) (*m.*), induction-motor. (57)

moteur de battage (Sondage) (*m.*), walking-beam drilling-machine. (58)

moteur de démarrage (*m.*), starting-motor. (59)

moteur de roulage (Mines) (*m.*), hauling-engine. (60)

moteur de roulage et d'extraction (*m.*), hauling and winding-engine. (61)

moteur de traction (*m.*), hauling-engine. (62)

moteur Diesel (*m.*), Diesel engine. (63)

moteur du fond (*m.*), underground engine. (64)

moteur électrique (*m.*), electric motor ; motor ; electromotor ; electromagnetic engine. (65)

moteur électrique de traction (*m.*), electric hauling-engine. (66)

moteur électromagnétique (*m.*). Same as **moteur électrique.**

moteur en dérivation (*m.*), shunt motor ; shunt-wound motor. (67)

moteur éolien (*m.*), wind-engine ; windmill. (68)

moteur fixe (*m.*), stationary engine. (69)

moteur-générateur [moteurs-générateurs *pl.*] (Élec.) (*n.m.*), motor-generator. (70)

moteur horizontal (*m.*), horizontal engine. (71)

moteur hydraulique (*m.*), hydraulic engine ; water-motor. (72)

moteur inanimé (*m.*). See **moteurs inanimés.**

moteur marin (*m.*), marine engine. (1)

moteur polycylindrique (*m.*), multicylinder engine. (2)

moteur sans condensation (*m.*), non-condensing engine. (3)

moteur sans détente (*m.*), non-expansion engine. (4)

moteur semi-Diesel (*m.*), semi-Diesel engine. (5)

moteur-série [**moteurs-série** *pl.*] (Élec.) (*n.m.*), series-motor ; series-wound motor. (6)

moteur-série-répulsion [**moteurs-série-répulsion** *pl.*] *ou* **moteur-série-compensé** [**moteurs-série-compensés** *pl.*] (*n.m.*), series-repulsion motor. (7)

moteur-shunt [**moteurs-shunt** *pl.*] (Élec.) (*n.m.*), shunt motor ; shunt-wound motor. (8)

moteur souterrain (*m.*), underground engine. (9)

moteur synchrone (Élec.) (*m.*), synchronous motor. (10)

moteur vertical (*m.*), vertical engine ; upright engine. (11)

moteurs animés (opp. **à moteurs inanimés**) (*m.pl.*), muscular power of men and animals ; hand or animal power : (12)

extraction par **moteurs animés** (*f.*), hand or animal-power hoisting. (13)

moteurs inanimés (*m.pl.*), mechanical power. (14)

motif de congé (d'un employé) (*m.*), reason for dismissal (of an employee). (15)

motif de départ (d'un emploi) (*m.*), reason for leaving (an employ). (16)

motte (*n.f.*), clod ; clump ; lump ; block ; ball ; turf : (17)

motte de terre, clod, clump, lump, of earth. (18)

motte de tourbe, block of peat. (19)

mottes de tannée, tan-balls ; tan-turf. (20)

motte (Moulage en mottes) (Fonderie) (*n.f.*), snap-mould. (21)

motte de recouvrement (Moulage en terre) (Fonderie) (*f.*), cake ; loam cake. (22)

mottramite (Minéral) (*n.f.*), mottramite. (23)

mou (*m.*) *ou* **mol** (*m.*) [devant un mot commençant par une voyelle], **molle** (*f.*) (*adj.*), soft : (24)

le plomb et l'étain sont des métaux mous, lead and tin are soft metals. (25)

cire molle (*f.*), soft wax. (26)

mou, molle (se dit des câbles, des courroies, des chaînes) (*adj.*), slack ; loose : (27)

câble mou (*m.*), slack rope ; loose rope. (28)

mou (d'un câble, d'une courroie, d'une chaîne) (*n.m.*), slack (of a rope, of a belt, of a chain). (29 *ou* 30 *ou* 31)

mouche (Méc.) (*n.f.*), sun-and-planet motion ; planet-gear ; planet-gearing. (32)

mouchette (opp. **à rabot rond**) (*n.f.*), hollow plane. (33) (opp. to **round plane**).

mouchette (pour lampes) (*n.f.*), pricker ; wick-trimmer. (34)

moudre (*v.t.*), to mill ; to grind : (35)

moudre du minerai, to mill, to grind, ore. (36)

moufette (*n.f.*). Same as **mofette**.

moufle (Métall.) (*n.m.*), muffle. (37)

moufle (assemblage de poulies dans une même chape) (*n.f.*), pulley-block ; tackle-block ; sheave-block ; sheave ; block ; pulley ; muffle. (38)

moufle (palan) (*rare en ce sens*) (*n.f.*), tackle ; tackle and fall. (39)

moufle à chaîne (*f.*), chain-block. (40)

moufle à corde (*f.*), rope-block. (41)

moufle à estrope double (*f.*), double-strapped block ; purchase-block. (42)

moufle à tendre (Télégr. & Téléph.) (*f.*), lines-men's block. (43)

moufle à trois poulies à crochet tournant (*f.*), three-sheave pulley-block with swivel-hook ; triple-sheaved pulley-block with swivelling hook ; threefold block with swivel-hook. (44)

moufle avec chape à mentonnet et goupille à chaînette (*f.*), snatch-block with pin and chain. (45)

moufle de White (*f.*), White's tackle. (46)

moufle double (*f.*), double block. (47)

moufle fixe (*f.*), standing block. (48)

moufle mobile (*f.*), running block ; runner ; hoisting-block. (49)

moufle modèle de Londres, à trois poulies (*f.*), London pattern sheave-block with three sheaves ; three-sheave London pattern pulley-block. (50)

moufler (pourvoir de moufles) (*v.t.*), to rig with pulley-blocks : (51)

moufler un monte-charge, to rig a hoist with pulley-blocks. (52)

mouillage *ou* **mouillement** (*n.m.*), wetting ; moistening. (53)

mouillé, -e (*adj.*), wet ; moist : (54)

périmètre mouillé (Hydraul.) (*m.*), wet per-imeter. (55)

meule mouillée (meule en émeri, etc.) (*f.*), wet wheel. (56)

mouiller (*v.t.*), to wet ; to moisten. (57)

moulage (action) (*n.m.*), moulding ; molding ; casting ; founding. (58)

moulage (pièce moulée) (*n.m.*), casting. (59) See also **pièce**.

moulage à découvert (*m.*), open sand molding. (60)

moulage à vert (*m.*), green-sand moulding. (61)

moulage au sable (*m.*), sand-moulding. (62)

moulage au trousseau *ou* **moulage à la trousse** *ou* **moulage au gabarit** (*m.*), strickle-mould-ing ; sweep-moulding. (63)

moulage d'acier (*m.*), steel casting. (64)

moulage de fonte (*m.*), iron casting. (65)

moulage des métaux (*m.*), metal-founding ; casting metals. (66)

moulage des pièces en fonte (*m.*), iron-founding. (67)

moulage en châssis (*m.*), flask-molding ; box-moulding. (68)

moulage en coquille (*m.*), chill-casting. (69)

moulage en fonte douce (*m.*), soft-iron casting. (70)

moulage en fonte trempée (action) (*m.*), chill-casting. (71)

moulage en fonte trempée (pièce coulée) (*m.*), chilled-iron casting ; chill-casting. (72)

moulage en mottes (*m.*), snap-moulding. (73)

moulage en pièces battues (*m.*), false-core moulding. (74)

moulage en sable sec (*m.*), dry-sand moulding ; dry casting. (75)

moulage en sable vert (*m.*), green-sand mould-ing. (76)

moulage en série (*m.*), repeat work ; repetition work. (77)

moulage en terre (*m.*), loam-moulding ; mould-ing in loam ; loam-work. (78)

moulage étuvé (*m.*), stoved moulding. (79)

moulage sain (m.), sound casting. (1)

moulage sur plaque-modèle (m.), plate-moulding ; card-moulding. (2)

moule (n.m.), mould ; mold. (3)

moule à découvert (m.), open-sand mould. (4)

moule à lingots (m.), ingot-mould. (5)

moule à saumons ou moule de gueuses (m.), pig-mold. (6)

moule à vert (m.), green-sand mould. (7)

moule de sable (m.), sand mould. (8)

moule de terre (m.), loam mold. (9)

moule externe (d'un fossile) (opp. à moule interne) (Paléon.)(m.), mould ; external mould. (10)

moule interne (Paléon.) (m.), cast ; internal mould. (11)

moule pour mordaches en plomb, pour étaux (m.), mould for lead jaws, for vices. (12)

mouler (v.t.), to mould ; to mold ; to cast ; to found : (13)

mouler à vert, to mould in green sand. (14)

mouler en sable sec, to dry-mould. (15)

mouler la fonte, to found, to cast, iron. (16)

mouler (se) (v.r.), to cast ; to be cast ; to be moulded : (17)

l'aluminium se moule bien (m.), aluminium casts well. (18)

moulerie (n.f.), moulding-floor. (19)

mouleur (pers.) (n.m.), moulder ; pattern-molder. (20)

mouleur à sable (m.), sand-moulder. (21)

mouleur en terre (m.), loam-moulder. (22)

moulière (n.f.), millstone grit. (23)

moulin (n.m.), mill. (24)

moulin (Glaciologie) (n.m.), moulin ; glacier-mill. (25)

moulin à bocards (m.), stamp-mill. (26)

moulin à boulets (m.), ball-mill ; ball-crusher. (27)

moulin à boulets pour le broyage par voie humide (m.), wet-grinding ball-mill. (28)

moulin à boulets pour le broyage par voie sèche (m.), dry-grinding ball-mill. (29)

moulin à bras (m.), hand-mill. (30)

moulin à cylindres (m.), crushing-rolls ; rolls. (31)

moulin à cylindres à grande vitesse ou moulin à cylindres à marche rapide (m.), high-speed rolls. (32)

moulin à cylindres oscillants (m.), oscillating rolls. (33)

moulin à eau (m.), water-mill. (34)

moulin à échantillons (m.), sampling-mill. (35)

moulin à force centrifuge (m.), centrifugal mill. (36)

moulin à galets (m.), pebble-mill. (37)

moulin à or (m.), gold-mill. (38)

moulin à poudre (m.), powder-mill. (39)

moulin à quartz (m.), quartz-mill ; quartz-battery. (40)

moulin à vent (m.), windmill. (41)

moulin chilien (m.), Chilian mill. (42)

moulin colombien (m.), Colombian mill. (43)

moulin d'amalgamation (m.), amalgamating-pan ; amalgamation-pan ; pan ; pan-mill ; grinding-pan ; grinder. (44)

moulin de prospection (m.), prospecting-mill. (45)

moulin dit aéromoteur (m.), windmill for pump-ing ; wind-engine. (46)

moulin glaciaire (Glaciologie) (m.), moulin ; glacier-mill. (47)

moulin hongrois (m.), Hungarian mill. (48)

moulin travaillant à façon (m.), customs mill. (49)

mouliné, -e (se dit du bois) (adj.), worm-eaten. (50)

moulinet (d'un moulin à vent) (n.m.), flier, flyer, (of a windmill). (51)

moulinet (d'un instrument tel qu'un radiomètre) (n.m.), fly ; vane. (52)

moulinet (Hydraul.) (n.m.), current-meter. (53)

moulinet (de cheminée) (n.m.), smoke-jack. (54)

moulineur (Mines) (pers.) (n.m.), trammer ; brakeman ; incline-braker. (55)

moulure (n.f.), moulding ; molding. (56)

moulure (latte à rainures pour pose de fils) (Élec.) (n.f.), moulding ; casing. (57)

moulurés (fer ou acier) (n.m.pl.), mouldings. (58 ou 59)

moulurière (n.f.), moulding-machine (wood-working). (60)

mourir (v.i.), to die : (61)

mourir des blessures reçues, to die of injuries received. (62)

mourir (s'effacer ; disparaître) (Mines) (v.i.), to peter out ; to pinch out ; to die out : (63)

filon qui meurt (m.), lode which peters out (or pinches out) (or dies out). (64)

mousqueton (n.m.), safety-hook ; spring-hook ; clip-hook ; clevis ; clevice. (65)

mousse (n.f.), moss. (66)

mousse d'or (f.), sponge gold ; gold-sponge. (67)

mousse de platine (f.), sponge platinium ; platinum sponge. (68)

mousson (n.f.), monsoon. (69)

mousson d'hiver (f.), dry monsoon. (70)

mousson pluvieuse d'été ou simplement mousson d'été (f.), wet monsoon ; the monsoon. (71)

moustiquaire (n.f.), mosquito-net. (72)

moustique (n.m.), mosquito. (73)

mouton (d'une sonnette) (n.m.), monkey, ram, ramming-block, tup, beetle-head, (of a pile-driver). (74)

mouton (d'un marteau-pilon) (n.m.), tup, ram, hammer-head, head, (of a power-hammer). (75)

mouton (d'une machine à essayer au choc) (n.m.), tup (of a drop-test machine). (76)

mouture (n.f.), grinding : (77)

mouture du minerai, grinding ore. (78)

mouvement (n.m.), motion ; movement. (79)

mouvement (circulation des trains) (Ch. de f.) (n.m.), traffic. (80)

mouvement (d'air) (n.m.), circulation (of air). (81)

mouvement à secousse (m.), shaking motion. (82)

mouvement accéléré (m.), accelerated motion. (83)

mouvement alternatif (Méc.) (m.), reciprocating motion ; alternating motion : (84)

mouvement alternatif d'une scie, reciprocat-ing, alternating, motion of a saw. (85)

mouvement alternatif d'un piston, reciprocat-ing motion of a piston. (86)

mouvement alternatif (à) (Méc.), reciprocating ; alternating : (87)

pompe à mouvement alternatif (f.), reciprocat-ing pump. (88)

mouvement ascendant et descendant (m.), up-and-down motion. (89)

mouvement basculaire (m.), swinging movement ; seesaw motion ; balancing motion. (90)

mouvement circulaire (m.), rotary motion. (91)

mouvement composé (*m.*), compound motion. (1)
mouvement curviligne (*m.*), curvilinear motion. (2)
mouvement d'avance et de recul (*m.*), forward-and-backward movement; back-and-forth motion. (3)
mouvement d'avancement (Méc.) (*m.*), feed-motion. (4)
mouvement d'enroulement (*m.*), winding motion; coiling motion. (5)
mouvement d'horlogerie (*m.*), clockwork-move-ment. (6)
mouvement d'horlogerie à réveille-matin (*m.*), alarm clockwork movement. (7)
mouvement de bascule (*m.*). Same as **mouve-ment basculaire.**
mouvement de chariotage (tour) (*m.*), traversing motion (lathe). (8)
mouvement de monte et baisse automatique (*m.*), automatic rising-and-falling movement; self-acting raising-and-lowering motion. (9)
mouvement de pression (Méc.) (*m.*), feed-motion. (10)
mouvement de rotation (*m.*), rotary motion. (11)
mouvement de sonnette *ou* mouvement à ailes (*m.*), bell-crank motion. (12)
mouvement de surfaçage (tour) (*m.*), surfacing motion (lathe). (13)
mouvement de translation (*m.*), translation; translatory motion. (14)
mouvement de va-et-vient (*m.*), to-and-fro motion; reciprocating motion; alternating motion. (15)
mouvement giratoire (*m.*), gyratory motion. (16)
mouvement irrotationnel (*m.*), irrotational motion. (17)
mouvement latéral (*m.*), side motion; lateral motion. (18)
mouvement perpétuel (*m.*), perpetual motion. (19)
mouvement planétaire (*m.*), sun-and-planet motion; planet-gear; planet-gearing. (20)
mouvement rectiligne (*m.*), rectilinear motion; straight-line motion. (21)
mouvement rectiligne uniformément accéléré (*m.*), uniformly accelerated rectilinear motion. (22)
mouvement retardé (*m.*), retarded motion. (23)
mouvement rétrograde (*m.*), backward move-ment. (24)
mouvement rotatif (*m.*), rotary motion. (25)
mouvement simple (*m.*), simple motion. (26)
mouvement tourbillonnaire (*m.*), vortical motion. (27)
mouvement transversal automatique de l'outil (*m.*), automatic transverse movement of the tool. (28)
mouvement uniformément varié (*m.*), uniformly variable motion. (29)
mouvement varié (*m.*), variable motion. (30)
mouvement vibratoire (*m.*), vibratory motion. (31)
mouver (*v.t.*), to stir. (32)
mouvoir (*v.t.*), to move: (33)
mouvoir une pierre, un meuble, to move a stone, a piece of furniture. (34 *ou* 35)
mouvoir (se) (*v.r.*), to move: (36)
le piston se meut en ligne droite dans le cylindre, the piston moves in a straight line in the cylinder. (37)
moyen, -enne (*adj.*), average; mean; middle; mid; middling; medium: (38)

teneur moyenne (*f.*), average grade; medium grade. (39)
moyenne latitude (*f.*), mean latitude; middle latitude. (40)
position moyenne du tiroir (*f.*), mid-position of the slide-valve. (41)
moyenne pression *ou* moyenne tension (*f.*), medium pressure; medium tension. (42)
moyen (*n.m.*). See moyens.
moyen de fortune (*m.*), makeshift. (43)
moyennant (*prép.*), in consideration of; in return for; on: (44)
avoir le droit de propriété moyennant rede-vance, to have the right of possession on royalty. (45)
moyenne (*n.f.*), average; mean: (46)
la mèche de sûreté brûle à raison d'un mètre par minute, en moyenne, safety-fuse burns at the rate of one metre per minute, on an average. (47)
moyenne annuelle des pluies (*f.*), average annual rainfall. (48)
moyenne proportionnelle (Math.) (*f.*), proportion. (49)
moyenne proportionnelle arithmétique (*f.*), arithmetical proportion. (50)
moyenne proportionnelle géométrique (*f.*), geometrical proportion. (51)
moyennement (*adv.*), moderately; middlingly; fairly; on an average; upon an average: (52)
être moyennement riche, to be fairly rich. (53)
objets vendus de deux à trois francs moyen-nement (*m.pl.*), articles sold from two to three francs, on an average. (54)
moyens (instrumentalité) (*n.m.pl.*), means: (55)
moyens d'accès dans la mine, means of access to the mine. (56)
moyens d'existence *ou* moyens de vivre, means of subsistence (or of living). (57)
moyeu (d'une roue) (*n.m.*), nave, hub, boss, (of a wheel). (58)
moyeu (d'une poulie) (*n.m.*), hub (of a pulley). (59)
moyeu (de la came) (bocard) (*n.m.*), hub (of the cam) (ore-stamp). (60)
moyeu (d'une manivelle) (*n.m.*), boss (of a crank). (61)
mulet (animal) (*n.m.*), mule. (62)
multiple (*adj.*), multiple: (63)
machine à multiple expansion (*f.*), multiple-expansion engine. (64)
multiplicateur (Élec.) (*n.m.*), multiplier. (65)
multiplicateur de pression (Hydraul.) (*m.*), intensifier. (66)
multiplication (rapport d'engrenages) (*n.f.*), gear-ratio. (67)
multipolaire (*adj.*), multipolar: (68)
dynamo à courant continu multipolaire (*f.*), multipolar direct-current dynamo. (69)
multitubulaire (*adj.*), multitubular: (70)
chaudière multitubulaire (*f.*), multitubular boiler. (71)
mundick (Minéral) (*n.m.*), mundic. (72)
munir (*v.t.*), to supply; to provide; to furnish; to equip; to fit: (73)
munir d'aubes une roue-turbine, to provide, to furnish, to fit, a turbine-wheel with blades; to blade a turbine-wheel. (74)
munir (se) de, to provide oneself with. (75)
mur (*n.m.*), wall. (76)

mur (d'un filon) (opp. au **toit**) (*n.m.*), foot-wall, foot, lying-wall, ledger-wall, ledger, (of a lode). (1) (opp. to the **hanging wall, hanging side, top wall, or roof**).

mur (d'une couche de houille) (opp. au **toit**) (*n.m.*), floor, bottom, pavement, (of a coal-seam). (2) (opp. to the **roof** or **top**).

mur biais (*m.*), skew wall. (3)

mur bombé *ou* **mur bouclé** (*m.*), bulged wall. (4)

mur d'appui (*m.*), breast-wall ; breast-high wall. (5)

mur de chute (d'une écluse) (*m.*), lift-wall (of a canal-lock). (6)

mur de cloison (*m.*), partition wall. (7)

mur de clôture (*m.*), enclosing wall. (8)

mur de face (*m.*), front wall. (9)

mur de garde *ou* **mur de barrage** (Hydraul.) (*m.*), fender-wall. (10)

mur de parapet (*m.*), breast-wall ; breast-high wall ; parapet. (11)

mur de refend (*m.*), cross-wall. (12)

mur de retenue *ou* **mur de revêtement** (*m.*). Same as **mur de soutènement**.

mur de soubassement (*m.*), basement wall ; base wall. (13)

mur de soutènement (*m.*), retaining wall ; retain-wall ; sustaining wall ; face-wall ; breast-wall. (14)

mur de soutènement avec fruit extérieur (*m.*), retaining wall battering away from the bank. (15)

mur de soutènement avec fruit intérieur (*m.*), retaining wall battering towards the bank. (16)

mur de soutènement avec retraites intérieures (*m.*), retaining wall with retreats on the inside. (17)

mur de terrasse (*m.*). Same as **mur de soutène-ment**.

mur déversé (*m.*), overhanging wall. (18)

mur en briques (*m.*), brick wall. (19)

mur en pierre (*m.*), stone wall. (20)

mur en pierres sèches (*m.*), dry wall ; dry-stone wall. (21)

mur en pierres sèches *ou* **mur de remblai** (Mines) (*m.*), pack ; pack-wall ; building. (22)

mur en retour (*m.*), return-wall ; flank wall. (23)

mur en surplomb *ou* **mur forjeté** (*m.*), over-hanging wall. (24)

mur latéral (*m.*), side wall. (25)

mur menaçant ruine (*m.*), falling wall. (26)

mur mitoyen (*m.*), party wall. (27)

mur orbe (*m.*), blank wall ; blind wall ; dead wall. (28)

mûr, -e (Géogr. phys.) (*adj.*), mature : (29)
vallée mûre (*f.*), mature valley. (30)

murage (*n.m.*), walling. (31)

muraille (*n.f.*), wall. (32)

muraille de rochers (*f.*), wall of rocks. (33)

muraille qui pousse (*f.*), bulging wall. (34)

muraillement (*n.m.*), walling : (35)
muraillement d'un puits de mine, walling a mine-shaft. (36)

muraillement en pierre sèche (*m.*), dry-walling. (37)

murailler (*v.t.*), to wall ; to wall in ; to wall up. (38)

mural, -e, -aux (*adj.*), wall (*used as adj.*) ; mural : (39)
console murale (*f.*), wall-bracket. (40)

murchisonite (Minéral) (*n.f.*), murchisonite. (41)

murer (*v.t.*), to wall ; to wall in ; to wall up : (42)
murer une porte, to wall up a door. (43)

muretin en pierres sèches *ou* **muretin de remblai** (Mines) (*m.*), pack ; pack-wall ; building. (44)

muschelkalk (Géol.) (*n.m.*), muschelkalk. (45)

muscovite (Minéral) (*n.f.*), muscovite. (46)

myriagramme (*n.m.*), myriagramme =22·046 lbs. (47)

mysorine (Minéral) (*n.f.*), mysorin. (48)

N

nacre (*n.f.*) *ou* **nacre de perle**, mother-of-pearl ; pearl ; nacre. (49)

nacré, -e (*adj.*), nacreous ; pearly ; pearlaceous ; perlaceous ; margaritaceous. (50)

nacrite (Minéral) (*n.f.*), nacrite. (51)

nagelfluh (Géol.) (*n.m.*), nagelfluh ; nagelflue. (52)

nager (*v.i.*), to swim ; to float : (53)
l'huile nage sur l'eau (*f.*), oil floats on water. (54)

nagyagite (Minéral) (*n.f.*), nagyagite. (55)

naissance (d'une voûte) (*n.f.*), spring, springing, (of an arch). (56)

naissance (culotte de chaudière) (*n.f.*), water-leg ; leg. (57)

naissant, -e (*adj.*), nascent : (58)
état naissant (Chim.) (*m.*), nascent state (*or* condition). (59)
rouge naissant (Métall.) (*m.*), nascent red. (60)

nantokite (Minéral) (*n.f.*), nantokite. (61)

naphtadil (Minéral) (*n.m.*), neftgil. (62)

naphtalène (*n.m.*) *ou* **naphtaline** (*n.f.*), naph-thalene ; naphthaline ; tar-camphor. (63)

naphtalique (*adj.*), naphthalic. (64)

naphte (*n.m.*), naphtha. (65)

naphte brut (*m.*), crude naphtha. (66)

naphte de charbon (*m.*), coal-naphtha ; coal-tar naphtha. (67)

naphte de pétrole (*m.*), petroleum naphtha. (68)

naphte de schistes (*m.*), shale-naphtha. (69)

naphte minéral *ou* **naphte natif** (*m.*), mineral naphtha ; petroleum ; rock-oil ; stone-oil ; earth-oil ; fossil oil ; coal-oil. (70)

nappe (vaste étendue) (*n.f.*), sheet. (71)

nappe (couche plane) (Géol.) (*n.f.*), sheet ; flat sheet ; blanket deposit ; blanket vein ; plate : (72)
mine de fer disposée en nappes (*f.*), iron-mine formed of flat sheets. (73)

nappe (de roche) (Géol.) (*n.f.*), rock-sheet. (74)

nappe (Hydraul.) (*n.f.*), sheet ; nappe : (75)
nappe d'eau au-dessus du seuil d'un déversoir, sheet, nappe, of water above the crest of a weir. (76)

nappe (Géom.) (*n.f.*), nappe. (77)

nappe adhérente (Hydraul.) (*f.*), adhering nappe. (78)

nappe aquifère (Géol.) (*f.*), water-bearing stratum ; water-bearing strata ; watery strata. (1)

nappe charriée (Géol.) (*f.*).   Same as **nappe de charriage**.

nappe d'eau (*f.*), sheet, body, of water. (2)

nappe d'eau souterraine (*f.*), underground sheet of water. (3)

nappe d'infiltration (Géol.) (*f.*), underground water. (4)

nappe de charriage (Géol.) (*f.*), translated rock-sheet ; travelled rock-sheet. (5)

nappe de feu (*f.*), sheet of flame. (6)

nappe de pétrole (*f.*). Same as **nappe pétrolifère**.

nappe déprimée (Hydraul.) (*f.*), depressed nappe. (7)

nappe éruptive (Géol.) (*f.*), eruptive rock-sheet ; sheet of eruptive rock. (8)

nappe ignée (Géol.) (*f.*), igneous rock-sheet. (9)

nappe libre (Hydraul.) (*f.*), free nappe. (10)

nappe noyée *ou* **nappe noyée en dessous** (Hydraul.) (*f.*), drowned nappe ; wetted nappe. (11)

nappe pétrolifère (*f.*), oil-sheet ; body of petroleum. (12)

nappe phréatique (Géol.) (*f.*), underground water. (13)

nappe salée *ou* **nappe saline** (*f.*), sheet of saline water. (14)

nappe superficielle (opp. à **nappe phréatique** *ou* **nappe d'infiltration**) (Géol.) (*f.*), ground-water ; bottom-water. (15)

narine (orifice de la crépine) (pompe) (*n.f.*), snore-hole. (16)

natif, -ive (*adj.*), native : (17)

or natif (*m.*), native gold. (18)

natrolite (Minéral) (*n.f.*), natrolite. (19)

natron *ou* **natrum** (Minéral) (*n.m.*), natron. (20)

natronitre (Minéral) (*n.m.*), soda-nitre ; nitratine. (21)

natte (*n.f.*), mat ; matting. (22)

natte de jonc (*f.*), rush matting. (23)

nature (puissance, force active) (*n.f.*), nature : (24)

les lois de la nature (*f.pl.*), the laws of nature. (25)

nature (tempérament) (*n.f.*), nature ; character : (26)

nature des couches traversées, character of the strata passed through. (27)

nature erratique des gîtes, erratic nature of the deposits. (28)

nature exceptionnelle du minerai, unusual nature of the ore. (29)

nature (sorte) (*n.f.*), nature ; kind ; sort. (30)

nature (grandeur naturelle ; vraie grandeur) (*n.f.*), full size : (31)

dessin nature (*m.*), full-size drawing. (32)

dessin 1/2 nature (*m.*), half-size drawing. (33)

gravure 1/4 nature (*f.*), quarter-size illustra-tion. (34)

naturel, -elle (*adj.*), natural : (35)

phénomènes naturels (*m.pl.*), natural phe-nomena. (36)

naturel, -elle (Minéralogie) (*adj.*), native : (37)

chlorure hydraté naturel d'alumine (*m.*), native hydrous aluminum chloride. (38)

naturellement (*adv.*), naturally. (39)

naumannite (Minéral) (*n.f.*), naumannite. (40)

navette (de menuisier) (*n.f.*), smoothing-plane with curved stock. (41)

navette (Sondage) (*n.f.*), bulldog casing-spear. (42)

navigabilité (*n.f.*), navigability : (43)

navigabilité d'un fleuve, navigability of a river. (44)

navigable (*adj.*), navigable : (45)

fleuve navigable (*m.*), navigable river. (46)

navigation (*n.f.*), navigation. (47)

navigation fluviale (*f.*), river navigation. (48)

naviguer (*v.t. & v.i.*), to navigate. (49)

navire (*n.m.*), ship ; vessel. (50)

navire à vapeur (*m.*), steamship ; steamer. (51)

navire à vapeur au long cours (*m.*), ocean-going steamer. (52)

navire à voiles (*m.*), sailing vessel ; sailing ship. (53)

navire charbonnier (*m.*), collier. (54)

navire-citerne [**navires-citernes** *pl.*] (*n.m.*), tank-ship ; tank-vessel ; tanker. (55)

navire long-courrier (*m.*), ocean-going ship. (56)

navire pétrolier (*m.*), oil-ship. (57)

nécessaire (*adj.*), necessary ; requisite ; needful ; required. (58)

nécessaire (trousse) (*n.m.*), outfit ; set. (59)

nécessaire à réparations (*m.*), repair-outfit. (60)

nécessaire à souder (*m.*), soldering-outfit ; soldering-set. (61)

nécessaire d'outils (*m.*), tool-pad ; pricker-pad. (62)

nécessairement (*adv.*), necessarily. (63)

nécessité (*n.f.*), necessity. (64)

nécessiter (*v.t.*), to necessitate ; to require : (65)

travail nécessitant une grande précision (*m.*), work requiring great precision. (66)

neck (cheminée d'ascension) (Géol.) (*n.m.*), neck. (67)

neftgil (Minéral) (*n.m.*), neftgil. (68)

négatif, -ive (*adj.*), negative. (69)

négatif (Photogr.) (*n.m.*), negative. (70) See also **cliché**.

négligeable (*adj.*), negligible ; unimportant : (71)

quantité négligeable (*f.*), negligible quantity. (72)

négligemment (*adv.*), negligently ; carelessly. (73)

négligence (*n.f.*), negligence ; carelessness ; neglect : (74)

négligence de la part des ouvriers, negligence, carelessness, on the part of the workmen. (75)

négligence de précautions convenables, neglect of proper precautions. (76)

négligent, -e (*adj.*), negligent ; careless ; neglect--ful. (77)

négliger (*v.t.*), to neglect : (78)

négliger l'effet de la pesanteur, la résistance de l'air, to neglect the effect of gravity, the resistance of the air. (79 *ou* 80)

nègre (pers.) (*n.m.*), negro ; black : (81)

les nègres africains, the African negroes. (82)

neige (*n.f.*), snow. (83) (Often used in the plural in French where singular is used in English ; as, enlèvement des neiges *or* enlèvement de la neige, clearing away snow ; removal of snow) :

neiges d'hiver *ou* neiges hivernales, winter snows. (84)

neiges perpétuelles *ou* neiges persistantes *ou* neiges éternelles, perpetual snow. (85)

neigé, -e (*adj.*), snowy : (86)

des cimes neigées (*f.pl.*), snowy summits. (87)

neiger (*v. impersonnel*), to snow : (88)

quand il neige, when it snows. (89)

neigeux, -euse (*adj.*), snowy : **(1)**
montagnes aux pics neigeux (*f.pl.*), mountains with snowy peaks. (2)

néo-ytterbium (Chim.) (*n.m.*), neoytterbium ; ytterbium ; aldebaranium. (3)

néocomien, -enne (Géol.) (*adj.*), Neocomian. (4)

néocomien (*n.m.*), Neocomian : (5)
le néocomien, the Neocomian. (6)

néodyme (Chim.) (*n.m.*), neodymium. **(7)**

néogène (Géol.) (*adj.*), Neogene. (8)

néogène (*n.m.*), Neogene. (9)

néojurassique (Géol.) (*adj.*), Neo-Jurassic. (10)

néojurassique (*n.m.*), Neo-Jurassic. (11)

néolithique (Archéologie & Géol.) (*adj.*), neolithic : (12)
âge néolithique (*m.*), neolithic age. (13) (frequently written with an initial capital letter ; as, the Neolithic age).

néon (Chim.) (*n.m.*), neon. (14)

néovolcanique (Géol.) (*adj.*), neovolcanic. (15)

néozoïque (Géol.) (*adj.*), Neozoic. (16)

néozoïque (*n.m.*), Neozoic. (17)

néphéline (Minéral) (*n.f.*), nephelite ; nepheline. (18)

néphélinique (*adj.*), nephelinic. (19)

néphélinite (Pétrol.) (*n.f.*), nephelinite. (20)

néphrite (Minéral) (*n.f.*), nephrite ; kidney-stone. (21)

neptunien, -enne (Géol.) (*adj.*), Neptunian. (22)

neptunisme (*n.m.*), Neptunianism. (23)

neptuniste (pers.) (*n.m.*), Neptunist. (24)

nerf (Métall.) (*n.m.*), fibre : (25)
cassure à nerf (*f.*), fibrous fracture. (26)

nerf (Géol.) (*n.m.*), horse ; rider : (27)
nerf de roche encaissante, horse of country-rock. (28)

nerveux, -euse (Métall.) (*adj.*), fibrous : (29)
fer nerveux (*m.*), fibrous iron. (30)
cassure nerveuse (*f.*), fibrous fracture. (31)

nervure (*n.f.*), rib : (32)
bâti à nervures (*m.*), ribbed frame. (33)

nervure d'entretoisage (du banc d'un tour) (*f.*), cross-girth, cross-bar, (of the bed of a lathe). (34)

nervurer (*v.t.*), to rib. (35)

net, nette (Commerce) (*adj.*), net ; nett : (36)
poids net (*m.*), net weight. (37)

net, nette (clair) (*adj.*), clear ; sharp : (38)
une image nette (Opt.), a clear (or a sharp) image. (39)

net, nette (propre) (*adj.*), clean : (40)
un chemin net, a clean path. (41)

net, nette (bien marqué ; point baveux) (*adj.*), clean : (42)
une cassure nette, a clean fracture. (43)

nettement (clairement) (*adv.*), clearly ; sharply : (44)
point nettement visible par tous les temps (*m.*), point clearly visible in all weathers. (45)
surface nettement couverte à pleine ouverture (de l'objectif) (Photogr.) (*f.*), size of plate sharply covered at full aperture (of the lens). (46)

netteté (d'une image) (Opt.) (*n.f.*), clearness, sharpness, (of an image). (47)

nettoyage *ou* nettoiement (*n.m.*), cleaning ; cleaning up ; clean-up ; cleansing ; clearing : (48)
nettoyage des lampes, cleaning the lamps ; lamp-cleaning. (49)

nettoyage des sluices, cleaning up the sluices ; clean-up of the sluices. (50)

nettoyage général (sluices ou bocards), general clean-up. (51)

nettoyage partiel, partial clean-up. (52)

nettoyer (*v.t.*), to clean ; to clean up ; to clean out ; to cleanse ; to clear : (53)
nettoyer la grille, to clean the fire ; to prick the fire-bars ; to slice the furnace-grate. (54)
nettoyer un puits, to clean out a well. (55)
nettoyer une pièce coulée, to clean, to dress, a casting. (56)

nettoyeur (pers.) (*n.m.*), cleaner. (57)

nettoyeur de machines (*m.*), engine-cleaner. (58)

nettoyeuse (*n.f.*), cleaning-machine. (59)

neuf, neuve (*adj.*), new : (60)
sable neuf (Fonderie) (*m.*), new sand. **(61)**

neutralement (*adv.*), neutrally. (62)

neutralisant, -e (*adj.*), neutralizing : (63)
agent neutralisant (*m.*), neutralizing agent ; neutralizing agency. (64)

neutralisant (*n.m.*), neutralizer. (65)

neutralisation (*n.f.*), neutralization ; neutraliz-ing : (66)
neutralisation des poussières de charbon, neutralizing coal-dust. (67)

neutraliser (*v.t.*), to neutralize : (68)
neutraliser les sels solubles, to neutralize the soluble salts. (69)

neutraliser (se) (*v.r.*), to neutralize each other. (70)

neutralité (Chim.) (*n.f.*), neutrality. **(71)**

neutre (*adj.*), neutral : (72)
sel neutre (Chim.) (*m.*), neutral salt. (73)
ligne neutre d'un barreau aimanté (Phys.) (*f.*), neutral line of a bar magnet. (74)
conducteur neutre (Élec.) (*m.*), neutral con-ductor. (75)
roches neutres (Géol.) (*f.pl.*), neutral rocks ; intermediate rocks. (76)

névadite (Pétrol.) (*n.f.*), nevadite ; nevadyte. (77)

névé (Géol.) (*n.m.*), névé ; glacier-snow ; firn : (78)
roches cristallines blanches de névés, de glaciers étincelant au soleil (*f.pl.*), crystalline rocks white with névés, with glaciers sparkling in the sun. (79)

névianskite *ou* newjanskite (Minéral) (*n.f.*), nevyanskite. (80)

nez (*n.m.*), nose. (81)

nez d'un ciseau (*m.*), nose of a chisel. (82)

nez de l'arbre porte-foret (*m.*), nose of the drilling-spindle. (83)

nez de marche (escalier) (*m.*), nosing of step. (84)

nez de marche (rabot) (*m.*), nosing-plane. (85)

nez de raccord (*m.*), nose-piece ; stud-union ; stud-coupling. (86)

nez du mandrin (tour) (*m.*), nose of the mandrel ; mandrel-nose ; spindle-nose (lathe). (87)

niche (pour recevoir un palier, or analogue) (Mach.) (*n.f.*), box ; housing. (88)

niche (*n.f.*) *ou* niche de refuge, shelter ; refuge ; refuge-hole ; manhole ; place of safety. (89)

niche murale (pour transmissions) (*f.*), wall-box (for shafting). (90)

nickel (*n.m.*), nickel. (91)

nickel antimonial (Minéral) (*m.*), antimonial nickel ; breithauptite. (92)

nickel arsenical (Minéral) (*m.*), Same as nickéline.

nickelage (*n.m.*), nickel-plating ; nickeling ; nickelling ; nickelizing ; nickelization ; nickelage. (1)

nickelé, -e (*adj.*), nickel-plated ; nickeled ; nickelled. (2)

nickeler (*v.t.*), to nickel-plate ; to nickel ; to nickelize. (3)

nickelgymnite (Minéral) (*n.f.*), nickel-gymnite ; genthite. (4)

nickélifère (*adj.*), nickeliferous ; niccoliferous ; nickel-bearing. (5)

nickéline (Minéral) (*n.f.*), niccolite ; nickelite ; nickeline ; arsenical nickel ; copper-nickel ; kupfernickel. (6)

nickélique (Chim.) (*adj.*), nickelic. (7)

nickélisage (*n.m.*), **nickéliser** (*v.t.*). Same as nickelage, nickeler.

nickelocre (Minéral) (*n.m.*), nickel-ochre ; nickel-bloom ; nickel-green ; annabergite. (8)

nickelure (*n.f.*), nickelage ; nickelure. (9)

nicol (*n.m.*), nicol ; Nicol prism ; Nicol's prism. (10)

nicols croisés (*m.pl.*), crossed nicols. (11)

nicols parallèles (*m.pl.*), parallel nicols. (12)

nicopyrite (Minéral) (*n.f.*), pentlandite. (13)

nid (de minerai) (*n.m.*), nest, pocket, bunny, (of ore). (14)

nid (de grisou) (*n.m.*), pocket (of fire-damp). (15)

niobate (Chim.) (*n.m.*), niobate ; columbate. (16)

niobite (Minéral) (*n.f.*), niobite. (17)

niobium (Chim.) (*n.m.*), niobium ; columbium. (18)

nitramite (Explosif) (*n.f.*), nitramite. (19)

nitrate (Chim.) (*n.m.*), nitrate. (20)

nitrate d'ammoniaque (*m.*), nitrate of ammonia ; ammonium nitrate. (21)

nitrate d'argent (*m.*), nitrate of silver ; silver nitrate. (22)

nitrate de potassium (*m.*), potassium nitrate ; nitrate of potash ; nitre ; saltpetre. (23)

nitrate de sodium *ou* **nitrate de soude** (*m.*), nitrate of sodium ; sodium nitrate ; nitrate of soda. (24)

nitratine (Minéral) (*n.f.*), nitratine ; soda-nitre. (25)

nitre (*n.m.*), nitre ; saltpetre ; saltpeter. (26)

nitré, -e (*adj.*), nitrate (*used as adj.*) : (27) explosifs nitrés (*m.pl.*), nitrate mixtures. (28)

nitreux, -euse (Chim.) (*adj.*), nitrous : (29) acide nitreux (*m.*), nitrous acid. (30)

nitrière (*n.f.*), nitrate-field ; nitrate-land ; nitre-bed ; saltpetre-bed. (31)

nitrifère (*adj.*), nitriferous ; nitrate-bearing. (32)

nitrificateur, -trice (*adj.*), nitrifying. (33)

nitrification (*n.f.*), nitrification. (34)

nitrifier (*v.t.*), to nitrify. (35)

nitrifier (se) (*v.r.*), to nitrify. (36)

nitrique (Chim.) (*adj.*), nitric : (37) acide nitrique (*m.*), nitric acid. (38)

nitrobenzène (*n.m.*) *ou* **nitrobenzine** (*n.f.*), nitro-benzene ; nitrobenzol. (39)

nitroferrite (Explosif) (*n.f.*), nitroferrite. (40)

nitrogène (*n.m.*), nitrogen. (41)

nitroglycérine (Explosif) (*n.f.*), nitroglycerine ; blasting-oil. (42)

nitromètre (*n.m.*), nitrometer. (43)

nitronaphtaline (*n.f.*), nitronaphthalene. (44)

niveau (élévation d'un point, d'une droite ou d'un plan au-dessus d'une surface horizontale de comparaison) (*n.m.*), level : (45)

le niveau d'une côte, d'un plateau, the level of a coast, of a plateau. (46 *ou* 47)

niveau (état de ce qui est horizontal) (*n.m.*), level : (48)

l'eau cherche constamment son niveau (*f.*), water tries to find its own level. (49)

niveau (Mines) (*n.m.*), level ; lift ; stage : (50)

les niveaux profonds d'une mine, the deep levels of a mine. (51)

le —ᵉ niveau, the —th level. (52)

le niveau de — mètres *ou* le niveau —, the —-metre level. (53)

un puits ou un flanc de montagne est divisé en étages ; chaque étage ou niveau est caractérisé par sa voie de fond, a shaft or mountainside is divided into stages, each stage, lift, or level, has its own bottom level (i.e., bottom roadway). (54)

niveau (instrument) (*n.m.*), level. (55)

niveau (tube indicateur de niveau) (*n.m.*), gauge-glass. (56)

niveau (de) *ou* **niveau (à)** (situé dans un même plan horizontal), level ; at grade ; grade (*used as adj.*) : (57)

surface de niveau (*f.*), level surface. (58)

poutre reposant librement par ses extrémités sur deux appuis de niveau (*f.*), beam resting freely by its ends on two level supports. (59)

passage à niveau *ou* passage de niveau (Ch. de f.) (*m.*), level crossing ; crossing at grade ; grade crossing. (60)

niveau à bulle d'air *ou* **niveau à alcool** (*m.*), spirit-level ; air-level ; bubble-level. (61)

niveau à bulle indépendante sur la lunette *ou* **niveau à cheval** (*m.*), striding-level ; axis-level. (62)

niveau à longue portée (*m.*), long-range level. (63)

niveau à lunette (*m.*), surveyors' level ; levelling-instrument. (64)

niveau à lunette (lunette fixe) (*m.*), dumpy level. (65)

niveau à lunette (lunette mobile reposant dans des étriers en forme de fourches) (*m.*), Y level. (66)

niveau à pinnules (*m.*), sighted level. (67)

niveau à plomb (*m.*), plumb-level ; plummet-level. (68)

niveau d'eau *ou* **niveau aquifère** *ou* *simplement* **niveau** (*n.m.*) (Géol.), water-bearing stratum ; water-bearing strata ; watery strata. (69)

niveau d'eau (plan d'eau) (*m.*), water-level : (70)

indicateur de niveau d'eau (*m.*), water-level indicator. (71)

niveau d'eau (instrument) (*m.*), water-level. (72)

niveau de base de l'érosion (Géogr. phys.) (*m.*), base-level of erosion : (73)

les fleuves débouchant dans l'océan ont pour niveau de base la surface des mers, leurs affluents le thalweg du fleuve principal à leur embouchure ; les cours d'eau débouchant dans un lac ont comme niveau de base la surface de ce lac, rivers debouching into the ocean have as base-level the surface of the seas, their tributaries the thalweg of the main river at their mouth ; streams debouch-ing into a lake have as base-level the surface of that lake. (74)

niveau de fond (Mines) (*m.*), bottom level. (75)

niveau de l'eau (plan d'eau) (*m.*), water-level. (76)

niveau de la charge (haut fourneau) (*m.*), stock-line (blast-furnace). (1)

niveau de la mer *ou* **niveau moyen de la mer** (*m.*), sea-level ; mean sea-level ; level of the sea : (2)

— mètres au-dessus du niveau de la mer, — metres above sea-level. (3)

le plan de comparaison auquel on rapporte les altitudes est généralement le niveau moyen de la mer, the datum-level to which heights are referred is generally the mean sea-level. (4)

niveau de maçon (*m.*), masons' level. (5)

niveau de pente (*m.*), slope-level ; gradiometer. (6)

niveau de poseur de voie (*m.*), plate-layers' level. (7)

niveau pétrolifère (Géol.) (*m.*), oil-horizon. (8)

niveau piézométrique (Hydraul.) (*m.*), hydraulic gradient. (9)

niveau piézométrique *ou* **niveau hydrostatique** (Géol.) (*m.*), water-table ; water-level. (10)

niveau triangulaire (*m.*), triangular level ; A level. (11)

nivelant, -e *ou* **nivelateur, -trice** (*adj.*), levelling ; leveling : (12)

l'action nivelante de la mer (*f.*), the levelling action of the sea. (13)

alidade nivelatrice (*f.*), levelling-alidade. (14)

nivelée (*n.f.*), observation (with a levelling-instrument) : (15)

mille nivelées, a thousand observations. (16)

niveler (rendre horizontal, plan) (*v.t.*), to level ; to make even : (17)

éminences nivelées par l'érosion (*f.pl.*), eminences leveled by erosion. (18)

niveler (mesurer au niveau) (*v.t.*), to level ; to bone : (19)

niveler le tracé probable d'une ligne de chemin de fer, to level the probable alignment of a railway-line. (20)

nivelette (*n.f.*), boning-rod ; boning-stick. (21)

nivelette à coulisse *ou* **nivelette à pied** (*f.*), boning-rod with standard ; boning-rod with driving-spur. (22)

niveleur (pers.) (*n.m.*), levelman. (23)

nivellement (action de rendre un plan uni, horizontal) (*n.m.*), levelling ; leveling : (24)

nivellement d'un terrain, levelling a piece of ground. (25)

nivellement (Arpent.) (*n.m.*), levelling ; leveling ; boning : (26)

nivellement d'une voie entre deux points, levelling. boning, a road between two points. (27)

nivellement barométrique (*m.*), barometric levelling. (28)

nivellement dans les mines (*m.*), mine-levelling. (29)

noble (se dit des pierres, des métaux) (*adj.*), noble ; precious. (30)

nodulaire (*adj.*), nodular : (31)

concrétion nodulaire (*f.*), nodular concretion. (32)

nodule (*n.m.*), nodule ; ball. (33)

nodules phosphatés (*m.pl.*), phosphatic nodules. (34)

noduleux, -euse (*adj.*), nodulous. (35)

nœud (cordage) (*n.m.*), knot. (36)

nœud (bois) (*n.m.*), knot. (37)

nœud (d'une penture) (*n.m.*), knuckle (of a hinge) (T or gate-hinge). (38)

nœud (du réseau) (distribution d'énergie électrique) (*n.m.*), feeding-point (of the system). (39)

nœud allemand (*m.*), figure-of-eight knot ; German knot. (40)

nœud coulant (*m.*), running knot. (41)

nœud d'ajust *ou* **nœud d'ajut** *ou* **nœud de vache** (*m.*), granny-knot ; granny's-knot ; granny's-bend. (42)

nœud d'empattement (Plomberie) (*m.*), T joint. (43)

nœud de jonction (Plomberie) (*m.*), butt-joint. (44)

nœud de soudure (Plomberie) (*m.*), wipe-joint ; wiped joint. (45)

nœud du tisserand (*m.*), weaver's knot ; weaver's hitch. (46)

nœud lâche (*m.*), loose knot. (47)

nœud plat (*m.*), square knot ; reef-knot. (48)

noir, -e (*adj.*), black. (49)

noir (couleur) (*n.m.*), black. (50)

noir (de fonderie) (*n.m.*), blacking ; blackening ; facing. (51)

noir (nègre) (pers.) (*n.m.*), black ; negro. (52)

noir (-e) comme du jais, jet-black. (53)

noir (-e) comme poix, pitch-dark ; pitch-black. (54)

noir d'étuve *ou* **noir de couche** *ou* **noir liquide** (Fonderie) (*m.*), blackwash ; wet blacking ; liquid facing ; paint. (55)

noir de fumée (*m.*), lampblack. (56)

noir minéral (Fonderie) (*m.*), mineral blacking. (57)

noirâtre (*adj.*), blackish. (58)

noircir (*v.t.* & *v.i.*), to blacken. (59)

noircir (se) (*v.r.*), to blacken : (60)

minéral qui se ternit et noircit à l'air (*m.*), mineral which tarnishes and blackens in the air. (61)

noircissement *ou* **noircissage** (*n.m.*), blackening ; blacking : (62)

noircissement de l'ampoule d'une lampe à incandescence, blackening of the bulb of a glow-lamp. (63)

noircissage d'un moule au noir d'étuve (Fonderie), blacking a mould with black-wash. (64)

noix (d'un robinet) (*n.f.*), plug (of a cock). (65)

noix (languette demi-circulaire) (Menuis.) (*n.f.*), cock-bead. (66)

noix (rainure ayant un fond arrondi en forme de demi-cercle) (Menuis.) (*n.f.*), groove (to receive a cock-bead). (67)

noix (pour chaîne calibrée) (*n.f.*), sheave (for pitched chain). (68)

nomade (*adj.*), nomad ; nomadic. (69)

nombre (*n.m.*), number : (70)

nombre d'heures de travail, number of working-hours. (71)

nombreux, -euse (*adj.*), numerous. (72)

nomenclature des pièces (*f.*), names of parts. (73)

nominal, -e, -aux (*adj.*), nominal : (74)

cheval nominal (*m.*), nominal horse-power. (75)

nommer un directeur, to appoint a manager. (76)

L'emploi du trait d'union entre la particule négative non *et le mot suivant est facultatif, tant en français qu'en anglais.*

non-achèvement (*n.m.*), non-completion : (77)

non-achèvement d'un chemin de fer, non-completion of a railway. (78)

non-adultéré, -e (*adj.*), unadulterated. (1)
non-aéré, -e (*adj.*), unventilated. (2)
non-affecté, -e (*adj.*), unaffected ; free : (3)
district non-affecté par des crises ouvrières
(*m.*), district unaffected by (*or* free from)
labour troubles. (4)
non-affûté, -e (*adj.*), unsharpened. (5)
non-altéré, -e (Géol.) (*adj.*), unweathered ;
unaltered : (6)
roche non-altérée (*f.*), unweathered rock. (7)
non-amalgamable (*adj.*), non-amalgamable : (8)
or non-amalgamable (*m.*), non-amalgamable
gold. (9)
non-amalgamé, -e (*adj.*), unamalgamated. (10)
non-arrivée (*n.f.*), non-arrival. (11)
non-azoté, -e (*adj.*), non-nitrogenous. (12)
non-bitumineux, -euse (*adj.*), non-bituminous.
(13)
non-boisé, -e (*adj.*), untimbered ; unwooded.
(14)
non-bouché, -e (*adj.*), unstoppered : (15)
flacon non-bouché (*m.*), unstoppered bottle.
(16)
non-broyé, -e (*adj.*), uncrushed. (17)
non-classé, -e *ou* **non-classifié, -e** (*adj.*), un-
-classified. (18)
non-clos, -e (*adj.*), unfenced : (19)
terrain non-clos (*m.*), unfenced ground. (20)
non-cokéfiant, -e (*adj.*), non-coking : (21)
combustible non-cokéfiant (*m.*), non-coking
fuel. (22)
non-collant, -e (se dit du charbon, de la houille)
(*adj.*), non-caking. (23)
non-combiné, -e (*adj.*), uncombined. (24)
non-conducteur, -trice (*adj.*), non-conducting ;
non-conductive : (25)
en réalité il n'y a pas de corps absolument
non-conducteurs, in reality there are no
absolutely non-conducting substances. (26)
non-conducteur (*n.m.*), non-conductor. (27)
non-corrosif, -ive (*adj.*), non-corrosive. (28)
non-criblé, -e (*adj.*), unscreened : (29)
charbon non-criblé (*m.*), unscreened coal. (30)
non-dissous, -oute (*adj.*), undissolved. (31)
non-distillé, -e (*adj.*), undistilled. (32)
non-drainé, -e (*adj.*), undrained. (33)
non-durci, -e (*adj.*), unhardened ; unindurated.
(34)
non-emballé, -e (*adj.*), unpacked. (35)
non-essayé, -e (*adj.*), untried ; untested. (36)
non-exécution (*n.f.*), non-execution ; non-
fulfilment ; non-completion. (37)
non-existant, -e (*adj.*), non-existent. (38)
non-existence (*n.f.*), non-existence. (39)
non-filtré, -e (*adj.*), unfiltered. (40)
non-fondu, -e (*adj.*), unmelted ; unfused. (41)
non-fossilifère (*adj.*), non-fossiliferous ; unfossil-
-iferous. (42)
non-fossilisé, -e (*adj.*), non-fossilized ; un-
-fossilized. (43)
non-galvanisé, -e (*adj.*), ungalvanized. (44)
non-gazeux, -euse (*adj.*), non-gaseous. (45)
non-grisouteux, -euse (*adj.*), non-gassy ; non-
gaseous ; non-fiery : (46)
mine non-grisouteuse (*f.*), non-gassy mine ;
non-fiery mine. (47)
non-imprégné, -e (*adj.*), unimpregnated ; non-
impregnated. (48)
non-isolé, -e (Élec.) (*adj.*), uninsulated. (49)
non-lavé, -e (*adj.*), unwashed : (50)
charbon non-lavé (*m.*), unwashed coal. (51)
non-livraison (*n.f.*), non-delivery. (52)

non-magnétique (*adj.*), non-magnetic. (53)
non-malléable (*adj.*), unmalleable. (54)
non-métallifère (*adj.*), non-metalliferous. (55)
non-métallique (*adj.*), non-metallic. (56)
non-muraillé, -e (*adj.*), unwalled. (57)
non-ouvré, -e (*adj.*), unwrought. (58)
non-oxydant, -e (*adj.*), non-oxidizing. (59)
non-payant, -e (*adj.*), unpayable ; non-payable ;
unremunerative. (60)
non-procurable (*adj.*), unprocurable ; unobtain-
-able. (61)
non-protégé, -e (*adj.*), unprotected ; unshielded.
(62)
non-reconnu, -e (*adj.*), unproved : (63)
territoire non-reconnu (*m.*), unproved territory.
(64)
non-recuit, -e (*adj.*), unannealed : (65)
verre non-recuit (*m.*), unannealed glass.
(66)
non-rémunérateur, -trice (*adj.*), unpayable ;
non-payable ; unremunerative. (67)
non-réussite (*n.f.*), insuccess ; failure. (68)
non-scientifique (*adj.*), unscientific. (69)
non-solidifié, -e (*adj.*), unsolidified. (70)
non-stratifié, -e (*adj.*), unstratified : (71)
roches non-stratifiées (*f.pl.*), unstratified
rocks. (72)
non-susceptible de preuve, incapable of proof.
(73)
non-systématique (*adj.*), unsystematic. (74)
non-taillé, -e (pierres de taille) (*adj.*), uncut ;
undressed ; unhewn. (75)
non-taillé, -e (pierres précieuses) (*adj.*), uncut.
(76)
non-tamisé, -e (*adj.*), unsifted. (77)
non-taraudé, -e (*adj.*), untapped : (78)
écrou non-taraudé (*m.*), untapped nut. (79)
non-travaillé, -e (*adj.*), unworked. (80)
non-trempé, -e (*adj.*), untempered ; unhardened.
(81)
non-triable (*adj.*), unsortable : (82)
minerai non-triable (*m.*), unsortable ore. (83)
non-usé, -e (*adj.*), unworn. (84)
non-valeurs (*n.f.pl.*), worthless objects ; waste.
(85)
non-ventilé, -e (*adj.*), unventilated. (86)
non-volatil, -e (*adj.*), non-volatile. (87)
non-volcanique (*adj.*), non-volcanic. (88)
nonne (Géol.) (*n.f.*), erosion column ; chimney-
rock. (89)
nontronite (Minéral) (*n.f.*), nontronite. (90)
nord (*adj.*), north ; northern ; northerly. (91)
nord (*n.m.*), north. (92)
nord-est (*adj.*), northeast ; northeastern ; north-
-easterly. (93)
nord-est (*n.m.*), northeast. (94)
nord magnétique (*m.*), magnetic north. (95)
nord-nord-est (*adj.* & *n.m.*), north-northeast.
(96)
nord-nord-ouest (*adj.* & *n.m.*), north-northwest.
(97)
nord-ouest (*adj.*), northwest ; northwestern ;
northwesterly. (98)
nord-ouest (*n.m.*), northwest. (99)
nord vrai (*m.*), true north. (100)
noria (*n.f.*) *ou* **noria à roue,** noria ; Persian wheel ;
flush-wheel. (101)
norite (Pétrol.) (*n.m.*), norite. (102)
normal, -e, -aux (*adj.*), normal : (103)
allure normale d'un haut fourneau (*f.*), normal
working of a blast-furnace. (104)
normale (Géom.) (*n.f.*), normal. (105)

normale commune des profils *ou* **normale de contact** (Engrenages) (*f.*), common normal to the curves of the teeth in contact; line of action (Gearing): (1)

il faut que la normale commune des profils passe par le point de contact des cercles primitifs, the common normal to the curves of the teeth in contact must pass through the pitch-point. (2)

normalement (*adv.*), normally. (3)

normanite (Explosif) (*n.f.*), normanite. (4)

noséane *ou* **nosiane** *ou* **noséite** *ou* **nosite** *ou* **nosélite** (Minéral) (*n.f.*), noselite; nosean; nosin; nosite. (5)

notation (*n.f.*), notation: (6)

notation chimique, chemical notation. (7)

notation cristallographique, crystallographic notation. (8)

noter (*v.t.*), to note; to record. (9)

notification (*n.f.*), notification. (10)

notifier (*v.t.*), to notify; to give notice. (11)

noue (endroit où se joignent deux combles en angle rentrant) (Arch.) (*n.f.*), valley. (12)

noue (arbalétrier) (*n.f.*), valley-rafter. (13)

noue (lame de plomb ou tuile creuse servant de rigole à la noue) (*n.f.*), valley-gutter. (14)

noueux, -euse (*adj.*), knotty: (15)

bois noueux (*m.*), knotty timber. (16)

nouméite (Minéral) (*n.f.*), noumeite; noumeaite; garnierite. (17)

nourrice des gueuses (mère-gueuse) (Métall.) (*f.*), sow. (18)

nourriture (*n.f.*), food; chop. (19)

nourriture pour chevaux (*f.*), horsefeed; feed for horses. (20)

nourriture pour les animaux (*f.*), provender for animals. (21)

nouveau (*m.*) *ou* **nouvel** (*m.*) [devant une voyelle ou un *h* muet], **nouvelle** (*f.*) (*adj.*), new; recent; fresh. (22)

nouveau venu [nouveaux venus *pl.*] (pers.) (*m.*), newcomer; greenhorn; tenderfoot; Johnny-come-lately. (23)

nouvelle (*n.f.*), news; piece of news; report: (24)

confirmer une nouvelle, to confirm a report. (25)

nouvellement (*adv.*), newly; recently; freshly. (26)

novaculite (Pétrol.) (*n.f.*), novaculite; novaculyte; Turkey stone; Turkey slate. (27)

novice (*adj.*), inexperienced; unskilled; raw. (28)

novice (pers.) (*n.m.*), raw hand; green hand. (29)

noyage (*n.m.*), flooding: (30)

noyage de la mine, flooding the mine. (31)

noyau (*n.m.*), core; centre; nucleus. (32)

noyau (Fonderie) (*n.m.*), core; nowel. (33)

noyau (Élec.) (*n.m.*), core. (34)

noyau (d'escalier tournant) (*n.m.*), newel (of winding staircase). (35)

noyau (d'une vis) (*n.m.*), core (of a screw). (36)

noyau (d'une vis d'Archimède) (*n.m.*), core, newel, (of an Archimedean screw). (37)

noyau (d'un cylindre broyeur) (*n.m.*), core (of a crushing-roll). (38)

noyau (d'un pli) (Géol.) (*n.m.*), core (of a fold). (39)

noyau chambré (Fonderie) (*m.*), chambered core. (40)

noyau creux *ou* **noyau évidé** (d'escalier) (*m.*), open newel; hollow newel. (41)

noyau de fer feuilleté (Élec.) (*m.*), laminated-iron core. (42)

noyau de l'induit [noyaux des induits *pl.*] (Élec.) (*m.*), armature-core. (43)

noyau en terre (Fonderie) (*m.*), loam core. (44)

noyau igné (Géol.) (*m.*), central fire. (45)

noyau plein (escalier) (*m.*), solid newel. (46)

noyau troussé (Fonderie) (*m.*), strickled core; struck up core; swept up core. (47)

noyautage (Fonderie) (*n.m.*), coring; coring out: (48)

noyautage en sable, en terre, coring in sand, in loam. (49 *ou* 50)

noyautage mécanique (*m.*), machine-coring. (51)

noyauter (Fonderie) (*v.t.*), to core; to core out: (52)

noyauter un moule, to core out a mould. (53)

noyauterie (Fonderie) (*n.f.*), core-making. (54)

noyauteur (Fonderie) (pers.) (*n.m.*), core-maker. (55)

noyé, -e (*adj.*), flooded; drowned; submerged; under water; sunken: (56)

mine noyée (*f.*), flooded mine. (57)

turbine noyée (*f.*), drowned turbine; sub-merged turbine. (58)

rail noyé (*m.*), sunken rail. (59)

noyer (arbre ou bois) (*n.m.*), walnut. (60)

noyer (*v.t.*), to flood; to drown; to submerge: (61)

noyer les travaux (Mines), to flood the workings. (62)

noyer l'aube d'une turbine, to drown the blade of a turbine. (63)

noyer le grisou (Mines), to baffle the fire-damp. (64)

nu, -e (opp. à recouvert, -e *ou* isolé, -e) (Élec.) (*adj.*), bare. (65) (opp. to **covered** *or* **insulated**):

fil nu (*m.*), bare wire. (66)

nu, -e (se dit des feux, des flammes de lampes) (*adj.*), naked; open: (67)

feu nu (*m.*), naked light; open light. (68)

nuage (*n.m.*), cloud. (69)

nuage de poussière (*m.*), cloud of dust. (70)

nuageux, -euse (*adj.*), cloudy; overcast: (71)

ciel nuageux (*m.*), cloudy sky. (72)

nuance (*n.f.*), shade (of colour). (73)

nucléus (*n m.*), nucleus. (74)

nugget (*n.m.*), nugget. (75)

nuisible (*adj.*), harmful; hurtful; injurious; detrimental; noxious: (76)

dans la mécanique, le frottement peut être utile, comme il peut être nuisible, in mechanics, friction can be useful, as it can be detrimental. (77)

eau nuisible (*f.*), noxious water. (78)

nuisiblement (*adv.*), harmfully; hurtfully; injuriously; detrimentally; noxiously. (79)

nuit (*n.f.*), night. (80)

numéro (*n.m.*), number. (81)

numéro d'ordre (*m.*), running number. (82)

numéro de fabrication (*m.*), shop-number. (83)

numéro de jauge (jauge pour fils métalliques) (*m.*), gauge-number (wire-gauge). (84)

numéro de référence *ou* **numéro de renvoi** (*m.*), reference-number. (85)

numéro de série (*m.*), serial number. (86)

numéroter (*v.t.*), to number. (87)

nystagmus des mineurs (Pathologie) (*m.*), miners' nystagmus. (88)

O

oaklite (Explosif) (*n.f.*), oaklite. (1)
obélisque (Géom.) (*n.m.*), obelisk. (2)
objectif (Opt.) (*n.m.*), objective ; object-glass ; lens : (3)
objectif d'un microscope, d'une lunette, objective, object-glass, lens, of a microscope, of a telescope. (4 *ou* 5)
objectif (Photogr.) (*n.m.*), lens : (6)
objectif photographique, photographic lens. (7)
objectif à court foyer (*m.*), short-focus lens. (8)
objectif à grande ouverture (*m.*), large-aperture lens ; lens of large aperture. (9)
objectif à immersion (*m.*), immersion-lens. (10)
objectif à immersion à eau (*m.*), water-immersion lens. (11)
objectif à immersion à huile (*m.*), oil-immersion lens. (12)
objectif à immersion homogène (*m.*), homo-geneous immersion lens. (13)
objectif à long foyer (*m.*), long-focus lens. (14)
objectif à mise au point fixe (*m.*), fixed-focus lens. (15)
objectif à portrait *ou* objectif à portraits (*m.*), portrait-lens. (16)
objectif à sec (opp. à objectif à immersion) (*m.*), dry lens. (17)
objectif achromatique (*m.*), achromatic lens. (18)
objectif anachromatique (*m.*), soft-focus lens ; diffused-focus lens ; anachromatic lens. (19)
objectif anastigmat *ou* objectif anastigmate *ou* objectif anastigmatique *ou* objectif stig--matique (*m.*), anastigmat lens ; anastig--matic lens ; anastigmat ; stigmatic lens. (20)
objectif aplanétique (*m.*), aplanatic lens ; aplanetic lens. (21)
objectif apochromatique *ou* objectif à spectre secondaire réduit (*m.*), apochromatic lens. (22)
objectif combinable (*m.*), combinable lens. (23)
objectif composé (*m.*), compound lens. (24)
objectif condensateur (*m.*), condensing lens. (25)
objectif dédoublable (*m.*), doubled lens ; doublet lens ; doublet. (26)
objectif demi-grand angle [objectifs demi-grands angles *pl.*] (*m.*), mid-angle lens ; medium-angle lens ; normal-angle lens. (27)
objectif grand angle [objectifs grands angles *pl.*] *ou* objectif à grand angle *ou* objectif grand angulaire [objectifs grands angulaires *pl.*] *ou* objectif panoramique (*m.*), wide-angle lens. (28)
objectif petit champ [objectifs petits champs *pl.*] *ou* objectif petit angulaire [objectifs petits angulaires *pl.*] (*m.*), narrow-angle lens. (29)
objectif rectiligne (*m.*), rectilinear lens. (30)
objectif simple (*m.*), single lens. (31)
objectif stéréoscopique (*m.*), stereoscopic lens. (32)
objectif symétrique (*m.*), symmetrical lens. (33)
objet (chose quelconque) (*n.m.*), object ; article. (34)
objets en fer-blanc (*m.pl.*), tinware. (35)
obligation (*n.f.*), obligation ; duty. (36)
obligatoire (*adj.*), obligatory. (37)
oblique (*adj.*), oblique ; skew ; slanting : (38)
ligne oblique (*f.*), oblique line. (39)

fer de rabot oblique (*m.*), skew plane-iron. (40)
coup de sonde oblique (*m.*), slanting bore-hole. (41)
obliquement (*adv.*), obliquely ; on the skew ; skew ; slantwise. (42)
obliquer (*v.i.*), to deviate ; to slant ; to oblique ; to skew. (43)
obliquité (*n.f.*), obliquity ; skew : (44)
obliquité des rayons solaires, obliquity of the sun's rays. (45)
obliquité de la bielle (*f.*), angularity of the connecting-rod. (46) See bielle infinie for example of use.
oblitérer (*v.t.*), to obliterate. (47)
oblong, -ongue (*adj.*), oblong : (48)
une caisse oblongue, an oblong box. (49)
obscur, -e (*adj.*), dark ; obscure : (50)
laboratoire obscur (Photogr.) (*m.*), dark room. (51)
obscurité (*n.f.*), dark ; obscurity : (52)
visées dans l'obscurité (*f.pl.*), sighting in the dark. (53)
obséquent, -e (Géol.) (*adj.*), obsequent : (54)
cours d'eau obséquent (*m.*), obsequent stream. (55)
observation (*n.f.*), observation ; observing : (56)
observation de la Polaire à la plus grande digression, d'une étoile au moment du passage, observing Polaris at extreme elongation, a star at the instant of culmina--tion. (57 *ou* 58)
observation du soleil, observation of the sun. (59)
observations sur le terrain, field-work ; observations in the field. (60)
observer (*v.t.*), to observe : (61)
observer le secret, to observe secrecy. (62)
obsidienne *ou* obsidiane (*n.f.*), obsidian ; volcanic glass. (63)
obsidienne porphyroïde (*f.*), porphyritic obsidian. (64)
obstacle (*n.m.*), obstacle : (65)
obstacle infranchissable *ou* obstacle insurmon--table, insurmountable obstacle ; unsur--mountable obstacle ; insuperable obstacle. (66)
obstacle surmontable, surmountable obstacle ; superable obstacle. (67)
obstruction (*n.f.*), obstruction ; stoppage ; choking ; blocking : (68)
obstruction de la cheminée (Mines), choking, clogging, of chute. (69)
obstruction du trou de sonde par des outils, par des débris, obstruction of the bore-hole by tools, by debris. (70 *ou* 71)
obstruer (*v.t.*), to obstruct ; to stop up ; to choke ; to choke up ; to block ; to block up : (72)
obstruer une rue, to obstruct, to block up, a street. (73)
obtenir (*v.t.*), to obtain ; to get ; to secure : (74)
obtenir une concession, to obtain, to get, to secure, a concession. (75)
obtenir de l'outillage le plus grand rendement possible, to get the most out of the plant. (76)
obtention (*n.f.*), obtaining ; getting ; securing. (77)

obturateur (d'entrée de vapeur dans le cylindre) (*n.m.*), cut-off. (1)

obturateur (Sondage) (*n.m.*), cap. (2)

obturateur (rondelle obturatrice) (*n.m.*), blind flange ; blind washer ; blank flange ; blank washer ; no-thoroughfare. (3)

oburateur (*n.m.*) *ou* obturateur d'objectif (Photogr.), shutter ; lens-shutter ; obturator. (4)

obturateur à rideau (*m.*), roller-blind shutter. (5)

obturateur à secteurs (*m.*), sector-shutter. (6)

obturateur de plaque (*m.*), focal-plane shutter. (7)

obturateur toujours armé, faisant la pose, la demi-pose et l'instantané (*ou* faisant la pose en un temps et deux temps et l'instantané), marchant au doigt ou au moyen du déclencheur métallique, donnant des vitesses variables de — " à — ' de seconde (*m.*), everset shutter, for time, bulb and instan-taneous exposures, operated by trigger or wire release (*or* worked by finger or cable release), giving speeds ranging from — th to — th of a second. (8)

obturation (*n.f.*). See examples. (9)

obturation d'un conduit, stopping up a pipe. (10)

obturation d'un puits de mine en cas d'incendie, sealing a mine-shaft in case of fire. (11)

obturation de l'admission de vapeur, cutting off, shutting off, the admission of steam. (12)

obturer (*v.t.*). See examples : (13)

obturer l'admission de la vapeur, to cut off, to shut off, the admission of steam. (14)

obturer un conduit, to stop up a pipe. (15)

obturer un puits de mine, to seal a mine-shaft. (16)

obturer un puits jaillissant (puits à pétrole), to cap a fountain. (17)

obtus, -e (Géom.) (*adj.*), obtuse : (18)

angle obtus (*m.*), obtuse angle. (19)

obtus, -e (émoussé ; arrondi) (*adj.*), dull ; blunt : (20)

pointe obtuse (*f.*), dull point ; blunt point. (21)

obtusangle *ou* obtusangulaire (Géom.) (*adj.*), obtuse-angled ; obtuse-angular ; obtus-angular : (22)

triangle obtusangle (*m.*), obtuse-angled triangle. (23)

obvier à une difficulté, to obviate a difficulty. (24)

occasion (*n.f.*), occasion. (25)

occasion (d'), second-hand : (26)

meubles d'occasion (*m.pl.*), second-hand furniture. (27)

occasionnel, -elle (*adj.*), occasional. (28)

occasionner (*v.t.*), to occasion ; to cause ; to bring about : to give rise to : (29)

occasionner des explosions, to cause, to give rise to, to bring about, explosions. (30)

occasionner une reprise d'intérêt dans les affaires minières, to cause a revival of interest in mining ; to bring about a revival of interest in mining business. (31)

occidental, -e, -aux (*adj.*), west ; western ; occidental : (32)

longitude occidentale (*f.*), west longitude. (33)

occidental, -e, -aux (de moindre valeur comme pierre gemme) (opp. à oriental) (*adj.*), occidental ; Occidental : (34)

turquoise occidentale (*f.*), occidental turquoise ; Occidental turquoise. (35)

occlus, -e (*adj.*), occluded : (36)

gaz occlus (*m.pl.*), occluded gases. (37)

occlusion (Chim.) (*n.f.*), occlusion : (38)

occlusion des gaz, occlusion of gases. (39)

occlusion (d'une conduite d'eau) (*n.f.*), stoppage, stopping up, of a water-pipe. (40)

occuper (*v.t.*), to occupy ; to employ ; to engage : (41)

occuper son temps, to occupy, to employ, one's time. (42)

occuper des ouvriers, to employ workmen. (43)

occuper (s') du boisage, to attend to the timber-ing. (44)

occurrence (*n.f.*), occurrence : (45)

une occurrence singulière, a singular occur-rence. (46)

océan (*n.m.*), ocean. (47)

océanique (*adj.*), oceanic ; ocean (*used as adj.*) : (48)

sédiments océaniques provenant de la démo-lition de la terre ferme (*m.pl.*), oceanic sediments derived from the waste of the land. (49)

bassin océanique (*m.*), ocean-basin. (50)

océanographie (*n.f.*), oceanography. (51)

océanographique (*adj.*), oceanographic ; oceano-graphical : (52)

recherche océanographique (*f.*), oceano-graphical research. (53)

ocre (*n.f.*), ochre ; ocher. (54)

ocre jaune (*f.*), yellow ochre. (55)

ocre rouge (*f.*), red ochre. (56)

ocreux, -euse (*adj.*), ochreous ; ocherous ; ochry. (57)

octaèdre *ou* octaédrique (*adj.*), octahedral ; octahedric ; octahedrous : (58)

cristal octaédrique (*m.*), octahedral crystal. (59)

octaèdre (*n.m.*), octahedron. (60)

octaédrite (Minéral) (*n.f.*), octahedrite ; anatase. (61)

octogonal, -e, -aux *ou* octogone (*adj.*), octag-onal : (62)

figure octogonale (*f.*), octagonal figure. (63)

octogone (*n.m.*), octagon. (64)

octogones (fer ou acier) (*n.m.pl.*), octagons. (65 *ou* 66)

octopolaire (Élec.) (*adj.*), eight-pole ; octopolar. (67)

octovalent, -e (Chim.) (*adj.*), octavalent ; octo-valent. (68)

oculaire (*adj.*), ocular ; eye (*used as adj.*) : (69)

les lentilles oculaires (*f.pl.*), the ocular lenses. (70)

anneau oculaire (*m.*), eye-ring. (71)

oculaire (Opt.) (*n.m.*), eyepiece ; ocular. (72)

oculaire à réticule (*m.*), webbed eyepiece ; eyepiece with cross-wires. (73)

oculaire composé (*m.*), compound eyepiece. (74)

oculaire de Huygens (*m.*), Huygenian eyepiece. (75)

oculaire de Ramsden (*m.*), Ramsden's eyepiece. (76)

oculaire-micromètre [oculaires-micromètres *pl.*] (*n.m.*), micrometer eyepiece. (77)

oculaire négatif (*m.*), negative eyepiece. (78)

oculaire positif (*m.*), positive eyepiece. (79)

odeur (*n.f.*), odour ; odor ; smell : (80)

liquide à odeur désagréable (*m.*), liquid with an unpleasant odour (*or* with a disagreeable smell). (81)

odite (Explosif) (*n.f.*), odite. (1)

odomètre (*n.m.*), odometer. (2)

odontographe (pour dents d'engrenage) (*n.m.*), odontograph (for gear-teeth). (3)

odontolite (Minéral) (*n.f.*), odontolite. (4)

œil [**yeux** *pl.*] (*n.m.*), eye : (5)
     œil d'une aiguille, d'un outil, d'une roue, eye of a needle, of a tool, of a wheel. (6 *ou* 7 *ou* 8)

œil (d'un cordage) (*n.m.*), eye, loop, bight, (of a rope). (9)

œil (d'un fourneau) (*n.m.*), tap-hole, tapping-hole, draw-hole, mouth, (of a furnace). (10)

œil (d'une penture) (*n.m.*), knuckle (of a hinge) (T or gate-hinge). (11)

œil (contre-rivure) (*n.m.*), burr ; riveting-burr ; rivet-washer. (12)

œil-de-chat [**œils-de-chat** *pl.*] (Minéral) (*n.m.*), cat's-eye. (13)

œil-de-poisson [**œils-de-poisson** *pl.*] (Minéral) (*n.m.*), moonstone. (14)

œil-de-tigre [**œils-de-tigre** *pl.*] (Minéral) (*n.m.*), tiger-eye ; tiger's-eye. (15)

œillard (d'un ventilateur d'une pompe centrifuge) (*n.m.*), intake, inlet, ear, (of a fan, of a cen-trifugal pump). (16 *ou* 17)

œillard (dans le centre d'une meule en émeri, ou similaire) (*n.m.*), hole (in the centre of an emery-wheel, or the like). (18 *ou* 19)

œillet (*n.m.*), eyelet ; eye. (20)

œillet de métal (*m.*), metal eyelet. (21)

œilleton de visée (d'un viseur direct) (Photogr.) (*m.*), eyepiece, magnifying-lens, (of a direct view-finder). (22)

œllacherite (Minéral) (*n.f.*), œllacherite. (23)

œrsted (unité C.G.S.) (Élec.) (*n.m.*), oersted. (24)

œsar [as *sing.*] (mot scandinave) (Géol.) (*n.m.pl.*), kames ; eskers. (25)

œtite (Minéral) (*n.f.*), aetites ; eaglestone. (26)

œuvre (*n.f.*), work : (27)
     la science est l'œuvre des siècles, science is the work of centuries. (28)

officiel, -elle (*adj.*), official. (29)

officiellement (*adv.*), officially. (30)

offre (*n.f.*), offer : (31)
     offre ferme d'une propriété, firm offer of a property. (32)
     offre labiale *ou* offre verbale, verbal offer. (33)

offre et la demande (l'), demand and supply ; supply and demand : (34)
     la loi de l'offre et de la demande, the law of demand and supply. (35)

offrir (*v.t.*), to offer ; to proffer : (36)
     offrir ses services, to offer, to proffer, one's services. (37)
     offrir une grande attraction au mineur, to offer great inducements to the miner. (38)

offrir une sécurité à toute épreuve, to be thoroughly reliable : (39)
     machine qui offre une sécurité à toute épreuve (*f.*), machine which is thoroughly reliable. (40)

ohm (unité électrique) (*n.m.*), ohm. (41)

ohm international (*m.*), international ohm (= the resistance offered to an unvarying electric current by a column of mercury at freezing-point 14·4521 grammes in mass, of a constant cross-sectional area, and of the length of 106·3 centimetres). (42)

ohmique (Elec.) (*adj.*), ohmic : (43)
     chute ohmique (*f.*), ohmic drop. (44)

ohmmètre (*n.m.*), ohmmeter. (45)

oisanite (épidote) (Minéral) (*n.f.*), a variety of epidote or delphinite. (46)

oisanite (TiO²) (Minéral) (*n.f.*), octahedrite ; anatase. (47)

okénite (Minéral) (*n.f.*), okenite ; dysclasite. (48)

oléagineux, -euse (*adj.*), oleaginous. (49)

oléfiant, -e *ou* **oléifiant, -e** (*adj.*), olefiant. (50)

oléomètre (*n.m.*), oleometer ; oilometer ; oil-gauge. (51)

oligiste (*adj.*), oligistic ; oligistical ; oligist (*used as adj.*) : (52)
     fer oligiste (*m.*), oligist iron. (53)

oligiste (Minéral) (*n.m.*), oligist ; oligist iron. (54)

oligocène (Géol.) (*adj.*), Oligocene. (55)

oligocène (*n.m.*), Oligocene : (56)
     l'oligocène, the Oligocene. (57)

oligoclase (Minéral) (*n.f.*), oligoclase. (58)

oligonite (Minéral) (*n.f.*), oligonite. (59)

olive *ou* **olive (à)** (en forme d'olive) (*adj.*), oval : (60)
     lime olive (*f.*), oval file. (61)
     bouton à olive (*m.*), oval knob. (62)

olive (bouton à olive) (Serrur.) (*n.f.*), oval knob. (63)

olivénite (Minéral) (*n.f.*), olivenite ; olive-ore. (64)

olivine (Minéral) (*n.f.*), olivine ; night-emerald. (65)

ombre (Opt.) (*n.f.*), shadow. (66)

ombre (terre d'ombre) (*n.f.*), umber. (67)

ombre électrique (*f.*), electric shadow. (68)

ombre portée (Opt.) (*f.*), umbra. (69)

omettre (*v.t.*), to omit ; to neglect ; to fail : (70)
     omettre de signaler un accident aux autorités, to fail to report an accident to the authorities. (71)

omission (*n.f.*), omission ; omitting ; failing ; failure : (72)
     omission de signaler un accident survenu à un employé, failing to report an accident to an employee. (73)

omphazite (Minéral) (*n.f.*), omphacite. (74)

once (avoirdupois) (unité anglaise de poids) (*n.f.*), ounce = 28·350 grammes. (75)

once (troy) (unité anglaise de poids) (*n.f.*), ounce = 31·1035 grammes. (76)

onctueux, -euse (Minéralogie) (*adj.*), unctuous ; greasy ; soapy : (77)
     le graphite est onctueux au toucher, graphite is unctuous to the touch. (78)
     couches onctueuses (*f.pl.*), soapy seams. (79)

onctuosité (*n.f.*), unctuousness ; greasiness ; soapiness. (80)

onde (*n.f.*), wave. (81)

onde calorifique (Phys.) (*f.*), heat-wave. (82)

ondes de la mer (*f.pl.*), waves of the sea. (83)

onde électrique *ou* **onde électromagnétique** (*f.*), electric wave ; electromagnetic wave. (84)

onde enveloppe (Phys.) (*f.*), wave-front. (85)

onde explosive (coups de feu) (Mines) (*f.*), explosion-wave ; detonation-wave. (86)

ondes hertziennes (*f.pl.*), Hertzian waves. (87)

onde lumineuse (*f.*), light-wave. (88)

onde prolongée (coups de feu) (Mines) (*f.*), collision-wave. (89)

onde réfléchie (coups de feu) (Mines) (*f.*), reflection-wave. (90)

onde rétrograde (coups de feu) (Mines) (*f.*), retonation-wave. (91)

onde secondaire (Phys.) (*f.*), secondary wave. (92)

onde sonore (*f.*), sound-wave. (93)

ondée (*n.f.*), shower ; shower of rain : (1)
ondée passagère, passing shower. (2)

ondographe (Élec.) (*n.m.*), ondograph. (3)

ondulation (*n.f.*), undulation ; wave ; corruga-
-tion : (4)
ondulations du sol, undulations of the ground.
(5)

ondulation (Phys.) (*n.f.*), undulation ; wave-
motion ; wave : (6)
le son se transmet par ondulation, sound is
transmitted by wave-motion. (7)

ondulatoire (*adj.*), undulatory : (8)
courant ondulatoire (Élec.) (*m.*), undulatory
current. (9)

ondulé, -e (*adj.*), undulating ; wavy ; corrugated :
(10)
terrain ondulé (*m.*), undulating ground. (11)
tôle ondulée (*f.*), corrugated iron. (12)

onduleux, -euse (*adj.*), wavy ; undulating : (13)
ligne onduleuse (*f.*), wavy line. (14)

onglet (angle de 45°) (*n.m.*), mitre ; miter : (15)
boîte à onglet (*f.*), mitre-box. (16)

onglet (Géom.) (*n.m.*), ungula. (17)

onglet d'encadrement (*m.*), mitre-joint ; bevel-
joint ; chamfered joint. (18)

onglet sphérique (Géom.) (*m.*), spherical ungula ;
spherical wedge. (19)

onyx (Minéral) (*n.m.*), onyx. (20)

onyx à fortifications (*m.*), fortification agate.
(21)

onyx d'Algérie *ou* **onyx calcaire** (*m.*), Algerian
onyx. (22)

onyx œillé (*m.*), eye agate. (23)

oolithe (calcaire) (*n.m.*), oolite ; eggstone ; roe-
stone. (24)

oolithe (Géol.) (*n.m.*), Oolite : (25)
l'oolithe, the Oolite. (26)

oolithe corallien (*m.*), coralline oolite. (27)

oolithique (*adj.*), oolitic : (28)
structure oolithique (*f.*), oolitic structure.
(29)

opacité (*n.f.*), opacity ; opaqueness. (30)

opale (Minéral) (*n.f.*), opal. (31)

opale cireuse (*f.*), wax opal. (32)

opale commune (*f.*), common opal. (33)

opale de feu *ou* **opale à flammes** *ou* **opale flam-
-boyante** (*f.*), fire-opal ; flame-opal ; sun-
opal. (34)

opale engagée dans sa gangue (*f.*), opal-matrix.
(35)

opale hydrophane (*f.*), hydrophanous opal. (36)

opale mélinite (*f.*), melinite opal. (37)

opale noble (*f.*), noble opal ; precious opal. (38)

opale xyloïde (*f.*), wood-opal. (39)

opalescence (*n.f.*), opalescence. (40)

opalescent, -e (*adj.*), opalescent : (41)
l'adulaire s'appelle pierre de lune lorsqu'elle
est opalescente (*f.*), adularia is called moon-
-stone when it is opalescent. (42)

opalin, -e (*adj.*), opaline ; opal (*used as adj.*) :
(43)
globe opalin (Verrerie) (*m.*), opal globe (Glass-
-ware). (44)

opalisé, -e (*adj.*), opalized : (45)
bois opalisé (*m.*), opalized wood. (46)

opaque (*adj.*), opaque : (47)
corps opaque (*m.*), opaque body. (48)

opérateur (pers.) (*n.m.*), operator : (49)
opérateur exercé, skilled operator. (50)

opération (*n.f.*), operation ; working : (51)
opérations de levé de plans et de nivellement,
surveying and levelling operations. (52)

opérer (*v.t. & v.i.*), to operate ; to work ; to carry
on operations : (53)
opérer d'une façon modeste, to carry on
operations in a modest way. (54)
opérer un sondage, to put down a bore-hole.
(55)

ophicalce (Pétrol.) (*n.f.*), ophicalcite. (56)

ophite (Pétrol.) (*n.m.*), ophite. (57)

ophitique (*adj.*), ophitic : (58)
structure ophitique (*f.*), ophitic structure.
(59)

opinion (*n.f.*), opinion : (60)
opinion des spécialistes miniers, opinion of
mining men. (61)

option (*n.f.*), option. (62)

optique (*adj.*), optic ; optical : (63)
l'axe optique d'une lunette (*m.*), the optic
axis of a telescope. (64)
les propriétés optiques des cristaux (*f.pl.*),
the optical properties of crystals. (65)

optique (*n.f.*), optics. (66)

optiquement (*adv.*), optically : (67)
minéral qui est cristallographiquement et
optiquement identique avec un autre (*m.*),
mineral which is crystallographically and
optically identical with another. (68)

or (*n.m.*), gold. (69)

or affiné (*m.*), refined gold. (70)

or alluvien *ou* **or alluvionnaire** *ou* **or alluvionnien**
(*m.*), alluvial gold ; placer-gold ; stream-
gold. (71)

or argental (Minéral) (*m.*), electrum. (72)

or au titre (*m.*), standard gold ; gold of standard
fineness. (73)

or cémentatoire (*m.*), cement-gold. (74)

or combiné (*m.*), combined gold. (75)

or d'alluvion (*m.*). Same as or alluvien.

or de coupelle (*m.*), cupelled gold. (76)

or de lavage (*m.*). Same as or alluvien.

or de Mannheim (*m.*), Mannheim gold. (77)

or des montagnes *ou* **or des roches** (*m.*). Same
as or filonien.

or en bain *ou* **or en fusion** (*m.*), molten gold.
(78)

or en barre *ou* **or en barres** (*m.*), bar gold. (79)

or en écailles (*m.*), scaly gold. (80)

or en pépites (*m.*), nuggety gold. (81)

or filonien (*m.*), reef-gold ; lode-gold ; vein-gold.
(82)

or fin (*m.*), fine gold. (83)

or flottant (*m.*), floating gold ; float-gold. (84)

or fulminant (*m.*), fulminating gold ; aurum
fulminans. (85)

or graphique (Minéral) (*m.*), graphic gold ;
graphic tellurium ; graphic ore ; sylvanite ;
sylvan. (86)

or gros (*m.*), coarse gold. (87)

or libre (*m.*), free gold. (88)

or natif (*m.*), native gold. (89)

or natif en grains (*m.*), granular native gold ;
granulated native gold. (90)

or natif en mousse (*m.*), moss-gold. (91)

or non-amalgamable (*m.*), non-amalgamable
gold. (92)

or rouillé (*m.*), rusty gold. (93)

or vierge (*m.*), virgin gold. (94)

orage (*n.m.*), storm ; thunderstorm. (95)

orage magnétique (*m.*), magnetic storm. (96)

orageux, -euse (*adj.*), stormy : (97)
temps orageux (*m.*), stormy weather. (98)

orange *ou* **orangé, -e** (*adj.*), orange. (99)

orange *ou* **orangé** (couleur) (*n.m.*), **orange. (100)**

orangite (Minéral) (*n.f.*), orangite. (1)
orbiculaire (*adj.*), orbicular : (2)
 diorite orbiculaire (*f.*), orbicular diorite. (3)
ordinaire (*adj.*), ordinary ; common. (4)
ordinairement (*adv.*), ordinarily ; commonly. (5)
ordonnance (*n.f.*), ordinance. (6)
ordonnée (Géom.) (*n.f.*), ordinate : (7)
 ordonnée d'un point, ordinate of a point. (8)
ordovicien, -enne (Géol.) (*adj.*), Ordovician. (9)
ordovicien (*n.m.*), Ordovician : (10)
 l'ordovicien, the Ordovician. (11)
ordre (*n.m.*), order : (12)
 ordre de chute des pilons, order of drop of stamps. (13)
 ordre de superposition des couches (Géol.), order of superposition of strata. (14)
oreille (Méc.) (*n.f.*), ear ; lug ; snug ; wing. (15)
oreille (d'un châssis de fonderie) (*n.f.*), lug, snug, (of a foundry-flask). (16)
oreille (d'une vis ailée) (*n.f.*), wing (of a thumb-screw). (17)
organes (Méc.) (*n.m.pl.*), gear ; mechanism ; parts ; organs : (18)
 organes d'une machine à vapeur, mechanism, parts, organs, of a steam-engine. (19)
organes accessoires de chaudières *ou* organes accessoires pour chaudières (*m.pl.*), boiler-accessories ; boiler-fittings. (20)
organes d'arrêt (*m.pl.*), stopping-gear ; stop-gear. (21)
organes de changement de marche (*m.pl.*), reversing-gear. (22)
organes de commande (*m.pl.*), driving-gear. (23)
organes de distribution de vapeur (*m.pl.*), valve-gear. (24)
organes de manœuvre (*m.pl.*), working-gear ; working mechanism : (25)
 les organes de manœuvre sont tous à l'avant et à portée de l'ouvrier, the working mechanism is all in front and within reach of the workman. (26)
organes de mise en mouvement (*m.pl.*), starting-gear. (27)
organes de roulement (*m.pl.*), running-gear. (28)
organes de transmission (*m.pl.*), transmission-gear. (29)
organique (*adj.*), organic ; organical : (30)
 chimie organique (*f.*), organic chemistry. (31)
organiquement (*adv.*), organically. (32)
organisation (*n.f.*), organization ; arrangement : (33)
 organisation du travail, organization, arrange-ment, of labour. (34)
organiser (*v.t.*), to organize ; to arrange : (35)
 organiser une surveillance effective, to arrange for efficient supervision. (36)
orgues basaltiques (Géol.) (*f.pl.*), prismatic columns of basalt. (37)
orient (perles) (*n.m.*), orient : (38)
 perle d'un bel orient (*f.*), pearl of a fine orient. (39)
oriental, -e, -aux (*adj.*), east ; eastern ; oriental : (40)
 longitude orientale (*f.*), east longitude. (41)
oriental, -e, -aux (se dit des pierres fines) (*adj.*), Oriental ; oriental ; true : (42)
 rubis oriental (*m.*), Oriental ruby ; true ruby. (43)
orientation (*n.f.*), orientation ; bearing ; bear-ings : (44)

l'orientation des particules cristallines, the orientation of crystalline particles. (45)
la sonorité de la phonolite est due à l'orientation des microlithes d'orthose, the sonorousness of phonolite is due to the orientation of the orthoclase microliths. (46)
orientation d'un filon, d'une galerie de mine, bearing of a lode, of a mine-level. (47 *ou* 48)
l'orientation se fait par l'observation des astres ou à l'aide de la boussole, bearings are taken by astronomical observations or with the compass. (49)
les orientations des diverses faces d'un édifice, the bearings of the several faces of a building. (50)
orientation quelconque (à), capable of being set (*or* swivelled) to any angle (*or* in any direction) : (51)
 porte-outil à revolver à quatre faces à diviseur et à orientation quelconque (*m.*), four-tool turret with base graduated in degrees so that tools may be set to any angle ; four-tool turret with graduated base capable of being swivelled to any angle. (52)
orienter (déterminer, marquer les points cardinaux de) (*v.t.*), to orient ; to orientate : (53)
 orienter un instrument de nivellement, to orient, to orientate, a levelling-instrument. (54)
 orienter un plan, to orient, to orientate, a plan. (55)
 orienter un anémomètre, to bring an ane-mometer into the wind. (56)
 orienter un moulinet dans le sens du courant, to point a current-meter in the direction of the current. (57)
orienter (s') (reconnaître l'orient et les autres points cardinaux) (*v.r.*), to orient ; to orientate : (58)
 un barreau aimanté suspendu par son centre de gravité s'oriente dans l'espace suivant une direction fixe voisine de la direction nord-sud, a magnetic bar suspended by its centre of gravity orients in space pointing in a fixed direction bordering on north-south. (59)
orienter (s') (reconnaître la situation des lieux où l'on est, pour se guider dans sa marche) (*v.r.*), to find one's bearings : (60)
 s'orienter à l'aide d'une boussole, to find one's bearings with a compass. (61)
orifice (*n.m.*), mouth ; port ; port-hole ; aperture ; opening ; orifice. (62)
orifice (*n.m.*) *ou* orifice de distribution (de cylindre à vapeur), port. (63)
orifice d'admission *ou* orifice d'entrée *ou* orifice d'arrivée *ou* orifice d'adduction (*m.*), inlet ; infall : (64)
 l'orifice d'arrivée d'un réservoir, the infall of a reservoir. (65)
orifice d'admission (de cylindre à vapeur) (*m.*), steam-port ; induction-port. (66)
orifice d'eau (*m.*), water-port. (67)
orifice d'échappement (de cylindre à vapeur) (*m.*), exhaust-port ; eduction-port. (68)
orifice d'introduction (*m.*), inlet. (69)
orifice d'introduction d'eau (*m.*), water-inlet. (70)
orifice d'introduction de vapeur (*m.*), steam-inlet. (71)
orifice de chauffe (d'un foyer) (*m.*), fire-hole (of a furnace). (72)

orifice de coulée du laitier (*m.*), slag-notch ; cinder-notch ; slag-eye ; slag-hole ; slag-tap ; slagging-hole ; breast-hole ; floss-hole ; floss ; monkey ; pee-pee. (1)

orifice de décharge (*m.*), outlet. (2)

orifice de grandes dimensions (Hydraul.) (*m.*), large orifice. (3)

orifice de la fosse (mine de charbon) (*m.*), pit-mouth ; mouth of the pit ; pit-head. (4)

orifice de passage (Aérage de mines) (*m.*), reduced opening. (5)

orifice de sortie (*m.*), outlet. (6)

orifice de sortie de l'eau (*m.*), water-outlet. (7)

orifice de sortie de vapeur (*m.*), steam-outlet. (8)

orifice de vapeur (*m.*), steam-port. (9)

orifice du puits (mine de charbon) (*m.*). Same as orifice de la fosse.

orifice du puits (mines métallifères) (*m.*), mouth of the shaft. (10)

orifice du puits (puits à pétrole, etc.) (*m.*), mouth of the well. (11)

orifice en mince paroi *ou* orifice percé en mince paroi (Hydraul.) (*m.*), sharp-edged orifice ; orifice in thin plate ; opening in thin partition ; standard orifice. (12)

orifice équivalent d'une mine (Aérage) (*m.*), equivalent orifice of a mine. (13)

orifice évasé (Hydraul.) (*m.*), rounded-approach orifice. (14)

orifice noyé (Hydraul.) (*m.*), submerged orifice ; orifice under water. (15)

orifice ouvert à sa partie supérieure (écoulement de l'eau par déversoir) (Hydraul.) (*m.*), notch. (16)

originaire (*adj.*), original ; primitive ; primary. (17)

originairement (*adv.*), originally. (18)

original, -e, -aux (*adj.*), original ; inventive. (19)

origine (*n.f.*), origin ; source : (20)
tracer l'origine d'une roche, to trace the origin of a rock. (21)

originel, -elle (*adj.*), original ; primitive ; primary. (22)

orme (arbre) (*n.m.*), elm. (23)

ornière (trace creusée dans le sol par les roues des voitures) (*n.f.*), rut. (24)

ornière (canal pour recevoir le boudin d'une roue à boudin) (*n.f.*), flangeway. (25)

orogénie *ou* orogénèse (*n.f.*), orogeny ; orogenesis. (26)

orogénique (*adj.*), orogenic ; orogenetic : (27)
l'effort orogénique auquel certaines îles doivent leur naissance (*m.*), the orogenic effort to which certain islands owe their birth. (28)
régions encore tourmentées par le volcanisme et les séismes, c'est-à-dire où le travail orogénique n'est pas achevé (*f.pl.*), regions still broken by volcanism and seismism, that is to say, where the orogenetic labour is not yet ended. (29)

orographe (*n.m.*), orograph. (30)

orographie (*n.f.*), orography. (31)

orographique (*adj.*), orographic ; orographical : (32)
les divisions orographiques du système alpin (*f.pl.*), the orographic divisions of the Alpine system. (33)

orographiquement (*adv.*), orographically. (34)

orohydrographie (*n.f.*), orohydrography. (35)

orohydrographique (*adj.*), orohydrographic ; orohydrographical. (36)

orologie (*n.f.*), orology. (37)

orologique (*adj.*), orological. (38)

orométrie (*n.f.*), orometry. (39)

orométrique (*adj.*), orometric. (40)

oropion (Minéral) (*n.m.*), mountain-soap ; rock-soap. (41)

orpaillage (*n.m.*), gold-washing ; alluvial digging. (42)

orpailleur (pers.) (*n.m.*), gold-washer ; gold-digger ; alluvial digger. (43)

orpailleur marron (*m.*), fossicker. (44)

orpiment (Minéral) (*n.m.*), orpiment. (45)

orthite (Minéral) (*n.f.*), orthite ; allanite ; cerine. (46)

orthochlorite (Minéral) (*n.f.*), orthochlorite. (47)

orthochromatique *souvent en abrév.* ortho. (*adj.*), orthochromatic ; ortho. : (48)
plaque orthochromatique (*f.*), orthochro-matic plate. (49)

orthochromatisation (*n.f.*), orthochromatization. (50)

orthochromatiser (*v.t.*), to orthochromatize. (51)

orthochromatisme (*n.m.*), orthochromatism. (52)

orthoclase (Minéral) (*n.f.*), orthoclase ; orthose. (53)

orthodiagonale (Cristall.) (*n.f.*), orthodiagonal. (54)

orthogneiss (Pétrol.) (*n.m.*), orthogneiss. (55)

orthophosphate (Chim.) (*n.m.*), orthophosphate. (56)

orthophyre (Pétrol.) (*n.m.*), orthophyre. (57)

orthophyrique (*adj.*), orthophyric. (58)

orthopinacoïde (Cristall.) (*n.m.*), orthopinacoid. (59)

orthoprisme (Cristall.) (*n.m.*), orthoprism. (60)

orthorhombique (*adj.*), orthorhombic. (61)

orthose (Minéral) (*n.m.*), orthoclase ; orthose. (62)

oscillant, -e (*adj.*), oscillating ; oscillatory : (63)
cylindre oscillant (*m.*), oscillating cylinder. (64)
décharge oscillante (Élec.) (*f.*), oscillatory discharge. (65)

oscillation (*n.f.*), oscillation ; swinging ; swing ; swaying : (66)
oscillations d'un pendule, swing, swinging, oscillations, of a pendulum. (67)
oscillations d'un ressort, oscillations of a spring. (68)
oscillations dans le niveau de la mer, oscilla-tions of the sea-level. (69)
oscillations des bennes non-guidées, swaying of unguided kibbles. (70)
oscillations des rivages (Géol.), oscillations of beaches : (71)
les oscillations des rivages sont faciles à constater par le recul ou l'invasion des eaux de la mer, oscillations of beaches are easy to establish by the recession or encroach-ment of the sea-water. (72)
oscillations électriques dans les conducteurs, electric oscillations in conductors. (73)

oscillatoire (*adj.*), oscillatory ; oscillating : (74)
vibrations oscillatoires (*f.pl.*), oscillatory vibrations. (75)
courant oscillatoire (Élec.) (*m.*), oscillating current. (76)

osciller (*v.i.*), to oscillate ; to swing ; to sway about : (77)
pendule qui oscille (*m.*), pendulum which swings. (78)
câble qui oscille (*m.*), rope which sways about. (79)

oscillographe (Élec.) (*n.m.*), oscillograph. (1)
osmieux (Chim.) (*adj.m.*), osmious. (2)
osmique (Chim.) (*adj.*), osmic. (3)
osmiridium (*n.m.*) *ou* **osmiure d'iridium** (*m.*) (Minéral), osmiridium ; iridosmine ; iridosmium. (4)
osmium (Chim.) (*n.m.*), osmium. (5)
osmondite (Métall.) (*n.f.*), osmondite. (6)
osmose (Phys.) (*n.f.*), osmose ; osmosis. (7)
osmose électrique (*f.*), electric osmose. (8)
osmotique (*adj.*), osmotic ; osmositic : (9)
    pression osmotique (*f.*), osmotic pressure. (10)
ossature (*n.f.*), framework ; ossature : (11)
    wagon à ossature métallique (*m.*), truck with metal framework. (12)
    l'ossature d'une voûte, the ossature of a vault. (13)
    les masses calcaires jurassiques constituant l'ossature des Alpes (*f.pl.*), the masses of Jurassic limestone constituting the ossature of the Alps. (14)
osseux,-euse (*adj.*), osseous ; bone (*used as adj.*) : (15)
    brèche osseuse (*f.*), bone breccia ; osseous breccia. (16)
ostéolite (Minéral) (*n.f.*), osteolite. (17)
ottrélite (Minéral) (*n.f.*), ottrelite. (18)
ouest (*adj.*), west ; western ; westerly. (19)
ouest (*n.m.*), west. (20)
ouest-nord-ouest (*n.m.*), west-northwest. (21)
ouest-sud-ouest (*n.m.*), west-southwest. (22)
ouïe (d'un ventilateur, d'une pompe centrifuge) (*n.f.*), intake, inlet, ear, (of a fan, of a centrifugal pump). (23 *ou* 24)
ouragan (*n.m.*), hurricane. (25)
ouralite (Minéral) (*n.f.*), uralite. (26)
ouralitisation (Minéralogie) (*n.f.*), uralitization. (27)
ouralitiser (Minéralogie) (*v.t.*), to uralitize. (28)
outil (*n.m.*), tool : (29)
    outils à charioter *ou* outils pour chariotage (opp. aux *outils à main*) (tour), slide-rest tools ; slide-rest turning-tools. (30)
    outils à grande vitesse, high-speed tools. (31)
    outils à main (tour), hand-tools ; hand turning-tools. (32)
    outils à mortaiser, slotting-tools ; slotter-tools. (33)
    outils à raboter, planing-tools ; planer-tools. (34)
    outils coupants, cutting-tools ; edge-tools. (35)
    outils d'étaux-limeurs *ou* outils de limeuses, shaper-tools. (36)
    outils de charpentier, carpenters' tools. (37)
    outils de curage (Sondage), cleaning-tools ; well-cleaning tools. (38)
    outils de forage *ou* outils de sondage *ou* outils foreurs, drilling-tools ; boring-tools. (39)
    outils de forge *ou* outils pour forgerons, smiths' tools ; blacksmiths' tools. (40)
    outils de mines, mining-tools. (41)
    outils de mortaiseuses, slotter-tools ; slotting-tools. (42)
    outils de raboteuses, planer-tools ; planing-tools. (43)
    outils de rechange, spare tools ; duplicate set of tools. (44)
    outils de repêchage *ou* outils de pêchage *ou* outils de secours (Sondage), fishing-tools. (45)
    outils de sculpteur, carving-tools. (46)

    outils de terrassement, earthworking tools. (47)
    outils de tour au cuivre, brassworking tools (lathe). (48)
    outils de tour au fer, ironworking tools (lathe). (49)
    outils de tourneurs sur bois, wood-turning tools. (50)
    outils de tours *ou* outils pour tours, lathe-tools ; turning-tools. (51)
    outils pour forage de puits, well-boring tools. (52)
    outils tranchants, edge-tools ; cutting-tools. (53)
    outils types, typical tools. (54)
outil à aléser (tour) (*m.*), boring-tool (lathe). (55)
outil à calibre (*m.*), forming-tool. (56)
outil à centrer (*m.*), bell centering-punch. (57)
outil à clavetage (machine à mortaiser) (*m.*), splining-tool (slotting-machine). (58)
outil à couteau (tour) (*m.*), knife-tool. (59)
outil à crochet (tour) (*m.*), cranked tool ; hang-ing tool ; hook tool. (60)
outil à dégrossir *ou* **outil à charioter** (tour) (*m.*), roughing-tool. (61)
outil à dresser (tour) (*m.*), facing-tool. (62)
outil à fileter (tour) (*m.*), threading-tool. (63)
outil à fileter extérieurement (*m.*), outside-threading tool. (64)
outil à fileter intérieurement (*m.*), inside-thread-ing tool. (65)
outil à galets (Sondage) (*m.*), roller-swedge. (66)
outil à grain-d'orge (tour) (*m.*), round-nose tool. (67)
outil à planer (tour) (*m.*), planishing-tool ; planisher. (68)
outil à pointe (tour) (*m.*), point-tool. (69)
outil à pointe de diamant (tour) (*m.*), diamond-point tool ; diamond point ; diamond tool. (70)
outil à raboter de côté (*m.*), side planer-tool. (71)
outil à rainer (*m.*), keyway-cutting tool. (72)
outil à saigner *ou* **outil à tronçonner** (tour) (*m.*), cutting-off tool ; parting-tool. (73)
outil coudé (tour) (*m.*), bent tool. (74)
outil de côté ; à droite, ou à gauche (tour) (*m.*), side-tool ; right-hand, or left-hand. (75 *ou* 76)
outil de face (tour) (*m.*), front tool. (77)
outil de forme *ou* **outil de reproduction** (tour ou autre machine-outil) (*m.*), forming-tool. (78 *ou* 79)
outillage (*n.m.*), tools ; plant ; machinery ; implements ; equipment ; outfit ; appliances : (80) See also **matériel**.
    l'ouvrier mouleur possède toujours son petit outillage, qui constitue son trousseau professionnel (*m.*), the moulder always owns his small tools, which constitute his trade kit. (81)
outillage d'extraction (Mines) (*m.*), hoisting-plant ; winding-plant. (82)
outillage de dragage (*m.*), dredge-plant ; dredg-ing-machinery. (83)
outillage de mines (*m.*), mining-plant ; mining-appliances ; mining-machinery. (84)
outillage de prospecteur (*m.*), prospectors' outfit. (85)
outillage des poseurs de voies (*m.*), plate-layers' tools ; track-laying tools. (86)

outillage fixe (*m.*), fixed plant ; stationary plant. (1)

outillage hydraulique à river, monté sur roues (*m.*), travelling hydraulic rivetting-plant. (2)

outillage mobile (*m.*), loose plant ; portable plant. (3)

outillage pour exploitation des alluvions (*m.*), alluvial mining-plant ; alluvial mining-machinery. (4)

outillage pour le forage de puits (*m.*), outfit of well-boring tools ; boring-plant. (5)

outillage pour travaux miniers (*m.*), plant, machinery, for mining purposes. (6)

outiller (*v.t.*), to equip ; to fit out ; to supply with tools ; to provide with machinery : (7)

outiller une usine, to equip, to fit out, a works. (8)

outilleur (pers.) (*n.m.*), tool-maker. (9)

outremer (Chim.) (*n.m.*), ultramarine. (10)

outremer (Minéral) (*n.m.*), lapis lazuli. (11)

ouvarovite (Minéral) (*n.f.*), ouvarovite ; uvaro-vite. (12)

ouvert, -e (*adj.*), open. (13)

ouverture (action) (*n.f.*), opening : (14)

ouverture d'une route, d'une voie ferrée, de la navigation, opening of a road, of a railway, of navigation. (15 *ou* 16 *ou* 17)

ouverture (d'un circuit) (Élec.) (*n.f.*), opening, breaking, disconnecting, (a circuit). (18)

ouverture (d'une lampe de sûreté) (*n.f.*), unlock-ing (a safety-lamp). (19)

ouverture (orifice) (*n.f.*), opening ; orifice ; aperture ; hole ; gap. (20)

ouverture (d'un objectif) (*n.f.*), aperture (of a lens) : (21)

surface nettement couverte à pleine ouverture (*f.*), size of plate sharply covered at full aperture. (22)

ouverture (d'une caverne) (*n.f.*), mouth (of a cave). (23)

ouverture (d'un pont, d'une voûte) (*n.f.*), span (of a bridge, of an arch) : (24 *ou* 25)

pont de — mètres d'ouverture (*m.*), bridge of — metres span. (26)

ouverture (d'une riveuse) (*n.f.*), gap (of a rivetter). (27)

ouverture (d'un palmer) (*n.f.*), capacity (of a micrometer). (28)

· ouverture aérante (*f.*), ventilation-hole. (29)

ouverture de chargement et de coulée (con-vertisseur Bessemer) (*f.*), pouring and charging-hole. (30)

ouverture de décrassage (gazogène) (*f.*), poking-hole. (31)

ouverture de fenêtre (*f.*), window-opening. (32)

ouverture de lumière (cylindre à vapeur) (*f.*), port-opening. (33)

ouverture en mince paroi, etc. (Hydraul.) (*f.*). Same as orifice en mince paroi, etc.

ouverture maximum de la lumière (*f.*), maxi-mum port-opening. (34)

ouverture relative (d'un objectif) (*f.*), relative aperture. (35)

ouverture utile (d'un objectif) (*f.*), working aperture ; effective aperture. (36)

ouvrage (*n.m.*), work : (37)

l'ouvrage d'un maçon, the work of a mason. (38)

un ouvrage inachevé, an unfinished work. (39)

ouvrage (d'un haut fourneau) (*n.m.*), hearth, crucible, well, (of a blast-furnace). (40)

Note.—ouvrage is the part between the tuyère-zone and the boshes ; distinguished from creuset, the part below the tuyère-zone. There is no such distinction in England, the whole being known as the hearth, crucible, or well.

ouvrage au rocher (Mines) (*m.*), rock-chute. (41)

ouvrage d'art (*m.*), permanent work ; permanent structure ; structure : (42)

la limite de charge d'une locomotive ne dépend pas seulement de la résistance de la voie, mais aussi de celle des ouvrages d'art, the loading-limit of a locomotive does not only depend on the strength of the permanent way, but also on that of the permanent works. (43)

pour traverser un chemin de fer, toute cana-lisation électrique doit, de préférence, emprunter un ouvrage d'art (passage supé-rieur ou passage inférieur) et, autant que possible, ne pas franchir cet ouvrage en diagonale, when crossing a railway, every electric transmission system should, pref-erably, make use of a permanent structure (overcrossing or undercrossing) and, as far as possible, not cross such structure diago-nally. (44)

un pont est un ouvrage d'art destiné à relier deux parties d'une voie de communication séparées par un cours d'eau, fleuve ou rivière, etc., a bridge is a structure intended to unite two parts of a way of communication separated by a stream or river, etc. (45)

ouvrer (*v.t.*), to work : (46)

la batte est un outil dont se servent les plombiers pour ouvrer le plomb, the dresser is a tool used by plumbers for working lead. (47)

ouvrer des métaux précieux, to work precious metals. (48)

ouvrer (*v.i.*), to work. (49)

ouvrier, -ère (*adj.*), working ; labouring ; operative : (50)

la classe ouvrière, the working class ; the labouring class. (51)

dans une filière, les coussinets sont la partie ouvrière, in a screw-stock, the dies are the working part. (52)

ouvrier (pers.) (*n.m.*), workman ; man ; hand. (53)

ouvrier à l'accrochage (Mines) (*m.*), platman ; hanger-on ; onsetter ; hooker ; hooker-on ; hitcher. (54)

ouvrier à l'exhaure (Mines) (*m.*), pumpman. (55)

ouvrier à l'extraction (Mines) (*m.*), hoistman. (56)

ouvrier à la manœuvre des portes d'aérage (Mines) (*m.*), trapper ; door-tender. (57)

ouvrier à la manœuvre des wagons (*m.*), car-handler ; wagon-shifter ; shunter. (58)

ouvrier à la recette du fond (Mines) (*m.*), bot-tomer ; bottomman. (59)

ouvrier à la recette du jour (Mines) (*m.*), banks-man ; lander ; braceman. (60)

ouvrier à la tâche (*m.*), piece-worker ; jobber. (61)

ouvrier à la veine (Mines) (*m.*), getter. (62)

ouvrier à marteau [ouvriers à marteau *pl.*] (*m.*), hammerman. (63)

ouvrier à tout faire (*m.*), handy man. (64)

ouvrier accrocheur [ouvriers accrocheurs *pl.*] (Mines) (*m.*). Same as ouvrier à l'accrochage.

ouvrier amalgamateur [ouvriers amalgamateurs *pl.*] (*m.*), amalgamator. (1)

ouvrier au treuil (*m.*), winchman. (2)

ouvrier aux bocards (*m.*), millman. (3)

ouvrier aux câbles (*m.*), bandsman. (4)

ouvrier aux grilles (*m.*), screenman; screener. (5)

ouvrier aux pièces (*m.*), piece-worker. (6)

ouvrier aux sonneries (*m.*), bell-tender. (7)

ouvrier aux ventilateurs (*m.*), fanman. (8)

ouvrier batteur d'or [ouvriers batteurs d'or *pl.*] (*m.*), gold-beater. (9)

ouvrier boiseur [ouvriers boiseurs *pl.*] (Mines) (*m.*), timberman. (10)

ouvrier boutefeux [ouvriers boutefeux *pl.*] (Mines) (*m.*), shotfirer; shotlighter; shotman; blaster. (11)

ouvrier cafre (*m.*), Kaffir labourer; Kaffir boy. (12)

ouvrier carrier [ouvriers carriers *pl.*] (*m.*), quarry--man; quarrier. (13)

ouvrier chinois (*m.*), Chinese labourer; Chinese boy. (14)

ouvrier cisailleur [ouvriers cisailleurs *pl.*] (*m.*), shearman. (15)

ouvrier d'à bas (Mines) (*m.*), bottomer; bottom--man. (16)

ouvrier d'à haut (Mines) (*m.*), banksman; lander; braceman. (17)

ouvrier d'accrochage (Mines) (*m.*). Same as ouvrier à l'accrochage.

ouvrier de choix (*m.*), picked man. (18)

ouvrier débourbeur [ouvriers débourbeurs *pl.*] (Mines) (*m.*), trunker; washerman. (19)

ouvrier dragueur [ouvriers dragueurs *pl.*] (*m.*), dredger; dredgerman. (20)

ouvrier du fond (Mines) (*m.*), underground man; underground hand. (21)

ouvrier du jour (Mines) (*m.*), surfaceman; surface-workman; surface-hand; above--ground hand. (22)

ouvrier exercé *ou* ouvrier expérimenté (*m.*), skilled workman. (23)

ouvrier fonceur [ouvriers fonceurs *pl.*] (Mines) (*m.*), sinker. (24)

ouvrier fondeur [ouvriers fondeurs *pl.*] (de fonderie) (*m.*), foundryman. (25)

ouvrier fondeur (de four métallurgique) (*m.*), furnaceman. (26)

ouvrier forgeur [ouvriers forgeurs *pl.*] (*m.*), forge--man; forger. (27)

ouvrier indigène (*m.*), native labourer; native workman; boy. (28)

ouvrier inexpérimenté (*m.*), unskilled workman; inexperienced workman; raw hand. (29)

ouvrier jeune (*m.*), green hand; green work--man. (30)

ouvrier maçon [ouvriers maçons *pl.*] (*m.*), brick--layer. (31)

ouvrier mécanicien [ouvriers mécaniciens *pl.*] (*m.*), mechanic. (32)

ouvrier mineur [ouvriers mineurs *pl.*] (*m.*), miner. (33)

ouvrier piocheur [ouvriers piocheurs *pl.*] (*m.*), pickman. (34)

ouvrier puddleur [ouvriers puddleurs *pl.*] (*m.*), puddler. (35)

ouvrier puisatier [ouvriers puisatiers *pl.*] (*m.*), pit-sinker. (36)

ouvrier serrurier [ouvriers serruriers *pl.*] (*m.*), metal-worker; ironworker; ironsmith. (37)

ouvrier sondeur [ouvriers sondeurs *pl.*] (*m.*), drillman. (38)

ouvrier syndicaliste (*m.*), trade-unionist. (39)

ouvrier terrassier [ouvriers terrassiers *pl.*] (*m.*), navvy. (40)

ouvrir (*v.t.*), to open; to open up: (41)

ouvrir dans le remblai des voies de roulage, to open haulage-roads in the goaf. (42)

ouvrir en plein, to turn full on: (43)

ouvrir le robinet d'eau en plein, to turn the water-tap full on. (44)

ouvrir la vapeur en plein, to turn the steam full on. (45)

ouvrir l'eau, to turn on the water; to turn the water on. (46)

ouvrir l'introduction (machine à vapeur), to open the throttle. (47)

ouvrir la terre au pic, to break the ground with a pick. (48)

ouvrir le courant (Élec.), to switch on the current. (49)

ouvrir le terrain, to open up the ground. (50)

ouvrir un circuit (Élec.), to open, to break, to disconnect, a circuit. (51)

ouvrir un nouveau champ d'investigation, to open up a new field of investigation. (52)

ouvrir un pays au commerce, to open, to open up, a country to trade. (53)

ouvrir une ferme à la prospection, to throw open a farm to prospecting. (54)

ouvrir une lampe de sûreté, to unlock a safety-lamp. (55)

ouvrir une porte, to open a door. (56)

il faut un temps assez long pour ouvrir une mine et la développer, a fairly long time is required to open a mine and develop it. (57)

ouvrir (*v.i.*), to open: (58)

magasin qui n'ouvre pas le dimanche (*m.*), shop which does not open on Sunday. (59)

ouvrir (s') (*v.r.*), to open: (60)

soupape qui s'ouvre et se ferme (*f.*), valve which opens and shuts. (61)

ouwarowite (Minéral) (*n.f.*), ouvarovite; uvaro--vite. (62)

ovale (*adj.*), oval. (63)

ovale (*n.m.*), oval. (64)

ovales (fer ou acier) (*n.m.pl.*), ovals. (65 *ou* 66)

ovoïde (*adj.*), ovoid. (67)

oxalate (Chim.) (*n.m.*), oxalate. (68)

oxhydrique (*adj.*), oxyhydrogen; oxyhydric: (69)

chalumeau oxhydrique (*m.*), oxyhydrogen blowpipe. (70)

oxydabilité (*n.f.*), oxidability. (71)

oxydable (*adj.*), oxidable; oxidizable; corrod--ible; corrosible. (72)

oxydant, -e (*adj.*), oxidizing: (73)

flamme oxydante (*f.*), oxidizing flame. (74)

oxydant (*n.m.*), oxidant. (75)

oxydation (*n.f.*), oxidation. (76)

oxydation par chauffage (Métall.) (*f.*), heat-tinting. (77)

oxyde (Chim.) (*n.m.*), oxide. (78)

oxyde cuivreux (*m.*), cuprous oxide. (79)

oxyde cuivrique (*m.*), cupric oxide. (80)

oxyde d'aluminium (*m.*), aluminium oxide; alumina. (81)

oxyde de carbone (CO) (*m.*), carbon monoxide; carbonic oxide; carbonic-oxide gas; white damp. (82)

oxyde de fer (*m.*), oxide of iron; iron oxide. (83)

oxyde ferreux (*m.*), ferrous oxide. (84)

oxyde ferrique (*m.*), ferric oxide. (85)

oxyde stanneux (*m.*), stannous oxide. (86)

oxyde stannique (*m.*), stannic oxide. (1)
oxyde sulfureux (*m.*), sulphur dioxide ; sul-
-phurous oxide ; sulphurous anhydride.
(2)
oxyder (*v.t.*), to oxidize. (3)
oxyder (s') (*v.r.*), to oxidize ; to become oxidized :
(4)
la fonte s'oxyde beaucoup moins rapidement
que le fer, cast iron oxidizes much less
rapidly than wrought iron. (5)
oxygénable (*adj.*), oxygenizable ; oxidable. (6)
oxygénant, -e (*adj.*), oxidizing. (7)

oxygénation (*n.f.*), oxygenation ; oxidation.
(8)
oxygène (Chim.) (*n.m.*), oxygen. (9)
oxygéné, -e (*adj.*), oxygenated ; oxidized. (10)
oxygéner (*v.t.*), to oxygenate ; to oxygenize ;
to oxidize. (11)
oxysel (Chim.) (*n.m.*), oxysalt ; oxisalt. (12)
ozocérite *ou* **ozokérite** (Minéral) (*n.f.*), ozocerite ;
ozokerite. (13)
ozone (*n.m.*), ozone. (14)
ozoneur *ou* **ozoniseur** (*n.m.*), ozonizer. (15)
ozoniser (*v.t.*), to ozonize. (16)

P

pachnolite (Minéral) (*n.f.*), pachnolite. (17)
packfond *ou* **packfong** (alliage) (*n.m.*), packfong ;
paktong. (18)
paddock (chantier à ciel ouvert) (Exploitation des
placers) (*n.m.*), paddock. (19)
pagodite (Minéral) (*n.f.*), pagodite ; pagoda
stone. (20)
pahage [Belgique] (Mines) (*n.m.*), lodge (a lower
level in a mine in which water collects). (21)
pahoéhoé (lave lisse) (opp. à aa) (Géol.) (*n.m.*),
pahoehoe. (22) (contrasted with **a-a** *or*
aa).
paie (des ouvriers) (*n.f.*), pay, wages, (of work-
-men). (23)
paiement (*n.m.*), payment. (24)
paiement à l'heure (*m.*), payment by time ;
payment on a time basis. (25)
paiement à la journée (*m.*), day-pay ; payment
by the day. (26)
paiement au poids (*m.*), payment by weight. (27)
**paiement proportionné à la valeur du minerai
extrait** (*m.*), payment by value of mineral
gotten. (28)
paillasse (d'un laboratoire) (*n.f.*), bench (of a
laboratory). (29)
paille (tige de graminée) (*n.f.*), straw : (30)
jaune de paille (*m.*), straw-yellow. (31)
paille (défaut dans une pierre précieuse) (*n.f.*),
flaw ; feather. (32)
paille (défaut dans un métal) (*n.f.*), flaw ; fault.
(33)
paille (*n.f.*) *ou* **paille de fer** (battitures), scale ;
iron-scale ; hammer-scale ; forge-scale ;
hammer-slag ; cinder ; anvil-dross ; nill.
(34)
paille de bois (*f.*), wood-wool. (35)
paillé, -e (*adj.*), flawy : (36)
fonte paillée (*f.*), flawy iron. (37)
diamant paillé (*m.*), flawy diamond. (38)
pailleteur (pers.) (*n.m.*), digger ; gold-digger ;
gold-washer ; alluvial digger. (39)
paillette (défaut dans une pierre précieuse) (*n.f.*),
flaw ; feather. (40)
paillettes d'or (*f.pl.*), float-gold ; floating gold.
(41)
paillettes de mica (*f.pl.*), flakes of mica. (42)
paillettes métalliques (*f.pl.*), float-mineral. (43)
pailleux, -euse (*adj.*), flawy : (44)
fer pailleux (*m.*), flawy iron. (45)
pain (Métallurgie du fer) (*n.m.*), bloom. (46)
pain précieux de l'industrie (terme employé au
figuré pour désigner la houille) (*m.*), black
diamonds. (47)
paire (*n.f.*), pair : (48)

une paire de ciseaux, de pinces, a pair of scissors,
of pliers. (49 *ou* 50)
palagonite (Pétrol.) (*n.f.*), palagonite. (51)
palagonitique (*adj.*), palagonitic. (52)
palan (Méc.) (*n.m.*), pulley-block ; pulley. (53)
palan (Marine) (*n.m.*), tackle ; tackle and fall ;
fall and tackle ; block and tackle ; purchase.
(54)
palan à chaîne (*m.*), chain pulley-block. (55)
palan à corde (*m.*), rope pulley-block. (56)
palan à croc (*m.*), tackle with hook-block. (57)
palan à émerillon (*m.*), tackle with swivel-
block. (58)
palan à engrenage à vis sans fin *ou* **palan à
vis sans fin** (*m.*), worm-geared pulley-
block ; worm pulley-block. (59)
palan à engrenage droit (*m.*), spur-geared
pulley-block. (60)
palan de chèvre (*m.*), gin-tackle. (61)
palan différentiel (*m.*), differential pulley-block ;
differential pulley ; differential tackle ;
differential purchase. (62)
palan différentiel Weston, à engrenage (*m.*),
Weston's differential pulley-block, with gear ;
Weston's differential geared pulley-block.
(63)
palan différentiel Weston, à volant (*m.*), Weston's
differential pulley-block, with sprocket-
wheel. (64)
palan différentiel Weston, avec guide-chaîne *ou*
palan différentiel Weston, à guide (*m.*),
Weston's differential pulley-block, with
chain-guide ; Weston pulley with guide. (65)
palan différentiel Weston, sans guide (*m.*),
Weston's differential pulley-block, without
guide. (66)
palan double (*m.*), double tackle ; twofold
tackle ; twofold purchase. (67)
palan roulant (*m.*), travelling pulley-block. (68)
palan simple (*m.*), single tackle ; luff-tackle. (69)
palan triple (*m.*), threefold tackle. (70)
palastre *ou* **palâtre** (d'une serrure) (la boîte)
(*n.m.*), case (of a lock). (71)
palastre *ou* **palâtre** (d'une serrure) (la platine)
(*n.m.*), plate (of a lock). (72)
pale (Hydraul.) (*n.f.*), shut-off ; shut ; paddle.
(73)
pâle (*adj.*), pale ; light : (74)
bleu pâle (*m.*), pale blue ; light blue. (75)
palée (d'un pont) (*n.f.*), pier (of a bridge). (76)
(a timber pier, distinguished from **pile**
(*n.f.*), a stone pier).
palefrenier (pers.) (*n.m.*), stableman ; horse-
-keeper. (77)

paléocène (Géol.) (*adj.*), Paleocene. (1)

paléogéographie (*n.f.*), paleogeography. (2)

paléolithique (Archéol. & Géol.) (*adj.*), paleolithic : (3)

âge paléolithique (*m.*), paleolithic age. (4) (frequently written with an initial capital letter ; as, the Paleolithic age).

paléophytique (Géol.) (*adj.*), paleophytic : (5) terrain houiller paléophytique (*m.*), paleo--phytic coal-measures. (6)

paléovolcanique (Géol.) (*adj.*), paleovolcanic. (7)

paléozoïque (Géol.) (*adj.*), Paleozoic ; Primary : (8) l'âge paléozoïque ou primaire, the Paleozoic or Primary age. (9)

palette (d'une roue hydraulique, d'un ventilateur) (*n.f.*), blade, paddle, (of a water-wheel, of a fan). (10 *ou* 11)

palette (d'une turbine) (*n.f.*), blade, vane, (of a turbine). (12)

palette (d'un transporteur à palettes) (*n.f.*), push-plate, flight, scraper, (of a push-plate, flight or scraper-conveyor). (13)

palette à forer (*f.*), breast-plate ; palette ; conscience. (14)

palette de forge (*f.*), blacksmiths' shovel ; scoven. (15)

palette, tisonnier, et ringard (de forgeron) [**palette** (*n.f.*) ; **tisonnier** (*n.m.*) ; **ringard** (*n.m.*)], shovel, poker, and rake (blacksmiths'). (16)

palier (partie horizontale d'une voie) (*n.m.*), level : (17) une ligne de chemin de fer présente des paliers, des rampes, et des pentes, a line of railway contains levels, rising gradients, and falling gradients. (18)

palier (en), on the level ; level : (19) locomotives roulant en palier (*f.pl.*), engines running on the level. (20) voie de roulage en palier (*f.*), level haulage--way. (21)

palier (Mach.) (*n.m.*), bearings ; bearing ; bearing-block ; plummer-block ; plumber-block ; pillow-block ; pillow ; pedestal ; journal-bearing ; journal-box ; plummer-box. (22)

palier (Muraillement d'un puits de mine) (*n.m.*), walling-scaffold ; walling-stage ; cradle. (23)

palier (d'escalier) (Constr.) (*n.m.*), landing (of stairs). (24)

palier à billes *ou* palier à coussinets à billes *ou* palier à roulement à billes (*m.*), ball-bearing plummer-block. (25)

palier à butée à double roulement (*m.*), double-thrust ball bearings. (26)

palier à coussinets lisses (*m.*), plain bearings. (27)

palier à coussinets rigides (*m.*), rigid bearings ; rigid-bearing plummer-block. (28)

palier à graissage automatique par bagues (*m.*), self-oiling plummer-block with rings ; ring-oiled bearings. (29)

palier à graissage automatique par bagues avec coussinets à rotule, tout fonte (*m.*), all-iron self-oiling swivelling plummer-block with rings. (30)

palier à graissage automatique par chaîne (*m.*), chain-oiled bearings. (31)

palier à graissage par mèche *ou* palier à rotins (*m.*), wick-oiled bearings. (32)

palier à potence (*m.*), bracket-hanger. (33)

palier à rotule (*m.*), swivel-bearings ; swivel plummer-block ; swivel-hanger ; plummer-block with swivel-bearing. (34)

palier à rouleaux (*m.*), roller bearings. (35)

palier d'arbre à cames (*m.*), cam-shaft box. (36)

palier de butée (*m.*), thrust-bearing ; thrust-block. (37)

palier de machine (*m.*), engine-bearing. (38)

palier de pied (*m.*), footstep ; footstep-bearing ; step-bearing ; step-box ; step ; shaft-step. (39)

palier de repos (d'escalier) (Constr.) (*m.*), landing (of stairs). (40)

palier de repos (Mines) (*m.*), platform ; sollar ; ladder-sollar ; stage of ladderway. (41)

palier de tête (*m.*), cap-bearing. (42)

palier de transmission (*m.*), shaft-bearing. (43)

palier-gabarit [paliers-gabarits *pl.*] (Mines) (*n.m.*), shaft-template. (44)

palier garni d'antifriction (*m.*), bearing, plummer-block, lined with antifriction metal ; babbitted bearings. (45)

palier glissant (*m.*), water-bearing. (46)

palier graisseur à bagues (*m.*). Same as **palier à graissage automatique par bagues**.

palier horizontal (*m.*), horizontal bearing. (47)

palier lisse (*m.*), plain bearings. (48)

palier-manivelle [paliers-manivelles *pl.*] (*n.m.*), crank-bearing. (49)

palier mural (transmission) (*m.*), wall-bearing ; wall-hanger. (50)

palier renforcé, semelle rabotée, portée double du diamètre (*m.*), heavy-type plummer-block, planed on sole, length of bearing equal to 2 diameters. (51)

palier vertical (*m.*), vertical bearing. (52)

paliers série légère (*m.pl.*), light-series plummer-blocks. (53)

palis (*n.m.*), pale ; paling. (54)

palissade (*n.f.*), palisade ; palisading ; fencing. (55)

palladeux (Chim.) (*adj.m.*), palladious ; palladous. (56)

palladique (Chim.) (*adj.*), palladic. (57)

palladium (Chim.) (*n.m.*), palladium. (58)

palle (Hydraul.) (*n.f.*), shut-off ; shut ; paddle. (59)

palmer (*n.m.*), micrometer ; micrometer-gauge ; micrometer-caliper ; micrometer-calipers : (60) palmer au —° de m/m, ouverture — m/m, avec bouton à friction *ou* palmer, donnant le --° de millimètre, ouvrant à — milli--mètres, avec tête à friction, micrometer, capacity 0—mm. × —¹— mm., with ratchet-stop ; micrometer-calipers, measuring all sizes less than — millimetres by — of a millimetre, with ratchet-stop. (61)

palmer d'extérieur (*m.*), external micrometer. (62)

palmer d'intérieur (*m.*), internal micrometer. (63)

palplanche (Boisage de mines) (*n.f.*), spill ; spile ; spiling-piece ; lagging-piece ; lath. (*In the plural*, **palplanches**, spills, spiles ; spilling ; spiling ; spiling-pieces ; lagging ; lagging-pieces ; laths ; lathing). (64)

palplanche (Constr. hydraul.) (*n.f.*), sheet pile ; sheeting pile ; flat pile ; pile-plank. (*In the plural*, **palplanches** *or* **palplanches jointives**, sheet piling ; close piling ; flat piling ; pile-planking). (65)

paludéen, -enne (*adj.*), paludal : (1) fièvre paludéenne (*f.*), paludal fever ; marsh-fever. (2)

paludisme (*n.m.*), paludism ; impaludism ; marsh-fever. (3)

palustre (*adj.*), palustral ; palustrian ; palustrine. (4)

pan (face ou côté) (*n.m.*), face ; side ; pane : (5)
les pans d'un écrou, the faces (*or* the sides) (*or* the panes) of a nut. (6)
écrou à six pans (*m.*), hexagonal nut ; six-sided nut ; six-paned nut. (7)
écrou à 8 pans (*m.*), octagonal nut. (8)
manche à pans (*m.*), polygon-sided handle. (9)
roue hydraulique qui s'emboîte entre deux pans de mur verticaux (*f.*), water-wheel which fits in between two vertical faces (*or* panes) of wall. (10)
les deux pans d'un comble à deux égouts, the two sides (*or* panes) of a span-roof. (11)

pan (batée) (Mines & Placers) (*n.m.*), pan. (12)

pan d'amalgamation (Préparation mécanique des minerais) (*m.*), amalgamating-pan ; amal-gamation-pan ; pan ; pan-mill ; grinding-pan ; grinder. (13)

pan de bois (Constr.) (*m.*), timber frame ; timber framing (for walls or partitions) : (14)
façade en pans de bois (*f.*), front of timber frames. (15)

pan de fer (Constr.) (*m.*), iron frame ; iron framing. (16)

pan de rocher (*m.*), pane, slab, of rock. (17)

panabase (Minéral) (*n.f.*), panabase. (18)

panchromatique (Photogr.) (*adj.*), panchromatic : (19)
plaque panchromatique (*f.*), panchromatic plate. (20)

pandermite (Minéral) (*n.f.*), pandermite. (21)

panier (*n.m.*), basket. (22)

panier à coke (*m.*), coke-basket. (23)

panier à minerai (*m.*), ore-basket. (24)

panier à patins (*m.*), sledge-basket. (25)

panier d'osier (*m.*), wicker basket. (26)

panique (*n.f.*), panic ; stampede. (27)

panne (arrêt accidentel) (*n.f.*), breakdown : (28)
avoir une panne, to have a breakdown. (29)
panne de moteur, engine breakdown. (30)

panne (*n.f.*) *ou* **panne filière** (Constr.), purlin ; purline ; side-timber ; side-waver. (31)

panne (d'un marteau) (*n.f.*), pane, pean, peen, (of a hammer). (32)

panne bombée (marteau) (*f.*), ball pane ; ball pean ; ball peen. (33)

panne de brisis (Constr.) (*f.*), curb-plate ; curb. (34)

panne du mouton (appareil à choc) (*f.*), striking face of the tup (drop-test machine). (35)

panne du pilon (de marteau-pilon) (opp. à tas *ou* tas inférieur) (*f.*), tup pallet, tup die, pallet for tup, die for tup, top pallet, top die, (of power-hammer). (36)

panne en long (marteau) (*f.*), straight pane ; straight pean ; straight peen. (37)

panne en travers (marteau) (*f.*), cross-pane ; cross-pean ; cross-peen. (38)

panne faîtière (Constr.) (*f.*), ridge-pole ; ridge-piece ; ridge-plate ; ridge-tree ; ridge-beam. (39)

panneau (Menuis., etc.) (*n.m.*), panel. (40)

panneau (grand pilier rectangulaire de charbon) (Exploitation) (*n.m.*), panel. (41)

panneau (chambres isolées) (Exploitation des mines de charbon) (*n.m.*), panel. (42)

panneau de porte (*m.*), door-panel. (43)

panner (*v.t.*), to peen ; to pean ; to pean-hammer : (44)
panner du cuivre, to peen copper. (45)

panneresse (brique ou pierre disposée de telle sorte que sa face la plus longue soit en parement) (opp. à **boutisse**) (*n.f.*), stretcher. (46)

panneton (d'une clef) (Serrur.) (*n.m.*), bit, web, (of a key). (47)

pantellérite (Pétrol.) (*n.f.*), pantellerite. (48)

pantographe (Dessin) (*n.m.*), pantograph. (49)

pantographe (machine à tailler les fraises de forme suivant gabarit) (*n.m.*), pantographic machine ; cutter-making machine for making formed milling-cutters according to copy. (50)

pantographie (*n.f.*), pantography. (51)

pantographique (*adj.*), pantographic ; panto-graphical : (52)
dessin pantographique (*m.*), pantographic drawing. (53)

pantographiquement (*adv.*), pantographically. (54)

papier (*n.m.*), paper. (55)

papier à calquer (*m.*), tracing-paper. (56)

papier à dessin (*m.*), drawing-paper. (57)

papier à image directe *ou* **papier à image apparente** *ou* **papier à** (*ou* par) **noircissement direct** (Photogr.) (*m.*), printing-out paper. (58)

papier à image latente (Photogr.) (*m.*), gaslight-paper ; gaslight-printing paper ; developing-paper ; developing-out paper. (59)

papier à réactif (Chim.) (*m.*), test-paper ; standard paper. (60)

papier au bromure *ou simplement* **papier bromure** (Photogr.) (*m.*), bromide paper. (61)

papier au charbon (Photogr.) (*m.*), carbon-paper ; carbon-tissue. (62)

papier au ferroprussiate *ou* **papier ferroprussiate** *ou* **papier prussiate** (*m.*), blue-paper ; blue-print paper ; blue-process paper ; ferro-prussiate paper. (63)

papier auto-vireur (Photogr.) (*m.*), self-toning paper. (64)

papier bitumé *ou* **papier bituminé** (*m.*), bitu-minized paper. (65)

papier-calque [**papiers-calques** *pl.*] (*n.m.*), tracing-paper. (66)

papier carbone *ou* **papier carboné** (*m.*), carbon-paper (for reproduction of a copy under-neath). (67)

papier contraste (Photogr.) (*m.*), contrasty paper. (68) See **contrasté, -e** for example.

papier d'amiante (*m.*), asbestos paper ; amian-thus. (69)

papier d'émeri (*m.*), emery-paper. (70)

papier de bois du Brésil (*m.*), Brazil wood paper. (71)

papier de curcuma (*m.*), turmeric-paper ; turcuma-paper. (72)

papier de montagne (Minéral) (*m.*), mountain-paper ; fossil paper. (73)

papier de silex (*m.*), flint-paper. (74)

papier de tournesol (*m.*), litmus paper. (75)

papier de tournesol bleu (*m.*), blue litmus paper. (76)

papier de tournesol neutre (*m.*), neutral litmus paper. (77)

papier de tournesol rouge (m.),   red litmus paper. (1)

papier de verre (m.), glass-paper ; sandpaper. (2)

papier doux (Photogr.) (m.), soft paper. (3)

papier-émeri [papiers émeri pl.] (n.m.) ou **papier émerisé** (m.), emery-paper. (4)

papier fossile (Minéral) (m.),   fossil paper ; mountain-paper. (5)

papier huilé ou **papier gras** (m.), oiled paper. (6)

papier normal (Photogr.) (m.), normal paper. (7)

papier par développement ou **papier gaslight** (Photogr.) (m.), gaslight-paper ; gaslight-printing paper ; developing-paper ; develop-ing-out paper. (8)

papier quadrillé (m.), sectional paper ; plotting-paper ; squared paper. (9)

papier sensible (m.), sensitive paper ; sensitized paper. (10)

papier sodé (m.), soda paper. (11)

papier translucide ou **papier végétal** (m.), tracing-paper. (12)

papier verré (m.), glass-paper ; sandpaper. (13)

papier vigoureux (Photogr.) (m.), vigorous paper. (14)

papillon (régulateur à papillon) (n.m.), butterfly-valve ; throttle-valve. (15)

papillon (bec papillon pour le gaz) (n.m.), butter-fly-burner. (16)

papillon (écrou à oreilles) (n.m.), wing-nut ; thumb-nut ; finger-nut ; fly-nut ; butterfly-nut ; hand-nut. (17)

papillonage ou **papillonnage** (Dragage) (n.m.), swinging across the face. (18)

papillonner (Dragage) (v.i.), to swing across the face. (19)

papyracé, -e (adj.), papyraceous : (20)
  'houille papyracée (f.), papyraceous coal. (21)

paquet (n.m.), packet ; bundle ; parcel : (22)
  paquet de diamants, parcel of diamonds. (23)

paquet (de minerai) (Mines) (n.m.),   patch, splash, dab, (of ore). (24)

paquet (de tourbe) (n.m.), block (of peat). (25)

paquet (Métall.) (n.m.), fagot ; faggot ; pile ; packet : (26)
  paquets pour fabriquer des poutres en double T, pour fabriquer des rails de chemin de fer, fagots for H beams, for railway-rails ; piles for making H beams, for making railway-rails. (27 ou 28)

paquet d'eau (quantité d'eau nécessaire pour écluser un bateau) (m.), feed. (29)

paquet exotique (Géol.) (m.), outlier. (30)

paquetage (Métall.) (n.m.), fagoting ; fagotting ; faggoting ; piling. (31)

paqueter (Métall.) (v.t.), to fagot ; to faggot ; to pile. (32)

par les temps humides, in wet weather. (33)

par places ou **par endroits,** in places : (34)
  une largeur atteignant par places un kilomètre, a breadth reaching a kilometre in places. (35)

par tête, per head ; per capita : (36)
  donner au moins — litres d'eau par tête d'habitant par jour, to allow at least — litres of water per head of population per day. (37)

par tous les temps, in all weathers : (38)
  point nettement visible par tous les temps (m.), point clearly visible in all weathers. (39)

par voie d'eau, by water ; water (used as adj.) : (40)

transport par voie d'eau (m.), transportation by water ; water-carriage. (41)

par voie de fer ou **par voie ferrée** ou **par chemin de fer,** by rail : (42)
  expédier des marchandises par chemin de fer, to send goods by rail. (43)

par voie de terre ou **par terre,** by road ; overland : (44)
  voyage par terre (m.), overland travel. (45)

parabole (n.f.), parabola. (46)

parabolique (adj.), parabolic : (47)
  miroir parabolique (m.), parabolic mirror. (48)

paraboliquement (adv.), parabolically. (49)

parachèvement (n.m.), finishing : (50)
  parachèvement des objets coulés,   finishing castings. (51)

parachever (v.t.), to finish. (52)

parachute (de cage de mine) (n.m.), parachute ; safety-catch. (53)

parachute (plans inclinés) (Mines) (n.m.),   set stopping device. (54)

parachute (pour arrêter une maîtresse-tige de pompe en cas de rupture) (Mines) (n.m.), horse-tree ; banging-beam. (55)

parachute à bras pointus (m.),   pointed-arm safety-catch. (56)

parachute à excentriques (m.), cam-parachute ; cam safety-catch. (57)

paraclase (fracture du sol accompagnée de rejet) (Géol.) (n.f.), paraclase ; fault. (58)

paradoxe (n.m.), paradox. (59)

paradoxe de Ferguson (Méc.) (m.), Ferguson's paradox. (60)

paradoxe hydrostatique (m.), hydrostatic paradox. (61)

paraffène (Chim.) (n.m.), paraffin series ; methane series. (62)

paraffinage (n.m.), paraffining. (63)

paraffine (n.f.), paraffin ; paraffin-wax. (64)

paraffiné, -e (adj.), paraffined. (65)

paraffiner (v.t.), to paraffin ; to paraffine. (66)

paraffinique (Chim.) (adj.), paraffinic. (67)

parafoudre (n.m.), lightning-arrester ; lightning-discharger ; lightning-protector. (68) See also **déchargeur.**

parafoudre à cornes (m.),   horn-type lightning-arrester. (69)

parafoudre à cuvettes d'aluminium (m.), electro-lytic lightning-arrester. (70)

parafoudre à soufflage magnétique (m.), magnetic-blowout lightning-arrester. (71)

paragneiss (Pétrol.) (n.m.), paragneiss. (72)

paragonite (Minéral) (n.f.), paragonite. (73)

paraître (v.i.), to appear. (74)

parallaxe (n.f.), parallax. (75)

parallèle (adj.), parallel : (76)
  lignes parallèles (f.pl.), parallel lines. (77)

parellèle (ligne parallèle à une autre) (n.f.), parallel. (78)

parallèlement (adv.), parallelly. (79)

parallélépipède (n.m.), parallelepiped ; parallele-pipedon. (80)

parallélépipédique (adj.), parallelepipedal ; parallelepipedic ; parallelepipedous ; paral-lelepipedonal. (81)

parallélisme (n.m.), parallelism : (82)
  parallélisme de deux plans,   parallelism of two planes. (83)

parallélogrammatique ou **parallélogrammique** (adj.), parallelogrammatic ; parallelogram-matical ;   parallelogrammic ;   parallelo-grammical. (84)

parallélogramme (*n.m.*), parallelogram. (1)

parallélogramme de Watt (*m.*), Watt's parallelo-gram. (2)

parallélogramme des forces (*m.*), parallelogram of forces. (3)

paramagnétique (*adj.*), paramagnetic. (4)

paramagnétiquement (*adv.*), paramagnetically. (5)

paramagnétisme (*n.m.*), paramagnetism. (6)

paramètre (Géom. & Cristall.) (*n.m.*), parameter. (7)

paramétrique (*adj.*), parametric ; parametrical ; parametral. (8)

parangon (perle, diamant sans défaut) (*n.m.*), parangon. (9)

paranthine (Minéral) (*n.f.*), paranthine. (10)

parapet (*n.m.*), parapet : (11)
parapet d'un pont, parapet of a bridge. (12)

parapluie (d'une cage de mine) (*n.m.*), bonnet. (13)

parasite (Minéralogie) (*n.m.*), parasite : (14)
parasites et leur hôte, parasites and their host. (15)

parasoleil d'objectif (Photogr.) (*m.*), lens-hood. (16)

paratonnerre (*n.m.*), lightning-conductor ; lightning-rod. (17)

parc à blooms (*m.*), bloom-yard. (18)

parc à charbon (*m.*), coal-yard. (19)

parc à châssis (Fonderie) (*m.*), box-yard. (20)

parc à fonte (Fonderie) (*m.*), pig-yard ; pig-iron yard. (21)

parc à matières (*m.*), stock-yard. (22)

parc à mitraille (*m.*), scrap-yard. (23)

parcelle (*n.f.*), particle : (24)
des parcelles minuscules d'or, minute particles of gold. (25)

parcelle (de terrain) (*n.f.*), parcel, plot, patch, (of ground). (26)

parcourir — kilomètres à l'heure, to travel at, to run at, — kilometres per hour. (27)

parcours (trajet) (*n.m.*), haul ; journey : (28)
parcours par voie ferrée, railroad-haul. (29)

pare-étincelles [pare-étincelles *pl.*] (locomotive) (*n.m.*), spark-catcher ; spark-arrester ; spark-consumer ; smoke-box netting ; spark-netting ; netting. (30)

pare-poussière [pare-poussière *pl.*] (*n.m.*), dust-guard. (31)

pare-ringard [pare-ringards *pl.*] (de porte de foyer) (*n.m.*), sill-plate (of fire-door). (32)

parement (*n.m.*), face ; facing : (33)
parement d'un talus, d'un mur, face of a slope, of a wall. (34 *ou* 35)

parer (*v.t.*), to dress ; to trim : (36)
parer une pierre, to dress a stone. (37)
parer une tôle, une barre de fer, une éprouvette, to trim a plate, an iron bar, a test-bar. (38 *ou* 39 *ou* 40)

parer à, to guard against ; to provide against : (41)
parer à la soudaineté des crues de certains fleuves torrentiels, to guard against the suddenness of the floods of certain torrential rivers. (42)

paresseux, -euse (*adj.*), lazy ; sluggish : (43)
cours d'eau paresseux (*m.*), lazy stream ; sluggish stream. (44)

parfait, -e (*adj.*), perfect : (45)
gaz parfait (*m.*), perfect gas. (46)

parfaitement (*adv.*), perfectly. (47)

pargasite (Minéral) (*n.f.*), pargasite. (48)

parleur (Téléph.) (*n.m.*), transmitter ; sender. (49)

paroi (*n.f.*), side ; interior surface ; wall : (50)
paroi d'un cylindre, wall of a cylinder. (51)
parois d'un filon, sides, walls, cheeks, of a lode. (52)
paroi d'un tuyau *ou* parois d'un tuyau *ou* paroi intérieure d'un tuyau (Hydraul.), interior surface of a pipe. (53)
parois d'un fossé, d'une vallée, sides of a ditch, of a valley. (54 *ou* 55)
parois d'une chambre, d'un réservoir, walls of a room, of a reservoir. (56 *ou* 57)
parois d'une chaudière, walls of a boiler. (58)
parois latérales d'une galerie de mine, sides, walls, of a mine-level ; side-walls. (59)

paroxysmal, -e, -aux (Géol.) (*adj.*), paroxysmal. (60)

paroxysme (Géol.) (*n.m.*), paroxysm : (61)
paroxysmes volcaniques, volcanic paroxysms. (62)

parpaing (Maçonn.) (*n.m.*), bonder ; binder ; bond-stone ; bond ; binding-stone ; through-stone ; through-binder ; perpend ; perpend-stone. (63)

parquet (*n.m.*), floor ; flooring. (64)

parquetage (*n.m.*), flooring. (65)

parqueter (*v.t.*), to floor. (66)

parsemer (*v.t.*), to strew ; to sprinkle ; to intersperse. (67)

part (*n.f.*), part ; share ; portion. (68)

partage (*n.m.*), division ; dividing. (69)

partage des eaux (Géogr. phys.) (*m.*), parting of the waters ; dividing of the waters : (70)
les plis ou axes anticlinaux déterminent le partage des eaux (*m.pl.*), anticlinal folds or axes decide the parting of the waters. (71)

partager (*v.t.*), to divide : (72)
partager une terre, to divide a piece of land. (73)

partager (se) (*v.r.*), to divide ; to share. (74)

partial, -e, -aux (*adj.*), partial ; biased ; biassed. (75)

partialement (*adv.*), partially. (76)

particularité (*n.f.*), peculiarity ; characteristic ; feature : (77)
particularité frappante, striking peculiarity (*or* feature). (78)

particule (*n.f.*), particle ; speck : (79)
particule de matière, particle of matter. (80)
particules d'or, specks, particles, of gold. (81)

particulier, -ère (opp. à **général**) (*adj.*), particular ; peculiar. (82)

particulier, -ère (privé) (*adj.*), private. (83)

particulièrement (*adv.*), particularly ; peculiarly ; especially. (84)

partie (*n.f.*), part ; portion ; section : (85)
diviser une droite en un nombre quelconque de parties égales, to divide a straight line into any number of equal parts. (86)
parties frottantes d'une machine, wearing parts of a machine. (87)
partie stérile d'un filon, barren, sterile, portion of a lode. (88)
dans une filière, les coussinets sont la partie ouvrière, in a screw-stock, the dies are the working part. (89)

partie (portion constituante) (*n.f.*), part : (90)
— parties de mercure alliées à une partie d'argent, — parts of mercury alloyed with one part of silver. (91)

alliage qui renferme de — à — parties d'argent pour 1 partie d'or (*m.*), alloy which contains from — to —parts of silver to 1 part of gold. (1)

partie (expédition) (*n.f.*), party : (2)
organiser une partie de recherches, to get up a prospecting party. (3)

partie (en), in part ; partly ; partially : (4)
en partie détruit (-e) par un incendie, partially destroyed by fire. (5)

partiel, -elle (*adj.*), partial ; incomplete : (6)
décarburation partielle de la fonte (*f.*), partial decarburization of cast iron. (7)

partiellement (*adv.*), partially ; incompletely. (8)

partir à la dérive (se dit d'un wagon) (Ch. de f.), to run back ; to run wild ; to break away : (9)
partir à la dérive sur un plan incliné, to run back down an incline. (10)

partschine (Minéral) (*n.f.*), partschinite. (11)

parvenir (*v.i.*), to reach ; to arrive ; to attain : (12)
parvenir au but de son voyage, to reach the end of one's journey. (13)

pas (distance comprise entre deux filets contigus d'une vis) (*n.m.*), pitch ; thread : (14)
vis à pas rapide (*f.*), quick-pitch screw. (15)

pas (filet de vis) (*n.m.*), thread : (16)
vis à pas carré (*f.*), square-thread screw. (17)

pas (d'un engrenage) (*n.m.*), pitch (of a gear). (18)

pas (d'une chaîne) (*n.m.*), pitch (of a chain). (19)

pas à droite (*m.*), right-hand thread. (20)

pas à faire *ou* **pas à exécuter** *ou* **pas à fileter** (Filetage) (*m.*), pitch to be cut (Screw-cutting). (21)

pas à gauche (*m.*), left-hand thread. (22)

pas allongé (*m.*), long pitch ; coarse pitch ; quick pitch : (23)
vis de pas allongé (*f.*), long-pitch screw ; coarse-pitch screw ; quick-pitch screw. (24)

pas américain système Sellers (*m.*), Sellers (U. S. standard) thread ; United States standard thread ; American standard thread ; Franklin Institute thread. (25)

pas anglais système Whitworth *ou* **pas Whit-worth** (*m.*), Whitworth (British standard) thread ; Whitworth's standard thread ; Whitworth thread. (26)

pas apparent (*m.*), apparent pitch ; divided pitch. (27)

pas bâtard (opp. à **pas exact**) (*m.*), bastard pitch ; odd pitch ; fractional pitch. (28)

pas carré (*m.*), square thread. (29)

pas circulaire *ou* **pas circonférentiel** (Engrenages) (*m.*), circular pitch ; circumferential pitch ; arc pitch. (30)

pas d'écrou (*m.*), inside thread. (31)

pas de cuffat (Mines) (*m.*), kibble-landing. (32)

pas de la vis mère (*m.*), pitch of lead-screw. (33)

pas de vis (*m.*), screw-pitch ; screw-thread. (34)

pas diamétral (Engrenages) (*m.*), diametral pitch ; diametrical pitch. (35)

pas du gaz (*m.*), gas-thread. (36)

pas exact (opp. à **pas bâtard**) (*m.*), even pitch. (37)

pas fin (*m.*), fine pitch : (38)
vis à pas fin (*f.*), fine-pitch screw. (39)

pas normal (*m.*), normal pitch. (40)

pas rapide (*m.*). Same as **pas allongé**.

pas réel (*m.*), true pitch ; total pitch. (41)

pas rond (*m.*), round thread. (42)

pas système français (S.F.) (*m.*), French standard thread. (43)

pas système international (S.I.) (*m.*), international standard thread. (44)

pas triangulaire (*m.*), V thread. (45)

passage (action) (*n.m.*), passage ; passing ; crossing : (46)
passage d'un courant électrique, passage of an electric current. (47)
passage d'un rayon lumineux à travers un prisme, passage of a ray of light through a prism. (48)
le passage d'un train, the passing of a train. (49)
le passage d'un fleuve, crossing a river. (50)

passage (*n.m.*) *ou* **passage au méridien** (Astron.), transit ; culmination : (51)
passage de la Polaire, transit, culmination, of Polaris. (52)
passage inférieur, lower transit ; lower culmination. (53)
passage supérieur, upper transit ; upper culmination. (54)

passage (d'une faille) (Exploitation de mines) (*n.m.*), negotiating (a fault). (55)

passage (action de passer une barre par le laminoir) (*n.m.*), pass : (56)
laminage en — passages (*m.*), rolling in—passes. (57)

passage (endroit où l'on passe) (*n.m.*), passage ; passageway ; thoroughfare : (58)
un passage souterrain, an underground passage. (59)
passage souterrain (Ch. de f., etc.), subway. (60)
passage interdit (avis), no thoroughfare. (61)

passage (endroit où l'on traverse une rivière dans un bac) (*n.m.*), ferry. (62)

passage (Signaux de ch. de f.) (opp. à **arrêt**), proceed. (63) (opp. to **stop**).

passage à niveau (Ch. de f.) (*m.*), level crossing ; grade crossing ; crossing at grade. (64)

passage à niveau portatif (Ch. de f.) (*m.*), port-able level crossing. (65)

passage au crible *ou* **passage à la claie** (*m.*), screening. (66)

passage au tamis (*m.*), sifting. (67)

passage au trommel (*m.*), trommeling. (68)

passage carrossable (*m.*), roadway. (69)

passage d'air (*m.*), air-passage. (70)

passage d'eau (voie d'eau d'un robinet, d'une soupape) (*m.*), waterway (of a tap, of a valve). (71 *ou* 72)

passage d'escalier (*m.*), stairway ; staircase. (73)

passage de niveau (Ch. de f.) (*m.*). Same as **passage à niveau**.

passage de prise de vapeur (*m.*), steam-passage to chest. (74)

passage en contrebande de diamants hors des camps de concentration (*m.*), smuggling away of diamonds from the compounds. (75)

passage en cul-de-sac (Mines) (*m.*), blind drift. (76)

passage en dessous *ou* **passage inférieur** (Ch. de f.) (*m.*), bridge under ; undercrossing. (77)

passage en dessus *ou* **passage supérieur** (Ch. de f.) (*m.*), bridge over ; overcrossing. (78)

passage pour les hommes (Mines) (*m.*), manway ; manhole. (79)

passager, -ère (*adj.*) passing : (80)
une ondée passagère, a passing shower. (81)

passe (passage navigable entre deux terres) (*n.f.*), channel ; passage : (1)
passe navigable, navigable channel (2)
passe profonde, deep-water channel. (3)
passe (opération)(*n.f.*), stage ; operation : (4)
préparation mécanique du minerai effectuée en une seule passe (*f.*), dressing of the ore effected in one stage. (5)
passe (action de passer une barre, etc., par le laminoir) (*n.f.*), pass : (6)
passe de dégrossissage, roughing-pass. (7)
passe de finissage, finishing-pass. (8)
passe (mouvement ou course d'une machine-outil ou analogue) (*n.f.*), going over ; cut : (9)
taraudage d'une seule passe (*m.*), screwing at once going over. (10)
filetage en plusieurs passes (*m.*), screwing in several cuts. (11)
prendre une faible, une forte passe, to take a light, a heavy cut. (12 *ou* 13)
abaissements successifs de l'élinde après chaque passe (*m.pl.*), successive lowerings of the dredging-ladder after each cut. (14)
passe-courroie [passe-courroie *pl.*] (*n.m.*), belt-shifter ; belt-shipper ; strike-gear ; striking-gear ; striker. (15)
passe-déversoir [passes-déversoirs *pl.*] (Hydraul.) (*n.f.*), spillway. (16)
passe-diable [passe-diables *pl.*] (Sondage) (*n.m.*), go-devil ; scraper ; tube-cleaner. (17)
passe-partout [passe-partout *pl.*] (clef) (*n.m.*), master-key. (18)
passe-partout (pour scier les gros arbres) (*n.m.*), lumberman's two-handed saw ; felling-saw. (19)
passe-partout (à pierres) (avec ou sans dents) (*n.m.*), stone-saw. (20 *ou* 21)
passe-partout (Photogr.) (*n.m.*), slip-in mount. (22) Cf. **carton pour coller les épreuves.**
passer (*v.t.* & *v.i.*). See examples : (23)
passer à — kilomètres d'une propriété, to pass within — kilometres of a property. (24)
passer à la filière (*v.t.*), to draw ; to draw through the drawplate ; to wiredraw : (25)
passer de l'or à la filière, to draw gold through the draw-plate. (26)
passer à travers *ou* passer au travers de, to pass through ; to run through : (27)
matières qui passent au travers des mailles d'un tamis (*f.pl.*), matter which passes through the meshes of a sieve. (28)
rivière qui passe à travers un champ (*f.*), river which runs through a field. (29)
passer au crible *ou* passer à la claie (*v.t.*), to screen ; to pass through the screen : (30)
passer le minerai au crible, to screen the ore ; to pass the ore through the screen. (31)
passer au delà d'un écriteau signalant qu'il y a danger, to pass beyond a danger-board. (32)
passer au tamis (*v.t.*), to sift. (33)
passer au trommel (*v.t.*), to trommel. (34)
passer aux bocards seulement le minerai à haute teneur, to put high-grade ore only through the mill. (35)
passer de la marche par engrenages à la marche à la volée, to change over from gear-drive to belt-drive. (36)
passer l'eau, to cross, to ferry across, a river. (37)

passer l'hiver sur la propriété, to pass the winter, to winter, on the property. (38)
passer le minerai dans un petit moulin de prospection, to put, to run, to pass, the ore through a small prospecting-mill. (39)
passer le niveau (Mines), to pass through, to negotiate, water-bearing strata. (40)
passer par-dessus ses berges (se dit d'une rivière), to overflow, to overrun, its banks. (41)
passer par un cours d'instruction pour travaux de sauvetage dans les mines, to go through a course of training for mine-rescue work. (42)
passer plusieurs jours à visiter une propriété, to spend several days examining a property. (43)
passer sous le remblai (Mines), to pass through the goaf. (44)
passer un contrat *ou* passer un traité, to enter into an agreement : (45)
passer un traité avec un entrepreneur, to enter into an agreement with a contractor. (46)
passer une commande pour des marchandises, to give an order for goods. (47)
passer une faille (Exploitation de mines), to negotiate a fault. (48)
passer une rivière, to cross a river. (49)
passer une rivière à gué, to ford a river. (50)
passer (se) de, to do without ; to dispense with : (51)
se passer de boisage par suite de la solidité du toit, to dispense with timbering owing to the firmness of the roof. (52)
passerelle (*n.f.*), foot-bridge. (53)
passeur (cribleur) (pers.) (*n.m.*), screener ; screenman. (54)
passeur (homme qui conduit un bac) (*n.m.*), ferryman : (55)
'héler le passeur, to hail the ferryman. (56)
passible d'une amende (être), to be liable to a fine. (57)
pastille (pièce métallique fixée dans la base du culot d'une lampe électrique à incandescence) (*n.f.*), contact. (58)
pâte (*n.f.*), paste. (59)
pâte (Géol.) (*n.f.*), paste. (60)
patère (pour recevoir une rosace, un interrupteur) (appareillage électrique) (*n.f.*), base, block, (to receive a rose, a switch). (61 *ou* 62)
pâteux, -euse (*adj.*), pasty : (63)
masse pâteuse (*f.*), pasty mass. (64)
patin (de rail de chemin de fer) (*n.m.*), flange (of railway-rail). (65)
patin (d'un palier) (*n.m.*), sole, sole-plate, sill-plate, base, (of a plummer-block). (66)
patin (du tiroir) (*n.m.*), face-flange (of the slide-valve). (67)
patin (d'une enclume) (*n.m.*), foot (of an anvil). (68)
patin (pièce de bois qui se place sous le pied d'un étai pour former empattement) (Constr.) (*n.m.*), sole-piece ; footing-block. (69)
patin (de la crosse du piston) (*n.m.*), shoe, slipper, gib, (of piston cross-head). (70)
patin (d'une benne à patins, ou analogue) (Mines) (*n.m.*), runner, sledge-runner, slipe, (of a sledge-kibble, or the like). (71 *ou* 72)
patin (d'une cuve d'amalgamation) (*n.m.*), muller, shoe, (of an amalgamating-pan). (73)

patin (d'une happe) (*n.m.*), toe (of a G cramp). (1)

patinage (*n.m.*), slipping : (2)

le patinage des cylindres d'un laminoir, des roues motrices d'une locomotive, the slipping of the rolls of a rolling-mill, of the driving-wheels of a locomotive. (3 *ou* 4)

patine (*n.f.*), patina ; green rust ; ærugo. (5)

patiner (*v.i.*), to slip : (6)

roue de locomotive qui patine au lieu de rouler (*f.*), engine-wheel which slips instead of rolling. (7)

patouillet (*n.m.*) *ou* **patouille** (*n.f.*) (Préparation mécanique des minerais), patouillet ; washer. (8)

patron (pers.) (*n.m.*), employer ; master. (9)

patron (modèle ; gabarit) (*n.m.*), template ; templet ; pattern ; former ; patron. (10)

patte (Méc.) (*n.f.*), cramp ; holdfast. (11)

patte (d'une pellicule film-pack) (*n.f.*), tab (of a pack-film). (12)

patte à crochet (pour embrasser les tuyaux de gaz, d'eau, etc.) (*f.*), pipe-strap. (13)

patte à futaille (*f.*), barrel-sling. (14)

patte d'attache (pour tubes) (appareillage électrique) (*f.*), saddle, clip, (for conduits). (15)

patte d'élingue (*f.*), can-hook ; sling-dog ; dog ; dog-hook. (16)

patte de lièvre (d'un croisement) (Ch. de f.) (*f.*), wing-rail (of a crossing or frog). (17)

patte de lièvre (d'un contre-rail) (Ch. de f.) (*f.*), wing (of a guard-rail). (18)

patte de sustentation (*f.*), bracket (for supporting a boiler). (19)

patte du câble (par opp. à l'enlevage) (Extraction) (Mines) (*f.*), end of the rope (attached to the cage). (20) (in contradistinction to the **lifting-point** [of the rope on the pit-head pulleys]) :

les coupages à la patte du câble ont l'avantage de changer l'enlevage, cutting off the end of the rope has the advantage of changing the lifting-point. (21)

pattes d'araignée (*f.pl.*), oil-grooves ; oil-channels. (22)

pattinsonage (*n.m.*), pattinsonation ; pattinsonization : (23)

la désargentation du plomb d'œuvre par pattinsonage, the desilverization of work-lead by pattinsonation. (24)

pattinsonner (*v.t.*), to pattinsonize. (25)

paumelle double (*f.*), H hinge. (26)

pauvre (*adj.*), poor : (27)

minerai pauvre (*m.*), poor ore ; poor stone. (28)

pauvrement (*adv.*), poorly. (29)

pauvreté (*n.f.*), poverty ; poorness ; baseness : (30)

pauvreté d'une mine, poverty of a mine. (31)

pauvreté du filon, poverty, poorness, of the lode. (32)

pauvreté du minerai, poorness, baseness, of the ore. (33)

pavage (action) (*n.m.*), paving. (34)

pavage (pavé) (*n.m.*), paving ; pavement. (35)

pavage en bois (*m.*), wood paving ; wood pavement. (36)

pavage en cailloux (*m.*), pebble paving. (37)

pavage en pierres (*m.*), stone paving. (38)

pavé (bloc de liais ou de grès dont on garnit les chaussées) (*n.m.*), paving-stone ; sett ; set. (39)

pavé (assemblage de pavés) (*n.m.*), pavement ; paving : (40)

le pavé de bois se généralise, wood paving is becoming general. (41)

pavé (partie pavée d'une rue) (*n.m.*), trackway ; pavé : (42)

le macadam et le pavé, the macadam and the trackway. (43)

pavé en dalles (*m.*), flagstone pavement ; flagging. (44)

pavement (action) (*n.m.*), paving. (45)

pavement (pavé) (*n.m.*), pavement ; paving. (46)

pavement de marbre (*m.*), marble pavement. (47)

paver (*v.t.*), to pave : (48)

paver une rue, to pave a street. (49)

paveur (pers.) (*n.m.*), pavior. (50)

pavillon (d'un entonnoir) (*n.m.*), flare (of a funnel). (51)

pavillon (d'un récepteur téléphonique) (*n.m.*), earpiece (of a telephone-receiver). (52)

payant, -e (*adj.*), payable ; paying ; pay (used as *adj.*) ; remunerative : (53)

minerai payant (*m.*), payable ore ; pay-ore. (54)

paye (des ouvriers) (*n.f.*), pay, wages, (of workmen). (55)

payement (*n.m.*). Same as **paiement**.

payer (*v.t.*), to pay ; to pay for : (56)

payer le salaire des ouvriers, to pay the men's wages. (57)

lorsqu'on est dans la nécessité de payer l'eau, de payer la casse, when one is obliged to pay for water, to pay for breakages. (58 *ou* 59)

pays (*n.m.*), country. (60) See also **contrée**.

pays accidenté (*m.*), undulating country ; uneven country ; rolling country. (61)

pays bien boisé (*m.*), well-wooded country. (62)

pays chaud (*m.*), hot country. (63)

pays de la soif (*m.*), thirsty country. (64)

pays de plaine *ou* **pays plat** (*m.*), plain country ; flat country. (65)

pays découvert (*m.*), open country. (66)

pays inhabité (*m.*), uninhabited country. (67)

pays montagneux *ou* **pays de montagne** *ou* **pays montueux** (*m.*), mountainous country ; hilly country. (68)

pays producteur d'or (*m.*), gold-producing country. (69)

pays stérile (*m.*), barren country. (70)

pays tourmenté (*m.*), broken country. (71)

peau (cuir détaché du corps de l'animal) (*n.f.*), skin ; hide. (72)

peau (croûte qui se forme à la surface des liquides, etc.) (*n.f.*), skin : (73)

la peau d'un lingot, the skin of an ingot. (74)

fonte à peau rugueuse, à peau lisse (*f.*), rough-skinned, smooth-skinned, pig iron. (75 *ou* 76)

peau crue *ou* **peau verte** (*f.*), raw hide ; green hide. (77)

peau de chamois (*f.*), wash-leather ; shammy ; chamois. (78)

pêchage des outils (Sondage) (*m.*), fishing up tools. (79)

pechblende (Minéral) (*n.f.*), pitchblende ; uraninite. (80)

pêche des perles (*f.*), pearl-fishing. (81)

pêche du corail (*f.*), coral-fishing. (82)

pêcher (*v.t.*), to fish ; to fish up. (83)

pêcherie de perles (f.), pearl-fishery. (1)
pêcheur de perles (pers.) (m.), pearl-diver ; pearl-fisher. (2)
pechkohle (Minéral) (n.f.), pitch-coal. (3)
pechstein (Pétrol.) (n.m.), pitchstone. (4)
péchurane (Minéral) (n.m.), pitchblende ; uraninite. (5)
peck (mesure anglaise de capacité) (n.m.), peck = 9·092 litres. (6)
pectolite (Minéral) (n.f.), pectolite. (7)
pédale (n.f.), treadle ; pedal. (8)
pédale (à), by foot power ; foot (used as adj.) ; treadle (used as adj.) ; pedal (used as adj.) : (9)
tour à pédale (m.), foot-lathe ; treadle-lathe ; pedal-lathe. (10)
pédale de calage ou pédale de sûreté (Ch. de f.) (f.), locking-bar ; detector-bar. (11)
pegmatite (Pétrol.) (n.f.), permatite ; pegmatyte. (12)
pegmatite graphique (f.), graphic pegmatite. (13)
pegmatitique (adj.), pegmatitic. (14)
pegmatoïde (adj.), pegmatoid : (15)
structure pegmatoïde (f.), pegmatoid structure. (16)
peignage (Méc.) (n.m.), chasing. (17)
peigne (n.m.) ou peigne à fileter, chaser ; screw-tool ; comb. (18)
peigne d'extérieur ou peigne à fileter pour l'extérieur (m.), chaser, screw-tool, comb, outside or external. (19)
peigne d'intérieur ou peigne à fileter pour l'intérieur (m.), chaser, screw-tool, comb, inside or internal. (20)
peigner (v.t.), to chase : (21)
peigner un filet de vis, to chase a screw-thread. (22)
peindre (v.t.), to paint : (23)
peindre les boiseries, to paint the woodwork. (24)
peinture (n.f.), paint ; colour ; color. (25)
peinture à l'huile (f.), oil-colour ; oil-paint. (26)
peinture vernissante (f.), enamel (for house decoration). (27)
péliom (Minéral) (n.m.), peliom. (28)
pelle (n.f.), shovel. (29)
pelle (Hydraul.) (n.f.), shut-off ; shut ; paddle. (30)
pelle à charbon (f.), coal-shovel. (31)
pelle à coke (f.), coke-shovel. (32)
pelle à échantillons (f.), sample-shovel. (33)
pelle à feu (f.), fire-shovel ; firing-shovel. (34)
pelle à grille (f.), pronged shovel. (35)
pelle à vanner (f.), vanning-shovel ; van. (36)
pelle à vapeur (Trav. publ.) (f.), steam-shovel ; steam-digger ; steam-excavator ; steam-navvy ; navvy. (37)
pelle de chauffeur (f.), firing-shovel ; fire-shovel. (38)
pelle de four (f.), furnace-shovel. (39)
pelle de mineur (f.), miners' shovel. (40)
pelle de mouleur (f.), moulders' shovel. (41)
pelle emmanchée (f.), handled shovel. (42)
pelle, tisonnier, et ratissette (de forgeron) [pelle (n.f.) ; tisonnier (n.m.) ; ratisette (n.f.)], shovel, poker, and rake (blacksmiths'). (43)
pelletage (n.m.), shovelling ; shovel-work : (44)
pelletage du minerai, shovelling the ore. (45)
pelletée (n.f.), shovelful : (46)
pelletée de terre, shovelful of earth. (47)
pelleter (v.t.), to shovel : (48)

pelleter le charbon, to shovel coal. (49)
pelleteur (pers.) (n.m.), shoveller. (50)
pelliculage (Photogr.) (n.m.), stripping : (51)
pelliculage des négatifs, stripping negatives. (52)
pellicule (membrane très mince) (n.f.), film ; pellicle : (53)
objet recouvert d'une pellicule de platine précipité (m.), article covered with a film of precipitated platinum. (54)
pellicule (Photogr.) (n.f.), film : (55)
pellicule en bobine [pellicules en bobines pl.] ou pellicule en rouleau [pellicules en rouleaux pl.], roll film. (56)
pellicules film-pack ou pellicules bloc-film, pack films. (57)
pelliculé, -e (adj.), filmy. (58)
pelliculer (Photogr.) (v.t.), to strip : (59)
pelliculer un cliché, to strip a negative. (60)
pellon (outil de mouleur) (Fonderie) (n.m.), pegging-rammer ; pegging-pean. (61)
peloton de ficelle (m.), ball of string. (62)
pembrite (Explosif) (n.f.), pembrite. (63)
pénalité pour fausse déclaration (f.), penalty for making false return. (64)
penchant, -e (hors d'aplomb) (adj.), leaning : (65)
muraille penchante (f.), leaning wall. (66)
penchant (pente) (n.m.), slope ; declivity : (67)
penchant d'une colline, slope of a hill. (68)
pencher (être hors de son aplomb) (v.i.), to lean ; to lean over : (69)
mur qui penche (m.), wall which leans. (70)
pendage (Géol.) (n.m.), dip : (71)
le pendage d'un filon, the dip of a lode. (72)
pendre (v.t.), to hang ; to hang up : (73)
pendre une lampe à un croc, to hang a lamp on a hook. (74)
pendre (v.i.), to hang. (75)
pendule (n.m.), pendulum. (76)
pendule compensateur ou pendule compensé (m.), compensation pendulum ; compensated pendulum ; compensating pendulum. (77)
pendule composé (m.), compound pendulum. (78)
pendule conique (m.), conical pendulum. (79)
pendule cycloïdal (m.), cycloidal pendulum. (80)
pendule électrique (m.), electric pendulum. (81)
pendule simple (m.), simple pendulum. (82)
pêne (n.m.) ou pêne de serrure, bolt ; lock-bolt. (83)
pêne à demi-tour ou pêne à ressort (m.), latch-bolt ; spring-bolt. (84)
pêne dormant (m.), dead bolt. (85)
pêne en biseau (m.), bevel-headed bolt. (86)
pénéplaine (Géol.) (n.f.), peneplain ; plain of erosion ; denudation-plain. (87)
pénétrabilité (n.f.), penetrability. (88)
pénétrable (adj.), penetrable. (89)
pénétrant, -e (adj.), penetrating ; penetrative. (90)
pénétration (n.f.), penetration ; permeation. (91)
pénétrer (v.t. & v.i.), to penetrate ; to permeate. (92)
pénible (adj.), painful ; laborious ; hard : (93)
travail pénible (m.), hard work. (94)
péniblement (adv.), painfully ; laboriously ; with difficulty. (95)
péniche (n.f.), lighter ; barge. (96)
péninsulaire (adj.), peninsular. (97)
péninsule (n.f.), peninsula. (98)
pénitent (Géol.) (n.m.), erosion column ; chimney-rock. (99)

pennine *ou* **penninite** (Minéral) (*n.f.*), penninite ; pennine. (1)
pennyweight (*en abrégé s'ecrit* **dwt**) (*n.m.*), penny-weight = 1·5552 grammes. (2)
pénombre (Opt.) (*n.f.*), penumbra. (3)
pénombré, -e (*adj.*), penumbral ; penumbrous. (4)
pentagonal, -e, -aux *ou* **pentagone** (*adj.*), pen-tagonal : (5)
prisme pentagonal (*m.*), pentagonal prism. (6)
pentagone (*n.m.*), pentagon. (7)
pentane (Chim.) (*n.m.*), pentane. (8)
pentavalence (Chim.) (*n.f.*), pentavalence ; pentavalency ; quinquevalence. (9)
pentavalent, -e (*adj.*), pentavalent ; quinque-valent. (10)
pente (*n.f.*), slope ; declivity ; falling gradient ; downward gradient ; gradient ; down grade ; grade ; incline ; pitch ; slant : (11)
une ligne de chemin de fer présente des paliers, des rampes, et des pentes, a line of railway contains levels, rising gradients, and falling gradients. (12)
parcourue en sens contraire, la rampe est une pente, traversed in the opposite direction, the up grade is a down grade. (13)
pente d'un comble (l'égout ou versant), slope of a roof. (14) Cf. **pente de comble**.
pente d'une colline, slope, pitch, of a hill. (15)
pente des sluices, slope of the sluices. (16)
pente douce, easy grade ; easy gradient ; gentle slope. (17)
pente faible, low gradient ; slight gradient. (18)
pente graduelle, gradual slope ; gradual incline. (19)
pente hydraulique, hydraulic slope. (20)
pente raide, steep gradient ; steep incline. (21)
pente (à) *ou* **pente (en)**, sloping ; slanting ; inclined : (22)
terrain en pente (*m.*), sloping ground. (23)
chemin à forte pente (*m.*), steeply inclined road. (24)
pente d'éboulis (Géol.) (*f.*), talus. (25)
pente de comble (l'inclinaison) (*f.*), roof-pitch ; pitch of roof. (26)
pente-limite (Géogr. phys.) (*n.f.*), grade : (27)
rivière qui atteint, qui possède, sa pente-limite (*f.*), river which reaches, which is at, grade. (28 *ou* 29)
pente naturelle de tassement (*f.*), natural slope ; angle of repose ; earth-slope. (30)
pentes et rampes (Ch. de f.) (*f.pl.*), gradients. (31)
pentlandite (Minéral) (*n.f.*), pentlandite. (32)
penture (*n.f.*), hinge ; ride : (33)
gond et penture, hook and hinge ; hook and ride. (34) See **gond et penture** for varieties of that class.
penture à T (*f.*), T hinge ; butt-and-strap hinge ; cross-garnet hinge ; cross-tail hinge ; cross-tailed hinge ; garnet-hinge ; garnet. (35)
pénurie (*n.f.*), scarcity ; want ; shortage ; shortness ; insufficiency ; lack ; dearth : (36)
pénurie d'eau, want, scarcity, lack, of water. (37)
pénurie de main-d'œuvre, scarcity, shortage, dearth, insufficiency, (of labour). (38)
péperin *ou* **peperino** (tuf) (*n.m.*), peperino ; peperine. (39)
pépite (*n.f.*), nugget ; pepita : (40)

pépite d'or, nugget of gold ; gold nugget. (41)
perçage (dans le métal) (*n.m.*), drilling : (42)
perçage à grande vitesse de petits trous, drilling small holes at high speed. (43)
perçage (dans le bois) (*n.m.*), boring : (44)
perçage des trous dans le bois, boring holes in wood. (45)
percée (*n.f.*) *ou* **percé** (*n.m.*), boring ; opening ; cutting ; passage : (46)
les crêtes de moraines frontales sont souvent trouées par d'anciennes percées de torrents glaciaires (*f.pl.*), the crests of frontal moraines are often holed by ancient borings of glacial torrents. (47)
faire une percée à travers une forêt, to make a cutting through a forest. (48)
percée à flanc de coteau, hillside opening ; hillside cutting. (49)
percée aux eaux accumulées dans de vieux chantiers, boring against water in old workings ; tapping water accumulated in old workings. (50)
percement (perforage) (*n.m.*), piercing. (51)
See verb for examples.
percement (dans le métal) (*n.m.*), drilling (in metal). (52)
percement (dans le bois) (*n.m.*), boring (in wood). (53)
percement aux eaux (Mines) (*m.*), boring against water ; tapping water. (54)
percement de descenderies (Mines) (*m.*), winzing. (55)
percement de galeries (Mines) (*m.*), driving levels. (56)
percement de galeries en direction (*m.*), drifting ; driving levels parallel in direction to the strike. (57)
percement de recoupes (Mines) (*m.*), cross-driving. (58)
percement de trous de sonde dans le toit (*m.*), putting bore-holes up in the roof. (59)
percement de trous de sonde pour diminuer la pression du gaz (*m.*), boring to relieve gas-pressure. (60)
percement de tunnels (*m.*), tunnelling. (61)
percement du massif (Mines) (*m.*), breaking through ; holing. (62)
percement en travers-banc (Mines) (*m.*), cross-cutting. (63)
perceptible (*adj.*), perceptible. (64)
perceptiblement (*adv.*), perceptibly. (65)
percer (perforer) (*v.t.*), to pierce : (66)
percer un trou avec un instrument pointu, un trou dans un mur, to pierce a hole with a pointed instrument, a hole in a wall. (67 *ou* 68)
percer (le métal) (*v.t.*), to drill : (69)
percer un trou dans un morceau de métal, to drill a hole in a piece of metal. (70)
percer (le bois) (*v.t.*), to bore : (71)
percer un trou dans un morceau de bois, to bore a hole in a piece of wood. (72)
percer (la roche avec la perforatrice) (Mines) (*v.t.*), to drill (rock with the rock-drill). (73)
percer (un trou de sonde) (*v.t.*), to bore, to put down, (a bore-hole). (74)
percer (une galerie de mine) (*v.t.*), to drive (a mine-level). (75)
percer aux eaux accumulées dans de vieux chantiers, to bore against water in old workings ; to tap water accumulated in old workings. (76)

percer dans une ancienne galerie, to hole into an old level. (1)

percer des galeries en direction, to drive levels along the strike ; to drift. (2)

percer en direction (Mines) (*v.i.*), to drift ; to drive. (3)

percer en montant (Mines) (*v.i.*), to rise ; to raise. (4)

percer en travers-banc (Mines) (*v.i.*), to cross-cut. (5)

percer un débouché, to make an outlet ; to make an exit ; to open a passage ; to make a way out. (6)

percer un tunnel, to drive a tunnel ; to tunnel. (7)

percer une remontée (Mines), to put up a raise ; to put in a rise. (8)

percer une voie dans le ferme (Mines), to drive a road in the solid. (9)

perceur (pers.) (*n.m.*), driller. (10)

perceuse (*n.f.*), drilling-machine ; drill ; driller. (11) See **machine à percer** for varieties.

perche (*n.f.*), pole. (12)

perche (monte-courroie) (*n.f.*), belt-mounter ; belt-shifter ; belt-shipper. (13)

perche (mesure agraire anglaise) (*n.f.*), perch =25·293 sq. metres. (14)

perche d'échafaud (*f.*), scaffold-pole ; standard. (15)

perche de trolley (tramway électrique) (*f.*), trolley-pole. (16)

perchlorate (Chim.) (*n.m.*), perchlorate : (17) perchlorate de potasse, perchlorate of potash. (18)

perçoir (de forgeron) (*n.m.*), bolster ; punching-tool (blacksmiths'). (19)

percolateur (*n.m.*), percolator. (20)

percussion (*n.f.*), percussion ; shock ; impact. (21)

percutant, -e (*adj.*), percussive ; percussion (*used as adj.*) : (22) perforatrice percutante (*f.*), percussive drill ; percussion-drill. (23)

percuter (*v.t.*), to percuss. (24)

percylite (Minéral) (*n.f.*), percylite. (25)

perd-fluide (d'un paratonnerre) (*n.m.*), roots (of a lightning-conductor). (26)

perdre (*v.t.*), to lose : (27) perdre de vue une considération importante, to lose sight of an important consideration. (28)

perdre l'équilibre, to lose one's balance ; to overbalance. (29)

méthode qui perd plutôt du terrain qu'elle n'en gagne (*f.*), method which is rather losing than gaining ground. (30)

perdre (*v.i.*), to lose. (31)

perdre (laisser fuir) (*v.i.*), to leak : (32) tonneau qui perd (*m.*), cask which leaks. (33)

perdre (se) (s'égarer) (*v.r.*), to lose oneself ; to lose one's way : (34) se perdre dans un bois, to lose oneself in a wood. (35)

perdre (se) (disparaître) (*v.r.*), to lose itself : (36) rivière qui se perd dans les sables (*f.*), river which loses itself in the sands. (37)

perdre (se) (se dit des eaux) (*v.r.*), to go to waste ; to run to waste. (38)

perdre (se) (tomber en désuétude) (*v.r.*), to fall into disuse : (39) usage qui se perd de jour en jour (*m.*), custom which falls into disuse from day to day. (40)

pérenne (*adj.*), perennial : (41) source pérenne (*f.*), perennial spring. (42)

perfection (*n.f.*), perfection. (43)

perfectionné, -e (*adj.*), perfected ; improved : (44) méthodes perfectionnées (*f.pl.*), improved methods. (45)

perfectionnement (*n.m.*), perfecting ; improve-ment : (46) perfectionnement d'un procédé, perfecting a process. (47) machine qui comporte tous les derniers perfectionnements (*f.*), machine which contains all the latest improvements. (48)

perfectionner (*v.t.*), to perfect ; to improve : (49) perfectionner l'invention d'un autre, to improve, to perfect, another's invention. (50)

perforage (*n.m.*), drilling ; perforating ; per-foration. (51)

perforateur, -trice *ou* **perforatif, -ive** (*adj.*), per-forating ; perforative ; drilling. (52)

perforateur (instrument qui sert à perforer) (*n.m.*), drill ; perforator. (53)

perforateur (machine à perforer les roches, le charbon, etc.) (*n.m.*). Same as **perforatrice**.

perforation (*n.f.*), drilling ; perforating ; per-foration : (54) perforation au diamant, diamond drilling. (55) perforation mécanique, machine-drilling. (56) perforation mécanique des trous de mine, drilling shot-holes by machine. (57)

perforatrice (Mines) (*n.f.*), drill ; rock-drill ; drilling-machine ; driller ; borer. (58)

perforatrice à air comprimé (*f.*), compressed-air drill ; air-drill ; pneumatic rock-drill. (59)

perforatrice à commande électrique (*f.*), electri-cally driven drill ; electric rock-drill. (60)

perforatrice à injection d'eau (*f.*), water-flush drill ; water-drill ; wash-drill. (61)

perforatrice à main (*f.*), hand-drill. (62)

perforatrice à percussion (*f.*). Same as per-foratrice percutante.

perforatrice à pointes de diamant (*f.*), diamond drill ; diamond borer. (63)

perforatrice à rotation (*f.*). Same as **perforatrice rotative**.

perforatrice à taquet (*f.*), tappet-valve drill. (64)

perforatrice américaine à diamants (*f.*), American diamond rock-drill. (65)

perforatrice au charbon (*f.*), coal-drill. (66)

perforatrice au rocher (*f.*), rock-drill ; rock-borer. (67)

perforatrice diamantée (*f.*), diamond drill. (68)

perforatrice mécanique (*f.*), machine-drill ; power-drill ; drilling-machine. (69)

perforatrice percutante (*f.*), percussion-drill ; percussive drill ; hammer-drill. (70)

perforatrice percutante à main (*f.*), percussive hand-drill ; hand hammer-drill ; plug-drill ; plugger. (71)

perforatrice pneumatique (*f.*). Same as per-foratrice à air comprimé.

perforatrice rotative (*f.*), rotary drill. (72)

perforatrice rotative à main (*f.*), rotary hand-drill. (73)

perforatrice rotative au diamant (*f.*), rotary diamond drill. (74)

perforer (*v.t.*), to drill ; to perforate. (75)

périclase (Minéral) (*n.f.*), periclasite ; periclase. (1)

périclinal, -e, -aux (Géol.) (*adj.*), periclinal. (2)

péricline (Minéral) (*n.m.*), pericline. (3)

péridot (Minéral) (*n.m.*), peridot. (4)

péridot de Ceylan (*m.*), Ceylonese chrysolite. (5)

péridotique *ou* péridoteux, -euse (*adj.*), peridotic. (6)

péridotite (Pétrol.) (*n.f.*), peridotite ; peridotyte ; olivine-rock. (7)

périer (outil de fondeur) (*n.m.*), tapping-bar. (8)

péril (*n.m.*), peril ; danger. (9)

périlleusement (*adv.*), perilously ; dangerously. (10)

périlleux, -euse (*adj.*), perilous ; dangerous. (11)

périmer (*v.i.*), to lapse : (12)

droits périmés (*m.pl.*), lapsed rights. (13)

périmètre (*n.m.*), perimeter. (14)

périmètre mouillé (Hydraul.) (*m.*), wet perim-eter. (15)

périmétrique (*adj.*), perimetric ; perimetrical. (16)

période (*n.f.*), period. (17)

période (Élec.) (*n.f.*), period. (18)

période carbonifère *ou* période carboniférienne *ou* période houillère (Géol.) (*f.*), Carbonif-erous period ; Carboniferous. (19)

période crétacée (Géol.) (*f.*), Cretaceous period ; Cretaceous ; chalk period. (20)

période d'admission (de la vapeur dans le cylindre) (*f.*), admission-period ; admission. (21)

période d'élasticité (Phys.) (*f.*), period of elasticity. (22)

période d'émission (vapeur) (*f.*), release-period ; release. (23)

période d'une vibration (Phys.) (*f.*), period of a vibration. (24)

période de compression (vapeur) (*f.*), compres-sion-period ; compression. (25)

période de décarburation *ou* période des flammes (Procédé Bessemer) (*f.*), boil period. (26)

période de détente (vapeur) (*f.*), expansion-period ; expansion ; cut-off. (27)

période de scorification *ou* période des étincelles (Procédé Bessemer) (*f.*), slag-formation period. (28)

période des fumées (Procédé Bessemer) (*f.*), fining period. (29)

période glaciaire *ou* période pluviaire (Géol.) (*f.*), Glacial period ; Glacial epoch ; ice period ; ice age ; boulder period ; Bowlder period. (30)

périodique (*adj.*), periodic ; periodical : (31)

loi périodique des éléments chimiques (*f.*), periodic law of chemical elements. (32)

source périodique (*f.*), periodical spring. (33)

périodiquement (*adv.*), periodically. (34)

périphérie (*n.f.*), periphery : (35)

la périphérie d'une roue, the periphery of a wheel. (36)

périphérique (*adj.*), peripheral ; peripheric : (37)

vitesse périphérique (*f.*), peripheral speed. (38)

périr (*v.i.*), to perish. (39)

périscopique (Opt.) (*adj.*), periscopic : (40)

lentille périscopique (*f.*) *ou* objectif péri-scopique (*m.*), periscopic lens. (41)

périssable (*adj.*), perishable. (42)

perlaire (*adj.*), pearly ; pearlaceous ; perlaceous ; nacreous ; margaritaceous. (43)

perle (*n.f.*), pearl : (44)

perles d'une belle eau, pearls of fine water. (45)

perles fines, fine pearls. (46)

perle baroque, baroque pearl. (47)

perle parangon, parangon pearl. (48)

perle vierge, virgin pearl. (49)

perle (Essais au chalumeau) (*n.f.*), bead : (50)

perle de borax, borax-bead. (51)

perle de carbonate de soude *ou* perle de soude, sodium-carbonate bead. (52)

perle de sel de phosphore, salt of phosphorus bead. (53)

perlé, -e (*adj.*), pearly ; pearlaceous ; perlaceous ; nacreous ; margaritaceous. (54)

perles de coke (traces laissées par les coups de poussières) (Mines) (*f.pl.*), globular coke (traces left by dust-explosions). (55)

perlier, -ère (*adj.*), pearl-bearing ; margaritif-erous ; pearly. (56)

perlière (huitre perlière) (*n.f.*), pearl-oyster. (57)

perlite (Pétrol.) (*n.f.*), perlite ; pearlstone. (58)

perlite (Métall.) (*n.f.*), perlite ; pearlite. (59)

perlitique (Géol.) (*adj.*), perlitic : (60)

structure perlitique (*f.*), perlitic structure. (61)

permanence (*n.f.*), permanence. (62)

permanent, -e (*adj.*), permanent : (63)

déformation permanente (Méc.) (*f.*), permanent set. (64)

permanganate (Chim.) (*n.m.*), permanganate : (65)

permanganate de potasse *ou* permanganate de potassium, permanganate of potash ; potassium permanganate. (66)

perméabilité (*n.f.*), permeability : (67)

la perméabilité des grès favorise le passage des eaux d'infiltration, the permeability of sandstones favours the passage of percolating water. (68)

perméabilité magnétique (*f.*), magnetic perme-ability : (69)

la perméabilité magnétique est une propriété inhérente à presque tous les minéraux, magnetic permeability is an inherent property in nearly all minerals. (70)

perméable (*adj.*), permeable ; porous ; pervious : (71)

roches perméables (*f.pl.*), permeable rocks ; porous rocks. (72)

permettre (*v.t.*), to permit. (73)

permien, -enne (Géol.) (*adj.*), Permian : (74)

le système permien, the Permian system. (75)

permien (*n.m.*), Permian : (76)

le permien, the Permian. (77)

permis (*n.m.*), permit ; licence. (78)

permis de circulation (Ch. de f.) (*m.*), pass (over a railroad). (79)

permis de recherche *ou* permis de recherches (*m.*), prospecting licence. (80)

permis de s'absenter (*m.*), leave of absence. (81)

permis exclusif de recherches (*m.*), exclusive prospecting licence ; E.P.L. (82)

permission (*n.f.*), permission ; authorization. (83)

permissionnaire (pers.) (*n.m ou f.*), licence-holder. (84)

permonite (Explosif) (*n.f.*), permonite. (85)

pérovskite *ou* pérowskite (Minéral) (*n.f.*), perof-skite ; perovskite ; perooskite. (86)

peroxyde (Chim.) (*n.m.*), peroxide. (87)

peroxyder (*v.t.*), to peroxide. (88)

perpendiculaire (*adj.*), perpendicular : (1)
ligne perpendiculaire (*f.*), perpendicular line. (2)
perpendiculaire (*n.f.*), perpendicular. (3)
perpendiculairement (*adv.*), perpendicularly. (4)
perpendicularité (*n.f.*), perpendicularity : (5)
perpendicularité d'une ligne sur un plan, perpendicularity of a line on a plane. (6)
perpétuel, -elle (*adj.*), perpetual : (7)
mouvement perpétuel (*m.*), perpetual motion. (8)
perpétuellement (*adv.*), perpetually. (9)
perpétuité (*n.f.*), perpetuity. (10)
parré (Arch. hydraul.) (*n.m.*), water-wing. (11)
perrier (pers.) (*n.m.*), quarryman ; slate-quarry--man. (12)
perrière (*n.f.*), quarry ; slate-quarry. (13)
persistance (*n.f.*), persistence ; persistency. (14)
persistant, -e (*adj.*), persistent ; perpetual : (15)
neiges persistantes (*f.pl.*), perpetual snow. (16)
personne (*n.f.*), person : (17)
personnes de couleur, coloured persons. (18)
personnel, -elle (*adj.*), personal : (19)
surveillance personnelle (*f.*), personal super--vision. (20)
personnel (*n.m.*), staff ; personnel. (21)
personnel au complet (*m.*), full force of men. (22)
personnel de service (*m.*), staff on duty. (23)
personnel du fond (Mines) (*m.*), underground staff. (24)
personnel du jour (*m.*), surface-staff. (25)
personnel ouvrier (*m.*), workmen ; hands. (26)
personnel ouvrier du jour (*m.*), surface-hands ; surfacemen ; aboveground hands. (27)
personnel réduit (*m.*), reduced staff. (28)
personnellement (*adv.*), personally. (29)
perspectif, -ive (*adj.*), perspective. (30)
perspective (Géom.) (*n.f.*), perspective : (31)
perspective cavalière *ou* perspective iso--métrique, isometric projection ; isometric perspective. (32)
perspective linéaire, linear perspective. (33)
perspective sphérique, spherical perspective. (34)
perspective (aspect ; espérance) (*n.f.*), outlook ; prospect ; prospects : (35)
les perspectives d'avenir d'une contrée, the future prospects of a country. (36)
perspective commerciale, trade prospects. (37)
perspective encourageante, encouraging out--look ; promising outlook ; bright prospect. (38)
persulfate (Chim.) (*n.m.*), persulphate. (39)
persulfure (Chim.) (*n.m.*), persulphide. (40)
perte (privation) (*n.f.*), loss : (41)
perte d'un outil (Sondage), loss of a tool. (42)
perte (dommage) (*n.f.*), loss : (43)
travailler à perte, to work at a loss. (44)
perte sèche (*f.*), dead loss. (45)
perte (mauvais emploi) (*n.f.*), loss ; waste : (46)
perte au traitement (minerai, etc.), loss in treatment (ore, etc.). (47)
perte d'énergie, loss of power. (48)
perte de temps, loss, waste, of time. (49)
perte (fuite) (*n.f.*), leak ; leakage : (50)
perte dans un tuyau, leak in a pipe. (51)
perte à la terre (Élec.) (*f.*), earth-leakage. (52)
perte au feu (Chim.) (*f.*), ignition loss. (53)

perte de charge (*f.*), loss of pressure : (54)
perte de charge de l'air comprimé dans les tuyaux de conduite, loss of pressure of compressed air in conduit-pipes. (55)
perte de charge (Hydraul.) (*f.*), loss of head ; loss of pressure ; pressure-drop ; pressure-loss : (56)
perte de charge dans les conduites d'eau, loss of head in water-pipes. (57)
perte de charge (Élec.) (*f.*), pressure-drop ; pressure-loss ; drop of potential ; fall of potential. (58)
perte de charge en volts (Élec.) (*f.*), voltage-drop ; volts lost. (59)
perte de puissance (Élec., etc.) (*f.*), loss of power ; power-loss ; power-drop. (60)
perte de travail (*f.*), lost motion : (61)
perte de travail résultant d'une détente incomplète de la vapeur, lost motion result--ing from incomplete expansion of steam. (62)
perte de vie (*f.*), loss of life. (63)
perte électrique (*f.*), electric leakage. (64)
perte par évaporation (*f.*), loss by evaporation. (65)
perte par hystérésis (*f.*), hysteretic loss. (66)
perthite (Minéral) (*n.f.*), perthite. (67)
perturbateur, -trice (*adj.*), disturbing ; perturb--ing : (68)
forces perturbatrices (*f.pl.*), disturbing forces. (69)
perturbation (*n.f.*), disturbance ; perturbation : (70)
perturbation atmosphérique, atmospheric disturbance. (71)
perturbation de l'aérage, disturbance of the ventilation. (72)
perturbation magnétique, magnetic perturba--tion ; magnetic disturbance. (73)
pesage (*n.m.*), weighing : (74)
pesage de l'or, weighing gold. (75)
pesamment (*adv.*), heavily : (76)
être pesamment chargé,-e, to be heavily loaded. (77)
pesant, -e (lourd) (*adj.*), heavy ; weighty : (78)
un pesant fardeau, a heavy burden. (79)
pesant, -e (doué de pesanteur) (*adj.*), ponderable : (80)
les gaz sont pesants, comme tous les autres corps (*m.pl.*), like all other bodies, gases are ponderable. (81)
pesanteur (*n.f.*), heaviness ; weight : (82)
pesanteur d'un fardeau, heaviness, weight, of a burden. (83)
pesanteur (Phys.) (*n.f.*), gravity : (84)
minerai qui descend avec l'aide de la pesanteur jusqu'à la voie de fond (*m.*), ore which falls by gravity to the bottom level. (85)
pesanteur spécifique (Phys.) (*f.*), specific gravity. (86)
pesée (action de peser) (*n.f.*), weighing : (87)
la pesée des tubes de chaudière est un moyen d'en apprécier l'épaisseur moyenne, weighing boiler-tubes is a means of estimating their average thickness. (88)
peser (*v.t.*), to weigh : (89)
peser un paquet, to weigh a package. (90)
peser (*v.i.*), to weigh : (91)
un litre d'air pèse 1 ^{gr}, 293, one litre of air weighs 1·293 grammes. (92)
peseur (pers.) (*n.m.*), weighman ; weigher. (93)
peson (*n.m.*), balance. (94)

peson à contrepoids (*m.*), bent-lever balance ; lever-balance. (1)

peson à ressort *ou* peson à hélice *ou* peson cylindrique (*m.*), spring-balance ; spiral balance. (2)

pestilentiel, -elle (*adj.*), pestilential. (3)

pétalite (Minéral) (*n.m.*), petalite. (4)

pétard (coup de mine) (*n.m.*), shot ; blast. (5)

pétard (Ch. de f.) (*n.m.*), detonator ; fog-signal ; fog-tin ; railroad-torpedo. (6)

pétardement (*n.m.*), blowing up ; blasting : (7) pétardement des roches, blasting rocks. (8)

pétarder (*v.t.*), to blow up ; to blast : (9) pétarder la roche, to blast the rock. (10)

petit, -e (*adj.*), little ; small : (11)
petit bois (d'un châssis de fenêtre) (*m.*), sash-bar (wood) ; window-bar. (12)
petit bois en fer (*m.*), sash-bar (iron) ; window-bar. (13)
petit cheval (*m.*), donkey-engine ; jack-engine. (14)
petit cheval alimentaire [petits chevaux alimentaires *pl.*] (*m.*), donkey-pump ; boiler-feed pump. (15)
petit cylindre (d'une machine à vapeur com-pound) (*m.*), high-pressure cylinder (of a compound steam-engine). (16)
petit laminoir *ou* petit train (*m.*), small bar-mill. (17)
petit outillage (*m.*), small tools. (18)
petit rayon (courbe de ch. de f.) (*m.*), inner radius. (19)
petite calorie (Phys.) (*f.*), small calory ; lesser calory ; gramme-calory ; gramme-centigrade heat-unit. (20)
petite dimension (*f.*), small size. (21)
petite forge (*f.*), portable forge. (22)
petite multiplication (Méc.) (*f.*), low gear. (23)
petite tête de bielle (*f.*), connecting-rod little end ; little end of connecting-rod ; small end of connecting-rod. (24)
petite vitesse (Méc.) (*f.*), low speed ; slow speed. (25)
petite vitesse (à) (Méc.), low-speed : (26)
moteur à petite vitesse (*m.*), low-speed motor. (27)
petites pièces (Fonderie) (*f.pl.*), light castings ; fine castings. (28)
petites réparations (*f.pl.*), small repairs ; minor repairs. (29)

pétrifiant, -e *ou* pétrificateur, -trice (qui change en pierre) (*adj.*), petrifying ; petrifactive ; petrescent. (30)

pétrifiant, -e (incrustant) (*adj.*), incrusting : (31) des eaux pétrifiantes (*f.pl.*), incrusting waters. (32)

pétrification (formation des pierres) (*n.f.*), petrifaction ; petrification ; petrescence ; petrescency. (33)

pétrification (fossilisation) (*n.f.*), petrifaction ; petrification ; petrescence ; fossilization : (34) la pétrifaction des bois, the petrifaction of woods. (35)

pétrification (incrustation) (*n.f.*), incrustation. (36)

pétrifié, -e (*adj.*), petrified : (37)
bois pétrifié (*m.*), petrified wood ; wood-stone. (38)

pétrifier (*v.t.*), to petrify. (39)

pétrifier (se) (*v.r.*), to petrify. (40)

pétrir (*v.t.*), to knead ; to work : (41)
pétrir de l'argile, to knead, to work, clay. (42)

pétrissage *ou* pétrissement (*n.m.*), kneading ; working. (43)

pétrographe (pers.) (*n.m.*), petrographer ; petrol--ogist. (44)

pétrographie (*n.f.*), petrography. (45)

pétrographique (*adj.*), petrographic ; petro--graphical : (46)
province pétrographique (*f.*), petrographical province. (47)

pétrographiquement (*adv.*), petrographically. (48)

pétrole (*n.m.*), petroleum ; oil ; rock-oil ; stone-oil ; earth-oil ; fossil oil ; coal-oil ; mineral naphtha. (49)

pétrole brut (*m.*), crude petroleum ; crude oil. (50)

pétrole combustible (*m.*), oil-fuel ; fuel-oil. (51)

pétrole du Caucase (*m.*), Caucasian petroleum. (52)

pétrole épuré *ou* pétrole raffiné (*m.*), refined petroleum ; refined oil. (53)

pétrole lampant (*m.*), lamp-oil ; paraffin. (54)

pétroléen, -enne (*adj.*), petroleum (*used as adj.*) ; oil (*used as adj.*) : (55)
industrie pétroléenne (*f.*), petroleum-industry ; oil-industry. (56)

pétrolerie (*n.f.*), petroleum-works ; petroleum-refinery ; oil-refinery. (57)

pétrolier, -ère (*adj.*), petroleum (*used as adj.*) ; oil (*used as adj.*) : (58)
industrie pétrolière (*f.*), petroleum-industry ; oil-industry. (59)
bateau pétrolier (*m.*), oil-ship. (60)

pétrolifère (*adj.*), petroliferous ; oil-bearing ; oil-producing ; petroleum-bearing ; petroleum-producing ; petroleum (*used as adj.*) ; oil (*used as adj.*) : (61)
région pétrolifère (*f.*), petroliferous region ; oil-bearing region ; oil-region. (62)

pétrologie (*n.f.*), petrology. (63)

pétrologique (*adj.*), petrologic ; petrological : (64)
examen pétrologique (*m.*), petrological examination. (65)

pétrologiquement (*adv.*), petrologically. (66)

pétrosilex (Pétrol.) (*n.m.*), petrosilex. (67)

pétrosiliceux, -euse (*adj.*), petrosiliceous : (68)
roche pétrosiliceuse (*f.*), petrosiliceous rock. (69)

petticoat (*n.m.*), petticoat ; petticoat-pipe ; draught-pipe : (70)
le petticoat est un tronçon de tuyau avec embouchure qui surmonte la tuyère d'échap--pement dans la boîte à fumée d'une loco--motive, the petticoat is a bell-mouthed piece of pipe over the exhaust-nozzle in the smoke-box of a locomotive. (71)

petzite (Minéral) (*n.f.*), petzite. (72)

peu abondant, -e, scarce : (73)
pays où le bois et la houille sont peu abondants (*m.pl.*), countries where wood and coal are scarce. (74)

peu connu, -e, little known : (75)
région peu connue (*f.*), little-known region. (76)

peu profond, -e, shallow : (77)
eau peu profonde (*f.*), shallow water. (78)

peuplé, -e (*adj.*), inhabited ; populated. (79)

peuplier (arbre) (*n.m.*), poplar. (80)

phacolite (Minéral) (*n.f.*), phacolite. (81)

pharmacie (coffret de pharmacie) (*n.f.*), medicine-chest. (82)

pharmacolite (Minéral) (*n.f.*), pharmacolite. (1)
pharmacosidérite (Minéral) (*n.f.*), pharmacosider
-ite ; cube-ore. (2)
phase (Phys., Elec., etc.) (*n.f.*), phase. (3)
phase de la distribution *ou* phase de la marche du
tiroir (*f.*), event of the cycle ; event of the
stroke ; critical event : (4)
les six phases de la distribution : avance à
l'admission, admission, détente, avance à
l'échappement, échappement, compression,
the six events of the cycle : preadmission,
admission, expansion, prerelease, release,
compression. (5)
phénacite *ou* phénakite (Minéral) (*n.f.*), phenacite.
(6)
phengite (Minéral) (*n.f.*), phengite ; fengite. (7)
phénocristal, -e, -aux (Pétrol.) (*adj.*), pheno-
-cristic. (8)
phénocristal (*n.m.*), phenocryst ; phenocrystal.
(9)
phénoménal, -e, -aux (*adj.*), phenomenal. (10)
phénoménalement (*adv.*), phenomenally. (11)
phénomène (*n.m.*), phenomenon : (12)
phénomènes thermiques dans les machines à
vapeur, thermic phenomena in steam-
engines. (13)
phillipsite (zéolite) (Minéral) (*n.f.*), phillipsite ;
christianite ; lime harmotome. (14)
phillipsite (érubescite) (Minéral) (*n.f.*), bornite ;
erubescite ; horse-flesh ore ; peacock ore ;
peacock copper ; purple copper ; variegated
copper ore ; variegated pyrites. (15)
phlogopite (Minéral) (*n.f.*), phlogopite. (16)
phœnicite *ou* phœnicochroïte (Minéral) (*n.f.*),
phœnicochroite ; phœnicite. (17)
phonolite (Pétrol.) (*n.f.*), phonolite ; phonolyte ;
clinkstone. (18)
phonolitique (*adj.*), phonolitic. (19)
phosgénite (Minéral) (*n.f.*), phosgenite ; horn-
lead ; corneous lead ; plumbum corneum.
(20)
phosphatation (*n.f.*), phosphatization. (21)
phosphate (Chim.) (*n.m.*), phosphate : (22)
phosphate de chaux, phosphate of lime. (23)
phosphaté, -e (*adj.*), phosphatic ; phosphated ;
phosphate (*used as adj.*) : (24)
craie phosphatée (*f.*), phosphatic chalk. (25)
roche phosphatée (*f.*), phosphate rock. (26)
phosphatisation (*n.f.*), phosphatization. (27)
phosphite (Chim.) (*n.m.*), phosphite. (28)
phosphore (Chim.) (*n.m.*), phosphorus ; phos-
-phor. (29)
phosphore blanc *ou* phosphore ordinaire (*m.*),
white phosphorus ; yellow phosphorus ;
ordinary phosphorus. (30)
phosphore rouge *ou* phosphore amorphe (*m.*),
red phosphorus ; amorphous phosphorus.
(31)
phosphorer (*v.t.*), to phosphorate. (32)
phosphorescence (*n.f.*), phosphorescence. (33)
phosphorescent, -e (*adj.*), phosphorescent : (34)
certaines variétés de blende sont phosphores-
-centes dans l'obscurité (*f.pl.*), certain
varieties of blende are phosphorescent in the
dark. (35)
phosphoreux (Chim.) (*adj.m.*), phosphorous. (36)
phosphorique (Chim.) (*adj.*), phosphoric : (37)
acide phosphorique (*m.*), phosphoric acid.
(38)
phosphoriser (*v.t.*), to phosphatize. (39)
phosphorite (Minéral) (*n.f.*), phosphorite. (40)
phosphoritique (*adj.*), phosphoritic. (41)

phosphorochalcite (Minéral) (*n.f.*), phosphoro-
-chalcite ; phosphochalcite ; pseudomal-
-achite. (42)
phosphoroscope (*n.m.*), phosphoroscope. (43)
phosphure (Chim.) (*n.m.*), phosphide. (44)
photo-jumelle [photo-jumelles *pl.*] (*n.f.*), bin-
-ocular camera. (45)
photogramme négatif (Photogr.) (*m.*), negative.
(46)
photogramme positif *ou simplement* photogramme
(*n.m.*) *ou* photocopie (*n.f.*) (Photogr.),
positive ; print. (47)
photographe (pers.) (*n.m.*), photographer. (48)
photographe amateur (*m.*), amateur photographer.
(49)
photographie *ou familièrement* photo (art) (*n.f.*),
photography. (50)
photographie *ou familièrement* photo (reproduction)
(*n.f.*), photograph ; photo. (51)
photographie des couleurs *ou* photo des couleurs
(*f.*), colour-photography ; photography
in colour. (52)
photographie en chambre *ou* photographie à
l'atelier (*f.*), indoor photography. (53)
photographie en plein air *ou* photographie au
dehors (*f.*), outdoor photography. (54)
photographie instantanée *ou* photo instantanée
(*f.*), instantaneous photography ; snap-
shotting. (55)
photographie sans objectif (*f.*), pinhole photog-
-raphy. (56)
photographie sur pellicules *ou* photo sur pellicules
(*f.*), film-photography. (57)
photographier (*v.t.*), to photograph. (58)
photographique (*adj.*), photographic ; photo-
-graphical : (59)
objectif photographique (*m.*), photographic
lens. (60)
photographiquement (*adv.*), photographically.
(61)
photomètre (*n.m.*), photometer. (62)
photométrie (*n.f.*), photometry. (63)
photométrique (*adj.*), photometric ; photo-
-metrical : (64)
unité photométrique (*f.*), photometric unit.
(65)
photométriquement (*adv.*), photometrically. (66)
phototype (Photogr.) (*n.m.*), negative. (67)
phtanite *ou* phthanite (Pétrol.) (*n.f.*), phthanite ;
chert ; hornstone ; rock-flint. (68)
phtisie (Pathologie) (*n.f.*), phthisis ; pulmonary
tuberculosis. (69)
phtisie des mineurs (*f.*), miners' phthisis. (70)
phtisique (*adj.*), phthisic ; phthisical. (71)
phyllite (Pétrol.) (*n.f.*), phyllite. (72)
physalite (Minéral) (*n.f.*), physalite. (73)
physicien (pers.) (*n.m.*), physicist. (74)
physico-chimique (*adj.*), physicochemical. (75)
physico-mécanique (*adj.*), physicomechanical.
(76)
physiographe (pers.) (*n.m.*), physiographer. (77)
physiographie (*n.f.*), physiography. (78)
physiographique (*adj.*), physiographic ; physio-
-graphical. (79)
physiographiquement (*adv.*), physiographically.
(80)
physique (*adj.*), physical : (81)
chimie physique (*f.*), physical chemistry. (82)
physique (*n.f.*), physics. (83)
physique expérimentale (*f.*), experimental physics.
(84)
physiquement (*adv.*), physically. (85)

pic (*n.m.*), pick. (1)

pic (Géogr. phys.) (*n.m.*), peak : (2)
les principaux pics du massif de la Bernina sont/le pic (*ou* piz) Bernina et le pic Palü, the /principal peaks of the Bernina massif are the Piz Bernina and the Piz Palü. (3)

pic (à) (*locution adverbiale*), perpendicularly ; precipitously ; sheer. (4)

pic à deux pointes (*m.*), double-pointed pick ; mandril ; mandrel ; maundrill. (5)

pic à tête (*m.*), poll-pick ; hammer-pick. (6)

pic au charbon *ou* pic à la veine (opp. à pic au rocher) (Exploitation du charbon) (*m.*), coal-pick. (7) (pick for coal-work, dis--tinguished from stone-pick or pick for stone-work) (Coal-mining).

pic au rocher *ou* pic à roc *ou* pic au mur (Exploita--tion du charbon) (*m.*), stone-pick. (8)

pic de mineur (*m.*), miners' pick. (9)

pic de tailleur de pierre (*m.*), stone-dressing pick. (10)

picot (Mines) (*n.m.*), wedge. (11)

picotage (*n.m.*), wedging : (12)
picotage d'un puits de mine, wedging a mine-shaft. (13)

picoter (*v.t.*), to wedge. (14)

picotite (Minéral) (*n.f.*), picotite. (15)

picoture (Photogr.) (*n.f.*), pinhole. (16)

picrate (Chim.) (*n.m.*), picrate : (17)
picrate de potassium, potassium picrate. (18)

picrique (Chim.) (*adj.*), picric : (19)
acide picrique (*m.*), picric acid. (20)

picrite (Pétrol.) (*n.f.*), picrite ; picryte. (21)

picrolite (Minéral) (*n.f.*), picrolite. (22)

picroméride *ou* picromérite (Minéral) (*n.f.*), pic--romerite. (23)

picrosmine (Minéral) (*n.f.*), picrosmine. (24)

pièce (*n.f.*), piece ; bit ; part ; member. (25)

pièce (Résistance des matériaux) (*n.f.*), bar : (26)
pièce prismatique, prismatic bar. (27)
For the phrases beginning pièce appuyée, pièce encastrée, etc., see poutre appuyée, poutre encastrée, etc., substituting the word *bar* for the word *beam*.

pièce (Fonderie) (*n.f.*), casting. (28) See pièce coulée. See also moulage.

pièce à demeure (*f.*). Same as pièce fixe.

pièce à noyaux (Fonderie) (*f.*), cored casting. (29)

pièce à vis (*f.*), screw-piece. (30)

pièce battue (Fonderie) (*f.*), false core ; draw--back. (31)

pièce chargée debout (Résistance des matériaux) (*f.*), strut (Strength of materials). (32)

pièce coulée *ou* pièce moulée *ou* pièce fondue *ou* pièce de fonte *ou* pièce venue de fonte *ou* pièce de fonderie *ou simplement* pièce (*n.f.*), casting : (33)
réparation des pièces coulées (*f.*), repairing castings. (34)
perçage des pièces fondues (*m.*), drilling castings. (35)
usiner une pièce venue de fonte, to machine a casting. (36)
pièces de fonderie brutes ou travaillées, rough or machined castings. (37)
une pièce saine, a sound casting. (38)

pièce coulée en coquille (*f.*), chill-casting. (39)

pièce d'acier forgé (*f.*), steel forging. (40)

pièce d'écartement (*f.*), distance-piece ; dis--stance-bar. (41)

pièce de bois (*f.*), piece, bit, of wood. (42)

pièce de bois (pièce longue et grande) (*f.*), piece, stick, of timber : (43)
un cadre complet (de boisage de galeries de mines) est formé de quatre pièces, à savoir : un chapeau, deux montants, et une semelle, a full set (of timbering for mine-levels) consists of four sticks, viz., a head-piece, two uprights, and a sill. (44)

pièce de bois en grume (*f.*), log. (45)

pièce de fonte *ou* pièce de fonderie (*f.*). See pièce coulée.

pièce de forge *ou* pièce forgée (*f.*), forging : (46)
meulage de grosses pièces de forge (*m.*), grinding heavy forgings. (47)
pièces de forge brutes, dégrossies, ou finies, de tous genres, poids, et dimensions, forgings, rough, roughed, or finished, of all kinds, weights, and sizes. (48)

pièce de métal (*f.*), piece, bit, of metal. (49)

pièce de rapport (*f.*). Same as pièce rapportée.

pièce de renfort (*f.*), stiffener. (50)

pièce de terre (*f.*), piece of land. (51)

pièce en acier moulé (*f.*), steel casting. (52)

pièce en fonte *ou* pièce en fonte de fer (*f.*), iron casting. (53)

pièce en fonte malléable (*f.*), malleable casting. (54)

pièce fixe *ou* pièce à demeure (*f.*), fixture ; fixed member : (55)
les pièces fixes d'une pompe, the fixtures of a pump. (56)

pièce fondue (*f.*), casting. (57) See also pièce coulée.

pièce fondue au sable (*f.*), sand-casting. (58)

pièce fondue en coquille (*f.*), chill-casting. (59)

pièce manquée (Fonderie) (*f.*), waster ; waster casting ; spoiled casting. (60)

pièce moulée (opp. à pièce troussée) (*f.*), mould--ed casting. (61) See also pièce coulée.

pièce moyenne (Fonderie) (*f.*), average casting (between light and heavy). (62)

pièce polaire (dynamo) (*f.*), pole-piece ; pole. (63)

pièce rapportée *ou* pièce de rapport (*f.*), loose piece : (64)
les modèles de fonderie s'établissent avec des portées, dont les empreintes dans les moules servent d'assise ou d'encastrement à des pièces rapportées appelées noyaux (*m.pl.*), foundry-patterns are made with prints, of which the impressions in the moulds serve as a bed or housing for loose pieces called cores. (65)

pièce rapportée (pièce battue) (Fonderie) (*f.*), drawback ; false core. (66)

pièce troussée (*f.*), struck-up casting ; strickled casting ; swept-up casting. (67)

pièce venue de forge (*f.*), forging. (68)

pièces d'automobiles (*f.pl.*), motor-car parts. (69)

pièces d'une charpente (*f.pl.*), members of a frame. (70)

pièces d'une machine (*f.pl.*), parts of a machine : (71)
machine à pièces interchangeables (*f.*), machine with interchangeable parts. (72)

pièces de rechange (*f.pl.*), spare parts ; spares ; duplicates. (73)

pièces détachées d'un tour (*f.pl.*), separate parts of a lathe. (74)

pièces pour raccords (pistolets) (Dessin) (*f.pl.*), French curves. (75)

pied (*n.m.*), foot ; base ; bottom : (76)
pied d'une colline, foot, base, bottom, of a hill. (77)

pied (support ; affût) (*n.m.*), stand ; standard ; support ; foot : (1)

pied pour théodolite, pour appareil photo--graphique, stand for theodolite, for camera. (2 *ou* 3) See also **pied à trois branches** *ou simplement* **pied**.

pieds d'un tour, standards of a lathe. (4)

pied d'une balance, support of a balance (scales). (5)

pied en forme de fer à cheval (microscope, etc.), horseshoe foot. (6)

pied (d'un étau à pied) (*n.m.*), leg (of a leg-vice). (7)

pied (d'une happe) (*n.m.*), heel (of a G cramp). (8)

pied (d'une dent d'engrenage) (*n.m.*), root, dedendum, (of a gear-tooth). (9)

pied (de la came) (bocard) (*n.m.*), root (of the cam) (ore-stamp). (10)

pied (mesure linéaire anglaise) (*n.m.*), foot= 0·30480 metre. (11)

pied carré, square foot=9·2903 square deci--metres. (12)

pied cube, cubic foot =0·028317 cubic metre. (13)

pied à coulisse (compas) (*m.*), sliding caliper-gauge ; sliding calipers ; slide-calipers ; caliper-gauge ; caliper-square ; beam-caliper ; beam caliper-gauge. (14)

pied à fourche (pour machine à percer) (*m.*), U-shaped base-plate (for drilling-machine). (15)

pied à trois branches *ou simplement* **pied** (*n.m.*), tripod ; tripod-stand ; stand : (16)

pied à calotte sphérique, facilitant la mise en station, quick-levelling tripod. (17)

pied à deux, à trois, à quatre, brisures, twofold, threefold, fourfold, tripod-stand. (18 *ou* 19 *ou* 20)

pied à 6 branches, split-leg tripod. (21)

pied à translation, tripod with centering head. (22)

pied à trois branches articulées, fold-over leg tripod. (23)

pied à trois branches coulissantes *ou* pied à coulisse, sliding-leg tripod-stand ; sliding tripod ; adjustable tripod ; telescopic tripod. (24)

pied à trois branches droites, rigid tripod. (25)

pied à trois branches extensibles, extension-tripod. (26)

pied-canne [pieds-cannes *pl.*] (*n.m.*) *ou* pied anglais, dit pied-canne, broomstick tripod ; light broomstick tripod. (27)

pied de campagne, field tripod-stand. (28)

pied photographique, camera-stand. (29)

pied-armoire [pieds-armoires *pl.*] (tour, etc.) (*n.m.*), cabinet-foot (lathe, etc.). (30)

pied d'œuvre (à), on site ; delivered site : (31)

prix des matériaux à pied d'œuvre (*m.pl.*), prices of materials on site (*or* of materials delivered site). (32)

pied-de-biche [pieds-de-biche *pl.*] (levier qui sert à arracher les crampons) (*n.m.*), claw-bar ; spike-lever ; spike-drawer ; spike-extractor ; clawed spike-lever. (33)

pied-de-biche (petit levier fendu qui sert à arracher les clous) (*n.m.*), claw ; nail-claw. (34)

pied-de-biche (organe de l'ancienne distribution, servant à obtenir le changement de marche) (*n.m.*), gab ; gab-hook ; eccentric-hook ; hook-gear : (35)

la coulisse supplanta le pied-de-biche, the link supplanted the gab-hook. (36)

pied de bielle (*m.*), connecting-rod little end ; little end of connecting-rod ; small end of connecting-rod. (37)

pied-de-bœuf [pieds-de-bœuf *pl.*] (sortie et rentrée de la sonde) (*n.m.*), rod-elevator ; rod-hoister ; lifting-dog. (38)

pied-de-chèvre [pieds-de-chèvre *pl.*] (*n.m.*). Same as **pied-de-biche**.

pied-droit [pied-droits *pl.*] (d'une voûte) (*n.m.*), pier, pillar, (of an arch). (39)

pied-droit (d'une galerie de mine) (*n.m.*), side, wall, side wall, (of a mine-level). (40)

pied-droit (de cadre) (Boisage des galeries de mines) (*n.m.*), upright, post, prop, arm, leg, leg-piece, (of frame or set) (Timbering mine-levels). (41)

piedroit [piedroits *pl.*] (*n.m.*). Same as **pied-droit**.

piémontite (Minéral) (*n.f.*), piedmontite. (42)

pierraille (*n.f.*), small stones : (43)

chemin ferré de pierraille (*m.*), road metaled with small stones. (44)

pierre (*n.f.*), stone. (45)

pierre à aiguiser (*f.*), oilstone ; whetstone ; hone ; honestone. (46)

pierre à aiguiser d'Arkansas (*f.*), Arkansas oilstone. (47)

pierre à aiguiser Washita (*f.*), Washita oilstone ; Ouachita oilstone. (48)

pierre à bâtir (*f.*), building-stone. (49)

pierre à chaux (*f.*), limestone. (50)

pierre à facettes (Géol.) (*f.*), glyptolith ; gibber ; dreikanter. (51)

pierre à feu (*f.*), fire-stone. (52)

pierre à filtrer (*f.*), filtering-stone. (53)

pierre à fusil (*f.*), flint ; flintstone. (54)

pierre à huile *ou* **pierre à l'huile** (*f.*), oilstone. (55)

pierre à huile montée *ou* **pierre à huile enchâssée** (*f.*), mounted oilstone. (56)

pierre à morfiler (*f.*), oilstone. (57)

pierre à morfiler du Levant (*f.*), Turkey oilstone. (58)

pierre à plâtre (*f.*), plaster-rock ; plaster-stone ; gypsum. (59)

pierre à rasoir (*f.*), razor-stone. (60)

pierre à statuettes (*f.*), figure-stone ; agalmato--lite ; bildstein. (61)

pierre bleue (Géol.) (*f.*), bluestone. (62)

pierre branlante (*f.*), rocking stone ; logan stone ; logan ; loggan stone ; loggan ; logging rock. (63)

pierre brute (pierre précieuse) (*f.*), rough stone ; rough. (64)

pierre calcaire (*f.*), limestone. (65)

pierre cassée (*f.*), broken stone. (66)

pierre d'aigle (*f.*), eaglestone ; aetites. (67)

pierre d'aimant (*f.*), loadstone ; lodestone ; natural magnet. (68)

pierre d'alun (*f.*), alum-stone ; alum-rock ; alunite. (69)

pierre d'Arkansas (*f.*), Arkansas stone. (70)

pierre d'attente *ou* **pierre d'arrachement** (Maçonn.) (*f.*), toothing-stone ; tooth. (71)

pierre d'azur (*f.*), azure-stone. (72)

pierre d'émeri (*f.*), emery-stone. (73)

pierre d'imitation (*f.*), imitation stone ; paste. (74)

pierre d'ornement *ou* **pierre d'ornementation** (*f.*), ornamental stone. (75)

pierre de Bologne (*f.*), Bologna stone ; Bolo--gnian stone. (1)

pierre de bornage (*f.*), boundary-stone. (2)

pierre de Caïrngorm (*f.*), cairngorm ; Cairngorm stone. (3)

pierre de chat (*f.*), fetid quartz ; stinkstone. (4)

pierre de corne (*f.*), hornstone. (5)

pierre de corne fusible (*f.*), petrosilex. (6)

pierre de corne infusible (*f.*), horn-flint. (7)

pierre de couleur (pierre précieuse colorée) (*f.*), coloured stone. (8)

pierre de croix (*f.*), cross-stone. (9)

pierre de fantaisie (*f.*), fancy stone ; fancy. (10)

pierre de grès (*f.*), grit-stone ; grit-rock. (11)

pierre de lard (*f.*), lardstone. (12)

pierre de limaçons (*f.*), lumachelle. (13)

pierre de lune (*f.*), moonstone. (14)

pierre de meule (*f.*), millstone grit. (15)

pierre de miel (*f.*), honeystone ; mellite. (16)

pierre de mine (*f.*), minestuff ; mine-stuff ; mine-stone ; stone ; ore ; mineral ; metal. (17)

pierre de Moka (*f.*), Mocha stone ; Mocha pebble. (18)

pierre de Portland (*f.*), Portland stone. (19)

pierre de savon (*f.*), soapstone. (20)

pierre de soleil (*f.*), sunstone ; aventurine feldspar. (21)

pierre de soude (*f.*), natrolite. (22)

pierre de taille (*f.*), cut-stone ; building-stone (stone which is cut, or intended to be cut, for building purposes). (23)

pierre de touche (*f.*), touchstone ; lydite ; Lydian stone (24)

pierre de tripes (*f.*), tripe-stone. (25)

pierre des Amazones (*f.*), Amazon stone ; amazonite. (26)

pierre du Levant (*f.*), Turkey stone ; Turkey slate ; novaculite. (27)

pierre façonnée par l'usure éolienne (Géol.) (*f.*), wind-worn stone. (28)

pierre factice *ou* pierre fausse (*f.*), imitation stone ; paste. (29)

pierre figurée (*f.*), figure-stone. (30)

pierre fine (opp. à pierre fausse) (*f.*), real stone : (31)

tromper l'acheteur sur la qualité d'une pierre fausse vendue pour fine, to deceive the buyer on the nature of an imitation stone sold as real. (32)

pierre fine (pierre de moindre valeur qu'une pierre précieuse proprement dite) (*f.*), semiprecious stone : (33)

les pierres précieuses se subdivisent en pierres précieuses proprement dites, et en pierres fines, precious stones are subdivided into precious stones proper, and semiprecious stones. (34)

pierre franche (*f.*), freestone ; liver-rock. (35)

pierre gélisse (*f.*), green stone (newly quarried). (36)

pierre gemme (*f.*), gem-stone ; jewel-stone ; gem. (37)

pierre hématite (*f.*), bloodstone ; hematite. (38)

pierre incolore (*f.*), colourless stone. (39)

pierre lithographique (*f.*), lithographic stone ; lithographic slate. (40)

pierre météorique (*f.*), meteoric stone. (41)

pierre meulière (*f.*), millstone grit. (42)

pierre non-taillée (pierre de taille) (*f.*), uncut stone ; undressed stone ; unhewn stone. (43)

pierre non-taillée (pierre précieuse) (*f.*), uncut stone ; uncut gem. (44)

pierre ollaire (*f.*), potstone ; lapis ollaris. (45)

pierre ponce (*f.*), pumice ; pumice-stone. (46)

pierre pour gouges (*f.*), gouge-slip. (47)

pierre précieuse (*f.*), precious stone ; gem ; gem-stone ; jewel-stone. (48)

pierre précieuse engagée dans sa gangue (*f.*), matrix-gem. (49)

pierre précieuse sans défaut (*f.*), flawless gem. (50)

pierre précieuse taillée (*f.*), cut gem ; cut stone. (51)

pierre rouge (*f.*), bloodstone ; hematite. (52)

pierre sèche (*f.*), dry stone : (53)

mur en pierres sèches (*m.*), dry-stone wall ; dry wall. (54)

pierre spéculaire (*f.*), specular stone. (55)

pierre taillée (pierre de taille) (*f.*), cut stone ; dressed stone ; hewn stone. (56)

pierre taillée (pierre précieuse) (*f.*), cut stone. (57)

pierre verte (*f.*), green stone (newly quarried). (58)

pierreries (*n.f.pl.*), precious stones ; stones ; gems : (59)

pierreries étincelantes, sparkling gems. (60)

pierreries orientales, oriental stones. (61)

pierreux, -euse (*adj.*), stony : (62)

terrain pierreux (*m.*), stony ground. (63)

rivière à fond pierreux (*f.*), river with stony bottom. (64)

pieu (*n.m.*), stake ; post ; pile ; spile : (65)

ficher un pieu en terre, to drive a stake into the ground. (66)

pieu à vis (*m.*), screw-pile. (67)

pieu de sécurité (derrick) (Sondage) (*m.*), headache post ; life preserver. (68)

piézo-électricité (*n.f.*), piezoelectricity. (69)

piézo-électrique (*adj.*), piezoelectric. (70)

piézomètre (*n.m.*), piezometer. (71)

piézométrie (*n.f.*), piezometry. (72)

piézométrique (*adj.*), piezometric ; piezometer (*used as adj.*) ; pressure (*used as adj.*) : (73)

tube piézométrique (Hydraul.) (*m.*), pressure-tube ; piezometer-tube ; piezometric tube. (74)

Cf. niveau piézométrique.

pigment (*n.m.*), pigment. (75)

pignon (roue dentée s'engrenant sur une roue plus grande) (*n.m.*), pinion. (76)

pignon (Arch.) (*n.m.*), gable. (77)

pignon de chaîne (*m.*) *ou* pignon de Galle (*m.*) *ou* pignon-Galle [pignons-Galle *pl.*] (*n.m.*) *ou* pignon Galle [pignons Galle *pl.*] (*m.*), sprocket-wheel ; rag-wheel. (78)

pignon de commande *ou* pignon d'attaque *ou* pignon d'entraînement (*m.*), driving-pinion. (79)

pignon en cuir vert *ou* pignon en cuir d'abatis (*m.*), raw-hide pinion. (80)

pignon engrenant sur la crémaillère (locomotive à crémaillère) (*m.*), rack-pinion. (81)

pignon et crémaillère [pignon (*n.m.*) ; crémaillère (*n.f.*)], rack and pinion. (82)

pignon satellite (*m.*), differential pinion. (83)

pilage (*n.m.*), pounding. (84)

pilastre (de rampe d'escalier) (*n.m.*), newel. (85)

pile (amas) (*n.f.*), pile ; heap : (86)

une pile de bois, a pile of wood. (87)

pile (massif de maçonnerie formant pilier) (*n.f.*), pier : (88)

les piles d'un pont, the piers of a bridge. (89)

pile (Élec.) (*n.f.*), battery; pile: (1)
une pile de — éléments, a battery of — cells. (2)

pile à auges (*f.*), trough-battery. (3)

pile à colonnes (*f.*). Same as **pile voltaïque.**

pile à deux liquides (*f.*), double-fluid battery. (4)

pile à gaz (*f.*), gas-battery. (5)

pile à un seul liquide (*f.*), single-fluid battery. (6)

pile au bichromate (*f.*), bichromate battery. (7)

pile de Bunsen (*f.*), Bunsen battery. (8)

pile de Callaud (*f.*), Callaud battery. (9)

pile de Daniell (*f.*), Daniell battery. (10)

pile de Grove (*f.*), Grove battery. (11)

pile de Leclanché (*f.*), Leclanché battery. (12)

pile de sonnerie (*f.*), bell-battery. (13)

pile de Volta (*f.*). Same as **pile voltaïque.**

pile électrique *ou simplement* **pile** (*n.f.*) (pile électrique quelconque), electric battery; battery; pile. (14)

pile électrique *ou simplement* **pile** (*n.f.*) (particulièrement,—appareil transformant en courant électrique l'énergie développée dans une réaction chimique), voltaic battery; galvanic battery; chemical battery; battery; pile. (15)

pile en batterie (*f.*), multiple battery; parallel battery. (16)

pile en série (*f.*), series-battery. (17)

pile étalon (*f.*), standard battery. (18)

pile galvanique (*f.*). Same as **pile voltaïque.**

pile humide (*f.*), wet battery. (19)

pile hydro-électrique (*f.*), water-battery. (20)

pile mixte *ou* **pile en séries parallèles** (*f.*), multiple-series battery; parallel-series battery. (21)

pile primaire (*f.*), primary battery. (22)

pile rectangulaire (Boisage de mines) (*f.*), square set; square timbering; Nevada timbering. (23)

pile sèche (*f.*), dry battery; dry pile. (24)

pile secondaire (*f.*), secondary battery; storage-battery; accumulator. (25)

pile thermo-électrique (*f.*), thermoelectric battery; pyroelectric battery; thermoelectric pile; thermopile; thermoelectric multiplier. (26)

pile voltaïque (*f.*), voltaic pile; Volta's pile; galvanic pile; column battery; electric column. (27)

piler (*v.t.*), to pound. (28)

pilier (*n.m.*), pillar; column. (29)

pilier (Mines) (*n.m.*), pillar; post; stoop: (30)
un pilier de charbon, de minerai, de stérile, a pillar of coal, of ore, of barren rock. (31 *ou* 32 *ou* 33)

pilier-abri [piliers-abris *pl.*] (Mines) (*n.m.*), sheet-pillar. (34)

pilier d'érosion (Géol.) (*m.*), erosion column; chimney-rock. (35)

pilier de protection de galerie (*m.*), chain-pillar. (36)

pilier de protection de puits (Mines) (*m.*), shaft-pillar; bottom pillar. (37)

pilier de soutènement (Mines) (*m.*), supporting-pillar. (38)

pilier de voûte (Mines) (*m.*), arch-pillar. (39)

pilier du mur (Mines) (*m.*), wall-pillar. (40)

pilon (instrument pour piler dans un mortier) (*n.m.*), pestle; pounder: (41)
mortier et pilon, pestle and mortar [mortier (*n.m.*)]. (42)

pilon (de bocard) (*n.m.*), stamp (of ore-stamp). (43)

pilon (dame) (*n.m.*), beetle; rammer; earth-rammer. (44)

pilon (d'un marteau-pilon) (le mouton) (*n.m.*), tup, ram, hammer-head, head, (of a power-hammer). (45)

pilon (marteau-pilon) (*n.m.*), hammer; power-hammer; forge-hammer; forging-hammer. (46) See also **marteau-pilon.**

pilon à chute libre (bocard) (*m.*), free-falling stamp; gravity-stamp; gravitation-stamp. (47)

pilon à vapeur (marteau-pilon) (*m.*), steam-hammer. (48)

pilonnage (damage) (action) (*n.m.*), ramming; tamping. (49)

pilonnage (damage) (ouvrage) (*n.m.*), tamp-work. (50)

pilonner (damer) (*v.t.*), to ram; to tamp: (51)
pilonner une allée, to ram, to tamp, a path. (52)

pilot (pieu de pilotis) (*n.m.*), pile; spile. (53)

pilot (chasse-bestiaux) (locomotive) (*n.m.*), pilot; cowcatcher. (54)

pilot à vis (*m.*), screw-pile. (55)

pilotage (action) (*n.m.*), pile-driving; driving piles; pilework; piling; spile-driving; spiling. (56)

pilotage (ouvrage) (*n.m.*), piling; spiling; pilework. (57)

pilote (Ch. de f.) (*n.m.*), pilot-engine. (58)

piloter (*v.t.*), to pile; to spile. (59)

piloter (*v.i.*), to drive in piles; to drive in spiles. (60)

pilotis (ensemble de pilots) (*n.m.*), piling; spiling; pilework. (61)

pilotis (pilot; chacun des pieux qui entrent dans un pilotis) (*n.m.*), pile; spile. (62)

pilotis à vis (*m.*), screw-pile. (63)

pimélite (Minéral) (*n.f.*), pimelite. (64)

pin (*n.m.*), pine; deal. (65)

pin sylvestre (*m.*), Scotch fir; red deal. (66)

pinacoïde (Cristall.) (*adj.*), pinacoid; pinacoidal. (67)

pinacoïde (*n.m.*), pinacoid. (68)

pince (tenailles) (*n.f.*) (s'emploie plutôt au pluriel), pliers; nippers; tongs. (69) See also **tenailles.**

pince (*n.f.*) *ou* **pince pour tubes** (tubes à caoutchouc), clip; clamp; tube-clip; clip for tubes; pinch-cock: (70)
burette avec pince (*f.*), burette with clip. (71)
pince pour burettes de Mohr *ou* pince de Mohr, Mohr pinch-cock; Mohr's clip. (72)

pince (*n.f.*) *ou* **pince à levier**, crowbar; crow. (73)

pince (de mineur) (*n.f.*), gad; ringer. (74)

pince (pied-de-biche pour arracher les clous) (*n.f.*), claw; nail-claw. (75)

pince (pour câble-guide) (Mines) (*n.f.*), gland (for guide-rope). (76)

pinces à becs plats, ronds, etc. Same as **pinces plates, rondes, etc.** (77)

pinces à boutonnières (*f.*), belt-punch. (78)

pinces à boutons (Essais) (*f.pl.*), button-pliers. (79)

pinces à brucelles (*f.pl.*), tweezers. (80)

pinces à combinaisons (*f.pl.*), combination pliers. (81)

pinces à cônes (*f.pl.*), cone-pliers. (82)

pinces à coupelles (*f.pl.*), cupel-tongs. (83)

pinces à creusets (*f.pl.*), crucible-tongs. (84)

pince à décocher (Fonderie) (f.), stripping-bar. (1)

pince à donner la voie aux scies (f.), plier saw-set. (2)

pinces à emporte-pièce à revolver (f.pl.), revolv-ing-head punch. (3)

pinces à emporte-pièce pour courroies (f.pl.), belt-punch. (4)

pinces à gaz (f.pl.), gas-pliers. (5)

pinces à genouillère (pinces articulées pour monter les pierres de taille) (f.pl.), self-adjusting stone-dogs ; block-clip ; lever-grip tongs. (6)

pince à loquet (Sondage) (f.), jar-latch ; boot-jack ; boot-latch. (7)

pince à pied-de-biche (f.), claw-bar ; clawed spike-lever ; spike-lever ; spike-drawer ; spike-extractor. (8)

pinces à plomber (f.pl.), sealing-pliers. (9)

pinces à rails ou pinces à soulever et à porter les rails (f.pl.), rail-tongs. (10)

pince à ressort (f.), spring-clip. (11)

pince à riper (pose de voie) (Ch. de f.) (f.), lining-bar ; straightening-bar ; lifting-bar ; slewing-bar. (12)

pince à scie (f.), plier saw-set. (13)

pince à talon (f.), pinch-bar ; pinching-bar ; pinch. (14)

pinces à télégraphie (f.pl.), telegraph-pliers. (15)

pinces à tirer (f.pl.), draw-tongs. (16)

pinces à tirer à la main (f.pl.), hand draw-tongs. (17)

pinces à tirer au banc (f.pl.), bench draw-tongs. (18)

pince à torsades (pour fils) (f.), joint-twisting pliers. (19)

pince à tourmaline (f.), tourmaline tongs. (20)

pinces à tuyaux (f.pl.), pipe-tongs. (21)

pince à vis, à lames parallèles à charnières (f.), clamp (or clip) with double centre-bar, the bottom part with hinge. (22)

pinces brucelles (f.pl.), tweezers. (23)

pince - cisaille [pinces-cisailles pl.] (n.f.) ou pince cisailleuse (f.), cutting-nippers. (24)

pinces coupantes (f.pl.), cutting-nippers. (25)

pinces coupantes sur bout, branches noires, mâchoires polies (f.pl.), end cutting-nippers, black handles, bright jaws. (26)

pinces coupantes sur côté (f.pl.), side cutting-nippers. (27)

pince d'accrochage (Roulage) (Mines) (f.), clip ; grip ; haulage-clip ; rope-clip ; jigger. (28)

pince d'arrêt (d'un instrument de nivellement, ou analogue) (f.), clamp : (29)
   triangle à 3 vis de calage avec pince d'arrêt et vis de rappel pour le mouvement lent (m.), tribrach with 3 levelling-screws with clamp and slow-motion adjusting-screw. (30)

pinces d'électricien, manche garni de matière isolante (f.pl.), electricians' pliers, with insulated handles. (31)

pince de débouchage du trou de coulée (Fonderie) (f.), tapping-bar ; tap-bar. (32)

pinces de forgeron (f.pl.), blacksmiths' tongs ; anvil-tongs. (33) See tenaille de forgeron for varieties.

pince de raccordement (Élec.) (f.), connector. (34)

pinces de serrage (f.pl.), nippers. (35)

pinces droites (de forgeron) (f.pl.), straight tongs. (36)

pince en bois, pour tubes à essais (f.), test-tube holder, wood. (37)

pinces parallèles (f.pl.), parallel pliers. (38)

pinces plates (f.pl.), flat-nose pliers ; flat-nosed pliers. (39)

pinces pour électriciens (f.pl.), electricians' pliers. (40)

pinces pour fours de fusion (f.pl.), furnace-tongs. (41)

pinces pour tubes (f.pl.), tube-tongs. (42)

pinces rondes (f.pl.), round-nose pliers ; round-nosed pliers. (43)

pinces serre-tubes à chaîne (f.pl.), chain pipe-wrench. (44)

pince thermo-électrique (f.), thermoelectric couple ; thermoelectric pair ; thermo-electric thermometer. (45)

pinces tourne-à-gauche (f.pl.), plier saw-set. (46)

pinces universelles (f.pl.), universal pliers. (47)

pinceau (n.m.), brush ; soft brush : (48)
   pinceau pour peintre en bâtiments, paint-brush. (49)

pinceau lumineux (Opt.) (m.), luminous pencil ; pencil of light. (50)

pincer (v.t.), to pinch ; to nip ; to grip ; to clip : (51)
   pincer des tiges dans un étau pour les empêcher de tourner quand on taraude, to grip rods in a vice to prevent them turning while being tapped. (52)

pincette (ressort à pincette) (n.f.), elliptic spring. (53)

pincettes (n.f.pl.), tongs ; fire-tongs. (54)

pinite (Minéral) (n.f.), pinite. (55)

pinnule (n.f.), sight ; sight-vane ; pinnule ; vane : (56)
   alidade à pinnules (f.), alidade with sights ; sighted alidade. (57)

pinnule à charnière (f.), folding sight. (58)

pinte (mesure anglaise de capacité) (n.f.), pint = 0·568 litre. (59)

pinte (mesure de capacité pour les liquides en usage dans les États-Unis) (n.f.), pint= 0·473164 litre. (60)

piochage (n.m.), picking ; picking up ; breaking with a pick. (61)

pioche (n.f.), pick. (62)

pioche à bec pointu et tête ou pioche à bec pointu et marteau (f.), poll-pick ; hammer-pick. (63)

pioche à bourrer ou pioche à tasser et pointue (Pose de voie) (Ch. de f.) (f.), beating-pick ; beater-pick ; beater ; packer ; tamping-pick. (64)

pioche à défricher (f.) ou pioche à défricher à hache (f.) ou pioche-hache [pioches-haches pl.] (n.f.), grubbing-mattock ; cutter-mattock ; mattock. (65)

pioche à long manche (f.), long-handled pick. (66)

pioche ordinaire ou pioche de terrassier, forme ordinaire ou pioche à bec plat et pointu (f.), pick-axe ; chisel-and-point pick. (67)

piochement (n.m.), picking ; picking up ; breaking with a pick. (68)

piocher (v.t.), to pick ; to pick up ; to break with a pick ; to pickaxe : (69)
   piocher la terre ou piocher le sol, to pick up the ground ; to break the ground with a pick. (70)

piocheur (pers.) (n.m.), pickman. (71)

piocheuse (*n.f.*), steam-shovel ; steam-digger ; steam-excavator ; steam-navvy ; navvy. (1)

piochon (bisaiguë) (*n.m.*), mortise-axe (double-ended). (2)

pionnier (pers.) (*n.m.*), pioneer. (3)

pipette (Verrerie de laboratoire) (*n.f.*), pipette. (4)

pipette compte-gouttes (*f.*), dropping-tube ; dropper ; dripping-tube ; pipette. (5)

pipette graduée (*f.*), graduated pipette. (6)

piquage (de chaudières) (enlevage du tartre) (*n.m.*), scaling (boilers). (7)

piquage (Exploitation des mines de houille) (*n.m.*), hewing ; digging : (8)

piquage de la houille, hewing, digging, coal. (9)

pique-feu [pique-feu *pl.*] (*n.m.*), poker ; pricker ; prick-bar. (10)

pique-mine [pique-mine *pl.*] (bocardeur) (pers.) (*n.m.*), millman. (11)

piquer (faire des petites blessures) (*v.t.*), to pit : (12)

l'eau acide pique les chaudières (*f.*), acid water pits boilers. (13)

piquer une chaudière (enlever le tartre), to scale a boiler ; to remove scale from a boiler. (14)

piquer une conduite sur une autre, to branch one pipe on another. (15)

piquer (se) (*v.r.*), to pit : (16)

tôle de chaudière qui se pique (*f.*), boiler-plate which pits. (17)

piquet (Géod., Ponts & Chaussées, Ch. de f.) (*n.m.*), stake ; picket. (18)

piquet (Dragage) (*n.m.*), spud. (19)

piquet de tente (*m.*), tent-peg ; tent-picket. (20)

piquetage (*n.m.*), staking ; staking off ; staking out. (21)

piqueter (*v.t.*), to stake ; to stake off ; to stake out : (22)

piqueter la direction générale du tracé à étudier, to stake out the general direction of the alignment to be surveyed. (23)

piqueur (Mines de houille) (pers.) (*n.m.*), hewer ; digger : (24)

le rendement du piqueur, the get of the hewer. (25)

piqueur à la veine (*m.*), getter. (26)

piqûre (*n.f.*), pitting : (27)

des piqûres qui se forment dans les tôles de chaudière, pitting which forms in boiler-plates. (28)

piqûre d'aiguille (Photogr.) (*f.*), pinhole. (29)

pis aller (chose à laquelle on se résout faute de mieux) (*m.*), makeshift. (30)

pisanite (Minéral) (*n.f.*), pisanite. (31)

pisiforme (*adj.*), pisiform ; pea (*used as adj.*): (32)

minerai pisiforme (*m.*), pisiform ore ; pea ore. (33)

pisolithe (Géol.) (*n.f.*), pisolite ; pea-grit ; pea-stone. (34)

pisolithique (*adj.*), pisolitic : (35)

minerai pisolithique (*m.*), pisolitic ore. (36)

pissasphalte (Minéral) (*n.m.*), pissasphalt ; piss-asphaltum ; maltha. (37)

pissée (*n.f.*), breakout (of molten slag from a blast-furnace). (38)

pistazite (Minéral) (*n.f.*), pistacite ; pistazite. (39)

piste (*n.f.*), track ; trail ; path ; road : way. (40)

piste à traîneaux (*f.*), sledge-road. (41)

piste de roulement (*f.*), roller-track ; roller-path ; runway. (42)

piste muletière *ou* piste à portage (*f.*), mule-track ; horse-trail ; pack-trail ; bridle-path ; bridle-road ; bridle-track ; bridleway. (43)

pistolet (du levier de changement de marche) (*n.m.*), latch (of reverse-lever). (44)

pistolet (fleuret de mineur) (*n.m.*), chisel ; drill ; borer. (45)

pistolet (Dessin) (*n.m.*), French curve. (46)

pistolet de mine (*m.*), Lens igniter. (47)

pistomésite (Minéral) (*n.f.*), pistomesite. (48)

piston (*n.m.*), piston. (49)

piston (*n.m.*) *ou* piston plongeur *ou* piston plein, ram ; plunger-piston ; plunger ; solid piston : (50)

pistons de haute et de basse pression, high and low-pressure rams. (51)

piston (de douille de lampe électrique à incandes-cence) (*n.m.*), plunger (of incandescent electric lamp-holder). (52)

piston à air (*m.*), air-piston. (53)

piston à contre-tige (*m.*), piston with tail-rod ; piston with extended rod. (54)

piston à eau (*m.*), water-piston. (55)

piston à plateau (*m.*), disc-piston. (56)

piston à segments (*m.*), ring-piston. (57)

piston à vapeur (*m.*), steam-piston. (58)

piston conique *ou* piston en chapeau chinois (*m.*), conical piston. (59)

piston creux *ou* piston à clapet (de pompe) (*m.*), bucket ; hollow piston ; valved piston. (60)

piston d'équilibre *ou* piston compensateur (turbine à réaction) (*m.*), balancing-piston. (61)

piston oscillant (*m.*), oscillating piston. (62)

piston rotatif (*m.*), rotary piston. (63)

pit à lingots (Métall.) (*m.*), soaking-pit. (64)

pitchpin (arbre) (*n.m.*), pitch-pine. (65)

pitite (Explosif) (*n.f.*), pitite. (66)

piton (pic) (Géogr. phys.) (*n.m.*), piton ; peak : (67)

les montagnes des Antilles françaises sont surmontées de hauts pitons (*f.pl.*), the mountains of the French Antilles are sur-mounted by high pitons (or peaks). (68)

piton (*n.m.*) *ou* piton à vis, screw-eye. (69)

pitticite (Minéral) (*n.f.*), pittizite ; pitticite. (70)

pivot (*n.m.*), pivot ; pin. (71)

pivot (d'une mèche à trois pointes) (*n.m.*), centre-point (of a centre-bit). (72)

pivot central *ou* pivot du bogie (*m.*), centre-pin ; main-pin ; bogie-pin ; king-pin ; king-bolt ; pintle. (73)

pivot central (d'une plaque tournante) (*m.*), centre-pin, centre-pivot, (of a turntable). (74)

pivot d'entraînement (du plateau-toc d'un tour) (*m.*), catch-pin (of the catch-plate of a lathe). (75)

pivotant, -e (*adj.*), pivoting ; swivelling ; swivel-ing ; swivel (*used as adj.*) : (76)

bogie pivotant (*m.*), swivelling bogie ; swivel-ing truck. (77)

support à chariot pivotant (*m.*), swivel slide-rest. (78)

pivotation (*n.f.*), pivoting ; swivelling ; swiveling. (79)

pivoter (*v.i.*), to pivot ; to swivel. (80)

placage (des métaux) (*n.m.*), plating (overlaying mechanically with a more precious metal). (81)

placage (Menuis.) (*n.m.*), veneering. (82)

place (emploi) (*n.f.*), place ; situation ; post ; position ; employment ; berth. (83)

place (endroit) (*n.f.*), place ; spot ; situation. (1)

place (en) (Géol.), in place ; in situ : (2) gisement en place (*m.*), deposit in situ ; in-place deposit. (3)

placer (*n.m.*), placer ; diggings ; alluvial diggings. (4)

placer (*v.t.*), to place ; to put : (5) placer les ventilateurs au fond de la mine, to place, to put, the fans underground. (6)

placer d'or *ou* placer aurifère (*m.*), gold-placer ; gold-diggings ; placer gold-field. (7)

placer de petite profondeur (*m.*), shallow placer. (8)

placer écrémé *ou* placer gaspillé (*m.*), wasted placer. (9)

placer fluvial (*m.*), river-diggings. (10)

placer public (*m.*), public diggings. (11)

placer sec *ou* placer dépourvu d'eau (*m.*), dry placer ; dry diggings. (12)

placer stannifère (*m.*), tin-placer. (13)

placérien, -enne (*adj.*), placer (*used as adj.*) : (14) exploitation placérienne (*f.*), placer-mining. (15)

placérien (pers.) (*n.m.*), placer-worker. (16)

plafond (Arch.) (*n.m.*), ceiling. (17)

plafond (d'une galerie de mine) (*n.m.*), roof (of a mine-level). (18)

plafond (d'un bassin, d'un réservoir) (*n.m.*), floor, bottom, (of a basin, of a reservoir). (19 *ou* 20)

plage (*n.f.*), beach ; sea-beach ; shore ; sea-shore : (21) une plage de sable, a sandy beach. (22)

plages soulevées (Géol.) (*f.pl.*), raised beaches. (23)

plagioclase (Minéralogie) (*adj.*), plagioclastic. (24)

plagioclase (*n.m.*), plagioclase. (25)

plaine (Géogr. phys.) (*n.f.*), plain ; flat : (26) pays de plaine (*m.*), flat country. (27)

plaine (outil) (*n.f.*), drawing-knife. (28)

plaine alcalifère (*f.*), alkali flat. (29)

plaine d'inondation *ou* plaine de débordement *ou* plaine alluviale *ou* plaine alluviale d'inondation (*f.*), flood-plain. (30)

plaine de comblement (*f.*), plain of accumulation. (31)

plaine de dénudation (*f.*), denudation-plain ; plain of erosion ; peneplain. (32)

plaine de lavage superficiel (Glaciologie) (*f.*), overwash plain ; outwash plain. (33)

plainte (*n.f.*), complaint : (34) plaintes de la clientèle, des ouvriers, com-plaints of customers, of workmen. (35 *ou* 36)

plan, -e (*adj.*), plane ; plain ; flat ; even ; level : (37) surface plane (*f.*), plane surface ; plain surface ; flat surface ; even surface ; level surface. (38)

plan (surface plane) (*n.m.*), plane ; plane surface. (39)

plan (tracé ; carte) (*n.m.*), plan ; map. (40)

plan (projet) (*n.m.*), plan ; project ; scheme. (41)

plan (plan incliné) (*n.m.*), plane ; inclined plane ; incline ; run ; plan. (42)

plan à double effet (*m.*). Same as plan incliné automoteur à double effet.

plan à simple effet (*m.*). Same as plan incliné automoteur à simple effet.

plan à traction mécanique (*m.*), engine-plane. (43)

plan au lavis (*m.*), wash-plan. (44)

plan automoteur (*m.*). Same as plan incliné automoteur.

plan axial (*m.*), axial plane ; axis-plane : (45) on représente les plis, dans les ouvrages de géologie, par des coupes perpendiculaires à leur plan axial, in works on geology, folds are represented by sections perpendicular to their axial plane. (46)

plan-concave (*adj.*), plano-concave : (47) lentilles plan-concaves (*f.pl.*), plano-concave lenses. (48)

plan-convexe (*adj.*), plano-convex ; convexo-plane : (49) verre plan-convexe (*m.*), plano-convex glass ; convexo-plane glass. (50)

plan coté (*m.*), dimensioned plan. (51)

plan d'agate (balance) (*m.*), agate plane (scales). (52)

plan d'eau (*m.*), water-level ; water-plane. (53)

plan d'ensemble (*m.*), general plan. (54)

plan d'essais (*m.*), assay-plan. (55)

plan d'hémitropie (Cristall.) (*m.*), twinning-plane. (56)

plan d'incidence (*m.*), plane of incidence. (57)

plan de cassure (Géol.) (*m.*), fracture-plane ; plane of fracture. (58)

plan de charriage (Géol.) (*m.*), thrust-plane. (59)

plan de clivage (*m.*), cleavage-plane. (60)

plan de comparaison (*m.*), datum-level ; datum-plane ; datum : (61) le plan de comparaison auquel on rapporte les altitudes est généralement le niveau moyen de la mer, the datum-level to which heights are referred is generally the mean sea-level. (62)

plan de conduite (*m.*), policy ; plan of action : (63) se tracer un plan de conduite, to map out a policy. (64)

plan de diaclase (Géol.) (*m.*), joint-plane. (65)

plan de faille (Géol.) (*m.*), fault-plane. (66)

plan de fondations (*m.*), foundation-plan. (67)

plan de fracture (Géol.) (*m.*), fracture-plane ; plane of fracture. (68)

plan de glissement (Cristall.) (*m.*), gliding-plane ; slip-plane. (69)

plan de jonction (*m.*), junction-plane. (70)

plan de la surface, des travaux souterrains (*m.*), plan of the surface, of underground works (71 *ou* 72)

plan de mine (*m.*), mine-plan. (73)

plan de montage (*m.*), erection-plan ; construc-tion-plan. (74)

plan de naissance (d'une voûte) (*m.*), spring, springing, (of an arch). (75)

plan de polarisation (*m.*), plane of polarization. (76)

plan de réflexion (Opt.) (*m.*), plane of reflection. (77)

plan de réfraction (Opt.) (*m.*), plane of refraction. (78)

plan de schistosité (Géol.) (*m.*), plane of schistosity ; foliation-plane. (79)

plan de séparation (Géol.) (*m.*), joint-plane. (80)

plan de stratification (Géol.) (*m.*), bedding-plane ; stratification-plane ; plane of stratification. (81)

plan de symétrie (*m.*), plane of symmetry ; symmetry-plane. (82)

plan de traçage (Mines) (*m.*), development-plan ; lines of development. (83)

plan des chantiers d'abatage (*m.*),　stope-plan. (1)

plan en croquis (*m.*), sketch-plan ; rough plan. (2)

plan et élévation d'un bâtiment [**plan** (*n.m.*) ; **élévation** (*n.f.*)], plan and elevation of a building. (3)

plan exécutable (*m.*), workable plan. (4)

plan focal (*m.*), focal plane. (5)

plan géométrique (*m.*), geometric plane ; ground-plane. (6)

plan graphique (*m.*), diagram. (7)

plan incliné *ou simplement* **plan** (*n.m.*), inclined plane ; incline ; plane ; run. (8)

plan incliné à trois rails (*m.*), three-rail incline. (9)

plan incliné automoteur *ou simplement* **plan incliné** (*m.*), self-acting incline ; self-acting plane ; gravity-incline ; gravity-road ; gravity-railroad ; gravity-plane ; gravity-plan ; incline ; brow. (10)

plan incliné automoteur à double effet avec câble sans fin (*m.*), endless-rope incline ; self-acting incline worked on the endless-rope principle. (11)

plan incliné automoteur à double effet avec chaîne à deux bouts (*m.*), cut-chain incline ; self-acting incline worked on the cut-chain principle. (12)

plan incliné automoteur à simple effet (*m.*), jig-brow ; jig-plane ; jig ; balance-incline ; balance-brae ; balance-brow ; jinny-road ; jinny ; self-acting incline worked on the balance-truck principle. (13)

plan moteur (*m.*), engine-plane. (14)

plan-relief [**plans-reliefs** *pl.*] (*n.m.*), relief-map. (15)

plan tangent (*m.*), tangent plane. (16)

planage (*n.m.*), planing ; planishing ; smoothing. (17)

planage (surfaçage) (Travail de tour) (*n.m.*), surfacing. (18)

planche (*n.f.*), board. (19)

planche (de charbon) (Dépilage en long avec planche protectrice) (*n.f.*), rib (of coal) (Rib-and-pillar method). (20)

planche à dessin (*f.*), drawing-board. (21)

planche à dessin en trois épaisseurs (*f.*), triplex drawing-board. (22)

planche à dresser (Charp.) (*f.*), shooting-board ; shoot-board. (23)

planche à noyaux (Fonderie) (*f.*), core-board. (24)

planche à trousser (Fonderie) (*f.*), strickle-board ; sweep-board ; striking-board ; loam-board. (25)

planche bouvetée (*f.*), match-board. (26)

planche d'échafaud (*f.*), scaffold-board. (27)

planche en cuivre rouge (*f.*), copper sheet. (28)

planchéiage (*n.m.*), boarding ; flooring. (29)

planchéier (*v.t.*), to board ; to floor. (30)

plancher (*n.m.*), floor ; flooring : (31)

　plancher d'un bâtiment, d'une maison, floor, flooring, of a building, of a house. (32 *ou* 33)

　plancher incombustible, fire-proof floor. (34)

　plancher simple *ou* plancher ordinaire, single floor. (35)

　plancher sur poutre, double floor (one binder). (36)

　plancher sur poutres, double floor (more than one binder). (37)

plancher sur poutre pan de bois, framed floor. (38)

plancher (plus particulièrement, assemblage de planches simplement posées jointives ; par opposition à **parquet**) (*n.m.*), square-jointed floor. (39) (opp. to **grooved-and-tongued floor**, or the like)

plancher (d'un pont) (*n.m.*), floor (of a bridge). (40)

plancher (d'une cage d'extraction) (Mines) (*n.m.*), deck (of a mining-cage). (41)

plancher de manœuvre (Mines) (*m.*), working-stage ; working-platform ; battery. (42)

plancher de manœuvre (à l'orifice supérieur d'un puits de mine) (*m.*), brace ; shaft-brace ; bracket. (43)

plancher de protection (puits de mine) (*m.*), pentice ; penthouse. (44)

plancher de repos (Mines) (*m.*), sollar ; platform ; ladder-sollar ; stage of a ladderway. (45)

plancher volant (Muraillement de puits de mine) (*m.*), walling-scaffold ; walling-stage ; cradle. (46)

planchette (petite planche) (*n.f.*), small board. (47)

planchette (plaque conscience) (*n.f.*), breast-plate ; conscience ; palette. (48)

planchette (Géod.) (*n.f.*), plane-table. (49)

planchette d'objectif (d'un appareil photo-graphique) (*f.*), lens-board (of a camera). (50)

plane (outil tranchant à deux poignées) (*n.f.*), drawing-knife. (51)

plane (outil de tour) (*n.f.*), planisher ; planish-ing-tool. (52)

planer (*v.t.*), to plane ; to planish ; to smooth : (53)

　planer une douve, to plane a stave. (54)

　planer le métal, to planish metal. (55)

planer (surfacer) (Travail de tour) (*v.t.*), to surface. (56)

　planer un moule (Fonderie), to smooth, to slick, to sleek, a mould. (57)

planeur (pers.) (*n.m.*), planisher. (58)

planimètre (*n.m.*), planimeter. (59)

planimétrie (*n.f.*), planimetry. (60)

planimétrique (*adj.*), planimetric ; planimetrical. (61)

plantation de poteaux télégraphiques (*f.*), plant-ing, setting, telegraph-poles. (62)

planter un jalon dans le sol, to set, to plant, a stake in the ground. (63)

plaque (*n.f.*), plate. (64)

plaque à glacer (Photogr.) (*f.*), glazing-pad ; glazer ; print-glazer. (65)

plaque à noyaux (Fonderie) (*f.*), core-plate. (66)

plaque à rebord (*f.*), flange-plate. (67)

plaque amalgamée (*f.*). Same as **plaque d'amal-gamation.**

plaque anti-halo (Photogr.) (*f.*), antihalation plate ; anti-halo plate ; non-halation plate. (68)

plaque argentée (*f.*), silvered plate. (69)

plaque argentée par électrolyse (*f.*), electro-silvered plate. (70)

plaque arrière de boîte à feu (*f.*), fire-box back-sheet ; door-sheet of fire-box. (71)

plaque avant de bo.te à feu (*f.*), fire-box front-sheet. (72)

plaque conscience (pour vilebrequins et drilles) (*f.*), breast-plate ; conscience ; palette. (73)

plaque d'accumulateur (*f.*), accumulator-plate. (1)

plaque d'amalgamation (*f.*), amalgamation-plate ; amalgamated plate ; amalgamated copper plate. (2)

plaque d'amalgamation extérieure (*f.*), outside amalgamation-plate. (3)

plaque d'amalgamation intérieure (*f.*), inside amalgamation-plate. (4)

plaque d'ancrage (*f.*), anchoring-plate. (5)

plaque d'assise (*f.*), bed-plate ; base-plate ; foundation-plate ; sole-plate. (6)

plaque d'autel (grille de foyer) (*f.*), bridge-plate (furnace-grate). (7)

plaque d'avant-foyer (grille de foyer) (*f.*), dead-plate ; coking-plate ; dumb-plate. (8)

plaque de base (*f.*). Same as plaque d'assise.

plaque de biscuit de porcelaine (Minéralogie) (*f.*), streak plate. (9)

plaque de blindage (*f.*), armour-plate. (10)

plaque de butée (*f.*), thrust-plate. (11)

plaque de contre-feu (d'une forge) (*f.*), hearth-back ; forge-back. (12)

plaque de croisement (Ch. de f.) (*f.*), crossing-plate. (13)

plaque de cuivre (*f.*), copper plate. (14)

plaque de cuvelage (Mines) (*f.*), tubbing-plate. (15)

plaque de détente (de tiroir de detente) (*f.*), expansion-plate, expansion-slide, cut-off plate, (of expansion-valve, of cut-off valve). (16)

plaque de devanture (de foyer de chaudière) (*f.*), front plate (of boiler-furnace). (17)

plaque de fer (*f.*), iron plate ; plate of iron (wrought). (18)

plaque de fond *ou* plaque de fondation (*f.*). Same as plaque d'assise.

plaque de fond (d'un moule en terre) (Fonderie) (*f.*), bed-plate, bottom plate, foundation-plate, loam-plate, (of a loam mould). (19)

plaque de fonte (*f.*), iron plate ; plate of iron (cast). (20)

plaque de friction *ou* plaque de frottement (*f.*), friction-plate ; chafing-plate ; wear-plate ; wearing-plate. (21)

plaque de garde (wagon ou locomotive de chemin de fer) (*f.*), horn-plate ; horn-block ; axle-guard ; pedestal ; guard-plate. (22 *ou* 23)

plaque de manœuvre (Ch. de f. de mines) (*f.*), switch-plate ; flat plate ; flat sheet ; turn-sheet ; tarantula. (24)

plaque de marbre (*f.*), slab of marble. (25)

plaque de numéro (*f.*), number-plate. (26)

plaque de pile (*f.*), battery-plate. (27)

plaque de pression (*f.*), pressure-plate. (28)

plaque de propreté (*f.*), finger-plate. (29)

plaque de recouvrement (Fonderie) (*f.*), covering-plate ; top plate. (30)

plaque de renfort (*f.*), stiffening-plate. (31)

plaque de revêtement (*f.*), lining-plate. (32)

plaque de scheidage (pour minerai) (*f.*), bucking-plate ; buck-plate (for ore). (33)

plaque de serrage (pour meules en émeri, ou analogues) (*f.*), side-plate (for emery-wheels, or the like). (34 *ou* 35)

plaque de trituration (*f.*), trituration-plate ; grinding-plate. (36)

plaque de trou d'homme (*f.*), manhole-plate ; manhole-door. (37)

plaque en cuivre amalgamée (*f.*). Same as plaque d'amalgamation.

plaque en verre extra-mince (Photogr.) (*f.*), extra-thin glass plate. (38)

plaque ferrotype pour le glaçage des épreuves (Photogr.) (*f.*), ferrotype print-glazer ; ferrotype glazing-pad. (39)

plaque indicatrice (Ch. de f.) (*f.*), quarter-mile post. (40)

plaque mince (Microscopie) (*f.*), section ; thin section ; thin plate : (41)

depuis l'emploi du microscope à l'étude des plaques minces de roches, certaines houilles peuvent être examinées par transparence et l'on peut y reconnaître facilement la nature des tissus, since the use of the microscope for the study of sections of rocks, certain coals can be examined by trans-parency and the nature of the tissues can be easily recognized. (42)

plaque-modèle [plaques-modèles *pl.*] (*n.f.*) *ou* plaque porte-modèles (Fonderie), pattern-plate ; match-plate ; carded patterns. (43)

plaque-modèle en bois (Fonderie) (*f.*), match-board ; joint-board. (44)

plaque négative (élément de pile) (*f.*), negative plate (battery-cell). (45)

plaque négative (Photogr.) (*f.*), negative plate. (46)

plaque orthochromatique (Photogr.) (*f.*), ortho-chromatic plate. (47)

plaque orthochromatique sans écran *ou en abrév.* plaque ortho. sans écran (*f.*), non-screen plate ; anti-screen plate ; self-screen plate ; autoscreen plate ; non-filter plate ; colour-screen plate ; screened chromatic plate. (48)

plaque panchromatique (Photogr.) (*f.*), pan-chromatic plate. (49)

plaque photographique (*f.*), photographic plate. (50)

plaque porte-objet [plaques porte-objets *pl.*] (Microscopie) (*f.*), slide ; slider. (51)

plaque positive (élément de pile) (*f.*), positive plate (battery-cell). (52)

plaque positive *ou* plaque pour positif [plaques pour positifs *pl.*] (Photogr.) (*f.*), lantern-plate ; lantern or transparency plate. (53)

plaque pour le glaçage des épreuves (Photogr.) (*f.*), print-glazer ; glazing-pad ; glazer. (54)

plaque rapide (Photogr.) (*f.*), rapid plate. (55)

plaque rotative (d'une machine à mouler à plaque rotative) (Fonderie) (*f.*), turn-over plate ; turn-over table. (56)

plaque sèche (Photogr.) (*f.*), dry plate. (57)

plaque sensible (Photogr.) (*f.*), sensitized plate ; sensitive plate. (58)

plaque striée (*f.*), chequer-plate ; chequered plate. (59)

plaque tournante (Ch. de f.) (*f.*), turntable ; turnplate ; turning-plate. (60)

plaque tubulaire (d'une chaudière) (*f.*), tube-sheet ; tube-plate ; flue-sheet ; flue-plate ; end-plate. (61)

plaque tubulaire de boîte à fumée (locomotive) (*f.*), front tube-sheet ; smoke-box flue-sheet. (62)

plaque tubulaire du foyer (*f.*), back tube-sheet ; fire-box flue-sheet. (63)

plaquer (le métal) (*v.t.*), to plate (to overlay mechanically with a more precious metal) : (64)

plaquer une feuille de cuivre, to plate a sheet of copper. (65)

plaquer (le bois) (*v.t.*), to veneer. (1)

plaqueur (pers.) (*n.m.*), plater. (2)

plasma (Minéral) (*n.m.*), plasma. (3)

plasticité (*n.f.*), plasticity : (4)

plasticité de l'argile, plasticity of clay. (5)

plastique (*adj.*), plastic : (6)

déformation plastique (*f.*), plastic deformation. (7)

plastron (*n.m.*), breast-plate ; conscience ; palette. (8)

plat, -e (*adj.*), flat ; level : (9)

terrain plat (*m.*), flat ground ; level ground. (10)

plat (la partie plate d'une chose) (*n.m.*), flat : (11)

le plat d'une lame, the flat of a blade. (12)

plat (fer ou acier) (*n.m.*) (*s'emploie généralement au pluriel*), flat. (13 *ou* 14)

plat (d'une balance) (*n.m.*), pan, scale, bowl, (of a balance or scale). (15)

plat-bord [**plats-bords** *pl.*] (Constr.) (*n.m.*), scaffold-board. (16)

plat-coin [**plats-coins** *pl.*] (Mines) (*n.m.*), plug ; feather (the internal wedge of a plug-and-feather wedge). (17)

plate-bande [**plates-bandes** *pl.*] (voûte plate) (*n.f.*), platband ; flat arch. (18)

plate-bande (d'une poutre composée) (*n.f.*), flange-plate, flange, boom-plate, boom, (of a built-up girder). (19)

plate-cuve [**plates-cuves** *pl.*] (dans un puits abandonné) (Mines) (*n.f.*), dam (in an abandoned shaft). (20)

plate-forme [**plates-formes** *pl.*] (*n.f.*), platform. (21)

plate-forme (d'une locomotive) (*n.f.*), foot-plate ; foot-board. (22)

plate-forme (wagon plat) (Ch. de f.) (*n.f.*), plat-form-wagon ; platform-car ; platform-carriage. (23)

plate-forme (sablière reposant sur le mur) (Constr.) (*n.f.*), wall-plate ; roof-plate ; plate. (24)

plate-forme (sablière reposant sur les extrémités des entraits) (Constr.) (*n.f.*), pole-plate. (25)

plate-forme (pièce de bois qui se place sous le pied d'un étai pour former empattement) (Constr.) (*n.f.*), sole-piece ; footing-block. (26)

plate-forme continentale (Océanogr.) (*f.*), continental platform ; continental shelf. (27) See **seuil continental** for illustrative example.

plate-forme corallienne (Géogr. phys.) (*f.*), coralline platform. (28)

plate-forme de chargement *ou* *simplement* **plate-forme** (*n.f.*) (four métallurgique), charging-platform ; charging-scaffold ; platform ; landing. (29)

plate-forme de coulée (haut fourneau) (*f.*), tapping-floor (blast-furnace). (30)

plate-forme de repos (puits de mine) (*f.*), platform ; sollar ; ladder-sollar ; stage of a ladderway. (31)

plate-forme des terrassements *ou* *simplement* **plate-forme** (*n.f.*) (Ch. de f. & Routes), forma-tion-level ; road-bed ; bed ; subgrade. (32)

plate-forme élévatrice hydraulique à action directe, à haute pression (*f.*), high-pressure hydraulic direct-acting platform lift. (33)

plate-forme littorale *ou* **plate-forme côtière** *ou* **plate-forme d'abrasion** (Géogr. phys.) (*f.*), shore-terrace. (34)

plate-forme littorale rocheuse *ou* **plate-forme côtière entaillée par les vagues** (opp. à talus littoral) (*f.*), wave-cut terrace. (35) (opp. to wave-built terrace).

plateau (*n.m.*), plate ; disc , table ; tray. (36)

plateau (d'une machine-outil, d'une machine à mouler, etc.) (*n.m.*), table, plate, platen, (of a machine-tool, of a moulding-machine, etc.). (37 *ou* 38)

plateau (d'une presse) (*n.m.*) (*s'emploie générale-ment au pluriel*), plate (of a press). (39)

plateau (d'un soufflet) (*n.m.*), board (of a bellows). (40)

plateau (d'une balance) (*n.m.*), pan, bowl, scale, (of a balance or scales). (41)

plateau (d'un tour) (*n.m.*), face-plate, chuck-plate, chuck, face-plate chuck, face-chuck, (of a lathe). (42) See also **mandrin**.

plateau (d'échafaud) (Constr.) (*n.m.*), scaffold-board. (43)

plateau (Géogr. phys.) (*n.m.*), plateau ; table-land. (44)

plateau à double coulisse, pour machines à percer (*m.*), compound table for drilling-machines. (45)

plateau à hauteur variable (*m.*), rising table ; table which can be raised or lowered : (46) support en plateau à hauteur variable (*m.*), stand with rising table. (47)

plateau à quatre griffes (*m.*), four-jaw chuck ; 4-jaw face-plate. (48)

plateau à quatre griffes indépendantes et réversibles (*m.*), independent four-jaw chuck, with reversible jaws ; four-jaw independent chuck, with reversible jaws. (49)

plateau à rainures, bien assis, et pouvant se mouvoir automatiquement dans les sens longitudinal, transversal, et circulaire (*m.*), slotted table, well bedded, and self-acting in the circular, transverse, and longitudinal motions. (50)

plateau à rainures hélicoïdales (*m.*), scroll-chuck. (51)

plateau à secousses (*m.*), shaking-tray ; bumping-tray. (52)

plateau à serrage concentrique à trois mors (*m.*), three-jaw concentric gripping-chuck. (53)

plateau à toc (*m.*). Same as **plateau-toc**.

plateau à trois griffes concentriques (*m.*), three-jaw concentric gripping-chuck. (54)

plateau à trous et étau parallèle montés sur collier tournant (machine à percer) (*m.*), plate with holes and parallel vice mounted on revolving table. (55)

plateau à trous, muni de quatre poupées à pompe (*m.*), face-plate, with four screw-dogs ; chuck face-plate with four dogs. (56)

plateau à trous nus (*m.*), face-plate with holes only (i.e., not including dogs). (57)

plateau arrière du cylindre (locomotive) (*m.*), back cylinder-head ; back cylinder-cover ; back-head of cylinder. (58)

plateau avant du cylindre (locomotive) (*m.*), front cylinder-head ; front cylinder-cover ; front-head of cylinder. (59)

plateau circulaire automatique (machine-outil) (*m.*), self-acting circular table. (60)

plateau conducteur (*m.*), wrist-plate : (61)

le plateau conducteur d'une distribution Corliss, the wrist-plate of a Corliss valve-gear. (62)

plateau continental (Géogr. phys.) (*m.*),   con-
-tinental plateau. (1)

plateau d'érosion (Géol.) (*m.*), plateau of erosion.
(2)

plateau de comblement (Géol.) (*m.*),   plateau of
accumulation. (3)

plateau de cylindre   (*m.*),   cylinder-head ;
cylinder-cover. (4)

plateau de dessous (*m.*), bottom plate. (5)

plateau de dessus (*m.*), top plate. (6)

plateau de division *ou* plateau diviseur (*m.*),
index-plate ;   index-dial ;   division-plate.
(7)

plateau de friction (*m.*), friction-disc. (8)

plateau de serrage (pour meules en émeri, ou
analogues) (*m.*),   side-plate (for emery-
wheels, or the like). (9 *ou* 10)

plateau-étau [plateaux-étaux *pl.*] (pour machines
à percer) (*n.m.*),   vice-plate ;   combined
parallel vice and circular table. (11)

plateau-manivelle [plateaux-manivelles *pl.*] (*n.m.*),
disc crank ; wheel crank. (12)

plateau muni de rigoles pour recevoir l'eau de
savon (*m.*), table provided with sud-channel.
(13)

plateau porte-coussinets (d'une machine à
tarauder les tiges ou les tubes) (*m.*), diehead,
screwing-head, screwing-chuck, (of a machine
for screwing bolts or tubes). (14)

plateau porte-pièce (*m.*),   work-plate ; work-
table. (15)

plateau pousse-toc   [plateaux   pousse-toc   *ou*
plateaux pousse-tocs *pl.*] (*m.*) *ou* plateau-
toc [plateaux-tocs *pl.*] (*n.m.*) (tour), catch-
plate ;   driver-plate ;   driver-chuck ;   dog-
chuck ;   point-chuck ;   take-about chuck.
(16)

plateau universel à quatre griffes indépendantes
(*m.*),   universal or independent four-jaw
chuck. (17)

platelage (d'un pont) (*n.m.*), flooring (of a bridge.)
(18)

plateure *ou* plateur (Houillères) (*n.f.*), flat seam ;
slightly-dipping seam. (19)
Note.—So-called flat seams usually have a
more or less decided dip, often rising to 3%,
in one direction or another.

platinage (*n.m.*), platinization. (20)

platine (Chim.) (*n.m.*), platinum ; platina. (21)

platine (d'une serrure) (*n.f.*),   plate (of a lock).
(22)

platine (d'un microscope) (*n.f.*),   stage (of a
microscope). (23)

platine iridé (*m.*),   platinum-iridium ; platino-
iridium ; iridioplatinum. (24)

platine natif (*m.*),   native platinum ;   native
platina. (25)

platine porte-objet [platines porte-objets   *pl.*]
(Microscope) (*f.*),   slide ; slider. (26)

platine tournante (de microscope) (*f.*),   revolving
stage. (27)

platiner (*v.t.*), to platinize ; to platinate : (28)
platiner du cuivre, to platinize copper. (29)

platineux, -euse (Chim.) (*adj.*),   platinous. (30)

platinides (*n.m.pl.*),   platinum metals ;   metals
of the platinum group. (31)

platinifère (*adj.*), platiniferous ; platinum (*used
as adj.*) : (32)
sables platinifères (*m.pl.*), platinum sands. (33)

platinique (Chim.) (*adj.*), platinic. (34)

platiniridium (*n.m.*),   iridioplatinum ;   platino-
iridium ; platinum-iridium. (35)

plâtrage (action) (*n.m.*), plastering. (36)

plâtrage (ouvrage) (*n.m.*), plaster-work ; plaster-
-ing. (37)

plâtre (*n.m.*), plaster. (38)

plâtre de moulage (*m.*), plaster of Paris. (39)

plâtre du puits (Mines) (*m.*),   pit-head works ;
heapstead. (40)

plâtrer (*v.t.*), to plaster : (41)
plâtrer un plafond, to plaster a ceiling. (42)

plâtrerie (ouvrage) (*n.f.*), plaster-work ; plaster-
-ing. (43)

plâtreux, -euse (*adj.*), plastery. (44)

plâtrière (carrière) (*n.f.*), gypsum-quarry. (45)

plâtrière (four à plâtre) (*n.f.*), gypsum-furnace.
(46)

plattnérite (Minéral) (*n.f.*), plattnerite. (47)

plein, -e (*adj.*), full : (48)
un tonneau plein, a full cask. (49)
un avenir plein de promesses, a future full of
promise. (50)
la pleine lune, the full moon. (51)
plein tuyau (Hydraul.) (*m.*),   short pipe ;
cylindrical   mouthpiece ;   orifice in thick
wall ;   opening in thick wall. (52) Cf.
écoulement.

pleine charge (Élec., etc.) (*f.*), full load. (53)

pleine charge d'eau (Hydraul.) (*f.*),   full head
of water. (54)

pleine marche arrière (conduite d'une loco-
-motive, etc.) (*f.*), full gear back ; full back-
-ward gear. (55)

pleine marche avant (*f.*),   full gear forward ;
full forward gear. (56)

pleine ouverture (d'un objectif) (Photogr.)
(*f.*),   full aperture. (57) See example
under ouverture.

pleins pouvoirs (*m.pl.*),   full powers ;   plenary
powers. (58)

plein, -e (opp. à creux) (*adj.*), solid. (59) (opp.
to hollow) :
brique pleine (*f.*), solid brick. (60)

plein (d'une vis) (opp. à écuelle, creux, *ou* vide)
(*n.m.*),   tooth (of a screw). (61) (opp. to
groove).

plein (d'une roue d'engrenage) (opp. à vide *ou*
creux) (*n.m.*),   tooth (of a gear-wheel). (62)
(opp. to space).

plein de l'eau (*m.*), high water. (63)

plein de l'été (*m.*), midsummer. (64)

plein de l'hiver (*m.*), midwinter. (65)

pleinement (*adv.*), fully ; entirely. (66)

pléistocène (Géol.) (*adj.*), Pleistocene ; Pleisto-
-cenic. (67)

pléistocène (*n.m.*),   Pleistocene ;   Pleistocenic :
(68)
le pléistocène, the Pleistocene. (69)

pléistosiste (*n.f.*), pleistosist. (70)

pléochroïque (*adj.*), pleochroic ; polychroic. (71)

pléochroïsme (*n.m.*), pleochroism ; polychroism.
(72)

pléonaste (Minéral) (*n.m.*), pleonaste. (73)

pléromorphe (Minéralogie) (*n.m.*),   pleromorphe.
(74)

pléromorphose (*n.f.*), pleromorphosis. (75)

pleurer (*v.i.*), to weep : (76)
robinet qui pleure (*m.*), tap which weeps. (77)

pleuvoir (*v. impersonnel*), to rain (*v.i.*) : (78)
il pleut très peu dans le Sahara, it rains very
little in the Sahara. (79)
pleuvoir à verse *ou* pleuvoir à seaux, to rain
cats and dogs ;   to rain pitchforks ;   to rain
in torrents. (80)

plexus d'injection (Géol.) (*m.*), injection plexus. (1)

pli (*n.m.*), fold. (2)

pli (épaisseur) (*n.m.*), ply : (3)
courroies à quatre plis *ou* courroies à 4 plis (*f.pl.*), four-ply belting ; 4-ply belting. (4)

pli (Géol.) (*n.m.*), fold ; flexure. (5)

pli anticlinal (*m.*), anticlinal fold ; anticlinal flexure. (6)

pli couché (opp. à **pli droit**) (*m.*), recumbent fold. (7)

pli déjeté (*m.*), fold with slightly inclined axial plane. (8)

pli déversé *ou* **pli renversé** (*m.*), inverted fold ; reversed fold ; overfold ; overturned fold ; overturned flexure. (9)

pli droit (*m.*), erect fold. (10)

pli dyssymétrique (*m.*), unsymmetrical fold ; asymmetrical fold. (11)

pli en éventail (*m.*), fan-fold. (12)

pli-faille [**plis-failles** *pl.*] (*n.m.*), flexure fault. (13)

pli-faille inverse (*m.*), overthrust-fault ; over-fault ; thrust-fault. (14)

pli isoclinal (*m.*), isoclinal fold ; isoclinic fold ; carinate fold. (15)

pli lâche (opp. à **pli serré**) (*m.*), open fold. (16)

pli monoclinal (*m.*), monoclinal fold ; uniclinal flexure. (17)

pli normal (*m.*), normal fold. (18)

pli retourné (*m.*), fold tilted below the horizontal. (19)

pli serré (*m.*), compressed fold. (20)

pli simple (*m.*), simple fold. (21)

pli symétrique (*m.*), symmetrical fold. (22)

pli synclinal (*m.*), synclinal fold ; synclinal flexure. (23)

pliable *ou* **pliant, -e** (*adj.*), pliable ; flexible ; supple ; pliant ; folding. (24)

pliage (*n.m.*), folding. (25)

plier (*v.t.*), to fold ; to fold up ; to bend : (26)
plier une feuille de papier, to fold a sheet of paper. (27)
plier l'osier, to bend osier. (28)

plier (*v.i.*), to fold ; to fold up ; to bend : (29)
le jonc plie aisément, rush bends easily. (30)

plier (se) (*v.r.*), to fold ; to fold up ; to bend ; to be bent : (31)
la tige du rivet doit se plier à froid à 180°, à bloc, comme l'indique la figure, sans criques à l'extérieur du pli (Cahier des charges unifié), the rivet-shank shall bend cold through 180 deg. flat on itself, as shown in the figure, without cracking on the outside of the bent portion (Standard specification). (32)
un bon rivet doit se plier en deux, à froid, a good rivet should bend double cold. (33)

plieuse (*n.f.*), folding-machine. (34)

plinthe (au bas d'un mur d'appartement) (*n.f.*), skirting ; skirting-board ; skirt-board ; base-board ; wash-board ; mop-board. (35)

pliocène (Géol.) (*adj.*), Pliocene ; Pleiocene ; Pliocenic. (36)
alluvion pliocène (*f.*), Pliocene alluvium. (37)

pliocène (*n.m.*), Pliocene ; Pleiocene : (38)
le pliocène, the Pliocene. (39)

plionnage (d'un puits de mine) (*n.m.*), basket-cribbing. (40)

plissé, -e (Géol.) (*adj.*), plicated ; folded : (41)
couche plissée (*f.*), plicated stratum. (42)

plissement (Géol.) (*n.m.*), folding ; flexuring ; plication : (43)
la forme élémentaire du plissement est le pli, qui se compose d'une partie convexe ou anticlinal et d'une partie concave ou synclinal, the elementary form of folding is the fold, which is composed of a convex part or anticlinal and of a concave part or synclinal. (44)

plissotement (Géol.) (*n.m.*), crumpling : (45)
plissotement de schistes, crumpling of schists. (46)

plistosiste (*n.f.*), pleistosist. (47)

pliure (*n.f.*), folding ; fold. (48)

plomb (métal) (*n.m.*), lead. (49)

plomb (petite masse de plomb, ou de tout autre métal, servant à lester le fil à plomb) (*n.m.*), plumb-bob ; plumb ; plummet ; bob. (50)

plomb (fil à plomb) (*n.m.*), plumb-line ; plummet-line ; plummet. (51) See **fil à plomb** for varieties.

plomb (sceau de plomb) (*n.m.*), lead seal ; leaden seal. (52)

plomb (à), plumb ; perpendicular : (53)
la maison se tient à plomb, the house stands plumb. (54)

plomb antimonié (*m.*), antimoniated lead. (55)

plomb argentifère (*m.*), silver lead ; argentif-erous lead. (56)

plomb blanc *ou* **plomb carbonaté** (Minéral) (*m.*), white lead ; lead carbonate ; cerusite. (57)

plomb chromaté (Minéral) (*m.*), lead chromate ; red lead ore ; crocoite. (58)

plomb corné (Minéral) (*m.*), horn-lead ; corneous lead ; plumbum corneum ; phosgenite. (59)

plomb d'œuvre (Métall.) (*m.*), work-lead. (60)

plomb désargenté (*m.*), desilverized lead. (61)

plomb en feuilles (*m.*), sheet lead. (62)

plomb en saumon (*m.*), pig lead. (63)

plomb fusible *ou* **plomb de sûreté** *ou* *simplement* **plomb** (*n.m.*) (Élec.), fuse ; fuse-wire ; safety-fuse : (64)
un plomb a fondu, a fuse has blown (*or* has gone). (65)

plomb fusible (chaudière) (*m.*), fusible plug ; safety-plug ; plug. (66)

plomb jaune (Minéral) (*m.*), yellow lead ore ; wulfenite. (67)

plomb pour essais (*m.*), assay lead. (68)

plomb rouge (Minéral) (*m.*), red lead ore ; lead chromate ; crocoite. (69)

plomb sulfuré (Minéral) (*m.*), lead-glance ; galena ; galenite. (70)

plombage (*n.m.*), plumbing ; leading ; sealing. (71)

plombagine (*n.f.*), plumbago ; black lead ; graphite. (72)

plombate (Chim.) (*n.m.*), plumbate. (73)

plomber (appliquer du plomb à) (*v.t.*), to plumb ; to lead : (74)
plomber un tuyau, to plumb a pipe. (75)

plomber (attacher un petit sceau de plomb à) (*v.t.*), to seal ; to plumb : (76)
plomber un colis, to seal, to plumb, a package. (77)

plomber (vérifier par le fil à plomb la verticalité de) (*v.t.*), to plumb ; to plumb-line : (78)
plomber un mur, to plumb, to plumb-line, a wall. (79)

plomberie (*n.f.*), plumbing ; plumbery. (80)

plombeux, -euse (*adj.*), plumbeous ; leady ; lead (*used as adj.*) : (81)
fusion plombeuse (Métall.) (*f.*), lead fusion. (82)

plombeux, -euse (Chim.) (*adj.*), plumbous. (1)

plombgomme (*n.m.*) *ou* **plombogummite** (*n.f.*) *ou* **plomborésinite** (*n.f.*) (Minéral), plumbo--gummite ; plumboresinite. (2)

plombier (pers.) (*n.m.*), plumber. (3)

plombifère (*adj.*), plumbiferous ; lead-bearing. (4)

plombique (Chim.) (*adj.*), plumbic. (5)

plongeant, -e (*adj.*), plunging ; dipping ; pitch--ing. (6)

plongée (de la lunette d'un théodolite) (*n.f.*), plunging (the telescope of a theodolite). (7)

plongée (*n.f.*) *ou* **plongement** (*n.m.*) (Géol.), dip : (8)

la plongée (*ou* le plongement) d'un filon, the dip of a lode. (9)

plonger (*v.t.*), to plunge ; to dip : (10)

plonger la lunette d'un théodolite, to plunge the telescope of a theodolite. (11)

plonger le fer dans l'eau (trempe), to plunge iron into water. (12)

plonger (*v.i.*), to plunge ; to dip. (13)

plonger (Géol.) (*v.i.*), to dip : (14)

apprécier de quel côté de la direction plonge le filon, to determine on which side of the strike the lode dips. (15)

une partie du bassin de Newcastle plonge sous la mer du Nord, a part of the Newcastle basin dips under the North Sea. (16)

plongeur (*n.m.*), ram ; plunger ; plunger-piston ; solid piston. (17)

plot (*n.m.*) *ou* **plot de contact** (Élec.), stud ; contact-stud ; contact. (18)

ployable (*adj.*), bendable ; pliable. (19)

ployer (*v.t.*), to bend ; to fold : (20)

ployer une branche, to bend a branch. (21)

ployer (*v.i.*), to bend ; to fold : (22)

poutre qui a ployé (*f.*), girder which has bent. (23)

ployer (se) (*v.r.*), to bend ; to fold ; to be bent. (24)

pluie (*n.f.*), rain. (25)

pluie *ou* **pluie ou vent** (Barométrie), rain : (26)

le baromètre est à la pluie (*ou* est à pluie ou vent), the barometer is at rain. (27)

pluie abondante (*f.*), heavy rain ; copious rain ; downpour ; rainpour. (28)

pluie artificielle (*f.*), artificial rain. (29)

pluie d'orage (*f.*), rain-storm. (30)

pluie de sable (*f.*), sand-storm ; simoom ; simoon. (31)

pluie de sang (*f.*), blood-rain ; red rain. (32)

pluie fine (*f.*), fine rain. (33)

pluie fossile (Géol.) (*f.*), rain-prints ; rain-pits. (34)

plume (*n.f.*), pen : (35)

dessin à la plume (*m.*), pen-drawing. (36)

plume de paon (agate) (*f.*), peacock-stone. (37)

plumosite (Minéral) (*n.f.*), plumosite ; feather-ore ; heteromorphite. (38)

plurivalence (Chim.) (*n.f.*), multivalence. (39)

plurivalent, -e (*adj.*), multivalent. (40)

plus grand diamètre admis *ou* **plus grand diamètre admissible** (tour, etc.) (*m.*), swing (lathe, etc.) : (41)

plus grand diamètre admis au-dessus des chariots, swing over saddle ; swing of the rest. (42)

plus grand diamètre admis au-dessus du banc, swing over bed ; swing of the bed. (43)

plus grand diamètre admis dans le rompu *ou* plus grand diamètre admissible dans la coupure, swing in gap. (44)

plus-value [**plus-values** *pl.*] (*n.f.*), appreciation ; increase of (*or* in) value ; extra : (45)

plus-value d'une propriété, appreciation, increase in the value, of a property. (46)

plus-value des terrains aux abords des grandes villes, increase in value of land in the vicinity of large towns. (47)

plus-value pour un renvoi de mouvement, extra for overhead motion ; additional price for counter-shaft and appurtenances. (48)

plutonique *ou* **plutonien, -enne** (qui se rapporte au plutonisme) (*adj.*), Plutonic ; Plutonian ; Vulcanian : (49)

théorie plutonienne (*f.*), Plutonic theory ; Vulcanian theory. (50)

plutonique *ou* **plutonien, -enne** (qui a été produit par l'action des feux souterrains) (*adj.*), plutonic ; plutonian : (51)

roches plutoniques (*ou* plutoniennes) (*f.pl.*), plutonic rocks. (52)

plutonisme (*n.m.*), Plutonism. (53)

plutoniste *ou* **plutonien** (pers.) (*n.m.*), Plutonist ; Plutonian. (54)

pluviaire (Géol.) (*adj.*), Glacial : (55)

période pluviaire (*f.*), Glacial period. (56)

pluvial, -e, -aux (*adj.*), rainy ; rain (*used as adj.*) ; pluvial : (57)

saison pluviale (*f.*), rainy season. (58)

érosion pluviale (*f.*), rain erosion. (59)

pluvieux, -euse (*adj.*), rainy ; pluvious. (60)

pluviographe *ou* **pluviomètre** (*n.m.*), rain-gauge ; pluviometer ; pluviameter ; udometer ; ombrometer. (61)

pluviomètre enregistreur (*m.*), self-recording rain-gauge ; pluviograph. (62)

pluviométrie (*n.f.*), pluviometry ; udometry. (63)

pluviométrique (*adj.*), pluviometric ; pluvio--metrical ; pluviametric ; pluviametrical ; udometric ; ombrometric ; ombrometrical. (64)

pluviométrographe (*n.m.*). Same as **pluviomètre enregistreur.**

pluvioscope (*n.m.*), pluvioscope. (65)

pluviosité (*n.f.*), raininess. (66)

pneumaticité (*n.f.*), pneumaticity. (67)

pneumatique (*adj.*), pneumatic : (68)

bandage pneumatique (*m.*), pneumatic tire. (69)

pneumatique *ou* **familièrement** **pneu** (bandage pneumatique) (*n.m.*), pneumatic tire ; pneumatic tyre. (70)

pneumatique (science) (*n.f.*), pneumatics. (71)

pneumatiquement (*adv.*), pneumatically. (72)

pneumatolyse (Géol.) (*n.f.*), pneumatolysis. (73)

pneumatolytique (Géol.) (*adj.*), pneumatolitic : (74)

formation pneumatolytique (*f.*), pneumato--lytic formation. (75)

pneumatophore (appareil respiratoire pour mines grisouteuses) (*n.m.*), pneumatophore. (76)

poche (*n.f.*), pocket. (77)

poche (Fonderie) (*n.f.*), ladle. (78)

poche à armature fixe et à anse démontable (*f.*), bull-ladle. (79)

poche à chariot (*f.*), car-ladle. (80)

poche à cristaux (Géol. & Mines) (*f.*), vug ; vugh ; geode ; druse. (81)

poche à engrenage (*f.*), geared ladle. (82)

poche à engrenage et vis sans fin (*f.*), worm-geared ladle. (83)

poche à fourche *ou* poche avec armature à fourche (*f.*), shank ; shank-ladle. (1)

poche à grisou *ou* poche de grisou (Mines) (*f.*), fire-damp pocket. (2)

poche à laitier (*f.*), slag-pot. (3)

poche à main (*f.*), hand-ladle. (4)

poche à main à double armature *ou* poche à main à double manche (*f.*), double-handled hand-shank. (5)

poche d'eau (Mines) (*f.*), feeder of water ; water feeder. (6)

poche de coulée *ou* poche de fonderie *ou* poche de fondeur *ou* poche à fonte *ou* *simplement* poche (*n.f.*), casting-ladle ; foundry-ladle ; ladle. (7)

poche de dissolution *ou* *simplement* poche (*n.f.*), washout. (8)

poche de gaz (*f.*), gas-pocket. (9)

poche de minerai (*f.*), pocket of ore ; ore-pocket ; bunch of ore. (10)

poche de vidange (*f.*), drain-cup ; drip-cup. (11)

poche montée sur wagonnet (*f.*), car-ladle. (12)

poche pour grues (*f.*), crane-ladle. (13)

poché, -e (Mines) (*adj.*), pockety ; bunchy : (14)
des alluvions pochées, c'est-à-dire formées par une série de lentilles souvent extrême--ment riches, séparées par des espaces stériles (*f.pl.*), pockety alluvials, that is to say, formed by a series of lenses, often extremely rich, separated by barren intervals. (15)

filon poché (*m.*), bunchy lode. (16)

pochette d'instruments de mathématiques (*f.*), pocket-case of mathematical instruments. (17)

pochette de papier photographique (*f.*), packet of photographic paper. (18)

pochoir (*n.m.*), stencil ; stencil-plate. (19)

podomètre (*n.m.*), pedometer ; paceometer ; passimeter ; passometer. (20)

poêle (de plombier) (*n.f.*), lead-pot ; melting-pot. (21)

poêle à flamber (Fonderie) (*m.*), drying-kettle ; kettle ; fire-lamp ; lamp ; lantern ; devil ; chauffer. (22)

poids (qualité d'un corps pesant) (*n.m.*), weight : (23)
le poids de l'air, the weight of the air. (24)

poids (morceau de métal d'une pesanteur déterminée, servant à peser d'autres corps) (*n.m.*), weight : (25)
une balance et ses poids, a scale and its weights. (26)

poids (*n.m.*) *ou* poids curseur (d'une balance romaine), weight, bob, (of a steelyard). (27)

poids à vide (*m.*), weight empty. (28)

poids adhérent (locomotive) (*m.*), weight on drivers ; weight on coupled wheels. (29)

poids atomique (*m.*), atomic weight. (30)

poids brut (*m.*), gross weight. (31)

poids de charge *ou* *simplement* poids (*n.m.pl.*) (Fonderie), weights : (32)
charger un moule avec des poids suffisants pour résister au soulèvement du châssis, to weight a mould sufficiently to resist the lifting of the box. (33)

poids de précision (*m.pl.*), precision weights. (34)

poids en cuivre (*m.pl.*), brass weights. (35)

poids en cuivre, forme godet (*m.pl.*), brass weights, cup form. (36)

poids en fonte (*m.pl.*), iron weights. (37)

poids en lame de cuivre (*m.pl.*), brass fractional weights ; brass fractions. (38)

poids en service (locomotive) (*m.*), weight in working order. (39)

poids et mesures métriques [poids (*n.m.pl.*) ; mesures (*n.f.pl.*)], metric weights and measures. (40)

poids faible (*m.*), light weight ; short weight ; scant weight ; lazy weight. (41)

poids fort (*m.*), overweight. (42)

poids juste (opp. à faux poids) (*m.*), just weight (opp. to unjust weight). (43)

poids moléculaire (*m.*), molecular weight. (44)

poids mort (poids d'un appareil qui absorbe une partie du travail utile) (*m.*), dead-weight : (45)
le poids mort des wagonnets de mines représente en moyenne 40 0/0 de la charge utile, the dead-weight of mine-cars represents on an average 40% of the live-weight. (46)

poids mort (Ponts et Chaussées) (*m.*), dead load : (47)
les ponts sont soumis à deux natures d'épreuves, l'une par poids mort, l'autre par poids roulant (*m.pl.*), bridges are subjected to two kinds of tests, the one by dead load, the other by live load. (48)

poids net (*m.*), net weight. (49)

poids roulant (Ponts et Chaussées) (*m.*), live load ; moving load ; rolling load. (50)
See poids mort for example of use.

poids spécifique (*m.*), specific gravity : (51)
le poids spécifique de l'eau à +4°=1, the specific gravity of water at +4° C.=1. (52)

poids utile (charge utile) (par opposition à poids mort) (*m.*), live weight. (53)

poignée (quantité que la main fermée peut empoigner ou contenir) (*n.f.*), handful : (54)
une poignée de sable, a handful of sand. (55)

poignée (d'un outil) (*n.f.*), handle, hand-grip, (of a tool). (56)

poignée (d'une varlope, d'une demi-varlope) (*n.f.*), toat, handle, (of a trying-plane, of a jack-plane). (57 *ou* 58)

poignée (d'un châssis de fonderie) (*n.f.*), handle (of a foundry-flask). (59)

poignée ansée (*f.*), basket-handle. (60)

poignée de porte (*f.*), door-handle. (61)

poignée montoire (locomotive, etc.) (*f.*), hand--hold ; grab-iron. (62)

poinçon (instrument sur lequel on frappe pour percer ou marquer) (*n.m.*), punch ; solid punch ; driver. (63)

poinçon (alène) (*n.m.*), awl. (64)

poinçon (d'une ferme de comble en bois) (*n.m.*), king-post, crown-post, joggle-post, joggle-piece, broach-post, middle post, (of a wooden roof-truss). (65)

poinçon (d'une ferme de comble métallique) (*n.m.*), king-rod (of an iron or steel roof-truss). (66)

poinçon à la machine (*m.*), machine-punch. (67)

poinçon à main (*m.*), hand-punch. (68)

poinçon à œil (*m.*), eyed punch. (69)

poinçon carré (*m.*), square punch. (70)

poinçon de forge (*m.*), smiths' punch. (71)

poinçon de garantie *ou* *simplement* poinçon (*n.m.*), hall-mark ; mark ; assayer's mark ; maker's mark. (72)

poinçon hydraulique pour trous d'homme (*m.*), hydraulic manhole-punch. (73)

poinçon rond (*m.*), round punch. (74)

poinçonnage *ou* poinçonnement (*n.m.*), punching ; stamping. (1)

poinçonner (percer avec un poinçon) (*v.t.*), to punch : (2)

poinçonner un trou dans une plaque de métal, to punch a hole in a plate of metal. (3)

poinçonner (marquer avec un poinçon) (*v.t.*), to stamp ; to mark ; to brand. (4)

poinçonneuse (*n.f.*), punching-machine. (5)

poinçonneuse à deux leviers (*f.*), two-lever punching-machine. (6)

poinçonneuse à levier (*f.*), lever punching-machine. (7)

poinçonneuse à main *ou* poinçonneuse à vis simple (*f.*), punching-bear ; screw punching-bear ; screw-bear ; bear. (8)

poinçonneuse-cisaille [poinçonneuses-cisailles *pl.*] (*n.f.*), punching and shearing-machine. (9)

poinçonneuse-cisaille à engrenages à bâti creux, à cisaille oblique pour longues barres et débrayage sur le devant, fonctionnant à bras ou au moteur (*f.*), geared punching and shearing-machine, hollow frame, skew shears for cutting long bars, stop-motion in front, for hand or power. (10)

poinçonneuse-cisaille à levier à bâti creux ; porte-matrice disposé pour poinçonner les fers plats ordinaires et les fers spéciaux (L, T, I, U) (*f.*), hollow-frame lever punching and shearing-machine ; die-holder set to punch ordinary flat iron and special sections (L, T, I, U). (11)

poinçonneuse-cisaille à levier, (1) avec cisaille sur le devant, ou (2) avec cisaille sur le derrière (*f.*), lever punching and shearing-machine, (1) with shear at front, or (2) with shear at back. (12 *ou* 13)

poinçonneuse-cisaille à levier ; cisaille en travers pour longues barres (*f.*), lever punching and shearing-machine ; shears set at an angle for cutting off long bars. (14)

poinçonneuse-cisaille à levier, poinçonnant en bas et cisaillant en haut (*f.*), lever punching and shearing-machine, punching below and shearing above. (15)

poinçonneuse-cisaille à levier, poinçonnant par devant et cisaillant sur le côté (*f.*), lever punching and shearing-machine ; punch in gap, shear at side. (16)

poinçonneuse-cisaille double, cisaillant d'un côté et poinçonnant de l'autre (*f.*), double-ended punching and shearing-machine, shearing at one end and punching the other. (17)

poinçonneuse duplex (*f.*), duplex punching-bear ; duplex lever-punch. (18)

poinçonneuse multiple (*f.*), multiple-punching machine ; gang-punch. (19)

point (*n.m.*), point. (20)

point (opp. à trait) (Télégr.) (*n.m.*), dot. (21) (opp. to dash).

point (au), in focus : (22)

image qui est au point (*f.*), image which is in focus. (23)

point cardinal (*m.*), cardinal point : (24)

les points cardinaux du compas, the cardinal points of the compass. (25)

point conséquent (Magnétisme) (*m.*). See points conséquents.

point critique (Phys.) (*m.*), critical point. (26)

point d'abatage (*m.*), mining-point. (27)

point d'application d'une force (*m.*), point of application of a force. (28)

point d'appui (d'un levier) (*m.*), fulcrum (of a lever) : (29)

le point d'appui du fléau d'une balance, the fulcrum of a balance-beam. (30)

point d'appui (Constr., etc.) (*m.*), point of support : (31)

répartition d'une charge entre deux points d'appui (*f.*), equalizing a load between two points of support. (32)

point d'attache du plateau d'une balance (*m.*), point of suspension of the scale-pan of a balance. (33)

point d'eau (Hydraul.) (*m.*), head-limit. (34)

point d'ébullition (*m.*), boiling-point ; tempera-ture of ebullition : (35)

100° C. (*ou* 212° F.) est le point d'ébullition de l'eau, 100° C. (*or* 212° F.) is the boiling-point of water. (36)

point d'éclair *ou* point d'inflammabilité (*m.*), flashing-point ; flash-point ; flash. (37)

point d'embarquement (*m.*), shipping-point. (38)

point d'équilibre (*m.*), point of equilibrium. (39)

point d'ignition *ou* point d'ignitibilité (*m.*), burn-ing-point. (40)

point d'intersection (*m.*), intersecting-point ; point of intersection. (41)

point de congélation (*m.*), freezing-point : (42)

point de congélation de l'eau, freezing-point of water. (43)

point de contact (*m.*), point of contact. (44)

point de contact (des cercles primitifs) (Engre-nages) (*m.*), pitch-point (Gearing). (45)

point de départ (*m.*), starting point ; point of departure. (46)

point de distance (Perspective) (*m.*), point of distance. (47)

point de fuite (Perspective) (*m.*), vanishing-point. (48)

point de fusion (*m.*), melting-point ; point of fusion ; fusion-point ; fusing-point. (49)

point de mire (*m.*), datum-point ; datum ; levelling-point. (50)

point de repère (*m.*), mark ; guiding-mark ; mark-point ; datum-point ; datum ; fixed point : (51)

0° et 100° sont des points de repère de l'échelle thermométrique centigrade, 0° and 100° are fixed points of the centrigrade thermo-metric scale. (52)

point de repère (marque faite sur un mur, sur un jalon, sur un terrain, pour qu'on puisse retrouver un alignement, un niveau) (*m.*), bench-mark. (53)

point de rupture (*m.*), breaking-point. (54)

point de station (Topogr.) (*m.*), station-point. (55)

point de suspension du fléau d'une balance, du plateau d'une balance (*m.*), point of suspension of the beam of a balance, of the scale-pan of a balance. (56 *ou* 57)

point de tangence (*m.*), point of tangency : (58)

point de tangence de deux cercles, point of tangency of two circles. (59)

point de vue (endroit où l'on se place pour voir un objet éloigné) (*m.*), point of view. (60)

point de vue (Perspective) (*m.*), point of sight ; station-point. (61)

point de vue (manière d'envisager les choses) (*m.*), point of view ; standpoint : (62)

examiner une affaire à différents points de vue, to examine a matter from different points of view ; to look at a thing from different standpoints. (1)

point donné (*m.*), given point. (2)

point fixe (*m.*), fixed point. (3)

point mort (Méc.) (*m.*), dead-centre ; dead-point : (4)

lorsque le piston est à une des extrémités de sa course, la manivelle est dite au point mort, when the piston is at one of the ends of its travel, the crank is said to be on the dead-centre. (5)

marche au point mort de la distribution (*f.*), running on the dead-point of the valve-gear. (6)

point mort matériel (*m.*), material point ; physical point. (7)

point perdu (opp. à **point fixe** *ou* **point de repère**) (Topogr.) (*m.*), turning-point. (8)

point principal (Perspective) (*m.*), centre of vision ; principal point ; point of sight. (9)

pointage (contrôle) (*n.m.*), tally ; timekeeping. (10)

pointe (*n.f.*), point (sharp end of a thing) : (11)

pointe d'une aiguille, d'un clou, d'une vis, d'un pieu, point of a needle, of a nail, of a screw, of a pile. (12 *ou* 13 *ou* 14 *ou* 15)

pointe (de l'aiguille) (Ch. de f.) (*n.f.*), point, nose, (of the switch). (16)

pointe (petit clou) (*n.f.*), brad ; nail ; point. (17)

pointe (d'un tour) (*n.f.*), centre (of a lathe) : (18)

'hauteur des pointes (*f.*), height of centres. (19)

pointe (pour dégelage du sol) (Exploitation des alluvions gelées) (*n.f.*), point ; steam-point. (20)

pointe à tracer *ou simplement* **pointe** (*n.f.*), scriber. (21)

pointe à trois dents (tour à bois) (*f.*), prong-centre ; fork-centre (wood-turning lathe). (22)

pointe au cône Morse (tour) (*f.*), Morse taper centre. (23)

pointe de cœur (d'un croisement de ch. de f.) (*f.*), point (of frog) ; frog-point ; tongue (of crossing). (24)

pointe de croisement *ou simplement* **pointe** (*n.f.*) (Ch. de f.), point-rail, tongue-rail, main point-rail, main point, of crossing : (25)

la pointe et la branche de pointe, the point-rail and the splice-rail ; the main point and the side point. (26)

pointe de diamant (diamant de vitrier) (*f.*), diamond point ; point ; diamond pencil ; diamond tool ; glaziers' diamond ; glass-cutters' diamond ; glass-cutter. (27)

pointe de diamant (outil de tour) (*f.*), diamond point ; diamond-point tool ; diamond tool (lathe-tool). (28)

pointe de la poupée fixe (tour) (*f.*), live centre. (29)

pointe de la poupée mobile (tour) (*f.*), dead centre ; back centre. (30)

pointe de mouleur (*f.*), moulders' nail ; moulders' point ; sprig. (31)

pointe de mouleur, à crochet (*f.*), moulders' nail, hook head. (32)

pointe de Paris (*f.*), French nail ; wire nail. (33)

pointe de terre (Géogr.) (*f.*), point ; promontory ; headland. (34)

pointe de tour (*f.*), lathe-centre. (35)

pointe de traçage (*f.*), scriber. (36)

pointe de traversée (Ch. de f.) (*f.*), point-rail of crossover. (37)

pointe mathématique du cœur (Ch. de f.) (*f.*), theoretical point of frog ; fine point of the crossing ; point of tongue. (38)

pointe réelle du cœur (Ch. de f.) (*f.*), actual point of the crossing ; point of frog ; nose of the crossing. (39)

pointeau (*n.m.*) *ou* **pointeau de mécanicien**, centre-punch ; prick-punch ; finder-point punch ; centre-pop ; bob-punch ; punch. (40)

pointeau (valve à vis-pointeau) (*n.m.*), needle-valve ; pin-valve. (41)

pointeau de débit visible (*m.*), sight-feed needle-valve. (42)

pointer (contrôler) (*v.t.*), to tally ; to check ; to tick ; to tick off. (43)

pointer une lunette (l'orienter), to point a telescope. (44)

pointerolle (*n.f.*), dresser (coal-mining pick). (45)

pointeur (pers.) (*n.m.*), tallyman ; timekeeper. (46)

points conséquents (Magnétisme) (*m.pl.*), con-sequent points ; consequent poles. (47)

pointu, -e (*adj.*), pointed ; sharp. (48)

poire (poulie étagée) (*n.f.*), stepped pulley ; stepped cone ; cone-pulley ; cone ; speed-pulley ; speed-cone ; speed-rigger ; speeder. (49)

poire (d'une balance romaine) (*n.f.*), weight, bob, (of a steelyard). (50)

poire (en caoutchouc pour l'obturateur d'un appareil photographique, etc.) (*n.f.*), ball ; blowing-ball ; bulb : (51)

poire avec tube pour déclenchement pneu-matique et raccord, ball with tubing for pneumatic release and teat. (52)

déclenche à la poire (*f.*), bulb release. (53)

poison (*n.m.*), poison. (54)

poison minéral (*m.*), mineral poison. (55)

poisser (*v.t.*), to pitch : (56)

poisser un cordage, to pitch a rope. (57)

poisseux, -euse (*adj.*), pitchy. (58)

poitrail (*n.m.*), breast-summer ; brestsummer ; bressomer. (59)

poix (provenant des conifères) (*n.f.*), pitch. (60)

poix (Minéralogie) (*n.f.*), pitch ; bitumen. (61)

poix de houille (*f.*), coal-tar pitch. (62)

poix de Judée (*f.*), Judean bitumen ; bitumen of Judea ; Jew's pitch ; Jews' pitch. (63)

poix liquide (*f.*), tar. (64)

polaire (Astron., Géogr., Phys. & Math.) (*adj.*), polar : (65)

régions polaires (*f.pl.*), polar regions. (66)

force polaire (*f.*), polar force. (67)

Polaire *ou* **la Polaire** (*n.f.*), Polaris ; pole-star ; polar star ; North Star : (68)

observation de la Polaire à la plus grande digression (*f.*), observing Polaris at extreme elongation. (69)

polarimètre (*n.m.*), polarimeter. (70)

polarisant, -e *ou* **polarisateur, -trice** (*adj.*), polar-izing. (71)

polarisateur (*n.m.*), polarizer. (72)

polarisation (*n.f.*), polarization. (73)

polarisation chromatique (*f.*), chromatic polariza-tion. (74)

polarisation rotatoire (*f.*), rotary polarization. (75)

polariscope (*n.m.*), polariscope. (1)
polariscopique (*adj.*), polariscopic : (2)
analyse polariscopique (*f.*), polariscopic analysis. (3)
polariser (*v.t.*), to polarize : (4)
polariser la lumière, to polarize light. (5)
polariseur (*n.m.*), polarizer. (6)
polarite (Explosif) (*n.f.*), polarite. (7)
polarité (*n.f.*), polarity : (8)
tous les corps sont susceptibles d'acquérir par influence la polarité magnétique (*m.pl.*), all bodies are capable of acquiring magnetic polarity by influence. (9)
pôle (Astron., Géogr., Phys. & Math.) (*n.m.*), pole. (10)
pôle (mesure anglaise de longueur) (*n.m.*), pole =5·0292 metres. (11)
pôle analogue (Cristall.) (*m.*), analogous pole. (12)
pôle antilogue (Cristall.) (*m.*), antilogous pole. (13)
pôle d'un cercle (Géom.) (*m.*), pole of a circle. (14)
pôle magnétique (*m.*), magnetic pole. (15)
pôle négatif (Électricité voltaïque) (*m.*), negative pole. (16)
pôle nord *ou* **pôle austral** (Magnétisme) (*m.*), north pole ; north-seeking pole ; austral pole ; positive pole ; red pole ; marked pole ; marked end. (17)
pôle nord *ou* **pôle boréal** *ou* **pôle septentrional** *ou* **pôle arctique** (Géogr.) (*m.*), north pole ; boreal pole ; arctic pole. (18)
pôle positif (Électricité voltaïque) (*m.*), positive pole. (19)
pôle sud *ou* **pôle boréal** (Magnétisme) (*m.*), south pole ; south-seeking pole ; boreal pole ; negative pole ; blue pole ; unmarked pole. (20)
pôle sud *ou* **pôle austral** *ou* **pôle méridional** *ou* **pôle antarctique** (Géogr.) (*m.*), south pole ; austral pole ; antarctic pole. (21)
pôles célestes (*m.pl.*), celestial poles ; poles of the heavens. (22)
pôles conséquents (Magnétisme) (*m.pl.*), con-sequent poles ; consequent points. (23)
pôles d'une pile, de l'aimant (*m.pl.*), poles of a battery, of the magnet. (24 *ou* 25)
pôles dissemblables *ou* **pôles de nom contraire** (Magnétisme) (*m.pl.*), unlike poles. (26)
pôles ennemis (Magnétisme) (*m.pl.*), opposite poles. (27)
pôles semblables *ou* **pôles de même nom** (Magné-tisme) (*m.pl.*), like poles. (28)
pôles terrestres (*m.pl.*), terrestrial poles. (29)
poli, -e (*adj.*), polished ; bright : (30)
marbre poli (*m.*), polished marble. (31)
outil à manche poli (*m.*), tool with bright handle. (32)
outil poli partout (*m.*), tool bright all over. (33)
poli (travail) (*n.m.*), polishing. (34)
poli (lustre) (*n.m.*), polish : (35)
pierre susceptible d'un beau poli (*f.*), stone capable of receiving a beautiful polish. (36)
poli éclatant, brilliant polish. (37)
poli glaciaire (Géol.), glacial polish. (38)
polianite (Minéral) (*n.f.*), polianite. (39)
polir (*v.t.*), to polish : (40)
polir les diamants, to polish diamonds. (41)
polir (se) (*v.r.*), to be polished ; to polish : (42)

les diamants se polissent avec de la poussière de diamant (*m.pl.*), diamonds are polished with diamond-dust. (43)
polissable (*adj.*), polishable : (44)
métal polissable (*m.*), polishable metal. (45)
polissage (*n.m.*), polishing : (46)
polissage du diamant, diamond polishing. (47)
polissage en bas-relief (Métall.) (*m.*), polish attack. (48)
polisseur (pers.) (*n.m.*), polisher. (49)
polissoir (touret) (*n.m.*), polishing-head ; polishing-lathe. (50)
politique (*n.f.*), policy : (51)
politique du conseil, policy of the board. (52)
polje [**polje** *pl.*] (Géol.) (*n.m.*), dolina ; sink-hole ; sink ; limestone sink ; swallow-hole ; swallow. (53)
pollux (*n.m.*) *ou* **pollucite** (*n.f.*) (Minéral), pollucite ; pollux. (54)
polonium (Chim.) (*n.m.*), polonium. (55)
polybasite (Minéral) (*n.f.*), polybasite. (56)
polychroïque (*adj.*), polychroic ; pleochroic. (57)
polychroïsme (*n.m.*), polychroism ; pleochroism. (58)
polycrase (Minéral) (*n.f.*), polycrase. (59)
polycylindrique (*adj.*), multicylinder ; multi-cylindered : (60)
moteur polycylindrique (*m.*), multicylinder engine. (61)
polyèdre *ou* **polyédrique** (*adj.*), polyhedral ; polyhedric ; polyhedrical ; polyhedrous : (62)
angle polyèdre (*m.*), polyhedral angle. (63)
polyèdre (*n.m.*), polyhedron. (64)
polygénique (Géol.) (*adj.*), polygenetic : (65)
roche polygénique (*f.*), polygenetic rock. (66)
polygonal, -e, -aux *ou* **polygone** (*adj.*), polygonal ; polygonous. (67)
polygone (*n.m.*), polygon. (68)
polygone concave (*m.*), concave polygon. (69)
polygone convexe (*m.*), convex polygon. (70)
polygone curviligne (*m.*), curvilinear polygon. (71)
polygone des forces (*m.*), polygon of forces. (72)
polygone étoilé (*m.*), star polygon. (73)
polygone funiculaire (*m.*), funicular polygon ; link polygon. (74)
polygone régulier (*m.*), regular polygon. (75)
polyhalite (Minéral) (*n.f.*), polyhalite. (76)
polymignite (Minéral) (*n.f.*), polymignite. (77)
polymorphe (*adj.*), polymorphous ; polymorphic. (78)
polymorphisme (*n.m.*), polymorphism. (79)
polyphasé, -e (Élec.) (*adj.*), polyphase : (80)
courant polyphasé (*m.*), polyphase current. (81)
polyphote (Élec.) (*adj.*), polyphote ; polyphotal : (82)
régulateur polyphote (*m.*), polyphote regulator. (83)
polysilicate (Chim.) (*n.m.*), polysilicate. (84)
polysilicique (Chim.) (*adj.*), polysilicic. (85)
polysynthétique (*adj.*), polysynthetic. (86)
polyvalence (Chim.) (*n.f.*), polyvalence ; poly-valency. (87)
polyvalent, -e (*adj.*), polyvalent. (88)
polyxène (Minéral) (*n.m.*), polyxene ; native platina. (89)
pomme (bouton) (*n.f.*), knob. (90)
pomme (d'un arrosoir) (*n.f.*), rose (of a watering-can). (91)

pompage (action de pomper) (*n.m.*), pumping ; pumping up ; pumping out. (1)

pompage (Fonderie) (*n.m.*), feeding ; pumping ; churning. (2) See **pomper**.

pompe (*n.f.*), pump. (3)

pompe (Fonderie) (*n.f.*), feeding-rod ; feeder. (4) See **pomper** for example of use.

pompe à action directe (*f.*), direct-acting pump ; direct-action pump. (5)

pompe à action indirecte (*f.*), indirect-action pump. (6)

pompe à aéromoteur (*f.*), wind-pump ; aero--motor-pump. (7)

pompe à air (*f.*), air-pump. (8)

pompe à boue (*f.*), mud-pump ; sludge-pump ; sludger. (9)

pompe à bras (*f.*), hand-pump. (10)

pompe à chapelet (*f.*), chain-pump ; chapelet ; paternoster pump. (11)

pompe à chute libre (*f.*), free-fall pump ; free-delivery pump. (12)

pompe à courroie horizontale à engrenage (*f.*), geared horizontal belt-pump. (13)

pompe à diaphragme (*f.*), diaphragm-pump. (14)

pompe à double effet (*f.*), double-acting pump ; double-action pump. (15)

pompe à eau (*f.*), water-pump. (16)

pompe à engrenage (*f.*), geared pump ; gear-pump. (17)

pompe à feu (*f.*), fire-engine ; fire-pump. (18)

pompe à godets (*f.*), chain-pump ; chapelet ; paternoster pump. (19)

pompe à grande vitesse (*f.*), high-speed pump. (20)

pompe à impulsion de vapeur (*f.*), pulsometer ; pulsator. (21)

pompe à incendie (*f.*), fire-engine ; fire-pump. (22)

pompe à main (*f.*), hand-pump. (23)

pompe à maîtresse-tige (Mines) (*f.*), bull-pump. (24)

pompe à mouvement alternatif (*f.*), reciprocat--ing pump. (25)

pompe à mouvement continu (*f.*), continuous-action pump. (26)

pompe à piston (*f.*), piston-pump. (27)

pompe à piston creux (*f.*), hollow-piston pump. (28)

pompe à piston plein (*f.*), solid-piston pump. (29)

pompe à piston plongeur *ou* **pompe à plongeur** (*f.*), ram-pump ; plunger-pump ; force-pump ; forcing-pump. (30)

pompe à puits à pétrole (*f.*), oil-well pump. (31)

pompe à quadruple effet (*f.*), quadruple-acting pump. (32)

pompe à quatre corps et à engrenage (*f.*), four-throw geared pump. (33)

pompe à sable (*f.*), sand-pump ; sludge-pump ; sludger. (34)

pompe à sangle (*f.*), hydraulic-belt pump (endless woolen band). (35)

pompe à simple effet (*f.*), single-acting pump ; single-action pump. (36)

pompe à trois corps (*f.*), three-throw pump. (37)

pompe à vapeur (*f.*), steam-pump. (38)

pompe à vapeur horizontale (*f.*), horizontal steam-pump. (39)

pompe à vapeur verticale (*f.*), vertical steam-pump. (40)

pompe à vide (*f.*), vacuum-pump. (41)

pompe à volant (*f.*), hand fly-wheel pump. (42)

pompe alimentaire (*f.*), donkey-pump ; feed-pump ; boiler-feed pump. (43)

pompe aspirante (*f.*), suction-pump ; sucking pump ; aspiring pump ; aspirating pump. (44)

pompe aspirante et foulante *ou* **pompe aspirante et refoulante** (*f.*), lift and force pump. (45)

pompe aspirante et soulevante *ou* **pompe aspirante et élévatoire** (*f.*), bucket-pump. (46)

pompe centrifuge (*f.*), centrifugal pump. (47)

pompe centrifuge à haute pression (*f.*), high-pressure centrifugal pump. (48)

pompe chinoise (*f.*), Chinese pump ; China pump. (49)

pompe compound à condensation (*f.*), compound condensing pump. (50)

pompe d'avaleresse (Mines) (*f.*), sinking-pump. (51)

pompe d'avaleresse à vapeur (*f.*), steam sinking-pump. (52)

pompe d'épreuve (*f.*), test-pump. (53)

pompe d'épuisement (*f.*), draining-pump ; drainage-pump ; mine-pump. (54)

pompe d'incendie (*f.*), fire-engine ; fire-pump. (55)

pompe de circulation (*f.*), circulating-pump. (56)

pompe de compression (*f.*), compression-pump. (57)

pompe de Cornouailles (*f.*), Cornish pump. (58)

pompe de dragage *ou* **pompe de drague** (*f.*), dredging-pump. (59)

pompe de fonçage (Mines) (*f.*), sinking-pump. (60)

pompe de lubrification et sa tuyauterie (tour, etc.) (*f.*), pump and connections ; lubricating-pump and pipe-connections ; oil-pump and circulating-pipes. (61)

pompe de mine (*f.*), mine-pump ; mining-pump. (62)

pompe de puits à pétrole (*f.*), oil-well pump. (63)

pompe de purge (*f.*), drip-pump. (64)

pompe de secours (*f.*), emergency pump. (65)

pompe de sondage (*f.*), bore-hole pump. (66)

pompe demi-rotative (*f.*), semi-rotary pump. (67)

pompe différentielle (*f.*), differential pump. (68)

pompe duplex (*f.*), duplex pump. (69)

pompe électrique (*f.*), electric pump. (70)

pompe élévatoire (*f.*), lift-pump. (71)

pompe foulante (*f.*). Same as **pompe refoulante**.

pompe hydraulique à main (*f.*), hand hydraulic pump. (72)

pompe pouvant fonctionner noyée (*f.*), sub--merged pump ; subaqueous pump. (73)

pompe refoulante (*f.*), ram-pump ; force-pump ; forcing-pump ; plunger-pump. (74)

pompe refoulante à eau de savon avec sa tuyauterie (*f.*), sud-pump and connections. (75)

pompe refoulante horizontale (*f.*), horizontal ram-pump. (76)

pompe refoulante verticale (*f.*), vertical ram-pump. (77)

pompe rotative (*f.*), rotary pump. (78)

pompe rotative à ailettes *ou* **pompe rotative à palettes** (*f.*), vane-pump. (79)

pompe semi-rotative (*f.*), semi-rotary pump. (80)

pompe soulevante (*f.*), lift-pump. (81)

pompe spirale (*f.*), spiral pump. (82)

pompe suceuse (*f.*), suction-pump ; sucking pump ; aspiring pump ; aspirating pump. (83)

pompe sur brouette, avec, ou sans, bâche (*f.*), barrow-pump, with, or without, water-box. (1 *ou* 2)

pompe turbine (*f.*),   turbine-pump ; turbo-pump. (3)

pompe verticale à commande par courroie, sans engrenage (*f.*), non-geared vertical belt-pump ; gearless vertical belt-driven pump. (4)

pompe volante *ou* **pompe suspendue** (Mines) (*f.*), suspended pump ; slung pump. (5)

pompe volante à bras (*f.*), portable hand-pump. (6)

pomper (*v.t.*), to pump ; to pump up ; to pump out : (7)

pomper l'air d'un récipient, de l'eau d'un puits, to pump the air out of a receiver, water out of (*or* from) a well. (8 *ou* 9)

pomper le sable, to pump up sand. (10)

pomper (Fonderie) (*v.t.*), to feed ; to pump ; to churn : (11)

les tassements sont combattus en pompant dans chaque évent avec une pompe (*m. pl.*), draws are combatted by feeding in each riser with a feeding-rod. (12)

pompier (exhaure) (pers.) (*n.m.*), pumpman. (13)

pompier (incendie) (pers.) (*n.m.*), fireman. (14)

ponçage (*n.m.*), pumicing. (15)

ponçage (avec le papier de verre) (*n.m.*), sand-papering. (16)

ponce (*n.f.*), pumice ; pumice-stone. (17)

ponceau (*n.m.*), culvert. (18)

poncelet (unité de force) (*n.m.*), poncelet. (19)

poncelet-heure [poncelets-heure *pl.*] (*n.m.*), poncelet-hour. (20)

poncer (*v.t.*), to pumice. (21)

poncer (avec le papier de verre) (*v.t.*), to sand-paper. (22)

ponceux, -euse (*adj.*), pumiceous ; pumicose. (23)

poncif (Fonderie) (*n.m.*), face-dust ; facing. (24)

pondérabilité (*n.f.*), ponderability : (25)

pondérabilité d'un gaz, ponderability of a gas. (26)

pondérable (*adj.*), ponderable : (27)

fluide pondérable (*m.*), ponderable fluid. (28)

poney [poneys *pl.*] *ou* **ponet** (*n.m.*), pony. (29)

poney de mine (*m.*), pit-pony. (30)

pont (*n.m.*), bridge. (31)

pont (d'un pont roulant) (*n.m.*), bridge (of a travelling crane). (32)

pont (d'un navire, d'une drague) (*n.m.*), deck (of a ship, of a dredge). (33)

pont (autel de four ou de foyer) (*n.m.*), fire-bridge ; bridge ; flame-bridge ; furnace-bridge ; bridge-wall ; fire-stop. (34 *ou* 35)

pont (sur le rompu) (tour) (*n.m.*), gap-piece ; gap-bridge ; bridge ; bridge-piece (lathe). (36)

pont (Élec.) (*n.m.*), bridge. (37)

pont à arches surbaissées (*m.*), bridge with diminished arches. (38)

pont à bascule (Ponts & Chaussées) (*m.*), bascule bridge ; balance bridge ; counterpoise bridge ; drawbridge. (39)

pont à bascule (machine à peser) (*m.*), weigh-bridge. (40)

pont à consoles (*m.*), cantilever bridge. (41)

pont à corde (Élec.) (*m.*), slide-bridge ; slide-wire bridge. (42)

pont à coulisse (*m.*), sliding bridge. (43)

pont à poutres (*m.*), girder bridge. (44)

pont à poutres à consoles (*m.*), cantilever bridge. (45)

pont à poutres armées (*m.*), truss-bridge. (46)

pont à tablier inférieur (*m.*), through bridge ; bottom-road bridge ; overgrade bridge. (47)

pont à tablier supérieur (*m.*), deck-bridge ; top-road bridge ; undergrade bridge. (48)

pont à travées égales (*m.*), bridge with equal bays. (49)

pont biais (*m.*), skew bridge ; oblique bridge. (50)

pont bow-string (*m.*), bowstring bridge ; tension bridge. (51)

pont cantilever [ponts cantilever *pl.*] (*m.*), canti-lever bridge. (52)

pont d'induction (Élec.) (*m.*), induction-bridge. (53)

pont de bateaux (*m.*), bridge of boats. (54)

pont de chargement (four métallurgique) (*m.*), charging-platform ; charging-scaffold ; plat-form ; landing. (55)

pont de chemin de fer (*m.*), railway-bridge. (56)

pont de chemin de fer à une travée et à voie normale (*m.*), single-span railway bridge for standard-gauge track. (57)

pont de chevalets (*m.*), trestle-bridge. (58)

pont de coulée (*m.*), ladle-crane ; hot-metal crane. (59)

pont de manutention pour parcs à matières (*m.*), stock-yard crane. (60)

pont de — mètres d'ouverture (*m.*), bridge of — metres span. (61)

pont de service (*m.*), service-bridge. (62)

pont de travail (Mines) (*m.*), stull. (63)

pont de Wheatstone (Élec.) (*m.*), Wheatstone's bridge ; Wheatstone's balance ; electrical bridge ; electric balance. (64)

pont dormant (ponceau) (*m.*), culvert. (65)

pont dormant (passerelle) (*m.*), foot-bridge. (66)

pont droit (opp. à pont biais) (*m.*), straight bridge. (67)

pont en arc (*m.*), arched-beam bridge. (68)

pont en béton armé (*m.*), reinforced-concrete bridge. (69)

pont en bois (*m.*), wooden bridge ; timber bridge. (70)

pont en charpente (*m.*), frame bridge. (71)

pont en encorbellement (*m.*), cantilever bridge. (72)

pont en fer (*m.*), iron bridge. (73)

pont en maçonnerie (*m.*), masonry bridge. (74)

pont en pierre (*m.*), stone bridge. (75)

pont en poutres (*m.*), girder bridge. (76)

pont en treillis (*m.*), lattice bridge. (77)

pont flottant (*m.*), floating bridge. (78)

pont-grue [ponts-grues *pl.*] (*n.m.*), bridge-crane. (79)

pont levant (*m.*), lifting-bridge ; lift-bridge ; drawbridge. (80)

pont métallique (*m.*), iron or steel bridge. (81)

pont mobile (*m.*), movable bridge. (82)

pont naturel (Géogr. phys.) (*m.*), natural bridge. (83)

pont par-dessous (Ch. de f.) (*m.*), bridge under. (84)

pont par-dessus (Ch. de f.) (*m.*), bridge over. (85)

pont pour routes (*m.*) *ou* **pont-route** [ponts-routes *pl.*] (*n.m.*), road-bridge ; highway-bridge. (86)

pont roulant (Ponts & Chaussées) (*m.*), roller-bridge ; rolling bridge ; running bridge ; drawbridge. (87)

pont roulant (grue roulante aérienne) (*m.*), travelling crane ; traveling crane ; traveller ; overhead travelling crane ; overhead crane ; overhead traveller. (1) See also **grue roulante.**

pont roulant (chariot transbordeur) (Ch. de f.) (*m.*), traverse-table ; traverse-carriage ; traverser ; slide-rail ; railway-slide. (2)

pont roulant à commande électrique (*m.*), electric overhead traveller. (3)

pont roulant à électro-aimant de levage *ou* **pont roulant à électro-aimant porteur** (*m.*), magnet-crane. (4)

pont roulant à main, à 2 treuils ; puissance — kilos, portée — mètres, hauteur de levage — mètres (*m.*), hand-power travelling crane, with two crabs ; to lift — kilos, span — metres, lift — metres. (5)

pont roulant complètement électrique (*m.*) all-electric travelling crane. (6)

pont roulant d'atelier (*m.*), shop-traveller. (7)

pont roulant électrique (*m.*), electric travelling crane. (8)

pont roulant semi-électrique (*m.*), semi-electric travelling crane. (9)

pont sur chevalets *ou* **pont sur tréteaux** (*m.*), trestle-bridge. (10)

pont suspendu (*m.*), suspension bridge ; hanging bridge. (11)

pont suspendu à chaînes (*m.*), chain bridge. (12)

pont tournant (*m.*), turning bridge ; turn-bridge ; swing-bridge ; swivel-bridge ; pivot-bridge ; drawbridge. (13)

pont transbordeur (*m.*), aerial ferry. (14)

pont tubulaire (*m.*), tubular bridge ; box-girder bridge. (15)

pont viaduc (*m.*), viaduct. (16)

pont volant (*m.*), flying bridge. (17)

ponte (d'un filon) (*n.f.*), wall (of a lode). (18)

ponton (*n.m.*), pontoon : (19)
le ponton d'une drague, the pontoon of a dredge. (20)

population (*n.f.*), population. (21)

porcelaine (*n.f.*), porcelain. (22)

porcelanique (*adj.*), porcelanic ; porcelainous. (23)

pore (*n.m.*), pore : (24)
les pores d'une roche, de la houille, the pores of a rock, of coal. (25 *ou* 26)

poreux, -euse (*adj.*), porous : (27)
couche poreuse (*f.*), porous bed. (28)

porion (Mines) (pers.) (*n.m.*), overman. (29)

porosité (*n.f.*), porosity : (30)
porosité de la pierre ponce, porosity of pumice-stone. (31)

porpezite (Minéral) (*n.f.*), porpesite ; palladium-gold. (32)

porphyre (Pétrol.) (*n.m.*), porphyry. (33)

porphyre microgranulitique (*m.*), microgranulitic porphyry. (34)

porphyre pétrosiliceux (*m.*), petrosiliceous porphyry. (35)

porphyre quartzifère (*m.*), quartz-porphyry. (36)

porphyre rouge antique (*m.*), porfido rosso antico. (37)

porphyre syénitique (*m.*), syenitic porphyry. (38)

porphyre vert antique (*m.*), verd-antique ; Oriental verd-antique. (39)

porphyrique *ou* **porphyroïde** (*adj.*), porphyritic ; porphyric ; porphyritical ; porphyraceous : (40)
roches porphyroïdes (*f.pl.*), porphyritic rocks. (41)

porphyrisation (*n.f.*), porphyrization ; pulveriza-tion ; comminution. (42)

porphyriser (*v.t.*), to porphyrize ; to pulverize ; to comminute. (43)

porphyrite (Pétrol.) (*n.f.*), porphyrite. (44)

porphyrite micacée (*f.*), mica-porphyrite. (45)

porphyroïde (*adj.*). Same as **porphyrique.**

porphyroïde (Pétrol.) (*n.m.*), porphyroid. (46)

port (*n.m.*), port ; harbour : (47)
port d'embarquement, port of shipment ; shipping-port. (48)
port de mer *ou* port maritime, seaport ; harbour. (49)

port (transport) (*n.m.*), carriage ; carrying : (50)
port dû, carriage forward : (51)
un envoi est fait franco, ou en port dû, suivant que le prix est payé d'avance par l'expéditeur, ou à l'arrivée par le destina-taire, goods are sent carriage free (*or* paid), or carriage forward, according as the money is paid in advance by the sender, or on arrival by the consignee. (52)
port payé, carriage paid. (53)

portage (transport) (*n.m.*), carriage ; carrying ; conveyance : (54)
portage à dos, carriage by persons. (55)
portage à dos du charbon, coal-backing. (56)

portatif, -ive (*adj.*), portable : (57)
machine à percer portative (*f.*), portable drilling-machine. (58)

porte (*n.f.*), door ; gate. (59) See also **portes.**

porte à charnières (*f.*), hinged door. (60)

porte à coulisse (*f.*), sliding door. (61)

porte-à-faux (*n.m.*), overhang : (62)
comme la longueur du truck est de 6 m. et la longueur de la caisse de 8 m., il existe un porte-à-faux considérable, as the length of the truck is 6 metres and the length of the body 8 m., there is a considerable overhang. (63)
bâti (d'une machine à raboter) d'une longueur suffisante pour réduire à son minimum le porte-à-faux de la table (*m.*), bed (of a planing-machine) of sufficient length to reduce the overhang of the platen to a minimum ; bed of sufficient length to ensure ample bearing of the table in all positions. (64)

porte-à-faux (en) *ou* **porte à faux (en)**, over-hanging ; overhung ; unsupported ; sus-pended : (65)
manivelle en porte-à-faux (*f.*), overhanging crank ; overhung crank. (66)
joints de rails placés en porte-à-faux, c'est-à-dire entre deux traverses (*m.pl.*), un-supported rail-joints (*or* suspended rail-joints), that is to say, joints between two sleepers. (67)

porte à guichet (Aérage de mines) (*f.*), regulat-ing-door ; gauge-door ; box-regulator. (68)

porte à rabat (*f.*), flap-door. (69)

porte automatique (*f.*), self-closing door ; auto-matic door. (70)

porte-balai [**porte-balais** *pl.*] (Élec.) (*n.m.*), brush-holder. (71)

Nota.: *Le pluriel de* **porte-balai**, *et des composés similaires, peut s'écrire sans l* s; *ainsi,* **porte-balais** *ou* **porte-balai** ; **porte-bougies** *ou* **porte-bougie,** *etc.*

porte-bougie [**porte-bougies** *pl.*] (*n.m.*), candle-holder. (72)

porte-chaîne [**porte-chaînes** *pl.*] (Arpent.) (pers.) (*n.m.*), chain-bearer. (73)

porte-charbon [porte-charbons *pl.*] (Élec.) (*n.m.*), carbon-holder. (1)

porte-cliché [porte-clichés *pl.*] (agrandisseur) (*n.m.*), negative-carrier. (2)

porte-coussinets [porte-coussinets *pl.*] (d'une machine à tarauder les tiges ou les tubes) (*n.m.*), diehead, screwing-head, screwing-chuck, (of a machine for screwing bolts or tubes). (3)

porte d'aérage (Mines) (*f.*), air-door; ventilat--ing-door; weather-door; trap-door. (4)

porte d'agrafe *ou simplement* **porte** (*n.f.*) (d'agrafe et porte), eye (of hook and eye). (5)

porte d'amont (d'une écluse) (*f.*), head-gate, crown-gate, (of a canal-lock). (6)

porte d'aval (d'une écluse) (*f.*), tail-gate; aft-gate. (7)

porte d'écluse *ou* **porte d'écluse de canal** (*f.*), lock-gate; canal lock-gate. (8)

porte de boîte à fumée (locomotive) (*f.*), smoke-arch door; smoke-box door. (9)

porte de cage (Mines) (*f.*), cage-gate. (10)

porte de chargement (four) (*f.*), charging-door. (11)

porte de foyer (*f.*), fire-door. (12)

porte de secours (*f.*), emergency door. (13)

porte de sûreté (*f.*), safety-door. (14)

porte de travail (d'un four) (*f.*), working-door (of a furnace). (15)

porte de trémie à minerai (*f.*), ore-bin gate. (16)

porte de vidange (d'une chaudière à vapeur) (*f.*), mud-plug (of a steam-boiler). (17)

porte de visite (*f.*), inspection-door; inspection-cover. (18)

porte du cendrier (*f.*), ash-pan dump; hopper-door. (19)

porte-écran [porte-écrans *pl.*] (Opt.) (*n.m.*), screen-holder. (20)

porte-électrode [porte-électrodes *pl.*] (four électrique) (*n.m.*), electrode-holder; ferrule. (21)

porte fausse *ou* **porte feinte** (*f.*), false door; blank door; blind door. (22)

porte-filière [porte-filières *pl.*] (*n.m.*), screw-stock; screwing-stock; die-stock; stock; screw-plate. (23) See **filière** for varieties.

porte-film [porte-films *pl.*] (portefeuille métal--lique pour films rigides) (*n.m.*), cut-film sheath. (24)

porte-fleuret [porte-fleurets *pl.*] (pour perfora--trices) (Mines) (*n.m.*), drill-holder (for rock-drills). (25)

porte-foret [porte-forets *pl.*] (*n.m.*), drill-holder; drill-stock; bit-holder; drill. (26)

porte-foret à archet (*m.*), bow-drill; piercer. (27)

porte-foret à conscience *ou* **porte-foret à engrenages à plaque conscience** (*m.*), breast-drill; breast-brace; breast drill-brace. (28)

porte-foret à main *ou* **porte-foret à engrenages** (*m.*), hand-drill. (29)

porte-foret ordinaire, garni de 6 forets (*m.*), drill-box, with 6 drills. (30)

porte-goutte [porte-gouttes *pl.*] (d'un fer à souder) (*n.m.*), bit, copper-bit, (of a soldering-iron). (31)

porte-lame [porte-lames *pl.*] (*n.m.*), cutter-bar. (32)

porte-lame pour aléser (*m.*), cutter-bar for boring; boring-bar. (33)

porte-lampe [porte-lampes *pl.*] (*n.m.*), lamp-holder. (34)

porte-lime [porte-limes *pl.*] (*n.m.*), file-carrier. (35)

porte-loupe *ou* **porte-loupes** [porte-loupe *ou* porte-loupes *pl.*] (*n.m.*), lens-holder. (36)

porte-lunette [porte-lunettes *pl.*] (d'une machine à tarauder les tiges ou les tubes) (*n.m.*), diehead, screwing-head, screwing-chuck, (of a machine for screwing bolts or tubes). (37)

porte-lunette revolver (*m.*), revolving diehead; capstan diehead. (38)

porte-mâchoire [porte-mâchoires *pl.*] (*n.m.*), jaw-holder. (39)

porte-matrice [porte-matrices *pl.*] (de poinçon--neuse, etc.) (*n.m.*), die-holder (of punching-machine, etc.). (40)

porte-mèche [porte-mèches *pl.*] (porte-foret) (Méc.) (*n.m.*). Same as **porte-foret.**

porte-mèche (lampes) (*n.m.*), wick-tube. (41)

porte-meule [porte-meules *pl.*] (*n.m.*), wheel-head; emery-wheel attachment. (42)

porte-mire [porte-mires *pl.*] (pers.) (*n.m.*), rod--man; rodsman; staff-holder. (43)

porte mobile (*f.*), removable door. (44)

porte-noyau [porte-noyaux *pl.*] (Fonderie) (*n.m.*). Same as **portée de noyau.**

porte-objectif [porte-objectifs *pl.*] (Photogr.) (*n.m.*), lens-front. (45) See syn. **avant porte-objectif** for a full specification.

porte-objet [porte-objets *pl.*] (Microscopie) (*n.m.*), slide; slider. (46)

porte-outil [porte-outils *pl.*] (*n.m.*), tool-holder. (47)

porte-outil (de machine à raboter) (*n.m.*), head (of planing-machine). (48)

porte-outil à base pivotante (*m.*), tool-holder on pivoting base. (49)

porte-outil à double coulisse (*m.*), compound tool-holder. (50)

porte-outil à tourelle carrée (tour) (*m.*). Same as **porte-outil revolver à quatre faces** (lathe).

porte-outil à tronçonner (tour) (*m.*), cutting-off tool holder. (51)

porte-outil articulé (*m.*), jointed tool-holder. (52)

porte-outil de reproduction (tour) (*m.*), forming-tool holder. (53)

porte-outil revolver (tour, etc.) (*m.*), capstan; capstan-head; turret; turret-head; monitor; revolving tool-holder. (54)

porte-outil revolver à quatre faces à diviseur et à orientation quelconque (*m.*), four-tool turret with base graduated in degrees so that tools may be set to any angle; four-stud tool-post with graduated base capable of being swivelled to any angle. (55)

porte-plaque(s) [porte-plaques *pl.*] (Photogr.) (*n.m.*), plate-holder; dark slide. (56 ou 57)

porte-poinçon [porte-poinçons *pl.*] (*n.m.*), punch-holder. (58)

porte roulante (*f.*), rolling door. (59)

porte-sabot [porte-sabots *pl.*] (*n.m.*), brake-head; brake-block. (60)

porte-scie [porte-scies *pl.*] (*n.m.*), saw-frame; saw-gate. (61)

porte-scie à découper *ou* **porte-scie à repercer** (*m.*), fret-saw frame; fret-saw. (62)

porte-scie à glissière (*m.*), adjustable hack-saw frame. (63)

porte-scie à métaux (*m.*), hack-saw; hack-saw frame. (64)

porte-tampon [porte-tampons *pl.*] (Fonderie) (*n.m.*), bott-stick. (65)

porte-taraud [porte-tarauds *pl.*] (*n.m.*), tap-holder. (1)

porte-tube [porte-tubes *pl.*] (Appareil de labora-toire) (*n.m.*), tube-holder. (2)

porte va-et-vient (*f.*), swing-door; swinging door. (3)

porte-vent [porte-vent *pl.*] (de tuyère) (*n.m.*), blowpipe; belly-pipe. (4)

porte vitrée (*f.*), glazed door; sash-door. (5)

portée (distance entre deux points d'appui) (*n.f.*), span: (6)
portée d'une voûte, d'un pont, span of an arch, of a bridge. (7 *ou* 8)
pont avec — mètres de portée (*m.*), bridge with a — -metre span. (9)
portée d'une poutre, d'un câble aérien, span of a girder, of an aerial ropeway. (10 *ou* 11)

portée (d'une grue) (*n.f.*), radius (of a crane). (12)

portée (d'une grue roulante, d'un pont roulant) (*n.f.*), span (of an overhead travelling crane). (13)

portée (d'une lunette) (*n.f.*), range (of a telescope). (14)

portée (saillie) (*n.f.*), projection; set-off; offset; scarcement: (15)
la portée d'une pierre, d'une gouttière au delà d'un mur, the projection of a stone, of a gutter beyond a wall. (16 *ou* 17)

portée (partie d'un essieu ou d'une fusée d'arbre en contact avec le coussinet) (*n.f.*), bearing: (18)
un arbre trempé ne sera pas rayé par un corps étranger s'introduisant dans la portée, a hardened shaft will not be scratched by a foreign substance entering the bearing. (19)
portée une fois et demie *ou* portée 1 f. ½ (palier), length of brasses equal to 1½ diameters; bearings 1½ diameters in length (plummer-block). (20)

portée (surélévation sur laquelle s'appuie un écrou de serrage, remplaçant une rondelle) (Méc.) (*n.f.*), boss. (21)

portée (*n.f.*) *ou* **portée de noyau** *ou* **portée de renmoulage** (Fonderie), print; core-print; seating; bearing; steady-pin: (22)
les modèles de fonderie s'établissent avec des portées, dont les empreintes dans les moules servent d'assise ou d'encastrement à des pièces rapportées appelées noyaux (*m.pl.*), foundry-patterns are made with prints, of which the impressions in the moulds serve as a bed or housing for loose pieces called cores. (23)

portée (volume d'eau débité dans un temps donné) (Hydraul.) (*n.f.*), discharge; flow. (24)

portée (Arpent.) (*n.f.*), chain (a land-measure of length, equal to the length of the chain employed). (25)

portée (à) *ou* **à la portée de la main**, at hand; within reach: (26)
levier placé bien à la portée de l'ouvrier (*m.*), lever placed well within reach (*or* within easy reach) of the workman; conveniently placed lever. (27)

portée de calage (d'un essieu) (*f.*), wheel-seat; axle-seat; wheel-fit; axle-fit. (28)

portée de la fusée (d'essieu) (*f.*), journal-bear-ing; journal-surface, (of axle). (29)

portée de la vue (*f.*), reach, range, of vision. (30)

porter (*v.t.*), to carry; to bear: (31)
porter un fardeau, des marchandises à un magasin, to carry a burden, goods to a storehouse. (32 *ou* 33)
tréteaux qui portent une poutre (*m.pl.*), trestles which bear a beam. (34)

porter (*v.i.*), to bear; to rest; to be carried: (35)
châssis qui porte sur des rouleaux (*m.*), frame which bears on rollers. (36)
bâtiment qui porte sur des colonnes (*m.*), building which rests (*or* is carried) on pillars. (37)

porter à faux, to overhang; to be unsupported; to be without support. (38) Cf. **porte-à-faux** (en).

portes battantes (*f.pl.*), folding doors. (39)

portes doubles (Aérage de mines) (*f.pl.*), double doors; set of doors; pair of doors. (40)

portes enclanchées *ou* **portes solidaires** (Aérage de mines) (*f.pl.*), interlocked air-doors. (41)

porteur (pers.) (*n.m.*), carrier. (42)

porteur (Boisage des puits de mines) (*n.m.*), bearer. (43)

porteur d'eau (*m.*), water-carrier. (44)

porteur de lampes (*m.*), lamp-carrier; light-carrier. (45)

porteuse (d'un cadre porteur) (Boisage des galeries de mine) (*n.f.*), bearer; bearer-crown; carrying-crown. (46)

portier (pers.) (*n.m.*), gatekeeper. (47)

portier (ouvrier à la manœuvre des portes d'aérage) (Mines) (*n.m.*), trapper; door-tender. (48)

portion (*n.f.*), portion; part: (49)
portions du gîte reconnues, proved parts of the deposit. (50)

portique (charpente en forme de portique) (*n.m.*), gantry; gantree; gauntry; gauntree: (51)
grue à portique (*f.*), gantry-crane. (52)

portlandien, -enne (Géol.) (*adj.*), Portlandian. (53)

portlandstone (*n.m.*), Portland stone. (54)

posage (*n.m.*), laying; fixing; setting: (55)
posage des conduites, des câbles, laying pipes, cables. (56 *ou* 57)

pose (*n.f.*), laying; fixing; setting: (58)
pose d'un câble électrique, laying an electric cable. (59)
pose d'une serrure, d'une gâche, fixing a lock, a keeper. (60 *ou* 61)
pose de bornes, setting landmarks (*or* boundaries). (62)
pose de bornes-signaux (Levé de plans), monumenting. (63)
pose de canalisations, laying pipe-lines. (64)
pose de fils, wiring; running wires. (65)
pose de rails, rail-laying; laying rails. (66)
pose de tuyaux, pipe-laying. (67)
pose de verres, glazing; setting glass. (68)
pose de voie *ou* pose des voies (Ch. de f.), plate-laying; track-laying. (69)
pose fixe (Ch. de f.), laying a permanent way. (70)
pose volante (Ch. de f.), laying a train-road, a service-track. (71)

pose (exposition) (Photogr.) (*n.f.*), exposure: (72)
bobine de — poses (*f.*), spool of — exposures. (73)

pose (temps, durée, de pose) (Photogr.) (*n.f.*), exposure; time exposure: (74)

une pose de plusieurs secondes, an exposure of several seconds. (1)

pose, demi-pose, et l'instantané *ou* pose en un et deux temps, et l'instantané, time, bulb, and instantaneous exposures. (2) See **obturateur toujours armé.**

pose exagérée, overexposure. (3)

pose insuffisante, underexposure. (4)

pose-mètre [pose-mètres *pl.*] (Photogr.) (*n.m.*), exposure-meter. (5)

poser (*v.t.*).   See examples :   (6)

poser des bornes-signaux sur (Levé de plans), to monument : (7)

poser des bornes-signaux sur la méridienne, to monument the meridian.   (8)

poser des fils dans une maison pour éclairage électrique,   to wire a house for electric lighting.   (9)

poser des lignes aériennes (Télégr., Téléph., etc.), to run overhead lines.   (10)

poser des planches de champ, to place boards edgeways ; to set boards edgewise.   (11)

poser des poteaux télégraphiques le long d'une ligne de chemin de fer,   to set, to plant, telegraph-posts along a railway-line.   (12)

poser des rails, to lay rails ; to lay " steel." (13)

poser des tuiles sur le comble d'un bâtiment, to tile the roof of a building.   (14)

poser des tuyaux (*ou* des conduites),   to lay pipes.   (15)

poser les fondements d'une maison,   to lay the foundations of a house.   (16)

poser un rayon, to put up a shelf.   (17)

poser un rivet, to set a rivet.   (18)

poser une canalisation,   to lay, to lay down, a pipe-line.   (19)

poser une pierre dans le sens de son lit de carrière,   to set a stone in the sense of its natural bed.   (20)

poser une pierre en délit, to set a stone bed out.   (21)

poser une plaque photographique, to expose a photographic plate.   (22)

poser une serrure, to fix a lock.   (23)

poser une voie ferrée, to lay a railway-track. (24)

poser (se) (*v.r.*),   to be laid ; to be placed ; to be set ; to be fixed :   (25)

les rails de chemin de fer ne se posent pas d'aplomb, mais sont inclinés de 1/20e vers l'intérieur de la voie (*m.pl.*),   railway-rails are not laid perpendicularly, but are canted $\frac{1}{20}$th inwardly to the road.   (26)

serrure qui se pose sur une porte (*f.*),   lock which is fixed on a door.   (27)

poseur (*n.m.*) *ou* **poseur de voie** (Ch. de f.) (pers.), plate-layer ; track-layer.   (28)

poseur de lignes (Télégr., etc.) (*m.*),   linesman ; lineman.   (29)

positif, -ive (*adj.*), positive.   (30)

positif (transparence) (Photogr.) (*n.m.*), positive ; slide.   (31)

positif (épreuve) (Photogr.) (*n.m.*),   positive ; print.   (32)

positif pour la projection (*m.*),   lantern-slide. (33)

positif pour la stéréoscopie *ou* **positif stéréoscopique** (*m.*), stereoscopic slide.   (34)

positif sur verre (*m.*),   positive on glass.   (35)

position (situation d'une chose) (*n.f.*),   position ; situation :   (36)

position moyenne du tiroir,   mid-position of slide-valve.   (37)

position (emploi) (*n.f.*),   position ; post ; situa-tion ; place ; employment ; berth.   (38)

positivement (*adv.*), positively.   (39)

posséder (*v.t.*),   to possess ; to own ;   to have ; to hold :   (40)

posséder une maison, de grands biens, to possess, to own, a house, great wealth. (41 *ou* 42)

posséder des particularités bien définies, to have well-marked peculiarities.   (43)

posséder son profil d'équilibre *ou* **posséder sa courbe de lit** *ou* **posséder sa pente-limite** (Géogr. phys.), to be at grade :   (44)

rivière qui possède son profil d'équilibre (*f.*), river which is at grade.   (45)

possesseur (pers.)   (*n.m.*),   owner ;   holder ; possessor.   (46)

possession (*n.f.*), owning ; possession ; holding. (47)

possibilité (*n.f.*),   possibility.   (48)

possible (*adj.*),   possible :   (49)

minerai possible (*m.*), possible ore.   (50)

post-glaciaire (Géol.) (*adj.*), postglacial :   (51)

alluvions post-glaciaires (*f.pl.*),   postglacial alluvia.   (52)

postpliocène (*adj.*), posttertiaire (*adj.*).   See below.

poste (station) (*n.m.*), station ; post.   (53)

poste (durée du service actif d'un mineur) (*n.m.*), shift.   (54)

poste (équipe ; groupe d'ouvriers) (*n.m.*), shift ; gang ; crew ; spell.   (55) See also **équipe.**

poste au charbon (opp. à poste d'entretien) (*m.*), working-shift.   (56)

poste d'entretien (Houillères) (*m.*),   repairing-shift.   (57)

poste d'explosion (Mines) (*m.*), firing-station. (58)

poste d'extraction (Mines) (*m.*),   drawing-gang. (59)

poste d'incendie *ou* **poste de pompiers** *ou* **poste d'eau** (*m.*),   fire-station.   (60)

poste de 8 heures (*m.*), 8-hour shift.   (61)

poste de jour (*m.*), day-shift.   (62)

poste de nuit (*m.*), night-shift.   (63)

poste de rallumage souterrain (*m.*),   underground lighting-station.   (64)

poste de sauvetage *ou* **poste de secours** (*m.*), rescue-station ; ambulance-station.   (65)

poste entrant (*m.*), oncoming-shift.   (66)

poste sortant (*m.*), outgoing-shift.   (67)

poste télégraphique (*m.*), telegraph-station.   (68)

poste téléphonique (*m.*), telephone-exchange. (69)

post-glaciaire (*adj.*).   See above.

postillon (clapet sphérique) (*n.m.*), ball valve. (70)

postpliocène (Géol.) (*adj.*),   Post-Pliocene.   (71)

posttertiaire (Géol.) (*adj.*),   Post-Tertiary.   (72)

pot (*n.m.*),   pot ; jar :   (73)

un pot de terre, an earthenware pot.   (74)

pot (Métall.) (*n.m.*),   pot ; crucible ; melting-pot ; melter.   (75)

pot (aven) (Géogr. phys.) (*n.m.*), limestone cave ; limestone cavern.   (76)

pot à colle (*m.*), glue-pot.   (77)

pot à suif (*m.*), tallow-pot.   (78)

pot d'échappement (*m.*), silencer ; muffler.   (79)

pot de pompe (*m.*), bucket.   (80)

potabilité (*n.f.*),   drinkableness ; drinkability ; potability.   (81)

potable (*adj.*), drinkable ; potable :   (82)

eau potable (*f.*), drinkable water.   **(83)**

potasse (Chim.) (*n.f.*), potash. (1)

potasse caustique (*f.*), caustic potash. (2)

potassique (Chim.) (*adj.*), potassic ; potash (*used as adj.*) ; potassium (*used as adj.*) : (3)

sel potassique (*m.*), potash salt ; potassium salt. (4)

potassium (Chim.) (*n.m.*), potassium. (5)

poteau (*n.m.*), post ; pole. (6)

poteau armé (Télégr., etc.) (*m.*), trussed pole. (7)

poteau battant (d'une porte) (*m.*), shutting-post, striking-post, (of a door). (8)

poteau busqué *ou* poteau battant (d'une porte d'écluse) (*m.*), mitre-post, meeting-post, (of a lock-gate). (9)

poteau cornier *ou* poteau d'angle (*m.*), corner-post ; corner-pillar. (10)

poteau d'ancrage (*m.*), anchor-post. (11)

poteau d'arrêt (Télégr., etc.) (*m.*), terminal pole. (12)

poteau d'huisserie (*m.*), jamb-post ; jamb ; door-post ; post. (13)

poteau de bornage (*m.*), boundary-post. (14)

poteau de ligne (Télégr., etc.) (*m.*), line-pole. (15)

poteau de vanne (*m.*), staple-post of sluice-gate. (16)

poteau indicateur (*m.*), sign-post ; guide-post ; finger-post. (17)

poteau indicateur de changement de pente (Ch. de f.) (*m.*), gradient-post ; grade-post ; indicator. (18)

poteau télégraphique (*m.*), telegraph-pole ; telegraph-post. (19)

poteau téléphonique (*m.*), telephone-pole. (20)

poteau tourillon (d'une porte d'écluse) (*m.*), heel-post, quoin-post, hanging-post, (of a lock-gate). (21)

poteau tubulaire (Télégr., etc.) (*m.*), tubular pole. (22)

poteaux couplés (Télégr., etc.) (*m.pl.*), H pole. (23)

poteaux dans les angles pour relier les cadres (Boisage de puits de mines) (*m.pl.*), studdles. (24)

potée (terre de fonderie) (*n.f.*), loam. (25)

potée d'émeri (*f.*), flour-emery ; flour of emery. (26)

potée d'étain (*f.*), putty-powder ; tin-putty ; putty ; jeweler putty. (27)

potelle (Boisage de galeries de mine) (*n.f.*), hitch ; post-hole : (28)

le pied de la butte est assujetti dans une potelle creusée préalablement dans le sol, the foot of the post is fixed in a hitch previously dug in the ground. (29)

potence (*n.f.*), bracket ; angle-bracket. (30)

potentiel, -elle (*adj.*), potential : (31)

énergie potentielle (*f.*), potential energy. (32)

potentiel (Phys.) (*n.m.*), potential : (33)

potentiel d'un conducteur électrique, d'une substance explosive, potential of an electric conductor, of an explosive substance. (34 *ou* 35)

potentiel constant (Élec.) (*m.*), constant potential. (36)

potentiellement (*adv.*), potentially. (37)

potentiomètre (Élec.) (*n.m.*), potentiometer. (?8)

poterie (vaisselle) (*n.f.*), pottery ; earthenware. (39)

poterie (fabrique) (*n.f.*), pottery. (40)

poterie de grès (*f.*), stoneware. (41)

poterie de terre (*f.*), earthenware. (42)

potier (pers.) (*n.m.*), potter. (43)

potille (d'une vanne) (*n.f.*), staple-post (of a sluice-gate). (44)

potstone (pierre ollaire) (*n.m.*), potstone. (45)

pouce (mesure linéaire anglaise) (*n.m.*), inch =25·400 millimetres. (46)

pouce carré (mesure anglaise) (*m.*), square inch =6·4516 square centimetres. (47)

pouce cube (mesure anglaise) (*m.*), cubic inch =16·387 cubic centimetres. (48)

pouce de mineur *ou* pouce d'eau *ou* pouce de fontainier (*m.*), miners' inch ; water-inch ; inch of water. (49)

poucier (d'un loquet) (*n.m.*), thumb-piece (of a thumb-latch). (50)

poudingue (Pétrol.) (*n.m.*), pudding-stone ; plum pudding-stone. (51)

poudre (*n.f.*), powder. (52)

poudre (Explosif) (*n.f.*), powder ; gunpowder (53)

poudre à combustion lente (*f.*), slow-burning powder ; low powder. (54)

poudre à grains fins (Explosif) (*f.*), fine-grained powder. (55)

poudre à gros grains (Explosif) (*f.*), coarse-grained powder. (56)

poudre à souder (*f.*), welding-powder. (57)

poudre au nitrate (*f.*), nitrate powder. (58)

poudre au nitrate d'ammoniaque (*f.*), nitrate of ammonium powder. (59)

poudre au nitrate de soude (*f.*), nitrate of sodium powder. (60)

poudre au picrate (*f.*), picric powder. (61)

poudre comprimée (*f.*), compressed gunpowder. (62)

poudre de chasse (*f.*), sporting powder ; sporting gunpowder. (63)

poudre de fusion (*f.*), smelting-powder. (64)

poudre de guerre (*f.*), gunpowder. (65)

poudre de mine *ou* poudre de mines (*f.*), blasting-powder. (66)

poudre de sûreté (*f.*), miners' safety-explosive. (67)

poudre détonante (*f.*), detonating powder. (68)

poudre-éclair [poudres-éclairs *pl.*] (Photogr.) (*n.f.*), flash-powder. (69)

poudre fulminante (*f.*), fulminating powder ; fulminating compound. (70)

poudre géante (*f.*), giant-powder ; dynamite. (71)

poudre lente (*f.*), slow powder ; slow-burning powder ; low powder. (72)

poudre noire (*f.*), black powder ; black gun-powder. (73)

poudre noire de mine (*f.*), black blasting-powder. (74)

poudre sans fumée (*f.*), smokeless powder. (75)

poudre vive (*f.*), hasty powder. (76)

poudrerie (*n.f.*), powder-mill ; powder-works. (77)

poudreux, -euse (*adj.*), dusty : (78)

route poudreuse (*f.*), dusty road. (79)

poudrière (*n.f.*), powder-magazine ; powder-house. (80)

poulie (de transmission) (*n.f.*), pulley ; wheel : (81)

poulies en fonte en une ou deux pièces, plates ou bombées, tournées et alésées, cast-iron whole or split pulleys, flat or crowned on face, turned and bored. (82)

poulie (réa) (*n.f.*), pulley ; sheave ; pulley-sheave ; pulley-wheel. (83)

poulie (moufle) (*n.f.*), pulley; block; sheave; pulley-block; sheave-block; tackle-block. (1)

poulie à câble (*f.*), rope-wheel. (2)

poulie à chape croisée (*f.*), gin-block; rubbish-pulley; rubbish-wheel; whip-gin. (3)

poulie à chape simple (*f.*), funnel-block. (4)

poulie à chicanes (*f.*), spider-wheel; sprocket-wheel; rag-wheel. (5)

poulie à corde *ou simplement* **poulie** (*n.f.*), V pulley; grooved wheel; pulley for gut (*or* round) band; sheave. (6)

poulie à courroie (*f.*), belt-pulley; band-pulley; band-wheel; flat-strap pulley. (7)

poulie à croc (*f.*), hook block. (8)

poulie à émerillon (*f.*), swivel-block; block with swivel-hook. (9)

poulie à empreintes (*f.*), pocket-wheel; cupped chain-sheave; indented wheel. (10)

poulie à engrenage à vis sans fin (*f.*), worm-geared pulley-block. (11)

poulie à engrenage droit (*f.*), spur-geared pulley-block. (12)

poulie à fouet (*f.*), tail-block. (13)

poulie à gorge (*f.*), grooved pulley; scored pulley; sheave. (14)

poulie à gradins (*f.*), stepped pulley; stepped cone; cone-pulley; cone; speed-pulley; speed-cone; speed-rigger; speeder. (15)

poulie à joues (*f.*), flange-pulley. (16)

poulie à mortaise (opp. à **poulie d'assemblage**) (*f.*), mortise-block. (17)

poulie à noix (pour chaîne calibrée) (*f.*), sheave (for pitched chain). (18)

poulie à rebords (*f.*), flange-pulley. (19)

poulie à violon (*f.*), fiddle block; thick-and-thin block. (20)

poulie à vis (*f.*), screw-pulley. (21)

poulie à vis sans fin (*f.*), worm pulley-block. (22)

poulie bombée (*f.*), round-faced pulley; crowned pulley; crowning pulley; crown-face pulley; rounding pulley. (23)

poulie conductrice (*f.*), driving-pulley; drive-pulley; driver. (24)

poulie conduite (*f.*), driven pulley; follower. (25)

poulie coupée (*f.*), snatch-block; return-block. (26)

poulie d'adhérence (voie à câbles) (*f.*), clip-pulley; grip-pulley. (27)

poulie d'assemblage (*f.*), made block. (28)

poulie d'excentrique (*f.*), eccentric-sheave; eccentric-disc. (29)

poulie de chevalement *ou simplement* **poulie** (*n.f.*) (Mines), pit-head pulley; winding-pulley; hoisting-pulley; pulley. (30)

poulie de chèvre (*f.*), gin-pulley. (31)

poulie de commande *ou* **poulie d'attaque** (*f.*), driving-pulley; drive-pulley; driver. (32)

poulie de frein (*f.*), brake-wheel. (33)

poulie de friction (*f.*), friction-pulley. (34)

poulie de moufle *ou simplement* **poulie** (*n.f.*), sheave; pulley-sheave; pulley-wheel; pulley. (35)

poulie de puits (*f.*), well-pulley. (36)

poulie de renvoi *ou* **poulie de retour** (*f.*), return-pulley; return-wheel; turn-pulley; tail-sheave. (37)

poulie de renvoi à chape double articulée (Serrur.) (*f.*), lazy pulley. (38)

poulie de renvoi à plat (Serrur.) (*f.*), side-pulley. (39)

poulie de renvoi à pont (Serrur.) (*f.*), axle-pulley. (40)

poulie de renvoi avec tige filetée (Serrur.) (*f.*), screw-pulley. (41)

poulie de renvoi sur champ (Serrur.) (*f.*), upright pulley. (42)

poulie de tension (*f.*), tightening-pulley; bind-ing-pulley; tension-pulley; tension-roller; idler; belt-idler; idler-pulley; idle pulley; jockey-pulley; jockey-roller; jockey-wheel; jockey; belt-tightener. (43)

poulie de transmission (*f.*), pulley; driving-pulley: (44)

poulies et arbres de transmission, pulleys and shafting. (45)

poulie démontable (*f.*), split pulley; parting pulley. (46)

poulie différentielle (*f.*), differential pulley; differential pulley-block. (47) See **palan différentiel** for varieties.

poulie double (*f.*), double block. (48)

poulie en bois (*f.*), wood pulley. (49)

poulie en bois en deux pièces (*f.*), wood split pulley. (50)

poulie en deux pièces (*f.*), split pulley; parting pulley. (51)

poulie en dos d'âne (*f.*). Same as **poulie bombée**.

poulie en fer en deux pièces (*f.*), wrought-iron split pulley. (52)

poulie en fonte en une seule pièce, tournée et alésée (*f.*), cast-iron whole pulley, turned and bored. (53)

poulie en une seule pièce *ou* **poulie en une pièce** (*f.*), whole pulley; solid pulley. (54)

poulie étagée (*f.*). Same as **poulie à gradins**.

poulie ferrée (*f.*), iron block; iron pulley-block. (55)

poulie fixe (opp. à **poulie folle**) (Méc.) (*f.*), fast pulley; tight pulley; runner. (56)

poulie fixe (de palan) (*f.*), standing block. (57)

poulie folle (*f.*), loose pulley; dead-pulley. (58)

poulie-guide [**poulies-guides** *pl.*] (*n.f.*), guide-pulley; guide; idler; belt-idler; idler-pulley; idle pulley; idler-wheel; leading-on pulley; runner. (59)

poulie guide-fils (*f.*), wire-carrier. (60)

poulie menante *ou* **poulie motrice** (*f.*), driving-pulley; drive-pulley; driver. (61)

poulie menée (*f.*), driven pulley; follower. (62)

poulie mobile (de palan) (*f.*), running block; runner; hoisting-block. (63)

poulie mouflée (*f.*), sheave (one of two or more grooved pulleys in the same block). (64)

poulie plate (*f.*), flat-faced pulley; flat pulley; straight-faced pulley. (65)

poulie porte-lame [**poulies porte-lame** *pl.*] (scierie à ruban) (*f.*), band-saw pulley; saw-pulley; band-wheel. (66)

poulie pour câble métallique (*f.*), wire-rope pulley. (67)

poulie-support [**poulies-supports** *pl.*] (pour câble flottant) (Mines) (*n.f.*), bearing-up pulley; hat-roller (for endless rope above wagons). (68)

poulie triple (*f.*), threefold block; three-sheave block. (69)

poulie-volant [**poulies-volants** *pl.*] (*n.f.*), fly-pulley. (70)

poulierie (*n.f.*), pulley-works. (71)

poulies-cônes (*n.f.pl.*), belt-speeder. (72)

poulies folle et fixe (*f.pl.*), fast and loose pulleys. (73)

poulieur (pers.) (*n.m.*), pulley-maker. (74)

poupée (de tour ou de fraiseuse horizontale) (*n.f.*), head-stock, headstock, poppet-head, puppet-head, head, (of lathe or horizontal miller). (1 *ou* 2)

poupée à coussinets coniques à rattrapage de jeu concentrique (*f.*), headstock with conical bearings to take up wear. (3)

poupée à frictions (*f.*), friction-headstock. (4)

poupée à griffes (pour plateaux de tours) (*f.*), face-plate dog ; face-plate jaw. (5)

poupée à monopoulie *ou* **poupée à monopoulie et boîte de vitesses** (*f.*), all-gear head ; all-geared headstock ; single-pulley head ; constant-speed belt head. (6)

poupée à pompe (pour plateaux de tours) (*f.*), screw-dog (for lathe face-plate). (7)

poupée blindée (tour à commande par mono--poulie et boîte de vitesses) (*f.*), ironclad headstock ; headstock completely enclosed ; totally enclosed headstock. (8)

poupée courante (*f.*). Same as **poupée mobile**.

poupée de lœss (Géol.) (*f.*), loess-kindchen. (9)

poupée de tour (*f.*), lathe-head. (10)

poupée diviseur et sa contre-pointe (pour machine à fraiser) (*f.*), dividing-heads (for milling-machine). (11)

poupée fixe (*f.*), fast head-stock ; headstock ; fast-head ; fast poppet-head ; live-head ; head. (12)

poupée fixe à cône *ou simplement* **poupée à cône** (*f.*), headstock with cone-pulley. (13)

poupée fixe à double harnais d'engrenages (*f.*), double-geared fast head-stock. (14)

poupée fixe à poulies (*f.*), headstock with fast and loose pulleys. (15)

poupée mobile (*f.*), loose headstock ; loose-head ; tail-stock ; tailstock ; loose poppet-head ; sliding poppet ; sliding headstock ; tail-block ; deadhead ; back-head ; back-puppet ; foot-stock. (16) See **contre-pointe** for varieties.

pour cent, per cent (*without full stop*) ; per cent. (*with full stop*) ; per centum : (17)
tant pour cent, so much per cent. (18)

pour mille, per mille ; per mill ; per thousand. (19)

pourcentage (*n.m.*), percentage : (20)
pourcentage de carbone fixe, percentage of fixed carbon. (21)
pourcentage élevé d'humidité, heavy percent--age of moisture. (22)

pourri, -e (*adj.*), rotten ; decayed : (23)
filon pourri (*m.*), rotten lode. (24)

pourrir (*v.t. & v.i.*), to rot ; to decay. (25)

pourrir (se) (*v.r.*), to rot ; to decay. (26)

pourriture du bois (*f.*), rotting of the timber ; timber-rot ; decay of timber. (27)

poursuites (Droit) (*n.f.pl.*), prosecution ; pro--ceedings : (28)
poursuites en vertu de la loi réglementant l'exploitation des mines de charbon, pro--secution under the coal-mines regulation act. (29)

poursuivre (*v.t.*), to pursue ; to follow up ; to carry on : (30)
poursuivre une affaire, les conséquences d'une théorie, to pursue, to follow up, a matter, the consequences of a theory. (31 *ou* 32)
poursuivre des travaux de sondage pendant l'hiver, to carry on drilling operations during the winter. (33)

poursuivre *ou* **poursuivre en justice** (*v.t.*), to prosecute. (34)

pourtour (*n.m.*), periphery ; rim ; circuit : (35)
pourtour d'une roue, periphery, rim, of a wheel. (36)

pourvoir (*v.t.*), to provide ; to supply ; to furnish ; to equip ; to fit : (37)
pourvoir à la sécurité des ouvriers, to provide for the safety of the workmen. (38)
pourvoir à un déficit, to make good a deficiency. (39)
pourvoir une place de vivres, to supply, to furnish, a place with provisions. (40)
se pourvoir de vivres, to provide, to supply, oneself with provisions. (41)

poussage (Boisage des galeries de mine) (*n.m.*), forepoling ; spilling ; spiling ; poling ; lagging. (42)

poussage au bouclier (Boisage des galeries de mine) (*m.*), forepoling with breast-boards. (43)

poussage par palplanches (puits de mine) (*m.*), sinking by piling. (44)

pousse (Mines) (*n.f.*), black damp ; choke-damp ; stythe. (45)

pousse-pointe [**pousse-pointe** *ou* **pousse-pointes** *pl.*] (*n.m.*), nail-set ; brad-punch ; nail-punch ; set. (46)

poussée (effort) (*n.f.*), thrust : (47)
poussée d'une voûte, thrust of an arch. (48)
poussée au vide, bulging ; bellying. (49)
Cf. **pousser au vide**.
poussée au vide (Génie civil), thrust passing outside material : (50)
éviter la poussée au vide (dans la construction des voûtes biaises, par exemple), to avoid the thrust passing outside the material (in the building of skew arches, for example). (51)
poussée des terres (Ponts et Chaussées), thrust of the ground. (52)
direction de la poussée orogénique (Géol.) (*f.*), direction of the orogenic thrust. (53)

poussée (pression) (*n.f.*), pressure : (54)
centre de poussée (Phys.) (*m.*), centre of pressure. (55)
les portes d'écluse doivent résister à la poussée des eaux (*f.pl.*), lock-gates have to resist the pressure of the water. (56)

poussée (Hydros.) (*n.f.*), buoyancy. (57)

poussée dans la production (*f.*), spurt in the production. (58)

pousser (déplacer, tendre à déplacer par un effort) (*v.t.*), to push ; to drive : (59)
pousser une voiture, un verrou, to push a cart, a bolt. (60 *ou* 61)
sable poussé par le vent (*m.*), wind-driven sand ; wind-blown sand ; blown sand. (62)

pousser au vide *ou* **pousser en dehors** (se dit d'un mur qui fait ventre), to bulge ; to belly ; to belly out : (63)
mur qui pousse au vide (*m.*), wall which bulges. (64)

pousser des palplanches sous les semelles, si la sole le réclame (Boisage de mines), to spill under the floor-sills, if necessary ; to drive in spills under the sills, if the pavement needs it. (65)

pousser la construction d'une voie ferrée, to urge forward, to push forward, the con--struction of a railway. (66)

pousser le traçage jusqu'aux limites du champ d'exploitation, avant de commencer le déhouillement, to push the development to the boundaries of the winning, before commencing to get out the coal from the pillars. (1)

pousser les travaux, to urge forward, to crowd, to push, to push on with, the work. (2)

pousser un chemin jusqu'au fleuve, to carry a road right up to the river. (3)

pousser un clou dans un mur, to drive a nail into a wall. (4)

pousser un mur à l'alignement, to throw back a wall into alignment. (5)

pousser une moulure, une baguette sur un joint pour dissimuler une fente (Menuis.), to run, to stick, a moulding, a bead on a joint to conceal a crack. (6 *ou* 7)

pousseur (Ch. de f.) (*n.m.*), push-jack ; pushing-jack ; car-starter. (8)

poussier (*n.m.*) (*s'emploie souvent au pluriel*), dust : (9)

poussier de charbon *ou* poussier de houille *ou simplement* poussier *ou* poussiers, coal-dust ; coal-screenings ; coal-dirt. (10)

poussier de charbon de bois, charcoal-dust ; powdered charcoal. (11)

poussier isolant *ou* poussier de sable brûlé (Fonderie), parting-dust ; parting-sand. (12)

poussiers de coke, coke-dust. (13)

poussiers toxiques d'arsenic, arsenic toxical dust. (14)

poussiers toxiques de cinabre, cinabar toxical dust. (15)

poussière (*n.f.*) (*s'emploie souvent au pluriel*), dust : (16)

poussière d'eau, spray. (17)

poussière d'or, gold-dust. (18)

poussière de charbon *ou* poussières de charbon *ou* poussières charbonneuses, coal-dust. (19)

poussière de charbon éminemment inflam--mable, highly inflammable coal-dust. (20)

poussière de charbon légèrement inflammable, slightly inflammable coal-dust. (21)

poussière de charbon non-inflammable, non-inflammable coal-dust. (22)

poussière de coke, coke-dust. (23)

poussière de diamant, diamond-dust. (24)

poussière produite par les coups de mine, dust created by blasting. (25)

poussières incombustibles (Schistification des galeries de mine), stone-dust ; rock-dust. (26)

poussiéreux, -euse (*adj.*), dusty : (27)

mine poussiéreuse (*f.*), dusty mine. (28)

poussoir (*n.m.*), push-button ; push ; button. (29)

pouteure (Mines) [Belgique] (*n.f.*), black damp ; choke-damp ; stythe. (30)

poutrage (*n.m.*) *ou* **poutraison** (*n.f.*), girderage. (31)

poutre (*n.f.*), girder ; beam ; baulk ; balk. (32)

poutre (d'un plancher sur poutre) (*n.f.*), binder (of a double floor). (33)

poutre à âme pleine (*f.*), plate-web girder. (34)

poutre à jour (*f.*). Same as **poutre en treillis.**

poutre à larges ailes (*f.*), broad-flange girder ; wide-flange girder. (35)

poutre à ventre de poisson (*f.*), fish-bellied girder. (36)

poutre appuyée (Résistance des matériaux) (*f.*), supported beam (Strength of materials). (37)

poutre appuyée aux extrémités et chargée en son milieu (*f.*), beam supported at the ends and loaded at the centre ; beam supported at the ends and loaded in the middle. (38)

poutre appuyée aux extrémités et chargée en un point quelconque de sa longueur (*f.*), beam supported at the ends and loaded at any intermediate point. (39)

poutre appuyée aux extrémités et chargée uniformément (*f.*), beam supported at the ends and loaded uniformly. (40)

poutre armée (*f.*), trussed beam ; truss-beam ; trussed girder ; truss-girder. (41)

poutre bow-string (*f.*), bowstring girder. (42)

poutre-caisson [**poutres-caissons** *pl.*] (*n.f.*), box girder ; box beam ; box-section girder ; tubular girder. (43)

poutre composée (*f.*), compound girder ; com--pound beam ; built-up girder. (44)

poutre continue (*f.*), continuous girder. (45)

poutre de plancher (*f.*), summer-beam ; summer-tree. (46)

poutre de rive (d'un pont) (*f.*), stringer (of a bridge). (47)

poutre en béton armé (*f.*), reinforced-concrete beam. (48)

poutre en bois sous-bandée (*f.*), trussed wooden beam (underbraced). (49)

poutre en double T (*f.*), H beam ; H girder ; I beam ; I girder. (50)

poutre en fer laminé (*f.*), rolled-iron girder. (51)

poutre en porte à faux (Résistance des matériaux) (*f.*), unsupported beam ; free beam. (52)

poutre en simple T (*f.*), T beam. (53)

poutre en tôle (*f.*), plate girder ; plate-iron girder. (54)

poutre en tôle, section rectangulaire (*f.*), box-section girder ; box girder ; box beam ; tubular girder. (55)

poutre en treillis *ou* **poutre évidée** (*f.*), lattice girder ; lattice truss ; lattice beam ; lattice frame ; open-web girder ; skeleton girder. (56)

poutre encastrée (Résistance des matériaux) (*f.*), fixed beam ; encastré beam. (57)

poutre encastrée à une extrémité et chargée à l'autre (*f.*), beam fixed at one end and loaded at the other ; cantilever loaded at free end. (58)

poutre sablière (reposant sur le mur) (*f.*), wall-plate ; roof-plate ; plate. (59)

poutre simple (*f.*), simple beam. (60)

poutre sous-bandée (*f.*), trussed beam (under--braced). (61)

poutre tubulaire (*f.*), tubular girder ; box girder ; box beam ; box-section girder. (62)

poutrelle (*n.f.*), small girder ; I girder ; H girder ; joist. (63)

poutrelle à croisillons (*f.*). Same as **poutre en treillis,** but of smaller dimensions. (64)

poutrelle en acier (*f.*), steel joist. (65)

pouvoir (*n.m.*), power. (66)

pouvoir absorbant d'un corps (*m.*), absorbent power of a substance. (67)

pouvoir brisant d'un explosif (*m.*), disruptive strength of an explosive. (68)

pouvoir calorifique (*m.*), calorific power ; heating-power. (69)

pouvoir débitant d'un ventilateur (*m.*), capacity of output of a fan ; volume yielded by a fan. (70)

pouvoir des pointes (Élec.) (*m.*), power of points. (1)

pouvoir diffusif (Phys.) (*m.*), diffusive power. (2)

pouvoir dispersif (Opt.) (*m.*), dispersive power : (3)

le flint-glass est doué de pouvoirs dispersif et réfringent plus considérables que ceux du verre ordinaire, flint glass is possessed of greater dispersive and refractive powers than ordinary glass. (4)

pouvoir éclairant (*m.*), illuminating power ; lighting-power : (5)

pouvoir éclairant du gaz, illuminating power of gas. (6)

pouvoir émissif (Phys.) (*m.*), emissive power ; emissivity. (7)

pouvoir émissif calorifique (*m.*), calorific emissivity. (8)

pouvoir grossissant d'une loupe (*m.*), magnifying power of a lens. (9)

pouvoir inducteur spécifique (*m.*), specific in- -ductive capacity ; dielectric capacity ; dielectric constant ; permittivity. (10)

pouvoir réflecteur (*m.*), reflective power. (11)

pouvoir réfractif *ou* pouvoir réfringent (*m.*), refractive power : (12)

pouvoir réfractif des pierres précieuses, refractive power of gems. (13)

pouvoir thermo-électrique d'un métal (*m.*), thermoelectric power of a metal. (14)

pouvoir vaporisateur d'un combustible (*m.*), evaporative power of a fuel. (15)

pouzzolane *ou* pozzolane (*n.f.*), pozzuolana ; pozzolana ; pozzuolan ; puzzolana. (16)

pouzzolanique (*adj.*), pozzuolanic ; pozzolanic : (17)

ciment pouzzolanique (*m.*), pozzolanic cement. (18)

prase (Minéral) (*n.m.*), prase. (19)

praséodyme (Chim.) (*n.m.*), praseodymium ; prasodymium. (20)

praséolite (Minéral) (*n.f.*), praseolite. (21)

praticabilité (*n.f.*), practicability. (22)

praticable (*adj.*), practicable. (23)

praticien (pers.) (*n.m.*), practician ; practical man. (24)

pratique (*adj.*), practical : (25)

un traité pratique de géologie, a practical treatise on geology. (26)

pratique (*n.f.*), practice : (27)

personnes ayant de longues années de pratique dans la fonderie (*f.pl.*), persons having long years of practice in foundry work. (28)

pratiquement (*adv.*), practically : (29)

pratiquement parlant, practically speaking. (30)

pratiquer (faire ; exécuter) (*v.t.*), to make ; to do. (31)

pratiquer des rouillures (Mines de charbon), to nick ; to shear. (32)

pratiquer des sondages pour se procurer de l'eau, to put down bore-holes for water. (33)

pratiquer la découverte *ou* pratiquer la découver- -ture *ou* pratiquer le découvert (Mines), to remove the overburden ; to unsoil ; to strip ; to untop : (34)

pratiquer le découvert d'un gisement, to remove the overburden from a deposit. (35)

pratiquer un canal pour évacuer l'eau d'un fossé, to make a channel to carry off the water from a ditch. (36)

pratiquer un trou, to make a hole ; to hole : (37)

pratiquer un trou dans une pièce de bois pour recevoir un tenon, to make a hole in a piece of wood to receive a tenon. (38)

pratiquer un trou de sonde, to put down a bore- hole ; to bore-hole ; to bore. (39)

pratiquer (se) (s'exécuter ; s'opérer) (*v.r.*), to be done : (40)

le moulage en terre se pratique sans modèle, loam-moulding is done without patterns. (41)

préalable (*adj.*), previous ; preliminary : (42)

reconnaissance préalable des quartiers (*f.*), preliminary exploration of the districts. (43)

préalablement (*adv.*), previously ; first. (44)

précambrien, -enne (Géol.) (*adj.*), Precambrian : (45)

le système précambrien est caractérisé par l'absence des fossiles, the Precambrian system is characterized by the absence of fossils. (46)

précambrien (*n.m.*), Precambrian : (47)

le précambrien, the Precambrian. (48)

précaution (*n.f.*), precaution ; care ; caution : (49)

précautions à prendre pour éviter les accidents, precautions to be taken to prevent accidents ; prevention of accidents. (50)

précautions conseillées pour le tirage des coups de mine, precautions recommended when firing blasts. (51)

précautions contre la poussière, precautions against dust. (52)

précautions d'hygiène, sanitary precautions, (53)

précédent, -e (*adj.*), preceding ; former : (54)

le précédent propriétaire, the former owner. (55)

précession des équinoxes (*f.*), precession of the equinoxes. (56)

précieux, -euse (*adj.*), precious ; valuable ; noble : (57)

métal précieux (*m.*), precious metal ; noble metal. (58)

précipice (*n.m.*), precipice : (59)

les précipices des Alpes, the precipices of the Alps. (60)

précipitable (*adj.*), precipitable : (61)

matière précipitable par les acides (*f.*), matter precipitable by acids. (62)

précipitant (*n.m.*), precipitant. (63)

précipitation (Chim.) (*n.f.*), precipitation : (64)

précipitation par le charbon, par le zinc, au moyen des acides, precipitation by char- -coal, by zinc, by means of acids. (65 *ou* 66 *ou* 67)

précipitation (*n.f.*) *ou* précipitation atmosphérique, precipitation ; atmospheric precipitation ; rainfall : (68)

la neige est la forme normale des précipitations atmosphériques dans les lieux élevés, snow is the normal form of atmospheric precipita- -tion in elevated places. (69)

la précipitation annuelle, the annual rainfall. (70)

précipité (Chim.) (*n.m.*), precipitate : (71)

précipité d'or, precipitate of gold. (72)

précipiter (Chim.) (*v.t.*), to precipitate : (73)

l'acide sulfurique précipite le chlorure d'or de sa solution concentrée (*m.*), the sulphuric acid precipitates the chloride of gold from its concentrated solution. (74)

le froid précipite l'humidité en pluie, cold precipitates moisture in rain. (1)

précipiter (*v.i.*), to precipitate. (2)

précipiter (se) (*v.r.*), to precipitate ; to be pre--cipitated : (3)

matière qui se précipite sous forme de poudre blanche (*f.*), matter which precipi--tates in the form of white powder. (4)

dans les hautes régions la pluie se précipite sous forme de neige, in the high regions rain precipitates in the form of snow. (5)

précipiteux, -euse (*adj.*), precipitous. (6)

précis, -e (*adj.*), precise ; exact : (7)

calcul précis (*m.*), precise calculation ; exact reckoning. (8)

précisément (*adv.*), precisely ; exactly. (9)

précision (*n.f.*), precision ; exactness ; exacti--tude : (10)

balance de précision (*f.*), balance of precision ; precision balance. (11)

précisions (détails ; renseignements) (*n.f.pl.*), particulars. (12)

prédominant, -e (*adj.*), predominant ; predomi--nating ; prevailing. (13)

prédominer (*v.i.*), to predominate : (14)

terrain où le sable prédomine (*m.*), ground in which sand predominates. (15)

préglaciaire (Géol.) (*adj.*), preglacial : (16)

alluvions préglaciaires (*f.pl.*), preglacial alluvium. (17)

préhistorique (*adj.*), prehistoric ; prehistorical : (18)

l'homme préhistorique, prehistoric man. (19)

prehnite (Minéral) (*n.f.*), prehnite. (20)

préjudice (dommage) (*n.m.*), damage ; injury : (21)

porter préjudice à quelqu'un, to cause someone damage. (22)

préjudiciable (*adj.*), injurious ; prejudicial. (23)

prélart (*n.m.*), tarpaulin. (24)

prélèvement d'échantillons (*m.*), sampling : (25)

prélèvement d'échantillons au hasard, random sampling. (26)

prélever un échantillon type du minerai sur la halde, to take a representative sample of the ore on the dump. (27)

préliminaire (*adj.*), preliminary : (28)

étude préliminaire d'un chemin de fer (*f.*), preliminary survey of a railway. (29)

prématuré, -e (*adj.*), premature : (30)

explosion prématurée d'un coup de mine (*f.*), premature explosion of a shot. (31)

prématurément (*adv.*), prematurely. (32)

premier, -ère (en abrégé **1ᵉʳ, 1ʳᵉ**) (*adj.*), first ; primary. (33)

premier arc (arc-en-ciel) (*m.*), primary bow ; primary rainbow. (34)

premier choix ou **1ᵉʳ choix** (*m.*), best quality ; finest quality. (35)

premier plan (*m.*), foreground : (36)

au premier plan, in the foreground. (37)

premier poste au charbon (*m.*), drawing-gang. (38)

premier taraud (*m.*), first tap ; taper tap ; entering tap. (39)

première occasion (à la), at the earliest oppor--tunity ; on the first occasion. (40)

première vue (à), at first glance ; at first sight. (41)

premièrement (*adv.*), firstly ; primarily. (42)

premiers soins ou **premiers secours** (aux blessés) (*m.pl.*), first aid : (43)

premiers soins à donner aux ouvriers blessés à la suite des explosions de grisou, first aid to workmen injured by explosions of fire-damp. (44)

premiers soins en cas d'accidents, first aid in case of accidents. (45)

prendre (*v.t.*), to take. (46)

prendre (se congeler) (*v.i.*), to congeal ; to co--agulate ; to freeze ; to freeze over ; to catch over : (47)

rivière qui prend rarement (*f.*), river which seldom freezes (or catches) over. (48)

prendre à bail, to take on lease ; to lease. (49)

prendre au passage ou **prendre au vol** (Photogr.), to snap ; to snap-shot. (50)

prendre des dimensions, to take measurements ; to take dimensions. (51)

prendre des dispositions pour éviter tous dégâts aux machines, to take steps to prevent damage being done to the machinery. (52)

prendre des dispositions pour la bonne marche des travaux pendant l'absence du directeur, to arrange for efficient conduct of the works during the manager's absence. (53)

prendre des informations sur quelqu'un, to make enquiries about someone. (54)

prendre du bois contre le fil, to work wood against the grain. (55)

prendre du jeu (Méc.), to work loose : (56)

clavetage qui a pris du jeu (*m.*), keying which has worked loose. (57)

prendre du lâche, to slack : (58)

câble qui prend du lâche (*m.*), rope which slacks. (59)

prendre en creux le moule du fond du trou de sonde, to take an impression of the bottom of the bore-hole. (60)

prendre feu (*v.i.*), to take fire ; to catch fire ; to become ignited ; to light : (61)

le phosphore prend feu à l'air libre à une température qui dépasse à peine son point de fusion, phosphorus takes fire in the open air at a temperature which hardly exceeds its melting-point. (62)

prendre intérêt aux affaires minières, to take interest in mining matters. (63)

prendre la direction d'une locomotive, to take charge of a locomotive. (64)

prendre la peine d'examiner une chose, to take the trouble to examine a thing. (65)

prendre le plus court, to take a short cut. (66)

prendre nature (Puddlage du fer), to come to nature. (67)

prendre sa source (se dit d'une rivière), to take its rise ; to rise. (68)

prendre soin de ne pas surcharger le câble, to take care not to overload the rope. (69)

prendre un beau poli (se dit d'une pierre), to take a beautiful polish. (70)

prendre un grand essor, to make great progress (or headway). (71)

prendre un (ou **l'**) **instantané** de (Photogr.), to snap ; to snap-shot : (72)

prendre un instantané d'un train qui passe, to snap, to snap-shot, a passing train. (73)

prendre un ouvrier, to engage a workman ; to take on a hand. (74)

prendre une empreinte du fond du trou de sonde, to take an impression of the bottom of the bore-hole. (75)

prendre une maison en location, to rent a house. (76)

prendre une photographie, une vue, to take a photograph, a view (*or* a picture). (1 *ou* 2)

prendre une tige dans l'étau, to grip a rod in the vice. (3)

prendre (se) (se congeler) (*v.r.*), to congeal ; to coagulate ; to freeze ; to freeze over ; to catch over : (4)

l'huile se prend facilement (*f.*), oil congeals readily. (5)

preneur, -euse (pers.) (*n.*), lessee ; leaseholder. (6)

préparatif (*n.m.*), preparative ; preparation : (7)

faire ses derniers préparatifs, to make one's final preparations. (8)

préparation (*n.f.*), preparation ; preparing : (9)

préparation du charbon pour la vente, prep--aration of coal for market. (10)

préparation d'un quartier (Mines), preparing a district ; setting a district in working-order. (11)

préparation mécanique (du minerai, de la houille) (*f.*), dressing : (12)

préparation mécanique des houilles *ou* pré--paration mécanique du charbon, coal-dressing. (13)

préparation mécanique des minerais *ou* pré--paration mécanique du minerai, ore-dressing. (14)

préparation mécanique du minerai d'étain, tin-dressing ; dressing tin-ore. (15)

préparatoire (*adj.*), preparatory ; preliminary : (16)

travaux préparatoires (*m.pl.*), preparatory work ; preliminary work. (17)

préparer (*v.t.*), to prepare : (18)

préparer un quartier (Mines), to prepare a district ; to set a district in working-order. (19)

préparer la carte d'un district, to prepare a map of a district. (20)

préparer mécaniquement, to dress : (21)

préparer mécaniquement du minerai, de la houille, to dress ore, coal. (22 *ou* 23)

préparer (se) (*v.r.*), to prepare ; to prepare oneself ; to get ready : (24)

se préparer à commencer l'abatage, to prepare to commence stoping. (25)

prépondérance (*n.f.*), preponderance. (26)

prépondérant, -e (*adj.*), preponderant ; pre--ponderating. (27)

prépondérer (*v.i.*), to preponderate. (28)

près (*adv.*), near. (29)

présence (*n.f.*), presence ; existence : (30)

la présence de la magnétite, de la pyrite de fer magnétique, etc., trouble les indications de la boussole, the presence of magnetite, magnetic iron pyrites, etc., disturbs the indications of the compass. (31)

présenter (*v.t.*), to present ; to show : (32)

présenter de grandes difficultés dans le traite--ment, to present great difficulties in treatment. (33)

présenter une réduction, to show a decline. (34)

présenter l'aspect, to look like ; to be like : (35)

roche qui présente l'aspect de granit (*f.*), rock which looks like granite. (36)

présenter un balourd, to be out of balance ; to be unbalanced : (37)

toutes les fois que la pièce présente un balourd, il faut rétablir l'équilibre au moyen de contrepoids, whenever the part is out of balance, equilibrium must be restored by means of counterweights. (38)

Cf. **balourd.**

présenter un rapport annuel, to submit an annual report. (39)

présenter un tenon à une mortaise, to try a tenon in a mortise (to see if it fits). (40)

présenter (se) (*v.r.*). See examples : (41)

se présenter à l'état natif, to occur in the native state. (42)

se présenter bien, to promise well ; to show well ; to show up very nicely : (43)

le filon se présente bien, the lode is showing up very nicely. (44)

se présenter en gisements, to occur in beds. (45)

se présenter en masses, to occur massive : (46)

minéral qui se présente en masses (*m.*), mineral which occurs massive. (47)

se présenter en masses compactes, en masses fibreuses, to occur in compact masses, in fibrous masses : (48 *ou* 49)

la cérusite se présente en cristaux, en masses compactes, en stalactites, etc., cerusite occurs in crystals, in compact masses, in stalactites, etc. (50)

se présenter sous forme de, to appear in the form of : (51)

l'anhydride arsénieux, condensé sur une paroi froide (d'une chambre en maçonnerie), se présente sous forme de poudre blanche cristallisée, the arsenious anhydride, condensed on a cold wall (of a masonry chamber), appears in the form of white crystallized powder. (52)

préservatif (*n.m.*), preservative. (53)

préservation (*n.f.*), preservation ; protection : (54)

préservation du bois, preservation of timber. (55)

préserver (*v.t.*), to preserve ; to protect : (56)

préserver la maçonnerie du travail et des dégradations incessantes des eaux, to protect the masonry from the continual action and wash of the water. (57)

présider à (régir), to govern : (58)

les principes fondamentaux qui président au choix de la disposition d'ensemble (*m.pl.*), the fundamental principles which govern the choice of the general arrangement. (59)

presqu'île [presqu'îles *pl.*] (*n.f.*), peninsula : (60)

la presqu'île des Balkans, the Balkan peninsula. (61)

presse (*n.f.*), press. (62)

presse à border (*f.*), flanging-press ; flanger ; flanging-machine. (63)

presse à cintrer et à dresser les rails (*f.*), rail-bender ; jim-crow. (64)

presse à coin (*f.*), wedge-press. (65)

presse à coller (*f.*). Same as **presse à vis** (pour menuisiers).

presse à cylindre (*f.*), cylinder-press ; rolling-press. (66)

presse à étirer les tubes (*f.*), tube-drawing press. (67)

presse à excentrique (*f.*), cam-press. (68)

presse à faire les moyeux des roues (*f.*), wheel-bossing press. (69)

presse à forger (*f.*), forging-press. (70)

presse à forger à quatre colonnes (*f.*), four-column forging-press. (71)

presse à levier (*f.*), lever-press. (72)

presse à macquer (pour fer puddlé) (*f.*), alligator squeezer; crocodile squeezer; squeezer; saurian squeezer. (1)

presse à refouler (*f.*), upsetting-press. (2)

presse à souder (*f.*), welding-press. (3)

presse à vis (*f.*), screw-press; vice-press. (4)

presse à vis (pour menuisiers et ébénistes) (*f.*), screw-clamp; C cramp; C clamp; G cramp. (5)

presse à vis à cintrer et redresser les rails *ou* presse à vis à deux bras, pour cintrer et redresser les rails (*f.*), jim-crow; rail-bender; screw rail-bender. (6)

presse à vis horizontale (*f.*), horizontal screw-press. (7)

presse à vis verticale (*f.*), vertical screw-press. (8)

presse de chaudronnier *ou* presse en fer, pour grosse chaudronnerie (*f.*), boiler-cramp; boiler-clamp. (9)

presse-étoupe [presse-étoupe *pl.*] *ou* presse-garniture [presse-garniture *pl.*] (*n.m.*), stuf-fing-box. (10)

presse hydraulique (*f.*), hydrostatic press; hydraulic press; Bramah press. (11)

presse hydraulique pour roues *ou* presse hydrauli-que pour le calage des roues (*f.*), hydraulic wheel-press; wheel-press. (12)

presser (serrer; comprimer) (*v.t.*), to press; to squeeze: (13)

presser sur un bouton, to press a button. (14)

presser une éponge, to squeeze a sponge. (15)

presser les travaux, to push on with, to hurry forward, the work. (16)

presser (se) (se serrer) (*v.r.*), to press. (17)

presser (se) (se hâter) (*v.r.*), to make haste; to hurry up. (18)

pression (*n.f.*), pressure. (19)

pression (de vapeur) (Méc.) (*n.f.*), pressure, tension, (of steam). (20)

pression (en) *ou* pression (sous) (en parlant d'une machine à vapeur), under steam: (21)

la visite des soupapes, quand la machine est en pression, présente quelque danger, the inspection of valves, when the engine is under steam, is somewhat dangerous. (22)

pression (Élec.) (*n.f.*), pressure; tension; voltage. (23)

pression (avancement; entraînement) (Méc.) (*n.f.*), feed. (24)

pression à la main par levier (*f.*), hand lever-feed. (25)

pression à levier (*f.*), lever-feed. (26)

pression à main par volant (*f.*), hand-feed by hand-wheel. (27)

pression à vis (*f.*), screw-feed. (28)

pression absolue (Méc.) (*f.*), absolute pressure. (29)

pression atmosphérique (*f.*), atmospheric pres-sure; air-pressure. (30)

pression automatique à débrayage instantané (*f.*), self-acting feed with instantaneous stop-motion; automatic feed with instan-taneous tripping. (31)

pression barométrique (*f.*), barometric pressure. (32)

pression critique (Phys.) (*f.*), critical pressure. (33)

pression d'air (*f.*), air-pressure. (34)

pression d'eau *ou simplement* pression (*n.f.*), water-pressure; pressure of water; head

of water; head; pressure-head; static head; pressure. (35)

pression d'épreuve (*f.*), test pressure. (36)

pression de — kilos par centimètre carré (*f.*), pressure of — kilos to the square centimetre. (37)

pression de marche *ou* pression de service *ou* pression de travail *ou* pression de régime (Mach.) (*f.*), working-pressure: (38)

la pression de marche d'une chaudière, the working-pressure of a boiler. (39)

pression de refoulement (*f.*), delivery-pressure. (40)

pression de vapeur (*f.*), steam-pressure. (41)

pression des gaz renfermés dans le charbon (*f.*), pressure of gases in coal. (42)

pression des liquides sur le fond des vases (*f.*), pressure of liquids on the bottom of vessels. (43)

pression du gaz (*f.*), gas-pressure. (44)

pression du vent (Météor.) (*f.*), wind-pressure; pressure of wind. (45)

pression du vent (haut fourneau, etc.) (*f.*), blast-pressure; pressure of blast (blast-furnace, etc.). (46)

pression effective (*f.*), effective pressure; active pressure. (47)

pression effective de vapeur (*f.*), effective steam-pressure. (48)

pression électrique (*f.*), electric pressure. (49)

pression électrostatique (*f.*), electrostatic pressure; electrostatic stress. (50)

pression en dessous par vis à tête ronde percée de trous pour l'emploi d'une broche (*f.*), feed from underneath by screw with round head pierced with holes for use with tommy. (51)

pression en un point (*f.*), pressure at a point. (52)

pression gazeuse (*f.*), gas-pressure. (53)

pression moyenne (*f.*), mean pressure; average pressure. (54)

pression osmotique (*f.*), osmotic pressure. (55)

pression produite par une explosion (*f.*), pressure of an explosion. (56)

pression relative (Méc.) (*f.*), relative pressure. (57)

prêt, -e (*adj.*), ready; prepared: (58)

prêt (-e) à l'usage *ou* prêt (-e) à servir, ready for use. (59)

être prêt (-e) à toute éventualité, to be ready for any emergency; to be prepared for any eventuality. (60)

prétendu, -e (*adj.*), pretended; alleged: (61)

contamination prétendue d'une rivière (*f.*), alleged pollution of a river. (62)

prétendu expert (*m.*), pretended expert; quasi-expert. (63)

preuve (*n.f.*), proof: (64)

preuve concluante, conclusive proof; positive proof. (65)

preuve convaincante, convincing proof. (66)

preuve indubitable de l'existence du pétrole, undoubted proof of the existence of oil. (67)

prévaloir (*v.i.*), to prevail: (68)

une situation normale des affaires prévaut en ce moment, a normal condition of affairs now prevails. (69)

prévaloir (se) de, to avail oneself of; to take advantage of: (70)

se prévaloir de la saison pluvieuse, to take advantage of the rainy season. (71)

prévenir (empêcher) (*v.t.*), to prevent : (1)
prévenir un accident, to prevent an accident.
(2)

prévenir (avertir) (*v.t.*), to warn : (3)
prévenir quelqu'un d'un danger, to warn
someone of a danger. (4)

prévenir un concurrent, to forestall a com-
-petitor. (5)

préventif, -ive (*adj.*), preventive. (6)

prévision (*n.f.*), forecast ; forecasting : (7)
prévision du temps, forecasting the weather ;
weather-forecast. (8)

prévoir (*v.t.*), to foresee ; to forecast ; to an-
-ticipate : (9)
prévoir le temps, to forecast the weather. (10)
prévoir des difficultés dans l'épuisement de la
mine, to anticipate trouble in unwatering
the mine. (11)

primage (*n.m.*), priming : (12)
en réalité, la vapeur entraîne toujours avec
elle une certaine quantité de liquide que les
bulles gazeuses projettent en crevant à
la surface : cet entraînement d'eau s'appelle
primage, in point of fact, steam always
entrains with it a certain quantity of liquid
which the gassy bubbles throw up when
bursting on the surface : this entrainment
of water is called priming. (13)

primaire (*adj.*), primary : (14)
pile primaire (*f.*), primary battery. (15)

primaire (Géol.) (*adj.*), Primary ; Paleozoic : (16)
l'âge primaire ou paléozoïque (*m.*), the
Primary or Paleozoic age. (17)

primaire (Élec.) (*n.m.*), primary. (18)

prime (récompense accordée aux ouvriers) (*n.f.*),
bonus ; premium : (19)
système à primes (*m.*), bonus system ; pre-
-mium system. (20)

primer (chaudières à vapeur) (*v.i.*), to prime : (21)
la vapeur produite dans une locomotive
entraîne dans les cylindres des gouttelettes
d'eau non-transformées en vapeur : on dit
que la chaudière prime, the steam produced
in a locomotive carries along into the
cylinders particles of water untransformed
into steam : the boiler is said to prime.
(22)

primitif, -ive (qui appartient au premier état des
choses) (*adj.*), primitive ; crude ; early : (23)
méthodes primitives (*f.pl.*), primitive methods ;
crude methods ; early methods. (24)

primitif, -ive (Géol.) (*adj.*), primitive ; primary :
(25)
roches primitives (*f.pl.*), primitive rocks ;
primary rocks. (26)

primitif, -ive (Phys.) (*adj.*), primary ; primitive :
(27)
les couleurs primitives du spectre (*f.pl.*), the
primary (*or* primitive) colors of the spectrum.
(28)

primitif, -ive (Engrenages) (*adj.*), pitch (*used as
adj.*) ; primitive (Gearing) : (29)
cercle primitif (*m.*), pitch-circle ; primitive
circle [*also called* pitch-line ; dividing-
circle *and* rolling circle]. (30)

primitif (Engrenages) (*n.m.*), pitch-circle ; pitch-
line ; primitive circle ; dividing-circle ;
rolling circle : (31)
épaisseur de la dent au primitif (*f.*), thickness
of tooth at pitch-circle. (32)

primordial, -e, -aux (*adj.*), primordial ; primitive.
(33)

principal, -e, -aux (*adj.*), principal ; main ;
chief ; master : (34)
les principales caractéristiques géologiques
(*f.pl.*), the principal (*or* the main) (*or* the
chief) geological characteristics (*or* features).
(35)
joint principal (Géol.) (*m.*), master-joint. (36)

principalement (*adv.*), principally ; mainly ;
chiefly. (37)

principe (*n.m.*), principle : (38)
principes de la mécanique, de la géométrie,
principles of mechanics, of geometry. (39 *ou*
40)

principe (Chim.) (*n.m.*), principle : (41)
substance qui contient un principe azoté (*f.*),
substance which contains a nitrogenic
principle. (42)

principe d'Archimède (Hydros.) (*m.*), Archi-
-medean principle ; Archimedes' principle.
(43)

principe des vitesses virtuelles (Méc.) (*m.*), prin-
-ciple of virtual velocities. (44)

printanier, -ère (*adj.*), spring : (45)
saison printanière (*f.*), spring season. (46)

printemps (*n.m.*), spring. (47)

prise (action de prendre) (*n.f.*), taking : (48)
prise d'une empreinte du fond du trou de sonde,
taking an impression of the bottom of the
bore-hole. (49)
See **prendre** for many other phrases.

prise (facilité de saisir) (*n.f.*), hold ; grip : (50)
tige carrée qui fournit une prise solide à une
clef (*f.*), square rod which affords a firm
hold for a spanner. (51)

prise (des chaux et ciments) (*n.f.*), setting (of
limes and cements) : (52)
ciment à prise lente (*m.*), slow-setting cement.
(53)
ciment à prise rapide, quick-setting cement.
(54)

prise (Engrenages) (*n.f.*), engagement ; meshing ;
pitching ; mesh ; pitch (Gearing) : (55)
la durée de la prise de deux dents, the duration
of the engagement (*or* the meshing) (*or* the
pitching) of two teeth. (56)
deux dents qui sont complètement en prise
(*f.pl.*), two teeth which are completely
in mesh (*or* which engage completely). (57)
arc de prise (*m.*), pitch-arc. (58)

prise d'air (détournement) (*f.*), intake of air.
(59)

prise d'air (échantillon) (*f.*), air sample ; sample
of air. (60)

prise d'air (entrée d'air) (*f.*), air-inlet. (61)

prise d'eau (détournement) (*f.*), tapping ;
intake of water : (62)
prises d'eau sur cours d'eau, tapping
streams. (63)

prise d'essai (*f.*), sample for analysis ; assay-
sample. (64)

prise de courant *ou* **prise du courant** (admission)
(Élec.) (*f.*), collecting current ; picking
up the current ; intake of current : (65)
différents systèmes de prise du courant (*m.pl.*),
different systems of collecting current. (66)
prise de courant par des frotteurs, picking
up current with plows. (67)

prise de courant (appareil mobile servant à
faire une prise de courant électrique) (*f.*),
switch (of the plug-and-socket type). (68)

prise de courant (balai ou trolley) (Élec.) (*f.*),
current-collector. (69)

prise de courant à deux broches (Élec.) (*f.*), two-pin plug-switch. (1)

prise de courant à fiche (Élec.) (*f.*), plug-switch ; peg-switch ; removable-key switch. (2)

prise de gaz latérale (haut fourneau) (*f.*), down--comer ; downtake (blast-furnace). (3)

prise de terre (Élec.) (*f.*), earth-plate ; ground-plate. (4)

prise de vapeur (admission) (*f.*), intake of steam ; input of steam. (5)

prise de vapeur (robinet de prise de vapeur) (*f.*), steam-cock. (6)

prise de vapeur (soupape de prise de vapeur) (*f.*), steam-valve. (7)

prise de vapeur (soupape régulatrice d'admission de vapeur) (locomotive) (*f.*), throttle-valve ; throttle. (8)

prise de vapeur de l'injecteur (*f.*), injector-throttle ; injection-cock. (9)

prise de vapeur de la pompe à air (*f.*), air-pump throttle. (10)

prise du courant (Élec.) (*f.*). Same as **prise de courant.**

priser (*v.t.*), to value ; to make a valuation of. (11)

priseur (pers.) (*n.m.*), valuer. (12)

prismatique (*adj.*), prismatic ; prismatical : (13)
couleurs prismatiques (*f.pl.*), prismatic colours. (14)
pièce prismatique (Résistance des matériaux) (*f.*), prismatic bar (Strength of materials). (15)

prisme (*n.m.*), prism. (16)

prisme à angle variable (*m.*), prism with variable angle. (17)

prisme à réflexion totale (*m.*), right-angled prism. (18)

prisme de Nicol (*m.*), Nicol prism ; Nicol's prism ; nicol. (19)

prisme de première espèce (Cristall.) (*m.*), prism of the first order ; first order prism ; unit prism ; protoprism. (20)

prisme de seconde espèce *ou* **prisme de deuxième espèce** (Cristall.) (*m.*), prism of the second order ; second order prism ; deuteroprism. (21)

prisme de troisième espèce (Cristall.) (*m.*), prism of the third order ; third order prism ; tritoprism. (22)

prisme de Wollaston (*m.*), Wollaston's prism. (23)

prisme droit (*m.*), right prism. (24)

prisme droit triangulaire (*m.*), triangular right prism. (25)

prisme oblique (*m.*), oblique prism. (26)

prisme orthorhombique (*m.*), orthorhombic prism. (27)

prisme pentagonal (*m.*), pentagonal prism. (28)

prisme quadrangulaire (*m.*), quadrangular prism ; four-sided prism. (29)

prisme tronqué (*m.*), truncated prism. (30)

prismé, -e (*adj.*), prismatic ; prismatical : (31)
cristal de quartz prismé (*m.*), prismatic quartz crystal. (32)

prismes jumeaux (*m.pl.*), twin prisms. (33)

prisonnier (boulon) (*n.m.*), stud-bolt ; stud ; standing-bolt. (34)

prisonnier à clavette (*m.*), cotter stud-bolt. (35)

prisonnier borgne (*m.*), blind stud-bolt. (36)

privation de droits (*f.*), confiscation of rights. (37)

privé, -e (*adj.*), private : (38)
industrie privée (*f.*), private industry. (39)

privé (-e) de tous approvisionnements pendant l'hiver (être), to be cut off from all supplies during the winter. (40)

privilège (*n.m.*), privilege ; right. (41)

prix (*n.m.*), price ; quotation : (42)
le prix d'une propriété, the price of a property. (43)
un prix pour matériel de sondage, a quotation for drilling-plant. (44)

prix à forfait *ou* **prix forfaitaire** (*m.*), contract-price. (45)

prix courant (prix réglé par la balance de l'offre et de la demande) (*m.*), current price ; market price. (46)

prix courant (bulletin) (*m.*), price-list ; price-current. (47)

prix d'achat *ou* **prix d'acquisition** (*m.*), purchase-price. (48)

prix de base (*m.*), basis price. (49)

prix de revient *ou* **prix coûtant** *ou* **prix de fabri--que** *ou* **prix de fabrication** (*m.*), cost ; cost price : (50)
les prix de revient des divers procédés de fabrication de l'acier, the costs of the different processes of steel-manufacture. (51)

prix de vente *ou* **prix marchand** (*m.*), sale-price ; selling-price. (52)

prix du fer (*m.*), price of old iron ; break-up price. (53)

prix le plus bas (*m.*), lowest price ; rock-bottom price ; bed-rock price. (54)

prix sauf variations (*m.pl.*), prices subject to alteration. (55)

prix sur demande (*m.pl.*), prices on application. (56)

prix unitaire (*m.*), unit price. (57)

probabilité (*n.f.*), probability : (58)
probabilité des erreurs (Math.), probability of errors. (59)

probable (*adj.*), probable : (60)
minerai probable (*m.*), probable ore. (61)

probant, -e (*adj.*), conclusive : (62)
essai probant (*m.*), conclusive test. (63)

problématique (*adj.*), problematic ; problematical. (64)

problématiquement (*adv.*), problematically. (65)

problème (*n.m.*), problem : (66)
un problème de géométrie, a geometrical problem. (67)
le problème de l'eau, the water problem. (68)
le problème des transports, the transport problem. (69)

procédé (*n.m.*), process ; way ; method. (70)
See also **méthode.**

procédé à la fonte et au minerai (*m.*), pig-and-ore process. (71)

procédé au carbone (Photogr.) (*m.*), carbon printing ; carbon process. (72)

procédé basique (acier) (*m.*), basic process ; dephosphorizing process ; Thomas-Gilchrist process. (73)

procédé Bessemer acide (*m.*), acid Bessemer process. (74)

procédé Bessemer basique (*m.*), basic Bessemer process. (75)

procédé d'analyse par la voie humide (*m.*), wet process of assaying ; moist method of assaying. (76)

procédé d'analyse par la voie sèche (*m.*), dry process of assaying. (77)

procédé d'exploitation du charbon (*m.*), method of working coal ; coal-mining method. (1)

procédé d'injection de ciment (*m.*), cement-injection process. (2)

procédé de chloruration (*m.*), chlorination-process. (3)

procédé de congélation (Fonçage des puits de mine) (*m.*), freezing process. (4)

procédé de cyanuration (*m.*), cyanide process. (5)

procédé de fonçage des puits Kind-Chaudron (*m.*), Kind-Chaudron shaft-sinking method. (6)

procédé direct de cyanuration (*m.*), direct cyanide process ; straight cyanide process. (7)

procédé du baril (minerais) (*m.*), barrel process. (8)

procédé électrolytique de cyanuration (*m.*), electrocyanide process. (9)

procédé électrolytique par voie humide (*m.*), wet electrolytic process. (10)

procédé électrolytique par voie ignée (*m.*), igneous electrolytic process. (11)

procédé Manhès d'affinage du cuivre (*m.*), Manhès copper process. (12)

procédé Martin *ou* **procédé Martin-Siemens** (*m.*), Siemens-Martin process ; open-hearth process. (13)

procédé par flottage (minerais) (*m.*), flotation-process. (14)

procéder (*v.i.*), to proceed : (15)
procéder par étapes successives, par tâtonne-ments, to proceed by successive stages, by trial and error. (16 *ou* 17)

procès-verbal [**procès-verbaux** *pl.*] (pièce émanée d'un fonctionnaire public) (*n.m.*), report : (18)
dresser procès-verbal d'un accident suivi de mort, to draw up a report, to report, on a fatal accident. (19)

processus [**processus** *pl.*] (*n.m.*), process : (20)
le processus qui aboutit à la formation de la houille, the process which has as a result the formation of coal. (21)

prochain, -e (*adj.*), next ; nearest. (22)

prochainement (*adv.*), soon ; at an early date. (23)

procurable (*adj.*), procurable ; obtainable. (24)

procurer (*v.t.*), to procure ; to obtain ; to get. (25)

procurer (se) (*v.r.*), to procure ; to obtain ; to get : (26)
difficulté à se procurer des provisions (*f.*), difficulty in obtaining provisions. (27)

producteur, -trice *ou* **productif, -ive** (*adj.*), pro-ducing ; productive : (28)
pays producteur (*ou* productif) de charbon (*m.*), coal-producing country. (29)

producteur (*n.m.*), producer. (30)

production (rendement) (*n.f.*), production ; out-put ; yield ; get. (31)

production augmentant de façon soutenue (*f.*), steadily increasing output. (32)

production d'une mine de charbon (*f.*), produc-tion, get, output, yield, of a coal-mine. (33)

production de vapeur *ou* **production de la vapeur d'eau** (*f.*), steam-raising ; generation of steam ; production of steam. (34)

production du courant (Élec.) (*f.*), current-production ; generation of current. (35)

production du vide dans une ampoule de lampe électrique à incandescence (*f.*), production,

creation, of a vacuum in an incandescent electric lamp bulb. (36)

production forcée (*f.*), forced production. (37)

production houillère en milliers de tonnes (*f.*), production of coal in thousands of tons. (38)

production journalière moyenne (*f.*), average daily output. (39)

production mondiale d'or (*f.*), world's production of gold. (40)

production synthétique du saphir (*f.*), synthetic production of the sapphire. (41)

productivité (*n.f.*), productivity ; productiveness. (42)

productrice (*n.f.*), producer. (43)

produire (en parlant d'un rendement) (*v.t.*), to produce ; to output : (44)
produire du minerai à raison de — tonnes par semaine, to produce, to output, ore at the rate of — tons per week. (45)
produire du pétrole sur une échelle industrielle, to produce petroleum on a commercial scale ; to produce oil in commercial quantities. (46)

produire (en parlant de l'énergie) (*v.t.*), to gener-ate ; to produce ; to raise : (47)
produire de la vapeur, to raise, to generate, to produce, steam. (48)
produire la force motrice, to generate, to produce, power. (49)
courant produit par une réaction chimique (*m.*), current generated by a chemical reaction. (50)

produire le vide dans un récipient, to produce, to create, a vacuum in a vessel. (51)

produire (se) (*v.r.*), to occur ; to take place : (52)
effondrements qui peuvent se produire inopinément et blesser ou tuer des ouvriers (*m.pl.*), falls which may occur unexpectedly and injure or kill workmen. (53)

produit (production) (*n.m.*), product ; yield ; produce : (54)
produit marchand, marketable product. (55)

produit (recette) (*n.m.*), proceeds ; yield : (56)
le produit d'une vente, de la vente d'une cargaison de minerai, the proceeds of a sale, of the sale of a cargo of ore. (57 *ou* 58)

produit (Chim.) (*n.m.*), product. (59)

produit chimique (*m.*), chemical : (60)
produits chimiques photographiques, photo-graphic chemicals. (61)
produits et appareils de laboratoires, chemical and scientific apparatus and chemicals. (62)

produit de la distillation (*m.*), distillate ; prod-uct of distillation : (63)
produit de la distillation du pétrole, distillate of petroleum. (64)

produit secondaire (*m.*), by-product : (65)
produits secondaires de la distillation de la houille, by-products of the distillation of coal. (66)

produits d'écumage (*m.pl.*), skimmings. (67)

produits de raffinage (*m.pl.*), refinery products. (68)

produits de rejet (*m.pl.*), waste products. (69)

produits dérivés du pétrole (*m.pl.*), products derived from petroleum. (70)

produits du lavage *ou* **produits lavés** (*m.pl.*), wash-ings ; wash. (71)

professionnel, -elle (*adj.*), professional ; trade (*used as adj.*) : (72)
le trousseau professionnel d'un ouvrier mouleur, the trade kit of a moulder. (73)

professionnel, -elle (pers.) (*n.*), professional : (1)
l'amateur et le professionnel (*m.*), the amateur and the professional. (2)

profil (*n.m.*), profile ; contour ; outline ; section ; cross-section : (3)
profil d'une dent d'engrenage, profile of a gear-tooth. (4)
profil du terrain entre la mine et le moulin, contour, profile, of ground between the mine and the mill. (5)
profil d'un rail, d'une poutre, section, cross-section, of a rail, of a girder. (6 *ou* 7)
profil en U, U section. (8)
profil terrestre, contour of the earth. (9)

profil d'équilibre (Géogr. phys.) (*m.*), grade : (10)
rivière qui atteint, qui possède, son profil d'équilibre (*f.*), river which reaches, which is at, grade. (11 *ou* 12)

profil de l'horizon (*m.*), sky-line ; summit-line. (13)

profil en long *ou* **profil longitudinal** (*m.*), longi-tudinal section ; lengthwise section : (14)
profil en long d'une route, longitudinal section of a road. (15)

profil en travers *ou* **profil transversal** (*m.*), cross-section ; transverse section : (16)
profil en travers d'une route, cross-section of a road. (17)

profil-limite de chargement [profils-limites de chargement *pl.*] (Ch. de f.) (*m.*), loading-gauge. (18)

profilage (*n.m.*), profiling. (19)

profilé, -e (*adj.*), sectional (consisting of special sections, as L, T, H, U) : (20)
fers profilés (*m.pl.*), sectional iron. (21)

profiler (*v.t.*), to profile. (22)

profilés *ou* **profils** (fer *ou* acier) (*n.m.pl.*), sections ; shapes : (23 *ou* 24)
profilés en acier, steel sections ; steel shapes. (25)
profils à vitrages, sash-bar sections. (26)
profils moulurés, mouldings. (27)

profit (*n.m.*), profit. (28)

profit (à), at a profit ; profitably : (29)
travailler à profit, to work at a profit. (30)

profitable (*adj.*), profitable. (31)

profitablement (*adv.*), profitably. (32)

profiter (*v.i.*), to profit ; to take advantage of : (33)
profiter d'une échappée de beau temps, to take advantage of a burst of fine weather ; to profit by a spell of fine weather. (34)

profond, -e (*adj.*), deep : (35)
eau profonde (*f.*), deep water. (36)

profondément (*adv.*), deeply. (37)

profondeur (*n.f.*), depth. (38)

profondeur (en) *ou* **profondeur** (de), in depth ; deep : (39)
un puits de — mètres en profondeur, a well — metres in depth ; a well — metres deep. (40)
le puits a — mètres de profondeur, the shaft is — metres deep. (41)

profondeur (partie enfoncée) (*n.f.*), depth ; deep : (42)
les profondeurs de l'océan, d'une caverne, de l'espace, the depths of the ocean, of a cave, of space. (43 *ou* 44 *ou* 45)
cheminée volcanique par où arrivent des profondeurs les matières minérales en fusion (*f.*), volcanic chimney through which the molten mineral matter comes from the depths. (46)

roches que l'on suppose avoir été apportées des profondeurs (*f.pl.*), rocks which it is supposed have been brought up from the deeps ; rocks which it is thought emanate from the recesses of the earth. (47)

profondeur d'eau (*f.*), depth of water. (48)

profondeur d'exploitation (Mines) (*f.*), depth of working. (49)

profondeur de champ (Photogr.) (*f.*), depth of field. (50)

profondeur de foyer (Photogr.) (*f.*), depth of focus ; depth of definition. (51)

profondeur de passe *ou* **profondeur de coupe** (Travail aux machines-outils) (*f.*), depth of cut. (52)

profondeur du creux (Engrenages) (*f.*), depth below pitch-line ; dedendum (Gearing). (53)

profondeur du dragage (*f.*), dredging-depth. (54)

profondeur du trou de sonde (*f.*), depth of the bore-hole. (55)

profondeur moyenne des mers (*f.*), average depth of the seas. (56)

progrès (*n.m.pl.*), progress ; headway : (57)
faire de grands progrès, to make great progress (*or* great headway). (58)
progrès d'une inondation, progress of a flood. (59)

progresser (*v.i.*), to progress ; to make headway. (60)

progressif, -ive (*adj.*), progressive : (61)
salaire progressif (*m.*), progressive wage. (62)

progression (*n.f.*), progression. (63)

progressivement (*adv.*), progressively. (64)

prohiber (*v.t.*), to prohibit. (65)

prohibitif, -ive (*adj.*), prohibitive : (66)
coût prohibitif (*m.*), prohibitive cost. (67)

prohibition (*n.f.*), prohibition. (68)

projecteur (*n.m.*), projector ; search-light. (69)

projection (*n.f.*), projection : (70)
projection latérale, lateral projection ; side projection. (71)
projection d'une image sur l'écran, projection of an image on the screen. (72)
projection de Mercator, Mercator's projection. (73)

projection (action de lancer un corps pesant) (*n.f.*), projection ; throwing up ; throwing out : (74)
projection de scories, de cendres, et de laves (volcan), projection, throwing up, of scoriæ, ashes, and lava. (75)

projection de sable (*f.*), sand-blasting : (76)
affûtage des limes par projection de sable (*m.*), sharpening files by sand-blasting. (77)

projet (*n.m.*), project ; scheme ; plan ; draft : (78)
projet de chemin de fer (dessein ; intention), railway project ; railway scheme. (79)
projet de chemin de fer (représentation graphique avec devis), plan and estimate for a railway. (80)
projet de marché de sondage pour recherche de houille, draft of boring-contract for prospecting for coal. (81)
projet de rapport, draft report ; draft of report. (82)
projet de route, plan of route. (83)
projet exécutable, workable plan ; feasible plan. (84)
projet inexécutable, unworkable plan. (85)

projeter (porter en avant) (*v.t.*), to project : (86)
corps qui projette son ombre (*m.*), body which projects its shadow. (87)

projeter (lancer) (*v.t.*), to project ; to throw out ; to hurl out ; to force out : (1)
projeter la bourre au dehors du trou (se dit d'un coup de mine qui fait canon), to force the tamping out of the hole (said of a shot which blows out). (2)

projeter (faire un projet) (*v.t.*), to plan ; to project ; to contemplate : (3)
projeter l'érection d'un nouveau bâtiment, to contemplate erecting, to plan the erection of, a new building. (4)
ligne de chemin de fer projetée (*f.*), projected railway-line. (5)

prolongation (délai) (*n.f.*), prolongation ; extension : (6)
prolongation du terme d'un bail, extension of the term of a lease. (7)

prolongement (accroissement de longueur) (*n.m.*), prolongation ; extension ; extending : (8)
prolongement des gîtes, extension of deposits. (9)

prolonger (*v.t.*), to prolong ; to extend : (10)
prolonger un bail, to extend a lease. (11)

promontoire (*n.m.*), promontory ; headland ; cape. (12)

prononcé, -e (*adj.*), pronounced ; marked : (13)
substance d'un éclat métallique très prononcé (*f.*), substance with a very pronounced metallic lustre. (14)

propagation (*n.f.*), propagation : (15)
propagation de la lumière, du son, des ondes électriques, d'une explosion, propagation of light, of sound, of electric waves, of an explosion. (16 *ou* 17 *ou* 18 *ou* 19)

propager (*v.t.*), to propagate ; to spread. (20)

propager (se) (*v.r.*), to propagate ; to spread : (21)
l'expérience montre que la lumière ne se propage pas instantanément, mais avec une vitesse finie (*f.*), experiment shows that light does not propagate instantaneously, but with a finite speed. (22)
incendie qui se propage (*m.*), fire which spreads. (23)

propice (*adj.*), propitious ; favourable ; favor-able : (24)
des vents propices (*m.pl.*), favourable winds. (25)

proportion (*n.f.*), proportion ; percentage : (26)
proportion exacte, exact proportion. (27)
proportion élevée d'humidité, heavy (*or* high) percentage of moisture. (28)

proportionnel, -elle (*adj.*), proportional : (29)
échelle proportionnelle (*f.*), proportional scale. (30)

proportionnellement (*adv.*), proportionally ; in proportion. (31)

proportionner (*v.t.*), to proportion : (32)
proportionner ses dépenses à ses ressources, to proportion one's expenses to one's resources. (33)

proposer (*v.t.*), to propose. (34)

proposer (se) (avoir l'intention) (*v.r.*), to intend ; to have in view. (35)

proposition (*n.f.*), proposition ; proposal. (36)

propre (convenable) (*adj.*), suitable ; suited ; proper : (37)
bois propre à la construction (*m.*), wood suit-able for building. (38)

propre (net) (*adj.*), clean : (39)
minerai propre (*m.*), clean ore. (40)

proprement (*adv.*), suitably ; properly ; cleanly. (41)

proprement dit, -e, proper (*commonly following the noun modified*) : (42)
les pierres précieuses se subdivisent en pierres précieuses proprement dites, et en pierres fines (*f.pl.*), precious stones are subdivided into precious stones proper, and semiprecious stones. (43)

propriétaire (pers.) (*n.m. ou f.*), proprietor ; owner ; holder. (44)

propriétaire de la surface (*m.*), surface-owner. (45)

propriétaire de mine(s) (*m.*), mine-owner. (46 *ou* 47)

propriétaire de mine(s) de charbon (*m.*), colliery-owner ; coalowner. (48 *ou* 49)

propriétaire en droit (*m.*), rightful owner. (50)

propriétaire foncier (*m.*), landowner ; landlord ; landholder. (51)

propriété (possession en propre ; bien-fonds) (*n.f.*), property ; estate : (52)
propriété de grande valeur, property of great value ; most valuable property. (53)
propriété de valeur, valuable property. (54)
propriété minière, mining-property. (55)

propriété (caractère propre ; vertu particulière) (*n.f.*), property : (56)
propriétés chimiques et physiques de l'or, chemical and physical properties of gold. (57)
propriétés mécaniques des aciers au nickel, mechanical properties of nickel steels. (58)
propriétés minéralisatrices du soufre, mineral-izing properties of sulphur. (59)
propriétés optiques des cristaux, optical properties of crystals. (60)

propulseur (*adj. m.*) *ou* **propulsif, -ive** (*adj.*), propelling ; propellent ; propulsive ; pro-pulsatory ; propulsory ; impelling ; impel-lent ; impulsive. (61)

propulseur (*n.m.*), propeller. (62)

propulsion (*n.f.*), propulsion ; propelling ; impulsion ; impelling. (63)

propylite (Pétrol.) (*n.f.*), propylite. (64)

prosopite (Minéral.) (*n.f.*), prosopite. (65)

prospecter (*v.t.*), to prospect : (66)
prospecter une concession, des alluvions, to prospect a claim, alluvials. (67 *ou* 68)

prospecteur (pers.) (*n.m.*), prospector : (69)
prospecteur d'or, gold-prospector. (70)

prospecteur (machine) (Mines) (*n.m.*), prospect-ing-machine. (71)

prospecteur à bras (*m.*), hand-power prospecting-machine. (72)

prospecteur à diamants (*m.*), diamond-drill prospecting-machine. (73)

prospectif, -ive (*adj.*), prospective. (74)

prospection (*n.f.*), prospection ; prospecting : (75)
prospection de terrains de dragage, prospect-ing dredging-ground. (76)
prospection pour diamants, prospecting for diamonds ; diamond-prospecting. (77)

protecteur, -trice (*adj.*), protecting ; protective. (78)

protecteur (*n.m.*), protector ; shield ; guard. (79)

protecteur de niveau d'eau en cristal armé *ou* **protecteur de niveau d'eau à fil de fer noyé dans le cristal** (*m.*), water-level protector, with wire-glass tube. (80)

protecteur en fil de fer (*m.*), wire guard. (81)

protection (*n.f.*), protection : (82)

protection contre les moustiques et contre la fièvre paludéenne, protection against mosquitoes and malaria. (1)

protéger (*v.t.*), to protect ; to shield ; to guard : (2)
protéger les machines contre les intempéries, to protect the machinery against the weather. (3)

protobastite (Minéral) (*n.f.*), enstatite. (4)

protogène(Géol.) (*adj.*), protogenic ; protogenetic : (5)
roches protogènes (*f.pl.*), protogenic rocks. (6)

protogine *ou* **protogyne** (Pétrol.) (*n.m.*), protogin ; protogene ; protogine. (7)

protomorphique (*adj.*), protomorphic. (8)

protoprisme (Cristall.) (*n.m.*), protoprism ; unit prism ; prism of the first order ; first order prism. (9)

protopyramide (Cristall.) (*n.f.*), protopyramid ; pyramid of the first order ; first order pyramid ; unit pyramid. (10)

protoxyde (Chim.) (*n.m.*), protoxide ; monoxide. (11)

protubérance (*n.f.*), protuberance ; boss. (12)

proustite (Minéral) (*n.f.*), proustite ; arsenical silver blende ; light-red silver ore ; light-ruby silver ore. (13)

prouver (*v.t.*), to prove : (14)
les faits prouvent plus que les raisonnements (*m.pl.*), facts prove more than arguments. (15)

provenance (*n.f.*), origin ; source ; source of supply : (16)
proximité de la provenance du combustible (*f.*), proximity to source of fuel-supply. (17)
provenance de l'eau (*f.*), source of the water (*or* of the water-supply). (18)

province (*n.f.*), province : (19)
province pétrographique, petrographical province. (20)

provision (*n.f.*), provision ; store ; supply : (21)
provision abondante de main-d'œuvre, plentiful supply of labour. (22)
provision d'air comprimé, supply of compressed air. (23)
provision d'eau assurée pour toute l'année, assured water-supply throughout the whole year. (24)

provisions (*n.f.pl.*) *ou* **provisions de bouche**, provisions ; food ; chop. (25)

provisoire (*adj.*), provisional ; temporary : (26)
barrage provisoire (*m.*), provisional dam ; temporary dam. (27)

provisoirement (*adv.*), provisionally ; temporarily. (28)

provoquer un tirage *ou* **provoquer un appel d'air**, to create a draught : (29)
l'échappement de vapeur des locomotives sert à provoquer un tirage forcé (*m.*), the exhaust from locomotives serves to create a forced draught. (30)
jet de vapeur qui provoque un appel d'air (*m.*), jet of steam which creates a draught. (31)

provoquer une explosion, to give rise to, to cause, to bring about, an explosion. (32)

proximité (*n.f.*), proximity ; nearness : (33)
proximité de la provenance du combustible, proximity to source of fuel-supply. (34)

prudence (*n.f.*), care ; carefulness ; prudence. (35)

prudent, -e (*adj.*), careful ; prudent. (36)

prussiate (Chim.) (*n.m.*), prussiate : (37)
prussiate de potasse, prussiate of potash. (38)

psammite (Pétrol.) (*n.m.*), psammite. (39)

psammitique (*adj.*), psammitic. (40)

psaturose *ou* **psathurose** (Minéral) (*n.f.*), psaturose ; stephanite ; brittle silver ore. (41)

pseudo-clivage (Géol.) (*n.m.*), false cleavage ; slip-cleavage. (42)

pseudo-malachite (Minéral) (*n.f.*), pseudo-malachite ; phosphorochalcite ; phospho-chalcite. (43)

pseudo-métallique (*adj.*), pseudometallic. (44)

pseudo-porphyrique (*adj.*), pseudoporphyritic. (45)

pseudo-rubis (*n.m.*), Bohemian ruby. (46)

pseudo-saphir (*n.m.*), sapphire-quartz ; blue quartz. (47)

pseudo-topaze (*n.f.*), false topaz ; citrine quartz ; citrine. (48)

pseudo-volcanique (*adj.*), pseudovolcanic. (49)

pseudomorphe *ou* **pseudomorphique** (*adj.*), pseudomorphous ; pseudomorphic : (50)
le quartz est souvent pseudomorphique d'autres minéraux, quartz is often pseudomorphic of other minerals. (51)

pseudomorphe (*n.m.*), pseudomorph. (52)

pseudomorphisme (*n.m.*), pseudomorphism. (53)

pseudomorphose (*n.f.*), pseudomorphosis. (54)

pseudomorphose de moulage (Minéralogie) (*f.*), pleromorphosis. (55)

psilomélane (Minéral) (*n.f.*), psilomelane ; manganese hydrate. (56)

psychromètre (*n.m.*), psychrometer. (57)

psychrométrie (*n.f.*), psychrometry. (58)

psychrométrique (*adj.*), psychrometric ; psychro-metrical. (59)

public, -ique (*adj.*), public : (60)
travaux publics (*m.pl.*), public works. (61)

publiquement (*adv.*), publicly. (62)

puddlage (Métall.) (*n.m.*), puddling. (63)

puddlage gras *ou* **puddlage chaud** *ou* **puddlage bouillant** (*m.*), wet puddling ; pig-boiling. (64)

puddlage sec *ou* **puddlage froid** *ou* **puddlage maigre** (*m.*), dry puddling. (65)

puddler (*v.t.*), to puddle : (66)
puddler le fer, to puddle iron. (67)

puddleur (pers.) (*n.m.*), puddler. (68)

pui (Géol. & Géogr.) (*n.m.*), puy. (69)

puisage (*n.m.*), drawing ; bailing ; fetching : (70)
puisage de l'eau d'un puits, drawing water from a well. (71)
puisage d'un puits à pétrole, bailing an oil-well. (72)

puisard (Mines) (*n.m.*), sump ; sink-hole ; standage. (73)

puisatier (fonceur de puits) (pers.) (*n.m.*), well-sinker. (74)

puisatier (exhaure) (Mines) (pers.) (*n.m.*), sump-man. (75)

puisement (*n.m.*). Same as **puisage**.

puiser (*v.t.*), to draw ; to fetch ; to bail : (76)
puiser de l'eau, to draw, to bail, water. (77)
puiser de l'eau à la rivière, to draw, to fetch, water from the river. (78)

puiser à la cuiller le liquide surnageant, to skim off the supernatant liquid with a ladle. (79)

puiser l'eau avec des pompes, to pump out the water. (80)

puissamment (*adv.*), powerfully. (81)

puissance (force) (*n.f.*), power ; strength : (1)
puissance absorbée par les machines-outils, power required by machine-tools. (2)
puissance d'un feuillet magnétique, strength of a magnetic shell. (3)
puissance d'une dynamo exprimée en watts, power of a dynamo expressed in watts. (4)
puissance d'une lentille, d'un microscope, d'une lunette, power of a lens, of a micro--scope, of a telescope. (5 *ou* 6 *ou* 7)
puissance du vent, strength of the wind. (8)
puissances mécaniques, mechanical powers. (9)
puissance (levier) (*n.f.*), power. (10) See **leviers des premier, second, et troisième genres.**
puissance (Géol. & Mines) (*n.f.*), thickness : (11)
la puissance d'une couche de houille, the thickness of a seam of coal. (12)
puissance apparente (Élec.) (*f.*), apparent power. (13)
puissance calorifique (*f.*), heating-power. (14)
puissance de levage *ou simplement* **puissance** (*n.f.*) (d'une grue, d'un pont roulant), to lift : (15)
grue d'une puissance de levage de — tonnes *ou* grue, puissance : — tonnes (*f.*), crane to lift — tons. (16)
puissance de propulsion (des explosifs) (*f.*), pro--pulsive strength (of explosives). (17)
puissance effective *ou* **puissance au frein** (*f.*), brake-power ; effective power ; actual power ; net power. (18)
puissance effective en chevaux *ou* **puissance en chevaux-vapeur effectifs** *ou* **puissance au frein en chevaux** (*f.*), brake horse-power ; effective horse-power ; actual horse-power ; net horse-power. (19)
puissance efficace (Élec.) (*f.*), true power. (20)
puissance électrique (*f.*), electric power. (21)
puissance en chevaux (*f.*), horse-power. (22)
See note under **cheval.**
puissance indiquée (Méc.) (*f.*), indicated power. (23)
puissance indiquée en chevaux (*f.*), indicated horse-power ; gross horse-power. (24)
puissance lumineuse en bougies (*f.*), candle-power : (25)
une lampe à acétylène, puissance lumineuse — bougies, a — -candle-power acetylene-lamp. (26)
puissance massique (Méc.) (*f.*), power-weight ratio. (27)
puissance motrice (*f.*), motive power. (28)
puissance nominale (Méc.) (*f.*), nominal power. (29)
puissance utile (Méc.) (*f.*), useful power. (30)
puissant, -e (*adj.*), powerful ; strong : (31)
une puissante machine, a powerful machine. (32)
un homme puissant, a strong man. (33)
puissant, -e (Géol. & Mines) (*adj.*), thick ; powerful ; strong : (34)
couches puissantes (*f.pl.*), thick beds. (35)
filon puissant (*m.*), powerful lode ; strong lode. (36)
puits (d'une mine) (*n.m.*), shaft (of a mine). (37)
puits (d'une mine de charbon) (*n.m.*), pit, shaft, (of a coal-mine). (38)
puits (pour en tirer de l'eau, du pétrole, du gaz, etc.) (*n.m.*), well ; hole. (39)
puits (fontaine ; source) (*n.m.*), well ; fountain. (40)

puits (Ponts & Chaussées) (*n.m.*), well. (41)
puits (d'un ascenseur) (*n.m.*), well (of a lift). (42)
puits à cabestan (*m.*), whim-shaft ; gin-pit. (43)
puits à eau (*m.*), water-well. (44)
puits à jaillissement périodique (*m.*), periodical fountain. (45)
puits à main (*m.*). Same as **puits creusé à la main.**
puits à pétrole (*m.*), oil-well ; petroleum-well. (46)
puits absorbant (*m.*), absorbing well ; drain-well. (47)
puits abyssinien (*m.*), driven well ; drive-well ; tube-well. (48)
puits artésien (*m.*), Artesian well. (49)
puits aux échelles (*m.*), ladderway ; ladder-shaft ; ladder-road ; footway-shaft. (50)
puits bétonné (*m.*), shaft lined with concrete. (51)
puits blindé (*m.*), metal-lined shaft ; ironclad shaft. (52)
puits boisé (*m.*), timbered shaft. (53)
puits capricieux (*m.*), capricious well ; fitful well ; freak well. (54)
puits caractérisé par une stabilité de production (Exploitation du pétrole) (*m.*), stayer : (55)
les puits sont caractérisés par une grande stabilité de production, the wells are good stayers. (56)
puits central (*m.*), main shaft ; central shaft. (57)
puits chinois (*m.*), China shaft. (58)
puits collecteur (*m.*), collecting-pit ; sump-shaft. (59)
puits coulant naturellement (opp. à **puits pompé**) (Exploitation du pétrole) (*m.*), flowing well. (60)
puits creusé à la main (Mines) (*m.*), hand-sunk shaft ; hand-dug surface-shaft. (61)
puits creusé à la main (Puits à pétrole, à eau, etc.) (*m.*), hand-dug well. (62)
puits d'aérage (*m.*), air-shaft ; ventilation-shaft. (63)
puits d'appel *ou* **puits d'appel d'air** (*m.*), down--cast ; downcast-shaft. (64)
puits d'embouage (*m.*), flushing-shaft. (65)
puits d'entrée *ou* **puits d'entrée d'air** (*m.*), down--cast ; downcast-shaft. (66)
puits d'épuisement *ou* **puits d'exhaure** (*m.*), pumping-shaft ; water-shaft ; sump-shaft. (67)
puits d'essai *ou* **puits d'exploration** (Mines) (*m.*), trial-shaft ; exploratory pit. (68)
puits d'essai *ou* **puits d'exploration** (pétrole, etc.) (*m.*), trial well ; trial hole ; exploratory well. (69)
puits d'extraction (*m.*), winding-shaft ; hoisting-shaft ; working-shaft ; mining-shaft ; work--ing-pit. (70)
puits d'extraction de saumure *ou* **puits d'extraction du sel** (*m.*), brine-well ; brine-pit ; salt-well. (71)
puits de drainage (*m.*), drain-shaft. (72)
puits de feu (*m.*), fire-well. (73)
puits de gaz *ou* **puits de gaz naturel** (*m.*), gas-well ; natural gas-well ; gasser. (74)
puits de mine (*m.*), mine-shaft ; mining-shaft. (75)
puits de pétrole (*m.*), oil-well ; petroleum-well. (76)
puits de recherches *ou* **puits de prospection** (Mines) (*m.*), prospecting-shaft ; discovery-shaft ; trial-shaft. (77)

puits de recherches (pétrole) (*m.*), exploratory well ; trial hole. (1)

puits de refoulement d'air (*m.*), downcast ; downcast-shaft. (2)

puits de remblai (*m.*), waste-shaft. (3)

puits de retour *ou* puits de retour d'air *ou* puits de sortie *ou* puits de sortie d'air (*m.*), upcast ; upcast-shaft ; uptake ; uptake-shaft ; out--take. (4)

puits de surface (*m.*), surface-shaft. (5)

puits de travail (*m.*). Same as puits d'extraction.

puits de ventilation (*m.*), air-shaft ; ventilation-shaft. (6)

puits des échelles (*m.*). Same as **puits aux échelles**.

puits des pompes (*m.*), pump-shaft. (7)

puits en production *ou* puits en rapport (*m.*), producing well. (8)

puits foré (*m.*), bored well ; drilled well ; Artesian well. (9)

puits foré à la main (*m.*), hand-drilled well. (10)

puits incliné (*m.*), incline-shaft ; inclined shaft ; incline. (11)

puits instantané (*m.*), driven well ; drive-well ; tube-well. (12)

puits intérieur (*m.*), internal shaft ; blind shaft ; blind pit ; staple ; staple-shaft ; staple-pit. (13)

puits intermittent (*m.*), intermittent well. (14)

puits jaillissant (pétrole) (*m.*), spouter ; gusher ; spout-well ; oil-fountain. (15)

puits jumeaux (*m.pl.*), twin shafts. (16)

puits muraillé (*m.*), walled shaft. (17)

puits naturel (*m.*), natural well. (18)

puits pompé (opp. à puits coulant naturellement) (Exploitation du pétrole) (*m.*), pumping well ; pumper. (19)

puits principal (*m.*), main shaft ; central shaft. (20)

puits stérile *ou* puits sec (Exploitation du pétrole) (*m.*), dry hole. (21)

puits tari (*m.*), dry well. (22)

pulpage (*n.m.*) *ou* pulpation (*n.f.*), pulping. (23)

pulpe (*n.f.*), pulp. (24)

pulper (*v.t.*), to pulp. (25)

pulsation (*n.f.*), pulsation : (26)

pulsation du courant (Élec.), pulsation of the current. (27)

pulsomètre (*n.m.*), pulsometer ; pulsator. (28)

pulvérin (poussière d'eau) (*n.m.*), spray. (29)

pulvérin (poussière de charbon)(*n.m.*), coal-dust. (30)

pulvérin schisteux *ou* pulvérin rocheux *ou* pulvérin incombustible (Schistification des galeries de mine) (*m.*), stone-dust ; rock-dust. (31)

pulvérisable (*adj.*), pulverizable. (32)

pulvérisateur (*n.m.*), pulverizer ; pulverizing-machine ; spray-producer ; spray ; water-sprayer ; atomizer ; vaporizer. (33)

pulvérisation (*n.f.*), pulverization ; pulverizing ; spraying. (34)

pulvériser (solides) (*v.t.*), to pulverize ; to powder ; to reduce to powder. (35)

pulvériser (liquides) (*v.t.*), to spray ; to vaporize ; to pulverize. (36)

pulvériser (se) (*v.r.*), to pulverize ; to powder ; to vaporize : (37)

substance qui se pulvérise très facilement (*f.*), substance which pulverizes very easily (*or* which powders very readily). (38)

pulvérulence (*n.f.*), pulverulence ; powderiness. (39)

pulvérulent, -e *ou* pulvifère (*adj.*), pulverulent ; powdery ; pulverous. (40)

pumiciforme (*adj.*), pumiciform : (41)

lave pumiciforme (*f.*), pumiciform lava. (42)

pumicite *ou* pumite (*n.f.*), pumice ; pumice-stone. (43)

pumiqueux, -euse (*adj.*), pumiceous ; pumicose : (44)

structure pumiqueuse (*f.*), pumiceous structure. (45)

punaise (de dessin) (*n.f.*), drawing-pin. (46)

punaise à tête plate, à tête en goutte-de-suif (*f.*), drawing-pin with flat head, with dome-shaped head. (47 *ou* 48)

pupitre à retouche, avec réflecteur (Photogr.) (*m.*), retouching-desk, with reflector. (49)

pur, -e (*adj.*), pure : (50)

cuivre pur (*m.*), pure copper. (51)

pur anthracite (opp. à *houille anthraciteuse*) (*m.*), pure anthracite. (52) (distinguished from *semianthracite*).

purbeckien, -enne (Géol.) (*adj.*), Purbeckian. (53)

purbeckien (*n.m.*), Purbeckian. (54)

pureau (partie visible d'une ardoise, d'une tuile sur la toiture) (*n.m.*), bare ; margin. (55)

purement (*adv.*), purely. (56)

pureté (*n.f.*), purity. (57)

purette (sable noir trouvé sur le bord de la mer) (*n.f.*), black sand. (58)

purge (d'un cylindre à vapeur) (*n.f.*), draining (a steam-cylinder). (59)

purger le fer par le cinglage, to purify iron by shingling. (60)

purger un cylindre, to drain a cylinder. (61)

purgeur (de cylindre à vapeur) (*n.m.*), drip-cock ; drain-cock ; cylinder-cock. (62)

purgeur à flotteur (*m.*), float-trap. (63)

purgeur automatique (*m.*), steam-trap. (64)

purifiant, -e *ou* purificateur, -trice (*adj.*), purify--ing. (65)

purifiant *ou* purificateur (*n.m.*), purifier. (66)

purificateur pour gaz naturel (*m.*), purifier for natural gas. (67)

purification (*n.f.*), purification. (68)

purifier (*v.t.*), to purify : (69)

purifier des métaux, des huiles, to purify metals, oils. (70 *ou* 71)

purifier (se) (*v.r.*), to purify. (72)

puy (Géol. & Géogr.) (*n.m.*), puy. (73)

pycnite (Minéral.) (*n.f.*), pycnite. (74)

pycnomètre (*n.m.*), pycnometer ; specific-gravity bottle ; specific-gravity flask. (75)

pylône (*n.m.*), tower ; lattice tower ; standard ; trellis-post ; pylon. (76)

pylône (échafaudage) (Constr.) (*n.m.*), tower. (77)

pyr (Photom.) (*n.m.*), pyr ; decimal candle ; bougie décimale : (78)

le pyr ou la bougie décimale vaut un vingtième de l'étalon Violle, the pyr or decimal candle has the value of one-twentieth of the Violle standard. (79) See **violle** in vocabulary.

pyramidal, -e, -aux (*adj.*), pyramidal ; pyramid-shaped : (80)

comble pyramidal (*m.*), pyramidal roof. (81)

pyramide (Géom.) (*n.f.*), pyramid. (82)

pyramide (quelque chose ayant la forme d'une pyramide) (*n.f.*), pyramid : (83)

la superbe pyramide rocheuse du mont Cervin, the superb rocky pyramid of the Matterhorn. (84)

pyramide coiffée ou **pyramide d'érosion** (Géol.) (f.), erosion column ; chimney-rock. (1)

pyramide de première espèce (Cristall.) (f.), pyramid of the first order ; first order pyramid ; unit pyramid ; protopyramid. (2)

pyramide de seconde espèce ou **pyramide de deuxième espèce** (Cristall.) (f.), pyramid of the second order ; second order pyramid ; deuteropyramid. (3)

pyramide de terre (Géol.) (f.), earth pillar : (4) les pyramides de terre de Bolzano, dans le Tyrol méridional, the earth pillars of Bolzano, in the Southern Tyrol. (5)

pyramide de troisième espèce (Cristall.) (f.), pyramid of the third order ; third order pyramid ; tritopyramid. (6)

pyramide droite (f.), right pyramid. (7)

pyramide hexagonale (f.), hexagonal pyramid. (8)

pyramide oblique (f.), oblique pyramid. (9)

pyramide régulière (f.), regular pyramid. (10)

pyramide tronquée (f.), truncated pyramid. (11)

pyrargyrite (Minéral) (n.f.), pyrargyrite ; argyrythrose ; dark-ruby silver ; dark-red silver ore. (12)

pyrgome (Minéral) (n.m.), pyrgom. (13)

pyrite (n.f.) ou **pyrites** (n.f.pl.) (sulfure métallique naturel), pyrites : (14) pyrites aurifères, auriferous pyrites ; gold-bearing pyrites ; sulphurets. (15)

pyrite (n.f.) ou **pyrite de fer** ou **pyrite jaune** ou **pyrite martiale** (Minéral), pyrite ; pyrites ; iron pyrites ; sulphur-ore ; brasses ; fool's gold. (16)

pyrite blanche ou **pyrite rhombique** (Minéral) (f.), white pyrites ; white iron pyrites ; marcasite. (17)

pyrite crêtée (Minéral) (f.), cockscomb pyrites. (18)

pyrite de cuivre (Minéral) (f.), copper pyrites ; yellow pyrites ; yellow copper ; yellow copper ore ; yellow ore ; fool's gold ; chalcopyrite. (19)

pyrite magnétique (Minéral) (f.), magnetic pyrites ; pyrrhotite. (20)

pyriteux, -euse (adj.), pyritic ; pyritical ; pyritous ; pyritose ; pyritaceous : (21) filon pyriteux (m.), pyritic lode. (22)

pyritifère (adj.), pyritiferous. (23)

pyritisation (n.f.), pyritization. (24)

pyritiser (v.t.), to pyritize : (25) pyritiser le fer, to pyritize iron. (26)

pyritoèdre (Cristall.) (adj.), pyritohedral. (27)

pyritoèdre (n.m.), pyritohedron. (28)

pyritologie (n.f.), pyritology. (29)

pyrochimique (adj.), pyrochemical. (30)

pyrochlore (Minéral) (n.m.), pyrochlore. (31)

pyrochroïte (Minéral) (n.f.), pyrochroite. (32)

pyroélectricité (n.f.), pyroelectricity. (33)

pyroélectrique (adj.), pyroelectric : (34) en raison de son hémimorphisme, la tourmaline est nettement pyroélectrique, by reason of its hemimorphism, tourmaline is distinctly pyroelectric. (35)

pyrogène (Géol.) (adj.), pyrogenous : (36) roches pyrogènes (f.pl.), pyrogenous rocks. (37)

pyrogénèse (n.f.), pyrogenesis ; pyrogenesia. (38)

pyrogénésique ou **pyrogénétique** (adj.), pyro-genetic. (39)

pyrognostique (adj.), pyrognostic : (40) analyse pyrognostique (f.), pyrognostic analysis. (41)

pyrologie (n.f.), pyrology. (42)

pyrologique (adj.), pyrological. (43)

pyrologiste (pers.) (n.m.), pyrologist. (44)

pyrolusite (Minéral) (n.f.), pyrolusite. (45)

pyromagnétique (adj.), pyromagnetic. (46)

pyromètre (n.m.), pyrometer. (47)

pyromètre électrique (m.), electric pyrometer. (48)

pyromètre optique (m.), optical pyrometer. (49)

pyrométrie (n.f.), pyrometry. (50)

pyrométrique (adj.), pyrometric ; pyrometrical. (51)

pyrométriquement (adv.), pyrometrically. (52)

pyromorphite (Minéral) (n.f.), pyromorphite. (53)

pyrope (Minéral) (n.m.), pyrope ; Bohemian garnet. (54)

pyrophyllite (Minéral) (n.f.), pyrophyllite ; pencil-stone. (55)

pyropneumatique (adj.), hot-air : (56) appareil pyropneumatique (m.), hot-air apparatus. (57)

pyrorthite (Minéral) (n.f.), pyrorthite. (58)

pyroschiste (Géol.) (n.m.), pyroschist. (59)

pyrosclérite (Minéral) (n.f.), pyrosclerite. (60)

pyroscope (n.m.), pyroscope. (61)

pyrosphère (Géol.) (n.f.), pyrosphere. (62)

pyrostat (n.m.), pyrostat. (63)

pyrostatique (adj.), pyrostatic. (64)

pyroxène (Minéral) (n.m.), pyroxene. (65)

pyroxéneux, -euse ou **pyroxénique** (adj.), pyroxenic. (66)

pyroxénite (Pétrol.) (n.f.), pyroxenite ; pyrox-enyte ; augite-rock. (67)

pyroxyle (n.m.) ou **pyroxyline** (n.f.) (Explosif), pyroxyline ; pyroxyle. (68)

pyrrhotine (Minéral) (n.f.), pyrrhotite ; pyrrhotine ; magnetic pyrites. (69)

Q

quadrangulaire (adj.), quadrangular ; quadri-lateral ; four-sided : (70) prisme quadrangulaire (m.), quadrangular prism ; four-sided prism. (71)

quadrant (Géom.) (n.m.), quadrant. (72)

quadratique (adj.), quadratic. (73)

quadrature (Math.) (n.f.), quadrature ; squaring : (74) quadrature du cercle, quadrature of the circle ; squaring the circle. (75)

quadrilatéral, -e, -aux ou **quadrilatère** (adj.), quadrilateral ; four-sided ; quadrangular ; tetragonal. (76)

quadrilatère (n.m.), quadrilateral ; quadrangle ; four-side ; tetragon. (77)

quadruple (adj.), quadruple : (78) machine à quadruple expansion (f.), quad-ruple-expansion engine. (79)

quai (n.m.) ou **quai de chargement** ou **quai de port**, quay ; wharf. (80)

quai de chemin de fer (*m.*), railway-platform. (1)

qualitatif, -ive *ou* **qualificatif, -ive** (*adj.*), qualita--tive : (2)

analyse qualitative (*ou* qualificative) (*f.*), qualitative analysis. (3)

qualitativement (*adv.*), qualitatively. (4)

qualité (*n.f.*), quality ; value : (5)

qualité industrielle d'une roche, commercial value of a rock. (6)

qualité inférieure, inferior quality ; poor quality. (7)

qualité médiocre, indifferent quality. (8)

qualité moyenne, middling quality : average quality. (9)

qualité ordinaire, ordinary quality ; common quality. (10)

qualité supérieure, superior quality ; better quality. (11)

quantitatif, -ive (*adj.*), quantitative : (12)

analyse quantitative (*f.*), quantitative analysis. (13)

quantitativement (*adv.*), quantitatively. (14)

quantité (*n.f.*), quantity : (15)

quantité d'eau nécessaire, de minerai extrait, quantity of water required, of mineral raised. (16 *ou* 17)

quantité d'électricité, quantity of electricity : (18)

le coulomb est l'unité pratique de mesure des quantités d'électricité, the coulomb is the practical unit of measurement of quantities of electricity. (19)

quantité d'eau tombée (précipitation atmosphé--rique) (*f.*), rainfall : (20)

quantité d'eau tombant annuellement, annual rainfall. (21)

quantité de mouvement (Méc.) (*f.*), momentum. (22)

quart (mesure anglaise de capacité) (*n.m.*), quart =1·136 litres. (23)

quart (mesure de capacité pour les liquides en usage dans les États-Unis) (*n.m.*), quart =·946327 litre. (24)

quart de cercle (instrument d'astronomie) (*m.*), quadrant. (25)

quart de cercle (d'un compas, d'un compas de calibre) (*m.*), quadrant, wing, (of a compass, of a calipers). (26 *ou* 27)

quart de rond (moulure) (Menuis.) (*m.*), quarter-round ; ovolo ; Roman ovolo. (28)

quart de rond (rabot) (Menuis.) (*m.*), quarter-round (plane). (29)

quartation (Métall.) (*n.f.*), quartation ; in--quartation : (30)

départ avec quartation d'argent, de cadmium (*m.*), parting with quartation of silver, of cadmium. (31 *ou* 32)

quartation (Échantillonnage du minerai) (*n.f.*), quartering. (33)

quarter (mesure anglaise de poids) (*n.m.*), quarter =12·70 kilogrammes. (34)

quarter (mesure anglaise de capacité) (*n.m.*), quarter=2·909 hectolitres. (35)

quartier (*n.m.*), quarter ; district. (36)

quartier général (*m.*), headquarters. (37)

quartier tournant (escalier) (*m.*), quarter space. (38) Cf. **escalier.**

quartz (*n.m.*), quartz. (39)

quartz à astéries (*m.*), asteriated quartz ; star quartz. (40)

quartz à or libre (*m.*), free-milling quartz. (41)

quartz aurifère (*m.*), auriferous quartz ; gold-bearing quartz ; gold quartz. (42)

quartz aventuriné (*m.*), aventurine quartz. (43)

quartz bleu (*m.*), blue quartz ; sapphire-quartz. (44)

quartz citrine (*m.*), citrine quartz ; citrine ; false topaz. (45)

quartz cristallisé (*m.*), hyaline-quartz. (46)

quartz enfumé (*m.*), smoky quartz. (47)

quartz hyalin (*m.*), hyaline-quartz. (48)

quartz hyalin bleu (*m.*), blue hyaline-quartz. (49)

quartz laiteux (*m.*), milky quartz ; greasy quartz. (50)

quartz lydien (*m.*), Lydian stone ; lydite ; touchstone. (51)

quartz œil-de-chat (*m.*), cat's-eye quartz. (52)

quartz saphirin (*m.*), sapphire-quartz ; blue quartz. (53)

quartzeux, -euse (*adj.*), quartzose ; quartzous ; quartzy : (54)

sable quartzeux (*m.*), quartzy sand. (55)

quartzifère (*adj.*), quartziferous. (56)

quartziforme (*adj.*), quartziform. (57)

quartzine (Minéral) (*n.f.*), quartzine. (58)

quartzique (*adj.*), quartzic. (59)

quartzite (Pétrol.) (*n.m.*), quartzite ; quartzyte ; quartz-rock. (60)

quartzitique (*adj.*), quartzitic. (61)

quaternaire (Géol.) (*adj.*), Quaternary : (62)

âge quaternaire (*m.*), Quaternary age. (63)

quaternaire (Cristall.) (*adj.*), quaternary : (64)

système quaternaire (*m.*), quaternary system. (65)

quatre roues (à), four-wheeled : (66)

wagon à quatre roues (*m.*), four-wheeled wagon. (67)

quelconque (*adj.*), any : (68)

prisme quelconque (*m.*), any prism. (69)

queue (*n.f.*), tail. (70)

queue (d'un foret) (*n.f.*), shank (of a drill). (71)

queue (d'une lime) (*n.f.*), tang (of a file) : (72)

lime à deux queues (*f.*), two-tanged file. (73)

queue (d'un moulin à vent) (*n.f.*), rudder, tail, (of a windmill). (74)

queue (de chambre noire) (*n.f.*), base-board, bed, (of camera). (75)

queue carrée (*f.*), taper square shank ; bit-stock shank : (76)

foret hélicoïdal à queue carrée (*m.*), twist-drill with taper square shank ; bit-stock twist-drill. (77)

queue conique (*f.*), taper shank. (78)

queue conique au cône Morse (*f.*), Morse taper shank. (79)

queue cylindrique (*f.*), parallel shank ; straight shank. (80)

queue-d'aronde [queues-d'aronde *pl.*] *ou* **queue-d'hironde** [queues d'hironde *pl.*] (Charp.) (*n.f.*), dovetail ; fantail : (81)

assemblage à queue d'aronde *ou* assemblage en queue d'hironde (*m.*), dovetail joint ; dovetailed joint ; fantail joint. (82)

queue-d'aronde cachée (*f.*), secret dovetail. (83)

queue d'orientation (d'un anémomètre, ou in--strument similaire) (*f.*), rudder ; tail ; vane to bring the instrument into the wind. (84 *ou* 85)

queue-de-cochon [queues-de-cochon *pl.*] (*n.f.*), auger-gimlet. (86)

queue de piston (*f.*), tail piston-rod. (87)

queue-de-rat [queues-de-rat *pl.*] (*n.f.*), rat-tail file ; rat-tailed file. (88)

queue de soupape (*f.*), valve-rod ; valve-stem. (1)

queue de vanne (Hydraul.) (*f.*), gate-stem. (2)

quille (coin de carrière) (*n.f.*), wedge ; stone-wedge ; plug. (3)

quincite *ou* **quincyte** (Minéral) (*n.f.*), quincite. (4)

quinconce (en), staggered ; zigzag ; cross (*used as adj.*) : (5)

rivure en quinconce (*f.*), staggered rivetting ; zigzag riveting ; cross-riveting. (6)

quintal (*n.m.*), centner=50 kilos. (7)

quintal métrique (*m.*), metric quintal ; metric centner=100 kilos *or* 1·968 cwt. (8)

quintuple (*adj.*), quintuple : (9)

machine à quintuple expansion (*f.*), quintuple-expansion engine. (10)

quinzaine (*n.f.*) *ou* **quinze jours** (*m.pl.*), fortnight. (11)

quitter les rails, to jump the metals ; to leave the metals ; to run off the rails. (12)

quitter son poste, to leave, to quit, one's post. (13)

quotidi n, -enne (*adj.*), daily : (14)

tâche quotidienne (*f.*), daily task. (15)

R

rabat-eau [rabat-eau *pl.*] *ou* **rabat-l'eau** [rabat-l'eau *pl.*] (pour une meule) (*n.m.*), splash-guard ; splash-board ; splash-wing ; splasher. (16)

rabattement des bords de tôles (*m.*), flanging plates. (17)

rabattement des collerettes de tubes (*m.*), flang-ing tubes. (18)

rabattoir (de plombier) (*n.m.*), dresser ; lead-dresser ; beater ; bat. (19)

rabattoir (d'ardoisier) (*n.m.*), slate-knife. (20)

rabattre la collerette de, les collerettes de (en parlant d'un tube, des tubes) (*v.t.*), to flange ; to bead over : (21)

rabattre la collerette d'un tube de chaudière, to flange, to bead over, a boiler-tube. (22)

rabattre le bord de, les bords de (en parlant d'une tôle, des tôles, etc.) (*v.t.*), to flange : (23)

rabattre le bord d'une tôle de chaudière, to flange a boiler-plate. (24)

rabattre un collet sur une tôle, to turn up a flange on a sheet. (25)

râble (tire-braise) (*n.m.*), rake. (26)

rabot (outil) (*n.m.*), plane. (27)

rabot à contre-fer *ou* **rabot à fer double** (*m.*), plane with break-iron ; plane with double iron. (28)

rabot à lumière de côté (*m.*), side-plane. (29)

rabot à semelle d'acier (*m.*), plane with steel-plated sole. (30)

rabot cintré (*m.*), compass-plane. (31)

rabot de menuisier, plat à navette (*m.*), smooth-ing-plane, smooth-plane, flat sole, curved stock. (32)

rabot de menuisier, plat carré (*m.*), smoothing-plane, smooth-plane, flat sole, square stock. (33)

rabot denté (*m.*), tooth-plane ; toothing-plane. (34)

rabot-racloir [rabots-racloirs *pl.*] (*n.m.*), scraping-plane. (35)

rabot rond (opp. à mouchette) (*m.*), round plane. (36) (opp. to hollow plane).

rabotage *ou* **rabotement** (*n.m.*), planing. (37)

rabotage glaciaire (Géogr. phys.) (*m.*), glacial planing. (38)

raboter (*v.t.*), to plane : (39)

raboter une planche, to plane a board. (40)

glacier qui rabote son lit (*m.*), glacier which planes its bed. (41)

raboter (se) (*v.r.*), to be planed : (42)

planche qui ne peut se raboter à cause de ses nœuds (*f.*), board which cannot be planed because of knots. (43)

raboteur (pers.) (*n.m.*), planer. (44)

raboteuse (*n.f.*), planer ; planing-machine. (45)

See **machine à raboter** for varieties.

raboteux, -euse (*adj.*), rough ; rugged ; uneven : (46)

bois raboteux (*m.*), rough wood. (47)

sentier raboteux (*m.*), rough path ; rugged track ; uneven trail. (48)

raccagnac (cliquet) (*n.m.*), ratchet ; ratchet-brace ; engineers' ratchet-brace ; brace. (49)

raccagnac à clef (*m.*), ratchet-brace with tommy-bar. (50)

raccommodable (*adj.*), mendable ; repairable. (51)

raccommodage (action) (*n.m.*), mending ; repair-ing. (52)

raccommodage (travail fait) (*n.m.*), repair. (53)

raccommoder (*v.t.*), to mend ; to repair. (54)

raccommoder (se) (*v.r.*), to be mended ; to be repaired. (55)

raccommodeur (pers.) (*n.m.*), repairer. (56)

raccord (accord, ajustement de deux parties d'abord séparées d'un ouvrage) (*n.m.*), con-nection ; connecting : (57)

le groupement en dérivation présente l'incon-vénient de nécessiter de nombreux raccords de fils avec les deux parties de la ligne, coupling in parallel has the disadvantage of necessitating numerous connections of wires with the two parts of the line. (58)

raccord (pièce métallique) (*n.m.*), coupling ; union ; connection ; coupler. (59)

raccord à culotte (*m.*), breeching-piece (Y pipe-connection). (60)

raccord à T (*m.*), tee-piece union ; union T. (61)

raccord coudé *ou* **raccord en équerre** (*m.*), elbow union. (62)

raccord d'air (*m.*), air-connection. (63)

raccord de pompe (*m.*), pump-connection. (64)

raccord de tuyaux (*m.*), pipe-coupling ; pipe-union ; pipe-connection ; hose-coupling ; hose-union. (65)

raccord pour tuyaux d'arrosage (*m.*), hose-coupling ; hose-union (for garden-hose, or the like). (66)

raccord universel (*m.*), universal coupling. (67)

raccordement (*n.m.*), coupling ; connecting ; connection ; union. (68)

raccordement (Ch. de f.) (*n.m.*), transition ; easement : (69)

introduire un raccordement après coup, to introduce a transition on an existing track. (70)

raccorder (*v.t.*), to connect; to couple; to con-
-nect up; to couple up; to join: (1)
raccorder des tuyaux, to connect pipes. (2)
raccorder une courbe à l'extrémité d'une ligne
droite, to join a curve to the end of a straight
line. (3)
raccorder (Ch. de f.) (*v.t.*), to transition: (4)
raccorder des courbes, to transition curves.
(5)
raccorder (se) (Ch. de f.) (*v.r.*), to be transitioned:
(6)
segments de chemins de fer qui se raccordent
par un arc de cercle (*m.pl.*), segments of
railway which are transitioned by a circular
arc. (7)
raccourcir (*v.t.*), to shorten: (8)
raccourcir la course d'un piston, to shorten
the stroke of a piston. (9)
raccourcir (*v.i.*), to shorten; to draw in: (10)
les nuits raccourcissent en été (*f.pl.*), the nights
shorten in summer. (11)
raccourcir (se) (*v.r.*). to shorten; to draw in: (12)
l'hiver, les jours se raccourcissent, in winter,
the days shorten (*or* draw in). (13)
raccourcissement (*n.m.*), shortening. (14)
racine (d'un arbre) (*n.f.*), root (of a tree). (15)
racine (Géol.) (*n.f.*), root: (16)
le mont Cervin et la Dent Blanche sont sans
racines, la nappe dont ils forment partie
ayant été charriée une cinquantaine de
kilomètres ou plus au delà de son lieu
d'origine en Piémont, the Matterhorn and
the Dent Blanche are without roots, the
rock-sheet of which they form part having
travelled some thirty miles or more from
its place of origin in Piedmont. (17)
racine (d'une dent d'engrenage) (*n.f.*), root,
dedendum, (of a gear-tooth). (18)
raclage (grattage) (*n.m.*), scraping. (19)
racler (gratter) (*v.t.*), to scrape. (20)
racler (ratisser) (*v.t.*), to rake. (21)
raclette (grattoir) (*n.f.*), scraper. (22)
raclette (rateau) (*n.f.*), rake. (23)
raclette (*n.f.*) *ou* **racloir** (*n.m.*) (de cantonnier),
squeegee; squilgee. (24)
raclette pour essorer les épreuves (Photogr.)
(*f.*), squeegee for drying prints (flat pattern).
(25) Cf. rouleau pour le collage des épreuves.
raclette pour tubes de chaudière (*f.*), tube-
scraper. (26)
racloir de menuisier (*m.*), joiners' scraper;
cabinet-scraper. (27)
radeau (*n.m.*), raft. (28)
radeau d'arbres (*m.*), log raft. (29)
radial, -e, -aux (*adj.*), radial: (30)
bras radial (*m.*), radial arm. (31)
radiale (*n.f.*), radial drilling-machine. (32) See
machine à percer radiale for varieties.
radialement (*adv.*), radially. (33)
radian (unité d'angle) (*n.m.*), radian. (34)
radiant, -e (*adj.*), radiant: (35)
chaleur radiante (*f.*), radiant heat. (36)
radiateur (*n.m.*), radiator: (37)
radiateur pour le chauffage d'un appartement,
radiator for warming a room. (38)
radiateur intégral (Phys.) (*m.*), perfect radiator;
black body; ideal black body. (39)
radiation (*n.f.*), radiation: (40)
radiations des corps minéraux, radiations of
mineral substances. (41)
radical, -e, -aux (*adj.*), radical. (42)
radical (Chim. & Math.) (*n.m.*), radical. (43)

radical composé (Chim.) (*m.*), compound radical.
(44)
radier (*n.m.*), floor, invert; inverted arch: (45)
radier d'un sas d'écluse, floor, invert, of a lock-
chamber. (46)
radier (revêtement) (Constr. hydraul.) (*n.m.*),
apron. (47)
radifère (*adj.*), radiferous. (48)
radio-actif, -ive (*adj.*), radioactive. (49)
radio-activité (*n.f.*), radioactivity; radioaction.
(50)
radioconducteur (Élec.) (*n.m.*), radioconductor.
(51)
radiogramme (*n.m.*), radiogram; radiograph.
(52)
radiographique (*adj.*), radiographic; radio-
-graphical: (53)
essai radiographique (*m.*), radiographic test.
(54)
radiolite (Minéral) (*n.f.*), radiolite (55)
radiologie (*n.f.*), radiology. (56)
radiologique (*adj.*), radiologic. (57)
radiomètre (*n.m.*), radiometer. (58)
radiométrie (*n.f.*), radiometry. (59)
radiométrique (*adj.*), radiometric. (60)
radiomicromètre (*n.m.*), radiomicrometer. (61)
radiotélégraphie (*n.f.*), radiotelegraphy; wireless
telegraphy; aerial telegraphy. (62)
radiotéléphonie (*n.f.*), radiotelephony; wireless
telephony. (63)
radium (Chim.) (*n.m.*), radium. (64)
rafale (*n.f.*), gust of wind. (65)
raffinage (*n.m.*), refining: (66)
raffinage à l'acide, refining by acid. (67)
raffinage des huiles, refining of oils. (68)
raffinage du plomb, lead-refining. (69)
raffinage du soufre, sulphur-refining. (70)
raffinage électrolytique du cuivre, electro-
-refining copper. (71)
raffiner (*v.t.*), to refine. (72)
raffiner (se) (*v.r.*), to refine; to be refined: (73)
substance qui se raffine par divers procédés (*f.*),
substance which is refined by various pro-
-cesses. (74)
raffinerie (*n.f.*), refinery. (75)
raffinerie d'huiles de schiste (*f.*), shale-oil
refinery. (76)
raffinerie de pétrole (*f.*), petroleum-refinery;
petroleum-works. (77)
raffinerie de sel (*f.*), salt-refinery; salt-works.
(78)
raffinerie de soufre (*f.*), sulphur-refinery; sul-
-phur-works. (79)
raffineur (pers.) (*n.m.*), refiner. (80)
raffûtage (*n.m.*), resharpening. (81)
raffûter (*v.t.*), to resharpen: (82)
raffûter un ciseau, to resharpen a chisel. (83)
rafraîchir (*v.t.*), to keep cool: (84)
rafraîchir un outil au moyen d'un courant
d'eau, to keep a tool cool by means of a
current of water. (85)
raide (rigide) (*adj.*), stiff; rigid. (86)
raide (abrupt) (*adj.*), steep; stiff: (87)
pente raide (*f.*), steep slope; stiff slope. (88)
raide (tendu) (se dit d'un câble) (*adj.*), taut;
tight; stiff: (89)
câble raide (*m.*), taut rope; tight rope. (90)
raideur (rigidité) (*n.f.*), stiffness; rigidity. (91)
raideur (rapidité d'une pente) (*n.f.*), steepness:
(92)
la raideur d'un escalier, the steepness of a
staircase. (93)

raideur (d'une corde) (*n.f.*), tautness, tightness, (of a rope). (1)

raidir (rendre rigide) (*v.t.*), to stiffen : (2)
raidir une plaque par des armatures, to stiffen a plate with straps. (3)

raidir un câble, to take up the slack of, to tighten, to tauten, a rope. (4)

raidissement (*n.m.*), stiffening ; tightening. (5)

raidisseur (pour fils) (*n.m.*), wire-stretcher ; wire-strainer ; strainer ; wire-straining ratchet ; raidisseur. (6)

raie (*n.f.*), line ; stroke ; streak. (7)

raie (Spectroscopie) (*n.f.*), line : (8)
la flamme de sodium est jaune et présente à l'analyse spectrale deux belles raies jaunes, the sodium flame is yellow and gives in spectrum analysis two beautiful yellow lines. (9)

raies brillantes *ou* raies lumineuses, bright lines. (10)

raies d'absorption, absorption lines. (11)

raies de Fraunhofer, Fraunhofer's lines. (12)

raies du spectre, lines of the spectrum ; spectral lines. (13)

raies sombres *ou* raies obscures, dark lines. (14)

raies telluriques, telluric lines ; atmospheric lines. (15)

rail (*n.m.*), rail. (16) (*Nota.—Le pluriel* **rails** *se traduit quelquefois par le terme* metals ; *ainsi*, le train a quitté les rails, the train jumped the metals ; the train left the rails).

rail à adhérence (opp. à **crémaillère**) (*m.*), adhesion-rail. (17) (opp. to **rack-rail**).

rail à champignon unique (*m.*), single-headed rail ; single-ended rail. (18)

rail à double champignon (*m.*), double-headed rail ; double-ended rail ; I rail. (19)

rail à double champignon dissymétrique (*m.*), bull-headed rail ; bull-head rail. (20)

rail à double champignon dissymétrique, type renforcé (*m.*), bull-headed rail, bull-head rail, heavy section. (21)

rail à double champignon symétrique (*m.*), double-headed reversible rail ; reversible rail. (22)

rail à gorge *ou* **rail à ornière** (*m.*), tram-rail ; tramway-rail ; street-railway rail ; grooved rail. (23)

rail à patin *ou* **rail à patin et à champignon unique** (*m.*), flange-rail ; flanged T-headed rail ; flat-bottomed rail ; Vignoles rail ; foot-rail. (24)

rail à pont (*m.*), bridge-rail. (25)

rail à simple champignon (*m.*), single-headed rail ; single-ended rail. (26)

rail à T (*m.*), T rail ; flange-rail. (27)

rail-bond [rails-bonds *pl.*] (Ch. de f. élec.) (*n.m.*), rail-bond ; bond. (28)

rail central (Ch. de f. élec.) (*m.*), middle rail. (29)

rail conducteur (Ch. de f. élec.) (*m.*), conductor-rail. (30)

rail contre-aiguille [rails contre-aiguilles *pl.*] *ou* **rail contraiguille** (*m.*), stock-rail ; main rail. (31)

rail contre-aiguille de la voie déviée (*m.*), stock-rail of diverging road ; stock-rail on the turnout side ; follower. (32)

rail contre-aiguille de la voie directe (*m.*), stock rail of through road ; through rail. (33)

rail coudé (*m.*), elbow-rail. (34)

rail creux (*m.*). Same as **rail à gorge**.

rail de champ (*m.*), edge-rail. (35)

rail de fer (*m.*), iron rail. (36)

rail de roulement *ou* **rail de la voie** (Ch. de f.) (*m.*), running-rail. (37)

rail de tramway (*m.*). Same as **rail à gorge**.

rail de translation (pour pont roulant) (*m.*), crane-rail ; tram-rail ; runway-rail ; running-rail. (38)

rail dissymétrique (*m.*). Same as **rail à double champignon dissymétrique**.

rail en acier (*m.*), steel rail. (39)

rail en aile (*m.*), wing-rail. (40)

rail en U (*m.*), bridge-rail. (41)

rail-guide [rails-guides *pl.*] (*n.m.*), guide-rail. (42)

rail léger (*m.*), light rail. (43)

rail méplat (*m.*), flat rail ; plate-rail ; strap-rail. (44)

rail noyé (opp. à **rail saillant**) (*m.*), sunken rail. (45)

rail-poutre [rails-poutres *pl.*] (*n.m.*), girder-rail. (46)

rail saillant (*m.*), raised rail. (47)

rail symétrique (*m.*), reversible rail ; double-headed reversible rail. (48)

rail Vignole (*m.*), Vignoles rail ; flange-rail ; flanged T-headed rail ; flat-bottomed rail. (49)

rails gras (cause de mauvaise adhérence) (*m.pl.*), greasy rails ; greasy metals. (50)

rainer (*v.t.*), to groove ; to slot ; to flute : (51)
rainer une planche, to groove a board. (52)

rainés (fer) (*n.m.pl.*), channel-iron ; channels (small sections). (53)

rainette de charpentier (*f.*), combined saw-set and marking-knife. (54)

rainurage (*n.m.*), grooving ; slotting ; fluting. (55)

rainure (*n.f.*), groove ; slot ; flute. (56)

rainure (à) *ou* **rainures (à)**, grooved ; slotted ; fluted : (57)
assemblage à rainure et languette (*m.*), grooved and tongued joint. (58)

table à rainures (*f.*), slotted table. (59)

foret à rainures droites (*m.*), fluted drill. (60)

rainure à T (*f.*), T slot. (61)

rainure de clavette *ou* **rainure de clavetage** *ou* **rainure de cale** (Méc.) (*f.*), keyway ; key-seat ; key-seating ; key-groove. (62)

rainure de graissage (*f.*), oil-groove ; oil-channel. (63)

rainureuse (*n.f.*), slot-drilling machine. (64)

rais [rais *pl.*] (d'une roue) (*n.m.*), spoke (of a wheel). (65)

raison (cause ; motif) (*n.f.*), reason : (66)
avoir de bonnes raisons pour changer, to have good reasons for changing. (67)

raison (Math.) (*n.f.*), ratio. (68)

raison de (à), at the rate of : (69)
la mèche de sûreté brûle à raison d'un mètre par minute, en moyenne, safety-fuse burns at the rate of one metre a minute, on an average. (70)

raison de l'équipage (*f.*), gear-ratio. (71)

raison directe (*f.*), direct ratio. (72)

raison géométrique (*f.*), geometrical ratio. (73)

raison inverse (*f.*), inverse ratio ; reciprocal ratio : (74)
l'intensité de la lumière est en raison inverse du carré de la distance du foyer (*f.*), the intensity of light is in inverse ratio to the square of the distance of the focus. (75)

dans les milieux autres que le vide, la lumière se propage avec une vitesse qui est en raison inverse de l'indice de réfraction, in media other than space, light propagates at a speed which is in inverse ratio to the index of refraction. (1)

raison inverse de (en), in inverse ratio to; inversely as : (2)

l'attraction ou la répulsion qui s'exerce entre deux pôles d'aimant varie en raison inverse du carré de leur distance (*f.*), the attraction or repulsion which is exercised between two magnet poles varies inversely as the square of their distance. (3)

raisonnable (*adj.*), reasonable : (4)

prix raisonnable (*m.*), reasonable price (*or* charge). (5)

raisonnablement (*adv.*), reasonably. (6)

rajeunir (Géogr. phys.) (*v.t.*), to rejuvenate : (7)

rivière rajeunie (*f.*), rejuvenated river. (8)

rajeunissement (*n.m.*), rejuvenation : (9)

la rajeunissement d'un réseau hydrographique vieilli, the rejuvenation of an aged hydro-graphical system. (10)

rajustement (*n.m.*), readjustment ; repairing ; refitting. (11)

rajuster (*v.t.*), to readjust ; to repair ; to refit. (12)

rajusteur (pers.) (*n.m.*), repairer. (13)

ralentir (*v.t.*) *ou* **ralentir la marche** (de) *ou* **ralentir l'allure** (de), to slow down ; to slacken ; to slacken speed ; to reduce the speed (of) : (14)

ralentir la marche d'un moteur, to slow down an engine. (15)

ralentir le feu, to slow down, to slacken, the fire. (16)

ralentir (*v.i.*) *ou* **ralentir sa marche**, to slow down ; to slacken ; to slacken speed ; to reduce speed : (17)

train qui ralentit sa marche aux bifurcations importantes (*m.*), train which slows down at important junctions. (18)

ralentissement (d'un train) (*n.m.*), slowing down (of a train). (19)

ralentissement (de commandes) (*n.m.*), falling off, slackening, of orders. (20)

ralentisseur automatique de vitesse (*m.*), auto-matic speed-checker. (21)

rallonge (*n.f.*), lengthening-piece ; lengthening-rod ; extension-piece. (22)

rallonge (tige de sonde) (*n.f.*), rod ; boring-rod ; bore-rod ; drill-rod ; drill-pole. (23)

rallongement (*n.m.*), lengthening. (24)

rallonger (*v.t.*), to lengthen. (25)

rallonger (se) (*v.r.*), to lengthen. (26)

rallumage (*n.m.*), relighting : (27)

rallumage intérieur des lampes de sûreté, internal relighting of safety-lamps. (28)

rallumer (*v.t.*), to relight. (29)

rallumer (se) (*v.r.*), to relight. (30)

rallumeur (lampes de sûreté) (*n.m.*), igniter ; relighter. (31)

rallumeur à amorce fulminante (lampes) (*m.*), fulminating-cap relighter. (32)

ralstonite (Minéral) (*n.f.*), ralstonite. (33)

ramassage (*n.m.*), gathering ; collecting ; picking up. (34)

ramassage à la pelle (*m.*), shovelling. (35)

ramasser (*v.t.*), to gather ; to collect ; to pick up. (36)

ramasser à la pelle, to shovel. (37)

rame (de wagons) (Mines) (*n.f.*), train, trip, set, (of cars). (38)

rameau (*n.m.*), branch ; spur : (39)

rameau d'un arbre, branch of a tree. (40)

rameau d'un filon, branch, spur, of a lode. (41)

rameau d'une montagne, spur of a mountain. (42)

rameaux de quartz, branches of quartz. (43)

ramenage à 0 (d'un instrument tel qu'un compte-tours) (*m.*), setting to zero ; resetting to zero. (44)

ramener à 0, to set, to reset, to zero. (45)

ramener la voie à son écartement (Ch. de f.), to bring back the road to gauge. (46)

ramification (*n.f.*), ramification ; branching. (47)

ramifier (*v.t.*), to ramify ; to branch. (48)

ramifier (se) (*v.r.*), to ramify ; to branch ; to branch off. (49)

ramollir (*v.t.*), to soften : (50)

la chaleur ramollit la cire, heat softens wax. (51)

ramollir le terrain gelé, to soften the frozen ground. (52)

ramollir (se) (*v.r.*), to soften ; to become soft : (53)

l'asphalte se ramollit sous un soleil ardent (*m.*), asphalt softens under a hot sun. (54)

ramollissement (*n.m.*), softening. (55)

ramonage (des tubes de chaudière) (*n.m.*), clean-ing (boiler-tubes). (56)

ramoner (*v.t.*), to clean ; to sweep : (57)

ramoner des tubes à fumée, to clean fire-tubes. (58)

ramoner une cheminée, to sweep a chimney. (59)

rampant, -e (Arch.) (*adj.*), rampant : (60)

arc rampant (*m.*), rampant arch. (61)

rampant (d'un comble) (*n.m.*), slope (of a roof). (62)

rampant (d'un four métallurgique) (*n.m.*), neck (of a metallurgical furnace). (63)

rampe (*n.f.*), rise ; slope ; bank ; acclivity ; ascent ; incline ; gradient ; grade ; rising gradient ; upward gradient ; up grade : (64)

rampe très rapide, very sudden rise. (65)

les rampes rocailleuses d'un coteau, the rocky slopes of a hill. (66)

une ligne de chemin de fer présente des paliers, des rampes, et des pentes, a line of railway contains levels, rising gradients, and falling gradients. (67)

parcourue en sens contraire, la rampe est une pente, traversed in the opposite direction, the up grade is a down grade. (68)

traction en rampe (*f.*), up-grade traction. (69)

rampe de 0^m, 00X, gradient of x in 1,000 ; grade of x $^0/_{00}$. (70)

ligne de chemin de fer avec rampe maximum de 0^m, 01, railway-line with a maximum gradient of 10 in 1,000 (*or* 1 in 100) ; rail-way-line with a maximum grade of 10 $^0/_{00}$ (*or* 1%). (71)

rampe (garde-fou) (*n.f.*), stair-rail ; hand-rail ; railing. (72)

rampe (instrument pour faire sortir ou remettre sur les rails un wagon) (*n.f.*), car-replacer ; wagon-replacer ; replacing-switch ; wreck-ing-frog ; jumper ; ramp. (73)

rampe (*n.f.*) *ou* **rampe d'escalier** (volée), flight ; flight of stairs. (74)

rampe de chevrons (Arch.) (*f.*), roof-pitch ; pitch of roof. (75)

rampe de graissage *ou* rampe de distribution (*f.*), lubricating-rack ; oil-distributor ; feed-rack. (1)

rampe de graissage à départs multiples (*f.*), multiple feed-rack. (2)

rampe de graissage à trois départs (*f.*), triple feed-rack. (3)

ramper dans un passage, to crawl through a passage. (4)

rancher (montant posé sur un wagon plat) (*n.m.*), stanchion ; stake : (5)
 wagon plat avec ranchers (*m.*), flat truck with stanchions. (6)

rang (classe) (*n.m.*), rank ; order ; class. (7)

rang (*n.m.*) *ou* rangée (*n. f.*), row : (8)
 un rang d'arbres, de rivets *ou* une rangée d'arbres, de rivets, a row of trees, of rivets. (9 *ou* 10)

ranimer (*v.t.*), to revive ; to resuscitate. (11)

ranimer (se) (*v.r.*), to revive. (12)

rankinisation des diagrammes (machine poly--cylindrique) (*f.*), combining the diagrams (multicylinder engine) : (13)
 rankinisation des diagrammes relevés sur une machine à double expansion, combining the diagrams taken on a double-expansion engine. (14)

rankiniser les diagrammes, to combine the dia--grams. (15)

râpage (*n.m.*), rasping. (16)

râpe (*n.f.*), rasp. (17)

râpe à maréchale (*f.*), horseshoe rasp. (18)

râpe à taille bâtarde (*f.*), bastard-cut rasp. (19)

râpe à taille demi-douce *ou* râpe à taille mi-douce (*f.*), second-cut rasp ; 2nd-cut rasp. (20)

râpe à taille douce (*f.*), smooth-cut rasp. (21)

râpe à taille moyenne (*f.*), middle-cut rasp. (22)

râpe à taille rude *ou* râpe à grosse taille (*f.*), rough-cut rasp. (23)

râpe demi-ronde (*f.*), half-round rasp. (24)

râpe plate-à-main [râpes plates-à-main *pl.*] (*f.*), flat rasp. (25)

râpe ronde (*f.*), round rasp. (26)

râper (*v.t.*), to rasp : (27)
 râper du bois, to rasp wood. (28)

rapide (*adj.*), rapid ; quick ; swift : (29)
 rivière rapide (*f.*), rapid river ; swift river. (30)
 mouvement rapide (*m.*), quick movement. (31)

rapide (très incliné) (*adj.*), steep : (32)
 pente rapide (*f.*), steep slope. (33)

rapide (courant d'un cours d'eau qui a une grande vitesse) (*n.m.*), rapid : (34)
 les rapides du Saint-Laurent, the rapids of the St Lawrence. (35)

rapidement (*adv.*), rapidly ; quickly ; swiftly. (36)

rapidité (*n.f.*), rapidity ; quickness ; swiftness ; speed : (37)
 rapidité d'un courant, rapidity, swiftness, of a current. (38)
 rapidité d'une pente, steepness of a gradient (*or* slope). (39)

rapiéçage *ou* rapiéçement (action) (*n.m.*), patch--ing ; patching up ; piecing ; mending. (40)

rapiéçage (résultat) (*n.m.*), patchwork. (41)

rapiécer (*v.t.*), to patch ; to patch up ; to piece ; to mend. (42)

rappel automatique à zéro (Élec.) (*m.*), no-voltage release. (43)

rappel de l'usure (Méc.) (*m*), taking up wear ; compensation for wear. (44)

rappel des mineurs (*m.*), knocking by entombed miners. (45)

rappeler l'usure (Méc.), to take up the wear ; to compensate for wear. (46)

rappeler un mineur à la vie, to resuscitate a miner. (47)

rapport (exposé) (*n.m.*), report : (48)
rapport collectif, joint report. (49)
rapport d'expert, expert's report. (50)
rapport de mines, mining-report. (51)
rapport de tout repos, reliable report. (52)
rapport hebdomadaire, weekly report. (53)
rapport de quinzaine, fortnightly report. (54)
rapport trimestriel, quarterly report. (55)

rapport (raison ; proportion) (*n.m.*), ratio ; proportion : (56)
 le rapport du travail utile au travail moteur absorbé, the ratio of useful work to the mechanical energy expended. (57)
 rapport d'engrenage, gear-ratio. (58)
 rapport de vitesse, speed-ratio. (59)
 rapport des vitesses angulaires, ratio of angular velocities. (60)

rapport (en), producing ; productive : (61)
 mine en rapport (*f.*), producing mine. (62)
 capital en rapport (*m.*), productive capital. (63)

rapport manométrique d'un ventilateur (*m.*), manometric efficiency of a fan. (64)

rapporté, -e (*adj.*) *ou* rapport (de), loose ; detach--able ; inserted ; added ; built. (65) (distinct and separate and added afterwards ; not made as part and parcel [opposed to such terms as venu de fonte (q.v.), venu de bois, etc.]:
 les modèles de fonderie s'établissent avec des portées, dont les empreintes dans les moules servent d'assise ou d'encastrement à des pièces rapportées (*ou* pièces de rapport) appelées noyaux (*m.pl.*), foundry-patterns are made with prints of which the impres--sions in the moulds serve as a bed or housing for loose pieces called cores. (66)
 languette rapportée (*f.*), loose tongue ; slip tongue. (67) [opposed to languette venue de bois (q.v.)].
 étau à mâchoires rapportées (*m.*), vice with detachable jaws ; vice with inserted jaws. (68)
 arbre à manivelle rapportée (*m.*), built crank-shaft. (69)
 Cf. rapporter (*v.t.*), pièce rapportée (pièce battue), alésoir à lames rapportées, and bielle motrice.

rapporter (ajouter à une chose pour la compléter) (*v.t.*), to add ; to join on ; to put in ; to insert : (70)
 rapporter un bout de planche à une étagère, to add, to join on, a bit of board to a shelf. (71)
 rapporter un morceau de bois pour dissimuler une fente dans un ouvrage de menuiserie, to put in a bit of wood to conceal a crack in a piece of joinery work. (72)

rapporter (tracer sur le papier) (*v.t.*), to lay down ; to lay off ; to plot ; to draw : (73)
 rapporter des angles, to lay down, to lay off, to plot, angles. (74)
 plan rapporté à l'échelle de 1 : — (*m.*), plan drawn to the scale of 1 in —. (75)

rapporter (référer) (*v.t.*), to refer : (76)
 le plan de comparaison auquel on rapporte les altitudes est généralement le niveau moyen de la mer, the datum-level to which heights

are referred is generally the mean sea-level. (1)
des vitesses rapportées à des repères fixes ou mobiles (*f.pl.*), velocities referred to fixed or movable datum-points. (2 *ou* 3)

rapporter un bénéfice, to pay ; to show a profit. (4)

rapporteur (Dessin) (*n.m.*), protractor. (5)

rapporteur cercle entier (*m.*), circular protractor. (6)

rapporteur demi-cercle (*m.*), semicircular pro--tractor. (7)

rapporteur rectangulaire (*m.*), rectangular pro--tractor. (8)

raquette (Travail à la poudre) (*n.f.*), squib ; german ; match. (9)

rare (*adj.*), rare ; scarce ; uncommon : (10)
terres rares (*f.pl.*), rare earths. (11)

raréfactif, -ive (*adj.*), rarefactive. (12)

raréfaction (*n.f.*), rarefaction ; scarcity : (13)
raréfaction de l'atmosphère, rarefaction of the atmosphere. (14)
raréfaction de la main-d'œuvre, scarcity of labour. (15)

raréfiable (*adj.*), rarefiable. (16)

raréfiant, -e (*adj.*), rarefactive ; rarefying. (17)

raréfier (*v.t.*), to rarefy : (18)
raréfier l'air contenu dans un récipient à l'aide de la machine pneumatique, to rarefy the air contained in a receiver by an air-pump. (19)

raréfier (se) (*v.r.*), to rarefy. (20)

rarement (*adv.*), rarely ; seldom. (21)

rareté (*n.f.*), rarity ; rareness ; scarcity : (22)
le taux du salaire se règle sur la rareté ou sur l'abondance du travail, the rate of wages depends on the scarcity or abundance of work. (23)

ras (*n.m.*). Same as **raz.**

ras de (à) ou ras de (au), level with : (24)
à ras de terre *ou* au ras de terre *ou* à rase terre, level with the ground. (25)

rassemblement (*n.m.*), collection ; gathering. (26)

rassembler (assembler de nouveau) (*v.t.*), to reassemble : (27)
rassembler les pièces d'une charpente, to reassemble the members of a frame. (28)

rassembler (faire amas) (*v.t.*), to collect ; to gather : (29)
rassembler des matériaux, des minéraux, to collect materials, minerals. (30 *ou* 31)

raté (*n.m.*) *ou* **raté d'allumage** (Travail aux explosifs, moteurs à combustion interne, etc.), misfire ; miss-fire. (32 *ou* 33)

râteau (*n.m.*), rake. (34)

râtelier pour tubes à essais (*m.*), rack for test-tubes. (35)

rater (Travail aux explosifs, moteurs à combustion interne, etc.) (*v.i.*), to misfire ; to miss fire ; to fail to explode : (36 *ou* 37)
coup de mine qui rate (*m.*), shot which misfires (*or* which misses fire). (38)

rationnel, -elle (opp. à **appliqué**) (*adj.*), pure ; abstract : (39)
mécanique rationnelle (*f.*), pure mechanics ; abstract mechanics. (40)

ratissette (de forgeron) (*n.f.*), rake (blacksmiths'). (41)

rattrapage de jeu (Méc.) (*m.*), taking up play. (42)

rattrapage de l'usure (*m.*), taking up the wear ; compensation for wear. (43)

rattraper l'usure (Méc.), to take up the wear ; to compensate for wear. (44)

rattraper le jeu (Méc.), to take up the play : (45)
rattraper le jeu provenant de l'usure, to take up the play resulting from wear. (46)

rattraper le temps perdu, to make up lost time. (47)

rattraper les tiges brisées, to recover broken rods. (48)

ravages occasionnés par une explosion (*m.pl.*), destruction, havoc, caused by an explosion. (49)

rave (*n.f.*), miners' lamp. (50)

ravin (*n.m.*) *ou* **ravine** (*n.f.*) *ou* **ravinée** (*n.f.*) (lit creusé par un torrent), ravine ; gully. (51)

ravine (torrent) (*n.f.*), torrent. (52)

ravinement (*n.m.*), gullying ; channeling ; ravinement. (53)

raviner (*v.t.*), to gully ; to channel ; to gutter : (54)
coteau que les pluies ont raviné (*m.*), hill that the rains have gullied. (55)

ravineux, -euse (*adj.*), gullied ; channeled. (56)

raviver l'intérêt relatif à un district, to revive interest in a district. (57)

raviver le feu, to brisk up, to brisk, the fire. (58)

raviver (se) (*v.r.*), to revive ; to break out again : (59)
incendie qui se ravive (*m.*), fire which breaks out again. (60)

rayable (*adj.*), that can be scratched : (61)
le plomb est rayable à l'ongle, lead can be scratched with the nail. (62)

rayage ou rayement (*n.m.*), scratching ; streak--ing. (63)

rayer (*v.t.*), to scratch ; to streak : (64)
le cristal de roche raye le verre, rock-crystal scratches glass. (65)
le rubis oriental est rayé seulement par le diamant, the true ruby is scratched only by the diamond. (66)
la craie de Briançon est assez tendre pour être aisément rayée par l'ongle ou coupée au couteau, French chalk is soft enough to be easily scratched by the nail or cut with a knife. (67)

rayon (Phys.) (*n.m.*), ray : (68)
rayons actiniques *ou* rayons chimiques, actinic rays ; chemical rays. (69)
rayons alpha *ou* rayons α, alpha rays ; α rays. (70)
rayons bêta *ou* rayons β, beta rays ; β rays. (71)
rayon calorifique, heat-ray. (72)
rayons-canaux de Goldstein, canal rays ; Goldstein rays. (73)
rayons cathodiques, cathode rays ; cathode streams ; kathode rays. (74)
rayons centraux, central rays. (75)
rayons convergents, converging rays. (76)
rayon de lumière, ray of light ; light-ray ; beam of light. (77)
rayon de lumière blanche, ray of white light. (78)
rayons du soleil, rays of the sun ; sun's rays. (79)
rayons émergents, emergent rays. (80)
rayon extraordinaire, extraordinary ray. (81)
rayons gamma *ou* rayons γ, gamma rays ; γ rays. (82)
rayons incidents, incident rays. (83)

rayons infra-rouges, infra-red rays; ultra-red rays. (1)

rayons lumineux, luminous rays; light-rays. (2)

rayons marginaux, marginal rays. (3)

rayon ordinaire, ordinary ray. (4)

rayon réfléchi, reflected ray. (5)

rayons réfléchis par un miroir, rays reflected by a mirror. (6)

rayons réfractés, refracted rays. (7)

rayons solaires, solar rays; sun's rays: (8) variations de température dues à l'obliquité croissante des rayons solaires lorsqu'on avance vers le pôle (f.pl.), fluctuations of temperature due to the obliquity of the sun's rays when advancing towards the pole. (9)

rayons ultra-rouges, ultra-red rays; infra-red rays. (10)

rayons ultra-violets, ultra-violet rays. (11)

rayons visuels, visual rays. (12)

rayons X, X-rays. (13)

rayon (Géom.) (n.m.), radius: (14) les rayons d'un cercle, the radii of a circle. (15)

à un rayon de — kilomètre(s) ou dans un rayon de — kilomètre(s), within a radius of — kilometre(s). (16)

rayon (d'une roue) (n.m.), spoke (of a wheel). (17)

rayon (tablette) (n.m.), shelf: (18) les rayons d'une armoire, the shelves of a cupboard. (19)

rayon de giration (m.), radius of gyration. (20)

rayon primitif (Engrenages) (m.), pitch-radius; primitive radius (Gearing). (21)

rayon principal ou rayon visuel principal (Per-spective) (m.), line of sight; line of direc-tion; direct radial; principal visual ray; principal ray. (22)

rayon visuel (Opt.) (m.), line of vision; line of sight. (23)

rayonnant, -e (adj.), radiating; radiant: (24) chaleur rayonnante (f.), radiant heat. (25)

rayonnement (n.m.), radiation: (26) rayonnement de la lumière, radiation of light. (27)

rayonner (v.t.), to radiate: (28) la quantité de chaleur rayonnée par une source, the quantity of heat radiated by a source. (29)

rayonner (v.i.), to radiate: (30) la lumière rayonne, light radiates. (31)

rayure (n.f.), scratch; streak. (32)

raz (n.m.) ou raz de marée, tidal wave; eager. (33)

raz de courant (m.), race; overfall. (34)

réa (n.m.), pulley-wheel; pulley; sheave; pulley-sheave. (35)

réabsorber (v.t.), to reabsorb. (36)

réabsorption (n.f.), reabsorption. (37)

réactance (Élec.) (n.f.), reactance. (38)

réactif, -e (adj.), reactive. (39)

réactif (Chim.) (n.m.), reagent. (40)

réaction (Chim., Phys. & Méc.) (n.f.), reaction. (41)

réaction d'appui (Résistance des matériaux) (f.), reaction of support (Strength of materials). (42)

réaction d'induit (Élec.) (f.), armature reaction; reaction of armature. (43)

réactions du haut fourneau (f.pl.), reactions in the blast-furnace. (44)

réadmission (n.f.), readmission. (45)

réaffûter (v.t.), to resharpen. (46)

réagir (v.i.), to react. (47)

réaimanter (v.t.), to remagnetize: (48) réaimanter l'aiguille d'une boussole, to remag-netize the needle of a compass. (49)

réaléser (Méc.) (v.t.), to rebore. (50)

réalgar (Minéral) (n.m.), realgar; ruby-arsenic; ruby-sulphur; ruby of arsenic; ruby of sulphur; red arsenic. (51)

réalisable (adj.), realizable: (52) valeur réalisable (f.), realizable value. (53)

réalisation (n.f.), realization. (54)

réaliser (v.t.), to realize. (55)

réamorçage d'un injecteur (m.), restarting of an injector. (56)

réamorcer un injecteur, to restart an injector. (57)

réanimer le feu, to brisk up, to brisk, the fire. (58)

réapparaître (v.i.), to reappear; to recur. (59)

réapparition (n.f.), reappearance; recurrence. (60)

réapprovisionnement (n.m.), restocking; fresh supply. (61)

réapprovisionner (v.t.), to restock. (62)

réargenter (v.t.), to resilver. (63)

réarpentage (n.m.), resurvey; remeasurement (of land). (64)

réarpenter (v.t.), to resurvey; to remeasure (land). (65)

réavalement de puits déjà creusés (m.), deepening pits already sunk. (66)

rebanchage du mur (Mines) (m.), sinking the floor; lifting the pavement. (67)

rebancher le mur (Mines), to sink the floor; to lift the pavement. (68)

rebâtir (v.t.), to rebuild. (69)

rebelle (adj.), rebellious; stubborn; refractory: (70) minerai rebelle (m.), rebellious ore; stubborn ore; refractory ore. (71)

reboisage (n.m.), retimbering. (72)

reboiser (v.t.), to retimber: (73) reboiser une galerie de mine, to retimber a mine-level. (74)

rebond (n.m.), rebound. (75)

rebondir (v.i.), to rebound. (76)

rebondissant, -e (adj.), rebounding. (77)

rebondissement (n.m.), rebounding. (78)

rebord (n.m.), edge; rim: (79) rebord d'une table, d'un fossé, edge of a table, of a ditch. (80 ou 81)

rebord (d'un tiroir, d'une lumière de cylindre) (n.m.), edge (of a slide-valve, of a cylinder-port): (82 ou 83) rebord extérieur, steam-edge; outer edge. (84) rebord intérieur, exhaust-edge; inner edge. (85)

rebord (bride; boudin) (n.m.), flange: (86) plaque à rebord (f.), flange-plate. (87)

rebord (d'une serrure) (n.m.), selvage; edge-plate, (of a lock). (88)

rebroussement (Géol.) (n.m.), upturning. (89)

rebroyage (n.m.), regrinding; recrushing. (90)

rebroyer (v.t.), to regrind; to recrush. (91)

rebut (n.m.) ou rebuts (n.m.pl.), waste; refuse; rubbish; rejects. (92)

rebut (pièce manquée) (Fonderie) (n.m.), waster. (93)

rebuts (Mines) (n.m.pl.), waste; attle; deads; chats. (94)

récalescence (Phys.) (*n.f.*), recalescence. (1)

recarbonisation (*n.f.*), recarbonization ; recar--burization. (2)

recarboniser *ou* **recarburer** (*v.t.*), to recarbonize ; to recarburize : (3)
recarburer l'acier, to recarbonize, to recar--burize, steel. (4)

récemment (*adv.*), recently ; newly. (5)

récent, -e (*adj.*), recent ; new. (6)

récent, -e (Géol.) (opp. à **ancien, -enne**) (*adj.*), recent ; Recent ; late : (7)
alluvions récentes (*f.pl.*), Recent alluvium ; alluvium of a late period. (8)
couches qui recouvrent des formations plus récentes (*f.pl.*), strata which overlap later formations. (9)

recepage *ou* **recépage** (*n.m.*), cutting off ; striking off : (10)
recepage des pieux *ou* recépage des pilotis, cutting off piles ; striking off the heads of piles. (11)

receper (*ou* **recéper**) **des pieux** (*ou* **des pilotis**), to cut off piles ; to strike off the heads of piles. (12)

réceptacle (*n.m.*), receptacle. (13)

réceptacle de la vapeur (*m.*), steam-dome ; dome. (14)

récepteur, -trice (*adj.*), receiving : (15)
station réceptrice (*f.*), receiving station. (16)

récepteur (*n.m.*), receiver. (17)

récepteur (*n.m.*) *ou* **réceptrice** (*n.f.*) (opp. à **source, générateur,** *ou* **génératrice**) (Elec.), load ; sink. (18) (opp. to **source** or **generator**). See **générateur** and **source** for examples of use.

récepteur (opp. à **source**) (Hydraul.) (*n.m.*), sink. (19) (opp. to **source**) :
les roues hydrauliques sont des récepteurs dont on se sert pour recueillir la force vive des eaux courantes, afin de la transformer en travail utile, water-wheels are sinks which are used to gather the living force of running water in order to convert it into useful work. (20)

récepteur télégraphique (*m.*), telegraph receiver. (21)

récepteur téléphonique (*m.*), telephone receiver. (22)

réception des cages (Mines) (*f.*), landing of the cages. (23)

recette (*n.f.*), receipts ; receipt ; proceeds ; yield : (24)
compter la recette et la dépense, to reckon up the receipts and expenditure. (25)
perdre la meilleure partie de la recette par un dépouillement insuffisant du fond (Dragage de rivières), to lose the best part of the yield by insufficient stripping of the bottom. (26)

recette (Mines) (*n.f.*), landing ; landing-station ; landing-stage ; station ; onsetting-station ; plat ; platt ; lodge. (27)

recette à eau (*f.*), lodge-room ; pound-room. (28)

recette de puits (*f.*), shaft-station ; pit-landing. (29)

recette du fond *ou* **recette d'à bas** (*f.*), bottom landing ; bottom onsetting-station ; bottom station ; bottom plat. (30)

recette du jour *ou* **recette d'à haut** (*f.*), bank ; top-landing. (31)

recette intérieure (*f.*), underground landing-station. (32)

recette intermédiaire (*f.*), intermediate landing-station. (33)

recette pour cuffats (*f.*), kibble-landing. (34)

receveur (Mines) (pers.) (*n.m.*), banksman ; lander. (35)

recevoir (*v.t.*), to receive : (36)
recevoir une communication du fond, to receive a communication from below. (37)

rechange (**de**), spare ; duplicate : (38)
outils de rechange (*m.pl.*), spare tools ; dupli--cate set of tools. (39)

rechargement (*n.m.*). See examples : (40)
rechargement d'un wagon, reloading a wagon ; refilling a truck. (41)
rechargement d'une pile électrique, recharging an electric battery. (42)
rechargement d'une route, remetalling a road. (43)
rechargement d'un chemin de fer, reballasting a railway. (44)

recharger (*v.t.*). See examples : (45)
recharger un wagon, to reload a wagon ; to refill a car. (46)
recharger un accumulateur (Élec.), to recharge an accumulator. (47)
recharger un four, to recharge a furnace. (48)
recharger un fourneau de mine, to recharge, to reload, a hole. (49)
recharger une route, to remetal a road. (50)
recharger la voie (Ch. de f.), to reballast the road. (51)

rechargeur (Élec.) (*n.m.*), replenisher. (52)

réchauffage *ou* **réchauffement** (*n.m.*), reheating : (53)
four de réchauffage (*m.*), reheating-furnace. (54)

réchauffer (*v.t.*), to reheat : (55)
réchauffer de l'eau, de la vapeur à basse pression, to reheat water, low-pressure steam. (56 *ou* 57)

réchauffeur (*n.m.*), reheater ; heater. (58)

réchauffeur d'air (*m.*), air-reheater. (59)

réchauffeur d'eau d'alimentation (*m.*), feed-water heater. (60)

recherche (*n.f.*) (*s'emploie souvent au pluriel*), search ; searching ; research ; prospecting ; prospection ; exploration ; location ; locat--ing ; finding : (61)
recherche de pépites d'or, nuggeting. (62)
recherche des dérangements (Élec.), locating faults. (63)
recherche des gîtes minéraux, des gîtes en profondeur, des sources, du charbon, du grisou, search, searching, for mineral deposits, for deposits in depth, for water, for coal, for fire-damp. (64 *ou* 65 *ou* 66 *ou* 67 *ou* 68)
recherche du centre de gravité d'un corps, finding the centre of gravity of a body. (69)
recherches filoniennes, prospecting for lodes. (70)
recherches métallurgiques, metallurgical research. (71)
recherches scientifiques, scientific research. (72)

rechercher (*v.t.*), to search ; to search for ; to locate ; to find : (73)
rechercher des pierres précieuses, to search for gems. (74)
rechercher la valeur de l'inconnue (Math.), to find the value of the unknown quantity. (75)
rechercher le grisou, to search for fire-damp. (76)

rechercher un dérangement (Élec.), to locate a fault. (1)

récif (n.m.), reef. (2)

récif-barrière [récifs-barrières pl.] (corail) (n.m.), barrier reef. (3)

récif corallien ou récif de corail (m.), coral reef. (4)

récif frangeant [récifs frangeants pl.] (m.) ou récif-bordure [récifs-bordures pl.] (n.m.), fringing reef. (5)

récipient (n.m.), receiver ; vessel : (6)
faire le vide dans un récipient, to create a vacuum in a receiver. (7)

récipient (d'un alambic) (n.m.), receiver (of a still). (8)

récipient à fond de cuivre (m.), copper-bottomed vessel. (9)

récipient à pétrole fermé (m.), closed cup. (10)

récipient à pétrole ouvert (m.), open cup. (11)

récipient cylindrique en tôle (m.), iron drum ; drum. (12)

récipient en forme de poire (m.), pear-shaped vessel. (13)

récipient florentin (m.), Florentine receiver. (14)

réciproque (adj.), reciprocal ; inverse : (15)
rapport réciproque (m.), reciprocal ratio ; inverse ratio. (16)

réciproque (n.f.), reciprocal ; inverse. (17)

réciproquement (adv.), reciprocally ; inversely. (18)

réclamation (n.f.), claim. (19)

réclamer (v.t.), to claim ; to call for ; to require : (20)
réclamer un rabais, to claim a rebate. (21)
minerai qui réclame un traitement spécial (m.), ore which calls for special treatment. (22)
machine qui réclame beaucoup de soins (f.), machine which requires a lot of attention. (23)

recoin (n.m.), corner ; recess : (24)
les recoins d'un moule, the recesses of a mould. (25)

récolte (résultat de recherches ; recette ; profit ; bénéfice) (Mines, etc.) (n.f.), winnings ; yield : (26)
la récolte de la journée, the day's winnings. (27)
perdre la meilleure partie de la récolte par un dépouillement insuffisant du fond, to lose the best part of the yield by insufficient stripping of the bottom. (28)

récolte de l'or (levée de la production) (f.), clean-up. (29)

récolter l'or des plaques, to clean up the gold from the plates. (30)

recombinaison (n.f.), recombination. (31)

recombiner (v.t.), to recombine. (32)

recombiner (se) (v.r.), to recombine. (33)

recommencer (v.t.), to recommence ; to restart ; to resume. (34)

récompense (n.f.), recompense ; reward ; repayment. (35)

récompenser (v.t.), to recompense ; to reward ; to repay. (36)

recomposer (v.t.), to recompose : (37)
recomposer la lumière blanche, to recompose white light. (38)

recomposition (n.f.), recomposition. (39)

recomposition (Chim.) (n.f.), synthesis. (40)

recondensation (n.f.), recondensation. (41)

recondenser (v.t.), to recondense. (42)

reconnaissable (adj.), recognizable ; identifiable : (43)
la turquoise est facilement reconnaissable à sa couleur bleu céleste, the turquoise is easily recognizable by its sky-blue colour. (44)
minéral qui est reconnaissable à sa cristallisation (m.), mineral which is identifiable by its crystallization. (45)

reconnaissance (n.f.), exploration ; examination ; inspection ; reconnaissance ; reconnoissance ; proving ; going over : (46)
reconnaissance préalable des quartiers, pre-liminary exploration of the districts. (47)
reconnaissance du terrain, examination, exploration, reconnaissance, reconnoissance, of the ground ; going over the ground. (48)
reconnaissance des gisements, proving the deposits. (49)

reconnaître (distinguer à certains caractères) (v.t.), to recognize ; to identify. (50)

reconnaître (parvenir à constater) (v.t.), to ascertain : (51)
reconnaître la valeur d'un lot de minerai, to ascertain the value of a parcel of ore. (52)

reconnaître (examiner) (v.t.), to examine ; to explore ; to reconnoitre ; to go over : (53)
reconnaître le terrain, to examine, to explore, to reconnoitre, to go over, the ground. (54)

reconnaître (constater la présence de) (v.t.), to prove ; to test ; to locate : (55)
reconnaître la nature du terrain avant de commencer les travaux souterrains, to prove the nature of the ground before commencing work underground. (56)
reconnaître une couche de houille par sondages, to prove a coal-seam by bore-holing. (57)
reconnaître l'emplacement d'un massif de minerai, to locate an ore-body. (58)
reconnaître le filon en profondeur, to prove, to test, the lode in depth. (59)

reconstruction (n.f.), reconstruction ; rebuilding ; reerection. (60)

reconstruire (v.t.), to reconstruct ; to rebuild ; to reerect. (61)

recoupage du toit (Mines) (m.), ripping, ripping down, brushing, the roof. (62)

recoupage du toit et du mur (Mines) (m.), brush-ing the roof and the floor. (63)

recoupe (galerie réunissant deux galeries princi-pales pendant les travaux de traçage) (Mines) (n.f.), drift ; cross-drift ; cross-drivage ; cross-drive ; cross-entry ; butt-entry ; offset. (64)

recoupe d'aérage (Mines) (f.), cross-heading. (65)
galeries jumelles réunies par des recoupes d'aérage (f.pl.), parallel entries connected by cross-headings. (66)

recoupement (intersection) (n.m.), intersection : (67)
méthode du recoupement (Levés) (f.), method of intersections (Surveying). (68)

recouper (couper de nouveau) (v.t.), to recut. (69)

recouper (intersecter) (v.t.), to intersect. (70)

recouper (Mines) (v.t.), to cross-drive. (71)

recouper le toit ou recouper le plafond (Mines), to rip, to rip down, to brush, the roof. (72)

recouper le toit et le mur (Mines), to brush the roof and the floor. (73)

recouper un filon, to cut, to tap, a lode : (74)

les puits inclinés suivent le pendage lorsqu'il est régulier ; sinon, ils s'écartent du filon pour le recouper en profondeur, incline-shafts follow the dip when it is regular ; if not, they deviate from the lode to cut it in depth. (1)

recoupeur (Mines) (pers.) (*n.m.*), ripper. (2)

recouvert, -e (opp. à **nu, -e**) (Élec.) (*adj.*), covered. (3) (opp. to **bare**) :
fil recouvert (*m.*), covered wire. (4)

recouvrable (*adj.*), recoverable ; extractable. (5)

recouvrant, -e (*adj.*), overlapping. (6)

recouvrement (action de recouvrer ce qui était perdu) (*n.m.*), recovery : (7)
recouvrement de cadavres, recovery of bodies. (8)

recouvrement (action de recouvrir) (*n.m.*), cover-ing. (9)

recouvrement (action de chevaucher) (*n.m.*), lapping ; overlapping. (10)

recouvrement (partie qui couvre un joint, une entaille) (*n.m.*), lap ; overlap : (11)
assemblage à recouvrement (*m.*), lap-joint ; overlap-joint. (12)

recouvrement (Toiture) (*n.m.*), lap : (13)
ardoises avec un recouvrement de — cm. (*f.pl.*), slates with a lap of — centimetres. (14)

recouvrement (du tiroir) (Méc.) (*n.m.*), lap, cover, (of the slide-valve) : (15)
recouvrement extérieur *ou* recouvrement à l'admission, outside lap ; steam-lap ; lap. (16)
recouvrement extérieur négatif, negative outside lap ; outside clearance. (17)
recouvrement intérieur *ou* recouvrement à l'échappement, inside lap ; exhaust-lap ; exhaust-cover ; inside cover. (18)
recouvrement intérieur négatif, negative inside lap ; inside clearance. (19)
recouvrement négatif, negative lap ; clearance. (20)

recouvrement (Géol.) (*n.m.*), overlap. (21)

recouvrement (en), overlapping. (22)

recouvrer (récupérer) (*v.t.*), to recover : (23)
recouvrer l'or, to recover the gold. (24)

recouvrir (couvrir de nouveau) (*v.t.*), to recover. (25)

recouvrir (couvrir complètement) (*v.t.*), to cover ; to overlay ; to line : (26)
neige recouvrant la terre (*f.*), snow covering the earth. (27)
recouvrir un fil (Élec.), to cover a wire. (28)
recouvrir d'antifriction un palier, to line a bearing-block with antifriction metal. (29)

recouvrir (chevaucher) (*v.t.*), to overlap : (30)
couches qui recouvrent des formations plus récentes (Géol.) (*f.pl.*), strata which overlap later formations. (31)

recouvrir (se) (chevaucher) (*v.r.*), to overlap. (32)

recreuser un puits obstrué, to resink a blocked up well. (33)

recreuser un puits trop peu profond, to deepen a too shallow well. (34)

recristallisation (*n.f.*), recrystallization. (35)

recroiser (*v.t.*), to recross. (36)

recrudescence (*n.f.*), recrudescence. (37)

recrutement du personnel ouvrier (*m.*), recruiting workmen. (38)

recruter la main-d'œuvre, to recruit labour. (39)

rectangle *ou* **rectangulaire** (*adj.*), rectangular ; right-angled : (40)

triangle rectangle (*m.*), right-angled triangle. (41)

rectangle (*n.m.*), rectangle ; right angle. (42)

rectangularité (*n.f.*), rectangularity. (43)

rectifiable (*adj.*), rectifiable. (44)

rectifiant (*n.m.*), rectifier. (45)

rectification (*n.f.*), rectifying ; truing ; straight-ening ; correction : (46)
rectification des limites, rectification of bound-aries. (47)

rectification (Chim.) (*n.f.*), rectification. (48)

rectifié (-e) et affûté (-e) mécaniquement après la trempe, trued and set mechanically after hardening. (49)

rectifier (*v.t.*), to rectify ; to true ; to straighten ; to correct : (50)
rectifier à la meule d'émeri une surface plane, to true, to rectify, a flat surface with an emery-wheel. (51)
rectifier le calage d'un instrument de nivelle-ment immédiatement avant et après la visée, to correct the setting of a levelling-instrument immediately before and after sighting. (52)
rectifier le tracé d'une route, to straighten the alignment of a road. (53)
rectifier les limites, to rectify the boundaries. (54)
rectifier une courbe (Géom.), to rectify a curve. (55)

rectifier (Chim.) (*v.t.*), to rectify. (56)

rectifier (se) (*v.r.*), to straighten. (57)

rectifieuse (*n.f.*), grinding-machine ; grinder (of the truing type). (58) See **machine à rectifier** for varieties.

rectiligne (*adj.*), rectilinear ; rectilineal ; straight-line ; true : (59)
mouvement rectiligne (*m.*), rectilinear motion ; straight-line motion. (60)
tenir le trou de sonde rectiligne, to keep the bore-hole true. (61)

rectilignement (*adv.*), rectilinearly. (62)

rectilignité (*n.f.*), rectilinearity. (63)

rectitude d'une ligne (*f.*), straightness of a line. (64)

recueillir (*v.t.*), to gather ; to collect ; to recover ; to reap : (65)
recueillir des données, to collect data. (66)
recueillir une bonne proportion de l'or, to recover a fair proportion of the gold. (67)
recueillir le bénéfice d'un développement systématique, to reap the benefit of system-atic development. (68)

recuire (*v.t.*), to anneal : (69)
recuire de l'acier, du verre, to anneal steel, glass. (70 *ou* 71)

recuire après trempe *ou simplement* **recuire** (*v.t.*), to draw the temper of ; to let down the temper of ; to temper : (72)
recuire un burin, to draw the temper of, to let down the temper of, to temper, a chisel. (73)

recuit, -e (*adj.*), annealed : (74)
fil recuit (*m.*), annealed wire. (75)

recuit (*n.m.*) *ou* **recuite** (*n.f.*), annealing. (76)

recuit après trempe *ou simplement* **recuit** (*n.m.*), drawing the temper ; letting down the temper ; tempering. (77)

recul (*n.m.*), recoil ; retreat ; recession ; retire-ment : (78)
recul d'une explosion, recoil of an explosion. (79)
recul d'un glacier, retreat, retirement, of a glacier. (80)

recul ou invasion des eaux de mer [recul (*n.m.*), invasion (*n.f.*)], retirement or invasion, recession or encroachment, of the sea-water. (1)

recul dans les affaires (*m.*), setback in business. (2)

reculer devant la dépense que devait entraîner un changement de système, to shrink from incurring the expense which a change of system would entail. (3)

reculer les bornes, to move back the boundaries. (4)

reculeur (Mines) (pers.) (*n.m.*), stower. (5)

reculoire (d'un chariot de mine) (*n.f.*), backstay, safety-dog, dog, drag, (of a mine-car). (6)

récupérable (*adj.*), recoverable. (7)

récupérateur (régénérateur) (*n.m.*), recuperator ; regenerator. (8)

récupération (*n.f.*), recovery ; recuperation ; winning : (9)

récupération d'or par tonne broyée, recovery of gold per ton milled. (10)

récupération de gazoline provenant de gaz naturels, recovery of gasoline from natural gases. (11)

récupération de chaleur, recuperation of heat. (12)

récupérer (*v.t.*), to recover ; to recuperate ; to win : (13)

récupérer le métal du minerai, to recover, to win, metal from ore. (14)

récupérer de la chaleur perdue, to recuperate waste heat. (15)

récurrence (*n.f.*), recurrence. (16)

recuvelage (*n.m.*), recasing. (17)

recuveler (*v.t.*), to recase : (18)

recuveler un puits, to recase a well. (19)

redentage (*n.m.*), retoothing : (20)

redentage d'une scie, retoothing a saw. (21)

redenter (*v.t.*), to retooth. (22)

redéposer (*v.t.*), to redeposit. (23)

redevance (*n.f.*), royalty : (24)

avoir le droit de propriété moyennant rede-vance, to have the right of possession on royalty. (25)

redevance à payer à l'État (*f.*), government royalty ; royalty payable to the State. (26)

redevance en nature (*f.*), royalty in kind. (27)

redevance tréfoncière à payer au propriétaire de la surface (*f.*), surface-royalty ; royalty payable to surface-owner. (28)

rédiger (*v.t.*), to draw up : (29)

rédiger un rapport, to draw up a report. (30)

rédintégration (Chim.) (*n.f.*), redintegration. (31)

redissoudre (*v.t.*), to redissolve : (32)

redissoudre un sel, to redissolve a salt. (33)

redistillation (*n.f.*), redistillation. (34)

redistiller (*v.t.*), to redistil. (35)

redoublement (augmentation) (*n.m.*), redoubling ; increasing ; increase : (36)

redoublement de précautions, redoubling of precautions ; increasing precautions. (37)

redoublement (action de remettre une doublure à) (*n.m.*), relining. (38)

redoublement d'engrenages (*m.*), back-gear ; back-speed. (39)

redoubler (augmenter) (*v.t.*), to redouble ; to increase. (40)

redoubler (remettre une doublure à) (*v.t.*), to reline : (41)

redoubler de fer-blanc une boîte, to reline a box with tin. (42)

redouter (*v.t.*), to dread ; to fear ; to apprehend danger from : (43)

mine où la présence de grisou est à redouter (*f.*), mine where the presence of fire-damp is to be feared. (44)

redouter des glissements de terrain, to appre-hend danger from landslides. (45)

redresse-tubes [redresse-tubes *pl.*] (Sondage) (*n.m.*). Same as **redresseur.**

redressement *ou* **redressage** (*n.m.*), straightening ; truing ; rectifying ; rectification : (46)

redressement d'une tige, straightening a rod. (47)

redressement de tubes faussés (*ou* gauchis), straightening buckled tubes. (48)

redressement du puits, truing the well. (49)

redresser (rendre droit) (*v.t.*), to straighten ; to true ; to rectify : (50)

redresser à la meule d'émeri une surface plane, to true, to rectify, a flat surface with an emery-wheel. (51)

redresser un courant alternatif (Élec.), to rectify an alternating current. (52)

redresser un essieu faussé, un puits de mine, to straighten a bent axle, a mine-shaft. (53 ou 54)

redresser une meule d'émeri, to true an emery-wheel. (55)

redresser les griefs des ouvriers, to redress the grievances of the men. (56)

redresser (se) (se remettre droit) (*v.r.*), to straighten. (57)

redresser (se) (se relever ; se tenir debout) (*v.r.*), to rear : (58)

couche de houille qui se redresse d'une manière prononcée (*f.*), seam of coal which rears in a pronounced way. (59)

redresseur (Élec.) (*n.m.*), rectifier. (60)

redresseur (Sondage) (*n.m.*), casing-swedge ; drift ; swage. (61)

redresseur à galets (*m.*), roller-swedge. (62)

redresseur à rainures (*m.*), casing-swedge with fluted watercourses. (63)

redresseur olive (*m.*), oval casing-swedge. (64)

réducteur, -trice (*adj.*), reducing ; reductive : (65)

agent réducteur (Chim.) (*m.*), reducing agent. (66)

réducteur (Chim.) (*n.m.*), reducer ; reducing agent. (67)

réducteur (de tuyau) (*n.m.*), reducer ; reducing pipe-fitting ; diminisher. (68)

réducteur de course (indicateur de pression) (*m.*), stroke-reducer (steam-reducer for steam indicator). (69)

réducteur de vitesse (*m.*), speed-reducer ; speed-controller ; speeder. (70)

réductibilité (*n.f.*), reductibility. (71)

réductible (*adj.*), reducible : (72)

métaux facilement réductibles (*m.pl.*), easily reducible metals. (73)

réductif, -ive (*adj.*), reductive ; reducing. (74)

réduction (diminution) (*n.f.*), reduction ; reduc-ing ; dimunition ; curtailment : (75)

réduction de température, reduction of temper-ature. (76)

réduction des salaires, reduction of wages. (77)

réduction des frais d'exploitation (Mines), reducing the cost of mining. (78)

réduction (Chim. et Métall.) (*n.f.*), reduction : (79)

réduction électrique des minerais de **fer,** electrical reduction of iron ores. (80)

réduction à 0° (Barométrie) (*f.*), correction for temperature. (1)
réduire (*v.t.*), to reduce : (2)
  réduire au minimum les chances d'incendie, to minimize the chances of fire ; to reduce the chances of fire to a minimum. (3)
  réduire en cendres, to reduce to ashes. (4)
  réduire en farine, to flour. (5)
  réduire en poudre, to reduce to powder ; to powder. (6)
  réduire l'acier (Métall.), to reduce the steel. (7)
  réduire un plan (le refaire en petit), to reduce a plan. (8)
  réduire un oxyde (Chim.), to reduce an oxide. (9)
réduire (se) en farine (se dit du mercure dans le procédé d'amalgamation), to flour. (10)
réduire (se) en poudre, to powder ; to be reduced to powder. (11)
réduisant, -e (Chim.) (*adj.*), reducing. (12)
reef (banc de quartz aurifère) (*n.m.*), reef. (13)
réel, -elle (*adj.*), real ; actual ; true. (14)
réellement (*adv.*), really ; actually ; truly. (15)
réembarquement (*n.m.*), reshipment ; reshipping. (16)
réembarquer (*v.t.*), to reship. (17)
réestimation *ou* **réévaluation** (*n.f.*), revaluation. (18)
réestimer *ou* **réévaluer** (*v.t.*), to reestimate ; to revalue. (19)
réexpédier (*v.t.*), to send back ; to send again ; to resend ; to reship. (20)
réexpédition (*n.f.*), sending back ; resending ; reshipping ; reshipment. (21)
réexporter (*v.t.*), to reexport. (22)
refaire (faire de nouveau) (*v.t.*), to remake ; to do again : (23)
  refaire un joint, to remake a joint. (24)
refaire (raccommoder) (*v.t.*), to repair ; to restore : (25)
  refaire un mur, to repair a wall. (26)
réfection (*n.f.*), remaking ; repairing ; repair ; restoration : (27)
  réfection de joints, remaking joints. (28)
  réfection d'une route, repairing a road ; repair of a road. (29)
refendre (*v.t.*), to split ; to cleave ; to rip : (30)
  refendre l'ardoise, to split slate. (31)
  refendre les piliers pour se procurer du charbon avant que le traçage ait atteint les limites du quartier, to rip the pillars to get coal before the first working has reached the boundaries of the district. (32)
  refendre un arbre en deux, to cleave, to split, a tree in two. (33)
  refendre un pilier en deux, en quatre (Mines), to spilt a pillar into two, into four. (34 *ou* 35)
  refendre une planche en longueur, to rip a board lengthways. (36)
refente (*n.f.*), splitting ; cleaving ; ripping : (37)
  refente d'ardoises, splitting slates ; slate-splitting. (38)
refiltrer (*v.t.*), to refilter. (39)
refixer (*v.t.*), to refix. (40)
réfléchir (*v.t.*), to reflect : (41)
  les miroirs réfléchissent la lumière (*m.pl.*), mirrors reflect light. (42)
réfléchir (se) (*v.r.*), to be reflected ; to reflect : (43)
  objets qui se réfléchissent dans une eau tranquille (*m.pl.*), objects which are reflected in still water. (44)

réfléchissant, -e (*adj.*), reflecting : (45)
  miroir réfléchissant (*m.*), reflecting mirror. (46)
réfléchissement (*n.m.*), reflection ; reflecting : (47)
  réfléchissement de la lumière, reflection of light. (48)
réflecteur, -trice (*adj.*), reflecting ; reflective : (49)
  miroir réflecteur (*m.*), reflecting mirror. (50)
réflecteur (*n.m.*), reflector. (51)
réflecteur de lampe à acétylène (*m.*), acetylene-lamp reflector. (52)
réflecteur parabolique (*m.*), parabolic reflector. (53)
réflecteur plaqué argent (*m.*), silver-plated reflector. (54)
réflectivité (*n.f.*), reflectivity. (55)
reflets changeants *ou* **reflets chatoyants** (*m.pl.*), changeable hues ; chatoyant lustre. (56)
reflets irisés (*m.pl.*), play of colours ; pavonine ; iris : (57)
  l'opale noble, utilisée dans la joaillerie, est remarquable par la beauté de ses reflets irisés (*f.*), the noble opal, used in jewelry, is remarkable for the beauty of its play of colours. (58)
réflexion (*n.f.*), reflection ; reflecting : (59)
  réflexion de la lumière, reflection of light. (60)
  réflexion totale, total reflection. (61)
reflux (de la marée) (*n.m.*), fall, ebb, reflux, (of the tide). (62)
refondre (*v.t.*), to remelt ; to resmelt ; to recast ; to refound. (63)
refonte (*n.f.*), remelting ; refusion : (64)
  les fontes écossaises peuvent subir cinq à six refontes sans blanchir (*f.pl.*), Scotch pig irons can stand five to six remeltings without whitening. (65)
reforer (*v.t.*), to redrill : (66)
  reforer un puits, to redrill a well. (67)
refouillement (*n.m.*), recess. (68)
refouiller (*v.t.*), to recess. (69)
refoulant, -e (*adj.*), force (*used as adj*) ; forcing : (70)
  pompe refoulante (*f.*), force-pump ; forcing-pump ; ram-pump ; plunger-pump. (71)
refoulement (opp. à aspiration) (Aérage de mines) (*n.m.*), forcing ; forcing down. (72) (in contradistinction to exhausting) (Mine-ventilation).
refoulement (renflement d'une des extrémités d'une barre, etc.) (*n.m.*), upsetting ; jumping ; battering ; stoving ; stoving up : (73)
  la tête d'un rivet est fabriquée par le refoulement du métal de l'extrémité de la barre, the head of a rivet is made by upsetting the metal at the end of the bar. (74)
refoulement (en parlant d'une pompe, ou analo-gue) (*n.m.*), delivery : (75)
  le refoulement d'eau dans une chaudière, the delivery of water into a boiler. (76)
  tuyau de refoulement (*m.*), delivery-pipe. (77)
refoulement (action de reculer) (conduite d'une locomotive) (*n.m.*), backing. (78)
refoulement de l'eau d'aval (Hydraul.) (*m.*), backing up of the tail-water. (79)
refoulement de l'eau de la chaudière (*m.*), back-flow of boiler-water : (80)
  le clapet de retenue est destiné à empêcher tout refoulement de l'eau de la chaudière dans l'appareil alimentaire, the back-pressure valve is intended to prevent any

back-flow of boiler-water into the feeding apparatus. (1)

refouler (faire entrer de force) (*v.t.*), to force *or* drive in, up, down, back *or* along : (2)

refouler de l'air dans un puits de mine, to force air into (*or* down) a mine-shaft. (3)

refouler des boulons, to drive in bolts. (4)

flot qui refoule le cours d'un fleuve (*m.*), wave which drives back the flow of a river. (5)

refouler (renfler une des extrémités de) (*v.t.*), to upset ; to jump ; to batter ; to stove ; to stove up : (6)

refouler la tête d'un boulon, to upset, to jump, to batter, to stove, to stove up, the head of a bolt. (7)

refouler un cercle de roue, to upset, tc jump, a wheel-tire. (8)

refouler (comprimer) (*v.t.*), to compress : (9)

refouler un gaz avec une pompe, to compress a gas with a pump. (10)

refouler (en parlant d'une pompe, ou analogue) (*v.t.*), to deliver : (11)

refouler l'eau d'un seul jet au moyen de pompes foulantes commandées par un moteur souterrain, to deliver the water in one lift by means of force-pumps driven by an engine underground. (12)

injecteur qui refoule — litres d'eau par minute dans une chaudière (*m.*), injector which delivers — litres of water per minute into a boiler. (13)

refouler (ne pas s'enfoncer sous les coups) (*v.i.*), to refuse : (14)

le pieu, le boulon, refoule, the pile, the bolt, refuses. (15 *ou* 16)

refouler la denture d'une scie, to set the teeth of a saw. (17)

refouler un rivet, to close a rivet. (18)

refouler une charge (Travail aux explosifs), to ram home a charge. (19)

refouleur (*n.m.*), upset. (20)

réfractaire (*adj.*), refractory ; fire (*used as adj.*) : (21)

argile réfractaire (*f.*), fire-clay ; refractory clay. (22)

réfractaire (*n.m.*), refractory. (23)

réfracter (*v.t.*), to refract : (24)

le prisme réfracte les rayons lumineux, the prism refracts luminous rays. (25)

réfracter (se) (*v.r.*), to be refracted : (26)

rayon lumineux qui se réfracte à travers un prisme (*m.*), light-ray which is refracted through a prism. (27)

réfractif, -ive (*adj.*), refractive ; refracting ; refringent : (28)

pouvoir réfractif (*m.*), refractive power. (29)

réfraction (*n.f.*), refraction : (30)

réfraction de la lumière, refraction of light. (31)

réfractomètre (*n.m.*), refractometer. (32)

réfractomètre interférentiel (*m.*), interference refractometer ; interferential refractometer. (33)

réfrigérant, -e (*adj.*), refrigerating : (34)

liquide réfrigérant (*m.*), refrigerating liquid. (35)

réfrigérant (d'un appareil frigorifique) (*n.m.*), refrigerator, cooler, (of a freezing-apparatus). (36)

réfrigérant (d'un alambic) (*n.m.*), condenser (of a still). (37)

réfrigérant à cheminée (*m.*), cooling-tower ; chimney-cooler. (38)

réfrigération (*n.f.*), refrigeration ; freezing ; cooling. (39)

réfrigérer (*v.t.*), to refrigerate ; to freeze ; to cool. (40)

réfringence (*n.f.*), refractivity ; refractiveness ; refringency ; refringence. (41)

réfringent, -e (*adj.*), refracting ; refractive ; refringent : (42)

angle réfringent d'un prisme (*m.*), refracting angle of a prism. (43)

refroidir (*v.t.*), to cool : (44)

refroidir par l'eau, to water-cool. (45)

refroidir (*v.i.*), to cool : (46)

les corps qui s'échauffent lentement refroidissent de même (*m.pl.*), bodies which heat slowly cool slowly. (47)

refroidir (se) (*v.r.*), to cool. (48)

refroidissement (*n.m.*), cooling : (49)

refroidissement de l'eau, de la lave, cooling of water, of lava. (50 *ou* 51)

refroidissement (d'un haut fourneau) (*n.m.*), slacking (of a blast-furnace). (52)

refroidissement d'air (à), air-cooled : (53)

moteur à refroidissement d'air (*m.*), air-cooled motor. (54)

refroidissement d'eau (à), water-cooled : (55)

piston à refroidissement d'eau (*m.*), water-cooled piston. (56)

refroidisseur (*n.m.*), cooler. (57)

refuge (*n.m.*), refuge ; refuge hole ; shelter ; manhole ; place of safety. (58)

réfugier (se) (*v.r.*), to take shelter. (59)

refus (*n.m.*), refusal : (60)

refus d'un pieu, refusal of a pile. (61)

refus (du crible) (*n.m.pl.*), oversize ; riddlings ; skimpings. (62)

refusé (-e) par le crible (être), to pass over the screen : (63)

matières refusées par le crible (*f.pl.*), stuff passing over the screen. (64)

refuser (*v.t.*), to refuse. (65)

refuser (ne pas s'enfoncer sous les coups) (*v.i.*), to refuse : (66)

le pilotis refuse, the pile refuses. (67)

regard (ouverture qui permet de visiter certains organes) (Mach.) (*n.m.*), peep-hole ; sight-hole ; spy-hole ; peek-hole ; hole ; door : (68)

regard servant à constater le jeu d'un mécanisme, peep-hole for watching the action of a mechanism ; door for the purpose of observing the movement of the parts of a mechanism. (69)

regard (d'un fourneau, d'un cubilot) (*n.m.*), sight-hole, eye, peep-hole, spy-hole, glory-hole, (of a furnace, of a cupola). (70 *ou* 71)

regard (d'un aqueduc, d'un égout) (*n.m.*), man-hole (of an aqueduct, of a sewer). (72 *ou* 73)

regard (fenêtre) (Géol.) (*n.m.*), inlier. (74)

regard (d'une faille) (Géol.) (*n.m.*), heaved side, upthrow side, upcast side, (of a fault). (75)

regard de lavage (*m.*), washout-hole ; hand-hole ; mud-door. (76)

regard de nettoyage (*m.*), cleaning-hole ; cleaning-door. (77)

regard de visite (*m.*), inspection-door ; inspection-cover. (78)

regarnir (*v.t.*), to refit ; to replenish. (79)

regarnir un piston, to repack a piston. (80)

regarnissage (*n.m.*), refitting ; replenishing. (81)

regarnissage de pistons (*m.*), repacking pistons. (1)

regel (Phys.) (*n.m.*), regelation. (2)

régénérateur, -trice (*adj.*), regenerative ; regener-atory. (3)

régénérateur (*n.m.*), regenerator ; recuperator. (4)

régénération (*n.f.*), regeneration : (5)
régénération de l'acier brûlé, regeneration of burnt steel. (6)

régénérer (*v.t.*), to regenerate : (7)
régénérer un métal, to regenerate a metal. (8)

régime (*n.m.*), conditions ; prevailing conditions : (9)
la diversité du relief du Massif central donne à chaque unité orographique un régime climatérique particulier, the diversity of the relief of the Central Plateau (a mountain-ous region in France) gives to each oro-graphic unit distinct climatic conditions. (10)
le régime lacustre est indiqué par la présence de mombreux mollusques d'eau douce et de petits crustacés, lacustrine conditions are indicated by the presence of numerous fresh-water molluscs and small crustaceans. (11)
le régime climatérique d'une vallée, the climatic conditions prevailing in a valley. (12)

régime (d'un fleuve) (*n.m.*), regimen (of a river) : (13)
la corrosion des rives des fleuves est souvent la principale cause des perturbations de leur régime, the erosion of the banks of rivers is often the cause of the disturbances of their regimen. (14)

régime (Méc.) (*n.m.*), normal working conditions ; working conditions ; normal conditions ; conditions : (15)
le régime d'une machine à vapeur, des lampes à incandescence, des lampes à arc voltaïque, the normal working conditions of a steam-engine, of incandescent lamps, of arc-lamps. (16 *ou* 17 *ou* 18)
un moteur marchant en régime uniforme, an engine running under uniform conditions. (19)
pour chaque puits il y a un régime de pompage qui procure le meilleur rendement en pétrole, et le moindre danger de le voir envahi par l'eau, for each well there are pumping conditions which secure the best yield of oil, and the least danger of having the well flooded with water. (20)

régime (de), normal ; working : (21)
courant de régime (Élec.) (*m.*), normal current. (22)
tension de régime (*f.*), normal voltage ; working-pressure. (23)

régime imbriqué (Géol.) (*m.*), imbricate structure. (24)

régime isoclinal (Géol.) (*m.*), isoclinal structure. (25)

région (*n.f.*), region ; district ; area ; field ; territory. (26)

région bien arrosée *ou* région bien pourvue d'eau (*f.*), well-watered district (*or* region). (27)

région carbonifère (*f.*), coal-field. (28)

région dépourvue de pluie (*f.*), rainless district. (29)

région désertique (*f.*), desert region. (30)

région effondrée (Mines) (*f.*), caved-in district. (31)

région forestière (*f.*), forest area. (32)

région mal arrosée (*f.*), badly watered district. (33)

région minière (*f.*), mining-region ; mining-district ; mine-field ; minery. (34)

région pétrolifère (*f.*), oil-field ; oil-region ; oil-producing territory ; oil-bearing country ; petroleum-territory. (35)

région productive de charbon (*f.*), coal-producing region. (36)

régional, -e, -aux (*adj.*), regional : (37)
glaciation régionale (*f.*), regional glaciation. (38)

régir (*v.t.*), to govern ; to rule : (39)
le circuit magnétique est régi par une loi, analogue à la loi d'Ohm, the magnetic circuit is governed by a law, analogous to Ohm's law. (40)

registre (d'un fourneau) (*n.m.*), damper, draught-regulator, register, (of a furnace). (41)

registre de réglage du vent (*m.*), blast-gate. (42)

registre de vapeur (*m.*), throttle ; throttle-valve. (43)

réglable (*adj.*), adjustable : (44)
coussinets réglables (*m.pl.*), adjustable bear-ings. (45)

réglable à volonté, adjustable at will. (46)

réglage (action ou manière de régler du papier) (*n.m.*), ruling. (47)

réglage (action de régler un mécanisme, d'en régulariser la marche) (*n.m.*), regulation ; regulating ; adjustment ; setting : (48)
réglage d'un chronomètre, regulation of a chronometer ; regulating a chronometer. (49)
réglage rapide de la course, même en marche, rapid adjustment of the stroke, without stopping the machine ; quick adjustment of the travel, even while running. (50)

règle (instrument pour tracer des lignes) (*n.f.*), rule ; ruler ; straight-edge. (51)

règle (usage) (*n.f.*), rule : (52)
en règle générale, as a general rule ; as a rule. (53)
règle invariable, invariable rule ; hard-and-fast rule. (54)

règle à calcul (*f.*), slide-rule, sliding-rule, sliding-scale, for calculation (55)
règle à calcul, à curseur, acajou, plaqué celluloïde, avec étui et instruction, slide-rule for calculation, with cursor, mahogany, faced with celluloid, with case and instruc-tions. (56)

règle à dévers *ou* règle à dévers et niveau de posage (Ch. de f.) (*f.*), superelevation-gauge ; track-level ; track level and gauge. (57)

règle à filet de cuivre (*f.*), brass-edged rule. (58)

règle à tracer des parallèles (*f.*), parallel rule ; parallel ruler. (59)

règle articulée (*f.*), jointed rule. (60)

règle cornière pour tracer les rainures de clavetage *ou* règle pour transmissions (*f.*), key-seat rule. (61)

règle de lord Kelvin (Élec.) (*f.*), Kelvin's law. (62)

règle des trois doigts de la main droite *ou* règle des générateurs *ou* règle des génératrices (Élec.) (*f.*), right-hand rule ; rule for generators. (63)

règle des trois doigts de la main gauche *ou* règle des moteurs (Élec.) (*f.*), left-hand rule ; rule for motors. (64)

règle divisée à la machine (*f.*), engine-divided scale. (1)

règle du tire-bouchon (Élec.) (*f.*), corkscrew rule. (2)

règle en acier, très soignée, dressée sur plat et sur champ (*f.*), steel straight-edge, accurately ground on the flat and on the edge. (3)

règle flexible en acier trempé (*f.*), tempered flexible steel rule. (4)

règle plate *ou* **règle plate à dessin** (*f.*), flat ruler. (5)

règle spéciale (*f.*). See **règles spéciales.**

règlement (action) (*n.m.*), regulating; regulation; adjustment; adjusting; settlement; settling: (6)

règlement d'une contestation, settlement of a dispute; settling a differ nce. (7)

règlement (ordonnance) (*n.m.*), regulation; by-law: (8)

règlements en vigueur, regulations in force. (9)

règlements pour l'exploitation des mines de charbon, coal-mines regulations. (10)

règlements relatifs aux mines métallifères, metalliferous-mines regulations. (11)

règlements sur les mines, mining regulations. (12)

réglementaire (*adj.*), prescribed: (13)

maintenir le niveau de l'eau dans la chaudière à la hauteur réglementaire, to keep the level of the water in the boiler at the pre--scribed height. (14)

réglementation (*n.f.*), regulating; regulation: (15)

réglementation de la production, regulating the output. (16)

réglementation des mines à grisou, regulation of fiery mines. (17)

régler (tirer des lignes sur) (*v.t.*), to rule: (18)

régler du papier, to rule paper. (19)

régler (diriger la marche de) (*v.t.*), to regulate; to adjust; to set: (20)

régler la tension d'un ressort, la marche d'une machine, l'admission de la vapeur aux tiroirs, un chronomètre, to regulate the tension of a spring, the running of an engine, the admission of steam to the slide-valves, a chronometer. (21 *ou* 22 *ou* 23 *ou* 24)

régler un instrument de nivellement, to adjust a levelling-instrument. (25)

régler un tiroir, une distribution par coulisse, to set a slide-valve, a link-gear. (26 *ou* 27)

régler l'itinéraire d'un voyage, to fix the itinerary of a journey; to settle the route of a pro--posed tour. (28)

régler les ouvriers, to pay off the men. (29)

régler une affaire en dispute, to settle a matter in dispute. (30)

règles spéciales (*f.pl.*), slide-rules for special purposes. (31)

réglette (carrelet) (Dessin) (*n.f.*), square ruler. (32)

réglette (d'une règle à calcul) (*n.f.*), slide, slider, (of a slide-rule): (33)

faire glisser la réglette, to slide the slide. (34)

réglette (règle portant les cavaliers d'une balance) (*n.f.*), graduated beam; divided beam. (35)

régnant, -e (dominant) (*adj.*), prevailing: (36)

vent régnant (*m.*), prevailing wind. (37)

règne minéral (*m.*), mineral kingdom; mineral world. (38)

régression (*n.f.*), regression. (39) Cf. **érosion régressive** and **marche régressive de l'érosion torrentielle.**

régularisation (*n.f.*), regularization. (40)

régularisation de pente *ou simplement* **régularisa--tion** (*n.f.*), grading. (41)

régulariser (*v.t.*), to regularize: (42)

régulariser le débit d'un cours d'eau, to regularize the flow of a stream. (43)

régulariser la pente de *ou simplement* **régulariser** (*v.t.*), to grade: (44)

régulariser la pente d'une route, to grade a road. (45)

rivière qui régularise son lit (*f.*), river which grades its bed. (46)

une surface régularisée, a graded surface. (47)

régularité (*n.f.*), regularity. (48)

régulateur, -trice (*adj.*), regulating. (49)

régulateur (Mach.) (*n.m.*), regulator; governor. (50)

régulateur (soupape régulatrice d'admission de vapeur) (*n.m.*), throttle; throttle-valve. (51)

régulateur (*n.m.*) *ou* **régulateur d'arc électrique** *ou* **régulateur à arc voltaïque** *ou* **régulateur de lumière électrique,** regulator; arc-regulator. (52)

régulateur à boules *ou* **régulateur à force centri--fuge** *ou* **régulateur de Watt** (*m.*), ball governor; centrifugal governor; fly-ball governor; simple governor; Watt's governor. (53)

régulateur à courant constant (Élec.) (*m.*), con--stant-current regulator. (54)

régulateur à masse centrale (*m.*), loaded governor. (55)

régulateur à papillon (*m.*), butterfly throttle-valve. (56)

régulateur à potentiel constant (Élec.) (*m.*), con--stant-potential regulator. (57)

régulateur à ressort (*m.*), spring-governor. (58)

régulateur à tiroir (*m.*), slide throttle-valve. (59)

régulateur à volant d'inertie (*m.*), inertia gover--nor. (60)

régulateur automatique (*m.*), automatic regu--lator; self-acting regulator. (61)

régulateur d'alimentation (*m.*), feed-box. (62)

régulateur de tension *ou* **régulateur d'intensité** (Élec.) (*m.*), voltage-regulator; pressure-regulator. (63)

régulateur de vapeur (*m.*), steam-governor. (64)

régulateur de vitesse (*m.*), governor; speed-governor. (65)

régulateur différentiel (Élec.) (*m.*), differential regulator. (66)

régulateur hydraulique (*m.*), hydraulic governor. (67)

régulateur isochrone (*m.*), isochronous governor. (68)

régulateur monophote (Élec.) (*m.*), monophote regulator. (69)

régulateur polyphote (Élec.) (*m.*), polyphote regulator. (70)

régulateur-série [**régulateurs-série** *pl.*] (Élec.) (*n.m.*), series-regulator. (71)

régulateur-shunt [**régulateurs-shunt** *pl.*] (Élec.) (*n.m.*), shunt regulator (72)

régule (Métall.) (*n.m.*), regulus. (73)

régule (métal antifriction) (*n.m.*), antifriction metal; white metal; babbitt metal; babbitt. (74)

régule d'antimoine (*m.*), antimony regulus : (1)
la facilité avec laquelle l'antimoine s'allie à l'or lui faisait attribuer par les alchimistes des qualités nobles. d'où son nom de régule (petit roi), the facility with which antimony alloys with gold caused noble qualities to be attributed to it by the alchemists, whence its name regulus (little king). (2)

régule de cuivre (*m.*), copper regulus. (3)

réguler (*v.t.*), to line with antifriction metal ; to babbitt : (4)
réguler un palier, to line a bearing-block with antifriction metal ; to babbitt a bearing. (5)

régulier, -ère (*adj.*), regular ; proper : (6)
pyramide régulière (*f.*), regular pyramid. (7)

régulièrement (*adv.*), regularly ; properly. (8)

régulin, -e (*adj.*), reguline. (9)

rehaussement (*n.m.*), raising ; heightening : (10)
rehaussement d'une muraille, raising, heightening, a wall. (11)

rehausser (*v.t.*), to raise ; to heighten. (12)

reillère (conduit qui amène l'eau sur une roue hydraulique) (*n.f.*), flume ; leat. (13)

rein (d'une voûte) (*n.m.*), haunch, hanch, hance, flank, flank, (of an arch). (14)

rejet (accident ; cran) (Géol. & Mines) (*n.m.*), throw ; jump ; leap ; check. (15)

rejet (déplacement vertical) (Géol. & Mines) (*n.m.*), throw. (16)

rejet en bas (*m.*), downthrow ; downcast. (17)

rejet en haut (*m.*), upthrow ; upcast. (18)

rejet en profondeur (*m.*), throw in depth. (19)

rejeter (Géol. & Mines) (*v.t.*), to throw : (20)
faille qui rejette un filon (*f.*), fault which throws a lode. (21)
on constate, presque toujours, que le filon croisé est rejeté par le filon croiseur, it is nearly always found that the intersected lode is thrown by the cross-lode. (22)

rejeter une offre, to reject an offer. (23)

relâchement (*n.m.*), loosening ; relaxing ; slackening ; slacking. (24)

relâcher (*v.t.*), to loosen ; to relax ; to slacken ; to slack : (25)
relâcher une corde, to slacken, to slack, to relax, a cord. (26)

relais (*n.m.*), relay ; spell ; shift. (27)

relais (Élec.) (*n.m.*), relay. (28)

relais (Géol.) (*n.m.*), warp ; silt. (29)

relais de chevaux (*m.*), relay of horses. (30)

relais-moteur [relais-moteurs *pl.*] (*n.m.*), relay ; servomotor. (31)

relais polarisé (Élec.) (*m.*), polarized relay. (32)

relatif, -ive (*adj.*), relative ; comparative : (33)
pression relative (Méc.) (*f.*), relative pressure. (34)

relation (*n.f.*), relation. (35)

relativement (*adv.*), relatively ; comparatively : (36)
roche relativement tendre (*f.*), comparatively soft rock. (37)

relaver (*v.t.*), to rewash. (38)

relayer (*v.t.*), to relay. (39)

relevage (*n.m.*), raising ; lifting : (40)
relevage de la voie (Ch. de f.), raising the road ; surfacing the track. (41)
relevage rapide de l'arbre porte-foret par volant à main, quick raising of the drilling-spindle by hand-wheel. (42)

relève-rail [relève-rail *pl.*] (*n.m.*), rail-jack ; railway-jack ; railway track-lifter. (43)

relevé d'un diagramme *ou* **relevé d'un diagramme d'indicateur** (*m.*), taking a diagram ; taking an indicator-card. (44)

relevé d'une empreinte du fond du trou de sonde (*m.*), taking an impression of the bottom of the bore-hole. (45)

relevé du terrain (Géod.) (*m.*), plotting ground. (46)

relevée (hauteur sur l'inclinaison) (opp. à **hauteur verticale**) (Mines) (*n.f.*), inclined depth. (47) (distinguished from **vertical depth**).

relèvement (*n.m.*), raising : (48)
relèvement d'un mur, raising a wall. (49)

relèvement (*ou* relevé) **de la dépense de l'année** (*m.*), statement of the year's expenses ; return of the annual expenditure. (50)

relèvement des galeries éboulées (Mines) (*m.*), ridding out the fallen roof ; clearing falls. (51)

relever (porter en haut) (*v.t.*), to raise ; to lift : (52)
relever l'arbre porte-foret, to raise the drilling-spindle. (53)
relever la coulisse (distribution par coulisse), to raise the link (link-gear). (54)
relever la voie (Ch. de f.), to raise the road ; to surface the track. (55)
relever le niveau d'un plancher, to raise the level of a floor. (56)
relever les eaux (Mines), to raise water. (57)
relever les terres, to raise the ground. (58)

relever (Géod.) (*v.t.*), to plot : (59)
relever les points du terrain, to plot the points of the ground. (60)

relever la teneur en grisou, to take the per-centage of fire-damp : (61)
la teneur en grisou des retours d'air est relevée quotidiennement dans les mines franchement grisouteuses, the percentage of fire-damp in the air-returns shall be taken daily in mines where much gas is present. (62)

relever un diagramme *ou* **relever un diagramme d'indicateur**, to take a diagram ; to take an indicator-card ; to indicate : (63)
rankinisation des diagrammes relevés sur une machine à double expansion (*f.*), combining diagrams taken (*or* indicated) on a double-expansion engine. (64)

relever une empreinte du fond du trou de sonde, to take an impression of the bottom of the bore-hole. (65)

relever une galerie éboulée (Mines), to rid out the fallen roof ; to clear a fall. (66)

reliage (*n.m.*), connection ; connecting ; coupling. (67)

relié (-e) à la terre *ou* **relié (-e) au sol** (Élec.), connected to earth ; earthed ; grounded. (68)

relief (Géogr. phys.) (*n.m.*), relief ; form : (69)
le relief de la surface terrestre, the relief (*or* the form) of the earth's surface. (70)

relier (*v.t.*), to connect ; to couple ; to couple up : (71)
relier des fils (Élec.), to connect wires. (72)

relier à la terre (Élec.), to connect to earth ; to earth ; to ground. (73)

réluctance (*n.f.*), reluctance ; magnetic resis-tance. (74)

réluctivité (Magnétisme) (*n.f.*), reluctivity ; specific reluctance. (75)

rémanence (*n.f.*), remanence ; remanent mag-netism ; residual magnetism. (76)

rembarquer (*v.t.*), to reship : (1)
rembarquer des marchandises, to reship goods. (2)

remblai (action) (*n.m.*). Same as remblayage.

remblai (talus) (*n.m.*), embankment : (3)
remblai d'un chemin de fer, embankment of a railway. (4)

remblai (Mines) (*n.m.*), gob ; goaf ; goave ; stowing ; filling ; rock-filling ; waste ; pack. (5)

remblai (crassier) (*n.m.*), slag-dump ; slag-tip ; slag-heap. (6)

remblai d'embouage (Mines) (*m.*), flush ; slush. (7) See remblayage par embouage.

remblai hydraulique *ou* remblai d'ensablage (Mines) (*m.*), water pack ; sand pack. (8)

remblai stérile (*m.*), waste filling. (9)

remblayage (d'un fossé) (*n.m.*), filling up (a ditch). (10)

remblayage (Mines) (*n.m.*), stowing ; filling ; gobbing ; gobbing up ; rock-filling ; pack-ing. (11)

remblayage hydraulique *ou* remblayage à l'eau *ou* remblayage par ensablage (*m.*), water-packing ; sand-packing ; sand-filling. (12)

remblayage hydraulique par embouage *ou* rem--blayage par embouage (*m.*), flushing ; slush--ing ; silting ; water-flush system (filling up the worked-out portions of a mine with refuse material brought in on streams of water). (13)

remblayer (*v.t.*), to fill up ; to fill : (14)
remblayer un fossé, to fill up a ditch. (15)

remblayer (Mines) (*v.t.*), to stow ; to fill ; to pack ; to gob : (16)
remblayer d'anciens chantiers, to stow, to fill, to pack, to gob, old workings. (17)

remblayeur (Mines) (pers.) (*n.m.*), stower. (18)

rembourrage *ou* rembourrement (*n.m.*), stuffing ; padding. (19)

rembourrer (*v.t.*), to stuff ; to pad. (20)

remède (*n.m.*), remedy. (21)

remédier (*v.t.*), to remedy ; to make good : (22)
remédier à un défaut, to remedy a defect : (23)
quelques défauts auxquels il importe de remédier, several defects which it is necessary to remedy. (24)
remédier à l'usure, to make good the wear and tear. (25)

remercier (congédier) (*v.t.*), to dismiss ; to discharge : (26)
remercier un employé, to dismiss an employee. (27)

remesurage (*n.m.*), remeasuring ; remeasure--ment. (28)

remesurer *ou* remétrer (*v.t.*), to remeasure. (29)

remettre à 0 (un instrument, tel qu'un compte-tours), to set to zero ; to reset to zero. (30)

remettre en état (*v.t.*), to repair. (31)

remettre en marche (*v.t.*), to restart : (32)
remettre les machines en marche, to restart the machinery. (33)

remise (Ch. de f.) (*n.f.*), covered siding. (34)

remise à 0 (*f.*), setting to zero ; resetting to zero. (35)

remise en état (*f.*), repairing. (36)

remontage (action de monter, d'assembler de nouveau les diverses pièces d'une machine) (*n.m.*), remounting ; reerection ; refitting. (37)

remontage (action de percer de bas en haut) (Mines) (*n.m.*), rising ; raising ; upraising. (38)

remontage (ouvrage montant—opp. à descenderie) (Mines) (*n.m.*), rise ; raise ; riser ; rising ; upraise. (39)

remontage (d'un fleuve) (*n.m.*), going up, ascend--ing, (a river). (40)

remontage (d'une montre) (*n.m.*), winding up (a watch). (41)

remonte (extraction) (*n.f.*), raising ; winding ; hoisting ; drawing ; lifting. (42)

remonte (ouvrage montant) (Mines) (*n.f.*), rise ; raise ; riser ; rising ; upraise. (43)

remonte (d'un fleuve) (*n.f.*), going up, ascend--ing, (a river) ; ascent (of a river). (44)

remontée (action) (*n.f.*), raising ; hoisting ; winding ; drawing ; lifting ; ascending : (45)
remontée des ouvriers, raising the men ; raising workmen ; bringing the workmen to the surface. (46)
remontée du minerai à la surface *ou* remontée du minerai au jour, raising ore to the surface ; bringing ore to grass. (47)

remontée (ouvrage montant) (Mines) (*n.f.*), rise ; raise ; riser ; rising ; upraise. (48)

remonter (aller contre le mouvement ; gravir) (*v.t.*), to ascend ; to mount ; to go up : (49)
remonter le courant d'une rivière *ou* remonter le cours d'un fleuve, to go up-river ; to go up-stream ; to go up a river ; to ascend a river. (50)
remonter une rampe, to mount a gradient ; to ascend a slope. (51)

remonter (extraire) (*v.t.*), to raise ; to hoist ; to wind ; to draw ; to lift : (52)
remonter le charbon de la mine, to raise, to hoist, to wind, to draw, coal from the mine. (53)
remonter le minerai à la surface *ou* remonter le minerai au jour, to raise the ore ; to lift ore to the surface ; to bring ore to grass ; to grass the ore. (54)
remonter le pétrole par air comprimé, to lift the oil by compressed air. (55)
remonter les ouvriers, to raise workmen ; to bring the men to the surface. (56)

remonter (exhausser) (*v.t.*), to raise : (57)
remonter un mur, to raise a wall. (58)

remonter (réassembler) (*v.t.*), to remount ; to reset. (59)

remonter (tendre de nouveau les ressorts) (*v.t.*), to wind ; to wind up : (60)
remonter un ressort, to wind up a spring. (61)

remonter (s'élever de bas en haut) (*v.i.*), to ascend ; to go up ; to mount ; to rise : (62)
remonter dans un puits, to ascend, to go up, a shaft. (63)

remonter (percer en montant) (Mines) (*v.i.*), to rise ; to raise ; to put up a rise ; to put in a raise. (64)

remorquage (*n.m.*), towing. (65)

remorque (traction exercée) (*n.f.*), tow : (66)
prendre un bateau à la remorque, to take a boat in tow. (67)

remorque (véhicule routier remorqué) (*n.f.*), trailer ; trail-car ; tow-car : (68)
atteler une remorque à un tracteur, to attach a trailer to a tractor. (69)

remorque (câble de remorque) (*n.f.*), tow-rope ; tow-line ; towing-rope : (70)
jeter la remorque, to throw the tow-rope. (71)

remorquer (*v.t.*), to tow ; to haul ; to draw : (72)
remorquer un vapeur, to tow a steamer. (73)

remorquer un train par locomotive électrique, to haul a train by an electric locomotive. (1)
locomotive qui remorque une lourde charge (*f.*), locomotive which hauls (*or* draws) a heavy load. (2)
remorqueur (*n.m.*), tug ; tug-boat. (3)
remoudre (*v.t.*), to regrind. (4)
remoulage (*n.m.*), remoulding ; remolding ; recasting ; refounding. (5)
remouler (*v.t.*), to remould ; to remold ; to recast ; to refound. (6)
remouleur (pers.) (*n.m.*), knife-grinder ; grinder. (7)
remous (*n.m.*), eddy. (8)
rempart morainique (*m.*), boulder-wall ; bowlder-wall. (9)
remplaçable (*adj.*), replaceable : (10)
machine difficilement remplaçable (*f.*), machine replaceable with difficulty. (11)
remplacement (*n.m.*), replacement ; replacing ; substitution : (12)
remplacement des pièces d'une machine, replacement of the parts of a machine. (13)
remplacer (*v.t.*), to replace ; to substitute ; to supersede : (14)
remplacer, dans un composé, un corps simple par un autre corps simple ou par un radical composé, to replace an element in a compound by another element or by a compound radical. (15)
remplacer l'emploi de la vapeur par l'électricité, to replace the use of steam by electricity. (16)
remplacer le charbon par le pétrole, to replace coal by oil-fuel ; to substitute oil for coal-fuel. (17)
remplacer le travail manuel, to supersede hand-labour. (18)
remplage (Constr.) (*n.m.*), filling. (19)
remplir (*v.t.*), to fill ; to fill up : (20)
remplir une citerne, une lampe, un tonneau vide, to fill a cistern, a lamp, an empty cask. (21 *ou* 22 *ou* 23)
remplir à nouveau, to refill. (24)
remplir (se) (*v.r.*), to fill ; to fill up ; to be filled. (25)
remplissage (*n.m.*), filling ; filling up : (26)
remplissage d'un fossé, filling up a ditch. (27)
remplissage filonien (Géol.), lode filling. (28)
remplissage (remplage) (Constr.) (*n.m.*), filling. (29)
remplissage à nouveau (*m.*), refilling. (30)
remuage (*n.m.*), agitation ; agitating ; shaking ; stirring ; stirring up. (31)
remuage à la pelle (*m.*), shovelling. (32)
remuer (*v.t.*), to agitate ; to shake ; to stir ; to stir up : (33)
remuer le feu, to stir, to stir up, the fire. (34)
remuer la terre *ou* remuer le sol, to turn up, to disturb, the ground. (35)
remuer à la pelle, to shovel. (36)
rémunérateur, -trice (*adj.*), payable ; paying ; pay (*used as adj.*) ; remunerative ; profit-able : (37)
minerai rémunérateur (*m.*), payable ore ; pay-ore. (38)
affaire rémunératrice (*f.*), paying concern ; profitable business ; remunerative under-taking. (39)
rémunérer (*v.t.*), to pay ; to pay for : (40)
rémunérer les dépenses, to pay expenses. (41)

rémunérer des services, to pay for services. (42)
renaissance de l'industrie minière (*f.*), revival of the mining industry. (43)
renard (pers.) (*n.m.*), blackleg ; scab ; strike-breaker. (44)
renchérir (*v.t.*), to raise the price of. (45)
renchérir (*v.i.*), to rise in price ; to rise : (46)
les loyers renchérissent (*m.pl.*), rents are rising. (47)
renchérissement (*n.m.*), rise in price ; rise in cost : (48)
renchérissement des matières premières, rise in the price of raw materials. (49)
renchérissement progressif de la main-d'œuvre, progressive rise in the cost of labour. (50)
rencontre (*n.f.*). See examples : (51)
rencontre d'or dans une veine, d'eau en pro-fondeur, occurrence of gold in a vein, of water in depth. (52 *ou* 53)
rencontre de pétrole, strike of oil ; striking oil. (54)
rencontre de deux trains, collision of two trains. (55)
rencontre des routes, de deux rivières, meeting of the roads, of two rivers. (56 *ou* 57)
rencontre (de), second-hand : (58)
meubles de rencontre (*m.pl.*), second-hand furniture. (59)
rencontrer (*v.t.*), to meet with ; to encounter ; to come upon ; to strike : (60)
rencontrer beaucoup d'obstacles, to meet with a lot of obstacles. (61)
rencontrer du minerai de nature à donner bon espoir, to come upon mineral of a promising character. (62)
rencontrer le fond solide (Mines & Placers), to bottom ; to strike bed-rock. (63)
rencontrer le filon, to strike the lode. (64)
rencontrer une nappe d'eau, to encounter, to strike, a body of water. (65)
rencontrer (tamponner ; se heurter à) (*v.t.*), to collide with ; to run into. (66)
rencontrer (se) (se trouver ; se présenter) (*v.r.*), to occur ; to be found ; to be met with ; to be present : (67)
se rencontrer à l'état natif, to occur in a native state ; to occur native ; to be found native : (68)
minéral qui se rencontre dans la nature à l'état natif (*m.*), mineral which occurs in nature in the native state ; mineral which is found native in nature. (69)
le bismuth se rencontre à l'état natif, bismuth occurs native. (70)
se rencontrer associé (-e) à, to be found associated with ; to be found in association with : (71)
espèce minérale qui se rencontre associée au grenat (*f.*), mineral species which is found associated with garnet. (72)
se rencontrer dans les eaux, to be present in the waters : (73)
le borax se rencontre dans les eaux, borax is present in the waters. (74)
se rencontrer en gisements, to occur in beds. (75)
se rencontrer en masses *ou* se rencontrer en amas, to occur massive ; to be found massive : (76)
la turquoise se rencontre en masses, toujours amorphes, turquoise is found massive, always amorphous. (77)

se rencontrer en masses fibreuses, to occur in fibrous masses. (1)

rencontrer (se) (entrer en collision) (*v.r.*), to collide ; to run into each other : (2)

deux trains qui se rencontrent (*m.pl.*), two trains which collide (*or* which run into each other). (3)

rencontrer (se) (confluer) (*v.r.*), to meet : (4)

la Seine et la Marne se rencontrent à Charenton, the Seine and the Marne meet at Charenton. (5)

rendement (ce que produit une chose en raison de la quantité) (*n.m.*), output ; production ; yield ; get ; returns ; result ; capacity : (6)

rendement à l'heure, output per hour. (7)

rendement augmentant régulièrement, steadily increasing output. (8)

rendement d'or pendant l'année, returns of gold during the year. (9)

rendement d'un puits à pétrole, yield, capacity, of an oil-well. (10)

rendement de la mine, mine-returns ; mine-results ; production, yield, output, get, of the mine. (11)

rendement des graviers, yield of the gravel. (12)

rendement donné par la batée (Mines), pan-ning ; pannings. (13)

rendement du bocard, mill-result. (14)

rendement du piqueur (Mines de houille), get of the hewer. (15)

rendement journalier moyen, average daily output. (16)

rendement (effet utile d'une machine, d'un appareil ; quantité de travail produit) (*n.m.*), output ; capacity ; duty ; perform-ance : (17)

rendement d'une dynamo exprimé en watts, output of a dynamo expressed in watts. (18)

rendement d'une locomotive, capacity of a locomotive. (19)

rendement (*n.m.*) *ou* **rendement industriel** (rapport existant entre le travail obtenu après la transformation par un système mécanique et la quantité d'énergie fournie à ce système), efficiency ; commercial coefficient ; coefficient of efficiency ; coefficient of useful effect : (20)

le rendement d'une machine est le rapport du travail utile produit au travail moteur absorbé, the efficiency of an engine is the ratio of the useful work performed to the mechanical energy expended. (21)

rendement calorifique, heat-efficiency ; ther-mal efficiency ; calorific efficiency : (22)

rendement calorifique total d'une chaudière, total heat-efficiency of a boiler. (23)

rendement d'un accumulateur, efficiency of an accumulator. (24)

rendement d'une chaudière, efficiency of a boiler. (25)

rendement du moteur, engine efficiency. (26)

rendement dynamique, dynamic efficiency. (27)

rendement électrique d'une dynamo, electric efficiency of a dynamo. (28)

rendement en volume, volumetric efficiency. (29)

rendement lumineux (*ou* rendement optique) d'une lampe, efficiency of a lamp. (30)

rendement manométrique d'un ventilateur, manometric efficiency of a fan. (31)

rendement mécanique, mechanical efficiency : (32)

rendement mécanique d'une pompe, d'un ventilateur de mine, mechanical efficiency of a pump, of a mine-fan. (33 *ou* 34)

rendement thermique. *Same as* rendement calorifique.

rendement volumétrique, volumetric efficiency. (35)

rendre (rapporter ; produire) (*v.t.*), to yield ; to produce ; to bring in ; to output : (36)

rendre un bénéfice de — par tonne, to yield a profit of — per ton. (37)

rendre à la batée (Mines), to pan ; to pan out. (38)

rendre une moyenne de, to yield an average of ; to average : (39)

le minerai rend une moyenne de — par tonne, the ore averages — per ton. (40)

rendre (faire devenir) (*v.t.*), to render ; to make : (41)

rendre un chenal navigable, to make a channel navigable. (42)

rendre compte de la valeur du minerai extrait, to account for the value of the ore extracted. (43)

rendre ininflammable des objets combustibles, to fireproof combustible articles. (44)

rendre service à quelqu'un, to render someone a service ; to be of use to someone. (45)

rendre (se) approximativement compte, d'après les affleurements, de la direction des filons souterrains, to ascertain approximately the trend of the lodes underground by following the outcrops ; to find out roughly from the outcrops the strike of the lodes beneath the surface. (46)

rendre (se) en voiture d'une ville à une autre, to drive from one town to another. (47)

rendre (se) maître des venues d'eau dans un puits, to cope with, to master, the water in a shaft. (48)

rénette de charpentier (*f.*), combined saw-set and marking-knife. (49)

renfermer (contenir) (*v.t.*), to contain : (50)

substance qui renferme de l'arsenic (*f.*), substance which contains arsenic. (51)

renflement (*n.m.*), swelling ; enlargement ; boss : (52)

renflement de la terre à l'équateur, swelling of the earth at the equator. (53)

renflement sur une pièce venue de fonte, boss on a casting. (54)

renflement des tubes (*m.*), expanding tubes. (55)

renfler (*v.t. & v.i.*), to swell. (56)

renfler un tube de chaudière, to expand a boiler-tube. (57)

renfler (se) (*v.r.*), to swell. (58)

renforçateur (opp. à affaiblisseur *ou* faiblisseur) (Photogr.) (*n.m.*), intensifier. (59)

renforcé, -e (*adj.*), reinforced ; reenforced ; strengthened ; braced ; strong ; for heavy duty ; of heavy type *or* section : (60)

tour renforcé (*m.*), strong lathe ; heavy-duty lathe. (61)

palier renforcé (*m.*), heavy-type plummer-block. (62)

rail dissymétrique, type renforcé (*m.*), heavy-section bull-headed rail. (63)

renforcement *ou* **renforçage** (*n.m.*), reinforce-ment ; reenforcement ; reinforcing ; strengthening ; bracing. (64)

renforcement *ou* renforçage (opp. à affaiblisse-
-ment *ou* atténuation *ou* baissage) (Photogr.)
(*n.m.*), intensification. (1)

renforcer (*v.t.*), to reinforce ; to reenforce ;
to strengthen ; to brace : (2)
renforcer le boisage, un barrage, to reinforce,
to strengthen, the timbering, a dam. (3 *ou* 4)
banc de tour renforcé par des nervures d'entre-
-toisage (*m.*), lathe-bed braced by cross-
girths. (5)

renforcer (Photogr.) (*v.t.*), to intensify : (6)
renforcer un cliché faible, to intensify a weak
negative. (7)

renfort (Charp.) (*n.m.*), tusk ; haunch. (8)

renfort carré (*m.*), square haunch. (9)

renfort en chaperon (*m.*), bevelled haunch. (10)

reniflard (pompe) (*n.m.*), strainer ; wind-bore ;
snore-piece ; tail-piece ; rose. (11)

reniflard (de chaudière à vapeur) (*n.m.*), snifting-
valve, snifter-valve, (of steam-boiler). (12)

renifler (se dit d'une pompe de mine lorsque
l'abaissement du niveau liquide dans le
puisard permette à l'air de s'introduire
par les narines supérieures du reniflard
qui termine en bas la colonne d'aspiration)
(*v.i.*), to be in fork ; to have the water in
fork. (13)

reniveler (*v t.*), to relevel. (14)

renivellement (*n.m.*), relevelling. (15)

renmoulage de moules (Fonderie) (*m.*), setting
up moulds. (16)

renmoulage de noyaux (Fonderie) (*m.*), coring
up ; coring ; setting cores. (17)

renmouler (*v.t.*), to set up ; to core ; to core up.
(18)

renom (*n.m.*), renown ; reputation. (19)

renommé, -e (*adj.*), famous ; renowned ; noted ;
celebrated ; well-known. (20)

renoncer (*v.i.*), to give up ; to renounce : (21)
renoncer à l'emploi de la dynamite, to give
up using dynamite. (22)

renouvelable (*adj.*), renewable. (23)

renouveler (*v.t.*), to renew ; to renovate : (24)
renouveler l'air dans les chantiers, to renew
the air in the working-places. (25)

renouvellement (*n.m.*), renewal ; renewing ;
renovation : (26)
le renouvellement des rails, des traverses, des
coussinets, the renewal of rails, of sleepers,
of chairs. (27 *ou* 28 *ou* 29)

renseignements (*n.m.pl.*), information ; enquiries;
inquiries ; particulars ; details : (30)
sujet sur lequel les renseignements font défaut
(*m.*), subject on which information is lack-
-ing. (31)
prendre des renseignements sur quelqu'un,
to make enquiries about someone. (32)

rentrant, -e (*adj.*), reentrant ; reentering ;
sunk : (33)
angle rentrant (*m.*), reentrant angle ; reenter-
-ing angle. (34)
monture rentrante (objectif photographique)
(*f.*), sunk mount. (35)

rentrant (Géom.) (*n.m.*), reentrant. (36)

rentrée dans un quartier barré (Mines) (*f.*), re-
-entering sealed workings. (37)

rentrée des cannelures (l'excédent de largeur
d'une cannelure sur celle qui la précède
immédiatement) (Laminage des métaux)
(*f.*), draught. (38)

rentrer dans, to reenter (*v.t.*) : (39)
rentrer dans la mine, to reenter the mine. (40)

renversé (-e) par la chute d'une pièce de bois
(être), to be knocked down by a piece of
falling timber. (41)

renversé (-e) par une locomotive (être), to be
run over by a locomotive. (42)

renversement (*n.m.*), reversing ; inversion : (43)
renversement de la vapeur, reversing steam.
(44)
renversement de marche (Mach.), reversing;
reversing the motion. (45)
renversement du courant d'air, de l'aérage,
reversing the air-current, the ventilation.
(46 *ou* 47)

renversement (d'un wagon) (*n.m.*), overturning
(of a truck). (48)

renversement (Géol.) (*n.m.*), inversion ; overturn.
(49)

renversement (Fonderie) (*n.m.*), rolling over. (50)

renverser (mettre dans un état contraire à celui
qui existait antérieurement) (*v.t.*), to reverse ;
to invert: (51)
renverser la marche (Mach.), to reverse ;
to reverse the motion. (52)
renverser la marche d'un moteur, to reverse
an engine. (53)
renverser la vapeur, to reverse ; to reverse
steam. (54)
renverser le courant d'air, to reverse the
air-current. (55)

renverser (verser, en parlant d'une voiture)
(*v.t. & v.i.*), to overturn ; to upset ; to
capsize : (56)
voiture qui renverse (*f.*), carriage which over-
-turns. (57)

renverser (se) (*v.r.*), to overturn ; to upset ;
to capsize. (58)

renvoi (*n.m.*), return ; sending back ; throwing
back : (59)
renvoi de marchandises, return of goods. (60)
renvoi des stériles à la fouille, throwing back
the tailings into the cut. (61)

renvoi (congé) (*n.m.*), dismissal ; discharge : (62)
motif de renvoi (*m.*), reason for dismissal.
(63)

renvoi (*n.m.*) *ou* renvoi de mouvement, counter-
shaft ; countershaft ; counter-motion ;
counter-gear ; counter-shaft and accessories ;
countershaft and appurtenances. (64)
Note.—Although the word counter-shaft used
alone in the sense of counter-shaft and
accessories is incomplete, it is the most
commonly used.

renvoi (levier de changement de marche) (*n.m.*),
reversing-lever ; reverse-lever. (65)

renvoi à chape double articulée (Serrur.) (*m.*),
lazy pulley. (66)

renvoi à plat (Serrur.) (*m.*), side-pulley. (67)

renvoi à pont (Serrur.) (*m.*), axle-pulley. (68)

renvoi adhérent avec débrayage par pédale (*m.*),
self-contained pedal-operated driving-
motion ; countershaft forming part of the
machine, belt shifted by pedal ; counter-
-shaft attachment operated by foot control.
(69)

renvoi avec tige filetée (Serrur.) (*m.*), screw-
pulley. (70)

renvoi de mouvement à double vitesse (*m.*),
two-speed counter-motion. (71)

renvoi de mouvement à simple vitesse (*m.*),
single-speed counter-motion. (72)

renvoi de mouvement adhérent au bâti (*m.*).
Same as **renvoi adhérent.**

renvoi de mouvement se fixant au plafond (*m.*), overhead motion ; overhead driving motion ; overhead counter-driving appa- -ratus ; ceiling countershaft and appurte- -nances. (1)

renvoi de mouvement se fixant indifféremment au sol ou au plafond (*m.*), countershaft for either floor or ceiling. (2)

renvoi de mouvement se fixant sur le sol *ou simplement* renvoi sur le sol (*m.*), under- -hand motion ; underhand counter-driving motion ; floor countershaft and appurte- -nances. (3)

renvoi de sonnette (*m.*), bell-crank. (4)

renvoi sur champ (Serrur.) (*m.*), upright pulley. (5)

renvoi tendeur (*m.*), tension-block. (6)

renvoyer des marchandises, to return goods ; to send goods back. (7)

renvoyer le mouvement (Méc.), to counter the motion. (8)

renvoyer les mixtes au classeur, to return the middlings to the classifier. (9)

renvoyer les ouvriers, to discharge the work- -men ; to dismiss the men. (10)

réorganisateur, -trice (*adj.*), reorganizing : (11) pouvoir réorganisateur (*m.*), reorganizing power. (12)

réorganisation (*n.f.*), reorganization. (13)

réorganiser (*v.t.*), to reorganize. (14)

réouverture (*n.f.*), reopening ; resumption : (15) réouverture de la mine, reopening the mine. (16) réouverture des communications télégraphi- -ques, resumption of telegraphic communi- -cation. (17)

réoxydation (*n.f.*), reoxidation ; reoxidizement. (18)

réoxyder (*v.t.*), to reoxidize ; to reoxygenize ; to reoxygenate. (19)

répandre (*v.t.*), to spread : (20) répandre du sable, to spread sand. (21) répandre une forte odeur (se dit de certaines substances lorsqu'on les chauffe), to give off a strong smell. (22)

répandu, -e (*adj.*), widespread ; widely dis- -tributed ; prevalent : (23) le titane est assez répandu dans la nature, titanium is fairly widely distributed in nature. (24)

réparable (*adj.*), repairable ; mendable. (25)

reparaître (*v.i.*), to reappear ; to recur. (26)

réparateur (pers.) (*n.m.*), repairer. (27)

réparation (action) (*n.f.*), repairing ; mending : (28) réparation des galeries (dans une mine), repairing, mending, roads. (29) réparation (travail fait) (*n.f.*), repair : (30) réparations au puits, repairs to shaft. (31) réparations peu importantes, minor repairs. (32)

réparer (*v.t.*), to repair ; to mend ; to make good : (33) réparer un bâtiment, to repair a building. (34) réparer l'usure, to make good the wear and tear. (35)

réparer (se) (*v.r.*), to be repaired ; to be mended. (36)

répareur (pers.) (*n.m.*), repairer. (37)

répartir (*v.t.*), to distribute. (38)

répartition d'air dans les chantiers (Mines) (*f.*), distribution of air in the workings. (39)

répartition d'une charge entre deux points d'appui (*f.*), equalizing a load between two points of support. (40)

répartition des eaux (Géol.) (*f.*), distribution of water. (41)

répartition des erreurs (Math.) (*f.*), frequency of errors. (42)

repassage de couteaux (*m.*), sharpening, whet- -ting, knives. (43)

repasser au crible, to rescreen. (44)

repasser aux sluices les vieux tailings, to resluice old tailings. (45)

repasser un couteau, to sharpen, to whet, a knife. (46)

repasser une rivière, to recross a river. (47)

repêchage des outils (Sondage) (*m.*), fishing up tools. (48)

repêcher des outils, to fish up tools. (49)

repêcher les tiges brisées, to fish up, to recover, broken rods. (50)

repérage (*n.m.*), marking. (51)

reperçage (*n.m.*), fret-cutting ; fret-sawing. (52)

repercer (*v.t.*), to fret-saw. (53)

répercussion d'une explosion (*f.*), backlash of an explosion ; backlashing. (54)

repère (*n.m.*), mark ; guiding-mark ; reference- mark ; datum ; datum-point : (55) tracer un repère sur un tube de verre, to scratch a guiding-mark on a glass tube. (56) des vitesses rapportées à des repères fixes ou mobiles (*f.pl.*), velocities referred to fixed or movable datum-points. (57) repère (marque faite sur un mur, sur un jalon, sur un terrain, pour qu'on puisse retrouver un alignement, un niveau) (*n.m.*), bench- mark. (58)

repère de profondeur (*m.*), depth-mark. (59)

repérer (*v.t.*), to mark : (60) repérer les degrés de température sur une échelle thermométrique, to mark the degrees of temperature on a thermometric scale. (61) pour repérer l'emplacement exact du dessus de châssis sur le sol de la fonderie, on plante sur chaque face du châssis, extérieure- -ment, deux piquets en fer, qui serviront de guide lors du renmoulage, in order to mark the exact location of the cope on the floor of the foundry, two iron stakes should be driven in against each face of the box, on the outside, to serve as guides when setting up the mould. (62)

répéter (*v.t.*), to repeat : (63) répéter une opération un certain nombre de fois, to repeat an operation a certain number of times. (64)

répétiteur (*adj. m.*), repeating : (65) instrument répétiteur (*m.*), repeating instru- -ment. (66)

répétiteur (Télégr.) (*n.m.*), repeater ; translator. (67)

répétition (*n.f.*), repetition. (68)

replacage (*n.m.*), replating. (69)

replanir (Charp.) (*v.t.*), to clean off : (70) replanir un parquet, to clean off a floor. (71)

replanissage *ou* replanissement (*n.m.*), cleaning off. (72)

replaquer (*v.t.*), to replate. (73)

replenisher (Élec.) (*n.m.*), replenisher. (74)

repolir (*v.t.*), to repolish. (75)

repolissage (*n.m.*), repolishing. (76)

répondre à l'attente, to come up to expectations. (77)

répondre à la formule, to comply with, to agree with, the formula : (1)
la cuprite répond à la formule Cu²O, cuprite complies with the formula Cu_2O. (2)
répondre au but, to answer the purpose. (3)
report sur un plan (*m.*), transfer to a plan. (4)
reporter (transférer) (*v.t.*), to transfer : (5)
mesurer la longuer d'une ligne avec la chaîne et la reporter sur le papier à l'échelle du plan, to measure the length of a line with the chain and transfer it to the paper to the scale of the plan. (6)
repos (*n.m.*), rest ; resting ; repose : (7)
corps en repos (*m.*), body at rest. (8)
angle de repos (*m.*), angle of repose. (9)
repos (d'escalier) (*n.m.*), landing (of stairs). (10)
repos (**de tout**), safe ; reliable : (11)
estimation de tout repos (*f.*), safe estimate ; reliable estimate. (12)
reposer (*v.t.*), to relay : (13)
reposer une voie ferrée, une voie de tramway, to relay a railway, a tram-track. (14 *ou* 15)
reposer (*v.i.*), to rest ; to lie : (16)
poutre reposant sur deux appuis (*f.*), girder resting on two supports. (17)
la couche reposant immédiatement au-dessus, the bed lying immediately above. (18)
repousser (Phys.) (*v.t.*), to repel : (19)
le mercure repousse le fer, mercury repels iron. (20)
repousser (**se**) (*v.r.*), to repel each other : (21)
les pôles de même nom se repoussent (*m.pl.*), like poles repel each other. (22)
repoussoir (*n.m.*), pin-punch ; pin-drift. (23)
reprendre dans les — jours de sa réception un instrument présentant des vices de con--struction, to take back within — days of receipt an instrument showing signs of faulty construction. (24)
reprendre le matériel et les approvisionnements à dire d'experts, to take over the plant and stores at a valuation. (25)
reprendre les travaux, to restart work ; to resume operations. (26)
reprendre un mur, to repair a wall. (27)
reprendre un mur en sous-œuvre *ou* **reprendre un mur par-dessous œuvre** *ou* **reprendre un mur sous-œuvre,** to underpin, to under--set, to underprop, a wall. (28)
reprendre une ancienne mine, to rework an old mine. (29)
représentant (pers.) (*n.m.*), representative ; agent. (30)
représentant de l'Administration (*m.*), official : (31)
un représentant de l'Administration des Mines, a mining official. (32)
représenter (*v.t. & v.i.*), to represent. (33)
reprise (défaut d'une pièce coulée) (Fonderie) (*n.f.*), cold shut ; cold shot. (34)
reprise d'activité (*f.*), renewal of activity. (35)
reprise d'anciennes mines (*f.*), reworking old mines. (36)
reprise d'une mine après une explosion (*f.*), recovering a mine after an explosion. (37)
reprise des affaires (*f.*), revival of business. (38)
reprise des piliers (Mines) (*f.*), drawing back the pillars. (39)
reprise des tailings (*f.*), retreatment of tailings. (40)
reprise des travaux (*f.*), resumption of work ; restarting work. (41)

reprise en sous-œuvre (*f.*), underpinning ; under--setting ; underpropping. (42)
reproducteur de charge (Élec.) (*m.*), replenisher. (43)
reproduction (*n.f.*), reproduction. (44)
reproduire (*v.t.*), to reproduce. (45)
reproduire (**se**) (*v.r.*), to recur. (46)
répulsion (Phys.) (*n.f.*), repulsion : (47)
la répulsion est le contraire à l'attraction, repulsion is the opposite to attraction. (48)
répulsion capillaire (*f.*), capillary repulsion. (49)
répulsion électrique (*f.*), electrical repulsion. (50)
réputation (*n.f.*), reputation. (51)
resabotage des traverses (Ch. de f.) (*m.*), rechair--ing sleepers. (52)
rescite (Explosif) (*n.f.*), rescite. (53)
réseau (*n.m.*), network ; system. (54)
réseau à deux fils, à trois fils, à quatre fils, à cinq fils (Élec.) (*m.*), two-wire, three-wire, four-wire ; five-wire system. (55 *ou* 56 *ou* 57 *ou* 58)
réseau aérien de fils téléphoniques (*m.*), over--head system of telephone-wires. (59)
réseau bifilaire (Élec.) (*m.*), two-wire system. (60)
réseau d'éclairage (*m.*), lighting system. (61)
réseau de canalisations, de conduites, de tubes servant à transporter un liquide (*m.*), system of pipe-lines, of pipes, of tubing for conveying a liquid. (62 *ou* 63 *ou* 64)
réseau de chemins de fer *ou* **réseau de voies ferrées** (*m.*), railway system ; network of railways : (65)
le réseau de chemins de fer de la Grande-Bre--tagne, the railway system of Great Britain. (66)
réseau de conducteurs électriques (*m.*), network, system, of electric conductors. (67)
réseau de distribution (*m.*), distribution system. (68)
réseau de failles, de veines (*m.*), network of faults, of veins. (69 *ou* 70)
réseau de fils (*m.*), network, system, of wires. (71)
réseau électrique aérien (*m.*), overhead electric system. (72)
réseau fluvial (*m.*), river system. (73)
réseau téléphonique (*m.*), telephone system. (74)
réseau trifilaire (Élec.) (*m.*), three-wire system. (75)
réseau triphasé (Élec.) (*m.*), three-phase system. (76)
réserve (*n.f.*), reserve ; reserves : (77)
réserve de minerai, reserves of ore ; ore-reserves. (78)
réserver (*v.t.*), to reserve : (79)
réserver quelque argent pour les cas imprévus, to reserve some money for eventualities. (80)
réservoir (*n.m.*), reservoir ; tank ; bunker ; bin. (81)
réservoir (pour le moteur d'une turbine) (*n.m.*), flume ; turbine-chamber ; turbine-pit ; forebay. (82)
réservoir à air (*m.*). Same as **réservoir d'air.**
réservoir à eau *ou* **réservoir d'eau** *ou simplement* **réservoir** (*n.m.*), cistern ; water-tank ; reservoir ; water-reservoir. (83)
réservoir à minerai (*m.*), ore-bin ; ore-bunker. (84)
réservoir à pétrole (*m.*), oil-tank ; petroleum-reservoir ; oil-bunker ; oilometer. (85)

réservoir d'air *ou* **réservoir d'air comprimé** (d'un compresseur d'air) (*m.*), air-receiver (of an air-compressor) ( )

réservoir d'air *ou* **réservoir d'air comprimé** (d'une locomotive ou d'un véhicule de ch. de f.) (*m.*), air-reservoir, air-drum, (of a loco-motive or of a railway-car). (2 *ou* 3)

réservoir d'air auxiliaire *ou* **réservoir auxiliaire** (de véhicule de ch. de f.) (*m.*), auxiliary air-reservoir ; auxiliary reservoir. (4)

réservoir d'air principal *ou* **réservoir principal** (de locomotive ou de voiture automotrice de ch. de f.) (*m.*), main air-reservoir, main reservoir, (of locomotive or railway motor-car). (5 *ou* 6)

réservoir d'eau (d'une chaudière) (*m.*), water-space (of a boiler). (7)

réservoir d'emmagasinage (*m.*), storage-tank ; storage-reservoir. (8)

réservoir de chasse (*m.*), flush-tank ; flush-box ; flushing-box. (9)

réservoir de grisou (*m.*), fire-damp feeder. (10)

réservoir de vapeur (d'une chaudière) (*m.*), steam-space (of a boiler). (11)

réservoir intermédiaire (d'une machine à détente) (*m.*), receiver (of an expansion-engine). (12)

réservoir réfrigérant intermédiaire (d'un com-presseur d'air) (*m.*), intercooler (of an air-compressor). (13)

réservoir sous le banc (tour) (*m.*), drop-pan, tray, trough, under the bed (lathe). (14)

résidu (*n.m.*), residue ; residuum. (15)

résiduaire *ou* **résiduel, -elle** (*adj.*), residuary ; residual. (16)

résidus (*n.m.pl.*), residues ; residuals ; residual products ; residua. (17)

résidus de pétrole (*m.pl.*), petroleum residues. (18)

résidus stériles *ou simplement* **résidus** (*n.m.pl.*) (Mines), tailings ; tails. (19)

résiliable (*adj.*), terminable. (20)

résiliation (*n.f.*), cancellation ; cancelling. (21)

résilience (Méc.) (*n.f.*), resilience. (22)

résilier (*v.t.*), to cancel : (23)

résilier un contrat, to cancel a contract. (24)

résine (*n.f.*), resin ; rosin. (25)

résine de Highgate (Minéral) (*f.*), Highgate resin ; fossil copal ; copalin ; copaline ; copalite. (26)

résine fossile (*f.*), fossil resin. (27)

résineux, -euse (*adj.*), resinous. (28)

résinifère (*adj.*), resiniferous. (29)

résinification (*n.f.*), resinification. (30)

résinifier (*v.t.*), to resinify. (31)

résinifier (se) (*v.r.*), to resinify. (32)

résiniforme (*adj.*), resiniform. (33)

résinite (Minéral) (*n.m.*), pitch-opal ; resin opal ; resinous opal. (34)

résinoïde (*adj.*), resinoid ; retinoid. (35)

résistance (*n.f.*), resistance ; strength ; tenacity ; toughness. (36)

résistance (force qui s'oppose au mouvement) (Méc.) (*n.f.*), resistance : (37)

vaincre la résistance d'une machine au moment de sa mise en marche, to overcome the resistance of a machine at the moment of starting. (38)

résistance (*n.f.*) *ou* **résistance de conductivité** (Élec.), resistance : (39)

résistance des conducteurs, resistance of con-ductors. (40)

résistance au passage d'un courant électrique, resistance to the passage of an electric current. (41)

l'unité de résistance électrique est l'ohm (*f.*), the unit of electrical resistance is the ohm. (42)

résistance (appareil) (Élec.) (*n.f.*), resistance : (43)

intercaler une résistance dans un circuit, to insert a resistance in a circuit. (44)

résistance (levier) (*n.f.*), resistance ; weight. (45) See **leviers du premier, second, & troisième genres.**

résistance à froid (Élec.) (*f.*), cold resistance. (46)

résistance à l'arrachement (*f.*), resistance to tearing. (47)

résistance à l'écrasement *ou* **résistance à la compression** (*f.*), crushing-strength ; com-pressive strength ; resistance to crushing ; reactive tenacity ; retroactive tenacity. (48)

résistance à la flexion (*f.*), resistance to bending ; deflective strength ; stiffness. (49)

résistance à la rupture (*f.*), breaking-strength ; ultimate strength ; resistance to rupture : (50)

dans la pratique, il est d'usage de faire travailler les câbles métalliques au sixième environ de leur résistance à la rupture, et même quel-quefois au dixième, suivant le degré de sécurité dont on veut s'entourer, in practice, it is usual to stress wire ropes to about one-sixth of their breaking-strength, some-times even to one-tenth, according to the degree of safety adopted. (51)

résistance à la rupture par compression (*f.*), ultimate crushing-strength. (52)

résistance à la rupture par flexion (*f.*), ultimate bending-strength. (53)

résistance à la rupture par traction (*f.*), ultimate tensile strength. (54)

résistance à la torsion (*f.*), resistance to twisting ; torsional strength ; torsional tenacity. (55)

résistance à la traction (*f.*), tensile strength ; resistance to tension ; absolute tenacity. (56)

résistance apparente (Élec.) (*f.*), apparent resistance. (57)

résistance au choc (*f.*), resistance to impact ; resistance to shock. (58)

résistance au cisaillement (*f.*), shearing-strength ; shearing-tenacity ; resistance to shearing. (59)

résistance au déchirement (*f.*), resistance to tear-ing. (60)

résistance au démarrage (*f.*), starting resis-tance. (61)

résistance au flambage (*f.*), resistance to buck-ling. (62)

résistance au glissement (*f.*), resistance to sliding. (63)

résistance au roulement (*f.*), resistance to rolling. (64)

résistance d'isolement (Phys.) (*f.*), insulation resistance. (65)

résistance de l'induit (Élec.) (*f.*), armature resistance. (66)

résistance de la mine (Aérage) (*f.*), mine-resistance ; resistance, drag, of the mine : (67)

la résistance de la mine est la résistance opposée par la mine au passage d'un courant d'air, the resistance (*or* drag) of the mine is the resistance offered by the mine to the passage of an air-current. (1)

résistance de mise à la terre (Élec.) (*f.*), earthing resistance. (2)

résistance des matériaux, des métaux, des poutres, des ressorts (*f.*), strength of materials, of metals, of beams, of springs. (3 *ou* 4 *ou* 5 *ou* 6)

résistance des trains (*f.*), train resistance. (7)

résistance extrême (*f.*), ultimate strength ; breaking-strength ; resistance to rupture. (8)

résistance limite au cisaillement (*f.*), ultimate shearing-strength. (9)

résistance magnétique (*f.*), magnetic resistance ; reluctance. (10)

résistance passive (Méc.) (*f.*), resistance to motion : (11)

   si le moteur marche à vide, il n'absorbe que la puissance nécessaire pour vaincre les frottements, la résistance de l'air, etc., en un mot, les résistances passives, if the motor runs light, it only absorbs the power necessary to overcome friction, the resistance of the air, etc., in a word, the resistances to motion. (12)

résistance relative (d'un assemblage) (Rivetage, etc.) (Méc.) (*f.*), efficiency (of a joint). (13)

résistance spécifique (Élec.) (*f.*), specific resistance ; resistivity ; volume resistivity. (14)

résistance utile (Méc.) (*f.*), useful resistance. (15)

résistance vive (Méc.) (*f.*), resilience. (16)

résistance vive de rupture (*f.*), breaking-resilience. (17)

résistance vive élastique (*f.*), elastic resilience. (18)

résistant, -e (*adj.*), resistant ; resisting ; strong ; tough. (19)

résister à, to resist ; to withstand ; to stand : (20)

   résister à l'action des agents atmosphériques *ou* résister à la destruction par les agents atmosphériques (se dit des roches, etc.) (Géol.), to resist weathering. (21)

   résister à la pression du toit, to withstand the pressure of the roof. (22)

   résister à des manipulations brutales, to stand rough handling. (23)

résistivité (*n.f.*), resistivity. (24)

résistivité (Élec.) (*n.f.*), resistivity ; specific resistance ; volume resistivity. (25)

résoluble (*adj.*), resoluble. (26)

résolution (Chim.) (*n.f.*), resolution : (27)

   résolution de l'eau en vapeur, resolution of water into steam. (28)

résoudre (Chim.) (*v.t.*), to resolve. (29)

résoudre un problème, to solve a problem. (30)

respirabilité (*n.f.*), respirability. (31)

respirable (*adj.*), respirable ; breathable : (32)

   air respirable (*m.*), respirable air ; breathable air. (33)

respirateur (appareil) (*n.m.*), respirator. (34)

respiration (*n.f.*), respiration ; breathing : (35)

   respiration des hommes et des chevaux (Mines), respiration, breathing, of men and horses. (36)

respiratoire (*adj.*), breathing ; respiratory : (37)

   appareil respiratoire (*m.*), breathing-apparatus. (38)

respirer (*v.t. & v.i.*), to breathe ; to respire. (39)

responsabilité (*n.f.*), responsibility ; liability : (40)

   responsabilité des patrons en cas d'accidents de personnes, responsibility of employers for personal accidents. (41)

responsable (*adj.*), responsible ; liable : (42)

   être responsable du dommage causé par quelqu'un, to be liable for the damage caused by someone. (43)

ressac (*n.m.*), surf. (44)

ressaut (saillie) (*n.m.*), projection ; set-off : offset ; scarcement : (45)

   ressaut d'une pierre au delà d'un mur, projection of a stone beyond a wall. (46)

ressaut (dans un cours d'eau) (Hydraul.) (*n.m.*), jump. (47)

ressemblance (*n.f.*), resemblance. (48)

ressembler à, to resemble ; to look like ; to be like : (49)

   la roche ressemble à du granit, the rock is like granite. (50)

resserrement (*n.m.*), contraction ; narrowing ; pinch ; pinching. (51)

resserrer (se) (*v.r.*), to contract ; to narrow ; to shrink ; to pinch. (52)

ressort (élasticité) (*n.m.*), spring ; elasticity (53)

   le ressort de l'air, de la vapeur d'eau comprimée, the elasticity of air, of compressed steam. (54 *ou* 55)

ressort (organe élastique) (*n.m.*), spring. (56)

ressort (à), spring (*used as adj.*) ; snap (*used as adj.*) : (57)

   compas à ressort (*m.*), spring-dividers. (58)

   commutateur à ressort (*m.*), snap-switch. (59)

ressort à boudin (*m.*), spiral spring. (60)

ressort à boudin cylindrique (*m.*), cylindrical spiral spring. (61)

ressort à demi-pincette (*m.*), half-elliptic spring ; semielliptic spring. (62)

ressort à lame (*m.*), plate-spring. (63)

ressort à lame en triangle (*m.*), triangular plate-spring. (64)

ressort à lame rectangulaire (*m.*), rectangular plate-spring. (65)

ressort à lame rectangulaire à profil paraboloïde (*m.*), rectangular plate-spring with end tapered in the form of a parabola. (66)

ressort à lames *ou* **ressort à lames étagées** *ou* **ressort à lames superposées** (*m.*), leaf-spring ; laminated spring ; blade-spring ; plate-spring. (67)

ressort à pincette (*m.*), elliptic spring. (68)

ressort compensateur (*m.*), equalizer-spring. (69)

ressort conique (*m.*), conical spring. (70)

ressort cylindrique (*m.*), cylindrical spring. (71)

ressort d'embrayage (*m.*), clutch-spring. (72)

ressort de choc (wagon de chemin de fer) (*m.*), buffer-spring. (73)

ressort de compression (opp. à ressort de rappel) (*m.*), open-coil spring ; open spiral spring ; compression-spring. (74)

ressort de flexion (*m.*), spring subjected to bending. (75)

ressort de friction (*m.*), friction-spring. (76)

ressort de l'essieu porteur à l'arrière (locomotive) (*m.*), trailing-spring. (77)

ressort de l'essieu porteur à l'avant (locomotive) (*m.*), leading-spring. (78)

ressort de la roue motrice (locomotive) (*m.*), driving-spring ; driver-spring. (79)

ressort de rappel (*m.*), close-coil spring; close spiral spring; drawback spring. (1)

ressort de rappel (de frein de wagon de ch. de f.) (*m.*), release-spring. (2)

ressort de sommier (*m.*), spool spring; hour-glass spring; sofa-spring; mattress-spring. (3)

ressort de soupape (*m.*), valve-spring. (4)

ressort de suspension (locomotive ou wagon de chemin de fer, etc.) (*m.*), bearing-spring; journal-spring; journal-box spring; bolster-spring. (5 ou 6)

ressort de torsion (*m.*), torsion-spring; torsional spring; spring subjected to torsion. (7)

ressort de traction (*m.*), draw-spring; draught-spring; draft-spring; drag-spring. (8)

ressort de voiture (*m.*), carriage-spring. (9)

ressort de wagon (*m.*), car-spring. (10)

ressort en C (*m.*), C spring. (11)

ressort en caoutchouc (*m.*), india-rubber spring; rubber spring. (12)

ressort en hélice (*m.*), helical spring. (13)

ressort en spirale (*m.*), coil spring; coiled spring; spiral coiled spring. (14)

ressort en volute (*m.*), volute spring. (15)

ressort plat (*m.*), flat spring. (16)

ressoudage (*n.m.*) *ou* **ressoudure** (*n.f.*), resolder-ing; rewelding. (17)

ressouder (*v.t.*), to resolder; to reweld. (18)

ressource (*n.f.*), resource: (19)

ressources d'un pays en chutes d'eau, resources of a country in waterfalls. (20)

ressources minérales d'un district, mineral resources of a district. (21)

ressuage (action d'un corps qui ressue) (*n.m.*), sweating. (22)

ressuage (Métall.) (*n.m.*), sweating. (23)

ressuer (*v.i.*), to sweat: (24)

les murs neufs ressuent pendant un certain temps (*m.pl.*), new walls sweat for a certain time. (25)

restapler (Mines) (*v.t.*), to stow; to gob. (26)

restapleur *ou* **restapeur** (Mines) (pers.) (*n.m.*), stower. (27)

reste (*n.m.*), rest; remainder; remains: (28)

le reste du matériel, the remainder of the plant. (29)

des restes fossiles, fossil remains. (30)

rester (se tenir en arrière) (*v.i.*), to remain: (31)

rester en dépôt *ou* rester en magasin, to remain in storage. (32)

rester entre ses repères (se dit de la bulle d'air dans la fiole d'un instrument de nivellement), to keep its centre. (33)

restreindre (*v.t.*), to restrict; to limit; to confine: (34)

restreindre son attention aux travaux de prospection, to confine one's attention to prospecting. (35)

restriction (*n.f.*), restriction: (36)

restrictions apportées au travail aux explosifs, restriction of blasting. (37)

résultant, -e (*adj.*), resultant. (38)

résultante (Méc.) (*n.f.*), resultant. (39)

résultat (*n.m.*), result; outcome: (40)

résultat d'une épreuve, result of a test; outcome of a trial. (41)

résulter (*v.i.*), to result. (42)

rétablir (*v.t.*), to reestablish; to restore: (43)

rétablir l'aérage, to reestablish, to restore, the ventilation. (44)

rétablissement (*n.m.*), reestablishment; restora-tion. (45)

retaillage *ou* **retaillement** (*n.m.*), recutting: (46)

retaillage de limes, recutting files. (47)

retailler (*v.t.*), to recut. (48)

rétamage (*n.m.*), retinning. (49)

rétamer (*v.t.*), to retin. (50)

retarauder (*v.t.*), to retap. (51)

retard (*n.m.*), delay: (52)

retard dans l'arrivée des machines, delay in the arrival of the machinery. (53)

retard à l'admission (tiroir) (*m.*), retarded admis--sion; late admission (slide-valve). (54)

retard à l'échappement (tiroir) (*m.*), retarded release; late release. (55)

retard d'aimantation (*m.*), lag of magnetization; magnetic lag; time-lag of magnetism; viscous hysteresis. (56)

retardataire *ou* **retardateur, -trice** (*adj.*), retarding; retardant: (57)

l'effet retardataire produit par la résistance de l'eau (*m.*), the retarding effect produced by the resistance of water. (58)

frottement retardateur (*m.*), retarding friction. (59)

retardateur (opp. à **accélérateur**) (Photogr.) (*n.m.*), restrainer; retarder. (60)

retarder (*v.t.*), to retard; to delay: (61)

retarder les travaux, to retard operations; to delay the work. (62)

retarder un mouvement (Méc.), to retard a motion. (63)

retassure (vide central dans un lingot d'acier) (*n.f.*), pipe. (64)

retenir (*v.t.*), to retain; to hold back; to keep back: (65)

retenir les services de quelqu'un, to retain someone's services. (66)

retenir le minerai jusqu'à nouvel ordre, to hold back the ore till further advice; to keep the ore back till further advised. (67)

réticulaire (*adj.*), reticular. (68)

réticule (*n.m.*) *ou* **réticule à croisée** (Opt.), web; cross-wires; cross-hairs; spider-lines; reticle; reticule. (69)

retient (Méc.) (*n.m.*), lock-plate: (70)

un retient fixé sur le couvercle d'un piston, a lock-plate fixed on the junk-ring of a piston. (71)

rétinalite (Minéral) (*n.f.*), retinalite. (72)

rétinasphalte (*n.m.*) *ou* **rétinellite** (*n.f.*) *ou* **rétinite** (*n.f.*) (Minéral), retinasphalt; retinasphaltum; retinellite; retinite. (73)

rétinite (Pétrol.) (*n.f.*), pitchstone. (74)

retire-goupille [retire-goupilles *pl.*] (*n.m.*), pin-extractor; split-pin extracting-tool. (75)

retirer (arracher) (*v.t.*), to draw; to withdraw: (76)

retirer les étais, to draw the props; to with-draw the props; to unprop. (77)

retirer (extraire) (*v.t.*), to extract; to collect: (78)

retirer de l'huile du schiste, de l'argent du plomb argentifère, de l'or de solutions cyanurées, to extract oil from shale, silver from argentiferous lead, gold from cyanide solutions. (79 ou 80 ou 81)

jadis on retirait le salpêtre des murs de caves, formerly saltpetre was collected from the walls of cellars. (82)

retirer (contracter) (*v.t.*), to shrink; to contract. (83)

retirer (se) (se contracter) (*v.r.*), to shrink; to contract: (84)

le drap se retire à l'eau, cloth shrinks in water. (1)

retirer (se) (*v.r.*) (se dit des eaux), to recede ; to subside ; to retire. (2)

retirer (se) à l'abri, to retire to a place of safety ; to withdraw to shelter. (3)

retirure (dans une pièce coulée) (Fonderie) (*n.f.*), draw (in a casting). (4) (a horizontal or oblique sinking in a casting, distinguished from **tassement**, a vertical sinking, i.e., one on the top face).

retombée *ou* **retombe** (d'une voûte) (*n.f.*), spring, springing, (of an arch). (5)

retorcher (*v.t.*), to fettle : (6)
retorcher les parties érodées dans l'intérieur d'un fourneau, to fettle the eroded parts inside a furnace. (7)

retouche d'une photographie (*f.*), retouching a photograph. (8)

retoucher un cliché photographique, to retouch a photographic negative. (9)

retour (*n.m.*), return : (10)
retour à vide de l'outil *ou* retour à blanc de l'outil (machine-outil, ou analogue), idle return, non-cutting return, of the tool. (11)
retour aux conditions normales, return to normal conditions. (12)
retour de marchandises, return of goods. (13)
retour du chariot à la main sur crémaillère en acier à denture taillée, hand-return of the carriage by steel rack with cut teeth. (14)
retour par la terre (Élec.), earth-return ; ground-return. (15)
retour rapide (machine-outil), quick return ; quick return motion. (16)
retours d'air (Mines), air-returns. (17)

retour d'eau (soupape de retenue) (*m.*), check-valve ; retaining-valve ; back-pressure valve ; non-return valve. (18)

retour de palan (*m.*), monkey-block. (19)

retournage de moules (Fonderie) (*m.*), turning over moulds. (20)

retourner (revenir) (*v.i.*), to return : (21)
retourner au travail, to return to work. (22)

retourner le sol, to disturb the ground. (23)

retourner un moule, un châssis de fonderie, to turn over a mould, a moulding-box. (24 *ou* 25)

retourner un rail sens dessus dessous, to turn a rail upside down. (26)

retrait (diminution de volume) (*n.m.*), shrinkage ; shrinking : (27)
retrait de l'argile, du ciment, shrinkage, shrink-ing, of clay, of cement. (28 *ou* 29)

retrait (Fonderie) (*n.m.*), shrinkage ; shrinking. (30)

retrait (dans un mur) (*n.m.*), recess, retreat, (in a wall). (31)

retrait (des eaux) (*n.m.*), retirement (of the waters). (32)

retrait (d'un glacier) (*n.m.*), retreat (of a glacier). (33)

retraite (action) (*n.f.*), retreat ; retirement : (34)
une retraite précipitée, a precipitate retreat. (35)
retraite des eaux, retirement of the waters. (36)

retraite (lieu de refuge) (*n.f.*), shelter ; refuge ; manhole ; recess. (37)

retraite (recul en arrière d'un alignement) (Constr.) (*n.f.*), retreat. (38)

retraiter (*v.t.*), to retreat : (39)
retraiter le minerai, to retreat the ore. (40)

rétrécir (*v.t.*), to narrow ; to shrink ; to contract. (41)

rétrécir (se) (*v.r.*), to narrow ; to shrink ; to contract ; to pinch. (42)

rétrécissement (*n.m.*), narrowing ; shrinking ; shrinkage ; contraction ; pinching : (43)
rétrécissement brusque de section (Hydraul.), sudden contraction of cross-section. (44)
rétrécissement d'une pièce de drap, shrinking of a piece of cloth. (45)

retrempe (*n.f.*), retempering ; rehardening. (46)

retremper (*v.t.*), to retemper ; to reharden. (47)

retrier (*v.t.*), to resort. (48)

rétrograde (*adj.*), retrograde ; backward ; reversed : (49)
mouvement rétrograde (*m.*), backward motion. (50)

retrouver (*v.t.*), to find again ; to relocate ; to pick up again : (51)
retrouver un filon, to find (*or* to pick up) a lode again ; to relocate a vein. (52)

réunion (action de rejoindre) (*n.f.*), reuniting ; uniting ; joining ; connection ; connecting : (53)
la réunion de deux parties au moyen de boulons, the joining of two parts by bolts. (54)

réunion (action de rassembler) (*n.f.*), collection ; collecting. (55)

réunir (rejoindre) (*v.t.*), to reunite ; to unite ; to join ; to connect : (56)
galeries jumelles réunies par des recoupes d'aérage (*f.pl.*), parallel entries connected by cross-headings. (57)

réunir (rassembler) (*v.t.*), to collect : (58)
réunir les eaux qui s'écoulent d'une montagne, to collect the water which runs down a mountain. (59)

réunir (se) (*v.r.*), to collect : (60)
eaux qui se réunissent dans le puisard (*f.pl.*), water which collects in the sump. (61)

réussi, -e (*adj.*), successful : (62)
entreprise bien réussie (*f.*), very successful enterprise. (63)
un type de grue des mieux réussis, a most successful type of crane. (64)

réussir (*v.i.*), to succeed. (65)

réussite (*n.f.*), success. (66)

révélateur (Photogr.) (*n.m.*), developer. (67) Cf. **développateur**.

révélateur à l'acide pyrogallique (*m.*), pyro-soda developer. (68)

révélateur à l'hydroquinone (*m.*), hydroquinone developer. (69)

révélateur à la métoquinone (*m.*), metol-quinol developer. (70)

révélateur au métol hydroquinone (*m.*), metol and hydroquinone developer. (71)

révéler (*v.t.*), to reveal ; to disclose : (72)
une forte proportion de silice révèle la présence de quartz ou tout au moins de silice libre dans la roche, a high percentage of silica reveals the presence of quartz or at any rate of free silica in the rock. (73)
ne révéler rien d'importance, to disclose nothing of value. (74)

révéler (se) (*v.r.*), to be revealed ; to be disclosed : (75)
structure qui se révèle au microscope (*f.*), structure which is revealed by the micro-scope. (76)

revendication (n.f.), claim. (1)

revendiquer (v.t.), to claim : (2)
revendiquer le droit de passage, to claim right of way. (3)
les avantages revendiqués pour un système (m.pl.), the advantages claimed for a system. (4)

revenir (v.i.), to come back ; to return ; to come again : (5)
revenir aux conditions normales, to return, to come back, to normal conditions. (6)
revenir entre ses repères (se dit de la bulle d'air dans la fiole d'un instrument de nivellement), to come back to the centre of its run ; to come to the centre of its run again. (7)

revenir au même, to come, to amount, to the same thing : (8)
chute des corps dans le vide ou, ce qui revient au même, en négligeant la résistance de l'air (f.), fall of bodies in vacuo or, what comes to the same thing, disregarding the resistance of the air. (9)

revenu (n.m.) ou revenu après trempe, drawing the temper ; letting down the temper ; tempering : (10)
la trempe et le revenu des outils, the hardening and the tempering of tools. (11) Cf. faire revenir.

réverbère (réflecteur) (n.m.), reflector. (12)

réversibilité (n.f.), reversibility. (13)

réversible (adj.), reversible : (14)
ventilateur réversible (m.), reversible fan. (15)

revêtement (n.m.), lining ; liner ; covering ; coating ; facing ; cleading ; revetment. (16)

revêtement de puits (m.), shaft-lining. (17)

revêtement en fascines (m.), revetment formed of fascines. (18)

revêtement en maçonnerie sur la surface d'un talus (m.), facing of masonry on the surface of a slope. (19)

revêtement en plaques de fonte (m.), plating ; lining of cast-iron plates. (20)

revêtement en tôle (m.), plating ; lining of plate-iron : (21)
revêtement en tôle d'un haut fourneau, plating of a blast-furnace. (22)

revêtement étanche (m.), water-tight lining. (23)

revêtir (v.t.), to line ; to cover ; to coat ; to face ; to revet. (24)

revient (recuit après trempe) (n.m.), drawing the temper ; letting down the temper ; temper-ing. (25)

révolution (n.f.), revolution ; turn : (26)
— révolutions à la minute ou — révolutions par minute, — revolutions a minute ; — turns per minute. (27)
révolution de la terre, revolution of the earth. (28)

révolution (Géom.) (n.f.), revolution : (29)
cylindre de révolution (m.), cylinder of revolu-tion. (30)

révolutionner (v.i.), to revolve ; to turn. (31)

révolutionner complètement (en parlant d'un cercle d'alignement, d'un théodolite à lunette centrale) (v.i.), to transit : (32)
lunette qui révolutionne complètement autour de son axe horizontal (f.), telescope which transits about its horizontal axis. (33)

révolu, -e (adj.), revolving. (34)

revolver (d'un tour) (n.m.), capstan, capstan-head, turret, turret-head, monitor, (of a lathe) : (35)

tour à revolver (m.), capstan-lathe ; turret-lathe ; monitor-lathe. (36)

revolver (de pinces à emporte-pièce à revolver) (n.m.), revolving head (of revolving-head punch). (37)

revolver porte-objectifs ou simplement revolver (n.m.) (d'un microscope), revolving nose-piece (of a microscope). (38)

rexite (Explosif) (n.f.), rexite. (39)

rhabillage (de meules en émeri) (n.m.), dressing (emery-wheels). (40)

rhabiller une meule d'émeri, to dress an emery-wheel. (41)

rhabilleur pour meules d'émeri, avec 2 garnitures de molettes (m.), emery-wheel dresser, with 2 sets of cutters. (42)

rhætizite (Minéral) (n.f.), rhætizite. (43)

rhéophore (Elec.) (n.m.), rheophore. (44)

rhéoscopique (adj.), rheoscopic. (45)

rhéostat (n.m.), rheostat. (46)

rhéostat d'arc (m.), arc-rheostat. (47)

rhéostat d'artère (m.), feeder-rheostat. (48)

rhéostat de champ ou rhéostat d'excitation (m.), field-rheostat ; exciting rheostat. (49)

rhéostat de démarrage ou rhéostat démarreur (m.), motor-starter ; starting-rheostat ; starting-resistance ; motor-starting resis-tance ; starting-box ; starter. (50)

rhéostat liquide (m.), liquid rheostat. (51)

rhéostatique (adj.), rheostatic. (52)

rhodite (Minéral) (n.f.), rhodite ; rhodium gold. (53)

rhodium (Chim.) (n.m.), rhodium. (54)

rhodizite (Minéral) (n.f.), rhodizite. (55)

rhodochrosite (Minéral) (n.f.), rhodochrosite ; rhodochroisite ; dialogite ; diallogite. (56)

rhodonite (Minéral) (n.f.), rhodonite ; manganese spar ; red manganese. (57)

rhombe ou rhombique (adj.), rhombic ; rhom-bical. (58)

rhombe (n.m.), rhomb ; rhombus ; lozenge. (59)

rhomboèdre (n.m.), rhombohedron. (60)

rhomboédrique (adj.), rhombohedral ; rhom-bohedric. (61)

rhomboïdal, -e, -aux (adj.), rhomboid ; rhom-boidal. (62)

rhomboïde (n.m.), rhomboid. (63)

rhyolite (Pétrol.) (n.f.), rhyolite ; rhyolyte. (64)

riblons (n.m.pl.), swarf (metal turnings). (65)

riblons d'acier (m.pl.), steel swarf. (66)

riche (adj.), rich : (67)
un minerai riche en or, an ore rich in gold. (68)

richesse (n.f.), richness ; wealth : (69)
richesse d'un sol, richness of a soil. (70)
richesse minérale d'un pays, mineral wealth of a country. (71)

richesse (Mines) (n.f.) (s'emploie souvent au pluriel), value ; strength : (72)
estimer la richesse moyenne d'un filon, to estimate the average value (or strength) of a lode. (73)
une estimation approximative des richesses sur lesquelles on peut compter dans un quartier reconnu, a rough estimate of the values on which one can reckon in a proved locality. (74)
les richesses contenues dans un minerai, the values contained in an ore. (75)

ride (sur la surface terrestre) (Géogr. phys.) (n.f.), wrinkle (on the earth's surface). (76)

rides (de l'eau) (n.f.pl.), ripples (of the water). (77)

rides éoliennes (dans le sable) (Géol.) (*f.pl.*), wind-ripples. (1)

riffle *ou* **rifle** (languette ou tasseau de bois fixé sur le fond d'un sluice) (*n.m.*), riffle. (2)

riffle à losanges (*m.*), diamond riffle. (3)

riflard (rabot) (*n.m.*), jack-plane. (4)

rifloir (lime) (*n.m.*), riffler. (5)

rigide (*adj.*), rigid ; stiff : (6)

palier à coussinets rigides (*m.*), rigid bearings. (7)

film rigide (Photogr.) (*m.*), stiff film. (8)

rigidement (*adv.*), rigidly. (9)

rigidité (*n.f.*), rigidity ; stiffness : (10)

rigidité d'une barre de fer, rigidity of a bar of iron. (11)

rigole (*n.f.*), channel ; ditch. (12)

rigole d'asséchement (*f.*), drainage-channel ; drainage-ditch ; gutter. (13)

rigole pour recevoir l'eau de savon (plateaux de machines à percer, etc.) (*f.*), sud-channel (drilling-machine tables, etc.) (14)

rigole pour recevoir l'huile (*f.*), oil-channel. (15)

rigoureux, -euse (*adj.*), severe ; rigorous : (16)

climat rigoureux (*m.*), severe (*or* rigorous) climate. (17)

rigueur du temps (*f.*), severity of the weather. (18)

rimaye (Glaciologie) (*n.f.*), bergschrund. (19)

rinçage (*n.m.*), rinsing. (20)

rincer (*v.t.*), to rinse : (21)

rincer des bouteilles, un cliché photographique, to rinse bottles, a photographic negative. (22 *ou* 23)

ringard (*n.m.*), prick-bar ; pricker ; poker. (24)

ringard (Puddlage) (*n.m.*), rabble ; rabbler ; staff : (25)

brasser le fer puddlé avec le ringard, to stir the puddled iron with the rabble. (26)

ringard à crochet (*m.*), rake. (27)

ringard à lance (*m.*), prick-bar ; pricker. (28)

ringard mécanique (Puddlage) (*m.*), mechanical rabble. (29)

ringarder (*v.t.*), to poke ; to prick. (30)

ripage de la voie (Ch. de f.) (*m.*), lining the road : (31)

le ripage de la voie se fait avec les pinces à riper, the lining of the road is done with lining-bars. (32)

riper la voie (Ch. de f.), to line the road. (33)

ripidolite (Minéral) (*n.f.*), ripidolite. (34)

rippite (Explosif) (*n.f.*), rippite. (35)

ripple-marks (Géol.) (*n.f.pl.*), ripple-marks : (36)

les ripple-marks, ces innombrables rides et ondulations qui recouvrent le sable des plages à marée basse, sont dues au clapotement, sous l'action du vent, des eaux peu profondes, ripple-marks, those innumerable ridges and undulations which cover the sand on the beach at low tide, are due to the plashing of shallow water, under the action of the wind. (37)

grès à ripple-marks (*m.*), ripple-marked sand-stone. (38)

risque (*n.m.*), risk : (39)

risque d'explosion de grisou, risk of fire-damp explosion. (40)

aux risques et périls des destinataires, at owner's risk : (41)

nos marchandises, quoique vendues franco, voyagent aux risques et périls des destinataires, et nous prions nos clients de les vérifier avant d'en prendre livraison, our goods, although sold carriage paid, are sent at owner's risk, and we beg our customers to examine them carefully before taking delivery. (42)

risquer (*v.t.*), to risk ; to venture. (43)

rivage (de la mer, d'un lac, d'un fleuve) (*n.m.*), shore, beach, of the sea, of a lake, of a river. (44 *ou* 45 *ou* 46)

rivage (Méc.) (*n.m.*), riveting ; rivetting. (47)

rive (*n.f.*), bank ; shore ; margin : (48)

rive d'un cours d'eau, bank of a stream. (49)

rive d'un fleuve, bank, shore, of a river. (50)

rive d'un glacier, margin of a glacier. (51)

rive de la mer, seashore. (52)

rive droite (d'un cours d'eau), right bank. (53)

rive gauche (d'un cours d'eau), left bank. (54)

rive (d'un four) (*n.f.*), lip (of a furnace). (55)

rive (d'une planche) (*n.f.*), thickness (of a board). (56)

rivé, -e (*adj.*), riveted ; rivetted : (57)

assemblage rivé (*m.*), riveted joint. (58)

rivelaine (Mines) (*n.f.*), holing-pick ; jadding-pick ; rivelaine. (59)

rivement (Méc.) (*n.m.*), riveting ; rivetting. (60)

river (*v.t.*), to rivet ; to clinch : (61)

river une tôle, to rivet a plate. (62)

river un clou, to clinch a nail. (63)

river (se) (*v.r.*), to be riveted ; to be clinched : (64)

les gazomètres se rivent à froid (*m.pl.*), gas-ometers are riveted cold. (65)

clous qui doivent se river (*m.pl.*), nails which should be clinched. (66)

riverain, -e (*adj.*), riverside ; riparian ; water-side : (67)

propriété riveraine (*f.*), riverside property. (68)

rivet (*n.m.*), rivet. (69)

rivet à tête conique *ou* **rivet à tête en pointe de diamant** (*m.*), steeple-head rivet ; cone-head rivet. (70)

rivet à tête cylindrique (*m.*), cheese-head rivet. (71)

rivet à tête en goutte-de-suif *ou* **rivet à tête en arc de cercle** *ou* **rivet à tête bombée** (*m.*), button-head rivet. (72)

rivet à tête fraisée (*m.*), countersunk rivet ; countersunk-head rivet. (73)

rivet à tête fraisée et goutte de suif (*m.*), raised-countersunk rivet ; bull-head rivet ; bull-headed rivet ; countersunk button-head rivet ; oval countersunk rivet. (74)

rivet à tête noyée *ou* **rivet à tête perdue** (*m.*), flush countersunk rivet ; flat countersunk rivet. (75)

rivet à tête plate (*m.*), flat-head rivet. (76)

rivet à tête ronde *ou* **rivet à tête hémisphérique** (*m.*), round-head rivet ; round-headed rivet ; cup-head rivet ; snap-head rivet. (77)

rivet à tête tronconique (*m.*), pan-head rivet ; smoke-pipe rivet ; cone-head rivet. (78)

rivet bifurqué (*m.*), bifurcated rivet ; slotted rivet ; slotted clinch rivet. (79)

rivet en cuivre rouge (*m.*), copper rivet. (80)

rivet pour chaudière (*m.*), boiler-rivet. (81)

rivet pour courroie (*m.*), belt-rivet. (82)

rivet tubulaire (*m.*), tubular rivet. (83)

rivetage (*n.m.*), riveting ; rivetting. (84)

rivetage à couvre-joints, à recouvrement, etc. (*m.*). Same as **rivure à couvre-joints, à recouvrement, etc.**

rivetage au marteau (*m.*), hammer-riveting. (85)

rivetage mécanique (*m.*), machine-riveting. (86)

riveter (*v.t.*), to rivet. (1)

riveur (pers.) (*n.m.*), riveter; rivetter. (2)

riveuse (*n.f.*), riveter; rivetter; riveting-machine. (3) See also **machine à river.**

riveuse hydraulique (*f.*), hydraulic riveter. (4)

riveuse pour river les dômes de chaudières (*f.*), dome-riveter. (5)

riveuse pour river les tôles des bouts des chaudières (*f.*), boiler-end rivetter. (6)

rivière (*n.f.*), river. (7) (a river discharging into a lake or into another stream : dis-tinguished from **fleuve**, a river discharging into the sea). See also **fleuve.**

rivière à fond mobile (*f.*), river with a shifting bottom. (8)

rivière décapitée (*f.*), beheaded river. (9)

rivière flottable (*f.*), floatable river. (10)

rivière franchissable (*f.*), passable river. (11)

rivière innavigable (*f.*), unnavigable river. (12)

rivière maîtresse (*f.*), master-river. (13)

rivière navigable *ou* **rivière marchande** (*f.*), navigable river. (14)

rivière rajeunie (*f.*), rejuvenated river. (15)

rivière souterraine (*f.*), underground river ; subterranean river. (16)

rivière subaérienne (*f.*), surficial river : (17) la rivière subaérienne naît souvent d'une rivière souterraine, the surficial river often originates from an underground river. (18)

rivoir (marteau) (*n.m.*), riveting-hammer. (19)

rivoir (machine) (*n.m.*), riveter ; riveting-machine. (20)

rivoir pneumatique (*m.*), pneumatic riveter. (21)

rivoire (*n.f.*) *ou* **rivois** (*n.m.*). Same as **rivoir.**

rivure (action de river) (*n.f.*), riveting ; rivetting : (22) rivure de chaudières, riveting boilers. (23)

rivure (assemblage rivé) (*n.f.*), riveted joint. (24)

rivure (tête que l'on fait à une broche en fer, pour l'assujettir dans un trou) (*n.f.*), head (the second head of a rivet or pin formed by battering or snapping). (25)

rivure (broche de charnière) (*n.f.*), hinge-pin ; pintle. (26)

rivure (contre-rivure) (*n.f.*), burr ; riveting-burr ; rivet-washer. (27)

rivure à couvre-joint *ou* **rivure à bande de recouvre-ment** *ou* **rivure à franc-bord** (*f.*), butt-riveting ; flush rivetting. (28)

rivure à deux rangs de rivets *ou* **rivure double** *ou* **rivure à double clouure** (*f.*), double-riveted joint. (29)

rivure à deux rangs de rivets à clin, avec rivets disposés en quinconce (*f.*), double-riveted lap-joint, with staggered rivets. (30)

rivure à deux rangs de rivets à simple couvre-joint *ou* **rivure double à une seule bande de recouvrement** *ou* **rivure à plat-joint** (*f.*), double-rivetted butt-joint with one welt. (31)

rivure à deux rangs de rivets et à double couvre-joint *ou* **rivure double à deux couvre-joints** (*f.*), double-riveted butt-joint with two welts. (32)

rivure à recouvrement *ou* **rivure à clin** (*f.*), lap-rivetting. (33)

rivure à trois rangs de rivets *ou* **rivure triple** *ou* **rivure à triple clouure** (*f.*), triple-riveted joint. (34)

rivure à trois rangs de rivets et à double couvre-joint *ou* **rivure triple à deux couvre-joints** (*f.*), triple-riveted double-welt butt-joint. (35)

rivure à un rang de rivets *ou* **rivure simple** *ou* **rivure à simple clouure** (*f.*), single-riveted joint. (36)

rivure à un rang de rivets à clin *ou* **rivure simple à recouvrement** (*f.*), single-rivetted lap-joint. (37)

rivure à un rang de rivets à simple couvre-joint (*f.*), single-riveted butt-joint with one welt. (38)

rivure à un rang de rivets et à double couvre-joint (*f.*), single-riveted butt-joint with two welts. (39)

rivure bouterollée (opp. à **rivure écrasée**) (*f.*), snap-head. (40)

rivure d'assemblage (*f.*), joint-riveting. (41)

rivure écrasée (*f.*), battered head. (42)

rivure en chaîne (*f.*), chain-rivetting. (43)

rivure en quinconce (*f.*), staggered rivetting ; zigzag riveting ; cross-riveting. (44)

rivure étanche (*f.*), tight riveting. (45)

rivure prisonnière (*f.*), countersunk riveting. (46)

rivure saillante (*f.*), raised head. (47)

roable *ou* **roatie** (tire-braise) (*n.m.*), rake. (48)

robine (Fonderie) (*n.f.*), casters' mallet. (49)

robinet (*n.m.*), cock ; tap ; faucet ; spigot. (50)

robinet à bec courbe (*m.*), bib-cock ; bib. (51)

robinet à bec droit (*m.*), straight-nose cock. (52)

robinet à carré (*m.*), cock with square head ; tap with square head. (53)

robinet à clef *ou* **robinet à boisseau conique et à clef** (*m.*), plug-cock. (54)

robinet à col de cygne (*m.*), swanneck cock. (55)

robinet à deux faces (*m.*), twin cock. (56)

robinet à deux voies *ou* **robinet à deux eaux** (*m.*), two-way cock. (57)

robinet à flotteur (*m.*), ball cock. (58)

robinet à gaz (*m.*), gas-tap ; gas-cock. (59)

robinet à quatre voies *ou* **robinet à quatre eaux** (*m.*), four-way cock. (60)

robinet à raccord, avec ou sans tubulure (*m.*), union-cock, cock with union, with or without tail-pipe. (61 *ou* 62)

robinet à repoussoir (*m.*), push-button faucet. (63)

robinet à ressort (*m.*), self-closing cock ; self-closing faucet ; spring-faucet. (64)

robinet à soupape (*m.*), screw-down valve. (65)

robinet à soupape à volant (*m.*), wheel-valve. (66)

robinet à tête (*m.*), cock with crutch key ; tap with crutch key. (67)

robinet à trois voies *ou* **robinet à trois eaux** (*m.*), three-way cock ; switch-cock. (68)

robinet à vis de pression (*m.*), screw-down cock. (69)

robinet-coffret [**robinets-coffrets** *pl.*] (*n.m.*), main-tap. (70)

robinet d'admission de l'air (*m.*), air-inlet cock. (71)

robinet d'air (*m.*), air-tap. (72)

robinet d'alimentation (*m.*), feed-cock. (73)

robinet d'amorçage (*m.*), priming-cock. (74)

robinet d'arrêt *ou* *simplement* **robinet** (*n.m.*), stop-cock ; stopcock : (75) le robinet d'une burette, the stop-cock of a burette. (76)

robinet d'arrêt (Mach.) (*m.*), stop-valve. (77)

robinet d'arrêt à soupape (*m.*), screw-down stop-valve. (78)

robinet d'arrêt de vapeur (*m.*), steam-stop valve. (79)

robinet d'arrosage (*m.*), jet-cock; cooler-cock; flood-cock. (1)

robinet d'isolement (*m.*), cut-out cock; isolating-valve. (2)

robinet de compression (*m.*), compression-cock. (3)

robinet de cylindre (*m.*), cylinder-cock. (4)

robinet de fermeture (*m.*), stop-cock; stopcock. (5)

robinet de gaz *ou* robinet de conduite de gaz (*m.*), gas-tap; gas-cock. (6)

robinet de hauteur d'eau *ou* robinet de jauge (*m.*), gauge-cock; test-cock; try-cock; trial-cock. (7)

robinet de prise d'eau (*m.*), water-valve. (8)

robinet de prise de vapeur (*m.*), steam-cock; steam-valve. (9)

robinet de purge (*m.*), drip-cock; drain-cock; cylinder-cock; waste-cock. (10)

robinet de sûreté (*m.*), safety-cock. (11)

robinet de vapeur (*m.*), steam-cock; steam-valve. (12)

robinet de vidange (*m.*), mud-cock; mud-plug; purge-cock; purging-cock; blow-off cock. (13)

robinet du souffleur (*m.*), blower-cock. (14)

robinet en cuivre (*m.*), brass cock; (Brass tap. (15)

robinet en bronze (*m.*), gun-metal cock. (16)

robinet graisseur (*m.*), grease-cock; oil-cock. (17)

robinet pour chaudière à vapeur (*m.*), boiler-cock. (18)

robinet pour eau *ou* robinet pour conduite d'eau *ou* robinet hydraulique (*m.*), water-cock; water-tap. (19)

robinet pour fûts (*m.*), butt-cock. (20)

robinet pour vapeur (*m.*), steam-valve. (21)

robinet purgeur (*m.*), drip-cock; drain-cock; cylinder-cock; waste-cock. (22)

robinet purgeur (d'un réservoir à air) (*m.*), bleeding-cock; bleeding-valve. (23)

robinet se refermant automatiquement *ou* robinet se refermant seul (*m.*), self-closing cock; self-closing faucet. (24)

robinet-valve [robinets-valves *pl.*] (*n.m.*) *ou* robinet-valve à soupape (*m.*), valve-cock. (25)

robinet-vanne [robinets-vannes *pl.*] (*n.m.*), sluice-valve; gate; gate-valve; water-gate. (26)

robinet-vanne pour vapeur (*m.*), gate-valve for steam. (27)

robinetier (pers.) (*n.m.*), brass-founder and finisher; brass-smith. (28)

robinetterie (fabrication) (*n.f.*), brass-founding and finishing. (29)

robinetterie (usine) (*n.f.*), brass-foundry. (30)

robinetterie (robinets et accessoires) (*n.f.*), cocks and fittings; cocks, taps, and fittings. (31)

roburite (Explosif) (*n.f.*), roburite. (32)

robuste (*adj.*), strong; robust: (33)

une machine robuste, a strong (*or* a robust) machine. (34)

robustesse (d'une machine) (*n.f.*), strength, robustness, (of a machine). (35)

roc (*n.m.*), rock; stone. (36)

Note.—Distinctions of meanings of roc, roche, and rocher,—roc imports hardness, solidity; as, une habitation creusée dans le roc, a dwelling cut in the rock; an abode hewn out of the solid rock; des deux côtés du port, un vaste roc s'avance, from either side of the harbour, a huge rock juts out.

roche is applied to the stone itself, considered from the point of view of its nature or quali-ties; as, il y a des roches dures et des roches tendres, there are hard rocks and soft rocks.

rocher is a steep, precipitous rock, or piece of rock, difficult or dangerous of access; as, les rochers anfractueux des côtes de la mer, the rugged rocks of the seacoast; des rochers surplombent le ravin, rocks over-hang the ravine; le rocher de Gibraltar, the rock of Gibraltar; eau qui cascade de rocher en rocher (*f.*), water which cascades from rock to rock; un chaos de rochers, a chaos of rocks.

When a detached piece of rock is considered, either rocher or roche can generally be used; as, un bloc perché ou roche perchée est un bloc de rocher juché sur son socle, etc., a perched block or perched rock is a lump of rock perched on its pedestal, etc.

roc vif (*m.*), living rock; living stone. (37)

rocailleux, -euse (*adj.*), stony: (38)

un chemin rocailleux, a stony path. (39)

rochage (de l'argent ou du platine en fusion) (Métall.) (*n.m.*), spitting; sprouting; vegetation. (40)

rochage (Brasage) (*n.m.*), fluxing; wetting *or* sprinkling with borax. (41)

roche (Géol.) (*n.f.*), rock; stone. (42)

roche (*n.f.*) *ou* roche mère, matrix; gangue; gangue-matter; mother-rock; parent-rock: (43)

roche d'émeraude, emerald matrix; mother-rock of emerald. (44)

roche de topaze, topaz matrix. (45)

roche mère du diamant, mother-rock of dia-mond; diamond matrix. (46)

roches à feldspaths (*f.pl.*), feldspathic rocks. (47)

roches abyssales (*f.pl.*), abyssal rocks; plutonic rocks. (48)

roches acides (*f.pl.*), acid rocks; acidic rocks. (49)

roches afeldspathiques (*f.pl.*), feldspar-free rocks. (50)

roches altérées (*f.pl.*), weathered rocks; altered rocks. (51)

roches amorphes (*f.pl.*), amorphous rocks; structureless rocks. (52)

roches archéennes (*f.pl.*), Archæan rocks. (53)

roche arénacée (*f.*), arenaceous rock. (54)

roche argileuse (*f.*), argillaceous rock. (55)

roches aschistes (*f.pl.*), aschistic rocks. (56)

roche asphaltique (*f.*), asphalt rock; asphalt stone. (57)

roches basaltiques (*f.pl.*), basaltic rocks. (58)

roches basiques (*f.pl.*), basic rocks. (59)

roche bitumineuse (*f.*), bituminous rock. (60)

roches brouillées (*f.pl.*), jumbled rocks. (61)

roche calcaire (*f.*), limestone-rock; calcareous rock. (62)

roches cataclastiques (*f.pl.*), cataclastic rocks. (63)

roches charbonneuses (*f.pl.*), carbonaceous rocks. (64)

roches chimiques (*f.pl.*), chemical rocks. (65)

roches cimentées (*f.pl.*), cemented rocks. (66)

roches clastiques (*f.pl.*), clastic rocks; detrital rocks; fragmental rocks; fragmentary rocks. (67)

roches complémentaires (*f.pl.*), complementary rocks. (68)

roches concrétionnées (*f.pl.*), concretionary rocks. (1)

roche corallienne (*f.*), coral rock; coralline rock. (2)

roches cristallines (*f.pl.*), crystalline rocks. (3)

roches cristallophylliennes (*f.pl.*), schistose crystalline rocks; foliated crystalline rocks; phyllocrystalline rocks. (4)

roche d'adieu (grès meulier) (Mines) (*f.*), fare-well-rock. (5)

roches d'effusion *ou* roches d'épanchement (*f.pl.*), effusive rocks; extrusive rocks. (6)

roche d'émeraude (*f.*), emerald matrix; mother-rock of emerald. (7)

roches d'intrusion (*f.pl.*), intrusive rocks; intruded rocks; deep-seated rocks. (8)

roches d'origine aqueuse (*f.pl.*), aqueous rocks. (9)

roches d'origine chimique (*f.pl.*), chemically formed rocks. (0)

roches d'origine externe (*f.pl.*), exogenous rocks. (11)

roches d'origine interne (*f.pl.*), endogenous rocks. (12)

roches d'origine mécanique (*f.pl.*), mechani-cally formed rocks. (13)

roches d'origine organique (*f.pl.*), organically derived rocks. (14)

roches de demi-profondeur (*f.pl.*), dyke rocks; hypabyssal rocks. (15)

roche de filon (*f.*), veinstone. (16)

roche de fond (Géol.) (*f.*), bed-rock. (17)

roches de profondeur (*f.pl.*), plutonic rocks; abyssal rocks. (18)

roches de surface (*f.pl.*), surface-rocks; surface-formed rocks. (19)

roche de topaze (*f.*), topaz matrix. (20)

roches de transition (*f.pl.*), transition rocks. (21)

roches dépourvues de feldspath (*f.pl.*), feldspar-free rocks. (22)

roches détritiques *ou* roches deutogènes (*f.pl.*), detrital rocks; clastic rocks; fragmental rocks; fragmentary rocks. (23)

roches diaschistes (*f.pl.*), diaschistic rocks. (24)

roches effusives (*f.pl.*), effusive rocks; extrusive rocks. (25)

roche encaissante (Géol. & Mines) (*f.*), country-rock; country; enclosing rock; wall-rock. (26)

roches endogènes (*f.pl.*), endogenous rocks. (27)

roches éoliennes (*f.pl.*), eolian rocks; æolian rocks. (28)

roches éruptives (*f.pl.*), eruptive rocks; eruptives. (29)

roches essentielles (*f.pl.*), essential rocks. (30)

roches euritiques (*f.pl.*), euritic rocks. (31)

roches exogènes (*f.pl.*), exogenous rocks. (32)

roches extra-terrestres (*f.pl.*), meteoric stone. (33)

roche feuilletée (*f.*), foliated rock. (34)

roche filonienne (*f.*), veinstone. (35)

roches fondamentales (*f.pl.*), fundamental rocks. (36)

roche franche (*f.*), freestone. (37)

roches glaciaires (*f.pl.*), glacial rocks. (38)

roche granitoïde (*f.*), granitoid rock. (39)

roches hétérogènes (*f.pl.*), heterogeneous rocks. (±0)

roches holocristallines (*f.pl.*), holocrystalline rocks. (41)

roches hypabyssales (*f.pl.*), hypabyssal rocks; dyke rocks. (42)

roches hypocristallines (*f.pl.*), hypocrystalline rocks. (43)

roches ignées (*f.pl.*), igneous rocks. (44)

roches ignées acides (*f.pl.*), acid igneous rocks. (45)

roches inaltérées (*f.pl.*), unweathered rocks; unaltered rocks. (46)

roches intermédiaires (*f.pl.*), intermediate rocks; neutral rocks (between acid and basic). (47)

roches interstratifiées (*f.pl.*), interbedded rocks; contemporaneous rocks. (48)

roches légères (*f.pl.*), acid rocks; acidic rocks. (49)

roches leucocrates (*f.pl.*), leucocratic rocks. (50)

roches lourdes (*f.pl.*), basic rocks. (51)

roche marmoréenne (*f.*), marmoraceous rock. (52)

roches mélanocrates (*f.pl.*), melanocratic rocks. (53)

roche mère (*f.*). See roche *ou* roche mère.

roches mésocrates (*f.pl.*), mesocratic rocks. (54)

roche métallifère (*f.*), metalliferous rock; ore-bearing rock; metal-bearing rock. (55)

roches métamorphiques (*f.pl.*), metamorphic rocks. (56)

roche micacée (*f.*), micaceous rock. (57)

roche minéralisée (*f.*), mineralized rock; mineralized stone. (58)

roches mixtes (*f.pl.*), heterogeneous rocks. (59)

roches monogéniques (*f.pl.*), monogenetic rocks (60)

roche morte (*f.*), dead rock. (61)

roches moutonnées (Géol.) (*f.pl.*), roches mou-tonnées; dressed rocks. (62)

roches neutres (*f.pl.*), neutral rocks; inter-mediate rocks (between acid and basic). (63)

roches non-altérées (*f.pl.*), unweathered rocks; unaltered rocks. (64)

roches non-stratifiées (*f.pl.*), unstratified rocks. (65)

roches organiques (*f.pl.*), organic rocks. (66)

roche perchée (Géol.) (*f.*), perched rock; perched block. (67)

roche pétrolifère (*f.*), oil-rock. (68)

roche phosphatée (*f.*), phosphate rock. (69)

roches plutoniques *ou* roches plutoniennes (*f.pl.*), plutonic rocks; abyssal rocks. (70)

roches polygéniques (*f.pl.*), polygenetic rocks. (71)

roches primitives (*f.pl.*), primitive rocks; primary rocks. (72)

roches protogènes (*f.pl.*), protogenic rocks. (73)

roche puante (*f.*), stinkstone; fetid sandstone. (74)

roches pyrogènes (*f.pl.*), pyrogenous rocks. (75)

roches sans feldspaths (*f.pl.*), feldspar-free rocks. (76)

roches sans quartz (*f.pl.*), quartzless rocks. (77)

roche schisteuse (*f.*), schistous rock; schist-rock; shaly rock. (78)

roches secondaires (*f.pl.*), secondary rocks. (79)

roches sédimentaires (*f.pl.*), sedimentary rocks. (80)

roches sédimentaires à grain fin (*f.pl.*), fine-grained sedimentary rocks. (81)

roche silicatée (*f.*), silicate rock. (82)

roche siliceuse (*f.*), siliceous rock; silicious rock. (83)

roche stérile (*f.*), barren rock; sterile rock; sterile stone; unproductive rock. (84)

roches sous-jacentes *ou* roches subjacentes (*f.pl.*), subjacent rocks; underlying rocks. (85)

roches stratifiées (*f.pl.*), stratified rocks ; bedded rocks. (1)

roches striées (*f.pl.*), striate rocks ; striated rocks. (2)

roches subaériennes (*f.pl.*), subaerial rocks. (3)

roche suintante (*f.*), weeping rock. (4)

roches surjacentes *ou* **roches susjacentes** (*f.pl.*), overlying rocks. (5)

roche tendre (*f.*), soft rock ; tender rock. (6)

roches thalassiques (*f.pl.*), thalassic rocks. (7)

roche trappéenne (*f.*), trap-rock. (8)

roches ultra-acides (*f.pl.*), ultra-acid rocks. (9)

roches ultra-basiques (*f.pl.*), ultrabasic rocks. (10)

roche verte (*f.*), greenstone. (11)

roches vitreuses (*f.pl.*), vitreous rocks. (12)

roche vive (*f.*), living rock ; living stone. (13)

roches volcaniques (*f.pl.*), volcanic rocks. (14)

rocher (*n.m.*), rock ; stone. (15) See note under **roc.**

rocher (roche encaissante, opp. **à charbon**) (Exploitation des mines de houille) (*n.m.*), rock ; stone ; metal ; dead ground : (16)

galerie au rocher *ou* **galerie en rocher** (*f.*), rock-drift ; stone-drift ; metal-drift ; gallery in dead ground. (17)

mineur au rocher (*m.*), stoneman ; metalman. (18)

rocher (saupoudrer, mouiller de borax) (Brasage) (*v.t.*), to flux ; to wet *or* sprinkle with borax. (19)

rocher (fusion de l'argent, du platine) (*v.i.*), to spit ; to sprout ; to vegetate. (20)

rocher branlant *ou* **rocher tremblant** (*m.*), rocking stone ; logan stone ; logan ; loggan stone ; loggan ; logging rock. (21)

rocher en surplomb (*m.*), overhanging rock. (22)

rochers anfractueux (*m.pl.*), rugged rocks ; craggy rocks ; crags. (23)

rochers caverneux (*m.pl.*), cavernous rocks. (24)

rocheux, -euse (*adj.*), rocky ; rock (*used as adj.*) : (25)

minerai rocheux (*m.*), rocky ore. (26)

masse rocheuse (*f.*), rock-mass. (27)

rodage (alésage à l'aide de la machine à roder, de la meule ou de la poudre émeri) (*n.m.*), grinding : (28)

rodage des cylindres, des soupapes, grinding cylinders, valves. (29 *ou* 30)

rodage (polissage, nettoyage avec le rodoir) (*n.m.*), lapping : (31)

rodage de l'intérieur d'un trou, d'un arbre de transmission, lapping the inside of a hole, a shaft. (32 *ou* 33)

rôdage (pivotation) (*n.m.*), rotation : (34)

sondage par rôdage (*m.*), boring by rotation. (35)

roder (aléser à l'aide de la machine à roder, de la meule ou de la poudre émeri) (*v.t.*), to grind : (36)

roder un bouchon de verre, to grind a glass stopper. (37)

les soupapes de sûreté doivent toujours être bien rodées sur leur siège (*f.pl.*), safety-valves should always be well ground on their seat. (38)

roder (polir, nettoyer avec le rodoir) (*v.t.*), to lap. (39)

rôder (accomplir sur un pivot un mouvement de rotation) (Méc.) (*v.i.*), to rotate. (40)

rôder autour d'un camp, to mooch about a camp. (41)

rôdeur (pers.) (*n.m.*), rubber-neck (one who prys about a mining settlement for information which may be of value). (42)

rodoir (*n.m.*), lap ; lapping-tool. (43)

rodoir en cuivre rouge (*m.*), copper lap. (44)

rodoir en plomb (*m.*), lead lap. (45)

rodoir pour transmissions (*m.*), shaft-lap. (46)

rogner les dépenses, to cut down, to curtail, to reduce, expenses. (47)

rognoir (*n.m.*), scraper. (48)

rognon (Géol.) (*n.m.*), kidney. (49)

rognon de minerai (*m.*), kidney of ore. (50)

rognon de turquoise (*m.*), kidney of turquoise. (51)

rognons de fer carbonaté (du terrain houiller) (*m.pl.*), ball-ironstone, ball-mine, (of the coal-measures). (52)

rognure (*n.f.*), cutting ; clipping ; paring ; shaving : (53)

rognures de zinc, zinc shavings ; zinc cuttings. (54)

rohwand (*n.m.*), ferruginous limestone (used as a flux). (55)

roide (*adj.*), **roideur** (*n.f.*), **roidir** (*v.t.*), **roidisse--ment** (*n.m.*), etc. Same as **raide, raideur, raidir, raidissement,** etc.

rôle (*n.m.*), rôle ; part (played or assumed) ; factor ; function : (56)

la pierre remplit, en géologie, le rôle de para--pluie et protège tout ce qu'elle recouvre, in geology, stone performs the rôle of an umbrella and protects everything it covers. (57)

le rôle joué par l'argent en affaires est con--sidérable, the part played by money in business is considerable ; in business, money is a most important factor. (58)

le cordon littoral exerce le rôle protecteur d'un brise-lames, the offshore bar performs the protective function of a breakwater. (59)

romaine (*n.f.*), steelyard ; scale-beam ; lever scales ; Roman balance. (60)

romaine-bascule [**romaines-bascules** *pl.*] (*n.f.*), suspended weighing-machine ; suspension-scales. (61)

roméine *ou* **roméite** (Minéral) (*n.f.*), romeite ; romeine. (62)

rompre (*v.t.*), to break : (63)

rompre un essieu, un attelage, to break an axle, a coupling. (64 *ou* 65)

rompre un circuit (Elec.), to break, to discon--nect, to open, a circuit. (66)

rompre charge, to break bulk. (67)

rompre l'équilibre, to upset the equilibrium. (68)

rompre (*v.i.*), to break. (69)

rompre (se) (*v.r.*), to break. (70)

rompre (se) tout à coup, to snap. (71)

rompu du banc de tour (*m.*), gap of lathe-bed. (72)

rompu en avant du plateau (*m.*), gap in front of face-plate. (73)

ronce (*n.f.*), barbed wire ; barb-wire. (74)

rond, -e (*adj.*), round. (75)

rondelle (*n.f.*), washer. (76)

rondelle de cuir (*f.*), leather washer. (77)

rondelle décolletée (*f.*), machine-made washer. (78)

rondelle fusible (pour chaudières) (*f.*), fusible plug ; safety-plug ; plug. (79)

rondelle obturatrice (*f.*), blind washer ; blind flange ; blank washer ; blank flange ; no-thoroughfare. (80)

rondelle tournée (*f.*), turned washer. (1)

rondeur (*n.f.*), roundness : (2)
rondeur d'une roue, roundness of a wheel. (3)

rondin (*n.m.*) *ou* **rondins** (*n.m.pl.*) *ou* **ronds** (*n.m.pl.*) (fer), round bar ; round bar iron ; round bars ; rounds. (4)

rondin (bois) (*n.m.*), round timber. (5)

ronger (entamer à petits coups ; corroder) (*v.t.*), to eat ; to eat away ; to eat into ; to eat up ; to wear away : (6)
le métal est rongé par l'acide, metal is eaten by acid. (7)
les vers rongent le bois (*m.pl.*), worms eat away wood. (8)
roches rongées par érosion (*f.pl.*), rocks worn away by erosion. (9)

ronger (miner) (*v.t.*), to undermine : (10)
la mer ronge les falaises, the sea undermines the cliffs. (11)

ronger (se) (*v.r.*), to be eaten ; to be eaten away ; to be worn away. (12)

rood (mesure agraire anglaise) (*n.m.*), rood = 10·117 ares. (13)

rosace de plafond (Élec.) (*f.*), ceiling-rose. (14)

rose (diamant) (*n.f.*), rose ; rose diamond ; rose-cut diamond. (15)

rose des vents *ou* **rose de boussole** *ou* **rose de compas** *ou* simplement **rose** (*n.f.*), compass-card ; compass-dial ; rose. (16)

rose mobile (boussole) (*f.*), floating dial. (17)

roseau (*n.m.*), reed. (18)

rosée (*n.f.*), dew. (19)

rosette (cuivre rosette) (*n.f.*), rose copper ; rosette copper. (20)

rosette (contre-rivure) (*n.f.*), burr ; riveting-burr ; rivet-washer. (21)

rotatif, -ive (*adj.*), rotary ; rotating ; rotative ; rotatory ; revolving. (22)

rotation (*n.f.*), rotation. (23)

rotation à droite (*f.*), right-hand rotation. (24)

rotation à gauche (*f.*), left-hand rotation. (25)

rotation dans le sens du mouvement des aiguilles d'une montre (*f.*), rotation clockwise ; clockwise rotation. (26)

rotation dans le sens inverse du mouvement des aiguilles d'une montre (*f.*), rotation reverse to clockwise ; counterclockwise rotation ; contraclockwise rotation ; anti-clockwise rotation. (27)

rotatoire (*adj.*), rotary ; rotating ; rotative ; rotatory ; revolving. (28)

rothoffite (Minéral) (*n.f.*), rothoffite. (29)

rotonde (Ch. de f.) (*n.f.*), roundhouse. (30)

rotor (Élec.) (*n.m.*), rotor. (31)

rotor à cage d'écureuil *ou* **rotor en court-circuit** (*m.*), squirrel-cage rotor. (32)

rotor à enroulement (*m.*), wound rotor. (33)

rotule (*n.f.*), swivel. (34)

rotule sphérique (*f.*), ball and socket. (35)

rouage (ensemble des roues d'une machine) (*n.m.*), wheels ; wheelwork : (36)
lubrifier le rouage d'une machine, to lubricate the wheels of a machine. (37)

rouage (roue dentée ; roues dentées) (*n.m.*), gear-wheel ; toothed wheel ; cog-wheel ; gearing ; mill-gearing ; gear ; gear-work. (38)

rouage (*au figuré*) (*n.m.*), machinery : (39)
le grand rouage des phénomènes météorologi-ques, the great machinery of meteorological phenomena. (40)

rouane (de charpentier) (*n.f.*), race-knife ; racer. (41)

roue (*n.f.*), wheel. (42)

roue (cadre porteur) (Fonçage des puits de mines) (*n.f.*), curb ; crib ; sinking-frame. (43)

roue à adhérence (opp. à pignon engrenant sur la crémaillère) (Ch. de f.) (*f.*), adhesion-wheel. (44) (opp. to **rack-pinion**).

roue à aubes (*f.*), paddle-wheel. (45)

roue à augets *ou* **roue à auges** (*f.*), bucket-wheel. (46)

roue à augets en dessus (*f.*), overshot wheel ; overshot water-wheel. (47)

roue à boudin *ou* **roue à bourrelet** (*f.*), flange-wheel. (48)

roue à bras (*f.*), spoke-wheel ; spoke-centre wheel. (49)

roue à cames (*f.*), cam-wheel ; wiper-wheel : (50)
la roue à cames d'un martinet, the wiper-wheel of a tilt-hammer. (51)

roue à centre plein (*f.*), disc-centre wheel ; plate-wheel. (52)

roue à chaîne (*f.*), chain-wheel ; chain-pulley ; chain-sheave. (53)

roue à chevilles (*f.*), treadwheel. (54)

roue à cliquet *ou* **roue à chien** (*f.*), ratchet-wheel ; ratchet ; ratch ; click-wheel ; dog-wheel. (55)

roue à cuiller (*f.*), horizontal water-wheel. (56)

roue à cuve (*f.*), tub-wheel. (57)

roue à dents (*f.*), toothed wheel ; cog-wheel ; rack-wheel ; gear-wheel ; gear. (58)

roue à dents creuses *ou* **roue à denture creuse** (Engrenage à vis sans fin) (*f.*), hollow-tooth wheel ; enveloping-tooth wheel (Worm-gearing). (59)

roue à dents droites *ou* **roue à denture droite** (Engrenages) (*f.*), straight-tooth wheel. (60)

roue à eau (*f.*), water-wheel ; hydraulic wheel. (61)

roue à empreintes (*f.*), pocket-wheel ; cupped chain-sheave ; indented wheel. (62)

roue à fileter (tour) (*f.*), change-wheel (lathe). (63)

roue à godets (*f.*), bucket-wheel. (64)

roue à gorge (*f.*), grooved wheel ; sheave. (65)

roue à joues (Engrenages) (*f.*), shrouded gear ; shrouding gear ; flanged gear. (66)

roue à marches (*f.*), treadwheel. (67)

roue à palettes (*f.*), paddle-wheel. (68)

roue à pignon (*f.*), pinion-wheel. (69)

roue à plateau plein (*f.*), disc-centre wheel ; plate-wheel. (70)

roue à rais *ou* **roue à rayons** (*f.*), spoke-wheel ; spoke-centre wheel. (71)

roue à réaction (*f.*), reaction-wheel ; reaction water-wheel. (72)

roue à rochet (*f.*), ratchet-wheel ; ratchet ; ratch ; click-wheel ; dog-wheel. (73)

roue à rochet du frein à main (*f.*), brake ratchet-wheel. (74)

roue à sabots (*f.*), noria ; Persian wheel ; flush-wheel. (75)

roue à tailings (Mines) (*f.*), tailings-wheel. (76)

roue à toile (*f.*), plate-wheel ; disc-centre wheel. (77)

roue à tuyaux (*f.*), reaction-wheel ; reaction water-wheel. (78)

roue à vis sans fin (*f.*), worm-wheel ; worm-gear. (79)

roue calée *ou* **roue fixe** (calée ou fixée sur l'essieu) (opp. à **roue folle**) (*f.*), fast wheel ; fixed wheel. (80)

roue conduite (Engrenages) (*f.*), follower. (1)

roue conique (*f.*), conical wheel ; cone-wheel. (2)

roue d'angle (*f.*), bevel-wheel. (3)

roue d'engrenage (*f.*), gear-wheel ; pitch-wheel. (4)

roue de Barlow (Élec.) (*f.*), Barlow's wheel. (5)

roue de chaîne (*f.*), chain-wheel ; chain-pulley ; chain-sheave. (6)

roue de champ (*f.*), crown-wheel ; face-wheel. (7)

roue de correction (tour à fileter) (*f.*), translat-ing-wheel (screw-cutting lathe). (8)

roue de côté (Hydraul.) (*f.*), breast-wheel. (9)

roue de division (*f.*), division-wheel ; dividing-wheel. (10)

roue de filetage (tour) (*f.*), change-wheel (lathe). (11)

roue de friction *ou* roue de frottement (*f.*), fric-tion-wheel. (12)

roue de poitrine (Hydraul.) (*f.*), breast-wheel. (13)

roue de Poncelet (Hydraul.) (*f.*), Poncelet wheel. (14)

roue de rechange (Engrenages) (*f.*), change-wheel. (15)

roue de tramway (*f.*), tram-wheel. (16)

roue de translation (pour pont roulant, etc.) (*f.*), travelling wheel ; running-wheel ; runner ; rail-wheel. (17)

roue de transmission (*f.*), driving-wheel. (18)

roue de voiture (*f.*), carriage-wheel ; cart-wheel ; vehicle-wheel. (19)

roue de volée (*f.*), man-power ; manual gear. (20)

roue de wagon (Ch. de f.) (*f.*), car-wheel ; carriage-wheel ; wagon-wheel ; truck-wheel. (21)

roue de wagon en fonte trempée (*f.*), chilled cast-iron car-wheel. (22)

roue dentée (*f.*), toothed wheel ; cog-wheel ; rack-wheel ; gear-wheel ; gear. (23)

roue dentée en partie (*f.*), mutilated gear. (24)

roue dépourvue de mentonnet (*f.*), flangeless wheel. (25)

roue directrice (*f.*), steering-wheel. (26)

roue diviseur (*f.*), dividing-wheel ; division-wheel. (27)

roue droite (roue d'engrenage) (*f.*), spur-wheel ; spur-gear. (28)

roue élévatoire (*f.*), elevating-wheel ; lifting-wheel. (29)

roue élévatrice de tailings (Mines) (*f.*), tailings-wheel. (30)

roue en dessous (*f.*), undershot wheel ; under-shot water-wheel. (31)

roue en dessus (*f.*), overshot wheel ; overshot water-wheel. (32)

roue épaulée (*f.*), Same as roue à joues.

roue fixe (d'une turbine) (*f.*), guide-ring (of a turbine). (33)

roue folle (folle sur l'essieu) (opp. à roue calée) (*f.*), loose wheel. (34)

roue hélicoïdale (*f.*), helicoidal wheel ; spiral wheel ; screw-wheel. (35)

roue hydraulique (*f.*), water-wheel ; hydraulic wheel. (36)

roue hyperboloïde (Engrenages) (*f.*) hyper-bolical wheel ; skew bevel wheel ; skew-wheel ; skew-gear. (37)

roue intermédiaire (Engrenages) (*f.*), idle wheel ; idler ; idler-wheel ; intermediate wheel ; stud-wheel. (38)

roue maîtresse (Méc.) (*f.*), master-wheel ; leader. (39)

roue menante (d'une locomotive) (*f.*), driving-wheel ; driver. (40)

roue menante (Engrenages) (*f.*), driver ; driving-wheel. (41)

roue menée (Engrenages) (*f.*), follower ; driven wheel. (42)

roue mobile (d'une turbine) (*f.*), runner, wheel, (of a turbine). (43)

roue mobile (d'une pompe centrifuge) (*f.*), impeller, runner, (of a centrifugal pump). (44)

roue motrice (*f.*), driving-wheel ; driver : (45) roue motrice d'une locomotive, driving-wheel, driver, of a locomotive. (46)

roue parasite (Engrenages) (*f.*), idle wheel ; idler ; idler-wheel ; intermediate wheel ; stud-wheel. (47)

roue Pelton (*f.*), Pelton wheel. (48)

roue planétaire (*f.*), planet-wheel ; planet-gear ; planetary pinion. (49)

roue pleine (*f.*), solid wheel. (50)

roue porteuse (d'une locomotive) (opp. à roue motrice *ou* roue menante) (*f.*), carrying-wheel. (51) (opposed to driving-wheel).

roue porteuse (de bogie de locomotive) (*f.*), truck-wheel ; engine-truck wheel. (52)

roue porteuse d'arrière (locomotive) (*f.*), trailing-wheel ; trailer ; back engine-truck wheel. (53)

roue porteuse d'avant (locomotive) (*f.*), leading-wheel ; pilot-wheel ; front engine-truck wheel. (54)

roue-turbine [roues-turbines *pl.*] (*n.f.*), turbine-wheel. (55)

roue-turbine à impulsion (*f.*), impulse wheel. (56)

rouer un câble, to coil a rope. (57)

roues couplées (locomotive) (*f.pl.*), coupled wheels. (58)

rouet (réa) (*n.m.*), pulley-wheel ; pulley ; sheave ; pulley-sheave. (59)

rouet (roue munie d'alluchons) (*n.m.*), cog-wheel (inserted wooden cogs) ; mortise-wheel. (60)

rouet (cadre porteur) (Fonçage des puits de mines) (*n.m.*), curb ; crib ; sinking-frame. (61)

rouet porteur (Mines) (*m.*), supporting-curb ; supporting-frame. (62)

rouge (*adj.*), red. (63)

rouge (devenu rouge au feu) (*adj.*), red-hot : (64) fer rouge (*m.*), red-hot iron. (65)

rouge (*n.m.*), red. (66)

rouge antique (marbre) (*m.*), rosso antico. (67)

rouge cerise (coloration du fer) (Métall.) (*m.*), cherry red. (68)

rouge clair *ou* rouge vif (Métall.) (*m.*), bright red. (69)

rouge d'Angleterre *ou* rouge à polir (*m.*), rouge. (70)

rouge feu (*m.*), fiery red : (71) pierre d'un beau rouge feu (*f.*), stone of a beautiful fiery red. (72)

rouge naissant (Métall.) (*m.*), nascent red ; black red. (73)

rouge sang (*m.*), blood red. (74)

rouge sombre (Métall.) (*m.*), sombre red ; dark red ; dull red. (75)

rougeâtre (*adj.*), reddish. (76)

rougir (*v.t.*), to redden : (77)

l'acide rougit la teinture bleue de tournesol (*m.*), acid reddens blue litmus solution. (1)

rougir (*v.i.*), to redden. (2)

rougir un fer au feu, to heat an iron to redness in the fire ; to make an iron red hot. (3)

rouille (*n.f.*), rust. (4)

rouille de cuivre (*f.*), verdigris. (5)

rouille de plomb (*f.*), white lead ; ceruse. (6)

rouillé, -e (*adj.*), rusty. (7)

rouiller (*v.t.*), to rust : (8)

l'humidité rouille le fer (*f.*), damp rusts iron. (9)

rouiller (**se**) (*v.r.*), to rust ; to become rusty ; to get rusty. (10)

rouilleuse (Exploitation du charbon) (*n.f.*), nicking-machine ; shearing-machine. (11)

rouilleuse à chaîne (*f.*), chain shearing-machine. (12)

rouilleuse à pic (*f.*), pick shearing-machine. (13)

rouillure (du fer) (*n.f.*), rustiness (of iron). (14)

rouillure (Exploitation du charbon) (*n.f.*), nicking ; nick ; shearing ; shear. (15)

roulage (*n.m.*), haulage ; hauling ; trucking ; tramming ; wheeling ; drawing ; putting : (16)

roulage du minerai des fronts de taille jusqu'au puits, haulage of the ore from the faces to the shaft. (17)

roulage à la cale (Mines) (*m.*), sprag-haulage. (18)

roulage en palier (*m.*), level haulage. (19)

roulage souterrain (*m.*), underground haulage. (20)

roulant, -e (*adj.*), rolling ; portable : (21)

matériel roulant (*m.*), rolling stock. (22)

étau roulant (*m.*), portable vice-stand. (23)

roulante (Géom.) (*n.f.*), rolling circle. (24)

roulé (-e) par les eaux (Géol.) (*adj.*), water-rolled : (25)

cailloux roulés par les eaux (*m.pl.*), water-rolled pebbles. (26)

rouleau *ou* **roule** (bâton cylindrique) (*n.m.*), roller : (27)

rouleau pour une carte géographique, roller for a map. (28)

rouleau (cylindre) (*n.m.*), roll ; roller. (29)

rouleau compresseur (pour chaussées) (*m.*), roller (for roads). (30)

rouleau de câble, de fil (*m.*), coil of rope, of wire. (31 *ou* 32)

rouleau de courroies de transmission (*m.*), roll of belting. (33)

rouleau de friction (*m.*), friction-roller. (34)

rouleau de renvoi (*m.*), return-roller. (35)

rouleau de tension (*m.*), tension-roller ; tension-pulley ; tightening-pulley ; binding-pulley ; idler ; belt-idler ; idler-pulley ; idle pulley ; jockey-roller ; jockey-pulley ; jockey-wheel ; jockey ; belt-tightener. (36)

rouleau entraîneur (*m.*), feed-roll ; feed-roller. (37)

rouleau-guide [rouleaux-guides *pl.*] (*n.m.*), guide-roller. (38)

rouleau pour le collage des épreuves (Photogr.) (*m.*), roller-squeegee. (39)

roulement (mouvement de ce qui roule) (*n.m.*), rolling : (40)

le roulement d'une bille, the rolling of a ball. (41)

roulement (opp. à glissement) (Méc.) (*n.m.*), rolling. (42) (opp. to sliding) :

frottement de roulement (*m.*), rolling friction. (43)

roulement (marche ; fonctionnement) (*n.m.*), running ; working : (44)

une voie ondulée est mauvaise pour le roulement des trains, a wavy road is bad for the running of trains. (45)

douceur de roulement d'une machine (*f.*), smoothness of running (*or* of working) of a machine. (46)

roulement (mécanisme permettant à certains appareils de rouler) (*n.m.*), running-gear ; carriage : (47)

les wagons roulent à une très grande vitesse : on soigne tout particulièrement le roulement afin d'éviter les accidents, the wagons run at a very high speed : particular attention is given to the running-gear in order to avoid accidents. (48)

roulement (des voitures) (*n.m.*), rumbling, noise, (of the traffic). (49)

roulement (durée de travail d'un fourneau) (*n.m.*), campaign. (50)

roulement (*n.m.*) *ou* **roulements** (*n.m.pl.*) (mécanisme permettant à certains appareils de rouler), bearing ; bearings. (51)

roulement à billes *ou* **roulement sur billes** (*m.*), ball bearing ; ball bearings. (52)

roulements à billes, type pour faible charge, ou type pour charge moyenne, ou type pour forte charge, à double rangée de billes, ball bearings, light type, or medium type, or heavy type, with double row of balls. (53 *ou* 54 *ou* 55)

le roulement à billes s'introduit de plus en plus dans la construction des paliers, ball bearings are coming more and more into use in the construction of bearings. (56)

rouler (faire avancer en tournant) (*v.t.*), to roll : (57)

rouler un tonneau, to roll a cask. (58)

rouler (plier en rond sur soi-même) (*v.t.*), to roll ; to roll up : (59)

rouler du papier, to roll, to roll up, paper. (60)

rouler (*v.t.*) *ou* **rouler en couronne**, to coil : (61)

les tuyaux en plomb du commerce sont roulés en couronne (*m.pl.*), commercial lead piping is coiled. (62)

rouler (herscher ; traîner) (Mines) (*v.t.*), to haul ; to wheel ; to tram ; to truck ; to draw ; to put : (63)

rouler le charbon du point d'abatage jusqu'à la galerie de roulage, to haul, to wheel, to tram, to put, the coal from the mining-point to the haulageway. (64)

rouler (avancer en tournant sur soi-même) (*v.i.*), to roll : (65)

boule qui roule (*f.*), ball which rolls. (66)

rouler (marcher ; circuler) (*v.i.*), to run ; to work : (67)

locomotives roulant en palier (*f.pl.*), engines running on the level. (68)

rouler (**se**) (*v.r.*) *ou* **rouler** (**se**) **en couronne**, to coil. (69)

roulette (petit rouleau) (*n.f.*), wheel ; roller (a small wheel under an article to aid in moving it about) : (70)

socle à roulettes (*m.*), stand on wheels. (71)

roulette (galet pivotant) (*n.f.*), caster ; castor. (72)

roulette (Géom.) (*n.f.*), roulette. (73)

roulette (mesure à ruban) (*n.f.*), tape ; measure ; measuring-tape (in circular case). (74)

roulette à manivelle (*f.*), wind-up measure. (75)

roulette à ressort (*f.*), spring-measure. (1)

roulette d'arpenteur (*f.*), surveyors' tape. (2)

roulette de la came (*f.*), cam-follower ; cam-roller. (3)

roulette de poche à ruban d'acier (*f.*), pocket steel tape. (4)

roulette du trolley (tramway électrique) (*f.*), trolley-wheel. (5)

rouleur (Mines) (pers.) (*n.m.*), trammer ; wheeler ; drawer ; roller ; putter ; haulage--man. (6)

rouleuse (*n.f.*), bending-machine ; bending-rollers. (7)

roulis (mouvement d'une locomotive) (*n.m.*), rolling. (8)

roulisse (cadre porteur) (Fonçage des puits de mines) (*n.f.*), curb ; crib ; sinking-frame. (9)

roulure (dans un tronc d'arbre) (*n.f.*), cupshake. (10)

round-buddle [round-buddles *pl.*] (Préparation mécanique des minerais) (*n.m.*), round buddle. (11)

roussier (*n.m.*), earthy ore ; muddy ore ; sandy ore. (12)

roussir (brûler légèrement) (*v.t. & v.i.*), to scorch. (13)

roussir (se) (*v.r.*), to scorch. (14)

route (voie de terre) (*n.f.*), road ; way. (15)

route (direction suivie) (*n.f.*), route ; path ; itinerary. (16)

route carrossable (*f.*), carriage-road ; carriage--way ; wagon-road : (17)

la seule route carrossable qui unisse la vallée de l'Inn et la vallée de l'Adda est le col de la Bernina, the only carriage-road which unites the Valley of the Inn and the Valley of the Adda is the Bernina Pass. (18)

route de mer (*f.*), sea-route. (19)

route de terre (*f.*), overland route ; land-route. (20)

route du soleil (*f.*), path of the sun. (21)

route empierrée *ou* **route ferrée** (*f.*), metalled road. (22)

route en lacet (*f.*), winding road. (23)

route la plus directe (*f.*), shortest route. (24)

route la plus rapide (*f.*), quickest route. (25)

route la plus sûre (*f.*), safest route. (26)

route macadamisée (*f.*), macadamized road. (27)

routier, -ère (*adj.*), route ; road (*used as adjs*) : (28)

carte routière (*f.*), route-map ; road-map. (29)

rouverain *ou* **rouverin** (se dit du fer) (*adj.m.*), hot-short ; red-short. (30)

rouvine (Géol.) (*n.f.*), erosion column ; chimney-rock. (31)

rouvrir (*v.t.*), to reopen : (32)

rouvrir une usine, to reopen a works. (33)

rubace *ou* **rubacelle** (Minéral) (*n.f.*), rubasse ; rubace ; Ancona ruby ; Mont Blanc ruby. (34)

ruban (*n.m.*), ribbon ; riband ; tape : (35)

un ruban d'acier, a ribbon of steel. (36)

ruban (roulette) (*n.m.*), tape ; measuring-tape ; ribbon ; riband : (37)

ruban d'acier (*m.*), steel tape ; steel measuring-tape ; steel ribbon. (38) See also **roulette**.

ruban caoutchouté (*m.*), rubber tape. (39)

ruban isolant *ou* **ruban isolateur** *ou* **ruban à isoler** (*m.*), insulating-tape ; electric tape. (40)

rubané, -e (*adj.*), banded ; ribbon : (41)

agate rubanée (*f.*), banded agate. (42)

jaspe rubané (*m.*), ribbon jasper. (43)

rubellite (Minéral) (*n.f.*), rubellite. (44)

rubicelle (Minéral) (*n.f.*), Same as **rubace**.

rubidium (Chim.) (*n.m.*), rubidium. (45)

rubis (*n.m.*), ruby. (46)

rubis-balais (*n.m.*), balas ruby ; balas. (47)

rubis blanc (*m.*), leucosapphire. (48)

rubis d'arsenic (Minéral) (*m.*), ruby-arsenic ; ruby-sulphur ; ruby of arsenic ; ruby of sulphur ; red arsenic ; realgar. (49)

rubis de Bohême (*m.*), Bohemian ruby. (50)

rubis du Brésil (*m.*), burnt topaz. (51)

rubis oriental (*m.*), Oriental ruby ; true ruby. (52)

rubis spinelle (*m.*), spinel ruby. (53)

rude (*adj.*), rough ; coarse : (54)

lime à taille rude (*f.*), rough-cut file. (55)

rudement (*adv.*), roughly ; coarsely. (56)

rudesse (*n.f.*), roughness ; coarseness : (57)

roche caractérisée par sa rudesse au toucher (*f.*), rock characterized by its roughness to the touch. (58)

rudimentaire (*adj.*), rudimentary. (59)

rue (*n.f.*), street. (60)

ruée aux champs aurifères (*f.*), rush to the gold-fields. (61)

rugosité (*n.f.*), roughness ; ruggedness : (62)

rugosité de la paroi intérieure des tuyaux (Hydraul.), roughness of the interior surface of pipes. (63)

rugueux, -euse (*adj.*), rough ; rugged : (64)

surface rugueuse (*f.*), rough surface. (65)

montagnes rugueuses (*f.pl.*), rugged mountains. (66)

ruine (*n.f.*), ruin. (67)

ruiner (*v.t.*), to ruin. (68)

ruisseau (*n.m.*), brook ; rivulet. (69)

ruisselant, -e (*adj.*), trickling : (70)

eaux ruisselantes (*f.pl.*), trickling water. (71)

ruisseler (*v.i.*), to run off over the surface ; to run down ; to trickle ; to stream ; to rill : (72)

eau de pluie qui ruisselle et arrive au thalweg (*f.*), rain-water which runs off over the surface and reaches the thalweg. (73)

la pluie ruisselle sur les toits, the rain runs down the roofs. (74)

ruisselet (*n.m.*), rivulet ; streamlet ; brooklet ; rill ; run. (75)

ruissellement (*n.m.*), running off over the surface ; running down ; trickling ; streaming ; rilling : (76)

un sous-sol imperméable favorise le ruisselle--ment des eaux, an impermeable subsoil favours the running off of water over the surface. (77)

l'observation du ruissellement des eaux de pluie sur une plage à marée basse offre l'exemple le plus simple de formation du réseau hydrographique (*f.*), the observation of the trickling of rain-water on a beach at low tide furnishes the simplest example of formation of the hydrographical system. (78)

traces de ruissellement (Géol.) (*f.pl.*), rill-marks. (79)

rumeur (*n.f.*), rumour ; rumor : (80)

rumeur sans aucun fondement d'aucune sorte, rumour without any foundation whatever. (81)

rupture (*n.f.*), breaking ; breakage ; fracture ; rupture : (82)

rupture d'un câble, d'une tige, breaking, breakage, of a rope, of a rod. (1 *ou* 2)

rupture du filon, break in the lode ; ore-break. (3)

rupture (Résistance des matériaux) (*n.f.*), break-ing ; fracture. (4)

rupture (d'un circuit) (Élec.) (*n.f.*), breaking, disconnecting, opening, a circuit. (5)

russélite (Explosif) (*n.f.*), russelite. (6)

ruthénium (Chim.) (*n.m.*), ruthenium. (7)

rutile (Minéral) (*n.m.*), rutile ; red schorl. (8)

S

S (*n.m.*), S : (9)
en cet endroit, la rivière décrit un S, here the river describes an S. (10)

S de suspension (*m.*), S hook ; S-shaped hook. (11)

sablage (*n.m.*), sanding. (12)

sable (*n.m.*), sand. (13)

sable à gaz (Géol.) (*m.*), gas-sand. (14)

sable à la houille (Fonderie) (*m.*), new sand ; facing-sand ; facing. (15)

sable à noyaux (Fonderie) (*m.*), core-sand. (16)

sable aurifère (*m.*), auriferous sand ; gold-bearing sand. (17)

sable boulant *ou* **sable bouillant** (*m.*), (*s'emploie généralement au pluriel*), quicksand. (18)

sable brûlé (Fonderie) (*m.*), burned sand ; dead sand. (19)

sable calcaire (*m.*), calcareous sand. (20)

sable de carrière (*m.*), pit-sand. (21)

sable de chantier (Fonderie) (*m.*), floor-sand ; heap-sand ; old sand ; black sand. (22)

sable de fonderie *ou* **sable de moulage** *ou* **simple-ment sable** (*n.m.*), foundry-sand ; founders' sand ; moulding-sand ; moulders' sand ; sand. (23)

sable de mer (*m.*), sea-sand. (24)

sable de plage (*m.*), beach-sand. (25)

sable de quartz (*m.*), quartz sand. (26)

sable de quartz à arêtes vives (*m.*), sharp quartz sand. (27)

sable de râperie (Fonderie) (*m.*), burned sand ; dead sand. (28)

sable de rivière (*m.*), river-sand. (29)

sable diamantifère (*m.*), diamondiferous sand ; diamond-bearing sand. (30)

sable étuvé *ou* **sable d'étuve** (Fonderie) (*m.*), stoved sand ; baked sand. (31)

sable femelle (*m.*), light-coloured sand. (32)

sable fluviatile (*m.*), river-sand ; fluviatile sand. (33)

sable fort (*m.*), strong sand. (34)

sable fossilifère (*m.*), fossiliferous sand. (35)

sable glauconifère (*m.*), glauconiferous sand ; glauconite. (36)

sable gris (Fonderie) (*m.*), burned sand ; dead sand. (37)

sable grossier (*m.*), coarse sand. (38)

sable maigre (*m.*), weak sand. (39)

sable mâle (*m.*), dark-coloured sand. (40)

sable marin (*m.*), sea-sand. (41)

sable micacé (*m.*), micaceous sand. (42)

sable monazité (*m.*), monazite sand. (43)

sable mouvant (*m.*), running sand ; quicksand. (44)

sable neuf (Fonderie) (*m.*), new sand ; facing-sand ; facing. (45)

sable noir ferrugineux (*m.*), black iron-sand. (46)

sable pétrolifère (*m.*), oil-sand. (47)

sable phosphaté (*m.*), phosphatic sand. (48)

sable platinifère (*m.*), platinum sand. (49)

sable pour grosses pièces (Fonderie) (*m.*), sand for heavy castings ; heavy sand. (50)

sable pour noyautage (Fonderie) (*m.*), core-sand. (51)

sable pour petites pièces (Fonderie) (*m.*), sand for fine (*or* light) castings ; light sand. (52)

sable pour pièces moyennes (Fonderie) (*m.*), sand for average castings ; medium sand. (53)

sable poussé par le vent (Géol.) (*m.*), wind-blown sand ; blown sand ; wind-driven sand. (54)

sable quartzeux (*m.*), quartz sand ; quartzous sand. (55)

sable recuit (Fonderie) (*m.*), stoved sand ; baked sand. (56)

sable réfractaire (*m.*), fire-sand ; refractory sand. (57)

sable sec *ou* **sable séché** (Fonderie) (*m.*), dry sand ; dried sand. (58)

sable siliceux (*m.*), siliceous sand ; silicious sand. (59)

sable silicocalcaire (*m.*), silicocalcareous sand ; silicicalcareous sand. (60)

sable terrein (*m.*), pit-sand. (61)

sable vert (Géol.) (*m.*), greensand. (62)

sable vert (Fonderie) (*m.*), green sand. (63)

sable vieux (Fonderie) (*m.*), old sand ; black sand ; floor-sand ; heap-sand. (64)

sabler (*v.t.*), to sand. (65)

sablerie (atelier) (Fonderie) (*n.f.*), sand-shop. (66)

sablerie (installation) (Fonderie) (*n.f.*), sand-preparation plant. (67)

sableuse (appareil à jet de sable) (*n.f.*), sand-blast machine. (68)

sableux, -euse (*adj.*), sandy ; arenaceous ; arenose : (69)
terre sableuse (*f.*), sandy earth. (70)

sablière (carrière) (*n.f.*), sand-pit. (71)

sablière (de locomotive) (*n.f.*), sand-box ; sander. (72)

sablière à air comprimé (*f.*), pneumatic sander. (73)

sablière basse (sablière d'huisserie d'un pan de bois) (opp. à sablière haute) (Constr.) (*f.*), groundsill ; sill. (74)

sablière de comble (poutre sablière reposant sur le mur) (Constr.) (*f.*), wall-plate ; roof-plate. (75)

sablière de comble (panne sablière reposant sur les entraits ou tirants) (*f.*), pole-plate. (76)

sablière haute (sablière d'huisserie d'un pan de bois) (*f.*), head-plate ; plate. (77)

sablon (*n.m.*), fine sand. (78)

sablonneux, -euse (*adj.*), sandy ; arenaceous ; arenose. (79)

sablonnière (*n.f.*), sand-pit. (80)

sabot (*n.m.*), shoe. (81)

sabot (pour voitures, dans les descentes) (*n.m.*), shoe ; skid-shoe ; skid-pan ; drag. (82)

sabot (frotteur) (chemin de fer électrique) (*n.m.*), shoe. (1)

sabot (pour soutenir le bout d'une solive) (Constr.) (*n.m.*), hanger. (2)

sabot (d'un étau-limeur, d'une machine à raboter) (*n.m.*), tool-box (of a shaping-machine, of a planing-machine). (3 *ou* 4)

sabot à quatre branches avec culot en fer (sabot de pieu) (*m.*), strap-shoe. (5)

sabot à vis (pour pieu à vis) (*m.*), screw-shoe. (6)

sabot de frein (*m.*), brake-shoe ; brake-block ; brake-rubber ; rubber. (7)

sabot de pieu (*m.*), pile-shoe. (8)

sabot de pilon (bocard) (*m.*), stamp-shoe (ore-stamp). (9)

sabot tranchant *ou* **sabot coupant** (Sondage) (*m.*), drive-shoe. (10)

sabotage de traverses (Ch. de f.) (*m.*), chairing sleepers ; sabotage. (11)

saboter (*v.t.*), to shoe : (12)

saboter un pieu, to shoe a pile. (13)

saboter une traverse (Ch. de f.), to chair a sleeper. (14)

sabre (d'une distribution Corliss, ou analogue) (*n.m.*), lifting-rod (of a Corliss valve-gear, or the like). (15 *ou* 16)

sac (*n.m.*), sack ; bag ; case. (17)

sac (cavité naturelle) (Géol. & Mines) (*n.m.*), pocket : bag. (18)

sac à charbon (*m.*), coal-sack. (19)

sac à échantillons (*m.*), sample-bag. (20)

sac à minerai (*m.*), ore-bag. (21)

sac à noir (de mouleur) (*m.*), blacking-bag. (22)

sac à poussiers (de mouleur) (*m.*), dusting-bag. (23)

sac de grisou (*m.*), pocket of fire-damp ; fire-damp pocket. (24)

sac de minerai (sac rempli de minerai) (*m.*), bag of ore. (25)

sac de minerai (cavité naturelle remplie de minerai) (*m.*), pocket of ore ; ore-pocket. (26)

sac de sable (*m.*), sand-bag. (27)

sac tout cuir avec courroie (*m.*), solid leather sling case. (28)

saccharoïde (Géol. & Minéralogie) (*adj.*), sac-charoid ; saccharoidal ; saccharine : (29)

cassure saccharoïde (*f.*), saccharoidal fracture. (30)

safflorite (Minéral) (*n.f.*), safflorite. (31)

sagénite (Minéral) (*n.f.*), sagenite ; sagenitic quartz ; rutilated quartz ; Venus's hair-stone ; Veneris crinis ; flèches d'amour. (32)

sahlite (Minéral) (*n.f.*), sahlite ; salite. (33)

saignée (tranchée) (*n.f.*), cut ; trench. (34)

saignée (entaille) (*n.f.*), cut ; nick : (35)

une saignée pratiquée avec le ciseau à froid, a nick made with the cold-chisel. (36)

saigner un fossé, to drain a ditch. (37)

saillant, -e (*adj.*), salient ; projecting ; jutting out ; raised : (38)

angle saillant (*m.*), salient angle. (39)

rail saillant (*m.*), raised rail. (40)

saillie (*n.f.*), projection ; jutting out ; ledge ; offset ; set-off ; scarcement. (41)

saillie (tête d'une dent d'engrenage) (*n.f.*), point, addendum, (of a gear-tooth). (42)

saillie (hauteur au-dessus du primitif) (Engre-nages) (*n.f.*), addendum ; height above pitch-line. (43)

saillie (d'une vis) (*n.f.*), depth of thread (of a screw). (44)

saillie (d'une mèche) (opp. à la **rainure**) (*n.f.*), land (of a drill). (45) (in contradistinction to the groove).

saillir (*v.i.*), to project ; to jut out ; to stand out. (46)

sain, -e (salubre) (*adj.*), healthy : (47)

air sain (*m.*), healthy air. (48)

sain, -e (qui n'est point gâté) (*adj.*), sound : (49)

bois sain (*m.*), sound wood. (50)

une pièce saine *ou* un moulage sain, a sound casting. (51)

saisir au passage *ou* **saisir au vol** (Photogr.), to snap ; to snap-shot : (52)

saisir au passage un train qui passe, to snap a passing train. (53)

saisir une tige dans l'étau, to grip a rod in the vice. (54)

saison (*n.f.*), season. (55)

saison d'été *ou* **saison estivale** (*f.*), summer season. (56)

saison d'hiver *ou* **saison hivernale** (*f.*), winter season. (57)

saison des hautes crues (*f.*), high-water season. (58)

saison humide (*f.*), wet season. (59)

saison pluvieuse *ou* **saison pluviale** *ou* **saison des pluies** (*f.*), rainy season. (60)

saison printanière (*f.*), spring season. (61)

saison sèche (*f.*), dry season. (62)

salaire (*n.m.*), wages ; wage ; pay : (63)

salaire des ouvriers, workmen's wages ; men's wages ; men's pay. (64)

salaire à la journée (*m.*), day-wage ; day-wages. (65)

salaire à la tâche *ou* **salaire aux pièces** *ou* **salaire à façon** (*m.*), piece-wage. (66)

salaire au temps (*m.*), time-wage. (67)

salaire progressif (*m.*), progressive wage. (68)

salaire proportionné à la production (*m.*), effi-ciency wages. (69)

salbande (*n.f.*) *ou* **salbande argileuse** (Géol.), selvage ; clay-selvage ; clay-course ; gouge ; dig ; pug ; flucan ; saalband ; salband ; selvedge ; selfedge. (70)

sale (*adj.*), dirty : (71)

eau sale (*f.*), dirty water. (72)

saler une mine, to salt, to plant, a mine. (73)

salicole (*adj.*), salt-producing ; salt (*used as adj.*) : (74)

industrie salicole (*f.*), salt industry. (75)

saliculture (*n.f.*), salt industry (the working of a salt-lagoon or salina). (76)

salifère (*adj.*), saliferous ; salt-bearing ; salt-producing ; saliniferous : (77)

terrain salifère (*m.*), saliferous ground ; salt-bearing ground. (78)

saliférien, -enne (Géol.) (*adj.*), saliferous : (79)

système saliférien (*m.*), saliferous system. (80)

salifiable (Chim.) (*adj.*), salifiable : (81)

base salifiable (*f.*), salifiable base. (82)

salification (*n.f.*), salification. (83)

salifier (*v.t.*), to salify. (84)

salin, -e (*adj.*), saline : (85)

eau saline (*f.*), saline water. (86)

saline (marais salant) (*n.f.*), salina ; salt-marsh. (87)

saline (mine de sel gemme) (*n.f.*), rock-salt mine. (88)

salinelle (Géol.) (*n.f.*), mud-volcano (volcanic) ; mud-cone ; salse. (89)

salinité (*n.f.*), salinity ; saltness : (90)

salinité des eaux de mer, salinity, saltness, of sea-water. (1)

salir (*v.t.*), to dirty ; to foul. (2)

salite (Minéral) (*n.f.*), salite ; sahlite. (3)

salle des chaudières *ou* **salle de chaufferie** (*f.*), boiler-room. (4)

salle des machines (*f.*), engine-room. (5)

salmiac (*n.m.*), sal ammoniac. (6)

salpêtre (*n.m.*), saltpetre ; saltpeter ; nitre. (7)

salpêtre du Chili *ou* **salpêtre du Pérou** *ou* **salpêtre des mers du Sud** (*m.*), Chile saltpetre ; Chilian saltpetre ; Peruvian saltpeter. (8)

salpêtre terreux (*m.*), wall saltpetre. (9)

salpêtrerie (*n.f.*), saltpetre-works. (10)

salpêtreux, -euse (*adj.*), salpetrous. (11)

salpêtrière (*n.f.*), saltpetre-bed ; nitre-bed. (12)

salse (volcan de boue) (*n.f.*), salse ; mud-volcano (volcanic) ; mud-cone. (13)

salubre (*adj.*), healthy : (14)
climat salubre (*m.*), healthy climate. (15)

salubrité (*n.f.*), healthiness ; health. (16)

salure (*n.f.*), saltness ; salinity : (17)
salure des eaux marines, saltness, salinity, of sea-water. (18)

samarium (Chim.) (*n.m.*), samarium. (19)

samarskite (Minéral) (*n.f.*), samarskite. (20)

samsonite (Explosif) (*n.f.*), samsonite. (21)

sandbergérite (Minéral) (*n.f.*), sandbergerite. (22)

sanguine (Minéral) (*n.f.*), bloodstone ; hematite. (23)

sanidine (Minéral) (*n.f.*), sanidine ; glassy feldspar. (24)

sanitaire (*adj.*), sanitary. (25)

sans câble, ropeless : (26)
appareil d'extraction sans câble (*m.*), ropeless hoisting-apparatus. (27)

sans clef, keyless : (28)
mécanisme sans clef (*m.*), keyless mechanism. (29)

sans condensation, non-condensing : (30)
machine sans condensation (*f.*), non-condensing engine. (31)

sans contrastes (Photogr.), without contrasts ; flat : (32)
un excès de pose donne un cliché gris sans contrastes, overexposure gives a grey negative without contrasts (*or* a flat gray negative). (33)

sans couture, seamless : (34)
tube en acier sans couture (*m.*), seamless-steel tube. (35)

sans défaut, faultless ; flawless. (36)

sans engrenage, gearless ; non-geared : (37)
pompe sans engrenage (*f.*), gearless pump ; non-geared pump. (38)

sans étage (opp. à étagé, -e), stageless : (39)
compression sans étage (d'air) (*f.*), stageless compression. (40)

sans feldspaths, feldspar-free : (41)
roches sans feldspaths (*f.pl.*), feldspar-free rocks. (42)

sans ferme, untrussed : (43)
comble sans ferme (*m.*), untrussed roof. (44)

sans fil, wireless : (45)
télégraphie sans fil (*f.*), wireless telegraphy. (46)

sans fin, endless ; perpetual : (47)
vis sans fin (*f.*), endless screw ; perpetual screw ; worm. (48)

sans frais, free of cost. (49)

sans fumée, smokeless : (50)
charbon sans fumée (*m.*), smokeless coal. (51)

sans joint, jointless : (52)
tube sans joint (*m.*), jointless tube. (53)

sans marée, tideless : (54)
mers sans marées (*f.pl.*), tideless seas. (55)

sans odeur, without smell ; odourless. (56)

sans pluie, rainless. (57)

sans quartz, quartzless : (58)
roches sans quartz (*f.pl.*), quartzless rocks. (59)

sans recouvrement *ou* **sans recouvrements**, lapless ; without lap : (60)
tiroir sans recouvrement (*m.*), lapless valve ; slide-valve without lap. (61)

sans ressort *ou* **sans ressorts**, springless : (62)
voiture sans ressorts (*f.*), springless carriage. (63)

sans soudure, weldless : (64)
tube en acier sans soudure (*m.*), weldless-steel tube. (65)

sans système, without system ; unsystematically. (66)

sans tête, headless : (67)
vis sans tête (*f.*), headless screw. (68)

sans valeur, valueless ; worthless : (69)
gangue sans valeur (*f.*), worthless gangue. (70)

santé (*n.f.*), health. (71)

sape (*n.f.*), hoe (a miners' shovel with the blade at right angles to the handle). (72)

sapement (*n.m.*), sapping ; undermining. (73)

saper (*v.t.*), to sap ; to undermine : (74)
saper les fondations, to sap, to undermine, the foundations. (75)

saphir (Minéral) (*n.m.*), sapphire. (76)

saphir blanc (*m.*), white sapphire ; leuco-sapphire. (77)

saphir d'eau (*m.*), water-sapphire. (78)

saphir du Brésil (*m.*), Brazilian sapphire. (79)

saphir femelle (*m.*), female sapphire ; pale blue sapphire. (80)

saphir mâle (*m.*), male sapphire ; indigo-sapphire. (81)

saphir occidental (*m.*), blue hyaline-quartz. (82)

saphir oriental (*m.*), blue sapphire. (83)

saphir spath (*m.*), transparent blue disthene. (84)

saphirin, -e (*adj.*), sapphirine ; sapphire : (85)
quartz saphirin (*m.*), sapphirine quartz ; sapphire-quartz. (86)

saphirine (Minéral) (*n.f.*), sapphirine. (87)

sapin (arbre) (*n.m.*), fir. (88)

sapine (Constr.) (*n.f.*), masons' scaffold. (89)

sapinette (arbre) (*n.f.*), spruce. (90)

saponite (Minéral) (*n.f.*), saponite. (91)

sarcolite (Minéral) (*n.f.*), sarcolite. (92)

sarde (Minéral) (*n.m.*), sard. (93)

sardoine (Minéral) (*n.f.*), sardonyx. (94)

sartorite (Minéral) (*n.f.*), sartorite ; scleroclase. (95)

sas (tamis) (*n.m.*), sieve. (96)

sas (*n.m.*) *ou* **sas d'écluse**, chamber ; lock-chamber ; coffer. (97)

sas à air (d'un caisson) (*m.*), air-lock (of a caisson). (98)

sassage *ou* **sassement** (*n.m.*), jigging. (99)

sasser (*v.t.*), to jig. (100)

sasseur (*m.*), jig ; jigger ; jigging-machine. (101)

sasseur à sec (*n.m.*), dry jigger. (102)

sassoline (Minéral) (*n.f.*), sassoline. (103)

satisfaire à, to satisfy ; to fulfil ; to meet : (104)
satisfaire à une condition, to fulfil a condition. (105)

satisfaire l'attente, to come up to expectations. (1)

saturabilité (n.f.), saturability. (2)

saturable (adj.), saturable : (3)
  substance saturable (f.), saturable substance. (4)

saturant, -e (adj.), saturant; saturating. (5)

saturateur (n.m.), saturater; saturator. (6)

saturation (n.f.), saturation : (7)
  aimanter une barre d'acier à saturation, to magnetize a bar of steel to saturation. (8)

saturé, -e (adj.), saturated : (9)
  solution saturée (f.), saturated solution. (10)

saturer (v.t.), to saturate : (11)
  saturer un acide, to saturate an acid. (12)

saturer (se) (v.r.), to become saturated; to be saturated. (13)

saturnisme (n.m.), saturnism; saturnismus; lead-poisoning. (14)

saucier (d'un cabestan) (n.m.), pawl-bitt (of a capstan). (15)

saucisson (fausse cartouche) (n.m.), dummy cartridge. (16)

sauf, sauve (adj.), safe. (17)

sauf de rares exceptions, with very few excep- -tions. (18)

sauf variations, subject to alteration : (19)
  prix sauf variations (m.pl.), prices subject to alteration. (20)

saumâtre (adj.), brackish : (21)
  eau saumâtre (f.), brackish water. (22)

saumon (Métall.) (n.m.), pig; ingot. (23)

saumon d'étain (m.), ingot of tin. (24)

saumon de fonte (m.), pig of iron. (25)

saumon de plomb (m.), pig of lead. (26)

saumure (n.f.), brine; saline water. (27)

saunerie (n.f.), salt-works; salt-refinery. (28)

saunier (pers.) (n.m.), salt-worker; salter. (29)

saupoudrage (n.m.), sprinkling; dusting. (30)

saupoudrer (v.t.), to sprinkle; to dust : (31)
  saupoudrer de menu gravier une chaussée macadamisée, to sprinkle a macadamized road with fine gravel. (32)
  saupoudrer de noir la surface d'un moule, to dust the surface of a mould with blacking. (33)

saussurite (Pétrol.) (n.f.), saussurite. (34)

saut (de rivière) (n.m.), fall; falls; waterfall; water-jump : (35)
  le saut du Niagara, the falls of Niagara. (36)

sautage (Travail aux explosifs) (n.m.), blasting; blowing up : (37)
  sautage des roches, blasting the rock. (38)

sautage (effet) (n.m.), blast : (39)
  dans les travaux de Panama, on a effectué des sautages de 30 000 mètres cubes de roche, in the operations at Panama, blasts of 30,000 cubic metres of rock were effected. (40)

sauter (être détruit par une explosion) (v.i.), to blow up : (41)
  poudrière qui saute (f.), powder-works which blows up. (42)

sauter (voler; être projeté à distance) (v.i.), to fly; to fly off : (43)
  tête de rivet qui saute lorsqu'on la mate (f.), rivet-head which flies off when it is jagged. (44)

sauter un fossé, to jump, to jump over, to leap over, a ditch. (45)

sauterelle (fausse équerre) (n.f.), bevel; bevel-square. (46)

sauterelle graduée (f.), bevel-protractor. (47)

sauteuse (scie à découper alternative) (n.f.), jig-saw; scroll-saw; fret-saw machine. (48)

sauvegarder la vie, to safeguard life. (49)

sauver (v.t.), to save; to rescue : (50)
  sauver la vie, to save life. (51)
  sauver les ouvriers qui manquent, to rescue the missing men. (52)

sauver (Mines) (v.t.), to save : (53)
  les Sibériens s'aident rarement de mercure pour sauver l'or fin qui se trouve dans la batée (m.pl.), the Siberians seldom use mercury to save the fine gold which is present in the pan. (54)

sauvetage (n.m.), saving; rescue; rescuing; rescue-work. (55)

sauvetage du métal précieux (Mines) (m.), saving the precious metal. (56)

sauveteur (pers.) (n.m.), rescuer. (57)

savant (pers.) (n.m.), scientist; scholar; savant. (58)

savon (n.m.), soap. (59)

savon de montagne ou savon minéral ou savon blanc (Minéral) (m.), mountain-soap; rock-soap. (60)

savonnette (boussole savonnette) (n.f.), pocket magnetic compass in hunter case. (61)

savonneux, -euse (adj.), soapy : (62)
  le talc est savonneux au toucher, talc is soapy to the touch. (63)
  eau savonneuse (f.), soapy water. (64)

saxonite (Pétrol.) (n.f.), saxonite. (65)

saxonite (Explosif) (n.f.), saxonite. (66)

scandium (Chim.) (n.m.), scandium. (67)

scapolite (Minéral) (n.f.), scapolite; wernerite. (68)

sceau (n.m.), seal. (69)

sceau de plomb (m.), lead seal; leaden seal. (70)

scellage (fermeture) (n.m.), sealing. (71)

scellement (fixation dans un trou) (n.m.), sealing. (72)

sceller (fermer) (v.t.), to seal : (73)
  sceller l'extrémité d'un tube en la fondant au chalumeau, to seal the end of a tube by melting it in the blowpipe. (74)

sceller (fixer dans un trou à l'aide d'une sub- -stance qu'on y coule et qui s'y durcit) (v.t.), to seal : (75)
  sceller un boulon dans une pierre, une poutre, un corbeau, dans un mur, to seal a bolt in a stone, a beam, a corbel, in a wall. (76 ou 77 ou 78)

scène d'un accident (f.), scene of an accident. (79)

scheelite (n.f.) ou schélin calcaire (m.) (Minéral), scheelite. (80)

scheelitine (Minéral) (n.f.), stolzite. (81)

scheerérite (Minéral) (n.f.), scheererite; scherer- -ite. (82)

schefférite (Minéral) (n.f.), schefferite. (83)

scheidage (Mines) (n.m.), cobbing; bucking; spalling; hammer-sorting; hand-sorting; sorting. (84)

scheidage préalable ou scheidage préliminaire ou scheidage d'épuration (m.), ragging. (85)

scheider (v.t.), to cob; to buck; to spall; to sort. (86)

scheideur (pers.) (n.m.), cobber; bucker; sorter; ore-sorter. (87)

schéma (Dressin) (n.m.), diagrammatic represen- -tation; diagram; plan; arrangement; scheme : (88)

schéma du chevalement et des machines, dia-
-grammatic representation of the head-gear
and machinery. (1)

schéma de la disposition des roues (locomotive)
(*m.*), wheel-plan ; wheel-arrangement. (2)

schéma de lavage (Préparation mécanique des
minerais) (*m.*), flow-sheet. (3)

schématique (*adj.*), diagrammatic ; schematic ;
schematical : (4)
coupe schématique (*f.*), diagrammatic section ;
schematic section. (5)

schématiquement (*adv.*), diagrammatically ;
diagramically ; schematically. (6)

schème (Dessin) (*n.m.*). Same as **schéma**.

schillerspath (Minéral) (*n.m.*), schiller-spar ;
schillerfels ; bastite. (7)

schiste (*n.m.*), schist ; shale ; slate-clay. (8)

schiste à calymènes (*m.*), calymene schist. (9)

schiste à kérosène (*m.*), kerosene shale. (10)

schiste à séricite (*m.*), sericite schist. (11)

schiste actinolithique (*m.*), actinolite schist. (12)

schiste aluneux *ou* **schiste alumineux** (*m.*), alum
schist ; alum shale ; alum slate. (13)

schiste ampéliteux (*m.*), ampelitic schist. (14)

schiste ardoisier (*m.*), clay-slate ; argillaceous
slate. (15)

schiste argileux (*m.*), argillaceous schist. (16)

schiste bitumineux (*m.*), oil-shale. (17)

schiste carton (*m.*), paper shale. (18)

schiste chloriteux (*m.*), chlorite schist ; chlorite
slate. (19)

schiste cristallin (*m.*), crystalline schist. (20)

schiste cuivreux (*m.*), copper-slate. (21)

schiste gaufré (*m.*), goffered schist. (22)

schiste houiller (*m.*), carboniferous shale. (23)

schiste lustré (*m.*), lustrous shale. (24)

schiste maclifère (*m.*), macled shale. (25)

schiste métamorphique (*m.*), metamorphic
schist. (26)

schiste micacé (*m.*), micaceous schist ; mica-
schist ; mica-slate. (27)

schiste noir (*m.*), black shale. (28)

schiste sériciteux (*m.*), sericitic schist. (29)

schisteux, -euse (*adj.*), schistose ; schistous ;
schistic ; schist (*used as adj.*) ; shaly ;
shale (*used as adj.*) : (30)
roche schisteuse (*f.*), schistose rock ; schist-
rock ; shaly rock ; shale-rock. (31)

schistification (Mines) (*n.f.*), stone-dusting ;
rock-dusting : (32)
schistification des galeries, stone-dusting,
rock-dusting, the roads. (33)

schistifier (*v.t.*), to stone-dust ; to rock-dust. (34)

schistoïde (*adj.*), schistoid : (35)
cassure schistoïde (Minéralogie) (*f.*), schistoid
fracture. (36)

schistosité (*n.f.*), schistosity ; foliation : (37)
schistosité des roches, schistosity of rocks.
(38)

schlamms (Préparation mécanique du minerai)
(*n.m.pl.*), slimes ; sludge. (39)

schlich (Métall.) (*n.m.*), schlich ; slick. (40)

schorl (Minéral) (*n.m.*), schorl ; shorl ; tour-
-maline. (41)

schorl rouge (*m.*), red schorl ; rutile. (42)

schorl vert (*m.*), epidote. (43)

schorlacé, -e (*adj.*), schorlaceous ; schorlous ;
schorly. (44)

schorlifère (*adj.*), schorliferous. (45)

schorliforme (*adj.*), schorliform. (46)

schorlite (Minéral) (*n.f.*), pycnite. (47)

schreibersite (Minéral) (*n.f.*), schreibersite. (48)

schwartzembergite (Minéral) (*n.f.*), schwartzem-
-bergite. (49)

schwatzite (Minéral) (*n.f.*), schwatzite. (50)

sciage (*n.m.*), sawing : (51)
sciage du bois, *ou* marbre, sawing wood,
marble. (52 *ou* 53)

sciage (bois de sciage) (*n.m.*), saw-timber. (54)

sciage du diamant (*m.*), diamond cleaving ;
diamond splitting. (55)

scialet (Géogr. phys.) (*n.m.*), limestone cave ;
limestone cavern. (56)

scie (*n.f.*), saw. (57)

scie à bois (*f.*), wood-saw. (58)

scie à bûches (*f.*), buck-saw ; wood-saw ;
wood-cutters' saw. (59)

scie à chaînette (*f.*), chain-saw. (60)

scie à chantourner (à main) (*f.*), turning-saw ;
bow saw ; sweep saw. (61)

scie à chantourner (mécanique) (*f.*), jig-saw ;
scroll-saw ; fret-saw machine. (62)

scie à châssis (*f.*), frame-saw. (63)

scie à chaud (*f.*), hot-saw. (64)

scie à couper en travers (*f.*), cross-cut saw ;
cross-cutting saw. (65)

scie à découper (*f.*), fret-saw ; fret-saw blade.
(66)

scie à dents affûtées et mises en voie *ou* **scie à
dents affûtées et avoyées** (*f.*), saw with teeth
sharpened and set. (67)

scie à dos (*f.*), back-saw ; carcass-saw ; tenon-
saw. (68)

scie à froid (*f.*), cold-saw. (69)

scie à guichet (*f.*), keyhole-saw ; compass-saw ;
lock-saw. (70)

scie à lame sans fin (*f.*), band-saw ; ribbon-
saw ; belt-saw. (71)

scie à lingots (*f.*), ingot-saw. (72)

scie à main (*f.*), hand-saw ; arm-saw. (73)

scie à marbre *ou* **scie à grès pour marbre** (*f.*),
marble-saw. (74)

scie à métaux (*f.*), metal-saw ; hack-saw. (75)

scie à pierre (*f.*), stone-saw. (76)

scie à placage (*f.*), veneer-saw. (77)

scie à plusieurs lames (*f.*), gang-saw. (78)

scie à refendre (*f.*), rip-saw ; ripper ; cleaving-
saw ; split saw. (79)

scie à ruban (*f.*), band-saw ; ribbon-saw ;
belt-saw. (80)

scie à ruban pour le débit des bois en grume (*f.*),
log band-mill. (81)

scie à tenon (*f.*), tenon-saw. (82)

scie alternative (opp. à scie continue, comme
une scie circulaire ou une scie à ruban) (*f.*),
reciprocating saw ; alternating saw. (83)

scie alternative à découper dite sauteuse (*f.*),
jig-saw ; scroll-saw ; fret-saw machine. (84)

scie alternative à main (pour métaux) (*f.*), hand-
power hack-saw. (85)

scie alternative à tronçonner (*f.*), log cross-cut-
-ting machine ; reciprocating cross-cut saw ;
drag-saw. (86)

scie alternative au moteur (pour métaux) (*f.*),
power hack-saw. (87)

scie articulée (*f.*), chain-saw. (88)

scie circulaire (*f.*), circular saw ; annular saw ;
buzz-saw. (89)

scie circulaire à balancier (*f.*), pendulum cross-
cut saw. (90)

scie circulaire à bois (*f.*), circular saw for wood.
(91)

scie circulaire à métaux (*f.*), circular saw for
metal. (92)

scie de carrier (*f.*), quarryman's saw. (1)
scie de long (*f.*), long saw ; pit-saw. (2)
scie de tailleur de pierres (*f.*), stonecutters' saw. (3)
scie de travers (*f.*), cross-cut saw ; cross-cutting saw. (4)
scie égoïne (*f.*), small hand-saw. (5)
scie hélicoïdale (Abatage en carrières) (*f.*), wire-saw. (6)
scie mécanique (*f.*), power-saw ; sawing-machine ; sawmill. (7)
scie ordinaire (*f.*), frame-saw ; span-saw. (8)
scie passe-partout (pour scier les gros arbres) (*f.*), lumberman's two-handed saw ; felling-saw. (9)
scie passe-partout (à pierres) (avec ou sans dents) (*f.*), stone-saw. (10 *ou* 11)
scie sans fin (*f.*), endless saw. (12)
scie verticale alternative à plusieurs lames pour débiter les madriers et autres pièces équarries (*f.*), deal and flitch frame. (13)
scie verticale alternative à plusieurs lames pour le débit des bois en grume et des fortes pièces équarries (*f.*), log-frame. (14)
scié, -e (*adj.*), sawn. (15)
science (*n.f.*), science. (16)
science appliquée (*f.*), applied science. (17)
science de l'ingénieur (*f.*), science of engineer--ing ; engineering. (18)
sciences mathématiques (*f.pl.*), mathematical sciences. (19)
sciences naturelles (*f.pl.*), natural sciences. (20)
sciences physiques (*f.pl.*), physical sciences. (21)
scientifique (*adj.*), scientific : (22)
    recherches scientifiques (*f.pl.*), scientific research. (23)
scientifiquement (*adv.*), scientifically. (24)
scier (*v.t.*), to saw : (25)
    scier du bois, du métal, to saw wood, metal. (26 *ou* 27)
scier en travers, to cross-cut : (28)
    scier du bois en travers, to cross-cut wood. (29)
scier (se) (*v.r.*), to saw : (30)
    le bois se scie aisément, the wood saws easily. (31)
scierie (usine) (*n.f.*), sawmill. (32)
scierie (machine) (*n.f.*), sawing-machine ; saw--mill ; saw-bench. (33)
scierie à ruban (*f.*), band-sawing machine. (34)
scierie à ruban à grumes (*f.*), band-sawing machine for logs. (35)
scierie à vapeur (*f.*), steam-sawmill. (36)
scierie circulaire à axe fixe (*f.*), fixed-spindle circular-saw bench. (37)
scierie circulaire à table mobile (*f.*), circular-saw bench, with rising and falling table. (38)
scierie circulaire à table pivotante, marchant au pied ou au moteur (*f.*), circular-saw bench, with tilting table, for foot or power. (39)
scierie mécanique (*f.*), power-saw ; sawing-machine ; saw-mill. (40)
scieur (pers.) (*n.m.*), sawyer. (41)
scieur de bois (*m.*), wood-sawyer. (42)
scieur de long (*m.*), pit-sawyer ; pitman. (43)
scieur de pierres (*m.*), stone-sawyer. (44)
scintillant, -e (*adj.*), scintillating ; sparkling. (45)
scintillation (*n.f.*) *ou* scintillement (*n.m.*), scintil--lation ; sparkling : (46)
    scintillement d'une pierre précieuse, sparkling of a gem. (47)

scintiller (*v.i.*), to scintillate ; to sparkle : (48)
    les diamants scintillent (*m.pl.*), diamonds sparkle. (49)
sciotte (*n.f.*), gauge-saw (for marble). (50)
sciure (*n.f.*) *ou* sciure de bois, sawdust. (51)
sciure de marbre (*f.*), marble-dust. (52)
scléroclase (Minéral) (*n.f.*), scleroclase ; sartorite. (53)
scléromètre (*n.m.*), sclerometer ; scleroscope. (54)
scloneur (Mines) (pers.) (*n m.*), trammer ; drawer ; wheeler ; putter ; haulageman. (55)
scolécite *ou* scolésite *ou* scolézite (Minéral) (*n.f.*), scolecite. (56)
scoriacé, -e (*adj.*), scoriaceous ; scorious ; scoriated ; slaggy. (57)
scorie (*n.f.*) *ou* scories (*n.f.pl.*) (Géol.), scoria ; scoriæ ; slag ; cinder. (58)
scorie (*n.f.*) *ou* scories (*n.f.pl.*) (Métall.), slag ; cinder ; scoria ; scoriæ. (59) Note.—Particularly, **scories** are silicates retaining metallic oxides, as distinguished from **laitiers**, which are silicates containing earthy bases.
scorie (mâchefer) (*n.f.*), clinker ; furnace-clinker. (60)
scories (crasses de fonte dans les poches de coulée ou dans les moules) (Fonderie) (*n.f.pl.*), sullage. (61)
scories basiques *ou* scories de déphosphoration (*f.pl.*), basic slag ; phosphatic slag ; Thomas's slag. (62)
scories bibasiques (*f.pl.*), dibasic slag ; bibasic slag. (63)
scories de convertisseur (*f.pl.*), converter-slags. (64)
scories de forge (*f.pl.*), forge-scale ; iron-scale ; hammer-scale ; scale ; hammer-slag ; cinder ; anvil-dross ; nill. (65)
scories de haut fourneau (*f.pl.*), blast-furnace slag. (66)
scories de laminoir (*f.pl.*), mill-scale ; roll-scale. (67)
scories enfermées (*f.pl.*) *ou* scorie renfermée (*f.*), slag inclusions. (68)
scories volcaniques (Géol.) (*f.pl.*), volcanic scoriæ ; volcanic scoria ; volcanic slag ; volcanic cinders. (69)
scorification (Métall.) (*n.f.*), scorification. (70)
scorificatoire (Métall.) (*n.m.*), scorifier. (71)
scorifier (*v.t.*), to scorify ; to slag : (72)
    scorifier les matières étrangères contenues dans un métal, to scorify, to slag, the foreign matter contained in a metal. (73)
scorifier (se), to scorify ; to slag. (74)
scoriforme (*adj.*), scoriform. (75)
scorodite (Minéral) (*n.f.*), scorodite. (76)
scoulérite (Minéral) (*n.f.*), pipe-clay. (77)
scraps de fonderie (*m.pl.*), foundry-scrap ; cast scrap. (78)
sculptage (*n.m.*), sculpturing ; carving. (79)
sculpter (*v.t.*), to sculpture ; to carve : (80)
    sculpter le marbre, to sculpture marble. (81)
    des roches profondément sculptées par les vents (*f.pl.*), rocks deeply sculptured by the winds. (82)
    sculpter le bois, to carve wood. (83)
sculpture (*n.f.*), sculpture ; carving : (84)
    sculpture éolienne (Géogr. phys.), wind-carving ; eolian sculpture. (85)
seau (*n.m.*), bucket ; pail. (86)
seau (contenu d'un seau) (*n.m.*), bucket ; bucket--ful ; pail ; pailful : (87)

un seau d'eau, a bucket (or a pailful) of water. (1)

seau d'incendie (*m.*), fire-bucket. (2)

seau de puits (*m.*), well-bucket. (3)

seau en toile (*m.*), canvas bucket. (4)

seau en tôle galvanisée (*m.*), galvanized-iron pail. (5)

sébile (Mines) (*n.f.*), pan ; dish ; batea. (6)

sec, sèche (*adj.*), dry : (7)

sable sec (*m.*), dry sand. (8)

pile sèche (*f.*), dry battery. (9)

sec (à), dry (in the sense of *when dry* or *while in a dry state*) : (10)

minerai qui est généralement broyé à sec (*m.*), ore which is generally crushed dry. (11)

sécable (*adj.*), sectile. (12)

sécant, -e (Géom.) (*adj.*), secant : (13)

ligne sécante (*f.*), secant line. (14)

sécante (*n.f.*), secant. (15)

séchage (*n.m.*), drying : (16)

séchage à l'étuve des noyaux, des moules, dry-ing cores, moulds, in the stove. (17 ou 18)

séché (-e) à l'air, air-dried. (19)

séché (-e) au four, kiln-dried. (20)

sécher (*v.t.*), to dry ; to dry up. (21)

sécher (*v.i.*), to dry ; to dry up : (22)

la rivière a séché, the river has dried up. (23)

sécher au four, to kiln-dry. (24)

sécher les bois, to season timber. (25)

sécher (se) (*v.r.*), to dry ; to dry up. (26)

sécheresse (*n.f.*), dryness : (27)

la sécheresse de l'air, the dryness of the air. (28)

sécheresse (disposition du temps) (*n.f.*), drought ; dryness : (29)

la sécheresse tarit les puits, the drought dries up the wells. (30)

sécheur, -euse (*adj.*), drying : (31)

appareil sécheur (*m.*), drying apparatus. (32)

séchoir (*n.m.*), dryer ; drier ; drying-house ; drying-chamber ; drying-room ; drying-stove. (33)

séchoir à air (*m.*), air-dryer ; air-drier. (34)

séchoir à air chaud (*m.*), hot-air dryer. (35)

séchoir à minerai (*m.*), ore-dryer. (36)

second, -e (*adj.*), second : (37)

second arc (arc-en-ciel) (*m.*), secondary bow ; secondary rainbow. (38)

second poste au charbon (Mines) (*m.*), back-shift. (39)

second taraud (*m.*), second tap ; plug tap. (40)

secondaire (*adj.*), secondary. (41)

secondaire (Géol.) (*adj.*), Secondary ; Mesozoic. (42)

secondaire (Géol.) (*n.m.*), Secondary ; Mesozoic : (43)

le secondaire, the Secondary ; the Mesozoic. (44)

secondaire (Élec.) (*n.m.*), secondary. (45)

seconde (unité de temps ou de mesure angulaire) (*n.f.*), second. (46)

secouer (*v.t.*), to shake. (47)

secourir (porter secours à) (*v.t.*), to rescue. (48)

secours (*n.m.*), aid ; help ; rescue : (49)

premiers secours, first aid. (50)

secours (de), emergency (*used as adj.*); relief (*used as adj.*) ; standby (*used as adj.*) ; standing by ; breakdown (*used as adj.*) : (51)

sortie de secours (*f.*), emergency exit. (52)

groupe de secours (Élec.) (*m.*), emergency set ; standby set ; breakdown set. (53)

locomotive de secours (*f.*), relief engine ; engine standing by. (54)

secousse (*n.f.*) *ou* **secousses** (*n.f.pl.*), shock ; jar ; jarring ; jerk ; shake ; shaking. (55)

secousse de tremblement de terre *ou* **secousse sismique** (*f.*), earthquake-shock ; seismic shock. (56)

secousse électrique (*f.*), electric shock. (57)

sécréter (*v.t.*), to secrete. (58)

sécrétion (*n.f.*), secretion : (59)

gîte de sécrétion (Géol.) (*m.*), secretion-vein. (60)

secteur (Géom.) (*n.m.*), sector. (61)

secteur denté *ou* **secteur crénelé** (*m.*), toothed sector ; toothed segment ; sector-gear ; sector-wheel ; segment-gear ; quadrant. (62)

secteur denté de l'appareil de mise en marche *ou* **secteur du changement de marche** *ou simple--ment* **secteur denté** *ou* **secteur crénelé** (machine à vapeur) (*m.*), quadrant. (63)

secteur sphérique (Géom.) (*m.*), spherical sector. (64)

sectile (*adj.*), sectile : (65)

minéral sectile (*m.*), sectile mineral. (66)

section (*n.f.*), section. (67)

section (Géom.) (*n.f.*), section : (68)

la section d'un cône, the section of a cone. (69)

section (d'un bâtiment) (*n.f.*), section, vertical section, (of a building). (70)

section (d'un rail, d'une poutre) (*n.f.*), section, cross-section, (of a rail, of a girder). (71 ou 72)

section (d'un cours d'eau, d'un canal, d'un canal découvert, d'une conduite) (*n.f.*), cross-section (of a stream, of a canal, of an open channel, of a pipe). (73 ou 74 ou 75 ou 76)

section carrée (*f.*), square section. (77)

section conique (*f.*), conic section. (78)

section contractée (Hydraul.) (*f.*), contracted section ; vena contracta. (79)

section d'égale résistance (Résistance des maté--riaux) (*f.*), section of uniform strength. (80)

section d'encastrement (d'une poutre, d'une colonne, ou analogue) (*f.*), embedment (of a beam, of a column, or the like). (81 ou 82 ou 83)

section dangereuse (Résistance des matériaux) (*f.*), section of maximum intensity of stress (Strength of materials). (84)

section en U (*f.*), U section. (85)

section plane (*f.*), plane section. (86)

section transversale (*f.*), cross-section ; trans--verse section : (87)

section transversale d'une pièce prismatique, cross-section of a prismatic bar. (88)

sectionner (*v.t.*), to sectionalize : (89)

sectionner un puits de mine, to sectionalize a mine-shaft. (90)

machine sectionnée pour le transport à dos de mulets (*f.*), machine sectionalized for mule-back transportation. (91)

séculaire (*adj.*), secular : (92)

variations séculaires de climat (*f.pl.*), secular variations of climate. (93)

sécurité (*n.f.*), security ; safety ; reliability : (94)

compromettre la sécurité des ouvriers, to endanger the safety of the workmen. (95)

la sécurité d'une soupape, the reliability of a valve. (96)

dispositif de sécurité (*m.*), safety-device. (97)

sécurité (Explosif) (*n.f.*), securite. (98)

sédiment (*n.m.*), sediment ; deposit ; settlings. (1)

sédiment (Géol.) (*n.m.*), sediment. (2)

sédimentaire (*adj.*), sedimentary : (3)
roches sédimentaires (*f.pl.*), sedimentary rocks. (4)

sédimentation (*n.f.*), sedimentation. (5)

sédimenteux, -euse (*adj.*), sedimental : (6)
dépôts sédimenteux (*m.pl.*), sedimental deposits. (7)

segment (*n.m.*), segment. (8)

segment de cercle (*m.*), segment of a circle. (9)

segment de cuvelage (pour puits de mines) (*m.*), segment of tubbing. (10)

segment de piston (*m.*), piston-ring ; piston packing-ring. (11)

segment sphérique (Géom.) (*m.*), spherical segment ; spherical cap. (12)

segmentaire (*adj.*), segmental ; segmentary. (13)

segmentation (*n.f.*), segmentation. (14)

segmenter (*v.t.*), to segment ; to cut into segments ; to divide into segments. (15)

segmenter (se) (*v.r.*), to segment ; to divide into segments. (16)

ségrégation (*n.f.*), segregation. (17)

seiche (Géogr. phys.) (*n.f.*), seiche. (18)

seille (*n.f.*), bucket. (19)

seille en bois (*f.*), wooden bucket. (20)

seille en toile (*f.*), canvas bucket. (21)

seilleau *ou* seillot (*n.m.*). Same as seille.

sein de la terre (*m.*), bosom of the earth. (22)

séisme (*n.m.*), séismique (*adj.*), séismologie (*n.f.*), etc. Same as sisme, sismique, sismologie, etc.

sel (*n.m.*), salt. (23)

sel acide (*m.*), acid salt. (24)

sel admirable de Glauber *ou* sel de Glauber (*m.*), Glauber's salt ; sal mirabile ; mirabilite. (25)

sel alcalin (*m.*), alkaline salt. (26)

sel ammoniac (*m.*), sal ammoniac ; chloride of ammonium. (27)

sel commun (*m.*), common salt. (28)

sel de radium (*m.*), radium salt. (29)

sel du commerce (*m.*), commercial salt. (30)

sel gemme (*m.*), rock salt ; fossil salt ; sal gemmæ ; halite. (31)

sel marin (*m.*), sea-salt. (32)

sel microcosmique *ou* sel de phosphore (*m.*), microcosmic salt ; salt of phosphorous ; sal microcosmicum. (33)

sel neutre (*m.*), neutral salt ; normal salt. (34)

sel potassique (*m.*), potash salt ; potassium salt. (35)

sel sodique (*m.*), soda salt ; sodium salt. (36)

séléniate (Chim.) (*n.m.*), selenate ; seleniate. (37)

sélénieux, -euse (Chim.) (*adj.*), selenious. (38)

sélénifère (*adj.*), seleniferous. (39)

sélénique (Chim.) (*adj.*), selenic. (40)

sélénite (Minéral) (*n.f.*), selenite. (41)

séléniteux, -euse (*adj.*), selenitic. (42)

sélénium (Chim.) (*n.m.*), selenium. (43)

séléniure (Chim.) (*n.m.*), selenide : (44)
séléniure d'argent, selenide of silver ; silver selenide. (45)

self-inductance (Élec.) (*n.f.*), self-inductance ; coefficient of self-induction. (46)

self-induction (Élec.) (*n.f.*), self-induction. (47)

sellaïte (Minéral) (*n.f.*), sellaite. (48)

selle (Sellerie) (*n.f.*), saddle (Saddlery). (49)

selle (pli anticlinal ; voûte) (opp. à auge *ou* fond de bateau) (Géol.) (*n.f.*), arch ; saddle. (50) (opp. to trough).

selle d'arrêt (semelle) (Ch. de f.) (*f.*), sole-plate ; tie-plate ; packing-piece. (51)

selle de joint (Ch. de f.) (*f.*), joint-chair. (52)

seller (se) (*v.r.*), to settle ; to sink : (53)
terre qui est sujette à se seller (*f.*), ground which is apt to sink. (54)

semaine (*n.f.*), week. (55)

sémaphore (*n.m.*), semaphore. (56)

sémaphorique (*adj.*), semaphoric ; semaphorical : (57)
signaux sémaphoriques (*m.pl.*), semaphoric signals. (58)

semblable (*adj.*), similar ; like : (59)
pôles semblables (*m.pl.*), like poles. (60)

semblablement (*adv.*), similarly ; likewise. (61)

séméline (Minéral) (*n.f.*), semelin ; semeline. (62)

semelle (Boisage de mines) (*n.f.*), sill ; sill-piece ; sole ; sole-piece ; sole-plate ; foot-piece ; groundsill. (63)

semelle (pièce de bois qui se place sous le pied d'un étai pour former empattement) (Constr.) (*n.f.*), sole-piece ; footing-block. (64)

semelle (sablière de comble) (Constr.) (*n.f.*), wall-plate ; roof-plate ; plate. (65)

semelle (plaque interposée entre la traverse et le patin du rail Vignole, ou entre la traverse et le coussinet) (*n.f.*), sole-plate ; tie-plate ; packing-piece. (66)

semelle (d'un rabot) (*n.f.*), sole (of a plane). (67)

semelle (d'un sabot de frein) (*n.f.*), sole (of a brake-shoe). (68)

semelle (d'un palier) (*n.f.*), sole-plate, sole, sill-plate, base, (of a plummer-block). (69)

semelle (d'une mine) (*n.f.*), sole, seat, (of a mine). (70)

semelle (d'une poutre) (*n.f.*), tread, table, (of a girder). (71)

semelle (de la crosse du piston) (*n.f.*), shoe, slipper, gib, (of piston cross-head). (72)

semelle (d'un bocard) (*n.f.*), die (of an ore-stamp). (73)

semelle (d'une cuve d'amalgamation) (*n.f.*), die (of an amalgamating-pan). (74)

semelle (d'un électro-aimant) (*n.f.*), spool (of an electromagnet). (75)

semelle (d'une contre-pointe se déplaçant sur sa semelle pour tourner conique) (tour) (*n.f.*), cross-slide (of a tail-stock which can be set over for taper turning) (lathe). (76)

semelle (pièce d'acier préparée pour faire une lime) (*n.f.*), file-blank. (77)

semelle de support à main (tour) (*f.*), hand-rest socket (lathe). (78)

semelles (d'une presse) (*n.f.pl.*), plates (of a press). (79)

semence de diamants (*f.*), diamond sparks. (80)

semence de perles (*f.*) *ou* semences de perles (*f.pl.*), seed-pearls. (81)

semestre (*n.m.*), half-year ; six months. (82)

semestriel, -elle (*adj.*), half-yearly ; six-monthly. (83)

semi-automatique (*adj.*), semiautomatic. (84)

semi-circulaire (*adj.*), semicircular. (85)

semi-conducteur (Élec.) (*n.m.*), semiconductor. (86)

semi-cristallin, -e (*adj.*), semicrystalline. (87)

semi-électrique (*adj.*), semielectric : (88)

pont roulant semi-électrique (*m.*), semielectric travelling crane. (1)

semi-fixe (*adj.*), semiportable : (2)
chaudière semi-fixe (*f.*), semiportable boiler. (3)

semi-fluide (*adj.*), semifluid ; semifluidic. (4)

semi-fluide (*n.m.*), semifluid. (5)

semi-hebdomadaire (*adj.*), half-weekly. (6)

semi-opale (Minéral) (*n.f.*), semiopal. (7)

semi-transparence (*n.f.*), semitransparency. (8)

semi-transparent, -e (*adj.*), semitransparent. (9)

semi-tubulaire (*adj.*), semitubular : (10)
chaudière semi-tubulaire (*f.*), semitubular boiler. (11)

sénarmontite (Minéral) (*n.f.*), senarmontite. (12)

senestrorsum (*adj. invar. & adv.*). Same as **sinistrorsum**.

sénile (Géogr. phys.) (*adj.*), senile : (13)
la pénéplaine est le type des formes séniles, the peneplain is the type of senile forms. (14)

sénilité (Géogr. phys.) (*n.f.*), senility. (15)

sens (direction) (*n.m.*), direction ; sense : (16)
inverser le sens d'un courant électrique, to reverse the direction of an electric current. (17)
arbre qui tourne dans le sens de la flèche (*m.*), shaft which turns in the direction of the arrow. (18)
sens d'un fleuve *ou* sens d'une rivière, direction of flow, drift, of a river : (19)
sur les cartes topographiques, le sens d'un fleuve se note par une flèche dessinée entre les rives, on topographical maps, the direction of flow of a river is shewn by an arrow drawn between the banks. (20)

sens (Géom.) (*n.m.*), sense : (21)
le sens de la force se distingue par les signes + et — , the sense of force is distinguished by the signs + and — . (22)

sens (opinion) (*n.m.*), opinion : (23)
à mon sens, in my opinion. (24)

sensibilisation (d'une plaque photographique) (*n.f.*), sensitization (of a photographic plate). (25)

sensibiliser (Photogr.) (*v.t.*), to sensitize. (26)

sensibilité (*n.f.*), sensitiveness ; sensibility ; delicacy : (27)
sensibilité d'une balance, sensitiveness, sensibility, delicacy, of a balance (*or* scale). (28)
sensibilité de la poussière de charbon, sensitive-ness of coal-dust. (29)

sensible (en parlant d'un instrument de précision) (*adj.*), sensitive ; sensible ; delicate : (30)
balance sensible (*f.*), sensitive balance ; sensible balance ; delicate scales. (31)
balance sensible au milligramme, balance sensible to 1 milligramme ; balance turning to 1 milligramme. (32)

sensible (Photogr.) (*adj.*), sensitive ; sensitized : (33)
papier sensible (*m.*), sensitive paper ; sensitized paper. (34)

sensitif, -ive (*adj.*), sensitive : (35)
machine à percer sensitive (*f.*), sensitive drill. (36)

sentier (*n.m.*), path ; footpath ; footway ; track ; trail. (37)

sentier muletier (*m.*), mule-track ; horse-trail ; pack-trail ; bridle-path ; bridle-road ; bridle-track ; bridleway. (38)

sentir (exhaler une odeur de) (*v.t.*), to smell of : (39)

sentir fortement le pétrole, to smell strongly of oil. (40)

séparabilité (*n.f.*), separability. (41)

séparable (*adj.*), separable. (42)

séparateur (*n.m.*), separator. (43)

séparateur à air (*m.*), air-separator. (44)

séparateur à huile (*m.*), oil-separator ; steam-separator. (45)

séparateur à minerai (*m.*), ore-separator. (46)

séparateur d'eau *ou* **séparateur d'eau et de vapeur** (*m.*), water-separator ; steam-separator. (47)

séparateur magnétique *ou* **séparateur électro-magnétique** (pour minerais) (*m.*), magnetic separator ; electromagnetic separator. (48)

séparation (action) (*n.f.*), separation ; separating : (49)
séparation du métal d'avec sa gangue, separa-tion of, separating, metal from its gangue. (50)

séparation (cloison) (*n.f.*), partition : (51)
une séparation de planches, a wood partition. (52)

séparation à sec par le vent (*f.*), dry-vanning process. (53)

séparation magnétique (*f.*), magnetic separation. (54)

séparation mécanique (*f.*), mechanical separation. (55)

séparation par densité (*f.*), separation by specific gravity. (56)

séparation par la voie humide (*f.*), wet separation. (57)

séparation par la voie sèche (*f.*), dry separation. (58)

séparé, -e (*adj.*), separate. (59)

séparément (*adv.*), separately. (60)

séparer (*v.t.*), to separate : (61)
séparer le métal d'avec sa gangue, l'or de ses combinaisons, to separate metal from its gangue, gold from its combinations. (62 *ou* 63)

séparer (se) (*v.r.*), to separate. (64)

sépiolite (Minéral) (*n.f.*), sepiolite ; meerschaum. (65)

septentrional, -e, -aux (*adj.*), north ; northern : (66)
la Sibérie septentrionale, Northern Siberia. (67)

septivalence (Chim.) (*n.f.*), septivalence ; septivalency. (68)

septivalent, -e (*adj.*), septivalent. (69)

sérac (Glaciologie) (*n.m.*), sérac. (70)

sergent (serre-joint) (*n.m.*), joiners' cramp ; sash-cramp ; sash-clamp. (71)

séricite (Minéral) (*n.f.*), sericite. (72)

sériciteux, -euse (*adj.*), sericitic : (73)
schiste sériciteux (*m.*), sericitic schist. (74)

série (*n.f.*), series : (75)
une série de profils à vitrages, a series of sash-bar sections. (76)

série (Élec.) (*n.f.*), series : (77)
enroulement en série (*m.*), series-winding. (78)

série carboniférienne (Géol.) (*f.*), coal-measures. (79)

série de couleurs, de dimensions (*f.*), range of colours, of sizes. (80 *ou* 81)

série de poids (*f.*), set of weights ; weights : (82)
balance avec série de poids (*f.*), scale with set of weights (*or* with weights). (83)

série de roues de filetage (tour) (*f.*), set of change-wheels (lathe). (84)

série de roues pour les variations de vitesse des avances (machine à engrenages quelconque) (*f.*), set of change-wheels. (1)

série éocène (Géol.) (*f.*), Eocene series. (2)

série hétérologue (Chim.) (*f.*), heterologous series. (3)

série homologue (Chim.) (*f.*), homologous series. (4)

sérieux, -euse (grave ; qui peut avoir des con-séquences fâcheuses) (*adj.*), serious : (5)
accident sérieux (*m.*), serious accident. (6)

sérieux, -euse (réel ; sincère) (*adj.*), bona-fide ; businesslike ; serious ; regular : (7)
maison qui peut fournir de sérieuses références (*f.*), firm which can give bona-fide references. (8)
proposition sérieuse (*f.*), businesslike proposal ; serious proposition. (9)
exploitation minière sérieuse (*f.*), regular mining ; serious mining. (10)

seringage *ou* **seringuement** (*n.m.*), syringing. (11)

seringue (*n.f.*), syringe. (12)

seringuer (*v.t.*), to syringe. (13)

serpenter (*v.i.*), to wind ; to meander : (14)
rivière qui serpente entre des collines boisées (*f.*), river which winds between wooded hills. (15)

serpentin (tuyau ou tige contourné en spirale) (*n.m.*), coil ; serpent coil ; worm : (16)
serpentin à grille, grid-coil. (17)

serpentin (d'un alambic) (*n.m.*), worm (of a still). (18)

serpentine (Minéral ou roche) (*n.f.*), serpentine : (19)
serpentine noble, precious serpentine ; noble serpentine. (20)

serpentineux, -euse (*adj.*), serpentinous ; serpentinic. (21)

serpentinisation (Géol.) (*n.f.*), serpentinization. (22)

serpentiniser (Géol.) (*v.t.*), to serpentinize. (23)

serrage (action de serrer) (*n.m.*), tightening : (24)
serrage d'un nœud, tightening a knot. (25)

serrage (d'un écrou) (*n.m.*), screwing up, tighten-ing, (a nut). (26)

serrage (d'une vis) (*n.m.*), driving, driving in, tightening, (a screw). (27)

serrage (d'un mandrin) (*n.m.*), grip, holding, (of a chuck) : (28)
vis auxiliare donnant un serrage énergique (*f.*), auxiliary screw giving a strong grip. (29)
diamètre de serrage (*m.*), holding diameter. (30)

serrage (du sable d'un moule) (*n.m.*), packing (the sand of a mould) : (31)
serrage du sable au fouloir, packing sand with the rammer. (32) See also **serrage du sable à bras** and **serrage du sable à pied.**

serrage (différence de diamètres à ménager à froid entre les parties à assembler) (Emman-chement à chaud) (*n.m.*), shrinkage allow-ance ; allowance for shrinkage ; shrinkage (Shrinking on) : (33)
un serrage de — dixième(s) de millimètre, a shrinkage allowance of — -tenth(s) of a millimetre. (34)

serrage (profondeur de passe) (Travail aux machines-outils) (*n.m.*), depth of cut. (35)

serrage (barrage) (Mines) (*n.m.*), dam (in a mine-level). (36)

serrage (Freinage) (Ch. de f.), brakes on. (37)

serrage à fond (Ch. de f.), brakes full on. (38)

serrage de l'échappement (*m.*), throttling the exhaust. (39)

serrage du frein *ou* simplement **serrage** (*n.m.*), braking ; putting on the brake ; applying the brake. (40)

serrage du sable à bras (Fonderie) (*m.*), tucking sand by hand. (41)

serrage du sable à pied (Fonderie) (*m.*), treading sand ; tramping sand. (42)

serrage progressif (*m.*), progressive braking. **(43)**

serre (petites pinces en fer ou acier) (*n.f.*), clip. (44)

serre (pour oreilles de châssis, etc.) (Fonderie) (*n.f.*), dog (for box-lugs, etc.). (45)

serre-câble [serre-câble *ou* serre-câbles *pl.*] (pour câble métallique) (*n.m.*), cappel ; capel ; wire-rope clamp. (46)

serre-câble (apparcillage électrique) (*n.m.*), cleat : (47)
serre-câble en porcelaine, à trois rainures, avec deux trous pour vis de fixation, three-groove porcelain cleat, with two holes for fixing-screws. (48)

serre-écrou [serre-écrou *pl.*] (*n.m.*), bicycle spanner ; bicycle wrench. (49)

serre-fils [serre-fils *pl.*] (appareillage électrique) (*n.m.*). Same as **serre-câble.**

serre-frein *ou* **serre-freins** [serre-frein *ou* serre-freins *pl.*] (pers.) (*n.m.*), brakeman ; brakes-man ; braker. (50)

serre-joint *ou* **serre-joints** [serre-joint *ou* serre-joints *pl.*] (pour menuisiers) (*n.m.*), joiners' cramp ; sash-cramp ; sash-clamp. (51)

serre-joint *ou* **serre-joints** (happe) (*n.m.*), cramp ; clamp ; G cramp ; G clamp ; C cramp. (52)

serre-rail *ou* **serre-rails** [serre-rail *ou* serre-rails *pl.*] (*n.m.*), rail-clip ; clip ; sleeper-clip. (53)

serre-tube *ou* **serre-tubes** [serre-tube *ou* serre-tubes *pl.*] (*n.m.*), tube-wrench ; pipe-wrench ; cylinder-wrench. (54)

serré, -e (dont les parties constituantes sont très rapprochées) (*adj.*), close ; tight : **(55)**
grain serré (*m.*), close grain. (56)
gravier serré (*m.*), tight gravel. (57)

serré, -e (étroit) (*adj.*), close ; narrow : **(58)**
un defilé long et serré, a long and narrow defile. (59)

serrée (*n.f.*) *ou* **serrement** (*n.m.*) (Géol. & Mines), pinch ; balk ; nip. (60)

serrement (action de serrer) (*n.m.*), tightening. (61)

serrement (barrage) (Mines) (*n.m.*), dam (in a mine-level) : (62)
serrement voûté en maçonnerie, masonry dam. (63)

serrer (*v.t.*). See examples : (64)
serrer — (se dit d'une happe, d'un mandrin, ou analogue), to take in — ; to hold — : (65)
'happe pouvant serrer — ᵐ/ₘ (*f.*), G cramp to take in — mm. (66)
mandrin porte-mèche serrant des mèches de 0 à — ᵐ/ₘ (*m.*), drill-chuck holding drills from 0 to — mm. (67)
serrer le frein, to put on the brake ; to apply the brake ; to brake. (68)
serrer le sable d'un moule, to pack the sand of a mould. (69)
serrer les freins à bloc, to put the brakes hard on ; to skid the wheels. (70)
serrer les freins à fond, to put the brakes full on. (71)

serrer les nappes aquifères (Mines), to shut off the water-bearing strata. (1)

serrer un écrou, to screw up, to tighten, a nut. (2)

serrer un écrou à fond, to screw a nut right home. (3)

serrer un nœud, to tighten a knot. (4)

serrer une tige dans l'étau, to grip a rod in the vice. (5)

serrer une tôle (de chaudière), to close a plate. (6)

serrer une vis, to drive, to drive in, to tighten, a screw. (7)

serrer une vis à bloc, to drive a screw right home ; to tighten a screw hard up. (8)

serrure (*n.f.*), lock. (9)

serrure à broche (*f.*), piped-key lock. (10)

serrure à combinaisons (*f.*), combination lock ; letter lock ; puzzle lock. (11)

serrure à droite (*f.*), right-hand lock. (12)

serrure à garnitures (*f.*), warded lock. (13)

serrure à gauche (*f.*), left-hand lock. (14)

serrure à gorge (*f.*), tumbler-lock. (15)

serrure à larder *ou* **serrure lardée** *ou* **serrure à mortaiser** (*f.*), mortise-lock ; mortice-lock. (16)

serrure à pêne à demi-tour *ou* **serrure à ressort** (*f.*), latch-lock ; spring-lock. (17)

serrure à pêne dormant (*f.*), dead lock ; dormant lock. (18)

serrure à pompe (*f.*), Bramah lock. (19)

serrure bénarde (*f.*), pinned-key lock. (20)

serrure de coffre (*f.*), box-lock ; chest-lock. (21)

serrure de sûreté (*f.*), safety-lock. (22)

serrure de sûreté à gorges mobiles (*f.*), tumbler-lock. (23)

serrure encastrée *ou* **serrure entaillée** (*f.*), flush lock. (24)

serrure encloisonnée (*f.*), rim-lock. (25)

serrurerie (fabrication des serrures) (*n.f.*), lock-smithing ; locksmithery. (26)

serrurerie (ouvrages en serrures) (*n.f.*), lock-work. (27)

serrurerie (fabrication des pièces en fer forgé, etc.) (*n.f.*), metal-working ; ironworking. (28)

serrurerie (ouvrages en fer forgé, etc.) (*n.f.*), metal work ; ironwork. (29)

serrurerie d'art (*f.*), art metal work (railings, gates, etc.). (30)

serrurerie de bâtiment (*f.*), builders' hardware. (31)

serrurerie de charronnage (*f.*), coachsmithing. (32)

serrurier (ouvrier qui fait des serrures) (*n.m.*), locksmith. (33)

serrurier (ouvrier qui fait des ouvrages en fer forgé, etc.) (*n.m.*), metal-worker ; ironworker ; ironsmith. (34)

serrurier charron (*m.*), coachsmith. (35)

serrurier d'art (*m.*), art metal worker. (36)

serrurier en bâtiments (*m.*), builders' hardware merchant. (37)

serrurier mécanicien (*m.*), mechanic ; artificer. (38)

sertir (*v.t.*), to set : (39)

sertir un diamant dans la couronne (Sondage), to set a diamond in the crown. (40)

serveur (Mines de charbon) (pers.) (*n.m.*), helper ; boy ; colliers' helper. (41)

service (*n.m.*), service. (42)

service (de) (Méc.), what is necessary for working a machine, or the like : (43)

clefs de service (*f.pl.*), screw-keys necessary for working the machine ; all necessary spanners. (44) Cf. tension de service.

service d'exhaure (*m.*), pumping service. (45)

service de l'entretien (Ch. de f.) (*m.*), maintenance department. (46)

service de la voie (Ch. de f.) (*m.*), permanent way department. (47)

service des eaux (*m.*), water-service ; water-supply. (48)

servir (*v.t.*), to serve : (49)

servir les intérêts de quelqu'un, to serve someone's purpose. (50)

servir (*v.i.*), to serve ; to be used ; to be of use : (51)

machine qui sert à élever l'eau (*f.*), machine which serves to raise water (*or* which is used to raise water). (52)

servir (se) de, to use ; to make use of ; to employ ; to avail oneself of : (53)

se servir d'une machine, to use a machine. (54)

se servir de la vapeur comme force motrice, to employ steam-power. (55)

ne pas se servir de crochets (avis), use no hooks. (56)

servo-moteur [servo-moteurs *pl.*] (*n.m.*), servo-motor ; relay. (57)

sesquioxyde (Chim.) (*n.m.*), sesquioxide : (58)

sesquioxyde de fer, sesquioxide of iron. (59)

settler (cuve de lavage) (Préparation mécanique des minerais) (*n.m.*), settler. (60)

settler (colon) (pers.) (*n.m.*), settler. (61)

seuil (*n.m.*), sill. (62)

seuil (d'un dormant de porte) (*n.m.*), groundsill (of a door-frame). (63)

seuil (sole gravière) (Constr. hydraul.) (*n.m.*), mudsill. (64)

seuil (d'un barrage, d'un déversoir) (*n.m.*), crest (of a dam, of a weir). (65 *ou* 66)

seuil continental (Océanogr.) (*m.*), continental shelf ; continental platform : (67)

la véritable limite entre les océans et les continents n'est pas en général le rivage actuel ; elle se trouve au large, où une dénivellation brusque sépare, des grands fonds qui se trouvent au delà, une plate-forme sous-marine plus ou moins étendue, dont la profondeur dépasse rarement 200 m. Cette plate-forme a reçu le nom de seuil continental, the true limit between the oceans and the continents is not generally the actual shore ; it is to be found out at sea, where a sudden drop separates, from the great depths beyond, a more or less extensive submarine platform, of which the depth rarely exceeds 100 fathoms. This platform is known as the continental shelf. (68)

seuil d'écluse (canal) (*m.*), lock-sill ; mitre-sill ; clap-sill. (69)

seuil de porte (*m.*), door-sill ; threshold. (70)

seul propriétaire (*m.*), sole proprietor ; sole owner. (71)

sévérite (Minéral) (*n.f.*), severite. (72)

sexivalence (Chim.) (*n.f.*), sexivalence. (73)

sexivalent, -e (*adj.*), sexivalent. (74)

sextant (*n.m.*), sextant. (75)

sextuple (*adj.*), sextuple. (76)

seybertite (Minéral) (*n.f.*), seybertite ; clintonite. (77)

shed (*n.m.*), shed-roof ; square-to roof ; saw-tooth roof. (78)

shunt (Élec.) (*n.m.*), shunt. (79)

shunt de galvanomètre (*m.*),   galvanometer-shunt. (1)
shunter (Élec.) (*v.t.*), to shunt : (2)
shunter un galvanomètre, to shunt a galva-nometer. (3)
si le temps le permet, weather permitting. (4)
sibérite (Minéral) (*n.f.*), siberite. (5)
siccatif, -ive (*adj.*), drying ; siccative : (6)
huile siccative (*f.*), drying oil. (7)
siccité (*n.f.*), dryness : (8)
évaporer à siccité (Chim.), to evaporate to dryness. (9)
sidérique (*adj.*), sideritic. (10)
sidérite (Minéral) (*n.f.*), siderite. (11)
sidérite (météorite de fer) (*n.f.*), siderite. (12)
sidérochrome (Minéral) (*n.m.*), chromite ; chrome iron ; chrome iron ore. (13)
sidérolithe (*n.f.*), iron ore. (14)
sidéromagnétique   (*adj.*),   sideromagnetic ; ferromagnetic. (15)
sidéromélane (Pétrol.) (*n.f.*), sideromelane. (16)
sidéroscope (*n.m.*), sideroscope. (17)
sidérose (Minéral) (*n.f.*), siderite. (18)
sidérotechnie (*n.f.*), siderotechny. (19)
sidérotechnique (*adj.*), siderotechnical. (20)
sidérurgie (*n.f.*), siderurgy. (21)
sidérurgique (*adj.*), siderurgic ; siderurgical ; iron : (22)
usine sidérurgique (*f.*), ironworks (metallurgical works). (23)
siège (Méc.) (*n.m.*), seat ; seating. (24)
siège d'exploitation *ou* **siège d'extraction** (Mines) (*m.*), working ; workings ; works ; winning-place : (25)
la généralité des mines françaises sont pourvues d'au moins deux communications avec le jour, pour chaque siège d'exploitation, most of the French mines are provided with at least two communications with the surface, for each working. (26)
un siège d'extraction à puits unique, a single-shaft working. (27)
siège de charbonnage (*m.*),   coal-works ; coal-workings : (28)
l'aménagement et l'installation d'un siège de charbonnage, the laying out and equipment of a coal-works (*or* of a coal-workings). (29)
siège de l'industrie (*m.*), seat of industry. (30)
siège de soupape (*m.*), valve-seat. (31)
siège du clapet (*m.*), seat of the clack. (32)
siège social *ou* *simplement* **siège** (*n.m.*), head office. (33)
sierra [**sierras** *pl.*] (Géogr.) (*n.f.*), sierra. (34)
sifflement (*n.m.*), whistling : (35)
sifflement des vents, whistling of the winds. (36)
sifflement de l'arc voltaïque (*m.*), singing of the voltaic arc. (37)
siffler (*v.i.*), to whistle : (38)
siffler au disque (Ch. de f.), to whistle for signals. (39)
sifflerie (*n.f.*), whistling. (40)
sifflet (*n.m.*), whistle. (41)
sifflet à vapeur (*m.*), steam-whistle. (42)
sifflet avertisseur *ou* **sifflet d'alarme** (*m.*), alarm-whistle. (43)
sifflet d'alarme à flotteur (*m.*), alarm-whistle with float ; low-water alarm, with float ; boiler-alarm. (44)
sifflet de locomotive (*m.*), locomotive-whistle. (45)
siffet (en) *ou* **sifflet (à)** (Charp.), splayed : (46)

assemblage en sifflet *ou* assemblage à sifflet (*m.*), splayed joint. (47)
sifflez (avis) (Ch. de f.), whistle : (48)
indicateur portant l'inscription :   sifflez, indicator bearing the notice : whistle. (49)
signal (*n.m.*), signal. (50)
signal à distance (Ch. de f.) (*m.*), distant signal ; distant block-signal ; distance-signal. (51)
signal acoustique (*m.*), audible signal. (52)
signal d'alarme *ou* **signal avertisseur** (*m.*), alarm-signal ; warning signal. (53)
signal d'arrêt (Ch. de f.) (*m.*), stop-signal ; danger-signal. (54)
signal d'arrêt absolu (*m.*), absolute stop-signal. (55)
signal de bifurcation (*m.*), junction signal. (56)
signal de cantonnement (*m.*), block-signal. (57)
signal de direction *ou* **signal indicateur de direction** (*m.*), direction-signal. (58)
signal de position (*m.*), position-target. (59)
signal-disque [**signaux-disques** *pl.*] (*n.m.*), disc signal. (60)
signal du fond (Mines) (*m.*), signal from below-ground ; bottom signal. (61)
signal effacé, open switch ; blind target. (62)
signal électrique (*m.*), electric signal. (63)
signal fermé, closed switch. (64)
signal indicateur de position des aiguilles (*m.*), switch-signal ; switch-target ; target. (65)
signal optique (*m.*), visible signal. (66)
signal rapproché (*m.*), home signal. (67)
signaler (*v.t.*), to signal : (68)
signaler un train, to signal a train. (69)
signaler à la surface de tenir bon (Mines), to signal to bank to hold. (70)
signaler un accident aux autorités, to report an accident to the authorities. (71)
signaleur (Ch. de f.) (pers.) (*n.m.*), signalman ; towerman. (72)
signaux dans le puits par cris (*m.pl.*), signaling in the shaft by shouting. (73)
signe (*n.m.*), sign ; indication : (74)
signe conventionnel, conventional sign. (75)
signe de pluie, sign of rain. (76)
signifier (notifier) (*v.t.*), to give notice. (77)
silencieux, -euse (*adj.*), silent ; noiseless ; quiet : (78)
marche silencieuse (*f.*), silent running. (79)
chaîne silencieuse (*f.*), silent chain ; noiseless chain. (80)
silencieux (pot d'échappement) (*n.m.*), silencer ; muffler. (81)
silex (*n.m.*), flint ; flintstone ; silex. (82)
silex corné (*m.*), horn-flint. (83)
silex meulier *ou* **silex molaire** (*m.*),   millstone grit. (84)
silex noir (*m.*), chert ; phthanite ; hornstone ; rock-flint. (85)
silex pyromaque (*m.*),   flint ; flintstone ; fire-stone. (86)
silex résinite (*m.*), pitch-opal ; resin opal. (87)
silex volcanique (*m.*), volcanic glass ; obsidian. (88)
silex xyloïde (*m.*), silicified wood. (89)
silicate (Chim.) (*n.m.*), silicate : (90)
silicate d'alumine, de chaux, de magnésie, silicate of alumina, of lime, of magnesia. (91 *ou* 92 *ou* 93)
silicaté, -e (*adj.*),   silicated ; silicate (*used as adj.*) : (94)
roche silicatée (*f.*), silicate rock. (95)
silice (Chim.) (*n.f.*), silica. (96)

silice (Minéralogie) (*n.f.*), silica : (1)
la famille de la silice, the family of the silicas. (2)
silicé, -e (*adj.*), siliceous ; silicious. (3)
silicéo-calcaire (*adj.*), silicocalcareous ; silici--calcareous. (4)
siliceux, -euse (*adj.*), siliceous ; silicious : (5)
sable siliceux (*m.*), siliceous sand ; silicious sand. (6)
silicifère (*adj.*), siliciferous. (7)
silicification (*n.f.*), silicification. (8)
silicifier (*v.t.*), to silicify. (9)
silicique (Chim.) (*adj.*), silicic : (10)
acide silicique (*m.*), silicic acid. (11)
silicium (Chim.) (*n.m.*), silicon ; silicium ; silicum. (12)
siliciure (Chim.) (*n.m.*), silicide. (13)
silicocalcaire (*adj.*), silicocalcareous ; silici--calcareous. (14)
sillimanite (Minéral) (*n.f.*), sillimanite. (15)
sillon (*n.m.*), furrow ; groove : (16)
les sillons qui se forment dans les tôles de chaudières, the grooves which form in boiler-plates. (17)
sillon houiller (*m.*), coal-belt : (18)
en France, le sillon houiller est limité entre deux murs, in France, the coal-belt lies between two walls. (19)
sillonner (*v.t.*), to furrow ; to groove : (20)
plateau sillonné de profondes gorges (*m.*), plateau furrowed with deep gorges. (21)
silurien, -enne (Géol.) (*adj.*), Silurian. (22)
silurien (*n.m.*), Silurian : (23)
le silurien, the Silurian. (24)
simbleau (*n.m.*), centring-bridge ; centering-bridge (for finding the centre of hollow cylindrical work). (25)
similaire (*adj.*), similar. (26)
similargent (*n.m.*), imitation silver. (27)
similarité *ou* **similitude** (*n.f.*), similarity. (28)
similibronze (*n.m.*), imitation bronze. (29)
similicuir (*n.m.*), leatherette ; leatheret ; imitation leather. (30)
similimarbre (*n.m.*), imitation marble. (31)
similipierre (*n.m.*), imitation stone. (32)
similor (*n.m.*), similor ; imitation gold. (33)
simoun (*n.m.*), simoom ; simoon ; sand-storm. (34)
simple (*adj.*), simple ; single : (35)
mouvement simple (Méc.) (*m.*), simple motion. (36)
simple harnais d'engrenages (*m.*), single gear. (37)
simple touche (Magnétisme) (*f.*), single touch. (38)
simple effet (à) (Méc.), single-acting ; single-action : (39)
machine à simple effet (*f.*), single-acting engine ; single-action engine. (40)
simplement (*adv.*), simply. (41)
simplicité (*n.f.*), simplicity : (42)
simplicité d'une substance, simplicity of a substance. (43)
simplicité de manœuvre, de réglage, simplicity of operation, of adjustment. (44 *ou* 45)
simplification (*n.f.*), simplification. (46)
simplifier (*v.t.*), to simplify. (47)
simultané, -e (*adj.*), simultaneous. (48)
simultanéité (*n.f.*), simultaneity. (49)
simultanément (*adv.*), simultaneously : (50)
le télégraphe duplex permet de faire passer simultanément sur un fil unique deux dépêches en sens inverse, the duplex telegraph allows two messages to be sent simultaneously over a single wire in opposite directions. (51)
singe (calibre reproducteur) (Méc.) (*n.m.*), former ; jig. (52)
singe (d'une chèvre) (*n.m.*), winch (of a derrick-crane). (53)
sinistre (*n.m.*), disaster ; accident ; fire : (54)
en cas de sinistre, briser la glace (avertisseur d'incendie), in case of fire, break the glass (fire-alarm). (55)
sinistrorsum (*adj. invar.*), sinistrorsal ; sinistrorse ; left-handed : (56)
hélice sinistrorsum (*f.*), sinistrorsal helix ; left-handed spiral. (57)
sinistrorsum (*adv.*), sinistrorsally ; sinistrorsely ; left-handedly : (58)
pas de vis enroulé sinistrorsum (*m.*), screw-thread wound sinistrorsally. (59)
sinopite (argile ferrugineuse) (*n.f.*), sinople ; sinopite. (60)
sinople (quartz ferrugineux) (*n.m.*), sinople ; sinopal. (61)
sinueux, -euse (*adj.*), winding ; sinuous : (62)
rivière sinueuse (*f.*), winding river. (63)
sinuosités d'un fleuve (*f.pl.*), windings of a river. (64)
sinus (Géom.) (*n.m.*), sine : (65)
sinus d'un angle, sine of an angle. (66)
sinus verse (*m.*), versed sine. (67)
sinusoïdal, -e, -aux (Géom. & Élec.) (*adj.*), sinusoidal : (68)
courant sinusoïdal (*m.*), sinusoidal current. (69)
sinusoïde (*n.f.*), sinusoid. (70)
siphon (tube recourbé à deux branches inégales, dont on se sert pour transvaser les liquides) (*n.m.*), siphon ; syphon. (71)
siphon (Plomberie sanitaire) (*n.m.*), trap ; stench-trap ; stink-trap ; drain-trap. (72)
siphon en S (Plomberie) (*m.*), S trap. (73)
siphon horizontal (Plomberie) (*m.*), running trap. (74)
siphon inverse (*m.*), inverted siphon. (75)
siphonal, -e, -aux (*adj.*), siphonal. (76)
siphonnement (*n.m.*), siphoning. (77)
siphonner (*v.t.*), to siphon. (78)
sirène (*n.f.*) *ou* **sirène d'alarme**, hooter ; siren. (79)
sirène à vapeur (*f.*), steam-hooter. (80)
sirène électrique (*f.*), electric hooter. (81)
sisme *ou* **séisme** (*n.m.*), earthquake. (82)
Note.—*The plural* sismes *or* séismes *can be translated by* seismism *when the meaning* earthquake shocks collectively *is intended* ; *as*, régions encore tourmentées par le volcanisme et les séismes (*f.pl.*), regions still broken by volcanism and seismism.
sismicité *ou* **séismicité** (*n.f.*), seismicity. (83)
sismique *ou* **séismique** *ou* **sismal, -e, -aux** *ou* **séismal, -e, -aux** (*adj.*), seismic ; seismical ; seismal ; earthquake (*used as adj.*) : (84)
secousse sismique (*f.*), seismic shock ; earth--quake shock. (85)
sismogramme *ou* **séismogramme** (*n.m.*), seis--mogram : (86)
sismogramme d'un tremblement de terre, seismogram of an earthquake. (87)
sismographe *ou* **séismographe** (*n.m.*), seismograph. (88)
sismographie *ou* **séismographie** (*n.f.*), seismog--raphy. (89)

sismographique *ou* séismographique (*adj.*), seismographic ; seismographical. (1)

sismologie *ou* séismologie (*n.f.*), seismology. (2)

sismologique *ou* séismologique (*adj.*), seismo--logical. (3)

situation (*n.f.*), situation ; position ; location : (4)
situation actuelle des affaires, present position of affairs. (5)
situation des travaux, situation, location, of workings. (6)

situer (*v.t.*), to place ; to locate. (7)

skip (caisse guidée) (Mines) (*n.m.*), skip ; gun--boat ; skep. (8)

skip à bascule (*m.*), dump-skip. (9)

skip à bascule automatique *ou* skip à culbutage automatique (*m.*), self-dumping skip. (10)

slimes (boues) (Métall.) (*n.f.pl.*), slimes ; sludge. (11)

sluice (canal de lavage) (Mines) (*n.m.*), sluice. (12)

sluice à losanges (*m.*), sluice with diamond-shaped riffles. (13)

sluice à or (*m.*), gold-sluice. (14)

sluice-box [sluice-boxes *pl.*] (*n.m.*), sluice-box. (15)

sluice de décharge (*m.*), tail-sluice. (16)

sluice de tête (*m.*), head-sluice. (17)

sluice garni de drap de laine (pour retenir l'or fin) (*m.*), blanket-sluice. (18)

sluice sibérien (*m.*), Siberian sluice. (19)

smaltine (Minéral) (*n.f.*), smaltite ; gray cobalt ; speiss cobalt ; speiskobalt. (20)

smaragdite (Minéral) (*n.f.*), smaragdite ; green diallage. (21)

smectite (terre à foulon) (*n.f.*), smectite ; fullers' earth ; fullers' chalk. (22)

smectite (variété d'halloysite) (*n.f.*), smectite. (23)

smélite (kaolin) (*n.f.*), smelite. (24)

smithsonite (Minéral) (*n.f.*), smithsonite ; zinc spar ; calamine. (25)

soc (*n.m.*), ploughshare ; plowshare. (26)

société (*n.f.*), company ; partnership ; society. (27)

société de mines *ou* société minière (*f.*), mining-company. (28)

sociétés de mines d'or (*f.pl.*), gold-mining companies. (29)

socle (*n.m.*), stand ; pedestal. (30)

socle à rainures (pour machines à percer, etc.) (*m.*), box table (for drilling-machines, etc.). (31)

socle à roulettes (*m.*), stand on wheels ; pedestal on wheels. (32)

soda (Minéral) (*n.m.*), natron. (33)

sodalite (Minéral) (*n.f.*), sodalite. (34)

sodé, -e (Chim.) (*adj.*), soda (*used as adj.*) : (35)
papier sodé (*m.*), soda paper. (36)

sodique (*adj.*), soda ; sodium (*used as adjs.*) : (37)
sel sodique (*m.*), soda salt ; sodium salt. (38)

sodium (Chim.) (*n.m.*), sodium. (39)

soffione [soffioni *pl.*] (Géol.) (*n.m.*), soffione ; suffione ; air-volcano ; mud-volcano (non-volcanic). (40)

soffite (Arch.) (*n.m.*), soffit. (41)

soi-disant expert (*m.*), self-styled expert ; pretended expert ; would-be expert ; quasi-expert ; so-called expert. (42)

soie (*n.f.*), silk : (43)
fil de cuivre recouvert de soie (*m.*), silk-covered copper wire. (44)

soie (d'un couteau, d'un ciseau, d'une lime) (*n.f.*), tang, tongue, fang, (of a knife, of a chisel, of a file). (45 *ou* 46 *ou* 47)

soie de manivelle (manivelle à main) (*f.*), crank-pin (hand-crank). (48)

soigneusement (*adv.*), carefully. (49)

soigneux, -euse (*adj.*), careful. (50)

soin (*n.m.*), care ; carefulness : (51)
travailler avec soin, to work with care. (52)
soins et traitement des animaux, care and treatment of animals. (53)

sol (*n.m.*), soil ; ground. (54)

sol (Élec.) (*n.m.*), ground ; earth. (55)

sol (au) (Élec.), grounded ; earthed : (56)
conducteur au sol (*m.*), grounded conductor ; earthed conductor. (57)

sol (d'une fonderie) (*n.m.*), floor : (58)
creuser une fouille dans le sol d'une fonderie, to dig a pit in the floor of a foundry. (59)

sol (d'un filon) (*n.m.*), foot-wall, ledger-wall, ledger, (of a lode). (60)

sol alluvial (*m.*), alluvial soil. (61)

sol argileux *ou* sol glaiseux (*m.*), clay soil. (62)

sol ferrifère (*m.*), iron-bearing ground. (63)

sol marécageux (*m.*), marshy soil ; marshy ground. (64)

sol perméable (*m.*), permeable soil ; leachy soil. (65)

sol primordial (Géol.) (*m.*), primitive strata ; primitive terrane. (66)

sol suintant (*m.*), oozy ground. (67)

sol vierge (*m.*), virgin soil ; virgin ground. (68)

solaire (*adj.*), solar ; sun (*used as adj.*) : (69)
spectre solaire (*m.*), solar spectrum. (70)
cadran solaire (*m.*), sun-dial. (71)

sole (Boisage de mines) (*n.f.*), sill ; sill-piece ; sole ; sole-piece ; sole-plate ; foot-piece ; groundsill. (72)

sole (pour une machine) (*n.f.*), sole-plate ; bed-plate ; base-plate. (73)

sole (pièce de bois qui se place sous le pied d'un étai pour former empattement) (*n.f.*), sole-piece ; footing-block. (74)

sole (support des barreaux de grille à l'avant) (*n.f.*), dead-plate ; coking-plate ; dumb-plate. (75)

sole (d'un four électrique, d'un four à réverbère) (*n.f.*), hearth, sole, (of an electric furnace, of a reverberatory furnace). (76 *ou* 77)

sole (du four Martin) (*n.f.*), open hearth (of the Siemens-Martin furnace). (78) **Cf. acier sur sole.**

sole (d'une galerie de mine) (*n.f.*), floor, pave--ment, sole, sill, (of a mine-level). (79)

sole du chantier (Mines) (*f.*), stope-floor. (80)

sole gravière (Constr. hydraul.) (*f.*), mudsill. (81)

soleil (*n.m.*), sun. (82)

soleil ardent (*m.*), burning sun ; hot sun ; blazing sun. (83)

solénoïde (Élec.) (*n.m.*), solenoid ; electro--magnetic cylinder. (84)

solfatare (Géol.) (*n.f.*), solfatara. (85)

solfatarien, -enne (*adj.*), solfataric. (86)

solidaire (Méc.) (*adj.*), integral ; solid : (87)
une tige solidaire d'un plateau, a rod integral (*or* solid) with a plate. (88)
pièces de machine rendues solidaires au moyen de clavettes (*f.pl.*), machine parts made integral by means of keys. (89)

solide (qui a de la consistance) (*adj.*), solid : (90)
les corps solides et les corps fluides, solid and fluid substances. (91)

solide (ferme) (*adj.*), solid ; firm : (92)
bâtiment solide (*m.*), solid building. (93)
fonçage en terrains solides (*m.*), sinking through firm ground. (94)

solide écrin en cuir à courroie (*m.*), solid leather sling case. (1)

sol!de (*n.m.*), solid. (2)

solide d'égale résistance (Méc.) (*m.*), solid of uniform strength. (3)

solidement (*adv.*), solidly; firmly: (4)
tête porte-outil solidement bloquée (*f.*), tool-head firmly clamped. (5)

solidifiable (*adj.*), solidifiable. (6)

solidification (*n.f.*), solidification: (7)
solidification de matières liquides, solidification of liquid matter. (8)

solidifier (*v.t.*), to solidify: (9)
solidifier l'eau en la congelant, to solidify water by freezing it. (10)

solidifier (se) (*v.r.*), to solidify; to become solid: (11)
se solidifier au contact de l'air, to solidify on contact with the air. (12)

solidité (*n.f.*), solidity; firmness: (13)
solidité d'un bâtiment, solidity of a building. (14)

solin (Constr.) (*n.m.*), fillet-gutter. (15)

solivage (*n.m.*), joisting; girderage. (16)

solive (Constr.) (*n.f.*), joist; girder. (17)

solive boiteuse *ou* solive bâtarde (*f.*), trimmed joist. (18)

solive d'enchevêtrure (*f.*), trimming-joist. (19)

solive de plafond (*f.*), ceiling-joist. (20)

solive de plancher (*f.*), floor-joist. (21)

solive en acier (*f.*), steel joist. (22)

solive en fer à double T (*f.*), H girder; I girder; H beam; I beam. (23)

soliveau (*n.m.*), small joist *or* girder. (24)

sollicité (-e) par la pesanteur, acted upon by gravity: (25)
les fragments de rocher qui se détachent sont sollicités par la pesanteur, et roulent ou glissent sur les pentes jusqu'à ce que la déclivité du terrain soit trop faible pour que ce mouvement se continue (*m.pl.*), the pieces of rock which become detached are acted upon by gravity, and roll or slide down the slopes until the declivity of the ground is too slight for this movement to continue. (26)

solliciter (*v.t.*), to ask; to invite; to solicit: (27)
solliciter la permission de visiter la mine, to ask for permission to visit the mine. (28)
solliciter quelqu'un à faire (*ou* de faire) quelque chose, to ask someone to do something. (29)
solliciter des soumissions pour un bâtiment, to invite tenders for a building. (30)

solorite (Explosif) (*n.f.*), solorite. (31)

solubiliser (*v.t.*), to solubilize. (32)

solubilité (*n.f.*), solubility: (33)
la solubilité d'un sel dans l'eau, de l'or dans le chlore, the solubility of a salt in water, of gold in chlorine. (34 *ou* 35)

soluble (*adj.*), soluble: (36)
substance soluble dans l'eau (*f.*), substance soluble in water. (37)

solution (*n.f.*), solution. (38)

solution aurifère (*f.*), gold-bearing solution. (39)

solution bromocyanurée (*f.*), bromocyanide solution. (40)

solution concentrée *ou* solution forte (*f.*), concentrated solution; strong solution; stock solution. (41)

solution cyanurée concentrée (*f.*), concentrated cyanide solution. (42)

solution d'une difficulté (*f.*), solution of a difficulty. (43)

solution de continuité (*f.*), break of continuity. (44)

solution de fixage (Photogr.) (*f.*), fixing-solution. (45)

solution de virage-fixage (Photogr.) (*f.*), toning-and-fixing solution. (46)

solution étendue *ou* solution faible (*f.*), diluted solution; weak solution. (47)

solution saturée (*f.*), saturated solution. (48)

solution sursaturée (*f.*), supersaturated solution. (49)

solution Thoulet (*f.*), Thoulet solution; Thoulet's solution. (50)

solvant (Chim.) (*n.m.*), solvent. (51)

sombre (*adj.*), sombre; dark; dull: (52)
rouge sombre (Métall.) (*m.*), sombre red; dark red; dull red. (53)
temps sombre (*m.*), dull weather. (54)

sombrer (se dit d'un bateau) (*v.i.*), to sink. (55)

somme (banc de sable) (*n.f.*), sand-bank (outside a port or the mouth of a river). (56)

sommet (d'une montagne) (*n.m.*), summit, top, peak, apex, (of a mountain): (57)
les sommets géants des Andes, the giant summits of the Andes. (58)

sommet (d'une colline) (*n.m.*), top (of a hill). (59)

sommet (d'un puits de mine) (*n.m.*), top (of a mine-shaft). (60)

sommet (d'un angle, d'un cône, d'une pyramide) (Géom.) (*n.m.*), vertex, apex, (of an angle, of a cone, of a pyramid). (61 *ou* 62 *ou* 63)

sommet (d'une voûte) (*n.m.*), vertex, crown, (of an arch). (64)

sommet du filon (*m.*), apex of the lode; ore-apex. (65)

sommier (premier claveau reposant à plat sur le pied-droit d'une voûte) (*n.m.*), springer; springing-stone; springing. (66)

sommier (pierre qui reçoit la retombée d'une voûte) (*n.m.*), impost. (67)

sommier (d'un plancher sur poutre) (*n.m.*), binder (of a double floor). (68)

sommier (d'un pont en bois) (*n.m.*), stringer (of a wooden bridge). (69)

sommier (support des barreaux de grille d'un foyer) (*n.m.*), bearer. (70)

sommier (d'une porte) (*n.m.*), transom, lintel, (of a door). (71)

sommier (d'une scie ordinaire) (*n.m.*), stretcher (of a frame-saw). (72)

sommier de caisse (Ch. de f.) (*m.*), body-bolster; bolster. (73)

sommier de voûte en briques (grille de foyer) (*m.*), brick-arch bearer (furnace-grate). (74)

sommier transversal (grille de foyer) (*m.*), cross-bearer. (75)

son (Phys.) (*n.m.*), sound. (76)

sondage (action de reconnaître la profondeur de, la nature de) (*n.m.*), sounding: (77)
sondage d'un lac, sounding a lake. (78)
sondage au marteau des pièces venues de fonte, sounding castings with the hammer. (79)

sondage (forage) (*n.m.*), boring; bore-holing; drilling: (80)
reconnaître une couche de houille par sondages, to prove a coal-seam by bore-holing. (81)

sondage (trou de sonde) (*n.m.*), bore-hole; bore; drill-hole. (82)

sondage à chute libre (m.), free-fall boring; free-fall drilling. (1)

sondage à grande profondeur (m.), deep drilling; deep boring. (2)

sondage à injection d'eau (m.), water-flush drilling. (3)

sondage à la corde ou sondage chinois ou sondage américain (m.), cable-drilling; rope-drilling; boring by percussion with rope. (4)

sondage à la grenaille d'acier (m.), shot-boring; boring by shot-drills. (5)

sondage à la tige (m.), boring by percussion with rods; pole-tool boring. (6)

sondage à main (m.), hand-drilling. (7)

sondage à sec (m.), dry boring. (8)

sondage au diamant (m.), diamond drilling. (9)

sondage au trépan (m.), boring with the bit. (10)

sondage avec curage hydraulique (m.), water-flush drilling. (11)

sondage de puits (m.), well-boring; well-drilling; boring wells; drilling wells. (12)

sondage de puits à grand diamètre (m.), boring shafts of large diameter. (13)

sondage de recherches ou sondage de reconnais--sance ou sondage d'exploration (action) (m.), experimental boring; exploratory drilling; trial-boring. (14)

sondage de recherches ou sondage de reconnais--sance ou sondage d'exploration (trou de sonde) (m.), trial-hole; proving-hole. (15)

sondage en rivière (m.), boring under water. (16)

sondage hydraulique par rôdage (m.), hydraulic rotary drilling. (17)

sondage par battage ou sondage percutant (m.), boring by percussion; percussion-drilling; churn-drilling. (18)

sondage par rôdage (m.), boring by rotation; drilling by rotation; rotary drilling. (19)

sondage pour recherches de pétrole (m.), boring for oil; drilling for petroleum. (20)

sondage profond (trou de sonde) (m.), deep bore-hole. (21)

sonde (la sonde proprement dite, c.-à-d., partie de l'ensemble de l'équipage de sonde qui s'enfonce dans le sein de la terre, comprenant la tête, les rallonges, et l'outil d'attaque) (n.f.), ground-rig. (22)

sonde (sondeuse) (n.f.), drill; driller; drilling-machine; borer; boring-machine. (23) See also sondeuse.

sonde à pointes de diamant et à tube carottier (f.), diamond core-drill. (24)

sonde de fontainier (f.), well-drill. (25)

sonde de prospection (f.), prospecting-drill; testing-drill. (26)

sonde percutante (f.), percussion-drill; churn-drill. (27)

sonde rotative (f.), rotary drill. (28)

sonder (v.t.), to sound: (29)
sonder un lac, le toit d'une galerie de mine, to sound a lake, the roof of a mine-level. (30 ou 31)

sondeur (pers.) (n.m.), driller; well-driller; well-borer. (32)

sondeur au diamant (m.), diamond-driller. (33)

sondeuse (n.f.), drill; drilling-machine; driller; borer; boring-machine. (34) See also sonde.

sondeuse à couronne (f.), annular borer. (35)

sondeuse à couronne à dents d'acier (f.), calyx drill. (36)

sondeuse à grenaille d'acier (f.), shot-drill; adamantine drill. (37)

sondeuse à main (f.), hand-drill. (38)

sondeuse au diamant (f.), diamond drill; diamond borer. (39)

sondeuse rotative à carottage continu (f.), rotary continuous-core drill. (40)

sondeuse rotative au diamant (f.), rotary diamond drill. (41)

sonner une cloche, to ring a bell. (42)

sonnerie (n.f.), bell; bells; system of bells. (43)

sonnerie d'appel (f.), call-bell. (44)

sonnerie électrique (f.), electric bell; magnetic bell. (45)

sonnerie électromagnétique (f.), magneto-bell. (46)

sonnette (clochette) (n.f.), bell; small bell: (47)
sonnette électrique, electric bell. (48)

sonnette (machine à enfoncer des pilotis) (n.f.), pile-driver; post-driver; spile-driver; gin. (49)

sonnette à déclic (f.), trip pile-driver; pile-driver with tripping-device. (50)

sonnette à main (f.), hand pile-driver. (51)

sonnette à tiraude (f.), ringing-engine. (52)

sonnette à vapeur (f.), steam pile-driver. (53)

sonore (adj.), sonorous; sound (used as adj.): (54)
onde sonore (Phys.) (f.), sound-wave. (55)

sonorité (n.f.), sonorousness; sonority. (56)

sorbite (Métall.) (n.f.), sorbite. (57)

sortage à la main (du minerai) (m.), hand-picking (ore). (58)

sorte (n.f.), sort; kind: (59)
sorte de matière, kind of matter. (60)

sortie (n.f.), exit; way out; outlet: (61)
sortie d'une mine, outlet of a mine. (62)

sortie (d'un modèle du moule) (n.f.), delivery: (63)
ébranler un modèle pour faciliter sa sortie du moule, to rap a pattern to facilitate its delivery from the mould. (64)

sortie d'air (orifice par où sort l'air) (f.), air-outlet. (65)

sortie de l'eau (f.), water-outlet. (66)

sortie de la vapeur (f.), steam-outlet. (67)

sortie de secours (f.), emergency exit. (68)

sortie du vent (machine soufflante) (f.), air-outlet (blower). (69)

sortir (Moulage de fonderie) (v.t.), to deliver: (70)
sortir un modèle du moule, to deliver a pattern from the mould. (71)

sortir (Fonderie) (v.i.), to deliver. (72)

sortir (faire saillie) (v.i.), to project; to stand out; to jut out: (73)
pierre qui sort du mur (f.), stone which projects from the wall. (74)

sortir de la voie ou sortir des rails (Ch. de f.), to jump the metals; to leave the metals; to run off the rails. (75)

soubassement (Arch.) (n.m.), base; basement. (76)

soubassement (d'une machine-outil, ou analogue) (n.m.), base, base-plate, (of a machine-tool, or the like). (77 ou 78)

souche (d'un arbre) (n.f.), stump, stub, (of a tree). (79)

souchevage, souchever, soucheveur. Same as souschevage, souschever, etc.

soudabilité (n.f.), weldability. (80) Cf. essai de soudabilité.

soudable (adj.), weldable. (81)

soudage (jonction des métaux par interposition d'une composition fusible) (n.m.), soldering. (82)

soudage (jonction des métaux par martelage, etc.) (*n.m.*), welding : (1)

soudage au marteau-pilon, welding by the power-hammer. (2)

See **soudure** for different kinds of welding and soldering.

soudain, -e (*adj.*), sudden : (3)

averse soudaine (*f.*), sudden shower. (4)

soudainement (*adv.*), suddenly. (5)

soudant, -e (*adj.*), soldering ; welding. (6)

soudant (*n.m.*), welding-heat ; sparkling heat. (7)

soude (*n.f.*), soda. (8)

soude caustique (*f.*), caustic soda. (9)

soude du commerce (*f.*), commercial soda. (10)

souder (joindre les métaux par l'interposition d'une composition fusible) (*v.t.*), to solder : (11)

souder un joint avec le fer à souder, au chalu-meau, to solder a joint with the soldering-iron, with the blowpipe. (12 ou 13)

souder (joindre les métaux par martelage, etc.) (*v.t.*), to weld : (14)

souder les mailles d'une chaîne-câble, to weld the links of a chain cable. (15)

chaîne soudée (*f.*), welded chain. (16)

souder à l'étain, to soft-solder ; to sweat. (17)

souder à rapprochement, to butt-weld. (18)

souder à recouvrement, to lapweld ; to lap-weld : (19)

tube soudé à recouvrement (*m.*), lapwelded tube. (20)

souder (se) (par composition fusible) (*v.r.*), to be soldered : (21)

les métaux se soudent à l'aide d'un métal différent (*m.pl.*), metals are soldered by means of a different metal. (22)

souder (se) (par martelage, etc.) (*v.r.*), to weld ; to be welded : (23)

tandis que le fer se soude, la fonte ne peut généralement pas se souder, whereas wrought iron welds, cast iron cannot generally be welded. (24)

soudeur (pers.) (*n.m.*), solderer ; welder. (25)

soudure (jonction des métaux par interposition d'une composition fusible) (*n.f.*), soldering. (26)

soudure (composition fusible) (*n.f.*), solder : (27)

un bâton de soudure, a stick of solder. (28)

soudure (jonction des métaux par martelage, etc.) (*n.f.*), welding. (29)

soudure (endroit soudé) (*n.f.*), weld. (30)

soudure à chaude portée (*f.*). See **soudure par amorces**.

soudure à gueule-de-loup (*f.*), split weld ; split welding ; cleft weld ; cleft welding ; tongue-weld ; tongue-welding. (31)

soudure à l'arc électrique (*f.*), electric-arc weld-ing ; arc-welding. (32)

soudure à l'étain (action) (*f.*), soft-soldering ; sweating. (33)

soudure à l'étain (composition) (*f.*). Same as **soudure d'étain**.

soudure à nœud (*f.*), wipe-joint ; wiped joint. (34)

soudure à recouvrement (*f.*), lapweld ; lapweld-ing ; lap-weld ; lap-welding ; scarf-weld ; scarf-welding. (35)

soudure aluminothermique (*f.*), aluminothermic welding. (36)

soudure au chalumeau (*f.*), blowpipe soldering ; welding with the blowpipe. (37)

soudure au cuivre (composition) (*f.*), brass-solder ; brazing-solder ; hard solder. (38)

soudure au fer à souder (*f.*), soldering with the soldering-iron. (39)

soudure au gaz (*f.*), gas-welding. (40)

soudure au gaz *ou* **soudure au gaz d'éclairage** (*f.*), gas-soldering. (41)

soudure au quart (*f.*), hard solder ; hard silver-solder (3 parts silver, 1 part brass wire). (42)

soudure au tiers (*f.*), hard pale solder (2 lead, 1 tin). (43)

soudure autogène (*f.*), autogenous welding ; autogenous soldering. (44)

soudure avec apport de fer-thermit (*f.*), thermit process intermediate welding. (45)

soudure d'aluminium (*f.*), aluminium-solder. (46)

soudure d'argent (*f.*), silver-solder. (47)

soudure d'étain (*f.*), tinmen's solder ; tin-solder. (48)

soudure d'étain en baguettes (*f.*), tinmen's solder in strips. (49)

soudure des plombiers (*f.*), plumbers' solder. (50)

soudure électrique (*f.*), electric welding. (51)

soudure en T (*f.*), T weld ; T welding. (52)

soudure grasse (*f.*), fine solder. (53)

soudure maigre (*f.*), coarse solder. (54)

soudure oxhydrique (*f.*), oxyhydrogen welding. (55)

soudure oxy-acétylénique (*f.*), oxyacetylene welding. (56)

soudure par amorces *ou* **soudure à chaude portée** (*f.*), scarf-weld ; scarf-welding ; lapweld ; lapwelding ; lap-weld ; lap-weld-ing. (57)

soudure par encollage *ou* **soudure par rapproche-ment** (*f.*), jump-weld ; jump-welding ; butt-weld ; butt-welding. (58)

soudure par la fonte liquide (*f.*), burning ; casting on. (59)

soudure sans apport de fer-thermit (*f.*), thermit process butt-welding. (60)

soudure tendre (*f.*), soft solder. (61)

soufflage (*n.m.*), blowing ; blow : (62)

fournir le vent nécessaire au soufflage, to provide the necessary blast for the blow. (63)

soufflage magnétique (*m.*), magnetic blowout. (64)

soufflante (*n.f.*), blowing-engine ; blower ; piston-blower ; blast-engine. (65)

soufflard (Géol.) (*n.m.*), air-volcano ; mud-volcano (non-volcanic) ; soffione ; suffione. (66)

soufflard de grisou (*m.*), blower ; feeder ; bleeder ; piper ; gas-vent ; fire-damp feeder. (67)

souffler (*v.t.*), to blow ; to blow out : (68)

souffler de l'air frais dans une mine, to blow fresh air into a mine. (69)

souffler le feu (avec le soufflet), to blow the fire (with the bellows). (70)

souffler un cubilot par un ventilateur, to blow a cupola with a fan. (71)

souffler une bougie, to blow out a candle. (72)

souffler (*v.i.*), to blow : (73)

soufflet qui ne souffle plus (*m.*), bellows which no longer blows. (74)

vent qui souffle en tempête (*m.*), wind which blows a gale. (75)

soufflerie (*n.f.*), blower ; blowers ; blowing-machinery ; bellows. (76)

soufflerie de forge (f.), forge-blower; forge-bellows. (1)

soufflet (n.m.), bellows. (2)

soufflet (machine soufflante de haut fourneau) (n.m.), blower; blowing-engine; blast-engine; piston-blower. (3)

soufflet cylindrique, à double vent (m.), double-blast circular bellows. (4)

soufflet cylindrique, à simple vent (m.), single-blast circular bellows. (5)

soufflet d'une chambre noire (m.), bellows of a camera. (6)

soufflet de forge (m.), smiths' bellows; black-smiths' bellows; forge-bellows. (7)

soufflet de mouleur (m.), moulders' bellows. (8)

souffleur (de locomotive, etc.) (n.m.), blower. (9)

souffleur (de grisou) (n.m.), blower; feeder; bleeder; piper; gas-vent; fire-damp feeder. (10)

souffleur magnétique (m.), magnetic blowout. (11)

soufflure (Fonderie, etc.) (n.f.), blow-hole; blow; air-hole; blister. (12)

souffrir de l'agitation ouvrière, to suffer from labour troubles. (13)

soufre (n.m.), sulphur; brimstone. (14)

soufre amorphe (m.), amorphous sulphur. (15)

soufre en canon (m.), roll-sulphur; stick-sulphur. (16)

soufrière (n.f.), sulphur-mine. (17)

soulager (v.t.), to relieve: (18)

soulager une soupape, to relieve a valve. (19)

soulève-rails [soulève-rails pl.] (n.m.), rail-lifter. (20)

soulèvement (n.m.), rising; lifting; heaving. (21)

soulèvement (Géol.) (n.m.), upheaval; upthrust; uplift: (22)

le soulèvement des montagnes représente le maximum d'effet produit par les contrac-tions de l'écorce terrestre, the upheaval of mountains represents the maximum of effect produced by the contractions of the earth's crust. (23)

le mont Cervin et le mont Rose représentent le soulèvement le plus puissant des Alpes, the Matterhorn and Monte Rosa represent the mightiest upheaval of the Alps. (24)

soulèvement des indigènes (m.), rising of the natives; native rising; native insurrection. (25)

soulever (v.t.), to raise; to lift: (26)

vent qui soulève des tourbillons de poussière (m.), wind which raises whirls of dust. (27)

soulever des rails au moyen d'un électro-aimant porteur, to lift rails by means of a lifting-magnet. (28)

soulever au moyen d'un levier ou soulever avec la pince, to lever; to lever up. (29)

soulever (Géol.) (v.t.), to upraise; to upheave; to uplift: (30)

couches soulevées (f.pl.), upraised beds; upheaved beds. (31)

soulever (se) (v.r.), to rise; to heave. (32)

soumettre (v.t.), to submit; to subject: (33)

soumettre à l'action de la chaleur, to subject to heat. (34)

soumettre un rapport annuel, to submit an annual report. (35)

soumission (pour une entreprise, une fourniture) (n.f.), tender. (36)

soumission cachetée (f.), sealed tender. (37)

soumissionnaire (pers.) (n.m.), tenderer. (38)

soumissionner (v.t.), to tender for; to put in a tender for: (39)

soumissionner une fourniture de marchandises, to tender for a supply of goods. (40)

soumissionner (v.i.), to tender; to put in a tender: (41)

soumissionner pour une fourniture de mar--chandises, to tender for a supply of goods. (42)

soupape (n.f.), valve. (43) See also valve and clapet.

soupape à boulet (f.), ball valve; globe valve. (44)

soupape à charnière (f.), hinged valve; leaf-valve. (45)

soupape à clapet (f.), clack-valve; flap-valve; clapper-valve. (46)

soupape à cloche (f.), bell-valve; Cornish valve. (47)

soupape à cône (f.), cone-valve. (48)

soupape à disque (f.), disc valve. (49)

soupape à double siège (f.), double-seated valve. (50)

soupape à flotteur (f.), float-valve; ball valve. (51)

soupape à levier (f.), lever-valve. (52)

soupape à levier et à ressort (f.), locomotive-balance (the spring which controls the safety-valve of a locomotive-boiler). (53)

soupape à manchon (f.), equilibrium-valve (sleeve pattern). (54)

soupape à ressort (f.), spring-valve; spring-loaded valve. (55)

soupape à siège conique (f.), valve with conical seat. (56)

soupape à simple siège (f.), single-seated valve. (57)

soupape à trois voies (f.), three-way valve; switch-valve. (58)

soupape annulaire (f.), annular valve. (59)

soupape commandée (f.), mechanically operated valve. (60)

soupape conique (f.), conical valve. (61)

soupape d'admission (f.), admission-valve; inlet-valve; intake-valve; induction-valve. (62)

soupape d'admission d'air ou soupape d'air (f.), air-valve. (63)

soupape d'admission de vapeur ou simplement soupape d'admission (f.), steam-valve. (64)

soupape d'arrêt (f.), stop-valve. (65)

soupape d'aspiration (f.), suction-valve. (66)

soupape d'échappement (f.), exhaust-valve; outlet-valve; eduction-valve. (67)

soupape d'équilibre (f.), equilibrium-valve. (68)

soupape d'évent (f.), snifting-valve; snifter-valve; blow-valve; air-valve; air-snifting valve; pet-valve. (69)

soupape de contrôle (f.), control-valve. (70)

soupape de Cornouailles (f.), Cornish valve; bell-valve. (71)

soupape de décharge (f.), discharge-valve. (72)

soupape de déviation (f.), deflecting-valve. (73)

soupape de pied (f.), foot-valve. (74)

soupape de prise de vapeur (f.), steam-valve. (75)

soupape de reflux (f.), reflux-valve. (76)

soupape de refoulement (f.), delivery-valve. (77)

soupape de rentrée d'air (f.), relief-valve; air-relief valve. (78)

soupape de réservoir (f.), tank-valve. (79)

soupape de retenue (*f.*), check-valve ; retaining valve ; back-pressure valve ; non-return valve. (1)

soupape de sûreté (*f.*), safety-valve ; relief-valve ; release-valve. (2)

soupape de trop-plein (*f.*), overflow-valve ; relief-valve. (3)

soupape de vapeur (*f.*), steam-valve. (4)

soupape équilibrée (*f.*), balanced valve. (5)

soupape hydraulique de manœuvre (*f.*), hydraulic working-valve. (6)

soupape soulevante (*f.*), lifting-valve ; poppet-valve. (7)

soupirail (*n.m.*), air-hole ; air-vent ; vent ; vent-hole. (8)

souple (*adj.*), supple ; flexible ; pliable ; pliant. (9)

souplesse (*n.f.*), suppleness ; flexibility ; plia-bility ; pliancy : (10)
souplesse d'un câble, flexibility of a rope. (11)

source (cause ; origine) (*n.f.*), source ; origin ; rise. (12)

source (fontaine) (*n.f.*), source ; spring ; foun-tain ; well ; fountainhead ; spring-head ; headspring. (13) See also **fontaine** and **puits**.

source (opp. à **récepteur**) (Hydraul.) (*n.f.*), source. (14) (opp. to **sink**).

source (opp. à **récepteur** *ou* **réceptrice**) (Élec.) (*n.f.*), source. (15) (opp. to **sink** *or* **load**) :
l'énergie électrique produite par une source (*f.*), the electrical energy produced by a source. (16)

source boueuse (Géol.) (*f.*), mud-geyser ; mud-pot. (17)

source calcaire (*f.*), calcareous spring. (18)

source capricieuse (*f.*), capricious spring ; fitful spring ; freak spring. (19)

source constante d'ennuis (*f.*), constant source of trouble. (20)

source d'eau (*f.*), spring of water. (21)

source d'eau thermale (*f.*), hot-water spring. (22)

source d'eau vive (*f.*), spring of running water. (23)

source d'énergie (*f.*), source of energy ; power-producer. (24)

source d'un fleuve (*ou* **d'une rivière**) (*f.*), source, head, rise, of a river. (25)

source de danger (*f.*), source of danger. (26)

source de feu *ou* **source inflammable** (Géol.) (*f.*), fire-well. (27)

source de pétrole (*f.*), oil-spring. (28)

source ferrugineuse (*f.*), ferruginous spring ; chalybeate spring. (29)

source incrustante (*f.*), incrusting spring. (30)

source intarissable (*f.*), unfailing spring. (31)

source intermittente (*f.*), intermittent spring. (32)

source lumineuse (Opt.) (*f.*), light-source ; source of light. (33)

source minérale (*f.*), mineral spring. (34)

source pérenne *ou* **source permanente** (*f.*), perennial spring. (35)

source périodique (*f.*), periodical spring. (36)

source salée *ou* **source saline** *ou* **source saumâtre** (*f.*), salt-spring ; saline spring ; brine-spring ; brine-pit. (37)

source suintante (*f.*), weeping spring. (38)

source thermale (*f.*), hot spring ; thermal spring. (39)

source vive (*f.*), running spring. (40)

sources vauclusiennes (Géol.) (*f.pl.*), very copious springs which issue from the foot of limestone cliffs, the type being realized in the *Fontaine de Vaucluse* in S.E. France. (41)

sourdre (*v.i.*), to spring ; to well ; to well up : (42)
eau qui sourd de la terre (*f.*), water which springs from the earth (*or* which wells from the ground). (43)
certaines sources ne se manifestent que par des suintements ; d'autres sourdent en abondantes cascades et donnent immédiate-ment naissance à des cours d'eau, certain springs only make their presence known by oozings ; others well up in copious cascades and immediately give birth to streams. (44)

souricière (repêchage des outils) (Sondage) (*n.f.*), mouse-trap. (45)

souris (employée pour indiquer la présence de gaz toxiques dans les mines) (*n.f.*), mouse. (46)

souris apprivoisées (*f.pl.*), tame mice. (47)

souris blanches (*f.pl.*), white mice. (48)

sous-âge (*n.m.*), subage. (49)

sous-bail [**sous-baux** *pl.*] (*n.m.*), sublease ; underlease. (50)

sous-bief *ou* **sous-biez** (d'un canal) (*n.m.*), back-water. (51)

sous-bois (*n.m.*), underwood. (52)

sous-cave (*n.f.*), **sous-cavé, -e** (*adj.*), **sous-caver** (*v.t.*). Same as **souschevage, souschevé, -e, souschever**.

sous-chef de gare (*m.*), assistant station-master. (53)

sous condition (Commerce), on approval. (54)

sous contrôle, under control. (55)

sous-cristallin, -e (*adj.*), subcrystalline. (56)

sous-directeur (*n.m.*), submanager ; under-manager ; assistant manager. (57)

sous-élément (*n.m.*), subelement ; secondary element. (58)

sous-estimation (*n.f.*), underestimation ; under-valuation. (59)

sous-estimer (*v.t.*), to underestimate ; to under-value. (60)

sous-étage (Géol.) (*n.m.*), substage. (61)

sous-étage (Mines) (*n.m.*), substage ; sublevel : (62)
partager un gîte en étages et sous-étages, to divide up a seam into stages and substages ; to divide a seam into levels and sublevels. (63)

sous-étampe (*n.f.*) *ou* **sous-étampe pour fers ronds**, bottom swage ; bottom rounding-tool. (64)

sous-évaluation (*n.f.*), undervaluation ; under-estimation. (65)

sous-évaluer (*v.t.*), to undervalue ; to under-estimate. (66)

sous-exposer (Photogr.) (*v.t.*), to underexpose. (67)

sous-exposition (Photogr.) (*n.f.*), underexposure. (68)

sous-famille (*n.f.*), subfamily. (69)

sous-fluvial, -e, -aux (*adj.*), subfluvial. (70)

sous forme de, in the form of : (71)
dans les hautes régions, la pluie se précipite sous forme de neige, in the high regions, rain precipitates in the form of snow. (72)

sous-genre (Minéralogie) (*n.m.*), subgroup : (73)
les sous-genres du genre clintonite, the sub-groups of the clintonite group. (74)

sous-glaciaire (*adj.*), subglacial ; infraglacial : (75)

cours d'eau sous-glaciaire (*m.*), subglacial stream ; infraglacial stream. (1)

sous-ingénieur (*n.m.*), assistant engineer. (2)

sous-ingénieur des mines (*m.*), assistant mining-engineer. (3)

sous-inspecteur (*n.m.*), subinspector ; assistant inspector. (4)

sous-jacent, -e (Géol.) (*adj.*), subjacent ; under-lying : (5)

roches sous-jacentes (*f.pl.*), subjacent rocks ; underlying rocks. (6)

sous-lacustre (*adj.*), sublacustrine. (7)

sous-louer (*v.t.*), to sublet ; to sublease. (8)

sous-marin, -e (*adj.*), submarine ; undersea : (9)

mine sous-marine (*f.*), submarine mine ; undersea mine. (10)

sous-métallique (*adj.*), submetallic : (11)

éclat sous-métallique (*m.*), submetallic lustre. (12)

sous-phosphate (*n.m.*), subphosphate. (13)

sous plomb (Isolement), leaded : (14)

câble sous plomb (*m.*), leaded cable. (15)

sous pression (en parlant d'une machine à vapeur), under steam : (16)

la visite des soupapes, quand la machine est sous pression, présente quelque danger, the inspection of valves, when the engine is under steam, is somewhat dangerous. (17)

sous-pression (Méc.) (*n.f.*), underpressure ; pressure exerted from beneath. (18)

sous-produit (*n.m.*), by-product : (19)

sous-produits de la distillation de la houille, by-products of the distillation of coal. (20)

sous réserve d'inspection et de rapport, subject to inspection and report. (21)

sous ruban (Isolement), taped : (22)

fil sous ruban (*m.*), taped wire. (23)

sous-sol (*n.m.*), subsoil ; undersoil. (24)

sous-station (*n.f.*), substation. (25)

sous-station de conversion (Élec.) (*f.*), sub-station for conversion of current. (26)

sous-tendre (Géom.) (*v.t.*), to subtend : (27)

on dit que la corde sous-tend l'arc ou que l'arc est sous-tendu par la corde, we say that the chord subtends the arc or that the arc is subtended by the chord. (28)

sous-traitant (pers.) (*n.m.*), subcontractor. (29)

sous-traité (*n.m.*), subcontract. (30)

sous-traiter (*v.t. & v.i.*), to sublet ; to subcon-tract. (31)

sous tresse (Isolement), braided : (32)

fil conducteur souple sous tresse soie paraffinée (*m.*), flexible conductor-wire silk braided and paraffined. (33)

sous-tropical, -e, -aux (*adj.*), subtropical : (34)

climat sous-tropical (*m.*), subtropical climate. (35)

sous-voltage (des lampes à incandescence) (Élec.) (*n.m.*), underrunning (glow-lamps). (36)

souschevage (action) (Mines) (*n.m.*), under-cutting ; holing ; underholing ; kirving ; kerving ; jadding. (37)

souschevage (entaille) (*n.m.*), undercut ; holing ; kerf ; jad. (38)

souschevé, -e (*adj.*), undercut ; holed ; under-holed. (39)

souschever (*v.t.*), to undercut ; to hole ; to kirve ; to kerve ; to jad. (40)

souscheveur (pers.) (*n.m.*), undercutter ; cutter ; holer. (41)

soute (*n.f.*), bunker. (42)

soute à charbon (*f.*), coal-bunker. (43)

soute à combustible (*f.*), fuel-bunker. (44)

soute à eau (de tender de locomotive) (*f.*), tank ; water-tank. (45)

soute du tender (*f.*), tender-tank. (46)

soutènement (*n.m.*), supporting ; propping ; support ; holding up : (47)

soutènement des chantiers, support of work-ings. (48)

soutenir (*v.t.*), to support ; to hold up : (49)

soutenir une maison avec des étais, to support a house with props. (50)

colonne qui soutient tout l'édifice (*f.*), column which holds up the whole building. (51)

soutenir la masse sous-cavée au moyen de cales, to sprag the holed coal. (52)

souterrain, -e (*adj.*), subterranean ; underground ; belowground : (53)

rivière souterraine (*f.*), subterranean river ; underground river. (54)

souterrain (*n.m.*), underground passage ; tunnel ; subway : (55)

les détours d'un souterrain, the windings of an underground passage. (56)

le souterrain du mont Cenis, the Mont Cenis tunnel. (57)

souterrainement (*adv.*), underground : (58)

une mine exploitée souterrainement, a mine worked underground. (59)

soutien (*n.m.*), support. (60)

soutirer (*v.t.*), to draw off : (61)

soutirer l'eau d'un tonneau, to draw off the water from a cask. (62)

soyeux, -euse (*adj.*), silky : (63)

éclat soyeux (*m.*), silky lustre. (64)

spacieux, -euse (*adj.*), spacious ; capacious. (65)

spalt (fondant) (*n.m.*), spalt. (66)

spalt (bitume de Judée) (*n.m.*), Judean bitumen ; Jew's pitch. (67)

spaniolite (Minéral) (*n.f.*), schwartzite. (68)

spartalite (Minéral) (*n.f.*), spartalite ; zincite. (69)

spath (Minéral) (*n.m.*), spar. (70)

spath brunissant (*m.*), brown spar. (71)

spath calcaire (*m.*), calcareous spar ; calc-spar. (72)

spath d'Islande (*m.*), Iceland spar ; Iceland crystal. (73)

spath fluor (*m.*), fluor spar ; fluor ; fluorite. (74)

spath perlé (*m.*), pearl spar. (75)

spath pesant (*m.*), heavy spar ; barite ; terra ponderosa. (76)

spath satiné (*m.*), satin-spar ; satin-stone. (77)

spathiforme (*adj.*), spathiform. (78)

spathique (*adj.*), spathic ; spathose ; sparry : (79)

fer spathique (*m.*), spathic iron ; sparry iron. (80)

spatule (*n.f.*), spatula. (81)

spatule en corne (*f.*), horn spatula. (82)

speaking-tube [speaking-tubes *pl.*] (tube acous-tique) (*n.m.*), speaking-tube. (83)

spécial, -e, -aux (*adj.*), special ; peculiar ; par-ticular : (84)

acier spécial (*m.*), special steel. (85)

spécialiste (pers.) (*n.m.*), specialist. (86)

spécialité (Commerce) (*n.f.*), specialty ; special-ity. (87)

spécification (*n.f.*), specification : (88)

spécification unifiée britannique pour les matériaux employés dans la construction du matériel roulant des chemins de fer,

British standard specification for material used in the construction of railway rolling stock. (1)

spécifier (*v.t.*), to specify. (2)

spécifique (*adj.*), specific. (3)

spécifiquement (*adv.*), specifically. (4)

spécimen (*adj.*), specimen (*used as adj.*) : (5)
cristal spécimen (*m.*), specimen crystal. (6)

spécimen [spécimens *pl.*] (*n.m.*), specimen. (7)

spectral, -e, -aux (Phys.) (*adj.*), spectral ; spec-trum (*used as adj.*) : (8)
couleurs spectrales (*f.pl.*), spectral colours. (9)
analyse spectrale (*f.*), spectrum analysis. (10)

spectre (*n.m.*), spectrum : (11)
spectre du radium, spectrum of radium. (12)
spectres caractéristiques des éléments chimi-ques, characteristic spectra of the chemical elements. (13)

spectre anormal (*m.*), abnormal spectrum. (14)

spectre calorifique (*m.*), heat spectrum ; thermal spectrum. (15)

spectre cannelé (*m.*), fluted spectrum ; channeled spectrum. (16)

spectre chimique (*m.*), chemical spectrum ; actinic spectrum. (17)

spectre continu (*m.*), continuous spectrum. (18)

spectre d'absorption (*m.*), absorption spectrum. (19)

spectre d'arc *ou* **spectre de l'arc** (*m.*), arc spectrum. (20)

spectre d'étincelle (*m.*), spark spectrum. (21)

spectre de bandes (*m.*), band spectrum. (22)

spectre de diffraction (*m.*), diffraction spectrum. (23)

spectre de flamme (*m.*), flame spectrum. (24)

spectre de lignes (*m.*), line spectrum. (25)

spectre discontinue (*m.*), discontinuous spectrum. (26)

spectre infra-rouge (*m.*), infra-red spectrum ; dark heat spectrum. (27)

spectre magnétique (*m.*), magnetic spectrum ; magnetic phantom ; magnetic fantom ; magnetic curves. (28)

spectre normal (*m.*), normal spectrum. (29)

spectre prismatique (*m.*), prismatic spectrum. (30)

spectre pur (*m.*), pure spectrum. (31)

spectre secondaire (Photogr.) (*m.*), ghost ; flare ; flare-spot ; central spot. (32)

spectre solaire (*m.*), solar spectrum. (33)

spectre ultra-violet (*m.*), ultra-violet spectrum. (34)

spectrographe (*n.m.*), spectrograph. (35)

spectromètre (*n.m.*), spectrometer. (36)

spectroscope (*n.m.*), spectroscope. (37)

spectroscope à vision directe (*m.*), direct-vision spectroscope. (38)

spectroscopie (*n.f.*), spectroscopy. (39)

spectroscopique (*adj.*), spectroscopic ; spectro-scopical : (40)
essai spectroscopique (*m.*), spectroscopic test. (41)

spectroscopiquement (*adv.*), spectroscopically. (42)

spéculaire (*adj.*), specular : (43)
fer spéculaire (*m.*), specular iron. (44)

spéculatif, -ive (*adj.*), speculative : (45)
science spéculative (*f.*), speculative science. (46)
affaire spéculative *ou* affaire de spéculation (*f.*), speculative enterprise. (47)

spéculation (*n.f.*), speculation ; venture. (48)

speerkies (Minéral) (*n.f.*), spear-pyrites. (49)

speiskobalt (Minéral) (*n.m.*), speiss cobalt ; speiskobalt ; gray cobalt ; smaltite. (50)

speiss (Métall.) (*n.m.*), speiss. (51)

spéléologie (*n.f.*), speleology. (52)

spéléologique (*adj.*), speleological. (53)

spéléologiste *ou* **spéléologue** (pers.) (*n.m.*), spele-ologist. (54)

sperkise (Minéral) (*n.f.*), spear-pyrites. (55)

spessartine (Minéral) (*n.f.*), spessartite ; spessart-ine. (56)

sphalérite (Minéral) (*n.f.*), sphalerite ; blende ; zinc blende ; false galena ; pseudogalena ; mock lead ; mock ore ; blackjack ; rosin-jack ; jack ; lead marcasite. (57)

sphène (Minéral) (*n.m.*), sphene. (58)

sphère (*n.f.*), sphere. (59)

sphère creuse (Géom.) (*f.*), hollow sphere. (60)

sphère pleine (Géom.) (*f.*), solid sphere. (61)

sphéricité (*n.f.*), sphericity : (62)
sphéricité de la terre, sphericity of the earth. (63)

sphérique (*adj.*), spherical ; spheric : (64)
miroir sphérique (*m.*), spherical mirror. (65)

sphériquement (*adv.*), spherically. (66)

sphéroïdal, -e, -aux (*adj.*), spheroidal : (67)
cristal sphéroïdal (*m.*), spheroidal crystal. (68)

sphéroïde (*n.m.*), spheroid. (69)

sphéroïdique (*adj.*), spheroidic ; spheroidical. (70)

sphérolithe (Pétrol.) (*n.m.*), spherulite. (71)

sphérolithique (*adj.*), spherulitic : (72)
structure sphérolithique (*f.*), spherulitic structure. (73)

sphéromètre (*n.m.*), spherometer. (74)

sphérophyre (Pétrol.) (*n.m.*), spherophyric rock. (75)

sphérosidérite (Minéral) (*n.f.*), spherosiderite. (76)

sphragide *ou* **sphragidite** (Minéral) (*n.f.*), sphrag-ide. (77)

spiegel *ou* **spiegeleisen** (*n.m.*), spiegeleisen ; spiegel iron ; spiegel ; specular cast iron. (78)

spilite (Pétrol.) (*n.f.*), spilite. (79)

spilosite (Pétrol.) (*n.f.*), spilosite. (80)

spinelle (*adj. invar.*), spinel (*used as adj.*) : (81)
rubis spinelle (*m.*), spinel ruby. (82)
le genre spinelle, the spinel group. (83)

spinelle (Minéral) (*n.m.*), spinel ; spinelle. (84)

spiral, -e, -aux (*adj.*), spiral : (85)
pompe spirale (*f.*), spiral pump. (86)

spirale (*n.f.*), spiral. (87)

spirale d'Archimède (*f.*), spiral of Archimedes. (88)

spirale hyperbolique (*f.*), hyperbolic spiral ; reciprocal spiral. (89)

spirale logarithmique (*f.*), logarithmic spiral. (90)

spiraloïde (*adj.*), spiral : (91)
tambour spiraloïde (*m.*), spiral drum. (92)

spire (*n.f.*), turn ; spire : (93)
bobine à fil fin et à grand nombre de spires (*f.*), coil with many turns of fine wire. (94)
les spires d'une hélice, the turns of a spiral ; the spires of a helix. (95)

spiritueux, -euse (*adj.*), spirituous. (96)

spiroïdal, -e, -aux (*adj.*), spiral. (97)

spitzkasten (Préparation mécanique des minerais) (*n.m.*), pointed box ; funnel-box ; V vat ; spitzkasten. (98)

spitzluttle (Préparation mécanique des minerais) (*n.m.*), spitzluttle. (99)

spodumène (Minéral) (*n.m.*), spodumene ; triphane. (1)

spongieux, -euse (*adj.*), spongy. (2)

spontané, -e (*adj.*), spontaneous : (3)
combustion spontanée (*f.*), spontaneous com- -bustion. (4)

spontanément (*adv.*), spontaneously : (5)
combustible qui est susceptible de s'échauffer spontanément (*m.*), fuel which is liable to heat spontaneously. (6)

spring-jack [spring-jacks *pl.*] (Élec. & Téléph.) (*n.m.*), spring-jack. (7)

squatter (pers.) (*n.m.*), squatter. (8)

squeezer (cylindre compresseur) (pour fer puddlé) (*n.m.*), rotary squeezer ; cam squeezer ; squeezer. (9)

squelette de cristal [squelettes de cristaux *pl.*] (Cristall.) (*m.*), skeleton crystal. (10)

stabilité (*n.f.*), stability : (11)
stabilité des barrages en maçonnerie, stability of masonry dams. (12)

stabilité de production d'un puits à pétrole (*f.*), staying qualities of an oil-well. (13)

stable (*adj.*), stable : (14)
équilibre stable (*m.*), stable equilibrium. (15)

stade (degré ; partie distincte d'un développe- -ment) (*n.m.*), stage : (16)
les rajeunissements des cours d'eau arrivés au stade de vieillesse (*m.pl.*), the rejuvenations of streams arrived at the stage of old age. (17)
les stades d'une évolution, the stages of an evolution. (18)
au stade primitif, in the early stage. (19)

stadia (*n.m.*), stadia ; stadium : (20)
levé des plans au stadia (*m.*), stadia-surveying. (21)

stadiomètre (*n.m.*), stadiometer ; stadimeter. (22)

stagnant, -e (*adj.*), stagnant : (23)
eau stagnante (*f.*), stagnant water. (24)

stagnation (*n.f.*), stagnation. (25)

stalactiforme (*adj.*), stalactiform. (26)

stalactite (*n.f.*), stalactite : (27)
les stalactites suspendues à la voûte des grottes, ou disposées en draperies contre leur parois, les stalagmites qui reposent sur le sol et s'élancent à la rencontre des stalactites, résultent de la dissolution des couches calcaires par les eaux d'infiltration, the stalactites pendent from the roofs of grottoes, or draping their walls, the stalagmites resting on the ground and rising to meet the stalactites, are the result of the dissolution of limestone beds by percolating water. (28)

stalactitite *ou* **stalactifère** (*adj.*), stalactitic ; stalactitical ; stalactic ; stalactical ; stalac- -tited : (29)
grotte stalactifère (*f.*), stalactited grotto. (30)

stalagmite (*n.f.*), stalagmite. (31)

stalagmitique (*adj.*), stalagmitic ; stalagmitical : (32)
concrétion stalagmitique (*f.*), stalagmitic concretion. (33)

stalagmomètre (*n.m.*), stalagmometer ; stactom- -eter. (34)

stannate (Chim.) (*n.m.*), stannate. (35)

stanneux, -euse (Chim.) (*adj.*), stannous : (36)
oxyde stanneux (*m.*), stannous oxide. (37)

stannifère (*adj.*), stanniferous ; tin-bearing ; tin (used as adj.) : (38)
terrain stannifère (*m.*), tin-bearing ground ; tin-ground. (39)

les gîtes stannifères de Cornouailles, the tin- deposits of Cornwall. (40)

stannine (Minéral) (*n.f.*), stannite ; stannine ; tin pyrites ; bell-metal ore. (41)

stannique (Chim.) (*adj.*), stannic : (42)
oxyde stannique (*m.*), stannic oxide. (43)

stappe (Mines) (*n.m.*), pillar. (44)

stassfurtite (Minéral) (*n.f.*), stassfurtite. (45)

statif (d'un microscope) (*n.m.*), stand (of a micro- -scope) : (46)
statif inclinable, sans objectifs ni oculaires, en boîte acajou fermant à clef, inclinable stand, without objectives or eyepieces, in mahog- -any lock-box. (47)

station (*n.f.*), station. (48)

station (Ch. de f.) (*n.f.*), station ; intermediate station ; way station. (49)

station (Géod.) (*n.f.*), station. (50)

station centrale d'électricité *ou* **station centrale d'énergie** *ou* **station centrale électrique** *ou* **station centrale génératrice** (*f.*), central electric station ; central power-station ; central generating-station. (51)

station d'angle (voie aérienne) (*f.*), angle-station. (52)

station d'épuisement *ou* **station de pompes** (*f.*), pumping-station ; pump-station. (53)

station d'essais (*f.*), testing-station. (54)

station de charge (accumulateurs) (Élec.) (*f.*), charging-station. (55)

station de chargement *ou* **station de départ** (câble aérien, etc.) (*f.*), loading-station. (56)

station de chemin de fer (*f.*), railway-station. (57)

station de culbutage (*f.*), dumping-station ; tippler-station. (58)

station de déchargement *ou* **station d'arrivée** (câble aérien) (*f.*), unloading-station ; discharging-station ; off-loading station ; knock-out station. (59)

station de distribution (*f.*), distributing-station. (60)

station de force motrice *ou* **station génératrice** *ou* **station génératrice d'énergie** (*f.*), power- station ; power-house ; generating-station ; power-generating station. (61)

station génératrice d'énergie électrique (*f.*), electric-power station. (62)

station grisoumétrique (*f.*), station for testing fire-damp ; fire-damp testing-station. (63)

station météorologique (*f.*), meteorological station. (64)

station réceptrice (*f.*), receiving station. (65)

station télégraphique (*f.*), telegraph-station. (66)

station terminus (*f.*), terminal station ; terminal ; terminus. (67)

station transmettrice *ou* **station émettrice** (*f.*), transmitting station. (68)

stationnement (*n.m.*), halt : (69)
le stationnement d'un glacier, the halt of a glacier. (70)

stationner (*v.i.*), to halt. (71)

statique (*adj.*), static ; statical : (72)
électricité statique (*f.*), static electricity. (73)
ventilateur statique (*m.*), statical fan. (74)

statique (*n.f.*), statics. (75)

statique graphique (*f.*), graphical statics. (76)

statistique (*adj.*), statistical. (77)

statistique (*n.f.*), statistics ; returns : (78)
statistique de la production mondiale d'or, statistics of the world's production of gold. (79)
statistique mensuelle, monthly statistics ; monthly returns. (80)

stator (Élec.) (*n.m.*), stator. (1)

stauroscope (Cristall.) (*n.m.*), stauroscope. (2)

staurotide *ou* staurolite (Minéral) (*n.f.*), staurolite ; staurotide. (3)

staurotique (Minéralogie) (*adj.*), staurolitic. (4)

steam-boat [steam boats *pl.*] *ou* steamboat (*n.m.*), steamboat. (5)

steam-loop [steam-loops *pl.*] (boucle de vapeur) (*n.m.*), steam-loop. (6)

steamer (*n.m.*), steamer ; steamship. (7)

steamer au long cours (*m.*), ocean-going steamer. (8)

steamer fluvial (*m.*), river-steamer. (9)

stéatite (Minéral) (*n.f.*), steatite. (10)

stéatiteux, -euse (*adj.*), steatitic. (11)

steinmannite (Minéral) (*n.f.*), steinmannite. (12)

stéphanite (Minéral) (*n.f.*), stephanite ; psaturose ; brittle silver ore. (13)

stercorite (Minéral) (*n.f.*), stercorite. (14)

stère (*n.m.*), stere = 1 cubic metre or 35·315 cubic feet. (15) (A French measure for firewood, analogous in use to the English word cord).

stéréométrie (*n.f.*), stereometry. (16)

stéréométrique (*adj.*), stereometric ; stereo-metrical. (17)

stéréoscope (*n.m.*), stereoscope. (18)

stéréoscope lenticulaire (*m.*), lenticular stereo-scope. (19)

stéréoscope réflecteur (*m.*), reflecting stereoscope. (20)

stéréoscopie (*n.f.*), stereoscopy. (21)

stéréoscopique (*adj.*), stereoscopic : (22)
   vue stéréoscopique (*f.*), stereoscopic view ; stereoscopic picture. (23)

stéréoscopiquement (*adv.*), stereoscopically. (24)

stérile (Mines, etc.) (*adj.*), barren ; sterile ; unproductive ; hungry ; mullocky : (25)
   terrain stérile (*m.*), barren ground. (26)

stérile (*n.m.*) *ou* stériles (*n.m.pl.*) (Mines), barren rock ; sterile rock ; deads ; attle ; chats ; waste ; leavings ; mullock : (27)
   pilier de stérile (*m.*), pillar of barren rock. (28)

sternbergite (Minéral) (*n.f.*), sternbergite. (29)

stibieux, -euse (Chim.) (*adj.*), stibious ; anti-monious. (30)

stibine *ou* stibnite (Minéral) (*n.f.*), stibnite ; antimony-glance ; gray antimony ; anti-monite. (31)

stibique (Chim.) (*adj.*), stibic ; antimonic. (32)

stiblite *ou* stibiconise (Minéral) (*n.f.*), stiblite ; stibiconite. (33)

stigmatique (Opt.) (*adj.*), stigmatic ; anastig-matic ; anastigmat. (34)

stilbite (Minéral) (*n.f.*), stilbite. (35)

stilpnomélane (Minéral) (*n.f.*), stilpnomelane. (36)

stimulant (*n.m.*), stimulus ; inducement ; in-centive. (37)

stimuler (*v.t.*), to stimulate : (38)
   stimuler l'intérêt concernant les affaires minières, to stimulate interest in mining. (39)

stinkal (*n.m.*), stinkstone. (40)

stock (existence) (*n.m.*), stock : (41)
   stock de charbon, de pièces de rechange, stock of coal, of spare parts. (42 *ou* 43)
   stock en magasin, stock in hand ; stock on hand. (44)

stock (dépôt) (*n.m.*), stock : (45)
   pièces de rechange toujours en stock (*f.pl.*), spare parts always in stock. (46)

stock (Mines) (*n.m.*). Same as **stot**.

stockwerk (Géol.) (*n.m.*), stockwork ; stock-werk. (47)

stolzite (Minéral) (*n.f.*), stolzite. (48)

stomonal (Explosif) (*n.m.*), stomonal. (49)

stone (mesure anglaise de poids) (*n.m.*), stone = 14 lbs *or* 6·350 kilogrammes. (50)

stot (dans une galerie de mine) (*n.m.*), arch (portion of lode left standing to support hanging wall). (51)

stot (dans un puits de mine) (*n.m.*), pentice (roof of rock or slice of ground left provisionally between the bottom of a shaft and its exten-sion beneath). (52)

stowite (Explosif) (*n.f.*), stowite. (53)

stratamètre (*n.m.*), stratameter. (54)

strate (*n.f.*), stratum : (55)
   strates à faible inclinaison, gently dipping strata. (56)
   strates affleurant au jour, outcropping strata. (57)
   strates favorables, kind strata ; kindly strata. (58)

stratification (Géol.) (*n.f.*), stratification ; bed-ding. (59)

stratification concordante *ou* stratification con-forme (*f.*), conformable stratification. (60)

stratification discordante (*f.*), unconformable stratification ; discordant stratification. (61)

stratification entrecroisée *ou* stratification tor-renticlle *ou* stratification oblique (*f.*), cross-bedding ; current-bedding ; false bedding ; cross-lamination ; oblique lamination ; diagonal stratification. (62)
   grès à stratification entrecroisée (*m.*), false-bedded sandstone. (63)

stratifié, -e (*adj.*), stratified ; bedded : (64)
   roches stratifiées (*f.pl.*), stratified rocks ; bedded rocks. (65)

stratifier (*v.t.*), to stratify. (66)

stratifier (se) (*v.r.*), to stratify. (67)

stratiforme (*adj.*), stratiform. (68)

stratigraphie (*n.f.*), stratigraphy. (69)

stratigraphique (*adj.*), stratigraphic ; strati-graphical : (70)
   géologie stratigraphique (*f.*), stratigraphic geology. (71)

stratigraphiquement (*adv.*), stratigraphically. (72)

strato-volcan [strato-volcans *pl.*] (*n.m.*), com-posite cone. (73)

striction (Résistance des matériaux) (*n.f.*), con-traction of area (Strength of materials). (74)

strie (*n.f.*), stria ; ridge ; corrugation ; chequer-ing ; checkering ; checking ; groove ; streak ; scratch : (75)
   stries d'une coquille, striæ of a shell. (76)
   stries glaciaires, glacial striæ ; glacial scratches: (77)
      les stries glaciaires sont dues au frottement des pierrailles enchâssées dans la glace, glacial striæ are due to the abrasion of stones set in the ice. (78)
   stries d'acier à vives arêtes, sharp ridges of steel. (79)
   stries sur le corps d'un pointeau, chequerings, checking, on the barrel of a centre-punch. (80)

strié, -e (*adj.*), ridged ; corrugated ; chequered ; checkered ; grooved ; streaked ; streaky : (81)
   mâchoires striées (*f.pl.*), corrugated jaws. (82)

tôle striée (*f.*), chequered plate. (1)

strié, -e (Géol.) (*adj.*), striated ; striate ; strial : (2)

  roches striées (*f.pl.*), striated rocks ; striate rocks. (3)

strier (*v.t.*), to striate ; to groove ; to corrugate ; to chequer ; to checker ; to streak : (4)

  glacier qui strie les roches (*m.*), glacier which striates the rocks. (5)

striure (état) (*n.f.*), striation ; corrugation ; chequering. (6)

striure (strie) (*n.f.*), stria ; corrugation. (7)

stromeyérite (Minéral) (*n.f.*), stromeyerite. (8)

strontianite (Minéral) (*n.f.*), strontianite. (9)

strontique (Chim.) (*adj.*), strontic ; strontitic. (10)

strontium (Chim.) (*n.m.*), strontium. (11)

stross (Mines) (*n.m.*), core (of rock to be taken out of the centre or side of a heading, when driving, or of a shaft, when sinking) : (12)

  abattre le stross central au moyen de coups de mine, ou au pic, to break down the centre core by means of shots, or with the pick. (13 *ou* 14)

structural, -e, -aux (*adj.*), structural : (15)

  surface structurale d'une contrée (Géol.) (*f.*), structural surface of a country. (16)

  montagnes structurales (*f.pl.*), structural mountains. (17)

  vallées structurales (*f.pl.*), structural valleys. (18)

structuralement (*adv.*), structurally. (19)

structure (*n.f.*), structure : (20)

  structure des corps cristallisés, structure of crystallized substances. (21)

  pont de structure massive (*m.*), bridge of massive structure. (22)

structure (construction ; bâtisse) (*n.f.*), structure : (23)

  structure en bois, wooden structure. (24)

structure (Géol.) (*n.f.*), structure : (25)

  structure géologique du pays, geological structure of the country. (26)

structure alvéolaire (usure éolienne, etc.) (*f.*), honeycomb structure. (27)

structure amygdaloïde (*f.*), amygdaloid structure. (28)

structure bréchiforme (*f.*), brecciated structure. (29)

structure cisaillée (*f.*), shear-structure. (30)

structure concrétionnée (*f.*), concretionary structure. (31)

structure en écailles (*f.*), imbricate structure ; imbricated structure. (32)

structure en éventail (*f.*), fan-shaped structure. (33)

structure en mortier (*f.*), mortar-structure. (34)

structure fluidale (*f.*), flow-structure ; fluidal structure ; fluxion-structure ; fluxional structure ; fluctuation structure ; flowage. (35)

structure foliacée (*f.*), foliaceous structure. (36)

structure granitique (*f.*), granitic structure. (37)

structure granitoïde (*f.*), granitoid structure. (38)

structure grenue *ou* structure granulaire (*f.*), granular structure. (39)

structure imbriquée (*f.*), imbricate structure ; imbricated structure. (40)

structure isoclinale (*f.*), isoclinal structure. (41)

structure microcristalline (*f.*), microcrystalline structure. (42)

structure microgrenue (*f.*), microgranular structure. (43)

structure microlithique (*f.*), microlithic structure ; microlitic structure. (44)

structure ophitique (*f.*), ophitic structure. (45)

structure rubanée *ou* structure zonaire *ou* structure zonée (*f.*), banded structure ; streaky structure. (46)

structure vacuolaire (*f.*), vesicular structure. (47)

structure vitreuse (*f.*), vitreous structure. (48)

struvite (Minéral) (*n.f.*), struvite. (49)

stuffing-box [stuffing-boxes *pl.*] (presse-étoupe) (*n.m.*), stuffing-box. (50)

stylolithe (Géol.) (*n.m.*), stylolite. (51)

stylolithique (*adj.*), stylolitic. (52)

suage (outil) (*n.m.*), swage. (53)

suager (*v.t.*), to swage. (54)

suant, -e (*adj.*), sweating : (55)

  murailles suantes (*f.pl.*), sweating walls. (56)

subaérien, -enne (Géol.) (*adj.*), subaerial ; sur--ficial ; superterranean : (57)

  érosion subaérienne (*f.*), subaerial erosion. (58)

  roches subaériennes (*f.pl.*), subaerial rocks. (59)

  la rivière subaérienne naît souvent d'une rivière souterraine, the surficial river often originates from an underground river. (60)

subalpin, -e (*adj.*), subalpine. (61)

subcarbonifère (Géol.) (*adj.*), subcarboniferous. (62)

subcristallin, -e (*adj.*), subcrystalline : (63)

  calcaire subcristallin (*m.*), subcrystalline limestone. (64)

subdiviser (*v.t.*), to subdivide ; to split : (65)

  subdiviser le courant d'air (Mines), to split the air-current. (66)

subdiviser (se) (*v.r.*), to subdivide ; to split. (67)

subdivision (*n.f.*), subdivision ; splitting : (68)

  subdivision du courant d'air (Mines), splitting the air-current. (69)

subfossile (*adj.*), subfossil. (70)

subfossile (*n.m.*), subfossil. (71)

subgranulaire *ou* subgranuleux, -euse (*adj.*), subgranular. (72)

subir (être soumis à) (*v.t.*), to undergo ; to go through ; to pass through ; to be subjected to ; to sustain ; to stand ; to suffer : (73)

  la chaleur fait subir à l'alun la fusion aqueuse, heat causes alum to undergo aqueous fusion. (74)

  résidus qui subissent plusieurs lavages (*m.pl.*), residues which are subjected to several washings. (75)

  les fontes écossaises peuvent subir cinq à six refontes sans blanchir (*f.pl.*), Scotch pig irons can stand five to six remeltings without whitening. (76)

subit, -e (*adj.*), sudden : (77)

  changement subit (*m.*), sudden change. (78)

subitement (*adv.*), suddenly. (79)

subjacent, -e (Géol.) (*adj.*), subjacent ; under--lying : (80)

  couche subjacente (*f.*), subjacent bed ; under--lying stratum. (81)

sublimable (*adj.*), sublimable. (82)

sublimation (Chim. & Géol.) (*n.f.*), sublimation. (83)

sublimatoire (*adj.*), sublimatory. (84)

sublimatoire (appareil) (*n.m.*), sublimer. (85)

sublimé, -e (*adj.*), sublimed ; sublimated : (86)

  arsenic sublimé (*m.*), sublimed arsenic. (87)

sublimé (*n.m.*), sublimate. (88)

sublimer (*v.t.*), to sublimate ; to sublime : (1) sublimer le soufre, to sublimate, to sublime, sulphur. (2)

sublimer (se) (*v.r.*), to sublime : (3) l'arsenic se sublime sans fondre lorsqu'on le chauffe à l'air libre vers 400° (*m.*), arsenic sublimes without melting when it is heated in the open air at about 400° C. (4)

submergé, -e (*adj.*), submerged ; flooded. (5)

submerger (*v.t.*), to submerge ; to flood : (6) submerger une mine (en cas d'incendie), to flood a mine (in case of fire). (7)

submersion (*n.f.*), submersion ; submergence. (8)

subséquent, -e (Géol.) (*adj.*), subsequent : (9) cours d'eau subséquent (*m.*), subsequent stream. (10)

substance (*n.f.*), substance. (11)

substance diamagnétique (*f.*), diamagnetic substance ; diamagnetic ; diamagnet : (12) le bismuth est le type des substances dia-magnétiques, bismuth is the type of dia-magnetics. (14)

substance étrangère (*f.*), foreign substance. (14)

substance magnétique (*f.*), magnetic substance ; magnetic. (15)

substance terreuse (*f.*), earthy substance. (16)

substantiel, -elle (*adj.*), substantial. (17)

substantiellement (*adv.*), substantially. (18)

substituer (*v.t.*), to substitute ; to replace : (19) substituer l'énergie de la vapeur à la puissance hydraulique, to substitute steam for water-power ; to replace water-power by steam. (20)

substituer (Chim.) (*v.t.*), to substitute : (21) substituer l'hydrogène par un métal, to substi-tute hydrogen by a metal. (22)

substitut (*n.m.*), substitute. (23)

substitution (*n.f.*), substitution ; replacement : (24) substitution de la traction électrique à la traction à vapeur, substitution of electric traction for steam-traction. (25)

substitution (Chim.) (*n.f.*), substitution ; meta-lepsy ; metalepsis ; metathesis : (26) substitution à l'hydrogène du chlore, substitu-tion of chlorine for hydrogen. (27)

substruction *ou* **substructure** (*n.f.*), substructure. (28)

subtranslucide (*adj.*), subtranslucent. (29)

subtropical, -e, -aux (*adj.*), subtropical : (30) climats subtropicaux (*m.pl.*), subtropical climates. (31)

subvention (*n.f.*), subsidy ; subvention. (32)

subventionner (*v.t.*), to subsidize ; to subvention. (33)

succès (*n.m.*), success. (34)

successif, -ive (*adj.*), successive : (35) procéder par étapes successives, to proceed by successive stages. (36)

succession (*n.f.*), succession. (37)

successivement (*adv.*), successively. (38)

succin (*n.m.*) *ou* **succinite** (*n.f.*) (ambre), succinite ; succin ; amber ; yellow amber. (39)

succinite (grenat) (*n.f.*), succinite. (40)

succion (*n.f.*), suction. (41)

sud (*adj.*), south ; southern ; southerly. (42)

sud (*n.m.*), south. (43)

sud-est (*adj.*), southeast ; southeastern ; south-easterly. (44)

sud-est (*n.m.*), southeast. (45)

sud-ouest (*adj.*), southwest ; southwestern ; southwesterly. (46)

sud-ouest (*n.m.*), southwest. (47)

sud-sud-est (*adj.* & *n.m.*), south-southeast. (48)

sud-sud-ouest (*adj.* & *n.m.*), south-southwest. (49)

suer (Métall.) (*v.t.*), to sweat : (50) suer le fer, to sweat iron. (51)

suer (*v.i.*), to sweat : (52) les murs suent dans les temps humides (*m.pl.*), walls sweat in wet weather. (53)

suffione [**suffioni** *pl.*] (Géol.) (*n.m.*), soffione ; suffione ; air-volcano ; mud-volcano (non-volcanic). (54)

suffocant, -e (*adj.*), suffocating. (55)

suffocation (*n.f.*), suffocation. (56)

suffoquer (*v.t.* & *v.i.*), to suffocate. (57)

suie (*n.f.*), soot. (58)

suie arsenicale (*f.*), arsenical soot. (59)

suif (*n.m.*), tallow : (60) chandelle de suif (*f.*), tallow candle. (61)

suif minéral (*m.*), mineral tallow ; mountain-tallow ; hatchettite. (62)

suintant, -e (*adj.*), oozy ; seepy ; sweating ; weeping : (63) sol suintant (*m.*), oozy ground. (64) roche suintante (*f.*), weeping rock. (65)

suintement (*n.m.*), oozing ; ooze ; seepage ; exudation ; sweating ; filtration ; per-colation ; weeping : (66) certaines sources ne se manifestent que par des suintements (*f.pl.*), certain springs only make their presence known by oozings. (67) suintement de pétrole, seepage of oil. (68) suintement d'une muraille, sweating of a wall. (69) suintement d'une roche, weeping of a rock. (70)

suinter (*v.i.*), to ooze ; to seep ; to exude ; to sweat ; to filter ; to percolate ; to weep : (71) eau qui suinte à travers le sol (*f.*), water which oozes through the ground. (72)

suiveur de rames (Mines) (pers.) (*m.*), trip-rider ; rider ; train-boy ; tram-boy. (73)

suivre (*v.t.*), to follow ; to follow up ; to carry out : (74) suivre le cours d'un fleuve, to follow the course of a river. (75) suivre une entreprise, to carry out, to follow up, an enterprise. (76)

suivre (*v.i.*), to follow. (77)

suivre à la trace, to trace ; to train : (78) suivre un filon à la trace, to trace, to train, a lode. (79)

suivre la direction de, to follow a course ; to run ; to strike : (80) le filon suit la direction du nord, the lode follows a northerly course ; the lode runs north ; the lode strikes north. (81)

suivre un cours d'instruction pour travaux de sauvetage dans les mines, to undergo a course of training for mine-rescue work. (82)

sujet, -ette (*adj.*), subject ; liable : (83) être sujet (-ette) à l'impôt foncier, to be subject to (*or* to be liable for) land-tax. (84)

sujet (*n.m.*), subject. (85)

sulfacide (*n.m.*), sulphur acid ; sulpho acid ; sulphacid ; sulfacid. (86)

sulfantimonieux (*adj. m.*), sulphantimonious ; sulfantimonious. (87)

sulfantimonique (*adj.*), sulphantimonic ; sulf-antimonic. (88)

sulfarséniate (*n.m.*), sulpharsenate ; sulfarsenate. (89)

sulfarsénieux (*adj. m.*), sulpharsenious ; sulf-arsenious. (1)

sulfarsénique (*adj.*), sulpharsenic ; sulfarsenic. (2)

sulfarséniure (*n.m.*), sulpharsenide ; sulfarsenide. (3)

sulfatation (*n.f.*), sulphatizing. (4)

sulfate (Chim.) (*n.m.*), sulphate ; sulfate : (5)

sulfate de chaux, sulphate of lime ; lime sulphate. (6)

sulfate de cuivre, sulphate of copper ; copper sulphate ; bluestone ; blue vitriol ; copper vitriol ; blue copperas. (7)

sulfaté, -e (Chim.) (*adj.*), sulphatic : (8)

eau minérale sulfatée (*f.*), sulphatic mineral water. (9)

chaux sulfatée (*f.*), sulphate of lime. (10)

sulfater (*v.t.*), to sulphate : (11)

sulfater des traverses de chemin de fer, to sulphate railway-sleepers. (12)

sulfatique (Chim.) (*adj.*), sulphatic. (13)

sulfhydrate (Chim.) (*n.m.*), sulphydrate ; sul-fohydrate ; sulfhydrate. (14)

sulfhydrique (Chim.) (*adj.*), sulphydric ; sulf-hydric. (15)

sulfite (Chim.) (*n.m.*), sulphite ; sulfite : (16)

sulfite de sodium, sulphite of sodium ; sodium sulphite. (17)

sulfobase (*n.f.*), sulphur base. (18)

sulfocarbonate (*n.m.*), sulphocarbonate ; sul-focarbonate. (19)

sulfocarbonique (*adj.*), sulphocarbonic ; sul-focarbonic. (20)

sulfochlorure (*n.m.*), sulphur chloride. (21)

sulfosel (*n.m.*), sulphur salt ; sulpho salt ; sul-fosalt ; sulfosel. (22)

sulfuration (*n.f.*), sulphuration ; sulfuration ; sulphurization : (23)

sulfuration directe du métal par le soufre, direct sulphuration of the metal by sulphur. (24)

sulfure (Chim.) (*n.m.*), sulphide ; sulfide : (25)

sulfure de carbone (CS²), carbon bisulphide ; carbon disulphide. (26)

sulfure de plomb, sulphide of lead ; lead sulphide. (27)

sulfures métalliques, metallic sulphides ; sulphurets. (28)

sulfuré, -e (Chim.) (*adj.*), sulphureted ; sul-phuretted : (29)

hydrogène sulfuré (*m.*), sulphureted hydrogen. (30)

sulfurer (*v.t.*), to sulphurate ; to sulfurate ; to sulphurize. (31)

sulfureux, -euse (*adj.*), sulphureous ; sulphurous ; sulfureous ; sulfurous ; sulphury : (32)

vapeurs sulfureuses (*f.pl.*), sulphurous fumes. (33)

sulfurifère (*adj.*), sulphur-bearing : (34)

calcaire sulfurifère (*m.*), sulphur-bearing lime-stone. (35)

sulfurique (*adj.*), sulphuric ; sulfuric : (36)

acide sulfurique (*m.*), sulphuric acid. (37)

sulfurisation (*n.f.*). Same as sulfuration.

superficie (*n.f.*), superficies ; area : (38)

superficie de la terre, superficies of the earth. (39)

superficie d'un champ, area of a field. (40)

superficiel, -elle (qui a rapport à la superficie) (*adj.*), superficial : (41)

étendue superficielle (*f.*), superficial extent. (42)

superficiel, -elle (situé à la surface) (*adj.*), super-ficial ; surface (*used as adj.*) ; aboveground : (43)

alluvion superficielle (*f.*), superficial alluvial. (44)

dynamitière superficielle (*f.*), surface dynamite-store ; aboveground dynamite-stores. (45)

superficiel, -elle (léger) (*adj.*), superficial : (46)

examen superficiel (*m.*), superficial examina-tion. (47)

superficiellement (*adv.*), superficially. (48)

superfin, -e (*adj.*), superfine. (49)

supérieur, -e (situé au-dessus) (*adj.*), upper ; top ; superior : (50)

les étages supérieurs d'un édifice (*m.pl.*), the upper stories (*or* the top stories) of a building. (51)

supérieur, -e (qui atteint un degré plus élevé) (*adj.*), higher ; superior : (52)

température de beaucoup supérieure à la normale (*f.*), much higher temperature than normal. (53)

supérieur, -e (qui dépasse les autres) (*adj.*), superior : (54)

pelles en acier fondu supérieur (*f.pl.*), superior cast-steel shovels. (55)

supériorité (*n.f.*), superiority : (56)

supériorité de niveau, superiority of level. (57)

superphosphate (Chem.) (*n.m.*), superphosphate : (58)

superphosphate de chaux, superphosphate of lime. (59)

superposer (*v.t.*), to superpose. (60)

superposition (*n.f.*), superposition ; overplace-ment : (61)

superposition des terrains, superposition of the strata. (62)

superstructure (*n.f.*), superstructure : (63)

superstructure d'un pont roulant, superstruc-ture of an overhead travelling crane. (64)

superstructure de la voie (Ch. de f.), super-structure of the permanent way. (65)

supplément (*n.m.*), supplement ; extra : (66)

supplément de dépense, extra expense ; further expense : (67)

supplément de dépense qui est largement compensé par la diminution des frais d'entretien, extra expense which is largely set off by the reduction of upkeep costs. (68)

supplémentaire (*adj.*), supplementary ; supple-mental ; extra. (69)

supplémentaire (Géom.) (*adj.*), supplementary ; supplemental : (70)

angles supplémentaires (*m.pl.*), supplementary angles ; supplemental angles. (71)

support (*n.m.*), support ; supporting : (72)

support d'une voûte, support of an arch. (73)

support de tuyaux dans le puits, supporting pipes in the shaft. (74)

support (dispositif) (*n.m.*), support ; rest ; stand ; holder. (75)

support (palier) (Méc.) (*n.m.*), pedestal ; pillow ; pillow-block. (76)

support (de noyaux) (Fonderie) (*n.m.*), chaplet (for cores). (77)

support à anneau (Appareil de laboratoire) (*m.*), stand with ring. (78)

support à chariot (tour) (*m.*), slide-rest ; slide-head ; turning-rest (lathe). (79)

support à chariot à double coulisse *ou* support à chariot à mouvements longitudinal et transversal (*m.*), compound slide-rest. (80)

support à chariot pivotant, à base graduée (*m.*), swivel slide-rest with graduated base. (1)
support à éventail (tour) (*m.*), T rest ; hand-rest. (2)
support à fourche (laboratoire) (*m.*), hook support. (3)
support à main (tour) (*m.*), hand-rest ; turning-rest. (4)
support à main pour tourner en l'air (tour à bois) (*m.*), pillar hand-rest (wood-turning lathe). (5)
support à noyaux (support pour dépôt de noyaux) (Fonderie) (*m.*), core-rack. (6)
support à pince (laboratoire) (*m.*), stand with clamp. (7)
support à trois pieds (*m.*), tripod stand. (8)
support d'arbre porte-fraise [supports d'arbres porte-fraise *pl.*] (machine à fraiser) (*m.*), arbor-support (milling-machine). (9)
support d'équerre (*m.*), angle-plate. (10)
support d'outil (*m.*), tool-rest. (11)
support de pied (d'un appareil photographique) (*m.*), supporting-leg (of a camera). (12)
support des glissières (glissières de la crosse de piston) (*m.*), guide-yoke ; guide-bearer ; guide-crosstie ; spectacle-plate ; motion-plate. (13)
support double (de noyaux) (Fonderie) (*m.*), stud ; single stud ; stud chaplet ; double-headed chaplet. (14)
support double étamé, plaque ronde ou carrée, single stud, tinned, round or square plate. (15 *ou* 16)
support double à platine (de noyaux) (*m.*), double stud. (17)
support en plateau à hauteur variable (*m.*), stand with rising table. (18)
support en V pour le traçage (*m.*), draughts-man's V block ; engineers' vee block : (19)
support en V, à une entaille, à — entailles, V block with one vee, with — vees. (20 *ou* 21)
support porte-fraise (machine à fraiser) (*m.*), overhanging arm (milling-machine). (22)
support pour burettes (*m.*), burette-stand. (23)
support pour entonnoir (*m.*), funnel-stand ; stand for funnel ; funnel-holder. (24)
support pour tubes à essais (*m.*), test-tube stand. (25)
support réglable (machines-outils) (*m.*), adjust-able rest. (26)
support simple (de noyaux) (Fonderie) (*m.*), stem chaplet ; staple : (27)
support simple, creux ou plat, stem chaplet, bent or flat. (28 *ou* 29)
support triple à platine (de noyaux) (Fonderie) (*m.*), treble stud ; triple stud chaplet. (30)
supporter (porter ; soutenir) (*v.t.*), to support ; to bear ; to carry : (31)
poutre supportée par des tréteaux (*f.*), beam (*or* girder) supported (*or* borne) (*or* carried) on trestles. (32)
supporter une galerie par des piliers, to support a gallery by pillars. (33)
supporter les frais d'un voyage, to bear the expenses of a journey. (34)
supporter (être à l'épreuve de) (*v.t.*), to stand ; to withstand : (35)
supporter une pression de — kilos par centi-mètre carré, to stand a pressure of — kilos to the square centimetre. (36)
supporter la pression du toit, to withstand the pressure of the roof. (37)

supposer (*v.t.*), to suppose : (38)
propriété supposée être sans valeur (*f.*), prop-erty supposed to be worthless. (39)
suppression des passages à niveau (Ch. de f.) (*f.*), abolishing level crossings ; eliminating grade crossings. (40)
supprimer (*v.t.*), to do away with ; to dispense with ; to abolish ; to eliminate ; to suppress : (41)
supprimer la main-d'œuvre, to do away with, to dispense with, hand-labour. (42)
supputation (*n.f.*), computation ; reckoning ; calculation. (43)
supracrétacé, -e (Géol.) (*adj.*), supercretaceous ; supracretaceous ; Upper Cretaceous : (44)
la flore supracrétacée, the supercretaceous flora ; the Upper Cretaceous flora. (45)
supracrétacé (*n.m.*), Upper Cretaceous : (46)
le supracrétacé, the Upper Cretaceous. (47)
suprajurassique (Géol.) (*n.m.*), White Jura. (48)
sur demande, on application : (49)
prix sur demande (*m.pl.*), prices on application. (50)
sur le carreau de la mine, at the pit's mouth : (51)
prix du charbon sur le carreau de la mine (*m.*), price of coal at the pit's mouth. (52)
sur place *ou* **sur les lieux,** locally ; on the spot ; on the ground : (53)
la tournure de zinc est fabriquée sur place, the zinc shavings are made on the spot. (54)
sur une base industrielle, on a commercial basis. (55)
sur une grande échelle, on a large scale ; extensively. (56)
sur une petite échelle, on a small scale ; in a small way. (57)
sur wagon, on rail : (58)
franco sur wagon, free on rail ; f.o.r. (59)
sûr, -e (*adj.*), sure ; certain ; safe ; secure ; reliable : (60)
profit sûr (*m.*), sure profit ; certain profit. (61)
port sûr (*m.*), safe port. (62)
la soupape la plus sûre qu'il soit possible de fabriquer, the most reliable valve it is possible to make. (63)
suraffinage (*n.m.*), overrefining. (64)
suraffiner (*v.t.*), to overrefine. (65)
suralésage (*n.m.*), counterbore. (66)
suraléser (*v.t.*), to counterbore. (67)
suralimentation (*n.f.*), overfeeding ; overfeed. (68)
suranné, -e (*adj.*), out of date ; antiquated ; obsolete. (69)
surcalciner (*v.t.*), to overcalcine. (70)
surcharge (excès de charge) (*n.f.*), overload ; overloading ; overstress ; overstressing ; overworking : (71)
les intensités de courant indiquées ne doivent être dépassées que pour des surcharges de très courte durée (*f.pl.*), the current-strengths shown should not be exceeded except for overloads of very short duration. (72)
surcharge d'épreuve (de chaudière) (*f.*), test pressure. (73)
surcharge de neige (Constr.) (*f.*), pressure of snow. (74)
surcharge de pilon (bocard) (*f.*), stamp-head ; stamp-boss (ore-stamp). (75)
surcharger (*v.t.*), to overload ; to overstress ; to overwork : (76)
surcharger un cheval, to overload a horse. (77)

surcharger un arbre, to overload, to overstress, a shaft. (1)

surcharger les ouvriers, to overwork the men. (2)

surchauffage (*n.m.*) *ou* **surchauffe** (*n.f.*), super-heating ; overheating : (3)

surchauffe de la vapeur, superheating steam. (4)

surchauffe des parois d'une chaudière, over-heating of the walls of a boiler. (5)

surchauffer (*v.t.*), to superheat ; to overheat. (6)

surchauffeur (*n.m.*) *ou* **surchauffeur de vapeur**, superheater ; steam-superheater. (7)

surcreusement (d'une vallée par l'érosion glaciaire) (*n.m.*), deepening (of a valley by glacial erosion). (8)

surécartement (Ch. de f.)(*n.m.*), gauge-clearance ; clearance space between the wheel-flanges and the rails. (9)

surélevé, -e (*adj.*), elevated ; overhead. (10)

surélévation (*n.f.*) *ou* **surélèvement** (*n.m.*), heightening ; raising. (11)

surélever (*v.t.*), to heighten ; to raise : (12)

surélever un mur, to heighten a wall. (13)

sûrement (*adv.*), surely ; securely ; safely ; reliably. (14)

surépaisseur pour usinage (Fonderie) (*f.*), machining allowance ; allowance for machining ; tooling allowance. (15)

surestarie (*n.f.*), demurrage. (16)

surestimation (*n.f.*), overestimation ; over-rating ; overvaluation. (17)

surestimer (*v.t.*), to overestimate ; to overrate ; to overvalue : (18)

surestimer la valeur d'une propriété, to overrate the value of a property. (19)

sûreté (*n.f.*), safety ; security : (20)

soupape de sûreté (*f.*), safety-valve. (21)

sûreté de fonctionnement (de machines, etc.), (*f.*), reliability. (22)

surévaluation (*n.f.*), overvaluation ; over-estimating ; overrating. (23)

surévaluer (*v.t.*), to overvalue ; to overestimate ; to overrate the value of : (24)

surévaluer une propriété, to overrate the value of a property. (25)

surexhausser (*v.t.*). Same as **surhausser**.

surexposer (Photogr.) (*v.t.*), to overexpose. (26)

surexposition (Photogr.) (*n.f.*), overexposure. (27)

surfaçage (*n.m.*), surfacing : (28)

surfaçage sur le tour, surfacing on the lathe. (29)

surface (partie extérieure ; dehors d'un corps) (*n.f.*), surface. (30)

surface (superficie) (*n.f.*), surface ; area ; space : (31)

unité de surface (*f.*), unit of surface ; unit of area. (32)

la masse surface d'un terrain, the total area of a piece of ground. (33)

surface (opp. à **fond**) (Mines) (*n.f.*), surface ; day ; grass. (34)

surface (à la) (Mines), at the surface ; on the surface ; aboveground ; at grass : (35)

minerai à la surface (*m.*), ore at grass. (36)

surface courbe (*f.*), curve surface. (37)

surface d'assise (*f.*), bedding-surface. (38)

surface d'onde (Phys.) (*f.*), wave-surface. (39)

surface de chauffe (*f.*), heating-surface ; fire-surface. (40)

surface de chauffe des tubes (chaudière) (*f.*), tube heating-surface. (41)

surface de chauffe du foyer (locomotive) (*f.*), fire-box heating-surface. (42)

surface de contact (*f.*), surface of contact. (43)

surface de friction *ou* **surface de glissement** (Géol.) (*f.*). See **surfaces de friction**.

surface de frottement (*f.*), rubbing-surface ; bearing-surface ; wearing-surface. (44)

surface de grille (*f.*), grate-area ; grate-surface. (45)

surface de niveau (Topogr.) (*f.*), level surface. (46)

surface de niveau (Phys.) (*f.*), level surface ; equipotential surface. (47)

surface de piston (*f.*), piston-area ; piston-surface. (48)

surface de portée (*f.*), bearing-surface. (49)

surface de révolution (Géom.) (*f.*), surface of revolution. (50)

surface de roulement (d'une roue) (*f.*), tread (of a wheel) : (51)

la conicité de la surface de roulement d'un bandage de roue de locomotive, the taper of the tread of a locomotive wheel-tire. (52)

surface de roulement (d'un rail) (*f.*), tread, table, (of a rail). (53)

surface de séparation (Géol.) (*f.*), joint-plane. (54)

surface de surchauffe (*f.*), superheater heating-surface. (55)

surface des étages (d'un bâtiment) (*f.*), floor-space (of a building). (56)

surface développable (Géom.) (*f.*), developable surface ; developable. (57)

surface du plancher (d'un wagon de ch. de f., ou analogue) (*f.*), floor-space. (58 *ou* 59)

surface équipotentielle (Phys.) (*f.*), equipotential surface ; level surface. (60)

surface frottante (*f.*), rubbing-surface ; bearing-surface ; wearing-surface. (61)

surface libre *ou* **surface dégagée** (Travail aux explosifs) (*f.*), free face ; free end. (62)

surface libre d'un liquide (*f.*), free surface of a liquid. (63)

surface lisse (*f.*), smooth surface ; plain surface ; even surface. (64)

surface piézométrique (Géol.) (*f.*), water-table ; water-level. (65)

surface plane (*f.*), plane surface ; plane ; flat surface ; even surface ; level surface. (66)

surface portante (*f.*), bearing-surface. (67)

surface primitive (Engrenages) (*f.*), pitch-surface (Gearing). (68)

surface réflectrice (*f.*), reflective surface. (69)

surface rugueuse (*f.*), rough surface. (70)

surface terrestre (*f.*), earth's surface : (71)

la convexité naturelle de la surface terrestre, the natural convexity of the earth's surface. (72)

surface unie (*f.*), even surface ; plain surface. (73)

surface utile (*f.*), useful surface. (74)

surfacer (Travail au tour) (*v.t.*), to surface. (75)

surfaces de friction *ou* **surfaces de glissement** (Géol.) (*f.pl.*), slickensides. (76)

surfondre (Phys.) (*v.t.*), to surfuse ; to supercool ; to undercool. (77)

surfondu, -e (*adj.*), surfused ; supercooled ; undercooled : (78)

liquide surfondu (*m.*), surfused liquid ; super-cooled liquid ; undercooled liquid. (79)

surfusibilité (*n.f.*), surfusibility. (1)

surfusible (*adj.*), surfusible. (2)

surfusion (*n.f.*), surfusion ; supercooling ; undercooling : (3)

surfusion du phosphore, surfusion, super--cooling, undercooling, of phosphorus. (4)

un corps en surfusion est en état d'équilibre instable et il suffit la plupart du temps de frapper un coup sur le vase qui le contient pour que le corps se prenne en masse solide, a body in surfusion is in a state of unstable equilibrium and it is generally sufficient merely to tap the vessel which contains it, to cause the body to congeal in a solid mass. (5)

surhaussement (*n.m.*), heightening ; raising. (6)

surhaussement (Ch. de f.) (*n.m.*), superelevation ; elevation of the outer rail ; cant : (7)

le surhaussement augmente à mesure que le rayon de la courbe diminue, the super--elevation increases as the radius of the curve diminishes. (8)

surhausser (*v.t.*), to heighten ; to raise : (9)

surhausser un mur, to heighten a wall. (10)

surimposé, -e (Géol.) (*adj.*), superimposed ; epigenetic : (11)

rivière surimposée (*f.*), superimposed river ; epigenetic river. (12)

surimposition (Géol.) (*n.f.*), superimposition. (13)

surjacent, -e (Géol.) (*adj.*), overlying : (14)

roches surjacentes (*f.pl.*), overlying rocks. (15)

surmenage (*n.m.*), overworking. (16)

surmener (*v.t.*), to overwork : (17)

surmener les ouvriers, to overwork the men. (18)

surmener un cheval, to overwork a horse. (19)

surmontable (*adj.*), surmountable ; superable : (20)

difficulté surmontable (*f.*), surmountable difficulty ; superable difficulty. (21)

surmonter (*v.t.*), to surmount ; to rise above : (22)

surmonter une colline, to surmount a hill. (23)

dans les inondations, l'eau quelquefois surmonte les maisons, in times of flood, the water sometimes rises above the houses. (24)

surmonter des difficultés, to overcome, to surmount, to get over, difficulties. (25)

surnageant, -e (*adj.*), supernatant ; floating on the top ; swimming on the top : (26)

transvaser le liquide surnageant, to decant the supernatant liquid. (27)

l'huile surnageante (*f.*), the oil floating on the top. (28)

surnager (*v.i.*), to float on the top ; to float ; to swim on the top ; to swim : (29)

enlever les scories qui surnagent, to remove the dross which floats on the top. (30)

le liège surnage dans l'eau, cork floats (*or* swims) in water. (31)

suroxyder (*v.t.*), to overoxidize. (32)

surplomb (*n.m.*), overhang : (33)

le surplomb d'un mur, d'un rocher, the over--hang of a wall, of a rock. (34 *ou* 35)

surplomb (en), overhanging ; overhead : (36)

rocher en surplomb (*m.*), overhanging rock. (37)

ascenseur électrique, mécanisme en surplomb (*m.*), electric passenger-lift, gear overhead. (38)

surplombement (*n.m.*), overhanging. (39)

surplomber (*v.t.*), to overhang : (40)

des rochers surplombent le ravin (*m.pl.*), rocks overhang the ravine. (41)

surplomber (*v.i.*), to overhang : (42)

mur qui surplombe (*m.*), wall which overhangs. (43)

surplus (*n.m.*), surplus. (44)

surproduction (*n.f.*), overproduction. (45)

surrection (Géol.) (*n.f.*), upheaval ; uplift : (46)

surrection des chaînes de montagnes, des fonds sous-marins, upheaval, uplift, of mountain-chains, of sea-floors. (47 *ou* 48)

sursaturation (*n.f.*), supersaturation. (49)

sursaturation magnétique (*f.*), magnetic super--saturation. (50)

sursaturé, -e (*adj.*), supersaturated : (51)

solution sursaturée (*f.*), supersaturated solution. (52)

sursaturer (*v.t.*), to supersaturate : (53)

sursaturer un liquide, to supersaturate a liquid. (54)

sursoufflage (Fabrication de l'acier) (*n.m.*), after-blow. (55)

surtendre (*v.t.*), to overstretch ; to overstrain. (56)

surtension (*n.f.*), overpressure ; excess pressure. (57)

surtension (Élec.) (*n.f.*), surge ; volt-rise ; pressure-rise. (58)

surturbrand (variété de lignite) (*n.m.*), surtur--brand. (59)

surveillance (*n.f.*), supervision ; surveillance ; superintendence : (60)

exercer une surveillance minutieuse, to exercise a close supervision. (61)

surveillant (pers.) (*n.m.*), overseer ; superinten--dent ; boss. (62)

surveiller (*v.t.*), to superintend ; to supervise. (63)

survenir (*v.i.*), to supervene ; to take place ; to happen ; to occur : (64)

accidents qui surviennent trop fréquemment dans les exploitations au jour (*m.pl.*), accidents which occur too frequently in open-cast mining. (65)

survivants (pers.) (*n.m.pl.*), survivors. (66)

survoltage (Élec.) (*n.m.*), boosting : (67)

survoltage d'un courant alternatif, boosting an alternating current. (68)

survoltage (des lampes à incandescence) (Élec.) (*n.m.*), overrunning (of glow lamps) : (69)

le survoltage, au delà de 6 0/0 du voltage normal, noircit très vite le verre de l'ampoule, overrunning, beyond 6% of the normal voltage, blackens the glass of the bulb very quickly. (70)

survolter (Élec.) (*v.t.*), to boost. (71)

survolter (lampes) (Élec.) (*v.t.*), to overrun : (72)

survolter une lampe, to overrun a lamp. (73)

survolteur (Élec.) (*n.m.*), booster. (74)

survolteur d'artère (*m.*), feeder-booster. (75)

survolteur d'induction (*m.*), induction-regulator. (76)

survolteur-dévolteur [survolteurs-dévolteurs *pl.*] (*n.m.*), reversible booster. (77)

susceptibilité (*n.f.*), susceptibility ; capability. (78)

susceptibilité magnétique (*f.*), magnetic suscepti--bility ; coefficient of induced magnetization. (79)

susceptible (*adj.*), susceptible ; capable. (80)

susceptible d'arriver *ou* susceptible de se produire, apt to occur. (81)

susceptible d'être traité (-e) à profit *ou* susceptible d'un traitement rémunérateur, amenable to profitable treatment. (1)

susceptible de preuve, capable of proof. (2)

susceptible de recevoir un beau poli *ou* susceptible d'un beau poli (se dit des pierres, etc.), capable of receiving a beautiful polish ; susceptible of a high polish ; susceptible to a fine polish. (3)

susjacent, -e (Géol.) (*adj.*), overlying : (4)

roches susjacentes (*f.pl.*), overlying rocks. (5)

suspendre (pendre) (*v.t.*), to suspend ; to hang ; to hang up : (6)

suspendre une aiguille aimantée sur un pivot, to suspend a magnetic needle on a pivot. (7)

suspendre une lampe à un croc, to hang, to hang up, to suspend, a lamp on a hook. (8)

suspendre les pilons d'un bocard au moyen des doigts de retenue, to hang up the stamps of a stamp-mill by means of the finger-bars. (9)

suspendre les travaux, to suspend, to cease, to stop, to discontinue, work ; to hold up the work. (10)

suspendre (se) (*v.r.*), to be suspended ; to be hung up ; to hang. (11)

suspension (action) (*n.f.*), suspension ; suspend-ing ; hanging ; hanging up; holding up ; stopping ; stoppage : (12)

suspension de la sonde, suspension of the rods ; holding up the rods. (13)

suspension de tous les travaux souterrains, suspension, stoppage, of all underground work. (14)

suspension des pilons (bocard), hanging up of the stamps. (15)

suspension (Chim.) (*n.f.*), suspension : (16)

fluide qui tient une substance en suspension (*m.*), fluid which holds a substance in suspension. (17)

suspension (dispositif par lequel est suspendue une pièce mobile) (*n.f.*), suspension : (18)

suspension bifilaire, bifilar suspension. (19)

suspension (ensemble des pièces par lesquelles est suspendu un châssis de locomotive, de wagon, etc.) (*n.f.*), suspension ; spring-gear ; spring-rigging : (20)

la suspension du bissel arrière d'une locomotive, the suspension of the trailing-truck of a locomotive. (21)

suspension à billes (*f.*), ball-bearing swivel-hook. (22)

suspension à la Cardan (*f.*), Cardanic suspension ; Cardan's suspension. (23)

suspension de tubage (Sondage) (*f.*), lowering-tee. (24)

sussultoire (Sismologie) (*adj.*), sussultatory : (25)

secousse sussultoire (*f.*), sussultatory shock. (26)

swalite (Explosif) (*n.f.*), swalite. (27)

syénite (Pétrol.) (*n.f.*), syenite. (28)

syénite à mica (*f.*), mica-syenite. (29)

syénite éléolitique (*f.*), elæolite-syenite ; eleolite-syenite; nephelite-syenite ; nephe-line-syenite. (30)

syénite rose d'Egypte (*f.*), Egyptian red syenite. (31)

syénite zirconienne (*f.*), zircon syenite. (32)

syénitique (*adj.*), syenitic : (33)

porphyre syénitique (*m.*), syenitic prophyry. (34)

sylvane (*n.m.*) *ou* sylvanite (Minéral) (*n.f.*), sylvanite ; sylvan ; graphic tellurium ; graphic gold ; graphic ore. (35)

sylvine *ou* sylvite (Minéral) (*n.f.*), sylvite ; sylvine ; sylvinite. (36)

symbole (*n.m.*), symbol : (37)

le symbole d'un corps simple, the symbol of an element. (38)

le symbole de l'or est Au, the symbol of gold is Au. (39)

symétrie (*n.f.*), symmetry. (40)

symétrie oblique (*f.*), skew symmetry. (41)

symétrie par rapport à un plan (*f.*), symmetry as to a plan. (42)

symétrique (*adj.*), symmetric ; symmetrical : (43)

aiguillage symétrique (*m.*), symmetrical switch. (44)

symétriquement (*adv.*), symmetrically. (45)

synchrone (*adj.*), synchronous : (46)

moteur synchrone (*m.*), synchronous motor. (47)

synchronisation (*n.f.*), synchronization : (48)

synchronisation des chronomètres, des moteurs électriques, synchronization of chronometers, of electric motors. (49 *ou* 50)

synchroniser (*v.t.*), to synchronize. (51)

synchronisme (*n.m.*), synchronism. (52)

synclinal, -e, -aux (opp. à anticlinal, -e, -aux) (Géol.) (*adj.*), synclinal. (53) (opp. to anticlinal) :

pli synclinal (*m.*), synclinal fold. (54)

synclinal (Géol.) (*n.m.*), synclinal ; syncline. (55)

syndicaliste (pers.) (*n.m.*), trade-unionist. (56)

syndicat (*n.m.*), syndicate. (57)

syndicat ouvrier (*m.*), trades-union. (58)

syngénite (Minéral) (*n.f.*), syngenite. (59)

synthèse (*n.f.*), synthesis : (60)

synthèse du rubis, synthesis of the ruby. (61)

synthétique (*adj.*), synthetic ; synthetical : (62)

production synthétique du saphir (*f.*), synthetic production of the sapphire. (63)

synthétiquement (*adv.*), synthetically. (64)

syntoniser (Phys.) (*v.t.*), to syntonize ; to tune : (65)

syntoniser un récepteur, to syntonize, to tune, a receiver. (66)

syntonisme (Phys.) (*n.m.*), syntony ; tuning. (67)

systématique (*adj.*), systematic ; systematical : (68)

boisage systématique (*m.*), systematic timber-ing. (69)

systématiquement (*adv.*), systematically. (70)

systématisation (*n.f.*), systematization. (71)

systématiser (*v.t.*), to systematize. (72)

système (*n.m.*), system ; method. (73)

système (Géol., Cristall., etc.) (*n.m.*), system. (74)

système à bloc enclenché (Ch. de f.) (*m.*), lock-and-block system ; controlled manual block system. (75)

système à plots (tramway électrique) (*m.*), stud system ; surface-contact system. (76)

système à primes (pour les ouvriers) (*m.*), bonus system ; premium system. (77)

système asymétrique *ou* système triclinique *ou* système clinoédrique *ou* système anorthique *ou* système doublement oblique (Cristall.) (*m.*), triclinic system ; asymmetric system ; clinorhomboidal system ; anorthic system ; doubly oblique system. (78)

système bielle et manivelle *ou* système bielle-manivelle (*m.*), crank and connecting-rod system. (1)

système binaire *ou* système monoclinique *ou* système clinorhombique (Cristall.) (*m.*), monoclinic system ; monosymmetric system ; clinorhombic system ; oblique system. (2)

système câble-tête et câble-queue (Roulage,—Mines) (*m.*), main and tail-rope system ; tail-rope system. (3)

système carbonifère *ou* système carboniférien (Géol.) (*m.*), Carboniferous system. (4)

système centimètre-gramme-seconde *ou* système C.G.S. (*m.*), centimetre-gramme-second system ; C.G.S. system : (5) dans le système centimètre-gramme-seconde, l'unité de longueur est le centimètre, l'unité de masse est la gramme, et l'unité de temps est la seconde, in the centimetre-gramme-second system, the unit of length is the centimetre, the unit of mass is the gramme, and the unit of time is the second. (6)

système d'aérage bien conçu (*m.*), well-designed system of ventilation. (7)

système d'exploitation à l'entreprise (*m.*), contract system of mining. (8)

système de caniveaux (tramway électrique) (*m.*), conduit system. (9)

système de distribution à trois fils (Élec.) (*m.*), three-wire distributing system. (10)

système de l'arbre (Élec.) (*m.*), tree-system. (11)

système de montagnes (*m.*), mountain system. (12)

système métrique (*m.*), metric system : (13) le système métrique est obligatoire en France, Allemagne, Autriche, Hongrie, Belgique, Italie, Espagne, Portugal, Hollande, Suisse, Suède, Norvège, Roumanie, Grèce, Égypte, le Mexique, et la plupart des républiques du sud. En Grande-Bretagne, au Canada, et aux États-Unis, le système métrique est légal, mais non obligatoire, the metric system is compulsory in France, Germany, Austria, Hungary, Belgium, Italy, Spain, Portugal, Holland, Switzerland, Sweden, Norway, Roumania, Greece, Egypt, Mexico, and most of the Southern republics. In Great Britain, Canada, and the United States, the metric system is legal, but not compulsory. (14)

système par fil aérien avec trolley (traction électrique) (*m.*), overhead-trolley system ; trolley system. (15)

système par troisième rail (ch. de f. électrique) (*m.*), third-rail system. (16)

système par voie souterraine (traction électrique) (*m.*), conduit system. (17)

système quaternaire *ou* système quadratique (Cristall.) (*m.*), tetragonal system ; dimetric system ; quadratic system ; pyramidal system ; quaternary system. (18)

système russe de forage à chute libre (*m.*), Russian free-fall boring method. (19)

système sénaire *ou* système hexagonal (Cristall.) (*m.*), hexagonal system. (20) See note under système ternaire.

système terbinaire *ou* système rhombique *ou* système orthorhombique (Cristall.) (*m.*), orthorhombic system ; orthosymmetric system ; trimetric system ; rhombic system ; prismatic system. (21)

système ternaire *ou* système rhomboédrique (Cristall.) (*m.*), rhombohedral system. (22) Note.—In the English system of crystallization the rhombohedral is a subdivision of the hexagonal.

système terquaternaire *ou* système cubique *ou* système régulier (Cristall.) (*m.*), isometric system ; monometric system ; regular system ; cubic system ; tesseral system ; tessular system. (23)

T

T (*n.m.*). Same as té.

tabergite (Minéral) (*n.f.*), tabergite. (24)

table (meuble) (*n.f.*), table. (25)

table (Préparation mécanique des minerais) (*n.f.*), table (Ore-dressing). (26)

table (d'une machine-outil, d'une machine à mouler, etc.) (*n.f.*), table, plate, platen, (of a machine-tool, of a moulding-machine, etc.). (27 *ou* 28)

table (d'une enclume) (*n.f.*), face, crown, (of an anvil). (29)

table (d'une poutre) (*n.f.*), table, tread, (of a girder). (30)

table (d'un cylindre de laminoir) (*n.f.*), body, barrel, (of a rolling-mill roll). (31)

table (surface d'une pierre fine de forme plane) (*n.f.*), table : (32) diamant en table (*m.*), table-cut diamond. (33)

table à bascule (*f.*), lift-up table. (34)

table à boues (minerais) (*f.*), slimes-table. (35)

table à dessin (*f.*), drawing-table. (36)

table à double coulisse (pour machines à percer, etc.) (*f.*), compound table ; table with compound slides (for drilling-machines, etc.). (37)

table à double équerre (machine-outil) (*f.*), table with top and side faces. (38)

table à hauteur variable (*f.*), rising table ; table which can be raised or lowered. (39)

table à percussion (minerais) (*f.*), percussion-table. (40)

table à rainures *ou* table à rainures de montage (*f.*), slotted table. (41)

table à riffles (minerais) (*f.*), riffled table. (42)

table à secousses (minerais) (*f.*), bumping-table ; shaking-table ; joggling-table. (43)

table carrée universelle (machine-outil) (*f.*), square universal table. (44)

table d'épandage (*f.*), spreading-table. (45)

table d'équerre à rainures (*f.*), slotted box table. (46)

table de concentration (minerais) (*f.*), concentrating-table. (47)

table de concentration à sec (*f.*), dry concentrating-table. (48)

table de diamant (*f.*), table diamond ; table-cut diamond. (49)

table de glacier (Géol.) (*f.*), glacier-table. (1)

table de renversement (d'une machine à démouler à renversement) (Fonderie) (*f.*), roll-over table. (2)

table de roulement (d'un rail) (*f.*), tread, table, (of a rail). (3)

table de triage (minerais) (*f.*), sorting-table ; picking-table. (4)

table des lumières (tiroir) (*f.*), port-face ; valve-seat (slide-valve). (5)

table dormante (minerais) (*f.*), sleeping-table ; frame ; framing-table. (6)

table hongroise (minerais) (*f.*), Hungarian table. (7)

table inclinable (*f.*), canting-table ; tilting-table. (8)

table inclinée à aire plane (minerais) (*f.*), plane-table. (9)

table porte-pièce (*f.*), work-table ; work-plate. (10)

table porte-pièce à rainures à double coulisse (*f.*), compound slotted work-table. (11)

table ronde (*f.*), round table ; circular table. (12)

table rotative *ou* table tournante (*f.*), revolving table. (13)

table s'inclinant sous tout angle (*f.*), table canting to any angle ; table tilting to any angle. (14)

tableau (Perspective) (*n.m.*), picture-plane ; perspective plane ; plane of delineation ; principal plane. (15)

tableau comparatif (*m.*), comparative table. (16)

tableau de commutateurs (Élec.) (*m.*), switch-board. (17)

tableau de distribution (Élec.) (*m.*), distribution-board ; distributing-board. (18)

tableau de vanne (*m.*), sluice-gate frame ; shut-frame. (19)

tableau des signaux (*m.*), list of the signals. (20)

tableau indicateur (sonnerie électrique) (*m.*), annunciator. (21)

tableau-marquoir [tableaux-marquoirs *pl.*] (*n.m.*), tally-board. (22)

tableau noir *ou simplement* tableau (*n.m.*), black--board. (23)

tableau synoptique (*m.*), synoptic table. (24)

tablette (rayon) (*n.f.*), shelf : (25)
les tablettes d'une armoire, the shelves of a cupboard. (26)

tablette (d'une balance) (*n.f.*), sole (of a balance *or* scales). (27)

tablier (d'un pont) (*n.m.*), floor, deck, (of a bridge). (28)

tablier (d'une forge) (*n.m.*), hearth (of a forge). (29)

tablier (de locomotive, de wagon) (*n.m.*), running-board ; foot-board. (30 *ou* 31)

tablier de forgeron (*m.*), blacksmiths' apron. (32)

tablier de vanne (Constr. hydraul.) (*m.*), shuttle-plate. (33)

tablier du chariot (tour) (*m.*), apron of the carriage (lathe). (34)

tablier releveur (de laminoir) (*m.*), lifting-table (of rolling-mill). (35)

tabulaire (*adj.*), tabular : (36)
cristal tabulaire (*m.*), tabular crystal. (37)
les vastes régions tabulaires entre la sierra Nevada et les montagnes Rocheuses (*f.pl.*), the vast tabular regions between the Sierra Nevada and the Rocky Mountains. (38)

tache (*n.f.*), spot ; stain : (39)

tache centrale (Photogr.), flare ; flare-spot ; ghost ; central spot. (40)

tache d'étain (dans le bronze), tin spot. (41)

tache de graisse, grease spot ; grease stain. (42)

tache de minerai, spot, splash, of ore. (43)

tache ferrugineuse, iron-stain. (44)

tache solaire (Astron.), sun-spot. (45)

tâche (*n.f.*), task ; job. (46)

tachéomètre (Levés de plans) (*n.m.*), tacheom--eter. (47)

tachéométrie (Levés) (*n.f.*), tacheometry. (48)

tacher (*v.t.*), to stain ; to spot. (49)

tacher (se) (*v.r.*), to stain ; to spot. (50)

tâcheron (pers.) (*n.m.*), piece-worker. (51)

tacheter (*v.t.*), to stain ; to spot. (52)

tachydrite *ou* tachhydrite (Minéral) (*n.f.*), tachy--drite ; tachhydrite. (53)

tachylyte *ou* tachylite (Pétrol.) (*n.f.*), tachylyte ; tachylite ; basalt glass. (54)

tachymètre (*n.m.*), tachometer ; speed-indicator ; revolution-indicator ; motion-indicator. (55)

tailings (résidus stériles) (*n.m.pl.*), tailings ; tails. (56)

tailings sableux (*m.pl.*), sand tailings. (57)

taillage (*n.m.*), cutting : (58)
taillage des engrenages à la fraise, cutting gears with milling-cutters. (59)

taillanderie (métier) (*n.f.*), edge-tool making ; tool-making. (60)

taillanderie (outils) (*n.f.*), edge-tools. (61)

taillandier (pers.) (*n.m.*), edge-tool maker ; tool-maker. (62)

taille (action) (*n.f.*), cutting : (63)
taille des dents des roues d'engrenage, cutting gear-wheel teeth. (64)
taille en série des crémaillères, cutting racks in gangs. (65)

taille (des pierres précieuses) (*n.f.*), cutting : (66)
taille des pierres précieuses, gem-cutting. (67)

taille (des pierres de taille) (*n.f.*), cutting ; dressing ; hewing : (68)
taille des pierres, stonecutting ; stone-dressing ; hewing stone. (69)

taille (façon) (*n.f.*), cut : (70)
la taille d'une pierre précieuse, d'une lime, the cut of a gem, of a file. (71 *ou* 72)

taille (Mines) (*n.f.*), room ; stall ; chamber ; breast. (73)

taille bâtarde (limes) (*f.*), bastard cut. (74)

taille brute (de pierres de taille) (*f.*), rough dressing ; spalling. (75)

taille chassante (Exploitation des mines de houille) (*f.*), forward stall. (76)

taille chassante (Exploitation des mines métallifères) (*f.*), drift-stope. (77)

taille croisée *ou* taille double (limes) (*f.*), cross-cut ; double cut. (78)

taille demi-douce *ou* taille mi-douce (limes) (*f.*), second cut ; 2nd cut. (79)

taille douce (limes) (*f.*), smooth cut. (80)

taille en râpe (*f.*), rasp-cut. (81)

taille grosse *ou* taille rude (limes) (*f.*), rough cut. (82)

taille montante (Exploitation des mines de houille) (*f.*), rise-workings. (83)

taille montante (Exploitation des mines métallifères) (*f.*), raise-stope. (84)

taille moyenne (limes) (*f.*), middle cut. (85)

taille simple (limes) (*f.*), single cut ; float cut. (86)

taille très douce (limes) (*f.*), dead-smooth cut. (1)

taillé, -e (pierres précieuses) (*adj.*), cut : (2) pierre taillée (*f.*), cut stone. (3)

taillé, -e (pierres de taille) (*adj.*), cut ; dressed ; hewn. (4)

taillé (-e) à la machine, machine-cut : (5) engrenages taillés à la machine (*m.pl.*), machine-cut gearing. (6)

taillé (-e) dans la masse, cut from the solid ; cut from solid blanks : (7) engrenages taillés dans la masse (*m.pl.*), gears cut from the solid ; gear-wheels cut from solid blanks. (8)

tailler (*v.t.*), to cut : (9) tailler une crémaillère, les dents d'une lime, to cut a rack, the teeth of a file. (10 *ou* 11)

tailler (pierres précieuses) (*v.t.*), to cut : (12) tailler des pierres précieuses, to cut gems. (13)

tailler (pierres de taille) (*v.t.*), to cut ; to dress ; to hew. (14)

tailler en pointe (*v.t.*), to point. (15)

tailler par éclats (*v.t.*), to chip : (16) tailler par éclats un bloc de marbre, to chip a block of marble. (17)

tailler (se) (*v.r.*), to cut : (18) le grès se taille facilement, sandstone cuts easily. (19)

tailleur (pers.) (*n.m.*), cutter ; dresser ; hewer. (20)

tailleur de limes (*m.*), file-cutter. (21)

tailleur de pierre (de taille) (*m.*), stone-cutter ; stone-dresser ; stone-mason ; stonemason ; jadder. (22)

tailleuse d'engrenages (*f.*), gear-cutter ; gear-cutting machine. (23)

taillis (*n.m.*), coppice ; copse ; underwood. (24)

tain (amalgame pour miroirs) (*n.m.*), quick-silvering ; silvering. (25)

talc (*n.m.*), talc. (26)

talc endurci (*m.*), indurated talc ; talc-slate. (27)

talcaire *ou* **talcique** (*adj.*), talcose ; talcous ; talcky ; talcy. (28)

talcite (Minéral) (*n.m.*), talcite. (29)

talcite (Lithologie) (*n.f.*), talcite ; talc-schist. (30)

talcitoïde (*adj.*), talcoid. (31)

talco-chloritique (*adj.*), talcochloritic : (32) schiste talco-chloritique (*m.*), talcochloritic schist. (33)

talco-micacé, -e (*adj.*), talcomicaceous. (34)

talcschiste (*n.m.*), talc-schist ; talcose schist. (35)

talon (partie d'un outil voisine du manche) (*n.m.*), heel : (36) le talon de la lame d'un couteau, the heel of the blade of a knife. (37)

talon (de l'aiguille) (Ch. de f.) (*n.m.*), heel (of the switch). (38)

talon (du croisement) (Ch. de f.) (*n.m.*), heel (of the frog). (39)

talon (de la came) (bocard) (*n.m.*), root (of the cam) (ore-stamp). (40)

talon (d'un essieu) (*n.m.*), collar (of an axle). (41)

talon (d'un barreau de grille) (*n.m.*), thickening piece (of a grate-bar). (42)

talonnable (Ch. de f.) (*adj.*), trailable : (43) voie, aiguille, qui est laissée talonnable (*f.*), track, switch, which is left trailable. (44 *ou* 45)

talonnement (Ch. de f.) (*n.m.*), trailing. (46)

talonner (Ch. de f.) (*v.i.*), to trail. (47)

talqueux, -euse (*adj.*), talcose ; talcous ; talcky ; talcy. (48)

talus (banquette) (*n.m.*), bank ; slope : (49) talus d'un canal, d'une rivière, d'une fouille, bank, slope, of a canal, of a river, of a cut. (50 *ou* 51 *ou* 52)

talus (pente) (*n.m.*), slope ; batter : (53) talus d'un remblai, slope, batter, of an embankment. (54) talus d'un mur, batter of a wall. (55)

talus (en), sloping ; battering ; inclined. (56)

talus (Géol.) (*n.m.*), talus ; slope. (57) Cf. talus d'éboulis.

talus continental (Océanogr.) (*m.*), continental slope. (58)

talus d'éboulis (Géol.) (*m.*), talus (sloping mass of fallen fragments at the foot of a mountain). (59)

talus de neige (Géol.) (*m.*), snow talus. (60)

talus littoral (opp. à **plate-forme côtière entaillée par les vagues** *ou* **plate-forme littorale rocheuse**) (*m.*), wave-built terrace. (61) (opp. to **wave-cut terrace**).

talus naturel (*m.*), natural slope ; angle of repose ; earth-slope. (62)

talutage (*n.m.*), sloping ; battering : (63) talutage d'un fossé, sloping, battering, a ditch. (64)

taluter (*v.t.*), to slope ; to batter : (65) taluter un remblai, to slope, to batter, an embankment. (66) taluter un mur, to batter a wall. (67)

tambour (*n.m.*), drum. (68)

tambour (d'une turbine à vapeur) (*n.m.*), drum (of a steam-turbine). (69)

tambour (d'un treuil) (*n.m.*), barrel, drum, (of a crab-winch). (70)

tambour (d'un baromètre anéroïde) (*n.m.*), box, vacuum-box, (of an aneroid barometer). (71)

tambour conique (*m.*), conical drum ; cone-drum. (72)

tambour conique (Extraction) (Mines) (*m.*), conical drum ; fusee-wheel. (73)

tambour cylindrique (Extraction) (Mines) (*m.*), cylindrical drum. (74)

tambour cylindro-conique (Extraction) (Mines) (*m.*), semiconical drum ; cylindroconical drum. (75)

tambour cylindro-spiraloïde (Extraction) (Mines) (*m.*), cylindrospiral drum. (76)

tambour d'enroulement (*m.*), winding-drum. (77)

tambour débourbeur *ou* **tambour laveur** (pour minerai) (*m.*), washing-drum ; drum-washer ; revolving washer. (78)

tambour dessableur (Fonderie) (*m.*), tumbling-drum ; tumbling-barrel ; tumbling-box ; tumbling-mill ; tumbler ; rattler ; foundry-rattler ; rattling-barrel ; cleaning-barrel ; rumbler ; shaking-machine ; shaking-barrel ; shaking-mill. (79)

tambour-queue [tambours-queues *pl.*] (Système câble-tête et câble-queue) (Mines) (*n.m.*), tail-rope drum (Main and tail-rope system). (80)

tambour spiraloïde (Extraction) (Mines) (*m.*), spiral drum. (81)

tambour-tête [tambours-têtes *pl.*] (Système câble-tête et câble-queue) (*n.m.*), main-rope drum. (82)

tamis (*n.m.*), sieve. (1)

tamis (d'un bocard à minerai) (*n.m.*), grate, screen, (of an ore-stamp). (2)

tamis (d'une lampe de sûreté, d'un entonnoir) (*n.m.*), gauze (of a safety-lamp, of a funnel). (3 *ou* 4)

tamis de mouleur (*m.*), moulders' sieve; foundry-riddle. (5)

tamis métallique (*m.*), wire gauze; gauze. (6)

tamis métallique *ou* **tamis de fil de fer** (pour cribler) (*m.*), wire sieve. (7)

tamis métallique pour entonnoir (*m.*), gauze strainer for funnel. (8)

tamis oscillant (*m.*), oscillating sieve. (9)

tamis roulant (*m.*), travelling belt-screen. (10)

tamisage (*n.m.*), sifting. (11)

tamiser (*v.t.*), to sift : (12)

tamiser du sable, to sift sand. (13)

tamiser l'air avant l'inspiration au moyen d'un respirateur, to sift the air before inspiration by means of a respirator. (14)

le brûleur est entouré d'un globe opalin, qui tamise la lumière, the burner is enclosed in an opal globe, which sifts the light. (15)

tamiser (*v.i.*), to sift. (16)

tamiseur (pers.) (*n.m.*), sifter. (17)

tampon (bouchon) (*n.m.*), plug. (18)

tampon (*n.m.*) *ou* **tampon de choc**, buffer; bumper : (19)

tampon de choc d'un wagon de chemin de fer, buffer, bumper, of a railway-truck. (20)

tampon (bouchon d'obturation de trou de coulée) (Fonderie) (*n.m.*), bott; plug; boat. (21)

tampon (calibre) (*n.m.*), plug; plug-gauge; internal cylindrical gauge : (22)

tampons et bagues, plugs and rings; plug and ring-gauges. (23)

tampon à ressort (*m.*), spring-buffer. (24)

tampon à tolérance (calibre) (*m.*), limit internal gauge. (25)

tampon de bois (bouchon) (*m.*), wooden plug. (26)

tampon de bois (tampon de choc) (*m.*), wooden bumper. (27)

tampon de lavage (*m.*), washout-plug. (28)

tampon de vidange (*m.*), emptying-plug. (29)

tampon obturateur (*m.*), sealing-plug; obturat-ing-plug. (30)

tampon sec (wagon de chemin de fer) (*m.*), dead-block; dead-wood. (31)

tamponnement (bouchage) (*n.m.*), plugging; plugging up; stopping; stopping up. (32)

tamponnement (rencontre) (*n.m.*), collision; colliding : (33)

le tamponnement de deux trains, the collision, the colliding, of two trains. (34)

wagons de chemin de fer qui se télescopent à la suite d'un tamponnement (*m.pl.*), railway-carriages which telescope as the result of a collision. (35)

tamponner (*v.t.*), to plug; to plug up; to stop; to stop up : (36)

tamponner un trou dans un puits, to plug a hole in a well. (37)

tamponner une voie d'eau, to stop a leak. (38)

tamponner (heurter avec des tampons) (*v.t.*), to collide with; to run into : (39)

tamponner un train qui n'a pas eu le temps de se garer, to collide with, to run into, a train which has not had time to shunt. (40)

tamponner (se) (*v.r.*), to collide; to run into each other : (41)

deux véhicules qui se tamponnent (*m.pl.*), two vehicles which collide (*or* which run into each other). (42)

tandem (en), tandem : (43)

cylindres en tandem (*m.pl.*), tandem cylinders. (44) (opp. à cylindres placés côte à côte, opp. to side-by-side cylinders).

tangage (mouvement d'une locomotive) (*n.m.*), pitching. (45)

tangence (*n.f.*), tangency; tangence. (46)

tangent, -e (*adj.*), tangent : (47)

une ligne tangentielle est une ligne qui est tangente à une courbe en un certain point, a tangent line is a line which is tangent to a curve at a certain point. (48)

tangente (*n.f.*), tangent. (49)

tangenter (*v.t.*), to tangent. (50)

tangentiel, -elle (*adj.*), tangential; tangent : (51)

force tangentielle (*f.*), tangential force. (52)

clavette tangentielle (*f.*), tangent key. (53)

tangentiellement (*adv.*), tangentially. (54)

tanguer (se dit d'un bateau) (*v.i.*), to pitch. (55)

tannée (*n.f.*), spent tan; spent tan-bark. (56)

tant pour cent, so much per cent. (57)

tantalate (Chim.) (*n.m.*), tantalate. (58)

tantale (Chim.) (*n.m.*), tantalum. (59)

tantaleux, -euse (Chim.) (*adj.*), tantalous. (60)

tantalique (Chim.) (*adj.*), tantalic. (61)

tantalite (Minéral) (*n.f.*), tantalite. (62)

tapage (Mach.) (*n.m.*), knocking. (63)

taper (Mach.) (*v.i.*), to knock : (64)

moteur qui tape (*m.*), engine which knocks. (65)

taper sur une porte, to knock at a door. (66)

tapiolite (Minéral) (*n.f.*), tapiolite. (67)

tapis en fibres de coco (*m.*), coco matting. (68)

tapisser (*v.t.*), to carpet; to line; to set around : (69)

cavité tapissée de cristaux (*f.*), cavity carpeted (*or* lined) (*or* set around) with crystals. (70)

tapisser le front d'une mine, to salt, to plant, a mine. (71)

tapure (Fonderie) (*n.f.*), shrinkage crack; contraction crack. (72)

taque (*n.f.*), plate (of cast iron). (73)

taque d'assise (*f.*), bed-plate; base-plate; foundation-plate; sole-plate; base. (74)

taquet (arrêt) (*n.m.*), stop; block : (75)

taquets réglables fixés à la table servant à débrayer la courroie au bout de course, adjustable stops fixed on the table for throwing the belt off at end of stroke. (76)

taquet (fermeture de porte, etc.) (*n.m.*), button. (77)

taquet (pièce de bois ou de métal servant à amarrer les cordages) (*n.m.*), cleat. (78)

taquet (clichage) (*n.m.*). See **taquets**.

taquet (d'un bocard) (*n.m.*), tappet (of an ore-stamp). (79)

taquet d'arrêt (Ch. de f.) (*m.*), stop-block. (80)

taquet de calage *ou* **taquet écluseur** (pour maintenir les wagons dans une cage de mine) (*m.*), car-stop. (81)

taquet de détente (de tiroir de détente) (*m.*), expansion-plate, expansion-slide, cut-off plate (of expansion-valve *or* cut-off valve). (82)

taquet de porte de boîte à fumée (*m.*), smoke-box door-clamp. (83)

taquet de sûreté (*m.*), safety-stop. (84)

taquets (clichages) (Mines) (*n.m.pl.*), keps ; keeps ; props ; rests ; cage-shuts ; cage-sheets ; landing-chairs ; landing-dogs ; dogs ; catches ; wings ; fangs. (1)

taquets à abaissement (Mines) (*m.pl.*), fallers. (2)

taquets automatiques (*m.pl.*), automatic keps. (3)

taquets de sûreté (*m.pl.*), safety-keps ; safety-chairs. (4)

taquets hydrauliques (*m.pl.*), hydraulic keps. (5)

tarage (*n.m.*), calibration ; calibrating ; scaling ; measurement ; measuring : (6)
   tarage d'un ressort, calibration of a spring ; scaling a spring. (7)
   un tarage fait avec une boîte de résistance (Élec.), a measurement made with a resis-tance-box. (8)

taraud (à fileter) (*n.m.*), tap. (9)

taraud à becs de gaz (*m.*), burner-tap. (10)

taraud à denture interrompue (*m.*), interrupted-tooth tap. (11)

taraud à droite (*m.*), right-hand tap. (12)

taraud à filet carré *ou* taraud à pas carré (*m.*), square-thread tap. (13)

taraud à gaz (*m.*), gas-tap. (14)

taraud à la machine (*m.*), machine-tap. (15)

taraud à main (*m.*), hand-tap. (16)

taraud à rainures hélicoïdales (*m.*), twist-tap. (17)

taraud accrocheur (Sondage) (*m.*), grabbing-tap ; recovering-tap. (18)

taraud aléseur (*m.*), reaming-tap ; reamer-tap. (19)

taraud aléseur à main à — rainures, fileté au pas système international (S.I.) (*m.*), hand reaming-tap with — grooves, international standard thread. (20)

taraud au pas Whitworth (*m.*), tap with Whit-worth thread. (21)

taraud conique *ou* taraud ébaucheur (*m.*), taper tap ; entering-tap ; first tap. (22)

taraud court à main (*m.*), short hand-tap. (23)

taraud cylindrique *ou* taraud finisseur (*m.*), bottoming-tap ; plug tap ; straight tap ; finishing-tap ; third tap. (24)

taraud dégagé mécaniquement *ou* taraud dépouillé mécaniquement (*m.*), machine-relieved tap. (25)

taraud demi-conique *ou* taraud intermédiaire (*m.*), second tap ; plug tap. (26)

taraud long pour machines (*m.*), long machine-tap. (27)

taraud mère *ou* taraud matrice (*m.*), master-tap ; hob ; hob-tap. (28)

taraud mère court (*m.*), short hob. (29)

taraud mère fileté au pas américain système Sellers (*m.*), Sellers hob. (30)

taraud mère fileté au pas anglais système Whit-worth (*m.*), master-tap, Whitworth thread. (31)

taraud mère long (*m.*), long hob. (32)

taraud pour entretoises (*m.*), stay-bolt tap. (33)

taraud pour machines (*m.*), machine-tap. (34)

taraud pour tuyauteries (*m.*), pipe-tap. (35)

taraudage (*n.m.*), tapping ; screwing : (36)
   taraudage d'une seule passe, screwing at once going over. (37)

tarauder (*v.t.*), to tap ; to screw : (38)
   tarauder un trou, un écrou, to tap, to screw, a hole, a nut. (39 *ou* 40)

taraudeuse (*n.f.*), tapping-machine ; screwing-machine ; screw-machine ; screwing and tapping-machine ; threader. (41) See machine à tarauder for varieties.

tare (poids des caisses, etc.) (*n.f.*), tare. (42)

tare du ressort (Dynamométrie) (*f.*), scale of spring. (43)

tarer (Commerce) (*v.t.*), to tare : (44)
   tarer un wagon, to tare a truck. (45)

tarer (Dynamométrie) (*v.t.*), to calibrate ; to scale ; to measure : (46)
   tarer un ressort, to calibrate, to scale, a spring. (47)
   Cf. tarage.

tarière (*n.f.*), auger. (48)

tarière à courant d'eau continu (Sondage) (*f.*), auger with water-flushing arrangement. (49)

tarière à cuiller *ou* tarière à cuillère (pour le bois) (*f.*), spoon-auger. (50)

tarière à double spire *ou* tarière à tire-bouchon (bois) (*f.*), twisted auger ; screw-auger. (51)

tarière à glaise (Sondage) (*f.*), surface-auger ; clay-auger ; earth-auger ; earth-boring auger ; mud-bit. (52)

tarière à gravier (*f.*), gravel-auger ; miser. (53)

tarière à main (*f.*), hand-auger. (54)

tarière à vrille (*f.*), gimlet auger. (55)

tarière rubanée *ou* tarière torse (pour le bois) (*f.*), screw-auger ; twisted auger. (56)

tarière rubanée (Sondage) (*f.*), auger-worm. (57)

tarif (*n.m.*), tariff ; rate ; rates ; price-list : (58)
   ce tarif annule les précédents, this price-list cancels all previous ones. (59)

tarif des chemins de fer (*m.*), railway rates. (60)

tarif des transports (*m.*), transport rates. (61)

tarif douanier (*m.*), customs tariff. (62)

tarir (*v.t.*), to dry up : (63)
   la sécheresse tarit les puits, the drought dries up the wells. (64)

tarir (*v.i.*), to dry up ; to run dry : (65)
   source qui a tari (*f.*), spring which has dried up (*or* has run dry). (66)
   la source est tarie, the spring is dried up (*or* is dry). (67)

tarissement (*n.m.*), drying up : (68)
   tarissement des sources, drying up of springs. (69)

tarmac (*n.m.*), tarmac. (70)

tartre (à l'intérieur des chaudières, des tubes, etc.) (*n.m.*), scale ; boiler-scale ; fur ; furring ; incrustation. (71)

tartrifuge (chaudières) (*n.m.*), disincrustant. (72)

tas (monceau) (*n.m.*), heap ; pile. (73)

tas (enclumette) (*n m.*), stake ; stake anvil. (74) See also tasseau.

tas (*n.m.*) *ou* tas inférieur (de marteau-pilon) (opp. à frappe *ou* frappe supérieure *ou* panne du pilon), anvil-pallet, anvil-die, pallet for anvil, die for anvil, bottom pallet, bottom die, (of power-hammer). (75)

tas à bouteroller *ou simplement* tas (*n.m.*), rivet-dolly ; dolly. (76)

tas à planer (*m.*), planishing-stake. (77)

tas à queue (*m.*), stake with tang ; stake with peg. (78)

tas de déchets *ou* tas de rejets (*m.*), heap of waste ; dump ; dump-heap. (79)

tas de grillage (fusion du cuivre) (*m.*), open heap. (80)

tas de mineral (*m.*), heap of ore ; pile of ore. (81)

tas de résidus *ou* tas de résidus stériles (*m.*), tailings-heap. (82)

tas-étampe [tas-étampes *pl.*] (*n.m.*), swage-block. (1)
tasmanite (Minéral) (*n.f.*), tasmanite. (2)
tassage à la main du sable dans les châssis de moulage (*m.*), tucking sand by hand in moulding-boxes. (3)
tasseau (placé dans la sous-cave) (Boisage de mines) (*n.m.*), sprag ; holing-prop ; punch-prop ; gib. (4)
tasseau (enclumette) (*n.m.*), stake ; stake anvil. (5) See also **tas.**
tasseau d'établi (*m.*), bench-stake. (6)
tasseau de bois (*m.*), strip of wood ; cleat. (7)
tasseau de fixation des glissières (glissières de la crosse de piston) (*m.*), guide-block ; slide-block ; slide-bar carrier. (8)
tasseau rond (*m.*), round-headed stake. (9)
tassement (de terrains) (*n.m.*), settling ; settle-ment ; sinking : (10)
le tassement de la voie (Ch. de f.), the settling of the road. (11)
tassement (du toit) (Mines) (*n.m.*), settling, settlement, sinking, squeeze, squeezing, (of the roof). (12)
tassement (dans une pièce moulée) (Fonderie) (*n.m.*), draw ; draw-down ; sinking (of casting). (13) (a vertical sinking, i.e., one on the top face of the casting : dis-tinguished from **retirure,** a horizontal or oblique sinking) :
les tassements sont combattus en pompant dans chaque évent avec une pompe, draws are combatted by feeding in each riser with a feeding-rod. (14)
tasser (Mines) (*v.i.*), to settle ; to squeeze ; to undergo squeeze : (15)
le toit tasse, the roof is settling (*or* is under-going squeeze). (16)
tasser (se) (terrains) (*v.r.*), to settle ; to sink : (17)
ces terrains se sont tassés (*m.pl.*), this ground has settled. (18)
tâtonnement (*n.m.*) *ou* **tâtonnements** (*n.m.pl.*), trial and error ; tentative : (19)
procéder par tâtonnements jusqu'à ce que le résultat désiré soit atteint, to proceed by trial and error until the desired result is obtained. (20)
tauriscite (Minéral) (*n.f.*), tauriscite. (21)
taux (*n.m.*), rate. (22)
taux de vitesse (*m.*), rate of speed. (23)
taux du salaire (*m.*), rate of wages. (24)
taxe (*n.f.*), tax. (25)
taylorite (Minéral) (*n.f.*), taylorite. (26)
té (pièce quelconque ayant la forme d'un T) (*n.m.*), tee ; T : (27)
fer à té *ou* fer en té (*m.*), T iron. (28)
té (tuyauterie) (*n.m.*), tee ; T. (29)
té (*n.m.*) *ou* **té de dessin,** T square : (30)
té en poirier, 1ᵉʳ choix, (1) à tête fixe (2) à tête tournante, filet cuivre, T square, pear wood, best quality, (1) with fixed head (2) with shifting head, brass edge. (31 *ou* 32)
té de raccordement (*m.*), T connection. (33)
té double (tuyauterie) (*m.*), cross. (34)
technicien (pers.) (*n.m.*), technical man ; techni-cist ; technician. (35)
technicité (*n.f.*), technicality. (36)
technique (*adj.*), technical : (37)
données techniques (*f.pl.*), technical data. (38)
difficultés d'ordre technique (*f.pl.*), technical difficulties. (39)
technique (*n.f.*), technics ; engineering. (40)

technique électrique (*f.*), electrical engineering. (41)
technique hydraulique (*f.*), hydraulic engineering. (42)
technique minière (*f.*), mining engineering. (43)
techniquement (*adv.*), technically. (44)
technologie (*n.f.*), technology. (45)
technologique (*adj.*), technologic ; technological. (46)
teck (arbre) (*n.m.*), teak. (47)
tectonique (Géol.) (*adj.*), tectonic : (48)
vallée tectonique (*f.*), tectonic valley. (49)
teinte (*n.f.*), tint ; tinge ; shade : (50)
une teinte grise, a gray shade. (51)
teinter (*v.t.*), to tint ; to tinge : (52)
pierre précieuse teintée par des oxydes métal-liques (*f.*), precious stone tinted by (*or* gem tinged with) metallic oxides. (53)
teinture de tournesol (*f.*), solution of litmus ; litmus solution. (54)
tek (arbre) (*n.m.*), teak. (55)
télé-objectif (Photogr.) (*n.m.*), telephoto lens ; telephotographic lens ; tele lens. (56)
télécommunication (*n.f.*), telecommunication. (57)
télédynamique (*adj.*), telodynamic ; teledynamic : (58)
câble télédynamique (*m.*), telodynamic rope ; teledynamic rope. (59)
télégraphe (*n.m.*), telegraph. (60)
télégraphe de menuisier (*m.*), joiners' bevel. (61)
télégraphe duplex (*m.*), duplex telegraph. (62)
télégraphie (*n.f.*), telegraphy. (63)
télégraphie avec fil (*f.*), wire-telegraphy. (64)
télégraphie sans fil (*f.*), wireless telegraphy ; aerial telegraphy ; radiotelegraphy. (65)
télégraphier (*v.t. & v.i.*), to telegraph. (66)
télégraphique (*adj.*), telegraphic ; telegraph (*used as adj.*) : (67)
code télégraphique (*m.*), telegraphic code. (68)
poteau télégraphique (*m.*), telegraph-pole. (69)
télégraphiquement (*adv.*), telegraphically. (70)
télégraphiste (pers.) (*n.m. ou f.*), telegraphist. (71)
télémécanique (*adj.*), telemechanical. (72)
télémécanique (*n.f.*), telemechanics. (73)
télémécanisme (*n.m.*), telemechanism. (74)
télémètre (*n.m.*), telemeter. (75)
télémétrie (*n.f.*), telemetry. (76)
télémétrique (*adj.*), telemetric ; telemetrical. (77)
télémétrographe (*n.m.*), telemetrograph. (78)
téléobjectif (Photogr.) (*n.m.*), telephoto lens ; tele lens ; telephotographic lens. (79)
téléphone (*n.m.*), telephone. (80)
téléphone à ficelle (*m.*), string telephone ; mechanical telephone. (81)
téléphone acoustique (*m.*), acoustic telephone ; mechanical telephone. (82)
téléphone haut-parleur (*m.*), intensifier tele-phone. (83)
téléphone magnétique (*m.*), magnetic telephone ; magneto-telephone. (84)
téléphoner (*v.t. & v.i.*), to telephone. (85)
téléphonie (*n.f.*), telephony. (86)
téléphonie avec fil (*f.*), wire-telephony. (87)
téléphonie sans fil (*f.*), wireless telephony ; radiotelephony. (88)
téléphonique (*adj.*), telephonic ; telephone (*used as adj.*) : (89)
appel téléphonique (*m.*), telephone-call. (90)
téléphoniquement (*adv.*), telephonically. (91)

téléphoniste (pers.) (*n.m. ou f.*), telephonist. (1)
téléphotographie (*n.f.*), telephotography. (2)
téléphotographique (*adj.*), telephotographic. (3)
télescopage (*n.m.*), telescoping. (4)
télescopage (à), telescoping : (5)
  mécanisme à télescopage (*m.*), telescoping mechanism. (6)
télescope (*n.m.*) *ou* **télescope à miroir** *ou* **téles-cope à réflexion**, telescope ; reflecting telescope : (7)
  le télescope diffère de la lunette par la sub-stitution d'un miroir concave à la lentille objective, the reflecting telescope differs from the refracting telescope by the substitution of a concave mirror for the object-lens. (8)
télescoper (se) (*v.r.*), to telescope : (9)
  wagons de chemin de fer qui se télescopent à la suite d'un tamponnement (*m.pl.*), railway-carriages which telescope as the result of a collision. (10)
télescopie (*n.f.*), telescopy. (11)
télescopiforme (*adj.*), telescopiform. (12)
télescopique (*adj.*), telescopic ; telescopical : (13)
  tubage télescopique (*m.*), telescopic casing. (14)
tellurate (Chim.) (*n.m.*), tellurate. (15)
tellure (Chim.) (*n.m.*), tellurium. (16)
tellure auroplombifère (Minéral) (*m.*), black tellurium ; foliated tellurium ; leaf tellu-rium ; nagyagite. (17)
tellure graphique (Minéral) (*m.*), graphic tellu-rium ; graphic gold ; graphic ore ; white tellurium ; yellow tellurium ; sylvanite. (18)
telluré, -e (Minéralogie) (*adj.*), telluride (*used as adj.*) : (19)
  minerai telluré (*m.*), telluride ore. (20)
tellureux, -euse (Chim.) (*adj.*), tellurous. (21)
tellurhydrique (Chim.) (*adj.*), tellurhydric. (22)
tellurien, -enne (qui provient de la terre) (*adj.*), tellurian : (23)
  émanations telluriennes (*f.pl.*), tellurian emanations. (24)
tellurifère (Minéralogie) (*adj.*), telluriferous. (25)
tellurine (Minéral) (*n.f.*), tellurite ; telluric ochre. (26)
tellurique (terrestre) (*adj.*), telluric : (27)
  fer tellurique (*m.*), telluric iron. (28)
tellurique (Chim.) (*adj.*), telluric : (29)
  acide tellurique (*m.*), telluric acid. (30)
tellurite (Minéral) (*n.f.*), tellurite ; telluric ochre. (31)
tellurite (Chim.) (*n.m.*), tellurite. (32)
tellurure (Chim.) (*n.m.*), telluride ; telluret : (33)
  tellurure d'or, d'argent, de plomb, telluride of gold, of silver, of lead. (34 *ou* 35 *ou* 36)
télodynamique (*adj.*). Same as **télédynamique**.
telphérage (*n.m.*), telpherage. (37)
témoin (butte-témoin) (opp. à **fenêtre**) (Géol.) (*n.m.*), outlier ; butte. (38) (opp. to **inlier**) :
  les monticules isolés qui échappent à la destruc-tion par ablation sont connus sous le nom de témoins (*ou* de buttes-témoins). Ils sont souvent le meilleur indice de l'ancienne extension d'une couche aujourd'hui presque entièrement disparue, the isolated mounds which escape destruction by ablation are known as outliers. They are often the best indication of the ancient extension of a bed which to-day has almost entirely dis-appeared. (39)
témoin (Coupellation) (*n.m.*), test ; test-button. (40)

témoin (Sondage) (*n.m.*), core : bore-core ; drill-core. (41)
témoin de profondeur (Moulage en terre) (Fonderie) (*m.*), depth-gauge (Loam-moulding). (42)
tempérament (*n.m.*), character : (43)
  tempérament chimique de la mine, chemical character of the mine. (44)
  tempérament du gisement, des alluvions, character of the deposit, of the alluvium. (45 *ou* 46)
température (*n.f.*), temperature. (47)
température absolue (Phys.) (*f.*), absolute tem-perature. (48)
température critique (Phys.) (*f.*), critical tem-perature. (49)
température d'ébullition (*f.*), boiling-point ; temperature of ebullition. (50)
température d'inflammation du grisou (*f.*), fire-damp ignition-temperature. (51)
température de beaucoup supérieure à la normale (*f.*), much higher temperature than usual. (52)
température de congélation (*f.*), freezing-point : (53)
  la température de congélation de l'eau sert de base à l'échelle thermométrique centrigrade, the freezing-point of water forms the base of the centigrade thermometric scale. (54)
  la température de congélation du mercure est de — 39,5° C. ou — 39° F., the freezing-point of mercury is — 39·5° C. or — 39° F. (55)
température de détonation (*f.*), detonating-point ; detonation-temperature. (56)
température de fusion (*f.*), melting-point ; fusing-point ; fusion-point ; point of fusion. (57)
température de la glace fondante (*f.*), ice-point ; freezing-point of water ; freezing-point ; temperature at which ice melts (*or* at which water freezes) : (58)
  0° C. ou 32° F. est la température de la glace fondante, 0° C. or 32° F. is the freezing-point of water. (59)
température de vaporisation (*f.*), heat of vapor-ization ; latent heat of vaporization ; vaporization latent temperature. (60)
température du rouge (*f.*), red heat. (61)
température du rouge cerise (*f.*), cherry-red heat. (62)
température du rouge clair (*f.*), bright-red heat. (63)
température du rouge sang (*f.*), blood-red heat. (64)
température du vent (machine soufflante) (*f.*), temperature of the blast (blowing-engine). (65)
température égale (*f.*), even temperature ; equable temperature ; equal temperature. (66)
température élevée (*f.*), high temperature. (67)
température estivale (*f.*), summer temperature. (68)
température hivernale (*f.*), winter temperature. (69)
température moyenne (*f.*), mean temperature ; average temperature. (70)
tempéré, -e (*adj.*), temperate : (71)
  climat tempéré (*m.*), temperate climate. (72)
tempête (*n.f.*), storm ; tempest ; gale : (73)
  tempête de neige, snow-storm. (74)
  tempête magnétique, magnetic storm. (75)

vent qui souffle en tempête (*m.*), wind which blows a gale. (1)

tempête (Barométrie), stormy. (2)

temporaire (*adj.*), temporary ; provisional. (3)

temporairement (*adv.*), temporarily ; provision--ally. (4)

temps (durée limitée) (*n.m.*), time : (5)
temps nécessaire pour mettre la mine en exploi--tation normale, time required to put the mine in working order. (6)
temps de chute d'un corps (Phys.), time required for a body to fall. (7)

temps (état de l'atmosphère) (*n.m.*) (*s'emploie quelquefois au pluriel*), weather : (8)
point nettement visible par (*ou* dans) tous les temps (*m.*), point clearly visible in all weathers. (9)

temps chaud (*m.*), hot weather. (10)

temps d'été (*m.*), summer weather. (11)

temps d'hiver (*m.*), winter weather. (12)

temps de brouillard (*m.*), foggy weather. (13)

temps de pose (Photogr.) (*m.*), exposure. (14)

temps fixe (*m.*), settled weather. (15)

temps froid (*m.*), cold weather. (16)

temps glacial (*m.*), frosty weather. (17)

temps hors de saison (*m.*), unseasonable weather. (18)

temps humide (*m.*), wet weather ; damp weather : (19)
dans les temps humides, in wet weather. (20)

temps orageux (*m.*), stormy weather. (21)

temps périodique (Élec.) (*m.*), period. (22)

temps pluvieux (*m.*), rainy weather. (23)

temps rigoureux (*m.*), severe weather ; wintry weather. (24)

temps sec (*m.*), dry weather. (25)

temps sombre (*m.*), dull weather. (26)

temps variable (*m.*), changeable weather. (27)

tenace (*adj.*), tenacious ; tough ; stiff. (28)

tenacement (*adv.*), tenaciously ; toughly ; stiffly. (29)

ténacité (*n.f.*), tenacity ; tenaciousness ; toughness ; stiffness : (30)
ténacité d'un métal, de la poix, tenacity, tenaciousness, toughness, of a metal, of pitch. (31 *ou* 32)

ténacité (Résistance des matériaux) (*n.f.*), tenac--ity ; strength. (33)

ténacité extrême (Résistance des matériaux) (*f.*), ultimate strength ; breaking-strength. (34)

tenaille (*n.f.*) *ou* **tenailles** (*n.f.pl.*), tongs ; pincers ; nippers. (35) See also **pince** and **pinces**.

tenaille à bec plat (de forgeron) (*f.*), flat-mouth tongs (blacksmiths'). (36)

tenaille à bec recourbé *ou* **tenaille à cornières** (de forgeron) (*f.*), side-mouth tongs ; crook-bit tongs ; elbow-tongs ; angle-jaw tongs ; duck-bill tongs. (37)

tenaille à bec rond (de forgeron) (*f.*), round-mouth tongs. (38)

tenaille à demi-coquille (de forgeron) (*f.*), flat-bar tongs. (39)

tenaille à forger (*f.*), smiths' pliers. (40)

tenaille à mors coupants, branches noires, mâchoires polies (*f.*), cutting-nippers, black handles, bright jaws. (41)

tenaille à rails (*f.*), rail-tongs. (42)

tenaille à rivets (*f.*), rivet-tongs. (43)

tenaille à vis (*f.*), hand-vice. (44)

tenaille buse (de forgeron) (*f.*), hollow-bit tongs. (45)

tenaille coupante (*f.*) *ou* **tenailles coupantes** (*f.pl.*), cutting-nippers. (46)

tenaille d'attelage (Roulage) (Mines) (*f.*), clip ; grip ; haulage-clip ; rope-clip ; jigger. (47)

tenaille d'attelage à desserrage automatique (*f.*), automatic detacher ; jockey. (48)

tenaille d'attelage à vis (*f.*), screw-clip ; screw-grip. (49)

tenaille de charpentier (*f.*), carpenters' pincers. (50)

tenaille de forge *ou* **tenaille de forgeron** (*f.*), blacksmiths' tongs ; anvil-tongs ; tue-irons. (51)

tenaille droite (de forgeron) (*f.*), straight-tip tongs. (52)

tenaille plate fermée *ou* **tenaille juste** (de forge--ron) (*f.*), close-mouth tongs. (53)

tenaille plate ouverte *ou* **tenaille goulue** (de forgeron) (*f.*), open-mouth tongs. (54)

tenaille pour boulons (*f.*), bolt-tongs. (55)

tenaille pour fer méplat (*f.*), flat-bar tongs. (56)

tenaille pour fer rond (*f.*), round-bar tongs. (57)

tenaille pour rivets (*f.*), rivet-tongs. (58)

tendance (*n.f.*), tendency ; inclination : (59)
tendance à s'améliorer, tendency to improve. (60)

tendard (Boisage de mines) (*n.m.*), stretcher-piece ; reacher. (61)

tender (*n.m.*) *ou* **tender de locomotive**, tender ; engine-tender. (62)

tendeur (*n.m.*), strainer ; stretcher ; tightener ; take-up. (63)

tendeur (barre d'attelage) (Ch. de f.) (*n.m.*), draw-bar ; draught-iron ; draft-iron ; draw-link ; drag-bar ; drag-link. (64)

tendeur à cliquet (*m.*), tension-ratchet. (65)

tendeur à pouce articulé *ou* **tendeur articulé pour fils** (*m.*), Dutch draw-tongs ; lion's claw ; devil's claw ; wire-dog. (66)

tendeur à vis (*m.*), straining-screw ; union-screw ; stretching-screw ; screw-tightener. (67)

tendeur pour courroie (*m.*), belt-tightener ; belt-stretcher. (68)

tendeur pour fil de fer (*m.*), wire-strainer. (69)

tendre (*adj.*), soft ; tender : (70)
roche tendre (*f.*), soft rock ; tender rock. (71)
bois tendre (*m.*), soft wood. (72)

tendre (*v.t.*), to stretch ; to tighten ; to strain ; to make tense ; to tense : (73)
tendre un cordeau d'une paroi à l'autre, to stretch a line from one wall to the other. (74)
tendre une courroie, to stretch, to tighten, a belt. (75)
toile tendue entre deux rouleaux (*f.*), cloth stretched between two rollers. (76)

tendre (se) (*v.r.*), to stretch. (77)

teneur (*n.f.*), grade ; content ; contents ; per--centage ; amount ; tenor ; tenure. (78)

teneur de minerai (*f.*), grade (*or* tenor) of ore. (79)

teneur de tas (Rivetage) (pers.) (*m.*), holder-up. (80)

teneur en cendres (*f.*), percentage of ash ; ash-content. (81)

teneur en grisou (*f.*), amount (*or* percentage) (*or* content) of fire-damp : (82)
la teneur en grisou des retours d'air est relevée quotidiennement dans les mines franche--ment grisouteuses, the percentage of fire-damp in the air-returns shall be taken daily in mines where much gas is present. (83)

teneur en humidité, en matières volatiles (*f.*), percentage of moisture, of volatile matter. (1 *ou* 2)

teneur en minerai (*f.*), ore-contents ; percentage of ore. (3)

teneur en or (*f.*), gold-content ; gold-contents ; tenure of gold. (4)

teneur en soufre du minerai (*f.*), percentage of sulphur in ore ; sulphur contents of the ore. (5)

teneur limite (*f.*), pay-line ; lowest grade at which working is payable. (6)

teneur moyenne (*f.*), average grade ; fair grade ; medium grade. (7)

teneur payante de minerai (*f.*), pay-grade of ore ; payable grade of ore. (8)

teneur pour cent en carbone (*f.*), percentage of carbon. (9)

tenir (*v.t.* & *v.i.*), to hold. (10)

tenir avec soin un carnet à jour du sondage, to keep a careful log of the bore-hole. (11)

tenir beaucoup de place, to take up a lot of room. (12)

tenir bon (*v.i.*), to hold ; to hold fast : (13)
le barrage tint bon, the dam held. (14)
signaler à la surface de tenir bon, to signal to bank to hold. (15)

tenir compte de la différence de poids, to make allowance for, to allow for, the difference in weight. (16)

tenir compte des considérations économiques, to have regard to the economic considerations ; to take the economic factors into account. (17)

tenir en suspension, to hold in suspension : (18)
tous les gaz tiennent en suspension une mul- -titude de poussières microscopiques (*m.pl.*), all gases hold in suspension large quantities of microscopical dust. (19)

tenir ferme (*v.i.*). Same as tenir bon.

tenir le boisage près du front de taille, to carry the timbering close to the face. (20)

tenir le coup (Rivetage) (*v.i.*), to hold up. (21)

tenir le trou de sonde rectiligne, to keep the bore-hole true. (22)

tenir — 0/0 d'hydrocarbures volatils, to contain — % of volatile hydrocarbons. (23)

tenir tête au débit d'un puits à pétrole, to keep pace with the flow of an oil-well. (24)

tenir (se) (*v.r.*), to hold ; to hold up : (25)
la roche se tient bien, the rock holds up well. (26)

tenir (se) prêt, -e, to hold oneself in readiness. (27)

tennantite (Minéral) (*n.f.*), tennantite. (28)

tenon (*n.m.*), tenon. (29)

tenon avec renfort carré (*m.*), haunched tenon. (30)

tenon bâtard (*m.*), barefaced tenon. (31)

tenon en queue d'aronde (*m.*), dovetailed tenon. (32)

tenon épaulé (*m.*), shouldered tenon. (33)

tenon invisible (*m.*), stub-tenon ; stub ; tooth. (34)

tenon oblique (*m.*), oblique tenon. (35)

tenon passant (*m.*), through-tenon. (36)

tenon renforcé *ou* tenon à renfort (*m.*), tusk- tenon. (37)

ténorite (Minéral) (*n.f.*), tenorite. (38)

tenseur (*n.m.*). Same as tendeur.

tension (*n.f.*), tension ; tightening ; strain ; stretching. (39)

tension (de vapeur) (Méc.) (*n.f.*), pressure, tension, (of steam). (40)

tension (*n.f.*) *ou* tension en volts (Élec.), tension ; pressure ; voltage. (41)

tension au départ (Élec.) (*f.*), station-pressure. (42)

tension d'un gaz, d'une vapeur (Phys.) (*f.*), tension of a gas, of a vapour. (43 *ou* 44)

tension d'une corde, d'un câble, d'un fil, d'une ligne aérienne, d'un ressort (*f.*), tension of a cord, of a rope, of a wire, of an overhead line, of a spring. (45 *ou* 46 *ou* 47 *ou* 48 *ou* 49)

tension de régime *ou* tension de régime en volts (Élec.) (*f.*), normal voltage. (50)

tension de service (Élec.) (*f.*), running voltage ; working-pressure. (51)

tension de vapeur (*f.*), steam-pressure. (52)

tension élastique pour lame de scie (*f.*), spring tension for saw-blade. (53)

tension moléculaire (*f.*), molecular strain : (54)
les pièces moulées viennent parfois à se briser sous l'effort de tensions moléculaires insoup- -çonnées (*f.pl.*), it happens sometimes that castings break under the stress of unsuspected molecular strains. (55)

tension superficielle (Phys.) (*f.*), surface-tension. (56)

tentative (*n.f.*), attempt. (57)

tente (*n.f.*), tent. (58)

tente conique (*f.*), bell tent. (59)

ténu, -e (*adj.*), thin ; slender ; tenuous : (60)
un fil ténu, a slender wire. (61)

tenue (manière de soigner) (*n.f.*), keeping ; care : (62)
tenue du carnet (Géod.), keeping notes. (63)
tenue des chaudières de locomotive, care of locomotive-boilers. (64)

tenue (des couches) (Mines) (*n.f.*), hold-up (of the seams). (65)

ténuement (*adv.*), thinly ; slenderly ; tenuously. (66)

ténuité (*n.f.*), thinness ; slenderness ; tenuity ; tenuousness : (67)
ténuité d'un liquide, d'une pellicule, tenuity, thinness, of a liquid, of a film. (68 *ou* 69)

téphrite (Pétrol.) (*n.f.*), tephrite. (70)

téphroïte (Minéral) (*n.f.*), tephroite. (71)

terbium (Chim.) (*n.m.*), terbium. (72)

térébenthine (*n.f.*), turpentine. (73)

terme (mot, expression technique) (*n.m.*), term : (74)
terme de chimie, chemical term. (75)

terminable (*adj.*), terminable. (76)

terminaison (*n.f.*), termination ; completion. (77)

terminer (*v.t.*), to terminate ; to finish ; to end ; to complete ; to put an end to. (78)

terminer (se) (*v.r.*), to terminate ; to finish ; to end. (79)

terminer (se) en pointe, to taper : (80)
le continent de l'Amérique du Sud se termine en pointe vers le sud, the South American continent tapers towards the south. (81)

terminus (*n.m.*), terminus : (82)
terminus d'un chemin de fer, terminus of a railway. (83)

ternaire (*adj.*), ternary : (84)
un alliage ternaire de cuivre, d'étain, et de zinc, a ternary alloy of copper, tin, and zinc. (85)

terne (*adj.*), lustreless ; dull. (86)

ternir (*v.t.*), to tarnish ; to dull ; to dim. (87)

ternir (se) (*v.r.*), to tarnish : (1)
  métal qui se ternit à l'air (*m.*), metal which
  tarnishes in the air. (2)
terra rossa (Géol.) (*f.*), terra rossa ; red earth.
  (3)
terrain (espace de terrain) (*n.m.*), ground ;
  land ; piece of ground ; plot of ground ;
  field : (4)
  le terrain se vend cher à Londres, ground is dear
  in London. (5)
  occuper un vaste terrain, to occupy a big piece
  of ground. (6)
  observations sur le terrain (*f.pl.*), observations
  in the field (*or* on the ground). (7)
  géologie sur le terrain (*f.*), field-geology. (8)
terrain (sol considéré au point de vue de sa nature,
  ou par rapport à l'état de sa surface) (*n.m.*),
  ground ; land ; soil. (9)
terrain (*n.m.*) *ou* **terrains** (*n.m.pl.*) (Géol.) (*n.m.*),
  ground ; stratum ; strata ; measures ;
  terrane ; terrain. (10)
terrain **à niveau** (*m.*), level ground. (11)
terrain **aquifère** (*m.*) *ou* **terrains aquifères** (*m.pl.*),
  water-bearing strata ; watery ground. (12)
terrain **attrayant** (pour une exploitation minière)
  (*m.*), attractive ground. (13)
terrain **aurifère** (*m.*), auriferous ground ;
  gold-bearing ground. (14)
terrain **carbonifère** (*m.*). Same as **terrain**
  houiller.
terrain **coulant** (*m.*), quick ground. (15)
terrain **d'alluvion** (*m.*) *ou* **terrains d'alluvion**
  (*m.pl.*), alluvial ground ; alluvial soil ;
  alluvial ; alluvium , placer-ground. (16)
terrains d'alluvions (exploitation) (*m.pl.*), alluvial
  diggings. (17)
terrains d'alluvions diamantifères (*m.pl.*), allu-
  -vial diamond-diggings. (18)
terrains de couverture *ou* **terrains de recouvre-**
  -ment (Géol. & Mines) (*m.pl.*), overburden ;
  overplacement ; burden ; soil-cap ; strip-
  -pings ; muck. (19)
terrains de dragage (*m.pl.*), dredging-ground.
  (20)
terrain **de niveau** (*m.*), level ground. (21)
terrains de transition (Géol.) (*m.pl.*), transition
  strata ; transition terrane ; transition
  terrain. (22)
terrain **ébouleux** (*m.*), running ground. (23)
terrain **en palier** (*m.*), level ground. (24)
terrain **en pente** (*m.*), sloping ground. (25)
terrain **encaissant** (Géol. & Mines) (*m.*), country-
  rock ; country ; enclosing rock ; wall-
  rock. (26)
terrain **erratique** (argile à blocaux) (Géol.) (*m.*),
  drift ; glacial drift ; till ; glacial till ;
  boulder-clay ; bowlder-clay. (27)
terrain **ferme** (*m.*), firm ground. (28)
terrain **forestier** (*m.*), forest land. (29)
terrain **houiller** (Géol.) (*m.*), coal-measures. (30)
terrain **houiller inférieur** (*m.*), Lower coal-
  measures. (31)
terrain **houiller moyen** (*m.*), Middle coal-
  measures. (32)
terrain **houiller productif** (*m.*), productive coal-
  measures. (33)
terrain **houiller stérile** (*m.*), barren coal-
  measures. (34)
terrain **houiller supérieur** (*m.*), Upper coal-
  measures. (35)
terrains imbibés d'eau (*m.pl.*), water-logged
  strata. (36)

terrain **inconsistant** (*m.*), loose ground. (37)
terrain **marécageux** (*m.*), marshy ground ;
  marsh-land. (38)
terrain **marin** (*m.*), marine deposit. (39)
terrain **métallifère** (*m.*), metalliferous ground ;
  ore-bearing ground. (40)
terrain **meuble** (*m.*), loose ground ; running
  ground. (41)
terrains modernes (Géol.) (*m.pl.*), modern strata.
  (42)
terrain **perméable** (*m.*), permeable ground ;
  leachy soil. (43)
terrain **pétrolifère** (*m.*), petroleum-land ; oil-
  land. (44)
terrain **pierreux** (*m.*), stony ground. (45)
terrain **plat** (*m.*), level ground ; even ground.
  (46)
terrains primitifs (Géol.) (*m.pl.*), primitive
  strata ; primitive terrane ; primitive
  terrain. (47)
terrains privés (*m.pl.*), private land. (48)
terrain **solide** (*m.*), firm ground : (49)
  fonçage en terrains solides (*m.*), sinking
  through firm ground. (50)
terrain **stérile** (*m.*), barren ground ; sterile
  ground ; hungry ground. (51)
terrain **uni** (*m.*), even ground. (52)
terrain **vague** (*m.*), waste land. (53)
terrasse (*n.f.*), terrace ; earthwork. (54)
terrasse (Géol.) (*n.f.*), terrace. (55)
terrasse (cavité remplie d'une matière terreuse)
  (Mines) (*n.f.*), horse of ground. (56)
terrasse (en), terraced ; terraciform : (57)
  dépôt en terrasse (Géol.) (*m.*), terraciform
  deposit. (58)
terrasse alluviale *ou* **terrasse d'alluvions** (Géol.)
  (*f.*), alluvial terrace ; flood-plain terrace ;
  built terrace ; terrace of construction. (59)
terrasse d'érosion (Géol.) (*f.*), cut terrace. (60)
terrasse dans le roc (Géol.) (*f.*), rock-terrace. (61)
terrasse de cours d'eau (Géol.) (*f.*), stream-
  terrace ; river-terrace. (62)
terrasse glaciaire (Géol.) (*f.*), glacial terrace. (63)
terrasse littorale (Géol.) (*f.*), shore-terrace. (64)
terrassement (action) (*n.m.*), earthwork : (65)
  terrassement relatif aux chemins de fer, earth-
  -work in connection with railways. (66)
terrassement (terres creusées) (*n.m.*), earthwork :
  (67)
  les terrassements d'une voie ferrée, the earth-
  -works of a railway. (68)
terrassement en remblai (*m.*), embankment ;
  earthwork embankment. (69)
terrasser (*v.t.*), to bank up with earth ; to bank
  with earth ; to bank up ; to bank ; to
  embank ; to navvy : (70)
  terrasser un mur, to bank up a wall with earth.
  (71)
terrasser (*v.i.*), to navvy. (72)
terrassier (pers.) (*n.m.*), navvy. (73)
terrassier à vapeur (*m.*), steam-navvy ; navvy ;
  steam-digger ; steam-excavator ; steam-
  shovel. (74)
terre (planète habitée par l'homme) (*n.f.*), earth :
  (75)
  la rotation de la terre, the rotation of the
  earth. (76)
terre (partie solide de la surface terrestre, par
  opposition à la mer, etc.) (*n.f.*), land ; earth :
  (77)
  où la terre finit, la mer commence, where the
  earth ends, the sea begins. (78)

terre (sol) (*n.f.*) (*s'emploie aussi au pluriel*), earth ; ground ; land ; soil : (1)
terre vierge, virgin ground. (2)
les grandes pluies affaissent les terres (*f.pl.*), heavy rains cause the earth to subside. (3)
terre (Chim.) (*n.f.*), earth : (4)
terres rares, rare earths. (5)
terre (contrée) (*n.f.*), land ; country : (6)
terres inconnues, unknown lands. (7)
terre (Élec.) (*n.f.*), earth ; ground : (8)
communication entre un conducteur et la terre (*f.*), connection between a conductor and the earth. (9)
employer la terre seule comme partie du circuit, to use the earth only as part of the circuit. (10)
terre (à la) (Élec.), earthed ; grounded : (11)
conducteur à la terre (*m.*), earthed conductor ; grounded conductor. (12)
terre (défaut dans une ligne de transmission électrique provoqué par une communication accidentelle du conducteur métallique avec la terre) (*n.f.*), earth ; ground : (13)
une terre, an earth ; a ground. (14)
terre trop faible pour faire jouer un fusible, earth too weak to blow out a fuse. (15)
terre (de fonderie) (*n.f.*), loam : (16)
moulage en terre (*m.*), loam-molding. (17)
terre à briques *ou simplement* terre (*n.f.*), brick-earth. (18)
terre à diatomées (*f.*), diatomaceous earth ; diatom-earth. (19)
terre à foulon (*f.*), fullers' earth ; fullers' chalk ; smectite. (20)
terre à infusoires (*f.*), infusorial earth ; in-fusorial silica. (21)
terre à porcelaine (*f.*), porcelain-clay ; china-clay ; kaolin ; china-stone. (22)
terre à poterie *ou* terre à poteries *ou simplement* terre (*n.f.*), potters' clay ; potters' earth ; pot-earth ; figuline. (23)
terre alcaline (*f.*), alkali soil. (24) Cf. terres alcalines.
terre alumineuse (*f.*), aluminous earth. (25)
terre amère (magnésie) (*f.*), bitter earth. (26)
terre aurifère (*f.*), auriferous ground ; gold-bearing gravel ; dirt. (27)
terre bleue (dans laquelle se trouve l'ambre) (*f.*), blue earth. (28)
terre d'infusoires (*f.*), infusorial earth ; in-fusorial silica. (29)
terre d'ombre *ou* terre de Nocera (*f.*), umber. (30)
terre d'ombre de Cologne (*f.*), Cologne earth ; Cologne umber ; Cologne brown ; German umber. (31)
terre d'os (*f.*), bone-phosphate. (32)
terre de couverture (*f.*). See terres de couverture.
terre de pipe (*f.*), pipe-clay. (33)
terre de Sienne (*f.*), sienna. (34)
terre des potiers (*f.*). Same as terre à poterie.
terre ferme (*f.*), terra firma ; dry land ; con-tinent ; mainland ; land : (35)
sédiments océaniques provenant de la démolition de la terre ferme (*m.pl.*), oceanic sediments derived from the waste of the land. (36)
terre glaise (*f.*), clay (potters' clay, tile-earth, or clay suitable for modeling purposes). (37)
terre jaune (*f.*), yellow earth. (38)
terre légère (*f.*), light soil. (39)
terre meuble (*f.*), loose earth. (40)

terre noire (Géol.) (Russie) (*f.*), black earth. (41)
terre pesante (Chim.) (*f.*), heavy earth ; baryta. (42)
terre pourrie d'Angleterre (*f.*), rottenstone. (43)
terre réfractaire (*f.*), refractory earth ; fire-clay. (44)
terre rouge (Géol.) (*f.*), red earth ; terra rossa. (45)
terre sableuse (*f.*), sandy earth. (46)
terre siliceuse (*f.*), siliceous earth ; silica. (47)
terre végétale (*f.*), soil ; mold earth. (48)
terre verte (Minéral) (*f.*), green earth. (49)
terre vierge (*f.*), virgin ground. (50)
terres alcalines (Chim.) (*f.pl.*), alkaline earths. (51)
métaux alcalino-terreux (*m.pl.*), metals of the alkaline earths. (52)
terres de couverture (Géol. & Mines) (*f.pl.*), overburden ; overplacement ; burden ; soil-cap ; strippings ; muck. (53)
terres rares (Chim.) (*f.pl.*), rare earths. (54)
terrestre (*adj.*), terrestrial ; earth's : (55)
les pôles terrestres, the terrestrial poles. (56)
la convexité naturelle de la surface terrestre, the natural convexity of the earth's surface. (57)
terreux, -euse (*adj.*), earthy : (58)
gangue terreuse (*f.*), earthy gangue. (59)
terrigène (*adj.*), terrigenous ; terrigene : (60)
dépôts terrigènes (*m.pl.*), terrigenous deposits. (61)
terris (Mines) (*n.m.*), dumping-ground ; dump ; tip. (62)
territoire (*n.m.*), territory. (63)
terrou (Mines) (*n.m.*), fire-damp ; gas ; pit-gas ; fire. (64)
tertiaire (Géol.) (*adj.*), Tertiary : (65)
l'âge tertiaire (*m.*), the Tertiary age. (66)
tertiaire (*n.m.*), Tertiary : (67)
le tertiaire, the Tertiary. (68)
tertre (*n.m.*), hillock ; mound ; knoll. (69)
tessélite (Minéral) (*n.f.*), tesselite. (70)
test-objet [test-objets *pl.*] (Microscopie) (*n.m.*), test-object ; test. (71)
tétartoèdre *ou* tétartoédrique (Cristall.) (*adj.*), tetartohedral ; tetartohedric ; tetartohedri-cal. (72)
tétartoèdre (*n.m.*), tetartohedron. (73)
tétartoédrie (*n.f.*), tetartohedrism ; tetartohedry. (74)
tête (*n.f.*), head. (75)
tête (d'une vis, d'un boulon, d'un clou) (*n.f.*), head (of a screw, of a bolt, of a nail). (76 *ou* 77 *ou* 78) For the various forms of screw-heads, boltheads, and nail-heads, see **vis, boulon, and clou** ; thus, **vis à tête à six pans**, hexagon-head screw.
tête (d'un rivet) (opp. à **rivure**) (*n.f.*), head (of a rivet) (the head formed on one end of a rivet before use). (79) For the various forms of rivet-heads, see **rivet**.
tête (d'un pieu) (opp. à la **pointe**) (*n.f.*), head (of a pile). (80) (opp. to the **point**).
tête (d'un marteau) (corps ou masse métallique) (opp. au **manche**) (*n.f.*), head (of a hammer). (81) (in contradistinction to the **handle**).
tête (d'un marteau) (opp. à la **panne**) (*n.f.*), face (of a hammer). (82) (in contra-distinction to the **peen**) :
marteau à devant à deux têtes (*m.*), double-faced sledge-hammer. (83)

tête (d'un coin) (n.f.), head (of a wedge *or* key). (1)

tête (d'un pic à tête) (n.f.), poll (of a poll-pick) ; hammer (of a hammer-pick). (2)

tête (d'une herminette à tête) (n.f.), poll (of a poll-adze). (3)

tête (d'une dent d'engrenage) (n.f.), point, addendum, (of a gear-tooth). (4)

tête (d'un cabestan) (n.f.), drumhead (of a capstan). (5)

tête (d'un gond) (n.f.), pintle (of a gate-hook). (6)

tête à friction (d'un palmer) (f.), ratchet-stop (of a micrometer). (7)

tête d'accouplement du frein (f.), brake-hose coupling. (8)

tête d'habitant (f.), head of population : (9) donner au moins — litres d'eau par tête d'habitant par jour, to allow at least — litres of water per head of population per day. (10)

tête de bélier (f.), ram's-horn ; ram's-horn crane-hook ; double crane-hook ; double hook ; change-hook. (11)

tête de bielle (f.), connecting-rod end ; rod-end ; pitman-head. (12)

tête de bielle (grosse tête de bielle,—opp. à pied de bielle) (f.), connecting-rod big end ; big end, large end, stub-end, butt, of connect-ing-rod. (13)

tête de bielle à cage fermée (f.), box connecting-rod end. (14)

tête de bielle à chape (f.), strap connecting-rod end. (15)

tête de bielle à chapeau *ou* tête de bielle à cage ouverte et à chapeau (f.), marine-type connecting-rod end. (16)

tête de bielle à fourche (f.), forkhead ; fork-head. (17)

tête de boulon (f.), bolthead. (18)

tête de chat [têtes de chat *pl.*] (Géol.) (f.), cat's-head ; cat-head ; cathead. (19)

tête de cheval (d'un tour à fileter) (f.), quadrant, quadrant-plate, radial arm, radial plate, swing-frame, tangent plate, (of a screw-cutting lathe). (20)

tête de hache (f.), axe-head. (21)

tête de la barre d'attelage (f.), draw-head. (22)

tête de ligne (Ch. de f.) (f.), terminus ; terminal : (23) la tête de ligne du réseau P.-L.-M., the terminus of the P.L.M. system. (24)

tête de palastre (serrure) (f.), selvage ; selvedge ; edge-plate (lock). (25)

tête de pilon (bocard) (f.), stamp-head ; stamp-boss (ore-stamp). (26)

tête de piston (f.), cross-head of piston-rod ; crosshead ; slipper-block. (27)

tête de piston guidée par quatre glissières (f.), cross-head with 4-bar guide. (28)

tête de sonde (f.), top rod ; swivel-rod ; swivel. (29)

tête de tube (Sondage) (f.), casing-head. (30)

tête de tube à vis (f.), screw casing-head. (31)

tête de voie (Ch. de f.) (f.), road-head. (32)

tête du gisement (Géol. & Mines) (f.), cap ; capping ; cap-rock ; rock capping. (33)

tête du pilon (bocard) (f.), stamp-head ; stamp-boss (ore-stamp). (34)

tête du tubage (Sondage) (f.), casing-head. (35)

tête porte-foret (f.), drill-head. (36)

tête porte-fraise (f.), cutter-head ; milling-head. (37)

tête porte-outil (d'un étau-limeur) (f.), tool-head (of a shaper). (38)

tête porte-outil à secteur denté (f.), tool-head fitted with quadrant ; tool-head with toothed sector. (39)

têtes (le minerai le plus pur obtenu par lavage) (n.f.pl.), heads ; headings : (40) les têtes riches, the rich heads. (41)

têtière (d'une serrure) (n.f.), selvage, selvedge, edge-plate, (of a lock). (42)

teton *ou* téton (Méc.) (n.m.), teat ; tit ; centre-point : (43) vis à téton (f.), teat-screw ; tit-screw. (44) le teton d'un foret à teton, the centre-point of a centre-bit. (45)

tétrabasique (Chim.) (adj.), tetrabasic. (46)

tétrabromure d'acétylène (m.), acetylene tetra-bromide. (47)

tétradymite (Minéral) (n.f.), tetradymite ; telluric bismuth. (48)

tétraédral, -e, -aux *ou* tétraèdre (adj.), tetrahedral. (49)

tétraèdre (n.m.), tetrahedron. (50)

tétraédrite *ou* tétrahédrite (Minéral) (n.f.), tetra-hedrite. (51)

tétragone (adj.), tetragonal ; quadrangular ; quadrilateral. (52)

tétragone (n.m.), tetragon ; quadrangle ; quadri-lateral. (53)

tétraphasé, -e (Élec.) (adj.), four-phase. (54)

tétrapolaire (Élec.) (adj.), four-pole : (55) moteur tétrapolaire (m.), four-pole motor. (56)

tétrasilicate (n.m.), tetrasilicate. (57)

tétrasilicique (adj.), tetrasilicic. (58)

tétratomicité (n.f.), tetratomicity. (59)

tétratomique (adj.), tetratomic. (60)

tétravalence (Chim.) (n.f.), tetravalence ; tetravalency ; quadrivalence. (61)

tétravalent, -e (adj.), tetravalent ; quadrivalent. (62)

têtu (n.m.), sledge ; stone-hammer. (63)

texasite (Minéral) (n.f.), zaratite ; emerald nickel. (64)

texture (n.f.), texture : (65) texture cellulaire de certaines roches, cellular texture of certain rocks. (66) texture d'un fer, d'un acier, du charbon, texture of an iron, of a steel, of the coal. (67 *ou* 68 *ou* 69) texture empâtée (Géol.), pasty texture. (70)

thalassique (adj.), thalassic : (71) roches thalassiques (f.pl.), thalassic rocks. (72)

thallieux, -euse (Chim.) (adj.), thallous ; thallious. (73)

thallique (Chim.) (adj.), thallic. (74)

thallium (Chim.) (n.m.), thallium. (75)

thalweg (Géol.) (n.m.), thalweg : (76) le ruissellement des eaux participe au creuse-ment des vallées en portant aux thalwegs les matériaux meubles des pentes, the run-ning off of water over the surface shares in the hollowing out of valleys by carrying down the loose material from the slopes to the thalwegs. (77)

thalweg d'amont (m.), upper thalweg. (78)

thalweg d'aval (m.), lower thalweg. (79)

théâtre des opérations (m.), scene of operations. (80)

thénardite (Minéral) (n.f.), thenardite. (81)

théodolite (n.m.), theodolite. (82)

théodolite à boussole (*m.*), transit-compass ; surveyors' transit ; transit. (1)

théodolite à boussole pour mines (*m.*), mine-transit ; mining-transit. (2)

théodolite à lunette centrale *ou* théodolite à lunette centrée (*m.*), transit-theodolite ; transit. (3)

théodolite altazimutal (*m.*), altazimuth theodolite. (4)

théodolite d'exploration (*m.*), mountain-theodolite. (5)

théodolite simplifié (*m.*), simple theodolite. (6)

théodolite souterrain (*m.*), miners' transit-instrument. (7)

théoricien (pers.) (*n.m.*), theorist. (8)

théorie (*n.f.*), theory ; hypothesis. (9)

théorie atomique (Chim.) (*f.*), atomic theory ; atomic hypothesis. (10)

théorie de l'infiltration (Géol.) (*f.*), infiltration theory. (11)

théorie de l'injection (Géol.) (*f.*), injection theory. (12)

théorie de la sublimation (Géol.) (*f.*), sublimation theory. (13)

théorie de la terre (*f.*), geology. (14)

théorie des ondulations (Phys.) (*f.*), undulatory theory ; wave theory of light. (15)

théorie du son, de l'électricité, de la lumière (*f.*), theory of sound, of electricity, of light. (16 *ou* 17 *ou* 18)

théorie glaciaire (*f.*), glacier theory ; glacier hypothesis ; glacial theory ; glacial hypothesis. (19)

théorie per ascensum (Géol.) (*f.*), ascensional theory. (20)

théorie per descensum (Géol.) (*f.*), descension theory ; descensional theory. (21)

théorie plutonienne *ou* théorie vulcanienne (Géol.) (*f.*), Plutonic theory ; Vulcanian theory. (22)

théorique (*adj.*), theoretic ; theoretical : (23) diagramme théorique (*m.*), theoretical diagram. (24)

théoriquement (*adv.*), theoretically. (25)

théoriser (*v.t.* & *v.i.*), to theorize. (26)

théoriste (pers.) (*n.m.*), theorist. (27)

thermal, -e, -aux (*adj.*), thermal ; thermic ; calorific ; hot : (28) source thermale (*f.*), thermal spring ; hot spring. (29)

thermalement (*adv.*), thermally. (30)

thermalité (*n.f.*), thermality. (31)

thermique (*adj.*), thermic ; thermal ; calorific ; hot : (32) les phénomènes thermiques dans les machines à vapeur (*m.pl.*), the thermic phenomena in steam-engines. (33) rendement thermique (*m.*), thermal efficiency ; calorific efficiency. (34)

thermite (Métall.) (*n.f.*), thermit ; thermite. (35)

thermo-électricité (*n.f.*), thermoelectricity. (36)

thermo-électrique (*adj.*), thermoelectric ; thermo-electrical : (37) le pouvoir thermo-électrique d'un métal, the thermoelectric power of a metal. (38)

thermochimie (*n.f.*), thermochemistry. (39)

thermochimique (*adj.*), thermochemic ; thermo-chemical. (40)

thermodynamique (*adj.*), thermodynamic. (41)

thermodynamique (*n.f.*), thermodynamics. (42)

thermographe (*n.m.*), thermograph. (43)

thermomètre (*n.m.*), thermometer. (44)

thermomètre à air (*m.*), air-thermometer. (45)

thermomètre à alcool (*m.*), spirit-thermometer ; alcoholic thermometer. (46)

thermomètre à ébullition (*m.*), boiling-point thermometer ; hypsometer ; hypsothermom-eter. (47)

thermomètre à éther de pétrole (*m.*), petroleum-ether thermometer. (48)

thermomètre à gaz (*m.*), gas-thermometer. (49)

thermomètre à maxima (*m.*), maximum ther-mometer. (50)

thermomètre à maxima et minima (*m.*), maxi-mum and minimum thermometer. (51)

thermomètre à mercure (*m.*), mercurial ther-mometer ; mercury-thermometer. (52)

thermomètre à minima (*m.*), minimum ther-mometer. (53)

thermomètre avertisseur (*m.*), alarm-thermom-eter ; thermal alarm. (54)

thermomètre centigrade (*m.*), centigrade ther-mometer. (55)

thermomètre différentiel (*m.*), differential ther-mometer. (56)

thermomètre enregistreur (*m.*), recording ther-mometer ; registering thermometer ; self-registering thermometer. (57)

thermomètre étalon (*m.*), standard thermometer. (58)

thermomètre Fahrenheit (*m.*), Fahrenheit ther-mometer. (59)

thermomètre-fronde [thermomètres-frondes *pl.*] (*n.m.*), sling thermometer ; whirled ther-mometer. (60)

thermomètre mouillé (*m.*), wet-bulb thermometer. (61)

thermomètre Réaumur (*m.*), Réaumur thermom-eter. (62)

thermomètre sec (*m.*), dry-bulb thermometer. (63)

thermométrie (*n.f.*), thermometry. (64)

thermométrique (*adj.*), thermometric ; thermo-metrical : (65) échelle thermométrique (*f.*), thermometric scale. (66)

thermométriquement (*adv.*), thermometrically. (67)

thermométrographe (*n.m.*), thermometrograph. (68)

thermominéral, -e, -aux (*adj.*), thermomineral : (69) émission thermominérale (*f.*), thermomineral emission. (70)

thermonatrite (Minéral) (*n.f.*), thermonatrite. (71)

thermoscope (*n.m.*), thermoscope. (72)

thermoscopique (*adj.*), thermoscopic ; thermo-scopical. (73)

thermoscopiquement (*adv.*), thermoscopically. (74)

thermostat (*n.m.*), thermostat. (75)

thermostatique (*adj.*), thermostatic. (76)

thermostatique (*n.f.*), thermostatics. (77)

thermostatiquement (*adv.*), thermostatically. (78)

thierne (galerie de mine) (*n.f.*), slant ; run. (79)

thiosulfate de sodium (*m.*), sodium thiosulphate ; hyposulphite of soda ; hypo. (80)

thomsénolite (Minéral) (*n.f.*), thomsenolite. (81)

thomsonite (Minéral) (*n.f.*), thomsonite. (82)

thorianite (Minéral) (*n.f.*), thorianite. (83)

thorine (Chim.) (*n.f.*), thoria. (84)

thorite (Minéral) (*n.f.*), thorite. (85)

thorium (Chim.) (*n.m.*), thorium. (1)
thulite (Minéral) (*n.f.*), thulite. (2)
thulium (Chim.) (*n.m.*), thulium. (3)
thundérite (Explosif) (*n.f.*), thunderite. (4)
thuringite (Minéral) (*n.f.*), thuringite. (5)
tiède (*adj.*), lukewarm ; tepid : (6)
eau tiède (*f.*), lukewarm water ; tepid water. (7)
tiemannite (Minéral) (*n.f.*), tiemannite. (8)
tierce (‴) (Math. & Astron.) (*n.f.*), third. (9)
tiers-point [tiers-points *pl.*] (*n.m.*), three-square file ; tri-square file ; triangular file. (10)
tiers-point double (*m.*), double-ended hand saw file. (11)
tiffanyite (Minéral) (*n.f.*), tiffanyite. (12)
tige (*n.f.*), rod ; stem ; shank. (13)
tige (d'un boulon, d'un rivet, d'un clou) (*n.f.*), shank (of a bolt, of a rivet, of a nail). (14 *ou* 15 *ou* 16)
tige (d'une vis) (*n.f.*), barrel, stem, (of a screw). (17)
tige (d'une clef) (*n.f.*), stem, shank, (of a key). (18)
tige (d'un thermomètre) (*n.f.*), stem (of a thermometer). (19)
tige (d'une pompe) (*n.f.*), rod, spear, (of a pump) : (20)
la maîtresse-tige d'une pompe, the main rod of a pump. (21)
tige à œil (*f.*), rod with eye ; eyed rod. (22)
tige creuse (*f.*), hollow rod. (23)
tige d'excentrique (*f.*), eccentric-rod. (24)
tige de commande du piston (tige de piston de pompe aspirante) (*f.*), sucker-rod. (25)
tige de fer (*f.*), iron rod. (26)
tige de jaugeage (haut fourneau) (*f.*), stock-indicator ; charge-gauge ; test-rod ; gauge. (27)
tige de manœuvre du cavalier (balance de précision) (*f.*), rider-slide. (28)
tige de piston (*f.*), piston-rod. (29)
tige de pompe (*f.*), pump-rod ; spear. (30)
tige de rallonge (*f.*), lengthening-piece ; ex-tension-piece. (31)
tige de sonde (la tige de sonde proprement dite, c.-à-d., la tête, les rallonges, et la coulisse) (*f.*), rods. (32)
tige de sonde (rallonge) (*f.*), boring-rod ; bore-rod ; drill-rod ; rod ; drill-pole. (33)
tige de sonde en bois (*f.*), wooden drill-rod. (34)
tige de sonde en fer (*f.*), iron boring-rod ; iron drill-pole. (35)
tige de soupape (*f.*), valve-rod ; valve-stem. (36)
tige de surcharge (sonde) (*f.*), sinker-bar. (37)
tige de suspension du ressort (locomotive ou wagon) (*f.*), spring-hanger ; swing-hanger. (38)
tige de tampon (*f.*), buffer-rod. (39)
tige de transmission (*f.*), flat rod ; string-rod ; transmission-rod. (40)
tige du pilon (*f.*), stamp-stem. (41)
tige du sabot de pilon (*f.*), shank of the stamp-shoe. (42)
tige du tiroir (*f.*), valve-stem ; valve-spindle ; stem of the slide-valve ; slide-rod. (43)
tige du trépan (Sondage) (*f.*), stem of the boring-chisel. (44)
tige du trolley (tramway électrique) (*f.*), trolley-pole. (45)
tige faussée *ou* **tige voilée** (*f.*), bent rod. (46)
tille de couvreur (*f.*), slaters' hammer. (47)

timbre (de chaudière à vapeur) (*n.m.*), test-plate, badge-plate, (of steam-boiler). (48)
By extension, in specifications of boilers, **timbre** *signifies* working-pressure. (49)
timbrer une chaudière, to badge a boiler : (50)
chaudière timbrée à — kg., boiler badged at — kilogrammes. (51)
timonerie des freins (*f.*), brake-gear. (52)
tincal *ou* **tinkal** (Minéral) (*n.m.*), tincal ; tinkal. (53)
tincalconite (Minéral) (*n.f.*), tincalconite. (54)
tir (Travail aux explosifs) (*n.m.*), firing ; blasting : (55)
tir électrique *ou* tir électrique des coups de mine, electric firing ; blasting by electricity. (56)
tir en volée, volley-firing ; volley-blasting. (57)
tirage (traction) (*n.m.*), drawing ; pulling ; pull ; dragging ; hauling ; traction : (58)
tirage d'un hauban, pull of a guy-rope. (59)
tirage (effort pour tirer quelque chose dans une montée) (*n.m.*), drag ; draught ; draft : (60)
il y a du tirage sur ce chemin, there is a heavy drag on this road. (61)
tirage (halage) (Navigation fluviale) (*n.m.*), towing : (62)
tirage d'un bateau sur un canal par des chevaux, towing a boat on a canal by horses. (63)
tirage (chemin de halage) (Navigation fluviale) (*n.m.*), tow-path ; towing-path. (64)
tirage (de métaux) (*n.m.*), drawing ; wire-drawing. (65)
tirage (Travail aux explosifs) (*n.m.*), firing ; blasting : (66)
tirage des coups de mine *ou* tirage des fourneaux *ou* tirage des pétards, shot-firing ; blasting. (67)
tirage en volées, volley-blasting ; volley-firing. (68)
tirage à la chaux, blasting with lime cartridges. (69)
tirage (Photogr.) (*n.m.*), printing : (70)
tirage des épreuves, printing positives. (71)
tirage à la lumière artificielle, gas or electric-light printing. (72)
tirage au jour, daylight printing. (73)
tirage par contact, contact-printing. (74)
tirage par agrandissement, enlargement-printing. (75)
tirage (du soufflet d'un appareil photographique) (*n.m.*), extension (of a camera-bellows) : (76)
chambre à double tirage (*f.*), camera with double extension. (77)
tirage (action par laquelle un foyer attire l'air pour la combustion) (*n.m.*), draught ; draft : (78)
tirage forcé, forced draught. (79)
tirage induit, induced draught. (80)
tirage naturel, natural draught. (81)
tirage (extraction des pierres d'une carrière) (*n.m.*), quarrying. (82)
tirage d'épaisseur (*m.*), thicknessing. (83)
tirage de l'air *ou* **tirage des trous d'air** (Fonderie) (*m.*), venting. (84)
tirant, -e (*adj.*), tractive ; pulling : (85)
force tirante (*f.*), tractive force. (86)
tirant (Constr., etc.) (*n.m.*), tie-rod ; tie-bar ; tie-bolt ; tie-beam ; tie ; stay ; tension-piece. (87)
tirant (d'une ferme de comble en bois) (*n.m.*), tie-beam, stringer, stretcher, binding-beam, (of a wooden roof-truss). (88)

tirant (d'une ferme de comble métallique) (*n.m.*), tie-rod, tie-bar, truss-rod, (of an iron or steel roof-truss). (1)

tirant (d'une chaudière, d'un foyer de locomotive) (*n.m.*), stay-bolt, stay-rod, stay, brace, (of a boiler, of a locomotive fire-box). (2 *ou* 3)

tirant (d'une grue) (*n.m.*), tie (of a crane). (4)

tirant articulé permettant la dilatation de la plaque tubulaire (*m.*), flexible stay-bolt ; expansion stay-bolt. (5)

tirant d'eau (d'un bateau) (*m.*), draught, draft, (of a boat). (6)

tirant de caisse (wagon de ch. de f.) (*m.*), body truss-rod. (7)

tirant de suspension (locomotive ou wagon) (*m.*), spring-hanger ; swing-hanger. (8 *ou* 9)

tire-bouchon [tire-bouchons *pl.*] (*n.m.*), cork--screw. (10)

tire-bourre [tire-bourre *pl.*] (Travail aux explosifs) (*n.m.*), spiral worm ; worm-screw ; wad-hook. (11)

tire-braise [tire-braise *pl.*] (*n.m.*), rake. (12)

tire-clou [tire-clous *pl.*] (*n.m.*), nail-puller ; nail-extractor. (13)

tire-fond [tire-fond *pl.*] (pour rails) (*n.m.*), coach-screw ; screw-spike ; lag-screw ; sleeper-screw. (14)

tire-goupille [tire-goupilles *pl.*] (*n.m.*), split-pin extracting-tool ; pin-extractor. (15)

tire-ligne [tire-lignes *pl.*] (*n.m.*), drawing-pen. (16)

tire-ligne à pointiller (*m.*), dotting-pen ; wheel-pen. (17)

tire-point [tire-points *pl.*] (*n.m.*), stabbing-awl. (18)

tire-sonde [tire-sondes *ou* **tire-sonde** *pl.*] (*n.m.*), drill-rod grab. (19)

tirefonner (*v.t.*), to fasten by coach-screws. (20)

tirer (mouvoir, amener vers soi, ou après soi) (*v.t.*), to draw ; to pull ; to drag ; to haul : (21)

tirer un verrou, to draw a bolt. (22)

tirer (passer par la filière) (*v.t.*), to draw ; to draw out ; to wiredraw : (23)

tirer des métaux, de l'or, to draw metals, gold. (24 *ou* 25)

tirer (se) (*v.r.*), to draw ; to draw out : (26)

l'or se tire en fils très déliés (*m.*), gold draws out into very fine wire. (27)

tirer (Travail aux explosifs) (*v.t.*), to fire : (28

tirer un coup de mine, to fire a shot (*or* a blast). (29)

tirer (Photogr.) (*v.t.*), to print : (30)

tirer une épreuve, to print a positive. (31)

tirer (aspirer) (*v.t.*), to draw ; to suck ; to exhaust ; to aspirate : (32)

piston qui tire de l'air (*m.*), piston which draws air. (33)

tirer (se dit d'une cheminée) (*v.i.*), to draw : (34)

cheminée qui ne tire pas (*f.*), chimney which does not draw. (35)

tirer à sa fin, to draw to a close. (36)

tirer au vide (se dit d'un mur qui fait ventre), to bulge ; to belly ; to belly out : (37)

mur qui tire au vide (*m.*), wall which bulges. (38)

tirer avantage de la saison pluvieuse, to take advantage of the rainy season. (39)

tirer bon parti de, to make the most of ; to use to the best advantage : (40)

pour tirer bon parti des locomotives, il importe de leur faire traîner des charges aussi lourdes que possible, in order to use locomotives to the best advantage, it is of moment to make them draw as heavy loads as possible. (41)

tirer d'épaisseur (*v.t.*), to thickness : (42)

tirer d'épaisseur une planche, to thickness a board. (43)

tirer d'épaisseur (Moulage en terre) (Fonderie) (*v.t.*), to thickness : (44)

tirer d'épaisseur un noyau, to thickness a core. (45)

tirer de grands bénéfices d'une entreprise, to draw big profits from a business ; to make large profits out of a business. (46)

tirer de l'air *ou* **tirer des trous d'air** (Fonderie), to vent ; to pierce vent-holes : (47)

tirer de l'air sur la face supérieure du dessus, to vent over the top face of the cope. (48)

tirer à l'aiguille, sur toute la surface du moule, de nombreux trous d'air, to pierce with the wire numerous vent-holes over the whole surface of the mould. (49)

tirer de l'eau d'un récipient, to draw water from a vessel. (50)

tirer de la pierre d'une carrière, to quarry stone. (51)

tirer de largeur, to reduce to width : (52)

tirer de largeur une planche, to reduce a board to width. (53)

tirer des approvisionnements de l'étranger, to draw supplies from abroad. (54)

tirer des briques d'un four, to draw, to take, bricks out of a kiln. (55)

tirer parti de la saison pluvieuse, to take advantage of the rainy season. (56)

tirer — pieds d'eau (se dit d'un bateau), to draw — feet of water. (57)

tirer sur le blanc, to incline to white (in colour) ; to be whitish ; to be somewhat white : (58)

un gris d'acier tirant légèrement sur le blanc d'argent, a steel-grey inclining slightly to silver-white. (59)

tirer sur un câble, to pull on, to pull, a rope. (60)

tirer une affaire au clair, to clear up a matter. (61)

tirer une courroie, to stretch a belt. (62)

tirer une déduction des sondages, to draw an inference from boring. (63)

tirer une ligne sur le papier, to draw a line on paper. (64)

tireur de coups de mine *ou* **tireur de mines** (pers.) (*m.*), shotfirer ; shotlighter ; blaster ; shot--man. (65)

tireuse (Photogr.) (*n.f.*), printing-machine. (66)

tiroir (meuble) (*n.m.*), drawer : (67)

les tiroirs d'une table, the drawers of a table. (68)

tiroir (coulisseau) (Méc.) (*n.m.*), slide : (69)

tiroir en fer mobile en tous sens, iron slide movable in any direction. (70)

tiroir (*n.m.*) *ou* **tiroir de distribution** (organe de distribution de vapeur), slide-valve ; sliding-valve ; slide ; valve. (71)

tiroir (pièce battue) (Moulage de fonderie) (*n.m.*), drawback ; false core. (72)

tiroir à canal *ou* **tiroir d'Allen** (*m.*), trick valve ; trick-ported valve ; Allen valve. (73)

tiroir à double coquille (*m.*), double-chambered slide-valve. (74)

tiroir à doubles orifices (*m.*), double-ported valve. (75)

tiroir à grille (*m.*), gridiron valve. (1)

tiroir à orifices **multiples** (*m.*), multiported valve. (2)

tiroir à piston *ou* tiroir cylindrique (*m.*), piston-valve ; circular slide-valve. (3)

tiroir à recouvrement *ou* tiroir à recouvrements (*m.*), lap-valve ; lapped valve ; valve with lap. (4)

tiroir de démarrage (locomotive compound) (*m.*), starting-valve. (5)

tiroir de détente *ou* tiroir à détente *ou* tiroir à tuile(s) de détente *ou* tiroir d'expansion (*m.*), cut-off valve ; expansion-valve ; riding cut-off valve ; independent cut-off valve ; auxiliary valve. (6)

tiroir en coquille (*m.*), slide-valve, ordinary pattern ; plain slide-valve. (7)

tiroir en **D** (*m.*), D valve. (8)

tiroir équilibré *ou* tiroir compensé (*m.*), balanced slide-valve. (9)

tiroir Meyer (*m.*), Meyer slide-valve ; Meyer valve. (10)

tiroir normal (sans recouvrements) (*m.*), normal slide-valve (lapless). (11)

tiroir oscillant *ou* tiroir rotatif (*m.*), rocking slide-valve ; rocking valve ; swinging valve. (12)

tiroir rond (*m.*), circular slide-valve ; piston-valve. (13)

tiroir sans recouvrement *ou* tiroir sans recouvre--ments (*m.*), lapless valve ; valve without lap. (14)

tisard *ou* tisart (*n.m.*), firing-door ; stoking-door ; stoke-hole. (15)

tisonner le feu, to poke, to stir, the fire. (16)

tisonnier (*n.m.*), poker. (17)

titan (*n.m.*), titan crane. (18)

titanate (Chim.) (*n.m.*), titanate. (19)

titane (Chim.) (*n.m.*), titanium. (20)

titane oxydé (Minéral) (*m.*), titanic oxide ; titanic schorl ; rutile. (21)

titane oxydé ferrifère (Minéral) (*m.*), titaniferous oxide of iron ; titaniferous iron ore ; titanic iron ore ; ilmenite. (22)

titane silicéo-calcaire (Minéral) (*m.*), calcium titano-silicate ; sphene. (23)

titaneux, -euse (Chim.) (*adj.*), titanous. (24)

titanifère *ou* titané, -e (*adj.*), titaniferous. (25)

titanique (Chim.) (*adj.*), titanic ; titanitic : (26)

acide titanique (*m.*), titanic acid. (27)

titanite (Minéral) (*n.f.*), titanite. (28)

titanium (Chim.) (*n.m.*), titanium. (29)

titrage (Chim.) (*n.m.*), standardization ; titration. (30) See titrer for example of use.

titre (acte) (*n.m.*), title : (31)

titre de propriété, title to property. (32)

titre à perpétuité, perpetual title. (33)

titre de concession (Mines), licence ; claim-licence. (34)

titre (d'une solution) (Chim.) (*n.m.*), strength (of a solution). (35)

titre (degré de fin ; teneur) (*n.m.*), fineness ; title ; tenure ; grade ; content ; contents ; tenor : (36)

titre de l'or, fineness, title, of gold. (37)

minerai à bas titre (*m.*), low-grade ore. (38)

titre en millièmes (*m.*) *ou simplement* millièmes (*n.m.pl.*), thousands fine : (39)

or, 900 millièmes *ou* or, 900/1000, gold, 900 thousands fine ; gold, ·900. (40)

titre en or, gold-contents ; content in gold ; tenure in gold : (41)

l'essai d'or a pour but de déterminer le titre en or du bouton obtenu par la coupellation (*m.*), the assaying of gold has as its object the determination of the gold-contents of the button obtained by cupellation. (42)

titre en or fin, content in fine gold. (43)

titre standard (matières d'or et d'argent) (Angleterre et États-Unis), standard fineness. (44)

au titre, standard ; of standard fineness : (45)

or au titre (*m.*), standard gold. (46)

titre (benne de mine) (*n.m.*), hoisting-bucket. (47)

titre d'essai (à), by way of trial ; trial (*used as adj.*) : (48)

envoi à titre d'essai (*m.*), trial lot ; trial shipment. (49)

titre de jauge (jauge pour fils métalliques) (*m.*), gauge-number (wire-gauge). (50)

titre de la vapeur (*m.*), quality of wet steam ; relative dryness of wet steam. (51)

titré, -e (Chim.) (*adj.*), standard : (52)

liqueur titrée (*f.*), standard solution. (53)

titrer (Chim.) (*v.t.*), to standardize ; to standardize by titration ; to titrate : (54)

titrer une liqueur, to standardize, to titrate, a solution : (55)

liqueur titrée d'iode (*f.*), solution standardized with iodine. (56)

titrer (Chimie minérale) (*v.i.*), to assay : (57)

un minerai titrant — 0/0 de métal, an ore assaying —% of metal. (58)

un gravier aurifère titrant — grammes au mètre cube, a gold-bearing gravel assaying — grammes to the cubic metre. (59)

toc (*n.m.*) *ou* toc de tour *ou* toc pour tours, lathe-dog ; dog ; lathe-carrier ; turning-carrier ; carrier ; driver. (60)

toc à coussinets (*m.*), jaw-dog ; lathe-carrier with 2 jaws. (61)

toc à tige coudée (*m.*), bent-tail dog. (62)

toc à tige droite (*m.*), straight-tail dog. (63)

toc à vis (*m.*), screw lathe-dog. (64)

toc d'entraînement (du plateau-toc d'un tour) (*m.*), catch-pin (of the catch-plate of a lathe). (65)

toc en acier estampé (*m.*), drop-forged steel dog. (66)

toile (*n.f.*), cloth ; canvas. (67)

toile (feuille de métal qui se produit entre deux pièces contiguës d'un moule) (Fonderie) (*n.f.*), flash. (68)

toile à bâches (*f.*), tarpaulin. (69)

toile à calquer *ou* toile d'architecte (*f.*), tracing-cloth. (70)

toile à sacs (*f.*), sackcloth ; sacking ; bagging. (71)

toile à voile (*f.*), sail-cloth ; canvas. (72)

toile cirée (*f.*), oilcloth. (73)

toile d'amiante (*f.*), asbestos cloth. (74)

toile d'émeri *ou* toile émerisée (*f.*), emery-cloth. (75)

toile de filtre *ou* toile filtrante (*f.*), filter-cloth. (76)

toile de klaubage *ou* toile de triage (*f.*), picking-belt ; sorting-belt ; canvas table. (77)

toile dite américaine (*f.*), American cloth. (78)

toile en fil de fer (*f.*), iron-wire gauze. (79)

toile en fil de laiton (*f.*), brass-wire gauze. (80)

toile métallique (*f.*), wire-gauze ; wire-cloth ; woven wire. (81)

toile transporteuse (*f.*), conveyor-belt ; travelling apron ; apron. (1)

toile vernie (*f.*), oil-skin. (2)

toile verrée (*f.*), glass-cloth. (3)

toiles (*n.f.pl.*) *ou* **toiles d'aérage** (Mines), brat--tice-cloth. (4)

toit (*n.m.*), roof : (5)

un toit de chaume, a thatched roof. (6)

See also **toiture** and **couverture.**

toit (d'une galerie de mine) (*n.m.*), roof (of a mine-level). (7)

toit (d'un filon) (opp. au **mur**) (*n.m.*), hanging wall, hanging side, top wall, roof, (of a lode). (8) (opp to the **foot-wall, foot, lying-wall, ledger-wall,** *or* **ledger**).

toit (d'une couche de houille) (opp. au **mur**) (*n.m.*), roof, top, (of a coal-seam). (9) (opp. to the **floor, pavement,** *or* **bottom**) : (10)

recoupage du toit et du mur (*m.*), brushing the roof and the floor. (11)

toiture (*n.f.*), roofing ; roof : (12)

toiture en toile, canvas roofing. (13)

toiture en tôle galvanisée, galvanized-iron roofing. (14)

tôle (*n.f.*) *ou* **tôle de fer,** sheet ; sheets ; plate ; plates ; sheet iron ; plate iron ; iron sheeting ; iron : (15)

du fer laminé en tôles, iron rolled into sheets. (16)

cisailler une tôle, to shear a plate. (17)

la fabrication de la tôle, the manufacture of sheets (*or* of sheet iron). (18)

Note.—In the English trade, a plate is properly ¼ inch thick or over ; a sheet is a plate less than ¼ inch thick, but some makers depart from this rule. In French, heavy plates are called grosses tôles *or* tôles fortes ; medium plates, tôles moyennes, and thin plates *or* sheets fines tôles, tôles fines, *or* tôles minces.

tôle (*n.f.*) *ou* **tôle d'acier,** sheet ; plate ; sheet steel. (19)

tôle d'enveloppe du corps du cylindre (*f.*), cylinder-casing. (20)

tôle d'enveloppe extérieure (chaudière) (*f.*), jacket ; boiler-jacket ; clothing-plate. (21)

tôle de chaudière (*f.*), boiler-plate. (22)

tôle de coup de feu *ou* **tôle du coup de feu** (*f.*), flame-plate. (23)

tôle de cuivre (*f.*), copper sheets. (24)

tôles douces (*f.pl.*), mild plates ; mild sheets. (25)

tôle du ciel (de foyer) (*f.*), crown-sheet (of fire-box). (26)

tôle émaillée (*f.*), enamelled iron. (27)

tôle étamée (*f.*), tinned iron ; tinned sheet iron. (28)

tôles extra-douces (*f.pl.*), extra-mild plates ; extra-mild sheets. (29)

tôles fines (*f.pl.*), sheets ; thin plates. (30)

tôles fortes (*f.pl.*), heavy plates. (31)

tôle galvanisée (*f.*), galvanized iron ; gal--vanized-iron sheets ; galvanized sheets. (32)

tôles minces (*f.pl.*), thin plates ; sheets. (33)

tôles moyennes (*f.pl.*), medium plates. (34)

tôles noires (*f.pl.*), black sheets. (35)

tôle ondulée (*f.*), corrugated iron ; corrugated sheet iron ; corrugated-iron sheet : (36)

la tôle ondulée, pour la garantir de la rouille, est galvanisée, corrugated iron, to keep it from rusting, is galvanized. (37)

les tôles ondulées se fabriquent au moyen de laminoirs particuliers, corrugated-iron sheets are made by means of special rolls. (38)

tôle perforée (*f.*), perforated plate ; perforated iron plate ; punched plate. (39)

tôles planes (*f.pl.*), flat sheets. (40)

tôles pour bordages (*f.pl.*), flange-plate. (41)

tôles pour chaudières (*f.pl.*), boiler-plate. (42)

tôles pour foyers (*f.pl.*), fire-box plate ; furnace-plate. (43)

tôles puddlées (*f.pl.*), puddled plates. (44)

tôles recuites (*f.pl.*), annealed sheets. (45)

tôles russes (*f.pl.*), Russia iron ; Russian sheet iron. (46)

tôle striée (*f.*), chequer-plate ; chequered plate. (47)

tôle zinguée (*f.*), galvanized iron ; galvanized-iron sheets ; galvanized sheets. (48)

tolérance (Méc.) (*n.f.*), tolerance ; margin ; limit : (49)

une tolérance de — ᵐ/ₘ accordée sur le diamètre, a tolerance of — mm. allowed on the diameter. (50)

calibre de tolérance (*m.*), limit-gauge. (51)

tolérance de laminage (*f.*), rolling-margin. (52)

tolérance en moins (*f.*), margin under. (53)

tolérance en plus (*f.*), margin over. (54)

tolérance sur l'épaisseur (*f.*), thickness margin. (55)

tolérance sur la longueur (*f.*), length margin. (56)

tolérance sur les dimensions (*f.*), size margin. (57)

tôlerie (art du tôlier) (*n.f.*), sheet-iron manu--facture ; plating. (58)

tôlerie (fabrique de tôle) (*n.f.*), sheet-iron works ; plate-works. (59)

tôlier (pers.) (*n.m.*), sheet-iron manufacturer. (60)

tombac (alliage) (*n.m.*), tombac ; tomback ; tombak ; tambac. (61)

tombée (*n.f.*), fall ; collapse ; subsidence ; caving ; caving in : (62)

tombée du toit (Mines), fall, collapse, subsidence, caving in, of the roof. (63)

tombée de pluie, fall of rain ; downpour : (64)

quelle tombée de pluie ! what a downpour ! (65)

tomber (*v.i.*), to fall ; to drop : (66)

les vitesses acquises par un corps tombant librement dans le vide (*f.pl.*), the velocities acquired by a body falling freely in a vacuum. (67)

tomber d'accord sur certaines conditions, to agree to certain conditions. (68)

tomber dans la mer (se dit des fleuves), to fall, to flow, into the sea. (69)

tomber dans le puits, to fall down (*or* to drop down) the shaft. (70)

tomber de sur une échelle, to fall off a ladder. (71)

tomber en déliquescence (*v.i.*), to deliquesce : (72)

sous une atmosphère humide, certains sels tombent en déliquescence, certain salts deliquesce in a damp atmosphere. (73)

tomber en désuétude, to fall into disuse. (74)

tomber en efflorescence (*v.i.*), to effloresce. (75)

tomber en poudre, to crumble to powder. (76)

tomber en ruine, to fall into ruins ; to go to rack and ruin. (77)

tombereau (*n.m.*), tip-cart ; tipping-cart ; tilting-cart ; dump-cart ; dumping-cart. (78)

tombolo (Géogr. phys.) (*n.m.*), loop. (79)

ton (Photogr.) (*n.m.*), tone : (1)
obtention de tons sépia avec les papiers au bromure (*f.*), obtaining sepia tones with bromide papers. (2)

tonalite (Pétrol.) (*n.f.*), tonalite. (3)

tonite (Explosif) (*n.f.*), tonite. (4)

tonnage (poids en tonnes) (*n.m.*), tonnage : (5)
tonnage approximatif de minerai en réserve, approximate tonnage of ore in reserve. (6)

tonnage (Marine) (*n.m.*), tonnage. (7)

tonnant, -e (moteurs à combustion interne) (*adj.*), explosive : (8)
mélange tonnant (*m.*), explosive mixture. (9)

tonne (unité de poids) (*n.f.*), ton. (10)

tonne (benne) (Mines) (*n.f.*), bucket ; hoisting-bucket ; tub. (11)

tonne courte (mesure anglaise) (*f.*), short ton ; net ton = 2,000 lbs avoirdupois or 907·184856 kg. (12)

tonne d'essai (mesure anglaise) (*f.*), assay ton : (13)
Note.—The assay ton contains as many milli--grammes as the *ton* contains troy ounces ; therefore, 1 milligramme of precious metal obtained from 1 assay ton represents 1 troy ounce in 1 *ton*.
The *ton* used in the calculation may be the *long ton* (2,240 lbs) (used in England and Australia), or the *short ton* (2,000 lbs) (used in North America and Africa), which are represented as follows :—
long ton by assay ton containing 32,666+ milligrammes (or 32·666 + grammes).
short ton by assay ton containing 29,166+ milligrammes (or 29·166 + grammes).
si une tonne d'essai d'un minerai a été prise, chaque milligramme de métal trouvé représente une once de métal par tonne de minerai : si deux tonnes d'essai de minerai ont été prises, chaque milligramme de métal trouvé représentera ½ once de métal à la tonne, et ainsi de suite, if one assay ton of an ore has been taken, each milligramme of metal found represents one ounce of metal per ton of ore : if two assay tons of ore have been taken, each milligramme of metal found will represent ½ ounce of metal to the ton, and so on. (14)

tonne forte (mesure anglaise) (*f.*), long ton ; gross ton=2,240 lbs avoirdupois or 1,016 kg. (15) Note.—This is the weight of what is ordinarily understood by a " ton " in England.

tonne kilométrique (Ch. de f.) (*f.*), kilometre-ton. (16) (=the transport of 1 *tonne* (1,000 kg.) of freight 1 kilometre : analogous to the English *ton-mile* (tonne *millénaire*)= transport of 1 ton (2,240 lbs) 1 mile).

tonne métrique ou *simplement* **tonne** (*n.f.*), metric ton ; tonne=1,000 kg. or 0·9842 ton or 2,204·6 lbs avoirdupois. (17)

tonneau (*n.m.*), cask ; barrel ; butt ; bin. (18)

tonneau (*n.m.*) ou **tonneau en fer,** drum ; iron drum ; bin : (19)
tonneau à carbure, carbide-drum ; carbide-bin. (20)

tonneau à eau (*m.*), water-butt. (21)

tonneau à malaxer le mortier ou **tonneau à mortier** (*m.*), mortar-mixer ; malaxator. (22)

tonneau-brouette [**tonneaux-brouettes** *pl.*] (*n.m.*), water-barrow. (23)

tonneau d'amalgamation (*m.*), amalgamating-barrel ; amalgamation-barrel. (24)

tonneau d'arrosage (*m.*), watering-cart ; water-cart. (25)

tonneau dessableur ou **tonneau à dessabler** (Fonderie) (*m.*), tumbling-barrel ; tumbling-box ; tumbling-drum ; tumbling-mill ; tumbler ; rattler ; foundry-rattler ; rattling-barrel ; cleaning-barrel ; rumbler ; shaking-machine ; shaking-barrel ; shaking-mill. (26)

tonneau étanche (*m.*), tight cask ; water-tight cask. (27)

tonnelet (*n.m.*). Same as **tonneau en fer,** but smaller in size. (28)

tonnerre (*n.m.*), thunder. (29)

topaze (*n.f.*), topaz. (30)

topaze brûlée (*f.*), burnt topaz. (31)

topaze d'Ecosse (*f.*), Scotch topaz. (32)

topaze de Saxe (*f.*), Saxon topaz. (33)

tapaze de Sibérie (*f.*), Siberian topaz. (34)

topaze du Brésil (*f.*), Brazilian topaz. (35)

topaze enfumée (*f.*), smoky topaz. (36)

topaze fausse (*f.*), false topaz ; citrine quartz ; citrine. (37)

topaze orientale (*f.*), Oriental topaz ; Indian topaz. (38)

topazolite (Minéral) (*n.f.*), topazolite. (39)

topographe (pers.) (*n.m.*), topographer. (40)

topographie (*n.f.*), topography : (41)
topographie d'un pays, topography of a country. (42)

topographique (*adj.*), topographic ; topographical : (43)
levé topographique (*m.*), topographical survey. (44)

topographiquement (*adv.*), topographically. (45)

toque-feu [**toque-feu** *pl.*] (*n.m.*), fire-pot hanging in the upcast shaft. (46)

torbanite (Minéral) (*n.f.*), torbanite ; Torbane Hill mineral. (47)

torbernite ou **torbérite** (Minéral) (*n.f.*), torbernite ; torberite ; chalcolite ; uranite ; copper uranite. (48)

torche (*n.f.*), torch. (49)

torche de foin (Fonderie) (*f.*), hay band ; hay rope. (50)

torche de paille (*f.*), straw band ; straw rope. (51)

tordre (*v.t.*), to twist : (52)
tordre du fil, to twist wire. (53)

tordre (se) (*v.r.*), to twist. (54)

tordu au quart (*m.*), quarter-twist. (55)

tore (Géom.) (*n.m.*), tore ; torus. (56)

tornade (*n.f.*) ou **tornado** (*n.m.*), tornado. (57)

toron (d'un cordage) (*n.m.*), strand : (58)
cordage composé de — torons (*m.*), rope made up of — strands. (59)

torpillage d'un puits à pétrole (*m.*), torpedoing, shooting, an oil-well. (60)

torpille (*n.f.*), torpedo ; shell. (61)

torpille à inflammateur (*f.*), torpedo with igniter. (62)

torpiller un puits, to torpedo, to shoot, a well. (63)

torréfacteur (*n.m.*), torrefier. (64)

torréfaction (*n.f.*), torrefaction ; roasting. (65)

torréfier (*v.t.*), to torrefy ; to roast. (66)

torrent (*n.m.*), torrent. (67)

torrent persistant (*m.*), constant torrent ; perpetual torrent ; perennial torrent. (68) (as opposed to a freshet).

torrentiel, -elle (*adj.*), torrential : (69)
apports torrentiels (*m.pl.*), torrential drift. (70)

torrentiellement (*adv.*), torrentially. (1)
torride (*adj.*), torrid : (2)
  zone torride (*f.*), torrid zone. (3)
tors, -e (*adj.*), twisted : (4)
  du fil tors, twisted wire. (5)
torsade (*n.f.*), twist. (6)
torse (tarière) (*n.f.*), screw-auger ; twisted auger. (7)
torsion (*n.f.*), torsion ; twisting ; twist : (8)
  fil raccourci par la torsion (*m.*), wire shortened by torsion. (9)
  torsion du câble, twisting of the rope. (10)
torsion magnétique (Phys.) (*f.*), magnetic twist. (11)
tortillement (action) (*n.m.*), kinking ; twisting : (12)
  tortillement du câble, kinking of the rope. (13)
tortillement (état) (*n.m.*), kink ; twist. (14)
tortiller (*v.t.*), to kink ; to twist : (15)
  tortiller une corde, to kink a cord. (16)
tortiller (se) (*v.r.*), to kink ; to twist ; to twine. (17)
tortu, -e (*adj.*), crooked ; tortuous. (18)
total, -e, -aux (*adj.*), total ; over-all : (19)
  éclipse totale (*f.*), total eclipse. (20)
  longueur totale (*f.*), total length ; over-all length. (21)
total (*n.m.*), total. (22)
totalement (*adv.*), totally. (23)
totalisation des diagrammes (machine poly-cylindrique) (*f.*), combining the diagrams (multicylinder engine) : (24)
  totalisation des diagrammes relevés sur une machine à double expansion, combining the diagrams taken on a double-expansion engine. (25)
totaliser les diagrammes (machine polycylin-drique), to combine the diagrams. (26)
touage (*n.m.*), towing (by submerged chain or cable). (27)
touchau ou **touchaud** (*n.m.*), touch-needle ; test-needle ; proof needle. (28)
touchaud à plusieurs alliages (*m.*), star for touchstone. (29)
touche (action de toucher) (*n.f.*), touch : (30)
  touche légère, light touch. (31)
touche (essai au moyen d'une pierre de touche) (*n.f.*), touch. (32)
touche (Élec.) (*n.f.*), contact ; contact-piece. (33)
touche (d'une machine à fraiser) (*n.f.*), profiling-roller (of a milling-machine). (34)
touche (d'une distribution Corliss, ou analogue) (*n.f.*), toe (of a Corliss valve-gear, or the like). (35 ou 36)
touche séparée (Magnétisme) (*f.*), separate touch. (37)
toucheau (*n.m.*). Same as **touchau**.
toucher (*n.m.*), touch : (38)
  certains minéraux (talc) sont onctueux et gras, tandis que d'autres sont simplement doux au toucher (amiante). Il y en a d'âpres au toucher (pierre ponce), certain minerals (talc) are unctuous and greasy, while others are simply soft to the touch (asbestos). There are some rough to the touch (pumice-stone). (39)
toucher (être en contact avec) (*v.t.*), to touch : (40)
  toucher un objet du doigt, to touch an object with the finger. (41)
toucher (percevoir) (*v.t.*), to draw (money) : (42)

toucher un salaire, ses honoraires, to draw a wage, one's fees. (43 ou 44)
toucher de l'or avec la pierre de touche, to test gold with the touchstone. (45)
toucher l'aiguille d'une boussole, to magnetize the needle of a compass. (46)
toucher (se) (*v.r.*), to touch ; to touch each other : (47)
  des sphères qui se touchent (*f.pl.*), spheres which touch. (48)
touer (*v.t.*), to tow (by submerged chain or cable). (49)
touffe (Mines) (*n.f.*), black damp ; choke-damp ; stythe. (50)
touffe d'arbres (*f.*), clump, tuft, of trees. (51)
toundra (*n.f.*), tundra. (52)
toupie (machine à moulurer) (*n.f.*), spindle moulding-machine. (53)
toupie (de plombier) (*n.f.*), turn-pin (plumbers'). (54)
toupie gyroscopique (*f.*), gyroscopic top. (55)
tour (révolution) (*n.m.*), revolution ; turn : (56)
  roue qui fait — tours à la minute (*f.*), wheel which makes — revolutions a minute ; wheel which does — turns a minute. (57)
tour (*n.m.*) ou **tour de spire**, turn ; spire : (58)
  les tours d'une spirale, the turns of a spiral. (59)
tour (d'un moulin à vent) (*n.f.*), tower (of a windmill). (60)
tour (Mach.) (*n.m.*), lathe ; turning-lathe. (61)
tour à archet (*m.*), turn. (62)
tour à banc rompu (*m.*), gap-lathe ; break-lathe. (63)
tour à bois (*m.*), wood-turning lathe : wood-turners' lathe. (64)
tour à charioter ou **tour à chariot** (*m.*), sliding-lathe ; slide-lathe. (65)
tour à charioter et surfacer, très perfectionné, avec poupée fixe à triple harnais d'engrenages (*m.*), high-class, treble-geared sliding and surfacing-lathe. (66)
tour à combinaisons (*m.*), combination lathe. (67)
tour à copier (*m.*), copying-lathe ; forming-lathe. (68)
tour à cuivre (*m.*), brass-finishers' lathe. (69)
tour à cylindrer (*m.*), plain-turning lathe. (70)
tour à cylindres (*m.*), roll-lathe. (71)
tour à décolleter (*m.*), cutting-off lathe. (72)
tour à dégager ou **tour à dépouiller** (*m.*), relieving-lathe. (73)
tour à dégager les fraises (*m.*), cutter-relieving lathe. (74)
tour à engrenages, marchant au pied (*m.*), back-geared foot-lathe ; back-geared treadle-lathe. (75)
tour à essieux (*m.*), axle-lathe. (76)
tour à facer (*m.*), facing-lathe ; face-lathe. (77)
tour à fileter (*m.*), screw-cutting lathe. (78)
tour à fileter, marchant au pied, avec poupée fixe à double harnais d'engrenages (*m.*), screw-cutting foot-lathe, with double-geared headstock. (79)
tour à marche rapide (*m.*), high-speed lathe ; speed-lathe. (80)
tour à noyaux (*m.*), core-lathe. (81)
tour à ovales (*m.*), oval lathe. (82)
tour à pédale (*m.*), treadle-lathe ; pedal-lathe ; foot-lathe. (83)
tour à plateau horizontal (*m.*), boring-mill ; turning-mill : boring and turning-mill. (84)

tour à plateau horizontal à commande électrique et à deux montants et deux porte-outils ; très robuste, permettant l'emploi des aciers à coupe rapide (*m.*), motor-driven boring and turning-mill, with two uprights and two tool-holders ; very strong, for use with high-speed cutting-steels. (1)

tour à pointes (*m.*), centre lathe ; centering-lathe ; pole-lathe. (2)

tour à poulies (*m.*), pulley-turning lathe ; pulley-lathe. (3)

tour à revolver (*m.*), turret-lathe ; capstan-lathe ; monitor-lathe. (4)

tour à revolver pour le décolletage et le façon-nage des pièces en série (*m.*), cutting-off and forming-lathe. (5)

tour à roues (*m.*), wheel-lathe ; wheel-turning lathe. (6)

tour à roues de locomotives (*m.*), locomotive-wheel lathe. (7)

tour à roues de wagons (*m.*), carriage-wheel lathe ; car-wheel lathe. (8)

tour à simple harnais d'engrenages (*m.*), single-geared lathe. (9)

tour à singer *ou* **tour à touche** (*m.*), copying-lathe ; forming-lathe. (10)

tour à tourelle hexagonale (*m.*), hexagon-turret lathe. (11)

tour à trains montés (*m.*), double-wheel lathe. (12)

tour à trousse coupante (Mines) (*f.*). Same as **tour de cuvelage.**

tour à volant à poignée pour la marche à bras (*m.*), throw ; hand-wheel lathe. (13)

tour avec porte-outil revolver *ou* **tour avec tourelle revolver** (*m.*). Same as **tour à revol-ver.**

tour avec poupée fixe à cône (*m.*), speed-lathe. (14)

tour d'amateur (*m.*), amateurs' lathe. (15)

tour d'outillage (*m.*), tool-makers' lathe. (16)

tour de cuvelage (Procédé de la trousse cou-pante) (Mines) (*f.*), drop-shaft ; open caisson ; caisson. (17)

tour de Gay-Lussac (*f.*), Gay-Lussac's tower ; Gay-Lussac tower. (18)

tour de Glover (*f.*), Glover's tower ; Glover tower. (19)

tour de — **HDP** (*m.*), — -millimetre centres lathe ; — $^{m}/_{m}$ lathe : (20)

tour de 200 HDP, 200 $^{m}/_{m}$ centres lathe. (21)

Note.—HDP stands for **hauteur des pointes**, height of centres.

tour de modeleur (*m.*), pattern-makers' lathe. (22)

tour de précision (*m.*), precision lathe. (23)

tour de robinetterie (*m.*), brass-finishers' lathe. (24)

tour de spire (*m.*). See **tour** *ou* **tour de spire.**

tour double à bois sur banc (banc droit en fonte, raboté et dressé, monté sur pieds) : colonne en fonte montée sur plaque de fondation à rainures, et munie d'un porte-outil à double coulisse pour tourner en l'air (*m.*), wood-turning bed-lathe, with extra face-plate on back end of live spindle, overhang-ing the bed (straight cast-iron bed, planed and finished, mounted on standards) : cast-iron pillar mounted on slotted base-plate, and provided with a compound tool-rest for face-plate turning. (25)

tour double à roues (*m.*), double-wheel lathe. (26)

tour en l'air *ou* **tour en l'air à plateau vertical** (*m.*), surfacing and boring-lathe ; face-lathe ; facing-lathe. (27) *Cf.* tourner en l'air.

tour en l'air à plateau horizontal (*m.*). Same as **tour à plateau horizontal.**

tour entre-pointes (*m.*), centre lathe ; centering-lathe ; pole-lathe. (28)

tour marchant à pédale *ou* **tour marchant au pied** (*m.*). Same as **tour à pédale.**

tour marchant au moteur (*m.*), power-lathe ; lathe for power ; turning-engine. (29)

tour ovale (*m.*), oval lathe. (30)

tour parellèle à commande électrique directe (*m.*), direct motor-driven lathe. (31)

tour parallèle à fileter, charioter, et surfacer (*m.*), sliding, surfacing, and screw-cutting lathe. (32)

tour parallèle à fileter, charioter, et surfacer automatiquement par la vis mère, marchant au moteur (*m.*), self-acting sliding, surfacing, and screw-cutting lathe, for power ; engine-lathe. (33)

tour parallèle renforcé, monté sur banc rompu, marchant au pied ou au moteur (*m.*), strong gap-lathe, for foot or power. (34)

tour parallèle très renforcé (*m.*), heavy-duty lathe ; heavy-service lathe ; heavy-cutting lathe. (35)

tour pour arbres de transmission (*m.*), shafting-lathe. (36)

tour pour bâtons ronds (*m.*), rounding-machine. (37)

tour pour l'emploi des aciers à coupe rapide, avec commande par monopoulie et boîte de vitesses (*m.*), all-gear single-pulley lathe, for use with high-speed cutting-steels. (38)

tour rapide (*m.*), high-speed lathe ; speed-lathe. (39)

tour-revolver [**tours-revolvers** *pl.*] (*n.m.*). Same as **tour à revolver.**

tour simple (*m.*), plain lathe ; plain ungeared lathe. (40)

tour simple à bois (*m.*), plain lathe for wood-turning. (41)

tour simple avec support à main (opp. **à tour à chariot**) (*m.*), hand-lathe. (42)

tour simple marchant à pédale (*m.*), plain treadle-lathe. (43)

tour simple sans banc (*m.*), plain lathe-heads. (44)

tour sur banc (*m.*), bed-lathe. (45)

tourbe (*n.f.*), peat : (46)

la tourbe est surtout utilisée comme com-bustible de ménage dans les pays où le bois et la houille sont peu abondants, peat is principally used as household fuel in countries where wood and coal are scarce. (47)

tourbe des marais (*f.*), bog-peat. (48)

tourbe mottière (*f.*), peat in blocks. (49)

tourbe mousseuse (*f.*), moss-peat. (50)

tourbeux, -euse *ou* **tourbier, -ère** (*adj.*), peaty : (51)

substance tourbeuse (*f.*), peaty substance. (52)

tourbière (*n.f.*), peat-bog ; peat-bed ; peat-moor ; peat-moss ; peatery ; turbary ; moss. (53)

tourbillon (vent) (*n.m.*), whirlwind. **(54)**

tourbillon (eau) (*n.m.*), whirlpool. (55)

tourbillon (Phys.) (*n.m.*), vortex. (56)

tourbillon (Mine de charbon) (*n.m.*), nest of stones in a coal-seam. (57)

tourbillon cyclonal (*m.*), cyclonal whirlwind. (58)

tourbillon de poussière (*m.*), dust-whirl. (59)

tourbillonnaire (Phys.) (*adj.*), vortical : (1) mouvement tourbillonnaire (*m.*), vortical motion. (2)

tourbillonnement (*n.m.*), whirling. (3)

tourbillonner (*v.i.*), to whirl. (4)

tourelle (*n.f.*) *ou* **tourelle revolver** (d'un tour), capstan, capstan-head, turret, turret-head, monitor, (of a lathe). (5)

tourelle hexagonale (*f.*), hexagon turret. (6)

tourelle revolver pour 6 outils (*f.*), six-tool capstan. (7)

touret (poulie à corde) (*n.m.*), pulley for gut band ; sheave. (8)

touret (dévidoir de corderie) (*n.m.*), reel ; winder. (9)

touret (porte-foret à archet) (*n.m.*), bow-drill ; piercer. (10)

touret de polisseur *ou* **touret pour polisseurs** (*m.*), polishing-head ; polishing-lathe. (11)

touret hydraulique (Sondage) (*m.*), water-swivel. (12)

tourie (*n.f.*), carboy. (13)

tourillon (*n.m.*), trunnion ; journal ; gudgeon ; stud ; pin. (14)

tourillon (d'une benne à bascule, d'un convertis- -seur, ou analogue) (*n.m.*), trunnion (of a tipping-bucket, of a converter, or the like). (15 *ou* 16 *ou* 17)

tourillon (d'un arbre) (*n.m.*), journal, gudgeon, (of a shaft). (18)

tourillon (d'un cylindre de laminoir) (*n.m.*), journal, neck, (of a rolling-mill roll). (19)

tourillon (d'un châssis de fonderie) (*n.m.*), trun- -nion, swivel, (of a foundry-flask). (20)

tourillon (d'une lanterne) (Fonderie) (*n.m.*), gudgeon (of a core-barrel). (21)

tourillon (d'un tambour de treuil) (*n.m.*), gudgeon (of a winch-barrel). (22)

tourillon (d'une grille) (*n.m.*), pivot (of an iron gate). (23)

tourillon creux (*m.*), hollow trunnion. (24)

tourillon de crosse *ou* **tourillon de la tête de piston** (*m.*), cross-head pin ; gudgeon-pin ; gudgeon ; wrist-pin ; wrist. (25)

tourillon de manivelle (*m.*), crank-pin ; wrist- pin ; wrist. (26)

tourillon de suspension de la coulisse (distri- -bution à coulisse) (*m.*), suspension-stud ; link-bearing (link-motion). (27)

tourmaline (Minéral) (*n.f.*), tourmaline ; schorl ; shorl. (28)

tourmaline bleue (*f.*), blue tourmaline. (29)

tourmaline brune (*f.*), brown tourmaline. (30)

tourmaline incolore (*f.*), colourless tourmaline ; white tourmaline. (31)

tourmaline jaune (*f.*), yellow tourmaline. (32)

tourmaline noire (*f.*), black tourmaline. (33)

tourmaline quartzeuse (Pétrol.) (*f.*), schorl- rock ; shorl-rock. (34)

tourmaline rose (*f.*), pink tourmaline. (35)

tourmaline rouge (*f.*), red tourmaline. (36)

tourmaline verte (*f.*), green tourmaline. (37)

tourmaline violet rougeâtre (*f.*), violet-red tour- -maline. (38)

tourmalinite (Pétrol.) (*n.f.*), schorl-rock ; shorl- rock. (39)

tourmenté, -e (*adj.*), distorted ; contorted ; broken : (40)
couche tourmentée (Géol.) (*f.*), distorted seam. (41)
les côtes du Pacifique vers la frontière des États-Unis et du Canada sont très

tourmentées et rocheuses : les plages se font rares, the Pacific coast towards the frontier of the United States and of Canada is very broken and rocky : beaches are of rare occurrence. (42)

tourmenter (se) (se dit du bois) (*v.r.*), to warp : (43)
ce bois était vert, il s'est tourmenté, this wood was green, it has warped. (44)

tournage (Travail au tour) (*n.m.*), turning : (45) tournage d'un arbre coudé, turning a crank- shaft. (46)
tournage en l'air, turning on the face-plate. (47) *See example under* tourner en l'air.
tournage entre pointes, turning between centres. (48)

tournant, -e (qui tourne) (*adj.*), turning ; revolv- -ing ; rotary ; rotatory ; rotating ; rota- -tive ; swiveling. (49)

tournant, -e (qui fait des détours) (*adj.*), winding : (50)
allée tournante (*f.*), winding path. (51)

tourne-à-gauche [**tourne-à-gauche** *pl.*] (*n.m.*), wrench. (52)

tourne-à-gauche (pour tarauds) (*n.m.*), tap- wrench. (53)

tourne-à-gauche (pour assembler ou désassembler les rallonges) (Sondage) (*n.m.*), hand- wrench ; hand-dog ; rod-wrench. (54)

tourne-à-gauche à carré variable (*m.*), adjustable tap-wrench. (55)

tourne-à-gauche à dévisser (Sondage) (*m.*), hand-wrench for screwing up boring-rods. (56)

tourne-à-gauche de manœuvre *ou simplement* **tourne-à-gauche** (*n.m.*) (Sondage), tiller ; turning-tool ; rod-turning tool ; pole- turner ; brace-head ; brace-key. (57)

tourne-à-gauche de support (Sondage) (*m.*), catch-wrench ; rod-support ; supporting- fork ; resting-fork ; lye-key ; tiger. (58)

tourne-à-gauche double (Sondage) (*m.*), double- ended hand-wrench. (59)

tourne-à-gauche pour donner la voie aux scies *ou simplement* **tourne-à-gauche** (*n.m.*), saw-set ; set ; jumper ; swage ; saw- jumper ; saw-swage ; saw-upsetter ; saw- wrest. (60)

tourne-à-gauche simple (Sondage) (*m.*), single- ended hand-wrench. (61)

tournée d'inspection (*f.*), tour of inspection. (62)

tourner (*v.t.*), to turn ; to rotate ; to swivel : (63)
tourner une manivelle, to turn a handle (*or* a crank). (64)
tourner sens dessus dessous, to turn upside down : (65)
tourner un rail sens dessus dessous, to turn a rail upside down. (66)

tourner (façonner au tour) (*v.t.*), to turn : (67) tourner du bois, to turn wood. (68)
tourner en l'air, to turn on the face-plate : (69) on dit que l'on tourne une pièce en l'air, lorsqu'elle est maintenue sur le plateau sans le secours des pointes, a part is said to be turned on the face-plate when it is kept on the face-plate without the aid of the centres. (70)
tourner entre pointes, to turn between centres. (71)

tourner (*v.i.*), to turn ; to revolve ; to rotate ; to run ; to swivel : (72)

la terre tourne sur elle-même autour du soleil, the earth turns on itself around the sun. (1)

mandrin tournant dans des coussinets en bronze (*m.*), mandrel running in gun-metal bearings. (2)

engrenages tournant dans un bain d'huile (*m.pl.*), gears running in an oil bath. (3)

tourner (avoir une issue) (*v.i.*), to turn out : (4) l'affaire a bien tourné (*f.*), the affair has turned out well. (5)

tournesol (*n.m.*), litmus. (6)

tourneur (pers.) (*n.m.*), turner. (7)

tourneur sur bois (*m.*), wood-turner. (8)

tournevis (*n.m.*), screw-driver ; screwdriver ; turn-screw ; turnscrew. (9)

tournevis au fût *ou* **tournevis pour vilebrequins** (*m.*), screwdriver-bit ; turnscrew-bit. (10)

tourniquet de cabestan (*m.*), roller. (11)

tourniquet de ventilateur (*m.*), fan-wheel. (12)

tourniquet électrique (*m.*), electric whirl ; electric vane. (13)

tournoiement (*n.m.*), spinning : (14) le guidage (*ou* guidonnage) est destiné à empê--cher le tournoiement et les rencontres des cages (Mines), guides are intended to prevent spinning and collision of cages. (15)

tournoyer (*v.i.*), to spin. (16)

tournure (déchets métalliques détachés d'une pièce pendant le tournage) (*n.f.*), turnings ; shavings : (17) la tournure d'acier, steel turnings. (18) tournure de zinc, zinc shavings. (19)

tourte (méthode d'amalgamation du minerai d'argent au " patio ") (*n.f.*), torta. (20)

tourte spongieuse (amalgame d'or) (*f.*), gold-sponge ; sponge gold. (21)

tourteau (d'une drague) (*n.m.*), tumbler (of a dredge). (22)

tourteau d'or (*m.*), gold-sponge ; sponge gold. (23)

tourteau inférieur (d'une drague) (*m.*), lower tumbler. (24)

tourteau supérieur *ou* **tourteau moteur** (d'une drague) (*m.*), upper tumbler. (25)

tout battant neuf (*m.*), **tout battant neuve** (*f.*), brand new. (26)

tout compte fait, taking everything into account. (27)

tout d'une pièce (*invar.*), in one piece ; all in one piece : (28) objet qui est tout d'une pièce (*m.*), article which is in one piece. (29)

tout neuf (*m.*), **toute neuve** (*f.*), quite new. (30)

tout premier choix (de), of the finest quality : (31) acier de tout premier choix (*m.*), finest quality steel. (32)

tout venant (*invar.*), unsorted ; ungraded : (33) carbure tout venant (*m.*), ungraded carbide. (34)

tout-venant (houille tout venant) (*n.m.*), run of mine ; mine-run ; run-of-mine coal ; through-and-through coal ; unsorted coal. (35)

tout-venant (minerai tout venant) (*n.m.*), run of mine ; mine-run ; run-of-mine ore ; unsorted ore. (36)

toute vapeur (à) *ou* **toute vitesse** (à), at full steam ; at full speed. (37)

toxicité (*n.f.*), poisonousness ; toxicity. (38)

toxique (*adj.*), poisonous ; foul ; toxic ; toxical : (39)

gaz toxique (*m.*), poisonous gas ; foul gas. (40)

poussiers toxiques de cinabre (*m.pl.*), cinnabar toxical dust. (41)

toxique (*n.m.*), poison ; toxicant : (42) toxiques minéraux, mineral poisons. (43)

traçage (action de tracer) (*n.m.*), laying out ; setting out ; drawing ; tracing ; plotting. (44) See the verb **tracer** for examples.

traçage (Exploitation des filons métallifères) (*n.m.*), development ; developing : (45) le traçage d'une mine, the development of a mine ; developing a mine. (46)

traçage (Exploitation des couches de houille) (*n.m.*), working in the whole ; whole working ; first working ; forewinning ; opening up the seam ; development ; developing. (47)

traçage au rocher (*m.*), developing by rock-chutes. (48)

traçage du gîte en massifs (*m.*), cutting up the bed into pillars. (49)

trace (signe ; marque) (*n.f.*), trace ; mark ; show ; print ; track : (50)

minerai contenant des traces d'or (*m.*), ore containing traces of gold. (51)

or qui ne contient pas de trace d'argent (*m.*), gold which contains no trace of silver. (52)

trace d'un filon, the neck or line of fracture connecting the two parts of a faulted lode. (53)

trace saillante, ridge : (54) une trace saillante laissée sur une pièce moulée, à l'endroit des joints du moule, a ridge left on a casting, at the junction of the parts of a mould. (55)

traces d'animaux (Géol.), animal-tracks. (56)

traces d'outil (Géol.), tool-marks. (57)

traces de courants (Géol.), current-marks. (58)

traces de gaz (Exploitation des gisements naturels), gas-shows. (59)

traces de pétrole, oil-shows. (60)

traces de pluie (Géol.), rain-prints ; rain-pits. (61)

traces de ruissellement (Géol.), rill-marks. (62)

traces de vagues *ou* traces du mouvement des vagues (Géol.), wave-marks. (63)

tracé (*n.m.*). See examples : (64)

tracé d'un assemblage, d'une courbe, d'une épure de distribution, du profil d'une dent d'engrenage (Dessin), layout of a joint, of a curve, of a valve-diagram, of the profile of a gear-tooth. (65 *ou* 66 *ou* 67 *ou* 68)

tracé d'un chemin de fer (ligne suivie ; par--cours), alignment of a railway : (69) baliser le tracé d'un chemin de fer, to beacon the alignment of a railway. (70)

tracé d'une courbe (action), laying out, setting out, plotting, a curve. (71)

tracé d'une courbe (ligne suivie), alignment of a curve : (72) rectifier le tracé des courbes (Ch. de f.), to rectify the alignment of curves. (73)

tracé d'un diagramme (action), plotting a diagram. (74)

tracé d'un fleuve à travers son bassin, course, track, of a river across its basin. (75)

tracé d'une ligne *ou* tracé d'un alignement (Topogr.), running a line : (76) le tracé d'alignements très longs nécessitant une grande précision se fait à l'aide d'instru--ments à lunette dits cercles d'alignement, running very long lines requiring great precision is done with telescope instruments called transits. (1)

tracé d'une route (ligne suivie), alignment of a road : (2)

rectifier le tracé d'une route, to straighten the alignment of a road. (3)

tracelet (*n.m.*). Same as **traceret**.

tracement (action de tracer) (*n.m.*). Same as **traçage**.

tracer (marquer le dessin de) (*v.t.*), to lay out ; to set out ; to draw ; to plot ; to trace : (4)

tracer un carré, to draw a square. (5)

tracer un diagramme, to plot a diagram. (6)

tracer un plan, to lay out, to draw, to trace, a plan. (7)

tracer une courbe, to lay out, to set out, to plot, a curve. (8)

tracer (suivre la piste de) (*v.t.*), to trace ; to train : (9)

tracer un filon jusqu'à son origine, to trace, to train, a lode to its head ; to chase. (10)

tracer (Exploitation des filons métallifères) (*v.t.*), to develop : (11)

tracer de nouveaux étages en profondeur, to develop new levels in depth. (12)

tracer la coupe d'une pierre, to line out a stone. (13)

tracer le chemin à quelqu'un, to show someone the way. (14)

tracer le gîte (Exploitation des couches de houille), to work in the whole ; to open up the seam ; to cut up the bed into pillars ; to develop the seam. (15)

tracer un itinéraire, to map out a route ; to mark out an itinerary. (16)

tracer un repère sur un tube de verre, to scratch a mark on a glass tube. (17)

tracer une ligne ou tracer un alignement (Topogr.), to run a line : (18)

tracer une ligne pour un chemin de fer, to run a line for a railway. (19)

tracer une ligne de conduite à quelqu'un, to map out, to outline, to trace, a policy for someone. (20)

tracer (se) un plan de conduite, to map out a policy (for oneself) ; to trace a policy to be followed. (21)

traceret (de menuisier, de charpentier) (*n.m.*), marking-awl ; scratch-awl ; scribing-awl ; scribe-awl ; scriber. (22)

traceret (d'un trusquin) (*n.m.*), scriber (of a surface-gauge). (23)

traceur (pers.) (*n.m.*), draughtsman. (24)

traceur mécanicien (*m.*), mechanical draughts-man. (25)

trachydolérite (Pétrol.) (*n.f.*), trachydolerite (26)

trachyte (Pétrol.) (*n.m.*), trachyte. (27)

trachytique (*adj.*), trachytic : (28)

lave trachytique (*f.*), trachytic lava. (29)

trachytoïde (*adj.*), trachytoid. (30)

traçoir (de menuisier, de charpentier) (*n.m.*), marking-awl ; scratch-awl ; scribing-awl ; scribe-awl ; scriber. (31)

traçoir (d'une mèche à trois pointes) (*n.m.*), scriber, nicker, (of a centre-bit). (32)

tracteur (*n.m.*), tractor ; traction-engine : (33)

atteler une remorque à un tracteur, to attach a trailer to a tractor. (34)

tractif, -ive (*adj.*), tractive ; pulling : (35)

force tractive (*f.*), tractive force. (36)

traction (*n.f.*), traction ; haulage ; hauling ; draught ; draft ; pulling ; pull. (37)

traction (Résistance des matériaux) (*n.f.*), tension (Strength of materials) : (38)

pièce prismatique qui travaille à la traction (*f.*), prismatic bar which is in tension. (39)

rivet qui est à la fois soumis à des efforts de traction et de cisaillement (*m.*), rivet which is subjected at the same time to tensile and shearing-stresses. (40)

traction à bras d'hommes (*f.*), manual haulage. (41)

traction à vapeur (*f.*), steam-traction. (42)

traction animale (*f.*), animal traction ; animal draught. (43)

traction électrique (*f.*), electric traction : (44)

il y a cinq systèmes de traction électrique : le système par accumulateurs, le système par fil aérien avec trolley, le système par voie souterraine, le système par troi-sième rail, et le système à plots, there are five systems of electric traction : the accumulator system, the overhead-trolley system, the conduit system, the third-rail system, and the stud system. (45)

traction en dessous (*f.*), underhaulage. (46)

traction en dessus (*f.*), overhaulage. (47)

traction en palier (*f.*), level traction ; level haulage. (48)

traction en rampe (*f.*), up-grade traction. (49)

traction mécanique sur voies de terre (*f.*), me-chanical road traction. (50)

traction par câble métallique (*f.*), wire-rope haulage ; haulage by wire rope. (51)

traction par chevaux (*f.*), horse-traction ; draught by horses. (52)

traction par machines fixes (*f.*), haulage by stationary engines. (53)

traction par moteurs animés (*f.*), hand or animal-power haulage. (54)

tractoire (*adj.*), tractional. (55)

trade-union [trade-unions *ou* trades-unions *pl.*] (*n.f.*), trade-union ; trades-union. (56)

tradition (*n.f.*), tradition : (57)

traditions de captage d'or dans l'ancien temps, traditions of gold recovered in the early days. (58)

trafic (négoce) (*n.m.*), traffic ; trade ; trading. (59)

trafic (mouvement) (*n.m.*), traffic : (60)

trafic de chemin de fer, railway-traffic. (61)

trafiquant (pers.) (*n.m.*), trader. (62)

trahir (*v.t.*), to betray : (63)

les roches striées trahissent l'action glaciaire (*f.pl.*), striated rocks betray glacial action. (64)

train (Ch. de f.) (*n.m.*), train. (65)

train (équipage de laminoir) (*n.m.*), train ; rolls ; mill. (66) See also **laminoir**.

train à billettes (laminoir) (*m.*), billeting-rolls. (67)

train à blooms *ou* **train blooming** (laminoir) (*m.*), blooming-mill ; cogging-mill. (68)

train à gros profilés (laminoir) (*m.*), heavy-section rolls. (69)

train à rails (laminoir) (*m.*), rail-train ; rail-mill. (70)

train à tôle (laminoir) (*m.*), plate-mill. (71)

train baladeur (Engrenages) (*m.*), sliding gear ; sliding gear-train ; balladeur train : (72)

arbre qui transmet trois vitesses par train baladeur (*m.*), shaft which transmits three speeds by a sliding gear-train. (73)

train d'échantillons moyens (laminoir) (*m.*), medium-sized bar-mill. (74)

train d'engrenages (*m.*), gear-train ; train of gearing ; multiple of gearing ; gear-work. (1)

train d'ouvriers (*m.*), workmen's train. (2)

train de bois (*m.*), raft ; timber raft ; lumber raft ; log raft. (3)

train de cylindres (laminoir) (*m.*), train of rolls ; set of rolls ; battery of rolls. (4)

train de gros fers marchands (laminoir) (*m.*), merchant train ; merchant-bar train ; merchant mill ; bar-mill. (5)

train de laminoir (*m.*), roll-train ; rolling-mill train ; rolling-mill rolls. (6)

train de loupes (laminoir) (*m.*), forge-train. (7)

train de machine *ou* **train de serpentage** *ou* **train à serpenter** (laminoir) (*m.*), rod-mill ; wire-mill ; looping-mill. (8)

train de marchandises (*m.*), goods-train ; freight-train. (9)

train de petits fers (laminoir) (*m.*), small bar-mill. (10)

train de puddlage (*m.*). Same as **train ébaucheur.**

train de roues (Méc.) (*m.*), train of wheels ; wheel-train. (11)

train de tôlerie (laminoir) (*m.*), plate-mill. (12)

train de voyageurs (*m.*), passenger-train. (13)

train de wagons (Mines) (*m.*), train, trip, set, of cars. (14)

train déraillé (Ch. de f.), train off line. (15)

train descendant (Ch. de f.) (*m.*), down train. (16)

train duo (laminoir) (*m.*), two-high mill ; two-high rolls ; two-high train. (17)

train ébaucheur (laminoir) (*m.*), muck-train ; puddle-train ; puddle-bar train ; puddle-rolls ; roughing-rolls. (18)

train épicycloïdal (Engrenages) (*m.*), epicyclic train. (19)

train finisseur (*m.*), finishing-mill ; finishing-rolls. (20)

train fixe (Engrenages) (opp. à **train baladeur**) (*m.*), fixed gear-train. (21)

train forgeur (laminoir) (*m.*), forge-train. (22)

train montant (Ch. de f.) (*m.*), up train. (23)

train réversible (laminoir) (*m.*), reversing mill. (24)

train trio (laminoir) (*m.*), three-high mill ; three-high rolls ; three-high train. (25)

traînage (transport sur la glace au moyen de traîneaux) (*n.m.*), sledging ; sleighing. (26)

traînage magnétique (*m.*), magnetic lag ; lag of magnetization ; time-lag of magnetism ; viscous hysteresis. (27)

traînant, -e (*adj.*), trailing : (28)
câble traînant (*m.*), trailing rope. (29)

traînard (d'un tour) (*n.m.*), saddle (of a lathe). (30)

traîneau (*n.m.*), sledge ; sleigh. (31)

traînée d'éboulis *ou* **traînée de cailloux roulés** (*f.*), stoneslide ; rockslide. (32)

traînée de débris sur la surface d'un glacier (*f.*), train of debris on the surface of a glacier. (33)

traînée de poudre (*f.*), train of powder. (34)

traînée houillère (*f.*), coal-belt : (35)
la traînée houillère qui traverse du Nord au Sud le Plateau central français, the coal-belt which crosses the French Central Plateau from North to South. (36)

traîner (*v.t.*), to draw ; to haul ; to drag : (37)
locomotive qui traîne une lourde charge (*f.*), locomotive which draws (*or* engine which hauls) a heavy load. (38)

traîneur (Mines) (pers.) (*n.m.*), drawer. (39)

trait (ligne tracée d'un seul coup) (*n.m.*), line ; stroke ; streak. (40)

trait (Minéralogie) (*n.m.*), streak : (41)
la molybdénite donne un trait bleu-verdâtre sur le biscuit, molybdenite gives a greenish blue streak on the streak plate. (42)

trait (tracé des opérations diverses relatives à la taille de la pierre et du bois) (*n.m.*), cutting-line ; line. (43)

trait (longe de corde ou de cuir avec laquelle les chevaux tirent) (*n.m.*), trace ; tug. (44)

trait (cordée) (Mines) (*n.m.*), lift ; journey ; trip. (45)

trait (opp. à point) (Télégr.) (*n.m.*), dash. (46) (opp. to **dot**).

trait de centre (*m.*), centre-line. (47)

trait de centre pour affûtage régulier (foret hélicoïdal) (*m.*), grinding-line (twist-drill). (48)

trait de division (cercle primitif) (Engrenages) (*m.*), pitch-line ; pitch-circle ; primitive circle ; dividing-circle ; rolling circle. (49)

trait de Jupiter (Charp.) (*m.*), splayed indent scarf. (50)

trait de Jupiter horizontal (*m.*), tabled scarf ; table-scarf. (51)

trait de lumière (*m.*), streak of light. (52)

trait de repère (*m.*), guiding-line. (53)

trait de scie (marque que l'on fait sur l'objet que l'on veut scier) (*m.*), cutting-line for sawing ; sawing-line. (54)

trait de scie (place que se fait la scie à mesure que son travail avance) (*m.*), kerf ; saw-kerf ; saw-cut. (55)

trait de stadia (*m.*), stadia-line. (56)

trait interrompu (*m.*), broken-section line. (57)

trait plein (opp. à **trait pointillé** *ou* **trait ponctué** *ou* **trait interrompu**) (*m.*), full line ; solid line. (58)

trait pointillé *ou* **trait ponctué** (*m.*), dotted line. (59)

trait zéro (d'une échelle) (*m.*), zero line (of a scale). (60)

traitable (*adj.*), treatable. (61)

traité (ouvrage où l'on traite d'une science) (*n.m.*), treatise : (62)
traité pratique de géologie, practical treatise on geology. (63)

traitement (manière d'opérer sur certaines matières qu'on veut transformer) (*n.m.*), treatment : (64)
traitement des minerais de fer au haut fourneau, treatment of iron ores in the blast-furnace. (65)
traitement à l'eau (du minerai), wet treatment. (66)
traitement par cyanuration, cyanide treatment. (67)
traitement par voie électrique, electric treatment. (68)
traitement par voie humide, wet treatment. (69)
traitement par voie sèche, dry treatment. (70)

traitement (émoluments) (*n.m.*), salary : (71)
traitement du directeur, salary of the manager. (72)

traiter (modifier au moyen de tel ou tel agent) (*v.t.*), to treat : (73)
traiter le minerai par le cyanure, to treat the ore with cyanide. (74)
traiter une liqueur pour argent, to treat a solution for silver. (75)

traiter (exposer ; développer) (*v.t.*), to treat : (1)
traiter un sujet théoriquement, to treat a subject theoretically. (2)

traiter des affaires, to transact business. (3)

trajet (*n.m.*), passage ; path ; journey ; way ; haul : (4)
trajet d'un rayon lumineux à travers un prisme, passage of a ray of light through a prism. (5)
trajet d'une ligne électrique, path of an electric line. (6)
trajet par mer, sea-passage. (7)
trajet par voie ferrée, railroad-haul. (8)

tramway [tramways *pl.*] (voie) (*n.m.*), tramway ; street-railroad ; street-railway. (9)

tramway (voiture) (*n.m.*), tram ; tram-car ; car ; street-car. (10)

tramway à accumulateurs (*m.*), accumulator-car. (11)

tramway à chevaux (*m.*), horse-tramway. (12)

tramway à la surface (Mines) (*m.*), surface-tramway. (13)

tramway à traction animale (*m.*), animal-traction tramway ; horse-tramway. (14)

tramway à trolley (*m.*), trolley-car (overhead system). (15)

tramway aérien (*m.*), aerial tramway. (16)

tramway automoteur (*m.*), gravity-tramway. (17)

tramway de mine (*m.*), mine-tramway ; tram-road ; rolleyway. (18)

tramway électrique (*m.*), electric tramway. (19)

tramway électrique à frotteur souterrain (*m.*), underground-trolley car. (20)

tranchage (*n.m.*), cutting ; cutting off ; slicing ; slicing off ; slabbing : (21)
tranchage des feuilles de placage, cutting veneers. (22)
tranchage d'un seul coup d'une pièce de fer, cutting off, slicing off, a piece of iron at one operation. (23)
tranchage du marbre, slabbing marble. (24)

tranchant, -e (*adj.*), cutting ; sharp ; keen : (25)
l'arête tranchante d'un outil (*f.*), the cutting edge of a tool. (26)
une herminette bien tranchante, a very sharp adze. (27)
un fil tranchant, a keen edge. (28)

tranchant (d'un outil) (*n.m.*), cutting edge, edge, (of a tool) : (29)
le tranchant d'un couteau, the edge of a knife. (30)

tranchant (d'un coin) (*n.m.*), point, edge, (of a wedge, of a key). (31)

tranche (morceau coupé mince) (*n.f.*), slice. (32)

tranche (de marbre, de pierre) (*n.f.*), slab (of marble, of stone). (33 ou 34)

tranche (Mines) (*n.f.*), slice ; lift ; jud. ; judd. (35)

tranche (bord mince ; faible épaisseur) (*n.f.*), edge : (36)
la tranche d'une planche, the edge of a board. (37)

tranche à chaud (outil de forgeron) (*f.*), hot-set ; hot-sate ; hot-cutter. (38)

tranche à froid (outil de forgeron) (*f.*), cold-set ; cold-sate ; cold-cutter. (39)

tranche à gouge (outil de forgeron) (*f.*), gouge. (40)

tranchée (*n.f.*), trench ; cutting ; cut ; surface-cut. (41)

tranchée (pour une voie ferrée) (*n.f.*), cutting (for a railway). (42)

tranchée de prospection *ou* **tranchée de recherches** (Mines) (*f.*), prospecting-trench ; costean-trench. (43)

trancher (*v.t.*), to cut ; to cut off ; to slice ; to slice off ; to slab. (44) See the noun **tranchage** for examples.

tranchet d'enclume (outil de forgeron) (*m.*), hardy ; hardie ; anvil-cutter. (45)

trancheur (Exploitation des carrières) (pers.) (*n.m.*), channeler ; channeller. (46)

trancheuse (*n.f.*), channeler ; channeling-machine ; quarrying-machine ; rock-chan-neller ; bar-channeler. (47)

trancheuse montée sur rails (*f.*), track-channeler. (48)

trancheuse pour le bois de bout (*f.*), wood-trimmer ; wood-trimming machine ; trim-mer ; paring-machine. (49)

trancheuse pour le bois de bout, fonctionnant à bras *ou* **trancheuse à levier** (*f.*), draw-stroke trimmer. (50)

transaction (*n.f.*), transaction : (51)
transactions commerciales, commercial trans-actions. (52)

transbordement (Marine) (*n.m.*), transshipment ; transhipment. (53)

transborder (Marine) (*v.t.*), to transship ; to tranship : (54)
transborder des marchandises, to transship goods. (55)
transborder une locomotive (au moyen d'un chariot de service), to traverse a locomotive. (56)

transbordeur (pont transbordeur) (*n.m.*), aerial ferry. (57)

transbordeur (chariot de service) (Ch. de f.) (*n.m.*), traverser ; traverse-table ; traverse-car-riage ; slide-rail ; railway-slide. (58)

transférer (*v.t.*), to transfer : (59)
transférer ses quartiers sur l'autre rive de la rivière, to transfer one's quarters to the other side of the river. (60)

transformateur (*n.m.*) *ou* **transformateur de ten-sion** (Élec.) (*n.m.*), transformer. (61)

transformateur à courant alternatif (*m.*), alternat-ing-current transformer. (62)

transformateur à cuirasse *ou* **transformateur cuirassé** (*m.*), shell transformer. (63)

transformateur à noyau (*m.*), core transformer. (64)

transformateur à rapport variable (*m.*), variable-ratio transformer. (65)

transformateur élévateur de tension (*m.*), step-up transformer. (66)

transformateur en série (*m.*), series-transformer. (67)

transformateur hérisson (*m.*), hedgehog trans-former. (68)

transformateur réducteur de tension (*m.*), step-down transformer. (69)

transformateur rotatif (*m.*), rotary transformer. (70)

transformateur statique *ou* **transformateur à action instantanée** (*m.*), static transformer ; stationary transformer. (71)

transformateur triphasé (*m.*), three-phase trans-former. (72)

transformateur uniphasé (*m.*), single-phase transformer. (73)

transformation (*n.f.*), transformation ; trans-forming ; conversion ; converting : (74)
transformation adiabatique (Phys.), adiabatic transformation. (75)

transformation de l'énergie calorifique en énergie mécanique, transformation of heat-energy into mechanical energy. (1)
transformation (Élec.) (*n.f.*), transformation ; transforming. (2)
transformation en coke (*f.*), coking. (3)
transformation en sulfate (*f.*), vitriolation. (4)
transformer (*v.t.*), to transform ; to convert ; to change : (5)
transformer la force motrice de l'eau en travail mécanique, to transform the motive force of water into mechanical energy. (6)
transformer le fer en acier par cémentation, to transform iron into steel by cementation ; to convert wrought iron into steel by cementation. (7)
transformer un mouvement alternatif rectiligne en mouvement circulaire continu au moyen d'une bielle, to transform a rectilinear reciprocating motion into continuous rotary motion by means of a connecting-rod. (8)
les machines électriques sont des appareils destinés à transformer l'énergie mécanique en énergie électrique : les piles transforment, au contraire, l'énergie chimique et calorifique en énergie électrique, electric machines are apparatuses intended to convert mechanical energy into electrical energy : batteries, on the contrary, convert chemical and heat-energy into electrical energy. (9)
transformer (Élec.) (*v.t.*), to transform : (10)
transformer les courants à haute tension en courants à basse tension, to transform high-tension currents into low-tension currents. (11)
transformer en coke (*v.t.*), to coke. (12)
transformer en sulfate (*v.t.*), to sulphatize ; to vitriolate : (13)
l'attaque par l'acide sulfurique des argiles ou silicates d'aluminium hydratés, conduit à les transformer en sulfate par élimination de la silice qu'ils contiennent (*f.*), the attack by sulphuric acid on clays or hydrous alumi-num silicates, tends to sulphatize them by elimination of the silica they contain. (14)
transformer (se) (*v.r.*), to transform ; to become transformed ; to be transformed ; to become changed ; to change : (15)
les névés se solidifient peu à peu et se trans-forment en glaciers (*m.pl.*), névés solidify gradually and transform to glaciers. (16)
de la chaleur qui se transforme en travail, heat which becomes transformed into work. (17)
se transformer sous l'influence de la chaleur, to become changed by the action of heat. (18)
transformer (se) en coke (*v.r.*), to coke. (19)
transgressif, -ive (Géol.) (*adj.*), transgressive (i.e., formed by encroachment of the sea) : (20)
dépôt transgressif (*m.*), transgressive deposit. (21)
transgression (envahissement par la mer) (Géol.) (*n.f.*), transgression ; encroachment. (22)
transit (*n.m.*), transit : (23)
marchandises en transit (*f.pl.*), goods in transit. (24)
transition (*n.f.*), transition ; change ; change over : (25)
la transition du chaud au froid, the transition from heat to cold. (26)
transition (Géol.) (*n.f.*), transition. (27)

translateur (Télégr.) (*n.m.*), translator ; repeater. (28)
translation (opp. à rotation) (Méc.) (*n.f.*), trans-lation. (29)
translation du chariot d'un pont roulant (*f.*), traversing the jenny of an overhead travel-ling crane ; racking the carriage of a travel-ler. (30)
la translation du chariot s'opère au moyen d'une poulie à chicanes actionnée par chaîne pendante sans fin, traversing the jenny is done by means of a spider-wheel and endless chain from below ; the carriage is racked along by a sprocket-wheel and dependent endless chain. (31)
translation du personnel (*f.*), transfer of the workmen (moving them from one place to another). (32)
translation du pont (d'un pont roulant) (*f.*), travelling, longitudinal motion, of the bridge, (of an overhead travelling crane) : (33)
la vitesse de translation du pont, the speed of travelling of the bridge. (34)
translucide (*adj.*), translucent : (35)
corps translucide (*m.*), translucent body. (36)
translucidité (*n.f.*), translucence ; translucency. (37)
transmetteur, -trice (*adj.*), transmitting : (38)
station transmettrice (*f.*), transmitting station. (39)
transmetteur (Télégr.) (*n.m.*), transmitter ; sender ; manipulator. (40)
transmetteur (Téléph.) (*n.m.*), transmitter ; sender : (41)
transmetteur téléphonique, telephone-trans-mitter. (42)
transmetteur d'ordres (*m.*), signalling-apparatus. (43)
transmettre (faire parvenir) (*v.t.*), to transmit ; to convey : (44)
transmettre le mouvement de l'arbre principal aux arbres secondaires, to transmit motion from the main shaft to the intermediate shafts. (45)
transmettre (Phys.) (*v.t.*), to transmit : (46)
transmettre la lumière, la chaleur, to transmit light, heat. (47 *ou* 48)
transmettre (se) (*v.r.*), to be transmitted : (49)
le son se transmet par ondulation, sound is transmitted by wave motion. (50)
transmission (action de transmettre) (*n.f.*), trans-mission ; transmitting ; conveyance ; con-veying : (51)
la transmission d'énergie électrique, the trans-mission of electrical energy ; the conveyance of electrical power. (52)
transmission (Phys.) (*n.f.*), transmission : (53)
la transmission du son, the transmission of sound. (54)
transmission (communication du mouvement d'un organe à un autre) (Méc.) (*n.f.*), trans-mission ; driving ; drive : (55)
la transmission par engrenages est le système adopté pour les trains de laminoirs, the gear-drive is the system adopted in rolling-mill trains. (56)
transmission (organe servant à transmettre le mouvement) (Méc.) (*n.f.*), driving-gear ; driving-motion ; driving-apparatus ; drive ; transmission-gear. (57)
transmission (système d'arbres de transmission) (*n.f.*), shafting ; driving-motion. (58)

transmission (l'arbre lui-même ; les arbres eux-mêmes) (*n.f.*), shaft ; shafting. (1)

transmission (d'un droit, d'une propriété) (*n.f.*), transfer (of a right, of a property). (2 *ou* 3)

transmission flexible (*f.*), flexible shaft ; flexible shafting. (4)

transmission intermédiaire (*f.*). Same as transmission secondaire.

transmission par balancier (*f.*), beam-drive. (5)

transmission par câbles (*f.*), rope-drive ; rope-transmission ; transmission of power by ropes. (6)

transmission par câbles métalliques (*f.*), wire-rope transmission. (7)

transmission par chaîne (*f.*), chain-drive. (8)

transmission par courroies (*f.*), belt-drive ; transmission of power by belts. (9)

transmission par engrenages (*f.*), gear-drive. (10)

transmission par frottement (*f.*), friction-gear ; friction-gearing. (11)

transmission principale (système) (*f.*), main drive. (12)

transmission principale (l'arbre ; les arbres) (*f.*), main shaft ; main shafting. (13)

transmission secondaire (système) (*f.*), counter-motion ; counter-shafting ; counterdriving motion ; counter-gear. (14)

transmission secondaire (l'arbre ou l'arbre et ses accessoires) (*f.*), counter-shaft ; counter-shaft and accessories ; counter-shafting. (15) See also **renvoi** *ou* **renvoi de mouvement.**

transmission secondaire à placer en l'air *ou* **transmission secondaire se fixant au plafond** (*f.*), overhead motion ; overhead driving-motion ; overhead counter-driving apparatus ; ceiling countershaft and appurtenances. (16)

transmission secondaire à placer sur le sol *ou* **transmission secondaire se fixant sur le sol** (*f.*), underhand motion ; underhand counter-driving motion ; floor countershaft and appurtenances. (17)

transmission secondaire se fixant indifféremment au sol ou au plafond (*f.*), countershaft for either floor or ceiling. (18)

transmission supérieure (disposition permettant au bras de faire un tour complet) (machine à percer radiale) (*f.*), top driving apparatus (arrangement enabling the arm to make a complete turn) (radial drilling-machine). (19)

transparence (*n.f.*), transparency : (20)
transparence de l'eau, transparency of water. (21)

transparent, -e (*adj.*), transparent : (22)
corps transparent (*m.*), transparent body. (23)

transport (*n.m.*), transport ; transportation ; conveyance ; conveying ; carriage ; carrying ; haulage ; hauling. (24)

transport à ciel ouvert *ou* **transport à la surface** (opp. à transport souterrain) (*m.*), surface-haulage ; aboveground conveyance. (25)

transport à dos de chameau (*m.*), camel-transport. (26)

transport à dos de mulets (*m.*), mule-transport ; mule-back transportation ; carriage by mules. (27)

transport d'énergie à grande distance (*m.*), long-distance conveyance of power. (28)

transport de force motrice (*m.*), conveyance of power ; conveying power. (29)

transport du pétrole par canalisation (*m.*), conveying oil by pipe-line ; pipage of oil. (30)

transport électrique de force (*m.*), electrical conveyance of power. (31)

transport gratuit, carriage free. (32)

transport outre-mer (*m.*), overseas transport ; shipment : (33)
l'emballage pour transport outre-mer est compté—0/0 du prix de la fourniture, packing for shipment is reckoned at — % of the cost of the goods. (34)

transport par chemin de fer (*m.*), railway-transport. (35)

transport par les eaux (Géol.) (*m.*), transportation by water. (36)

transport par roulage (*m.*), haulage ; haulage by road ; carting ; cartage. (37)

transport par voie d'eau (*m.*), transportation by water ; water-carriage. (38)

transport par voie de fer (*m.*), transportation by rail. (39)

transport par voie de terre (*m.*), transportation by land ; land-carriage. (40)

transport par voiture (*m.*), carting ; cartage. (41)

transport routier (*m.*), road-transport. (42)

transport souterrain (*m.*), underground haulage ; underground conveyance. (43)

transportable (*adj.*), transportable. (44)

transporter (*v.t.*), to transport ; to convey ; to carry ; to remove : (45)
transporter la force motrice, to convey power. (46)
transporter le camp en amont de la rivière, to remove the camp up-stream. (47)
transporter le minerai des chantiers aux puits, de la mine au moulin par tramway, to carry the ore from the workings to the shafts, from the mine to the mill by tramway. (48 *ou* 49)
transporter un liquide au moyen d'un tuyau, to convey a liquid by means of a pipe. (50)

transporteur (*n.m.*), conveyer ; conveyor ; transporter. (51) See also **convoyeur.**

transporteur à chaîne (*m.*), chain-conveyor. (52)

transporteur à courroie *ou* **transporteur à toile sans fin** (*m.*), belt-conveyor ; endless-belt conveyer ; band-conveyor. (53)

transporteur à godets (*m.*), bucket-conveyor. (54)

transporteur à palettes (*m.*), push-plate conveyer ; flight-conveyer ; scraper-conveyer ; trough-conveyer. (55)

transporteur à sec (pour matières sèches) (*m.*), dry conveyer. (56)

transporteur à vis sans fin (*m.*), screw-conveyer ; spiral conveyer. (57)

transporteur aérien (*m.*), aerial transporter ; aerial tramway. (58)

transporteur aérien à voie unique (*m.*), cableway. (59)

transporteur-trembleur [transporteurs-trembleurs *pl.*] (*n.m.*), shaker-conveyor. (60)

transporteuse (*n.f.*). Same as **transporteur.**

transvasement (*n.m.*), decantation ; transvasation. (61)

transvaser (*v.t.*), to decant ; to transvase ; to transvasate : (62)
transvaser un liquide d'un récipient, to decant a liquid from a vessel. (63)

transversal, -e, -aux (*adj.*), transverse ; cross ; transversal ; latitudinal. (64)

transversalement (*adv.*), transversally ; latitudinally. (65)

trapèze (Géom.) (*n.m.*), trapezium. (1)
trapézoïdal, -e, -aux *ou* trapézoïde (*adj.*), trape-
-zoidal ; trapezoid : (2)
section trapézoïdale (*f.*), trapezoidal section.
(3)
trapézoïde (*n.m.*), trapezoid. (4)
trapp (Pétrol.) (*n.m.*), trap ; trap-rock. (5)
trappe (*n.f.*), trap-door ; trap ; flap ; gate. (6)
trappe d'expansion (*f.*), expansion-trap. (7)
trappe de cave (*f.*), cellar-flap. (8)
trappe de cheminée (Mines) (*f.*), chute-gate. (9)
trappéen, -enne (Géol.) (*adj.*), trappean ; trap-
-poid. (10)
trass (Géol.) (*n.m.*), trass. (11)
travail (labeur) (*n.m.*) (*s'emploie souvent au
pluriel*), work ; working ; labour ; operation.
(12)
travail (ouvrage fait ou à faire) (*n.m.*), work ;
piece of work ; job : (13)
  distribuer le travail aux ouvriers, to distribute
  the work amongst the men. (14)
  un travail délicat, a delicate piece of work.
  (15)
  un travail de longue haleine, a long job. (16)
travail (occupation ; emploi) (*n.m.*), work ;
employment : (17)
  le taux du salaire se règle sur la rareté ou sur
  l'abondance du travail, the rate of wages
  depends on the scarcity or abundance of
  work. (18)
travail (façon) (*n.m.*), workmanship ; execution :
(19)
  un instrument d'un beau travail, an instrument
  of fine workmanship ; a beautifully made
  instrument. (20)
travail (Méc. et Phys.) (*n.m.*), work ; energy :
(21)
  de la chaleur qui se transforme en travail,
  heat which transforms to work. (22)
  emmagasiner du travail, to store up energy.
  (23)
travail (Résistance des matériaux) (*n.m.*), stress
  (the reaction of the interior parts of a solid
  body against forces tending to deform it)
  (Strength of materials) : (24)
  on appelle travail du métal la résistance que
  les molécules de ce métal opposent à l'effort
  qui agit sur eux, the stress of the metal is
  the resistance which the molecules of the
  metal offer to the stress which acts on
  them. (25)
travaux (chantiers) (Mines, etc.) (*n.m.pl.*), work-
-ings : (26)
  vieux travaux, old workings. (27)
travaux à flanc de coteau (Mines, etc.) (*m.pl.*),
work on hillside ; hillside work. (28)
travail à l'eau (*m.*), hydraulic mining ; hydrau-
-licking ; piping ; spatter-work. (29)
travail à l'entreprise (*m.*), contract work ;
bargain-work. (30)
travail à l'étroit (Mines) (*m.*), narrow work ;
straitwork ; straight work. (31)
travail à l'heure (*m.*), working by the hour.
(32)
travail à la compression *ou* travail à l'écrasement
(*m.*), compressive stress ; crushing-stress ;
positive stress. (33)
travail à la flexion (*m.*), bending-stress. (34)
travail à la journée (*m.*), day-work ; working
by the day. (35)
travail à la main (*m.*), hand-work ; hand-
working ; manual labour. (36)

travail à la pelle (*m.*), shovelling ; shovel-work.
(37)
travail à la poudre (*m.*), blasting. (38)
travail à la surface (Mines, Carrières, etc.) (*m.*),
surface-work ; surface-working ; surface-
labour ; grass-work. (39)
travail à la tâche (*m.*), job-work ; piece-work.
(40)
travail à la traction *ou* travail à l'extension (*m.*),
tensile stress. (41)
travaux à mi-pente (Mines) (*m.pl.*), mid-workings.
(42)
travail ardu (*m.*), hard work ; arduous work ;
up-hill work. (43)
travail au cisaillement (*m.*), shearing-stress ;
shear-stress. (44)
travail au dehors *ou* travail au grand air (*m.*),
outdoor work ; out-of-door work ; outside
work ; open-air work. (45)
travail au feu (Mines) (*m.*), fire-setting. (46)
travail au jour (Mines, etc.) (*m.*), surface-work ;
surface-working ; surface-labour ; grass-
work. (47)
travail au rocher (*m.*) *ou* travaux au rocher
(*m.pl.*) (Mines), stonework ; mullocking. (48)
travail au temps (*m.*), time-work. (49)
travail au tour (*m.*), lathe-work. (50)
travail aux explosifs (*m.*), blasting. (51)
travaux d'art (*m.pl.*), permanent works ; per-
-manent structures ; structures. (52) See
ouvrage d'art (synonymous expression) for
examples.
travaux d'exploitation (Mines) (*m.pl.*), mining-
work ; mining. (53)
travaux d'exploration (*m.pl.*), exploring work ;
exploration-work ; exploratory work ;
discovery-work ; search-work. (54)
travaux de construction (*m.pl.*), construction-
work ; constructional work. (55)
travaux de déblayement (*m.pl.*), clearing opera-
-tions. (56)
travaux de mines (*m.pl.*), mining ; mining-work.
(57)
travail de nuit (*m.*), night-work ; night-labour.
(58)
travaux de premier établissement (*m.pl.*), prelimi-
-nary work ; preparatory work ; dead-
work. (59)
travaux de prospection *ou* travaux de recherches
(*m.pl.*), prospecting work ; exploration-
work ; search-work ; experimental work.
(60)
travaux de relèvement (Mines) (*m.pl.*), clearing
operations. (61)
travaux de sauvetage (*m.pl.*), rescue-work. (62)
travaux de terrassement *ou* travaux de terrasse
(*m.pl.*), earthwork : (63)
  travaux de terrassement relatifs aux chemins
  de fer et aux routes, earthwork in connection
  with railways and roads. (64)
travail de tour (*m.*), lathe-work. (65)
travaux de traçage (Exploitation des couches
de houille) (*m.pl.*), working in the whole ;
whole working ; first working ; opening
operations ; forewinning operations ; de-
-velopment-work. (66)
travaux de traçage (Exploitation des filons
métallifères) (*m.pl.*), development-work.
(67)
travail des eaux (*m.*), action of water : (68)
  dispositif qui protège le fond d'une rivière
  contre le travail des eaux (*m.*), device which

protects the bottom of a river against the action of the water. (1)

travail des entretoises (*m.*), working of the stay-bolts. (2)

travail des femmes (*m.*), woman-labour. (3)

travail des métaux (*m.*), metal-working. (4)

travail des placers (*m.*), placer-working ; placer-work ; placer-mining. (5)

travail des résistances passives (Méc.) (*m.*), loss due to friction. (6)

travaux du fond (Mines) (*m.pl.*), underground workings ; underground work. (7)

travaux du jour (*m.pl.*), surface-working ; surface-work. (8)

travail électrique (*m.*), electric work : (9) unité de travail électrique (*f.*), unit of electric work. (10)

travail emmagasiné (*m.*), stored-up energy. (11)

travaux en cloche (Mines) (*m.pl.*), bell-shaped workings. (12)

travail en couche (*m.*) *ou* travaux en couche (*m.pl.*) (Mines), seam-work (13) (i.e., work in coal-seams: distinguished from **travaux en roche**).

travail en couche grisouteuse (*m.*), working in fiery seams. (14)

travaux en cul-de-sac (Mines) (*m.pl.*), blind workings ; dead end. (15)

travail en main (*m.*), work in hand. (16)

travaux en profondeur (Mines) (*m.pl.*), work in depth ; deep workings. (17)

travaux en progrès (*m.pl.*), work in progress ; works in progress. (18)

travaux en roche (Mines) (*m.pl.*), rock-work. (19)

travaux en tranchées (*m.pl.*), trench-work. (20)

travaux étriqués (Mines) (*m.pl.*), narrow work ; straitwork ; straight work. (21)

travail extérieur (*m.*), outdoor work ; out-of-door work ; outside work. (22)

travail grossier (*m.*), rough work. (23)

travail manuel (*m.*), manual labour ; hand-work ; hand-labour. (24)

travail mécanique (Méc.) (*m.*), mechanical energy ; energy : (25) transformer la force motrice de l'eau en travail mécanique, to transform the motive force of water into mechanical energy. (26)

travail mécanique (travail fait à la machine) (*m.*), machine-work. (27)

travaux miniers (*m.pl.*), mining ; mining-work. (28)

travaux miniers proprement dits (placers) (*m.pl.*), drift-mining. (29)

travail moteur (Méc.) (*m.*), mechanical energy ; energy. (30) See **travail utile** for example of use.

travail négatif (Méc.) (*m.*), negative work. (31)

travail nuisible (Méc.) (*m.*), loss (i.e., power or energy wasted in a machine). (32)

travail pénible (*m.*), hard work ; arduous work ; up-hill work. (33)

travail préparatoire (*m.*) *ou* travaux préparatoires *ou* travaux préliminaires (*m.pl.*), prepara-tory work ; preliminary operations ; dead-work. (34)

travaux publics (*m.pl.*), public works. (35)

travail résistant (Méc.) (*m.*), resistance energy. (36)

travaux souterrains (*m.pl.*), underground work. (37)

travaux sur l'affleurement (Mines) (*m.pl.*), work-ing the outcrop. (38)

travail sur le terrain (*m.*), field-work. (39)

travail utile (Méc.) (*m.*), useful work ; useful energy ; power : (40) le rendement d'une machine est le rapport du travail utile produit au travail moteur absorbé, the efficiency of an engine is the ratio of the useful work performed to the mechanical energy expended. (41) le frottement occasionne une perte notable de travail utile, friction causes a considerable loss of power ; friction occasions a note-worthy loss of useful energy. (42)

travaillante (d'une pompe de mine) (*n.f.*), working-barrel (of a mine-pump). (43)

travailler (*v.t.* & *v.i.*), to work. (44) Examples in vocabulary.

travailler (usiner) (Méc.) (*v.t.*), to machine ; to tool : (45) pièces de fonderie brutes ou travaillées (*f.pl.*), rough or machined castings. (46)

travailler (Mach.) (*v.i.*), to work ; to run : (47) machine qui travaille à haute pression (*f.*), engine which works (*or* runs) at high pressure. (48)

travailler (Résistance des matériaux) (*v.i.*), to be in stress ; to be stressed : (49) poutre qui travaille à — kilogrammes par millimètre carré (*f.*), beam which is in stress of — kilogrammes per square millimetre ; girder which is stressed to — kilogrammes per square millimetre. (50) Cf. travailler à la traction, etc.

travailler à ciel ouvert (Mines), to work open-cast. (51)

travailler à force, to work hard. (52)

travailler à forfait *ou* travailler à l'entreprise, to work on contract ; to do work on contract. (53)

travailler à grande échelle, to work, to operate, on a large scale. (54)

travailler à la traction, à la compression, au cisaillement, etc. (Résistance des matériaux), to be in tension, in compression, in shear, etc. (Strength of materials) : (55 *ou* 56 *ou* 57) rivet qui travaille au double cisaillement (*m.*), rivet which is in double shear. (58)

travailler à niveau plein (Mines), to sink through watery strata without any pumping. (59)

travailler à niveau vide (Mines), to sink through watery strata with continuous pumping. (60)

travailler à plein rendement *ou* travailler à pleine charge, to work to full capacity. (61)

travailler à profit, à perte, to work at a profit, at a loss. (62 *ou* 63)

travailler au fond (Mines), to work underground. (64)

travailler dans le grisou (Mines), to work in gassy places. (65)

travailler du bois à contre-fil, to work wood against the grain. (66)

travailler en bon mineur, to work in a minerlike way. (67)

travailler en butte (Dragage), to work, to dredge, the banks. (68)

travailler en grand, to work, to operate, on a large scale. (69)

travailler en remontant vers la surface (Mines), to work upwards to the surface. (70)

travailler le marbre (le façonner), to work marble. (71)

travailler sur une assez grande échelle, to work, to operate, on a fairly large scale. (72)

travailler (se) (*v.r.*), to work ; to be worked : (1) le fer ne se travaille pas aisément, iron does not work easily. (2)

travailleur (pers.) (*n.m.*), labourer ; workman ; worker. (3)

travailleur à la surface (Mines) (*m.*), surface-labourer. (4)

travailleur du fond (*m.*), underground workman. (5)

trave (Charp.) (*n.f.*), single notch ; single notch joint. (6)

travée (Arch., Ponts, Télégr., etc.) (*n.f.*), bay ; span : (7)
pont à travées égales (*m.*), bridge with equal bays. (8)
pont de chemin de fer à une travée et à voie normale (*m.*), single-span railway-bridge for standard-gauge track. (9)
travée de comble, span of roof. (10)
travée de ligne aérienne, bay, span, of overhead line. (11)

travée (Mines) (*n.f.*), lift ; stage : (12)
'hauteur fractionnée en plusieurs travées (*f.*), height divided into several lifts (*or* stages). (13)

travers-banc [travers-bancs *pl.*] (Mines) (*n.m.*), cross-cut ; cross-cutting. (14)

travers-banc au rocher (*m.*), stone-drift. (15)

traverse (Boisage) (*n.f.*), cross-bar ; cross-beam ; crosspiece ; cross-tie ; traverse. (16)

traverse (d'un châssis de charpente) (*n.f.*), transom, traverse, lintel, (of a frame *or* skeleton). (17)

traverse (d'un bâti de porte, d'un châssis de fenêtre) (*n.f.*), rail (of a door-frame, of a window-sash). (18 *ou* 19)

traverse (d'un châssis de fonderie) (*n.f.*), cross-bar, bar, stay, (of a foundry-flask). (20)

traverse (d'un poteau télégraphique, d'un poteau téléphonique, ou analogue) (*n.f.*), cross-arm (of a telegraph-post, of a telephone-pole, or the like). (21 *ou* 22 *ou* 23)

traverse (d'une machine à raboter) (*n.f.*), cross-rail, cross-slide, (of a planing-machine). (24)

traverse (seuil) (*n.f.*), sill. (25)

traverse (Ch. de f.) (*n.f.*), sleeper ; tie ; cross-tie. (26)

traverse (galerie transversale) (Mines) (*n.f.*), cross-heading. (27)

traverse basse (porte ou fenêtre) (*f.*), bottom rail. (28 *ou* 29)

traverse d'arrière (locomotive) (*f.*), tail-piece ; back bumper. (30)

traverse d'avant (locomotive) (*f.*), buffer-beam ; front bumper ; bumper ; bumper-beam. (31)

traverse dansante (bogie) (Ch. de f.) (*f.*), truck-bolster ; bolster. (32)

traverse danseuse (Entretien des voies) (Ch. de f.) (*f.*), yielding sleeper : (33)
les traverses danseuses sont des traverses qui ballottent au passage des trains, yielding sleepers are sleepers which jump up and down on the passing of trains. (34)

traverse de chemin de fer (bois ou acier) (*f.*), railway-sleeper ; railroad-tie (wood or steel). (35 *ou* 36)

traverse de fondation (*f.*), mudsill. (37)

traverse de nez (*f.*), nose-sill. (: 8)

traverse de tête (d'une locomotive) (*f.*), buffer-beam ; front bumper ; bumper ; bumper-beam. (39)

traverse de tête (d'un wagon de chemin de fer) (*f.*), end-sill. (40)

traverse du bas (*f.*). Same as traverse basse.

traverse du haut (*f.*). Same as traverse haute.

traverse du haut du derrick (*f.*), derrick crown-block. (41)

traverse du milieu (d'une porte) (*f.*), middle rail, lock-rail, (of a door). (42)

traverse haute (porte ou fenêtre) (*f.*), top rail. (43 *ou* 44)

traverse porte-outil (d'une machine à raboter) (*f.*), cross-rail, cross-slide, (of a planing-machine). (45)

traversée (*n.f.*), crossing : (46)
la traversée des vallées se fait sur des remblais ou des viaducs, the crossing of valleys is effected on embankments or viaducts. (47)
la traversée d'un fleuve, crossing a river. (48)

traversée (d'une faille) (Mines) (*n.f.*), negotiating (a fault). (49)

traversée (*n.f.*) *ou* **traversée de voie** (Ch. de f.), crossing. (50) (two railway-tracks crossing one another : not to be confounded with crossing, a railway-frog, which is in French, croisement *or* croisement de voie).

traversée-bretelle [traversées-bretelles *pl.*] (Ch. de f.) (*n.f.*), double crossover road ; double crossover ; scissors crossing ; diamond switch. (51)

traversée-jonction [traversées-jonctions *pl.*] (Ch. de f.) (*n.f.*), slip-switch ; slip-points ; slip-road ; compound switch. (52)

traversée-jonction double (*f.*), double slip-switch ; double slip-points ; double slip-road ; double compound switch. (53)

traversée-jonction simple (*f.*), single slip-switch ; single slip-points ; single slip-road ; single compound switch. (54)

traversée oblique (Ch. de f.) (*f.*), diamond crossing. (55)

traversée rectangulaire *ou* **traversée de voie à angle droit** (Ch. de f.) (*f.*), right-angle crossing ; four-way crossing. (56)

traverser (*v.t.*), to cross ; to traverse ; to pass through ; to run through : (57)
traverser une rivière, une montagne, to cross a river, a mountain. (58 *ou* 59)
chemin de fer qui traverse une route (*m.*), railway which crosses a road. (60)
les rayons qui traversent une lentille (*m.pl.*), the rays which traverse (*or* pass through) a lens. (61)
les matières traversant les mailles d'un tamis (*f.pl.*), the matter passing through the meshes of a sieve. (62)
étude géologique des terrains traversés (*f.*), geological survey of the strata passed through. (63)
la galerie traverse actuellement une roche dure, the level is at present running through hard rock. (64)

traverser les plans de quelqu'un, to upset some-one's plans. (65)

traverser une planche avec le rabot, to traverse a board with the plane. (66)

traverser une rivière à gué, to ford a river. (67)

travertin (calcaire) (*n.m.*), travertine ; travertin ; calc-sinter ; calcareous sinter ; sinter. (68)

travertin (siliceux) (*n.m.*), siliceous sinter ; quartz sinter ; sinter. (69)

travertin (ferrugineux) (*n.m.*), ferruginous sinter. (1)

trébucher (faire pencher la balance) (*v.i.*), to turn the scale : (2)

quand on pèse une monnaie d'or, il faut qu'elle trébuche, when one weighs a gold coin, it should turn the scale. (3)

trébuchet (balance) (*n.m.*), physical balance ; balance (light scales on stand for delicate work, such as chemical analysis, assaying, physical experiments, etc.) : (4)

trébuchet de précision, sur socle acajou, à tiroir, precision balance, on mahogany stand, with drawer. (5)

trébuchet (fourchette qui soulève le fléau d'une balance) (*n.m.*), beam-support. (6)

trébuchet (fourchette qui soulève les plateaux d'une balance) (*n.m.*), pan-support. (7)

tréfilage (*n.m.*), wiredrawing. (8)

tréfiler (*v.t.*), to wiredraw ; to draw (metal) into wire ; to draw (metals) through the draw-plate : (9)

tréfiler du métal ductile, to wiredraw ductile metal. (10)

tréfilerie (action) (*n.f.*), wiredrawing. (11)

tréfilerie (usine) (*n.f.*), wire-works ; wire-drawing works. (12)

tréfileur (pers.) (*n.m.*), wiredrawer. (13)

trèfle (d'un cylindre de laminoir) (*n.m.*), wobbler (of a rolling-mill roll). (14)

treillage (*n.m.*), trellis ; lattice ; trellis-work ; lattice-work. (15)

treillage métallique en losange (*m.*), diamond wire lattice. (16)

treillis (*n.m.*), netting ; network ; lattice ; lattice-work ; trellis ; trellis-work : (17)

pont en treillis (*m.*), lattice bridge. (18)

treillis métallique *ou simplement* **treillis** (*n.m.*) (pour clôtures), wire netting. (19)

tremblement (*n.m.*), trembling ; shaking : (20)

tremblement d'un pont suspendu, trembling of a suspension bridge. (21)

tremblement de terre (*m.*), earthquake. (22)

trembler (*v.i.*), to tremble ; to shake. (23)

trembleur (Élec.) (*n.m.*), trembler. (24)

trémie (auge en forme d'entonnoir) (*n.f.*), hopper. (25)

trémie (Géol.) (*n.f.*), pan. (26)

trémie à minerai (*f.*), ore-bin ; ore-hopper. (27)

trémie d'alimentation *ou* **trémie de chargement** (*f.*), feed-hopper. (28)

trémie de cheminée (Constr.) (*f.*), chimney-opening. (29)

trémie de sel (Géol.) (*f.*), salt-pan ; salt bottom. (30)

trémolite (Minéral.) (*n.f.*), tremolite ; grammatite. (31)

trempe (action de tremper le fer, l'acier, etc.) (*n.f.*), hardening ; tempering : (32)

alésoirs affûtés mécaniquement après la trempe (*m.pl.*), reamers ground mechanically after hardening. (33)

la trempe à l'eau froide rend le fer cassant, hardening (*or* tempering) in cold water renders iron brittle. (34)

trempe de l'acier, hardening steel ; temper-ing of steel. (35)

trempe du bronze, hardening gun-metal ; tempering of bronze. (36)

trempe du verre, tempering glass. (37)

trempe à l'eau, water-hardening ; water-tempering. (38)

trempe à l'huile, oil-hardening ; oil-temper-ing. (39)

trempe en coquille, chilling ; chill-hardening. (40)

trempe (qualité que le fer contracte quand on le trempe) (*n.f.*), temper : (41)

éviter une trempe trop dure et cassante, to avoid too hard and brittle a temper. (42)

trempe (plongement dans un liquide) (*n.f.*), dipping ; quenching : (43)

essai de pliage après trempe (*m.*), bend-test after quenching. (44)

trempé, -e (Métall.) (*adj.*), hardened ; tempered : (45)

coussinets en acier trempé (*m.pl.*), hardened steel bearings. (46)

trempé (immersion) (*n.m.*), dipping : (47)

dorure au trempé (*f.*), gilding by dipping. (48)

tremper (Métall, etc.) (*v.t.*), to harden ; to temper : (49)

tremper de l'acier, to harden, to temper, steel. (50)

tremper une lame, to temper a blade. (51)

pièce qui doit être trempée et revenue jaune paille (*f.*), part which should be hardened and tempered straw yellow. (52)

billes en acier, trempées à cœur (*f.pl.*), steel balls, hardened throughout. (53)

tremper à l'eau, to water-harden ; to water-temper. (54)

tremper à l'huile, to oil-temper. (55)

tremper en coquille *ou simplement* tremper, to chill ; to chill-harden. (56)

tremper (plonger dans un liquide) (*v.t.*), to dip ; to quench : (57)

tremper de la tôle dans un bain de zinc fondu pour la galvaniser, to dip sheet iron in a bath of molten zinc to galvanize it. (58)

l'éprouvette est chauffée au rouge vif et trempée dans l'eau ayant une température qui ne dépasse pas 27° C. (*f.*), the test-piece to be heated to a bright red and quenched in water of a temperature not exceeding 27° C. (59)

tremper (mouiller) (*v.t. & v.i.*), to soak ; to steep ; to drench ; to wet. (60)

tremper (recouvrir avec du métal par immersion) (*v.t*), to dip. (61)

tremper (se) (Métall., etc.) (*v.r.*), to be tempered ; to be hardened ; to harden : (62)

les très petits objets se trempent en paquets après avoir été chauffés en vase clos (*m.pl.*), very small articles are tempered in bundles after having been heated in a closed vessel. (63)

trenail (cheville en bois) (*n.m.*), treenail ; trenail. (64)

trépan (Sondage) (*n.m.*), bit ; chisel ; boring-chisel ; boring-bit ; bore-bit ; drilling-bit. (65)

trépan à biseau (*m.*), chopping-bit ; cutting-bit ; chisel-bit. (66)

trépan à chute libre (*m.*), free-falling bit. (67)

trépan à joues (*m.*), fluted bit. (68)

trépan à tranchant en croix (*m.*), star bit ; X chisel. (69)

trépan batteur (*m.*), churn-drill bit. (70)

trépan carottier *ou* **trépan découpeur** (*m.*), core-bit ; core-cutter. (71)

trépan composé (*m.*), boring-head ; composite borer with chisels fixed in sockets. (72)

trépan contondant (opp. à trépan tranchant) (*m.*), blunt chisel. (73)

trépan élargisseur (*m.*), underreamer. (1)

trépan en double marteau *ou* trépan à double marteau (Procédé Kind-Chaudron) (*m.*), trepan (in the form of a double Y). (2)

trépan excentrique (*m.*), eccentric bit ; eccentric chisel. (3)

trépan plat *ou* trépan simple (*m.*), flat chisel ; plain bit. (4)

trépan plat à tranchant droit (*m.*), straight-edged chopping-bit. (5)

trépan plat à tranchant en pointe de diamant (*m.*), V-shaped chisel. (6)

trépan tranchant (*m.*), chopping-bit ; cutting-bit ; chisel-bit. (7)

trépidation (*n.f.*), tremor ; vibration : (8)
  machine qui marche sans trépidation (*f.*), machine which runs without tremor. (9)
  trépidations occasionnées par le passage d'un train, tremor caused by the passing of a train. (10)
  trépidation des vitres d'une voiture, vibration of the windows of a carriage. (11)

trépider (*v.i.*), to vibrate. (12)

trépied (*n.m.*), tripod. (13) See also **pied**.

trépied à coulisse (*m.*), sliding tripod ; sliding-leg tripod-stand ; adjustable tripod ; tele--scopic tripod. (14)

trépied en forme de triangle, muni de 3 vis de calage, avec pince d'arrêt et vis de rappel pour le mouvement lent (*m.*), tribrach with 3 levelling-screws, with clamp and slow-motion adjusting-screw. (15)

trépied extensible (*m.*), extension-tripod. (16)

très sec (Barométrie), very dry. (17)

tresse (bourrage) (Méc.) (*n.f.*), gasket ; packing : (18)
  tresse en chanvre *ou* tresse en étoupe de chanvre, hemp gasket ; gasket ; hemp packing. (19)

tresse (Isolement des fils électriques) (*n.f.*), braid : (20)
  tresse de coton, cotton braid. (21)
  fil conducteur souple sous tresse soie paraffinée (*m.*), flexible conductor-wire silk braided and paraffined. (22)

tréteau (*n.m.*), trestle ; horse. (23)

tréteau de noyauteur (Fonderie) (*m.*), core-trestle. (24)

tréteau de scieur de long (*m.*), pit-saw horse ; pit-sawyers' trestle. (25)

tréteau volant (*m.*), flying trestle. (26)

treuil (*n.m.*), winch ; windlass ; crab ; hoist ; hoisting-crab. (27)

treuil à air comprimé (*m.*), air-hoist ; compressed-air hoist ; pneumatic hoist. (28)

treuil à bras *ou* treuil à manivelle *ou simplement* treuil (*n.m.*), hand-winch ; crab-winch ; winch with crank-handle. (29)

treuil à chaîne (*m.*), chain-hoist. (30)

treuil à cheval (*m.*), horse-power ; horse-gear. (31)

treuil à double engrenage (*m.*), double-purchase crab-winch ; double-purchase hoisting-crab. (32)

treuil à manège (Mines) (*m.*), horse-whim ; whim ; whin ; gin. (33)

treuil à simple engrenage, avec frein (*m.*), single-purchase crab, with brake. (34)

treuil à vapeur (*m.*), steam-winch ; steam-hoist. (35)

treuil chinois (*m.*), Chinese windlass ; differential windlass. (36)

treuil d'applique (*m.*), bracket-crab ; wall-crab ; gipsy-winch ; gypsy-winch. (37)

treuil d'extraction (Mines) (*m.*), hoisting-crab ; windlass (*or* winch) for raising ore ; tackle. (38)

treuil d'extraction à manège (*m.*). Same as treuil à manège.

treuil de curage (Sondage) (*m.*), sand-reel. (39)

treuil de forage *ou* treuil de manœuvre (*m.*), hoist ; draw-works ; drilling-winch. (40)

treuil de levage (*m.*). Same as treuil d'extraction.

treuil de puits (*m.*), windlass for wells ; well-bucket windlass. (41)

treuil de remonte des tringles de sondage (*m.*), winch for raising drill-rods. (42)

treuil des carriers (*m.*), treadwheel. (43)

treuil différentiel (*m.*), differential windlass ; Chinese windlass. (44)

treuil électrique (*m.*), electric hoist. (45)

treuil hydraulique (*m.*), hydraulic hoist. (46)

treuil pour lampes à arc (*m.*), arc-lamp winch ; arc-lamp hoist. (47)

treuil roulant (*m.*), travelling crab. (48)

tri-compound (*adj. invar.*), triple-expansion : (49)
  machine tri-compound (*f.*), triple-expansion engine. (50)

triable (*adj.*), sortable. (51)

triage (*n.m.*), sorting. (52)

triage à l'eau *ou* triage par la voie humide (*m.*), wet sorting. (53)

triage à la main (*m.*), hand-sorting. (54)

triage à sec *ou* triage par la voie sèche (*m.*), dry sorting. (55)

triage au marteau (minerai) (*m.*), hammer-sorting. (56)

triage des wagons de chemin de fer (*m.*), marshal-ling railway-trucks ; drilling railway-cars. (57)

triage du minerai dans les chantiers (*m.*), sorting ore in the stopes. (58)

triangle (Géom.) (*n.m.*), triangle. (59)

triangle (système triphasé) (Élec.) (*n.m.*), delta ; mesh : (60)
  montage en triangle (*m.*) *ou* groupement en triangle (*m.*), delta-connection ; mesh-grouping. (61)

triangle à 3 vis de calage avec pince d'arrêt et vis de rappel pour le mouvement lent (*m.*), tribrach with 3 levelling-screws, with clamp and slow-motion adjusting-screw. (62)

triangle acutangle (*m.*), acute-angled triangle. (63)

triangle curviligne (*m.*), curvilinear triangle. (64)

triangle équiangle (*m.*), equiangular triangle. (65)

triangle équilatéral (*m.*), equilateral triangle. (66)

triangle isocèle *ou* triangle isoscèle (*m.*), isosceles triangle. (67)

triangle mixtiligne (*m.*), mixtilinear triangle ; mixtilineal triangle. (68)

triangle obliquangle (*m.*), oblique-angled triangle. (69)

triangle obtusangle (*m.*), obtuse-angled triangle. (70)

triangle rectangle (*m.*), right-angled triangle. (71)

triangle rectiligne *ou simplement* triangle (*n.m.*), plane triangle ; triangle. (72)

triangle scalène (*m.*), scalene triangle. (73)

triangle sphérique (*m.*), spherical triangle. (74)

triangulaire (*adj.*), triangular : (1)
prisme droit triangulaire (*m.*), triangular right prism. (2)
triangulairement (*adv.*), triangularly. (3)
triangulation (*n.f.*), triangulation. (4)
trianguler (*v.t.*), to triangulate : (5)
trianguler un pays, to triangulate a country. (6)
trias (Géol.) (*n.m.*), Triassic ; Trias : (7)
le trias, the Triassic ; the Trias. (8)
triasique (*adj.*), Triassic : (9)
le système triasique, the Triassic system. (10)
triatomique (Chim.) (*adj.*), triatomic. (11)
tribasique (Chim.) (*adj.*), tribasic. (12)
tribu (*n.f.*), tribe. (13)
tributaire (*adj.*), tributary ; affluent : (14)
un cours d'eau tributaire, a tributary stream. (15)
la Marne est tributaire de la Seine, the Marne is tributary to the Seine. (16)
tributaire de (être), to be dependent on *or* upon ; to depend on *or* upon : (17)
être tributaire de l'étranger, to be dependent upon, to depend on, foreign supplies. (18)
être tributaire pour la matière première des pays d'extrême Orient, to be dependent upon the Far East for raw material. (19)
trichite (Pétrol.) (*n.f.*), trichite. (20)
trichopyrite (Minéral) (*n.f.*), millerite ; hair-pyrites ; capillary pyrites. (21)
trichroïque (Phys.) (*adj.*), trichroic. (22)
trichroïsme (Phys.) (*n.m.*), trichroism. (23)
triclinique (Cristall.) (*adj.*), triclinic. (24)
tricoises (*n.f.pl.*), pincers. (25)
tridymite (Minéral) (*n.f.*), tridymite. (26)
trièdre (Géom.) (*adj.*), trihedral. (27)
trièdre (*n.m.*), trihedron. (28)
trier (*v.t.*), to sort : (29)
trier du minerai à la main, to sort ore by hand. (30)
trier au marteau (minerai), to sort with the hammer. (31)
trier des wagons (Ch. de f.), to marshal trucks ; to drill cars. (32)
trieur (pour minerais) (Mach.) (*n.m.*), separator. (33)
trieur, -euse (pers.) (*n.*), sorter. (34)
trieur de minerai (pers.) (*m.*), ore-sorter. (35)
trieur électromagnétique *ou* trieur magnétique (pour minerais) (*m.*), magnetic separator ; electromagnetic separator. (36)
trifilaire (*adj.*), three-wire : (37)
réseau trifilaire (Élec.) (*m.*), three-wire system. (38)
trifurcation (*n.f.*), trifurcation. (39)
trifurquer (se) (*v.r.*), to trifurcate. (40)
trigonométrie (*n.f.*), trigonometry. (41)
trigonométrie rectiligne (*f.*), plane trigonometry. (42)
trigonométrie sphérique (*f.*), spherical trigonom-etry. (43)
trigonométrique (*adj.*), trigonometric ; trigono-metrical : (44)
levé trigonométrique (*m.*), trigonometrical survey. (45)
trigonométriquement (*adv.*), trigonometrically. (46)
trimestriel, -elle (*adj.*), quarterly : (47)
locataire trimestriel (*m.*), quarterly tenant. (48)
trimétrique (*adj.*), trimetric ; trimetrical. (49)
tringle (verge métallique) (*n.f.*), rod. (50)

tringle (marque faite par le cordeau blanchi ou rougi) (*n.f.*), chalk-line. (51)
tringle d'écartement (aiguillage) (Ch. de f.) (*f.*), bridle-rod ; stretcher-rod. (52)
tringle de fer (*f.*), iron rod. (53)
tringle de manœuvre *ou* tringle de connexion (aiguillage) (Ch. de f.) (*f.*), switch-rod ; switch-connecting rod ; connecting-rod ; switch-bar ; pull-rod. (54)
tringle de manœuvre du régulateur (*f.*), throttle-rod ; throttle reach-rod ; throttle-stem. (55)
tringle de suspension (suspension de cadre porteur) (Fonçage de puits de mines) (*f.*), hanging rod. (56)
tringle de tirage (*f.*), pull-rod. (57)
tringle de tirage du frein (wagon de ch. de f.) (*f.*), brake connecting-rod ; brake-rod. (58)
tringle du régulateur (*f.*), governor-rod. (59)
tringler (tracer une ligne droite avec le cordeau frotté de craie) (*v.t.*), to line out : (60)
tringler un morceau de bois, to line out a piece of wood. (61)
tringler (Métall.) (*v.t.*), to shingle ; to knobble. (62)
trio (laminoir) (*n.m.*), three-high mill ; three-high rolls ; three-high train. (63)
triphane (Minéral) (*n.m.*), triphane ; spodumene. (64)
triphasé, -e (Élec.) (*adj.*), three-phase ; triphase : (65)
courant triphasé (*m.*), three-phase current. (66)
triphyline (Minéral) (*n.f.*), triphylite ; triphyline. (67)
triple (*adj.*), triple ; treble ; threefold : (68)
triple harnais d'engrenage (*m.*), triple gear. (69)
triple tirage (appareil photographique) (*m.*), triple extension. (70)
triple valve (frein à air comprimé) (*f.*), triple valve. (71)
palan triple (*m.*), threefold purchase. (72)
à triple effet (Méc.), triple-acting ; triple-action. (73)
triplite (Minéral) (*n.f.*), triplite. (74)
tripoli (*n.m.*), tripoli ; rottenstone. (75)
tripolir *ou* tripolisser (*v.t.*), to rottenstone. (76)
tripolite (Pétrol.) (*n.f.*), tripolite. (77)
triqueballe (pour transporter les grosses pièces de bois) (*n.m.*), timber-cart ; timber-wagon ; timber-wain. (78)
trisilicate (Chim.) (*n.m.*), trisilicate. (79)
trisilicique (Chim.) (*adj.*), trisilicic. (80)
tritoprisme (Cristall.) (*n.m.*), tritoprism ; prism of the third order ; third order prism. (81)
tritopyramide (Cristall.) (*n.f.*), tritopyramid ; pyramid of the third order ; third order pyramid. (82)
triturable (*adj.*), triturable. (83)
triturant, -e, *ou* triturateur (*adj.*), triturating. (84)
tirturateur (Mach.) (*n.m.*), triturator. (85)
trituration (*n.f.*), trituration ; grinding : (86)
trituration d'une substance dans un mortier, trituration of a substance in a mortar. (87)
trituration des masses continentales (Géol.), trituration of continental masses. (88)
triturer (*v.t.*), to triturate ; to grind : (89)
triturer l'amalgame, to triturate, to grind, the amalgam. (90)
trivalence (Chim.) (*n.f.*), trivalence ; trivalency. (91)

trivalent, -e (*adj.*), trivalent. (1)

troctolite (Pétrol.) (*n.f.*), troctolite ; troctolyte. (2)

trögérite (Minéral) (*n.f.*), trögerite ; troegerite. (3)

troïlite (Minéral) (*n.f.*), troilite. (4)

trois-carrés *ou* **trois-quarts** (*n.m.*), three-square file ; tri-square file ; triangular file. (5)

trois-pieds (*n.m.*), tripod. (6)

troisième *ou* **3ᵉ** (*adj.*), third ; 3rd : (7)
troisième rail (Élec.) (*m.*), third rail. (8)
troisième taraud (*m.*), third tap ; bottoming tap ; plug tap ; straight tap ; finishing-tap. (9)

trolley [**trolleys** *pl.*] (tramway électrique) (*n.m.*), trolley ; trolly. (10)

trolley (câble aérien) (*n.m.*), trolley ; trolly ; runner ; truck. (11)

trolley à archet (tramway électrique) (*m.*), bow-type trolley. (12)

trombe (Météor.) (*n.f.*), waterspout. (13)

trommel (Préparation mécanique des minerais) (*n.m.*), trommel ; revolving screen. (14)

trommel classeur (*m.*), sizing-trommel. (15)

trommel débourbeur (*m.*), washing-trommel. (16)

trommel des fins (*m.*), fine trommel. (17)

trommel des gros (*m.*), coarse trommel. (18)

trommelage (*n.m.*), trommeling. (19)

trommeler (*v.t.*), to trommel. (20)

trompe (de fourneau, de forge catalane) (*n.f.*), trompe ; tromp ; waterfall-blower. (21)

trompe (*n.f.*) *ou* **trompe à eau** *ou* **trompe hydraulique** (Mines), water-blast. (22)

trompe d'alarme (*f.*), hooter ; siren. (23)

trompe évasée (d'un ventilateur de mine) (*f.*), évasé chimney (of a mine-fan). (24)

trompette (d'un étau-limeur) (*n.f.*), ram (of a shaper). (25)

trompeur, -euse (*adj.*), deceiving ; deceptive ; misleading. (26)

trona (Minéral) (*n.m.*), trona. (27)

tronc (Géom.) (*n.m.*), frustum. (28)

tronc d'arbre (*m.*), tree-trunk. (29)

tronc de cône (Géom.) (*m.*), frustum of a cone ; frustum of cone. (30)

tronc de pyramide (Géom.) (*m.*), frustum of a pyramid ; frustum of pyramid. (31)

troncature (Géom. & Cristall.) (*n.f.*), truncation. (32)

tronçon (*n.m.*), piece ; section (a piece cut off or broken off from some object having more length than width or diameter) : (33)
tronçon de tube, piece of tube ; section of tubing. (34)

tronçon inférieur (d'un cours d'eau) (Géogr. phys.) (*m.*), lower track ; plain-track. (35)

tronçon moyen (d'un cours d'eau) (*m.*), middle track ; valley-track. (36)

tronçon supérieur (d'un cours d'eau) (*m.*), upper track ; torrent-track ; mountain-track. (37)

tronconique (*adj.*), in the shape of a truncated cone. (38)

tronçonnage *ou* **tronçonnement** (*n.m.*), cutting up ; cutting up into sections ; cutting off : (39)
tronçonnage d'un tube, cutting up a tube into sections ; cutting off pieces from a tube. (40)

tronçonner (*v.t.*), to cut up ; to cut up into sections ; to cut off : (41)
tronçonner un arbre, to cut up a tree. (42)

outil à tronçonner (tour) (*m.*), cutting-off tool (lathe). (43)

tronqué, -e (*adj.*), truncated : (44)
cône tronqué (Géom.) (*m.*), truncated cone. (45)
arête tronquée (Cristall.) (*f.*), truncated edge. (46)

tronquer (*v.t.*), to truncate : (47)
tronquer une pyramide, to truncate a pyramid. (48)

troostite (Minéral) (*n.f.*), troostite. (49)

troostite (Métall.) (*n.f.*), troostite. (50)

trop-plein [**trop-pleins** *pl.*] (quantité d'eau, etc., en excès) (*n.m.*), overflow ; surplus water ; waste water : (51)
tuyau par où s'échappe le trop-plein (*m.*), pipe through which the overflow escapes. (52)
le trop-plein d'un bassin, the overflow of a basin. (53)

trop-plein (orifice ; tuyau) (*n.m.*), overflow ; overflow-pipe ; waste-pipe : (54)
eau qui s'échappe par le trop-plein (*f.*), water which escapes by the overflow. (55)

trop-plein (déversoir) (*n.m.*), weir ; waste-weir. (56)

trop-plein (d'un injecteur) (*n.m.*), overflow (of an injector). (57)

tropical, -e, -aux (*adj.*), tropical : (58)
climat tropical (*m.*), tropical climate. (59)

tropique (*n.m*), tropic. (60)

tropique du Cancer (*m.*), tropic of Cancer. (61)

tropique du Capricorne (*m.*), tropic of Capricorn. (62)

trottoir (*n.m.*), path ; footpath ; footway ; sidewalk. (63)

trottoir roulant (*m.*), moving platform ; moving sidewalk ; travelling platform ; travelling sidewalk. (64)

trou (*n.m.*), hole. (65)

trou (dans le centre d'une meule en émeri, ou similaire) (*n.m.*), hole (in the centre of an emery-wheel, or the like). (66 ou 67)

trou à laitier *ou* **trou à crasse** (de four métallur-gique) (*m.*), slag-notch ; cinder-notch ; slag-eye ; slag-hole ; slag-tap ; slagging-hole ; breast-hole ; floss-hole ; floss ; monkey ; pee-pee. (68)

trou à main (*m.*), Same as **trou de bras**.

trou à tarauder (*m.*), tap-sized hole. (69)

trou au cône Morse (*m.*), Morse taper hole. (70)

trou borgne (*m.*), blind hole ; no-thoroughfare hole. (71)

trou d'aiguille (Photogr.) (*m.*), pinhole. (72)

trou d'air (d'un moule) (Fonderie) (*m.*), vent, vent-hole, sand-vent, (of a mould). (73)

trou d'évent (canal pour renouveler l'air) (*m.*), vent-hole ; vent ; air-vent ; air-hole. (74)

trou d'évent (dans un moule) (Fonderie) (*m.*), riser ; rising-head. (75)

trou d'exploration (*m.*), Same as **trou de prospection**.

trou d'homme (*m.*), manhole. (76)

trou de boulon (*m.*), bolt-hole. (77)

trou de bras (chaudière) (*m.*), hand-hole ; wash-out-hole ; mud-door. (78)

trou de centrage (*m.*), centre-hole. (79)

trou de cheville (de tenon) (*m.*), draw-bore. (80)

trou de clef (d'un robinet) (*m.*), plug-hole (of a cock *or* tap). (81)

trou de coulée (fourneau) (*m.*), tap-hole ; tapping-hole ; draw-hole ; mouth. (82)

trou de coulée (dans un moule) (Fonderie) (*m.*), runner; gate; git; geat; sprue; cast-gate; pouring-gate; running-gate; pouring-hole; down-runner; down-gate; upright runner; sprue-hole; funnel; jet. (1)

trou de fleuret (Mines) (*m.*), hand-drilled hole. (2)

trou de graissage (*m.*), oil-hole. (3)

trou de gueuse (fourneau) (*m.*). Same as **trou de coulée**.

trou de la scorie (*m.*). Same as **trou à laitier**.

trou de mine (*m.*), blast-hole; blasting-hole; shot-hole; mine-chamber; mining-hole; powder-mine. (4)

trou de passage (pour un boulon fileté) (*m.*), clearance-hole (for a threaded bolt). (5)

trou de prospection (*m.*), prospect-hole; pros-pecting-pit; trial-pit; test-pit; costean-pit. (6)

trou de sable (dans une pièce venue de fonte) (*m.*), sand-hole (in a casting). (7)

trou de sel (*m.*). Same as **trou de bras**.

trou de sonde (*m.*), bore-hole; drill-hole. (8)

trou de sonde dévié (*m.*), bore-hole out of true; crooked hole. (9)

trou de sonde en montage (*m.*), bore-hole up in the roof. (10)

trou de sonde pratiqué à sec (*m.*), dry hole. (11)

trou en cul-de-sac (*m.*), no-thoroughfare hole; blind hole. (12)

trou foré à la main (*m.*), hand-drilled hole. (13)

trou graisseur (*m.*), oil-hole. (14)

trou percé de part en part *ou* **trou traversant la pièce de part en part** *ou* **trou traversant la pièce d'outre en outre** (opp. à **trou en cul-de-sac** *ou* **trou borgne**) (*m.*), thoroughfare hole. (15)

trouble (*adj.*), cloudy; troubled: (16)

eau trouble (*f.*), cloudy water. (17)

trouble (*n.m.*), trouble: (18)

troubles ouvriers, labour troubles; labour disturbances. (19)

trouer (*v.t.*), to hole: (20)

les crêtes de moraines frontales sont souvent trouées par d'anciennes percées de torrents glaciaires (*f.pl.*), the crests of frontal moraines are often holed by ancient borings of glacial torrents. (21)

trousquin (*n.m.*). Same as **trusquin**.

troussage (Fonderie) (*n.m.*), strickling; striking; striking up; sweeping; sweeping up: (22)

troussage d'un noyau en terre, strickling, striking up, sweeping up, a loam core. (23)

troussage (Aérage de mines) (*n.m.*), airway. (24)

trousse (faisceau d'objets de même nature) (*n.f.*), set; kit; outfit; series. (25)

trousse (étui ou portefeuille divisé en comparti-ments) (*n.f.*), case; wallet-case; leather case; roll. (26)

trousse (Fonçage des puits de mines) (*n.f.*), drum-curb; curb; crib. (27)

trousse (Fonderie) (*n.f.*), strickle; strike; sweep; former; template; templet. (28)

trousse à réparations (*f.*), repair-outfit. (29)

trousse coupante (Mines) (*f.*), sinking-drum; sinking drive-pipe. (30)

trousse d'explorateur, consistant en un baromètre de nivellement, une boussole, et un thermo-mètre: le tout en écrin (*f.*), explorers' set, consisting of one surveying-barometer, one compass, and one thermometer: the whole in a case. (31)

trousse d'objectifs (Photogr.) (*f.*), series, set, outfit, case, casket-case, of supplementary lenses; supplementary lenses. (32)

trousse d'outils (*f.*), kit, outfit, of tools. (33)

trousse de clefs (*f.*), bunch of keys. (34)

trousse de clefs à douille (*f.*), wallet-case of box-spanners; set of box-spanners in leather case; kit of socket-wrenches in leather case; case of socket-spanners. (35)

trousse en fonte (Mines) (*f.*), cast-iron crib. (36)

trousse en toile, avec — outils (*f.*), canvas roll, with — tools. (37)

trousse picotée (Mines) (*f.*), wedging-curb; wedging-crib. (38)

trousseau (faisceau d'objets de même nature) (*n.m.*), kit; outfit: (39)

le trousseau professionnel d'un ouvrier mouleur, the trade kit of a moulder. (40)

trousseau (Fonderie) (*n.m.*), spindle and sweep; rig for loam-work. (41)

trousseau à potence murale (Fonderie) (*m.*), loam-moulders' horse. (42)

trousseau transportable (Fonderie) (*m.*), loam-moulders' gig. (43)

troussequin (*n.m.*), **troussequiner** (*v.t.*). Same as **trusquin**, **trusquiner**.

trousser (Fonderie) (*v.t.*), to strickle; to strike; to strike up; to sweep; to sweep up: (44)

trousser un noyau en terre, to strickle (*or* to sweep up) a loam core. (45)

trousser (se) (Fonderie) (*v.r.*), to be strickled; to be struck up; to be swept up. (46)

trouvaille (*n.f.*), find; discovery: (47)

trouvaille accidentelle, accidental discovery; adventitious find. (48)

trouvaille de diamants, find of diamonds. (49)

trouvaille encourageante, promising find. (50)

trouvé (-e) associé (-e) avec, found in association with: (51)

bismuth trouvé associé avec la molybdénite (*m.*), bismuth found in association with molybdenite. (52)

trouver (*v.t.*), to find: (53)

trouver une pépite, to find a nugget. (54)

trouver l'azimut du soleil, l'heure du passage supérieur de la Polaire, to find the azimuth of the sun, the time of upper culmination of Polaris. (55 *ou* 56)

trouver un écoulement facile, to find a ready market; to command a ready sale. (57)

trouver (se) (se rencontrer) (*v.r.*), to be found; to be met with; to be present; to occur: (58)

minéral qui se trouve dans la nature à l'état natif (*m.*), mineral which is found (*or* which occurs) in nature in a native state; mineral which is found native in nature. (59)

en règle générale, les fossiles les mieux conservés se trouvent dans les roches sédimentaires à grain fin, as a rule, the best preserved fossils are met with in fine-grained sedimentary rocks. (60)

trouver (se) en présence d'une difficulté, to be confronted (*or* to be faced) with a difficulty. (61)

trouver (se) pris (-e) entre un wagon et le toit, to get caught between a wagon and the roof. (62)

truck *ou* **truc** (*n.m.*), truck. (63)

truck à bogie (*m.*), bogie-truck; bogie. (64)

truck à deux essieux (*m.*), four-wheeled truck. (65)

truck à un seul essieu (bissel de locomotive) (*m.*), two-wheeled truck ; pony truck. (1)
truck de locomotive (*m.*), engine-truck. (2)
truck de mine (*m.*), mine-truck ; mining-truck ; tram. (3)
truck de wagon à bogie(*m.*), bogie-wagon truck. (4)
truck porteur (Mines) (*m.*), carriage for tub on inclines ; slope-carriage. (5)
truelle (*n.f.*), trowel. (6)
truelle à cœur (de mouleur) (Fonderie) (*f.*), heart trowel. (7)
truelle à feuille de laurier (de mouleur) (*f.*), leaf ; leaf-shaped trowel. (8)
truelle à mortier (*f.*), brick-trowel. (9)
truelle à plâtre (*f.*), plastering-trowel. (10)
truelle carrée (de mouleur) (*f.*), square trowel. (11)
trusquin (*n.m.*) *ou* **trusquin à pointe** (de menuisier, de charpentier), marking-gauge ; scribing-gauge ; carpenters' gauge ; joiners' gauge. (12)
trusquin (*n.m.*) *ou* **trusquin à marbre** *ou* **trusquin à marbre, ordinaire** *ou* **trusquin debout, pour marbres**, surface-gauge ; scribing-block. (13)
trusquin à combinaisons (*m.*), combination surface-gauge. (14)
trusquin à couper (de menuisier) (*m.*), cutting-gauge (joiners'). (15)
trusquin à double traçoir (de charpentier, de menuisier) (*m.*), mortise-gauge. (16)
trusquin à main *ou* **trusquin de côté** (de mécanicien) (*m.*), scratch-gauge. (17)
trusquin à main, carré (*m.*), scratch-gauge, with square beam. (18)
trusquin à main, rond (*m.*), scratch-gauge, with round beam. (19)
trusquin debout, pour marbres, à vis de rappel (*m.*), surface-gauge, with adjusting-screw. (20)
trusquin en V *ou* **trusquin d'alésage** (*m.*), surface-gauge with V-shaped groove in bottom. (21)
trusquin porte-planche (d'un appareil à trousser) (Fonderie) (*m.*), spindle-arm (of a rig for loam-work). (22)
trusquiner (*v.t.*), to mark (with the marking-gauge) ; to scribe (with the scribing-gauge or surface-gauge). (23)
tscheffkinite (Minéral) (*n.f.*), tscheffkinite ; tschewkinite. (24)
tschermigite (Minéral) (*n.f.*), tschermigite. (25)
tubage (action) (*n.m.*), casing ; tubing : (26)
tubage d'un puits à pétrole, casing, tubing, an oil-well. (27)
tubage d'un trou de sonde, tubing a borehole. (28)
tubage (tubes) (*n.m.*), tubing ; casing. (29)
tubage à joint à insertion (*m.*), inserted-joint casing. (30)
tubage d'extraction *ou* **tubage de puits** (*m.*), tubing ; well-tubing. (31) (the inside **tubing** as distinguished from the outside **casing**).
tubage d'isolement *ou* **tubage de puits** (*m.*), casing ; well-casing. (32) (the outside **casing** as distinguished from the inside **tubing**).
tubage de puits artésien (*m.*), Artesian-well casing. (33)
tubage télescopique (*m.*), telescopic casing. (34)
tube (*n.m.*), tube ; tubing. (35)

tube à ailettes extérieures (*m.*), gilled tube. (36)
tube à ailettes intérieures (*m.*), Serve tube. (37)
tube à boule (Verrerie de laboratoire) (*m.*), safety-funnel with bulb. (38)
tube à boulet (Extraction du pétrole) (*m.*), bailer. (39)
tube à condensation (*m.*), condenser-tube. (40)
tube à — courbures (*m.*), tube with — bends. (41)
tube à dégagement (*m.*), by-pass. (42)
tube à entonnoir (Verrerie de laboratoire) (*m.*), funnel-tube. (43)
tube à essai *ou* **tube à essais** (*m.*), test-tube. (44)
tube à fumée (*m.*), fire-tube. (45)
tube à noyaux (Fonderie) (*m.*), core-tube. (46)
tube à potasse (Verrerie de laboratoire) (*m.*), potash-bulbs. (47)
tube à robinet (Verrerie de laboratoire) (*m.*), tube with stop-cock. (48)
tube à sable (Sondage) (*m.*), sand-pump ; shell-pump ; sludge-pump. (49)
tube à tirage (d'un instrument d'optique) (*m.*), draw-slide, draw-tube, sliding-tube, (of an optical instrument). (50)
tube à vide (*m.*), vacuum-tube. (51)
tube abducteur (Verrerie de laboratoire) (*m.*), leading-tube. (52)
tube abducteur de gaz (*m.*), gas-leading tube. (53)
tube acoustique (*m.*), speaking-tube. (54)
tube bagué (tube de chaudière) (*m.*), ferruled tube (boiler-tube). (55)
tube broyeur (*n.m.*). Same as **tube finisseur**.
tube brûleur (d'une lampe à braser) (*m.*), flame-tube (of a brazing-lamp). (56)
tube capillaire (*m.*), capillary tube. (57)
tube carottier (Sondage) (*m.*), core-tube ; core-barrel. (58)
tube condenseur (*m.*), condenser-tube. (59)
tube conducteur (Sondage) (*m.*), guide-tube ; conductor-pipe ; drill-pipe. (60)
tube congélateur (*m.*), freezing-tube. (61)
tube coudé (*m.*), bent tube. (62)
tube-cuiller [tubes-cuillers *pl.*] (Sondage) (*n.m.*), shell-pump ; sand-pump ; sludger ; sludge-pump. (63)
tube-cuiller à clapet (*m.*), sand-pump, sludger, with clack-valve. (64)
tube d'affluence (Verrerie de laboratoire) (*m.*), filling-tube. (65)
tube d'eau (chaudière) (*m.*), water-tube (boiler). (66)
tube d'équilibre (Procédé Kind-Chaudron) (*m.*), equilibrium-tube. (67)
tube de chaudière (*m.*), boiler-tube. (68)
tube de couleur (*m.*), tube of colour (paint). (69)
tube de Crookes (*m.*), Crookes tube. (70)
tube de dégagement (*m.*), by-pass. (71)
tube de dégagement de la vapeur (*m.*), steam-pipe. (72)
tube de force (Phys.) (*m.*), tube of force. (73)
tube de fumée *ou* **tube de flammes** (*m.*), fire-tube : (74)
chaudière à tubes de fumée (*ou* à tubes de flammes) (*f.*), fire-tube boiler. (75)
tube de Geissler (*m.*), Geissler tube. (76)
tube de Liebig à potasse (*m.*), Liebig bulb. (77)
tube de Moore (*m.*), Moore tube. (78)
tube de niveau d'eau (*m.*), gauge-glass. (79)
tube de Pitot (Hydraul.) (*m.*), Pitot tube. (80)
tube de Plücker (*m.*), Plücker tube. (81)
tube de retour de fumée (*m.*), return-flue ; return-tube. (82)

tube de succion (turbine) (*m.*), draught-tube ; draft-box. (1)

tube de sûreté (Verrerie de laboratoire) (*m.*), safety-funnel ; safety-tube. (2)

tube de Torricelli (Phys.) (*m.*), Torricellian tube. (3)

tube de verre (*m.*), glass tube. (4)

tube de verre gradué (*m.*), graduated glass tube. (5)

tube de Welter (Chim.) (*m.*), Welter's safety-funnel. (6)

tube effilé (en verre) (*m.*), tube with drawn end. (7)

tube en acier (*m.*), steel tube. (8)

tube en acier étiré sans soudure (*m.*), solid-drawn steel tube. (9)

tube en acier sans couture (*m.*), seamless-steel tube. (10)

tube en acier sans soudure (*m.*), weldless-steel tube. (11)

tube en acier sans soudure, étiré à froid [tubes en acier sans soudure, étirés à froid *pl.*] (*m.*), cold-drawn weldless-steel tube. (12)

tube en cuivre rouge (*m.*), copper tube. (13)

tube en fer (*m.*), wrought-iron tube. (14)

tube en S, en T, en U, en V, en Y (*m.*), S tube, T tube, U tube, V tube, Y tube. (15 *ou* 16 *ou* 17 *ou* 18 *ou* 19)

tube en verre (*m.*), glass tube. (20)

tube en verre de débit visible (graisseur) (*m.*), sight-feed glass (lubricator). (21)

tube en verre soufflé (*m.*), blown-glass tube. (22)

tube en W avec deux boucles (Verrerie de labo-ratoire) (*m.*), W-form tube with two bulbs. (23)

tube étiré (*m.*), drawn tube. (24)

tube fermé (*m.*), closed tube : (25)

essai au tube fermé (Essais au chalumeau) (*m.*), closed-tube test (Blowpipe analysis). (26)

tube fermé par un bout *ou* tube fermé à une de ses extrémités (*m.*), tube closed at one end. (27)

tube finisseur (pour minerai) (*m.*), tube-mill. (28)

tube finisseur à galets *ou* tube finisseur à galets en silex (*m.*), pebble-mill ; flint-mill. (29)

tube finisseur pour le broyage par voie humide (*m.*), wet-grinding tube-mill. (30)

tube finisseur pour le broyage par voie sèche (*m.*), dry-grinding tube-mill. (31)

tube-foyer [tubes-foyers *pl.*] (*n.m.*), flue ; cylin-drical flue (as of Cornish or Lancashire boiler). (32)

tube frigorifique (*m.*), freezing-tube. (33)

tube Galloway (chaudière) (*m.*), Galloway tube. (34)

tube-guide [tubes-guides *pl.*] (Sondage) (*n.m.*), guide-tube ; conductor-pipe ; drill-pipe. (35)

tube indicateur de niveau (*m.*), gauge-glass. (36)

tube jaugeur (*m.*), gauge-tube. (37)

tube laveur (Verrerie de laboratoire) (*m.*), wash-ing-tube. (38)

tube lisse (*m.*), plain tube. (39)

tube ouvert (*m.*), open tube : (40)

essai au tube ouvert (Essais au chalumeau) (*m.*), open-tube test (Blowpipe analysis). (41)

tube perforateur (Sondage) (*m.*), drive-pipe ; drive-tube. (42)

tube piézométrique (Hydraul.) (*m.*), pressure-tube ; piezometer-tube ; piezometric tube. (43)

tube porte-objectif (d'une lunette) (*m.*), outer tube (of a telescope). (44)

tube porte-oculaire (d'une lunette) (*m.*), eye-tube. (45)

tube porte-réticule (d'une lunette) (*m.*), inner tube. (46)

tube pour distillation fractionnée (*m.*), fractional-distillation tube. (47)

tube pour dosage (Chim.) (*m.*), measuring-tube. (48)

tube pour essais (*m.*), test-tube. (49)

tube pour thermomètre (*m.*), tube, tubing, for thermometer. (50)

tube sans joint (*m.*), jointless tube. (51)

tube scellé (*m.*), sealed tube. (52)

tube Serve (*m.*), Serve tube. (53)

tube siphonal (*m.*), siphon-tube. (54)

tube soudé à rapprochement (*m.*), butt-welded tube. (55)

tube soudé à recouvrement (*m.*), lapwelded tube. (56)

tube surchauffeur (*m.*), superheater-pipe. (57)

tube-tirant [tubes-tirants *pl.*] (chaudière) (*n.m.*), stay-tube. (58)

tube unité (Phys.) (*m.*), unit tube. (59)

tuber (*v.t.*), to tube ; to case : (60)

tuber un puits, to tube, to case, a well. (61)

tuber un trou de sonde, to tube a bore-hole. (62)

tuberculose pulmonaire (Pathologie) (*f.*), pulmo-nary tuberculosis ; phthisis. (63)

tubiforme (*adj.*), tubiform. (64)

tubulaire (*adj.*), tubular : (65)

chaudière tubulaire (*f.*), tubular boiler. (66)

tubule (*n.m.*), small tube. (67)

tubulé, -e (*adj.*), tubulated : (68)

flacon tubulé (*m.*), tubulated bottle. (69)

tubulure (*n.f.*), tubulure ; tubulature ; pipe : (70)

flacon à deux tubulures (*m.*), bottle with two tubulures. (71)

tubulure (Géol.) (*n.f.*), pipe ; sand-pipe ; sand-gall. (72)

tubulure (d'un robinet) (*n.f.*), tail-pipe (of a cock). (73)

tubulure d'aspiration (pompe) (*f.*), suction-pipe ; tail-pipe. (74)

tubulure de prise de vapeur (*f.*), steamway ; steam-passage to chest. (75)

tubulure de refoulement (pompe) (*f.*), delivery-pipe ; head-pipe. (76)

tuer (*v.t.*), to kill : (77)

être tué (-e) dans un accident, to be killed in an accident. (78)

être tué (-e) par un éboulement, to be killed by a fall of ground. (79)

tuf (Géol.) (*n.m.*), tufa ; tuff. (80) See following entries for distinctions.

Note.—Many geologists prefer to apply the term tuff to the volcanic rock, restricting tufa to the calcareous deposit.

tuf basaltique (volcanique) (*m.*), basalt tuff ; basaltic tuff. (81)

tuf calcaire *ou* simplement tuf (*n.m.*), calcareous tufa ; calc-tufa ; calc-tuff ; tufa. (82)

tuf palagonitique (volcanique) (*m.*), palagonite tuff. (83)

tuf ponceux (*m.*), pumiceous tuff. (84)

tuf volcanique *ou* simplement tuf (*n.m.*), tuff ; volcanic tufa ; tufa. (85)

tufacé, -e (calcaire) (*adj.*), tufaceous : (86)

calcaire tufacé (*m.*), tufaceous limestone. (87)

tufacé, -e (volcanique) (*adj.*), tuffaceous. (88)

tufier, -ère (*adj.*), tufous. (89)

tuile (*n.f.*), tile. (1)

tuile à rebord (*f.*), flange-tile. (2)

tuile arêtière (*f.*), hip-tile. (3)

tuile cornière (*f.*), corner-tile. (4)

tuile creuse *ou* **tuile canal** (*f.*), arched tile; crown-tile. (5)

tuile de détente (de tiroir de détente) (*f.*), expan-sion-plate, expansion-slide, cut-off plate, (of expansion-valve *or* cut-off valve). (6)

tuile en dos d'âne (*f.*), saddle-tile. (7)

tuile en S (*f.*), pantile. (8)

tuile faîtière (*f.*), ridge-tile; crest-tile. (9)

tuile plate (*f.*), flat tile; plain tile; plane tile; crown-tile. (10)

tuileau de détente (*m.*). Same as **tuile de détente**.

tuilerie (*n.f.*), tilery. (11)

tuilier (pers.) (*n.m.*), tiler. (12)

tungstate (Chim.) (*n.m.*), tungstate; wolframate. (13)

tungstaté, -e (*adj.*), tungstenic. (14)

tungstène [Tu *ou* W] (Chim.) (*n.m.*), tungsten; wolfram. (15)

tungsténifère (*adj.*), tungsteniferous. (16)

tungsteux (Chim.) (*adj.m.*), tungstous. (17)

tungstique (Chim.) (*adj.*), tungstic; wolframic. (18)

tungstite [WO³] (Minéral) (*n.f.*), tungstite; tungstic ocher; wolfram ochre. (19)

tungstosilicate (Chim.) (*n.m.*), tungstosilicate. (20)

tungstosilicique (Chim.) (*adj.*), tungstosilicic. (21)

tunnel (*n.m.*), tunnel : (22)

le tunnel du Saint-Gothard, the St Gothard tunnel. (23)

tunnel d'assèchement (*m.*), drainage-tunnel. (24)

tunnel dans le bed-rock (*m.*), bed-rock tunnel. (25)

tunnel de chemin de fer (*m.*), railway-tunnel. (26)

tunnel de prospection *ou* **tunnel de recherches** (*m.*), prospecting-tunnel; prospect-tunnel; discovery-tunnel. (27)

turbine (*n.f.*), turbine. (28)

turbine à air *ou* **turbine atmosphérique** (*f.*), air-turbine. (29)

turbine à axe horizontal *ou* **turbine à arbre horizontal** (*f.*), horizontal-shaft turbine. (30)

turbine à axe vertical *ou* **turbine à arbre vertical** (*f.*), vertical-shaft turbine. (31)

turbine à eau *ou* **turbine hydraulique** (*f.*), water-turbine; hydraulic turbine. (32)

turbine à étages de pression (*f.*), pressure-stage turbine. (33)

turbine à étages de vitesse (*f.*), velocity-stage turbine. (34)

turbine à gaz (*f.*), gas-turbine. (35)

turbine à impulsion *ou* **turbine à action** *ou* **turbine à libre déviation** (*f.*), impulse turbine; action-turbine; turbine with free deviation. (36)

turbine à injection partielle (*f.*), partial-injection turbine. (37)

turbine à injection totale *ou* **turbine à pleine injection** (*f.*), full-injection turbine. (38)

turbine à réaction *ou* **turbine à pression** (*f.*), reaction-turbine; pressure-turbine. (39)

turbine à réaction nulle *ou* **turbine limite** *ou* **turbine à aubes garnies** *ou* **turbine à veine moulée** (*f.*), limit-turbine. (40)

turbine à vapeur (*f.*), steam-turbine. (41)

turbine axiale *ou* **turbine parallèle** *ou* **turbine hélico.de** (*f.*), axial-flow turbine; parallel-flow turbine; journal-turbine. (42)

turbine centrifuge (*f.*), outward-flow turbine. (43)

turbine centripète (*f.*), inward-flow turbine. (44)

turbine d'action (*f.*). Same as **turbine à action**.

turbine d'action-réaction (*f.*), reaction-and-impulse turbine. (45)

turbine mixte *ou* **turbine hélico-centripète** *ou* **turbine américaine** (*f.*), mixed-flow turbine; combined-flow turbine. (46)

turbine multiple *ou* **turbine compound** (*f.*), mul-tistage turbine. (47)

turbine noyée (*f.*), drowned turbine; sub-merged turbine. (48)

turbine radiale (*f.*), radial-flow turbine. (49)

turbine radiale centrifuge (*f.*), radial outward-flow turbine. (50)

turbine radiale centripète (*f.*), radial inward-flow turbine. (51)

turbine simple (*f.*), single-stage turbine. (52)

turbine tangentielle (*f.*), tangential-flow turbine. (53)

turbo-alternateur [turbo-alternateurs *pl.*] (*n.m.*), turbo-alternator. (54)

turbo-compresseur [turbo-compresseurs *pl.*] (*n.m.*), turbo-compressor. (55)

turbo-pompe [turbo-pompes *pl.*] (*n.f.*), turbo-pump; turbine-pump. (56)

turbo-ventilateur [turbo-ventilateurs *pl.*] (*n.m.*), turbo-ventilator; turbo-fan. (57)

turc, à tête inclinable (*m.*), screw packing-jack, with head made to swivel; jack-screw with self-adjusting head. (58)

turck (*n.m.*). Same as **turc**.

turgite *ou* **turjite** (Minéral) (*n.f.*), turgite. (59)

turquoise (Minéral) (*n.f.*), turquoise. (60)

turquoise engagée dans sa gangue (*f.*), turquoise-matrix. (61)

turquoise occidentale (*f.*), occidental turquoise; Occidental turquoise. (62)

turquoise odontolite *ou* **turquoise osseuse** (*f.*), odontolite turquoise; bone turquoise. (63)

turquoise orientale (*f.*), Oriental turquoise; true turquoise. (64)

tutol (Explosif) (*n.m.*), tutol. (65)

tuyau (*n.m.*), pipe. (66) (*plural* pipes, *or* col-*lective plural* piping; thus, tuyaux de plomb, lead pipes; lead piping). See also **conduit** and **conduite**.

tuyau (tuyau flexible en caoutchouc, en toile caoutchoutée, en cuir, etc.) (*n.m.*), hose. (67)

Note.—*The word* hose *remaining unaltered in the plural, must be rendered by* tuyaux *in plural senses, thus*: rubber hose for washing carriages, tuyaux en caoutchouc pour lavage de voitures; hose-coupling, raccord de tuyaux (*m.*).

tuyau (un des tuyaux d'une colonne d'exhaure) (Mines) (*n.m.*), pipe; stock; tree (one of the pipes in a rising-main). (68)

tuyau à bride (*m.*), flange-pipe. (69)

tuyau à emboitement (*m.*), socket-pipe. (70)

tuyau à incendie (*m.*), fire-hose. (71)

tuyau à joint lisse (*m.*), flush-joint pipe. (72)

tuyau à joints de caoutchouc (*m.*), pipe with india-rubber joints. (73)

tuyau à sable (locomotive) (*m.*), sand-pipe; sand-box pipe. (74)

tuyau acoustique (*m.*), speaking-tube. (75)

tuyau adducteur (Hydraul.) (*m.*), flow-pipe. (76)

tuyau alimentaire (*m.*). Same as tuyau d'alimen-
-tation.

tuyau avissé (*m.*), roll-joint pipe. (1)

tuyau bifurqué (*m.*), forked pipe. (2)

tuyau Bunsen (*m.*), Bunsen tube. (3)

tuyau coudé (*m.*), bent pipe. (4)

tuyau d'accouplement du frein (*m.*), air-brake
hose. (5)

tuyau d'admission (*m.*), induction-pipe. (6)

tuyau d'air (*m.*), air-pipe ; air-hose. (7)

tuyau d'alimentation (*m.*), feed-pipe ; feeding-
pipe. (8)

tuyau d'alimentation de chaudière (*m.*), boiler-
feed pipe. (9)

tuyau d'amenée (*m.*), supply-pipe ; head-pipe.
(10)

tuyau d'amenée d'eau (*m.*), water supply-pipe.
(11)

tuyau d'amenée d'eau fraiche (*m.*), fresh-water
supply-pipe. (12)

tuyau d'arrivée (*m.*), inlet-pipe. (13)

tuyau d'arrivée de l'air (*m.*), air-inlet pipe. (14)

tuyau d'arrosage (*m.*), garden-hose ; hose (for
spraying or watering purposes); squirt
hose. (15)

tuyau d'aspiration (*m.*), suction-pipe ; suction-
hose. (16)

tuyau d'aspiration, en caoutchouc (*m.*), rubber
suction-hose. (17)

tuyau d'eau (*m.*), water-pipe. (18)

tuyau d'échappement (*m.*), exhaust-pipe. (19)

tuyau d'échappement de la pompe à air (*m.*), air-
pump exhaust-pipe. (20)

tuyau d'écoulement du trop-plein (*m.*), over-
-flow-pipe ; waste-pipe. (21)

tuyau d'évacuation (*m.*), discharge-pipe. (22)

tuyau d'incendie (*m.*), fire-hose. (23)

tuyau d'injection (*m.*), injection-pipe. (24)

tuyau de caoutchouc entoilé (*m.*), rubber and
canvas hose. (25)

tuyau de cheminée (*m.*), chimney-flue. (26)

tuyau de conduite (*m.*), conduit-pipe ; conduit.
(27)

tuyau de débit *ou* tuyau de décharge (*m.*),
discharge-pipe. (28)

tuyau de descente *ou* tuyau de descente des
eaux pluviales (*m.*), downpipe ; rain-water
downpipe ; downcomer ; leader ; stack-
pipe. (29)

tuyau de drainage (*m.*), drain-pipe. (30)

tuyau de fer (*m.*), wrought-iron pipe ; iron pipe.
(31)

tuyau de fonte (*m.*), cast-iron pipe ; iron pipe.
(32)

tuyau de fonte à bride (*m.*), flanged cast-iron
pipe. (33)

tuyau de gaz (*m.*), gas-pipe. (34)

tuyau de graissage (*m.*), oil-pipe. (35)

tuyau de la sablière (locomotive) (*m.*), sand-
box pipe ; sand-pipe. (36)

tuyau de plomb (*m.*), lead pipe ; leaden pipe.
(37)

tuyau de prise d'eau (*m.*), water-pipe ; feed-
pipe. (38)

tuyau de prise de vapeur (*m.*), steam-pipe ;
steam supply-pipe. (39)

tuyau de refoulement (pompe) (*m.*), delivery-
pipe ; discharge-pipe. (40)

tuyau de refoulement, en caoutchouc (*m.*),
delivery-hose ; rubber delivery-hose. (41)

tuyau de trop-plein (*m.*), overflow-pipe ; waste-
pipe. (42)

tuyau de vapeur (*m.*), steam-pipe ; steam-hose.
(43)

tuyau de vidange des escarbilles (*m.*), cinder-
chute ; cinder-pocket. (44)

tuyau distributeur (*m.*), distributing-pipe. (45)

tuyau élévatoire (*m.*), lift-pipe. (46)

tuyau en caoutchouc (*m.*), india-rubber hose.
(47)

tuyau en caoutchouc à gaine métallique (*m.*),
wire-bound rubber hose. (48)

tuyau en cuivre cloué (*m.*), rivetted copper pipe.
(49)

tuyau en fer, en fonte, en plomb, etc. (*m.*). Same
as tuyau de fer, de fonte, de plomb, etc.

tuyau en terre (*m.*), earthenware pipe. (50)

tuyau en tôle (*m.*), sheet-iron pipe. (51)

tuyau flexible (*m.*), hose. (52)

tuyau flexible de la prise d'eau (*m.*), feed-pipe
hose. (53)

tuyau intérieur de prise de vapeur (chaudière)
(*m.*), dry-pipe. (54)

tuyau porte-vent (de tuyère) (*m.*), blowpipe ;
belly-pipe. (55)

tuyau principal de vapeur (*m.*), main steam-pipe.
(56)

tuyau protégé (*m.*), armoured hose. (57)

tuyau sableur (locomotive) (*m.*), sand pipe ;
sand-box pipe. (58)

tuyautage (*n.m.*), piping ; pipes ; tubing : (59)
tuyautage d'une machine à vapeur, tubing of a
steam-engine. (60)
tuyautage de fer, iron piping ; iron pipes.
(61)

tuyauterie (fabrique) (*n.f.*), pipe-works ; tube-
works. (62)

tuyauterie (ensemble de tuyaux) (*n.f.*), piping ;
pipes ; tubing ; pipes and fittings ; con-
-nections ; pipe-connections : (63)
tuyauterie d'échappement, exhaust-piping ;
exhaust-pipes. (64)
tuyauterie d'une machine à vapeur, tubing of
a steam-engine. (65)
tuyauterie de plomb, lead piping. (66)
pompe de lubrification et sa tuyauterie (*f.*),
oil-pump and connections ; lubricating-
pump and pipe-connections. (67)

tuyauteur (pers.) (*n.m.*), pipe-fitter. (68)

tuyère (de fourneau, de forge, etc.) (*n.f.*), tuyère ;
tweer ; twere ; twyer ; tewel ; tue-iron ;
blast-nozzle ; blast-orifice ; nozzle ; nozle.
(69 *ou* 70)

tuyère (d'un haut fourneau, d'un cubilot) (*n.f.*),
tuyère (of a blast-furnace, of a cupola).
(71 *ou* 72)

tuyère (d'une forge portative, d'un bâti de forge)
(*n.f.*), tue-iron (of a portable forge, of a
smiths' hearth). (73 *ou* 74)

tuyère (de soufflet, ou analogue) (*n.f.*), nose,
nose-piece, nozzle, nozle, snout, (of bellows,
or the like). (75 *ou* 76)

tuyère à paroi creuse (*f.*), water-tuyère, jacket
type. (77)

tuyère à serpentin (*f.*), water-tuyère, serpent-
coil type. (78)

tuyère à vapeur (d'un injecteur) (*f.*), steam-
nozzle, steam-cone, (of an injector). (79)

tuyère convergente (d'un injecteur) (*f.*), combin-
-ing-tube, combining-nozzle, combining-
cone, (of an injector). (80)

tuyère d'échappement (d'une locomotive) (*f.*),
exhaust-nozzle, blast-nozzle, (of a locomo-
-tive). (81)

tuyère divergente (d'un injecteur) (*f.*), diverging tube, diverging nozzle, diverging cone, delivery-tube, delivery-nozzle, delivery-cone, (of an injector). (1)

tympan (Arch.) (*n.m.*), spandrel. (2)

tympan (Hydraul.) (*n.m.*), tympanum; scoop-wheel; scoop water-wheel. (3)

tympe (haut fourneau) (*n.f.*), tymp (blast-furnace). (4)

type (*n.m.*), type: (5)

type de chaudière, type of boiler. (6)

typique (*adj.*) *ou* type (*n.m. employé adjectivement*), typical; standard: (7)

exemple typique (*m.*), typical example. (8)

outils types (*m.pl.*), typical tools. (9)

lampe type (*f.*), standard lamp. (10)

Note.—*Sometimes* type *is hyphened to the noun qualified, as* éprouvette-type [éprouvette-types *pl.*] (*n.f.*), standard test-piece.

U

udomètre (*n.m.*), **rain-gauge**; udometer. (11)

ulexite (Minéral) (*n.f.*), ulexite. (12)

ullmannite (Minéral) (*n.f.*), ullmannite. (13)

ultra-acide (*adj.*), ultra-acid; highly acid; strongly acid: (14)

roches ultra-acides (*f.pl.*), ultra-acid rocks. (15)

laitier ultra-acide (*m.*), highly acid slag; strongly acid slag. (16)

ultra-basique (*adj.*), ultrabasic; highly basic; strongly basic: (17)

roches ultra-basiques (*f.pl.*), ultrabasic rocks. (18)

laitier ultra-basique (*m.*), highly basic slag; strongly basic slag. (19)

ultra-rouge (*adj.*), ultra-red; infra-red: (20)

rayons ultra-rouges (*m.pl.*), ultra-red rays; infra-red rays. (21)

ultra-violet, -ette (*adj.*), ultra-violet: (22)

rayons ultra-violets (*m.pl.*), ultra-violet rays. (23)

uni, -e (*adj.*), even; level; smooth; plain: (24)

surface unie (*f.*), even surface; level surface; smooth surface; plain surface. (25)

uniaxe (*adj.*), uniaxial; uniaxal; monoaxal: (26)

cristal uniaxe (*m.*), uniaxial crystal; mono-axal crystal. (27)

unification (*n.f.*), standardization: (28)

unification des méthodes d'essai, des filetages, des écrous, des têtes de boulons, standard-ization of test methods, of screw-threads, of nuts, of bolt-heads. (29 *ou* 30 *ou* 31 *ou* 32)

unifier (*v.t.*), to standardize: (33)

unifier des cahiers des charges, to standardize specifications. (34)

See also **cahier des charges** and **spécification** for further examples.

uniforme (*adj.*), uniform; even: (35)

accélération uniforme (*f.*), uniform accelera-tion. (36)

uniformément (*adv.*), uniformly; evenly: (37)

mouvement rectiligne uniformément accéléré (*m.*), uniformly accelerated rectilinear motion. (38)

uniformité (*n.f.*), uniformity; evenness: (39)

uniformité de texture, uniformity of texture. (40)

unilatéral, -e, -aux (*adj.*), unilateral; one-sided. (41)

uniment (*adv.*), evenly; smoothly. (42)

union (*n.f.*), union; junction; joining. (43)

uniphasé, -e (Élec.) (*adj.*), single-phase; one-phase; uniphase; monophase: (44)

courant uniphasé (*m.*), single-phase current; one-phase current. (45)

unipolaire (Élec.) (*adj.*), unipolar; homopolar: (46)

dynamo unipolaire (*f.*), unipolar dynamo; homopolar dynamo. (47)

unique (*adj.*), single; sole; only: (48)

ligne à voie unique (*f.*), single-track line. (49)

unir (joindre) (*v.t.*), to unite; to join: (50)

canal qui unit deux lacs (*m.*), canal which joins two lakes. (51)

unir ensemble (*v.t.*), to join; to couple. (52)

unir (aplanir) (*v.t.*), to smooth; to level: (53)

unir une planche avec un rabot, to smooth a board with a plane. (54)

unir une allée, to level a path. (55)

unir (s') (s'associer) (*v.r.*), to unite; to join; to be joined: (56)

l'or a une grande tendance à s'unir au mercure (*m.*), gold has a great tendency to unite with mercury. (57)

uniréfringence (Opt.) (*n.f.*), monorefringence. (58)

uniréfringent, -e (*adj.*), monorefringent: (59)

cristal uniréfringent (*m.*), monorefringent crystal. (60)

unité (*n.f.*), unit. (61)

unité C.G.S. (*f.*), C.G.S. unit: (62)

dans le système centimètre-gramme-seconde (*ou* système C.G.S.), l'unité de longueur est le centimètre, l'unité de masse est la gramme, et l'unité de temps est la seconde, in the centimetre-gramme-second system (or C.G.S. system), the unit of length is the centimetre, the unit of mass is the gramme, and the unit of time is the second. (63)

unité calorimétrique (*f.*), calorimetric unit. (64)

unité d'accélération (*f.*), unit of acceleration. (65)

unité d'aire (*f.*), unit of area; unit of surface. (66)

unité d'angle (*f.*), unit of angle. (67)

unité d'éclairement (Photom.) (*f.*), unit of illumination. (68)

unité d'énergie (*f.*), unit of energy; unit of work. (69)

unité d'intensité de champ magnétique (*f.*), unit of intensity of magnetic field. (70)

unité d'intensité lumineuse (*f.*), unit of light; light-unit: (71)

la bougie décimale est l'unité française d'intensité lumineuse, the bougie décimale (or decimal candle) is the French light-unit. (72)

unité de capacité (Élec.) (*f.*), unit of capacity. (73)

unité de chaleur (*f.*), unit of heat ; heat-unit ; thermal unit. (1)

unité de charge (*f.*), unit of stress ; stress-unit : (2)

prendre le kilogramme pour unité de charge, to take the kilogramme as the unit of stress. (3)

unité de courant (Élec.) (*f.*), unit of current. (4)

unité de flux lumineux (*f.*), unit of light-flux ; unit of luminous flux. (5)

unité de flux magnétique *ou* **unité de flux de force** (*f.*), unit of magnetic flux. (6)

unité de force (*f.*), unit of force ; force-unit. (7)

unité de force électromotrice (*f.*), unit of electro- -motive force. (8)

unité de longueur (*f.*), unit of length. (9)

unité de masse (*f.*), unit of mass. (10)

unité de poids (*f.*), unit of weight ; weight-unit. (11)

unité de pression (*f.*), unit of pressure ; pressure- unit. (12)

unité de puissance (*f.*), **unit of power** ; power- unit : (13)

l'unité pratique de puissance électrique est le watt, the practical unit of electrical power is the watt. (14)

en Angleterre, l'unité vulgaire de puissance s'appelle horse-power, in England, the common power-unit is called horse-power. (15)

unité de quantité (*f.*), unit of quantity : (16)

le coulomb est l'unité pratique de mesure des quantités d'électricité, the coulomb is the practical unit of measurement of quantities of electricity. (17)

unité de résistance (Élec.) (*f.*), unit of resistance. (18)

unité de surface (*f.*), unit of area ; unit of surface. (19)

unité de temps (*f.*), unit of time. (20)

unité de travail (*f.*), unit of work ; unit of energy. (21)

unité de vitesse (*f.*), unit of velocity. (22)

unité de vitesse angulaire (*f.*), unit of angular velocity. (23)

unité de volume (*f.*), unit of volume. (24)

unité dérivée (opp. à unité fondamentale) (*f.*), derived unit. (25)

unité électrique (*f.*), electrical unit. (26)

unité électromagnétique (*f.*), electromagnetic unit. (27)

unité électrostatique (*f.*), electrostatic unit. (28)

unité fondamentale (*f.*), fundamental unit : (29)

dans la physique, les unités fondamentales sont au nombre de trois : la longueur, la masse, et le temps, in physics, the fundamental units are three in number : length, mass, and time. (30)

unité géométrique (*f.*), geometrical unit. (31)

unité magnétique (*f.*), magnetic unit. (32)

unité mécanique (*f.*), mechanical unit. (33)

unité optique (*f.*), optical unit. (34)

unité pratique (*f.*), practical unit. (35)

unité thermique (*f.*), thermal unit ; heat-unit ; unit of heat. (36)

univalence (Chim.)(*n.f.*),univalence ; univalency ; monovalence ; monovalency. (37)

univalent, -e (*adj.*), univalent ; monovalent. (38)

universel, -elle (*adj.*), universal : (39)

machine à percer radiale universelle (*f.*), universal radial drill. (40)

universellement (*adv.*), universally. **(41)**

uranate (Chim.) (*n.m.*), uranate. (42)

urane (*n.m.*), uranium oxide. (43)

uraneux (Chim.) (*adj.m.*), uranous. (44)

uranifère (*adj.*), uraniferous ; uranium-bearing. (45)

uranine *ou* **uraninite** (Minéral) (*n.f.*), uraninite ; pitchblende. (46)

uranique (Chim.) (*adj.*), uranic. (47)

uranite (Minéral) (*n.f.*), uranite ; lime uranite ; autunite. (48)

uranium (Chim.) (*n.m.*), uranium. (49)

uranyle (Chim.) (*n.m.*), uranyl. (50)

urao (Minéral) (*n.m.*), urao. (51)

urgemment (*adv.*), urgently. (52)

urgence (*n.f.*), urgency ; emergency : (53)

en cas d'urgence, in case of emergency. (54)

frein d'urgence (*m.*), emergency brake. (55)

urgent, -e (*adj.*), urgent. (56)

usage (*n.m.*), usage ; custom ; practice : (57)

usages du pays, custom of the country. (58)

usages miniers, mining practice. (59)

usage rend maître, practice makes perfect. (60)

usé, -e (*adj.*), worn ; worn out ; worn off ; worn away ; worn down ; abraded : (61)

lime usée (*f.*), worn file. (62)

usé (-e) par l'eau, water-worn : (63)

gravier usé par l'eau (*m.*), water-worn gravel. (64)

user (détériorer, diminuer par le frottement) (*v.t.*), to wear ; to wear out ; to wear off ; to wear away ; to wear down ; to abrade : (65)

le grès use le fer, sandstone wears away (*or* abrades) iron. (66)

user la pointe d'un outil, to wear down the point of a tool. (67)

user (s') (*v.r.*), to wear ; to wear out, away, off, *or* down : (68)

les foyers s'usent de diverses manières (*m.pl.*), fire-boxes wear in different ways. (69)

usinage (Méc.) (*n.m.*), machining ; tooling : (70)

usinage des pièces venues de fonte, machining, tooling, castings. (71)

usinage en série (*m.*), gang-machining ; multiple machining. (72)

usine (*n.f.*), works ; station. (73)

usine à gaz (*f.*), gas-works. (74)

usine à pétrole (*f.*), petroleum-works ; oil- refinery. (75)

usine à plomb (*f.*), lead-works. (76)

usine centrale (Élec.) (*f.*), central power-station. (77)

usine d'affinage (métaux) (*f.*), refining-works ; refinery. (78)

usine d'affinage de métaux (*f.*), metal-refinery. (79)

usine de chloruration (*f.*), chlorination-works. (80)

usine de cyanuration (*f.*), cyanide-works. (81)

usine de force hydraulique (*f.*), hydraulic-power station. (82)

usine de force motrice (*f.*), power-station ; power-generating station. (83)

usine de traction (*f.*), traction-works. (84)

usine génératrice *ou* **usine génératrice de courant électrique** (*f.*), generating-station ; power- station ; electricity-works. (85)

usine hydraulique (*f.*), water-works. (86)

usine métallurgique (usine où l'on fond ou raffine les métaux) (*f.*), smelting-works ; smeltery ; smelter ; smelting and refining works. (1)

usine métallurgique d'affinage (*f.*), metal-refinery. (2)

usine sidérurgique (*f.*), ironworks. (3)

usiner (Méc.) (*v.t.*), to machine ; to tool : (4)
usiner partout une pièce fondue *ou* usiner entièrement une pièce venue de fonte, to machine, to tool, a casting all over. (5)

ustensile (*n.m.*), utensil : (6)
ustensiles de ménage, household utensils. (7)

usuel, -elle (*adj.*), usual : (8)
termes usuels (*m.pl.*), usual quarter-days. (9)

usuellement (*adv.*), usually. (10)

usure (*n.f.*), wear ; wearing : (11)

usure des rails, wear on the rails ; wearing of the rails. (12)

usure normale *ou simplement* **usure**, wear and tear. (13)

utile (*adj.*), useful : (14)
travail utile (*m.*), useful work. (15)

utilement (*adv.*), usefully. (16)

utilisable (*adj.*), serviceable ; utilizable. (17)

utilisation (*n.f.*), utilization ; utilizing : (18)
utilisation des gaz sortant des puits, utilization of gases escaping from the wells. (19)

utiliser (*v.t.*), to utilize : (20)
utiliser la force motrice d'une chute d'eau, to utilize the motive power of a waterfall. (21)

utilité (*n.f.*), utility ; usefulness. (22)

uwarowite *ou* **uvarovite** (Minéral) (*n.f.*), uvarovite ; ouvarovite. (23)

V

V de mécanicien *ou simplement* **V** (*n.m.*), engineers' V block ; V block ; vee block : (24)
V à — entaille(s), V block with — vee(s). (25)

va-et-vient (*adj. invar.*), to-and-fro ; reciprocating ; alternating ; oscillating ; swinging ; swing : (26)
mouvement va-et-vient (*m.*), to-and-fro motion ; reciprocating motion ; alternating motion. (27)
pivot va-et-vient (*m.*), oscillating pivot. (28)
porte va-et-vient (*f.*), swing-door ; swinging door. (29)

va-et-vient [va-et-vient *pl.*] (mouvement alternatif) (*n.m.*), to-and-fro ; reciprocating motion ; alternating motion ; oscillation ; swing ; swinging : (30)
va-et-vient d'un piston, reciprocating motion of a piston. (31)
va-et-vient d'un pendule, oscillations of a pendulum. (32)

va-et-vient (organe de machine qui est doué d'un mouvement alternatif rectiligne) (*n.m.*), reciprocating-device. (33)

va-et-vient (bac servant à passer un cours d'eau) (*n.m.*), trail-bridge ; ferry-boat. (34)

va-et-vient (transporteur aérien) (*n.m.*), reversible cable-tramway ; jig-back. (35)

vache (soufflet de forge) (*n.f.*), smiths' bellows ; blacksmiths' bellows ; forge-bellows. (36)

vacuolaire (Géol.) (*adj.*), vesicular : (37)
structure vacuolaire (*f.*), vesicular structure. (38)

vacuole (Géol.) (*n.f.*), vesicle. (39)

vacuum [vacuums *pl.*] (Phys.) (*n.m.*), vacuum. (40)

vadeux, -euse (Géol.) (*adj.*), vadose. (41)

vagabonder dans la campagne, to ramble in the country. (42)

vagon, vagon-citerne, vagonnet, etc. Same as **wagon, wagon-citerne, wagonnet, etc.**

vague (*n.f.*), wave : (43)
vagues qui se brisent contre les rochers, waves which break against the rocks. (44)

vague à basse fréquence (Élec.) (*f.*), low-frequency wave. (45)

vague de chaleur (Météor.) (*f.*), heat-wave. (46)

vague de fond (*f.*), tidal wave. (47)

vague de froid (Météor.) (*f.*), cold-wave. (48)

vaincre (*v.t.*), to overcome ; to surmount ; to master ; to get over : (49)
vaincre des difficultés, to overcome, to surmount, to get over, difficulties. (50)
vaincre la résistance d'une machine au moment de sa mise en marche, to overcome the resistance of a machine at the moment of starting. (51)

vaisseau (récipient) (*n.m.*), vessel ; recipient. (52)

vaisseau (navire) (*n.m.*), vessel ; ship. (53)

val [vaux, *quelquefois* vals, *pl.*] (Géol.) (*n.m.*), synclinal valley. (54)

valable (*adj.*), valid ; good : (55)
permis de recherches valable pour — années (*m.*), prospecting-licence valid (*or* good) for — years. (56)

valence (Chim.) (*n.f.*), valence. (57)

valentinite (Minéral) (*n.f.*), valentinite ; antimony-bloom. (58)

valet (de la platine d'un microscope) (*n.m.*), clip (of the stage of a microscope). (59)

valet d'arrêt (d'une plaque tournante) (*m.*), latch (of a turntable). (60)

valet d'établi (de menuisier) (*m.*), bench-holdfast (joiners'). (61)

valet d'établi, avec vis de pression (*m.*), bench-holdfast, with screw and handle. (62)

valet de laboratoire, en bois, en paille, etc. (*m.*), stand for laboratory use, consisting of a block of wood, a straw ring, etc. (63 *ou* 64)

valeur (*n.f.*), value. (65)

valeur apparente de la résistance d'un conducteur (Élec.) (*f.*), apparent value of the resistance of a conductor. (66)

valeur d'usage (*f.*), value as a going concern. (67)

valeur en puissance (*ou* valeur virtuelle) **d'un champ pétrolifère** (*f.*), potential value of an oil-field. (68)

valeur industrielle d'une roche (*f.*), commercial value of a rock. (69)

valeur intrinsèque (*f.*), intrinsic value. (70)

valeur marchande (*f.*), market value ; sale value. (71)

valeur ohmique de la résistance d'un conducteur (Élec.) (*f.*), ohmic value of the resistance of a conductor. (72)

validité (*n.f.*), validity : (1)
validité d'un contrat, validity of a contract.
(2)

vallée (*n.f.*), valley. (3)

vallée (Mines) (*n.f.*), dip-head. (4)

vallée anticlinale (Géol.) (*f.*), anticlinal valley.
(5)

vallée en direction (*f.*), strike valley. (6)

vallée épigénétique *ou* **vallée surimposée** (*f.*),
epigenetic valley ; superimposed valley. (7)

vallée longitudinale (*f.*), longitudinal valley.
(8)

vallée monoclinale (*f.*), monoclinal valley. (9)

vallée sèche *ou* **vallée morte** (*f.*), dry valley.
(10)

vallée suspendue (*f.*), hanging valley. (11)

vallée synclinale (*f.*), synclinal valley. (12)

vallée transversale (*f.*), transverse valley ;
cross-valley. (13)

valleuse (Géogr. phys.) (*n.f.*), hanging valley.
(14)

vallon (*n.m.*), small valley; glen ; dale. (15)

valoir (*v.i.*), to be worth : (16)
valoir bien la peine de plus ample prospection,
to be well worth further prospecting.
(17)

valve (*n.f.*), valve. (18) See also **soupape** and
clapet.

valve à gaz (appareil à air chaud) (haut fourneau)
(*f.*), gas-valve (hot-blast stove) (blast-
furnace). (19)

valve à vent chaud (appareil à air chaud) (haut
fourneau) (*f.*), hot-blast valve. (20)

valve à vent froid (appareil à air chaud) (haut
fourneau) (*f.*), cold-blast valve. (21)

valve d'interception (locomotive compound) (*f.*),
intercepting-valve. (22)

valve de démarrage (locomotive compound) (*f.*),
starting-valve. (23)

valve de la cheminée (appareil à air chaud)
(haut fourneau) (*f.*), chimney-valve. (24)

valve de pneumatique (*f.*), pneumatic-tyre valve.
(25)

valve de réduction (*f.*), reducing-valve. (26)

valve de régulateur (*f.*), governor-valve. (27)

valve de renversement (régénérateur) (*f.*),
reversing-valve. (28)

valve oscillante (*f.*), rocking slide-valve ; rocking
valve ; swinging valve. (29)

valve-tiroir [valves-tiroirs *pl.*] (*n.f.*), slide-valve ;
sliding valve ; slide ; valve. (30)

van (pelle à vanner) (*n.m.*), van ; vanning-
shovel. (31)

vanadate (Chim.) (*n.m.*), vanadate ; vanadiate.
(32)

vanadeux, -euse (Chim.) (*adj.*), vanadious ;
vanadous. (33)

vanadifère (*adj.*), vanadiferous. (34)

vanadine (Minéral) (*n.f.*), vanadic ocher. (35)

vanadinite (Minéral) (*n.f.*), vanadinite. (36)

vanadiolite (Minéral) (*n.f.*), vanadiolite. (37)

vanadique (Chim.) (*adj.*), vanadic. (38)

vanadite (Chim.) (*n.m.*), vanadite. (39)

vanadium (Chim.) (*n.m.*), vanadium. (40)

vannage (Préparation mécanique des minerais)
(*n.m.*), vanning. (41)

vannage (d'une turbine, etc.) (*n.m.*), gating (of a
turbine, etc.). (42)

vannage à sec (*m.*), dry vanning ; dry blowing.
(43)

vannage à sec des alluvions (*m.*), alluvium dry
vanning. (44)

vanne (Constr. hydraul.) (*n.f.*), gate; sluice-
gate; water-gate; draw-gate; shut-off;
shut; shutter; shuttle. (45)

vanne (d'une turbine) (*n.f.*), gate (of a turbine).
(46)

vanne (d'un ventilateur) (*n.f.*), shutter (of a
fan). (47)

vanne à coulisse (Hydraul.) (*f.*), sash-gate. (48)

vanne de caisson à minerai (*f.*), ore-bin gate.
(49)

vanne de chasse (Hydraul.) (*f.*), flush-gate. (50)

vanne de décharge *ou* **vanne de passe** (*f.*),
flood-gate : (51)
aqueducs fermés à l'amont par des vannes de
décharge (*ou* de passe) qui pourraient être
ouvertes en cas de crue anormale (*m.pl.*),
aqueducts closed upstream by flood-gates
which can be opened in case of abnormal
flood. (52)

vanne de réglage du vent (fourneau) (*f.*), blast-
gate. (53)

vanne lançoire *ou* **vanne motrice** *ou* **vanne de
travail** (Hydraul.) (*f.*), head-gate ;
regulator-gate. (54)

vanne plongeante (*f.*), falling-sluice ; flood-gate :
(55)
vanne plongeante disposée de façon à s'abaisser
pour laisser passer l'eau par-dessus, falling-
sluice (*or* flood-gate) arranged to lie down
so as to allow the water to pass over. (56)

vannelle (petite vanne)(Hydraul.) (*n.f.*), paddle :
(57)
la vannelle d'une porte d'écluse, the paddle of
a lock-gate. (58)

vannelle (valve) (Hydraul.) (*n.f.*), sluice-valve.
(59)

vanner (Préparation mécanique des minerais)
(*v.t.*), to van. (60)

vanner (garnir de vannes) (*v.t.*), to gate : (61)
vanner une turbine, to gate a turbine. (62)

vanneur *ou* **vanner** (Préparation mécanique des
minerais) (*n.m.*), vanner ; vanning-machine.
(63)

vanneur (pers.) (*n.m.*), vanner. (64)

vanneur à sec (*m.*), dry vanner ; dry blower.
(65)

vanneur Frue *ou* **Frue vanner** (*m.*), Frue vanner.
(66)

vannoir *ou* **vanoir** (*n.m.*). Same as vanneur.

vantail [vantaux *pl.*] (d'une porte) (*n.m.*), leaf
(of a door) : (67)
les deux vantaux d'une porte d'écluse, the two
leaves of a lock-gate. (68)

vantelle (*n.f.*). Same as vannelle.

vapeur (exhalaison quelconque de forme gazeuse)
(*n.f.*), vapour ; vapor ; fume : (69)
vapeur de mercure *ou* vapeur mercurielle,
mercury vapour ; mercurial vapour. (70)
vapeur de pétrole, petroleum vapour. (71)
vapeurs de soufre, fumes of sulphur. (72)
vapeurs sulfureuses, sulphurous fumes. (73)

vapeur (exhalaison gazeuse) (Phys.) (*n.f.*),
vapour ; vapor. (74)

vapeur (vapeur d'eau employée comme force
motrice) (*n.f.*), steam : (75)
machine à vapeur (*f.*), steam-engine. (76)

vapeur (Météor.) (*n.f.*), mist : (77)
les vapeurs du matin, the morning mists.
(78)

vapeur (bateau mû par la vapeur) (*n.m.*), steam-
-boat ; steamship ; steamer : (79)
un vapeur fluvial, a river-steamer. (80)

vapeur d'eau (Phys.) (*f.*), vapour of water ; water vapor ; aqueous vapour : (1)
la force élastique de la vapeur d'eau, the tension of vapour of water. (2)
la vapeur d'eau répandue dans l'air est absorbée par un grand nombre de corps solides, the aqueous vapour diffused in the air is absorbed by a great number of solid bodies. (3)

vapeur d'eau (Méc.) (*f.*), steam : (4)
l'inventeur qui, le premier, eut l'idée d'utiliser la pression de la vapeur d'eau comme moteur industriel (*m.*), the inventor who first had the idea of utilizing the pressure of steam as an industrial motor. (5)

vapeur d'eau surchauffée (*f.*). Same as **vapeur surchauffée.**

vapeur d'échappement (*f.*), exhaust-steam ; dead steam ; spent steam. (6)

vapeur fraîche (*f.*), live steam. (7)

vapeur laminée (*f.*), wiredrawn steam. (8)

vapeur mouillée *ou* **vapeur humide** *ou* **vapeur aqueuse** *ou* **vapeur globuleuse** *ou* **vapeur vésiculaire** (*f.*), wet steam. (9)

vapeur non-saturante (Phys.) (*f.*), unsaturated vapour ; superheated vapor. (10)

vapeur saturante (Phys.) (*f.*), saturated vapour. (11)

vapeur sèche *ou* **vapeur saturée** (*f.*), dry steam ; anhydrous steam ; saturated steam. (12)

vapeur surchauffée *ou* **vapeur désaturée** *ou* **vapeur non-saturée** (*f.*), superheated steam ; sur-charged steam. (13)

vapeur vive (*f.*), live steam. (14)

vapo-hydraulique (*adj.*), steam-hydraulic : (15)
presse vapo-hydraulique (*f.*), steam-hydraulic press. (16)

vaporeux, -euse (*adj.*), vaporous ; misty ; hazy : (17)
émanations vaporeuses (*f.pl.*), vaporous emanations. (18)
atmosphère vaporeuse (*f.*), misty atmosphere. (19)
ciel vaporeux (*m.*), hazy sky. (20)

vaporisable (*adj.*), vaporizable. (21)

vaporiseur (*n.m.*), vaporizer ; atomizer ; pulverizer ; spray ; sprayer. (22)

vaporisation (*n.f.*), vaporization ; spraying. (23)

vaporiser (faire passer à l'état de vapeur) (*v.t.*), to vaporize : (24)
la chaleur vaporise l'eau, heat vaporizes water. (25)
vaporiser le mercure, to vaporize mercury. (26)

vaporiser (disperser en gouttelettes fines ; pul-vériser) (*v.t.*), to spray. (27)

vaporiser (se) (*v.r.*), to vaporize. (28)

variabilité (*n.f.*), variability ; variableness ; changeability ; changeableness. (29)

variable (*adj.*), variable ; changeable : (30)
transformateur à rapport variable (*m.*), vari-able-ratio transformer. (31)
temps variable (*m.*), changeable weather. (32)

variable (Barométrie) (*n.m.*), change : (33)
le baromètre est au variable, the barometer is at change. (34)

variation (*n.f.*), variation ; change ; fluctuation : (35)
variation d'allure, de puissance, d'un filon, vari-ation in character, in thickness, of a lode. (36 *ou* 37)
variation du temps, change in the weather. (38)

variations accidentelles (Barométrie), acciden-tal variations. (39)

variations brusques de la température, sudden changes, sudden fluctuations, sudden vari-ations, of temperature. (40)

variations en largeur, variations in breadth ; changes in width. (41)

variations horaires (Barométrie), daily vari-ations. (42)

variations magnétiques, magnetic variations. (43)

varié, -e (*adj.*), varied ; various ; variable ; different : (44)
vitesse variée (Méc.) (*f.*), variable velocity. (45)
couleurs variées (*f.pl.*), different colours. (46)

varier (*v.t.*), to vary ; to change. (47)

varier (*v.i.*), to vary ; to change ; to differ ; to range : (48)
varier de — à — centimètres en largeur, to vary from — to — centimetres in width ; to range from — to — centimetres in breadth. (49)

variété (*n.f.*), variety : (50)
variété à grain fin de granit, fine-grained variety of granite. (51)

variolite (Pétrol.) (*n.f.*), variolite. (52)

variolitique (*adj.*), variolitic. (53)

variscite (Minéral) (*n.f.*), variscite. (54)

varlet (d'une pompe de mine) (*n.m.*), quadrant ; angle-bob ; V bob. (55)

varlopage (Charp.) (*n.m.*), trying ; trying up. (56)

varlope (*n.f.*), trying-plane ; try-plane ; joint-ing-plane ; jointer. (57)

varloper (*v.t.*), to try ; to try up : (58)
varloper une planche, to try, to try up, a board. (59)

varlopeuse (*n.f.*), trying-up machine. (60)

vase (récipient) (*n.m.*), vessel : (61)
vases communicants (Phys.), communicating vessels ; vessels in communication. (62)

vase (boue) (*n.f.*), mud ; silt ; slime ; ooze : (63)
vase inconsistante, mud lacking consistency ; inconsistent silt. (64)

vase à filtrations chaudes en verre d'Iéna, avec bec (*m.*), beaker, Jena glass, with spout (*or* lipped). (65)

vase à filtrations chaudes en verre de Bohême, sans bec (*m.*), beaker, Bohemian glass, with-out spout (*or* plain). (66)

vase clos (*m.*), closed vessel. (67)

vase clos (Distillation) (*m.*), retort. (68)

vase de Mariotte (Phys.) (*m.*), Mariotte's bottle ; Marriotte's flask. (69)

vaseline (*n.f.*), vaseline. (70)

vaseux, -euse (*adj.*), muddy ; silty ; slimy ; oozy : (71)
fond vaseux (*m.*), muddy bottom. (72)

vastringue (*n.f.*), spokeshave. (73)

vauquelinite (Minéral) (*n.f.*), vauquelinite. (74)

vecteur (Math.) (*n.m.*), vector. (75)

vectoriel, -elle (*adj.*), vectorial. (76)

végétation (*n.f.*), vegetation ; growth : (77)
végétation luxuriante, luxuriant vegetation. (78)
végétation tropicale, tropical vegetation ; tropical growth. (79)

véhicule (moyen de transport) (*n.m.*), vehicle ; car : (80)
véhicule de mine, mine-car. (81)

véhicule (ce qui sert à transmettre) (*n.m.*), vehicle : (82)

l'air est le véhicule du son (*m.*), air is the vehicle of sound. (1)

veiller à l'observation des règlements sur les mines, to see that the mining regulations are carried out. (2)

veiller à la sécurité des ouvriers, to look after the safety of the men. (3)

veilleur de nuit (*m.*), night-watchman. (4)

veilleuse (pour gaz) (*n.f.*), by-pass : (5)
bec à incandescence par le gaz avec veilleuse (*m.*), incandescent-gas burner with by-pass. (6)

veine (partie longue et étroite dans le bois et les pierres dures) (*n.f.*), vein : (7)
marbre qui a de très belles veines (*m.*), marble which has very beautiful veins. (8)

veine (partie longue et étroite qui, dans une roche, diffère des parties voisines) (Géol.) (*n.f.*), vein : (9)
une veine de marbre, a vein of marble. (10)

veine (*n.f.*) *ou* **veine fluide** *ou* **veine liquide** (Phys.), jet ; water-vein : (11)
contraction de la veine *ou* contraction de la veine liquide (*f.*), contraction of the jet ; contraction of the water-vein. (12)

veine (filon) (Géol. & Mines) (*n.f.*), vein ; lode : (13) See also **filon**.

veine centrale, main lode. (14)
veine d'intrusion, intrusive vein. (15)
veine de contact, contact-vein. (16)
veine de quartz, quartz-vein. (17)
veine exploitable, workable vein. (18)
veine métallifère, metalliferous vein. (19)
veine rubanée, banded vein. (20)
veine stratifiée, bedded vein. (21)
veines réticulées, linked veins. (22)

veine (du placer) (*n.f.*), pay-streak ; pay-lead ; pay-channel ; run of gold ; gutter. (23)

veiné, -e (*adj.*), veined : (24)
marbre veiné (*m.*), veined marble. (25)

veineux, -euse (*adj.*), veiny. (26)

veinule (*n.f.*), veinlet ; small vein ; venule ; veinule ; stringer. (27)

vêlage (Glaciologie) (*n.m.*), calving. (28)

vêler (Glaciologie) (*v.i.*), to calve. (29)

vélocipède (pour visiter la voie) (Ch. de f.) (*n.m.*), velocipede ; velocipede-car. (30)

venasquite (Minéral) (*n.f.*), venasquite. (31)

vendable (*adj.*), saleable; marketable; merchant-able. (32)

vendeur, -euse (pers.) (*n.*), seller. (33)

vendeur, -eresse (pers.) (*n.*), vendor. (34)

vendre (*v.t.*), to sell : (35)
vendre une maison à l'enchère (*ou* aux enchères), to sell a house by (*or* at) auction. (36)

vendre (à), for sale. (37)

vendre (se) (*v.r.*), to sell. (38)

venir de fonte, to be cast : (39)
le palier est une pièce qui vient de fonte avec le bâti ou qui est fixée au moyen de boulons, the bearing-block is a part which is cast with the frame or which is secured by means of bolts. (40)
See also examples under **venu (-e) de fonte.**

venir en aide à quelqu'un, to come to someone's assistance. (41)

venir en contact avec, to come in contact with ; to impinge *followed by* upon, on, *or* against : (42)
en sortant du générateur, la vapeur se détend avant de venir en contact avec les aubes de la roue-turbine, on leaving the boiler, the steam expands before impinging upon the vanes of the turbine-wheel. (43)

vent (Météor.) (*n.m.*), wind : (44)
entendre souffler le vent, to hear the wind blow. (45)

vent (air agité par un moyen quelconque) (*n.m.*), wind ; blast ; air-blast ; draught : (46)
le vent d'un soufflet (*m.*), the wind of a bellows. (47)
four à vent chaud (*m.*), hot-blast furnace. (48)
valve à vent froid (*f.*), cold-blast valve. (49)
fournir le vent nécessaire au soufflage, to provide the necessary blast for the blow. (50)
faire du vent avec un ventilateur, to make a draught (*or* an air-blast) with a fan. (51)

vent (air en général) (*n.m.*), air ; wind : (52)
ballon plein de vent (*m.*), balloon full of air. (53)

vent à rafales (*m.*), gusty wind. (54)

vent d'est *ou* **vent est** (*m.*), east wind. (55)

vent d'ouest *ou* **vent ouest** (*m.*), west wind. (56)

vent du midi *ou* **vent du sud** *ou* **vent sud** (*m.*), south wind. (57)

vent du nord *ou* **vent nord** (*m.*), north wind. (58)

vent glacial (*m.*), icy wind. (59)

vent régnant (*m.*), prevailing wind. (60)

vente (*n.f.*), sale ; selling : (61)
vente de l'or à l'État, sale of gold to the Government. (62)

ventilateur (*n.m.*), fan ; blower ; fan-blower ventilator ; ventilating-fan. (63)

ventilateur (Constr.) (*n.m.*), stench-pipe. (64)

ventilateur à ailettes (*m.*), wing-fan. (65)

ventilateur à commande directe (*m.*), direct-driven fan. (66)

ventilateur à commande électrique (*m.*), electrically driven fan ; motor-driven fan. (67)

ventilateur à commande par câble (*m.*), rope-driven fan. (68)

ventilateur à commande par courroie (*m.*), belt-driven fan. (69)

ventilateur à deux ouïes (*m.*), double-inlet fan. (70)

ventilateur à hélice (*m.*), propeller-fan. (71)

ventilateur à la surface (Mines) (*m.*), surface ventilating-fan ; aboveground ventilating-fan. (72)

ventilateur à une seule ouïe (*m.*), single-inlet fan. (73)

ventilateur aspirant *ou* **ventilateur négatif** (*m.*), exhaust-fan ; exhauster ; suction-fan ; induced-draught fan ; vacuum-fan ; vacuum-ventilator ; negative blower. (74)

ventilateur aspirateur mural [ventilateurs aspirateurs muraux *pl.*] (*m.*), port-hole fan. (75)

ventilateur de fonderie (*m.*), foundry-blower. (76)

ventilateur de mine (*m.*), mine-fan ; mine-ventilating fan. (77)

ventilateur de secours (*m.*), emergency fan. (78)

ventilateur de table (*m.*), table-fan. (79)

ventilateur déprimogène *ou* **ventilateur dynamique** *ou* **ventilateur centrifuge** (opp. à **ventilateur volumogène** *ou* **ventilateur statique** *ou* **ventilateur déplaceur**) (*m.*), dynamical ventilator ; centrifugal fan ; cased fan. (80)

ventilateur foulant (*m.*). See **ventilateur soufflant**.

ventilateur mécanique (*m.*), power-fan. (81)

ventilateur négatif (*m.*). See **ventilateur aspirant**.

ventilateur plafonnier (*m.*), ceiling-fan. (82)

ventilateur portatif (*m.*), portable fan. (83)

ventilateur positif (*m.*). See **ventilateur soufflant**.

ventilateur réversible (*m.*), reversible fan. (84)

ventilateur rotatif (*m.*), rotary fan. (1)

ventilateur silencieux (*m.*), silent fan ; noiseless fan. (2)

ventilateur soufflant *ou* **ventilateur foulant** *ou* **ventilateur positif** (*m.*), force-fan ; pressure-fan ; blower ; positive blower ; compress-ing-fan ; plenum fan ; plenum ventilator. (3)

ventilateur soufflant pour fonderie (*m.*), foundry-blower. (4)

ventilateur souterrain (*m.*), underground ventilat-ing-fan. (5)

ventilateur volumogène *ou* **ventilateur statique** *ou* **ventilateur déplaceur** (*m.*), statical ventilator ; propulsion-fan ; displacement-fan. (6)

ventilation (*n.f.*), ventilation ; ventilating ; airing : (7) See also **aérage**.

ventilation des mines, mine-ventilation ; ventilating mines. (8)

ventilation artificielle (*f.*), artificial ventilation. (9)

ventilation mécanique (*f.*), mechanical ventila-tion. (10)

ventilation mécanique par aspiration (*f.*), vacu-um method of ventilation ; exhaust-draught ; induced draught. (11)

ventilation mécanique par insufflation (*f.*), plenum method of ventilation ; forced draught ; pressure-draught. (12)

ventilation naturelle (*f.*), natural ventilation. (13)

ventilation par foyers (Mines) (*f.*), furnace-ventilation. (14)

ventiler (*v.t.*), to ventilate ; to air : (15)

ventiler une mine, to ventilate a mine. (16)

ventouse (Constr.) (*n.f.*), ventiduct. (17)

ventouse (d'un foyer) (*n.f.*), draught-hole (of a furnace). (18)

ventre (*n.m.*), belly ; bulge : (19)

ventre d'un flacon, d'un filon, d'un haut four-neau, belly of a flask , of a lode, of a blast-furnace. (20 *ou* 21 *ou* 22)

ventre d'un mur, belly, bulge, of a wall. (23)

ventru, -e (*adj.*), bellied : (24)

noyau ventru (Fonderie) (*m.*), bellied core. (25)

venu (-e) de fonte *ou* **venu (-e) à la coulée**, cast : (26)

une roue venue de fonte d'un seul morceau, a wheel cast in one piece. (27)

une console venue de fonte avec le banc, a bracket cast in one piece with the bed. (28)

ces paliers sont venus de fonte (*m.pl.*), these bearing-blocks are cast ; these plummer-blocks are castings. (29)

See also example under **venir de fonte**.

venu (-e) de forge *ou* **venu (-e) à la forge**, forged : (30)

manivelle qui est venue de forge avec l'arbre (*f.*), crank which is forged in one piece with the shaft. (31)

venu (-e) de laminage *ou* **venu (-e) au laminage**, rolled : (32)

dans les rails à gorge, le champignon comporte un canal venu au laminage, de sorte que le rail et le contre-rail ne forment qu'une seule et même pièce, in tram-rails, the head contains a rolled groove, so that the rail and the guard-rail form one and the same piece. (33)

venu (-e) de tour, turned ; turned in the lathe ; turned on the lathe : (34)

poinçon qui possède un petit teton venu de tour destiné à venir se placer dans le coup de pointeau (*m.*), punch which has a small teat turned in the lathe for placing in the punch-mark. (35)

venue (rencontre) (Géol.) (*n.f.*), occurrence : (36)

les grandes venues de cuivre se sont placées à l'époque permienne et surtout dans le terrain permien supérieur, the great occur-rences of copper are placed at the Permian epoch and especially in the Upper Permian strata. (37)

accidents géologiques marqués par des venues d'eaux thermales (*m.pl.*), geological acci-dents marked by occurrences of thermal water. (38)

la venue au jour de produits d'origine interne, the occurrence on the surface of internally derived products. (39)

la venue aurifère de l'Afrique du Sud, the occur-rence of gold in South Africa. (4 ')

venue irrégulière du pétrole, irregular occur-rence of the oil. (41)

venue d'eau (Mines) (*f.*), inrush of water ; influx of water ; irruption of water ; inflow of water ; incoming of water ; make of water ; advent of water : (42)

venue d'eau dans le puits à pétrole, advent of water in the oil-well. (43)

venues d'eaux souterraines, underground inflows. (44)

venue de l'image (Photogr.) (*f.*), appearance of the image. (45)

venue du vent (*f.*), draught ; indraught : (46)

régler la venue du vent d'une machine soufflante, to regulate the draught of a blower. (47)

verdâtre (*adj.*), greenish ; greeny. (48)

verge (tringle de métal) (*n.f.*), rod : (49)

verge de fer, rod of iron ; iron rod. (50)

verge de balance (*f.*), balance-beam ; scale-beam. (51)

verge de piston (*f.*), piston-rod. (52)

verglas (*n.m.*), glazed frost : (53)

le verglas est une couche de glace, unie et transparente, qui se dépose sur la surface du sol. Il se produit lorsque, la température du sol étant au-dessous de 0° C., après quelques jours d'un froid continu, il vienne à tomber un peu de pluie, qui se congèle aussitôt. Il se produit aussi quand les gouttelettes de pluie arrivent à la surface du sol en état de surfusion, glazed frost is an even and transparent sheet of ice, which is deposited on the surface of the ground. It is produced when, the temperature of the ground being below 32° F., after several days of continuous cold, slight rain falls, which freezes immediately. It is also produced when the rain-drops arrive at the surface of the ground in a state of surfusion. (54)

vérificateur des poids et mesures (pers.) (*m.*), inspector of weights and measures. (55)

vérifier (*v.t.*), to verify ; to check : (56)

vérifier les mesures prises, to check the measurements made. (57)

vérin (*n.m.*) *ou* **vérin à vis**, jack ; screw-jack ; jack-screw ; screw lifting-jack ; lifting-screw ; lifting-jack ; hoisting-jack ; handscrew. (58) See also **cric**.

vérin à bouteille (*m.*), bottle jack. (59)

vérin à bouteille avec cliquet (*m.*), ratchet bottle-jack ; ratchet-jack. (1)

vérin à chariot (*m.*), traversing-jack. (2)

vérin à chariot, à bouteille (*m.*), bottle travers--ing jack. (3)

vérin à chariot, à 4 colonnes *ou* **vérin à chariot avec fût à colonnes** (*m.*), leg traversing jack. (4)

vérin à trépied (*m.*), tripod jack. (5)

vérin de calage avec tête à rotule (*m.*), screw packing-jack with head made to swivel ; jack-screw with self-adjusting head. (6)

vérin hydraulique (*m.*), hydraulic jack ; hydrau--lic lifting-jack ; hydrostatic jack. (7)

vérin multiple (*m.*), lift, pull, and push jack, and cramp. (8)

vérin télescope *ou* **vérin télescopique** (*m.*), tele--scope-jack ; telescopic jack. (9)

vérin télescopique à chariot (*m.*), telescope traversing jack. (10)

vermiculite (Minéral) (*n.f.*), vermiculite. (11)

vermoulu, -e (*adj.*), worm-eaten : (12)
bois vermoulu (*m.*), worm-eaten timber. (13)

vernaille (Minéral) (*n.f.*), corundum. (14)

vernier (*n.m.*), vernier. (15)

vernier au dixième *ou* **vernier au 1/10ᵉ** (*m.*), vernier divided into ten equal parts. (16)

vernir (*v.t.*), to varnish ; to lacquer ; to japan : (17)
verrou de porte, verni noir (*m.*), door-bolt, japanned black. (18)

vernis (*n.m.*), varnish ; lacquer ; japan. (19)

vernissage (*n.m.*), varnishing ; lacquering ; japanning. (20)

verre (*n.m.*), glass. (21)

verre à fil de fer noyé (*m.*), wire-glass ; ferro--glass : (22)
protecteur de niveau d'eau à fil de fer noyé dans le cristal (*m.*), water-level protector with wire-glass tube. (23)

verre à vitres (*m.*), window-glass. (24)

verre ardent (Phys.) (*m.*), burning-glass ; sun-glass. (25)

verre argenté (*m.*), silvered glass. (26)

verre armé (*m.*), wire-glass ; ferroglass. (27)

verre cannelé (*m.*), corrugated glass. (28)

verre clair (opp. à verre dépoli, etc.) (*m.*), clear glass. (29)

verre conique, gradué (gobelet gradué) (Verrerie de laboratoire) (*m.*), graduated measure, conical form (30)

verre d'Iéna (*m.*), Jena glass. (31)

verre d'œil (Opt.) (*m.*), eye-glass ; eye-lens : (32)
le verre d'œil d'un oculaire, the eye-glass of an eyepiece. (33)

verre d'optique (*m.*), optical glass. (34)

verre de Bohême (*m.*), Bohemian glass. (35)

verre de champ (opp. à verre d'œil) (Opt.) (*m.*), field-lens ; field-glass. (36)

verre de couleur (*m.*), coloured glass. (37)

verre de lampe (*m.*), lamp-glass ; lamp-chimney. (38)

verre de montre (*m.*), watch-glass. (39)

verre de Moscovie (Minéral) (*m.*), Muscovy glass ; mica. (40)

verre dépoli (*m.*), ground glass ; frosted glass. (41)

verre des volcans (Minéral) (*m.*), volcanic glass ; obsidian. (42)

verre durci (*m.*), hardened glass ; tempered glass ; toughened glass. (43)

verre épais (*m.*), stout glass : (44)
bouteille en verre épais (*f.*), bottle of stout glass. (45)

verre fumé (*m.*), smoked glass. (46)

verre grossissant (*m.*), magnifying-glass ; lens. (47)

verre moulé (*m.*), moulded glass. (48)

verre non-recuit (*m.*), unannealed glass. (49)

verre objectif (Opt.) (*m.*), object-glass ; objective. (50)

verre oculaire (Opt.) (*m.*). Same as verre d'œil.

verre recuit (*m.*), annealed glass. (51)

verre rouge *ou* **verre rubis** (pour laboratoire) (Photogr.) (*m.*), ruby glass. (52)

verre soufflé (*m.*), blown glass. (53)

verre strié (*m.*), corrugated glass. (54)

verre trempé (*m.*). Same as verre durci.

verre volcanique (Minéral) (*m.*), volcanic glass ; obsidian. (55)

verrerie (*n.f.*), glassware. (56)

verrerie de laboratoire (*f.*), laboratory glass--ware. (57)

verrou (Serrur.) (*n.m.*), bolt. (58)

verrou à coquille (*m.*), barrel-bolt. (59)

verrou à la capucine (*m.*), tower bolt. (60)

verrou à ressort (*m.*), spring-bolt. (61)

verrou à ressort (du levier de changement de marche) (*m.*), latch (of reverse-lever). (62)

verrou de fermeture (*m.*), locking-bolt. (63)

verrou de porte (*m.*), door-bolt. (64)

verrouillage (*n.m.*), bolting. (65)

verrouillage des aiguilles (Ch. de f.) (*m.*), locking of points. (66)

verrouiller (*v.t.*), to bolt : (67)
verrouiller une porte, to bolt a door. (68)

verrouiller une aiguille (Ch. de f.), to lock a switch. (69)

verrucano (Géol.) (*n.m.*), verrucano. (70)

versage (vidage) (*n.m.*), tipping : (71)
versage des wagons dans les bennes, tipping the wagons into the kibbles. (72)

versant, -e (*adj.*), liable to overturn ; liable to upset ; liable to capsize : (73)
voiture qui est très versante (*f.*), carriage which is very liable to overturn. (74)

versant (*n.m.*), side ; slope (the side or slope of a hill or mountain considered as opposed to another side or slope over the ridge) : (75)
les Pyrénées ont un versant français et un versant espagnol (*f.pl.*), the Pyrenees have a French side and a Spanish side. (76)
le versant le plus abrupt d'une montagne, the most abrupt slope of a mountain. (77)

versant (d'un comble) (*n.m.*), slope (of a roof.) (78)

versant de colline (*m.*), hillside. (79)

verser (faire couler) (*v.t.*), to pour ; to pour out : (80)
verser du métal en fusion dans un moule, to pour molten metal into a mould. (81)

verser (vider, en parlant des wagons, des bennes, etc.) (*v.t.*), to tip. (82)

verser (faire tomber, en parlant d'une voiture) (*v.t.*), to overturn ; to upset ; to capsize. (83)

verser (tomber sur le côté) (*v.i.*), to overturn ; to upset ; to capsize : (84)
wagon sujet à verser (*m.*), wagon liable to upset. (85)

verseur (*n.m.*), tip ; tipper ; tipple ; tippler. (86)

versicolore (*adj.*), versicolour. (87)

vert, -e (*adj.*), green. (88)

vert (*n.m.*), green. (89)

vert antique (marbre) (*m.*), verd-antique. (1)

vert de cuivre (Minéral) (*m.*), green mineral; malachite. (2)

vert-de-gris (*n.m.*), verdigris. (3)

vert-de-grisé, -e (*adj.*), covered with verdigris. (4)

vertical, -e, -aux (*adj.*), vertical; upright: (5) chaudière verticale (*f.*), vertical boiler; upright boiler. (6)

verticale (*n.f.*), vertical. (7)

verticalement (*adv.*), vertically. (8)

verticalité (*n.f.*), verticality; verticalness: (9) verticalité du puits, verticality of the shaft. (10)

vestiaire (*n.m.*), change-room; changing-room; change-house. (11)

vestiges (*n.m.pl.*), vestiges; remains: (12) vestiges d'une ancienne voie, remains of an old road. (13) vestiges fossiles, fossil remains. (14)

vésuvienne *ou* vésuvianite (Minéral) (*n.f.*), vesuvianite; vesuvian; idocrase. (15)

viaduc (*n.m.*), viaduct. (16)

viaille (galerie perpendiculaire à la direction) (Mines) (*n.f.*), cross-drift; cross-drivage; cross-drive; cross-entry; butt-entry; drift; offset. (17)

vibration (*n.f.*), vibration. (18)

vibration complète (Phys.) (*f.*), complete vibra-tion. (19)

vibration longitudinale (Phys.) (*f.*), longitudinal vibration. (20)

vibration sonore (*f.*), sound-vibration. (21)

vibration transversale (Phys.) (*f.*), transverse vibration. (22)

vibratoire (*adj.*), vibratory: (23) mouvement vibratoire (*m.*), vibratory motion. (24)

vibrer (*v.i.*), to vibrate. (25)

vibro-classeur [vibro-classeurs *pl.*] (*n.m.*), vibro-classifier. (26)

vice caché (*m.*), hidden defect. (27)

vice de construction (*m.*), constructional defect; faulty construction: (28) instrument présentant des vices de construc-tion (*m.*), instrument showing signs of faulty construction. (29)

vicier (*v.t.*), to vitiate; to foul: (30) vicier l'air, to vitiate the air. (31)

victimes de l'incendie d'une fosse (*f.pl.*), victims of a pit-fire. (32)

victorite (Explosif) (*n.f.*), victorite. (33)

vidage (*n.m.*) *ou* vidange (*n.f.*), emptying; dis-charge: (34) vidange automatique des cuffats d'épuisement (Mines), automatic discharge of the water-buckets. (35)

vide (*adj.*), empty; vacant; void: (36) un tonneau vide, an empty cask. (37)

vide (wagon vide, caisse vide, etc.) (*n.m.*), empty. (38 *ou* 39)

vide (espace, milieu sans bornes) (*n.m.*), space; void: (40) la lumière est susceptible de se transmettre à travers le vide, light is capable of being transmitted across space. (41)

vide (Phys.) (*n.m.*), vacuum: (42) faire le vide dans un récipient, to create a vacuum in a receiver. (43)

vide (d'une vis) (opp. à plein) (*n.m.*), groove (of a screw). (44) (opp. to tooth).

vide (opp. à plein *ou* dent) (Engrenages) (*n.m.*), space. (45) (opp. to tooth) (Gearing): largeur du vide (*f.*), width of space. (46)

vide (cuve de fourneau) (*n.m.*), tunnel; fire-room; shaft. (47)

vide (*.*), empty (in the sense of *when empty*): (48) poids à vide (*m.*), weight empty. (49)

vide (*.*) (marche des moteurs, des machines-outils, etc.) (opp. à en charge *ou* chargé, -e), light; on no load; empty; idle; non-cutting: (50) moteur qui marche à vide (*m.*), engine which runs light; motor which runs on no load. (51) machine marchant à vide (*f.*), empty machine; machine running light. (52) course à vide (*f.*), idle stroke; non-cutting stroke. (53)
Cf. charge (en) *ou* chargé, -e *and* marcher à vide.

vide central (dans un lingot d'acier) (*m.*), pipe (in a steel ingot). (54)

vide imparfait (Phys.) (*m.*), low vacuum; partial vacuum. (55)

vide parfait *ou* vide absolu (Phys.) (*m.*), absolute vacuum. (56)

vide presque parfait (Phys.) (*m.*), high vacuum. (57)

vider (*v.t.*), to empty; to discharge: (58) vider un tonneau, to empty a cask. (59)

vider (se) (*v.r.*), to empty; to empty itself. (60)

vie (*n.f.*), life; lifetime: (61) vie d'une mine, life, lifetime, of a mine. (62)

vie (aliments) (*n.f.*), living: (63) la vie est chère, living is dear. (64)

vieilles fontes (*f.pl.*), scrap; cast scrap; foundry-scrap. (65)

vieilles fontes de bâtiment (*f.pl.*), heavy scrap. (66)

vieilles fontes de poterie (*f.pl.*), hardware scrap; pot-scrap; muck-scrap. (67)

vieilles fontes mécaniques (*f.pl.*), machine-scrap. (68)

vieillesse (Géogr. phys.) (*n.f.*), old age. (69)

vieillir (Phys.) (*v.t.*), to age: (70) vieillir un aimant, to age a magnet. (71)

vieillir (*v.i.*), to age. (72)

vieillissement (Phys.) (*n.m.*), aging; ageing. (73)

vierge (*adj.*), virgin: (74) forêt vierge (*f.*), virgin forest. (75) or vierge (*m.*), virgin gold. (76)

vierge de, free from: (77) il y a bien peu de gîtes minéraux affleurant au jour vierges de vieux travaux témoignant du passage des anciens, there are very few mineral deposits outcropping on the surface free from old workings testifying to the passage of the ancients. (78)

vieux (*m.*) [vieil (*m.*) *devant une voyelle ou* h *muet*] vieille (*f.*) (*adj.*), old. (79) See also vieille.

vieux fer (*m.*) *ou* vieux fers (*m.pl.*), old iron. (80)

vieux grès rouge (Géol.) (*m.*), Old Red Sand-stone. (81)

vieux sable (Fonderie) (*m.*), old sand; black sand; floor-sand; heap-sand. (82)

vieux travaux (Mines) (*m.pl.*), old workings. (83)

vif, vive (*adj.*), live; living; quick; sharp; bright: (84) vapeur vive (*f.*), live steam. (85) roche vive (*f.*), living rock. (86) combustion vive (*f.*), quick combustion. (87) l'arête vive du couteau d'une balance (*f.*), the sharp edge of the knife-edge of a balance. (88)

couleur vive (*f.*), bright colour. (1)

vif-argent (*n.m.*), quicksilver ; mercury. (2)

vif de l'eau (*m.*) *ou* **vives eaux** (*f.pl.*), spring-tide. (3)

vigoureux, -euse (*adj.*), vigorous ; strong ; plucky : (4)

cliché vigoureux (Photogr.) (*m.*), vigorous negative ; strong negative ; plucky negative. (5)

vilebrequin (outil) (*n.m.*), brace ; bit-stock ; bit-brace ; stock. (6)

vilebrequin (pour forerie) (*n.m.*), crank-brace ; drill-brace ; drill-crank. (7)

vilebrequin (arbre à manivelle) (*n.m.*), crank-shaft. (8)

vilebrequin à cliquet (*m.*), ratchet-brace ; ratchet. (9)

vilebrequin à coudes multiples (*m.*), multithrow crank-shaft. (10)

vilebrequin à engrenage (*m.*), geared ratchet-brace. (11)

vilebrequin à vis de pression (pour forerie) (*m.*), crank-brace with screw ; drill-brace with screw. (12)

vilebrequin d'angle (*m.*), corner-brace. (13)

vilebrequin porte-forets à engrenages et conscience (*m.*), breast-drill ; breast-drill brace ; breast-brace. (14)

vilebrequin universel à genouillères (*m.*), universal angular bit-stock. (15)

village (*n.m.*), village. (16)

ville (*n.f.*), town. (17)

violation (*n.f.*), violation. (18)

violation de propriété (*f.*), trespass. (19)

viole (*n.f.*), timber-jack. (20)

violemment (*adv.*), violently. (21)

violence (*n.f.*), violence : (22)

violence d'un choc, violence of a shock (*or* of an impact). (23)

violent, -e (*adj.*), violent : (24)

éruption violente (*f.*), violent eruption. (25)

violer (*v.t.*), to violate. (26)

violer une propriété, to trespass upon a property. (27)

violet, -ette (*adj.*), violet. (28)

violet (*n.m.*), violet : (29)

le violet est une couleur du spectre solaire, violet is a colour of the solar spectrum. (30)

violle *ou* **violle-étalon** (Photom.) (*n.m.*), violle ; Violle standard ; Violle's standard : (31)

le violle correspond à la lumière émise par un centimètre carré de platine à la température de solidification. Il équivaut à environ 18½ bougies anglaises, the violle corresponds to the light afforded by a square centimetre of platinum at solidification-point. It is equivalent to about 18½ English candles. (32)

violon (poulie à violon) (*n.m.*), fiddle-block ; thick-and-thin block. (33)

violon (plaque conscience) (*n.m.*), breast-plate ; conscience ; palette. (34)

virage (action de tourner) (*n.m.*), turning ; turning round ; slewing round ; swinging round : (35)

virage d'une manivelle, turning a handle (*or* a crank). (36)

virage d'une grue, turning, swinging round, slewing round, of a crane. (37)

virage au rouge, au bleu, etc., turning red, blue, etc. : (38 *ou* 39)

l'acide sulfurique contenu dans une burette graduée (alcalimètre) est versé goutte à goutte dans la liqueur alcaline, préalablement colorée en bleu par du tournesol, jusqu' à virage au rouge (*m.*), the sulphuric acid contained in a graduated burette (alkalim-eter) is poured drop by drop into the alkaline solution, previously coloured blue by litmus, until it turns red. (40)

virage (action) (Photogr.) (*n.m.*), toning : (41)

virage à l'or et fixage séparés, separate gold toning and fixing. (42)

virage au platine, platinum toning. (43)

virage (substance, solution) (Photogr.) (*n.m.*), toning ; toning-solution. (44)

virage-fixage (action) (Photogr.) (*n.m.*), toning and fixing (combined). (45)

virage-fixage [virages-fixages *pl.*] (*n.m.*) *ou* **virage-fixage combiné** [virages-fixages com-binés *pl.*] (substance, solution), toning and fixing ; toning-and-fixing solution ; combined toning and fixing. (46)

virer (Photogr.) (*v.t.*), to tone : (47)

virer une épreuve, to tone a print. **(48)**

virer (Photogr.) (*v.i.*), to tone. (49)

virer (tourner) (*v.i.*), to turn : (50)

virer un cercle complet, to turn a complete circle. (51)

virer au rouge, au bleu, etc., to turn red, blue, etc. : (52 *ou* 53)

l'acide vire au rouge la teinture bleue de tournesol (*m.*), acid turns blue litmus solution red. (54) *Cf.* virage au rouge.

virgule (Méc.) (*n.f.*), wiper ; wipe ; lifter ; cam : (55)

la virgule et la touche d'une distribution Corliss, the wiper (*or* lifter) (*or* cam) and the toe of a Corliss valve-gear. (56)

virite (Explosif) (*n.f.*), virite. (57)

virolage (*n.m.*), ferruling. (58)

virole (*n.f.*), ferrule ; thimble ; clip : (59)

virole de manche d'outil, ferrule of tool-handle. (60)

virole pour tubes de chaudière, ferrule, thimble, for boiler-tubes. (61)

virole à air (d'un brûleur à gaz) (*f.*), air-regulator, air-clip, (of a gas-burner). (62)

virole de boîte à fumée (locomotive) (*f.*), smoke-box shell. (63)

virole de corps cylindrique (de chaudière) (*f.*), shell-plate, strake, of barrel (of boiler). (64)

viroler (*v.t.*), to ferrule : (65)

viroler le manche d'un outil, les tubes d'une chaudière, to ferrule the handle of a tool, the tubes of a boiler. (66 *ou* 67)

virtuel, -elle (*adj.*), virtual : (68)

image virtuelle (Opt.) (*f.*), virtual image. (69)

vitesse virtuelle (Méc.) (*f.*), virtual velocity. (70)

vis (*n.f.*), screw. (71)

vis à bois (*f.*), wood-screw ; screw for wood (72)

vis à bois à tête carrée (*f.*), coach-screw ; square-head pointed coach-screw ; lag-screw. (73)

vis à bois en fer, à tête plate (*f.*), iron countersunk wood-screw. (74)

vis à bois en laiton, à tête ronde (*f.*), brass round-head wood-screw. (75)

vis à deux filets (*f.*), double-threaded screw. (76)

vis à droite (*f.*), right-hand screw ; right-handed screw. (77)

vis à filet carré *ou* **vis à filet rectangulaire** (*f.*), square-threaded screw ; square-thread screw (78)

vis à filet droite et gauche (*f.*), right-and-left screw. (1)

vis à filet rond (*f.*), round-threaded screw. (2)

vis à filet triangulaire (*f.*), V-threaded screw. (3)

vis à filets convergents (*f.*). Same as **vis globique**.

vis à gauche (*f.*), left-hand screw ; left-handed screw. (4)

vis à jour (*f.*), open-newel stair ; hollow-newel stair ; open-well stair. (5)

vis à métaux (opp. à **vis à bois**) (*f.*), metal-screw ; screw for metal ; machine-screw. (6)

vis à mouvement lent de rotation (*f.*), slow-motion screw. (7)

vis à noyau plein (*f.*), solid-newel stair. (8)

vis à œil (*f.*), capstan screw ; capstan-headed screw. (9)

vis à oreilles (*f.*), thumb-screw ; wing-screw ; winged screw ; butterfly-screw. (10)

vis à pas allongé *ou* **vis à pas rapide** (*f.*), long-pitch screw ; coarse-pitch screw ; quick-pitch screw. (11)

vis à pas carré, à pas triangulaire, etc. (*f.*). Same as **vis à filet carré**, etc.

vis à pas contraires (*f.*), right-and-left screw; compound screw. (12)

vis à pas couché de haute précision (*f.*), finely threaded micrometer-screw. (13)

vis à pas fin (*f.*), fine-pitch screw. (14)

vis à plusieurs filets (*f.*), multiple-threaded screw. (15)

vis à pointeau sans tête (*f.*), cone-pointed grub screw ; cone-point headless screw. (16)

vis à pompe (*f.*), shrouded screw. (17)

vis à tête à six pans (*f.*), hexagon-head screw. (18)

vis à tête carrée (*f.*), square-headed screw ; square-head screw. (19)

vis à tête cylindrique (*f.*), cheese-head screw ; fillister-head screw. (20)

vis à tête en goutte-de-suif *ou* **vis à tête saillante** (*f.*), raised-head screw. (21)

vis à tête fraisée (*f.*), countersunk-head screw. (22)

vis à tête plate (*f.*), flat-headed screw ; flat-head screw. (23)

vis à tête romaine (*f.*), capstan screw ; capstan-headed screw. (24)

vis à tête ronde (*f.*), round-headed screw ; round-head screw. (25)

vis à teton *ou* **vis à téton** (*f.*), tit-screw ; teat-screw. (26)

vis à trois filets (*f.*), three-threaded screw. (27)

vis à un filet (*f.*), single-threaded screw. (28)

vis ailée (*f.*), thumb-screw ; wing-screw ; winged screw ; butterfly-screw. (29)

vis arrêtoir (*f.*), levelling-screw ; foot-screw. (30)

vis au pas Whitworth (*f.*), screw with Whitworth thread. (31)

vis cachée (d'un étau, ou analogue) (*f.*), protected screw (of a vice, or the like). (32 *ou* 33)

vis calante (*f.*), levelling-screw ; foot-screw. (34)

vis casse-joint (*f.*), break-joint screw. (35)

vis creuse (*f.*), female screw ; internal screw ; interior screw ; inside screw ; companion screw. (36)

vis d'Archimède (*f.*), Archimedean screw ; Archimedes' screw. (37)

vis d'arrêt *ou* **vis de blocage** (*f.*), clamp-screw ; clamping-screw. (38)

vis d'assemblage (*f.*), connecting-screw. (39)

vis d'établi de menuisier (*f.*), joiners' bench-screw. (40)

vis de calage *ou* **vis de bride** (*f.*), levelling-screw ; foot-screw. (41)

vis de changement de marche (*f.*), reversing-screw ; reverse-screw. (42)

vis de fin calage (*f.*), fine-adjustment screw. (43)

vis de fixation (*f.*), fixing-screw. (44)

vis de frein de chariot [**vis de freins de chariots** *pl.*] *ou* **vis de mécanique de voiture** [**vis de mécaniques de voitures** *pl.*] (*f.*), brake-screw. (45)

vis de pas allongé (*f.*). Same as **vis à pas allongé**.

vis de pression (*f.*), set-screw ; pressure-screw. (46)

vis de rallonge (du câble de forage) (Sondage) (*f.*), temper-screw. (47)

vis de rappel (*f.*), adjusting-screw ; temper-screw ; draw-screw ; drag-screw. (48)

vis de rappel pour le mouvement lent (*f.*), slow-motion adjusting-screw. (49)

vis de réglage (*f.*), regulating-screw. (50)

vis de réglage de la bulle (d'un instrument de nivellement) (*f.*), bubble-nut (of a levelling-instrument). (51)

vis de serrage (*f.*), tightening-screw. (52)

vis de tension (*f.*), tension-screw ; straining-screw ; stretching-screw ; union-screw. (53)

vis décolletées (*f.pl.*), machine-made screws ; screws cut from bar. (54)

vis différentielle (*f.*), differential screw ; compound screw. (55)

vis en blanc (*f.*), screw-blank ; blank for screw. (56)

vis femelle (*f.*), female screw ; internal screw ; interior screw ; inside screw ; companion screw. (57)

vis fraise (*f.*), hob (for cutting the teeth of worm-wheels). (58)

vis globique *ou* **vis globoïde** (*f.*), hour-glass screw ; Hindley's screw ; Hindley's worm ; globoid worm ; curved worm. (59)

vis mâle (*f.*), male screw ; external screw ; exterior screw ; outside screw. (60)

vis mécanique (*f.*), machine-screw. (61)

vis mère [**vis mères** *pl.*] (*f.*) *ou* **vis-mère** [**vis-mères** *pl.*] (*n.f.*) (tour), lead-screw ; leading-screw ; feed-screw ; guide-screw (lathe). (62)

vis micrométrique (*f.*), micrometer-screw ; micrometric screw. (63)

vis monte-et-baisse (*f.*), elevating-screw. (64)

vis multiple (*f.*), multiple-threaded screw. (65)

vis noyée (*f.*), flush screw. (66)

vis pleine (*f.*), male screw ; external screw ; exterior screw ; outside screw. (67)

vis pneumatique (Aérage de mines) (*f.*), screw-fan. (68)

vis sans fin (*f.*), worm ; endless screw ; perpetual screw. (69)

vis sans fin globique (*f.*). Same as **vis globique**.

vis sans tête (*f.*), grub screw ; headless screw. (70)

vis tangente (*f.*), tangent screw. (71)

vis transporteuse (*f.*), screw-conveyor ; spiral conveyor. (72)

viscosimètre (*n.m.*), viscosimeter ; viscometer. (73)

viscosité (*n.f.*), viscosity : (74)

viscosité des liquides, viscosity of liquids. (1)

visée (action)(*n.f.*), sighting ; observation : (2)
rectifier le calage d'un instrument de nivelle-
-ment immédiatement avant et après la
visée, to correct the setting of a levelling-
instrument immediately before and after
sighting. (3)
visées dans l'obscurité, sighting (*or* observa-
-tions) in the dark. (4)
visées à longue portée, long-range sightings.
(5)

visée (direction de la vue vers un but) (*n.f.*),
sight ; observation : (6)
la ligne de visée d'une lunette, the line of sight
of a telescope. (7)

viser (*v.t.*), to sight : (8)
viser un astre situé dans le plan du méridien
magnétique, to sight a star situated in the
plane of the magnetic meridian. (9)
viser sur la ligne de séparation des couleurs du
voyant d'une nivelette, to sight on the line
dividing the colours on the target of a boning-
rod. (10)

viseur (Photogr.) (*n.m.*), view-finder ; finder. (11)
viseur à lunette (*m.*), telescopic view-finder.
(12)
viseur brillant *ou* **viseur clair** (opp. à **viseur
dépoli**) (*m.*), brilliant view-finder. (13)
viseur dépoli (*m.*), ground-glass view-finder.
(14)
viseur direct *ou* **viseur clair** (opp. à **viseur indirect**
ou **obscur**) (*m.*), direct-vision view-finder ;
direct finder. (15)
viseur direct avec œilleton de visée (*m.*), direct
finder with eyepiece (*or* with magnifying-
lens). (16)
viseur de mine (*m.*), lining-sight. (17)
viseur iconomètre *ou* **viseur iconométrique** *ou*
viseur à cadre (*m.*), frame view-finder ;
iconometer finder. (18)
viseur indirect *ou* **viseur obscur** *ou* **viseur à
chambre noire** (*m.*), indirect view-finder ;
reflecting view-finder. (19)
viseur pliant à réticule (*ou* **réticulé**) **et aiguille
de mire** (*m.*), folding view-finder with cross-
lines and sight-bar (*or* sighter) (*or* sighting-
pin) ; folding graphic sight-finder. (20)
viseur redresseur (*m.*), collapsible finder ; self-
erecting and closing view-finder. (21)
viseur réversible (*m.*), reversible view-finder.
(22)

visible (*adj.*), visible : (23)
cristaux qui sont visibles à l'œil nu (*m.pl.*),
crystals which are visible to the naked eye.
(24)

visiblement (*adv.*), visibly. (25)

visière de cheminée (locomotive) (*f.*), chimney-
lip ; smoke-stack lip. (26)

visite (tournée d'inspection) (*n.f.*), visit. (27)
visite (examen détaillé) (*n.f.*), inspection ;
examination : (28)
visite de la chaudière, boiler-inspection. (29)
visite de la voie (Ch. de f.), railroad inspection.
(30)
visite fréquente et minutieuse du câble,
frequent minute examination of the rope.
(31)

visiter (aller voir) (*v.t.*), to visit. (32)
visiter (inspecter) (*v.t.*), to inspect ; to examine :
(33)
visiter l'intérieur d'une chaudière, to examine
the inside of a boiler. (34)

visiteur (inspecteur) (pers.) (*n.m.*), inspector ;
examiner. (35)
visiteur de chaudières (*m.*), boiler-inspector. (36)
visiteur des wagons (Ch. de f.) (*m.*), carriage-
examiner ; car-inspector. (37)

visqueux, -euse (*adj.*), viscous ; viscid. (38)

vissage (*n.m.*), screwing ; screwing on ; screwing
down ; screwing up. (39)

vissé, -e (*adj.*), screwed : (40)
manchon vissé (*m.*), screwed coupling. (41)

visser (*v.t.*), to screw ; to screw on ; to screw
down ; to screw up : (42)
visser les tiges les unes au bout des autres,
to screw the rods the ones to the ends of the
others ; to screw the rods together. (43)
visser un écrou à bloc, to screw a nut right
home. (44)
visser une serrure, to screw on a lock. (45)
visser le couvercle d'une boîte, to screw down
the lid of a box. (46)

visser (se) (*v.r.*), to screw : (47)
écran qui se visse à l'intérieur du barillet
(Photogr.) (*m.*), screen which screws into
the tube. (48)

visserie (vis et articles analogues, comme écrous
et boulons) (*n.f.*), screws, etc. ; bolts and
nuts. (49)
visserie (usine) (*n.f.*), screw-works ; bolt-and-
nut works. (50)

visuel, -elle (*adj.*), visual : (51)
angle visuel (*m.*), visual angle. (52)

vite (*adj.*), quick ; fast. (53)

vite (*adv.*), quickly ; quick ; fast. (54)

vitesse (*n.f.*), speed ; velocity ; swiftness. (55)
vitesse absolue (Méc.) (*f.*), absolute velocity.
(56)
vitesse absolue de sortie (turbines) (*f.*), absolute
exit-velocity. (57)
vitesse accélérée (Méc.) (*f.*), accelerated velocity.
(58)
vitesse angulaire (Méc.) (*f.*), angular velocity.
(59)
vitesse commerciale (d'un train) (Ch. de f.) (*f.*).
average speed including stoppages. (60)
vitesse critique (Phys.) (*f.*), critical velocity.
(61)
vitesse d'arrivée (Hydraul.) (*f.*), velocity of
approach. (62)
vitesse d'ascension du pétrole (*f.*), ascensional
speed of the oil. (63)
vitesse d'avancement (Mines) (*f.*), heading speed.
(64)
vitesse d'écoulement des eaux (*f.*), velocity of
flow of the water. (65)
vitesse d'entrée (turbines) (*f.*), inlet-velocity.
(66)
vitesse d'un courant (*f.*), velocity, speed, drift,
of a current. (67)
vitesse de combustion des mèches (*f.*), burning-
speed of fuses. (68)
vitesse de coupe (*f.*), cutting-speed. (69)
vitesse de l'air (*f.*), velocity of the air. (70)
**vitesse de propagation de la lumière, du son,
de la flamme d'une explosion** (*f.*), velocity
of propagation of light, of sound, of the
flame of an explosion. (71 *ou* 72 *ou* 73)
vitesse de régime (Mach.) (*f.*), normal working-
speed ; working-speed. (74) Cf. régime.
vitesse de sortie (turbines) (*f.*), exit-velocity. (75)
vitesse des trains (*f.*), speed of trains. (76)
vitesse du renvoi (tours) (*f.*), speed of counter-
-shaft (revolutions). (77)

vitesse effective de marche (Ch. de f.) (*f.*), actual running speed : (1)
la vitesse effective de marche dépasse à certains moments la vitesse moyenne, the actual running speed exceeds at certain moments the average speed. (2)

vitesse initiale (Méc.) (*f.*), initial velocity. (3)

vitesse linéaire (*f.*), linear speed : (4)
vitesse linéaire du piston, linear speed of the piston. (5)

vitesse moyenne de marche (d'un train d'une station à la suivante) (*f.*), average running speed (of a train from one station to the following). (6)

vitesse optimum (*f.*), speed of greatest energy ; speed of maximum power. (7)

vitesse relative (Méc.) (*f.*), relative velocity ; comparative velocity. (8)

vitesse relative d'entrée (turbines) (*f.*), relative inlet-velocity. (9)

vitesse relative de sortie (turbines) (*f.*), relative exit-velocity. (10)

vitesse retardée (Méc.) (*f.*), retarded velocity. (11)

vitesse uniforme (Méc.) (*f.*), uniform velocity. (12)

vitesse uniformément accélérée (Méc.) (*f.*), uniformly accelerated velocity. (13)

vitesse uniformément retardée (Méc.) (*f.*), uniformly retarded velocity. (14)

vitesse variée (Méc.) (*f.*), variable velocity. (15)

vitesse virtuelle (Méc.) (*f.*), virtual velocity. (16)

vitrage (*n.m.*), glazing. (17)

vitrer (*v.t.*), to glaze : (18)
vitrer un châssis de fenêtre, to glaze a window-sash. (19)

vitreux, -euse (*adj.*), vitreous ; glassy ; glass (*used as adj.*) : (20)
cassure vitreuse (*f.*), vitreous fracture. (21)
feldspath vitreux (*m.*), glassy feldspar. (22)
inclusion vitreuse (Minéralogie) (*f.*), vitreous inclusion ; glass-inclusion. (23)

vitrier (pers.) (*n.m.*), glazier ; glass-cutter. (24)

vitrifiabilité (*n.f.*), vitrifiability. (25)

vitrifiable (*adj.*), vitrifiable. (26)

vitrification (*n.f.*), vitrification ; vitrifaction : (27)
vitrification du sable, vitrification of sand. (28)

vitrifier (*v.t.*), to vitrify. (29)

vitrifier (se) (*v.r.*), to vitrify. (30)

vitriol (acide sulfurique) (*n.m.*), vitriol ; oil of vitriol ; vitrolic acid ; sulphuric acid. (31)

vitriol blanc (*m.*), white vitriol ; salt of vitriol ; zinc vitriol ; sulphate of zinc ; zinc sulfate. (32)

vitriol bleu (*m.*), blue vitriol ; bluestone ; copper vitriol ; sulphate of copper ; copper sulphate ; blue copperas. (33)

vitriol vert *ou* vitriol martial (*m.*), green vitriol ; martial vitriol ; ferrous sulphate. (34)

vitriolique (*adj.*), vitriolic. (35)

vitriolisation (*n.f.*), vitriolization. (36)

vitrioliser (*v.t.*), to vitriolize. (37)

vitrophyre (Pétrol.) (*n.m.*), vitrophyre. (38)

vitrose (Photogr.) (*n.f.*), cut film ; flat film ; flat cut film ; stiff film. (39)

vitrosité (*n.f.*), vitreosity. (40)

vives eaux (*f.pl.*), spring-tide. (41)

vivianite (Minéral) (*n.f.*), vivianite ; blue iron earth ; blue ochre ; iron phosphate. (42)

vivres (*n.m.pl.*), provisions ; food ; living : (43)

les vivres sont chers, living is dear. (44)

vogésite (Pétrol.) (*n.f.*), vogesite ; vogesyte. (45)

voie (*n.f.*), way. (46)

voie (chemin ; route) (*n.f.*), way ; road ; route. (47)

voie (Mines) (*n.f.*), way ; road ; roadway ; level. (48) See also galerie.

voie (Ch. de f.) (*n.f.*), permanent way ; road ; way ; track ; trackway ; line. (49)

voie (écartement ; intervalle qui sépare les rails d'une voie ferrée) (Ch. de f.) (*n.f.*), gauge ; gage ; railway-gauge ; track-width ; distance between rails. (50)

voie (distance entre les roues d'un véhicule) (*n.f.*), gauge ; gage. (51)

voie (ornières, traces laissées sur le sol par les roues d'une voiture, etc.) (*n.f.*), track : (52)
la voie d'une roue, d'un traîneau, the track of a wheel, of a sleigh. (53 *ou* 54)

voie (d'une scie) (*n.f.*), set (of a saw). (55)

voie (Chim.) (*n.f.*), way ; method ; process. (56)

voie à câble aérien *ou* voie à câbles (*f.*), aerial ropeway ; ropeway ; cableway ; rope-railway ; wire-ropeway ; wire-tramway ; wireway. (57)

voie à coussinets (Ch. de f.) (*f.*), chaired road. (58)

voie à écartement normal (Ch. de f.) (*f.*), standard-gauge track. (59)

voie aérienne (*f.*), overhead track ; aerial trackway. (60)

voie barrée (Mines) (*f.*), fenced-off road. (61)

voie courante (Ch. de f.) (*f.*), running line. (62)

voie courbe (Ch. de f.) (*f.*), curve track. (63)

voie d'accès (*f.*), approach : (64)
les voies d'accès au Simplon, the approaches to the Simplon. (65)

voie d'air (Mines) (*f.*), airway ; windway ; air-heading ; air-head ; air-course. (66)

voie d'eau (passage d'eau d'un robinet, d'une soupape) (*f.*), waterway. (67)

voie d'eau (fissure par laquelle l'eau s'échappe) (*f.*), leak : (68)
les bateaux en bois ne sont jamais complète-ment exempts des voies d'eau (*m.pl.*), wooden boats are never completely free from leaks. (69)

voie d'écoulement (Mines) (*f.*), drainage-level ; water-level ; watercourse-level ; water-adit ; offtake. (70)

voie d'évitement (Ch. de f.) (*f.*), passing-place ; passing-track ; turnout ; pass-by ; passing ; shunting-loop. (71)

voie d'extraction du minéral (*f.*), ore-level ; ore-road. (72)

voie d'un mètre *ou* voie de 1 mètre (Ch. de f.) (*f.*), metre gauge. (73)

voie de chemin de fer (*f.*), railway-track ; set of rails : (74)
grue à portique chevauchant — voies de chemin de fer (*f.*), gantry-crane spanning — sets of rails. (75)

voie de classement (Ch. de f.) (*f.*), marshalling-line ; assembling-track ; drill-track. (76)

voie de dérivation (Ch. de f.) (*f.*), loop-line ; loop. (77)

voie de droite (Ch. de f.) (*f.*), down road ; out-bound track. (78) Note.—In France the down road is on the right going from Paris ; in England it is on the left going from London.

voie de fond (Mines) (*f.*), bottom level. (1)

voie de garage (Ch. de f.) (*f.*), siding; side-track; turnout; side-tracking; lieby; liebye; lie. (2)

voie de gauche (Ch. de f.) (*f.*), up road; inbound track. (3) Note.—In France the up road is on the left looking from Paris; in England it is on the right looking from London.

voie de jonction (Ch. de f.) (*f.*), crossover-road; crossover. (4)

voie de manœuvre (Ch. de f.) (*f.*), shunting-line; switching-track. (5)

voie de niveau (Mines) (*f.*), level. (6)

voie de raccordement (Ch. de f.) (*f.*), transition-road. (7)

voie de remisage (Ch. de f.) (*f.*), covered siding. (8)

voie de retour d'air (Mines) (*f.*), return airway. (9)

voie de roulage (Mines) (*f.*), haulageway; haulage-road; haulage-level; tram-level; wagonway; wagon-road; rolleyway; hutch-road. (10)

voie de roulement (*f.*), runway; race; roller-track; roller-path. (11)

voie de roulement pour billes (*f.*), ball-race. (12)

voie de terre (*f.*), overland route; land-route; (13)
prendre la voie de terre, to take the overland route. (14)

voie de traçage (Mines) (*f.*), development-heading. (15)

voie de traction par chevaux (*f.*), horse-road. (16)

voie de tramway (*f.*), tram-track. (17)

voie de translation (d'un pont roulant) (*f.*), runway (of an overhead travelling crane). (18)

voie descendante (Ch. de f.) (*f.*). Same as **voie de droite.**

voie déviée (opp. à **voie directe**) (Ch. de f.) (*f.*), diverging road. (19)

voie diagonale (Ch. de f.) (*f.*), crossover; cross-over-road; diagonal road. (20)

voie directe (Ch. de f.) (*f.*), through road. (21)

voie double (Ch. de f.) (*f.*), double line; double track. (22)

voie droite *ou* **voie en ligne droite** (Ch. de f.) (*f.*), straight track. (23)

voie en courbe (Ch. de f.) (*f.*), curve track. (24)

voie en déblai *ou* **voie en tranchée** (Ch. de f.) (*f.*), cutting; cut; railway-cutting. (25)

voie en remblai (Ch. de f.) (*f.*), embankment; railway-embankment. (26)

voie étroite (Ch. de f.) (*f.*), narrow gauge; (27)
chemin de fer à voie étroite (*m.*), narrow-gauge railway. (28)

voie ferrée (*f.*), railway; railroad. (29)

voie ferrée à traction par cheval (*f.*), horse-railway; horse-railroad. (30)

voie ferrée aérienne (*f.*), aerial railway; aerial railroad; runway; overhead runway; overhead track. (31)

voie fixe (Ch. de f.) (*f.*), permanent way. (32)

voie humide (Chim.) (*f.*), wet way; wet method; wet process; humid way; (33)
cristallisation par voie humide (*f.*), crystallization by the wet method. (34)

voie interdite (Mines) (*f.*), fenced-off road. (35)

voie large (Ch. de f.) (*f.*), broad gauge; wide gauge; (36)
chemin de fer à voie large (*m.*), broad-gauge railway; wide-gauge railway. (37)

voie libre (Signaux de ch. de f.), line clear; clear: (38)
donner voie libre, to give line clear. (39)
signal à voie libre, signal at clear. (40)

voie lourde (Ch. de f.) (*f.*), heavy line. (41)

voie maîtresse (Mines) (*f.*), main road; main-way; main level; gangway; entry. (42)

voie montante (Ch. de f.) (*f.*). Same as **voie de gauche.**

voie navigable (*f.*), waterway; water-route. (43)

voie normale (Ch. de f.) (*f.*), standard gauge; normal gauge: (44)
chemin de fer à voie normale (*m.*), standard-gauge railway. (45)

voie occupée (Signaux de ch. de f.), line occupied. (46)

voie principale (Ch. de f.) (*f.*), main line; main road; main track. (47)

voie principale (Mines) (*f.*), main road; main-way; main level; gangway; entry. (48)

voie principale d'air (Mines) (*f.*), main airway. (49)

voie principale d'entrée d'air (*f.*), main intake-airway. (50)

voie principale de retour d'air (*f.*), main return-airway. (51)

voie publique (*f.*), public road; highway. (52)

voie sèche (Chim.) (*f.*), dry way; dry method; dry process. (53)

voie secondaire (Ch. de f.) (*f.*), branch-line; branch-road; branch; side line. (54)

voie souterraine (*f.*), underground road. (55)

voie thierne (galerie de mine) (*f.*), slant; run. (56)

voie unique (Ch. de f.) (*f.*), single line; single track: (57)
circulation à voie unique (*f.*), single-line working. (58)
ligne à voie unique (*f.*), single-track line. (59)

voies et moyens [voies (*n.f.pl.*); moyens (*n.m.pl.*)] ways and means. (60)

voies jumelles *ou* **voies conjuguées** (Mines de houille) (*f.pl.*), parallel entries. (61)

voies jumelles doubles *ou* **voies conjuguées doubles** (*f.pl.*), double entries. (62)

voies jumelles triples *ou* **voies conjuguées triples** (*f.pl.*), triple entries. (63)

voile (d'un moulin à vent) (*n.f.*), sail (of a wind-mill). (64)

voile (gauchissement) (*n.m.*), buckle: (65)
la rectification a pour but d'enlever le voile, the object of truing is to remove the buckle. (66)

voile (Photogr.) (*n.m.*), fog. (67)

voile chimique (*m.*), chemical fog. (68)

voile dichroïte (*m.*), red fog. (69)

voile noir pour la mise au point (Photogr.) (*m.*), focussing-cloth. (70)

voilé, -e (courbé; faussé) (*adj.*), buckled; bent: (71)
roue voilée (*f.*), buckled wheel. (72)

voilé, -e (déjeté) (*adj.*), warped: (73)
planche voilée (*f.*), warped board. (74)

voilé, -e (Photogr.) (*adj.*), fogged: (75)
cliché voilé (*m.*), fogged negative. (76)

voiler (courber; fausser) (*v.t.*), to buckle; to bend: (77)
voiler la tige (Sondage), to buckle the rods. (78)

voiler (Photogr.) (*v.t.*), to fog. (79)

voiler (se courber, en parlant du métal) (*v.i.*), to buckle; to bend. (80)

voiler (se déjeter) (*v.i.*), to warp : (1)
planche qui voile (*f.*), board which warps. (2)
voiler (se) (en parlant du métal) (*v.r.*), to buckle ;
to bend. (3)
voiler (se) (en parlant du bois) (*v.r.*), to warp :
(4)
planche qui se voile (*f.*), board which warps.
(5)
voiler (se) (Photogr.) (*v.r.*), to fog. (6)
voilure (*n.f.*), buckling ; bending ; warping :
(7)
voilure d'une tôle de chaudière, buckling of
a boiler-plate. (8)
voilure d'une planche, warping of a board.
(9)
voisin, -e (*adj.*), neighbouring ; adjoining ;
near ; bordering. (10)
voisinage (*n.m.*), neighbourhood ; vicinity. (11)
voiture (véhicule routier) (*n.f.*), carriage ; cart ;
vehicle ; van ; wagon ; waggon. (12)
voiture (wagon de chemin de fer) (*n.f.*), carriage ;
coach ; car ; wagon ; waggon. (13) See
also **wagon** (*n.m.*).
voiture (transport) (*n.f.*), transport ; carriage :
(14)
voiture par mulets, transport by mules. (15)
voiture à bogies (Ch. de f.) (*f.*), bogie-carriage.
(16)
voiture à bras (*f.*), hand-cart. (17)
voiture à quatre roues (*f.*), four-wheeled carriage.
(18)
voiture à vapeur (*f.*), traction-engine ; road-
engine. (19)
voiture à voyageurs (Ch. de f.) (*f.*), passenger-
carriage ; passenger-coach ; passenger-car.
(20)
voiture automotrice (Tramway électrique) (*f.*),
motor-car. (21)
voiture automotrice (Ch. de f.) (*f.*), locomotive-
car ; motor-car ; rail motor-car. (22)
voiture d'arrosage (*f.*), watering-cart ; water-
cart. (23)
voiture de livraison (*f.*), delivery-van ; parcel-
van ; van. (24)
voiture de remorque (Ch. de f. ou Tramway)
(opp. à voiture automotrice) (*f.*), trail-car ;
trailer. (25 *ou* 26)
voiture de secours (Ch. de f.) (*f.*), breakdown
van ; wrecking-car. (27)
voiture de tramway (*f.*), tram-car ; tram. (28)
voiture de transport (*f.*), transport-wagon. (29)
voiturier (pers.) (*n.m.*), carrier ; carman ; carter.
(30)
vol d'oiseau (à) (directement), as the crow flies.
(31)
volant, -e (*adj.*), flying : (32)
pont volant (*m.*), flying bridge. (33)
volant, -e (mobile ; portatif) (*adj.*), portable :
(34)
forge volante (*f.*), portable forge. (35)
volant (*n.m.*) *ou* **volant de chasse** *ou* **volant
d'entrainement** *ou* **volant de commande**
(roue pesante qui sert à maintenir l'uniformité
du mouvement d'une machine), fly-wheel ;
fly ; driving-wheel. (36)
volant (arbre garni de palettes qui sert à modérer
la rapidité d'un mouvement circulaire)
(*n.m.*), fly ; fly-governor. (37)
volant (d'un moulin à vent) (*n.m.*), wing (of a
windmill). (38)
volant (distance entre deux points d'appui) (*n.m.*),
span : (39)

volant d'une poutre, d'une voûte, span of a
girder, of an arch. (40 *ou* 41)
volant à cloison pour éviter la ventilation (*m.*),
fly-wheel enclosed to prevent draught. (42)
volant à main *ou simplement* **volant** (*n.m.*), hand-
wheel (without handle). (43)
volant à manivelle *ou* **volant à poignée** (*m.*),
hand-wheel (with handle). (44)
volant à poignées radiales (*m.*), pilot-wheel ;
capstan-wheel. (45)
volant de changement de marche (*m.*), hand-
wheel of reversing-gear. (46)
volant de machine (*m.*), engine fly-wheel. (47)
volant de manœuvre (*m.*), operating hand-wheel.
(48)
volant de manœuvre du frein *ou* **volant du frein
à main** (*m.*), brake wheel. (49)
volant porte-lame (*m.*), band-saw pulley ; saw-
pulley ; band-wheel. (50)
volant pour commande à bras (*m.*), hand-wheel
for driving (a *machine*) by hand. (51)
volatil, -e (*adj.*), volatile : (52)
huile volatile (*f.*), volatile oil. (53)
volatilisable (*adj.*), volatilizable : (54)
minéraux volatilisables (*m.pl.*), volatilizable
minerals. (55)
volatilisation (*n.f.*), volatilization : (56)
volatilisation du mercure, volatilization of
mercury. (57)
volatiliser (*v.t.*), to volatilize : (58)
volatiliser du soufre, to volatilize sulphur.
(59)
volatiliser (se) (*v.r.*), to volatilize : (60)
substance qui se volatilise à une certaine
température (*f.*), substance which volatilizes
at a certain temperature. (61)
métal qui se volatilise au four électrique (*m.*),
metal which volatilizes in the electric furnace.
(62)
volatilité (*n.f.*), volatility ; volatileness : (63)
volatilité de l'éther, volatility of ether. (64)
volborthite (Minéral) (*n.f.*), volborthite. (65)
volcan (*n.m.*), volcano. (66)
volcan actif *ou* **volcan en activité** *ou* **volcan
agissant** (*m.*), active volcano. (67)
volcan d'air *ou* **volcan de boue** (*m.*), mud-volcano
(volcanic) ; mud-cone ; salse ; maccaluba.
(68)
volcan en repos (*m.*), dormant volcano. (69)
volcan éteint (*m.*), extinct volcano. (70)
volcan sous-marin (*m.*), submarine volcano. (71)
volcanicité (*n.f.*), volcanicity ; volcanism ;
vulcanicity ; vulcanism. (72)
volcanique (*adj.*), volcanic : (73)
éruption volcanique (*f.*), volcanic eruption.
(74)
volcaniquement (*adv.*), volcanically. (75)
volcanisation (*n.f.*), volcanization : (76)
volcanisation des basaltes, volcanization of
basalts. (77)
volcaniser (*v.t.*), to volcanize. (78)
volcaniser (se) (*v.r.*), to volcanize. (79)
volcanisme (*n.m.*), volcanism ; vulcanism ;
volcanicity ; vulcanicity : (80)
une région tourmentée par le volcanisme,
a region broken by volcanism. (81)
volcaniste (pers.) (*n.m.*), volcanist. (82)
volée (Travail aux explosifs) (*n.f.*), volley : (83)
une volée de coups de mine, a volley of shots.
(84)
volée (d'une grue) (*n.f.*), jib, boom, (of a crane)
(85)

volée (galerie perpendiculaire à la direction) (Mines) (*n.f.*), cross-drift; cross-drivage; cross-drive; cross-entry; butt-entry; drift; offset. (1)

volée (*n.f.*) *ou* volée d'escalier, flight; flight of stairs. (2)

volée droite (escalier) (*f.*), straight flight. (3)

voler en éclats (*v.i.*), to splinter; to fly; to fly into fragments: (4)
les meules d'émeri se brisent parfois et volent en éclats (*f.pl.*), emery-wheels sometimes break and fly into fragments. (5)

volet (*n.m.*), shutter; screen; guard; shield: (6)
volet en bois pour protéger l'ouvrier contre les accidents, wood guard to protect the workman against accidents. (7)

volet (d'une roue hydraulique) (*n.m.*), paddle, paddle-board, float, floatboard, (of a water-wheel). (8)

voleur de rivières (Géogr. phys.) (*m.*), stream robber. (9)

volige *ou* volice (*n.f.*), lath; slate-lath; batten. (10)

volige chanlattée (*f.*), eaves-board; eaves-catch; eaves-lath. (11)

voligeage (*n.m.*), lathing; battening. (12)

voliger (*v.t.*), to lath; to batten: (13)
voliger un toit, to lath, to batten, a roof. (14)

volt (Élec.) (*n.m.*), volt. (15)

volt-ampère [volts-ampères *pl.*] (*n.m.*), volt-ampere; ampere-volt; watt. (16)

volt-coulomb [volts-coulombs *pl.*] (*n.m.*), volt-coulomb; coulomb-volt; joule. (17)

voltage (Élec.) (*n.m.*), voltage; pressure; tension. (18)

voltage primaire (*m.*), primary voltage. (19)

voltage secondaire (*m.*), secondary voltage. (20)

voltaïque (Phys.) (*adj.*), voltaic; galvanic: (21)
pile voltaïque (*f.*), voltaic pile; galvanic pile. (22)

voltaïquement (*adv.*), galvanically: (23)
électricité développée voltaïquement (*f.*), electricity developed galvanically. (24)

voltaïsme (Phys.) (*n.m.*), voltaism. (25)

voltamètre (Phys.) (*n.m.*), voltameter. (26)

voltampère (Élec.) (*n.m.*), volt-ampere; ampere-volt; watt. (27)

voltampèremètre (Élec.) (*n.m.*), volt-ammeter; wattmeter. (28)

voltmètre (Élec.) (*n.m.*), voltmeter. (29)

volume (*n.m.*), volume; bulk; mass; capacity: (30)
volume d'un cône, d'une pyramide, volume of a cone, of a pyramid. (31 *ou* 32)
volume d'un cours d'eau, volume of a stream. (33)
volume d'un cylindre, volume, capacity, of a cylinder. (34)
volume d'une pierre, volume, bulk, mass, of a stone. (35)

volume critique (Phys.) (*m.*), critical volume. (36)

voluménomètre (Phys.) (*n.m.*), volumenometer. (37)

volumètre (Phys.) (*n.m.*), volumeter. (38)

volumétrique (Phys.) (*adj.*), volumetric; volu-metrical (39)
analyse volumétrique (*f.*), volumetric analysis. (40)

volumétriquement (*adv.*), volumetrically. (41)

vorscheidage (du minerai) (*n.m.*), ragging (ore). (42)

vorscheider (*v.t.*), to rag. (43)

voussoir *ou* vousseau (*n.m.*), arch-stone; quoin; voussoir. (44)

voussoir de clé (*m.*), keystone; key. (45)

voûte (Arch.) (*n.f.*), arch; vault. (46) Cf. arc (Arch.) (*n.m.*).

voûte (d'un tunnel, d'une caverne) (*n.f.*), roof (of a tunnel, of a cave). (47 *ou* 48)

voûte (d'un fourneau) (*n.f.*), arch, crown, dome, (of a furnace). (49)

voûte (pli anticlinal; selle) (opp. à auge *ou* fond de bateau) (Géol.) (*n.f.*), arch; saddle. (5.) (opp. to trough).

voûte à trois centres (Arch.) (*f.*), three-centred arch. (51)

voûte aplatie *ou* voûte à quatre centres (*f.*), four-centred arch. (52)

voûte biaise (*f.*), skew arch; skewed arch; oblique arch. (53)

voûte d'arête (*f.*), groined vault. (54)

voûte elliptique *ou* voûte en ellipse (*f.*), elliptical arch. (55)

voûte en anse de panier (*f.*), basket-handle arch. (56)

voûte en arc de cercle *ou* voûte bombée (*f.*), segmental arch; segment arch (flat curve). (57)

voûte en arc de cloître (*f.*), groined vault. (58)

voûte en berceau (*f.*), cradle vault; barrel vault; wagon vault; tunnel vault. (59)

voûte en briques (de foyer de chaudière) (*f.*), brick arch (of boiler-furnace). (60)

voûte en fer à cheval *ou* voûte outrepassée (*f.*), horseshoe arch. (61)

voûte en maçonnerie (*f.*), masonry arch. (62)

voûte en plein cintre (*f.*), round arch; semi-circular arch. (63)

voûte extradossée (*f.*), extradosed arch. (64)

voûte plate (*f.*), flat arch; platband. (65)

voûte rampante (*f.*), rampant arch; rising arch; rampant vault. (66)

voûte surbaissée (opp. à voûte surhaussée) (*f.*), segmental arch; segment arch; imperfect arch; diminished arch; scheme arch; skene arch. (67)

voûte surhaussée *ou* voûte surmontée *ou* voûte exhaussée *ou* voûte surélevée (*f.*), stilted arch. (68)

voûte triangulaire (*f.*), triangular arch. (69)

voûté, -e (*adj.*), arched; vaulted. (70)

voûter (*v.t.*), to arch; to vault: (71)
voûter une cave, to vault a cellar. (72)

voûter (se) (*v.r.*), to arch. (73)

voyage (*n.m.*), voyage; journey; travel; travelling; traveling; trip; tour. (74)

voyage à cheval (*m.*), journey on horseback; travelling on horseback. (75)

voyage au long cours *ou* voyage de long cours (*m.*), ocean voyage; ocean travel. (76)

voyage de prospection (*m.*), prospecting-tour. (77)

voyage en automobile (*m.*), motor-car trip. (78)

voyage par mer (*m.*), sea voyage. (79)

voyage par terre (*m.*), journey overland; over-land travel. (80)

voyager (*v.i.*), to travel; to journey: (81)
voyager par mer, to travel by sea. (82)
voyager par terre, to travel overland. (83)

voyageur (pers.) (*n.m.*), traveller; traveler; passenger: (84)
train à voyageurs (*m.*), passenger-train. (85)

voyant (Géod.) (*n.m.*), target; vane: (1)
mire à voyant (*f.*), target levelling-rod. (2)
vrac (en) *ou* **vrague (en)**, in bulk: (3)
expédier en vrac, to ship in bulk. (4)
vrai, -e (*adj.*), true; real: (5)
nord vrai (*m.*), true north. (6)
une vraie pierre précieuse, a real stone. (7)
vraisemblable (*adj.*), likely; probable. (8)
vraisemblablement (*adv.*), likely; probably. (9)
vrillage (*n.m.*), corkscrewing: (10)
le vrillage d'un fil, the corkscrewing of a wire.
(11)
vrille (*n.f.*), gimlet. (12)
vrille à torsade (*f.*), twist-gimlet. (13)
vrille façon suisse (*f.*), shell gimlet; Swiss bit.
(14)
vriller (percer avec une vrille) (*v.t.*), to gimlet;
to bore: (15)
vriller une planche, to gimlet a board. (16)
vriller (se tordre en se rétrécissant) (*v.i.*), to cork-
-screw: (17)
fil qui vrille (*m.*), wire which corkscrews. (18)
vrillerie (*n.f.*), small tools. (19)
vue (manière dont les objets se présentent **aux**
regards) (*n.f.*), view. (20)
vue d'ensemble (*f.*), general view. (21)
vue de bas en haut (*f.*), bottom view. (22)
vue de côté (*f.*), side view. (23)
vue de derrière (*f.*), back view. (24)
vue de face (*f.*), front view. (25)
vue de haut en bas (*f.*), top view. (26)

vue de projection (Photogr.) (*f.*), lantern-slide.
(27)
vue en bout (*f.*), end view. (28)
vue en coupe (*f.*), sectional view. **(29)**
vue en plan (*f.*), plan-view. (30)
vue instantanée (Photogr.) (*f.*), snap-shot;
snap. (31)
vue stéréoscopique (*f.*), stereoscopic view;
stereoscopic picture; stereoscopic slide. (32)
vues fondantes (*f.pl.*), dissolving views. (33)
vulcanicité (*n.f.*), vulcanisme (*n.m.*) (Géol.).
Same as **volcanicité, volcanisme.**
vulcanien, -enne (Géol.) (*adj.*), Vulcanian;
Plutonic: (34)
hypothèse vulcanienne (*f.*), Vulcanian theory;
Plutonic theory. (35)
vulcanisation (*n.f.*), vulcanization. (36)
vulcanisé, -e (*adj.*), vulcanized: (37)
fibre vulcanisée (*f.*), vulcanized fibre. **(38)**
vulcaniser (*v.t.*), to vulcanize: (39)
vulcaniser le caoutchouc, to vulcanize india-
rubber. (40)
vulcanite (*n.f.*), vulcanite; ebonite; hard
rubber. (41)
vulgaire (communément reçu) (*adj.*), common:
(42)
en Angleterre, l'unité vulgaire de puissance
s'appelle horse-power, in England, the
common unit of power is called horse-power.
(43)
vulgairement (*adv.*), commonly. **(44)**

W

wabstringue (*n.f.*), spokeshave. (45)
wacke (Pétrol.) (*n.f.*), wacke; wacky. (46)
wad (Minéral) (*n.m.*), wad; wadd. (47)
wagnérite (Minéral) (*n.f.*), wagnerite. (48)
wagon (Ch. de f.) (*n.m.*), wagon; truck; car;
carriage. (49) See note under **wagonnet.**
wagon (contenu, charge d'un wagon) (*n.m.*),
wagon; wagon-load; car; car-load; truck-
load; truck: (50)
un wagon de houille, a truck of coal. (51)
wagon à bascule (*m.*), dump-wagon; dump-
car; dumping-car; dumper; tipping-
wagon; tip-truck; tip-car; tip. (52)
wagon à bestiaux *ou* **wagon à bétail** (*m.*), cattle-
truck; cattle-wagon. (53)
wagon à charbons (*m.*), coal-wagon; coal-
truck. (54)
wagon à coke (*m.*), coke-wagon; coke-truck;
coke-car. (55)
wagon à couloir (*m.*), corridor-carriage. (56)
wagon à culbutage automatique (*m.*), self-
dumping wagon; self-tipping car. (57)
wagon à culbutage latéral (*m.*), side-tipping
truck; side-dump car; side-dumper. (58)
wagon à déchargement par bout (*m.*), end-
tipping wagon; endwise-tipping wagon. (59)
wagon à double bascule (*m.*), wagon to tip either
side; double side-tipping car. (60)
wagon à fond mobile (*m.*), drop-bottom wagon;
bottom-dump car. (61)
wagon à guérite (*m.*), lookout-car; cupola-car.
(62)

wagon à haussettes, à côtés rabattants et frein
à vis *ou* **wagon à haussettes, à côtés**
tombants et frein à vis (*m.*), low-sided wagon,
with sides hinged at bottom and screw-brake.
(63)
wagon à houille (*m.*), coal-wagon; coal-truck.
(64)
wagon à marchandises (*m.*), goods-wagon;
goods-truck; goods-van; freight-car. (65)
wagon à marchandises à hauts bords (*m.*), high-
sided goods-wagon; box car. (66)
wagon à minerai (*m.*), ore-wagon; ore-car.
(67)
wagon à ressorts (*m.*), spring-wagon; wagon
with springs. (68)
wagon à trémie (*m.*), hopper-car; hopper-
wagon. (69)
wagon-basculeur [**wagons-basculeurs** *pl.*] (*n.m.*).
Same as **wagon à bascule.**
wagon-boxe pour le transport de chevaux [**wagons-**
boxes *pl.*] (*m.*), horse-box. (70)
wagon chargé (*m.*), loaded wagon. (71)
wagon-citerne [**wagons-citernes** *pl.*] (*n.m.*), tank-
car. (72)
wagon-citerne pour le transport de pétrole (*m.*),
oil tank-car; petroleum-car. (73)
wagon couvert (*m.*), covered wagon; covered
truck; box wagon; box car; house-car.
(74)
wagon couvert à marchandises *ou* **wagon couvert**
pour transport de marchandises (*m.*), covered
goods-wagon. (75)

wagon culbutant dans tous les sens (*m.*), all-round-dumping wagon ; universal tipping wagon. (1)

wagon culbutant dans un sens (*m.*), wagon to tip one side. (2)

wagon-culbuteur [wagons-culbuteurs *pl.*] (*n.m.*). Same as **wagon à bascule.**

wagon de grande capacité (*m.*), high-capacity wagon. (3)

wagon de marchandises (*m.*). Same as **wagon à marchandises.**

wagon de — tonnes (*m.*), — -ton wagon ; — -ton truck : (4)
   wagon à charbons de 10 tonnes, 10-ton coal-wagon ; 10-ton coal-truck. (5)

wagon découvert (*m.*), open wagon ; open truck. (6)

wagon découvert pour transport de marchandises (*m.*), open goods-wagon. (7)

wagon détaché (Ch. de f.) (*m.*), slip-carriage. (8)

wagon fermé (*m.*). Same as **wagon couvert.**

wagon plat (*m.*), flat truck. (9)

wagon plat avec ranchers (*m.*), flat truck with stanchions. (10)

wagon plat-gondole à double trémie (*m.*), double-hopper gondola car. (11)

wagon plat pour le transport de bois (*m.*), timber-wagon ; timber-truck. (12)

wagon plate-forme [wagons plates-formes *pl.*] (*m.*), platform wagon ; platform car ; flat truck. (13)

wagon plate-forme, sur bogies (*m.*), bogie plat-form-wagon. (14)

wagon plein (*m.*), loaded wagon. (15)

wagon pour plan incliné (*m.*), incline-car. (16)

wagon pour transport de ballast, à 4 portes latérales rabattantes (*m.*), ballast-car, with 4 hinged side doors. (17)

wagon pour transport de rails (*m.*), rail-car. (18)

wagon-réservoir [wagons-réservoirs *pl.*] (*n.m.*). Same as **wagon-citerne.**

wagon sans ressorts (*m.*), springless wagon. (19)

wagon se vidant de côté ou par bout (*m.*), end-and-side-tip wagon. (20)

wagon se vidant sur le devant (*m.*), front-dis-charge wagon. (21)

wagon vide (*m.*), empty wagon ; empty truck ; empty. (22)

wagonnet (*n.m.*), wagon ; truck ; car. (23)
   Note.—The word **wagonnet**, being merely a diminutive of **wagon**, can be used in many of the descriptions given under that word where a small wagon is in point.

wagonnet à bec (*m.*), scoop wagon. (24)

wagonnet à fond mobile (*m.*), drop-bottom wagon ; bottom-dump car. (25)

wagonnet de mine (*m.*), mine-wagon ; mine-car ; mining-wagon ; mining-truck ; tub. (26)

wagonnet-porteur [wagonnets-porteurs *pl.*] (*n.m.*), trolley ; trolly. (27)

walchowite (Minéral) (*n.f.*), walchowite. (28)

warocquère (Mines) (*n.f.*), man-engine ; man-machine ; mining-engine ; travelling ladder-way ; movable ladder. (29)

water-jacket [water-jackets *pl.*] (chemise d'eau) (*n.m.*), water-jacket. (30)

water-jacket à cuivre (*m.*), copper water-jacket. (31)

watt (Élec.) (*n.m.*), watt ; volt-ampere ; ampere-volt. (32)

watt-heure [watts-heure *pl.*] (*n.m.*), watt-hour. (33)

watt-minute [watts-minute *pl.*] (*n.m.*), watt-minute. (34)

watt-seconde [watts-seconde *pl.*] (*n.m.*), watt-second. (35)

wattman [wattmen *pl.*] (pers.) (*n.m.*), motorman. (36)

wattmètre (Élec.) (*n.m.*), wattmeter. (37)

wavellite (Minéral) (*n.f.*), wavellite. (38)

wavellitique (*adj.*), wavellitic. (39)

wealdien, -enne (Géol.) (*adj.*), Wealden. (40)

wealdien (*n.m.*), Wealden : (41)
   le wealdien, the Wealden. (42)

webstérite (Minéral) (*n.f.*), websterite ; aluminite. (43)

wernérite (Minéral) (*n.f.*), wernerite ; scapolite. (44)

westfalite (Explosif) (*n.f.*), westfalite. (45)

wharf [wharfs *pl.*] (quai) (*n.m.*), wharf. (46)

whewellite (Minéral) (*n.f.*), whewellite. (47)

whitneyite (Minéral) (*n.f.*), whitneyite. (48)

willémite (Minéral) (*n.f.*), willemite. (49)

williamsite (Minéral) (*n.f.*), williamsite. (50)

wiluite (Minéral) (*n.f.*), wiluite. (51)

withamite (Minéral) (*n.f.*), withamite. (52)

withérite (Minéral) (*n.f.*), witherite. (53)

wittichénite *ou* wittichite (Minéral) (*n.f.*), witti-chenite. (54)

wöhlérite (Minéral) (*n.f.*), woehlerite. (55)

wolfram [W *ou* Tu] (Chim.) (*n.m.*), wolfram ; tungsten. (56)

wolfram [(Fe, Mn) WO^4] (Minéral) (*n.m.*), wolf-ramite ; wolfram. (57)

wolframiate (Chim.) (*n.m.*), wolframate ; tung-state. (58)

wolframine [WO^3] (Minéral) (*n.f.*), wolfram ocher ; tungstic ochre ; tungstite. (59)

wolframite [(Fe, Mn) WO^4] (Minéral) (*n.f.*), wolf-ramite ; wolfram. (60)

wolframocre [WO^3] (Minéral) (*n.m.*), wolfram ocher ; tungstic ochre ; tungstite. (61)

wolfsbergite (Minéral) (*n.f.*), wolfsbergite ; chalcostibite. (62)

wollastonite (Minéral) (*n.f.*), wollastonite ; tabular spar. (63)

wulfénite (Minéral) (*n.f.*), wulfenite ; yellow lead ore. (64)

wurtzite (Minéral) (*n.f.*), wurtzite. (65)

X

xanthophyllite (Minéral) (*n.f.*), xanthophyllite. (66)

xénomorphe (Pétrol.) (*adj.*), xenomorphic ; allotriomorphic. (67)

xénon (Chim.) (*n.m.*), xenon. (68)

xénotime (Minéral) (*n.f.*), xenotime. (69)

xylonite (*n.f.*), xylonite ; zylonite. (70)

xylotile (Minéral) (*n.m.*), xylotile. (71)

Y

yard (*y* aspiré) (mesure linéaire anglaise) (*n.m.*), yard = 0·914399 metre. (1)

yard carré (mesure anglaise) (*m.*), square yard = 0·836126 square metre. (2)

yard cube (mesure anglaise) (*m.*), cubic yard = 0·764553 cubic metre. (3)

ytterbium (Chim.) (*n.m.*), ytterbium ; neo-ytterbium ; aldebaranium. (4)

yttreux (Chim.) (*adj.m.*), yttrious. (5)

yttria (Chim.) (*n.m.*), yttria. (6)

yttrifère (*adj.*), yttriferous. (7)

yttrique (Chim.) (*adj.*), yttric. (8)

yttrite (Minéral) (*n.f.*), gadolinite. (9)

yttrium (Chim.) (*n.m.*), yttrium. (10)

yttrocalcite *ou* **yttrocérite** (Minéral) (*n.f.*), yttrocerite. (11)

yttrotantale (*n.m.*) *ou* **yttrotantalite** (*n.f.*) (Minéral), yttrotantalite. (12)

yttrotitanite (Minéral) (*n.f.*), yttrotitanite ; keilhauite. (13)

Z

zaratite (Minéral) (*n.f.*), zaratite ; emerald nickel. (14)

zénith (*n.m.*), zenith. (15)

zénithal, -e, -aux (*adj.*), zenithal : (16)

angles zénithaux (*m.pl.*), zenithal angles. (17)

zéolite (Minéral) (*n.f.*), zeolite. (18)

zéolitiforme (*adj.*), zeolitiform. (19)

zéolitique (*adj.*), zeolitic. (20)

zéro (*n.m.*), zero : (21)

le zéro d'une échelle, the zero of a scale. (22)

aiguille qui reste au zéro (*f.*), needle which remains at zero. (23)

zéro (Phys.) (*n.m.*), zero : (24)

le thermomètre est à zéro, à — degrés au-dessous de zéro, the thermometer is at zero, at — degrees below zero. (25 *ou* 26) Inasmuch, however, as the centigrade scale is used in France and the Fahrenheit scale in England, there is a considerable difference in the values of the terms. 0° C. = 32 F. 0° F. = —17·8 C.

zéro absolu (Phys.) (*m.*), absolute zero (= —273·13° C. *or* —459·64° F.). (27)

zeunérite (Minéral) (*n.f.*), zeunerite. (28)

zigzag (*n.m.*), zigzag. (29)

zigzag (en), zigzag : (30)

marcher en zigzag, to walk zigzag. (31)

zigzaguer (*v.i.*), to zigzag. (32)

zinc (*n.m.*), zinc ; spelter. (33)

zinc carbonaté (Minéral) (*m.*), zinc-spar ; smithsonite ; calamine. (34)

zincage (*n.m.*). Same as zingage.

zincico- (*préfixe*), zinco (*prefix*) ; zinc (*used as adj.*) : (35)

sel zincico-potassique (*m.*), zincopotassic salt ; zinc-potash salt. (36)

zincifère (*adj.*), zinciferous ; zinckiferous ; zinc-bearing : (37)

gîte zincifère (*m.*), zinc-bearing deposit. (38)

zincique (Chim.) (*adj.*), zincic. (39)

zincite (Minéral) (*n.f.*), zincite ; red zinc ore ; spartalite. (40)

zinckénite (Minéral) (*n.f.*), zinkenite. (41)

zinconise *ou* **zinconite** (Minéral) (*n.f.*), hydro-zincite ; zinc-bloom. (42)

zincosite (Minéral) (*n.f.*), zinkosite. (43)

zingage (*n.m.*), zinking ; zincing ; zinkification ; galvanization ; galvanizing : (44)

zingage de la tôle, zincing sheet iron ; galvan-ization of sheet iron. (45)

zinguer (*v.t.*), to zinc ; to zincify ; to galvanize : (46)

zinguer du fer (le galvaniser), to galvanize, to zincify, to zinc, iron. (47)

zinguer un toit (le couvrir avec du zinc), to zinc a roof. (48)

zinguerie (atelier) (*n.f.*), zinc-works. (49)

zingueur (pers.) (*n.m.*), zinc-worker. (50)

zingueux, -euse (Chim.) (*adj.*), zincous. (51)

zinnwaldite (Minéral) (*n.f.*), zinnwaldite. (52)

zircon (Minéral) (*n.m.*), zircon. (53)

zirconate (Chim.) (*n.m.*), zirconate. (54)

zircone (Chim.) (*n.f.*), zirconia ; zircona. (55)

zirconico- (*préfixe*), zirco ; zircono : (56)

sel zirconico-potassique (*m.*), zircopotassic salt. (57)

zirconien, -enne (*adj.*), zirconiferous ; zircon (*used as adj.*) : (58)

syénite zirconienne (*f.*), zircon syenite. (59)

zirconique (Chim.) (*adj.*), zirconic ; zirconian. (60)

zirconium (Chim.) (*n.m.*), zirconium. (61)

zoïsite (Minéral) (*n.f.*), zoisite. (62)

zonaire *ou* **zoné, -e** (*adj.*), zoned ; banded ; streaky : (63)

agate zonaire (*f.*), zoned agate ; banded agate. (64)

structure zonaire *ou* structure zonée (*f.*), banded structure ; streaky structure. (65)

zonal, -e, -aux (*adj.*), zonal. (66)

zone (Géogr., Géol., Géom., etc.) (*n.f.*), zone. (67)

zone (espace de pays long et étroit, caractérisé par quelque circonstance particulière ; étendue de pays formant une division) (*n.f.*), zone ; belt ; area. (68)

zone d'oxydation *ou* **zone oxydée** *ou* **zone d'altération** (Géol.) (*f.*), zone of oxidation ; oxidized zone ; zone of weathering ; weathering-zone : (69)

la zone d'altération est la couche superficielle de l'écorce terrestre située au-dessus du niveau hydrostatique, the weathering-zone is the top bed of the earth's crust situated above the water-table. (70)

zone d'oxydation *ou* **zone oxydante** (Métall.) (*f.*), zone of oxidation ; oxidizing zone. (71)

zone de cémentation (*f.*), cementation-zone. (72)

zone de combustion (Métall.) (*f.*), zone of combustion ; combustion zone. (1)

zone de contrée granitique (*f.*), belt of granitic country. (2)

zone de fissuration (*f.*), fissure-zone. (3)

zone de fracture de l'écorce terrestre (*f.*), zone of fracture of the earth's crust. (4)

zone de fusion (Métall.) (*f.*), melting zone ; zone of fusion. (5)

zone de plissement (Géol.) (*f.*), zone of folds. (6)

zone de réduction *ou* **zone réductrice** (Métall.) (*f.*), zone of reduction ; reducing zone. (7)

zone forestière (*f.*), belt of forest. (8)

zone inoxydée *ou* **zone inaltérée** (Géol.) (*f.*), unoxidized zone ; unweathered zone. (9)

zone minéralisée (*f.*), mineralized zone ; mineral-bearing zone ; mineral belt. (10)

zone non-reconnue (*f.*), unproved area. (11)

zone oxydée (Géol.) (*f.*). See **zone d'oxydation**.

zone pétrolifère (*f.*), oil-horizon ; petroleum-zone. (12)

zone sphérique (Géom.) (*f.*), spherical zone. (13)

zone stannifère (*f.*), tin-belt. (14)

zone torride (Géogr.) (*f.*), torrid zone. (15)

zones de l'onyx (*f.pl.*), zones, bands, of onyx. (16)

zones glaciales (Géogr.) (*f.pl.*), frigid zones. (17)

zones tempérées (Géogr.) (*f.pl.*), temperate zones ; variable zones. (18)

zorgite (Minéral) (*n.f.*), zorgite. (19)

zunyite (Minéral) (*n.f.*), zunyite. (20)

SUPPLEMENT
FRENCH-ENGLISH

A

abaque (graphique) (*n.m.*), chart; graph.
abaque de centrage (aviation) (*n.m.*), weight and balance chart.
abattée sur une aile (aviation) (*n.f.*), roll-off.
abattre (un angle) (*v.t.*), to chamfer; to remove sharp edges.
abcisse (*n.m.*), abcissa; X-axis.
abrégé (*n.m.*), abstract; abridgment; summary.
abrité, -e (construction des machines) (*adj.*), drip-proof.
abrité-grillagé, -e, -e (construction des machines) (*adj.*), drip-proof screen protected.
accélérateur (chim.) (*n.m.*), activator.
accélération arrêt (*n.f.*), acceleration stop.
accélération centrifuge (*n.f.*), centrifugal acceleration.
accélération centripète (*n.f.*), centripetal acceleration.
accélération normale de la pesanteur (*n.f.*), standard gravitation acceleration.
accéléromètre (*n.m.*), accelerometer.
accéléromètre latéral (*n.m.*), yaw-axis accelerometer.
accéléromètre transversal (*n.m.*), lateral accelerometer.
accès libre (*n.m.*), free access.
accessoire (*n.m.*), component; fitting.
accessoire d'installation (*m.*), installation accessory.
accessoiriste (*n.m.*), component manufacturer *or* stockist.
accord décalé (électronique) (*n.m.*), stagger tuning.
accostage (soudure) (*n.m.*), squeeze.
accotement (routes) (*n.m.*), verge.
accoudoir (*n.m.*), arm-rest.
accouplement à cannelures (*n.m.*), splined coupling.
accouplement à cardan (*n.m.*), cardan.
accouplement à cliquet (*n.m.*), pawl coupling.
accouplement à coquilles (*n.m.*), split coupling.
accouplement à déclic (*n.m.*), impulse coupling.
accouplement à douille (*n.m.*), sleeve coupling.
accouplement à friction (*n.m.*), friction clutch coupling.
accouplement à glissement (*n.m.*), slipping clutch.
accouplement à griffe (*n.m.*), dog coupling.
accouplement à ruban (*n.m.*), band coupling.
accouplement à segments extensibles (*n.m.*), spring ring coupling.

accouplement à surcharge (*n.m.*), torque limiting clutch.
accouplement articulé (*n.m.*), flexible shaft coupling.
accouplement de palier à cannelures (*n.m.*), bearing splined coupling.
accouplement de transmission arrière (aviation) (*n.m.*), tail drive shaft coupling.
accouplement électromagnétique (*n.m.*), electromagnetic coupling.
accouplement flexible (*n.m.*), flexible coupling.
accrochage (*n.m.*), catch.
accrochage de la charge (*n.m.*), load hook-up.
accrochage des trappes de train (aviation) (*n.m.*), landing-gear door latch.
accrochage sur un faisceau (aviation) (*n.m.*), lock-on.
accrochage train bas (aviation) (*n.m.*), landing-gear down latch.
accrochage train rentré (aviation) (*n.m.*), landing-gear up latch.
accroissement d'une fonction (*n.m.*), function increase.
accumètre (*n.m.*), accumulator capacity indicator.
accumulateur inversable (*n.m.*), non-spill battery.
accumulation de la glace (*n.f.*), ice accretion.
accusé de réception (*n.m.*), acknowledgment (of a document).
acéré, -e (*adj.*), sharp edged.
acétage de cellulose (plast.) (*n.m.*), cellulose acetate.
acétobutyrate de cellulose (plast.) (*n.m.*), cellulose acetate butyrate.
acétone (chim.) (*n.m.*), acetone.
acide acétique concentré (chim.) (*n.m.*), glacial *or* concentrated acetic acid (CH_3 CO OH).
acide sélénieux (chim.) (*n.m.*), selenious acid.
acide sulfurique fumant (chim.) (*n.m.*), fuming sulphuric acid.
acidité libre (chim.) (*n.f.*), free acidity.
acier allié (*n.m.*), alloy steel.
acier austénitique (*n.m.*), austenitic steel.
acier basique (*n.m.*), basic steel.
acier bleu (*n.m.*), spring steel.
acier brut (*n.m.*), crude steel.
acier calmé (*n.m.*), killed steel.
acier cémenté (*n.m.*), case-hardened steel.
acier de décolletage (*n.m.*), free-cutting steel.
acier demi-doux (*n.m.*), low carbon steel.
acier effervescent (*n.m.*), rimming steel.

acier eutectoïde (*n.m.*), eutectoid steel.
acier extra-doux (*n.m.*), dead soft steel.
acier ferreux (*n.m.*), ferrous steel.
acier martensitique (*n.m.*), martensitic steel.
acier matricé (*n.m.*), drop-forged steel.
acier moulé (*n.m.*), cast steel.
acier non allié (*n.m.*), plain carbon steel.
acier plaqué (*n.m.*), plated steel.
acier poli (*n.m.*), bright steel.
acier réfractaire (*n.m.*), heat-resisting steel.
acier stubs (*n.m.*), silver steel.
acier trempant à l'air (*n.m.*), air-hardening steel.
acquisition (électronique) (*n.f.*), acquisition of a beam.
acrylique (plast.) (*adj.*), acrylic.
actionneur (*n.m.*), actuator.
activant (*n.m.*), dope (aircraft construction).
activateur (électronique) (*n.m.*), activator
actuel, -elle (*adj.*), current.
adaptateur à démontage rapide (*n.m.*), quick release adapter.
adaptateur de tuyère (*n.m.*), nozzle adapter.
adaptateur pour plateaux magnétiques (*n.m.*), additional top plate for magnetic chucks.
adaptation-client (*n.f.*), customization.
adaptation plastique (*n.f.*), plastic flow properties.
additif antimousse (*n.m.*), defoaming agent.
additionneur (*n.m.*), adder.
additive (*n.m.*), additive.
adduction d'air (*n.f.*), air supply; air ducting.
adhésif (*n.m.*), adhesive.
adjonction (*n.f.*), addition.
adjuvant (*n.m.*), additive.
admissible (*adj.*), permissible.
admission (*n.f.*), inlet.
admission (auto) (*n.f.*), induction.
admittance (élect.) (*n.f.*), admittance.
adoucissement à la flamme (*n.f.*), flame annealing.
adoucissement de l'eau (*n.m.*), water softening.
adoucisseur d'eau (*n.m.*), water softener.
adresse absolue (électronique) (*n.f.*), absolute address.
adsorption (*n.f.*), adsorption.
advection (*n.f.*), advection.
aérateur (*n.m.*), ventilator; aerator.
aérien, -ienne (*adj.*), aerial.
aérodrome (*n.m.*), airfield; airport.
aérodrome de dégagement (*n.m.*), alternate airport.
aérodynamique (*n.f.*), aerodynamics.
aérodyne à voilure tournante (*n.m.*), rotary-wing aircraft, helicopter.
aéro-élasticité (*n.f.*), aero-elasticity.
aérofrein (*n.m.*), air brake.
aéromètre (*n.m.*), aerometer.
aéronef (*n.m.*), aircraft (in general).
aéroport (*n.m.*), airport; aerodrome.
aéroporté, -e (*adj.*), airborne.
aérothermodynamique (*adj.*), aerothermo-dynamics.
aérosol (*n.m.*), aerosol.
affaiblissement (élec.) (*n.m.*), attenuation; decay; fading.
affichage (lecture) (*n.m.*), display.
affichage (pétrole) (*n.m.*), posting.
affichage (réglage) (*n.m.*), setting; setting up.

affichage inverse (*n.m.*), reciprocal setting.
afficher (*v.t.*), to set; to select; to set up.
agencement (*n.m.*), layout.
agent de conservation (*n.m.*), preservative.
agent donnant du collant (plast.) (*n.m.*), tackifier.
agent mouilleur (*n.m.*), wetting agent.
agent retardateur (plast.) (*n.m.*), retarder.
agglomérant (trav. pub.) (*n.m.*), binder; matrix; cement.
agglomération (plast.) (*n.f.*), pelletizing.
agrafe (*n.f.*), clip.
agrandissement (photo) (*n.m.*), blow-up; enlargement.
agréments (*n.m.pl.*), amenities.
aigu, -e (*adj.*), high-pitched.
aiguille (d'un compteur) (*n.f.*), pointer.
aile (aviation) (*n.f.*), wing.
aile de cornière (*n.f.*), extrusion flange.
aile en flèche (aviation) (*n.f.*), sweptback wing.
aile rotor à cycle chaud (aviation) (*n.f.*), hot cycle rotor wing.
aileron (aviation) (*n.m.*), aileron.
aileron compensé (aviation) (*n.m.*), balanced aileron.
ailette (*n.f.*), lug.
ailettes (aviation) (*n.f.pl.*), sponsons.
ailettes (de refroidissement) (*n.f.pl.*), fins.
aimant de concentration (faisceau électro--nique) (*n.m.*), focussing magnet.
aimant de signalisation (ch. de fer) (*n.m.*), magnetic needle (for instruments).
aimant électro-permanent (*n.m.*), electro permanent magnet.
aimant permanent (magnétisme) (*n.m.*), per--manent magnet.
air ambiant (*n.m.*), surrounding air.
air de dégivrage (*n.m.*), de-icing air.
air de dilution (moteur) (*n.m.*), (engine) by-pass air.
air de réchauffage (*n.m.*), heating air.
air de refroidissement (*n.m.*), cooling air.
air forcé (*n.m.*), air pressure in excess of ambient; supercharged air.
aire de compensation (compas) (*n.f.*), compass compensation base.
aire de stationnement (aviation) (*n.f.*), park--ing area; apron.
ajustage (*n.m.*), alignment.
ajustement à chaud (*n.m.*), hot shrink fit.
ajustement à force (*n.m.*), drive fit.
ajustement à froid (*n.m.*), cold shrink fit.
ajustement à frottement dur (*n.m.*), force fit.
ajustement à la presse (*n.m.*) press fit.
ajustement auto-calibré (*n.m.*), shear fit.
ajustement avec serrage (*n.m.*), interference fit.
ajustement doux (*n.m.*), slip fit.
ajustement glissant (*n.m.*), slide fit.
ajustement glissant juste (*n.m.*), close slid-ing fit.
ajustement gras (*n.m.*), push fit.
ajustement libre (*n.m.*), loose fit.
ajustement serré (*n.m.*) tight fit.
ajustement tournant (*n.m.*), running fit.
ajutage de Pitot (*n.m.*), Pitot tube.
alarme (*n.f.*), warning signal.
alarme lumineuse (*n.f.*), warning light.
alarme sonore (*n.f.*), audible warning signal.
alarme sonore monocoup (*n.f.*), single stroke alarm bell.

alcool éthylique (*n.m.*), ethyl-alcohol.
alcool polyvinylique (plast.) (*n.m.*), polyvinyl alcohol.
alésage calibré (*n.m.*), reamed bore.
alésoir ébaucheur (*n.m.*), roughing reamer.
alésoir en bout (*n.m.*), chucking reamer.
aligné, -e (*adj.*), in line with.
alignement de descente (aviation) (*n.m.*), glide path.
alignement des pales (aviation) (*n.m.*), blade tracking.
alimentation (élec.) (*n.f.*), power supply.
alimentation générale (élec.) (*n.f.*), common power supply.
alimentation secteur (elec.) (*n.f.*), mains supply.
alimenté (-e) en parallèle (*adj.*), parallel fed.
alimenter (élec.) (*v.t.*), to supply; to energize.
alios ferrugineux (géol.) (*n.m.*), iron pan.
alkyde (plast.) (*n.f.*), alkyd resin; glyptal resin
alliage binaire (*n.f.*), binary alloy.
alliage coulé au sable (*n.m.*), sand cast alloy.
alliage coulé en coquille (*n.m.*), chill cast alloy.
alliage eutectique (*n.m.*), eutectic alloy.
alliage léger (*n.m.*), light alloy.
alliage non-ferreux (*n.m.*), non-ferrous alloy.
allite (sol.) (*adj.*), allitic.
allongement d'une aile (aviation) (*n.m.*), wing aspect ratio.
allongement de la pale (aviation) (*n.m.*), blade aspect ratio.
allongement en % (*n.m.*), strain.
allumage continu (*n.m.*), continual relight.
allumage prématuré (auto) (*n.m.*), backfire.
allumé, -e (adj.), lit; on.
allumer (s') (*v.r.*), to light up; to come on.
allumeur (auto) (*n.m.*), distributor.
allumeur-torche (*n.m.*), torch-igniter.
allure (d'un moteur) (*n.f.*), rate.
alodinage (*n.m.*), alodinizing.
alternance (*n.f.*), cycle.
alternat (*n.m.*), microphone push-to-talk button.
alternateur à fer tournant (élec.) (*n.m.*), inductor alternator.
alternateur à réaction (élec.) (*n.m.*), reaction alternator.
alternateur asynchrone (élec.) (*n.m.*), asyn-chron alternator.
alternateur auto-excitateur à induit tournant (élec.) (*n.m.*), self-excited alternat-ing current generator with revolving armature.
alternateur de synchronisation (*n.m.*), synchronizing alternator.
alternateur synchrone (élec.) (*n.m.*), syn-chronous generator; alternator.
alternatif brut (*n.m.*), unrectified AC.
altimètre à tambour (aviation) (*n.m.*), drum altimeter.
altimètre de cabine (aviation) (*n.m.*), cabin altimeter.
altitude cabine (aviation) (*n.f.*), cabin altitude.
altitude d'attente (aviation) (*n.f.*), holding altitude.
altitude de croisière (aviation) (*n.f.*), cruising altitude.
altitude de rétablissement (aviation) (*n.f.*), critical altitude, height.
altitude de sécurité (*n.f.*), safe altitude, height.
altitude densité (*n.f.*), height density.

altitude du terrain (aviation) (*n.f.*), airport altitude, height.
altitude fictive (aviation) (*n.f.*), cabin altitude, height.
altitude géométrique (*n.f.*), true height.
altitude maximum en exploitation (avia-tion) (*n.f.*), maximum operating altitude, height.
altitude nominale (aviation) (*n.f.*), rated alti-tude, height.
altitude pression (*n.f.*), pressure height.
alumel (*n.m.*), alumel.
alumilitage (*n.m.*), aluminizing.
aluminothermie (*n.f.*), thermit; Goldschmidt process.
alvéolaire (*adj.*), cellular; honeycomb structure.
alvéole (*n.f.*), recess.
amagnétique (*adj.*), non magnetic.
amarrage (*n.m.*), stowing.
ambiance (*n.f.*), environment.
ambiant, -e (*adj.*), ambient; environmental.
âme (alésage) (*n.f.*), bore.
âme (voile) (*n.f.*), web.
âme d'acier (*n.f.*), steel strip insertion.
âme de longeron (aviation) (*n.f.*), spar web.
âme de poutre de fuselage (aviation) (*n.f.*), fuselage box beam wall.
amenage rapide (machines-outils) (*n.m.*), quick (*or* rapid) traverse.
amenage rapide automatique (machines-outils) (*n.m.*), power rapid traverse.
aménagement (*n.m.*), accommodation; furnishings.
aménagement fixe de soutes (aviation) (*n.m.*), cargo compartment equipment.
aménagement office (*n.m.*), galley furnishings.
aménagements commerciaux (*n.m.pl.*), equipment; furnishings.
amenée d'air (*n.f.*), air duct.
amenée de courant (*n.f.*), electrical lead.
américium (*n.m.*), americium.
amerrissage (aviation) (*n.m.*), landing on water.
amerrissage forcé (aviation) (*n.m.*), ditching.
amidon (*n.m.*), starch.
aminoplaste (plast.) (*n.m.*), aminoplastic resin.
amorçage (formation d'un arc) (*n.m.*), arcing; flashover.
amorçage acoustique (*n.m.*), microphonics; Larsen effect.
amorçage d'arc (élec.) (*n.m.*), arc ignition; arcing.
amorçage d'un relais (*n.m.*), energizing relay.
amorçage d'une pompe (*n.m.*), priming.
amorce (élec.) (*n.f.*), starting bath.
amorce de crique (*n.f.*), incipient crack.
amortir (*v.t.*), to absorb; to damp.
amortissement (*n.m.*), absorption; damping; choking.
amortissement (de l'équipement, du matériel) (*n.m.*), depreciation.
amortissement interne (*n.m.*), internal damping.
amortisseur à friction (*n.m.*), friction damper.
amortisseur à ressort (*n.m.*), spring-type shock absorber.
amortisseur auxiliaire (*n.m.*), secondary shock strut.
amortisseur caoutchouc (*n.m.*), bumper.
amortisseur de commande (aviation) (*n.m.*), control damper.
amortisseur de direction (aviation) (*n.m.*),

rudder damper.

amortisseur de lacet (aviation) (*n.m.*), yaw damper.

amortisseur de pale (aviation) (*n.m.*), rotor blade damper.

amortisseur de palonnier (aviation) (*n.m.*), pedal damper assembly.

amortisseur de sabot (aviation) (*n.m.*), tail bumper oleo-pneumatic shock absorber.

amortisseur de shimmy (aviation) (*n.m.*), shimmy damper.

amortisseur de tangage (aviation) (*n.m.*), pitch damper.

amortisseur de traînée (aviation) (*n.m.*), drag damper.

amortisseur de vibrations (*n.m.*), shock mount.

amortisseur des sautes de pression (*n.m.*), surge damper.

amortisseur électrique (*n.m.*), electrical damper.

amortisseur hydraulique (*n.m.*), dashpot.

amortisseur oléopneumatique (*n.m.*), oleo-pneumatic shock absorber.

amortisseur orientable (*n.m.*), steerable shock absorber.

amortisseur pneumatique (*n.m.*), air dashpot.

amortisseur principal (*n.m.*), landing gear main shock strut.

amovible (*adj.*), interchangeable.

ampérage (élec.) (*n.m.*), amperage.

ampère-tour (élec.) (*n.m.*), ampere-turn.

ampèremètre (*n.m.*), ammeter.

amplificateur (élec.) (*n.m.*), amplifier.

amplificateur à contre réaction (radio) (*n.m.*), negative-feedback amplifier.

amplificateur basse fréquence (radio) (*n.m.*), audio-amplifier.

amplificateur d'asservissement (électro-nique) (*n.m.*), slaving amplifier.

amplificateur d'effacement (aviation) (*n.m.*), washout amplifier.

amplificateur d'intégration (aviation) (*n.m.*), integrator amplifier.

amplificateur d'interphone (*n.m.*), inter-phone amplifier.

amplificateur d'isolement (*n.m.*), isolating amplifier.

amplificateur de faisceau et d'altitude (aviation) (*n.m.*), beam and altitude amplifier.

amplificateur de gyro (*n.m.*), gyro amplifier.

amplificateur de puissance (*n.m.*), power amplifier.

amplificateur de régulation de tempé-rature (*n.m.*), temperature control amplifier.

amplificateur de sécurité (*n.m.*), shut-down amplifier.

amplificateur de sonorisation (*n.m.*), public address amplifier.

amplificateur de synchro d'erreur de cap (aviation) (*n.m.*), heading error synchronizer amplifier.

amplificateur en cascade (*n.m.*), cascade amplifier.

amplificateur haute fréquence (radio) (*n.m.*), radio frequency amplifier.

amplificateur linéaire (*n.m.*), linear amplifier.

amplificateur magnétique (*n.m.*), magnetic amplifier.

amplificateur principal (*n.m.*), main amplifier.

amplitude maximum de l'intensité de rafale (aviation) (*n.f.*), derived gust velocity.

ampoule électrique (*n.f.*), bulb; lamp.

ampoule grillée (*n.f.*), burnt-out bulb.

analogique (électronique) (*adj.*), analogous.

analyse volumétrique (*n.f.*), volumetric analysis.

anéchoïde (*adj.*), anechoic.

angle d'attaque (*n.m.*), angle of attack.

angle d'attaque de pale (aviation) (*n.m.*), blade pitch angle; feathering angle.

angle d'atterrissage (aviation) (*n.m.*), ground angle.

angle d'enroulement de la courroie (*n.m.*), arc of contact of the belt.

angle d'incidence (*n.m.*), incidence angle.

angle d'incidence de pale (aviation) (*n.m.*), blade angle of attack.

angle de basculement du disque balayé (aviation) (*n.m.*), rotor disc tilt angle.

angle de battement (aviation) (*n.m.*), flapping angle.

angle de battement cyclique (aviation) (*n.m.*), cyclic flapping angle.

angle de biellette de pas (aviation) (*n.m.*), pitch control rod angle.

angle de braquage des gouvernes (avia-tion) (*n.m.*), control surface angle.

angle de braquage du train avant (avia-tion) (*n.m.*), steering angle.

angle de calage (*n.m.*), setting angle.

angle de calage de l'aile (aviation) (*n.m.*), angle of wing setting angle.

angle de calage de pale (aviation) (*n.m.*), blade setting angle.

angle de calage des balais (élec.) (*n.m.*), brush displacement.

angle de carrossage (auto) (*n.m.*), camber angle.

angle de chasse (auto) (*n.m.*), castor angle.

angle de cintrage (*n.m.*), bend angle.

angle de conicité (aviation) (*n.m.*), coning angle.

angle de couplage (élec.) (*n.m.*), circuit angle.

angle de crabe (aviation) (*n.m.*), crab angle.

angle de déclination (*n.m.*), declination angle.

angle de décrochage (*n.m.*), angle of stall.

angle de déflexion des filets d'air vers le bas (aviation) (*n.m.*), angle of downwash.

angle de déflexion des filets d'air vers le haut (aviation) (*n.m.*), angle of upwash.

angle de déflexion du sillage rotor (avia-tion) (*n.m.*), rotor wake skew angle.

angle de dépouille (affûtage) (*n.m.*), relief angle; rake angle.

angle de dérapage (aviation) (*n.m.*), sideslip angle.

angle de dérive (aviation) (*n.m.*), drift angle.

angle de filet (*n.m.*), lead angle.

angle de flèche (*n.m.*), sweep angle.

angle de lacet (aviation) (*n.m.*), yaw angle.

angle de levée de pale (aviation) (*n.m.*), flap-ping angle.

angle de pas (hélicoptère) (*n.m.*), pitch angle; blade angle.

angle de pas de la pale (hélicoptère) (*n.m.*), blade pitch angle.

angle de pente (aviation) (*n.m.*), angle of glide.

angle de phase (*n.m.*), phase angle.

angle de plané (aviation) (*n.m.*), gliding angle.

angle de portance nulle (*n.m.*), zero lift angle of attack.

angle de rayonnement (radio) (*n.m.*), radia--tion angle; wave angle.

angle de retard (aviation) (*n.m.*), lag angle.

angle de route (aviation) (*n.m.*), course angle.

angle de [*ou* du] sillage rotor (hélicoptère) (*n.m.*), rotor wake angle.

angle de site (*n.m.*), elevation angle.

angle de tangage (aviation) (*n.m.*), pitch angle.

angle de traînée (aviation) (*n.m.*), drag angle.

angle entre la vitesse d'un point de pale et la vitesse relative (*n.m.*), inflow angle.

angle induit (*n.m.*), inflow angle.

angle induit (aviation) (*n.m.*), induced attack angle.

angle mort (*n.m.*), blind angle.

angle ouvert (n.m.), open angle.

angle plat (à 180°) (*n.m.*), flat angle.

angle polyèdre (*n.m.*), solid angle.

angle vif (*n.m.*), sharp edge.

angledozer (*n.m.*), angledozer.

Ångström (mesure des longueurs d'ondes) (*n.m.*), Ångström unit; tenth metre.

aniline-formol (plast.) (*n.m.*), aniline formaldehyde.

animateur (radio) (*n.m.*), disc-jockey.

anneau (construction) (*n.m.*), loop.

anneau collecteur (*n.m.*), collector ring.

anneau d'amarrage au sol (*n.m.*), mooring ring.

anneau d'amarrage chargement (*n.m.*), stowing ring.

anneau d'attelage (*n.m.*), towing ring.

anneau d'azimuth extérieur (*n.m.*), outer axis gimbal.

anneau d'entrée d'air (*n.m.*), nacelle intake ring.

anneau de butée (*n.m.*), stop ring.

anneau de débit (*n.m.*), flow straightener ring.

anneau de distributeur de turbine (*n.m.*), turbine nozzle shroud.

anneau de hissage (*n.m.*), hoisting ring.

anneau de levage (*n.m.*), hoisting ring.

anneau de retenue (*n.m.*), retainer ring.

anneau de roue libre (*n.m.*), freewheel driven head.

anneau de suspension (*n.m.*), axis gimbal.

anneau entretoise (*n.m.*), spacer sleeve; spacer ring.

anneau réciproque (*n.m.*), droop restraining ring.

année budgétaire (*n.f.*), financial year.

annexe (*n.f.*), enclosure; appendix.

annulateur de frottement (*n.m.*), vibration damper.

annulateur de manche (*n.m.*), stick canceller.

annuler et remplacer (*v.t.*), to supersede.

anode d'entretien (élec.) (*n.f.*), excitation anode.

anode de shuntage (élec.) (*n.f.*), by-pass anode.

anode de transfert (élec.) (*n.f.*), transition anode.

anodisation (traitement des métaux) (*n.f.*), anodizing.

anomalie (*n.f.*), discrepancy.

antenne (radio) (*n.f.*), aerial; antenna (U.S.A.).

antenne à feeder coaxial (radio) (*n.f.*), coax--ial aerial; coaxial antenna (U.S.A.).

antenne anémométrique (radio) (*n.f.*), pilot static tube.

antenne artificielle (radio) (*n.f.*), mute aerial.

antenne bipolaire [*ou* dipole] (radio) (*n.f.*), dipole aerial.

antenne cadre (*n.f.*), loop aerial.

antenne d'incidence (*n.f.*), incidence probe.

antenne de balise (*n.f.*), marker aerial.

antenne de lever de doute (*n.f.*) sense aerial.

antenne de pente (*n.f.*), glide aerial.

antenne directive (*n.f.*), directional aerial.

antenne en losange (*n.f.*), rhombic aerial.

antenne encastrée (*n.f.*), flush aerial.

antenne fouet (*n.f.*), whip aerial.

antenne omnidirectionnelle (*n.f.*), non-directional aerial.

antenne pendante (*n.f.*), trailing aerial.

antenne quart d'onde (*n.f.*), quarter wave aerial.

antenne sabre (*n.f.*), blade aerial.

antenne tournante (radar) (*n.f.*), scanner.

anti-brouillage (*n.m.*), anti-jamming.

anti-brouillard (phare) (*n.m.*), fog light.

anti-brouilleur (*n.m.*), anti-jammer.

anti-bruit (*n.m.*), anti-chatter.

anti-buée (*n.m.*), demister.

anti-corrosif (*n.m.*), anti-corrosive.

anti-déflagrant, -e (*adj.*), flame-proof; explo--sion-proof (U.S.A.). (La norme américaine est assez semblable à la norme française, mais dans la majorité des cas *flame-proof* et *explosion-proof* décrivent souvent des articles de con--struction complètement différente.).

anti-éblouissant, -e (*adj.*), anti-dazzling; glare free.

anti-givrage (*n.m.*), anti-icing.

anti-mousse (*n.m.*), anti-froth.

anti-parasitage (radio) (*n.m.*), noise suppression.

anti-parasite (radio) (*n.m.*), suppressor.

anti-parasité, -e (radio) (*adj.*), shielding.

anti-retour de flamme (*n.m.*), flame trap.

anti-rouille (*n.m.*), anti-rust.

anti-shimmy (*n.m.*), shimmy damper.

anti-vibratoire (*adj.*), shock-proof.

antibiotique (*n.m.*), antibiotics.

antibourrage (électronique) (*n.m.*), anti-blocking.

antidérapant, -e (*adj.*), non-skidding; skid-proof.

antidétonnant, -e (essence) (*adj.*), anti-knock.

antifading (radio) (*n.m.*), automatic volume control.

antifriction (*n.m.*), anti-friction.

antimonage (*n.m.*), antimony plating.

antiprotos (*n.m.*), anti-protos.

aplati, -e (*adj.*), flattened.

aplatissement (*n.m.*), collapsing.

aplatissement du pneumatique (*n.m.*), flat--tening of the tyre.

aplomb (*n.m.*), indent number.

appareil (aviation) (*n.m.*), aircraft.

appareil (mécanique) (*n.m.*), instrument; set; device.

appareil à distillation discontinue (raffi--nerie) (*n.m.*), batch still.

appareil à fraiser les angles et les axes (*n.m.*), cherrying attachment.

appareil à fraiser les crémaillères (*n.m.*), rack milling attachment.

appareil à fraiser les filets (machines-outils) (*n.m.*), thread milling attachment.

appareil contre la surdité (*n.m.*), hearing-

appareil de commande (*n.m.*), control gear.
appareil de correction auditive (*n.m.*), hearing aid.
appareil de mesure des épaisseurs à rayons bêta (*n.m.*), beta gauge.
appareil de prise de vues (cinéma) (*n.m.*), camera; cine-camera; movie-camera.
appareil humidifiant (*n.m.*), humidifier.
appareillage d'essais (*n.m.*), test equipment.
appareillage de commutation (*n.m.*), switchgear.
appareils de radio-guidage (*n.m.pl.*), aids to navigation; radio-control.
apparier (*v.t.*), to match.
apparition graduelle (*n.f.*), fade-in.
appel codé (*n.m.*), code tone.
appel commandant de bord (aviation) (*n.m.*), captain call.
appel sélectif (*n.m.*), selective call.
appendice (*n.m.*), appendix.
applique de raccord (*n.f.*), wall connector.
apprêt (peinture) (*n.m.*), primer.
approbation (*n.f.*), acceptance; approval.
approche asymptotique (*n.f.*), level *or* asymptotic approach.
approche automatique (aviation) (*n.f.*), autoland approach.
approche directe (aviation) (*n.f.*), straight-in approach.
approche finale (aviation) (*n.f.*), final approach.
approche initiale (aviation) (*n.f.*), initial approach.
approche sur radar de précision (*n.f.*), precision approach.
approvisionnement cyclique (*n.m.*), routine placing of orders.
approximation (*n.f.*), approximation.
appui de cric (auto) (*n.m.*), jacking pad.
appui-tête (auto) (*n.m.*), head rest.
appui vertical (*n.m.*). prop.
aptitude à la remise en état (*n.f.*), serviceability.
araignée (hélicoptère) (*n.f.*), star; spider.
araignée de changement de pas (hélicoptère) (*n.f.*), pitch change spider.
araser (*v.t.*), to machine flush.
arborescence (*n.f.*), arborescence.
arbre à casser (*n.m.*), shear shaft.
arbre à collerette (*n.m.*), flanged shaft.
arbre à rotule (*n.m.*), swivel end shaft.
arbre bridé (*n.m.*), flanged shaft.
arbre cannelé (*n.m.*), splined shaft.
arbre creux (*n.m.*), hollow shaft.
arbre d'articulation (*n.m.*), hinge shaft.
arbre d'entraînement (*n.m.*), drive shaft.
arbre d'entrée (*n.m.*), input shaft.
arbre de montage pour fraises (machines-outils) (*n.m.*), cutter adaptor.
arbre de montage pour mandrin (machines-outils) (*n.m.*), chuck adaptor.
arbre de pignon (*n.m.*), gear shaft.
arbre de relais (*n.m.*), connecting shaft.
arbre de renvoi (*n.m.*), layshaft.
arbre de repérage (*n.m.*), locating arbor.
arbre de rotor (hélicoptère) (*n.m.*), main rotor shaft.
arbre de sortie (*n.m.*), output shaft.
arbre de torsion (*n.m.*), torque shaft.
arbre de transmission (auto) (*n.m.*), trans-mission shaft.

arbre de transmission dans pylône (hélicoptère) (*n.m.*), pylon drive shaft.
arbre de transmission principale (hélicoptère) (*n.m.*), main drive shaft.
arbre de transmission rotor arrière (hélicoptère) (*n.m.*), tail rotor drive shaft.
arbre flottant (*n.m.*), free shaft.
arbre intermédiaire (*n.m.*), layshaft.
arbre lisse (*n.m.*), arbor.
arbre porte-galet (*n.m.*), main shaft.
arbre primaire (*n.m.*), main shaft.
arbre secondaire (*n.m.*), layshaft.
arbre téléscopique (*n.m.*), telescopic shaft.
arc de lecture (d'un instrument) (*n.m.*), swing.
arc en retour (élec.) (*n.m.*), flash-back.
arceau support de réacteur (*n.m.*), engine support arch.
archet porte-scie (*n.m.*), saw frame.
archives (*n.f.pl.*), records; files.
arête de dérive (aviation) (*n.f.*), fin leading edge.
arête dorsale fuselage (aviation) (*n.f.*), fuselage dorsal fin.
arêtier (*n.m.*), ledge.
argentan (*n.m.*), German silver; nickel silver.
armature de balai d'essuie glace (*n.f.*), wiper blade backing strip.
armature de câble (élec.) (*n.f.*), cable sheath.
armer (un mécanisme) (*v.t.*), to set; to cock; to arm.
armoire (*n.f.*), cabinet.
armoire de rangement (*n.f.*), stowage equipment.
armoire électrique (*n.m.*), electrical equipment cabinet.
arrache-aiguille (*n.m.*), pointer jack; puller.
arrache-moyeu (*n.m.*), hub extractor.
arrachement de métal (*n.m.*), metal pick-up.
arrangement (*n.m.*), array.
arrêt (*n.m.*), "off" (*adj.*).
arrêt d'un moteur en vol (aviation) (*n.m.*), engine shut-down in flight.
arrêt de câble (*n.m.*), cable stop.
arrêt de gaine (*n.m.*), sleeve stop.
arrêter (moteur) (*v.t.*), to shut down; to stop.
arrêtoir (*n.m.*), retainer.
arrêtoir de joint (*n.m.*), seal retainer.
arrière (*n.f.*), aft.
arrimage (*n.m.*), stowing, mooring (boat or aircraft).
arrondi, -e (*adj.*), rounded off (figures).
arrosage (machines-outils) (*n.m.*), cooling.
article (*n.m.*), item.
articulation (*n.f.*), hinge; connection; link.
articulation à rotule (*n.f.*), ball joint.
articulation de battement (hélicoptère) (*n.f.*), flapping hinge.
articulation de battement excentrée (hélicoptère) (*n.f.*), offset flapping hinge.
articulation de pas (hélicoptère) (*n.f.*), blade pitch change hinge.
articulation de repliage des pales (hélicoptère) (*n.f.*), blade folding hinge.
articulation de repliage du pylône (hélicoptère) (*n.f.*), pylon folding hinge.
articulation de traînée (hélicoptère) (*n.f.*), drag hinge.
articulation rotor (hélicoptère) (*n.f.*), rotor hingeing.
ascensionnel, -elle (*adj.*), up-; climbing.

courant ascensionnel (*n.m.*), updraft.
vitesse ascensionnelle (*n.f.*), climbing speed.
aspirateur de poussière (*n.m.*), vacuum cleaner.
aspiration (*n.f.*), intake.
assemblage (*n.m.*), assembly.
assemblage à baïonnette (*n.m.*), bayonet joint.
assemblage en fausse coupe (*n.m.*), bevel joint.
assemblé (-e) et équipé (-e) (*adj.*), complete assembly.
asservi, -e (électronique) (*adj.*), slaved.
asservissement (électronique) (*n.m.*), control device; automatic guidance system; servo system.
assiette (aviation) (*n.f.*), attitude.
assiette de vol (aviation) (*n.f.*), flight attitude.
assiette longitudinale (aviation) (*n.f.*), pitch attitude.
assiette transversale (aviation) (*n.f.*), roll attitude.
assistance technique (*n.f.*), technical support.
assortir (*v.t.*), to match.
astate (*n.m.*), astatine.
asymptote (*n.m.*), asymptote.
-ate (suffixe chimique), -ate.
atelier de montage (*n.m.*), assembly shop.
atmosphère agitée (*n.f.*), turbulent air.
atmosphériques (*n.f. pl.*), statics; atmos--pherics.
atomisation (*n.f.*), atomisation.
attache (*n.f.*), fastener; clamp.
attache rapide (*n.f.*), quick connection.
attaque acide (*n.f.*), etching.
attaque d'un faisceau (*n.f.*), beam interception.
attaque électrolytique (*n.f.*), electrolytic corrosion.
attaquer (*v.t.*), to actuate; to engage.
attente (aviation) (*n.f.*), stand-off; holding.
attente, en (*adj., adv.*), stand-by.
atténuateur (*n.m.*), damper.
atténuation (*n.f.*), attenuation; damping.
atténuer (*v.t.*), to weaken.
atterrissage (aviation) (*n.m.*), landing.
atterrissage aux instruments (*n.m.*), instrument landing.
atterrissage brutal (*n.m.*), rough landing.
atterrissage en autorotation (hélicoptère) (*n.m.*), autorotation landing.
atterrissage en cabré (*n.m.*), tail landing.
atterrissage en flare (*n.m.*), flared landing.
atterrissage en surcharge (*n.m.*), over land--ing (U.S.A.).
atterrissage forcé (*n.m.*), forced landing.
atterrissage manqué (*n.m.*), balked landing.
atterrissage moteur réduit (*n.m.*), glide landing.
atterrissage par vent de travers (*n.m.*), cross wind landing.
atterrissage sans visibilité (*n.m.*), blind landing.
atterrissage sur trois moteurs (*n.m.*), three-engine landing.
atterrissage tout-temps (*n.m.*), all-weather landing.
atterrissage train rentré (*n.m.*), belly landing.
atterrissage trop court (*n.m.*), undershoot.
atterrissage trop long (*n.m.*), overshoot.

atterrissage vent arrière (*n.m.*), tail wind landing.
atterrissage vertical (hélicoptère) (*n.m.*), ver--tical landing.
atterrisseur (*n.m.*), landing gear.
atterrisseur à flotteurs (hélicoptère) (*n.m.*), floating gear; seaplane floats.
atterrisseur à patins (hélicoptère) (*n.m.*), skid-type landing gear.
atterrisseur auxiliaire (hélicoptère) (*n.m.*), auxiliary landing gear.
atterrisseur avant (hélicoptère) (*n.m.*), nose gear.
atterrisseur principal (hélicoptère) (*n.m.*), main landing gear.
attitude (*n.f.*), attitude.
attraction (magnétisme) (*n.f.*), magnetic attraction.
attraction moléculaire (*n.f.*), cohesive force.
attraction universelle (*n.f.*), gravitation.
attraits (*n.m.pl.*), amenities.
attrition (*n.f.*), attrition.
au droit de (*adv.*), in line with; level with.
aubage (d'un réacteur) (*n.m.*), shroud.
aube caisson (*n.f.*), shrouded blade.
aube de distributeur de turbine (*n.f.*), noz--zle guide vane.
aube de pré-rotation (*n.f.*), intake guide vane.
aube de turbulence (*n.f.*), swirl vane.
aube directrice d'entrée (*n.f.*), intake guide vane.
aube fixe de stator (*n.f.*), fixed stator vane.
aube mobile de rotor (hélicoptère) (*n.f.*), mov--able rotor blade.
aube renforcée de turbine (hélicoptère) (*n.f.*), shrouded blade.
audibilité de l'émission (radio) (*n.f.*), clarity of transmission.
audiomètre (*n.m.*), audiometer; sound meter.
auditif, -ive (*adj.*), aural.
augmentation de l'incidence à l'extrémité de l'aile (*n.f.*), wash-in.
augmentation de pas (hélicoptère) (*n.f.*), pitch increase.
auto-alignement (*n.m.*), self-aligning.
auto-allumage (*n.m.*), compression ignition.
auto-cabrage (*n.m.*), pitch-up.
auto-étanche (adj.), self-sealing.
auto-excitation (*n.f.*), self-energization.
auto-freineur, -euse (*adj.*), self-locking.
auto-induction (*n.f.*), self-induction.
auto-obturateur, -trice (*adj.*), self-sealing.
auto-réglable (*adj.*), self-adjusting.
auto-trempant, -e (*adj.*), self-hardening.
autochenille (*n.f.*), half-track vehicle.
autodémarrage (*n.m.*), self-starting.
autogire (*n.m.*), autogyro; gyroplane.
autoguidage actif (*n.m.*), homing active guidance.
autoguidage passif (*n.m.*), homing passive guidance.
autoguidage semi-actif (*n.m.*), homing semi-active guidance.
automanette (*n.f.*), autothrottle.
automate (électronique) (*n.m.*), automation.
automatisation (*n.f.*), automation.
automatisme (*n.m.*), automatic device.
autonome (*adj.*), self-contained.
autonomie (distance) (*n.f.*), range.
autonomie (temps) (*n.f.*), endurance.
autorefroidi, -e (*adj.*), self-cooled.

autorisation de vol (*n.f.*), flight clearance.
autoriser l'excitation (*v.*), to close the ener-
-gizing circuit.
autorotation (*n.f.*), autorotation.
autorotation du réacteur (hélicoptère) (*n.f.*),
engine windmilling.
autosyn (*n.f.*), autosyn.
autotransformateur (*n.m.*), autotransformer.
autour de l'axe (*adv.*), about the axis.
auvent (auto) (*n.m.*), cowl.
auvent d'éclairage (*n.m.*), glare shield.
auxiliaire (*adj.*), auxiliary; secondary.
avachissement des ressorts (*n.m.*), slack-
-ening of the springs.
aval (*n.m.*), endorsement.
avance (*n.f.*), advance.
avance automatique (machines-outils) (*n.f.*),
power feed.
avance de la broche (machines-outils) (*n.f.*),
spindle feed.
avance de la table (machines-outils) (*n.f.*),
table feed.
avance de la tête (machines-outils) (*n.f.*), head
feed.
avance descendante (machines-outils) (*n.f.*),
down feed.
avance en plongée (machines-outils) (*n.f.*),
in-feed.
avance lente (machines-outils) (*n.f.*), slow-feed.
avance longitudinale (machines-outils) (*n.f.*),
traverse feed.
avance manuelle (machines-outils) (*n.f.*), hand
feed.
avance mécanique (machines-outils) (*n.f.*),
power-feed.
avance rapide (machines-outils) (*n.f.*), coarse
feed; rapid feed.
avance transversale (machines-outils) (*n.f.*),
cross feed; traverse feed.
avant (aviation) (*n.m.*), forward.
avant-trou (*n.m.*), pilot-hole.
avenant (*n.m.*), amendment.
avertissement (*n.m.*), caution; alarm; warning;
notification.
avertisseur à son grave (*n.m.*), low-tone
horn.
avertisseur de décrochage (aviation) (*n.m.*),
stall warning indicator.
avertisseur de givrage (aviation) (*n.m.*), ice
detector.
avertisseur de Mach (aviation) (*n.m.*), audible
machmeter.
avertisseur sonore (*n.m.*), warning bell.
avion (*n.m.*), aircraft; airplane; plane.
avion à voilure basculante (*n.m.*), tilt-wing
aircraft.
avion combiné (*n.m.*), composite aircraft.
avion convertible (*n.m.*), convertible plane;
vertiplane.
avion d'affaires (*n.m.*), executive aircraft.
avion de pré-série (*n.m.*), pre-production
aircraft.
avion de série (*n.m.*), production aircraft.

avion en service (*n.m.*), aircraft in service.
avion lisse (*n.m.*), clean configuration.
avion prototype (*n.m.*), prototype aircraft.
avionneur (*n.m.*), aircraft manufacturer.
avis de modification (*n.m.*), notice of
modification.
axe (*n.m.*), centre line; axis; axle; spindle.
axe balisé (radio) (*n.m.*), beam.
axe cannelé (*n.m.*), splined pin.
axe creux (*n.m.*), hollow pin.
axe creux fendu (*n.m.*), spring pin.
axe d'arrêt (*n.m.*), stop pin.
axe d'articulation (*n.m.*), hinge pin.
axe d'articulation de train principal
(aviation) (*n.m.*), landing gear shaft.
axe d'entraînement (*n.m.*), drive pin.
axe d'incidence (hélicoptère) (*n.m.*), pitch-
change axis.
axe d'instrument (*n.m.*), staff.
axe de battement (hélicoptère) (*n.m.*), flapping
hinge pin.
axe de changement de pas (hélicoptère)
(*n.m.*), pitch change rod.
axe de chape (*n.m.*), clevis pin.
axe de charnière (*n.m.*), hinge pin.
axe de fixation (*n.m.*), fitting bolt.
axe de galet (*n.m.*), roller pin.
axe de guidage (*n.m.*), guide pin.
axe de la pale (hélicoptère) (*n.m.*), blade span
axis.
axe de lacet (aviation) (*n.m.*), yaw axis.
axe de levée de pale (hélicoptère) (*n.m.*), flap-
-ping hinge pin.
axe de piston (auto) (*n.m.*), gudgeon-pin (G.B.);
piston pin, wrist pin (U.S.A.).
axe de pivotement (*n.m.*), fulcrum pin.
axe de profondeur (*n.m.*), pitch axis.
axe de référence (*n.m.*), datum line.
axe de repliage (hélicoptère) (*n.m.*), folding
axis.
axe de roues (*n.m.*), wheel axle.
axe de roulis (*n.m.*), roll axis.
axe de tangage (*n.m.*), pitch axis.
axe de torsion (*n.m.*), torque shaft.
axe de traînée (hélicoptère) (*n.m.*), drag hinge
pin.
axe du croisillon du joint universel (*n.m.*),
universal joint cross pin.
axe épaulé (*n.m.*), shoulder pin; stepped pin.
axe expansible (*n.m.*), expansion pin.
axe fileté (*n.m.*), threaded pin.
axe fût balancier (*n.m.*), shock compensating
rocker beam shaft.
axe géométrique (*n.m.*), axis; centre line.
axe latéral (*n.m.*), lateral.
axe lisse (*n.m.*), shear pin.
axe longitudinal (*n.m.*), longitudinal axis.
axe normal (*n.m.*), normal axis.
axe principal (*n.m.*), centre line.
axe transversal (*n.m.*), lateral axis.
axial, -e (*adj.*), axial.
azoture (*n.m.*), nitride.

B

bac (*n.m.*), pan; tank.
bac d'accumulateur (*n.m.*), battery tray.

bac d'ionisation (*n.m.*), anodizing tank.
bac de décantation d'acide (pétr.) (*n.m.*),

acid drum.
bac de récupération (*n.m.*), drain pan.
bâche de protection (*n.f.*), cover.
bactéricide (*n.m.*), bactericide.
bactérie des nodosités (*n.f.*), nodule bacteria.
badin (*n.m.*), air speed indicator.
bagage (*n.m.*), baggage.
baguage (*n.m.*), bush setting.
bague (*n.f.*), sleeve; ring.
bague butée (*n.f.*), thrust ring.
bague cannelée (*n.f.*), splined bushing.
bague collectrice (élec.) (*n.f.*), slip ring; collec--tor ring.
bague d'appui (*n.f.*), thrust bushing.
bague d'arrêt (*n.f.*), stop ring.
bague d'entraînement (*n.f.*), coupling ring.
bague de blocage (*n.f.*), lock bushing.
bague de brochage (*n.f.*), rigging bush.
bague de centrage (*n.f.*), centering bush.
bague de friction (*n.f.*), thrust ring.
bague de guidage (*n.f.*), guide bush.
bague de projection (*n.f.*), oil slinger.
bague de réglage (*n.f.*), setting ring.
bague de retenue (*n.f.*), retainer.
bague de roulement (*n.f.*), bearing race.
bague de serrage (*n.f.*), collet.
bague de sertissage (*n.f.*), crimping bush.
bague entretoise (*n.f.*), spacer bushing.
bague épaulée (*n.f.*), shouldered bush.
bague étroite (*n.f.*), ring.
bague excentrée (*n.f.*), eccentric bush.
bague extérieure de roulement (*n.f.*), outer race.
bague fendue (*n.f.*), split bush.
bague filetée (*n.f.*), threaded bush.
bague flottante (*n.f.*), floating bush; floating ring.
bague intérieure de roulement (*n.f.*), inner race.
bague jauge (*n.f.*), ring gauge.
bague large (*n.f.*), sleeve.
bague sphérique (*n.f.*), spherical bush.
baguette de métal d'apport (soudure) (*n.f.*), filler rod.
baguette de moulure (*n.f.*), moulding.
baguette de soudure (*n.f.*), filler rod.
baillement (*n.m.*), gap.
bain de fusion (soudure) (*n.m.*), molten pool.
bain électrolytique (*n.m.*), electrolytic solution.
baïonnette (*n.f.*), bayonet.
baisse de débit (*n.f.*), drop in the rate of flow.
baisse de niveau (*n.f.*), reduction in level.
baisse de pression (*n.f.*), pressure drop.
baisse de tension (*n.f.*), voltage drop.
bakélite (*n.f.*), bakelite.
baladeuse (*n.f.*), inspection lamp.
balai d'essuie-glace (*n.m.*), wiper blade.
balance aérodynamique (*n.f.*), aerodynamic balance.
balancier (commande du volet obturateur de fente) (aviation) (*n.m.*), balance arm.
balancier (outillage) (*n.m.*), screw press.
balancier (train) (aviation) (*n.m.*), shock compensating rocker beam (landing gear).
balancier (trappes de train) (aviation) (*n.m.*), rocker arm of the landing gear sliding doors.
balancier, rotor en (hélicoptère) (*n.m.*), seesaw-type rotor.
balancine (aviation) (*n.f.*), outrigger.
balata (*n.m.*), balata.

balayage (radar) (*n.m.*), scanning.
balayage (moteur diesel) (*n.m.*), scavenging.
balayage écran radar (*n.m.*), radar scope sweep.
balayage en hélice (radar) (*n.m.*), helical scanning.
balayer (radar) (*v.t.*), to scan; to sweep.
balisage (aviation) (*n.m.*), ground lighting.
balise (*n.f.*), marker.
balise à faisceau dirigé oblique (*n.f.*), fan marker.
balise d'approche (radio) (*n.f.*), marker.
balise d'entrée de piste (radio) (*n.f.*), landing strip marker.
balise d'extrémité (aviation) (*n.f.*), boundary light.
balise de piste (aviation) (*n.f.*), runway light.
balise extérieure (radio) (*n.f.*), outer marker.
balise intérieure (radio) (*n.f.*), middle marker.
ballon de levage (aviation) (*n.m.*), lifting bag.
ballon-sonde (météo.) (*n.m.*), pilot balloon.
ballonnet de stabilization (aviation) (*n.m.*), stabilizing float.
ballonnets de secours gonflables (*n.m.pl.*), inflatable pontoons.
balourd (*n.m.*), unbalance.
banc (*n.m.*), stand.
banc d'équilibrage dynamique rotor (hélicoptère) (*n.m.*), rotor dynamic balancing stand.
banc d'essai (*n.m.*), test bench.
banc d'essai réacteur (aviation) (*n.m.*), engine test stand.
banc d'essai volant (aviation) (*n.m.*), flying test bench.
banc de charge (*n.m.*), charging bench.
banc vibrant (trav. publics) (*n.m.*), vibrating table.
bandage de roue (*n.m.*), solid tyre.
bande (*n.f.*), tape; channel (radio).
bande à écrous prisonniers (*n.f.pl.*), gang channel.
bande à maroufler (*n.f.*), winding tape.
bande anti-érosion (*n.f.*), erosion strip.
bande crantée (*n.f.*), notched band.
bande d'atterrissage (aviation) (*n.f.*), landing strip.
bande d'usure (volets de courbure) (aviation) (*n.f.*), rub strip.
bande de fréquences (élec.) (*n.f.*), frequency band.
bande de frottement (*n.f.*), chafing strip.
bande de métallisation (*n.f.*), bonding strip.
bande de modulation (*n.f.*), modulation band.
bande de protection de collier de ser--rage (*n.f.*), clamping band; cushion.
bande de retenue (*n.f.*), retaining band.
bande élastique (*n.f.*), elastic band.
bande latérale inférieure (radio) (*n.f.*), lower sideband.
bande latérale supérieure (radio) (*n.f.*), upper sideband.
bande latérale unique (radio) (*n.f.*), single sideband.
bande magnétique (*n.f.*), magnetic tape.
bande passante (*n.f.*), pass band.
bande renfort (*n.f.*), backing band.
bande stratifiée (*n.f.*), laminated strip.
bande-vidéo (*n.f.*), video-tape.
bandeau (*n.m.*), band.
banjo (*n.m.*), banjo union.

banquette du poste de pilotage (aviation) (*n.f.*), console.
baquet (aviation) (*n.m.*), bottom structure.
baquet de siège (*n.m.*), seat pan.
bar (mesure de pression = hectopièze) (*n.m.*), bar (1 bar = 1 hpz. = 1,02 kg./cm² = 14,5 lb./sq. in.).
barbe (tranchant d'un outil) (*n.f.*), cutting edge.
barbotage (*n.m.*), splashing.
barbotage (d'un gaz) (*n.m.*), bubbling.
barboteur (laboratoire) (*n.m.*), wash bottle.
bardeau (building) (*n.m.*), shingle.
barillet de sécurité (*n.m.*), safety lock.
barographe (*n.m.*), barograph.
barographe enregistreur (*n.m.*), recording barograph.
barostat (*n.m.*), barostat.
barque (hélicoptère) (*n.f.*), bottom structure.
barrage à contreforts (*n.m.*), buttress dam.
barrage en enrochement (*n.m.*), rock-fill dam.
barrage en terre cylindrée (*n.m.*), rolled earth dam.
barrage poids (*n.m.*), gravity dam.
barrage poids déversoir (*n.m.*), gravity spill--way dam.
barrage-voûte (*n.m.*), arch dam.
barre (élec.) (*n.f.*), busbar.
barre batterie (*n.f.*), battery bar.
barre d'accouplement (direction auto) (*n.f.*), track rod.
barre d'écartement (*n.f.*), spreader bar.
barre d'induit (élec.) (*n.f.*), armature bar.
barre de commande de direction (auto) (*n.f.*), steering drag.
barre de frein (*n.f.*), brake bar.
barre de guidage (*n.f.*), steering bar.
barre de guidage en l'air (tour) (*n.f.*), over--head pilot bar.
barre de masse (aviation) (*n.f.*), grounding bar.
barre de relevage pour roues (aviation) (*n.f.*), wheel retraction bar.
barre de remorquage (*n.f.*), towing bar.
barre de rappel (*n.f.*), drag link.
barre de sécurité (électronique) (*n.f.*), stand-by battery bar.
barre de torsion (auto) (*n.f.*), torsion bar; tor--que shaft.
barre délestable (électronique) (*n.f.*), shedder bar.
barre dérivée (*n.f.*), secondary bar.
barre essentielle (*n.f.*), essential bar.
barre génératrice (*n.f.*), generator bar.
barre-levier (*n.f.*), jimmy bar.
barre négative (élec.) (*n.f.*), negative busbar.
barre omnibus (élec.) (*n.f.*), busbar.
barre principale (élec.) (*n.f.*), main bar.
barre secondaire (élec.) (*n.f.*), secondary bar.
barreau de traction (*n.m.*), tensile test bar.
barres de tendence (aviation) (*n.f.pl.*), hori--zontal and vertical bars of flight director.
barrette (roue) (*n.f.*), wheel drive block.
barrette à bornes (élec.) (*n.f.*), terminal bar; terminal strip.
barrette à cosses (élec.) (*n.f.*), tag block.
barrette à plots (élec.) (*n.f.*), contact stud bar.
barrette de connexions (élec.) (*n.f.*), con--necting strap.
barrette de masse (élec.) (*n.f.*), grounding strip.
barrette de raccordement (élec.) (*n.f.*), con-

-nection strip.
barrière de décrochage (*n.f.*), wing fence.
barrière thermique (*n.f.*), thermal barrier.
bas-volet (aviation) (*n.m.*), tab.
basane (*n.f.*), sheepskin.
basculant, -e (*adj.*), rocking.
bascule (*n.f.*), weighing cell.
bascule (élec.) (*n.f.*), rocking lever.
basculer (*v.t.*), to overturn.
basculeur (*n.m.*), rocker.
basculeur de train principal (aviation) (*n.m.*), main gear axle beam.
base de temps (*n.f.*), time base.
basse fréquence (radio) (*n.f.*), audio frequency.
basse fréquence (élec.) (*n.f.*), low frequency.
basse tension (élec.) (*n.f.*), low voltage.
bassin versant (*n.m.*), catchment basin.
bâti (*n.m.*), mount.
bâti d'assemblage (*n.m.*), assembly jig.
bâti d'équilibrage (*n.m.*), balance jig.
bâti d'essai (*n.m.*), test jig.
bâti de réception (*n.m.*), engine stand.
bâti de réparation (*n.m.*), repair jig.
bâti de reprise (*n.m.*), out of jig cradle.
bâti de soudage (*n.m.*), welding jig.
bâti mécanique (hélicoptère) (*n.m.*), main gear-box support.
bâti moteur (*n.m.*), engine mount.
bâti tournant (*n.m.*), rotating jig.
battement (élec.) (*n.m.*), beat.
battement (hélicoptère) (*n.m.*), flapping.
battement d'une aiguille (*n.m.*), hunting.
battement de vibration (*n.m.*), buffeting.
battement régulier (*n.m.*), beat.
battement vertical (hélicoptère) (*n.m.*), flapping.
batterie (groupe) (*n.f.*), bank.
batterie à électrolyte libre (*n.f.*), dry charged battery.
batterie au plomb (*n.f.*), lead-acid battery.
batterie de bord (*n.f.*), main battery (on an aircraft).
batterie de condensateurs (élec.) (*n.f.*), bank of capacitors.
batterie de polarisation de grille (radio) (*n.f.*), grid polarisation battery.
batterie de réserve (élec.) (*n.f.*), emergency battery.
batterie de secours (élec.) (*n.f.*), emergency battery.
batterie de sécurité (*n.f.*), stand-by battery.
batterie de tension anodique (radio) (*n.f.*), B-battery.
batterie tampon (élec.) (*n.f.*), buffer battery.
bec d'attaque (aviation) (*n.m.*), leading edge.
bec de compensation (*n.m.*), balance horn.
bec de dérive (aviation) (*n.m.*), vertical stabi--lizer leading edge.
bec de descente (*n.m.*), downspout.
bec de nervure (aviation) (*n.m.*), leading edge rib.
bécher (*n.m.*), beaker.
bélière (*n.f.*), suspension brace.
bélinographie (*n.f.*), facsimile telegraphy.
béquille (moto) (*n.f.*), stand.
béquille arceau (hélicoptère) (*n.f.*), tail rotor.
béquille arrière (hélicoptère) (*n.f.*), tail skid.
béquille de combinateur (hélicoptère) (*n.f.*), mixer rod.
béquille de queue (aviation) (*n.f.*), supporting strut.

béquille de sécurité (aviation) (*n.f.*), tail prop.

berceau (outillage) (*n.m.*), cradle.

berceau moteur (*n.m.*), engine mounting.

bêtatron (*n.m.*), betatron.

béton précontraint (*n.m.*), prestressed concrete.

biais (*n.m.*), bias.

bicarré, -e (*adj.*), biquadratic (the 4th power of a number).

bichromate (*n.m.*), dichromate.

bicône (*n.m.*), cone union body.

bielle (*n.f.*), link.

bielle à contact (*n.f.*), force link.

bielle à ressort (*n.f.*), spring rod.

bielle à rotule (*n.f.*), spherical head rod.

bielle d'attaque (*n.f.*), actuating rod.

bielle de clapet (*n.f.*), valve rod.

bielle de commande (*n.f.*), control rod.

bielle de compensation (aviation) (*n.f.*), land-ing gear compensation rod.

bielle de contreventement (*n.f.*), strut.

bielle de liaison (*n.f.*), link rod; connecting rod.

bielle de recul (*n.f.*), drag brace.

bielle de suspension (*n.f.*), suspension rod.

bielle de triangulation (*n.f.*), bracing truss.

bielle diagonale (aviation) (*n.f.*), landing gear diagonal truss.

bielle double (*n.f.*), dual rod.

bielle entretoise (*n.f.*), intermediate rod.

bielle tournante (*n.f.*), rotating control rod.

bielle va-et-vient (*n.f.*), push-pull rod.

bielle vérin (*n.f.*), actuating rod.

biellette (*n.f.*), link; rod.

biellette coupe-feu (aviation) (*n.f.*), fuel shut off cock control link.

biellette de commande de pas (hélicoptère) (*n.f.*), blade pitch-change rod.

biellette de liaison (*n.f.*), link.

biellette double (aviation) (*n.f.*), landing gear fork rod.

biellette sélectrice (*n.f.*), sloppy link.

biellette vitesse (aviation) (*n.f.*), governor control link.

bilame (*n.m.*), bimetallic strip; (élec.) thermal switch.

bilan énergétique d'une réaction nucléaire (*n.m.*), Q-value.

bilan thermique d'une réaction chimique (*n.m.*), thermal value of a chemical reaction.

bille-aiguille (*n.f.*), turn and bank indicator.

billion (*n.f.*), billion (U.S.A. = 1 thousand million; G.B. = 1 million million).

biphasé, -e (*adj.*), two-phase.

bipolaire à deux directions (élec.) (*n.m.*), double-pole double-throw.

bipolaire à une direction (élec.) (*n.m.*), double-pole single-throw.

bi-réacteur (aviation) (*n.m.*), twin jet.

blanc d'Espagne (*n.m.*), whiting.

blanchiment (décoloration) (*n.m.*), bleaching.

blanchir (décolorer) (*v.t.*), to bleach.

bleuissage (*n.m.*), bluing of metal.

blindage (*n.m.*), shield.

blindage d'induit (*n.m.*), armature casing.

blindé, -e (*adj.*), armoured; armour-clad; shielded.

bloc (*n.m.*), unit; pack.

bloc à contact (*n.m.*), contact block.

bloc à fusible (élec.) (*n.m.*), fuse block.

bloc à shunt (élec.) (*n.m.*), shunt block.

bloc cylindres (auto) (*n.m.*), power unit.

bloc d'accessoires radar (*n.m.*), radar acces--sory unit.

bloc d'alimentation (*n.m.*), power unit supply.

bloc d'alimentation (tube fluorescent) (*n.m.*), ballast.

bloc de commande (manettes) (*n.m.*), quadrant.

bloc de serrage (outillage) (*n.m.*), wrench adaptor.

bloc de transformation (*n.m.*), matching transformer unit.

bloc électrodes hélice (*n.m.*), solenoid valve block assembly.

bloc frein (*n.m.*), brake assembly.

bloc moteur (auto) (*n.m.*), power unit.

bloc pompe/régulateur de carburant (*n.m.*), fuel pump and control unit.

bloc raccord hydraulique (*n.m.*), connecting block for hydraulics.

bloc selfs de filtrage (*n.m.*), filter choke unit.

blocage (*n.m.*), binding; jamming; seizing.

blocage de gouverne (aviation) (*n.m.*), control surface locking.

blocage des gyros (aviation) (*n.m.*), gyro caging.

blocage des manettes des gaz (*n.m.*), throt--tle friction lock.

blocage hydraulique (*n.m.*), hydraulic lock.

blocage réception (*n.m.*), sidetone dead-stop point.

bloqué, -e (*adj.*), blocked; jammed.

bloquer (radio) (*v.t.*), to cut off.

bloquer à refus (un écrou) (*v.*), to screw up a nut spanner tight.

bobinage (élec.) (*n.m.*), coiling.

bobinage imbriqué (élec.) (*n.m.*), lap winding.

bobine à ruban (*n.f.*), spool.

bobine d'accord (*n.f.*), tuning coil.

bobine d'allumage (*n.f.*), ignition coil.

bobine d'amortissement (*n.f.*), choke.

bobine d'arrêt (*n.f.*), choke.

bobine d'articulation (*n.f.*), hinge shaft.

bobine d'équilibrage (élec.) (*n.f.*), arc suppres--sion coil; balancing coil.

bobine d'excitation (*n.f.*), field coil.

bobine d'induit (*n.f.*), armature coil.

bobine de champ (*n.f.*), field coil.

bobine de compensation (*n.f.*), bucking coil.

bobine de concentration (faisceau électro--nique) (*n.f.*), focussing coil.

bobine de départ (*n.f.*), booster coil.

bobine de filtrage (*n.f.*), smoothing coil.

bobine de liaison (*n.f.*), spacer.

bobine de maintien (*n.f.*), holding coil.

bobine de polarisation (*n.f.*), bias coil.

bobine de réactance (*n.f.*), reactor.

bobine de réaction (*n.f.*), feedback coil.

bobine de recopie (*n.f.*), output multiplier.

bobine de relais (*n.f.*), trip coil.

bobine de soufflage (élec.) (*n.f.*), blow-out coil.

bobine de verrou (*n.f.*), lock bush.

bobine dérouleuse (*n.f.*), take-off spool.

bobine en fond de panier (élec.) (*n.f.*), basket coil.

bobine en nid d'abeilles (élec.) (*n.f.*), lattice coil.

bobine enrouleuse (*n.f.*), take-up spool.

bobine entretoise (*n.f.*), spacer.

bobine exploratrice (élec.) (*n.f.*), search coil.

bobine guide-câble (*n.f.*), fairlead.

bobine interchangeable (élec.) (*n.f.*), plug-in coil.

bobine mobile (*n.f.*), moving coil.
bogie porte-galets (*n.f.*), roller carriage.
boisseau (*n.m.*), sleeve.
boîte à bornes (*n.f.*), terminal.
boîte à clapets (*n.f.*), valve box.
boîte à détritus (*n.f.*), waste-bin.
boîte à documents (*n.f.*), document box.
boîte à fusibles (*n.f.*), fuse box.
boîte à graisse (*n.f.*), grease packing gland.
boîte à Jacks (*n.f.*), Jack box.
boîte à papier (*n.f.*), waste paper basket.
boîte à pharmacie (*n.f.*), first-aid box.
boîte à relais (*n.f.*), relay box.
boîte à résistance (*n.f.*), resistor box.
boîte accord d'antenne (*n.f.*), aerial tuner.
boîte d'accord (*n.f.*), tuner.
boîte d'accord automatique d'antenne (*n.f.*), automatic aerial tuner.
boîte d'alarme (*n.f.*), warning box.
boîte d'alimentation (*n.f.*), power supply unit.
boîte d'allumage (*n.f.*), ignition box.
boîte d'amplification (*n.f.*), amplifier.
boîte d'appel (*n.f.*), call-box.
boîte d'écoute (radio) (*n.f.*), audio control panel.
boîte d'expansion carburant (saumon d'aîle) (aviation) (*n.f.*), fuel ullage box.
boîte de claquage (*n.f.*), flash tester.
boîte de commande (*n.f.*), control box; control unit.
boîte de commande à distance (*n.f.*), remote control box.
boîte de commande d'accessoires (aviation) (*n.f.*), accessory gearbox.
boîte de commande de pilote automatique (aviation) (*n.f.*), auto-pilot control unit.
boîte de commande train (aviation) (*n.f.*), landing gear control unit.
boîte de commande V.H.F. (*n.f.*), V.H.F. control unit.
boîte de commutation gyro (aviation) (*n.f.*), gyro data switching control.
boîte de décades (*n.f.*), decade box.
boîte de démarrage réacteur (aviation) (*n.f.*), engine starting control box.
boîte de dérivation (élec.) (*n.f.*), junction box.
boîte de détection pour pilote automatique (aviation) (*n.f.*), auto-pilot gyro sensor unit.
boîte de gyromètre pour pilote automatique (aviation) (*n.f.*), angular three-axis rate sensor.
boîte de jonction coupleur (*n.f.*), aerial tuner transfer unit.
boîte de mesure (*n.f.*), meter.
boîte de minirupteur (*n.f.*), microswitch box.
boîte de raccordement (élec.) (*n.f.*), connection box.
boîte de recopie de cap (élec.) (*n.f.*), heading synchro-transmitter.
boîte de relais d'alimentation et d'interdiction de pilote automatique (aviation) (*n.f.*), power and interlock relay box.
boîte de relais débitmètre (*n.f.*), flowmeter instrumentation unit.
boîte de répartition (conditionnement d'air) (*n.f.*), distribution chamber.
boîte de tarage pyromètre (*n.f.*), thermocouple calibration unit.
boîte de temporisation (*n.f.*), timer.
boîte de transmission (*n.f.*), transmission gearbox.
boîte de transmission arrière (hélicoptère) (*n.f.*), tail rotor gearbox.
boîte de transmission intermédiaire (hélicoptère) (*n.f.*), intermediate gearbox.
boîte de transmission principale (hélicoptère) (*n.f.*), main gearbox.
boîte des avances (machines-outils) (*n.f.*), feed-box.
boîte des changements d'avance (machines-outils) (*n.f.*), change-feed box.
boîte noire (*n.f.*), black box.
boîtier (*n.m.*), casing; case; unit.
boîtier à cellules redresseuses (*n.m.*), rectifier box.
boîtier à inertie (*n.m.*), inertia reel (safety belt).
boîtier d'accrochage train rentré (aviation) (*n.m.*), landing gear up-lock box.
boîtier d'accrochage trappes de train (aviation) (*n.m.*), landing gear door latching box.
boîtier d'éclairage (*n.m.*), box-type lamp.
boîtier d'étanchéité (*n.m.*), seal housing.
boîtier de cloche (hélicoptère) (*n.f.*), bell housing.
boîtier de crépine (*n.m.*), filter housing.
boîtier de démarrage (*n.m.*), automatic starting unit.
boîtier de détection de givrage (*n.m.*), ice probe.
boîtier de pré-affichage carburant (aviation) (*n.m.*), fuel level pre-setting controls.
boîtier de verrou (*n.m.*), lock casing.
boîtier relais d'hélice (*n.m.*), propeller relay unit.
boîtier tangentiel (*n.m.*), tangential gearbox.
boîtier téléphonique (*n.m.*), telephone box.
bolomètre (thermomètre à résistance) (*n.m.*), bolometer.
bombe à hydrogène (*n.f.*), H bomb.
bombe atomique (*n.f.*), atomic bomb.
bomber (*v.t.*), to dish.
bombonne (*n.f.*), carboy.
bon de commande (*n.m.*), order.
bon de travail (*n.m.*), work order.
bonhomme de verrouillage (boîte de vitesse) (*n.m.*), locking plunger.
bonnette à trèfle (*n.f.*), dimmer cap.
bord à bord (*adj.*, *adv.*), edge-to-edge.
bord croqué (*n.m.*), pinked edge; crimped edge.
bord d'attaque (*n.m.*), leading edge.
bord d'attaque de pale (hélicoptère) (*n.m.*), blade leading edge.
bord de fuite (aviation) (*n.m.*), trailing edge.
bord de fuite de pale (hélicoptère) (*n.m.*), blade trailing edge.
bord marginal (*n.m.*), tip edge.
bord tombé (*n.m.*), bent over *or* turned over edge; flanged edge.
bordereau d'envoi (*n.m.*), release note.
bordereau de livraison (*n.m.*), consignment note.)
bordure (*n.f.*), trimming.
bordure d'encadrement (*n.f.*), frame edging.
bordure de marche (*n.f.*), step edging.
borne (*n.f.*), terminal.
borne de batterie (n.f.), battery terminal.
borne de masse (élec.) (*n.f.*), earth terminal.
borne de métallisation (*n.f.*), bonding stud.
borne de piquage (élec.) (*n.f.*), tap terminal.
bossage de came (*n.f.*), cam lobe.

bossage de montage (*n.f.*), mounting pad.
bosseler (*v.t.*), to dent.
bosselure (*n.f.*), dent.
bossette (*n.f.*), boss.
bossette d'entrée d'air (*n.f.*), air scoop.
bouchardage (*n.m.*), roughening.
bouche de sortie d'oxygène (*n.f.*), oxygen outlet.
bouche de soufflage (*n.f.*), aerator.
bouché, -e (*adj.*), clogged.
bouchon à baguer (*n.m.*), bushing blank.
bouchon à baïonnette (*n.m.*), bayonet plug.
bouchon à vis (*n.m.*), threaded plug.
bouchon anti-poussière (*n.m.*), dust cap.
bouchon atomiseur (*n.m.*), swirl plug.
bouchon d'étanchéité (*n.m.*), sealing plug.
bouchon d'expedition (*n.m.*), shipping plug.
bouchon de protection (*n.m.*), cap.
bouchon de purge (*n.m.*), bleed plug.
bouchon de remplissage par gravité (*n.m.*), gravity filler plug.
bouchon de stockage (*n.m.*), storage plug.
bouchon de tuyauterie (*n.m.*), stopper.
bouchon filtre (*n.m.*), filter plug.
bouchon magnétique (*n.m.*), magnetic drain plug.
bouchon obturateur (*n.m.*), blanking cover.
boucle d'asservissement (*n.f.*), feedback loop; minor loop.
boucle de commande (*n.f.*), control loop; major loop.
boucle fermée (aviation) (*n.f.*), closed loop.
boucle ouverte (aviation) (*n.f.*), open loop.
boudin d'étanchéité (*n.m.*), sealing bead.
boudin de dégivrage (*n.m.*), de-icer boot.
boudin de mastic (*n.m.*), sealing bead.
boudinage (plast.) (*n.m.*), extrusion.
boudineuse (plast.) (*n.f.*), extrusion machine.
bouée radio et sonore (marine) (*n.f.*), anchored radio sono-buoy.
bougie (auto) (*n.f.*), ignition plug; sparking plug; plug.
bougie à incandescence (*n.f.*), glow plug.
bougie d'allumage (réacteur) (*n.f.*), spark igniter.
boulon à encoches (*n.m.*), slotted bolt.
boulon à oeil (*n.m.*), eye bolt.
boulon à six pans (*n.m.*), hexagonal headed bolt.
boulon à tête bouterollée (*n.m.*), snap-head bolt.
boulon à tête creuse (*n.m.*), socket head bolt.
boulon à tête plate (*n.m.*), flat head bolt.
boulon à tête rectangulaire (*n.m.*), tee-head bolt.
boulon à tête ronde à un pan (*n.m.*), dee head bolt.
boulon à tige cônique (*n.m.*), taper shank bolt.
boulon carrossier (*n.m.*), coach bolt.
boulon creux de raccord (n.m.), banjo bolt.
boulon d'arrêt (*n.m.*), stop bolt.
boulon de butée (*n.m.*), stop bolt.
boulon de chape (*n.m.*), clevis bolt.
boulon de montage (*n.m.*), mounting bolt.
boulon de sécurité (*n.m.*), safety bolt.
boulon épaulé (*n.m.*), shoulder bolt.
boulon mécanique (*n.m.*), machine bolt.
boulon rectifié (*n.m.*), machined bolt.
bourrage (de cartes mécanographiques) (*n.m.*), jamming of cards.
bourre (*n.f.*) wadding.
bourrelet (*n.m.*), bead.

bourrelet (soudure) (*n.m.*), weld upset.
bout (*n.m.*), tip.
bout de pale (hélicoptère) (*n.m.*), blade tip.
bout mort (bobinage) (*n.m.*), dead end.
bouteille d'oxygène (*n.f.*), oxygen cylinder.
bouteille extincteur (*n.f.*), fire extinguisher.
bouteillon (*n.m.*), thermos flask.
bouterolle (*n.f.*), die bar; riveting tool.
bouteur (travaux publics) (*n.m.*), bulldozer.
bouteur à pneus (travaux publics) (*n.m.*), tournadozer.
bouteur biais (travaux publics) (*n.m.*), angledozer.
bouteur inclinable (travaux publics) (*n.m.*), tilt dozer.
bouton à index (*n.m.*), pointer knob.
bouton cranté (*n.m.*), dented knob.
bouton d'affichage (*n.m.*), set knob.
bouton de débrayage rapide (aviation) (*n.m.*), autopilot disengage push button.
bouton de percussion (*n.m.*), discharge button.
bouton de réallumage réacteur (aviation) (*n.m.*), engine relight push button.
bouton de réglage de friction de manche (hélicoptère) (*n.m.*), stick friction knob.
bouton de sélection de cap (aviation) (*n.m.*), set HDG knob.
bouton de virage (aviation) (*n.m.*), autopilot turn knob.
bouton flèche (*n.m.*), pointer knob.
bouton moleté (*n.m.*), knurled knob.
bouton poignée (*n.m.*), knob.
bouton-poussoir (*n.m.*), press-button; push-button.
bouton-poussoir à verrouillage magné-tique (*n.m.*), magnetic hold-down push-button.
bouton-pression (*n.m.*), press-stud.
bouton sphérique (*n.m.*), ball knob.
bouton test (*n.m.*), push-to-test button.
bouton tirette (*n.m.*), push-and-hold button.
boutonnière (*n.f.*), slot.
brai (*n.m.*), tar; filler; (élec.) cable compound.
brai de pétrole fluxé (*n.m.*), cut-back.
brame (aciers) (*n.f.*), bloom.
branche de compas femelle (hélicoptère) (*n.f.*), upper link of rotor shaft.
branche de compas mâle (hélicoptère) (*n.f.*), lower link of rotor shaft.
branche de pont de Wheatstone (*n.f.*), Wheatstone bridge arm.
branche vent arrière (aviation) (*n.f.*), down wind leg.
branche vent debout (aviation) (*n.f.*), up wind leg.
branché, -e (*adj.*), connected.
branchement (*n.m.*), connection.
branchement-débranchement (élec.) (*n.m.*), make-and-break.
braquage ailerons (aviation) (*n.m.*), aileron deflection.
braquage direction (aviation) (*n.m.*), rudder deflection.
braquage du train avant (aviation) (*n.m.*), nose gear steering.
braquage hypersustentateurs (aviation) (*n.m.*), wing flap deflection.
braquage profondeur (aviation) (*n.m.*), ele-vator deflection.
bras d'essuie-glace (*n.m.*), windshield wiper

arm.

bras d'un treuil (hélicoptère) (*n.m.*), jib.

bras de commande de pas (hélicoptère) (*n.m.*), pitch control arm.

bras de flotteur (*n.m.*), float arm.

bras de levier (*n.m.*), moment arm.

bras de réglage (*n.m.*), adjusting arm.

bras mobile (*n.m.*), support arm.

bras profilé (aviation) (*n.m.*), support strut of engine.

bras support (fraiseuse) (*n.m.*), overarm.

brasage à l'arc (*n.m.*), arc brazing.

brassage (moteur à piston) (*n.m.*), cranking.

bretelle (fraiseuse) (*n.f.*), brace.

bretelle (aviation) (*n.f.*), feeder line.

bretelles (aviation) (*n.f.pl.*), shoulder harness.

brève impulsion (magnétisme) (*n.f.*), short impulse.

bridage magnétique (*n.m.*), magnetic holding.

bride (collier) (*n.f.*), clamp.

bride (sangle) (*n.f.*), strap; bridle.

bride à moyeu cannelé (*n.f.*), splined hub flange.

bride à moyeu dentelé (*n.f.*), serrated hub flange.

bride à moyeu lisse (*n.f.*), plain hub flange.

bride de raccordement (*n.f.*), connecting flange.

brillance (électronique) (*n.f.*), brilliance.

brillantage (*n.f.*), brightening.

brise-béton (travaux publics) (*n.m.*), concrete breaker.

brise-copeau meulé (*n.m.*), ground-in cutting rake.

brisure (*n.f.*), offset.

brisure de bord d'attaque (aviation) (*n.f.*), leading edge glove.

brochage (usinage) (*n.m.*), broaching.

brochage (mécanique) (*n.m.*), pin setting.

broche (de tubes électroniques) (*n.f.*), pin.

broche (de machine-outil) (*n.f.*), broach.

broche à canneler (machine-outil) (*n.f.*), spline broach.

broche à denture rapportée (machine-outil) (*n.f.*), inserted tooth broach.

broche à denture taillée (machine-outil) (*n.f.*), solid tooth broach.

broche à ergot (aviation) (*n.f.*), landing gear lock pin.

broche à rainurer (machine-outil) (*n.f.*), slot-broach.

broche à strier (machine-outil) (*n.f.*), serration broach.

broche d'entraînement (*n.f.*), spindle.

broche d'usinage (*n.f.*), broach.

broche de charnière (*n.f.*), hinge pin.

broche de contact électrique (*n.f.*), contact pin.

broche de réglage (*n.f.*), rigging pin.

broche de verrouillage (*n.f.*), locking pin.

broche type à fourreau (machines-outils) (*n.f.*), sleeve-type spindle.

broche type à moyeu (machines-outils) (*n.f.*), hub-type spindle.

broche type à pince (machines-outils) (*n.f.*), collet-type spindle.

broche type à quille (machines-outils) (*n.f.*), quill-type spindle.

brochure (*n.f.*), booklet; leaflet.

brosse métallique (*n.f.*), wire brush.

brouette motorisée (*n.f.*), powered barrow.

brouillage (radio) (*n.m.*), interference.

brouillard (*n.m.*), haze; mist.

brouillard salin (*n.m.*), salt spray.

bruine (*n.f.*), drizzle.

bruinissage (*n.m.*), black finishing.

bruit de fond (*n.m.*), background noise.

brut, -e (*adj.*), crude; rough; in bulk.

buée (*n.f.*), mist.

buffeting à grande vitesse (aviation) (*n.m.*), high-speed buffeting.

buffeting avertisseur de décrochage (aviation) (*n.m.*), stall buffeting.

bulbe (*n.f.*), bubble.

bulldozer (travaux publics) (*n.m.*), bulldozer.

bulle (couleur) (*adj.*), buff.

bulletin météorologique (*n.m.*), weather report.

bureau des dépêches (journaux et radio) (*n.m.*), news desk.

bureaux d'études (*n.m.pl.*), engineering and design departments.

buse d'entrée (*n.f.*), bellmouth.

buse mobile (*n.f.*), two-position nozzle.

buse variable (*n.f.*), adjustable nozzle.

butane (gaz) (*n.m.*), butane.

butée (d'un micromètre) (*n.f.*), anvil.

butée à billes (aviation) (*n.f.*), ball stop unit.

butée à double effet (*n.f.*), two-way thrust bearing.

butée anti-cône (hélicoptère) (*n.f.*), anti-cone stop.

butée centrifuge (hélicoptère) (*n.f.*), anti-flapping restrainer.

butée d'accrochage (*n.f.*), lock stop.

butée d'affaissement (*n.f.*), droop restrainer.

butée d'extrémité (*n.f.*), end stop.

butée d'inclinaison de moyeu (hélicoptère) (*n.f.*), hub tilt-stop.

butée de battement (hélicoptère) (*n.f.*), anti-flapping restrainer.

butée de fin de course (*n.f.*), stroke end stop; limit trip.

butée de moyeu (hélicoptère) (*n.f.*), hub spacer.

butée de pale (hélicoptère) (*n.f.*), blade stop.

butée de pas (hélicoptère) (*n.f.*), pitch stop.

butée de secteur (aviation) (*n.f.*), throttle gate.

butée de signalisation (*n.f.*), indicating stop.

butée de traînée (hélicoptère) (*n.f.*), drag stop.

butée double (pédale) (*n.f.*), stop yoke.

butée élastique du train (aviation) (*n.f.*), landing gear bumper.

butée électrique (*n.f.*), electrical limit.

butée fixe (*n.f.*), fixed stop.

butée grand pas (*n.f.*), high pitch stop.

butée mécanique (*n.f.*), mechanical stop unit.

butée mobile (*n.f.*), movable stop.

butée petit pas (hélicoptère) (*n.f.*), low pitch stop.

butée plein gaz (aviation) (*n.f.*), full open throttle.

butée ralentie (aviation) (*n.f.*), idle throttle stop.

butée réacteur éteint (aviation) (*n.f.*), engine shut off stop.

butée régulateur (aviation) (*n.f.*), governor control stop.

butée statique (hélicoptère) (*n.f.*), droop restrainer.

C

cabestan à roue dentée (*n.m.*), sprocket capstan.

cabestan électrique (aviation) (*n.m.*), electrical disconnect capstan.

câblage (élec.) (*n.m.*), wiring; cabling.

câblage pré-assemblé (*n.m.*), cable loom.

câble armé (élec.) (*n.m.*), sheathed cable; shielded cable; screened cable; armoured cable.

câble coaxial (*n.m.*), coaxial cable.

câble d'alimentation (élec.) (*n.m.*), feeder.

câble d'énergie (élec.) (*n.m.*), power cable.

câble de commande (*n.m.*), control cable.

câble de tierçage (hélicoptère) (*n.m.*), spacing cable.

câble krarupisé (élec.) (*n.m.*), continuously loaded cable.

câble prolongateur (*n.m.*), extension cord.

câble pupinisé (élec.) (*n.m.*), coil loaded cable.

câbliste (*n.m.*), cableman.

cabochon (*n.m.*), dimmer cap.

cabrage (hélicoptère) (*n.m.*), rotation.

cabré, en (aviation) (*adj., adv.*), in a nose-up attitude.

cabrer (aviation) (*v.t.*), to pull up.

cache (*n.m.*), mask; cover.

cache-borne (élec.) (*n.m.*), terminal cover.

cache de protection (*n.m.*), cover.

cache de sécurité (*n.m.*), guard.

cache pour roulements de train (aviation) (*n.m.*), wheel bearing cover plate.

cache poussière (*n.m.*), dust cover.

cadence (*n.f.*), rate.

cadmiage (*n.m.*), cadmium plating.

cadmiage au tonneau (*n.m.*), barrel cadmium plating.

cadran ajouré (*n.m.*), cut-out dial.

cadre (radio) (*n.m.*), frame aerial; frame antenna (U.S.A.).

cadre antenne (radio) (*n.m.*), loop aerial.

cadrer (cinéma) (*v.t.*), to centre.

cadreur (cinéma) (*n.m.*), cameraman.

cage d'ascenseur (*n.f.*), lift-shaft.

cage d'écrou (*n.f.*), nut cage.

cage de galets (*n.f.*), roller retainer.

cage de rotule (*n.f.*), ball joint cage.

cage de roulement (*n.f.*), bearing cage.

cagoule de soudeur (*n.f.*), welding helmet.

caillebotis (aviation, marine) (*n.m.*), walkway.

caisse (*n.f.*), crate; case.

caisse d'outillage (*n.f.*), tool-box.

caisson (*n.m.*), box-type structure.

caisson d'altitude (*n.m.*), altitude chamber.

caisson de bordure (aviation) (*n.m.*), edge box member.

caisson de dérive (aviation) (*n.m.*), fin spar box.

caisson de fuselage (aviation) (*n.m.*), fuselage box.

caisson de pale (hélicoptère) (*n.m.*), blade pocket.

caisson de voilure (aviation) (*n.m.*), wing spar box.

caisson raidisseur (aviation) (*n.m.*), box-type stiffener.

caisson résistant (aviation) (*n.m.*), stressed box structure.

calage (*n.m.*), setting.

calage altimétrique (*n.m.*), altimeter setting.

calage de l'allumage (*n.m.*), ignition timing.

calage de la voilure (*n.m.*), wing setting.

calaminage (*n.m.*), carbonization.

calandre de radiateur (auto) (*n.f.*), radiator grille.

calciné, -e (*adj.*), burned; charred.

calculateur analogique (électronique) (*n.m.*), analogue computer.

calculateur d'arrondi (électronique) (*n.m.*), flare-out computer.

calculateur de poussée (aviation) (*n.m.*), thrust computer (autothrottle).

calculateur de vitesse (aviation) (*n.m.*), velocity computer.

calculateur de vol (aviation) (*n.m.*), flight computer.

calculateur électronique (*n.m.*), electronic computer; electronic brain.

calculateur latéral (électronique) (*n.m.*), lateral computer.

calculateur longitudinal (électronique) (*n.m.*), longitudinal computer.

cale biaise (*n.f.*), wedge.

cale d'épaisseur (*n.f.*), shim.

cale de blocage (*n.f.*), wedge.

cale de blocage de pales rotor arrière (hélicoptère) (*n.f.*), tail rotor blade locking clamp.

cale de centrage (*n.f.*), positioning block.

cale de forme (*n.f.*), form shim.

cale de montage (*n.f.*), saddle plate.

cale de réglage (*n.f.*), setting shim.

cale de roue (*n.f.*), wheel chock.

cale lamellée (*n.f.*), peel shim.

caler le moteur (auto) (*v.*), to stall the engine.

calfdozer (trav. pub.) (*n.m.*), calfdozer.

calfeutrage (*n.m.*), caulking.

calibre (*n.m.*), gauge.

calibre à lame (*n.m.*), feeler gauge.

calibre d'épaisseur (*n.m.*), thickness gauge.

calibre de contrôle (*n.m.*), inspection gauge.

calibre "entrant" (*n.m.*), "go" gauge.

calibre "n'entrant pas" (*m.*), "no go" gauge.

calorifuge (*adj.*), heat-proof; heat-resistant.

calorifugé, -e (*adj.*), insulated against heat.

calorifugeage (*n.m.*), thermal insulation; heat shielding.

calorimétrique (*adj.*), heat sensing.

calotte avant (aviation) (*n.f.*), detachable nose cone.

calotte de bouton pression (*n.f.*), press button.

cambouis (*n.m.*), sludge.

came de blocage (*n.f.*), locking cam.

came double (*n.f.*), two-lobe cam.

came sélectrice (*n.f.*), selector cam.

camion à benne basculante (*n.m.*), dump truck (U.S.A.); tipping lorry (G.B.).

camion à six roues motrices (*n.m.*), six-wheeler truck (three-axle drive); six-wheeler lorry.

camion citerne (*n.m.*), road tanker.

camionnette (carrosserie) (*n.f.*), van (G.B.); delivery truck (U.S.A.).

camionnette bâchée (*n.f.*), light lorry (G.B.); pick-up truck (U.S.A.).

campement (*n.m.*), parking.
canaux de ventilation (*n.m.pl.*), cooling ducts.
candélabre (aviation) (*n.m.*), landing gear hinge beam fitting.
candella (unité d'intensité lumineuse) (*n.f.*), candella; new candle.
caniveau de soudure (*n.m.*), undercut of a weld.
cannelure (*n.f.*), keyway; spline.
cannelures à développantes (*n.f.pl.*), involute serrations.
canon (outil) (*n.m.*), guide.
canon à électrons (tube cathodique) (*n.m.*), electron gun.
canon de centrage (*n.m.*), centering bush.
canon de guidage (*n.m.*), guide bush.
canon nylon (*n.m.*), nylon bush.
canot de sauvetage (*n.m.*), lifeboat.
canot pneumatique (*n.m.*), inflatable dinghy.
caoutchouc mousse (*n.m.*), foam rubber.
caoutchouc régénéré (*n.m.*), reclaimed rubber.
caoutchouc spongieux (*n.m.*), sponge rubber.
cap (*n.m.*), heading.
  prendre un cap, to set a course.
  suivre un cap, to steer a course.
cap compas (*n.m.*), compass heading.
cap d'éloignement (*n.m.*), outbound heading.
cap d'une piste (*n.m.*), localizer beam heading.
cap inverse (*n.m.*), reciprocal heading.
cap magnétique (*n.m.*), magnetic heading.
cap retour (*n.m.*), inbound heading.
cap vrai (*n.m.*), true heading.
capacimètre (*n.m.*), capacitance meter.
capacité (*n.f.*), capacity.
capacité calorifique (*n.f.*), heat storage capacity.
capacité d'un condensateur (élec.) (*n.f.*), capacity of a capacitor.
capacité d'un conducteur (élec.) (*n.f.*), capacity of a conductor.
capacité de rétention au champ (géol.) (*n.f.*), field capacity.
capacité de rétention en eau (géol.) (*n.f.*), water-holding capacity.
capacité des réservoirs de carburant (*n.f.*), fuel tankage.
capacité électrique (*n.f.*), absorbing power; electric capacity.
capacité limite (*n.f.*), load capacity; cut-out capacity.
capacité parasite (*n.f.*), stray capacity.
capacité propre (*n.f.*), self-capacitance.
capacité utile (*n.f.*), service capacity.
capot (auto) (*n.m.*), bonnet (G.B.); engine hood (U.S.A.).
capot (machines-outils) (*n.m.*), guard.
capot (couvercle) (*n.m.*), cover.
capot couvre-broche (machines-outils) (*n.m.*), spindle guard.
capot couvre-courroie (machines-outils) (*n.m.*), belt-guard.
capot couvre-mandrin (machines-outils) (*n.m.*), spindle-guard.
capot couvre-meule (machines-outils) (*n.m.*), wheel-guard.
capot d'instrument (*n.m.*), hood.
capot de buse réacteur (aviation) (*n.m.*), nozzle cowl.
capot de carénage (aviation) (*n.m.*), fairing.
capot de protection (*n.m.*), guard.

capot moteur (*n.m.*), cowl.
capotage (*n.m.*), cowling.
capotage démarreur (*n.m.*), starter fairing.
capsule anéroïde manométrique (*n.f.*), aneroid.
capsule bathymétrique Sonar (*n.f.*), sonar transducer.
capter (*v.t.*), to pick up.
capteur (*n.m.*), pick-up; pick-off.
capteur de vibrations (*n.m.*), vibration pick-up.
capture d'un faisceau (*n.f.*), beam capture.
capuchon (*n.m.*), cap.
capuchon de sécurité (*n.m.*), safety cover.
caractéristique (*adj.*), typical.
caractéristiques (*n.f.pl.*), characteristics; data.
caractéristiques à vide (*n.f.pl.*), no-load characteristics.
caractéristiques aérodynamiques (*n.f.pl.*), airflow characteristics.
caractéristiques de persistance d'écran (élec.) (*n.f.pl.*), decay characteristics.
caractéristiques en charge (*n.f.pl.*), load characteristics.
caractéristiques mécaniques (*n.f.pl.*), mechanical properties.
carbonitruration (*n.f.*), carbonitriding.
carburant (*n.m.*), fuel.
carburant embarqué (*n.m.*), fuel load.
carburant inutilisable (*n.m.*), unusable fuel.
carburant non utilisable récupérable (*n.m.*), drainable unusable fuel.
carburant pour réacteur (*n.m.*), turbine fuel (G.B.); jet fuel (U.S.A.).
carburant résiduel (*n.m.*), trapped fuel.
carburant utilisable (*n.m.*), usable fuel.
carburateur inversé (*n.m.*), down draught carburettor (G.B.); down draft carburetor (U.S.A.).
carburéacteur (aviation) (*n.m.*), jet engine fuel.
carcasse lisse, à (*adj.*), plain surface.
carcasse ventilée, à (*adj.*), ventilated frame.
cardan (*n.m.*), gimbal joint.
carénage (aviation) (*n.m.*), fairing.
carénage de bout de pale (aviation) (*n.m.*), blade tip fairing.
caréné, -e (aviation) (*adj.*), faired.
carotte (moule pour plast.) (*n.f.*), insert.
carré, au (*adj.*), square.
carré d'entraînement (outillage) (*n.m.*), square bit drive.
carrossage (automobile) (*n.m.*), wheel camber.
carrosserie (auto) (*n.f.*), car body.
carte (*n.f.*), card.
carte d'entretien (*n.f.*), maintenance data card.
carte d'incident (*n.f.*), failure data card.
carte d'ouverture (*n.f.*), leader card.
carte de fermeture (ordinateur) (*n.f.*), trailer card.
carte de rechanges (*n.f.*), spare part data card.
carte maîtresse (*n.f.*), master card.
carte mécanographique (*n.f.*), key punched card.
carte perforée (*n.f.*), punch card.
carter (mécanique) (*n.m.*), housing.
carter (auto) (*n.m.*), crankcase; oil sump (G.B.); oil pan (U.S.A.).
carter d'entrée (*n.m.*), inlet case.
carter de distribution (auto) (*n.m.*), timing gear case.

carter de jonction (*n.m.*), junction case.
carter de sécurité (*n.m.*), guard.
carter de sortie (*n.m.*), exhaust case.
carter du volant (auto) (*n.m.*), flywheel housing.
carter externe (*n.m.*), external wheel case.
carter intermédiaire (*n.m.*), intermediate case.
carter interne (*n.m.*), internal wheel case.
carter réacteur (aviation) (*n.m.*), case.
carter turbine (aviation) (*n.m.*), turbine box.
carton d'amiante (*n.m.*), sheet asbestos.
cartouche annexe (*n.f.*), annex block.
cartouche d'un dessin (*n.f.*), engineering drawing block.
cartouche de filtre (*n.f.*), filter cartridge.
case (*n.f.*), compartment.
casier (*n.m.*), rack.
casque d'écoute (radio) (*n.m.*), head-set.
casquette (hélicoptère) (*n.f.*), visor.
cassant, -e (*adj.*), short.
casserole d'hélice (aviation) (*n.f.*), spinner.
caténaire (ch. de f. électriques) (*n.m.*), trolley wire.
catergol (fusée) (*n.m.*), catalyte propellant.
cathode à arc (élec.) (*n.f.*), arc cathode.
cathode à décharge luminescente (élec.) (*n.f.*), glow discharge cathode.
cathode chauffée (élec.) (*n.f.*), hot cathode.
cathode compensée (*n.f.*), dispenser cathode.
cathode froide (*n.f.*), cold cathode.
cathode liquide (élec.) (*n.f.*), pool cathode.
cavalier (*n.m.*), jumper.
cavitation (*n.f.*), cavitation.
cavité (*n.f.*), recess.
cavité résonante (*n.f.*), tuned cavity.
cé de réglage (outillage) (*n.m.*), "c" spacer.
ceinture (*n.f.*), belt.
ceinture de sécurité (auto) (*n.f.*), safety belt.
célérité (*n.f.*), velocity.
cellule (aviation) (*n.f.*), airframe.
cellule de batterie (*n.f.*), battery cell.
cellule photoélectrique (*n.f.*), photo-electric cell; electric eye.
centistoke (*n.m.*), centistoke.
centrage (aviation) (*n.m.*), aircraft balance.
centrage des disques de frein (auto) (*n.m.*), brake disc alignment.
centrale à béton (*n.f.*), concrete batching and mixing plant.
centrale aérodynamique (*n.f.*), air data computer.
centrale baro-altimétrique (*n.f.*), barome-tric altitude controller.
centrale bi-gyroscopique (*n.f.*), dual platform.
centrale d'assiette (aviation) (*n.f.*), attitude data generator.
centrale d'enrobage (trav. pub.) (*n.f.*), gravel coating plant.
centrale de cap (aviation) (*n.f.*), heading data generator.
centrale de verticale (aviation) (*n.f.*), control gyro and amplifier.
centrale de vol (aviation) (*n.f.*), three-axis data generator.
centrale gyroscopique (aviation) (*n.f.*), gyroscopic platform.
centrale hydraulique (*n.f.*), hydraulic generator.
centrale téléphonique automatique (*n.f.*), automatic exchange.
centraliser (*v.t.*), to centralize; to repeat.
centre d'essai en vol (*n.m.*), flight test centre.
centre de poussée (aviation) (*n.m.*), blade centre of pressure.
centre de sustension (*n.m.*), lift centre.
centré, -e (*adj.*), zeroed.
centreur (*n.m.*), centering pin.
centreur de forêt (perceuse) (*n.m.*), drill locator.
cerce oscillante (*n.f.*), vibrating tamper.
cercle circonscrit (*n.f.*), circumscribed circle.
cercle de perçage (*n.m.*), bolt circle.
cercle inscrit (*n.m.*), inscribed circle.
cerf-volant (aviation) (*n.m.*), kite.
cermet (*n.m.*), ceramal.
certificat d'homologation (*n.m.*), approval certificate.
certificat de navigabilité (aviation) (*n.m.*), certificate of airworthiness.
chaîne (radio) (*n.f.*), channel.
chaîne (textiles) (*n.f.*), weft.
chaîne cinématique (aviation) (*n.f.*), trans-mission system mechanism.
chaîne de lacet (aviation) (*n.f.*), yaw channel.
chaîne de montage (*n.f.*), assembly line.
chaîne de roulis (aviation) (*n.f.*), roll channel.
chaîne de tangage (aviation) (*n.f.*), pitch channel.
chaîne dynamique (aviation) (*n.f.*), power train.
chaînette à boules (*n.f.*), beaded chain.
chaleur d'humectation (géol.) (*n.f.*), heat of wetting.
chaleur latente de vaporisation (*n.f.*), latent vaporization heat.
chaleur rémanente résiduelle (*n.f.*), after-heat.
chalumeau (*n.m.*), gas welding torch.
chalumeau à découper (*n.m.*), cutting torch.
chambrage (*n.m.*), counterbore; recess.
chambre à air (auto) (*n.f.*), inner tube.
chambre annulaire (*n.f.*), annulus.
chambre d'asservissement (*n.f.*), pilot pres-sure chamber.
chambre de carburation (auto) (*n.f.*), mixing chamber.
chambre de combustion à écoulement direct (*n.f.*), straight flow combustion chamber.
chambre de combustion à écoulement inversé (*n.f.*), return flow combustion chamber.
chambre de combustion à turbulence (*n.f.*), swirl-type combustion chamber.
chambre de combustion turbo-annulaire (*n.f.*), cannular combustion chamber.
chambre de détente (*n.f.*), expansion chamber.
chambre de pression statique (*n.f.*), static pressure chamber.
chambre de sédimentation (*n.f.*), sediment chamber.
chambre de tranquillisation (*n.f.*), plenum chamber.
chambrer (*v.t.*), to counterbore.
champ alternatif (élec.) (*n.m.*), alternating field.
champ d'audibilité (*n.m.*), range of hearing.
champ inducteur (*n.m.*), induction field.
champ irrotationnel (élec.) (*n.m.*), irrotational field.

champ rotationnel (élec.) (*n.m.*), curl field.

champ tourbillonnaire (élec.) (*n.m.*), curl field.

champignon (*n.f.*), poppet valve.

chandelle (*n.f.*), positioning screw jack.

chanfrein d'entrée (*n.m.*), leading chamfer.

chanfrein de brasage (*n.m.*), brazing chamfer.

changement de vitesse (auto) (*n.m.*), gear-shift (U.S.A.); gear change (G.B.).

chantournage à la fraise (*n.m.*), form milling.

chape (*n.f.*), clevis; fork; yoke.

chape à queue (*n.f.*), turnbuckle fork.

chape de commande de pale (aviation) (*n.f.*), trunnion yoke.

chape de jonction demi-voilure (aviation) (*n.f.*), wing root fitting.

chape double (*n.f.*), twin clevis; twin fork; twin yoke.

chape sur moyeu (*n.f.*), hub grip.

chapeau de couple (*n.m.*), frame cap.

chapeau de moyeu (aviation) (*n.m.*), hub cover plate.

chapeau de palier (*n.m.*), bearing cap.

chapelle de structure (aviation) (*n.f.*), fin stub frame.

chapiteau (aviation) (*n.m.*), landing gear hinge beam.

charbon actif (*n.m.*), activated carbon.

charge (plast.) (*n.f.*), filler.

charge à l'élingue (*n.f.*), underslung load.

charge admissible (*n.f.*), safe load.

charge aérodynamique (aviation) (*n.f.*), aerodynamic load.

charge alaire (aviation) (*n.f.*), wing load.

charge au cheval (*n.f.*), horsepower loading.

charge au flambage (*n.f.*), buckling load.

charge d'épreuve (*n.f.*), test load.

charge d'étalonnage (ressort) (*n.f.*), preload.

charge de mise en rotation des roues (*n.f.*), wheel spin-up load.

charge de pale (aviation) (*n.f.*), blade loading.

charge de pointe (élec.) (*n.f.*), peak load.

charge de rafale limite (*n.f.*), gust load limit.

charge de rupture (*n.f.*), ultimate strength.

charge du disque balayé (aviation) (*n.f.*), disc loading.

charge du disque rotor (aviation) (*n.f.*), rotor disc loading.

charge électrique (*n.f.*), charge.

charge équilibrée (élec.) (*n.f.*), balanced load.

charge extrême (*n.f.*), ultimate load.

charge fictive (*n.f.*), dummy load.

charge limite (*n.f.*), maximum load; safe work-ing load.

charge limite de pale (aviation) (*n.f.*), limit load of blade.

charge marchande (*n.f.*), payload.

charge marchande au maximum de capacité (*n.f.*), payload to maximum capacity.

charge marchande limite (*n.f.*), weight lim-ited payload.

charge marchande maximum (*n.f.*), maxi-mum payload.

charge marchande normale estimée (*n.f.*), estimated normal payload.

charge maximum de réaction élastique (*n.f.*), maximum spring back load.

charge payante (*n.f.*), payload.

charge soulevée (*n.f.*), lifted load.

charge théorique (*n.f.*), design load.

charge unitaire (*n.f.*), basic load.

charge utile (*n.f.*), useful load.

charge verticale maximum sur les roues (*n.f.*), maximum vertical load wheel.

chargeur (magasin) (*n.m.*), magazine.

chargeur de batterie (*n.m.*), battery charger.

chargeuse (génie civil) (*n.f.*), loader.

chargeuse-pelleteuse (génie civil) (*n.f.*), back hoe loader.

chariot (tour) (*n.m.*), slide.

chariot à tronçonner (*n.m.*), cutting-off slide.

chariot de guidage volets (aviation) (*n.m.*), flap roller carriage.

chariot de hissage (*n.m.*), hoisting carriage.

chariot de manutention (*n.m.*), dolly.

chariot de transport réacteur (*n.m.*), engine trolley.

chariot élévateur à fourche (*n.m.*), fork-lift truck.   (00)

chariot transversal (*n.m.*), cross slide.

chariotage (machines-outils) (*n.m.*), turning.

charpente (*n.f.*), structure.

chasse d'une roue (*n.f.*), castor action of a wheel.

chasse-goupille (*n.m.*), pin drift.

châssis (de montage) (*n.m.*), rack.

châssis-caisson (auto) (*n.m.*), box-type frame.

chatterton (élec.) (*n.m.*), adhesive tape; insulat-ing tape.

chaudronnerie (*n.f.*), sheet metal work.

chauffage au mazout (*n.m.*), oil-firing.

chauffage par induction (*n.m.*), induction heating.

chauffe-eau (*n.m.*), water-heater.

chauffe-eau à accumulation (*n.m.*), storage water-heater.

chauffe-eau à immersion (*n.m.*), immersion water-heater.

chauffer à blanc (*v.t.*), to bright anneal.

chemin optique (*n.m.*), optical path.

cheminée (hélicoptère) (*n.f.*), message chute.

cheminement (*n.m.*), routing.

chemise de cylindre (auto) (*n.f.*), cylinder liner.

chenille (de masque) (*n.f.*), mask pipe.

chercheur (*n.m.*), cat's-whisker; whisker.

cheval de bois (*n.m.*), ground loop.

cheville d'assemblage (*n.f.*), drift pin.

chevron (*n.m.*), herringbone (teeth of wheel).

chevrons, à (*adj.*), herringbone.

chicane (*n.f.*), baffle.

chiffon (*n.f.*), cloth; rag; duster.

chiffon non pelucheux (*n.m.*), lintless cloth.

chiffre (*n.m.*), cipher; digit.

chignole (*n.f.*), hand-brace.

chlorure de polyvinyle (plast.) (*n.m.*), poly-vinyl chloride.

chlorure de vinyl (plast.) (*n.m.*), vinyl chloride.

chromage (*n.m.*), chromium plating.

chromel (*n.m.*), chromel.

chronométrage (*n.m.*), timing.

chronomètre (*n.m.*), stop-watch.

chronoscope électronique (*n.m.*), electronic chronoscope.

chrysocale (*n.m.*), ormolu.

chute de pression (*n.f.*), pressure drop.

chute de tension (*n.f.*), voltage drop.)

chute libre (*n.f.*), free fall.

chuter (*v.i.*), to drop.

cinématique (*n.f.*), kinematics; sequence of

operation.

cinémomètre (n.m.), speedometer.

cinéscope (élec.) (n.m.), television picture tube.

cintre (n.m.), former.

circlip (n.m.), circlip.

circonférence (n.f.), circumference.

circuit (n.m.), line; system.

circuit à déclenchement périodique (radio) (n.m.), gate.

circuit à distorsion (n.m.), shaping circuit.

circuit anémométrique (n.m.), pitot static system.

circuit anodique (élec.) (n.m.), plate circuit.

circuit bouchon (n.m.), anti-resonant circuit.

circuit d'amorçage (élec.) (n.m.), triggering circuit.

circuit d'attente (en vol) (aviation) (n.m.), flight holding pattern.

circuit d'effacement (aviation) (n.m.), wash-out network

circuit d'entrée (élec.) (n.m.), input circuit.

circuit de balayage (élec.) (n.m.), sweeping circuit.

circuit de graissage (n.m.), lubricating system.

circuit de maintien (n.m.), hold circuit.

circuit de retour par les essieux (ch. de fer) (n.m.), axle circuit.

circuit de secours (n.m.), emergency system.

circuit de sortie (élec.) (n.m.), output circuit.

circuit excitation (n.m.), energizing circuit.

circuit fermé (n.m.), closed circuit.

circuit hydraulique (n.m.), hydraulic system.

circuit imprimé (électronique) (n.m.), printed circuit.

circuit iso-contour (électronique) (n.m.), iso-contour circuit.

circuit ouvert (n.m.), open circuit.

circuit stabilisation (électronique) (n.m.), sta-bilization circuit.

circuit synchrone (aviation) (n.m.), synchro loop.

circuit tronqueur (électronique) (n.m.), clamp-ing circuit.

circuit vidéo (électronique) (n.m.), video circuit.

cisaille de câble de treuil (hélicoptère) (n.f.), hoist cable cutter.

citerne (n.f.), tank.

civière (n.f.), stretcher.

clapet à bille (n.m.), ball valve.

clapet à flotteur (n.m.), float valve.

clapet anti-retour (n.m.), check valve; non-return valve.

clapet battant sur nervure (aviation) (n.m.), rib flap valve.

clapet commandé (aérofreins) (aviation) (n.m.), lock out valve.

clapet d'expiration (masque) (n.m.), exhala-tion valve.

clapet d'intercommunication (du carburant) (n.m.), fuel cross feed valve.

clapet d'intercommunication d'air (n.m.), air cross bleed valve.

clapet d'interdiction (n.m.), safety valve.

clapet d'isolement sur nervure (n.m.), rib crabpot valve.

clapet de décharge (n.m.), discharge valve.

clapet de dégazage (n.m.), vapour relief valve.

clapet de dépression (bâche hydraulique) (n.m.), depressurization valve.

clapet de dépression (conduite d'air) (n.m.), vacuum relief valve.

clapet de dérivation (n.m.), by-pass valve.

clapet de détente (n.m.), pressure reducing valve.

clapet de gonflage (n.m.), air charging valve.

clapet de laminage (n.m.), restrictor.

clapet de mise à l'air libre (n.m.), air vent valve.

clapet de mise à l'air libre étanche au carburant (aviation) (n.m.), air-no-fuel vent valve.

clapet de purge (n.m.), bleed valve.

clapet de remplissage (n.m.), filling valve.

clapet de remplissage carburant (n.m.), refuelling valve.

clapet de surpression (n.m.), pressure relief valve.

clapet de verrouillage aérofrein (aviation) (n.m.), speed brake lock out valve.

clapet de vidange (n.m.), drain valve.

clapet navette (n.m.), shuttle valve.

clapet pilote (n.m.), pilot valve.

clapet plat (n.m.), flat valve.

clapet préférentiel (hydraulique) (n.m.), prior-ity valve.

clapet taré (n.m.), calibrated valve.

claquage (élec.) (n.m.), insulation breakdown.

clause (n.f.), clause; provision.

clauses techniques (n.f.pl.), type specifica-tion.

clavette à disque (n.f.), disc key.

clavette à talon (n.f.), gib head key.

clavette conique (n.f.), taper pin.

clavette demi-lune (n.f.), Woodruff key.

clavier (n.m.), keyboard.

clé à cadran (n.f.), torque wrench.

clé à col de cygne (n.f.), goose-neck wrench.

clé à ergot (n.f.), pin wrench.

clé à griffe (n.f.), hook spanner.

clé anglaise (n.f.), monkey wrench.

clé coudée (n.f.), offset wrench.

clé crocodile (n.f.), alligator wrench.

clé d'âme de longeron (aviation) (n.f.), spar web wedge.

clé d'écoute (n.f.), audio switch.

clé de frein (n.f.), brake wedge.

clé de serrure (n.f.), lock key.

clé dynamométrique (n.f.), torque wrench.

clé plate ouverte (n.f.), thin spanner.

clé universelle (n.f.), monkey wrench.

cliché simili (photo) (n.m.), half-tone block.

clignoter (v.i.), to blink; to flash on and off.

clignoteur (n.m.), indicator; flasher.

climatisation (n.m.), air-conditioning.

cliquet de verrouillage (n.m.), catch.

cliquetis (n.m.), chatter.

cloche à vide (n.f.), vacuum chamber.

cloche d'accouplement (n.f.), ring gear.

cloche de mesure (instrument) (n.f.), drag cup.

cloche de pompe carburant (n.f.), fuel pump shroud.

cloche dentée (n.f.), splined coupling.

cloison anti-ballast (n.f.), anti-surge baffle.

cloison de décrochage (aviation) (n.f.), wing fence.

cloison de sangles (n.f.), webbing.

cloison de séparation (n.f.), partition.

cloison fine (n.f.), septum.

cloison pare-feu (n.f.), fire wall.

cloisonnement (n.m.), partitioning.)

cloque de revêtement (n.f.), blister; skin.

coaxial, -e, (élec.) (*adj.*), coaxial.
cocher (*v.t.*), to tick; to check.
code (phares d'auto) (*n.m.*), dipped beam.
coefficient aérodynamique (*n.m.*), aerody-namic factor.
coefficient d'atténuation de rafale (avia-tion) (*n.m.*), gust alleviation factor.
coefficient de charge (*n.m.*), load factor.
coefficient de couple (*n.m.*), torque coefficient.
coefficient de dilatation (*n.m.*), expansion factor.
coefficient de force latérale (*n.m.*), lateral force coefficient.
coefficient de moment (*n.m.*), moment coefficient.
coefficient de Poisson (phys.) (*n.m.*), Poisson's ratio.
coefficient de portance (*n.m.*), lift coefficient.
coefficient de proportionalité (*n.m.*), pro-portionality factor.
coefficient de remplissage (*n.m.*), occupation factor.
coefficient de sécurité (*n.m.*), safety factor; factor of ignorance.
coefficient de stabilité de la structure (géol.) (*n.m.*), structure index.
coefficient de traînée (aviation) (*n.m.*), drag coefficient.
coefficient numérique forfaitaire (*n.m.*), empirical operation factor.
coeur (planification) (*n.m.*), core.
coeur électrique (*n.m.*), electrical master box.
coffret de servitude (électronique) (*n.m.*), servo-module.
coiffe de clapet (*n.f.*), valve retainer.
coiffe mobile (aviation) (*n.f.*), landing gear slid-ing valve.
col (d'un tube cathodique) (*n.m.*), neck.
col-de-cygne (machines-outils) (*n.m.*), throat; swan-neck.
colisage (*n.m.*), preparation for despatch; packing specification.
colle (*n.f.*), adhesive.
collecteur (élec.) (*n.m.*), slip ring.
collecteur (radiateur) (auto) (*n.m.*), header-tank.
collecteur annulaire (*n.m.*), annulus.
collecteur d'échappement (auto) (*n.m.*), exhaust manifold.
collecteur d'huile (*n.m.*), oil cup.
collecteur de génératrice (élec.) (*n.m.*), gen-erator commutator.
collecteur de pompe carburant (auto) (*n.m.*), fuel pump manifold.
collecteur de pressurisation (*n.m.*), pres-surizing manifold.
coller (*v.t.*), to bond.
collier (ressort) (*n.m.*), spring clamp.
collier à garniture de protection (*n.m.*), lined clamp.
collier de jonction (*n.m.*), coupling sleeve.
collier de serrage (*n.m.*), clamp.
collier Marman (*n.m.*), V-band clamp.
collier porte-balais (élec.) (*n.m.*), brush-rocker.
colmatage (*n.m.*), filling-up; blocking; clogging.
colonne à plateau (distillation) (*n.f.*), coffey still (U.S.A.); distillation column (G.B.).
colonne d'eau (*n.f.*), head of water.
colonnette d'assemblage (*n.f.*), assembly post.
colonnette de soutien (*n.f.*), stanchion.

colorant (textiles) (*n.m.*), dye.
colorant à l'aniline (textiles) (*n.m.*), aniline dye.
colorant à la cuve (textiles) (*n.m.*), vat dye.
colorant acide (textiles) (*n.m.*), acid dye.
colorant anthracénique (*n.m.*), coal-tar dye.
colorant azoïque (*n.m.*), azo-dye.
colorant basique (*n.m.*), basic dye.
colorant substantif (*n.m.*), substantive dye.
combinateur de pas (hélicoptère) (*n.m.*), mix-ing unit.
combinateur de pas général et cyclique (hélicoptère) (*n.m.*), collective pitch synchronizer.
combiné, -e (*adj.*), compound.
hélicoptère combiné (*n.m.*), compound helicopter; gyrodyne.
combiné téléphonique (*n.m.*), telephone handset.
comburant (*n.m.*), oxydizer.
commandant de bord (aviation) (*n.m.*), pilot.
commande (*n.f.*), control; operation; actuation.
commande à distance (*n.f.*), remote control.
commande cyclique (aviation) (*n.f.*), azimu-thal control.
commande d'interdiction (*n.f.*), interlock control.
commande de direction (aviation) (*n.f.*), rud-der control; (hélicoptère) tail rotor control system.
commande de direction (roue avant) (avia-tion) (*n.f.*), steering control for nose wheel.
commande de gauchissement (aviation) (*n.f.*), aileron control; (hélicoptère) roll control.
commande de mélange (*n.f.*), mixture control.
commande de pas (hélicoptère) (*n.f.*), pitch control.
commande de pas cyclique (hél.) (*n.f.*), cyclic pitch control.
commande de pas général (hél.) (*n.f.*), col-lective pitch control.
commande de profondeur (aviation) (*n.f.*), elevated control; (hélicoptère) pitch control.
commande de puissance (aviation) (*n.f.*), throttle control.
commande de richesse du carburant (*n.f.*), fuel control.
commande de secours (*n.f.*), emergency control.
commande directe (*n.f.*), direct control.
commande flexible (*n.f.*), flexible control.
commande manuelle (*n.f.*), manual control.
commande manuelle à distance (*n.f.*), man-ual remote control.
commande numérique (*n.f.*), digital control.
commande servo-actionnée (*n.f.*), power-operated control.
commande servo-assistée (*n.f.*), power-assisted control.
commande téléflex (*n.f.*), control via flexible control at a distance.
commande transparente (aviation) (*n.f.*), override control.
commandes conjuguées (*n.f.pl.*), intercon-nected controls.
commandes de vol (aviation) (*n.f.pl.*), flight controls.
commandes et contrôles (*n.m.pl.*), controls and feedback.
commodités (génie civil) (*n.f.pl.*), utilities.
communication bi-latérale (*n.f.*), two-way

communication.

commutateur (électricité) (*n.m.*), selector switch; lever switch.

commutateur à galette (électricité) (*n.m.*), wafer switch.

commutateur à plots (électricité) (*n.m.*), step switch.

commutateur barométrique (*n.m.*), baro- -metric switch.

commutateur de gammes d'ondes (radio) (*n.m.*), band selector.

commutateur sélecteur de phase (électri- -cité) (*n.m.*), phase selector switch.

commutation (électricité) (*n.f.*), switching.

commutation de voies (*n.f.*), channel switching.

commutatrice (électricité) (*n.f.*), current inverter; rotary converter.

commuter (*v.t.*), to switch over.

compactage (*n.m.*), compaction.

compacter (*v.t.*), to roller; to compact.

compagnie d'affrètement (*n.f.*), charter company.

comparateur à cadran (*n.m.*), dial gauge.

comparateur de phase (*n.m.*), phase comparator.

comparateur de signaux (aviation) (*n.m.*), three-axis signal comparator.

compartiment de rangement (*n.m.*), storage space.

compas (aviation) (atterrisseur) (*n.m.*), torque link (in the landing gear).

compas (hélicoptère) (*n.m.*), scissors.

compas de secours (*n.m.*), stand-by compass.

compas fixe (hélicoptère) (*n.m.*), stationary scissors.

compas gyrosyn (hélicoptère) (*n.m.*), gyrosyn compass.

compas mobile (hélicoptère) (*n.m.*), rotating scissors.

compensateur (hélicoptère) (*n.m.*), trim cylinder.

compensateur aérodynamique (aviation) (*n.m.*), trim tab.

compensateur de centrage (aviation) (*n.m.*), trimmer.

compensateur de commande de pas (hélicoptère) (*n.m.*), blade pitch control compensator.

compensateur de couple (*n.m.*), anti-torque device.

compensateur de profondeur (aviation) (*n.m.*), pitch trim.

compensateur pneumatique (hélicoptère) (*n.m.*), pneumatic cylinder.

compensateur synchrone (élec.) (*n.m.*), syn- -chronous capacitor.

compensation (aviation) (*n.f.*), counterbal- -ance; trim.

compensation de compas (*n.f.*), compass compensating.

compensation de direction (*n.f.*), directional trim.

compensation de gauchissement (aviation) (*n.f.*), lateral trim.

compensation de l'hélicoptère (*n.f.*), heli- -copter behaviour.

compensation de pas (hélicoptère) (*n.f.*), pitch compensation.

compléter (le plein) (*v.t.*), to top up a tank.

completion (*n.f.*), completion.

complexe (*adj.*), intricate; complex; complicated.

compliqué, -e (*adj.*), complicated; intricate.

comportement (*n.m.*), behaviour; performance.

composant (*n.m.*), component.

composante de portance (aviation) (*n.f.*), lift component.

composante de poussée (aviation) (*n.f.*), thrust component.

composante de traînée (aviation) (*n.f.*), drag component.

composition granulométrique (géol.) (*n.f.*), granular content.

compresseur (auto) (*n.m.*), supercharger.

compresseur à plusieurs étages (*n.m.*), multi-stage compressor.

compresseur à un seul étage (*n.m.*), single- stage compressor.

compresseur axial (*n.m.*), axial compressor.

compresseur basse pression (*n.m.*), low- pressure compressor.

compresseur centrifuge (*n.m.*), centrifugal compressor.

compresseur de moteur à piston (*n.m.*), supercharger.

compresseur double (*n.m.*), dual compressor; twin compressor.

compresseur-expanseur (*n.m.*), com- -pounder.

compresseur haute pression (*n.m.*), high- pressure compressor.

compresseur simple (*n.m.*), single compressor.

compression, en (*adj.*), compressed.

comprimer (*v.t.*), to press.

compte-tours (*n.m.*), tachometer.

compteur d'impulsions (*n.m.*), pulse meter.

compteur de vitesse (auto) (*n.m.*), speedometer.

compteur électronique (*n.m.*), electronic rate counter.

compteur Geiger-Müller (*n.m.*), Geiger counter.

compteur-intégrateur (élec.) (*n.m.*), integrat- -ing meter.

concave (outil) (*adj.*), hollow ground.

concentration (faisceau électronique) (*n.f.*), focussing.

concordance (*n.f.*), coherence; agreement.

condensateur (élec.) (*n.m.*), capacitor.

condensateur variable (élec.) (*n.m.*), vari- -able capacitor.

conditionnement d'air (*n.m.*), air- conditioning.

conditionneur d'air (*n.m.*), air-conditioner.

conditions nominales de fonctionnement (*n.f.*), rating.

conducteur (élec.) (*n.m.*), lead.

conduit collecteur (génie civil) (*n.m.*), shunt.

conduit de pale (hélicoptère) (*n.m.*), blade duct.

conduite (d'air) (*n.f.*), duct.

configuration transitoire (aviation) (*n.f.*), transient configuration.

cône (d'un tube cathodique) (*n.m.*), cone.

cône arrière (aviation) (*n.m.*), tail cone.

cône avant (aviation) (*n.m.*), nose cone.

cône d'échappement (aviation) (*n.m.*), exhaust cone.

cône de butée (aviation) (*n.m.*), thrust cone.

cône de confusion (aviation) (*n.m.*), confusion cone.

cône de queue (aviation) (*n.m.*), tail cone.

cône de serrage (*n.m.*), tightening cone.

cône de silence (*n.m.*), cone of silence.

configuration (*n.f.*), outline.
configuration lisse (*n.f.*), clean configuration.
conformité (*n.f.*), accordance.
en conformité de (*adv.*), in accordance with.
conicité (pale) (*n.f.*), blade taper ratio.
conicité (d'un pignon) (*n.f.*), gear cone angle.
conicité (hélicoptère) (*n.f.*), coning angle.
conicité rotor (hélicoptère) (*n.f.*), rotor coning.
conique (*adj.*), tapered.
conjoncteur-disjoncteur (élec.) (*n.m.*), circuit-breaker.
conjoncteur-disjoncteur différentiel (élec.) (*n.m.*), reverse current relay.
conjugaison (*n.f.*), combination; inter--connection.
connexion d'électrode (*n.f.*), lead-in wire.
conservateur d'huile (transformateur) (*n.m.*), oil conservator.
conservateur de cap (aviation) (*n.m.*), course indicator.
consigne d'utilisation (*n.f.*), operating procedure.
console (*n.f.*), console.
consommable (*adj.*), expendable.
consommation kilométrique (*n.f.*), specific fuel consumption.
constante de temps (*n.f.*), time constant.
contact à déclic (*n.m.*), snap-action contact.
contact à deux directions avec che--vauchement (élec.) (*n.m.*), two-way make-before-break contact.
contact à deux directions sans che--vauchement (élec.) (*n.m.*), two-waybreak-before-make contact.
contact à soufflage magnétique (*n.m.*), magnetic arc blow-out contact.
contact bimétallique (élec.) (*n.m.*), bimetal switch.
contact de fin de course (*n.m.*), limit switch; end-of-travel switch.
contact électrique (*n.m.*), electrical contact.
contacts platinés (*n.m.pl.*), breaker points.
contacts repos (*n.m.pl.*), normally closed contacts.
contacts travail (*n.m.pl.*), normally open contacts.
contacteur (*n.m.*), contactor.
contacteur à flotteur (*n.m.*), float switch.
contacteur à inertie (*n.m.*), inertia switch.
contacteur à palette (*n.m.*), paddle switch.
contacteur barométrique (*n.m.*), pressure switch.
contacteur centrifuge (*n.m.*), centrifugal switch.
contacteur de liaison entre barres (élec--tronique) (*n.m.*), tie-in relay between busbars.
contacteur tripolaire (*n.m.*), three-pole switch.
contacteur variométrique (*n.m.*), pressure rate-of-change switch.
container (pour le transport) (*n.m.*), container.
contenance (*n.f.*), capacity.
continuité de masse (*n.f.*), earth bonding.
continuité électrique (*n.f.*), electrical continuity.
contournement (élec.) (*n.m.*), flashover.
contournement d'isolant (élec.) (*n.m.*), insu--lation breakdown.
contrainte (*n.f.*), strain.
contrainte de cisaillement (*n.f.*), shear stress.

contrainte transitoire (*n.f.*), transient stress.
contraste (radar) (*n.m.*), brightness ratio.
contrat (*n.m.*), agreement.
contre-bouterolle (*n.f.*), backing bar.
contre-bride (*n.f.*), adapter.
contre-butée (*n.f.*), thrust ring.
contre-clavette (*n.f.*), nose key.
contre-essai (*n.m.*), counter-test.
contre-plaque (*n.f.*), backing plate; counterplate.
contre-plaqué (bois) (*n.m.*), plywood.
contre-réaction (radio) (*n.f.*), negative feedback.
contre-type (photo) (*n.m.*), duplicate.
contrefiche de couple (*n.f.*), frame cross-member.
contrefiche de potence (aviation) (*n.f.*), upper support arm (landing gear).
contrefiche de train principal (aviation) (*n.f.*), main landing gear brace strut.
contrefiche horizontale (*n.f.*), horizontal strut.
contrefiche télescopique (*n.f.*), telescopic strut.
contrôle automatique d'approche (avia--tion) (*n.m.*), automatic approach control.
contrôle automatique de gain (radio) (*n.m.*), automatic gain control.
contrôle d'aspect (*n.m.*), visual inspection.
contrôle d'écoute (radio) (*n.m.*), side-tone output test.
contrôle de fatigue (*n.m.*), fatigue inspection.
contrôle de qualité (*n.m.*), quality control.
contrôle fluoroscopique (*n.m.*), fluoroscopic inspection.
contrôle magnétoscopique (*n.m.*), magnetic particle inspection; magnetic crack detection.
contrôle radiographique (*n.m.*), X-ray inspection.
contrôle statistique (*n.m.*), statistical control.
contrôle systématique (*n.m.*), routine inspection.
contrôler (surveiller) (*v.t.*), to monitor.
contrôleur d'accélération (aviation) (*n.m.*), acceleration control unit.
contrôleur d'altitude (aviation) (*n.m.*), alti--tude controller.
contrôleur de vol (aviation) (*n.m.*), flight controller.
convertisseur (élec.) (*n.m.*), inverter.
convertisseur d'ozone (*n.m.*), de-ozonizer.
convertisseur de couple (*n.m.*), torque converter.
convertisseur de secours (*n.m.*), stand-by inverter.
convertisseur radio (*n.m.*), dynamotor.
convertisseur statique (*n.m.*), static inverter.
convoyage (aviation) (*n.m.*), ferry flight.
copie en clair (*n.f.*), hard copy.
co-pilote (*n.m.*), co-pilot.
copolymère (plast.) (*n.m.*), copolymer; cross polymer.
coque (auto) (*n.f.*), shell.
corde aérodynamique moyenne (aviation) (*n.f.*), mean aerodynamic chord.
corde d'évacuation (*n.f.*), escape rope.
corde d'extrémité (*n.f.*), tip chord.
corde de la pale (hélicoptère) (*n.f.*), blade chord.
corde de piston (auto) (*n.m.*), piston land.
corde de profil (aviation) (*n.f.*), airfoil chord.

corde géométrique moyenne (*n.f.*), mean geometric chord.

corde moyenne de la gouverne (aviation) (*n.f.*), mean chord of the control surface.

cordon (*n.m.*), cord.

cordon de soudure (*n.m.*), weld bead.

cordon détecteur d'incendie (*n.m.*), fire-wire.

cordon tire-feu (*n.m.*), lanyard.

corne (de tubes électroniques) (*n.f.*), top-cap; side contact.

corne de gouverne (aviation) (*n.f.*), horn.

cornière à boudin (*n.f.*), beaded extrusion.

cornière de raidissement (*n.f.*), stiffener.

cornière de reprise (*n.f.*), pick-up angle.

cornière en tôle pliée (*n.f.*), bent section.

corps de moyeu (hélicoptère) (*n.m.*), hub.

corps de poignée (*n.m.*), hand grip.

correcteur d'effort (aviation) (*n.m.*), pitch cor-rector unit.

correcteur de Mach (aviation) (*n.m.*), Mach compensator.

correction d'altitude (*n.f.*), altitude correction.

corrélation (*n.f.*), correlation.

corrosion de contact (*n.f.*), fretting corrosion.

corrosion intergranulaire (*n.f.*), intergran-ular corrosion.

corset (outillage) (*n.m.*), collar.

cosse (d'un câble électrique) (*n.f.*), terminal con-nector; terminal lug.

cosse à river (élec.) (*n.f.*), rivet terminal lug.

cosse à sertir (élec.) (*n.f.*), crimp terminal lug.

cosse à souder (élec.) (*n.f.*), solder terminal lug.

cosse circulaire (élec.) (*n.f.*), ring terminal lug.

cotation (*n.f.*), quotation.

cote de réalésage (*n.f.*), reaming allowance.

cote de trusquinage (*n.f.*), edge distance.

cote nominale (*n.f.*), nominal size.

cote sur plat (*n.f.*), dimension across flats.

cote théorique (*n.f.*), theorical dimension.

cotes d'encombrement (*n.f.pl.*), overall dimensions.

côté, de (*adj.,adv.*), askew.

côté de refoulement (*n.m.*), pressure side.

côté droit (d'une voiture) (*n.m.*) (s'il s'agit de la France ou d'autres pays où l'on conduit à droite) near side; (s'il s'agit de la Grande-Bre-tagne ou d'autres pays où l'on conduit à gauche) off side.

côté gauche (d'une voiture) (*n.m.*) (s'il s'agit de la France ou d'autres pays où l'on conduit à droite) off side; (s'il s'agit de la Grande-Bre-tagne ou d'autres pays où l'on conduit à gauche) near side.

couche (*n.f.*), coat; layer; course.

couche d'accrochage (peinture) (*n.f.*), priming coat.

couche d'apprêt (peinture) (*n.f.*), priming coat.

couche d'usure (route) (*n.f.*), wearing coat.

couche de finition (peinture) (*n.f.*), top coat.

couche de fond (peinture) (*n.f.*), undercoat.

couche de liaison (route) (*n.f.*), binder course.

couche de roulement (route) (*n.f.*), carpet.

couche de scellement (peinture) (*n.f.*), sealing coat.

couche électronique (électronique) (*n.f.*), electron-shell.

couche inférieure (route) (*n.f.*), base coat.

couche ionisée (troposphère) (*n.f.*), ionosphere.

couche limite (aviation) (*n.f.*), boundary layer.

couche supérieure (peinture) (*n.f.*), top coat.

coude de piquage (*n.m.*), bleed elbow.

coude orientable (*n.m.*), swivelling elbow.

coulée à la cire perdue (*n.f.*), lost wax casting.

coulée au sable (*n.f.*), sand casting.

coulée en coquille (*n.f.*), chill casting.

coulée en matrice (*n.f.*), die casting.

coulée sous pression (*n.f.*), pressure casting.

coulis (trav. pub.) (*n.m.*), grouting.

coulis réfractaire (*n.m.*), refractory wash.

coup, à- (*n.m.*), impact load.

coup de bélier (*n.m.*), pressure surge.

coupage à l'arc (*n.m.*), arc cutting.

coupe (*n.f.*), cross-section.

coupe (d'un joint) (*n.f.*), gap.

coupe-feu (robinet) (*n.m.*), fuel shut-off cock.

coupe négative (*n.f.*), negative rake.

coupe positive (*n.f.*), positive rake.

coupe tout (*n.m.*), master switch.

coupé, -e (élec.) (*adj.*), off.

coupé (-e) à la longueur (*adj.*), cut to length.

coupelle (d'une colonne à plateaux) (raffinerie) (*n.f.*), bell cap.

coupelle (d'un plateau) (raffinerie) (*n.f.*), bubble cap.

coupelle d'appui (*n.f.*), saddle.

coupelle de protection (*n.f.*), protective cap.

coupelle de soufflet (*n.f.*), boot retainer.

coupelle serre-gaine (*n.f.*), ferrule.

couper (*v.t.*), to turn out; to turn off.

couper un réacteur (aviation) (*v.*), to shut down an engine.

couplage à double voie (élec.) (*n.m.*), double-way connection.

couplage à simple voie (élec.) (*n.m.*), single-way connection.

couplage capacitif (élec.) (*n.m.*), capacitive coupling.

couplage d'antenne (radio) (*n.m.*), aerial cou-pling; antenna coupling (U.S.A.).

couplage d'un transformateur (élec.) (*n.m.*), vestor group of a transformer.

couplage en double zig-zag (élec.) (*n.m.*), fork connection.

couplage en fourche (élec.) (*n.m.*), fork connection.

couple (*n.m.*), frame.

couple antagoniste (*n.m.*), antagonistic torque.

couple conique d'entraînement (*n.m.*), bevel gear drive.

couple courant (*n.m.*), secondary frame.

couple d'accrochage (élec.) (*n.m.*), pull-in torque.

couple d'amortissement (*n.m.*), damping torque.

couple de décrochage (élec.) (*n.m.*), pull-out torque.

couple de freinage (*n.m.*), braking torque.

couple de serrage (*n.m.*), tightening torque.

couple de serrage à sec (*n.m.*), dry thread torque.

couple de serrage avec lubrifiant (*n.m.*), lubricated thread torque.

couple étanche (*n.m.*), pressure bulkhead.

couple fort (*n.m.*), main frame.

couple longeron (*n.m.*), spar frame.

couple maximum (*n.m.*), stall torque.

couple maximum constant à vide (*n.m.*), pull-out torque.

couple maximum constant en charge (*n.m.*), pull-in torque.

couple minimum au démarrage (*n.m.*), pull-up torque.

couple mobile (*n.m.*), removable frame.

couple résiduel (*n.m.*), residual torque.

couple résiduel d'une vis (*n.m.*), thread friction torque.

couple rotor (hélicoptère) (*n.m.*), rotor torque.

couple thermoélectrique (élec.) (*n.m.*), thermo-couple.

couplé (-e) en étoile (*adj.*), star connected.

couplemètre (*n.m.*), torquemeter.

coupler (*v.t.*), to link; to gang.

coupleur (*n.m.*), coupler.

coupleur automatique d'antenne (radio) (*n.m.*), automatic aerial tuner.

coupleur d'automanette (aviation) (*n.m.*), thrust computer.

coupleur de faisceau latéral (aviation) (*n.m.*), lateral beam coupler.

coupleur de faisceau longitudinal (aviation) (*n.m.*), longitudinal beam coupler.

coupleur de vol stationnaire (hélicoptère) (*n.m.*), hover flight coupler.

coupleur latéral (aviation) (*n.m.*), lateral computer.

coupleur longitudinal (aviation) (*n.m.*), longitudinal computer.

coupleur sonar (radio) (*n.m.*), sonar coupler.

coupole (*n.f.*), blister.

coupure (élec.) (*n.f.*), cutting-off a circuit.

coupure (radio) (*n.f.*), radio set out of order.

courant anodique (élec.) (*n.m.*), anode current.

courant ascensionnel (aviation) (*n.m.*), updraft.

courant d'excitation (élec.) (*n.m.*), energizing current.

courant d'induit (élec.) (*n.m.*), armature current.

courant de chauffage (élec.) (*n.m.*), filament current.

courant de collage (électronique) (*n.m.*), cut-in current.

courant de décharge (élec.) (*n.m.*), discharge current.

courant de fuite (élec.) (*n.m.*), leakage current.

courant de grille (élec.) (*n.m.*), grid current.

courant de repos (électron.) (*n.m.*), quiescent current.

courant demi-alternance (élec.) (*n.m.*), half-wave current.

courant force (élec.) (*n.m.*), power; power-mains.

courbe de manche cyclique (hélicoptère) (*n.f.*), stick displacement curve.

courbe de température (*n.f.*), temperature chart.

courbe isopoids (*n.f.*), iso-weight curve.

courbes de Wöhler (*n.f.pl.*), Wöhler curves.

couronne de butée basse (*n.f.*), droop restraining ring.

couronne de liaison de train (*n.f.*), ring nut for landing gear attachment.

couronne fixe de réducteur (*n.f.*), fixed ring gear.

couronne mobile de réducteur (*n.f.*), first stage sun gear.

couronne porte-balais (élec.) (*n.f.*), brush-rocker.

couronne principale (engrenages) (*n.f.*), bull gear.

courroie à créneaux (*n.f.*), cog belt.

courroie d'entraînement (*n.f.*), drive belt.

courroie de fixation (*n.f.*), strap.

courroie trapézoïdale (*n.f.*), v-belt; vee-belt.

course à l'atterrissage (aviation) (*n.f.*), landing run.

course ascendante (machines-outils) (*n.f.*), upward stroke.

course au décollage (aviation) (*n.f.*), take-off run.

course de détente (piston) (*n.f.*), expansion stroke.

course descendante (piston) (*n.f.*), downstroke.

coussin d'air (*n.m.*), air cushion.

coussinet autolubrifiant (*n.m.*), impregnated bearing; journal bearing.

coussinet sphérique (*n.m.*), ball cup.

couteau (machine à tailler les engrenages) (*n.m.*), cutter.

couteau à queue (machine à tailler les engrenages) (*n.m.*), shank-type cutter.

couteau de shaving (machine à tailler les engrenages) (*n.m.*), shaving cutter.

couteau droit (machine à tailler les engrenages) (*n.m.*), spur cutter.

couteau générateur (machine à tailler les engrenages) (*n.m.*), generating cutter.

couvercle antiturbulence pour compresseur de réacteur (*n.m.*), windage cover.

couvercle bombé (*n.m.*), welch plug.

couvercle d'expédition (*n.m.*), shipping cover.

couvercle porte-gicleur (aviation) (*n.m.*), fuel jet support cover.

couvert, -e (*adj.*), covered.

couvre-joint (*n.m.*), cover strip.

crabot (engrenages) (*n.m.*), dog.

crabotage (*n.m.*), dog clutch.

cranté, -e (*adj.*), notched; serrated.

crapaudine (*n.f.*), socket.

craquage (raffinerie) (*n.m.*), cracking.

craquage avec fixation du coke (*n.m.*), coking cracking.

craquage catalytique (*n.m.*), catcracking.

craquage en présence d'hydrogène (*n.m.*), hydro-cracking.

craquelage (*n.m.*), crazing.

craquelé, -e (peinture) (*adj.*), wrinkle finish.

craquelures (*n.f.pl.*), fine cracks.

crayon électrique (*n.m.*), electric pencil.

crémaillère d'avance (machines-outils) (*n.f.*), feed rack.

crémone (*n.f.*), lock bolt.

crête de tension (élec.) (*n.f.*), peak voltage.

creux (déformation) (*n.m.*), dent.

crever (*v.t.*), to puncture.

critère (*n.m.*), criterion.

crochet à ressort (*n.m.*), snap hook.

crochet commutateur de combiné téléphonique (*n.m.*), telephone switch hook.

crochet d'interdiction (*n.m.*), safety pawl.

crochet de verrouillage (*n.m.*), catch.

crochet délesteur (hélicoptère) (*n.m.*), release hook.

crochet délesteur de fret (hélicoptère) (*n.m.*), cargo release hook.

crochetage d'un faisceau (aviation) (*n.m.*), beam interception.

crocheter un faisceau (aviation) (*v.*), to cap-
-ture a beam.
croisière (*n.f.*), cruise.
croisière ascendante (aviation) (*n.f.*), climb
cruise.
croisillon (*n.m.*), cross-brace; spider.
croisillon d'induit (élec.) (*n.m.*), armature
spider.
croisillon de changement de pas (hélicop-
-tère) (*n.m.*), pitch change spider.
croisillon du joint universel (*n.m.*), universal
joint spider.
croisillon inducteur (élec.) (*n.m.*), field spider;
magnet wheel.
croisillonné, -e (*adj.*), cross-braced.
croix de Malte (*n.f.*), Geneva wheel.
croquer (*v.t.*), to pink.
croquis (*n.m.*), sketch.
croquis de mise au point (*n.m.*), revision
drawing.
crosse de dragage (de mines) (*n.f.*), minesweeping boom; tow hook.
croûte (métallurgie) (*n.f.*), scale.
cryoscopie (*n.f.*), cryoscopy.
cuissard (*n.m.*), thigh strap.
cuisson (caoutchouc-plastique) (*n.f.*), curing.
"cuit-vite" (caoutchouc-plastique) (*adj.*), fast-
curing.
cuivrage (*n.m.*), copper plating.

cuivré, -e (*adj.*), copper plated.
culasse (auto) (*n.f.*), cylinder head.
culasse (arme) (*n.f.*), breech.
culbuteur (auto) (*n.m.*), rocker.
culot (de tube électronique) (*n.m.*), base.
culot Edison (*n.m.*), screw base.
culot octal (*n.m.*), octal base.
culotte de tuyère (réacteur) (*n.f.*), exhaust noz-
-zle breeches.
cunéiforme (*adj.*), wedge-shaped.
curie (mesure de l'activité d'une matière radio-
-active) (*n.f.*), curie.
curium (élément) (*n.m.*), curium.
curseur de potentiomètre (*n.m.*), potentio-
-meter wiper.
cut-back (travaux publics) (*n.m.*), cut-back.
cuve à huile (transformateur) (*n.f.*), oil tank.
cuvette (*n.f.*), bowl.
cuvette d'arrêt (*n.f.*), retainer.
cuvette de clapet (*n.f.*), valve flange.
cuvette de levage (*n.f.*), socket pad.
cuvette de poignée (*n.f.*), handle plate.
cuvette de verrou (*n.f.*), lock cup.
cyanuration (*n.f.*), cyanide hardening.
cybernétique (*adj.*), cybernetics.
cycle d'accord (*n.m.*), tuning cycle.
cycle de marche (radio) (*n.m.*), duty cycle.
cycle des visites (*n.m.*), inspection cycle.
cyclotron (élec.) (*n.m.*), cyclotron.

D

dalot ((*n.m.*), scupper.
danger d'incendie (*n.m.*), fire hazard.
dard (support de maquette) (*n.m.*), sting.
de front (*adv.*), abreast.
dé (*n.m.*), thimble.
dé d'entraînement (*n.m.*), drive bit.
dé de cardan (*n.m.*), trunnion.
dé du joint universel (*n.m.*), universal joint
block.
déaérateur (*n.m.*), deaerator.
débattement (*n.m.*), deflection; displacement;
travel.)
débattement des pédales (aviation) (*n.m.*),
rudder pedal travel.
débit (*n.m.*), delivery.
débit de carburant (*n.m.*), mass fuel rate of
flow.
débit masse (air) (*n.m.*), mass air flow.
débit massique (*n.m.*), mass flow.
débitmètre (*n.m.*), flowmeter.
débitmètre instantané (*n.m.*), instant
flowmeter.
débitmètre totalisateur (*n.m.*), integrating
flowmeter.
déblocage (*n.m.*), unlocking.
déblocage réception (radio) (*n.m.*), sidetone
starting point.
débloquer (*v.t.*), to loosen.
déboîter (*v.t.*), to disengage; to separate.
débordement (*n.m.*), spillage.
debout (*adv.*), upright; vertical.
déboutonner (soudure) (*v.t.*), to peel off.
débrancher (*v.t.*), to disconnect.
débrayage (auto) (*n.m.*), declutching;
disengagement.
débrayage du pilote automatique (avia-

-tion) (*n.m.*), autopilot disengagement.
débrayage du rappel de manche (hélicop-
-tère) (*n.m.*), stick trim release.
débrayage rapide (*n.m.*), quick
disengagement.
débroussailleur (trav. pub.) (*n.m.*), grubber.
début de rupture (*n.m.*), starting point of a
fatigue crack.
décalage (*n.m.*), shifting; stagger.
décalage de pale (hél.) (*n.m.*), out-of-pitch
blade.
décalage de temps (*n.m.*), time lag.
décalaminage (*n.m.*), decarbonizing; descaling.
décalaminant (métal.) (*n.m.*), descaler.
décalcomanie (*n.f.*), transfer.
décalé, -e (*adj.*), offset; displaced; staggered;
shifted.
décaleur de phase (élec.) (*n.m.*), phase shifter.
décantation (*n.f.*), decanting
décanteur (*n.m.*), water trap.
décapage (*n.m.*), grading; scraping; etching;
scouring; stripping.
décapant (*n.m.*), etching agent.
décapant de soudure (*n.m.*), flux.
décapant pour peinture (*n.m.*), paint
remover.
décaper (*v.t.*), to grade; to scrape; to etch; to
scour; to strip.
décapeuse (trav. pub.) (*n.f.*), grader; scraper.
décapeuse-chargeuse (trav. pub.) (*n.f.*),
scraper-loader.
décélération (*n.f.*), deceleration.
décentré, -e (*adj.*), off-centre.
décharge autonome (électronique) (*n.f.*), self-
maintained discharge.
décharge disruptive (dans un gaz) (*n.f.*),

breakdown.

décharge non autonome (électronique) (*n.f.*), non-self-maintained discharge.

décharge semi-autonome (électronique) (*n.f.*), semi-self-maintained discharge.

déchargé, -e (*adj.*), discharged; run down.

déchiffrage (*n.m.*), decoding.

déchirement (*n.m.*), ripping.

déchirure (*n.f.*), rip; tear.

décibel (unité d'intensité sonore) (*n.m.*), decibel.

décibelmètre (*n.m.*), noise level meter.

décinormal, -e (*adj.*), decinormal.

déclenchement indépendant (élec.) (*n.m.*), inter-tripping.

déclenchement par bobine en dérivation (élec.) (*n.m.*), shunt tripping.

déclenchement par bobine en série (élec.) (*n.m.*), series tripping.

déclenchement par manque de tension (*n.m.*), under voltage tripping.

déclencher (*v.t.*), to initiate; to set off; to trigger; to disconnect.

déclencheur (mécanisme) (*n.m.*), trip; initiator.

déclencheur à action différée (*n.m.*), delayed action trip.

déclencheur à action instantanée (*n.m.*), fast acting trip.

déclencheur à maxima (*n.m.*), overcurrent trip.

déclencheur à minima (*n.m.*), undercurrent trip; no-volt trip.

déclencheur de fin de course (*n.m.*), limit trip.

déclin (*n.m.*), decay.

déclinaison (*n.f.*), grid declination.

déclivité de la piste (aviation) (*n.f.*), runway slope.

déclivité vers le bas (*n.f.*), downhill slope.

déclivité vers le haut (*n.f.*), uphill slope.

décodeur (*n.m.*), decoder.

décollage (aviation) (*n.m.*), take-off.

décollage avec vitesse initiale (aviation) (*n.m.*), rolling take-off.

décollage roulé (aviation) (*n.m.*), running take-off.

décollage sauté (hélicoptère) (*n.m.*), jump take-off.

décollage vertical (hélicoptère) (*n.m.*), vertical take-off.

décollement d'un relais (*n.m.*), tripping off of a relay.

décollement d'une soupape (*n.m.*), cracking open of a valve.

décollement des filets d'air (aviation) (*n.m.*), airstream separation.

décoller (aviation) (*v.i.*), to take off.

décolletage (*n.m.*), undercut; undercutting.

décompresseur (*n.m.*), pressure reducer.

décontamination (*n.f.*), decontamination.

découpage à la presse (*n.m.*), blanking; press cutting.

découpage au chalumeau (*n.m.*), flame-cutting; torch cutting.

découpe (*n.f.*), breakdown.

découplage (*n.m.*), uncoupling.

décraber (aviation) (*v.*), to decrab.

décrément (élec.) (*n.m.*), decay factor; decrement.

décrochage (aviation) (*n.m.*), stall.

décrochage (magnétisme) (*n.m.*), wipe off.

décrochage (mécanique) (*n.m.*), release; unlock-ing; unlatching.

décrochage (élec.) (*n.m.*), cut-out; de-energization.

décrochage (compresseur) (*n.m.*), engine stall.

décrochage de bout de pale (aviation) (*n.m.*), blade tip stall.

décrochage de la sonde altimétrique (*n.m.*), radio altimeter cut-out.

décrochage des pales (aviation) (*n.m.*), blade stall.

décrochage des trappes de train (aviation) (*n.m.*), landing gear door unlatching.

décrochage du train (aviation) (*n.m.*), landing gear unlocking.

décrochage latéral (aviation) (*n.m.*), roll-off

décrochage manuel du train (aviation) (*n.m.*), landing gear manual release.

décrochage tournant (aviation) (*n.m.*), rotat-ing stall.

décrocher (aviation) (*v.i.*), to stall.

décrocher (mécanique) (*v.t.*), to unlatch; to unlock.

défaillance intrinsèque (*n.f.*), basic failure.

défaillant, -e (*adj.*), faulty.

défaut d'alignement (*n.m.*), misalignment.

défense anti-sous-marine (*n.f.*), anti-submarine defence.

déflecteur (*n.m.*), baffle.

déflecteur d'huile (*n.m.*), oil slinger.

déflexion (*n.f.*), deflection; wash.

déflexion vers le bas (aviation) (*n.f.*), downwash.

déflexion vers le haut (aviation) (*n.f.*), upwash.

défocalisation (tube cathodique) (*n.f.*), deflec-tion; defocussing.

défonceuse (trav. pub.) (*n.f.*), ripper.

défonceuse tractée (trav. pub.) (*n.f.*), rooter.

déformation (*n.f.*), distortion.

déformation d'un pneu (*n.f.*), deformation of a tyre.

déformation de la pale (aviation) (*n.f.*), blade distortion.

déformation permanente (*n.f.*), permanent set.

défreiner (*v.t.i.*), to take off the brake; to un-safety (U.S.A.).

dégagement (mécanique) (*n.m.*), undercut; waist.

dégagement de chaleur (*n.m.*), heat release.

dégagement gazeux (*n.m.*), gassing.

dégazage (électronique) (*n.m.*), degassing.

dégazage (métallurgie) (*n.m.*), baking.

dégazage (chimie) (*n.m.*), vapour relief.

dégazeur (électronique) (*n.m.*), getter.

dégazeur (chimie) (*n.m.*), vapour relief valve.

dégivrage planeur (aviation) (*n.m.*), airfoil de-icing.

dégivrage réacteur (aviation) (*n.m.*), engine de-icing.

dégivreur (*n.m.*), de-icer; defroster.

dégommage (moteur) (*n.m.*), priming; cranking.

dégonfler (*v.t.*), to deflate.

dégoupiller (*v.t.*), to remove a cotter pin.

dégradation (*n.f.*), degradation.

dégraissage (*n.m.*), degreasing.

dégraissage électrolytique (*n.m.*), electro-lytic cleaning.

dégraissant (poterie) (*n.m.*), grog.

dégraisser (*v.t.*), to degrease.

degré de dureté Brinell (*n.m.*), Brinell hard-ness number.

dégrossissement (mécanique) (*n.m.*), rough-machining; roughing out.
dégroupage (*n.m.*), disassembly.
délai (*n.m.*), lead time.
délai de passage en autorotation (hélicop-tère) (*n.m.*), autorotation transition time.
délamination (*n.f.*), delamination.
délestage (électricité) (*n.m.*), load shielding.
délestage (aviation) (*n.m.*), lift-off.
délesteur de charge (hélicoptère) (*n.m.*), cargo sling.
délimiter (*v.t.*), to locate.
déloqueter (*v.t.*), to unlatch.
démagnétisateur (*n.m.*), demagnetizer.
démagnétiser (*v.t.*), to demagnetize; to degauss.
demande, à la (*adj.*, *adv.*), as required.
démarcation (*n.f.*), separation line; demarcation.
démarrage autonome (*n.m.*), engine starting under its own power.
démarrage avec surchauffe (*n.m.*), hot start.
démarrage étoile-triangle (élec.) (*n.m.*), star-delta starting.
démarrage par self (élec.) (*n.m.*), reactor starting.
démarreur à combustion (*n.m.*), combustion starter.
démarreur pneumatique (*n.m.*), air starter.
demi-acier ferreux (*n.m.*), half steel.
demi-additionneur (électronique) (*n.m.*), half adder.
demi-alternance (*n.f.*), half cycle.
demi-bague (*n.f.*), half bushing.
demi-cage de roulement (*n.f.*), ball bearing retainer.
demi-charnière (*n.f.*), half hinge.
demi-collerette (*n.f.*), half flange.
demi-collier (*n.m.*), half clamp.
demi-coquille (*n.f.*), half shell.
demi-cuvette (*n.f.*), half cup.
demi-écrou (*n.m.*), half-nut.
demi-étrier (*n.m.*), half stirrup.
demi-nervure (*n.f.*), rib section.
demi-peigne (*n.m.*), half cleat.
demi-période (élec.) (*n.f.*), half cycle.
demi-période de radioactivité (atom) (*n.f.*), half cycle; half-life.
demi produit (*n.m.*), semi-finished product.
demi-raccord (*n.m.*), half union.
demi-rondelle (*n.f.*), half washer.
demi-teinte (*n.f.*), half tone.
demi-tour en vol (aviation) (*n.m.*), return to base.
demi-voilure (*n.f.*), wing.
démontable (*adj.*), detachable.
démontage (*n.m.*), disassembly; removal; taking down; dismantling.
démonter (*v.t.*), to dismantle; to disassemble; to take down; to remove.
démoulage (fonderie) (*n.m.*), shake-out.
démultiplicateur (*n.m.*), reduction gear.
dénoyage des pompes (*n.m.*), pump draining.
densité spectrale d'énergie (*n.f.*), power spectrum density.
dent (*n.f.*), prong.
dent d'arrêt (*n.f.*), ratchet tooth.
dent de repère (*n.f.*), guide tooth; master spline.
denté, -e (*adj.*), indented.
dentelé, -e (*adj.*), serrated.
dénudé, -e (*adj.*), stripped.
dépannage (*n.m.*), repair, breakdown service.

dépanner (*v.t.*), to repair.
départ (*n.m.*), departure.
départ usine (*adj.*), ex-works.
dépassement (*n.m.*), overshooting.
dépassement de course (aviation) (*n.m.*), overrun.
dépasser (*v.t.*), to protrude.
déperditeur de potentiel (*n.m.*), static discharger.
déphasage (élec.) (*n.m.*), angle of phase devia-tion; phase splitting; phase shift.
déphasé, -e (élec.) (*adj.*), out of phase.
déphasé (-e) en arrière (élec.) (*adj.*), lagging.
déphasé (-e) en avant (élec.) (*adj.*), leading.
déphaser (*v.*), to split phase; to shift phase.
déphaseur (élec.) (*n.m.*), phase converter.
déphaseur (hélicoptère) (*n.m.*), phasing unit.
déplacement (*n.m.*), relocation; shift.
déplacer (*v.t.*), to relocate.
déplomber (*v.t.*), to unseal.
déployer (*v.t.*), to spread out.
déport (*n.m.*), mismatch.
déport dans l'avancement (aviation) (*n.m.*), blade tilt.
déport dans le plan de rotation (aviation) (*n.m.*), blade sweep.
déporté, -e (*adj.*), offset.
dépose (*n.f.*), removal.
dépôt (*n.m.*), residue; sediment.
dépôt morainique (géol.) (*n.m.*), boulder clay.
dépoussiérage (*n.m.*), dust removal.
dépression (*n.f.*), negative pressure.
dérapage (auto) (*n.m.*), skidding.
déréglage des pales en rotation (hélicop-tère) (*n.m.*), out-of-track.
déréglage en pas (hélicoptère) (*n.m.*), out-of-pitch.
dérivateur (radio) (*n.m.*), rat race.
dérivation (construction) (*n.f.*), by-pass.
dérive (aviation) (*n.f.*), vertical stabilizer.
dérive de fréquence (*n.f.*), frequency shift.
dérive en vol (aviation) (*n.f.*), drift in flight.
dérivée (mathématiques) (*n.f.*), derivative.
dériver (*v.t.*), to by-pass; to shunt; to tap.
dérivomètre (*n.m.*), driftmeter.
dérochage (*n.m.*), pickling.
dérouillage (*n.m.*), derusting.
dérouler (*v.t.*), to spring back; to unroll.
déroutement (*n.m.*), diversion.
désaccord gyro (*n.m.*), gyro unbalance.
désamorçage d'une pompe (*n.m.*), pump unloading.
désaxé, -e (*adj.*), off centre.
descente (aviation) (*n.f.*), let down.
descente automatique (machines-outils) (*n.f.*), automatic down feed.
descente d'antenne (radio) (*n.f.*), aerial lead; antenna lead (U.S.A.).
descente du train (aviation) (*n.f.*), landing gear extension.
descente rapide (aviation) (*n.f.*), emergency descent.
désembourbage (*n.m.*), towing out of the mud; debogging.
déséquilibre (*n.m.*), unbalance.
déséquilibre du rotor (hélicoptère) (*n.m.*), rotor unbalance.
déséquiper un réacteur (aviation) (*v.t.*), to tear down engine equipment.
désétamage (*n.m.*), detinning.
désexciter (*v.t.*), to de-activate; to de-energize.

déshabillage (*n.m.*), disassembly; stripping.
déshydratant (*n.m.*), desiccant.
déshydratation (*n.f.*), dewatering.
désignation (*n.f.*), description; nomenclature.
désolidariser (*v.t.*), to free from.
dessangler (*v.t.*), to unstrap.
dessécheur d'air (transformateurs à l'huile) (*n.m.*), breather.
desserrage (*n.m.*), untightening.
dessertir (*v.t.*), to uncrimp.
dessin animé (cinéma) (*n.m.*), cartoon.
dessin d'étude (*n.m.*), engineering drawing.
dessin de fabrication (*n.m.*), engineering drawing.
dessoucheur (trav. pub.) (*n.m.*), rooter; stumper.
dessouder (*v.t.*), to unsolder.
dessus (*n.m.*), top.
destinataire (*n.m.*), addressee.
détalonnage (machines-outils) (*n.m.*), backing-off; relief; undercut.
détartrage (*n.m.*), descaling.
détecter (élec.) (*v.t.*), to demodulate.
détecteur (*n.m.*), detector; pick-off; sensor.
détecteur (élec.) (*n.m.*), demodulator.
détecteur à fil continu (*n.m.*), closed circuit sensor.
détecteur angulaire à trois axes (aviation) (*n.m.*), angular three-axis rate sensor.
détecteur d'accélération (*n.m.*), acceleration detector.
détecteur d'écart (aviation) (*n.m.*), deviation detector.
détecteur d'incidence (*n.m.*), incidence probe.
détecteur de défaut d'alimentation (élec.) (*n.m.*), line fault detector.
détecteur de déplacement (*n.m.*), follow-up.
détecteur de fissures (*n.m.*), crack detector.
détecteur de flux (*n.m.*), flux valve.
détecteur de fréquence (*n.m.*), frequency detector.
détecteur de givrage (aviation) (*n.m.*), ice detector.
détecteur de panne de sensation musculaire (*n.m.*), artificial feel-failure detector.
détecteur de vitesse angulaire (aviation) (*n.m.*), angular velocity rate sensor.
détecteur embrayable (aviation) (*n.m.*), clutched pick-off.
détecteur gravimétrique (aviation) (*n.m.*), levelling unit.
détecteur ponctuel de surchauffe (*n.m.*), local overheat detector.
détection d'avions en vol (radar) (*n.f.*), aircraft interception.
détendeur (*n.m.*), pressure reducing valve.
détendre (*v.t.*), to expand; to relax; to release; to relieve.
détente (*n.f.*), expansion; relaxation; release; stress relief.
détente adiabatique (*n.f.*), adiabatic pressure drop.
détergent (*n.m.*), detergent.
détérioration (*n.f.*), degradation.
détermination du centrage (*n.f.*), computation of central gravity.
détersif (*n.m.*), cleaner; detergent.
détonation (dans un moteur) (*n.f.*), pinking.
détourage (*n.m.*), routine.
détourer (machines-outils) (*v.t.*), to cut out; to rout.

détoureuse (machines-outils) (*n.f.*), routing machine; router.
détrempe (peinture) (*n.f.*), distemper.
détremper (métallurgie) (*v.t.*), to anneal.
détresse (aviation) (*n.f.*), last emergency action.
détrompeur (*n.m.*), locating pin.
deutéron (*n.m.*), deuteron.
deuton (*n.m.*), deuteron.
développement (photo) (*n.m.*), processing; development.
déverrouillage (*n.m.*), unlocking; unlatching.
déviateur de jet (*n.m.*), thrust reverser.
déviation (électronique) (*n.f.*), deflection.
devis de masse (*n.m.*), weight breakdown.
dévolteur (élec.) (*n.m.*), stepdown transformer.
déwatté, -e (élec.) (*adj.*), watless.
diabolo (*n.m.*), bogie.
diagramme d'écoulement (*n.m.*), flow pattern.
diagramme de manoeuvres et de rafales (aviation) (*n.m.*), flight envelope.
diagramme de rayonnement (*n.m.*), radiation pattern.
diagramme logique (*n.m.*), logical diagram.
diamètre de centrage (*n.m.*), snap diameter.
diamètre de cercle primitif (*n.m.*), pitch circle diameter.
diamètre de fond de filets (*n.m.*), minor diameter.
diamètre de perçage (*n.m.*), pitch centre diameter.
diamètre nominal (*n.m.*), nominal diameter.
diamètre sur flancs de filets (*n.m.*), pitch diameter.
diamètre sur plats (*n.m.*), diameter across flats.
diapason (*n.m.*), tuning fork.
diaphonie (radio) (*n.f.*), cross-talk.
diaphragme (photographie) (*n.m.*), aperture.
diaphragme de réduction (*n.m.*), restrictor.
diaphragme rotatif (*n.m.*), rotary diaphragm.
diapositive (photo) (*n.f.*), transparency.
diastase (chimie) (*n.f.*), diastase.
dièdre négatif (*n.m.*), anhedral.
diélectrique (élec.) (*adj.*), non-conductive.
différence de pression (*n.f.*), pressure difference.
diffraction côtière (navigation) (*n.f.*), shore effect.
diffuseur d'ordres (*n.m.*), public address system.
diffusion (*n.f.*), distribution; broadcasting.
diluant (*n.m.*), diluent; solvent; thinner.
dilution (*n.f.*), dilution.
dimensions hors-tout (*n.f.pl.*), overall dimensions.
diode (électronique) (*n.f.*), diode.
diode à crystal (électronique) (*n.f.*), crystal diode.
diode de souffle (électronique) (*n.f.*), noise diode.
dioptrie (*n.f.*), diopter.
directeur de vol (*n.m.*), flight director.
direction à vis et doigt (auto) (*n.f.*), worm and peg steering.
direction d'une force (*n.f.*), line of action of a force.
direction des traces d'usinage (*n.f.*), lay.
directrice (*n.f.*), directing line.
discontinuité de mouillage d'eau (*n.f.*), water break.
discriminant (*n.m.*), discriminant.

discriminateur (*n.m.*), discriminator.

discriminateur de fréquence (radio) (*n.m.*), frequency discriminator.

discriminateur de phase (électronique) (*n.m.*), phase discriminator.

disjoncteur à maxima (élec.) (*n.m.*), maximum cut-out.

disjoncteur à réenclenchement automatique (élec.) (*n.m.*), auto-release circuit-breaker.

disjoncteur de sécurité (élec.) (*n.m.*), limit-switch; safety cut-out.

disjoncteur directionnel (élec.) (*n.m.*), reverse-current relay.

disjoncteur principal de génératrice (élec.) (*n.m.*), generator main circuit-breaker.

disjoncteur thermique (élec.) (*n.m.*), thermal circuit-breaker.

dispatching (*n.m.*), dispatching.

dispersion (*n.f.*), stray.

disponibilité intrinsèque (*n.f.*), inherent availability.

disponibilité opérationnelle (*n.f.*), operational availability.

disponibilité réalisable (*n.f.*), achievable availability.

dispositif anti-couple (*n.m.*), anti-torque device.

dispositif anti-dérapant (*n.m.*), anti-skid unit.

dispositif anti-pompage (élec.) (*n.m.*), surge guard.

dispositif avertisseur de décrochage (aviation) (*n.m.*), stall warning device.

dispositif compensateur (*n.m.*), compensating device.

dispositif d'amarrage rapide (*n.m.*), quick mooring gear.

dispositif d'escamotage (*n.m.*), retracting device installation.

dispositif de blocage (*n.m.*), locking device.

dispositif de commande de pale (hélicoptère) (*n.m.*), blade control system.

dispositif de contrôle optique du train (aviation) (*n.m.*), landing gear downlock visual check.

dispositif de délestage (élec.) (*n.m.*), automatic load limitation.

dispositif de filetage (machines-outils) (*n.m.*), threading attachment.

dispositif de fraisage (machines-outils) (*n.m.*), milling attachment.

dispositif de hissage des volets de profondeur (aviation) (*n.m.*), elevator hoist.

dispositif de hissage turbine (aviation) (*n.m.*), turbine hoist.

dispositif de mise en tension des barres de torsion (*n.m.*), torque bar stressing fixture.

dispositif de perçage (machines-outils) (*n.m.*), drilling attachment.

dispositif de rainurage (machines-outils) (*n.m.*), slotting attachment.

dispositif de rappel dans l'axe (aviation) (*n.m.*), wheel centering device.

dispositif de réarmement (relais) (*n.m.*), resetting device.

dispositif de repérage (*n.m.*), locating device.

dispositif de repliage automatique des pales (hélicoptère) (*n.m.*), automatic blade-folding system.

dispositif de reproduction (machines-outils) (*n.m.*), copying attachment.

dispositif de roue libre (*n.m.*), free-wheel mechanism.

dispositif de sécurité (*n.m.*), safety device.

dispositif de surfaçage (machines-outils) (*n.m.*), facing attachment.

dispositif de taraudage (machines-outils) (*n.m.*), tapping attachment.

dispositif de verrouillage de diabolo train avant (aviation) (*n.m.*), nose gear steer lock.

dispositif de verrouillage de pas (hélicoptère) (*n.m.*), pitch locking system.

dispositif pour centrage des disques de frein (*n.m.*), brake disc alignment jig.

disposition (schéma) (*n.f.*), layout.

disque à friction (frein) (*n.m.*), rotor disc.

disque à polir (*n.m.*), buffing wheel.

disque d'entraînement (*n.m.*), driving disc.

disque d'étanchéité (*n.m.*), sealing disc.

disque de frein (*n.m.*), brake disc.

disque de turbine (*n.m.*), turbine disc.

disque fixe (sur frein) (*n.m.*), stator disc.

disque menant (*n.m.*), driving disc.

disque mené (*n.m.*), driven disc.

disque parabolique (radar) (*n.m.*), parabolical deflector.

disque phonographique (*n.m.*), gramophone record.

disque rotor (hélicoptère) (*n.m.*), rotor disc.

disque sustentateur (*n.m.*), actuator disc.

dissolution (caoutchouc) (*n.f.*), rubber solution.

distance, à (*adj.*), remote.

distance accélération-arrêt (auto) (*n.f.*), accelerate-stop distance.

distance d'accélération (auto) (*n.f.*), acceleration distance.

distance d'arrêt (auto) (*n.f.*), stopping distance.

distance d'atterrissage (*n.f.*), landing distance.

distance de décollage (aviation) (*n.f.*), take-off distance.

distance de mise en vitesse (aviation) (*n.f.*), distance needed to achieve (take-off) speed.

distance de plané (aviation) (*n.f.*), gliding distance.

distance disruptive (élec.) (*n.f.*), break distance.

distance focale (photo) (*n.f.*), focal length.

distance franchissable (aviation) (*n.f.*), range.

distance parcourue (*n.f.*), distance covered.

distance spécifique (aviation) (*n.f.*), specific range.

distillateur (*n.m.*), still.

distributeur (*n.m.*), dealer.

distributeur à tiroir (*n.m.*), slide valve distributor.

distributeur d'ordonnancement (*n.m.*), sequence valve.

distributeur de freinage (*n.m.*), brake valve.

distributeur de savon (*n.m.*), liquid soap dispenser.

distributeur de servodyne (aviation) (*n.m.*), actuator control valve.

distributeur de turbine (*n.m.*), turbine nozzle.

distributeur quadruple de freins (aviation) (*n.m.*), landing gear master brake cylinder.

distribution (artistique) (cinéma) (*n.f.*), casting.

distribution (auto) (*n.f.*), timing.
diviseur de tension (élec.) (*n.m.*), voltage divider.
documentation technique (*n.f.*), documen--tation; technical data.
doigt d'accrochage (*n.m.*), catch.
doigt d'entraînement (*n.m.*), driving pin.
doigt de blocage (*n.m.*), stop.
doigt de butée d'affaissement (*n.m.*), droop restraining shaft.
doigt de verrouillage (*n.m.*), locking pin.
domaine de vol (aviation) (*n.m.*), flight envelope.
dôme radar (radar) (*n.m.*), radome.
dôme sonar (radar) (*n.m.*), sonar dome.
dope (*n.m.*), dope.
doppler (*n.m.*), doppler.
dosage (*n.m.*), solution analysis.
dosage des gaz dans les métaux (*n.m.*), gas determination in metals.
dosage du volume (radio) (*n.m.*), volume adjustment.
dosage potentiométrique (*n.m.*), electro--metric titration.
doser (*v.t.*), to meter.
doser (radio) (*v.t.*), to adjust reception output volume.
doseur pondéral (ciment) (*n.m.*), weighing batcher.
doseur volumétrique (ciment) (*n.m.*), volu--metric batcher.
dosseret (*n.m.*), backing.
dossier (document) (*n.m.*), folder; file; record.
dossier de calcul (*n.m.*), engineering calcula--tions record.
double (*adj.*), dual.
double chape (*n.f.*), dual clevis.
double commande (*n.f.*), dual control.
double coupure (élec.) (*n.f.*), dual cut-out.
double diabolo (*n.m.*), four-wheel bogie.
double enveloppe (*n.f.*), double-casing.
double sensibilité (aviation) (*n.f.*), auto-pilot pitch sensitivity system.
doublet (*n.m.*), dipole.
doublet quart d'onde (*n.m.*), quarter-wave dipole.
douille à baïonnette (élec.) (*n.f.*), bayonet lamp-holder.
douille de brochage (*n.f.*), rigging bushing.
douille de fixation de crépine (*n.f.*), strainer retaining bush.
douille de serrage (*n.f.*), collet.
douille de verrou (*n.f.*), lock bush.

douille électrique (*n.f.*), socket.
douille entretoise (*n.f.*), spacer bushing.
douille filetée (*n.f.*), screw bushing.
douille lisse (*n.f.*), bushing.
douille tendeur (*n.f.*), turnbuckle barrel.
douille voleuse (élec.) (*n.f.*), lamp adaptor.
dragage de mines (*n.m.*), minesweeping.
drainage à la charrue taupe (*n.m.*), mole drainage.
drapeau d'alarme (sur instrument) (*n.m.*), warning flag.
drapeau d'avertissement (sur instrument) (*n.m.*), warning flag.
dressage (machines-outils) (*n.m.*), surfacing.
dresser (*v.t.*), to set up; to straighten; to arrange; to hold upright.
drisse (*n.f.*), halyard.
dumper (*n.m.*), dumper.
duplexeur (*n.m.*), duplexer.
duplicata (*n.m.*), duplicate; copy.
duraluminium (*n.m.*), duralumin.
durcissement (plast.) (*n.m.*), curing.
durcissement par précipitation (*n.m.*), pre--cipitation hardening.
durcissement par revenue (*n.m.*), temper hardening.
durcissement par trempe (*n.m.*), quench hardening.
durcissement par vieillissement (*n.m.*), age hardening.
durcissement superficiel (*n.m.*), case hardening.
durcisseur (*n.m.*), hardener.
durée (*n.f.*), period; time.
durée d'autorotation d'un réacteur (avia--tion) (*n.f.*), engine coasting down time.
durée d'échauffement (d'un moteur) (*n.f.*), warm-up time.
durée d'établissement (radio) (*n.f.*), build-up time.
durée d'utilisation (*n.f.*), operating time.
durée de stockage (*n.f.*), shelf life.
durée de vie (*n.f.*), service life.
durée de vie de la pale (aviation) (*n.f.*), blade life.
dureté sclérométrique (*n.f.*), abrasive hardness.
duromètre (*n.m.*), hardness tester.
dynamo-démarreur (*n.f.*), generator starter.
dynamoteur (*n.m.*), dynamotor.
dyne (*n.f.*), dyne.
dynode (électronique) (*n.f.*), dynode.

E

eau douce (*n.f.*), soft water.
eau dure (*n.f.*), hard water.
eau lourde (*n.f.*), heavy water.
eau oxygénée (*n.f.*), hydrogen peroxide.
eau pelliculaire (géol.) (*n.f.*), film water.
eau usée (*n.f.*), waste water.
eaux résiduaires (*n.f.pl.*), residual water.
ébauchage (usinage) (*n.f.*), rough machining.
ébauche (*n.f.*), blank.
ébauche coulée (*n.f.*), rough casting.
ébauche d'engrenage (*n.f.*), gear blank.

ébauche forgée (*n.f.*), rough forging.
ébauche matricée (*n.f.*), drop forging.
ébavurage (*n.m.*), deburring.
ébavurage au tonneau (*n.m.*), tumbling.
ébrèchement (*n.m.*), chipping.
écart (*n.m.*), deviation.
écart angulaire (*n.m.*), angular displacement.
écart moyen quadratique (*n.m.*), standard deviation.
écarté, -e (*adj.*), separated.
écartement (*n.m.*), clearance; gap.

échancrer (*v.t.*), to indent.

échancrure (*n.f.*), indentation.

échange standard (*n.m.*), standard replacement.

échangeur air/air (*n.m.*), air-to-air heat exchanger.

échangeur de chaleur (*n.m.*), heat exchanger.

échangeur huile/kérosène (*n.m.*), fuel-coolant/oil-cooler heat exchanger.

échangeur primaire (*n.m.*), primary heat exchanger.

échantillon (*n.m.*), specimen; sample; pattern.

échappement (horlogerie) (*n.m.*), escapement mechanism.

échauffement (*n.m.*), overheating.

échauffement cinétique (*n.m.*), kinetic heating.

échauffement en régime transitoire (*n.m.*), transient heating.

échelle (*n.f.*), ladder.

échelle accès au poste de pilotage (avia--tion) (*n.f.*), flight compartment access stairway.

échelle anémométrique (*n.f.*), scale of wind forces.

échelle de corde (*n.f.*), cable ladder.

échelle grandeur (*n.f.*), actual size; full scale.

échelon de commande cyclique (hélicop--tère) (*n.m.*), cyclic control step.

écho radar (*n.m.*), radar echo.

éclairage commercial (*n.m.*), domestic lights.

éclairage de veille (*n.m.*), night lighting system.

éclairage du poste de pilotage (*n.m.*), flight compartment lights.

éclairage fluorescent (*n.m.*), fluorescent lighting.

éclairage indirect (*n.m.*), concealed lighting.

éclairage interne (*n.m.*), internal lighting.

éclairage par la tranche (*n.m.*), edge lighting.

éclairage ultraviolet (*n.m.*), ultra-violet lighting.

éclatement en éventail (*n.m.*), fanning out.

éclateur (élec.) (*n.m.*), arrester.

éclateur sous tube (élec.) (*n.m.*), glass sealed discharge gap.

éclissage (*n.m.*), splicing.

éclisse (*n.m.*), batten.

éclisse magnétique réglable (magnétisme) (*n.f.*), magnetic adjustable link.

écope (*n.f.*), scoop.

écorner (*v.t.*), to chamfer; to smooth corners.

écoulement axial (*n.m.*), axial flow.

écoulement d'air (*n.m.*), airflow.

écoulement laminaire (*n.m.*), laminar flow.

écoulement plastique (*n.m.*), plastic flow.

écoulement turbulent (*n.m.*), turbulent flow.

écoute (*n.f.*), reception; listening-in; output.

écoute sonar (*n.f.*), sonar search.

écouteur (radio-téléphone) (*n.m.*), earphone; headphone.

écran (d'un tube cathodique) (*n.m.*), screen.

écran absorbant (électronique) (*n.m.*), dark trace screen.

écran acoustique (radio) (*n.m.*), baffle board.

écran antiarc (élec.) (*n.m.*), flash barrier.

écran luminescent (électronique) (*n.m.*), lumi--nescent screen.

écran magasin (radio) (*n.m.*), display storage tube.

écran radar (*n.m.*), radarscope.

écran thermique (de cathode) (*n.m.*), heat shield.

écrasement (*n.m.*), collapsing; flattening.

écrasement (d'un joint) (*n.m.*), compression (of a seal).

écrêteur (*n.m.*), chopper; clipper, amplitude limiter.

écrou à auto-freinage (*n.m.*), self-locking nut.

écrou à collet (*n.m.*), flanged nut.

écrou à cuvette (*n.m.*), spherical nut.

écrou à encoches (*n.m.*), slotted nut.

écrou à face (*n.m.*), shear nut.

écrou à fente (n.m.), split nut.

écrou à oreilles (*n.m.*), thumb screw.

écrou à pattes (*n.m.*), tab nut.

écrou à portée sphérique (*n.m.*), acorn nut.

écrou à river (à patte) (*n.m.*), anchor nut.

écrou à rotule (*n.m.*), knuckle nut.

écrou à six pans (*n.m.*), hexagonal nut.

écrou auto-freiné (*n.m.*), elastic stop nut.

écrou baladeur (*n.m.*), sliding nut.

écrou borgne (*n.m.*), capnut.

écrou de blocage (*n.m.*), lock nut.

écrou de centrage (*n.m.*), centering nut.

écrou de raccord de tuyauterie (*n.m.*), union nut.

écrou de serrage (*n.m.*), retaining nut.

écrou de vis mère (*n.m.*), clasp nut.

écrou élastique (*n.m.*), elastic stop nut.

écrou étanche (*n.m.*), seal nut.

écrou fendu (*n.m.*), split nut.

écrou flottant (*n.m.*), floating anchor nut.

écrou frein (*n.m.*), lock nut.

écrou guide (*n.m.*), guide nut.

écrou haut (*n.m.*), heavy nut; thick nut.

écrou indesserrable (*n.m.*), elastic stop nut.

écrou normal (*n.m.*), ordinary hexagonal nut.

écrou nylstop (*n.m.*), nylstop self-locking nut.

écrou pinacle (*n.m.*), self-locking nut.

écrou prisonnier (*n.m.*), captive nut.

écrou sphérique (*n.m.*), spherical nut.

écroui, -e (métallurgie) (*adj.*), work-hardened.

écrouissage (métallurgie) (*n.m.*), cold rolling; work-hardening.

effacement (aviation) (*n.m.*), wash-out.

effacement de la sécurité (aviation) (*n.m.*), landing gear safety override.

effacer (enregistrement sur bande) (*v.t.*), to erase.

effacer (une interdiction) (*v.t.*), to override.

effacer (une alarme) (*v.t.*), to cancel (a warning).

effacer la piste (aviation) (*v.*), to overshoot the runway.

effacer (s') (*v.t.*), to fade out.

effectif (*n.m.*), personnel; staff.

effet béquille de l'escalier (aviation) (*n.m.*), stairway propping effect.

effet Brinell (*n.m.*), Brinell effect.

effet capillaire (*n.m.*), capillary action.

effet d'antenne (*n.m.*), aerial effect.

effet d'aspiration (*n.m.*), syphoning; suction effect.

effet d'échelle (*n.m.*), scale effect.

effet d'entaille (*n.m.*), notch effect.

effet d'hydroglisseur (aviation) (*n.m.*), aquaplaning.

effet de compressibilité en bout de pale (aviation) (*n.m.*), tip blade compressibility effect.

effet de Coriolis (*n.m.*), Coriolis effect.

effet de pincement (élec.) (*n.m.*), pinch effect.
effet de pression dynamique (*n.m.*), ram effect.
effet de rivage (*n.m.*), shore effect.
effet de sol (*n.m.*), ground effect.
effet de vague (dans un réservoir) (*n.m.*), surge effect (in a tank).
effet drapeau (*n.m.*), feathering effect.
effet différentiel (*n.m.*), differential effect.
effet Doppler (*n.m.*), Doppler effect.
effet dynamique (*n.m.*), ram effect.
effet Larsen (*n.m.*), Larsen effect.
effet local (*n.m.*), sidetone effect.
effet microphonique (radio) (*n.m.*), micro--phonics.
effet pelliculaire (*n.m.*), fringe; skin effect.
effet phugoïde (*n.m.*), phugoid effect.
effet tranchant mobile (*n.m.*), running shear load.
efficacité (*n.f.*), effectiveness.
effilement (*n.m.*), taper ratio.
effilochage (*n.m.*), fraying.
effluve (élec.) (*n.f.*), brush discharge; glow discharge.
effort (*n.m.*), load.
effort au manche (aviation) (*n.m.*), stick force; stick load.
effort de cisaillement (*n.m.*), shearing stress.
effort de commande de pas (hélicoptère) (*n.m.*), pitch control load.
effort de rupture (*n.m.*), breaking stress.
effort longitudinal au manche cyclique (hélicoptère) (*n.m.*), longitudinal cyclic stick load.
effort tranchant (*n.m.*), shearing stress.
éjecter (*v.t.*), to expel; to jettison.
élasticité (*n.f.*), yield.
élastomère (plast.) (*n.m.*), elastomer.
électro-aimant (*n.m.*), electro-magnet.
électro-formage (*n.m.*), electro-forming.
électrocinétique (élec.) (*n.f.*), electrokinetics.
électrode d'amorçage (élec.) (*n.f.*), starting electrode.
électrode d'entrée (élec.) (*n.f.*), input electrode.
électrode de commande (élec.) (*n.f.*), control electrode.
électrode de sortie (élec.) (*n.f.*), output electrode.
électrodéposition (*n.f.*), electroplating.
électrodéposition rapide et superficielle (*n.f.*), flash plating.
électron germe (*n.m.*), initiating electron.
électron interne (*n.m.*), inner-shell electron.
électron libre (*n.m.*), free electron.
électron lié (*n.m.*), bound electron.
électron périphérique (*n.m.*), outer-shell electron.
électron-volt (*n.m.*), electron-volt.
électronique (*adj.*), electronic.
électronique (*n.f.*), electronics.
électrophone (*n.m.*), electric gramophone.
électroplastie (*n.f.*), electroplating.
électrovalve (*n.f.*), electrovalve.
électrovalve de remplissage (carburant) (*n.f.*), refuelling valve.
électrovanne (*n.f.*), electrovalve.
élément (*n.m.*), item.
élément chargé (élec.) (*n.m.*), active cell.
élément chauffant (*n.m.*), heating element.
élément consommable (*n.m.*), expendable item.

élément d'accouplement à roue libre (*n.m.*), free wheel and clutch unit.
élément d'accumulateur (élec.) (*n.m.*), bat--tery cell.
élément de fabrication (*n.m.*), spare part.
élément détecteur (*n.m.*), sensing element.
élément dynamique (*n.m.*), dynamic component.
élément filtrant (*n.m.*), filter element.
élément longitudinal (*n.m.*), longitudinal member.
élément menant (*n.m.*), driving member.
élément mené (*n.m.*), driven member.
élément pare-feu (*n.m.*), heat shield.
élément principal (*n.m.*), main unit.
élément récupérable (*n.m.*), recoverable item.
élément simple (*n.m.*), part.
élément transversal (*n.m.*), transverse member.
élévateur à fourche (manutention) (*n.m.*), fork-lift truck.
élévateur de cuve (transformateur) (*n.m.*), tank-lifter.
élévation (*n.f.*), front elevation.
élevon (*n.m.*), elevon.
élimé, -e (*adj.*), frayed.
éliminateur de bruit de fond (radio) (*n.m.*), squelch.
éliminateur de quadrature (élec.) (*n.m.*), quadrature stripper.
élingue balançoire (hélicoptère) (*n.f.*), cargo swing.
élingue de chargement (*n.f.*), cargo sling.
élingue de fret (*n.f.*), cargo sling.
élingue de hissage (*n.f.*), hoisting sling.
émail au four (*n.m.*), baking enamel.
emballage (*n.m.*), packaging; wrapper.
emballement (moteur, rotor) (*n.m.*), overspeed.
embardée (aviation) (*n.f.*), gust.
embarqué, -e (aviation) (*adj.*), ship-borne (aircraft).
embarquer (aviation) (*v.i.*), to swerve off the runway.
embase (*n.f.*), base.
embase de raccord (*n.f.*), pipe-union.
embiellage (*n.m.*), link mechanism; linkage.
emboîter (*v.t.*), to engage.
embouchure (*n.f.*), opening; orifice.
embout (*n.m.*), end; fitting.
embout à oeil (*n.m.*), eye end.
embout à rotule (*n.m.*), ball bearing rod end.
embout adapteur (*n.m.*), adaptor.
embout arrêtoir (*n.m.*), end stop.
embout cannelé (*n.m.*), splined end.
embout d'articulation (*n.m.*), hinge fitting.
embout de bielle (*n.m.*), rod end.
embout de câble (*n.m.*), cable end.
embout de clé (*n.m.*), wrench adaptor.
embout de contrefiche (*n.m.*), strut end fitting.
embout de drain (*n.m.*), drain nipple.
embout de flexible (*n.m.*), flexible shaft adaptor.
embout de levage (*n.m.*), jacking pad.
embout de levage mâle (*n.m.*), dome pad.
embout de palier (*n.m.*), bearing end.
embout de réglage (*n.m.*), setting head.
embout de tube (*n.m.*), tube end.
embout de tuyauterie (*n.m.*), nipple.
embout dynamométrique (*n.m.*), torque wrench adaptor.

embout fileté (*n.m.*), threaded end.
embout rotulaire (*n.m.*), swivel end.
embout sphérique (*n.m.*), ball end.
embout taraudé (*n.m.*), tapped end.
embouti de hublot (navire) (*n.m.*), window frame.
embrasse de rideau (*n.f.*), curtain loop.
embrayage à friction (*n.m.*), slip clutch.
embrayage à poudre magnétique (*n.m.*), magnetic powder clutch.
embrayage à roue libre (*n.m.*), overrunning clutch.
embrayage hydraulique (*n.m.*), hydraulic clutch.
embrayage hydromécanique (*n.m.*), hydro-mechanical clutch.
embrayage pneumatique (*n.m.*), air clutch.
embrèvement (*n.m.*), dimpling.
embrèvement à chaud (*n.m.*), hot dimpling process.
embrèvement à froid (*n.m.*), cold dimpling process.
émeri potée (*n.f.*), emery powder.
émetteur (radio) (*n.m.*), transmitter.
émetteur d'asservissement (électronique) (*n.m.*), servo transmitter.
émetteur omnidirectionnel (radio) (*n.m.*), omnirange transmitter.
émetteur-récepteur (radio) (*n.m.*), trans-ceiver.
émetteur-répondeur (radio) (*n.m.*), trans-ponder.
émettre (radio) (*v.t.*), to transmit.
émettre (*v.t.*), to radiate.
émission (radio) (*n.f.*), transmission; broadcast.
émission météorologique (*n.f.*), weather broadcast.
émission sur bande latérale unique (*n.f.*), single sideband transmission.
emmanchement dur (*n.m.*), tight fit.
emmancher à froid (*v.t.*), to shrink in.
emmêler (*v.t.*), to entangle.
empennage horizontal (aviation) (*n.m.*), hor-izontal stabilizer.
empennage vertical (aviation) (*n.m.*), vertical stabilizer.
empennages (aviation) (*n.m.pl.*), tail unit.
empiriquement (*adv.*), by rule of thumb.
emplanture de l'aile (aviation) (*n.f.*), wing root.
emplanture de pale (hélicoptère) (*n.f.*), blade root.
empochement (*n.m.*), recess; bay.
empoilage (*n.m.*), bristles.
empois d'amidon (*n.m.*), starch paste.
empreinte (*n.f.*), dent; indentation; tyre thread.
empreinte de billes (*n.f.*), brinelling.
émulsifiant (*n.m.*), emulsifying agent.
émulsionner (*v.t.*), to emulsify.
en-tête (d'une lettre) (*n.m.*), heading.
encadrement de porte (*n.m.*), door frame.
encaisser un effort (*v.*), to withstand stress.
encart (*n.m.*), insert.
encastré, -e (*adj.*), built-in; inserted; embedded.
encastrement (*n.m.*), tail-in.
encastrer (*v.t.*), to build-in; to recess; to mount flush.
enclaver (*v.t.*), to encompass.
enclenchement (*n.m.*), cutting-in; engagement.
enclencher (*v.t.*), to actuate; to engage.
encoche (élec.) (*n.f.*), slot; recess.

encodeur (radio) (*n.m.*), encoder.
encombrant, -e (*adj.*), bulky.
encrassé, -e (*adj.*), fouled.
encrassement de la meule (*n.m.*), wheel loading.
endoscope (*n.m.*), endoscope.
endroit de stationnement (route) (*n.m.*), lay-by.
enduire (*v.t.*), to smear.
enduit (*n.m.*), coating; compound.
enduit superficiel (trav. pub.) (*n.m.*), sealing coat; surface dressing.
endurance (*n.f.*), fatigue strength; life.
énergie d'entrée (*n.f.*), input.
énergie hydraulique (*n.f.*), hydraulic power.
énergie totale (*n.f.*), total energy.
enfermé, -e (*adj.*), enclosed.
enfiler (*v.t.*), to insert; to slip on.
enfoncement du train (aviation) (*n.m.*), land-ing gear shock strut compression.
enfoncer (*v.i.*), to sink.
engagement en roulis (aviation) (*n.m.*), lat-eral divergence.
engagement longitudinal (*n.m.*), longitudi-nal divergence; pitching.
engrenage (*n.m.*), pinion.
engrenage à bride (*n.m.*), flanged gear.
engrenage à crémaillère (*n.m.*), rack gear.
engrenage à plateau (*n.m.*), flanged gear.
engrenage conique à denture hélicoï-dale (*n.m.*), spiral bevel gear.
engrenage conique plat (*n.m.*), face gear.
engrenage cylindrique (*n.m.*), spur gear.
engrenage de commande (machines-outils) (*n.m.*), bull gear.
engrenage de distribution (*n.m.*), timing gear.
engrenage de prise directe (*n.m.*), direct drive; high-speed gear.
engrenage de vis sans fin (*n.m.*), worm gear.
engrenage excentrique (*n.m.*), eccentric gear.
engrenage hypoïde (*n.m.*), hypoid gear.
enjoliveur (*n.m.*), trim.
enquête (*n.f.*), investigation; survey.
enregistrement (*n.m.*), record.
enregistrement automatique (*n.m.*), logger.
enregistrement fractionné (audio-visuel) (*n.m.*), duoplay; multi-playback.
enregistrer (*v.t.*), to log; to record.
enregistreur à bande magnétique (*n.m.*), tape-recorder.
enregistreur de flèche de pont (*n.m.*), bridge sagging recorder.
enregistreur de pointe (élec.) (*n.m.*), demand meter.
enregistreur de vol (*n.m.*), flight recorder.
enregistreur sur fil magnétique (*n.m.*), wire recorder.
enregistreur sur ruban (*n.m.*), tape-recorder.
enrobage (*n.m.*), coating.
enrobage par coulée (*n.m.*), potting.
enroulement (spire) (*n.m.*), convolution.
enroulement à double cage d'écureuil (élec.) (*n.m.*), double squirrel cage winding.
enroulement à fils tirés (élec.) (*n.m.*), pull-through winding.
enroulement à nombre entier d'encoches par pôle et par phase (élec.) (*n.m.*), integral slot winding.

enroulement à nombre fractionnaire d'encoches par pôle et par phase (élec.) (*n.m.*), fractional slot winding.

enroulement à pas allongé (élec.) (*n.m.*), long pitch winding.

enroulement à pas brouillé (élec.) (*n.m.*), stepped winding; split winding.

enroulement à pas diamétral (élec.) (*n.m.*), full pitch winding.

enroulement à pas partiel (élec.) (*n.m.*), fractional pitch winding.

enroulement à pas raccourci (élec.) (*n.m.*), short pitch winding.

enroulement d'induit (élec.) (*n.m.*), armature winding.

enroulement de polarisation (élec.) (*n.m.*), bias winding.

enroulement en anneaux (élec.) (*n.m.*), ring winding.

enroulement en disque (élec.) (*n.m.*), disc winding.

enroulement en losange (élec.) (*n.m.*), diamond winding.

enroulement en pattes de grenouille (élec.) (*n.m.*), frog-leg winding.

enroulement en tambour (élec.) (*n.m.*), drum winding.

enroulement enchevêtré (élec.) (*n.m.*), concentric winding.

enroulement imbriqué (élec.) (*n.m.*), lap winding.

enroulement inducteur (élec.) (*n.m.*), field winding.

enroulement ondulé (élec.) (*n.m.*), wave winding.

enroulement primaire (élec.) (*n.m.*), primary winding.

enroulements alternés (transformateur) (*n.m.pl.*), sandwich windings.

enrouleur (*n.m.*), reel; jockey pulley.

enrouleur à inertie (*n.m.*), shoulder harness reel.

ensemble (*n.m.*), assembly.

ensemble autonome (*n.m.*), self-contained unit.

ensemble cyclique (hélicoptère) (*n.m.*), swash-plate assembly.

ensemble de propulsion (*n.m.*), drive system.

ensemble supérieur (*n.m.*), next higher assembly.

ensemble symétrique (*n.m.*), handed assembly.

ensemble transmission mécanique (*n.m.*), transmission system.

ensemencement (*n.m.*), seed sowing.

entablement (*n.m.*), windshield frame section.

entoilage (*n.m.*), fabric.

entortillé, -e (*adj.*), twisted.

entraînement (*n.m.*), drive.

entraînement (instruction) (*n.m.*), instruction; training.

entraînement à cliquet (*n.m.*), pawl and ratchet drive.

entraînement à tournevis (*n.m.*), blade and slot drive.

entraînement à vitesse constante (*n.m.*), constant speed drive.

entraînement du manche (aviation) (*n.m.*), control column whip.

entraînement mécanique (*n.m.*), mechanical drive.

entraînement par croix de Malte (*n.m.*). Geneva mechanism.

entraînement par galet (*n.m.*), rim drive.

entraînement par hélice propulsive (*n.m.*), pusher propeller propulsion.

entraînement par réacteurs en bout de pale (aviation) (*n.m.*), tip blade jet propulsion.

entraîner (*v.t.*), to drive.

entrée (*n.f.*), input; intake.

entrée d'air fuseau réacteur (*n.f.*), nacelle intake ring.

entrée d'air réacteur (*n.f.*), engine air intake.

entrée de filetage (*n.f.*), thread lead in.

entrée de piste (aviation) (*n.f.*), approach end of the runway.

entrée micro (radio) (*n.f.*), audio input.

entrefer (magnétisme) (*n.m.*), gap.

entrer en vigueur (*v.i.*), to become effective.

entretien (élec.) (*n.m.*), maintenance.

entretien courant (*n.m.*), servicing.

entretoise (pièce d'épaisseur) (*n.f.*), spacer.

entretoise calibrée (*n.f.*), close tolerance spacer.

entretoise conique (*n.f.*), dishpan spacer.

enveloppe (*n.f.*), envelope.

enveloppe (emballage) (*n.f.*), wrapping material.

enveloppe annulaire de la chambre de combustion (*n.f.*), combustion chamber annular case.

enveloppe de manoeuvre (aviation) (*n.f.*), manoeuvring envelope.

enveloppe de pneumatique (auto) (*n.f.*), tyre.

enveloppe de rafale (aviation) (*n.f.*), gust envelope.

enveloppe de rallumage (*n.f.*), relight envelope.

envergure (*n.f.*), span.

envol (aviation) (*n.m.*), take-off.

envoyer un faisceau (*v.*), to radiate a beam.

épaisseur curviligne (engrenages) (*n.f.*), circular thickness.

épaisseur de déplacement (*n.f.*), displacement thickness.

épaisseur de la couche limite (*n.f.*), thickness of boundary layer.

épaisseur de serrage (*n.f.*), grip.

épaisseur rectiligne (engrenages) (*n.f.*), chordal thickness.

épandeur-régleur (trav. pub.) (*n.m.*), spreader-grader.

épandeuse (trav. pub.) (*n.f.*), spreader.

épanouir (*v.t.*), to make something open *or* expand.

épinglage (avant rivetage) (*n.m.*), fastening.

épinglage (avant soudage) (*n.m.*), tack welding.

épingle de freinage (*n.f.*), lockpin.

épingle de sécurité (*n.f.*), safety pin.

épingler (*v.t.*), to pin.

épointé, -e (*adj.*), blunt.

épreuve (*n.f.*), test run.

épreuve d'homologation (*n.f.*), certification test.

épreuves (de tournage) (cinéma) (*n.f.pl.*), rushes.

éprouvette (*n.f.*), test specimen.

éprouvette de résilience (*n.f.*), resilience test specimen.

épurateur d'air (*n.m.*), air cleaner.

épure (*n.f.*), draught (technical drawing).

épure de déplacement du manche (avia-

-tion) (n.f.), stick displacement pattern.

équerre (n.f.), angle bracket.

équerre de métallisation (n.f.), bonding angle.

équidistant, -e (adj.), equally spaced.

équilibrage (n.m.), balancing; equalizing.

équilibrage aérodynamique (aviation) (n.m.), aerodynamic balance.

équilibrage des pales (aviation) (n.m.), blade balance.

équilibrage dynamique (aviation) (n.m.), dynamic balancing.

équilibrage statique (aviation) (n.m.), static balancing.

équipage (marine, aviation) (n.m.), crew.

équipage mobile (électricité) (n.m.), moving coil; moving contact.

équipement commercial (n.m.), equipment; furnishings.

équipement de stabilisation automa--tique (aviation) (n.m.), automatic stabiliza--tion equipment.

équipement de tête (n.m.), headgear.

équipement facultatif (n.m.), optional equipment.

équipement permanent (n.m.), permanent equipment.

équipement variable (n.m.), alternative equipment.

équivalent de vitesse (aviation) (n.m.), equi--valent airspeed.

éraflure (n.f.), scratch.

ergot (n.m.), spigot.

ergot conique (n.m.), taper spigot.

ergot d'entraînement (n.m.), drive pin.

erreur de tri (n.f.), missorting.

erreur due à l'installation (n.f.), installation error.

erreur due à l'instrument (n.f.), instrument error.

erreur quadrantale (n.f.), quadrantal error.

éruption (pétrole) (n.f.), blow-out.

escabeau (n.m.), step-ladder.

escamotable (adj.), retractable.

escamotage (n.m.), retraction; folding back.

espace (n.m.), gap.

espace aérien (n.m.), air space.

espace aérien contrôlé (n.m.), controlled air space.

espacé, -e (adj.), apart; spaced.

espacement de fréquence (radio) (n.m.), fre--quency separation.

essai (n.m.), test; test run.

essai à l'entrave (n.m.), tethered trial.

essai à la bille *ou* **essai Brinell** (n.m.), Brinell test; hardness testing.

essai à la flexion sur éprouvette en--taillée (n.m.), notch bending test.

essai à la goutte (n.m.), drop test.

essai au brouillard salin (n.m.), salt spray test.

essai au cisaillement (n.m.), shearing test.

essai d'adhérance (n.m.), bonding test.

essai d'affaissement (trav. pub.) (n.m.), slump test.

essai d'arrachement (n.m.), peeling test.

essai d'échauffement d'acide (pétrole) (n.m.), acid heat test.

essai d'endurance (n.m.), endurance test.

essai d'entraînement (trav. pub.) (n.m.), elu--triation test.

essai d'étalonnage (n.m.), calibration test.

essai d'étanchéité (n.m.), leakage test.

essai d'homologation (n.m.), prototype test.

essai d'interrupteur (n.m.), make-and-break check.

essai d'isolement (n.m.), dielectric test; flash test.

essai d'usure à la meule (n.m.), abrasion test.

essai d'usure par frottement réciproque (n.m.), attrition test.

essai de chute limite (aviation) (n.m.), landing gear drop test.

essai de cisaillement par compression (n.m.), block shear test.

essai de dépression (n.m.), vacuum test.

essai de dureté par striage (n.m.), scratch test.

essai de fatigue (n.m.), fatigue test.

essai de fissibilité (n.m.), crack test.

essai de flexion au choc (n.m.), blow bending test.

essai de fluage (n.m.), creep test.

essai de fonctionnement (n.m.), operational test.

essai de la perle de borax (métal.) (n.m.), borax bead test.

essai de masse (n.m.), earth test.

essai de mouillage d'eau (n.m.), water break test.

essai de moussage (n.m.), foaming test.

essai de résilience (n.m.), impact test.

essai de résonance (n.m.), resonance test.

essai de traction (n.m.), tensile test.

essai de vibration (n.m.), vibration test.

essai destructif (n.m.), destructive test.

essai en cage (radio) (n.m.), screen room test.

essai en piscine (n.m.), water tank test.

essai en pression (n.m.), pressure test.

essai en soufflerie (n.m.), wind tunnel test.

essai en vol (n.m.), flight test.

essai mécanique (n.m.), physical test.

essai non destructif (n.m.), non destructive test.

essai par rupture (n.m.), fracture test.

essai partiel de fatigue (n.m.), sectional test.

essai statique (n.m.), static test.

essai sur prélèvement (n.m.), sampling test.

essai téléphonique (n.m.), voice test.

essais par lots (n.m.pl.), batch test; sampling.

essais tropicaux (n.m.pl.), tropical tests.

essence minérale (n.f.), white spirit.

essuie-glace (auto) (n.m.), windscreen wiper; wiper.

estimation (n.f.), assessment.

estime, navigation à, (n.f.), dead reckoning.

établir le courant (élec.) (v.), to switch on.

établissement de la pression (n.m.), pres--sure build-up.

étage basse fréquence (radio) (n.m.), low frequency stage.

étage de commande (n.m.), control stage.

étage de compresseur (n.m.), compressor stage.

étage de turbine (n.m.), turbine stage.

étage haute fréquence (radio) (n.m.), high frequency stage.

étage intermédiaire (n.m.), buffer stage.

étage moyenne fréquence (radio) (n.m.), intermediate frequency stage.

étage séparateur (n.m.), buffer stage.

étagère (n.f.), shelf.

étagère de meuble radio (*n.f.*), radio rack shelf.

étai (*n.f.*), brace.

étalement de bande (radio) (*n.m.*), band-spread.

étalon (*n.m.*), master; standard.

étalonnage (*n.m.*), calibration.

étalonné, -e (*adj.*), calibrated.

étanchage (*n.m.*), sealing.

étanche à l'air (*adj.*), air-tight.

étanche au gaz (*adj.*), gas-proof.

étanche aux vapeurs (*adj.*), vapour-proof.

état (document) (*n.m.*), record.

état de chargement (*n.m.*), load distribution manifest.

état de surface (*n.m.*), surface condition.

état métastable (*n.m.*), metastable state.

étau à base pivotante (*n.m.*), swivelling base vice.

étau-limeur d'établi (*n.m.*), bench-type shap-ing machine.

étayage (*n.m.*), buttressing.

étayé, -e (*adj.*), buttressed.

éteindre (*v.t.*), to switch off the lights.

étiquette (*n.f.*), label; tag.

étiquette sur panneaux (*n.f.*), placard.

étirage (*n.m.*), stretching.

étirer (*v.t.*), to stretch.

étoile (hélicoptère) (*n.f.*), star.

étoile fixe (hélicoptère) (*n.f.*), stationary star.

étouffoir (auto) (*n.m.*), choke.

étoupe (*n.f.*), packing; stuffing.

étoupille (*n.f.*), squib.

étranglé, -e (*adj.*), restricted.

étranglement (*n.m.*), constriction; neck restriction.

étrésillon (*n.m.*), drag link.

étrier (*n.m.*), bracket.

étude (*n.f.*), analysis.

étude d'exploitation (*n.f.*), operation analysis.

étuve (*n.f.*), drying furnace.

étuve de séchage à infra-rouge (peinture) (*n.f.*), infra-red oven.

eutectique (*adj.*), eutectic.

évacuer (*v.t.*), to dissipate.

évaluer (*v.t.*), to evaluate.

évanouissement (radio) (*n.m.*), fading.

évidement (*n.m.*), recess.

évidement (détourage) (*n.m.*), cut-out.

évidement de noyautage (*n.m.*), cored passage.

évier (*n.m.*), sink.

évolutif, -ive (*adj.*), tapered.

évolution (*n.f.*), evolution.

évolution au sol (aviation) (*n.f.*), ground manoeuvre.

évolution en vol (aviation) (*n.f.*), flight manoeuvre.

examen au métalloscope (*n.m.*), eddy current inspection.

examiner (*v.t.*), to inspect.

excentrage (*n.m.*), off-setting.

excentré, -e (*adj.*), off-set.

excentricité de battement nulle (aviation) (*n.f.*), zero flapping eccentricity.

excité, -e (relais élec.) (*adj.*), energized.

excitron (élec.) (*n.m.*), excitron.

exclusivité (audio-visuel) (*n.f.*), scoop.

extérieur (*n.m.*), outboard.

extincteur (électronique) (*n.m.*), quencher.

extincteur à la mousse (*n.m.*), foam extinguisher.

extinction auditive (*n.f.*), aural nulling.

extinction du réacteur (aviation) (*n.f.*), engine flame-out.

extinction par excès d'air (aviation) (*n.f.*), lean die out.

extinction par excès de carburant (*n.f.*), rich blow-out.

extra-courant de rupture (radio) (*n.m.*), break impulse; doubling effect.

extracteur (*n.m.*), puller.

extracteur à vis (*n.m.*), jack-off screw.

extraction électrolytique (métal.) (*n.f.*), electro-extraction; electro-winning.

extrados (aviation) (*n.m.*), wing upper surface.

extrados (pale) (*n.m.*), suction face.

extrados de la pale (*n.m.*), blade upper surface.

extrafort (*n.m.*), fabric tape.

extraire (*v.t.*), to extract.

extrême (*n.m.*), outer.

extrémité de l'aile (aviation) (*n.f.*), wing tip.

extrémité de pale (aviation) (*n.f.*), blade tip.

extrémité de pied de pale (aviation) (*n.f.*), blade shank.

F

fabricant de moteurs (*n.m.*), engine manufacturer.

fabrication (*n.f.*), fabrication; manufacturing.

fabrication en série (*n.f.*), mass production; series production.

face arrière (*n.f.*), rear panel.

face au vent (aviation, navigation) (*adv.*), into the wind.

face avant (*n.f.*), front face; front panel.

face d'appui (machines-outils) (*n.f.*), chuck face.

face de roulement (engrenages) (*n.f.*), tooth face.

face magnétique (*n.f.*), magnetic face.

facetté, -e (d'un outil) (*adj.*), chamfered nose.

facilité d'entretien (*n.f.*), maintainability.

façonnage (*n.m.*), manufacturing; turning.

façonnier (*n.m.*), job contractor.

facteur d'accroissement (*n.m.*), growth factor.

facteur d'amortissement (*n.m.*), break-even factor.

facteur d'amplification (élec.) (*n.m.*), ampli-fication factor.

facteur d'atténuation (*n.m.*), alleviation fac-tor; attenuation factor.

facteur d'efficacité de la pale (hélicoptère) (*n.m.*), blade efficiency factor.

facteur de charge (*n.m.*), load factor.

facteur de charge de rafales (aviation) (*n.m.*), gust load factor.

facteur de charge extrême (*n.m.*), ultimate load factor.

facteur de charge limite (*n.m.*), limit load factor.

facteur de correction de la traînée induite (aviation) (*n.m.*), correction factor for induced drag.

facteur de flexibilité (*n.m.*), flexibility factor.

facteur de permittivité (elec.) (*n.m.*), relative permittivity.

facteur de perte en bout de pale (hélicop--tère) (*n.m.*), blade tip loss factor.

factice (*adj.*), dummy (of bottles, boxes); fictitious.

facultatif, -ive (*adj.*), optional.

faible encombrement, de (*adj.*), compact.

faire coïncider (*v.t.*), to align; to match.

faisceau (échangeur de chaleur) (*n.m.*), core (heat exchanger).

faisceau d'éloignement (*n.m.*), outbound beam.

faisceau de balayage (radar) (*n.m.*), scanning sweep.

faisceau de câble (*n.m.*), cable bundle.

faisceau de détection incendie (*n.m.*), fire detecting wire.

faisceau de la trajectoire d'atterrissage (*n.m.*), glide path beam.

faisceau de lames (*n.m.*), laminated torsion bar.

faisceau de radiophare de balisage (avia--tion) (*n.m.*), localizer beam.

faisceau de thermocouple (*n.m.*), thermo--couple harness.

faisceau explorateur (radar) (*n.m.*), scanning beam.

faisceau radio (*n.m.*), radio beam.

farine (trav. pub.) (*n.f.*), filler.

faucille (aviation) (*n.f.*), engine support arch.

faussage (*n.m.*), buckling; warping.

fausse alarme (*n.f.*), false warning.

fausse bague de roulement (*n.f.*), dummy bearing race.

fausse nervure (*n.f.*), false rib.

faux (factice) (*n.m.*), dummy.

faux alignement (*n.m.*), mismatch.

faux cardan à membrane (*n.m.*), sliding joint.

faux châssis (*n.m.*), sub-frame.

faux contact (élec.) (*n.m.*), bad contact.

faux couple (*n.m.*), false frame.

faux équerrage (*n.m.*), out-of-square.

faux longeron (*n.m.*), false spar.

faux parallélisme (*n.m.*), out-of-parallelism.

faux rond (*n.m.*), out-of-round.

fêlure (*n.f.*), split.

fendillement (*n.m.*), crazing.

fenestron (hélicoptère) (*n.m.*), shrouded tail fan.

fenêtre sur instrument (*n.f.*), flag window.

fente en haricot (*n.f.*), kidney-shaped slot.

fer d'induit (*n.m.*), armature iron.

fer fritté (*n.m.*), sintered iron.

fermé, -e (*adj.*), closed.

fermé, -e (construction de machines) (*adj.*), totally enclosed.

fermeture (contact) (*n.f.*), make.

fermeture éclair (*n.f.*), zip fastener.

ferrailler (*v.t.*), to scrap.

ferrure d'accrochage (*n.f.*), catch fitting.

ferrure d'articulation (*n.f.*), hinge fitting.

ferrure d'attache de pale (hélicoptère) (*n.f.*), blade attachment fitting.

ferrure de fixation (*n.f.*), attachment fitting.

ferrure de hissage (*n.f.*), hoist fitting.

ferrure de pied de pale (hélicoptère) (*n.f.*), blade cuff.

ferrure de reprise (*n.f.*), pick-up fitting.

ferrure de verrouillage (*n.f.*), lock fitting.

ferrure en delta (*n.f.*), delta fitting.

ferrure forgée d'attache voilure-fuse--lage (aviation) (*n.f.*), forged wing attachment.

ferrure support (*n.f.*), bracket.

feston (*n.m.*), scallop.

feu anti-collision (*n.m.*), anti-collision light.

feu clignotant (auto) (*n.m.*), flashing light.

feu de cockpit (aviation) (*n.m.*), cockpit light.

feu de position (auto) (*n.m.*), parking light; side light.

feu de position (aviation) (*n.m.*), navigation light.

feu électrique (*n.m.*), electrical fire.

feu fixe (*n.m.*), steady light.

feuille volante (*n.f.*), loose leaf.

feuillet magnétique (élec.) (*n.m.*), magnetic shell.

feuillure de glace (pare-brise) (auto) (*n.f.*), pane rabbet (windscreen).

fiabilité (*n.f.*), reliability.

fibrage (métallurgie) (*n.m.*), grain flow.

fibrane (textile) (*n.f.*), spun rayon.

fibre de verre (*n.f.*), glass wool; fibreglass.

fibre synthétique (*n.f.*), man-made fibre.

fibre vulcanisée (*n.f.*), vulcanized fibre.

fiche (document) (*n.f.*), card; record sheet.

fiche (de tournage) (audio-visuel) (*n.f.*), dope sheet.

fiche à deux broches (élec.) (*n.f.*), two-pin plug.

fiche à trois broches (élec.) (*n.f.*), three-pin plug.

fiche banane (radio) (*n.f.*), banana plug.

fiche d'entretien élémentaire (*n.f.*), elemen--tary servicing sheet.

fiche d'essai (*n.f.*), test sheet.

fiche de câble d'allumage (auto) (*n.f.*), igni--tion cable terminal.

fiche de contrôle (*n.f.*), inspection card.

fichier (*n.m.*), card index; filing cabinet.

fichier électrique (*n.m.*), electrical socket panel.

fictif, -ive (*adj.*), dummy.

fidélité de modulation (électronique) (*n.f.*), audio frequency response.

fil à casser (*n.m.*), snap wire.

fil à freiner (*n.m.*), lockwire.

fil à ligaturer (*n.m.*), lacing cord.

fil de bougie (auto) (*n.m.*), ignition cable.

fil de lin (*n.m.*), linen thread.

fil de masse (*n.m.*), grounding lead.

fil de métallisation (aviation) (*n.m.*), airframe bonding lead.

fil de sonnerie (élec.) (*n.m.*), bell wire.

fil électrique (*n.m.*), lead.

fil guipé (élec.) (*n.m.*), braided wire.

fil métallique (*n.m.*), wire.

fil nylon (*n.m.*), nylon thread.

fil plombé (*n.m.*), lead seal wire.

fil repéré (*n.m.*), tracer.

fil textile (*n.m.*), thread.

fil tréfilé (*n.m.*), drawn wire.

fil tressé (*n.m.*), braided wire.

filage (*n.m.*), extrusion.

filament chauffant (électronique) (*n.m.*), heater.

filet conique (outillage) (*n.m.*), taper thread.

filet d'amarrage (aviation) (*n.m.*), mooring harness.
filet d'atterrissage sur neige (aviation) (*n.m.*), snow landing net.
filet extérieur (outillage) (*n.m.*), external thread; male thread.
filet intérieur (outillage) (*n.m.*), internal thread; female thread.
filet laminé (*n.m.*), rolled thread.
filet rapporté (*n.m.*), thread insert.
filetage (*n.m.*), thread cutting; screw thread.
filetage femelle (*n.m.*), female thread; internal thread.
filetage mâle (*n.m.*), external thread; male thread.
fileté, -e (*adj.*), threaded.
filière (*n.f.*), threading *or* chasing die.
fillerisé, -e (trav. pub.) (*adj.*), filled.
filtrat (*n.m.*), filtrate.
filtre à accord décalé (radio) (*n.m.*), stagger tuned filter.
filtre à cartouche (*n.m.*), cartridge filter.
filtre anti-parasite (radio) (*n.m.*), radio noise filter.
filtre carburant basse pression (*n.m.*), low pressure fuel filter.
filtre de bande (radio) (*n.m.*), radio range filter.
filtre de combustible (*n.m.*), fuel filter.
filtre de courant redressé (élec.) (*n.m.*), smoothing filter.
filtre de dérivation (*n.m.*), by-pass filter.
filtre magnétique (*n.m.*), magnetic filter.
filtre métallique (*n.m.*), wire gauze filter.
filtre micronique (*n.m.*), micronic filter.
filtre passe-bas (*n.m.*), low stop filter.
filtre passe-haut (*n.m.*), high stop filter.
filtre passe-tout (élec.) (*n.m.*), all-pass filter.
filtre piézo-électrique (*n.m.*), crystal filter.
filtrer (électricité) (*v.t.*), to smooth.
fin de bande (*n.f.*), run-out.
fin de course (machines-outils) (*n.f.*), stroke end; travel end.
fin de course, en (*adj.*), at the end of travel position.
fin de déplacement (machines-outils) (*n.f.*), travel end.
fin de filetage (*n.f.*), thread run-out.
fines (construction) (*n.f.pl.*), filler.
finesse (aviation) (*n.f.*), aerodynamic efficiency.
fini de surface (*n.m.*), surface finish.
fini granité (*n.m.*), crinkled finish.
fini mat (*n.m.*), matt finish.
fini vermiculé (*n.m.*), crackled finish.
finisseur (trav. pub.) (*n.m.*), finisher.
finition (*n.f.*), finish.
finition (peinture) (*n.f.*), top coat.
fissure par retrait du métal (*n.f.*), shrinkage crack.
fixation (*n.f.*), attachment.
flambage local (*n.m.*), wrinkling.
flamestat (*n.m.*), flamestat.
flamme (*n.f.*), pennant.
flan (*n.m.*), blank.
flanc (d'un pneu) (*n.m.*), side wall.
flasque (*n.f.*), flange; end plate.
flèche (d'un ressort) (*n.f.*), compression of a spring.
flèche de remorquage (*n.f.*), tow bar.
fléchissement (d'une aile) (aviation) (*n.f.*), wing droop.
flector (*n.m.*), coupling.
flettner (aviation) (*n.m.*), trim tab.

fleur (trav. pub.) (*n.f.*), flower.
flexible d'entraînement (*n.m.*), flexible drive shaft.
flocon (de neige) (*n.m.*), flake.
flocon (plastique) (*n.m.*), flock.
floculation (*n.f.*), floculation.
flottabilité (*n.f.*), buoyancy.
flottement (aviation) (*n.m.*), buffeting; flutter.
flottement des roues (auto) (*n.m.*), wheel wobble.
flotteur escamotable (hélicoptère) (*n.m.*), retractable float.
fluage (métallurgie) (*n.m.*), creep.
fluage (plastique) (*n.m.*), flow.
fluatation (trav. pub.) (*n.f.*), impregnation (with fluosilicates).
fluide de refroidissement (dans les frigori-fiques, etc.) (*n.m.*), coolant.
Fluon (matière plastique) (*n.m.*), Fluon (G.B.); Teflon (U.S.A.).
fluoroscopie (*n.f.*), fluoroscopy.
fluorure (*n.m.*), fluoride.
fluotournage (*n.m.*), flow-spinning.
flûte marine (pétrole) (*n.f.*), streamer.
flux de chaleur (*n.m.*), heat transfer rate.
flux décapant (soudure) (*n.m.*), welding flux.
flux induit (*n.m.*), inflow.
foirage (*n.m.*), stripping.
fonctionnement (*n.m.*), operation.
fonctionnement défectueux (*n.m.*), faulty operation.
fonctionnement en autogyre (hélicoptère) (*n.m.*), autorotative flight.
fonctionnement en moulinet (hélicoptère) (*n.m.*), windmilling.
fonctionner (*v.i.*), to operate.
fond (*n.m.*), base; background.
fond (d'un tube cathodique) (*n.m.*), face-plate.
fond de filet (outillage) (*n.m.*), thread-root.
fond tournant (*n.m.*), revolving centre.
fondamental, -e (*adj.*), basic.
fondoir-réchauffeur (trav. pub.) (*n.m.*), heating-melter.
fondu (cinéma) (*n.m.*), fading.
forçage (*n.m.*), interference.
force antagoniste (*n.f.*), counteracting force.
force contre-électromotrice (*n.f.*), back electromotive force.
force d'impulsion (*n.f.*), impetus.
force d'inertie (*n.f.*), inertia effect.
force de champ (*n.f.*), field strength.
force électromotrice (*n.f.*), electromotive force.
force répulsive (*n.f.*), repelling power.
force résultante (*n.f.*), resultant force.
force vive (*n.f.*), momentum.
forgeage mécanique (*n.m.*), drop forging.
formage (*n.m.*), forming.
format (*n.m.*), format.
formation (soupapes électriques) (*n.f.*), degassing.
formation catalytique (*n.f.*), catforming.
formation de frisures (peinture) (*n.f.*), alligatoring.
forme (*n.f.*), line; pattern.
forme en saillie (*n.f.*), projection.
formulaire (*n.m.*), document.
fosse de bâti (*n.f.*), jig pit.
fou (*n.m.*), idler (wheel or pulley).
four à cloche (*n.m.*), bell furnace.
four à réverbère (*n.m.*), air furnace; reverba-tory furnace.

four électrique à baguette rayonnante (*n.m.*), electric resistance furnace.

four électrique à induction (*n.m.*), electric induction furnace.

four tunnel (*n.m.*), tunnel furnace.

fourche d'articulation (aviation) (*n.f.*), hinge fork.

fourche de direction (aviation) (*n.f.*), nose--wheel steering bar.

fourche du joint universel (*n.f.*), universal joint yoke.

fourchettage (d'un faisceau) (*n.m.*), bracketing of a beam.

fourchette (relais élec.) (*n.f.*), dead zone.

fourchette (butée mécanique) (*n.f.*), fork arm.

fourneau (machines-outils) (*n.m.*), quill.

fournir (*v.t.*), to deliver.

fourreau d'avance (machines-outils) (*n.m.*), feed sleeve.

fourreau de broche (machines-outils) (*n.m.*), spindle sleeve.

fourrure (*n.f.*), liner.

foyer (*n.m.*), focal point.

foyer d'un profil (*n.m.*), aerodynamic centre.

foyer de la pale (hélicoptère) (*n.m.*), blade aero--dynamic centre.

foyer lumineux (*n.m.*), light source.

fractionnement (*n.m.*), breakdown.

fragilisation (*n.f.*), embrittlement.

fragilité de revenu (*n.f.*), temper brittleness.

frais d'exploitation (*n.m.pl.*), operating costs; running costs.

fraisage à la vis-mère (*n.m.*), hobbing.

fraisage chimique (*n.m.*), chemical milling.

fraisage conique (*n.m.*), taper milling.

fraisage en avalant (*n.m.*), climb milling.

fraisage en monte et baisse (*n.m.*), rise and fall milling.

fraisage en roulant (*n.m.*), out milling; up milling.

fraisage hélicoïdal (*n.m.*), spiral milling.

fraisage par reproduction (*n.m.*), copy milling.

fraise à champignon (*n.f.*), rose countersink.

fraise à détourer (*n.f.*), routing cutter.

fraise à rainurer (*n.f.*), slot milling cutter.

fraise à rouler (*n.f.*), plain milling cutter.

fraise à surfacer (*n.f.*), face milling cutter.

fraise angulaire (*n.f.*), cone countersink.

fraise conique (*n.f.*), rose countersink.

fraise-mère (machine à tailler les engrenages) (*n.f.*), hob.

fraise-scie (*n.f.*), slitting saw.

fraiseuse à reproduire (*n.f.*), copy milling machine.

fraisure (*n.f.*), countersunk hole.

francium (élément) (*n.m.*), francium.

franklin (unité électrostatique de charge élec--trique) (*n.m.*), franklin.

frapper (*v.t.*), to hit.

frein à disque (*n.m.*), disc brake.

frein à ergot (*n.m.*), tab washer.

frein à friction (*n.m.*), friction brake.

frein à pattes (*n.m.*), tab washer.

frein aérodynamique (*n.m.*), speed brake.

frein d'axe (*n.m.*), circlip.

frein d'écrou à languette (*n.m.*), tab washer.

frein de détresse (*n.m.*), emergency brake.

frein de gouverne (*n.m.*), gust lock.

frein de parc (*n.m.*), parking brake.

frein de vis (*n.m.*), screw retainer.

frein magnétique (élec.) (*n.m.*), eddy current brake.

frein plat (*n.m.*), lockplate.

frein rotor (hélicoptère) (*n.m.*), rotor brake.

freinage d'un écrou (*n.m.*), lockwiring; safe--tying of a nut.

freinage de la visserie (*n.m.*), screw locking; safetying.

freinage différentiel (*n.m.*), differential braking.

freinage par coups de pointeau (*n.m.*), staking.

freiner (*v.t.*), to stake (aviation) (U.S.A.).

freiner avec fil (*v.t.*), to lockwire.

fréon (*n.m.*), freon.

fréquence acoustique (radio) (*n.f.*), audio-frequency.

fréquence d'appel (radio) (*n.f.*), call frequency.

fréquence d'attaque (radio) (*n.f.*), injection frequency.

fréquence d'écoute (radio) (*n.f.*), listening frequency.

fréquence de coupure (radio) (*n.f.*), cut-off frequency.

fréquence de ligne (*n.f.*), line frequency.

fréquence de récurrence (*n.f.*), pulse recur--rence frequency.

fréquence de travail (*n.f.*), operating frequency.

fréquence étalon (*n.f.*), standard frequency.

fréquence pilotée par quartz (radio) (*n.f.*), crystal controlled frequency.

fréquence porteuse (radio) (*n.f.*), carrier frequency.

fréquence propre (*n.f.*), natural frequency.

fréquence résultante (*n.f.*), sum frequency.

fréquence surveillée (*n.f.*), monitored frequency.

fréquencemètre (*n.m.*), frequency meter.

fret (*n.m.*), cargo.

fréter (*v.t.*), to charter.

frettage (*n.m.*), shrinking.

fretté, -e (*adj.*), shrunk.

frigorie (*n.f.*), French unit of cold.

frittage (métal.) (*n.m.*), sintering.

fritté, -e (métal.) (*adj.*), sintered.

friture (*n.f.*), radio interference.

front (*n.m.*), front.

front avant d'une impulsion (*n.m.*), leading edge of a pulse.

front, de (*adv.*), abreast.

frottement onctueux (graissage) (*n.m.*), boundary friction.

frotteur (élec.) (*n.m.*), sliding contact.

fuel de soute (pétrole) (*n.m.*), bunker C.

fuite d'aileron (*n.f.*), snape.

fungicide (*n.m.*), fungicide.

furaldéhyde (plast.) (*n.f.*), fural; fufural; fufuraldehyde.

fuseau horaire (*n.m.*), time zone.

fuseau réacteur (*n.m.*), engine nacelle.

fusée (moteur) (*n.f.*), rocket.

fusée (*n.f.*), spindle.

fusée d'essieu (auto) (*n.f.*), stub axle.

fusée de pale (hélicoptère) (*n.f.*), blade spindle.

fusée de roue (*n.f.*), wheel axle.

fuselage (aviation) (*n.m.*), fuselage.

fuselage étanche (aviation) (*n.m.*), sealed fuselage.

fuselé, -e (*adj.*), streamlined.

fusible (*n.m.*), fuse.

fuse.

fusible à cartouche (*n.m.*), cartridge fuse.
fusible d'éclatement (aviation) (*n.m.*), landing gear wheel rim fusible plug.
fusible de fort calibre (*n.m.*), heavy duty

fusible temporisé (*n.m.*), delayed action fuse.
fût de cylindre (*n.m.*), cylinder barrel.
fût de l'atterrisseur (*n.m.*), landing gear leg.

G

gabarit (*n.m.*), jig.
gabarit de fraisage (*n.m.*), milling jig.
gabarit de perçage (*n.m.*), drilling jig.
gabarit de rainurage (*n.m.*), slotting jig.
gabarit de réglage (*n.m.*), rigging template.
gabarit de traçage (*n.m.*), template.
gabarit de vérification (*n.m.*), inspection gauge.
gâchée (de béton) (*n.f.*), batch.
gâcher (*v.t.*), to mix.
gâcher le mortier (*v.*), to mix mortar.
gain (*n.m.*), amplification; increase.
gaine de dégivrage (aviation) (*n.f.*), de-icing duct.
gaine de faisceau de câbles (*n.f.*), conduit.
gaine de parachute de queue (aviation) (*n.f.*), drag-chute cover.
gaine isolante (*n.f.*), insulating sleeve.
gaine métallique rigide (*n.f.*), rigid metal conduit.
gaine métallique souple (*n.f.*), flexible metal conduit.
gaine repère de câbles (*n.f.*), cable identification sleeve.
gaine tressée (*n.f.*), braiding.
gal (unité du système C.G.S.) (*n.m.*), gal; 1 gal = 1 cm./s./s.
galber (*v.t.*), to curve.
galerie (*n.f.*), manifold; level; drift; heading (in mining).
galet suiveur (*n.m.*), cam follower.
galopin de tension (*n.m.*), jockey pulley.
galvanomètre à cadre (*n.m.*), loop galvanometer.
galvanomètre à cadre mobile (*n.m.*), moving coil galvanometer.
gammagraphie (*n.f.*), gamma radiography.
gamme (*n.f.*), band; channel; range.
gamme d'ondes (radio) (*n.f.*), waveband.
gamme d'opérations (*n.f.*), sequence of operations.
gamme de fréquence (radio) (*n.f.*), frequency range.
gamme de travail (*n.f.*), detailed work-sheet.
garde (*n.f.*), clearance.
garde au sol (auto) (*n.m.*), ground clearance.
garde-boue (auto) (*n.m.*), mudguard; wing; fender (U.S.A.).
garnir de graisse (*v.t.*), to pack with grease.
garniture (décoration) (*n.f.*), trimming.
garniture (de frein) (*n.f.*), lining.
garniture anti-friction (*n.f.*), antifriction lining.
garniture d'étanchéité (*n.f.*), packing.
garniture de disque d'embrayage (*n.f.*), clutch lining.
garniture de frein (*n.f.*), brake lining.
garniture de joint (*n.f.*), seal packing.
garniture de siège (*n.f.*), covering.
gas-oil (pétrole) (*n.m.*), gas-oil.

gauche (navigation) (*n.f.*), port.
gauchissement (aviation) (*n.m.*), aileron control; roll control.
gauchissement négatif (aviation) (*n.m.*), wash-out.
gauchissement positif (aviation) (*n.m.*), wash-in.
gaufrage (*n.m.*), corrugation; embossment.
gaufré, -e (*adj.*), embossed.
gaz d'éclairage (*n.m.*), coal-gas.
gaz de carneau (*n.m.*), flue gas.
gaz réduits (moteur) (*n.m.pl.*), idling.
gaze (*n.f.*), gauze.
gel de silice (*n.m.*), silica gel.
générateur d'allumage (*n.m.*), ignition generator.
générateur de tourbillons (*n.m.*), vortex generator.
générateur haute-tension (*n.m.*), high energy unit.
générateur hydraulique (*n.m.*), hydroelectric generator.
générateur toute basse fréquence (*n.m.*), very low frequency generator.
génération (*n.f.*), generation.
génération alternative (*n.f.*), AC generation system.
génération continue (*n.f.*), DC generation system.
génératrice couplée aux essieux (ch. de fer) (*n.f.*), axle generator.
génératrice de pression hydraulique (*n.f.*), hydraulic pressure source.
génératrice de profil (*n.f.*), generatrix.
génératrice polymorphique (élec.) (*n.f.*), multiple-current generator.
génératrice tachymétrique (*n.f.*), velocity generator.
génératrice tachymétrique (réacteur) (*n.f.*), engine tachometer generator.
genouillère (*n.f.*), knuckle joint; toggle joint.
gicler (*v.t.*), to squirt.
gicleur (de carburateur) (auto) (*n.m.*), jet.
gicleur calibré (*n.m.*), metering jet.
gicleur d'huile (*n.m.*), oil jet.
gicleur de dégivrage pare-brise (aviation) (*n.m.*), windshield de-icing sprayer.
gicleur de démarrage (*n.m.*), starting jet.
gicleur de ralenti (*n.m.*), idle jet; slow running.
gicleur principal (*n.m.*), main jet.
giravion (*n.m.*), rotorcraft.
gisement (navigation) (*n.m.*), bearing.
givrage (*n.m.*), icing.
glace chauffante (*n.f.*), heated windshield pane.
glace coulissante (*n.f.*), sliding panel.
glace de pare-brise (*n.f.*), windshield panel.
glace latérale (*n.f.*), side window.
glacé, -e (*adj.*), glossy (of paper).
gland (*n.m.*), glans.
glissade (*n.f.*), sideslip.

glissement de fréquence (radio) (*n.m.*), fre-
-quency shift.
glissières de tour (machines-outils) (*n.f.pl.*),
lathe bed.
globulation (métal.) (*n.f.*), spherodizing.
glyptal (plast.) (*n.m.*), glyptal resin.
gobelet (*n.m.*), cup.
gommage (*n.m.*), sticking.
gommé, -e (*adj.*), sticky.
gomme-laque (*n.f.*), shellac.
gonflage (*n.m.*), inflation; charging.
gonflé, -e (*adj.*), inflated; charged.
gond (*n.m.*), hinge.
gong (*n.m.*), gong.
goniométrie (*n.f.*), direction finding.
gorge (*n.f.*), throat.
gorge d'étanchéité (*n.f.*), seal groove.
gorge de dégagement de filet (*n.f.*), thread
root draft.
gougeage (soudage) (*n.m.*), gouging.
goujon repère (*n.m.*), locating pin.
goujure (*n.f.*), gash.
goujure d'usure (*n.f.*), gouging.
goupille cannelée (*n.f.*), grooved pin.
goupille cylindrique (*n.f.*), cylindrical pin.
goupille d'arrêt (*n.f.*), stop pin.
goupille de sécurité (*n.f.*), safety pin.
goupille élastique (*n.f.*), spring pin.
goupille filetée (*n.f.*), threaded pin.
goupille frein (*n.f.*), locking pin.
gousset en équerre (*n.m.*), angle plate.
gouttelette (*n.f.*), small drop.
gouttière d'éclairage (*n.f.*), channel.
gouttière de protection (*n.f.*), shield.
gouttière de volet (aviation) (*n.f.*), wing flap
eaves.
gouttière pour câbles (*n.f.*), cable trough.
gouvernail de direction (aviation) (*n.m.*),
rudder.
gouverne (aviation) (*n.f.*), control surface.
gouverne compensée (aviation) (*n.f.*), bal-
-anced control surface.
gouverne de direction (aviation) (*n.f.*),
rudder.
gouverne de profondeur (aviation) (*n.f.*),
elevator.
grade (unité d'angle) (*n.m.*), grade.
gradient (math.) (*n.m.*), gradient (of a
function).
grain (abrasif) (*n.m.*), grit.
grain (météor.) (*n.m.*), rainstorm; squall.
grain de câble (*n.m.*), cable end bead.
graissage à compte-gouttes (*n.m.*), drip-
feed lubrication.
graissage à film d'huile (*n.m.*), boundary
lubrication.
graissage par barbotage (*n.m.*), splash
lubrication.
graissage sous pression (*n.m.*), force feed
lubrication.
graisse courante No. 1 (*n.f.*), low temperature
grease.
graisse courante No. 2 (*n.f.*), general service
grease.
graisse graphitée (*n.f.*), graphited grease.
graisse très consistante (*n.f.*), block grease.
graisseur (auto) (*n.m.*), grease nipple; lubricator.
graisseur à pression (outil) (*n.m.*), grease gun.
grand pas (hélicoptère) (*n.m.*), high pitch.
grand pas (taraudage) (*n.m.*), coarse pitch.
grande puissance (*n.f.*), heavy duty.

grande visite (aviation) (*n.f.*), major inspection.
grandeur nature (*n.f.*), full scale.
grandeur oscillante (élec.) (*n.f.*), oscillating
quantity.
grandeur pulsatoire (élec.) (*n.f.*), pulsating
quantity.
grandeur scalaire (géol.) (*n.f.*), scalar
quantity.
grandeur sinusoïdale (élec.) (*n.f.*), sinusoidal
quantity.
grandeur sinusoïdale amortie (élec.) (*n.f.*),
damped sinusoidal quantity.
granité, -e (fini) (*adj.*), crinkled (finish).
granulé (*n.m.*), granule.
granulométrie (métallurgie) (*n.f.*), grain size
structure.
graphique (*n.m.*), plot
gras (*n.m.*), stock allowance.
gravé, -e (*adj.*), engraved; etched.
graver au crayon électrique (*v.t.*), to
electro-engrave.
gravillon (trav. pub.) (*n.m.*), grit.
gravillonnage (trav. pub.) (*n.m.*), gritting;
spreading of grit or chippings (on a road
surface).
gravité spécifique (*n.f.*), density.
gravure (*n.f.*), engraving; etching.
grenade sous-marine (*n.f.*), depth charge.
grenaillage (*n.m.*), sand blasting with steel shot;
shot peening.
griffe (*n.f.*), dog; clamp.
grillagé, -e (construction des machines) (*adj.*),
screen protected.
grille (élec.) (*n.f.*), grid.
grille à aube (dégivrage) (*n.f.*), deflector.
grille antigivre (*n.f.*), ice guard.
grille d'arrêt (élec.) (*n.f.*), suppressor grid.
grille d'attaque (*n.f.*), injection grid.
grille d'aubes (*n.f.*), cascade blades; cascade
vanes.
grille de charge d'espace (*n.f.*), space-charge
grid.
grille de commande (élec.) (*n.f.*), control grid.
grille de désionisation (élec.) (*n.f.*), de-
ionization grid.
grille de protection d'entrée (*n.f.*), guard
against debris.
grille de protection de haut-parleur
(radio) (*n.f.*), cover grille.
grille directrice d'entrée (*n.f.*), inducer.
grille directrice de sortie (*n.f.*), exducer.
grille-écran (radio) (*n.f.*), screen-grid.
grille-écran (*n.f.*), louvre.
grille flottante (élec.) (*n.f.*), floating grid.
grillé, -e (élec.) (*adj.*), burnt-out.
grippage (d'un moteur) (*n.m.*), binding; jamming;
seizing.
grippé, -e (*adj.*), jammed; seized.
gros grain (textile) (*n.m.*), petersham.
gros plan (cinéma) (*n.m.*), close-up.
groupe (*n.m.*), bank; unit; cluster.
groupe (cinéma) (*n.m.*), pool.
groupe auxiliaire au sol (aviation) (*n.m.*),
ground power unit.
groupe auxiliaire de bord (aviation) (*n.m.*),
airborne auxiliary power unit.
groupe d'injecteurs (aviation) (*n.m.*), engine
nozzle cluster.
groupe de démarrage au sol (aviation)
(*n.m.*), ground power unit.
groupe de quartz (radio) (*n.m.*), bank of

crystals.
groupe de réfrigération (*n.m.*), cooling unit.
groupe de soudage à postes multiples (*n.m.*), multiple operated welding set.
groupe électrogène (*n.m.*), electrical ground power unit.
groupe en cascade (élec.) (*n.m.*), cascade set.
groupe hydraulique (élec.) (*n.m.*), hydro-electric generating set.
groupe moteur (*n.m.*), power plant.
groupe moto-sustentateur (*n.m.*), rotor transmission and power plant group.
groupe thermique (élec.) (*n.m.*), thermo-electric generating set.
groupe turbo-alternateur (*n.m.*), steam tur-bine set.
groupe turbo-générateur (élec.) (*n.m.*), tur-bine driven set.
groupe turbo-moteur (élec.) (*n.m.*), turbo shaft power plant.
groupe turbo-réacteur (*n.m.*), power plant.
groupement (*n.m.*), pool.
grue volante (hélicoptère) (*n.f.*), rotor crane.
guidage par radio vers station émettrice (radio) (*n.m.*), homing.
guidage vertical (aviation) (*n.m.*), elevation guidance.
guide-câble (*n.m.*), fairlead.
guide d'ondes (radio) (*n.m.*), wave guide.
guide d'ondes cloisonné (radio) (*n.m.*), ridge wave guide.
guide d'ondes flexible (radio) (*n.m.*), flexible wave guide.
guide d'ondes hétérogènes (radio) (*n.m.*), composite wave guide.
guide d'ondes rigide (radio) (*n.m.*), rigid wave guide.
guide d'ondes vrillé (radio) (*n.m.*), twisted wave guide.
guide de clapet (*n.m.*), valve guide.

guide de ressort (*n.m.*), spring guide.
guignol (*n.m.*), bellcrank.
guignol additionneur (hélicoptère) (*n.m.*), collective bellcrank.
guignol d'angle (hélicoptère) (*n.m.*), bellcrank.
guignol de décrochage des trappes et trains (hélicoptère) (*n.m.*), door unlatch and landing gear unlock control bellcrank.
guignol de pas général (hélicoptère) (*n.m.*), collective bellcrank.
guignol droit (hélicoptère) (*n.m.*), lever (reciprocal).
guignol porte-galet (*n.m.*), roller carrying bellcrank.
guillotine (d'un robinet) (*n.f.*), slide gate of a cut-off valve.
guipage (élec.) (*n.m.*), braiding; cover band.
guipage coton, à (fil élec.) (*adj.*), cotton covered.
guiper un câble (*v.*), to wrap a cable.
gyro directionnel (aviation) (*n.m.*), directional gyro.
gyro vertical (aviation) (*n.m.*), vertical gyro.
gyrodyne (aviation) (*n.m.*), compound helicopter; aerogyro; gyrodyne.
gyromètre (aviation) (*n.m.*), rate gyro.
gyromètre à contact de virage (aviation) (*n.m.*), switching rate gyro.
gyromètre a trois axes (aviation) (*n.m.*), three-axis rate sensor.
gyromètre de lacet (aviation) (*n.m.*), yaw rate gyro.
gyromètre de roulis (aviation) (*n.m.*), roll rate gyro.
gyromètre longitudinal (aviation) (*n.m.*), pitch rate gyro.
gyroscope de commande (hélicoptère) (*n.m.*), control gyro.
gyroscope libre (aviation) (*n.m.*), free gyroscope.

H

habillage (*n.m.*), covering.
'hachure (*n.f.*), cross hatching; hatching.
'hachuré, -e (*adj.*), hatched.
'halo (dans un tube cathodique) (*n.m.*), halation.
halogénure (chimie) (*n.m.*), halide.
'hampe (outillage) (*n.f.*), pole.
'hangar (*n.m.*), hangar.
'harnais pyrométrique (*n.m.*), fire detection harness.
'harpon (hélicoptère) (*n.m.*), quick mooring gear.
'haut-parleur (radio) (*n.m.*), loudspeaker.
'haut-parleur à aimant permanent (*n.m.*), permanent magnet loudspeaker.
'haut-parleur électro-dynamique (*n.m.*), moving coil loudspeaker.
'haut-parleur électrostatique (radio) (*n.m.*), capacitor loudspeaker; condenser loudspeaker.
'haut-parleur piézo-électrique (radio) (*n.m.*), crystal loudspeaker.
'haute fréquence (radio) (*n.f.*), radio frequency.
'hauteur de chute libre (*n.f.*), free drop height.
'hauteur des pointes (machines-outils) (*n.f.*), swing over bed.

'hauteur sous traverse (*n.f.*), under rail height.
'hauteur théorique de filet (*n.f.*), theoretical thread height.
hectopièze (mesure de pression) (*n.m.*), hecto-pieze (1 hpz. = 1 bar = $1,02$ kg./cm^2).
hélibus à cabine amovible (aviation) (*n.m.*), skylounge.
hélice anti-couple (hélicoptère) (*n.f.*), tail rotor.
hélice d'avion (*n.f.*), propeller.
hélice de queue carénée (hélicoptère) (*n.f.*), ring tail.
hélicoplane (*n.m.*), cyclogyro.
hélicoptère (*n.m.*), helicopter.
hélicoptère à cycle chaud (*n.m.*), hot cycle helicopter.
hélicoptère à missions multiples (*n.m.*), multi-purpose helicopter.
hélicoptère à réaction (*n.m.*), jet helicopter.
hélicoptère à rotors en tandem (*n.m.*), tan-dem rotor helicopter.
hélicoptère à turbine (*n.m.*), turbine powered helicopter.

hélicoptère amphibie (*n.m.*), amphibian helicopter.

hélicoptère anti-sous-marin (*n.m.*), anti-submarine helicopter.

hélicoptère armé (*n.m.*), combat helicopter.

hélicoptère bi-rotor (*n.m.*), dual rotor helicopter.

hélicoptère combiné (*n.m.*), compound heli--copter; gyrodyne.

hélicoptère d'affaires (*n.m.*), executive helicopter.

hélicoptère de manoevre (*n.m.*), tactical helicopter.

hélicoptère de transport à réaction (*n.m.*), transport jet helicopter.

hélicoptère embarqué (*n.m.*), carrier-borne helicopter.

hélicoptère expérimental (*n.m.*), experimen--tal helicopter.

hélicoptère grue (*n.m.*), crane helicopter.

hélicoptère léger d'observation (*n.m.*), light observation helicopter.

hélicoptère monorotor (*n.m.*), single rotor helicopter.

hélicoptère multimoteur (*n.m.*), multi-engine helicopter.

hélicoptère piloté (*n.m.*), manned helicopter.

hélicoptère polyvalent (*n.m.*), multi-purpose helicopter; utility helicopter.

hélicoptère radio-commandé (*n.m.*), radio controlled helicopter.

hélicoptère ravitailleur (*n.m.*), tanker helicopter.

hélicoptère téléguidé (*n.m.*), drone helicopter.

hélicoptère tout temps (*n.m.*), all weather helicopter.

héligrue (*n.f.*), rotor crane.

hélion (atome) (*n.m.*), alpha particle.

héliport (*n.m.*), heliport.

hélium (*n.m.*), helium.

heptode (*n.f.*), heptode.

'hérisson (trav. pub.) (*n.m.*), stone foundation (of a road).

hermétique (*adj.*), watertight.

'herse (*n.f.*), cable rack.

'hertz (unité de fréquence) (*n.m.*), hertz (1 hertz = 1 c./s.).

hétérodyne (radio) (*adj.*), heterodyne.

hétérodyne (radio) (*n.f.*), heterodyne.

heure locale (*n.f.*), local time.

heures de fonctionnement (*n.f.pl.*), operat--ing hours.

heures de vol (*n.f.pl.*), flying hours.

'heurter (*v.t.*), to hit; to run into.

hexode (*n.f.*), hexode.

hiloire de protection (*n.m.*), coaming.

homme-heure directe (*n.f.*), direct man-hour.

homme-heure indirecte (*n.f.*), indirect man-hour.

homologation (*n.f.*), approval; certification.

homologué, -e (*adj.*), approved; certified.

horaire (*n.m.*), timetable.

horizon argileux compact (géol.) (*n.m.*), clay pan.

horizon artificiel (*n.m.*), artificial horizon.

horizon d'accumulation calcaire (géol.) (*n.m.*), lime pan.

horizon de secours (*n.m.*), stand-by horizon.

horizon gyroscopique (*n.m.*), gyro horizon.

horizon indicateur de vol (aviation) (*n.m.*), horizon flight director.

'hors d'usage (*adj.*), unserviceable.

'hors tolérance (*adj.*), out of tolerance; beyond tolerance.

'hors tout (*adj.*), overall.

'hors trim (*adj.*), out of trim.

'houle (navigation) (*n.f.*), swell.

'housse (*n.f.*), cover.

'hublot (navire, avion) (*n.m.*), window; porthole.

'hublot largable (*n.m.*), jettisonable window.

huile compoundée (*n.f.*), compound oil.

huile de coupe (machines-outils) (*n.f.*), coolant; cutting oil.

huile de dégrippage (*n.f.*), penetrating oil.

huile de protection (*n.f.*), anti-corrosion oil.

huile épaisse (*n.f.*), thick oil.

huile fluide (*n.f.*), thin oil.

huile hydrogénée (*n.f.*), hardened oil.

huile neutre (*n.f.*), neutral oil.

huile régénérée (*n.f.*), reclaimed oil.

huile siccative (peinture) (*n.f.*), drying oil.

huile soufflée (*n.f.*), blown oil.

huiler (*v.t.*), to lubricate.

humidificateur (*n.m.*), humidifier.

humidité absolue (*n.f.*), absolute humidity.

humidité relative (*n.f.*), relative humidity.

hydrofuge (*n.m.*), water repellent.

hydrogène lourd (atome) (*n.m.*), deuteron; heavy hydrogen.

hydromécanique (*adj.*), hydromechanical.

hydrométrie (mesure de la dureté de l'eau) (*n.f.*), hydrometry.

hyperfréquence (radio) (*n.f.*), ultra-high frequency.

hypergol (fusées) (*n.m.*), bipropellant.

hyperluminosité du spot (élec.) (*n.f.*), blooming.

hypersustentateurs (*n.m.pl.*), high lift devices.

hypertrempé, -e (métal.) (*adj.*), re-annealing.

hypoxie (*n.f.*), hypoxia.

hystérésis (*n.f.*), magnetic lag.

'hz (hertz) (*n.m.*), hertz; cycle per second.

I

identification (*n.f.*), identification.

ignifugé, -e (*adj.*), fire resistant; flame resistant.

igniteur (élec.) (*n.m.*), ignitor.

ignitron (électron.) (*n.m.*), ignitron.

image d'écoulement (*n.f.*), flow pattern.

image de fond (radar) (*n.f.*), clutter.

imbiber (*v.t.*), to dampen.

immatriculation (*n.f.*), registration.

immobiliser (*v.t.*), to ground; to immobilize; to secure.

impact (*n.m.*), impingement.

impact à l'atterrissage (aviation) (*n.m.*), touch-down.

impair, -e (nombre) (*adj.*), odd (of number).

impédance de sortie (*n.f.*), output impedance.
impédance propre (*n.f.*), self-impedance.
impédance réelle (*n.f.*), normal impedance.
impédance variable (*n.f.*), bleeder.
impératif, -ive (*adj.*), imperative; mandatory.
implantation (*n.f.*), layout; location.
imposte (*n.m.*), transom.
imprécis, -e (*adj.*), inaccurate.
imprimante (machine) (*n.f.*), printer.
imprimé (*n.m.*), print; printed form; (*pl.*) printed matter.
impulseur (*n.m.*), pulsating system.
impulsion à bord raide (*n.f.*), steep slope pulse.
impulsion d'accord (*n.f.*), tuning pulse.
impulsion électrique (*n.f.*), pulse.
impulsion modulée (*n.f.*), modulated pulse.
impuretés (*n.f.pl.*), foreign matter.
inaltérable (*adj.*), weather resistant.
inattaquable par des acides (*adj.*), acid-proof.
incandescence résiduelle (métal.) (*n.f.*), after-glow.
incassable (*adj.*), shatterproof.
incidence (*n.f.*), attack angle.
incidence de décrochage (aviation) (*n.f.*), stall angle.
incidence de la pale (aviation) (*n.f.*), blade attack angle.
incidence de portance nulle (aviation) (*n.f.*), zero-lift angle.
incident (*n.m.*), trouble.
inclinaison (*n.f.*), cant; tilt.
inclinaison du rotor (hélicoptère) (*n.f.*), rotor tilt.
inclinaison longitudinale (aviation) (*n.f.*), pitch.
inclinaison transversale (aviation) (*n.f.*), bank.
incliné, -e (*adj.*), inclined; slanted; tilted.
incombustible (*adj.*), flame-resistant.
incombustible (plast.) (*adj.*), self-extinguishing.
incorporé, -e (*adj.*), built-in; embodied; incorporated.
incorrect, -e (*adj.*), erroneous; false.
indélébile (*adj.*), indelible.
indentation (*n.f.*), dent.
indentation à bord vif (*n.f.*), nick.
indentation d'usure de roulement (*n.f.*), brinelling.
independant, -e (*adj.*), independent; separate; unrelated.
indesserrable (*adj.*), self-locking.
index (*n.m.*), index.
index d'instrument (*n.m.*), marker.
index de l'abaque de centrage (*n.m.*), weight and balance chart basic index.
index en V (*n.m.*), V pointer.
indicateur à comparaison de deux résistances (*n.m.*), ratiometer-type indicator.
indicateur d'angle de câble (sonar) (*n.m.*), cable angle indicator.
indicateur d'incidence (*n.m.*), angle of attack indicator.
indicateur de changement de direction (auto) (*n.m.*), trafficator.
indicateur de charge (élec.) (*n.m.*), loadmeter.
indicateur de débit (*n.m.*), flowmeter.
indicateur de décharge (*n.m.*), discharge indicator.

indicateur de déviation (*n.m.*), deviation indicator.
indicateur de frottement de turbine (aviation) (*n.m.*), rub indicator.
indicateur de jaugeur de carburant (aviation) (*n.m.*), fuel gauge indicator.
indicateur de niveau (*n.m.*), level indicator.
indicateur de pas (hélicoptère) (*n.m.*), pitch indicator.
indicateur de pas des pales (hélicoptère) (*n.m.*), blade pitch indicator.
indicateur de pas général (hélicoptère) (*n.m.*), collective pitch indicator.
indicateur de pente (aviation) (*n.m.*), glide path localizer.
indicateur de perte de poussée (aviation) (*n.m.*), thrust loss indicator.
indicateur de pointe (élec.) (*n.m.*), maximum demand indicator; peak indicator.
indicateur de position de train (aviation) (*n.m.*), landing gear position indicator.
indicateur de position gauchissement (aviation) (*n.m.*), aileron position indicator.
indicateur de position gouverne direc-tion (aviation) (*n.m.*), rudder position indicator.
indicateur de position gouverne profon-deur (aviation) (*n.m.*), elevator position indicator.
indicateur de poussée (aviation) (*n.m.*), thrust indicator.
indicateur de pré-affichage (*n.m.*), presetting indicator.
indicateur de pression (*n.m.*), pressure gauge.
indicateur de recopie de cap (aviation) (*n.m.*), heading remote indicator.
indicateur de relèvement (aviation) (*n.m.*), omni-bearing indicator.
indicateur de taux d'ondes stationnaires (radio) (*n.m.*), standing wave ratio meter.
indicateur de température (*n.m.*), temper-ature indicator.
indicateur de température d'huile (aviation) (*n.m.*), oil temperature indicator.
indicateur de température de l'air dégi-vrage (aviation) (*n.m.*), de-icing air temp-erature indicator.
indicateur de température de tuyère (*n.m.*), nozzle temperature indicator.
indicateur de température du poste pi-lote (aviation) (*n.m.*), cockpit temperature indicator.
indicateur de température extérieure (aviation) (*n.m.*), outside air temperature indicator.
indicateur de T4 (aviation) (*n.m.*), exhaust gas temperature indicator.
indicateur de trim (aviation) (*n.m.*), trim indicator.
indicateur de virage-glissade (aviation) (*n.m.*), turn and bank indicator.
indicateur de volets (aviation) (*n.m.*), wing flap position indicator.
indicateur double (*n.m.*), dual indicator.
indicateur gyro compas (aviation) (*n.m.*), gyrosyn compass indicator.
indicateur magnétique (*n.m.*), magnetic indicator.
indicateur omnidirectionnel (*n.m.*), omni-range indicator.
indicateur principal (*n.m.*), master indicator.

indicateur radar (aviation) (*n.m.*), radar-scope.

indicateur radiomagnétique (*n.m.*), radiomagnetic indicator.

indicateur sélecteur de course (aviation) (*n.m.*), course indicator selector.

indicateur sphérique (aviation) (*n.m.*), three-axis attitude indicator.

indicateur totalisateur (carburant) (aviation) (*n.m.*), integrating flowmeter.

indicatif d'appel (*n.m.*), call signal.

indice (chiffre) (*n.m.*), dash number.

indice (lettre) (*n.m.*), suffix letter.

indice d'octane (pétr.) (*n.m.*), octane; fuel grade.

indice de détonation (pétr.) (*n.m.*), knock-rating.

indice de dureté (*n.m.*), hardness number.

indice de glissement amont (magnétron) (*n.m.*), pushing figure.

indice de pulsation (élec.) (*n.m.*), pulse number.

indice horaire de couplage (élec.) (*n.m.*), clock hour figure of the vector group.

indice inférieur (*n.m.*), subscript.

indice numérique de couplage (élec.) (*n.m.*), numerical index of the vector group.

indice supérieur (*n.m.*), superscript.

indisponible (*adj.*), not available.

inducteur série (*n.m.*), series inductor.

inducteur shunt (*n.m.*), shunt inductor.

industrie du spectacle (audio-visuel) (*n.f.*), show business.

inférieur, -e (*adj.*), sub-standard.

infime (*adj.*), minute.

infini amont (aviation) (*n.m.*), upstream infinite.

inflammabilité (*n.f.*), flammability.

information (*n.f.*), data; information.

information cathodique (radio) (*n.f.*), cathode-ray tube display.

information de cap (aviation) (*n.f.*), heading information.

information de roulis (aviation) (*n.f.*), roll information.

information de tangage (aviation) (*n.f.*), pitch information.

information technique (*n.f.*), technical information.

infrastructure (*n.f.*), ground installation.

ingéniérie (*n.f.*), engineering.

ingrédients (*n.m.pl.*), products.

inhalateur (*n.m.*), oxygen mask.

inhibiteur (plast.) (*n.m.*), inhibitor.

ininflammable (*adj.*), non-flammable.

initiateur (radar) (*n.m.*), magneto trigger generator.

injecteur à téton (moteur diesel) (*n.m.*), pintle-type nozzle.

injecteur d'huile (*n.m.*), oil jet.

injecteur de brûleur (aviation) (*n.m.*), burner nozzle.

injecteur de ventilation (fuseau réacteur) (aviation) (*n.m.*), ventilation nozzle (nacelle).

injection (*n.f.*), priming.

injection d'eau (*n.f.*), water injection.

inscription (*n.f.*), entry; mark; marking.

insonore (*adj.*), sound-proof.

insonorisation (*n.f.*), sound-proofing.

inspection périodique (*n.f.*), periodic inspection.

instabilité du type oscillatoire (*n.f.*), oscil-latory mode instability.

instabilité spirale (aviation) (*n.f.*), spiral instability.

instable (*adj.*), unsteady.

installation de pulvérisation (*n.f.*), spraying installation.

installation de transport sanitaire (*n.f.*), ambulance installation.

instruction (*n.f.*), instruction.

instruction technique (*n.f.*), technical training.

instruments de vol (aviation) (*n.m.pl.*), flight instruments.

instruments de vol sans visibilité (avia-tion) (*n.m.pl.*), blind flying instruments.

instruments gyroscopiques (aviation) (*n.m.pl.*), gyro instruments.

instruments moteur (aviation) (*n.m.pl.*), engine instruments.

intégrateur (*n.m.*), integrator.

intégrateur d'erreur de cap (aviation) (*n.m.*), heading error integrator.

intégrateur de trajectoire transver-sale (*n.m.*), lateral path integrator.

intégrateur synchroniseur (*n.m.*), heading synchronizer and lateral path integrator.

intempestif, -ive (*adj.*), untimely.

intensité (élec.) (*n.f.*), amperage.

intensité d'ouverture (élec.) (*n.f.*), drop-out current.

intensité de crête (élec.) (*n.f.*), peak current.

intensité de fermeture (élec.) (*n.f.*), pick-up current.

intensité de rafale (aviation) (*n.f.*), gust intensity.

intensité de rupture (*n.f.*), interruption capacity.

intensité efficace (élec.) (*n.f.*), effective current.

intensité nominale (élec.) (*n.f.*), contact rating; rated intensity.

intensité sonore (radio) (*n.f.*), loudness; sound volume.

interaction (*n.f.*), interference.

interaction électrolytique (métal.) (*n.f.*), electrolytic action.

intercalaire (*n.m.*), divider.

interception d'un faisceau (navigation) (*n.f.*), interception of a radio beam.

interchangeabilité (*n.f.*), interchangeability.

interclasseuse (*n.f.*), collator.

intercommunication (*n.f.*), cross-feed; inter-communication; interconnection.

interdiction (*n.f.*), interlock.

interdiction de vol (aviation) (*n.f.*), grounding of aircraft.

interface (*n.f.*), interface.

interférence (*n.f.*), cross-talk.

interféromètre (optique) (*n.m.*), inter-ferometer.

interligne (*n.f.*), spacing line.

interphone (*n.m.*), intercom.

interpolation (*n.f.*), interpolation.

interpoler (*v.t.*), to interpolate.

interrégional, -e (téléphone) (*adj.*), toll.

interrogateur (*n.m.*), interrogator.

interrompre le courant (*v.*), to switch off.

interrupteur à bascule (*n.m.*), tumbler switch.

interrupteur à couteau (*n.m.*), knife switch.

interrupteur à plots (*n.m.*), step switch.

interrupteur d'allumage (*n.m.*), ignition switch.

interrupteur de choc (*n.m.*), impact switch.

interrupteur de crash (n.m.), crash switch.
interrupteur de débrayage de servo-moteur (n.m.), servo actuator disconnect switch.
interrupteur de sécurité (n.m.), safety switch.
interrupteur général (n.m.), master switch.
interrupteur marche-arrêt (n.m.), "on"-"off" switch.
interrupteur pas-à-pas (sur marche cyclique) (n.m.), beep switch (on cycle stick).
interrupteur temporisé (n.m.), time delay switch.
interrupteur thermique (n.m.), thermo--switch.
interrupteur tripolaire (n.m.), three-pole switch.
interurbain, -e (téléphone) (adj.), trunk; long distance.
intervallomètre (n.m.), intervalometer.
intervention (n.f.), action; interference.
interverrouillage (n.m.), interlocking.
intervis (n.m.), screw thread insert.
intrados (aviation) (n.m.), lower surface; under surface.
intrados de la pale (n.m.), blade lower surface.
invariable (adj.), invariable.
inversable (adj.), non-spill.
inverseur (élec.) (n.m.), toggle switch.
inverseur coupe-tout (élec.) (n.m.), master switch.
inverseur de commande (aviation) (n.m.), control reversal.
inverseur de commande génératrice (élec.) (n.m.), generator control switch.
inverseur de poussée (aviation) (n.m.), thrust reverser.
inverseur de prise statique (aviation) (n.m.), static air selector.
inverseur de réenclenchement (élec.) (n.m.), reset switch.
inverseur de sélection démarrage (avia--tion) (n.m.), starter selection switch.
inverseur de sélection pompes (aviation) (n.m.), fuel pump selector switch.
inverseur de sélection réservoirs (avia--tion) (n.m.), fuel tank selector switch.
inverseur démarrage/ventilation (mo--teur) (n.m.), engine starter/crank switch.
inverseur marche-arrêt (n.m.), "on"-"off" switch.
ion (n.m.), ion.
ion-gramme (n.m.), gram-ion.
ionisation cumulative (électronique) (n.f.), cumulative ionization.
ionisation multiple (électronique) (n.f.), multiple ionization.
ionisation par choc (électronique) (n.f.), colli--sion ionization.
ionisation par rayonnement (électronique) (n.f.), radiation ionization.
ionisation thermique (électronique) (n.f.), thermal ionization.
irrégulier, -ière (adj.), erratic.
irréparable (adj.), beyond repair.
irrétrécissable (adj.), unshrinkable.
irréversible (adj.), irreversible.
iso-contrainte (n.f.), iso-stress.
isocèle (adj.), isosceles.
isolation acoustique (n.f.), sound-proofing.
isolation thermique (n.f.), thermal insulation.
isolé, -e (élec.) (adj.), insulated.
isolement (n.m.), cut-out; isolation; shut-off.
isoler (v.t.), to cut-out.
isosurpression (n.f.), constant differential pressure.
isotope (n.m.), isotope.
isotope radioactif (n.m.), radioactive isotope.
issue de secours (n.f.), emergency exit.
-ite (suffixe chimique), -ite.

J

jambe (n.f.), leg.
jambe de train avant (aviation) (n.f.), nose gear leg.
jambe oléo-pneumatique (n.f.), oleo-strut.
jambe oscillante (n.f.), swivel arm.
jante de pneu (n.f.), tyre bead.
jarret d'une courbe (n.m.), knee of a curve.
jarretière de masse (n.f.), bonding strip.
jauge à fil résistant (n.f.), strain gauge.
jauge à vis (n.f.), screw gauge.
jauge d'enfoncement (aviation) (n.f.), bottom--ing indicator.
jauge de contrainte (n.f.), strain gauge.
jauge "entre" (n.f.), "go" gauge.
jauge "n'entre pas" (n.f.), "no go" gauge.
jaugeur (n.m.), capacity or quantity gauge.
jaugeur à canne (n.m.), dripstick.
jaugeur à capacité (n.m.), capacitor gauge.
jaugeur à flotteur (élec.) (n.m.), float-type transmitter.
jaugeur à niveau visible (n.m.), sight gauge.
jaugeur d'huile (n.m.), oil gauge.
jaugeur d'intrados (aviation) (n.m.), dripstick.
jaugeur de réservoir de carburant (n.m.), fuel gauge transmitter.
jaugeur manuel (n.m.), dipstick.
jet d'eau (auto) (n.m.), drip moulding.
jet d'extraction (outil) (n.m.), drift.
jet des réacteurs (aviation) (n.m.), jet wash.
jeu (écart) (n.m.), gap.
jeu axial (n.m.), end play.
jeu d'usure (n.m.), play.
jeu de cales (n.m.), feeler gauges.
jeu de denture (n.m.), backlash.
jeu de modification (n.m.), modification kit.
jeu de tolérance (n.m.), clearance.
jeu léger (n.m.), wink.
jeux de montage (n.m.pl.), fits and clearances.
joint à brides à emboîtement (n.m.), recessed flange joint.
joint à enture (n.m.), scarf joint.
joint à lèvre (n.m.), lip seal.
joint à olive (n.m.), block joint.
joint à oulice (n.m.), triangular tenon joint.
joint à rivure encastrée (n.m.), recessed riv--eted joint.
joint à rotule (n.m.), ball and socket joint.
joint articulé (n.m.), swivel joint.

joint biseauté (*n.m.*), bevel joint.
joint d'about (soudure) (*n.m.*), butt joint.
joint d'about sans espace (soudure) (*n.m.*), closed butt joint.
joint d'étanchéité (*n.m.*), gasket.
joint d'huile (*n.m.*), oil seal.
joint de chambre (*n.m.*), gasket.
joint de culasse (auto) (*n.m.*), cylinder head gasket.
joint de glace (auto) (*n.m.*), window seal.
joint de réglage (*n.m.*), calibrated spacer.
joint en bout (soudure) (*n.m.*), butt joint.
joint en caoutchouc (*n.m.*), rubber seal.
joint flexible (*n.m.*), flexible joint.
joint ignifuge (*n.m.*), fireseal.
joint labyrinthe (*n.m.*), labyrinth seal.
joint métallo-plastic (*n.m.*), copper asbestos gasket.
joint moulé en néoprène (*n.m.*), neoprene moulded seal.
joint papier (*n.m.*), paper gasket.
joint plat (*n.m.*), gasket.

joint torique (*n.m.*), "O" ring.
joint tubulaire (*n.m.*), seal tube.
jonc (*n.m.*), retaining ring.
jonc à ergot (*n.m.*), snap ring.
jonction (*n.f.*), joint; junction.
jonction de revêtement (*n.f.*), skin joint.
jonction des demi-voilures (aviation) (*n.f.*), wing centre attachment.
jonction différentielle (radio) (*n.f.*), hybrid junction.
joue (*n.f.*), web.
joue de manivelle (*n.f.*), crankshaft.
jumelage (magnétron) (*n.m.*), strapping.
jumelé, -e (*adj.*), dual; split; tandem; twin.
jumelle (*n.f.*), yoke.
jumelle d'arrêt (*n.f.*), stop yoke.
jumelle de liaison (*n.f.*), connecting twin yoke.
jumelle de ressort (auto) (*n.f.*), spring shackle.
jupe de cloche (aviation) (*n.f.*), drag cup skirt.
jupe de piston (auto) (*n.f.*), piston skirt.

K

karman (*n.m.*), fillet.
karman d'empennage (*n.m.*), stabilizer fillet.
karman de mât réacteur (*n.m.*), stub fillet.
karman de voilure (*n.m.*), wing fillet.
kelvinomètre (*n.m.*), kelvinometer; thermo--colourimeter.
kénotron (*n.m.*), kenotron.

kilocycle (radio) (unité de fréquence) (*n.m.*), kilocycle.
kinétoscope (*n.m.*), kinetoscope.
klaxon (auto) (*n.m.*), horn.
klystron (*n.m.*) klystron.
krarupisation (élec.) (*n.m.*), continuous coil loading.

L

labyrinthe d'étanchéité (*n.m.*), labyrinth seal.
lacet (aviation) (*n.m.*), yaw.
lâcher des freins (*n.m.*), brake release.
lacune-trou (d'électron) (*n.f.*), hole (electron hole).
laine de verre (*n.f.*), fibreglass; glass wool.
lamage (*n.m.*), spotfacing.
lamage avec fraise à coupe négative (*n.m.*), back spotfacing.
lambert (unité de luminescence photoélectrique) (*n.m.*), lambert (1 lambert = $1/\pi$ candle/cm^2).
lame de retenue de pale (hélicoptère) (*n.f.*), blade retention strap.
lame ressort (*n.m.*), leafspring.
laminage (*n.m.*), laminating; rolling (steel, etc.).
laminé, -e (*adj.*), rolled.
laminer (*v.t.*), to laminate; to roll.
laminer un fluide (*v.*), to throttle a fluid.
lampe à vapeur de sodium (*n.f.*), sodium vapour lamp.
lampe d'appel (*n.f.*), call lamp.
lampe d'éclairage (*n.f.*), bulb; electric light; lamp.
lampe de poche (*n.f.*), flashlight; torch.
lampe de signalisation (*n.f.*), signal light.
lampe fluorescente (*n.f.*), fluorescent lamp.
lampe radio (*n.f.*), tube.

lampe torche (*n.f.*), flashlight; torch.
lancement du rotor (hélicoptère) (*n.m.*), rotor spinning.
lancer un réacteur (*v.*), to crank; to start.
languette (*n.f.*), tab.
languette de masse (*n.f.*), grounding lug.
lanière (*n.f.*), cord; lanyard.
laps de temps (*n.m.*), space of time.
laque (*n.f.*), lacquer.
lardon conique (machines-outils) (*n.m.*), tapered gib.
lardon de guidage (machines-outils) (*n.m.*), gib.
lardon de guidage en coin (machines-outils) (*n.m.*), taper gib.
lardon de réglage (machines-outils) (*n.m.*), adjusting gib.
lardon réglable (machines-outils) (*n.m.*), adjustable gib.
largage de la charge (aviation) (*n.m.*), load release.
largage par corde lisse (aviation) (*n.m.*), roping.
large (*adj.*), broad; wide.
largeur de bande (*n.f.*), bandwidth.
largeur hors-tout (*n.f.*), overall width.
larmier (*n.m.*), gutter.
laryngophone (*n.m.*), throat telephone.

latence (*n.f.*), latency.

latéralement (*adv.*), sideways: sidewise.

lave-glace (*n.m.*), windshield cleaner.

lecteur de bande (*n.m.*), tape reader.

lecture directe (instruments) (*n.f.*), direct reading.

légende des couleurs (*n.f.*), colour code.

lent, -e (*adj.*), sluggish.

lessive de potasse (chimie) (*n.f.*), potassium hydroxide.

lest (*n.m.*), ballast.

lesté, -e (*adj.*), ballasted; weighted.

levage (*n.m.*), jacking.

levage par hélicoptère (*n.m.*), heli-lifting.

lève-tôles (*n.m.*), plate-hoist.

lever (avec une grue) (*v.t.*), to hoist (with a crane).

lever de doute (*n.m.*), to-from sensor.

levier à déclic (*n.m.*), tripping lever.

levier à main (*n.m.*), hand lever.

levier à poussoir (*n.m.*), latched lever.

levier à rallonge (*n.m.*), extension lever.

levier articulé (suspension de voitures) (*n.m.*), wishbone; suspension control arm (U.S.A.).

levier coudé (*n.m.*), bellcrank.

levier d'attaque (*n.m.*), actuating lever.

levier d'attaque de fusée (auto) *n.m.*), steering arm.

levier d'intention (*n.m.*), setting lever.

levier de commande (direction d'auto) (*n.m.*), drop arm.

levier de commande (en général) (*n.m.*), control lever.

levier de commande de pas (hélicoptère) (*n.m.*), pitch control lever.

levier de pas général (hélicoptère) (*n.m.*), collective pitch lever.

levier de serrage (*n.m.*), clamping lever.

levier double (*n.m.*), compound lever.

levier gaz-pas (*n.m.*), pitch-throttle synchronizer.

lèvre de bossette (*n.f.*), gill.

lèvre de clapet (*n.f.*), valve lip.

liaison (*n.f.*), liaison.

liaison navette par hélicoptère (*n.f.*), helicopter shuttle service.

liaison radio maritime (*n.f.*), ship-to-shore communication.

liaison voilure/fuselage (*n.f.*) wing-to-fuselage attachment.

liaisons électriques (*n.f.pl.*), electrical connections.

liant (*n.m.*), bond; binder.

liant activé (*n.m.*), doped binder.

liant fillerisé (*n.m.*), filled binder.

liant hydrocarboné (*n.m.*), hydrocarbon binder.

liasse de plans (*n.f.*), set of drawings.

libre (*adj.*), free; vacant.

libre parcours moyen (électron.) (*n.m.*), mean free path.

lié, -e (*adj.*), bound; connected; joined; linked.

lien (*n.m.*), link.

lieu (*n.m.*), place; spot.

lieu géométrique (*n.m.*), locus.

ligature de fil frein (*n.f.*), lockwire twist.

ligne (*n.f.*), cable; feeder; lead.

ligne bifilaire (*n.f.*), two-wire line.

ligne d'action (engrenages) (*n.f.*), line of contact.

ligne de charge (*n.f.*), load line.

ligne de cote (plans) (*n.f.*), dimension line.

ligne de départ de cote (plans) (*n.f.*), extension line.

ligne de dispersion (élec.) (*n.f.*), stray line.

ligne de flottaison (*n.f.*), water line.

ligne de foi (*n.f.*), lubber line.

ligne de force électrique (*n.f.*), electric flux.

ligne de rappel de cote (plans) (*n.f.*), leader line.

ligne de référence (*n.f.*), datum line.

limaille (*n.f.*), metal particles.

limite d'utilisation (*n.f.*), operating limitations.

limite de décrochage (hélicoptère) (*n.f.*), stall limit.

limite de décrochage sur le disque (hélicoptère) (*n.f.*), rotor disc stall limit.

limite de fonctionnement, à la (*adj.*), time-expired.

limite de rupture (*n.f.*), ultimate strength.

limite de vie (*n.f.*), service life.

limite élastique (*n.f.*), yield strength.

limiteur (machines-outils) (*n.m.*), limiter; trip.

limiteur d'avance (*n.m.*), feed limiter.

limiteur d'ouverture de valve (*n.m.*), valve adjuster.

limiteur de couple (*n.m.*), torque limiter.

limiteur de courant (élec.) (*n.m.*), current limiter.

limiteur de course (*n.m.*), stroke limiter; travel stop.

limiteur de débit (*n.m.*), flow limiter.

limiteur de parasites (radio) (*n.m.*), noise suppressor.

limiteur de puissance (*n.m.*), power limiter.

limiteur de vitesse (*n.m.*), speed limiter.

linéarité (*n.f.*), linearity.

liquide hydraulique (*n.m.*), hydraulic fluid.

liquide réfrigérant (*n.m.*), coolant.

lisse (aviation) (*n.f.*), stringer.

lisse clarinette (aviation) (*n.f.*), tapered stringer.

liste de vérification (*n.f.*), check-list.

litière du sol (*n.f.*), litter.

livres par heure (*n.f.pl.*), pounds per hour.

livres par pouce carré (*n.f.pl.*), pounds per square inch (p.s.i.).

livret moteur (auto) (*n.m.*), engine log book.

localisation horaire (*n.f.*), clock-face reference.

localiser (*v.t.*), to locate; to pinpoint.

logement (*n.m.*), compartment; groove; recess.

logement de train (aviation) (*n.m.*), landing gear well.

loger (*v.t.*), to accommodate; to place.

loi des aires (aviation) (*n.f.*), area rule.

longeron (auto) (*n.m.*), side member; side rail (U.S.A.).

longeron (aviation) (*n.m.*), spar.

longeron de pale (hélicoptère) (*n.m.*), blade spar.

longeron de treillis (*n.m.*), chord member of a truss.

longeronnet (aviation) (*n.m.*), small spar.

longévité (*n.f.*), durability.

longueur-accélération-arrêt (*n.f.*), accelerate-stop distance.

longueur d'onde (radio) (*n.f.*), wavelength.

longueur de décollage (aviation) (*n.f.*), take-off distance.

longueur de filetage (outils) (*n.f.*), threaded length.

longueur de piste équivalente (aviation) (*n.f.*), balanced field strength.

longueur de piste minimum (aviation) (*n.f.*), minimum runway length.

longueur de roulement à l'atterrissage (aviation) (*n.f.*), landing run.

longueur de roulement au décollage (aviation) (*n.f.*), take-off run.

longueur de trajectoire d'envol (aviation) (*n.f.*), take-off distance.

longueur dépassante (*n.f.*), projection length.

longueur hors-tout (*n.f.*), overall length.

longueur libre d'un ressort (*n.f.*), free length of a spring.

loqueteau (*n.m.*), latch.

loqueteau de fermeture (*n.m.*), lock.

lot (*n.m.*), batch; kit; set.

lot de modification (*n.m.*), modification kit.

lueur (*n.f.*), glare; glow.

lumière (ouverture) (*n.f.*), aperture; opening; sight.

lumière noire (*n.f.*), black light.

luminosité (*n.f.*), brightness.

lunette à coussinets (machines-outils) (*n.f.*), jaw steady rest.

lunette à galets (machines-outils) (*n.f.*), roller steady rest.

lunette arrière (auto) (*n.f.*), rear window.

lunette d'instrument (*n.f.*), bezel.

lunette de visée (*n.f.*), sight tube.

lunette équipée (outils) (*n.f.*), sight assembly.

lustrage (*n.m.*), glazing; glossing.

lyre (commandes de vol) (aviation) (*n.f.*), flight controls lyre-shaped bellcrank.

lyre d'accrochage (*n.f.*), finger grip clip.

M

Mach (M.) (*n.m.*), Mach number (M.).

Mach à ne jamais dépasser (aviation), (*n.m.*), Mach number not to be exceeded.

Mach maximum en utilisation normale (aviation) (*n.m.*), maximum operating limit Mach number.

machine à aléser à deux broches angulaires (*n.f.*), two-way boring machine.

machine à aléser à deux broches opposées (*n.f.*), duplex boring machine.

machine à aléser et à pointer (*n.f.*), jig boring machine.

machine à brocher (*n.f.*), broaching machine.

machine à brocher à plusieurs coulisseaux (*n.f.*), multiple ram broaching machine.

machine à brocher type à coulisseau (*n.f.*), ram-type broaching machine.

machine à brocher type par poussée (*n.f.*), push-type broaching machine.

machine à brocher type par traction (*n.f.*), pull-type broaching machine.

machine à brocher verticale type presse (*n.f.*), press-type vertical broaching machine.

machine à calculer (*n.f.*), calculating machine; adding machine; adder; calculator.

machine à calculer à analogie (*n.f.*), ana-logic computer.

machine à calculer électronique (*n.f.*), electronic computer.

machine à découper les bandes de tôles (*n.f.*), slitting machine.

machine à fileter (*n.f.*), threading machine.

machine à fileter à la filière (*n.f.*), die threading machine.

machine à fileter à la fraise (*n.f.*), thread milling machine.

machine à fileter à la meule (*n.f.*), thread grinding machine.

machine à fileter au peigne (*n.f.*), chasing machine.

machine à fraiser à console (*n.f.*), knee-type milling machine.

machine à fraiser à table coulissante (*n.f.*), sliding table milling machine.

machine à fraiser à tête coulissante (*n.f.*), sliding head milling machine.

machine à fraiser les matrices et les moules (*n.f.*), die sinking machine.

machine à mortaiser (*n.f.*). *Voir* mor-taiseuse (*n.f.*).

machine à mouler en coquilles (*n.f.*), die-casting machine.

machine à mouler par compression (plast.) (*n.f.*), compression moulding machine.

machine à percer sur colonne (*n.f.*), column-type drilling machine.

machine à pointer (*n.f.*), jib boring machine.

machine à raboter (*n.f.*). *Voir* raboteuse (*n.f.*).

machine à rectifier à meule à 90° (*n.f.*), face grinder.

machine à régler le sol (trav. pub.) (*n.f.*), grader; road grader.

machine à rétreindre (*n.f.*), swagging machine.

machine à roder (avec pierres abrasives rota-tives) (*n.f.*), honing machine (with rotating abrasive stones).

machine à sabler (*n.f.*), sand blasting machine.

machine à shaver les engrenages (*n.f.*), gear shaving machine.

machine à souder à la molette (*n.f.*), seam welding machine.

machine à souder par points (*n.f.*), spot welding machine.

machine à souder par rapprochement (*n.f.*), butt welding machine.

machine à souffler les noyaux (*n.f.*), core blowing machine.

machine à tailler les crémaillères (*n.f.*), rack cutting machine.

machine à tailler les engrenages par couteau (*n.f.*), gear shaping machine.

machine à tailler les engrenages par fraise-disque (*n.f.*), gear milling machine.

machine à tailler les engrenages par fraise-mère (*n.f.*), gear hobbing machine.

machine à tréfiler (*n.f.*), wire-drawing machine.

machine pour atmosphère grisouteuse (*n.f.*), fire-damp proof machine.

machine transfert (*n.f.*), transfer machine.

machine ventilée en circuit fermé (*n.f.*),

machine with closed circuit ventilation.

machine ventilée en circuit ouvert (*n.f.*), machine with open-circuit ventilation.

machmètre (aviation) (*n.m.*), machmeter.

machmètre audible (aviation) (*n.m.*), audible machmeter.

magasin (chargeur) (*n.m.*), magazine.

magasin d'alimentation (machines-outils) (*n.m.*), hopper.

magasin en douanes (*n.m.*), bonded warehouse.

magma (soudure) (*n.m.*), puddle.

magnesyn (*n.m.*), magnesyn.

magnétiser (*v.t.*), to magnetize.

magnétophone (*n.m.*), magnetic recorder.

magnétron (*n.m.*), magnetron.

magnétron à anode à segments multi- -ples (*n.m.*), multi-segment magnetron.

magnétron à anode fendue (*n.m.*), split- anode magnetron.

magnétron à cavité (*n.m.*), cavity magnetron.

maillage (aviation) (*n.m.*), riveting pattern.

maille de structure (*n.f.*), area between rivet lines.

maillechort (*n.m.*), argentan.

maillet de cuir (*n.m.*), hide faced mallet.

main d'essieu (auto) (*n.f.*), axle casing.

main d'oeuvre (*n.f.*), manpower.

main de ressort (auto) (*n.f.*), spring bracket.

maintenabilité (*n.f.*), maintainability.

maintenance (*n.f.*), maintenance.

maintener (trav. pub.) (*n.m.*), maintener.

maintien (électricité) (*n.m.*), maintenance; secur- -ing device; holder.

maintien à température (*n.m.*), soaking.

maître-cylindre (frein) (*n.m.*), master-cylinder.

malléable (*adj.*), ductile.

management (*n.m.*), management.

manche à air (*n.f.*), air scoop.

manche à balai (aviation) (*n.m.*), control column.

manche cyclique (hélicoptère) (*n.m.*), cyclic pitch stick.

manche d'accouplement (*n.m.*), coupling sleeve.

manche d'aspiration d'air (*n.f.*), air duct.

manche d'évacuation (aviation) (*n.f.*), escape chute.

manche de commande de pas cyclique (hélicoptère) (*n.m.*), cyclic pitch control stick.

manche de pas général (hélicoptère) (*n.m.*), collective pitch lever.

manche en butée (hélicoptère) (*n.m.*), stick limit position.

manche treuilliste (hélicoptère) (*n.m.*), moist lever.

manchon coulissant (*n.m.*), sliding sleeve.

manchon de pale (hélicoptère) (*n.m.*), blade sleeve.

manchon de protection (*n.m.*), protection sleeve.

manchon de raccordement (*n.m.*), coupling sleeve.

manchon de remplissage (*n.m.*), filler neck.

manchon fileté (*n.m.*), sleeve nut.

manchon incombustible (*n.m.*), heatshield.

manchon repère (*n.m.*), identification sleeve.

mandrin à commande pneumatique (machines-outils) (*n.m.*), air-chuck.

mandrin à mors (machines-outils) (*n.m.*), jaw- chuck.

mandrin de serrage élastique (*n.m.*), draw- in spring chuck.

mandrin magnétique circulaire (*n.m.*), cir- -cular magnetic chuck.

mandrin pince-barre (machines-outils) (*n.m.*), bar-chuck; collet-chuck.

mandrin pneumatique (*n.m.*), air-operated chuck.

mandrin type à genouillères (machines- outils) (*n.m.*), toggle chuck.

mandrin type rentrant (machines-outils) (*n.m.*), draw-in chuck.

mandrin type sans retrait (machines-outils) (*n.m.*), dead-length-type chuck.

maniabilité (*n.f.*), controllability; handling; manoeuvrability.

manipulateur (radio) (*n.m.*), morse key.

manipulation (radio) (*n.f.*), keying.

manocontact (élec.) (*n.m.*), pressure switch.

manocontact différentiel (*n.m.*), differential pressure switch.

manocontacteur (élec.) (*n.m.*), pressure switch.

manocontacteur anémométrique (*n.m.*), ram pressure switch.

mano-détendeur (*n.m.*), pressure reducing valve.

manoeuvrabilité (aviation) (*n.f.*), maniability.

manoeuvre (*n.f.*), manoeuvre.

manoeuvre au sol (*n.f.*), ground operation.

manoeuvre en vol (*n.f.*), flight manoeuvre.

manoeuvres transparentes (aviation) (*n.f.pl.*), limited authority manoeuvres.

manomètre (*n.m.*), pressure indicator.

manomètre à eau (*n.m.*), water gauge.

manomètre double de pression kérosène (aviation) (*n.m.*), dual fuel pressure gauge.

manomètre triple (*n.m.*), triple reading gauge.

mantisse (*n.f.*), mantissa.

manuel d'entretien (*n.m.*), maintenance manual.

manuel d'instruction (*n.m.*), instruction manual.

manuel d'utilisation (*n.m.*), operating instructions.

manuel de l'équipage (aviation) (*n.m.*), crew operating manual.

manuel de révision (*n.m.*), overhaul manual.

manuel de schémas de câblage (*n.m.*), wir- -ing diagram manual.

manuel de vol (aviation) (*n.m.*), flight manual.

manuel des études (*n.m.*), engineering standards.

maquette (*n.f.*), mock-up; model.

maquette d'entraînement au sol (aviation) (*n.f.*), link trainer.

marbré, -e (*adj.*), veined.

marche, en (*adj.*), operating.

marche à suivre (*n.f.*), procedure.

marche à vide (*n.f.*), no load operation; off-load operation.

marche altimétrique (aviation) (*n.f.*), fixed error on radio altimeter.

marche arrêt (*adj.*), "on"-"off".

marchepied (auto) (*n.m.*), footstep; running board.

marge de sécurité (*n.f.*), safety margin.

marouflage (*n.m.*), taping.

marquage (*n.m.*), marking.

marquage au burin (*n.m.*), chisel marking.

marquage au poinçon (*n.m.*), punch marking.

marque (*n.f.*), make; model; stamp.

marque déposée (*n.f.*), trade mark.
marquer (*v.t.*), to stamp.
marqueur d'étalonnage (radar) (*n.m.*), range mark.
marronnage (trav. pub.) (*n.m.*), crazing; crocodiling.
marteau (mécanisme) (*n.m.*), striker.
marteau à chûte libre (*n.m.*), drop hammer.
marteau à mater (*n.m.*), caulking hammer; riveting hammer.
marteau-batte (*n.m.*), beating hammer.
marteau postillon (*n.m.*), creasing hammer; seam hammer.
marteau-riveur (*n.m.*), rivet gun.
martelage (*n.m.*), peening.
martinet (*n.m.*), power hammer.
masque (barrage) (*n.m.*), facing.
masque (à porter) (*n.m.*), mask.
masque à oxygène (*n.m.*), oxygen mask.
masque de soudure (*n.m.*), face shield.
masquer (*v.t.*), to blank; to mask.
masse (électricité) (*n.f.*), earth; ground.
masse (poids) (*n.f.*), bulk; weight.
masse à l'atterrissage (aviation) (*n.f.*), land-ing weight.
masse à sec (*n.f.*), dry weight.
masse à vide (*n.f.*), empty weight.
masse au cheval (*n.f.*), power loading; weight per horsepower.
masse au décollage (aviation) (*n.f.*), take-off weight.
masse brute (*n.f.*), gross weight.
masse brute initiale (*n.f.*), initial gross weight.
masse d'équilibrage (aviation) (*n.f.*), balance weight.
masse d'équilibrage de pale (aviation) (*n.f.*), blade balance weight.
masse de calcul (*n.f.*), design weight.
masse de calcul à l'atterrissage (aviation) (*n.f.*), design landing weight.
masse de calcul de l'avion en vol (avia-tion) (*n.f.*), design flight weight.
masse de calcul de roulage (aviation) (*n.f.*), design taxi weight.
masse de l'avion (*n.f.*), aircraft weight.
masse de tiérçage (aviation) (*n.f.*), cable weight.
masse et centrage (*n.m.*), weight and balance.
masse flottille (*n.f.*), fleet weight.
masse garantie (*n.f.*), guaranteed weight.
masse maximale (*n.f.*), gross weight.
masse polaire (*n.f.*), pole piece.
masse polaire mobile (*n.f.*), mobile pole core.
masse spécifique (*n.f.*), density; specific mass; specific weight.
masse totale (aviation) (*n.f.*), certification weight.
masse volumétrique (*n.f.*), density.
masselotte (*n.f.*), flyweight.
masselotte (fonderie) (*n.f.*), runner.
mastic (*n.m.*), cement; putty; chemical filler.
mastic d'étanchéité (*n.m.*), sealing compound.
mât de compensation (aviation) (*n.m.*), rudder balancing arm.
mât réacteur (*n.m.*), engine nacelle stub.
mât rotor (hélicoptère) (*n.m.*), main rotor shaft.
matage (*n.m.*), peening.
maté, -e (*adj.*), peened.
matelas d'air (*n.m.*), air cushion.
matelas d'étanchéité (*n.m.*), sealing pad.
matelas d'isolation (*n.m.*), insulation blanket.

matelassage (*n.m.*), padding.
matériau de remplissage (*n.m.*), core; filler.
matière interstellaire (*n.f.*), cosmic dust.
matière plastique (*n.f.*), plastic.
matriçage (*n.m.*), die forging.
matricé, -e (*adj.*), drop forged.
maturation (*n.f.*), ageing.
mauvais, -e (*adj.*), faulty.
mauvais fonctionnement (*n.m.*), faulty oper--ation; malfunctioning.
maximum d'une courbe (*n.m.*), peak of a curve.
mazout (*n.m.*), fuel-oil.
mécanicien navigant (*n.m.*), flight engineer.
mécanique ondulatoire (*n.f.*), wave mechanics.
mécanisme de roue libre (*n.m.*), free wheel mechanism.
mèche de déperditeur (*n.f.*), static discharger wick.
méga- (préfixe), mega-.
mégohmmètre (*n.m.*), megger.
mélange (*n.m.*), compound.
mélange pauvre (carburant) (*n.m.*), lean mixture.
mélange riche (carburant) (*n.m.*), rich mixture.
mélangeage sur chantier (trav. pub.) (*n.m.*), mix in-place.
mélanine (plast.) (*n.f.*), melanine formaldehyde resin.
membrane (pompe) (*n.f.*), diaphragm.
membrane à cardan (*n.f.*), universal joint.
membre (*n.m.*), member.
membrure (*n.f.*), member.
membrure de triangulation (*n.f.*), brace strut.
mémoire (ordinateur) (*n.f.*), storage.
mémoire, pour, for reference.
mémoires de contrôle (*n.m.pl.*), acceptance test sheets.
mendélévium (élément) at.) (*n.m.*), mendelevium.
ménisque (*n.m.*), meniscus.
mentionnet (*n.m.*), stop of a rotating part.
mentonnet (*n.m.*), wiper.
méson (*n.m.*), meson.
mésopause (couche de l'atmosphère) (*n.f.*), mesopause.
mésosphère (couche de l'atmosphère) (*n.f.*), mesosphere.
message publicitaire (audio-visuel) (*n.m.*), spot.
métacentre (*n.m.*), metacentre.
métal centrifugé (*n.m.*), spun metal.
métal d'apport (soudure) (*n.m.*), filler metal.
métal de base (*n.m.*), base metal.
métal déployé (*n.m.*), expanded metal.
métal non-ferreux (*n.m.*), non-ferrous metal.
métallisation (*n.f.*), bonding; metal coating.
métallisation sous vide (*n.f.*), vacuum metal coating.
métallisé, -e (*adj.*), metal coated.
métallo-plastique (*adj.*), copper-asbestos; metal-plastic.
métalloscope (*n.m.*), magnetic crack detector.
méthode (*n.f.*), procedure.
méthode empirique (*n.f.*), rule of thumb; trial and error method.
méthyl-éthylcétone (*n.m.*), methyl-ethyl-ketone.
mettre (*v.t.*), to lay; to place; to set.

mettre à jour (*v.t.*), to bring up to date.
mettre à la masse (élec.) (*v.t.*), to earth; to ground.
mettre au point (un projet) (*v.t.*), to develop.
mettre en circuit (élec.) (*v.t.*), to switch on; to turn on.
mettre en oeuvre (*v.t.*), to implement.
mettre en palier (aviation) (*v.t.*), to level out.
mettre en pression (*v.t.*), to pressurize; to set under pressure.
mettre hors circuit (élec.) (*v.t.*), to switch off; to turn off.
mettre hors service (*v.t.*), to de-activate.
mettre les gaz (aviation) (*v.*), to advance the throttle.
mettre sous tension (élec.) (*v.t.*), to apply the voltage; to energize.
meuble radio (*n.m.*), radio rack.
meule à tronçonner (*n.f.*), cutting-off wheel.
meule à un embrèvement (*n.f.*), recessed one side wheel.
meule abrasive (*n.f.*), abrasive wheel.
meule embrevée (*n.f.*), recessed wheel.
meule encrassée (*n.f.*), loaded wheel.
mi-course (*n.f.*), mid travel.
mi-dur, -e (*adj.*), medium hard.
micro (radio) (*n.m.*), microphone.
micro bouton-poussoir (*n.m.*), push microswitch.
micro-contact (élec.) (*n.m.*), microswitch.
micro masque (radio) (*n.m.*), mask microphone.
micro-micron (unité de longueur) (*n.m.*), micromicron (= 10^{-12} m.).
micro-onde (*n.f.*), microwave.
microfilm (*n.m.*), microfilm.
micromètre pneumatique (*n.m.*), air gauge.
microphone (*n.m.*), microphone.
microphone à ampli incorporé (radio) (*n.m.*), dynamic microphone with self-contained amplifier.
microphone à carbone (*n.m.*), carbon microphone.
microphone à main (*n.m.*), hand microphone.
micropompe (*n.f.*), micropump.
microrupteur (*n.m.*), microswitch.
microscope électronique (*n.m.*), electron microscope.
microsillon (disque) (*adj.*), long playing.
microsillon (*n.m.*), long playing record.
milieu ambiant (*n.m.*), environment.
mille marin (*n.m.*), nautical mile.
mille terrestre (*n.m.*), statute mile.
milliampèremètre (élec.) (*n.m.*), milliammeter.
millicron (unité de longueur) (*n.m.*), millicron; micronmillimetre (= 10^{-9}m.).
mini-contact (*n.m.*), microswitch.
mini-rupteur (*n.m.*), microswitch.
miniature (*adj.*), miniaturized.
miniaturiser (*v.t.*), to miniaturize.
minimum d'atterrissage (aviation) (*n.m.*), minimum weather conditions.
minuterie (élec.) (*n.f.*), time-switch.
mise au repos (élec.) (*n.f.*), drop out.
mise au travail (élec.) (*n.f.*), pick-up.
mise en cône (hélicoptère) (*n.f.*), coning.
mise en mémoire (*n.f.*), memorizing.
mise rapportée, à (outil) (*adj.*), tipped.
mise sur vérin (*n.f.*), jacking.
mixage (*n.m.*), mixing.
mixte (*adj.*), combined.

mobile axial (hélicoptère) (*n.m.*), axial compressor rotor.
mobile de compression (hélicoptère) (*n.m.*), compressor rotor.
mobilier (*n.m.*), furnishing; furniture.
mode d'emploi (*n.m.*), instructions (for use).
mode opératoire (*n.m.*), procedure.
mode vibratoire (*n.m.*), vibration mode.
modèle (*n.m.*), mark.
modification (*n.f.*), change.
modification facultative (*n.f.*), recommended modification.
modulaire (*adj.*), modular.
modulateur (*n.m.*), modulator.
modulation (*n.f.*), modulation.
modulation d'amplitude (*n.f.*), amplitude modulation.
module de Coulomb (module de rigidité) (*n.m.*), modulus of transverse elasticity; modulus of rigidity.
module de dent (*n.m.*), tooth pitch.
module de masse (*n.m.*), bulk modulus.
module de réglage (aviation) (*n.m.*), calibration module.
modulé, -e (radio) (*adj.*), modulated.
moduler (*v.t.*), to modulate.
moignon (*n.m.*), stub.
moins (prép.), less; minus.
mole (*n.f.*), gram-molecule.
molécule-gramme (*n.m.*), gram-molecule.
moleté, -e (*adj.*), knurled.
molette à dresser (par roulage) (*n.f.*), crusher.
molette à dresser (par taillage) (*n.f.*), dresser-cutter.
molikotage (*n.m.*), molibonding.
moment cabreur (aviation) (*n.m.*), stalling moment.
moment cinétique (*n.m.*), anular momentum.
moment d'amortissement (*n.m.*), damping moment.
moment d'inertie de la pale (hélicoptère) (*n.m.*), blade moment of inertia.
moment de battement (hélicoptère) (*n.m.*), flapping moment.
moment de charnière (hélicoptère) (*n.m.*), hinge moment.
moment de flexion (*n.m.*), bending moment.
moment de lacet (aviation) (*n.m.*), yawing moment.
moment de torsion de la pale (hélicoptère) (*n.m.*), blade twisting moment.
moment de traînée (aviation) (*n.m.*), drag moment.
monergol (fusée) (*n.m.*), monopropellant.
monophase (élec.) (*adj.*), single phase.
monorotor (hélicoptère) (*adj.*), single rotor.
montage (machines-outils) (*n.m.*), fixture.
montage de bridage (machines-outils) (*n.m.*), clamping fixture.
montage de rodage (machines-outils) (*n.m.*), lapping fixture.
montage de vérification (machines-outils) (*n.m.*), checking fixture.
montage mural (*n.m.*), wall mounting.
montage sur châssis (*n.m.*), rack mounting.
montage sur panneau (*n.m.*), panel mounting.
montage sur poteau (*n.m.*), pole mounting.
montant (raboteuse) (*n.m.*), housing.
monte-jus à l'acide (raffinerie) (*n.m.*), acid blow case.
monté (-e) en parallèle (*adj.*), parallel-

connected; (élec.) wired in parallel.

monté (-e) en série (*adj.*), series-connected.

montée (*n.f.*), climb.

montée en caisson (*n.f.*), chamber ascent.

montée en chandelle (aviation) (*n.f.*), zooming up.

montée verticale (aviation) (*n.f.*), vertical climb.

monter (*v.t.*), to assemble; to install.

monter en rattrapage (*v.t.*), to retrofit.

montre (*n.f.*), clock.

mordançage (*n.m.*), chromate finishing.

mors de serrage (*n.m.*), clamping tool.

mortaiseuse à bielle et manivelle (*n.f.*), crank-drive slotting machine.

mortaiseuse à crémaillère (*n.f.*), rack-drive slotting machine.

mortaiseuse à excentrique (*n.f.*), eccentric drive slotting machine.

morue (aviation) (*n.f.*), horizontal stabilizer centre fishplate.

moteur à bagues (élec.) (*n.m.*), slip-ring motor.

moteur à cage (élec.) (*n.m.*), squirrel-cage motor.

moteur à combustion interne (*n.m.*), compression ignition engine.

moteur à condensateur (élec.) (*n.m.*), capac--itor motor.

moteur à démarrage par self (élec.) (*n.m.*), reactor-start motor.

moteur à induit bobiné (élec.) (*n.m.*), wound rotor motor.

moteur à réaction (*n.m.*), jet engine; reaction engine.

moteur à répulsion à enroulement aux--iliaire (élec.) (*n.m.*), split phase motor.

moteur aérobie (*n.m.*), air breathing engine.

moteur asynchrone synchronisé (élec.) (*n.m.*), synchronous induction motor.

moteur couple (*n.m.*), torque motor.

moteur d'érection (*n.m.*), levelling motor.

moteur des volets de courbure (aviation) (*n.m.*), wing flap motor.

moteur hors-bord (*n.m.*), outboard motor.

moteur fractionnaire (*n.m.*), fractional h.p. motor.

moteur pneumatique (*n.m.*), pneumatic motor; air motor.

moteur primaire (*n.m.*), prime mover.

motoriste (*n.m.*), engine manufacturer; car driver.

moufle (*n.f.*), chain block; hoisting block.

mouillage de mines (marine) (*n.m.*), mine laying.

mouillant (*n.m.*), wetting agent.

moulage à la barbotine (*n.m.*), slip casting.

moulage à la cire perdue (*n.m.*), investment casting.

moulage centrifuge (*n.m.*), centrifugal casting.

moulage par compression (plast.) (*n.m.*), compression moulding.

moulage par formage (plast.) (*n.m.*), pulp preforming.

moulage par injection (plast.) (*n.m.*), injection moulding.

moulage par transfert (plast.) (*n.m.*), transfer moulding.

moule (*n.m.*), die; matrix.

moule à empreintes multiples (plast.) (*n.m.*), multi-impression mould.

moule de transfert (plast.) (*n.m.*), transfer mould.

moulé, -e (*adj.*), cast; moulded.

moulinet (*n.m.*), spinner.

moulinet, fonctionnement en (hélicoptère) (*n.m.*), windmilling.

moulure (*n.f.*), trimming.

mousqueton (*n.m.*), snap hook.

mouvement de traînée (hélicoptère) (*n.m.*), hunting; lagging.

mouvement ondulatoire (*n.m.*), wave motion.

mouvement pendulaire (*n.m.*), simple har--monic motion.

mouvement sinusoïdal (*n.m.*), simple har--monic motion.

moyen courrier (aviation) (*n.m.*), medium range airliner.

moyenne fréquence (radio) (*n.f.*), intermediate frequency.

moyenne quadratique (*n.f.*), root mean square value (R.M.S.).

moyeu à charnière (*n.m.*), door hinge hub.

moyeu de rotor arrière (hélicoptère) (*n.m.*), tail rotor hub.

moyeu de rotor principal (hélicoptère) (*n.m.*), main rotor hub.

moyeu flexible (*n.m.*), flex hub.

moyeu flottant (*n.m.*), free floating hub.

multibroche (élec.) (*adj.*), multi-pin.

multifilaire (élec.) (*adj.*), multi-wire.

multilisse (*n.m.*), multistringer.

multimètre (*n.m.*), multimeter.

multiplicande (*adj.*), multiplicand.

multiplicateur (radio) (*n.m.*), amplifier.

multiplicateur d'électrons (*n.m.*), electron multiplier.

mur de la chaleur (*n.m.*), thermal barrier.

mur du son (*n.m.*), sound barrier.

mur thermique (*n.m.*), thermal barrier.

myria- (préfixe), myria- (= 10^4).

N

nacelle (aviation) (*n.f.*), nacelle.

nacelle amovible (hélicoptère) (*n.f.*), detachable pod.

nageoires (*n.f.pl.*), sponsons.

nappe de rayons (dans une roue à rayons) (*n.f.*), row of spokes.

narrateur tactique (aviation) (*n.m.*), tactical plotter.

navette (hélicoptère) (*adj.*), shuttle.

navette (hélicoptère) (*n.f.*), shuttle service helicopter.

navette (*n.f.*), shuttle valve.

navigabilité (*n.f.*), airworthiness.

navigation à l'estime (marine) (*n.f.*), dead reckoning.

navire-citerne (pétrole) (*n.m.*), tanker.

nébulisation (*n.f.*), fogging.
nébulosité (*n.f.*), cloudiness.
négatron (élec.) (*n.m.*), negatron.
neige carbonique (*n.f.*), carbonic acid snow.
néoprène (*n.m.*), neoprene.
néphoscope (météor.) (*n.m.*), nephoscope.
neptunium (élément) (*n.m.*), neptunium.
nervure courante (aviation) (*n.f.*), secondary rib.
nervure d'encastrement (aviation) (*n.f.*), root rib.
nervure d'extrémité (aviation) (*n.f.*), end rib.
nervure de bec de volet (aviation) (*n.f.*), wing flap nose rib.
nervure de fermeture de chapelle (avia-tion) (*n.f.*), fin stub top rib.
nervure de rive (aviation) (*n.f.*), end rib.
nervure en arc (aviation) (*n.f.*), crescent shaped rib.
nervure en treillis (aviation) (*n.f.*), lattice rib.
nervure étanche (aviation) (*n.f.*), sealed rib.
nervure forte (aviation) (*n.f.*), main rib.
nervure glissière (aviation) (*n.f.*), flap track rib.
nervuré, -e (*adj.*), webbed.
nervures non ventilées, à (*adj.*), ribbed surface.
nervures ventilées, à (*adj.*), ventilated ribbed surface.
neutralisation (*n.f.*), neutralization.
neutraliser (*v.t.*), to de-activate.
neutron (*n.m.*), neutron.
neutron lent (*n.m.*), slow neutron.
neutron rapide (*n.m.*), fast neutron.
névasse (*n.f.*), slush.
nez d'arbre (*n.m.*), extension shaft.
niche (*n.f.*), recess.
niche d'aspiration (conditionnement d'air) (*n.f.*), suction box.
nid d'abeilles (*n.m.*), honeycomb.
nid de poule (trav. pub.) (*n.m.*), pot hole (in roads).
nit (unité de luminance) (*n.m.*), nit.
nitomètre-luxmètre (*n.m.*), nitometer-luxmeter.
nitrate de cellulose (plast.) (*n.m.*), cellulose nitrate.
nitruration (*n.f.*), nitriding.
niveau d'artillerie (*n.m.*), clinometer.
niveau d'entrée (radio) (*n.m.*), input level.
niveau de bruit (*n.m.*), noise level.
niveau de bruit perçu (*n.m.*), recorded noise decibels.
niveau de la mer (*n.m.*), sea level.
niveau de retenue (barrage) (*n.m.*), top water level.

niveau de sortie (radio) (*n.m.*), output level.
niveau énergétique (électron.) (*n.m.*), energy level of a particle.
niveau interfacial (*n.m.*), interface level.
niveau sonore (*n.m.*), sound level.
niveau Vinot (aviation) (*n.m.*), side-slip indicator.
niveleuse (trav. pub.) (*n.f.*), levelling machine; grader; drag-line; motor-grader.
nocif, -ive (*adj.*), noxious.
nodulation (*n.f.*), coring.
noeud (marine) (*n.m.*) knot (1 nautical mile per hour).
noeud (élec.) (*n.m.*), node.
noeud (soudure) (*n.m.*), cluster; junction.
noir de fumée (*n.m.*), lamp-black.
noix d'articulation (*n.f.*), hinge yoke.
noix de cardan (*n.f.*), yoke.
noix de démarreur (*n.f.*), starter jaw.
noix de serrage (*n.f.*), tightening yoke.
nombre (*n.m.*), figure.
nombre d'impulsions (électronique) (*n.m.*), pulse repetition frequency.
nombre de Mach (*n.m.*), Mach number.
nombre de Reynolds (*n.m.*), Reynolds number.
nombre quantique interne (*n.m.*), total angular momentum quantum number.
nombre quantique principal (*n.m.*), main quantum number.
nombre quantique secondaire (*n.m.*), orbital quantum number.
nombres quantiques d'un atome (*n.m.pl.*), quantum numbers of an atom.
nominal, -e (*adj.*), indicated; rated.
nomogramme (*n.m.*), nomogram.
non-congelable (*adj.*), anti-freezing.
non-fonctionnement (*n.m.*), failure.
non-pelucheux, -euse (*adj.*), lint free.
non-travaillant, -e (*adj.*), unstressed.
normal, -e (*adj.*), perpendicular.
normalisation (*n.f.*), standardization.
norme (*n.f.*), specification; standard.
norme britannique (*n.f.*), British standard specification (B.S.S.).
normograph (*n.m.*), graph stencil.
nota (*n.m.*), note; remark.
notice (*n.f.*), notice.
notice technique (*n.f.*), technical manual.
nourrice (aviation) (*n.f.*), feeder tank.
nourrice de secours (aviation) (*n.f.*), emer-gency supply tank.
noyau (planification) (*n.m.*), core.
noyau atomique (*n.m.*), atomic nucleus.
noyau magnétique (*n.m.*), magnetic core.
nucléaire (*adj.*), nuclear.
nucléon (*n.m.*), nucleon.
numéroté, -e (*adj.*), numbered.

O

obscurcisseur de voyant (*n.m.*), dimmer.
observateur (aviation) (*n.m.*), observer.
obstruer (*v.t.*), to clog.
obturateur (pétrole) (*n.m.*), blow-out preventer.
obturateur d'entrée d'air dérivée (*n.m.*), blanking cover for fin air scoop.
obturateur d'évacuation air refroidis-seur (*n.m.*), blanking cover for air cooling unit outlet.
obturateur de prise statique (aviation) (*n.m.*), static port stopper.
obturateur partiel (*n.m.*), partial shutter.
obturer (*v.t.*), to blank off; to close up.
obus (outil) (*n.m.*), thimble.

octa (unité de mesure de la nébulosité) (*n.m.*), okta.
octane (*n.m.*), octane.
octode (*n.f.*), octode.
oeil à queue filetée (*n.m.*), eyebolt.
oeil de câble (*n.m.*), thimble.
oeil de levage (*n.m.*), hoisting eye.
oeil de tendeur (*n.m.*), turnbuckle eye.
oeil magnétique (radio) (*n.m.*), electron-ray tube.
ogive de guidage (outil) (*n.f.*), guide bush.
ogive de montant (*n.f.*), post staff.
ohm réciproque (unité de conductance) (*n.m.*), reciprocal ohm.
oléosoluble (plast.) (*adj.*), oil-soluble.
oligo-élément (*n.m.*), trace element.
olive (*n.f.*), olive.
olive de butée (*n.f.*), stop olive.
ombre aérodynamique (*n.f.*), blanking effect.
onde amortie (radio) (*n.f.*), damped wave.
onde carrée (radio) (*n.f.*), square wave.
onde courte (radio) (*n.f.*), short wave.
onde de choc (*n.f.*), shock wave.
onde de choc amont (*n.f.*), bow wave.
onde de choc aval (*n.f.*), tail wave.
onde de jour (*n.f.*), day wave.
onde de nuit (*n.f.*), night wave.
onde de tension (élec.) (*n.f.*), voltage wave.
onde diurne (*n.f.*), day wave.
onde électromagnétique (*n.f.*), electromag
-netic wave.
onde entretenue (*n.f.*), sustained wave.
onde ionosphérique (*n.f.*), sky wave; space wave.
onde moyenne (radio) (*n.f.*), standard waveband.
onde nocturne (*n.f.*), night wave.
onde plane (élec.) (*n.f.*), plane wave.
onde polarisée horizontalement (*n.f.*), hor
-izontally polarized wave.
onde porteuse (radio) (*n.f.*), carrier wave.
onde porteuse intermédiaire (*n.f.*), subcarrier.
onde stationnaire (*n.f.*), standing wave.
ondemètre (radio) (*n.m.*), wavelength meter.
ondemètre à absorption (*n.m.*), absorption frequency meter.
ondulateur (élec.) (*n.m.*), inverter.
ondulation de courant (*n.f.*), current ripple.
ondulation de surface (*n.f.*), waviness of surface.
ondulé, -e (*adj.*), channelled; corrugated.
onglet (*n.m.*), tab marker.
opaque (*adj.*), non-transparent.
opérations mineures et majeures (*n.f.pl.*), minor and major servicing operations.
ordinateur (*n.m.*), computer.
ordinogramme (*n.m.*), flow chart.
ordonnée (*n.f.*), Y-axis.
ordre (*n.m.*), sequence; signal.
ordre de fabrication (*n.m.*), work order.
ordre de phase (*n.m.*), phase sequence.
ordre inverse (*n.f.*), reverse order.
organe (*n.m.*), component.
organigramme (*n.m.*), organization chart.
organisme (*n.m.*), agency; organization.
organisme officiel (*n.m.*), authority.
orientable (*adj.*), swivelling.
orientation (*n.f.*), direction; turning; heading.
orientation train avant (aviation) (*n.f.*), nosewheel steering.
orienter (*v.t.*), to direct; to position; to steer.
orifice calibré (*n.m.*), metering hole.

orifice de mise à l'air libre (aviation) (*n.m.*), vent hole.
orifice de raccordement (*n.m.*), connection port.
orifice de retour (*n.m.*), spill orifice.
oscillateur (*n.m.*), oscillator.
oscillateur à battement de fréquence (*n.m.*), beat frequency oscillator.
oscillateur à quartz (*n.m.*), crystal oscillator.
oscillateur de balayage (*n.m.*), sweep oscillator.
oscillation (*n.f.*), pendulum motion; vibration.
oscillation amortie (*n.f.*), stable oscillation.
oscillation amplifiée (*n.f.*), instable oscillation.
oscillation d'incidence (*n.f.*), incidence oscillation.
oscillation de lacet (aviation) (*n.f.*), snaking.
oscillation de rotation (aviation) (*n.f.*), rotary oscillation.
oscillation en dents de scie (radio) (*n.f.*), saw-tooth wave.
oscillation entretenue (*n.f.*), sustained oscillation.
oscillation phugoïde (*n.f.*), phugoid oscillation.
oscillation propre (*n.f.*), natural oscillation.
oscilloscope cathodique (*n.f.*), cathode ray oscilloscope.
ossature (aviation) (*n.f.*), structure.
ouate (*n.f.*), cotton wool.
ouïe d'entrée d'air (*n.f.*), air scoop.
ouïes de capot (auto) (*n.f.pl.*), louvres.
outil à aléser, section carrée (*n.m.*), square shank boring tool.
outil à charioter (*n.m.*), light roughing tool; bar turning tool.
outil à charioter et à dresser (*n.m.*), turning and facing tool.
outil à dresser (*n.m.*), inside corner tool.
outil à dresser d'angle (*n.m.*), cranked fin
-ishing tool.
outil à dresser les faces (*n.m.*), facing tool.
outil à dresser les faces, dégrossissage (*n.m.*), facing tool for roughing.
outil à saigner (*n.m.*), parting tool.
outil à tronçonner pour porte-lame (*n.m.*), parting-off blade for tool-holder.
outil au carbure (*n.m.*), carbide tool.
outil au diamant (*n.m.*), diamond tool.
outil coudé à charioter (*n.m.*), cranked turn
-ing and facing tool.
outil coudé à charioter arrondi (*n.m.*), cranked round nose turning tool.
outil coudé à dresser les faces (*n.m.*), cranked facing tool.
outil couteau (*n.m.*), knife tool.
outil couteau coudé (*n.m.*), cranked knife tool.
**outil couteau coudé pour tours à repro
-duire** (*n.m.*), cranked knife tool for copy turning.
outil d'aléseuse à dégrossir (*n.m.*), boring tool for roughing.
outil d'évasement de tube (*n.m.*), tube flaring tool.
outil de coupe (*n.m.*), cutting tool.
outil de finition droit (*n.m.*), straight finishing tool.
outil de tour à reproduire (*n.m.*), copying lathe tool.
outil de tour tangentiel (*n.m.*), tangential turning tool.

outil droit (*n.m.*), straight tool.
outil droit à charioter (*n.m.*), bar turning tool.
outil droit à charioter, arrondi (*n.m.*), straight round-nosed turning tool.
outil pelle (*n.m.*), recessing tool.
outil pour charioter, dresser et chan-freiner (*n.m.*), bar turning, facing and chamfering tool.
outil pour gorges de courroies trapézoï-dales (*n.m.*), pulley grooving tool.
outil sérié (*n.m.*), gang tool.
outil spécial (*n.m.*), special tool.
outillage d'étalonnage (*n.m.*), calibration tool.
outillage de fixation (*n.m.*), holding fixture.
outillage de sertissage (*n.m.*), crimping tool.
ouvert, -e (construction de machines) (*adj.*), open-type.

ouverture d'un circuit (élec.) (*n.f.*), breaking of a circuit.
overture du faisceau d'antenne (radio) (*n.f.*), aerial beam width.
ouverture du moule (presse) (*n.f.*), daylight.
ouverture quart de tour (*n.f.*), quarter turn open type.
ouvrabilité (*n.f.*), workability.
ovalisation (*n.f.*), elongation.
ovalisé, -e (*adj.*), elongated; out of true.
oxy-coupage (*n.m.*), gas cutting; oxygen-arc cutting.
oxy-coupage automatique (*n.m.*), automatic oxygen cutting.
oxydation anodique (*n.f.*), anodizing; anodic oxidation.
oxydé, -e (*adj.*), oxided; oxidized.
oxygénation (*n.f.*), passivation.
ozonation (*n.f.*), ozonation.

P

pair (nombre) (*adj.*), even.
paladium (*n.m.*), paladium.
pale (hélicoptère) (*n.f.*), blade.
pale articulée (hélicoptère) (*n.f.*), articulated blade.
pale avançante (hélicoptère) (*n.f.*), advancing blade.
pale décalée (hélicoptère) (*n.f.*), offset blade.
pale principale (hélicoptère) (*n.f.*), main rotor blade.
pale reculante (hélicoptère) (*n.f.*), retreating blade.
pale repliable (hélicoptère) (*n.f.*), folding blade.
pale rigide (hélicoptère) (*n.f.*), rigid blade.
pale traînante (hélicoptère) (*n.f.*), hunting blade.
palet d'accouplement (*n.m.*), coupling buffer.
palette (man. méc.) (*n.f.*), pallet; stillage.
palette à ressort (*n.f.*), spring flap.
palette caisse (*n.f.*), box pallet.
palette de chargement (*n.f.*), pallet.
palette de contrôle optique (aviation) (*n.f.*), landing gear downlock optical inspection flap.
palette de manoeuvre (aviation) (*n.f.*), actuating plate.
palier (*n.m.*), bearing surface.
palier (aviation) (*n.m.*), level flight.
palier à alignement (*n.m.*), align reaming box.
palier à bague (*n.m.*), sleeve bearing.
palier à serrage (*n.m.*), friction-type bearing.
palier d'essai (*n.m.*), test point.
palier de guidage (*n.m.*), flanged guide.
palier de tête de bielle (*n.m.*), big end bearing.
palier lisse (*n.m.*), journal bearing.
palier réacteur (aviation) (*n.m.*), engine shaft bearing.
palonnier (*n.m.*), cross bar.
palonnier (hélicoptère) (*n.m.*), tail rotor control pedals.
palonnier (commandes de vol) (*n.m.*), rudder bar.
palonnier de mise en place reverse (avia-tion) (*n.m.*), thrust reverser hoist.
palonnier orientation train avant (avia-tion) (*n.m.*), nose gear steering base post.

palpeur (raboteuse) (*n.m.*), follower.
pan (d'une clé) (*n.m.*), spanner inner spline.
panne (*n.f.*), failure.
panne d'un fer à souder (*n.f.*), soldering bit.
panne franche (*n.f.*), straight failure.
panne intermittente (*n.f.*), intermittent failure.
panne permanente (*n.f.*), permanent failure.
panne subite (*n.f.*), sudden failure.
panneau alvéolé (*n.m.*), waffle panel.
panneau arrière (*n.m.*), back-end plate; rear panel.
panneau avant (*n.m.*), front panel.
panneau central (*n.m.*), centre panel.
panneau d'accès (*n.m.*), access panel.
panneau d'extrémité (*n.m.*), end panel.
panneau d'insonorisation (*n.m.*), sound-proofing panel.
panneau de bordure (*n.m.*), edging panel.
panneau de commande (*n.m.*), control panel.
panneau de custode (auto) (*n.m.*), quarter panel.
panneau de fermeture (*n.m.*), closing panel.
panneau de plafond (*n.m.*), overhead panel.
panneau de plancher (*n.m.*), floor panel.
panneau de voûte (*n.m.*), arch panel.
panneau latéral (*n.m.*), side panel.
panneau mobile (*n.m.*), sliding panel.
panneau ouvrant (*n.m.*), hinged panel.
panneau pare-gouttes (machines-outils) (*n.m.*), splash guard.
panneau service (*n.m.*), servicing panel.
papier bulle (*n.m.*), manila paper.
papier d'aluminium (*n.m.*), aluminium foil.
papier de soie (*n.m.*), tissue paper.
papier indicateur (*n.m.*), litmus paper.
papier millimétré (*n.m.*), graph paper.
papier paraffiné (*n.m.*), waxed paper.
papier sulfurisé (*n.m.*), greaseproof paper.
parachor (chimie) (*n.m.*), parachor.
parachute de queue (aviation) (*n.m.*), drag chute.
parafoudre à peigne (élec.) (*n.m.*), gap arrester.

parafouille (barrage) (*n.m.*), cut-off trench; cut-off wall.

parallélisme des roues (auto) (*n.m.*), wheel alignment.

paramètre d'avancement (aviation) (*n.m.*), tip speed ratio.

paramètre de flux axial (aviation) (*n.m.*), inflow ratio.

paramètre de glissement (aviation) (*n.m.*), tip speed ratio.

paramètre de masse de l'avion (*n.m.*), air-plane mass ratio.

paramètre de traînée rotor (*n.m.*), rotor drag parameter.

parasite (*adj.*), spurious.

parasite (radio) (*n.m.*), interference; noise; statics.

parasites atmosphériques (radio) (*n.m.*), atmospherics.

parcours (*n.m.*), course; route.

parcours à l'atterrissage (aviation) (*n.m.*), landing run.

parcours au décollage (aviation) (*n.m.*), take-off run.

pare-brise (auto) (*n.m.*), windscreen (G.B.); windshield (U.S.A.).

pare-choc (auto) (*n.m.*), bumper.

pare-étincelles (machines-outils) (*n.m.*), spark guard.

pare-gouttes (*n.m.*), splash-guard.

pare-soleil (*n.m.*), glare shield.

parkérisation (*n.f.*), parkerizing.

paroi de calorifugeage (*n.f.*), heat insulating wall.

paroi latérale (*n.f.*), side-wall.

parsec (unité astronomique de longueur) (*n.m.*), parsec (= 3,26 années-lumières = 30,8 × 10^{12} km. = 19,16 × 10^{12} miles = 206,265 astro-nomical units).

particule alpha (atome) (*n.f.*), alpha particle.

particule béta (atome) (*n.f.*), beta particle.

partie chaude (d'un réacteur) (*n.f.*), hot section (of an engine).

partie froide (d'un réacteur) (*n.f.*), cold section (of an engine).

partie lisse (d'une visse) (*n.f.*), plain length (of a screw).

partie non étanche d'un fuselage (aviation) (*n.f.*), fuselage non-pressurized section.

partiel, -ielle (*adj.*), fractional.

pas collectif (hélicoptère) (*n.m.*), collective pitch.

pas cyclique (hélicoptère) (*n.m.*), cyclic pitch.

pas cyclique latéral (hélicopre) (*n.m.*), lat-eral cyclic pitch.

pas cyclique longitudinal (hélicoptère) (*n.m.*), fore-and-aft cyclic pitch.

pas de dent (*n.m.*), tooth pitch.

pas de freinage (hélicoptère) (*n.m.*), braking pitch.

pas diamétral (*n.m.*), diametral pitch.

pas du rotor arrière (hélicoptère) (*n.m.*), tail rotor pitch.

pas général (hélicoptère) (*n.m.*), collective pitch.

pas géométrique (*n.m.*), geometric pitch.

pas large (visserie) (*n.m.*), coarse thread.

pas moyen (*n.m.*), mean pitch angle.

pas multicyclique (hélicoptère) (*n.m.*), high order cyclic pitch.

pas négatif (hélicoptère) (*n.m.*), reverse pitch.

passage au vol en translation (aviation) (*n.m.*), transition to forward flight.

passage calibré (*n.m.*), metering hole.

passage dur (machines-outils) (*n.m.*), friction point.

passage en palier (aviation) (*n.m.*), transition to level flight.

passage machine (*n.m.*), machine run.

passant de courroie (*n.m.*), strap loop.

passe-bas (*n.m.*), low-pass.

passe de soutien (soudage) (*n.f.*), backing-pass.

passe-fil (*n.m.*), grommet.

passe-haut (*n.m.*), high-pass.

passe-poil (*n.m.*), edging.

passe sur l'envers (*n.f.*), back weld.

passer au bleu (*v.t.*), to blue.

passerelle de travail (*n.f.*), servicing platform.

passivation (*n.f.*), passivation.

pastille (plast.) (*n.f.*), pellet.

pastille d'écouteur (*n.f.*), earphone inset.

pastille de carbone (*n.f.*), carbon contact.

pastille de serrage (*n.f.*), tightening disc.

pastille de positionnement (*n.f.*), locating disc.

pastille radioactive (*n.f.*), luminescent nugget.

pâte à caractères (*n.f.*), typewriter type cleaner.

pâte à joint (*n.f.*), sealant; sealing compound.

pâte à roder (*n.f.*), grinding paste; lapping compound.

pâte à souder (*n.f.*), soldering flux.

patin (d'un frein) (*n.m.*), brake friction pad.

patin de sabot arrière (*n.m.*), rear bumper skid.

patinage (*n.m.*), skidding.

patte d'araignée (*n.f.*), oil groove.

patte de métallisation (*n.f.*), bonding tab.

pavillon (toit de voiture) (*n.m.*), top-roof.

pavillon (d'écouteur téléphonique) (*n.m.*), tele-phone earpiece.

pavillon microphone (téléphone) (*n.m.*), tele-phone mouthpiece.

pédale d'alternat (radio) (*n.f.*), press-to-talk bar.

pédale de commande de direction (avia-tion) (*n.f.*), rudder control pedal.

pédale de palonnier (aviation) (*n.f.*), rudder pedal.

peigne (*n.m.*), cleat.

peigne (rampe) (*n.m.*), rake.

peigne à fileter par roulage (*n.m.*), thread rolling die.

peinture (*n.f.*), paint.

peinture au pistolet (*n.f.*), spraying.

peinture cellulosique (*n.f.*), cellulose paint.

peinture-émulsion (*n.f.*), emulsion paint.

peinture frein (*n.f.*), screw stop varnish.

peinture fungicide (*n.f.*), fungicid paint.

peler une rondelle (*v.*), to peel the laminations off a shim washer.

pelle-décapeuse (trav. pub.) (*n.f.*), scraper.

pelleteuse (trav. pub.) (*n.f.*), power shovel.

pelleteuse-chargeuse (trav. pub.) (*n.f.*), load-ing shovel.

peluche (*n.f.*), lint.

pène de serrure (*n.m.*), locking bolt.

pente aérodynamique (*n.f.*), climb gradient.

pente de la piste (aviation) (*n.f.*), runway slope.

pente de montée (aviation) (*n.f.*), climb gradient.

pente normale (aviation) (*n.f.*), normal descent angle.

pente latérale (aviation) (n.f.), bank.
pente longitudinale (aviation) (n.f.), pitch.
pente radiogoniométrique (n.f.), glide slope.
pentode (n.f.), pentode.
perçage à la lance (n.m.), oxygen lancing.
perçage d'arrêt (n.m.), stop drilling.
perçage débouchant (n.m.), through hole.
percée (aviation) (n.f.), let-down.
percée en G.C.A. (aviation) (n.f.), ground con-trolled approach.
percée radio-compas (aviation) (n.f.), ADF letdown.
perchiste (cinéma) (n.m.), perchman.
percussion de bouteille (n.f.), extinguisher percussion.
percuteur (n.m.), striking pin.
percuteur d'extincteur (n.m.), extinguisher striker.
percuteur électrique (n.m.), initiator.
péremption (n.f.), cure date.
perfectionnement (n.m.), development.
perforateur (n.m.), punch.
perforation (élec.) (n.f.), puncture.
perforation en grille (n.f.), lace punching.
perforé, -e (adj.), perforated.
périmé, -e (adj.), obsolete.
périmètre de sécurité (d'un avion) (n.m.), danger area.
période (élec.) (n.f.), frequency.
période (sens général) (n.f.), cycle.
période (temps) (n.f.), period.
période de régulation (helicoptère) (n.f.), cycling.
période de stockage sans contrôle (n.f.), shelf life.
périodicité (n.f.), frequency; recurrence.
périphérie (n.f.), edge; rim.
perlé, -e (adj.), beaded.
permis, -e (adj.), allowed; authorized; permissible.
permittivité (élec.) (n.f.), permittivity.
peroxyde d'azote (n.m.), nitrogen dioxide.
persistance d'écran (électron.) (n.f.), afterglow; after-image.
persistance des impressions lumineuses (n.f.), persistance of vision.
personnel au sol (aviation) (n.m.), ground staff.
perspective cavalière (n.f.), isometric projection.
perte de pression (n.f.), pressure drop.
perte de vitesse (moteur) (n.f.), stall.
perte de vitesse marginale (aviation) (n.f.), tip stall.
perte Joule (n.f.), ohmic loss.
pertinence (n.f.), dependability; reliability.
perturbation (n.f.), interference.
perturbé, -e (adj.), disturbed.
pèse-acide (n.f.), acidimeter.
pèse-sirop (n.m.), saccharometer.
peson (n.m.), spring scale; weight indicator.
petit moteur (élec.) (n.m.), fractional h.p. motor.
petite visite (aviation) (n.f.), minor base check.
pétrole lampant (n.m.), kerosene.
pH mètre (n.m.), pH meter.
phare (n.m.), light; lighthouse.
phare anti-brouillard (auto) (n.m.), fog light.
phare d'atterrissage (aviation) (n.m.), land-ing light.
phare d'éclairage voilure (aviation) (n.m.), wing light.

phare de roulement (aviation) (n.m.), taxi light.
phare portatif (n.m.), portable light.
phase, en (élec.) (adj.), in phase; in step; in synchronism.
phénomène transitoire (n.m.), transient.
phénoplaste (plast.) (n.m.), phenolic plastic.
phone (unité de puissance sonore) (n.m.), phon.
phosphorescence (n.f.), after-glow.
phot (unité d'éclairement) (n.m.), phot.
photo-conducteur, -trice (adj.), light-positive.
photo-émissif, -ive (adj.), photo-emissive.
photo-résistant, -e (adj.), light-negative.
photo-sensible (adj.), photo-sensitive.
photon (phys.) (n.m..), photon.
phugoïde (aviation) (adj.), phugoid.
physique nucléaire (n.f.), nuclear physics; nucleonics.
pick-up piézo-électrique (radio) (n.m.), crys-tal pick-up
pièce brute (usinage) (n.f.), blank.
pièce d'arrêt (n.f.), stopper.
pièce de fixation (n.f.), attaching part.
pièce de fonderie (n.f.), casting.
pièce de rechange (n.f.), spare part.
pièce de retenue (n.f.), retainer.
pièce détachée (n.f.), spare part.
pièce en T (n.f.), tee.
pièce factice (n.f.), dummy part.
pièce forgée (n.f.), forging.
pièce omnibus (n.f.), workhorse part.
pièce permutable (n.f.), interchangeable part.
pièce rapportée (n.f.), insert.
pièce récupérée (n.f.), salvaged part.
pièce symétrique (n.f.), LH or RH part.
pied (de tubes électroniques) (n.m.), pinch.
pied avant (auto) (n.m.), front post.
pied d'aube (n.m.), blade root.
pied d'une courbe (n.m.), toe of a curve.
pied d'une dent d'engrenage (n.m.), root of a gear tooth.
pied de positionnement (n.m.), locator.
pied support (n.m.), stand.
piétage (hélicoptère) (n.m.), cover strip of root rib.
piétage femelle (n.m.), locating hole.
piétage mâle (n.m.), locating stud.
piéter (v.t.), to dowel.
pièze (unité de pression) (n.f.), pieze (approx. 0.145 p.s.i.).
piézo-électrique (adj.), piezo-electric.
pige (n.f.), feeler gauge.
pignon (n.m.), gear.
pignon à chevrons (n.m.), herringbone gear.
pignon à ergot (n.m.), pick-up gear.
pignon à queue (n.m.), shaft gear.
pignon cloche (n.m.), bell gear.
pignon conique (n.m.), bevel gear.
pignon d'attaque (n.m.), driving gear; drive pinion.
pignon d'entraînement à chaîne (n.m.), sprocket.
pignon de distribution (auto) (n.m.), timing sprocket.
pignon différentiel (n.m.), differential gear.
pignon droit (n.m.), spur gear.
pignon excentrique (n.m.), eccentric gear.
pignon fou (n.m.), idler gear.
pignon hélicoïdal (n.m.), helical gear.
pignon principal d'engrenage épicycloï-

-dal (*n.m.*), sun gear.
pignon réducteur (*n.m.*), reducing gear.
pignon satellite (*n.m.*), planet gear.
pignon silencieux (*n.m.*), quiet gear.
pile à dépolarisation par l'air (élec.) (*n.f.*), air cell.
pile atomique (*n.f.*), atomic pile; nuclear reactor.
pile atomique transportable (*n.f.*), aspatron.
pile de carbone (*n.f.*), carbon pile.
pile de polarisation (*n.f.*), bias cell.
pile électrique (*n.f.*), cell.
pilonnage (*n.m.*), drop forging; stamping.
pilotage automatique transparent (aviation) (*n.m.*), limited authority autopilot.
pilotage piézo-électrique (aviation) (*n.m.*), crystal control.
pilote (aviation, marine) (*n.m.*), pilot.
pilote automatique (aviation) (*n.m.*), autopilot.
pince (machines-outils) (*n.f.*), collet.
pince à dénuder (élec.) (*n.f.*), cable stripper.
pince à dessin (*n.f.*), spring clip.
pince à long bec (*n.f.*), long-nose pliers.
pince à sertir (*n.f.*), crimping pliers.
pince américaine (*n.f.*), collet.
pince bec-de-corbin (*n.f.*), bent-nose pliers.
pince crocodile (outil) (*n.f.*), alligator clip; alligator wrench.
pince de serrage (machines-outils) (*n.f.*), holding collet.
pince élastique (machines-outils) (*n.f.*), spring collet.
pince fendue (machines-outils) (*n.f.*), split collet.
pince multiprise (*n.f.*), multi-grip pliers.
pincement (*n.m.*), pinching.
pincement (roues avant d'une auto) (*n.m.*), toe-in.
pion (*n.m.*), spigot.
pion d'entraînement (*n.m.*), drive pin.
pion de butée (*n.m.*), stop pin.
pion de centrage (*n.m.*), guide pin.
pion de guidage (*n.m.*), stud.
pion de verrouillage (*n.m.*), locking pin.
pipe cache-borne caoutchouc (élec.) (*n.f.*), rubber terminal cap.
pipe d'échappement (*n.f.*), exhaust pipe.
pipeline (pétrole) (*n.f.*), pipe-line.
pipette de prélèvement (*n.f.*), pipette.
piquage (*n.m.*), pitting.
piquage d'un circuit (radio) (*n.m.*), tapping of a circuit.
piqué, -e (*adj.*), pitted.
piqué en vol (aviation) (*n.m.*), dive.
piratage de pièces (sur une auto) (*n.m.*), cannibalizing.
piste (aviation) (*n.f.*), landing strip; runway.
piste d'atterrissage (aviation) (*n.f.*), landing strip; runway.
piste d'envoi (aviation) (*n.f.*), runway.
piste de circulation (aviation) (*n.f.*), taxiway.
piste de dégagement (aviation) (*n.f.*), secondary runway.
piste périphérique (aviation) (*n.f.*), perimeter track.
piste sonore (cinéma) (*n.f.*), sound-track.
pistolet (pour peinture) (*n.m.*), spray-gun; sprayer.
pistolet à dessin (*n.m.*), French curve.
pistolet à extrusion (*n.m.*), extrusion gun.
piston de commande (*n.m.*), control piston.

piston de vérin (*n.m.*), cylinder piston rod.
piston isodrome (*n.m.*), single-action piston.
piston plongeur (*n.m.*), plunger.
piston séparateur (*n.m.*), separator piston.
pistonnage (extraction de pétrole) (*n.m.*), swabbing.
piton (*n.m.*), stud.
pitot (*n.m.*), pitot.
placer (*v.t.*), to locate.
placette (planification) (*n.f.*), piazzetta.
plafond absolu (aviation) (*n.m.*), absolute ceiling.
plafond opérationnel (aviation) (*n.m.*), operational ceiling.
plafond pratique (aviation) (*n.m.*), service ceiling.
plafonnier (*n.m.*), overhead light.
plage d'utilisation (*n.f.*), operational range.
plage de vitesse (*n.f.*), speed range.
plage graduée d'un instrument (*n.f.*), scale of an instrument.
plan (*n.m.*), layout.
plan d'archives (cinéma) (*n.m.*), stock shot.
plan d'ensemble (*n.m.*), assembly drawing.
plan de construction (*n.m.*), shaft plane.
plan de détails (*n.m.*), detailed drawing.
plan de forme (*n.m.*), outline drawing.
plan de joint (*n.m.*), mating surfaces.
plan de joint (en fonderie) (*n.m.*), parting line.
plan de pilotage (aviation) (*n.m.*), control plane.
plan de référence structure (aviation) (*n.m.*), airframe reference plane.
plan de rotation (*n.m.*), rotation plane.
plan de symétrie (*n.m.*), symmetry plane.
plan de vol (aviation) (*n.m.*), flight plan.
plan des supports (cinéma) (*n.m.*), media planning.
plan du disque (aviation) (*n.m.*), tip path plane.
plan du disque rotor (aviation) (*n.m.*), rotor disc plane.
plan fixe horizontal (aviation) (*n.m.*), horizontal stabilizer.
plan fixe vertical (aviation) (*n.m.*), vertical stabilizer.
plan incliné (*n.m.*), slanted plane.
plan mobile (aviation) (*n.m.*), stabilator.
plan-paquet (cinéma) (*n.m.*), pack shot.
plan rapproché (cinéma) (*n.m.*), close-up.
plan serré (cinéma) (*n.m.*), close-up.
plan trois-vues (*n.m.*), three view drawing.
planche d'ambiance (*n.f.*), general layout drawing.
planche de bord (aviation) (*n.f.*), instrument panel.
plancher mécanique (hélicoptère) (*n.m.*), transmission support platform.
plancher pressurisé (aviation) (*n.m.*), pressurized floor.
planchette (*n.f.*), slide table.
planchette à borne (élec.) (*n.f.*), terminal strip.
planéité (*n.f.*), flatness.
planétaire primaire (hélicoptère) (*n.m.*), first stage planet gear.
planeur (aviation) (*n.m.*), glider.
plante indicatrice (*n.f.*), indicator plant.
plaque (radio) (*n.f.*), grid.
plaque à fusibles (élec.) (*n.f.*), fuse base-plate.
plaque à plots (élec.) (*n.f.*), contact plate.
plaque à résistance (élec.) (*n.f.*), resistor plate.
plaque à serre-fils (élec.) (*n.f.*), terminal strip.

plaque d'accrochage (n.f.), catch plate.
plaque d'appui en deux pièces (n.f.), split removal plate.
plaque d'appui pour extraction (n.f.), push-out plate.
plaque d'arrêt (n.f.), stop plate.
plaque d'extracteur (n.f.), drift plate.
plaque d'identification (n.f.), name plate.
plaque d'usure de patin (hélicoptère) (n.f.), skid plate.
plaque de butée (n.f.), stop plate.
plaque de couple (sur freins) (n.f.), torque plate.
plaque de déviation (n.f.), deflection plate.
plaque de fermeture (n.f.), cover plate.
plaque de flexion (hélicoptère) (n.f.), rotor flexbeam.
plaque de maintien (n.f.), support plate.
plaque de métallisation (n.f.), earth bonding plate.
plaque de poussée (sur freins) (n.f.), thrust plate.
plaque de projection (n.f.), splash guard.
plaque de raccordement à bornes (élec.) (n.f.), terminal block.
plaque de renfort (n.f.), reinforcement plate.
plaque frein (n.f.), locking plate.
plaque inférieure (n.f.), base plate.
plaque obturatrice (n.f.), blanking plate.
plaque signalétique (n.f.), rating plate; name plate.
plaque supérieure (n.f.), top plate.
plaqué, -e (adj.), clad.
plaquette (n.f.), small plate.
plaquette à braser (n.f.), tip for brazing.
plaquette à jeter (n.f.), cutting insert.
plaquette butée (n.f.), stop plate.
plaquette d'arrêt (n.f.), lock plate.
plaquette d'alarme (n.f.), warning label.
plaquette d'identification (n.f.), name plate.
plaquette d'instructions (n.f.), instruction plate.
plaquette d'obturation (n.f.), blanking plate.
plaquette de métallisation (n.f.), earth bonding plate.
plaquette frein (n.f.), lock-plate.
plaquette isolante (élec.) (n.f.), insulating plate.
plaquette joint (n.f.), seal plate.
plaquette lumineuse (n.f.), warning label.
plaquette porte-écrou (n.f.), gang channel.
plaquette racleuse (n.f.), scraper insert.
plaquette serre-câble (n.f.), cable grip plate.
plaquette sur panneau (n.f.), placard.
plastifiant (plast.) (n.m.), plasticizer.
plastifier (v.t.), to plasticize.
plastique (n.m.), plastic.
plastique armé (n.m.), reinforced plastic.
plastique expansé (n.m.), expanded plastic.
plastique mousse (n.m.), expanded plastic; foamed plastic.
plateau (d'une presse) (n.m.), platen.
plateau à aimants permanents (n.m.), per-manent magnet chuck.
plateau cyclique (hélicoptère) (n.m.), swashplate.
plateau cyclique fixe (hélicoptère) (n.m.), non-rotating star.
plateau cyclique mobile (hélicoptère) (n.m.), rotating star.
plateau de barbotage (raffinerie) (n.m.), bub-ble tray.

plateau de commande (hélicoptère) (n.m.), spider.
plateau de commande du rotor arrière (hélicoptère) (n.m.), pitch change spider.
plateau de manutention (outillage) (n.m.), dolly.
plateau de moyeu rotor (hélicoptère) (n.m.), rotor hub plate.
plateau de pression d'embrayage (n.m.), clutch pressure plate.
plateau fixe (hélicoptère) (n.m.), stationary star.
plateau flottant (presse) (n.m.), floating platen.
plateau mobile (hélicoptère) (n.m.), rotating star.
plateau oscillant (hélicoptère) (n.m.), swashplate.
plateau porte-charge (hélicoptère) (n.m.), cargo carrier support.
plateau porte-chariot à aléser (machines-outils) (n.m.), boring head.
plateau porte-chariot à avance continue (machines-outils) (n.m.), continuous-feed head.
plateau porte-chariot à fraiser (machines-outils) (n.m.), milling head.
plateau porte-chariot à surfacer (machines-outils) (n.m.), facing head.
plateau rectangulaire type lourd (n.m.), heavy duty rectangular magnetic chuck.
plateau tournant (hélicoptère) (n.m.), rotating star.
plateforme à inertie (n.f.), inertial platform.
plateforme d'accès (n.f.), servicing platform.
plateforme de levage (n.f.), lifting platform.
plateforme de manutention (n.f.), pallet.
plateforme mécanicien (n.f.), servicing platform.
plateforme roulante élévatrice (n.f.), port-able hoisting platform.
platine (n.f.), support plate; lock plate; module.
platine support (n.f.), mounting plate.
plats, sur (adj.), across flats.
plein, compléter le (auto, aviation) (v.), to top up with fuel.
plein gaz (adj.), full throttle.
pleine charge (n.f.), full load.
plénitude (hélicoptère) (n.f.), solidity.
pleurage (radio) (n.m.), wow.
plexiglas (n.m.), Lucite (U.S.A.); Perspex (G.B.).
plexiglas de saumon d'aile (n.m.), plexiglass fairing.
pli (n.m.), wrinkle.
pliage (pièce) (n.m.), bent section.
pliage (opération) (n.m.), bending.
pliage à la presse (n.m.), brake press forming.
pliage accordéon (n.m.), fanfold.
plissement (n.m.), wrinkling.
plomb (élec.) (n.m.), fuse.
plomb tétraéthyle (n.m.), lead tetraethyl.
plot magnétique avec extracteur (n.m.), magnetic holdfast.
plots à soufflage magnétique (n.m.pl.), magnetic blow-out contacts.
plutonium (élément) (n.m.), plutonium.
pneu (auto) (n.m.), tyre.
pneu à bavette (n.m.), chine tyre.
pneu laminé (auto) (n.m.), fabric laminated tread tyre.
pneu sans chambre à air (auto) (n.m.), tube-less tyre.
poche d'air (aviation) (n.f.), air lock.
pochette (n.f.), small bag, envelope.

poignée amovible (*n.f.*), detachable handle.
poignée de largage de parachute (avia-
-tion) (*n.f.*), parachute release handle.
poignée de verrouillage (*n.f.*), locking
handle.
poignée escamotable (*n.f.*), retractable
handle.
poignée tournante (*n.f.*), twist grip.
point (*n.m.*), stitch.
point (navigation) (*n.m.*), fix.
point, faire le (cinéma) (*v.*), to round up.
point d'adhésiveté (géol.) (*n.m.*), sticky
point.
point d'appui (*n.m.*), bearing point; leverage.
point d'arrêt (aviation) (*n.m.*), stagnation point.
point d'attente (aviation) (*n.m.*), holding point.
point d'équilibre (*n.m.*), balance point.
point d'inflammabilité (*n.m.*), flash-point.
point de blocage (radio) (*n.m.*), cut-off point.
point de déblocage (radio) (*n.m.*), sidetone
starting.
point de décollement (*n.m.*), separation point.
point de démontage (*n.m.*), disconnect point.
point de flétrissement (géol.) (*n.m.*), wilting
point.
point de levage (*n.m.*), jacking point.
point de mesure (*n.m.*), test point.
point de montage (*n.m.*), division point.
point de nuage (huile) (*n.m.*), clouding point.
point de référence (aviation) (*n.m.*), station.
point de rosée (aviation) (*n.m.*), dew point.
point de soudure (*n.m.*), weld spot.
point de transition (*n.m.*), transition point.
point dur (*n.m.*), friction point.
point estimé (radar) (*n.m.*), dead reckoning.
point milieu (aviation) (*n.m.*), point of no return.
point mort bas (auto) (*n.m.*), bottom dead
centre.
point mort haut (auto) (*n.m.*), top dead centre.
point par point, step by step.
point remarquable d'une courbe (*n.m.*),
characteristic point of a curve.
pointage (soudure) (*n.m.*), tack welding.
pointe (élec.) (*n.f.*), making demand; peak period;
peak value.
pointe (de tube électronique) (*n.f.*), tip.
pointe à tracer (*n.f.*), scriber.
pointe arrière du fuselage (aviation) (*n.f.*),
tail cone.
pointe avant du fuselage (aviation) (*n.f.*),
nose cone.
pointe de contact (*n.f.*), prod.
pointe de contrainte (*n.f.*), stress peak.
pointe de contrainte en battement (héli-
-coptère) (*n.f.*), flapping stress peak.
pointeau de soupape (*n.m.*), valve needle.
pointillés (*n.m.pl.*), dotted line.
poise (unité de viscosité) (*n.f.*), poise (= 1
dyne/s./cm²).
poix (*n.f.*), resin.
polarisation (*n.f.*), bias.
polarisation automatique (*n.f.*), self bias.
polarisation de coupure (*n.f.*), cut-off bias.
polarité (*n.f.*), polarity.
pôle d'un noyau (*n.m.*), node of a core.
pôle négatif (élec.) (*n.m.*), cathode.
pôle positif (élec.) (*n.m.*), anode.
pôle saillant (*n.m.*), salient pole.
poli, -e (*adj.*), polished; smooth.
polir (*v.t.*), to buff.
polissage (*n.m.*), honing.

polissage au tonneau (*n.m.*), barrel finishing.
polissage électrolytique (*n.m.*), electro-
polishing.
pollution (*n.f.*), contamination.
polyamide (plast.) (*n.m.*), polyamide.
polyester (plast.) (*n.m.*), polyester.
polyéthylène (plast.) (*n.m.*), polyethylene;
polythene.
polymère (*n.m.*), polymer.
polymère linéaire (plast.) (*n.m.*), linear
polymer.
polymérisation (*n.f.*), polymerization.
polymériser (*v.t.*), to cure.
polymètre (*n.m.*), multipurpose meter.
polystyrolène (plast.) (*n.m.*), polystyrene.
polyvalence (*n.f.*), versability.
polyvalent, -e (hélicoptère) (*adj.*), utility; multi-
purpose.
pompage (moteur) (*n.m.*), surging.
pompage (élec.) (*n.m.*), hunting.
pompe à barillet (*n.f.*), wobble pump.
pompe à carburant (*n.f.*), fuel pump.
pompe à clapets (*n.f.*), flap valve pump.
pompe à débit constant (*n.f.*), constant flow
pump.
pompe à débit variable (*n.f.*), adjustable dis-
-charge pump; variable flow pump.
pompe à graisse (*n.f.*), grease gun.
pompe à plateau oscillant (*n.f.*), wobble
pump.
pompe à vide élevé (*n.f.*), low vacuum pump.
pompe à vide préliminaire (*n.f.*), absolute
pressure vacuum pump.
pompe annulaire (*n.p.*), ring pump.
pompe autorégulatrice (*n.f.*), self-regulating
pump.
pompe carburant haute pression (*n.f.*),
fuel high pressure pump.
pompe d'alimentation (*n.f.*), supply pump.
pompe d'amorçage (*n.f.*), primer pump.
pompe d'appoint (*n.f.*), booster pump.
pompe d'arrosage (machines-outils) (*n.f.*),
coolant pump.
pompe d'injection (moteur diesel) (*n.f.*), fuel
injection pump.
pompe de gavage (*n.f.*), booster pump.
pompe de pression d'huile (*n.f.*), oil pressure
pump.
pompe de récupération (*n.f.*), scavenge pump.
pompe de refoulement (*n.f.*), pressure pump.
pompe de reprise (auto) (*n.f.*), accelerating
pump.
pompe de servo-commande (*n.f.*), servo-
pump.
pompe de suralimentation (*n.f.*), booster
pump.
pompe de transfert (carburant) (*n.f.*), transfer
pump.
pompe de treuil (hélicoptère) (*n.f.*), hoist pump.
pompe double d'essuie-glace (*n.f.*), wind-
-shield wiper twin cylinder pump.
pompe entraînée par le réacteur (aviation)
(*n.f.*), engine driven pump.
pompe éthyl-glycol (aviation) (*n.f.*), de-icing
pump.
pompe noyée (*n.f.*), submerged pump.
pompe volumétrique (*n.f.*), positive displace-
-ment pump.
poncer (*v.t.*), to sand.
pont-arrière (*n.m.*), rear-axle.
pont de mesure (*n.m.*), measuring bridge.

pont de redressement (*n.m.*), rectifier bridge.

pont quotientométrique (*n.m.*), ratiometric bridge.

portance (aviation) (*n.f.*), lift.

portance de la pale (hélicoptère) (*n.f.*), blade lift.

portance maximum (aviation) (*n.f.*), maximum lift.

portance nulle (aviation) (*n.f.*), zero lift.

porte à faux (*n.m.*), cantilever.

porte à fermeture rapide (*n.f.*), snap door.

porte-balais alternateur (*n.m.*), alternator brush-holder.

porte-balais d'essuie-glace (*n.m.*), wind-shield wiper arm assembly.

porte cargo (*n.f.*), loading door.

porte-contacts (*n.m.*), contact-holder.

porte coulissante (*n.f.*), sliding door.

porte d'accès (*n.f.*), access door.

porte de soute (aviation) (*n.f.*), cargo compart-ment door.

porte du train avant (aviation) (*n.f.*), nose gear door.

porte du train principal (aviation) (*n.f.*), main landing gear door.

porte escalier (aviation) (*n.f.*), stairway door.

porte-étiquette (*n.m.*), label-holder.

porte fusible (*n.m.*), fuse-holder.

porte-garniture (*n.m.*), packing retainer.

porte-joint (*n.m.*), seal holder.

porte-outil d'alésage (*n.m.*), tool-holder for boring.

porte-outil pour ébaucher (*n.m.*), tool-holder for external roughing.

porte-outil tangentiel (*n.m.*), tangential tool-holder.

porte-protecteur (aviation) (*n.m.*), landing gear boot retainer.

porte-satellites (hélicoptère) (*n.m.*), planet pinion cage.

porte vissée (*n.f.*), plug door.

portée (*n.f.*), bearing surface.

portée (performance) (*n.f.*), range.

portée d'arbre sur palier (*n.f.*), journal.

portée diurne (aviation) (*n.f.*), day range.

portée du radar (*n.f.*), radar range.

portée lisse (*n.f.*), journal bearing.

portée nocturne (aviation) (*n.f.*), night range.

portée radio (*n.f.*), radio range.

porteur électrisé (*n.m.*), charged particle.

porteuse (radio) (*n.f.*), carrier.

portillon (*n.m.*), flap.

pose (*n.f.*), installation.

pose-tubes (trav. pub.) (*n.m.*), side-boom.

poser (*v.t.*), to lay; to place; to set.

position (*n.f.*), location; setting.

position de repos (*n.f.*), off-position.

position de travail (élec.) (*n.f.*), on-position.

position écoute (radio) (*n.f.*), listen position.

position émission (radio) (*n.f.*), transmit position.

position repos (*n.f.*), rest contacts.

position travail (*n.f.*), operative contacts.

positionnement (*n.m.*), location.

positionner (*v.t.*), to locate; to position.

positionneur (*n.m.*), positioner.

positon (*n.m.*), positron.

possibilité (*n.f.*), capability.

possibilité de vol stationnaire (hélicoptère) (*n.f.*), hovering capability.

post-chauffage (*n.m.*), post heating.

post-combustion (*n.f.*), afterburner.

post-formable (plast.) (*adj.*), post-forming.

poste (avion) (*n.m.*), station.

poste (nomenclature) (*n.m.*), item.

poste à galène (radio) (*n.m.*), crystal receiver; crystal set.

poste d'équipage (aviation) (*n.m.*), cockpit; flight compartment; flight deck.

poste de pilotage (aviation) (*n.m.*), flight compartment.

poste de soudage oxy-acétylénique (*n.m.*), oxy-acetylene welding equipment.

poste de T.S.F. (radio) (*n.m.*), radio set.

poste radio à modulation d'amplitude (radio) (*n.m.*), amplitude modulated radio.

postenquête (cinéma) (*n.m.*), post testing.

postsonorisation (audio-visuel) (*n.f.*), playback.

pot d'équilibrage (*n.m.*), plenum chamber.

pot de condensation (*n.m.*), water trap.

pot de décantation (*n.m.*), spill chamber.

pot de détente (*n.m.*), dash pot.

pot de transfert (plast.) (*n.m.*), transfer chamber.

potence (aviation—train d'atterrissage) (*n.f.*), support beam.

potence (hélicoptère) (*n.f.*), hoist arm.

potence de hissage (*n.f.*), portable crane.

potence de manutention (hélicoptère) (*n.f.*), derrick.

potence de treuil (manutention) (*n.f.*), hoist arm.

potentiel (élec.) (*n.m.*), voltage.

potentiel capillaire (géol.) (*n.m.*), capillary potential.

potentiel de sécurité (*n.m.*), safe life.

potentiomètre d'éclairage (*n.m.*), lighting rheostat.

potentiomètre de détection (aviation) (*n.m.*), sensing potentiometer.

potentiomètre de gain (aviation) (*n.m.*), gain potentiometer.

potentiomètre de pré-affichage (*n.m.*), pre-setting potentiometer.

potentiomètre de réglage (*n.m.*), adjusting potentiometer.

poudrage de récolte (agriculture) (*n.m.*), crop dusting.

poudre à mouler (plast.) (*n.f.*), moulding powder.

poulie d'entraînement (*n.f.*), driving pulley.

poulie folle (*n.f.*), idle pulley.

poulie ovale (*n.f.*), oval pulley.

poupée de tour (machines-outils) (*n.f.*), lathe headstock.

poursuite automatique (radar) (*n.f.*), auto-matic tracking.

pourtour (*n.m.*), circumference; periphery.

pousse-toc (machines-outils) (*n.m*), driving dog.

poussée au décollage (aviation) (*n.f.*), take-off thrust.

poussée au point fixe (aviation) (*n.f.*), static thrust.

poussée brute (*n.f.*), gross thrust.

poussée du rotor (hélicoptère) (*n.f.*), rotor thrust.

poussée du rotor arrière (hélicoptère) (*n.f.*), tail rotor thrust.

poussée garantie (aviation) (*n.f.*), guaranteed thrust.

poussée hydrostatique (n.f.), buoyancy.
poussée inversée (n.f.), reverse thrust.
poussée nette (n.f.), net thrust.
poussée nominale (n.f.), rated thrust.
poussée normale (n.f.), forward thrust.
poussée statique du rotor (hélicoptère) (n.f.), rotor static thrust.
poussoir (auto) (n.m.), tappet.
poussoir à tige (auto) (n.m.), tappet.
poussoir à vis (n.m.), screw plunger.
poussoir bi-polaire (élec.) (n.m.), two-pole push button.
poussoir de contact (élec.) (n.m.), contact button.
poutre centrale du fuselage (aviation) (n.f.), fuselage centre box.
poutre de queue (hélicoptère) (n.f.), tail boom.
poutre de treuil (hélicoptère) (n.f.), hoist boom.
poutre en treillis (hélicoptère) (n.f.), truss boom.
poutrelle en U (n.f.), channel section.
pouvoir de coupure (élec.) (n.m.), interrupting capacity.
pouvoir de séparation (n.m.), resolving power.
pouvoir émissif (n.m.), radiating power.
pouvoir séparateur (optique) (n.m.), resolving power.
pré-affichage (n.m.), pre-setting.
pré-allumage (n.m.), pre-ignition.
pré-amplification (n.f.), pre-amplification.
précession (n.f.), precession.
préchauffage (plast.) (n.m.), preheating.
précipitation artificielle (n.f.), cloud seeding.
précis, -e (adj.), accurate.
précision (n.f.), accuracy.
précontraint, -e (béton) (adj.), prestressed.
préenquête (cinéma) (n.f.), pre-testing.
préfabriqué, -e (adj.), prefabricated.
préforme (plast.) (n.f.), preform.
prélèvement d'air sur un compresseur (n.m.), compressor air-bleed.
prélèvement d'échantillonnage (n.m.), sampling.
prélèvement par lot (inspection) (n.m.), batch sampling.
premier vol (aviation) (n.m.), maiden flight.
pré-rodage (n.m.), honing; preliminary seating.
présentation (vol) (n.f.), approach.
présentation (texte) (n.f.), format; layout.
présenter (v.t.), to position.
préséparateur d'huile par congélation (n.m.), oil-chiller.
pré-série (n.f.), preproduction.
présonorisation (audio-visuel) (n.f.), playback.
presse à balancier (n.f.), fly press; friction screw.
presse à cintrer (n.f.), bending press.
presse à découper (n.f.), blanking press; punch press.
presse à ébavurer (n.f.), trimming press.
presse à injection (plast.) (n.f.), injection moulding press.
presse à mandriner (n.f.), arbor press.
presse à poinçons sériés (n.f.), gang press.
presse à sertir (n.f.), flanging press.
presse-étoupe (n.m.), packing gland.
presse-garniture (n.m.), gland.
presse-glace (n.m.), window pane retainer.
pression à débit nul (n.f.), zero flow pressure.
pression absolue relative (n.f.), relative absolute pressure.
pression d'admission d'air (n.f.), air intake pressure.
pression d'arrêt (aviation) (n.f.), impact pressure.
pression d'éclatement (n.f.), bursting pressure.
pression d'épreuve (n.f.), proof pressure.
pression de décharge (n.f.), discharge pressure.
pression de déclenchement (n.f.), calibration pressure.
pression de référence (n.f.), reference pressure.
pression détendue (n.f.), reduced pressure.
pression différentielle (n.f.), differential pressure.
pression dynamique (n.f.), dynamic pressure; impact pressure.
pression motrice (n.f.), driving pressure.
pression régulée (n.f.), controlled pressure.
pression statique (n.f.), static pressure.
pressurisation (n.f.), pressurization.
primaire (peinture) (adj.), primer.
principe de fonctionnement (n.m.), operational principle.
priorité (n.f.), priority.
prise (élec.) (n.f.), connector.
prise coaxiale (n.f.), coaxial connector.
prise d'air dynamique (n.f.), ram air inlet.
prise d'air sur compresseur (n.f.), compressor air bleed port.
prise d'entraînement des accessoires (n.f.), accessory drive.
prise de baladeuse (élec.) (n.f.), utility outlet.
prise de compte-tours (n.f.), tachometer drive.
prise de coque (aviation) (n.f.), fuselage ground connection.
prise de courant (élec.) (n.f.), plug.
prise de courant force (élec.) (n.f.), power plug.
prise de faisceau (aviation) (n.f.), beam capture.
prise de mouvement (n.f.), power take-off.
prise de mouvement de l'arbre d'entraînement (n.f.), shaft drive.
prise de parc (élec.) (n.f.), ground power receptacle.
prise de pression (conditionnement d'air) (n.f.), pressure inlet.
prise de remplissage carburant (n.f.), refuelling connector.
prise de remplissage eau (n.f.), water connector.
prise de rinçage (n.f.), flushing connector.
prise de vidange (n.f.), drain connector.
prise directe (n.f.), direct drive.
prise électrique (élec.) (n.f.), electrical connector; plug.
prise fixe de repos (hélicoptère) (n.f.), stowage connector.
prise mobile (n.f.), movable connector.
prise murale (élec.) (n.f.), wall outlet.
prise statique (n.f.), static port.
prise tourne-disque (audio-visuel) (n.f.), pick-up attachment.
prix affiché (pétrole) (n.m.), posted price.
procédure d'approche finale (aviation) (n.f.), landing procedure.
procédure d'attente (aviation) (n.f.), holding

procedure.
procédure de secours (*n.f.*), emergency procedure.
profil aérodynamique (*n.m.*), aerofoil.
profil d'une came (*n.m.*), cam contour.
profil de référence (aviation) (*n.m.*), wing reference profile.
profil têtard (aviation) (*n.m.*), tadpole airfoil.
profilé (*n.m.*), extrusion; formed section.
profilé d'étanchéité (*n.m.*), extruded seal.
profilé de bordure (*n.m.*), edging.
profilé en caoutchouc (*n.m.*), rubber extrusion.
profilé en L (*n.m.*), L section.
profilé en U (*n.m.*), channel section.
profilé filé (*n.m.*), extrusion.
profilé plié (*n.m.*), bent section.
profilé stratifié (*n.m.*), laminated section.
profilomètre (*n.m.*), contour follower.
profondeur, commande de (aviation) (*n.f.*), pitch control.
profondeur, gouverne de (aviation) (*n.f.*), elevator.
profondeur d'une aile (aviation) (*n.f.*), chord.
profondeur de pale (aviation) (*n.f.*), blade depth.
programme (cinéma) (*n.m.*), planning.
projecteur (*n.m.*), floodlight.
prométhéum (élément) (*n.m.*), prometheum.
propergol (fusées) (*n.m.*), propellant.
propriétés (*n.f.pl.*), characteristics; properties.
propriétés mécaniques (*n.f.pl.*), mechanical properties.
propriétés physiques (*n.f.pl.*), physical properties.
propulsion, ensemble de (*n.m.*), drive system.
propulsion à réaction (*n.f.*), jet propulsion.
protecteur de mandrin (*n.m.*), chuck guard.
protection (*n.f.*), finishing; protection coating.
protection thermique (*n.f.*), heat shield.
protège chaîne (*n.m.*), chain guard.
protégé (-e) contre l'humidité (*adj.*), moisture-proof.
protégé (-e) contre le vent (*adj.*), windproof.
protégé (-e) contre les chocs (*adj.*), shock-resistant.
protégé (-e) contre les corps étrangers (*adj.*), splash-proof.
protégé (-e) contre les intempéries (*adj.*), weather-proof.
protégé (-e) contre les jets d'eau (*adj.*), hose-proof; watertight.
protégé (-e) contre les moisissures (*adj.*), fungus-proof.
protégé (-e) contre les poussières (*adj.*), dust-proof.

protégé (-e) contre les projections d'eau latérales (*adj.*), splash-proof.
protoactinium (élément) (*n.m.*), protoactinium.
proton (*n.m.*), particle; proton.
prototype (*n.m.*), prototype.
protoxyde d'azote (*n.m.*), nitrous oxide.
provision (*n.f.*), allowance.
provisoire (*adj.*), interim.
psychromètre (*n.m.*), wet and dry bulb hygrometer.
publipostage (audio-visuel) (*n.m.*), mailing.
puisard d'eau (*n.m.*), water sump.
puisard d'huile (*n.m.*), oil-sump.
puissance au frein (*n.f.*), brake horsepower.
puissance d'émission (radio) (*n.f.*), radiating power.
puissance d'entrée (*n.f.*), input power.
puissance d'excitation de grille (électron.) (*n.f.*), grid-driving power.
puissance d'un feuillet (élec.) (*n.f.*), strength of shell.
puissance de décollage (aviation) (*n.f.*), take-off power.
puissance de pointe (élec.) (*n.f.*), peak output power.
puissance de sortie (élec.) (*n.f.*), output power.
puissance disponible (*n.f.*), available power.
puissance équivalente sur arbre (*n.f.*), equivalent shaft horsepower.
puissance inférieure à 1 CV (élec.) (*n.f.*), fractional horsepower.
puissance nominale (*n.f.*), rated power.
puissance nominale en chevaux (*n.f.*), indicated horsepower.
puissance utile (*n.f.*), effective power.
puissance utilisable (*n.f.*), available power.
puits de chaleur (*n.m.*), heat sink.
pulsation (élec.) (*n.f.*), angular frequency.
pulso-réacteur (moteur) (*n.m.*), pulsejet engine.
pulvérisation (*n.f.*), atomization.
pulvérisation agricole (hélicoptère) (*n.f.*), crop spraying.
pupinisation (élec.) (*n.f.*), coil loading.
pupitre (*n.m.*), control panel.
pupitre de commande (élec.) (*n.m.*), control desk; control pedestal.
purge (*n.f.*), purge drain.
purger (chasser l'air d'une canalisation) (*v.t.*), to bleed; to purge.
push-pull (radio) (*n.m.*), push-pull.
pylône (hélicoptère) (*n.m.*), pylon.
pylône central (*n.m.*), control pedestal.
pylône repliable (hélicoptère) (*n.m.*), folding pylon.
pyroélectricité (élec.) (*n.f.*), pyro-electricity.
pyromètre (*n.m.*), pyrometer.

Q

qualité (*n.f.*), grade.
qualité de fabrication (*n.f.*), workmanship.
quantité (*n.f.*), amount.
quantum (*n.m.*) [*pl.* **quanta**], quantum (quanta).
quartz (*n.m.*), crystal.
queue de distillation (pétr.) (*n.f.*), after-running; tail.

queue de l'avion (*n.f.*), aircraft tail unit.
queue de pignon conique d'attaque (hélicoptère) (*n.f.*), input bevel pinion shaft.
queue de poussée (aviation) (*n.f.*), thrust tail off.
queue de roue conique (hélicoptère) (*n.f.*), bevel ring flanged shaft.
quille (machines-outils) (*n.f.*), quill.

quille (hélicoptère) (*n.f.*), hull.
quille d'angle (*n.f.*), chine.
quille d'échouage (aviation) (*n.f.*), keel.
quille de poussée (*n.f.*), thrust member.

quille de voilure (aviation) (*n.f.*), wing bolster beam.
quincaillerie (*n.f.*), hardware.
quotient (*n.m.*), quotient.

R

rabattable (*adj.*), folding; hinged.
rabattre (*v.t.*), to fold back; to fold down; to fold over.
raboteuse à deux montants (*n.f.*), double housing planing machine.
raboteuse à un montant (*n.f.*), open-side planing machine.
raccord à démontage rapide (*n.m.*), quick disconnect.
raccord banjo (*n.m.*), banjo union.
raccord coaxial (*n.m.*), coaxial connector.
raccord de remplissage (*n.m.*), filler adapter.
raccord droit (*n.m.*), straight union.
raccord électrique (*n.m.*), connector.
raccord en croix (*n.m.*), cross union.
raccord flexible (*n.m.*), flexible coupling.
raccord orientable (*n.m.*), swivel connector.
raccord presse-étoupe (*n.m.*), gland.
raccord réducteur (*n.m.*), restricting union.
raccord universel (*n.m.*), union; universal adaptor.
raccordement (*n.m.*), junction.
raccordement d'entrée d'air réacteur (*n.m.*), engine air intake extension.
racleur (*n.m.*), scraper; wiper.
racleur à feutre (*n.m.*), felt wiper.
racleur d'huile (auto) (*n.m.*), oil wiper.
radar (*n.m.*), radar.
radar d'obstacle (*n.m.*), terrain avoidance radar.
radar de tenue de poste (*n.m.*), station keep-ing radar.
radar météo (*n.m.*), weather radar.
radiateur d'électrode (*n.m.*), electrode radiator.
radiateur d'huile (*n.m.*), oil cooler.
radiateur ventilé, à (*adj.*), ventilated radiator.
radio balise (*n.f.*), beacon.
radio compas (*n.m.*), automatic direction-finder; radio compass.
radio-sonde (*n.f.*), radio altimeter.
radio-téléphonie (*n.f.*), voice communication.
radiogoniomètre (marine) (*n.m.*), direction-finder.
radiographie (*n.f.*), radiography.
radionavigation (*n.f.*), navigational aids; radio aids to navigation.
radiophare (*n.m.*), beacon.
radiophare d'aéroport (*n.m.*), airport beacon; aerodrome beacon (U.S.A.).
radiophare d'alignement (*n.m.*), radio range beacon.
radiophare omnidirectionnel (*n.m.*), omni-range radio; visual omni-range.
radioralliement (radar) (*n.m.*), homing.
radôme (*n.m.*), radome.
radon (élément) (*n.m.*), radon.
rafale (aviation) (*n.f.*), gust.
rafale, tir en (*n.m.*), burst of firing.
ralenti, -e (moteur) (*adj.*), idle.

rampage (élec.) (*n.m.*), crawling.
rampe (*n.f.*), ramp.
rampe d'allumage (*n.f.*), ignition harness.
rampe d'éclairage (*n.f.*), light ramp.
rampe d'escalier (*n.f.*), staircase hand-rail.
rampe d'extincteur (*n.f.*), extinguisher dis-charge tube.
rampe de chargement (*n.f.*), loading ramp.
rampe de distribution (*n.f.*), manifold.
rampe de guidage (*n.f.*), guide ramp.
rampe de répandage (trav. pub.) (*n.f.*), spray-ing bank.
rampe de soufflage (*n.f.*), hot air gallery.
rampe hélicoïdale (*n.f.*), helical groove.
rapide (*adj.*), fast.
rapidité (*n.f.*), velocity.
rappel (*n.m.*), return.
rappel à plat (hélicoptère) (*n.m.*), zero pitch return.
rappel de manche (hélicoptère) (*n.m.*), beeper trim; stick trim device.
rapport d'amortissement (*n.m.*), damping ratio.
rapport de compression (*n.m.*), compression ratio.
rapport de démultiplication (*n.m.*), step-down ratio.
rapport de finesse (*n.m.*), fineness ratio.
rapport de multiplication (*n.m.*), step-up ratio.
rapport de transformation (élec.) (*n.m.*), voltage ratio.
rapport de vitesse (*n.m.*), gear ratio.
rapport pondéral (*n.m.*), mass ratio.
rapport signal/bruit (*n.m.*), signal-to-noise ratio.
raté (*n.m.*) misfiring.
rattrapage (en chaîne) (*n.m.*), retrofit.
ravinement (*n.m.*), guttering.
rayé, -e (*adj.*), scored.
rayon (élec.) (*n.m.*), beam.
rayon à fond de filet (*n.m.*), root radius.
rayon admis (machines-outils) (*n.m.*), swing radius.
rayon d'action (aviation) (*n.m.*), operating range.
rayon de braquage (auto) (*n.m.*), lock; turning radius.
rayon de cintrage (*n.m.*), bend radius.
rayon de pied de denture (engrenage) (*n.m.*), tooth fillet.
rayon de pliage (*n.m.*), bending radius.
rayon de raccordement (engrenage) (*n.m.*), fillet.
rayon de virage (*n.m.*), turn radius.
rayon du circle primitif (*n.m.*), pitch radius.
rayon du rotor (hélicoptère) (*n.m.*), rotor radius.
rayonne (textile) (*n.f.*), rayon. (The use of the term "artificial silk" (soie artificielle) is for-bidden in France by law.)

rayons alpha (*n.m.pl.*), alpha rays.
rayons bêta (*n.m.pl.*), beta rays.
rayons canaux (électron.) (*n.m.pl.*), canal rays.
rayons cosmiques (*n.m.pl.*), cosmic rays.
rayons delta (*n.m.pl.*), delta rays.
rayons gamma (*n.m.pl.*), gamma rays.
rayure (*n.f.*), score; scoring.
réactance de capacité (*n.f.*), capacitive reactance.
réactance inductive (*n.f.*), inductive reactance.
réacteur (électron.) (*n.m.*), reactor.
réacteur (moteur) (*n.m.*), jet engine.
réacteur à double flux (*n.m.*), dual flow jet engine.
réacteur à simple flux (*n.m.*), single flow jet engine.
réacteur en bout de pale (*n.m.*), tip jet.
réacteur nucléaire (*n.m.*), nuclear reactor.
réacteur pilote (*n.m.*), master engine.
réacteur sec (*n.m.*), dry engine.
réaction (*n.f.*), feedback; interaction; response.
réaction en chaîne (*n.f.*), chain reaction.
réaléser (*v.t.*), to counterbore.
réarmer (*v.t.*), to reset; to recock.
rebord (*n.m.*), bordering; ledge.
rebord d'une jante (*n.m.*), rim bead.
rebut (*n.m.*), scrap.
rebuter (*v.t.*), to reject; to discard; to scrap.
recalage des gyros (*n.m.*), gyro resetting.
recalage rapide (aviation) (*n.m.*), fast slaving.
récepteur (instrument) (*n.m.*), indicator.
récepteur de balise (*n.m.*), marker beacon receiver.
récepteur de garde (radio) (*n.m.*), stand-by frequency receiver.
récepteur de poche (radio) (*n.m.*), pocket radio.
récepteur double (*n.m.*), dual indicator.
récepteur triple (*n.m.*), threefold indicator.
réception de matériel (*n.f.*), acceptance.
réception radio (*n.f.*), radio reception.
rechange (*n.m.*), replacement.
rechange, pièce de (*n.f.*), spare part.
rechapage (pneumatiques) (*n.m.*), retreading.
rechaper (pneumatiques) (*v.t.*), to retread.
recharge (*n.f.*), filler.
rechargé (-e) en dur (*adj.*), hard faced.
rechargement (*n.m.*), refacing.
réchauffage (*n.m.*), reheating.
réchauffage (radio) (*n.m.*), warming-up.
réchauffe (réacteur) (*n.f.*), reheat.
réchauffeur de carburant (*n.m.*), fuel heater.
recherche (*n.f.*), research; investigation.
réciprocité (*n.f.*), reciprocity.
réciproque (*adj.*), converse.
réciproquement (*adv.*), conversely.
reconditionnement (*n.m.*), work-over.
rectificatif (*n.m.*), amendment; correction.
rectification à l'enfilade (machines-outils) (*n.f.*), straight-through precision grinding.
rectification à sec (machines-outils) (*n.f.*), dry precision grinding.
rectification avec arrosage (machines-outils) (*n.f.*), wet precision grinding.
rectification de précision (machines-outils) (*n.f.*), precision grinding.
rectification plane (machines-outils) (*n.f.*), surface precision grinding.
rectification sans centre (machines-outils) (*n.f.*), centreless precision grinding.

rectifier (*v.t.*), to amend.
rectifier (usinage) (*v.t.*), to grind; to straighten; to true.
recuit bleu (*n.m.*), blue annealing.
recuit brillant *ou* **blanc** (*n.m.*), bright annealing.
recuit complet (*n.m.*), full annealing.
recuit de trempe (*n.m.*), quench annealing.
recuit noir (*n.m.*), black annealing.
recuit partiel (*n.m.*), half hard process annealing.
reculante, pale (*n.f.*), retreating blade.
récupération (*n.f.*), recovery.
redan (coque) (*n.m.*), hull step.
redevance (*n.f.*), royalty.
redressement (aviation) (*n.m.*), recovery.
redresseur au sélénium (*n.m.*), selenium rectifier.
redresseur demi-onde (élec.) (*n.m.*), half-wave rectifier.
redresseur pleine onde (élec.) (*n.m.*), full-wave rectifier.
réducteur (engrenage) (*n.m.*), reduction gearbox; gear head.
réducteur (débit) (*n.m.*), restricter.
réduire les gaz (aviation) (*v.*), to throttle down.
réenclencher (*v.t.*), to reset.
refaire les portées (*v.*), to reseat; to rebed.
référence (*n.f.*), reference.
référence fuselage (aviation) (*n.f.*), fuselage datum line.
réflectomètre (*n.m.*), reflectometer.
reformage (pétrole) (*n.m.*), reforming.
refoulage (métal.) (*n.m.*), upsetting.
refoulement (*n.m.*), discharge; output.
refoulement, côté de (*n.m.*), pressure side.
refoulement sous pression (*n.m.*), pressure delivery.
refouler (*v.t.*), to supply.
réfractaire (*adj.*), heat resisting; refractory.
réfrigérant (*n.m.*), coolant.
réfrigérateur à compresseur (*n.m.*), compression refrigerator.
refroidissement forcé (*n.m.*), forced cooling.
refroidissement forcé à l'air (*n.m.*), forced air cooling.
refroidissement forcé à l'eau (*n.m.*), forced water cooling.
refroidissement forcé à l'huile (*n.m.*), forced oil cooling.
refroidissement par air (élec.) (*n.m.*), air-cooling.
refroidissement par l'huile (*n.m.*), oil-cooling.
régalage (trav. pub.) (*n.m.*), grading; levelling.
regard (*n.m.*), inspection hole.
régénération d'huile (*n.f.*), oil reclaiming.
régime (élec.) (*n.m.*), operating conditions; power.
régime plein (aviation) (*n.m.*), full throttle.
régime d'anneau tourbillonnaire (aviation) (*n.m.*), vortex ring state.
régime de croisière (aviation) (*n.m.*), cruising power.
régime de décollage (aviation) (*n.m.*), take-off rating.
régime de montée (aviation) (*n.m.*), climb setting.
régime économique (aviation) (*n.m.*), economical cruising conditions.
régime maximum rotor (aviation) (*n.m.*), maximum rotor speed.

régime moteur (*n.m.*), engine speed.
régime nominal (élec.) (*n.m.*), rating; rate.
régime rotor (hélicoptère) (*n.m.*), rotor speed.
réglage (radio) (*n.m.*), tuning.
réglage altimétrique (*n.m.*), altimeter setting.
réglage automatique (*n.m.*), automatic control.
réglage automatique de fréquence (radio) (*n.m.*), automatic frequency control.
réglage automatique de volume (radio) (*n.m.*), automatic volume control.
réglage de pale (aviation) (*n.m.*), blade setting.
réglage de précision (*n.m.*), precision setting.
réglage de puissance (radio) (*n.m.*), gain control; volume control.
réglage du pas des pales (hélicoptère) (*n.m.*), blade pitch setting.
réglage du zéro (*n.m.*), zero adjustment.
règle porte-niveau (*n.f.*), spirit level rest.
règle-support magnétique (*n.f.*), magnetic rack.
réglet (*n.m.*), rule.
réglette (*n.f.*), connection bar.
réglette d'éclairage (élec.) (*n.f.*), light strip.
régulateur à force centrifuge et dépres-sion (moteur diesel) (*n.m.*), centrifugal and vacuum governor.
régulateur altimétrique (*n.m.*), barometric controller.
régulateur de carburant (*n.m.*), fuel control unit.
régulateur de débit (*n.m.*), flow regulator.
régulateur de fréquence (*n.m.*), frequency regulator.
régulateur de pression (*n.m.*), pressure regulator.
régulateur de température (*n.m.*), temper-ature regulator.
régulateur par tout ou rien (*n.m.*), hit or miss governor.
régulation (*n.f.*), control; regulation.
régulation automatique (*n.f.*), automatic control.
régulation variométrique (*n.f.*), pressure rate-of-change regulating.
rehausse (*n.f.*), raising block.
rehaut (*n.m.*), bezel ring.
relâcher (*v.t.*), to release.
relais à deux directions (*n.m.*), throw-over relay.
relais à deux seuils (*n.m.*), two-step relay.
relais à fiches (*n.m.*), plug-in relay.
relais à maximum (élec.) (*n.m.*), over-voltage relay; overload relay.
relais à maximum et minimum (*n.m.*), over- and under-voltage relay.
relais à minimum (*n.m.*), under-voltage relay.
relais à prise (élec.) (*n.m.*), plug-in-type relay.
relais à retenue (*n.m.*), biased relay.
relais à seuil de fréquence (*n.m.*), underfre-quency relay.
relais accordé (*n.m.*), resonant relay.
relais avertisseur de givrage (aviation) (*n.m.*), ice detector relay.
relais commutateur (*n.m.*), centre zero relay.
relais d'autorisation (*n.m.*), gate relay.
relais d'excitation (*n.m.*), exciter control relay.
relais de chauffage (*n.m.*), heat demand relay.
relais de coupure (*n.m.*), cut-out relay.
relais de délestage (*n.m.*), load shedder relay.
relais de démarrage (*n.m.*), starting relay.

relais de fin de course (*n.m.*), end of travel relay.
relais de fréquence (élec.) (*n.m.*), frequency relay.
relais de protection (élec.) (*n.m.*), protective relay.
relais de ralenti (*n.m.*), idling relay.
relais de recalage rapide (*n.m.*), fast slaving relay.
relais de refroidissement (*n.m.*), cold demand relay.
relais de réglage (*n.m.*), regulating relay.
relais de sécurité (*n.m.*), safety relay.
relais de sélection (*n.m.*), selective relay.
relais de signalisation (*n.m.*), warning relay.
relais de surtension (*n.m.*), over-voltage relay.
relais de transfert (*n.m.*), transfer relay.
relais de verrouillage (élec.) (*n.m.*), blocking element.
relais différentiel (*n.m.*), differential relay.
relais hydraulique (*n.m.*), anti-surge valve.
relais intégrateur d'impulsions (*n.m.*), notching relay.
relais manométrique (*n.m.*), anti-surge valve.
relais mécanique (*n.m.*), bellcrank block.
relais pilote (*n.m.*), pilot relay.
relais pneumatique (*n.m.*), differential pres-sure relay.
relais polarisé (*n.m.*), biased relay.
relais temporisé (*n.m.*), time lag relay.
relais thermique (*n.m.*), thermic relay.
relais thermique à maximum (élec.) (*n.m.*), temperature overload relay.
relais thermique de temporisation (*n.m.*), thermal timer relay.
relaxation (*n.f.*), stress relief.
relevé (navigation) (*n.m.*), plot.
relèvement (navigation) (*n.m.*), bearing.
réluctance (magnétisme) (*n.f.*), reluctance.
rémanence (élec.) (*n.f.*), residual induction.
rémanence (radar) (*n.f.*), image retention.
remblayeuse (construction) (*n.f.*), back-filler.
rembourrage (*n.m.*), wadding.
remettre en état (*v.t.*), to recondition.
remise de pas (hélicoptère) (*n.f.*), pitch increase.
remise en état (*n.f.*), rectification.
remontage (*n.m.*), re-assembly.
remorque (*n.f.*), trailer; trolley.
remplacement (*n.m.*), exchange.
remplissage (*n.m.*), topping-up.
remplissage par gravité (*n.m.*), gravity refuelling.
remplissage sous pression (*n.m.*), pressure refuelling.
rendement (*n.m.*), efficiency; output.
rendement à l'antenne (radio) (*n.m.*), aerial efficiency; antenna efficiency (U.S.A.).
rendement de sustension (hélicoptère) (*n.m.*), lifting efficiency.
rendement du rotor (hélicoptère) (*n.m.*), rotor efficiency.
rendement quantique (électron.) (*n.m.*), quan-tum yield.
rendre la main (aviation) (*v.*), to ease the stick forward.
renflé, -e (*adj.*), bulged; expanded.
renflement (*n.m.*), bulge.
renforcement (*n.m.*), backing.
renfort (*n.m.*), doubler; reinforcement plate.
reniflard (auto) (*n.m.*), crankcase breather.

renouvellement de l'air (*n.m.*), air renewal.
renseignements techniques (*n.m.pl.*), tech-
-nical data.
rentrée (du train d'atterrissage) (aviation) (*n.f.*),
retraction.
renversement (*n.m.*), reversal.
renvoi (*n.m.*), lever.
renvoi à une planche (*n.m.*), cross-reference
of a drawing.
renvoi d'angle (*n.m.*), angle gearbox.
renvoi de chaîne (*n.m.*), chain sprocket.
renvoi de combinateur (hélicoptère) (*n.m.*),
mixer bellcrank.
renvoi intermédiaire (*n.m.*), intermediate
gearbox.
répandeur finisseur (trav. pub.) (*n.m.*),
spreader-finisher.
répandeuse (trav. pub.) (*n.f.*), spreader; spread-
-ing machine.
répartiteur (plateau magnétique) (*n.m.*), adaptor
plate.
répartiteur (*n.m.*), distributor.
répartiteur (pétrole) (*n.m.*), dispatcher.
répartition des charges (*n.f.*), load
distribution.
répartition du rivetage (*n.f.*), riveting
pattern.
repasser les filets (*v.*), to chase threads.
repérage (*n.m.*), identification.
repérage de l'objectif (aviation) (*n.m.*), target
acquisition.
repérage sous-marin par ultrasons
(*n.m.*), asdic; sonar (U.S.A.).
répercussion technique (*n.f.*), technical
effect.
repère (*n.m.*), index; item number; station.
repère de niveau horizontal (aviation)
(*n.m.*), levelling mark.
repérer (*v.t.*), to find; to spot.
répertoire (*n.m.*), index.
répétiteur de cap (navigation) (*n.m.*), heading
repeater.
répétiteur de jaugeur (*n.m.*), gauge repeater.
repliage des pales (hélicoptère) (*n.m.*), blade
folding.
repliage du pylône (hélicoptère) (*n.m.*), pylon
folding.
réponse (*n.f.*), answer; response.
reporteur d'images (cinéma) (*n.m.*), reporter-
cameraman.
repos, au (*adj.*), at rest; non-operating.
repose-pied (*n.m.*), foot-rest.
repoussage (*n.m.*), spinning.
reprise (fixation) (*n.f.*), pick-up.
reprise (usinage) (*n.f.*), rework.
reproductible (*n.m.*), master.
réseau (de distribution électrique) (grande distance généralement à 132.000 volts) (*n.m.*),
grid; mains.
réseau (pile atomique) (*n.m.*), lattice.
réseau d'antennes directionnelles (radio)
(*n.m.*), aerial array; antenna array (U.S.A.).
réseau d'effacement (aviation) (*n.m.*), wash-
-out network.
réseau d'intégration (aviation) (*n.m.*), inte-
-grating network.
réseau de bord (aviation) (*n.m.*), aircraft
mains.
réseau dipôle (élec.) (*n.m.*), two-terminal
network.
réseau quadripôle (élec.) (*n.m.*), four-terminal

network.
réserve de combustible (*n.f.*), fuel reserve.
réserve de liquide hydraulique (*n.f.*),
hydraulic fluid reserve.
réservoir d'eau usée (*n.m.*), waste water tank.
réservoir d'équilibrage (*n.m.*), trim tank.
réservoir d'huile (*n.m.*), oil tank.
réservoir de carburant (*n.m.*), fuel tank.
réservoir hydraulique (*n.m.*), hydraulic
reservoir.
réservoir structural (*n.m.*), built-in tank.
résidu (élec.) (*n.m.*), harmonic content.
résilience (*n.f.*), impact strength.
résine d'urée-formol (plast.) (*n.f.*), urea-
formaldehyde resin; U.F. resin.
résine de coumarone (plast.) (*n.f.*), coumarone
resin.
résine de crésol (plast.) (*n.f.*), cresol resin.
résine de phénol (plast.) (*n.f.*), phenolic resin.
résistance (élec.) (*n.f.*), resistor.
résistance à la fatigue (*n.f.*), fatigue strength.
résistance au fluage (*n.f.*), creep strength.
résistance chauffante (*n.f.*), heating resistor.
résistance chutrice (élec.) (*n.f.*), ballast
resistance.
résistance de charge (élec.) (*n.f.*), load
resistor.
résistance de fuite (*n.f.*), bleeder.
résistance de polarisation (élec.) (*n.f.*), bias
resistor.
résistance de rupture (*n.f.*), ultimate
strength.
résistance diélectrique (*n.f.*), dielectric
strength.
résistance permanente à la traction (*n.f.*),
endurance tensile strength.
résonance (*n.f.*), resonance.
résonance au sol (*n.f.*), ground resonance.
résonance de bielle de commande (héli-
-coptère) (*n.f.*), control rod resonance.
résonateur (élec.) (*n.m.*), resonator.
ressort à lames (auto) (*n.m.*), leaf spring.
ressort de flexion (*n.m.*), flexion spring.
ressort de rappel (*n.m.*), return spring.
ressort de traction (*n.m.*), extension spring.
ressort en épingle (*n.m.*), hairpin spring.
ressort spiral (*n.m.*), hair spring.
ressource (aviation) (*n.f.*), pull-out; recovery.
ressuage (*n.m.*), dye penetrant inspection.
rétablissement (aviation) (*n.m.*), pull-out.
retard de phase (élec.) (*n.m.*), phase lag.
retardateur (*n.m.*), delay timer.
réticulaire (structure) (*adj.*), network-like.
réticule (*n.m.*), reticle.
retient-graisse (*n.m.*), grease retainer; oil seal.
retouche (*n.f.*), rectification in situ.
retouche de fabrication (*n.f.*), rework.
retouche de soudure (*n.f.*), touch-up.
retour bâche (*n.m.*), reservoir return line.
retour d'asservissement (*n.m.*), feedback.
retour de flamme (*n.m.*), back-fire.
retour en arrière (cinéma) (*n.m.*), flash-back.
retrait (*n.m.*), contraction.
rétreindre (*v.t.*), to shrink (in swaging).
rétreint (*n.m.*), necking; throat (in swaging).
rétroaction (radio) (*n.f.*), feedback.
rétrochargeuse (construction) (*n.f.*), back-
loader.
rétroviseur (auto) (*n.m.*), driving mirror.
réusiner (*v.t.*), to rework.
revêtement (cellule) (*n.m.*), skin.

revêtement électrolytique à cuve rota-
-tive (*n.m.*), barrel plating.
revêtement par immersion (*n.m.*), dip-
coating; dip-plating.
révision (d'un moteur) (*n.f.*), overhaul.
rhénium (élément) (*n.m.*), rhenium.
rhéostat de champ (élec.) (*n.m.*), field rheostat.
rhéostriction (élec.) (*n.f.*), rheostriction.
rideau (*n.m.*), curtain.
ridelle (d'une camion) (*n.f.*), ledge (of a lorry).
ridoir (*n.m.*), turnbuckle.
rigide (*adj.*), steady; stiff.
rigidité diélectrique (*n.f.*), disruptive
strength.
rigidité du moyen en battement (hélicop-
-tère) (*n.f.*), hub flapping stiffness.
rinçage (*n.m.*), flushing.
risque (*n.m.*), hazard.
rivet à clin (*n.m.*), clinch rivet.
rivet à tête affleurie (*n.m.*), flush-head rivet.
rivet à tête cylindrique (*n.m.*), cylin-
-drical head rivet.
rivet à tête goutte de suif (*n.m.*), brazier
head rivet.
rivet à tête réduite (*n.m.*), small head rivet.
rivet bouterollé (*n.m.*), snap-head rivet.
rivet explosif (*n.m.*), "pop" rivet.
rivet fendu (*n.m.*), split rivet.
rivet plein (*n.m.*), solid rivet.
rivet tête chaudronnée (*n.m.*), brazier head
rivet.
rivet tubulaire (*n.m.*), hollow rivet.
riveter à chaud (*v.t.*), to hot rivet.
riveter à froid (*v.t.*), to cold rivet.
riveuse à martelage (*n.f.*), percussion riveter.
robe (d'une tube cathodique) (*n.f.*), cone.
robinet à guillotine (*n.m.*), gate valve.
robinet à passage intégral (*n.m.*), non-
restricted valve.
robinet à pointeau (*n.m.*), needle valve.
robinet carburant (*n.m.*), fuel cock.
robinet coupe-feu (*n.m.*), fuel shut-off cock.
robinet de chute de pression (*n.m.*), pressure
releasing valve.
robinet de débit (*n.m.*), throttle valve.
robinet de décantation (*n.m.*), sediment drain
cock.
robinet de drainage (*n.m.*), drain valve.
robinet de purge (*n.m.*), purge tap.
robinet de trop-plein (*n.m.*), overflow valve.
robinet de vidange (*n.m.*), drain cock.
robinet électrique (*n.m.*), electric valve.
robinet trois-voies (*n.m.*), three-way valve.
rochet (*n.m.*), ratchet.
rodage (d'un moteur) (*n.m.*), running-in.
rodage (d'un surface) (*n.m.*), honing.
roder (*v.t.*), to run in.
rogner (*v.t.*), to nibble.
rompu, -e (circuit électrique) (*adj.*), broken.
rond (*n.m.*), round bar.
rond rainuré (*n.m.*) slotted bead.
rondelle à collerette (*n.f.*), flanged washer.
rondelle à crans (*n.f.*), toothed washer.
rondelle à ergots (*n.f.*), tab washer.
rondelle amiante (*n.f.*), asbestos washer.
rondelle autofreineuse (*n.f.*), stop washer.
rondelle autolubrifiante (*n.f.*), oil ring.
rondelle biseautée (*n.f.*), bevelled washer.
rondelle bombée (*n.f.*), disher washer.
rondelle chambrée (*n.f.*), recessed washer.
rondelle contre-rivure (*n.f.*), rivet washer.

rondelle cuvette (*n.f.*), cup washer.
rondelle d'appui (*n.f.*), thrust washer.
rondelle d'arrêt (*n.f.*), stop washer.
rondelle d'épaisseur (*n.f.*), thickness washer.
rondelle d'équilibrage (*n.f.*), balance washer.
rondelle de butée (*n.f.*), stop washer.
rondelle de calage (*n.f.*), set ring.
rondelle de feutre (*n.f.*), felt washer.
rondelle de frottement (*n.f.*), friction ring.
rondelle de liège (*n.f.*), cork washer.
rondelle de poussée (*n.f.*), thrust washer.
rondelle de réglage (*n.f.*), adjusting washer;
shim washer.
rondelle de retenue (*n.f.*), retainer.
rondelle de sécurité (*n.f.*), safety washer.
rondelle de sertissage (*n.f.*), crimping
washer.
rondelle décolletée (*n.f.*), stepped washer.
rondelle élastique (*n.f.*), spring washer.
rondelle embase (*n.f.*), base washer.
rondelle entretoise (*n.f.*), spacing washer.
rondelle éventail (*n.f.*), lock washer.
rondelle fendue (*n.f.*), split washer.
rondelle filetée (*n.f.*), threaded washer.
rondelle frein (*n.f.*), lock washer.
rondelle Grower (*n.f.*), lock washer; spring
washer.
rondelle Grower à rebord (*n.f.*), bend-up
lock washer.
rondelle isolante (*n.f.*), insulating washer.
rondelle joint (*n.f.*), gasket.
rondelle plate (*n.f.*), flat washer.
rondelle rectangulaire (*n.f.*), lock plate.
rondelle ressort (*n.f.*), spring washer.
rondelle sphérique (*n.f.*), spherical washer.
rondelle stratifiée (*n.f.*), laminated washer.
rondelle striée (*n.f.*), serrated washer.
ronflement (d'un moteur) (*n.m.*), hum.
ronfleur (*n.m.*), buzzer.
ronronner (d'un moteur) (*v.i.*), to hum.
rotamètre (*n.m.*), rotameter.
rotatif, -ive (*adj.*), rotary; revolving; rotative.
rotation journalière (*n.f.*), daily use.
rotationnel (d'un vecteur) (élec.) (*n.m.*), rotation;
curl (of a vector).
rotodyne (*n.m.*), rotodyne.
rotor anti-couple (hélicoptère) (*n.m.*), tail
rotor.
rotor arrière (hélicoptère) (*n.m.*), tail rotor.
rotor articulé (hélicoptère) (*n.m.*), articulated
rotor.
rotor auxiliaire (hélicoptère) (*n.m.*), auxiliary
rotor.
rotor balancier (hélicoptère) (*n.m.*), see-saw-
type rotor.
rotor basculant (hélicoptère) (*n.m.*), tilting
rotor.
rotor escamotable (hélicoptère) (*n.m.*),
retractable rotor.
rotor inclinable (hélicoptère) (*n.m.*), tilting
rotor.
rotor libre (hélicoptère) (*n.m.*), free rotor.
rotor pivotant (hélicoptère) (*n.m.*), swinging
rotor.
rotor principal (hélicoptère) (*n.m.*), main rotor.
rotor rigide (hélicoptère) (*n.m.*), rigid rotor.
rotor soufflé (hélicoptère) (*n.m.*), jet flapped
rotor.
rotor sustentateur (hélicoptère) (*n.m.*), lifting
rotor.
rotule (*n.f.*), ball joint.

rotule à queue (*n.f.*), ball end.
rotule d'articulation (*n.f.*), hinge and ball joint.
rotule d'embout (*n.f.*), ball end.
rotule de friction (*n.f.*), friction ball.
roue à aubes (*n.f.*), impeller.
roue avant (avion) (*n.f.*), nose gear wheel.
roue caoutchoutée (*n.f.*), rubber treaded wheel.
roue conique (*n.f.*), turbine bevel ring.
roue de turbine (*n.f.*), turbine wheel.
roue de vis sans fin (*n.f.*), worm wheel.
roue droite (*n.f.*), spur gear.
roue libre (*n.f.*), free wheel.
roue soufflante (*n.f.*), blower wheel.
roues en diabolo (*n.f.pl.*), twin wheels.
roues jumelées (*n.f.pl.*), twin wheels.
rouet d'antenne (radio) (*n.m.*), aerial reel; aerial wind; antenna reel, antenna wind (U.S.A.).
roulage au sol (aviation) (*n.m.*), taxiing.
roulage des filets (*n.m.*), thread rolling.
rouleau à pieds de mouton (trav. pub.) (*n.m.*), sheep's foot roller.
rouleau compresseur vibrateur (trav. pub.) (*n.m.*), vibrating roller.
rouleau de câble (*n.m.*), cable bundle.
roulement à aiguilles (*n.m.*), needle bearing.
roulement à billes à gorge profonde (*n.m.*), deep groove ball bearing.
roulement à deux rangées de billes (*n.m.*), double row ball bearing.
roulement à galets (*n.m.*), roller bearing.
roulement à l'atterrissage (aviation) (*n.m.*), landing run.
roulement à rotule (*n.m.*), self-aligning bearing.
roulement à rouleaux (*n.m.*), roller bearing.
roulement à segments (*n.m.*), slipper bearing.
roulement à tige fileté (*n.m.*), rod end bearing.
roulement à une rangée de billes (*n.m.*), single row ball bearing.
roulement au sol (aviation) (*n.m.*), taxiing.
roulement de butée (*n.m.*), thrust bearing.
roulement orientable (*n.m.*), self-aligning bearing.
roulement oscillant (*n.m.*), self-aligning bearing.
roulette de queue (aviation) (*n.f.*), tail wheel.
roulis (navigation) (*n.m.*), roll.
route (navigation) (*n.f.*), course.
route aérienne (*n.f.*), airway.
route magnétique (*n.f.*), magnetic heading.
route vraie (*n.f.*), true course.
ruban à masquer (*n.m.*), masking tape.
ruban adhésif (*n.m.*), adhesive tape.
rubis (*n.m.*), jewel (for instrument bearings).
rupteur (auto) (*n.m.*), contact breaker.
rupteur thermique (*n.m.*), thermal switch.
rupture (*n.f.*), breakdown; failure.

S

sablage au jet (*n.m.*), sand blasting.
sablage humide (*n.m.*), vapour blasting.
sabot arrière (aviation) (*n.m.*), tail skid.
sabot de balai d'essuie-glace (*n.m.*), wiper blade shoe.
sacoche (*n.f.*), bag.
saignée (*n.f.*), groove.
saillie (*n.f.*), protrusion.
sain, -e (*adj.*), sound.
salle de séjour (construction) (*n.f.*), living room.
salle des dépêches (journaux et radio) (*n.f.*), news desk.
sandow (*n.m.*), bungee cord.
sangle (*n.f.*), strap.
satellite (*n.m.*), planet gear.
saturation du sol par l'eau (*n.f.*), waterlogging.
saumon d'aile (aviation) (*n.m.*), wing tip.
saumon de dérive (aviation) (*n.m.*), vertical stabilizer tip.
saumon de pale (aviation) (*n.m.*), blade tip cap.
saumon de plan fixe (*n.m.*), horizontal stabilizer tip.
saupoudrage de culture (agriculture) (*n.m.*), crop dusting.
saute de vent (navigation) (*n.f.*), veer of the wind.
sauterelle (*n.f.*), snap fastener.
savoir-faire (*n.m.*), know-how.
scarificatrice (trav. pub.) (*n.f.*), scarifier.
schéma (*n.m.*), layout.
schéma de câblage (*n.m.*), wiring diagram.
schéma de principe (*n.m.*), functional diagram.
schéma simplifié (*n.m.*), block diagram.
schoopage (*n.m.*), metal spraying.
scléroscope (*n.m.*), scleroscope.
scripte (cinéma) (*n.m.*), script.
second régime d'un moteur (*n.m.*), engine second rating.
secteur (élec.) (*n.m.*), mains.
secteur de manette (*n.m.*), control lever quadrant.
secteur de poulie (*n.m.*), pulley sector.
section d'écoulement libre (*n.f.*), unrestricted passage area.
section de la pale (aviation) (*n.f.*), blade cross-section.
section évolutive (*n.f.*), tapered section.
sectionneur (élec.) (*n.m.*), isolating switch.
sécurité électrique (*n.f.*), electrical safety.
sécurité en vol (*n.f.*), safety in flight.
segment (auto) (*n.m.*), cylinder ring; piston ring.
segment d'étanchéité (*n.m.*), sealing ring; compression ring.
segment de couronne (*n.m.*), ring segment.
segment de feu (auto) (*n.m.*), top ring.
segment racleur (auto) (*n.m.*), oil scraper ring.
ségrégation de solidification (*n.f.*), freezing.
sélecteur (*n.m.*), selector.
sélecteur à trois voies (*n.m.*), three-way selector.
sélecteur d'écoute (radio) (*n.m.*), audio switch.
sélecteur de cap (aviation) (*n.m.*), heading selector.
sélecteur de démarrage (auto) (*n.m.*), starter selector switch.

sélecteur de niveau de carburant (avia-
-tion) (*n.m.*), fuel level selector.
sélecteur de route (aviation) (*n.m.*), course
selector.
sélecteur de téléphone (*n.m.*), phone switch.
sélecteur de vol (aviation) (*n.m.*), mode selector
switch.
sélecteur omnidirectionnel (aviation)
(*n.m.*), omni-bearing selector.
sélecteur prise statique (*n.m.*), static air
selector.
sélecteur "tout ou rien" (*n.m.*), "hit or miss"
selector valve.
sélectivité (*n.f.*), selectivity.
self de filtrage (*n.m.*), filter choke.
self-obturateur (*n.m.*), self-sealing.
semelle de longeron (aviation) (*n.f.*), spar
flange.
semelle de longeron rapportée (aviation)
(*n.f.*), spar cap.
semi-conducteur (élec.) (*n.m.*), semi-conductor.
semi-conducteur extrinsèque (électron.)
(*n.m.*), extrinsic semi-conductor.
semi-conducteur intrinsèque (électron.)
(*n.m.*), intrinsic semi-conductor.
semi-noyé, -e (*adj.*), semi-flush.
sens de la fibre (*n.m.*), grain flow.
sens de rotation (*n.m.*), sense of rotation.
sens de rotation des phases (*n.m.*), phase
sequence.
sens de rotation du rotor (aviation) (*n.m.*),
rotor rotation sense.
sens des aiguilles d'une montre (*n.m.*),
clockwise.
sens inverse des aiguilles d'une montre
(*n.m.*), anti-clockwise.
sensibilisateur (électron.) (*n.m.*), sensitizer.
séquence (*n.f.*), sequence.
séquence à pas de pélerin (soudure) (*n.f.*),
backstep sequence.
serpentin réfrigérant (*n.m.*), cooling coil.
serre-câbles (*n.m.*), cable grip.
serre-fils (*n.m.*), clamp.
serre-flan (*n.m.*), draw ring.
serre-glace (*n.m.*), window pane retainer.
serre-joint (*n.m.*), screw clamp.
sertir par gaufrage (*v.t.*), to swage.
sertir par replis (*v.t.*), to crimp.
sertissage (*n.m.*), crimping; swaging.
servante (machines-outils) (*n.f.*), work stand.
service intermittent (élec.) (*n.m.*), intermit-
-tent duty.
service navette (*n.m.*), shuttle service.
service permanent (élec.) (*n.m.*), permanent
duty.
service technique (*n.m.*), engineering
department.
servitudes au sol (aviation) (*n.f.pl.*), ground
handling services.
servitudes avion (aviation) (*n.f.pl.*), equipment
services.
servo-actionneur de gouverne (aviation)
(*n.m.*), servo rotary actuator.
servo-actionneur de trim (aviation) (*n.m.*),
trim actuator.
servo-amplificateur (aviation) (*n.m.*), servo
amplifier.
servo-amplificateur de trim (aviation)
(*n.m.*), trim servo amplifier.
servo-commande (*n.f.*), servo control.
servo-commande auxiliaire (*n.f.*), auxiliary

servo control.
servo-commande principale (*n.f.*), main
servo control.
servo régulateur du pas cyclique (héli-
-coptère) (*n.m.*), cyclic pitch servo trim.
servo-rotor (aviation) (*n.m.*), control rotor.
servodyne (*n.f.*), servodyne unit.
servodyne double (*n.f.*), tandem servodyne.
seuil (*n.m.*), threshold.
seuil d'audibilité (acoust.) (*n.m.*), threshold of
hearing.
seuil de blocage réception (radio) (*n.m.*),
squelch threshold.
shunt de masse (*n.m.*), grounding shunt.
shunt universel (élec.) (*n.m.*), Ayrton shunt.
siallite (géol.) (*adj.*), siallitic.
siccatif, -ive (*adj.*), air drying.
siège de bonde (*n.m.*), plug seat.
siège de clapet (*n.m.*), clapper seat.
siège de rotule (*n.m.*), ball socket seat.
signal continu (radio) *n.m.*), steady-rate signal.
signal d'écart (aviation) (*n.m.*), deviation signal.
signal d'entrée (radio) (*n.m.*), input signal.
signal d'erreur (aviation) (*n.m.*), error signal.
signal de sortie (radio) (*n.m.*), output signal.
signal de stabilisation (aviation) (*n.m.*), sta-
-bilization signal.
signal vidéo (*n.m.*), video signal.
signalisation (*n.f.*), indicating.
silicone (plast.) (*n.m.*), silicone.
silencieux de piste (aviation) (*n.m.*), jet noise
suppressor.
silencieux de radio (*n.m.*), squelch.
silencieux de soufflage (*n.m.*), aerator
silencer.
silentbloc (*n.m.*), flexible mounting.
silicagel (*n.m.*), silica gel.
silicium (élément) (*n.m.*), silicon.
sillage (aviation) (*n.m.*), stream; wake.
sillage du rotor (hélicoptère) (*n.m.*), rotor
slipstream.
sillonnement d'usure (*n.m.*), grooving.
similitude (*n.f.*), similarity.
simulateur de pilotage (aviation) (*n.m.*), link
trainer.
simulateur de vol (aviation) (*n.m.*), flight
simulator.
site (*n.m.*), elevation.
socle (*n.m.*), base.
socle de fixation (*n.m.*), mounting base.
soie artificielle (*n.f.*), celanese.
solénoïde de blocage (*n.m.*), snubber.
solénoïde de remplissage (*n.m.*), refuelling
solenoid.
solénoïde de vidange (*n.m.*), drain solenoid.
solidaire (*adj.*), embodied.
sollicitation (*n.f.*), stress.
solution tampon (chrome) (*n.f.*), buffer solution.
solvant (peinture) (*n.m.*), thinner.
somme (*n.f.*), sum.
sommet (*n.m.*), vertex.
son (*n.m.*), sound.
son grave (*n.m.*), low tone.
sondage de recherche (extraction de pétrole)
(*n.m.*), core drilling.
sondage ultrasonore (marine) (*n.m.*), echo-
sounding.
sonde (électron.) (*n.f.*), probe.
sonde altimétrique (*n.f.*), radio altimeter.
sonde d'ambiance (conditionnement d'air)
(*n.f.*), ambient thermoresistor.

sonde de cyclage (n.f.), cycling thermo-resistor.

sonde de givrage (n.f.), icing probe.

sonde de surchauffe (n.f.), overheat thermoresistor.

sonde de température carburant (n.f.), fuel temperature probe.

sonde de température d'huile (n.f.), oil temperature probe.

sonde de température extérieure (aviation) (n.f.), outside air temperature probe.

sonde magnétométrique (n.f.), fluxvalve.

sonde thermométrique (n.f.), temperature probe.

sonner un circuit (v.), to check the circuit continuity.

sonnerie continue (n.f.), continuous ring bell.

sonnerie d'alarme (n.f.), alarm bell.

sonnerie monocoup (n.f.), single ring bell.

sortie (élec.) (n.f.), outlet.

sortie asservie (élec.) (n.f.), controlled outlet.

sortie d'air dégivrage (n.f.), de-icing air outlet.

sortie du train d'atterrissage (aviation) (n.f.), landing gear extension.

soudage (d'une enveloppe plastique) (n.m.), heat sealing.

soudage à droite (n.m.), backward welding.

soudage à gauche (n.m.), forward welding.

soudage à l'arc à électrodes de carbone (n.m.), carbon arc welding.

soudage à l'arc à électrodes métalliques enrobées (n.m.), shielded metal arc welding.

soudage à l'arc à électrodes métalliques nues (n.m.), unshielded metal arc welding.

soudage à l'arc en atmosphère inerte (n.m.), inert gas welding.

soudage à l'argon (n.m.), argon-arc welding.

soudage à l'hydrogène atomique (n.m.), atomic hydrogen welding.

soudage à la forge (n.m.), forge welding.

soudage à la molette (n.m.), seam welding.

soudage autogène (n.m.), fusion welding.

soudage bout à bout (n.m.), butt seam welding.

soudage des goujons (n.m.), stud welding.

soudage en arrière (n.m.), backhand welding.

soudage en bout par résistance (n.m.), resistance butt welding.

soudage oxy-acétylénique (n.m.), oxy-acetylene welding.

soudage par étincelage (n.m.), flash welding.

soudage par fusion (n.m.), fusion welding.

soudage par induction (n.m.), induction welding.

soudage par points (n.m.), spot welding.

soudage par points à impulsions (n.m.), pulsation welding; woodpecker point welding.

soudage par pression (n.m.), pressure welding.

soudage par projection (n.m.), projection welding.

soudage par rapprochement (n.m.), butt welding.

soudage par recouvrement continu (n.m.), seam welding.

soudage par résistance (n.m.), resistance welding.

soude caustique (chimie) (n.f.), sodium hydroxide.

soudo-brazure (n.f.), brazing.

soudure à décapant incorporé (n.f.), rosin cored solder.

soudure à l'étain (n.f.), tin solder.

soudure autodécapante (n.f.), self-fluxing solder.

soudure de pointage (n.f.), tack weld.

soudure de soutien (n.f.), back weld.

soudure de thermocouple (n.f.), hot junction.

soudure en bout (n.f.), butt weld.

soudure non traitée thermiquement (n.f.), as welded.

soudure par points à la molette (n.f.), roll spot weld.

soudure par points continus (n.f.), stitch weld.

soudure par recouvrement (n.f.), lap weld.

souffle (n.m.), blast.

souffle d'antenne (radio) (n.m.), aerial noise; antenna pick-up (U.S.A.).

souffle des réacteurs (aviation) (n.m.), engine jet wash.

souffle du récepteur (radio) (n.m.), receiver tube noise.

souffle du rotor (hélicoptère) (n.m.), rotor stream.

souffle tourbillonnaire du rotor (n.m.), rotor vortex stream.

souffle vertical (n.m.), vertical stream.

soufflerie à arc bref (n.f.), hot shot wind tunnel.

soufflerie aérodynamique (n.f.), wind tunnel.

soufflet de gonflage (n.m.), air pump.

soufflet de protection (n.m.), boot.

soufflure (n.f.), blow hole.

soulever (v.t.), to lift.

soupape à arc (élec.) (n.f.), arc rectifier.

soupape à cathode liquide (élec.) (n.f.), pool rectifier.

soupape à décharge luminescente (élec.) (n.f.), glow-discharge rectifier.

soupape à gaz (élec.) (n.f.), gas-filled rectifier.

soupape à semi-conducteur (élec.) (n.f.), semi-conductor rectifier.

soupape à solénoïde (n.f.), solenoid valve.

soupape à tulipe (n.f.), poppet valve.

soupape à va-et-vient (n.f.), shuttle valve.

soupape à vapeur de mercure (élec.) (n.f.), mercury arc rectifier.

soupape à vide entretenu (élec.) (n.f.), pumped rectifier.

soupape champignon (n.f.), poppet valve.

soupape de décompression (n.f.), depressurization valve.

soupape de dosage (n.f.), metering valve.

soupape de prélèvement d'air (n.f.), air bleed valve.

soupape de purge (n.f.), purge valve.

soupape de surpression (n.f.), pressure relief valve.

soupape de vidange (n.f.), drain valve.

soupape électrique (élec.) (n.f.), rectifier.

soupape électrolytique (élec.) (n.f.), electrolytic rectifier.

soupape ionique (élec.) (n.f.), gas filled rectifier.

soupape monoanodique (élec.) (n.f.), single-anode rectifier.

soupape polyanodique (élec.) (n.f.), multi-anode rectifier.

soupape thermionique (élec.) (n.f.), thermionic rectifier.

souplisseau (n.m.), spaghetti tubing.

source d'alimentation (n.f.), power supply.

source d'alimentation extérieure (avia-
-tion) (*n.f.*), ground power supply.
source électrique (*n.f.*), electrical power
supply.
source hydraulique (*n.f.*), hydraulic pressure
supply.
source pneumatique (*n.f.*), air pressure supply.
sous-ensemble (*n.m.*), sub-assembly.
sous tension (élec.) (*adj.*), alive; energized; live.
soute (*n.f.*), cargo compartment.
soutien logistique (*n.m.*), logistic support.
soyage (*n.m.*), joggling.
soyé, -e (*adj.*), joggled.
spatula (*n.f.*), scraper.
spectre d'énergie (*n.m.*), power spectrum.
spectre de vol (*n.m.*), flight spectrum.
spiral (*n.m.*), helix.
spire (*n.m.*), coil.
spires jointives (*n.f.pl.*), solid length of a spring.
spoiler (*n.m.*), spoiler.
spot (dans un tube cathodique) (*n.m.*), spot.
stabilisant (plast.) (*n.m.*), stabilizer.
stabilisateur (aviation) (*n.m.*), stabilizer.
stabilisateur de flamme (aviation) (*n.m.*),
flame holder.
stabilisation (*n.f.*), stabilization.
stabilisation de cap (aviation) (*n.f.*), heading
tool.
stabilisation du sol (trav. pub.) (*n.f.*), soil
stabilizing.
stabilité dynamique en lacet et en roulis
(aviation) (*n.f.*), dynamic directional and lat-
-eral stability.
stabilité latérale statique (aviation) (*n.f.*),
static lateral stability.
stabilité longitudinale dynamique (avia-
-tion) (*n.f.*), dynamic longitudinal stability.
stabilité longitudinale statique (aviation)
(*n.f.*) static longitudinal stability.
stabilité manche bloqué (aviation) (*n.f.*),
fixed stick stability.
stage (étape) (*n.m.*), stage.
stage d'instruction (*n.m.*), training course.
stagiaire (*n.m.*), trainee.
stall-vane (aviation) (*n.f.*), wing fence.
stalle (construction) (*n.f.*), box.
standard, -e (*adj.*), standard.
starter (auto) (*n.m.*), choke.
station radio (*n.f.*), ground radio station.
station service (*n.f.*), service station.
stationnement (*n.m.*), parking.
stator compresseur (*n.m.*), compressor stator.
statoréacteur (moteur) (*n.m.*), ramjet.
stéradian (math.) (unité d'angle solide) (*n.m.*),
steradian.
stilb (unité de brillance) (C.G.S.) (*n.m.*), stilb.
stockage (*n.m.*), storing.
stratifié, -e (*adj.*), laminated.
stratifié (plast.) (*n.m.*), laminate; laminated
plastics.
stratopause (couche de l'atmosphère) (*n.f.*),
stratopause.
stratosphère (couche de l'atmosphère) (*n.f.*),
stratosphere.
strié, -e (*adj.*), serrated.
stries (*n.f.pl.*), scratches; serrations; scores.
stroboscope (*n.m.*), stroboscope.
stroboscopie (*n.f.*), stroboscopy.
structure caisson (*n.f.*), box-type structure.
structure en nids d'abeilles (*n.f.*), honey-
-comb structure.

structure grumeleuse (*n.f.*), crumb structure.
structure résistante (*n.f.*), primary structure.
structure secondaire (*n.f.*), secondary
structure.
structure soudée (*n.f.*), welded structure.
suint (*n.m.*), wool grease.
super-polissage (*n.m.*), mirror finish.
supercarburant (*n.m.*), premium grade petrol.
superficie totale des pales (aviation) (*n.f.*)
total blade area.
superhétérodyne (radio) (*n.m.*), super-
-heterodyne.
superposé, -e (*adj.*), superimposed.
supersonique (*adj.*), supersonic.
supplémentaire (*adj.*), additional.
support (*n.m.*), bracket.
support à rotule (*n.m.*), swivel fitting.
support anti-vibratoire (radio) (*n.m.*), shock
mount.
support bornes (*n.m.*), terminal board.
support de couple (*n.m.*), torque pin.
support de démarreur (*n.m.*), starter support.
support de jambe de train (aviation) (*n.m.*),
landing gear leg support.
support de train avant (aviation) (*n.m.*), nose
gear saddle.
support de tube (*n.m.*), tube socket.
support de turbine (aviation) (*n.m.*), engine
mount.
support gauchissement (hélicoptère) (*n.m.*),
lateral cyclic control support.
support prismatique (*n.m.*), vee block.
support profondeur (hélicoptère) (*n.m.*), fore
and aft cyclic control support.
support reproductible (plans) (*n.m.*), master
copy.
suppresseur (*n.m.*), eliminator.
suppresseur de bruit de fond (radio) (*n.m.*),
squelch.
surbau (*n.m.*), coaming.
surdiamétrage (*n.m.*), oversizing.
surdimensionné, -e (*adj.*), oversize.
surélevé, -e (*adj.*), raised.
surépaisseur de soudure (*n.f.*), weld
reinforcement.
surfaçage (machines-outils) (*n.m.*), facing.
surface alaire (aviation) (*n.f.*), design wing area.
surface conductrice (*n.f.*), conducting surface.
surface d'appui (*n.f.*), bearing surface.
surface de fond de creux de denture
(engrenage) (*n.f.*), tooth bottom land.
surface de freinage (*n.f.*), brake area.
surface de tête de denture (engrenage) (*n.f.*),
tooth top land.
surface du disque balayé (hélicoptère) (*n.f.*),
disc swept area.
surface du disque rotor (hélicoptère) (*n.f.*),
rotor disc area.
surface frontale (*n.f.*), frontal area.
surface mouillée (*n.f.*), wetted area.
surface nette de l'aile (*n.f.*), net wing area.
surface projetée (*n.f.*), projected area.
surface totale de l'aile (*n.f.*), wing total area.
surface verticale de l'empennage (avia-
-tion) (*n.f.*), vertical tail area.
surfacer dur (*v.t.*), to hard face.
surintensité (élec.) (*n.f.*), overcurrent; overload.
surjeu (audio-visuel) (*n.m.*), playback.
surpression (*n.f.*), overpressure.
surtension (élec.) (*n.f.*), over-voltage.
surveillance (*n.f.*), monitoring.

survitesse (*n.f.*), overspeed.
suspension à la cardan (*n.f.*), gimbal suspension.
suspension articulée (*n.f.*), hinged suspension.
suspension caténaire (ch. de f. élec.) (*n.f.*), catenary suspension.
sustension (*n.f.*), lift.
sustension de translation (hélicoptère) (*n.f.*), translation lift.
symbole de couplage d'un transforma-
-teur (élec.) (*n.m.*), vector group symbol of a transformer.
symptôme (*n.m.*), symptom.
synchro d'érection (aviation) (*n.m.*), gyro lev-elling synchro.
synchro de maintien quadrangulaire (aviation) (*n.m.*), quadrantal synchro.
synchro de recalage (aviation) (*n.m.*), slaving synchro.
synchro détecteur de roulis (aviation) (*n.m.*), roll detector synchro.
synchro détecteur de tangage (aviation) (*n.m.*), pitch detector synchro.
synchro répétiteur (aviation) (*n.m.*), synchro transmitter.
synchroniseur (*n.m.*), synchronizer.
synchroniseur de cap (aviation) (*n.m.*), head-ing synchronizer.
synchroniseur de tangage (aviation) (*n.m.*), pitch synchro.
synchroniseur intégrateur (aviation) (*n.m.*), heading synchronizer and lateral pitch integrator.
synchroniseur par gaz (hélicoptère) (*n.m.*), throttle synchronizer.

synchronisme, en (*adj.*), synchronized.
synchrotron (atom.) (*n.m.*), synchrotron.
système aérien d'appui feu avancé (*n.m.*), advanced airborne fire-support system.
système autopositionnement (*n.m.*), auto positioning unit.
système d'approche contrôlé au sol (*n.m.*), ground controlled approach.
système d'armes intégré (aviation) (*n.m.*), integrated weapon system.
système d'articulation rotor (hélicoptère) (*n.m.*), rotor hingeing.
système d'asservissement du gyroscope directionnel (aviation) (*n.m.*), slueing assembly.
système d'atterrissage radiogoniomét-
-rique (aviation) (*n.m.*), instrument landing system.
système d'augmentation de stabilité (aviation) (*n.m.*), stability augmentation system.
système de commande (*n.m.*), control system.
système de contrôle de vol automatique (aviation) (*n.m.*), automatic flight-control system.
système de flottabilité de secours (avia-tion) (*n.m.*), emergency flotation gear.
système mètre, kilogramme, seconde, ampère (système MKSA) (*n.m.*), metre, kil-ogram, second, ampere system; Giorgi system.
système MKSA (système mètre, kilogramme, seconde, ampère) (*n.m.*), Giorgi system; metre, kilogram, second, ampere system.
système retardateur (*n.m.*), lagging system.
système réticulé (*n.m.*), reticulated system.

T

tab à ressort (*n.m.*), spring tab.
tab de réglage d'incidence de pale avia-tion) (*n.m.*), blade trim tab.
table des matières (*n.f.*), contents list; table of contents.
table tactique (mil.) (*n.f.*), tactical plotting system.
tableau (*n.m.*), chart; diagram; panel.
tableau de bord (auto) (*n.m.*), dashboard; instrument panel.
tableau de dépannage (*n.m.*), fault finding chart.
tableau lumineux des alarmes (*n.m.*), gen-eral warning panel.
tablette (*n.f.*), tablet.
tablier (auto) (*n.m.*), dash.
tabulatrice (électronique) (*n.f.*), tabulator.
tache cathodique (élec.) (*n.f.*), cathode spot.
tachymètre de haute précision (*n.m.*), high sensitivity tachometer.
taillage par fraise mère (*n.m.*), hob cutting.
talon d'épaulement (*n.m.*), shoulder.
talon d'une aube (*n.m.*), blade root.
talon de pneu (*n.m.*), tyre bead.
talonnage du sabot de queue (aviation) (*n.m.*), tail drag.
talus d'éboulement (aluminium) (*n.m.*), angle of repose.
tambour à inertie (*n.m.*), inertia reel.

tambour d'embrayage (*n.m.*), clutch drum.
tambour de frein (auto) (*n.m.*), brake drum.
tambour de treuil (hélicoptère) (*n.m.*), winch drum.
tampon de contrôle (*n.m.*), inspection stamp.
tampon-jauge (*n.m.*), plug gauge.
tamponnage (*n.m.*), plugging (of a wall).
tangage (navigation) (*n.m.*), pitch.
tapoter (*v.t.*), to tap gently.
taquet (*n.m.*), bracket.
tarage (*n.m.*), setting.
tarage nominal (*n.m.*), nominal setting.
taraud à rainures hélicoïdales (*n.m.*), spiral flute tap.
tasseau (*n.m.*), cleat.
taupe (trav. pub.) (*n.f.*), mole.
taux (*n.m.*), ratio.
taux d'amplitude (*n.m.*), peak-to-valley ratio.
taux d'appoint (*n.m.*), make-up rate.
taux d'ondes stationnaires (radio) (*n.m.*), standing wave ratio.
taux d'ondulations résiduelles (*n.m.*), rip-ple ratio.
taux de compression (*n.m.*), compression ratio.
taux de pannes (*n.m.*), failure rate.
taux de plané (aviation) (*n.m.*), glide ratio.
taux de réluctances étagées[1] (*n.m.*), grad-uated reluctance.

té de commande de pas (hélicoptère) (*n.m.*), pitch-change beam.

technétium (élément) (*n.m.*), technetium.

télécommande (*n.f.*), remote control.

téléflex (*n.m.*), flexible control.

télémétrie au son (*n.f.*), echo-ranging.

téléscopable (*adj.*), collapsible.

téléscripteur (*n.m.*), teleprinter.

téléviseur (*n.m.*), television set.

téléviseur avec filtre optique (*n.m.*), black screen television set.

télévision (*n.f.*), video (U.S.A.); television (G.B.).

télévision commerciale (*n.f.*), commercial television.

témoin (*n.m.*), signal light.

température absolue relative (*n.f.*), relative absolute temperature.

température ambiante (*n.f.*), ambient temperature.

température extérieure ambiante (*n.f.*), outside ambient temperature.

température tuyère (aviation) (*n.f.*), jet pipe temperature.

temporisateur (élec.) (*n.m.*), timer.

temporisateur électro-mécanique (*n.m.*), repeat cycle timer.

temporisateur thermique (*n.m.*), thermal timer.

temporisation (élec.) (*n.f.*), time-lag; timing.

temps (*n.m.*), duration.

temps bloc (*n.m.*), block time.

temps d'intégration (élec.) (*n.m.*), integration time.

temps d'ouverture (électron.) (*n.m.*), "on" period.

temps de blocage du courant (électron.) (*n.m.*), "off" period.

temps de cuisson (plast.) (*n.m.*), curing time.

temps de réaction (*n.m.*), response time.

temps de réglage (*n.m.*), setting up time.

temps de régulation (*n.m.*), pressure switch cut-out period.

temps de réponse (*n.m.*), response time.

temps de vol (aviation) (*n.m.*), airborne time.

tendance (*n.f.*), trend.

tendance à cabrer (aviation) (*n.f.*), tail heaviness.

tendance à piquer (aviation) (*n.f.*), nose heaviness.

tendance au flambage (aviation) (*n.f.*), elastic instability.

tendance de retour du manche au neutre (aviation) (*n.f.*), stick tendency to neutral.

tendeur à vis (*n.m.*), turnbuckle.

tendeur automatique de câble (*n.m.*), cable tension compensator.

teneur en eau (*n.f.*), moisture content; water content.

tension à l'arc (élec.) (*n.f.*), input voltage.

tension à vide (élec.) (*n.f.*), off-load voltage.

tension-actif, -ive (*adj.*), surface active.

tension aux bornes (élec.) (*n.f.*), terminal voltage.

tension carrée (élec.) (*n.f.*), square wave voltage.

tension d'alimentation (élec.) (*n.f.*), supply voltage.

tension d'asservissement (élec.) (*n.f.*), feed-back voltage.

tension d'éclatement (élec.) (*n.f.*), disruptive voltage.

tension d'ondulation (élec.) (*n.f.*), ripple voltage.

tension de blocage (d'un écran luminescent) (*n.f.*), blocking voltage.

tension de chauffage (élec.) (*n.f.*), heater voltage.

tension de claquage (élec.) (*n.f.*), breakdown voltage.

tension de collage (élec.) (*n.f.*), operating voltage.

tension de court-circuit d'un transformateur (élec.) (*n.f.*), impedance voltage of a transformer.

tension de crête (élec.) (*n.f.*), peak voltage.

tension de décollage (élec.) (*n.f.*), release voltage.

tension de déséquilibre (élec.) (*n.f.*), unbalance voltage.

tension de l'eau (géol.) (*n.f.*), soil-moisture content.

tension de percement (élec.) (*n.f.*), puncture voltage.

tension de polarisation (élec.) (*n.f.*), bias voltage.

tension de repos (élec.) (*n.f.*), rest potential.

tension de retour (élec.) (*n.f.*), feedback voltage.

tension de rupture (élec.) (*n.f.*), breakdown.

tension de secteur (élec.) (*n.f.*), mains voltage.

tension de serrage (élec.) (*n.f.*), torque value.

tension de sortie (élec.) (*n.f.*), output voltage.

tension disruptive (élec.) (*n.f.*), breakdown voltage.

tension efficace (élec.) (*n.f.*), effective voltage.

tension nominale (élec.) (*n.f.*), rated voltage.

tension résiduelle (élec.) (*n.f.*), residual stress.

tenue (*n.f.*), behaviour.

tenue de cap (navigation) (*n.f.*), heading hold.

tenue de stationnaire (hélicoptère) (*n.f.*), hover control.

téra- (préfixe utilisé en électronique), tera- (= 10^{12}).

terrasse de colature (*n.f.*), drainage terrace.

terre arable (*n.f.*), top soil; arable land.

test (*n.m.*), test.

tête d'avertisseur de givrage (aviation) (*n.f.*), ice probe.

tête d'axe (*n.f.*), driving head.

tête d'effacement (enregistrement sur bande) (*n.f.*), erasing head.

tête d'essuie-glace (auto) (*n.f.*), wiper head.

tête de rotor (hélicoptère) (*n.f.*), rotor head.

tête de rotor principal (hélicoptère) (*n.f.*), main rotor head.

tête double de percussion (*n.f.*), dual discharge head.

tête du manche pilote (aviation) (*n.f.*), control column boss.

tête inclinable (machines-outils) (*n.f.*), tilting head.

tête pivotante (machines-outils) (*n.f.*), swivelling head.

tête porte-outil inclinable (machines-outils) (*n.f.*), tilting tool-head.

tête porte-outil orientable (machines-outils) (*n.f.*), swivelling tool-head.

téton (*n.m.*), stud.

téton de blocage (*n.m.*), locking stud.

téton de centrage (*n.m.*), centering pin; locating spigot.

tétrachlorure de carbone (chimie) (*n.m.*), carbon tetrachloride.

tétrode (*n.f.*), tetrode.
texte (cinéma) (*n.m.*), script.
thermie (unité de quantité de chaleur du système MTS) (*n.f.*), therm.
thermistor (*n.m.*), thermistor.
thermocontact (*n.m.*), thermal switch.
thermocouple (*n.m.*), thermocouple.
thermodurcissable (plast.) (*adj.*), thermo--setting.
thermoélectronique (*adj.*), thermionic.
thermoplastique (plast.) (*adj.*), thermoplastic.
thermo-plongeur (*n.m.*), immersion heater.
thermosoudable (plast.) (*adj.*), heat-sealable.
thermosphère (couche de l'atmosphère) (*n.f.*), thermosphere.
thermostable (*adj.*), heat stable.
thyratron (électronique) (*n.m.*), thyratron.
tiédeur (*n.f.*), tepidity.
tierçage (aviation) (*n.m.*), blade spacing system.
tige coulissante (*n.f.*), sliding rod.
tige de centrage (*n.f.*), centering rod.
tige de crémaillère (pompe d'injection) (*n.f.*), rack link.
tige de culbuteur (auto) (*n.f.*), push rod.
tige de rappel (machines-outils) (*n.f.*), drawbar.
tige de réglage (*n.f.*), adjusting rod.
tige de repérage (*n.f.*), centering pin.
tige de vérin (*n.f.*), cylinder rod.
tige filetée (*n.f.*), threaded rod.
timon de remorquage (*n.m.*), towing bar.
timonerie de direction (auto) (*n.f.*), steering linkage.
tir en rafale (*n.m.*), burst of firing.
tirage (impression) (*n.m.*), blueprint.
tirage (radio) (*n.m.*), homing.
tirant (*n.m.*), tie rod.
tirant d'amarrage (*n.m.*), mooring ring.
tire-aiguille (*n.m.*), extractor.
tirette (*n.f.*), pull knob.
tiroir rotatif (*n.m.*), throttle barrel.
tissu de verre (*n.m.*), fibreglass.
tissu de verre imprégné (*n.m.*), bonded glass cloth.
toile (*n.f.*), fabric.
toile de pignon (*n.f.*), pinion web.
toile émeri fine (*n.f.*), crocus cloth.
toile filtrante (*n.f.*), gauze.
toiture à redents (construction) (*n.f.*), shed.
tôlage (*n.m.*), sheet metal planting.
tôle à doigts (aviation) (*n.f.*), skin doubler.
tôle bleue (*n.f.*), spring sheet metal.
tôle de fermeture (*n.f.*), cover plate.
tôle de liaison (*n.f.*), junction panel.
tôle de revêtement (*n.f.*), skin panel.
tôle emboutie (*n.f.*), stamping.
tôle pare-feu (*n.f.*), fire wall.
tôle pliée (*n.f.*), brake formed sheet.
tôle roulée (*n.f.*), roll-formed sheet.
tolérance (*n.f.*), allowance; clearance.
tolérance sur brut (*n.f.*), general tolerance.
tomber un bord (*v.t.*), to flange.
tombereau (trav. pub.) (*n.m.*), dumper.
torche de soudure (*n.f.*), welding torch.
tordu, -e (*adj.*), distorted.
toron (*n.m.*), strand; cable form.
totalisateur de débit (*n.m.*), integrating flowmeter.
touche (d'un clavier) (*n.f.*), key.
touffe du déperditeur (*n.f.*), static discharged wick.

toupie à rainure (*n.f.*), router.
toupilleuse (*n.f.*), routing machine.
tour à banc coulissant (*n.m.*), sliding-bed lathe.
tour à banc droit (*n.m.*), plain-bed lathe.
tour à barre prismatique (*n.m.*), gantry lathe.
tour à dégrossir (*n.m.*), roughing-down lathe.
tour à détalonner (*n.m.*), relieving lathe; backing-off lathe.
tour à écrouter (*n.m.*), peeling lathe; roughing lathe.
tour à fileter et à charioter (*n.m.*), screw and sliding cutting lathe.
tour à outils multiples (*n.m.*), multi-tool lathe.
tour à outils multiples à un chariot (*n.m.*), single-slide multi-tool lathe.
tour à repousser (*n.m.*), spinning lathe; chasing lathe.
tour à reproduire (*n.m.*), copy milling lathe.
tour à tronçonner (*n.m.*), cutting-off lathe.
tour automatique à mandrin (*n.m.*), auto--matic chucking lathe.
tour d'absorption (pétr.) (*n.m.*), absorber.
tour de contrôle (aviation) (*n.f.*), control tower.
tour de forage (pétrole) (*n.f.*), derrick.
tour de fractionnement (raffinerie) (*n.f.*), bubble tower.
tour de piste (courses) (*n.m.*), lap.
tour de reprise (*n.m.*), second operation lathe.
tour en l'air sur banc (*n.m.*), bed-type surfac--ing and boring lathe.
tour en l'air sur taque (*n.m.*), floor-type sur--facing and boring lathe.
tour/minute (*n.m.*), r.p.m.
tour parallèle (*n.m.*), slide lathe.
tourbillon (*n.m.*), eddy.
tourbillon d'extrémité de pale (hélicoptère) (*n.m.*), blade tip vortex.
tourbillon en bout d'aile (aviation) (*n.m.*), tip vortex.
tourbillon marginal (aviation) (*n.m.*), tip vortex.
tourbillonnement (*n.m.*), vorticity.
tourbillonnement (réacteur) (*n.m.*), windage.
tourillonner (*v.i.*), to rotate.
tournette (*n.f.*), circular diamond-cutting apparatus.
tourniquet de verrouillage (*n.m.*), latch.
tournoiement (*n.m.*), swirling; whirling.
tout ou rien, hit or miss.
trace du spot (dans un tube cathodique) (*n.f.*), trace.
tracer (*v.t.*), to mark; to scribe.
traceur de route (aviation) (*n.m.*), course tracer.
tracking (*n.m.*), alignment; tracking.
tracteur (*n.m.*) **et semi-remorque** (*n.f.*), tractor and semi-trailer; articulated vehicle.
traction (*n.f.*), towing.
traction avant (auto) (*n.f.*), front-wheel drive.
train à patins à voie large (hélicoptère) (*n.m.*), wide-track skid landing gear.
train automoteur (aviation) (*n.m.*), self-propelling landing gear.
train avant (aviation) (*n.m.*), nose gear.
train d'atterrissage (aviation) (*n.m.*), landing gear.
train d'engrenages (*n.m.*), gear train.
train épicycloïdal (*n.m.*), epicyclic gear train;

sun and planet gear.

train planétaire (engrenage) (*n.m.*), planetary gear train.

train principal (aviation) (*n.m.*), main landing gear.

train tricycle (aviation) (*n.m.*), tricycle landing gear.

traînée (aviation) (*n.f.*), drag.

traînée d'équilibrage (aviation) (*n.f.*), bal-ancing drag.

traînée de forme (aviation) (*n.f.*), form drag.

traînée de frottement (aviation) (*n.f.*), fric-tion drag.

traînée de fuite (aviation) (*n.f.*), spillage drag.

traînée de pression (aviation) (*n.f.*), pressure drag.

traînée de profil (aviation) (*n.f.*), profile drag.

traînée induite (aviation) (*n.f.*), induced drag.

traînée parasite (aviation) (*n.f.*), parasitic drag.

trait continu (*n.m.*), continuous line.

traitement (*n.m.*), process.

traitement anodique (*n.m.*), anodizing; anodic treatment; anodic etching.

traitement de détente (*n.m.*), stress relief treatment.

traitement de mise en solution (*n.m.*), solution heat treatment.

traitement de surface (*n.m.*), surface treatment.

traitement mécanographique de l'information (*n.m.*), data processing.

traitement thermique (*n.m.*), heat treatment.

traitement thermique de précipitation (*n.m.*), precipitation heat treatment.

trajectoire (*n.f.*), path; trajectory.

trajectoire d'atterrissage (aviation) (*n.f.*), glide path.

trajectoire de décollage (aviation) (*n.f.*), take-off path.

trajectoire de vol (aviation) (*n.f.*), flight path.

trajectoire garantie (aviation) (*n.f.*), guar-anteed flight path.

trame (imprimerie) (*n.f.*), screening.

tranche (*n.f.*), section.

trancheuse (*n.f.*), digger.

transducteur magnétique (élec.) (*n.m.*), transductor.

transfert (*n.m.*), transfer.

transfert de carburant (*n.m.*), fuel transfer.

transfert de température (*n.m.*), heat transfer.

transformateur à champ tournant (élec.) (*n.m.*), rotating field transformer.

transformateur à colonnes (élec.) (*n.m.*), core-type transformer.

transformateur cuirassé (élec.) (*n.m.*), shell-type transformer.

transformateur d'intensité (élec.) (*n.m.*), current transformer.

transformateur de sonnerie (élec.) (*n.m.*), bell transformer.

transformateur redresseur (élec.) (*n.m.*), transformer rectifier.

transformateur suceur (élec.) (*n.m.*), drain-ing transformer.

transformateur survolteur ou dévolteur (élec.) (*n.m.*), booster transformer.

transistor (électronique) (*n.m.*), transistor.

transistor à effet photo-électrique (élec.) (*n.m.*), photistor.

transistor à pointes (électronique) (*n.m.*), point contact transistor.

transition en vol (aviation) (*n.f.*), in-flight transition.

transitron (*n.m.*), transitron.

transmetteur de débit (*n.m.*), flow transmitter.

transmetteur de direction (*n.m.*), rudder follow-up.

transmetteur de flux magnétique (*n.m.*), flux gate transmitter.

transmetteur de gauchissement (aviation) (*n.m.*), aileron follow-up.

transmetteur de jaugeur (*n.m.*), fuel level transmitter.

transmetteur de pas (hélicoptère) (*n.m.*), blade pitch transmitter.

transmetteur de position de volets (avia-tion) (*n.m.*), wing flap follow-up.

transmetteur de poussée (aviation) (*n.m.*), thrust transmitter.

transmetteur de pression (*n.m.*), pressure transmitter.

transmetteur de profondeur (aviation) (*n.m.*), elevator follow-up.

transmission à distance (*n.f.*), remote control.

transmission arrière (hélicoptère) (*n.f.*), tail rotor drive.

transmission au rotor principal (hélicop-tère) (*n.f.*), main drive shaft.

transmission oblique (hélicoptère) (*n.f.*), inclined drive shaft.

transmission rotor arrière (hélicoptère) (*n.f.*), tail rotor drive shaft.

transmodulation (radio) (*n.f.*), monkey chatter.

transport de charge par élingue (hélicop-tère) (*n.m.*), external load carrying.

transport tactique (mil.) (*n.m.*), tactical lift.

transport urbain (*n.m.*), town transport.

trappe (*n.f.*), hatch.

trappe de plancher (aviation) (*n.f.*), floor hatch.

trappe sous train (aviation) (*n.f.*), main gear sliding door.

traverse (auto) (*n.f.*), cross member (G.B.); cross rail (U.S.A.).

traverse de couple (*n.f.*), frame cross-beam.

traverse support de plancher (*n.f.*), floor beam.

traversée étanche (*n.f.*), pressure seal.

traversier (*n.m.*), through bolt.

trèfle occultable (*n.m.*), dimmer cap.

trembleur (élec.) (*n.m.*), vibrator.

trempe au brouillard (métall.) (*n.f.*), spray quenching.

trempe au chalumeau (métall.) (*n.f.*), flame hardening.

trempe fraîche (métall.) (*n.f.*), solution-treated (*adj.*).

trempe locale (métall.) (*n.f.*), selective hardening.

trempe structurale (métall.) (*n.f.*), precipita-tion hardening.

trépan-benne (*n.f.*), hammer grab.

tresse de métallisation (*n.f.*), bonding jumper.

tresse de mise à la masse (élec.) (*n.f.*), earthing strip; bonding jumper.

treuil à tambour (*n.m.*), winch.

treuil de levage (*n.m.*), hoist.

treuil de sauvetage (hélicoptère) (*n.m.*), rescue hoist.

treuil pneumatique (*n.m.*), air hoist.

treuil réacteur (*n.m.*), engine hoist.

treuilliste (*n.m.*), hoist operator.

tri (*n.m.*), sorting.

triangle à chape (*n.m.*), clevis link.

triangulation du train (aviation) (*n.f.*), land-ing gear bracing installation.

triangulation voilure (aviation) (*n.f.*), wing bracing installation.

trichloréthylène (chimie) (*n.m.*), trichlor-ethylene.

trièdre de référence (aviation) (*n.m.*), pitch, roll and yaw axes.

trieuse (*n.f.*), sorter.

trigatron (*n.m.*), trigatron.

trillion (*n.m.*), trillion (France, U.S.A. = 10^{12}; G.B. = 10^{18}).

trim (aviation) (*n.m.*), trim.

trim de profondeur (aviation) (*n.m.*), elevator trim.

tringlerie (*n.f.*), linkage.

triode (électronique) (*n.f.*), triode.

triode au germanium (électronique) (*n.f.*), transistor.

tritureuse (*n.f.*), pulvi-mixer.

trompe (*n.f.*), nozzle.

trompe à vide (*n.f.*), filter pump.

trompe de dépression (*n.f.*), syphon.

trompe de soufflage (*n.f.*), syphon.

trompe de ventilation (*n.f.*), ventilation nozzle.

trompette de gouvernail de direction (aviation) (*n.f.*), rudder arm.

trompette de roue conique (hélicoptère) (*n.f.*), bevel gear housing; bevel ring flared stub shaft.

trop-plein de batterie (*n.m.*), battery overflow.

trop-plein de réservoir (*n.m.*), overflow port.

tropicalisation (*n.f.*), tropic-proofing.

tropicalisé, -e (*adj.*), tropic-proof.

tropopause (couche de l'atmosphère) (*n.f.*), tropopause.

troposphère (couche de l'atmosphère) (*n.f.*), troposphere.

trou à bord tombé (*n.m.*), flanged hole.

trou chambré (*n.m.*), counterbored hole.

trou d'allégement (*n.m.*), lightening hole.

trou d'arrêt (*n.m.*), stop hole.

trou de fabrication (*n.m.*), tooling hole.

trou de piétage (*n.m.*), locating hole.

trou de rivet (*n.m.*), rivet hole.

trou embrevé (*n.m.*), dimpled hole.

trou ovalisé (*n.m.*), elongated hole.

trou pilote (*n.m.*), pilot hole.

trou radial (*n.m.*), radial hole.

trou taraudé (*n.m.*), tapped hole.

trou traversant (*n.m.*), through hole.

trouée dégagée (*n.f.*), clearway.

trousse à outil (*n.f.*), tool kit.

trusquin à tracer (*n.m.*), marking gauge.

tube à disques scellés (électron.) (*n.m.*), disc-seal valve.

tube à écran luminescent (*n.m.*), luminescent screen tube.

tube à faisceau électronique (*n.m.*), elec-tron beam valve.

tube à flamme (*n.m.*), liner.

tube à gaz (électronique) (*n.m.*), gas-filled valve.

tube à grilles alignées (électronique) (*n.m.*), aligned-grid valve.

tube à grilles multiples (électronique) (*n.m.*), multi-electrode valve.

tube à mémoire (électronique) (*n.m.*), storage tube.

tube à pente réglable (électronique) (*n.m.*), variable-mu valve.

tube à vide (électronique) (*n.m.*), vacuum valve.

tube cathodique (radio) (*n.m.*), cathode-ray tube.

tube cathodique à écran absorbant (*n.m.*), dark-trace tube.

tube d'amenée d'air chaud (*n.m.*), hot air duct.

tube d'emmanchement (*n.m.*), drift tube.

tube de gonflage (*n.m.*), inflating tube.

tube de guidage (*n.m.*), guide tube.

tube de maturation (raffinerie) (*n.m.*), soaker tube.

tube de mise à l'air libre (*n.m.*), vent tube.

tube de puissance à faisceau électro-nique (*n.m.*), beam power valve.

tube de raccordement (*n.m.*), coupling tube. (00

tube de refroidissement de génératrice (*n.m.*), generator blast tube.

tube de torsion (*n.m.*), torque tube.

tube de visée (*n.m.*), sight tube.

tube électromètre (*n.m.*), electrometer valve.

tube électronique (*n.m*), electronic tube; elec-tronic valve; electron tube (U.S.A.); ther-mionic tube.

tube entretoise (*n.m.*), spacer tube.

tube épanoui (*n.m.*), flared tube.

tube indicateur à néon (électronique) (*n.m.*), neon indicator.

tube mélangeur (radar) (*n.m.*), mixer tube.

tube multiple (électronique) (*n.m.*), multiple unit valve.

tube photo-électronique (*n.m.*), photovalve.

tube push-pull (radar) (*n.m.*), push-pull tube.

tube régulateur (électronique) (*n.m.*), regulator tube.

tube sans soudure (*n.m.*), seamless tube.

tubes analyseurs de télévision (électro-nique) (*n.m.pl.*), camera tubes.

tubes hyperfréquence (*n.m.pl.*), microwave tubes.

tubes micro-ondes (*n.m.pl.*), microwave tubes.

tubulure (*n.f.*), tube.

tubulure d'admission (auto) (*n.f.*), induction manifold.

tulipage (*n.m.*), gyro caging.

turbidimètre (*n.m.*), turbidimeter.

turbine à trois étages débit axial (*n.f.*), three-stage axial flow turbine.

turbine de détente (*n.f.*), expansion turbine.

turbine de réfrigération (*n.f.*), cooling turbine.

turbine libre (*n.f.*), free turbine.

turbo-alternateur (élec.) (*n.m.*), turbo-generator.

turbo-moteur (*n.m.*), gas turbine engine; turbo shaft engine.

turbo-propulseur (moteur) (*n.m.*), turbo-prop engine.

turbo-réacteur (moteur) (*n.m.*), turbo-jet engine.

turbo-réacteur à écoulement axial (*n.m.*),

jet engine with axial exhaust.

turbo-réacteur à soufflante canalisée (*n.m.*), turbo cooler unit.

turbo-réfrigérateur (*n.m.*), turbine cooler unit.

turbulence (*n.f.*), eddy; turbulence.

tuyauterie souple (*n.f.*), flexible pipe.

tuyère (moteur à réaction) (*n.f.*), expansion nozzle.

tuyère à deux positions (*n.f.*), two-position

nozzle.

tuyère d'éjection des gaz (*n.f.*), exhaust nozzle.

tuyère d'extrémité de pale (aviation) (*n.f.*), blade tip nozzle.

tuyère de lorin (*n.f.*), aero-thermodynamic duct; athodyd.

tuyère en régime sonique (*n.f.*), choked nozzle.

U

U (*n.m.*), channel section.

ultra-sons (*n.m.pl.*), ultrasonics.

unipolaire à deux directions (*n.m.*), single pole double throw.

unipolaire à une direction (*n.m.*), single pole single throw.

unité avionique d'hélicoptère (aviation) (*n.f.*), helicopter avionics package.

universel, -elle (élec.) (*adj.*), AC-DC; all-purpose; all-mains.

-ure (suffixe chimique), -ide.

urée (*n.f.*), urea.

urée-formol (plast.) (*n.f.*), urea formaldehyde.

usager (*n.m.*), user.

usinabilité à chaud (*n.f.*), hot strength.

usinage chimique (*n.m.*), chemical milling.

usine (*n.f.*), factory; plant; shop.

usiné, -e (*adj.*), machined.

usure (*n.f.*), attrition.

usure par frottement (*n.f.*), fretting; galling.

utile (*adj.*), effective; serviceable.

utilisateur (*n.m.*), operator; user.

utilisation (*n.f.*), use; operating instructions.

utiliser (*v.t.*), to use.

utilité (*n.f.*), use.

V

vacuomètre (*n.m.*), vacuum gauge.

valeur de crête (élec.) (*n.f.*), peak value.

valeur de pas indiqué (hélicoptère) (*n.f.*), indicated pitch angle.

valeur de pointe (élec.) (*n.f.*), peak value.

valeur de puissance (*n.f.*), power rating.

valeur efficace (élec.) (*n.f.*), root mean square value.

valeur moyenne (*n.f.*), mean value.

valeur propre (*n.f.*), Eigen value.

valide (*adj.*), applicable; effective; valid.

validité avion (*n.f.*), aircraft effectivity.

valve à flotteur (*n.f.*), float valve.

valve à pointeau (*n.f.*), needle valve.

valve anti-retour (*n.f.*), check valve.

valve de mise en pression (*n.f.*), pressurizing valve.

valve de remplissage (*n.f.*), filler valve.

valve de séquence (*n.f.*), sequence valve.

valve pneumatique (*n.f.*), pneumatic valve.

vanne (*n.f.*), valve.

vanne à commande électrique (*n.f.*), electrovalve.

vanne à papillon (*n.f.*), butterfly-type valve.

vanne d'admission d'air (*n.f.*), air intake valve.

vanne d'air (*n.f.*), airflow.

vanne d'air chaud (*n.f.*), hot air valve.

vanne de climatisation (*n.f.*), air-conditioning master valve.

vanne de dégivrage planeur (aviation) (*n.f.*), airfoil de-icing valve.

vanne de dégivrage réacteur (aviation) (*n.f.*), engine anti-icing gate valve.

vanne de dépressurisation (aviation) (*n.f.*), depressurization valve.

vanne de direction (aviation) (*n.f.*), yaw con-trol valve.

vanne de flux (aviation) (*n.f.*), flux valve.

vanne de prélèvement (*n.f.*), bleed valve.

vanne de régulation température (*n.f.*), temperature regulating valve.

vanne de répartition (*n.f.*), distribution valve.

vanne de transfert (*n.f.*), transfer valve.

vanne de ventilation (*n.f.*), ventilation valve.

vanne pneumatique (*n.f.*), pneumatic valve.

vanne régulatrice de débit (*n.f.*), flow reg-ulating valve.

vanne wagon (barrages) (*n.f.*), roller sluice.

vaporisé, -e (*adj.*), atomized.

varia (cinéma) (*n.m.*), feature.

variation cyclique principale (hélicoptère) (*n.f.*), primary cyclic variation.

variation du pas de la pale (hélicoptère) (*n.f.*), blade pitch variation.

variomètre (aviation) (*n.m.*), indicator of the rate of climb.

vaseline liquide (*n.f.*), liquid paraffin.

vaseline neutre (*n.f.*), petroleum jelly.

vé (*n.m.*), vee.

veille (radio) (*n.f.*), listening.

veilleur (*n.m.*), stand-by watch.

veilleuse (gaz) (*n.f.*), pilot-light.

veine (aviation) (*n.f.*), airflow.

vent arrière (navigation) (*n.m.*), down wind; tail wind.

vent debout (navigation) (*n.m.*), head wind.

vent latéral (navigation) (*n.m.*), cross-wind.

vent mesuré (navigation) (*n.m.*), reported wind.
vent nul (navigation) (*n.m.*), zero wind.
ventilateur de désembuage (*n.m.*), defog-ging fan.
ventilateur de queue caréné (hélicoptère) (*n.m.*), shrouded tail fan.
ventilateur-réchauffeur (*n.m.*), heater-blower.
ventilation air frais (*n.f.*), air cooling.
ventilation du réacteur (aviation) (*n.f.*), cranking.
ventilation forcée (*n.f.*), air blowing.
ventilé, -e (*adj.*), aired; ventilated.
ventouse (*n.f.*), suction cup.
venturi (*n.m.*), venturi.
vérificateur d'angle de pale (hélicoptère) (*n.m.*), blade angle check gauge.
vérificateur positionnement axe train (aviation) (*n.m.*), tool for checking landing gear alignment.
vérification (*n.f.*), check.
vérification de la sensibilité (*n.f.*), accuracy test.
vérin (*n.m.*), actuator; cylinder; jack; ram.
vérin à bille des volets (aviation) (*n.m.*), flap jack.
vérin à simple effet (*n.m.*), single action cylinder.
vérin à vis (*n.m.*), screw jack.
vérin correcteur d'effort (aviation) (*n.m.*), pitch corrector unit.
vérin de compensation (*n.m.*), compensating cylinder.
vérin de décrochage (aviation) (*n.m.*), pitch corrector unit.
vérin de déverrouillage (*n.m.*), unlocking cylinder.
vérin de levage hydraulique (*n.m.*), hydraulic jack.
vérin de rappel dans l'axe (*n.m.*), compen-sating cylinder.
vérin des aubes de guidage (*n.m.*) intake guide vane ram.
vérin électrique (*n.m.*), electrical actuator.
vérin hydraulique (*n.m.*), hydraulic actuating cylinder.
vérin mobile (*n.m.*), mobile jack.
vernis à tracer (*n.m.*), layout dye.
verre à trois foyers (*n.m.*), tri-focal lens.
verre bi-focal (*n.m.*), bi-focal lens.
verre bi-focal à segment rond (*n.m.*), bi-focal lens with round insert.
verre coloré (*n.m.*), tinted glass.
verre de contact (*n.m.*), contact lens.
verre de sûreté (*n.m.*), safety glass.
verre ponctuel (*n.m.*), pin-point lens.
verrou (*n.m.*), lock.
verrou à ressort (*n.m.*), spring lock.
verrou d'interdiction de relevage (avia-tion) (*n.m.*), landing gear safety lock.
verrou électromagnétique (*n.m.*), snubber.
verrou train rentré (aviation) (*n.m.*), landing gear uplock.
verrouillage (*n.m.*), locking.
verrouillage à action rapide (*n.m.*), quick acting locking device.
verrouillage hydraulique (*n.m.*), hydraulic locking.
verrouillage mécanique (*n.m.*), mechanical locking.
verrouillé, -e (*adj.*), locked.

version (*n.f.*), type; version.
vésicant, -e (*adj.*), blistering.
vestiaire (*n.m.*), dressing room.
vibrateur (*n.m.*), vibrator.
vibration aérolastique (*n.f.*), flutter.
vibreur anti-friction (*n.m.*), vibrator.
vibreur de signalisation (*n.m.*), buzzer.
vibro-réacteur (*n.m.*), vibrator.
vidange (*n.f.*), drainage.
vide, à (*adj.*), no-load; empty; light running.
vide-vite (aviation) (*n.m.*), fuel jettison.
vieillissement (plast.) (*n.m.*), ageing.
vieillissement artificiel (*n.m.*), artificial ageing.
vieillissement étagé (*n.m.*), interrupted ageing.
vieillissement naturel (*n.m.*), natural ageing.
vieillissement progressif (*n.m.*), progressive ageing.
vilebrequin à main (outillage) (*n.m.*), breast drill.
vilebrequin démonte-roue (auto) (*n.m.*), wheel-brace.
virage (*n.m.*), turn.
virage de procédure (*n.m.*), landing pattern turn.
virage du moteur (*n.m.*), belting-in.
virole de cône arrière (aviation) (*n.f.*), shroud.
virole de raccordement (*n.f.*), coupling ring.
virole intérieur de réacteur (aviation) (*n.f.*), inner shroud.
vis à bride (*n.f.*), tab screw.
vis à collerette (*n.f.*), flanged bolt.
vis à côte réparation (*n.f.*), oversize screw.
vis à filets contraires (*n.f.*), translating screw.
vis à six pans (*n.f.*), Allen screw.
vis à tête creuse (*n.f.*), socket head screw.
vis à tête cylindrique bombée (*n.f.*), fillister head machine screw.
vis à tête cylindrique forée (*n.f.*), drilled fillister head screw.
vis à tête fraisée bombée (*n.f.*), oval head screw.
vis à tête moletée (*n.f.*), thumb screw.
vis ajustée (*n.f.*), tight fitting screw.
vis borne (*n.f.*), terminal screw.
vis butée (*n.f.*), stop screw.
vis creuse (*n.f.*), hollow bolt.
vis d'arrêt (*n.f.*), set screw.
vis d'épinglage (*n.f.*), locating screw.
vis de blocage (*n.f.*), lock screw.
vis de centrage (*n.f.*), centering screw.
vis de décolletage (*n.f.*), jack screw.
vis de fixation (*n.f.*), attaching screw.
vis de purge (*n.f.*), bleed screw.
vis de réglage (*n.f.*), adjusting screw.
vis épaulée (*n.f.*), washer head screw.
vis-grain (*n.f.*), screw bushing.
vis imperdable (*n.f.*), captive screw.
vis Parker (*n.f.*), self-tapping screw.
vis pointeau (*n.f.*), set screw.
vis profilée (*n.f.*), round head screw.
vis purgeur (*n.f.*), bleeder screw.
vis taraudeuse (*n.f.*), self-tapping screw.
vis taraudeuse à bout plat (*n.f.*), flat end self-tapping screw.
viscosité cinématique (*n.f.*), kinematic viscosity.
visière (*n.f.*), visor.
visite (*n.f.*), check.
visite après-vol (aviation) (*n.f.*), post-flight

vissé (-e) à fond (*adj.*), screwed home.
visser (*v.t.*), to screw in.
visseuse (*n.f.*), screwing machine.
vitesse angulaire de lacet (aviation) (*n.f.*), angular yaw rate.
vitesse angulaire de roulis (aviation) (*n.f.*), angular roll rate.
vitesse angulaire de tangage (aviation) (*n.f.*), angular pitch rate.
vitesse ascensionnelle (aviation) (*n.f.*), rate of climb.
vitesse badin (aviation) (*n.f.*), indicated air speed.
vitesse cale à cale (*n.f.*), block speed.
vitesse corrigée (aviation) (*n.f.*), calibrated air speed.
vitesse d'impact (aviation) (*n.f.*), touchdown speed.
vitesse de calcul d'atterrissage (aviation) (*n.f.*), design landing speed.
vitesse de croisière (aviation) (*n.f.*), cruising speed.
vitesse de débit (*n.f.*), rate of flow.
vitesse de décrochage (*n.f.*), stalling speed.
vitesse de descente verticale (aviation) (*n.f.*), sinking speed.
vitesse de montée (aviation) (*n.f.*), climbing speed.
vitesse de rotation (*n.f.*), rotation speed.
vitesse décollage (aviation) (*n.f.*), take-off speed.
vitesse descensionnelle (aviation) (*n.f.*), rate of descent.
vitesse limite d'impact (aviation) (*n.f.*), limit rate of descent at touchdown.
vitesse limite de piqué (aviation) (*n.f.*), design diving speed.
vitesse nominale de rafale (aviation) (*n.f.*), nominal gust velocity.
vitesse périphérique (*n.f.*), circumferential speed.
vitesse sol (aviation) (*n.f.*), ground speed.
vitesse sonique (*n.f.*), acoustic velocity.
vitesse subsonique (*n.f.*), subsonic speed.
vitesse supersonique (*n.f.*), supersonic speed.
vitesse surmultipliée (auto) (*n.f.*), over-drive.
vitesse synchronisée (auto) (*n.f.*), synchro-mesh gear.
vitesse vraie (aviation) (*n.f.*), true air speed.
vitrifié, -e (*adj.*), enamelled; vitreous.
voie (auto) (*n.f.*), track.
voie du train d'atterrissage (aviation) (*n.f.*), landing gear track.
voile (*n.m.*), coat of paint.
voilé, -e, (*adj.*), out of true.

voilure (aviation) (*n.f.*), wings.
voilure de parachute (*n.f.*), canopy.
voilure en porte à faux (aviation) (*n.f.*), can-tilever wing.
voilure tournante (hélicoptère) (*n.f.*), rotary wing. (00
voix dans le champ (*n.f.*), voice in.
voix hors du champ (*n.f.*), voice off.
vol (*n.m.*), flight.
vol aux instruments (*n.m.*), instrument flying.
vol d'essai (*n.m.*), test flight.
vol de convoyage (*n.m.*), ferry flight.
vol de nuit (*n.m.*), night flight.
vol en autorotation (*n.m.*), autorotation flight.
vol en dérapage (*n.m.*), drifting flight.
vol en palier (*n.m.*), level flight.
vol horizontal (*n.m.*), level flight.
vol latéral (*n.m.*), sideways flight.
vol plané (*n.m.*), gliding flight.
vol rectiligne (*n.m.*), straight and level flight.
vol sans escale (*n.m.*), non-stop flight.
vol sans visibilité (*n.m.*), blind flight.
vol stationnaire (*n.m.*), hovering.
vol sur un réacteur (*n.m.*), single-engine flight.
vol vers l'arrière (*n.m.*), backward flight.
vol vers l'avant (*n.m.*), forward flight.
volant d'aileron (aviation) (*n.m.*), aileron con-trol wheel.
volant de commande de trim (aviation) (*n.m.*), trim control wheel.
volant de direction train avant (aviation) (*n.m.*), nosewheel steering control wheel.
volant de serrage (*n.m.*), friction wheel.
volet de fermeture (*n.m.*), shutter.
volet obturateur de fente (*n.m.*), slot flap.
voyant à test (*n.m.*), push-to-test light.
voyant alarme (*n.m.*), warning light.
voyant de baisse de niveau (*n.m.*), low level warning light.
voyant de fonctionnement (*n.m.*), pilot light.
voyant électrique (*n.m.*), indicator light.
voyant lumineux (*n.m.*), indicator light.
voyant magnétique (*n.m.*), magnetic indicator.
vraie grandeur (*n.f.*), actual size; full scale.
vrillage (*n.m.*), twisting.
vrille (aviation) (*n.f.*), spin.
vrille à plat (aviation) (*n.f.*), flat spin.
vrille serrée (aviation) (*n.f.*), steep spin.
VU-mètre (*n.m.*), volume meter.
vue coupée (*n.f.*), cut-away view.
vue de face (*n.f.*), front elevation.
vue de profil (*n.f.*), side view.
vue dégagée (*n.f.*), unobstructed view.
vue éclatée (*n.f.*), exploded view.
vulcanisation (*n.f.*), vulcanization.

W

weber (unité de flux magnétique MKSA) (*n.m.*), weber (= 10 maxwells).

white spirit (essence minérale) (*n.m.*), white spirit.

Z

zéro (*n.m.*), nought (G.B.); zero (U.S.A.).
zéro reader (*n.m.*), zero reader.
zicral (*n.m.*), zicral alloy.
ziglo (*n.m.*), dye-penetrant inspection.
zonage (planification) (*n.m.*), zoning.

zone (*n.f.*), field.
zone d'utilisation (aviation) (*n.f.*), operating range.
zone de balayage (électronique) (*n.f.*), sweeping zone.

SIGNES CONVENTIONNELS

La SÉPARATION DES DÉCIMALES, indiquée en français par une virgule ; ainsi **0,005 1,005** est indiquée **en** anglais par un point ; ainsi **·005** *ou* **0·005 1·005**
En Amérique, le point est ordinairement placé sur la ligne ; ainsi **.005** *ou* **0.005 1.005**

Les TRANCHES DE TROIS CHIFFRES, séparées en français, ou par des espaces ; ainsi **1 005 1 000 000** ou par des points (*viellissant*) ; ainsi **1.005 1.000.000** sont séparées en anglais par des virgules ; ainsi **1,005 1,000,000**
La séparation par des espaces s'emploie aussi en Amérique.

La DÉNOMINATION DES POIDS ET MESURES, qui est généralement placée en français entre le nombre entier et la fraction ; ainsi **1^m,25 0^m,25** suit les chiffres en anglais ; ainsi **1·25m. ·25m.** *ou* **0·25m.**

Les FRACTIONS et le SIGNE DU POUR CENT, communément imprimés en français en gros caractères ; ainsi **2 1/2 0/0** sont généralement imprimés en anglais en petits caractères ; ainsi **2$\frac{1}{2}$%**

Les INDICES DES SYMBOLES CHIMIQUES, placés en haut en français ; ainsi **AL^{2}O^3** sont généralement placés en bas en anglais ; ainsi **AL$_2$O$_3$**